MW00843696

Wastewater Treatment and Reuse

Theory and Design Examples

Volume 1: Principles and Basic Treatment

Wastewater Treatment and Reuse

Theory and Design Examples

Volume 1: Principles and Basic Treatment

Syed R. Qasim
The University of Texas at Arlington

Guang Zhu
CP&Y, Inc.

In Cooperation with

CP&Y, Inc.
Consulting Engineers · Planners · Project Managers
Dallas, Texas

CRC Press
Taylor & Francis Group
Boca Raton London New York

CRC Press is an imprint of the
Taylor & Francis Group, an **informa** business

CRC Press
Taylor & Francis Group
6000 Broken Sound Parkway NW, Suite 300
Boca Raton, FL 33487-2742

Printed on acid-free paper

International Standard Book Number-13: 978-1-138-30089-7 (Hardback)

Visit the Taylor & Francis Web site at
http://www.taylorandfrancis.com

and the CRC Press Web site at
http://www.crcpress.com

Printed and bound in the United States of America by Sheridan

Contents

VOLUME 2 Post-Treatment, Reuse, and Disposal

Preface

Over the last decade there have been rapid developments and changes in the field of wastewater treatment. The emphasis has been on identification, detection, and removal of specific constituents; computer simulation and modeling; membrane processes; renovation and reuse of wastewater effluent; nutrients recovery, and reduction and utilization of biosolids; energy conservation; greater understanding of theory and principles of treatment processes; and application of these fundamentals into facility design. Environmental engineers have many responsibilities. One of the most demanding yet satisfying of these are the design of wastewater treatment and reuse facilities. There are several books that discuss the fundamentals, scientific principles, and concepts and methodologies of wastewater treatment. The actual design calculation steps in numerical examples with intense focus on practical application of theory and principles into process and facility design are not fully covered in these publications. The intent of the authors writing this book is threefold: *first*, to present briefly the theory involved in specific wastewater treatment processes; *second*, to define the important design parameters involved in the process, and provide typical design values of these parameters for ready reference; and *third*, to provide a design approach by providing numerical applications and step-by-step design calculation procedure in the solved examples. Over 700 illustrative example problems and solutions have been worked out to cover the complete spectrum of wastewater treatment and reuse from fundamentals through advanced technology applied to primary, secondary and advanced treatment, reuse of effluent, by-product recovery and reuse of biosolids. These examples and solutions enhance the readers' comprehension and deeper understanding of the basic concepts. They also serve as a good source of information for more experienced engineers, and also aid in the formal design training and instruction of engineering students. Equipment selection and design procedures are the key functions of engineers and should be emphasized in engineering curricula. Many practice problems with step-by-step solution provide skills to engineering students and professionals of diverse background for learning, and to master the problem-solving techniques needed for professional engineering (PE) exams. Also, these solved examples can be applied by the plant designers to design various components and select equipment for the treatment facilities. Thus, the book is a consolidated resource of valuable quick-and-easy access to a myriad of theory and practice information and solved examples on wastewater treatment processes and reuse.

This work is divided into two volumes. Principles and basic treatment processes are covered in Volume 1, which includes Chapters 1 through 10. Volume 2 contains Chapters 11 through 15 to cover post-treatment processes, reuse, and solids disposal.

Volume 1: Principles and Basic Treatment. Chapter 1 is an overview of wastewater treatment: past, present, and future directions. Chapters 2 and 3 cover the stoichiometry, reaction kinetics, mass balance, theory of reactors, and flow and mass equalization. Sources of municipal wastewater and flow rates and characteristics are provided in Chapters 4 and 5. Chapter 6 provides an in-depth coverage of wastewater treatment objectives, design considerations, and treatment processes and process diagrams. The preliminary treatment processes are covered in Chapters 7 and 8. These unit processes are screening and grit removal. Chapter 9 deals with primary treatment with plain and chemically enhanced sedimentation. Chapter 10 provides an in-depth coverage of biological waste treatment and nutrients removal processes.

Volume 2: Post-Treatment, Reuse, and Disposal. Chapter 11 covers major processes for effluent disinfection, while Chapter 12 deals with effluent disposal and reuse. Chapter 13 is devoted to residuals management, recovery of resources, and biosolids reuse. The plant layout, yard piping, plant hydraulics, and instrumentation and controls are covered in Chapter 14. Upgrading of secondary treatment facility, land application, wetlands, filtration, carbon adsorption, BNR and MBR; and advanced wastewater treatment processes such as ion exchange, membrane processes, and distillation for demineralization are covered in Chapter 15.

This book will serve the needs of students, teachers, consulting engineers, equipment manufacturers, and technical personnel in city, state, and federal organizations who are involved with the review of designs, permitting, and enforcement. To maximize the usefulness of the book, the technical information is summarized in many tables that have been developed from a variety of sources. To further increase the utility of this book six appendices have been included. These appendices contain (a) abbreviations and symbols, basic information about elements, useful constants, common chemicals used in water and wastewater treatment, and U.S. standard sieves and size of openings; (b) physical constants and properties of water, solubility of dissolved gases in water, and important constants for solubility and sodicity of water; (c) minor head loss coefficients for pressure conduits and open channels, normal commercial pipe sizes, and design information of Parshall fumes; (d) unit conversions; (e) design parameters for wastewater treatment processes; and (f) list of examples presented and solved in this book. These appendices are included in both volumes. The numerical examples are integrated with the key words in the subject index. This gives additional benefit to the users of this book to identify and locate the solved examples that deal with the step-by-step calculations on the specific subject matter.

Enough material is presented in this textbook that cover supplemental material for a water treatment course, and a variety of wastewater treatment courses that can be developed and taught from this title. The supplemental material for a water treatment course include components of municipal water demand (Section 4.3), rapid mix, coagulation, flocculation, and sedimentation (Sections 9.6, 9.7, and 10.9), filtration (Section 15.4.6), carbon adsorption (Section 15.4.8), chlorine and ozone disinfection (Sections 11.6 and 11.8), demineralization by ion exchange and membrane processes (Sections 15.4.9 and 15.4.10), and residuals management (Sections 13.4.1 through 13.4.3, 13.5 through 13.8, and 13.11.6). At least *three* one-semester, and *one* two-semester sequential wastewater treatment courses at undergraduate or graduate levels can be developed and taught from this book. The specific topics to be covered will depend on time available, depth of coverage, and the course objectives. The suggested wastewater treatment and reuse courses are:

Course A: A one-semester introductory course on wastewater treatment and reuse
Course B: A sequential two-semester advance course on wastewater treatment and reuse
Course C: A one-semester course on physical and chemical unit operations and processes
Course D: A one-semester course on biological wastewater treatment

The suggested course outlines of these courses are provided in the tables below. The information in these tables is organized under three columns: topic, chapter, and sections. The examples are not included in these tables. It is expected that the instructor of the course will select the examples to achieve the depth of coverage required.

Course A: Suggested course contents of a one-semester introductory course on wastewater treatment and reuse

Topic	Chapter	Section
Overview of wastewater treatment	1	All
Stoichiometry and reaction kinetics	2	2.1 and 2.2

Continued

Topic	Chapter	Section
Mass balance, reactors, and equalization	3	3.1 to 3.3, and 3.4.1 to 3.4.3
Sources and flow rates of wastewater	4	All
Characteristics of municipal wastewater	5	5.1 to 5.6, 5.7.1, 5.8, and 5.9
Treatment and design objectives, and processes	6	All
Screening	7	All
Grit removal	8	8.1 to 8.3, 8.4.1 to 8.4.5, 8.5, and 8.6
Conventional and chemically enhanced primary sedimentation	9	9.1 to 9.6, 9.7.1, and 9.7.2
Biological waste treatment: basics, oxygen transfer, fixed film attached growth processes, anaerobic treatment, biological nitrogen removal, and final clarifier	10	10.1, 10.2, 10.3.1, 10.3.2, 10.3.4 to 10.3.8, 10.3.10, 10.3.11, 10.4 to 10.6, 10.7.1 to 10.7.3, 10.8, and 10.9
Effluent disinfection	11	11.1 to 11.7
Effluent reuse and disposal	12	12.1, 12.2, 12.5, and 12.6
Residuals processing, reuse, and disposal	13	13.1 to 13.8, and 13.11
Plant layout, piping, hydraulics, and instrumentation and control	14	All
Advanced wastewater treatment and upgrading secondary treatment facility	15	15.1 to 15.3, 15.4.5, 15.4.6, and 15.4.8 to 15.4.10

Course B: Suggested course contents of a sequential two-semester advanced course on wastewater treatment and reuse

Topic	Chapter	Section
First Semester		
Overview of wastewater treatment	1	All
Stoichiometry and reaction kinetics	2	All
Mass balance, reactors, and equalization	3	All
Sources and flow rates of wastewater	4	All
Characteristics of municipal wastewater	5	All
Treatment objectives, design considerations, and treatment processes	6	All
Screening	7	All
Grit removal	8	All
Primary and enhanced sedimentation	9	All
Biological waste treatment: fundamentals and types	10	10.1 and 10.2
Second Semester		
Biological waste treatment (cont'd): suspended, attached, aerobic, anaerobic kinetics, oxygen transfer, biological nutrient removal (BNR), computer	10	10.3 to 10.10
application, and final clarifiers		
Disinfection and kinetics	11	All
Effluent reuse and disposal	12	All

Continued

Topic	Chapter	Section
Residuals processing, reuse, and disposal	13	All
Plant layout, piping, hydraulics, and instrumentation and control	14	All
Advanced wastewater treatment facilities	15	All

Course C: Suggested course contents of a one-semester course on physical and chemical unit operations and processes

Topic	Chapter	Sections
Overview of wastewater treatment	1	All
Stoichiometry and reaction kinetics	2	All
Mass balance, reactors, and equalization	3	All
Sources and flow rates of wastewater	4	4.4 and 4.5
Characteristics of municipal wastewater	5	5.1 to 5.4
Wastewater treatment processes	6	6.3.5
Screening: coarse and fine screens	7	7.1, and 7.2.1 to 7.2.4
Discrete settling and grit removal	8	8.1, 8.3, 8.4.2, and 8.4.4
Flocculant settling, rapid mixing, flocculation, and sedimentation	9	9.1, 9.2, 9.5.5, 9.6.5, 9.6.6, and 9.7.2
Zone or hindered settling	10	10.9.2
Disinfection kinetics, chlorination, dechlorination, ozonation, and UV radiation	11	11.4, 11.5, 11.6.1, 11.6.2, 11.7.1, 11.8.6, and 11.9.4 to 11.9.6
Compression settling, dissolved air flotation, anaerobic digestion, conditioning, and dewatering	13	13.4.1, 13.4.2, 13.5.1 to 13.5.3, 13.6.1, 13.6.2, 13.7.1, 13.8.1, and 13.8.2
Air stripping, filtration, carbon adsorption, ion exchange, and membrane processes	15	15.4.5, 15.4.6 , 15.4.8, 15.4.9, and 15.4.10

Course D: Suggested course contents of a one-semester course on biological wastewater treatment

Topic	Chapter	Section
Overview of wastewater treatment	1	All
Stoichiometry and reaction kinetics	2	All
Mass balance, reactors, and equalization	3	All
Sources and flow rates of wastewater	4	All
Characteristics of municipal wastewater	5	All
Wastewater treatment processes	6	6.3.5
Biological waste treatment, biological nutrient removal (BNR), and final clarifier	10	All
Pathogens reduction in treatment processes and natural die-off kinetics	11	11.2.1, and 11.5.1
Anaerobic and aerobic digestion of sludge, material mass balance, and composting	13	13.6.1, 13.6.2, 13.9, and 13.11.1
Aquatic treatment systems, and membrane processes	15	15.3.2, and 15.4.10

In the solutions of examples, full expressions are provided to demonstrate step-by-step calculations. Many process and hydraulic parameters are involved in these expressions. To be more efficient, these parameters are represented by symbols. Sometimes, in the same example, parameters are applied multiple times to different streams or reactors. Therefore, subscripted notations are also used to identify these parameters. Each symbol is fully defined when it appears for the first time in the solution of the example. After that this symbol is repeated in the entire solution. This approach is helpful in (1) saving space by replacing lengthy descriptions of a parameter, and (2) providing an identification of the numerical value used or obtained in the expression. Additionally, these symbols provide the designers a ready reference in their design calculations while using Mathcad or spreadsheet software.

The International System of Units (SI) is used in this book. This is consistent with the teaching practices in most universities in the United States and around the world. Most tables in the book have dual units and include conversion from SI to U.S. customary units in footnotes. Useful conversion data and major treatment process design parameters are provided in Appendices D and E.

Acknowledgment

A project of this magnitude requires the cooperation and collaboration of many people and organizations. We are indebted to many professionals, faculty members, students, and friends who have helped and provided constructive suggestions. We must acknowledge the support, encouragement, and stimulating discussion by Michael Morrison, W. Walter Chiang, and Pete K. Patel throughout this project. CP&Y, Inc., a multidisciplinary consulting engineering firm in Dallas, Texas provided the technical support. We gratefully appreciate the support and assistance provided by Michael F. Graves, Marisa T. Vergara, Gregory W. Johnson, Ellen C. Carpenter, Barbara E. Vincent, Megan E. Martin, Gil W. Barnett, and Dario B. Sanchez. Many students also assisted with typing, artwork, literature search, and proofreading. Among them are Bernard D'Souza, Rajeshwar Kamidi, Neelesh Sule, Richa Karanjekar, Gautam Eapi, and Olimatou Ceesay.

Kelcy Warren established Syed Qasim Endowed Professorship in Environmental Engineering in the Department of Civil Engineering at The University of Texas at Arlington. Funds from this endowment helped to support students. The support of the Department of Civil Engineering at The University of Texas at Arlington is greatly appreciated. In particular, we thank the support of Dr. Ali Abolmaali, and tireless support of Sara Ridenour in making departmental resources available.

Finally, we must acknowledge with deep appreciation the support, encouragement, and patience of our families.

Although the portions of this book have been reviewed by professionals and students, the real test will not come until this book is used in classes, and by professionals in design of wastewater treatment facilities. We shall appreciate it very much if all who use this book will let us know of any errors and changes they believe would improve its usefulness.

Syed R. Qasim and Guang Zhu
Arlington, Texas

Authors

Syed R. Qasim is a professor emeritus in the Department of Civil Engineering at the University of Texas at Arlington. Dr. Qasim earned PhD and MSCE from West Virginia University, and BSCE from India. He served on the faculty of Polytechnic University, Brooklyn, New York, and on the faculty of the University of Texas at Arlington, Texas, from 1973 till his retirement in 2007. Dr. Qasim has conducted full-time research with Battelle Memorial Institute, Columbus Laboratories, and has worked as a design engineer with a consulting engineering firm in Columbus, Ohio. He has over 47 years of experience as an educator, researcher, and practitioner in the related fields of environmental engineering. His principal research and teaching interests include water and wastewater treatment processes and plant design, industrial waste treatment, and solid and hazardous waste management. He served nationally and internationally as a consultant with governmental agencies and private concerns. Dr. Qasim has written 3 books, and he is the author or coauthor of over 150 technical papers, encyclopedia and book chapters, and research reports. His papers, seminars, and short courses have been presented nationally and internationally. He is a *life member* and *Fellow* of American Society of Civil Engineers, and a *life member* of Water Environment Federation; a member of Association of Environmental Engineering and Science Professors, American Water Works Association, and other professional and honor societies. He is a registered professional engineer in the state of Texas.

Guang Zhu is an associate and senior water and wastewater engineering director with CP&Y, Inc., a multidisciplinary consulting engineering firm headquartered in Dallas, Texas. Dr. Zhu has over 30 years of consulting experience in planning, process evaluation, pilot testing, design, and commissioning of numeral conventional and advanced water and wastewater treatment plants in the United States and China. He had 10 years of consulting experience with Beijing General Municipal Engineering Design and Research Institute, Beijing, China. Dr. Zhu has coauthored one textbook and many technical papers and has translated two water and wastewater books in Chinese. He has taught several design courses as an adjunct assistant professor at the University of Texas at Arlington. He is a registered professional engineer in the state of Texas.

1

Introduction to Wastewater Treatment: An Overview

1.1 Historical Development

As societies moved from nomadic cultures to building more permanent settlements, the waste disposal became an important concern. Odors, nuisance, and health risks primarily from infectious diseases became apparent. The advent of the water carriage system was the major development toward removal of human waste from the residential areas. Now the human waste could be diluted with large quantities of water, carried through underground pipes to a centralized location outside the population centers and disposed of in a large and suitable water body.

As the population grew with increased urbanization and concentration in major population centers, discharging untreated wastewater in natural waters created serious water pollution problems. Using the water for drinking with some or no treatment still caused serious health risk. By the mid-nineteenth century, the engineers, public health officials, and the public were searching for treatment methods. How to treat the wastewater before discharging continued to be an unsolved problem until early to the mid-twentieth century.

The standards for discharge loading and treatment were investigated for Chicago in 1886. In 1887, the first intermittent sand filter was installed in Medford, Massachusetts. In 1899, first federal regulation of Sewage, River and Harbors Appropriations (Refuse Act) prohibited discharge of solids to navigable waters without permit from U.S. Army Corps of Engineers. Sewered population increased from 1 million in 1860 to 25 million in 1900. Various types of wastewater treatment technologies were developed and used: trickling filter in 1901, Imhoff tank in 1909, liquid chlorine for disinfection in 1914, and activated sludge in 1916. Increased wastewater treatment produced large quantities of residuals. Heated sludge digesters and use of gas were introduced in 1921, and mechanical dewatering by vacuum filters and centrifuge were developed soon. In the early 1930s, sludge drying and incineration were utilized in Chicago. By 1948, wastewater treatment plant served some 45 million Americans out of a total population of 145 million.[1–3]

1.2 Current Status

In 1956, Congress enacted the *Federal Water Pollution Control Act* (FWPCA), which established the *Federal Construction Grants Program* (FCGP). Under the 1972 Amendments to FWPCA (Public Law (PL) 92–500) and *Clean Water Act* (CWA) of 1977 (PL 95–217), thousands of municipal wastewater treatment facilities were built or expanded across the nation to control or prevent water pollution.[4,5] The law established the *National Pollutant Discharge Elimination System* (NPDES), which calls for limitation on the amount or quality of effluent from point sources, and requires all municipal and industrial dischargers to obtain permits. The interim goal is to achieve fishable/swimmable water quality of nation's waters.

The CWA authorized each state to develop a *water quality management* (WQM) plan to identify bodies of water as either *effluent-limited* or *water-quality limited* based on secondary level of treatment. The Minimum National Standards for Secondary Treatment as defined by U.S. EPA are given in Table 6.1. The *General Pretreatment Regulations* were established for the industrial users of municipal sewers to control contaminants that may pass through or interfere with treatment processes or which may concentrate in the municipal sludge. The *total maximum daily load* (TMDL) rule was promulgated in 2000 to protect ambient water quality of surface waters. The rule required to limit TMDL of pollutants that the water body can receive and still maintain a margin of safety in meeting the *established water quality standards.* The TMDL includes total waste load from all point source and nonpoint source (NPS) discharges, and the natural background levels. Under this rule, an integrated planning effort is needed to include drainage from agriculture, urban runoff, and discharges from all point source and NPS. Under the TMDL requirement, advanced wastewater treatment facilities may be needed for point source discharges in order to meet the specific water quality objectives for specific watershed.[6,7]

The law authorized billions of dollars for construction grants. The FCGP was later converted to *State Revolving Fund* (SRF) which provided loans to municipalities for construction of wastewater treatment facilities, until it was phased out as a method of financing publicly owned treatment works (POTW).

The early wastewater treatment objectives were based primarily on total suspended solids (TSS), biochemical oxygen demand (BOD), and pathogenic organisms. Subsequently, removal of nutrients nitrogen and phosphorus was addressed. Increased understanding of environmental and health effects of effluent discharge, and indirect and expected direct potable reuse (DPR) have shifted the attention toward identification and removal of specific constituents in wastewater. Many of these constituents can cause long-term (chronic) health effects. Even today, while the early treatment objectives remain valid, the required degree of treatment has increased significantly, and additional treatment objectives and goal have been added.

Enhanced treatment of liquid waste results in increased quantity of sludge. The national standards for handling and disposal of beneficial reuse of biosolids include pathogens and heavy metals. The objective is to protect human health and environment where biosolids are applied beneficially over land.[8]

The U.S. EPA *Clean Watershed Needs Survey (CWNS) 2012 Report to Congress* reported that in 2012 there were a total of 14,748 wastewater treatment facilities serving a total population of 238.2 million. The number of treatment facilities and expected population served in 2012 and 2032 at different wastewater treatment levels are summarized in Table 1.1.[9]

The expected total documented water quality capita investment needs for the nation over a period up to 2032 are $271.0 billion as dollars on January 1, 2012. This is equivalent to a national average need of $868 per capita. This figure represents capital needs for publicly owned wastewater collection systems, treatment facilities, combined sewer overflows (CSO) correction, storm water management, and recycled water distribution. The documented capital investment needs in different categories are given in Table 1.2.[9]

TABLE 1.1 Improvement in Wastewater Treatment Levels and Population Served

Level of Treatment	2012		2032	
	Number of Facilities	Population Served, Millions	Number of Facilities	Population Served, Millions
Less than secondary	57[a]	4.1	38[a]	4.5
Secondary	7374	90.4	6670	88.7
Greater than secondary	5036	127.7	6111	174.9
No discharge	2281	16.0	2461	26.7
Total	14,748	238.2	15,280	294.9

[a] Includes the partial treatment facilities.
Source: Adapted in part from Reference 9.

TABLE 1.2 Expected Total Documented Capital Investment Needs up to 2032 in Different Categories

Investment Category	Investment Needs over a Period up to 2032	
	Billion January 2012 Dollar	% of Total
Wastewater treatment systems	102.0	37.6
Collection system repair and new collection systems	95.7	35.3
Combined sewer overflow (CSO) correction	48.0	17.7
Stormwater management program	19.2	7.1
Recycled water distribution	6.1	2.3
Total	271.0	100

Source: Adapted in part from Reference 9.

1.3 Future Directions

U.S. EPA in its recently released *Blueprint for Integrating Technology Innovation into National Water Program* stated that over the past 40 years, great progress has been made in protecting the nation's water resources through an array of efforts by federal, state, and local governments, and the private sector under the auspices of the CWA and *Safe Drinking Water Act* (SDWA). But the United States is facing serious challenges to its water resources, including deteriorating infrastructure, continued population growth and development, impacts of climate change, emerging contaminants, widespread nutrient pollution, and strains on water supply. We need to discover new energy-efficient and cost-effective ways to meet today's demands and tomorrow's challenges.[10] New directions and concerns are evident in many areas. For this reason, there will be some shift in strategy in new plants design, and retrofitting and upgrading the existing facilities. Future directions are expected in many interrelated areas of wastewater treatment fields: (1) health and environmental concerns; (2) improved wastewater characterization and sidestreams; (3) rehabilitation of aging infrastructure; (4) energy reduction and recovery from wastewater; (5) retrofitting and upgrading POTWs; (6) treatment selection, performance reliability, and resiliency; (7) reduction in sludge quantity, nutrients recovery, and reuse of biosolids; (8) effluent disposal and reuse; (9) control of CSO and storm water management; (10) decentralized and on-site treatment and disposal; and (11) technology assessment and implementation.[3,6,9–11]

1.3.1 Health and Environmental Concerns

The health and environmental concerns are in two major areas: contaminants in effluent discharge and in air emissions. It is estimated that some 10,000 new organic compounds are manufactured each year. New and improved analytical techniques and sophisticated instrumentation make their detection and monitoring possible. Many of these compounds are detected in wastewater effluents, receiving waters, and drinking water supply. Among the concerned chemicals are priority pollutants such as pesticides, herbicides, and metabolites; chlorinated hydrocarbons, disinfection by-products (DBPs), trichloroethylene (TCE), polychlorinated biphenyls (PCBs), N-nitrosodimethylamine (NDMA), methyl tertiary butyl ether (MTBE), and more. A list of 129 EPA priority pollutants may be found in Reference 3. Most of the priority pollutants are very persistent, and are known or suspected to have carcinogenic, mutagenic, teratogenic, or high acute toxic properties. The toxicity and biomonitoring tests are covered in Sections 5.7 and 12.3. Pharmaceutical and personal care products (PPCPs) are highly bioactive and occur at trace concentrations. They are endocrine disruptors and interfere with reproduction, development, and behavior in many aquatic animals. Conventional treatment technologies are incapable of removing these trace contaminants. New technologies need to be developed, tested, and applied to achieve the acceptable levels in the effluent. Information on PPCPs may be found in Section 12.2.1. Additionally, enforcement of industrial pretreatment regulations, control of CSOs, stormwater management, and public education program are needed to reduce their entry into wastewater treatment plants.

Release of many volatile organic compounds (VOCs) and volatile toxic organic compounds (VTOCs) from wastewater collection and treatment facilities are of great concern. These compounds are odorous, corrosive, and have serious health implications. For this reason, selection of proper plant site and treatment processes is needed. Compact layout, covered and multistory process designs, and even underground facilities are needed to reduce the atmospheric exposure of wastewater. Provision should be made for collection and treatment of air from treatment facilities prior to release in the atmosphere. This will minimize the adverse environmental impacts, and objections by the neighborhood residents. Compact and modular wastewater treatment facilities are discussed in Section 14.2.3.

1.3.2 Improved Wastewater Characterization and Sidestreams

Characteristics of municipal wastewater are changing because of technical advancements in manufacturing processes, and use of newly developed organic compounds in raw material (Table 5.1). Advancements in analytical technologies make detection of these compounds possible even in extremely low concentrations in the influent, effluent, and biosolids. Biological kinetic and stoichiometric constants are widely used in process modeling and computer simulation for optimization of treatment processes and plant design. For this reason, fractionation of total nitrogen, phosphorous, chemical oxygen demand (COD), and carbon into organic, inorganic, soluble and particulate constituents is needed. Wastewater characterization for computer modeling of biological nutrient removal (BNR) facility is given in Example 10.161. Also, available concentrations of short-chain volatile fatty acids (SCVFAs) in the influent are needed for design of biological phosphorous removal. The biochemical reactions involving SCVFAs are discussed in Section 10.7.1. To identify the active mass in biological treatment, techniques from the microbiological science such as RNA and DNA typing may be needed.[6,7] The sidestreams from solids processing areas are normally returned to the plant for treatment. These streams contain high nitrogen and phosphorus; suspended, colloidal and dissolved organic and inorganic solids; and heavy metals. The side streams if mixed at the headworks may totally change the characteristics of incoming wastewater, increase hydraulic, organic, solids, and nutrients loadings. The concentrations of many contaminants may build up due to recirculation. The impact of side streams on influent quality is illustrated through material mass balance analysis in Section 13.9. The most significant development in wastewater treatment technology is to keep these sidestreams separate from the main plant. Specialized treatment processes are needed to remove specific contaminants. These specific processes may include (1) dissolved air floatation for nonsettleable solids, (2) steam stripping for ammonia removal, (3) high rate or chemically enhanced sedimentation for difficult-to-settle colloidal material, (4) chemical precipitation for removal of heavy metals, (5) biological treatment, and (6) combination of these processes. The selection of specific treatment processes may depend upon the constituents that may adversely affect the performance of the main plant.

1.3.3 Rehabilitation of Aging Infrastructure

The rehabilitation and renewal of aging wastewater collection and treatment infrastructures is the top national priority. The wastewater collection systems particularly in parts of older cities have combined sewers. Most of the sewers and appurtenances are old, leaky, and undersized. Excessive infiltration and inflow enter the sanitary sewers, and the storm runoff enters the combined sewers during wet weather. Whenever the design capacity of the treatment plant is exceeded, the excess untreated flow is diverted into the receiving waters. This causes serious water quality impairment primarily due to nutrients, sediments, and bacteria. Municipal wastewater flow and infiltration and inflow are presented in Section 4.5. Exfiltration causes untreated wastewater to enter the groundwater and nearby surface waters. It is estimated in the 2012 CWNS that nearly $95.7 billion (January 2012 dollars) will be needed up to 2032 to rehabilitate and replace the deteriorated pipes and appurtenances in the collection system, and build new ones (Table 1.2). The U.S. EPA's goals include encouraging most cost-effective leak detection and rehabilitation techniques.[9]

1.3.4 Energy Reduction and Recovery from Wastewater

Most conventionally run wastewater treatment plants require a large amount of energy for operation. There is an urgent need to conserve energy in all aspects of wastewater collection, transport, treatment, and reuse. Selection of energy-efficient processes and equipment is needed. Nearly half of the entire plant electricity usage is for aeration. Efficient aeration devices are needed. In-depth discussions on aeration devices and system design are provided in Sections 10.3.9 and 10.3.10. Additionally, the BNR facilities use less chemicals, reduce waste sludge production, lower energy consumption, and achieve nutrients removal. For these benefits, every effort should be made to retrofit the conventional activated sludge plants by BNR processes. Discussion on BNR processes may be found in Sections 10.6 through 10.8. Enhanced anaerobic digestion would increase biogas generation and reduce the quantity of biosolids. Anaerobic sludge digestion and quantity of biogas generation are discussed in Sections 13.6.1 and 13.10.2. Several successful alternative energy generation technologies are in development phases. These are solar–microbial device that uses wastewater and sunlight to generate clean energy, and electricity-producing microbes that treat wastewater and provide power source.[11,12] The U.S. EPA plans to encourage wastewater treatment facilities to take energy-saving measures and utilize alternative energy sources by using new technology.

1.3.5 Building, Retrofitting, and Upgrading POTWs

A large number of wastewater treatment plants were built under FCGP in 1975–1985. Most of these facilities are in need for retrofitting and upgrading to meet the stricter permit requirements, and to handle increased flow and loadings. It is expected that a great deal of effort is needed to apply new and improved processes to modify, improve performance, and expand the existing plants. It is reported in the 2012 CWNS that the total investment needs for wastewater treatment plants improvement and new construction from 2012 to 2032 is $102 billion of January 2012 dollars (Table 1.2).[9]

1.3.6 Process Selection, Performance, Reliability, and Resiliency

Innovative treatment processes should be applied for retrofitting and upgrading of existing plants and building new plants. The future goal is to produce high-quality treated effluent suitable for indirect and direct reuse, and minimize the environmental impacts. Among the innovative treatment methods are fine screens (Sections 7.2.3, 7.2.4, and 9.7), vortex degritters (Section 8.4.4), chemically enhanced primary treatment (CEPT) (Section 9.5.2), high rate clarification (Section 9.6), integrated fixed-film activated sludge (IFAS) and moving bed biofilm reactor (MBBR) (Section 10.4.3), enhanced biological phosphorus removal (EBPR) and BNR processes (Sections 10.7 and 10.8), membrane bioreactors (MBR) (Chapter 10 and Section 15.4.10), advanced oxidative processes (AOP) (Section 11.8), and innovative design and application of UV lamps for disinfection (Section 11.9). Wastewater treatment and sludge management processes are summarized in Tables 6.8 and 6.10. Many proprietary processes are commercially available that utilize innovative processes. Equipment and process performance and reliability are essential for consistently meeting the permit requirements and effluent reuse criteria. The performance of a plant is measured in terms of conventional parameters, toxicity and microbiological quality. The reliability is measured as probability that an equipment and process meet the performance criteria over an extended period of time.

In 2010, Superstorm Sandy revealed just how big of an impact a natural disaster can have on water and wastewater infrastructure. It is perhaps the ultimate wake-up call that gives a sense of vulnerability of our infrastructure and need for improvements. Technological innovations and higher safety factors may be utilized to achieve greater resiliency against damage caused by the raging nature.[11,13]

1.3.7 Reduction in Sludge Quantity, Nutrients Recovery, and Biosolids Reuse

Safe and cost-effective disposal of wastewater sludge is one of the biggest challenges currently facing the industry. Innovative technologies are being investigated to reduce the quantity of sludge. Among these are pretreatment of sludge (Section 13.10.1), enhanced anaerobic digestion (Section 13.10.2), and resource recovery (Section 13.10.3). The partial nitritation/Anammox (PN/A) process is also important for side-stream treatment of municipal wastewater (Section 13.10.4).

Hydrolysis of waste activated sludge (WAS) by alkaline treatment followed by ultrasonic treatment improves the subsequent biotransformation of organics into fatty acids. Hydrolyzed sludge undergoes improved volatile solids destruction, enhanced methane generation, and reduced quantity of digested sludge. Other development is temperature-phased anaerobic digestion (TPAnD) which also achieves similar results.[6,7,14]

Protein recovery from WAS has great potential for animal feed, and quantity of sludge solids reduction. The biosolids are hydrolyzed by alkali treatment separately or in combination with ultrasonication. The extracted protein contains all the essential amino acids and is suitable for animal feed supplement.[15,16]

Phosphorus removal from the wastewater is aggressively pursued by the regulatory agencies even at great cost. Phosphorus along with nitrogen and potassium are the essential component of (N-P-K) fertilizer. It is now realized that phosphorus is a limited resource. Current practice is not sustainable, and phosphorus must be recovered and reused. Phosphorus recovery from wastewater sludge is in the form of magnesium ammonium phosphate ($MgNH_4PO_4 \cdot 6H_2O$) called struvite. The rich source of phosphorus recovery is the wasted biomass from EBPR process. Under anaerobic environment, the phosphorus is released from the biomass and recovered as struvite by precipitation, electrochemical processes (ECPs), or bioelectrochemical systems (BESs).[17–19] Another source of phosphorus recovery is sidestream. The phosphorus is recovered as struvite in a fluidized-bed reactor.[20] Beneficial reuse of biosolids has increased in recent years because of environmental constraints and increasing costs of many other disposal alternatives (Sections 13.4.4 and 13.11.5). There has been technical advancement in sludge processing and stabilization to produce Class A biosolids that meets the heavy metals and pathogen requirements. It is suitable for application over farm lands, home gardens, nurseries, and golf courses. Class B biosolids is of lower quality and is applied over agricultural land, forests, and reclamation sites. Still there are public concerns over many beneficial uses of biosolids. Finding better methods for biosolids processing, reuse, and disposal will remain highest priority for the future.

1.3.8 Effluent Disposal and Reuse

Communities worldwide are facing water supply challenges, the increasing water demands combined with drought make water scarcity a real issue. Wastewater renovation and reuse as a resource is U.S. EPA's top promotional goal for the future. Wastewater effluent projects for nonpotable end uses are a common practice worldwide. Documented uses are irrigation in urban areas, agricultural (food and nonfood crops), industrial, recreational, and groundwater recharge. These uses are discussed in detail in Section 12.5. It is estimated in the 2012 CWNS that about \$6.1 billion (January 2012 dollars) will be needed between 2012 and 2032 for the improvements in recycled water distribution systems (Table 1.2).[9]

In many regions, the utilities are water-stressed and will sooner or later be compelled to investigate water reuse with two options: indirect potable reuse (IPR) or DPR.[21–24] IPR involves releasing reclaimed water into groundwater or surface water resource then reclaiming it and treating it to meet the drinking water standards. IPR water undergoes environmental purification before treatment. It has positive public perception but not 100% backed by communities, and is more expensive because same water is treated twice. The DPR is purified water created from treated wastewater and is introduced directly into municipal water supply system without any environmental buffer. Sometimes it is referred to as "pipe-to-pipe" reuse. The concept is difficult to sell to the public, although it provides cost savings, reduced carbon footprint, and is less vulnerable to natural or manmade disasters. However, the DPR does have some safety concerns, face negative public perception, require large storage buffer before distribution, and require extensive monitoring program. Because of natural buffer, the IPR will be more acceptable than DPR.

Numerous toxicological and epidemiological studies support that IPR is reliable, safe, and sustainable urban resource. There is, however, some perceived reservations amongst the communities about the safety and quality of recycled water particularly DPR, and should be considered an option of last resort.[23,24] Discussion on health risk analysis of water reuse is presented in Section 12.3.

Over the next decade, it is expected that many more epidemiological studies will be conducted with IPR and DPR, development of newer treatment and monitoring technologies, building and maintaining public trust, and ensuring a fair and sound decision-making process and outcome.

1.3.9 Control of Combined Sewer Overflows and Stormwater Management

The discharges from CSO, sanitary sewer overflow (SSO), stormwater runoff, and other NPS during wet weather have been recognized as a serious and difficult national problem. The sources of SSO are surcharge from stormwater entry, blockage, or structural, mechanical, or electrical failures. Pollution resulting from these sources often has prevented the attainment of mandated water quality standards, and closures of recreational beaches and shellfish beds. To control pollution from CSO and SSO, many techniques are applied. Some of these are: construction of new separate sewer and storm drainage systems, treatment at the overflow or discharge outlet, storage followed by treatment under dry weather conditions. All these techniques are expensive to build and operate. It is reported in the 2012 CWNS that $48 billion (January 2012 dollars) will be needed up to 2032 for correction of CSOs (Table 1.2).

On December 8, 1999, The U.S. Congress authorized states to implement the NPDE Stormwater Program (Stormwater Phase II rule).[25] It regulates point and nonpoint stormwater discharges. The peak flow, volume, and quality are tied intimately to the land use. Potential sources of point sources are: municipal separate storm sewer systems (MS4s), construction and industrial activities. Permit is required for these discharges.

The NPS pollution is drainage from diffused sources such as roads, buildings, lawns, drainage pipes, landscapes, agriculture, forests, and other undeveloped lands. The diffused runoffs carry pollutant into drainage ditches, ponds, lake, wetlands, rivers, bays, and aquifers. The federal CWA requires states to minimize NPS pollution from land use activities such as agriculture, urban development, forestry, recreational activities, and wetlands. To control runoff in both urban and suburban areas, buffer strips (grassy barrier, retention ponds, and porous pavements) may be considered. Restoration methods such as constructed wetlands are provided to slow down the runoff and absorb contamination. U.S. EPA has taken a number of actions to practice sustainable stormwater management and use low impact development (LID) techniques and natural wet weather green infrastructure. Effective public education program is essential to involve state and local governments, volunteer groups, water quality professionals, and general public to participate in watershed management efforts.[25–28] It is estimated in the 2012 CWNS that nearly $19.2 billion (January 2012 dollars) will be needed up to 2032 for stormwater management program (Table 1.2).[9] This is a priority area in national water quality improvement efforts, and rapid growth is expected in this sector of the industry.

1.3.10 Decentralized and On-Site Treatment and Disposal

Early thinking during the construction grants program was given to areawide wastewater planning and management. Centralized regional facilities were built to apply best practicable waste treatment technology. Experience has shown that regional facilities require pumping wastewater from service areas located in different watersheds. For this reason, such facilities became energy and resource intensive. Also, serious environmental problems developed as a result of processing large quantities of wastewater and sludge at one location.

There is renewed interest in the concept of satellite or decentralized wastewater treatment systems in the overall context of economics, innovative energy-efficient technology, and reuse of effluent and sludge. Small communities in rural areas find construction and maintenance of sewerage systems very expensive and in many cases impractical. For this reason, the planning, design, construction, and management of on-site systems for individual and cluster of homes and small establishments appear more practical and less expensive. Many innovative processes have been developed for such applications, and continued growth in this area is expected.

1.3.11 Technology Assessment and Implementation

Implementing new technologies in wastewater industries has always been a slow process. There are complex federal, state, and local requirements that slow down the acceptance, adaption, and implementation. One of the priorities of U.S. EPA is to break this cycle. Testing and evaluation of these technologies by independent third party will be promoted and adapted. Such effort will certainly enhance the pace for implementation. Additionally, smart sensor technology, telemetry, and remote sensing of water quality data will be developed and applied. More cost-effective leak detection and rehabilitation techniques and more emphasis on green infrastructure for stormwater management can be implemented. Also, new technologies and operational controls will be applied to lower the cost of maintaining the small wastewater systems with greater reliability.

1.4 Wastewater Treatment Plants

Planning, design, and process analysis of wastewater treatment plants involves understanding of service areas, sources and characteristics of wastewater and conveyance system, and treatment processes for liquid and residues. Many factors such as legal issues, regulatory constraints, public participation, effluent discharge or reuse, and sludge disposal may influence planning and design. Also, the design period for the facility may be 20 years or more. During this period, technology may change, new laws may be passed and effluent standards may become tighter. Engineers must consider these possibilities and select the processes that are flexible to remain useful and can be upgraded. Also, it is important to recognize that a treatment plant has many processes arranged sequentially in a process train. The performances of individual processes affect each other and the overall quality of the plant effluent.

Nowadays, many computer programs are commercially available and are used extensively in design and operation of wastewater facility. The designer can compare the alternative processes and process trains with a speed that was not possible in the past. The supervisory control and data acquisition (SCADA) system and expert system can provide operators with accurate process control variables and operation and maintenance records. The computational fluid dynamics (CFD) simulations are applied to design new systems, and optimize the operation of many processes. Also, the dynamic models can be integrated with process control system to optimize ongoing operation. Many commercially available computer programs for process modeling and CFD analysis are presented in Sections 10.8.4, 10.9.3, and 11.10.

In recent years, many wastewater treatment facilities are privatized. Privatization is public–private partnership. Normally, the private partner arranges the financing, design-builds, and operates the treatment facility. In some cases, the private partner may own the facility. The private financing is the main reason for privatization, but loss of local control is the major concern against privatization. It is expected that privatization trend may continue.

1.5 Scope of This Book

The objective of this book is to present theory and design concepts of wastewater treatment and reuse. It is a consolidated resource of valuable quick-and-easy access to theory and design examples and step-by-step solutions for complete spectrum of wastewater treatment from fundamentals to advanced technology. These numerical examples and solutions enhance the reader's comprehension, deeper understanding of the basic concepts, and master the problem-solving techniques. The subject matter covered in this book includes reaction kinetics, mass balance and reactors; equalization; preliminary, primary, secondary treatment, BNR and advanced treatment; and effluent and biosolids reuse. It is expected that a cross-section of simple to advanced problem will benefit undergraduate and graduate students, researchers, educators, and professionals.

Discussion Topics and Review Problems

1.1 What parameters are included in the national standard for secondary treatment and their maximum 30-day average allowable concentrations?

1.2 What is chronic toxicity and what class of compounds are associated by chronic toxicity? What are the health effects of many of these compounds resulting from long-term exposure?

1.3 With reference to chronic toxicity, define the following terms: biomonitoring, TRE, dose–response curve, NOAEL, LOAEL, and ADI.

1.4 Review Example (10.161), list the fractionation components of total nitrogen that is used in computer simulation.

1.5 What are the sources of sidestreams? If side streams are returned to the head of the plant, how will they change the influent quality to the plant? List the treatment processes commonly used to treat the sidestream.

1.6 Define infiltration and inflow (I/I). What is the normal I/I allowance for design of laterals and submains, and I/I allowance for mains and trunk sewers.

1.7 Describe anaerobic process and biogas generation. Based on rule of thumb, what is the rate of gas generation in m^3 per kg of volatile solids stabilized?

1.8 Review various treatment processes listed in Table 6.8. These processes are used for treatment of bulk liquid. Describe conventional activated sludge, trickling filter processes, and membrane bioreactor.

1.9 Review various treatment processes listed in Table 6.10. These processes are used for treatment of sludge. Describe gravity thickener and belt filter press.

1.10 Define IPR and DPR. List your concerns about IPR as a water supply source in your community.

1.11 Review sections on Public Education and Public Involvement in Reference 28 As a concerned citizen, what actions would you take for watershed management to reduce contaminants from non-point source discharges?

References

1. U.S. Environmental Protection Agency, *Building for Clean Water*, Office for Public Affairs (A-107), Washington, DC, August 1975.
2. Task Force of the Water Environment Federation and the American Society of Civil Engineers/Environmental and Water Resources Institute, *Design of Municipal Wastewater Treatment Plant*,

WEF Manual of Practice No. 8 and ASCE Manuals and Reports on Engineering Practice No. 76, 5th ed., McGraw-Hill, New York, NY, 2009.
3. Qasim, S. R., *Wastewater Treatment Plants: Planning, Design, and Operation*, 2nd ed., CRC Press, Boca Raton, FL, 1999.
4. *Federal Water Pollution Control Act Amendments of 1972* (P.L. 92-500), 92nd Congress, October 18, 1972.
5. *Clean Water Act* (P.L. 95-217), 95th Congress, December 27, 1977.
6. Metcalf & Eddy | AECOM, *Wastewater Engineering: Treatment and Resource Recovery*, 5th ed., McGraw-Hill, New York, NY, 2014.
7. Metcalf & Eddy, Inc., *Wastewater Engineering: Treatment and Reuse*, 4th ed., McGraw-Hill Companies, New York, NY, 2003.
8. U.S. Environmental Protection Agency, 40 CFR Part 503—Standards for the Use or Disposal of Sewage Sludge, 58 *FR* 9247, February 19, 1993, and as amended later.
9. U.S. Environmental Protection Agency, *Clean Watershed Needs Survey 2012 Report to Congress*, EPA-830-R-15005, Office of Wastewater Management, Washington, D.C., January 2016.
10. U.S. Environmental Protection Agency, *Blueprint for Integrating Technology Innovation into the National Water Program (Version 1.0)*, Office of Water, Washington, D.C., March 27, 2013.
11. Martin, L., EPA's top 10 technology needs for water, *Water Online*, December 23, 2013. http://www.wateronline.com/doc/epa-s-top-technology-needs-for-water-0001 (accessed on November 18, 2016).
12. Min, B. and B. E. Logan, Continuous electricity generation from domestic wastewater and organic substrates in a flat plate microbial fuel cell, *Environmental Science and Technology*, 38(21), 2004, 5809–5814.
13. Westerling, K., Sandy's aftermath: Is there a silver lining? *Water Online*, November 5, 2012. http://www.wateronline.com/doc/sandy-s-aftermath-is-there-a-silver-lining-0001? (accessed on November 18, 2016).
14. Chiu, Y., C. Chang, J. Lin, and S. Huang, Alkaline and ultrasonic pretreatment of sludge before anaerobic digestion, *Water Science and Technology*, 36(11), 1997, 155–162.
15. Chishti, S. S., S. N. Hasnain, and M. A. Khan, Studies on the recovery of sludge protein, *Water Research*, 26(2), 1992, 241–248.
16. Hwang, J., L. Zhang, S. Seo, Y. Lee, and D. Jahng, Protein recovery from excess sludge for its use as animal feed, *Bioresource Technology*, 99(18), 2008, 8945–8954.
17. Westerling, K., Save nutrients, save the world, *Water Online the Magazine*, October 2013, pp. 8–11.
18. Zhang, Y., E. Desmidt, A. V. Lover, L. Pinoy, B. Meesschaert, and B. Van der Bruggen, Phosphate separation and recovery from wastewater by novel electrodialysis, *Environmental Science and Technology*, 47(11), 2013, 5888–5895.
19. Wang, X., Y. Wang, X. Zhang, H. Feng, C. Li, and T. Xu, Phosphorus recovery from excess sludge by conventional electrodialysis (ECD) and electrodialysis with bipolar membranes (EDBM), *Industrial & Engineering Chemistry Research*, 52(45), 2013, 15896–15904.
20. Battistoni, P., P. Pavan, F. Cecchi, and J. Mata-Alvarez, Phosphate removal in real anaerobic supernatants: Modelling and performance of a fluidized bed reactor, *Water Science and Technology*, 38(1), 1998, 275–283.
21. Westerling, K., Meet the new water, same as the old water, *Water Online*, November 5, 2012. http://www.wateronline.com/doc/meet-the-new-water-same-as-the-old-water-0001 (accessed on November 18, 2016).
22. Martin, L., Direct potable reuse vs. indirect: Weighing the pros and cons, *Water Online*, November 4, 2013. http://www.wateronline.com/doc/direct-potable-reuse-vs-indirect-weighing-the-pros-and-cons-0001 (accessed on November 18, 2016).

23. Rodriguez, C., P. V. Buynder, R. Lugg, P. Blair, B. Devine, A. Cook, and P. Weinstein, Indirect potable reuse: A sustainable water supply alternative, *International Journal of Environmental Research and Public Health*, 6(3), 2009, 1174–1209.
24. National Research Council, *Issues in Potable Reuse: The Viability of Augmenting Drinking Water Supplies with Reclaimed Water*, National Academic Press, Washington, DC, 1998.
25. U.S. Environmental Protection Agency, *Stormwater Phase II Final Rule Fact Sheet Series, National Pollutant Discharge Elimination System (NPDES)*, https://www.epa.gov/npdes/stormwater-phase-ii-final-rule-fact-sheet-series (accessed on November 18, 2016).
26. Reese, A. J. and the Water Environment Federation, Ten reasons managing stormwater is different from wastewater, *Water Online*, February 13, 2013. http://www.wateronline.com/doc/ten-reasons-managing-stormwater-different-from-wastewater-0001 (accessed on November 18, 2016).
27. U.S. Environmental Protection Agency, *Handbook for Developing Watershed Plans to Restore and Protect Our Waters*, EPA 841-B-08-002, Office of Water, Washington, DC, March 2008.
28. U.S. Environmental Protection agency, *National Menu of Best Management Practices (BMPs) for Stormwater, National Pollutant Discharge Elimination System (NPDES)*, https://www.epa.gov/npdes/national-menu-best-management-practices-bmps-stormwater#edu (accessed on November 18, 2016).

2

Stoichiometry and Reaction Kinetics

2.1 Chapter Objectives

Stoichiometry deals with chemical reactions and the numerical relationships between the reactants and the products. Chemical reactions are classified based on their reaction kinetics. Reaction kinetics is the study of the rates of chemical reactions at which reactants are consumed, or products are formed in a stoichiometric reaction. The objectives of this chapter are to review:

- Stoichiometric homogeneous and heterogeneous reactions
- Reaction rates: zero-order, first-order, second-order, and saturation-type and enzymatic reactions
- Effect of temperature on reaction rate
- Reaction order data analysis and design
- Nitrification reactions, equations and solutions, and graphical representation
- Parallel irreversible reactions
- Derivation and application of type II of second-order reactions

2.2 Stoichiometry

A typical stoichiometry reaction is expressed by Equation 2.1.

$$aA + bB \leftrightarrow cC + dD \tag{2.1}$$

where

A and B = reactants
C and D = products
a and b = stoichiometric coefficients of reactants A and B, mole
c and d = stoichiometric coefficients of products C and D, mole

The stoichiometric reactions are used to (1) determine the number of moles of reactants entering into the reaction, and (2) the number of moles of products produced.

There are two principal types of reactions: (a) *homogeneous* and (b) *heterogeneous* (or *nonhomogeneous*).

2.2.1 Homogeneous Reactions

In homogeneous reactions, the reactants are distributed uniformly throughout the fluid. As a result, the potential for reaction at any point within the fluid is the same. These reactions may be *irreversible* or *reversible.* The irreversible reaction represented by one arrow continues to occur in the forward direction until equilibrium is reached. For reversible reaction, double arrow is employed. This means that the reverse reaction is significant. When the rate of forward reaction equals the reverse reaction, the system is in a state of *equilibrium*. In fact, both forward and reverse reactions may be occurring, but their rates are balanced, so there is no overall change in the concentrations with time. Examples of these reactions are expressed by Equations 2.2 and 2.3.

Irreversible Reactions: Irreversible reactions are single and multiple.

Single reactions are characterized by reactants that produce only one product as shown below:

$$A \rightarrow P \tag{2.2a}$$

$$A + A \rightarrow P \tag{2.2b}$$

$$A + B \rightarrow P \tag{2.2c}$$

where P = product

Multiple reactions can be either parallel or consecutive reactions.

Parallel Reactions: In parallel reaction, a reactant produces two products in a competitive reaction (Example 2.27).

$$A \begin{array}{l} \nearrow P_1 \\ \searrow P_2 \end{array} \tag{2.2d}$$

where P_1 and P_2 = products

Consecutive or Series Reactions: In these reactions, the product of one step becomes the reactant of the subsequent reaction steps (Example 2.25).

$$A \rightarrow P_1 \rightarrow P_2 \tag{2.2e}$$

Reversible Reactions: Reversible reactions are characterized by forward as well as reverse reactions occurring simultaneously (Example 2.9).

$$A \leftrightarrow P \tag{2.3a}$$

$$A + B \leftrightarrow C + D \tag{2.3b}$$

2.2.2 Heterogeneous Reactions

Heterogeneous reactions take place on specific sites such as those on an adsorbent and exchange resins. The reaction may involve (1) transport of reactants from bulk solution to solid–liquid interface, and transport of products from interior sites to outer surface, (2) adsorption at interior sites, and (3) chemical reaction of adsorbed reactants and desorbed products.[1]

EXAMPLE 2.1: QUANTITY OF BASE TO NEUTRALIZE ACID

In an accident, 500 lb of hydrochloric acid spilled into a stagnant pool. Calculate the quantity of 95% Ca $(OH)_2$ required to neutralize the acid. Also, demonstrate that mass is conserved (in the stoichiometric equation, mass of reactants (negative sign) + mass of product (positive sign) = 0).

Solution

1. Write the stoichiometric reaction (irreversible).

   $2\ HCl + Ca(OH)_2 \rightarrow CaCl_2 + 2\ H_2O$

   2 moles of HCl react with one mole of $Ca(OH)_2$.

2. Calculate the moles of HCl neutralized (n_{HCl}).

$$n_{HCl} = 500\,\text{lb HCl} \times 453.6\,\text{g/lb} \times \frac{1}{36.5\,\text{g/mole HCl}} = 6214\,\text{mole HCl}$$

3. Calculate the moles of $Ca(OH)_2$ required for neutralization.

$$n_{Ca(OH)_2} = 6214\,\text{mole HCl} \times \frac{1\,\text{mole Ca(OH)}_2}{2\,\text{mole HCl}} = 3107\,\text{mole Ca(OH)}_2$$

4. Calculate the quantity of 95% lime required.

Quantity of lime required

$$= 3107\,\text{mole Ca(OH)}_2 \times 74\,\text{g/mole Ca(OH)}_2 \times \frac{1}{0.95\,\text{(purity)}} \times \frac{1}{453.6\,\text{g/lb}} = 534\,\text{lb lime}$$

5. Demonstrate that the mass is conserved.

$$\begin{aligned}\text{Mass of reactants} &= -2\,\text{moles of HCl} \times 36.5\,\text{g/mole HCl} - 1\,\text{mole Ca(OH)}_2 \times 74\,\text{g/mole Ca(OH)}_2\\ &= -147\,\text{g}\\ \text{Mass of products} &= 1\,\text{mole CaCl}_2 \times 111\,\text{g/mole CaCl}_2 + 2\,\text{mole H}_2\text{O} \times 18\,\text{g/mole H}_2\text{O}\\ &= 147\,\text{g}\\ \text{Mass of reactants} + \text{mass of products} &= -147\,\text{g} + 147\,\text{g} = 0\end{aligned}$$

EXAMPLE 2.2: QUANTITY OF GASEOUS PRODUCT

$CaCO_3$ is heated to produce anhydrous lime (CaO). Calculate the quantity of CaO and CO_2 produced for each ton of $CaCO_3$ heated.

Solution

1. Write the stoichiometric reaction equation.

$$CaCO_3 \xrightarrow[\text{Heat}]{} CaO + CO_2$$

1 mole of $CaCO_3$ produces 1 mole of CaO and 1 mole of CO_2.

2. Calculate the number of moles of $CaCO_3$ in one ton of $CaCO_3$ heated.

$$n_{CaCO_3} = 1\,\text{ton CaCO}_3 \times 2000\,\text{lb/ton} \times 453.6\,\frac{\text{g}}{\text{lb}} \times \frac{1}{100\,\text{g/mole CaCO}_3} = 9072\,\text{mole CaCO}_3$$

3. Calculate the number of moles of CaO and CO_2 produced.

$$n_{CaO} = 9072\,\text{mole CaCO}_3 \times \frac{1\,\text{mole CaO}}{1\,\text{mole CaCO}_3} = 9072\,\text{mole CaO}$$

$$n_{CO_2} = 9072\,\text{mole CaCO}_3 \times \frac{1\,\text{mole CO}_2}{1\,\text{mole CaCO}_3} = 9072\,\text{mole CO}_2$$

4. Calculate the quantity of CaO and CO_2 produced.

$$CaO = 9072\,\text{mole CaO} \times 56\,\text{g/mole CaO} \times \frac{1}{453.6\,\text{g/lb}} = 1120\,\text{lb CaO}$$

$$CO_2 = 9072\,\text{mole } CO_2 \times 44\,\text{g/mole } CO_2 \times \frac{1}{453.6\,\text{g/lb}} = 880\,\text{lb } CO_2$$

5. Demonstrate that the mass is conserved.

$$-2000\,\text{lb } CaCO_3 + 1120\,\text{lb CaO} + 880\,\text{lb } CO_2 = 0$$

EXAMPLE 2.3: OXYGEN CONSUMPTION

Calculate the theoretical amount of oxygen required to completely burn 50 lb of sugar ($C_6H_{12}O_6$).

Solution

1. Write the stoichiometric reaction.

$$C_6H_{12}O_6 + 6\,O_2 \rightarrow 6\,CO_2 + 6\,H_2O$$

1 mole of sugar reacts with 6 moles of O_2.

2. Calculate the mole of sugar incinerated.

$$n_{\text{sugar}} = 50\,\text{lb sugar} \times 453.6\,\text{g/lb} \times \frac{1}{180\,\text{g/mole sugar}} = 126\,\text{mole sugar}$$

3. Calculate the theoretical amount of oxygen (O_2) required.

$$n_{O_2} = 126\,\text{mole sugar} \times \frac{6\,\text{mole } O_2}{1\,\text{mole sugar}} = 756\,\text{mole } O_2$$

$$\text{Theoretical amount of oxygen, } O_2 = 756\,\text{mole } O_2 \times 32\,\text{g/mole } O_2 \times \frac{1}{453.6\,\text{g/lb}} = 53.3\,\text{lb } O_2$$

2.3 Reaction Rates and Order of Reaction

The reaction rate is a mathematical expression describing the rate at which the mass or volume of a substance changes with time. Reactants have negative rates of reaction, and products have positive rates of reaction.

The controlling stoichiometry and the rate of the reaction are of principal concern in process selection and design. Many reactions have rates that are proportional to the concentration of one, two, or more of the reactant power. In this section, many examples and relationships between the rates of reaction, reactants, and products are presented.[1–4]

The generic chemical reaction is expressed by Equation 2.4 and the rate law describing the decrease in concentration of chemical A with respect to time is written by Equation 2.5.

$$A \rightarrow \text{Product} \tag{2.4}$$

$$r = -\frac{d[A]}{dt} = k[A]^n \tag{2.5}$$

where

A = reactant A that is converted to some unknown product

r = overall rate of reaction, mole/L·t. The time (t) may be in an unit of s, min, h, or d.

[A] = molar concentration of A at any time t, mole/L

k = reaction rate constant, $(\text{mole/L})^{1-n} \cdot \text{t}^{-1}$

n = an exponent that is typically determined through an experiment and used to define the order of the reaction with respect to the concentration of reactant(s), dimensionless

The order of reaction in chemical kinetics is the power to which its concentration term in the rate equation is raised. Depending upon the condition, the reaction rate may be:

$n = 0$ for the zero-order reactions,
$n = 1$ for the first-order reactions, and
$n = 2$ for the second-order reactions.

2.3.1 Reaction Rates

The definition and basic information on irreversible and reversible reactions are presented in Section 2.2.1. The reaction rates of different order reactions are presented in Examples 2.4 through 2.9.

EXAMPLE 2.4: RATE OF SINGLE REACTION

A generalized stoichiometric single irreversible reaction is given by the expression: $a\,\text{A} + b\,\text{B} \rightarrow c\,\text{C} + d\,\text{D}$ (Equation 2.1). Express (a) overall rate of reaction, and define all related terms, (b) relationships between the rates of overall reaction and each individual reaction, and (c) relationships between the reaction rates of reactants and products.

Solution

1. The overall reaction rate is defined by Equation 2.6.

$$r = k[\text{A}]^{\alpha}[\text{B}]^{\beta} \tag{2.6}$$

where

r = overall rate of reaction, mole/L·t

k = reaction rate constant, $(\text{mole/L})^{1-(\alpha+\beta)} \cdot \text{t}^{-1}$

[A] and [B] = molar concentrations of reactants A and B, mole/L

α and β = empirical exponents that are used to define the order of reaction with respect to reactants, dimensionless. The exponents α and β are usually 0, 1 or 2.

2. Express the relationships between the rates of overall reaction and each individual reaction.

$$r = -\frac{r_\text{A}}{a} = -\frac{r_\text{B}}{b} = \frac{r_\text{C}}{c} = \frac{r_\text{D}}{d} \quad \text{or} \quad r = \frac{|r_\text{A}|}{a} = \frac{|r_\text{B}|}{b} = \frac{r_\text{C}}{c} = \frac{r_\text{D}}{d}$$

3. Express the relationships between the reaction rates of reactants and products.

Using the reactant A and product C as an example, the ratio of the rates must equal to the ratio of the stoichiometric coefficient of reactant A to that of product C.

$$-\frac{r_\text{A}}{r_\text{C}} = \frac{a}{c} \quad \text{or} \quad \frac{|r_\text{A}|}{r_\text{C}} = \frac{a}{c}$$

Similarly, the ratios of reaction rates between any other reactants and products can also be established.

EXAMPLE 2.5: REACTION ORDER AND UNITS OF REACTION RATE CONSTANT

The overall rate of reaction of a single reversible reaction is given by the equation: $r = k\,[A][B]^2$. Determine (a) the order of reaction, and (b) units of reaction rate constant.

Solution

1. Determine the order of reaction.
   The overall reaction $r = k\,[A][B]^2$.
   The reaction is the first order with respect to reactant A.
   The reaction is the second order with respect to reactant B.
   The reaction is the third order with respect to overall reaction.
2. Determine the units of reaction rate constants.
   The reaction rates of and units for individual reactants can be written as follows:

$$\frac{d[A]}{dt} = r_A = k_A[A] \text{ (first-order reaction)}$$

$$\text{Unit for } k_A = \frac{r_A}{[A]} = \frac{\text{mole/L·t}}{\text{mole/L}} = \text{t}^{-1}$$

$$\frac{d[B]}{dt} = r_B = k_B[B]^2 \text{ (second-order reaction)}$$

$$\text{Unit for } k_B = \frac{r_B}{[B]^2} = \frac{\text{mole/L·t}}{(\text{mole/L})^2} = \frac{\text{L}}{\text{mole·t}}$$

$$r = k[A][B]^2 \text{ (third-order reaction)}$$

$$\text{Unit for } k = \frac{r}{[A][B]^2} = \frac{\text{mole/L·t}}{(\text{mole/L})(\text{mole/L})^2} = \frac{\text{L}^2}{\text{mole}^2\text{·t}}$$

EXAMPLE 2.6: REACTION RATE FOR SINGLE REACTION

A single irreversible first-order reaction has only one reaction step. A generalized stoichiometric single reaction is given below. Determine the rates of reaction for the reactant and each product as a function of reaction rate constant k and molar concentration of A.

$$3\,A \rightarrow 2\,B + C$$

Solution

1. Express the overall rate of reaction.

$$r = k[A]$$

2. Determine the rates of reaction of individual reactants.

$$r = -\frac{r_A}{a} = \frac{r_B}{b} = \frac{r_C}{c} = k[A]$$

$$r_A = -ak[A] = -3k[A]$$

$$r_B = bk[A] = 2k[A]$$

$$r_C = ck[A] = k[A]$$

Note: The r_A has negative sign because the concentration of [A] is reduced.

EXAMPLE 2.7: REACTION RATE OF CONSECUTIVE REACTION

In a consecutive first-order reaction, the product of one step becomes the reactant of the subsequent reaction steps. A consecutive irreversible reaction is given below. Determine the stoichiometric reaction relationships.

$$a\,\text{A} \xrightarrow{k_1} b\,\text{B} \xrightarrow{k_2} c\,\text{C}$$

Solution

1. Express the overall rates of reactions, assuming the first-order reaction.

$$r_1 = -\frac{r_\text{A}}{a} = \frac{r_{\text{B}_1}}{b} = k_1[\text{A}]$$

$$r_2 = -\frac{r_{\text{B}_2}}{b} = \frac{r_\text{C}}{c} = k_2[\text{B}]$$

Note: The r_{B_1} represents generation of B; and the rate r_{B_2} represents reduction in B and has negative sign.

2. Express the rates of reaction of individual reactants with respect to overall rate.

$$r_\text{A} = -ar_1$$

$$r_\text{B} = r_{\text{B}_1} + r_{\text{B}_2} = br_1 - br_2$$

$$r_\text{C} = -cr_2$$

3. Express the rates of reaction of individual reactants with respect to concentration remaining.

$$r_\text{A} = -ak_1[\text{A}]$$

$$r_\text{B} = bk_1[\text{A}] - bk_2[\text{B}]$$

$$r_\text{C} = -ck_2[\text{B}]$$

Proper signs should be used in the above reaction rates.

EXAMPLE 2.8: NITRIFICATION OF AMMONIA, A CONSECUTIVE REACTION

A classic example of consecutive reaction in environmental engineering is nitrification of ammonia expressed by the following expression. Develop the stoichiometric reaction relationships.

$$NH_3 \xrightarrow{O_2\ \&\ \textit{Nitrosomonas},\ k_1} NO_2^- \xrightarrow{O_2\ \&\ \textit{Nitrobacter},\ k_2} NO_3^-$$

Solution

1. Express the overall rates of reactions, assuming first-order reaction.

$$r_1 = -r_{NH_3} = r_{(NO_2^-)_1} = k_1[NH_3]$$

$$r_2 = -r_{(NO_2^-)_2} = r_{NO_3^-} = k_2\left[NO_2^-\right]$$

2. Express the rates of reaction of NH_3, NO_2^-, and NO_3^-.

$$r_{NH_3} = -k_1[NH_3]$$

$$r_{NO_2^-} = r_{(NO_2^-)_1} + r_{(NO_2^-)_2}$$

$$r_{NO_2^-} = k_1[NH_3] - k_2\left[NO_2^-\right]$$

$$r_{NO_3^-} = k_2\left[NO_2^-\right]$$

EXAMPLE 2.9: SINGLE REVERSIBLE REACTION

A single reversible first-order reaction is given below. Express the reaction rates of reactants.

$$a\,\mathrm{A} \underset{k_2}{\overset{k_1}{\longleftrightarrow}} b\,\mathrm{B}$$

Solution

1. Express the overall reaction rates.

$$r_1 = -\frac{r_{A_1}}{a} = \frac{r_{B_1}}{b} = k_1[\mathrm{A}],$$

$$r_2 = -\frac{r_{B_2}}{b} = \frac{r_{A_2}}{a} = k_2[\mathrm{B}],$$

where

r_{A_1} = expresses decrease in concentration of A due to forward reaction
r_{A_2} = expresses increase in concentration of A due to reverse reaction
r_{B_1} = expresses increase in concentration of B due to forward reaction
r_{B_2} = expresses decrease in concentration of B due to reverse reaction

2. Express the individual rates of reaction in terms of overall reaction rate

$$r_A = r_{A_1} + r_{A_2} = -ar_1 + ar_2$$
$$r_B = r_{B_1} + r_{B_2} = br_1 - br_2$$

3. Express the individual rates of reactions in terms of concentration.

$$r_A = -ak_1[\mathrm{A}] + ak_2[\mathrm{B}]$$
$$r_B = bk_1[\mathrm{A}] - bk_2[\mathrm{B}]$$

2.3.2 Saturation-Type or Enzymatic Reactions

The saturation-type reactions reach a maximum rate. After reaching the maximum rate, the reaction becomes independent of the concentration of the reactants. The reaction rate constant of a simplified saturation-type reaction $a\,\mathrm{A} \rightarrow b\,\mathrm{B}$ is given by Equation 2.7a.

$$r = \frac{k[\mathrm{A}]}{K_s + [\mathrm{A}]} \tag{2.7a}$$

where

r = overall reaction rate of saturation reaction, mole/L·t
k = maximum reaction rate, mole/L·t
K_s = half-saturation constant or substrate concentration at one-half the maximum reaction rate, mole/L

The relationship between the constants k and K_s are indicated in Figure 2.1.

In a saturation-type reaction, the reaction rate r reaches the maximum rate of reaction k and the half-saturation constant K_s is equal to the substrate concentration at which the reaction rate r is one-half of maximum ($r = (1/2)k$).

Equation 2.7a may be modified to develop a linear relationship to obtain the coefficients k and K_s. The procedure involves inversing Equation 2.7a to develop a linear relationship that is expressed by Equation 2.7b.

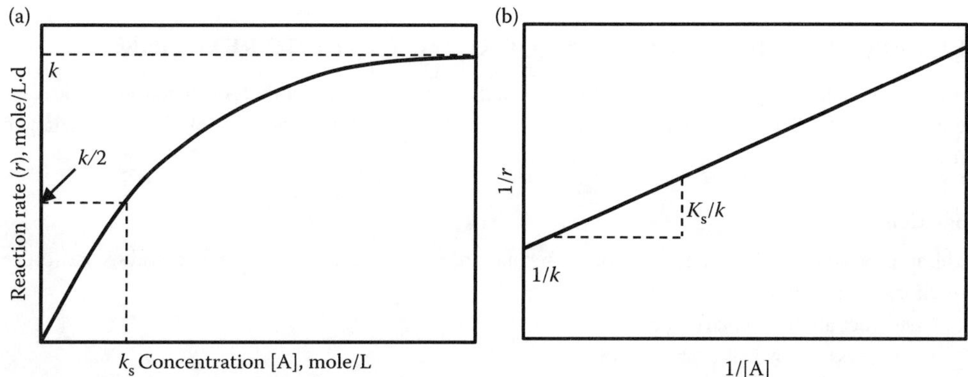

FIGURE 2.1 The relationship between substrate concentration and reaction rate in saturation-type reaction: (a) reaction rate r reaching to a maximum rate of k and (b) linear representation of the data.

$$\frac{1}{r} = \frac{K_s}{k}\frac{1}{[A]} + \frac{1}{k} \tag{2.7b}$$

A plot of $1/r$ versus $1/[A]$ gives a linear relationship as shown in Figure 2.1b. The slope of the line is K_s/k and the intercept is $1/k$. From these relationships, the coefficients k and K_s can be obtained.

More complex saturation reactions are used to express the specific substrate utilization rate, and specific biomass growth rate with respect to substrate concentration remaining and kinetic coefficients. This topic is covered in detail in Chapter 10, Section 10.3.2, Examples 10.28 through 10.30.

EXAMPLE 2.10: TWO LIMITING CASES OF SATURATION-TYPE REACTION

Two limiting cases of enzymatic reaction are commonly developed from Equation 2.7a. Express these limiting cases.

Solution

Two limiting cases of enzymatic reactions are dependent upon the substrate concentration. These are described below:

1. Expression for the substrate in excess.

   When the substrate concentration is much greater $[A] >> K_s$ and $K_s + [A] \approx [A]$, the reaction rate is approaching the maximum rate and is independent of the concentration of [A]. It is derived from Equation 2.7a and is expressed by the equation:

   $$r = \frac{k[A]}{K_s + [A]} \approx \frac{k[A]}{[A]}, \quad \text{or} \quad r = k \qquad \text{(This is a pseudo zero-order reaction.)}$$

2. Expression for the limited substrate.

   When the substrate concentration is small $[A] << K_s$ and $K_s + [A] \approx K_s$ (or a constant), the rate of enzymatic reaction is derived from Equation 2.7a and is expressed by the equation:

   $$r = \frac{k[A]}{K_s + [A]} \approx \frac{k[A]}{K_s}, \quad \text{or} \quad r = k'[A] \quad \text{and} \quad k' = \frac{k}{K_s} \quad \text{(This is a pseudo first-order reaction.)}$$

   Many enzymatic reactions are function of the product. Examples are the number of microorganisms that increases in proportion to the number present. These reactions can be first order, second order, or saturation type.

EXAMPLE 2.11: REACTION RATES OF AUTOCATALYTIC REACTIONS

A reaction is autocatalytic if the rate is a function of the product concentration. A generalized single reaction is given by: $a\,\mathrm{A} \rightarrow b\,\mathrm{B}$. Express the stoichiometric reaction relationships of these autocatalytic reactions.

Solution

The autocatalytic reactions can be first order, second order or saturation type. The first- and second-order reactions are as follows:

1. First-order autocatalytic reaction.
   a. Express the overall reaction rate.

   $$r = k[\mathrm{B}]$$

   b. Express the individual reactant and product reaction rates.

   $$r_\mathrm{A} = -ar = -ak[\mathrm{B}]$$
   $$r_\mathrm{B} = br = bk[\mathrm{B}]$$

2. Second-order autocatalytic reaction or partially autocatalytic reaction.
   a. Express the overall reaction rate.

   $$r = k[\mathrm{A}][\mathrm{B}]$$

   b. Express the individual reactant and product reaction rates.

   $$r_\mathrm{A} = -ar = ak[\mathrm{A}][\mathrm{B}]$$
   $$r_\mathrm{B} = br = bk[\mathrm{A}][\mathrm{B}]$$

   Other examples of pseudo first-order reaction are when the concentration of one component remains constant during the reaction. Example of such reactions are (1) the initial concentration of one reactant is much higher than that of the other, (2) concentration of one reactant is buffered such as alkalinity remaining unchanged by dissolution of $CaCO_3$, and (3) one reactant is supplied continuously. In these situations, a pseudo first-order reaction is used.

EXAMPLE 2.12: PSEUDO FIRST-ORDER REACTION RATE

Consider an irreversible elementary reaction. Assume $[\mathrm{A}] \approx [\mathrm{A_0}]$ and $[\mathrm{A_0}] \gg [\mathrm{B_0}]$, where $\mathrm{A_0}$ and $\mathrm{B_0}$ are initial concentrations. Express the modified pseudo first-order overall reaction rate.

Solution

1. Express the irreversible elementary reaction.

   $$a\,\mathrm{A} + b\,\mathrm{B} \rightarrow c\,\mathrm{C} + d\,\mathrm{D}$$

2. Express the rate law for the reaction.

   $$r = k[\mathrm{A}]^a[\mathrm{B}]^b$$

3. If concentration of A does not change significantly during the reaction, the concentration of [A] remains essentially constant. Assume $[\mathrm{A}]^a \approx K$, the overall reaction rate is expressed below.

   $$r = kK[\mathrm{B}]^b = k'[\mathrm{B}]^b \text{ and } k' = k[\mathrm{A}]$$

2.4 Effect of Temperature on Reaction Rate

Reaction rates are strongly affected by temperature change, but the reaction stoichiometry is not affected. The temperature effects are expressed by Equation 2.8. This is a modified form of the *van't Hoff–Arrhenius* relationship.[1]

$$k_T = Ae^{-E/RT} \tag{2.8}$$

where
k_T = reaction rate constant, variable unit
A = *van't Hoff–Arrhenius* coefficient, d^{-1}
E = activation energy, J/mole
R = universal gas constant, 8.314 J/mole·°K
T = temperature, °K

EXAMPLE 2.13: SIMPLIFICATION OF VAN'T HOFF–ARRHENIUS EQUATION

Equation 2.8 is the generalized *van't Hoff–Arrhenius* equation. Simplify this equation for water and wastewater applications.

Solution

The ratio of k at two temperatures T_1 and T_2 may be written as follows:

$$\frac{k_{T_2}}{k_{T_1}} = \frac{Ae^{-E/RT_2}}{Ae^{-E/RT_1}} = \exp\left[\frac{E}{RT_1} - \frac{E}{RT_2}\right] = \exp\left[\frac{E}{RT_1T_2}(T_2 - T_1)\right]$$

$$k_{T_2} = k_{T_1} \exp\left[\frac{E}{RT_1T_2}(T_2 - T_1)\right] \tag{2.9}$$

In the range of 0–35°C the exp $[E/RT_1T_2]$ is considered approximately constant and Equation 2.9 is reduced to Equation 2.10.

$$k_{T_2} = k_{T_1}\theta^{(T_2-T_1)} \tag{2.10}$$

where θ = temperature coefficient, dimensionless

The value of θ varies slightly with temperature. The value of θ in different temperature ranges are:

Temperature, °C	θ
4–20	1.135
20–35	1.047

EXAMPLE 2.14: REACTION RATE CONSTANT AT DIFFERENT TEMPERATURES

The standard value of reaction rate constant is given at the standard temperature of 20°C. The typical value of k_{20} (base e) for wastewater is 0.20 d^{-1}. Determine the value of the reaction rate constant at 25°C. Assume θ = 1.047.

Solution

The value of k_{25} is obtained from Equation 2.10.

$$k_{T_2} = k_{T_1}\theta^{(T_2-T_1)}$$

$$k_{25} = k_{20}(1.047)^{(25-20)} = 0.20\,\text{d}^{-1} \times (1.047)^5 = 0.20\,\text{d}^{-1} \times 1.258 = 0.25\,\text{d}^{-1}$$

2.5 Reaction Order Data Analysis and Design

For design of water and wastewater treatment facilities, the physical, chemical and biological processes are used. Most commonly, the laboratory experiments are conducted to model the performance of reactors. Valuable information such as order of reaction, reaction rate constants, and range and limits of these reactions are developed through analysis of experimental data. Numerous examples on zero-, first-, and second-order reactions are provided below to describe the procedures used in these analyses.

2.5.1 Zero-Order Reaction

If $n = 0$, the generic Equation 2.5 is reduced to Equation 2.11 with respect to the concentration of reactant.

$$-\frac{dC}{dt} = k \tag{2.11}$$

where

C = concentration remaining of reactant after reaction time t, mg/L
t = reaction time, variable time unit (t)
k = zero-order reaction rate constant, mg/L·t

This is the rate law that describes the zero-order reaction. In the zero-order system, the reaction proceeds at a rate independent of the concentration of the reactants. This is often the case when the reactant is present in high concentration. Rearranging Equation 2.11 and integrating it, the concentration of C can be expressed by Equation 2.12.

$$dC = -kdt$$

$$\int_{C_0}^{C} dC = -k\int_0^t dt$$

$$C = C_0 - kt \tag{2.12}$$

where C_0 = initial concentration or reactant at $t = 0$, mg/L

Equation 2.12 is a linear relationship. If concentration data are plotted with respect to time, the result is a straight line. The slope of the line is the zero-order rate constant k, which has unit of concentration/time (e.g., mole/L·d). It is often useful to express a reaction in terms of *half-life*. The half-life, $t_{1/2}$, is defined as the time required for the concentration to decrease to one-half or $C = (1/2)C_0$. Substituting this relationship in Equation 2.12, it is reduced to Equation 2.13.

$$\frac{1}{2}C_0O = C_0 - kt_{1/2}$$

$$t_{1/2} = \frac{1}{2}\frac{C_0}{k} \tag{2.13}$$

EXAMPLE 2.15: REACTION ORDER, AND RATE CONSTANT

In a batch reactor, an industrial waste is exposed to UV radiation in the presence of a catalyst. The initial concentration of an organic dye in the industrial waste was 3000 mg/L. Samples were withdrawn from the reactor at different time intervals and the concentrations of organic dye remaining were measured. Following results were obtained. Determine (a) the order of the reaction, (b) calculate the reaction rate constant, (c) half-life, and (d) dye concentration remaining after 24 h of UV exposure.

Time, h	Dye Concentration Remaining, mg/L
0	3000
4	2800
8	2590
12	2380
16	2150
20	1990

Solution

1. Plot the concentration of dye remaining with time; draw the best-fit line and indicate the reaction order. The graphical relationship is shown in Figure 2.2.

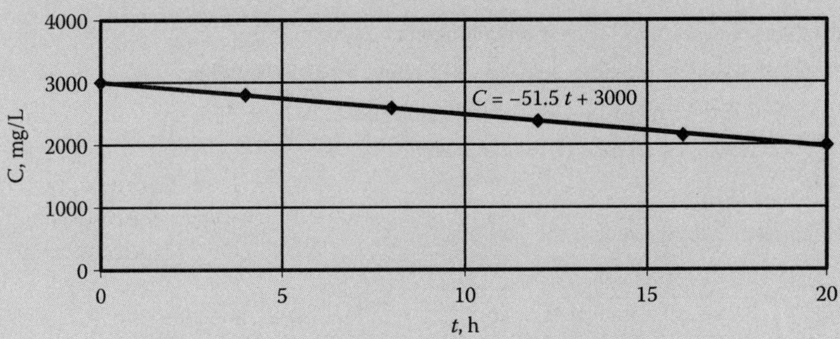

FIGURE 2.2 Zero-order destruction of organic dye (Example 2.15).

Since the dye concentration remaining represents a linear relationship, the reaction is zero order.

2. Determine the slope and k.

   The slope of the line is the reaction rate constant.

   Slope $= k = -51.5$ mg/L·h

3. Determine the half-life.

   $C_0 = 3000$ mg/L

   Use Equation 2.13 to calculate the half-life.

$$t_{1/2} = \frac{1}{2} \times \frac{3000\text{ mg/L}}{51.5\text{ mg/L·h}} = 29.1\text{ h}$$

4. Determine the concentration of dye remaining after 24 h of UV exposure from Equation 2.12.
   Use Equation 2.12 to calculate the concentration of dye remaining.

   $C = C_0 - kt = 3000\,\text{mg/L} - 51.5\,\text{mg/L·h} \times 24\,\text{h} = 1764\,\text{mg/L}$

EXAMPLE 2.16: CONCENTRATION REMAINING

In the fermentation of sugar, an enzymatic reaction is utilized. Initial sugar solution was 0.15 mole/L. After 4 hours of reaction, 10% sugar was converted into alcohol. Assuming no side reactions and zero-order reaction rate, calculate (a) reaction rate constant, (b) concentration remaining after 12 h, and (c) half-life.

Solution

1. Calculate the concentration of sugar remaining after 4 h.

   $C = (1 - 0.1) \times 0.15\,\text{mole/L} = 0.135\,\text{mole/L}$

2. Calculate the reaction rate constant.
   Use Equation 2.12 to calculate the reaction rate constant.

   $C = C_0 - kt$

   $0.135\,\text{mole/L} = 0.15\,\text{mole/L} - k \times 4\,\text{h}$

   $$k = \frac{1}{4\,\text{h}} \times (0.15\,\text{mole/L} - 0.135\,\text{mole/L}) = 3.75 \times 10^{-3}\,\text{mole/L·h}$$

3. Calculate the concentration of sugar remaining after 24 h.
   Use Equation 2.12 to calculate the concentration of sugar remaining.

   $$\begin{aligned} C &= C_0 - kt = 0.15\,\text{mole/L} - 3.75 \times 10^{-3}\,\text{mole/L·h} \times 24\,\text{h} \\ &= 0.15\,\text{mole/L} - 0.09\,\text{mole/L} = 0.06\,\text{mole/L} \end{aligned}$$

4. Calculate the half-life.
   Use Equation 2.13 to calculate the half-life.

   $$t_{1/2} = \frac{1}{2} \times \frac{0.15\,\text{mole/L}}{3.75 \times 10^{-3}\,\text{mole/L·h}} = 20\,\text{h}$$

EXAMPLE 2.17: REACTION RATES OF REACTANTS

Gaseous ammonia is produced in a catalytic experiment utilizing the Haber process: $N_2 + 3\,H_2 \rightarrow 2\,NH_3$. The measured generation rate of ammonia (dC_{NH_3}/dt) was 2.0×10^{-4} mole/L·s. If there were no side reactions, calculate the rate of reaction of (a) N_2 and (b) H_2 expressed in terms of NH_3.

Solution

1. Calculate the reaction rate of N_2.
   The balanced equation $N_2 + 3\,H_2 \rightarrow 2\,NH_3$ indicates that ammonia is produced at twice the utilization rate of N_2. This means that the reaction rate of N_2 is half that of NH_3.

   $$-\frac{dC_{N_2}}{dt}O = \frac{1}{2}\frac{dC_{NH_3}}{dt} = \frac{1}{2} \times 2.0 \times 10^{-4}\,\text{mole/L·s} = 1.0 \times 10^{-4}\,\text{mole/L·s}$$

   $$\frac{dC_{N_2}}{dt} = -1.0 \times 10^{-4}\,\text{mole/L·s}$$

2. Calculate the reaction rate of H_2.

$$-\frac{dC_{H_2}}{dt} = \frac{3}{2}\frac{dC_{NH_3}}{dt} = \frac{3}{2} \times 2.0 \times 10^{-4}\ \text{mole/L·s} = 3.0 \times 10^{-4}\ \text{mole/L·s}$$

$$\frac{dC_{H_2}}{dt} = -3.0 \times 10^{-4}\ \text{mole/L·s}$$

2.5.2 First-Order Reaction

If $n = 1$, the generic Equation 2.5 becomes Equation 2.14.

$$-\frac{dC}{dt} = kC \tag{2.14}$$

This is the rate law for the first-order reaction. This suggests that the reaction rate is dependent upon the concentration of the chemical remaining. As reaction continues, the concentration C remaining decreases with time. Therefore, the first-order reaction rate slows down over time. Equation 2.14 is rearranged and integrated with respect to time to give Equations 2.15a and 2.15b.

$$\frac{1}{C}\,dC = -kdt$$

$$\int_{C_0}^{C} \frac{1}{C}dC = -k\int_{0}^{t} dt$$

$$\ln\left(\frac{C}{C_0}\right) = -kt$$

$$C = C_0 e^{-kt} \tag{2.15a}$$

$$\ln(C) = \ln(C_0) - kt \quad \text{or} \quad C = C_0 10^{-kt} \tag{2.15b}$$

where

k = the first-order reaction rate constant to the base e, t^{-1}

K = the first-order reaction rate constant to the base 10, t^{-1}

Equations 2.15a and 2.15b express an exponential decay. The semi-log plot of this equation will give a linear relationship, and slope of the line is the reaction rate constant. The half-life for the first-order relationship is expressed by Equations 2.16a and 2.16b.

$$t_{1/2} = \frac{0.693}{k} \tag{2.16a}$$

$$t_{1/2} = \frac{0.301}{K} \tag{2.16b}$$

EXAMPLE 2.18: DETERMINATION OF FIRST-ORDER REACTION RATE CONSTANT FROM GRAPHICS

The following COD data have been collected from a batch biological reactor study with an acclimated culture of microorganisms. Prepare arithmetic and semi-log plot of the data. Determine the reaction rate constants to the base 10, and to the base e.

t, h	0	3	5	7	10	12	15
C, mg/L COD	350	178	110	75	35	20	13

Solution

1. Prepare the arithmetic and semi-log plot of COD data.

   The arithmetic and semi-log plots of COD data are illustrated in Figure 2.3a and b.

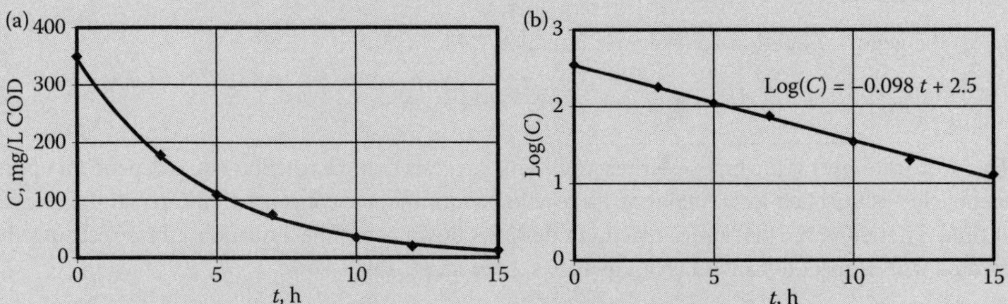

FIGURE 2.3 Graphical presentation of COD data: (a) arithmetic plot and (b) semi-log plot (base 10) (Example 2.18).

2. Determine the reaction order.

   The arithmetic plot is nonlinear while the semi-log plot shows a linear relationship. Therefore, the data represents the first-order reaction rate.
3. Determine the slope from the semi-log plot.

   Slope $= -0.098\ \text{h}^{-1}$
4. Determine the reaction rate constants to the base 10 and to the base *e*.

   The slope of the line is the reaction rate constant K (base 10) $= 0.098\ \text{h}^{-1}$.

   Based on the relationship given in Equations 2.15a and 2.15b, k (base e) is calculated.

$$10^{-Kt} = e^{-kt}$$

$$-Kt\log(10) = -kt\log(e)$$

$$-Kt \times 1 = -kt \times 0.434 \quad \text{or} \quad K = 0.434\,k \quad \text{or} \quad k = 2.03\,K$$

$$k = \frac{K}{0.434} = \frac{0.098}{0.434} = 0.23\ \text{h}^{-1}$$

EXAMPLE 2.19: DETERMINATION OF REACTION RATE CONSTANT BY CALCULATION

Determine the reaction rate constant from the data in Example 2.18.

Solution

1. Select the reaction rate equation.

   Assume first-order reaction rate. Determine k (base e) from Equation 2.15a.

$$C = C_0 e^{-kt}$$

$$\ln\left(\frac{C}{C_0}\right) = -kt$$

$$k = -\frac{1}{t}\ln\left(\frac{C}{C_0}\right)$$

2. Apply the data of Example 2.18.

t, h	C, mg/L	$\frac{C}{C_0}$	$\ln\left(\frac{C}{C_0}\right)$	$-\frac{1}{t}\ln\left(\frac{C}{C_0}\right)$, h^{-1}
0	350	1.000	0	–
3	178	0.509	−0.676	0.23
5	110	0.314	−1.157	0.23
7	75	0.214	−1.540	0.22
10	35	0.100	−2.303	0.23
12	20	0.057	−2.862	0.24
15	13	0.037	−3.293	0.22
				Average = 0.23

3. The reaction rate constants for every data point are approximately the same. Therefore, the plot is linear, and the assumption for the first-order reaction rate is correct.
4. The average value of k (base e) = 0.23 h^{-1}.

EXAMPLE 2.20: CALCULATIONS FOR HALF-LIFE

A storage basin contains toluene and dieldrin. The first-order removal constants k for toluene and dieldrin are 0.067 h^{-1} and 2.67 × 10^{-5} h^{-1}, respectively. Calculate the storage time for both chemicals to reduce their concentrations to one-half of their initial values.

Solution

1. Determine the half-life of toluene from Equation 2.16a.

$$k_{\text{toulene}} = 0.067\,\text{h}^{-1}$$

$$t_{\text{toulene},\,1/2} = \frac{0.693}{k_{\text{toulene}}} = \frac{0.693}{0.067\,\text{h}^{-1}} = 10.3\,\text{h}$$

2. Determine the half-life of dieldrin from Equation 2.16a.

$$k_{\text{dieldrin}} = 2.67 \times 10^{-5}\,\text{h}^{-1}$$

$$t_{\text{dieldrin},\,1/2} = \frac{0.693}{k_{\text{dieldrin}}} = \frac{0.693}{2.67 \times 10^{-5}\,\text{h}^{-1}} = 26{,}000\,\text{h} \quad \text{or} \quad 26{,}000\,\text{h} \times \frac{1}{24\,\text{h/d} \times 365\,\text{d/yr}} = 3\,\text{yr}$$

EXAMPLE 2.21: DETERMINATION OF INITIAL CONCENTRATION

An industrial sludge contains benzene. The allowable benzene limit for disposal of sludge in a landfill is 15 ppb. A 30-day-old sludge had benzene concentration of 1.1 ppb. If the first-order removal rate constant k for benzene is 0.00345 h^{-1}, calculate if fresh sludge would have met the allowable concentration for disposal in the landfill.

Solution

1. Calculate the concentration of benzene in the fresh sludge from Equation 2.15a.

$$C = C_0 e^{-kt}$$

$$\ln\left(\frac{C}{C_0}\right) = -0.00345\,\text{h}^{-1} \times (30\,\text{d} \times 24\,\text{h/d}) = -2.48$$

$$\frac{C}{C_0} = 0.0837$$

$$C_0 = \frac{C}{0.0837} = \frac{1.1\,\text{ppb}}{0.0837} = 13.1\,\text{ppb}$$

2. Compare the limit.

   The initial concentration of benzene in fresh sludge is 13.1 ppb, which is lower than 15 ppb. Therefore, the fresh sludge would have met the allowable limit for disposal in the landfill.

EXAMPLE 2.22: CONCENTRATION RATIO REMAINING, AND CONCENTRATION RATIO REMOVED

A hazardous waste has half-life ($t_{1/2}$) of 12 h. Calculate the time to reach 40% and 80% concentrations remaining, and the time to achieve 40% and 80% decay of the original hazardous waste. Plot the percent decay and percent concentration remaining versus time.

Solution

1. Calculate the reaction rate constant k from Equation 2.16a.

$$t_{1/2} = \frac{0.693}{k}$$

$$k = \frac{0.693}{t_{1/2}} = \frac{0.693}{12\,\text{h}} = 0.058\,\text{h}^{-1}$$

2. Calculate the time to achieve 40% and 80% concentration remaining from Equation 2.15a.

$$C = C_0 e^{-kt} \quad \text{or} \quad \frac{C}{C_0} = e^{-kt}$$

$$0.4 = e^{-0.058\,\text{h}^{-1} \times t_{40\%\,\text{removal}}}$$

$$\ln(0.4) = -0.058\,\text{h}^{-1} \times t_{40\%\,\text{removal}}$$

$$t_{40\%\,\text{removal}} = \frac{\ln(0.4)}{-0.058\,\text{h}^{-1}} = \frac{-0.916}{-0.058\,\text{h}^{-1}} = 15.8\,\text{h}$$

   The time for 40% concentration remaining is 15.8 h.

$$0.8 = e^{-0.058\,\text{h}^{-1} \times t_{80\%\,\text{removal}}}$$
$$\ln(0.8) = -0.058\,\text{h}^{-1} \times t_{80\%\,\text{removal}}$$

$$t_{80\%\,\text{removal}} = \frac{\ln(0.8)}{-0.058\,\text{h}^{-1}} = \frac{-0.223}{-0.058\,\text{h}^{-1}} = 3.8\,\text{h}$$

   The time for 80% concentration remaining is 3.8 h.

3. Calculate the time to achieve 40% and 80% decay.

   The decay or destruction = $(1 - C/C_0)$

   To achieve 40% decay, $0.4 = (1 - C/C_0)$

$C/C_0 = 0.6$ (60% concentration remaining)

$$t_{40\%\,\text{decay}} = t_{60\%\,\text{removal}} = \frac{\ln(0.6)}{-0.058\,\text{h}^{-1}} = \frac{-0.511}{-0.058\,\text{h}^{-1}} = 8.8\,\text{h}$$

The time for 40% concentration decay is 8.8 h.
To achieve 80% decay, $0.8 = (1 - C/C_0)$
$C/C_0 = 0.2$ (20% concentration remaining)

$$t_{80\%\,\text{decay}} = t_{20\%\,\text{removal}} = \frac{\ln(0.2)}{-0.058\,\text{h}^{-1}} = \frac{-1.61}{-0.058\,\text{h}^{-1}} = 27.8\,\text{h}$$

The time for 80% concentration decay is 27.8 h.

4. Tabulate the concentration ratio of decay $(1 - C/C_0)$, and concentration ratio of remaining (C/C_0) with time.

   The concentration ratios of decay and remaining are derived from each other. These ratios are:

$(1 - C/C_0)$ (decay)	0.2	0.4	0.6	0.8
C/C_0 (remaining)	0.8	0.6	0.4	0.2
Time, h	3.8	8.8	15.8	27.8

5. Plot the percent decay and percent remaining concentration.

   The plot of percent concentration destroyed and percent concentration remaining are shown in Figure 2.4.

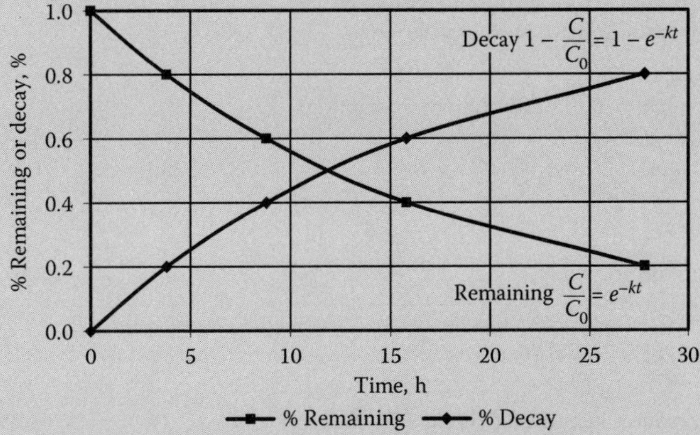

FIGURE 2.4 The graphical representation of percent destruction and percent remaining concentrations of a hazardous waste (Example 2.22).

EXAMPLE 2.23: REACTION RATE CONSTANT OBTAINED FROM HALF-LIFE

Chemical contamination of groundwater aquifer has been reported. The contaminants are benzene, trichloroethane (TEC), and toluene. The half-lives of these chemicals are 69, 231, and 12 days, respectively. Calculate the first-order reaction rate constants of all these chemicals.

Solution

The reaction rate constants are calculated from Equation 2.16a.

$$t_{1/2} = \frac{0.693}{k}$$

$$k_{\text{benzene}} = \frac{0.693}{69\,\text{d}} = 0.01\,\text{d}^{-1}$$

$$k_{TCE} = \frac{0.693}{231\,\text{d}} = 0.003\,\text{d}^{-1}$$

$$k_{\text{toluene}} = \frac{0.693}{12\,\text{d}} = 0.058\,\text{d}^{-1}$$

EXAMPLE 2.24: RADIOACTIVE DECAY

Strontium-90 (^{90}Sr) is a radioactive nuclide that has a half-life of 28 years. Like most radioactive nuclides, it exhibits first-order decay. If a groundwater source is contaminated by ^{90}Sr, how long will it take to reduce the radioactivity by 99%.

Solution

1. Calculate the first-order reaction rate from Equation 2.16a.

$$k = \frac{0.693}{28\,\text{yr}} = 0.0248\,\text{yr}^{-1}$$

2. Calculate the reaction period from Equation 2.15a.
   To reduce the radioactivity by 99%, $0.99 = (1 - C/C_0)$

$$C/C_0 = 1 - 0.99 = 0.01 \text{ (radioactivity remaining)}$$

$$C = C_0 e^{-kt} \quad \text{or} \quad \frac{C}{C_0} = e^{-kt}$$

$$0.01 = e^{-0.0248\,\text{yr}^{-1} \times t}$$

$$\ln(0.01) = -0.0248\,\text{yr}^{-1} \times t$$

$$t = \frac{\ln(0.01)}{-0.0248\,\text{yr}^{-1}} = \frac{-4.61}{-0.0248\,\text{yr}^{-1}} = 186\,\text{yr}$$

The time to reduce the radioactivity by 99% is 186 years.

EXAMPLE 2.25: NITRIFICATION REACTION

The simplified consecutive reaction that describes the biological nitrification is given by equation:

$$NH_3\text{-N} \underset{k_1}{\overset{O_2}{\longrightarrow}} NO_2\text{-N} \underset{k_2}{\overset{O_2}{\longrightarrow}} NO_3\text{-N}.$$

Write the decomposition equations that describe the rates of decomposition and formation of the reactants and products. Also, give the solution expressing the concentrations of each constituent with respect to time.

Solution

1. Develop the differential equations.

   Assume that the rate of consecutive reactions are first order. The differential equations expressing the concentrations (as NH_4-N) of NH_3-N, NO_2-N, and NO_3-N are given by Equations 2.17 through 2.19.

$$\frac{dC_{NH_3\text{-N}}}{dt} = -k_1 C_{NH_3\text{-N}} \tag{2.17}$$

$$\frac{dC_{NO_2\text{-N}}}{dt} = k_1 C_{NH_3\text{-N}} - k_2 C_{NO_2\text{-N}} \tag{2.18}$$

$$\frac{dC_{NO_3\text{-N}}}{dt} = k_2 C_{NO_2\text{-N}} \tag{2.19}$$

2. Express the solutions of the differential equations.

   It is assumed that at $t = 0$, the initial concentration of NH_3-N, NO_2-N, and NO_3-N are, respectively, $C^0_{NH_3\text{-N}}$, $C^0_{NO_2\text{-N}}$, and $C^0_{NO_3\text{-N}}$.

   The solutions of the differential equations are given by Equations 2.20 through 2.22.

$$C_{NH_3\text{-N}} = C^0_{NH_3\text{-N}} \cdot e^{-k_1 t} \tag{2.20}$$

$$C_{NO_2\text{-N}} = \frac{k_1 C^0_{NH_3\text{-N}}}{k_2 - k_1}\left(e^{-k_1 t} - e^{-k_2 t}\right) + C^0_{NO_2\text{-N}} \cdot e^{-k_2 t} \tag{2.21}$$

$$C_{NO_3\text{-N}} = C^0_{NH_3\text{-N}}\left(1 - \frac{k_2 \cdot e^{-k_1 t} - k_1 \cdot e^{-k_2 t}}{k_2 - k_1}\right) + C^0_{NO_2\text{-N}}\left(1 - e^{-k_2 t}\right) + C^0_{NO_3\text{-N}} \tag{2.22}$$

EXAMPLE 2.26: GRAPHICAL REPRESENTATION OF NITRIFICATION REACTIONS

In a wastewater treatment plant, nitrification is carried out by oxidation of NH_3. The initial concentration of ammonia nitrogen (NH_3-N) is 25 mg/L, and the initial concentration of nitrite nitrogen (NO_2-N) and nitrate nitrogen (NO_3-N) are zero. Plot the concentration profiles of NH_3-N, NO_2-N, and NO_3-N with respect to time in days. The reaction rate constants k_1 and k_2 at 20°C are 0.6 d^{-1} and 0.3 d^{-1}, respectively. The operating temperature in the nitrification facility is 15°C. Assume $\theta = 1.047$.

Solution

1. Determine the reaction rate constants at 15°C from Equation 2.10.

$$(k_1)_{15} = 0.6\,d^{-1} \times (1.047)^{(15-20)} = 0.48\,d^{-1}$$

$$(k_2)_{15} = 0.3\,d^{-1} \times (1.047)^{(15-20)} = 0.24\,d^{-1}$$

2. Calculate the concentrations of $C_{NH_3\text{-N}}$, $C_{NO_2\text{-N}}$, and $C_{NO_3\text{-N}}$ at different time intervals t in days.

   Calculate these concentrations from Equations 2.20 through 2.22.

   At $t = 0$, NH_3-N = 25 mg/L, and NO_2-N and NO_3-N = 0.

The calculation steps at $t = 1$ day are given below.

$$C_{NH_3\text{-}N} = C^0_{NH_3\text{-}N} \cdot e^{-k_1 t} = 25\,\text{mg/L} \times e^{-(0.48\,\text{d}^{-1}) \times 1\,\text{d}} = 15.5\,\text{mg/L}$$

$$C_{NO_2\text{-}N} = \frac{k_1 C^0_{NH_3\text{-}N}}{k_2 - k_1}\left(e^{-k_1 t} - e^{-k_2 t}\right) + C^0_{NO_2\text{-}N} \cdot e^{-k_2 t}$$

$$= \frac{0.48\,\text{d}^{-1} \times 25\,\text{mg/L}}{(0.24 - 0.48)\text{d}^{-1}}\left(e^{-(0.48\,\text{d}) \times 1\,\text{d}} - e^{-(0.24\,\text{d}) \times 1\,\text{d}}\right) + 0$$

$$= -50\,\text{mg/L} \times (0.619 - 0.787)$$

$$= 8.4\,\text{mg/L}$$

$$C_{NO_3\text{-}N} = C^0_{NH_3\text{-}N}\left(1 - \frac{k_2 \cdot e^{-k_1 t} - k_1 \cdot e^{-k_2 t}}{k_2 - k_1}\right) + C^0_{NO_2\text{-}N}\left(1 - e^{-k_2 t}\right) + C^0_{NO_3\text{-}N}$$

$$= 25\,\text{mg/L} \times \left(1 - \frac{0.24\,\text{d}^{-1} \times e^{-(0.48\,\text{d}) \times 1\,\text{d}} - 0.48\,\text{d}^{-1} \times e^{-(0.24\,\text{d}) \times 1\,\text{d}}}{(0.24 - 0.48)\text{d}^{-1}}\right) + 0 + 0$$

$$= -25\,\text{mg/L} \times \left(1 - \frac{0.15 - 0.38}{0.24 - 0.28}\right)$$

$$= 1.1\,\text{mg/L}$$

It may be noted that the sum of the concentrations of NH_3-N, NO_2-N, and NO_3-N at all the points is 25 mg/L. This is because all concentrations are expressed in mg/L as NH_4-N.

3. Assume different values of t and calculate the concentrations of NH_3-N, NO_2-N, and NO_3-N at different time intervals. The calculated values are summarized below.

t, d	NH_3-N, mg/L	NO_2-N, mg/L	NO_3-N, mg/L
0	25.0	0	0
1	15.5	8.4	1.1
2	9.6	11.8	3.6
3	5.9	12.5	6.6
4	3.7	11.8	9.5
5	2.3	10.5	12.2
6	1.4	9.0	14.6
7	0.9	7.6	16.5
8	0.5	6.3	18.2
9	0.3	5.1	19.6
10	0.2	4.1	20.7
11	0.1	3.3	21.6
12	0.1	2.6	22.3
13	0.0	2.1	22.8
14	0.0	1.7	23.3

4. Plot the concentration profiles.

The concentration profiles of NH_3-N, NO_2-N, and NO_3-N with respect to time at 20°C are shown in Figure 2.5.

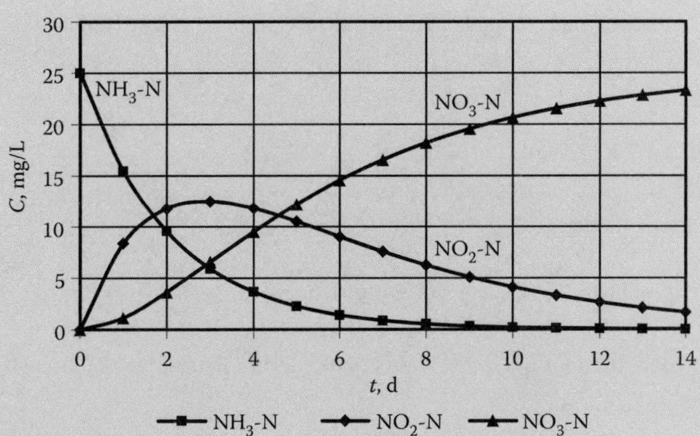

FIGURE 2.5 Concentration profiles of NH_3-N, NO_2-N, and NO_3-N with time (Example 2.26).

EXAMPLE 2.27: PARALLEL IRREVERSIBLE FIRST-ORDER REACTIONS

In a batch reactor, a reactant A produces two products B and C in a competitive reaction. Both reactions follow first-order kinetics with respect to A. Also, the concentrations of B and C at any time are expressed as A. Develop the rate equations and solve the differential equations. Plot the concentration profiles. The initial concentration of A is 45 mg/L and that of B and C are zero. Assume that the reaction coefficients k_1 and k_2 at 20°C are 0.35 and 0.21 per day when all concentrations of reactant and products are expressed in mg/L as reactant A. The reaction vessel temperature is 15°C and $\theta = 1.135$. The parallel reaction is as follows.

$$A \xrightarrow{k_1} B, \quad A \xrightarrow{k_2} C$$

Solution

1. Develop the rate equations.

$$-\frac{dC_A}{dt} = k_1 C_A + k_2 C_A = C_A(k_1 + k_2)$$

$$\frac{dC_B}{dt} = k_1 C_A$$

$$\frac{dC_C}{dt} = k_2 C_A$$

2. Integrate the rate equation and develop concentration equations Equations 2.23 through 2.25.

$$C_A = C_{A_0} e^{-(k_1+k_2)t} \tag{2.23}$$

$$C_B = \frac{k_1}{k_1 + k_2} C_{A_0}\left(1 - e^{-(k_1+k_2)t}\right) \tag{2.24}$$

$$C_C = \frac{k_2}{k_1 + k_2} C_{A_0}\left(1 - e^{-(k_1+k_2)t}\right) \tag{2.25}$$

3. Determine the reaction rate constants k_1 and k_2 at 15°C from Equation 2.10.

$$(k_1)_{15} = 0.35\,\text{d}^{-1} \times (1.135)^{(15-20)} = 0.19\,\text{d}^{-1}$$

$$(k_2)_{15} = 0.21\text{d}^{-1} \times (1.135)^{(15-20)} = 0.11\,\text{d}^{-1}$$

4. Set up the initial conditions.

$$\text{At}\, t = 0,\, C_{A_0} = 45\,\text{mg/L},\, C_{B_0} = 0,\, \text{and}\, C_{C_0} = 0.$$

5. Calculate the concentrations of C_A, C_B, and C_C at 0.5 d from Equations 2.23 through 2.25.

$$C_A = 45\,\text{mg/L} \times e^{-(0.19+0.11)\text{d}^{-1}\times 0.5\text{d}} = 45\,\text{mg/L} \times 0.86 = 38.7\,\text{mg/L}$$

$$C_B = \frac{0.19\,\text{d}^{-1}}{(0.19+0.11)\text{d}^{-1}} 45\,\text{mg/L} \times \left(1 - e^{-(0.19+0.11)\text{d}^{-1}\times 0.5\text{d}}\right)$$
$$= 0.63 \times 45\,\text{mg/L} \times (1 - 0.86) = 4.0\,\text{mg/L}$$

$$C_C = \frac{0.11\,\text{d}^{-1}}{(0.19+0.11)\,\text{d}^{-1}} 45\,\text{mg/L} \times \left(1 - e^{-(0.19+0.11)\text{d}^{-1}\times 0.5\,\text{d}}\right)$$
$$= 0.37 \times 45\,\text{mg/L} \times (1 - 0.86) = 2.3\,\text{mg/L}$$

6. Assume different values of t and tabulate the concentrations of C_A, C_B, and C_C with respect to time. The concentrations of C_B and C_C are expressed as C_A.

t, d	C_A, mg/L	C_B, mg/L	C_C, mg/L
0	45	0.0	0
0.5	38.7	4.0	2.3
1	33.3	7.4	4.3
2	24.7	12.9	7.4
3	18.3	16.9	9.8
4	13.6	19.9	11.5
5	10.0	22.1	12.8
6	7.4	23.8	13.8
7	5.5	25.0	14.5
8	4.1	25.9	15.0
9	3.0	26.6	15.4
10	2.2	27.1	15.7
11	1.7	27.4	15.9
12	1.2	27.7	16
13	0.9	27.9	16.2
14	0.7	28.1	16.3

Note: The sum of concentrations of C_A, C_B, and C_C at any time t is 45 mg/L since all concentrations are expressed in mg/L as reactant A.

7. Plot the concentration profiles of C_A, C_B, and C_C (Figure 2.6).

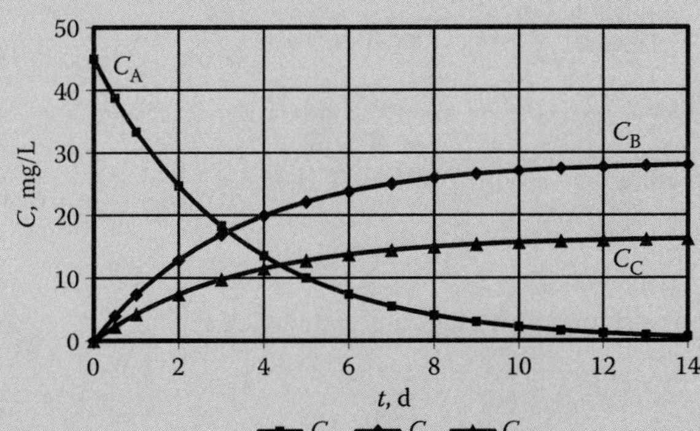

FIGURE 2.6 Time-dependent concentrations of reactant and products in parallel irreversible reaction (Example 2.27).

EXAMPLE 2.28: PSEUDO FIRST-ORDER OR AUTO-CATALYTIC REACTION

A waste stabilization pond has hydrogen sulfide (H_2S) concentration of 17 mg/L. It was found experimentally that the hydrogen sulfide odors are completely eliminated at a residual concentration of 0.034 mg/L. A decision therefore was made to aerate the lake so that H_2S is oxidized to nonodorous sulfate ion in accordance with the following oxidation reaction: $H_2S + 2\ O_2 \rightarrow SO_4^{2-} + 2H^+$. Experimentally it was determined that the overall reaction follows second-order kinetics with respect to both oxygen and hydrogen sulfide concentrations, $d[H_2S]/dt = -k[O_2][H_2S]$. It is expected that 2 mg/L oxygen concentration will be maintained in the lake. Therefore, the reaction can be modified to pseudo first order. The rate constant k for the reaction was determined experimentally and was 1000 L/mole·d. If the aeration completely inhibits the anaerobic respiration, how long would it take for H_2S concentration to reach the desired residual concentration for odor control.

Solution

1. Determine the reaction rate equation.

   Since the concentration of oxygen is maintained constant, the reaction can be modified to pseudo first order. The modified equation is expressed by Equation 2.14 and the H_2S concentration is obtained from Equation 2.15a.

$$\frac{d[H_2S]}{dt} = -k'[H_2S]$$

$$[H_2S] = [H_2S]_0 e^{-k't}$$

   where

   k' = modified reaction rate constant to express $k' = k[O_2]$, L/mole·d

   $[H_2S]$ = equilibrium concentration of H_2S, mole/L

   $[H_2S]_0$ = initial concentration of H_2S, mole/L

2. Calculate the value of the modified reaction rate constant k'.

$$k' = k[O_2] = 1000\frac{\text{L}}{\text{mole·d}} \times 2\,\text{mg/L}\,O_2 \times \frac{1\,\text{g}}{1000\,\text{mg}} \times \frac{1}{32\,\text{g/mole}\,O_2} = 0.0625\,\text{d}^{-1}$$

3. Calculate the molar concentration of initial and equilibrium H_2S concentrations.

$$\text{Initial concentration, } [H_2S]_0 = 17\,\text{mg/L}\,H_2S \times \frac{1\,\text{g}}{1000\,\text{mg}} \times \frac{1}{34\,\text{g/mole}\,H_2S} = 5 \times 10^{-4}\,\text{mole/L}\,H_2S$$

$$\text{Equilibrium concentration, } [H_2S] = 0.034\,\text{mg/L}\,H_2S \times \frac{1\,\text{g}}{1000\,\text{mg}} \times \frac{1}{34\,\text{g/mole}\,H_2S} = 10^{-6}\,\text{mole/L}\,H_2S$$

4. Calculate the time to reach the equilibrium concentration of H_2S.

$$10^{-6}\,\text{mole/L} = 5 \times 10^{-4}\,\text{mole/L} \times e^{-0.0625\,\text{d}^{-1} \times t}$$

$$e^{-0.0625\,\text{d}^{-1} \times t} = 0.002$$

$$t = -\frac{1}{0.0625\,\text{d}^{-1}} \ln(0.002) = -\frac{-6.21}{0.0625\,\text{d}^{-1}} = 99.4\,\text{d} \approx 100\,\text{d}$$

2.5.3 Second-Order Reaction

The second-order reactions are of two types:

a. Type I reactions ($2\,A \rightarrow P$) are expressed by the following equation:

$$-\frac{d[A]}{dt} = k[A]^2$$

b. Type II reactions ($A + B \rightarrow P$) are expressed by the following equation:

$$-\frac{d[A]}{dt} = k[A][B] \quad \text{or} \quad -\frac{d[B]}{dt} = k[A][B],$$

where k = reaction rate constant, L/mole·t

Several examples on both types of second-order reactions are given below.

EXAMPLE 2.29: SECOND-ORDER TYPE I EQUATION DERIVATION AND REACTION RATE CONSTANT

Consider the following reaction carried out in a batch reactor: $A + A \rightarrow P$.

Develop the following: (a) equation expressing the concentration with respect to time, (b) equation expressing half-life, and (c) determination of kinetic coefficients k and the initial concentration C_A.

t, h	0	1	2	3	4	5	6
C_A, mg/L	100	80	67	57	50	44	40
$1/C_A$, L/mg	0.01	0.0125	0.0149	0.0175	0.02	0.0227	0.025

Solution

1. Develop the concentration equation.

$$r = -\frac{dC_A}{dt} = kC_A^2$$

where C_A = concentration of A at time *t*, mg/L

$$\frac{dC_A}{C_A^2} = -k\,dt$$

Integrate and solve for C_A. This is represented by Equation 2.26.

$$\int_{C_{A_0}}^{C_A} \frac{dC_A}{C_A^2} = \int_0^t -k\,dt$$

$$\frac{1}{C_A} - \frac{1}{C_{A_0}} = kt \tag{2.26}$$

2. Develop the half-life expression.

The half-life expression (Equation 2.27) is obtained by substituting $C_A = (1/2)C_{A_0}$.

$$t_{1/2} = \frac{1}{kC_{A_0}} \tag{2.27}$$

3. Determine *k* and C_{A_0}.

A plot of $1/C_A$ versus *t* gives a linear relationship as shown in Figure 2.7.

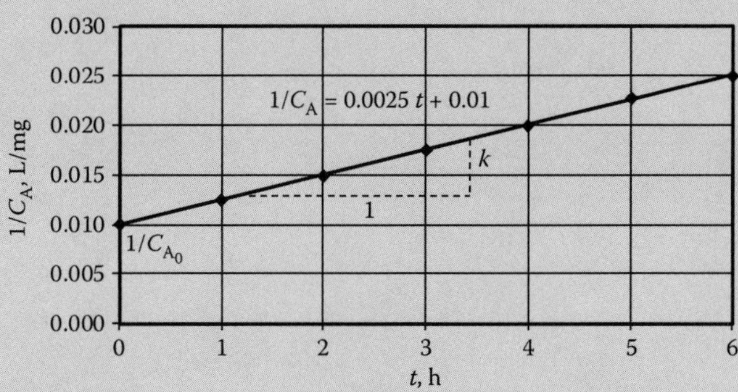

FIGURE 2.7 Determination of reaction rate constant of second-order Type I reaction (Example 2.29).

The slope of the line is *k* and the intercept is $1/C_{A_0}$.

$$C_{A_0} = \frac{1}{0.01\,\text{L/mg}} = 100\,\text{mg/L}$$

$$k = 0.0025\,\text{L/mg·h}$$

4. Determine the half-life from Equation 2.27.

$$t_{1/2} = \frac{1}{0.0025\,\text{L/mg·h} \times 100\,\text{mg/L}} = 4\text{h}$$

EXAMPLE 2.30: SECOND-ORDER DECAY

Benzene concentration in an industrial sludge is 10 ppb. What will be the concentration remaining after 90 days storage? Assume second-order decay, and reaction rate constant is 0.00085 L/μg·d.

Solution

Determine the concentration remaining from Equation 2.26.

$$\frac{1}{C_A} = -\frac{1}{C_{A_0}} + kt$$

$$\frac{1}{C_A} = \frac{1}{10\,\mu\text{g/L}} + 0.00085\frac{\text{L}}{\mu\text{g}\cdot\text{d}} \times 90\,\text{d} = (0.1 + 0.0765)\frac{\text{L}}{\mu\text{g}} = 0.1765\frac{\text{L}}{\mu\text{g}}$$

$$C_A = 5.7\,\mu\text{g}/L\ (\text{ppb})$$

EXAMPLE 2.31: REACTION ORDER AND REACTION RATE CONSTANT

A chemical reaction was carried out in a batch reactor. The experimental data are given below. Determine the order of reaction and the reaction rate constant.

Time, min	0	1	2	3	4	5	6
C_A, mg/L	100	80	67	57	50	44	40

Solution

1. Use Equation 2.5 for two data points.

$$-\frac{dC_{A_1}}{dt} = kC_{A_1}^n$$

$$-\frac{dC_{A_2}}{dt} = kC_{A_2}^n$$

2. Take log on both sides.

$$\log\left[-\frac{dC_{A_1}}{dt}\right] = \log(k) + n\log(C_{A_1})$$

$$\log\left[-\frac{dC_{A_2}}{dt}\right] = \log(k) + n\log(C_{A_2})$$

3. Equate log k in both equations to solve for n.

$$n = \frac{\log\left[-\frac{dC_{A_1}}{dt}\right] - \log\left[-\frac{dC_{A_2}}{dt}\right]}{\log(C_{A_1}) - \log(C_{A_2})}$$

4. Select two data points $t = 2$ and $t = 4$ to determine the differentials.

$$-\frac{dC_{t=2}}{dt} = -\frac{C_{t=2+1} - C_{t=2-1}}{t_{2+1} - t_{2-1}} = -\frac{57 - 80}{3 - 1} = 11.5$$

$$-\frac{dC_{t=4}}{dt} = -\frac{C_{t=4+1} - C_{t=4-1}}{t_{4+1} - t_{4-1}} = -\frac{44 - 57}{5 - 3} = 6.5$$

5. Calculate n and reaction order.

$$n = \frac{\log\left[-\frac{dC_{t=2}}{dt}\right] - \log\left[-\frac{dC_{t=4}}{dt}\right]}{\log(C_{t=2}) - \log(C_{t=4})} = \frac{\log(11.5) - \log(6.5)}{\log(67) - \log(50)} = \frac{1.061 - 0.813}{1.826 - 1.699} = \frac{0.248}{0.127} = 1.95 \approx 2$$

The reaction is Type I second-order and Equation 2.5 with $n = 2$ is valid.

EXAMPLE 2.32: EFFECT OF REACTION ON SPECIES CONCENTRATION

Many photochemical reactions occur in the atmosphere that are related with smog formation. Following are three example reactions:

$$NO_2 + OH \xrightarrow{k_1} HNO_3 \qquad \text{(Reaction 1)}$$

$$2\,HO_2 \xrightarrow{k_2} H_2O_2 + O_2 \qquad \text{(Reaction 2)}$$

$$NO_2 + O_3 \xrightarrow{k_3} NO_3^- + O_2 \qquad \text{(Reaction 3)}$$

Write differential equations to describe the net effect of these three reactions on the mole fractions of nitrogen dioxide (NO_2) and hydroperoxy radical ($HO_2^\bullet$). Assume that reactions occur under well-mixed condition and are second order.

Solution

1. Write the rate law expressions.

$$r_1 = k_1 C_{NO_2} C_{OH^\bullet}$$
$$r_2 = k_2 \left(C_{HO_2^\bullet}\right)^2$$
$$r_3 = k_3 C_{NO_2} C_{O_3}$$

2. Write the differential equations of NO_2 in Reaction 1.

   NO_2 and $OH^\bullet$ are consumed, and nitric acid (HNO_3) is produced.

$$\frac{dC_{NO_2}}{dt} = -r_1 = -k_1 C_{NO_2} \times C_{OH^\bullet}$$

3. Write the differential equation of $HO_2^\bullet$ in Reaction 2.

   $HO_2^\bullet$ is consumed and converted to hydrogen peroxide (H_2O_2) and oxygen (O_2).

$$\frac{dC_{HO_2^\bullet}}{dt} = -2r_2 = -2k_2 \left(C_{HO_2^\bullet}\right)^2$$

In this reaction, two moles of $HO_2^\bullet$ are consumed. The exponent 2 expresses second-order reaction of $HO_2^\bullet$.

4. Write the differential equation of NO_2 in Reaction 3.

   NO_2 reacts with ozone (O_3) to produce nitrate (NO_3^-) and oxygen (O_2).

$$\frac{dC_{NO_2}}{dt} = -r_3 = -k_3 C_{NO_3^-} \times C_{O_3}$$

5. Write the differential equation of NO_2 in Reactions 1 and 3.

   NO_2 is consumed in both Reactions 1 and 3. Therefore, the effects of each reaction are added.

$$\frac{dC_{NO_2}}{dt} = -r_1 - r_3 = -k_1 C_{NO_2} C_{OH^\bullet} - k_3 C_{NO_2} C_{O_3}$$

EXAMPLE 2.33: TYPE II SECOND-ORDER REACTION, DERIVATION OF EQUATION

Consider the following reaction is carried out in a batch reactor. $A + B \xrightarrow{k} P$. Develop the equation expressing the concentration of reactants A and B with respect to time.[1]

Solution

1. Write the Type II second-order reaction.

$$-\frac{dC_A}{dt} = -\frac{dC_B}{dt} = kC_A C_B$$

2. Use the stoichiometric relationships.

$$dC_A = dC_B$$

$$\int_{C_{A_0}}^{C_A} dC_A = \int_{C_{B_0}}^{C_B} dC_B$$

$$C_{A_0} - C_A = C_{B_0} - C_B$$

   where C_{A_0} and C_{B_0} are initial concentrations.

$$C_B = C_{B_0} - C_{A_0} + C_A$$

$$-\frac{dC_A}{dt} = kC_A(C_{B_0} - C_{A_0} + C_A)$$

$$-\frac{dC_A}{C_A(C_{B_0} - C_{A_0} + C_A)} = kdt$$

3. Using the method of partial fraction, integrate the equation.

$$\frac{1}{C_A(C_{B_0} - C_{A_0} + C_A)} \equiv \frac{p}{C_A} + \frac{q}{C_{B_0} - C_{A_0} + C_A},$$

   where p and q are constants, dimensionless

   Note: The identity symbol (≡) is used to indicate an equality between the two sides of the equation for all values of C_A.

$$\frac{C_A(C_{B_0} - C_{A_0} + C_A)}{C_A(C_{B_0} - C_{A_0} + C_A)} \equiv \frac{p}{C_A}\left(\frac{C_A(C_{B_0} - C_{A_0} + C_A)}{1}\right) + q\left(\frac{(C_{B_0} - C_{A_0} + C_A)C_A}{(C_{B_0} - C_{A_0} + C_A)}\right)$$

$$1 \equiv p(C_{B_0} - C_{A_0} - C_A) + qC_A$$

Rearrange the equation.

$$1 + 0 \equiv p(C_{B_0} - C_{A_0}) + C_A(p + q)$$

Equate constant terms.

$$1 = p\,(C_{B_0} - C_{A_0})$$

Equate coefficient of C_A.

$$0 = p + q$$

Solve these equations for q and p.

$$p = \frac{1}{(C_{B_0} - C_{A_0})}$$

$$q = -\frac{1}{(C_{B_0} - C_{A_0})}$$

Substitute these values in the equation $\left(-\dfrac{dC_A}{C_A(C_{B_0} - C_{A_0} + C_A)} = kdt\right)$.

$$\left(\frac{p}{C_A} + \frac{q}{C_{B_0} - C_{A_0} + C_A}\right)(-\,dC_A) = kdt$$

$$-\frac{p\,dC_A}{C_A} + \frac{-q\,dC_A}{C_{B_0} - C_{A_0} + C_A} = kdt$$

$$-\frac{dC_A}{C_A(C_{B_0} - C_{A_0})} + \frac{dC_A}{(C_{B_0} - C_{A_0})(C_{B_0} - C_{A_0} + C_A)} = kdt$$

Integrating the above equation to obtain Equation 2.28.

$$\frac{1}{C_{B_0} - C_{A_0}}\ \ln\left(\frac{C_{A_0}(C_{B_0} - C_{A_0} + C_A)}{C_{B_0}C_A}\right) = kt \qquad (2.28)$$

4. Draw the linear relationship.
   The above equation presents a linear relationship. The plot is shown in Figure 2.8. Slope of the line is the reaction rate constant k.

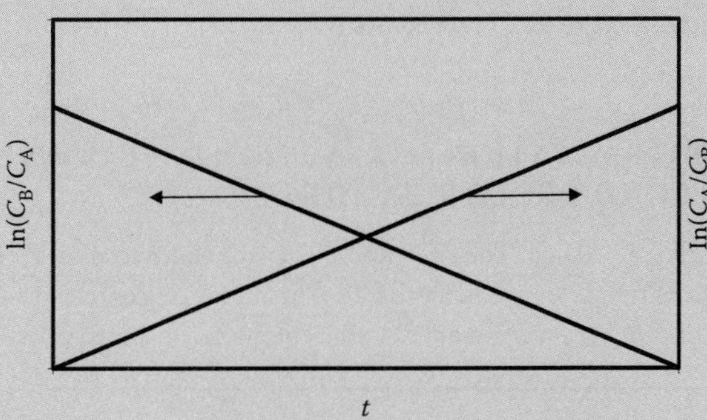

FIGURE 2.8 Linear plot of second-order Type II reaction (Example 2.33).

EXAMPLE 2.34: SECOND-ORDER TYPE II REACTION WITH A VARIABLE CONCENTRATION

The second-order Type II reaction rate can be simplified by expressing the result in terms of a variable concentration C_X of the reactants consumed. Develop the kinetic equation.

Solution

1. State the Type II second-order reaction.

$$A + B \rightarrow P$$

2. Write the equation to express the variable C_X representing the decrease in concentration of the reactants in a given time.

$$\frac{dC_X}{dt} = kC_A C_B = k(C_{A_0} - C_X)(C_{B_0} - C_X)$$

3. Integrate the equation and express variable concentration C_X with time by Equation 2.29.

$$\frac{1}{C_{B_0} - C_{A_0}} \ln\left(\frac{C_{A_0}(C_{B_0} - C_X)}{C_{B_0}(C_{A_0} - C_X)}\right) = kt \tag{2.29}$$

4. Plot the linear relationship.
   The linear relationship between variable concentration C_X and time is shown in Figure 2.9.

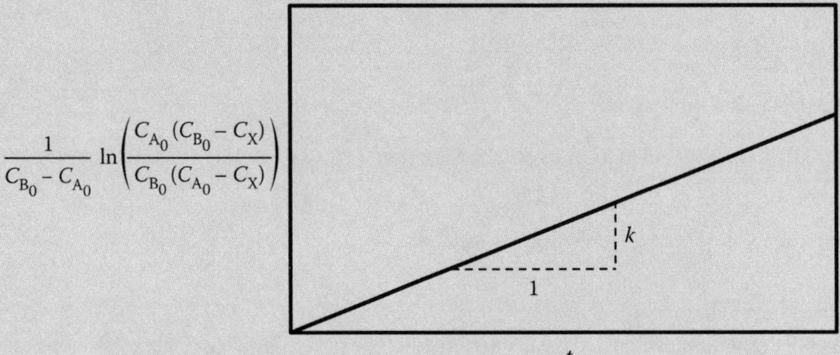

FIGURE 2.9 Linear plot of Type II second-order reaction with decreasing concentrations (Example 2.34).

The slope of the line is the reaction rate constant k.

EXAMPLE 2.35: DETERMINATION OF REACTION RATE CONSTANT OF TYPE II SECOND-ORDER REACTION

Reactants A and B produce a product. The decrease in reactant concentration C_x is variable with time. Determine the kinetic coefficients k in Equation 2.29. The initial concentrations of reactants A and B are 100 and 98 mg/L. The experimental values of C_x concentration consumed with time t are given below.

t, s	4	8	16	20
C_x, mg/L	90.3	94.6	96.7	97.1

Solution

1. Tabulate calculation results in the table below.

$C_{A_0} = 100\,mg/L$ and $C_{B_0} = 98\,mg/L$

t, s	0	4	8	16	20
C_X, mg/L	0	90.3	94.6	96.7	97.1
$C_{A_0} - C_X$, mg/L	100	9.7	5.4	3.3	2.9
$C_{B_0} - C_X$, mg/L	98	7.7	3.4	1.3	0.9
$C_{A_0}(C_{B_0} - C_X)$, mg²/L²	9800	770	340	130	90
$C_{B_0}(C_{A_0} - C_X)$, mg²/L²	9800	951	529	323	284
$\ln\left(\frac{C_{A_0}(C_{B_0} - C_X)}{C_{B_0}(C_{A_0} - C_X)}\right)$, dimensionless	0.000	−0.211	−0.442	−0.911	−1.15
$\frac{1}{C_{B_0} - C_{A_0}}\ln\left(\frac{C_{A_0}(C_{B_0} - C_X)}{C_{B_0}(C_{A_0} - C_X)}\right)$, L/mg	0.000	0.105	0.221	0.456	0.575
$k = \frac{1}{C_{B_0} - C_{A_0}}\frac{1}{t}\ln\left(\frac{C_{A_0}(C_{B_0} - C_X)}{C_{B_0}(C_{A_0} - C_X)}\right)$, L/mg·s	–	0.0263	0.0277	0.0285	0.0287
					Average = 0.028

2. Plotting the values (Figure 2.10).

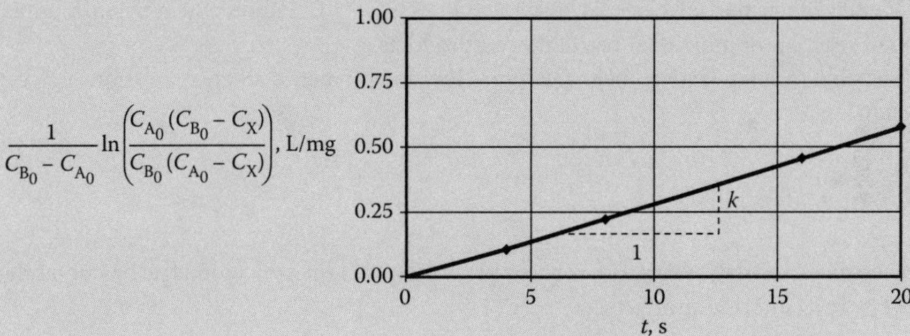

FIGURE 2.10 A linear plot of $\frac{1}{C_{B_0} - C_{A_0}}\ln\left(\frac{C_{A_0}(C_{B_0} - C_X)}{C_{B_0}(C_{A_0} - C_X)}\right)$ versus t (Example 2.35).

Equation 2.29 is a linear relationship and the slope of the line gives the kinetic coefficient k.

3. Determine the kinetic coefficient k.

The slope of the line is kinetic coefficient k.

The value of k can be obtained from either the calculations in the table or the plot in Figure 2.10.

$k = 0.028\,L/mg{\cdot}s$

Discussion Topics and Review Problems

2.1 A photoprocessing industry discharges 1700 m^3/d waste that contains 130 mg/L $AgNO_3$. Silver is precipitated by adding NaCl. The solubility of AgCl is very small.

$$AgNO_3 + NaCl \rightarrow AgCl\downarrow + NaNO_3$$

Calculate (a) the theoretical quantity of NaCl in kg/d, and (b) kg/d silver recovered.

2.2 Well water has carbonate hardness due to calcium. If 160 mg/L hardness as $CaCO_3$ is precipitated by adding lime, calculate lime dose in mg/L.

2.3 Recarbonation is necessary after water softening to solubilize unsettled $CaCO_3$ and $Mg(OH)_2$. The reactions are:

$$CaCO_3 + CO_2 + H_2O \rightarrow Ca(HCO_3)_2$$

$$Mg(OH)_2 + CO_2 \rightarrow MgCO_3 + H_2O$$

If 15 mg/L $CaCO_3$ and 17 mg/L $Mg(OH)_2$ concentrations are solubilized, calculate the stoichiometric dose of CO_2.

2.4 A single irreversible reaction is expressed by the following balanced reaction. Express the reaction rates of A and B in terms of C.

$$2\text{ A} + 3\text{ B} \rightarrow 3\text{ C}$$

2.5 A single irreversible reaction is given below. Determine the ratio of reaction rates.

$$3\text{ A} \rightarrow 2\text{ P}$$

2.6 A single reactant is being converted to a single product by the following reaction: $2\text{ A} \rightarrow \text{P}$. If C_A is the concentration of A at any time t, express (a) the disappearance of A with respect to time, and (b) the order of reaction.

2.7 A generalized reaction is expressed by $2\text{ A} + 3\text{ B} \rightarrow 3\text{ C} + 2\text{ D}$. Express the relationships between the rates of reaction of individual reactants and products.

2.8 A reversible reaction is given below. Express the stoichiometric reaction relationships between reactants.

$$2\text{ A} \underset{k_1}{\overset{k_2}{\longleftrightarrow}} 3\text{ B}$$

2.9 A reversible consecutive reaction is given below. Express the stoichiometric reaction relationships between reactants and products.

$$a\text{ A} \underset{k_1'}{\overset{k_1}{\longleftrightarrow}} b\text{ B} \underset{k_2'}{\overset{k_2}{\longleftrightarrow}} c\text{ C}$$

2.10 An organic waste is treated in a batch reactor. Concentration of organic waste is measured at different time intervals. The results are given below. Determine (a) the order of the reaction, and (b) the values of kinetic coefficients, (c) the kinetic equation, and (d) half-life.

t, h	0	1	2	3	5	10	15
Concentration, mg/L	500	475	450	425	375	250	120

2.11 A batch reactor is used to develop the reaction rate constant. The reactor is operated with an organic substrate. The concentration of the substrate is measured on time interval. The experimental data are tabulated below. Determine (a) reaction order, (b) reaction rate constants by numerical and graphical methods, and (c) half-life.

t, min	0	2	4	6	10	15	20
Concentration, mg/L	180	105	61	37	12	3	1

2.12 Biological studies were conducted at different substrate loadings to determine the reaction rate constant k. Following data were developed. Determine the value of half-saturation constant K_s.

Substrate concentration S, mg/L	5	10	15	20	25	30	35	45
K, d^{-1}	0.7	1.30	1.75	2.00	2.20	2.30	2.40	2.47

2.13 The reaction rate constants were developed for an enzymatic saturation reaction. The reaction rate constant k was 1800 mg/L.d and half saturation constant was 800 mg/L. Calculate the product formation rate if 2000 mg/L reactant concentration is maintained in the reactor.

2.14 The reaction rate of an autocatalytic reaction A → B is expressed by

$$-\frac{dC_A}{dt} = k(C_A)^{\alpha}.$$

Determine the concentration of C_A at 2nd min for zero-order, first-order, and second-order reactions. The initial concentration (C_{A_0}) is 50 mg/L. The reaction rate constants k for zero-, first-, and second-order reactions are 0.5 mg/L·min, 0.5 min^{-1}, and 0.5 L/mg·min.

2.15 A batch-reactor study gave the following result for the reaction A → P. Determine (a) the reaction order, (b) rate constant, and (c) the reaction time to achieve 90% stabilization.

t, min	0	1	2	3	5	10	20
C_A, mg/L	100	6.3	3.2	2.2	1.3	0.7	0.3

2.16 A chemical reaction is carried out in a batch reactor. The reaction is expressed by equation

$$-\frac{dC_A}{dt} = k(C_{A_0} - C_A)^2.$$

Calculate the value of reaction rate constant k. $C_{A_0} = 150$ mg/L and C_A after 6 h is 20 mg/L.

2.17 In Problem (2.16) if the reaction is expressed by

$$-\frac{dC_A}{dt} = k(C_{A_0} - C_A),$$

determine the value of k. Use the data given in the problem.

References

1. Tchobanoglous, G. and E. D. Schroeder, *Water Quality Characteristics, Modeling and Modification*, Addison-Wesley Publishing Co., Reading, MA, 1985.
2. Metcalf & Eddy, Inc., *Wastewater Engineering: Treatment, and Reuse*, 4th ed., McGraw-Hill Book Co., New York, NY, 2003.
3. Reynolds, T. D. and P. A. Richards, *Unit Operations and Processes in Environmental Engineering*, 2nd ed., PWS Publishing Co., Boston, MA, 1996.
4. Atkins, P. W. and J. De Paula, *Physical Chemistry for the Life Science*, W. H. Freeman & Co., New York, NY, 2006.

3

Mass Balance and Reactors

3.1 Chapter Objectives

Mass balance analysis, and type of reactors and their behavior are used to design treatment plants and evaluate process performance. The objectives of this chapter are to present:

- The principles of mass balance analysis in wastewater treatment
- Flow regime and types of reactors: batch, continuous-flow stirred-tank reactors (CFSTRs) and plug flow reactors (PFRs)
- Comparative performance of CFSTRs and PFRs
- Performance of CFSTERs in series, and performance of PFRs with dispersion and conversion
- Equalization of flow and mass loadings

3.2 Mass Balance Analysis

The law of conservation of mass states that mass can neither be created nor destroyed. Mass balance analyses are routinely used in environmental engineering. To apply mass balance analysis it is necessary to establish a system boundary, which is an imaginary barrier drawn around the system. The system boundary may surround a node (junction), reactor, container, or a process diagram. Proper selection of system boundary is extremely important to identify all flows and masses into and out of the system.[1,2] The generalized mass balance is expressed by the following statement and equations (Equations 3.1 and 3.2a).

$$\begin{bmatrix}\text{Accumulation}\\ \text{rate}\end{bmatrix} = \begin{bmatrix}\text{Input}\\ \text{rate}\end{bmatrix} - \begin{bmatrix}\text{Output}\\ \text{rate}\end{bmatrix} + \begin{bmatrix}\text{Utilization or}\\ \text{conversion rate}\end{bmatrix} \tag{3.1}$$

$$\frac{dm_A}{dt} = V\frac{dC_{Ai}}{dt} = \sum_{i=1}^{n}(Q_i C_{Ai}) - \sum_{j=1}^{m}(Q_j C_{Aj}) + r_A V \tag{3.2a}$$

where

m_A = mass of species A, mass

Q_i = volumetric flow rate of the species entering the system through line i ($i = 0, 1, \ldots, n$), volume/time

C_{Ai} = concentration of species A entering the system through line i ($i = 0, 1, \ldots, n$), mass/volume

Q_j = volumetric flow rate of the species leaving the system through line j ($j = 0, 1, \ldots, m$), volume/time

C_{Aj} = concentration of species A leaving the system through line j ($j = 0, 1, \ldots, m$), mass/volume

r_A = reaction rate of species A, mass/volume·time

V = volume of reactor, volume

For a system at steady state with no accumulation, the time-dependent term (dC_{Ai}/dt) goes to 0, and Equation 3.2a reduces to Equation 3.2b.

$$-r_A V = \sum_{i=1}^{n} (Q_i C_{Ai}) - \sum_{j=1}^{m} (Q_j C_{Aj}) \tag{3.2b}$$

3.2.1 Procedure for Mass Balance Analysis

Compounds with no chemical formation or loss within the control volume are termed *conservative* (i.e., mass is truly conserved). These compounds are not affected by chemical or biological reactions. Examples are chloride, fluoride, sodium, and tracer dyes. *Nonconservative* compounds such as BOD, COD, TKN undergo consumption or generation. To apply mass balance analysis around the system boundary, the following steps must be followed:

1. Draw system boundary, and identify the volumetric flow rate into and out of the system by arrows. All mass flows that are known or to be calculated must cross the system boundary.
2. Assume that the liquid volume within the system does not change.
3. Assume that the control volume is well mixed.
4. Determine whether the compound being balanced is conservative ($r_A = 0$) or nonconservative (r_A must be determined based on reaction kinetics).
5. Determine whether the process is steady state ($dC/dt = 0$) or nonsteady state ($dC/dt \neq 0$).
6. Solve the problem. This will require solution of a differential equation for nonsteady-state condition, and algebraic solution for steady-state condition.

EXAMPLE 3.1: SIMPLIFIED MASS BALANCE EQUATION

Simplify Equation 3.2a for a single stream entering and leaving the reaction vessel. Assume first-order reaction rate ($r_A = -kC$).

Solution

1. Draw the reactor and system boundary (Figure 3.1).

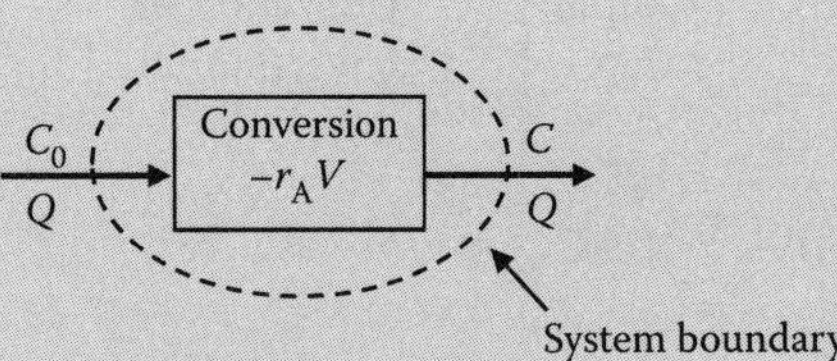

FIGURE 3.1 Reactor and system boundary (Example 3.1).

2. Write the mass balance (Equation 3.3).

$$V\frac{dC}{dt} = QC_0 - QC + V(-kC) \tag{3.3}$$

where k = reaction rate constant (first order), time^{-1}.

Other variables are defined earlier, and are shown in Figure 3.1.

EXAMPLE 3.2: FLUID-FLOW SYSTEM

Fluid enters and leaves a reactor. Assuming no accumulation of fluid $(d\bar{\rho}V/dt) = 0$, and fluid is neither produced nor lost in the system $(r_m = 0)$. Prove that $Q_{in} = Q_{out}$, if the fluid is incompressible.

Solution

1. Draw the reactor and system boundary (Figure 3.2).

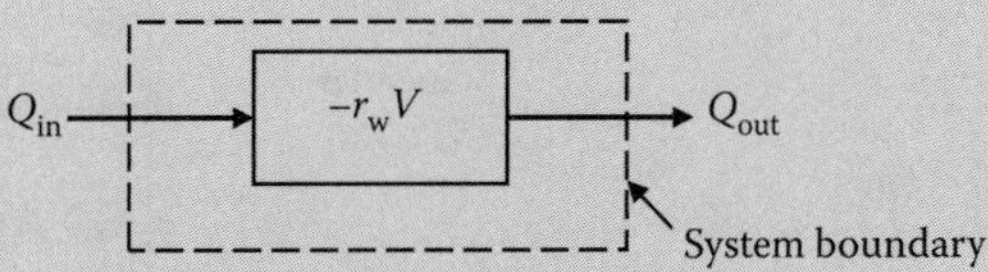

FIGURE 3.2 Reactor and system boundary (Example 3.2).

2. Write the mass balance equation.

$$\frac{d(\bar{\rho}V)}{dt} = Q_{in}\rho_1 - Q_{out}\rho_2 + r_m V$$

where

$\bar{\rho}$ = mean density of fluid in the control reactor, kg/m^3
ρ_1 and ρ_2 = density of the fluid entering and leaving the control reactor, kg/m^3
Q_{in} and Q_{out} = volumetric flow rate in and out of the control volume, m^3/s
V = volume of control reactor, m^3
r_m = mass rate of generation, $g/m^3\ s$

3. Given $d(\bar{\rho}V)/dt = 0$ and $r_w = 0$; write simplified equation.

$$Q_{in}\rho_1 = Q_{out}\rho_2$$

Since the fluid is incompressible, $\rho_1 = \rho_2$

$$Q_{in} = Q_{out}$$

3.2.2 Combining Flow Streams of a Single Material

A system boundary may receive several influent lines, and may have one or more effluent lines. Flow lines presenting one material may be combined in Equation 3.4.

$$\begin{bmatrix}\text{Rate of flow}\\ \text{change}\end{bmatrix} = \begin{bmatrix}\text{Rate of}\\ \text{flow in}\end{bmatrix} - \begin{bmatrix}\text{Rate of}\\ \text{flow out}\end{bmatrix} + \begin{bmatrix}\text{Rate of volume}\\ \text{increased}\end{bmatrix} - \begin{bmatrix}\text{Rate of volume}\\ \text{reduced}\end{bmatrix} \quad (3.4)$$

$$0 = \text{Flow in} - \text{Flow out} + 0 - 0$$

$$\text{Flow in} = \text{Flow out}$$

EXAMPLE 3.3: SEWERS FLOW AND JUNCTION BOXES

Intercepting sewers receive flows from several laterals and then discharges into a trunk line. The flow lines are shown in Figure 3.3. Determine the flow in the final trunk line.

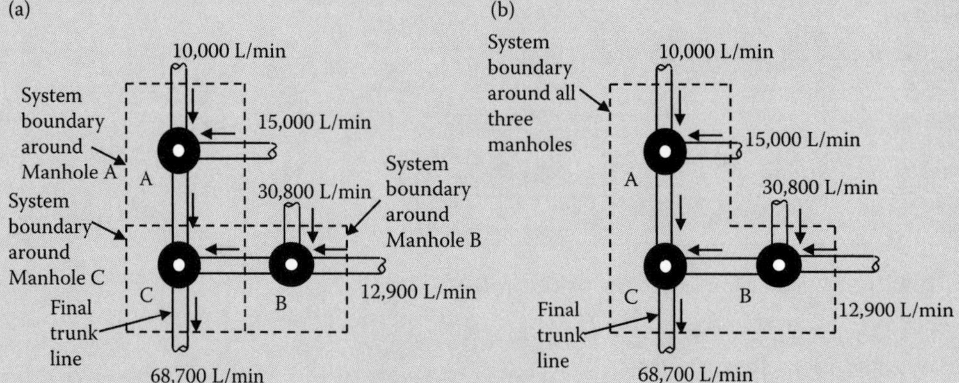

FIGURE 3.3 Sewer layout and system boundary: (a) Method 1 and (b) Method 2 (Example 3.3).

Solution

Method 1.

Draw three system boundaries.

1. Conduct flow balance at Manhole A and find flow in outgoing intercepting sewer.

$$\text{Flow out into the interceptor} = (10{,}000 + 15{,}000)\ \text{L/min} = 25{,}000\ \text{L/min}$$

2. Conduct flow balance at Manhole B and find flow in outgoing intercepting sewer.

$$\text{Flow out into the interceptor} = (30{,}800 + 12{,}900)\ \text{L/min} = 43{,}700\ \text{L/min}$$

3. Conduct flow balance at Manhole C and determine flow in the final trunk sewer.

$$\text{Flow in trunk line} = (25{,}000 + 43{,}000)\ \text{L/min} = 68{,}700\ \text{L/min}$$

Method 2.

Draw one system boundary around three manholes.

$$\text{Flow in the final trunk line} = (10{,}000 + 15{,}000 + 30{,}800 + 12{,}900)\ \text{L/min} = 68{,}700\ \text{L/min}$$

EXAMPLE 3.4: HYDROLOGICAL CYCLE AND GROUNDWATER RECHARGE

A 400 ha farm receives 100 cm precipitation per year. It is estimated that 40% precipitation returns into the atmosphere by evaporation, and 20% reaches the nearest watercourse as runoff. The remaining precipitation is percolated into the aquifer. The water is withdrawn from the aquifer throughout the year for irrigation purposes. Approximately 80% of the withdrawn groundwater is eventually lost as evapotranspiration. Calculate the amount of groundwater recharge in m^3 that can be withdrawn annually for irrigation without depleting the groundwater reservoir. Also, draw the hydrological cycle with water components.

Solution

1. Determine the annual precipitation.

$$\text{Precipitation volume} \quad = \frac{100\ \text{cm/year}}{100\ \text{cm/m}} \times 400\ \text{ha} \times 10{,}000\ \text{m}^2/\text{ha} = 4 \times 10^6\ \text{m}^3/\text{year}$$

2. Determine the components of water budget.

$$\text{Runoff} \quad = 0.2 \times (4 \times 10^6\ \text{m}^3/\text{year}) = 0.8 \times 10^6\ \text{m}^3/\text{year}$$

$$\text{Evaporation of precipitation} \quad = 0.4 \times (4 \times 10^6\ \text{m}^3/\text{year}) = 1.6 \times 10^6\ \text{m}^3/\text{year}$$

$$\text{Assume pumping rate for irrigation} = Q\ \text{m}^3/\text{year}$$

$$\text{Evaporation of irrigation water} \quad = 0.8\ Q\ \text{m}^3/\text{year}$$

3. Conduct the volume balance.

$$[\text{Accumulation}] = [\text{Precipitation}] - [(\text{Evaporation of precipitation}) + (\text{Runoff})] - [\text{Evaporation of irrigation water}]$$

$$= 4 \times 10^6 \text{m}^3/\text{year} - (1.6 \times 10^6 \text{m}^3/\text{year} + 0.8 \times 10^6\ \text{m}^3/\text{year}) - 0.8\,Q$$

Since annual withdrawal is without depletion of groundwater reservoir, net accumulation = 0.

$$0 \quad = 1.6 \times 10^6\ \text{m}^3/\text{year} - 0.8\,Q$$

$$Q \quad = 2 \times 10^6\ \text{m}^3/\text{year}$$

4. Draw the hydrological cycle showing water components in Figure 3.4.

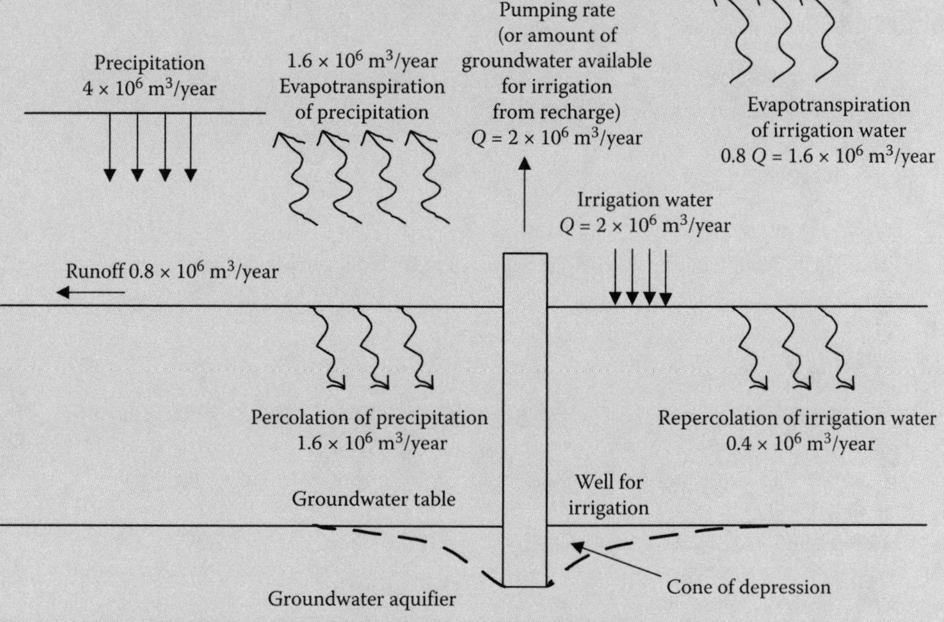

FIGURE 3.4 Hydrological cycle showing water budget components (Example 3.4).

3.2.3 Combining a Conservative Substance and Flow

If a system boundary receives flow streams that also contain a conservative material, a mass balance analysis will include flows and concentrations.

EXAMPLE 3.5: DISCHARGING A CONSERVATIVE SUBSTANCE IN A NATURAL STREAM

An industry is discharging waste brine into a stream. The concentration of total dissolved solids (TDS) in the industrial brine is 15,600 mg/L. The allowable concentration of TDS in the stream is 500 mg/L. The discharge of the stream under drought conditions is 8500 m^3/d and background TDS concentration in the stream is 210 mg/L. Calculate the permissible discharge of industrial brine into the stream.

Solution

1. Draw the system boundary (Figure 3.5).

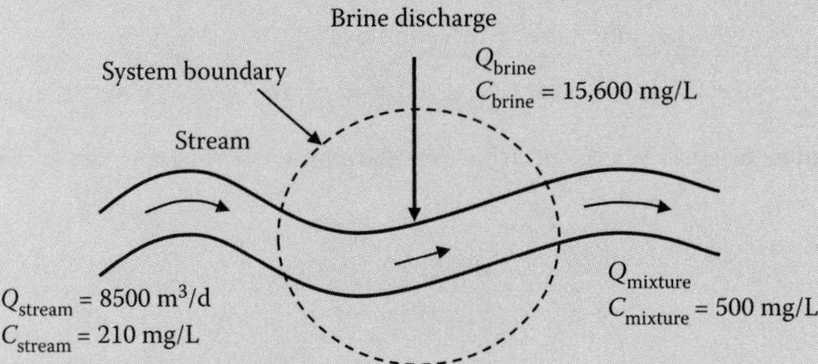

FIGURE 3.5 System boundary (Example 3.5).

2. Apply the mass balance equation (Equation 3.2a).

$$V\frac{dC}{dt} = \sum QC_{in} - \sum QC_{out} + r_A V$$

There is no accumulation $V(dC/dt) = 0$ and there is no conversion ($rV = 0$).

$$Q_{stream}C_{stream} + Q_{brine}C_{brine} - Q_{mixture}C_{mixture} = 0$$

$$8500\ m^3/d \times 210\ g/m^3 + Q_{brine} \times 15{,}600\ g/m^3 - (8500\ m^3/d + Q_{brine}) \times 500\ g/m^3 = 0$$

$$17.85 \times 10^5\ g/d + 15{,}600\ g/m^3 \times Q_{brine} - 42.5 \times 10^5\ g/d - 500\ g/m^3 \times Q_{brine} = 0$$

$$15{,}100\ g/m^3 \times Q_{brine} = 24.65 \times 10^5\ g/d$$

$$Q_{brine} = 163\ m^3/d$$

The permissible brine discharge = 163 m^3/d

EXAMPLE 3.6: MLSS CONCENTRATION IN AN ACTIVATED SLUDGE PLANT

The biomass concentration in an activated sludge process is maintained by returning the sludge. The mixed liquor suspended solid (MLSS) concentration in the aeration basin is 2500 mg/L and TSS concentration (TSS_{ras}) in return activated sludge (RAS) is 10,000 mg/L. TSS concentration (TSS_{inf}) in the influent is small and ignored. The process diagram is shown below. Calculate (a) the return rate of sludge (Q_{ras}), and (b) the flow ratio of RAS to influent.

Solution

1. Draw the process diagram and system boundary.

   The system boundary is drawn around Point A (Figure 3.6).

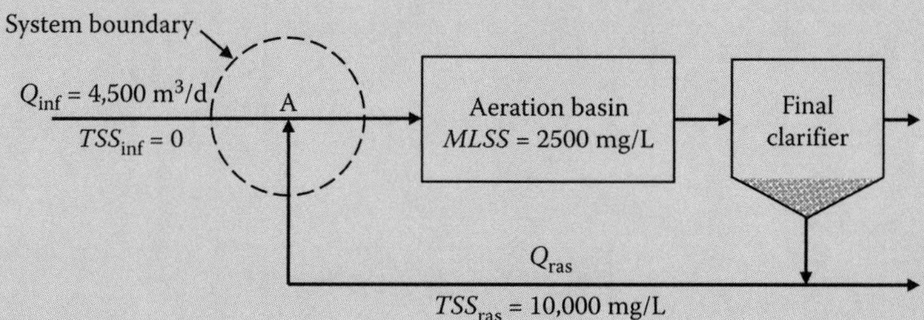

FIGURE 3.6 Process diagram and system boundary (Example 3.6).

2. Conduct a mass balance at Point A.

$$V\frac{dC_{TSS}}{dt} = Q_{inf}TSS_{inf} - Q_{ras}TSS_{ras} + r_{TSS}V$$

There is no accumulation $V(dC_{TSS}/dt) = 0$, and mass conversion in the connecting system is small and assumed zero ($r_{TSS}V = 0$).

$$0 = (4500\ m^3/d \times 0 + Q_{ras} \times 10{,}000\ g/m^3) - (4500\ m^3/d + Q_{ras}) \times 2500\ g/m^3 + 0$$

$$Q_{ras} = \frac{(2500 \times 4500)\ g/d}{(10{,}000 - 2500)\ g/m^3} = 1500\ m^3/d$$

3. Determine the return flow ratio (R_{ras}) of Q_{ras} to Q_{inf}.

$$R_{ras} = \frac{Q_{ras}}{Q_{inf}} = \frac{1500\ m^3/d}{4500\ m^3/d} = 0.33$$

The RAS flow is one-third of the influent flow.

EXAMPLE 3.7: INDUSTRIAL SEWER SURVEY

An industry uses electroplating process in several shops at the plant site. Wastewater from each shop is collected in an equalization basin, and discharged in individual sewer lines over a 24-h period. The characteristics of waste stream from each shop are given below in Figure 3.7. Determine the flow and pollutant concentrations in the combined waste stream.

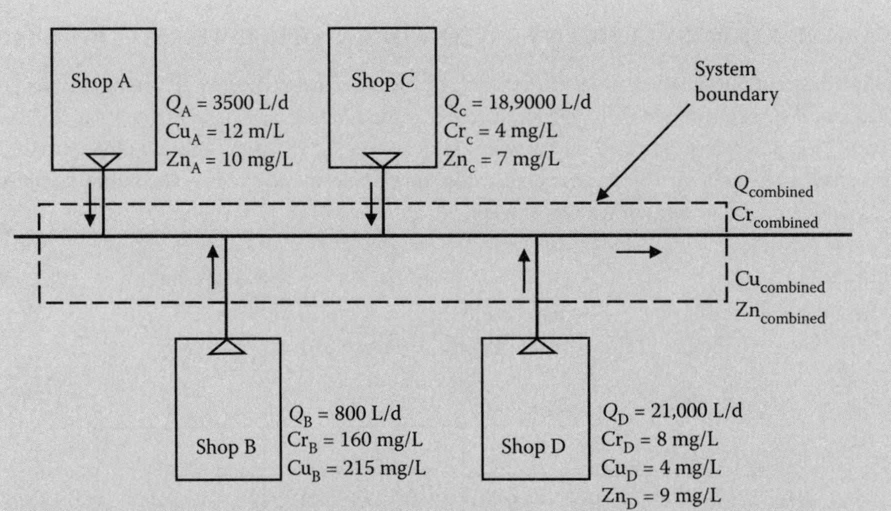

FIGURE 3.7 Industrial sewer plan (Example 3.7).

Solution

1. Draw the overall system boundary (Figure 3.7).
2. Conduct flow balance.

$$0 = (Q_A + Q_B + Q_C + Q_D) - Q_{combined}$$

$$Q_{combined} = (3500 + 800 + 18{,}900 + 21{,}000)\ \text{L/d} = 44{,}200\ \text{L/d}$$

3. Conduct a mass balance for each contaminant.

   Since there is no accumulation and all heavy metals are conservative substances, simplified mass balance analysis will give the result.

   a. Material balance for Cr.

$$0 = (Q_B \times Cr_B + Q_C \times Cr_C + Q_D \times Cr_D) - (Q_{combined} \times Cr_{combined})$$

$$Cr_{combined} = \frac{800\ \text{L/d} \times 160\ \text{mg/L} + 18{,}900\ \text{L/d} \times 4\ \text{mg/L} + 21{,}000\ \text{L/d} \times 8\ \text{mg/L}}{44{,}200\ \text{L/d}}$$

$$= \frac{371{,}600\ \text{mg/d}}{44{,}200\ \text{L/d}} = 8.4\ \text{mg/L}$$

   b. Material balance for Cu.

$$0 = (Q_A \times Cu_A + Q_B \times Cu_B + Q_D \times Cu_D) - (Q_{combined} \times Cu_{combined})$$

$$Cu_{combined} = \frac{3500\ \text{L/d} \times 12\ \text{mg/L} + 800\ \text{L/d} \times 215\ \text{mg/L} + 21{,}000\ \text{L/d} \times 4\ \text{mg/L}}{44{,}200\ \text{L/d}}$$

$$= \frac{298{,}000\ \text{mg/d}}{44{,}200\ \text{L/d}} = 6.7\ \text{mg/L}$$

c. Material balance for Zn.

$$0 = (Q_A \times Zn_A + Q_C \times Zn_C + Q_D \times Zn_D) - (Q_{combined} \times Zn_{combined})$$

$$Zn_{combined} = \frac{3500\,L/d \times 10\,mg/L + 18{,}900\,L/d \times 7\,mg/L + 21{,}000\,L/d \times 9\,mg/L}{44{,}200\,L/d}$$

$$= \frac{356{,}300\,g/d}{44{,}200\,L/d} = 8.1\,mg/L$$

EXAMPLE 3.8: SPLIT TREATMENT OF INDUSTRIAL WATER

An industrial water treatment plant produces 100 m^3/min finished water. The hardness in raw water is 15 mg/L as $CaCO_3$. The plant uses split treatment. Partial flow is treated by zeolite softener. Softened water has hardness of 0.02 mg/L as $CaCO_3$. A small stream is bypassed around the softener and then mixed with the softened water. The upper limit of hardness in treated water is 0.9 mg/L as $CaCO_3$. Calculate (a) hardness capture rate in the softener, (b) partial flow to the softener, and (c) flow bypassed around the softener.

Solution

1. Draw the process diagram and the system boundaries around the water treatment plant, softener, and flow splitter (Figure 3.8).

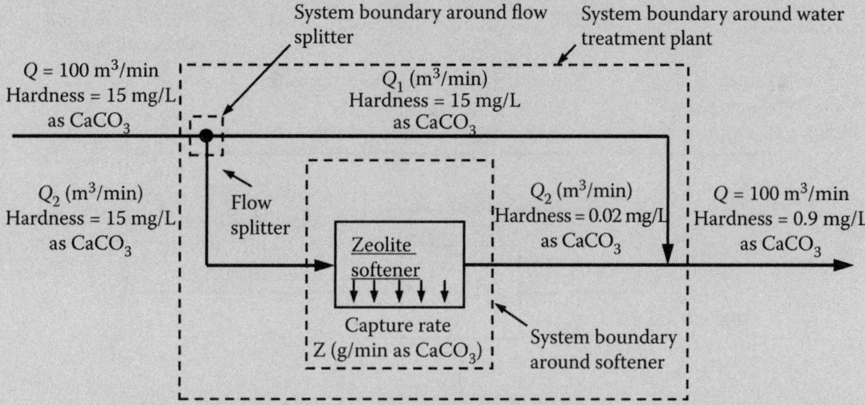

FIGURE 3.8 Process diagram of water treatment plant and system boundaries (Example 3.8).

2. Conduct a mass balance around the water treatment plant to determine hardness capture rate Z in the softener.

$$Z = 15\,g/m^3 \text{ as } CaCO_3 \times 100\,m^3/min - 0.9\,g/m^3 \text{ as } CaCO_3 \times 100\,m^3/min$$

$$Z = 1410\,g/min \text{ as } CaCO_3$$

3. Conduct a mass balance around the softener to determine filtration rate Q_2.

$$15\,g/m^3 \text{ as } CaCO_3 \times Q_2 = 1410\,g/min \text{ as } CaCO_3 + 0.02\,g/m^3 \text{as } CaCO_3 \times Q_2$$

$$Q_2 = \frac{1410\,g/min \text{ as } CaCO_3}{14.98\,g/m^3 \text{ as } CaCO_3} = 94.1\,m^3/min$$

4. Conduct a mass balance around the splitter to determine bypass flow Q_1 around the softener.

$$100\ \text{m}^3/\text{min} = 94.1\ \text{m}^3/\text{min} + Q_1$$

$$Q_1 = (100 - 94.1)\ \text{m}^3/\text{min} = 5.9\ \text{m}^3/\text{min}$$

EXAMPLE 3.9: PARTICULATE REMOVAL IN A BAGHOUSE

An air pollution facility is using a baghouse to remove dust from an air exhaust stream flowing at a rate of 200 m^3/min. The dirty air contains 10 g/m^3 of particulate, while the clean air from the baghouse contain 0.02 g/m^3 particulate. The operating permit allows the exhaust stream to contain as much as 0.9 g/m^3 of particulate matter. The industry wishes to use split treatment by bypassing some of the dirty air around the baghouse and mixing it back into the clean air so that the total exhaust stream meets the permit limit. Assume that there is no air leakage, and there is negligible change in pressure or temperature of air though the process. Calculate the flow rate of air through the baghouse and kilogram of dust collected per day at the baghouse.

Solution

1. Draw the process diagram of air pollution control facility.
   The process diagram and system boundaries are shown in Figure 3.9.

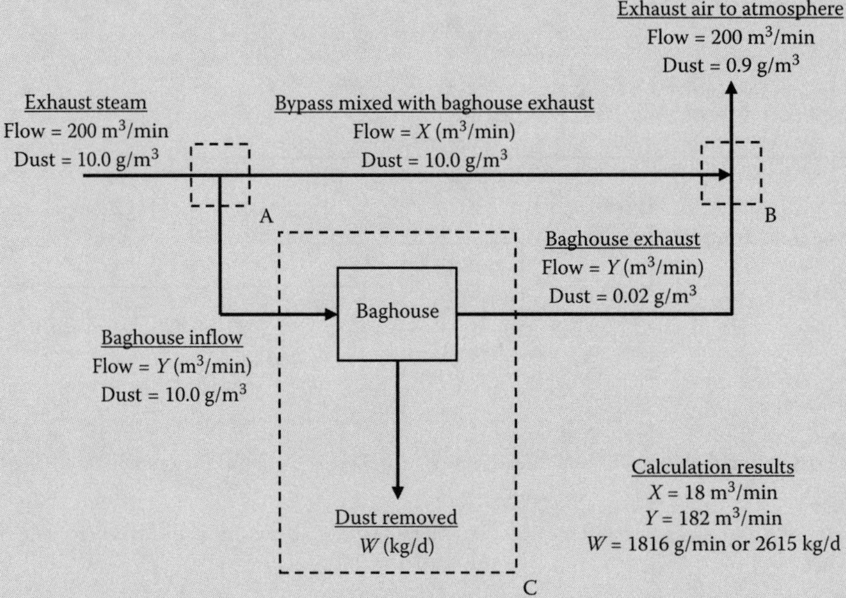

FIGURE 3.9 Process diagram of air pollution control facility (Example 3.9).

2. Conduct a flow balance at system boundary A.

$$200\ \text{m}^3/\text{min} = X + Y$$

$$X = 200\ \text{m}^3/\text{min} - Y$$

3. Conduct a mass balance at system boundary B and calculate gas flow through baghouse.

$$\begin{aligned} X \times 10.0\,\text{g/m}^3 + Y \times 0.02\,\text{g/m}^3 &= 200\,\text{m}^3\text{/min} \times 0.9\,\text{g/m}^3 \\ (200\,\text{m}^3\text{/min} - Y) \times 10\,\text{g/m}^3 + 0.02\,\text{g/m}^3 \times Y &= 180\,\text{g/min} \\ 2000\,\text{g/min} - 10\,\text{g/m}^3 \times Y + 0.02\,\text{g/m}^3 \times Y &= 180\,\text{g/min} \\ 9.98\,\text{g/m}^3 \times Y &= 1820\,\text{g/min} \\ Y &= 182\,\text{m}^3\text{/min} \end{aligned}$$

4. Determine the flow X bypassed the baghouse at system boundary A.

$$X = (200 - 182)\,\text{m}^3\text{/min} = 18\,\text{m}^3\text{/min}$$

5. Conduct a mass balance at system boundary C, and calculate dust collected in the baghouse.

$$Y \times 10\,\text{g/m}^3 - Y \times 0.02\,\text{g/m}^3 = W$$

$$182\,\text{m}^3\text{/min} \times (10.0 - 0.02)\,\text{g/m}^3 = W$$

$$W = 182\,\text{m}^3\text{/min} \times 9.98\,\text{g/m}^3 = 1816\,\text{g/min}$$

$$\text{Dust collected in the baghouse} = 1816\,\text{g/min} \times \frac{60\,\text{min/h} \times 24\,\text{h/d}}{1000\,\text{g/kg}} = 2615\,\text{kg/d}$$

EXAMPLE 3.10: SLUDGE SOLIDS CONCENTRATION IN THE FILTRATE

A sludge dewatering filter press receives thickened sludge that has 3% solids and specific gravity of 1.02. The dry solids in thickened sludge are 1800 kg/d. The solids capture efficiency of dewatering facility is 85%. The dewatered sludge cake has solids content of 30% and specific gravity of 1.04. Calculate (a) volumetric flow rate of sludge cake, m^3/d and (b) volumetric flow rate and TSS in the filtrate.

Solution

1. Draw the process diagram and system boundary (Figure 3.10).
2. Determine the volumetric flow rate of thickened sludge (Stream A).

$$Q_{\text{sludge}} = 1800\,\text{kg dry solids/d} \times \frac{100\,\text{kg wet sludge}}{3\,\text{kg dry solids}} \times \frac{1}{1020\,\text{kg/m}^3\text{ wet sludge}} = 58.8\,\text{m}^3\text{/d}$$

3. Calculate the solids in the sludge cake (Stream B) and filtrate (Stream C).

Solids capture efficiency of dewatering facility is 85%.

Solids captured in sludge cake (Stream B) = 1800 kg/d dry solids × 0.85 = 1530 kg/d dry solids

Solids in filtrate (Stream C) = 1800 kg/d dry solids × (1 − 0.85) = 270 kg/d dry solids

4. Calculate the volumetric flow rate of sludge cake (Stream B).

$$Q_{\text{cake}} = 1530\,\text{kg dry solids/d} \times \frac{100\,\text{kg sludge cake}}{30\,\text{kg dry solids}} \times \frac{1}{1040\,\text{kg/m}^3} = 4.9\,\text{m}^3\text{/d}$$

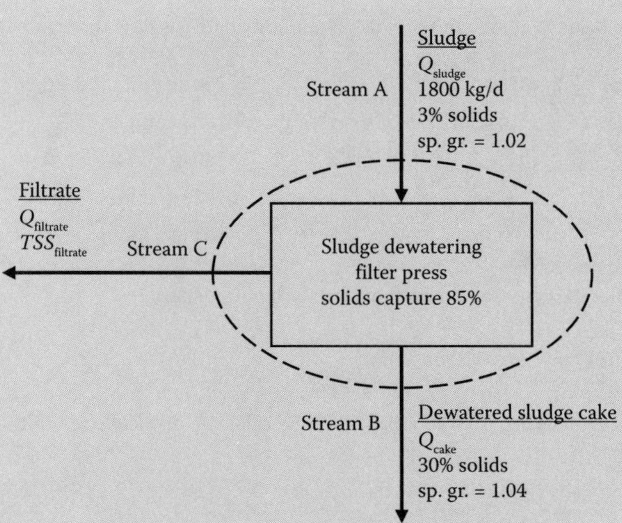

FIGURE 3.10 Process diagram and system boundary (Example 3.10).

5. Calculate the volume and TSS concentration of filtrate (Stream C).

$$Q_{\text{filtrate}} = 58.8\,\text{m}^3/\text{d} - 4.9\,\text{m}^3/\text{d} = 53.9\,\text{m}^3/\text{d}$$

$$TSS_{\text{filtrate}} = \frac{270\,\text{kg/d}}{53.9\,\text{m}^3/\text{d}} \times 1000\,\text{g/kg} = 5009\,\text{g/m}^3 \text{ or } 5009\text{ mg/L}$$

EXAMPLE 3.11: SOLIDS CONCENTRATION IN THICKENER OVERFLOW

Total quantity of sludge collected in a secondary wastewater treatment plant is 8500 lb per day. Primary sludge is 60% of combined sludge by weight, and has solids concentration of 3% by weight, and sp. gr. of 1.02. The secondary sludge has solids concentration of 0.8% and sp. gr. is 1. The combined sludge is thickened in a gravity thickener, and the supernatant is returned to the head of the plant. The solids capture efficiency of thickener is 90%. The thickened sludge has solids content of 8% and sp. gr. of 1.04. Calculate (a) the volume of the thickened sludge and (b) TSS in thickener supernatant.

Solution

1. Draw the process diagram and system boundary (Figure 3.11)
2. Calculate the dry solids and volumetric flow rate of primary sludge.

$$\text{TSS (dry solids) in primary sludge} = 8500\,\text{lb/d(dry solids)} \times 0.6 = 5100\,\text{lb/d}$$

$$Q_{\text{primary}} = 5100\,\text{lb dry solids/d} \times \frac{100\,\text{lb}}{3\,\text{lb}} \times \frac{1}{62.4\,\text{lb/ft}^3 \times 1.02} = 2671\,\text{ft}^3/\text{d}$$

3. Calculate the dry solids and volumetric flow rate in secondary sludge.

$$\text{TSS (dry solids) in secondary sludge} = 8500\,\text{lb/d(drysolids)} \times 0.4 = 3400\,\text{lb/d}$$

$$Q_{\text{secondary}} = \frac{3400\,\text{lb dry solids}}{\text{d}} \times \frac{100\,\text{lb}}{0.8\,\text{lb}} \times \frac{1}{62.4\,\text{lb/ft}^3 \times 1.00} = 6811\,\text{ft}^3/\text{d}$$

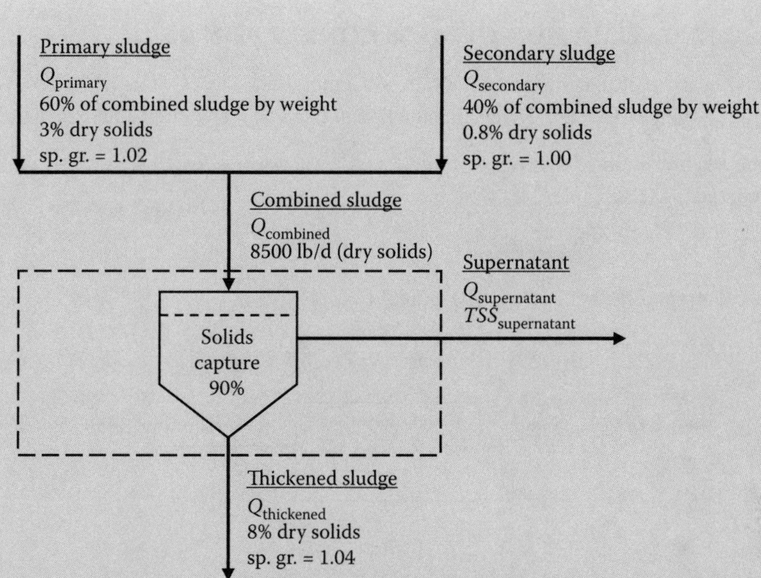

FIGURE 3.11 Process diagram and system boundary (Example 3.11).

4. Determine the volumetric flow rate of combined sludge.

TSS (dry solids) in combined sludge = 8500 lb/d

$$Q_{\text{combined}} = 2671\,\text{ft}^3/\text{d} + 6811\,\text{ft}^3/\text{d} = 9482\,\text{ft}^3/\text{d}$$

5. Determine the volumetric flow rate of thickened sludge.

TSS (dry solids) in thickened sludge = 8500 lb/d × 0.9 = 7650 lb/d

$$Q_{\text{thickened}} = \frac{7650\,\text{lb dry solids}}{\text{d}} \times \frac{100\,\text{lb}}{8\,\text{lb}} \times \frac{1}{62.4\,\text{lb/ft}^3 \times 1.04} = 1474\,\text{ft}^3/\text{d}$$

6. Determine the TSS concentration in supernatant.

$$Q_{\text{supernatant}} = (9482 - 1474)\,\text{ft}^3/\text{d} = 8008\,\text{ft}^3/\text{d}$$

TSS (dry solids) in supernatant = (8500 − 7650) lb/d = 850 lb/d

$$TSS_{\text{supernaqant}} = \frac{850\,\text{lb/d}}{8008\,\text{ft}^3/\text{d}} \times \frac{453.6\,\text{g/lb}}{0.0283\,\text{m}^3/\text{ft}^3} = 1700\,\text{g/m}^3 \text{ or } 1700\,\text{mg/L}$$

3.2.4 Mass or Concentration of Nonconservative Substances in Reactors

A nonconservative substance may undergo decay or generation; therefore, the mass balance analysis must also include conversion reactions.

EXAMPLE 3.12: ESTIMATION OF MASS CONVERSION RATE

A reactor receives a chemical compound at a rate of 0.2 mole/L·h. The exit rate of the same compound from the reactor is 0.001 mole/L·h. Determine the rate of mass conversion for the following conditions:

1. There is no accumulation in the reactor, and
2. The accumulation is 0.08 mole/L·h.

Solution

1. Draw the reactor and system boundary (Figure 3.12).

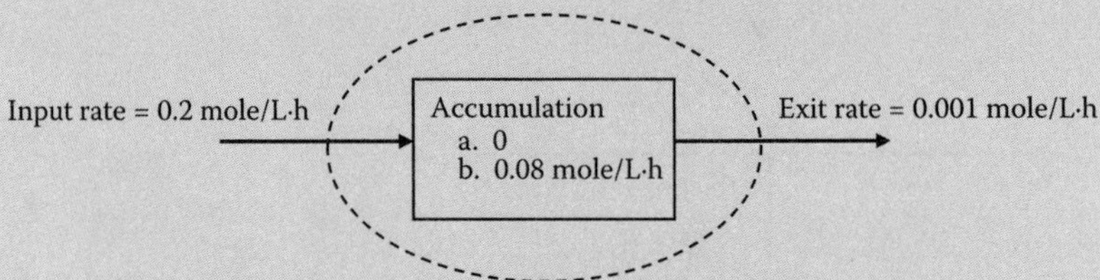

FIGURE 3.12 Reactor and system boundary (Example 3.12).

2. Calculate the conversion rate.
   a. Accumulation = 0

$$0 = \text{inflow} - \text{outflow} + \text{conversion}$$

$$0 = 0.2\ \text{mole/L·h} - 0.001\ \text{mole/L·h} - \text{Conversion}$$

$$\text{Conversion} = (0.2 - 0.001)\ \text{mole/L·h} = 0.199\ \text{mole/L·h}$$

   b. Accumulation = 0.08 mole/L·h

$$0.08\ \text{mole/L·h} = 0.2\ \text{mole/L·h} - 0.001\ \text{mole/L·h} - \text{Conversion}$$

$$\text{Conversion} = (0.2 - 0.001 - 0.08)\ \text{mole/L·h} = 0.119\ \text{mole/L}$$

EXAMPLE 3.13: SLUDGE STABILIZATION IN AN ANAEROBIC DIGESTER

The thickened sludge is pumped into an anaerobic digester for stabilization. The volatile matter is partly converted into gaseous products (CH_4 and CO_2). The digested sludge is pumped out daily and the supernatant from the digester is returned to the aeration basin. The information regarding the raw sludge and input rate, and digester performance is given below. Determine the quantity of solids (kg/d) and flow rate (m^3/d) of digested sludge and supernatant.

Thickened wet sludge input rate, $Q_{\text{thickened sludge}}$	$= 132\ m^3/d$
Thickened dry solids input rate, $W_{\text{thickened sludge}}$	$= 8180$ kg/d
VSS/TSS ratio in thickened sludge solids	= 71%
VS reduction (VSR) in anaerobic digester	= 52%
Solids in supernatant and density	= 0.4% and 1000 kg/m^3
Digested sludge solids and density	= 5% and 1030 kg/m^3

Solution

1. Draw the process diagram and system boundary (Figure 3.13).

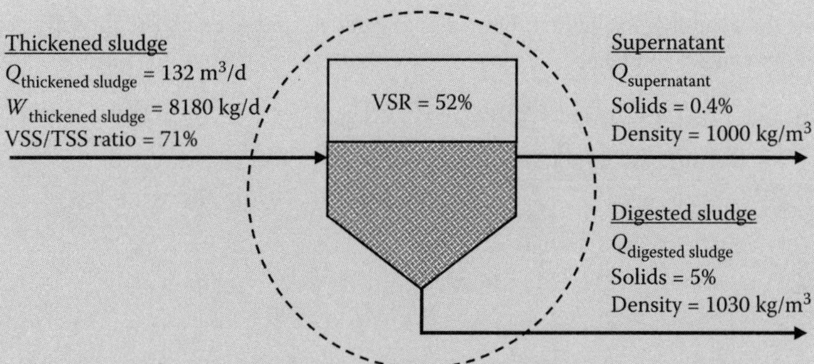

FIGURE 3.13 Mass balance around anaerobic digester (Example 3.13).

2. Conduct the flow balance.
   Assuming the loss of moisture is negligible, Equation 3.5a is developed.

$$\begin{bmatrix}\text{Rate of flow}\\ \text{accumulation}\end{bmatrix} = Q_{\text{thickened sludge}} - \left[Q_{\text{digested sludge}} + Q_{\text{supernatant}}\right] + \begin{bmatrix}\text{Volume lost}\\ \text{or consumed}\end{bmatrix} \quad (3.5a)$$

   Since both the rate of accumulation of flow, and volume lost or consumed = 0, relationship between weight of supernatant and weight of digested sludge is established by Equation 3.5b.

$$Q_{\text{thickened sludge}} = Q_{\text{digested sludge}} + Q_{\text{supernatant}}$$

$$132\,\text{m}^3/\text{d} = \frac{W_{\text{digested sludge}}}{0.05 \times 1030\,\text{kg/m}^3} + \frac{W_{\text{supernatant}}}{0.004 \times 1000\,\text{kg/m}^3}$$

$$0.0194\,\text{m}^3/\text{kg} \times W_{\text{digested sludge}} + 0.25\,\text{m}^3/\text{kg} \times W_{\text{supernatant}} = 132\,\text{m}^3/\text{d}$$

$$W_{\text{supernatant}} = 528\,\text{kg/d} - 0.0776 \times \text{W}_{\text{digested sludge}} \quad (3.5b)$$

3. Conduct a mass balance of volatile solids in the digester.
   From mass balance to develop Equation 3.6a.

$$\begin{bmatrix}\text{Rate of solids}\\ \text{accumulation}\end{bmatrix} = \begin{bmatrix}\text{Rate of thickened}\\ \text{solids input}\end{bmatrix} - \left[\begin{pmatrix}\text{Rate of digested}\\ \text{solids withdrawn}\end{pmatrix} + \begin{pmatrix}\text{Rate of solids}\\ \text{lost in supernatant}\end{pmatrix}\right] + \left[-\begin{pmatrix}\text{Rate of volatile}\\ \text{solids stabilizaion}\end{pmatrix}\right] \quad (3.6a)$$

   Substitute the values of solids mass in Equation 3.6a to establish Equation 3.6b.

$$\text{Volatile solids in input sludge} = 8180\,\text{kg/d} \times 0.71 = 5808\,\text{kg/d}$$

$$\text{Volatile solids reduced} = 5808\,\text{kg/d} \times 0.52 = 3020\,\text{kg/d}$$

$$0\,\text{kg/d} = 8180\,\text{kg/d} - [W_{\text{digested sludge}}] - [W_{\text{supernatant}}] - 3020\,\text{kg/d}$$

$$W_{\text{supernatant}} = 5160\,\text{kg/d} - W_{\text{digested sludge}} \quad (3.6b)$$

4. Determine the quantity of solids and flow rates of digested sludge, and those for supernatant.
   Solve Equation 3.5b and 3.6b.

$$W_{\text{digested sludge}} = 5022\,\text{kg/d and } W_{\text{supernatant}} = 138\,\text{kg/d}$$

$$Q_{\text{digestered sludge}} = \frac{5022\,\text{kg dry solids}}{\text{d}} \times \frac{100\,\text{kg wet solids}}{5\,\text{kg dry solids}} \times \frac{1}{1030\,\text{kg/m}^3\text{ wet solids}}$$

$$= 98\,\text{m}^3/\text{d wet solids}$$

$$Q_{\text{supernatant}} = (132 - 98)\,\text{m}^3/\text{d} = 34\,\text{m}^3/\text{d}$$

3.3 Flow Regime

Hydraulic flow models are useful in evaluating the effect of residence time, flow rate, and reactor performance. There are three flow regimes that may occur in a continuous flow system (a) *ideal completely mixed flow*, (b) *ideal plug flow*, and (c) *dispersed plug flow*.

In the ideal completely mixed flow regime, the elements of fluid upon entry into the system are dispersed immediately throughout the system. In the ideal plug flow regime, the elements of fluid pass through the system in the same sequence in which they enter. The particles retain their identity and remain in the reactor for a time equal to the theoretical detention time. It should be realized that the ideal plug flow and completely mixed flow regimes are the two limiting cases. The actual flow regimes will lie between the two ideal conditions. The dispersed plug flow regime will encounter any degree of partial mixing between plug flow and complete mixing.

3.4 Types of Reactors

Mixing of chemicals, and biological and chemical reactions are carried out in containers, vessels, or tanks. These are commonly called reactors. Homogenous reactors are (1) batch reactor, (2) continuous-flow stirred-tank reactor (CFSTR), or complete mix reactor, (3) the plug flow, tubular flow, or piston-flow reactor, and (4) dispersed plug flow, arbitrary-flow, plug flow with longitudinal mixing, or intermediate-mixed flow reactor.[3-5]

Heterogeneous reactions involve more than one phase (such as solid and liquid). Such reactions are normally carried out in (1) fixed-bed or packed-bed, (2) moving-bed, and (3) fluidized-bed reactors. The principal types of homogenous reactors are discussed below.

3.4.1 Batch Reactor

In a batch reactor, the flow is neither entering nor leaving the reactor. The contents are well mixed for desired time for conversion to occur, and then the resultant mixture is discharged. This is a nonsteady-state process where the composition changes with time. Since the composition is uniform throughout the reactor, a mass balance analysis can be applied at any time.

EXAMPLE 3.14: MASS BALANCE AND DECAY EQUATION IN A BATCH REACTOR

In a batch reactor, a substance is undergoing first-order decay. Sketch the batch reactor, and develop the equation representing the change in composition with reaction time, t.

Solution

1. Draw the definition sketch.
   The definition sketch of batch reactor is shown in Figure 3.14.

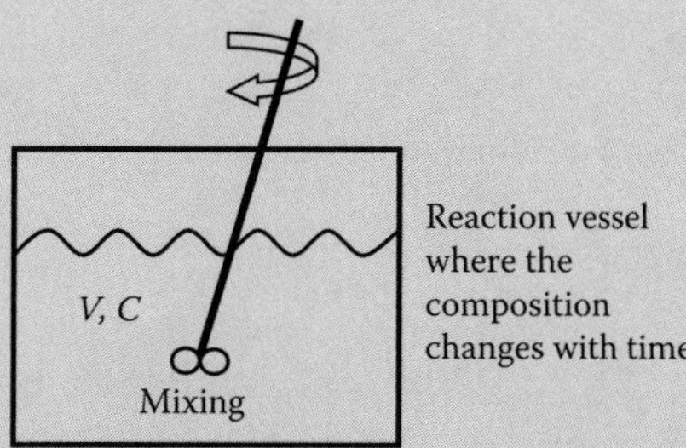

FIGURE 3.14 Definition of batch reactor (Example 3.14).

2. Write the generalized mass balance equation.
   The generalized mass balance equation is expressed by Equation 3.1.

   [Accumulation rate] = [Input rate] − [Output rate] + [Utilization or conversion rate]

   Since the input and output rates are zero, the equation is written as Equation 3.7.

   [Accumulation] = [Decrease due to conversion]

$$r_A = \frac{dC_A}{dt} = -kC_A \tag{3.7}$$

3. Rearrange and integrate Equation 3.7 within the limits to obtain Equations 3.8a through 3.8c.

$$\int_{C_0}^{C} \frac{dC_A}{C_A} = -k\int_0^t dt$$

$$\ln\left(\frac{C}{C_0}\right) = -kt$$

$$t = -\frac{1}{k}\ln\left(\frac{C}{C_0}\right) \tag{3.8a}$$

$$C = C_0 e^{-kt} \tag{3.8b}$$

$$\frac{C}{C_0} = e^{-kt} \tag{3.8c}$$

where

C_0 = initial concentration of a substance at $t = 0$, mg/L

C = concentration of a substance at time t, mg/L

EXAMPLE 3.15: MASS CONVERSION IN A BATCH REACTOR

A batch reactor is designed for removal or conversion of a substance. If the first-order reaction rate constant at 20°C is 0.21 h^{-1} and required removal at 10°C is 90%, determine the reaction time. Assume $\theta_T = 1.047$. Also, calculate the concentration ratio C/C_0 at time $t = 4$, 10, and 20 h.

Solution

1. Calculate the reaction rate from Equation 2.10 at 10°C.

$$k_{10} = k_{20}\,(1.047)^{T-20} = 0.21\,\text{h}^{-1} \times (1.047)^{10-20} = 0.21\,\text{h}^{-1} \times (1.047)^{-10} = 0.13\,\text{h}^{-1}$$

2. Calculate the time required to achieve 90% conversion from Equation 3.8a.

For 90%, removed C/C_0 remaining ratio $= (1 - 0.9) = 0.1$.

Substitute the remaining ratio in Equation 3.8a to determine t.

$$\frac{C}{C_0} = 0.1, \quad t = -\frac{1}{0.13\,\text{h}^{-1}} \ln(0.1) = 17.7\,\text{h}$$

3. Calculate the C to C_0 ratio at different time intervals from Equation 3.8c.

$$t = 4\,\text{h}, \quad \frac{C}{C_0} = e^{-0.13\,\text{h}^{-1} \times 4\,\text{h}} = 0.6$$

$$t = 10\,\text{h}, \quad \frac{C}{C_0} = e^{-0.13\,\text{h}^{-1} \times 10\,\text{h}} = 0.27$$

$$t = 20\,\text{h}, \quad \frac{C}{C_0} = e^{-0.13\,\text{h}^{-1} \times 20\,\text{h}} = 0.07$$

3.4.2 Continuous-Flow Stirred-Tank Reactor

In a CFSTR, material entering is dispersed instantly throughout the reactor. As a result, the concentration of material leaving the reactor is same as that at any point in the reactor. Mixing in a CFSTR is extremely important. A round, square, or slightly rectangular reactor may be used. Baffles may be necessary to control vortexing.

The flow schematic of a CFSTR is shown in Figure 3.15.

Conservative Tracer Response in a CFSTR: Tracer studies are conducted in a CFSTR to determine the reactor response and flow regime. A nonreactive conservative tracer is normally injected at the influent zone, and its concentration is measured with respect to time. The tracer may be applied as a slug input or as a step (continuous) feed. In a slug input, a known volume of stock solution containing known

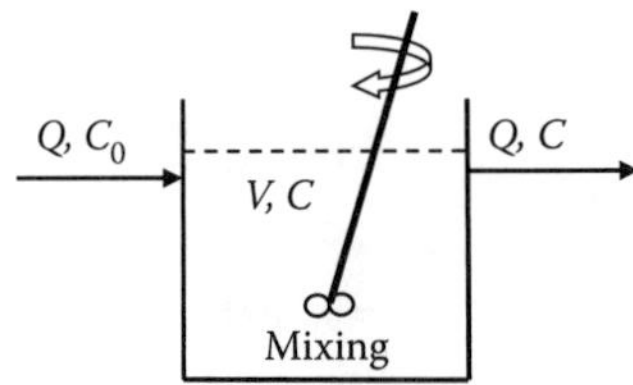

FIGURE 3.15 Flow schematic of a CFSTR.

quantity of tracer is released into the influent zone. A constant flow is maintained through the reactor. The tracer concentration in the effluent is measured with respect to time. A step-feed tracer study utilizes continuous feed of known flow and concentration of tracer at the influent, and its concentration is monitored in the effluent.[4,6]

Conversion of Nonconservative (Reactive) Substance in a CFSTR: A nonconservative substance undergoes conversion reaction in a CFSTR. The reaction may be zero order, first order, or second order. The concentration equation at any time t can be developed from mass balance relationship. The solution of mass balance equations has been presented earlier. The concentration equation with first-order decay under steady-state and nonsteady-state conditions are expressed by Equation 3.3.

EXAMPLE 3.16: C/C_0 RATIO OF A CONSERVATIVE TRACER IN A CFSTR

A conservative tracer is injected into a CFSTR. Develop the generalized concentration equations for the effluent from a CFSTR receiving (a) slug tracer input and (b) step (continuous) tracer feed.

Solution

1. Write the generalized mass balance equations (Equations 3.1 and 3.3).

$$\begin{bmatrix}\text{Accumulation}\\ \text{rate}\end{bmatrix} = \begin{bmatrix}\text{Input}\\ \text{rate}\end{bmatrix} - \begin{bmatrix}\text{Output}\\ \text{rate}\end{bmatrix} + \begin{bmatrix}\text{Utilization or}\\ \text{conversion rate}\end{bmatrix}$$

$$V\frac{dC}{dt} = QC_0 - QC + V(-kC)$$

For a conservative substance, the conversion term $-kC = 0$. Equation 3.3 is simplified by Equation 3.9a.

$$V\frac{dC}{dt} = QC_0 - QC \quad (3.9a)$$

where

Q = flow rate through the reactor, m^3/s
V = volume of reactor, m^3
C_0 = tracer input concentration, mg/L
C = tracer output concentration, mg/L

For slug input, $C_{0,\,slug}$ = initial concentration of tracer in the reactor volume right after the tracer is fed at $t = 0$, mg/L. It is determined by Equation 3.9b.

$$C_{0,slug} = \frac{V_{tracer}}{V}C_{tracer} \quad \text{or} \quad C_{0,slug} = \frac{w_{tracer}}{V} \quad (3.9b)$$

where

V_{tracer} = volume of slug tracer, m^3
C_{tracer} = concentration of trace in the slug feed, mg/L (g/m^3)
$w_{reactor}$ = weight of reactor, g

For step feed, $C_{0,step}$ = feed concentration of tracer in the influent into the reactor at $t > 0$. It is expressed by Equation 3.9c.

$$C_{0,step} = \frac{Q_{tracer}}{Q + Q_{tracer}}C_{tracer} \quad (3.9c)$$

When $Q \gg Q_{tracer}$, $(Q + Q_{tracer}) \approx Q$. Equation 3.9c is simplified to Equation 3.9d.

$$C_{0,step} = \frac{Q_{tracer}}{Q} C_{tracer} \tag{3.9d}$$

where

Q_{tracer} = tracer flow rate fed into influent, m^3/s

C_{tracer} = concentration of trace in the tracer flow, mg/L

2. Develop equations for slug input.

   Apply the condition of slug input to Equation 3.9a. For a slug input, $C_{0,slug} = 0$ at $t > 0$ after slug is added. Rearrange Equation 3.9a at $QC_{0,slug} = 0$.

$$\frac{dC}{C} = -\frac{Q}{V}dt$$

   Use the integration limits: $C_{0,slug}$ at $t = 0$, and C at t, integrate the equation to obtain a remaining concentration equation (Equation 3.10a) and a ratio equation (Equation 3.10b).

$$\int_{C_{0,slug}}^{C} \frac{dC}{C} = -\frac{Q}{V}\int_{0}^{t} dt$$

$$\ln\left(\frac{C}{C_{0,slug}}\right) = -\frac{Q}{V}t = \frac{-t}{\left(\frac{V}{Q}\right)} = -\frac{t}{\theta}$$

$$C = C_{0,slug}\, e^{-t/\theta} \tag{3.10a}$$

$$\frac{C}{C_{0,slug}} = e^{-t/\theta} \tag{3.10b}$$

   where θ = theoretical detention time, h. It is expressed by Equation 3.10c.

$$\theta = \frac{V}{Q} \tag{3.10c}$$

3. Develop the equations for step feed.

   Apply the condition of step feed to Equation 3.9a. For a step feed, $C_{0,step} \neq 0$ at $t > 0$ after tracer injection begins. Rearrange Equation 3.9a.

$$\frac{dC}{C_{0,step} - C} = \frac{Q}{V}dt$$

   Use the integration limits: $C_{0,step} = 0$ at $t = 0$, and C at t, integrate the equation to obtain a remaining concentration equation (Equation 3.10d) and a ratio equation (Equation 3.10e).

$$\int_{0}^{C} \frac{dC}{(C_{0,step} - C)} = \frac{Q}{V}\int_{0}^{t} dt$$

$$\ln\left(\frac{C_{0,\text{step}} - C}{C_{0,\text{step}}}\right) = -t\left(\frac{Q}{V}\right) = -t/\theta$$

$$\frac{C_{0,\text{step}} - C}{C_{0,\text{step}}} = e^{-t/\theta} \quad \text{or} \quad 1 - \frac{C}{C_{0,\text{step}}} = e^{-t/\theta}$$

$$C = C_{0,\text{step}}(1 - e^{-t/\theta}) \tag{3.10d}$$

$$\frac{C}{C_{0,\text{step}}} = 1 - e^{-t/\theta} \tag{3.10e}$$

EXAMPLE 3.17: CONSERVATIVE TRACER PROFILE FOR SLUG AND STEP FEED IN A CFSTR

A CFSTR has a volume of 35 m^3. The constant flow in the reactor is 200 m^3/d. The reactor flow regime was established by (a) releasing in the influent zones a slug of 20 L stock solution containing 3.5 g/L tracer and (b) feeding continuously 3.5 g/L stock tracer solution in the influent line at a rate of 100 L per day. Determine the following for both test conditions (a) expected tracer concentration in the effluent at 2 and 6 h after the tracer injection, and (b) tracer profile as a function of C/C_0 versus t/θ.

Solution

1. Determine the theoretical detention time θ of the CFSTR from Equation 3.10c.

$$\theta = \frac{V}{Q} = \frac{35\,\text{m}^3}{200\,\text{m}^3/\text{d}} \times 24\,\text{h/d} = 4.2\,\text{h}$$

2. Determine initial tracer concentration C_0 and concentration C at time 2 h (C_2) and 6 h (C_6) since tracer injection. Also, calculate dimensionless ratios C/C_0 and t/θ.

   a. Slug input.

   Calculate the initial concentration $C_{0,\text{slug}}$ from Equation 3.9b.

$$C_{0,\text{slug}} = \frac{3.5\,\text{g/L} \times 20\,\text{L}}{35\,\text{m}^3} = 2\,\text{g/m}^3$$

   Calculate the concentration C_2 at $t_2 = 2$ h from Equation 3.10a.

$$C_2 = C_{0,\text{slug}} e^{-t_2/\theta} = 2\,\text{g/m}^3 \times e^{-2\,\text{h}/4.2\,\text{h}} = 1.24\,\text{g/m}^3$$

   Calculate the concentration ratio $C_2/C_{0,\text{slug}}$ from Equation 3.10b.

$$\frac{C_2}{C_{0,\text{slug}}} = e^{-t_2/\theta} = e^{-2\,\text{h}/4.2\,\text{h}} = 0.62$$

$$\frac{t_2}{\theta} = \frac{2\,\text{h}}{4.2\,\text{h}} = 0.48$$

Repeat the calculations for $t_6 = 6$ h.

$$C_6 = C_{0,\text{slug}} e^{-t_6/\theta} = 2\,\text{g/m}^3 \times e^{-6\,\text{h}/4.2\,\text{h}} = 0.48\,\text{g/m}^3$$

$$\frac{C_6}{C_{0,\text{slug}}} = e^{-t_6/\theta} = e^{-6\,\text{h}/4.2\,\text{h}} = 0.24$$

$$\frac{t_6}{\theta} = \frac{6\,\text{h}}{4.2\,\text{h}} = 1.43$$

b. Step feed.
Calculate the initial concentration $C_{0,\text{step}}$ from Equation 3.9d.

$$C_{0,\text{step}} = \frac{100\,\text{L/d}}{200\,\text{m}^3/\text{d}} \times 3.5\ \text{g/L} = 1.75\,\text{g/m}^3$$

Calculate the concentration C_2 at $t_2 = 2$ h from Equation 3.10d.

$$C_2 = C_{0,\text{step}}(1 - e^{-t_2/\theta}) = 1.75\,\text{g/m}^3 \times (1 - e^{-2\,\text{h}/4.2\,\text{h}}) = 0.66\,\text{g/m}^3$$

Calculate the concentration ratio $(C_2/C_{0,\text{step}})$ from Equation 3.10e.

$$\frac{C_2}{C_{0,\text{slug}}} = 1 - e^{-t_2/\theta} = 1 - e^{-2\,\text{h}/4.2\,\text{h}} = 0.38$$

Repeat the calculations for $t_6 = 6$ h.

$$C_6 = C_{0,\text{step}}(1 - e^{-t_6/\theta}) = 1.75\,\text{g/m}^3 \times (1 - e^{-6\,\text{h}/4.2\,\text{h}}) = 1.33\,\text{g/m}^3$$

$$\frac{C_6}{C_{0,\text{slug}}} = 1 - e^{-t_6/\theta} = 1 - e^{-6\text{h}/4.2\text{h}} = 0.76$$

3. Draw the tracer profiles.
Assume a series of time intervals (t) after the slug is added or since the tracer injection began and calculate t/θ, $(C/C_{0,\text{slug}})$, and $(C/C_{0,\text{step}})$ ratios. The tracer profile data for the slug input and step feed are summarized in Table 3.1.
The profiles are shown for (a) $(C/C_{0,\text{slug}})$ versus t/θ, and (b) $(C/C_{0,\text{step}})$ versus t/θ in Figure 3.16.

TABLE 3.1 Tracer Profile Data for Slug and Step Feed (Example 3.17)

Time Step (t), h	t/θ	Slug Input		Step Feed	
		C, g/m^3	$C/C_{0,\text{slug}}$	C, g/m^3	$C/C_{0,\text{step}}$
0	0	2.00	1.00	0.00	0.00
0.5	0.12	1.77	0.89	0.20	0.11
1	0.23	1.59	0.80	0.36	0.21
2	0.48	1.24	0.62	0.66	0.38
4	0.95	0.77	0.39	1.07	0.61
6	1.43	0.49	0.25	1.33	0.76
10	2.38	0.19	0.10	1.59	0.91
20	4.76	0.02	0.01	1.74	0.99

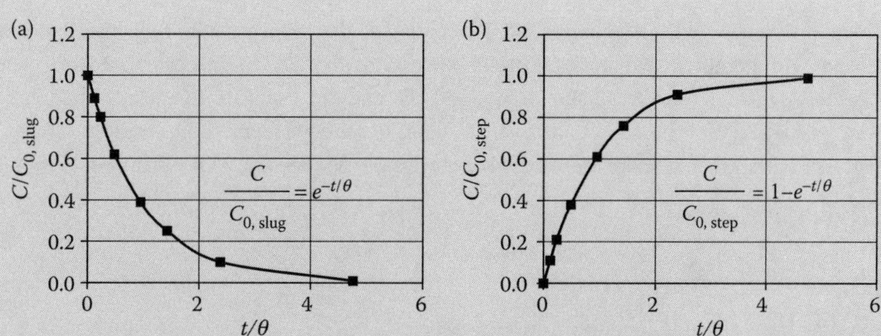

FIGURE 3.16 Tracer profile in CFSTR: (a) slug input, and (b) step feed (Example 3.17).

EXAMPLE 3.18: STEADY-STATE OPERATION OF A NONCONSERVATIVE SUBSTANCE WITH FIRST-ORDER DECAY

A reactor receives industrial material for product conversion. The reactor volume is 500 m^3, and influent and effluent flow rates are 50 m^3/d. The concentration of feed material is 650 mg/L, and it is consumed according to first-order kinetics with $k = 0.28\ \text{day}^{-1}$. Develop the kinetic equation, and determine the exit concentration of the material. The process is operating under steady-state condition.

Solution

Use Equation 3.3, and apply steady-state condition ($V(dC/dt) = 0$).

The solution is expressed by Equations 3.11a through 3.11c.

$$0 = QC_0 - QC - VkC$$

$$C = \frac{C_0}{1 + k\dfrac{V}{Q}} \tag{3.11a}$$

$$C = \frac{C_0}{1 + k\theta} \tag{3.11b}$$

$$\frac{C}{C_0} = \frac{1}{1 + k\theta} \tag{3.11c}$$

Substitute the data in Equation 3.11a to determine the value of C.

$$C = \frac{650\ \text{g/m}^3}{1 + 0.28\ \text{d}^{-1} \times \dfrac{500\ \text{m}^3}{50\ \text{m}^3/\text{d}}} = \frac{650\ \text{g/m}^3}{1 + 2.8} = 171\ \text{g/m}^3$$

EXAMPLE 3.19: TIME-DEPENDENT CONCENTRATION OF A NONCONSERVATIVE SUBSTANCE

A nonconservative substance has an influent concentration of C_0. It undergoes conversion reaction A → B which is known to be first order ($r = -kC$). Conduct mass balance analysis, and (a) develop time-dependent concentration equation, (b) determine the steady-state concentration equation when $t = \infty$ and compare the concentration with Equation 3.11b, (c) calculate the concentration of the substance at time $t = 1$ and 3 day, and ∞, and (d) calculate the steady-state concentration from Equation 3.11b and compare it with time-dependent concentration from time-dependent equation when $t = \infty$. Use the following data: $V = 480\ \text{m}^3$, $Q = 30\ \text{m}^3/\text{d}$, $C_0 = 40\ \text{g/m}^3$, and $k = 0.18\ \text{d}^{-1}$.

Solution

1. Develop the time-dependent concentration equation.
   a. Develop the mass balance equation from Equation 3.3.

$$V\frac{dC}{dt} = QC_0 - QC + V(-kC)$$

   b. Simplify and rearrange the above equation.

$$\frac{dC}{dt} = \frac{1}{V}[-C(Q + kV) + QC_0]$$

$$= -C\left(\frac{Q}{V} + k\right) + C_0\frac{Q}{V}$$

$$\frac{dC}{dt} + C\left(\frac{Q}{V} + k\right) = C_0\frac{Q}{V}$$

   c. Integrate the above equation.
      i. Substitute $(Q/V) + k$ by the following equation (Equation 3.12).

$$\beta = \frac{Q}{V} + k \quad \text{or} \quad \beta = \frac{1}{\theta} + k = \frac{1 + k\theta}{\theta} \tag{3.12}$$

$$\frac{dC}{dt} + \beta C = C_0\frac{Q}{V}$$

      ii. Multiply both sides by the integrating factor $e^{\beta t}$.

$$\left[\frac{dC}{dt} + \beta C\right]e^{\beta t} = \left(C_0\frac{Q}{V}\right)e^{\beta t}$$

      iii. The left side of the equation is written as a differential.

$$\frac{d}{dt}(Ce^{\beta t}) = C_0\frac{Q}{V}e^{\beta t}$$

$$d(Ce^{\beta t}) = C_0\frac{Q}{V}e^{\beta t}\,dt$$

$$Ce^{\beta t} = \frac{Q}{V}C_0\int e^{\beta t}\,dt$$

      iv. Integration yields the following (Equation 3.13a).

$$Ce^{\beta t} = \frac{Q}{V}\frac{C_0}{\beta}e^{\beta t} + k \tag{3.13a}$$

when $t = 0$, $C = C_0$

$$k = C_0 - \frac{Q}{V}\frac{C_0}{\beta}$$

v. Substitute the value of k in Equation 3.13a to obtain the nonsteady-state solution of time-dependent concentration expressed by Equation 3.13b.

$$C = \frac{Q}{V}\frac{C_0}{\beta}(1 - e^{-\beta t}) + C_0 e^{-\beta t} \tag{3.13b}$$

2. Determine the effluent concentration C when $t = \infty$.

a. Substitute $t = \infty$ and solve.

$$e^{-\beta t} = e^{-\beta\infty} = \frac{1}{e^{\infty}} \to 0$$

$$C = \frac{Q}{V}\frac{C_0}{\beta}(1 - e^{-\beta\infty}) + C_0 e^{-\beta\infty}$$

$$C = \frac{Q}{V}\frac{C_0}{\beta}$$

$$C = \frac{Q}{V}\frac{C_0}{\left(\frac{Q}{V} + k\right)} = \frac{Q}{V}\frac{C_0}{\frac{Q}{V}\left(1 + k\frac{V}{Q}\right)}$$

$$C = \frac{C_0}{\left(1 + k\frac{V}{Q}\right)} = \frac{C_0}{1 + k\theta}$$

b. Compare the above equation with the steady-state equation (Equation 3.11b). The steady-state equation Equations 3.11b and 3.13b are identical when $t = \infty$.

3. Calculate the concentration of nonconservative waste in the effluent when $t = 1$ day, 3 days, and ∞.

a. Calculate β from Equation 3.12.

$$\beta = \frac{Q}{V} + k = \frac{30\,\text{m}^3/\text{d}}{480\,\text{m}^3} + 0.18\,\text{d}^{-1} = 0.243\,\text{d}^{-1}$$

b. Calculate C for $t = 1$ day, 3 days, and ∞ from Equation 3.13b.

$t = 1$ day

$$C = \frac{Q}{V}\frac{C_0}{\beta}(1 - e^{-\beta t}) + C_0 e^{-\beta t}$$

$$= \frac{30\,\text{m}^3/\text{d}}{480\,\text{m}^3} \times \frac{40\,\text{g/m}^3}{0.243\,\text{d}^{-1}} \times (1 - e^{-0.243\text{d}^{-1}\times 1\text{d}}) + 40\,\text{g/m}^3 \times e^{-0.243\text{d}^{-1}\times 1\text{d}}$$

$$= 10.29\,\text{g/m}^3 \times (1 - 0.784) + 40\,\text{g/m}^3 \times 0.784$$

$$= (2.22 + 31.36)\,\text{g/m}^3 = 33.6\,\text{g/m}^3$$

$t = 3$ day

$$C = 10.29\,\text{g/m}^3 \times (1 - e^{-0.243\text{d}^{-1}\times 3\text{d}}) + 40\,\text{g/m}^3 \times e^{-0.243\text{d}^{-1}\times 3\text{d}}$$
$$= 10.29\,\text{g/m}^3 \times (1 - 0.482) + 40\,\text{g/m}^3 \times 0.482$$
$$= (5.33 + 19.28)\,\text{g/m}^3 = 24.6\,\text{g/m}^3$$

$t = \infty$

$$C = 10.29\,\text{g/m}^3(1 - 0) + 40\,\text{g/m}^3 \times 0 = 10.3\,\text{g/m}^3$$

4. Calculate the steady-state concentration C from Equation 3.11b and compare it with the concentration obtained from Equation 3.13b when $t = \infty$.

$$\text{Steady-state concentration } C = \frac{C_0}{1 + k\theta} = \frac{40\,\text{g/m}^3}{1 + 0.18\,\text{d}^{-1} \times \dfrac{480\,\text{m}^3}{30\,\text{m}^3/\text{d}}} = 10.3\,\text{g/m}^3$$

Concentration $C = 10.3\,\text{g/m}^3$ when $t = \infty$ (see Step 3b above)

This comparison clearly shows that the time-dependent nonsteady-state equation yields steady-state result when t is very large.

EXAMPLE 3.20: STEADY-STATE CONVERSION OF A REACTIVE SUBSTANCE IN A CFSTR

A CFSTR receives a reactive substance. The first-order reaction rate constant $k = 0.30\,\text{h}^{-1}$. What is the residence time θ to achieve 90% conversion under steady-state condition? Plot the fraction remaining (C/C_0), and fraction removed ($1 - C/C_0$) with respect to corresponding residence time θ. Also, derive linear relationships that can be used to determine the reaction rate constant k based on the experimental data.

Solution

1. Plot fraction remaining and removal curves.
   a. Select the conversion equation.
      The first-order conversion relationship under steady-state condition is expressed by Equation 3.11b.
      $$C = \frac{C_0}{1 + k\theta}$$
   b. Rearrange Equation 3.11b to solve for the residence time θ.
      $$\theta = \frac{C_0 - C}{kC}$$
   c. Equation 3.11c gives a relationship between C/C_0 and θ.
      $$\frac{C}{C_0} = \frac{1}{1 + k\theta}$$
   d. Rearrange Equation 3.11b to obtain the relationship between $(1 - C/C_0)$ and θ.
      $$1 - \frac{C}{C_0} = \frac{k\theta}{1 + k\theta}$$
   e. Develop the data for plotting the above two relationships.
      Use 90% conversion as an example.

$C = C_0 - 0.9 \times C_0 = 0.1C_0$ (fraction remaining)

$C/C_0 = 0.1$

$1 - C/C_0 = 0.9$ (fraction removed)

$$\theta = \frac{C_0 - C}{kC} = \frac{C_0 - 0.1C_0}{k \times (0.1C_0)} = \frac{0.9}{0.30\,\text{h}^{-1} \times 0.1} = 30\,\text{h}$$

Assume different values of conversions and repeat the calculations. The results are tabulated in Table 3.2.

Note: If the experimental data are utilized, the relationships between the fraction remaining (C/C_0), and fraction removed ($1 - C/C_0$) with respect to corresponding residence time θ are developed and summarized as in Table 3.2.

TABLE 3.2 Fraction Remaining and Removed of a Reactive Substance in a CFSTR with Respect to Residence Time (Example 3.20)

Parameter	Conversion, %							
	100	90	80	60	40	20	10	0
C/C_0, fraction remaining	0	0.1	0.2	0.4	0.6	0.8	0.9	1
$1 - C/C_0$, fraction removed	1	0.9	0.8	0.6	0.4	0.2	0.1	0
θ, h	∞	30	13.3	5	2.22	0.83	0.37	0
Relationship between C_0/C and θ								
θ, h	∞	30	13.3	5	2.22	0.83	0.37	0
C_0/C	∞	10	5	2.5	1.67	1.25	1.11	1
Relationship between $C_0/(C_0 - C)$ and $1/\theta$								
$1/\theta$, h^{-1}	0	0.033	0.075	0.2	0.45	1.2	2.7	∞
$C_0/(C_0 - C)$	1	1.11	1.25	1.67	2.5	5	10	∞

f. Plot the curves.

The relationships between the fraction remaining (C/C_0) and fraction removed ($1 - C/C_0$) with respect to corresponding residence time θ are plotted in Figure 3.17a.

2. Derive linear relationships for determining k.

a. Equation 3.11b is rearranged as follows.

$$\frac{C_0}{C} = 1 + k\theta$$

A linear relationship between C_0/C (a reverse of fraction remaining) and θ is developed. The first-order reaction rate constant k can be determined from the slope of the linear relationship from experimental data (Figure 3.17b).

b. Equation 3.11b can be rearranged into the following relationship.

$$\frac{C_0}{C_0 - C} = 1 + \frac{1}{k} \times \frac{1}{\theta}$$

A linear relationship between $C_0/(C_0 - C)$ (a reverse of fraction removed) and $1/\theta$ (a reverse of residence time) is developed. The first-order reaction rate constant k can be determined by reversing the slope of the linear relationship obtained from experimental data (Figure 3.17c).

c. Develop the data for plotting above two linear relationships.

Use 90% conversion as an example.

$C_0/C = 1 \div 0.1 = 10$

$C_0/(C_0 - C) = 1 \div 0.9 = 1.1$

$1/\theta = 1 \div 30\,\text{h} = 0.033\,\text{h}^{-1}$

Assume different values of conversions and repeat the calculations. The results are also provided in Table 3.2.

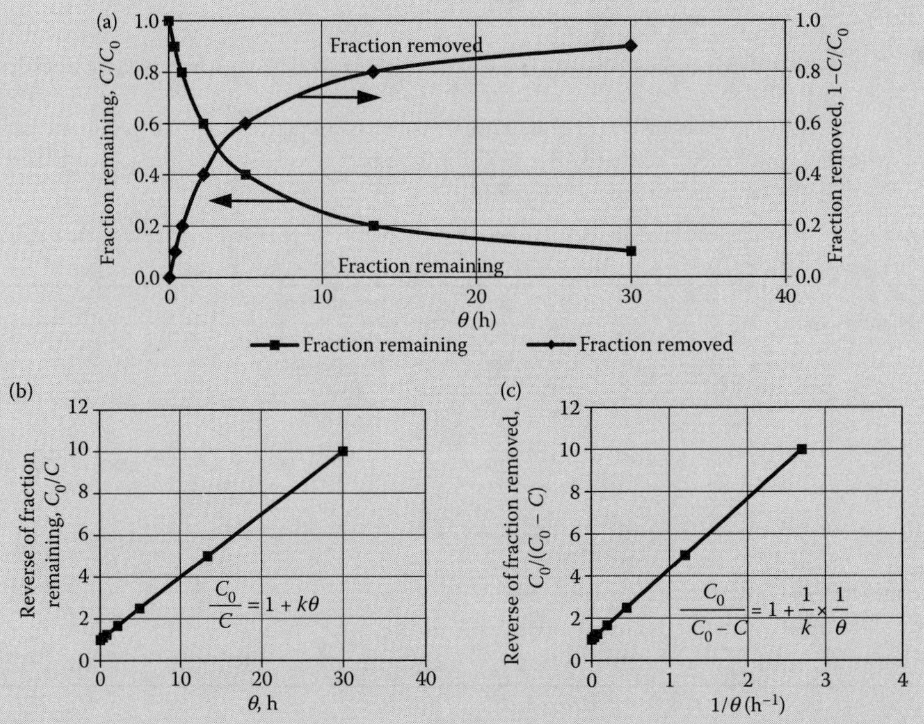

FIGURE 3.17 Concentration profiles and linear relationships of reactive substance in the effluent under steady-state conditions: (a) C/C_0 and $(1 - C/C_0)$ versus θ, (b) C_0/C versus θ, and (c) $C_0/(C_0 - C)$ versus $1/\theta$ (Example 3.20).

d. Determine k from linear plots.

The linear relationships are also plotted in Figure 3.17b and c.

e. Validate k from Figure 3.17b and c.

i. From Figure 3.17b.

$$k = \frac{(10 - 1)}{30\,\text{h}} = 0.3\,\text{h}^{-1}$$

ii. From Figure 3.17c.

$$\frac{1}{k} = \frac{(10 - 1)}{2.7\,\text{h}^{-1}} = 3.33\,\text{h}$$

$$k = \frac{1}{3.33\,\text{h}} = 0.3\,\text{h}^{-1}$$

iii. Compare the results with the given k.

The k value determined from either Figure 3.17b or c is exactly same of that given in the example statement.

EXAMPLE 3.21: NONSTEADY-STATE CONVERSION OF A REACTIVE SUBSTANCE IN A CFSTR

A reactor receives industrial waste for destruction of hazardous substance. The influent flow and concentration vary considerably. The reactor data are summarized below:

Reactor volume, $V = 480\ \text{m}^3$	Average waste concentration, $C_0 = 40\ \text{mg/L}$
Average flow, $Q = 30\ \text{m}^3/\text{d}$	First-order reaction kinetic coefficient, $k = 0.18\ \text{d}^{-1}$

The reactor is operating under nonsteady-state condition. Using Equation 3.13b develop the following plots: (a) concentration profile C versus reaction time t; and (b) Dimensionless parameters C/C_0 versus t/θ.

Also, indicate steady-state concentration line on these plots.

Solution

1. Calculate θ from Equation 3.10c.

$$\theta = \frac{V}{Q} = \frac{480\ \text{m}^3}{30\ \text{m}^3/\text{d}} = 16\ \text{d}$$

2. Calculate β from Equation 3.12, and steady-state concentration of hazardous substance.

$$\beta = \frac{Q}{V} + k = \frac{30\ \text{m}^3/\text{d}}{480\ \text{m}^3} + 0.18\ \text{d}^{-1} = 0.243\ \text{d}^{-1}$$

$$\frac{Q}{V} \times \frac{C_0}{\beta} = \frac{1}{16\ \text{d}} \times \frac{40\ \text{g/m}^3}{0.243\ \text{d}^{-1}} = 10.29\ \text{g/m}^3 \quad \text{(steady-state concentration)}$$

3. Substitute the above values in Equation 3.13b and simplify to obtain Equation 3.14.

$$C = \frac{Q}{V} \times \frac{C_0}{\beta}(1 - e^{-\beta t}) + C_0 e^{-\beta t}$$

$$C = 10.29\ \text{g/m}^3 \times (1 - e^{-0.243\ \text{d}^{-1} \times t}) + 40\ \text{g/m}^3 \times e^{-0.243\ \text{d}^{-1} \times t} \tag{3.14}$$

4. Tabulate the data for plotting the concentration profile curves (Table 3.3).

TABLE 3.3 Nonsteady-State Conversion of a Reactive Substance in a CFSTR (Example 3.21)

t, d	$e^{-0.243\ \text{d}^{-1} t}$	$(1 - e^{-0.243\ \text{d}^{-1} t})$	$10.29 \times (1 - e^{-0.243\ \text{d}^{-1} t})$	$40 \times e^{-0.243\ \text{d}^{-1} t}$	C, g/m^3	t/θ	C/C_0
0	1	0	0	40	40	0	1
0.5	0.886	0.114	1.18	35.4	36.6	0.0313	0.92
1	0.784	0.216	2.22	31.4	33.6	0.0625	0.84
1.5	0.695	0.305	3.14	27.8	30.9	0.0938	0.77
2	0.615	0.385	3.96	24.6	28.6	0.125	0.71
3	0.482	0.518	5.33	19.3	24.6	0.188	0.62
6	0.233	0.767	7.89	9.31	17.2	0.375	0.43
10	0.0880	0.912	9.38	3.52	12.9	0.625	0.32
15	0.0261	0.974	10.0	1.04	11.1	0.938	0.28
20	0.00775	0.992	10.2	0.310	10.5	1.25	0.26

5. Plot the concentration profile C versus t from the data tabulated above.

   The values are plotted in Figure 3.18a. The steady-state concentration of 10.29 g/m^3 obtained from Equation 3.11b is also shown in Figure 3.18a.

6. Plot the dimensionless values C/C_0 versus t/θ (Figure 3.18b).

   The steady-state concentration ratio $C/C_0 = (10.29\ \text{g/m}^3)/(40\ \text{g/m}^3) = 0.26$ (Figure 3.18b).

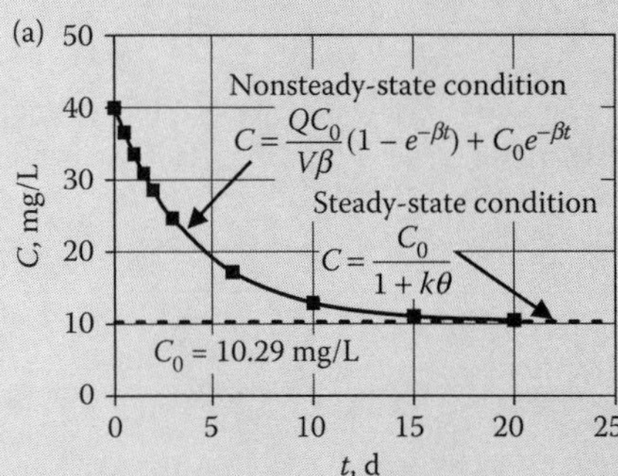

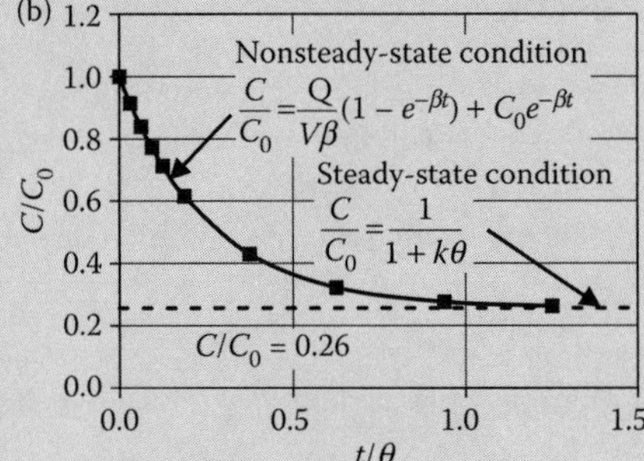

FIGURE 3.18 Concentration profile in the effluent under steady-state and nonsteady-state conditions: (a) C versus t, and (b) dimensionless parameters C/C_0 versus t/θ (Example 3.21).

EXAMPLE 3.22: STEP FEED OF A REACTIVE SUBSTANCE IN A CFSTR

A constant flow is maintained through a CFSTR. At $t = 0$, a reactive substance is added in the influent stream at a constant rate. The conversion reaction is of the first order. Determine (a) the output concentration equation as a function of time, (b) output concentration when $t = \infty$, (c) output concentration as g/m^3 at $t = 1$ and 4 h, and ∞ from the start of step feed, and (d) output concentration profile as g/m^3 and as dimensionless parameters C/C_0 and t/θ for both the reactive and conservative substances. The operational conditions of the reactor are: $V = 20\ \text{m}^3$, $Q = 200\ \text{m}^3/\text{d}$, C_0 in step feed $= 120\ \text{g/m}^3$, and $k = 0.2\ \text{h}^{-1}$.

Solution

1. Development of the output concentration equation of a reactive (nonconservative) substance.

   Write the mass balance equation for the reactor (Equation 3.3).

$$V\frac{dC}{dt} = QC_0 - QC + V(-kC)$$

   Rearrange the equation to obtain Equation 3.15a.

$$\frac{dC}{dt} = \frac{Q}{V}C_0 - \frac{Q}{V}C - kC$$

$$\frac{dC}{dt} = \frac{Q}{V}C_0 - \frac{Q}{V}C\left(1 + \frac{V}{Q}k\right) = \frac{1}{\theta}[C_0 - C(1 + \theta k)] \quad (3.15a)$$

   Rearrange and integrate Equation 3.15a to obtain Equations 3.15b and 3.15c.

$$\int_{C_0}^{C} \frac{dC}{C_0 - C(1 + k\theta)} = \frac{1}{\theta}\int_0^t dt - \left(\frac{1}{1 + k\theta}\right)\ln\left[\frac{C_0 - (1 + k\theta)C}{C_0}\right] = \frac{t}{\theta}$$

$$C = \frac{C_0(1 - e^{-(1+k\theta)t/\theta})}{1 + k\theta} \quad (3.15b)$$

$$\frac{C}{C_0} = \frac{(1 - e^{-(1+k\theta)t/\theta})}{1 + k\theta} \quad (3.15c)$$

2. Develop steady-state equation when $t = \infty$.
   Substitute $t = \infty$ in Equation 3.15b and solve. Steady-state equation $C = C_0/(1 + k\theta)$ (Equation 3.11b) is reached.
3. Determine output concentration in g/m^3.
   Calculate θ from Equation 3.10c.

$$\theta = \frac{V}{Q} = \frac{20\,\text{m}^3}{200\,\text{m}^3/\text{d}} \times 24\,\text{h/d} = 2.4\,\text{h}$$

Calculate C values from Equation 3.15b when $t = 1$, 4 h and ∞.

$$\text{when } t = 1\ h,\ C = \frac{120\,\text{g/m}^3 \times (1 - e^{-(1+0.2\,\text{h}^{-1}\times 2.4\,\text{h})\times(1\,\text{h}/2.4\,\text{h})})}{(1 + 0.2\text{h}^{-1} \times 2.4\text{h})}$$

$$= \frac{120\,\text{g/m}^3 \times (1 - e^{-0.61})}{1.48}$$

$$= 37.3\,\text{g/m}^3$$

$$\text{when } t = 4\ h,\ C = \frac{120\,\text{g/m}^3 \times (1 - e^{-(1+0.2\,\text{h}^{-1}\times 2.4\,\text{h})\times(4\,\text{h}/2.4\,\text{h})})}{(1 + 0.2\,\text{h}^{-1} \times 2.4\,\text{h})} = 74.2\,\text{g/m}^3$$

$$\text{when } t = \infty,\ C = \frac{120\,\text{g/m}^3 \times (1 - e^{-\infty})}{1.48} = \frac{120\,\text{g/m}^3}{1.48} = 81.1\,\text{g/m}^3$$

4. Develop the output time response calculation table.
   The calculations are developed from Equation 3.15b for reactive substance. The data for conservative substances utilize Equation 3.10d. These calculations are summarized in Table 3.4.

TABLE 3.4 Output Concentrations of Reactive and Conservative Substance in a CFSTR (Example 3.22)

		Reactive Substance		Conservative Substance	
t, h	t/θ	C from Equation 3.15b, g/m^3	C/C_0 from Equation 3.15c	C from Equation 3.10d, g/m^3	C/C_0 from Equation 3.10e
0	0	0	0	0	0
0.5	0.21	21.5	0.18	22.6	0.19
1	0.42	37.3	0.31	40.9	0.34
2	0.83	57.5	0.48	67.8	0.57
4	1.67	74.2	0.62	97.3	0.81
10	4.17	80.9	0.67	118	0.98
15	6.25	81.1	0.68	120	1.00
20	8.33	81.1	0.68	120	1.00

5. Plot concentration profiles.

   The output time response curve of reactive (nonconservative) substance in g/m^3 versus reaction time t is shown in Figure 3.19a. The output time response curves for reactive and conservative substances as dimensionless parameters C/C_0 versus t/θ are shown in Figure 3.19b.

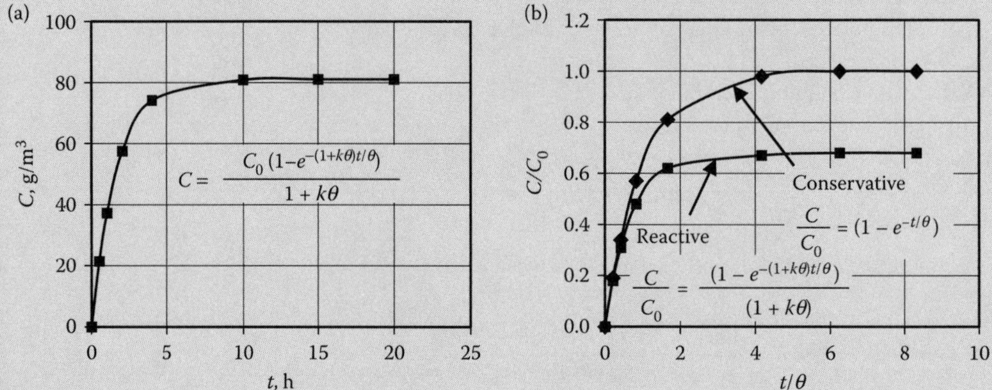

FIGURE 3.19 Concentration profiles: (a) output concentration of reactive (nonconservative) substance, and (b) comparison of concentration profiles of reactive (nonconservative) and nonreactive (conservative) substances in dimensionless parameters (Example 3.22).

EXAMPLE 3.23: NONSTEADY STATE REACTIVE PURGING OF A REACTOR WITH FIRST-ORDER DECAY

A manufacturing plant stabilizes its waste in a reactor before discharging the effluent into a sanitary sewer. The raw waste concentration is 250 mg/L. The reactor volume is 750 m^3, and volumetric flow rate is 30 m^3/d. The waste is stabilized under steady-state condition exhibiting first-order reaction kinetic ($k = 0.37\ d^{-1}$). For annual maintenance, the waste input into the reactor is terminated at $t = 0$, while the flow of clean liquid is continued for reactive purging of the reactor.

a. Determine the steady-state concentration of waste before purging started.
b. Develop the generalized equation to express the reactive purging of the reactor. Check the boundary conditions ($t = 0$ and $t = \infty$).
c. Determine the concentration of waste in the reactor after 2 days since waste input is terminated and purging started, at $t = 0$.
d. Draw the concentration profile in the reactor for steady-state condition until the reactor purging is complete.

Solution

1. Determine the steady-state concentration of waste in the reactor before purging started.

   Calculate θ from Equation 3.10c.

$$\theta = \frac{V}{Q} = \frac{750\ \text{m}^3}{30\ \text{m}^3/\text{d}} = 25\ \text{d}$$

Substitute the data in Equation 3.11b to determine the steady-state concentration in the reactor during continuous feed.

$$C = \frac{C_0}{1 + k\theta} = \frac{250\,\text{mg/L}}{1 + 0.37\,\text{d}^{-1} \times 25\,\text{d}} = \frac{250\,\text{mg/L}}{10.25} = 24.4\,\text{mg/L}$$

2. Develop the generalized equation to express the reactive purging of the reactor before purging started. Simplify Equation 3.3 to determine C as a function of time.

$$V\frac{dC}{dt} = QC_0 - QC - VkC$$

Since waste input is terminated, $C_0 = 0$.

$$V\frac{dC}{dt} = -C(Q + Vk)$$

Integrate the differential equation to solve for C.

$$\int_{C_0}^{C} \frac{dC}{C} = \int_{0}^{t} -(Q/V + k)\,dt$$

where

$t = 0$, exponential term $e^0 = 1$
$C = C_0$, steady-state concentration in the reactor is reached.
$t = \infty$, exponential term $e^{-\infty} = 0$
$C = 0$, the reactor purging is complete.

$$\ln(C) - \ln(C_0) = -(Q/V + k)t$$

$$\ln\left(\frac{C}{C_0}\right) = -(Q/V + k)$$

$$C = C_0 e^{-(Q/V+k)t}$$

$$C = C_0 e^{-(1+k\theta)t/\theta}$$

$$\frac{C}{C_0} = e^{-(1+k\theta)t/\theta}$$

3. Determine the concentration of waste in the effluent after 2 days ($t = 2$) since purging began.

$$C = C_0 e^{-(1+k\theta)t/\theta}$$

$$C = 24.4\,\text{mg/L} \times e^{-(1+0.37\,\text{d}^{-1}\times 25\,\text{d})\times(2\,\text{d}/25\,\text{d})}$$

$$C = 24.4\,\text{mg/L} \times e^{-0.82} = 24.4\,\text{mg/L} \times 0.44 = 10.7\,\text{mg/L}$$

4. Draw the concentration profile in the effluent before and after reactive purging began.
   a. Under steady-state condition, $C_0 = 24.4$ mg/L.
   b. Assume different values of t and repeat the calculations to determine values of C. The results are provided in Table 3.5.
5. Data plot is shown in Figure 3.20.

TABLE 3.5 Nonsteady State Reactive Purging Data of a Reactor (Example 3.23)

t, d	Reactive Substance		
	$(1+k\theta)\frac{t}{\theta}$	$e^{-(1+k\theta)\frac{t}{\theta}}$	C, mg/L
0	0	1	24.4
0.5	0.205	0.815	19.9
1	0.410	0.664	16.2
1.5	0.615	0.541	13.2
2	0.820	0.440	10.7
2.5	1.03	0.359	8.8
3.5	1.44	0.238	5.8
5	2.05	0.129	3.1
7	2.87	0.057	1.4
9	3.69	0.025	0.6
12	4.92	0.00730	0.2

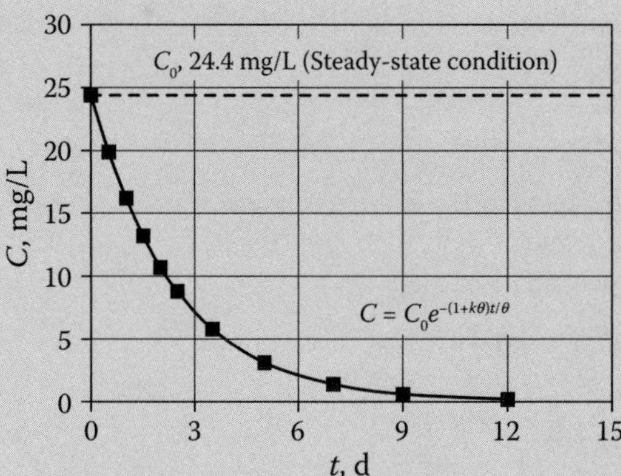

FIGURE 3.20 Concentration profile of industrial waste undergoing reactive purging from steady-state operation (Example 3.23).

3.4.3 Plug Flow Reactor

In an ideal plug flow reactor, the elements of fluid pass through the reactor and are discharged in the same sequence in which they enter the reactor. The flow regimes are characterized by the following (1) each fluid particle remains in the reactor for a time period equal to the theoretical detention time, and (2) there is no longitudinal dispersion or mixing of the fluid elements as they move through the system. Plug flow regime is approached in systems that have large length-to-width ratio for rectangular basin or large length-to-diameter ratios for circular pipes. At a length-to-diameter ratio of 50:1, the flow approaches plug flow regime if the velocity is not excessive.[4,5]

Conservative Tracer Response with Slug Input: If a slug of dye tracer is released in an ideal plug flow reactor, it will move and form a band. As the flow continues, the band will move through the reactor. It will emerge as a band in the effluent at the theoretical detention time θ (Equation 3.10c). The generalized tracer band movement is shown in Figure 3.21a.

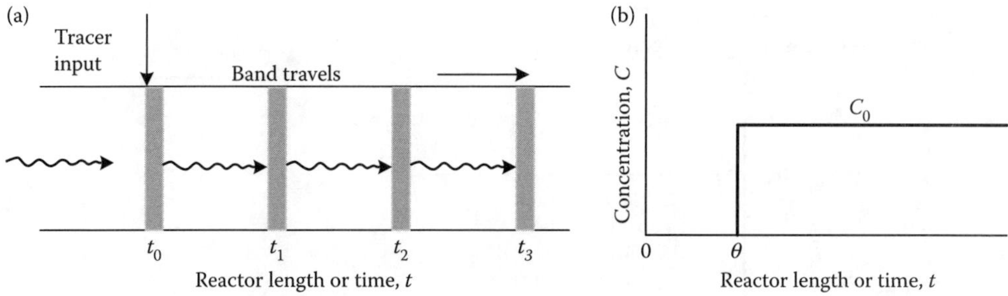

FIGURE 3.21 Movement of tracer in an ideal plug flow reactor: (a) slug band and (b) step-feed profile.

Conservative Tracer Response with Step Feed: In a plug flow reactor, the continuous dye tracer feed results in a continuous discharge of concentration C_0 after the theoretical detention time θ. Prior to θ, no dye will be detected in the effluent. The step-feed tracer profile is shown Figure 3.21b.

Stabilization of Nonconservative (Reactive) Substance in a PFR: A nonconservative substance undergoes zero-, first-, or second-order decay in a PFR. The concentration of reactive substance in the reactor varies from point to point along the flow path. The concentration equations for steady-state and nonsteady-state conditions can be developed from mass balance procedure.[4,5]

Saturation-Type Reaction: The residence time equations of CFSTR and PFR for saturation-type reaction under steady-state condition ($r = -kC/[K_s + C]$) were presented in Section 2.3.2. The saturation-type reaction requires values of C_0, K_s, and k. The final results are quite sensitive to these values. Readers may refer to Equation 2.7a, and Examples 2.10, 3.32, and 3.33 for more information.[3,4,7]

EXAMPLE 3.24: SLUG TRACER INPUT IN A PLUG FLOW REACTOR

An ideal plug flow reactor is a long channel. The channel volume is 38 m^3 and flow is 500 m^3/d. The tracer solution contains 4 g dye/L. A slug of 19 L dye solution is released into the influent zone. The tracer band has a concentration of 8.4 times C_0. Determine (a) average expected tracer concentration C_0 if it is completely mixed in the basin, (b) theoretical detention time θ, (c) tracer concentration in the band, and (d) time for emergence of the band in the effluent. Draw the tracer profile.

Solution

1. Calculate C_0 from Equation 3.9b.

$$C_{0,\text{slug}} = \frac{V_{\text{tracer}}}{V} C_{\text{tracer}} = \frac{19\,\text{L}}{38\,\text{m}^3} \times 4\,\text{g/L} = 2\,\text{g/m}^3 \text{ or } 2\,\text{mg/L}$$

2. Calculate the theoretical detention time θ from Equation 3.10c.

$$\theta = \frac{V}{Q} = \frac{38\,\text{m}^3}{500\,\text{m}^3/\text{d}} = 0.076\,\text{d}$$

or $\theta = 0.076\,\text{d} \times 24\,\text{h/d} = 1.8\,\text{h}$

3. Calculate the tracer concentration in the band, and time for the band to exit in the effluent.

Tracer concentration in the band or piston $= 2\,\text{g/m}^3 \times 8.4 = 16.8\,\text{g/m}^3$

Time for dye concentration band to reach effluent is θ or 1.8 h.

4. The tracer profile is shown in Figure 3.22.

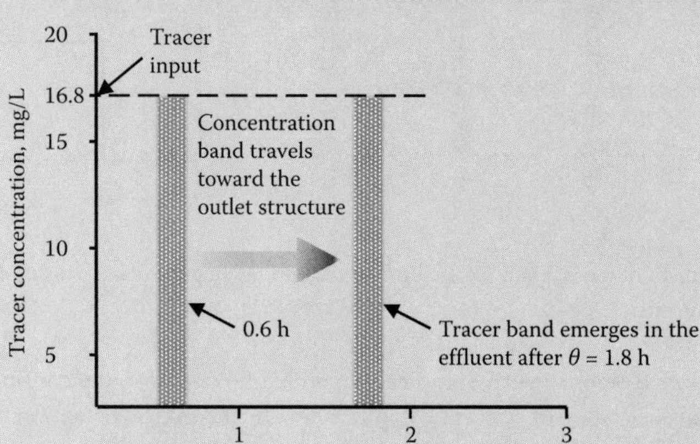

FIGURE 3.22 Slug conservative tracer band moving in an ideal plug flow reactor (Example 3.24).

EXAMPLE 3.25: STEP FEED OF A CONSERVATIVE TRACER IN A PLUG FLOW REACTOR

The flow in a force main 1000 m long and 92 cm diameter represents an ideal plug flow regime. The velocity in the force main is 1 m/s. A 2 g/L tracer solution is continuously injected into the force main at a rate of 0.5 L/s. Determine the following (a) flow through the force main, (b) theoretical detention time, (c) average tracer concentration in the force main, and (d) time and concentration of emergence of tracer at the outlet point. Draw the tracer profile.

Solution

1. Calculate the flow through the force main.

$$A = \frac{\pi}{4}D^2 = \frac{\pi}{4}(0.92\,\text{m})^2 = 0.665\,\text{m}^2$$

$$Q = v \times A = 1\,\text{m/s} \times 0.665\,\text{m}^2 = 0.665\,\text{m}^3/\text{s}$$

2. Calculate the theoretical detention time θ.

$$\theta = \frac{V}{Q} = \frac{L \times A}{v \times A} = \frac{L}{v} = \frac{1000\,\text{m}}{1\,\text{m/s}} = 1000\,\text{s}$$

$$\text{or } \theta = 1000\,\text{s} \times \frac{1\,\text{min}}{60\,\text{s}} = 16.7\,\text{min}$$

3. Calculate the average tracer concentration C_0 in the force main.

$$\text{Injection rate of tracer in the force main} = 2\,\text{g/L} \times 0.5\,\text{L/s} = 1\,\text{g/s}$$

$$\text{Concentration of tracer in the force main} = \frac{1\,\text{g/s}}{0.665\,\text{m}^3/\text{s}} = 1.5\,\text{g/m}^3$$

4. Determine the concentration and time of tracer emergence at the outlet zone.

Concentration of dye tracer at the outlet point = 1.5 g/m^3

Time of emergence of the dye tracer at the outlet point = 16.7 min

5. The tracer movement and concentration profile are shown in Figure 3.23a and b, respectively.

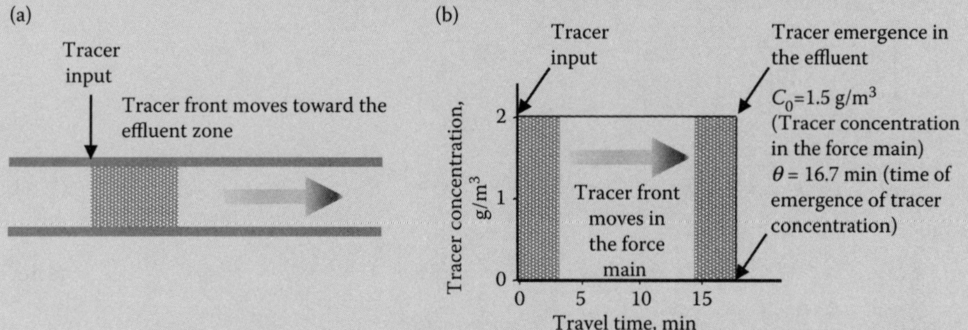

FIGURE 3.23 Profile of step feed of conservative tracer in an ideal plug flow reactor: (a) tracer front movement in the force main, and (b) tracer concentration in the effluent and travel time (Example 3.25).

EXAMPLE 3.26: CONVERSION EQUATION OF REACTIVE SUBSTANCE IN A PFR

Develop the effluent concentration equation of a reactive substance treated in a PFR. The influent concentration is C_0 and first-order reaction constant is k.

Solution

1. Draw the definition sketch in Figure 3.24 for analysis of a PFR.

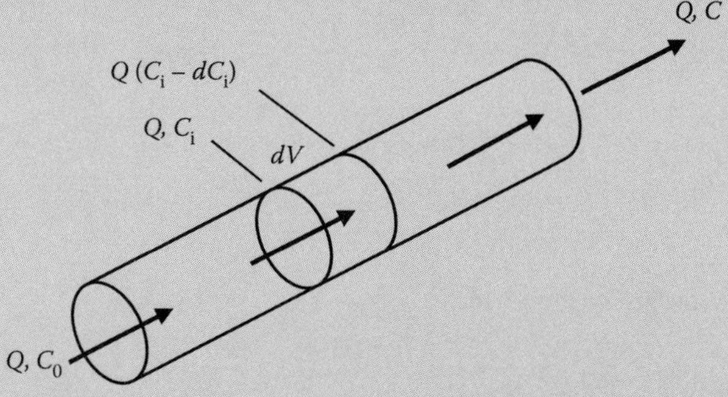

FIGURE 3.24 Definition sketch for analysis of nonconservative substance in a PFR (Example 3.26).

2. Apply the mass balance equations at the element dV (Equation 3.16a).

[Input] = [Output] + [Decrease due to reaction]

$QC_i = Q(C_i - dC_i) + rdV$

$$Q\,dC_i = r\,dV \tag{3.16a}$$

First-order reaction rate $r = -kC_i$

$$Q\,dC_i = -kC_i\,dV$$

3. Rearrange and integrate to obtain Equations 3.16b through 3.16d.

$$\int_{C_0}^{C} \frac{dC_i}{C_i} = -\frac{k}{Q}\int_{0}^{V} dV$$

$$\ln\left(\frac{C}{C_0}\right) = -\frac{k}{Q}V = -k\theta \tag{3.16b}$$

$$C = C_0 e^{-k\theta} \tag{3.16c}$$

$$\frac{C}{C_0} = e^{-k\theta} \tag{3.16d}$$

It may be noted that the concentration equations for a PFR (Equations 3.16c and 3.16d) are the same as those for the batch rector (Equations 3.8b and 3.8c). Therefore, for a reaction time equal to residence time the performance of a PFR is same as that of a batch reactor when $t = \theta$ (the theoretical detention time).

EXAMPLE 3.27: RESIDENCE TIME IN A PFR FOR A GIVEN CONVERSION

A PFR is designed to treat a waste that follows a first-order reaction kinetic. What is the residence time to achieve 90% conversion? The first-order kinetic coefficient at the operating temperature is $0.30\ \text{h}^{-1}$.

Solution

1. Apply Equation 3.16c.

For 90% conversion, $C = C_0\,(1 - 0.9) = 0.1C_0$

or $\dfrac{C}{C_0} = 0.1$

Substitute the value in Equation 3.16c.

$$\frac{C}{C_0} = 0.1 = e^{(-0.30\,\text{h}^{-1})\times\theta}$$

2. Determine the residence time θ.

Solve the above equation.

$$(-0.30\,\text{h}^{-1}) \times \theta = \ln(0.1)$$

$$\theta = \frac{\ln(0.1)}{-0.30\,\text{h}^{-1}} = \frac{-2.3}{-0.30\,\text{h}^{-1}} = 7.7\,\text{h}$$

3.4.4 Comparative Performance of a PFR and CFSTR

It is well recognized that in wastewater treatment, the performance of a PFR is significantly higher than that of a CFSTR. The comparison can easily be made by (1) comparing the hydraulic retention time θ for a given degree of removal or (2) comparing the degree of removal for a given θ. In both cases, same value of reaction rate constant and same reaction order must be used.[4]

First-order Reaction: The residence time equations of CFSTR and PFR are easily developed under steady-state condition ($r = -kC$). Readers may refer to Section 2.5.2, and Examples 2.18 through 2.28, and 3.18 through 3.23 as this information is developed earlier.

Second-order Reaction: The residence time equations of CFSTR and PFR for simple Type I second-order reaction under steady-state condition can be easily developed. Readers may refer to Section 2.5.3, and Examples 2.29 through 2.35.

EXAMPLE 3.28: RESIDENCE TIME EQUATIONS FOR CFSTR AND PFR

Express the hydraulic residence time equations of CFSTR and PFR for first-order reaction under steady-state condition. Also,calculate (a) residence times for PFR and CFSTR if removal is 90%, and (b) compare the performance of both reactors. Assume k is 0.28 h^{-1}.

Solution

1. Express the hydraulic residence time equations for both reactors ($r = -kC$).

   Equation 3.17 for CFSTR is developed from Equation 3.11b.

$$\theta_{CFSTR} = \frac{1}{k}\left(\frac{C_0}{C_{CFSTR}} - 1\right) \quad (3.17)$$

   The following equation for PFR (Equation 3.18) is developed from Equation 3.16b.

$$\theta_{PFR} = -\frac{1}{k}\ln\left(\frac{C_{PFR}}{C_0}\right) \quad (3.18)$$

2. Calculate the reaction time for both reactors to achieve 90% percent removal from Equations 3.17 and 3.18.

   For 90% removal, $C = (C_0 - 0.9C_0) = 0.1C_0$

$$\text{or } \frac{C}{C_0} = 0.1$$

$$\theta_{CFSTR} = \frac{1}{0.28\,h^{-1}} \times \left(\frac{1}{0.1} - 1\right) = 32.1\,h$$

$$\theta_{PFR} = -\frac{1}{0.28\,h^{-1}} \times \ln(0.1) = 8.2\,h$$

3. Compare the performance of both reactors to achieve 90% removal.

   To achieve 90% removal, the residence time required for the CFSTR is 32.1 h, while that for a PFR is 8.2 h. The ratio of ($\theta_{CFSTR}/\theta_{PFR}$) is 3.9. This shows that PFR is almost four times more efficient than a CFSTR in achieving 90% removal efficiency.

EXAMPLE 3.29: PERFORMANCE EVALUATION OF FIRST-ORDER REACTION IN CFSTR AND PFR FOR A GIVEN RESIDENCE TIME

A CFSTR and PFR are operating under identical condition receiving same waste. The volumes of both reactors are 35 m^3. The flow through each reactor is 100 m^3/d. The first-order reaction kinetic coefficient is 0.28 h^{-1}, and the influent concentration of waste is 280 g/m^3. Determine the following (a) the ratio of effluent and influent concentrations from both reactors and (b) percent removal in each reactor. Compare the performance of both reactors.

Solution

1. Calculate the hydraulic residence time in both reactors using Equation 3.10c.

$$\theta = \frac{V}{Q} = \frac{35\,\text{m}^3}{100\,\text{m}^3/\text{d}} = 0.35\,\text{h}$$

or $\theta = 0.35\,\text{d} \times 24\,\text{h/d} = 8.4\,\text{h}$

2. Calculate the effluent concentration from the CFSTR using Equation 3.17.

$$8.4\,\text{h} = \frac{1}{0.28\,\text{h}^{-1}} \times \left(\frac{280\,\text{g/m}^3}{C_{\text{CFSTR}}} - 1\right)$$

$$C_{\text{CFSTR}} = 83.5\,\text{g/m}^3$$

The CFSTR effluent concentration can also be calculated directly using Equation 3.11b.

$$C_{\text{CFSTR}} = \frac{1}{1 + k\theta} = \frac{280\,\text{g/m}^3}{1 + 0.28\,\text{h}^{-1} \times 8.4\,\text{h}} = 83.5\,\text{g/m}^3$$

3. Calculate the effluent concentration from the PFR from Equation 3.18.

$$8.4\,\text{h} = -\frac{1}{0.28\,\text{h}^{-1}} \times \ln\left(\frac{C_{\text{PFR}}}{280\,\text{g/m}^3}\right)$$

$$C_{\text{PFR}} = 26.7\,\text{g/m}^3$$

The PFR effluent concentration can also be calculated directly from Equation 3.16c.

$$C_{\text{PFR}} = C_0 e^{-k\theta} = 280\,\text{g/m}^3 \times e^{-0.28\,\text{h}^{-1} \times 8.4\,\text{h}} = 26.7\,\text{g/m}^3$$

4. Calculate the effluent and influent ratio for each reactor.

$$\frac{C_{\text{CFSTR}}}{C_0} = \frac{83.5\,\text{g/m}^3}{280\,\text{g/m}^3} = 0.30$$

$$\frac{C_{\text{PFR}}}{C_0} = \frac{26.7\,\text{g/m}^3}{280\,\text{g/m}^3} = 0.10$$

5. Calculate the percent removal in each reactor.

$$\text{Percent removal in CFSTR} = \frac{(280 - 83.5)\,\text{g/m}^3}{280\,\text{g/m}^3} \times 100\% = 70\%$$

$$\text{Percent removal in PFR} = \frac{(280 - 26.7)\,\text{g/m}^3}{280\,\text{g/m}^3} \times 100\% = 90\%$$

6. Compare the performance of both reactors under steady state and first-order reaction kinetics.
   For the given hydraulic residence time of 8.4 h, the performance of PFR is 90% while that of CFSTR is 70%.

EXAMPLE 3.30: SECOND-ORDER RESIDENCE TIME EQUATIONS OF CFSTR AND PFR

Express the hydraulic residence time equations for CFSTR and PFR for the Type I second-order reaction under steady-state condition. Also, calculate (a) residence times of CFSTR and PFR if removal is 90%, and (b) compare the performance of both reactors. Assume k is 0.09 $m^3/g{\cdot}h$ and $C_0 = 100\ g/m^3$.

Solution

1. Develop the generalized equation to express the Type I second-order reaction equation.

   Write the mass balance equation for the reactor (Equation 3.3).

$$V\frac{dC}{dt} = QC_0 - QC + V(-kC) \qquad (3.19a)$$

For the Type I second-order reaction, $r = -kC.^2$ The reaction is expressed by Equation 3.19.

$$V\frac{dC}{dt} = QC_0 - QC - VkC^2 \qquad (3.19)$$

2. Express the hydraulic residence time equation for the Type I second-order reaction in a CFSTR.

   Under steady state in a CFSTR $(dC/dt) = 0$. Equation 3.19 is expressed by Equation 3.20a.

$$C_0 - C = \frac{V}{Q}kC \qquad (3.20a)$$

Rearrange the equation to obtain Equation 3.20b.

$$\theta_{\text{CFSTR}} = \frac{C_0 - C}{kC^2} \qquad (3.20b)$$

3. Express the hydraulic residence time equation for the Type I second-order reaction in a PFR.

   Under steady-state condition $C_0 = C$. Equation 3.19 is expressed by Equation 3.20c.

$$\frac{dC}{dt} = -kC^2 \qquad (3.20c)$$

Integrate and solve within the limits (C_0 at $t = 0$ and C at $t = \theta_{\text{PFR}}$) to obtain the following (Equation 3.20d):

$$\int_{C_0}^{C} \frac{dC}{C^2} = -k \int_{0}^{\theta_{\text{PFR}}} dt$$

$$\left(\frac{1}{C_0} - \frac{1}{C}\right) = -k\theta_{\text{PFR}}$$

$$\theta_{\text{PFR}} = \frac{C_0 - C}{kC_0C} \qquad (3.20d)$$

4. Calculate the hydraulic residence time for both reactors to achieve 90% removal from Equations 3.20b and 3.20d.

   For 90% removal, $C = (C_0 - 0.9C_0) = 0.1C_0$

$$\theta_{CFSTR} = \frac{(C_0 - 0.1C_0)}{k(0.1C_0)^2} = \frac{0.9}{0.09\,\text{m}^3/\text{g·h} \times 0.01 \times 100\,\text{g/m}^3} = 10\,\text{h}$$

$$\theta_{PFR} = \frac{C_0 - 0.1C_0}{kC_0 \times 0.1C_0} = \frac{0.9}{0.09\,\text{m}^3/\text{g·h} \times 0.1 \times 100\,\text{g/m}^3} = 1\,\text{h}$$

5. Compare the performance of both reactors.

   To achieve 90% removal, the hydraulic residence time required for the CFSTR is 10 h, while that for a PFR is 1 h. The ratio of $(\theta_{CFSTR}/\theta_{PFR})$ is 10. This shows that PFR is significantly more efficient.

EXAMPLE 3.31: PERFORMANCE EVALUATION OF SECOND-ORDER REACTION IN A CFSTR AND PFR FOR A GIVEN RESIDENCE TIME

The volumes of a CFSTR and PFR are 35 m^3 each. The flow through each reactor is 100 m^3/d. The second-order kinetic coefficient is 0.09 m^3/g·h and $C_0 = 100$ g/m^3. Determine the following (a) the ratio of effluent and influent concentrations from both reactors, (b) percent removal in each reactor, and (c) compare the performance of both reactors.

Solution

1. Calculate the hydraulic residence time of both reactors from Equation 3.10c.

$$\theta = \frac{V}{Q} = \frac{35\,\text{m}^3}{100\,\text{m}^3/\text{d}} = 0.35\,\text{d}$$

   or $\theta = 0.35\,\text{d} \times 24\,\text{h/d} = 8.4\,\text{h}$

2. Calculate the effluent concentration from the CFSTR using Equation 3.20b.

$$8.4\,\text{h} = \frac{1}{0.09\,\text{m}^3/\text{g·h}}\left(\frac{100\,\text{g/m}^3 - C}{C^2}\right)$$

$$0.756\,\text{m}^3/\text{g} = \frac{100\,\text{g/m}^3 - C}{C^2}$$

$$0.756\,\text{m}^3/\text{g}C^2 = 100\,\text{g/m}^3 - C$$

$$C^2 + (1.323\,\text{g/m}^3)C = 132.3\,(\text{g/m}^3)^2$$

$$C = 10.8\,\text{g/m}^3$$

3. Calculate the effluent concentration from PFR using Equation 3.20d.

$$8.4\,\text{h} = \frac{100\,\text{g/m}^3 - C}{0.09\,\text{m}^3/\text{g·h} \times 100\,\text{g/m}^3 \times C} = \frac{100\,\text{g/m}^3 - C}{9\,\text{h}^{-1} \times C}$$

$$C = 1.3\,\text{g/m}^3$$

4. Calculate the ratio of effluent to influent C/C_0 in each reactor.

$$\frac{C}{C_0} \text{ in CFSTR} = \frac{10.8\,\text{g/m}^3}{100\,\text{g/m}^3} = 0.11$$

$$\frac{C}{C_0} \text{ in PFR} = \frac{1.3\,\text{g/m}^3}{100\,\text{g/m}^3} = 0.01$$

5. Calculate the percent removal in each reactor.

$$\text{Percent removal in CFSTR} = \frac{(100 - 10.8)\,\text{g/m}^3}{100\,\text{g/m}^3} = 89\%$$

$$\text{Percent removal in PFR} = \frac{(100 - 1.3)\,\text{g/m}^3}{100\,\text{g/m}^3} = 99\%$$

6. Compare the performance of both reactors.

For the given hydraulic residence time of 8.4 h, the removal efficiency of the PFR is 99%, while that of CFSTR is 89% for second-order reaction.

EXAMPLE 3.32: COMPARISON OF RESIDENCE-TIMES OF CFSTR AND PFR FOR A GIVEN REMOVAL BY SATURATION-TYPE REACTION

Express the hydraulic residence time equations of CFSTR and PFR for saturation-type reactions under steady-state condition. Also, calculate (a) residence time of CFSTR and PFR if removal is 80%, and (b) compare the performance of both reactors. Assume k, K_s, and C_0 are $0.19\,\text{g/m}^3{\cdot}\text{h}$, $10\,\text{g/m}^3$, and $100\,\text{g/m}^3$, respectively.

Solution

1. Express the saturation equation for both reactors ($r = kC/[K_s + C]$).

The θ for both reactors are expressed by Equations 3.21 and 3.22.[3]

$$\theta_{\text{CFSTR}} = \frac{(C_0 - C)(K_s + C)}{kC} \tag{3.21}$$

$$\theta_{\text{PFR}} = \frac{1}{k}\left(K_s \ln\left(\frac{C_0}{C}\right) + C_0 - C\right) \tag{3.22}$$

2. Calculate the residence time for both reactors to achieve 80% removal.

$$C = C_0 - 0.8\,C_0 = 0.2\,C_0 = 0.2 \times 100\,\text{g/m}^3 = 20\,\text{g/m}^3$$

$$\theta_{\text{CFSTR}} = \frac{(C_0 - 0.2\,C_0) \times (10\,\text{g/m}^3 + 0.2 \times 100\,\text{g/m}^3)}{0.19\,\text{g/m}^3{\cdot}\text{h} \times 0.2\,C_0} = \frac{0.8 \times (10 + 20)\,\text{g/m}^3}{0.19\,\text{g/m}^3{\cdot}\text{h} \times 0.2} = 632\,\text{h}$$

$$\theta_{\text{PFR}} = \frac{1}{0.19\,\text{g/m}^3{\cdot}\text{h}}\left(10\,\text{g/m}^3 \times \ln\left(\frac{100\,\text{g/m}^3}{20\,\text{g/m}^3}\right) + 100\,\text{g/m}^3 - 20\,\text{g/m}^3\right)$$

$$= \frac{1}{0.19\,\text{g/m}^3{\cdot}\text{h}}(10\,\text{g/m}^3 \times 1.61 + 100\,\text{g/m}^3 - 20\,\text{g/m}^3) = 506\,\text{h}$$

3. Compare the performance of both reactors.

$$\frac{\theta_{\text{CFSTR}}}{\theta_{\text{PFR}}} = \frac{632\,\text{h}}{506\,\text{h}} = 1.25$$

The required hydraulic residence time of CFSTR is 25% longer than that of PFR to achieve 80% removal. Therefore, PFR is more efficient than CFSTR.

EXAMPLE 3.33: PERFORMANCE EVALUATION OF SATURATION-TYPE REACTION IN A CFSTR AND PFR FOR A GIVEN HYDRAULIC RESIDENCE TIME

The volumes of a CFSTR and PFR for saturation-type reactions under steady-state condition are 60 m^3 each. The flow through each reactor is 100 m^3/d. The reaction rate kinetic coefficient $k = 0.85\ g/m^3{\cdot}h$, $C_0 = 100\ g/m^3$, and $K_s = 10\ g/m^3$. Determine the following (a) the ratio of effluent and influent concentrations from both reactors, (b) percent removal in each reactor, and (c) compare the performance of both reactors.

Solution

1. Calculate the hydraulic residence time of both reactors from Equation 3.10c.

$$\theta = \frac{V}{Q} = \frac{60\ m^3}{100\ m^3/d} = 0.6\ d$$

or $\theta = 0.6\ d \times 24\ h/d = 14.4\ h$

2. Calculate the effluent concentration from CFSTR using Equation 3.21.

$$14.4\ h = \frac{(100\ g/m^3 - C)(10\ g/m^3 + C)}{0.85\ g/m^3{\cdot}h \times C}$$

Solve the equation for C.

$C = 89\ g/m^3$

3. Calculate the effluent concentration from PFR using Equation 3.22.

$$14.4\ h = \frac{1}{0.85\ g/m^3{\cdot}h}\left(10\ g/m^3 \times \ln\left(\frac{100\ g/m^3}{C}\right) + 100\ g/m^3 - C\right)$$

Solve the equation for C.

$C = 89\ g/m^3$

4. Compare the performance of both reactors.
Both reactors have equal performance under saturation-type reaction.

3.4.5 Performance of CFSTRs in Series (Cascade Arrangement)

Two or more CFSTRs when arranged in series will give higher removal efficiency than a single reactor of combined volume. The performance increases with increasing number of reactors. A series of 10 CFSTRs will have a performance approaching to that of a plug flow reactor, which has the residence time equal to the combined residence time of 10 CFSTRs.[3,4,7]

EXAMPLE 3.34: SLUG CONSERVATIVE DYE TRACER IN SERIES CFSTRS

Total volume V_T is divided into n equal size of CFSTRs such that volume of each reactor, $V = V_T/n$. A slug of conservative dye is applied in the first reactor. If the concentration of dye in the first reactor after mixing is C_0, derive the expression for effluent concentration from the *second* and the *i*th reactor in terms of (a) $\theta = V/Q$ (based on divided volume V), and (b) $\theta_T = V_T/Q = n\theta$ (based on the total volume V_T and n).

Solution

1. Draw the flow schematic (Figure 3.25).

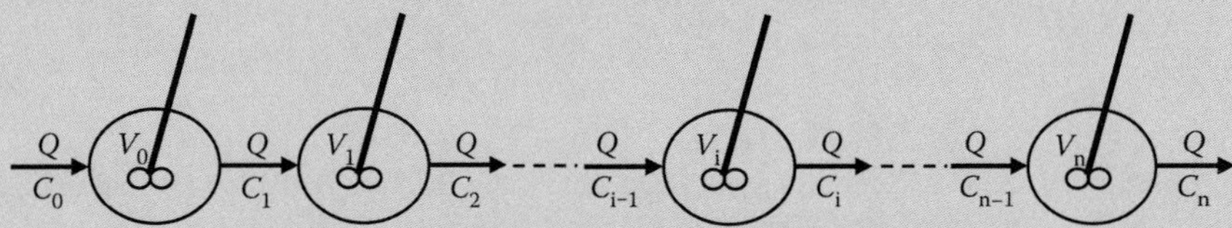

FIGURE 3.25 Process diagram with n identical CFSTRs in series (Example 3.34).

2. Write the word statement for conservative substance. There is no conversion in the reactors.

Accumulation = inflow − outflow

3. Develop the relationship in terms of divided volume V in series, $\theta = V/Q$.
   a. Write the symbolic equation for effluent from the second reactor (C_2).

$$V\frac{dC_2}{dt} = QC_1 - QC_2$$

$$\frac{dC_2}{dt} + \frac{Q}{V}C_2 = \frac{Q}{V}C_1$$

Using Equation 3.10a to determine the effluent concentration from the first reactor (C_1) with respect to C_0 which is the initial concentration in the first reactor.

$$C_1 = C_0 e^{-t/\theta}$$

Substitute C_1 in the above equation to obtain the following (Equation 3.23):

$$\frac{dC_2}{dt} + \frac{Q}{V}C_2 = \frac{Q}{V}C_0 e^{-t/\theta} \quad (3.23)$$

   b. Integrate Equation 3.23 to obtain the effluent concentration C_2 from the second reactor. The integration procedure by substitution is given in Example 3.19. Using the same procedure, the desired results are obtained.

Substitute $\theta = V/Q$ in Equation 3.23. The effluent concentration from second reactor in series is expressed by Equation 3.24.

$$C_2 = C_0\left(\frac{t}{\theta}\right)e^{-t/\theta} \quad (3.24)$$

The effluent concentration from ith ($i \geq 2$) reactor in series is given by the following Equation 3.25.

$$C_i = \frac{C_0}{(i-1)!}\left(\frac{t}{\theta}\right)^{(i-1)} e^{-t/\theta} \quad (3.25)$$

4. Develop the relationship in terms of total volume V_T divided in n reactors in series, $\theta_T = V_T/Q = n\theta$.
   a. Apply Equation 3.3 (without reactive term) and write the symbolic equation for effluent concentration from the second reactor (C_2).

$$\frac{V_T}{n}\frac{dC_2}{dt} = QC_1 - QC_2$$

$$\frac{dC_2}{dt} + n\frac{Q}{V_T}C_2 = n\frac{Q}{V_T}C_1$$

Substitute $C_1 = C_0 e^{-t/\theta}$ in the equation to obtain the following equation (Equation 3.26):

$$\frac{dC_2}{dt} + n\frac{Q}{V_T}C_2 = n\frac{Q}{V_T}C_0 e^{-t/\theta} = n\frac{Q}{V_T}C_0 e^{-nt/\theta_T} \tag{3.26}$$

   b. Integrate Equation 3.26 using substitution method.
   The effluent concentration from the second reactor is expressed by Equation 3.27.

$$C_2 = C_0 n\left(\frac{t}{\theta_T}\right)e^{-nt/\theta_T} \tag{3.27}$$

The effluent concentration from ith ($i \geq 2$) reactor is expressed by Equation 3.28.

$$C_i = \frac{C_0}{(i-1)!}\left[n\left(\frac{t}{\theta_T}\right)\right]^{(i-1)} e^{-nt/\theta_T} \tag{3.28}$$

EXAMPLE 3.35: EFFLUENT CONCENTRATION OF A CONSERVATIVE TRACER THROUGH THREE CFSTRs IN SERIES

The volume of a CFSTR is 60 m^3. It receives a flow of 240 m^3/d. A conservative dye slug of 120 g is placed into the reactor. Calculate the concentration of dye in the effluent after $t = 1.5$ h. If the reactor volume is divided into three equal CFSTRs, determine the dye concentration in the effluent from the first, second, and third reactors if the total hydraulic residence time remains at 1.5 h in three reactors.

Solution

1. Single CFSTR reactor ($n = 1$) having the combined volume $V_T = 60\ m^3$.
   a. Calculate the hydraulic residence time in the CFSTR.

$$\theta_T = \frac{V_T}{Q} = \frac{60\ m^3}{240\ m^3/d} = 0.25\ d$$

$$\text{or } \theta_T = 0.25\ d \times 24\ h/d = 6\ h$$

   b. Calculate the initial dye concentration C_0 in the CFSTR after mixing from Equation 3.9b.

$$C_0 = \frac{w_{tracer}}{V_T} = \frac{120\ g}{60\ m^3} = 2\ g/m^3.$$

c. Calculate the dye concentration C in the effluent from a single CFSTR of combined volume after $t = 1.5$ h (Equation 3.10a).

$$C = C_0 e^{-t/\theta_T} = 2\,\text{g/m}^3 \times e^{-\frac{1.5\,\text{h}}{6\,\text{h}}} = 2\,\text{g/m}^3 \times e^{-0.25} = 1.56\,\text{g/m}^3$$

2. Three CFSTR reactors ($n = 3$) in series of volume $V = 20\ \text{m}^3$ each.

a. Calculate the hydraulic residence time in each CFSTR.

$$\theta = \frac{V}{Q} = \frac{20\,\text{m}^3}{240\,\text{m}^3/\text{d}} = 0.083\,\text{d}$$

$$\text{or } \theta = 0.083\,\text{d} \times 24\,\text{h/d} = 2\,\text{h}$$

$$\theta_T = n\theta = 3 \times 2\,\text{h} = 6\,\text{h}$$

b. Calculate the initial dye concentration C_0 in the first CFSTR after mixing from Equation 3.9b.

$$C_0 = \frac{w_{\text{tracer}}}{V} = \frac{120\,\text{g}}{20\,\text{m}^3} = 6\,\text{g/m}^3$$

c. Calculate the dye concentration C_1 in the effluent from the first CFSTR after $t = 1.5$ h (Equation 3.10a).

$$C_1 = C_0 e^{-t/\theta} = 6\,\text{g/m}^3 \times e^{-\frac{1.5\,\text{h}}{2\,\text{h}}} = 6\,\text{g/m}^3 \times e^{-0.75} = 2.83\,\text{g/m}^3$$

d. Calculate the dye concentration C_2 in the effluent from the second CFSTR after $t = 1.5$ h (Equation 3.24).

$$C_2 = C_0\left(\frac{t}{\theta}\right)e^{-t/\theta} = 6\,\text{g/m}^3 \times \left(\frac{1.5\,\text{h}}{2\,\text{h}}\right) \times \text{e}^{-\frac{1.5\,\text{h}}{2\,\text{h}}} = 6\,\text{g/m}^3 \times 0.75 \times \text{e}^{-0.75} = 2.13\,\text{g/m}^3$$

e. Calculate the dye concentration C_3 in the effluent from the third CFSTR ($i = 3$) after $t = 1.5$ h (Equation 3.25).

$$C_3 = \frac{C_0}{(i-1)!}\left(\frac{t}{\theta}\right)^{(i-1)} e^{-t/\theta} = \frac{C_0}{(3-1)\times(2-1)}\left(\frac{t}{\theta}\right)^{(3-1)} e^{-t/\theta} = \frac{C_0}{2}\left(\frac{t}{\theta}\right)^2 e^{-t/\theta}$$

$$C_3 = \frac{6\,\text{g/m}^3}{2} \times \left(\frac{1.5\,\text{h}}{2\,\text{h}}\right)^2 \times \text{e}^{-\frac{1.5\,\text{h}}{2\,\text{h}}} = \frac{6\,\text{g/m}^3}{2} \times 0.75^2 \times \text{e}^{-0.75} = 0.80\,\text{g/m}^3$$

Note: The effluent quality from the last CFSTR in series is superior to that from a single CFSTR of volume equal to total volume of CFSTRs in series.

EXAMPLE 3.36: CONSERVATIVE TRACER CONCENTRATION PROFILE FROM THREE CFSTRs IN SERIES

Three CFSTRs are arranged in series and used in a slug input dye study. Plot the effluent concentration ratio profile (C/C_0) versus time dependent (t/θ) from each reactor.

Solution

1. Calculate the ratios of C/C_0 from the first, second, and third reactors at $t/\theta = 0.2$.

In the first reactor, from Equation 3.10b: $\frac{C_1}{C_0} = e^{-t/\theta} = e^{-0.2} = 0.82$

In the second reactor, from rearranged Equation 3.24: $\frac{C_2}{C_0} = \frac{t}{\theta} e^{-t/\theta} = 0.2 \times e^{-0.2} = 0.16$

In the third reactor, from rearranged Equation 3.25: $\frac{C_3}{C_0} = \frac{1}{2}\left(\frac{t}{\theta}\right)^2 e^{-t/\theta} = \frac{1}{2} \times 0.2^2 \times e^{-0.2} = 0.016$

2. Develop data to plot C/C_0 versus t/θ curve.

From the different values of t/θ, the calculated effluent to influent ratio (C/C_0) for these three reactors are summarized in Table 3.6.

TABLE 3.6 Conservative Tracer Data for Three CFSTRs in Series (Example 3.36)

t/θ	C_1/C_0 from the first reactor (Equation 3.10b)	C_2/C_0 from the second reactor (Equation 3.24)	C_3/C_0 from the third reactor (Equation 3.25)
0	1	0	0
0.2	0.82	0.16	0.016
0.4	0.67	0.27	0.054
0.6	0.55	0.33	0.10
0.8	0.45	0.36	0.14
1	0.37	0.37	0.18
1.2	0.30	0.36	0.22
1.4	0.25	0.35	0.24
1.6	0.20	0.32	0.26
1.8	0.17	0.30	0.27
2	0.14	0.27	0.27
2.2	0.11	0.24	0.27
2.4	0.091	0.22	0.26
2.6	0.074	0.19	0.25
2.8	0.061	0.17	0.24
3	0.050	0.15	0.22
3.5	0.030	0.11	0.18
4	0.018	0.073	0.15
4.5	0.011	0.050	0.11
5	0.0067	0.034	0.084

3. Plot the ratios of C_i/C_0 versus t/θ for three reactors (Figure 3.26).

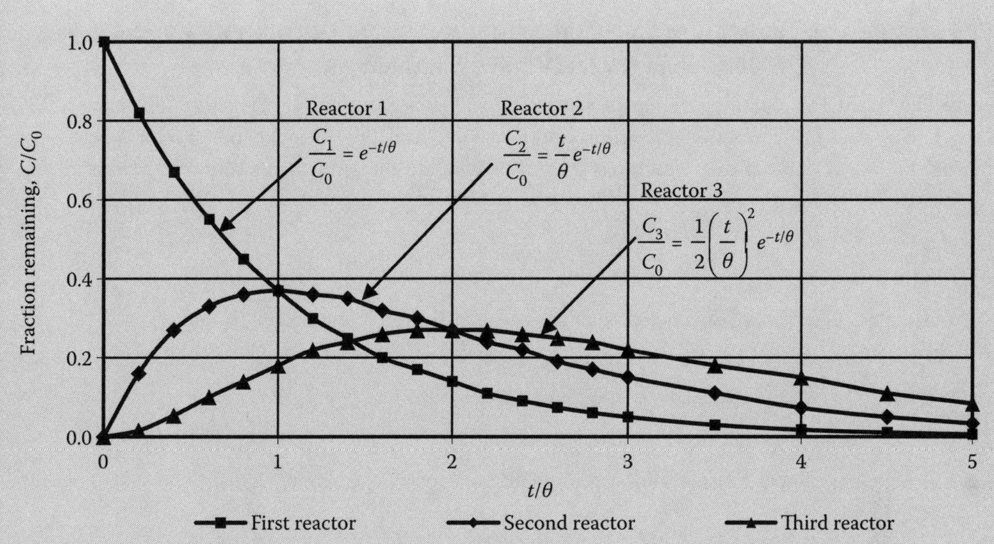

FIGURE 3.26 Time-related effluent concentration profiles from reactors 1, 2, and 3 in three identical CFSTRs operating in series (Example 3.36).

EXAMPLE 3.37: EFFLUENT QUALITY FROM CFSTRs IN SERIES WITH FIRST-ORDER REACTION AND UNDER STEADY STATE

Several CFSTRs of equal volume are operating in series. Develop the generalized equation expressing the effluent concentration from the nth reactor. The reactors are operating under the steady-state condition. The first-order reaction constant is k.

Solution

Develop the equations (Equations 3.29a through 3.29c) from Equation 3.11b for the effluent concentrations from the reactors in series.

$$C_1 = \frac{C_0}{1 + k\theta} \quad \text{(first reactor)} \tag{3.29a}$$

$$C_2 = \frac{C_1}{1 + k\theta} = \frac{C_0}{(1 + k\theta)^2} \quad \text{(second reactor)} \tag{3.29b}$$

$$C_n = \frac{C_{n-1}}{1 + k\theta} = \frac{C_0}{(1 + k\theta)^n} \quad (n\text{th reactor}) \tag{3.29c}$$

where

n = number of reactors in series
θ = residence time for each reactor
C_0 = initial concentration

EXAMPLE 3.38: PERFORMANCE OF FIVE CFSTRs IN SERIES TREATING A NONCONSERVATIVE SUBSTANCE

Five identical CFSTRs are operating in series. Influent concentration of a reactive substance is 180 mg/L. The first-order reaction rate constant is $0.83\,h^{-1}$ and flow through the reactor assembly is $35\,m^3/h$. The volume of each reactor is $24.5\,m^3$. Calculate the effluent concentration from all five reactors.

Solution

1. Calculate the hydraulic residence time of each reactor from Equation 3.10c.

$$\theta = \frac{V}{Q} = \frac{24.5\,\text{m}^3}{35\,\text{m}^3/\text{h}} = 0.7\,\text{h}$$

2. Calculate the effluent concentration from the first reactor from Equation 3.29a.

$$C_1 = \frac{C_0}{1+k\theta} = \frac{180\,\text{mg/L}}{1+0.83\,\text{h}^{-1}\times 0.7\,\text{h}} = \frac{180\,\text{mg/L}}{1.58} = 114\,\text{mg/L}$$

3. Calculate the effluent concentration from second, third, fourth, and fifth reactors from Equations 3.29b and 3.29c.

$$C_2 = \frac{180\,\text{mg/L}}{(1.58)^2} \quad \text{or} \quad \frac{114\,\text{mg/L}}{1.58} = 72\,\text{mg/L}$$

$$C_3 = \frac{180\,\text{mg/L}}{(1.58)^3} \quad \text{or} \quad \frac{72\,\text{mg/L}}{1.58} = 46\,\text{mg/L}$$

$$C_4 = \frac{180\,\text{mg/L}}{(1.58)^4} \quad \text{or} \quad \frac{46\,\text{mg/L}}{1.58} = 29\,\text{mg/L}$$

$$C_5 = \frac{180\,\text{mg/L}}{(1.58)^5} \quad \text{or} \quad \frac{29\,\text{mg/L}}{1.58} = 18\,\text{mg/L}$$

EXAMPLE 3.39: TOTAL VOLUME OF CFSTRs IN SERIES TO ACHIEVE A GIVEN REMOVAL RATIO OF A NONCONSERVATIVE SUBSTANCE

Develop a generalized equation to express total residence time θ of n reactors in series to achieve a given effluent to influent ratio (C/C_0). First-order reaction rate constant k is $0.85\,h^{-1}$. Also, calculate total volume of three CFSTRs to achieve 90% removal.

Solution

1. Express the C/C_0 ratio in terms of k and θ for n CFSTRs in series (Equation 3.29c).

$$\frac{C_n}{C_0} = \frac{1}{(1+k\theta)^n}$$

2. Express θ_T in terms of the C/C_0 ratio, and k for n CFSTRs in series (Equation 3.30).

$$\theta_T = n\theta$$

$$\frac{C_n}{C_0} = \frac{1}{\left[1 + \frac{k\theta_T}{n}\right]^n}$$

$$\theta_T = \frac{n}{k}\left[\left(\frac{C_0}{C_n}\right)^{\frac{1}{n}} - 1\right] \tag{3.30}$$

3. Calculate the total θ_T and θ for each reactor for 90% removal in three CFSTRs in series.

$$1 - \frac{C_n}{C_0} = 0.9$$

$$\frac{C_n}{C_0} = 0.1 \text{ or } \frac{C_0}{C_n} = 10$$

$$\theta_T = \frac{3}{0.85\,\text{h}^{-1}}\left[(10)^{\frac{1}{3}} - 1\right] = 3.53\,\text{h} \times (2.15 - 1) = 4.1\,\text{h}$$

$$\theta = \frac{\theta_T}{n} = \frac{4.1\,\text{h}}{3} = 1.4\,\text{h}$$

Check from Equation 13.29c: $\frac{C_n}{C_0} = \frac{1}{(1 + 0.85\,\text{h}^{-1} \times 1.4\,\text{h})^3} = 0.1$

EXAMPLE 3.40: COMPARISON OF HRTs IN A SINGLE AND THREE EQUAL CFSTRs WITH THAT OF A PFR

Compare the hydraulic residence times of a single CFSTR, and three equal volume CFSTRs with that of a PFR. In all cases, the removal efficiency is 90%, C_0 of reactive substance is 180 mg/L, and $k = 0.85\,\text{h}^{-1}$.

Solution

1. Calculate the C/C_0 ratio for 90% removal.

$$\left(1 - \frac{C}{C_0}\right) = 0.9$$

$$\frac{C}{C_0} = 0.1 \quad \text{or} \quad \frac{C_0}{C} = 10$$

2. Calculate the θ_T value of single CFSTR that has 90% removal efficiency from Equation 3.11c.

$$\frac{C}{C_0} = \frac{1}{1 + k\theta} \quad \text{or} \quad \theta = \frac{C_0}{C} \times \frac{1}{k}\left(1 - \frac{C}{C_0}\right)$$

$$\theta_T = 10 \times \frac{1}{0.85\,\text{h}^{-1}} \times (1 - 0.1)$$

$$\theta_T = 10.6\,\text{h}$$

3. Calculate the total retention time of three reactors arranged in series from Equation 3.30.

$$\theta_T = \frac{3}{0.85\,\text{h}^{-1}}\left[(10)^{\frac{1}{3}} - 1\right] = 4.1\,\text{h}$$

4. Calculate the retention time of each reactor ($n = 3$).

$$\theta = \frac{\theta_T}{n} = \frac{4.1\,\text{h}}{3} = 1.4\,\text{h}$$

5. Calculate the retention time of a single PFR from Equation 3.18.

$$\theta = -\frac{1}{k}\ln\left(\frac{C}{C_0}\right) = -\frac{1}{0.85\,\text{h}^{-1}}\ln(0.1) = 2.7\,\text{h}$$

6. Compare the θ_T values of different reactors.
   The calculated values for three reactors for 90% removal are compared below:

Reactor	θ_T, h
CFSTR	10.6
Three CFSTRs	4.1 (θ for each CFSTR in series = 1.4 h)
PFR	2.7

The θ_T of single CFSTR is 3.9 times (10.6 h ÷ 2.7 h) that of a PFR.
The total θ_T for three CFSTRs in series is 1.5 times (4.1 h ÷ 2.7 h) that of a PFR.

3.4.6 Graphical Solution of Series CFSTRs

The algebraic solution of CFSTRs in series can become very complex if reaction times and reaction rates vary in different reactors. A graphical solution can be easily performed for any number of reactors having different reaction rates and reaction times. The results are reasonably accurate.[5]

EXAMPLE 3.41: GRAPHICAL SOLUTION PROCEDURE OF CFSTRs IN SERIES

Describe the procedure of graphical solution of four CFSTRs operating in series. Use the following information:

C_0 = influent concentration of a reactive substance, g/m^3
C_1, C_2, C_3, and C_4 = effluent concentrations from reactors 1, 2, 3, and 4, g/m^3
θ_1, θ_2, θ_3, and θ_4 = hydraulic residence times of reactors 1, 2, 3, and 4, h
k = reaction rate constant

The rate law equation (Equation 2.5) is generalized to express the rate of reaction:
r or $-r_A = kC^n$, where n is reaction order ($n = 1$ and 2 for first- and second-order reactions).

Solution

1. Plot the reaction rate equation ($r = kC^n$).
   Assume different values of C and calculate r. Tabulate the data.

Plot on the y-axis the values of r against assumed values of C to obtain line r (Figure 3.27). This line represents equation $r = kC^n$ (the reaction rate equation).

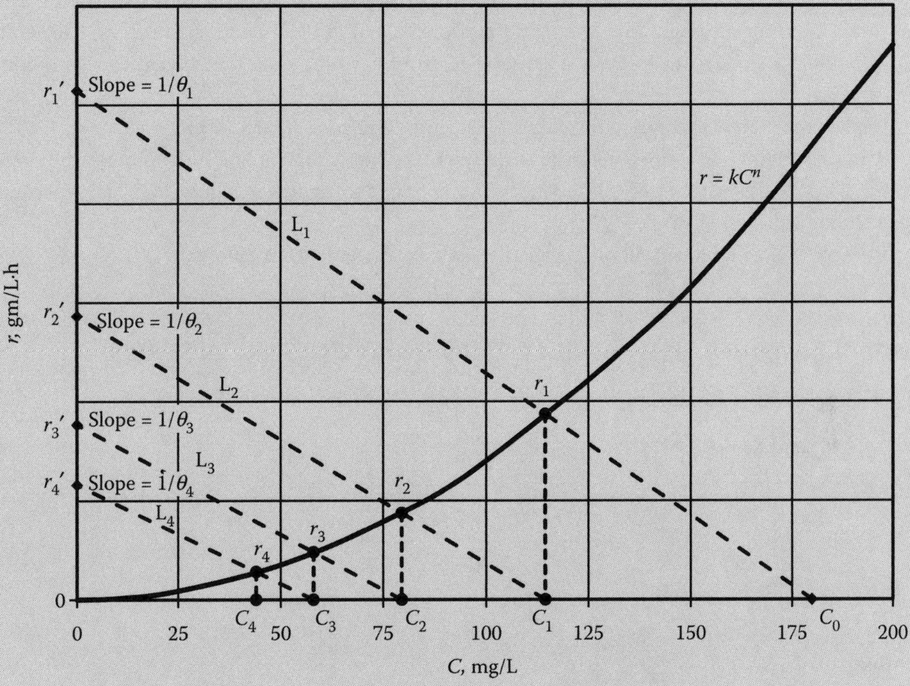

FIGURE 3.27 Graphical procedure for determining effluent concentration from four CFSTRs in series (Example 3.41).

2. Develop the generalized mass balance equation from Equation 3.3.

$$V\frac{dC}{dt} = QC_0 - QC + r_A V$$

Since there is no accumulation $(dC/dt) = 0$ and $r = -r_A$, the mass balance equation is expressed by Equation 3.31a.

$$0 = QC_0 - QC - rV$$

$$C = C_0 - \frac{V}{Q}r \quad \text{or} \quad C = C_0 - \theta r \tag{3.31a}$$

3. Obtain the slope equation (Equation 3.31b) by rearranging Equation 3.31a.

$$r = \frac{1}{\theta}(C_0 - C) \tag{3.31b}$$

The plot of r versus $(C_0 - C)$ gives a linear relationship and slope of the line is $1/\theta$. When $C = 0$, the equation is simplified to Equation 3.31c.

$$r' = \frac{C_0}{\theta} \tag{3.31c}$$

The intercept of the line on y-axis (r') is C_0/θ.

4. Determine the effluent concentration C_1 from the first reactor.
   The hydraulic residence time of reactor 1 is θ_1 and slope of the straight line L_1 is $1/\theta_1$. To draw this line, assume $C_1' = 0$ and calculate $r_1' = C_0/\theta_1$ from Equation 3.31c. Plot point $(0, C_0/\theta_1)$ on y-axis in Figure 3.27. Draw a straight line L_1 connecting from $(C_0, 0)$ to $(0, r_1')$ and crossing line r at point r_1. This line has a slope of $1/\theta_1$. Draw a vertical line from r_1 which gives the effluent concentration C_1 on x-axis.
5. Complete the graphical solution for effluent concentration from other CFSTRs.
   Draw a straight line connecting from $(C_1, 0)$ to $(0, r_2')$ and crossing line A at point r_2. The slope of the line is $1/\theta_2$. Repeat the procedure. The lines and points are shown in Figure 3.27. The procedure is explained further in Example 3.42.
   Note: The slope of lines will not be parallel if θ_1, θ_2, θ_3, and θ_4 are different.

EXAMPLE 3.42: GRAPHICAL SOLUTION OF FIVE CFSTRs IN SERIES

Solve Example 3.38 using the graphical solution. The problem data are given below:

Number of CFSTRs in series	$= 5$
C_0	$= 180$ mg/L
k	$= 0.83\ \text{h}^{-1}$
θ for each reactor	$= 0.7$ h
r or $-r_A$	$= kC$

Solution

1. Develop the rate versus concentration data.

   $r = -r_A = kC.$

   Assume different values of C and calculate $r = kC$.

   kC values for assumed C as summarized below:

C, mg/L	$r = kC$, mg/L·h
10	8.30
25	20.8
50	41.5
75	62.3
100	83.0
150	125
200	166

2. Prepare the rate line r as a graph between assumed C and calculated r.
   The rate line r is shown in Figure 3.28.
3. Draw the line to obtain the effluent concentration (C_1) from the first reactor.
   Calculate r_1' from Equation 3.31c.

$$r_1' = \frac{C_0}{\theta} = \frac{180\ \text{mg/L}}{0.7\ \text{h}} = 257\ \text{mg/L·h}$$

Plot a straight line L_1 from (180 mg/L, 0) to (0, 257 mg/L·h). Determine the reaction rate r_1 at the point, where L_1 is intercepting the rate line r. Draw a perpendicular from r_1 to intercept x-axis at C_1.
$C_1 = 114$ mg/L
Calculate the reaction rate r_1 from Equation 3.31b.

$$r_1 = \frac{C_0 - C_1}{\theta} = \frac{180\,\text{mg/L} - 114\,\text{mg/L}}{0.7\,\text{h}} = 94\,\text{mg/L·h}$$

4. Determine the effluent concentration from other reactors.
   Since all CFSTRs are identical, θ remains the same for all reactors and all slope lines will be parallel to have the same slope of $1/\theta$. Draw a straight line L_2 from (114 mg/L, 0) parallel to L_1. Determine r_2 by the intercepting point of straight lines L_2 and the rate line r and then C_2 on x-axis. Repeat the procedure to obtain effluent concentration from each of the other reactors. The procedure is also shown in Figure 3.28.

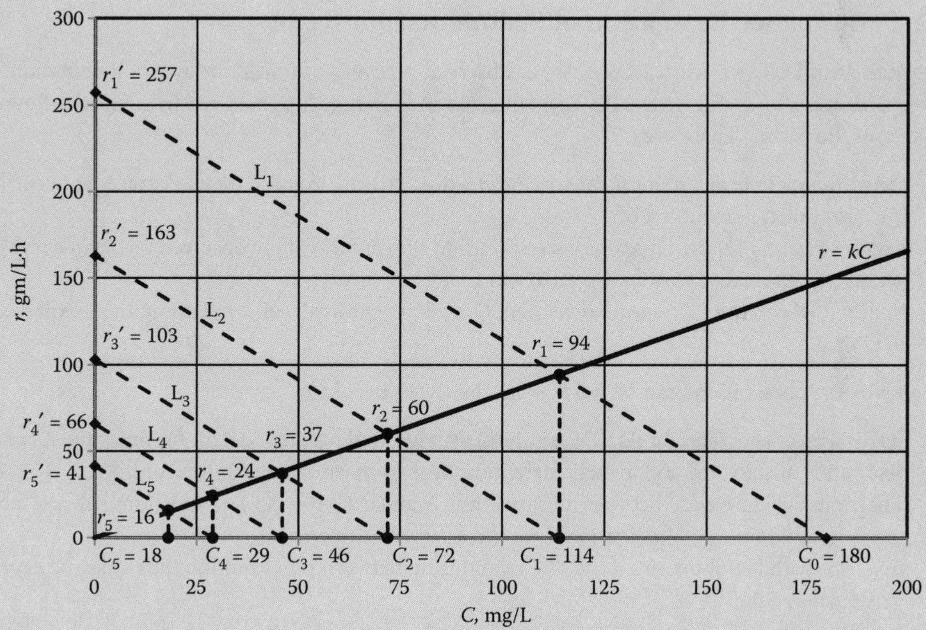

FIGURE 3.28 Graphical solution of five identical CFSTRs in series (Example 3.42).

5. Summary of the effluent concentrations from all five reactors.
   The results are summarized below:

$C_1 = 114$ mg/L	$r_1 = 94$ mg/L·h
$C_2 = 72$ mg/L	$r_2 = 60$ mg/L·h
$C_3 = 46$ mg/L	$r_3 = 37$ mg/L·h
$C_4 = 29$ mg/L	$r_4 = 24$ mg/L·h
$C_5 = 18$ mg/L	$r_5 = 16$ mg/L·h

3.5 Plug Flow Reactors with Dispersion and Conversion

Plug flow with dispersion is an intermediate flow regime between the ideal plug flow and ideal completely mixed flow. It is also referred to as *intermediate-mixed flow, arbitrary flow, nonideal flow, or flow with axial dispersion*. All dispersion problems are three dimensional, and dispersion coefficient varies with direction and degree of turbulence. To simplify the analysis, one-dimensional dispersion with longitudinal mixing (axial mixing) is usually assumed.

3.5.1 Flow Regime and Dispersion of Tracer

If a slug of conservative tracer is released into a rector near the inlet, a front is formed. The front is not straight because of longitudinal transport of material due to turbulence and molecular diffusion. This initial effect on front formation is shown in Figure 3.29a. As the tracer moves through the reactor, the mixing and dispersion lengthen the zone of tracer (Figure 3.29b). Finally, the tracer exits from the effluent zone (Figure 3.29c).

3.5.2 Performance Evaluation of Sedimentation Basin

Tracer Exit Profile: The tracer exit profile (or observed recovery of tracer) from a sedimentation basin with dispersion is shown in Figure 3.30. Several terms and relationships are used to define the flow characteristics of the basin.[8] These are:

1. *Flow through time or standard detention time.* It is the time to reach the peak tracer concentration. It is also called *modal* time θ_{SD}
2. *Average detention time.* This time corresponds to the *centroid* of the observed recovery curve of the tracer; it is also called *median* time (θ_{AD}).
3. *Nominal, theoretical, or mean detention time.* It is the hydraulic retention time expressed by V/Q or θ.

The following observations can be made from the tracer profile:

- In absence of short-circuiting, the standard, average, and nominal detention times must coincide.
- The ratios of standard and average detention time to mean detention time are <1.
- The ratios of difference between nominal and standard, $(\theta - \theta_{SD})/\theta$ and nominal and average $(\theta - \theta_{AD})/\theta$ to nominal detention time increase with dispersion.
- In basins with low short circuiting, the relationship between the detention times may be expressed by Equation 3.32.[9,10]

FIGURE 3.29 Progress of slug tracer input in a plug flow reactor with dispersion: (a) tracer front formation, (b) tracer front movement, and (c) tracer exit.

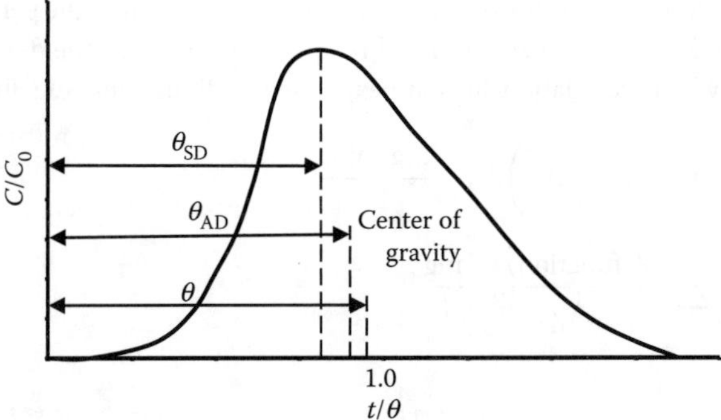

FIGURE 3.30 Slug tracer exit curve for plug flow with dispersion.

$$\theta_{SD} = \theta_{AD} - 3(\theta - \theta_{AD}), \tag{3.32}$$

where
θ_{SD} = standard detention time, min
θ_{AD} = average detention time, min
θ = nominal detention time, min

The actual shape of tracer exit profile can only be determined by tracer studies because the shape is a function of multiple parameters that may include (1) geometry and relative dimensions of the reactor, (2) mixing intensity, (3) dispersion coefficient, (4) dead volume, (5) short circuiting, and (6) density stratification.

Most reactors used in water and wastewater applications exhibit flow regimes that are plug flow with dispersion. Engineers prefer to design them as a plug flow system and apply a correction factor to simulate the dispersed plug flow condition.[5] The correction factor may vary from 0.1 for unbaffled low length-to-width ratio, to 1.0 for very high length-to-width ratio (pipeline flow) basins. The standard detention time for a well-designed tank is expected to be larger than 30% of nominal detention time.

Mostly tracer tests are utilized to estimate the actual efficiency factor. The slug dye tracer profiles for several types of sedimentation basins are shown in Figure 3.31.[11]

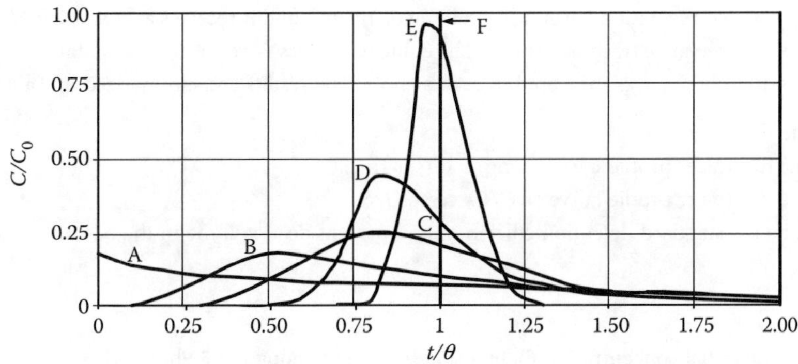

FIGURE 3.31 Typical slug dye tracer curves for several tanks.

Note: *Curve A* is for ideal CFSTR (complete dispersion); *Curve B* is for radial flow circular tank (large dispersion); *Curve C* is for wide rectangular tank with relatively shallow depth (medium to large dispersion); *Curve D* is for long narrow rectangular tank (medium or intermediate dispersion); *Curve E* is for around-the-end baffled tank (small dispersion); and *Curve F* is for ideal PF tank (no dispersion).

Performance Evaluation: The tracer exit profile is useful in determining the performance of a sedimentation basin. Fielder and Fitch developed empirical relationships between the dye tracer test data and the sedimentation efficiency.[8] These relationships are expressed by Equations 3.33 through 3.39.

$$\sum(Y\text{-function}) = \sum\left(C - \frac{1}{2}\Delta C\right)\Delta t, \quad \frac{\text{mg·min}}{\text{L}} \tag{3.33}$$

$$\sum(Z\text{-function}) = \sum\frac{(Y\text{-function})}{t - \frac{1}{2}\Delta t}, \quad \frac{\text{mg}}{\text{L}} \tag{3.34}$$

$$\sum(W\text{-function}) = \sum(Y\text{-function})\left(t - \frac{1}{2}\Delta t\right), \quad \frac{\text{mg·min}^2}{\text{L}} \tag{3.35}$$

$$\theta_{\text{SD}} = \frac{\Sigma(Y\text{-function})}{\Sigma(Z\text{-function})} \tag{3.36}$$

$$\theta_{\text{AD}} = \frac{\Sigma(W\text{-function})}{\Sigma(Y\text{-function})} \tag{3.37}$$

$$SDE = \frac{\theta_{\text{SD}}}{\theta} \times 100\% \tag{3.38}$$

$$PDV = \frac{\theta - \theta_{\text{AD}}}{\theta} \times 100\% \tag{3.39}$$

where

C = dye concentration, mg/L
ΔC = change in dye concentration, mg/L
Δt = time increment, min
SDE = standard detention efficiency, %
PDV = percent dead volume, %

EXAMPLE 3.43: PERFORMANCE OF A SEDIMENTATION BASIN WITH CONSERVATIVE DYE TRACER

A relatively long sedimentation basin has a volume of 100 m^3, and it receives a flow of 2 m^3/min. A 320 gram slug of a conservative tracer is applied. The effluent samples were collected at suitable time intervals and concentration of dye was measured in each sample. The results are summarized in Table 3.7.

a. Calculate C_0 and θ.
b. Draw the dye tracer profile curve as mg/L versus time.
c. Draw the dye tracer profile curve as C/C_0 versus t/θ.
d. Determine the standard detention efficiency, and percent dead volume in the basin.

Solution

1. Calculate the initial concentration C_0 of a slug input from Equation 3.9b.

   C_0 is the average theoretical initial concentration of dye if it is completely mixed in the basin.

$$C_0 = \frac{w_{\text{tracer}}}{V} = \frac{320\,\text{g}}{100\,\text{m}^3} = 3.2\,\text{g/m}^3 \text{ or mg/L}$$

TABLE 3.7 Tracer Profile Data and Calculated Values of Parameters (Example 3.43)

t, min	C, mg/L	$\frac{t}{\theta}$	$\frac{C}{C_0}$	Δt, min	ΔC, mg/L	$C - 1/2\Delta C$, mg/L	Y-function, mg·min/L	$t - 1/2\Delta t$, min	Z-function, mg/L	W-function, mg·min^2/L
(1)	(2)	(3)	(4)	(5)	(6)	(7)	(8)	(9)	(10)	(11)
0	0	0	0	–	–	–	–	–	–	–
4	0	0.08	0	4	–	–	–	–	–	–
7	0.10	0.14[a]	0.03[b]	3[c]	0.10[d]	0.05[e]	0.15[f]	5.50[g]	0.027[h]	0.83[i]
10	0.25	0.20	0.08	3	0.15	0.18	0.53	8.50	0.062	4.46
13	0.83	0.26	0.26	3	0.58	0.54	1.62	11.5	0.141	18.6
16	1.48	0.32	0.46	3	0.65	1.16	3.47	14.5	0.239	50.2
20	2.45	0.40	0.77	4	0.97	1.97	7.86	18.0	0.437	141
25	2.85	0.50	0.89	5	0.40	2.65	13.3	22.5	0.589	298
30	2.32	0.60	0.73	5	−0.53	2.59	12.9	27.5	0.470	355
35	1.75	0.70	0.55	5	−0.57	2.04	10.2	32.5	0.313	331
40	1.40	0.80	0.44	5	−0.35	1.58	7.88	37.5	0.210	295
50	0.92	1.00	0.29	10	−0.48	1.16	11.6	45.0	0.258	522
60	0.62	1.20	0.19	10	−0.30	0.77	7.70	55.0	0.140	424
70	0.41	1.40	0.13	10	−0.21	0.52	5.15	65.0	0.079	335
80	0.28	1.60	0.09	10	−0.13	0.35	3.45	75.0	0.046	259
90	0.18	1.80	0.06	10	−0.10	0.23	2.30	85.0	0.027	196
100	0.12	2.00	0.04	10	−0.06	0.15	1.50	95.0	0.016	143
110	0.03	2.20	0.01	10	−0.09	0.08	0.75	105	0.007	78.8
118	0	2.36	0.00	8	−0.03	0.02	0.12	114	0.001	13.7
							ΣY-function = 90.4		ΣZ-function = 3.06	ΣW-function = 3465

[a] $\frac{7\text{ min}}{50\text{ min}} = 0.14$

[b] $\frac{0.10\text{ mg/L}}{3\text{ mg/L}} = 0.13$

[c] 7 min − 4 min = 3 min

[d] 0.10 mg/L − 0 mg/L = 0.10 mg/L

[e] $0.10\text{ mg/L} - \frac{1}{2} \times 0.1\text{ mg/L} = 0.05\text{ mg/L}$

[f] 0.05 mg/L × 3 min = 0.15 mg·min/L

[g] $7\text{ min} - \frac{1}{2} \times 3\text{ min} = 5.5\text{ min}$

[h] $\frac{0.15\text{ mg·min/L}}{5.50\text{ min}} = 0.027\text{ mg/L}$

[i] 0.15 mg·min/L × 5.5 min = 0.83 mg·min$^{2/L}$

2. Calculate the nominal detention time θ from Equation 3.10c.

$$\theta = \frac{V}{Q} = \frac{100\text{ m}^3}{2\text{ m}^3/\text{min}} = 50\text{ min}$$

3. Plot the tracer profile of C, mg/L versus time t, min.

   The tracer concentration in the effluent with respect to time of sampling t and other parameters are calculated in Table 3.7. These values are plotted in Figure 3.32a.

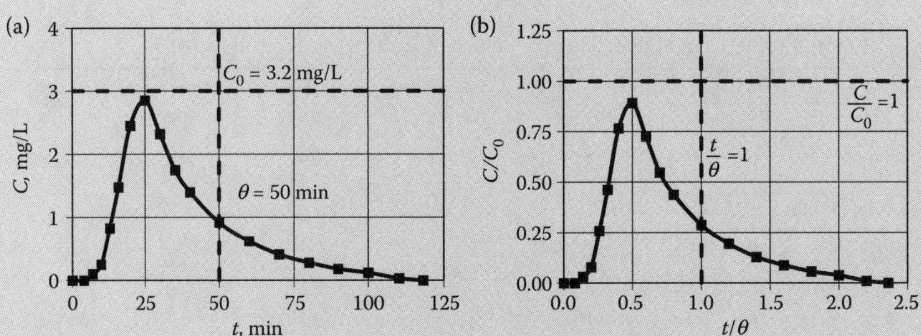

FIGURE 3.32 Tracer profile in the effluent from a plug flow reactor with dispersion during a slug tracer test: (a) C versus t, and (b) C/C_0 versus t/θ (Example 3.34).

4. Plot the tracer profile C/C_0 versus t/θ.

   The value of C/C_0 and t/θ at different time intervals are calculated in Columns (3) and (4) of Table 3.7. The results are plotted in Figure 3.32b.

5. Calculate the standard detention efficiency and fraction dead volume.

   The procedure for calculating Y-function, Z-function, and W-function from Equations 3.33 through 3.35 is tabulated in Columns (5) to (11) of Table 3.7.

   Standard detention time from Equation 3.36,

$$\theta_{SD} = \frac{\sum(Y\text{-function})}{\sum(Z\text{-function})} = \frac{90.4\ \text{mg·min/L}}{3.06\ \text{mg/L}} = 29.5\ \text{min}$$

   Average detention time from Equation 3.37,

$$\theta_{AD} = \frac{\sum(W\text{-function})}{\sum(Y\text{-function})} = \frac{3465\ \text{mg·min}^2/\text{L}}{90.4\ \text{mg·min/L}} = 38.3\ \text{min}$$

   Standard detention efficiency Equation 3.38, $SDE = \frac{\theta_{SD}}{\theta} \times 100\% = \frac{29.5\ \text{min}}{50.0\ \text{min}} \times 100\% = 59\%$

   Percent dead volume, $PDV = \frac{\theta - \theta_{AD}}{\theta} \times 100\% = \frac{50 - 38.5\ \text{min}}{50\ \text{min}} \times 100\% = 23\%$

EXAMPLE 3.44: DEAD VOLUME IN A CFSTR

A tracer study was conducted in a CFSTR. The tracer concentrations in the effluent at different time intervals are given below. The volume of the basin, the average flow, and influent concentration C_0 to the basin are 10 m^3, 0.04 m^3/s, and 2.2 mg/L, respectively. Determine the dead volume.

t, s	50	100	200	300	400	500
C, mg/L	1.78	1.42	0.88	0.48	0.20	0.10

Solution

1. Calculate the ratio of tracer concentration C/C_0 in an ideal CFSTR from the experimental data.

   The ratio of tracer concentration in an ideal CFSTR is expressed by Equation 3.10b.

$$\frac{C}{C_0} = e^{-t/\theta}$$

   C/C_0 is calculated from actual tracer concentration and tabulated in Column (3) of Table 3.8.

TABLE 3.8 Experimental and Theoretical Concentrations of Tracer in Effluent from a CFSTR (Example 3.44)

t, s	C, mg/L	C/C_0	θ', s	C_T, mg/L
(1)	(2)	(3)	(4)	(5)
50	1.78	0.81	236[a]	1.80
100	1.42	0.65	228	1.47
200	0.88	0.40	218	0.99
300	0.48	0.22	197	0.66
400	0.20	0.09	167	0.44
500	0.10	0.05	162	0.30
			$\sum\theta' = 1208$	

[a] $\dfrac{-50\text{ min}}{\ln\left(\dfrac{1.78\text{ mg/L}}{2.2\text{ mg/L}}\right)} = 236\text{ min}$

2. Determine the actual average experimental detention time θ from experimental tracer data.

   Rearrange Equation 3.10b to obtain the relationship.

$$\theta' = \frac{-t}{\ln(C/C_0)}$$

   Substitute C/C_0 and t for each data point to determine θ'. These values are summarized in Column (4) of Table 3.8.

3. Calculate the nominal detention time θ from Equation 3.10c.

$$\theta = \frac{V}{Q} = \frac{10\text{ m}^3}{0.04\text{ m}^3/\text{s}} = 250\text{ s}$$

4. Calculate the arithmetic average detention time from experimental data in Column (4).

$$\theta_{AD} = \frac{\Sigma\theta'}{n} = \frac{1208\text{ s}}{6} = 201\text{ s}$$

5. Calculate the theoretical tracer concentration C_T from nominal detention time θ using Equation 3.10a.

$$C_T = C_0 e^{-t/\theta} = 2.2\text{ mg/L} \times e^{-t/(250\text{ s})}$$

   Calculate C_T for different t values. These values are tabulated in Column (5) of Table 3.8.

6. Plot the actual and theoretical tracer concentrations in the effluent from the CFSTR.

   The measured concentrations in Column (2) and theoretical concentrations in Column (5) are plotted against t in Figure 3.33.

7. Compare the results.

   The plotted values show that dead volume exists because the tracer concentrations in the experimental curve are lower than that in the theoretical curve. This is due to tracer washout.

8. Determine the dead volume.

   The nominal detention time is 250 s. The average detention time is 201 s. Clearly, there is tracer washout.

$$\text{Percent dead volume, } PDV = \frac{\theta - \theta_{AD}}{\theta} \times 100\% = \frac{250\text{ s} - 201\text{ s}}{250\text{ s}} \times 100\% = 20\%$$

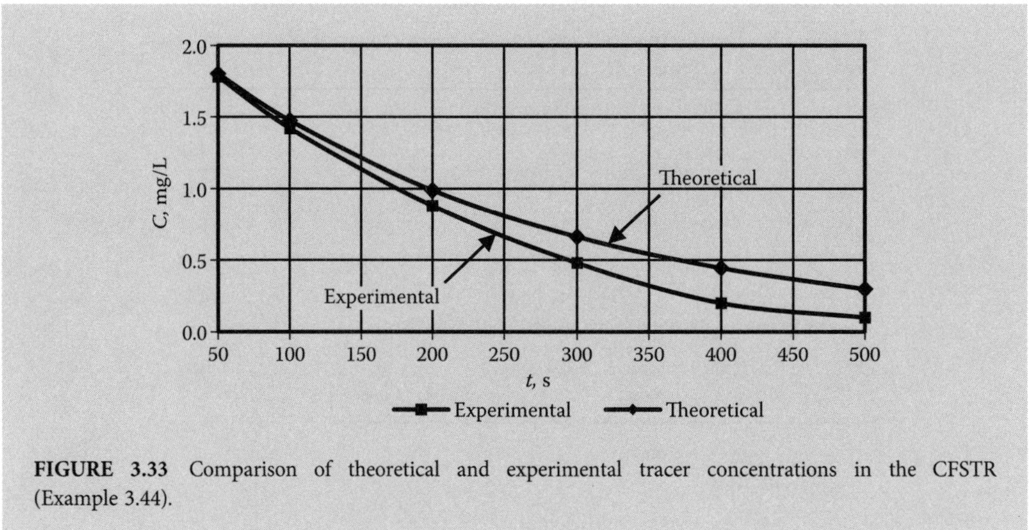

FIGURE 3.33 Comparison of theoretical and experimental tracer concentrations in the CFSTR (Example 3.44).

3.5.3 Dispersion with Conversion

The fundamental approach to understanding nonideal flow in reactors was proposed by Donckwerts.[12] Wehner and Wilhelm developed steady-state solution for the first-order reaction ($r = kC$). The solution is expressed by Equation 3.40, which is independent of inlet and outlet conditions, and depends upon dispersion number (Equation 3.41).[13]

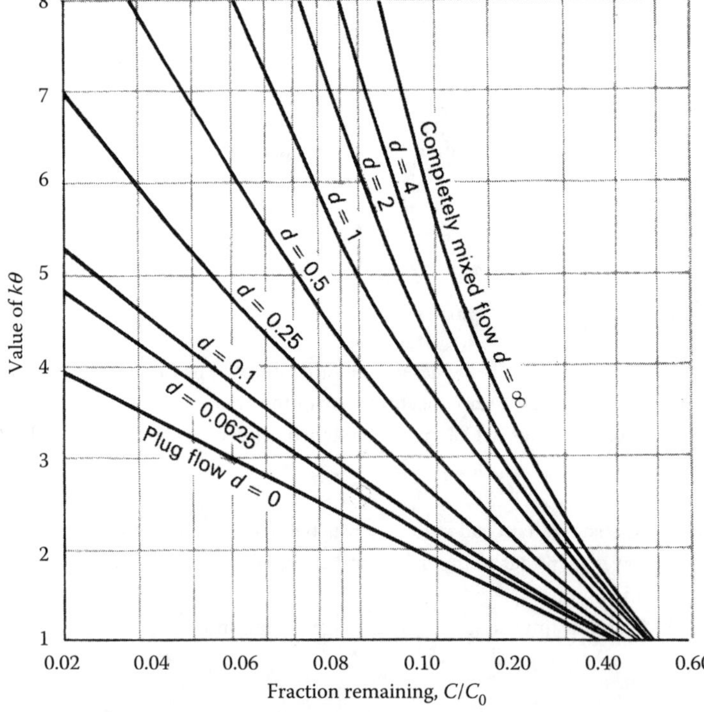

FIGURE 3.34 Relationship between k_θ and fraction of concentration remaining for different dispersion numbers. (From Reference 14 used with permission of American Society of Civil Engineers).

TABLE 3.9 Typical Dispersion Numbers for Various Treatment Facilities

Treatment Facility	Dispersion Number
Waste stabilization pond	
Single pond	1–4
Multiple ponds in series	0.1–1
Aerated lagoon	
Long rectangular shape	1–4
Square shaped	3–4
Rectangular sedimentation basin	0.2–2
Aeration basin	
Long plug flow	0.1–1
Complete mix	3–4
Oxidation ditch	3–4
Chlorine contact basin	0.02–0.08

Source: Adapted in part from Reference 3.

$$\frac{C}{C_0} = \frac{4ae^{1/2d}}{(1+a)^2 e^{a/2d} - (1-a)^2 e^{-a/2d}} \tag{3.40}$$

$$d = \frac{D}{vL} \quad \text{or} \quad d = \frac{D\theta}{L^2} \tag{3.41}$$

where

a = coefficient, dimensionless

$$a = \sqrt{1 + 4k\theta d}$$

d = dispersion number, dimensionless
D = longitudinal axial dispersion coefficient, m^2/h (ft^2/h)
v = fluid axial velocity, m/h (ft/h)
L = reactor length, m (ft)
θ = hydraulic retention time, h

Thirumurthi developed Figure 3.34 to facilitate the solution of Equation 3.40.[14] In this figure, the dimensionless term $k\theta$ is plotted against percent C/C_0 (remaining) for dispersion number d varying from 0 for ideal PFR to infinity (∞) for an ideal CFSTR. The dispersion number d varies for different reactors. Typical values of dispersion number d for some wastewater treatment facilities are summarized in Table 3.9.

EXAMPLE 3.45: EFFLUENT QUALITY FROM A REACTOR WITH DISPERSION

A stabilization pond is operating at 90% BOD_5 removal efficiency. The dispersion number d of the pond is 0.1. The average flow and first-order reaction rate constant are 1000 m^3/d and 0.4 d^{-1}, respectively. Calculate the volume of the pond.

Solution

1. Determine the C/C_0 ratio.

Removal efficiency = 90%

$(1 - C/C_0) = 0.9$

$C/C_0 = (1 - 0.9) = 0.1$

2. Determine $k\theta$.

   Read the value of $k\theta$ for $C/C_0 = 0.1$ and $d = 0.1$ from Figure 3.34.

   $k\theta = 2.75$

3. Calculate the volume of reactor.

$$\theta = \frac{2.75}{k} = \frac{2.75}{0.4\,\text{d}^{-1}} = 6.9\,\text{d}$$

$$V = Q\,\theta = 1000\,\text{m}^3/\text{d} \times 6.9\,\text{d} = 6900\,\text{m}^3$$

EXAMPLE 3.46: DISPERSION COEFFICIENT

A UV disinfection facility is designed for reduction of coliform count in the secondary effluent of a wastewater treatment plant. Total length of UV exposure is 300 cm. The velocity through the channel is 22.5 cm/s. Performance of UV disinfection facility is high under plug flow condition. At a dispersion coefficient $D = 200\,\text{cm}^2/\text{s}$ low to moderate dispersion exists. Calculate the dispersion number d (Equation 3.41). Also, calculate (a) N/N_o from Equation 3.40 and compare it with the value obtained from Figure 3.34, (b) coliform number remaining and percent reduction from UV radiation. Because of high rate of kill of coliform organism by UV radiation the first-order reaction rate k is assumed $0.29\,\text{s}^{-1}$. $N_0 = 10^4$ organisms/100 mL.

Solution

1. Calculate the dispersion number d for $D = 200\,\text{cm}^2/\text{s}$ from Equation 3.41.

$$d = \frac{D}{vL} = \frac{200\,\text{cm}^2/\text{s}}{22.5\,\text{cm/s} \times 300\,\text{cm}} = 0.03$$

2. Calculate the θ value.

$$\theta = \frac{L\,(\text{reactor length})}{v\,(\text{velocity})} = \frac{300\,\text{cm}}{22.5\,\text{cm/s}} = 13.3\,\text{s}$$

3. Calculate the dimensionless factor $k\theta$.

$$k\theta = 0.29\,\text{s}^{-1} \times 13.3\,\text{s} = 3.86$$

4. Calculate the dimensionless factor a.

$$a = \sqrt{1 + 4k\theta d} = \sqrt{1 + 4 \times 3.86 \times 0.03} = 1.21$$

5. Calculate the dimensionless ratio N/N_0 from Equation 3.40.

$$\frac{N}{N_0} = \frac{4ae^{1/2d}}{(1+a)^2 e^{a/2d} - (1-a)^2 e^{-a/2d}} = \frac{4 \times 1.21 \times e^{\frac{1}{2\times 0.03}}}{(1+1.21)^2 e^{\frac{1.21}{2\times 0.03}} - (1-1.21)^2 e^{-\frac{1.21}{2\times 0.03}}}$$

$$= \frac{4.84e^{16.7}}{4.88 \times e^{20.2} - 0.044 \times e^{-20.2}} = \frac{8.4 \times 10^7}{2.8 \times 10^9 - 7.7 \times 10^{-11}} = 0.03$$

6. Estimate the value of N/N_0 from Figure 3.34.

Estimate the value of N/N_0 for $k\theta = 3.86$ and $d = 0.03$.

$N/N_0 \approx 0.03$

7. Compare the calculated and estimated values.
   The calculated and estimated values are the same.
8. Calculate the coliform number remaining and percent reduction.

$N = 0.03 \times 10^4$ organisms/100 mL = 300 organisms per 100 mL

$$\text{Percent reduction} = \frac{(10^4 - 300)\text{ organisms per 100 mL}}{10^4\text{ organisms per 100 mL}} \times 100\% = 97\%$$

EXAMPLE 3.47: OBSERVED *E.COLI* DIE-OFF IN A SERIES OF STABILIZATION PONDS

Two stabilization ponds are operating in series. The operational data on both ponds are summarized below. Estimate the *E. coli* number in the effluent from the second pond. The average flow and influent *E. coli* counts are 4000 m^3/d and 10^7 organisms/100 mL, respectively.

Operational Data	First Pond	Second Pond
Volume, ha·m	2	5
Dispersion number, *d*	0.25	0.1
Die-off coefficient (*k*), d^{-1}	1.3	0.4

Solution

1. Calculate the θ values for both ponds using Equation 3.10c.

In the first pond, $\theta_1 = \frac{V_1}{Q} = \frac{2\text{ ha·m} \times 10{,}000\text{ m}^3/\text{ha}}{4000\text{ m}^3/\text{d}} = 5\text{ d}$

In the second pond, $\theta_2 = \frac{V_2}{Q} = \frac{5\text{ ha·m} \times 10{,}000\text{ m}^3/\text{ha}}{4000\text{ m}^3/\text{d}} = 12.5\text{ d}$

2. Calculate $k\theta$ for both ponds.

In the first pond, $k_1\theta_1 = 1.3\text{ d}^{-1} \times 5\text{ d} = 6.5$

In the second pond, $k_2\theta_2 = 0.4\,\text{d}^{-1} \times 12.5\,\text{d} = 5$

3. Determine N/N_0 for both ponds using Figure 3.34.

For $d_1 = 0.25$ in the first pond, $(N/N_0)_1 = 0.023$ and $k_1\theta_1 = 6.5$

For $d_2 = 0.1$ in the second pond, $(N/N_0)_2 = 0.022$ and $k_2\theta_2 = 5$

4. Estimate *E.coli* number in the effluent from the second pond.

$(N/N_0)_{\text{overall}} = (N/N_0)_1 \times (N/N_0)_1 = 0.023 \times 0.022 = 0.00051 = 5.1 \times 10^{-4}$

N in the effluent from the second pond

$= N_0 \times (N/N_0)_{\text{overall}} = 10^7$ organisms/100 mL $\times\ 5.1 \times 10^{-4} = 5100$ organisms/100 mL

EXAMPLE 3.48: DISPERSION FACTOR AND PERFORMANCE OF A CHLORINE CONTACT BASIN

A chlorine contact basin is 3 m wide and has three-pass baffle arrangement as shown in Figure 3.35. At a flow of 5000 m^3/d, the depth in the basin is 1 m. The dispersion number d and reaction rate constant are 0.1 and 2.9 h^{-1}. Determine (a) contact time θ, (b) dispersion coefficient D, (c) N/N_0 from Equation 3.40, and (d) N/N_0, from Figure 3.34. The baffle wall thickness is 0.3 m.

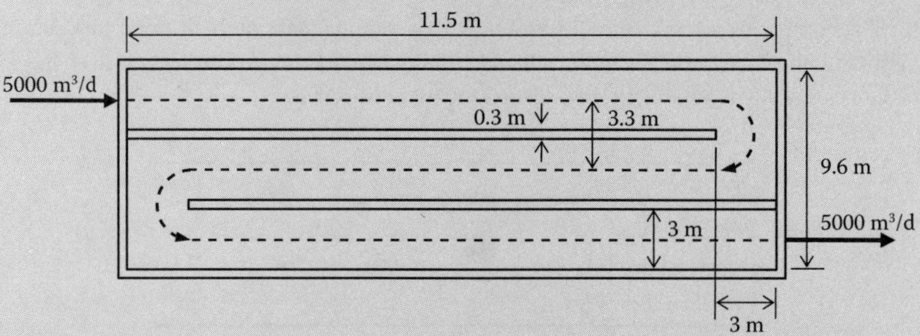

FIGURE 3.35 Chlorine contact basin layout (Example 3.48).

Solution

1. Calculate the contact time θ.

$$\begin{aligned}\text{Volume of basin, } V &= (\text{length} \times \text{width} \times \text{depth}) - (\text{volume of baffle walls})\\ &= (11.5\,\text{m} \times 9.6\,\text{m} \times 1\,\text{m}) - 2 \times (11.5 - 3)\,\text{m} \times 0.3\,\text{m} \times 1\,\text{m} = 105\,\text{m}^3\end{aligned}$$

Calculate θ from Equation 3.10c.

$$\theta = \frac{V}{Q} = \frac{105\,\text{m}^3}{5000\,\text{m}^3/\text{d}} = 0.021\,\text{d or } \theta = 0.021\,\text{d} \times 24\,\text{h/d} = 0.5\,\text{h}$$

2. Calculate the dispersion coefficient D.

$$\text{Velocity } v = \frac{Q}{A} = \frac{5000\,\text{m}^3/\text{d}}{3\,\text{m} \times 1\,\text{m}} = 1667\,\text{m/d or } v = \frac{1667\,\text{m/d}}{24\,\text{h/d}} = 69.5\,\text{m/h}$$

$$\begin{aligned}\text{Total approximate length of flow pass} &= 2 \times (11.5 - 1.5)\,\text{m} + 2 \times (3 + 0.3)\,\text{m} + (11.5 - 3)\,\text{m} \\ &= 35.1\,\text{m}\end{aligned}$$

Rearrange Equation 3.41 and calculate dispersion coefficient D.

$$D = d\,v\,L = 0.1 \times 69.5\,\text{m/h} \times 35.1\,\text{m} = 224\,\text{m}^2/\text{h}$$

3. Calculate the ratio of N/N_0 from Equation 3.40.

$$k\theta = 2.9\,\text{h}^{-1} \times 0.5\,\text{h} = 1.45$$

$$a = \sqrt{1 + 4k\theta d} = \sqrt{1 + 4 \times 1.45 \times 0.1} = 1.26$$

$$\frac{N}{N_0} = \frac{4a\,e^{1/2d}}{(1+a)^2 e^{a/2d} - (1-a)^2 e^{-a/2d}} = \frac{4 \times 1.26 \times e^{\frac{1}{2\times 0.1}}}{(1+1.26)^2 e^{\frac{1.26}{2\times 0.1}} - (1-1.26)^2 e^{-\frac{1.26}{2\times 0.1}}}$$

$$= \frac{5.04e^5}{5.11 \times e^{6.3} - 0.068 \times e^{-6.3}} = \frac{748}{2783 - 1.25 \times 10^{-4}} = 0.27$$

4. Determine the ratio of N/N_0 from Figure 3.34.
   Read the value of N/N_0 for $k\theta = 1.45$ and $d = 0.1$.

$$N/N_0 = 0.27$$

The results obtained from Equation 3.40 and Figure 3.34 are compatible.

EXAMPLE 3.49: COMPARISON OF RESIDENCE TIMES FOR IDEAL AND DISPERSED PFRs TREATING A NONCONSERVATIVE SUBSTANCE

An ideal first-order PFR receives a nonconservative substance that undergoes 90% first-order conversion. The residence time θ is 6 h. Determine the residence time of a dispersed PFR that has dispersion number $d = 1.0$ and achieves the same removal.

Solution

1. Determine the C/C_0 ratio for 90% removal.

$$C/C_0 = (1 - 0.90) = 0.1$$

2. Determine the k value of an ideal PFR with $\theta = 6$ h.
   In an ideal PFR, the C/C_0 ratio is expressed by Equation 3.16d.

$$\frac{C}{C_0} = e^{-k\theta}$$

Substitute C/C_0 ratio = 0.1 in the equation and solve for $k\theta$.

$$0.1 = e^{-k\theta}$$

$$k\theta = -\ln(0.1) = 2.3$$

$$k = \frac{2.3}{\theta} = \frac{2.3}{6\,\text{h}} = 0.38\,\text{h}^{-1}$$

3. Determine the θ value of a dispersed PFR with $k = 0.38\,\text{h}^{-1}$.
   Read the value of $k\theta$ for $C/C_0 = 0.1$ and $d = 1.0$ in a dispersed PFR from Figure 3.34.
   $k\theta = 5$

$$\theta = \frac{5}{k} = \frac{5}{0.38\,\text{h}^{-1}} = 13.2\,\text{h}$$

4. Comparison of residence times of ideal and dispersed PFRs.
   For achieving a given conversion rate of 90%, the required residence time in a dispersed PFR is more than double of that in an ideal PFR.

EXAMPLE 3.50: COMPARISON OF FRACTION REMAINING OF A NONCONSERVATIVE SUBSTANCE IN IDEAL AND DISPERSED PFRs

An ideal PFR has $k = 0.35\,\text{h}^{-1}$ and $\theta = 8$ h. Compare the fraction of a substance remaining in an ideal PFR and in a dispersed PFR having $d = 1.0$.

Solution

1. Determine the C/C_0 ratio for an ideal PFR from Figure 3.34.

$$k\theta = 0.35\,\text{h}^{-1} \times 8\,\text{h} = 2.8$$

   Read $C/C_0 = 0.06$ for $k\theta = 2.8$ and $d = 0$ in an ideal PFR.

2. Determine the C/C_0 ratio for a dispersed PFR with $d = 1.0$ from Figure 3.34.

   Read $C/C_0 = 0.2$ for $k\theta = 2.8$ and $d = 1.0$.

3. Compare the fraction remaining.
   Fraction remaining in an ideal PFR is 0.06, and fraction remaining in a dispersed PFR is 0.2. The fraction remaining in a dispersed PFR is more than three times that of an ideal PFR.

EXAMPLE 3.51: REACTORS IN SERIES TO ACHIEVE A GIVEN REMOVAL OF A NONCONSERVATIVE SUBSTANCE

A reactor has total volume of 50 m^3. It receives a flow of 150 m^3/d. The first-order reaction rate constant for removal of a species A is 0.5 h^{-1}. Determine the following:

a. The fraction of A remaining if the reactor is an ideal PFR, and

b. Number of equal volume reactors in series with dispersion number d of 0.1 each. The total volume of all reactors is 50 m^3 and overall efficiency being the same as that for the ideal PFR in (a).

Solution

1. Determine the fraction A remaining in the effluent from the ideal PFR.
   Calculate θ from Equation 3.10c.

$$\theta_{PFR} = \frac{V_{PFR}}{Q} = \frac{50\,m^3}{150\,m^3/d} \times 24\,h/d = 8\,h$$

C/C_0 ratio is expressed by Equation 3.16d for an ideal PFR.

$$\frac{C}{C_0} = e^{-k\theta_{PFR}} = e^{-0.5\,h^{-1} \times 8\,h} = 0.018$$

2. Determine the number of reactors in series with dispersion number d of 0.1 that give $C/C_0 = 0.018$.
   Assume the number of reactors $n = 3$.

Volume of each reactor $V'_{PFR} = \dfrac{V_{PFR}}{n} = \dfrac{50\,m^3}{3} = 16.7\,m^3$

Hydraulic residence time in each reactor $\theta'_{PFR} = \dfrac{16.7\,m^3}{150\,m^3/d} \times 24\,h/d = 2.7\,h$

$$k\theta'_{PFR} = 0.5\,h^{-1} \times 2.7\,h = 1.4$$

Obtain C/C_0 ratio of 0.26 for each reactor from Figure 3.34 at $k\theta'_{PFR} = 1.4$ and $d = 0.1$.
Overall C/C_0 ratios for three reactors in series $= (0.26)^3 = 0.018$

Three reactors each of volume 16.7 m^3 and dispersion number of 0.1 in series will produce an effluent quality approximately equal to that of a single ideal PFR with a total volume of 50 m^3. The ideal PFR has $d = 0$.

3.6 Equalization of Flow and Mass Loadings

3.6.1 Need and Types

All wastewater treatment facilities experience hourly, daily, and seasonal flow and strength variations. Flow equalization is damping of the flow rate variations so that a relatively constant flow rate may be maintained through the downstream facilities. Depending upon the situation, flow equalization facilities therefore are provided to: (1) overcome the operational problems caused by the flow variations, (2) reduce the surge through the units, (3) equalize the strength, and dilute toxic inhibitory constituents, and maintain uniform concentrations and pH, (4) improve the performance of the process, and (5) reduce the size and cost of treatment facilities.[15,16]

The flow equalization may be of two types (1) in-line, and (2) off-line. In the in-line system, the entire flow is passed through the equalization basin. A constant flow is pumped from the basin. In an off-line system, only the excess flow rate is diverted to the equalization basin and is rerouted through the plant under low flow situations. In-line equalization basins offer better damping of the constituent mass loading, while only slight damping of constituents is achieved with off-line equalization basins.[15,16]

3.6.2 Design Considerations

The design of equalization basin should utilize mixing and aeration to prevent solids deposition and odor problems. Location below grit removal or primary sedimentation basin is desirable. Rectangular, square, and circular shapes with 4–6 m liquid depth have been adapted by the designers. Controlled flow pumping is necessary from the equalization basin.

EXAMPLE 3.52: PROCESS TRAIN WITH EQUALIZATION BASINS

Equalization basin is provided between grit and primary sedimentation basins. Draw the process trains to show in-line and off-line arrangements of the equalization basins.

Solution

The process trains for in-line and off-line flow equalization basins, and flow patterns before and after equalization are shown in Figure 3.36.

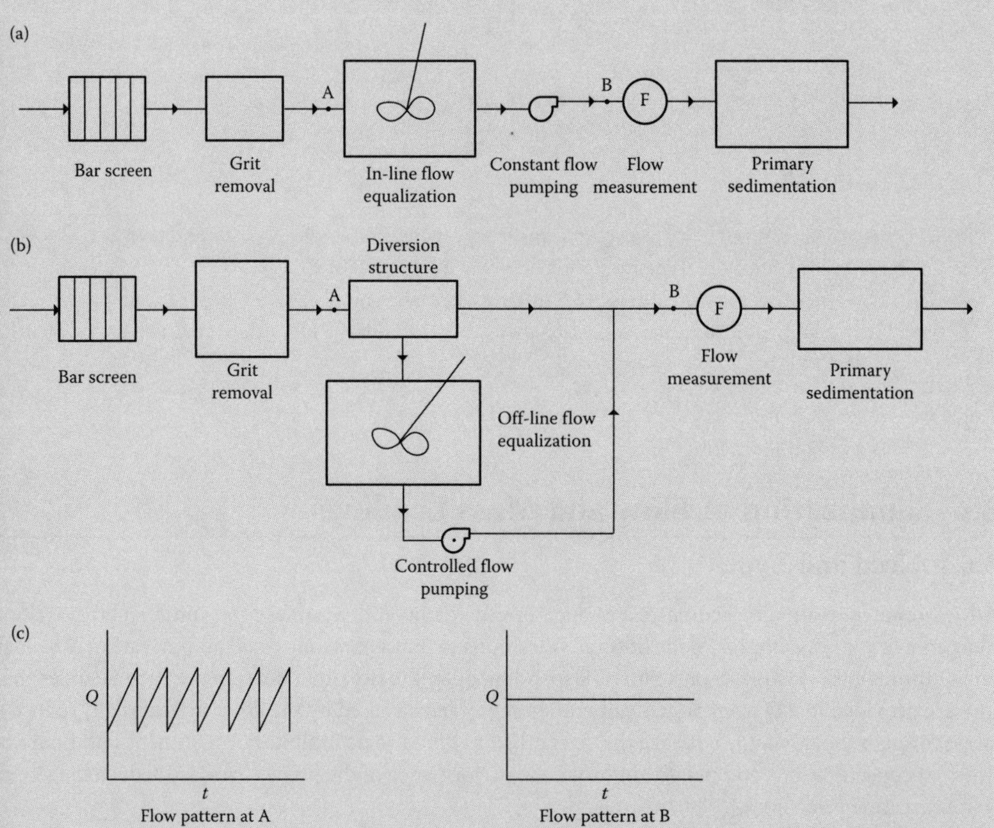

FIGURE 3.36 Process trains and flow pattern with equalization basins: (a) in-line, (b) off-line, and (c) flow patterns (Example 3.52).

3.6.3 Design Volume of Equalization Basin

The volume of an equalization basin is established for (a) flow equalization and (b) mass loading equalization. Basin sizing for both conditions are given below:

Flow Equalization: The volume or storage capacity of an equalization basin is determined by either graphical or analytical technique. The graphical technique requires preparation of a mass diagram in which cumulative inflow volume and cumulative average day volume are plotted on the same plot against the time of the day.[17,18] The analytical technique is based on cumulative change in volume deficiency between inflow and outflow as expressed by Equations 3.42a and 3.42b.

$$\Delta V = V_{in} - V_{out} \tag{3.42a}$$

$$\Delta V = (Q_{in} - Q_{out})\Delta t \tag{3.42b}$$

where

ΔV = change in storage volume during a specified time interval, volume
V_{in} = total inflow volume during the specified time interval, volume
V_{out} = total outflow volume during the specified time interval, volume
Q_{in} = total inflow, volume/time
Q_{out} = total outflow, volume/time
Δt = time interval, time

Mass Loading Equalization: In an in-line equalization basin, the variable mass loading rate is also equalized along with the flow rate. The degree of equalization depends upon the size of the basin. The basic expression used to calculate the effluent concentration is usually expressed by the general mass balance relationship of a completely mixed reactor without conversion. This relationship for an equalization basin is expressed by Equation 3.43.[19]

$$\underset{\text{(accumulation)}}{VdC} = \underset{\text{(input)}}{C_a q_a dt} - \underset{\text{(output)}}{C q_a dt} \tag{3.43}$$

where

V = volume of the tank, m^3
dC = change in concentration in the tank, mg/L or g/m^3
q_a = average flow rate over small time interval, m^3/s
dt = small time increment of time, s
C_a = average concentration of material in the influent, mg/L or g/m^3
C = concentration of material in the effluent (variable), mg/L or g/m^3

$$VdC = q_a(C_a - C)dt$$

$$dt = \frac{V}{q_a}\frac{dC}{(C_a - C)}$$

Using the integration limits C_1 at t_1 and C_2 at t_2, integrate the equation to obtain Equation 3.44.

$$\int_{t_1}^{t_2} dt = \frac{V}{q_a}\int_{C_1}^{C_2}\frac{dC}{(C_a - C)}$$

$$(t_2 - t_1) = \frac{V}{q_a}\ln\left(\frac{C_a - C_1}{C_a - C_2}\right) \tag{3.44}$$

where

t_1 = time at the start of the time increment dt, s
t_2 = time at the end of the time increment dt, s
C_1 = concentration in the effluent at the time t_1, mg/L or g/m^3
C_2 = concentration in the effluent at the time t_2, mg/L or g/m^3

EXAMPLE 3.53: STORAGE VOLUME BASED ON FLOW DIAGRAM AND CHANGE IN VOLUME

The hourly flow pattern of an industrial process is given below. The treatment facility is designed for a constant flow rate of 1500 m^3/h. Determine the volume of equalization basin by flow diagram and change in volume calculations. Also, draw the line representing the volume remaining in the basin.

Time Period	Midnight–1:00 a.m.	1–2	2–3	3–4	4–5	5–6	6–7	7–8	8–9	9–10	10–11	11:00 a.m.–noon
Flow, m^3/h	1050	960	930	930	975	1080	1170	1350	1570	1800	2100	2235
Time Period	Noon–1:00 p.m.	1–2	2–3	3–4	4–5	5–6	6–7	7–8	8–9	9–10	10–11	11:00 p.m.–midnight
Flow, m^3/h	2250	2235	2100	1920	1680	1380	1230	1350	1574	1574	1383	1170

Solution

1. Develop the calculation table for flow diagram and change in volume.
   The calculations are provided in Table 3.10.
2. Develop the flow diagram.
   a. Plot cumulative inflow and outflow with respect to time of the day. This plot is shown in Figure 3.37.
   b. Draw two tangents at points A and B on cumulative inflow line. Points A and B are at the valley and hump of the curvatures. The tangents are parallel to cumulative outflow or pumping line.
3. Determine the required volume of the equalization basin using change in volume calculations.
   The change in volume calculations are summarized in Table 3.11. The following information may be drawn:
   a. The basin is empty at 8:00 a.m.
   b. The basin starts filling after 8:00 a.m. by the incoming flow to the plant.
   c. It is completely full to its maximum volume of 4390 m^3 at 5:00 p.m.
   d. The basin starts emptying after 5:00 p.m.
   e. The basin is totally empty at 8:00 a.m on the next day.
   f. This cycle repeats again.
4. Determine the required volume of equalization basin from flow diagram.
   The vertical distance between the two tangents is the required theoretical volume of the equalization basin. This volume is 4390 m^3.
5. Draw the line representing the basin volume remaining based on cumulative change in volume (Figure 3.37).
   a. From 8:00 a.m. to 5:00 p.m., is the filling cycle. The basin is filled in 9 h. From 5:00 p.m. to 8:00 a.m., is the emptying cycle. The basin is emptied in 15 h.
   b. During the filling and emptying cycles, a constant flow of 1500 m^3/h is withdrawn from the basin.
   c. The maximum volume of 4390 m^3 on the curve represents the required theoretical volume of the equalization basin.

TABLE 3.10 Calculations for Determination of Volume of Equalization Basin (Example 3.53)

Time	Δt, h	Q_{in}, m^3/h	Q_{out}, m^3/h	$\sum \Delta V_{in}$, m^3	$\sum \Delta V_{out}$, m^3	ΔV, m^3	$\sum \Delta V$, m^3
(1)	(2)	(3)	(4)	(5)	(6)	(7)	(8)
Midnight	–	–	–	–	–	–	–
1:00 a.m.	1	1050	1500	1050	1500	−450	3105
2	1	960	1500	2010	3000	−540	2565
3	1	930	1500	2940	4500	−570	1995
4	1	930	1500	3870	6000	−570	1425
5	1	975	1500	4845	7500	−525	900
6	1	1080	1500	5925	9000	−420	480
7	1	1170	1500	7095	10,500	−330	150
8	1	1350	1500	8445	12,000	−150	0[a]
9	1	1570	1500	10,015	13,500	70[b]	70
10	1	1800	1500	11,815	15,000	300	370
11	1	2100	1500	13,915	16,500	600	970
Noon	1	2235	1500	16,150	18,000	735	1705
1:00 p.m.	1	2250	1500	18,400	19,500	750	2455
2	1	2235	1500	20,635	21,000	735	3190
3	1	2100	1500	22,735	22,500	600	3790
4	1	1920	1500	24,655	24,000	420	4210
5	1	1680	1500	26,335	25,500	180	4390[c]
6	1	1380	1500	27,715	27,000	−120[d]	4270
7	1	1230	1500	28,945	28,500	−270	4000
8	1	1350	1500	30,295	30,000	−150	3850
9	1	1574	1500	31,869	31,500	74	3924
10	1	1574	1500	33,443	33,000	74	3998
11	1	1383	1500	34,826	34,500	−117	3881
Midnight	1	1174	1500	36,000	36,000	−326	3555

[a] Basin is empty.
[b] Basin starts filling.
[c] Basin full. Required theoretical capacity of the basin is reached.
[d] Basin starts emptying.

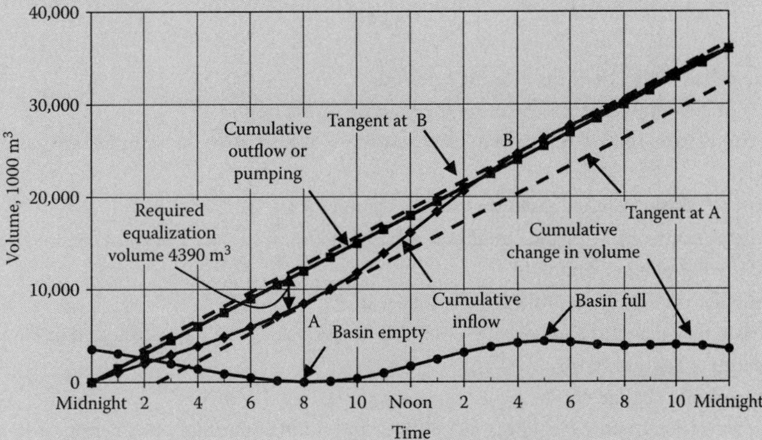

FIGURE 3.37 Mass diagram for determination of the capacity of the equalization basin and volume remaining (Example 3.53).

EXAMPLE 3.54: VOLUME OF AN IN-LINE EQUALIZATION BASIN

An in-line equalization basin receives variable flow. Determine the volume of an equalization basin to withdraw a constant flow of 652 m^3/min into a biological wastewater treatment plant. Use volume change calculations and flow diagram to determine the volume. Indicate what time of the day the basin will be full and empty.

Time Period	Midnight–1:00 a.m.	1–2	2–3	3–4	4–5	5–6	6–7	7–8	8–9	9–10	10–11	11:00 a.m.–noon
Average flow rate during time period, m^3/min	582	468	348	276	222	210	252	432	750	870	900	912
Time Period	Noon–1:00 p.m.	1–2	2–3	3–4	4–5	5–6	6–7	7–8	8–9	9–10	10–11	11:00 p.m.–midnight
Average flow rate during time period, m^3/min	900	858	817	744	690	690	696	774	864	846	804	743

Solution

1. Determine the constant withdrawal rate from the equalization basin during each hour of pumping.

   Constant flow rate $= 652\ m^3/min \times 60\ min/h = 39.1 \times 10^3\ m^3/h$

2. Determine the average variable flow during each hour time period.

   Change in volume due to flow into the basin during midnight and 1:00 a.m. $= 582\ m^3/min \times 60\ min/h \times 1\ h$ (duration)

   $= 34.9 \times 10^3\ m^3$

   Repeat calculations for each hour time period.
3. Prepare a calculation table to determine the volume of the basin.

   The cumulative inflow and cumulative outflow, and change in volume are summarized in Table 3.11.
4. Determine the basin volume based on volume change.

   The theoretical volume of the equalization basin is $146 \times 10^3\ m^3$. The basin will be empty at 8:00 a.m. and it will be full at midnight.
5. Determine the basin volume from the flow diagram.

   The basin flow diagram and volume remaining at any time of the day are plotted in Figure 3.38. The basin volume when full is $146 \times 10^3\ m^3$.
6. Draw the volume change curve.

   The equalization basin starts filling at 8:00 a.m. and is full at midnight in a period of 16 h. The basin starts emptying at midnight and becomes empty at 8:00 a.m. in a period of 8 h. The volume change curve is also shown in Figure 3.38.

TABLE 3.11 Calculations to Determine Equalization Basin Volume (Example 3.54)

Time of Day	Q_{in}, m^3/min	V_{in}, 10^3 m^3	$\sum \Delta V_{in}$, 10^3 m^3	V_{out}, 10^3 m^3	$\sum \Delta V_{out}$, 10^3 m^3	ΔV, 10^3 m^3	$\sum \Delta V$, 10^3 m^3
Midnight	–	–	–	–	–	–	–
1:00 a.m.	582	34.9	34.9	39.1	39.1	−4.20[a]	141
2	468	28.1	63.0	39.1	78.2	−11.0	130
3	348	20.9	83.9	39.1	117	−18.2	112
4	276	16.6	100	39.1	156	−22.6	89.5
5	222	13.3	114	39.1	196	−25.8	63.7
6	210	12.6	126	39.1	235	−26.5	37.2
7	252	15.1	141	39.1	274	−24.0	13.2
8	432	25.9	167	39.1	313	−13.2	0.00[b]
9	750	45.0	212	39.1	352	5.88[c]	5.88
10	870	52.2	265	39.1	391	13.1	19.0
11	900	54.0	319	39.1	430	14.9	33.8
Noon	912	54.7	373	39.1	469	15.6	49.4
1:00 p.m.	900	54.0	427	39.1	509	14.9	64.3
2	858	51.5	479	39.1	548	12.4	76.7
3	817	49.0	528	39.1	587	9.90	86.5
4	744	44.6	572	39.1	626	5.52	92.1
5	690	41.4	614	39.1	665	2.28	94.3
6	690	41.4	655	39.1	704	2.28	96.6
7	696	41.8	697	39.1	743	2.64	99.2
8	774	46.4	743	39.1	782	7.32	107
9	864	51.8	795	39.1	822	12.7	119
10	846	50.8	846	39.1	861	11.6	131
11	804	48.2	894	39.1	900	9.12	140
Midnight	745	44.7	939	39.1	939	5.58	146[d]

[a] Basin starts emptying cycle.
[b] Basin is empty.
[c] Basin starts filling.
[d] Basin is full. Required theoretical capacity of the basin is reached.

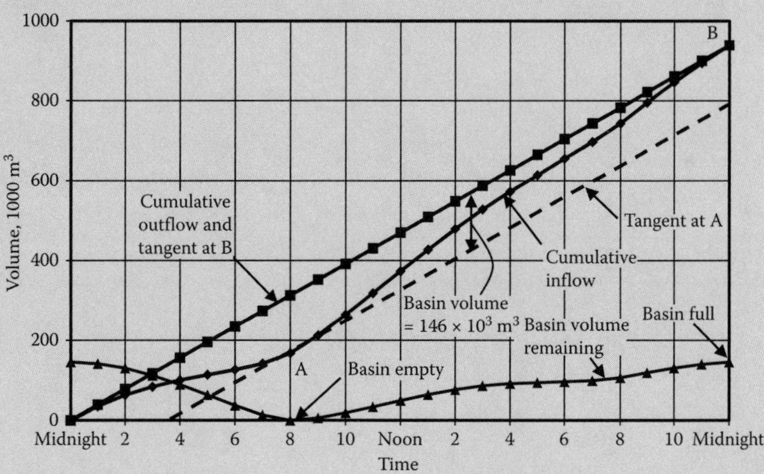

FIGURE 3.38 Flow diagram for determination of the capacity of equalization basin and volume remaining (Example 3.54).

EXAMPLE 3.55: EQUALIZATION OF MASS LOADING

An equalization basin has a volume of 1 million liters (ML). The influent flow and BOD_5 data for several time periods are given below. Calculate the BOD_5 concentration in the basin effluent. It is given that the BOD_5 concentration in the effluent at 8:00 a.m. is 160 mg/L.

Time	Influent Flow (MLD[a])	Influent BOD_5 (mg/L)
8:00 a.m.	0.2	150
9:00 a.m.	0.3	250
10:00 a.m.	0.4	180
11:00 a.m.	0.5	350
12:00 a.m. (Noon)	0.3	200

[a] MLD: million liters per day.

Solution

1. Calculate q_a and C_a at $t_1 = 8{:}00$ a.m., and $t_2 = 9{:}00$ a.m.

$$q_a = \frac{(0.2+0.3)\,\text{MLD}}{2} = 0.25\,\text{MLD or } q_a = 0.25\,\text{MLD} \times \frac{\text{d}}{24\,\text{h}} = 0.0104\,\text{ML/h}$$

$$C_a = \frac{(150+250)\,\text{mg/L}}{2} = 200\,\text{mg/L}$$

2. Calculate the effluent BOD_5 concentration (C_9) at 9:00 a.m.

Use Equation 3.44 to calculate effluent concentration. It is given that the effluent BOD_5 concentration at 8:00 a.m. is 160 mg/L.

$$(t_2 - t_1) = \frac{V}{q_a}\ln\left(\frac{C_a - C_8}{C_a - C_9}\right)$$

$$(9-8)\,\text{h} = \frac{1\,\text{ML}}{0.0104\,\text{ML/h}} \times \ln\left(\frac{(200-160)\,\text{mg/L}}{(200-C_9)\,\text{mg/L}}\right)$$

$$\ln\left(\frac{40}{200-C_9}\right) = 0.0104$$

$$\frac{40}{200-C_9} = e^{0.0104} = 1.01$$

Solving $C_9 = 160$ mg/L

3. Calculate the effluent BOD_5 concentration C_{10} at 10:00 a.m.

$$q_a = \frac{(0.3+0.4)\,\text{MLD}}{2} = 0.35\,\text{MLD or } q_a = 0.35\,\text{MLD} \times \frac{\text{d}}{24\,\text{h}} = 0.0146\,\text{ML/h}$$

$$C_a = \frac{(250+180)\,\text{mg/L}}{2} = 215\,\text{mg/L}$$

$$(10-9)\,\text{h} = \frac{1\,\text{ML}}{0.0146\,\text{ML/h}} \times \ln\left(\frac{(215-160)\,\text{mg/L}}{(215-C_{10})\,\text{mg/L}}\right)$$

$$\ln\left(\frac{55}{215-C_{10}}\right) = 0.0146$$

$$\frac{55}{215-C_{10}} = e^{0.0146} = 1.01$$

Solving $C_{10} = 161$ mg/L

4. Calculate the effluent BOD_5 concentration C_{11} at 11:00 a.m.

$$q_a = \frac{(0.4 + 0.5)\,\text{MLD}}{2} = 0.45\,\text{MLD or } q_a = 0.45\,\text{MLD} \times \frac{\text{d}}{24\,\text{h}} = 0.0188\,\text{ML/h}$$

$$C_a = \frac{(180 + 350)\,\text{mg/L}}{2} = 265\,\text{mg/L}$$

$$(11 - 10)\,\text{h} = \frac{1\,\text{ML}}{0.0188\,\text{ML/h}} \times \ln\left(\frac{(265 - 161)\,\text{mg/L}}{(265 - C_{11})\,\text{mg/L}}\right)$$

$$\ln\left(\frac{104}{265 - C_{11}}\right) = 0.0188$$

$$\frac{104}{265 - C_{11}} = e^{0.0188} = 1.02$$

Solving $C_{11} = 163$ mg/L

5. Calculate the effluent BOD_5 concentration C_{12} at noon.

$$q_a = \frac{(0.5 + 0.3)\,\text{MLD}}{2} = 0.4\,\text{MLD or } q_a = 0.4\,\text{MLD} \times \frac{\text{d}}{24\,\text{h}} = 0.0167\,\text{ML/h}$$

$$C_a = \frac{(350 + 200)\,\text{mg/L}}{2} = 275\,\text{mg/L}$$

$$(12 - 11)\,\text{h} = \frac{1\,\text{ML}}{0.0167\,\text{ML/h}} \times \ln\left(\frac{(275 - 163)\,\text{mg/L}}{(275 - C_{12})\,\text{mg/L}}\right)$$

$$\ln\left(\frac{112}{275 - C_{12}}\right) = 0.0167$$

$$\frac{112}{275 - C_{12}} = e^{0.0167} = 1.02$$

Solving $C_{12} = 165$ mg/L

6. Summarize the results.

Influent flows and BOD_5 concentrations, and the effluent BOD_5 concentrations are summarized in Table 3.12. It may be noted that the effluent concentrations at different times are based on the given or assumed value of effluent BOD_5 of 160 mg/L at 8:00 a.m. If the calculations are continued over 24-h period, the concentration of BOD_5 at 8:00 a.m. after 24-h cycle should match 160 mg/L. If it does not match, that means the given or assumed value of 160 mg/L of BOD_5 at 8:00 a.m. is not reliable enough. The procedure should be carried out over 24 h for several iterations until a stable value is reached.

TABLE 3.12 Effluent BOD_5 from Equalization Basin (Example 3.55)

Time	Influent Flow, MLD	Influent BOD_5, mg/L	q_a, MLD	C_a, mg/L	Effluent BOD_5, mg/L
8:00 a.m.	0.2	150	–	–	160 (given)
9:00 a.m.	0.3	250	0.25	200	160
10:00 a.m.	0.4	180	0.35	215	161
11:00 a.m.	0.5	350	0.45	265	163
Noon	0.3	200	0.40	275	165

EXAMPLE 3.56: EQUALIZATION OF MASS LOADING OVER 24-h CYCLE

An equalization basin receives flow from an industrial plant. The hourly flow and COD concentration data of the influent to the equalization basin are given below. Determine: (a) average flow rate over 24-h period, (b) the theoretical volume of the equalization basin in ML that will provide constant flow to the treatment plant, and (c) hourly concentration of COD in the effluent from the equalization basin that has a volume equal to the theoretical volume, 0.35 and 0.7 ML. Draw the influent and effluent flow profiles, and influent and effluent COD concentration profiles at all three volumes of the equalization basin.

Time	7:00 a.m.	8	9	10	11	Noon	1:00 p.m.	2	3	4	5	6
Influent flow, MLD	0.40	0.50	0.80	1.20	1.30	1.40	1.30	0.80	0.70	0.60	0.50	0.50
Influent COD, mg/L	100	115	125	150	200	280	300	300	150	125	100	100
Time	7:00 p.m.	8	9	10	11	Midnight	1:00 a.m.	2	3	4	5	6
Influent flow, MLD	0.60	0.80	1.00	0.90	0.60	0.40	0.30	0.25	0.20	0.20	0.25	0.30
Influent COD, mg/L	150	175	225	225	175	150	100	90	80	70	80	90

Solution

1. Determine the theoretical volume of the equalization basin by change in volume method.
   a. Prepare the calculation table to determine constant flow and the theoretical volume of the basin (Table 3.13).
   b. Determine the constant withdrawal flow from the basin.

      There are 24 hourly observations (7:00 a.m. to 6:00 a.m.). The average withdrawal flow based on 24 observations is 0.658 MLD.

$$\text{Mean constant withdrawal flow based on 24 observations} = \frac{15.8\,\text{MLD}}{24} = 0.658\,\text{MLD}$$

$$\text{Average flow per hour} = 0.658\,\text{MLD} \times \frac{\text{d}}{24\,\text{h}} \times \frac{10^6\,\text{L}}{\text{ML}} = 27.4 \times 10^3\,\text{L/h}$$

2. Determine the theoretical volume of the equalization basin.

   The theoretical volume of the equalization basin is obtained from the maximum value of the changes in volume (Column [7], Table 3.13). The procedure to calculate the required theoretical volume of a flow equalization basin is shown in Examples 3.53 and 3.54.

$$\text{Theoretical volume} = 133 \times 10^3\,\text{L or } 0.133 \times 10^6\,\text{L (0.133 ML)}$$

3. Plot the influent and effluent flow profiles from the equalization basins.

   The influent flow and constant effluent flow profiles from an equalization basin with the theoretical volume of 0.133 ML are shown in Figure 3.39. If there is no flow accumulation in the basin, the influent and effluent flow profiles are independent of the equalization basin volume. Therefore, the influent and

TABLE 3.13 Calculation Table for Determination of Theoretical Volume of the Basin, and COD Concentration in the Effluent from Basin Volumes of 0.133, 0.35, and 0.7 MG (Example 3.56)

t	Q_{in}, MLD	C_{in}, mg/L	V_{in}, 10^3 L	V_{out}, 10^3 L	ΔV, 10^3 L	$\sum \Delta V$, 10^3 L	q_a, 10^3 L/h	C_a, mg/L	C_{out} for Different Bain Volume, mg/L		
									$V = 0.133$ ML	$V = 0.35$ ML	$V = 0.7$ ML
(1)	(2)	(3)	(4)	(5)	(6)	(7)	(8)	(9)	(10)	(11)	(12)
7 a.m.	0.40	100	16.7[a]	–	–	–	–	–	146[b]	169	176
8	0.50	115	20.8	27.4	−6.6	0	18.8[c]	108[d]	141[e]	166	174
9	0.80	125	33.3	27.4	5.9	5.90	27.1	120	137	163	172
10	1.20	150	50.0	27.4	22.6	28.5	41.7	138	137	160	170
11	1.30	200	54.2	27.4	26.7	55.2	52.1	175	149	162	171
Noon	1.40	280	58.3	27.4	30.9	86.1	56.3	240	181	174	176
1 p.m.	1.30	300	54.2	27.4	26.7	113	56.3	290	218	191	185
2	0.80	300	33.3	27.4	5.9	119	43.8	300	241	204	192
3	0.70	150	29.2	27.4	1.7	120	31.3	225	238	205	193
4	0.60	125	25.0	27.4	−2.4	118	27.1	138	219	200	191
5	0.50	100	20.8	27.4	−6.6	111	22.9	113	202	195	189
6	0.50	100	20.8	27.4	−6.6	105	20.8	100	188	189	186
7	0.60	150	25.0	27.4	−2.4	102	22.9	125	178	185	184
8	0.80	175	33.3	27.4	5.9	108	29.2	163	175	183	183
9	1.00	225	41.7	27.4	14.2	123	37.5	200	181	185	184
10	0.90	225	37.5	27.4	10.1	133[f]	39.6	225	192	189	186
11	0.60	175	25.0	27.4	−2.4	130	31.3	200	194	190	187
Midnight	0.40	150	16.7	27.4	−10.8	119	20.8	163	189	189	186
1 a.m.	0.30	100	12.5	27.4	−14.9	105	14.6	125	183	186	185
2	0.25	90	10.4	27.4	−17.0	87.5	11.5	95	175	183	183
3	0.20	80	8.3	27.4	−19.1	68.4	9.4	85	169	181	182
4	0.20	70	8.3	27.4	−19.1	49.3	8.3	75	164	178	181

(Continued)

TABLE 3.13 (*Continued*) Calculation Table for Determination of Theoretical Volume of the Basin, and COD Concentration in the Effluent from Basin Volumes of 0.133, 0.35, and 0.7 MG (Example 3.56)

t	Q_{in}, MLD	C_{in}, mg/L	V_{in}, 10^3 L	V_{out}, 10^3 L	ΔV, 10^3 L	$\sum \Delta V$, 10^3 L	q_a, 10^3 L/h	C_a, mg/L	C_{out} for Different Bain Volume, mg/L		
									$V=0.133$ ML	$V=0.35$ ML	$V=0.7$ ML
5	0.25	80	10.4	27.4	−17.0	32.3	9.4	75	158	175	179
6	0.30	90	12.5	27.4	−14.9	17.4	11.5	85	152	172	178
7	0.40	100	16.7	27.4	−10.8	7.12	14.6	95	146[g]	169	176
Total	–	–	658	658	–	–	–	–	–	–	–
Average	0.658	152	27.4	27.4			27.4	152			

[a] $0.4\,\text{MLD} \times \frac{10^6\,\text{L}}{\text{ML}} \times \frac{\text{d}}{24\,\text{h}} \times 1\,\text{h} = 16.7 \times 10^3\,\text{L}$

[b] Assume an effluent concentration $C_1 = 146$ mg/L to start the calculation.

[c] $q_a = \frac{(0.4 + 0.5) \times \text{MLD}}{2} = 0.45\,\text{MLD}$ or $q_a = 0.45\,\text{MLD} \times \frac{10^6\,\text{L}}{\text{ML}} \times \frac{\text{d}}{24\,\text{h}} = 18.8 \times 10^3\,\text{ML/h}$

[d] $C_a = \frac{(100 + 115)\,\text{mg/L}}{2} = 108\,\text{mg/L}$

[e] Rearrange Equation 3.44 to calculate C_2.

$$C_2 = C_a - (C_a - C_1)e^{-\frac{q_a}{V} \times (t_2 - t_1)} = 108\,\text{mg/L} - (108\,\text{mg/L} - 146\,\text{mg/L}) \times e^{-\frac{18.8\times10^3\,\text{ML/h}}{133\,\text{ML}} \times (8-7)\,\text{h}} = 141\,\text{mg/L}$$

[f] Theoretical volume of the basin.

[g] The calculated effluent concentration at 7:00 a.m. is same as that assumed in the first row.

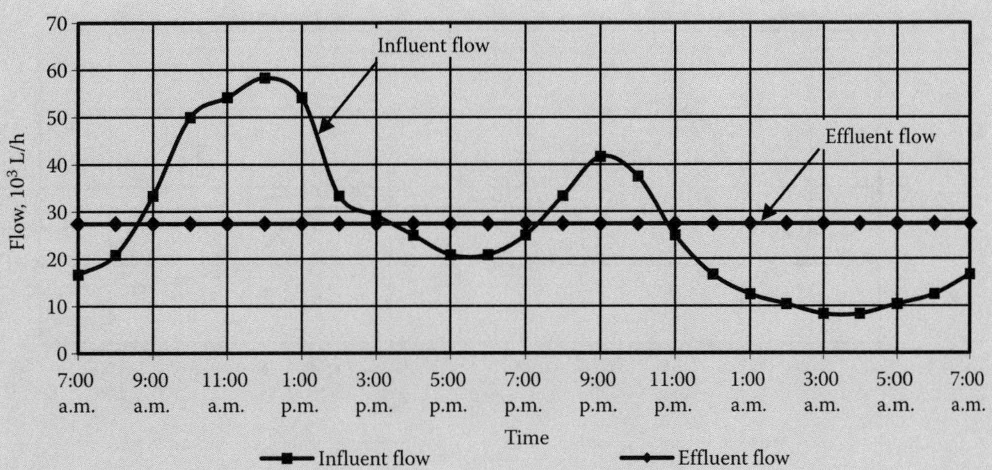

FIGURE 3.39 Influent and effluent flow profiles at theoretical volume of 0.133 ML (Example 3.56).
Note: The profiles for basin volumes of 0.35 and 0.7 ML will be the same as that for the theoretical volume of 0.133 ML.

effluent flow profiles at basin volumes of 0.35 and 0.7 ML will also be the same as those of 0.133 ML (Figure 3.39).

4. Determine the effluent concentration of COD from the equalization basin.

   An iterative procedure is used to determine the COD concentration in the effluent.

   a. Calculate average flow rate q_a between the consecutive time intervals.
   b. Calculate average concentration C_a in the effluent between the consecutive time intervals.
   c. Assume that the effluent COD concentration C_1 at the start of time (7:00 a.m.) for a given basin volume V. For example, $C_1 = 146$ mg/L for $V = 0.133$ ML.
   d. Calculate effluent COD concentration C_2 at the next time period for the given basin volume. In this case, the next time period is 8:00 a.m. Use the procedure shown in Example 3.55 to calculate C_2.
   e. Calculate effluent COD concentration for different time periods C_3, C_4, ... C_{24}. After a 24-h cycle, the calculated effluent concentration at 7:00 a.m. in the last row should match the assumed value in the first row. If not, repeat the procedure until the values fall within an acceptable limit. The entire procedure for effluent COD concentration from 0.133 ML basin is summarized in Table 3.13.
   f. Apply the similar procedure to determine the effluent COD concentrations from the basin with a volume of 0.35 and 0.7 ML, respectively. The results are summarized in Table 3.13.

5. Draw the influent and effluent COD concentration profiles.

   The influent and effluent COD concentration profiles from the equalization basin with a total volume of 0.133, 0.35, and 0.7 ML are respectively shown in Figure 3.40a through c. It may be noted that the effluent concentration is more uniform from the basin with a larger volume.

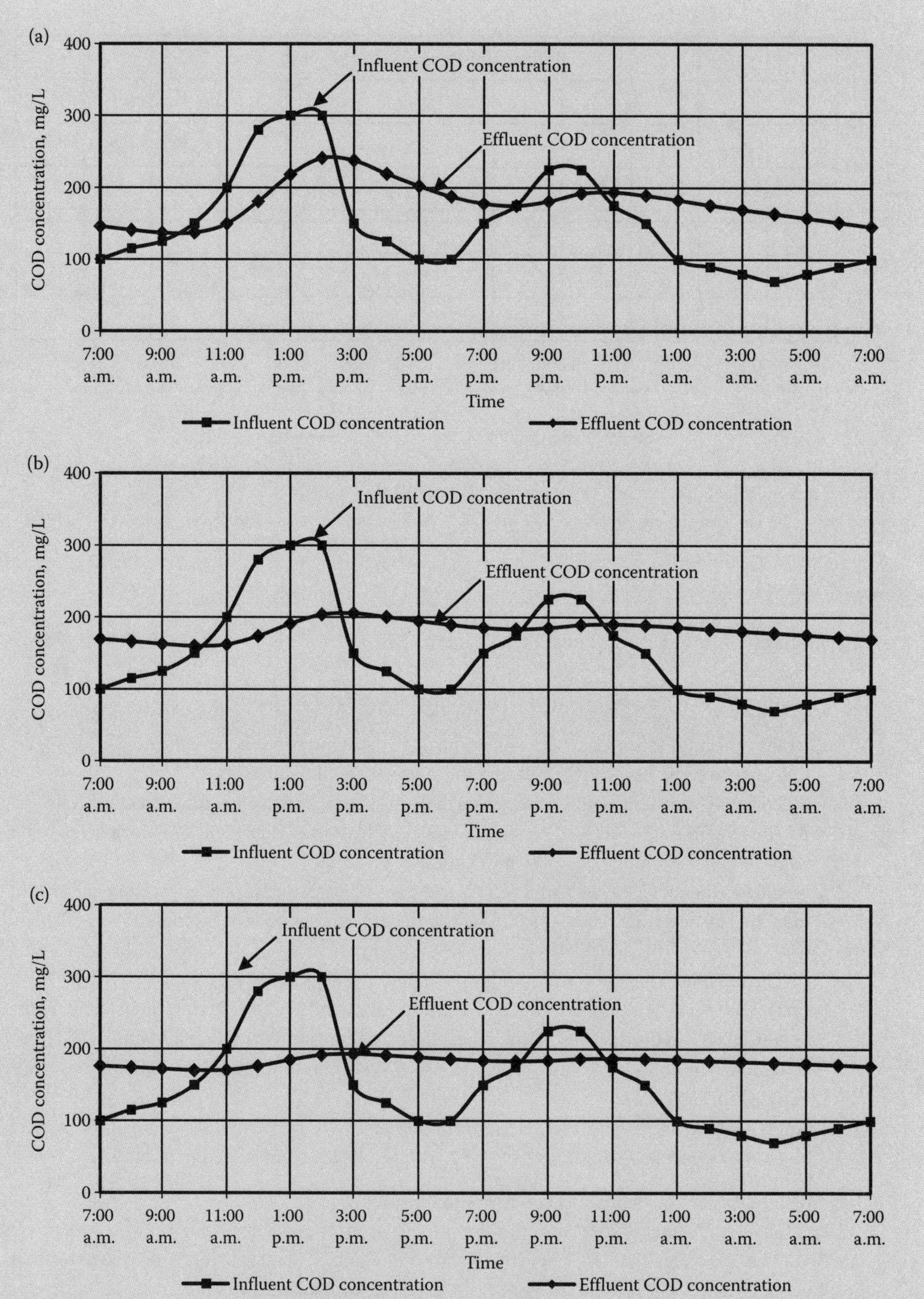

FIGURE 3.40 Influent and effluent COD concentration profiles: (a) basin volume of 0.133 ML, (b) basin volume of 0.35 MG, and (c) basin volume of 0.7 MG (Example 3.56).

Discussion Topics and Review Problems

3.1 Wastewater flows from manhole 1–2. Manhole 1 is connected to three sewer lines that bring average flow of 4.5 L/s, 80 gal/min, and 38 L/min. Manhole 2 receives additional flow of 35 L/s. Determine the flow in the outgoing sewer.

3.2 An industrial sewer survey was conducted to establish the concentrations of heavy metals in the intercepting sewer. Average flow and concentrations of heavy metals in different streams are given below. Determine the concentrations of various heavy metals in the flow from the intercepting sewer. Manhole (1) $Q = 3$ L/s, Cu = 4 mg/L, Zn = 10 mg/L; $Q = 15$ L/s, Cd = 15 mg/L; $Q = 8$ L/s, Cd = 8 mg/L, Ni = 8 mg/L, Manhole (2) $Q = 10$ L/s, Cu = 8 mg/L; $Q = 5$ L/s, Ni = 28 mg/L, Manhole (3) $Q = 8$ L/s, Cd = 8 mg/L, Ni = 8 mg/L; $Q = 10$ L/s, Cu = 8 mg/L, Co = 2 mg/L; $Q =$ 5 L/s, Zn = 6 mg/L. Manhole (4) receives flows from three manholes.

3.3 A domestic water softener produces water with hardness of 5 mg/L as $CaCO_3$. The raw water has hardness of 225 mg/L. The average finished water demand is 1600 gpd with hardness of 80 mg/L as $CaCO_3$. Determine the flow through water softener.

3.4 In an ideal process, a chemical reaction with two liquids F_1 and F_2 produces one liquid product P and one gaseous product G. The gas is produced at a mass ratio of 1 unit of gas for average 1000 units of liquid product. The recycle stream $R = 0.5\ P$. Determine the weight and volume of liquid and gaseous products generated, and recycle stream. Use the following data: the flow rates of streams F_1 and F_2 are 1000 L/min and 6000 L/min; the densities of streams F_1 and F_2 are 1.5 and 1.2 kg/L; and the densities of the products P and G are 1.3 kg/L and 1.0 kg/m^3.

3.5 A primary sedimentation basin receives an average wastewater flow of 1500 m^3/d. The suspended solids in the raw wastewater is 240 mg/L. The solids removal efficiency of the basin is 60 %. The sludge has 4% solids and specific gravity is 1.03. Determine (a) TSS in the primary settled wastewater, and (b) average quantity of sludge withdrawn from the basin in kg/d and m^3/d.

3.6 A CFSTR is receiving an average flow of 240 m^3/d. The volume of the reactor is 48 m^3. A slug of 25-L stock solution containing 6.0 g/L dye tracer is released at the influent channel of the reactor. Calculate the dye concentration in the effluent at time intervals of 0.5, 1.0, 2.0, 4.0, 5.0, 20.0, and 0.0 h from the time of release of the slug. Draw the tracer profiles as a function of (a) tracer concentration versus time, and (b) dimensionless functions C/C_0 versus t/θ.

3.7 The volume of a CFSTR is 45 m^3 and average constant flow in the reactor is 300 m^3/d. The reactor flow regime was established by feeding continuously 6 g/L stock tracer solution in the influent line at a rate of 125 L per day. Determine the theoretical tracer concentration in the effluent at 3 and 10 h after the tracer injection. Plot the tracer profile as a function of (a) concentration versus t, and (b) C/C_0 versus t/θ.

3.8 A CFSTR receives 150 g/m^3 reactive material that undergoes first-order conversion reaction with $k = 0.20\ \text{h}^{-1}$. The volume of the reactor is 400 m^3, and the influent flow rate is 150 m^3/d. Determine the exit concentration of the material and percent stabilization if the process is operating under steady state condition. What is the value of t in days to achieve 95% conversion?

3.9 A CFSTR receives industrial material for product conversion. The operational data of the reactor are: $V = 450\ \text{m}^3$, $Q = 35\ \text{m}^3/\text{d}$, $C_0 = 100\ \text{g/m}^3$, and $k = 0.25\ \text{d}^{-1}$. The reactor is operating under nonsteady-state conditions. Calculate (a) the concentration of the material at 0.5 d, 1 d, 2 d, 4 d, 6 d, 8 d, 10 d, 15 d, and 20 d, and (b) plot the graph of concentration remaining versus time.

3.10 A constant flow is maintained through a CFSTR. At $t = 0$ a reactive substance is added into the influent stream at a constant rate. The conversion reaction is of the first order. Calculate the output concentration as g/m^3 for the following time intervals: 0.5 h, 1.0 h, 1.5 h, 3 h, 6 h, 8 h, and 10 h. The reactor data are given below: $V = 30\ \text{m}^3$, $Q = 250\ \text{m}^3/\text{d}$, $C_0 = 100\ \text{g/m}^3$, and $k = 0.3\ \text{h}^{-1}$.

3.11 A reactive substance with influent concentration of 75 g/m^3 is treated in a plug flow reactor. The first-order reaction constant is 0.15/h. Calculate the effluent concentration. The volume of the reactor is 50 m^3 and average constant flow through the reactor is 150 m^3/d.

3.12 Ammonia nitrogen is oxidized to nitrite, and then to nitrate nitrogen in a nitrification facility which consists of two identical CFSTRs in series. The hydraulic retention time of each reactor is 60 h. Oxygen is transferred by means of aerators at a rate which depends on the oxygen deficit. The DO transfer rate is expressed by the equation $r_{aer} = k_2\,(C_{DO_s} - C_{DO_t})$. The temperature is 25°C.

where

r_{aer} = oxygen transfer rate mg/L·d
k_2 = reaction rate constant = 70 d^{-1}
C_{DO_s} = DO saturation concentration = 8.24 mg/L at 25°C

The influent NH_3-N concentration is 40 mg/L. The NH_3-N conversion rate is given by Equation 3.45.

$$C_{NH_3\text{-}N} = \frac{-k(C_{NH_3\text{-}N})}{K_s + (C_{NH_3\text{-}N})} \tag{3.45}$$

where

k = 0.12 g/m^3·h
K_s = 0.74 g/m^3

The DO concentration in the influent is zero. Calculate the dissolved oxygen and ammonia nitrogen concentrations in (a) reactor 1 and (b) reactor 2. Assume steady-state conditions. Assume that there is no carbonaceous oxygen demand and oxygen consumption is due to only nitrification.

3.13 One CFSTR is arranged in between two PFRs in series (Figure 3.41). The arrangement is shown below. The influent BOD_5 is 200 g/m^3. The wastewater flow is 200 m^3/d. Calculate the final effluent BOD_5. The first-order reaction rate constant is 0.08 h^{-1}. Assume that the system is at steady state.

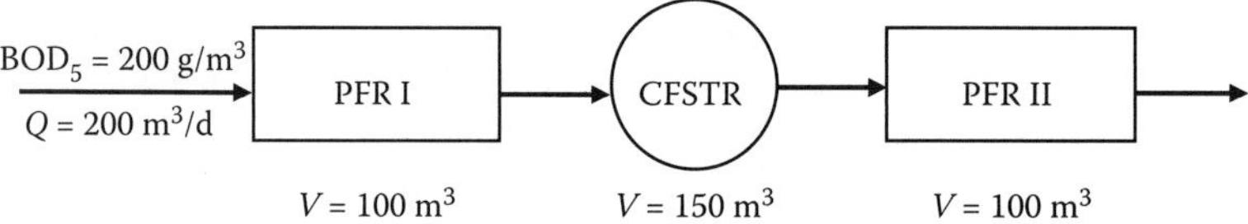

FIGURE 3.41 Definition sketch of one CFSTR between two PFRs (Problem 3.13).

3.14 Four identical CFSTRs are operating in series. The influent concentration of the reactive substance is 250 mg/L. The first-order reaction rate constant for each reactor is 0.74 h^{-1}. Volumes of reactors 1, 2, 3, and 4 are 20 m^3, 25 m^3, 30 m^3, and 35 m^3 respectively. The flow through the assembly is 50 m^3/h. Calculate the effluent concentration from all four reactors by using analytical and graphical method.

3.15 Calculate the first-order reaction rate constant in days for the given data. The number of CFSTRs in series are 5. The influent BOD_5 is 160 g/m^3 and final effluent BOD_5 is 10 g/m^3. The hydraulic residence time of each reactor is 1 h. Determine the detention time of a PFR that will produce same effluent quality as that from the series of reactors.

3.16 Derive the expression for PFR treating a reactive substance under steady state. The second-order reaction kinetics apply. The reaction rates are: (a) second-order reaction rate $r = -kC_1^2$ and (b) $r = -kC_1^{1.5}$.

3.17 A sedimentation basin receives an average flow of 4.8 m^3/min. The volume of the basin is 205 m^3. A 5-L slug of dye tracer solution containing 50,000 mg/L of dye was introduced at the influent zone of the basin. The effluent samples were collected at different time intervals and dye concentrations were measured. The sampling time and dye concentrations are given below.

Time t (min)	0	2	5	10	15	20	25	30	35	40	45	50	90
Dye Concentration C, mg/L	0.075	0.075	0.225	1.225	2.10	2.40	2.48	2.10	1.48	1.20	1.16	1.00	0.00

a. Calculate C_0 and nominal detention time θ
b. Draw the dye tracer profile C versus t
c. Draw the dye tracer profile C/C_0 versus t/θ
d. Determine standard detention time, standard detention efficiency, and fraction dead volume

3.18 A slug of dye tracer is applied at the influent channel of a CFSTR. The volume of the reactor is 28 m^3 and the flow is 0.08 m^3/s. The initial concentration C_0 after mixing the dye in the reactor content is 2.50 mg/L. The tracer concentration in the effluent sample collected at 4 min is 1.05 mg/L. Calculate (a) the average detention time and (b) percent dead volume.

3.19 A long and narrow channel is used as chlorine contact basin. The channel is 2-m wide and 30-m long, and 1.5-m liquid depth. The peak flow through the basin is 6480 m^3/d. The dispersion number d and reaction rate constants are 0.08 and 3.0 h^{-1}. Determine (a) contact time θ, (b) dispersion coefficient D, and (c) C/C_0.

3.20 A stabilization pond has dispersion number d of 0.25. The average flow and first-order reaction rate constants are 800 m^3/d and 0.3 d^{-1}, respectively. Calculate the volume of the pond to achieve 90% soluble BOD_5 removal.

3.21 A reactor is used for oxidation of a chemical compound. The reaction rate constant k is 0.8 h^{-1} and HRT is 4 h. The oxidation efficiency is 96%. Determine the flow regime in the reactor.

3.22 An equalization basin is designed for a small treatment plant. The hourly average flow and COD concentration data over a 24-h period are summarized below.

Time of Day	1 a.m.	2	3	4	5	6	7	8	9	10	11	Noon
Flow(m^3/h)	10	10	11	12	14	16	20	30	35	40	45	50
COD(mg/L)	120	120	130	135	130	135	140	145	200	250	300	300
Time of Day	1 p.m.	2	3	4	5	6	7	8	9	10	11	Midnight
Flow(m^3/h)	80	80	100	80	60	40	35	30	20	16	12	10
COD(mg/L)	300	350	400	400	300	250	250	200	150	150	120	120

Determine (a) the theoretical volume of the equalization basin that will provide a constant outflow, (b) the times when the basin will be empty and full, and (c) draw the influent and effluent COD concentrations from an equalization basin that is in 1200 m^3 volume.

References

1. Davis, M. L. and D. A. Cornwell, *Introduction to Environmental Engineering*, 2nd ed., McGraw-Hill, Inc., New York, 1991.
2. Vesilind, A. P., *Introduction to Environmental Engineering*, PWS Publishing Co., Boston, MA, 1997.
3. Metcalf & Eddy, Inc., *Wastewater Engineering: Treatment and Reuse*, 4th ed., McGraw-Hill Book Companies, Inc., New York, 2003.
4. Tchobanaglous, G. and E. D. Schroeder, *Water Quality: Characteristics, Modeling, and Modification*, Addison-Wesley Publishing Co., Reading, MA, 1985.
5. Reynolds, T. P. and P. A. Richards, *Unit Operations and Processes in Environmental Engineering*, 2nd ed., PWS Publishing Co., Boston, MA, 1996.

6. Qasim, S. R., E. M. Motley, and G. Zhu, *Water Works Engineering: Planning, Design, and Operation*, Prentice-Hall, Inc., Upper Saddle River, NJ, 2000.
7. Sawyer, C. N. and P. L. McCarty, *Chemistry for Environmental Engineering*, McGraw-Hill Book Company, New York, 1978.
8. Fiedler, R. A. and E. B. Fitch, Appraising basin performance from dye test results, *Sewage and Industrial Wastes*, 31(9), 1959, 1016–1021.
9. APHA, AWWA, and WEF, *Standard Methods for Examination of Water and Wastewater*, 9th ed., American Public Health Association, Washington, DC, 1995.
10. Spiegel, M. P. and L. J. Stephens, *Theory and Problems of Statistics*, 3rd ed., Schaum's Outlines Series, McGraw-Hill, New York, 1999.
11. Camp, T. R., Studies of sedimentation basin design, *Sewage and Industrial Wastes*, 25(1), 1953, 1–12.
12. Danckwerts, P. V., Continuous flow systems, *Chemical Engineering Science*, 2(1), 1953, 1–13.
13. Wehner, J. F. and R. F. Wilhelm, Boundary conditions of flow reactor, *Chemical Engineering Science*, 6(2), 1956, 89–93.
14. Thirumurthi, D., Design principles of waste stabilization ponds, *Journal of the Sanitary Engineering Division, ASCE*, 95(SA2), 1969, 311–330.
15. U.S. Environmental Protection Agency, *Flow Equalization*, Office of Technology Transfer, Washington D.C., 1974.
16. MacInnes, C. D., K. Adamowaski, and A. C. Middleton, Stochastic design of flow equalization basin, *Journal of the Environmental Engineering Division, ASCE*, 104(EE6), 1978, 1275–1289.
17. Qasim, S. R., *Wastewater Treatment Plants: Planning, Design, and Operation*, 2nd ed., CRC Press LLC, Boca Raton, FL, 1999.
18. Viessman, W.and M. J. Hammer, *Water Supply and Pollution Control*, Addison-Wesley, Menlo Park, CA, 1998.
19. Rudolfs, W. and J. N. Millar, A method for accelerated equalization of industrial wastes, *Sewage Works Journal*, 18(4), 1946, 686–689.

4

Sources and Flow Rates of Municipal Wastewater

4.1 Chapter Objectives

Municipal wastewater is the general term applied to the liquid wastes collected from residential, commercial, and industrial areas of a city. In addition to these waste streams, some undesired groundwater and surface runoff may also enter the collection system. All these flows are conveyed by means of a sewerage system to a central location for treatment. There is a wide variation in seasonal, daily, and hourly flow rates of the municipal wastewater, and several flow terms are used to identify these flow variations and characterize the municipal wastewaters.

The purpose of this chapter is to present the flow characteristics of municipal wastewater. The material is arranged under the following sections:

- Relationship between municipal water demand and wastewater flow
- Sources and components of water usage and wastewater generation
- Wastewater flow variations, infiltration and inflow, and definition of common terms
- Diurnal flows, sustained flows, and peaking factors
- Flow projections and forecasting design flow rates

4.2 Relationship between Municipal Water Demand and Wastewater Flow

Under dry weather conditions, municipal wastewater is derived largely from municipal water supply. A close relationship exists between water usage and wastewater flow in a community. Following is the interdependence between water usage and wastewater flow[1–5]:

1. Under dry weather conditions, the daily water demand and wastewater flow show a diurnal flow pattern.
2. The diurnal wastewater flow curve closely parallels that of water demand with a lag of several hours. This relationship is shown in Figure 4.1.
3. The fluctuations in wastewater flow are less than those of water supply.
4. A considerable portion of the water supply may not reach the sewers. This includes (a) lawn sprinkling, (b) street washing and fire fighting, (c) leakage from the water mains and service pipes, (d) water consumed in products and manufacturing processes, and (e) homes and establishments that may use city water but utilize on-site wastewater treatment and disposal systems.
5. Establishments using private water supply may increase average wastewater flow.

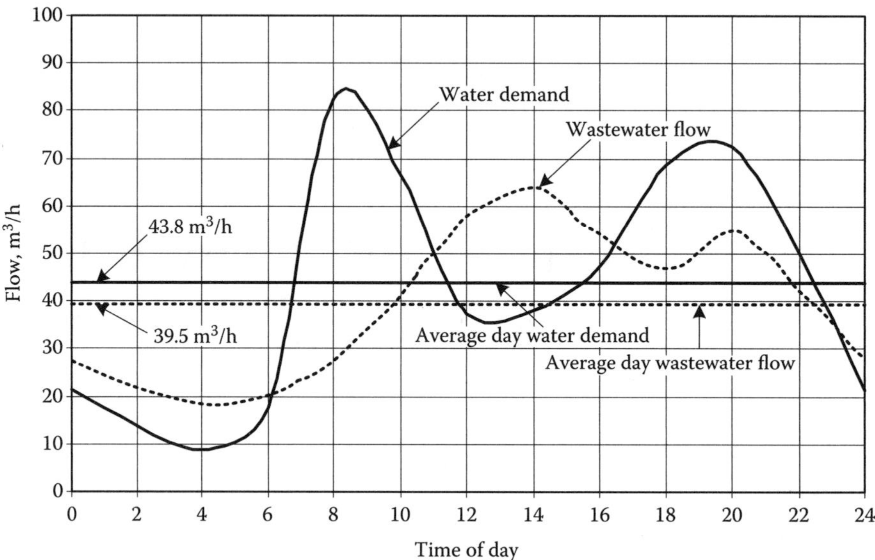

FIGURE 4.1 Typical diurnal municipal water demand and wastewater flow.
Source: Adapted in part from References 1 and 2.

6. Infiltration/inflow (I/I) may enter the sewers during wet weather conditions and may increase the flow substantially over a short period.
7. Exfiltration may occur during dry weather causing a reduction in average wastewater flow.
8. Annual average wastewater flow may vary from 60% to 130% of the annual average water demand. Designers frequently assume that annual average wastewater flow equals 80–100% of annual average water usage in a community.

EXAMPLE 4.1: RELATIONSHIP BETWEEN DIURNAL WATER DEMAND AND WASTEWATER FLOW

The diurnal wastewater flow pattern measured at a wastewater treatment plant is similar to that of water demand, but exhibits less fluctuation, and shows a lag of few hours. Why does this happen? Explain.

Solution

The diurnal wastewater flow has a lag of few hours due to:

a. The water lines are under pressure. Any usage in the distribution system transmits a pressure wave to the supply point. The pressure wave travels at the speed of sound. As a result, the water demand is almost instantly recorded.
b. The sewer lines flow under gravity.
c. The flow from different parts of the city take different time to reach a central location (treatment plant or lift station).
d. As the flow increases, the depth of flow in the lines also increases. As a result, the increase in velocity in sewer is relatively small. This results in a built-in storage capacity within the sewer lines, and has a damping effect upon the peaking factor.
e. As the length of collection system increases, these effects further reduce the peaking factor.

EXAMPLE 4.2: DETERMINATION OF AVERAGE FLOW FROM DIURNAL FLOW DATA

The diurnal flow pattern of municipal water demand and wastewater flow for a typical dry day is given in Figure 4.1. Calculate the following: (a) average-day water demand, (b) average-day wastewater flow, (c) ratio of average wastewater flow and water use, (d) ratio of peak and average wastewater flow, and (e) ratio of peak and average water demand.

Solution

1. The trapezoidal rule is generally used to calculate the areas under the demand curves. The trapezoidal rule is given by Equation 4.1.

$$A = \frac{1}{2} \times \Delta T\,[q_1 + 2(q_2 + q_3 + q_4 + \cdots + q_{n-1}) + q_n] \tag{4.1}$$

where

A = area under the curve, m^3

ΔT = constant time increment, h

$q_1, q_2, \ldots, q_n$ = flows at various time intervals, m^3/h

2. Calculate the daily average water demand from the area under water demand curve.

$$A_1 = \frac{1}{2} \times 2\,\text{h} \times [21\,\text{m}^3/\text{h} + 2 \times (12 + 8 + 18 + 85 + 65 + 36 + 39 + 48 + 69 + 74 + 50)\,\text{m}^3/\text{h} + 21\,\text{m}^3/\text{h}]$$

$$= 1050\,\text{m}^3 \text{ total in 24 h or } 43.8\,\text{m}^3/\text{h in daily average}$$

3. Calculate the daily average wastewater flow from the area under wastewater flow curve.

$$A_2 = \frac{1}{2} \times 2\,\text{h} \times [28\,\text{m}^3/\text{h} + 2 \times (21 + 18 + 20 + 27 + 41 + 58 + 64 + 53 + 46 + 55 + 43)\,\text{m}^3/\text{h} + 28\,\text{m}^3/\text{h}]$$

$$= 948\,\text{m}^3 \text{ total in 24 h or } 39.5\,\text{m}^3/\text{h in daily average}$$

4. Calculate the daily average flows and various ratios.

a. Daily average water demand $= 43.8\,\text{m}^3/\text{h}$

b. Daily average wastewater flow $= 39.5\,\text{m}^3/\text{h}$

c. Ratio of average wastewater flow to water demand $= \dfrac{39.5\,\text{m}^3/\text{h}}{43.8\,\text{m}^3/\text{h}} = 0.90$

d. Ratio of peak to average water demand $= \dfrac{85\,\text{m}^3/\text{h}}{43.8\,\text{m}^3/\text{h}} = 1.94$

e. Ratio of peak to average wastewater flow $= \dfrac{64\,\text{m}^3/\text{h}}{39.5\,\text{m}^3/\text{h}} = 1.62$

4.3 Components of Municipal Water Demand

Municipal water demand in the United States varies from 300 to1800 liters per capita per day (Lpcd).[1,5–7] Various components of municipal water uses are: (1) residential or domestic, (2) commercial, (3) industrial, (4) institutional and public, and (5) water lost or unaccounted for.[1]

The relative proportions of these uses with respect to overall municipal water demand are summarized in Table 4.1. General discussion is presented below.

TABLE 4.1 Components of Municipal Water Demand and Percent Distribution

Components of Municipal Water Demand	Percent of Total
Residential or domestic	35–50
Commercial	10–25
Industrial	15–30
Institutional and public	5–20
Water lost or unaccounted for	5–10

Source: Adapted in part from References 1, 2, 4, and 6.

4.3.1 Residential or Domestic Water Use

The residential or domestic water uses include toilet flush, bathing and washing, cooking and drinking, lawn watering, and others. Typical breakdown of residential water uses and flow rates are summarized in Table 4.2. The average residential water demand varies from 300 to 450 Lpcd. With flow reduction devices, the residential water demand for many types of residential establishments is decreasing.

4.3.2 Commercial Water Use

Commercial establishments include motels, hotels, office buildings, shopping centers, service stations, movie houses, airports, and the like. The water uses in commercial establishments vary greatly depending upon the size of operation. The commercial water demand may be estimated from unit loading or floor area. Typical unit loadings of residential and commercial establishments are provided in Table 4.3.

4.3.3 Institutional and Public Water Use

Water uses in public buildings and institutional establishments (city halls, prisons, hospitals, schools, etc.) as well as water used for public services (fire protection, street washing, park irrigation, and the like) are considered public water uses. Some residential, commercial, and institutional water uses are provided in Table 4.3.

TABLE 4.2 Typical Breakdown and Flow Rates of Residential Water Uses

Types of Water Use	Non Conserving Home Usage, %	Flow Rate
Toilet flush, including toilet leakage	33	Tank type 19–27 L/use
		Valve type 90–110 L/min
Shower and bathing		
Shower	20	Shower head 90–110 L/use or 19–40 L/min
Bathing	8	Tub bath 60–90 L/use
Wash basin	11	4–8 L/use
Kitchen	9	
Drinking, cooking	2–6	Kitchen sink
Dishwashing	3–5	15–30 L/dishwasher load
Garbage disposal	0–6	6000–7500 L/week, 4–8 L/person·d
Laundry and washing machine	16	110–200 L/load
Lawn	3	Sprinkler system

Source: Adapted in part from References 1, 2, 4, and 8.

TABLE 4.3 Average Water Demand in Residential, Institutional, Commercial, and Industrial Establishments

Source	Unit	Unit Flow, m^3/unit·d
	Residential	
Single-family detached, low income to high income	Person	0.25–0.38
Apartment	Person	0.18–0.23
Trailer park	Person	0.15
	Commercial	
Country club		
Resident	Member	0.38
Nonresident	Member	0.10
Hotel/Motel	Unit guest	0.38
Resort	Unit guest	0.19
Restaurant	Customer	0.03
Bar	Customer	0.08
Store	Toilet room	1.52
	Employee	0.04
Department store and shopping center	Employee	0.04
	Per m^2 floor area	0.001–0.002
Office building and complex	Employee	0.065
	Per m^2 floor area	0.015
Movie	Seat	0.008
Laundromat	Machine	2.5
Barber shop	Chair	0.065
Beauty salons	Station	1.026
Service station	First bay	3.8
	Additional bays	1.9
	Industrial	
Industrial building	Employee	0.055
Factories		
With shower	Employee-shift	0.133
Without shower	Employee-shift	0.095
Light industrial zone	ha	9–14
Medium industrial zone	ha	14–30
Heavy industrial zone	ha	30–100
Industrial products		
Cattle	Head	0.04–0.05
Dairy	Head	0.07–0.08
Chicken	Head	0.03–0.04
Canning	Metric ton/d	30–60
Dairy, milk	Metric ton/d	2–3
Meat packing	Metric ton/d	15–25
Pulp and paper	Metric ton/d	200–800
Steel	Metric ton/d	260–300

(*Continued*)

TABLE 4.3 (*Continued*) Average Water Demand in Residential, Institutional, Commercial, and Industrial Establishments

Source	Unit	Unit Flow, m^3/unit·d
Tannery	Metric ton of raw hides processed/d	60–70
	Institutional and Public Uses	
Nursing homes	Bed	0.38
Hospital	Bed	0.95
Prison	Inmate	0.45
School		
Boarding	Student	0.3
Day	Student	0.076

Note: $m^3 \times 264.5 =$ gallon
Source: Adapted in part from References, 1, 2, 4, 6, and 8 through 10.

4.3.4 Industrial Water Use

Industrial water demands in the United States are very large. Generally, large industries develop their own water supply systems. Only small industries purchase water and therefore impose demand on local municipal system. The industrial water demand may be estimated on the basis of proposed industrial zoning and unit loadings for specific industries. Unit loadings for many industries are summarized in Table 4.3.

4.3.5 Water Unaccounted for or Lost

Major sources of water unaccounted for are leaks from mains, faulty meters, and unauthorized connections. This loss mostly depends upon the condition of water distribution system. In many developing nations where individual water connections are unmetered, as much as 30–50% water supplied may be lost or unaccounted for due to old and leaky distribution system and wasted from unmetered connections. To reduce these losses, water is supplied only for a few hours in the mornings and in the evenings. This practice does save water but may cause serious contamination of water supply from the entry of polluted surface and groundwater into the distribution system during the periods when the supply is interrupted.

4.3.6 Factors Affecting Water Use

The water usage in a community depends upon many factors. Many of these factors are:

1. *Geographical location and climate:* More water is used in hot and dry climate or region than that in wet, humid, and cold climate or region.
2. *Size, population density, and economic conditions of the community:* Wealthier and sparsely populated communities have higher water demand.
3. *Industrialization:* Industries use large amounts of water especially seasonal industries such as vegetable canneries and others.
4. *Metered water supply and cost:* Less water is used where water supply is individually metered. Also, water usage is closely related to cost of water.
5. *Water pressure:* Higher supply pressure may cause unnecessary waste due to spurts from faucets, leaks, and drips.

6. *Water conservation:* Efforts of general public and regulatory agencies in water conservation have resulted in 20–30% savings. These savings are achieved by pressure reducing valves, faucet aerators, flow limiting shower heads, shallow trap water closet, level-controlled washing machines, and education of consumers.[1–3,7–9]

EXAMPLE 4.3: MUNICIPAL WATER DEMAND AND COMPONENTS

A city has a population of 30,000 residents. The residential water demand is 210 Lpcd. The commercial, industrial, institutional and public water uses, and water lost or unaccounted for are 18%, 23%, 10%, and 8%, respectively, of total municipal water demand. Calculate total municipal water demand in m^3/d and MGD, and the demand for each component.

Solution

1. Calculate the residential water demand.

$$\text{Residential water demand} = 30{,}000 \text{ residents} \times \frac{210\text{ L}}{\text{person·d}} \times \frac{\text{m}^3}{1000\text{ L}} = 6300\text{ m}^3/\text{d or } 1.66\text{ MGD}$$

2. Calculate the total municipal water demand.

Total demand for commercial, industrial, institutional and public, and water lost or unaccounted for

$$= (18 + 23 + 10 + 8)\%$$

$$= 59\% \text{ of total municipal water demand}$$

$$\text{Residential water demand} = (100-59)\% = 41\% \text{ of municipal water demand}$$

$$\text{Total municipal water demand} = \frac{6300\text{ m}^3/\text{d residential}}{0.41\text{ residential/total municipal}} = 15{,}366\text{ m}^3/\text{d}$$

3. Calculate the water demand for each component.

Water demand for each component is tabulated below:

Water Use	Percent of Total	Water Demand	
		m^3/d	MGD
Residential	41	6300	1.66
Commercial	18	2766	0.73
Industrial	23	3534	0.93
Institutional and public use	10	1537	0.40
Water lost or unaccounted for	8	1229	0.32
Total	100	15,366	4.04

EXAMPLE 4.4: WATER DEMAND AND WASTEWATER FLOW

A subdivision of a suburban city is being developed. The ultimate zoning plan shows the following residential, commercial, industrial, and institutional establishments. Assume that the water lost is small and is ignorable. Using the average values given in Table 4.3, estimate the following:

a. Annual average water demand for the entire subdivision in m^3/d
b. Residential, commercial, industrial, and institutional water demands and their percentages with respect to total municipal water demand

c. Annual average wastewater flow if 85% average water supply is returned into the municipal sewers.

Water-Using Establishment	Source of Water Demand
Residential	
Single-family detached residential population	1000 residents
Apartments	1650 residents
Trailer park	400 residents
Commercial	
One hotel/motel	250 units
Five restaurants, total seating	300 customers
Shopping centers	250 employees
Office buildings	500 employees
Office complexes	600 m^2
One movie theater	200 seats
One commercial laundry	40 machines
Barbershops	10 chairs
Beauty salons	20 stations
One service station	6 bays
Industrial	
Factories with shower	100 employee-shifts
Light industrial zone	2 ha
One feedlot	1000 heads
One canning plant	4 metric tons
One meat packaging plant	3 metric tons
Institutional	
One nursing home	50 beds
One hospital	200 beds
Two daytime schools	1200 students

Solution

1. Estimate the residential population of the subdivision.

Single-family detached housing residents	= 1000
Apartment residents	= 1650
Trailer park residents	= 400
Total	= 3050 residents

2. Estimate the residential water demand.

Single-family detached housing	= 1000 residents × 0.315 m^3/person·d	= 315 m^3/d
Apartments	= 1650 residents × 0.205 m^3/person·d	= 338 m^3/d
Trailer park	= 400 residents × 0.15 m^3/person·d	= 60 m^3/d
Total residential demand		= 713 m^3/d

3. Estimate the commercial water demand.

Hotel/motel	= 250 units × 0.38 m^3/unit·d	= 95 m^3/d
Restaurant	= 300 customers × 0.03 m^3/customer·d	= 9 m^3/d
Shopping centers	= 250 employees × 0.04 m^3/employee·d	= 10 m^3/d
Office building	= 500 employees × 0.065 m^3/employee·d	= 32.5 m^3/d
Office complexes	= 600 m^2 × 0.015 m^3/employee·d	= 9 m^3/d
Movie theater	= 200 seats × 0.008 m^3/seat·d	= 1.6 m^3/d
Laundromat	= 40 machines × 2.5 m^3/machine·d	= 100 m^3/d
Barber shops	= 10 chairs × 0.065 m^3/chair·d	= 0.65 m^3/d
Beauty salons	= 20 stations × 1.026 m^3/station·d	= 20.5 m^3/d
Service station		
Fist bay	= 1 bay × 3.8 m^3/bay·d	= 3.8 m^3/d
Additional bays	= 5 bays × 1.9 m^3/bay·d	= 9.5 m^3/d
Total commercial demand		= 292 m^3/d

4. Estimate the industrial water demand.

Factories with shower	= 100 employee-shifts × 0.133 m^3/employee-shift	= 13 m^3/d
Light industrial zone	= 2 ha × 11.5 m^3/ha·d	= 23 m^3/d
Feedlot	= 1000 heads × 0.045 m^3/head·d	= 45 m^3/d
Canning plant	= 4 metric tons/d × 45 m^3/metric ton	= 180 m^3/d
Meat packaging plant	= 3 metric tons/d × 20 m^3/metric ton	= 60 m^3/d
Total industrial water demand		= 321 m^3/d

5. Estimate the institutional water demand.

Nursing home	= 50 beds × 0.38 m^3/bed·d	= 19 m^3/d
Hospital	= 100 beds × 0.95 m^3/bed·d	= 95 m^3/d
School	= 1200 students × 0.076 m^3/student·d	= 91.2 m^3/d
Total institutional water demand		= 205 m^3/d

6. Estimate the annual municipal water demand and their percentages.

Components	Demand	Percentage
Residential	= 713 m^3/d	47
Commercial	= 292 m^3/d	19
Industrial	= 321 m^3/d	21
Institutional	= 205 m^3/d	13
Total	= 1531 m^3/d	100

$$\text{Average municipal water demand} = 1531\ \text{m}^3/\text{d} \times \frac{1000\ \text{L}}{\text{m}^3} \times \frac{1}{3050\ \text{residents}} = 502\ \text{Lpcd (133 gpcd)}$$

Note: 1 gallon = 3.8 L

7. Estimate the annual average wastewater flow.

Annual wastewater flow = 0.85 × 1531 m^3/d = 1301 m^3/d or 427 Lpcd (112 gpcd)

EXAMPLE 4.5: WATER SAVING FROM SUPPLY PRESSURE REDUCTION

In a community of 50,000 residents, the average water demand is 630 Lpcd and supply pressure is 414 kN/m^2. Average water lost through cracks at this pressure is 15%. How much saving would be achieved if the supply pressure is reduced to 276 kN/m^2? Assume cracks behave like orifices.

Solution

1. Estimate the average water demand, and average water lost through the cracks at the pressure of 414 kN/m^2.

$$\text{Average water demand} \quad = 50{,}000\,\text{people} \times \frac{630\ \text{L}}{\text{person}\cdot\text{d}} \times \frac{\text{m}^3}{1000\ \text{L}} = 31{,}500\ \text{m}^3/\text{d}$$

$$\text{Water lost through cracks} = 0.15 \times 31{,}500\ \text{m}^3/\text{d} = 4725\ \text{m}^3/\text{d}$$

2. Select the orifice equation.

Velocity and discharge through an orifice are expressed by Equations 4.2 and 4.3.

$$V = \sqrt{2gh} \tag{4.2}$$

$$q = C_d AV = C_d A\sqrt{2gh} \tag{4.3}$$

where

V = velocity through an orifice, m/s
q = discharge through an orifice, m^3/s
C_d = coefficient of discharge. Typical value for orifice is 0.61.
A = area of orifice, m^2
g = acceleration due to gravity, m/s^2
h = water head over the orifice, m

3. Calculate the water heads at pressures of 414 and 276 kN/m^2.

There is a direct conversion between the pressure head in kN/m^2 and water head in m based on specific weight of water. At 4°C (39°F), the specific weight for water is 9.81 kN/m^3 (62.4 lb/ft^3) and the static head of water in meter is expressed below:

$$\text{Head in meter} = \frac{\text{Pressure in kN/m}^2}{9.81\ \text{kN/m}^3}$$

$$\text{Water head at pressure of 414 kN/m}^2,\ h_{414} = \frac{414\ \text{kN/m}^2}{9.81\ \text{kN/m}^3} = 42.2\ \text{m}$$

$$\text{Water head at pressure of 276 kN/m}^2,\ h_{276} = \frac{276\ \text{kN/m}^2}{9.81\ \text{kN/m}^3} = 28.1\ \text{m}$$

4. Calculate the loss of water through the cracks at pressures 414 and 276 kN/m^2.

$$q_{414} = C_d A\sqrt{2gh_{414}}$$

$$q_{276} = C_d A\sqrt{2gh_{276}}$$

$$\frac{q_{414}}{q_{276}} = \frac{C_d A\sqrt{2gh_{414}}}{C_d A\sqrt{2gh_{276}}} = \sqrt{\frac{h_{414}}{h_{276}}} = \sqrt{\frac{42.2\ \text{m}}{28.1\ \text{m}}} = 1.23$$

$$\text{Water lost through the cracks at 414 kN/m}^2,\ q_{414} = 4725\ \text{m}^3/\text{d}$$

$$\text{Water lost through the cracks at 276 kN/m}^2,\ q_{276} = \frac{q_{414}}{1.23} = \frac{4725\ \text{m}^3/\text{d}}{1.23} = 3841\ \text{m}^3/\text{d}$$

5. Calculate the saving.

Saving achieved by reduction of supply pressure from 414 to 276 kN/m^2

$= (4725 - 3841)\ m^3/d = 884\ m^3/d$ or 17.7 Lpcd

Note: The average water loss through the cracks is reduced by about 19% when the supply pressure is lowered. It counts about 2.8% of the average daily water demand.

EXAMPLE 4.6: WATER SAVING FROM PRESSURE REDUCING VALVE (PRV)

A home has installed a PRV at the point of entry into the home. The PRV reduces supply pressure from 500 to 300 kN/m^2. If water usage before PRV is 350 Lpcd, calculate typical water saving. Use the information provided in Table 4.2.

Solution

1. Select from Table 4.2 the residential water uses that are affected by PRV.

   The toilet flush tank has a fixed volume, and the washing machine and dishwasher have level controller. Therefore, installation of the PRV will not reduce water uses in these categories. Only those uses will be affected that are directly from the faucet. These uses are: (a) shower 20% (bathing 8% not affected), (b) wash basin 11%, (c) kitchen assume 4% (out of 9%), and (d) lawn sprinkling 3%.

2. Calculate the water uses in home directly affected by PRV.

   Total percentage affected $= (20 + 11 + 4 + 3)\% = 38\%$

   Total usage affected at 500 kN/m^2, $q_{500} = 0.38 \times 350\ \text{Lpcd} = 133\ \text{Lpcd}$

3. Calculate the ratio of usage before and after installing the PRV.

$$\frac{q_{500}}{q_{300}} = \frac{C_d A\sqrt{2g\,h_{500}}}{C_d A\sqrt{2g\,h_{300}}} = \sqrt{\frac{\left(\dfrac{P_{500}}{9.81\ \text{kN/m}^3}\right)}{\left(\dfrac{P_{300}}{9.81\ \text{kN/m}^3}\right)}} = \sqrt{\frac{P_{500}}{P_{300}}} = \sqrt{\frac{500\ \text{kN/m}^2}{300\ \text{kN/m}^2}} = 1.29$$

4. Calculate the water saving due to PRV.

   Total usage at 300 kN/m^2, $q_{300} = \dfrac{q_{500}}{1.29} = \dfrac{133\ \text{Lpcd}}{1.29} = 103\ \text{Lpcd}$

   Total water saving, $\Delta q = q_{500} - q_{300} = 133\ \text{Lpcd} - 103\ \text{Lpcd} = 30\ \text{Lpcd}$

   Overall water saving based on average water demand $= \dfrac{\Delta q}{q_{500}} = \dfrac{30\ \text{Lpcd}}{350\ \text{Lpcd}} \times 100\% = 8.6\%$

EXAMPLE 4.7: WATER CONSERVATION IN HOMES

A home has installed water conservation devices. These devices and expectedpercentage savings over conventional devices are given in Table 4.4. Calculate total water saving in Lpcd and overall percent saving. The average water usage in the home before installation of these devices was 400 Lpcd.

TABLE 4.4 Water Saving Devices and Expected Savings over Conventional Devices (Example 4.7)

Device	Expected Saving, %
Shallow-trap water closet (toilet)	35
Sink faucet aerator	2
Limiting shower head	12
Water efficient dishwasher	7
Level-controlled washing machine	4

Solution

1. Select from Table 4.2 the residential water uses that are affected by new devices.
   The residential water uses that are affected by new devices are (a) toilet 33%, (b) wash basin 11%, (c) shower use is 20% (bathing 8% not affected), (d) dishwasher 4%, and (e) washing machine 16%.
2. Calculate the savings achieved by new devices.
   The savings achieved from different new devices are summarized below:

Water Uses	Usage		Saving	
	%	Lpcd	%	Lpcd
Water closet (toilet)	33	132[a]	35	46.2[b]
Wash basins and sinks	11	44	2	0.9
Shower	20	80	12	9.6
Dishwasher	4	16	7	1.1
Washing machine	16	64	4	2.6
Total	84	332	–	60 Lpcd

[a] 400 Lpcd × 0.33 = 132 Lpcd
[b] 132 Lpcd × 0.35 = 46.2 Lpcd

Overall savings from water conservation devices = 60 Lpcd

$$\text{Percent overall saving} = \frac{60\text{ Lpcd}}{400\text{ Lpcd}} = 15\%$$

4.4 Wastewater Flow

Municipal wastewater is derived largely from the water supply. A considerable portion of the water supply, however, does not reach the sewers. This includes water used for street washing, lawn sprinkling, fire fighting, and leakages from water mains and service pipes. A small portion of water may also be consumed in products and manufacturing processes. Also, many homes and other establishments that are not served by a sewerage system may use the city water supply but utilize on-site wastewater treatment and disposal. On the other hand, I/I, and water used by industries and residences that is obtained from privately owned sources may increase the quantity of wastewater larger than the public water supply. In general, the average wastewater flow may vary from 60% to 130% of the water used in the community. Many designers frequently assume that the average rate of wastewater flow, including a moderate allowance for I/I, equals the average rate of water consumption. Average wastewater flows from residential, commercial, industrial, institutional, and other sources may be obtained from careful consideration of local water consumption data or by using wastewater generation rates for other similar cities in that region.

4.5 Wastewater Flow Variation

4.5.1 Dry Weather Flow

Like water demand, wastewater flows also vary with respect to time of the day, day of the week, weather condition, and season of the year. Under dry weather conditions, the daily wastewater flow shows a diurnal pattern exhibiting a peak and a minimum in 24-h period. The diurnal wastewater flow parallels that of water demand with a lag of few hours (Figure 4.1).

Hourly peak and minimum dry weather flows are generally estimated from several equations and graphical relationships developed from case studies. The ratio of hourly peak to daily average, and hourly minimum to daily average flows depend upon the contributing population. Commonly used equations are given by (Equations 4.4 through 4.8).[1,4,5,11]

$$M_{max} = 1 + \frac{14}{4+\sqrt{P}} \tag{4.4}$$

$$M_{max} = \frac{5}{P^{0.167}} \tag{4.5}$$

$$Q'_{max} = 3.2\,(Q'_{avg})^{5/6} \tag{4.6}$$

$$\log(M_{max}) = -0.19\,\log(P') + 0.74 \tag{4.7}$$

$$\log(M_{min}) = 0.142\,\log(P') - 0.682 \tag{4.8}$$

where

M_{max} = peaking factor, or ratio of hourly maximum to daily average flows
P = population in thousands
Q'_{max} = hourly maximum dry weather flow, MGD
Q'_{avg} = daily average dry weather flow, MGD
M_{min} = ratio of hourly minimum to daily average flows
P' = population in thousands, range $1 < P' < 10{,}000$

EXAMPLE 4.8: PEAK DRY WEATHER WASTEWATER FLOW CALCULATION FROM DIFFERENT EQUATIONS

A city has a population of 20,000 residents. The average wastewater flow is 380 Lpcd. Calculate (1) daily average and hourly peak dry weather flows in m^3/d and MGD, (2) hourly minimum flow in m^3/d, and (3) compare the results.

Solution

1. Calculate the daily average wastewater flow.

$$\text{Daily average wastewater flow} = \frac{380\ \text{L}}{\text{person·d}} \times 20{,}000\ \text{persons} \times \frac{\text{m}^3}{1000\ \text{L}} = 7600\ \text{m}^3/\text{d or } 2.0\ \text{MGD}$$

2. Calculate the hourly peak dry weather flows from Equations 4.4 through 4.7.

$$\text{Peaking factor from Equation 4.4, } M_{max} = 1 + \frac{14}{4+\sqrt{P}} = 1 + \frac{14}{4+\sqrt{20}} = 2.65$$

$$\text{Hourly peak dry weather flow} = 2.65 \times 7600\ \text{m}^3/\text{d} = 20{,}140\ \text{m}^3/\text{d or } 5.30\ \text{MGD}$$

Peaking factor from Equation 4.5, $M_{max} = \frac{5}{P^{0.167}} = \frac{5}{(20)^{0.167}} = 3.03$

Hourly peak dry weather flow $= 3.03 \times 7600\ m^3/d = 23{,}028\ m^3/d$ or 6.06 MGD

Hourly peak dry weather flow Equation 4.6,

$$Q'_{max} = 3.2\,(Q'_{avg})^{5/6} = 3.2\,(2.0)^{5/6} = 5.70 \text{ MGD or } 21{,}575\ m^3/d$$

Peaking factor from Equation 4.7, $\log(M_{max}) = -0.19 \log(P') + 0.74 = -0.19 \log(20) + 0.74$

$$M_{max} = 3.11$$

Hourly peak dry weather flow $= 3.11 \times 7600\ m^3/d = 23{,}636\ m^3/d$ or 6.22 MGD

3. Calculate the hourly minimum flow.
   Ratio of hourly minimum to daily average flows from Equation 4.8,

$$\log(M_{min}) = 0.142 \log(P') - 0.682 = 0.142 \times \log(20) - 0.682 = -0.497$$
$$M_{min} = 0.32$$

Hourly minimum dry weather flow $= 0.32 \times 7600\ m^3/d = 2432\ m^3/d$ or 0.64 MGD

4. Compare the results.

Flow Condition	m^3/d	MGD	Ratio to Average Flow
Hourly peak dry weather flow from Equation 4.4	20,140	5.30	2.65
Hourly peak dry weather flow from Equation 4.5	23,028	6.06	3.03
Hourly peak dry weather flow from Equation 4.6	21,575	5.70	2.85
Hourly peak dry weather flow from Equation 4.7	23,636	6.22	3.11
Hourly minimum dry weather flow from Equation 4.8	2432	0.64	0.32

The results of hourly peak dry weather flow from four equations are comparable. The results indicate that the ratio of hourly peak and minimum dry weather flows is approximately 10:1.

4.5.2 Infiltration and Inflow

Infiltration is the groundwater that enters the sewers through service connections, cracked pipes, defective joints, and cracked manhole walls. Inflow is the direct flow due to surface runoff that may enter through a manhole cover, roof area drains, and cross connection from storm sewers and combined sewers. Flow from cellar and foundation drains, cooling water discharges, and drainage from springs and other steady flows are included with the infiltration. The conditions that apply to I/I are listed below:

1. The amount of I/I reaching a sewer system depends upon
   a. The length and age of the sewer
   b. The construction material, methods, and workmanship
   c. Number of roof or drain connections
   d. The groundwater table relative to sewer position
   e. Type of soil, ground cover, and topographic conditions

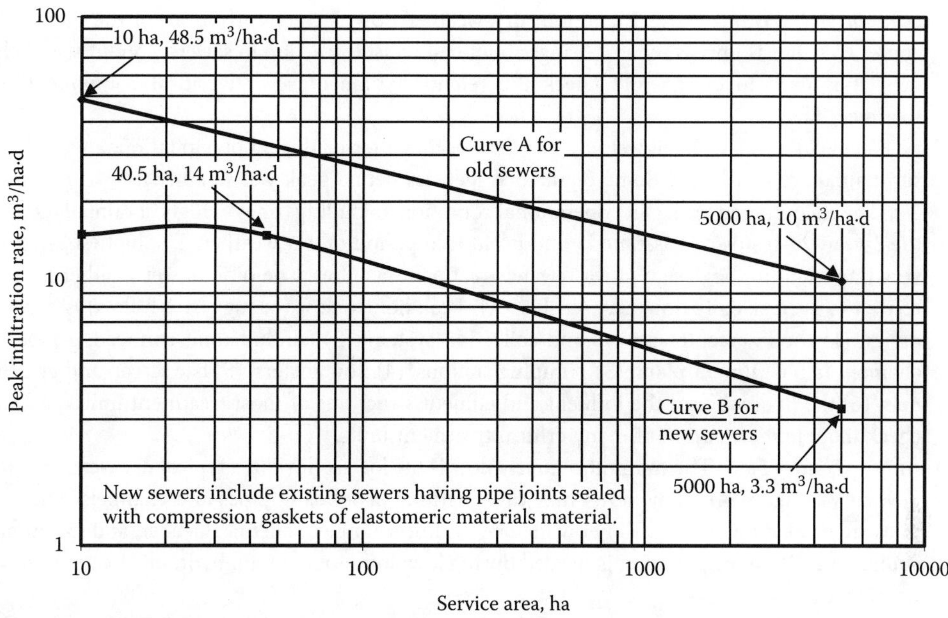

Note: ha × 2.4711 = acre; m^3/ ha·d × 106.9 = gal/acre·d.

FIGURE 4.2 Average allowance for infiltration rate in new sewers.
Source: Adapted in part from Reference 4.

2. Permissible I/I allowance is 780 Lpcd (205 gpcd) as long as there is no operational problems such as surcharges, bypasses, or poor treatment performance resulting from hydraulic overloading of publically owned treatment work (POTW) during storm events.
3. In older sewers, infiltration is high because of deteriorated joints and masonry mortar. Newer sewers use joints sealed with rubber gaskets or synthetic material, and use precast manholes. Also, the I/I rate is significantly smaller.
4. When designing sewers, allowance must be made for old and new sewers. Average values of I/I allowance are 94–9400 L/cm (diameter)·km (linear length)·d (100–10,000 gpd/in·mile) or 200–28,000 L/ha·d (20–3000 gpd/acre).
5. In the absence of flow records, many states specify 1500 Lpcd (400 gpcd) for design of laterals and submains, and 1300 Lpcd (350 gpcd) for mains and trunk sewers. These design flows include normal I/I allowance.
6. For new sewers or recently constructed sewer systems having precast manholes and pipe joints made with gaskets of rubber or rubber like material, the average infiltration allowance may be found from Figure 4.2.

4.5.3 Common Terms Used to Express Flow Variations

Several terms are commonly used to express the flow variations in the overall context of planning, design, and operation of wastewater treatment facilities. These terms and their significance are briefly presented below.

Daily Average Flow: The daily average flow (also called the annual average day flow) is based on annual flow rate data. It is the average flow occurring over 24-h period under dry weather condition. This flow is very important and used to (1) evaluate treatment plant capacity, (2) develop flow rate ratios, (3) size many treatment units, (4) calculate organic loadings, (5) estimate sludge solids, (6) estimate chemical needs, and (7) calculate pumping and treatment unit costs.

Maximum Dry Weather Flow: Maximum dry weather flow is the hourly peak of the diurnal flow curve. This flow is important as it brings about daily flushing action in sewers, conduits, and channels. It is used to design and check the retention time of several critical components of a treatment facility.

Peak Wet Weather Flow: Peak wet weather flow occurs after or during precipitation, and includes a substantial amount of I/I. Some regulatory agencies define peak wet weather flow as the highest 2-h flow encountered under any operational condition, including times of highest rainfall (generally the 2-year, 24-h storm is assumed), and prolonged period of wet weather. The highest 2-h flow is very important in the design of wastewater treatment facilities. The ratio of wet weather flow with respect to average daily flow may reach 3–5:1. Peak hourly flow (or highest 2-h flow) is needed to design (1) the interceptor sewers and collection system, (2) conduits and connecting pipes and channels in a treatment plant, (3) pumping stations, (4) flow meters, (5) bar screen and grit channels, (6) the hydraulics of the influent and effluent structures of most treatment units, and (7) to check the retention period of many critical treatment units.

Minimum Hourly Flow: The minimum hourly flow is the lowest flow on a typical dry weather diurnal flow curve (Figure 4.1). Low flows may cause settling of solids in pipes and channels. These flow rates are needed for sizing of (1) flow meters, (2) lower range of chemical feeder, and (3) pumping equipment. Often recirculation is needed during low flows to maintain hydraulic loadings in some treatment units.

Sustained Flows: Sustained flows are the flows that persist over several days. Sustained flows could be on the low or high sides. As an example, extraordinarily dry or hot weather may cause sustained low extremes. Likewise, prolong wet weathers and special events in a community such as fairs, conventions, games, exhibitions may be a source of population surges and could all cause high sustained flows. Data on sustained flow rates may be needed in sizing equalization basins and other plant hydraulic components, and to check the design of many units under extreme conditions. The sustained-flow envelope should be developed from the longest available period of record. The typical sustained flow envelope at a treatment plant is shown in Figure 4.3.

Emergency Flow Diversion: Many POTWs are equipped with a flow diversion facility to hold the excess or entire flow under emergency situations. Such conditions may develop if (1) power failure occurs, (2) the combined sewer overflows (CSOs) are directed, and (3) excessive I/I is diverted from the

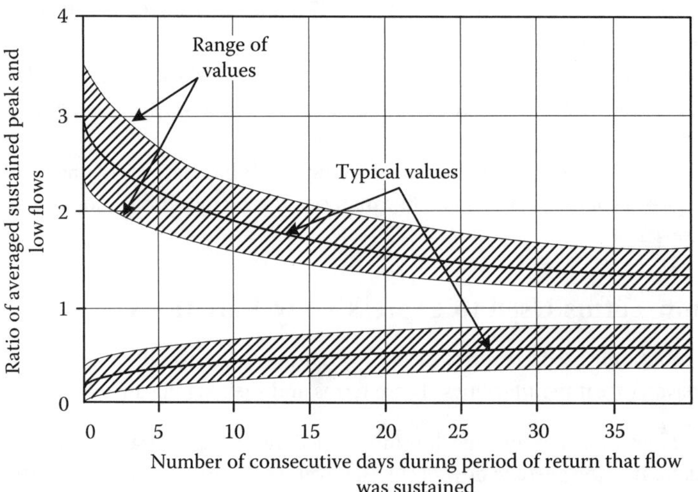

FIGURE 4.3 The typical sustained flow envelope at a wastewater treatment plant.
Source: Adapted in part from References 1, and 3 through 5.

sanitary sewers. The diverted flow is temporarily stored in a basin, and rerouted through the plant after restoration of power, or when the plant is able to handle the stored flows. Various types of flow-regulating devices are used to achieve proper flow diversion. Common diversion devices are side weir, transverse weir, leaping weir, orifice, relief siphon, tripping plate, and hydrobrake regulators. References 1, 3, 4, 11, and 12 should be consulted for design information on these diversion devices.

Design Flows: The design flows are the ultimate flows when the facility is expected to reach its full design capacity. Design flows occur at the end of the design period, which may be 10–20 years after initial construction is completed. The design flows are peak wet weather, average day, and minimum flows.

EXAMPLE 4.9: PEAK I/I

A city subdivision has a population density of 3500 residents per km^2. During wet weather condition, the expected average I/I from this subdivision is 3800 L/ha·d and the peaking factor is 1.5. Estimate average and peak I/I in Lpcd.

Solution

1. Estimate the population per ha.

$$\text{Population density} = \frac{3500 \text{ persons}}{\text{km}^2} \times \frac{\text{km}^2}{10^6 \text{ m}^2} \times \frac{10{,}000 \text{ m}^2}{\text{ha}} = 35 \text{ persons/ha}$$

2. Estimate the average I/I.

$$\text{Average I/I} = \frac{3800 \text{ L}}{\text{ha·d}}$$

$$\text{Average I/I per person} = \frac{3800 \text{ L}}{\text{ha·d}} \times \frac{1}{35 \text{ person/ha}} = 109 \text{ Lpcd}$$

3. Estimate the peak I/I.

$$\text{Peak I/I} = 1.5 \times 109 \text{ Lpcd} = 164 \text{ Lpcd}$$

EXAMPLE 4.10: PERMISSIBLE I/I FOR NEW DEVELOPMENT

A new housing subdivision is planned. The total land area that will be developed is 2470 acres. Using Figure 4.2, estimate the average I/I allowance for design of intercepting sewer.

Solution

1. Determine the area served in ha.

$$\text{Land area} = \frac{\text{ha}}{2.4711 \text{ acre}} \times 2470 \text{ acres} = 1000 \text{ ha}$$

2. Estimate from Figure 4.2 the I/I allowance for new sewer system.

$$\text{I/I allowance} = 5.7 \text{ m}^3\text{/ha·d}$$

3. Determine the average I/I.

$$\text{Average I/I} = 5.7 \text{ m}^3\text{/ha·d} \times 1000 \text{ ha} = 5700 \text{ m}^3\text{/d}$$

EXAMPLE 4.11: PERMISSIBLE I/I ALLOWANCE IN A SEWER AND PEAK WET WEATHER FLOW

The population of a city is 40,000. Average wastewater flow is 400 Lpcd, and permissible average I/I allowance is 890 L/cm·km·d (961 gpd/in·mile). The average length of sanitary sewer is 8 m per capita. The sewer distribution by pipe size is as follows: 15 cm dia. = 4%, 20 cm dia. = 79%, 31 cm dia. = 10%, 46 cm dia. = 5%, and 61 cm dia. = 2%. Estimate I/I in Lpcd and peak wet weather flow in m^3/d. Assume the peaking factor for I/I is 1.6.

Solution

1. Calculate the daily average and hourly peak dry weather flows.

$$\text{Daily average flow} = \frac{400\ \text{L}}{\text{person·d}} \times 40{,}000\ \text{persons} \times \frac{\text{m}^3}{1000\ \text{L}} = 16{,}000\ \text{m}^3/\text{d}$$

$$\text{Peaking factor from Equation 4.4, } M_{\max} = 1 + \frac{14}{4+\sqrt{P}} = 1 + \frac{14}{4+\sqrt{40}} = 2.36$$

$$\text{Hourly peak dry weather flow} = 2.36 \times 16{,}000\ \text{m}^3/\text{d} = 37{,}760\text{m}^3/\text{d}$$

$$\text{Hourly peak dry weather flow in Lpcd} = 2.36 \times 400\ \text{Lpcd} = 944\ \text{Lpcd}$$

2. Calculate the length and equivalent sewer diameter in the collection system.

$$\text{Total length of sanitary sewer} = \frac{8\ \text{m}}{\text{person}} \times 40{,}000\ \text{persons} \times \frac{\text{km}}{1000\ \text{m}} = 320\ \text{km}$$

Equivalent diameter (weighted average)

$$= \frac{0.04 \times 15\ \text{cm} + 0.79 \times 20\ \text{cm} + 0.1 \times 31\ \text{cm} + 0.05 \times 46\ \text{cm} + 0.02 \times 61\ \text{cm}}{0.04 + 0.79 + 0.1 + 0.05 + 0.02}$$

$$= \frac{23.02\ \text{cm}}{1} = 23\ \text{cm}$$

3. Calculate the peak I/I.

$$\text{Daily average I/I allowance} = \frac{890\ \text{L}}{\text{cm·km·d}} \times (23\ \text{cm} \times 320\ \text{km}) \times \frac{\text{m}^3}{1000\ \text{L}} = 6550\ \text{m}^3/\text{d}$$

$$\text{Daily average I/I allowance in Lpcd} = 6550\ \text{m}^3/\text{d} \times \frac{1000\ \text{L}}{\text{m}^3} \times \frac{1}{40{,}000\ \text{persons}} = 164\ \text{Lpcd}$$

$$\text{Hourly peak I/I allowance} = 1.6 \times 164\ \text{Lpcd} = 262\ \text{Lpcd}$$

$$\text{Hourly peak I/I allowance} = 40{,}000\ \text{persons} \times \frac{262\ \text{L}}{\text{person·d}} \times \frac{\text{m}^3}{1000\ \text{L}} = 10{,}480\ \text{m}^3/\text{d}$$

4. Calculate the hourly peak wet weather flow.

Total hourly peak wet weather flow to the plant

$$= \text{hourly peak dry weather flow} + \text{hourly peak I/I allowance}$$
$$= (37{,}760 + 10{,}480)\ \text{m}^3/\text{d} = 48{,}240\ \text{m}^3/\text{d}$$

$$\text{Total hourly peak wet weather flow in Lpcd} = (944 + 262)\ \text{Lpcd} = 1206\ \text{Lpcd}$$

EXAMPLE 4.12: I/I AND DIURNAL FLOW CURVE

A sewer evaluation study of a service area was conducted to determine the infiltration and inflow reaching the collection system and to estimate the peak wet weather flow. A flow measurement device was installed in the interceptor that carried the entire flow from the service area. The hourly flow measurements on a typical dry day gave the diurnal weather flow pattern. The diurnal flow pattern of a typical dry day is shown in Figure 4.4. After a long dry spell, the first high-intensity storm caused heavy downpour in the service area. Since the groundwater table was low, the infiltration was negligible, and the excess flow over the dry weather flow was the inflow. The measured hourly flow under this condition is illustrated in Figure 4.4. After this downpour, high-intensity storms and intermittent showers continued to occur over a 10-day period. Under this wet weather condition, the ground was completely saturated. The recorded flow data are a measure of combined infiltration and diurnal dry weather flow in the interceptor. It is also shown in Figure 4.4. Determine (a) daily average dry weather flow, (b) hourly peak dry weather flow, (c) daily average infiltration, (d) 2-h peak inflow, (e) 2-h peak wet weather flow, and (f) 2-h peaking factor, which is the ratio of 2-h peak wet weather flow and daily average dry weather flow.

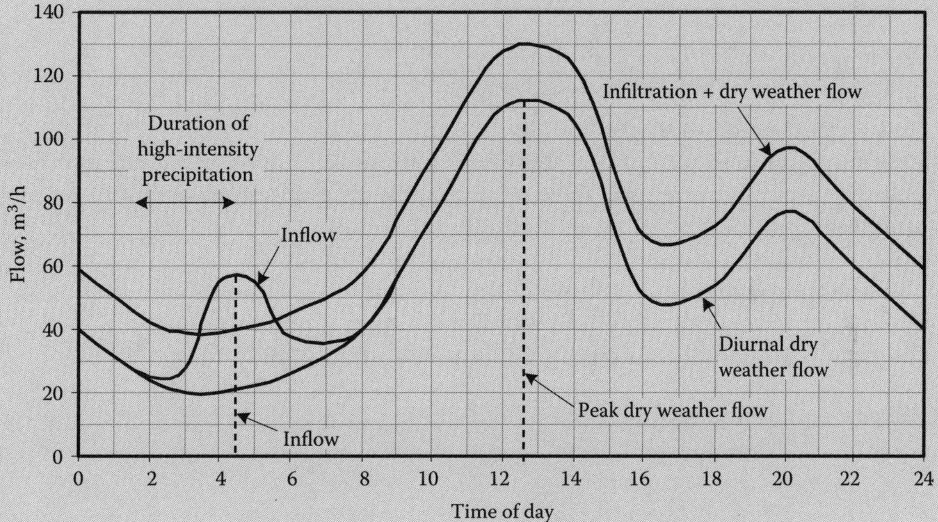

FIGURE 4.4 Diurnal dry weather flow, infiltration and inflow recorded in an interceptor (Example 4.12).

Solution

1. Calculate the daily average and hourly peak dry weather flows.

The daily average dry weather flow is calculated from the diurnal dry weather flow curve given in Figure 4.4.

Apply the trapezoidal rule (Equation 4.1).

Daily average dry weather flow

$$= \frac{1}{2} \times 2\text{ h} \times [40\text{ m}^3/\text{h} + 2 \times (24 + 20 + 26 + 40 + 75 + 110 + 105 + 51 + 54 + 77$$

$$+ 60)\text{ m}^3/\text{h} + 40\text{ m}^3/\text{h}]$$

$$= 1364\text{ m}^3 \text{ total in 24 h or } 56.8\text{ m}^3/\text{h as daily average}$$

Hourly peak dry weather flow as read between 12:00 and 1:00 p.m. in Figure 4.4 = 111 m^3/h

2. Calculate the daily average infiltration when soil is saturated completely.
   The combined infiltration and dry weather flow curve is also shown in Figure 4.4.
   The average infiltration is also calculated from the curve using the trapezoidal rule (Equation 4.1).
   Area under infiltration flow curve (includes dry weather flow)

$$
\begin{aligned}
&= \frac{1}{2} \times 2\,\text{h} \times [59\ \text{m}^3/\text{h} + 2 \times (42 + 39 + 45 + 58 + 94 + 128 + 123 + 70 + 73 \\
&\quad + 97 + 79)\ \text{m}^3/\text{h} + 59\ \text{m}^3/\text{h}] \\
&= 1814\ \text{m}^3 \text{ total in 24 h}
\end{aligned}
$$

$$\text{Daily average infiltration plus dry weather flow} = \frac{1814\ \text{m}^3}{24\ \text{h}} = 75.6\ \text{m}^3/\text{h}$$

   Daily average infiltration flow $= (75.6 - 56.8)\ \text{m}^3/\text{h} = 18.8\ \text{m}^3/\text{h}$

3. Estimate the 2-h peak inflow.

   The inflow is estimated from the inflow curve between 3:30 and 5:30 a.m. in Figure 4.4.
   2-h peak inflow plus base dry weather flow $= 54\ \text{m}^3/\text{h}$
   2-h base dry weather flow $= 21\ \text{m}^3/\text{h}$
   2-h peak inflow $= (54 - 21)\ \text{m}^3/\text{h} = 33\ \text{m}^3/\text{h}$

4. Estimate the 2-h peak wet weather flow.
   In the worst case scenario, assume peak inflow occurs between 12:00 and 2:00 p.m. when hourly peak dry weather flow is reached. To be conservative, hourly peak dry weather flow is used in the estimation.

$$
\begin{aligned}
\text{2-h peak wet weather flow} &= \text{hourly peak dry weather flow} + \text{daily average infiltration} \\
&\quad + \text{2-h peak inflow} \\
&= (111 + 18.8 + 33)\ \text{m}^3/\text{h} = 162.8\ \text{m}^3/\text{h}
\end{aligned}
$$

5. Calculate the 2-h peaking factor.

$$\text{2-h peaking factor} = \frac{162.8\ \text{m}^3/\text{h}}{56.8\ \text{m}^3/\text{h}} = 2.9$$

EXAMPLE 4.13: PEAKING FACTOR IN A CITY DEVELOPMENT PROJECT

The future land use plan of a city development project has residential, commercial, and industrial areas, and a day school. Data on the expected saturation population densities, daily average wastewater flows, and peaking factors for various types of land use plan of the development project are given in Table 4.5. The day school has an enrollment of 2000 students. Estimate the daily average and peak wastewater flow rates and the overall peaking factor.

Solution

1. Calculate the population and flows.
   Set up a computational table for estimating the population and flows. The summary of computations is provided in Table 4.6.

TABLE 4.5 Expected Saturation Population Densities, Wastewater Flows, and Peaking Factors for Various Land Use Plan of the City Development Project (Example 4.13)

Land Use	Area, ha	Expected Population Density per ha	Unit	Unit Loading	Peaking Factor
Single family dwellings	170	35	Lpcd	350	Equation 4.4
Mixed residential dwellings	120	45	Lpcd	260	Equation 4.4
Apartments	150	130	Lpcd	220	Equation 4.4
School	40	–	Lpcd	76	4.8
Commercial	150	–	m^3/ha·d	30	1.8
Industrial	110	–	m^3/ha·d	60	2.2
Infiltration/inflow	Total area	–	m^3/ha·d	5	6.0

TABLE 4.6 Computation Table for Population and Flow Estimation (Example 4.13)

Land Use	Area, ha	Expected Population Density per ha	Population	Unit	Unit Loading	Daily Average Flow, m^3/d	Peaking Factor	Peak Flow, m^3/d
Single family dwellings	170	35	5950	Lpcd	350	2083	3.2[a]	6666
Mixed residential dwellings	120	45	5400	Lpcd	260	1404	3.2[a]	4493
Apartments	150	130	19,500	Lpcd	220	4290	2.7[a]	11,583
School	40	–	2000	Lpcd	76	152	4.8	730
Commercial	150	–	–	m^3/ha·d	30	4500	1.8	8100
Industrial	110	–	–	m^3/ha·d	60	6600	2.2	14,520
Subtotal	740	–	32,850	–	–	19,029	2.4[b]	46,092
I/I	740	–	–	m^3/ha·d	5	3700	6.0	22,200
Total/overall	740	–	32,850	–	–	22,729	3.0[c]	68,292

[a] Values calculated from Equation 4.4.

[b] $\dfrac{46{,}092 m^3/d}{19{,}029 m^3/d} = 2.4.$

[c] See calculation in Step 4.

2. Estimate the daily average flows.

   Daily average dry weather flow $= 19{,}029\ m^3/d$

   Daily average I/I flow $= 3700\ m^3/d$

   Daily average flow including average I/I $= 22{,}729\ m^3/d$

3. Estimate the peak flows.

   Peak dry weather flow $= 46{,}092\ m^3/d$

   Peak I/I flow $= 22{,}200\ m^3/d$

   Peak wet weather flow $= 68{,}292\ m^3/d$

4. Estimate the overall peaking factor.

$$\text{Overall peaking factor} = \frac{68{,}292\ m^3/d}{22{,}729\ m^3/d} = 3.0$$

EXAMPLE 4.14: SUSTAINED AVERAGE FLOW RATIO (SAFR) ENVELOPE

Describe the procedure for developing sustained average flow ratio (*SAFR*) envelope. Daily average flow (*DAF*) data during 30 consecutive days at a wastewater treatment plant is summarized in Table 4.7. The annual average day flow (*AADF*) at the plant is 0.58 m^3/s. Develop the maximum and minimum *SAFR* envelope up to a 10-day period.

TABLE 4.7 Average Flow (AF) Data during 30 Consecutive Days (Example 4.14)

Day	*DAF*, m^3/s	Day	*DAF*, m^3/s	Day	*DAF*, m^3/s
1	0.25	11	0.65	21	0.83
2	0.16	12	0.66	22	0.80
3	0.15	13	0.67	23	0.72
4	0.13	14	0.68	24	0.66
5	0.14	15	0.69	25	0.66
6	0.25	16	0.70	26	0.65
7	0.35	17	0.75	27	0.50
8	0.48	18	1.12	28	0.32
9	0.52	19	1.20	29	0.20
10	0.60	20	1.15	30	0.18

Solution

1. Describe the procedure for developing the *SAFR* envelope.

   a. Select the longest available period of consecutive days (m) with *DAF* data.
   b. Determine the *AADF*.
   c. Calculate the $SAFR_i$ for each i consecutive days, $i = 1, 2, 3, \ldots, n$, and $n < m$.
   d. Identify the maximum and minimum values of $SAFR_i$ ($i = 1, 2, 3, \ldots, n$).
   e. Summarize the maximum and minimum *SAFR* data.
   f. Plot the maximum and minimum *SAFR* data as well as the *AADF* ratio with respect to i consecutive days the flow rate is sustained. The *AADF* line will be horizontal and pass from the ratio of 1. The maximum and the minimum *SAFR* curves will be above and below the *AADF* line, respectively.

2. Develop the maximum and minimum *SAFR* envelope for a 10-day period ($n = 10$).

   Calculation results are tabulated in Table 4.8. A brief description is provided below for each step of the calculations.

   a. In this problem, a 30-day period ($m = 30$) is selected in Column 1 and the consecutive *DAF* data is summarized in Column 2.
   b. The *AADF* is 0.58 m^3/s at the plant.
   c. $SAFR_i$ is calculated for each i consecutive days ($i = 1, 2, 3, \ldots, 10$). The calculation results are summarized in Columns 4–12.
   d. The maximum and minimum values of $SAFR_i$ are identified in bold numbers in Columns 3–12 for each i consecutive days over the 30-day period.

TABLE 4.8 Summary of Calculations of Maximum and Minimum of SAFRs (Example 4.14)

Day	*DAF*, m^3/s	$SAFR_i$									
		$i=1$	$i=2$	$i=3$	$i=4$	$i=5$	$i=6$	$i=7$	$i=8$	$i=9$	$i=10$
(1)	(2)	(3)	(4)	(5)	(6)	(7)	(8)	(9)	(10)	(11)	(12)
1	0.25	0.43[a]	–	–	–	–	–	–	–	–	–
2	0.16	0.28[a]	0.35[b]	–	–	–	–	–	–	–	–
3	0.15	0.26	0.27[b]	0.32[c]	–	–	–	–	–	–	–
4	0.13	**0.22**[c]	0.24	0.25[d]	0.30	–	–	–	–	–	–
5	0.14	0.24	**0.23**[e]	**0.24**[f]	**0.25**	**0.29**	–	–	–	–	–
6	0.25	0.43	0.34	0.30	0.29	0.29	**0.31**	–	–	–	–
7	0.35	0.60	0.52	0.43	0.38	0.35	0.34	**0.35**	–	–	–
8	0.48	0.83	0.72	0.62	0.53	0.47	0.43	0.41	**0.41**	–	–
9	0.52	0.90	0.86	0.78	0.69	0.60	0.54	0.50	0.47	**0.47**	–
10	0.60	1.03	0.97	0.92	0.84	0.76	0.67	0.61	0.56	0.53	**0.52**
11	0.65	1.12	1.08	1.02	0.97	0.90	0.82	0.74	0.67	0.63	0.59
12	0.66	1.14	1.13	1.10	1.05	1.00	0.94	0.86	0.79	0.72	0.68
13	0.67	1.16	1.15	1.14	1.11	1.07	1.03	0.97	0.90	0.83	0.77
14	0.68	1.17	1.16	1.16	1.15	1.12	1.09	1.05	0.99	0.93	0.86
15	0.69	1.19	1.18	1.17	1.16	1.16	1.14	1.10	1.07	1.02	0.96
16	0.70	1.21	1.20	1.19	1.18	1.17	1.16	1.15	1.11	1.08	1.03
17	0.75	1.29	1.25	1.23	1.22	1.20	1.19	1.18	1.16	1.13	1.10
18	1.12	1.93	1.61	1.48	1.41	1.36	1.32	1.30	1.28	1.25	1.21
19	1.20	**2.07**[c]	2.00	1.76	1.63	1.54	1.48	1.43	1.39	1.36	1.33
20	1.15	1.98	**2.03**[e]	**1.99**[f]	1.82	1.70	1.61	1.55	1.50	1.46	1.43
21	0.83	1.43	1.71	1.83	**1.85**	1.74	1.65	1.59	1.53	1.49	1.46
22	0.80	1.38	1.41	1.60	1.72	**1.76**	**1.68**	1.61	1.56	1.52	1.48
23	0.72	1.24	1.31	1.35	1.51	1.62	1.67	**1.62**	**1.57**	**1.52**	1.49
24	0.66	1.14	1.19	1.25	1.30	1.43	1.54	1.60	1.56	1.52	**1.49**
25	0.66	1.14	1.14	1.17	1.22	1.27	1.39	1.48	1.54	1.51	1.48
26	0.65	1.12	1.13	1.13	1.16	1.20	1.24	1.35	1.44	1.49	1.47
27	0.50	0.86	0.99	1.04	1.06	1.10	1.15	1.19	1.29	1.37	1.43
28	0.32	0.55	0.71	0.84	0.92	0.96	1.01	1.06	1.11	1.20	1.29
29	0.20	0.34	0.45	0.59	0.72	0.80	0.86	0.91	0.97	1.02	1.12
30	0.18	0.31	0.33	0.40	0.52	0.64	0.72	0.78	0.84	0.90	0.95

[a] The value of $SAFR_1$ is separately calculated for each day in Column 3, Column(3) = Column (2)/($0.58 m^3/s$).

For example, $SAFR_1 = \frac{DAF}{AADF} = \frac{0.25\ m^3/s}{0.58\ m^3/s} = 0.43$ for day 1, and $SAFR_1 = \frac{0.16\ m^3/s}{0.58\ m^3/s} = 0.28$ for day 2.

[b] The value of $SAFR_2$ is separately calculated for each day in Column 4, Column (4) = (1/2) × (Sum of two consecutive values in Column (3)).

For example, $SAFR_2 = (1/2) \times (0.43 + 0.28) = 0.35$ during the first two days, and $SAFR_2 = (1/2) \times (0.28 + 0.26) = 0.27$ during the second and third days.

[c] The maximum and minimum values of 2.07 and 0.22 are identified for $SAFR_1$ and highlighted in Column 3.

[d] The value of $SAFR_3$ is separately calculated for each day in Column 5, Column (5) = (1/3) × (Sum of three consecutive values in Column (3)).

For example, $SAFR_3 = (1/3) \times (0.43 + 0.28 + 0.26) = 0.32$ during the first three days, and $SAFR_3 = (1/3) \times (0.28 + 0.26 + 0.22) = 0.25$ between the second and fourth days. Similarly, values of $SAFR_i$ ($i = 4, 5, \ldots, 10$) are calculated and summarized in Columns 6–10.

[e] The maximum and minimum values of 2.03 and 0.23 are identified for $SAFR_2$ and highlighted in Column 4.

[f] The maximum and minimum values of 1.99 and 0.24 are identified for $SAFR_2$ and highlighted in Column 5. Similarly, the maximum and minimum values are identified for all $SAFR_i$ ($i = 4, 5, \ldots, 10$) and highlighted in Columns 6–10.

3. Summarize the maximum and minimum *SAFR* data.

The maximum and minimum *SAFR* data is summarized below.

Parameter	$SAFR_i$									
	$i=1$	$i=2$	$i=3$	$i=4$	$i=5$	$i=6$	$i=7$	$i=8$	$i=9$	$i=10$
Maximum	2.07	2.03	1.99	1.85	1.76	1.68	1.62	1.57	1.52	1.49
Minimum	0.22	0.23	0.24	0.25	0.29	0.31	0.35	0.41	0.47	0.52

4. Plot the maximum and minimum values of *SAFR* with respect to the number of consecutive days during the 10-day period.

These plotted lines form an envelope. As the number of consecutive days is increased, the envelope will tend to converge to the *AADF* ratio passing through 1. The *SAFR* envelope for 10 consecutive days is shown in Figure 4.5.

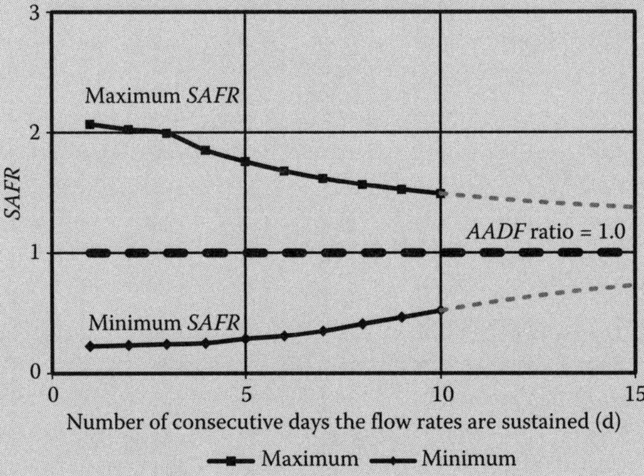

FIGURE 4.5 Sustained average flow ratio (SAFR) envelop (Example 4.14).

EXAMPLE 4.15: FORECASTING DESIGN FLOW RATES

A city has a current total population of 40,000 residents. The estimated design population is 55,000 at the end of a design period of 20 years. Determine the design average, peak, and minimum flow rates. Use the following data.

Current daily average flow to the wastewater treatment plant	$= 18{,}400\ m^3/d$
Daily average I/I flow	= 80 Lpcd
Hourly peak I/I flow	= 180 Lpcd

A new junior college will attract 1500 nonresident students per day. The daily average wastewater flow is 50 Lpcd. A new industry will add a daily average flow of 900 m^3/d for a 24 h/d operation. An hourly peak flow of 1500 m^3/d will occur during the day shift. The plant will be shut down one day a week. It is expected that 10% per capita reduction in wastewater flow will occur with the future population growth. The minimum to average flow ratio is 0.4 for the current dry weather wastewater flow.

Solution

1. Calculate the current wastewater flow in Lpcd.

$$\text{Daily average I/I flow} = \frac{80\ \text{L}}{\text{person}\cdot\text{d}} \times 40{,}000\ \text{people} \times \frac{\text{m}^3}{1000\ \text{L}} = 3200\ \text{m}^3/\text{d}$$

Daily average dry weather wastewater flow $= (18{,}400 - 3200)\ \text{m}^3/\text{d} = 15{,}200\ \text{m}^3/\text{d}$

Daily average dry weather flow in Lpcd $= 15{,}200\ \text{m}^3/\text{d} \times \dfrac{1000\ \text{L}}{\text{m}^3} \times \dfrac{1}{40{,}000\ \text{persons}} = 380\ \text{Lpcd}$

2. Calculate the expected future wastewater flow in Lpcd.

Future daily average flow after 10% reduction $= (1 - 0.1) \times 380\ \text{Lpcd} = 342\ \text{Lpcd}$

3. Calculate the design daily average wastewater flow.

Current daily average dry weather flow $= 15{,}200\ \text{m}^3/\text{d}$

Future addition daily average residential flow $= \dfrac{342\ \text{L}}{\text{person·d}} \times 15{,}000\ \text{people} \times \dfrac{\text{m}^3}{1000\ \text{L}} = 5310\ \text{m}^3/\text{d}$

Daily average students' contribution $= \dfrac{50\ \text{L}}{\text{student·d}} \times 1500\ \text{people} \times \dfrac{\text{m}^3}{1000\ \text{L}} = 75\ \text{m}^3/\text{d}$

Subtotal daily average residential flow $= (15{,}200 + 5130 + 75)\ \text{m}^3/\text{d} = 20{,}405\ \text{m}^3/\text{d}$

Daily average industrial flow $= 900\ \text{m}^3/\text{d}$

Daily average I/I flow $= \dfrac{80\ \text{L}}{\text{person·d}} \times 55{,}000\ \text{people} \times \dfrac{\text{m}^3}{1000\ \text{L}} = 4400\ \text{m}^3/\text{d}$

Design daily average flow $= (20{,}405 + 900 + 4400)\ \text{m}^3/\text{d} = 27{,}705\ \text{m}^3/\text{d}$

4. Calculate the design hourly peak flow.

Peaking factor from Equation 4.4, $= 1 + \dfrac{14}{4 + \sqrt{P}} = 1 + \dfrac{14}{4 + \sqrt{55}} = 2.23$

Hourly peak dry weather residential flow $= 2.23 \times 20{,}405\ \text{m}^3/\text{d} = 45{,}500\ \text{m}^3/\text{d}$

Hourly peak industrial flow $= 1500\ \text{m}^3/\text{d}$

Hourly peak I/I flow $= \dfrac{180\ \text{L}}{\text{person·d}} \times 55{,}000\ \text{people} \times \dfrac{\text{m}^3}{1000\ \text{L}} = 9900\ \text{m}^3/\text{d}$

Design hourly peak flow $= (45{,}500 + 1500 + 9900)\ \text{m}^3/\text{d} = 56{,}900\ \text{m}^3/\text{d}$

5. Calculate the design minimum flow.

Minimum dry weather flow $= 0.4 \times 15{,}200\ \text{m}^3/\text{d} = 6080\ \text{m}^3/\text{d}$

Minimum industrial flow $= 0$ (during the plant shut down)

Total design minimum flow $= 6080\ \text{m}^3/\text{d}$

Discussion Topics and Review Problems

4.1 The hourly diurnal water demand and wastewater flow data of a city is given below. Calculate the average water demand and average wastewater flow. What percentage of the water is returned to the wastewater treatment facility? Draw the diurnal water demand and wastewater flow curves.

Time	a.m.							p.m.					
	12	2	4	6	8	10	12	2	4	6	8	10	12
Water demand, m^3/h	14.3	9.3	5.9	11.7	55.3	44.4	25.1	26.0	31.8	46.0	48.6	33.5	14.3
Wastewater flow, m^3/h	18.5	14.3	12.1	13.5	18.5	27.7	38.5	42.8	36.0	31.2	26.5	28.5	18.4

4.2 A 200-home subdivision is proposed. Calculate the annual water saving that can be achieved if instead of conventional plumbing all homes are installed with faucet aerators, water efficient washing machines and dishwashers, and shallow trap water closets. Assume average nonconserving water consumption is 380 Lpcd, and there are at an average 3.5 residents per home. Make use of the information provided in Table 4.2. Average savings from conservation devices are given in Example 4.7.

4.3 A home has installed a pressure reducing value (PRV) at the point of entry into the home. If water usage is 350 Lpcd and there are four residents, calculate percent water saving. Typical breakdown of residential water usage is given in Table 4.2. Average savings from PRV is assumed 25%.

4.4 In a home, total average water loss caused by toilet leaks, faucet drips, and leakage from the lawn sprinkler is ~4% of total water consumed. The water supply pressure is 345 kPa. A PRV is installed which drops the supply pressure in the home to 110 kPa. Assuming water consumption is 304 Lpcd and all leaks behave like orifices, calculate the water saving that will be achieved from these sources after PRV is installed.

4.5 A wastewater treatment plant receives at an average 87% of water supply during dry weather conditions. Average, expected municipal water demand is 340 Lcpd. Calculate average and maximum dry weather flows. Population of the town is 45,000.

4.6 A small subdivision of a city is being developed. The ultimate zoning plan shows the following: residential, institutional, and commercial establishments. The single family population in low-income, medium-income, and high-income units are 500, 800, and 1500.

One apartment complex	500 residents
One hotel/motel	500 units
One hospital	300 beds
One nursing home	150 beds
One boarding school	1500 students
Five restaurants	1900 customers
One bar	150 customers
Shopping centers	250 employees
Three barber shops	30 chairs
Two beauty salons	20 stations
One commercial laundry	20 machines
One service station	bays
One movie theater	seats

Calculate the residential water demand, and institutional and commercial water demand. What is the average water demand for the entire community and Lpcd? Use the average unit loadings in Table 4.3.

4.7 A city of 60,000 residents has an average water demand of 350 Lpcd. The institutional and commercial, and industrial average areas in the city are 200 and 300 ha, and water demand expected is 20 and 23 m^3/ha·d. The public water use and water unaccounted for are 10% and 6% of total municipal water demand, respectively. Calculate total municipal demand, and that of each component as a percent of total municipal water demand.

4.8 A municipal service area is 1650 ha. Average estimated infiltration/inflow allowance is 3500 L/ha·d. Calculate the I/I quantity per capita per day. Assume that the population density of the service area is 25 persons per ha.

4.9 The total length of a sewerage system of a city is 200 km. The weighted equivalent diameter of the sewers is 21 cm. Calculate the I/I flow per capita per day that may reach the wastewater treatment plant if the I/I allowance is 1400 L/cm·km·d. Average sewerage length per capita is 8 m.

4.10 The population of a city is 68,000. Average wastewater flow and length of wastewater collection system are 380 Lpcd, and 10 m per capita. The distribution of collection system is as follows.

Sewer diameter, cm	15	20	31	46	61
Distribution, %	4	75	12	6	3

The permissible I/I allowance is 1200 L/cm·km·d (1300 gpd/in·mile). The peaking factor for I/I is 1.5. Determine I/I flow in Lpcd, and peak wet weather flow in m^3/d.

4.11 The population of a city is 25,000 residents. The design average flow to a wastewater treatment plant is 100 gpcd. Determine the average annual sustained one day and 5 consecutive days peak and low flows in MGD. Use the information given in Figure 4.3.

4.12 The design population of a city in 15 years is estimated to be 35,000 residents. The expected average per capita wastewater flow and peak I/I allowances are 400 and 150 Lpcd. The estimated peak industrial flow is 3500 m^3/d. Calculate the peak design flow.

References

1. Qasim, S. R., *Wastewater Treatment Plants: Planning, Design, and Operation*, 2nd ed., CRC Press, Boca Raton, FL, 1999.
2. Qasim, S. R., E. M. Motley, and G. Zhu, *Water Works Engineering: Planning, Design, and Operation*, Prentice Hall PTR, Upper Saddle River, NJ, 2000.
3. Metcalf & Eddy, Inc., *Wastewater Engineering: Treatment and Reuse*, 4th ed., McGraw-Hill, New York, 2003.
4. Tchobanoglous, G. and E. D. Shroeder, *Water Quality: Characteristics, Modeling, and Modification*, Addison-Wesley Publishing Co., Reading, MA, 1985.
5. Metcalf & Eddy, Inc., *Wastewater Engineering: Collection and Pumping of Wastewater*, McGraw - Hill Book Company, New York, 1981.
6. Murrary, C. R. and E. B. Reeves, *Estimated Use of Water in the United States in 1970*, U.S. Geological Survey, Department of Interior, Geological Survey Circular 676, Washington, DC, 1972.
7. Dufor, C. N. and E. Becker, *Public Water Supplies of the 100 Largest Cities in the United States*, U.S. Geological Survey, Water Supply Paper 1812, 1964, p. 35.
8. Bailey, J. R. et al., *A Study of Flow Reduction and Treatment of Wastewater from Households*, U.S. EPA Water Pollution Control Research Series Report 11050 FKE 12/69, December 1969.
9. U.S. Department of Housing and Urban Development, *Residential Water Conservation*, Summary Report, June 1984.
10. U.S. Department of Housing and Urban Development, *Water Saved by Low-Flush Toilets and Low-Flow Shower Heads*, March 1984.
11. Joint Committee of the Water Pollution Control Federation and American Society of Civil Engineers, *Design and Construction of Sanitary and Storm Sewer*, WPCF Manual of Practice No. 9, Water Pollution Control Federation, Washington, DC, 1970.
12. U.S. Environmental Protection Agency, *Handbook for Sewer System Evaluation and Rehabilitation*, Municipal Construction Division, Office of Water Program Operations, Washington, DC, 1975.

5 Characteristics of Municipal Wastewater

5.1 Chapter Objectives

Municipal wastewater contains ~99.9% water. The remaining constituents include suspended (settleable and nonsettleable) and dissolved solids that have organic and inorganic components, as well as microorganisms. These constituents give physical, chemical, and biological qualities that are characteristics of residential, commercial, and industrial wastewaters.

The objectives of this chapter are to present the physical, chemical, and biological characteristics of municipal wastewater. The material is arranged under the following sections:

- Physical quality such as temperature, color, turbidity, odor and solids
- Chemical quality
- Measurement of organic matter and organic strength
- Microbiological quality
- Priority pollutants, toxicity, and biomonitoring
- Unit waste loading and population equivalent (PE)

5.2 Physical Quality

The physical quality of municipal wastewater is generally reported in terms of temperature, color, turbidity, odor, and suspended (settleable and nonsettleable) and dissolved solids. The significance of these parameters is briefly summarized below. Measurement procedures may be found in References 1–3.

5.2.1 Temperature

The temperature of municipal wastewater is slightly higher than that of the water supply, and stays in the range of 10–21°C. The average temperature varies slightly with the season, and is higher than the average air temperature most of the year except during the hot summer months. The wastewater temperature has significant effect upon the microbiological activity, treatability, solubility of gases, density, and viscosity.

5.2.2 Color, Turbidity, and Odor

The color of fresh wastewater is slightly gray. Stale or septic wastewater is dark gray or black. Turbidity is due to suspended and colloidal particles. In general, stronger wastewaters have higher turbidity.

Fresh wastewater has soapy or oily odor that is somewhat disagreeable. Stale wastewater has putrid odor due to hydrogen sulfide, indol, and skatol, and other decomposition products. The intensity of odor is expressed by odor test. The *threshold odor number* (*TON*) corresponds to the greatest dilution ratio at which the odor is just perceptible. *TON* is expressed by Equation 5.1:

$$TON = \frac{A + B}{A} \tag{5.1}$$

where

TON = threshold odor number
A = volume of sample, mL
B = volume of odor-free water, mL. The recommended volume of A + B is 200 mL

A series of dilutions and blanks (no sample) are made and maintained at 60°C. Panel members are requested to give their response. Geometric mean of TON is developed from the response of panel members. Larger the TON value, more odorous is the sample.

EXAMPLE 5.1: TON TEST

TON study was conducted on a wastewater sample. A series of dilutions were made in odor-free water such that the total volume of diluted sample in each case was 200 mL. The results of a three-member panel are summarized below. Determine TON.

Panel Member	Panel Response to Sample Volume Diluted to 200 mL									
	0.5	1.0	B	1.5	2.0	2.5	B	3.0	3.5	4.0
1	−	−	−	−	+	+	−	+	+	+
2	−	−	−	−	−	+	−	+	+	+
3	−	−	−	+	+	+	−	+	+	+

B = blank (dilution water only)
\+ = odor detected
− = No odor perception

Solution

1. Determine the TON values based on results of each panel member.

Panel member 1, $\text{TON}_1 = \frac{200}{2} = 100$

Panel member 2, $\text{TON}_2 = \frac{200}{2.5} = 80$

Panel member 3, $\text{TON}_3 = \frac{200}{1.5} = 133$

2. Determine the mean TON values.

a. Arithmetic mean: $\text{TON} = \frac{1}{n}(\text{TON}_1 + \text{TON}_2 + \ldots + \text{TON}_n)$

$$= \frac{1}{3}(100 + 80 + 133) = 104$$

b. Geometric mean

Method 1 $\text{TON} = (\text{TON}_1 \times \text{TON}_2 \times \ldots \times \text{TON}_n)^{\frac{1}{n}} = (100 \times 80 \times 133)^{\frac{1}{3}} = 102$

Method 2 $m = \frac{1}{n}[\log(\text{TON}_1) + \log(\text{TON}_2) + \ldots \log(\text{TON}_n)]$

$$= \frac{1}{3}(\log 100 + \log 80 + \log 133) = \frac{1}{3}(2.00 + 1.90 + 2.12) = 2.01$$

$$\text{TON} = 10^m = 10^{2.01} = 102$$

5.2.3 Settleable and Suspended (Nonfilterable), Dissolved (Filterable), Volatile, and Fixed Solids

The settleable solids are organic and inorganic. They settle under low velocity, and may block the channels and pipes. Organic content of settleable solids will undergo decomposition and cause odors.

Solids in municipal wastewater contain 50–80% volatile, and 20–50% fixed solids, although this ratio may vary greatly. The solids settleability test is conducted by settling the wastewater in an *Imhoff cone* for 1 h. It is also obtained from the results of total suspended solids (mg/L) minus nonsettleable solids (mg/L). The suspended or nonfilterable solids are determined by filtration of sample through a glass fiber filter with a nominal pore size of about 1.2 μm (1 μm = 10^{-6} m). The dissolved or filterable solids are determined by evaporation of a filtered sample in a steam bath. The volatile solids in each category are determined by ignition of dry solids in a muffle furnace at 550 ± 50°C. Procedures for chemical analysis may be found in the Standard Methods.[1]

EXAMPLE 5.2: SETTLEABLE SOLIDS IN IMHOFF CONE

A 1-L sample of raw municipal wastewater is settled for 1 h in an *Imhoff* cone (Figure 5.1). The volume of settled solids in the bottom of the cone is 15 mL. Express the result in mL/L and mg/L. Assume bulk density of settled solids in the cone is 1015 kg/m^3.

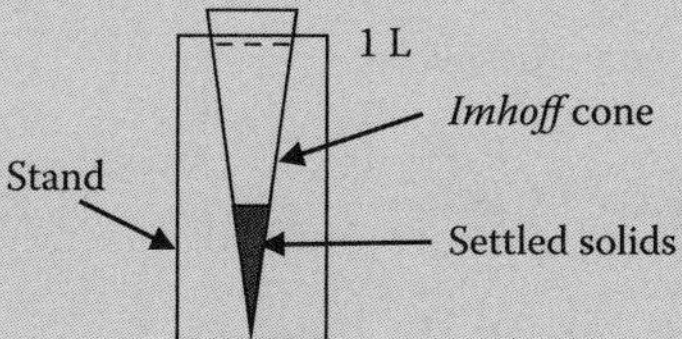

FIGURE 5.1 Determination of settleable solids in an *Imhoff* cone (Example 5.2).

Solution

1. Express the result as mL/L.
   Since the volume of sample in the *Imhoff* cone is 1 L, the volume of settleable solids = 15 mL/L
2. Express the result as mg/L
   Based on bulk density, the concentration of settled solids in the *Imhoff* cone

$$= 15\,\text{mL/L} \times \frac{\text{m}^3}{10^6\,\text{mL}} \times 1015\,\text{kg/m}^3 \times \frac{10^6\,\text{mg}}{\text{kg}} = 15{,}225\,\text{mg/L}$$

EXAMPLE 5.3: SETTLEABLE, NONSETTLEABLE, AND DISSOLVED (FILTERABLE) SOLIDS

Twenty-five milliliters of raw wastewater sample was filtered through a dried and preweighed glass-fiber filter to determine the total suspended solids (TSS). One liter of raw wastewater sample was settled in an *Imhoff* cone for 1 h to determine the volume of settleable suspended solids (SS). Twenty-five milliliters of settled sample was filtered through another predried and weighed glass-fiber filter to determine the non-settleable SS. Fifty milliliters of filtrate was evaporated in a predried and weighed petri dish over a steam bath to determine the total dissolved or filterable solids. Calculate (1) TSS, (2) nonsettleable (or nonfilterable) SS, (3) settleable SS, (4) total dissolved (filterable) solids (TDS), and (5) the concentration of solids in the settled sludge in the *Imhoff* cone.

The results of the tests are given below:

TSS, including settleable and nonsettleable SS

Weight of dried filter paper in an aluminum dish = 12.3478 g

Weight of dried filter paper in the aluminum dish and TSS from raw wastewater = 12.3534 g

Nonsettleable SS

Weight of dried filter paper in an aluminum dish = 12.3480 g

Weight of dried filter paper in the aluminum dish, and nonsettleable suspended solids = 12.3508 g

Settleable SS

Volume of settled SS in the *Imhoff* cone = 10 mL/L

TDS

Weight of dried petri dish = 65.4711 g

Weight of dried petri dish and filterable residue = 65.5080 g

Solution

1. Calculate the concentration of TSS in raw wastewater sample, including settleable and nonsettleable SS.

$$\text{Concentration of TSS} = \frac{(12.3534 - 12.3478)\,\text{g}}{25\,\text{mL}} \times \frac{10^3\,\text{mg}}{\text{g}} \times \frac{10^3\,\text{mL}}{\text{L}} = 244\,\text{mg/L}$$

2. Calculate the concentration of nonsettleable SS.

$$\text{Concentration of nonsettleable SS} = \frac{(12.3508 - 12.3480)\,\text{g}}{25\,\text{mL}} \times \frac{10^3\,\text{mg}}{\text{g}} \times \frac{10^3\,\text{mL}}{\text{L}} = 112\,\text{mg/L}$$

3. Determine the concentration of settleable SS.

$$\text{Concentration of settleable SS} = \text{TSS} - \text{nonsettleable SS} = (244 - 112)\,\text{mg/L} = 132\,\text{mg/L}$$

4. Calculate the concentration of TDS.

$$\text{Concentration of TDS} = \frac{(65.5080 - 65.4711)\,\text{g}}{50\,\text{mL}} \times \frac{10^3\,\text{mg}}{\text{g}} \times \frac{10^3\,\text{mL}}{\text{L}} = 738\,\text{mg/L}$$

5. Calculate the concentration of settled solids in the *Imhoff* cone.

   Volume of settled SS in 1-L *Imhoff* cone = 10 mL/L

$$\text{Concentration of solids in the sludge} = \frac{132\,\text{mg/L}}{10\,\text{mL/L}} \times \frac{10^3\,\text{mL}}{\text{L}} = 13{,}200\,\text{mg/L}$$

EXAMPLE 5.4: SLUDGE QUANTITY AND PUMPING RATE

Municipal wastewater is settled in a primary sedimentation basin. The settled solids are pumped out intermittently from the basin. The average wastewater flow to the basin is 0.25 m^3/s. A constant speed sludge pump runs 15 min per h of cycle time. The average pumping rate is 0.321 m^3/min. The concentration of total solids in raw and settled wastewater is 250 and 150 mg/L, respectively. The average solids content in the sludge is 1.85%. Calculate the following: (a) total volume of wet sludge, (b) total quantity of settleable solids in the sludge, (c) density and specific gravity of liquid sludge, and (d) the volumetric concentration expected in the *Imhoff* cone.

Solution

1. Calculate the volume of sludge pumped per day.

Pump running time $= \frac{15\,\text{min}}{\text{h}} \times \frac{24\,\text{h}}{\text{d}} = 360\,\text{min/d}$

Total volume of wet sludge pumped $= \frac{0.321\,\text{m}^3}{\text{min}} \times \frac{360\,\text{min}}{\text{d}} = 116\,\text{m}^3/\text{d}$

2. Calculate the quantity of settleable SS.

Concentration of settleable SS in wastewater $= (250 - 150)\,\text{mg/L} = 100\,\text{mg/L}$

Wastewater flow rate $= \frac{0.25\,\text{m}^3}{\text{s}} \times \frac{(60 \times 60 \times 24)\,\text{s}}{\text{d}} = 21{,}600\,\text{m}^3/\text{d}$

Quantity of settleable dry sludge $= 100\,\text{mg/L} \times \frac{\text{kg}}{10^6\,\text{mg}} \times 21{,}600\,\text{m}^3/\text{d} \times \frac{10^3\,\text{L}}{\text{m}^3}$

$= 2160\,\text{kg/d}$

3. Calculate the density and specific gravity of wet sludge.

Quantity of wet sludge $= 2160\,\text{kg/d dry solids} \times \frac{100\,\text{kg of wet sludge}}{1.85\,\text{kg dry solids}}$

$= 117{,}000\,\text{kg/d wet sludge}$

Density of wet sludge $= \frac{117{,}000\,\text{kg/d}}{116\,\text{m}^3/\text{d}} = 1010\,\text{kg/m}^3$

Specific gravity $= \frac{1010\,\text{kg/m}^3\,\text{sludge}}{1000\,\text{kg/m}^3\,\text{water}} = 1.01$

4. Calculate the expected volumetric concentration in the *Imhoff* cone.

The volumetric concentration in the 1 − L *Imhoff* cone $= \dfrac{116\,\text{m}^3/\text{d} \times \frac{10^6\,\text{mL}}{\text{m}^3}}{21{,}600\,\text{m}^3/\text{d} \times \frac{10^3\,\text{L}}{\text{m}^3}} = 5.4\,\text{mL/L}$

5. Summarize the final results.

a. Total volume of wet sludge $= 116\,\text{m}^3/\text{d}$

b. Total quantity of settleable solids in the sludge $= 2160\,\text{kg/d}$

c. Density and specific gravity of wet sludge $= 1010\,\text{kg/m}^3$ and 1.01

d. Volumetric concentration expected in *Imhoff* cone $= 5.4\,\text{mL/L}$

EXAMPLE 5.5: DETERMINATION OF VOLATILE AND FIXED SOLIDS

Twenty-five milliliters of municipal wastewater sample was filtered through a dried preweighed glass-fiber filter. The filter after filtration of solids was dried and weighed. Fifty milliliters of the sample was evaporated in a preweighed petri dish. The dish was dried and weighed. Both the filter paper and the dish were ignited in a muffle furnace for 15–20 min, cooled, and weighed. Determine (a) nonsettleable SS, and fixed and volatile components, (b) filterable and nonfilterable solids, and fixed and volatile components, and (c) dissolved or filterable solids, fixed and volatile components.

The experimental values are given below:

Weight of dried filter paper in the aluminum dish	= 12.3469 g
Weight of aluminum dish, filter paper, and suspended solids after drying	= 12.3540 g
Weight of aluminum dish and ash after ignition	= 12.3488 g
Weight of ignited petri dish	= 65.3821 g
Weight of petri dish, and total (filterable and nonfilterable) solids after drying	= 65.4221 g
Weight of petri dish and ash after ignition	= 65.3920 g

Solution

1. Determine the total suspended solids (TSS), fixed suspended solids (FSS), and volatile suspended solids (VSS).

$$\text{TSS} = \frac{(12.3540 - 12.3469)\,\text{g}}{25\,\text{mL}} \times \frac{10^3\,\text{mg}}{\text{g}} \times \frac{10^3\,\text{mL}}{\text{L}} = 284\,\text{mg/L}$$

$$\text{FSS} = \frac{(12.3488 - 12.3469)\,\text{g}}{25\,\text{mL}} \times \frac{10^3\,\text{mg}}{\text{g}} \times \frac{10^3\,\text{mL}}{\text{L}} = 76\,\text{mg/L}$$

$$\text{VSS} = \frac{(12.3540 - 12.3488)\,\text{g}}{25\,\text{mL}} \times \frac{10^3\,\text{mg}}{\text{g}} \times \frac{10^3\,\text{mL}}{\text{L}} = 208\,\text{mg/L}$$

2. Determine the total solids (TS), total fixed solids (TFS), and total volatile solids (TVS).

$$\text{TS} = \frac{(65.4221 - 65.3821)\,\text{g}}{50\,\text{mL}} \times \frac{10^3\,\text{mg}}{\text{g}} \times \frac{10^3\,\text{mL}}{\text{L}} = 800\,\text{mg/L}$$

$$\text{TFS} = \frac{(65.3920 - 65.3821)\,\text{g}}{50\,\text{mL}} \times \frac{10^3\,\text{mg}}{\text{g}} \times \frac{10^3\,\text{mL}}{\text{L}} = 198\,\text{mg/L}$$

$$\text{TVS} = \frac{(65.4221 - 65.3920)\,\text{g}}{50\,\text{mL}} \times \frac{10^3\,\text{mg}}{\text{g}} \times \frac{10^3\,\text{mL}}{\text{L}} = 602\,\text{mg/L}$$

3. Determine the total dissolved or filterable solids (TDS), fixed dissolved solids (FDS), and volatile dissolved solids (VDS).

TDS	= TS − TSS = (800 − 284) mg/L = 516 mg/L
Fixed dissolved solids (FDS)	= TFS − FSS = (198 − 76) mg/L = 122 mg/L
Volatile dissolved solids (VDS)	= TVS − VSS = (602 − 208) mg/L = 394 mg/L

EXAMPLE 5.6: BLOCK PRESENTATION OF SUSPENDED, DISSOLVED, VOLATILE, AND FIXED SOLIDS

Draw a block diagram. Indicate the concentrations of volatile and fixed solids of suspended and dissolved solids in typical municipal wastewater.

Solution

The block representation of various components of suspended and dissolved solids in municipal wastewater are given in Figure 5.2. The typical values are given in parentheses.

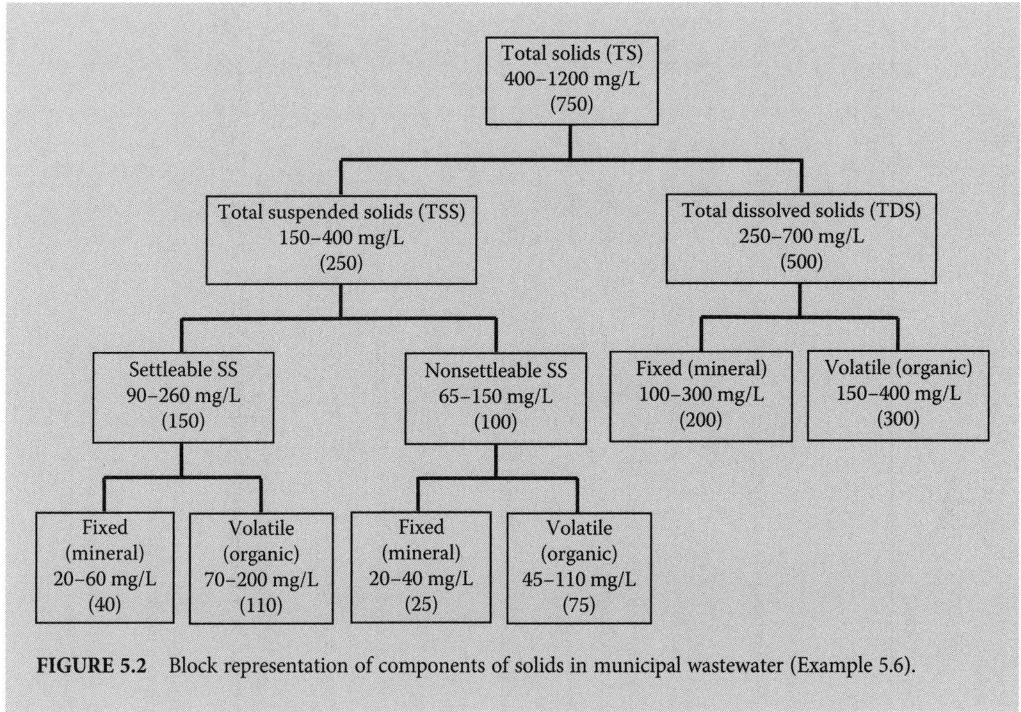

FIGURE 5.2 Block representation of components of solids in municipal wastewater (Example 5.6).

5.3 Chemical Quality

The chemical quality of wastewater is expressed in terms of organic and inorganic constituents. The organic components may be biodegradable, nonbiodegradable, or toxic. The inorganic constituents may be essential, nonessential, or toxic. The concentration of many common chemical quality constituents of municipal wastewater and their significance are summarized in Table 5.1. Brief discussion is provided in several sections.

5.4 Measurement of Organic Matter and Organic Strength

Major components of organic matter in municipal wastewater are proteins, carbohydrates, and fats, which are ~50%, 40%, and 10%, respectively. Proteins and carbohydrates are readily biodegradable. Fats are not easily decomposed by bacteria. Many surfactants, phenolic, and refractory organic compounds resist breakdown by biological means. The volatile organic compounds (VOCs) are emitted into the atmosphere, which cause health risks and lead to formation of photochemical oxidants. Nonbiodegradable and toxic organic compounds are presented in Section 5.4.8.

Most methods used for measurement of organic matter in wastewater are indirect measures or nonspecific. Specific measurements are seldom necessary in wastewater characterization. Many specific methods are presented in Section 5.4.7. Several nonspecific measurement methods are presented below.

5.4.1 Biochemical Oxygen Demand

Measurement: Most prevalent method of nonspecific measurement of organic strength of wastewater is carbonaceous biochemical oxygen demand (CBOD) or simply BOD. The standard BOD test is defined as the amount of oxygen utilized in 5 days and at 20°C by a mixed population of microorganisms under

TABLE 5.1 Concentration of Many Physical and Chemical Quality Constituents of Municipal Wastewater

Water Quality Parameters	Description	Concentration	
		Range	Typical
Total solids (TS), mg/L	Organic and inorganic, suspended and dissolved matter.	400–1200	750
Total suspended solids (TSS), mg/L	The total suspended solids in raw wastewater include settleable and nonsettleable (suspended) organic and inorganic solids.	150–400	250
Fixed, mg/L	Noncombustible or mineral components of TSS	40–100	65
Volatile,[a] mg/L	Combustible or organic components of TSS	115–310	185
Settleable suspended solids (SS), mL/L (mg/L)	Portion of organic and inorganic solids that settles in 1 h in an *Imhoff* cone. These solids are approximate measure of sludge that is removed in a sedimentation basin.	5–20 90–260	10 150
Fixed, mg/L	Noncombustible or mineral components of settleable SS	20–60	40
Volatile,[a] mg/L	Combustible or organic components of settleable SS	70–200	110
Nonsettleable SS, mg/L	Colloidal solids that do not settle in a sedimentation basin. They are removed by filtration through a glass fiber or membrane filter.	65–150	100
Fixed, mg/L	Noncombustible or mineral portion of nonsetteable SS	20–40	25
Volatile, mg/L	Combustible or organic fraction of non setteable SS	45–110	75
Total dissolved or filterable solids (TDS), mg/L	Portion of organic and inorganic solids that are not filterable. Solids smaller than 1 millimicron (μm)[b] fall in this category.	250–700	500
Fixed, mg/L Volatile, mg/L	Noncombustible or mineral components of TDS Combustible or organic fraction of TDS	100–300 150–400	200–300
BOD_5, mg/L	Biochemical oxygen demand (5 d at 20°C). It represents the biodegradable portion of organic component. It is a measure of dissolved oxygen required by microorganisms to stabilize the organic matter in 5 days at 20°C.	110–400	210
COD, mg/L	Chemical oxygen demand. It is a measure of organic matter and represents the amount of oxygen required to oxidize the organic matter by strong oxidizing chemicals (potassium dichromate) under acidic condition.	200–780	400
TOC, mg/L	Total organic carbon is a measure of organic matter. TOC is determined by converting organic carbon to carbon dioxide. It is done in a high-temperature furnace in the presence of a catalyst. Carbon dioxide is quantitatively measured.	80–290	150

(*Continued*)

TABLE 5.1 (*Continued*) Concentration of Many Physical and Chemical Quality Constituents of Municipal Wastewater

Water Quality Parameters	Description	Concentration	
		Range	Typical
Total nitrogen (TN)[c], mg/L as N	Total nitrogen includes organic, ammonia, nitrite, and nitrate nitrogen. Nitrogen and phosphorus along with carbon and other trace elements serve as nutrients. They accelerate the aquatic plant growth in natural waters.	20–85	40
Organic nitrogen (ON), mg/L as N	It is bound nitrogen into protein, amino acid, and urea.	8–30	15
Ammonia nitrogen (AN), mg/L as N	Ammonia nitrogen is produced as first stage of decomposition of organic nitrogen	12–50	30
Nitrite and nitrate nitrogen, mg/L as N	Nitrite and nitrate nitrogen are the higher oxidized forms of ammonia nitrogen. Both forms of nitrogen are absent in raw domestic wastewater.	0–small	0
Total phosphorus (TP)[d], mg/L as P	Total phosphorus exists in organic and inorganic form. Phosphorus in natural water is a source of eutrophication.	4–8	6
Organic, mg/L as P	Organic phosphorus is bound in proteins and amino acids.	1–3	2
Inorganic[e], mg/L as P	Inorganic form of phosphorous exists as orthophosphate and polyphosphate.	3–6	4
pH	pH is indication of acidic or basic nature of wastewater. A solution is neutral at pH 7.	6.7–7.5	7.0
Alkalinity, mg/L as $CaCO_3$	Alkalinity in wastewater is due to presence of bicarbonate, carbonate, and hydroxide ion.	80–350	220
Hardness, mg/L as $CaCO_3$	Hardness in wastewater is primarily due to calcium and magnesium ions. Hardness in wastewater depends on the hardness of water supply.	120–350	200
Chloride, mg/L	Chloride in wastewater comes from water supply, human wastes, and domestic water softeners.	30–100	50
Fats, oils, and grease (FOG), mg/L	These are soluble portion of organic matter in hexane. Their primary sources are fats and oils used in foods.	50–150	100

[a] Volatile fraction is obtained after ignition in a muffle furnace at $550 \pm 50°C$.

[b] Micrometer (μm) $= 10^{-6}$ m (also referred as micron [μ]). Other measures are nanometer (nm) $= 10^{-9}$ also referred to as millimicron (mμ); angstrom (Å) $= 10^{-10}$ m.

[c] TN = ON + AN + NO_2-N + NO_3-N. Total Kjeldahl nitrogen (TKN) = ON + AN.

[d] The concentration of phosphorous in municipal wastewaters in the United States has, in general, been falling over the past decade. In late 1960s, typical total P concentrations in raw wastewater was 10–12 mg/L. Currently, concentrations are usually in the range of 4–8 mg/L where phosphorous-based detergents are regulated. It is expected that in future the concentration of TP in municipal wastewater will decrease.

[e] The predominate form of inorganic phosphorus in raw wastewater is ortho-phosphate (ortho-P).

Source: Adapted in part from References 4 through 7.

aerobic condition to stabilize the organic matter in the sample. It is mostly expressed as BOD_5. Basic information about the BOD_5 test is summarized below.

1. The BOD test is conducted by placing a measured amount of wastewater sample in a 300-mL standard BOD bottle.
2. The remainder volume is filled with dilution water that contains the essential nutrients and is saturated with dissolved oxygen.
3. Well-acclimated microbial seed may be necessary for industrial wastes. Municipal wastewater is naturally well seeded and seeding is not needed.
4. The dissolved oxygen is measured initially and after incubation for 5 days at 20°C.
5. The BOD of wastewater is not a single point, rather a time-dependent variable. Also, biochemical reaction rate varies with temperature. Therefore, 5-d incubation at 20°C has become a standard practice. The microbial oxidation of organic matter and BOD relationships are expressed by Equations 5.2 through 5.4.

$$\text{Organic matter} + O_2 \xrightarrow{\text{Heterotrophic organisms}} CO_2 + \text{biomass} + \text{energy} \quad (5.2)$$

$$BOD_5 = \frac{D_1 - D_2}{P} \quad \text{(unseeded sample)} \quad (5.3)$$

$$BOD_5 = \frac{(D_1 - D_2) - (B_1 - B_2)f}{P} \quad \text{(seeded sample)} \quad (5.4)$$

where

BOD_5 = BOD_5 concentration, mg/L

D_1 and D_2 = dissolved oxygen of diluted samples immediately after preparation, and after 5-d incubation at 20°C, mg/L

B_1 and B_2 = dissolved oxygen of seeded control before and after incubation, mg/L

f = ratio of seed in diluted sample to seed in control

P = decimal volumetric fraction of sample used in the bottle

Ultimate BOD and Reaction Rate Constant: BOD is a time-dependent variable. It follows a first-order reaction rate, reaching an ultimate value in ~20 days. At this point, most of the biodegradable carbonaceous organic matter is stabilized. The ultimate BOD (L_0) is also called ultimate or total carbonaceous oxygen demand, or first-stage BOD. For municipal wastewater, the 5-d BOD is 60–70% of ultimate BOD. The value of reaction rate constant varies with type of waste. Typical value of reaction rate constant k (base e) for municipal wastewater is 0.23 d^{-1} (K [base 10] = 0.1 d^{-1}). The range of k for industrial wastes may be 0.05–0.3 d^{-1} (base e) (K = 0.02–0.13 d^{-1} [base 10]). The reaction rate constant k is temperature dependent. The time-dependent reaction rate equations are given below (Equations 5.5 through 5.7).[4–6]

$$-\frac{dL_t}{dt} = kL_t \quad \text{or} \quad -\frac{dL_t}{dt} = KL_t \quad (5.5)$$

$$L_t = L_0\, e^{-kt} \quad \text{or} \quad L_t = L_0\, 10^{-Kt} \quad (5.6)$$

$$y_t = L_0(1 - e^{-kt}) \quad \text{or} \quad y_t = L_0(1 - 10^{-Kt}) \quad (5.7)$$

$$k_T = k_{20}\theta_T^{T-20} \quad \text{or} \quad K_T = K_{20}\theta_T^{T-20} \quad (5.8)$$

where

L_t = BOD remaining at any time t, mg/L

L_0 = ultimate carbonaceous BOD (or first-stage BOD) at $t = 0$, mg/L. For municipal wastewater, $BOD_5 = (2/3)L_0$

y_t = BOD exerted at any time t, mg/L
k = reaction rate constant (base e), d^{-1}. For municipal wastewater, $k_{20} = 0.2$–$0.3\ d^{-1}$
K = reaction rate constant (base 10), d^{-1}. $K = k/2.3$
k_T, k_{20}, K_T, K_{20} = reaction rates at any temperature T and at 20°C
θ_T = temperature correction coefficient, dimensionless. The value of θ_T may be 1.03–1.09 depending upon wastewater characteristics, operating condition, and temperature range. Typical value of θ_T is 1.047.

Determination of Parameters k (or K) and L_0: Several methods are available to determine k (or K) and L_0 for given wastewater. In all cases, a BOD-time curve is prepared either by a series of BOD measurements or by using a laboratory respirometer. Once the BOD time data is obtained, any of the following methods can be used to determine k (or K) and L_0.[5,6,8]

- Thomas graphical method
- Least squares method
- Fujimoto method
- Method of moments
- Daily difference method
- Rapid-ratio method

Estimation of BOD_5 Exerted by Suspended and Dissolved Organic Matter: Organic matter exerts oxygen demand. BOD_5 exerted by organic matter including biomass can be estimated from the biodegradable fraction of the organic matter. Discussions on the related topics may be found in Chapter 10, Examples 10.7 through 10.11 and Example 10.70.

EXAMPLE 5.7: BOD DEFINITION

If a wastewater sample has BOD_5 of 120 mg/L, what does it mean?

Solution

BOD_5 of 120 mg/L means that microorganisms will consume under aerobic condition 120 mg of oxygen in 5 days and at 20°C to stabilize the organic matter contained in 1 L of wastewater. The oxygen consumed is therefore, an indirect measurement of organic content in the wastewater.

EXAMPLE 5.8: DO MEASUREMENT

Describe briefly the methods of DO measurement.

Solution

Analytical as well as electronic measurements of DO are commonly used.[1]

Analytical method is a modified Winkler procedure that involves sequential addition of manganese sulfate and alkali-iodide-azide reagents, and concentrated sulfuric acid in a BOD bottle containing wastewater sample. Complex series of reactions release an amount of iodine equivalent to the amount of oxygen originally present. The liberated iodine is titrated with standard sodium thiosulfate solution, or the color intensity is measured by an absorption spectrophotometer.

The electronic method utilizes membrane electrodes and is based on the rate of diffusion of molecular oxygen across the membrane. The instrument gives direct DO reading.

EXAMPLE 5.9: SOLUBILITY OF OXYGEN

List the factors on which the solubility of oxygen in water depends. Tabulate the solubility of oxygen in fresh water at temperatures 0°C, 10°C, 20°C, 30°C, and 40°C.

Solution

The solubility of oxygen in water depends upon:

1. Temperature (lesser solubility at higher temperature)
2. Pressure (higher solubility at higher pressure)
3. Salt content (lesser solubility at higher salt content)

The saturation value of oxygen in distilled water at different temperatures is given in Table 5.2. Discussion on solubility of oxygen and other gases may be found in Chapter 10 and Appendix B.

TABLE 5.2 Solubility of Oxygen at Different Temperatures in Distilled Water (Example 5.9)

Temperature, °C	Solubility of Oxygen, mg/L
0	14.6
10	11.3
20	9.2
30	7.6
40	6.6

EXAMPLE 5.10: PURPOSES OF DILUTION WATER

What is the purpose of dilution water in a BOD test, and what does it contain?

Solution

The purpose of dilution water in BOD test is to dilute the wastewater sample in the BOD bottle. Without dilution, the wastewater sample may deplete the oxygen in the bottle. As a result, after the incubation period of 5 days there may not be sufficient oxygen left in the bottle to give a reliable result.

The dilution water is prepared from distilled water. The following chemicals are added: (1) phosphate buffer, (2) magnesium sulfate, (3) calcium chloride, and (4) ferric chloride solutions. These chemicals provide needed nutrients for microbial growth. The dilution water is saturated with oxygen and free of organic contaminants. Often microbial seed is added in the dilution water. Thus, the use of dilution water (1) reduces the sample volume in the BOD bottles, (2) increases the available dissolved oxygen, (3) provides needed nutrients, and (4) with seed, provides microbial population.

EXAMPLE 5.11: PREPARATION OF DILUTION CHART

A sample was diluted into a 300-mL BOD bottle. The volume of sample depended upon the expected BOD_5 of the sample. The selected volume is expected to cause a DO depletion of at least 2 mg/L or give a residual DO of at least 1 mg/L. Prepare a dilution table that has range of BOD for different dilutions expressed as mL in 300-mL BOD bottle and percent volume. Assume that the initial DO of all dilutions is 8.5 mg/L.

Solution

1. Set up range of BOD for a given dilution.
   a. Minimum value of BOD_5 range is obtained when DO depletion is at least 2 mg/L, or DO remaining after 5 days = (8.5 − 2.0) mg/L = 6.5 mg/L.

$$\text{Minimum BOD}_5\ \text{value} = \frac{(8.5 - 6.5)\ \text{mg/L}}{P} = \frac{2\ \text{mg/L}}{P}$$

   b. Maximum value of BOD_5 range is obtained when minimum DO remaining after 5 days = 1 mg/L.

$$\text{Maximum BOD}_5\ \text{value} = \frac{(8.5 - 1)\ \text{mg/L}}{P} = \frac{7.5\ \text{mg/L}}{P}$$

2. Prepare a table with different values of P and mL sample used in 300-mL BOD bottle (Table 5.3).

TABLE 5.3 BOD Dilution Chart for Different Range of Values (Example 5.11)

% Mixture	P	Sample Volume in 300 mL Bottle, mL	BOD_5 Range, mg/L
0.01	0.0001[a]	0.03[b]	20,000–75,000[c]
0.05	0.0005	0.15	4000–15,000
0.10	0.001	0.30	2000–7500
0.20	0.002	0.60	1000–3700
0.50	0.005	1.50	400–1500
1.00	0.01	3.00	200–750
5.00	0.05	15.00	40–150
10.00	0.1	30.00	20–75
50.00	0.5	150.00	4–15
100.00	1	300.00	0–7

[a] For 0.01% mixture, $P = 0.0001$.
[b] $0.0001 \times 300\ \text{mL} = 0.03\ \text{mL}$
[c] $\text{Min BOD}_5 = \frac{2\ \text{mg/L}}{P} = 20{,}000\ \text{mg/L}$; and $\text{Max BOD}_5 = \frac{7.5\ \text{mg/L}}{P} = 75{,}000\ \text{mg/L}$

EXAMPLE 5.12: SINGLE DILUTION BOD TEST WITHOUT SEED

Five milliliters of a municipal wastewater sample was placed in a 300-mL BOD bottle. No seeding was needed. The BOD bottle was filled with dilution water that was free of organic matter, saturated with oxygen, and contained necessary nutrients. The initial DO in the diluted sample was 8.6 mg/L, and DO after incubation for 5 days at 20°C was 4.2 mg/L. Calculate BOD_5.

Solution

1. Select the applicable equation for calculation of BOD_5.
   Since the sample is not seeded, use Equation 5.3.

$$BOD_5 = \frac{D_1 - D_2}{P}$$

2. Calculate decimal volumetric fraction of sample.

$$P = \frac{5\ \text{mL}}{300\ \text{mL}} = 0.01667$$

3. Calculate BOD_5 from Equation 5.3.

$$BOD_5 = \frac{(8.6 - 4.2)\ \text{mg/L}}{0.01667} = 264\ \text{mg/L}$$

EXAMPLE 5.13: MULTIPLE DILUTION BOD TEST DATA WITHOUT SEED

Replicate BOD_5 tests were run on a municipal wastewater sample. Four sample dilutions were prepared without seeding. The results of measured DO in the initial and 5-d incubated dilution bottles are summarized below. Determine the average BOD_5.

Bottle No.	Sample Volume Used, mL	Initial DO (D_1), mg/L	DO after 5-days Incubation at 20°C (D_2), mg/L
1	1	8.9	8.1
2	4	8.8	5.5
3	6	8.8	3.8
4	10	8.7	0.7

Solution

1. Calculate BOD_5 from each dilution from Equation 5.3.

$$BOD_5 = \frac{D_1 - D_2}{P}$$

2. Set up the calculation table below.

Bottle No.	Sample Volume Used, mL	P	DO Depletion ($D_1 - D_2$), mg/L	Calculated BOD_5, mg/L
1	1	0.0033	0.8	242
2	4	0.0133	3.3	248
3	6	0.020	5.0	250
4	10	0.033	8.0	240

3. Select the BOD_5 results for averaging.
   a. In multiple dilution test, the dilution showing a residual DO of at least 1 mg/L, and a DO depletion of at least 2 mg/L provide most reliable result.
   b. The results show that dilution bottle 1 has DO depletion of 0.8 mg/L ($<$ 2 mg/L), and dilution bottle 4 has residual DO of 0.7 mg/L ($<$1 mg/L). Therefore, both results should not be used.
   c. The average results of dilution bottles 2 and 3 should be used (the DO depletion is $>$2 mg/L and residual DO is $>$1 mg/L).
4. Determine the average BOD_5.
   The average BOD_5 is 249 mg/L (the average values of dilution bottles 2 and 3).

EXAMPLE 5.14: BOD_5 DETERMINATION WITH SEED

An industrial wastewater sample was tested for BOD_5. The sample dilutions were seeded with a well-acclimated bacterial culture. The dilution procedure and test results are provided below. Calculate BOD_5 at 20°C.

Bottle No.	Volume of Sample, mL	Volume of Seed, mL	DO Initial, mg/L	DO after 5-d Incubation at 20°C, mg/L
1	2	1	8.8	–
2	2	1	–	4.3
3[a]	0	0.8	8.9	–
4[a]	0	0.8	–	8.7

[a] Seeded control.

Solution

1. Select the applicable equation for calculation of BOD_5.
   Since the sample is seeded, use Equation 5.4.

$$BOD_5 = \frac{(D_1 - D_2) - (B_1 - B_2)f}{P}$$

2. Determine the variables in Equation 5.4 from the experimental data.
   $D_1 = 8.8$ mg/L, $D_2 = 4.3$ mg/L, $B_1 = 8.9$ mg/L, $B_2 = 8.7$ mg/L

$$f = \frac{1\,\text{mL}}{0.8\,\text{mL}} = 1.25$$

$$P = \frac{2\,\text{mL}}{300\,\text{mL}} = 0.00667$$

3. Calculate BOD_5 from Equation 5.4.

$$BOD_5 = \frac{(8.8 - 4.3)\,\text{mg/L} - (8.9 - 8.7)\,\text{mg/L} \times 1.25}{0.00667} = \frac{(4.5 - 0.25)\,\text{mg/L}}{0.00667} = 637\,\text{mg/L}$$

EXAMPLE 5.15: DETERMINATION OF FIRST-ORDER REACTION KINETIC COEFFICIENTS

Develop the BOD remaining and BOD exerted relationship with respect to reaction rate constant and ultimate BOD.

Solution

1. Formulate the first-order reaction equation.
   The first-order reaction is expressed by Equation 5.5.

$$-\frac{dL_t}{dt} = kL_t$$

2. Develop the relationship between L_t (BOD remaining at any time t, mg/L) and L_0 (initial BOD at $t = 0$, mg/L).

Integrate Equation 5.5 to obtain the relationship.

$$\int_{L_0}^{L_t} \frac{dL_t}{L_t} = -k \int_0^t dt$$

$$\ln L_t|_{L_0}^{L_t} = -kt|_0^t$$

$$\ln \frac{L_t}{L_0} = -kt \quad \text{or} \quad \frac{L_t}{L_0} = e^{-kt}$$

$$L_t = L_0 e^{-kt}$$

Note: This equation is same as that expressed by Equation 3.16c.

3. Develop the relationship between y_t (BOD exerted at time t, mg/L) and L_0.

$$y_t = L_0 - L_t = L_0 - L_0 e^{-kt} = L_0(1 - e^{-kt})$$

Similarly, $y_t = L_0(1 - 10^{-Kt})$

4. Plot y_t and L_0 with respect to time t.

The BOD exerted (y_t) and BOD remaining (L_t) relationships with respect to time t are shown in Figurc 5.3.

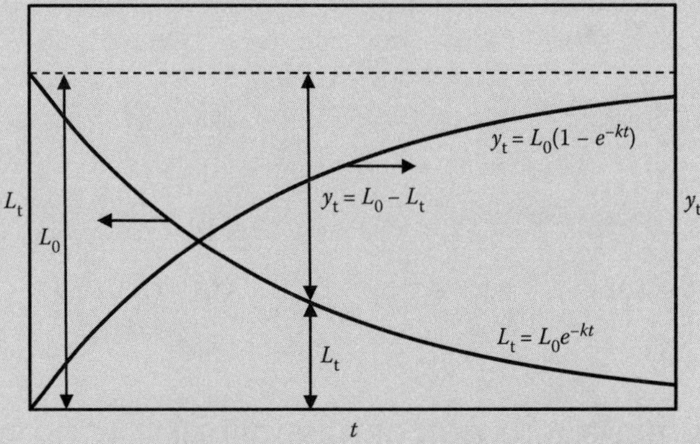

FIGURE 5.3 Time-dependent BOD remaining and BOD exerted curves (Example 5.15).

EXAMPLE 5.16: ULTIMATE CARBONACEOUS BOD

The 5-day BOD of a sample at 20°C is 180 mg/L. Calculate the ultimate carbonaceous BOD. The reaction rate constant k (base e) at 20°C = 0.25 d^{-1}.

Solution

1. Select the reaction rate equation.

The ultimate BOD is calculated from Equation 5.7.

$$y_t = L_0(1 - e^{-kt})$$

2. Substitute $y_5 = 180$ mg/L, $k = 0.25\ \text{d}^{-1}$, and $t = 5$ d in Equation 5.7 and calculate L_0.

$$180\ \text{mg/L} = L_0(1 - e^{-0.25\text{d}^{-1} \times 5\text{d}}) = L_0(1 - e^{-1.25}) = L_0(1 - 0.287) = 0.713L_0$$

$$L_0 = \frac{180\ \text{mg/L}}{0.713} = 252\ \text{mg/L}$$

EXAMPLE 5.17: RELATIONSHIP BETWEEN *k* AND *K*

What is the relationship between the reaction rate constant k (base e) and K (base 10)?

Solution

1. Select the reaction rate equation.
   Use Equation 5.7.

$$y_t = L_0(1 - e^{-kt})$$

$$y_t = L_0(1 - 10^{-Kt})$$

2. Equate the two equations and solve for Equation 5.9.

$$L_0(1 - e^{-kt}) = L_0(1 - 10^{-Kt})$$

$$e^{-kt} = 10^{-Kt}$$

$$\ln(e^{-kt}) = \ln(10^{-Kt})$$

$$-kt = -Kt \ln(10)$$

$$k = K \ln(10) = 2.3\ K \quad \text{or} \quad K = 0.43\ k \tag{5.9}$$

EXAMPLE 5.18: RATIO OF 5-DAY BOD AND ULTIMATE BOD

What percent of BOD reaction is completed in 5-d BOD test? Assume reaction rate $k = 0.23\ \text{d}^{-1}$.

Solution

1. Select the reaction rate equation.
   Use Equation 5.7.

$$y_t = L_0(1 - e^{-kt})$$

2. Substitute $y_5 = \text{BOD}_5$, $k = 0.23\ \text{d}^{-1}$, and $t = 5$ d in Equation 5.7.

$$\text{BOD}_5 = L_0(1 - e^{-0.23\,\text{d}^{-1} \times 5\,\text{d}}) = L_0(1 - e^{-1.15}) = L_0(1 - 0.32) = 0.68L_0$$

$$\frac{\text{BOD}_5}{L_0} = 0.68$$

BOD_5 is 68% of the ultimate BOD (or $\text{BOD}_5 \approx (2/3)L_0$).

EXAMPLE 5.19: BOD REMAINING AND BOD EXERTED CURVES

A wastewater sample has a BOD_5 of 120 mg/L. Draw the BOD remaining (L_t) and BOD exerted (y_t) curves (mg/L and %) with respect to time. The reaction rate constant (base e) is $0.22\,d^{-1}$.

Solution

1. Select the reaction rate equation.

   BOD remaining (L_t) is expressed by Equation 5.6.

   $$L_t = L_0\, e^{-kt}$$

   BOD exerted (y_t) is expressed by Equation 5.7.

   $$y_t = L_0(1 - e^{-kt})$$

2. Calculate the ultimate BOD.

   Substitute the data in Equation 5.7 and calculate L_0.

   $$120\,\text{mg/L} = L_0(1 - e^{-0.22\,d^{-1} \times 5\,d}) = L_0(1 - e^{-1.1}) = L_0(1 - 0.333) = 0.667 L_0$$

   $$L_0 = \frac{120\,\text{mg/L}}{0.667} = 180\,\text{mg/L}$$

3. Calculate the BOD remaining values.

   Select serial incubation periods of $t = 1, 2, \ldots, 12$ days and calculate BOD remaining L_t from Equation 5.6. Tabulate L_t for $t = 1$–12 days in Table 5.4, Row 1. Calculate percent BOD remaining for $t = 1$–12 days and provide in Table 5.4, Row 2.

TABLE 5.4 BOD Remaining and Exerted Data over a 12-d Time Period (Example 5.19)

Parameter	Incubation Period (t), d												
	0	1	2	3	4	5	6	7	8	9	10	11	12
BOD remaining (L_t), mg/L	180	144	116	93	75	60	48	39	31	25	20	16	13
BOD remaining (L_t), %	100%	80%	64%	52%	41%	33%	27%	21%	17%	14%	11%	9%	7%
BOD exerted (y_t), mg/L	0.0	36	64	87	105	120	132	141	149	155	160	164	167
BOD exerted (y_t), %	0%	20%	35%	48%	59%	67%	73%	79%	83%	86%	89%	91%	93%

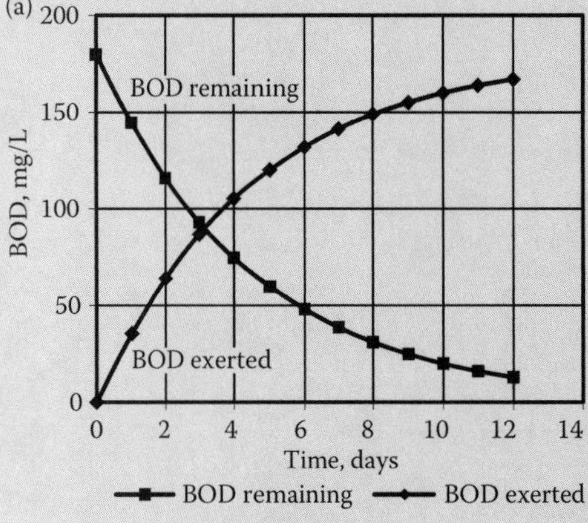

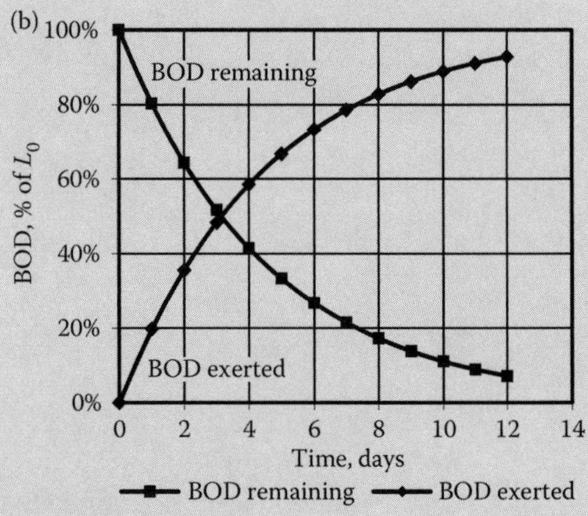

FIGURE 5.4 BOD exerted and remaining concentrations: (a) concentrations in mg/L and (b) concentrations in percent of L_0 (Example 5.19).

4. Calculated BOD exerted values.

   Use the same serial incubation periods and calculate BOD exerted y_t from Equation 5.7 for $t = 1$–12 days. Also, calculate percent BOD exerted for $t = 1$–12 days. Arrange the calculation results in Table 5.4, Rows 3 and 4.

   It may be noted that the sum of BOD remaining and exerted values are 180 mg/L or 100% at any time t during the time periods.

5. Prepare BOD remaining and exerted curves.

   Plot BOD exerted (mg/L) and BOD remaining (mg/L) versus incubation period. These plots are shown in Figure 5.4a. Plot percent BOD exerted and BOD remaining versus incubation period. These plots are shown in Figure 5.4b.

EXAMPLE 5.20: BOD CALCULATION AT DIFFERENT TIMES AND TEMPERATURES

The 7-d BOD of a sample at 22°C is 80 mg/L. Calculate (a) Ultimate BOD, and (b) 5-d BOD at 20°C. The reaction rate constant k (base e) = 0.23 d^{-1} at 20°C. Use the typical value of $\theta_T = 1.047$.

Solution

1. Calculate the reaction rate constant k at 22°C from Equation 5.8.

$$k_{22} = k_{20}\theta_T^{T-20} = 0.23\,\text{d}^{-1} \times (1.047)^{(22-20)} = 0.25\,\text{d}^{-1}$$

2. Calculate the ultimate BOD from Equation 5.7.

$$80\,\text{mg/L} = L_0(1 - e^{-0.25\text{d}^{-1}\times 7\text{d}}) = L_0(1 - e^{-1.75}) = L_0(1 - 0.174) = 0.826L_0$$

$$L_0 = \frac{80\,\text{mg/L}}{0.826} = 96.8\,\text{mg/L}$$

3. Calculate the BOD_5 concentration at 20°C.

$$\text{BOD}_5 = 96.8\,\text{mg/L} \times (1 - e^{-0.23\text{d}^{-1}\times 5\text{d}}) = 96.8\,\text{mg/L} \times (1 - e^{-1.15}) = 96.8\,\text{mg/L} \times 0.683 = 66.1\,\text{mg/L}$$

EXAMPLE 5.21: BOD CURVE AT DIFFERENT TIMES AND TEMPERATURES

Draw the BOD remaining and BOD exerted curves at temperatures 10, 20, and 30°C. The 5-d BOD is 120 mg/L and k at 20°C is 0.23 d^{-1}. Use the typical value of $\theta_T = 1.047$.

Solution

1. Determine the ultimate BOD by substituting the given values of BOD_5 and k_{20} in Equation 5.6.

$$120\,\text{mg/L} = L_0(1 - e^{-0.23\,\text{d}^{-1}\times 5\,\text{d}}) = L_0(1 - e^{-1.15}) = L_0(1 - 0.317) = 0.683L_0$$

$$L_0 = \frac{120\,\text{mg/L}}{0.683} = 176\,\text{mg/L}$$

2. Calculate k_T at $T = 10$ and 30°C from Equation 5.8.

$$k_{10} = k_{20}\theta_T^{T-20} = 0.23\,\text{d}^{-1} \times (1.047)^{(10-20)} = 0.15\,\text{d}^{-1}$$

$$k_{30} = k_{20}\theta_T^{T-20} = 0.23\,\text{d}^{-1} \times (1.047)^{(30-20)} = 0.36\,\text{d}^{-1}$$

3. Develop the BOD curves at different times.

   Calculate BOD from Equation 5.7 at different time intervals, using k_T at temperatures 10°C, 20°C, and 30°C. These values are summarized in Table 5.5.

TABLE 5.5 BOD Values at Different Times and at 10°C, 20°C, and 30°C (Example 5.21)

t, d	BOD, °C		
	10	20	30
2	46	65	90
4	79	106	134
6	104	131	155
8	123	148	166
10	136	158	171
12	147	164	173
14	154	169	174
16	160	171	175
18	164	173	175
20	167	174	175

4. Plot the BOD curves at 10°C, 20°C, and 30°C.

Plot the BOD values using k_T for temperatures 10°C, 20°C, and 30°C and at different time intervals. This plot is shown in Figure 5.5. It may be noted that all curves converge at ultimate BOD of 176 mg/L.

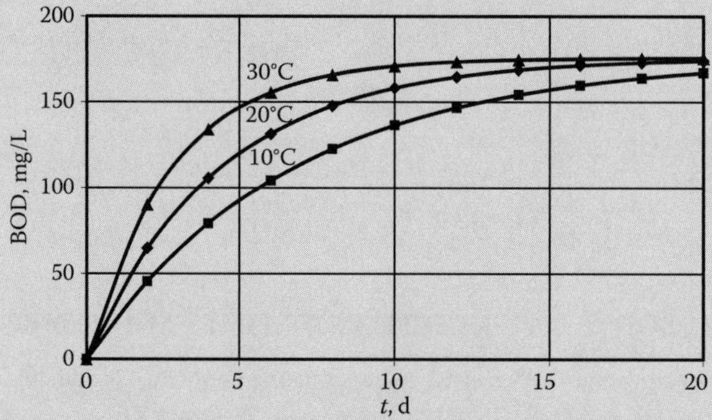

FIGURE 5.5 BOD curves at different times and temperatures (Example 5.21).

EXAMPLE 5.22: DETERMINATION OF REACTION RATE CONSTANT *k*, AND ULTIMATE CARBONACEOUS BOD, L_0

The BOD test results of an industrial wastewater are given below. Calculate *k* (base e) and L_0 using the following methods, (a) Thomas, (b) Least squares, and (c) Fujimoto methods.[5,8,9] Compare the results.

t, d	0	2	4	6	8	10
BOD, mg/L	0	86	138	175	193	210

Solution

A. Thomas Method.

1. Review the procedure.

   In Thomas method, a best-fit straight line is drawn between $(t/y_t)^{1/3}$ and time t. The k and L_0 are calculated using Equations 5.10 and 5.11.

$$K(\text{base } 10) = 2.61\frac{S}{I} \tag{5.10}$$

$$L_0 = \frac{1}{2.3\ KI^3} \quad \text{or} \quad L_0 = \frac{1}{kI^3} \tag{5.11}$$

where
S = slope
I = intercept

2. Calculate the values of $(t/y_t)^{1/3}$ with respect to time (t).

t, d	0	2	4	6	8	10
t/y_t	–	0.023	0.029	0.034	0.041	0.048
$(t/y_t)^{1/3}$	–	0.29	0.31	0.32	0.35	0.36

3. Plot the values of $(t/y_t)^{1/3}$ versus t in Figure 5.6.

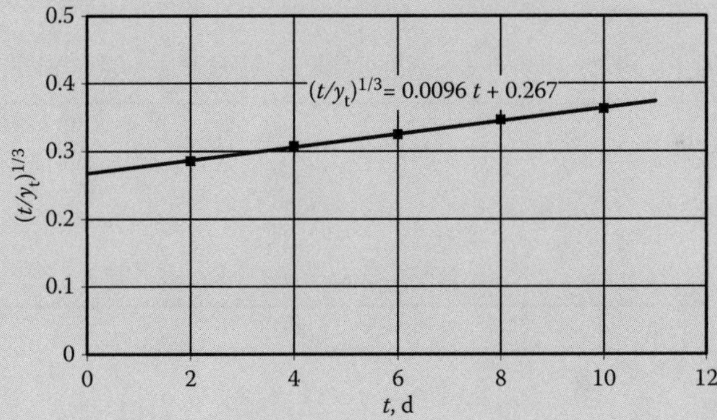

FIGURE 5.6 Linear plot for determination of k and L_0 by Thomas method (Example 5.22).

4. Determine the slope S and intercept I.

   Slope S = 0.0096

   Intercept I = 0.267

5. Calculate k (base e).

   Calculate K (base 10) from Equation 5.10.

$$K = 2.61\frac{S}{I} = 2.61 \times \frac{0.0096}{0.267} = 0.0938\ \text{d}^{-1}$$

   Calculate k (base e) from Equation 5.9.

$$k = 2.3K = 2.3 \times 0.0938 = 0.22\ \text{d}^{-1}$$

6. Calculate L_0.

   Calculate L_0 from Equation 5.11.

   $$L_0 = \frac{1}{k\,I^3} = \frac{1}{0.22 \times (0.267)^3} = 239\ \text{mg/L}$$

B. Least Squares Method.

1. Review the procedure.

   The least squares method involves fitting a curve through a set of data points, so that the sum of the squares and the residuals are minimal. The residuals are the differences between the observed value and the value of the fitted curve. The general relationships are expressed by Equations 5.12 and 5.13.

   $$na + b\sum y - \sum y' = 0 \tag{5.12}$$

   $$a\sum y + b\sum y^2 - \sum yy' = 0 \tag{5.13}$$

   where

   y = BOD, mg/L
   n = number of observations
   y' = $(dy/dx) = (y_{n+1} - y_{n-1})/2 \times \Delta t$
   a = $-b\,L_0$
   k (base e) = $-b$
   Δt = time interval, d

   Summarize the given values and calculate various values in a summary table below.

t, d	0	2	4	6	8	10	$\sum$
y	0	86	138	175	193	210	$\sum y = 592$[a]
y^2	–	7396	19,044	30,625	37,249	–	$\sum y^2 = 94{,}314$
y'	–	34.5[b]	22.3[c]	13.8[d]	8.75[e]	–	$\sum y' = 79.3$
yy'	–	2967	3071	2406	1689	–	$\sum yy' = 10{,}133$

[a] Does not include $y_{10} = 210$ mg/L.

[b] $y' = \frac{138 - 0}{2 \times 2} = 34.5.$

[c] $y' = \frac{175 - 86}{2 \times 2} = 22.3.$

[d] $y' = \frac{193 - 138}{2 \times 2} = 13.8.$

[e] $y' = \frac{210 - 175}{2 \times 2} = 8.75$

2. Substitute the values in Equations 5.12 and 5.13.

   $$4\,a + 592\,b - 79.3 = 0$$

   $$592\,a + 94{,}314\,b - 10{,}133 = 0$$

3. Solve for a and b.

   $$a + 148\,b - 19.8 = 0$$

   $$a + 159\,b - 17.1 = 0$$

$b = -0.24$

$4\ a = 79.3 - 592 \times (-0.24)$

$a = 55.3$

4. Determine k, L_0.

$k = -\ (-0.24) = 0.24\ \text{d}^{-1}$

$55.3 = -\ (-0.24)\ L_0$

$L_0 = 230\ \text{mg/L}$

C. Fujimoto Method.

1. Review the procedure.

The method involves preparation of an arithmetic plot of BOD_{t+1} versus BOD_t. The value at intersection of the line with the slope of unity corresponds to L_0. The value of k is calculated using Equation 5.7 and any pair of BOD and t values given in the data set.[9]

2. Plot the linear relationship between BOD $_{t+1}$ and BOD_t (Figure 5.7).

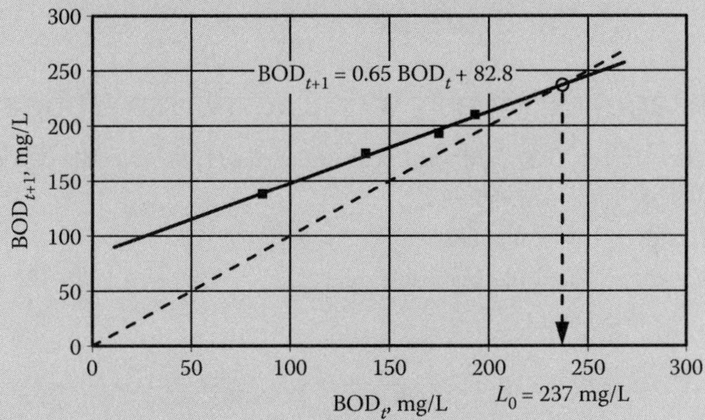

FIGURE 5.7 Linear plot for determination of k and L_0 by Fujimoto method (Example 5.22).

3. Determine the values of L_0.

L_0 is obtained from the intersection in Figure 5.6. $L_0 = 237$ mg/L

4. Determine the value of k.

k is calculated from the given data. In this example, only the BOD of 175 mg/L at the sixth day ($t = 6$ d) is used. Substitute this value and $L_0 = 237$ mg/L in Equation 5.7, and calculate k.

$175\ \text{mg/L} = 237\ \text{mg/L} \times (1 - e^{-k \times 6\,\text{d}})$

$k = 0.22\ \text{d}^{-1}$

A desirable procedure is to calculate k from several data points and use the average value.

5. Compare the values of k and L_0 obtained from three methods.

The values of k and L_0 calculated from three methods are compared below.

Method	k, d^{-1}	L_0, mg/L
Thomas	0.22	239
Least square	0.24	230
Fujimoto	0.22	237

EXAMPLE 5.23: SOLUBLE AND SUSPENDED BOD_5 IN PLANT EFFLUENT

An activated sludge plant is producing effluent that has TSS and total BOD_5 concentrations of 10 and 15 mg/L, respectively. Calculate suspended and soluble BOD_5 in the effluent. Assume (a) biodegradable fraction in TSS is 0.65, (b) each g of biodegradable fraction of organic matter exerts 1.42 g ultimate BOD (L_0), and (c) ratio of BOD_5 to L_0 is 0.68.

Solution

1. Calculate the BOD_5 exerted by TSS.

   Biodegradable solids in TSS = 10 mg/L TSS × 0.65 g biodegradable solid/g TSS
   = 6.5 mg/L biodegradable solids

   Ultimate BOD, L_0 = 6.5 mg/L biodegradable solids × 1.42 gL_0/g biodegradable solids
   = 9.23 mg/L·L_0

   BOD_5 exerted by TSS = 9.23 mg/L·L_0 × 0.68 g BOD_5/gL_0 = 6.3 mg/L BOD_5

2. Determine the suspended and soluble BOD_5 in the effluent.

   Suspended BOD_5 = 6.3 mg/L

   Soluble BOD_5 = 15 mg/L total BOD_5 − 6.3 mg/L suspended BOD_5
   = 8.7 mg/L

EXAMPLE 5.24: BOD_5 EXERTED BY WASTE ACTIVATED SLUDGE

An anaerobic digester receives 1200 kg/d waste activated sludge (TSS). Calculate BOD_5 (kg/d) reaching the digester. Use the following information.

Biodegradable solids = 0.64 × Biomass

Ultimate BOD, L_0 = 1.42 × Biodegradable solids

BOD_5 = 0.68 L_0

Solution

Biodegradable solids reaching the digester = 0.64 × 1200 kg/d = 768 kg/d

Ultimate BOD, L_0 reaching the digester = 1.42 × 768 kg/d = 1091 kg/d

BOD_5 reaching the digester = 0.68 × 1091 kg/d = 742 kg/d

5.4.2 Nitrogenous Oxygen Demand

Nitrogen and phosphorus are nutrients required for growth of living systems. The sources of these nutrients in wastewater are proteins, amino acids, and urea. Nitrogen exists in various states: organic, ammonia, nitrite, and nitrate nitrogen. Microorganisms bring about these conversions as expressed by Equations 5.14 through 5.16.

$$\text{Proteins (or organic nitrogen)} \xrightarrow{\text{Heterotrophes}} NH_3 + H_2O \quad (5.14)$$

$$NH_3 + \frac{3}{2}O_2 \xrightarrow{\text{Nitrosomonas}} HNO_2 + H_2O \quad (5.15)$$

$$HNO_2 + \frac{1}{2}O_2 \xrightarrow{\text{Nitrobacter}} HNO_3 \quad (5.16)$$

The oxygen required for oxidation of ammonia to nitrite and then to nitrate is called nitrogenous biochemical oxygen demand (NBOD), simply nitrogenous oxygen demand (NOD), or second-stage

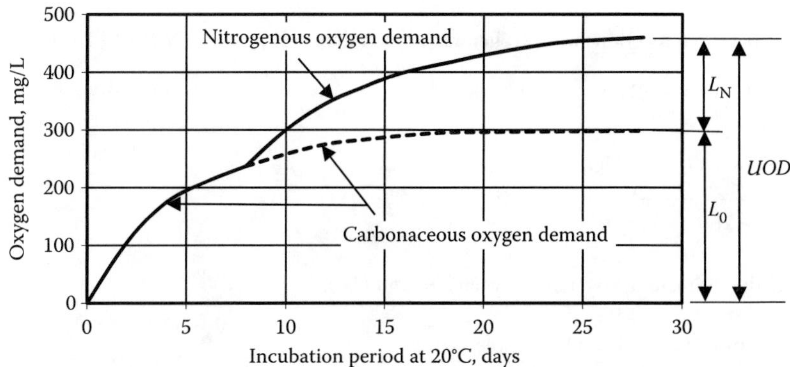

FIGURE 5.8 Typical BOD curve for domestic wastewater showing carbonaceous and nitrogenous oxygen demands.

BOD. In a BOD bottle, nitrification normally starts after 6 to 10 days. Nitrification may start much earlier if the population of nitrifies is initially high. This will cause serious interference with the BOD test. Many inhibitory chemicals are added in the BOD bottle to suppress the nitrification reaction.[1,5,10] A typical BOD curve with nitrification reaction is shown in Figure 5.8. The ultimate oxygen demand (UOD) is the total oxygen demand due to carbonaceous BOD (L_0) and nitrogenous BOD (L_N). This is expressed by Equation 5.17.

$$UOD = L_0 + L_N \tag{5.17}$$

where
UOD = ultimate oxygen demand, mg/L
L_0 = ultimate carbonaceous oxygen demand, or first-stage BOD, mg/L
L_N = ultimate nitrogenous oxygen demand, or second-stage BOD, mg/L

EXAMPLE 5.25: CARBONACEOUS AND NITROGENOUS OXYGEN DEMAND

An organic waste sample has molecular formula $C_5H_7O_2N$. Its concentration in a waste stream is 250 mg/L. Calculate carbonaceous, nitrogenous, and ultimate oxygen demands in mg/L.

Solution

1. Write the balanced reaction for carbonaceous oxygen demand.

$$C_5H_7O_2N + 5\,O_2 \rightarrow NH_3 + 5\,CO_2 + 2\,H_2O$$

2. Calculate the molecular weight of the waste material.

$$C_5H_7O_2N = (5 \times 12 + 7 \times 1 + 2 \times 16 + 1 \times 14)\text{ g/mole} = 113\text{ g/mole waste}$$

3. Calculate the ultimate carbonaceous oxygen demand (L_0).

$$L_0 = \frac{5\text{ mole }O_2}{\text{mole waste}} \times \frac{32\text{ g }O_2}{\text{mole }O_2} \times \frac{\text{mole waste}}{113\text{ g waste}} \times 250\text{ mg/L waste} = 354\text{ mg/L}O_2$$

4. Write the balanced reactions for nitrogenous oxygen demand.

$$NH_3 + \frac{3}{2}O_2 \rightarrow HNO_2 + H_2O$$

$$HNO_2 + \frac{1}{2}O_2 \rightarrow HNO_3$$

$$\overline{NH_3 + 2O_2 \rightarrow HNO_3 + H_2O}$$

5. Calculate the ultimate nitrogenous oxygen demand (L_N).

$$L_N = \frac{2\text{ mole }O_2}{\text{mole waste}} \times \frac{32\text{ g }O_2}{\text{mole }O_2} \times \frac{\text{mole waste}}{113\text{ g waste}} \times 250\text{ mg/L waste} = 142\text{ mg/L }O_2$$

6. Calculate the ultimate oxygen demand UOD.

$UOD = (354 + 142)\text{ mg/L} = 496\text{ mg/L}$

EXAMPLE 5.26: GENERAL UOD EQUATION FOR MUNICIPAL WASTEWATER

The UOD of a wastewater is expressed by Equation 5.18a. Calculate the values of constants A and B. The reaction rate constant k (base e) is 0.23 d^{-1}.

$$UOD = A(BOD_5) + B(NH_3\text{-}N) \tag{5.18a}$$

Solution

1. Determine the value of constant A.

Carbonaceous oxygen demand, $L_0 = A\ (BOD_5)$ or $A = L_0/BOD_5$

Use Equation 5.7.

$y_t = L_0(1 - e^{-kt})$

Substitute $y_5 = BOD_5$, $k = 0.23\text{ d}^{-1}$, and $t = 5\text{ d}$ in Equation 5.7.

$$BOD_5 = L_0\,(1 - e^{-0.23\text{d}^{-1}\times 5\text{d}}) = L_0\,(1 - e^{-1.15}) = L_0\,(1 - 0.32) = 0.68\,L_0$$

$$A = \frac{L_0}{BOD_5} = \frac{1}{0.68} = 1.5\text{ g }O_2/\text{g }BOD_5$$

$$L_0 = 1.5\,(BOD_5) \tag{5.18b}$$

2. Determine the value of constant B.

Nitrogenous oxygen demand, $L_N = B\ (NH_3\text{-}N)$ or $B = L_N/NH_3\text{-}N$

$$NH_3 + 2\,O_2 \rightarrow HNO_3 + H_2O$$

$$B = \frac{L_N}{NH_3\text{-}N} = \frac{2 \times 32\text{ g }O_2/\text{mole}}{14\text{ g }NH_3\text{-N/mole}} = 4.57\text{ g }O_2/\text{g }NH_3\text{-N}$$

$$L_N = 4.57\,(NH_3\text{-}N) \tag{5.18c}$$

3. Express the generalized UOD equation for municipal wastewater.

The generalized equation is expressed in Equation 5.18d.

$$UOD = 1.5\,(BOD_5) + 4.57\,(NH_3\text{-}N) \tag{5.18d}$$

EXAMPLE 5.27: UOD OF MUNICIPAL WASTEWATER

The BOD_5 and NH_3-N concentrations in a municipal wastewater are 210 and 30 mg/L, respectively. Determine UOD.

Solution

The UOD of municipal wastewater is calculated from Equation 5.18d.

$$UOD = 1.5(BOD_5) + 4.57(NH_3\text{-}N) = 1.5 \times 210\,\text{mg/L} + 4.57 \times 30\,\text{mg/L} = (315 + 137)\,\text{mg/L} = 452\,\text{mg/L}$$

EXAMPLE 5.28: INTERFERENCE OF NITRIFICATION IN BOD_5 TEST

The electrolytic respirometer test data on an industrial wastewater sample at 20°C are given in Figure 5.9. Determine (a) L_0, and k for carbonaceous BOD, and BOD5, and (b) ultimate nitrogenous oxygen demand L_n, and *UOD*.

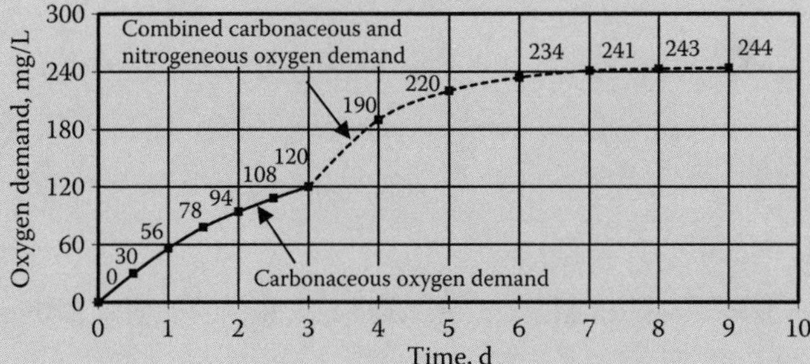

FIGURE 5.9 Respirometer test results showing nitrification from the third day (Examples 5.28 and 5.29).

Solution

1. Review of respirometer test data.

   As shown in Figure 5.9, the respirometer test results clearly indicate that the first-order carbonaceous BOD curve (portion as solid line). However, there was serious interference due to nitrification from the third day (portion as dash line). The reaction rate constants and ultimate oxygen demands values for carbonaceous and nitrogenous oxygen are developed using these two portions of the curve.

2. Tabulate the oxygen demand data within the first 3 days.

   Tabulate 0–3-d carbonaceous oxygen demand values at 0.5-d intervals.

Respirometer Test Time, day	Carbonaceous Oxygen Demand (y), mg/L
0	0
0.5	30
1	56
1.5	78
2	94
2.5	108
3	120

3. Plot the carbonaceous oxygen demand (y) data.

   Use Fujimoto method to draw a line between y_t and y_{t+1} values. Also, draw a line with slope = 1 from the origin in Figure 5.10.

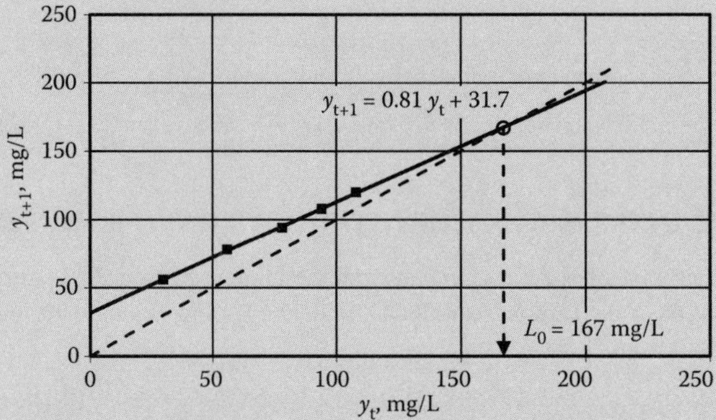

FIGURE 5.10 Determination of k and L_0 for carbonaceous oxygen demand (Example 5.28).

4. Determine L_0, and k for carbonaceous oxygen demand, and BOD_5.

   L_0 from Figure 5.10 is 167 mg/L.

   Use Equation 5.7.

   $y_t = L_0(1 - e^{-kt})$

   Use the third day oxygen demand value to calculate k. Substitute $y_3 = 120$ mg/L and $t = 3$ d in Equation 5.7 and solve for k.

$$120\text{ mg/L} = 167\text{ mg/L} \times (1 - e^{-k \times 3\text{ d}})$$

$$k = 0.42\text{ d}^{-1}$$

$$BOD_5 = 167\text{ mg/L} \times (1 - e^{-0.42\text{ d}^{-1} \times 5\text{ d}}) = 147\text{ mg/L}$$

The BOD_5 and L_0 are illustrated in Figure 5.11.

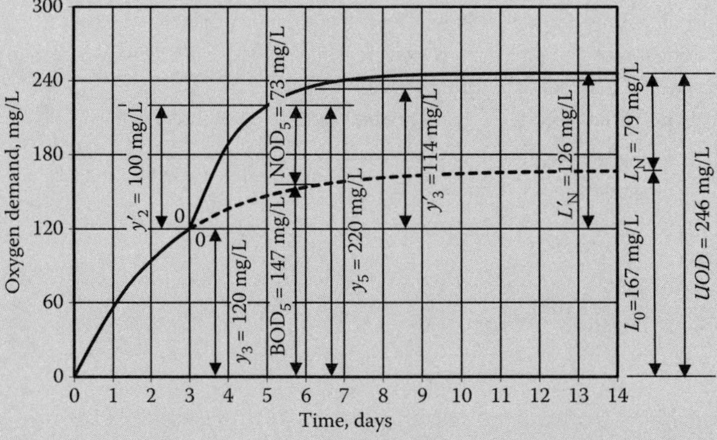

FIGURE 5.11 Illustration of carbonaceous and nitrogenous oxygen demands with BOD_5, L_0, L_N, and UOD (Example 5.28 and 5.29).

5. Tabulate the oxygen demand data beyond 3 days.

   Tabulate the combined carbonaceous and nitrogenous oxygen demand beyond 3 days. Shift origin to day 3 and tabulate oxygen demand data at 1-day intervals.

Time, day		Combined Carbonaceous and Nitrogenous Oxygen Demand, mg/L	
Duration of Test	Beyond 3 Days	Overall (y)	Beyond 120 mg/L (y')
3	$3-3=0$	120	$120-120=0$
4	$4-3=1$	190	$190-120=70$
5	$5-3=2$	220	$220-120=100$
6	$6-3=3$	234	$234-120=114$
7	$7-3=4$	241	$241-120=121$
8	$8-3=5$	243	$243-120=123$
9	$9-3=6$	244	$244-120=124$

6. Plot combined carbonaceous and nitrogenous oxygen demand (y') data.

   Use Fujimoto method to draw a line between y'_t and y'_{t+1} values from the shifted origin, and a line with slope = 1 in Figure 5.12.

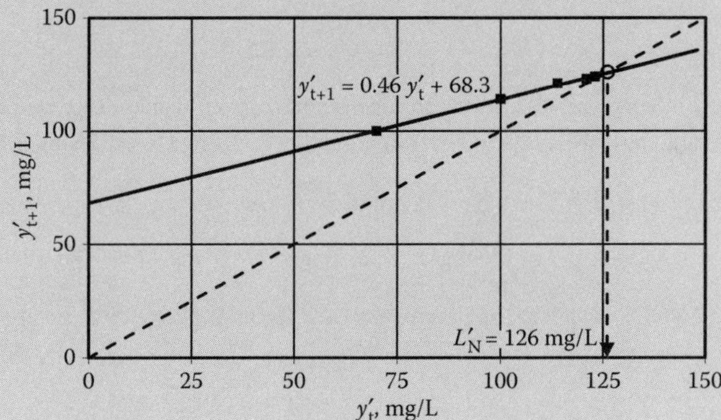

FIGURE 5.12 Determination of k' and L'_N for combined carbonaceous and nitrogenous oxygen demand (Example 5.28).

7. Determine L'_N and k' for combined carbonaceous and nitrogenous oxygen demand beyond 120 mg/L.

   From Figure 5.12, $L'_N = 126$ mg/L.

   Use Equation 5.7.

   $$y_t = L_0(1 - e^{-kt})$$

   Use the third day oxygen demand value (y'_3) to calculate k', which is the reaction rate constant for combined nitrification, and carbonaceous oxygen demand beyond 120 mg/L. This will be the sixth day of the respirometric data, and is 114 mg/L.

   Substitute $y'_3 = 114$ mg/L and $t = 3$ d in Equation 5.7 and solve for k'.

   $$114\text{ mg/L} = 126\text{ mg/L} \times (1 - e^{-k' \times 3\text{ d}})$$

   $$k' = 0.78\text{ d}^{-1}$$

8. Determine the ultimate nitrogenous oxygen demand L_N and UOD.

   L'_N includes the combined nitrogenous and carbonaceous oxygen demands beyond the first 3 days. Therefore,

   $$L_N = (120 + L'_N) - L_0 = (120 + 126 - 167)\ \text{mg/L} = 79\ \text{mg/L}$$

   $$UOD = L_0 + L_N = (167 + 79)\ \text{mg/L} = 246\ \text{mg/L}$$

   The L_N and UOD are also illustrated in Figure 5.11.

EXAMPLE 5.29: CARBONACEOUS AND NITROGENOUS OXYGEN DEMANDS ON A GIVEN DAY

Calculate BOD_5 and nitrogenous oxygen demand exerted on the fifth day (NOD_5) for the waste sample presented in Example 5.28. Use the reaction rate constants, and ultimate carbonaceous oxygen demand and nitrogenous oxygen demand values developed in Example 5.28.

Solution

1. Calculate BOD_5.

   Substitute $L_0 = 167$ mg/L, $k = 0.42\ \text{d}^{-1}$, and $t = 5$ d in Equation 5.7 to calculation BOD_5.

   $$BOD_5 = 167\ \text{mg/L} \times (1 - e^{-0.42\,\text{d}^{-1} \times 5\,\text{d}}) = 147\ \text{mg/L}$$

2. Calculate NOD_5.

   Calculate the combined carbonaceous and nitrogenous oxygen demands exerted beyond 120 mg/L between the third and the fifth days ($t = 2$ d) from Equation 5.7 using $L'_N = 126$ mg/L and $k' = 0.78\ \text{d}^{-1}$.

   $$y'_2 = 126\ \text{mg/L} \times (1 - e^{-0.78\,\text{d}^{-1} \times 2\,\text{d}}) = 100\ \text{mg/L}$$

   $$NOD_5 = y'_2 - (BOD_5 - y_3) = 100\ \text{mg/L} - (147\ \text{mg/L} - 120\ \text{mg/L}) = 73\ \text{mg/L}$$

   The BOD_5 and NOD_5 as well as the parameters used in the above calculation are also illustrated in Figure 5.11.

5.4.3 Chemical Oxygen Demand

Chemical oxygen demand (COD) represents the amount of oxygen required to oxidize the organic matter by a strong oxidizing chemical in acid solution. There is generally a linear relationship between COD and BOD_5 results. This relationship depends entirely on the composition of the wastewater. Standard potassium dichromate and potassium permanganate oxidation tests are in common use.

Standard Dichromate Oxidation: This test is performed by digesting for 2 h a measured sample. The digestion reagent contains a known excess potassium dichromate ($K_2Cr_2O_7$) solution and sulfuric acid. The organic matter is oxidized and potassium dichromate is reduced. The remaining potassium dichromate is titrated with standard ferrous ammonium sulfate ($Fe(NH_4)_2 \cdot (SO_4)_2 \cdot 6H_2O$) reagent (FAS).[1,11] Additional information on COD consumption and biomass growth may be found in Chapter 10, Examples 10.7 through 10.11, and 10.70.

Permanganate Oxidation: This test measures the easily oxidizable organic matter. The wastewater sample is oxidized for 10 min with excess potassium permanganate ($KMnO_4$) in a basic solution. The excess $KMnO_4$ is neutralized by excess ferrous ammonium sulfate solution. The excess ferrous ammonium sulfate is then titrated with $KMnO_4$ solution until pink color returns.

EXAMPLE 5.30: COD DETERMINATION

COD test was conducted on a wastewater sample. Ten milliliters sample, 6-mL potassium dichromate solution, and 14 mL of sulfuric acid digestion reagent were mixed in a flask. A blank was prepared with 10 mL distilled water, and all other reagents in same amounts as in the sample. The sample and blank flasks were refluxed for 2 h. After cooling to room temperature, the prepared samples were titrated with 0.1 M ferrous ammonium sulfate (FAS) titrant using ferroin indicator. The endpoint was reached by sharp color change from blue-green to reddish brown. Amounts of FAS standard used for blank and sample were 9.1 and 3.8 mL. Determine the COD of the sample in mg/L.

Solution

1. Use Equation 5.19 to calculate COD.

$$\mathrm{COD} = \frac{(A - B) \times M \times 8\,\mathrm{g\ O_2/mole\ FAS} \times 10^3\,\mathrm{mg/g}}{V} \tag{5.19}$$

where

COD = COD concentration in the sample ($\mathrm{mg{\cdot}O_2/L}$)
A and B = volume of FAS used for blank and sample (mL)
M = molarity of FAS (mole/L)
V = sample volume (mL)

2. Calculate COD.

$$\mathrm{COD} = \frac{(9.1 - 3.8)\,\mathrm{mL} \times 0.1\,\mathrm{mole/L\ FAS} \times 8\,\mathrm{g\,O_2/mole\ FAS} \times 10^3\,\mathrm{mg/g}}{10\,\mathrm{mL\ sample}} = 424\,\mathrm{mg/L}$$

EXAMPLE 5.31: RELATIONSHIP BETWEEN COD AND BOD_5 OF RAW MUNICIPAL WASTEWATER

BOD_5 and COD tests were conducted on 10 samples of same raw wastewater obtained on different days. The results are summarized below. Determine the COD/BOD_5 ratio and comment on result.

BOD_5, mg/L	191	135	140	120	229	193	150	102	238	212
COD, mg/L	418	298	309	266	499	422	330	228	518	462

Solution

1. Plot the data.

   Plot the COD versus BOD_5 values in Figure 5.13.

2. Develop the relationship.

   The slope and intercept of the linear relationship are obtained.

   Slope or $\left(\frac{\mathrm{COD}}{\mathrm{BOD_5}}\ \mathrm{ratio}\right) = 2.13$ and intercept $= 10.1$ mg/L

   The linear relationship is expressed by the following equation:

   $$\mathrm{COD} = 2.13\ \mathrm{BOD_5} + 10.1$$

   Correlation ratio, $r^2 = 1$

3. Comments on the result.

   In raw wastewater, a significant portion of organic matter is biodegradable. The ratio of COD to BOD_5 is slightly >2. Also, at $BOD_5 = 0$ mg/L, the concentration of COD is about 10 mg/L. It indicates the residual organic matter exerts COD.

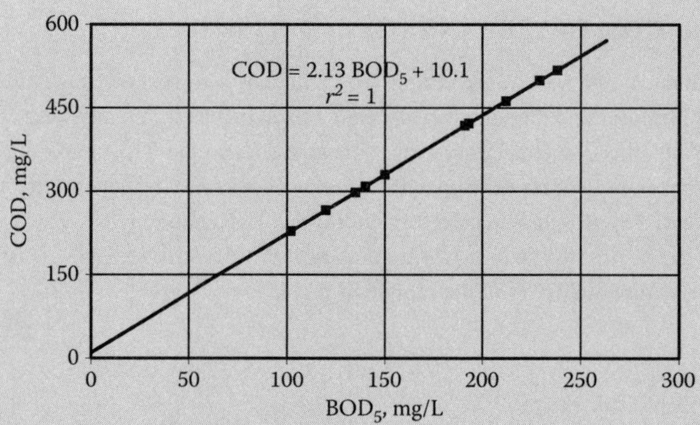

FIGURE 5.13 BOD_5 and COD relationship in raw municipal wastewater (Example 5.31).

EXAMPLE 5.32: RELATIONSHIP BETWEEN BOD_5 AND COD TESTS OF SECONDARY EFFLUENT

Secondary effluent contains low biodegradable and high nonbiodegradable organic matters. As a result, the BOD is low while COD is high, and COD to BOD_5 ratio is also high. A secondary wastewater treatment plant is achieving nitrification. Six samples of well-nitrified effluent on different dates were analyzed for total BOD_5 and COD. The results are summarized below. Develop a statistical linear relationship and ratio of COD and BOD_5. Comment on COD value in the effluent when BOD_5 is very low. Also, compare the COD to BOD_5 ratios in influent (Example 5.31) and effluent.

BOD_5, mg/L	10.2	11.0	7.5	6.5	5.5	8.2
COD, mg/L	55.0	57.7	45.8	40.1	39.0	48.3

Solution

1. Plot the COD versus BOD_5 values in Figure 5.14.

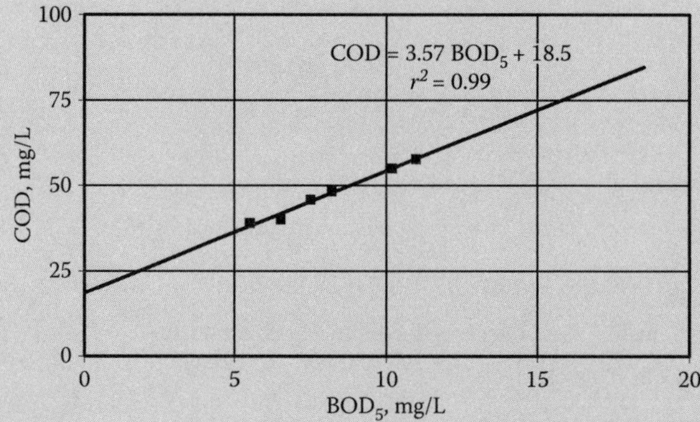

FIGURE 5.14 BOD_5 and COD relationship in secondary effluent (Example 5.32).

2. Develop the slope and intercept.

$$\text{Slope or } \left(\frac{\text{COD}}{\text{BOD}_5}\text{ ratio}\right) = 3.57 \text{ and intercept} = 18.5\text{ mg/L}$$

3. Determine the linear relationship.

   The linear relationship is expressed by equation: $\text{COD} = 3.57\ \text{BOD}_5 + 18.5$.

   Correlation ratio, $r^2 = 0.99$

4. Comment on the COD value in the effluent when BOD_5 is low.

   At low BOD_5 concentrations, most of the organic matter in the effluent is nonbiodegradable organics which exert COD. When $BOD_5 = 0$, the entire organic matter is nonbiodegradable and exerts a COD of 18.5 mg/L.

5. Compare the ratios of COD to BOD_5 in influent (Example 5.31) and effluent.

   The ratio of COD to BOD_5 in the influent is approximately 2.2–2.3 (Figure 5.13), and that in the effluent ranges from 2.5 to 7.1 (see Figure 5.14). A higher ratio means more nonbiodegradable organics. This is true as effluent has very little biodegradable organics remaining.

5.4.4 Total Organic Carbon

The total organic carbon (TOC) test is performed very rapidly and is becoming a popular test for small concentrations of organic matter.[12] It is measured by injecting a known quantity of sample into a high-temperature furnace or by wet oxidation. The carbon dioxide that is produced is measured quantitatively. Another measure of organic matter is theoretical organic carbon (ThOC).[1,5,6,11]

5.4.5 Total Oxygen Demand and Theoretical Oxygen Demand

Total oxygen demand (TOD) measurement is based on conversion of organic compounds into stable oxides in a platinum-catalyzed combustion chamber. The measurement can be completed in 3 min.

Both ThOD and ThOC are the measurements based on stoichiometric relationships. This is possible only if the chemical formula of the organic compound is known. Due to unknown chemicals in the wastewater, use of these parameters is very limited.[5,6,8,11,13]

5.4.6 Relationship between BOD_5 and Other Tests Used for Organic Content

COD, TOC, and TOD tests are used for rapid measurement of organic strength of wastewater. For this reason, these tests are gaining popularity. Similar to Examples 5.31 and 5.32, a statistical relationship or ratio between BOD_5 and any of these tests can be established by laboratory testing. This ratio will vary greatly depending upon the characteristics of wastewater and the degree of treatment the wastewater may have undergone. Therefore, statistical relationships must be developed for each case.[8,11,13] Generalized ratios of these rapid tests and BOD_5 results for raw wastewater are given in Examples 5.33 through 5.35.

EXAMPLE 5.33: CALCULATION OF ThOD, ThOC, AND ThOD/ThOC RATIO

An industrial waste stream contains 54 mg/L glycine ($CH_2(NH_2)COOH$). Calculate ThOD, ThOC, and ThOD/ThOC ratio.

Solution

1. Write the balanced equation for complete oxidation of glycine in Equation 5.20.

$$CH_2(NH_2)COOH + \frac{7}{2}O_2 \rightarrow HNO_3 + 2\,CO_2 + 2H_2O \tag{5.20}$$

2. Determine the molecular weight of glycine.

   The molecular weight of glycine = 75

3. Calculate ThOD.

$$\text{Oxygen consumed per mole of glycine} = \frac{3.5\text{ mole } O_2}{\text{mole glycine}} \times \frac{32\,g\,O_2}{\text{mole } O_2} = 112\,g\,O_2/\text{mole glycine}$$

$$\text{ThOD} = \frac{112\,g\,O_2}{\text{mole glycine}} \times \frac{\text{mole glycine}}{75\,g\text{ glycine}} \times 54\text{ mg/L glycine} = 81\text{ mg/L } O_2$$

4. Calculate the theoretical organic carbon (ThOC).

$$\text{ThOC} = \frac{2 \times 12\,g\,C}{75\,g\text{ glycine}} \times 54\text{ mg/L glycine} = 17\text{ mg/L C}$$

5. Calculate the ratio of ThOD to ThOC.

$$\frac{\text{ThOD}}{\text{ThOC}} = \frac{81\text{ mg/L}O_2}{17\text{ mg/L C}} = 4.8\,g\,O_2/g\text{ C}$$

EXAMPLE 5.34: RELATIONSHIP BETWEEN VARIOUS PARAMETERS AND BOD_5 TEST FOR RAW MUNICIPAL WASTEWATER

A number of studies have been conducted to establish the relationship between BOD_5 and many other tests commonly used for municipal wastewater. Tabulate these results.

Solution

The relationship between ThOD, UOD, TOD, COD, TOC, and BOD_5 are obtained from many sources. These relationships are summarized in Table 5.6.

TABLE 5.6 Concentrations of ThOD, UOD, TOD, COD, TOC and BOD_5 in Raw Municipal Wastewater and Their BOD_5 Ratios (Example 5.34)

Test	Concentration, mg/L		BOD_5 Ratio
	Range	Typical	
ThOD	400–430	410	1.95
UOD	390–405	400	1.90
TOD	370–390	380	1.81
COD (standard dichromate)	330–350	340	1.62
COD (rapid $KMnO_4$)	280–295	285	1.36
TOC	190–280	250	1.19
BOD (ultimate)	286–330	300	1.43
BOD_5	190–220	210	1.0

Source: Adapted in part from Reference 13.

EXAMPLE 5.35: ThOD AND TOC RELATIONSHIPS FOR PURE CHEMICALS

Tabulate the ratios of calculated ThOD/ThOC and measured COD/TOC ratios of some commonly used industrial chemicals.

Solution

The ratios of calculated ThOD and ThOC and measured COD/TOC for many substances are summarized in Table 5.7.

TABLE 5.7 Calculated and Measured Ratios of ThOD/ThOC and COD/TOC of Different Compounds (Example 5.35)

Substance	ThOD/ThOC (Calculated)	COD/TOC (Measured)
Acetone	3.56	2.44
Ethanol	4.00	3.35
Phenol	3.12	2.96
Benzene	3.34	0.84
Pyridine	3.33	–
Salicylic acid	2.86	2.86
Methanol	4.00	3.89
Benzoic acid	2.86	2.90
Sucrose	2.67	2.44

5.4.7 Other Nonspecific and Specific Tests for Organic Contents

Other common nonspecific tests for organics are UV absorbance and fluorescence. Many specific tests utilize gas chromatography and mass spectroscopy, and are used to identify different organic compounds. These methods are summarized in Table 5.8.[1,11,14]

TABLE 5.8 Some Nonspecific and Specific Analytical Methods for Measurement and Identification of Organic Compounds

Measurement	Description
	Nonspecific
Color	Most organic matters in water cause color. Therefore, color measurement is used to quantify organics. Colorimetric analysis is performed to measure the color.
UV absorbance	Ultraviolet absorbance at a specific wave length (UV 254 nm) is used to quantify groups of organic compounds such as aliphatic, aromatic, complex multiaromatic, and multiconjugated humic substances.
Fluorescence	Some organic compounds absorb UV energy and then release energy at some longer wavelength. This phenomenon provides a basis for measurement of organics.
	Specific
Gas chromatography	The sample is vaporized and swept by a carrier gas through a chromatographic column. The emergence of the compound is detected and measured.
Mass spectrometry	The sample is vaporized and the compounds are separated by gas chromatography. The bombardment of the organic molecules by the rapidly moving electrons breaks the organic molecule into charged fragments. The mass-to-charge ratio of each fragment provides quantitative analysis.
High-pressure liquid chromatography	The carrier stream is composed of a solvent or mixture of solvents maintained under high pressure. The compounds are separated in a solid or liquid stationary phase and measured.

5.4.8 Nonbiodegradable and Toxic Compounds

Most inorganic and organic compounds in municipal wastewater come from water supply and human wastes. Nitrogen and phosphorus are macronutrients. Many heavy metals, cyanide, and asbestos are toxic inorganics.

Many organic compounds are nonbiodegradable and toxic. Examples of these compounds are solvents, pesticides, and herbicides. These compounds are discussed under *priority pollutants* in Section 5.6.

5.5 Microbiological Quality

The municipal wastewater contains microorganisms that play an important role in biological waste treatment and public health and safety. The principal groups of microorganisms of significance in wastewater treatment include bacteria, fungi, protozoa, and algae.

5.5.1 Basic Concepts

The principal groups of organisms found in water and wastewater are *eukaryotes* and *prokaryotes*. *Eukaryotes* include algae, fungi, protozoa, and multicellular plants and animals. Each cell has a clearly defined nucleus containing nucleic acid (deoxyribonucleic acid [DNA]). The DNA has genetic information that is vital for the cell reproduction. *Prokaryotes* (mainly bacteria and blue-green algae) have poorly defined nuclei. Viruses fall between living and nonliving. They are not complete organisms, being made up of protein-protective coating surrounding a strand of nucleic acid.

TABLE 5.9 Basic Description of Organisms in Water and Wastewater

Types of Organisms	Description
Bacteria	Bacteria play very important role in wastewater treatment. They are single-celled prokaryotic eubacteria. Based on morphology bacteria are (1) spherical (*cocci*) 1–3 μm, (2) rod shaped (*bacilli*) 0.3–1.5 μm (dia.) and 1–10 μm long, (3) curved rod-shaped (*vibros*) 0.6–1 μm (in diameter) and 1–6 μm long, (4) spiral (*spirilla*) 50 μm long, and (5) filamentous 100 μm long. The pathogenic bacteria that are excreted by humans may cause many diseases. Bacteria-caused common water borne diseases are gastroenteritis, typhoid fever, dysentery, diarrhea, cholera.
Fungi	Fungi are aerobic, single to multicellular, nonphotosynthetic, heterotrophic *eukaryotic protists*. Most fungi and molds are *saprophytes* (obtain food from dead organic matter). Fungi, along with bacteria, play principal role in waste treatment. Fungi can compete and perform better than bacteria at lower pH, nutrients, and moisture in wastes.
Algae	Algae are autotrophic and photosynthetic, and contain chlorophyll. Algae play an important role in waste stabilization ponds.
Protozoa	Protozoa are a group of unicellular, nonphotosynthetic, aerobic organisms. Many protozoa cause disease. *Entamoeba histolytica* (amoebic dysentery), *Giardia lamblia* (giardiasis), and *Cryptosporidium parvum* (cryptosporidiosis) are transmitted by drinking water, while *Naegleria fowleri* may enter by nasal inhalation exposure from swimming in polluted water and causes amoebic meningoencephalitis. Protozoa feed upon bacteria and their presence in wastewater treatment plants indicates healthy operation.
Virus	Viruses are obligate parasites and, as such, require a host. Common virus-caused diseases are infectious hepatitis, gastroenteritis, and respiratory diseases.
Plants and animals	Plants and animals ranging in size from microscopic to larger are important in waste treatment and effluent quality control. Larger aquatic plants are used in constructed wetlands. Many smaller animals also play an important role in natural purification and food chain. Rotifers and crustacean are lower order animals that prey on bacteria, protozoa, and algae. They help to maintain a balance in population of primary producers, and become a part of the food chain. Sludge worms and other insect larvae feed on sludge and organic debris. Parasitic worms (*helminths*) cause many waterborne diseases such as intestinal roundworm, Guinea worm, lung fluke, and schistosomiasis.

TABLE 5.10 Basic Classification of Organisms Based on Nutritional and Environmental Requirements

Classification and Type Based on Nutritional and Environmental Requirements	Description
	Carbon Source
Heterotrophic	Organisms that use organic carbon for cell synthesis
Autotrophic	Organisms that derive their cell carbon from carbon dioxide
	Energy Source
Phototrophic	Organisms that derive energy for cell synthesis from sunlight
Chemotrophic	Organisms that derive energy from chemical reactions
	Oxygen Requirements
Aerobic	Organisms that require molecular oxygen (O_2) for their metabolism
Anaerobic	Microorganisms that grow in absence of molecular oxygen. For strict anaerobes, molecular oxygen may be toxic
Facultative	Microorganisms grow in presence or absence of molecular oxygen
	Temperature Requirement
Psychrophilic or cryophilic	Optimum temperature range 12–18°C
Mesophilic	Optimum temperature range 25–40°C
Thermophilic	Optimum temperature range 55–65°C

Brief discussion of bacteria, fungi, algae, protozoa, virus, and plants and animals is provided in Table 5.9. Different organisms have different nutritional and environmental requirements. These requirements include carbon source, energy source, oxygen, temperature, and others. Basic classifications of organisms based on carbon source, energy source, and oxygen and temperature requirement are summarized in Table 5.10.[4–6,15,16]

EXAMPLE 5.36: COMMON WATERBORNE DISEASES AND CAUSATIVE AGENTS

Two lists are given below. Match the terms.

	List A		List B
i	Cholera	a	Virus
ii	Typhoid fever	b	Helminths
iii	Dysentery	c	*G. lamblia*
iv	Hepatitis	d	Bacteria
v	Schistosomiasis	e	*Entamoeba histolytica*
vi	Giardiasis	f	Continued fever

EXAMPLE 5.37: CLASSIFICATION BASED ON NUTRITIONAL AND ENVIRONMENTAL REQUIREMENTS OF MICROORGANISMS

Match the terms in lists A and B.

	List A		List B
i	Eukaryotic cell	a	Need oxygen
ii	Phototrophs	b	12–18°C
iii	Prokaryotic cell	c	Do not need oxygen
iv	Aerobic	d	Need organic carbon
v	Heterotrophic	e	Genetic information
vi	Anaerobic	f	Nucleus clearly defined
vii	Psychrotrophic	g	Sunlight
viii	Mesophilic	h	Nucleus poorly defined
ix	Actinomycetes	i	25–40°C
x	Protozoa	j	Filamentous
xi	DNA	k	*Guardia lamblia*

Solution

The matching terms are (i–f), (ii–g), (iii–h), (iv–a), (v–d), (vi–c), (vii–b), (viii–i), (ix–j), (x–k), (xi–e).

5.5.2 Indicator Organisms

Water has long served as a mode of transmission of diseases. The most important of the waterborne diseases are those of the intestinal tract. Pathogens (disease-causing bacteria, viruses, protozoa, and parasitic helminths) are excreted in the feces of patients and carriers. In order to determine the presence of pathogenic organisms in water, the microbiologist must have a reliable measurement technique. Unfortunately, the analytical procedures for detection of pathogenic organisms are not clear-cut. Hence, rather than looking for the specific pathogens, there is a need to find a group of indicator organisms that can measure the potential of a water to transmit disease.

Characteristics of indicator organisms: The ideal indicator organism should have the following characteristics:[4–6]

1. Detection should be quick, simple, and reproducible.
2. Results should be applicable to all waters, that is, the numbers should correlate with the degree of pollution (higher numbers in sewage, less in polluted waters, and none in unpolluted waters).
3. The organism should have greater or equal survival time in nature than the pathogens and be present in larger numbers.
4. The organisms should not grow in nature.
5. It should be harmless to humans.

Coliform organisms: Neither organism nor group of organisms possesses all characteristics of indicator organisms. However, the coliform organisms currently used as indicator organisms have many of these qualities. These are nonpathogenic bacteria whose origin is in fecal matter. The presence of these bacteria in water is an indication of fecal contamination and probably unsafe water. The coliform bacteria are gram-negative, nonspore-forming *bacilli* capable of fermenting lactose with the production of acids and gases. The coliform bacteria are members of the *Enterobacteriaceae* family and include genera *Escherichia, Klebsiella, Citrobacter,* and *Enterobacter*. The *Escherichia coli* (*E. coli*) species appears to be the most representative of fecal contamination. Fecal *Streptococci* (FS) and *Enterococci* are also used as indicator organisms under specific conditions.

The coliform organisms were originally believed to be entirely of fecal origin, but it has been shown that certain genera can grow in soil. Therefore, the presence of coliform organisms in surface water may be due to fecal wastes from human and animal sources and from soil erosion. To separate *E. coli* species from possible soil type, special tests are generally run. Frequently, for this purpose, FS are also used as indicator organisms because they also originate from the intestinal tract of warm-blooded animals.

Number of coliform organisms: Each person discharges from 100 to 400 billion coliform organisms per day, in addition to many other types of bacteria. The numbers of specific indicator organisms found in untreated municipal wastewater are: total coliform (TC) 10^5–10^8, fecal coliform (FC) 10^4–10^5, FS 10^3–10^4, and *Enterococci* 10^2–10^3 per mL. Generally, the ratio of FC to FS in a sample can be used to show whether the suspected contamination derives from human or from animal wastes. As a general rule, the FC/FS ratio for domestic animals is <1, whereas this ratio for human beings is >4.

EXAMPLE 5.38: COLIFORM NUMBER IN RAW WASTEWATER

It is estimated that each person discharges 150 billion total coliform organisms per day. If the average wastewater flow is 120 gpcd, what is the average coliform density per mL in raw fresh wastewater?

Solution

$$\text{Average total coliform density} = \frac{150 \times 10^9 \text{ coliform organisms/person·d}}{120 \text{ gallons/person·d} \times 3.8 \text{ L/gallon}}$$

$$= 3.3 \times 10^8 \text{ coliform organisms/L or } 3.3 \times 10^5 \text{ coliform organisms/mL}$$

5.5.3 Measurement Techniques

The standard techniques used to enumerate the coliform organisms are (1) multiple-tube fermentation, and (2) the membrane filter. Both techniques are described below.

Multiple-Tube Fermentation Test: The multiple-tube fermentation technique uses lactose broth media, which is fermented by coliform group. Gas bubbles collected inside an inverted inner vial are an indication of gas formation. The test may be carried out to presumptive, confirmed, or completed test levels. For most routine water and wastewater analyses, only presumptive test is performed. Series of dilutions in multiple tubes are used for statistical enumeration. The result is reported as most probable number per 100 mL (MPN/100 mL). Probability equations or standard MPN index tables are used to determine MPN. The 95% confidence limits for five tubes each at dilutions of 10, 1, and 0.1 mL are provided in Table 5.11.

Membrane Filter Technique: Membrane filter technique is used to quantify the coliform organisms present in water and wastewater samples. This technique is faster, gives the actual number rather than the most probable number, uses larger volume of sample, requires less laboratory apparatus, and has higher degree of reproducibility than the standard multiple-tube technique. Using a standard filtration apparatus, a dilute sample is filtered through a membrane filter having a rated pore diameter of 0.45 μm. The filter is placed over an absorbent pad containing Endo-type selective media in a dish with tight cover, and is incubated at 35°C. The colonies with a golden-green metallic sheen are counted. The procedure requires media, dilution water, and fully sterilized filtration and other apparatus. The procedure is given in the Standard Methods.[1]

It is necessary that an appropriate volume of sample should be selected to give 20–80 colonies per filter for best results. Colonies >80 become too numerous to count accurately. Colonies <20 raise reliability issues. Membrane filtration results are often reported as coliform density that is number of colonies per mL of sample.

Based on the results of many studies, the sample sizes that have been proposed for different water sources are provided in Table 5.12.

TABLE 5.11 Most Probable Numbers (MPNs) of Coliform Bacteria per 100-mL Sample (Tubes Showing Gas of Five Tubes with Three Serial Dilutions, 10-, 1-, and 0.1-mL Samples)

10	1	0.1	MPN	10	1	0.1	MPN	10	1	0.1	MPN	10	1	0.1	MPN	10	1	0.1	MPN	10	1	0.1	MPN
0	0	0	0	1	0	0	2	2	0	0	4.5	3	0	0	7.8	4	0	0	13	5	0	0	23
0	0	1	1.8	1	0	1	4	2	0	1	6.8	3	0	1	11	4	0	1	17	5	0	1	31
0	0	2	3.6	1	0	2	6	2	0	2	9.1	3	0	2	13	4	0	2	21	5	0	2	43
0	0	3	5.4	1	0	3	8	2	0	3	12	3	0	3	16	4	0	3	25	5	0	3	58
0	0	4	7.2	1	0	4	10	2	0	4	14	3	0	4	20	4	0	4	30	5	0	4	76
0	0	5	9	1	0	5	12	2	0	5	16	3	0	5	23	4	0	5	36	5	0	5	95
0	1	0	1.8	1	1	0	4	2	1	0	6.8	3	1	0	11	4	1	0	17	5	1	0	33
0	1	1	3.6	1	1	1	6.1	2	1	1	9.2	3	1	1	14	4	1	1	21	5	1	1	46
0	1	2	5.5	1	1	2	8.1	2	1	2	12	3	1	2	17	4	1	2	26	5	1	2	64
0	1	3	7.3	1	1	3	10	2	1	3	14	3	1	3	20	4	1	3	31	5	1	3	84
0	1	4	9.1	1	1	4	12	2	1	4	17	3	1	4	23	4	1	4	36	5	1	4	110
0	1	5	11	1	1	5	14	2	1	5	19	3	1	5	27	4	1	5	42	5	1	5	130
0	2	0	3.7	1	2	0	6.1	2	2	0	9.3	3	2	0	14	4	2	0	22	5	2	0	49
0	2	1	5.5	1	2	1	8.2	2	2	1	12	3	2	1	17	4	2	1	26	5	2	1	70
0	2	2	7.4	1	2	2	10	2	2	2	14	3	2	2	20	4	2	2	32	5	2	2	95
0	2	3	9.2	1	2	3	12	2	2	3	17	3	2	3	24	4	2	3	38	5	2	3	120
0	2	4	11	1	2	4	15	2	2	4	19	3	2	4	27	4	2	4	44	5	2	4	150
0	2	5	13	1	2	5	17	2	2	5	22	3	2	5	31	4	2	5	50	5	2	5	180
0	3	0	5.6	1	3	0	8.3	2	3	0	12	3	3	0	17	4	3	0	27	5	3	0	79
0	3	1	7.4	1	3	1	10	2	3	1	14	3	3	1	21	4	3	1	33	5	3	1	110
0	3	2	9.3	1	3	2	13	2	3	2	17	3	3	2	24	4	3	2	39	5	3	2	140
0	3	3	11	1	3	3	15	2	3	3	20	3	3	3	28	4	3	3	45	5	3	3	180
0	3	4	13	1	3	4	17	2	3	4	22	3	3	4	31	4	3	4	52	5	3	4	210
0	3	5	15	1	3	5	19	2	3	5	25	3	3	5	35	4	3	5	59	5	3	5	250

(*Continued*)

TABLE 5.11 (*Continued*) Most Probable Numbers (MPNs) of Coliform Bacteria per 100-mL Sample (Tubes Showing Gas of Five Tubes with Three Serial Dilutions, 10-, 1-, and 0.1-mL Samples)

10	1	0.1	MPN	10	1	0.1	MPN	10	1	0.1	MPN	10	1	0.1	MPN	10	1	0.1	MPN	10	1	0.1	MPN
0	4	0	7.5	1	4	0	11	2	4	0	15	3	4	0	21	4	4	0	34	5	4	0	130
0	4	1	9.4	1	4	1	13	2	4	1	17	3	4	1	24	4	4	1	40	5	4	1	170
0	4	2	11	1	4	2	15	2	4	2	20	3	4	2	28	4	4	2	47	5	4	2	220
0	4	3	13	1	4	3	17	2	4	3	23	3	4	3	32	4	4	3	54	5	4	3	280
0	4	4	15	1	4	4	19	2	4	4	25	3	4	4	36	4	4	4	62	5	4	4	350
0	4	5	17	1	4	5	22	2	4	5	28	3	4	5	40	4	4	5	69	5	4	5	430
0	5	0	9.4	1	5	0	13	2	5	0	17	3	5	0	25	4	5	0	41	5	5	0	240
0	5	1	11	1	5	1	15	2	5	1	20	3	5	1	29	4	5	1	48	5	5	1	350
0	5	2	13	1	5	2	17	2	5	2	23	3	5	2	32	4	5	2	56	5	5	2	540
0	5	3	15	1	5	3	19	2	5	3	26	3	5	3	37	4	5	3	64	5	5	3	920
0	5	4	17	1	5	4	22	2	5	4	29	3	5	4	41	4	5	4	72	5	5	4	1600
0	5	5	19	1	5	5	24	2	5	5	32	3	5	5	45	4	5	5	81	5	5	5	N/A

TABLE 5.12 Suggested Range of Sample Sizes That Have Been Proposed for Filtration to Give Coliform Colonies in the Range of 20–80 per Filter

Water Source	Sample Size, mL
Drinking water, swimming pools, wells, and springs	Larger than 100
Lakes and reservoirs	50–100
Bathing beaches	1–10
Rivers	0.1–10
Chlorinated municipal wastewater	0.001–0.1
Raw municipal wastewater	0.00001–0.0001

Note: See Example 5.49 for common error in membrane filtration test.

EXAMPLE 5.39: SERIES DILUTION TO OBTAIN A DESIRED VOLUME OF DILUTED SAMPLE

A wastewater sample is expected to have a high coliform count. Sample dilution is necessary for the coliform test. Prepare a dilution scheme such that 10 mL of diluted sample gives 0.00001 mL of the original sample.

Solution

1. Describe the dilution technique.

   Sample dilution is mostly achieved by transferring 1 mL sample into a series of dilution bottles that contains 99 mL of autoclaved (sterilized) distilled water. The pipets used are also sterilized.

2. Set up the dilution scheme.

   The dilution scheme is shown in Figure 5.15.

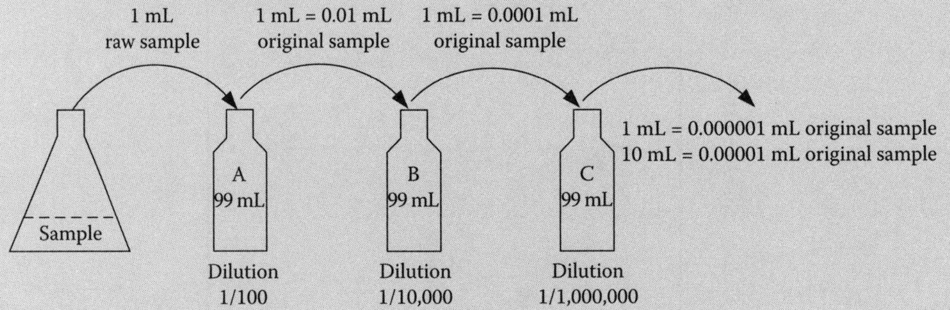

FIGURE 5.15 Series dilution to obtain the desired sample volume (Example 5.39).

3. Determine the sample size.

   The required original sample size of 0.00001 mL is obtained by withdrawing samples from bottles C.

   10 mL of diluted sample from bottle C × ($1/10^6$ dilution factor) = 0.00001 mL original sample.

   Also, 0.1 mL diluted sample from bottle B × ($1/10^4$ dilution factor) = 0.00001 mL original sample.

EXAMPLE 5.40: DILUTION SCHEME TO ACHIEVE A REQUIRED NUMBER OF ORGANISMS

A wastewater sample is expected to have 10^9 coliform organisms per 100 mL. Prepare a dilution sequence, and calculate the volume of diluted sample that will give 100 coliform organisms.

Solution

1. Prepare the dilution scheme using 99 mL dilution bottles

   The dilution scheme is shown in Figure 5.16.

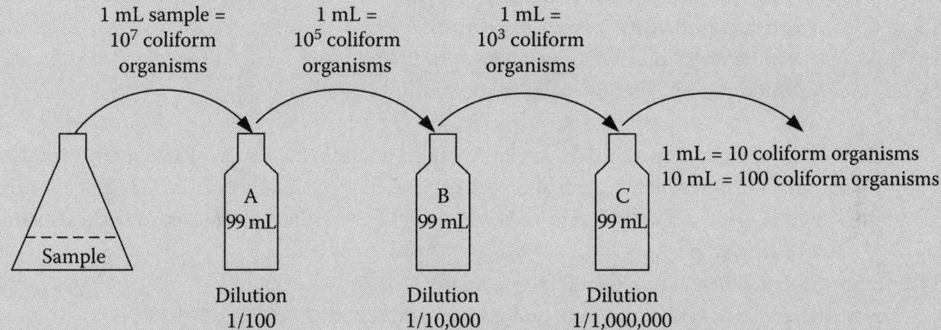

FIGURE 5.16 Series dilution to obtain a desired number of organisms (Example 5.40).

2. Determine the required sample size.

   Three dilutions of original sample are used. The dilutions are such that certain value of diluted sample withdrawn from bottles B or C will give 100 coliform organisms.

   a. Bottle C

$$\text{Sample volume required from bottle C} = \frac{100 \text{ organisms required}}{10^7 \text{ organisms/mL}} \times \frac{10^6}{1} \text{ dilution factor} = 10 \text{ mL}$$

   b. Bottle B

$$\text{Sample volume required from bottle B} = \frac{100 \text{ organisms required}}{10^7 \text{ organisms/mL}} \times \frac{10^4}{1} \text{ dilution factor} = 0.1 \text{ mL}$$

EXAMPLE 5.41: MPN USING PROBABILITY FORMULA (POISSON DISTRIBUTION)

A set of five lactose-broth fermentation tubes at three dilutions of 1 mL, 0.1 mL, and 0.01 mL (a total of 15 fermentation tubes) were prepared from a wastewater effluent. After incubation at 35°C ± 0.5°C for 24 ± 2 h, the following gas production results were obtained.

1 mL dilution—5 positive gas production tubes (5/5)
0.1 mL dilution—4 positive gas production tubes (4/5)
0.01 mL dilution—2 positive gas production tubes (2/5)

Calculate MPN/100 mL using probability formula based on Poisson distribution

Solution

1. State the probability equation.

The probability formula based on Poisson distribution is used for calculating MPN. The result for three dilutions is expressed by Equation 5.21.[17]

$$y = \frac{1}{k}\left[\left(1 - e^{-n_1\lambda}\right)^{p_1}\left(e^{-n_1\lambda}\right)^{q_1}\right]\left[\left(1 - e^{-n_2\lambda}\right)^{p_2}\left(e^{-n_2\lambda}\right)^{q_2}\right]\left[\left(1 - e^{-n_3\lambda}\right)^{p_3}\left(e^{-n_3\lambda}\right)^{q_3}\right] \quad (5.21)$$

where

y = 95% probability of occurrence of a given result
k = constant for a given set of conditions
n_1, n_2, n_3 = sample size in each dilution, mL
λ = coliform density, organisms per mL
p_1, p_2, p_3 = number of positive tubes in each sample dilution
q_1, q_2, q_3 = number of negative tubes in each sample dilution

It should be noted that Equation 5.21 can be expanded to utilize the results of any numbers of dilutions. The number of factorials in the equation will depend upon the number of dilutions. For example, with three dilutions, the factorials are three. With one or four dilutions, the factorials in the equation will be one and four, respectively.

2. Substitute the experimental results into the equation.

The multiple tube fermentation results are generally expressed as 5/5, 4/5, 2/5, or 5-4-2. Substitute the following values in the probability equation.

$p_1 = 5, p_2 = 4, p_3 = 2, q_1 = 0, q_2 = 1, q_3 = 3$

The probability equation after substitution becomes

$$y \cdot k = \left[\left(1 - e^{-1\lambda}\right)^5\left(e^{-1\lambda}\right)^0\right]\left[\left(1 - e^{-0.1\lambda}\right)^4\left(e^{-0.1\lambda}\right)^1\right]\left[\left(1 - e^{-0.01\lambda}\right)^2\left(e^{-0.01\lambda}\right)^3\right]$$

3. Determine the coliform density λ, organisms per mL.

Substitute different values of λ and solve for $y \cdot k$ in the above equation. The results are tabulated below.

λ, organisms per mL	$y \cdot k$
22.09	1.3963287×10^{-3}
22.10	1.3963315×10^{-3}
22.11	1.3963330×10^{-3}
22.12	1.3963331×10^{-3}
22.13	1.3963319×10^{-3}
22.14	1.3963293×10^{-3}
22.15	1.3963255×10^{-3}

The maximum value of $y \cdot k$ is reached when $\lambda = 22.12$ organisms per mL.
MPN = 22.12 organisms per mL
Coliform density is normally expressed as MPN/100 mL.
MPN = 2212 organisms per 100 mL

EXAMPLE 5.42: MPN USING THOMAS EQUATION

Lactose-broth multiple-tube fermentation technique was used to determine MPN value of a water sample in Example 5.41. The test was run with five-tube dilutions of 1, 0.1, and 0.01 mL. The positive gas bubble results are 5/5, 4/5, and 2/5. Determine MPN by using Thomas equation.

Solution

1. State Thomas equation.

   Thomas proposed a simpler equation, which is often used to estimate the coliform density. It is expressed by Equation 5.22.[17]

$$\text{MPN}/100\,\text{mL} = \frac{\text{No. of positive tubes} \times 100}{(V_1 V_2)^{1/2}} \tag{5.22}$$

   where

   V_1 = volume of sample in negative tubes, mL
   V_2 = volume of sample in all tubes, mL

2. Substitute the experimental values.

   $V_1 = (0.1 + 3 \times 0.01)\,\text{mL} = (0.1 + 0.03)\,\text{mL} = 0.13\,\text{mL}$
   $V_2 = (5 \times 1 + 5 \times 0.1 + 5 \times 0.01)\,\text{mL} = 5.55\,\text{mL}$
   Number of positive tubes = 11

3. Calculate MPN.

$$\text{MPN} = \frac{11 \times 100}{(0.13 \times 5.55)^{1/2}} = 1295\ \text{organisms per 100 mL}$$

EXAMPLE 5.43: MPN FROM PROBABILITY TABLE

The lactose-broth multiple-tube test results for dilutions of 1, 0.1, and 0.01 mL are 5/5, 4/5, and 2/5 (Example 5.41). Determine the coliform density using probability table (Table 5.11).

Solution

1. Determine the MPN value from the probability table.

   Refer to probability table (Table 5.11). The value in the table for dilutions of 10, 1, and 0.1 mL and the test result of 5-4-2 is 220 per 100 mL.

2. Determine the MPN density for the dilutions used.

   Since the dilutions used are 1, 0.1, and 0.01, the coliform density is calculated below.

$$\text{MPN} = \frac{220\ \text{coliforms}}{100\,\text{mL}} \times \frac{10\,\text{mL highest dilution in probability table}}{1\,\text{mL highest dilution used for the sample}}$$
$$= 2200\ \text{coliforms per100 mL sample}$$

EXAMPLE 5.44: MPN TEST RESULTS WITH MORE THAN THREE DILUTIONS

Often it is necessary to run more than three dilutions. From these dilutions, it is possible to obtain more than one result. A rule is applied to select the three dilutions and positive results for determination of the MPN value. Following are the MPN test results on six samples of water and wastewater. Select the three dilution and positive results. Also, determine the coliform density using the probability table.

Sample Source	Dilution Series and Positive Gas					
	10	1	0.1	0.01	0.001	0.0001
Clarifier effluent	5	5	5	3	2	0
Primary effluent	5	5	5	5	3	2
Chlorinated primary effluent (a)	5	5	5	3	1	1
Chlorinated primary effluent (b)	5	5	5	3	2	1
River water	5	5	0	3	2	1
Filtered water	0	1	0	0	0	0

Solution

Select the dilution and positive results for each sample. State the rule for this selection and determine the MPN value using the probability table. This information for each sample is provided in Table 5.13.

TABLE 5.13 Rules for Selection of Positive Results from More Than Three Dilutions (Example 5.44)

Sample Source	Highest Value of Selected Dilution	Selected Positive Result	Explanation of Rule for Selection	MPN, Coliform Density per 100 mL
Clarifier effluent	0.1	5-3-2	Select the highest dilution that has positive result in all five tubes. No positive result in lowest dilution.	14,000[a]
Primary effluent	0.01	5-3-2	Select the highest dilution that has positive result in all five tubes. No lower dilution remaining.	140,000
Chlorinated primary effluent (a)	0.1	5-3-(1 + 1) or 5-3-2	Select the highest dilution in lowest series that has positive result in all five tubes. Incorporate the result of next lower dilution into the higher dilution.	14,000
Chlorinated primary effluent (b)	0.1	5-3-(2 + 1) or 5-3-3	Select the highest dilution in lowest series that has highest positive result. Incorporate the result of lower dilution into the higher dilution.	18,000
River water	1	5-0-(3 + 2) or 5-0-5	Select the highest dilution that has positive result in all five tubes. Incorporate the result of next lower dilution into the higher dilution up to a total of five.	950
Filtered water	10	0-1-0	The three dilutions are used to include the positive result in the middle dilution.	2

[a] $140 \times (10/0.1) = 14{,}000$.

EXAMPLE 5.45: ARITHMETIC AND GEOMETRIC MEANS AND THE MEDIAN

Quite often for a given sample, a series of dilutions are run. The geometric mean, the arithmetic mean, or the median value is used to give a single MPN value from the results of series of dilutions. If unlikely combinations occur with a frequency >1%, it is an indication that the technique may be faulty or that the statistical assumptions underlying the MPN estimate are not being fulfilled. Using this principle, calculate the MPN/100 mL of a sample for which a series of multiple-tube dilutions are run. Use multiple combinations of the dilutions and obtain the arithmetic and geometric means, and the median value. Also, determine MPN using the rules outlined in Table 5.13. The experimental results are given below:

Dilutions, mL	10	1	0.1	0.01	0.001	0.0001
Positive Tubes	5/5	5/5	4/5	1/5	0/5	0/5

Solution

1. Select the multiple combinations for the series of dilutions.
   The series of dilutions, MPN value, multiplying factor, and MPN density are summarized below.

Multiple Combinations	MPN Value from the Table	Multiplying Factor	MPN Density, MPN/100 mL
5-5-4	1600	1	1600
5-4-1	170	10	1700
4-1-0	17	100	1700
1-0-0	2	1000	2000

2. Select the MPN values.
   The results are consistent for all dilutions. Mean and median values are calculated from the results.
3. Determine the MPN using the arithmetic and geometric means, and median value.

$$\text{Arithmetic mean} = \frac{1}{4} \times (1600 + 1700 + 1700 + 2000) = 1750 \text{ MPN/100 mL}$$

$$\text{Geometric mean} = (1600 \times 1700 \times 1700 \times 2000)^{1/4} = 1743 \text{ MPN/100 mL}$$

$$\text{Median value} = 1700 \text{ MPN/100 mL}$$

Based on rule for selection of positive results in Table 5.13, the selected combination is 5-4-1. This is the highest dilution that has positive result in all five tubes, and no positive result in lower dilution. The MPN value is 1700 per 100 mL.

EXAMPLE 5.46: SELECTION OF PROPER SAMPLE SIZE FOR MEMBRANE FILTRATION

A wastewater sample is expected to have 10^5 coliform organisms per mL. Determine the sample size that should be used to give 50 coliform colonies over the membrane filter. Also, develop dilution scheme.

Solution

1. Develop the dilution scheme.
   The dilution scheme is shown in Figure 5.17.
2. Estimate the volume of sample to be filtered from bottle B to give 50 colonies.

$$\text{Volume of raw sample to be filtered} = \frac{50 \text{ colonies needed}}{10^5 \text{ colonies/mL} - \text{raw sample}} = 0.0005 \text{ mL} - \text{raw sample}$$

$$\text{Volume of sample from dilution bottle B} = \frac{0.0005 \text{ mL} - \text{raw sample}}{0.0001 \text{ mL} - \text{raw sample/mL} - \text{sample B}} = 5 \text{ mL} - \text{sample B}$$

3. Check the number of colonies expected over membrane filter.

$$5 \text{ mL} - \text{sample B} \times 0.0001 \text{ mL} - \text{raw sample/mL} - \text{sample B} \times 10^5 \text{ colonies/mL} - \text{raw sample} = 50 \text{ colonies}$$

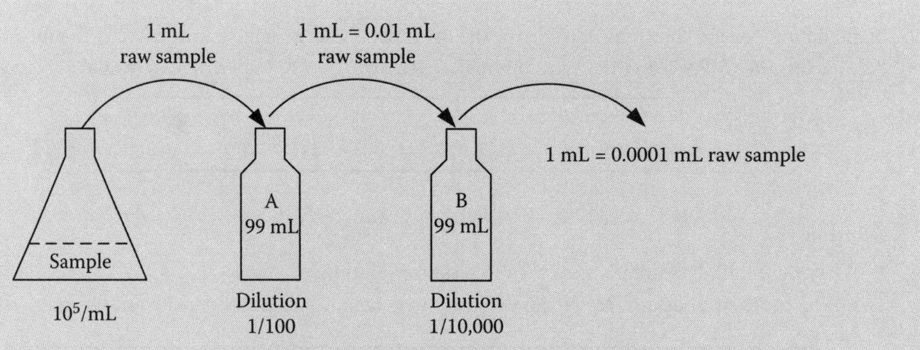

FIGURE 5.17 Series dilution to obtain a desired volume of sample (Example 5.46).

EXAMPLE 5.47: DILUTION TECHNIQUE TO OBTAIN PROPER SAMPLE VOLUME

A wastewater sample is to be filtered that results in a desired number of colonies over the membrane filter. Set up a procedure to obtain 0.005 and 0.0005 mL samples for filtration.

Solution

1. Set up the dilution scheme using 99-mL sterilized dilution water bottles (Figure 5.18).

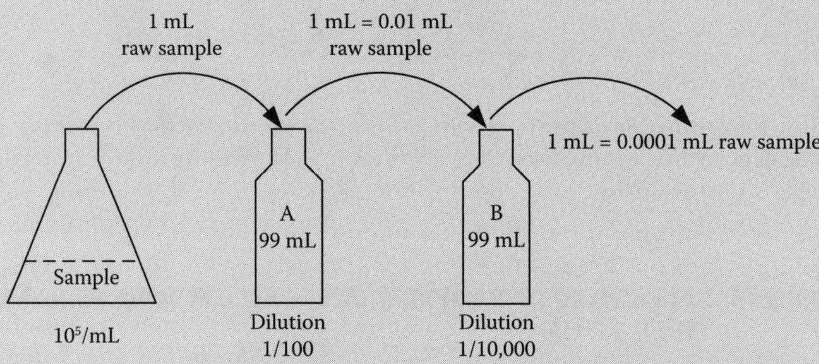

FIGURE 5.18 Series dilution to obtain a desired volume of sample (Example 5.47).

2. Withdraw samples from bottles to give 0.005 and 0.0005 mL sample volume.

   Withdraw 0.5 mL diluted sample from bottle A, or 50 mL diluted sample from bottle B, to obtain 0.005 mL original sample volume. Withdraw 0.5 mL diluted sample from bottle B to obtain 0.0005 mL original sample.

EXAMPLE 5.48: COLIFORM DENSITY

Five milliliters of river water was filtered through the membrane filter. Thirty coliform colonies were counted. Determine the coliform count per 100 mL and give membrane filter results as coliform density.

Solution

$$\text{Coliform count per mL of river water} = \frac{30 \text{ coliform colonies}}{5\,\text{mL}} = 6 \text{ coliforms/mL}$$

$$\text{Coliform count per 100 mL of river water} = \frac{6 \text{ coliforms}}{\text{mL}} \times 100\,\text{mL} = 600 \text{ coliforms/100 mL}$$

Coliform density = 6

EXAMPLE 5.49: COMMON ERROR IN MEMBRANE FILTRATION TEST

If a small sample volume or diluted sample is applied over membrane filter, an experimental error results. Describe the error and suggest methods to avoid it.

Solution

As soon as a small volume of sample or diluted sample ($<$5 mL) is dropped over a membrane filter, it is absorbed instantly at a spot. As a result, crowded colonies grow at a small area. The correct count of colonies is not possible. To avoid this problem, it is suggested that 50–100 mL of sterilized dilution water is poured into the filtration flask without turning on the vacuum pump. Then, the small volume of sample is dropped over the standing water and the vacuum pump is turned on for filtration. After that, additional 50–100 mL of sterilized water is filtered to rinse the apparatus. This procedure will give well-dispersed colonies over the entire filter area.

EXAMPLE 5.50: MULTIPLE DILUTION RESULTS

Multiple sample volumes of three different wastewater samples were tested for coliform density using membrane filtration technique. Calculate the coliform count from the following results.

Sample	Volumes Filtered	Respective Coliform Colonies Counted
A	50 mL, 50 mL	21, 23
B	50 mL, 20 mL, 1 mL	35, 40, none
C	10 mL, 1 mL, 0.1 mL	40, 6, none

Solution

Determine the total colonies/100 mL of each sample.

$$\text{Sample A, coliform count} = \frac{(21 + 23) \text{ colonies} \times 100\,\text{mL}}{(50 + 50)\,\text{mL}} = 44 \text{ colonies/100 mL}$$

$$\text{Sample B, coliform count} = \frac{35 \text{ colonies} \times 100\,\text{mL}}{50\,\text{mL}} = 70 \text{ colonies/100 mL}$$

It may be noted that the coliform counts for sample sizes 50, 20, and 1 mL are 70, 200, and 0 per 100 mL. Two hundred and 0 colonies are a wide variation from 70. Therefore, only one value of 35 colonies for 50 mL sample is expected.

$$\text{Sample C, coliform count} = \frac{(40 + 6) \text{ colonies} \times 100\,\text{mL}}{(10 + 1)\,\text{mL}} = 418 \text{ colonies/100 mL}$$

5.6 Priority Pollutants

Priority pollutants include both organic and inorganic compounds. They are selected on the basis of their known or suspected carcinogenicity, mutagenicity, teratogenicity, or high acute toxicity. These pollutants may interfere with the treatment processes, pass through unchanged, transformed, generated, or accumulated in the sludge. The U.S. EPA has identified approximately 129 pollutants in 65 classes to be regulated. It is anticipated that this list will continue to expand in the future. Under the National Pollutant Discharge Elimination System (NPDES), effluent as well as industrial pretreatment standards are being developed. Under the pretreatment regulations, two types of federal pretreatment standards are established: (1) prohibited discharges and (2) categorical standards. The prohibited discharges are those that cause fire or explosion hazard, corrosion, obstruction, slug discharges, and heated discharge to sewers or publicly owned treatment works (POTWs). The categorical standards are developed for priority pollutants to limit their discharge into natural waters.[18]

EXAMPLE 5.51: INORGANIC PRIORITY POLLUTANTS

List the inorganic priority pollutants.

Solution

The inorganic priority pollutants are 13 toxic metals and nonmetals, fibrous asbestos, and total cyanide. The nonmetals and metals are antimony, arsenic, beryllium, cadmium, chromium, copper, lead, mercury, nickel, selenium, silver, thallium, and zinc.

EXAMPLE 5.52: EXAMPLES OF ORGANIC PRIORITY POLLUTANTS

List the representative priority organic pollutants.

Solution

The 114 representative priority organic compounds can be subdivided into aliphatics (36), aromatics (59), and pesticides (19). Approximately 30 of these organic compounds can be considered volatile, and 69 contain chlorine.[18] Examples of common organic priority pollutants are: benzene, carbon tetrachloride, dichloroethane, vinyl chloride, bromoform, aldrin, dieldrin, polychlorinated biphenyl (PCB), and many more.

5.7 Toxicity and Biomonitoring

On March 9, 1984, the U.S. EPA published a new national policy based on 1972 Clean Water Act, prohibiting the discharge of toxic pollutants in toxic amounts into natural waters. As a result of these guidelines, toxicity tests have been instituted to (1) assess the toxicity of wastewater effluent to natural aquatic life, (2) assess effectiveness and degree of wastewater treatment to meet water pollution control requirements, and (3) determine compliance with federal and state water quality standards and water quality criteria associated with NPDES permits.

The effluent from municipal wastewater treatment facility may contain numerous toxic chemicals that may cause synergistic effects. For this reason, instead of specific chemical tests, biomonitoring was introduced.

5.7.1 Toxicity Test

Biomonitoring requires appropriate bioassay organisms to determine the level of toxicity in the effluent. The tests are run in effluent samples with varying degrees of dilution, and records are kept on observations of death, deformities, reproduction, and growth of test organisms. The tests may be short term, intermediate, and/or long term. The test may also be static, recirculation, renewal, or flow-through. Common freshwater species are *Ceriodaphnia dubia* (water flea, daphnid shrimp), and *Pimephales promelas* (fathead minnow). The marine organisms included in the biomonitoring are *Champia parvula* (the red algae), *Mysidopsis bahia* (the mysid shrimp), *Menidia beryllina* (the island silversides), and *Cyrinidon variegalus* (the sheephead minnow). The biomonitoring protocols have been published and commented on extensively in the literature. The results of acute toxicity tests are reported in 48 or 96 h. The LC_{50} is the concentration of effluent in dilution water that causes 50% mortality of the test organisms. The chronic toxicity is measured over a long period or generations. The results are based on mortality, reduced growth, or reproduction behavior.[19–25]

5.7.2 Toxicity Test Evaluation

Toxicity test results are expressed in terms of *toxic units (TU)*

TU Acute (TU_a) is the reciprocal of the effluent dilution that causes the acute effect by the end of the exposure period. This is expressed by Equation 5.23.[26]

$$TU_a = \frac{100}{LC_{50}} \tag{5.23}$$

where LC_{50} = lethal concentration of the toxin required to kill half the members of a tested population after a specified test duration

TU Chronic (TU_c) is the reciprocal of the maximum effluent dilution that causes no unacceptable effect on test organisms by the end of the chronic exposure period that is expressed by Equation 5.24.

$$TU_c = \frac{100}{NOEC} \tag{5.24}$$

where $NOEC$ = no observable effect concentration

The acute toxicity criterion is based on the Criterion Maximum Concentration (*CMC*), and is expressed by Equation 5.25.

$$CMC = \frac{TU_a}{CID} \leq 0.3\ TU_a \tag{5.25}$$

where CID = critical initial dilution

The chronic toxicity criterion is based on the Criterion Continuous Concentration (*CCC*), and is expressed by Equation 5.26.

$$CCC = \frac{TU_c}{CID} \leq 1.0 TU_c \tag{5.26}$$

Based on toxicity test evaluation, if the toxicity level is found and verified, *toxicity reduction evaluation (TRE)* is required. These studies involve *toxicity investigation examination (TIE)* and *effluent toxicity treatability (ETT)*. The objective is to determine the best method to reduce or eliminate the toxicity of the effluent. The plan established by *TIE* or *ETT* is carried out under corrective action. The plan may include process modification, chemical change, construction, or industrial pretreatment. After reduction or elimination of toxicity, the plant is considered in compliance by the U.S. EPA.

EXAMPLE 5.53: DETERMINATION OF LC_{50} OF AN INDUSTRIAL CHEMICAL

The acute toxicity level of an industrial chemical is measured over a 96-h exposure time in a static bioassay test. Four concentrations of the chemical were tested with 20 organisms in each concentration. The results over 96-h of exposure time are summarized below. Determine LC_{50} by using (a) log concentration and probability scales and (b) log concentration and arithmetic scales.

Control Concentration, μg/L	Organisms Surviving after 96 h	Mortality after 96 h	% Mortality
0	20	0	0
500	16	4	20
800	11	9	45
1300	6	14	70
2000	2	18	90

Solution

1. Determine the LC_{50} value by using log-probability scale.
   a. Plot the concentration of chemical (log scale) versus percent mortality of test organisms (probability scale). The plot is shown in Figure 5.19.
   b. Fit the data points by visual observation giving close consideration in the range of 16% and 84% mortality.
   c. Draw a vertical line from 50% mortality.
   d. Read the estimated LC_{50} value.

   $LC_{50} = 890$ mg/L

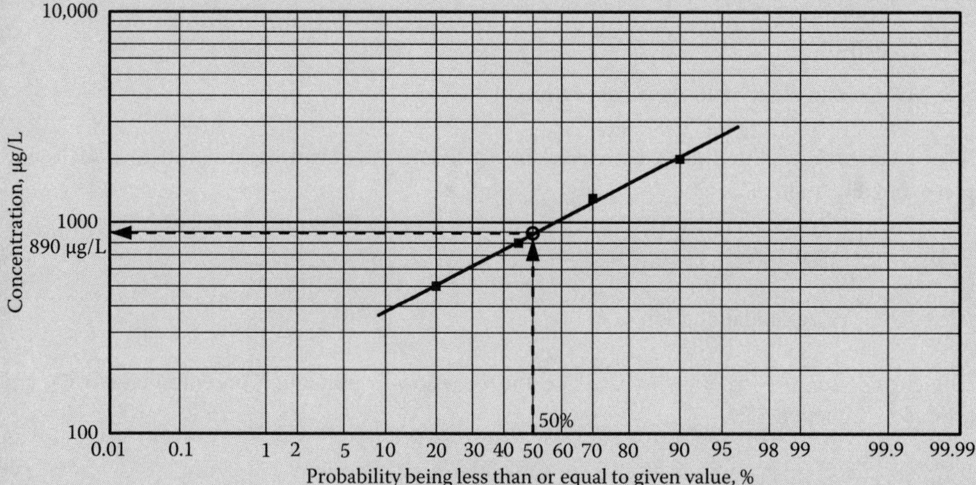

FIGURE 5.19 Determination of LC_{50} using log and probability scales (Example 5.53).

2. Determine the LC_{50} value by using log concentration and percent survival (both in arithmetic scale).
   a. Arrange the experimental data as log concentration and percent survival.

Control Concentration (μg/L)	Log Concentration	% Mortality	% Survival
0	N/A	0	100
500	2.70	20	80
800	2.90	45	55
1300	3.11	70	30
2000	3.30	90	10

b. Plot the log concentration of chemical (arithmetic scale) versus percent survival of test organisms (arithmetic scale). The plot is shown in Figure 5.20.

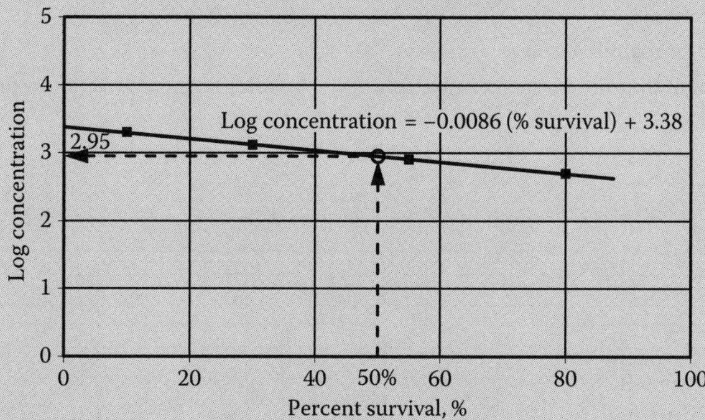

FIGURE 5.20 Determination of LC_{50} using log and arithmetic scales (Example 5.53).

c. Develop linear relationship between log concentration and percent survival.

$$\text{Log Concentration} = -0.0086\ (\%\ \text{survival}) + 3.38$$

d. Estimated the $\log(LC_{50})$ value from 50% survival.

$$\log(LC_{50}) = -0.0086 \times 50 + 3.38 = 2.95$$

Note: A vertical line from 50% mobility will also give $\log(LC_{50}) = 2.95$.

e. Convert the $\log(LC_{50})$ to LC_{50} value.

$$LC_{50} = 10^{2.95} = 891\ \mu\text{g/L}$$

EXAMPLE 5.54: EFFLUENT TOXICITY

The effluent toxicity of a wastewater treatment plant was examined by a rapid bioassay test. A series of effluent dilutions were prepared in samples from the receiving water. A group of 20 *Ceriodaphnia dubia* was tested in each dilution. The results of 96 h of testing are summarized below. Estimate LC_{50} of the 96 h using graphical method.

Effluent Dilutions, % by Volume	No. of Test Organisms	No. of Test Organisms Dead after 96 h	Percent Mortality
30	20	18	90
20	20	15	75
10	20	8	40
7	20	4	20
5	20	2	10
3	20	1	5

Solution

1. Use the log and probability scales.

   Plot the effluent dilutions in percent by volume (log scale) versus mortality in percent (probability scale) as in Figure 5.21.

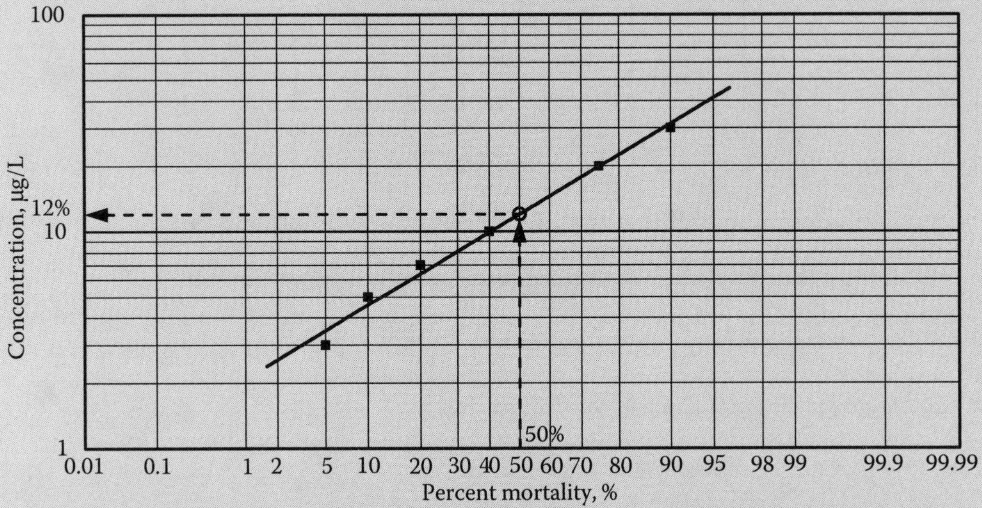

FIGURE 5.21 Determination of LC_{50} of effluent dilution using log and probability scale (Example 5.54).

2. Develop the best fit line (visual observation) giving close consideration in the range of 16%–18% mortality.
3. Draw a vertical line from 50% mortality and read the LC_{50} value.

$$LC_{50} = 12\%$$

EXAMPLE 5.55: PROBIT ANALYSIS

Solve Example 5.54 using the Probit analysis.

Solution

1. Describe the Probit analysis.

   The Probit value (or probability unit) provides a basis for transforming the sigmoidal curve of dose–response relationship to a complete linear transformation making linear regression and statistical analysis straightforward. Thus, LC_{50}, slope, and confidence limits can be established. For most analyses, a Probit table (Table 5.14) is used to transform raw dose-response data into Probit values.

TABLE 5.14 Transformation from Percentages to Probit Values (Example 5.55)

	1	2	3	4	5	6	7	8	9	10	11
1	%	0	1	2	3	4	5	6	7	8	9
2	0	–	2.67	2.95	3.12	3.25	3.36	3.45	3.52	3.59	3.66
3	10	3.72	3.77	3.82	3.87	3.92	3.96	4.01	4.05	4.08	4.12
4	20	4.16	4.19	4.23	4.26	4.29	4.33	4.36	4.39	4.42	4.45
5	30	4.48	4.50	4.53	4.56	4.59	4.61	4.64	4.67	4.69	4.72
6	40	4.75	4.77	4.80	4.82	4.85	4.87	4.90	4.92	4.95	4.97
7	50	5.00	5.03	5.05	5.08	5.10	5.13	5.15	5.18	5.20	5.23
8	60	5.25	5.28	5.31	5.33	5.36	5.39	5.41	5.44	5.47	5.50
9	70	5.52	5.55	5.58	5.61	5.64	5.67	5.71	5.74	5.77	5.81
10	80	5.84	5.88	5.92	5.95	5.99	6.04	6.08	6.13	6.18	6.23
11	90	6.28	6.34	6.41	6.48	6.55	6.64	6.75	6.88	7.05	7.33
12	99	7.33	7.37	7.41	7.46	7.51	7.58	7.65	7.75	7.88	8.09

Source: Adapted in part from Reference 27.

2. Convert the data of Example 5.54 to Probit value.

Effluent Dilution, % by Volume	% Mortality	Probit Value	log % Effluent Dilution
30	90	6.28[a]	1.48[b]
20	75	5.67[c]	1.30
10	40	4.75	1.00
7	20	4.16	0.85
5	10	3.72	0.70
3	5	3.36[d]	0.48

[a] 6.28 is obtained from Table 5.14 (Row 11, Column 2). Row 11 and Column 2 correspond to 90%.
[b] log 30 = 1.48.
[c] 5.67 is obtained from Table 5.14 (Row 9, Column 7). Row 9 and Column 7 correspond to 75%.
[d] 3.36 is obtained from Table 5.14 (Row 2, Column 7). Row 2 and column 7 corresponds to 5%.

3. Plot the Probit value as a function of log % effluent dilution. The plot is shown in Figure 5.22.

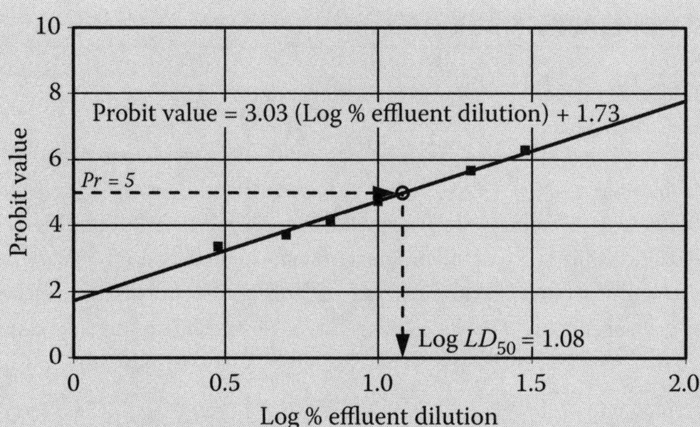

FIGURE 5.22 Probit value as a function of log % effluent dilution (Example 5.55).

4. Identify the Probit equation as expressed by Equation 5.27.

$$\text{Pr} = a + b\log(\%V) \tag{5.27}$$

where

P_r = the Probit value
$\%V$ = percent effluent dilution
a and b = intercept and slope

5. Develop the linear relationship between the Probit value and log % effluent dilution, and determine intercept and slope in the Probit equation.

$a = 1.73$ (intercept) and $b = 3.03$ (slope)

$P_r = 1.73 + 3.03\log(LD_{50})$

6. Determine the LD_{50} value.

LD_{50} is determined at the Probit value P_r of 5 that is corresponding to 50% (see Row 7 and Column 2 in Table 5.14). Substitute this value in Equation 5.27 and solve for LD_{50}.

$5 = 1.73 + 3.03\log(LD_{50})$, or $\log(LD_{50}) = 1.08$; $LD_{50} = 10^{1.08} = 12\%$

EXAMPLE 5.56: TOXICITY TEST EVALUATION RESULTS

A municipal wastewater treatment plant is discharging secondary effluent into a river. The dilution received at the boundary of the mixing zone at minimum 3-year dry weather flow condition is 50:1 (or $CID = 50$). Acute toxicity tests and chronic toxicity tests were conducted on three freshwater species to eliminate the effect of sensitivities of test species to the plant effluent. The acute toxicity tests were conducted over a 96-h period. The chronic toxicity tests utilized a series of effluent dilutions in continuous flow-through systems. *Ceriodaphnia dubia* was the test organism. The acute toxicity LC_{50} was based on mortality, and the chronic endpoint was measured on reduced reproduction and reported as *NOEC*. The test results are summarized below. Determine the toxicity compliance.

Acute toxicity:

Exposure period = 96 h; control survival % = 100; $LC_{50} = 5.7$

Chronic end point:

Exposure period = 10 d; control survival % = 98; $NOEC = 1.7$

Solution

1. Determine the TU_a and TU_c values from Equations 5.23 and 5.24.

$$TU_a = \frac{100}{LC_{50}} = \frac{100}{5.7} = 17.5 \quad \text{(acute)}$$

$$TU_c = \frac{100}{NOEC} = \frac{100}{1.7} = 58.8 \quad \text{(chronic)}$$

2. Determine the CMC value from Equation 5.25 and compliance value.

$$CMC = \frac{TU_a}{CID} = \frac{17.5}{50} = 0.35 \quad \text{(acute)}$$

The compliance criteria = 0.3 $TU_a = 0.3 \times 17.5 = 5.3$.

The CMC value of 0.35 is considerably <0.3 TU_a or 5.3. Therefore, acute toxicity is within the compliance criteria.

3. Determine the CCC value from Equation 5.26 and compliance value.

$$CCC = \frac{TU_c}{CID} = \frac{58.5}{50} = 1.2 \quad \text{(chronic)}$$

The compliance criteria = 1.0 $TU_c = 1 \times 58.8 = 58.8$.

The CCC value of 1.2 is considerably less than the compliance value of 1.0 TU_c or 58.8. Therefore, chronic toxicity is within the compliance criteria. *TRE* evaluation is not required.

5.8 Unit Waste Loading and Population Equivalent

Loadings of many constituents in municipal wastewater on a *per capita* basis remain relatively uniform. The variation in constituent loadings per capita per day may be due to industries served, usage of garbage grinders, domestic water softeners, and discharge of septage. In small treatment facilities, their effects may be significant. *Unit waste loadings* in municipal wastewater may be developed from flow rate (liters per capita per day) and concentrations of various constituents in mg/L. These unit loadings are used to estimate the population equivalent (PE) of industrial wastes. PE may be developed on the basis of flow, TSS, BOD_5, COD, nitrogen, and phosphorus.[4,11]

EXAMPLE 5.57: UNIT BOD_5 AND TSS LOADINGS

A municipal wastewater treatment plant receives mostly residential wastewater. The average wastewater flow is 440 Lpcd. The average BOD_5 and TSS concentrations are 220 and 240 mg/L. Calculate unit BOD_5 and TSS loadings as g/capita·d.

Solution

1. Calculate the unit BOD_5 loading.

$$BOD_5 = 220\,\text{mg/L} \times \frac{\text{g}}{10^3\,\text{mg}} \times 440\,\text{Lpcd} = 97\,\text{g/capita}\cdot\text{d}$$

2. Calculate the unit TSS loading.

$$TSS = 240\,\text{mg/L} \times \frac{\text{g}}{10^3\,\text{mg}} \times 440\,\text{Lpcd} = 106\,\text{g/capita}\cdot\text{d}$$

EXAMPLE 5.58: PE OF AN INDUSTRY

A small dairy industry discharges 500 m^3/d wastewater that has an average BOD_5 of 1500 mg/L. If the unit BOD_5 loading is 97 g/capita·d, calculate the population equivalent.

Solution

1. Calculate the BOD_5 load of the dairy plant.

$$BOD_5 = 1500\,mg/L \times \frac{kg}{10^6\,mg} \times \frac{10^3\,L}{m^3} \times 500\,m^3/d = 750\,kg/d$$

2. Calculate the PE value of dairy industry.

$$P.E. = 750\,kg/d \times \frac{10^3\,g}{kg} \times \frac{1}{97\,g/capita \cdot d} \times 500\,m^3/d = 7732\text{ persons}$$

EXAMPLE 5.59: UNIT LOADINGS BASED ON TSS, BOD_5, COD, NITROGEN, AND PHOSPHORUS

The typical concentrations of most common constituents of municipal wastewater are given in Table 5.1. If average unit flow to the plant is 450 Lpcd, determine the unit waste loadings on the basis of BOD_5, COD, TSS, total nitrogen (as N), organic nitrogen (as N), ammonia nitrogen (as N), and total phosphorus (as P). The concentrations of these constituents are provided in Table 5.1.

Solution

1. Calculate the unit waste loadings (g/capita·d) for each of the above constituents.

   The average flow rate is 450 Lpcd. The typical waste loadings (g/capita·d) are calculated from the typical concentrations given in Table 5.1.
2. Tabulate the unit mass loadings.

   The typical unit waste loadings (g/capita·d) for each constituent is summarized in Table 5.15.

TABLE 5.15 Typical Unit Waste Loadings Derived from Table 5.1 (Example 5.59)

Constituent	Typical Concentration, mg/L	Typical Unit Waste Loadings, g/capita·d
BOD_5	210	95[a]
COD	400	180
TSS	250	113
Total nitrogen (TN), as N	40	18
Organic nitrogen (ON), as N	15	7
Ammonia nitrogen (AN), as N	30	14
Total phosphorus (TP), as P	6	3

[a] $210\,g/m^3 \times 10^{-3}\,m^3/L \times 450\,L/capita{\cdot}d = 95\,g/capita{\cdot}d$.

EXAMPLE 5.60: PE CALCULATION BASED ON UNIT WASTE LOADINGS

A canning industry generates 1500 m^3/d waste stream. The average concentrations of $BOD_5 = 1200$ mg/L, COD = 2400 mg/L, TSS = 1800 mg/L, TN = 60 mg/L as N, ON = 44 mg/L as N, AN = 16 mg/L as N, and TP = 18 mg/L as P. Using the unit waste loadings in Table 5.15, calculate PE using each constituent. Select the highest PE.

Solution

1. Tabulate the industrial waste loadings.

   Use unit waste loading in Table 5.15 to obtain the industrial load of each constituent. These industrial loads are summarized below.

Constituent	Industrial Waste			PE
	Unit Waste Loadings, g/capita·d	Concentration, mg/L	Loading, kg/d	
BOD_5	95	1200	1800	18,947
COD	180	2400	3600	20,000
TSS	113	1800	2700	23,894
TN, as N	18	60	90	5000
ON, as N	7	44	66	9429
AN, as N	14	16	24	1714
TP, as P	3	18	27	9000

2. Select the highest PE value.

   The highest PE is obtained from TSS result and is 23,894 or 24,000. The TSS load from the canning industry is equivalent to a population of 24,000.

EXAMPLE 5.61: EFFECT OF GARBAGE GRINDER ON WASTEWATER CHARACTERISTICS

A wastewater treatment plant is serving 15,000 residents of a community that has garbage grinders to grind the kitchen wastes. The average wastewater flow is 7500 m^3/d. The average BOD_5 and TSS concentrations are 230 and 280 mg/L, respectively. Assume that the unit waste loadings in Table 5.15 apply for community without garbage grinders. Calculate the per capita BOD_5 and TSS load from the garbage grinder.

Solution

1. Calculate the unit BOD_5 and TSS loadings for the community with garbage grinders.

$$BOD_5 = 230\,\text{mg/L} \times \frac{\text{g}}{10^3\,\text{mg}} \times \frac{10^3\,\text{L}}{\text{m}^3} \times 7500\,\text{m}^3/\text{d} \times \frac{1}{15{,}000\text{ people}} = 115\,\text{g/capita·d}$$

$$TSS = 280\,\text{mg/L} \times \frac{\text{g}}{10^3\,\text{mg}} \times \frac{10^3\,\text{L}}{\text{m}^3} \times 7500\,\text{m}^3/\text{d} \times \frac{1}{15{,}000\text{ people}} = 140\,\text{g/capita·d}$$

2. Determine the unit waste loadings derived from Table 5.15.

$$BOD_5\text{ loading} = 95\,\text{g/capita·d}$$

$$TSS\text{ loading} = 113\,\text{g/capita·d}$$

3. Determine the per capita contributions of the garbage grinder.

$$BOD_5 = (115 - 95)\,\text{g/capita·d} = 20\,\text{g/capita·d}$$

$$TSS = (140 - 113)\,\text{g/capita·d} = 27\,\text{g/capita·d}$$

5.9 Mass Loadings and Sustained Mass Loadings

Mass Loadings: The analysis of wastewater data involves the determination of flow rates and mass loadings. Both wastewater flow and concentrations of contaminants have hourly, daily, and seasonal variations. Facility designs based on average flow rate and concentrations may have serious deficiencies during high mass loading. For example, during early stages of infiltration/inflow, high flows are encountered. Concentrations of various contaminants including BOD, TSS, nitrogen, and phosphorus may also be high due to "first-flush-effect." As a result, a surge of mass loading may pass through the facility over several hours. The mass loading is calculated from Equation 5.28.

$$C_m = \frac{\sum_{i=1}^{n} c_i q_i}{\sum_{i=1}^{n=1} q_i} \tag{5.28}$$

where

C_m = flow-weighted average concentration of contaminant, mg/L or g/m^3
c_i = average concentration of contaminant during the ith time period, mg/L or g/m^3
n = number of observations
q_i = average flow rate during ith time period, m^3/d

Sustained Mass Loading: The critical components of a wastewater treatment facility should be checked for expected sustained peak mass loadings of key constituents. Some of these critical components are aeration capacity, sludge zone and clarification zone in the final clarifier, digestion period, sludge dewatering capacity, and effluent quality. The peaking factors for sustained peak and low mass loadings are developed with respect to the number of consecutive days during the period of record that high and low mass loadings are sustained. The procedure for developing sustained mass loading envelop is similar to that of sustained peak and low flow rates (Section 4.5.3). The general procedure for developing the sustained mass loading envelope is briefly presented below.

1. Select the available daily mass loading record during a sustained period of time.
2. Determine the average day mass loading.
3. Calculate the ratio of the daily- to average-mass loading.
4. Calculate the running average values of the ratios over a period 1, 2, 3, ..., n consecutive days.
5. Identify and plot the maximum and minimum running average ratios with respect to each period 1, 2, 3, ..., n consecutive days.

EXAMPLE 5.62: MASS LOADINGS FOR DIURNAL WASTEWATER

At a wastewater treatment plant, the hourly diurnal flow and BOD$_5$ concentration values were recorded. These values are tabulated below. Calculate the BOD$_5$ diurnal mass-loading values. Plot the diurnal flow, BOD concentration and BOD mass-loading curves.

Time Period	Midnight–1 a.m.	1–2	2–3	3–4	4–5	5–6	6–7	7–8	8–9	9–10	10–11	11 a.m.–Noon
Flow, m^3/h	72	56	41	34	31	33	74	80	86	88	86	84
BOD$_5$, mg/L	160	132	93	63	40	40	59	110	139	180	203	211
Time Period	**Noon–1 p.m.**	**1–2**	**2–3**	**3–4**	**4–5**	**5–6**	**6–7**	**7–8**	**8–9**	**9–10**	**10–11**	**11 p.m.–Midnight**
Flow, m^3/h	86	84	77	72	67	65	67	72	80	82	88	77
BOD$_5$, mg/L	218	230	240	220	180	160	150	179	200	280	305	250

Solution

1. Set up a calculation table.

   Tabulate the 24-h individual data point. The flow input is the hourly average rate.

Time	Flow (q), m^3/h	BOD_5 Concentration (C), mg/L	BOD_5 Mass Loading ($C \times q$), kg/h
Midnight–1:00 a.m.	72	160	11.5
1–2 a.m.	56	132	7.39
2–3	41	93	3.81
3–4	34	63	2.14
4–5	31	40	1.24
5–6	33	40	1.32
6–7	74	59	4.37
7–8	80	110	8.80
8–9	86	139	12.0
9–10	88	180	15.8
10–11	86	203	17.5
11–12	84	211	17.7
Noon–1:00 p.m.	86	218	18.7
1–2 p.m.	84	230	19.3
2–3	77	240	18.5
3–4	72	220	15.8
4–5	67	180	12.1
5–6	65	160	10.4
6–7	67	150	10.1
7–8	72	179	12.9
8–9	80	200	16.0
9–10	82	280	23.0
10–11	88	305	26.8
11–12	77	250	19.3
Total	1682	4042	306

2. Calculate the daily average flow, BOD_5, and BOD_5 mass loading.

$$\text{Daily average flow} = \frac{1682\,\text{m}^3/\text{h} \times 1\,\text{h}}{24\,\text{h}} = 70\,\text{m}^3/\text{h}$$

$$\text{Daily average BOD}_5\text{concentration} = \frac{4042\,\text{mg/L} \times 1\,\text{h}}{24\,\text{h}} = 168\,\text{mg/L}$$

$$\text{Daily average BOD}_5\text{mass loading} = \frac{306\,\text{kg/h} \times 1\,\text{h}}{24\,\text{h}} = 12.8\,\text{kg/h}$$

3. Plot the diurnal flow, BOD_5 concentration, and BOD_5 mass loading.

   The average and hourly variation of flow, BOD_5 concentration, and BOD_5 mass loading plots are shown in Figure 5.23.

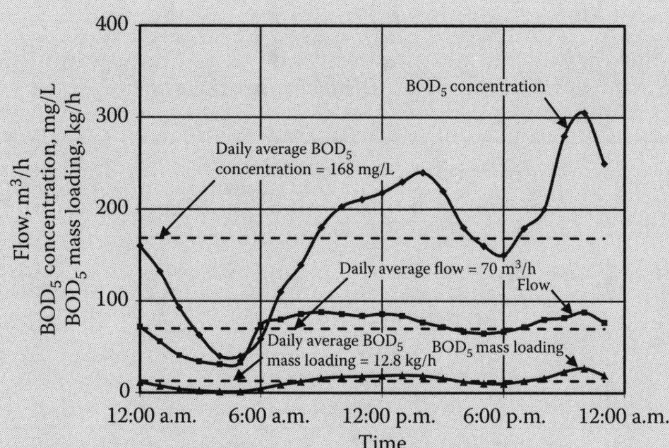

FIGURE 5.23 Plot of diurnal flow, BOD_5 concentration, and BOD_5 mass loading curves (Example 5.62).

EXAMPLE 5.63: RATIOS OF SUSTAINED AVERAGE PEAK AND MINIMUM LOADING CURVES OF BOD_5, TSS, NITROGEN, AND PHOSPHORUS

The typical sustained average peak and low mass loading ratios of BOD_5, TSS, nitrogen, and phosphorus with respect to average loadings and the number of consecutive days the mass loadings are maintained are provided in Table 5.16. Draw the typical maximum and minimum ratio curves.

TABLE 5.16 Ratio of Sustained Average Peak and Low Mass Loadings to Average Mass Loading (Example 5.63)

No. of Consecutive Days That the Average Mass Loading Was Sustained	Ratio of Sustained Average Peak and Low Mass Loading to Average Mass Loading									
	BOD_5		TSS		TKN		NH_3-N		Total P	
	Max	Min	Max	Min	Max	Min	Max	Min	Max	Min
1	2.48	0.13	2.58	0.19	2.13	0.19	1.50	0.50	1.73	0.31
2	2.13	0.37	2.31	0.31	1.85	0.44	1.31	0.70	1.44	0.56
4	1.85	0.50	1.93	0.50	1.60	0.65	1.24	0.81	1.24	0.79
7	1.56	0.63	1.56	0.63	1.50	0.75	1.23	0.85	1.23	0.88
10	1.40	0.65	1.38	0.65	1.44	0.80	1.20	0.86	1.20	0.90
15	1.28	0.73	1.25	0.71	1.38	0.88	1.20	0.86	1.20	0.91
20	1.25	0.75	1.25	0.79	1.29	0.88	1.20	0.86	1.20	0.92
25	1.23	0.75	1.23	0.79	1.25	1.00	1.20	0.86	1.20	0.93
30	1.20	0.75	1.19	0.80	1.23	1.00	1.20	0.86	1.20	0.94

Solution

The maximum and minimum consecutive days of sustained mass loading ratios with respect to average day mass loading for the period of record are provided in Table 5.16. The procedure for developing the sustained average flow ratio (SAFR) envelope is given in Example 4.14. Similar procedure is applied for developing the sustained peak and low mass loading ratio envelope. This procedure is not repeated here. Readers are referred to Example 4.14 to review the procedure for developing mass loading ratio envelope.

Plot the sustained peak and low mass loading ratio envelope.

The ratios of averaged sustained peak and low mass loadings to average mass loading for BOD, TSS, TKN, NH_3-N and total P with respect to number of consecutive days these mass loadings are sustained are plotted in Figure 5.24.

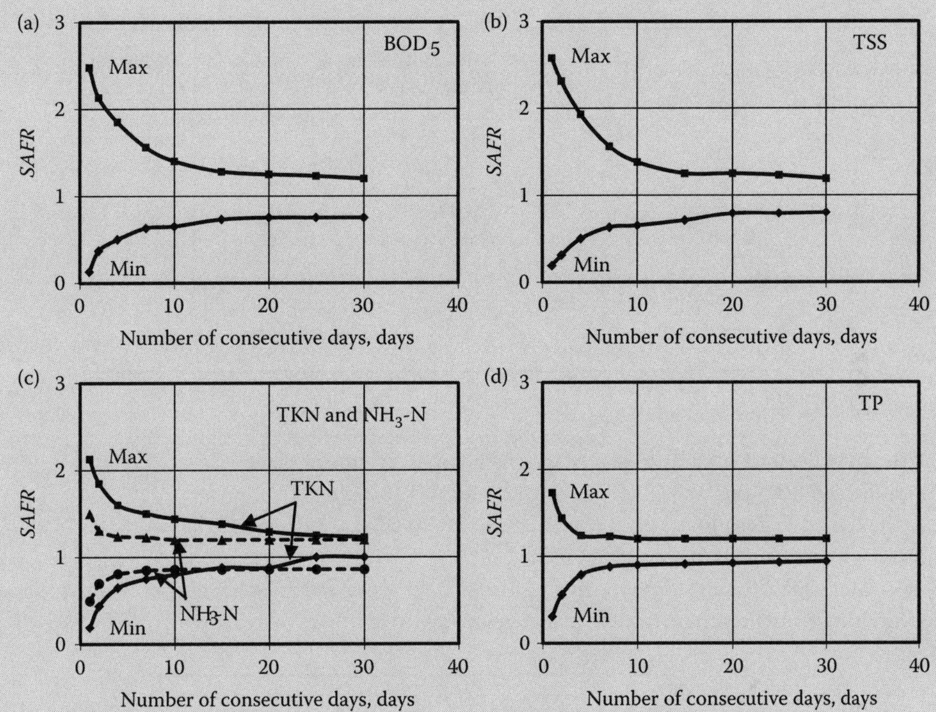

FIGURE 5.24 Sustained average ratios of maximum and minimum mass loadings: (a) BOD_5, (b) TSS, (c) TKN and NH_3-N, and (d) TP (Example 5.63).

EXAMPLE 5.64: APPLICATION OF SUSTAINED MASS LOADING FOR DESIGN AND OPERATION OF A WASTEWATER TREATMENT PLANT

A wastewater treatment plant has a design annual average wastewater flow of 38,000 m^3/d (10 MGD), and annual average BOD_5 concentration of 190 mg/L. The peaking factors and number of consecutive days of sustained peak loading are given below. Develop total mass loading curve for BOD_5 that the treatment plant must handle under different days of consecutive sustained peak loadings. What is the total amount of BOD_5 that is expected to reach the treatment plant during 5 days of sustained loading period? What is the significance of this sustained mass loading for plant design and operation?

No. of Consecutive Days of Sustained Peak	1	2	3	5	10	20	30
Peaking Factor	2.48	2.13	1.90	1.70	1.40	1.25	1.20

Solution

1. Determine the average daily BOD_5 mass loading.

$$\text{Daily BOD}_5 \text{ mass loading} = 190\,\text{mg/L} \times \frac{\text{kg}}{10^6\,\text{mg}} \times \frac{10^3\,\text{L}}{\text{m}^3} \times 38{,}000\,\text{m}^3/\text{d} = 7220\,\text{kg/d}$$

2. Set up a computational table to calculate the sustained peak BOD_5 mass loading
   a. Use the peaking factors for different lengths of sustained peak periods.
   b. Calculate peak BOD mass loading.
   c. Calculate total mass loading for different lengths of sustained peak loadings.

No. of Consecutive Days of Sustained Peak Loading, d	Peaking Factor	Sustained BOD_5 Mass Loading, kg/d	Total Mass Loading for Sustained Period, kg
1	2.48	17,906[a]	17,906
2	2.13	15,379	30,758[b]
3	1.90	13,718	41,154
5	1.70	12,274	61,370
10	1.40	10,108	101,080
20	1.25	9025	180,500
30	1.20	8664	259,920

[a] 7220 kg/d × 2.48 = 17,906 kg/d.
[b] 15,379 kg/d × 2 d = 30,758 kg.

3. Determine the total sustained 5-day mass loading to the treatment plant.
   Sustained BOD_5 mass loading = 1.70 × 7220 kg/d – 12,274 kg/d
   Total mass loading to the plant over 5 days = 12,274 kg/d × 5 d = 61,370 kg
4. Develop the total mass load curve for BOD_5.
   The total mass loading curve for BOD_5 that the treatment plant must handle for different consecutive days of sustained peak loading is shown in Figure 5.25.

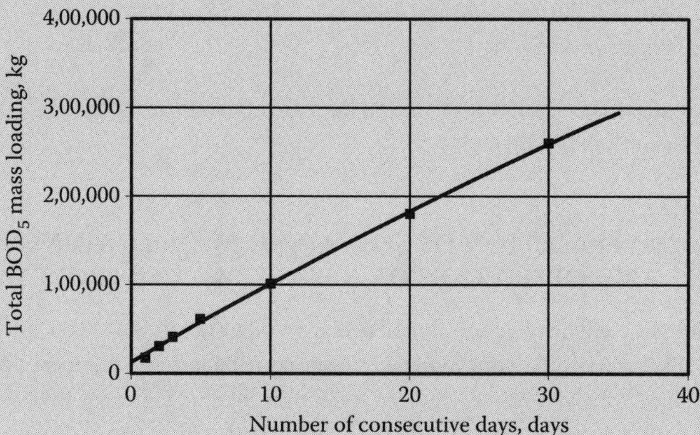

FIGURE 5.25 Sustained BOD_5 mass loading during different consecutive days at the plant (Example 5.64).

5. Significance of the sustained mass BOD_5 loading.
   The treatment plant will receive a total of 61,370 kg of sustained peak BOD_5 loading over a period of 5 days. In comparison, the average mass loading over 5 days is only 36,100 kg (5 d × 7220 kg/d). The plant components should be large enough to supply additional aeration, return activated sludge and waste-activated sludge to handle 5-day sustained peak condition. The clarifier should be checked to handle increased solids loading, have sufficiently large sludge zone and clarification zone. A desirable situation is to provide sufficient return sludge capacity to move solids from the clarifier into the aeration basin. This will reduce significantly the depth of sludge blanket in the clarifier. Additionally, the sludge digestion and dewatering facilities should be designed to handle the excess load during the peaking periods.

Discussion Topics and Review Problems

5.1 Define TON. A water sample at 2 mL in standard dilution test gave a perceptible odor. Calculate TON.

5.2 An odor test panel of five members was used in an odor evaluation study. The results are given below. Determine the geometric mean of TON.

Panel Member	Sample Volume Diluted to 200 mL									
	0.05	0.5	B	1.0	2.0	2.5	B	3.0	4.0	5.0
A	−	−	−	+	+	+	−	+	+	+
B	−	−	−	−	+	+	−	+	+	+
C	−	−	−	−	+	+	−	+	+	+
D	−	−	−	+	+	+	−	+	+	+
E	−	−	−	−	−	−	−	+	+	+

5.3 Settling test was performed in an *Imhoff* cone. The following results were obtained.

Total suspended solids in raw wastewater = 350 mg/L
Total settleable solids = 10 mL/L
Solids in sludge = 2%
Total suspended solids in effluent = 150 mg/L
A sedimentation basin is designed to treat 10,000 m^3/d.

Determine the following:

a. Volume of sludge, m^3/d
b. Density of sludge, kg/m^3

5.4 Suspended solids test was performed on a wastewater sample. Determine TSS, VSS, and FSS. Use the following data

Volume of sample filtered = 30 mL
Weight of dried filter paper in aluminum dish = 12.7324 g
Weight of dried filter paper + dish + TSS = 12.7369 g
Weight of dish + ash after ignition = 12.7357 g

5.5 A wastewater sample has total solids of 960 mg/L. TSS is 35% and TVSS is 70% of TSS. TDS has 45% volatile fraction. Calculate the concentrations of TSS, TVSS, TFSS, TDS, TVDSS, and TFDS.

5.6 Sludge production in a sedimentation basin is 185 m^3/d. The pumping rate of sludge is 200 L/min. Determine pump running time and idle time per hour.

5.7 Five milliliters municipal wastewater sample was diluted in 300 mL BOD bottle. The DO of initial diluted sample and after 5-d of incubation were 8.7 and 4.3 mg/L. Calculate BOD_5.

5.8 The selected dilution for BOD_5 test should be such that the residual DO is at least 1 mg/L, and DO depletion should be at least 2 mg/L. Assuming initial DO of 8.6 mg/L, calculate the range of BOD_5 that can be measured by a dilution of 1 mL sample in a 300-mL BOD bottle.

5.9 Determine BOD_5 of an industrial wastewater sample from the following laboratory data.

Volume of sample in BOD bottle = 3 mL
Initial DO of diluted sample = 8.5 mg/L
DO of diluted sample after 5-day incubation = 4.4 mg/L
Initial DO of seeded control = 8.5 mg/L
DO of seeded control after incubation = 8.3 mg/L
Volume of seed in diluted sample and control = 1 mL

5.10 Calculate the carbonaceous BOD of a sample after 7 days and at 30°C. BOD_5 at 20°C is 120 mg/L, and k at 20°C is 0.25 d^{-1}.

5.11 A wastewater sample has BOD_5 of 210 mg/L. The reaction rate constant k_{20} (base e) is 0.23 d^{-1}. Draw the time-dependent BOD remaining and BOD exerted curves.

5.12 The time series BOD data of an industrial waste sample is given below. Determine L_0 and k (base e) using (a) least squares, (b) Thomas method, and (c) Fujimoto method.

Time, d	0	2	4	6	8
BOD, mg/L	0	11	18	22	24

5.13 Calculate total theoretical oxygen demand (ThOD), total nitrogenous oxygen demand (TNOD), total carbonaceous oxygen demand (TCOD), and total organic carbon (TOC) of an industrial waste. The waste is represented by the chemical formula $C_6N_2H_6O_2$. Assume that nitrogen is converted to ammonia and then to nitrate. The average concentration of waste is 210 mg/L. Also, determine various ratios with respect to TCOD.

5.14 The electrolytic respirometer test results of a wastewater sample at 21°C are given in Figure 5.26. Determine reaction rate constants K and ultimate BOD for carbonaceous and nitrogenous oxygen demands. Also, determine UOD.

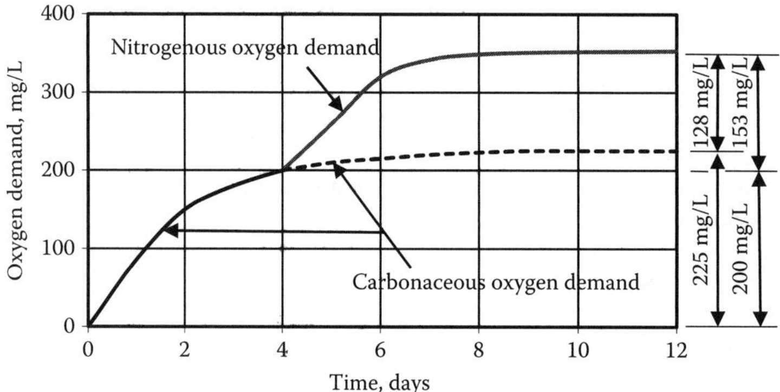

FIGURE 5.26 Electrolytic respirometer test results (Problem 5.14).

5.15 The ultimate oxygen demand (UOD) of a waste is expressed by the following equation:

$$\text{UOD (mg/L)} = A \times BOD_5 + B \times (NH_3\text{-N})$$

Calculate the factors A and B if the reaction rate k (base e) is 0.18 d^{-1}, and the ammonia nitrogen is fully converted to NO_3-N.

5.16 The BOD_5 and TOC test data of a sampling point in a process train over 7-day data collection period are given below. Develop the linear relationship and average BOD_5/TOC ratio.

Sampling Day	1	2	3	4	5	6	7
BOD_5, mg/L	122	97	215	250	100	139	144
TOC, mg/L	102	80	180	210	85	115	120

5.17 Two lists are given below. Match the terms.

List A	List B
i. Phototroph	a. Oxygen must be absent
ii. Autotroph	b. *G. lamblia*
iii. Anaerobes	c. 12–18°C
iv. Giardiasis	d. Poorly defined nucleus
v. Cryophilic	e. Derive energy from sunlight
vi. Prokaryotes	f. Schistosomiasis
vii. Helminthes	g. Derive cell carbon from CO_2

5.18 Identify the characteristics that apply to coliform organisms.

a. Pathogenic
b. Gram negative
c. Members of *Enterobateriaceae* family
d. Present in wastewater
e. Nonspore-forming cells
f. Include genera *Klebsiella*
g. Incapable of fermenting lactose
h. Its presence means safe water

5.19 The expected coliform count of a wastewater sample is 10^7 per mL. Indicate dilution series to achieve a dilution such that 1 mL = 0.000001 mL original sample. What is the expected coliform count in 1 mL final dilution?

5.20 Multiple dilution test with five lactose broth fermentation tubes at three dilutions of 0.1 mL, 0.01 mL, and 0.001 mL (total 15 fermentation tubes) were performed with primary settled effluent. After incubation, the positive gas production results were 5/5, 3/5, and 10/5. Determine the MPN using (a) probability table, (b) probability equation, and (c) Thomas equation.

5.21 Multiple dilution tests with five lactose broth fermentation tubes at four dilutions of 1, 0.1, 0.01, and 0.001 mL were performed with primary settled effluent. After incubation, the following positive gas production results were obtained 5/5, 3/5, 2/5, and 0/5. Determine the MPN using probability equation.

5.22 Series of dilutions with five tubes each were prepared from several wastewater samples to determine MPN from probability table (Table 5.11). The results are given below. Select three dilutions for each sample to determine MPN. Use the selection procedure given in Table 5.10.

Sampling Day	Dilution Series and Positive Gas Results					
	10	1	0.1	0.01	0.001	0.0001
1	5/5	5/5	5/5	3/5	2/5	1/5
2	5/5	5/5	0/5	3/5	2/5	1/5
3	5/5	5/5	5/5	4/5	3/5	1/5
4	5/5	5/5	5/5	5/5	3/5	2/5

5.23 Coliform density was determined by membrane filtration (Figure 5.27). The dilution is given below. 20 mL of series dilution was filtered through the membrane filter. A total of 43 colonies were counted. Calculate coliform density per 100 mL sample.

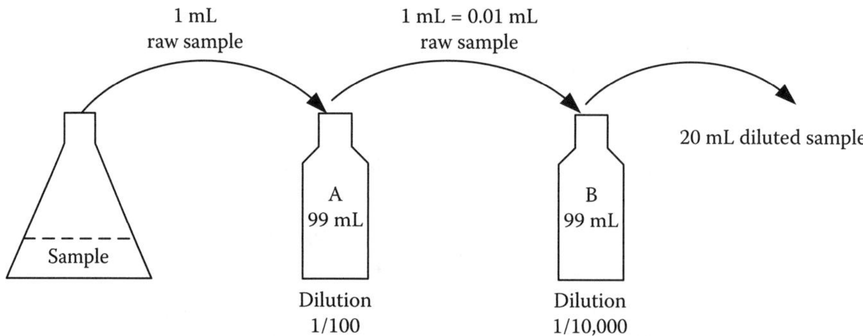

FIGURE 5.27 Series dilution to obtain a desired volume of sample [Problem 5.23].

5.24 Membrane filtration technique was used to determine coliform count in a wastewater sample. Several volumes of samples were filtered and coliform colonies were counted. The results are given below. Determine the coliform count.

Sample Volume, mL	100	10	5	2	1	0.5
Coliform Colonies	Too many to count	78	45	16	9	2

5.25 Acute toxicity test was conducted on an industrial chemical. Ten young *Ceriodaphnia dubia* test organisms in five concentrations of chemicals and in control were used in the acute toxicity test. The concentrations and mortality data are given below. Determine LD_{50} using (a) arithmetic plot, (b) arithmetic and log plot, (c) semilog scale, (d) probability plot, and (e) Probit equation.

Concentration, μg/L	0	370	750	1000	1600	2000
No. of Organisms Died	0	1	4	6	8	9

5.26 A municipal wastewater treatment plant is discharging secondary effluent into a creek. At low flow, the expected dilution is 10:1. The acute toxicity test was conducted with *C. dubia* over 96-h exposure. The following results were obtained. Determine the toxicity compliance. Control % survival = 100, $LC_{50} = 4.8$.

5.27 An industrial effluent was tested for chronic toxicity using *C. dubia*. The critical effluent dilution in the receiving water was 5:1. The test procedure utilized a series of effluent dilutions in continuous flow-through systems. The results of chronic test were based on reduced reproduction and are reported as *NOEC*. Following results were obtained. Determine the toxicity compliance. Exposure period = 10 days, control survived = 98%, and NOEC = 2.7%.

5.28 An industry is discharging 2500 m^3/d wastewater into a sanitary sewer. The concentrations of BOD_5, TSS, TN, and TP are 200, 280, 35, 10 mg/L, respectively. Calculate the population equivalent of the industry based on flow, BOD_5, TSS, TN, and TP.

5.29 A community has garbage grinders installed in homes. Food and kitchen wastes are ground and discharged into the sanitary sewers. Data from other communities with garbage grinders shows that average BOD_5 and TSS contributions are 0.24 and 0.28 lb per capita per day. The wastewater flow is 110 gpcd. Calculate: (a) concentration of BOD_5 and TSS in the wastewater; and (b) percent BOD_5 and TSS increase due to garbage grinder. Assume that the unit loadings in Table 5.15 are without garbage grinder.

5.30 The diurnal flow rate and BOD_5 concentrations were measured every 2 h at a wastewater treatment plant. The results are tabulated below. Determine average flow, BOD_5 concentration, and BOD_5 mass loading. Plot the diurnal flow, BOD_5, and mass loading curves.

	a.m.						p.m.					
Time of Day	12	2	4	6	8	10	12	2	4	6	8	10
Flow, m^3/h	75	38	36	74	89	88	88	73	63	63	83	89
BOD_5, mg/L	165	92	40	60	140	205	214	210	180	145	220	250

5.31 The peaking factors of sustained mass loadings are given in Figure 5.19. Determine total consecutive sustained 7-d BOD_5 mass loading at a treatment plant. The average design flow and annual average BOD_5 concentration for which the plant is designed are 4.0 MGD and 180 mg/L. Calculate the total sustained BOD_5 mass loading that the plant is expected to receive over the sustained mass loading period.

References

1. APHA, AWWA, and WEF, *Standard Methods for Examination of Water and Wastewater*, 22nd ed., American Public Health Association, Washington, DC, 2012.
2. Faust, D. D. and O. M. Aly, *Chemistry of Water Treatment*, 2nd ed., Lewis Publishers, Boca Raton, FL, 1997.
3. Perry, H. S., D. R. Rowe, and G. Tchobanoglous, *Environmental Engineering*, McGraw-Hill Book Co., New York, NY, 1985.
4. Qasim, S. R., *Wastewater Treatment Plants: Planning, Design, and Operation*, 2nd ed., CRC Press, Boca Raton, FL, 1999.
5. Metcalf & Eddy, Inc., *Wastewater Engineering: Treatment, Disposal, and Reuse*, 4th ed., McGraw-Hill, Inc., New York, 1991.
6. Tchobanoglous, G. and E. D. Shroeder, *Water Quality: Characteristics, Modeling, and Modification*, Addison-Wesley Publishing Co., Reading, MA, 1985.
7. Montgomery Watson Harza (MWH), *Water Treatment, Principles and Design*, 2nd ed., John Wiley & Sons, Inc., Hoboken, NJ, 2005.
8. Ramaldho, R. S., *Introduction to Wastewater Treatment Processes*, 2nd ed., Academic Press, New York, 1983.
9. Fujimoto, Y. Graphical use of first-stage BOD equation, *Journal Water Pollution Control Federation*, 36 (1), 1964, 69–72.
10. Young, J. C., Chemical methods for nitrification control, *Journal Water Pollution Control Federation*, 45 (4), 1973, 637.
11. U.S. Environmental Protection Agency, *Methods for Chemical Analysis of Water and Wastes*, EPA/600/4-79/020, Office of Research and Development, Washington, DC, 1983.
12. Niedercorn, J. G., S. Kaufman, and H. Senn, Rapid procedure for estimating organic materials in industrial wastes, *Journal Sewage and Industrial Wastes*, 25 (8), 1953, 950.
13. U.S. Environmental Protection Agency, *Handbook for Monitoring Industrial Wastewater*, Technology Transfer, August 1973.
14. Sawyer, C. N. and P. N. McCarty, *Chemistry for Environmental Engineering*, McGraw-Hill Book Co., New York, 1978.
15. Qasim, S. R., E. Motley, and G. Zhu, *Water Works Engineering: Planning, Design, and Operation*, Prentice Hall PTR, Upper Saddle River, NJ 2000.
16. Gaudy, A. F. and E. T. Gaudy, *Microbiology for Environmental Scientists and Engineers*, McGraw-Hill Book Co., New York, 1980.
17. Thomas, H. A., Bacterial densities from fermentation tube tests, *Journal American Water Works Association*, 34 (4), 1942, 572–576.

18. Miller, L. A., R. S. Taylor, and L. A. Monk, *NPDES Permit Handbook*, Government Institutes, Inc., Swidler & Berlin Chartered, Washington DC, May 1991.
19. McIntyre, D., M. DeGraere, and J. Fava, Monitoring effluent for toxicity, *Proceedings of the Specialty Conference on Toxicity Based Permits for NPDES Compliance and Laboratory Techniques*, Water Pollution Control Federation, April 16–19, 1989.
20. U.S. Environmental Protection Agency, *Technical Support Document for Water Quality-Based Toxics Control*, EPA/505/2-90-001, Office of Water, Washington, DC, 1991.
21. Peltier, W. and C. I. Weber, *Methods for Measuring the Acute Toxicity of Effluents to Aquatic Organisms*. 3rd ed., EPA-600/4-85-013, Office of Research and Development, Cincinnati, Ohio, 1992.
22. U.S. Environmental Protection Agency, *Short-Term Methods for Estimating Chronic Toxicity of Effluents and Receiving Waters to Freshwater Organisms*, 5th ed., EPA-821-R-02-013, Office of Water, Washington, DC, 2002.
23. U.S. Environmental Protection Agency, *Methods for Measuring the Acute Toxicity of Effluents and Receiving Waters to Freshwater and Marine Organisms*, 5th ed., EPA-821-R-02-012, Office of Water, Washington, DC, 2002.
24. U.S. Environmental Protection Agency, *User's Guide to the Conduct and Interpretation of Complex Effluent Toxicity Tests at Estuarine/Marine Sites*, EPA-600/X-86/224, Washington, DC, 1985.
25. U.S. Environmental Protection Agency, *Short-Term Methods for Estimating the Chronic Toxicity of Effluents and Receiving Waters to Marine and Estuarine Organisms*, 3rd ed., EPA-821-R-02-014, Office of Water, Washington, DC, 2002.
26. Stephen, C. E., "Methods for Calculating LC_{50}," In *Aquatic Toxicology and Hazard Evaluation* (F. L. Mayer and J. L. Hamelink, eds.), ASTM STP 634, American Society for Testing and Materials, Philadelphia, PA, 1984, pp. 65–84.
27. Louvar, J. F. and B. D. Louvar (eds.), *Health and Environmental Risk Analysis: Fundamentals with Applications*, Prentice Hall PTR, Upper Saddle River, NJ, 1998.

6

Wastewater Treatment Objectives, Design Considerations, and Treatment Processes

6.1 Chapter Objectives

This is an introductory chapter to provide prospective on wastewater treatment objectives and regulations, basic design considerations, and processes used for wastewater treatment. The following topics are covered:

- Wastewater treatment objectives and regulations
- Basic design considerations to include current and future trends in population growth, design period, wastewater characteristics and degree of treatment, treatment processes selection and combinations, and other design factors
- Service area, treatment plant site selection, regulatory requirements, and effluent discharge limitations
- Residuals production, handling and processing technology selection and combinations, and final disposal and reuse of biosolids
- Plant layout, plant hydraulics, flow through conduits, pumping, and flow measurement
- Energy and resource requirements, plant economics, and environmental impact assessment
- General considerations in plant planning, design, and construction management

6.2 Treatment Objectives and Regulations

6.2.1 Objectives

The objectives of a wastewater treatment plant are to meet the effluent quality standards established by the federal, state, and regional regulatory authorities; and to prevent many adverse environmental conditions that may develop due to inadequate wastewater treatment. Some of these adverse environmental conditions are (1) unsightliness, nuisance, and obnoxious odors at the plant and at the disposal site; (2) contamination of water supply sources; (3) destruction of fish, shellfish, and other aquatic life; (4) impairment of beneficial uses of receiving waters; and (5) decline in land values causing restriction in community growth and development. Ideally, a wastewater treatment plant must encourage the beneficial uses of effluent and residuals, and enhance community growth and development.

6.2.2 Regulations

The passage of the Federal Water Pollution Control Amendments of 1972 (PL 92-500), and subsequent amendments brought a significant change in water pollution control philosophy.[1,2] The specific national water quality goals and regulations are to eliminate the discharge of pollutants into all surface waters, and restore and maintain the physical, chemical, and biological integrity of the Nation's waters. The highlights of these regulations are presented below:

1. Under the National Pollutant Discharge Elimination System (NPDES) permitting program, effluent limitations are established for conventional, nonconventional, toxic, and hazardous pollutants based upon their levels of concentration in the effluent.[1]
2. Industrial pretreatment standards are developed for all pollutants that "interfere with, pass through, concentrate in sludge, or otherwise incompatible" with publicly owned treatment works (POTWs).[1]
3. The U.S. Environmental Protection Agency (U.S. EPA) published its definition of secondary treatment. The current definition is given in Table 6.1.[3]
4. Under the Clean Water Acts, each state agency is required to develop a *water quality management* (WQM) plan.[1,2] Bodies of state waters are classified as either *effluent-limited* or *water-quality limited.* The state agency established total maximum daily load (TMDL) waste allocation for all surface waters throughout the state. The water body is classified as effluent-limited if states water quality standards are met by discharge of secondary effluent. Water-quality limited bodies of water require discharge of higher effluent quality. The revised TMDL waste allocation includes (1) point sources, (2) nonpoint sources, (3) natural background levels, and (4) a margin of safety.[1,4]
5. Biomonitoring requirement is established to determine the toxicity in the wastewater effluent to control the discharge of toxic pollutants in toxic amounts.[5]
6. Construction grant program with federal share of 75% was initiated for construction of POTWs. The federal share was reduced to 55%. Subsequently, it was changed to loans under State Revolving Fund (SRF).[6–8]
7. The regulations that affect design of POTWs include those for the treatment, disposal and beneficial use of biosolids. National standards are set for pathogens and heavy metals contents and for safe handling and use of biosolids. The rule promotes the development of "clean sludge" that can be used as biosolids over cultivated land.[9]
8. Health and environmental concerns in design of POTWs are emphasized. These are (1) release of volatile organic compounds (VOCs) and toxic air contaminants (TACs), (2) odors, and (3) disinfection byproducts (DBPs).
9. Greater emphasis is on effluent reuses including potable water reuse. There is however much concern regarding detection, monitoring, and health risks of many contaminants. Chemicals of greatest concern are gasoline additives, solvents, phenolic compounds, and the pharmaceuticals and personal care products (PPCPs). Many of these compounds are *endocrine-disruptors*, meaning

TABLE 6.1 Minimum National Standards for Secondary Treatment as Defined by the U.S. EPA

Effluent Parameter	30-Day Average		7-day Average, Maximum Concentration, mg/L
	Maximum Concentration, mg/L	Minimum Removal, %	
BOD_5	30	85	45
$CBOD_5$[a]	25	85	40
TSS[b]	30	85	45
pH[c]		Between 6.0 and 9.0	

[a] CBOD = carbonaceous biochemical oxygen demand. $CBOD_5$ may be substituted for BOD_5 by the permitting authority.

[b] TSS = total suspended solids.

[c] The pH limit may be exceeded if POTW demonstrates that (1) inorganic chemicals are not added to waste stream as part of treatment process, and (2) contributions from industrial sources do not cause the pH of the effluent to be < 6 and >9.

Note: Special considerations may apply to combined sewers, certain industrial categories, waste stabilization ponds, and less concentrated influent wastewater for separate sewers.

they may enhance the hormones production, exaggerate response, or block the effect of hormone on the body. Some reported findings include problems with development, behavior, and reproduction of many aquatic animals. Decline in male sperm count has also been reported. Increases in certain cancers have been blamed on endocrine-disruptive chemicals.[10–13]

10. The U.S. EPA is required to report to congress the *Need Survey* about the progress and the status of the program and detailed capital cost investment additionally needed to comply with the requirements of the Acts. The ultimate goal of these Acts is to eliminate the discharge of pollutants into all surface waters.[14]

EXAMPLE 6.1: SECONDARY EFFLUENT COMPLIANCE

Grab samples were collected daily for 7 days from the secondary effluent pipe. The BOD_5 and TSS concentrations are tabulated below. Determine 7-day average BOD_5 and TSS values. Compare the effluent quality with national standards for secondary treatment.

Parameter	Concentration, mg/L						
	Monday	Tuesday	Wednesday	Thursday	Friday	Saturday	Sunday
BOD_5	20	15	15	50	54	40	30
TSS	28	25	30	65	68	45	25

Solution

1. Determine 7-d average BOD_5 and TSS.

$$BOD_5 = (20 + 15 + 15 + 50 + 54 + 40 + 30)\ \text{mg/L}/7 = (224\ \text{mg/L})/7 = 32\ \text{mg/L}$$

$$TSS = (28 + 25 + 30 + 65 + 68 + 45 + 25)\ \text{mg/L}/7 = (286\ \text{mg/L})/7 = 41\ \text{mg/L}$$

2. Compare the effluent quality with the minimum national standards for secondary treatment in Table 6.1.

   The minimum national secondary treatment standards for 7-d average BOD_5 and TSS are 45 mg/L each. Therefore, the effluent meets the secondary effluent quality standards.

6.3 Basic Design Considerations

There are many important design factors that must be considered during the initial planning and design stages of a wastewater treatment project. Basic design factors are:[15]

1. Initial and design years and design population
2. Service area and treatment plant site selection
3. Regulatory requirements and effluent limitations
4. Characteristics of wastewater and degree of treatment
5. Selection of treatment processes, equipment, and process train
6. Plant layout
7. Plant hydraulic conditions
8. Plant hydraulic profile
9. Energy and resource requirements and plant economics
10. Environmental impact assessment

Most of these factors are covered in great detail in several chapters; the information presented below is an introduction to wastewater project planning.

6.3.1 Initial and Design Years and Design Population

Initial and Design Years: It generally takes several years to plan, design, and construct a wastewater treatment facility. Accordingly, most of the plant's components are made large enough to satisfy the community needs for several years in the future. The initial year is the year when the construction is completed and the initial operation begins. The design or planning year is the year when the facility is expected to reach its full designed capacity. Selecting the design year requires sound judgments and skills in developing future population growth estimates from the past social and economic trends of a community. Design periods are generally chosen with the following factors in mind:

1. Useful life of treatment units
2. Ease or difficulty in expansion
3. Performance of the treatment facility during the initial years
4. Future growth in the population, commercial and industrial developments, water demands, and wastewater characteristics
5. Cost of present and future construction, and availability of funds

The design periods of different components of a treatment facility may also vary. As an example, main conduits, channels, and appurtenances that cannot be expanded readily are designed for periods up to 50 years in the future. On the other hand, treatment units, process equipment, pumps, and sludge processing and disposal facilities are constructed for shorter periods to avoid construction of oversized units. In such cases, adequate space is left at the plant site for expansion of the facility at different staging periods. According to the guidelines of the construction grants program, the design period may be divided into several staging periods (10, 15, and 20 years). The number and length of staging period depend on the ratio of wastewater flow expected at the design and initial years. The length of staging period is determined from Table 6.2.

Design Population: The volume of wastewater generated in a community depends on the population and per capita contribution of wastewater. It is therefore important to estimate the population to be served at the design year. Accurate population prediction is quite difficult because many factors influence the growth of a city. Among the important factors are industrial growth; state of development of the surrounding area; location with regard to transportation sources; availability of raw material, land, and water resources; local taxes and government activities; migration trends; and so on.

Sources of Population Data: The population data can be obtained from several sources. The U.S. Bureau of Census (Department of Commerce) publishes 10-year census data. For the interim periods, reliable data can usually be obtained from local census bureaus; the state, county, or local planning commissions; the chambers of commerce; voters registration lists; the post office; newspapers; and public utilities (telephone, electric, gas and water, etc.). It is important that the design engineer becomes familiar with the population data sources and the type of information that can be obtained from these sources.

TABLE 6.2 Staging Periods for Plant Expansion over the Design Period

Flow Growth Factor, Ratio of Flow at Initial and Design Years	Staging Period, Years
Less than 1.3	20
1.3–1.8	15
Greater than 1.8	10

Source: Adapted in part from Reference 16.

Method of Population Forecasting: There are many mathematical and graphical methods that are used to project past population data to the design year. Widely employed methods are:

1. Arithmetic growth
2. Geometric growth
3. Decreasing rate of increase
4. Mathematical or logistic curve fitting
5. Graphical comparison with similar cities
6. Ratio method
7. Employment forecast
8. Birth cohort

All these methods utilize different assumptions and therefore give different results. Selection of any method depends on the amount and type of data available and whether the projections are made for the short or long term. The arithmetic, geometric, decreasing rate of increase, and logistic curve-fitting methods are summarized in Table 6.3 (Equations 6.1 through 6.10). The remaining methods are presented in Table 6.4.

TABLE 6.3 Population Projections by Using Arithmetic, Geometric, Decreasing Rate of Increase, and Mathematical Curve Fitting

Method	Description	Basic Equations
Arithmetic method	Population is assumed to increase at a constant rate. The method is used for short-term estimates (1-5 year)	$\frac{dY}{dt} = K_a; \quad Y_t = Y_2 + K_a - T_2)$ (6.1) $K_a = \frac{Y_2 - Y_1}{T_2 - T_1}$ (6.2)
Geometric method	Population is assumed to increase in proportion to the number present. The method is commonly used for short-term estimates (1-5 year).	$\frac{dY}{dt} = K_p Y; \quad \ln(Y_t) = \ln(Y_2) + K_p(T - T_2)$ (6.3) $K_p = \frac{\ln(Y_2) - \ln(Y_1)}{T_2 - T_1}$ (6.4)
Decreasing rate of increase method	Population is assumed to reach some limiting value or saturation point. The method is normally applied for long-term estimates.	$\frac{dY}{dt} = K_d(Z - Y);$ $Y_t = Y_2 + (Z - Y_2)\left(1 - e^{-K_d(T-T_2)}\right)$ (6.5) $Z = \frac{2Y_0 Y_1 Y_2 - Y_1^2(Y_0 + Y_2)}{Y_0 Y_2 - Y_1^2}$ (6.6) $K_d = -\frac{1}{T_2 - T_1}\ln\left(\frac{Z - Y_2}{Z - Y_1}\right)$ (6.7)
Mathematical or logistic curve fitting	It is assumed that the population growth follows a logistic mathematical relationship. Most common relationship is an S-shaped curve.	$Y_t = \frac{Z}{1 + ae^{b(T-T_0)}}$ (6.8) $a = \frac{Z - Y_0}{Y_0}$ (6.9) $b = \frac{1}{n}\ln\left(\frac{Y_0(Z - Y_1)}{Y_1(Z - Y_0)}\right)$ (6.10)

Note: dY/dt = rate of changes in population with respect to time; Y_0, Y_1, and Y_2 = populations at time T_0, T_1, and T_2; Y_t = estimated population of the year of interest; Z = saturation population; K_a, K_p, and K_d = population constants for arithmetic, geometric, and decreasing rate of increase; a and b = constants; n = constant interval between T_0, T_1, and T_2 (generally 10 years).

Source: Adapted in part from References 15 and 17.

TABLE 6.4 Population Projections by Using Graphical Comparison, Ratio Method, Employment Forecast, and Birth Cohort

Method	Description
Graphical comparison	The procedure involves the graphical projection of the past population data for the city being studied. The population data of other similar but larger cities are also plotted in such a manner that all the curves are coincident at the present population of the city being studied. These curves are used as a guide in future population projections.
Ratio and correlation	In this method, the population of the city in question is assumed to follow the same trends as that of the region, county, or state. From the population records of a series of census years, the ratio is plotted and then projected to the year of interest. From the available estimated population of the region, county, or state, and the projected ratio, the population of the concerned city is obtained.
Employment forecast or other utility connections forecast	The population is estimated using the employment forecast. From the past data of population and employment, the ratio is plotted and population is obtained from the projected employment forecast. Procedure is similar to that of the ratio method. Similar procedure can be utilized from the forecast of various utility service connections such as telephone, electric, gas, water and sewers, etc. Utility companies conduct studies and develop reliable forecasts on the future connections. Forecasts of postal and newspaper service points have also been used in population estimates.
Birth cohort	A birth cohort is defined by demographers as a group of people born in a given year or period.[a] The existing populations of males and females in different age groups are determined from the past records. From birth and death rates of each group and population migration data, the net increase in each group is calculated. The population data are then shifted from one group to the other until the design period is reached.

[a] Demography is that branch of anthropology that deals with the statistical study of the characteristics of human population with reference to total size, density, number of deaths, births, migration, etc.

Source: Adapted in part from References 15 and 17.

Population Density: The average population density for the entire city rarely exceeds 7500–10,000 per km^2 (30–40 per acre). Often it is important to know the population density in different parts of the city in order to estimate the wastewater flows, and to design the collection network. Density varies widely within a city depending on the land use. The average population densities based on land use characteristics are summarized in Table 6.5.

TABLE 6.5 Range of Population Densities in Various Sections of a City

Land Use	Population Range	
	Persons per km^2	Persons per Acre
Residential areas		
Single-family dwellings, large lots	1250–3700	5–15
Single-family dwellings, small lots	3700–8700	15–35
Multiple-family dwellings, small lots	8700–25,000	35–100
Apartment or tenement houses	25,000–250,000	100–1000 or more
Commercial areas	3700–7500	15–30
Industrial areas	1250–3700	5–15
Total, exclusive of parks, playgrounds, and cemeteries	2500–12,500	10–50

Note: km^2 = acre × 247.1.

Source: Adapted in part from References 15 and 17.

EXAMPLE 6.2: PLANNING AND DESIGN PERIOD

A wastewater treatment plant is currently overloaded and will not meet the future effluent standards. A capital improvement plan is under consideration. The project planning schedule is given below. Determine (a) planning and design period, (b) initial year, (c) design year, (d) staging period, and (e) design period.

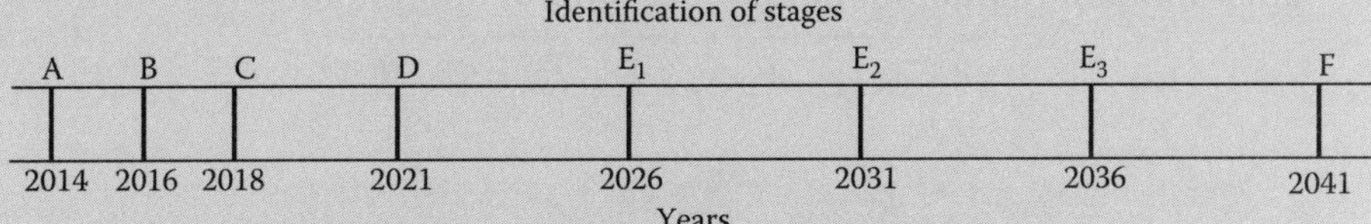

A = Problem identified and planning started
B = Completion of facility plan (phase I report or engineering report)
C = Completion of design, plans, specifications, and construction bid package
D = Completion of constructions and operation begins
D–E_1, E_1–E_2, and E_2–E_3 = Staging periods
F = Plant reaches full capacity.

Solution

a. Planning and design period is 4 years between 2014 (A) and 2018 (C).
b. Initial year is in 2021 (D).
c. Design year is in 2041 (F).
d. There are three staging periods, each of 5 years: 2021–2026 (D–E_1), 2026–2031 (E_1–E_2), and 2031–2036 (E_2–E_3).
e. Design period is 20 years between 2021 (D) and 2041 (F).

EXAMPLE 6.3: STAGING PERIOD

A wastewater treatment plant has initial year population of 38,000 and wastewater flow of 330 Lpcd. The estimated design year population and wastewater flow are 85,000 and 310 Lpcd. Calculate the staging period.

Solution

1. Calculate initial and design year flows.

$$\text{Initial year flow to the plant} = 330\ \text{L/person·d} \times 38{,}000\ \text{persons} \times 10^{-3}\ \text{m}^3/\text{L}$$
$$= 12{,}540\ \text{m}^3/\text{d}$$
$$\text{Design year capacity of the plant} = 310\ \text{L/person·d} \times 85{,}000\ \text{persons} \times 10^{-3}\ \text{m}^3/\text{L}$$
$$= 26{,}350\ \text{m}^3/\text{d}$$

2. Calculate the ratio of flow rates in design to initial year.

$$\frac{\text{Design year flow}}{\text{Initial year flow}} = \frac{26{,}350\ \text{m}^3/\text{d}}{12{,}540\ \text{m}^3/\text{d}} = 2.1$$

3. Determine staging period from Table 6.2.
   The staging period is 10 years or less.

EXAMPLE 6.4: POPULATION ESTIMATES FROM ARITHMETIC, GEOMETRIC, DECREASING RATE OF INCREASE, AND LOGISTIC CURVE FITTING METHODS

Estimate the population of a city using arithmetic, geometric, decreasing rate of increase, and logistic curve fitting methods. Three census years' data are given below: The design year for treatment plant expansion is 2030.

Year	Population, Person
1990	5000
2000	10,000
2010	15,000
2020	18,000

Solution

1. Apply the following data from the table.

   $T_0 = 2000$, $Y_0 = 10{,}000$ persons; $T_1 = 2010$, $Y_1 = 15{,}000$ persons; $T_2 = 2020$, $Y_2 = 18{,}000$ persons; $T = 2030$, and $n = 10$ years.

2. Estimate population using arithmetic growth from Equations 6.1 and 6.2.
   Most recent census data in 2020 is used for population projections.

$$K_a = \frac{(18{,}000 - 15{,}000)\text{ persons}}{(2020 - 2010)\text{ year}} = 300\text{ persons/year}$$

$$Y_{2030} = 18{,}000\text{ persons} + 300\text{ persons/year} \times (2030 - 2020)\text{ years} = 21{,}000\text{ persons}$$

3. Estimate population using geometric growth from Equations 6.3 and 6.4.

$$K_p = \frac{\ln(18{,}000) - \ln(15{,}000)}{(2020 - 2010)\text{ year}} = \frac{9.798 - 9.616}{10\text{ year}} = 0.0182\text{ year}^{-1}$$

$$\ln(Y_{2030}) = \ln(18{,}000) + 0.0182\text{ year}^{-1} \times (2030 - 2020)\text{ year} = 9.798 + 0.182 = 9.98$$

$$Y_{2030} = e^{9.98} = 21{,}600\text{ persons}$$

4. Estimate population using decreasing rate of increase method from Equations 6.5 through 6.7.

$$Z = \frac{2 \times (10{,}000 \times 15{,}000 \times 18{,}000)\text{ persons}^3 - (15{,}000\text{ persons})^2 \times (10{,}000 + 18{,}000)\text{ persons}}{(10{,}000 \times 18{,}000)\text{ persons}^2 - (15{,}000\text{ persons})^2}$$

$$= 20{,}000\text{ persons}$$

$$K_d = -\frac{1}{(2020 - 2010)\text{ years}} \times \ln\left(\frac{(20{,}000 - 18{,}000)\text{ persons}}{(20{,}000 - 15{,}000)\text{ persons}}\right) = -0.1\text{ year}^{-1} \times (-0.916)$$

$$= 0.0916\text{ year}^{-1}$$

$$Y_{2030} = 18{,}000\text{ persons} + (20{,}000 - 18{,}000)\text{ persons} \times \left(1 - e^{-0.0916\text{ year}^{-1} \times (2030-2020)\text{ year}}\right)$$

$$= 18{,}000\text{ persons} + 1200\text{ persons} = 19{,}200\text{ persons}$$

5. Estimate population using logistic curve fitting from Equations 6.8 through 6.10.

$$Z = 20{,}000\ \text{persons}$$

$$a = \frac{(20{,}000 - 10{,}000)\ \text{persons}}{10{,}000\ \text{persons}} = 1$$

$$b = \frac{1}{10\,\text{years}} \times \ln\left(\frac{10{,}000 \times (20{,}000 - 15{,}000)\,\text{persons}^2}{15{,}000 \times (20{,}000 - 10{,}000)\,\text{persons}^2}\right) = \frac{1}{10\,\text{years}} \times \ln 0.333 = -0.1099\,\text{year}^{-1}$$

$$Y_{2030} = \frac{20{,}000\ \text{persons}}{1 + 1 \times e^{-0.1099\,\text{year}^{-1} \times (2030-2000)\,\text{year}}} = \frac{20{,}000\ \text{persons}}{1 + 0.0370} = 19{,}290\ \text{persons}$$

EXAMPLE 6.5: GRAPHICAL COMPARISON METHOD FOR POPULATION PROJECTION

Estimate the population of City A by using graphical comparison with populations of Cities B and C. The design year is 2040.

Year	Population, 10^3 Persons		
	City A	City B	City C
1980	8.0	18.0	16.0
1990	11.0	20.3	20.0
2000	14.2	22.0	23.5
2010	18.0	23.2	26.0

Solution

1. Determine the years for Cities B and C so that their curves are coincident at the population of City A in the year of 2010.

   City A population in 2010 = 18,000 persons

   City B population in 1980 = 18,000 persons

   City C reaches a population of 18,000 persons between the years 1980 and 1990.

   Using arithmetic rate of increase method (Equations 6.1 and 6.2), calculate the year T when the City C reaches a population of 18,000 persons.

$$K_a = \frac{(20{,}000 - 16{,}000)\ \text{persons}}{(1990 - 1980)\ \text{year}} = 400\ \text{persons/year}$$

$$18{,}000\ \text{persons} = 16{,}000\ \text{persons} + \frac{400\ \text{persons}}{\text{year}}(T - 1980)\ \text{years}$$

$$T = 1980 + \frac{18{,}000\ \text{persons} - 16{,}000\ \text{persons}}{400\ \text{persons/year}} = 1985$$

2. Plot the census populations of the City A.

   Shift the year scale for Cities B and C so that their population curves start from 18,000 persons which is the year 2010 census population of City A (Point P). The graphical presentation of these populations is shown in Figure 6.1.

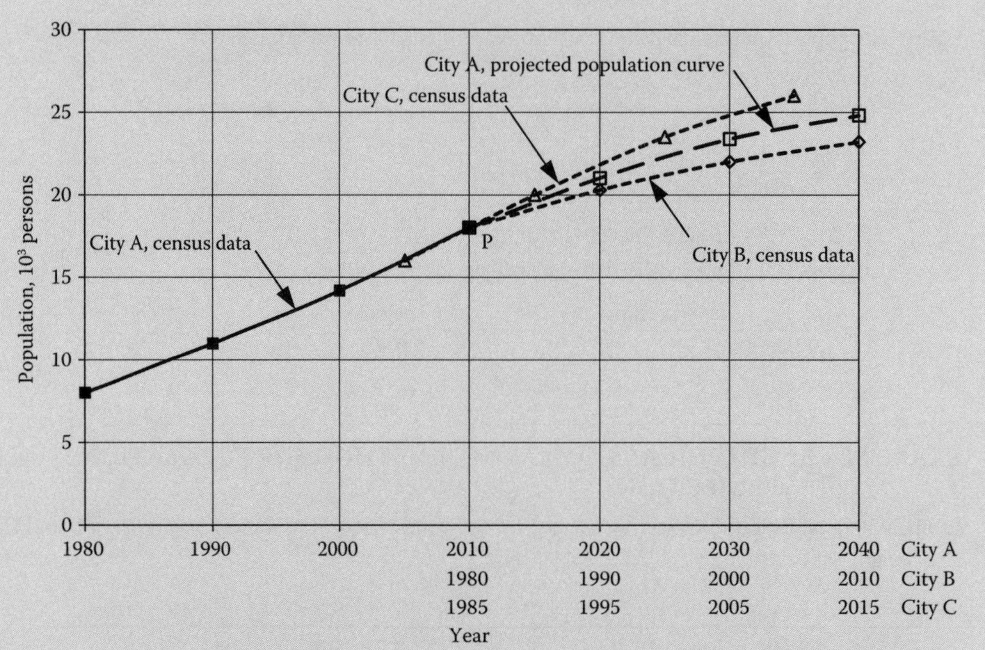

FIGURE 6.1 Population estimate by graphical comparison (Example 6.5).
Note: The time scales of Cities B and C have been shifted to start curves from point P (18,000 persons).

3. Project the population of the City A using shifted population curves of the Cities B and C as a guide.
4. The graphically estimated population of City A = 24,800 in the year 2040.

EXAMPLE 6.6: RATIO METHOD FOR POPULATION PROJECTION

Estimate the population of the City A using the ratio method. The design year is 2030. The estimated population of the region in the year 2030 is obtained from the State Planning Commission. It is 988,000 persons.

Year	Population, 10^3 Persons	
	City A	Region
1980	50	455
1990	61	623
2000	72	766
2010	77	850

Solution

1. Develop the ratio of city population and region population.

Year	City A Population, 10^3 Persons	Region Population, 10^3 Persons	Ratio
1980	50	455	0.110
1990	61	623	0.098
2000	72	766	0.094
2010	77	850	0.091
2030	—	988 (estimated)	—

2. Plot the population ratio with year.
   The population ratio is plotted in Figure 6.2.

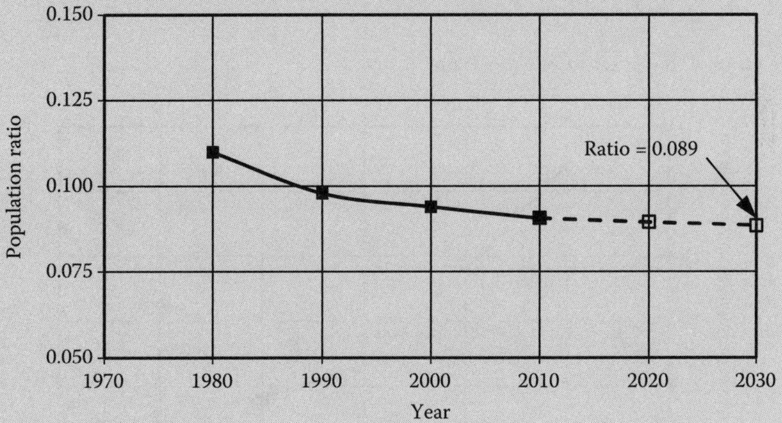

FIGURE 6.2 Population estimate by ratio method (Example 6.6).

3. Estimate the population ratio for the year 2030 and the population of the City A in the year 2030.
   Extend the population ratio graph in dotted line to the year 2030.
   Population ratio = 0.088
   The estimated population of the City A in the year 2030 = 0.089 × 988,000 persons = 88,000 persons.

EXAMPLE 6.7: EMPLOYMENT FORECAST

Estimate the population of a city using employment forecast. The design year is 2030. Use the following data. The employment forecast for the city is obtained from the Regional Planning Commission. For the year 2030, it is 21,300 persons.

Year	Population and Employment, 10^3 Persons	
	Population of the City	Employment
1980	20	6.80
1990	30	10.79
2000	39	14.77
2010	46	17.83
2030	—	21.30

Solution

1. Calculate the ratio of population to employment for the census years.

Year	Population of the City, 10^3 Persons	Employment in the City, 10^3 Persons	Ratio of Population to Employment
1980	20	6.80	2.94
1990	30	10.79	2.78
2000	39	14.77	2.64
2010	46	17.83	2.58
2030	—	21.30	—

2. Plot the ratios with respect to years in Figure 6.3.

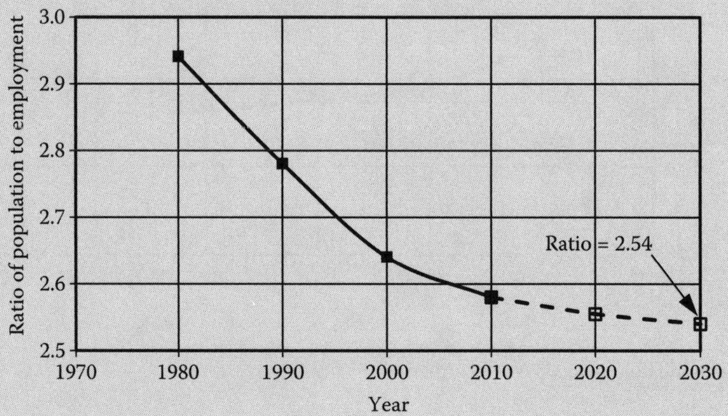

FIGURE 6.3 Population estimate by employment forecast (Example 6.7).

3. Estimate the ratio of population to employment for the year 2030 and the population of the city.
   Extend the dotted line of the ratios to the year 2030.
   The ratio population to employment for the year 2030 = 2.54
   Population of the city in the year 2030 = 2.54 × 21,300 persons = 54,100 persons

EXAMPLE 6.8: POPULATION ESTIMATE BASED ON POPULATION DENSITY

A city is rapidly growing and is expected to reach saturation population in the year 2030. The land-use plan for the year 2030 is developed by the City Planning Commission. The areas for various sections of the city in land-use plan are provided below. The average ultimate population densities for different land use are given below. Estimate (a) the saturation population of the city, and (b) average population density for the entire city.

Type of Land Use	Single-Family Large Lots	Single-Family Small Lots	Multiple-Family Small Lots	Apartments	Commercial	Industrial	Parks, Playground, and Cemeteries
Area, acres	380	250	120	20	560	875	680
Population density, persons/acre	10	25	68	550	23	10	30

Solution

1. Tabulate the land use and population.
   The land use average population densities and population are tabulated below.

Type of Land Use	Area, acre	Population Density, persons/acre	Population, Person
Single-family, large lots	380	10	3800
Single-family, small lots	250	25	6250
Multiple-family, small lots	120	68	8160
Apartments	20	550	11,000
Commercial	560	23	12,880
Industrial	875	10	8750
Parks, playground, and cemeteries	680	30	20,400
Total	2885	–	71,240

2. Determine the saturation population of the city, and average population density of the entire city.
   a. Saturation population of the city = 71,240 persons
   b. Average population density of the entire city = 71,240 persons ÷ 2885 acres = 25 persons/acre

6.3.2 Service Area and Treatment Plant Site Selection

Service Area: Service area (also called sewer district) is defined as the total land area that will be eventually served by the proposed wastewater treatment facility. The area may be based on natural drainage, political boundaries, or both. Site visits, and engineering data on topography, geology, hydrology, climate, ecological elements, and social and economic conditions, and land use or zoning plans should be studied. These factors may affect both developed and undeveloped lands. Such efforts should be carefully coordinated with the state, regional, and local planning agencies, and should be in conformance with the development and implementation of the area-wide waste management plan.

Site Selection: Site selection of a wastewater treatment facility should be based on careful consideration of the land use and development patterns in the region, as well as social, environmental, and engineering constraints. It is important to remember that the selection of a site for a wastewater treatment plant will have long-lasting social, economic, and political repercussions on the affected community and neighborhood. Therefore, public involvement in decision-making is crucial. The basic principles that must be considered during site evaluation are (a) low elevation to permit gravity flow, (b) isolated site from presently buildup areas and future developments, (c) large land area to maintain isolation (buffer land),

(d) opportunity for local disposal of residuals, (e) above flood zone, in low-lying areas proper flood protection measures must be taken, (f) year round, all-weather access roads, (g) alternate source of electric power, (h) effluent reuse and disposal potential, (i) geology to permit foundation stability, moderate slope for locating units in their normal sequence, and (j) absence of archaeological and historical sites, and critical habitats for endangered or threatened species of flora or fauna.

The project planning team must investigate topography, drainage, surface and groundwater, soil type, prevailing winds, temperature, precipitation, seasonal solar angles, wildlife habitats, ecosystems, regional and local land use and zoning, transportation, archaeological and historical features, and other factors.

EXAMPLE 6.9: SERVICE AREA OF A WASTEWATER TREATMENT PLANT

A city has one wastewater treatment plant. The treatment plant capacity and expansion is planned to take place in several stages. Show the service areas.

Solution

The service area proposed build out of wastewater facilities for the city is shown in the Figure 6.4.

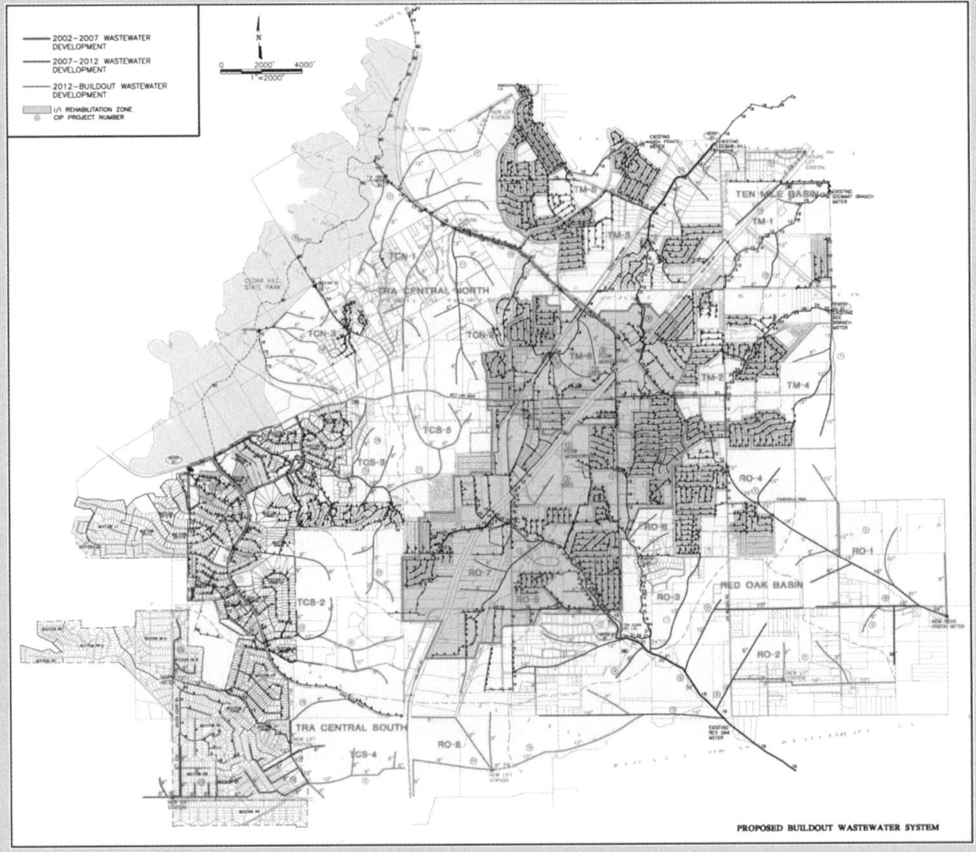

FIGURE 6.4 Example of a service area of a wastewater treatment plant project (Courtesy Freese and Nichols, Inc., Engineers, Architects, Planners, Fort Worth, Texas). (Example 6.9).

6.3.3 Regulatory Requirements and Effluent Limitations

Regulations are always subject to change as more information becomes available regarding wastewater characteristics, effectiveness of treatment processes, and environmental effects. The regulatory trends and minimum national standards for secondary treatment may also change. Many waste pollution control legislations in the United States have been passed in the past 40 years. Some of these are listed in Example 6.10.

EXAMPLE 6.10: WATER POLLUTION CONTROL LEGISLATIONS AND REGULATIONS

Provide (1) a chronological listing of major water pollution control legislation in the United States, and (2) a list of major federal water quality regulations.

Solution

1. The chronological listing of major water pollution control legislations are listed below:[1,2,18–20]
   1948: Federal Water Pollution Control Act (P.L. 80-845)
   1956: Water Pollution Control Act of 1956 (P.L. 84-660)
   1961: Federal Water Pollution Control Act Amendments (P.L. 87-88)
   1965: Water Quality Act of 1965 (P.L. 89-234)
   1966: Clean Water Restoration Act (P.L. 89-753)
   1969: National Environmental Policy Act (P.L. 91-190)
   1970: Water Quality Improvement Act of 1970 (P.L. 91-224)
   1972: Federal Water Pollution Control Act Amendments (P.L. 92-500)
   1977: Clean Water Act (CWA) of 1977 (P.L. 95-217)
   1981: Municipal Wastewater Treatment Construction Grants Amendments (P.L. 97-117)
   1987: Water Quality Act of 1987 (P.L. 100-4)
2. The major federal water quality regulations under the authorization of the CWA are selected and summarized below:[3,9,21–32]
   a. *40 CFR Part 122—EPA Administrated Permit Programs: The National Pollutant Discharge Elimination System (NPDES) (1983):* Establish regulatory requirements for general permit programs in NPDES, pretreatment programs, stormwater discharge, water transfers, and concentrated animal feeding operations (CAFOs).
   b. *40 CFR Part 123—State Program Requirements (1983):* Specify the procedures for the U.S. EPA in reviewing, approving, and withdrawing State programs, and the requirements State programs must meet to be approved by the U.S. EPA.
   c. *40 CFR Part 125—Criteria and Standards for the National Pollutant Discharge Elimination System (NPDES) (1979):* Establish criteria and standards for the treatment requirements and effluent discharge limits in the NPDES permits under.
   d. *40 CFR Part 130—Water Quality Planning and Management (1985):* Establish policies and program requirements for water quality planning and management under Section 303 of the CWA, including water quality standards (WQS), water quality monitoring, water quality management plans, water quality report, total maximum daily loadings (TMDL), and individual water quality-based effluent limitations.
   e. *40 CFR Part 131—Water Quality Standards (1983):* Describe the requirements and procedures for establishing water quality standards (WQS) under Section 303(c) of the CWA.
   f. *40 CFR Part 132—Water Quality Guidance for the Great Lakes System (1995):* Establish the water quality guidance for the Great Lakes system.
   g. *40 CFR Part 133—Secondary Treatment Regulation (1984):* Provide information on the level of effluent quality attainable through the application of secondary or equivalent treatment.

h. *40 CFR Part 136—Guidelines Establishing Test Procedures for the Analysis of Pollutants (1973):* Provide approved methods, alternative testing procedures, and method modifications, and analytical requirements for Permit Applications and Reporting under National Pollutant Discharge Elimination System (NPDES), including Whole Effluent Toxicity Test Methods.
i. *40 CFR Part 403—General Pretreatment Regulations for Existing and New Sources of Pollution (1981):* Establish responsibilities to implement National Pretreatment Standards to control pollutants which may interfere with the treatment processes in publicly owned treatment works (POTWs) or contaminate sewage sludge.
j. *40 CFR Part 412—Concentrated Animal Feeding Operations (CAFO) Point Source Category (2003):* Establish regulatory requirements for discharges of manure, litter, and/or process wastewater from applicable CAFOs.
k. *40 CFR Part 449—Airport Deicing Point Source Category (2012):* Establish regulatory requirements for discharges of pollutants from deicing operations at Primary Airport.
l. *40 CFR Part 450—Construction and Development Point Source Category (2009):* Establish regulatory requirements for discharges from construction activity required to obtain NPDES permit coverage.
m. *40 CFR Part 503—Standards for the Use or Disposal of Sewage Sludge (1993):* Establish standards, including general requirements, pollutant limits, management practices, and operational standards, for the final use or disposal of sewage sludge generated during the treatment of domestic sewage in a treatment works.
n. *Combined Sewer Overflow (CSO) Control Policy (1994):* Provide a national framework for control of CSOs through the NPDES permitting program (*59 FR 18688, April 19, 1994*).

6.3.4 Characteristics of Wastewater and Degree of Treatment

Reliable estimates of wastewater characteristics are important to design the treatment facilities. The characteristics of wastewater are developed in terms of flow conditions, and physical, chemical, and biological quality. The wastewater characteristics data are needed for the initial year and for the design year. The data includes minimum, average, and maximum dry weather flows; peak wet weather flows; sustained maximum flows; and chemical parameters such as BOD_5, TSS, pH, TDS, nitrogen, phosphorus, and toxic chemicals. Chapters 4 and 5 are devoted exclusively to develop wastewater characteristics.

The degree of treatment is determined based on the influent characteristics to the plant and the effluent quality required. If the effluent is discharged into the natural water, it should comply with NPDES permit requirements. If used for irrigation, the plant effluent must also satisfy the health regulations governing the types of crops that are irrigated. Other effluent uses such as recreational lakes, agricultures, industrial, and municipal may dictate the effluent quality and thus the degree of treatment.

EXAMPLE 6.11: WASTEWATER CHARACTERISTICS AND DEGREE OF TREATMENT

The wastewater characteristics data of a canning industry is given in Example 5.60. The pretreatment regulations require that the industrial wastewater discharged into municipal sewer must comply with the following standards: $BOD_5 \leq 180$ mg/L, COD ≤ 400 mg/L, TSS ≤ 220 mg/L, TN ≤ 40 mg/L, ON ≤ 20 mg/L, AN ≤ 20 mg/L, and TP ≤ 6 mg/L. Calculate the degree of treatment required in mg/L, percent, and kg/d. Average flow = 1500 m^3/d.

Solution

Tabulate the influent characteristics of the canning industry (Example 5.60), the pretreatment standards, and degree of treatment in mg/L, percent, and kg/d.

Constituents	Influent Concentration, mg/L	Pretreatment Standards, mg/L	Degree of Pretreatment Required		
			Concentration Removed, mg/L	Removal Efficiency, %	Loading Rate, kg/d
BOD_5	1200	180	1020[a]	85[b]	1530[c]
COD	2400	400	2000	83	3000
TSS	1800	220	1580	88	2370
TN	60	40	20	33	30
ON	44	20	24	55	36
AN	16	20	0	0	0
TP	18	6	12	67	18

[a] (1200−180) mg/L = 1020 mg/L.
[b] (1020 mg/L)/(1200 mg/L) × 100 = 85%.
[c] 1020 g/m^3 × 1500 m^3/d × 10^{-3} kg/g = 1530 kg/d.

6.3.5 Treatment Processes, Process Diagrams, and Equipment

Unit Operations and Processes: Municipal wastewaters contain approximately 99.9% water. Wastewater treatment facilities are designed to process liquid and solids streams. The small fraction of solids in the bulk liquid includes organic and inorganic suspended and dissolved solids.

Wastewater treatment units for bulk liquid and sludge generally fall into two broad divisions: *unit operations* and *unit processes.* In the unit operations, the treatment or removal of contaminants is brought about by the physical forces. In the unit processes, however, the treatment occurs predominantly due to chemical and biological reactions. Often the terms "unit operations" and "unit processes" are used interchangeably because many processes are integrated combinations of operations serving a single primary purpose. As an example, activated sludge combines mixing, gas transfer, flocculation, and biological phenomena to remove biodegradable organics.

Process Diagram: Wastewater treatment plants utilize a number of unit operations and processes to achieve the desired degree of treatment. The collective treatment schematic is called a *flow scheme, process diagram, flow sheet, process trains*, or *flow schematic.* Many different process trains can be developed from various unit operations and processes for the desired degree of treatment. However, the most desirable process train is the one that is most cost-effective.

Equipment Selection: Every wastewater treatment facility utilizes manufactured equipment or materials. In fact, many design details are often governed by the dimensions and installation requirement of the selected equipment. The design engineer must review the design standards, design procedure, and design assumptions; conduct preliminary design calculations; and study the manufacturers' online catalogs. It is necessary for the design engineer to work closely with the equipment supplier to ensure that the equipment selection is best for the specific application. Treatment processes and process diagrams are further covered in subsequent sections of this chapter.

Reactor Types Used in Wastewater Treatment: Wastewater treatment is usually achieved in tanks or reactors in which physical, chemical, and biological changes occur. The principal types of reactors used for wastewater treatment are (1) batch reactor (BR), (2) tubular-flow or plug-flow reactor (PFR), (3) complete-mixed or continuous-flow stirred-tank reactor (CFSTR), and (4) arbitrary-flow reactor.[33] Each of these reactors and numerous theory and design examples have been presented in Chapter 3.

Connecting Conduits, Pumping Units, Flow Measurement, and Flow Equalization: Wastewater treatment plants utilize connecting conduits, pumping units, flow measurement, and flow equalization. Basic information about these components is provided in Table 6.6. Discussion of plant hydraulic conditions is presented in Section 6.3.7 while development of plant hydraulic profile is described in Section 6.3.8.

TABLE 6.6 Connecting Conduits, Pumping Units, Flow Measurement, and Flow Equalization

Component	Description	Reference
Connecting conduits	Many types of pipes and channels are provided at a treatment plant to convey wastewater from upstream to downstream units. These conveyance systems are mostly open channels, partially flowing circular sewers, and pressure pipes.	Section 6.3.7
Pumping station (PS)	Treatment plants are normally located at a low point in order to provide gravity flow into the collection systems. At the plant site, the wastewater is pumped to an adequate height to achieve flow by gravity through the various treatment units. Pumping is also needed to remove grit; primary sludge; waste and return sludge, thickened and digested sludge; and deliver chemical solutions.	Section 6.3.7
Flow measurement (FM)	Measurement of wastewater flow, sludge, and chemical solutions at wastewater treatment facilities is essential for plant operation, process control, and record keeping. The flow measurement devices may be located in the interceptor sewer, after the pumping station, or at any other location within a plant.	Section 6.3.7
Flow and mass equalization (EQ)	Flow and mass equalization is simply the damping of the flow rate and mass-loading variations. With flow equalization, the plants can be designed and operated under a nearly constant ideal flow and mass-loading condition. This minimizes shock and achieves maximum utilization of the facilities.	Section 6.3.7

Treatment Levels of Bulk Liquid: At the present time, most common practice is to group several unit operations and processes to achieve the desired levels of treatment. These levels of treatment are (1) preliminary, (2) primary, (3) secondary, and (4) advanced. These levels of treatment are summarized in Table 6.7.

Bulk Liquid Treatment Systems: A number of treatment processes are used in combination to achieve various levels of treatment. A summary of different unit operations and processes considered in the design of preliminary, primary, secondary, and advanced treatment facilities is provided in Table 6.8.

Sludge Processing and Disposal: Safe handling and disposal of residues produced at a wastewater treatment plant is of equal importance. The residues include screenings, grit, scum, primary and secondary

TABLE 6.7 Levels of Wastewater Treatment

Treatment Level	Description	Reference
Preliminary	Gross solids (large objects) such as rags, floatables, trash, and grit that may cause equipment damage or operation and maintenance problems are removed in preliminary treatment. Unit operations such as screening and grit removal are part of preliminary treatment.	Chapters 7 and 8
Primary		
Conventional primary treatment (or primary clarifier)	In conventional primary treatment, the settleable organic and inorganic solids are settled, and floatable matter is skimmed from the surface.	Chapter 9
Chemically enhanced primary treatment (CEPT)	Chemicals are added before sedimentation to enhance the removal of settleable and suspended solids.	Chapter 9
Secondary		
Conventional	Dissolved and suspended biodegradable organics are removed. Disinfection is normally included at this level of treatment.	Chapters 10 and 11
Conventional with nutrient removal	Dissolved and suspended biodegradable organic matter and nutrients (nitrogen or phosphorus or both) are removed. Disinfection is normally included at this level of treatment.	Chapters 10 and 11
Advanced	Dissolved and suspended organic and inorganic materials including nutrients are removed. These constituents are normally not removed in the secondary treatment. This level of treatment is normally required where reuse of wastewater effluent becomes an important factor in water resource planning.	Chapter 15

TABLE 6.8 Major Unit Operations and Processes Applied to Bulk Liquid Treatment

Unit No.	Unit Operation and Process (UO and UP)	Principal Application	Removal Achieved, %	Reference
A	Screening	Racks or bar screens are normally the first step in wastewater treatment. They are used to remove large objects for pretreatment. Fine screens are also provided for removal of smaller objects including grit and organic matter for pretreatment and/or primary treatment.	Ignorable removal	Chapter 7 and Section 9.7
B	Grit removal	As a pretreatment process, grit removal facility removes heavy materials such as gravel, sand, cinder, eggshell, and like.	Small removal	Chapter 8
C	Primary clarifier (or sedimentation)	The main purpose of primary sedimentation is to remove settleable and floatable inorganic and organic solids. Fine screens are also used in lieu of primary clarifier, especially prior to a membrane bioreactor (MBR) process.	BOD_5 and COD = 20–40 TSS = 50–70 TP = 10–20 ON = 20–30 AN = 0	Chapter 9
D	Coagulation and chemical precipitation	Coagulation and chemical precipitation involves addition of chemicals, and rapid mixing (RM) or dispersion, followed by flocculation (FLOC) and sedimentation. The process is used for enhanced removal of TSS, BOD_5, and TP. Commonly used chemicals are alum, iron salts, polymers, and lime in one or two stages.	BOD_5 and COD = 40–70 TSS = 50–80 TP = 70–90 ON = 70–95 AN = 0	Chapters 9 and Section 15.4.2
E	Conventional suspended growth aerobic biological reactor (activated sludge)	The process is used to remove dissolved organics. Air is supplied for oxygen transfer into the liquid phase. Clarification of effluent is needed to separate biomass from the effluent and return sludge to maintain a desired level of mixed liquor suspended solids (MLSS) concentration in the reactor. Principal variation is activated sludge process.	BOD_5 and COD = 85–95 TSS = 80–90 TP = 10–25 ON = 60–85 AN = 8–15 High NO_3-N in the effluent	Section 10.3
F	Attached growth aerobic biological reactor (trickling filter)	The process is used to remove dissolved organics. The wastewater is trickled over a fixed media which contains the biomass. Biofilm removes organics. Effluent is clarified to remove biomass. Principal variation is a trickling filter.	BOD_5 and COD = 60–80 TSS = 60–85 TP = 8–12 ON = 60–80 AN = 8–15 High NO_3-N in the effluent	Section 10.4.1
G	Rotating biological contactor (RBC)	RBC also called biodiscs consists of circular plastic plates or disks mounted over a shaft that slowly rotates, and partially submerged in a tank that alternatively substrate and air. The biological growth occurs over the disks. Effluent is clarified.	BOD_5 and COD = 60–80 TSS = 60–85 TP = 8–12 ON = 60–80 AN = 8–15 High NO_3-N in the effluent	Section 10.4.1

(*Continued*)

TABLE 6.8 (*Continued*) Major Unit Operations and Processes Applied to Bulk Liquid Treatment

Unit No.	Unit Operation and Process (UO and UP)	Principal Application	Removal Achieved, %	Reference
H	Combined attached and suspended growth process	Combined attached and suspended growth process is developed by integrating an attached growth process into a conventional suspended growth aerobic reactor. The overall biological treatment performance is improved. The primary variations are trickling filter/solids contact (TF/SC), activated biofilter (ABF), integrated fixed-film activated sludge (IFAS), and moving bed bioreactor (MBBR). Single- or multistage clarifier may be required for clarification and solids return.	BOD_5 and COD = 80–95 TSS = 80–90 TP = 10–25 ON = 60–85 AN = 8–15 High NO_3-N in the effluent	Sections 10.4.2 and 10.4.3
I	Suspended growth anaerobic biological reactor	The suspended growth anaerobic treatment process is carried out in an airtight and mixed reactor, with or without heating. The process is suitable for treatment of medium to high strength industrial wastes. Biogas is produced. The principal variations include (1) anaerobic contact process (ACP) followed by solids separation and return, and (2) upflow anaerobic sludge blanket (UASB) process with an integrated gas–solids–liquid separation zone on top of the reactor.	BOD_5 and COD = 75–90 TSS = 80–90 TP = 8–12 ON = 50–70 High AN in the effluent	Section 10.5
J	Nitrification	Nitrification process converts most of AN to nitrate nitrogen. It is achieved in suspended or attached growth biological reactors. Nitrification can be carried out in a single-stage aerobic reactor in conjunction with carbonaceous BOD removal, or in a separate stage aerobic reactor after BOD removal. Ammonia-oxidizing bacteria (AOB) oxidize ammonia to nitrite, while nitrite-oxidizing bacteria (NOB) oxidize nitrite to nitrate.	ON = 75–85 AN = 85–95 High NO_3-N in the effluent	Section 10.6.1
K	Denitrification	Nitrate and nitrite are reduced to nitrogen by microorganisms under anoxic condition, and in presence of a suitable carbon source such as ready biodegradable organics in the influent or methanol as supplement an external carbon source.	TP = 5–10 NO_3-N and NO_2-N = 90–100	Section 10.6.2
L	Enhanced biological phosphorus removal (EBPR)	Removal of phosphorus is achieved by polyphosphate accumulating organisms (PAO) in a specially arranged configuration. Phosphate is released in an anaerobic zone, followed by uptake aeration basin.	TP = 70–90	Section 10.7

(*Continued*)

TABLE 6.8 (*Continued*) Major Unit Operations and Processes Applied to Bulk Liquid Treatment

Unit No.	Unit Operation and Process (UO and UP)	Principal Application	Removal Achieved, %	Reference
M	Continuous flow biological nutrient removal (BNR) using combined anaerobic, anoxic and aerobic reactors	The process uses multiple reactors utilizing anaerobic, anoxic, and aerobic sequence, sedimentation, and return sludge and internal recirculation lines. The process has capabilities to enhance biological phosphorus and nitrogen removal.	BOD_5 and COD = 90–95 TSS = 90–95 TP = 70–90 ON = 70–95 AN = 90–99 NO_3-N = 4–8 mg/L in the effluent	Sections 10.8 and 15.4.1
N	Sequential batch reactor (SBR) for biological nutrient removal (BNR)	This is a fill-and-draw or continuous flow-activated sludge process. It utilizes essentially the same principle as in a plug flow continuous system. The filling, decanting, and refilling operations are arranged so that anaerobic, anoxic, and aerobic conditions develop for enhanced biological phosphorus and nitrogen removal along with BOD_5 and TSS removals. The process is capable of achieving significantly higher effluent quality than that is achieved in a secondary treatment plant.	BOD_5 and COD = 90–95 TSS = 90–95 TP = 70–90 ON = 70–95 AN = 90–98 NO_3-N = 4–8 mg/L in the effluent	Section 10.8
O	Secondary (or final) clarifier	Secondary or final clarifier is used in conjunction with biological or chemical treatment processes to remove the biological and chemical floc.	It is typically included as an integral part of a suspended growth biological reactor system.	Section 10.9
P	Disinfection	Disinfection is used to destroy water borne pathogens in the effluent. Common methods of disinfection are UV radiation, ozonation, chlorination, and other chemicals. Chlorination is the most common method of disinfection. Chlorine residual in effluent causes toxicity in the receiving water. For this reason, dechlorination is used to destroy the chlorine residual in the effluent.	No removal of chemical constituents. Pathogens and coliform organisms are destroyed.	Chapter 11
Q	Natural systems	Many natural systems such as pond processes, land treatment systems, and natural and constructed wetlands are utilized for complete treatment or for polishing of final effluent from secondary treatment plants.	BOD_5 = 50–90 COD = 40–90 TSS = 50–80 TP = 60–90 ON = 60–90 AN = 85–95	Section 15.3
R	Ammonia stripping	Ammonia gas is stripped from the wastewater in a stripping tower. High pH is necessary for stripping of ammonia. Other volatile organic compounds are also removed with ammonia stripping.	AN = 60–95	Section 15.4.5

(*Continued*)

TABLE 6.8 (*Continued*) Major Unit Operations and Processes Applied to Bulk Liquid Treatment

Unit No.	Unit Operation and Process (UO and UP)	Principal Application	Removal Achieved, %	Reference
S	Filtration and microstraining	Filtration is used to polish the secondary effluent by removing TSS and turbidity as tertiary treatment, and for effluent reuse. Microstrainers are also used for this purpose, and to remove algae from stabilization pond effluent.	$BOD_5 = 30–60$ COD = 0–50 TSS = 60–80 TP = 20–30 ON = 50–70 AN = 0	Sections 15.4.6 and 15.4.7
T	Carbon adsorption	Carbon adsorption is used to remove refractory and other nonbiodegradable organics from wastewater effluent. Both powdered activated carbon (PAC) and granular activated carbon (GAC) columns are used.	$BOD_5 = 50–80$ COD = 50–85 TSS = 40–80 TP = 10–30 ON = 30–50 AN = 0	Section 15.4.8
U	Ion exchange	Ion exchange is a demineralization process to remove dissolved solids from the effluent. It is also used to selectively remove ammonia in a bed of *clinoptilolite*, a zeolite resin.	TDS = 70–95 BOD_5, COD, TSS, TP, and ON removal is small. AN = 85–95 in clinoptilolite bed	Section 15.4.9
V	Microfiltration (MF) and ultrafiltration (UF)	These are membrane processes that are primarily used to achieve high quality solids separation. Clarification and/or filtration of feed flow is not required.	TSS = 90–95	Section 15.4.10
W	Membrane bioreactor (MBR)	Membrane bioreactor (MBR) process is a modified suspended growth biological system. In this process, clarification replacing clarifier with an integrated MF/UF membrane module is used for direct solids separation inside the aeration basin.	$BOD_5 = 95–99$ TSS = 90–95 TP = 90–95 ON = 85–95 AN = 30–50	Section 15.4.10
X	Nanofiltration (NF)	As a partial demineralization process, this membrane process is normally used for softening or reduction of TDS.	TDS = 40–70	Section 15.4.10
Y	Reverse osmosis (RO)	This is a demineralization process capable of producing high quality reclaimed water from the effluent. The water is permeated through semipermeable membranes at high pressure. The reject stream or brine contains high concentration of dissolved solids.	TDS = 80–95 BOD_5, COD, TSS, TP, and ON removal = 90–100 AN = 60–90	Section 15.4.10
Z	Electrodialysis (ED)	It is a demineralization process. Electrical potential is used to transfer the cations and anions through ion selective membranes. An effluent stream that is low in dissolved solids, and brine stream containing high salt content are thus produced.	TDS = 70–95 BOD_5 and COD = 20–60 TP = 90–100 ON = 80–95 AN = 30–50	Section 15.4.10

Source: Adapted in part from Reference 15.

sludges, solids generated from coagulation and chemical precipitation, backwash wastes of filtration, and concentrates from membrane processes. The screenings, grit, and scum are generally disposed of by land filling. The sludge solids offer complex processing and disposal problems.

Quantity of Residues: The quantities of residues produced in different treatment processes at a wastewater treatment plant are summarized in Table 6.9.

TABLE 6.9 Quantity of Residues Produced at a Secondary Wastewater Treatment Plant

Residue	Quantity	Description	Reference
(a) Screenings	0.02–0.075 $m^3/10^3\ m^3$	Coarse solids. Disposal by land filling. Often comminuted.	Chapter 7
(b) Grit	0.005–0.05 $m^3/10^3\ m^3$	Heavy inorganic solids. Disposal by land filling.	Chapter 8
(c) Primary sludge	100–165 g/m^3 (135 g/m^3)	1–6% solids. Offensive odors. Needs processing.	Chapter 9
(d) Scum	8 g/m^3	Odorous. Disposal by land filling.	Chapter 9
(e) Chemical precipitation sludge	80–300 g/m^3 (200 g/m^3)	3–4% solids. Needs processing.	Chapter 9
(f) Single-stage lime sludge	500–600 g/m^3	2–5% solids. Needs processing.	Chapter 9
(g) Two-stage lime sludge	800–1000 g/m^3	4–5% solids. Needs processing.	Chapter 9
(h) Waste activated sludge (WAS)	50–100 g/m^3 (70 g/m^3)	0.3–2.0% solids. Needs processing.	Chapter 10
(i) Single-stage nitrification sludge	60–100 g/m^3	0.8% solids. Needs processing.	Chapter 10
(j) Separate-stage nitrification sludge	10–12 g/m^3	0.8–1% solids. Needs processing.	Chapter 10
(k) Separate-stage denitrification sludge	10–12 g/m^3	0.8–1% solids. Needs processing.	Chapter 10
(l) Chemical-biological sludge	100–150 g/m^3 (120 g/m^3)	0.8–1% solids. Needs processing.	Chapter 15
(m) Filter backwash waste	0.02–0.05 m^3/m^3	Needs processing.	Chapter 15
(n) Granular activated carbon (GAC) backwash waste	0.01–0.03 m^3/m^3	Needs processing.	Chapter 15
(o) MF/UF membrane backwash waste	0.05–0.15 m^3/m^3	Needs processing.	Chapter 15
(p) NF/RO membrane brine (or concentrate or reject)	0.1–0.4 m^3/m^3	Needs processing.	Chapter 15

Note: $g/m^3 \times 8.35 = lb/MG$; $m^3/10^3\ m^3 \times 134 = ft^3/MG$; and $m^3/m^3 \times 134{,}000 = ft^3/MG$.

Source: Adapted in part from Reference 15.

EXAMPLE 6.12: IDENTIFICATION OF UNIT OPERATIONS AND UNIT PROCESSES

Table 6.8 contains basic information on 26 unit operations (UOs) and unit processes (UPs) used for treatment of bulk liquid at municipal wastewater treatment plants. List separately these UOs and UPs.

Solution

Arrange separately UOs and UPs.

Using unit numbers in the Table 6.8 list separately the unit operations and unit processes.

UOs	UPs
A B C O	D E F G
R S V	H I J K
X Y	L M N P
Z	Q T U W

EXAMPLE 6.13: SECONDARY PLANT PROCESS DIAGRAM

Draw the typical process diagram of a secondary wastewater treatment plant. Indicate all unit operations and processes and residuals generated at each unit. Use the system or unit numbers in Tables 6.6, 6.8, and 6.9.

Solution

1. Draw the process diagram (Figure 6.5).

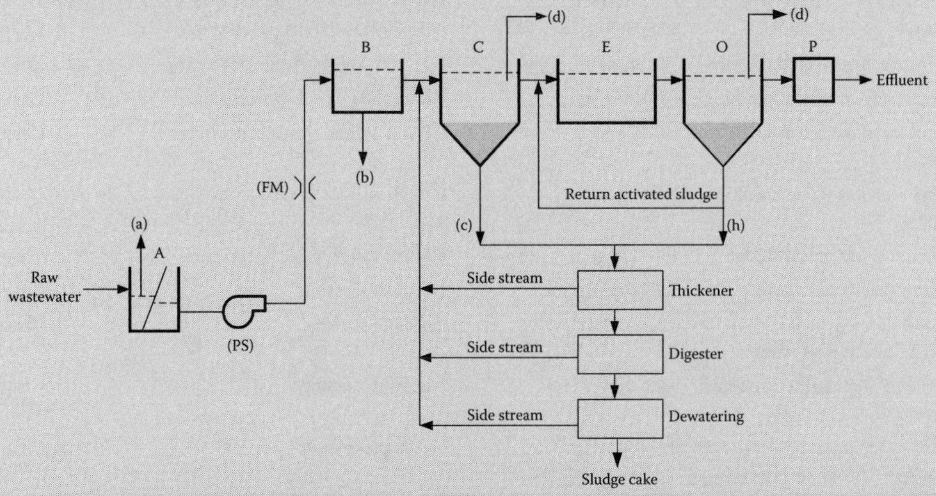

FIGURE 6.5 Process diagram of a secondary wastewater treatment plant (Example 6.13).

2. Indicate the unit operations and processes, and residuals streams.

The unit operations and processes for liquid treatment are named in accordance with the numbering system used in Tables 6.7 and 6.8. The residual streams are indicated by lower case alphabets using Table 6.9.

A = Bar screen, and (a) = screenings (large object)
(PS) = Pumping station
(FM) = Flow Measurement
B = Grit removal, and (b) = grit or heavy objects
C = Primary clarifier, (c) = primary sludge, and (d) = scum
E = Conventional suspended growth biological reactor
O = Final clarifier, (d) = scum, and (h) = WAS
P = Disinfection by chlorination and dechlorination

EXAMPLE 6.14: EFFLUENT QUALITY FROM A SECONDARY WASTEWATER TREATMENT PLANT

Municipal wastewater is treated at a secondary wastewater treatment plant. The raw wastewater has the following characteristics: BOD_5 = 210 mg/L, COD = 400 mg/L, TSS = 230 mg/L, TP = 6 mg/L, ON = 15 mg/L, and AN = 30 mg/L. The performance of various units in terms of key constituents is given in Table 6.8. Use the midpoint removal efficiency to estimate the potential effluent quality.

Solution

1. Tabulate the midpoint removal efficiencies given in Table 6.8 of different units of a secondary wastewater treatment plant.

Unit	Percent Removal, %					
	BOD_5	COD	TSS	TP	ON	AN
Bar screen	0	0	0	0	0	0
Grit removal	0	0	0	0	0	0
Primary sedimentation	33	30	60	15	25	0
Activated sludge	90	90	85	18	73	12
Disinfection	0	0	0	0	0	0

2. Determine the effluent quality.

The calculation procedure for estimating BOD_5 in the effluent is given below as an example:
Primary sedimentation BOD_5 removal = 30%

$$BOD_5 \text{ in the effluent from primary sedimentation facility} = (1-0.3)\times 210\text{ mg/L} = 0.7\times 210\text{ mg/L} = 147\text{ mg/L}$$

$$BOD_5 \text{ in the effluent from activated sludge facility} = (1-0.9)\times 147\text{ mg/L} = 0.1\times 147\text{ mg/L} = 15\text{ mg/L}$$

The effluent quality in terms of other parameter is summarized below:

$BOD_5 = 15\text{ mg/L}$
$COD = (1-0.3)\times(1-0.9)\times 400\text{ mg/L} = 0.07\times 400\text{ mg/L} = 28\text{ mg/L}$
$TSS = (1-0.6)\times(1-0.85)\times 230\text{ mg/L} = 0.06\times 230\text{ mg/L} = 14\text{ mg/L}$
$TP = (1-0.15)\times(1-0.18)\times 6\text{ mg/L} = 0.70\times 6\text{ mg/L} = 4\text{ mg/L}$
$ON = (1-0.25)\times(1-0.73)\times 15\text{ mg/L} = 0.20\times 15\text{ mg/L} = 3\text{ mg/L}$
$AN = (1-0)\times(1-0.12)\times 30\text{ mg/L} = 0.88\times 30\text{ mg/L} = 26\text{ mg/L}$

EXAMPLE 6.15: SEQUENCING BATCH REACTOR PROCESS FOR BIOLOGICAL NUTRIENT REMOVAL (BNR)

A wastewater treatment plant is designed without a primary sedimentation basin. It has a sequencing batch reactor process for BNR. Draw the process train and estimate the potential effluent quality. The raw wastewater quality is the same as that given in Example 6.14.

Solution

1. Draw the process diagram (Figure 6.6).
2. Identify the treatment units and residuals streams using the system or unit numbers in Tables 6.6, 6.8, and 6.9.

A = Bar and fine screens, and (a) = screenings
(PS) = Pumping station
(FM) = Flow measurement
B = Grit removal, and (b) = grit
N = Sequencing batch reactor for BNR, (d) = scum, and (h) = WAS
P = Disinfection

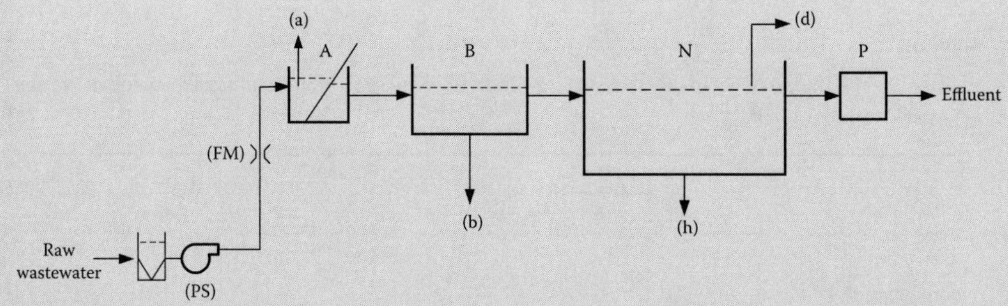

FIGURE 6.6 Process diagram of a sequencing batch reactor process for biological nutrient removal (Example 6.15).

3. Determine the effluent quality.

$$\begin{aligned}
\text{BOD}_5 &= (1 - 0.93) \times 210\ \text{mg/L} = 0.07 \times 210\ \text{mg/L} = 15\ \text{mg/L}\\
\text{COD} &= (1 - 0.93) \times 400\ \text{mg/L} = 0.07 \times 400\ \text{mg/L} = 28\ \text{mg/L}\\
\text{TSS} &= (1 - 0.93) \times 230\ \text{mg/L} = 0.07 \times 230\ \text{mg/L} = 16\ \text{mg/L}\\
\text{TP} &= (1 - 0.8) \times 6\ \text{mg/L} = 0.2 \times 6\ \text{mg/L} = 1\ \text{mg/L}\\
\text{ON} &= (1 - 0.83) \times 15\ \text{mg/L} = 0.17 \times 15\ \text{mg/L} = 3\ \text{mg/L}\\
\text{AN} &= (1 - 0.94) \times 30\ \text{mg/L} = 0.06 \times 30\ \text{mg/L} = 2\ \text{mg/L}
\end{aligned}$$

EXAMPLE 6.16: SECONDARY TREATMENT WITH EFFLUENT POLISHING FOR INDUSTRIAL REUSE

A secondary wastewater treatment plant is designed for effluent polishing for industrial reuse. Gravity filtration and carbon adsorption units are provided to polish the effluent from the secondary plant. Draw the process diagram, and estimate the effluent quality after post filtration and disinfection. The raw wastewater quality is the same as given in Example 6.14.

Solution

1. Draw the process diagram (Figure 6.7).
2. Identify the units and residuals streams using the system or unit numbers in Tables 6.6, 6.8, and 6.9.

A = Screen, and (a) = screenings
(PS) = Pumping station

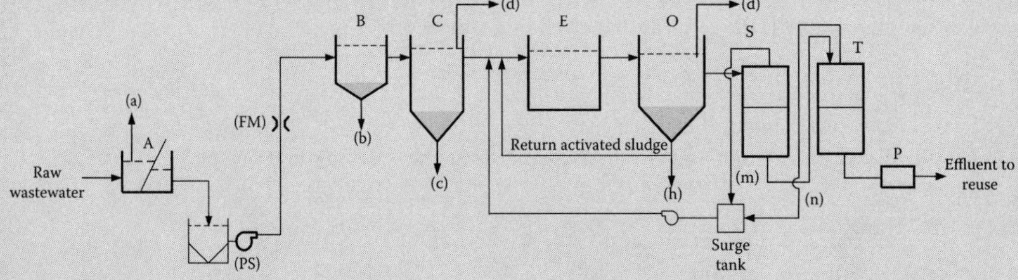

FIGURE 6.7 Gravity filtration and carbon adsorption to polish secondary treated effluent for industrial reuse (Example 6.16).

(FM) = Flow measurement
B = Grit removal, and (b) = grit
C = Primary clarifier, (c) = primary sludge, and (d) = scum
E = Conventional suspended growth biological reactor
O = Final clarifier, (d) = scum, and (h) = WAS
S = Gravity filter, and (m) = filter backwash waste
T = GAC, and (n) = GAC backwash wash
P = Disinfection

Note: Large flows are generated over a short period during the backwash of gravity filter and GAC column. A surge (or balancing) tank is necessary to reduce the slug flows, and return a uniform flow to the plant.

3. Determine the effluent quality.

The effluent quality is established from the removal efficiency of each unit.

$BOD_5 = (1-0.3) \times (1-0.9) \times (1-0.45) \times (1-0.65) \times 210\ mg/L = 0.013 \times 210\ mg/L = 3\ mg/L$

$COD = (1-0.3) \times (1-0.9) \times (1-0.25) \times (1-0.68) \times 400\ mg/L = 0.017 \times 400\ mg/L = 7\ mg/L$

$TSS = (1-0.6) \times (1-0.85) \times (1-0.7) \times (1-0.6) \times 230\ mg/L = 0.007 \times 230\ mg/L = 2\ mg/L$

$TP = (1-0.15) \times (1-0.18) \times (1-0.25) \times (1-0.2) \times 6\ mg/L = 0.42 \times 6\ mg/L = 3\ mg/L$

$ON = (1-0.25) \times (1-0.73) \times (1-0.6) \times (1-0.4) \times 15\ mg/L = 0.049 \times 15\ mg/L = 1\ mg/L$

$AN = (1-0) \times (1-0.12) \times (1-0) \times (1-0) \times 30\ mg/L = 0.88 \times 30\ mg/L = 26\ mg/L$

EXAMPLE 6.17: COAGULATION AND CHEMICAL PRECIPITATION

Chemically enhanced primary treatment (CEPT) is applied to remove key constituents prior to additional treatment processes at a wastewater treatment plant. Draw the process train and describe the processes utilized.

Solution

1. Draw the process train of the physical–chemical treatment facility (Figure 6.8).
2. Identify the treatment units using the system or unit numbers in Tables 6.6 and 6.8.

A = Bar screen, and (a) = screenings
(PS) = Pumping station
(FM) = Flow measurement
B = Grit removal, and (b) = grit
(RM) = Rapid mix
(FLOC) = Flocculation
C = Primary clarifier, (c) = primary sludge, (d) = scum, and (e) chemical precipitation sludge
D = Coagulation and chemical precipitation

Note: (RM), (FLOC), and C may also be collectively identified as unit process D in Table 6.8.

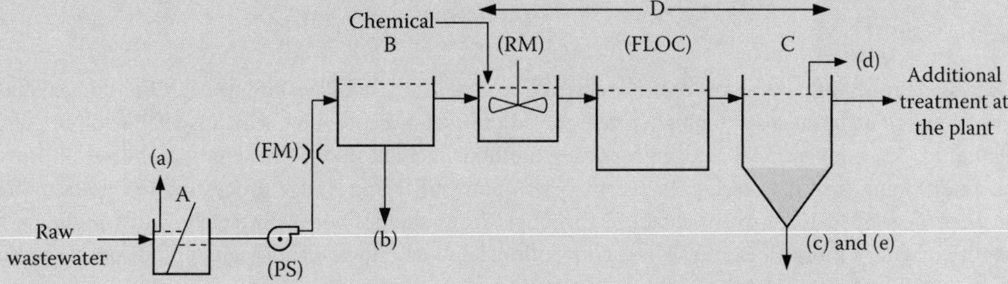

FIGURE 6.8 Process train of a chemically enhanced primary treatment (CEPT) facility (Example 6.17).

3. Describe the process diagram.
   a. Bar screen removes the large objects. The wastewater is pumped into a grit removal facility to remove settleable heavy solids. A flow measurement device is used to record influent flow.
   b. The chemical addition is achieved in a rapid mix unit. High speed mixer disperses the chemicals in 1–2 min. Pinhead floc develops in this unit. In the flocculation basin, the contents are mixed slowly. The pinhead floc grows to a large and settleable size.
   c. The floc is settled in the primary clarifier and effluent is sent to additional treatment at the plant.
   d. Enhanced BOD_5, COD, TSS, TP, and ON removals occurs. Removals expected are: BOD_5 and COD = 40–70%, TSS = 50–80%, TP = 70–90%, ON = 50–90%, and AN = 0.
   e. Due to chemical precipitation, large quantity of sludge is produced (see Table 6.9). Sludge with chemical precipitates is usually difficult to handle.

EXAMPLE 6.18: HIGH QUALITY EFFLUENT FOR REUSE

An industrial plant is planning to use wastewater effluent for process water. Effluent from a secondary wastewater treatment plant is further treated by a membrane process to achieve partial demineralization. Describe the treatment processes.

Solution

1. Describe pretreatment process train using the unit numbers from Table 6.8.

   The secondary wastewater treatment plant utilizes bar screen (A), grit removal (B), primary clarifier (C), aeration basin and final clarifier (E and O) to remove the dissolved organic maters and most suspended solids. The effluent is filtered (S) before the combined UF/NF membrane process (V and X) is applied.
2. Describe the membrane process.

   Membrane processes are used to remove contaminants by sieving, straining, rejection and exclusion, and diffusion. Based on pore size, membrane processes are classified as nanofiltration (NF), reverse osmosis (RO), and electrodialysis (ED).

   MF and UF are only used for high-quality solids separation without demineralization. They are also the required pretreatment before NF or RO is applied.

   NF is typically used for modest level of demineralization, including softening and reduction of TDS.

   RO has the highest material rejection capacity. It is mostly used for demineralization for industrial water supply, and/or for production of potable water from sea or brackish water.

   ED is also an effective process for demineralization and production of potable water from brackish water.

   Note: Theory and design information of membrane processes is presented in Chapter 15.

Residue Processing: Many processes are used to concentrate, stabilize, and dewater the sludge. Sludge processing and disposal costs are high; often exceed the cost of processing, which is 99.9% volume of the bulk liquid. The conventional sludge processing methods include sludge thickening, stabilization, chemical conditioning, and dewatering. In recent years, many emerging sludge processing technologies have also been developed to recover valuable byproducts from sludge more completely, while reducing the quantity of sludge more efficiently. These conventional and emerging technologies and disposal methods are described in Table 6.10.

Process Diagram for Sludge Management: The process diagram of a sludge management facility generally has thickener, digester, dewatering, and disposal of sludge cake. Proper selection of the sludge processing equipment is important for efficient operation of a wastewater treatment facility. Sludge is quite odorous and may cause serious environmental problems.

The liquid streams (also called side stream) that are generated at each stage of concentration of solids must also be treated. Normally, all side streams are returned to the head of the plant. The side streams usually contain high concentrations of suspended solids and BOD. Often, equalization facilities are provided to distribute the hydraulic and material loading over 24 h of operation. To predict the incremental loadings due to returned flows, material mass balance at average design flow is performed to determine the final loadings to the plant.

TABLE 6.10 Unit Operations and Processes Used for Sludge Processing and Disposal

Unit No.	UO and UP	Description	Reference
		Conventional Technologies for Sludge Processing	
A	Sludge thickening	Thickening of sludge is done to concentrate solids and reduce volume.	Section 13.5
A1	Gravity thickening	Used to thicken the primary, secondary, and combined sludges. The solids concentration in raw and thickened sludge is 1–7% and 5–10%. Solids capture is 85–95%.	Section 13.5.1
A2	Dissolved air flotation (DAF)	Used to thicken secondary and combined sludges. Chemical conditioning is needed. The solids concentrations in raw and thickened sludge are 0.2–3% and 4–6%. Solids capture is 85–95%.	Section 13.5.2
A3	Centrifuge	Used to thicken primary, secondary, and combined sludges. Chemical conditioning is needed. The solids concentrations in raw and thickened sludge are 0.2–4% and 3–5%. Solids capture is 85–95%.	Section 13.5.3
A4	Gravity belt	Used for thickening of secondary sludge. The sludge is held over a porous horizontal belt. Free water drains by gravity. Chemical conditioning is needed. The solids concentrations in raw and thickened sludge are 0.3–4% and 3–5%. Solids capture is 80–90%.	Section 13.5.4
A5	Rotary drum	Used for thickening of secondary and combined sludge. They consist of media covered drum that rotates slowly. Free water drains. Chemical conditioning is needed. The solids concentrations in raw and thickened sludge are 0.2–3% and 3–5%. Solids capture is 85–95%.	Section 13.5.5
B	Sludge stabilization	Used to reduce pathogens and offensive odors, and condition the solids for dewatering.	Section 13.6
B1	Anaerobic digestion	Sludge is digested under anaerobic condition. Methane is recovered as an energy source. Volatile solids are reduced by 48–60%.	Section 13.6.1
B2	Aerobic digestion	Sludge is aerated for extended period (10–15 days). Small plants use this option. Volatile solids are reduced by 40–50%.	Section 13.6.2
B3	Chemical	Oxidative chemicals such as chlorine, ozone, or hydrogen peroxide are used. Excess lime is also used to raise the pH to 12. There is no destruction in organic matter. Solids content in chemically treated sludge is increased particularly with lime.	Section 13.6.3
B4	Heat or thermal	High temperature condition is provided to destroy gel structure and coagulate solids.	Section 13.6.4

(Continued)

TABLE 6.10 (*Continued*) Unit Operations and Processes Used for Sludge Processing and Disposal

Unit No.	UO and UP	Description	Reference
C	Sludge conditioning	Sludge is conditioned to improve its dewatering characteristics.	Section 13.7
C1	Chemical	Alum, iron salts, lime, and polymers are the chemicals of choice. There is no reduction in volatile matter. Solids concentration in sludge is increased after conditioning.	Section 13.7.1
C2	Physical	Sludge can be destabilized by heat treatment, freeze and thaw, and elutriation processes.	Section 13.7.2
D	Sludge dewatering	The process is used to remove water so that sludge cake can be transported in trucks and applied over land.	Section 13.8
D1	Drying beds	These are shallow beds of sand. The liquid is removed by an underdrain system and decanting. There are many variations of drying beds. The solids content in sludge cake is 20–25%.	Section 13.8.1
D2	Belt-filter press	The conditioned sludge is pressed between horizontally or vertically mounted continuous belts. The sludge cake of 20–25% solids content is produced.	Section 13.8.2
D3	Filter press	Conditioned sludge is pressed between the filter sacks held vertically in a frame. This is a batch process. Solids content of dewatered sludge is 25–35%.	Section 13.8.2
D4	Centrifuge	Sticky chemical sludge is dewatered using centrifuge. Solids content in sludge cake is 15–20%.	Section 13.8.2
D5	Vacuum filter	The water is removed from a rotary drum. The drum rotates slowly, while partly dipped in conditioned sludge. The suction created inside the drum pulls the moisture. The sludge cake has solids content of 20–25%.	Section 13.8.2
	Emerging Technologies to Enhance Performance of Conventional Methods and By-products Recovery		
E	Advanced sludge processing and by-product recovery	Sludge is processed to recover phosphorus and protein from the organics, enhance biogas generation in digestion process, and reduce the quantity of solids for disposal. These processes are also used for efficient sidestream treatment.	Section 13.10
E1	Pretreatment of sludge	Solubilization of biosolids is achieved by destroying the cellular wall of microorganisms to release the liquid cell contents. Ozone or ultrasonic destruction methods are commonly used. Enzyme and thermal hydrolysis processes are also effective for breaking down the complex organic molecules into simple and ready oxidizable molecules.	Section 13.10.1
E2	Enhanced anaerobic digestion	Pretreatment of sludge by thermal hydrolysis process (THP) allows microorganisms to consume organics more completely in anaerobic digesters. Thus, biogas generation is enhanced and the quantity of digested solids is significantly reduced. Temperature-phased anaerobic digestion (TPAD) process is also efficient for increasing gas production.	Section 13.10.2
E3	Nutrient recovery	It is now realized that phosphorus is a limited resource. Phosphorus, nitrogen, and potassium are the essential components of fertilizer. Phosphorus is recovered from sludge as magnesium–ammonium phosphate also called struvite ($MgNH_4PO_4{\cdot}6H_2O$). After phosphorus recovery, the quantity of sludge solids for disposal is reduced.	Section 13.10.3

(*Continued*)

TABLE 6.10 (*Continued*) Unit Operations and Processes Used for Sludge Processing and Disposal

Unit No.	UO and UP	Description	Reference
E4	Protein recovery	Recovery of protein is achieved by solubilization of municipal sludge at a pH value around 12.5, followed by extraction by protein precipitating agents. The nutrient content in recovered protein is comparable with protein in commercial animal feed. Reduction in sludge quantity after protein recovery is very significant.	Section 13.10.3
E5	Partial nitrification/ anammox (PN/A)	The single-step deammnonification process is a breakthrough process for replacing. This is an effective process for nitrogen removal from sidestreams containing high ammonia concentration.	Section 13.10.4
		Sludge Disposal and Reuse	
F	Disposal	The sludge is finally disposed off by applying it over land, landfilling, or by incineration.	Section 13.11
F1	Composting	Sludge cake is composted and then used as a soil conditioner. The compost is prepared by aerobic biological process or by drying the cake at a temperature of ~37°C.	Section 13.11.1
F2	Heat drying	Biosolids is directly or indirectly heated to remove moisture to produce quality sludge pellets for beneficial reuse.	Section 13.11.2
F3	Incineration	Incineration involves drying of sludge cake followed by complete combustion of organic matter. Wet oxidation is also combustion in liquid phase at high temperature (200–300°C) and high pressure (5–20 mega·N/m^2).	Section 13.11.3
F4	Pyrolysis	For energy recovery purpose, the organic matter in the sludge can be converted to combustible gases, oil and tar, and charcoal in oxygen-free or oxygen-starved atmosphere at high temperature.	Section 13.11.3
F5	Wet air oxidation	The organic matter is destroyed under high temperature and high pressure conditions, and residues are physically separated.	Section 13.11.3
F6	Recalcination	Recalcination process is used to reduce sludge volume, while recovering lime from sludge that has been chemically treated at a high lime dosage.	Section 13.11.4
F7	Biosolids land application	The digested sludge or sludge cake is applied over farmland. The nutrients are taken up by growing plants.	Section 13.11.5
F8	Landfilling	Raw or digested sludge cake is buried in cells or trenches, and covered by soil. Daily cover is 15–30 cm, and final cover is not <60 cm of compacted soil. Proper base construction is necessary to protect the groundwater from contamination.	Section 13.11.6
G	Concentration of waste brine	The waste brine produced from NF, RO, ED, and ion exchange processes create major disposal problem. Proper concentration methods must be applied before disposal.	Section 15.4.11
G1	Concentration by multiple-stage RO	Multiple-stage RO system is applied to concentrate the waste brine to a TDS level of 40–50 kg/m^3.	Section 15.4.11
G2	Solar evaporation	The waste brine is evaporated in open or covered basins to a desired TDS level.	Section 15.4.11
G3	Controlled thermal distillation	Multiple-effect distillation with vapor compression and other methods is used to increase the solids content in the waste brine.	Section 15.4.11

(*Continued*)

TABLE 6.10 (*Continued*) Unit Operations and Processes Used for Sludge Processing and Disposal

Unit No.	UO and UP	Description	Reference
H	Disposal of concentrated waste brine	Disposal of concentrated waste brine is a major disposal problem. Many land- or water-based methods are used.	Section 15.4.11
H1	Disposal in wastewater collection system	The disposal of concentrated or unconcentrated waste brine into the municipal sewer is governed by local ordinance.	Section 15.4.11
H2	Disposal in surface water	Discharge of small quantities of waste brine may be allowable in inland waters under NPDES permit. Coastal facilities use extensively the ocean or saline water. Large facilities may require deep ocean outfall. NPDES permit requirements must be met.	Section 15.4.11
H3	Deep-well injection	Many oil producing states allow deep-well injection of brine. Extensive monitoring of injection facilities is required.	Section 15.4.11
H4	Evaporation and land disposal	Various types of evaporation devices are used to completely dry the concentrated brine. The dry residue is sent to land disposal facilities.	Section 15.4.11

Source: Adapted in part from References 15 and 33.

EXAMPLE 6.19: IDENTIFICATION OF UO AND UP IN SLUDGE PROCESSING

Table 6.10 contains basic information of 36 unit operations (UOs) and unit processes (UPs) for sludge and brine processing and disposal. List separately these unit operations and processes.

Solution

The unit operations and unit processes presented in Table 6.10 are separately listed below.

UOs	UPs
A1, A2, A3, A4, A5, B4, C2,	B1, B2, B3, C1, E2, E3,
D1, D2, D3, D4, D5, E1, F2,	E4, E5, F1, F3, F4, F5,
G1, H1, H2, H3, H4	F6, F7, F8, G2, G3

EXAMPLE 6.20: GENERALIZED PROCESS DIAGRAM FOR SLUDGE MANAGEMENT

The generalized conventional process diagram of sludge management includes thickening, stabilization, chemical conditioning, dewatering, and disposal of sludge cake. Draw the process diagram and indicate alternative process options under each stage.

Solution

Draw the process diagram.

The process diagram along with the alternative process option is shown in Figure 6.9.

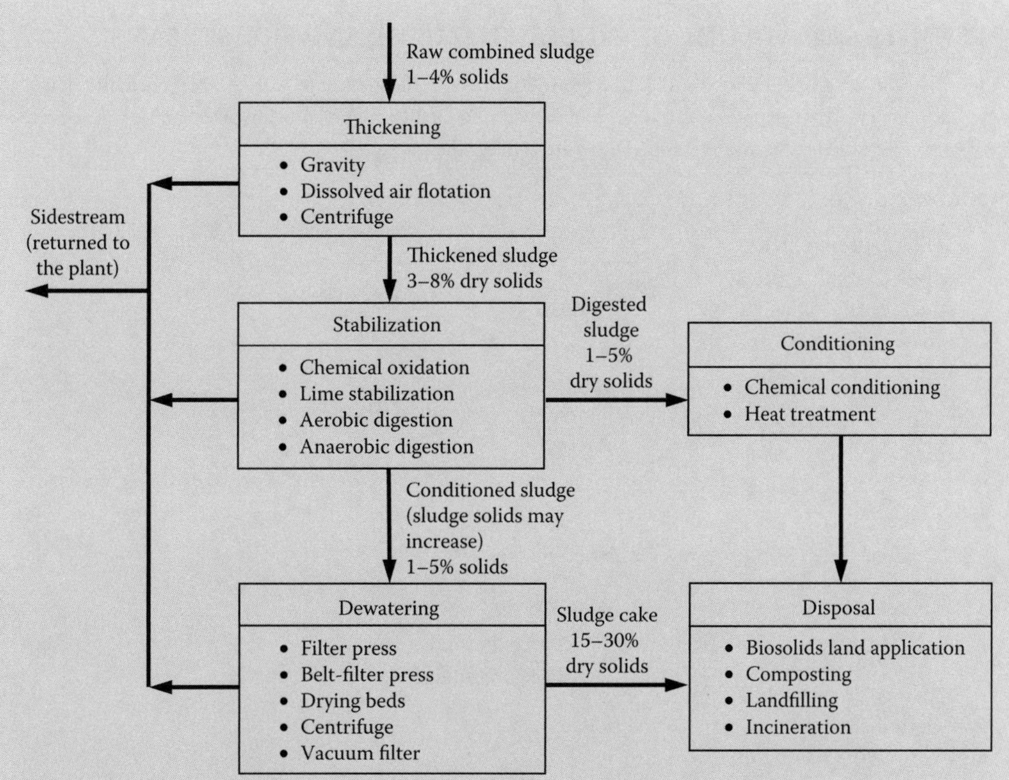

FIGURE 6.9 Process diagram with treatment options for sludge processing and disposal (Example 6.20).

EXAMPLE 6.21: QUANTITY OF RESIDUALS

A conventional suspended growth biological wastewater treatment plant is designed. The average design capacity is 110,000 m^3/d, and the treatment plant is expected to receive wastewater of medium strength. Estimate the quantities of screenings, grit, scum, and primary sludge. Use the typical values of residues given in Table 6.9. What will be the volume of combined sludge produced per day if solids content is 3% and specific gravity of wet sludge is 1.01.

Solution

1. Determine the quantities of residues using the typical values given in Table 6.9.

$$\text{Screenings} = 20 \times 10^{-3}\ m^3/1000\ m^3 \times 110{,}000\ m^3/d = 2.2\ m^3/d$$

$$\text{Grit} = 30 \times 10^{-3}\ m^3/1000\ m^3 \times 110{,}000\ m^3/d = 3.3\ m^3/d$$

$$\text{Scum} = 8\ g/m^3 \times 110{,}000\ m^3/d \times 10^{-3}\ kg/g = 880\ kg/d$$

$$\text{Primary sludge} = 135\ g/m^3 \times 110{,}000\ m^3/d \times 10^{-3}\ kg/g = 14{,}850\ kg/d$$

$$\text{Biological sludge} = 70\ g/m^3 \times 110{,}000\ m^3/d \times 10^{-3}\ kg/g = 7700\ kg/d$$

$$\text{Total dry solids in combined sludge} = (14{,}850 + 7700)\ kg/d = 22{,}550\ kg/d$$

2. Determine the volume of combined wet sludge.

$$\text{Volume with 3\% dry solids} = 22{,}550\ \text{kg/d dry solids} \times \frac{100\ \text{kg wet sludge}}{3\ \text{kg dry solids}} \times \frac{1}{1010\ kg/m^3} = 744\ m^3/d$$

EXAMPLE 6.22: VOLUME OF PROCESSED SLUDGE AND SIDE STREAMS

In total, 500 m^3/d of combined sludge at 2.5% solids and specific gravity of 1.01 are thickened and dewatered. The performance data of thickener and dewatering facility are given below. Determine the volume of processed sludge and side streams. Draw the process diagram.

Thickener

Solids capture efficiency = 85%
Solids content of thickened sludge = 6%
sp. gr. of thickened sludge = 1.015

Dewatering facility

Solids capture efficiency = 90%
Solids content of sludge cake = 25%
sp. gr. of sludge cake = 1.02

Solution

1. Determine the dry solids in raw combined sludge.

$$\text{Volume of raw sludge} = 500\ \text{m}^3/\text{d}$$

$$\text{Dry solids} = 500\ \text{m}^3/\text{d} \times \frac{2.5\ \text{kg dry solids}}{100\ \text{kg wet sludge}} \times 1010\ \text{kg/m}^3\ \text{wet sludge} = 12{,}625\ \text{kg/d}$$

2. Determine the volume of thickened sludge.

$$\text{Dry solids captured by sludge thickener} = 12{,}625\ \text{kg/d} \times 0.85 = 10{,}730\ \text{kg/d}$$

$$\text{Volume of thickened sludge} = 10{,}730\ \text{kg/d dry solids} \times \frac{100\ \text{kg wet sludge}}{6\ \text{kg dry solids}} \times \frac{1}{1015\ \text{kg/m}^3\ \text{wet sludge}} = 176\ \text{m}^3/\text{d}$$

3. Determine the volume of thickener overflow.

$$\text{Thickener overflow} = (500 - 176)\ \text{m}^3/\text{d} = 324\ \text{m}^3/\text{d}$$

4. Determine the volume of sludge cake.

$$\text{Dry solids captured by sludge dewatering facility} = 10{,}730\ \text{kg/d} \times 0.9 = 9660\ \text{kg/d}$$

$$\text{Volume of sludge cake} = 9660\ \text{kg/d dry solids} \times \frac{100\ \text{kg wet sludge}}{25\ \text{kg dry solids}} \times \frac{1}{1020\ \text{kg/m}^3\ \text{wet sludge cake}} = 38\ \text{m}^3/\text{d}$$

5. Determine the volume of side stream from sludge dewatering facility.

$$\text{Volume} = (176 - 38)\ \text{m}^3/\text{d} = 138\ \text{m}^3/\text{d}$$

6. Draw the process diagram.
   The process diagram with solids capture and side streams are shown in Figure 6.10.

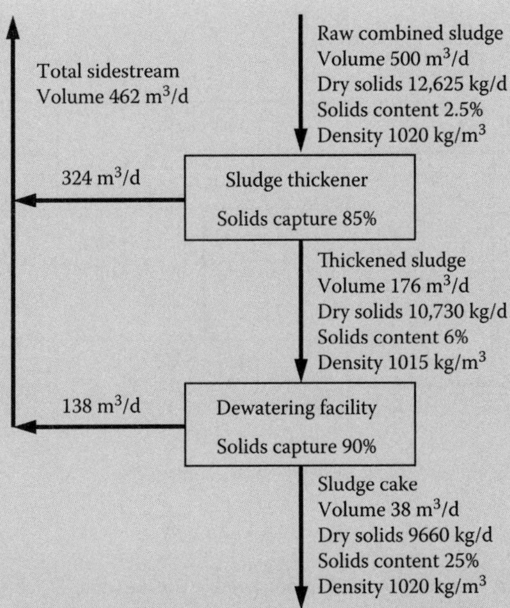

FIGURE 6.10 Process diagram of sludge thickening and dewatering facility (Example 6.22).

EXAMPLE 6.23: MATERIAL MASS BALANCE ANALYSIS

A primary wastewater treatment facility uses sludge drying beds for dewatering of raw primary sludge. The raw wastewater contains 250 TSS and 210 mg/L BOD_5. The average daily flow to the plant is 2784 m^3/d. In primary clarifier, TSS removal is 65%, and BOD_5 removal is 35% of incoming flow. The primary sludge has 3% solids. The solids and BOD_5 capture efficiency of the drying beds is 85% each, and the moisture content of the sludge cake is 72%. Conduct material mass balance analysis and determine the flow, TSS, and BOD_5 in the line that contains the mixture of influent and side streams. Also, determine the TSS and BOD_5 in the effluent from the primary treatment facility. Assume that the specific gravities of the primary sludge and sludge cake are 1.01 and 1.06, respectively.

Solution

1. Draw the process diagram.

   The process diagram with the given data is shown in Figure 6.11.

2. The material mass balance involves iterative process because the side stream (*Line B*) is returned prior to the primary clarifier. It will change the quality of raw wastewater. As a result, the influent characteristics to primary clarifier (*Line C*) will change. The iterative process is given below:

   A. First Iteration

   i. Calculate BOD_5 and TSS in raw or clarifier influent (*Line A* or *C*).

$$\text{BOD}_5 \text{ in raw or clarifier influent} = 2784\,\text{m}^3/\text{d} \times 210\,\text{mg/L} \times 10^{-6}\,\text{kg/mg} \times 10^3\,\text{L/m}^3 = 585\,\text{kg/d}$$

$$\text{TSS in raw or clarifier influent} = 2784\,\text{m}^3/\text{d} \times 250\,\text{mg/L} \times 10^{-6}\,\text{kg/mg} \times 10^3\,\text{L/m}^3 = 696\,\text{kg/d}$$

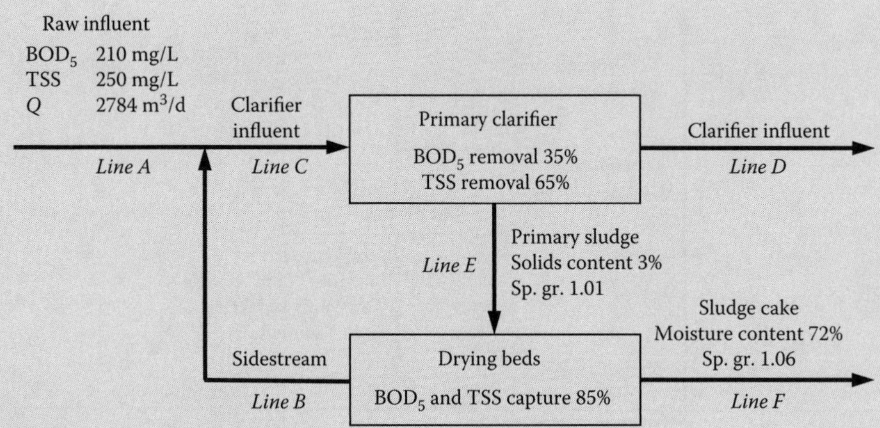

FIGURE 6.11 Process diagram of sludge dewatering (Example 6.23).

ii. Calculate BOD_5 and TSS in primary sludge (*Line E*).

$$\text{BOD}_5 \text{ in primary sludge} = 0.35 \times 585\,\text{kg/d} = 205\,\text{kg/d}$$

$$\text{TSS in primary sludge} = 0.65 \times 696\,\text{kg/d} \times 0.65 = 452\,\text{kg/d}$$

$$\text{Flow rate of primary sludge} = 452\,\text{kg/d dry solids} \times \frac{100\,\text{kg wet sludge}}{3\,\text{kg dry solids}} \times \frac{1}{1010\,\text{kg/m}^3\,\text{wet sludge}}$$

$$= 14.9\,\text{m}^3/\text{d}$$

iii. Calculate solids and flow rate of sludge cake (*Line F*).
Solids captured in drying beds are 85%.

$$\text{Dry solids in sludge cake} = 0.85 \times 452\,\text{kg/d} = 384\,\text{kg/d}$$

72% moisture content gives 28% dry solids content in the sludge cake.

$$\text{Volume of sludge cake} = 384\,\text{kg/d dry solids} \times \frac{100\,\text{kg sludge cake}}{28\,\text{kg dry solids}} \times \frac{1}{1060\,\text{kg/m}^3\,\text{sludge cake}}$$

$$= 1.3\,\text{m}^3/\text{d}$$

iv. Calculate flow, BOD_5, and TSS concentrations in side stream from drying beds (*Line B*).

$$\text{Flow rate of the side stream} = (14.9 - 1.3)\,\text{m}^3/\text{d} = 13.6\,\text{m}^3/\text{d}$$

BOD_5 in the side stream

$$(85\%\ \text{BOD}_5 \text{ is captured in the drying beds}) = (1 - 0.85) \times 205\,\text{kg/d} = 31\,\text{kg/d}$$

$$\text{BOD}_5 \text{ concentration in the side stream} = \frac{31\,\text{kg/d dry solids} \times 10^6\,\text{mg/kg}}{13.6\,\text{m}^3/\text{d} \times 10^3\,\text{L/m}^3} = 2279\,\text{mg/L}$$

$$\text{TSS in the side stream} = (425 - 384)\,\text{kg/d} = 68\,\text{kg/d}$$

$$\text{TSS concentration in the side stream} = \frac{68\,\text{kg/d dry solids} \times 10^6\,\text{mg/kg}}{13.6\,\text{m}^3/\text{d} \times 10^3\,\text{L/m}^3} = 5000\,\text{mg/L}$$

v. Calculate flow rate, BOD_5, and TSS concentrations in primary clarifier influent (*Line C*).

Flow rate in the influent $= (2784 + 13.6)\ m^3/d = 2798\ m^3/d$

BOD_5 in the influent $= (585 + 31)\ kg/d = 616\ kg/d$

BOD_5 concentration in the influent $= \dfrac{616\,kg/d \times 10^6\,mg/kg}{2798\,m^3/d \times 10^3\,L/m^3} = 220\,mg/L$

TSS in the influent $= (696 + 68)\ kg/d = 764\ kg/d$

TSS concentration in the influent $= \dfrac{764\,kg/d \times 10^6\,mg/kg}{2798\,m^3/d \times 10^3\,L/m^3} = 273\,mg/L$

B. Second Iteration

i. Calculate BOD_5 and TSS in clarifier influent (*Line C*).

BOD_5 in clarifier influent $= 616\ kg/d$

TSS in clarifier influent $= 764\ kg/d$

ii. Calculate BOD_5 and TSS in primary sludge (*Line E*).

BOD_5 in primary sludge $= 0.35 \times 616\ kg/d = 216\ kg/d$

TSS in primary sludge $= 0.65 \times 764\ kg/d = 497\ kg/d$

$$\text{Flow rate of primary sludge} = 497\,kg/d\,\text{dry solids} \times \frac{100\,kg\,\text{wet sludge}}{3\,kg\,\text{dry solids}} \times \frac{1}{1010\,kg/m^3\,\text{wet sludge}} = 16\,m^3/d$$

iii. Calculate solids and flow rate of sludge cake (*Line F*).

Solids capture in drying beds is 85%.

Dry solids in sludge cake $= 0.85 \times 497\ kg/d = 422\ kg/d$

Volume of sludge cake

(72% moisture content gives 28% beds content)

$$= 422\,kg/d\ \text{dry solids} \times \frac{100\,kg\ \text{sludge cake}}{28\,kg\ \text{dry solids}} \times \frac{1}{1060\,kg/m^3\ \text{sludge cake}} = 1.4\,m^3/d$$

iv. Calculate flow, BOD_5, and TSS concentrations in side stream from drying beds (*Line B*).

Flow rate of the side stream $= (16 - 1.4)\ m^3/d = 15\ m^3/d$

BOD_5 in the side stream

(85% BOD_5 is captured in the drying beds) $= (1 - 0.85) \times 216\ kg/d = 32\ kg/d$

BOD_5 concentration in the side stream $= \dfrac{32\,kg/d\ \text{dry solids} \times 10^6\,mg/kg}{15\,m^3/d \times 10^3\,L/m^3} = 2133\,mg/L$

TSS in the side stream

(85% TSS is captured in the drying beds) $= (1 - 0.85) \times 497\,\text{kg/d} = 75\,\text{kg/d}$

$$\text{TSS concentration in the side stream} = \frac{75\,\text{kg/d dry solids} \times 10^6\,\text{mg/kg}}{15\,\text{m}^3/\text{d} \times 10^3\,\text{L/m}^3} = 5000\,\text{mg/L}$$

v. Calculate flow rate, BOD_5, and TSS concentrations in primary clarifier influent (*Line C*).

Flow rate in the influent $= (2784 + 15)\,\text{m}^3/\text{d} = 2799\,\text{m}^3/\text{d}$

BOD_5 in the influent $= (585 + 32)\,\text{kg/d} = 617\,\text{kg/d}$

$$BOD_5 \text{ concentration in the influent} = \frac{617\,\text{kg/d} \times 10^6\,\text{mg/kg}}{2799\,\text{m}^3/\text{d} \times 10^3\,\text{L/m}^3} = 220\,\text{mg/L}$$

TSS in the influent $= (696 + 75)\,\text{kg/d} = 771\,\text{kg/d}$

$$\text{TSS concentration in the influent} = \frac{771\,\text{kg/d} \times 10^6\,\text{mg/kg}}{2799\,\text{m}^3/\text{d} \times 10^3\,\text{L/m}^3} = 275\,\text{mg/L}$$

3. Compare the results of first and second iterations.
The flow rate and the concentrations of BOD_5 and TSS in *Lines A* and *C* of first and second iterations are compared below:

Parameter	Raw Influent (*Line A*)	Side Stream (*Line C*)	
		First Iteration	Second Iteration
Flow rate, m^3/d	2784	2798	2799
BOD_5 concentration, mg/L	210	220	220
TSS concentration, mg/L	250	273	275

The results are close enough so that the third iteration may not be necessary.

4. Calculate the effluent quality from the primary treatment facility.

Flow rate of primary clarifier effluent $= (2799 - 16)\,\text{m}^3/\text{d} = 2783\,\text{m}^3/\text{d}$

BOD_5 in the effluent $= (617 - 216)\,\text{kg/d} = 401\,\text{kg/d}$

$$BOD_5 \text{ concentration in the influent} = \frac{401\,\text{kg/d} \times 10^6\,\text{mg/kg}}{2783\,\text{m}^3/\text{d} \times 10^3\,\text{L/m}^3} = 144\,\text{mg/L}$$

TSS in the influent $= (771 - 497)\,\text{kg/d} = 274\,\text{kg/d}$

$$\text{TSS concentration in the influent} = \frac{274\,\text{kg/d} \times 10^6\,\text{mg/kg}}{2783\,\text{m}^3/\text{d} \times 10^3\,\text{L/m}^3} = 98\,\text{mg/L}$$

6.3.6 Plant Layout

Plant layout refers to arrangement of treatment units, piping, and buildings over the selected site. A compact and modular layout is desirable. Important factors that are considered are topography, soil condition, accessibility, future expansion, hydraulics, aesthetics, and environmental control. Detailed discussion on plant layout is given in Chapter 14.

EXAMPLE 6.24: PLANT LAYOUT

Show the physical layout and aerial photograph of a medium-sized wastewater treatment plant.

Solution

The physical layout and aerial photograph of a compact regional wastewater treatment plant is given in Figure 6.12. The design flow is 13,250 m^3/d (3.5 MGD).

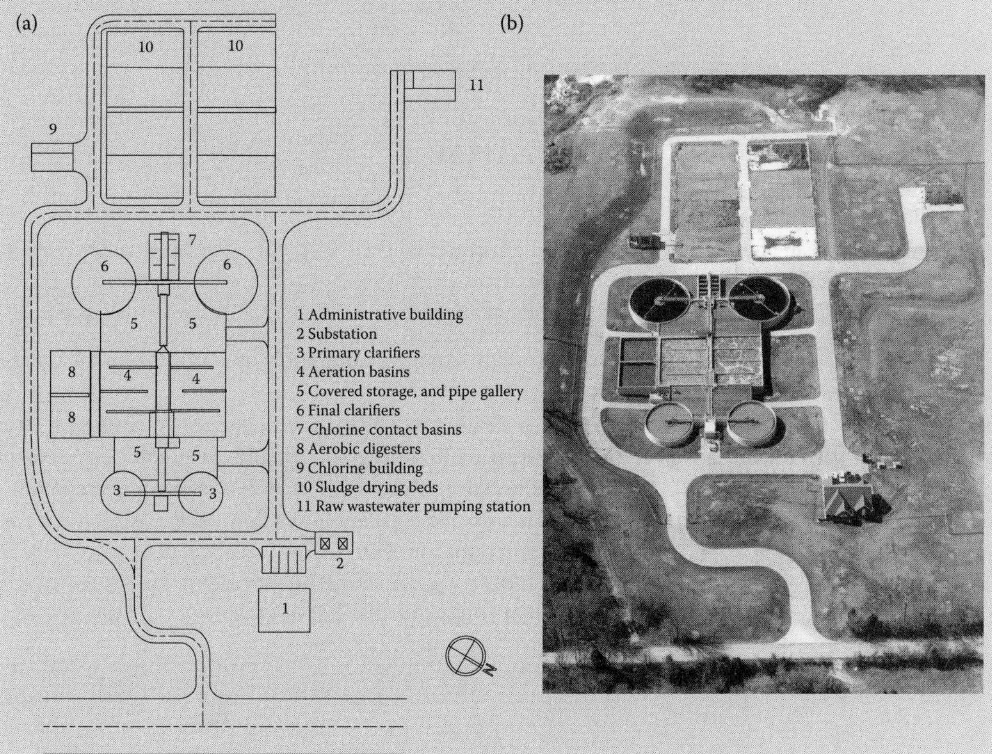

FIGURE 6.12 A regional wastewater treatment plant: (a) plant layout, and (b) aerial photograph (Courtesy CP&Y, Inc., Dallas, Texas). (Example 6.24).

6.3.7 Plant Hydraulic Conditions

Wastewater treatment plants utilize pipes and channels for connecting treatment units, pumping, flow measurement, and sometimes flow equalization. Although these components do not provide any direct treatment, they are considered an integral part of the overall process design for both bulk liquid and sludge streams. Basic information about these components is provided below.[15,33] The minor head loss constants for pressure pipes and open channels are provided in Appendix C.

Connecting Conduits: Many types of connecting conduits are provided to convey wastewater from upstream to downstream units at a treatment plant. These conveyance systems are mostly open channels, partially flowing circular sewers, and pressure pipes (force mains).

The Manning equation has received the most widespread application for open channels. The Manning equation in various forms is expressed below in Equations 6.11a through 6.11d.

$$V = \frac{1}{n} R^{2/3} S^{1/2} \qquad \text{(SI unit)} \tag{6.11a}$$

$$Q = \frac{0.312}{n} D^{8/3} S^{1/2} \qquad \text{(Circular pipe flowing full, SI unit)} \tag{6.11b}$$

$$V = \frac{1.486}{n} R^{2/3} S^{1/2} \qquad \text{(U.S. customary unit)} \tag{6.11c}$$

$$Q = \frac{0.464}{n} D^{8/3} S^{1/2} \qquad \text{(Circular pipe flowing full, U.S. customary unit)} \tag{6.11d}$$

where

V = velocity in a conduit (pipe or channel), m/s (ft/s)
Q = flow in pipe flowing full, m^3/s (ft^3/s)
D = diameter, m (ft)
R = hydraulic mean radius, m (ft). R = area/length of wetted perimeter. For a pipe flowing full, $R = D/4$.
S = slope of energy grade line or invert slope, m/m (ft/ft)
n = coefficient of roughness used in Manning equation

The value of n depends on the material and age of the conduit. Commonly used values of n for concrete and cast iron pipes are in the range of 0.013–0.015.

Sanitary sewers are primarily designed to flow partially full. The hydraulic element equations for circular pipe flowing partially full are given by Equations 6.12a through 6.12d, and are graphically shown in Figure 6.13.[15] The central angle is θ. It may be noted that the value of n decreases with the depth of flow (Figure 6.13b). However, in most designs, n is assumed constant for all flow depths. Also, it is a common practice to use d, v, q, a, and p (lowercase) notations for depth of flow, velocity, discharge, area, and wetted perimeter under partial flow condition, while D, V, Q, A, and P (uppercase) notations are used for sewer flowing full. As an example, a sewer line that is flowing 80% full has a d/D ratio of 0.8.

$$\cos\left(\frac{\theta}{2}\right) = \left(1 - 2\frac{d}{D}\right) \tag{6.12a}$$

$$a = \frac{D^2}{4}\left(\frac{\pi\theta}{360^\circ} - \frac{\sin\theta}{2}\right) \tag{6.12b}$$

$$p = \frac{\pi D\theta}{360^\circ} \tag{6.12c}$$

$$r = \frac{D}{4}\left(1 - \frac{360^\circ \sin\theta}{2\pi\theta}\right) \tag{6.12d}$$

Friction head loss in a pressure pipe (force main) can be calculated from Darcy–Weisbach or Hazen–Williams equation (Equation 6.13a or 6.13b).

$$h_f = \frac{fL}{D_h}\frac{V^2}{2g} \quad \text{or} \quad h_f = \frac{fL}{4R}\frac{V^2}{2g} \qquad \text{(Darcy–Weisbach)} \tag{6.13a}$$

$$h_f = 6.82\left(\frac{V}{C}\right)^{1.85} \times \frac{L}{D^{1.167}} \qquad \text{(Hazen–Williams, SI units)} \tag{6.13b}$$

The Hazen–Williams equation is most commonly used for force mains. This equation in various forms is given by Equations 6.14a through 6.14d.

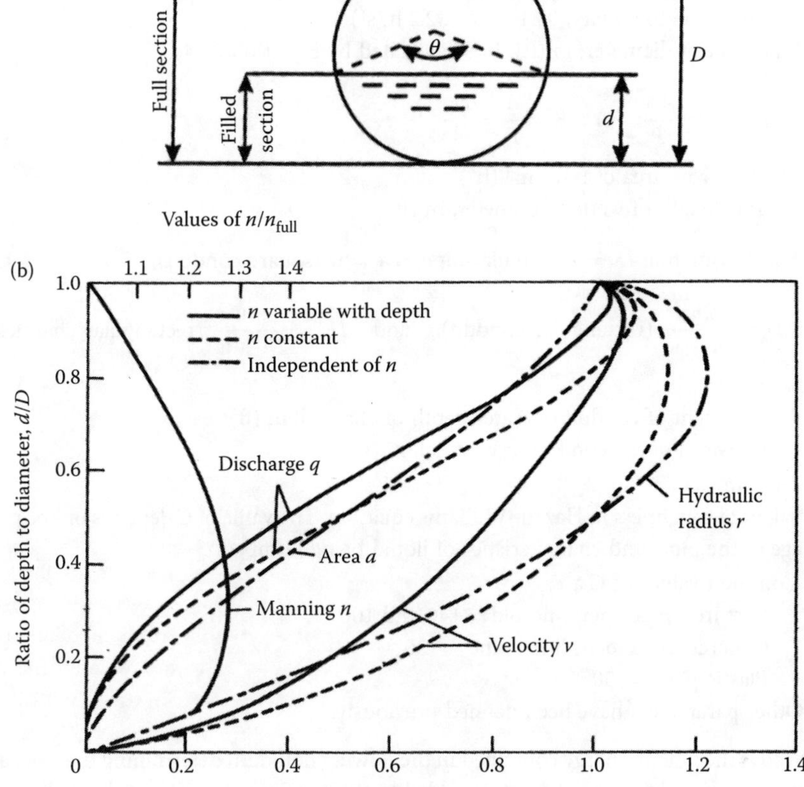

FIGURE 6.13 Hydraulic properties of circular sewer: (a) hydraulic elements and (b) partial flow condition. (Adapted in part from Reference 15.)

$$V = 0.355CD^{0.63}\left(\frac{h_f}{L}\right)^{0.54} \quad \text{(SI units)} \tag{6.14a}$$

$$V = 0.550CD^{0.63}\left(\frac{h_f}{L}\right)^{0.54} \quad \text{(U.S. customary units)} \tag{6.14b}$$

$$Q = 0.278CD^{2.63}\left(\frac{h_f}{L}\right)^{0.54} \quad \text{(SI units)} \tag{6.14c}$$

$$Q = 0.432CD^{2.63}\left(\frac{h_f}{L}\right)^{0.54} \quad \text{(U.S. customary units)} \tag{6.14d}$$

where

h_f = total friction head loss in pipe, m (ft)
V = velocity in pipe, m/s (ft/s)
Q = flow rate in pipe, m^3/s (ft^3/s)
L = length of pipe, m (ft)

f = coefficient of friction in Darcy–Weisbach equation. The value of f depends on the Reynolds number and the relative roughness and diameter of the pipe. It may range from 0.01 to 0.10.
g = acceleration due to gravity, 9.81 m/s^2 (32.2 ft/s^2)
D_h = hydraulic mean diameter, m (ft). It is calculated by Equation 6.14e.

$$D_h = 4R = \frac{4A}{P} \quad \text{or} \quad R = \frac{A}{P} = \frac{D_h}{4} \tag{6.14e}$$

A = cross area of flow, m^2 (ft^2)
P = length of wetted perimeter, m (ft)

For flowing full, $D_h = D$ (circular pipe), $D_h = w$ (square conduit),

$$D_h = \frac{2hw}{h+w} \text{ (rectangular conduit), and } D_h = \frac{4hw}{2h+w} \text{ (rectangular channel).}$$

where
h = height of conduit or water depth of channel, m (ft)
w = width of the conduit or channel, m (ft)

D = diameter, m (ft)
C = coefficient of roughness in Hazen–Williams equation. The value of C depends on the pipe material and age of the pipe, and characteristics of liquid being pumped.
Common values of C are:
Cast iron pipes new and old = 140 and 100
Concrete or cement lined pipe = 120
Plastic pipe = 150

Note: Other parameters have been defined previously.

Velocity head is the kinetic energy contained in the flowing liquid. In determining the head at any point in a piping system, the velocity head must be added to the gauge reading. The velocity head is given by Equation 6.15a. The minor head losses are produced because of fittings, valves, bends, entrance, exit, etc., and are normally calculated as a function of velocity head. The minor losses are expressed by Equation 6.15b. The velocity head is often added to the minor head losses.

$$h_v = \frac{V^2}{2g} \tag{6.15a}$$

$$h_m = K\frac{V^2}{2g} \quad \text{or} \quad h_m = Kh_v \tag{6.15b}$$

where
h_m = minor losses, m (ft)
h_v = velocity head, m (ft)
K = head loss coefficient (Appendix C)

Pumping: Treatment plants are normally located at a low point in order to receive gravity flow from the collection systems. At the plant site, the wastewater is pumped to an adequate elevation to achieve flow by gravity through the various treatment units. Pumping is also needed to remove grit and scum, transfer sludge, and deliver chemical solutions.

All pumps may be classified as *kinetic energy* or *positive displacement.* Brief descriptions and applications of many types of pumps in these two classes are provided in Table 6.11. The kinetic energy pumps are most commonly used in wastewater applications. In depth discussion on these pumps may be found in References 15 and 17.

The total dynamic head (*TDH*) is head against which the pump works to move the liquid. It is the total energy barrier that must be overcome before the water can be lifted at a given flow rate. It is determined by

TABLE 6.11 Pump Types and Major Applications in Wastewater

Major Classifications	Pump Type	Brief Description	Major Pumping Applications
Kinetic	Centrifugal	Consists of an impeller enclosed in a casing with inlet and discharge connections. The head is developed principally by centrifugal force.	Raw wastewater, settled primary and secondary sludge, return and waste sludge, thickened sludge, and effluent
	Peripheral (torque-flow or vortex)	Consists of a recessed impeller on the side of the casing entirely out of the flow stream. A pumping vortex is set up by viscous drag.	Scum, grit, sludge, and raw wastewater
Positive displacement	Rotary	Consists of a fixed casing containing gears, vanes, pistons, cams, screws, etc., operating with minimum clearance. The rotating element pushes the liquid around the enclosed casing into the discharge pipe.	Lubricating oils, chemical solutions, small flows of water, and wastewater
	Screw	Uses a spiral screw operating in an enclosed casing	Grit, settled primary and secondary sludge, thickened sludge, and raw wastewater
	Progressive cavity	The pump is composed of single-threaded rotor that operates with a minimum of clearance in a double-threaded helix stator made of rubber.	Used for pumping large objects of heavy solids concentration. Common use is for pumping thickened and digested sludge.
	Diaphragm	Uses flexible diaphragm or disk fastened over edges of a cylinder	Chemical solutions
	Plunger	Uses a piston or plunger that operates in a cylinder. The pump is self-priming, and discharges liquid during piston or plunger movement through each stroke	Scum, settled primary and secondary sludge, and chemical solutions
	High-pressure piston	Uses separate power pistons or membranes to separate the drive mechanism from contact with sludge	Used for thickened and digested sludge
	Airlift	Air is bubbled into a vertical tube partly submerged in water. The air bubbles reduce the unit weight of the fluid in the tube. The higher unit weight fluid displaces the low unit weight fluid, forcing it up into the tube.	Return and circulation of secondary sludge
	Pneumatic ejector	Air is forced into the receiving chamber which ejects the wastewater from the receiving chamber.	Raw wastewater at small installations (100–600 L/min)

Source: Adapted in part from References 15 and 17.

considering the static head, friction and minor head losses in suction and discharge pipings, and velocity head in the discharge pipe. Equation 6.16a is used to determine *TDH*.

$$TDH = H_{stat} + h_{fs} + \Sigma h_{ms} + h_{fd} + \Sigma h_{md} + \frac{V_d^2}{2g} \tag{6.16a}$$

where

TDH = total dynamic head, m (ft)

H_{stat} = static head, m (ft)

h_{fs} and h_{fd} = friction head losses in suction and discharge pipings, m (ft)

Σh_{ms} and Σh_{md} = sum of minor losses in fittings and valves in suction and discharge pipes, m (ft). Entrance loss in the suction bell may be included in the minor losses in suction piping.

V_d = velocity in discharge pipe, m/s (ft/s)

The Bernoulli's energy equation (Equation 6.16b) is also applied to determine *TDH*. The energy equation is written between the suction bell and the discharge nozzle of the pump. Distances above and below the datum are considered positive and negative, respectively. The datum is normally referenced to the centerline of the pump impeller. The *TDH* is expressed by Equation 6.16c.[15,17]

$$Z_d + \frac{P_d}{\gamma} + \frac{V_d^2}{2g} = Z_s + \frac{P_s}{\gamma} + \frac{V_s^2}{2g} + TDH \tag{6.16b}$$

$$TDH = \left(\frac{P_d}{\gamma} + \frac{V_d^2}{2g} + Z_d\right) - \left(\frac{P_s}{\gamma} + \frac{V_s^2}{2g} + Z_s\right) \tag{6.16c}$$

where

P_d and P_s = gauge pressures on discharge and suction nozzles of the pump, kN/m^2 (lb/ft^2)
γ = specific weight of the liquid pumped, kN/m^3 (lb/ft^3)
V_d and V_s = velocity in discharge and suction nozzles of the pump, m/s (ft/s)
Z_d and Z_s = elevation of gauges on discharge and suction nozzles of the pump above datum, m (ft)

The work done by a pump is proportional to the product of the specific weight of the fluid being discharged and the total head against which the flow is moved. The pump efficiency is the ratio of the useful pump output power to the input power. Pump power output, and pump and motor efficiencies are expressed by Equations 6.17a through 6.17c.[15,17]

$$P_w = K'Q(TDH)\gamma \tag{6.17a}$$

$$E_p = \frac{P_w}{P_p} \quad \text{or} \quad P_p = \frac{P_w}{E_p} \tag{6.17b}$$

$$E_m = \frac{P_p}{P_m} \quad \text{or} \quad P_m = \frac{P_w}{E_m} \tag{6.17c}$$

where

P_w = power output of the pump (water power), kW (horse power of hp)
P_p = power input to the pump (brake power), kW (hp)
P_m = power input to the motor (electrical energy or wire power), kW (hp)
Q = capacity, discharge, or flow rate, m^3/s (ft^3/s)
E_p = pump efficiency, usually 70–90%
E_m = motor efficiency, usually 90–98%
K' = constant depending on the units of expression

($TDH = m$, $Q = m^3/s$, $\gamma = 9.81\ kN/m^3$, $P_w = kW$, $K' = 1\ kW/[kN{\cdot}m/s]$)

($TDH = ft$, $Q = ft^3/s$, $\gamma = 62.4\ lb/ft^3$, $P_w = hp$, $K' = 1\ hp/(550\ ft{\cdot}lb/s)$)

Note: Power: 1 kN·m/s = 1 kW (SI unit)
1 hp = 550 ft·lb/s (U.S. customary unit)

Flow Measurement: Measurement of wastewater flow, sludge and chemical solutions at wastewater treatment facilities is essential for plant operation, process control, and record keeping. The flow measurement devices may be located in the interceptor sewer, after the pump station, or at any other location within a plant. Flow measurement systems are composed of primary and secondary elements. Primary elements produce a head, pressure, electrical current, or other measurable parameters that are proportional to the flow. The secondary element measures the parameter produced by the primary element and provides an indication of the flow rate. These indicators may be visual, such as an analog or digital readout, or a telemetry signal. Secondary elements are an integral part of the control system. In general, the primary flow measurement systems are applicable to either of the two major conditions: (1) pressure pipes, and (2) open channels. Some systems however, are applicable to both. Various types of primary flow elements and their principle of flow measurement are provided in Table 6.12. The

TABLE 6.12 Types of Flow Measurement Devices Available for Determining Liquid Discharge

Conduit	Flow Measurement Device	Principle of Flow Measurement
Pressure pipes	Venturi meter	The differential pressure is measured.
	Flow nozzle meter[a]	The differential pressure is measured.
	Orifice meter[a]	The differential pressure is measured.
	Pitot tube	The differential pressure is measured.
	Electromagnetic meter[a]	Magnetic field is induced, and voltage is measured.
	Rotameter	The rise of float in a tapered tube is measured.
	Turbine meter[a]	A velocity driven rotational element (turbine, vane, or wheel) is used.
	Ultrasonic velocity[a]	The ultrasonic transducers send and receive ultrasonic pressure pulses.
	Ultrasonic doppler[a]	The transducers transmit beams that are reflected to a receiver by suspended solids or gas bubbles.
	Elbow meter	The differential pressure is measured around a bend.
Open channels	Flumes (Parshall, Palmer-Bowlus)[a]	Critical depth is measured at the flume.
	Weirs[a]	Head is measured over a weir.
	Current meter	Rotational element is used to measure velocity.
	Pitot tube	The differential pressure is measured.
	Depth measurement[a]	Float is used to obtain the depth of flow.
	Sonic level meter[a]	The transducer emits and receives the beam reflected from the liquid surface.
Freely discharge from a pipe flowing full[b]	Nozzles and orifices	The water jet data is recorded.
	Vertical open-end flow	The vertical height of water jet is recorded.
Freely discharge from a pipe partly flowing full[b]	Open flow nozzle[a] (Kennison nozzle or California pipe method)	The depth of flow at free-falling end is determined.
Miscellaneous methods	Dilution method	The concentration of a constant flow of a dye tracer is measured.
	Bucket and stopwatch	A calibrated bucket is used and time to fill it is noted.
	Measuring level change in a tank	Change in level in a given time is obtained.
	Calculation from water meter readings	Water meter readings over a given time period give average wastewater flow.
	Pumping rate	Constant pumping rate and pumping duration.

[a] Commonly used devices for wastewater flow measurement.
[b] Flow is calculated from measurement at the end of pipe.
Source: Adapted in part from Reference 15.

applicable range of flow, accuracy, repeatability, and many selection criteria are summarized in Table 6.13.[15]

Flow and Mass Equalization: Flow and mass equalization is simply the damping of the flow rate and mass-loading variations. With flow equalization, the plants can be designed and operated under a nearly constant ideal flow and mass-loading condition. This minimizes shock and achieves maximum utilization of the facilities. Discussion on flow and mass equalization is given in Section 3.6.

TABLE 6.13 Evaluation of Various Types of Devices Commonly Used for Wastewater Flow Measurement

Type of Device	Pressure Flow	Open Channel	Range	Accuracy Maximum Flow, %	Repeatability, Percent of Full Scale	Effects of Solids in Wastewater	Head Loss	Power Requirement	Simplicity and Reliability	Maintenance Requirement	Ease of Calibration	Cost	Portability
Venturimeter	Y	N	10:1	±0.5[a]	±0.5	H[b]	L	L	G	M	G	H	N
Flow Nozzle meter	Y	N	4:1	±0.3	±0.5	H	M	L	G	L	G	M	N
Orifice meter	Y	N	4:1	±1	±1	H	H	L	G	H	G	L	Y
Electromagnetic meter	Y	N	10:1	±1–2	±0.5	S	L	M	F	M	G	H	N
Turbine meter[c]	Y	N	15:1	±0.25	±0.05	H	M	L	F	H	G	H	N
Ultrasonic velocity	Y	N	10:1	±1–2	±1	M	L	M	F	M	G	H	N
Ultrasonic doppler	Y	N	10:1	±1–2	±1	M	L	M	F	M	G	H	N
Parshall flume	N	Y	20:1	±5	±0.5	S	L	L	G	L	G	M	Y
Palmer-Bowlus flume	N	Y	20:1	±10	±0.5	S	L	L	G	L	G	L	Y
Weirs	N	Y	20:1	±0.5	±0.5	H	H	L	G	M	G	L	Y
Depth measurement	N	Y	10:1	±50		M	L	L	G	L	P	L	Y
Open flow nozzle	N	Y	20:1	±1	±0.5	S	H	L	G	M	F	L	Y

[a] Based on full scale reading.
[b] Effect of solids is substantially smaller if solids bearing or continuous flushing-type venturimeter is used.
[c] Positive displacement type.

Note: F = fair; G = good; H = high; M = medium; N = no, P = poor; S = slight; Y = yes.

Source: Adapted in part from Reference 15.

EXAMPLE 6.25: HEAD LOSSES IN A DISCHARGE PIPE

A submersible pump is pumping primary settled wastewater from a wet well to an aeration basin. There is no suction piping. The diameter of the suction bell is 30 cm, and $K = 0.04$. Calculate the components of head losses in the force main. The diameter and length of force main are 20 cm and 120 m, respectively. There is one enlarger $K = 0.25$, one check valve $K = 2.5$, one plug valve $K = 1.0$, one gate valve $K = 0.19$, three 45° elbows $K = 0.2$, and two 90° elbows $K = 0.3$. The pump discharge is $0.08\ \text{m}^3/\text{s}$ and Darcy–Weisbach coefficient of roughness $f = 0.02$.

Solution

1. Calculate the velocity head in the force main.

Area of the force main, $A = \dfrac{\pi}{4} \times (0.2\ \text{m})^2 = 0.0314\ \text{m}^2$

Velocity in the force main, $V = \dfrac{Q}{A} = \dfrac{0.08\ \text{m}^3/\text{s}}{0.0314\ \text{m}^2} = 2.55\ \text{m/s}$

Apply Equation 6.15a to calculate velocity head, $h_v = \dfrac{V^2}{2g} = \dfrac{(2.55\ \text{m})^2}{2 \times 9.81\ \text{m/s}^2} = 0.33\ \text{m}$

2. Calculate the total friction head loss in the force main from Darcy–Weisbach equation.

ApplyEquation6.13atocalculatetotalfrictionheadloss,

$$h_f = \frac{fL}{D_h} \times \frac{V^2}{2g} = \frac{0.02 \times 120\ \text{m}}{0.2\ \text{m}} \times 0.33\ \text{m} = 3.96\ \text{m}$$

3. Calculate the minor losses from Equation 6.15b.

Velocity through suction bell $= V_{\text{bell}} = \dfrac{0.08\ \text{m}^3/\text{s}}{\dfrac{\pi}{4} \times (0.3\ \text{m})^2} = 1.13\ \text{m/s}$

Head loss through suction bell $= 0.04 \times \dfrac{(1.13\ \text{m/s})^2}{2 \times 9.81\ \text{m/s}^2} = 0.003\ \text{m}$

Head loss through enlarger $= 0.25 \times 0.33\ \text{m} = 0.08\ \text{m}$

Head loss through check valve $= 2.5 \times 0.33\ \text{m} = 0.83\ \text{m}$

Head loss through plug valve $= 1.0 \times 0.33\ \text{m} = 0.33\ \text{m}$

Head loss through gate valve $= 0.19 \times 0.33\ \text{m} = 0.06\ \text{m}$

Head loss through 45° elbows $= 3 \times 0.2 \times 0.33\ \text{m} = 0.20\ \text{m}$

Head loss through 90° elbows $= 2 \times 0.3 \times 0.33\ \text{m} = 0.20\ \text{m}$

$\Sigma h_m = 1.70\ \text{m}$

4. Calculate the total head loss.

Total head loss $h_L = h_f + \Sigma h_m + h_v = (3.96 + 1.70 + 0.33)\ \text{m} = 5.99\ \text{m}$

EXAMPLE 6.26: DETERMINATION OF TDH FROM BERNOULLI'S EQUATION

Wastewater is pumped from a wet well to a primary sedimentation basin. The pump discharge is $0.5\ \text{m}^3/\text{s}$. The diameter of the discharge pipe is 35 cm, and the reading on the discharge gauge located at the centerline of the pump discharge pipe is $125\ \text{kN/m}^2$. The diameter of the suction pipe is 40 cm, and reading on the suction gauge located 0.6 m below the centerline of the discharge pipe is $10\ \text{kN/m}^2$. Determine the head on the pump using the Bernoulli's equation. Use $\gamma = 9.81\ \text{kN/m}^3$.

Solution

1. Determine the velocity in discharge and suction pipes.

$$V_\text{d} = \frac{0.5\ \text{m}^3/\text{s}}{\frac{\pi}{4} \times (0.35\ \text{m})^2} = 5.20\ \text{m/s}$$

$$V_\text{s} = \frac{0.5\ \text{m}^3/\text{s}}{\frac{\pi}{4} \times (0.4\ \text{m})^2} = 3.98\ \text{m/s}$$

2. Determine the pressure head in discharge and suction pipes.

$$h_\text{p} = \frac{P_\text{d}}{\gamma} = \frac{125\ \text{kN/m}^2}{9.81\ \text{kN/m}^3} = 12.7\ \text{m}$$

$$h_\text{s} = \frac{P_\text{s}}{\gamma} = \frac{10\ \text{kN/m}^2}{9.81\ \text{kN/m}^3} = 1.02\ \text{m}$$

3. Apply Bernoulli's equation (Equation 6.16c).

$$\begin{aligned} TDH &= \left(\frac{P_\text{d}}{\gamma} + \frac{V_\text{d}^2}{2g} + Z_\text{d}\right) - \left(\frac{P_\text{s}}{\gamma} + \frac{V_\text{s}^2}{2g} + Z_\text{s}\right) \\ &= \left(12.7\ \text{m} + \frac{(5.20\ \text{m/s})^2}{2 \times 9.81\ \text{m/s}^2} + 0\ \text{m}\right) - \left(1.02\ \text{m} + \frac{(3.98\ \text{m/s})^2}{2 \times 9.81\ \text{m/s}^2} + (-0.6\ \text{m})\right) \\ &= (12.7 + 1.38 + 0 - 1.02 - 0.81 + 0.6)\ \text{m} \\ &= 12.9\ \text{m} \end{aligned}$$

EXAMPLE 6.27: WIRE POWER AND EFFICIENCY

A pump is delivering 2000 gpm at a *TDH* of 64 ft. The pump efficiency is 85% and motor efficiency is 92%. Calculate the wire power or electrical energy input to the motor. Use $\gamma = 62.4\ \text{lb/ft}^3$.

Solution

1. Convert the pump discharge into cubic feet per second (ft^3/s or cfs).

$$Q = 2000\ \text{gpm} \times \frac{1}{60\ \text{s/min}} \times \frac{1}{7.48\ \text{gal/ft}^3} = 4.46\ \text{cfs}$$

2. Calculate the water power or power output of pump from Equation 6.17a.

$$P_\text{w} = K'Q(TDH)\gamma$$

$$P_\text{w} = \frac{1\ \text{hp}}{550\ \text{ft·lb/s}} \times 4.46\frac{\text{ft}^3}{\text{s}} \times 64\ \text{ft} \times 62.4\frac{\text{lb}}{\text{ft}^3} = 32.4\ \text{hp}$$

3. Calculate the power input to the pump from Equation 6.17b.

$$P_p = \frac{P_w}{E_p} = \frac{32.4\,\text{hp}}{0.85} = 38.1\,\text{hp}$$

4. Calculate the wire power or power input to motor from Equation 6.17c.

$$P_m = \frac{P_w}{E_m} = \frac{38.1\,\text{hp}}{0.92} = 41.4\,\text{hp}$$

EXAMPLE 6.28: DEPTH AND VELOCITY CALCULATED FROM HYDRAULIC ELEMENTS EQUATIONS AND NOMOGRAPH

A sewer is flowing 35% full. The sewer diameter and slope are 18 in and 0.0018, respectively. Determine depth, velocity, and discharge under partial full condition using hydraulic elements equations. $n = 0.013$. Also, check partial flow velocity and discharge using the monograph (Figure 6.13).

Solution

1. Calculations based on hydraulic elements equations.
   a. Calculate the hydraulic elements of the sewer running partially full.
      Apply Equations 6.12a through 6.12d when $d/D = 0.35$.

$$\cos\left(\frac{\theta}{2}\right) = \left(1 - 2\frac{d}{D}\right) = (1 - 2 \times 0.35) = 0.3$$

$$\theta = 145^\circ$$

$$a = \frac{(1.5\,\text{ft})^2}{4} \times \left(\frac{\pi \times 145^\circ}{360^\circ} - \frac{\sin(145^\circ)}{2}\right) = 0.55\,\text{ft}^2$$

$$p = \frac{\pi \times 1.5\,\text{ft} \times 145^\circ}{360^\circ} = 1.90\,\text{ft}$$

$$r = \frac{1.5\,\text{ft}}{4}\left(1 - \frac{360^\circ \times \sin 145^\circ}{2\pi \times 145^\circ}\right) = 0.29\,\text{ft}$$

   b. Calculate the velocity and discharge.

      From Equation 6.11c, $v = \frac{1.486}{0.013} \times (0.29\,\text{ft})^{2/3} \times (0.0018)^{1/2} = 2.12\,\text{ft/s}$

$$q = va = 2.12\,\text{ft/s} \times 0.55\,\text{ft}^2 = 1.17\,\text{ft}^3/\text{s}$$

2. Calculations based on the nomograph.
   a. Determine the velocity and discharge when the sewer is flowing full.

      Apply Equation 6.11c to calculate the velocity, $V = \frac{1.486}{0.013} \times \left(\frac{1.5\,\text{ft}}{4}\right)^{2/3} \times (0.0018\,\text{ft/ft})^{1/2}$
      $= 2.5\,\text{ft/s}$

$$Q = VA = 2.5\,\text{ft/s} \times \frac{\pi}{4} \times (1.5\,\text{ft})^2 = 4.4\,\text{ft}^3/\text{s}$$

   b. Determine v and q from Figure 6.13.

Read the ratios v/V and q/Q using given $d/D = 0.35$.

$$\frac{v}{V} = 0.82 \quad \text{and} \quad \frac{q}{Q} = 0.27$$

c. Calculate the velocity and discharge.

$$v = 0.82 \times 2.5\,\text{ft/s} = 2.1\,\text{ft/s}$$

$$Q = 0.27 \times 4.4\,\text{ft}^3/\text{s} = 1.2\,\text{ft}^3/\text{s}$$

6.3.8 Plant Hydraulic Profile

Hydraulic profile is the graphical representation of hydraulic grade line through the treatment plant. Hydraulic profile is prepared to ensure that (1) adequate hydraulic gradient exists for wastewater to flow by gravity, (2) pumps deliver adequate head, and (3) treatment units are not flooded or backed up during periods of peak flow. Connecting pipes, and collection and division boxes are provided to transmit flow from one unit to the other and to isolate some units from the process train for maintenance. Hydraulic profile is an important topic and it is covered in detail in Chapter 14.

EXAMPLE 6.29: HYDRAULIC PROFILE THROUGH A SEWER MANHOLE

An intercepting sewer is 61 cm in diameter and has a slope of 0.0013. It carries a flow of 113 L/s. The sewer enters a standard manhole. The outlet sewer is also 61 cm in diameter and has a slope of 0.0014. Using $n = 0.015$, draw the hydraulic profile and determine the invert elevation of the incoming sewer. Assume that the minor loss coefficients of exit and entrance are 0.4 and 0.1, respectively. The manhole invert is in line with that of the outlet sewer at the entrance point that is 0.00 m as the datum.

Solution

1. Calculate the velocity and discharge in the upstream sewer when the sewer is flowing full.

Apply Equation 6.11a to calculate the velocity, $V_1 = \dfrac{1}{0.015} \times \left(\dfrac{0.61\,\text{m}}{4}\right)^{2/3} \times (0.0013)^{1/2} = 0.686\,\text{m/s}$

Apply Equation 6.11b to calculate the discharge, $Q_1 = \dfrac{0.312}{0.015} \times (0.61\,\text{m})^{8/3} \times (0.0013)^{1/2} = 0.201\,\text{m}^3/\text{s}$

2. Calculate the depth of flow and velocity in the upstream sewer.

Calculate the flow ratio, $\dfrac{q_1}{Q_1} = \dfrac{113\,\text{L/s} \times 10^{-3}\,\text{m}^3/\text{L}}{0.201\,\text{m}^3/\text{s}} = 0.56$

Read the ratios d_1/D_1 and v_1/V_1 from Figure 6.13 when $q_1/Q_1 = 0.57$.

$$\frac{d_1}{D_1} = 0.54 \quad \text{and} \quad \frac{v_1}{V_1} = 1.04$$

$$d_1 = 0.54 \times 0.61\,\text{m} = 0.329\,\text{m}$$

$$v_1 = 1.04 \times 0.686\,\text{m/s} = 0.713\,\text{m/s}$$

3. Calculate the velocity and discharge in the downstream sewer when the sewer is flowing full.

Apply Equation 6.11a to calculate the velocity, $V_2 = \dfrac{1}{0.015} \times \left(\dfrac{0.61\,\text{m}}{4}\right)^{2/3} \times (0.0014)^{1/2} = 0.712\,\text{m/s}$

Apply Equation 6.11b to calculate the discharge, $Q_2 = \dfrac{0.312}{0.015} \times (0.61\,\text{m})^{8/3} \times (0.0014)^{1/2} = 0.208\,\text{m}^3/\text{s}$

4. Calculate the depth of flow and velocity in the downstream sewer.

Calculate the flow ratio, $\dfrac{q_2}{Q_2} = \dfrac{113\,\text{L/s} \times 10^{-3}\,\text{m}^3/\text{L}}{0.208\,\text{m}^3/\text{s}} = 0.54$

Read the ratios d_2/D_2 and v_2/V_2 from Figure 6.13 when $q_2/Q_2 = 0.54$.

$$\frac{d_2}{D_2} = 0.53 \quad \text{and} \quad \frac{v_2}{V_2} = 1.03$$

$$d_2 = 0.53 \times 0.61\,\text{m} = 0.323\,\text{m}$$

$$v_2 = 1.03 \times 0.712\,\text{m/s} = 0.733\,\text{m/s}$$

5. Calculate the head loss at entrance and exit of the manhole from Equation 6.15b.

$$h_{\text{L,entrance}} = 0.1 \times \frac{(0.733\,\text{m}^3/\text{s})^2}{2 \times 9.81\,\text{m/s}^2} = 0.003\,\text{m}$$

$$h_{\text{L,exit}} = 0.4 \times \frac{(0.713\,\text{m}^3/\text{s})^2}{2 \times 9.81\,\text{m/s}^2} = 0.010\,\text{m}$$

6. Compute the water surface elevations (WSELs) for preparing the hydraulic profile.

Invert elevation of the downstream sewer at the entrance point	= 0.00 m
Water depth in the downstream sewer line	= 0.323 m
WSEL in the downstream sewer line at the entrance point	= (0.00 + 0.323) m
	= 0.323 m
Entrance loss in the outlet sewer at the manhole	= 0.003 m
WSEL in the manhole prior to the entrance	= (0.323 + 0.003) m
	= 0.326 m
Exit loss in the inlet sewer at the manhole	= 0.010 m
WSEL in the upstream sewer line prior to the exit	= (0.326 + 0.010) m
	= 0.336 m
Water depth in the upstream sewer line	= 0.329 m
Invert elevation of the upstream sewer line	= (0.336 − 0.329) m
	= 0.007 m

7. Draw the hydraulic profile (Figure 6.14).

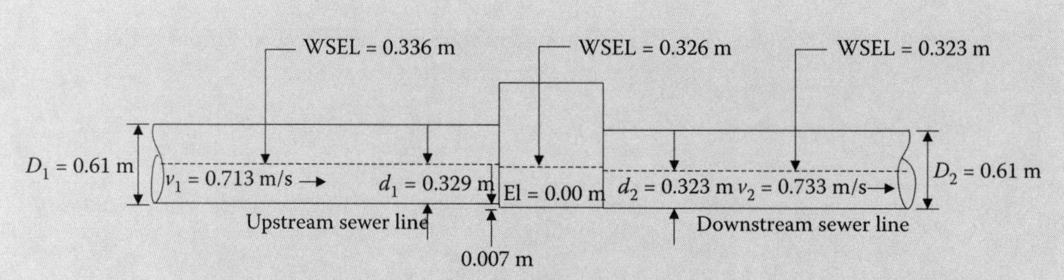

FIGURE 6.14 Hydraulic profile through a sewer manhole (Example 6.29).

EXAMPLE 6.30: HYDRAULIC PROFILE THROUGH A WASTEWATER TREATMENT PLANT

Show the hydraulic profile of a secondary wastewater treatment plant. Indicate the wastewater elevations at different treatment units of the plant.

Solution

1. Draw the hydraulic profile.

   The hydraulic profile of a regional wastewater treatment plant is shown in Figure 6.15.

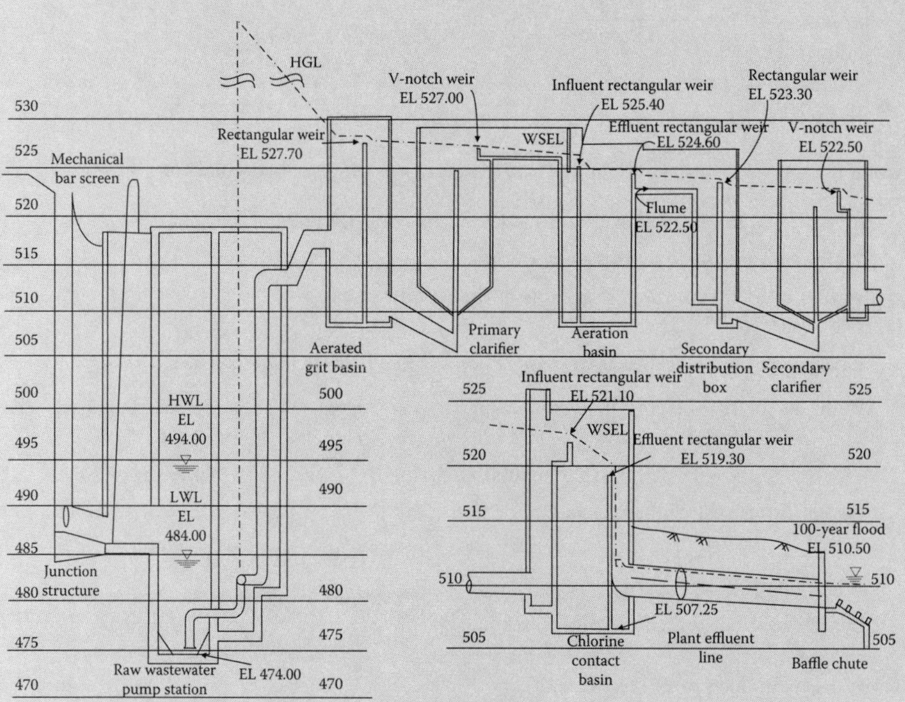

FIGURE 6.15 Hydraulic profile through a regional wastewater treatment plant (Example 6.30).

2. Determine the water surface elevation at different treatment units.

Unit	Elevation, ft
Low water level in wet well	484.00
High water level in wet well	494.00
Aerated grit basin, rectangular weir	527.70
Primary clarifier, V-notch weir	527.00
Aeration basin, influent rectangular weir	525.40
Aeration basin, effluent rectangular weir	524.60
Secondary distribution box, rectangular weir	523.30
Secondary clarifier, V-notch weir	522.50
Chlorine contact basin, influent rectangular weir	521.10
Chlorine contact basin, effluent rectangular weir	519.30
100-year flood elevation	510.50

6.3.9 Energy and Resource Requirements and Plant Economics

Primary energy is the energy used in plant operation, while secondary energy is needed to manufacture chemicals, other consumable materials, and construction materials. Resource requirements include land, equipment, instrumentation, and labor. The plant economics is therefore not only based on the initial construction costs but also on annual operation and maintenance (O&M) costs. Under the Clean Water Act of 1977, it is required that the designers utilize treatment alternatives that substantially conserve energy. Additionally, a cost-benefit analysis is necessary to ensure that the construction and O&M costs are reasonable and appropriate for the planned level of treatment.

6.3.10 Environmental Impact Assessment

The National Environmental Policy Act of 1969 (NEPA) was enacted to ensure that federal agencies consider environmental factors in the decision-making process, and utilize an interdisciplinary approach in evaluating these issues.[34,35] The environment impact assessment must evaluate all impacts—beneficial and adverse, and primary and secondary that may result from the construction of a wastewater treatment facility. The primary impacts are those directly associated with construction and operation of the treatment works. For example, changes in water quality and odors resulting from the plant operation are the primary impacts. The secondary impacts result from the growth or change in land use induced or facilitated by the construction of the plant or its associated sewers. To address these federal environmental considerations, the design engineer should work closely with the federal, state, and local regulatory agencies that have responsibilities for planning, design, and operation of the wastewater treatment facilities.

EXAMPLE 6.31: PLANT ECONOMICS

A biological nutrient removal plant is serving a design population of 80,000. Average wastewater flow treated is 38,000 m^3/d (10 MGD). The influent and effluent quality is provided below. The total current capital cost of the wastewater treatment plant is 34.296 million dollars. Current total annual O&M cost is 1.895 million dollars. The interest rate is 7.652%, and the planning period is 15 years. Determine the following: (a) present worth, (b) equivalent annual cost, (c) unit cost $/m³ and $/1000 gal, and (d) charges per family per month. Assume 3.5 members per family.

Solution

1. Describe the present worth.

   The present worth may be thought of as the sum which, if invested now at a given rate of interest, would provide exactly the funds required at the design period while making all necessary expenditures during the planning period.

2. Calculate the present worth of annual O&M cost.

   The value of annuity is expressed by Equation 6.18.

$$\text{Value of annuity} = \frac{[1-(1+i)^{-n}]}{i} \tag{6.18}$$

where

i = interest rate

n = planning period, yr

$$\text{Value of annuity} = \frac{[1-(1+0.07652)^{-15}]}{0.07652} = 8.744$$

Present worth of annual O&M cost = 1.895 million dollars × 8.744 = 16.570 million dollars

Present worth of capital and O&M cost = (34.296 + 16.570) million dollars = 50.866 million dollars

3. Describe the equivalent annual cost.

   The equivalent annual capital cost is the expression of nonuniform series of expenditures used as a uniform annual amount to simplify calculations of present worth.

4. Calculate the capital recovery factor.

   The capital recovery factor is expressed by Equation 6.19.

$$\text{Capital recovery factor} = \frac{i(1+\mathrm{i})^n}{[(1+i)-1]} \tag{6.19}$$

$$\text{Value of annuity} = \frac{0.07652 \times (1+0.07652)^{15}}{[(1+0.07652)^{15}-1]} = 0.1144/\text{year}$$

5. Calculate the equivalent annual cost.

   The equivalent annual cost is calculated in two ways.

   a. The equivalent capital annual cost = 34.296 million dollars × 0.1144/year = 3.923 million dollars/year

      It may be noted that an annual payment of 3.923 million dollars over 15-year period will pay back the loan principal and the interest on the loan.

$$\begin{aligned}\text{Total equivalent annual cost} &= [3.923 + 1.895\ (\text{annual O\&M cost})]\ \text{million dollars/year}\\ &= 5.818\ \text{million dollars/year}\end{aligned}$$

   b. Total equivalent annual cost = 50.866 × 0.1144 = 5.819 million dollars/year

      An annual payment of 5.819 million dollars per year for 15 years will pay back the loan principal, and interest, and O&M costs.

6. Calculate the unit treatment cost.

Total equivalent annual cost = $5.819 × 10^6/year

$$\text{Average wastewater treated in a year} = 38{,}000\ \text{m}^3/\text{d} \times 365\ \text{d/year}$$
$$= 13.87 \times 10^6\ \text{m}^3/\text{year}\ (3664 \times 10^6\ \text{gallon/year})$$

$$\text{Unit cost of treatment (SI unit)} = \frac{\$5.819 \times 10^6/\text{year}}{13.87 \times 10^6\ \text{m}^3/\text{year}} = \$0.42/\text{m}^3$$

$$\text{Unit cost of treatment (US customary unit)} = \frac{\$5.819 \times 10^6/\text{year}}{3664 \times 10^6\ \text{gallon/year}} \times 1000 = \$1.59/10^3\ \text{gallons}$$

7. Calculate the monthly charge per family.

$$\text{Number of families} = \frac{80{,}000\ \text{persons}}{3.5\ \text{persons/family}} = 22{,}900\ \text{families}$$

$$\text{Monthly charge per family} = \frac{\$5.819 \times 10^6/\text{year}}{22{,}900\ \text{families} \times 12\ \text{month/year}} = \$21.18\ \text{per family per month}$$

EXAMPLE 6.32: CONSTRUCTION COST ADJUSTED TO A DIFFERENT TIME PERIOD

The estimated construction cost of a wastewater treatment plant in 2008 was 27.7 million dollars. Estimate the construction cost in 2013 dollars.

Solution

The capital costs are adjusted to any period by using the Engineering News-Record Construction Cost Index (ENRCCI), or the U.S. EPA Cost Index. The project construction cost is estimated from Equation 6.20.

Projected construction cost

$$= \frac{\text{Projected future value of index}}{\text{Value of index at time of estimate}} \times \text{Estimated cost at the time of estimate} \tag{6.20}$$

1. Determine the ENRCCI for 1996 and 2008 from the ENR website.

ENRCCI for 2008 = 8362

ENRCCI for 2013 = 9183

2. Determine the construction cost for 2013.

$$\text{Construction cost for 2013} = \frac{9183}{8362} \times 27.7\ \text{million dollars} = 30.4\ \text{million dollars}$$

EXAMPLE 6.33: OPERATION AND MAINTENANCE COST

The annual O&M costs are important factors in the evaluation of alternative treatment processes. The principal element of O&M costs are labor, energy, chemicals, and materials and supplies. The total O&M cost of a 10 MGD diffused aeration basin for 2008 was $409,000. Various cost components and unit costs for aeration basin are provided below. Estimate the total annual O&M costs for 2013.

Category	O&M Cost, % of Total Cost	Unit Energy Rate, $/kWh		ENR Index	
		2008	2013	2008	2013
Labor	20	–	–	7861	9183
Energy	62	$0.045	$0.068	–	–
Materials and supplies	18	–	–	2822	2969
Total	100	–	–	–	–

Solution

1. Calculate the costs of labor, and materials and supplies from the ENR indices.

$$\text{The annual cost of labor in 2013} = (\$409{,}000/\text{year} \times 0.20) \times \frac{9183}{7861} = \$95{,}600/\text{year}$$

$$\text{The annual cost of materials and supplies in 2013} = (\$409{,}000/\text{year} \times 0.18) \times \frac{2969}{2822} = \$77{,}500/\text{year}$$

2. Calculate the cost of energy from the different unit energy rates.

$$\text{The annual cost of energy in 2013} = (\$409{,}000/\text{year} \times 0.62) \times \frac{\$0.068/\text{kWh}}{\$0.045/\text{kWh}} = \$383{,}200/\text{year}$$

3. Calculate the estimated total O&M costs in 2013.

$$\text{The total annual O\&M cost in 2013} = (\$95{,}600 + \$77{,}500 + \$383{,}200)/\text{year} = \$556{,}300/\text{year}$$

EXAMPLE 6.34: ENVIRONMENTAL IMPACT ASSESSMENT

A wastewater treatment project involves construction of interceptors, and expansion of existing wastewater treatment and disposal facilities. List the major primary and secondary beneficial and adverse impacts of the project.

Solution

1. List the beneficial impacts.

   The beneficial impacts are:

   a. Provides compliance with effluent standards
   b. Enhances surface water quality in the receiving water
   c. Encourages many beneficial uses of effluent and receiving waters (agriculture, recreation, commerce, industry, and water supply)
   d. Eliminates groundwater pollution due to septic tanks and absorption fields
   e. Reduces health hazards due to enhanced effluent quality
   f. Provides opportunity for biosolids reuse
   g. Improves shoreline and waterfront for recreation
   h. Encourage growth or change in land use induced by the construction of the plant and sewer lines

2. List the adverse impacts.

   Listed below are possible adverse impacts that may be minimized by modern design techniques and efficient plant operation.

The adverse impacts are:

a. Possible occasional odors emanating from the facility, and lower air quality
b. May reduce the value of the adjacent property
c. Disrupt the environment and inconveniences to the citizens during construction of interceptors
d. Traffic disruptions and higher noise level during construction
e. Require energy, and chemicals for plant operation
f. Lost construction materials and land permanently for the treatment facility
g. It is possible that some adverse impacts may occur on plant and animal communities, ecosystem, scenic views and aesthetics, and community growth pattern and land use.

EXAMPLE 6.35: REVIEWING AGENCIES OF ENVIRONMENTAL IMPACT ASSESSMENT REPORT

Many federal, state and local agencies, and private and public groups may review and provide comments on an environmental impact assessment report. List some of these potential agencies.

Solution

The Federal and state agencies that review and comment on environmental impact assessment reports are: U.S. Environmental Protection Agency, U.S. Fish and Wildlife Service, U.S. Army Corps of Engineers, State Environmental Quality Department, State Air Control Board, State Parks and Wildlife Department, State Water Development Board, State Historic Preservation Office, and concerned city departments.

Various appointed environmental advisory committees and concerned citizens groups should be notified. Some of the citizen groups are: League of Women Voters, Audubon Society, Conservation Societies, Historic preservation societies, Sierra Club, and others. Relevant documents should be placed in the designated depository for public review and comments. The fact sheet to highlight major elements of the project and notices of public hearing should be sent to local news paper, posted in libraries and city hall notice boards, and sent to the mailing list of concerned parties.

6.4 Wastewater Facility Planning, Design, and Management

The task of wastewater treatment plant planning and design involves understanding of service area, sources and characteristics of wastewater, plant site, collection system, treatment processes for liquid and residuals, legal issues, and regulatory constraints. For most wastewater treatment plant design projects, the engineering services are performed in three steps: (a) engineering report or facility plan, (b) preparation of design plans, specifications, cost estimates, and contractual documents, and (c) construction and construction management.

6.4.1 Facility Planning

A facility plan is prepared to identify the water pollution problems in a specific area, evaluate alternatives, and recommend a solution. Through the facility plan, the consultants make many decisions that are subsequently used in preparation of the detailed plans and specifications for the wastewater treatment facility.

EXAMPLE 6.36: MAJOR TOPICS COVERED IN A FACILITY PLAN

List the major topics covered, and the final outcome of a facility plan.

Solution

1. The major topics covered in a facility plan are:
   a. Project description, need, and service area
   b. Design period
   c. Effluent limitations
   d. Existing and future conditions
   e. Population projections, industrial growths, and unit flow rates
   f. Wastewater characteristics: flows and concentrations; I/I; minimum, average, and peak flow rates; mass-loadings and sustained conditions
   g. Forecast, and evaluation and selection of design flow rates and mass-loadings
   h. Wastewater treatment process alternatives and evaluation
   i. Description of selected alternative
   j. Elements of conceptual process design
   k. Operation and maintenance of the facility and emergency operation
   l. Preliminary costs and financial status
   m. Environmental impact assessment report and public participation
   n. Implementation plan including financial arrangements and time schedule for design and construction, and project milestone
2. Describe the outcome of the facility plan.

   At the completion of the facility plan, the project elements are fully defined so that preparation of detailed plans and specifications can proceed expeditiously. The completed facility plan in the prescribed format is submitted to the regional or state clearing house, EPA, and regulatory agency for review and comments, and public hearings. The plan is then revised or amended as necessary. The outcome of a facility plan therefore, is a well-defined, cost-effective, and environmentally sound. The approved project is capable of being implemented and acceptable to the taxpayers, regulatory authorities, and general public.

6.4.2 Design Plans, Specifications, Cost Estimates, and Support Documents

Once a facility plan and project application of a community is approved by the state water quality agency, the design phase of the project is initiated. This phase of the project deals with preparation of detailed engineering design plans, specifications, and cost estimates. The plans and specifications become the official document on which contractors base their bids for construction of the facilities. Based on the project milestones established in this phase, the construction managers or administrators may hold the contractor responsible for completion of the project.

EXAMPLE 6.37: PRELIMINARY INFORMATION AND DOCUMENTS

Many documents and some basic information are needed before the task of developing the design plans, specifications and cost estimates can be started. Provide a list of needed important documents and preliminary information.

Solution

Following is a list of the preliminary information and important documents needed before starting the design plans, specifications, cost estimates, and support documents.

1. Application to prepare design plans and specifications is necessary in specified forms
2. Statement regarding sources of local share of project cost (general taxes, sewer revenue funds, etc.)
3. A copy of resolution authorizing the official representative (mayor, council member, or others) to act on behalf of the applicant
4. Statement regarding availability of the proposed site
5. Proposed contracts or explanation for selection of consulting engineers
6. User's charge or resolution that a user's charge system will be developed
7. Letter of agreement from industries participating in cost recovery system
8. Copy of existing sewer use ordinance or intent to develop one
9. Assurance of compliance with Civil Rights Act, Uniform Relocation of Land Acquisition Policy Act, equal employment opportunity, and others

EXAMPLE 6.38: PREPARATION OF DESIGN PLANS AND SPECIFICATIONS

The most common approach for preparation of the design plans and specifications consists of the following tasks: (a) conceptual design, (b) preliminary design, (c) special studies, and (d) final design and specifications. Describe each task.

Solution

The successful completion of design and construction phase of a project largely depends upon the quality of information developed during the early stages of the project.

1. List the purpose and basic steps conducted for conceptual design.
   The conceptual design is conducted to develop framework for completion of the preliminary design. The tasks accomplished are:
   a. Make the principal engineering decisions
   b. Finalize the preliminary design criteria used in the facility plan.
   c. Select equipment and establish the preliminary facilities layout
   d. Define the necessary field investigations, topographic surveys, geological and hydrological studies, soil borings, etc.
   e. Prepare the process flow diagram, connecting pipings, and hydraulic profile
   f. Define operation and control strategies
2. List the purpose and basic steps conducted for preliminary design.
   The preliminary design stage is an expansion of the conceptual design. Major design decisions are completed in this phase. The tasks accomplished are:
   a. The site plan is finalized, and equipment and piping arrangements are completed.
   b. Mitigation measures are utilized to reduce or lessen unavoidable environmental impacts.
   c. Space and utilities requirements are finalized.
   d. Architectural concepts are developed.
   e. Preliminary cost estimates are developed.
   f. For large projects the *value engineering* should be well along at this stage.
3. List the information developed during the special studies.
   Special studies are often needed to develop specialized information and to refine the design criteria. The specialized information developed is:
   a. Bench-scale and pilot plant testing of processes and equipment to refine the design criteria

b. Dispersion studies for outfall siting
c. Odor studies to develop base-line data
d. The special studies must be completed before the start of the final design to avoid costly design changes.

4. List the information that is developed for final design.

The detailed design plans consists of plan views, elevations, sections, and supplementary views that together with the specifications and general layouts provide the working information for the contract and construction facilities. The design plans specifically contains the following information:

a. Dimensions and location of units and type of equipment, location and size of pipings and appurtenances and elevations
b. Final layout of treatment units, buildings and roads, elevation of all units and structures, and ground elevations
c. Hydraulic profiles at maximum, average and peak hourly flows, and elevation of high and low water levels in the receiving waters
d. Technical specifications for construction of sewers, pumping station, treatment plant, and appurtenances must accompany the design plans.
e. The specification shall include all construction information not shown on the drawings. Information such as quality of material, workmanship, fabrication of the equipment, complete requirements for all mechanical and electrical equipment including machinery, valves, pipings and jointing of pipes, electrical apparatus, wiring, instrumentation, and meters should be included.
f. The consultant is to prepare the construction cost estimates based on the scope of work covered in the project plans and specifications. This estimate is used to judge the reasonableness of the bids received.
g. For the construction contracts, the specifications must also include construction schedule and project milestones, bid bond, performance bond and payment bond, fire and extended coverage, workmen's compensation, public liability and property damage, and "all risk" insurance as required by local and state law, and flood insurance as required during and after construction.
h. Well-prepared design plans and specifications provide the following information (1) allow contractor submit bids with small allowances for unknowns, (2) utilize high quality material for construction, (3) complete work in timely manner, (4) integrate new facilities with existing units under operation, and (5) require minimum changes during construction.

EXAMPLE 6.39: VALUE ENGINEERING

What is value engineering (VE)? Describe the purpose and procedure to conduct the VE.

Solution

Value engineering (VE) is an intensive review of the project elements to utilize specialized cost control techniques. Value engineering is conducted by a review team. The purpose is to obtain the best project without sacrificing quality or reliability. The U. S. EPA has mandated that all projects having the total construction cost over \$10 million dollars and receiving federal funding are subjected to VE analysis. The procedure is as follows:

1. The VE team members are senior professionals who are not involved with the design of the project.
2. The number of VE teams and the number of review sessions depends upon the size and complexity of the project. It may vary from one team and one review session, to multiple teams and multiple sessions.
3. For medium to large projects, one team and two review sessions lasting 1 week each are needed. The first session is conducted at approximately 20–30% and second session at 60–70% completion of the design.

6.4.3 Construction and Construction Management

After the construction bid is approved and contractual agreements are completed, the construction phase of the project is initiated. The construction contract must define clearly integration of new construction with existing facility so that there is no violation of permit requirements, and construction does not create any safety hazards to the treatment plant personnel.

EXAMPLE 6.40: CONSTRUCTION MANAGEMENT TASKS

The construction management techniques are used for timely construction of the project in accordance with the design plans and specification. List the major construction management tasks.

Solution

The major construction and management tasks are:

1. Verify that the contractor has adequate technical and equipment resources onsite that is compatible with the project needs before construction begins.
2. Review the contractor's operation to ensure that the requirements of design plans and specifications are fully implemented.
3. Review and verify the construction schedule and progress in meeting the project milestones.
4. Control change orders and possible construction claims.

Discussion Topics and Review Problems

6.1 Effluent sampling data was collected over a 30-day period at a wastewater treatment plant that has a primary sedimentation basin and a trickling filter. The frequency distributions of 30-day BOD_5 results are given below. Determine if 30-day average BOD_5 data is in compliance with the Minimum National Standards for Secondary Treatment as defined by the U.S. EPA.

BOD_5 Concentration, mg/L	20	22	28	29	31	34	40
Frequency (or number of days), d	2	4	5	6	7	4	2

6.2 The initial and design years of a wastewater treatment facility are 2013 and 2028. The estimated populations of these years are 35,000 and 76,000, respectively. The expected flow rates during these periods are 600 Lpcd. What should the staging period be?

6.3 Estimate the year 2020 population of a community by using arithmetic, geometric, decreasing rate of increase, and logistic curve fitting methods. Use the following census data.

Year	Population (Thousands)
1980	31.6
1990	36.9
2010	42.3

6.4 The population data of a city are given below. Estimate the year 2020 population if the employment projection for the year 2020 is 9200.

Year	Population, Person	Employment, Person
1990	20,000	7500
2000	21,000	8000
2010	23,000	8800

6.5 Population of City A is estimated by graphical comparison. The population of Cities B and C are used for comparison. These cities are larger and had reached the year 2010 population of City A in the past. Estimate the population of City A in the year 2040. Use the following census data.

Year	Population, Person		
	City A	City B	City C
1980	52,000	90,000	85,000
1990	70,000	120,000	95,000
2000	80,000	128,000	105,000
2010	90,000	140,000	115,000

6.6 Draw a generalized process diagram of a secondary wastewater treatment plant using activated sludge process. The process diagram should include processes for both liquid and sludge streams.

6.7 A wastewater treatment plant has the following process train: bar screen, grit chamber, primary sedimentation, trickling filter (high rate), final clarifier, gravity filtration, and chlorine contact basin. Using the average percent removal efficiencies given for various units, estimate the effluent quality in terms of BOD_5, COD, TSS, TP, ON, and NH_3-N. The influent quality after mixing with the return flows from the sludge processing areas is as follows: $BOD_5 = 220$, $COD = 450$, $TSS = 225$, $TP = 9$, $ON = 8$, $NH_3\text{-}N = 21$ and $NO_3\text{-}N = 0$. All units are in mg/L.

6.8 The effluent discharge permit of a state is 5/5/10/2 (BOD_5/TSS/TN/TP). The average influent BOD_5, TSS, TN, and TP are 200, 240, 44, and 8 mg/L, respectively (assume TN = 50% ON + 50% AN). The selected process train uses bar screen, grit removal, primary clarifier, BNR facility, gravity filters, and chlorination/dechlorination. The selected processes and percent removal efficiencies of these processes are given below. Assume no removal through bar screen and chlorination/dechlorination process. Determine the effluent quality from each process, and overall plant effluent quality.

Removal Efficiency, %	Grit Chamber	Primary Sedimentation	BNR Facility	Gravity Filters
BOD_5	0	35	95	45
TSS	0	58	95	70
ON	0	30	83	60
NH_3-N	0	0	95	0
TP	0	15	80	35

6.9 A wastewater treatment plant utilizes chemically enhanced sedimentation followed by filtration, carbon adsorption, and disinfection. Draw the process diagram and determine the effluent quality in terms of BOD_5, COD, TSS, ON, AN, and TP. The influent quality is as follows: $BOD_5 = 220$ mg/L, $COD = 450$ mg/L, $TSS = 255$ mg/L, $ON = 25$ mg/L, $NH_3\text{-}N = 21$ mg/L, and $TP = 9$ mg/L. Use the average removal efficiency given in Table 6.8 for each process.

6.10 The effluent discharge permit of a state is 10/10/1 ($BOD_5/TSS/NH_3$-H). The average influent BOD_5, TSS, and TN are 200, 240, and 44 mg/L, respectively (assume TN = 50% ON + 50% AN). Select a process diagram that provides effluent quality within the state permit.

6.11 A secondary wastewater treatment plant is designed for an average flow of 19,100 m^3/d. Calculate the average quantities of screenings, grit, scum, primary, and secondary sludge. The unit quantities of residuals produced is given in Table 6.9.

6.12 The quantities of dry solids in primary and secondary sludge are 2600 and 1600 kg/d. the solids content and specific gravities of primary and secondary sludge are 3.5% and 0.3%, and 1.02 and 1.00, respectively. Calculate the dry solids, percent solids, and volume of combined sludge. The specific gravity of combines sludge is 1.01.

6.13 A gravity thickener receives combined primary and secondary sludge. The combined sludge is 500 m^3/d and contains 1% solids. Assume that the solids capture efficiency of the thickener is 90% and the thickened sludge has 6% solids. Calculate the average volumes of thickened sludge and supernatant, and average concentration of TSS in the supernatant. The specific gravities of combined and thickened sludges are 1.00 and 1.03, respectively.

6.14 The liquid elevations in the wet well and influent channel of grit chamber are 147.48 m and 158.23 m, respectively. The elevation of centerline of the pump impeller is 150.00 m. Determine (a) static suction head, (b) static discharge head, and (c) total static head.

6.15 A submersible pump is used for pumping wastewater at a treatment plant. The diameter of the suction bell is 40 cm and $K = 0.05$. Calculate the components of head losses and total head loss in the force main. The diameter and length of a force main are 25 cm and 150 m, respectively. There is one enlarger from 15 to 25 cm with $K = 0.2$, one check valve with $K = 2.5$, one plug valve with $K = 1.0$, two gate valves with $K = 0.19$ each, five 45° elbows with $K = 0.2$ each, and two 90° elbows with $K = 0.3$ each. The pump discharge rate is 0.10 m^3/s and Darcy–Weisbach coefficient of roughness $f = 0.02$.

6.16 A pump is delivering a flow of 1850 gpm at a *TDH* of 60 ft. The pump efficiency is 85% and motor efficiency is 92%. Calculate the wire power or electrical energy input.

6.17 An intercepting sewer is 500 mm in diameter and is serving a population of 6500 residents. The average wastewater flow is 450 Lpcd. Calculate the depth of flow and velocity at minimum flow. The slope of the line is 0.0025. Assume $n = 0.015$ and minimum flow is one-third of the average flow.

6.18 A trunk sewer has a diameter of 21 in. It has a slope of 0.001 and is flowing 40% full. Determine the depth, velocity, and discharge at partial flow condition using hydraulic element equations. Assume $n = 0.015$.

6.19 An intercepting sewer is carrying a flow of 0.12 m^3/s. It has a diameter of 61 cm and a slope of 0.001. The sewer enters a manhole that has an outlet sewer diameter of 61 cm at a slope of 0.0012. Draw the hydraulic profile and determine the invert elevation of the incoming sewer. Assume $n = 0.013$ and coefficients of exit and entrance are 0.42 and 0.14, respectively.

6.20 A wastewater treatment plant is designed to serve a population of 20,000 residents. The designed flow is 7600 m^3/d (2.0 MGD). The construction cost of the facility is $ 8.0 million. The annual O&M cost is 300,000. Calculate (a) present worth (b) equivalent annual cost (c) the unit treatment costs in $/$m^3$ and in $/1000 gallons, and (d) $/family per year. Amortize the construction cost over 20 years of useful life at an interest rate of 7% .

6.21 Review the model facility plan in Chapter 6 of Reference 15. Describe the purpose of a facility plan, and provide the contents of a facility plan that is developed for construction of a wastewater treatment facility.

References

1. *Federal Water Pollution Control Act Amendments of 1972* (P.L. 92-500), 92nd Congress, October 18, 1972.

2. *Clean Water Act* (P.L. 95-217), 95th Congress, December 27, 1977.
3. U.S. Environmental Protection Agency, *40 CFR Part 133—Secondary Treatment Regulation*, 49 *FR* 37006, September 20, 1984, and as amended later.
4. U.S. Environmental Protection Agency, *Total Maximum Daily Load (TDML)*, F-00-009, Washington, DC, 2000.
5. U.S. Environmental Protection Agency, *Biomonitoring to Achieve Control of Toxic Effluents*, Technology Transfer, EPA/625/8-87/013, Duluth, MN, September 1987.
6. U.S. Environmental Protection Agency, *Facilities Planning 1981, Municipal Wastewater Treatment, Construction Grants Program*, 430/9-81-002 FRD-20, Washington, DC, March 1981.
7. U.S. Environmental Protection Agency, *Construction Grants 1985 (CG-85)*, EPA 430/9-84-004, Office of Water Program Operations (WH-546), Washington, DC, July 1984.
8. U.S. Environmental Protection Agency, *The Clean Water State Revolving Fund: Financing America's Environmental Infrastructure—A Report of Progress*, Office of Water, Washington, DC, 1995.
9. U.S. Environmental Protection Agency, 40 CFR Part 503—Standards for the Use or Disposal of Sewage Sludge, 58 *FR* 9247, February 19, 1993, and as amended later.
10. U.S. Environmental Protection Agency, Contaminants of Emerging Concern Including Pharmaceuticals and Personal Care Products, website at https://www.epa.gov/wqc/contaminants-emerging-concern-including-pharmaceuticals-and-personal-care-products, accessed on August 11, 2017.
11. Daughton, C. G. and T. L. Jones-Lepp (editors), *Pharmaceuticals and Personal Care Products in the Environment: Scientific and Regulatory Issues*, ACS, June 2001.
12. Dietrich, D., S. Webb, and T. Petry, *Hot Spot Pollutants: Pharmaceuticals in the Environment*, Elsevier Science Ltd, September 2003.
13. Kummerer, K., *Pharmaceuticals in the Environment: Source, Fate, Effects and Risks*, Springer, June 2001.
14. U.S. Environmental Protection Agency, *Assessment of Needed Publicly Owned Wastewater Treatment Facilities, Correction of Combined Sewer Overflows, and Management of Storm Water and Non-Point Source Pollution in the United States,* Report to Congress, 1992 Needs Survey, EPA832-p-03-002, 1993.
15. Qasim, S. R., *Wastewater Treatment Plants: Planning, Design, and Operation*, 2nd ed., CRC Press, Boca Raton, FL, 1999.
16. U.S. Environmental Protection Agency, 40 CFR Part 35, Subpart E—Grants for Construction of Treatment Works, 43 *FR* 44049, September 27, 1978, and as amended later.
17. Qasim, S. R., E. Motley, and G. Zhu, *Water Works Engineering: Planning, Design, and Operation*, Prentice Hall PTR, Upper Saddle River, NJ, 2000.
18. Congressional Research Service, *Summaries of Major Statutes Administered by the Environmental Protection Agency*, RL30798, December 20, 2013.
19. Congressional Research Service, *Federal Pollution Control Laws: How Are They Enforced?*, RL34384, December 16, 2013.
20. U.S. Environmental Protection Agency, *Summary of the Clean Water Act*, website at https://www.epa.gov/laws-regulations/summary-clean-water-act, accessed on August 11, 2017.
21. U.S. Environmental Protection Agency, 40 CFR Part 122—EPA Administrated Permit Programs, 48 *FR* 14153, April 1, 1983, and as amended later.
22. U.S. Environmental Protection Agency, 40 CFR Part 123—State Program Requirements, 48 *FR* 14178, April 1, 1983, and as amended later.
23. U.S. Environmental Protection Agency, 40 CFR Part 125—Criteria and Standards for the National Pollutant Discharge Elimination System (NPDES), 44 *FR* 32948, June 7, 1979, and as amended later.
24. U.S. Environmental Protection Agency, 40 CFR Part 130—Water Quality Planning and Management, 50 *FR* 1779, January 11, 1985, and as amended later.

25. U.S. Environmental Protection Agency, 40 CFR Part 131—Water Quality Standards, 48 *FR* 51405, November 8, 1983, and as amended later.
26. U.S. Environmental Protection Agency, 40 CFR Part 132—Water Quality Guidance for the Great Lakes System, 60 *FR* 15387, March 23, 1995, and as amended later.
27. U.S. Environmental Protection Agency, 40 CFR Part 136—Guidelines Establishing Test Procedures for the Analysis of Pollutants, 38 *FR* 28758, October 16, 1973, and as amended later.
28. U.S. Environmental Protection Agency, 40 CFR Part 403—General Pretreatment Regulations for Existing and New Sources of Pollution, 46 *FR* 9439, January 28, 1981, and as amended later.
29. U.S. Environmental Protection Agency, 40 CFR Part 412—Concentrated Animal Feeding Operations (CAFO) Point Source Category, 68 *FR* 7269, February 12, 2003, and as amended later.
30. U.S. Environmental Protection Agency, 40 CFR Part 449—Airport Deicing Point Source Category, 77 *FR* 29203, May 16, 2012, and as amended later.
31. U.S. Environmental Protection Agency, 40 CFR Part 450—Construction and Development Point Source Category, 74 *FR* 63057, December 1, 2009, and as amended later.
32. U.S. Environmental Protection Agency, Combined Sewer Overflow (CSO) Control Policy, 59 *FR* 18688, April 19, 1994, and as amended later.
33. Metcalf & Eddy, Inc., *Wastewater Engineering: Treatment and Reuse*, 4th ed., McGraw-Hill, New York, 2003.
34. Ross, W. A., Evaluating environmental impact statements, *Journal of Environmental Management*, 25, 1987, 137–147.
35. Canter, L. W., *Environmental Impact Assessment*, 2nd ed., McGraw-Hill, Inc., New York, 1996.

7 Screening

7.1 Chapter Objectives

Screening is normally the first unit operation utilized at a wastewater treatment plant. The purpose of screens is to remove large objects that can damage equipment, block valves, nozzles, channels, pipelines, and appurtenances. This creates serious plant operation and maintenance problems. Fine screens are gaining popularity as a substitute for preliminary and primary treatment to remove suspended solids and BOD. Applications of microscreens have also been developed for polishing secondary effluent or tertiary treatment. The objectives of this chapter are to present:

- Types of screening devices and their applications
- Equipment description of coarse screens, design considerations, and installations
- Design criteria of coarse screens, and design examples
- Types of fine screens, design considerations, design criteria, and examples for preliminary and primary treatment
- Special screens for treatment of combined sewer overflows (CSOs)
- Quantity, characteristics, handling, grinding and comminution, and disposal of screenings

7.2 Screening Devices

Screening devices are widely used for wastewater treatment. They are broadly classified as coarse screens, fine screens, and microscreens. Coarse screens remove large objects for preliminary treatment only, while fine screens remove small particulates for either preliminary or primary treatment. Microscreens are actually filtration devices that are used for either primary or tertiary treatment.[1–3] Brief information about these devices is provided in Table 7.1.

7.2.1 Coarse Screens

Coarse screens (also called bar rack or bar screen) remove large objects such as rags, paper, plastics, cans, tree branches, and like. The screening element may consist of parallel bars (or rods) or perforated plates. The openings may be circular, rectangular, or square shapes. The opening sizes typically range from 6 to 75 mm (0.25–3 in). Coarse screens with larger opening sizes up to 100–150 mm (4–6 in) are also available and mainly used at intake structures of water supply or power generation facilities. The types of coarse screen commonly used in wastewater treatment are identified in Figure 7.1. These screens are either manually or mechanically cleaned. The manually cleaned coarse screens are used at small wastewater treatment plants. These screens are normally inclined parallel bar racks. Manually or mechanically cleaned parabolic screens are also available for such applications. The mechanically cleaned coarse screens use either fixed bars or moving screening elements. Except those in heavy-duty design, many of these screens are enclosed for odor control. The fixed bars are continuously or intermittently cleaned by multiple scrapers or a single

TABLE 7.1 Summary of Screens Used for Wastewater Treatment

Screen Classification	Opening Size, mm (in)	Commonly Used Screen Material				Application	References
		Bars or Rods	Perforated Plate	Wedge Wire	Wire or Fabric Mesh		
Coarse screens	25–75 (1–3)	X				Preliminary treatment	Sections 7.2.1 and 7.2.2
	6–25 (0.25–1)	X	X			Preliminary treatment	Sections 7.2.1 and 7.2.2
Fine screens	3–6 (0.12–0.25)	X	X			Preliminary treatment	Sections 7.2.3 and 7.2.4
	1–3 (0.04–0.12)		X	X		Preliminary treatment	Sections 7.2.3 and 7.2.4
	0.5–1 (0.02–0.04)			X	X	Primary treatment	Section 9.7
Micro screens	0.25–0.5 (0.01–0.02)			X	X	Primary treatment	Section 9.7
	0.1–0.25 (0.004–0.01)				X	Primary treatment	Section 9.7
	0.005–0.1 (0.0002–0.004)				X	Tertiary treatment	Section 15.4.7

arm rake. During recent years, many types of fine screens are modified for use in preliminary treatment. Most of these self-cleaning screens consist of moving screening elements with large openings for continuous removal of trapped screenings. Brief information on each type of coarse screens is developed based on available online information from the manufacturers. This information on different type of screens is summarized in Table 7.2.[1–25] The major components of these screens are illustrated in Figure 7.2.[14,17,20,24,25]

7.2.2 Design Considerations of Coarse Screens and Installations

The design of coarse screens involves (1) location, (2) screen chamber and screen arrangement, (3) velocity and head loss, and (4) control system.[1]

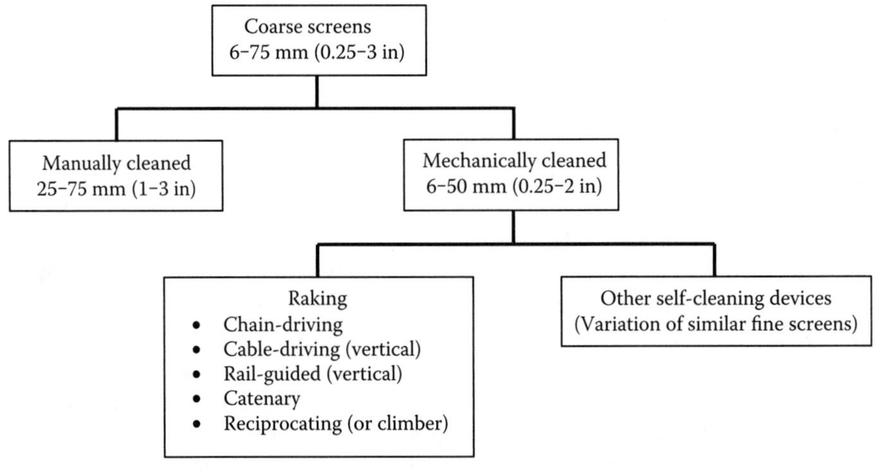

FIGURE 7.1 Types of coarse screens.

TABLE 7.2 General Information about Coarse Screens Used for Preliminary Treatment

Types of Screen	Range	Description
Manually cleaned screen	Opening size: 25–75 mm (1–3 in) Channel width: 1–1.5 m (3–5 ft) Channel depth: 1–1.2 m (3–4 ft) Incline angle: 45–60°	The parallel bar racks are widely used in small channels at small wastewater treatment plants (Figure 7.2a). The length of the bars is restricted to 3 m (10 ft) to facilitate manual raking. A drainage plate at the top temporarily stores the rakings for drainage. Manually cleaned parabolic screens with stainless-steel wedge-shaped bars may also be used for pretreatment (see fine screens Table 7.6 and Figure 7.20a).
Mechanically cleaned screens	Opening size: 6–50 mm (0.25–2 in)	The mechanically cleaned coarse screens are continuously or intermittently cleaned by cleaning device. The two basic types of cleaning methods are (1) raking and (2) other self-cleaning devices.
Raking	Opening size: 6–50 mm (0.25–2 in)	Raking is commonly used by mechanically cleaned bar and parabolic screens. It may use multiple scrapers or a single rake to pull the debris similar to the manual raking operation. Raking is performed from the front or back of the screens in three basic types of arrangements: (a) front clean/rear return, (b) front clean/front return, and (c) back clean/rear return. All types offer advantages and disadvantages with effectiveness of cleaning, protection against jamming, and vulnerability to fouling and breakage. The most common types of raking devices are (1) chain-driven, (2) cable-driven, (3) rail-guided, (4) catenary, and (5) reciprocating (or climber). Rotating arm raking is used for parabolic coarse screens.[4–24]
Chain-driven	Opening size: 6–50 mm (0.25–2 in) Channel width: 0.3–6 m (1–20 ft) Max. channel depth: 20 m (65 ft) Incline angle: 60–80°	Continuously moving scrapers clean the screen from the front or the back (Figure 7.2b). This type is available for special applications in deep channels up to 90 m (300 ft). The maintenance requirements are usually medium. This type of raking mechanism is also available for fine screens (Table 7.6).[4,7–14]
Cable-driven (vertical)	Opening size: 15–50 mm (0.6–2 in) Channel width: 0.6–6 m (2–20 ft) Channel depth: up to 20 m (65 ft)	The bar rack is cleaned from the front by a single rake moving up and down by multiple cables. They are usually in heavy-duty design. It can be used in deep channels up to 75 m (250 ft). Vertical design requires small footprint and can be retrofitted into existing channels. The maintenance requirements are usually low to medium (Figure 7.2c).[5,6,15,16]
Rail-guided (vertical)	Opening size: 10–50 mm (0.4–2 in) Channel width: 0.4–2.5 m (1.3–8 ft) Channel depth: up to 20 m (65 ft)	The screens are usually back cleaned/back return. The operation of this type of screens is very similar to that for cable-driven screens. The vertical structure reduces space requirements and allows for easy retrofit. The maintenance requirements are usually low to medium (Figure 7.2d).[4,17,18]
Catenary	Opening size: 6–40 mm (0.25–1.5 in) Channel width: up to 5 m (16 ft) Channel depth: up to 20 m (65 ft) Incline angle: 45–75°	This type of screen uses front cleaned/front return arrangement. It is suitable for use in wide channels up to 9 m (30 ft). The weight of chain holds the rake against the moderately inclined rack. The required headroom is relatively low, but need large floor area. The sprockets are not submerged, and most maintenance can be done above the operating floor. The heavy-duty design is suitable for the unattended facilities where loads may vary significantly. Periodical manual cleaning is required because the rakes pass over heavy objects at the base to avoid jamming. They require medium maintenance (Figure 7.2e).[6,19,20]

(*Continued*)

TABLE 7.2 (*Continued*) General Information about Coarse Screens Used for Preliminary Treatment

Types of Screen	Range	Description
Reciprocating (or climber)	Opening size: 6–50 mm (0.25–2 in) Channel width: 0.6–4 m (2–13 ft) Channel depth: up to 15 m (50 ft) Incline angle: 75–90°	The heavy-duty screens are cleaned from the front or back of the screens. Usually, a cogwheel-driven carriage with a pivot cleaning rake arm travels up and down along a looped guide track. The rake engages the screen at the bottom, cleans trapped materials while moving upward, and discharges the screenings through a chute into a container at the top. It is can be used in wide channels up to 9 m (30 ft). When it is used in deep channels, more headroom than other types is needed. The screen requires small floor area due to a steep incline. The maintenance requirements are low to medium since all moving parts are above the water surface (Figure 7.2f).[6,7,21–24]
Other self-cleaning devices	Opening size: 6–50 mm (0.25–2 in) Single unit capacity: 90–24,000 m^3/h (0.6–150 MGD)	As a variation of similar fine screens (Table 7.6), these coarse screens can be used for preliminary treatment. The screen elements are normally from stainless steel or high-strength plastics. These screens include: • Parabolic with rotating rakes: curved bars or wedge wires with openings 6–12 mm (0.25–0.5 in) (Figure 7.2g)[25] • Spiral (or basket): perforated plates with openings 6–10 mm (0.25–0.4 in) • Continuous moving elements: perforated plates, wedge-shaped bars, molded links, or patterned meshes or grids with openings 6–30 mm (0.25–1.2 in) • Inclined rotary drum: wedge-shaped bars with openings 6–10 mm (0.25–0.4 in) • Rotary shear drum: perforated plates with openings 6–50 mm (0.25–2 in) These screens require medium to high maintenance (Table 7.6).

Screen Location: The coarse screens should be located ahead of pumps or grit removal facility. They provide protection against large objects that may wrap around moving parts, jam equipment, and bock channels, valves, and piping.

Screen Chamber and Screen Arrangement: The screen chamber is designed to prevent accumulation of grit and other heavy materials. It is a rectangular channel having a flat (horizontal) or mild slope. It should provide a straight approach, perpendicular to the screen for uniform distribution of flow and screenings over the entire screen area. At least two bar racks, each designed to carry the peak flow, must be provided for continued protection in case one unit is out of service. Arrangements for stopping the flow and draining the channel should be made for routine maintenance. The entrance structure should have a smooth transition or divergence in order to minimize the entrance losses. The effluent structure should also have smooth transition. Rectangular weir to control channel depth should be avoided as heavy material will settle in the channel. If the screen is located ahead of the grit chamber, the depth and velocity in the channel should be controlled by the water surface elevation in a grit chamber. It is however required that the flushing velocity is developed under normal operating condition. If the screen is ahead of a wet well, a head control device such as *proportional weir* or *Parshall flume* should be provided so that low depth and high velocity do not occur at the screen due to drawdown resulting from a free fall.[1,26–28] The bar rack chambers, and influent and effluent arrangements are shown in Figure 7.3.[1]

Velocity and Head Loss: The desirable upstream channel length is usually 2–4 times of the width of the channel. The ideal range of approach velocity in the upstream channel is usually 0.4–0.45 m/s (1.25–1.5 ft/s) to prevent solids deposition and avoid strong turbulence in the channel. The velocity through the clean screen should not exceed 0.6 and 0.9 m/s (2 and 3 ft/s) at design average and peak flow, respectively. Higher velocity will push debris through the screen, while lower velocity accumulates larger quantity of screenings. The allowable head loss through the clogged bar racks is 150 mm (6 in). The maximum design head loss may range from 250 to 600 mm (10–24 in) for the continuous belt screens with perforated plates. For design purposes, the head loss through the screen may be calculated from Equations 7.1a or 7.1b.[1,2,26] The actual head loss shall be validated with the screen manufacturers.

Equation 7.1a is used to calculate head loss through clean or partly clogged bars, while Equation 7.1b is used to calculate head loss through clean screen only. The head loss across the entire screen chamber should also include head loss due to entrance, bends, expansion, contraction, and exit.

$$h_L = \frac{V^2 - v^2}{2g} \times \left(\frac{1}{C_d}\right) \tag{7.1a}$$

$$h_L = \beta \times \left(\frac{W}{b}\right)^{4/3} \times h_v \times \sin\theta \tag{7.1b}$$

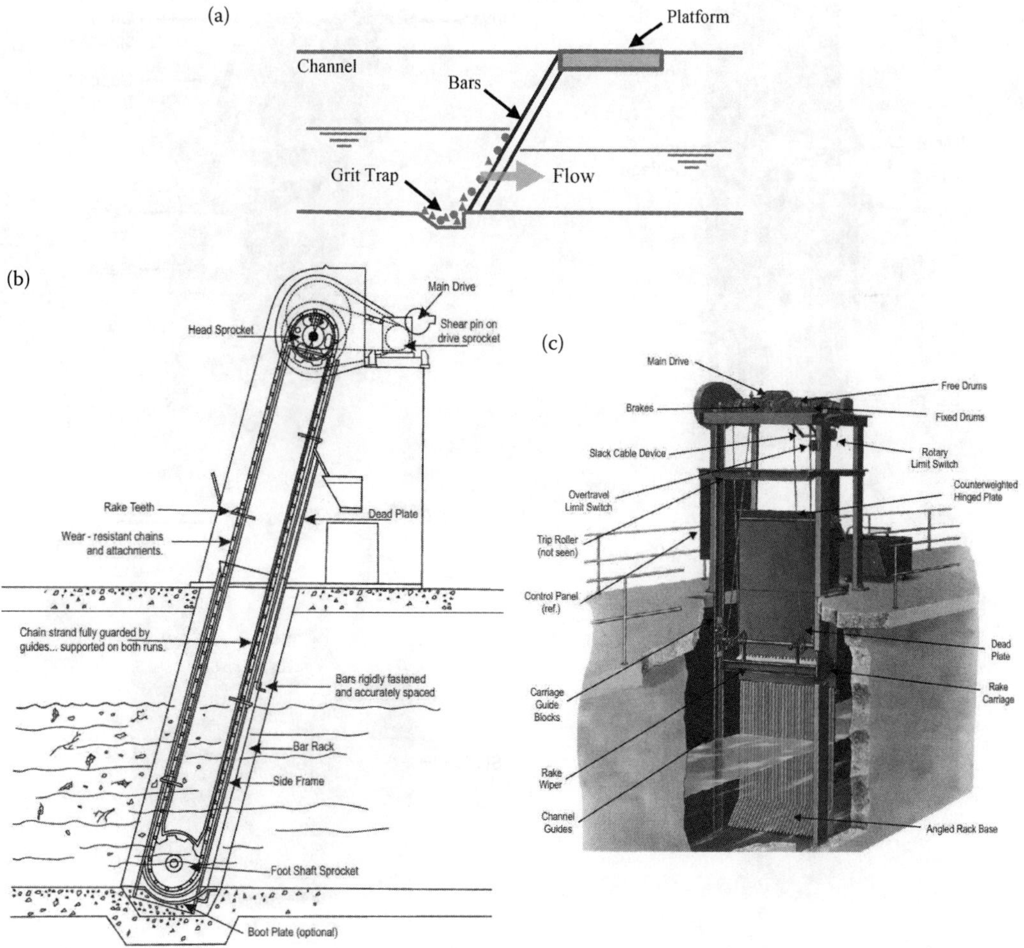

FIGURE 7.2 Design details of manually and mechanically cleaned coarse screens: (a) manually cleaned bars, (b) chain-driven (Courtesy Triveni Engineering and Industries Ltd.), and (c) cable-driven (Courtesy Triveni Engineering and Industries Ltd.). *(Continued)*

where

h_L = head loss across the screen, m (ft)

V and v = velocity through the screen, and approach velocity in the channel upstream of the screen, m/s (ft/s)

g = acceleration due to gravity, 9.81 m/s^2 (32.2 ft/s^2)

W = total cross-sectional width of the bars facing the direction of flow, m (ft)

b = total clear spacing of bars, m (ft)

h_v = velocity head of the flow approaching the bars, m

θ = angle of bars with respect to the horizontal, degree

Q = discharge through screen, m^3/s, (ft^3/s)

C_d = coefficient of discharge, dimensionless. $C_d = 0.60$ for clogged screen, and 0.70 for clean screen.

β = bar shape factor, dimensionless

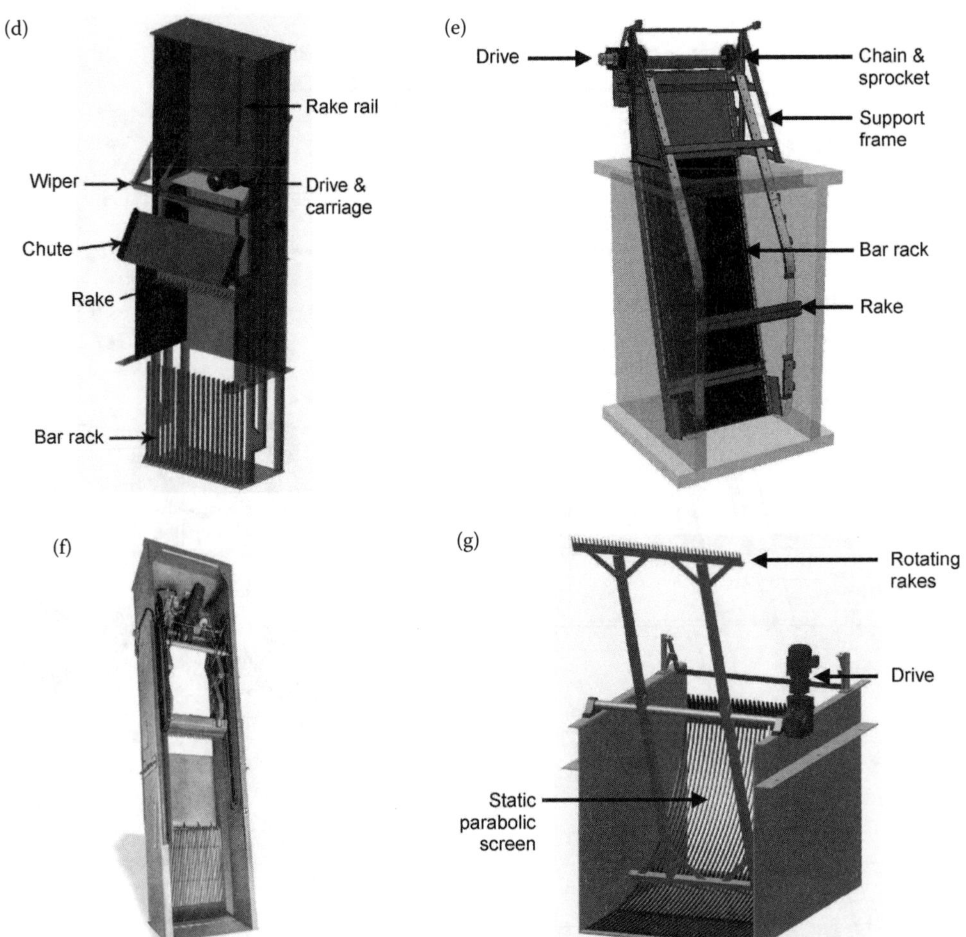

FIGURE 7.2 (Continued) Design details of manually and mechanically cleaned coarse screens: (d) rail-guided (Courtesy John Meunier Products/Veolia Water Technologies Canada), (e) catenary (Courtesy Mabarex, Inc.), (f) reciprocating (Courtesy Vulcan Industries, Inc.), and (g) parabolic (Courtesy John Meunier Products/Veolia Water Technologies Canada).

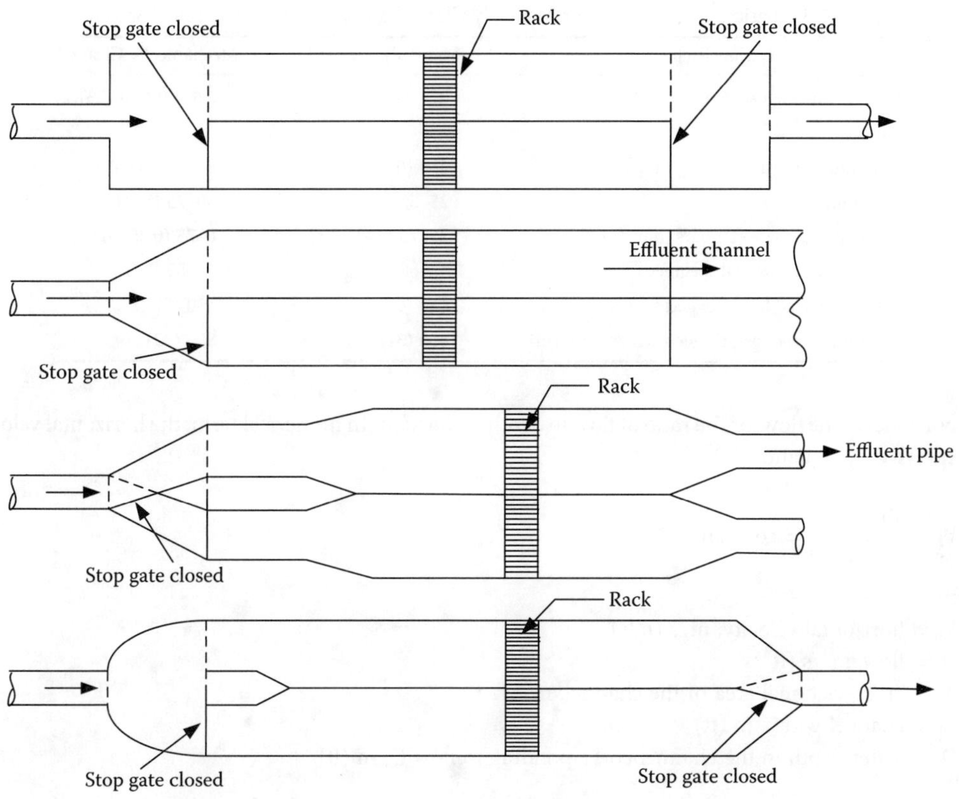

FIGURE 7.3 Double chamber bar rack, and influent and effluent arrangement.

The β values of bar shape factors for clean screen are summarized as follows:

Bar Type	β
Sharp-edged rectangular	2.42
Rectangular with semicircular upstream face	1.83
Circular	1.79
Rectangular with semicircular upstream and downstream faces	1.67
Tear shape	0.76

The basic design factors for manually and mechanically cleaned bar screen are presented in Table 7.3.[1,2]

Control Systems: The mechanically cleaned screens operate by an independent, motor-driven time clock adjustable to give cycle time ranging from 0 to 15 min. Additionally, the raking device has an automatic override if a preset high water level upstream of the screen or head loss across the screen is reached.

Velocity Control Requirements: The horizontal velocity through the screen chamber should remain relatively constant within the desired range under variable flow condition. Unfortunately, the velocity in a channel is typically and easily exceeding 0.3–1.0 m/s (1–3.3 ft/s) when the channel floor is flat at the downstream end with a free-fall condition. Therefore, a proper method must be applied to achieve proper velocity control purpose. The common velocity control methods include (a) sutro or proportional weir, (b) Parshall flume, (c) raised floor bottom, or (d) automatic controlled weir gate.

Velocity Control in a Rectangular Channel: In a rectangular channel since the channel width does not change, a constant horizontal velocity is achieved if the water depth in the channel is maintained

TABLE 7.3 Basic Design Factors for Manually Cleaned and Mechanically Cleaned Screens

Design Factor	Manually Cleaned	Mechanically Cleaned
Velocity through screen, m/s (ft/s)	0.3–0.6 (1–2)	0.6–1.0 (2.0–3.3)
Bar Size		
Width, mm (in)	4–8 (0.16–0.32)	8–10 (0.32–0.4)
Depth, mm (in)	25–50 (1–2)	50–75 (2–3)
Clear spacing between bars, mm (in)	25–75 (1–3)	6–75 (0.25–3)
Slope from horizontal, degrees	45–60	75–85
Allowable head loss, clogged screen, mm (in)	150 (6)	150 (6)
Maximum head loss, clogged screen, mm (in)	800 (32)	800 (32)

proportional to the flow, or the ratio of flow to depth is constant. In numerical term, the horizontal velocity is expressed by Equation 7.2.

$$V_{\mathrm{h}} = \frac{Q}{A} = \frac{Q}{WD} = \text{constant} \tag{7.2}$$

where

V_{h} = horizontal velocity, m/s (ft/s)
Q = flow, m^3/s (ft^3/s)
A = cross-sectional area of the channel, m^2 (ft^2)
W = channel width, m (ft)
D = water depth in the channel corresponding to flow Q, m (ft)

The *sutro* and *proportional* weirs are installed at the outlet section of the channel, and designed to maintain water depth in the channel proportional to the flow. These weirs are a combination of a straight weir crest and a well-shaped opening or orifice through which the discharge occurs. The details of sutro and proportional weir are shown in Figure 7.4. Ideally, the length needs to be extended to infinity at the theoretical weir datum elevation. In reality, it is impossible to achieve this condition in a channel. Therefore,

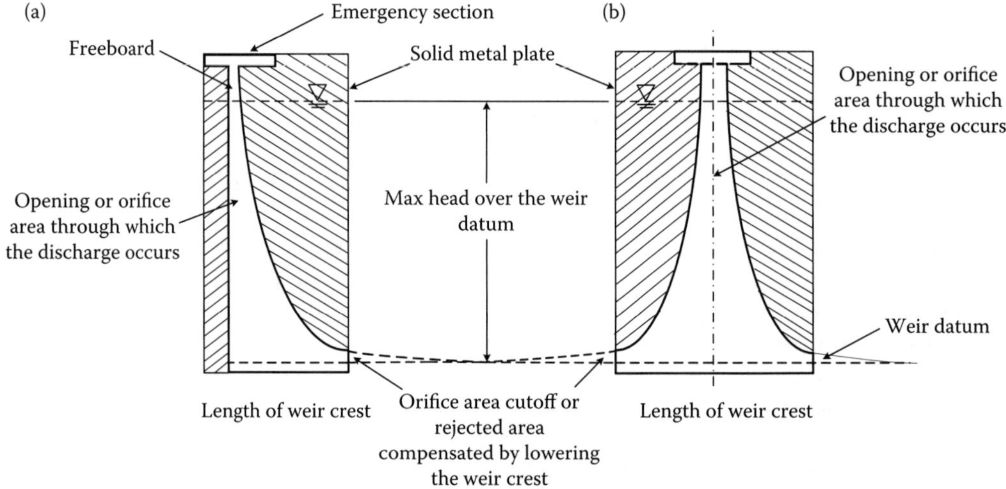

FIGURE 7.4 Definition sketch of constant velocity control section in a rectangular channel: (a) sutro weir, and (b) proportional weir.

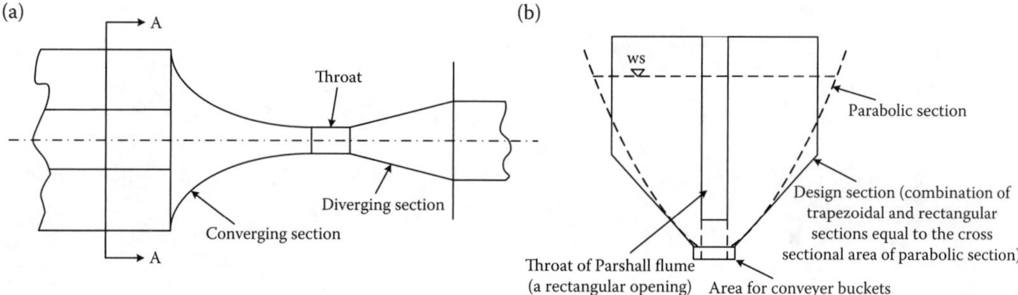

FIGURE 7.5 Definition sketch of a Parshall flume in a trapezoidal channel represented by a parabola: (a) plan, and (b) Section AA.

the curved section is typically cut off beyond a desired weir crest length that is less than or equal to the channel width. When the ideal weir shape is applied at a finite weir length, the actual velocity in the channel may vary slightly with the water depth because of such nonideal conditions. However, the velocity variation is usually within an acceptable range. In some designs, a compensation of the rejected area may also be considered by lowering the weir crest from the weir datum. Readers may review Examples 7.6 and 7.7 for details on this subject.

Velocity Control in a Parabolic or Trapezoidal Channel: In a channel with varying cross section such as parabolic, trapezoidal, or a combination of trapezoidal and rectangular channel, the control section is a *Parshall flume*. The throat or control section of Parshall flume is rectangular and maintains a constant velocity in the channel. The dimensions of standard Parshall flumes and design information are provided in Appendix C. True parabolic channel is difficult to construct. Therefore, it is normally replaced by a trapezoidal–rectangular channel of cross-sectional area. The definition sketches of parabolic and trapezoidal–rectangular channels with Parshall flume at the control section are shown in Figure 7.5. The design example of a Parshall flume in a parabolic channel is presented in Chapter 8.

EXAMPLE 7.1: RATIO OF CLEAR AREA TO TOTAL AREA

A bar rack has 30 parallel bars equally spaced in a chamber. The clear spacing is 25 mm. The width of the bars is 12 mm. The water depth in the channel upstream of the rack is 1.5 m. Calculate the width of the chamber and the ratio of the clear area to total screen area.

Solution

1. Determine the total clear opening area.

   There are 30 parallel bars. Total number of clear spaces = 31 spaces

   $$\text{Total width of clear opening} = 31\ \text{spaces} \times 25\ \text{mm/space} \times \frac{\text{m}}{10^3\ \text{mm}} = 0.775\ \text{m}$$

   $$\text{Total clear opening area} = 0.775\ \text{m} \times 1.5\ \text{m} = 1.16\ \text{m}^2$$

2. Determine total area of the screen.

   $$\text{Total width of all bars} = 30\ \text{bars} \times 12\ \text{mm/bar} \times \frac{\text{m}}{10^3\ \text{mm}} = 0.36\ \text{m}$$

   $$\begin{aligned}\text{Total width of the channel} &= \text{total width of clear opening} + \text{total width of all bars}\\ &= (0.775 + 0.36)\ \text{m} = 1.135\ \text{m}\end{aligned}$$

   $$\text{Total area of the screen} = 1.135\ \text{m} \times 1.5\ \text{m} = 1.70\ \text{m}^2$$

   This is the screen area that is submerged under water in the channel.

3. Calculate the ratio of total clear opening and total screen area.

$$\text{The ratio} = \frac{1.16\,\text{m}^2}{1.70\,\text{m}^2} = 0.68$$

This ratio is called screen efficiency factor. The manufacturer normally provides this ratio for their standard screens.

EXAMPLE 7.2: HEAD LOSS ACROSS A BAR SCREEN

A mechanically cleaned bar screen is installed ahead of a velocity controlled grit channel. There are 28 parallel sharp-edged rectangular bars 10 mm wide placed at a clear spacing of 20 mm. The flow through the screen is 0.75 m^3/s, and the depth of flow is 1.45 m upstream of the screen. The angle of inclination of bars is 80° from the horizontal. Calculate, (1) head loss across clean screen, $C_d = 0.7$, and (2) head loss across the screen when the screen is 50% clogged, $C_d = 0.6$.

Solution

1. Calculate the clear area of the screen.

$$\text{Total width of clear opening} = 29\,\text{spaces} \times 20\,\text{mm/space} \times \frac{\text{m}}{10^3\,\text{mm}} = 0.58\,\text{m}$$

$$\text{Total clear opening area} = 0.58\,\text{m} \times 1.45\,\text{m} = 0.84\,\text{m}^2$$

2. Calculate the cross-sectional area of the channel.

$$\text{Total width of all bars} = 28\,\text{bars} \times 10\,\text{mm/bar} \times \frac{\text{m}}{10^3\,\text{mm}} = 0.28\,\text{m}$$

$$\text{Total width of the channel} = (0.58 + 0.28)\,\text{m} = 0.86\,\text{m}$$

$$\text{Total area of the screen} = 0.86\,\text{m} \times 1.45\,\text{m} = 1.25\,\text{m}^2$$

3. Calculate the approach velocity in the channel, and velocity through clean screen openings.

$$\text{Approach velocity} = \frac{0.75\,\text{m}^3/\text{s}}{1.25\,\text{m}^2} = 0.60\,\text{m/s}$$

Clear screen opening area perpendicular to the direction of flow = total clear opening area = 0.84 m^2

$$\text{Velocity through screen openings} = \frac{0.75\,\text{m}^3/\text{s}}{0.84\,\text{m}^2} = 0.89\,\text{m/s}$$

4. Calculate the head loss through clean screen.
   Calculate the head loss h_L from Equation 7.1a with $C_d = 0.7$.

$$h_L = \frac{(0.89\,\text{m/s})^2 - (0.60\,\text{m/s})^2}{2 \times 9.81\,\text{m/s}^2} \times \frac{1}{0.7} = 0.031\,\text{m}$$

Calculate the head loss h_L from Equation 7.1b with a bar shape factor $\beta = 2.42$.

$$h_L = 2.42 \times \left(\frac{0.28\,\text{m}}{0.58\,\text{m}}\right)^{4/3} \times \frac{(0.60\,\text{m/s})^2}{2 \times 9.81\,\text{m/s}^2} \times \sin 80 = 0.017\,\text{m}$$

The actual head loss through clean screen may be between 0.02 and 0.03 m.

5. Calculate the head loss through 50% clogged screen.

Calculate the head loss through 50% clogged screen h_{L50} from Equation 7.1a with $C_d = 0.6$. At 50% clogged screen the area is reduced by half and velocity through screen is doubled.

$$h_{L50} = \frac{(2 \times 0.89\,\text{m/s})^2 - (0.60\,\text{m/s})^2}{2 \times 9.81\,\text{m/s}^2} \times \frac{1}{0.6} = 0.24\,\text{m}$$

The head loss through clogged screen is about 10 times of that when it is clean.

EXAMPLE 7.3: WATER DEPTH AND VELOCITY IN THE SCREEN CHAMBER UPSTREAM OF A BAR SCREEN

An interceptor discharges into a bar screen chamber. The velocity and depth of flow in the interceptor are 1 m/s and 1.2 m, respectively. The width of the screen chamber is 1.5 m, and coefficient of expansion $K_e =$ 0.5. The floor of screen chamber is horizontal, and the invert is 0.06 m below the invert of the interceptor. Calculate the velocity and depth of flow in the screen chamber upstream of the bar rack. The flow in the interceptor is 1.1 m^3/s.

Solution

1. Draw the problem definition sketch (Figure 7.6).

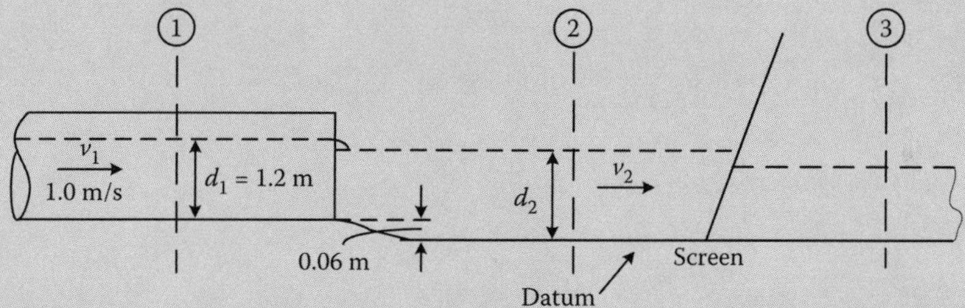

FIGURE 7.6 Definition sketch of screen chamber (Example 7.3).

2. Select equations for the calculations.

Select Bernoulli's energy equation that is expressed by Equation 7.3a at sections (1) and (2).

$$Z_1 + d_1 + \frac{v_1^2}{2g} = Z_2 + d_2 + \frac{v_2^2}{2g} + h_L \tag{7.3a}$$

where

Z_1 and Z_2 = chamber bottom elevations at sections (1) and (2) above datum, m (ft)
d_1 and d_2 = water depths at sections (1) and (2), m (ft)
v_1 and v_2 = velocities at sections (1) and (2), m/s (ft/s)
h_L = head loss between sections (1) and (2), m (ft)

The head loss h_L is calculated from the velocity head differential between sections (1) and (2) and expressed by Equations 7.3b or 7.3c. Either equation is modified from minor head loss equation (Equation 6.15b).

$$h_L = \frac{K_e}{2g}\left(v_1^2 - v_2^2\right) \quad \text{(for expansion, } v_1 > v_2\text{)} \tag{7.3b}$$

$$h_L = \frac{K_c}{2g}\left(v_2^2 - v_1^2\right) \quad \text{(for reduction, } v_2 > v_1\text{)} \tag{7.3c}$$

where K_e and K_c = coefficient of expansion or reduction, dimensionless

Typical value of K_e or K_c is between 0.3 and 0.5 (see Appendix C).

3. Determine the depth of flow (d_2) and velocity (v_2) in the screen chamber.
   Assume datum is at the chamber bottom, $Z_2 = 0$ m.
   Apply Equations 7.3a and 7.3b at sections (1) and (2) with $K_e = 0.5$.

$$0.06\,\text{m} + 1.2\,\text{m} + \frac{(1.0\,\text{m/s})^2}{2 \times 9.81\,\text{m/s}^2} = 0\,\text{m} + d_2 + \frac{v_2^2}{2 \times 9.81\,\text{m/s}^2} + \frac{0.5}{2 \times 9.81\,\text{m/s}^2}\left((1.0\,\text{m/s})^2 - v_2^2\right)$$

$$1.285\,\text{m} = d_2 + \frac{0.5 \times v_2^2}{2 \times 9.81\,\text{m/s}^2}$$

$$\text{Substitute } v_2 = \frac{Q}{d_2 \times W} = \frac{1.1\,\text{m}^3/\text{s}}{d_2 \times 1.5\,\text{m}} = \frac{0.733\,\text{m}^2/\text{s}}{d_2}$$

$$1.285\,\text{m} = d_2 + \frac{0.5}{2 \times 9.81\,\text{m/s}^2} \times \left(\frac{0.733\,\text{m}^2/\text{s}}{d_2}\right)^2$$

$$1.285\,\text{m} = d_2 + \frac{0.0137\,\text{m}^3}{d_2^2}$$

$$d_2^3 - (1.285\,\text{m}) \times d_2^2 + 0.0137\,\text{m}^3 = 0$$

Solving by trial and error procedure, $d_2 = 1.28$ m

$$v_2 = \frac{1.1\,\text{m}^3/\text{s}}{1.28\,\text{m} \times 1.5\,\text{m}} = 0.57\,\text{m/s} \ < v_1$$

EXAMPLE 7.4: DEPTH OF FLOW, AND VELOCITY IN THE SCREEN CHAMBER OF THE BAR RACK

A rectangular channel contains a bar rack. The width of the chamber downstream is 1.25 m and the flow through the chamber is 1 m^3/s. The water depth in the chamber upstream of the bar rack is 1.4 m. The velocity through the clear rack openings is 0.9 m/s and $C_d = 0.7$ for clean screen. Calculate (a) approach velocity, and (b) depth and velocity in the channel down stream of the bar rack.

Solution

1. Draw the definition sketch (Figure 7.7).
2. Calculate head loss through bar rack.

$$\text{Approach velocity, } v_2 = \frac{1\,\text{m}^3/\text{s}}{1.25\,\text{m} \times 1.4\,\text{m}} = 0.57\,\text{m/s}$$

Calculate h_L from Equation 7.1a with $C_d = 0.7$.

$$h_L = \frac{(0.9\,\text{m/s})^2 - (0.57\,\text{m/s})^2}{2 \times 9.81\,\text{m/s}^2} \times \frac{1}{0.7} = 0.035\,\text{m}$$

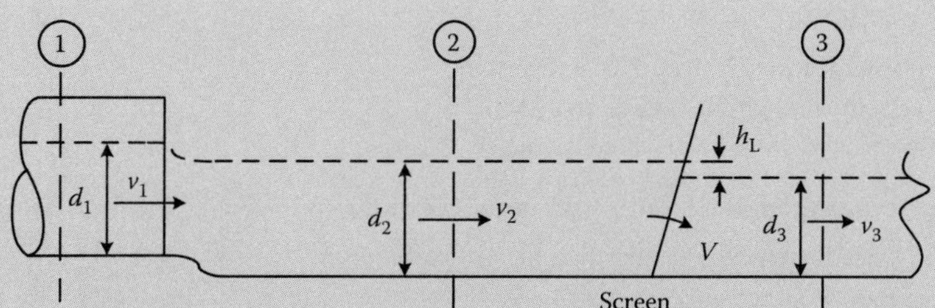

FIGURE 7.7 Definition sketch of screen chamber (Example 7.4).

3. Calculate the depth of flow in the channel downstream of the rack.

$$\text{Downstream velocity, } v_3 = \frac{Q}{d_3 \times W} = \frac{1\,\text{m}^3/\text{s}}{d_3 \times 1.25\,\text{m}} = \frac{0.8\,\text{m}^2/\text{s}}{d_3}$$

Apply Equation 7.3a at sections (2) and (3), $Z_2 = Z_3 = 0$

$$0\,\text{m} + 1.4\,\text{m} + \frac{(0.57\,\text{m/s})^2}{2 \times 9.81\,\text{m/s}^2} = 0\,\text{m} + d_3 + \frac{1}{2 \times 9.81\,\text{m/s}^2}\left(\frac{0.8\,\text{m}^2/\text{s}}{d_3}\right)^2 + 0.035\,\text{m}$$

$$1.382\,\text{m} = d_3 + \frac{0.0326\,\text{m}^3}{d_3^2}$$

$$d_3^3 - (1.382\,\text{m}) \times d_3^2 + 0.0326\,\text{m}^3 = 0$$

Solving by trial and error procedure, $d_3 = 1.36$ m

$$v_3 = \frac{1\,\text{m}^3/\text{s}}{1.36\,\text{m} \times 1.25\,\text{m}} = 0.59\,\text{m/s}$$

Note: The depth of d_3 may also be estimated from d_2 and h_L: $d_3 = d_2 - h_L = (1.4 - 0.035)\,\text{m} = 1.37$ m. This approach is much simplified and practically acceptable.

EXAMPLE 7.5: EQUATIONS FOR THE PROPORTIONAL WEIR DEVELOPED FROM ORIFICE EQUATION

Sutro and proportional weirs are devices that are constructed at the downstream section of a rectangular channel. The purpose is to maintain a constant velocity in the channel. Develop the orifice equations, and develop the generalized equations for discharge through ideal sutro and proportional weirs. Assumed weir crest, which is also the channel bottom, is set at the weir datum.

Solution

1. Write the orifice equations.
   The discharge through an orifice is expressed by Equations 7.4a and 7.4b.

$$Q = C_d A\sqrt{2gH} \tag{7.4a}$$

$$V = C_d\sqrt{2gH} \tag{7.4b}$$

where

Q = discharge through the orifice, m^3/s (ft^3/s)

V = velocity through the orifice, m/s (ft/s)

C_d = coefficient of discharge, dimensionless

Plate thickness is smaller than the orifice diameter	$C_d = 0.6$ (typical)
Plate thickness is greater than the orifice diameter	$C_d = 0.8$
Round edge orifice	$C_d = 0.92$

A = area of the orifice, m^2 (ft^2)

H = effective head over the orifice, m (ft)

For a free outfall, it is measured from the center of the orifice to the headwater (upstream water) surface (Figure 7.8a).

For a submerged orifice, it is the head loss through the orifice that is measured as the vertical difference between the headwater and tailwater (downstream water) surfaces (Figure 7.8b).

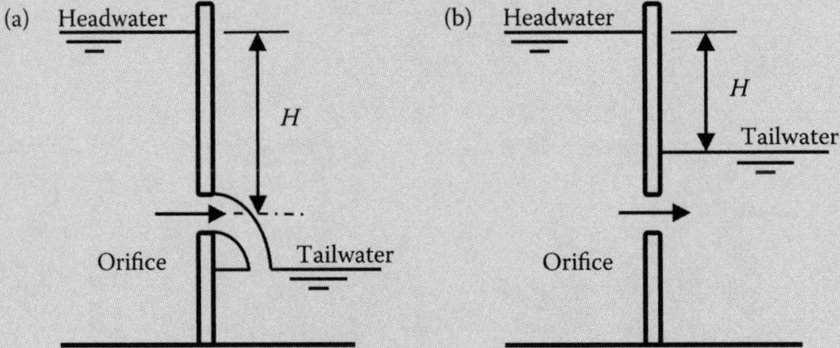

FIGURE 7.8 Effective head: (a) free outfall orifice, and (b) submerged orifice (Example 7.5).

2. Draw the definition sketches of sutro and proportional weirs (Figure 7.9).

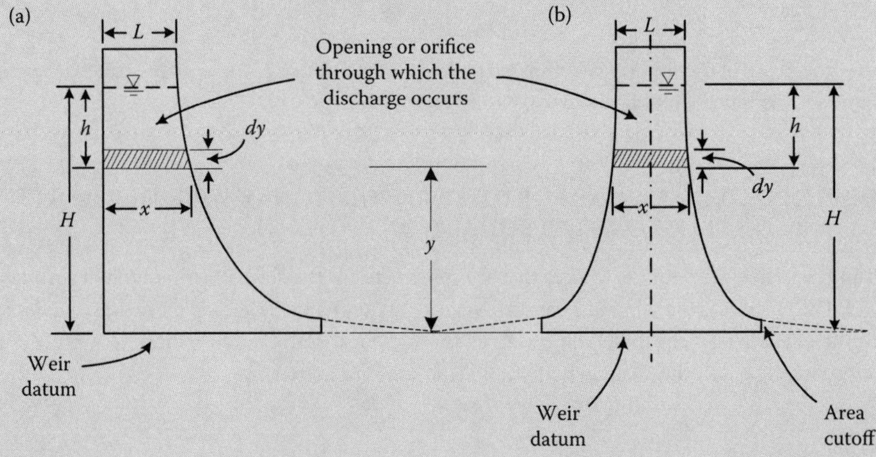

FIGURE 7.9 Definition sketch of velocity control structures: (a) Sutro weir, and (b) proportional weir (Example 7.5).

Notes: y = height of the elemental section above the weir datum (or crest), m (ft); x = width of the elemental section (weir opening) at height y, m (ft); dy = depth of the elemental section, m (ft); H = maximum water head over the weir datum at design flow, m (ft); h = head over the elemental section, $h = H - y$, m (ft).

Based on the definition sketch, the generalized equations for the weir through which discharge occurs are developed below.[29–31]

3. Write the flow equation through the elemental section (Equation 7.5a).

$$dQ = C_d\sqrt{2gh}(dA) = C_d\sqrt{2g(H-y)}(x\,dy) \tag{7.5a}$$

where

dQ = flow through the elemental area, m^3/s (ft^3/s)

dA = area of elemental section, $dA = x\,dy$, m^2 (ft^2)

C_d = coefficient of discharge, dimensionless

4. Write the integral equation of total flow through the proportional weir opening.

Assume the total flow through the weir opening is Q at the maximum water head H.

The integration of Equation 7.5a from $y = 0$ to $y = H$ should give the total flow Q by Equations 7.5b and 7.5c.

$$Q = \int_0^Q dQ = \int_0^H C_d\sqrt{2g}\sqrt{(H-y)}x\,dy \tag{7.5b}$$

$$Q = C_1\int_0^H \sqrt{(H-y)}x\,dy \tag{7.5c}$$

where Q = total flow through the weir opening (orifice), m^3/s (ft^3/s)

C_1 = constant that is expressed by Equation 7.5d.

$$C_1 = C_d\sqrt{2g} \tag{7.5d}$$

5. Write the constant horizontal velocity equation in the rectangular channel.

The horizontal velocity is derived from Equation 7.2 at the maximum head of H or depth D (Equation 7.6a).

$$v_h = \frac{Q}{A} = \frac{Q}{WH} = \frac{Q}{WD} \tag{7.6a}$$

where

v_h = constant horizontal velocity through the channel, m/s (ft/s)

A = cross-sectional area of the channel, $A = WH = WD$, m^2 (ft^2)

W = constant width of the rectangular channel, m (ft)

D = maximum water depth of the rectangular channel, m (ft)

The maximum water depth D in the channel is also equal to H when the weir datum is set at the channel bottom.

Since W is constant for a rectangular channel, v_h is kept constant in the channel if Q is proportional to H (Equation 7.6b).

$$Q = v_h WH = C_2 H \text{ (}Q\text{ is proportional to }H.\text{)} \tag{7.6b}$$

where C_2 = constant, $C_2 = v_h W$ (7.6c)

6. Find a function for x and y such that Equation 7.5c satisfies Equation 7.6b.

Assume that the openings (or orifice shape) of sutro or proportional weir is designed to fit the following conditions in Equation 7.7a:

$$x = Ky^{-1/2} \quad \text{or} \quad K = xy^{1/2} \tag{7.7a}$$

where K = constant

Substitute Equation 7.7a in Equation 7.5c to obtain Equation 7.7b.

$$Q = C_1 K \int_0^H \sqrt{\frac{(H-y)}{y}} dy \tag{7.7b}$$

Assume $z = \frac{y}{H}$ or $dy = Hdz$

At $y = 0$, $z = 0$,

At $y = H$, $z = 1$

Substitute theses values in the integral of Equation 7.7b.

$$Q = C_1 KH \int_0^1 \sqrt{\frac{(1-z)}{z}} dz$$

This function is integrated by trigonometric substitution, or found from integration table. The integrated value is given by Equation 7.7c.

$$Q = \frac{\pi}{2} C_d \sqrt{2g} KH \quad \text{or} \quad Q = C_3 H \, (Q \text{ is proportional to } H.) \tag{7.7c}$$

where C_3 = constant that is expressed by Equation 7.7d.

$$C_3 = \frac{\pi}{2} C_1 K \quad \text{or} \quad C_3 = \frac{\pi}{2} C_d \sqrt{2g} K \tag{7.7d}$$

7. Develop the general equations for obtaining the ideal profile of weir opening.

The generalized equations for the discharge through the proportional weir are developed from Equation 7.7c.

In accordance with Equation 7.7c, the weir opening profile must be developed in such a way as expressed by Equation 7.8a.

$$xy^{1/2} = x_1 y_1^{1/2} = x_2 y_2^{1/2} = \cdots = x_i y_i^{1/2} = \cdots = LH^{1/2} = K(\text{constant}) \tag{7.8a}$$

where L = top length of weir opening at the maximum head H over the weir crest, m (ft)

Substituting $K = LH^{1/2}$ in Equation 7.7c, the equation is expressed by Equation 7.8b.

$$Q = \frac{\pi}{2} C_d \sqrt{2g} (LH^{1/2}) H \quad \text{or} \quad Q = \frac{\pi}{2} C_d \sqrt{2g} LH^{3/2} \tag{7.8b}$$

where C_d = coefficient of discharge, dimensionless. Due to slime growth and obstructions, the commonly used value is 0.6 for applications of proportional weir in raw wastewater.

Appling $C_d = 0.6$, Equation 7.8b is simplified to Equations 7.8c through 7.8e.[31]

$$Q = 1.57 C_d \sqrt{2g} KH \quad \text{or} \quad Q = 1.57 C_d \sqrt{2g} LH^{3/2} \tag{7.8c}$$

$$Q = 4.17 KH \quad \text{or} \quad Q = 4.17 LH^{3/2} \quad \text{(SI units)} \tag{7.8d}$$

$$Q = 7.56 KH \quad \text{or} \quad Q = 7.56 LH^{3/2} \quad \text{(U.S. customary units)} \tag{7.8e}$$

As defined in Equation 7.8a, these equations should be applicable to any x and y. The generalized equations are expressed by Equations 7.8f through 7.8h.

$$q = 1.57 C_d \sqrt{2g} Ky \quad \text{or} \quad q = 1.57 C_d \sqrt{2g} xy^{3/2} \tag{7.8f}$$

$$q = 4.17Ky \quad \text{or} \quad q = 4.17xy^{3/2} \qquad \text{(SI units)} \tag{7.8g}$$

$$q = 7.56Ky \quad \text{or} \quad q = 7.56xy^{3/2} \qquad \text{(U.S. customary units)} \tag{7.8h}$$

The weir opening profiles (or orifice shapes) of ideal sutro and proportional weirs are illustrated in Figure 7.10. In reality, the curved weir section is typically cut off at a desired weir crest length b. For a practical velocity control purpose, the ideal equations can still be used to develop the approximate weir profile.

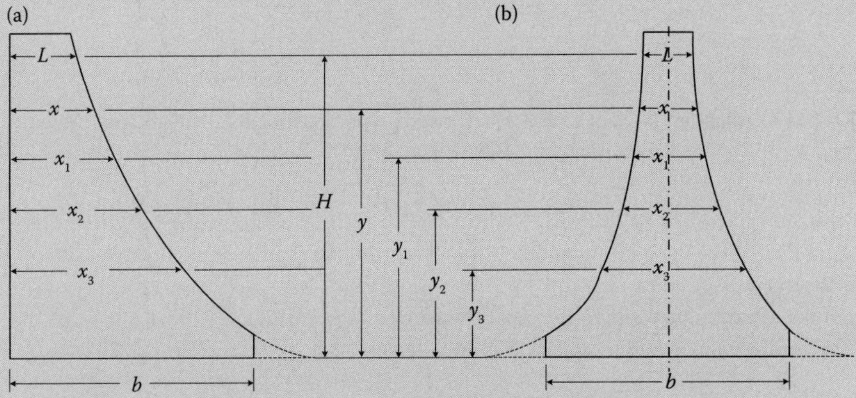

FIGURE 7.10 Weir opening profiles: (a) Sutro weir, and (b) proportional weir (Example 7.5).
Note: b = length of the actual weir crest, m (ft).

EXAMPLE 7.6: PROPORTIONAL WEIR DESIGN

A bar rack is provided ahead of a pumping station. The screened wastewater has a free fall into the wet well. A proportional weir is provided at the free fall to maintain the normal depth into the screen chamber. Design the proportional weir to maintain 0.61 m/s velocity in the channel downstream of the bar rack. The peak discharge through the rack is 1.25 m^3/s. The width of the screen chamber W = 1.6 m, the weir length is less than the chamber width, the bottom of the chamber is horizontal, and the weir crest is flushed with the channel bottom. Calculate at peak discharge (a) maximum depth of flow in the channel downstream of the bar rack, (b) develop the profile of ideal proportional weir and determine the weir length, (c) head loss through the bar rack if the velocity through the screen at peak discharge is 0.89 m/s, and (d) depth of flow and velocity in the channel upstream of bar rack. The weir crest is at the channel bottom. Assume it is also the weir datum.

Solution

1. Draw the definition sketch below (Figure 7.11).
2. Determine the maximum depth d_3 at flow Q in the channel upstream of the proportional weir.

$$\text{Cross area of flow } Q \text{ in the channel, } A_3 = \frac{Q}{v_3} = \frac{1.25\,\text{m}^3/\text{s}}{0.61\,\text{m/s}} = 2.05\,\text{m}^2$$

$$\text{Maximum depth in the channel, } d_3 = \frac{2.05\,\text{m}^2}{1.6\,\text{m}} = 1.28\,\text{m}$$

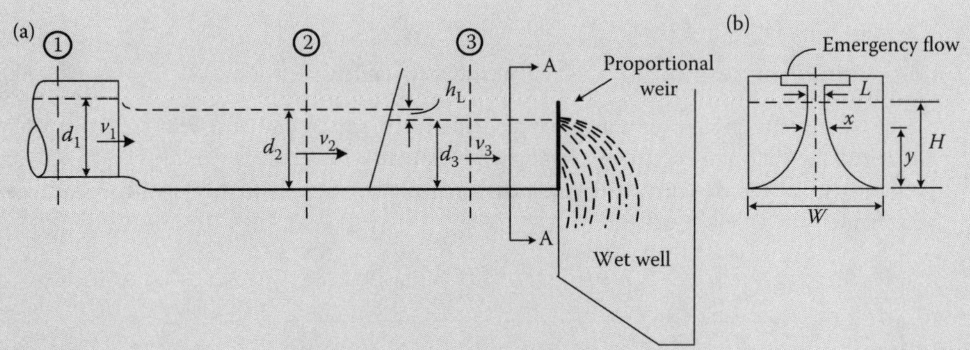

FIGURE 7.11 Definition sketch of screen chamber and proportional weir: (a) longitudinal section of screen chamber, and (b) Section AA showing proportional weir (Example 7.6).

Since the crest of the proportional weir is in level with the channel bottom, the maximum head H over the weir is also 1.28 m.

3. Determine the opening width of the proportional weir at $H = 1.28$ m.

   The ideal opening width is calculated from Equation 7.8d.

$$L = \frac{Q}{4.17H^{3/2}} = \frac{1.25\,\text{m}^3/\text{s}}{4.17 \times (1.28\,\text{m})^{3/2}} = 0.207\,\text{m} \quad \text{or} \quad 0.21\,\text{m}$$

4. Develop the ideal orifice profile of the proportional weir.

   In order to maintain a constant velocity in the channel, K is determined from Equation 7.8a.

   $K = LH^{1/2} = 0.207\,\text{m} \times (1.28\,\text{m})^{1/2} = 0.234\,\text{m}^{3/2}$

   The proportional weir calculations are tabulated below. The ideal weir profile is shown in Figure 7.12.

Assumed Flow q, m^3/s	$y = q/4.17K$ (Equation 7.8g), m	$x = Ky^{-1/2}$ (Equation 7.7a), m	Water Depth in the Channel $d_3 = y$, m	Velocity in the Channel $v_3 = Q/(d_3 \times W)$, m/s
1.25 (Q)	1.28	0.21	1.28	0.61
1.00	1.02	0.23	1.02	0.61
0.75	0.77	0.27	0.77	0.61
0.50	0.51	0.33	0.51	0.61
0.25	0.26	0.46	0.26	0.61
0.049[a]	0.05	1.05	0.05	0.61

[a] This flow is assumed to determine the length of weir crest of 1.05 m.

Note: Although flow varies, but the velocity in the channel remains constant.

5. Compute the depth of flow d_2 and velocity v_2 upstream of the bar rack.

   Apply Bernoulli's energy equation (Equation 7.3a) upstream (section [2]) and downstream of the bar rack (section [3]) with $Z_2 = Z_3 = 0$.

$$d_2 + \frac{v_2^2}{2g} = d_3 + \frac{v_3^2}{2g} + h_L$$

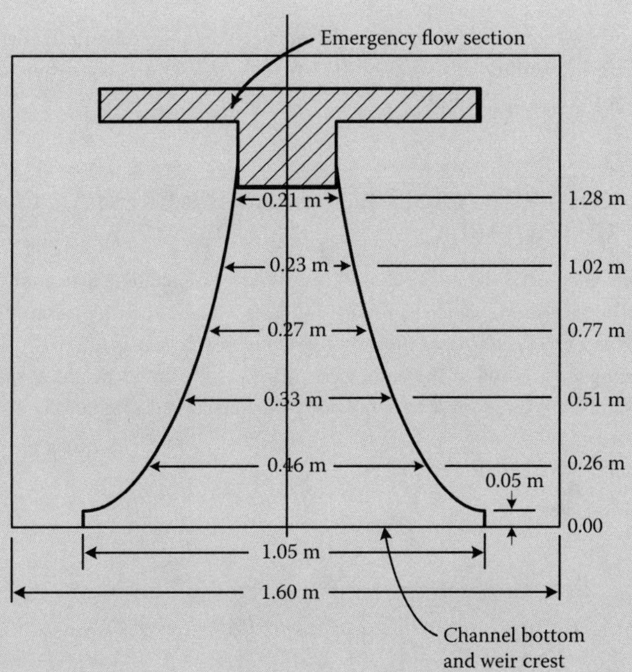

FIGURE 7.12 Design of proportional weir at the effluent structure (Example 7.6).

Head loss through screen is calculated from Equation 7.1a.

$$h_L = \frac{V^2 - v_2^2}{2g} \times \left(\frac{1}{C_d}\right)$$

Substitute, $d_3 = 1.28$ m, $v_3 = 0.61$ m/s, $V = 0.89$ m/s, and $C_d = 0.7$.

$$d_2 + \frac{v_2^2}{2 \times 9.81\,\text{m/s}^2} = 1.28\,\text{m} + \frac{(0.61\,\text{m/s})^2}{2 \times 9.81\,\text{m/s}^2} + \frac{(0.89\,\text{m/s})^2 - v_2^2}{2 \times 9.81\,\text{m/s}^2} \times \frac{1}{0.7}$$

Velocity of flow upstream of the bar rack, $v_2 = \frac{Q}{d_2 \times W} = \frac{1.25\,\text{m}^3/\text{s}}{d_2 \times 1.6\,\text{m}} = \frac{0.781\,\text{m}^2/\text{s}}{d_2}$

$$d_2^3 - 1.36d_2^2 + 0.0755 = 0$$

Solve by trial and error procedure, $d_2 = 1.31$ m

$$v_2 = \frac{1.25\,\text{m}^3/\text{s}}{1.31\,\text{m} \times 1.6\,\text{m}} = 0.57\,\text{m/s} = 0.60\,\text{m/s}$$

6. Calculate the head loss through the bar rack from Equation 7.1a with $C_d = 0.7$.

$$h_L = \frac{(0.89\,\text{m/s})^2 - (0.60\,\text{m/s})^2}{2 \times 9.81\,\text{m/s}^2} \times \frac{1}{0.7} = 0.031\,\text{m}$$

Note: The head loss across the bar rack is also equal to the difference in the depth of water in the chamber upstream (d_2) and downstream of the bar rack (d_3). This is consistent with the comments as a note given at the end of Example 7.4.

EXAMPLE 7.7: LOWERING PROPORTIONAL WEIR CREST TO COMPENSATE CUTOFF AREA

A proportional weir is constructed at the downstream end of a rectangular channel. The width of the channel is 1.4 m. The maximum water depth at peak design flow is 1.5 m, and the weir opening at the top is 0.30 m. The weir crest spans over the entire channel width; and is lowered from the weir datum to compensate for the areas cutoff at the ends. Determine (1) the cutoff height of the weir at the ends, and (2) the depth of lowered weir crest to compensate for the curved area cutoff.

Solution

1. Draw the definition sketch in Figure 7.13.

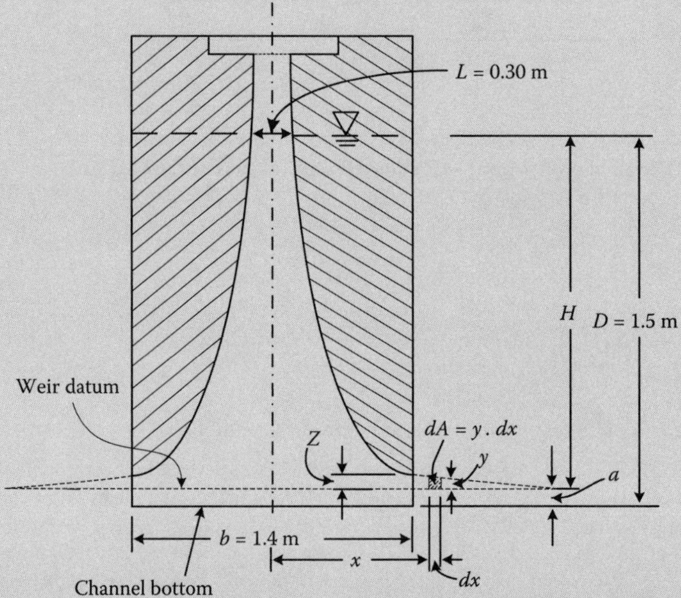

FIGURE 7.13 Definition sketch of proportional weir (Example 7.7).
Notes: a = depth of lowered weir crest to compensate for the curved area cutoff, m (ft); b = length of the weir crest, m (ft); and Z = cutoff height of the weir at the ends, m (ft).

2. Explain the purpose for lowering the weir crest.

   The relationship $x = K/y^{1/2}$ has a limitation that when $y = 0$, $x = \infty$. For this reason, the bottom portion of the weir is usually modified. A realistic length of weir crest is adapted, and the curved areas at both ends are cutoff. The weir crest is lowered to compensate for the rejected areas by a rectangular section ($a \times b$).

3. Calculate the cutoff height of the proportional weir that excludes the curved areas through trial-and-error procedure.

   a. First iteration.

Determine the water head over the weir crest H.

Since a is unknown, assume $H = D - a \approx D = 1.5$ m to initiate the procedure.

Determine the constant K.

The constant K in Equation 7.7a is calculated from the data given in the example, $L = 0.30$ m and $K = LH^{1/2} = 0.30 \text{ m} \times (1.5 \text{ m})^{1/2} = 0.367 \text{ m}^{3/2}$

Determine vertical portion Z.

The vertical portion Z, which is the cutoff height at the ends of the proportional weir, is also calculated from Equation 7.7a.

$$Z = \left(\frac{K}{b}\right)^2 = \left(\frac{0.367 \text{ m}^{3/2}}{1.4 \text{ m}}\right)^2 = 0.0687 \text{ m}$$

Calculate the areas cutoffs on both ends.

$$y = \frac{K^2}{x^2}$$

$$dA = y\,dx$$

where

dA = elemental area in the cutoff section, $dA = y\,dx$, m^2 (ft^2)

x = horizontal distance of the elemental area dA from the center line of the weir, m (ft)

dx = width of the elemental area, m (ft)

y = height of the elemental area, m (ft)

$$dA = \frac{K^2}{x^2}dx$$

Integrate the equation.

$$\int_0^A dA = \int_{b/2}^{\infty} \frac{K^2}{x^2}dx$$

$$A = \left[\frac{-K^2}{x}\right]_{b/2}^{\infty} = \frac{K^2}{b/2} = \frac{(0.367 \text{ m}^{3/2})^2}{0.7 \text{ m}} = 0.192 \text{ m}^2$$

Calculate the depth of lowered weir crest.

Total compensated area (both sides) $2A = 2 \times 0.192 \text{ m}^2 = 0.384 \text{ m}^2$

Height of the compensated area, $a = \dfrac{0.384 \text{ m}^2}{1.4 \text{ m}} = 0.274 \text{ m}$

b. Second iteration.

Determine H.

$$H = D - a = (1.5 - 0.274) \text{ m} = 1.23 \text{ m}$$

Determine the constant K from Equation 7.7a.

$$K = 0.30 \text{ m} \times (1.23 \text{ m})^{1/2} = 0.333 \text{ m}^{3/2}$$

Determine vertical portion Z from Equation 7.7a.

$$Z = \left(\frac{0.333 \text{ m}^{3/2}}{1.4 \text{ m}}\right)^2 = 0.0566 \text{ m}$$

Calculate the areas cutoffs on both ends.

$$A = \frac{(0.333\,\text{m}^{3/2})^2}{0.7\,\text{m}} = 0.158\,\text{m}^2$$

Calculate the depth of lowered weir crest.

$$2A = 2 \times 0.158\,\text{m}^2 = 0.316\,\text{m}^2$$

$$a = \frac{0.316\,\text{m}^2}{1.4\,\text{m}} = 0.226\,\text{m}$$

c. Third through six iterations.

After repeating the trial-and-error procedures six times, the final results are obtained.

$$H = 1.27\,\text{m}$$

$$K = 0.338\,\text{m}^{2/3}$$

$$Z = 0.058\,\text{m}$$

$$A = 0.163\,\text{m}^2$$

$$a = 0.23\,\text{m}$$

Therefore, the weir crest will be lowered by 0.23 m to compensate for the areas cutoff on both sides.

The vertical portion of the proportional weir $(Z + a) = (0.058 + 0.23)\,\text{m} = 0.29\,\text{m}$ (0.95 ft)

EXAMPLE 7.8: PARSHALL FLUME AT THE EFFLUENT STRUCTURE

A bar rack precedes a pumping station. The screened wastewater has a free fall into the wet well. A Parshall flume is provided before the outfall to regulate water depth in the screen chamber. The velocity through the bar screen and approach velocity are 0.9 and 0.6 m/s. Assume $C_d = 0.7$ for clean screen. The width of the channel is 1.5 m and the throat width of the Parshall flume is 0.6 m (2 ft). The peak flow through the screen is 1.25 m^3/s. Design the effluent structure and determine the water depth and velocity in the channel upstream and downstream of the bar rack, and head loss through the screen. Ignore the small head loss at entrance of the Parshall flume.

Solution

1. Compute the depth of flow in the channel upstream of the screen.

$$\text{Depth } d_2 = \frac{Q}{v_2 \times W} = \frac{1.25\,\text{m}^3/\text{s}}{0.6\,\text{m/s} \times 1.5\,\text{m}} = 1.39\,\text{m}$$

2. Compute head loss through the screen.

   Calculate h_L from Equation 7.1a with $C_d = 0.7$.

$$h_L = \frac{(0.9\,\text{m/s})^2 - (0.6\,\text{m/s})^2}{2 \times 9.81\,\text{m/s}^2} \times \frac{1}{0.7} = 0.033\,\text{m}$$

3. Compute the depth of flow and velocity in the channel after the bar rack.

$$\text{Velocity after bar screen } v_3 = \frac{Q}{d_3 \times W} = \frac{1.25\,\text{m}^3/\text{s}}{d_3 \times 1.5\,\text{m}} = \frac{0.833\,\text{m}^2/\text{s}}{d_3}$$

Apply Bernoulli's energy equation (Equation 7.3a) at sections (2) and (3), $Z_2 = Z_3 = 0$

$$0\,\text{m} + 1.39\,\text{m} + \frac{(0.6\,\text{m/s})^2}{2 \times 9.81\,\text{m/s}^2} = 0\,\text{m} + d_3 + \frac{1}{2 \times 9.81\,\text{m/s}^2}\left(\frac{0.833\,\text{m}^2/\text{s}}{d_3}\right)^2 + 0.033\,\text{m}$$

$$1.287\,\text{m} = d_3 + \frac{0.0354\,\text{m}^3}{d_3^2}$$

$$d_3^3 - (1.375\,\text{m}) \times d_3^2 + 0.0354\,\text{m}^3 = 0$$

Solving by trial and error procedure, $d_3 = 1.36\,\text{m}$

$$v_3 = \frac{1.25\,\text{m}^3/\text{s}}{1.36\,\text{m} \times 1.5\,\text{m}} = 0.61\,\text{m/s}$$

4. Determine the dimensions of the Parshall flume.
   a. Calculate water depth H_a at the throat.

      Provide a standard Parshall flume with throat width $W = 2$ ft (~0.6 m).

      Flow through a Parshall flume under free-flow condition is expressed by Equation 7.9.

$$Q = CH_a^n \tag{7.9}$$

where

Q = flow through the Parshall flume, ft^3/s

C and n = coefficient and exponent for design of the flume. C and n are given for various standard throat widths in Table C.2 of Appendix C.

H_a = water depth at the throat, ft

Apply all units in U.S. customary units in Equation 7.9.

Convert flow from m^3/s to ft^3/s.

$$Q = 1.25\,\text{m}^3/\text{s} \times \frac{\text{ft}^3/\text{s}}{0.0283\,\text{m}^3/\text{s}} = 44.2\,\text{ft}^3/\text{s}$$

For a Parshall flume throat width $W = 2$ ft, $C = 8$ and $n = 1.55$ are obtained from Table C.2 in Appendix C.

Calculate H_a from Equation 7.9.

$$44.2 = 8 \times H_a^{1.55}$$

$$H_a^{1.55} = \frac{44.2}{8} = 5.53$$

$$H_a = 3.0\,\text{ft} \quad \text{or} \quad H_a = 3.0\,\text{ft} \times \frac{0.305\,\text{m}}{\text{ft}} = 0.92\,\text{m}$$

It is required to keep the submergence (H_b) <0.7 of H_a for a free flow at the flume (Table C.4). In this example, the wet well design shall meet this requirement under the design peak flow condition. Also, see Figure C.1 for the locations for H_a and H_b.

   b. Calculate throat elevation T of Parshall flume.

$$T = d_3 - H_a = (1.36 - 0.92)\,\text{m} = 0.44\,\text{m}$$

   c. Draw the longitudinal section through the screen chamber.

      Determine the standard dimensions of a 2-ft Parshall flume from Table C.3 in Appendix C. The longitudinal section of screen chamber, Parshall flume dimensions, and water surface elevations are shown in Figure 7.14.

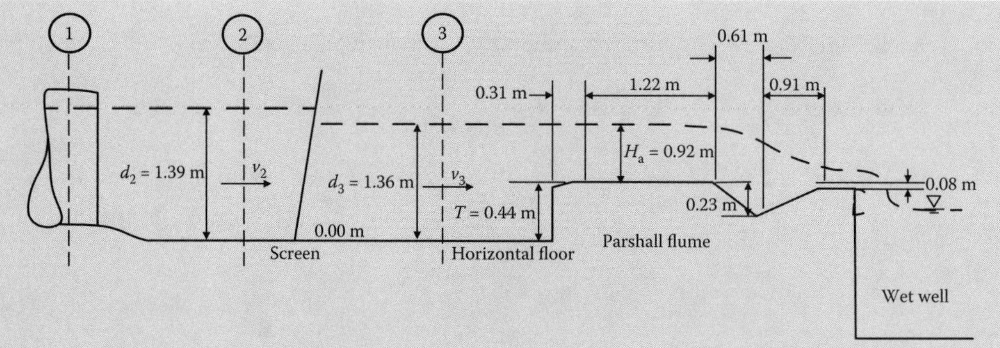

FIGURE 7.14 Longitudinal section of screen chamber and Parshall flume (Example 7.8).

The raised channel bottom at the throat may encourage settling of large grit particles at low velocities. Normally grit will not settle if the velocity in the channel is >0.3 m/s. In this case, the velocity in the channel is 0.6 m/s.

EXAMPLE 7.9: RAISED CHANNEL BOTTOM AT OUTFALL

A rectangular chamber of a bar rack is 1.74 m wide. The desired water depth is 1.25 m downstream of the bar rack. The wastewater drops into a wet well. The chamber bottom is raised at the outfall to maintain a nearly constant depth and velocity in the chamber. The bar screen is designed for a flow of 1.32 m^3/s. Assume weir coefficient $C_w = 1.0$ and coefficient of reduction $K_c = 0.5$. Calculate the height of the raised floor above the bottom of the channel. Draw the longitudinal section.

Solution

1. Calculate the critical depth at the raised section near the outfall point.

   The channel bottom is raised at the outfall. This presents an obstruction in the entire width of the channel similar to a broad-crested weir. Assume the critical depth and velocity are achieved at the outfall point.

   Critical flow is expressed by Equation 7.10.

$$Q = C_w A_c \sqrt{g d_c} = C_w b \sqrt{g} d_c^{3/2} \tag{7.10}$$

where

Q = critical flow, m^3/s (ft^3/s)
A_c = area of cross section at critical flow, m^2 (ft^2)
b = channel width at critical flow, m (ft)
d_c = critical depth, m (ft)
C_w = weir coefficient, dimensionless

Calculate d_c from Equation 7.10.

$$d_c = \left(\frac{1.32\,m^3/s}{1.74\,m \times \sqrt{9.81\,m/s^2}} \right)^{2/3} = 0.39\,m$$

Calculate velocity of critical flow $v_c = \dfrac{1.32\,m^3/s}{1.74\,m \times 0.39\,m} = 1.95\,m/s$

2. Calculate the velocity in the channel prior to the raised section.

$$\text{Velocity in the channel} = v_3 = \frac{1.32\,\text{m}^3/\text{s}}{1.74\,\text{m} \times 1.25\,\text{m}} = 0.61\,\text{m/s}$$

Note: If the floor at the outfall is not raised, the velocity in downstream of the channel will be 1.95 m/s under free-fall condition. It is more than three times of the desired velocity of 0.61 m/s in the channel.

3. Calculate the height of the raised bottom.

The approximate elevation of the channel bottom near the free fall into the wet well is calculated by applying Bernoulli's energy equation (Equation 7.3a) at sections (3) and (4), $Z_3 = 0$, and $v_4 = v_c = 1.95$ m/s.

Assume that the height of the raised floor above the channel bottom is Z_c.

$$0\,\text{m} + 1.25\,\text{m} + \frac{(0.61\,\text{m/s})^2}{2 \times 9.81\,\text{m/s}^2} = Z_c + 0.39\,\text{m} + \frac{(1.95\,\text{m/s})^2}{2 \times 9.81\,\text{m/s}^2} + h_L$$

$$Z_c = 0.69\,\text{m} - h_L$$

The head loss h_L due to reduction on cross section of flow is estimated from Equation 7.3c with $K_c = 0.5$.

$$h_L = \frac{K}{2g}(v_4^2 - v_3^2) = \frac{0.5}{2 \times 9.81\,\text{m/s}^2}((1.95\,\text{m/s})^2 - (0.61\,\text{m/s})^2) = 0.09\,\text{m}$$

$$Z_c = (0.69\ - 0.09)\,\text{m} = 0.6\,\text{m}$$

The floor of the chamber should therefore be raised by 0.6 m to maintain the desired water depth of 1.25 m in the chamber.

4. Draw the longitudinal section through the screen chamber.

The longitudinal section through the screen chamber and the raised floor is shown Figure 7.15.

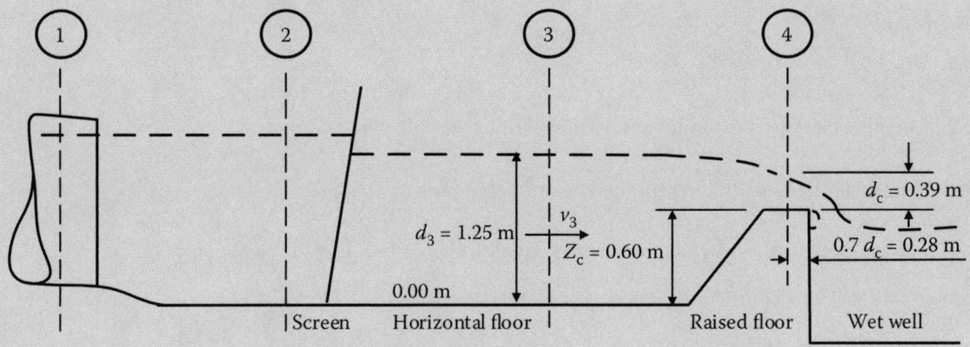

FIGURE 7.15 Longitudinal section of screen channel with a raised floor at the end (Example 7.9).

EXAMPLE 7.10: DESIGN OF A BAR RACK FACILITY

Design a bar rack facility at a wastewater treatment plant. The bar racks are located upstream of a wet well, and screened flow falls freely into the wet well. In each channel, a proportional weir is provided at the outfall structure to maintain a constant velocity in the upstream channel. The crest of the proportional weir is flushed with the channel floor. The weir length is less than the channel width. The design criteria for the bar rack is given below.

Flow:	Design peak wet weather flow	$= 1.32\ m^3/s$
	Design maximum dry weather flow	$= 0.92\ m^3/s$
	Design average dry weather flow	$= 0.44\ m^3/s$
	Design minimum dry weather flow	$= 0.09\ m^3/s$
Screens:	Provide two identical mechanically cleaned bar rack screens each capable of handling peak wet weather flow	
	Clear bar spacing	= 2.5 cm
	Bar width	= 1 cm
	Velocity through bar rack at design peak flow	= 0.8–0.9 m/s
	Angle of inclination of bars from horizontal	= 85°
	C_d	= 0.70 for clean screen
		= 0.60 for clogged screen
Incoming conduit:	Diameter	= 1.53 m
	Velocity at exit	= 0.87 m/s
	Depth of flow at exit	= 1.18 m (when the screen is clean)
	The coefficient of expansion K_e	= 0.30

The invert elevation of the conduit at exit is 0.08 m higher than the screen chamber floor.

The design should include (a) dimensions of bar rack and screen chamber, (b) ideal geometry of the proportional weir by assuming the weir crest is same as the weir datum, (c) head loss when the screen is clean and 50% clogged, (d) the velocity through the screen, and depth of flow and velocity in the channel upstream and downstream of the rack at maximum dry weather flow, and (e) hydraulic profile at peak wet weather flow.

Solution

1. Draw the bar rack arrangement.

   The plan and section of bar rack chamber are shown in Figure 7.16.
2. Compute bar spacing and the dimensions of the bar rack and chamber.

Assume the velocity through bar rack at design peak flow $V = 0.9$ m/s.

$$\text{Clear area of the bar rack openings perpendicular to the flow} = \frac{1.32\ m^3/s}{0.9\ m/s} = 1.47\ m^2$$

Assume the depth of flow in the rack chamber = 1.2 m.

$$\text{Total clear width of the openings required at the rack} = \frac{1.47\ m^2}{1.2\ m} = 1.22\ m$$

Provide 49 parallel bars with clear spacing of 25 mm or 0.025 m.

Total number of clear spaces = 50 spaces

$$\text{Total width of clear openings provided at the rack} = 50\ \text{spaces} \times 25\,\text{mm/space} \times \frac{\text{m}}{10^3\,\text{mm}} = 1.25\,\text{m}$$

$$\text{Total width of all bars} = 49\ \text{bars} \times 10\,\text{mm/bar} \times \frac{\text{m}}{10^3\,\text{mm}} = 0.49\,\text{m}$$

$$\begin{aligned}\text{Total width of the channel} &= \text{total width of clear opening} + \text{total width of all bars}\\ &= (1.25 + 0.49)\,\text{m} = 1.74\,\text{m}\end{aligned}$$

The design details of the bar screen and the channel are shown in Figure 7.16a.

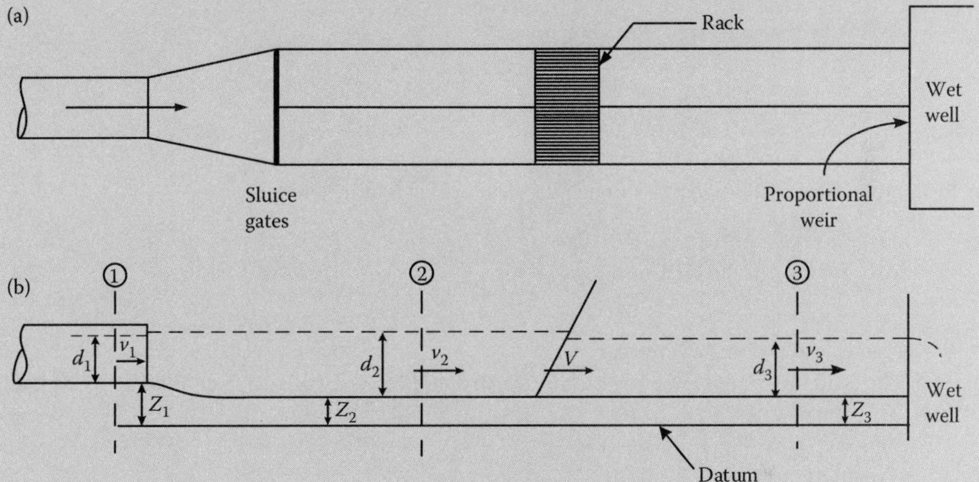

FIGURE 7.16 The bar rack arrangement: (a) plan, and (b) longitudinal section through the rack chamber (Example 7.10).

Note: This conceptual design example is provided for illustration and teaching purposed. The equipment manufacturers shall be consulted for the actual dimensions of their screens and associated screen chamber.

3. Calculate the efficiency coefficient.

$$\text{Efficiency coefficient} = \frac{\text{Total width of clear opening}}{\text{Total width of the channel}} = \frac{1.25\,\text{m}}{1.74\,\text{m}} = 0.72$$

4. Compute the actual depth of flow and velocity in the bar rack chamber at design peak flow.

$$\text{Velocity in the channel prior to the bar rack, } v_2 = \frac{1.32\,\text{m}^2/\text{s}}{d_2 \times 1.74\,\text{m}} = \frac{0.759\,\text{m}^2/\text{s}}{d_2}$$

Calculate h_L from Equation 7.3b, with $v_1 = 0.87$ m/s and $K_e = 0.30$ (given).

$$h_L = \frac{0.30}{2 \times 9.81\,\text{m/s}^2}\left[(0.87\,\text{m/s})^2 - \left(\frac{0.759\,\text{m}^2/\text{s}}{d_2}\right)^2\right] = 0.0116\,\text{m} - \frac{0.0088\,\text{m}^3}{d_2^2}$$

Assume the elevation at the floor of the rack chamber is the datum (0.00 m), $Z_2 = 0$ m. The invert elevation of the conduit $Z_1 = 0.08$ m.

Apply Bernoulli's energy equation (Equation 7.3a) at sections (1) and (2), $d_1 = 1.18$ m.

$$0.08\,\text{m} + 1.18\,\text{m} + \frac{(0.87\,\text{m/s})^2}{2 \times 9.81\,\text{m/s}^2} = 0\,\text{m} + d_2 + \frac{1}{2 \times 9.81\,\text{m/s}^2}\left(\frac{0.759\,\text{m}^2/\text{s}}{d_2}\right)^2$$

$$+\left(0.0116\,\text{m} - \frac{0.0088\,\text{m}^3}{d_2^2}\right)$$

$$1.287\,\text{m} = d_2 + \frac{0.0206\,\text{m}^3}{d_2^2}$$

$$d_2^3 - (1.287\,\text{m}) \times d_2^2 + 0.0206\,\text{m}^3 = 0$$

Solving by trial and error procedure, $d_2 = 1.28$ m

$$v_2 = \frac{1.32\,\text{m}^3/\text{s}}{1.28\,\text{m} \times 1.74\,\text{m}} = 0.59\,\text{m/s}$$

5. Compute the actual velocity V through the clear openings at the bar rack.

Velocity through the clear openings, $V = \dfrac{1.32\,\text{m}^3/\text{s}}{1.25\,\text{m} \times 1.28\,\text{m}} = 0.83\,\text{m/s}$

6. Compute head loss through the bar rack.
The head loss through the bar rack for clean condition is calculated from Equation 7.1a with $C_d = 0.7$.

$$h_L = \frac{(0.83\,\text{m/s})^2 - (0.59\,\text{m/s})^2}{2 \times 9.81\,\text{m/s}^2} \times \frac{1}{0.7} = 0.025\,\text{m}$$

7. Compute the depth of flow and velocity in the chamber after the bar rack.

Velocity in the channel after the bar rack, $v_3 = \dfrac{1.32\,\text{m}^3/\text{s}}{d_3 \times 1.74\,\text{m}} = \dfrac{0.759\,\text{m}^2/\text{s}}{d_3}$

Apply Bernoulli's energy equation (Equation 7.3a) at sections (2) and (3), $Z_2 = Z_3 = 0$ m.

$$0\,\text{m} + 1.28\,\text{m} + \frac{(0.59\,\text{m/s})^2}{2 \times 9.81\,\text{m/s}^2} = 0\,\text{m} + d_3 + \frac{1}{2 \times 9.81\,\text{m/s}^2}\left(\frac{0.759\,\text{m}^2/\text{s}}{d_3}\right)^2 + 0.025\,\text{m}$$

$$1.273\,\text{m} = d_3 + \frac{0.0294\,\text{m}^3}{d_3^2}$$

$$d_3^3 - (1.273\,\text{m}) \times d_3^2 + 0.0294\,\text{m}^3 = 0$$

Solving by trial and error procedure, $d_3 = 1.25$ m

$$v_3 = \frac{1.32\,\text{m}^3/\text{s}}{1.25\,\text{m} \times 1.74\,\text{m}} = 0.61\,\text{m/s}$$

8. Compute the head loss through the bar rack at 50% clogging.

At 50% clogging of the rack, the clear area through the rack is reduced to half and the velocity through clear openings is doubled.

$$V_{50} = 2 \times 0.83\,\text{m/s} = 1.66\,\text{m/s}$$

Calculate head loss from Equation 7.1a with $C_d = 0.6$.

$$h_{L,50} = \frac{(1.66\,\text{m/s})^2 - (0.59\,\text{m/s})^2}{2 \times 9.81\,\text{m/s}^2} \times \frac{1}{0.6} = 0.21\,\text{m}$$

9. Compute the depth of flow and velocity in the channel at 50% clogging.

Depth upstream of bar screen, $d_{2,50} = (1.25 + 0.21)\,\text{m} = 1.46\,\text{m}$

Velocity upstream of bar screen, $v_{2,50} = \dfrac{1.32\,\text{m}^3/\text{s}}{1.46\,\text{m} \times 1.74\,\text{m}} = 0.52\,\text{m/s}$

A summary of depth of flow, velocity, and head loss through the bar rack under clean and 50% clogging is given in Table 7.4.

TABLE 7.4 Summary of Depth of Flow, Velocity, and Head Loss through the Bar Rack at Design Peak Flow (Example 7.10)

Conditions	Upstream Channel		Bar Rack		Downstream Channel	
	Depth, m	Velocity, m/s	Velocity, m/s	Head Loss, m	Depth, m	Velocity, m/s
Clean rack	1.28	0.59	0.83	0.025	1.25	0.61
50% clogged rack	1.46	0.52	1.66	0.21	1.25	0.61

10. Design the proportional weir.

The proportional weir at the effluent control section maintains nearly constant velocity in the channel under variable flow condition. The head over weir is proportional to flow.

The flow through the proportional weir is given by Equation 7.8d.

$$Q = 4.17LH^{3/2}$$

Calculate the opening width L of the proportional weir from Equation 7.8d at the maximum water depth $d_3 = 1.25$ m.

$$L = \frac{Q}{4.17H^{3/2}} = \frac{1.32\,\text{m}^3/\text{s}}{4.17 \times (1.25\,\text{m})^{3/2}} = 0.227\,\text{m} \quad \text{or} \quad 0.23\,\text{m}$$

11. Develop the profile of the proportional weir and validate the velocity in the channel.

Determine the constant K from Equation 7.8a.

$$K = LH^{1/2} = 0.227\,\text{m} \times (1.25\,\text{m})^{1/2} = 0.254\,\text{m}^{3/2}$$

The proportional weir calculations are tabulated below. The weir profile is shown in Figure 7.17b.

TABLE 7.5 Design Calculations of Proportional Weir, and Depth and Velocity at Different Flow Conditions (Example 7.10)

Given and Assumed Flow Condition	Flow q, m^3/s	Head over Weir or Depth in Channel $d_3 = y = (q/4.17K)$ (Equation 7.8g), m	Weir Length $l_3 = x = Ky^{-1/2}$ (Equation 7.7a), m	Velocity in the Channel $v_3 = Q/(d_3 \times W)$, m/s
Design peak wet weather flow (given)	1.32	1.25	0.23	0.61[a]
Design maximum dry weather flow (given)	0.92	0.87[b]	0.27[c]	0.61[d]
Design average dry weather flow (given)	0.44	0.42	0.39	0.61
Assumed	0.22	0.21	0.56	0.61
Assumed	0.15	0.14	0.67	0.61
Design minimum dry weather flow (given)	0.090	0.085	0.87	0.61
Assumed to determine the length of weir crest	0.053	0.050	1.13	0.61

[a] $v_3 = \dfrac{1.32\,m^3/s}{1.25\,m \times 1.74\,m} = 0.61\,m/s.$

[b] $d_3 = \dfrac{0.92\,m^3/s}{4.17 \times 0.254\,m^{3/2}} = 0.87\,m.$

[c] $l_3 = 0.254\,m^{3/2} \times (0.869\,m)^{-1/2} = 0.27\,m.$

[d] $v_3 = \dfrac{0.92\,m^3/s}{1.74\,m \times 0.869\,m} = 0.61\,m/s.$

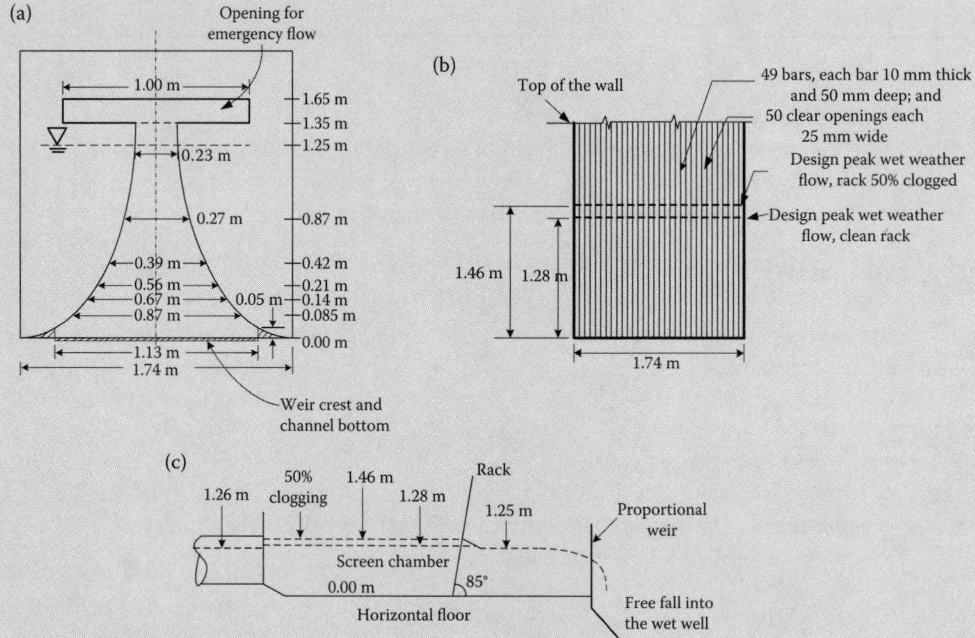

FIGURE 7.17 Bar rack design details of bar rack: (a) bar rack and channel, (b) proportional weir, and (c) hydraulic profile through the bar rack at peak design flow when rack is clean and at 50% clogging (Example 7.10).

The calculations to determine the depth of flow in the channel, head over the proportional weir, the opening width of the proportional weir, and the velocity in the channel at different flow conditions are summarized in the Table 7.5.

12. Sketch the design details of the proportional weir, bar rack, and the hydraulic profile.

    The hydraulic profile is shown in Figure 7.17c. The design details and equipment layout of the bar rack are shown in Figure 7.18.

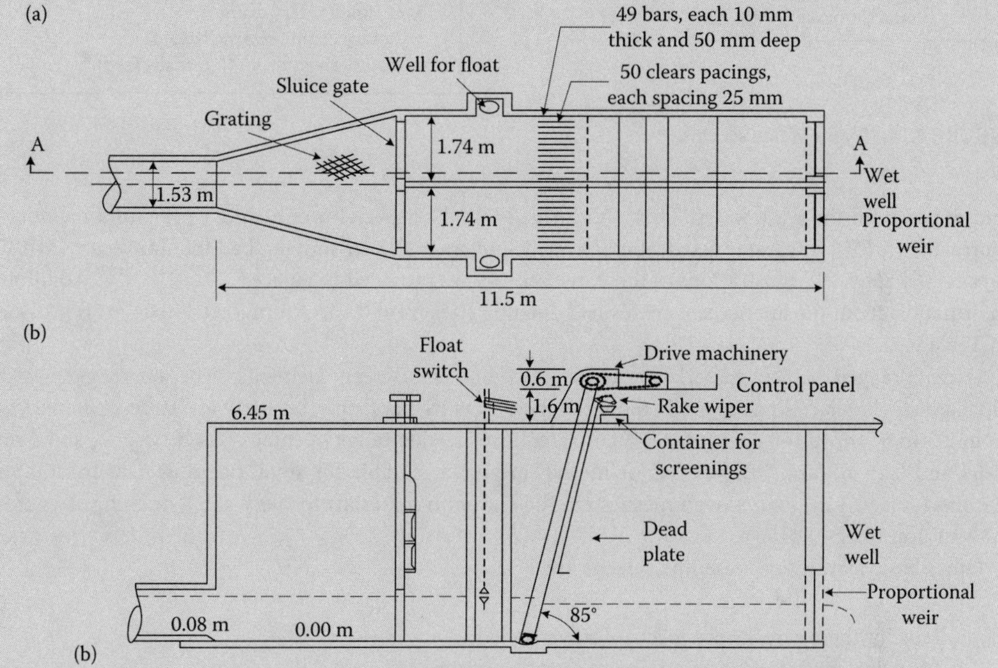

FIGURE 7.18 Design and equipment details of bar rack: (a) plan, and (b) Section AA (Example 7.10).
Note: All elevations are with respect to the datum (0.00 m) at the floor of the rack chamber.

7.2.3 Fine Screens

Application of Fine Screens: Fine screens are gaining popularity in wastewater application. The purpose is to provide preliminary or primary treatment as a substitute for primary clarifiers. They are mostly mechanically cleaned and preceded by coarse screens. In recent years, the use of fine screens has been extended to remove solids equivalent of primary treatment (Section 9.7). Various types of microscreens have also been developed to remove algae from stabilization pond effluent, and to upgrade secondary effluent for reuse purpose in tertiary treatment (Section 15.4.7). Fine screens may also be used for treatment of stormwater and CSO.

Types of Fine Screens: Many types of fine screens have been developed for different applications. The types of fine screen commonly used in wastewater treatment are identified in Figure 7.19.[1–7,32–34] In general, these fine screens can be categorized into: stationary and mobile screen types.

Basic Features: The opening sizes of fine screens typically range from 0.5 to 6 mm (0.02–0.25 in). General information about the fine screens with medium to large opening sizes from 1 to 6 mm (0.04–0.25 in) is developed based on available online information from the manufacturers and summarized in Table 7.6.[1–7,32–87] These fine screens are usually used for preliminary treatment. The fine screens with opening sizes from 1 to 3 mm (0.04–0.12 in) are not designed to remove a significant amount of TSS

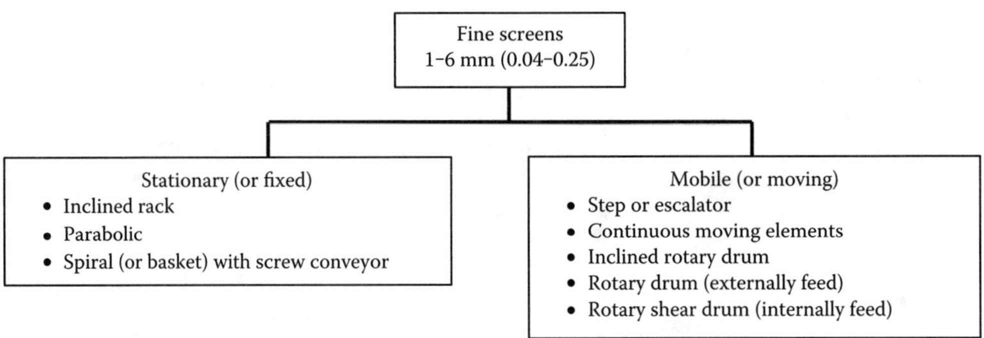

FIGURE 7.19 Types of fine screens.

and BOD_5 from the wastewater. They are normally used as special pretreatment prior to a membrane bioreactor (MBR), integrated fixed-film activated sludge (IFAS), or moving bed biofilm rector (MBBR) process (Chapter 10). Illustrations of these fine screens are provided in Figure 7.20.[35,44,60,71,87] Additional information about the fine screens with small opening sizes from 0.1 to 1 mm (0.004–0.04 in) is provided in Table 9.30.

Widely ranged screen materials can be used for fine screen elements. The openings may be circular, rectangular, square, or diamond shapes. Parallel bars may be used for large opening sizes from 2.5 to 6 mm (0.1–0.25 in); perforated plates cover a wide range opening sizes between 1 and 6 mm (0.04 and 0.25 in); and wedge wires or meshes are more suitable for small opening sizes from 0.5 to 2.5 mm (0.02–0.1 in), and woven meshes are the common materials for very small opening sizes from 0.5 to 1 mm (0.02–0.04 in).

Other basic features of these fine screens are:

1. Most of these screens are mechanically cleaned.
2. Most screens are of enclosed design for odor control.
3. The most commonly used screen material is stainless steel.
4. Mobile screens require more power than stationary screens.
5. The head loss through the screens may range from 0.25 to 1.5 m, although mobile screens exhibit less head loss than the stationary screens.`
6. Solids trapped on the fine screen create a "filter mat" that enhances the solids removal performance.
7. They also remove grit, grease, and increase dissolved oxygen (DO).

7.2.4 Design of Fine Screens

Design considerations: Many important design considerations must be evaluated during the design of fine screens. These may include: (1) plant capacity (large or small), (2) influent quality (mainly the solids data), (3) desired separation efficiency (preliminary or special requirement by the downstream processes), (4) type of improvement (new or retrofitting existing), (5) head loss allowance (high or low), (6) site and arrangement requirements (channel installation or package system), (7) available space, (8) final disposal of screenings (weight, volume, and dry solids contents), (9) concern and control of odors, and (10) operating staff skill and O&M intensity. The results from such evaluation provide a basis for selection of type of fine screen and the size of openings, expected performance, and development of hydraulic profile through the facility. The screen manufacturer shall be consulted to confirm the actual capacity of the fine screen since it may vary significantly with the channel width, upstream water depth, the expected raw influent, and desired effluent qualities.

TABLE 7.6 General Information about Fine Screens Used for Preliminary Treatment

Type of Screen	Range	Description
Stationary (or fixed) screens	Opening size: 1–6 mm (0.04–0.25 in)	These screens are usually used at small wastewater treatment plants. Rectangular or wedge-shaped bars and perforated plates are the most commonly used materials for static fine screens. There are three basic types of static fine screens: inclined rack, parabolic, or spiral (or basket) screw conveyor.[4,7,25,35–45]
Inclined rack	Opening size: 1–6 mm (0.04–0.25 in) Channel width: 1–6 m (3–20 ft) Channel depth: up to 20 m (65 ft) Incline angle: 60–80°	Inclined fine screens use stainless-steel bar racks or wedge-shaped wires. Chain-driven multiple rakes are the common cleaning mechanism for these fine screens. See Table 7.2 and Figure 7.2b for general information about the chain-driven screens.
Parabolic	Opening size: 1–6 mm (0.04–0.25 in) Wedge wires: 1–6 mm (0.04–0.25 in) Perforated plates: 1.5–6 mm (0.06–0.25 in) Manually cleaned: Width: 0.6–3 m (2–10 ft) Height: 1.2–3 m (4–10 ft) Mechanically cleaned: Width: 0.4–4.5 m (1.3–15 ft) Discharge height: 1.2–2 m (4–7 ft) Single unit capacity: 16–320 m^3/h (0.1–2 MGD)	Static parabolic screens are usually used in shallow channels or as freestanding units above the channels. The surface loading rate may be 400–1200 $L/m^2 \cdot min$ (10–30 $gal/ft^2 \cdot min$) of screen area. The head loss may be high up to 1–2 m (3–7 ft) when a freestand screen is installed on top of the channel and operated in gravity flow pattern. The manually cleaned fine screens use the screen medium of stainless-steel wedge-shaped bars. The screenings are washed down and discharged from the bottom. Cleaning is done once or twice a day with high-pressure hot water, steam, or degreaser. Perforated plates are normally used in mechanically cleaned parabolic fine screens. Multiple brushes wipe the retained solids (Figure 7.20a).[4,7,25,35–37]
Spiral (or basket) with screw conveyor	Opening size: 1–6 mm (0.04–0.25 in) Wedge wires or meshes: 1–6 mm (0.04–0.25 in) Perforated plates: 2–6 mm (0.08–0.25 in) Channel width: 0.2–1 m (0.7–3.5 ft) Channel depth: up to 1.5 m (5 ft) Upstream water depth: up to 0.75 m (2.5 ft) Basket diameter: 300–900 m (12–36 in) Incline angle: 35–48° Discharge height: 2–2.5 m (7–8.5 ft) Single unit capacity: 90–1600 m^3/h (0.6–10 MGD) Screening disposal: Washing efficiency: up to 90% Volume reduction: 40–50% Solids content: 30–40% as dry solids	This type of screen is typically used in shallow and narrow channels. It includes an inclined (35–45°) stationary screening "basket," cleaning "spiral" brushes, and a central screw conveyer. The wastewater flows into the semicircular basket and solids are retained. As the liquid level in the basket rises to a predetermined level, the spiral begins to rotate and cleans basket. The brushed solids drop into the conveyor that moves the solids up to a top discharge chute. Usually, the screened materials is washed, compacted, and dewatered by an integrated disposal device. A pivoting support option may be included for easy access to the screen elements (Figure 7.20b).[4,7,33,38–45]
Mobile (or moving) screens	Opening size: 1–6 mm (0.04–0.25 in)	These fine screens are widely used in medium to large plants. These screens are all mechanically cleaned. The most commonly used material are perforated plates or wedge wire meshes. These screens require medium to high maintenance. The basic types of mobile fine screens are (1) step or escalator, (2) continuous moving band or belt, (3) rotating belt filter, (4) inclined rotary drum, (5) rotary drum, and (6) rotary shear drum.[4–7,33,34,46–87]

(*Continued*)

TABLE 7.6 (*Continued*) General Information about Fine Screens Used for Preliminary Treatment

Type of Screen	Range	Description
Step or escalator	Opening size: 1–6 mm (0.04–0.25 in) Laminae elements: 1–6 mm (0.04–0.25 in) Perforated plates: 2–6 mm (0.08–0.25 in) Channel width: 0.4–2 m (1.3–7 ft) Channel depth: up to 3 m (10 ft) Upstream water depth: up to 2.3 m (7.5 ft) Incline angle: 40–75° Discharge height: up to 6 m (20 ft) Single unit capacity: 160–9600 m^3/h (1–60 MGD)	These screens have step-shaped laminae where every other lamina is connected to one fixed and one moveable part on which a carpet of solids is formed. This gives a high sieving efficiency. The rotating laminae convey the solids upwards step-by-step to the discharge point (Figure 7.20c). Escalator screens operate similar to step screens but use perforated plates to provide circular openings instead of slotted ones. In addition to wastewater screening, these screens are also used for removal of solids from septage, primary sludge, and/or digested sludge.[4,5,7,33,46–52]
Continuous moving elements	Opening size: 1–6 mm (0.04–0.25 in) Molded links, or patterned meshes or grids: 1–6 mm (0.04–0.25 in) Perforated plates: 2–6 mm (0.08–0.25 in) Wedge wires or meshes: 1–3 mm (0.04–0.12 in) Channel width: 0.3–5 m (1–16 ft) Channel depth: up to 11 m (36 ft) Incline angle: 45–90° Discharge height: up to 15 m (50 ft) Single unit capacity: 640–9600 m^3/h (4–60 MGD)	The continuous moving elements fine screens consist of an endless screen band or belt that passes over upper and lower sprockets. Chain-driven circulating sickle-shaped screen disks are the latest innovative design. The retained screenings are lifted up and removed by a continuous self-cleaning device at the top of the screen. The cleaning device may include rotating brush, wiping paddles, water spray nozzles, or a combination of these methods. There are three basic flow patterns available for these screens: (1) through flow, (2) inside-to-outside flow, and (3) outside-to-inside flow. A single- or dual-screening process can be used in the through flow pattern, while only single screening is applicable in the other two flow patterns. Each of these configurations has advantages and disadvantages in developing screening mat, head loss through the screen elements, cleaning efficiency, solids accumulation, and solids carryover. The screen elements are normally made of wedge wires, perforated plates, or patterned meshes or grids of stainless steel or high-strength plastics. These screens are inclined at 50–90°. The vertical design option requires small space and allows for easy retrofitting existing channels (Figure 7.20d).[4–7,33,34,53–67]
Inclined rotary drum	Opening size: 1–6 mm (0.04–0.25 in) Perforated plates: 1–6 mm (0.04–0.25 in) Wedge wires or meshes: 1–6 mm (0.04–0.25 in) Channel width: 1–3.5 m (3–11.5 ft) Channel depth: up to 4 m (13 ft) Incline angle: 30–35° Basket diameter: 0.6–3 m (2–10 ft) Single unit capacity: 320–7200 m^3/h (2–45 MGD) Screening disposal: Washing efficiency: up to 80% Volume reduction: ~50% Solids content: ~40% as dry solids	This type of screen is usually installed into the channel at an angle of 30–35°. It consists of a drum-shaped screen (stainless-steel wedge wires or meshes, or perforated plates), a central screw conveyer, and an integrated disposal device. The flow enters the open end of the inclined drum and the solids are retained by the lower part of the drum. The rotation of drum is activated at a preset water level. The screenings are carried to the upper part of the drum and dropped into the conveying screw. High-pressure water spray, scrapper and/or brushes flush down and clear the solids off the screen openings. The screw conveys the solids up to the integrated disposal device. Pivoting support provides easy access to the screen elements for maintenance purpose (Figure 7.20e).[4–6,68–75]

(*Continued*)

TABLE 7.6 (*Continued*) General Information about Fine Screens Used for Preliminary Treatment

Type of Screen	Range	Description
Rotary drum (externally feed)	Opening size: 1–2.5 mm (0.04–0.1 in) Perforated plates: 1–2 mm (0.04–0.08 in) Wedge wires or meshes: 1–2.5 mm (0.04–0.1 in) Drum dimensions: Diameter: 0.6–1 m (2–3 ft) Length: 0.3–3 m (1–10 ft) Single unit capacity: 16–1600 m^3/h (0.1–10 MGD)	The screen consists of a drum or cylinder-shape screen of stainless-steel wedge wires or meshes, or perforated plates. The drum is installed horizontally in either partially or nonsubmerged condition. The influent is evenly distributed over the entire length of the drum. As an externally feed system, the screenings are retained on the drum outside surface and carried to the discharge zone by the rotation movement. External scrapers or blades with internal high-pressure water spray are used to dislodge the solids into a screw conveyer/compactor. As a dual-screening process, the flow falling from the upper screen passes through the lower part of the drum again and is collected in an underneath trough. It is a compact, stand-alone, and fully enclosed unit. The head loss is high in gravity flow pattern under nonsubmerged condition.[7,34,76–78]
Rotary shear drum (internally feed)	Opening size: 1–6 mm (0.04–0.25 in) Perforated plates: 1–6 mm (0.04–0.25 in) Wedge wires or meshes: 1–3 mm (0.04–0.12 in) Drum diameter: 0.5–2 m (1.5–7 ft) Enclosure dimensions: Height: 1–2.7 m (3–9 ft) Width: 0.6–3 m (2–10 ft) Length: 1.4–6.5 m (4–21 ft) Single unit capacity: up to 3200 m^3/h (20 MGD) (up to 1420–14,200 m^3/h (9–90 MGD) for in-channel option)	This screen is very similar to the rotary drum screen except it is an internally feed system using single-screening process. The influent is fed internally from a distribution box near one end and the solids are retained inside as the flow passes through the screen. A series of diverters move solids to the other end of the drum for removal. Internal rotating brushes with external water spray are used to removal the solids. The flow passing through the screen is collected in a trough. The screen is usually a stand-alone unit, while in-channel design option is also available. The head loss is lower than the rotary drum screen in submerged flow pattern (Figure 7.20f).[5,7,33,79–87]

Head loss: The head loss in a fine screen depends upon (1) effective clear area of the screen and (2) the extent of filter mat clogging. In general, the head loss is calculated from Equations 7.11a and 7.11b. These equations are developed based on head loss calculated across an orifice. The actual head loss shall be validated with the screen manufacturers.

$$h_L = \frac{1}{2g} \times \left(\frac{Q}{C_d \times A_e}\right)^2 \tag{7.11a}$$

$$A_e = E_c \times A_T \tag{7.11b}$$

where

h_L = head loss, m (ft)
Q = discharge through screen, m^3/s (ft^3/s)
A_e = effective wetted screen area, m^2 (ft^2)
A_T = total wetted screen area, m^2 (ft^2)
C_d = coefficient of discharge, dimensionless (0.6–0.8 for clean screen)
E_C = efficiency factor which is the ratio of effective area to total area of the screen (manufacturers provide this factor)

FIGURE 7.20 Details of fine screens: (a) static parabolic (Courtesy Parkson Corporation), (b) spiral (Courtesy Huber Technology, Inc.), (c) step (Courtesy Huber Technology, Inc.), (d) continuous moving elements (Courtesy Kusters Water Div. of Kusters Zima Corp.), (e) inclined rotary drum, (Courtesy Lakeside Equipment Corporation.), and (f) rotary shear drum (Courtesy Lackeby Products).

The effective screen area A_e is the clear area through which the flow occurs. It depends upon the shape and configuration, milling slots, and wire diameter and weave.[2] The effective area can be calculated from the total screen area based on the screen *efficiency factor* (area reduction factor) that is usually available from the equipment manufacturers. The manufacturer's efficiency factor for fine screen may vary from 0.5 to 0.6.[1]

The fine screens should be preceded by coarse screen. A minimum of two fine screens in parallel (similar to coarse screen) should be provided in case one unit is out of service. If course and fine screens are provided in series, one velocity control device (proportional weir or Parshall flume) may be provided downstream of the fine screen to control the water depth in both screens.

EXAMPLE 7.11: HEAD LOSS AND VELOCITY THROUGH A CONTINUOUS BELT SCREEN

A stainless-steel fine band screen is installed downstream of a bar rack. As the screen revolves, the debris is lifted by upward travel of the screen. At the top of the screen, the debris is removed by powerful water jet sprays and brush action. The debris is washed into a trough and collected in a container. The flow through the screen chamber is 1.45 m^3/s, and the depth and width of the screen channel are 1.35 and 1.20 m, respectively. The manufacturer's recommended efficiency factor is 0.56, and coefficient of discharge is 0.6. Calculate (a) head loss and velocity through the clean screen, and (b) head loss and velocity through the screen at 20% clogging.

Solution

1. Calculate the head loss and velocity through the clean screen.
   Effective area A_e is calculated from Equation 7.11b.

$$A_e = 0.56 \times 1.35\,\text{m} \times 1.20\,\text{m} = 0.91\,\text{m}^2$$

Calculate the head loss h_L from Equation 7.11a.

$$h_L = \frac{1}{2g} \times \left(\frac{Q}{C_d \times A_e}\right)^2 = \frac{1}{2 \times 9.81\,\text{m/s}^2} \times \left(\frac{1.45\,\text{m}^3/\text{s}}{0.6 \times 0.91\,\text{m}^2}\right)^2 = 0.36\,\text{m}$$

Calculate velocity through the screen.

$$v = \frac{Q}{A_e} = \frac{1.45\,\text{m}^3/\text{s}}{0.91\,\text{m}^2} = 1.59\,\text{m/s}$$

2. Calculate the head loss and velocity through 20% clogged screen.
   Calculate effective area of 20% clogged screen.

$$A_{e,\text{clogged}} = 0.91\,\text{m}^2\,(1 - 0.2) = 0.73\,\text{m}^2$$

Calculate head loss $h_{L,\text{clogged}}$ through 20% clogged screen.

$$h_{L,\text{clogged}} = \frac{1}{2 \times 9.81\,\text{m/s}^2} \times \left(\frac{1.45\,\text{m}^3/\text{s}}{0.6 \times 0.73\,\text{m}^2}\right)^2 = 0.56\,\text{m}$$

Calculate velocity through 20% clogged screen.

$$v_{\text{clogged}} = \frac{1.45\,\text{m}^3/\text{s}}{0.73\,\text{m}^2} = 1.99\,\text{m/s}$$

EXAMPLE 7.12: LENGTH OF A ROTARY DRUM SCREEN

A rotary drum screen is designed to provide primary treatment. A flow of 0.4 m^3/s enters the inside of a slowly rotating drum that has a diameter of 1.5 m. The drum is covered with stainless-steel wedge-wire screening medium. The screen is cleaned by a high-pressure water jet applied at the top. The differential head between the inside and outside drum is 80 mm, and submergence outside drum is 0.5 m. The expected efficiency factor after partial clogging and discharge coefficient are 0.18 and 0.60, respectively. Calculate the length of the drum, velocity through the screen and hydraulic loading on the screening medium.

Solution

1. Draw the definition sketch (Figure 7.21).

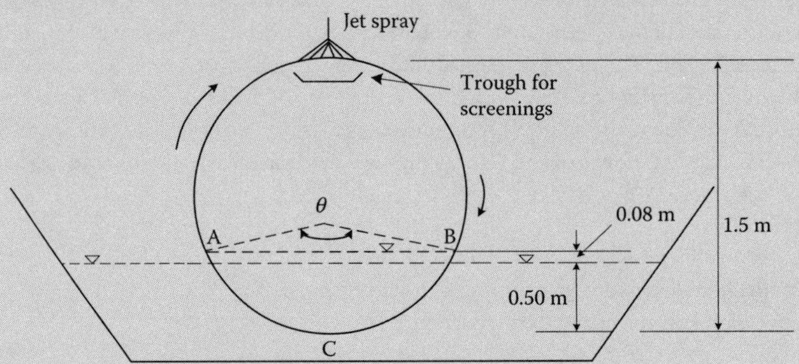

FIGURE 7.21 Definition sketch of rotary shear drum screen (Example 7.12).

2. Calculate the surface area of the screening medium.

The area of the screen through which the flow occurs is calculated from Equation 7.11a, $h_L = 80\text{ mm} = 0.08\text{ m}$, $A_e = 0.18 \times A_T$, and $C_d = 0.6$.

$$0.08\text{ m} = \frac{1}{2 \times 9.81\text{ m/s}^2} \times \left(\frac{0.4\text{ m}^3/\text{s}}{0.6 \times 0.18 \times A_T}\right)^2$$

$$A_T = 2.96\text{ m}^2$$

3. Calculate the circumferential length ACB of the rotary drum through which screening occurs.

The submerged depth inside the drum, $d = (0.50 + 0.08)\text{ m} = 0.58\text{ m}$

$$\text{Submergence to diameter ratio} = \frac{d}{D} \times 100\% = \frac{0.58\text{ m}}{1.5\text{ m}} \times 100\% = 39\%$$

The angle θ is calculated from Equation 6.12a, $D = 1.5$ m and $d = 0.58$ m.

$$\cos\left(\frac{\theta}{2}\right) = 1 - \frac{2d}{D} = 1 - \frac{2 \times 0.58\text{ m}}{1.5\text{ m}}$$

$$\theta = 154^\circ$$

The length of wetted perimeter P is obtained from Equation 6.12c.

$$\text{Length of perimeter ACB, } P = \frac{\pi D\theta}{360^\circ} = \frac{\pi \times 1.5\text{ m} \times 154^\circ}{360^\circ} = 2.02\text{ m}$$

This is also the circumferential length through which screening occurs.

4. Calculate ratios related to filtering drum area.

$$\text{Filtering area ratio} = \frac{\text{filtering area}}{\text{circumferential area}} = \frac{2.02\text{ m} \times \text{length of drum}}{\pi \times 1.5\text{ m} \times \text{lenght of drum}} \times 100\% = 43\%$$

5. Calculate the length of the rotary drum.

The unit filtering area per linear length of drum is $2.02\ m^2/m$.

$$\text{Length of rotary drum} = \frac{A_T}{P} = \frac{2.96\ m^2}{2.02\ m^2/m} = 1.47\ m$$

6. Calculate velocity through screen.

$$\text{Velocity through screen, } v = \frac{0.4\ m^3/s}{0.18 \times 2.96\ m^2} = 0.75\ m/s$$

7. Calculate hydraulic loading on the screen based on the total wetted area.

$$\text{Hydraulic loading} = \frac{Q}{A_T} = \frac{0.4\ m^3/s}{2.96\ m^2} \times \frac{60\ s}{min} = 8.1\ m^3/m^2 \cdot min \text{ or } 8100\ L/m^2\ min$$

EXAMPLE 7.13: DESIGN OF A ROTARY DRUM MICROSCREEN

A microscreen is used to polish secondary effluent for reuse as industrial cooling water. The manufacturer provided the following information about their product. The microscreen has stainless-steel screen. The recommended efficiency factor and discharge coefficient are 0.53 and 0.60, respectively. The acceptable hydraulic loading is $10\ m^3/m^2 \cdot min$ at the inside water depth at 60% of drum diameter. Design the microscreen system for average filtration rate of $1.5\ m^3/s$. The applied backwash pressure is 100 kPa.

Solution

1. Calculate the wetted surface area of the screening medium.

$$\text{Total wetted area} = \frac{1.5\ m^3/s \times 60\ s/min}{10\ m^3/m^2 \cdot min} = 9\ m^2$$

Provide two microscreens, each submerged fabric area $= 0.5 \times 9\ m^2 = 4.5\ m^2$

Flow through each microscreen $= 0.5 \times 1.5\ m^3/s = 0.75\ m^3/s$

2. Calculate head loss through screen from Equation 7.11a.

$C_d = 0.6$

$$h_L = \frac{1}{2 \times 9.81\ m/s^2} \times \left(\frac{0.75\ m^3/s}{0.6 \times 0.53 \times 4.5\ m^2}\right)^2 = 0.014\ m$$

3. Select the drum length and diameter D.
   At inside water depth of 60% of D, the θ value is obtained from Equation 6.12a.

$$\cos\left(\frac{\theta}{2}\right) = 1 - \frac{2d}{D} = 1 - \frac{2 \times 0.6D}{D} = -0.2$$

$\theta = 203°$

The length of wetted perimeter is obtained from Equation 6.12c.

$$\text{Length of wetted perimeter} = \frac{\pi D\theta}{360^\circ} = \frac{\pi \times D \times 203^\circ}{360^\circ} = 1.77D$$

Select 2.75 m standard length of the drum.

$$2.75\,\text{m} \times (1.77D) = 4.5\,\text{m}^2$$

$$D = \frac{4.5\,\text{m}^2}{1.77 \times 2.75\,\text{m}} = 0.92\,\text{m}$$

Select 1-m diameter standard drum.

4. Calculate velocity through screen.

$$\text{Total submerged area of screen} = 1.77 \times 1\,\text{m} \times 2.75\,\text{m} = 4.87\,\text{m}^2$$

$$\text{Velocity through screen opening} = \frac{0.75\,\text{m}^3/\text{s}}{0.53 \times 4.87\,\text{m}^2} = 0.29\,\text{m/s}$$

5. Check the hydraulic loading.

$$\text{Hydraulic loading} = \frac{0.75\,\text{m}^3/\text{s}}{4.87\,\text{m}^2} \times \frac{60\,\text{s}}{\text{min}} = 9.2\,\text{m}^3/\text{m}^2\cdot\text{min} < 10\,\text{m}^3/\text{m}^2\cdot\text{min}$$

7.2.5 Special Screens

The discharge of storm runoff and CSOs is a significant contributor to the surface water quality impairment of the nation's waters. In recent years, there has been significant development in pollution abatement technology from these sources. Coarse screens are normally used for minimal treatment of stormwater and CSOs. Use of fine screens may be required to meet more stringent quality requirements. Special screens and many other devices are also manufactured to remove floating debris, suspended solids, and grits specifically from stormwater and CSOs. These fixed or retractable devices can be used at manholes, junction boxes, weirs, wet wells at lift stations, or outfalls. Some special screens used for such applications are illustrated in Figure 7.22.[88,89]

(a)

(b)

FIGURE 7.22 Use of special screens for treatment of stormwater or combined sewer overflows: (a) retractable bar screen with trash basket (Courtesy Huber Technology, Inc.), and (b) overflow weir screen (Courtesy John Meunier Products/Veolia Water Technologies Canada).

7.3 Quantity, Characteristics, and Disposal of Screeings

7.3.1 Quantity and Characteristics

Screenings are residuals retained over coarse and fine screens. The quantity of screenings depends on type of wastewater, geographic location, weather, and type and size of screens. Larger quantity of screenings is retained over screens with smaller openings. The screenings contain ~60–80% moisture contents and 700–1000 kg/m^3 (44–65 lb/ft^3) in densities.[2,3] The screenings collected over coarse screens consist of large debris such as rags, paper and plastics, leaves, tree branches and roots, and lumber. The screens retained over fine screens include small rags, paper, plastics, grit, food waste, and feces. The average quantities of screenings collected over coarse screen are provided in Figure 7.23.[1] The average quantity of screenings retained over screens with opening size of 12.5 mm (0.5 in) are 44–110 $m^3/10^6\ m^3$. An average quantity of 30–60 $m^3/10^6\ m^3$ screenings is retained over rotary drum of opening size 6–7.5 mm (0.25–0.3 in) following coarse screen.[2,3]

7.3.2 Processing and Disposal of Screenings

Conveyance: The screenings are very odorous and attract flies. They should be transported over covered belt conveyor. Screenings should also be stored in containers with tight covers. Screening compactors are screw or hydraulic ram type. They are used to reduce the water content and the volume of the screenings.

Grinding and Comminution: Grinders or comminutors (macerators) are used to grind or cut up the screenings. They utilize cutting teeth or shredding devices on a rotating or oscillating drum that passes through stationary combs, screen, or disks. Large objects are shredded to pass through 6–10 cm (2.5–4 in) openings or slots. Manufacturers' rating tables are available for different capacity ranges, channel dimensions, submergence, and power requirements. Provision to bypass the device is always made. Comminutor installation is shown in Figure 7.24.[90–93]

Disposal of Screening: The most common method of disposal of screenings are (a) landfilling on plant site or off-site, (b) codisposal with municipal solid wastes, (c) incineration either alone or in combination with grit and scum, and (d) discharge to the head of the plant after grinding or comminution.

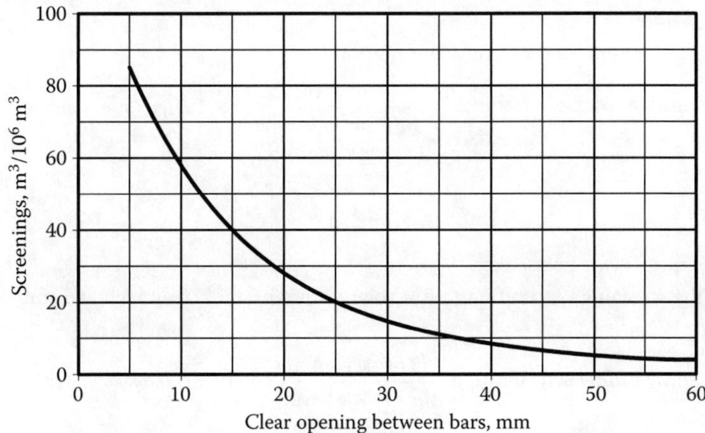

FIGURE 7.23 Quantity of screenings retained on coarse screens of different opening sizes.

FIGURE 7.24 Comminutor assemblies (Courtesy Franklin Miller, Inc.): (a) open channel comminutor and typical installation; and (b) duplex twin-shaft grinder and its application prior to a screw screening system.

EXAMPLE 7.14: QUANTITY OF SCREENINGS

The average design flow at a wastewater treatment plant is 1.8 m^3/s. The screening facility has coarse screen that has 15-mm clear bar spacing followed by a rotary drum fine screening. The size of rotary drum screen openings are 6 mm. Estimate the quantity of screenings collected per day.

Solution

1. Estimate the quantity of coarse screenings.

   The screenings generation rate at a coarse screen having 15-mm clear bar spacing is determined from Figure 7.14.

   Screening generation rate $= 40\ m^3/10^6\ m^3$

$$\text{Total quantity of coarse screenings} = \frac{40\ m^3 \text{ screenings}}{10^6\ m^3 \text{ wastewater}} \times 1.8\ m^3/s \text{ wastewater} \times \frac{(60 \times 60 \times 24)\ s}{d} = 6.2\ m^3/d$$

2. Estimate the quantity of fine screen.

   Assume the screenings generation rate at rotary drum of fine screen having 6-mm screen opening size $= 45\ m^3/10^6\ m^3$.

$$\text{Total quantity of fine screenings} = \frac{45\ m^3 \text{ screenings}}{10^6\ m^3 \text{ wastewater}} \times 1.8\ m^3/s \text{ wastewater} \times \frac{(60 \times 60 \times 24)\ s}{d} = 7.0\ m^3/d$$

3. Estimate total quality of coarse and fine screenings.

   Total quantity of screenings $= (6.2 + 7.0)\ m^3/d = 13.2\ m^3/d$

EXAMPLE 7.15: LANDFILL AREA FOR DISPOSAL OF SCREENINGS

A wastewater treatment plant produces 9 m^3/d of screenings, grit, and scum. These residues are land filled at the plant property. Calculate (a) the annual volume of raw residues, (b) the annual volume of raw daily cover materials required, (c) the annual volume of compacted fill, and (d) the land area required per year for final disposal of the residues. Assume the depth of landfill excluding the final cover is 3 m, and the daily cover is ~15% of the volume of residues. The compaction in landfill is 35% by volume.

Solution

1. Calculate the annual volume of raw residues.

$$\text{Volume of raw residue per year} = 9\ \text{m}^3/\text{d} \times 365\ \text{d/year} = 3290\ \text{m}^3/\text{year}$$

2. Calculate the annual volume of daily cover materials.

$$\text{Volume of daily cover} = 0.15 \times 9\ \text{m}^3/\text{d} = 1.35\ \text{m}^3/\text{d}$$

$$\text{Volume of daily cover per year} = 1.35\ \text{m}^3/\text{d} \times 365\ \text{d/year} = 490\ \text{m}^3/\text{year}$$

3. Calculate the annual volume of compacted fill.
   The total compacted fill include both residues and daily cover materials.

$$\text{Total volume of raw residue and daily cover} = (9\ + 1.35)\ \text{m}^3/\text{d} = 10.35\ \text{m}^3/\text{d}$$

$$\text{Volume of compacted fill} = (1 - 0.35) \times 10.35\ \text{m}^3/\text{d} = 6.73\ \text{m}^3/\text{d}$$

$$\text{Volume of compacted fill per year} = 6.73\ \text{m}^3/\text{d} \times 365\ \text{d/year} = 2460\ \text{m}^3/\text{year}$$

4. Calculate the required annual area of landfill.

$$\text{Available volume for compacted fill per m}^2 = 1\ \text{m} \times 1\ \text{m} \times 3\ \text{m} = 3\ \text{m}^3/\text{m}^2$$

$$\text{Area of required landfill per year} = \frac{2640\ \text{m}^3}{\text{year}} \times \frac{1}{3\ \text{m}^3/\text{m}^2} = 820\ \text{m}^2/\text{year}$$

The final cover will be on the top of compacted fill, and will not occupy additional land area.

Discussion Topics and Review Problems

7.1 A bar rack has 45 parallel bars equally spaced in a chamber. The clear spacing is 20 mm. The width of the bars is 10 mm. The water depth in the channel upstream of the rack is 1.6 m. Calculate the width of the chamber and the ratio of the clear area to total screen area.

7.2 A mechanically cleaned bar screen has 30 parallel sharp-edged rectangular bars 12 mm wide placed at a clear spacing of 20 mm. The facility is provided as a preliminary treatment unit at a wastewater treatment facility. The water depth and flow through the screen are 1.5 m and 0.85 m^3/s, respectively. The angle of inclination of bars is 10° from the vertical. Calculate (1) head loss across clean screen, $C_d = 0.7$, and (2) head loss across the screen at 50% clogging. $C_d = 0.6$.

7.3 An interceptor discharges 1.2 m^3/s wastewater into a bar screen chamber. The velocity and depth of flow in the interceptor are 1 m/s and 1.5 m, respectively. The screen chamber has a width of 1.75 m, and coefficient of expansion at the entrance is 0.5. The floor of screen chamber is horizontal, and its invert is 0.05 m below the invert of the interceptor. Calculate the velocity and depth of flow in the screen chamber upstream of bar rack.

7.4 A rectangular channel contains a bar rack. The chamber has a width of 1.5 m and receives a flow of 1.2 m^3/s. The depth of the chamber upstream of the bar rack is 1.9 m. The velocity through the clear rack openings is 0.9 m/s. Calculate (a) approach velocity, and (b) depth and velocity in the channel down stream of the bar rack.

7.5 A bar rack is provided ahead of a pumping station. The screened wastewater has a free fall into the wet well. A proportional weir is provided at the free fall to maintain the normal depth into the screen chamber. Design the proportional weir to maintain 0.61 m/s velocity in the channel downstream of the bar rack. The peak discharge through the rack is 1.5 m^3/s and the width of the screen chamber W is 1.8 m. The bottom slope of the chamber is horizontal and the weir crest is flushed with the channel bottom. Determine (a) depth of flow in the channel downstream of the bar rack, (b) head loss through the bar rack if velocity through the screen at given flow is 0.92 m/s, and (c) depth of flow and velocity in the channel upstream of the bar rack.

7.6 A bar rack proceeds pumping station. The screened wastewater has a free fall into the wet well. A Parshall flume is provided before the outfall to regulate water depth in the screen chamber. The Parshall flume has a throat width of 0.6 m (2 ft) and the average width of the trapezoidal chamber is 1.75 m. The flow velocity through the bar openings and approach velocity through the chamber are 1 and 0.7 m/s, respectively. The peak flow through the screen is 1.5 m^3/s. Design the effluent structure and determine the water depth and velocity in the channel upstream and downstream of the bar rack, and head loss through the screen.

7.7 A bar rack is installed in a rectangular channel. The width of the channel is 1.85 m, and it has a water depth of 1.5 m at the downstream of the rack. The wastewater has a free fall into a wet well. The chamber bottom is raised to maintain a nearly constant depth near the outfall. The bar screen is designed for a peak flow of 1.52 m^3/s. Calculate the height of the raised floor above the bottom of the channel. Draw the longitudinal section of the screen chamber.

7.8 Design a bar rack at a wastewater treatment plant. The bar rack is located upstream of a wet well, and screened flow falls freely into the wet well. A proportional weir is provided at the outfall structure to maintain a constant velocity in the upstream channel. The crest of the proportional weir is flushed with the channel floor. The design criteria for the bar rack is given below.

Flow:	Peak wet weather flow	= 1.52 m^3/s
	Maximum dry weather flow	= 1.12 m^3/s
	Average design dry weather flow	= 0.66 m^3/s
Screen:	Two identical mechanically cleaned bar racks each capable of handling peak wet weather flow.	
	Clear bar spacing	= 2 cm
	Bar width	= 1.2 cm
	The coefficient of expansion K_e	= 0.30
	The design velocity through rack at peak wet weather flow	= 0.9–1.2 m/s
	Angle of inclination of bars from vertical	= 15°
Incoming conduit:	Diameter	= 1.83 m
	Velocity v_1	= 0.89 m/s
	Depth of flow	= 1.25 m
	The invert elevation of the conduit is 0.1 m higher than the chamber floor.	

The design should include:

a. Dimensions of bar rack, and screen chamber,
b. Geometry of theproportional weir,

c. Head loss when the screen is clean, and at 50% clogging,
d. The velocity through the screen, and depth of flow and velocity in the channel upstream and downstream of the rack at maximum wet weather flow, and
e. Hydraulic profile at peak wet weather flow.

7.9 A stainless-steel fine band screen is installed downstream of a bar rack. The flow in the screen chamber is 1.75 m^3/s, and the depth and width of the screen channel are 1.5 and 1.3 m, respectively. The manufacturer's recommended efficiency factor is 0.56, and coefficient of discharge is 0.6. Calculate (a) head loss and velocity through the clean screen, and (b) head loss and velocity through the screen at 20% clogging.

7.10 A rotary drum screen is designed to provide primary treatment. A flow of 3 m^3/s enters the inside of a slowly rotating drum that has a diameter of 1.75 m. The drum is covered with a stainless-steel wedge-wire screening medium. The screen is cleaned by a high-pressure water jet applied at the top. The differential head between the inside and outside drum is 0.45 m, and drum submergence is 0.6 m. The manufacturer's recommended efficiency factor and discharge coefficient are 0.51 and 0.6, respectively. Calculate the length of the drum, velocity through the screen, and hydraulic loading on the screening medium.

7.11 Secondary effluent is considered for use as industrial cooling water. A microscreen is provided to polish the secondary effluent. The following information is available about the microscreen. It has stainless-steel fabric. The recommended efficiency factor and discharge coefficient are 0.58 and 0.60, respectively. The acceptable hydraulic loading is 15 $m^3/m^2 \cdot min$. The inside depth of water is 65% of drum diameter. Design the microscreen system for average filtration rate of 2.5 m^3/s. The applied backwash pressure is 100 kPa.

7.12 Calculate the quantity of screenings collected per day from a wastewater treatment plant that has an average design flow of 2.0 m^3/s. The treatment plant has coarse screens followed by a rotary drum fine screen. The width of clear bar spacings is 1.2 cm, and the rotary drum fine screen has openings of 6 mm.

7.13 The total residuals generated at a wastewater treatment is 10 m^3/d of screenings, grit, and scum. These residues are land filled. Calculate the land area required per year. The depth of landfilling excluding the final cover is 2.5 m. The daily cover is ~17% of the volume of residues, and compaction in landfill is 30%.

References

1. Qasim, S. R., *Wastewater Treatment Plants: Planning, Design, and Operation*, 2nd ed., CRC Press, Boca Raton, FL, 1999.
2. Metcalf & Eddy, Inc., *Wastewater Engineering: Treatment and Reuse*, 4th ed., McGraw-Hill, New York, 2003.
3. Water Environment Federation, *Design of Municipal Wastewater Treatment Plants*, vol. 2, WEF Manual of Practice No. 8, ASCE Manual and Report on Engineering Practice No. 76, Water Environment Federation, Alexandria, VA, 1998.
4. John Meunier Products/Veolia Water Technologies Canada, *Pretreatment, Complete Line of Products and Solutions*, http://www.veoliawatertech.com/en/markets/municipal-solutions/headworks/
5. Johnson Screens, *Contra-Shear® Wastewater Screening Products*, http://www.johnsonscreens.com, 2009.
6. Fairfield Service Company, *General Catalog*, http://www.fairfieldservice.com
7. Passavant-Geiger/Bilfinger Water Technologies, *Products & Service*, http://www.water-passavant.bilfinger.com

8. Huber Technology, Inc., *Multi-Rake Bar Screen RakeMax®*, http://www.huber-technology.com, 2011.
9. Huber Technology, Inc., *HUBER RakeMax®-hf (High Flow) Multi-Rake Bar Screen*, http://www.huber.de, 2012.
10. John Meunier Products/Veolia Water Technologies Canada, *CONT-FLO® TYPE ER/M-L Multi-Rake Bar Screen*, http://www.veoliawatertech.com/en/markets/municipal-solutions/headworks/
11. Kusters Water Div. of Kusters Zima Corp., *Multi-Rake Bar Screen*, http://www.kusterswater.com
12. Passavant-Geiger/Bilfinger Water Technologies, *Noggerath Revolving Chain Screen KLR*, http://www.water-passavant.bilfinger.com
13. Lackeby Products, *Roto-Sieve® Drum Screen*, http://lackeby.com
14. Triveni Engineering and Industries Ltd., *Mechanically Cleaned Bar Screens*, http://www.trivenigroup.com/water
15. Fairfield Service Company, *THE CAF Cable Operated Bar Screen*, http://www.fairfieldservice.com
16. Johnson Screens, *Automatic Raked Bar Screen*, http://www.johnsonscreens.com, 2009.
17. John Meunier Products/Veolia Water Technologies Canada, *John Meunier CONT-FLO® TYPE CF Vertical Bar Screen*, http://www.veoliawatertech.com/en/markets/municipal-solutions/headworks/
18. Aqualitec Corp., *SCREENTEC Vertical Bar Screen*, http://www.aqualitec.com
19. Fairfield Service Company, *CANTENARY Chain Operated Bar Screen*, http://www.fairfieldservice.com
20. Mabarex, Inc., *Cat-Rex™ Catenary Bar Screen*, http://www.mabarex.com
21. SUEZ Treatment Solutions, Inc., *Climber Screen®*, http://www.suez-na.com, 2010.
22. Fairfield Service Company, *THE CLAW Climber Bar Screen*, http://www.fairfieldservice.com
23. Passavant-Geiger/Bilfinger Water Technologies, *Geiger Climber Screen*, http://www.water-passavant.bilfinger.com
24. Vulcan Industries, Inc., *Mensch Severe Duty™ Bar Screen, Product Information Guide*, http://vulcanindustries.com.
25. John Meunier Products/Veolia Water Technologies Canada, *John Meunier ROTARC® Arc Bar Screens & Brush Fine Screens*, http://www.veoliawatertech.com/en/markets/municipal-solutions/headworks/
26. Metcalf & Eddy, Inc., *Wastewater Engineering: Treatment, Disposal and Reuse*, 3rd ed., McGraw-Hill, New York, 1991.
27. Pankratz, T., *Screening Equipment Handbook for Industrial and Municipal Water and Wastewater Treatment*, 2nd ed., Technomic Book Co., Lancaster, PA, 1995.
28. Rex Chain Belt Co., *Weir of Special Design for Grit Channel Velocity Control*, Binder No. 315, vol. 1, September 1963.
29. Droste, R. L., *Theory and Practice of Water and Wastewater Treatment*, John Wiley & Sons, Inc., New York, NY, 1997.
30. Reynolds, T. D. and P. A. Richards, *Unit Operations and Processes in Environmental Engineering*, 2nd ed., PWS Publishing Company, Boston, MA, 1996.
31. Babbitt, H. E. and E. R. Baumann, *Sewerage and Sewage Treatment*, John Wiley & Sons, Inc., New York, 1958.
32. Laughlin, J. E. and W. C. Roming, Design of rotary fine screen facilities in wastewater treatment, *Public Works*, vol. 124, no. 4, April 1993, pp. 47–50.
33. WesTech Engineering, Inc., *Headworks Process Equipment*, http://www.westech-inc.com, 2012.
34. Andritz Separation, Inc., *Andritz Separation*, http://www.andritz.com, 2013.
35. Parkson Corporation, *Hydroscreen Static Screen*, http://www.parkson.com
36. Kusters Water Div. of Kusters Zima Corp., *Static Screens*, http://www.kusterswater.com
37. Passavant-Geiger/Bilfinger Water Technologies, *Noggerath HYdroscreen HS High-Capacity Static Screen*, http://www.water-passavant.bilfinger.com

38. Smith and Loveless Inc., *OBEX™ Spiral Fine Screen*, http://www.smithandloveless.com, 2012.
39. John Meunier Products/Veolia Water Technologies Canada, *ROTARC® TYPE SB Shaftless Spiral Fine Screens*, http://www.veoliawatertech.com/en/markets/municipal-solutions/headworks/
40. WesTech Engineering, Inc., *CleanFlo™ Spiral*, http://www.westech-inc.com, 2005.
41. Kusters Water Div. of Kusters Zima Corp., *Spiral Screen*, http://www.kusterswater.com
42. Passavant-Geiger/Bilfinger Water Technologies, *Noggerath Spiral Sieves NSI*, http://www.water-passavant.bilfinger.com
43. Aqualitec Corp., *APIRALTEC Inclined Cylindrical Screen*, http://www.aqualitec.com
44. Huber Technology, Inc., *ROTAMAT® Micro Strainer Ro 9*, http://www.huber.de, 2012.
45. FSM Frankenberger GmbH & Co. KG, *Screw Screen*, http://www.fsm-umwelt.de
46. WesTech Engineering, Inc., *Step Screening and Dewatering*, http://www.westech-inc.com, 2009.
47. Johnson Screens, *Contra-Shear® Screenmat Step Screen*, http://www.johnsonscreens.com, 2009.
48. Huber Technology, Inc., *STEP SCREEN® Flexible SSF*, http://www.huber-technology.com, 2012.
49. Huber Technology, Inc., *STEP SCREEN® Vertical SSV*, http://www.huber-technology.com, 2012.
50. Passavant-Geiger/Bilfinger Water Technologies, *Noggerath Nogco-Step® Step-Screen*, http://www.water-passavant.bilfinger.com
51. FSM Frankenberger GmbH & Co. KG, *Filter Step Screen*, http://www.fsm-umwelt.de
52. John Meunier Products/Veolia Water Technologies Canada, *ESCALATOR® Fine Screen*, http://www.veoliawatertech.com/en/markets/municipal-solutions/headworks/
53. Andritz Separation, Inc., *Aqua-Guard Self-Washing Continuous Fine Screen*, http://www.andritz.com
54. Andritz Separation, Inc., *Aqua-Screen Perforated Plate Fine Screen*, http://www.andritz.com
55. WesTech Engineering, Inc., *CleanFlo™ Element*, http://www.westech-inc.com, 2005.
56. Parkson Corporation, *Aqua Guard® UltraClean™ Self-Cleaning Moving Media Channel Screen*, http://www.parkson.com
57. Parkson Corporation, *Aqua Guard® The Original Self-Cleaning In-Channel Moving Media Screen*, http://www.parkson.com
58. Johnson Screens, *Centre-Flo™ Screen*, http://www.johnsonscreens.com, 2009.
59. Fairfield Service Company, *Stream Guard*, http://www.fairfieldservice.com
60. Kusters Water Div. of Kusters Zima Corp., *Perforated Plate Filter Screen*, http://www.kusterswater.com
61. Huber Technology, Inc., *HUBER Belt Screen EscaMax® Perforated Plate Screen*, http://www.huber-technology.com, 2012.
62. FSM Frankenberger GmbH & Co. KG, *Center Flow Screen*, http://www.fsm-umwelt.de
63. FSM Frankenberger GmbH & Co. KG, *Dual Flow Screen*, http://www.fsm-umwelt.de
64. FSM Frankenberger GmbH & Co. KG, *Filterscreen*, http://www.fsm-umwelt.de
65. Passavant-Geiger/Bilfinger Water Technologies, *Noggerath Nogco-Guard® NT/ST Fine Screen*, http://www.water-passavant.bilfinger.com
66. Passavant-Geiger/Bilfinger Water Technologies, *Travelling Band Screen*, http://www.water-passavant.bilfinger.com
67. Passavant-Geiger/Bilfinger Water Technologies, *Geiger MultiDisc® Screen*, http://www.water-passavant.bilfinger.com
68. Johnson Screens, *Inclined Rotary Screen—VERSA™*, http://www.johnsonscreens.com, 2011.
69. Parkson Corporation, *Hycor® Helisieve In-Channel Fine Screen*, http://www.parkson.com
70. John Meunier Products/Veolia Water Technologies Canada, *ROTARC® TYPE SD Rotary Drum Fine Screen*, http://www.veoliawatertech.com/en/markets/municipal-solutions/headworks/
71. Lakeside Equipment Corporation, *Raptor® Rotating Drum Screen, Bulletin #2316*, http://www.lakeside-equipment.com, 2005.
72. Huber Technology, Inc., *ROTAMAT® Fine Screen Ro 1*, http://www.huber.de, 2010.
73. Huber Technology, Inc., *Rotary Drum Fine Screen ROTAMAT® Ro 2*, http://www.huber.de, 2012.

74. Kusters Water Div. of Kusters Zima Corp., *In Channel Drum Screen*, http://www.kusterswater.com
75. FSM Frankenberger GmbH & Co. KG, *Rotary Drum Screen*, http://www.fsm-umwelt.de
76. Andritz Separation, Inc., *Girapac Rotating Drum Screen with Screw Compactor*, http://www.andritz.com
77. Parkson Corporation, *Hycor® Rotostrainer® Automatic Wedgewire Fine Screen*, http://www.parkson.com
78. Passavant-Geiger/Bilfinger Water Technologies, *Noggerath ROTOPASS® Drum Sieve System*, http://www.water-passavant.bilfinger.com
79. Johnson Screens, *Contra-Shear®Milliscreen™*, http://www.johnsonscreens.com, 2009.
80. Johnson Screens, *Contra-Shear®Suboscreen®*, http://www.johnsonscreens.com, 2009.
81. WesTech Engineering, Inc., *CleanFlo™ Shear Internally Fed Rotary Drum Screen*, http://www.westech-inc.com, 2006.
82. Parkson Corporation, *Rotoshear® EZ-CARE™ Internally-Fed Rotating Drum Screen*, http://www.parkson.com
83. Parkson Corporation, *Hycor® Rotomesh Fine Screen*, http://www.parkson.com
84. Aqualitec Corp., *DRUMTEC Internally Fed Rotary Drum Screen*, http://www.aqualitec.com
85. Kusters Water Div. of Kusters Zima Corp., *Internally Fed Drum Screen*, http://www.kusterswater.com
86. Passavant-Geiger/Bilfinger Water Technologies, *Noggerath Multi-Drum® Drum Sieve System*, http://www.water-passavant.bilfinger.com
87. Lackeby Products, *Roto-Sieve® Drum Screen*, http://lackebyproducts.com
88. Huber Technology, Inc., *Basket Screen S16*, http://www.huber.de
89. John Meunier Products/Veolia Water Technologies Canada, *John Meunier StormGuard® Overflow Fine Screen*, http://www.veoliawatertech.com/en/markets/municipal-solutions/combined-sewer-overflow/
90. Franklin Miller, Inc., *Turbo Dimminutor® Automatic Channel Screening and Grinding of Water-Borne Solids*, http://www.franklinmiller.com, 2010.
91. Franklin Miller, Inc., *Taskmaster® 8500 Voracious Twin Shaft Grinder*, http://www.franklinmiller.com, 2009.
92. Franklin Miller, Inc., *Taskmaster® TM1600 Heavy-Duty Twin-Shaft Shredder for Gravity or Pressure Systems*, http://www.franklinmiller.com, 2009.
93. Franklin Miller, Inc., *Super Shredder® Voracious In-Line Disintegrator*, http://www.franklinmiller.com, 2009.

8 Grit Removal

8.1 Chapter Objectives

Grits are inert, dense, and highly abrasive material. They include pebble, sand, silt, broken glass, bone chips, and egg shells. Grit accumulates in pipes, corners, and bends, reducing flow capacity, and ultimately clogging pipes and channels. It also accumulates in the treatment units resulting in the loss of treatment capacity. The objectives of this chapter are to present theory and design of grit removal facilities. More specifically, this chapter covers:

- Need and location of grit removal
- Settling behavior of grit and types of gravity settling
- Discrete settling
- Types of grit removal facilities
- Design criteria of grit removal facilities and design examples
- Quantity, handling, and disposal of grit

8.2 Need and Location of Grit Removal Facility

Grit is a nonputrescible material and is much heavier than water. As a result, it makes heavy deposits in pipes and channels. Being inert and heavy, it accumulates in the aeration basin, and has cementing effect on the bottom of the sludge digesters. This results in the loss of usable volume in both situations. Abrasive nature of grit is associated with some abnormal wear to pumps and other equipment. For these reasons, grit removal is considered essential at most treatment plants.[1,2]

Grit removal may normally be achieved at three locations: (1) ahead of the raw wastewater pumps, (2) below the pumping station, or (3) degritting in conjunction with primary sludge. All locations have advantages and disadvantages. Some of these are stated below.[3]

Ahead of the raw wastewater pumps

- Maximum protection of pumping equipment is achieved.
- Frequently they are deep in the ground associated with high construction, operation, and maintenance costs. They are not easily accessible, and it is difficult to raise the grit to the ground level.

Below the pumping station

- Some abnormal wear to pumping equipment may occur. Abrasive resistance pumps may be needed.
- The structure is at ground level; thus, they are accessible and easy to operate and maintain.

Degritter in conjunction with primary sludge

- Pumping equipment is not adequately protected.
- The capital, operation, and maintenance costs are usually low.

- Separation of grit from primary sludge is needed.
- Fine screens are now replacing the primary clarifiers. As a result, grit removal prior to fine screens is needed for their protection.

8.3 Gravity Settling

Grit is removed from wastewater by gravity settling or by centrifugal force. Grit removal facilities are presented latter in this chapter. Since gravity settling is encountered in many treatment processes, an introductory discussion of gravity settling is presented below.

8.3.1 Types of Gravity Settling

Depending on the concentration and the tendency of particles to interact, four types of settling can occur in an aqueous solution. These are (1) discrete or Type I; (2) flocculant or Type II; (3) hindered, zone, or Type III; and (4) compression or Type IV. Each type of settling, their behavior, and occurrence in different processes and reference chapters in this book are summarized in Table 8.1. Discrete settling applies to grit particles. Therefore, detailed discussion on discrete settling is presented below.

8.3.2 Discrete Settling (Type I)

The settling particles have constant velocity during the fall in a fluid. The velocity depends upon the shape, size, and density of the particles; and temperature and viscosity of the fluid. Several equations are used to

TABLE 8.1 Types of Settling, Description, and Occurrence in Treatment Processes

Types of Settling	Description	Application	References
Discrete or Type I settling	Discrete settling occurs in suspension of low concentration. The particles settle as individual entities. There is no significant interference between the settling particles. The velocity remains constant throughout the fall. Both Newton's and Stokes' equations apply depending upon the size of particles.	• Grit removal channel or chamber • Settling of filter media after backwash	Section 8.3.2 Chapter 15
Flocculant or Type II settling	Flocculant settling occurs in dilute suspension. The particles coalesce, flocculate, or agglomerate. As a result, the particle size increases and settling velocity also increases.	• Primary sedimentation basin • Chemically enhanced sedimentation basin • Upper zone of secondary clarifier	Chapters 9 and 10
Zone, hindered or Type III settling	Hindered settling occurs in suspension of intermediate concentration. The solids settle as a "mass" or "blanket." The particle remains in fixed position with respect to each other. A solid–liquid interface develops at the top of settling mass leaving a clear water zone on the top, which gradually increases as settling continues. A sludge blanket builds up at the bottom.	• Settling zone of secondary clarifier • Upper zone of gravity thickners	Chapters 10 and 13
Compression or Type IV settling	Compression settling occurs at high concentration of solids that remain supported on top of each other; further settling is possible only by compression of the structure since solids are constantly added on the top. The compression of sludge blanket takes place slowly.	• Lower zone of final clarifier • Settling zone of gravity thickeners	Chapters 10 and 13

calculate the settling velocity of the particles. These equations and definition of the variables are presented below.

Newton's and Stokes' Law. The classic Newton's law yields the terminal velocity of a spherical particle by equating the gravitational force of the particle to the frictional resistance or drag. The settling velocity is expressed by Equations 8.1a and 8.1b.

$$v_s = \left[\frac{4g(\rho_s - \rho)d}{3C_D\rho}\right]^{1/2} \quad \text{or} \quad v_s = \left[\frac{4g(S_s - 1)d}{3C_D}\right]^{1/2} \tag{8.1a}$$

$$v_s = \left[\frac{4g(\rho_s - \rho)d}{3C_D\rho\phi}\right]^{1/2} \quad \text{or} \quad v_s = \left[\frac{4g(S_s - 1)d}{3C_D\phi}\right]^{1/2} \tag{8.1b}$$

$$C_D = \frac{24}{N_R} + \frac{3}{\sqrt{N_R}} + 0.34 \tag{8.2}$$

$$N_R = \frac{v_s\rho d}{\mu} \quad \text{or} \quad N_R = \frac{v_s d}{\nu} \tag{8.3}$$

where

C_D = drag coefficient that is expressed by Equation 8.2, dimensionless
d = particle diameter, m (ft)
g = acceleration caused by gravity, m/s^2 (ft/s^2)
N_R = Reynolds number that is expressed by Equation 8.3, dimensionless
S_s = specific gravity of particle $\left(S_s = \frac{\rho_s}{\rho}\right)$, dimensionless
t = settling time, s
v_s = settling velocity of particle, m/s (ft/s)
ρ = density of fluid, kg/m^3 ($slug/ft^3$)
ρ_s = density of particle, kg/m^3 ($slug/ft^3$)
μ = dynamic viscosity, $N{\cdot}s/m^2$
ν = kinematic viscosity, m^2/s (ft^2/s)
ϕ = shape factor, dimensionless

This factor is commonly used to account for the effects of nonspherical shapes on their settling velocity. The adjustment is achieved by multiplying the drag coefficient (C_D) by ϕ (Equation 8.1b). Typical ranges of ϕ for settling particles with different shapes are summarized below:[1,2,4]

	Spherical	Rounded Grits	Angular Grits	Raw Wastewater/Flocculated
ϕ	1.0–1.1	1.1–1.7	1.7–2.3	2.3–25

See Section 9.2 for the use of ϕ in calculations of settling velocities of nonspherical particles.

Note: Several useful unit conversions are:

$\nu = \mu/\rho$ and $\rho = \gamma/g$
N = mass × acceleration, $kg{\cdot}m/s^2$ ($lb_f{\cdot}ft/s^2$)
γ = specific weight of fluid, N/m^3 (lb_f/ft^3)

Equation 8.1a applies for spherical particles ($\phi = 1$) and is called Newton's law, while Equation 8.1b applies for nonspherical particles ($\phi > 1$). The drag coefficient C_D is calculated from Equation 8.2. For

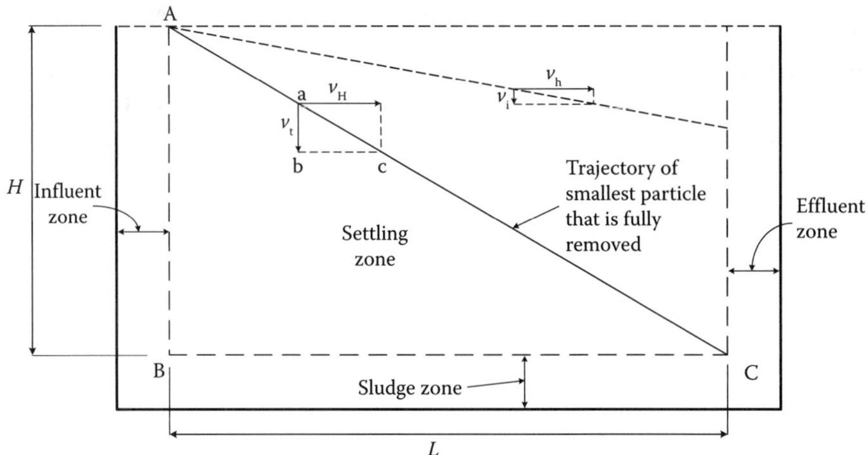

FIGURE 8.1 Definition sketch of terminal velocity and overflow rate.

laminar flow, N_R is <1. The first term in Equation 8.2 predominates and $C_D \approx 24/N_R$. Substituting this value in Equation 8.1a yields the Stokes' law (Equation 8.4).[2,5]

$$v_s = \frac{g(\rho_s - \rho)d^2}{18\mu} \quad \text{or} \quad v_s = \frac{g(S_s - 1)d^2}{18\nu} \quad \text{(Stokes' equation, } N_R < 1, \text{ laminar flow).} \tag{8.4}$$

A Reynolds number $N_R > 10^4$ indicates turbulent flow, and the third term in Equation 8.2 predominates ($C_D \approx 0.34$). Substituting $C_D = 0.34$, Equation 8.1 is transformed into Equation 8.5.

$$v_s = \left[3.92 \times \frac{g(\rho_s - \rho)d}{\rho}\right]^{1/2} \quad \text{or } v_s = \left[3.92g(S_s - 1)d\right]^{1/2} \ (N_R > 10^4, \text{turbulent flow}) \tag{8.5}$$

If the Reynolds number is >1 and $<10^4$, the settling is in the transition range, and the settling velocity of a particle from Equation 8.1a is determined by a trial-and-error solution. The values of C_D and N_R are calculated from Equations 8.2 and 8.3, respectively.

Terminal velocity, Overflow Rate, Hydraulic Loading, or Surface Loading: In the design of a settling basin, the settling velocity v_t (also called terminal velocity) of target particle is selected. The basin is designed such that all particles having a settling velocity equal to and greater than the terminal velocity are fully removed (Figure 8.1). The designed terminal velocity is equivalent to the overflow rate, hydraulic loading, or surface loading. These terms are numerically equal to the flow per unit area of the basin ($m^3/m^2 \cdot d$ or gpd/ft^2). These relationships are expressed by Equation 8.6. It may be noted from Equation 8.6b that the design capacity of the basin is independent of the depth.

$$v_t = \frac{H}{\theta} \tag{8.6a}$$

$$v_t = \frac{Q}{A} \tag{8.6b}$$

$$\theta = \frac{V}{Q} \tag{8.6c}$$

where

v_t = design terminal velocity, overflow rate, hydraulic loading, or surface loading rate, m/s, $m^3/m^2{\cdot}d$ (gpd/ft^2)

H = side water depth, m (ft)

θ = hydraulic retention time, h

Q = flow rate, m^3/s (ft^3/s)

A = surface area of the basin (length × width), m^2 (ft^2)

V = basin volume, m^3 (ft^3)

The performance of an idealized settling basin is reduced significantly due to (1) turbulence at the influent, effluent, and sludge zones, (2) short-circuiting, and (3) volume occupied by the sludge and sludge removal equipment. The design factors therefore are adjusted to account for these factors. These design considerations are discussed later.

Fraction removed at a given overflow rate: A grit removal facility is normally designed for a given terminal velocity or overflow rate. The particles having a terminal velocity equal to and greater than the given over flow rate v_t are fully removed. The particles of settling velocity v_i less than v_t are partially removed. If the particles of all sizes are uniformly distributed over the entire depth of the influent zone, the particles with settling velocity v_i less than v_t (Figure 8.2a) will be partially removed (Equation 8.7).

$$X_r = \frac{v_i}{v_t} \tag{8.7}$$

where X_r = the fraction of the particles removed with settling velocity v_i, dimensionless

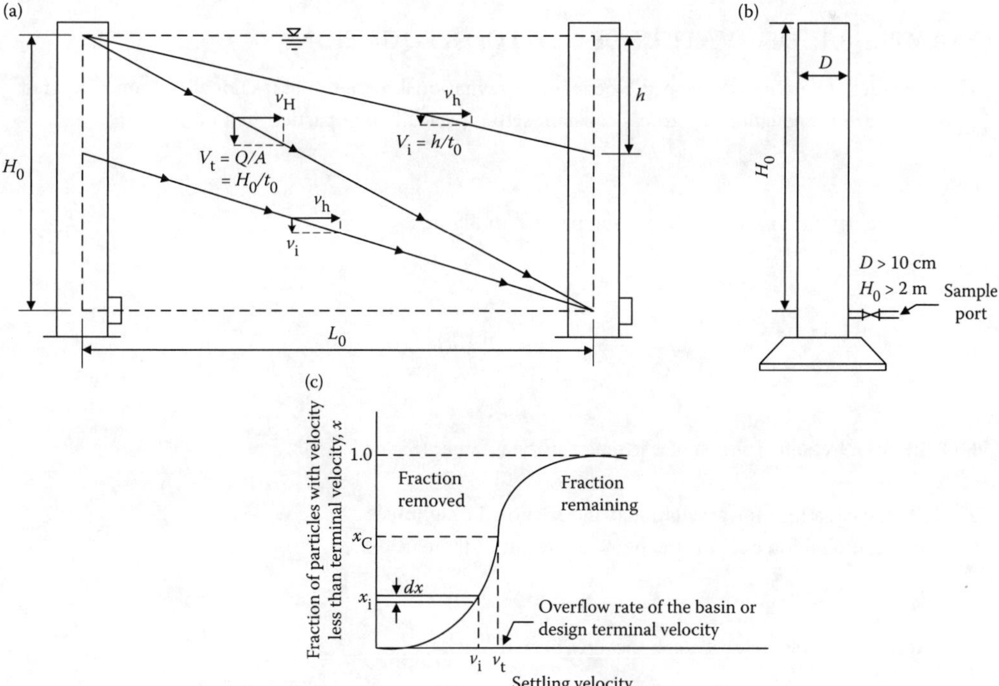

FIGURE 8.2 Settling behavior of discrete particles: (a) settling trajectory of particles $v_i < v_t$, (b) column for settling test, and (c) fraction removal curve.

The grit suspension has particles of different sizes. The size distribution is determined from the standard sieve analysis. The velocity of each fraction is obtained from a column test, or by using Newton's or Stokes' equations. A column used in such test is shown in Figure 8.2b.

A typical settling curve of discrete particles from a column analysis test is shown in Figure 8.2c. The fraction removed in a settling basin having a terminal velocity of v_t is given by Equation 8.8.

$$F = (1 - X_c) + \int_0^{X_C} \frac{v_i}{v_t} dx \approx (1 - X_c) + \sum_{i=1}^{i=n} \frac{v_i}{v_t} \Delta x_i \quad (8.8a)$$

$$v_t = \frac{H_0}{t_0} \quad (8.8b)$$

where

F	= fraction removed, dimensionless
X_c or $\sum_{i=1}^{i=n} \Delta x_i$	= fraction of particles with velocity v_i less than v_t, m/s (ft/s)
$(1 - X_c)$ or $\left(1 - \sum_{i=1}^{i=n} \Delta x_i\right)$	= fraction of particles removed with settling velocity v_i equal to and greater than v_t, dimensionless
$\int_0^{X_C} \frac{v_i}{v_t} dx$ or $\sum_{i=1}^{i=n} \frac{v_i}{v_t} \Delta x_i$	= fraction of particles removed with v_i less than v_t, dimensionless
v_i	= settling velocity of the particles v_i less than v_t, m/s (ft/s)
H_0	= effective depth of the column or tank, m (ft)
t_0	= time to fall the depth H_0 or detention time, s

The concept of fraction removal is difficult, and may require careful review of the following solved examples to fully comprehend the text material.

EXAMPLE 8.1: DERIVATION OF NEWTON'S EQUATION

The Newton's equation is developed by equating gravitational force to the frictional resistance or drag. Develop Newton's equation to express constant settling velocity of a particle in a fluid.

Solution

1. Draw the definition sketch of a settling particle in Figure 8.3.

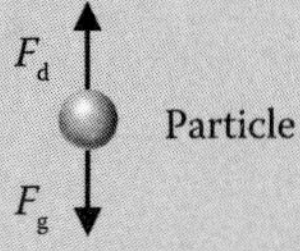

FIGURE 8.3 Definition sketch of a settling particle in water (Example 8.1).

2. Write the equations for gravitational and frictional drag forces.

Gravitational force (F_g) is the buoyant weight of the particle.

$$F_g = (\rho_s - \rho) g V$$

where V = volume of the spherical particle, m^3 (ft^3)

$$V = \frac{\pi}{6} d^3$$

Frictional (or drag) force (F_d) is determined by the interface between the particle and the fluid.

$$F_d = \frac{C_D A \rho v_s^2}{2}$$

where A = cross-sectional or projected area of the particle at right angle to the direction of velocity, m^2 (ft^2)

$$A = \frac{\pi}{4} d^2$$

3. Equate the two forces and simplify.
   At a constant settling velocity of v_s, the gravitational force must equal to the frictional force: $F_g = F_d$.

$$(\rho_s - \rho) g V = \frac{C_D A \rho v_s^2}{2}$$

$$(\rho_s - \rho) g \frac{\pi}{6} d^3 = \frac{C_D \frac{\pi}{4} d^2 \rho v_s^2}{2}$$

$$v_s^2 = \frac{4g(\rho_s - \rho)d}{3\rho C_D}$$

$$v_s = \left[\frac{4g(\rho_s - \rho)d}{3\rho C_D}\right]^{1/2} = \left[\frac{4g(S_s - 1)d}{3C_D}\right]^{1/2}$$

The Newton's equation (Equation 8.1a) is obtained.

EXAMPLE 8.2: DERIVATION OF STOKES' EQUATION

Stokes' equation applies for laminar condition when N_R is <1. Develop the Stokes' equation from Newton's equation.

Solution

1. Obtain the value of C_D for $N_R < 1$.
   C_D is obtained from Equation 8.2.
   For $N_R < 1$, the sum of the last two terms in Equation 8.2 is relatively much smaller than the first term. Therefore, the equation is simplified to $C_D = \frac{24}{N_R}$.
2. Develop the Stokes' equation.
   Substitute the value of $C_D = \frac{24}{N_R}$ and $N_R = \frac{v_s \rho d}{\mu}$ in Equation 8.1a.

$$v_s = \left[\frac{4g(\rho_s - \rho)d}{3\rho \times \frac{24}{N_R}}\right]^{1/2} = \left[\frac{g(\rho_s - \rho)d}{18\rho} \times N_R\right]^{1/2} = \left[\frac{g(\rho_s - \rho)d}{18\rho} \times \frac{v_s \rho d}{\mu}\right]^{1/2} = \left[\frac{g(\rho - \rho)v_s d^2}{18\mu}\right]^{1/2}$$

$$v_s^2 = \frac{g(\rho_s - \rho)v_s d^2}{18\mu} \quad \text{or} \quad v_s = \frac{g(\rho_s - \rho)d^2}{18\mu} = \frac{g(S_s - 1)d^2}{18\nu}$$

Therefore, the Stokes' equation (Equation 8.4) is obtained.

EXAMPLE 8.3: SETTLING EQUATION UNDER TURBULENT CONDITION

Settling behavior of large diameter particles is such that turbulent condition is created, and N_R is $>10^4$. Develop the Newton's equation for settling velocity under turbulent condition.

Solution

1. Determine C_D under turbulent condition.

   C_D is obtained from Equation 8.2. Under turbulent condition, $N_R > 10^4$. The sum of the first two terms in Equation 8.2 is significantly smaller than the third term. Therefore, these terms are ignored, and $C_D = 0.34$.
2. Develop the equation.

   Substitute $C_D = 0.34$ in Equation 8.1a and obtain the simplified settling equation (Equation 8.5) under turbulent condition.

$$v_s = \left[\frac{4g(\rho_s - \rho)d}{3\rho \times 0.34}\right]^{1/2} = \left[3.92g\frac{(\rho_s - \rho)}{\rho}d\right]^{1/2} = [3.92g(S_s - 1)d]^{1/2}$$

EXAMPLE 8.4: SETTLING VELOCITY OF PARTICLES IN THE TRANSITION RANGE

In a grit chamber, the sandy material is removed. Calculate the terminal settling velocity of the selected smallest particle that is fully removed in a grit chamber. The specific gravity and mean diameter of the selected particle are 2.7 and 0.3 mm, respectively. Wastewater temperature is 10°C.

Solution

1. Calculate the terminal settling velocity v_t.

   The kinematic viscosity v of wastewater at 10°C is obtained from Table B.2 in Appendix B.

   $\nu = 1.307 \times 10^{-6}\,\text{m}^2/\text{s}$

   Assume that Stokes' equation (Equation 8.4) is applicable, and calculate the terminal settling velocity.

$$v_t = \frac{g(S_s - 1)d^2}{18\nu} = \frac{9.81\,\text{m/s}^2 \times (2.7 - 1) \times (0.3 \times 10^{-3}\,\text{m})^2}{18 \times (1.307 \times 10^{-6}\,\text{m}^2/\text{s})} = 0.064\,\text{m/s}$$

2. Verify the Reynolds number.

   Calculated Reynolds number from Equation 8.3.

$$N_R = \frac{v_t d}{\nu} = \frac{0.064\,\text{m/s} \times 0.3 \times 10^{-3}\,\text{m}}{1.307 \times 10^{-6}\,\text{m}^2/\text{s}} = 14.7$$

   N_R is in transition range $(1.0 < N_R < 10^4)$ and Newton's equation is applicable. Use trial-and-error procedure.
3. Calculate v_t in *Trial 1*.

   Calculate C_D from Equation 8.2.

$$C_D = \frac{24}{N_R} + \frac{3}{\sqrt{N_R}} + 0.34 = \frac{24}{14.7} + \frac{3}{\sqrt{14.7}} + 0.34 = 2.76$$

Calculate terminal settling velocity v_t from Equation 8.1a.

$$v_t = \sqrt{\frac{4 \times 9.81\,\text{m/s}^2 \times (2.7 - 1) \times (0.3 \times 10^{-3}\,\text{m})}{3 \times 2.76}} = 0.049\,\text{m/s}$$

This is smaller than $v_t = 0.064$ m/s calculated previously from Stokes' equation (Equation 8.4).

4. Calculate v_t in *Trial 2*.

$$N_R = \frac{0.049\,\text{m/s} \times 0.3 \times 10^{-3}\,\text{m}}{1.307 \times 10^{-6}\,\text{m}^2/\text{s}} = 11.2$$

$$C_D = \frac{24}{11.2} + \frac{3}{\sqrt{11.2}} + 0.34 = 3.38$$

$$v_t = \sqrt{\frac{4 \times 9.81\,\text{m/s}^2 \times (2.7 - 1) \times (0.3 \times 10^{-3}\,\text{m})}{3 \times 3.38}} = 0.044\,\text{m/s}$$

This is smaller than $v_t = 0.049$ m/s calculated in *Trial 1*.

5. Repeat the trial-and-error procedure.

The calculation procedure for terminal settling velocity v_t is repeated until v_t does not change from that obtained in the previous trial. Finally, this objective is achieved after Trial 5.

$$N_R = 9.70$$
$$C_D = 3.78$$
$$v_t = 0.042\,\text{m/s}$$

Therefore, the terminal settling velocity of the particle is 0.042 m/s.

EXAMPLE 8.5: SETTLING VELOCITY AND REYNOLDS NUMBER OF PARTICLES OF DIFFERENT DIAMETER

A grit chamber receives a mixture of sandy material of different sizes. Sieve analysis showed the following average diameter of the particles: 0.07, 0.2, 0.3, 0.5, 1, 5, 10, and 15 mm. The specific gravity of sand particles and kinematic viscosity of the fluid are 2.7 and $1.306 \times 10^{-6}\,\text{m}^2/\text{s}$, respectively. Plot the settling velocity v_s and Reynold's number N_R values of each fraction with respect to average diameter. Assume spherical shape.

Solution

1. Calculate v_s and N_R of smallest particle ($d = 0.07$ mm $= 0.00007$ m).

Assume laminar condition and Stokes' law (Equation 8.4) is applicable.

Substitute $S_s = 2.7$ and $v = 1.306 \times 10^{-6}\,\text{m}^2/\text{s}$ in Equation 8.4.

$$v_s = \frac{9.81\,\text{m/s}^2 \times (2.7 - 1) \times (0.00007\,\text{m})^2}{18 \times (1.306 \times 10^{-6}\,\text{m}^2/\text{s})} = 0.0035\,\text{m/s}$$

Check N_R from Equation 8.3.

$$N_R = \frac{0.0035\,\text{m/s} \times 0.00007\,\text{m}}{1.306 \times 10^{-6}\,\text{m}^2/\text{s}} = 0.2$$

The N_R value is <1; therefore, the Stokes' law holds.

2. Calculate v_s and N_R of 0.2-mm diameter particles.

Assume transition condition is applicable and apply Newton's law (Equation 8.1a) through trial-and-error procedure for solutions.

a. Initial validation.

Apply Stokes' equation (Equation 8.4).

$$v_s = \frac{9.81\,\text{m/s}^2 \times (2.7 - 1) \times (0.0002\,\text{m})^2}{18 \times (1.306 \times 10^{-6}\,\text{m}^2/\text{s})} = 0.0284\,\text{m/s}$$

Check N_R.

$$N_R = \frac{0.0284\,\text{m/s} \times 0.0002\,\text{m}}{1.306 \times 10^{-6}\,\text{m}^2/\text{s}} = 4.35$$

Since $1 < N_R < 10^4$, settling velocity is in transition zone. Apply trial-and-error procedure.

b. First trial.

Calculate C_D from Equation 8.2.

$$C_D = \frac{24}{4.35} + \frac{3}{\sqrt{4.35}} + 0.34 = 7.30$$

Calculate terminal settling velocity v_t from Equation 8.1a.

$$v_s = \sqrt{\frac{4 \times 9.81\,\text{m/s}^2 \times (2.7 - 1) \times 0.0002\,\text{m}}{3 \times 7.30}} = 0.0247\,\text{m/s}$$

Check N_R.

$$N_R = \frac{0.0247\,\text{m/s} \times 0.0002\,\text{m}}{1.306 \times 10^{-6}\,\text{m}^2/\text{s}} = 3.78$$

c. Second trial.

$$C_D = \frac{24}{3.78} + \frac{3}{\sqrt{3.78}} + 0.34 = 8.23$$

Calculate terminal settling velocity v_t from Equation 8.1a.

$$v_s = \sqrt{\frac{4 \times 9.81\,\text{m/s}^2 \times (2.7 - 1) \times 0.0002\,\text{m}}{3 \times 8.23}} = 0.0232\,\text{m/s}$$

Check N_R.

$$N_R = \frac{0.0232\,\text{m/s} \times 0.0002\,\text{m}}{1.306 \times 10^{-6}\,\text{m}^2/\text{s}} = 3.55$$

Repeat the procedure until v_s does not change from that obtained in the previous trial. Finally, this objective is achieved after Trial 5. The final results are: $v_s = 0.022$ m/s and $N_R = 3.4$.

3. Calculate v_s and N_R of particles of 0.3-, 0.5-, 1-, 5-, and 10-mm diameter particles.

Use the trial-and-error procedure as that in Step 2 above. The results indicate the settling velocities are all within transition zone.

4. Calculate v_s and N_R of 15-mm diameter particles.

It is assumed that the turbulent condition exists and settling velocity is calculated by Equation 8.5.

$$v_s = \sqrt{3.92 \times 9.81\,\text{m/s}^2 \times (2.7 - 1) \times 0.015\,\text{m}} = 0.99\,\text{m/s}$$

$$N_R = \frac{0.99\,\text{m/s} \times 0.015\,\text{m}}{1.306 \times 10^{-6}\,\text{m}^2/\text{s}} = 1.1 \times 10^4$$

The N_R value is $>10^4$; therefore, the turbulent condition exists, and $v_s = 0.99$ m/s.

5. Summarize v_s and N_R of particles of different diameters.

The settling velocity v_s and Reynolds number N_R of particles of different diameters are summarized in Table 8.2.

TABLE 8.2 The Settling Velocity v and Reynolds Number N_R of Different Size Particles (Example 8.5)

Particle Diameter (d), mm	Settling Regime	Terminal Velocity (v_s), m/s	Reynolds Number (N_R)
0.07	Laminar	0.0035	0.2
0.2	Transition	0.022	3.4
0.3	Transition	0.042	10
0.5	Transition	0.083	32
1	Transition	0.17	129
5	Transition	0.51	1971
10	Transition	0.76	5831
15	Turbulent	0.99	1.1×10^4

6. Plot the settling behavior (v_s and N_R) with respect to particle diameter.

The settling velocity v_s and Reynolds number N_R of particles of different diameter are plotted in Figure 8.4.

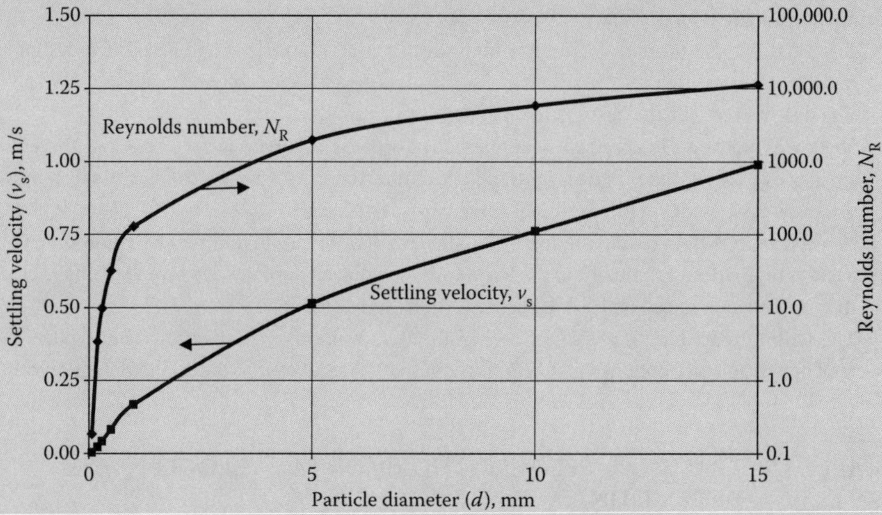

FIGURE 8.4 Settling behavior (settling velocity and Reynolds number) of different diameter particles (Example 8.5).

EXAMPLE 8.6: TERMINAL VELOCITY AND OVERFLOW RATE

A settling basin receives a suspension of discrete particles. The settling velocity of the smallest particle that is fully removed is the design terminal velocity v_t. Develop the following relationships and information (a) terminal velocity is equal to the design overflow rate (hydraulic or surface loading rate) and (b) hydraulic retention time (V/Q). Comment on these relationships.

Solution

1. Express the terminal velocity v_t as overflow rate.

   From similar triangles ABC and abc in Figure 8.1,

$$\frac{v_t}{v_H} = \frac{H}{L}$$

$$v_t = v_H \times \frac{H}{L}$$

   where v_H = horizontal velocity, m/s (ft/s). It is also expressed as:

$$v_H = \frac{\text{Flow}}{\text{Area perpendicular to flow}} = \frac{Q}{H \times W\,(\text{basin width})}$$

$$v_t = \left(\frac{Q}{H \times W}\right) \times \frac{H}{L} = \frac{Q}{L \times W} = \frac{Q}{A\,(\text{surface area})}$$

   Therefore, terminal velocity v_t is same as the overflow rate (hydraulic or surface loading rate).

2. Express the hydraulic retention time θ as V/Q.

   Equate v_t in Equations 8.6a and 8.6b.

$$v_t = \frac{H}{\theta} = \frac{Q}{A}$$

$$\theta = \frac{A \times H}{Q} = \frac{L \times W \times H}{Q} = \frac{V}{Q}$$

3. Comment on these relationships.
   a. The overflow rate (surface loading or hydraulic loading rate) is common basis for design of discrete particles.
   b. The overflow rate of the basin is independent of the depth.
   c. Shallower basins are theoretically more efficient because of larger surface area and therefore smaller overflow rate at a given hydraulic retention time. As a result, the basin will remove the suspension with smaller terminal velocity or particle size.
   d. In a continuous-flow sedimentation basin, the length of the basin and the detention time (the average time a particle remains in the basin) are such that all particles of a given terminal velocity v_t must reach to the bottom of the basin to be fully removed.
   e. The settling zone dimensions L, W, and H are used in the above derivations. In practice, the tank dimensions are obtained by using conservative values of overflow rate and detention time.

EXAMPLE 8.7: TERMINAL VELOCITY, HORIZONTAL VELOCITY, AND DETENTION TIME

A grit removal facility is designed to treat a flow of 18,800 m^3/d. The length, width, and depth of the channel are 18, 0.73, and 1.0 m, respectively. Calculate (a) overflow rate, (b) terminal velocity v_t of the smallest particle that will be fully removed, (c) horizontal velocity, and (d) detention time.

Solution

1. Calculate the overflow rate and terminal settling velocity of smallest particle that will be removed.
   Surface area of the channel $A = L \times W = 18\text{ m} \times 0.73\text{ m} = 13.14\text{ m}^2$

$$\text{Overflow rate} = \frac{Q}{A} = \frac{18{,}800\text{ m}^3/\text{d}}{13.14\text{ m}^2} = 1431\text{ m}^3/\text{m}^2{\cdot}\text{d}$$

2. Calculate the terminal velocity.

$$\text{Terminal velocity, } v_t = 1431\frac{\text{m}^3}{\text{m}^2{\cdot}\text{d}} \times \frac{\text{d}}{(24 \times 60)\text{ min}} = 1\text{ m/min or }0.017\text{ m/s}$$

3. Calculate the horizontal velocity.

$$\text{Flow, } Q = 18{,}800\frac{\text{m}^3}{\text{d}} \times \frac{\text{d}}{(24 \times 60 \times 60)\text{ s}} = 0.22\text{ m}^3/\text{s}$$

$$\text{Horizontal velocity, } v_H = \frac{Q}{H \times W} = \frac{0.22\text{ m}^3/\text{s}}{1.0\text{ m} \times 0.73\text{ m}} = 0.30\text{ m/s}$$

4. Calculate the hydraulic detention time.

$$\text{Basin volume, } V = L \times W \times H = 18\text{ m} \times 0.73\text{ m} \times 1.0\text{ m} = 13.14\text{ m}^3$$

$$\text{Detention time, } \theta = \frac{V}{Q} = \frac{13.14\text{ m}^3}{0.22\text{ m}^3/\text{s}} = 60\text{ s or }1\text{ min}$$

EXAMPLE 8.8: FRACTION REMAINING AND SETTLING VELOCITY

A column test was performed on a mixed sample of discrete particles. The column was filled with a mixed sample to a height of 3 m. The initial concentration of suspended solids at time zero is 285 mg/L. The concentration remaining after time 2 min of settling is 125 mg/L. Calculate (a) weight fraction remaining, (b) weight fraction settled, and (c) settling velocity, terminal velocity of the smallest particle that is fully removed in the weight fraction, or overflow rate.

Solution

1. Calculate the weight fractions remaining and settled.

$$\text{Weight fraction remaining in suspension} = \frac{C}{C_0} = \frac{125\text{ mg/L}}{285\text{ mg/L}} = 0.44$$

$$\text{Weight fraction settled} = \frac{C_0 - C}{C} = \frac{(285 - 125)\text{ mg/L}}{285\text{ mg/L}} = 0.56$$

2. Calculate the settling velocity or the terminal velocity of the smallest particle in the weight fraction that is fully removed.

$$\text{Settling velocity or } v_t = \frac{H}{\theta} = \frac{3\text{ m}}{2\text{ min}} = 1.5\text{ m/min}$$

$$\text{Overflow rate} = 1.5\text{ m/min} \times \frac{(24 \times 60)\text{ min}}{\text{d}} = 2160\text{ m/d or }2160\text{ m}^3/\text{m}^2{\cdot}\text{d}$$

EXAMPLE 8.9: WEIGHT FRACTION REMOVED FROM A GIVEN SETTLING VELOCITY DATA

A settling basin is designed to remove a discrete suspension. The results of a column test are given below. The settling basin has an overflow rate of 51,700 gpd/ft^2. Calculate percent weight fraction of solids removed.

Settling velocity, ft/min	8.0	4.0	2.0	1.0	0.7	0.5
Weight fraction remaining	0.56	0.48	0.37	0.19	0.05	0.02

Solution

1. Calculate the design terminal velocity of the smallest particle that is fully removed at the given overflow rate.

$$v_t = \frac{51{,}700\ \text{gal/d·ft}^2}{7.48\ \text{gal/ft}^3} \times \frac{\text{d}}{(24 \times 60)\ \text{min}} = 4.8\ \text{ft/min or } 4.8\ \text{ft}^3/\text{min·ft}^2$$

2. Plot the velocity curve as a function of the fraction of particles remaining.

   The data are plotted in Figure 8.5. A smooth curve is drawn through the data points.

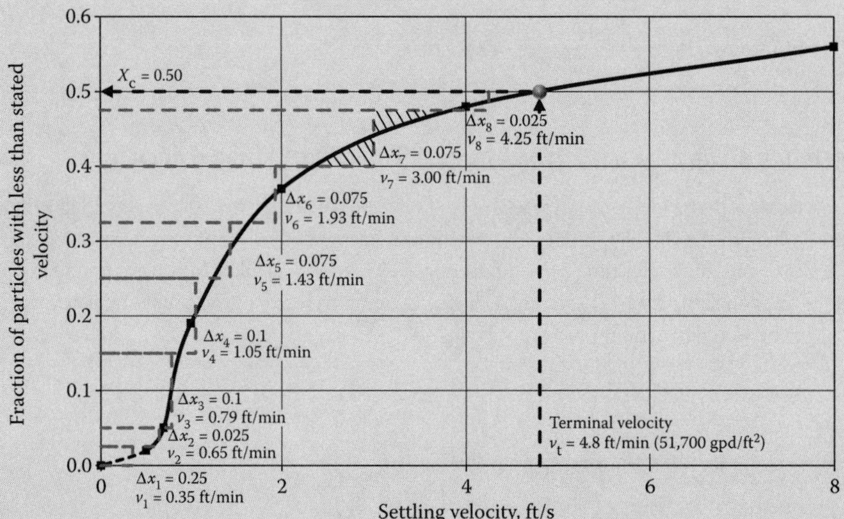

FIGURE 8.5 Fraction of particles remaining that have settling velocity equal to and less than the design terminal velocity, and graphical integration (Example 8.9).

3. Draw a vertical line from the terminal velocity v_t of 4.8 ft/min.

   The vertical line intersects the curve at 0.5 fraction of the suspension ($X_c = 0.5$). This fraction has a settling velocity less than the terminal velocity $v_t = 4.8$ ft/min.

4. Apply the graphical integration between the limit $x = 0$ and $x = 0.5$ to determine the fraction removed.

   Divide the area into eight small parallel strips. The selected velocity v_i should be such that the triangular area lost above the curve is gained in the strip. Tabulate each fraction and its settling velocity.

Δx_i	v_i, ft/min	$\frac{v_i}{v_t} \times \Delta x_i$
0.025	0.35	0.0018
0.025	0.65	0.0034
0.100	0.79	0.0165
0.100	1.05	0.0219
0.075	1.43	0.0223
0.075	1.93	0.0302
0.075	3.00	0.0469
0.025	4.25	0.0221
$X_c = \sum_{i=1}^{i=8} \Delta x_i = 0.50$		$\sum_{i=1}^{i=8} \frac{v_i}{v_t} \Delta dx_i = 0.17$

5. Apply Equation 8.8a to calculate the fraction removed.

$$F = (1 - X_c) + \int_0^{X_C} \frac{v_i}{v_t} dx \approx (1 - X_c) + \sum_{i=1}^{i=8} \frac{v_i}{v_t} \Delta x_i = (1 - 0.5) + 0.17 = 0.67 \quad \text{or} \quad 67\%$$

EXAMPLE 8.10: WEIGHT FRACTION REMOVAL FROM COLUMN ANALYSIS

Settling column analysis was performed on a grit sample. The column depth and initial concentration of TSS were 2.0 m and 380 mg/L, respectively. The test results are summarized below. Calculate the weight fraction removed in a grit basin that has an overflow rate of 25 m^3/m^2·d.

Time, min	0	60	80	100	130	200	240	420
TSS concentration, mg/L	380	239	228	213	198	141	99	34

Solution

1. Calculate the weight fraction remaining in suspension, and the settling rate of each fraction.

$$\text{Settling velocity at 60 min} = \frac{2.0\text{ m}}{60\text{ min}} = 3.3 \times 10^{-2}\text{ m/min}$$

$$\text{Fraction remaining at 60 min} = \frac{239\text{ mg/L}}{380\text{ mg/L}} = 0.63$$

Similarly calculate and summarize the settling velocity and fraction remaining at all other sampling periods.

Time, min	0	60	80	100	130	200	240	420
Settling velocity v_i, 10^{-2} m/min	NA	3.30	2.50	2.00	1.53	1.00	0.83	0.48
Fraction remaining	NA	0.63	0.60	0.56	0.52	0.37	0.26	0.09

2. Plot the settling curve of weight fraction remaining and the corresponding settling velocity.
   The data points are plotted in Figure 8.6 and a smooth curve is drawn through these points.

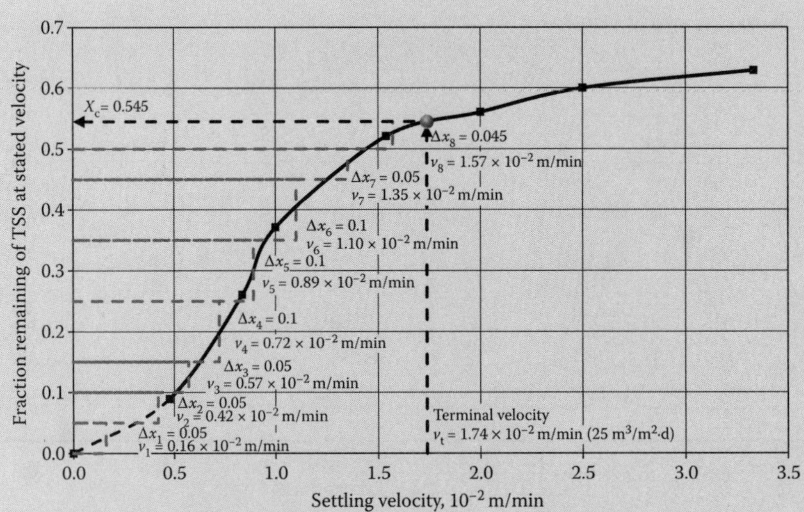

FIGURE 8.6 Fraction remaining that has settling velocity less than or equal to the design terminal velocity and graphical integration (Example 8.10).

3. Calculate the terminal velocity of the smallest particle that is fully removed in the grit basin.

$$v_t = 25\,\text{m}^3/\text{m}^2{\cdot}\text{d} \times \frac{\text{d}}{(24 \times 60)\,\text{min}} = 0.0174\,\text{m/min} \quad \text{or} \quad 1.74 \times 10^{-2}\,\text{m/min}$$

4. Draw a vertical line from $v_t = 1.74 \times 10^{-2}$ m/min.

   Fraction remaining $X_c = 0.545$. This fraction has settling velocity $v_i < 1.74 \times 10^{-2}$ m/min and will be partly removed. Divide area under X_c into a series of strips and draw vertical v_i lines such that triangular area lost above the curve is gained in the strip.
5. Determine the fraction removal.

   Determine v_i and Δx_i through graphical integration.

Δx_i	v_i, 10^{-2} m/min	$\frac{v_i}{v_t} \times \Delta x_i$
0.050	0.16	0.0046
0.050	0.42	0.0121
0.050	0.57	0.0164
0.100	0.72	0.0415
0.100	0.89	0.0513
0.100	1.10	0.0634
0.050	1.35	0.0389
0.045	1.57	0.0407
$X_c = \sum_{i=1}^{i=8} \Delta x_i = 0.545$		$\sum_{i=1}^{i=8} \frac{v_i}{v_t} \Delta dx_i = 0.269$

6. Calculate the overall removal efficiency of a given suspension in a basin that has surface loading rate of 25 m³/m²·d ($v_t = 1.74 \times 10^{-2}$ m/min) from Equation 8.8a.

$$F = (1 - X_c) + \int_0^{X_C} \frac{v_i}{v_t} dx \approx (1 - X_c) + \sum_{i=1}^{i=8} \frac{v_i}{v_t} \Delta x_i = (1 - 0.545) + 0.269 = 0.72 \quad \text{or} \quad 72\%$$

EXAMPLE 8.11: WEIGHT FRACTION REMOVAL CALCULATED FROM STANDARD SIEVE ANALYSIS

Standard sieve analysis was performed on a grit sample. The results of the analysis are tabulated below. Calculate the weight fraction removed in a grit chamber that has an overflow rate of 100 gpm/ft^2. The average wastewater temperature is 70°F, and average specific gravity of discrete particle is 2.65.

Standard sieve number	16	20	30	40	65	80	100	120	140	200
Weight fraction retained	0	0.06	0.17	0.17	0.40	0.10	0.05	0.03	0.01	0.01

Solution

1. Calculate the design terminal velocity v_t of the smallest particle that will be totally removed.

$$v_t = 100 \frac{\text{gal}}{\text{min}\cdot\text{ft}^2} \times \frac{\text{ft}^3}{7.48\,\text{gal}} \times \frac{\text{min}}{60\,\text{s}} = 0.22\,\text{ft/s}$$

2. Calculate the weight fraction passing and settling velocity of each fraction retained.

   The average theoretical settling velocity of each fraction retained is calculated from either Newton's or Stokes' equation, whichever applies. The procedure is as follows:

   a. Tabulate from sieve analysis data the weight fraction retained and passing for each standard sieve size.
   b. Determine from an engineering handbook the average diameter of the openings for each standard sieve.[4]
   c. Calculate the geometric mean diameter of the particles of each fraction. The diameters of the sieve through which the fraction is passing and retained should be used.
   d. Calculate the settling velocity of each fraction. Use either Newton's or Stokes' equation that may apply. Specific gravity = 2.65 and kinematic viscosity $\nu = 1.059 \times 10^{-5}\,\text{ft}^2/\text{s}$ at 70°F (Table B.3 in Appendix B). These values are summarized in Table 8.3.
   e. Plot standard curve between the weight fraction of particles passing with less than the stated velocity and settling velocity v_s of the particles. These results are shown in Figure 8.7.

3. Determine weight fraction removed by graphical integration of area under the terminal velocity v_t of 0.22 ft/s from Equation 8.8a.

Δx_i	v_i, ft/s	$\frac{v_i}{v_t} \times \Delta x_i$
0.050	0.049	0.0110
0.050	0.072	0.0162
0.075	0.123	0.0414
0.075	0.163	0.0549
0.075	0.191	0.0643
0.060	0.213	0.0574
$X_c = \sum_{i=1}^{i=6} \Delta x_i = 0.385$		$\sum_{i=1}^{i=6} \frac{v_i}{v_t} \Delta dx_i = 0.245$

$$F = (1 - X_c) + \int_0^{X_C} \frac{v_i}{v_t} dx \approx (1 - X_c) + \sum_{i=1}^{i=6} \frac{v_i}{v_t} \Delta x_i = (1 - 0.385) + 0.245 = 0.86 \quad \text{or} \quad 86\%$$

TABLE 8.3 Sieve Analysis and Calculated Theoretical Settling Velocity Data (Example 8.11)

Standard Sieve Number	Weight Distribution of Particles		Standard Sieve Opening, ft	Geometric Mean Diameter (d), ft	Theoretical Velocity (v_s), ft/s	N_R	Settling Regime
	Fraction Retained	Fraction Passing					
(1)	(2)	(3)	(4)	(5)	(6)	(7)	(8)
16[a]	0.00[b]	1.00	0.00391[d]	–	–	–	–
20	0.06	0.94[c]	0.00276	0.00329[e]	0.577[f]	179[g]	Transition
30	0.17	0.77	0.00195	0.00232	0.426	93	Transition
40	0.17	0.60	0.00138	0.00164	0.300	46	Transition
65	0.40	0.20	0.00069	0.00098	0.158	15	Transition
80	0.10	0.10	0.00058	0.00063	0.083	5	Transition
100	0.05	0.05	0.00049	0.00053	0.063	3	Transition
120	0.03	0.02	0.00041	0.00045	0.047	2	Transition
140	0.01	0.01	0.00034	0.00037	0.034	1.2	Transition
200	0.01	0.00	0.00024	0.00029	0.023[h]	0.6	Laminar

[a] U.S. Standard sieve number.

[b] Given sieve analysis data on grit sample.

[c] Weight fraction passing = 1.00 (in Column [3]) − 0.06 (in Column [2]) = 0.94. This fraction has velocity less than the velocity v_s of 0.577 ft/s (in Column [6]).

[d] Standard sieve diameter opening.

[e] Geometric mean $d = \sqrt{0.00391 \times 0.00276} = 0.00329$ ft.

[f] Theoretical settling velocity v_s is calculated using Newton's equation (Equation 8.1a) under transition settling condition.

[g] The Reynolds number N_R is calculated from Equation 8.3.

[h] Theoretical settling velocity v_s is calculated using Stokes' equation (Equation 8.4) under laminar settling condition.

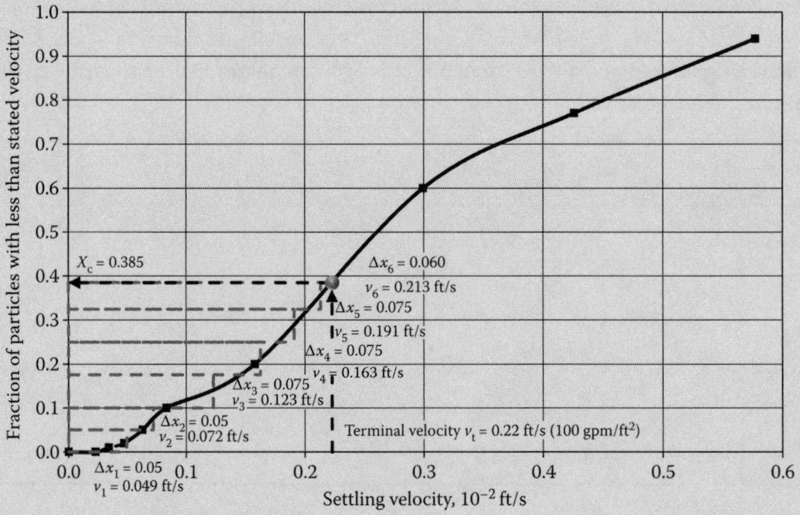

FIGURE 8.7 Fraction of particles that have settling velocity less than the stated velocity v_i, and graphical integration (Example 8.11).

8.4 Types of Grit Removal Facilities

Several types of grit removal facilities are available for selection. The choice of facility and equipment needs careful consideration. The factors considered for selection are: (1) quantity and quality of grit, (2) types of treatment units and equipment, connecting conduits, valves and fittings, and instrumentation provided at different parts of the plant, (3) head loss and space constraints, and (4) odors, VOCs, and environmental control issues.

There are three general types of grit removal facilities: (1) horizontal flow, (2) aerated, and (3) vortex type. Each type is discussed below.

8.4.1 Horizontal-Flow Grit Chamber

The horizontal-flow grit chambers utilize channel or basins. The flow through the chamber is in a horizontal direction. The velocity is controlled by the chamber dimensions, and influent and effluent structures. An optimum horizontal velocity of 0.3 m/s (1 ft/s) is normally used at a detention time of 45–90 s. Two types of horizontal-flow grit chambers are in common use. These are velocity-controlled grit channel, and detritus or square tank.

Velocity-Controlled Grit Channel: The grit in wastewater has a specific gravity in the range of 1.5–2.7. The organic matter in the wastewater has a specific gravity around 1.02. Therefore, differential sedimentation is a successful mechanism for separation of grit from organic matter. Also, the grit exhibits discrete settling, whereas organic matter settles as flocculant solids.

The velocity-controlled grit chamber is a shallow long and narrow sedimentation basin in which velocity is controlled. The velocity control is achieved by the use of a control sections at the outlet end of the channel. The control sections may be a *proportional weir*, *Sutro weir*, *Parshall flume*, *parabolic flume*, or like. These control sections maintain constant velocity in the channel at a wide range of flows. The grit channel may be manually cleaned or mechanically cleaned. Manually cleaned channels are used only at small plants. The channels have hoppers at the bottom for grit storage. The mechanically cleaned grit channel utilizes a conveyor with scrapers, bucket, or plows to move the grit to a sump. The grit is removed from the sump by screw conveyors or bucket elevators.[1–3,6–8]

The velocity-controlled grit channels are designed for peak wet weather flow and multiple units are provided. The typical design parameters are provided in Table 8.4. The components of velocity-controlled grit channel are shown in Figure 8.8.[7,8]

Detritus or Square Horizontal-Flow Grit Chamber: Detritus or square horizontal-flow grit chambers are essentially square sedimentation basins in which grit and organics are removed collectively. The influent is released on one side of the basin through a series of ports located near the bottom. The wastewater flows up across the basin to a rectangular weir, and has a free fall into the effluent trough. The

TABLE 8.4 Typical Design Parameters of Velocity-Controlled Channel

Design Parameter	Range	Typical
Detention time, s	45–90	60
Horizontal velocity, m/s (ft/s)	0.25–0.4 (0.8–1.3)	0.3 (1)
Settling velocity, m/min (ft/ min)		
0.21 mm (70-mesh)	1–1.3 (3.3–4.3)	1.2 (4)
0.15 mm (100-mesh)	0.6–0.9 (2–3)	0.75 (2.5)
Head loss, % of depth	30–40	36
Added length allowance for inlet and outlet turbulence, % of length	25–50	30

(a)

(b)

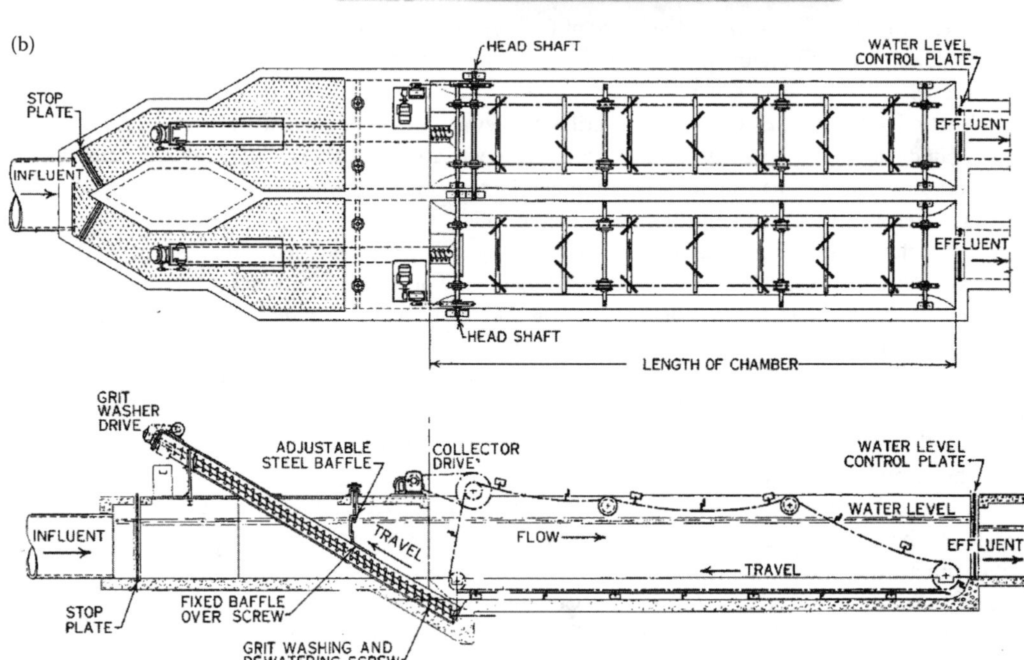

FIGURE 8.8 Details of velocity-controlled grit chamber: (a) velocity-controlled grit channel (Courtesy Fairfield Service Company), and (b) plan and longitudinal section of a double-channel grit collector (Adapted from Reference 8; reprinted with permission of the Water Environment Federation and the American Society of Civil Engineers).

hydraulic loading may vary, depending upon the specific gravity and diameter of the particles, and the temperature of the wastewater. The side water depth of the basin may be 2–3 m. The basin diameter can be in a range from 2.5 to 7 m (8–23 ft) for capacity from 8600 to 235,000 m^3/d (2.3–62 MGD). The hydraulic loadings for different diameter particles of specific gravity 2.65 and at different temperatures are given in Table 8.5.[2,3] These basins are designed for peak design flow, and at least two units are provided. The settled solids in the basin are moved by a rotating raking mechanism into a sump. The head loss through the unit is a minimum. The solids may build up on four sloping corners unless corner sweep raking mechanism is used to sweep the corners. Organic matter also settles with grit. As a result,

TABLE 8.5 The Design Hydraulic Loadings for Detritus Chamber at Different Temperatures for Removal of Different Grit Particle Sizes Having Specific Gravity of 2.65

Particle Size, mm	Hydraulic Loading[a], $m^3/m^2{\cdot}d$		
	10°C	15°C	20°C
0.21 (70-mesh)	800	950	1100
0.15 (100-mesh)	520	590	670
0.11 (140-mesh)	330	380	430

[a] To account for inlet and outlet turbulence and hydraulic inefficiencies, a safety factor of 2 has been applied.

Note: $m^3/m^2{\cdot}d \times 24.5 = gpd/ft^2$.

degritting or separation of organic matter from settled solids is essential. Degritting is achieved simultaneously by moving the settled grit up an incline over a reciprocating rake mechanism. The organic solids are separated from the grit and fall back into the basin while moving the grit up the incline. Degritting may also be achieved by an external hydrocyclone or a centrifuge. The detritus tank grit removal system installation is shown in Figure 8.9.[9,10]

(a)

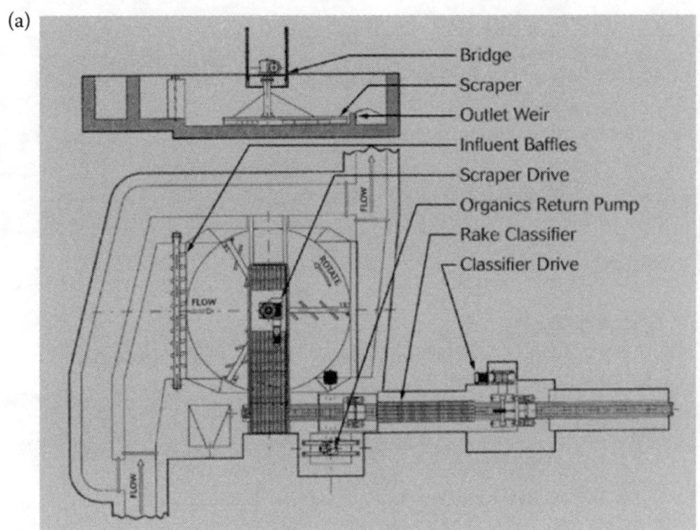

(b)

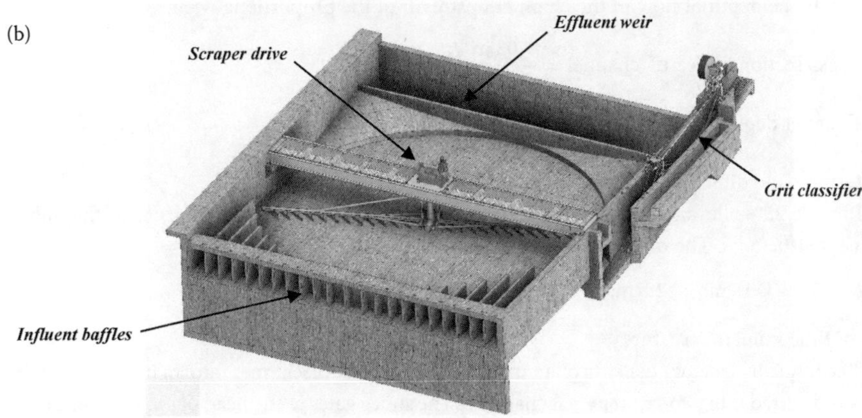

FIGURE 8.9 Detritus tank and degritting equipment: (a) components of detritus tank system (Courtesy Ovivo), and (b) bird view of detritus tank (Courtesy Hubert Stavoren BV).

EXAMPLE 8.12: DESIGN OF A VELOCITY-CONTROLLED RECTANGULAR GRIT CHANNEL

Design a velocity-controlled rectangular grit removal facility for peak design flow. The design average flow, peaking factor, and ratio of minimum to average flow are 0.35 m^3/s, 3.0, and 1/3, respectively. The horizontal and terminal velocity, and detention time are 0.30 m/s, 1.2 m/min, and 60 s, respectively. At this terminal velocity, the minimum discrete particle of 0.18 mm diameter (80-mesh) will be fully removed. The flow from a common influent chamber is diverted by two sluice gates to two identical channels. Each channel is capable of handling the design peak flow. A proportional weir is provided at the downstream of each channel to maintain a constant horizontal velocity at a wide flow variation. Assume that the weir crest is set at the weir datum, and the weir length is less than the channel width. The grit removal system consists of a conveyor that raises the grit and washes it to separate the organics.

Solution

1. Calculate the design peak and minimum flow per basin.

$$\text{Design peak flow} = 3.0 \times 0.35\,\text{m}^3/\text{s} = 1.05\,\text{m}^3/\text{s}$$

$$\text{Design minimum flow} = \frac{1}{3} \times 0.35\,\text{m}^3/\text{s} = 0.12\,\text{m}^3/\text{s}$$

2. Determine the theoretical length and width of the grit channel.

$$\text{Overflow rate} = 1.2\,\text{m/min} \times \frac{\text{m}^2}{\text{m}^2} = 1.2\,\text{m}^3/\text{m}^2{\cdot}\text{min}$$

$$\text{Surface area, } A = \frac{1.05\,\text{m}^3/\text{s}}{1.2\,\text{m}^3/\text{m}^2{\cdot}\text{min}} \times \frac{60\,\text{s}}{\text{min}} = 52.5\,\text{m}^2$$

Detention time, $\theta = 60$ s

Theoretical length of channel, $L = 0.3\,\text{m/s} \times 60\,\text{s} = 18\,\text{m}$

$$\text{Width of basin, } W = \frac{52.5\,\text{m}^2}{18\,\text{m}} = 2.9\,\text{m}$$

Provide a width W of 3.0 m to accommodate standard equipment.

3. Calculate the depth of flow in the channel upstream of the proportional weir.

$$\text{Cross-sectional area of channel} = \frac{1.05\,\text{m}^3/\text{s}}{0.30\,\text{m/s}} = 3.5\,\text{m}^2$$

$$\text{Channel depth, } D = \frac{3.5\,\text{m}^2}{3.0\,\text{m}} = 1.2\,\text{m}$$

4. Determine the dimensions of the channel.
   Provide 25% allowance in the length of the channel to account for turbulence at the influent and effluent structures. The dimensions of the channel are:

$$L = 1.25 \times 18\,\text{m} = 22.5\,\text{m},\ W = 3.0\,\text{m},\ \text{and}\ D = 1.2\,\text{m}$$

5. Layout the influent structure.
   The influent structure consists of an influent conduit that discharges into an influent chamber. The flow is diverted into two grit removal channels. The sluice gates at the head of each channel are used to remove one unit out of service. The flow divides equally when both gates are open. The channel layout is shown in Figure 8.10a.

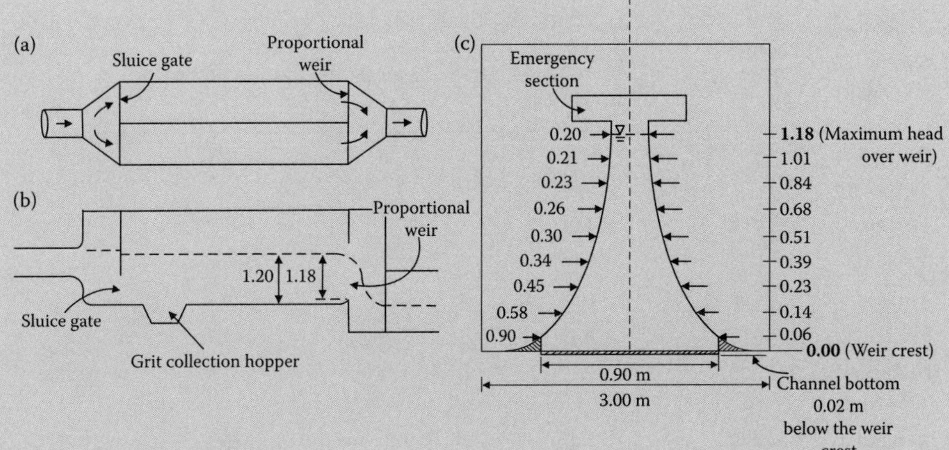

FIGURE 8.10 Layout of grit removal facility, (a) plan, (b) longitudinal section, and (c) design of proportional weir at the effluent structure (Example 8.12).

6. Design the effluent structure.

   The effluent structure is designed to handle the design peak flow when one channel is in service. It consists of a proportional weir to maintain a constant velocity of 0.3 m/s in the channel over a wide flow range.

   a. Determine the head over the proportional weir.

      Water depth D in the channel at design peak flow is 1.20 m. The weir crest is generally raised to 1–3 cm from the channel bottom so that the rolling grit over the channel bottom may be contained. In this design, raise the weir crest by 2 cm (0.02 m).

      Maximum head over the weir at peak design flow, $H = (1.2 - 0.02)\text{ m} = 1.18\text{ m}$. The conceptual longitudinal section view of the channel is shown in Figure 8.10b.

      $$Q = 4.17LH^{3/2}$$

   b. Determine the opening width of the proportional weir at the maximum water depth.

      Calculate the opening width L of the proportional weir from Equation 7.8d at the maximum water depth $H = 1.18$ m.

      $$L = \frac{Q}{4.17H^{3/2}} = \frac{1.05\text{ m}^3/\text{s}}{4.17 \times (1.18\text{ m})^{3/2}} = 0.196\text{ m} \quad \text{or} \quad 0.20\text{ m}$$

   c. Develop the ideal profile (shape of the opening) of the proportional weir and validate the velocity in the channel.

      In order to maintain a constant velocity in the channel, the factor ($LH^{1/2}$) in Equation 7.8a must remain constant (K).

      Determine the constant K from Equation 7.8a.

      $$K = LH^{1/2} = 0.196\text{ m} \times (1.18\text{ m})^{1/2} = 0.213\text{ m}^{3/2}$$

      The proportional weir calculations are tabulated. The weir profile is shown in Figure 8.10c.

Given and Assumed Flow Condition	Flow (q), m^3/s	Head Over Weir, $y = \frac{q}{4.17K}$ (Equation 7.8g), m	Weir Length $x = Ky^{-1/2}$ (Equation 7.7a), m	Depth in the Channel (d), m	Velocity in the Channel $v = \frac{Q}{d \times W}$, m/s
Design peak flow (given)	1.05	1.18	0.20	1.20	0.29
Assumed[a]	0.90	1.01	0.21	1.03	0.29
Assumed[a]	0.75	0.84	0.23	0.86	0.29
Assumed[a]	0.60	0.68	0.26	0.70	0.29
Assumed[b]	0.525	0.49	0.30	0.61	0.29
Design average flow (given)	0.35	0.39	0.34	0.41	0.28
Assumed	0.20	0.23	0.45	0.25	0.27
Design minimum flow (given)	0.12	0.14	0.58	0.16	0.26
Assumed[c]	0.05	0.06	0.90	0.08	0.22

[a] These flows are assumed to develop the ideal weir profile.
[b] This is the design peak flow when both channels are in service.
[c] This flow is assumed to determine the weir length of 0.90 m.

Note: Although the flow varies greatly, the velocity in the channel remains nearly constant. Moderate deviations occur at flows below the design average flow.

EXAMPLE 8.13: DESIGN OF A PARABOLIC GRIT CHANNEL AND PARSHALL FLUME

Design a velocity-controlled parabolic grit channel to treat municipal wastewater. The design average flow is 0.15 m^3/s. The design peak flow is three times the design average flow. The design horizontal velocity in the channel is 0.3 m/s, and the settling velocity is 19 mm/s. A Parshall flume with throat width of 15 cm (6 in) is selected for velocity control in the channel. The head losses from the end of grit chamber to the entrance of the Parshall flume (converging section) are small and ignorable because of the low velocity in the channel. Determine (a) the head H_a (at the throat) at the design peak flow, (b) the cross-sectional dimensions of the parabolic section, (c) the settling time, and (d) the theoretical and design lengths of the chamber. The design length of the channel is 1.35 times the theoretical length.

Solution

1. Describe the basic design considerations of Parshall flume as a velocity control device.

   The basic design considerations and design examples of Parshall flume for flow measurement are presented in Chapter 6 and Appendix C. The standard dimensions and discharge relationships for throat widths of Parshall flumes are provided in Figure C.1, and Tables C.2 and Table C.3 of Appendix C, respectively.

   A small Parshall flume throat width of 6 in (0.15 m) is selected to maintain an acceptable water depth in the channel. Match the head H_a at the throat with the channel depth at design peak flow. Keep the submergence (H_b) <0.6 of H_a to achieve free flow at the flume (Table C.4). Also see Figure C.1 for the locations for H_a and H_b.
2. Describe the design considerations of the parabolic channel.

The channel is designed to have horizontal velocity of 0.23–0.38 m/s (0.75–1.25 ft/s). Since a parabolic section is difficult to construct, in practice a combination of trapezoidal and rectangular shapes are used to approximate the cross-sectional area of the parabola (Figure 7.5b). The design width of the channel is 1.2–1.5 times the theoretical width to even out the turbulence at the entrance and exit of the channel.

3. Convert the design flow in the U.S. customary unit.

   The flow through a Parshall flume is provided in the U.S. customary unit in Equation 7.9.

   Calculate the design peak flow.

   $$Q = 3 \times 0.15\,\text{m}^3/\text{s} = 0.45\,\text{m}^3/\text{s}$$

   Convert the flow from m^3/s to ft^3/s.

   $$Q = 0.45\,\text{m}^3/\text{s} \times \frac{\text{ft}^3/\text{s}}{0.0283\,\text{m}^3/\text{s}} = 15.9\,\text{ft}^3/\text{s}$$

4. Calculate the head H_a at the throat.

   The head H_a for the standard 6-in throat is calculated from Equation 7.9.

   $$Q = CH_a^n$$

   where Q is in ft^3/s, and H_a is in ft

   For a Parshall flume throat width of 6 in (0.15 m), $C = 2.06$ and $n = 1.58$ are obtained from Table C.2.

   Calculate H_a.

   $$15.9 = 2.06 \times H_a^{1.58}$$

   $$H_a = \left(\frac{15.9}{2.06}\right)^{1/1.58} = 3.65\,\text{ft} \quad \text{or} \quad H_a = 3.65\,\text{ft} \times \frac{0.305\,\text{m}}{\text{ft}} = 1.11\,\text{m}$$

5. Calculate the approach velocity at the entrance of the Parshall flume.

   The dimension D (Figure C.1) of the standard 6 in (15 cm) Parshall flume in Table C.3 is 1 ft 3–5/8 in (1.3 ft) or 0.397 m.

   The approach velocity at the approach section, $v = \dfrac{Q}{A} = \dfrac{0.45\,\text{m}^3/\text{s}}{0.397\,\text{m} \times 1.11\,\text{m}} = 1.02\,\text{m/s}$

6. Calculate the water depth in the parabolic channel.

   Assume a minor loss coefficient $K_m = 0.1$ for the entrance.

   $$h_L = K_m\left(\frac{v^2}{2g} - \frac{v_h^2}{2g}\right)$$

   Apply Bernoulli's energy equation (Equation 7.3a) at the entrance of the flume, and assume the bottom elevation of the parabolic channel (Z) is the same as that at the entrance of the Parshall flume (z).

   $$Z + h + \frac{v_h^2}{2g} = z + H_a + \frac{v^2}{2g} + h_L = z + H_a + \frac{v^2}{2g} + K_m\left(\frac{v^2}{2g} - \frac{v_h^2}{2g}\right)$$

   $$Z = z = 0$$

   $$h + \frac{(0.30\,\text{m/s})^2}{2 \times 9.81\,\text{m/s}^2} = 1.11\,\text{m} + \frac{(1.02\,\text{m/s})^2}{2 \times 9.81\,\text{m/s}^2} + 0.1 \times \left(\frac{(1.02\,\text{m/s})^2}{2 \times 9.81\,\text{m/s}^2} - \frac{(0.30\,\text{m/s})^2}{2 \times 9.81\,\text{m/s}^2}\right)$$

   $$0.0046\,\text{m} + h = (0.053 + 1.11 + 0.0048)\,\text{m}$$

   $$h = 1.164\,\text{m}$$

7. Calculate the cross-sectional area A of the parabolic channel.

$$A = \frac{Q}{v_h} = \frac{0.45\,\text{m}^3/\text{s}}{0.30\,\text{m/s}} = 1.5\,\text{m}^2$$

8. Calculate the top width W of the parabolic channel.

The top width W of the parabola, $W = \frac{3A}{2h} = \frac{3 \times 1.5\,\text{m}^2}{2 \times 1.164\,\text{m}} = 1.93\,\text{m}$

9. Develop the cross-sectional profile of the parabolic channel.
   The cross-sectional profile is developed as follows:
   a. Select a series of head H'_a at the throat.
   b. Calculate the discharge q' through the flume from Equation 7.9.
   c. Perform steps 5–8 to calculate v', h', A', and W' at the assumed H'_a. The calculated values are summarized in Table 8.6.

TABLE 8.6 Tabulate Data to Determine the Cross-Sectional Profile of the Parabolic Channel (Example 8.13)

H'_a		q'		v', m/s	h', m	A', m^2	W', m
ft	m	ft^3/s	m^3/s				
0.50[a]	0.152[b]	0.689[c]	0.0195[d]	0.323[e]	0.153[f]	0.065[g]	0.64[h]
1.00	0.305	2.06	0.0583	0.482	0.313	0.194	0.93
1.50	0.457	3.91	0.111	0.610	0.473	0.369	1.17
2.00	0.610	6.16	0.174	0.721	0.634	0.581	1.38
2.50	0.762	8.76	0.248	0.820	0.795	0.827	1.56
3.00	0.914	11.7	0.331	0.912	0.956	1.10	1.73
3.50	1.067	14.9	0.422	0.997	1.118	1.41	1.89
3.65	1.111	15.9	0.450	1.02	1.164	1.50	1.93

[a] Assumed value

[b] $0.50\,\text{ft} \times \frac{0.3048\,\text{m}}{\text{ft}} = 0.152\,\text{m}$

[c] $q' = 2.06 \times 0.50^{1.58} = 0.689\,\text{ft}^3/\text{s}$

[d] $q' = 0.689\,\text{ft}^3/\text{s} \times \frac{0.0283\,\text{m}^3/\text{s}}{\text{ft}^3/\text{s}} = 0.0195\,\text{m}^3/\text{s}$

[e] $v' = \frac{0.0195\,\text{m}^3/\text{s}}{0.397\,\text{m} \times 0.152\,\text{m}} = 0.323\,\text{m/s}$

[f] $h' = \frac{(0.323\,\text{m/s})^2}{2 \times 9.81\,\text{m/s}^2} + 0.152\,\text{m} + 0.1 \times \left(\frac{(0.323\,\text{m/s})^2}{2 \times 9.81\,\text{m/s}^2} - \frac{(0.30\,\text{m/s})^2}{2 \times 9.81\,\text{m/s}^2}\right) - \frac{(0.30\,\text{m/s})^2}{2 \times 9.81\,\text{m/s}^2} = 0.153\,\text{m}$

[g] $A' = \frac{0.0195\,\text{m}^3/\text{s}}{0.30\,\text{m}} = 0.065\,\text{m}^2$

[h] $W' = \frac{3 \times 0.065\,\text{m}^2}{2 \times 0.153\,\text{m}} = 0.64\,\text{m}$

10. Calculate settling time θ.

$$\theta = \frac{h}{v_s} = \frac{1.164\,\text{m}}{19\,\text{mm/s}} \times \frac{10^3\,\text{mm}}{\text{m}} = 61.3\,\text{s}$$

11. Calculate length of parabolic channel.

Theoretical length $= v_h \times \theta = 0.30\,\text{m/s} \times 61.3\,\text{s} = 18.4\,\text{m}$

The design length $= 1.35 \times$ Theoretical length $= 1.35\,\text{m} \times 18.4\,\text{m} = 24.8\,\text{m}^2$

12. Select the parabolic section of the channel.

    The parabolic section of the channel is selected by plotting water depth in the channel h' versus channel width W'. These values are adopted from Table 8.6 and are summarized below.

h, m	0.15	0.31	0.47	0.63	0.80	0.96	1.12	1.16
W, m	0.64	0.93	1.17	1.38	1.56	1.73	1.89	1.93

    The cross section of the parabolic channel is shown in Figure 8.11.

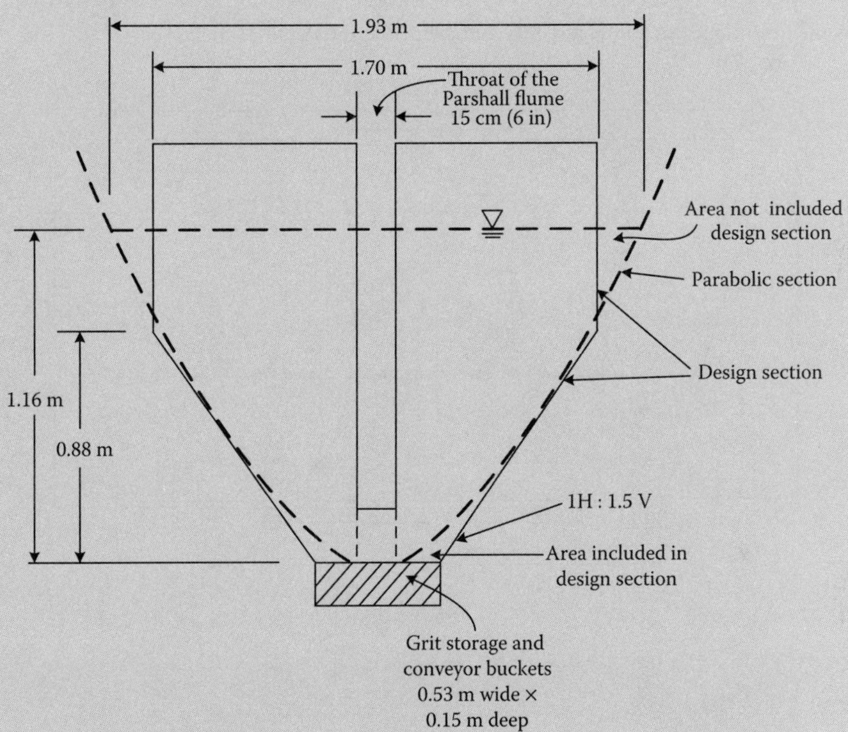

FIGURE 8.11 Cross section of the parabolic channel and the design section with rectangular base for grit collection and removal, and trapezoidal and rectangular sections of the channel (Example 8.13).

13. Select the design section of the channel.

    The parabolic channel is difficult to form and build. Therefore, for the ease of construction, a combination of trapezoidal and rectangular cross section is selected. The dimensions are such that the cross-sectional areas of both shapes are nearly equal. Also, at the bottom of trapezoidal section, a rectangular channel is provided for the grit storage and movement of the conveyor buckets. The channel is normally 15–30 cm (6–12 in) deep, and width depends upon the standard size of the equipment. For this design, select a 15-cm-deep and 0.53-m (21 in)-wide rectangular channel at the base of trapezoidal section. The widths of the trapezoidal section are 0.53 m (bottom) and 1.70 m (top). The depth is 0.88 m. The rectangular section of the channel is 1.70 m (wide) × 0.28 m (deep). The design details of the parabolic channel and Parshall flume are shown in Figure 8.11.

14. Check the cross-sectional area of the design channel section.

    Area of trapezoidal section $= (1/2) \times (0.53 + 1.70)\ \text{m} \times 0.88\ \text{m} = 0.98\ \text{m}^2$

    Area of rectangular section $= 1.70\ \text{m} \times 0.28\ \text{m} = 0.48\ \text{m}^2$

    Total cross-sectional area $= (0.98 + 0.48)\ \text{m}^2 = 1.46\ \text{m}^2$

    Note: This is close to 1.50 m^2 of the required area of parabolic section calculated in Step 7.

EXAMPLE 8.14: EQUATIONS FOR VELOCITY CONTROL BY A PROPORTIONAL WEIR IN A TRAPEZOIDAL CHANNEL

Constant velocity in a trapezoidal channel is achieved by designing a proportional weir of parabolic shape. Develop the generalized equation of a proportional weir of parabolic shape for a trapezoidal channel. For practical purpose, assume that the water head above the weir datum is approximately the same as the water depth in the channel.

Solution

1. Draw the definition sketch in Figure 8.12 and define the variables.[11,12]

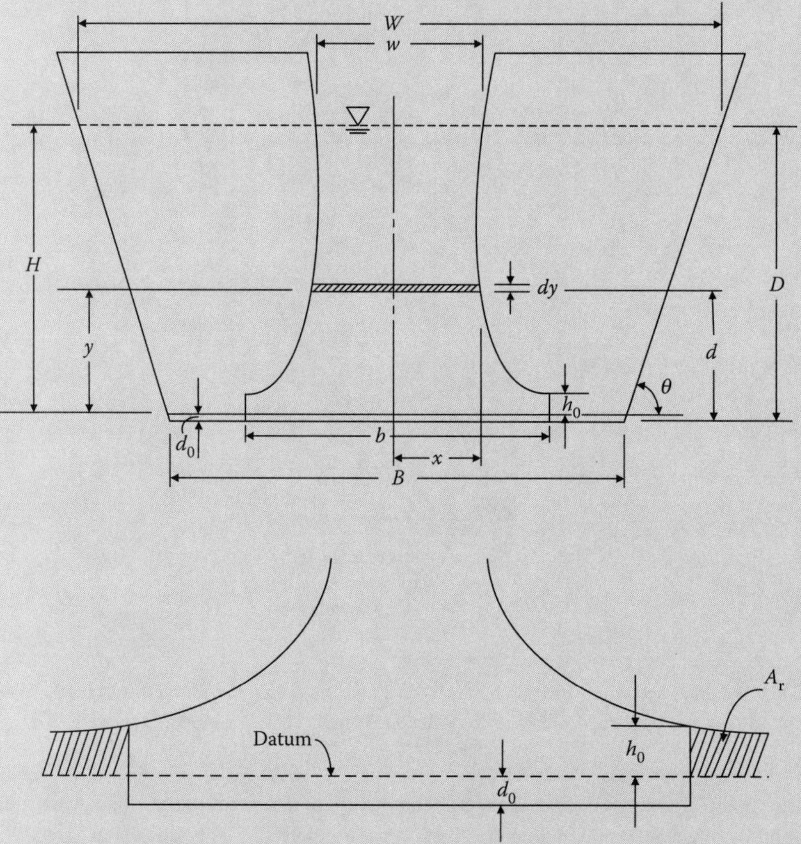

FIGURE 8.12 Definition sketch of velocity control in a trapezoidal channel by a proportional weir (Example 8.14).

Define the variables used in the development of the flow equations.

v = uniform velocity in the channel, m/s (ft/s)
B = bottom width of the channel, m (ft)
θ = angle of side slope, degree
H = maximum water head over weir datum, m (ft)
d_0 = bottom correction to be added below the weir datum, m (ft)
D = maximum water depth in the channel, which is approximately equal to the maximum water head over weir datum H, $D = H + d_0 \approx H$, m (ft)

w = weir width at the maximum water head H, m (ft)
W = channel width at the maximum water depth D, m (ft)
A = wetted cross-sectional area of the channel at the maximum water depth D, m^2 (ft^2)
Q_c = maximum channel flow at the maximum water depth d, m^3/s (cfs)
Q_w = total weir discharge at height y, m^3/s (cfs)
y = water head (or height of weir element) above weir datum, m (ft)
d = water depth in the channel, which is approximately equal to the height of weir element above weir datum y, $d = y + d_0 \approx y$, m (ft)
x = half width of weir at height y, m (ft)
q_c = channel flow at water depth d, m^3/s (cfs)
q_w = weir discharge at water head y, m^3/s (cfs)
q = generalized flow through either the weir or the channel at the approximate water head y, m^3/s (cfs)
A' = wetted cross-sectional area of the channel at water depth d, m^2 (ft^2)
C_d = weir discharge coefficient
k = first weir constant
a = second weir constant
k_1 = constant used during equation development
c_1, c_2 = variables used during equation development
h_0 = height above weir datum at cutoff, m (ft)
b = bottom width of the weir, m (ft)
A_r = rejected area on one side, m^2 (ft^2)

2. Develop the general channel flow equations (Equations 8.9a through 8.9f).

Maximum area of channel $A = BD + 2 \times \left(\frac{1}{2} \times D \times (D \cot\theta)\right) = BD + \cot\theta D^2$ (8.9a)

Maximum width of channel $W = B + 2 \times D\cot\theta$ (8.9b)

Maximum channel flow $Q_c = vA = (vB)D + (v\cot\theta)D^2$ (8.9c)

Assume $c_1 = (vB)$ and $c_2 = v\cot\theta$, Equation 8.9c is simplified by

$$Q_c = c_1 D + c_2 D^2 \quad (8.9d)$$

At a water depth d, the generalized channel flow from Equation 8.9d is expressed by

$$q_c = c_1 d + c_2 d^2 \quad (8.9e)$$

Area of channel from Equation 8.9a at water depth d, $A' = Bd + \cot\theta d^2$ (8.9f)

3. Develop the weir discharge equations.
Apply proportional weir equation (generalized Equation 7.8f) at the height of weir element y and the width of weir element $2x$ to obtain Equation 8.9g.

$$q_w = k_1 2xy^{3/2} \quad (8.9g)$$

Since the flow through the channel must be equal to the flow through the proportional weir ($q_c = q_w$), equate Equations 8.9e and 8.9g and rearrange the relationship to obtain Equation 8.9h.

$$k_1 2xy^{3/2} = c_1 d + c_2 d^2$$

It is given that $d = y + d_0 \approx y$ and gives Equation 8.9h.

$$k_1 2xy^{3/2} = c_1 y + c_2 y^2$$

$$2x = \frac{k}{\sqrt{y}} + a\sqrt{y} \tag{8.9h}$$

where $k = \frac{c_1}{k_1}$ and $a = \frac{c_2}{k_1}$

Discharge dq through the element dy at the height of weir element y is expressed by Equation 8.9i.

$$dq = C_d\sqrt{2g}\sqrt{(H-y)}(2x)dy \tag{8.9i}$$

Substitute the value of $2x$ in Equation 8.9i and expand it to obtain Equation 8.9j.

$$Q_w = C_d\sqrt{2g}\int_0^H \sqrt{(H-y)}\left(\frac{k}{\sqrt{y}} + a\sqrt{y}\right)dy$$

$$Q_w = C_d\sqrt{2g}\int_0^H \sqrt{(H-y)}\frac{k}{\sqrt{y}}dy + C_d\sqrt{2g}\int_0^H \sqrt{(H-y)}a\sqrt{y}dy \tag{8.9j}$$

The integral of Equation 8.9j is given by Equation 8.9k.

$$Q_w = C_d k\sqrt{2g}\left(\frac{\pi}{2}H\right) + C_d a\sqrt{2g}\left(\frac{\pi}{8}H^2\right) \tag{8.9k}$$

Substitute $C_d = 0.6$, Equation 8.9k is simplified to Equations 8.9l through 8.9n.

$$Q_w = 1.57C_d\sqrt{2g}kH + 0.393C_d\sqrt{2g}aH^2 \tag{8.9l}$$

$$Q_w = 4.17kH + 1.04aH^2 \quad \text{(SI units)} \tag{8.9m}$$

$$Q_w = 7.56kH + 1.89aH^2 \quad \text{(U.S. customary units)} \tag{8.9n}$$

Note that these equalities are used to determine the maximum flow rate Q_w through the weir at the maximum water head of h.

The generalized equations are expressed by Equations 8.9o through 8.9q.

$$q = 1.57C_d\sqrt{2g}ky + 0.393C_d\sqrt{2g}ay^2 \tag{8.9o}$$

$$q = 4.17ky + 1.04ay^2 \quad \text{(SI units)} \tag{8.9p}$$

$$q = 7.56ky + 1.89ay^2 \quad \text{(U.S. customary units)} \tag{8.9q}$$

Note that these equalities are used to obtain the generalized flow rate q through the weir or the channel at a water head of y above the weir datum. These equations are of the same form as Equation 8.9e.

4. Develop the general equations for obtaining constants k and a.

Since the flow through the proportional weir is equal to the flow through the channel, there must be $Q_w = Q_c$. Therefore, Equations 8.9c and 8.9l can be equated to obtain expression by Equation 8.9r.

$$1.57C_d\sqrt{2g}kH + 0.393C_d\sqrt{2g}aH^2 = (vB)D + (v\cot\theta)D^2$$

It is given that $D = H + d_0 \approx H$.

$$1.57C_d\sqrt{2g}kH + 0.393C_d\sqrt{2g}aH^2 = (vB)H + (v\cot\theta)H^2 \quad (8.9r)$$

Equate the coefficients on both sides in Equation 8.9r and substitute $C_d = 0.6$ to obtain expressions for k and a by Equations 8.9s through 8.9v.

$$vB = 1.57C_d\sqrt{2g}k = 4.17k \quad \text{or} \quad k = \frac{vB}{4.17} \quad \text{(SI units)} \quad (8.9s)$$

$$v\cot\theta = 0.393C_d\sqrt{2g}a = 1.04a \quad \text{or} \quad a = \frac{v\cot\theta}{1.04} \quad \text{(SI units)} \quad (8.9t)$$

$$vB = 1.57C_d\sqrt{2g}k = 7.56k \quad \text{or} \quad k = \frac{vB}{7.56} \quad \text{(U.S. customary units)} \quad (8.9u)$$

$$v\cot\theta = 0.393C_d\sqrt{2g}a = 1.89a \quad \text{or} \quad a = \frac{v\cot\theta}{1.89} \quad \text{(U.S. customary units)} \quad (8.9v)$$

When the design values of v, B and θ are known, and the values of k and a will be then determined. Using Equation 8.9h when values of k and a are known, the profile of the proportional weir (x and y) can be plotted.

5. Determine the equation for rejected area.

The cutoff end areas of the proportional weir are added by a rectangular area. The rejected area A_r on one side of the channel is expressed by Equation 8.9w.

$$A_r = \frac{1}{2}\int_0^{h_0}\left(\frac{k}{\sqrt{y}} + a\sqrt{y}\right)dy - \frac{b}{2}h_0 \quad (8.9w)$$

Integration of this equation gives Equation 8.9x.

$$A_r = k\sqrt{h_0} + \frac{1}{3}ah_0^{3/2} - \frac{b}{2}h_0 \quad (8.9x)$$

The rectangular area under the datum should have the same area as the rejected areas on both end of the proportional weir. This is expressed by Equation 8.9y.

$$d_0 b = 2A_r = 2 \times \left(k\sqrt{h_0} + \frac{1}{3}ah_0^{3/2} - \frac{b}{2}h_0\right) = 2k\sqrt{h_0} + \frac{2}{3}ah_0^{3/2} - bh_0 \quad (8.9y)$$

The rectangular area for the balanced discharge may vary from 90% to 99% of the area given by Equation 8.9w due to fluctuations in channel depth. It is suggested to use a coefficient of 0.96 for bottom correction to be added below the weir datum. Value of d_0 is therefore expressed by Equation 8.9z.

$$d_0 = 0.96\left[\frac{2k\sqrt{h_0} + \frac{2}{3}ah_0^{3/2} - bh_0}{b}\right] \quad (8.9z)$$

6. Describe the application of parabolic proportional weir.
   Equation 8.9o represents the generalized flow through a proportional weir of parabolic shape expressed for a trapezoidal channel. Note the following:
   a. If $a = 0$ in Equation 8.9o, the channel sides become vertical ($\theta = 90°$) and Equation 8.9o is of the same form Equation 7.8f. The design of the proportional weir for a rectangular channel has been presented earlier in Example 8.12.
   b. If $k = 0$ in Equation 8.9o, the channel becomes triangular ($B = 0$), and a parabolic weir maintains a constant velocity in a hypothetical triangular channel.

EXAMPLE 8.15: VELOCITY CONTROL WEIR IN A TRAPEZOIDAL CHANNEL

A trapezoidal channel is designed to carry a maximum flow of 7.44 cfs (ft^3/s). The design velocity in the channel is 1.0 fps (ft/s). The bottom width B is 2 ft and $\theta = 49°$. Draw the profile of the velocity control section.

Solution

1. Calculate the constants k and a.
   Calculate k and a using Equations 8.9u and 8.9v.

$$k = \frac{vB}{7.56} = \frac{1 \times 2}{7.56} = 0.265$$

$$a = \frac{v \cot \theta}{1.89} = \frac{1 \times \cot(49°)}{1.89} = \frac{1 \times 0.869}{1.89} = 0.460$$

2. Calculate the maximum water head H above the weir datum.
   The maximum water head H is calculated from Equation 8.9n at $Q_w = 7.44$ cfs.

$$7.44 = 7.56 \times 0.265H + 1.89 \times 0.460H^2$$

$$H^2 + 2.30H - 8.56 = 0$$

   Solve the equation and obtain $H = 1.99$ ft.
3. Calculate the weir width W at the maximum water head H.
   At $y = H = 1.99$ ft, calculate $2x = w$ from Equation 8.9h.

$$w = \frac{0.265}{\sqrt{1.99}} + 0.460\sqrt{1.99} = 0.84 \text{ ft}$$

4. Calculate the bottom width of the weir b.
   Assume $y = h_0 = 1.5$ in (0.125 ft), calculate $2x = b$ from Equation 8.9h.

$$b = \frac{0.265}{\sqrt{0.125}} + 0.460\sqrt{0.125} = 0.91 \text{ ft}$$

5. Develop the profile data of velocity control weir.
   Equation 8.9h is used to develop the profile of the proportional weir (x and y) with $k = 0.265$ and $a = 0.460$.

$$2x = \frac{0.265}{\sqrt{y}} + 0.460\sqrt{y}$$

Assume different water head of y. At a constant interval $Dy = 0.25$ ft, select water heads between $h_0 = 0.125$ ft (1.50 in) and $H = 1.99$ ft. Calculate the weir width ($2x$) from Equation 8.9h at each water head. The profile is given in Table 8.7.

TABLE 8.7 Tabulate Data to Determine the Profile of Velocity Control Weir (Example 8.15)

Assumed Water Head (y), ft	Weir Width (x) at Water Head (y) from Equation 8.9h, ft
1.99 (H)	0.84 (w)
1.75	0.81
1.50	0.78
1.25	0.75
1.00	0.72
0.75	0.70
0.50	0.70
0.25	0.76
0.15	0.86
0.125 (h_0)	0.91 (b)

6. Calculate the bottom correction d_0.

   At $h_0 = 0.125$ ft and $b = 0.91$ ft, calculate d_0 from Equation 8.9z.

$$d_0 = 0.96 \times \left[\frac{2 \times 0.265\sqrt{0.125} + \frac{2}{3} \times 0.460 \times (0.125)^{3/2} - 0.91 \times 0.125}{0.91}\right] = 0.092\,\text{ft} \quad \text{or} \quad 1.10\,\text{in}$$

7. Calculate the maximum water depth D in the channel.

   Calculate the maximum water depth from $H = 1.99$ ft and $d_0 = 0.092$ ft.

$$D = H + d_0 = (1.99 + 0.092)\,\text{ft} = 2.08\,\text{ft}$$

TABLE 8.8 Validation of Velocities in the Channel (Example 8.15)

Assumed Water Head (y), ft	Weir (or channel) Flow (q) at the Water Head (y) from Equation 8.9q, cfs	Water Depth (d), ft	Channel Area (A') at the Water Head (d) from Equation 8.9f, ft^2	Velocity in the Channel (v) at the Water Depth (d), fps
1.99 (H)	7.44 (Q_c or Q_w)	2.08 (D)	7.95	0.94
1.75	6.16	1.84	6.63	0.93
1.50	4.96	1.59	5.38	0.92
1.25	3.86	1.34	4.25	0.91
1.00	2.87	1.09	3.22	0.89
0.75	1.99	0.84	2.30	0.87
0.50	1.22	0.59	1.49	0.82
0.25	0.55	0.34	0.78	0.71
0.15	0.32	0.24	0.53	0.60
0.125 (h_0)	0.32	0.22	0.47	0.56

Note: Although the flow varies greatly, the velocity in the channel remains nearly constant and slightly below the design velocity of 1.0 fps. Moderate deviations occur at flows below 1.99 cfs.

8. Calculate the channel width W at the maximum water depth D.
   Calculate the maximum channel width from Equation 8.9b at $D = 2.08$ ft.

$$W = B + 2 \times D\cot\theta = 2\,\text{ft} + 2 \times 2.08\,\text{ft} \times \cot(49°) = 2\,\text{ft} + 2 \times 2.08\,\text{ft} \times 0.869 = 5.62\,\text{ft}$$

9. Verify the velocity at different water depth in the channel.
   At each water depth selected in Table 8.7, calculate the weir (or channel) flow, water depth, channel area, and velocity. The calculation results are summarized in Table 8.8.
10. Draw the profile of velocity control weir and the cross section of trapezoidal channel in Figure 8.13.

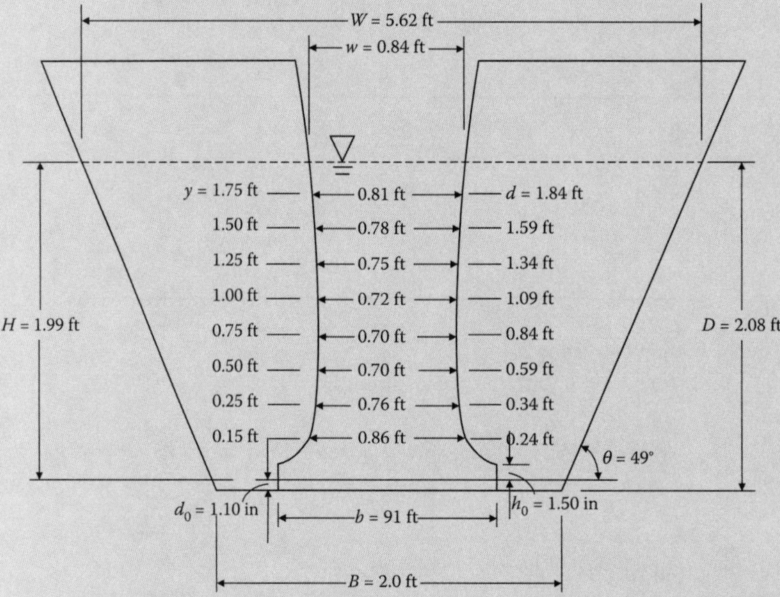

FIGURE 8.13 Proportional weir profile for velocity control in a trapezoidal channel (Example 8.15).

EXAMPLE 8.16: DESIGN OF A SQUARE HORIZONTAL-FLOW DEGRITTER

Design a square degritter tank for a flow of 0.56 m^3/s. The design overflow rate is 950 $m^3/m^2{\cdot}d$ to remove 0.21-mm (70-mesh) particles at 15°C. The detention time is 2 min. The design should include dimension of the basin and the influent and effluent structures.

Solution

1. Calculate the dimensions of square degritter tank.

$$\text{Design flow} = 0.56\,\text{m}^3/\text{s} \times \frac{(60 \times 60 \times 24)\,\text{s}}{\text{d}} = 48{,}384\,\text{m}^3/\text{d}$$

$$\text{Surface area} = \frac{48{,}384\,\text{m}^3/\text{d}}{950\,\text{m}^3/\text{m}^2{\cdot}\text{d}} = 50.9\,\text{m}^2$$

For a square tank, the required length of each side, $L = \sqrt{A} = \sqrt{50.9\,\text{m}^2} = 7.1\,\text{m}$
Provide a design length of 7 m.

Actual surface area, $7\text{ m} \times 7\text{ m} = 49\text{ m}^2$

$$\text{Settling velocity in tank} = \frac{0.56\text{ m}^3/\text{s}}{49\text{ m}^2} = 0.0114\text{ m/s}$$

At a detention time of 2 min, the required depth of tank $= 2\text{ min} \times \frac{60\text{ s}}{\text{min}} \times 0.0114\text{ m/s} = 1.37\text{ m}$

Provide a design side water depth of 1.4 m.

Consider an additional depth of 300 mm (0.3 m) for the bottom slope and raking mechanism.

Total depth water in the tank $= (1.4 + 0.3)\text{ m} = 1.7\text{ m}$

The dimensions of the basin are: 7 m × 7 m × 1.7 m (deep).

2. Design the influent structure.

The influent structure consists of an influent box 1 m × 1 m, and a tapering channel that has a maximum width of 1 m. There are 15 submerged 0.3-m diameter orifices distributed along the length of the channel. These orifices discharge the influent into the tank 0.5 m below the water surface in the tank. The design details are shown in Figure 8.14.

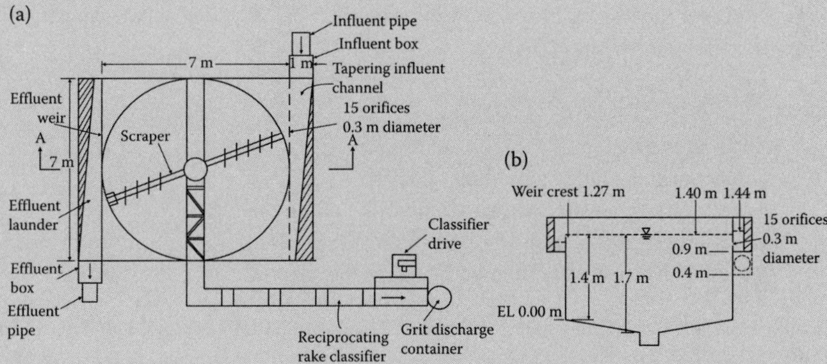

FIGURE 8.14 Design details of the detritus tank: (a) plan view and (b) sectional view (Section AA) (Example 8.16).

a. Calculate total area of the influent orifices.

$$\text{Area of each orifice} = \frac{\pi}{4}(0.3\text{ m})^2 = 0.0707\text{ m}^2$$

Total area of 15 orifices $= 15 \times 0.0707\text{ m}^2 = 1.06\text{ m}^2$

b. Calculate velocity through the influent orifices.

$$\text{Velocity} = \frac{0.56\text{ m}^3/\text{s}}{1.06\text{ m}^2} = 0.53\text{ m/s} \quad \text{or} \quad 1.7\text{ ft/s}$$

At this velocity, the flow will be distributed uniformly along the length of the influent channel.

c. Calculate head loss through the influent orifices.

The head loss is calculated from orifice equation (Equation 7.4a) and assume $C_d = 0.6$.

$$H = \frac{1}{2g}\left(\frac{Q}{C_d A}\right)^2 = \frac{1}{2 \times 9.81\text{ m/s}^2} \times \left(\frac{0.56\text{ m}^3/\text{s}}{0.6 \times 1.06\text{ m}^2}\right)^2 = 0.04\text{ m}$$

Head loss through the influent orifices $H = 0.04$ m.

3. Design the effluent structure.

   The effluent structure consists of a rectangular weir across the basin. The weir elevation maintains the desired water depth in the chamber. The flow over the weir has a free fall into a rectangular effluent trough that collects the effluent and carries it to an outfall channel (Figure 8.14). The design of effluent trough is complex and is discussed in details in the next section.

   a. Calculate the head over effluent weir.

      The effluent weir is placed along the opposite side of the in fluent channel.

      Head over the weir is calculated from the rectangular weir equation as expressed in Equation 8.10.

$$Q = \frac{2}{3} \times C_d L' \sqrt{2gH^3} \tag{8.10}$$

where

Q = flow over the weir, m^3/s (ft^3/s)

C_d = weir discharge coefficient, dimensionless. The value of C_d may vary between 0.6 and 0.95

H = head over the weir corrected to include the effect of velocity of approaching, m (ft)

L' = effective weir length, m (ft)

$L' = L - 0.1nH$

where

L = length of the weir, m (ft)

n = number of end contractions, dimensionless

$n = 0$ for weir that stretches across the width of the channel

$n = 1$ for weir that is flush ed on one side of the channel

$n = 2$ for weir that away from both ends of the channel

Calculated the head over the effluent weir from Equation 8.10 at $C_d = 0.6$, $L = 7$ m, and $n = 0$.

$$L' = L - 0.1nH = L = 7\,\text{m}$$

$$H = \left(\frac{3}{2} \times \frac{Q}{C_d L' \sqrt{2g}}\right)^{2/3} = \left(\frac{3}{2} \times \frac{0.56\,\text{m}^3/\text{s}}{0.6 \times 7\,\text{m} \times \sqrt{2 \times 9.81\,\text{m/s}^2}}\right)^{2/3} = 0.13\,\text{m}$$

Head over the effluent weir $H = 0.13$ m.

4. Describe the grit handling process.

   The settleable solids settle to the bottom of the tank. They are moved up an incline over a reciprocating rake mechanism for degritting. The organic solids are separated from the grit and fall back into the basin and carried in the effluent.

5. Draw the hydraulic profile.

   The hydraulic profile through the detritus tank is shown in Figure 8.14b.

8.4.2 Design of Effluent Trough

Like the influent zone, the *outlet zone* or *effluent structure* has a significant influence on the flow pattern and settling behavior of particles in a settling basin.

When a trough, launder, or flume with a free-falling weir is used to collect effluent, the flow rate varies throughout its entire length. The flow is maximum at the discharge point, while zero at the upper most point. The water surface profile is shown in Figure 8.15. Two methods have been developed to deal with the variable water depth in the trough and they are summarized below.[13–16]

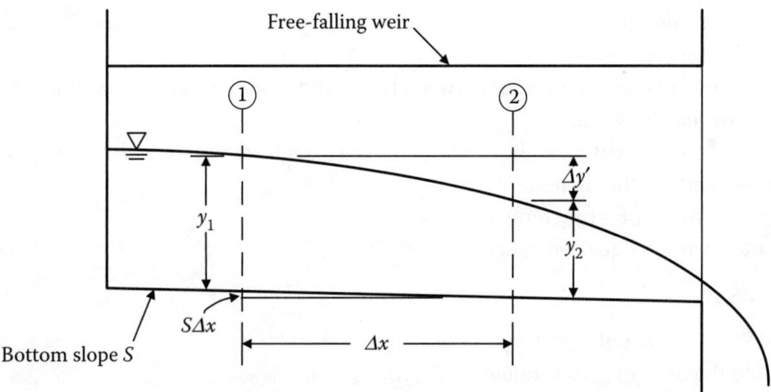

FIGURE 8.15 Water surface elevation in flume receiving weir discharge.

Increment Method to Calculate the Depth of Effluent Launder: For uniform velocity distribution, Chow expressed the drop in the water surface elevation between sections (1) and (2) by Equation 8.11a.[13]

$$\Delta y' = \frac{q_1 v_{avg}}{g q_{avg}}\left[\Delta v + \frac{v_2}{q_1}\Delta q\right] + (S_E)_{avg}\Delta x \tag{8.11a}$$

where

$\Delta y'$ = drop in water surface elevation between sections (1) and (2) ($\Delta y' = y_1 - y_2$), m (ft)
y_1 and y_2 = depths of flow at sections (1) and (2), respectively, m (ft)
q_1 and q_2 = discharges at sections (1) and (2), respectively, m^3/s (ft^3/s)
v_1 and v_2 = velocities at sections (1) and (2), respectively, m/s (ft/s)
v_{avg} = average velocity $\left(v_{avg} = \frac{1}{2}(v_1 + v_2)\right)$, m/s (ft/s)
q_{avg} = average discharge $\left(q_{avg} = \frac{1}{2}(q_1 + q_2)\right)$, m^3/s (ft^3/s)
Δv = velocity differential between sections (1) and (2) ($\Delta v = v_2 - v_1$), m/s (ft/s)
Δq = flow differential between sections (1) and (2) ($\Delta v = q_2 - q_1$), m^3/s (ft^3/s)
Δx = horizontal distance between sections (1) and (2), m (ft)
g = acceleration due to gravity, m/s^2 (ft/s^2)
$(S_E)_{avg}$ = average slope of the energy line, m/m (ft/ft). The value of $(S_E)_{avg}$ is obtained from Equation 8.11b or 8.11c

$$(S_E)_{avg} = \frac{n^2(v_{avg})^2}{(R_{avg})^{4/3}} \quad \text{(SI units)} \tag{8.11b}$$

$$(S_E)_{avg} = \frac{n^2(v_{avg})^2}{2.21(R_{avg})^{4/3}} \quad \text{(U.S. customary units)} \tag{8.11c}$$

where

n = Coefficient of roughness used in Manning equation
R_{avg} = average hydraulic mean radius $\left(R_{avg} = \frac{1}{2}(R_1 + R_2)\right)$, m (ft)

Note: The terms n and R were defined previously in Section 6.3.7.

The computational procedure for obtaining the depth of flow in the trough at the upstream section is given below.[8]

1. Determine the depth of flow (y_2) at the lower end of trough which is generally fixed by the downstream control conditions.
2. Select an incremental distance Δx between two sections for computational purpose. Δx may be a constant or may be varied.
3. Take the first increment Δx at the lower end of the trough having water depths of y_2 and y_1 at lower and upper ends of the increment.
4. Assume $\Delta y'$ for the first increment.
5. Compute Δy from Equation 8.12a.

$$\Delta y = S\Delta x - \Delta y' \tag{8.12a}$$

   where S = channel bottom slope, m/m
6. Compute the depth y_1 from Equation 8.12b.

$$y_1 = y_2 - Dy \tag{8.12b}$$

7. Determine discharge q_2 and q_1 below and above the selected incremental distance Δx.
8. Compute velocity v_2 and v_1.
9. Use Equation 8.11a to compute $\Delta y'$.
10. If the difference between the values of $\Delta y'$ assumed in Step 4 and computed in Step 9 is not within certain tolerance level set by the designer, repeat Steps 4 through 9.
11. After balancing the two sections, repeat Steps 3 through 10. At this time, the computed depth y_1 at the upper end of the previous selected increment becomes the water depth y_2 at the lower end of the newly selected increment.

Benefield et al. provided a computational scheme and solution of water surface profile for a lateral spillway channel receiving uniformly distributed flow along the entire channel length.[13] A similar solution using Equations 8.11a through 8.12b for point load discharges along the effluent channel is presented in Example 8.17. A computer program is also given in Appendix B of Reference 15.

Approximate Method to Calculate the Depth of Effluent Launder: The use of above procedure by the increment method is tedious and time consuming. Widely used practice by the designers is to utilize an approximate solution given by Equation 8.13a. This equation was originally developed for flumes with level inverts and parallel sides; channel friction is neglected; and the draw-down curve is assumed parabolic.[1,5] Equation 8.13b also gives approximate solution but the channel friction loss is included.[5,13,16]

$$y_1 = \sqrt{y_2^2 + \frac{2Q^2}{gb^2y_2}} \tag{8.13a}$$

$$y_1 = \sqrt{y_2^2 + \frac{2Q^2}{gb^2y_2} + \frac{2}{3}\frac{n^2LQ^2}{c^2b^2R_{\text{mean}}^{4/3}y_{\text{mean}}}} \tag{8.13b}$$

where

y_1 = water depth at the upstream end, m (ft)
y_2 = water depth in the trough at a distance L from the upstream end, m (ft)
Q = total discharge at the downstream section, m^3/s (ft^3/s)
b = width of the launder, m (ft)
g = acceleration due to gravity, m/s^2 (ft/s^2)
L = channel length, m (ft)
c = constant
c = 1.0 $m^{1/3}/s$ for SI units and c = 1.49 $ft^{1/3}/s$ for the U.S. customary units
n = coefficient of roughness used in Manning equation
y_{mean} = mean water depth, m (ft)

y_{mean} is calculated from Equation 8.13c.

$$y_{mean} = y_1 - \frac{1}{3}(y_1 - y_2) \tag{8.13c}$$

R_{mean} = mean hydraulic radius, m (ft)

R_{mean} is calculated from Equation 8.13d.

$$R_{mean} = \frac{b y_{mean}}{b + 2y_{mean}} \tag{8.13d}$$

EXAMPLE 8.17: DEPTH OF EFFLUENT TROUGH CALCULATED FROM BACKWATER EQUATIONS

The depth of effluent trough is calculated to assure that the effluent structure is not inundated under peak flow. A horizontal effluent launder is 7 m long and 1 m wide. It receives a design peak flow of 0.56 m^3/s from 90° V-notches located on both sides of the trough. The V-notches are 0.3 m above the maximum expected water surface in the trough. Calculate (a) the water depth at the upper end of the trough and (b) the height of the V-notches above the trough bottom. Show the hydraulic profile. Assume the bottom slope of the trough is ignorable ($S \approx 0$) and Manning's coefficient $n = 0.013$. Use a tolerance limit of 0.01 m in the water surface profile calculations.

Solution

1. Calculate the critical depth at the free fall of effluent trough.

   Critical depth y_c is calculated from Equation 7.10 with $C_w = 1.0$ at $Q = 0.56\ m^3/s$ and width of trough, $b = 1$ m.

   $$Q = C_w b \sqrt{g} d_c^{3/2}$$

   $$0.56\ m^3/s = 1.0 \times 1\ m \times \sqrt{9.81\ m/s^2} \cdot d_c^{3/2}$$

   $$d_c = 0.32\ m$$

   Critical depth d_c occurs near the outfall end and assume $y_2 = d_c$.

2. Calculate the unit flow over the effluent weir.

   $$\text{Flow per unit length } q' = \frac{0.56\ m^3/s}{7\ m} = 0.08\ m^3/m{\cdot}s$$

3. Calculate the upstream water depth of the selected first increment from the outfall end.

   a. Determine the depth of flow y_2 at the lower end of trough.

      Select the critical depth at the free fall as the depth at the lower end of trough y_2.

      $$y_2 = d_c = 0.32\ m$$

   b. Select a distance for the first increment.

      Assume x is the length from the upper to the lower end of the trough, and then $x_2 = 7$ m. Select $\Delta x = 1.5$ m and then $x_1 = x_2 - \Delta x = (7 - 1.5)\ m = 5.5$ m.

   c. Assume $\Delta y'$ for the first increment.

      Assume $\Delta y' = 0.16$ m.

      Note: The assumed value will be checked later with that calculated from Equation 8.11a to ensure the tolerance limit is not exceeded.

   d. Calculate Δy from Equation 8.12a.

      $$\Delta y = S\Delta x - \Delta y' = 0 \times 1.5\ m - 0.16\ m = -0.16\ m$$

e. Compute the depth y_1 from Equation 8.12b.

$$y_1 = y_2 - \Delta y = 0.32\,\text{m} - (-0.16\,\text{m}) = 0.48\,\text{m}.$$

f. Calculate the discharge q_2, q_1, q_{avg}, and Δq.

$$q_2 = Q = 0.56\,\text{m}^3/\text{s}$$

$$q_1 = q_2 - q'\Delta x = 0.56\,\text{m}^3/\text{s} - 0.08\,\text{m}^3/\text{m·s} \times 1.5\,\text{m} = 0.44\,\text{m}^3/\text{s}$$

$$q_{avg} = \frac{1}{2}(q_1 + q_2) = \frac{1}{2} \times (0.44 + 0.56)\,\text{m}^3/\text{s} = 0.50\,\text{m}^3/\text{s}$$

$$\Delta q = q_2 - q_1 = (0.56 - 0.44)\,\text{m}^3/\text{s} = 0.12\,\text{m}^3/\text{s}$$

g. Calculate the velocity v_2, v_1, v_{avg}, and Δv.

$$v_2 = \frac{q_2}{y_2 b} = \frac{0.56\,\text{m}^3/\text{s}}{0.32\,\text{m} \times 1.0\,\text{m}} = 1.75\,\text{m/s}$$

$$v_1 = \frac{q_1}{y_1 b} = \frac{0.44\,\text{m}^3/\text{s}}{0.48\,\text{m} \times 1.0\,\text{m}} = 0.92\,\text{m/s}$$

$$v_{avg} = \frac{1}{2}(v_1 + v_2) = \frac{1}{2}(0.92\,\text{m/s} + 1.75\,\text{m/s}) = 1.34\,\text{m/s}$$

$$\Delta v = v_2 - v_1 = (1.75 - 0.92)\,\text{m/s} = 0.83\,\text{m/s}$$

h. Calculate the hydraulic mean radius R_2, R_1, and R_{avg}.

$$R_2 = \frac{by_2}{b + 2y_2} = \frac{1.0\,\text{m} \times 0.32\,\text{m}}{1.0\,\text{m} + 2 \times 0.32\,\text{m}} = 0.20\,\text{m}$$

$$R_1 = \frac{by_1}{b + 2y_1} = \frac{1.0\,\text{m} \times 0.48\,\text{m}}{1.0\,\text{m} + 2 \times 0.48\,\text{m}} = 0.24\,\text{m}$$

$$R_{avg} = \frac{1}{2}(R_1 + R_2) = \frac{1}{2}(0.20\,\text{m} + 0.24\,\text{m}) = 0.22\,\text{m}$$

i. Calculate $(S_E)_{avg}$ from Equation 8.11b.

$$(S_E)_{avg} = \frac{n^2(v_{avg})^2}{(R_{avg})^{4/3}} = \frac{(0.013)^2(1.34\,\text{m/s})^2}{(0.22\,\text{m})^{4/3}} = 0.0023\,\text{m/m} \quad \text{or} \quad 2.3\,\text{mm/m}$$

j. Calculate $\Delta y'$ from Equation 8.11a.

$$\Delta y' = \frac{q_1 v_{avg}}{g q_{avg}}\left[\Delta v + \frac{v_2}{q_1}\Delta q\right] + (S_E)_{avg}\Delta x$$

$$= \frac{0.44\,\text{m}^3/\text{s} \times 1.34\,\text{m/s}}{9.81\,\text{m/s}^2 \times 0.50\,\text{m}^3/\text{s}}\left[0.83\,\text{m/s} + \frac{1.75\,\text{m/s}}{0.44\,\text{m}^3/\text{s}} \times 0.12\,\text{m}^3/\text{s}\right] + 0.0023\,\text{m/m} \times 1.5\,\text{m}$$

$$= 0.161\,\text{m}$$

k. Check the tolerance limit.

$$\Delta y'(\text{assumed}) - \Delta y'(\text{calculated}) = (0.16 - 0.161)\,\text{m} = -0.001\,\text{m} < 0.01\,\text{m}\,(\text{tolerance limit})$$

The assumed $\Delta y'$ of 0.16 m is within the tolerance limit of 0.01 m. Therefore, the water depth y_1 of 0.46 m at the upper end of the first increment is selected and used as the depth y_2 at the lower end of the second increment. The calculations are repeated for the remaining increments and the results are summarized in Table 8.9.

TABLE 8.9 Summary of Calculation Results to Determine the Water Surface Profile in the Effluent Trough (Example 8.17)

Parameter	Increment				Equation Used
	1	2	3	4	
x_2, m	7.00	5.50	4.00	2.50	
Δx, m	1.50	1.50	1.50	2.45	
x_1, m	5.50	4.00	2.50	0.05	
y_2, m	0.32	0.48	0.53	0.56	
$\Delta y'$ (assumed), m	0.16	0.050	0.030	0.015	
Δy, m	−0.16	−0.05	−0.03	−0.02	Equation 8.12a
y_1, m	0.48	0.53	0.56	0.57	Equation 8.12b
q_2, m^3/s	0.56	0.44	0.32	0.20	
q_1, m^3/s	0.44	0.32	0.20	0.00	
q_{avg}, m^3/s	0.50	0.38	0.26	0.10	
Δq, m^3/s	0.12	0.12	0.12	0.20	
v_2, m/s	1.76	0.92	0.61	0.36	
v_1, m/s	0.92	0.61	0.36	0.0070	
v_{avg}, m/s	1.34	0.76	0.48	0.18	
Δv, m/s	0.83	0.31	0.25	0.35	
R_1, m	0.24	0.26	0.26	0.27	
R_2, m	0.20	0.24	0.26	0.26	
R_{avg}, m	0.22	0.25	0.26	0.27	
$(S_E)_{avg}$, mm/m	2.3	0.63	0.24	0.033	Equation 8.11b
$\Delta y'$ (calculated), m	0.161	0.044	0.024	0.013	Equation 8.11a
$\Delta y'$ (assumed) − $\Delta y'$ (calculated)[a], m	−0.001	0.006	0.006	0.002	

[a] The difference between calculated and assumed values in each increment is less than the tolerance limit of 0.01 m.

4. Estimate the water depth at the upper end of the trough.
   The calculated water depth y_1 at the upper end of the trough is 0.57 m.
5. Determine the height of V-notches above the bottom of effluent trough.
   Assume a free fall allowance of 0.3 m above the water surface in the trough for all V-notches.
   The bottom height of V-notches $= y_1 +$ free fall $= (0.57 + 0.3)$ m $= 0.87$ m
6. Draw the hydraulic profile of free water surface in the effluent trough.
   The design details of the effluent trough are shown in Figure 8.16.

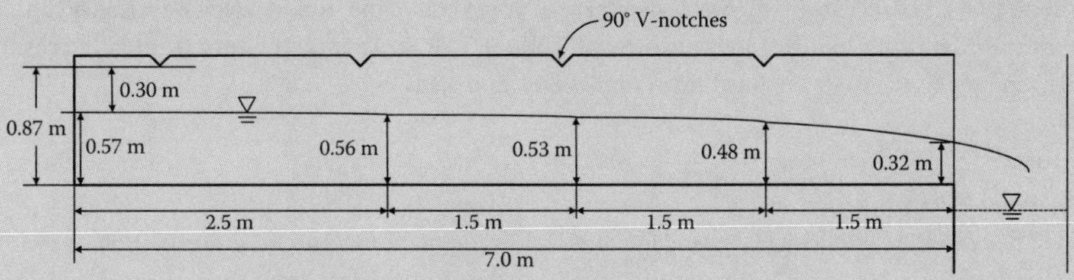

FIGURE 8.16 Design details of effluent trough (Example 8.17).

EXAMPLE 8.18: DEPTH OF EFFLUENT LAUNDER CALCULATED FROM EQUATION 8.12a

The procedure to determine the depth of effluent launder from variable flow equations is tedious and time consuming. The approximate solution using Equation 8.13a is most commonly used by the designers. Apply the data of Example 8.17 and calculate (a) the water depth at the upper end of the launder using Equation 8.13a and (b) the height of the V-notches above the trough bottom. Compare the result obtained in this example with that in Example 8.17. Since Equation 8.13a does not include head loss due to channel friction, add 4% additional depth to the calculated water depth at the upper end for additional loss due to friction and turbulence.

Solution

1. Calculate the critical depth at the discharge point of the effluent trough.

   The critical depth d_c is calculated in Step 1 of Example 8.17. The water depth y_2 at the outfall section is also equal to critical depth.

   $y_2 = d_c = 0.32\,\text{m}$

2. Calculate the water depth at the upper end of the trough (without channel friction loss).

   Use Equation 8.13a to obtain the water depth y_1 at the upper end.

$$y_1 = \sqrt{y_2^2 + \frac{2Q^2}{gb^2y_2}} = \sqrt{(0.32\,\text{m})^2 + \frac{2 \times (0.56\,\text{m}^3/\text{s})^2}{9.81\,\text{m/s}^2 \times (1\,\text{m})^2 \times 0.32\,\text{m}}} = 0.55\,\text{m}$$

3. Calculate the water depth with channel friction loss.

   The water depth y_1 calculated from Equation 8.13a is without additional losses due to friction and turbulence. Depending upon channel length, width, and configuration, the head loss may be 0%–15% of y_1 calculated from Equation 8.13a. The additional loss given in this problem is 4%.

   Water depth at the upper end $= (0.55 + 0.04 \times 0.55)\,\text{m} = 0.57\,\text{m}$

4. Calculate the height of V-notch above the bottom of effluent launder.

   Use the same free fall allowance of 0.3 m that is assumed in Example 8.17.

   The height of V-notches $= y_1 + \text{free fall} = (0.57 + 0.3)\,\text{m} = 0.87\,\text{m}$

5. Compare the result with that in Example 8.17.

   The height of V-notches above the launder bottom calculated in this example is the same as that obtained in Example 8.17.

EXAMPLE 8.19: DEPTH OF THE EFFLUENT LAUNDER CALCULATED FROM EQUATION 8.12b

Equation 8.13b gives approximate result but includes the channel friction loss. Apply the data of Examples 8.17 and 8.18 and calculate (a) the water depth at the upper end of the launder using Equation 8.13b, and (b) the height of the V-notches above the trough bottom. Use a tolerance limit of 0.01 m in the calculations. Compare the results obtained in Examples 8.17 and 8.18.

Solution

1. State the calculation procedure.

   The procedure to calculate y_1 using Equation 8.13b constitutes a trial-and-error solution. In the first trial, the friction loss term is ignored and water depth y_1 is calculated. From the calculated value of y_1, the mean hydraulic radius R_{avg} and mean depth y_{avg} are calculated. In the second trial, the friction loss term is included for calculation of y_1 from Equation 8.13b. The procedure is repeated until a stable value is reached.

2. Calculate y_1 by ignoring the friction loss term.

   Equation 8.13a and 8.13b are identical if friction term is ignored. Therefore, y_1 calculated in Example 8.18 is the one desired in this calculation step.

   $y_1 = 0.55\,\text{m}$

3. Calculate y1 including the friction loss term.

   Calculate mean depth y_{mean} from Equation 8.13c.

$$y_{mean} = y_1 - \frac{1}{3}(y_1 - y_2) = 0.55\,\text{m} - \frac{1}{3} \times (0.55 - 0.32)\,\text{m} = 0.473\,\text{m}$$

   Calculate mean hydraulic radius R_{mean} from Equation 8.13d.

$$R_{mean} = \frac{by_{mean}}{b + 2y_{mean}} = \frac{1\,\text{m} \times 0.473\,\text{m}}{1\,\text{m} + 2 \times 0.473\,\text{m}} = 0.243\,\text{m}$$

   Use Equation 8.13b to obtain y_1.

$$y_1 = \sqrt{y_2^2 + \frac{2Q^2}{gb^2y_2} + \frac{2}{3}\frac{n^2LQ^2}{c^2b^2R_{mean}^{4/3}y_{mean}}}$$

$$= \sqrt{(0.32\text{m})^2 + \frac{2 \times (0.56\text{m}^3/\text{s})^2}{9.81\text{m/s}^2 \times (1\text{m})^2 \times 0.32\text{m}} + \frac{2}{3} \times \frac{(0.013)^2 \times 7\text{m} \times (0.56\text{m}^3/\text{s})^2}{(1\text{m}^{1/3}/\text{s})^2 \times (1\text{m})^2 \times (0.243\text{m})^{4/3} \times 0.473\text{m}}}$$

$$= 0.553\text{m}$$

   The difference between the values of y_1 calculated from Equation 8.13a and 8.13b is 0.003 m, which is less than the tolerance limit of 0.01 m. Therefore, the water depth $y_1 = 0.55$ m at the upper end of the launder.

4. Calculate the height of V-notch above the bottom of effluent launder.

   Use the same free fall allowance of 0.3 m that is assumed in Example 8.17.

   The height of V-notches $= y_1 +$ free fall $= (0.55 + 0.3)$ m $= 0.85$ m

5. Compare the results.

   The head loss due to channel friction obtained from Equation 8.13b is small and is neglected.

   Therefore, the height of the V-notches above the launder bottom is lower than those obtained in Examples 8.17 and 8.18. For a short launder, turbulence is a more important factor than friction for the additional head loss.

8.4.3 Aerated Grit Chamber

Traditionally, aerated grit chambers utilize diffused air along one longitudinal side of the rectangular basin. A spiral rolling action within the chamber similar to that in a standard spiral-flow aeration basin is created. The air rate is adjusted to create a low lateral velocity near the bottom to settle the grit and roll it over the sloping bottom into a grit collection channel. The lighter organic particles are carried with the roll and eventually out of the basin. The unit is designed to target particles of specific gravity 2.5 and diameter 0.21 mm (70-mesh), although smaller particles may also be removed effectively by reducing the air supply.

In recent years, different configurations have also been seen in aerated grit removal system design. Integration of a calm settling chamber or zone has been made when effective removal of grease is also desired.

An innovative educator tube aerator may also be used to create the rolling action in a square-shaped grit chamber. The helical rolling action and cross sections in different types of aerated grit removal systems are shown in Figure 8.17.[2,17–19]

Advantages and Disadvantages: The aerated grit chamber offers many advantages and disadvantages over horizontal grit chambers. Some of the advantages are: (1) these chambers can be used for chemical addition, mixing, and flocculation ahead of primary treatment; (2) low to medium head loss occurs across the chamber; (3) grease removal can be achieved if skimming is provided; (4) grit of desired size may be removed by varying the air supply; (5) scouring action of air separates organics, and grit of low putrescible organic content can be produced; and (6) air introduces oxygen, and wastewater is freshened up. One of the major disadvantages of utilizing aerated grit chamber is the release of serious and objectionable odorous compounds, and VOCs due to stripping action of air. Therefore, it is essential that this unit be contained in a building and proper air pollution control devices be installed to purify the exhaust air.

Design Factors: Many factors are considered for planning and design of aerated grit chamber. Some of these include (1) type of grit and other solids, (2) odors potential, (3) detention time, (4) air supply, (5)

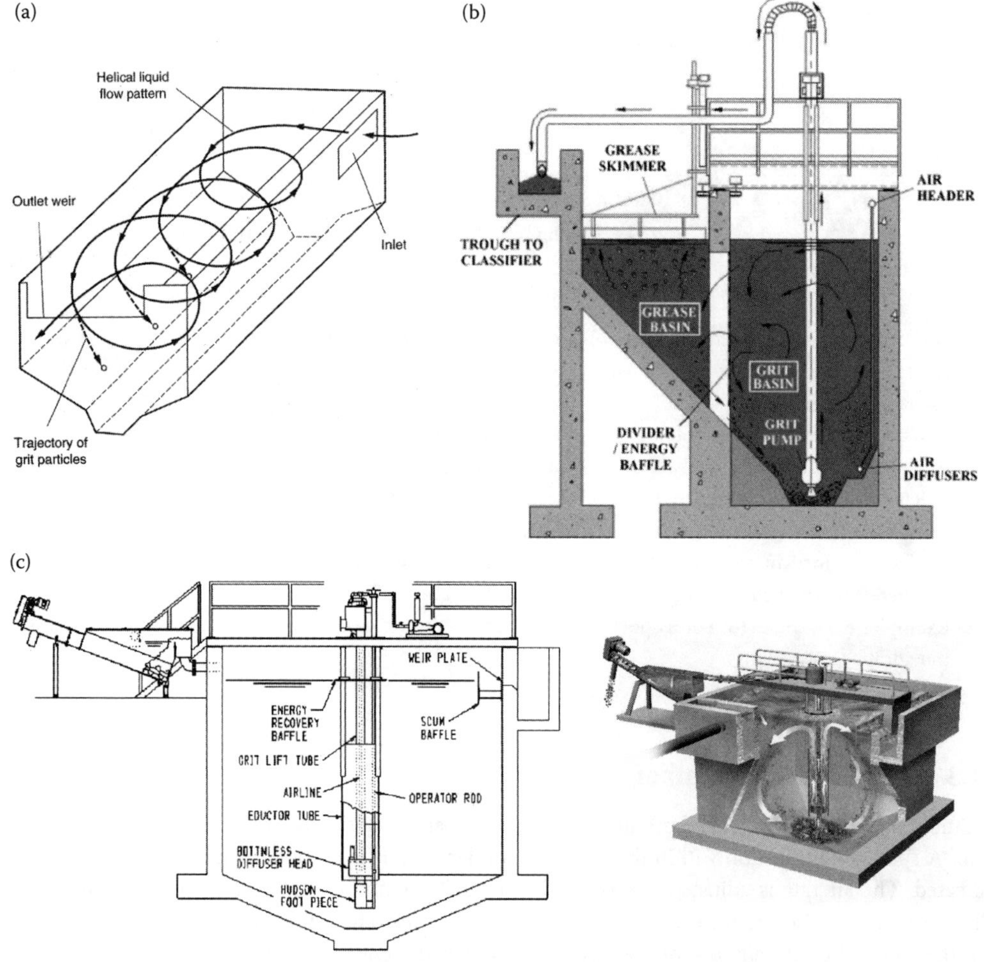

FIGURE 8.17 Details of aerated grit chamber: (a) conceptual helical-flow pattern in a traditional aerated grit chamber (Adapted from Reference 2; reprinted with permission of McGraw-Hill Education), (b) integrated grit and grease removal system using traveling bridge grit collector (Courtesy Kusters Water Div. of Kusters Zima Corp.), and (c) square aerated grit chamber using an educator tube (Courtesy Walker Process Equipment).

TABLE 8.10 Design Factors and Typical Design Values of Aerated Grit Chamber

Design Parameter	Range	Typical	Comment
Dimensions			The width of the basin is limited to provide roll action in the tank
Depth, m (ft)	2–5 (7–16)	3.5 (12)	
Length, m (ft)	7.5–20 (25–65)	12 (40)	
Width, m (ft)	2.5–7 (8–25)	4 (13)	
Width-to-depth ratio	1:1–5:1	1.5:1	
Length to width ratio	2.5:1–5:1	4:1	
Transverse velocity at surface, m/s (ft/s)	0.6–0.8 (2–2.5)	0.7 (2.3)	The velocity at the floor of the tank is 75% of the surface velocity. A velocity of 0.23 m/s is required to move 0.21 mm (70-mesh) sand particles along the tank bottom.
Detention time at peak flow, min	2–5	3	A longer detention time may be required for effective removal of grits <0.15 mm (100-mesh)
Air supply, L/s·m (scfm/ft)	4.5–12.5 (3–8)	9 (6)	Higher air rate should be used for wider and deeper tanks. Provisions should be made to vary the air flow. An air flow rate of 4.6–8 L/s·m in a 3.5–5-m-wide and 4.5-m-deep tank gives a surface velocity of approximately 0.5–0.7 m/s.
Removal efficiencies, %			Aerated grit chamber is typically designed to remove particles 0.21 mm (70-mesh) or larger. The removal efficiencies are reduced significantly for the removal of grits <0.15 mm (100-mesh).
≥ 0.25 mm (60-mesh)	92–98	95	
≥ 0.21 mm (70-mesh)	82–88	85	
≥ 0.15 mm (100-mesh)	75–85	80	
≥ 0.11 mm (140-mesh)	50–70	60	

Source: Adapted in part from References 1 through 3, 17, 20, and 21.

inlet and outlet structures, (6) dead spaces, (7) tank geometry, and (8) baffle arrangements. The typical values of many of these design factors are summarized in Table 8.10. Some factors are briefly discussed below.

Inlet and Outlet structures: The inlet and outlet structures should prevent short-circuiting and turbulence. Influent and effluent baffles are used for this purpose. Inlet to the chamber should induce the circulation pattern. Outlet should be at a right angle to the inlet. The inlet and outlet are sized and built in such a way that the velocity exceeds 0.3 m/s under all flow conditions to minimize the deposits.

Baffles: Longitudinal and transverse baffles improve grit removal efficiency. If the grit chamber is much longer than the width, a transverse baffle should be considered. The baffles offer obstruction to the flow in the chamber. The momentum equation (Equation 8.14) is used to calculate the head loss due to baffles.[22,23]

$$h_L = C_D \frac{v_H^2}{2g} \frac{A_b}{A} \tag{8.14}$$

where

h_L = head loss due to baffle, m (ft)
v_H = horizontal velocity in the chamber through the unobstructed area, m/s (ft/s)
C_D = coefficient of drag dimensionless. The value of C_D for flat plates is ~1.9.[1,2,12,13]
A_b = vertical projection of the area of the baffle, m^2 (ft^2)
A = cross-sectional area of the chamber, m^2 (ft^2)

Chamber Geometry: Location of air diffusers, sloping chamber bottom, grit hopper, and accommodation of grit collection and removal equipment should all be given consideration in chamber geometry. The diffusers are normally located 0.45–0.6 m (1.5–2 ft) above the sloping chamber bottom. The hopper for grit collection is provided beneath the air diffusers. In an integrated grit and grease removal system design, either a designed calm chamber is provided on the opposite side of the basin from the air diffusers or a slower spiral action zone is created to allow grease to float to the top of the chamber for easy separation.

EXAMPLE 8.20: DESIGN OF AN AERATED GRIT CHAMBER

Design an aerated grit removal facility. The design average flow and peaking factor are 0.46 m^3/s and 2.91, respectively. Provide two units with spiral circulation capable of removing grit particles of 0.21 mm (70-mesh) and larger. Air supply per linear tank length shall be 7.8 L/s·m, with provision to meet 150% air capacity for peaking purposes. The retention time at design peak flow when both units are in operation is 3.9 min. The influent and effluent structures shall be designed to handle the emergency flow conditions when one unit is out of service. The influent and effluent structures shall be sized to assure that the unit is not inundated under peak design flow. The influent and effluent baffles each occupy 70% of the cross-sectional area of the chamber with $C_D = 2$.

Solution

1. Calculate the design peak flow for each chamber.

Design peak flow when one chamber is in operation $= 0.46\ \text{m}^3/\text{s} \times 2.91 = 1.34\ \text{m}^3/\text{s}$

Design peak flow per chamber when both chambers are in operation $= \frac{1}{2} \times (1.34\ \text{m}^3/\text{s}) = 0.67\ \text{m}^3/\text{s}$

2. Calculate the dimensions of the chambers.

$$\text{Chamber volume} = 3.9\ \text{min} \times \frac{60\ \text{s}}{\text{min}} \times 0.67\ \text{m}^3/\text{s} = 157\ \text{m}^3$$

The chamber floor shall slope from the mid-width at a ratio of 1 (horizontal) to 1 (vertical).

Assume that the water depth (D) from the chamber bottom is 1.125 times of the width (W) when both chambers are operating.

$$\text{Cross-sectional area } DW - \frac{1}{2} \times \left(\frac{W}{2}\right)^2 = 1.125W^2 - 0.125W^2 = W^2$$

Use a length to width ($L{:}W$) ratio of 4:1.

Volume of the basin $= LW^2 = 4W \times W^2 = 4W^3 = 157\ \text{m}^3$

$$W = \sqrt[3]{\frac{157\ \text{m}^3}{4}} = 3.4\ \text{m}$$

Select $W = 3.5$ m standard width.
Length $L = 4 \times 3.5\ \text{m} = 14\ \text{m}$
Depth $D = 1.125 \times 3.5\ \text{m} = 3.94\ \text{m}$
The design dimensions of the chamber are:
$L = 14$ m, $W = 3.5$ m, and $D = 3.95$ m
Surface area $= 14\ \text{m} \times 3.5\ \text{m} = 49\ \text{m}^2$

$$\text{Cross-sectional area} = 3.95\,\text{m} \times 3.5\,\text{m} - \frac{1}{2} \times \left(\frac{3.5\,\text{m}}{2}\right)^2 = (13.83 - 1.53)\,\text{m}^2 = 12.3\,\text{m}^2$$

Volume $= 14\ \text{m} \times 12.3\ \text{m}^2 = 172\ \text{m}^3$

3. Check the detention times.
   a. Calculate detention time at design peak flow with both chambers operating.

$$\text{Detention time} = \frac{172\ \text{m}^3}{0.67\ \text{m}^3/\text{s}} \times \frac{\text{min}}{60\ \text{s}} = 4.3\ \text{min}$$

b. Calculate detention time at design peak flow with one chamber out of service.

$$\text{Detention time} = \frac{172\ \text{m}^3}{1.34\ \text{m}^3/\text{s}} \times \frac{\text{min}}{60\ \text{s}} = 2.1\ \text{min}$$

4. Check the surface loading rates.

a. Calculate the surface loading (or overflow) rate at design peak flow with both chambers operating.

$$\text{Surface loading rates} = \frac{0.67\ \text{m}^3/\text{s}}{49\ \text{m}^2} \times \frac{(60 \times 60 \times 24)\ \text{s}}{\text{d}} = 1181\ \text{m}^3/\text{m}^2{\cdot}\text{d} \quad \text{or} \quad 28{,}990\ \text{gpd/ft}^2$$

Note: This value is considerably less than that for velocity-controlled grit chamber (Table 8.4), and larger than that of square horizontal-flow chamber (Table 8.5).

b. Calculate the surface loading rate at design peak flow with one chamber out of service.

$$\text{Surface loading rates} = \frac{1.34\ \text{m}^3/\text{s}}{49\ \text{m}^2} \times \frac{(60 \times 60 \times 24)\ \text{s}}{\text{d}} = 2363\ \text{m}^3/\text{m}^2{\cdot}\text{d} \quad \text{or} \quad 58{,}000\ \text{gpd/ft}^2$$

5. Design the air supply and aeration systems.

a. Determine the air supply rate.

$$\text{Theoretical air required per chamber} = \frac{8.3\ \text{L/s}}{\text{m}} \times 14\ \text{m} = 116\ \text{L/s}$$

Provide 150% capacity for maximum air supply purposes.

Total capacity of air supply per chamber $= 1.5 \times 116\ \text{L/s} = 174\ \text{L/s}$ or $0.174\ \text{m}^3/\text{s}$

Total air supply capacity from blowers for both chambers $= 2 \times 174\ \text{L/s} \times \frac{60\ \text{s}}{\text{min}} \times \frac{\text{m}^3}{10^3\ \text{L}} =$ 20.9 standard m^3/min (sm^3/min) or 740 standard ft^3/min (scfm)

Provide two blowers each 21 sm^3/min each with one blower being standby unit. Air piping shall deliver 0.174 m^3/s air to each chamber. Control valves and flow meters shall be provided on all branch lines to balance air flow rate.

b. Describe the diffuser arrangement.

Locate diffusers along the length of the chamber on one side, and place them 0.6 m above the bottom. The upward draft of the air near the wall will create a spiral roll action of the liquid in the chamber. The chamber bottom is sloped toward a collection channel located on the same side of the air diffusers. Grit is swept into the grit collection channel due to rolling action of the liquid. A traveling bridge grit removal system is provided for both chambers. There are two bridge-mounted grit pumps, one for each grit collection channel to lift the grit to a grit discharge trough along the outside wall of one chamber. The pumped flow provides a cleaning action to sweep the grit into a grit classifier.

6. Design the influent structure.

a. Describe the arrangement of influent structure.

Providea 1-m-wide submerged influent channel that diverts the flow into two grit chambers. Each channel has one sluice gate (orifice 1 m × 1 m) that discharges the flow near the diffuser area. Provide a baffle at the influent to divert the flow transversally to follow the circulation pattern. Sluice gates are provided to remove one chamber from service for maintenance purposes. The details of the influent structure are shown in Figure 8.18.

b. Calculate the head loss through the influent structure.

The head loss is calculated from the orifice equation (Equation 7.4a) and assume $C_d = 0.6$.

$$Q = C_d A \sqrt{2gH}$$

1. Calculate the head loss when both units are operating.

$$0.67\,\text{m}^3/\text{s} = 0.6 \times 1\,\text{m}^2 \times \sqrt{2 \times 9.81\,\text{m/s}^2 \times H}$$

$$H = 0.06\,\text{m}$$

2. Calculate the head loss when one unit is out of service.

$$1.34\,\text{m}^3/\text{s} = 0.6 \times 1\,\text{m}^2 \times \sqrt{2 \times 9.81\,\text{m/s}^2 \times H}$$

$$H = 0.25\,\text{m}$$

7. Design the effluent structure.

a. Describe the arrangement of effluent structure.

The effluent structure consists of a rectangular weir, an effluent launder, an effluent box, and an outlet pipe. The effluent weir is 3 m long, and the effluent launder is 3 m long ×1 m wide. The effluent box is common to both chambers and is 1 m × 1 m. Slot logs are provided at the downstream end of the effluent launder to isolate the effluent launder of the chamber that is removed from. The outlet pipe carries flow to a collection–division box. Details of the effluent structure and design details of aerated grit chamber are shown in Figure 8.18.

b. Calculate the head over the effluent weir.

Calculate the head over the weir from Equation 8.10 at $n = 1$, and assume $C_d = 0.6$.

1. Calculate the head over the weir when both units are operating.

Assume $L' = 2.98$ m.

$$0.67\,\text{m}^3/\text{s} = \frac{2}{3} \times 0.6 \times 2.98\,\text{m}\sqrt{2 \times 9.81\,\text{m/s}^2} \times H^{3/2}$$

$$H = 0.25\,\text{m}$$

Check: $L' = 3\,\text{m} - 0.1 \times 1 \times 0.25\,\text{m} = 2.98\,\text{m}$ (It is same as the initial assumption.)

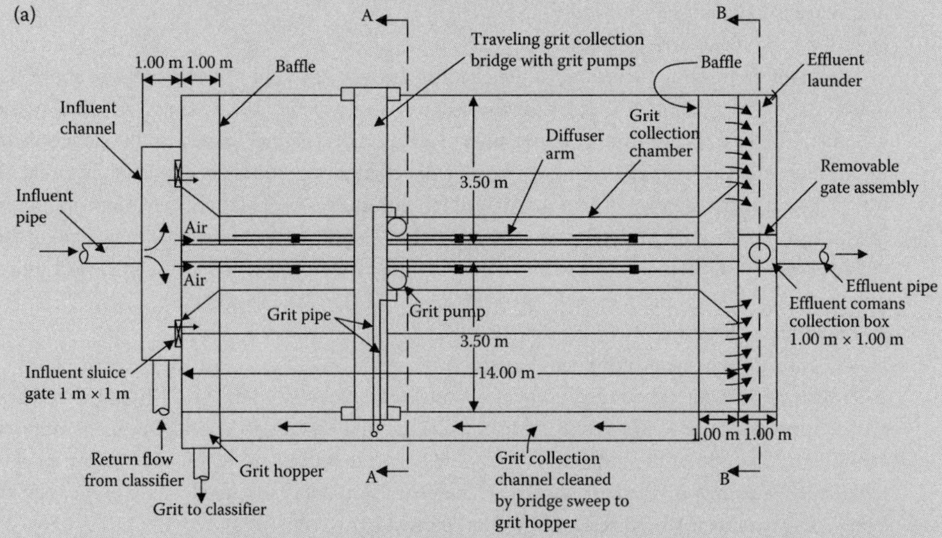

FIGURE 8.18 Design details and hydraulic profile of aerated grit chamber: (a) plan (Example 8.20).
(*Continued*)

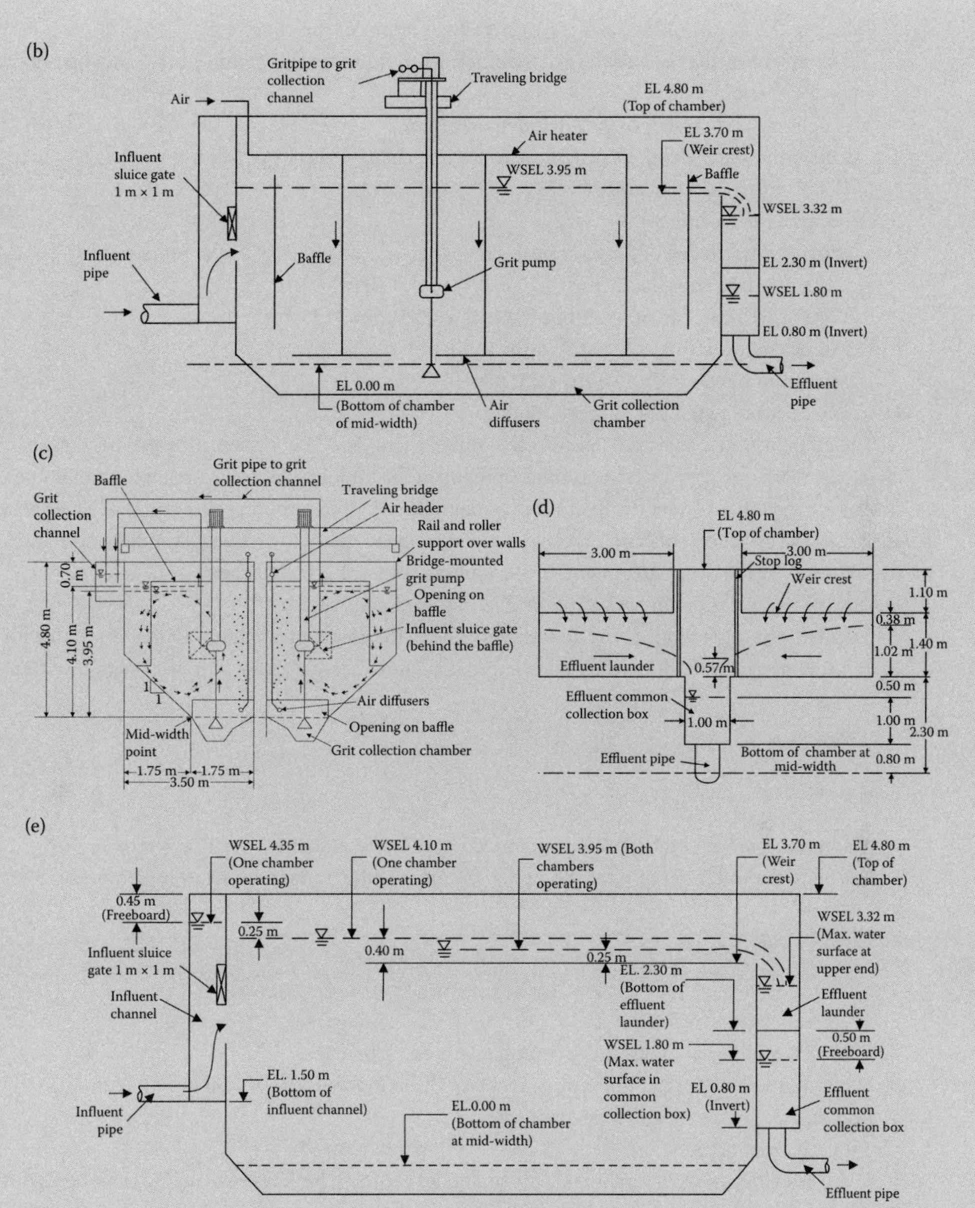

FIGURE 8.18 (Continued) Design details and hydraulic profile of aerated grit chamber: (b) longitudinal section, (c) Section AA, (d) Section BB, and (e) hydraulic profile (Example 8.20).

2. Calculate the head over the weir when one chamber is out of service.
   Assume $L' = 2.96$ m.

$$1.34\,\text{m}^3/\text{s} = \frac{2}{3} \times 0.6 \times 2.96\,\text{m}\sqrt{2 \times 9.81\,\text{m/s}^2} \times H^{3/2}$$

$$H = 0.40\,\text{m}$$

Check: $L' = 3\,\text{m} - 0.1 \times 1 \times 0.40\,\text{m} = 2.96\,\text{m}$ (It is same as the initial assumption.)

3. Calculate the height of the weir crest above the bottom of the chamber.

   The height of the weir crest is determined with the water depth and head loss when both units are operating.

   Height of weir crest = (3.95 − 0.25) m = 3.70 m

4. Calculate the water depth at peak flow when one chamber is out of service.

   Water depth in the chamber = (3.70 + 0.40) m = 4.10 m

5. Determine the basin depth.

   Provide a freeboard of 0.7 m at peak flow when one chamber is out of service.

   Basin depth in the chamber = (4.10 + 0.7) m = 4.80 m

   Water surface in the influent box = (4.10 + 0.25) m = 4.35 m

   Check the minimum freeboard in the influent box.

   Minimum freeboard = (4.80 − 4.35) m = 0.45 m

c. Describe the arrangement of effluent launder.

   The effluent weir has a free fall into the effluent launder. The effluent launders of both grit chambers discharge into a common collection box in the middle. The downstream hydraulic conditions are such that a water depth of 1.0 m is maintained in the common collection box. The water surface in the common collection box is 0.5 m below the invert of the effluent launder, and critical depth occurs at the downstream end of the launder for each grit chamber.

d. Determine the water depth and water profile in the effluent launder.

   Calculate critical depth y_c at the downstream end of the effluent trough from Equation 7.10 with $C_w = 1.0$ at design peak flow when one chamber is out of service and width of trough, $b = 1$ m.

$$1.34\,\text{m}^3/\text{s} = 1.0 \times 1\,\text{m} \times \sqrt{9.81\,\text{m/s}^2}\cdot d_c^{3/2}$$

$$d_c = 0.57\,\text{m}$$

   Assume the water depth at the downstream end of the effluent trough $y_2 = d_c = 0.57$ m.
   Calculate the water depth y_1 at the upper end of the effluent launder from Equation 8.13a.

$$y_1 = \sqrt{y_2^2 + \frac{2Q^2}{gb^2 y_2}} = \sqrt{(0.57\,\text{m})^2 + \frac{2 \times (1.34\,\text{m}^3/\text{s})^2}{9.81\,\text{m/s}^2 \times (1\,\text{m})^2 \times 0.57\,\text{m}}} = 0.98\,\text{m}$$

   Allow 4% additional depth for friction losses, and turbulence.

   Maximum water depth at the upstream end of the launder = 1.04 × 0.98 m = 1.02 m

   Add 0.33 m for free fall after the effluent weir.

   Total depth of effluent launder = (1.02 + 0.38) m = 1.40 m

8. Calculate the head loss due to influent and effluent baffles at design peak flow when one chamber is out of service.

   Apply Equation 8.14 to calculate the head loss through the baffles.

$$h_L = C_D \frac{v_H^2}{2g}\frac{A_b}{A}$$

   Water depth at mid-width point = 4.1 m

   Width of the chamber = 3.5 m

   Cross section area (A),

$$A = 4.1\,\text{m} \times 3.5\,\text{m} - \frac{1}{2} \times \left(\frac{3.5\,\text{m}}{2}\right)^2 = (14.35 - 1.53)\,\text{m}^2 = 12.8\,\text{m}^2$$

Calculate the horizontal velocity v through the unobstructed area.

$$v_H = \frac{1.34\,\text{m}^3/\text{s}}{12.8\,\text{m}^2} = 0.10\,\text{m/s}$$

Calculate the head loss through each baffle.

$$h_L = 2 \times \frac{(0.10\,\text{m/s})^2}{2 \times 9.81\,\text{m/s}^2} \times \frac{0.7 \times A}{A} = 0.0007\,\text{m}$$

The total head loss through two baffles is 0.0014 m. It is small head loss and can be ignored.

9. Show the design details and hydraulic profile through the aerated grit chamber.
   The design details and hydraulic profile are shown in Figure 8.18.

8.4.4 Vortex-Type Grit Chambers

The vortex-type grit chambers (also called accelerated grit separation devices) utilize both gravitational and centrifugal forces for the separation of grit. This type of unit is designed to remove grits of specific gravity 2.5 or larger with sizes ranging from 0.15 (100-mesh) to 2 mm (10-mesh). They offer very compact, totally enclosed, and low VOC emission facilities. These types of devices have been gaining popularity during recent years. Manufacturers provide rating tables or curves to select the units for desired applications. The selection criteria are based on peak design flow, grit size and percent removal, and head loss constraints.

Free Vortex Grit Chamber: In a free vortex grit chamber, a dominant and strong vortex is created by gravitational and swirling forces to separate inorganic solid from organic solids and water. The action is similar to that of a hydrocyclone. The unit consists of a cylindrical section on top of a conical section. The influent enters tangentially around the upper middle section of the cylinder. The high centrifugal forces retain particles near the center of the chamber. The retention depends upon the size of the particles, their density and drag within the free vortex. Grit and sand particles settle by gravity to the bottom of the unit, while organics and degritted wastewater exit the center, or "eye" of the fluid on the top of the unit. The effluent discharge may be under gravity or pressurized. The grit is discharged from the bottom for further treatment through a belt grit dewatering escalator. Examples of free vortex grit chamber include TeaCup® and Grit King® grit removal units. The basic components of these systems are illustrated in Figure 8.19.[24,25]

The head loss in the unit is a function of the particle size and their required removal rate. The head loss increases for higher removal and for smaller particles. These grit separators are generally designed to achieve 95% removal of 75 μm (200-mesh) to 106 μm (140-mesh) grits. The head required to achieve this removal is a few meters. For 95% removal of 25-μm grit, the head requirement is 5–7 m. Pumping into these grit removal units may be needed to exert such head in the unit above the grade. For gravity flow, installation deep in the ground is required. The inlet and outlet orientation of the unit is flexible to fit different piping configurations. This type of unit also features a free vortex vessel without moving parts. Due to its simple hydraulic design, it may be sensitive to flow variations. The available manufacturer design and performance information regarding this type of vortex units are summarized in Table 8.11.

Induced Vortex Grit Chamber: The *Mectan*® or other similar grit removal separator operates on the principle of mechanically induced vortex.[26–32] In general, a cylindrical section is on top of a central

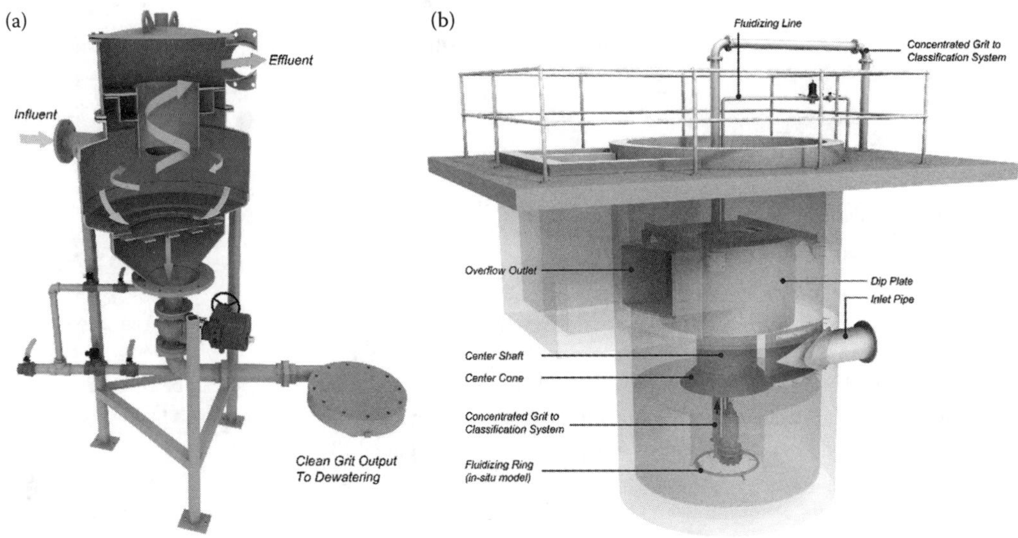

FIGURE 8.19 Free vortex grit chambers (Courtesy Hydro International): (a) flow regime through a TeaCup® unit, and (b) major components in a Grit King® unit.

grit hopper at the bottom. The influent is surface fed tangentially into the sloped inlet flume that minimizes turbulence. At the end of the inlet flume, the grit reaches the chamber floor. The rotating action of the adjustable pitch blades produces a toroidal-flow path for the grit particles and also promotes the separation of organics from the grit. In recent design, optional baffles are provided to improve grit removal efficiency under low flow conditions. The flow continues to move circumferentially, and the grit is propelled along the bottom toward the center, while lighter organics are lifted and carried in the effluent. The grit drops into the center storage hopper. Mechanical or other device is also installed near the bottom of the hopper to fluidize grit bed for grit removal by an airlift or grit pump for grit processing. The general principle of operation in the induced vortex grit separator is illustrated in Figure 8.20.[26–28]

TABLE 8.11 Design and Performance Information of Free Vortex Grit Chamber

Design Parameters	Range
Available unit capacity range	
Minimum, L/min (gpm)	265 (70)
Maximum, m^3/d (MGD)	0.35 m^3/s (8)
Turndown ratio of peak to average daily flow	3–4:1
Diameter, m (ft)	0.6–2.4 (2–8)
Performance	
Removal of 75 μm or greater particles at design flow, %	95
VSS removal, %	<15–20
TS removal, %	>60
Solids content, %	Up to 1.5%
Head loss	
At peak flow, m (ft)	0.3–1.5 (1–5)
At average flow, m (ft)	<0.15 (0.5)

Source: Developed from information in References 24 and 25.

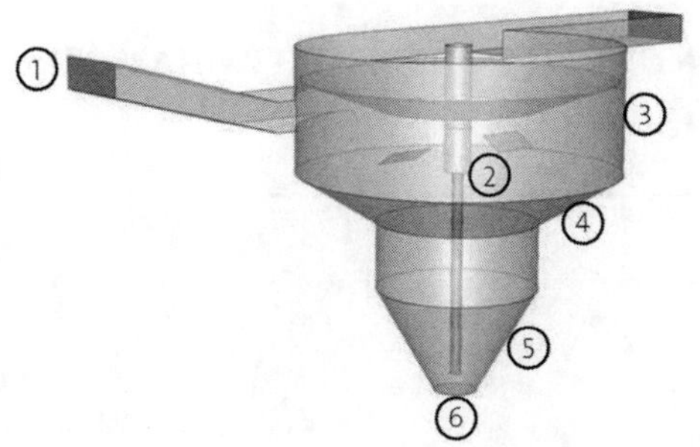

FIGURE 8.20 General principle of operation in John Meunier Mectan® induced vortex grit chamber (Courtesy John Meunier Products/Veolia Water Technologies Canada).

Sizing of the proprietary vortex device is based on recommended dimensions of the manufacturers. The available manufacturer design and performance information regarding to this type of vortex units are summarized in Table 8.12.

Special requirements are usually required in design of the inlet and outlet channels to ensure an ideal performance of the grit chambers. For an induced vortex grit chamber, the inlet and outlet channels must be arranged in either 360° or 270° configuration. In general, it is required to achieve inlet channel velocities of 0.6–0.9 m/s (2–3 ft/s) within the design flow range. The initial minimum inlet channel velocity must exceed 0.15 m/s (0.5 ft/s). A flush velocity of 0.6 m/s (2 ft/s) must reach during peak flows to wash the accumulated grits in the channel under low flow conditions into the grit chamber. A straight run for a smooth laminar-type flow with little turbulence is usually desired in both the inlet and outlet channels. The outlet channel is usually a free-flowing flume with a constant elevation and without bends or any other narrowed sections. In some applications, a submerged weir located in the discharge channel may be used to control the level in the grit chamber. Use of flow control baffles may reduce or eliminate some of these design requirements in inlet and/or outlet channels.[26–28]

TABLE 8.12 Typical Design and Performance Information of Induced Vortex Grit Chambers

Design Parameter	Range
Available unit capacity range, m^3/s (MGD)	0.022–4.4 (0.5–100)
Diameter	
Upper chamber, m (ft)	1.8–9.8 (6–32)
Lower hopper, m (ft)	0.9–2.4 (3–8)
Detention time, s	20–30
Height, m (ft)	2.6–6.9 (8.7–22.7)
Removal efficiencies, %	
$\geq$ 0.30 mm (50-mesh)	95–96
$\geq$ 0.21 mm (70-mesh)	85–90
$\geq$ 0.15 mm (100-mesh)	65–75
$\geq$ 0.11 mm (140-mesh)	40–70
Maximum head loss (without baffles), mm (in)	6.35 (0.25)
Power supply, kW (Hp)	0.56–1.5 (0.75–2)
Grit hopper storage volume, m^3 (ft^3)	0.9–9.5 (32–335)

Source: Developed from information in References 1 through 3, and 26 through 32.

EXAMPLE 8.21: DETENTION TIMES OF A INDUCED VORTEX GRIT CHAMBER

An induced vortex grit chamber has the following dimensions:

Upper cylindrical chamber	
Diameter, D_{upper}	1.8 m
Height, H_{upper}	1.0 m
Hopper frustum	
Height, H_{hopper}	0.5 m
Lower cylindrical section connecting the hopper and the grit transfer cone	
Diameter, D_{lower}	1.0 m
Height, H_{lower}	0.8 m

The design capacity of the grit chamber is 13,250 m^3/d (3.5 MGD). Calculate (a) the total detention time in the unit, and (b) the retention time in the upper cylindrical chamber. Assume that the grit transfer cone does not provide any flow retention.

Solution

1. Draw the definition sketch (Figure 8.21).

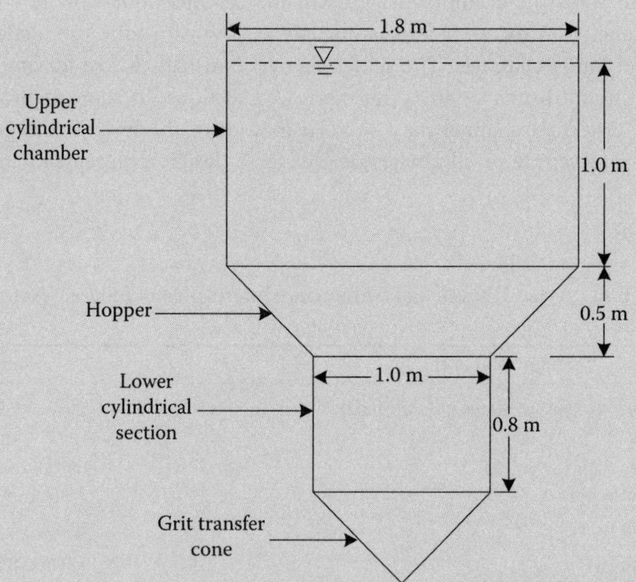

FIGURE 8.21 Definition sketch of centrifugal vortex-induced grit chamber (Example 8.21).

2. Calculate the volume of the upper cylindrical chamber.

$$\text{Area, } A_{upper} = \frac{\pi}{4}(D_{upper})^2 = \frac{\pi}{4}(1.8\text{ m})^2 = 2.54\text{ m}^2$$

$$\text{Volume, } V_{upper} = A_{upper}H_{upper} = 2.54\text{ m}^2 \times 1.0\text{ m} = 2.54\text{ m}^3$$

3. Calculate the volume of the hopper.

Volume of the hopper is calculated from Equation 8.15.

$$V = \frac{1}{3}H\left[A_1 + A_2 + \sqrt{A_1 A_2}\right] \quad (8.15)$$

where

V = volume of frustum of a cone, m^3 (ft^3)
H = height of the cone frustum, m (ft)
A_1 = area of the upper portion of the cone frustum, m^2 (ft^2)
A_2 = area of the lower portion of the cone frustum, m^2 (ft^2)

$$A_1 = A_{upper} = 2.54\,m^2$$

$$A_2 = \frac{\pi}{4}(D_{lower})^2 = \frac{\pi}{4}(1.0\,m)^2 = 0.79\,m^2$$

$$\text{Volume, } V_{hopper} = \frac{1}{3}H_{hopper}\left[A_1 + A_2 + \sqrt{A_1 A_2}\right]$$

$$= \frac{1}{3} \times 0.5\,m \times \left[2.54\,m^2 + 0.79\,m^2 + \sqrt{2.54\,m^2 \times 0.79\,m^2}\right]$$

$$= 0.79\,m^3$$

4. Calculate the volume of the lower cylindrical section connecting the hopper and the grit transfer cone.

Area, $A_{lower} = A_2 = 0.79\,m^2$
Volume, $V_{lower} = A_{lower}H_{lower} = 0.79\,m^2 \times 0.8\,m = 0.63\,m^3$

5. Determine the total volume of the grit chamber.

$$\text{Volume, } V_{total} = V_{upper} + V_{hopper} + V_{lower} = (2.54 + 0.79 + 0.63)\,m^3 = 3.96\,m^3$$

6. Calculate the detention times at the design capacity.

$$\text{Design capacity, } Q = 13{,}250\,m^3/d \times \frac{1\,d}{86{,}400\,s} = 0.153\,m^3/s$$

Total detention time in the grit chamber,

$$\theta_{total} = \frac{V_{total}}{Q} = \frac{3.96\,m^3}{0.153\,m^3/s} = 26\,s$$

Detention time in the upper cylindrical chamber,

$$\theta_{upper} = \frac{V_{upper}}{Q} = \frac{2.54\,m^3}{0.153\,m^3/s} = 17\,s$$

Note: The detention time in the upper cylindrical chamber is about two-third of the total detention time in the grit chamber.

8.4.5 Sludge Degritting

In many plants, grit chambers are not provided. Instead, grit is removed in primary sedimentation basin along with settleable organic matter. Therefore, degritting of primary sludge is necessary to separate grit from organic solids. The devices used are cyclones (hydrocyclones), centrifuges, and reciprocating rake similar to that used in detritus tanks.

Cyclone (Hydrocyclone) Degritter: The cyclones are cone-shaped devices into which dilute primary sludge is pumped tangentially. As the fluid spirals inward, centrifugal forces push the grit toward the wall. The grit slides spirally down to the apex of the cone and then discharged through the orifice. The effluent containing organic solids is discharged from the top. The action of a cyclone is similar to that of a vortex-type grit separator.

Centrifugal Degritter: Centrifuge exerts centrifugal force to cause selection and separation of grit from organic matter. Various types of continuous-flow centrifuges are available for such application. Discussion on centrifuges may be found in Chapter 13.

Reciprocating Rake Degritter: The sludge containing grit is moved up an incline by a reciprocating rake mechanism. The organic solids are separated and removed from the grit. This device is used with detritus tank to simultaneously degrit the settled sludge (Section 8.4.1, Figure 8.9, and Example 8.16).

8.4.6 Grit Collection and Removal

Grit collection equipment is required in horizontal-flow and aerated grit chambers. In general, it moves the grit to a central storage channel or to a hopper for the removal from the grit chamber. Depending upon the type of facility, the commonly used equipment for grit collection are chain-and-sprocket scraper or plows, chain-and-bucket conveyor, tabular conveyor, and screw collectors that run along the full length of the channel floor or storage hopper.

Traveling bridge type of grit collectors are developed for removing grits from aerated grit chambers (Figure 8.17b). A submersible grit pump is mounted on the bridge. The grits are lifted from the bottom of the chamber by the pump and discharged into a channel where they are further cleaned through classifier and washer prior to final disposal. A special skimmer mechanism may also be used for collection of grease. The equipment is fully automated and can be operated continuously or intermittently.[1–3,17,18]

In a vortex-type grit chamber, an airlift or grit pump is typically used to transfer the grits from the center storage hopper to a classifier or concentrator for grit processing. Direct discharge of grits by gravity may also be possible in an elevated grit chamber.[24–32]

8.5 Grit Characteristics and Quality

Grit Characteristics: Grit is predominantly inert material. It is heavy, very abrasive, and has cementing effect on the basin floor and channel bottom. At high organic contents, grit is highly odorous and attracts insects and rodents. Grit composition and characteristics vary greatly. Basic information is summarized in Table 8.13.[1–3]

Grit Quantity: The quantity of grit varies greatly, depending on (1) type of collection system (separate or combined), (2) climatic conditions, (3) soil type, (4) condition of sewers and grades, (5) types of

TABLE 8.13 Physical Characteristics of Grit from POTWS

Item	Range	Comment
Grit quantity in sanitary sewage[a], $m^3/10^3 m^3$ (ft^3/Mgal)	0.005–0.05 (0.7–7)[b]	The grit quantity in combined sewers is significantly higher
Specific gravity	1.3–2.7	Organic matter reduces specific gravity
Bulk density, kg/m^3 (lb/ft^3)	1300–1900 (80–120)[a]	Depends greatly on moisture content and organic matter
Moisture content, %	13–65	Moisture content may depend upon drainage provided
Volatile solids, %	1–56	Grit separators and washers remove organic solids, and make grit less odorous
Particle size, mm (mesh)	0.10 (140)–0.21 (70)	Variation due to the type of collection system and grit removal efficiency

[a] $kg/m^3 \times 0.0624 = lb/ft^3$
[b] $m^3/1000\ m^3 \times 134.2 = ft^3$/Mgal
Source: Adapted in part from References 1 through 3.

industrial wastes, (6) relative use of garbage grinders, and (7) proximity to the sandy bathing beaches. The grit quantity may range from 0.005 to 0.05 $m^3/10^3$ m^3 (0.7–7 ft^3/Mgal). Typical value is 0.03 $m^3/10^3$ m^3 (4 ft^3/Mgal) (see also Table 6.9 and Example 6.21).

8.6 Grit Processing, Reuse, and Disposal

Unwashed grit may contain 50% or more organic matter, has serious odors, and may attract insects and rodents. Grit processing may significantly reduce putrescible organic contents to <3%.[1]

Grit Processing: Effective grit processing is a part of complete grit removal system. Multiple functions are usually needed in grit processing prior to final disposal. These functions include grit separation, classification,

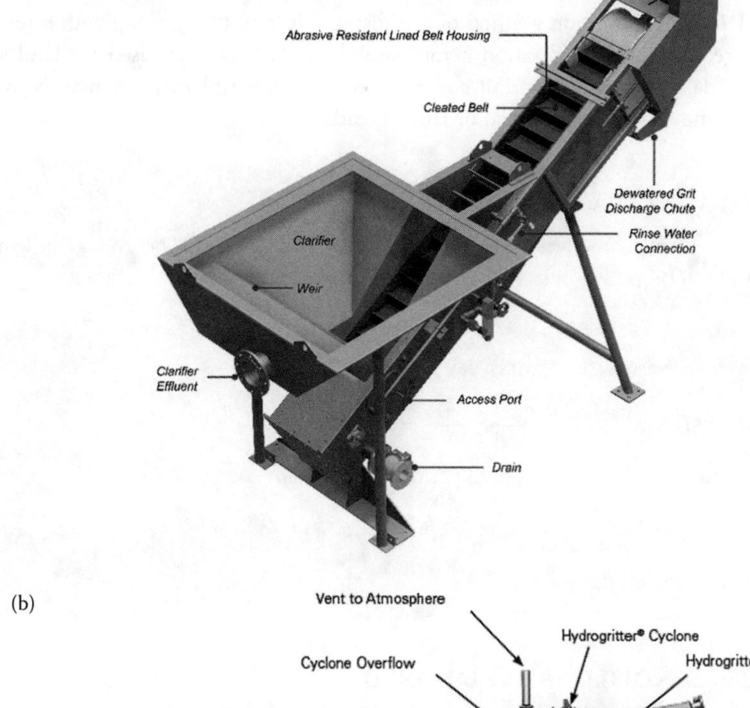

FIGURE 8.22 Grit processing equipment and system (a) Grit Snail® cleated belt grit dewatering escalator (Courtesy Hydro International), and (b) an integrated grit removal and processing system (Courtesy Weir Specialty Pumps).

concentration, washing/dewatering, and/or transportation.[33–41] Two major types of grit processing devices are (1) inclined submerged screw conveyor and (2) cleated belt escalator. These devices are shown in Figure 8.22.[33,34] Both types include a large setline hopper for effective settling of the grit at a low overflow rate, utilize wash and/or spray water for cleaning action to separate organic contents from the grit, dewater the grit to have low water content, and raise the grit up to the discharge level that is suitable for transporting to final disposal. A grit pump is usually required to remove the grit from the grit chamber to the grit processing devices and a hydrocyclone may also be used ahead of these devices to improve organic separation.

The multiple functions needed in grit processing can be achieved either by one or multiple devices. For example, a vortex-type grit chamber (Figure 8.19 or 8.20) can be combined with grit processing equipment (Figure 8.22a) to achieve the desired effects. Such a combination of equipment is commercially available.[25,26,28,33–41] An integrated grit removal and processing system is illustrated in Figure 8.22b.[34]

After a proper grit processing, the grit is usually clean with $<5\%$ putrescible organic matters and dry with $<10\%$ water content.[33–41]

Grit Reuse and Disposal: Common method of grit disposal is land filling along with screenings and grease. Some states require lime stabilization before landfilling. Grit is often used for land spreading, and also mixed with daily municipal landfill cover material. In large plants, grit may be incinerated with other solids, and the residue is landfilled or land spread.

EXAMPLE 8.22: GRIT QUANTITY REMOVED

A wastewater treatment plant receives an average daily flow of 0.68 m^3/s. The average grit quantity removal rate is 0.03 $m^3/10^3\ m^3$. Calculate the average grit quantity removed.

Solution

1. Calculate the average wastewater flow per day.

$$\text{Average flow} = 0.68\,\text{m}^3/\text{s} \times \frac{(60 \times 60 \times 24)\,\text{s}}{\text{d}} = 58{,}752\,\text{m}^3/\text{d}$$

2. Calculate the average grit quantity removed per day.

$$\text{Grit quantity removed} = \frac{0.03\,\text{m}^3}{1000\,\text{m}^3} \times 58{,}752\,\text{m}^3/\text{d} = 1.76\,\text{m}^3/\text{d}$$

Note: See also Example 6.21.

EXAMPLE 8.23: LANDFILL AREA REQUIRED

At a wastewater treatment plant, the total quantity of screenings and grit removal per day are 3.8 and 2.1 m^3/d. The combined residuals are disposed off in a permitted landfill on the plant site. The average depth of residuals in the landfill is 3 m. Calculate the land area required per year. Ignore the slight reduction in residuals volume by compaction during landfill operation.

Solution

1. Calculate the quality of residuals produced per year.

$$\text{Quantity of screenings and grit per day} = (3.8 + 2.1)\,\text{m}^3/\text{d} = 5.9\,\text{m}^3/\text{d}$$

$$\text{Quantity of residuals per year} = 5.9\,\text{m}^3/\text{d} \times \frac{365\,\text{d}}{\text{yr}} = 2154\,\text{m}^3/\text{yr}$$

2. Calculate the land area required per year.

$$\text{Land area required per year} = \frac{2154\,\text{m}^3}{\text{yr}} \times \frac{1}{3\,\text{m}} = 718\,\text{m}^2/\text{yr}$$

Discussion Topics and Review Problems

8.1 A grit chamber at a wastewater treatment plant is used to remove sandy material from raw wastewater. The specific gravity and mean diameter of the particles are 2.60 and 0.25 mm, respectively. Wastewater temperature is 25°C. Calculate the N_R value, and terminal velocity of the particles in m/s.

8.2 A grit chamber is designed to remove a mixture of sandy material of different sizes from municipal wastewater. Sieve analysis showed the following average diameter of the particles: 0.06, 0.09, 0.25, 0.35, 0.45, 1.25, 4.5, 10, and 15 mm. The temperature of the wastewater is 20°C. The specific gravity of sandy particles is 2.65. Calculate (1) the N_R value, (2) the terminal velocity, and (3) the settling regime of each fraction with respect to average diameter.

8.3 A grit removal facility is designed to treat a flow of 20,000 m^3/d. The grit chamber has length, width, and depth of 20, 0.85, and 1.1 m, respectively. Calculate the following: (a) overflow rate, (b) terminal velocity v_t of the smallest particle that will be removed, (c) horizontal velocity, and (d) detention time.

8.4 A column test is conducted to determine the percent removal of discrete suspended particles in a clarifier. The results of a column test are given below. The clarifier has an overflow rate of 56,010 gpd/ft^2. Calculate the performance of the clarifier in terms of weight fraction and percent solids removal.

Settling velocity, ft/min	8	4	2	1	0.7	0.5
Weight fraction remaining	0.52	0.50	0.34	0.16	0.04	0.02

8.5 Settling column analysis was performed on a grit sample. The column is 2.5 m deep. The test results are summarized below. Calculate the weight fraction removal of this material in a basin that has an overflow rate of 20 $m^3/m^2 \cdot d$.

Time, min	0	60	80	100	120	180	240	420
Total suspended solids Concentration, mg/L	518	445	401	352	297	208	147	71

8.6 Standard sieve analysis was performed on a grit sample. The results of the analysis are tabulated below. Calculate the weight fraction removed in a grit chamber that has an overflow rate of 120 gpm/ft^2. The average wastewater temperature is 68°F, and average specific gravity of discrete particle is 2.65. Tabulate the settling velocity and N_R value of each fraction retained, and settling regime.

Standard sieve number	16	20	30	40	65	80	100	120	140	200
Weight fraction retained	0	0.05	0.20	0.18	0.37	0.08	0.05	0.04	0.02	0.01

8.7 A wastewater treatment plant is designed to treat an average flow of 1.0 m^3/s. The peak flow is the 2.7 times the average flow. The horizontal and terminal velocities, and detention time of the grit chamber are 0.25 m/s, 1.5 m/min, and 70 s, respectively. At this terminal velocity, the minimum discrete particle of 0.21 mm (70-mesh) diameter will be fully removed. The flow to the influent channel is regulated by stop gates to divide the flow into two identical channels. Each channel is capable of handling half peak design flow. The effluent structure consists of a proportional weir to maintain a constant horizontal velocity in the channel at a wide flow variation. The grit removal system consists of a conveyor that moves the grit into a hopper, and raises the grit while washing it to separate the organics. Design the velocity-controlled grit removal facility.

8.8 A wastewater treatment plant has two identical horizontal-flow square degritting tanks. The flow to each basin is 38,000 m^3/d. The design overflow rate and side water depth are 1100 $m^3/m^2{\cdot}d$ and 15 m. Determine the dimensions of the basin and detention time. Design the influent and effluent structures, and draw the hydraulic profile through the unit. The influent channel has 10 square orifices 0.3 m × 0.3 m. The effluent trough is 0.85 m wide, and has free fall into an effluent box that is common to both units. The thickness of the common wall is 0.2 m, and the dimensions of common effluent box are 1.2 m × 1.2 m. Draw the plan view and hydraulic profile.

8.9 A wastewater treatment plant treats a flow of 0.65 m^3/s. The peak flow is 2.8 times the average flow treated. Design two identical aerated grit removal units that have spiral circulation capable of removing grit particles of 0.21 mm (70-mesh) and larger. The influent and effluent structures shall be designed to handle the emergency flow conditions when one unit is out of service. The effluent trough is 0.5 m higher than the water surface elevation in the common effluent box. Provide detention time of 5 min when both units are in operation. Air supply shall be 8 L/s·m of tank length, with provision to meet 150% air capacity for peaking purposes. The length to width ratio is 4:1. The inlet and outlet structures shall be sized to provide a minimum velocity of 0.25 mps.

8.10 A wastewater treatment plant receives an average daily flow of 0.75 m^3/s. The average grit quantity removal rate is 0.05 $m^3/10^3\ m^3$. Calculate the average grit quantity removed.

8.11 A wastewater treatment plant generates 4.5 m^3/d screenings and 2.7 m^3/d grit. These combined residuals are disposed off in a permitted landfill at the plant site. The average depth of residuals in the landfill is 4.3 m. Calculate the area of land used per year.

References

1. Qasim, S. R., *Wastewater Treatment Plants: Planning, Design, and Operation*, 2nd ed., CRC Press, Boca Raton, FL, 1999.
2. Metcalf & Eddy, Inc., *Wastewater Engineering: Treatment and Reuse*, 4th ed., McGraw-Hill, New York, 2003.
3. Water Environment Federation, *Design of Municipal Wastewater Treatment Plants*, vol. 2, WEF Manual of Practice No. 8, ASCE Manual and Report on Engineering Practice No. 76, Water Environment Federation, Alexandria, VA, 1998.
4. Reynolds, T. D. and P. A. Richards, *Unit Operation and Processes in Environmental Engineering*, 2nd ed., PWS Publishing Company, Boston, MA, 1996.
5. Droste, R. L., *Theory and Practice of Water and Wastewater Treatment*, John Wiley & Sons, Inc., New York, 1977.
6. WesTech Engineering, Inc., *Headworks Process Equipment*, westech-inc.com, 2012.
7. Fairfield Service Company, *Grit Collector*, http://www.fairfieldservice.com
8. Joint Committee of the Water Pollution Control Federation and the American Society of Civil Engineers, *Water Treatment Plant Design*, Manual of Practice No. 8, Water Pollution Control Federation, Washington, D.C., 1977.
9. Ovivo, *J+A CrossflowTM Grit Removal*, http://www.ovivowater.com, 2010.
10. Hubert Stavoren BV, *Hubert Grit Removal System*, http://www.hubert.nl
11. Gorden, R. T., *Weir of Special Design for Grit Removal, Product Manual, Sanitation Equipment Division*, Rex Chain Belt Company, Milwaukee, Wisconsin, 1963.
12. Rao, N. S. and D. Chandrasekaran, Outlet weirs for trapezoidal grit chamber, *Journal Water Pollution Control Federation*, 44(3), 1972, 459–469.
13. Benefield, L. O., J. F. Judkins, and A. D. Parr, *Treatment Plant Hydraulics for Environmental Engineers*, Prentice-Hall, Eaglewood Cliffs, NJ, 1984.

14. Chow, V. T., *Open-Channel HYDRAULICS*, McGraw-Hill Book Co., New York, 1959.
15. Qasim, S. R., *Wastewater Treatment Plants: Planning, Design, and Operation*, Technomic Publishing Company, Inc., Lancaster, PA, 1994.
16. Fair, G. M., J. C. Geyer, and J. C. Morris, *Water Supply and Wastewater Disposal*, McGraw-Hill Book Co., New York, 1954.
17. Kusters Water Div. of Kusters Zima Corp., *Traveling Bridge Grit & Grease Removal System*, http://www.kusterswater.com
18. SUEZ Treatment Solutions, Inc., *Suez Traveling Bridge Grit and Grease Removal System*, http://www.suez-na.com, 2009.
19. Walker Process Equipment, Rolling Grit–Grit Removal System, http://www.walker-process.com
20. Neighbor, J. B. and T. W. Cooper, Design and operation criteria for aerated grit chamber, *Water and Sewage Works*, 112(12), 1965, 448.
21. Albrecht, A. E., Aerated grit chamber operation and design, *Water and Sewage Works*, 114(9), 1967, 331.
22. Rouse, H., *Engineering Hydraulics*, Wiley & Sons, New York, NY, 1950.
23. Hwang, N.H.C., *Fundamentals of Hydraulic Engineering Systems*, Prentice-Hall, Englewood Cliffs, NJ, 1981.
24. Hydro International, *TeaCup®, Advanced Grit Removal and Classification*, http://www.hydro-int.com, 2017.
25. Hydro International, *Grit King®, All Hydraulic Grit Separation*, http://www.hydro-int.com, 2017.
26. John Meunier Products/Veolia Water Technologies Canada, *John Meunier Mectan® Induced Vortex Grit Chamber from Veolia Water Technologies Canada*, http://www.veoliawatertech.com-/en/markets/municipal-solutions/ headworks/
27. Smith and Loveless Inc., *Pista® Grit Removal System*, http://www.smithandloveless.com, 2012.
28. Smith and Loveless Inc., *Engineering Data: Pista® Grit Removal System Notes on Design*, http://www.smithandloveless.com, 2012.
29. SUEZ Treatment Solutions, Inc., *Vortex® Grit Remover*, http://www.suez-na.com, 2010.
30. Huber Technology Inc., *Huber Vortex Grit Chamber VORMAX*, http://www.huber-technology.com, 2010.
31. Napier-Reid Ltd., *Napier Reid Vortex Grit Chamber*, http://www.napier-reid.com, 2007.
32. Kusters Water Div. of Kusters Zima Corp., *XGT™ Vortex Grit Removal System*, http://www.kusterswater.com
33. Hydro International, *Grit Snail®, High Performance Grit Dewatering Escalator*, http://www.hydro-int.com, 2017.
34. Weir Specialty Pumps, *WEMCO® Hydrogritter® De-Gritting Machines*, http://www.global.weir/businesses/weir-specialty-pumps, 2012.
35. Smith and Loveless Inc., *Pista® Turbo™ Grit Washer*, http://www.smithandloveless.com, 2012.
36. Smith and Loveless Inc., *Pista® Duralyte™ Grit Concentrator*, http://www.smithandloveless.com, 2012.
37. John Meunier Products/Veolia Water Technologies Canada, *Sam® TYPE GFW Grit Washer with Shaftless Screw*, http://www.veoliawatertech.com/en/markets/municipal-solutions/headworks/
38. John Meunier Products/Veolia Water Technologies Canada, *Sam® Grit Dewatering Screw*, http://www.veoliawatertech.com/en/markets/municipal-solutions/headworks/
39. Huber Technology Inc., *COANDA Grit Classifier RoSF 3*, http://www.huber-technology.com, 2005.
40. Huber Technology Inc., *ROTAMAT® Grit Washer RoSF 4/t*, http://www.huber-technology.com, 2010.
41. Kusters Water Div. of Kusters Zima Corp., *Grit Washer*, http://www.kusterswater.com

9

Primary and Enhanced Sedimentation

9.1 Chapter Objectives

The purpose of a primary sedimentation basin is to remove settleable organic solids. It is achieved in large basins under relatively quiescent conditions. The settled solids are collected by mechanical scrapers into a hopper, from which they are pumped to a sludge processing area. Oil, grease, and other floating materials are skimmed off from the surface. The effluent is discharged over weirs into a collection trough.

Often, chemicals are added to enhance the removal efficiency of a plain sedimentation basin. The colloidal particles are flocculated and removed with the settling floc. These particles would otherwise not be removed in plain sedimentation. Chemicals may also be added to precipitate soluble organics, phosphorous, and heavy metals.

The objectives of this chapter are to present the basic design and selection criteria for plain and enhanced primary sedimentation basins. Secondary clarifiers are covered in Chapter 10. The specific objectives of this chapter are to present:

- Theory of flocculent settling
- Types of sedimentation basins
- Basic design factors and basin performance
- Theory of coagulation and flocculation
- Enhanced primary sedimentation
- High-rate clarification
- Design of basin components and example problems

9.2 Flocculent Settling (Type II)

The settling behavior of solids in primary and enhanced sedimentation basins is of a flocculent type as particles coalesce, mass increases and particles settle faster. The curvilinear particles trajectory is shown in Figure 9.1a. As described in Chapter 8, mathematical equations have been well developed for discrete settling (Type I). However, no mathematical model has been developed because of unknown settling behavior of flocculent particles.[1–3] Also, unlike discrete settling, detention time is an important parameter in the design of a flocculent settling basin. Design parameters (surface overflow rate at a detention time and water depth) are developed either by a batch-type settling column test or through experience with existing plants treating similar wastewater.[2]

Batch Settling Column: The batch flocculent settling column tests are performed in the laboratory. The settling column is 15–20 cm (6–8 in) in diameter and 2–5 m (6.5–16 ft) tall. Sampling ports are provided at

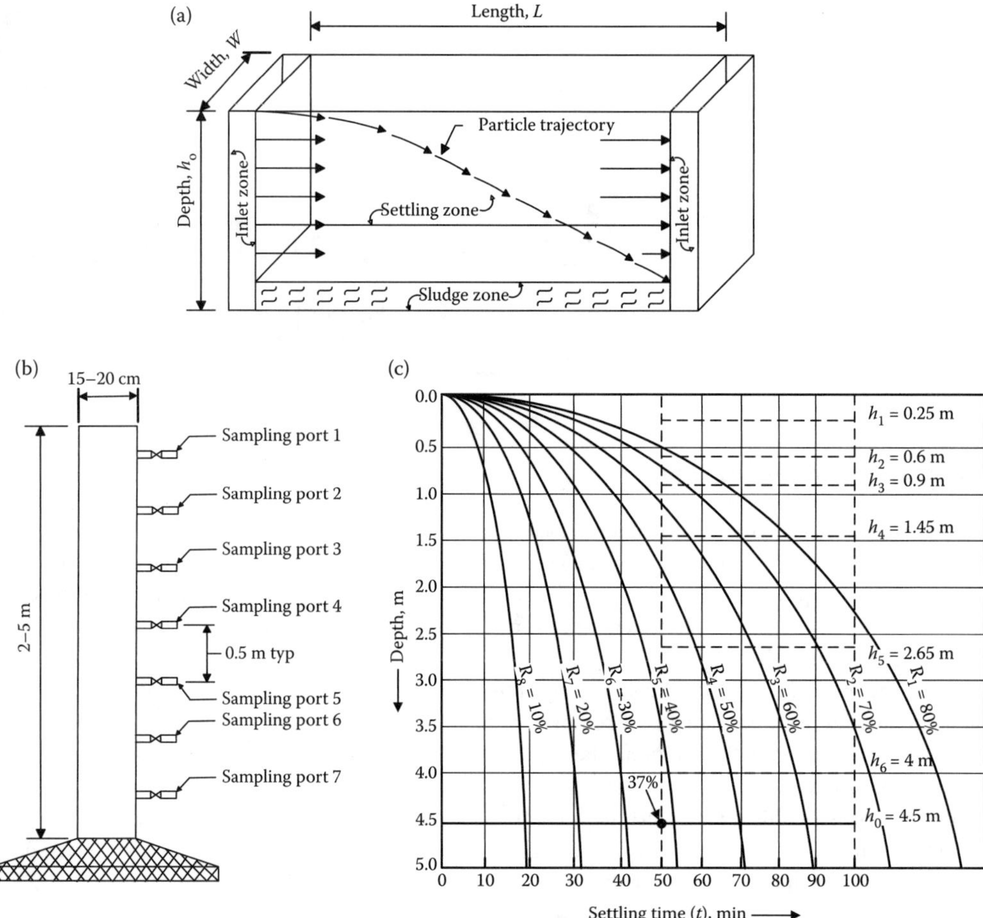

FIGURE 9.1 Flocculant settling: (a) settling trajectories of flocculant particle in a sedimentation basin, (b) flocculant column for batch settling test, and (c) graph showing results of percent removal and isoremoval lines.

uniform depths from top. The column details are shown in Figure 9.1b.[2–4] Ideally, the height should be equal to the proposed side water depth of the basin. The suspension is thoroughly mixed, then placed into the column to the desired depth.

Test Procedure: Samples are withdrawn usually at initial intervals of 5–10 min simultaneously from all ports. Later the frequency of sampling is increased. A test with duration of 1–3 h should yield sufficient data to develop the design parameters. Ideally, the tests should be conducted twice to ensure the repeatability of the results. The total suspended solids (TSS) concentration is determined for each sample, preferably in duplicate.

Test Results: The TSS results are reduced to yield percent removals. A summary table with reduced results is generated, and a grid showing percent hypothetical removal of TSS at each port and at different time intervals is plotted. Lines or contours of equal percentage removal or *isoremoval* are drawn (Figure 9.1c). These lines also trace the maximum trajectories of particles' settling paths for specific concentrations in a flocculent suspension.[1,4]

The overall percent removal of solids at a given detention time and depth of the column is calculated from Equation 9.1a or Equation 9.1b.[1–4]

$$\text{Overall percent removal} = \frac{h_1}{h_0}(100 - R_1) + \frac{h_2}{h_0}(R_1 - R_2) + \cdots + \frac{h_{n-1}}{h_0}(R_{n-1} - R_n) + R_n \tag{9.1a}$$

$$\text{Overall Percent removal} = \frac{\Delta h_1}{h_0}\left(\frac{100 + R_1}{2}\right) + \frac{\Delta h_2}{h_0}\left(\frac{R_1 + R_2}{2}\right) + \cdots + \frac{\Delta h_n}{h_0}\left(\frac{R_{n-1} + R_n}{2}\right) \tag{9.1b}$$

where

$h_1, h_2,\ldots,h_n$ = vertical distance from the top of the settling column to the midpoint between two consecutive particle isoremoval curves at the desired detention time (Figure 9.1c), m (ft)

h_0 = desired side water depth that is less than or equal to the depth of settling column (Figure 9.1c), m (ft)

$R_1, R_2,\ldots, R_n$ = consecutive particle isoremoval curves, percent removal

$\Delta h_1, \Delta h_2,\ldots, \Delta h_n$ = vertical distance between two consecutive particle isoremoval curves at the desired detention time (Figure 9.1c), m (ft)

The theoretical detention time and surface overflow rate are obtained from the percent particle removal efficiency curves. To account for less than optimum conditions encountered in the field, the design values are typically obtained by multiplying the theoretical values of surface overflow rate by a factor 0.65–0.85, and detention time by a factor 1.25–1.5.[2–4] The use of particle isoremoval curves and these equations, and procedure for determining the design values of the surface overflow rate and detention time from a settling column test are covered in several examples below.

EXAMPLE 9.1: PERCENT REMOVAL FOR A GIVEN COLUMN DEPTH AND DETENTION TIME

A flocculant settling column study was conducted on an industrial wastewater. The column depth was 4.5 m, and initial TSS concentration of the sample was 300 mg/L. The particle isoremoval graph is shown in Figure 9.1c. Determine (a) overall percent TSS removal at 50-min detention time and desired water depth of 4.5 m from Equations 9.1a and 9.1b, (b) surface overflow rate ($m^3/m^2{\cdot}d$, gpd/ft^2) corresponding to 50-min detention time and desired water depth of 4.5 m, (c) the percent hypothetical removal of particles, or the maximum trajectories or particles' settling path of 180 mg/L TSS concentration in the effluent, (d) percent removal of particles at a water depth of 3 m and 62 min detention time, (e) detention time for 30% removal of particles at a water depth of 2 m, and (f) side water depth for 70% removal of particles at a detention time of 80 min.

Solution

1. Determine the percent hypothetical removal of particles.

   The desire water depth (from top) is equal to the column depth, $h_0 = 4.5$ m.

   Draw a vertical line from detention time of 50 min in Figure 9.1c. A 36% hypothetical removal of particles is obtained on the vertical line at the desire water depth of 4.5 m.
2. Determine the midpoint depths between the consecutive isoremoval curves.

   Read the midpoint depths between two consecutive isoremoval curves on the vertical line at 50-min detention time in Figure 9.1c.

   $h_1 = 1/2 \times (0.5\text{ m}) = 0.25$ m (between 80% and 100%)
   $h_2 = 1/2 \times (0.5\text{ m} + 0.7\text{ m}) = 0.6$ m (between 70% and 80%)
   $h_3 = 1/2 \times (0.7\text{ m} + 1.1\text{ m}) = 0.9$ m (between 60% and 70%)
   $h_4 = 1/2 \times (1.1\text{ m} \times 1.8\text{ m}) = 1.45$ m (between 50% and 60%)
   $h_5 = 1/2 \times (1.8\text{ m} + 3.5\text{ m}) = 2.65$ m (between 40% and 50%)
   $h_6 = 1/2 \times (3.5\text{ m} + 4.5\text{ m}) = 4$ m (between 37% and 40%)

3. Calculate from Equation 9.1a the overall percent TSS removal at 50-min detention time and at the desire water depth of 4.5 m.

$$\text{Overall percent TSS removal} = \frac{0.25\,\text{m}}{4.5\,\text{m}} \times (100-80)\% + \frac{0.6\,\text{m}}{4.5\,\text{m}} \times (80-70)\%$$
$$+\frac{0.9\,\text{m}}{4.5\,\text{m}} \times (70-60)\% + \frac{1.45\,\text{m}}{4.5\,\text{m}} \times (60-50)\%$$
$$+\frac{2.65\,\text{m}}{4.5\,\text{m}} \times (50-40)\% + \frac{4\,\text{m}}{4.5\,\text{m}} \times (40-37)\% + 37\%$$
$$= (1.11 + 1.33 + 2 + 3.22 + 5.89 + 2.67 + 37)\%$$
$$= 53.2\%$$

4. Determine the depths of the isoremoval curves.

Read the incremental depths between two consecutive isoremoval curves on the vertical line at 50-min detention time in Figure 9.1c.

$\Delta h_1 = 0.5$ m (between 80% and 100%)
$\Delta h_2 = 0.7 - 0.5$ m $= 0.2$ m (between 70% and 80%)
$\Delta h_3 = 1.1 - 0.7$ m $= 0.4$ m (between 60% and 70%)
$\Delta h_4 = 1.8 - 1.1$ m $= 0.7$ m (between 50% and 60%)
$\Delta h_5 = 3.5 - 1.8$ m $= 1.7$ m (between 40% and 50%)
$\Delta h_6 = 4.5 - 3.5$ m $= 1$ m (between 37% and 40%)

5. Calculate from Equation 9.1b the overall percent TSS removal at 50-min detention time at a column depth of 4.5 m. Use the vertical distance between the consecutive curves of percent removal.

$$\text{Overall percent TSS removal} = \frac{0.5\,\text{m}}{4.5\,\text{m}} \times \left(\frac{100+80}{2}\right)\% + \frac{0.2\,\text{m}}{4.5\,\text{m}} \times \left(\frac{80+70}{2}\right)\%$$
$$+\frac{0.4\,\text{m}}{4.5\,\text{m}} \times \left(\frac{70+60}{2}\right)\% + \frac{0.7\,\text{m}}{4.5\,\text{m}} \times \left(\frac{60+50}{2}\right)\%$$
$$+\frac{1.7\,\text{m}}{4.5\,\text{m}} \times \left(\frac{50+40}{2}\right)\% + \frac{1\,\text{m}}{4.5\,\text{m}} \times \left(\frac{40+37}{2}\right)\%$$
$$= (10 + 3.33 + 5.78 + 8.56 + 17 + 8.56)\%$$
$$= 53.2\%$$

6. Calculate the surface overflow rate (*SOR*) for a depth of fall of 4.5 m in 50 min.

$$SOR = \frac{4.5\,\text{m}}{50\,\text{min}} = 0.09\,\text{m/min}$$

$$SOR = \frac{0.09\,\text{m}}{\text{min}} \times \frac{\text{m}^2}{\text{m}^2} \times \frac{(60 \times 24)\,\text{min}}{\text{d}} = 130\,\text{m}^3/\text{m}^2\cdot\text{d}$$

$$SOR = \frac{130\,\text{m}^3}{\text{m}^2\cdot\text{d}} \times \frac{10^3\,\text{L}}{\text{m}^3} \times \frac{\text{gal}}{3.79\,\text{L}} \times \frac{\text{m}^2}{10.8\,\text{ft}^2} = 3180\,\text{gpd/ft}^2$$

7. Calculate the percent removal of particles that have the maximum trajectories or settling path that gives 180 mg/L TSS concentration in the effluent.

The maximum trajectories of the settling path of particles that give 180 mg/L TSS concentration in the effluent is the percent removal of TSS.

Percent removal curve that gives 180 mg/L TSS concentration in the effluent

$$= \frac{(300-180)\,\text{mg/L}}{300\,\text{mg/L}} 100\% = 40\%$$

The maximum trajectories or particles' settling path that gives 180 mg/L TSS in the effluent is the isopercent removal curve of 40% in Figure 9.1c. This means that at the detention time of 53 min, 40% of particles as 120 mg/L TSS will reach the 4.5 m depth of the basin or 60% of particles as 180 mg/L TSS will remain in the effluent.

8. Determine the percent removal of particles, detention time, and water depth on the settling paths from Figure 9.1c.
   a. Percent removal of particles at a water depth of 3 m and 62-min detention time = 50% of particles.
   b. Detention time for 30% removal of particles at a water depth of 2 m = 32 min.
   c. Side water depth for 70% removal of particles at a detention time of 80 min = 1.95 m.

EXAMPLE 9.2: FLOCCULENT SETTLING COLUMN TEST FOR DETENTION TIME AND SURFACE OVERFLOW RATE

A chemically enhanced primary treatment (CEPT) study was conducted on a coagulated industrial wastewater sample. The objective was to design a basin to remove 80% TSS at a side water depth of 3.5 m. A batch column test was performed in a 4-m-tall and 15-cm-diameter column. The initial concentration of TSS in the coagulated sample was 200 mg/L. The sampling ports were 0.5 m apart, the lowest port being 3.5 m below the water surface to simulate the design side water depth of the basin. The samples were withdrawn from each port at 10-min interval. TSS was determined in each sample and the data matrix is given in Table 9.1. Provide the following information (a) write the step-by-step experimental procedure and (b) determine the design values of detention time and overflow rate for 80% removal.

TABLE 9.1 TSS Results in Samples Withdrawn from Various Ports (Example 9.2)

Port Number	Depth, m	TSS Concentration at Different Sampling Time (min), mg/L								
		10	20	30	40	50	60	70	80	90
1	0.5	134	76	52	20	–	–	–	–	–
2	1.0	158	118	70	59	40	22	20	–	–
3	1.5	168	128	82	66	52	38	28	18	–
4	2.0	166	134	88	72	58	44	36	24	18
5	2.5	172	136	92	72	60	44	36	30	24
6	3.0	172	140	96	74	62	50	38	34	30
7	3.5	176	140	98	80	62	52	40	34	32

Solution

1. Write the step-by-step procedure of batch flocculant column analysis.
   a. Mix the flocculent sample and run TSS test in triplicate.
   b. Fill the settling column with flocculent suspension.
   c. Draw samples from all ports at 10-min intervals.
   d. Determine TSS in each sample in duplicate, and record the average value (Table 9.1).
   e. Reduce the TSS results to percent removal for each port and prepare a summary table (Table 9.2).
   f. Plot a grid showing percent TSS removal at each port at different time intervals. This grid is shown in Figure 9.2.
   g. Draw lines of equal percentage removal (*isoremoval*) of particles. These lines are drawn similar to the contour lines (Figure 9.2).

TABLE 9.2 Reduced Batch Settling Test Results to Percent Removal at Various Ports (Example 9.2)

Port Number	Depth, m	Particle Removed at Different Sampling Time (min), % Removal as TSS								
		10	20	30	40	50	60	70	80	90
1	0.5	33[a]	62	74	90	–	–	–	–	–
2	1.0	21	41	65	71	80	89	90	–	–
3	1.5	16	36	59	67	74	81	86	91	–
4	2.0	17	33	56	64	71	78	82	88	91
5	2.5	14	32	54	64	70	78	82	85	88
6	3.0	14	30	52	63	69	75	81	83	85
7	3.5	12	30	51	60	69	74	80	83	84

[a] 33% of initial TSS removed at sampling time of 10 min since start, $\frac{(200-134)\ \text{mg/L}}{200\ \text{mg/L}} \times 100\% = 33\%$.

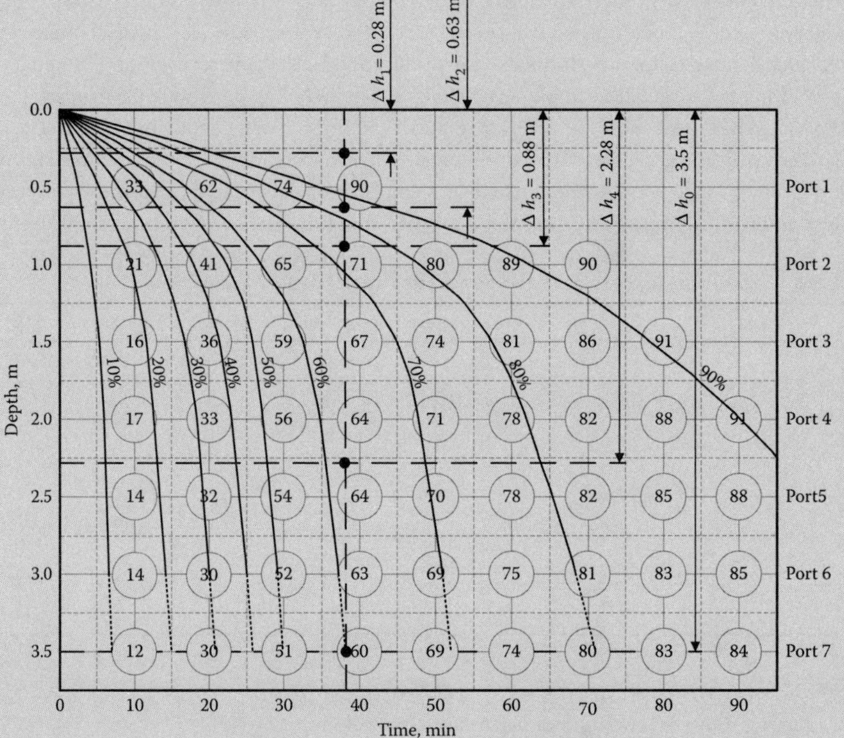

FIGURE 9.2 Data grid and isoremoval lines developed from batch settling test data (Example 9.2).

h. Draw a vertical line at each point of intersection of particle isoremoval curve at 3.5 m depth of the column. For example, the $R = 60\%$ particle isoremoval curve intercepts the 3.5 m column depth at 38 min. The intersection point represents 60% particles that have settled 3.5 m at a detention time t of 38 min.

2. Calculate the overall percent removal of TSS for each intersection point. The percent TSS removal is calculated from Equation 9.1a.

   a. Illustrate the overall percent removal calculations for 60% particle isoremoval curves.

The 60% particle isoremoval curve meets 3.5 m column depth at 38-min detention time. $h_0 =$ 3.5 m, and h_1, h_2, h_3 , and h_4 values are 0.28, 0.63, 0.88, and 2.28 m, respectively. These values are indicated in Figure 9.2.

$$\text{Overall percent TSS removal} = \frac{0.28\text{ m}}{3.5\text{ m}} \times (100-90)\% + \frac{0.63\text{ m}}{3.5\text{ m}} \times (90-80)\%$$

$$+ \frac{0.88\text{ m}}{3.5\text{ m}} \times (80-70)\% + \frac{2.28\text{ m}}{3.5\text{ m}}(70-60)\% + 60\%$$

$$= (0.79 + 1.79 + 2.5 + 6.5 + 60)\% = 71.6\%$$

b. Calculate the overall percent removal for other detention times.

The above steps are repeated to determine the percent removals at 7-, 15-, 21-, 26-, 30-, 38-, 52-, and 71-min detention times. These detention times include the intersection points of percent particle isoremoval curves of 10%, 20%, 30%, 40%, 50%, 60%, 70%, and 80% and at 3.5 m column depth. The calculated percent removals are summarized in Table 9.3.

TABLE 9.3 Detention Time, Settling Velocity, Overall Percent TSS Removal, and Surface Overflow Rate (Example 9.2)

Detention Time, min	Settling Velocity, m/min	Surface Overflow Rate, $m^3/m^2{\cdot}d$	Overall Percent TSS Removal, %
7[a]	0.50[b]	720[c]	21.3[d]
15	0.23	336	35.5
21	0.17	240	46.7
26	0.13	194	56.1
30	0.12	168	63.2
38	0.092	133	71.6
52	0.067	97	80.5
71	0.049	71	88.5

[a] 7-min detention time is obtained from the intersection of 10% particle isoremoval curve at the water depth of 3.5 m.

[b] Settling velocity $= \frac{3.5\text{ m}}{7\text{ min}} \times 100\% = 0.5\text{ m/min}$.

[c] Surface overflow rate $= \frac{0.5\text{ m}}{\text{min}} \times \frac{\text{m}^2}{\text{m}^2} \times \frac{(60 \times 24)\text{ min}}{\text{d}} = 720\text{ m}^3/\text{m}^2{\cdot}\text{d}$

[d] Overall percent TSS removal is obtained from Equation 9.1a. The procedure for detention time of 38 min (corresponds to 60% particle isoremoval curve) is shown in Step 2a.

3. Calculate the settling velocity and overflow rates.

The settling velocity of the suspension at the column depth of 3.5 m, detention time, and overflow rates for different percent removal are summarized in Table 9.3.

4. Draw the overall percent TSS removal curves with respect to detention time and surface overflow rate.

The overall percent TSS removal curves with respect to detention time and surface overflow rate are developed from data given in Table 9.3. These plots are shown in Figure 9.3a and b.

5. Determine the theoretical values of detention time and surface overflow rate for 80% TSS removal.

Read theoretical values of detention time and surface overflow rate for 80% TSS removal from the overall percent TSS removal curves in Figure 9.3.

Detention time from Figure 9.3a = 50 min

Surface overflow rate from Figure 9.3b $= 101\text{ m}^3/\text{m}^2{\cdot}\text{d}$

6. Determine the design values of detention time and surface overflow rate.

The batch settling test in the laboratory represents ideal settling conditions. The ideal conditions are rarely present in a full-scale continuous-flow sedimentation basin. To account for effects such as

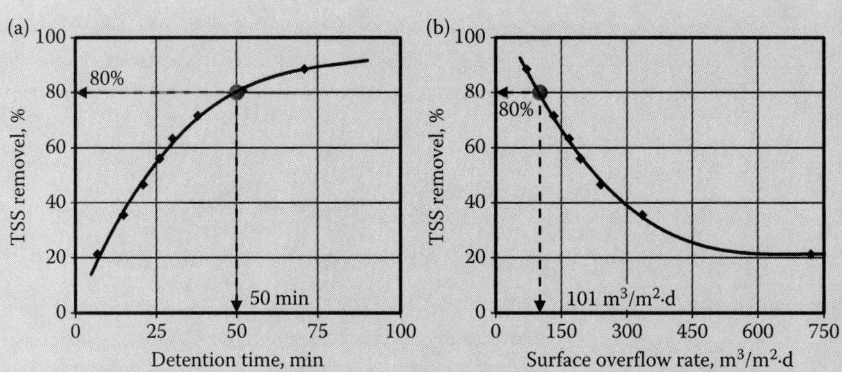

FIGURE 9.3 Percent TSS removal curves with respect to (a) detention time and (b) surface overflow rate (Example 9.2).

density currents, temperature currents, wind effect, and uneven flow distributions, correction factors are applied to the experimental results. The test results are multiplied by the safety factors to yield the design parameters. Use the multiplying factors of 1.6 for detention time, and 0.7 for surface overflow rates.

Design detention time = 1.6 × 50 min = 80 min

Design surface overflow rate = 0.7 × 101 $m^3/m^2{\cdot}d$ = 71 $m^3/m^2{\cdot}d$ or 1740 gpd/ft^2

9.3 Influent Quality of Primary Sedimentation Basin

Besides degritted raw wastewater, the primary basins may also receive *side streams* that are recycled within the plant. These side streams may include rejects from sludge processing areas, filter backwash return, and waste-activated sludge. These streams increase solids, BOD_5, nitrogen and phosphorus loadings to the primary basins by 10–30%. A material mass balance analysis should be conducted to determine the impact of these streams.

The side streams from sludge processing areas may include supernatant from thickeners and digesters, and filtrate from dewatering facilities. If the plant has gravity filters or microstrainers for effluent polishing, the solids-bearing streams may also be returned from the filter backwash recovery system.

The waste-activated sludge normally contains solids that are <1% (10,000 mg/L). At many small plants, the waste-activated sludge stream is also returned to the primary basin. The purpose is to concentrate these solids. The combined primary and secondary sludge removed from primary basins normally has 3–4% solids.

EXAMPLE 9.3: IMPACT OF SIDE STREAMS ON INFLUENT QUALITY TO PRIMARY BASIN

A wastewater treatment plant is designed for a flow of 38,000 m^3/d. The degritted stream has TSS and BOD_5 concentrations of 240 and 180 mg/L, respectively. The side streams from gravity thickener, sludge digester, and filter press areas are characterized as follows Examples 6.22 and 6.23.

Parameter	Gravity Thickener	Sludge Digester	Filter Press
Q, m^3/L	1750	30	300
TSS, mg/L	800	4000	650
BOD_5, mg/L	500	3000	1740

Calculate the characteristics of influent to the primary sedimentation basin, and percent TSS and BOD_5 increase due to these returned side streams.

Solution

1. Draw the definition sketch in Figure 9.4.

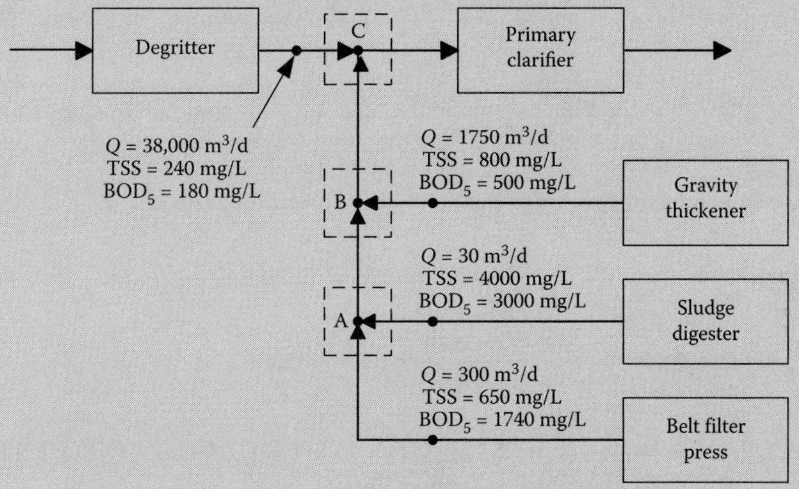

FIGURE 9.4 Definition sketch of degritter effluent and side streams prior to primary sedimentation basin (Example 9.3).

2. Determine the characteristics of combined side streams at point A.
   The system boundary is drawn around the point A.
   Determine the characteristics at point A based on side streams from sludge digester and filter press.

$$Q = (300 + 30)\,\text{m}^3/\text{d} = 30\,\text{m}^3/\text{d}$$

$$\text{TSS} = \frac{300\,\text{m}^3/\text{d} \times 650\,\text{g/m}^3 + 30\,\text{m}^3/\text{d} \times 4000\,\text{g/m}^3}{330\,\text{m}^3/\text{d}} = 955\,\text{g/m}^3 \quad \text{or} \quad 955\,\text{mg/L}$$

$$\text{BOD}_5 = \frac{300\,\text{m}^3/\text{d} \times 1740\,\text{g/m}^3 + 30\,\text{m}^3/\text{d} \times 3000\,\text{g/m}^3}{330\,\text{m}^3/\text{d}} = 1855\,\text{g/m}^3 \quad \text{or} \quad 1855\,\text{mg/L}$$

Note: Concentration units of mg/L and g/m^3 are identical.

3. Determine the characteristics of combined side streams at point B.
   The system boundary is drawn around the point B.
   Determine the characteristics at point B based on the combined side streams from point A and gravity thickener.

$$Q = (330 + 1750)\,\text{m}^3/\text{d} = 2080\,\text{m}^3/\text{d}$$

$$\text{TSS} = \frac{330\,\text{m}^3/\text{d} \times 955\,\text{g/m}^3 + 1750\,\text{m}^3/\text{d} \times 800\,\text{g/m}^3}{2080\,\text{m}^3/\text{d}} = 825\,\text{g/m}^3 \quad \text{or} \quad 825\,\text{mg/L}$$

$$\text{BOD}_5 = \frac{330\,\text{m}^3/\text{d} \times 1855\,\text{g/m}^3 + 1750\,\text{m}^3/\text{d} \times 500\,\text{g/m}^3}{2080\,\text{m}^3/\text{d}} = 715\,\text{g/m}^3 \quad \text{or} \quad 715\,\text{mg/L}$$

4. Determine the characteristics of influent to the primary sedimentation basin at point C.

The system boundary is drawn around the point C.

Determine the characteristics at point C based on the combined side streams from point B and effluent from degritter.

$$Q = (2080 + 38{,}000)\,\text{m}^3/\text{d} = 40{,}080\,\text{m}^3/\text{d}$$

$$\text{TSS} = \frac{2080\,\text{m}^3/\text{d} \times 825\,\text{g/m}^3 + 38{,}000\,\text{m}^3/\text{d} \times 240\,\text{g/m}^3}{40{,}080\,\text{m}^3/\text{d}} = 270\,\text{g/m}^3 \quad \text{or} \quad 270\,\text{mg/L}$$

$$\text{BOD}_5 = \frac{2080\,\text{m}^3/\text{d} \times 715\,\text{g/m}^3 + 38{,}000\,\text{m}^3/\text{d} \times 180\,\text{g/m}^3}{40{,}080\,\text{m}^3/\text{d}} = 208\,\text{g/m}^3 \quad \text{or} \quad 208\,\text{mg/L}$$

5. Calculate the percent increase in TSS and BOD_5 due to returned side streams.

$$\text{Percent TSS increase} = \frac{(270 - 240)\,\text{mg/L}}{240\,\text{mg/L}} \times 100\% = 12.5\%$$

$$\text{Percent BOD}_5\text{ increase} = \frac{(208 - 180)\,\text{mg/L}}{180\,\text{mg/L}} \times 100\% = 15.6\%$$

EXAMPLE 9.4: IMPACT OF WASTE-ACTIVATED SLUDGE ON INFLUENT QUALITY TO PRIMARY BASIN

An activated sludge plant is treating an average flow of 40,000 m^3/d. The TSS and BOD_5 of degritted raw wastewater are 250 and 200 mg/L, respectively. The biological solids are wasted from the aeration basin at a rate of 2300 kg/d. The waste-activated sludge (WAS) is pumped to the head of primary sedimentation basin at a rate of 250 m^3/d. The ratio of BOD_5 and TSS in waste-activated sludge is 0.43. Calculate the characteristics of combined influent stream reaching the primary basin. Also calculate the percent increase in TSS and BOD_5.

Solution

1. Draw the definition sketch in Figure 9.5.

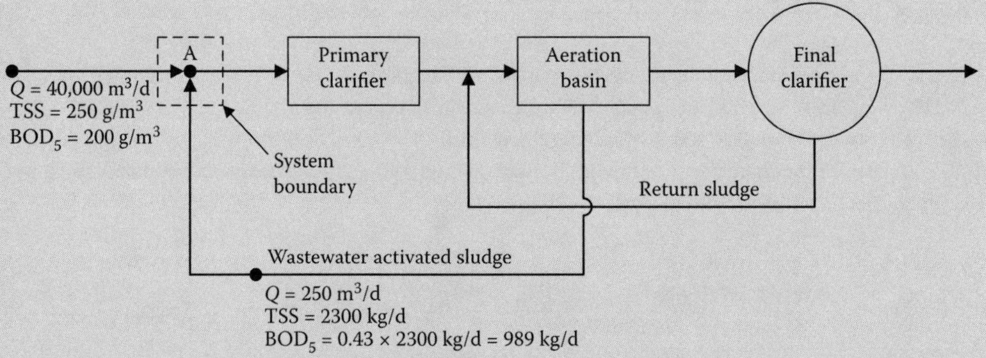

FIGURE 9.5 Definition sketch of the activated sludge plant (Example 9.4).

2. Calculate the combined flow, TSS, and BOD_5 reaching the primary sedimentation basin.

The system boundary is drawn around the point A.

$$\text{Combined flow} = (40{,}000 + 250)\,\text{m}^3/\text{d} = 40{,}250\,\text{m}^3/\text{d}$$

$$\text{TSS in combined flow} = \frac{40{,}000\,\text{m}^3/\text{d} \times 250\,\text{g/m}^3 + 2300\,\text{kg/d} \times \dfrac{10^3\,\text{g}}{\text{kg}}}{40{,}250\,\text{m}^3/\text{d}}$$

$$= 306\,\text{g/m}^3 \quad \text{or} \quad 306\,\text{mg/L}$$

$$\text{BOD}_5 \text{ in biosolids} = 0.43 \times 2300\,\text{kg/d} = 989\,\text{kg/d}$$

$$\text{BOD}_5 \text{ in combined flow} = \frac{40{,}000\,\text{m}^3/\text{d} \times 200\,\text{g/m}^3 + 989\,\text{kg/d} \times 10^3\,\text{g/kg}}{40{,}250\,\text{m}^3/\text{d}}$$

$$= 223\,\text{g/m}^3 \quad \text{or} \quad 223\,\text{mg/L}$$

3. Calculate the percent increase in TSS and BOD_5.

$$\text{Increase in TSS} = \frac{(306 - 250)\,\text{mg/L}}{250\,\text{mg/L}} \times 100\% = 22.4\%$$

$$\text{Increase in BOD}_5 = \frac{(223 - 200)\,\text{mg/L}}{200\,\text{mg/L}} \times 100\% = 11.5\%$$

9.4 Types of Primary Sedimentation Basins

The primary sedimentation basins are broadly divided into *plain* or *conventional*, *chemically enhanced*, and *high-rate* sedimentation basins. Most plain primary sedimentation basins are horizontal-flow type. The velocity gradient is primarily in the horizontal direction, and solids settle by gravity. The common types of horizontal-flow clarifiers are: rectangular, square, or circular. In some cases, stacked (multilevel) basins have also been designed. The selection of the type of basin for a given application depends upon (1) size of the facility, (2) design criteria or preference of local regulatory authorities, (3) experience and judgment of the design engineer, (4) preference of the operation and maintenance personnel, (5) local site considerations, and (6) economics. Two or more tanks should be provided for operational flexibility and maintenance needs. Brief description of different types of plain primary sedimentation basins is given below.

High-rate clarification basins are designed in conjunction with coagulation and flocculation. The sedimentation basin may utilize solids contact effect, inclined plate or tube settler, or ballasted flocculation for enhanced clarification. Most of these systems utilize proprietary equipment and are sold under various trade names. Discussion on this topic is provided in Section 9.6.

9.4.1 Rectangular Basin

Description: The dimensions of a rectangular basin are selected to accommodate standard-size equipment. Rectangular basins offer larger flow path than the circular basins. The flow distribution in rectangular basins is critical. Therefore, full-width influent channel with multiple submerged ports, and a baffle extending over the entire width, and 300 mm (12 in) in front of the submerged ports are provided. The horizontal component of the velocity should be maintained in the range of 3–9 m/min (10–30 ft/min). Multiple effluent weirs with V-notches are provided in the basin. The dimensions of the rectangular basins are given in Table 9.4.[2–7] The rectangular sedimentation basins are illustrated in Figure 9.6.

TABLE 9.4 Dimensions of a Rectangular Sedimentation Basin

Parameter	Range	Typical
Length, m (ft)	15–90 (50–300)	30–60 (100–200)
Width[a,b], m (ft)	6–24 (20–80)	6–9 (20–30)
Side water depth, m (ft)	3–5 (10–16)	3.7–4 (12–13)
Length-to-width ratio	1.5–15:1	4–5:1
Length-to-depth ratio	5–25:1	10–18:1
Bottom slope, percent	6–15	8

[a] Most manufacturers build equipment in a standard increment of 61 cm (2 ft) in width.

[b] The normal width of a single flight is 6 m (20 ft). If the width is >6 m (20 ft), use of multiple bays may be necessary for sludge collection.

Source: Adapted in part from References 2 through 7.

Advantages and Disadvantages: The advantages of rectangular basins are (1) less area requirement with multiple units, (2) cost-effective for using common walls, (3) easy to cover for odor control, (4) long travel distance for settling to occur, (5) low risk of short-circuiting, (6) low inlet–outlet losses, and (7) low power consumption for sludge collection mechanisms. The *disadvantages* are (1) possible dead spaces, (2) sensitive to flow surges, (3) restricted in width by collection equipment, (4) requirement of multirow weirs to achieve low weir loading rates, and (5) high upkeep and maintenance cost of sprockets, chains, and flights used for sludge collection.

Multiple rectangular tanks are constructed with equipment galleries to house the sludge pumps, piping, and maintenance equipment. Galleries are also connected with pipe tunnels for access to other units.

Scum removal: The scum is normally pushed toward the effluent end by flights of sludge mechanism in its return travel. All effluent weirs have baffles to stop the loss of scum into the effluent. The scum may be scrapped manually or mechanically up an inclined apron. In small installations, a hand-tilt slotted pipe with a lever or screw is commonly used. In large installations, a transverse-rotating helical wiper attached to a shaft moves the scum over an incline apron to a cross-collecting scum trough. A scum sump is provided outside the tank, and a scum pump is provided to transfer scum to the disposal facility. The details of the scum collection and removal systems are shown in Figure 9.7.[8]

FIGURE 9.6 Rectangular primary sedimentation basins (Courtesy CP&Y, Inc., Dallas, Texas).

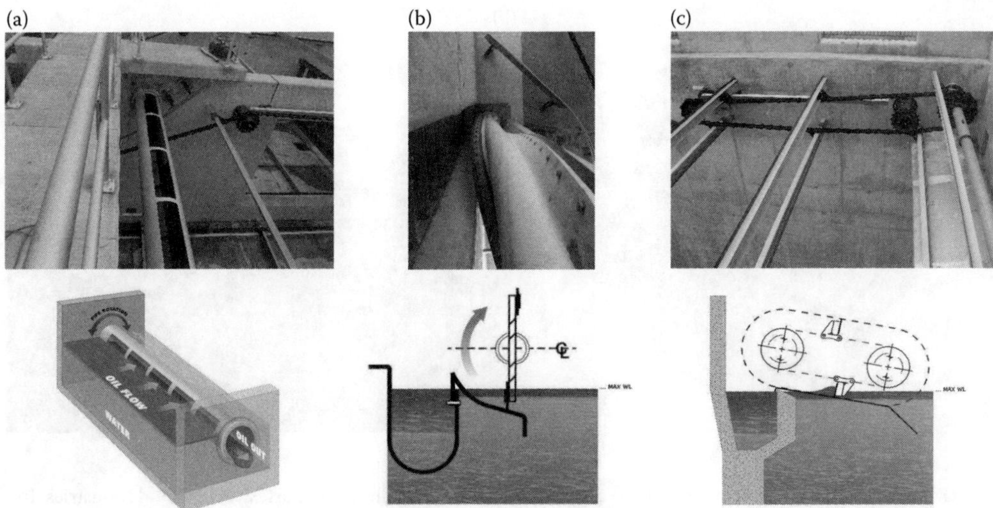

FIGURE 9.7 Scum collection and removal systems (Courtesy Brentwood Industries, Inc.): (a) rotary scum skimming trough to tilt the open-top of the slotted pipe to collect the scum into the trough, (b) rotating helical skimmer that turns the rubber wiper blades to push the scum up pass a curved beach into a scum trough, and (c) power skimmer that uses a chain and flight system to remove scrum from the water surface into a trough.

Sludge Collection: The sludge collection equipment consists of (1) a pair of endless conveyor chains running over sprockets attached to the shaft or (2) moving-bridge collector with a scraper to push the sludge into the hopper located on the influent end. The sludge is removed from the hopper by means of a pump. The bridge drive sludge suction arrangement is not used for primary basins. This type of sludge collector is typically used to withdraw very light sludge such as chemical or biological solids from the basin (see details in Section 10.9.3). System description and design details of sludge collection equipment are summarized below.

Conveyor Chain: One endless conveyor chain is connected to a sprocket, shaft, and drive unit. The linear speed is 0.6–1.2 m/min (2–4 ft/min). In small basins, the settled solids are scraped manually into a hopper. In large basins, an auto-transverse trough is used. The scrapper system consists of cross-wood flights that extend to full width of the tank. The flights are spaced at 3-m (10-ft) interval, and are 6 m in length. Multiple pairs of chains are used in tanks wider than 6 m. The transverse troughs are equipped with cross collectors which move the solids to the hopper. In very long units (over 50 m), two collection systems can be used to move solids in a transverse trough near the middle length. The bottom of tank is sloped toward the hopper typically at a rate of one vertical to eight horizontal. The floating material is pushed in opposite direction of sludge, and is collected in a scum collection box.

The conveyor chain arrangement is simple to install, low in power consumption, and efficient for scum collection. It is suitable for heavy sludge. The major problems of this system are high maintenance cost of chain and flight, and sludge removal mechanism. The tank must be dewatered to repair gears or chain. Light sludge may also resuspend due to turbulences caused by scrappers. The conveyor chain system is shown in Figure 9.8a.[9]

Moving-Bridge Drive Scrappers: The moving-bridge-type sludge collectors travel along the length of the tank. Standard traveling beam bridges are used for spans up to 13 m (40 ft), and truss bridges for spans over 13 m. The bridge travel is accomplished by gear motor. The wheels run on rails that are attached to the footing walls along each side wall of the basin. One or more scraper blades or rakes are hung from the top carriage. These scrappers push the sludge into the cross collector. The scraper

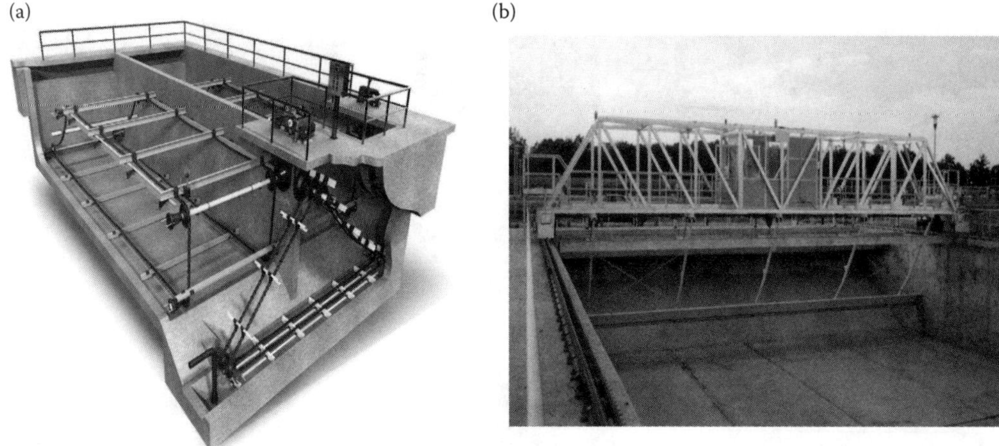

FIGURE 9.8 Sludge collector equipment: (a) conveyor chain sludge collector (Courtesy Brentwood Industries, Inc.), and (b) moving-bridge sludge collector (Courtesy Transdynamics Engineering Ltd.).

blades are lifted above the sludge blanket on return travel. Separate blades are also provided on the top to move the scum.

The moving-bridge drive scrappers offer benefits such as (1) all moving mechanisms are above water (no underwater bearings); (2) standard designs permit scraper repair or replacement without tank dewatering; (3) tank width is not restricted as that with conveyor chain type, and (4) there is long operation life and low maintenance cost in low-span bridges. The disadvantages of the system are high power requirement to move the bridge. The unit will not operate with ice-covered tanks; and in long-span bridges, the wind action may cause the wheel to climb over rails causing breakdowns. The details of a moving-bridge drive scrapper are shown in Figure 9.8b.[10]

Sludge Removal: Sludge removal involves pumping of solids from the hopper to sludge handling facilities. The following design considerations are given to the sludge removal system:

1. Each sludge hopper should have an individual sludge pump. The discharge pipe should be at least 15 cm (6 in) in diameter. Time-control pumping cycle may be necessary to achieve self-cleaning velocity of 0.6–0.9 m/s (2–3 ft/s).
2. The sludge pump must be self-priming centrifugal or submersible type. It is desirable to locate pumping station close to the collection hopper. One sludge pumping station with two or more pumps can serve two rectangular clarifiers. The sludge pumps should discharge into a common manifold.
3. Screw conveyors for sludge removal are also used.
4. The multiple hoppers pose operational difficulties and the use of cross collectors is preferred. Sludge accumulates in corners and slopes, and arches over the sludge draw-off piping. Cross collectors provide withdrawal of more uniform and concentrated sludge.

9.4.2 Circular Basins

Description: Circular basins have radial-flow pattern, in contrast to rectangular basins that have velocity gradient predominantly in the horizontal direction. Multiple tanks are normally arranged in groups of two or four. A central-flow splitter box and sludge pumping station are typically provided. The normally used dimensions of circular basins are given in Table 9.5 and design details are shown in Figure 9.9.[2–5,11–13]

TABLE 9.5 Components and Dimensions of Circular Basin

Parameter	Range	Typical
Diameter, m (ft)	3–60 (10–200)	12–45 (40–150)
Side water depth, m (ft)	3–5 (10–16)	4–4.5 (13–15)
Bottom slope, m/m (ft/ft)	1/16–1/6	1/12
Center well		
Diameter, % of basin diameter	15–20	18
Depth, % of basin diameter	25–50	35–40
Bottom edge of central well	1 m (3 ft) below the energy-dissipating inlet (EDI) port, or at least half the depth of the tank	1.7 (5.6)
Max downward velocity in central well, m/s (ft/s)	≤0.75 (2.5)	–
Freeboard, m (ft)	0.5–0.7 (1.5–2)	–
Sludge collector		
Travel speed, revolutions per minute (rpm)	0.02–0.06	0.03
Tip speed, m/min (ft/min)	≤10 (30)	3 (10)

Source: Adapted in part from References 2 through 5.

Center or peripheral feed basin: A circular tank can be either a center or peripheral feed basin. In a *center feed clarifier,* the inlet is at the center, and the outlet is along the periphery. A concentric baffle distributes the flow equally in radial direction. In a *peripheral feed clarifier,* the flow enters along the periphery. Although clarifiers of both flow configurations perform well, the center feed type are more commonly used for primary treatment. Peripheral feed tanks are generally used for secondary clarification (see details in Section 10.9.3).

Advantages and Disadvantages: The walls of circular tanks act as tension rings, which permit thinner walls than those for rectangular basins. As a result, the circular tanks have a lower capital cost per unit surface area than the rectangular tanks. The circular tanks require more yard piping than rectangular tanks. Other advantages of circular clarifier are low upkeep cost and ease of design and construction. The disadvantages include short-circuiting, low hydraulic detention efficiency, high risk of short-circuiting, lack of scum control, and loss of sludge into the effluent.

Scum Removal: In circular basins, the scum is moved by a radial arm that rotates on the surface with the sludge removal equipment. The radial arm pushes the scum over an inclined apron then into a sump. Water sprays are also used to move the scum.

Sludge Collection: The circular tanks utilize sludge scraping mechanisms installed with radial arm and plows set at an angle. The bottom of the tank is sloped around 1 vertical to 12 horizontal to form an inverted cone. The radial racking arm travels at speed of 0.02–0.06 rpm, while the scrapers push the solids to a relatively small hopper located near the center of the tank. Tanks in excess of 10-m diameter utilize central pier that support the mechanism and are reached by a walkway or bridge. The basins with small diameter utilize beam to support the radial arm. The equipment details of circular basins are also shown in Figure 9.9.

9.4.3 Square Tanks

Square tanks are the modification of circular tanks. They are hydraulically similar to circular tanks. Typically they utilize same primary sludge removal equipment as do the circular tanks. The radial racking arm and plows are supported on central pier or on beam across the basin. The removal of settled solids from the corners can be a problem. Corner sweeping equipment and rounded inside corners are used to prevent solids accumulation. The square clarifiers utilize common walls in multiple units, but require thicker walls

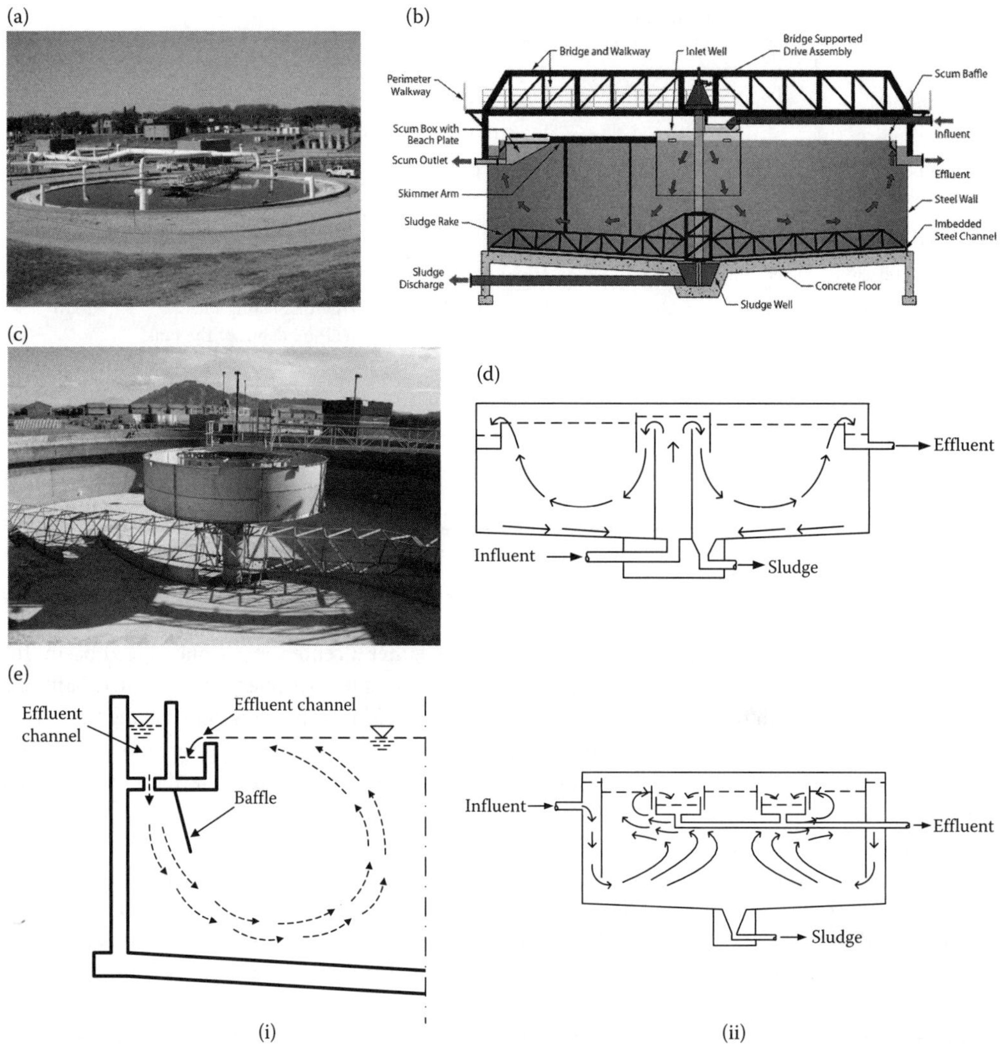

FIGURE 9.9 Circular sedimentation basins: (a) circular primary basin with covered effluent trough for odor control; (b) sectional view of a circular clarifier with center feed (Courtesy Monroe Environmental); (c) construction details of center feed well and sludge collection mechanism of a COP™ clarifier (Courtesy WesTech Engineering, Inc.); (d) flow pattern through a center feed clarifier; and (e) peripheral feed clarifier: (i) effluent and influent channels separated by a skirt baffle, and (ii) flow pattern in a basin with the effluent weirs near the center.

than circular units. Square tanks are rarely used as primary clarifier. Layout and sectional details of a square tank are shown in Figure 9.10.

9.4.4 Stacked, Multilevel, or Multitray Tanks

In areas where land is not available, stacked or multilevel rectangular tanks for both primary and secondary clarification have been used. Two- and three-level tanks have been designed to increase the treatment capacity per unit surface area. Compact, multistory, or underground treatment plants are receiving much interest these days.[14–18] The design criteria are similar to those of conventional primary tanks and are given in Section 9.4.1.

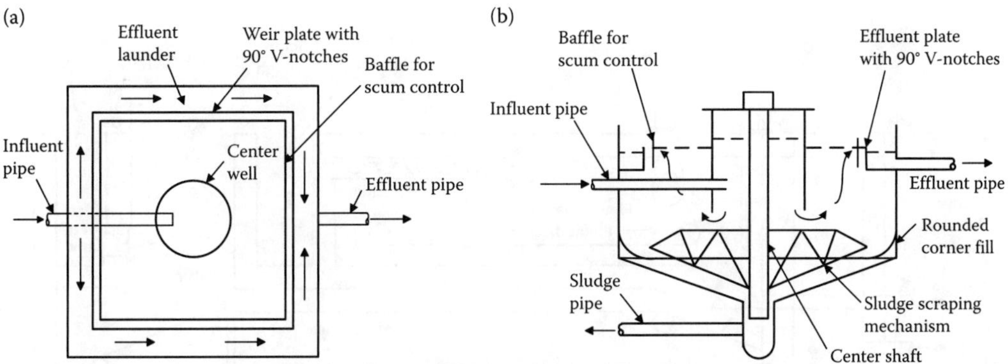

FIGURE 9.10 Square sedimentation tanks: (a) plan view and (b) cross-sectional view.

Series- or parallel-flow basin: The stacked tank can be either a series- or parallel-flow basin. In a *series-flow basin*, the influent enters the lower tray, goes up into the secondary tray on the far end, and travels in the opposite direction. The effluent exits from the upper tray. Baffles straighten the flow paths and minimize the turbulence at the influent point in the lower tray and at the turn around on the top tray. The sludge from both trays is collected in a common hopper. System components and flow schematics of a two-tray series-flow basin are shown in Figure 9.11a. In the *parallel-flow basin*, the split influent enters separately in the upper and lower trays at the same end and moves longitudinally. The influent baffles in both trays straighten the flow and minimize the turbulence. Effluent is removed from both trays by a common longitudinal effluent weir and launder along the top tray. The sludge is pumped from a common hopper. The parallel two-tray unit is most commonly used for primary treatment. System components and flow schematics of a two-tray parallel-flow basin are shown in Figure 9.11b.

Advantages and Disadvantages: The advantages of stacked basins are (1) compact and save space, (2) less piping and pumping requirements, and (3) better control of odors and volatile organic compounds emission due to less exposed surface area. The disadvantages include (1) higher construction cost, (2) complex structural design, (3) difficulties in observation of lower tray, and (4) high operation and maintenance costs.

9.4.5 Basic Design and Performance Criteria

The primary sedimentation basins are designed to provide sufficient time under quiescent condition for maximum settling to occur. The important design and performance criteria therefore are (1) surface overflow rate, (2) detention time, and (3) weir loading rate. Typical values of these parameters for sedimentation basins under different applications are summarized in Table 9.6.

The solids and BOD_5 removal efficiency of primary sedimentation basins is reduced by (1) eddy currents induced by the inertia of the incoming fluid, (2) surface currents induced by wind action, (3) vertical currents induced by the temperature difference between the influent and tank contents, (4) density currents that cause colder water (which is heavier) to under run a basin, and warmer water to flow across the surface, (5) vertical current induced by outlet structure, and (6) currents induced by the sludge scraper and sludge removal equipment. Factors that affect the performance of a basin are discussed below.

Surface Overflow Rate or Terminal Velocity: The surface overflow rate, surface loading rate, or hydraulic loading rate is expressed as flow divided by surface area of the tank (Q/A, $m^3/m^2{\cdot}d$). The surface overflow rate also represents the terminal velocity discussed in detail in Chapter 8. For a given

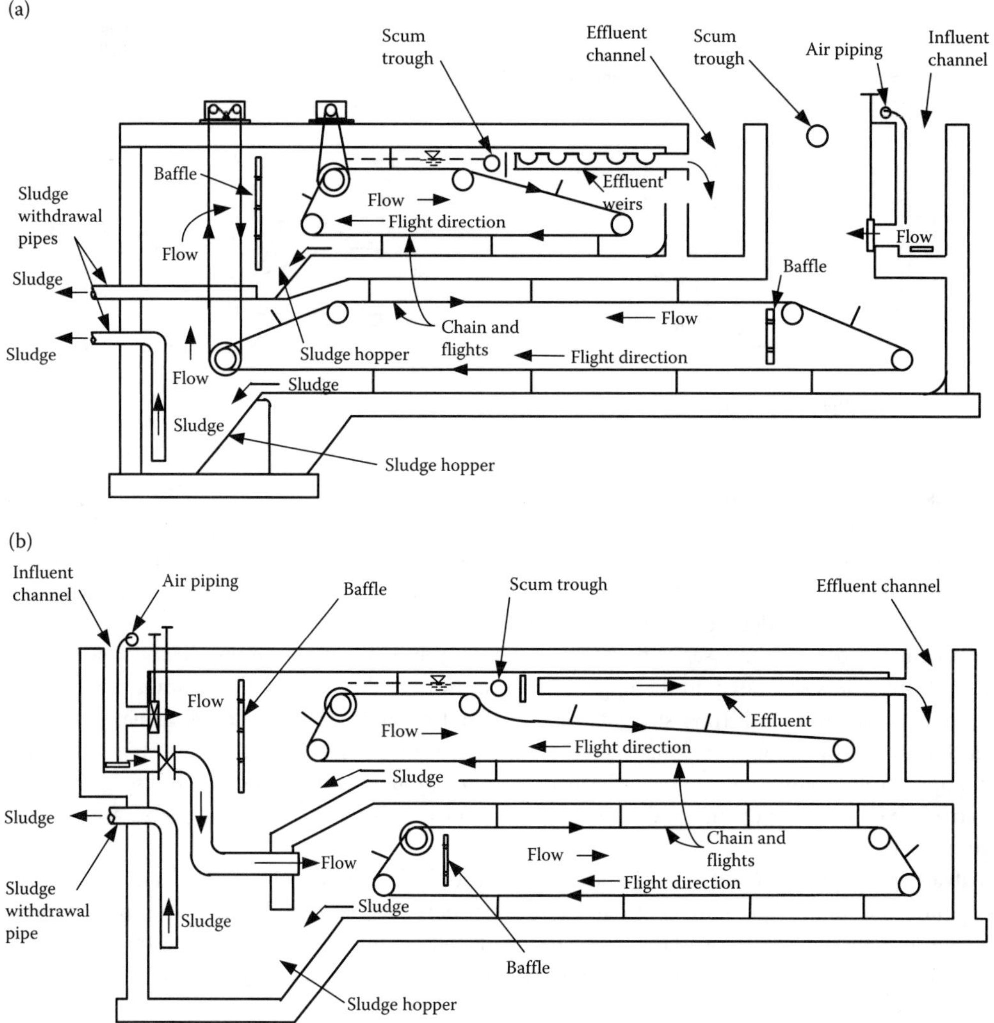

FIGURE 9.11 Stacked or two-tray sedimentation basins: (a) series flow and (b) parallel flow.

volume, shallower basins have larger surface area, and lower surface overflow rate, or lower terminal velocity. Thus, smaller particles are removed and the basin is more efficient. In very shallow tanks however, wind-induced waves near the surface and turbulence caused by sludge removing equipment near the bottom reduce the basin performance. For this reason, a side water depth of 3.7–4.5 m (12–15 ft) is typically used in design. Typical removal trends of BOD_5 and TSS are illustrated as functions of surface overflow rate in Figure 9.12a. In general, the surface overflow rate should be small enough to ensure satisfactory performance at peak design flow.[1–5] The range and typical design surface overflow rates for primary sedimentation basins are given in Table 9.6. Primary sedimentation basins for municipal wastewater applications are usually designed for an average surface overflow rate of $40\,m^3/m^2{\cdot}d$ ($1000\,gpd/ft^2$).

Detention Time: Sufficient contact time between the solids in the primary sedimentation basin is necessary for the coagulation of fine particles for effective settling. Some states have set limits (high and low) for detention time. Detention times >1.5 h without continuous sludge withdrawal may result in septic condition that will cause septic condition resulting in odors and solubilization of organic matter. Typical

TABLE 9.6 Typical Design and Performance Criteria for Primary Sedimentation Basins under Different Applications

Application	Parameter	Design Criteria	
		Range	Typical
Primary clarifier followed by secondary treatment	Surface overflow rate, m^3/m^2·d (gpd/ft^2)		
	At average flow	32–48 (800–1200)	40 (1000)
	At peak 2-h flow	60–100 (1500–2500)	72 (1800)
	Detention time (h)	1.5–3.5	2
	Weir loading rate, m^3/m·d (gpd/ft)	125–375 (10,000–30,000)	250 (20,000)
Primary clarifier receiving waste-activated sludge	Surface overflow rate, m^3/m^2·d (gpd/ft^2)		
	At average flow	24–32 (600–800)	28 (700)
	At peak 2-h flow	40–80 (1000–2000)	50 (1200)
	Detention time (h)	2–4	2.5
	Weir loading rate, m^3/m·d (gpd/ft)	125–375 (10,000–30,000)	250 (20,000)
Stacked multilevel or multitray primary sedimentation tanks	Surface overflow rate, m^3/m^2·d (gpd/ft^2)	16–40 (400–1000)	28 (700)
	Detention time, h	0.75–1.25	1
	Weir loading rate, m^3/m·d (gpd/ft)	85–170 (7000–14,000)	125 (10,000)

Note: 1 m^3/m^2·d = 24.5 gpd/ft^2 and 1 m^3/m·d = 80.5 gpd/ft.
Source: Adapted in part from References 2 through 5, and 19.

removals of BOD_5 and TSS with respect to detention time in a well-designed and -operated primary sedimentation basin are shown in Figure 9.12b. Equations 9.2a and 9.2b provide the curvilinear relationship between percent removal of BOD_5 and TSS with respect to detention time.[1–5]

$$R_{BOD_5} = \frac{\theta}{a_1 + b_1\theta} \quad (9.2a)$$

$$R_{TSS} = \frac{\theta}{a_2 + b_2\theta} \quad (9.2b)$$

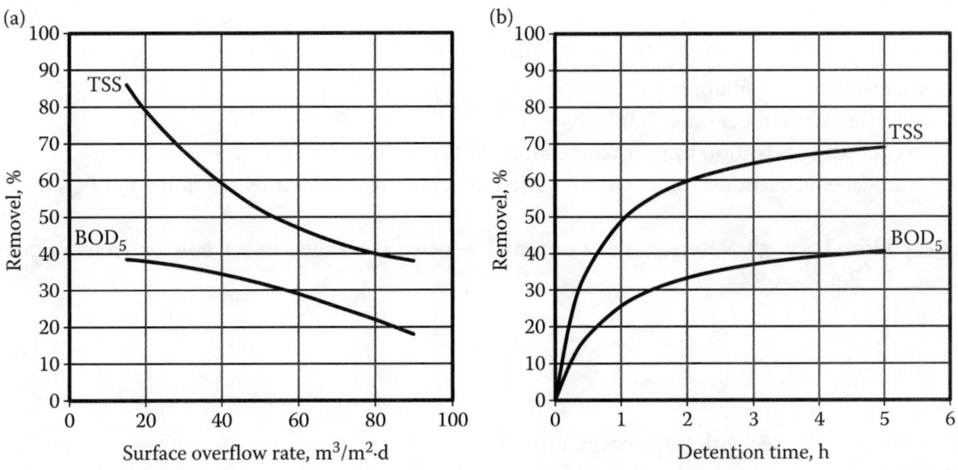

FIGURE 9.12 Typical percent removals of BOD_5 and TSS in primary sedimentation basin with respect to (a) surface overflow rate and (b) detention time.

where

R_{BOD_5} or R_{TSS} = expected removal efficiency of BOD_5 or TSS, percent
θ = detention time, h
a_1 and b_1 = empirical constants for BOD_5 removal. Typical range of values are $a_1 = 0.015-0.02$ and $b_1 = 0.015-0.025$
a_2 and b_2 = empirical constants for TSS removal. Typical range of values are $a_2 = 0.005-0.015$ and $b_2 = 0.01-0.02$

Weir Loading Rate: Weir loading rates have little effect on the removal efficiency of sedimentation basins with side water depths in excess of 3.7 m (12 ft). Primary sedimentation basins are generally designed for a weir loading rate <375 m^3 per meter length of the weir per day (30,000 gpd/ft). The Ten-States Standards recommend the following weir loading rates:[4,19]

- Weir loading rate of 250 m^3/m·d (20,000 gpd/ft) for plants designed for peak flows of 44 L/s (1 MGD) or less.
- Weir loading rate of 375 m^3/m·d (30,000 gpd/ft) for plants designed for peak design flows in excess of 44 L/s (1 MGD).

Weather Condition: Weather condition may affect the basin performance. Wind may cause the water surface on the leeward side higher than the windward side. This may result in uneven weir loading. Also, the scum may be pushed in windward direction. Wind sweeps over primary sedimentation basins are serious odor sources. Design considerations for wind mitigation include orientation of tanks, installation of wind breakers, increased tank freeboard, reduction in size by providing multiple units, and covering of basins.

Cold climate and freezing may cause serious operational problems. Freezing of pipes, weirs, scum collection equipments, and surface sprays are common in extreme cold climate. Protection techniques include pipe insulation, deeper underground pipes and channels, and covering of tanks. High viscosity of wastewater at low temperatures results in decreased settling velocity of particles and poor basin performance. A multiplier (Equation 9.3a) is used to increase the design detention time in cold regions where wastewater temperature drops below 20°C.[3,4] The use of the multiplier is expressed by Equation 9.3b.

$$M = 1.82e^{-0.03T} \tag{9.3a}$$

$$\theta_{T_2} = M\,\theta_{T_1} \tag{9.3b}$$

where

M = detention time multiplier
T = wastewater temperature, °C
θ_{T_1} = equivalent detention time at temperature T_1 that is typically 20°C, h
θ_{T_2} = actual equivalent detention time required or achieved at field at temperature T_2 ($T_2 < T_1$), h

Linear Flow-Through Velocity: The horizontal velocity also called linear flow or scour velocity is expressed by Equation 9.4a.

$$V_h = \frac{Q}{DW} \tag{9.4a}$$

where

V_h = horizontal velocity through basin, m/s (ft/s)
D = side water depth, m (ft)
W = basin width, m (ft)

The horizontal velocity should be kept low enough to prevent resuspension of settled particles. A critical scour velocity is calculated from Equation 9.4b and used in primary sedimentation basin design.[3,4,20]

$$V_c = \sqrt{\frac{8k(S-1)\,gd}{f}} \tag{9.4b}$$

where

V_c = critical horizontal scour velocity, m/s (ft/s)

k = cohesion constant for the type of scoured particles. Typical values are 0.04 for unigranular material, and 0.06 for sticky and interlocking material.

S = specific gravity of scoured particles, dimensionless. Typical value of S for primary settled solids is 1.1−1.7.

g = gravitational acceleration, 9.81 m/s^2 (32.2 ft/s^2)

d = average diameter of scoured particles, m (ft). Typical values for primary settled sludge solids 0.08−0.15 mm (0.003−0.006 in).

f = Darcy–Weisbach friction factor, dimensionless. Typical values range between 0.02 and 0.03. These values are a number and characteristics of function of Reynolds particles.

To prevent the resuspension of settled solids, horizontal velocity should be less than the critical scour velocity, $V_h < V_c$.

Influent Structure: The purpose of an influent structure is to (1) dissipate energy of incoming flow by means of baffles or stilling basins, (2) distribute the flow equally along the basin width, (3) prevent short-circuiting by disturbing the thermal or density stratification, (4) promote flocculation, and (5) keep low head loss.

In rectangular basins the influent flow is usually distributed across the width by (1) an inlet channel with submerged ports or an overflow weir, or (2) a circular manifold or pipe with laterals. The inlet channel has a minimum velocity of 0.3 m/s (1 ft/s) at 50% of design flow to prevent settling of solids. The submerged ports or orifices have velocities between 4.5 and 9 m/min (15 and 30 ft/min) at design average flow. The spacing between the ports is normally 1–2 m (3–6 ft) with a maximum spacing of 3 m (10 ft). A perforated baffle is typically installed 0.6–0.9 m (2–3 ft) away from the inlet ports and the lower end is about 15–30 cm (6–12 in) below the inlet ports. The top of the baffle is kept below the average water surface to allow scum to pass over the top. The most common influent structure is a tapering channel with submerged orifices. Different types of influent structures are shown in Figure 9.13.

In circular basins, there may be four different flow patterns depending upon location of influent and effluent structures. These flow patterns are (1) center feed with center withdrawal, (2) center feed with peripheral withdrawal, (3) peripheral feed with central withdrawal, and (4) peripheral feed with peripheral withdrawal. Baffles are usually provided to minimize the short-circuiting. The flow patterns in circular basins are shown in Figure 9.14.

Effluent Structure: The performance of a sedimentation basin greatly depends upon the location and operation of the effluent structure. Effluent should be uniformly withdrawn to prevent localized high-velocity gradients and short-circuiting. The minimum distance between the influent and effluent structures should be 3 m (10 ft), unless the tank includes special provision to prevent short-circuiting. If the approach velocity reaches the scour velocity, the settled particles may be swept into the effluent. Density gradient may also cause solids carryover.

The effluent structures are (a) overflow weir type or (b) submerged orifice type. The weir-type effluent structure consists of a baffle in front of an overflow weir, an effluent launder or channel, and an outlet box. The baffle stops the floating matter from escaping into the effluent. The overflow weir may be straight edged or V-notched. If the straight-edge weir is provided, the head calculations using the classical weir equation give extremely small head over the weir (1–2 mm). At such a small head, the capillary clinging effects at the weir crest are significant, and the use of classical weir equations does not provide satisfactory

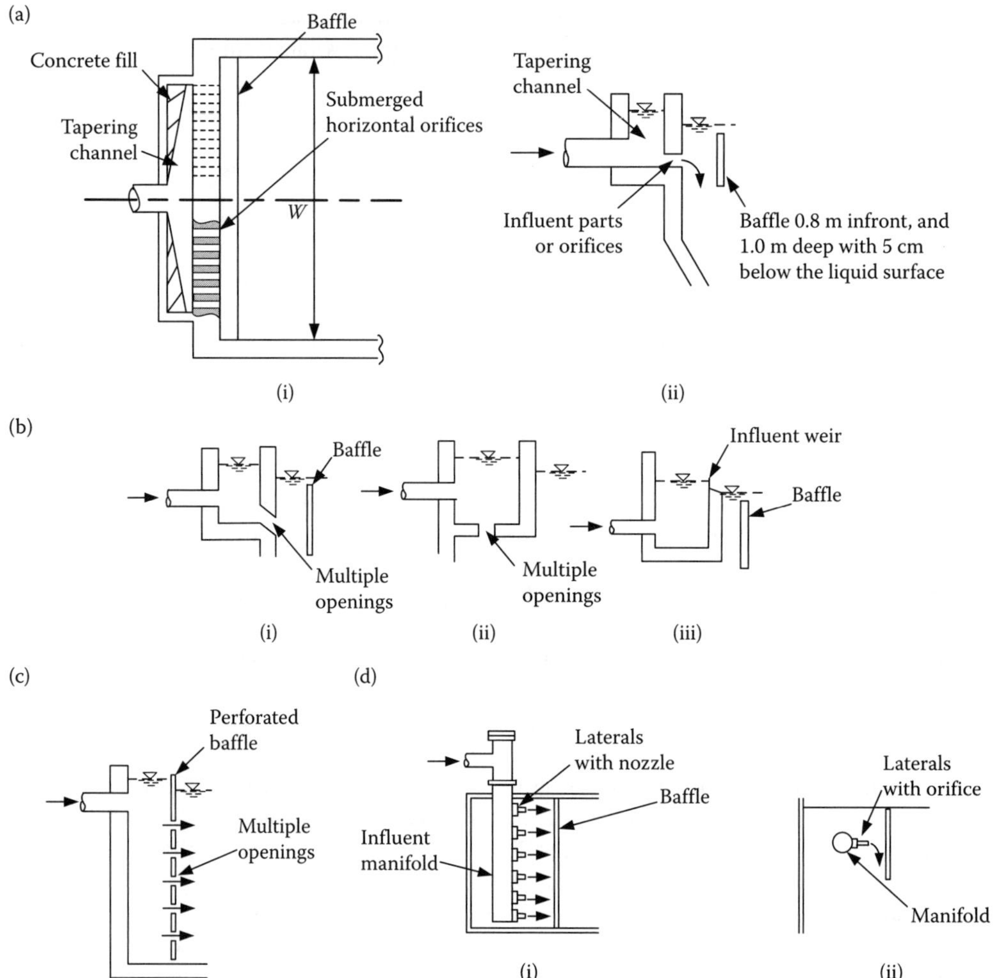

FIGURE 9.13 Typical design of influent structures for rectangular basins: (a) tapered influent channel and submerged horizontal ports with baffle wall: (i) plan and (ii) L-Section; (b) types of influent ports: (i) sloping ports in the distribution channel and a baffle, (ii) bottom ports in the distribution channel, and (iii) influent weir in the channel and baffle; (c) perforated baffle; and (d) influent manifold and laterals for flow distribution: (i) plan and (ii) L-section.

head calculations.[21,22] At small heads, the solids and slime tend to accumulate at the weir crest. Also, ice may form in cold climate. Additionally, if the weir is not perfectly levelled, the weir loading may be nonuniform. Due to these reasons, the designers prefer 90° V-notch weir. The effluent launder collects the free-falling flow from the weir and carry it to the effluent outlet box. The weir-type effluent structure is the most common effluent structure used in both rectangular and circular basins. This type of effluent structure for rectangular basins are shown in Figure 9.15a.

Another type of effluent structure consists of a partly submerged manifold or pipe with orifices discharging into the manifold (Figure 9.15b). The collected effluent into the manifold enters into the outlet channel or common manifold. Orifices are sized for uniform-flow distribution. Submerged launders have the advantage of reducing the stripping and release of entrained odorous gases. The major disadvantage is the flow distribution problems at peak design flow. Also, scum collection and removal may require special consideration. This type of effluent structure may be seen in rectangular basins but rarely be used in circular basins.

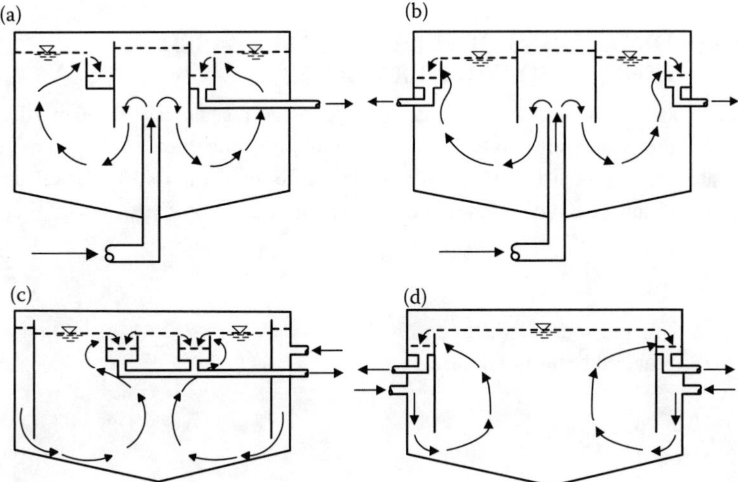

FIGURE 9.14 Flow pattern in circular basins: (a) center feed with central withdrawal, (b) center feed with peripheral withdrawal, (c) peripheral feed with central withdrawal, and (d) peripheral feed with peripheral withdrawal.

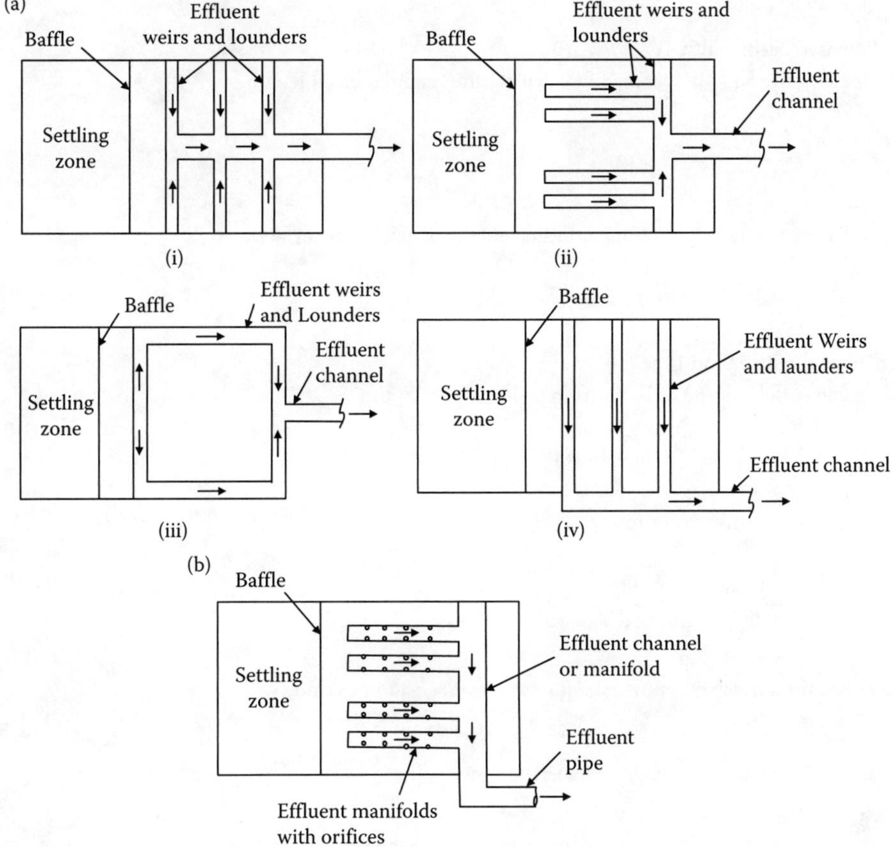

FIGURE 9.15 Effluent weirs and launders in rectangular basins: (a) multiple weir and launder configuration: (i) weirs on parallel launders discharging into central launder, (ii) finger weir launders discharging into an outlet channel then into an exit channel, (iii) rectangular launder and weirs, and (iv) parallel launders and weirs discharging into an external outlet channel; and (b) multiple manifolds with orifices discharging into an outlet channel or manifold.

EXAMPLE 9.5: DIMENSIONS OF PRIMARY BASIN BASED ON LENGTH-TO-WIDTH RATIO

A rectangular primary sedimentation basin is designed for an average flow of 10,000 m^3/d. Calculate the length and width of the basin. Use the following information. Surface overflow rate is $\leq 40\ m^3/m^2{\cdot}d$ at average flow, side water depth is equal to 4 m, and the basin length-to-width ratio is 3:1. Also calculate detention time and surface overflow rate at design peak flow. The peaking factor is 2.5:1.

Solution

1. Calculate the length and width of the basin.
   Calculate the required basin surface area.

$$A' = \frac{10{,}000\ m^3/d}{40\ m^3/m^2{\cdot}d} = 250\ m^2$$

Calculate the required basin width from a basin length-to-width ratio is 3:1.

$$L' = 3W' \quad and \quad A' = L'W' = (3W')W' = 3W'^2$$

Required basin width $W' = \sqrt{A'/3} = \sqrt{250\ m^2/3} = 9.1\ m$
Use a basin width $W = 9$ m, and calculate the required basin length.

$$L' = \frac{A'}{W} = \frac{250\ m^2}{9\ m} = 27.8\ m$$

Use a basin length $L = 28$ m, and calculate the actual basin surface area.

$$A = LW = 28\ m \times 9\ m = 252\ m^2$$

2. Calculate the detention time θ.
   Calculate the volume of basin from a side water depth $D = 4$ m.

$$V = AD = 252\ m^2 \times 4\ m = 1008\ m^3$$

Calculate the detention time.

$$\theta = \frac{V}{Q} = \frac{1008\ m^3}{10{,}000\ m^3/d} \times \frac{24\ h}{d} = 2.4\ h$$

3. Calculate the surface overflow rate at design average and peak flows.

$$\text{Surface overflow rate at design average flow} = \frac{Q}{A} = \frac{10{,}000\ m^3/d}{252\ m^2} = 40\ m^3/m^2{\cdot}d$$

Design peak flow $Q_{peak} = 2.5 \times 10{,}000\ m^3/d = 25{,}000\ m^3/d$

$$\text{Surface overflow at design peak flow} = \frac{Q_{peak}}{A} = \frac{25{,}000\ m^3/d}{252\ m^2} = 100\ m^3/m^2{\cdot}d$$

EXAMPLE 9.6: DIMENSIONS OF A PRIMARY SEDIMENTATION BASIN BASED ON LENGTH, WIDTH, AND DEPTH RATIOS

A rectangular primary sedimentation basin has length-to-width ratio of 3:1, and width-to-depth ratio of 2.5:1. Calculate the dimensions of the basin and detention time. The design average flow is 4 MGD and design average surface overflow rate is 1000 gpd/ft^2.

Solution

1. Calculate the dimensions of the basin.

Calculate the required surface area at the design average flow of 4 MGD and design average surface overflow rate of 1000 gpd/ft^2.

$$A' = \frac{4\,\text{MGD}}{1000\,\text{gpd/ft}^2} \times \frac{10^6\,\text{gal}}{\text{Mgal}} = 4000\,\text{ft}^2$$

Calculate the required basin width from a basin length-to-width ratio is 3:1.

$$L' = 3W' \text{ and } A' = L'W' = (3W')W' = 3W'^2$$

$$\text{Required basin width} \quad W' = \sqrt{\frac{A'}{3}} = \sqrt{\frac{4000\,\text{ft}^2}{3}} = 36.5\,\text{ft}$$

Use a standard basin width $W = 36$ ft, and calculate the required basin length.

$$L' = \frac{A'}{W} = \frac{4000\,\text{ft}^2}{36\,\text{ft}} = 111\,\text{ft}$$

Use a basin length $L = 112$ ft, and calculate the required side water depth from a basin width-to-depth ratio is 2.5:1.

$$D' = \frac{W}{2.5} = \frac{36\,\text{ft}}{2.5} = 14.4\,\text{ft}$$

Use a side water depth $D = 15$ ft, and calculate the actual surface overflow rate.

Calculate the actual basin surface area.

$$A = LW = 112\,\text{ft} \times 36\,\text{ft} = 4032\,\text{ft}^2$$

$$\text{Actualsurfaceoverflow rate} = \frac{Q}{A} = \frac{4\,\text{MGD}}{4032\,\text{ft}^2} \times \frac{10^6\,\text{gal}}{\text{Mgal}} = 992\,\text{gpd/ft}^2 < 1000\,\text{gpd/ft}^2$$

2. Calculate the detention time θ.

Calculate the actual volume of the basin.

$$V = AD = 4032\,\text{ft}^2 \times 15\,\text{ft} = 60{,}480\,\text{ft}^3$$

Calculate the detention time.

$$\theta = \frac{V}{Q} = \frac{60{,}480\,\text{ft}^3 \times \dfrac{7.48\,\text{gal}}{\text{ft}^3}}{4\,\text{MGD} \times \dfrac{10^6\,\text{gal}}{\text{Mgal}}} \times \frac{24\,\text{h}}{\text{d}} = 2.7\,\text{h}$$

EXAMPLE 9.7: LENGTH OF EFFLUENT WEIR PLATE AND NUMBER OF V-NOTCHES

A rectangular sedimentation basin has width of 12 m. The effluent weir plates are on both sides of the effluent launders that extend over the entire width of the basin. The width of the launders is 0.45 m, the design peak flow is 25,000 m^3/d, and the design weir loading rate is 350 $m^3/m{\cdot}d$. The weir plates have 90° V-notches with $C_d = 0.6$. Calculate the length of effluent weir plate, number of effluent launders, total number of V-notches, and depth of effluent launders. Assume that the effluent launders have free fall into the effluent exit channel that is located on one side of the basin, head loss due to friction and turbulence is 7% of the maximum calculate water depth in the launder, and $C_w = 1.0$. Provide a free fall of 0.20 m below the bottom of the V-notch and above the high water surface in the launder, a freeboard of 3 cm above the head over the notch, and an additional freeboard of 0.10 m about the weir plate to the top of the effluent launder. Draw the weir layout plan and sections.

Solution

1. Determine the effluent troughs with weir plates.

   Calculate the total length of effluent weir required at the design weir loading rate of 350 $m^3/m{\cdot}d$.

$$L_{\text{total}} = \frac{25{,}000\ \text{m}^3/\text{d}}{350\ \text{m}^3/\text{m·d}} = 71.4\ \text{m}$$

   Calculate the total number of weir plates required at a length of effluent weir plate $L_{\text{plate}} = 12$ m/plate.

$$N_{\text{plate}} = \frac{L_{\text{total}}}{L_{\text{plate}}} \frac{71.4\ \text{m}}{12\ \text{m/plate}} = 6\ \text{weir plates}$$

   Provide three troughs ($N_{\text{launder}} = 3$ launders) with weir plates on both sides of the trough. The weir arrangement is shown in Figure 9.16a.

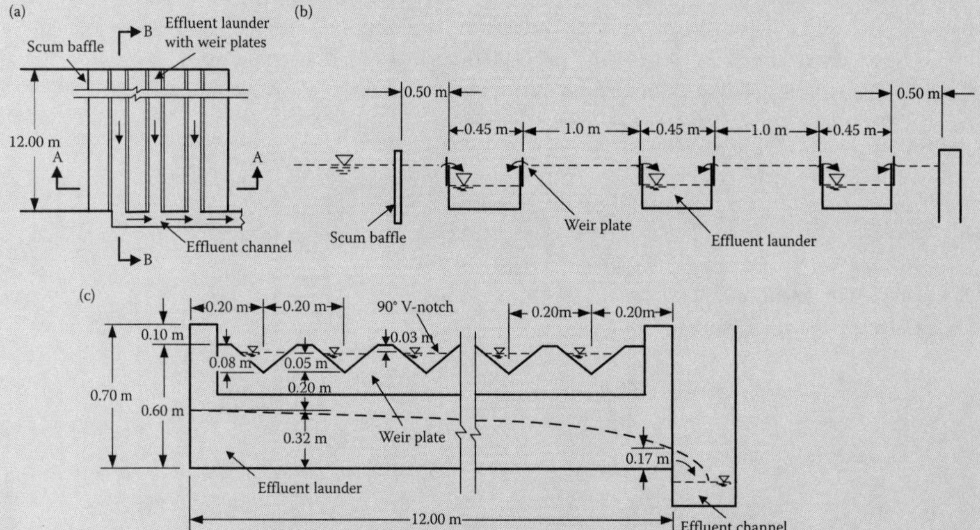

FIGURE 9.16 Design details of effluent structure: (a) layout plan of effluent launders, (b) effluent launders details (Section AA), and (c) water profile in the effluent launder (Section BB) (Example 9.7).

2. Calculate the total number of 90° V-notches.

   The head over a V-notch is calculated from Equation 9.5a.

$$q = \frac{8}{15} C_d \sqrt{2g} \tan\frac{\theta}{2} H^{5/2} \tag{9.5a}$$

where

q = flow per notch, m^3/s (ft^3/s)

C_d = coefficient of discharge

A C_d of 0.6 is typically used for a V-notch.

θ = angle of the V-notch, degree

H = head over the V-notch, m (ft)

For a 90° V-notch, $\theta = 90°$, and $\tan(\theta/2) = \tan(90°/2) = \tan(45°) = 1$. Equation 9.5a is therefore simplified to Equation 9.5b.

$$q = \frac{8}{15} C_d \sqrt{2g}\, H^{5/2} \tag{9.5b}$$

Provide notches 20-cm center-to-center and from each end of the weir plate to center of the nearest V-notch.

Calculate the number of 90° V-notches on each weir plate at $L_{space} = 0.20$ m.

$$N_{notches/plate} = \frac{L_{plate}}{L_{space}} - 1 = \frac{12\text{ m/plate}}{0.20\text{ m}} - 1 = 59\text{ notches/plate}$$

Calculate the total number of 90° V-notches.

$$N_{notch} = N_{plate} \times N_{notch/plate} = 6\text{ weir plates} \times 59\text{ notches/plate} = 354\text{ notches}$$

$$\text{Design flow } Q = 25{,}000\text{ m}^3/\text{d} \times \frac{\text{d}}{(60 \times 60 \times 24)\text{ s}} = 0.289\text{ m}^3/\text{s}$$

Calculate flow per 90° V-notch at design flow of 0.289 m^3/s.

$$q = \frac{Q}{N_{notch}} = \frac{0.289\text{ m}^3/\text{s}}{354\text{ notches}} = 0.00082\text{ m}^3/\text{s}$$

Calculate head over 90° V-notch from Equation 9.5b.

$$H = \left(\frac{15}{8} \times \frac{0.00082\text{ m}^3/\text{s}}{0.6 \times \sqrt{2 \times 9.81\text{ m/s}^2}}\right)^{2/5} = 0.05\text{ m} \quad \text{or} \quad 5\text{ cm}$$

3. Calculate the depth of effluent launders.

   Calculate flow per launder.

$$Q_{launder} = \frac{Q}{N_{launder}} = \frac{0.289\text{ m}^3/\text{s}}{3\text{ launders}} = 0.0963\text{ m}^3/\text{s}$$

Calculate the critical depth y_c at the free fall of effluent launder from Equation 7.10 with $C_w = 1.0$ and width of launder, $b = 0.45$ m.

$$0.0963\,\text{m}^3/\text{s} = 1.0 \times 0.45\,\text{m} \times \sqrt{9.81\,\text{m/s}^2}\cdot y_c^{3/2}$$

Critical depth is calculated.

$$y_c = \left(\frac{0.0963\,\text{m}^3/\text{s}}{1.0 \times 0.45\,\text{m} \times \sqrt{9.81\,\text{m/s}^2}}\right)^{2/3} = 0.17\,\text{m}$$

For a free fall into the effluent exit channel, assume the water depth $y_2 = y_c = 0.17$ m at the lower end of the launder.

Calculate the water depth y_1 at the upper end of the launder from Equation 8.12a.

$$y_1 = \sqrt{y_2^2 + \frac{2(Q_{\text{launder}})^2}{gb^2y_2}} = \sqrt{(0.17\,\text{m})^2 + \frac{2 \times (0.0963\,\text{m}^3/\text{s})^2}{9.81\,\text{m/s}^2 \times (0.45\,\text{m})^2 \times 0.17\,\text{m}}} = 0.29\,\text{m}$$

Add 10% additional depth for friction and turbulence losses.

Total water depth at the upper end = 1.1 × 0.29 m = 0.32 m

Add the height of 0.20 m for free fall below the V-notches, the head of 0.05 m over the V-notches, the freeboard of 0.03 m above the head at the notches, and the additional freeboard of 0.10 m above the weir plate to the top of the launder.

Total depth of effluent launder = (0.32 + 0.20 + 0.05 + 0.03 + 0.10) m = 0.70 m

4. Draw the plan and show the design details.

   The plan of effluent structure and design details are shown in Figure 9.16.

EXAMPLE 9.8: CIRCULAR BASIN AND EFFLUENT WEIR PLATE

A circular primary sedimentation basin is designed for a design average flow of 4 MGD. The design surface overflow rate is 1000 gal/ft²·d. The average side water depth is 14 ft. The effluent launder is 2 ft wide and is constructed as an integral part of the basin. The effluent notches are cut on the weir plate that is installed on the internal wall of the launder. Calculate the diameter and detention time of the basin, and weir loading rate. Show the design details.

Solution

1. Calculate the diameter of the basin.

   Calculate the required surface area at the design average flow of 4 MGD and design average surface overflow rate of 1000 gpd/ft².

$$A' = \frac{4\,\text{MGD}}{1000\,\text{gpd/ft}^2} \times \frac{10^6\,\text{gal}}{\text{Mgal}} = 4000\,\text{ft}^2$$

Calculate the required diameter of the basin.

$$\text{Required basin diameter } D' = \sqrt{\frac{4}{\pi}A'} = \sqrt{\frac{4}{\pi} \times 4000\,\text{ft}^2} = 71.4\,\text{ft}$$

Provide a standard diameter of basin $D = 72$ ft .

2. Calculate the detention time θ.
   Calculate the actual volume of the basin with a side water depth $SWD = 14$ ft.

$$V = \frac{\pi}{4} D^2 SWD = \frac{\pi}{4} \times (72\,\text{ft})^2 \times 14\,\text{ft} = 57{,}000\,\text{ft}^3$$

   Calculate the detention time.

$$\theta = \frac{V}{Q} = \frac{57{,}000\,\text{ft}^3 \times 7.48\,\text{gal/ft}^3}{4\,\text{MGD} \times 10^6\,\text{gal/Mgal}} \times \frac{24\,\text{h}}{\text{d}} = 2.6\,\text{h}$$

3. Calculate the weir loading.
   Calculate the diameter of the effluent weir plate for an effluent launder width $b = 2$ ft.

$$D_{\text{plate}} = D - 2b = 72\,\text{ft} - 2 \times 2\,\text{ft} = 68\,\text{ft}$$

   Calculate the total length of the effluent weir plate.

$$L_{\text{plate}} = \pi D_{\text{plate}} = \pi \times 68\,\text{ft} = 214\,\text{ft}$$

   Calculate the effluent weir loading.

$$\text{Weir loading} = \frac{Q}{L_{\text{plant}}} = \frac{4\,\text{MGD}}{214\,\text{ft}} \times \frac{10^6\,\text{gal}}{\text{Mgal}} = 18{,}700\,\text{gpd/ft}$$

4. Draw the plan and longitudinal section.
   The design details of the circular sedimentation basin are shown in Figure 9.17.

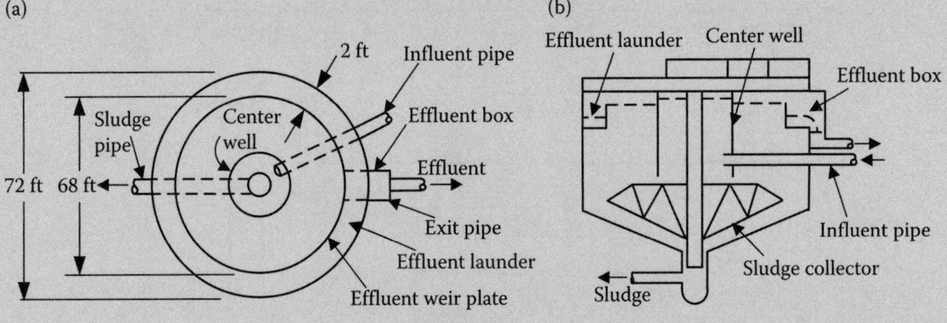

FIGURE 9.17 Details of a circular basin: (a) plan and (b) sectional view (Example 9.8).

EXAMPLE 9.9: FLOW PATTERN IN RECTANGULAR BASINS AND EFFLUENT STRUCTURES

Types of effluent structures have significant effect on the flow pattern through rectangular basins. Draw ideal-flow pattern through rectangular sedimentation basins of three different types of effluent structures (a) V-notch weir, (b) perforated baffle, and (c) perforated effluent manifold.

Solution

The flow pattern through a basin depends greatly upon the influent and effluent conditions. The idealized flow pattern through the basin for three different types of effluent structure is shown in Figure 9.18. All

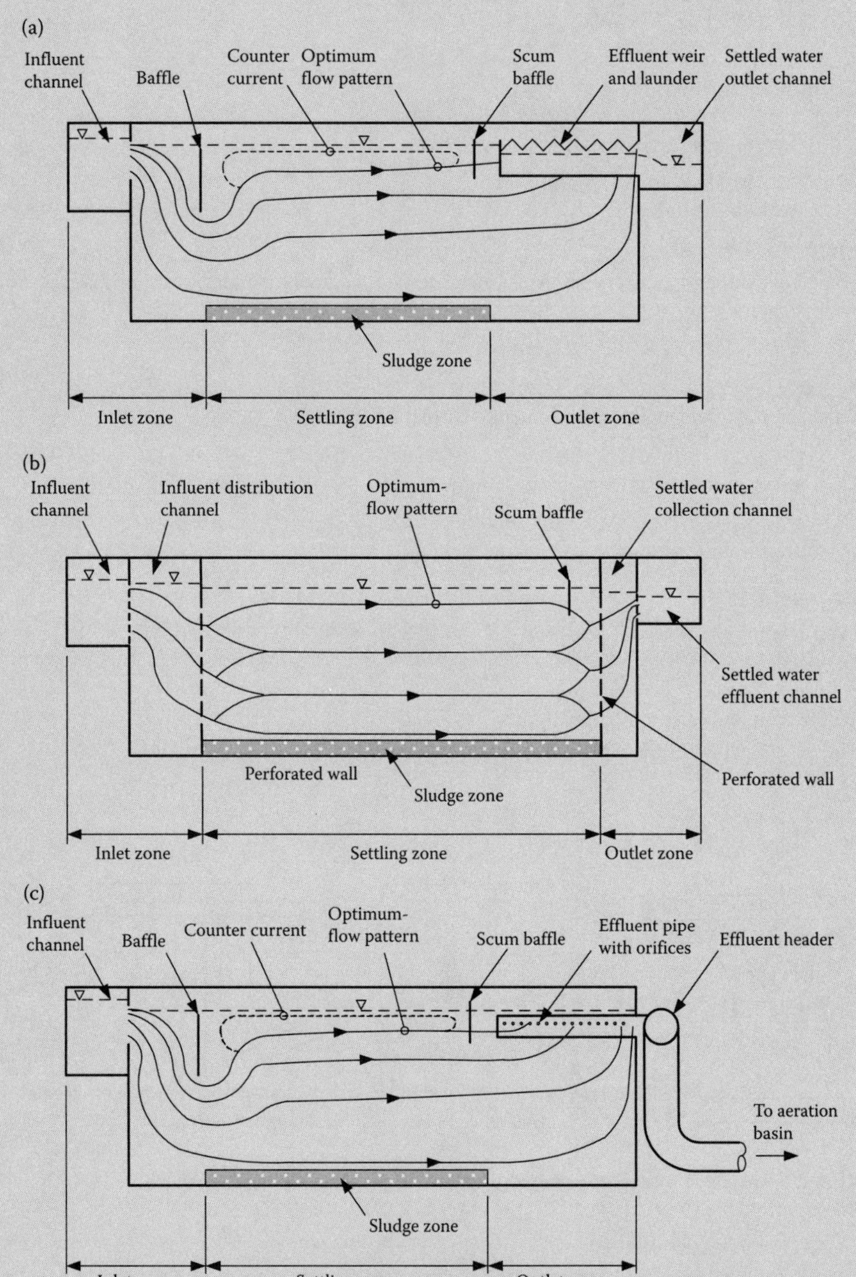

FIGURE 9.18 Effects of influent and effluent structures on flow patterns through sedimentation basins: (a) conventional horizontal-flow sedimentation basin with influent baffle and effluent V-notch weir, (b) sedimentation basin with perforated influent and effluent baffle walls, and (c) sedimentation basin with influent baffle and perforated effluent manifold (Example 9.9).

three idealized flow patterns will produce excellent settling behavior of solids. Basins with influent and effluent perforated baffles (Figure 9.18b) provide the best distribution of flow, and reduce density and thermal gradients both vertically and horizontally. Also, short-circuiting is minimized. Basins with perforated effluent manifold reduce the stripping of odorous gases for odor control.

EXAMPLE 9.10: FLOW PATTERN IN A RECTANGULAR BASIN WITH SHORT-CIRCUITING

In a rectangular sedimentation basin, the density gradient due to temperature difference between the influent and tank content may cause serious short-circuiting. The wind-driven circulation may also cause dead zones. Draw the flow pattern in a rectangular basin due to the density gradient and wind-driven circulation.

Solution

Dead zones are developed because of temperature or density effect (cooler water is denser). Wind circulation may also leave dead pockets in the tank. These dead zones in a basin cause short-circuiting and loss in performance efficiency due to reduced detention time. The flow pattern and dead spaces are illustrated in Figure 9.19.

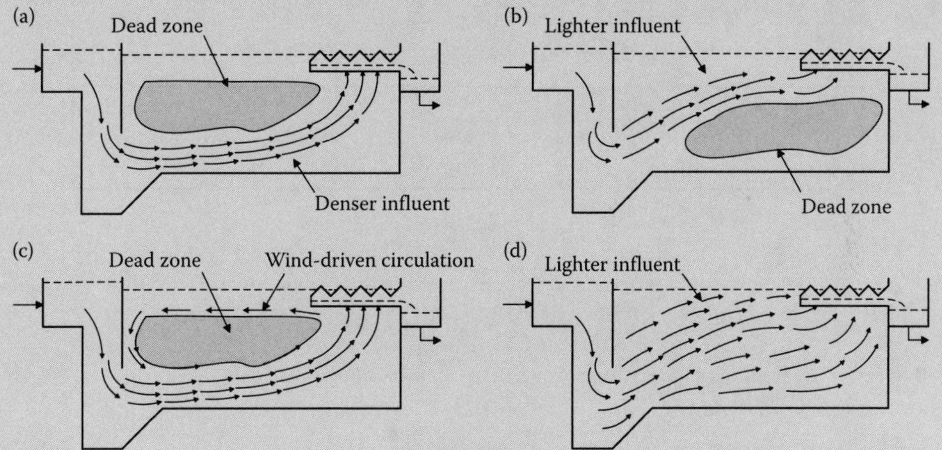

FIGURE 9.19 Short-circuiting in a horizontal-flow rectangular basin: (a) denser (cooler) influent, (b) lighter (warmer) influent, and (c) wind-driven circulation, and (d) ideal-flow pattern (Example 9.10).

EXAMPLE 9.11: DESIGN OF A DIVISION BOX WITH FLOW MEASUREMENT

A division box is designed to divide a flow of 20,000 m^3/d equally between two basins. Use of rectangular weirs is desired for flow split. Prepare the calibration curve for measurement of the divided flow in each basin. Determine the head over the weir when flow is equally distributed.

Solution

1. Describe the design details of the division box.

   The division box consists of a central chamber 2.3 m × 2.0 m, and two rectangular weirs, each having a length of 1.0 m. Both weirs have identical crest elevations and dimensions. The flow going over each identical weir has free fall into separate chambers each 1.0 m × 1.0 m. The exit pipes carry the distributed flow to separate sedimentation basins.

2. Draw the definition sketch.

   The plan of the division box is shown in Figure 9.20a.

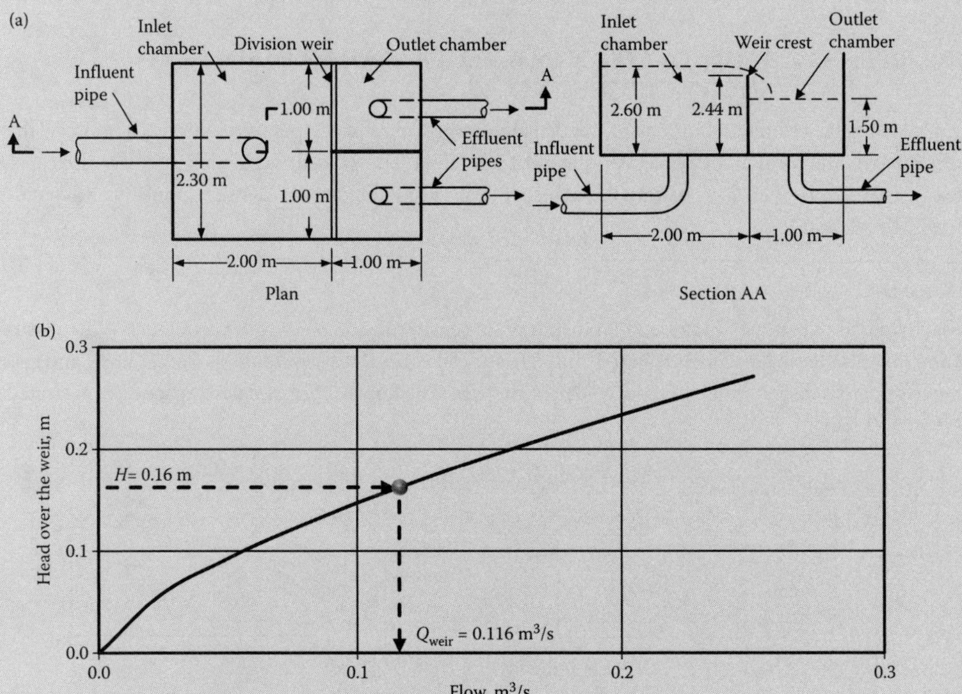

FIGURE 9.20 Definition sketch: (a) plan and Section AA, and (b) calibration curve to measure the distributed flow discharging over the weirs of division box (Example 9.11).

3. Determine the head over the rectangular weirs for an even flow split.

   Calculate flow per weir.

$$Q_{weir} = \frac{Q}{N_{weir}} = \frac{20{,}000\ \text{m}^3/\text{d}}{2\ \text{weirs}} \times \frac{\text{d}}{(60 \times 60 \times 24)\text{s}} = 0.116\ \text{m}^3/\text{s}$$

   The length of straight weir in each box $L = 1$ m.

   Calculate the head over the rectangular weir from Equation 8.10 with $C_d = 0.6$, and $L' = L = 1$ m at $n = 0$.

$$0.116\ \text{m}^3/\text{s} = \frac{2}{3} \times 0.6 \times 1\ \text{m}\sqrt{2 \times 9.81\ \text{m/s}^2} \times H^{3/2}$$

   $H = 0.16$ m

4. Draw the L-section of the weir.

   The water depth in the outbox is regulated by the exit pipe and outlet condition. Assume that the water depth in the outlet chamber is 1.5 m. Provide the weir height of 2.44 m to achieve free fall. The water depth in influent chamber = (2.44 + 0.16) m = 2.60 m. The L-section of the weir is shown in Figure 9.20a.

5. Draw the calibration curve and read.

   The head over the weir is a measure of flow distribution. The calibration curve in terms of head over the weir versus discharge over the weir is developed by assuming different values of H and calculating the corresponding Q_{weir} from Equation 8.14. The calibration curve is shown in Figure 9.20b. The head $H = 0.16$ m over the weir is read from the calibration curve when $Q_{weir} = 0.116\ m^3/s$.

EXAMPLE 9.12: DESIGN OF AN INFLUENT CHANNEL WITH MULTIPLE PORTS

An influent structure consists of a 1.0-m-wide tapered influent channel that runs across the entire width of the tank. Average velocity in the channel is 1.0 m/s. Eight submerged orifices 34 cm (13 in) square each are provided in the exit wall of the channel to distribute the flow over the entire width of the basin. The total discharge into the basin is $0.70\ m^3/s$, and the width of the basin is 12 m. Determine the head loss across the influent structure, and draw the plan and L-section.

Solution

1. Calculate the discharge through each orifice.

   There are eight orifices in the basin, $N_{orifice} = 8$ orifices.

   Calculate the discharge through each orifice.

$$q = \frac{Q}{N_{orifice}} = \frac{0.70\ m^3/s}{8} = 0.0875\ m^3/s$$

2. Calculate the water depth D in the influent channel.

   Calculate the water depth at the velocity $v = 1$ m/s and channel width $b = 1$ m.

$$D = \frac{q}{vb} = \frac{0.70\ m^3/s}{1\ m/s \times 1\ m} = 0.70\ m$$

3. Calculate the head loss across the influent structure.

   Calculate head loss through the influent orifices from orifice equation in Equation 7.4a with $A = (0.34\ m)^2$ and assuming $C_d = 0.6$.

$$0.0875\ m^3/s = 0.6 \times (0.34\ m)^2 \times \sqrt{2 \times 9.81\ m/s^2 \times \Delta H}$$

   $\Delta H = 0.08$ m

   Head loss through the influent orifices is 0.08 m.

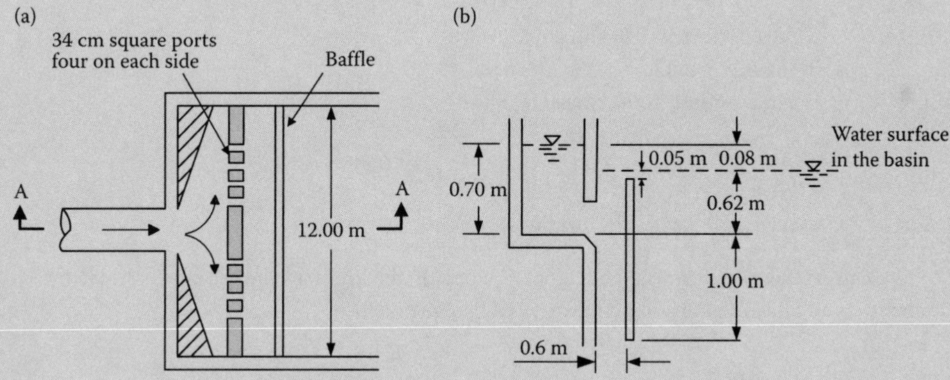

FIGURE 9.21 Details of influent structure, (a) plan, and (b) L-section (Example 9.12).

4. Describe the influent baffle wall.

   The influent baffle wall is 0.6 m in front, and 1.0 m below the invert of influent channel. The top of the baffle is 5 cm (0.05 m) below the water surface in the basin.

5. Draw the plan and L-section.

   The plan and L-section of the influent structures are shown in Figure 9.21.

EXAMPLE 9.13: FLOW DISTRIBUTION IN AN INFLUENT MANIFOLD AND LATERALS

Develop the variable flow distribution equations for discharge through multiple laterals. The laterals are arranged in a manifold that is designed for distribution of degritted influent into a primary sedimentation basin. Also describe the computational procedure to establish the flow from different laterals.

Solution

1. List the head losses in the manifold and laterals.

   The head losses in a manifold upstream of a lateral to the point of lateral discharge consists of (1) friction loss in the manifold and in the lateral, (2) entrance loss to the lateral, and (3) exit loss from the lateral. The friction losses are small and are usually ignored.

2. Draw the schematic of manifold and laterals.

   In Figure 9.22, the definition sketch of the manifold shows flow distribution in N laterals equally spaced over the entire length of the manifold.

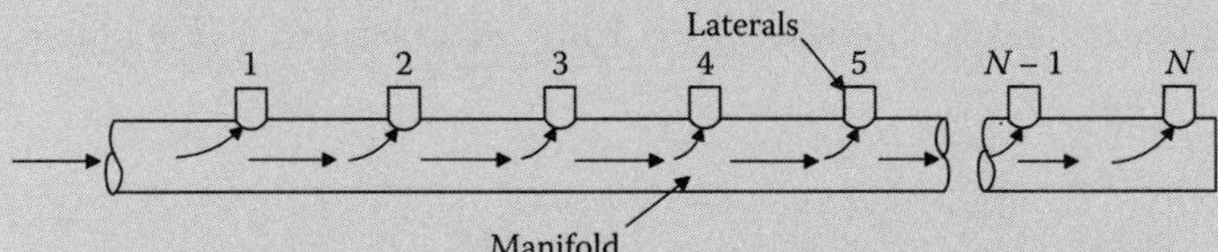

FIGURE 9.22 Definition sketch of a manifold and laterals (Example 9.13).

3. Establish the head loss equation in the laterals.

   The sum of head losses h_i in the ith lateral ($i = 1, 2, \ldots, N$) is given by Equation 9.6.

$$h_i = h_{i\,\text{entrance}} + h_{i\,\text{exit}} \text{ or } h_i = h_{\text{entrance}} + \frac{v_i^2}{2g} \tag{9.6}$$

where

h_i = total head loss in the ith lateral, m (ft)

$h_{i\,\text{entrance}}$ = entrance head loss in the ith lateral, m (ft)

$h_{i\,\text{exit}}$ = exit head loss in the ith lateral, m (ft)

$$h_{i\,\text{exit}} = k_{i\,\text{exit}}\frac{v_i^2}{2g} = \frac{v_i^2}{2g} \text{ since } \quad k_{i\,exit} = 1.0$$

v_i = velocity in lateral (i), m/s (ft/s)

The entrance loss in a lateral is a function of velocity in the manifold and velocity in the lateral. The entrance head loss in the ith lateral is expressed by Equation 9.7.[22,23]

$$h_{i\,\text{entrance}} = \left[\phi\left(\frac{v_{mi}}{v_i}\right)^2 + \theta\right]\frac{v_i^2}{2g} \tag{9.7}$$

where

v_{mi} = velocity in the manifold prior to the *i*th lateral, m (ft)

ϕ and θ = entrance loss constants

The constants ϕ and θ are determined experimentally, and depend upon the ratio of length (l_i) to diameter (d_i) of the lateral. These values for long and short laterals are given in Table 9.7. Short laterals have length less than three times the diameter, and long laterals have length substantially greater than three times the diameter.

TABLE 9.7 Entrance Head Loss Constants[22–24] (Example 9.13)

Length of Lateral	Constant	
	ϕ	θ
Short, $l_i < 3\ d_i$	1.67	0.70
Long, $l_i \geq 3\ d_i$	0.90	0.40

The equation for head loss through the lateral (i) is given by Equations 9.8 and 9.9.

$$h_i = \left[\phi\left(\frac{v_{mi}}{v_i}\right)^2 + \theta\right]\frac{v_i^2}{2g} + \frac{v_i^2}{2g} \quad \text{or} \quad h_i = \left[\phi\left(\frac{v_{mi}}{v_i}\right)^2 + \theta + 1\right]\frac{v_i^2}{2g} \tag{9.8}$$

$$h_i = \beta_i \frac{v_i^2}{2g} \tag{9.9}$$

where β_i is expressed by Equation 9.10.

$$\beta_i = \phi\left(\frac{v_{mi}}{v_i}\right)^2 + \theta + 1 \tag{9.10}$$

4. Establish equations for flow in subsequent laterals.

If an ideal-flow distribution from each lateral is assumed, then Equations 9.11 and 9.12 will apply.

$$\beta_1 \frac{v_1^2}{2g} = \beta_2 \frac{v_2^2}{2g} = \cdots = \beta_N \frac{v_N^2}{2g} = \text{constant} \tag{9.11}$$

$$v_i = v_1 \sqrt{\frac{\beta_1}{\beta_i}} \tag{9.12}$$

If all laterals are of the same diameter, the discharge and velocity relationships in the manifold and laterals are expressed by Equations 9.13 and 9.14.[23–25]

$$\sum_1^N q_i = Q \tag{9.13}$$

$$q_i = a v_i \tag{9.14a}$$

$$q_{mi} = A v_{mi} \tag{9.14b}$$

where

Q = total flow, m^3/s (ft^3/s)

q_i = discharge from the *i*th lateral, m^3/s (ft^3/s)

q_{mi} = flow in the manifold prior to the ith lateral, m^3/s (ft^3/s)
a and A = area of the lateral and manifold, m^2 (ft^2)
N = number of laterals

Substitute Equations 9.12 and 9.14 into Equation 9.13, the velocity in the first lateral is expressed by Equation 9.15.[23–25]

$$v_1 = \frac{Q}{a\sqrt{\beta_1}}\left(\sum_{1}^{N}\sqrt{\frac{1}{\beta_i}}\right)^{-1} \quad (j = 1, 2, \ldots, N) \tag{9.15}$$

5. Develop the computational procedure for flow distribution.
   Equations 9.6 through 9.15 are iteratively applied to determine the actual flow distribution in the laterals. In the first iteration, it is assumed that the flow is equally distributed in all laterals. The velocities in the lateral and manifold for a section are calculated from Equations 9.14a and 9.14b, respectively. The flow in the manifold at the section is calculated after subtracting the discharges in all laterals upstream of the section. In the second iteration Equation 9.15 is applied to calculate the velocity v_1, in the first lateral. Equation 9.12 is then used to calculate the velocities in the ith lateral using β_i from the previous iteration. These steps are repeated until the results converge within a tolerance value. The calculation steps are illustrated in Example 9.14.

EXAMPLE 9.14: FLOW DISTRIBUTION FROM AN INFLUENT MANIFOLD AND LATERALS

A 40-cm diameter manifold is designed to distribute an influent flow of 0.3 m^3/s into a primary sedimentation basin. The manifold has four 13-cm diameter short laterals with diffusers. Assume that the friction losses in the manifold and laterals are small. Calculate the flow discharged by each lateral.

Solution

1. Draw the definition sketch in Figure 9.23.

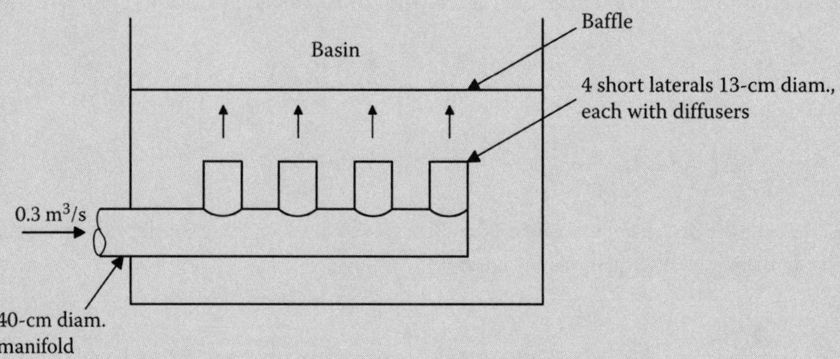

FIGURE 9.23 Definition sketch of influent manifold and laterals (Example 9.14).

2. Calculate the cross-sectional areas of the manifold and laterals.
   Calculate the cross area of manifold with a diameter $D = 40$ cm (0.4 m).

$$A = \frac{\pi}{4}D^2 = \frac{\pi}{4} \times (0.40\,\text{m})^2 = 0.126\,\text{m}^2$$

Calculate the cross area of lateral with a diameter $d = 13$ cm (0.13 m).

$$a = \frac{\pi}{4}d^2 = \frac{\pi}{4} \times (0.13\,\text{m})^2 = 0.0133\,\text{m}^2$$

3. Calculate various entries for the first iteration.
   a. First lateral.
      Assume an equal flow in all laterals ($N = 4$ laterals) and calculate flow in the first lateral.

$$q_1 = \frac{Q}{N} = \frac{0.30\,\text{m}^3/\text{s}}{4} = 0.075\,\text{m}^3/\text{s}$$

Calculate velocity in the first lateral from Equation 9.14a.

$$v_1 = \frac{q_1}{a} = \frac{0.075\,\text{m}^3/\text{s}}{0.0133\,\text{m}^2} = 5.64\,\text{m/s}$$

Flow in the section of manifold prior to the first lateral is equal to the total flow.

$$q_{\text{m1}} = Q = 0.30\,\text{m}^3/\text{s}$$

Calculate velocity in the manifold prior to the first lateral from Equation 9.14b.

$$v_{\text{m1}} = \frac{q_{\text{m1}}}{A} = \frac{0.30\,\text{m}^3/\text{s}}{0.126\,\text{m}^2} = 2.38\,\text{m/s}$$

Calculate the square of the ratio of $v_{\text{m1}}{:}v_1$.

$$\left(\frac{v_{\text{ml}}}{v_1}\right)^2 = \left(\frac{2.38\,\text{m/s}}{5.64\,\text{m/s}}\right)^2 = 0.178$$

For short laterals, the constants $\phi = 1.67$ and $\theta = 0.70$ are obtained from Table 9.7. Calculate β_1 from Equation 9.10.

$$\beta_1 = \phi\left(\frac{v_{\text{m1}}}{v_1}\right)^2 + \theta + 1 = 1.67 \times 0.178 + 0.70 + 1 = 2.00$$

$$\sqrt{\frac{1}{\beta_1}} = \sqrt{\frac{1}{2.00}} = 0.708$$

   b. Second lateral.

$$q_2 = q_1 = 0.075\,\text{m}^3/\text{s}$$

$$v_2 = v_1 = 5.64\,\text{m/s}$$

Calculate flow in the section of manifold prior to the second lateral.

$$q_{\text{m2}} = Q - q_{\text{m1}} = 0.30\,\text{m}^3/\text{s} - 0.075\,\text{m}^3/\text{s} = 0.225\,\text{m}^3/\text{s}$$

$$v_{m2} = \frac{q_{m2}}{A} = \frac{0.225\ m^3/s}{0.126\ m^2} = 1.79\ m/s$$

$$\left(\frac{v_{m2}}{v_2}\right)^2 = \left(\frac{1.79\ m/s}{5.64\ m/s}\right)^2 = 0.100$$

$$\beta_2 = 1.67 \times 0.100 + 0.70 + 1 = 1.87$$

$$\sqrt{\frac{1}{\beta_2}} = \sqrt{\frac{1}{1.87}} = 0.732$$

c. Third and fourth laterals.

Repeat these steps to obtain all values for the third and fourth laterals in the first iteration. These values are summarized in Table 9.8.

TABLE 9.8 Calculation Steps of Flow Distribution from a Manifold into Laterals (Example 9.14)

Lateral Number	Values						
	q_i, m^3/s	v_i, m/s	q_{mi}, m^3/s	v_{mi}, m/s	$(v_{mi}/v_i)^2$	β_i	$\sqrt{1/\beta_i}$
1	2	3	4	5	6	7	8
First Iteration							
1	0.075	5.64	0.300	2.38	0.178	2.00	0.708
2	0.075	5.64	0.225	1.79	0.100	1.87	0.732
3	0.075	5.64	0.150	1.19	0.0446	1.77	0.751
4	0.075	5.64	0.075	0.595	0.0111	1.72	0.763
$\sum_1^4$	0.300	–	–	–	–	–	2.95
Second Iteration							
1	0.0719	5.42	0.30	2.38	0.193	2.02	0.703
2	0.0748	5.64	0.228	1.81	0.103	1.87	0.731
3	0.0767	5.78	0.153	1.22	0.0443	1.77	0.751
4	0.0780	5.88	0.0766	0.608	0.0107	1.72	0.763
$\sum_1^4$	0.301	–	–	–	–	–	2.95
Third Iteration							
1	0.0716	5.39	0.30	2.38	0.195	2.03	0.703
2	0.0744	5.61	0.228	1.81	0.105	1.87	0.730
3	0.0765	5.76	0.154	1.22	0.0450	1.78	0.751
4	0.0777	5.85	0.0775	0.615	0.0110	1.72	0.763
$\sum_1^4$	0.300	–	–	–	–	–	2.95
Fourth Iteration							
1	0.0715	5.39	0.30	2.38	0.195	2.03	0.703
2	0.0744	5.60	0.228	1.81	0.105	1.87	0.730
3	0.0764	5.76	0.154	1.22	0.0451	1.78	0.751
4	0.0777	5.85	0.0777	0.616	0.0111	1.72	0.763
$\sum_1^4$	0.300	–	–	–	–	–	2.95

d. Final result for the first iteration

Calculate the sum of $\sqrt{1/\beta_i}$ for the first iteration.

$$\sum_1^4 \sqrt{\frac{1}{\beta_i}} = 2.95$$

4. Calculate various entries for the second iteration.

a. First lateral

Calculate velocity in the first lateral from Equation 9.15.

$$v_1 = \frac{Q}{a\sqrt{\beta_1}}\left(\sum_1^4 \sqrt{\frac{1}{\beta_i}}\right)^{-1} = \frac{0.30\,\text{m}^3/\text{s}}{(0.0133\,\text{m}^2) \times \sqrt{2.00}} \times (2.95)^{-1} = 5.42\,\text{m/s}$$

Calculate flow in the first lateral from Equation 9.14a.

$$q_1 = av_1 = (0.0133\,\text{m}^2) \times 5.42\,\text{m/s} = 0.0719\,\text{m}^3/\text{s}$$

$$v_{m1} = 2.38\,\text{m/s}$$

$$\left(\frac{v_{m1}}{v_1}\right)^2 = \left(\frac{2.38\,\text{m/s}}{5.42\,\text{m/s}}\right)^2 = 0.193$$

$$\beta_1 = 1.67 \times 0.193 + 0.70 + 1 = 2.02$$

$$\sqrt{\frac{1}{\beta_1}} = \sqrt{\frac{1}{2.02}} = 0.703$$

b. Second, third, and fourth laterals

Calculate velocity in the second lateral from Equation 9.12 using $\beta_2 = 1.87$ obtained in the first iteration.

$$v_2 = v_1\sqrt{\frac{\beta_1}{\beta_2}} = (5.42\,\text{m/s}) \times \sqrt{\frac{2.02}{1.87}} = 5.64\,\text{m/s}$$

Complete calculations for all values of the second, third, and fourth laterals. These values are summarized in Table 9.8.

5. Calculate various entries for the third and fourth iterations.

The procedure is repeated to calculate all values for the third and fourth iterations. These values are summarized in Table 9.8.

6. Comments on the discharge calculations of each lateral.

a. There is only ignorable change in discharge values of all laterals calculated in the third and fourth iterations. Therefore, no further iterations are needed.

b. The final discharge through each lateral is given below:

Lateral Number	1	2	3	4
Flow, m^3/s	0.072	0.074	0.076	0.078

c. Discharge increases slightly in the laterals that are further away.
d. The discharge of 0.078 m^3/s in the fourth lateral is 1.08 times the flow of 0.072 m^3/s in the first lateral.
e. The sum of discharges in all laterals = 0.30 m^3/s. This is equal to total flow entering the manifold.

EXAMPLE 9.15: BOD AND TSS REMOVAL

A sedimentation basin has detention time of 1.5 h. The empirical constants for BOD_5 removal are a_1 = 0.018 and b_1 = 0.021, and constants for TSS removal are a_2 = 0.0075 and b_2 = 0.013. Determine BOD_5 and TSS removals using Figure 9.12b. Compare these results with those obtained from Equations 9.2a and 9.2b.

Solution

1. Determine the BOD_5 and TSS removals.
   Read the removal efficiencies at θ = 1.5 h from Figure 9.12b.
   BOD_5 removal = 30%
   TSS removal = 56%
2. Calculate the BOD_5 removal.
   Calculate BOD_5 removal from Equation 9.2a at θ = 1.5 h.

$$R_{BOD_5} = \frac{\theta}{a_1 + b_1\theta} = \frac{1.5\,\text{h}}{0.018 + 0.021 \times 1.5\,\text{h}} = 30\%$$

3. Calculate the TSS removal.
   Calculate TSS removal from Equation 9.2b at θ = 1.5 h.

$$R_{TSS} = \frac{\theta}{a_2 + b_2\theta} = \frac{1.5\,\text{h}}{0.0075 + 0.013 \times 1.5\,\text{h}} = 56\%$$

4. Compare the results.
   The BOD_5 and TSS removals obtained from Figure 9.12b and Equations 9.2a and 9.2b are the same.

EXAMPLE 9.16: TEMPERATURE EFFECT ON BASIN PERFORMANCE

The performance of a primary sedimentation basin is lowered in cold climate. A sedimentation basin was designed for operation at average wastewater temperature above 20°C. The detention time of the basin is 2 h at 20°C. Due to excessively cold weather, the wastewater temperature dropped to 10°C. Estimate the BOD_5 and TSS removals at 10°C, and loss in removal efficiencies.

Solution

1. Determine the standard BOD_5 and TSS removal efficiencies at 20°C.
   Read the removal efficiencies at θ = 2 h from Figure 9.12b.
   BOD_5 removal = 33%
   TSS removal = 60%

2. Calculate the detention time at 20°C that is equivalent to the detention time of 2 h at 10°C.
   Calculate the multiplier for temperature correction from Equation 9.3a.

$$M = 1.82e^{-0.03T} = 1.82e^{-0.03\times 10^\circ\text{C}} = 1.35$$

   Calculate the equivalent detention time θ at 20°C from Equaiton 9.3b when the actual detention time $\theta_{T_{10}} = 2\,\text{h}$ is achieved at 10°C.

$$\theta_{T_{10}} = M\theta$$

$$\theta = \frac{\theta_{T_{10}}}{M} = \frac{2\,\text{h}}{1.35} = 1.5\,\text{h}$$

3. Determine the reduced BOD_5 and TSS removal efficiencies at 10°C.
   Read the removal efficiencies at the equivalent detention time $\theta = 1.5\,\text{h}$ from Figure 9.12b.
   BOD_5 removal = 30%
   TSS removal = 56%
4. Estimate the loss in performance efficiencies at 10°C.

$$\text{Loss in BOD}_5\text{ removal efficiency at }10^\circ\text{C} = \frac{(33-30)\%}{33\%} \times 100\% = 9\%$$

$$\text{Loss in TSS removal efficiency at }10^\circ\text{C} = \frac{(60-56)\%}{60\%} \times 100\% = 7\%$$

EXAMPLE 9.17: SCOUR VELOCITY AND CHECK FOR RESUSPENSION OF SETTLED SOLIDS

A primary sedimentation basin has width and depth of 10 and 4.5 m, respectively. Calculate the scour velocity and determine whether the settled material will resuspend at design peak flow of 50,000 m^3/d. Use the following standard values of coefficients in Equation 9.4b. Cohesion constant $k = 0.05$, specific gravity of settled particles $S = 1.25$, equivalent diameter of particles $d = 100\,\mu\text{m}$, and Darcy–Weisbach factor $f = 0.025$.

Solution

1. Calculate the design peak flow.

$$Q_{\text{peak}} = 50{,}000\,\text{m}^3/\text{d} \times \frac{\text{d}}{(60\times 60\times 24)\,\text{s}} = 0.579\,\text{m}^3/\text{s}$$

2. Calculate the horizontal velocity.
   Calculate the horizontal velocity V_h through the basin from Equation.9.4a at $D = 4.5$ m and $W =$ 10 m.

$$V_h = \frac{Q}{DW} = \frac{0.579\,\text{m}^3/\text{s}}{4.5\,\text{m} \times 10\,\text{m}} = 0.013\,\text{m/s}$$

3. Calculate the critical scour velocity.
   Calculate the critical scour velocity V_c through the basin from Equation 9.4b at $d = 100\,\mu\text{m}$ (100×10^{-6} m).

$$V_c = \sqrt{\frac{8k(S-1)gd}{f}} = \sqrt{\frac{8 \times 0.05 \times (1.25-1) \times 9.81\ \text{m/s}^2 \times (100 \times 10^{-6}\ \text{m})}{0.025}} = 0.063\ \text{m/s}$$

4. Comment on the resuspension of settled solids.

   To resuspend the settled solids, the horizontal velocity in the basin should be greater than the critical scour velocity. In this case, the horizontal velocity of 0.013 m/s is significantly less than the critical scour velocity of 0.063 m/s. Therefore, resuspension of settled solids will be unexpected at design peak flow.

EXAMPLE 9.18: EFFLUENT QUALITY FROM PRIMARY SEDIMENTATION BASIN

A primary sedimentation basin receives degritted wastewater. The design surface overflow rate is 40 $m^3/m^2{\cdot}d$ and influent quality is as follows: $BOD_5 = 200$ mg/L, TSS = 240 mg/L, total phosphorous (TP) = 6 mg/L, total nitrogen (TN) = 40 mg/L, and ammonia nitrogen (AN) = 30 mg/L. Determine the quantity of the primary settled effluent.

Solution

1. Determine the BOD_5 and TSS removals.

   Read the removal efficiencies at surface overflow rate of 40 $m^3/m^2{\cdot}d$ from Figure 9.12a.

   BOD_5 removal = 35%

   TSS removal = 59%

2. Determine the TP, TN, and AN removal.

   The typical removal ranges of TP, TN, and AN in primary sedimentation basin are provided in Table 6.8.

   The average removal values are TP = 15%, ON = 25%, and AN = 0%.

3. Determine the ON concentration in the influent.

   Total nitrogen consists of organic nitrogen (ON), ammonia nitrogen (AN), nitrite (NO_2-N), and nitrate (NO_3-N). In primary influent, NO_2-N and NO_3-N concentrations are very low and typically ignorable. Therefore, TN is mainly in the form of ON and AN, which is total Kjeldahl nitrogen (TKN).

   $$\text{ON} \approx \text{TN} - \text{AN} = (40 - 30)\ \text{mg/L} = 10\ \text{mg/L}$$

4. Determine the effluent quality from primary sedimentation basin.

   $BOD_5 = (1 - 0.35) \times 200\ \text{mg/L} = 130\ \text{mg/L}$

   $\text{TSS} = (1 - 0.59) \times 240\ \text{mg/L} = 98\ \text{mg/L}$

   $\text{TP} = (1 - 0.15) \times 6\ \text{mg/L} = 5.1\ \text{mg/L}$

   $\text{ON} = (1 - 0.25) \times 10\ \text{mg/L} = 7.5\ \text{mg/L}$

   $\text{AN} = (1 - 0.00) \times 30\ \text{mg/L} = 30\ \text{mg/L}$

   $\text{TN} \approx \text{TKN} = \text{ON} + \text{AN} = (7.5 + 30)\ \text{mg/L} = 37.5\ \text{mg/L}$

EXAMPLE 9.19: QUANTITY OF PRIMARY SLUDGE AND SCUM REMOVED IN A PRIMARY SEDIMENTATION BASIN

A primary sedimentation basin receives a degritted flow of 3.5 MGD. The TSS concentration in the influent is 230 mg/L, and TSS removal efficiency is 65%. The scum removal is 67 lb/Mgal (8 g/m^3). Calculate the primary sludge quantity in lb/d and gal/d. The solids content in the sludge is 4% and bulk specific gravity is 1.015. Also calculate the quantity of scum in lb/d.

Solution

1. Calculate the quantity of sludge produced.

$$\text{TSS removed} = 0.65 \times 230\,\text{mg/L} = 150\,\text{mg/L}$$

$$\text{Dry solids removed} = 150\,\text{mg/L} \times \frac{8.34\,\text{lb}}{\text{mg/L}\cdot\text{Mgal}} \times 3.5\,\text{MGD} = 4380\,\text{lb/d}$$

$$\text{or} \quad 4380\,\text{lb/d} \times \frac{\text{ton}}{2000\,\text{lb}} = 2.2\,\text{tons/d}$$

$$\text{Sludge volume removed} = 4380\ \text{lb/d drysolids} \times \frac{100\,\text{lb wet sludge}}{4\,\text{lb dry solids}} \times \frac{1}{(8.34 \times 1.015)\ \text{lb/gal as wet sludge}}$$

$$= 12{,}940\,\text{gal/d (gpd) wet sludge}$$

2. Calculate the quantity of scum produced.

$$\text{Quantity of scum removed} = \frac{67\,\text{lb}}{\text{Mgal}} \times 3.5\,\text{MGD} = 235\,\text{lb/d}$$

Note: General information about residual generation from wastewater treatment processes is summarized in Table 6.9.

EXAMPLE 9.20: SLUDGE PUMPING RATE AND CYCLE TIME

The volume of sludge removed from a primary sedimentation basin is 38,400 gal/d. The sludge pumping rate is 100 gpm, and the pump operates at a set time cycle. The pumping cycle is 4 min. Calculate the idle time and number of cycles per hour Example 5.4.

Solution

1. Determine the number of pumping cycles per hour.

$$\text{Sludge pumped per cycle} = \frac{100\,\text{gal}}{\text{min}} \times \frac{4\,\text{min}}{\text{cycle}} = 400\,\text{gal/cycle}$$

$$\text{Sludge pumped per hour} = \frac{38{,}400\,\text{gal}}{\text{d}} \times \frac{\text{d}}{24\,\text{h}} = 1600\,\text{gal/h}$$

$$\text{Number of pumping cycles per hour} = \frac{1600\,\text{gal}}{\text{h}} \times \frac{\text{cycle}}{400\,\text{gal}} = 4\,\text{cycles/h}$$

2. Calculate the idle time per cycle.

$$\text{Total time per cycle} = \frac{60\,\text{min}}{\text{h}} \times \frac{\text{h}}{4\,\text{cycles}} = 15\,\text{min/cycle}$$

Idle time per cycle = Total time per cycle − pumping time per cycle
= (15 − 4) min/cycle = 11 min/cycle

3. Describe the pumping operation.

The pump runs for 4 min at a rate of 100 gal/min. After each running cycle, there is an idle cycle of 11 min. Thus, the pump is kicked on every 15 min, runs for 4 min then stops for 11 min.

EXAMPLE 9.21: DESIGN OF A PRIMARY SEDIMENTATION BASIN

A rectangular primary sedimentation basin is designed for a design average flow of 0.22 m^3/s. The peaking factor is 2.8. The influent BOD_5 and TSS are 180 mg/L and 220 mg/L, respectively. The design criteria are given below:

1. The design surface overflow rate at design average flow shall be <36 m^3/m^2·d.
2. The detention time at design average flow shall be >1.5 h.
3. The water depth at mid-length and at design average flow shall be 4.0 m.
4. The slope of the basin floor shall be at 6%.
5. The length-to-width (L:W) ratio shall be 4:1.
6. The influent and effluent structures shall be designed for peak wet weather flow.
   a. Velocity in the influent channel shall not be <0.3 m/s to prevent accumulation of sediments.
   b. The influent channel shall have 0.38 m diameter submerged circular exit orifices. The velocity through influent orifices shall be <0.5 m/s to reduce potential turbulence caused by the jet flows in the inlet zone.
   c. The weir loading shall not be >340 m^3/m·d.

Solution

1. Determine the basin geometry and dimensions.

Calculate the design average flow.

$$Q = 0.22\,\text{m}^3/\text{s} \times \frac{(60 \times 60 \times 24)\,\text{s}}{\text{d}} = 19{,}000\,\text{m}^3/\text{d}$$

Calculate the required basin surface area at the design surface overflow rate of 36 m^3/m^2·d.

$$A' = \frac{19{,}000\,\text{m}^3/\text{d}}{36\,\text{m}^3/\text{m}^2\cdot\text{d}} = 528\,\text{m}^2$$

Calculate the required basin width from a basin length-to-width ratio is 4:1.

$$L' = 4W' \text{ and } A' = L'W' = (4W')W' = 4W'^2$$

$$\text{Required basin width } W' = \sqrt{\frac{A'}{4}} = \sqrt{\frac{528\,\text{m}^2}{4}} = 11.5\,\text{m}$$

Select standard basin width W = 11.58 m (38 ft in the U.S. customary unit) and calculate the required basin length.

$$L' = 4 \times 11.58\,\text{m} = 46.32\,\text{m}$$

Select standard basin length $L = 46.33$ m (152 ft in the U.S. customary unit) and check the actual length-to-width ratio.

$$L{:}W = \frac{46.33\text{ m}}{11.58\text{ m}} = 4.0 \quad \text{(It meets the design requirement of } L{:}W = 4{:}1.)$$

Calculate the actual basin surface area.

$$A = LW = 46.33\text{ m} \times 11.58\text{ m} = 537\text{ m}^2$$

Calculate basin volume at an average side water depth $SWD_{\text{mid}} = 4.0$ m.

$$V = SWD_{\text{mid}}\,A = 4.0\text{ m} \times 537\text{ m}^2 = 2146\text{ m}^2$$

Calculate the side water depths on the influent and effluent sides at the design slope $S_{\text{basin}} = 0.06$ (6%).

$$SWD_{\text{inf}} = SWD_{\text{mid}} + S_{\text{basin}}\frac{L}{2} = 4.0\text{ m} + 0.06 \times \frac{46.33\text{ m}}{2} = (4.0\text{ m} + 1.4)\text{ m} = 5.4\text{ m}$$

$$SWD_{\text{eff}} = SWD_{\text{mid}} - S_{\text{basin}}\frac{L}{2} = (4.0 - 1.4)\text{ m} = 2.6\text{ m}$$

The design details of the primary sedimentation basin are shown in Figure 9.24.

2. Check the surface overflow rates.
   Calculate design peak wet weather flow from the design peaking factor $PF = 2.8$.

   $$\text{Designpeak flow, } Q_{\text{peak}} = PF\;Q = 2.8 \times 19{,}000\text{ m}^3/\text{d} = 53{,}200\text{ m}^3/\text{d}$$

   Check surface overflow rates at design average and peak flows.

   $$\text{Surface overflow rate, } SOR = \frac{Q}{A} = \frac{19{,}000\text{ m}^3/\text{d}}{537\text{ m}^2} = 35\text{ m}^3/\text{m}^2{\cdot}\text{d}$$

   (It is less than the design requirement of $36\text{ m}^3/\text{m}^2$ d.)

   $$\text{Peak overflow rate, } SOR_{\text{peak}} = \frac{Q_{\text{peak}}}{A} = \frac{53{,}200\text{ m}^3/\text{d}}{537\text{ m}^2} = 99\text{ m}^3/\text{m}^2{\cdot}\text{d}$$

   (It is within the acceptable range; see Table 9.6.)

3. Check the detention times.
   Check detention times at design average and peak flows.

   $$\theta = \frac{V}{Q} = \frac{2146\text{ m}^3}{19{,}000\text{ m}^3/\text{d}} \times \frac{24\text{ h}}{\text{d}} = 2.7\text{ h} \qquad \text{(It is greater than the design requirement of 1.5 h.)}$$

   $$\theta_{\text{peak}} = \frac{V}{Q_{\text{peak}}} = \frac{2146\text{ m}^3}{53{,}200\text{ m}^3/\text{d}} \times \frac{24\text{h}}{\text{d}} = 1.0\text{ h} \quad \text{(It is greater than the typical requirement of 0.9 h.)}$$

4. Design the influent structure.
   a. Select the arrangement.

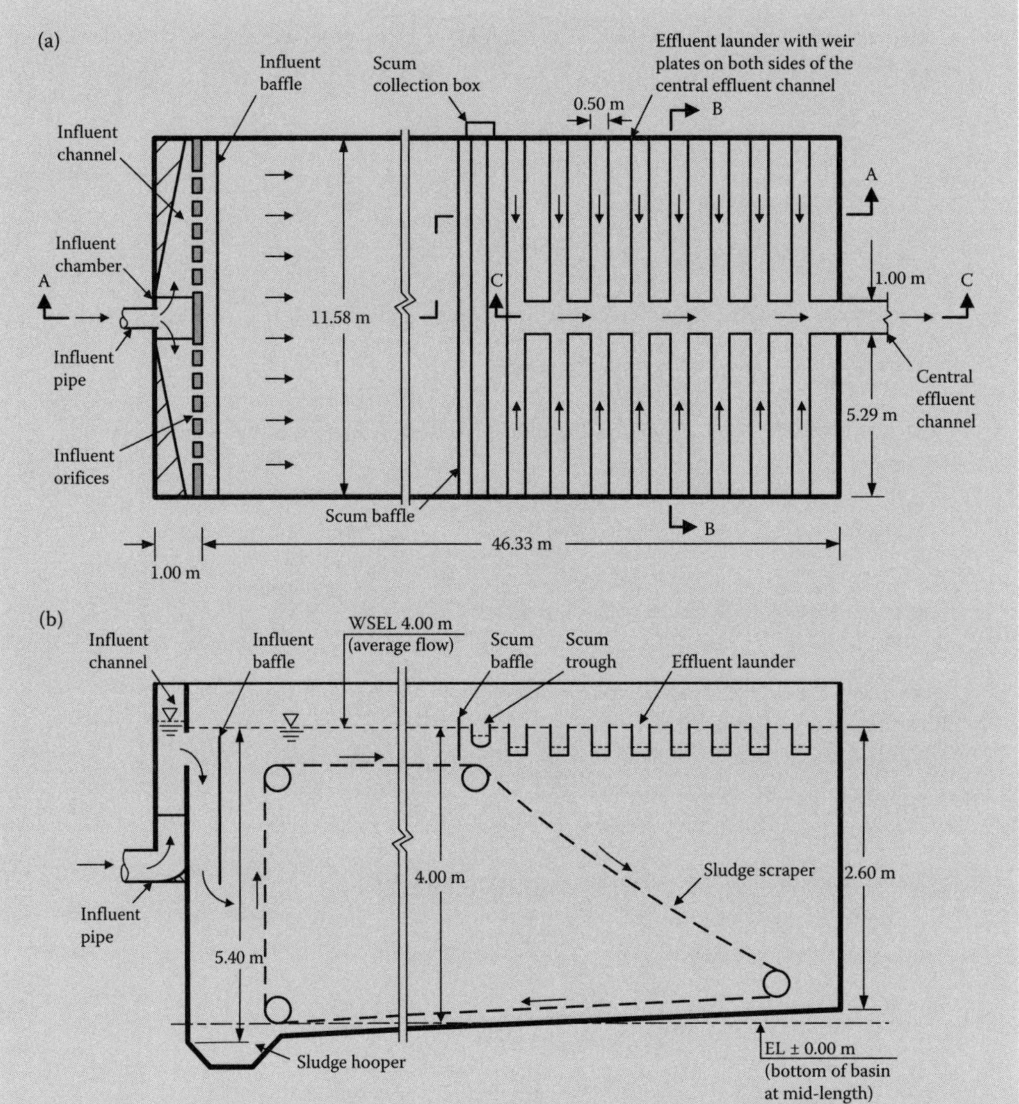

FIGURE 9.24 Design details of primary sedimentation basin: (a) plan, and (b) Section AA (Example 9.21). (*Continued*)

The influent structure is designed for peak wet weather flow. The influent structure consists of an influent pipe in the middle of a tapering 1-m-wide channel. The flow divides on both sides of the channel and exists from 12 submerged influent orifices; 6 on each half of the basin.

b. Calculate the velocity in the channel at the entrance.

Set the invert elevation of the channel 1.0 m below the water surface at the design peak flow. This will ensure a depth of flow <1.0 m in the influent channel at design average flow.

Calculate the design peak flow on each side of the channel.

$$q_{\text{channel}} = \frac{1}{2}(PF\ Q) = \frac{1}{2} \times 2.8 \times 0.22\,\text{m}^3/\text{s} = 0.31\,\text{m}^3/\text{s}$$

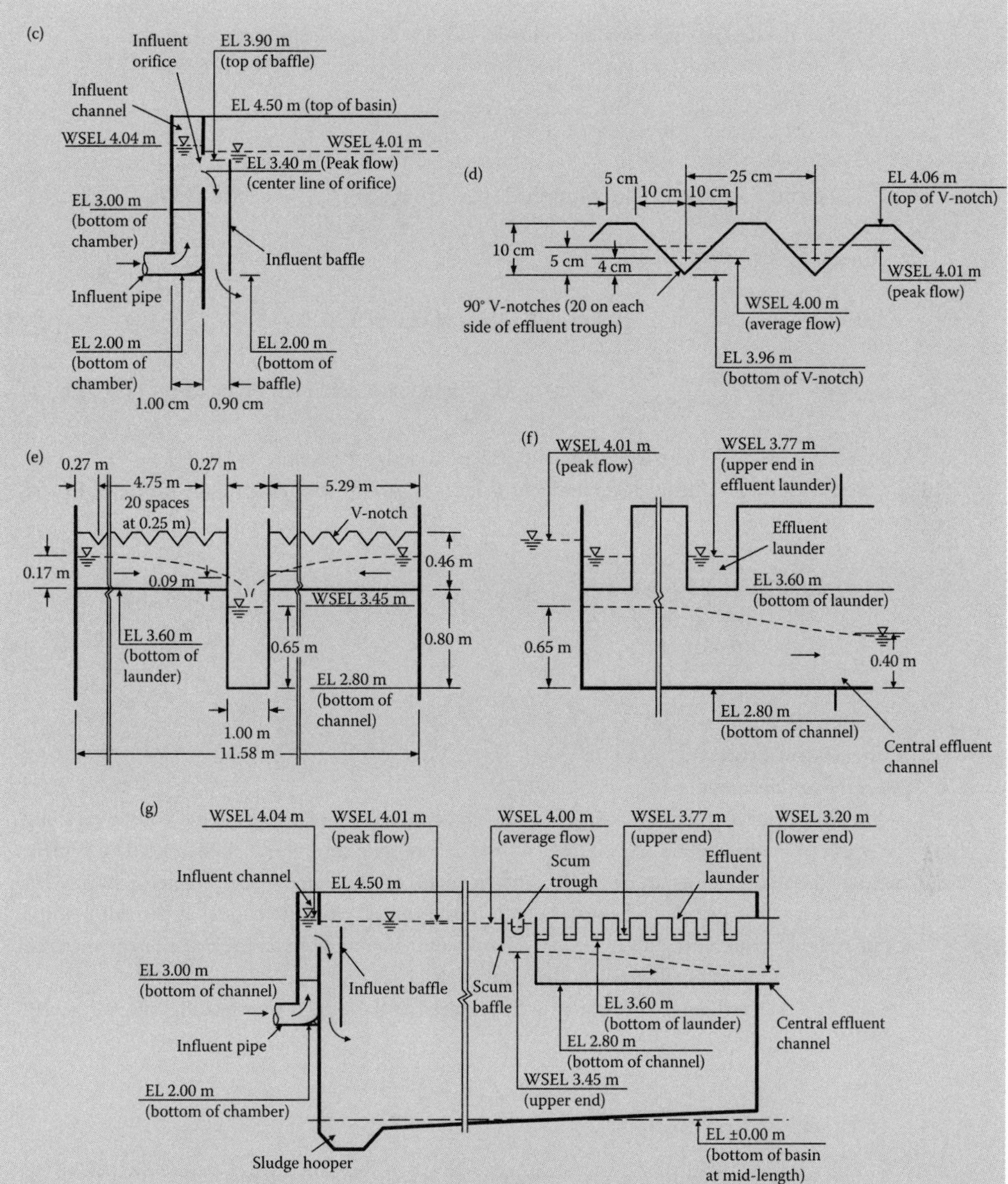

FIGURE 9.24 (Continued) Design details of primary sedimentation basin: (c) details of influent structure, (d) details of V-notches, (e) Section BB (details of effluent launder), (f) Section CC (details of central effluent channel), and (g) hydraulic profile through the basin (Example 9.21).

Calculate the velocity in the channel at design peak flow in a worst case scenario: $b_{\text{inf}} = 1$ m (maximum width of channel) and $h_{\text{inf}} = 1$ m (maximum depth of flow).

$$v_{\text{channel}} = \frac{q_{\text{channel}}}{h_{\text{inf}} b_{\text{inf}}} = \frac{0.31\,\text{m}^3/\text{s}}{1\,\text{m} \times 1\,\text{m}} = 0.31\,\text{m/s} \quad \text{(It is greater than the design requirement of 0.3 m/s.)}$$

c. Calculate the velocity through the influent orifices.

Calculate the design peak flow per influent orifice at $N_{orifice} = 6$ orifices/channel.

$$q_{orifice} = \frac{q_{channel}}{N_{orifice}} = \frac{0.31\ \text{m}^3/\text{s}}{6\ \text{orifices/channel}} = 0.051\ \text{m}^3/\text{s}$$

Calculate the area of each influent orifice at the diameter $d_{orifice} = 0.38$ m (15 in).

$$A_{orifice} = \frac{\pi}{4} \times (0.38\ \text{m})^2 = 0.113\ \text{m}^2$$

Calculate the velocity through influent orifices at design peak flow.

$$v_{orifice} = \frac{q_{orifice}}{A_{orifice}} = \frac{0.051\ \text{m}^3/\text{s}}{0.113\ \text{m}^2} = 0.45\ \text{m/s} \quad \text{(It is less than the design requirement of 0.5 m/s.)}$$

d. Calculate the head loss across the influent orifices at design peak flow.
Calculate head loss through the influent orifices from orifice equation in Equation 7.4b and assuming $C_d = 0.6$.

$$0.45\ \text{m/s} = 0.6 \times \sqrt{2 \times 9.81\ \text{m/s}^2 \times \Delta H}$$

$$\Delta H = \left(\frac{0.45\ \text{m/s}}{0.6 \times \sqrt{2 \times 9.81\ \text{m}^2/\text{s}}}\right)^2 = 0.03\ \text{m}$$

5. Design the effluent structure.
   a. Select the arrangement.
   The effluent structure consists of 0.5-m-wide multiple effluent launders that are installed along the width of the basin, and have a free fall into a 1-m-wide central outlet channel. The effluent launders have 90° V-notches over effluent weir plates on both sides. The V-notches have 0.19-m free fall into the launder. The water depth in the central effluent channel at the exit point is 0.4 m at design peak flow, and is regulated by the outfall conditions in the exit effluent channel.
   b. Calculate the number of effluent launders with weir plates.
   Calculate the total length of effluent weir required at the design weir loading rate $WL = 340$ m^3/m·d.

$$L'_{total} = \frac{Q_{peak}}{WL} = \frac{53{,}200\ \text{m}^3/\text{d}}{340\ \text{m}^3/\text{m·d}} = 156\ \text{m}$$

Calculate the maximum potential length of each weir from the basin width $W = 11.58$ m and the central outlet channel width $b_{eff} = 1$ m.

$$L'_{plate} = \frac{1}{2}(W - b_{eff}) = \frac{1}{2} \times (11.58 - 1)\ \text{m} = 5.29\ \text{m/plate}$$

Use an effluent weir plate length $L_{plate} = 5$ m/plate by considering additional spaces for construction.
Calculate the total number of weir plates required.

$$N_{plate} = \frac{L'_{total}}{L_{plate}} = \frac{156\ \text{m}}{5\ \text{m/plate}} \approx 32\ \text{weir plates}$$

Provide 16 launders ($N_{launder} = 16$ launders) on both sides of the central channel.

Number of launders on each side of the central effluent channel

$$= \frac{1}{2} N_{\text{launder}} = \frac{1}{2} \times 16 \text{ launders} = 8 \text{ launders}$$

Weir plates are installed on both sides of the launder. Calculate the total length of the weir plates.

$L_{\text{plate}} = 2\text{N}_{\text{launder}}\ \text{L}_{\text{plate}} = 2$ plates/launder × 16 launders × 5 m/plate = 160 m

Check the weir loading at design peak flow.

$$WL = \frac{Q_{\text{peak}}}{L_{\text{plate}}} = \frac{53{,}200\,\text{m}^3/\text{d}}{160\,\text{m}} = 333\,\text{m}^3/\text{m·d} \qquad \text{(It is less than the design requirement of } 340\,\text{m}^3/\text{m·d.)}$$

c. Calculate the total number of 90° V-notches.

Calculate the number of 90° V-notches on each weir plate at $L_{\text{space}} = 0.25$ m.

$$N_{\text{notches/plate}} = \frac{L_{\text{plate}}}{L_{\text{space}}} = \frac{5\,\text{m/plate}}{0.25\,\text{m}} = 20 \text{ notches/plate}$$

Provide notches 25-cm center-to-center and 27 cm from each end of the effluent trough to center of the nearest V-notch.

Calculate the total number of 90° V-notches.

$$N_{\text{notch}} = N_{\text{plate}} \times N_{\text{notch/plate}} = 32 \text{ weir plates} \times 20 \text{ notches/plate} = 640 \text{ notches}$$

Calculate design peak flow.

$$\text{Design peak flow,} \quad Q_{\text{peak}} = 53{,}200\,\text{m}^3/\text{d} \times \frac{\text{d}}{(60 \times 60 \times 24)\,\text{s}} = 0.616\,\text{m}^3/\text{s}$$

Calculate flow per 90° V-notch at design peak flow of $0.616\ \text{m}^3/\text{s}$.

$$q_{\text{peak}} = \frac{Q_{\text{peak}}}{N_{\text{notch}}} = \frac{0.616\,\text{m}^3/\text{s}}{640 \text{ notches}} = 0.00096\,\text{m}^3/\text{s per V-notch}$$

Calculate head over 90° V-notch from Equation 9.5b.

$$H_{\text{peak}} = \left(\frac{15}{8} \times \frac{0.00096\,\text{m}^3/\text{s}}{0.6 \times \sqrt{2 \times 9.81\,\text{m/s}^2}}\right)^{2/5} = 0.05\,\text{m} \quad \text{or} \quad 5\,\text{cm}$$

Provide 5-cm freeboard over weir head at design peak flow.

Total depth of V-notch weir plate $D_{\text{plate}} = 5 + 5$ cm (Freeboard) = 10 cm.

d. Calculate the head over 90° V-notches at design average flow.

Calculate flow per 90° V-notch at design average flow of $0.22\ \text{m}^3/\text{s}$.

$$q = \frac{Q}{N_{\text{notch}}} = \frac{0.22\,\text{m}^3/\text{s}}{640 \text{ notches}} = 0.00034\,\text{m}^3/\text{s}$$

Calculate head over 90° V-notch from Equation 9.5b.

$$H = \left(\frac{15}{8} \times \frac{0.00034\,\text{m}^3/\text{s}}{0.6 \times \sqrt{2 \times 9.81\,\text{m/s}^2}}\right)^{2/5} = 0.04\,\text{m} \quad \text{or} \quad 4\,\text{cm}$$

Freeboard over weir head at design average flow $D_{\text{plate}} - H = (10 - 4)$ cm $= 6$ cm (Freeboard).
The design details of effluent structure are shown in Figure 9.24.

e. Calculate the depth of flow in the effluent launder.

The effluent launders have increasing flow from upper to lower end, and have a free fall into central outlet channel. Critical depth will develop at the point of free fall.

Calculate flow per launder.

$$q_{\text{launder}} = \frac{Q_{\text{peak}}}{N_{\text{launder}}} = \frac{0.616\,\text{m}^3/\text{s}}{16\ \text{launders}} = 0.0385\,\text{m}^3/\text{s}$$

Calculate the critical depth y_c at the free fall of effluent launder from Equation 7.10 with $C_w =$ 1.0 and width of launder, $b_{\text{launder}} = 0.5$ m.

$$0.0385\,\text{m}^3/\text{s} = 1.0 \times 0.5\,\text{m} \times \sqrt{9.81\,\text{m/s}^2} \cdot y_c^{3/2}$$

Critical depth is calculated.

$$y_c = \left(\frac{0.0385\,\text{m}^3/\text{s}}{1.0 \times 0.5\,\text{m} \times \sqrt{9.81\,\text{m/s}^2}}\right)^{2/3} = 0.09\,\text{m}$$

For a free fall into the effluent exit channel, assume the water depth $y_2 = y_c = 0.09$ m at the lower end of the launder.

Calculate the water depth y_1 at the upper end of the launder from Equation 8.12a.

$$y_1 = \sqrt{y_2^2 + \frac{2(q_{\text{launder}})^2}{g(b_{\text{launder}})^2 y_2}} = \sqrt{(0.09\,\text{m})^2 + \frac{2 \times (0.0385\,\text{m}^3/\text{s})^2}{9.81\,\text{m/s}^2 \times (0.5\,\text{m})^2 \times 0.09\,\text{m}}} = 0.15\,\text{m}$$

Add 10% additional depth to account for friction losses and turbulence.
Total water depth at the upper end $y_{\text{launder}} = 1.1 \times 0.15\,\text{m} = 0.17\,\text{m}$
Add clearance of 0.19 m for free fall after the V-notches, and depth of V-notch of 0.10 m.
Total depth of effluent launder $= (0.17 + 0.19 + 0.10)\,\text{m} = 0.46\,\text{m}$

f. Calculate the depth of flow in the central outlet channel.

The central outlet channel receives flow from 16 effluent launders (8 on each side). At the peak design flow, the water depth y_2 in the central outlet channel is 0.4 m at the exit point. Calculate the water depth y_1 at the upper end of the central outlet channel from Equation 8.12a at the design peak flow $Q_{\text{peak}} = 0.616\,\text{m}^3/\text{s}$ and channel width $b_{\text{eff}} = 1.0$ m.

$$y_1 = \sqrt{(0.40\,\text{m})^2 + \frac{2 \times (0.616\,\text{m}^3/\text{s})^2}{9.81\,\text{m/s}^2 \times (1.0\,\text{m})^2 \times 0.40\,\text{m}}} = 0.59\,\text{m}$$

Add 10% additional depth to account for friction losses and turbulence.
Total water depth at the upper end $y_{\text{launder}} = 1.1 \times 0.59\,\text{m} = 0.65\,\text{m}$

Add clearance of 0.15 m below the bottom of effluent launder.

Total depth of central outlet channel = (0.65 + 0.15) m = 0.80 m

6. Check the possible resuspension of settled solids by horizontal velocity.
   a. Calculate the horizontal velocity.

      To be conservative, horizontal velocity is calculated using the shallowest side water depth on the effluent side of the basin and full design peak flow.

      Calculate the horizontal velocity V_h from Equation 9.4a at $SWD_{eff} = 2.6$ m, $W = 11.58$ m, and $Q_{peak} = 0.616\ m^3/s$.

$$V_h = \frac{Q_{peak}}{SWD_{eff}\,W} = \frac{0.616\,m^3/s}{2.6\,m \times 11.58\,m} = 0.020\,m/s$$

   b. Calculate the critical scour velocity.

      Calculate the critical scour velocity V_c through the basin from Equation 9.4b based on the following assumptions:

      Cohesion constant, $k = 0.05$

      Specific gravity, $S = 1.25$

      Average diameter of particles $d = 95\ \mu m$ (95×10^{-6} m)

      Darcy–Weisbach friction factor, $f = 0.025$

$$V_c = \sqrt{\frac{8\,k(S-1)gd}{f}} = \sqrt{\frac{8 \times 0.05 \times (1.25-1) \times 9.81\,m/s^2 \times (95 \times 10^{-6})m}{0.025}}$$

$$= 0.061\,m/s$$

   c. Comment on the resuspension of settled solids.

      In this case, the horizontal velocity is calculated at the extreme location and flow conditions. The actual velocity in most part of the basin is less than 0.2 m/s and significantly less than the critical scour velocity of 0.061 m/s. Therefore, resuspension of settled solids will not occur.

7. Determine the process performance and effluent quality.
   a. Determine the BOD_5 and TSS removal efficiencies.

      The BOD_5 and TSS removal efficiencies are evaluated at the design average flow based on (1) readings from Figure 9.12a and b and (2) calculation results of using Equations 9.2a and 9.2b. The evaluation results are summarized below.

Parameter	Reading from Figure 9.12a at $SOR = 35$ $m^3/m^2{\cdot}d$	Reading from Figure 9.12b at $\theta = 2.7$ h	Use of Equations 9.2a and 9.2b at $\theta = 2.7$ h	Recommendation for Design
		Removal, %		
BOD_5	35	36	36[a]	35
TSS	64	63	63[b]	60

[a] $R_{BOD_5} = \dfrac{\theta}{a_1 + b_1\theta} = \dfrac{2.7\,h}{0.018 + 0.021 \times 2.7\,h} = 36\%$

[b] $R_{TSS} = \dfrac{\theta}{a_2 + b_2\theta} = \dfrac{2.7\,h}{0.0075 + 0.013 \times 2.7\,h} = 63\%$

   b. Calculate BOD_5 and TSS in the primary effluent.

      Calculate BOD_5 and TSS concentrations in the primary effluent based on the recommended removal efficiencies.

$$BOD_5 = (1 - 0.35) \times 180\,mg/L = 117\,mg/L$$

$$TSS = (1 - 0.60) \times 220\,mg/L = 88\,mg/L$$

8. Determine the primary solids generation and handling.
   a. Calculate the generation of primary sludge.
      Calculate the quantity of dry primary sludge removed at TSS removal efficiency of 60%.

$$\text{Dry solids removed} = 0.60 \times 220\,\text{mg/L} \times \frac{10^3\,\text{L}}{\text{m}^3} \times \frac{\text{kg}}{10^6\,\text{mg}} \times 19{,}000\,\text{m}^3/\text{d} = 2510\,\text{kg/d}$$

      Calculate volume of bulk primary sludge produced using typical solids content of 4.5% and specific gravity of 1.015.

$$\text{Sludge volume} = 2510\,\text{kg/d dry solids} \times \frac{100\,\text{kg wet sludge}}{4.5\,\text{kg dry solids}} \times \frac{1}{(1.015 \times 10^3)\text{kg/m}^3\text{ as wet sludge}}$$

$$= 55\,\text{m}^3/\text{d wet sludge}$$

   b. Calculate the sludge pumping cycle.
      Select a sludge pumping rate of 500 L/min (0.5 m^3/min), and a pumping cycle 5 per hour.

$$\text{Daily pumping capacity} = \frac{5\,\text{min}}{\text{h}} \times \frac{0.5\,\text{m}^3}{\text{min}} \times \frac{24\,\text{h}}{\text{d}} = 60\,\text{m}^3/\text{d} > 55\,\text{m}^3/\text{d wet sludge}$$

   c. Calculate the generation of scum.
      Calculate scum generation using typical scum generation rate of 8 g/m^3 (Table 6.9).

$$\text{Scum quantity} = 8\,\text{g/m}^3 \times \frac{\text{kg}}{10^3\,\text{g}} \times 19{,}000\,\text{m}^3/\text{d} = 152\,\text{kg/d}$$

9. Draw the design details.
   The plan, longitudinal section, and details of the influent and effluent structures, and design water surface elevations are shown in Figure 9.24.
   Note: All elevations are with respect to water surface elevation in the basin at average design flow.

9.5 Enhanced Primary Treatment

The performance of a primary sedimentation basin is enhanced by preaeration or chemical coagulation and precipitation. Both methods are presented below.

9.5.1 Preaeration

Preaeration of wastewater increases the TSS and BOD_5 removal efficiency by ~7–8%. Preaeration promotes flocculation of solids, and uniform distribution of flow into the basin. Approximately 20–30 min of aeration at a minimum air supply of 0.82 L/L (0.11 ft^3/gal) is considered sufficient.[4] An aerated grit chamber also improves grit separation, and enhances the performance of a primary sedimentation basin. Other benefits of preaeration are addition of DO and prevention of septicity in sedimentation basin.

Preaeration causes scrubbing of VOCs and intensifies odor problems. Collection of gases from preaeration channel, and scrubbing of exhaust gases is an effective method of controlling odors at the primary sedimentation facility. Preaeration may also have negative impact on biological nutrient removal (BNR) processes by creating unfavorable conditions for the formations of (1) anaerobic or anoxic conditions and/or (2) short-chain volatile fatty acids (SCVFAs) for phosphorus release and removed are the critical carbon sours for achieving effective BNR processes.

9.5.2 Chemical Coagulation and Precipitation

Chemical coagulation promotes flocculation of colloidal particles into a more readily settleable floc in a chemically enhanced primary treatment (CEPT) process. Some chemicals cause precipitation by chemically or physically modifying the solubility of many dissolved solids. Thus, the removal efficiencies of BOD_5, TSS, and total phosphorous (TP) are increased. The typical removals of these constituents are summarized in Table 9.9 (see also Table 6.8).

The *advantages* of chemically enhanced sedimentation are (a) greater removal efficiencies, (b) the ability to use higher surface overflow rates for sedimentation basin, and (c) more consistent basin performance. The *disadvantages* of the process include (a) increased sludge production, (b) sludge is difficult to thicken and dewater, and (c) increased operation and maintenance cost.

Process Description of Coagulation and Flocculation: The colloidal particles have a very large surface-area-to-mass ratio. As a result, the gravitational force has little effect upon their behavior. Additionally, most colloids are electrically charged, and therefore remain in suspension and in motion*. This is called stability of colloids. Coagulants destabilize the colloids by a combination of three mechanisms: (a) compression of the double layer†, (b) interparticle bridging, and (c) enmeshment in a precipitate. The chemically destabilized colloidal particles after coagulation are gently stirred or flocculated to promote the growth of the pinhead flocs produced.

Traditionally, lime and metal salts such as alum (aluminum sulfate), ferric chloride, ferric sulfate, ferrous chloride, and ferrous sulfate have been utilized as coagulants. Physical and chemical properties of these chemicals are provided in Table 9.10. Standard jar test apparatus is utilized to evaluate the performance and to optimize the chemical dosages. In recent years, *polymers* (long-molecular-chain organic compounds) have been used in conjunction with or in lieu of metal salts to enhance the coagulation and flocculation process. Additionally, alum polymers such as *polyaluminum chloride*, *polyaluminum chlorohydrate*, and *polyaluminum sulfate* are gaining popularity. The stoichiometric chemical reactions between alkalinity, and lime and metal salts are expressed by Equations 9.16 through 9.19.[3,5,26–34]

TABLE 9.9 Typical Removal Efficiencies of CEPT

Constituent	Removal, %	
	Range	Typical
TSS	50–80	70
BOD_5	40–70	55
COD	30–60	50
ON	70–95	80
TP	70–90	85

* Electrostatic forces are due to similar charges that repel the colloids. The van der Waals forces are due to attraction between two masses.

† The surface charge on colloids attracts ions of opposite charge known as counter ions (hydrogen or other cations). This forms a dense layer or stern layer adjacent to the particle, and a diffused layer due to asymmetric electrical charge of water molecules. The two layers are referred to as the double layer.[3,4]

TABLE 9.10 Commonly Used Chemicals for Enhanced Sedimentation and Chemical Precipitation of Phosphorus

Chemical Name	Synonyms (Appearance)	Chemical Formula (Molecular Weight)	Commercial Grade Qualities				
			Bulk Density, kg/m^3 (Specific Gravity)	Solubility in Water, kg/m^3 (%)	Chemical Content, % w/w	Water Content, % w/w	pH
Calcium oxide	Lime, quick lime (off-white powder or lump)	CaO (56)	880–960 (–)	~1.2 at 25°C (–)	CaO: 75–95 (high calcium) 72–74 (dolomitic)	–	~12.5 (0.12% solution)
Calcium Hydroxide	Hydrated lime (off-white faint slurry)	$Ca(OH)_2$ (74)	400–560 (–)	~1.6 at 25°C (–)	CaO: 72–74 (high calcium) 46–48 (normal dolomitic) 40–42 (pressure dolomitic)	23–24[a] (high calcium) 15–17[a] (normal dolomitic) 25–27[a] (pressure dolomitic)	~12.5 (0.16% solution)
Aluminum sulfate	Aluminum sulfate hydrate, dry alum (white or pale green powder)	$Al_2(SO_4)_3{\cdot}14\ H_2O$ (594)	1010–1140 (–)	~500 at 0°C (–)	Al: ~9	–	~3.5 (1% solution)
	Aluminum sulfate octadecahydrate, alum, alum salt (white or pale green lump)	$Al_2(SO_4)_3{\cdot}18\ H_2O$ (666)	1610–1690 (–)	~870 at 0°C (–)	Al: 7.9–8.1	–	–
	Liquid alum, (clear, light green or amber liquid)	$Al_2(SO_4)_3{\cdot}49\ H_2O$[b] (1225)	– (1.30–1.34)	– (100)	Al: 4.1–5.0	45–55[c]	1.9–2.3
Ferric chloride	Iron (III) chloride anhydrous, ferric trichloride (green-black to dark gray or brown power)	$FeCl_3$ (162)	690–960 (–)	~740 at 0°C (–)	Fe: 31–34	–	~2 (1% solution)
	Iron trichloride, ferric chloride hexahydrate (yellow to brown lump)	$FeCl_3{\cdot}6\ H_2O$ (270)	960–1030 (–)	~920 at 20°C (–)	Fe: 20–21	–	–
	Ferric chloride solution (Reddish-brown syrupy liquid)	$FeCl_3{\cdot}13.1\ H_2O$[d] (398)	– (1.26–1.48)	– (100)	Fe: 8.6–16.3	53–75	< 2
Ferric sulfate	Iron(III) sulfate, ferric sulfate anhydrous (brown to yellow powder)	$Fe_2(SO_4)_3$ (400)	200 (–)	Soluble (–)	Fe: ~28	–	~2–3 (1% solution)
	Ferric sulfate nonahydrate (red-brown crystal lump)	$Fe_2(SO_4)_3{\cdot}9\ H_2O$ (562)	980–1150 (–)	Miscible (–)	Fe: ~20	–	–

(Continued)

TABLE 9.10 (*Continued*) Commonly Used Chemicals for Enhanced Sedimentation and Chemical Precipitation of Phosphorus

Chemical Name	Synonyms (Appearance)	Chemical Formula (Molecular Weight)	Commercial Grade Qualities				
			Bulk Density, kg/m^3 (Specific Gravity)	Solubility in Water, kg/m^3 (%)	Chemical Content, % w/w	Water Content, % w/w	pH
	Ferric sulfate solution (reddish-brown syrupy liquid)	$Fe_2(SO_4)_3{\cdot}30.1\ H_2O$[e] (940)	– (1.38–1.62)	– (100)	Fe: 10–14	28–50[c]	<2
Ferrous chloride	Iron(II) chloride tetrahydrate, ferrous dichloride tetrahydrate, ferrous chloride tetrahydrate (light green crystal lump)	$FeCl_2{\cdot}4\ H_2O$ (199)	900 (–)	~1600 at 10°C (–)	Fe: 27–28	–	<7 (1% solution)
	Ferrous chloride solution (Pale green liquid)	$FeCl_2{\cdot}12\ H_2O$[f] (343)	– (1.20 –1.40)	– (100)	Fe: 7–17	55–84	<2
Ferrous sulfate	Copperas, ferrous sulfate heptahydrate (blue green crystal lump)	$FeSO_4{\cdot}7\ H_2O$ (278)	700–1200 (–)	~260 at 20°C (–)	Fe: ~20	–	<7 (1% solution)
	Ferrous sulfate solution (green liquid)	$FeSO_4{\cdot}36\ H_2O$[g] (800)	– (1.14 –1.30)	– (100)	Fe: 4–11	70–89	<4

[a] Chemically combined water only.
[b] It is determined based on a typical commercial concentration of 48.5% as $Al_2(SO_4)_3{\cdot}14\ H_2O$ by weight.
[c] Not including chemically combined water.
[d] It is determined based on a typical commercial concentration of 41% as $FeCl_3$ by weight.
[e] It is determined based on a typical commercial concentration of 60% as $Fe_2(SO_4)_3{\cdot}9\ H_2O$ by weight.
[f] It is determined based on a typical commercial concentration of 37% as $FeCl_2$ by weight.
[g] It is determined based on a typical commercial concentration of 19% as $FeSO_4$ by weight.

Note: $kg/m^3 \times 0.0624 = lb/ft^3$ and specific gravity × 8.345 = lb/gal.

Source: Adapted in part from References 2, 5, 26, and 32 through 34.

Lime

$$\underset{56}{CaO} + \underset{2\times 18}{H_2O} \rightarrow \underset{74}{Ca(OH)_2} \tag{9.16a}$$

$$\underset{74}{Ca(OH)_2} + \underset{\substack{100\\(\text{as } CaCO_3)}}{Ca(HCO_3)_2} \rightarrow \underset{2\times 100}{2\ CaCO_3(s)} + \underset{2\times 18}{2\ H_2O} \tag{9.16b}$$

$$\underset{74}{Ca(OH)_2} + \underset{\substack{100\\(\text{as } CaCO_3)}}{Mg(HCO_3)_2} \rightarrow \underset{\substack{100\\(\text{as } CaCO_3)}}{MgCO_3} + \underset{100}{CaCO_3(s)} + \underset{2\times 18}{2\ H_2O} \tag{9.16c}$$

$$\underset{74}{Ca(OH)_2} + \underset{\substack{100\\(\text{as } CaCO_3)}}{MgCO_3} \rightarrow \underset{58}{Mg(OH)_2(s)} + \underset{100}{CaCO_3(s)} \tag{9.16d}$$

$$\underset{2\times 74}{2\ Ca(OH)_2} + \underset{\substack{100\\(\text{as } CaCO_3)}}{Mg(HCO_3)_2} \rightarrow \underset{58}{Mg(OH)_2(s)} + \underset{2\times 100}{2\ CaCO_3(s)} + \underset{2\times 18}{2\ H_2O} \tag{9.16e}$$

Alum

$$\underset{594}{Al_2(SO_4)_3\cdot 14\ H_2O} + \underset{\substack{3\times 100\\(\text{as } CaCO_3)}}{3\ Ca(HCO_3)_2} \rightarrow \underset{2\times 78}{2\ Al(OH)_3(s)} + \underset{3\times 136}{3\ CaSO_4} + \underset{6\times 44}{6\ CO_2} + \underset{14\times 18}{14\ H_2O} \tag{9.17a}$$

$$\underset{666}{Al_2(SO_4)_3\cdot 18\ H_2O} + \underset{\substack{3\times 100\\(\text{as } CaCO_3)}}{3\ Ca(HCO_3)_2} \rightarrow \underset{2\times 78}{2\ Al(OH)_3(s)} + \underset{3\times 136}{3\ CaSO_4} + \underset{6\times 44}{6\ CO_2} + \underset{18\times 18}{18\ H_2O} \tag{9.17b}$$

$$\underset{1236}{Al_2(SO_4)_3\cdot 49.6\ H_2O} + \underset{\substack{3\times 100\\(\text{as } CaCO_3)}}{3\ Ca(HCO_3)_2} \rightarrow \underset{2\times 78}{2\ Al(OH)_3(s)} + \underset{3\times 136}{3\ CaSO_4} + \underset{6\times 44}{6\ CO_2} + \underset{49.6\times 18}{49.6\ H_2O} \tag{9.17c}$$

Ferric Chloride

$$\underset{2\times 162}{2\ FeCl_3} + \underset{\substack{3\times 100\\(\text{as } CaCO_3)}}{3\ Ca(HCO_3)_2} \rightarrow \underset{2\times 107}{2\ Fe(OH)_3(s)} + \underset{3\times 111}{3\ CaCl_2} + \underset{6\times 44}{6\ CO_2} \tag{9.18a}$$

$$\underset{2\times 270}{2\ FeCl_3\cdot 6\ H_2O} + \underset{\substack{3\times 100\\(\text{as } CaCO_3)}}{3\ Ca(HCO_3)_2} \rightarrow \underset{2\times 107}{2\ Fe(OH)_3(s)} + \underset{3\times 111}{3\ CaCl_2} + \underset{6\times 44}{6\ CO_2} + \underset{12\times 18}{12\ H_2O} \tag{9.18b}$$

$$\underset{2\times 398}{2\ FeCl_3\cdot 13.1\ H_2O} + \underset{\substack{3\times 100\\(\text{as } CaCO_3)}}{3\ Ca(HCO_3)_2} \rightarrow \underset{2\times 107}{2\ Fe(OH)_3(s)} + \underset{3\times 111}{3\ CaCl_2} + \underset{6\times 44}{6\ CO_2} + \underset{13.1\times 18}{13.1\ H_2O} \tag{9.18c}$$

Ferric Sulfate

$$\underset{400}{Fe_2(SO_4)_3} + \underset{\substack{3\times100 \\ (\text{as } CaCO_3)}}{3\,Ca(HCO_3)_2} \rightarrow \underset{2\times107}{2\,Fe(OH)_3(s)} + \underset{3\times136}{3\,CaSO_4} + \underset{6\times44}{6\,CO_2} \tag{9.18d}$$

$$\underset{562}{Fe_2(SO_4)_3\cdot 9\,H_2O} + \underset{\substack{3\times100 \\ (\text{as } CaCO_3)}}{3\,Ca(HCO_3)_2} \rightarrow \underset{2\times107}{2\,Fe(OH)_3(s)} + \underset{3\times136}{3\,CaSO_4} + \underset{6\times44}{6\,CO_2} + \underset{9\times18}{9\,H_2O} \tag{9.18e}$$

$$\underset{940}{Fe_2(SO_4)_3\cdot 30.1\,H_2O} + \underset{\substack{3\times100 \\ (\text{as } CaCO_3)}}{3\,Ca(HCO_3)_2} \rightarrow \underset{2\times107}{2\,Fe(OH)_3(s)} + \underset{3\times136}{3\,CaSO_4} + \underset{6\times44}{6\,CO_2} + \underset{30.1\times18}{30.1\,H_2O} \tag{9.18f}$$

Ferrous Chloride

$$\underset{199}{FeCl_2\cdot 4\,H_2O} + \underset{\substack{100 \\ (\text{as } CaCO_3)}}{Ca(HCO_3)_2} \rightarrow \underset{90}{Fe(OH)_2(s)} + \underset{111}{CaCl_2} + \underset{2\times44}{2\,CO_2} + \underset{4\times18}{4\,H_2O} \tag{9.19a}$$

$$\underset{343}{FeCl_2\cdot 12\,H_2O} + \underset{\substack{100 \\ (\text{as } CaCO_3)}}{Ca(HCO_3)_2} \rightarrow \underset{90}{Fe(OH)_2(s)} + \underset{111}{CaCl_2} + \underset{2\times44}{2\,CO_2} + \underset{12\times18}{12\,H_2O} \tag{9.19b}$$

Ferrous Sulfate

$$\underset{278}{FeSO_4\cdot 7\,H_2O} + \underset{\substack{100 \\ (\text{as } CaCO_3)}}{Ca(HCO_3)_2} \rightarrow \underset{90}{Fe(OH)_2(s)} + \underset{136}{CaSO_4} + \underset{2\times44}{2\,CO_2} + \underset{7\times18}{7\,H_2O} \tag{9.19c}$$

$$\underset{800}{FeSO_4\cdot 36\,H_2O} + \underset{\substack{100 \\ (\text{as } CaCO_3)}}{Ca(HCO_3)_2} \rightarrow \underset{90}{Fe(OH)_2(s)} + \underset{136}{CaSO_4} + \underset{2\times44}{2\,CO_2} + \underset{36\times18}{36\,H_2O} \tag{9.19d}$$

The following important observations can be made from the above coagulation reactions:

1. The number of bound water molecules with coagulants varies.
2. The alkalinity in water is expressed as $CaCO_3$, which has a molecular weight of 100 g/mole (or mg/mmole). Therefore, the molecular weight of $Ca(HCO_3)_2$ if expressed as $CaCO_3$ is 100 g/mole (or mg/mmole).
3. In all these reactions, the alkalinity is consumed. The alkalinity consumption rate is dependent upon the coagulant dosage. The alkalinity consumption rates (mg/L as $CaCO_3$) per mg/L coagulant dose are summarized in Table 9.11.
4. Insoluble precipitates produced, such as $CaCO_3$ and metal hydroxides (indicated by [s]) destabilize the colloids and become nucleus for floc formation. They are a part of the sludge from coagulation process. The stoichiometric equations may be used to estimate the sludge production.
5. Most wastewaters contain sufficient natural alkalinity (50–200 mg/L as $CaCO_3$) for coagulation reaction to occur. If sufficient alkalinity is not present, proper floc will not produce. Lime or soda ash is commonly added for supplementation of alkalinity.
6. The reaction is analogous when the alkalinity-causing compound is magnesium bicarbonate.
7. The hardness is not removed; it simply changes from carbonate hardness to noncarbonate hardness.
8. CO_2 is produced, and pH is slightly or moderately lowered after coagulation.
9. The coagulation reactions are complex. The actual amounts of coagulant required for destabilization of colloids may depend not only on the reaction stoichiometry but also on other operational

TABLE 9.11 Alkalinity Consumptions in Coagulation Reactions

Coagulant	Alkalinity Consumed	Alkalinity Consumption, mg as $CaCO_3$ per mg Coagulant Dose as Chemical	Reference
Lime			
CaO	$Ca(HCO_3)_2$	(100 mg/mmole) ÷ (56 mg/mmole) = 1.79	Equations 9.16a and 9.16b
	$Mg(HCO_3)_2$	(100 mg/mmole) ÷ (2 × 56 mg/mmole) = 0.89	Equations 9.16a and 9.16e
$Ca(OH)_2$	$Ca(HCO_3)_2$	(100 mg/mmole) ÷ (74 mg/mmole) = 1.35	Equation 9.16b
	$Mg(HCO_3)_2$	(100 mg/mmole) ÷ (2 × 74 mg/mmole) = 0.68	Equation 9.16e
Alum			
$Al_2(SO_4)_3{\cdot}14\ H_2O$	$Ca(HCO_3)_2$	(3 × 100 mg/mmole) ÷ (594 mg/mmole) = 0.51	Equation 9.17a
$Al_2(SO_4)_3{\cdot}18\ H_2O$	$Ca(HCO_3)_2$	(3 × 100 mg/mmole) ÷ (666 mg/mmole) = 0.45	Equation 9.17b
$Al_2(SO_4)_3{\cdot}49.6\ H_2O$	$Ca(HCO_3)_2$	(3 × 100 mg/mmole) ÷ (1236 mg/mmole) = 0.24	Equation 9.17c
Ferric chloride			
$FeCl_3$	$Ca(HCO_3)_2$	(3 × 100 mg/mmole) ÷ (2 × 162 mg/mmole) = 0.93	Equation 9.18a
$FeCl_3{\cdot}6\ H_2O$	$Ca(HCO_3)_2$	(3 × 100 mg/mmole) ÷ (2 × 270 mg/mmole) = 0.56	Equation 9.18b
$FeCl_3{\cdot}13.1\ H_2O$	$Ca(HCO_3)_2$	(3 × 100 mg/mmole) ÷ (2 × 398 mg/mmole) = 0.38	Equation 9.18c
Ferric sulfate			
$Fe_2(SO_4)_3$	$Ca(HCO_3)_2$	(3 × 100 mg/mmole) ÷ (400 mg/mmole) = 0.75	Equation 9.18d
$Fe_2(SO_4)_3{\cdot}9\ H_2O$	$Ca(HCO_3)_2$	(3 × 100 mg/mmole) ÷ (562 mg/mmole) = 0.53	Equation 9.18e
$Fe_2(SO_4)_3{\cdot}30.1\ H_2O$	$Ca(HCO_3)_2$	(3 × 100 mg/mmole) ÷ (940 mg/mmole) = 0.32	Equation 9.18f
Ferrous chloride			
$FeCl_2{\cdot}4\ H_2O$	$Ca(HCO_3)_2$	(100 mg/mmole) ÷ (199 mg/mmole) = 0.50	Equation 9.19a
$FeCl_2{\cdot}12\ H_2O$	$Ca(HCO_3)_2$	(100 mg/mmole) ÷ (343 mg/mmole) = 0.29	Equation 9.19b
Ferrous sulfate			
$FeSO_4{\cdot}7\ H_2O$	$Ca(HCO_3)_2$	(100 mg/mmole) ÷ (278 mg/mmole) = 0.36	Equation 9.19c
$FeSO_4{\cdot}36\ H_2O$	$Ca(HCO_3)_2$	(100 mg/mmole) ÷ (800 mg/mmole) = 0.13	Equation 9.19d

Note: Alkalinity is consumed when the softening achieved by adding sufficient lime.

conditions such as ionic species, pH, temperature, type and properties of particles, mixing-energy input, and effective content of metal ions in the coagulant solution.

Process Description of Chemical Precipitation: Chemical precipitation involves the addition of chemicals to decrease the solubility of a targeted constituent so that the precipitate can be removed by flocculation and sedimentation. In wastewater treatment, the most common use of precipitation is for (1) the removal of bivalent metal cations and (2) chemical removal of phosphorus.

Chemical Precipitation of Metals: In industrial wastewater pretreatment, the cations of interest for precipitation include arsenic (As), barium (Ba), cadmium (Cd), calcium (Ca), copper (Cu), magnesium (Mg), manganese (Mn), mercury (Hg), nickel (Ni), selenium (Se), and zinc (Zn). Most of these metals can be precipitated as hydroxides or sulfides. These cations are bivalent and cause hardness. Their precipitation removes hardness. Solubility product (K_{sp}) is used to calculate the solubility of an ionic compound, and is expressed by Equation 9.20.

$$MX \leftrightarrow M^+ + X^- \tag{9.20a}$$

$$K_{sp} \leftrightarrow [M^+][X^-] \tag{9.20b}$$

where

K_{sp} = solubility product constant of the compound
$[M^+]$ = concentration of the cation, mole/L
$[X^-]$ = concentration of anion, mole/L

TABLE 9.12 Typical Solubility Product Constants of Compounds Used In Wastewater Treatment

Equilibrium Equation		K_{sp} at 25°C	Significance in Wastewater Treatment
$CaCO_3$	$\leftrightarrow Ca^{2+} + CO_3^{-}$	5×10^{-9}	Hardness removal, phosphorus removal, scaling issue
$Ca(OH)_2$	$\leftrightarrow Ca^{2+} + 2\ OH^-$	8×10^{-6}	Hardness removal
$Ca_3(PO_4)_2$	$\leftrightarrow 3\ Ca^{2+} + 2\ PO_4^{3-}$	1×10^{-27}	Phosphate precipitation
$CaHPO_4$	$\leftrightarrow Ca^{2+} + HPO_4^{2-}$	3×10^{-7}	Phosphate precipitation
$Ca_5OH(PO_4)_3$	$\leftrightarrow 5\ Ca^{2+} + 3\ PO_4^{3-} + OH^-$	1×10^{-56}	Phosphate precipitation
$Al(OH)_3$	$\leftrightarrow Al^{3+} + 3\ OH^-$	1×10^{-32}	Coagulation, phosphorus removal
$AlPO_4$	$\leftrightarrow Al^{3+} + PO_4^{3-}$	1×10^{-22}	Phosphorus precipitation
$Fe(OH)_3$	$\leftrightarrow Fe^{3+} + 3\ OH^-$	6×10^{-38}	Coagulation, iron removal, phosphorus removal, corrosion issue
$Fe(OH)_2$	$\leftrightarrow Fe^{2+} + 2\ OH^-$	1×10^{-15}	Coagulation, iron removal, phosphorus removal, corrosion issue
$FePO_4$	$\leftrightarrow Fe^{3+} + PO_4^{3-}$	1×10^{-22}	Phosphorus precipitation
$Fe_3(PO_4)_2$	$\leftrightarrow 3\ Fe^{3+} + 2\ PO_4^{3-}$	1×10^{-30}	Phosphorus precipitation
$MgCO_3$	$\leftrightarrow Mg^{2+} + CO_3^{-}$	4×10^{-5}	Hardness removal, scaling issue
$Mg(OH)_2$	$\leftrightarrow Mg^{2+} + 2\ OH^-$	7×10^{-12}	Hardness removal, phosphorus removal, scaling issue
$MgNH_4PO_4$	$\leftrightarrow Mg^{2+} + NH_4^+ + PO_4^{3-}$	3×10^{-13}	Phosphorus recovery, scaling issue
$Mn(OH)_3$	$\leftrightarrow Mn^{3+} + 3\ OH^-$	1×10^{-36}	Manganese removal
$Mn(OH)_2$	$\leftrightarrow Mn^{2+} + 2\ OH^-$	1×10^{-14}	Manganese removal, phosphorus removal

Note: Names of many compounds in the first column may be found in Table 9.17

Source: Adapted in part from References 3, 4, 26, 27, 34, and 35.

The solubility products of various compounds of interest in wastewater treatment are provided in Table 9.12.[3,4,26,27,34,35]

Chemical Precipitation of Phosphorus: Precipitation of phosphorus can be achieved by adding chemicals at different locations through the wastewater processes: (1) *preprecipitation*: in raw wastewater followed by removal in primary basin; (2) *coprecipitation*: in primary settled effluent, or effluent from biological reactor before secondary clarifier; (3) *postprecipitation*: after secondary clarifier followed by removal in tertiary clarifier and/or filter, or (4) *multi-precipitations*: a combination of any locations in (1) through (3). Preprecipitation process is discussed below. Additional discussions about the phosphorus removal by other chemical precipitation processes are presented in Chapters 10 and 15. Chemical precipitations have also been involved considerably in solids handling, side stream treatment, and nutrient recovery processes that are presented in Chapter 13.

In general, phosphorus must be in the soluble form of orthophosphate (PO_4^{3-} or ortho-P) and then can be removed through two major mechanisms: (1) formation of phosphate precipitates and (2) adsorption on the metal hydroxide flocs. Addition of salts of multivalent metal ions is needed to stimulate the precipitations. The common metal ions are Ca^{2+}, Al^{3+}, Fe^{3+}, and Fe^{2+}. Polymers have also been used effectively in conjunction with metals. The precipitation reactions with these metal ions are expressed by Equations 9.21 through 9.24.[1,3,5,26–34]

Calcium (lime), Ca^{2+}

$$5\,Ca^{2+} + 3\,PO_4^{3-} + OH^- \leftrightarrow Ca_5OH(PO_4)_3(s) \tag{9.21a}$$

$$Ca^{2+} + CO_3^{2-} \leftrightarrow CaCO_3(s) \tag{9.21b}$$

$$Mg^{2+} + 2\,OH^- \leftrightarrow Mg(OH)_2(s) \tag{9.21c}$$

Aluminum (alum), Al^{3+}

$$Al^{3+} + PO_4^{3-} \leftrightarrow AlPO_4(s) \tag{9.22a}$$

$$Al^{3+} + 3\ OH^- \leftrightarrow Al(OH)_3(s) \tag{9.22b}$$

Iron, Fe^{3+}

$$Fe^{3+} + PO_4^{3-} \leftrightarrow FePO_4(s) \tag{9.23a}$$

$$Fe^{3+} + 3\ OH^- \leftrightarrow Fe(OH)_3(s) \tag{9.23b}$$

Iron, Fe^{2+}

$$3\ Fe^{2+} + 2\ PO_4^{3-} \leftrightarrow Fe_3(PO_4)_2(s) \tag{9.24a}$$

$$Fe^{2+} + 2\ OH^- \leftrightarrow Fe(OH)_2(s) \tag{9.24b}$$

Theoretically 1.67 mole of Ca^{2+}, 1 mole of Al^{3+} or Fe^{3+}, and 1.5 mole of Fe^{2+} will precipitate 1 mole of PO^{4+} (or P). In reality, the chemistry of phosphate precipitate formation is complex because of complexes formed between (a) phosphate and metals, (b) metals and liquid in the wastewater, (c) pH effect, and (d) side reactions of metals with alkalinity to form other hydroxide precipitates. The actual amount of phosphorus removed at different conditions can be assessed by the *standard jar test*. The dose of coagulant required is a function of pH, temperature, influent characteristics, and chemicals. The selection of chemicals for phosphorus precipitation is based on the following factors: (1) influent quality (concentrations of phosphorus, TSS, and alkalinity), (2) cost of chemicals and reliability of supply, (3) sludge handling and disposal, and (4) compatibility with other processes. Typical coagulants used for phosphorus precipitation are lime, alum, ferric chloride, ferric sulfate, ferrous chloride, and ferrous sulfate. Basic requirements of phosphorus precipitation with these chemicals are summarized in Table 9.13.[1,3–5,31–34,36–41]

9.5.3 Effluent Neutralization

Wastewater effluents having excessively high or low pH may require neutralization before discharge or reuse. A number of chemical are used for pH adjustment. The choice of chemical may depend upon (1) availability, (2) reliability of supply, (3) particular application, (4) economics, and (5) safety concerns. The chemicals most commonly used for wastewater application are summarized in Table 9.14.

9.5.4 Sludge Production

Mineral salt addition for enhanced sedimentation and phosphorus precipitation significantly increase the quantity of sludge due to (1) improved TSS and BOD removal, (2) production of metal hydroxides, (3) production of metal–phosphate precipitate, and (4) lime treatment produces $CaCO_3$ precipitate, in addition to phosphorous precipitate. Metal salt addition may also increase the volume of sludge due to potentially lower solids concentration in the sludge.

Bench- or pilot-scale testing will provide precise data on optimum chemical dosages and sludge production. Stoichiometric reactions with lime, alum, and ferric ions are also used to estimate the quantity of chemical sludge produced. Stoichiometric quantities of chemical sludge produced from different

TABLE 9.13 Phosphorus Precipitation from Raw Municipal Wastewater with Lime, Alum, and Iron Salts Addition

Parameter		Metal Ion			
		Lime, Ca^{2+}	Alum, Al^{3+}	Iron, Fe^{3+}	Iron, Fe^{2+}
Chemical used	Name	Calcium oxide or calcium Hydroxide	Aluminum sulfate	Ferric chloride or ferric sulfate	Ferrous chloride or ferrous sulfate
	Formulas	CaO, or $Ca(OH)_2$	$Al_2(SO_4)_3 \cdot 14\ H_2O$, $Al_2(SO_4) \cdot 18\ H_2O$', or $Al_2(SO_4) \cdot 49.6\ H_2O$	$FeCl_3$, $FeCl_3 \cdot 6\ H_2O$, $FeCl_3 \cdot 13.1\ H_2O$, $Fe_2(SO_4)_3$, $Fe_2(SO_4)_3 \cdot 9\ H_2O$, or $Fe_2(SO_4)_3 \cdot 30.1\ H_2O$	$FeCl_2 \cdot 4\ H_2O$, $FeCl_2 \cdot 12\ H_2O$, $FeSO_4 \cdot 7\ H_2O$, or $FeSO_4 \cdot 36\ H_2O$
Reaction equations		Equations 9.21a through 9.21c	Equations 9.22a and 9.22b	Equations 9.23a and 9.23b	Equations 9.24a and 9.24b
Dosage requirements	Ratio of metal ion to phosphorus	Ca:P	Al:P	Fe:P	Fe:P
	Theoretical molar ratio	1.67:1	1:1	1:1	1.5:1
	Theoretical weight ratio	2.2:1	0.87:1	1.8:1	2.7:1
	Practical molar ratio	1.3:1–2:1 Higher dose of lime is required because apatite precipitate varies	1:1–2.5:1 Higher dose of Al is required because of side reactions involving alkalinity and organic matter	1.5:1–3.5:1 Higher dose of Fe is required to satisfy the side reactions with alkalinity to produce $Fe(OH)_3$. Effective phosphorus removal may require at least additional 10 mg/L of iron for hydroxide formation. Typical iron requirement is 15–30 mg/L Fe to reduce influent P by 85–90%.	1.5:1–1.7:1 The excess ferrous iron forms ferrous hydroxide floc to entrap the finely-divided phosphorus precipitate and improve the overall settleability of the solids

(*Continued*)

TABLE 9.13 (*Continued*) Phosphorus Precipitation from Raw Municipal Wastewater with Lime, Alum, and Iron Salts Addition

Parameter		Metal Ion							
		Lime, Ca^{2+}		Alum, Al^{3+}		Iron, Fe^{3+}		Iron, Fe^{2+}	
Optimum pH		9.5–11.5 Solubility of hydroxyl apatite decreases with increasing pH, and phosphorus precipitation also increases. Typically lime requirement to precipitate phosphorus in wastewater is typically about 1.4–1.5 times the total alkalinity expressed as $CaCO_3$. Removal of carbonate hardness due to Ca^{2+} and Mg^{2+} (Equation 9.21b and 9.21c enhances the phosphorus removal at higher pH.		5.5–6.5 The solubility of aluminum phosphate ($AlPO_4$) is a function of pH (lowest solubility of 0.01 mg/L at pH 6). Addition of alum depresses the pH to the range of minimum $AlPO_4$ solubility.		4.5–5 pH is lowered by addition of ferric salts. Significant P removal occurs at pH 7 or above.		~8 Effective P removal can be achieved between pH of 7 and 8. A lime dose of 1.6–1.9 g of $Ca(OH)_2$ per gram of Fe^{2+} may be used to achieve the optimum pH. Some study results indicate P removal may be improved if Fe^{2+} is oxidized to Fe^{3+}.	
Performance efficiencies	P removal(%)	$Ca(OH)_2$ dosage (mg/L)	pH	Typical alum dosage (range), ratio of Al:P in influent		Typical ferric chloride dosage (range), ratio of Fe:P in influent		Typical ferrous chloride dosage, ratio of Fe:P in influent	
				Molar basis	Weight basis	Molar basis	Weight basis	Molar basis	Weight basis
	75	63	8.7	1.4:1 (1.3:1–1.5:1)	1.2:1 (1.1:1–1.3:1)	0.8:1 (0.5:1–0.9:1)	1.5:1 (0.9:1–1.7:1)	–	–
	80	–	–	–	–	–	–	1.1:1	3.1:1
	85	85	9.4	1.7:1 (1.6:1–1.9:1)	1.5:1 (1.4:1–1.7:1)	1.1:1 (0.8:1–1.3:1)	2.0:1 (1.4:1–2.3:1)	–	–
	95	175	10.5	2.3:1 (2.1:1–2.6:1)	2.0:1 (1.8:1–2.3:1)	1.7:1 (1.3:1–1.9:1)	3.1:1 (2.4:1–3.5:1)	–	–

(*Continued*)

TABLE 9.13 (*Continued*) Phosphorus Precipitation from Raw Municipal Wastewater with Lime, Alum, and Iron Salts Addition

Parameter	Metal Ion			
	Lime, Ca^{2+}	Alum, Al^{3+}	Iron, Fe^{3+}	Iron, Fe^{2+}
Comments	pH of effluent is high and may need neutralization. Considerable amount of BOD_5 and TSS along with phosphorus are also removed after clarification. Removal efficiencies up to 80% and 60% may be achievable for TSS and COD, respectively.	BOD_5 and TSS removals are respectively 50–70% and 80–90%. Phosphorus removal depends upon applied dose until solubility limit is reached. Concentration of dissolved solids in the effluent is increased.	BOD_5, TSS, and P removals are increased. TDS levels are similar to that of alum. Solids capture is poor. Ferric iron may impart slight reddish color.	Removal efficiencies of 60%, 60%, and 55% may be achievable for TSS, BOD_5, and COD, respectively. Total iron residual of 10 mg/L as Fe may be seen in the effluent but 90% of the effluent iron is insoluble and removable through filtration.

Source: Adapted in part from References 1, 3 through 5, 31 through 34, and 36 through 41.

TABLE 9.14 Common Chemicals used for Effluent Neutralization

Chemical	Formula	Molecular Weight	Equivalent Weight	Appearance	Percent Purity
Chemicals for lowering pH					
Carbonic acid	H_2CO_3	62	31	CO_2 (gaseous)	100
Sulfuric acid	H_2SO_4	98	49	Liquid	93–98
Hydrochloric acid	HCl	36.5	36.5	Liquid	26–35
Chemicals for raising pH					
Calcium hydroxide (hydrated lime)	$Ca(OH)_2$	74	37	Powder, granules	40–74
Calcium oxide (quick lime)	CaO	56	28	Powder, lump, pebble	72–95
Sodium hydroxide (caustic soda)	NaOH	40	40	Powder, flake, solution	50–98
Chemicals for supplementing alkalinity					
Sodium bicarbonate	$NaHCO_3$	84	84	Powder, lump	99
Sodium carbonate (soda ash)	Na_2CO_3	106	53	Powder	99
Calcium carbonate	$CaCO_3$	100	50	Powder, granules	96–99

TABLE 9.15 Stoichiometric Quantity of Chemical Sludge Produced in Various Coagulation Precipitation Reactions

Coagulant	Precipitate Formed	Chemical Sludge Produced, mg Solids per mg Coagulant Dose as Chemical	Reference
Lime			
CaO	$CaCO_3$	$(2 \times 100\,mg/mmole) \div (56\,mg/mmole) = 3.57$	Equations 9.16a and 9.16b
	$Mg(OH)_2$ and $CaCO_3$	$((58 + 2 \times 100)\,mg/mmole) \div (2 \times 56\,mg/mmole) = 2.30$	Equations 9.16a and 9.16e
$Ca(OH)_2$	$CaCO_3$	$(2 \times 100\,mg/mmole) \div (74\,mg/mmole) = 2.70$	Equation 9.16b
	$Mg(OH)_2$ and $CaCO_3$	$((58 + 2 \times 100)\,mg/mmole) \div (2 \times 74\,mg/mmole) = 1.74$	Equation 9.16e
Alum			
$Al_2(SO_4)_3{\cdot}14\ H_2O$	$Al(OH)_3$	$(2 \times 78\,mg/mmole) \div (594\,mg/mmole) = 0.26$	Equation 9.17a
$Al_2(SO_4)_3{\cdot}18\ H_2O$	$Al(OH)_3$	$(2 \times 78\,mg/mmole) \div (666\,mg/mmole) = 0.23$	Equation 9.17b
$Al_2(SO_4)_3{\cdot}49.6\ H_2O$	$Al(OH)_3$	$(2 \times 78\,mg/mmole) \div (1236\,mg/mmole) = 0.13$	Equation 9.17c
Ferric chloride			
$FeCl_3$	$Fe(OH)_3$	$(2 \times 107\,mg/mmole) \div (2 \times 162\,mg/mmole) = 0.66$	Equation 9.18a
$FeCl_3{\cdot}6\ H_2O$	$Fe(OH)_3$	$(2 \times 107\,mg/mmole) \div (2 \times 270\,mg/mmole) = 0.40$	Equation 9.18b
$FeCl_3{\cdot}13.1\ H_2O$	$Fe(OH)_3$	$(2 \times 107\,mg/mmole) \div (2 \times 398\,mg/mmole) = 0.27$	Equation 9.18c
Ferric sulfate			
$Fe_2(SO_4)_3$	$Fe(OH)_3$	$(2 \times 107\,mg/mmole) \div (400\,mg/mmole) = 0.54$	Equation 9.18d
$Fe_2(SO_4)_3{\cdot}9\ H_2O$	$Fe(OH)_3$	$(2 \times 107\,mg/mmole) \div (562\,mg/mmole) = 0.38$	Equation 9.18e
$Fe_2(SO_4)_3{\cdot}30.1\ H_2O$	$Fe(OH)_3$	$(2 \times 107\,mg/mmole) \div (940\,mg/mmole) = 0.23$	Equation 9.18f
Ferrous chloride			
$FeCl_2{\cdot}4\ H_2O$	$Fe(OH)_2$	$(90\,mg/mmole) \div (199\,mg/mmole) = 0.45$	Equation 9.19a
$FeCl_2{\cdot}12\ H_2O$	$Fe(OH)_2$	$(90\,mg/mmole) \div (343\,mg/mmole) = 0.26$	Equation 9.19b
Ferrous sulfate			
$FeSO_4{\cdot}7\ H_2O$	$Fe(OH)_2$	$(90\,mg/mmole) \div (278\,mg/mmole) = 0.32$	Equation 9.19c
$FeSO_4{\cdot}36\ H_2O$	$Fe(OH)_2$	$(90\,mg/mmole) \div (800\,mg/mmole) = 0.11$	Equation 9.19d

TABLE 9.16 Stoichiometric Quantity of Chemical Sludge Produced in Various Phosphorus Precipitation Reactions

Metal Ion	Precipitate Formed	Chemical Sludge Produced, mg Solids per mg Metal Ion	Reference
Calcium			
Ca^{2+}	$Ca_5OH(PO_4)_3$	(502 mg/mmole) ÷ (40 mg/mmole) = 12.6	Equation 9.21a
Alum			
Al^{3+}	$AlPO_4$	(122 mg/mmole) ÷ (27 mg/mmole) = 4.52	Equation 9.22a
Iron			
Fe^{3+}	$FePO_4$	(151 mg/mmole) ÷ (56 mg/mmole) = 2.70	Equation 9.23a
Fe^{2+}	$Fe_3(PO_4)_2$	(357 mg/mmole) ÷ (56 mg/mmole) = 6.38	Equation 9.24a

coagulation reactions are summarized in Table 9.15. Chemical sludge production information for chemical phosphorus precipitation reactions is provided in Table 9.16.

EXAMPLE 9.22: ALKALINITY REMOVAL AND HARDNESS CONVERSIONS FROM METAL SALT

An untreated water sample has total alkalinity (TA) and total hardness (TH) of 95 and 145 mg/L as $CaCO_3$, respectively. Determine TA, and carbonate hardness (CH) and noncarbonated hardness (NCH) of treated water after coagulation with (a) 35 mg/L of alum ($Al_2(SO_4)_3 \cdot 14\ H_2O$), and (b) 40 mg/L of ferric chloride ($FeCl_3$). Assume there is no magnesium hardness.

Solution

1. Define the carbonate and noncarbonate hardness.

   Carbonate hardness (CH) is due to bivalent cations combined with alkalinity causing anions, such as $Ca(HCO_3)_2$, $MgCO_3$, or $Ca(OH)_2$. Noncarbonate hardness (NCH) is due to bivalent cations combined with nonalkalinity-causing anions, such as $CaSO_4$, $MgCl_2$, or $Ca(NO_3)_2$.
2. Determine the carbonate and noncarbonated hardness components of untreated water.

   The hardness-causing cations combine first with alkalinity; any remaining hardness-causing cations combine with nonalkalinity-causing anions such as SO_4^{2-}, Cl^-, NO_3^-, etc.

   Since TA < TH and no magnesium hardness, there are only CH and NCH.

   $CH = TA = 95$ mg/L as $CaCO_3$

   $NCH = TH - TA = (145 - 95)$ mg/L as $CaCO_3 = 50$ mg/L as $CaCO_3$
3. Calculate the alkalinity and hardness concentrations of treated water with 35 mg/L alum (Equation 9.17a).

   Alkalinity consumption by coagulation $= \dfrac{3 \times 100 \text{ mg/mmole as } CaCO_3}{594 \text{ mg/mmole as } Al_2(SO_4)_3 \cdot 14H_2O} \times 35 \text{ mg/L as } Al_2(SO_4)_3 \cdot 14H_2O$

   $= 18$ mg/L as $CaCO_3$

   TA concentration after coagulation $= (95 - 18)$ mg/L as $CaCO_3 = 77$ mg/L as $CaCO_3$

   CH concentration after coagulation $=$ TA concentration $= 77$ mg/L as $CaCO_3$

   NCH concentration after coagulation $= TH - CH = (145 - 77)$ mg/L as $CaCO_3 = 68$ mg/L as $CaCO_3$
4. Calculate the alkalinity and hardness concentrations of treated water with 40 mg/L ferric chloride (Equation 9.18a).

Alkalinity consumption by coagulation $= \frac{3 \times 100 \text{ mg/mmole as } CaCO_3}{2 \times 162 \text{ mg/mmole as } FeCl_3} \times 40 \text{ mg/L as } FeCl_3$

$= 37$ mg/L as $CaCO_3$

TA concentration after coagulation $= (95 - 37)$ mg/L as $CaCO_3 = 58$ mg/L as $CaCO_3$

CH concentration after coagulation = TA concentration = 58 mg/L as $CaCO_3$

NCH concentration after coagulation = TH − CH = (145 − 58) mg/L as $CaCO_3 = 87$ mg/L as $CaCO_3$

5. Summarize the TA, CH, NCH, and TH concentrations before and after the treatments.

   The TA, CH, NCH, and TH concentrations before and after the coagulation process are summarized below. It should be noticed that CH is converted to NCH in coagulation process but the TH concentration remains the same. Therefore, coagulation process cannot remove hardness.

Sampling Location	Concentrations, mg/L as $CaCO_3$			
	TA	CH	NCH	TH
Untreated water	95	95	50	145
Treated water				
Alum dosage of 35 mg/L	77	77	68	145
Ferric chloride dosage of 40 mg/L	58	58	87	145

EXAMPLE 9.23: LIME DOSE FOR HARDNESS REMOVAL

Hydrated lime ($Ca(OH)_2$) is used to precipitate carbonate hardness in a softening process. Prepare a curve between lime dose and carbonate and total hardness remaining in the treated water. Initial carbonate and noncarbonate hardness concentrations are 180 and 50 mg/L as $CaCO_3$, respectively. Assume there is no magnesium hardness. Ignore the small lime dosage required to raise pH value for effective removal of hardness.

Solution

1. Calculate the carbonate hardness concentration of softened water at lime dosage of 10 mg/L (Equation 9.16b).

CH removed (or TA consumed) $= \frac{100 \text{ mg/mmole as } CaCO_3}{74 \text{ mg/mmole as } Ca(OH)_2} \times 10 \text{ mg/L as } Ca(OH)_2$

$= 14$ mg/L as $CaCO_3$

CH concentration after softening $= (180 - 14)$ mg/L as $CaCO_3 = 166$ mg/L as $CaCO_3$

NCH concentration after softening = 50 mg/L as $CaCO_3$

TH concentration after softening = CH + NCH = (166 + 50) mg/L as $CaCO_3 = 216$ mg/L as $CaCO_3$

2. Summarize the carbonate hardness remaining at different doses of lime.

   The CH removed and hardness concentrations after the softening process are summarized below. CH is removed due to precipitation of $CaCO_3$ in the lime-softening process. Therefore, the TH concentration decreases with increase in lime dosage.

Lime Dosage, mg/L as Ca $(OH)_2$	CH Removed, mg/L as $CaCO_3$	Hardness Concentration, mg/L as $CaCO_3$		
		CH	NCH	TH
0	0	180	50	230
10	14	166	50	216
30	41	139	50	189
50	68	112	50	162
70	95	85	50	135
90	122	58	50	108
110	149	31	50	81
128	173	7.0	50	57

3. Prepare a curve to express lime dose (mg/L $Ca(OH)_2$) versus CH and TH concentrations (mg/L as $CaCO_3$).

   The tabulated data above are plotted in Figure 9.25. The lowest CH concentration in this example is 7 mg/L, which is the solubility of $CaCO_3$ (Example 9.24).

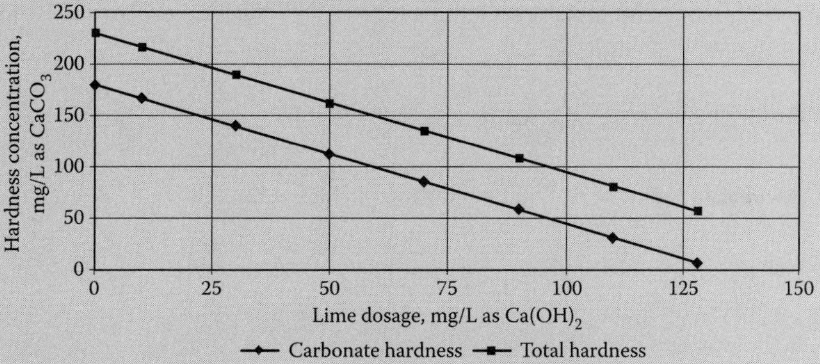

FIGURE 9.25 Carbonate and total hardness concentrations after removal of carbonate hardness due to precipitation of $CaCO_3$ at different lime dosages (Example 9.23).

EXAMPLE 9.24: SOLUBILITY OF $CaCO_3$ AND $Mg(OH)_2$

Hardness is removed by precipitating Ca^{2+} and Mg^{2+} as $CaCO_3$ and $Mg(OH)_2$. Calculate the concentrations of soluble $CaCO_3$ and $Mg(OH)_2$ remaining in the solution after water-softening process. Also calculate the equivalent alkalinity caused by the soluble $CaCO_3$ and $Mg(OH)_2$.

Solution

1. Calculate the concentration of soluble $CaCO_3$.

   The concentration of soluble $CaCO_3$ is calculated from the solubility product constant given in Table 9.12.

   The chemical equation for dissolution of $CaCO_3$ is expressed below.

$$CaCO_3 \leftrightarrow Ca^{2+} + CO_3^{2-}$$

Assume X is the number of moles per liter (mole/L) of Ca^{2+} and CO_3^{2-} resulting from dissolution of $CaCO_3$. At the equilibrium, $K_{sp} = [Ca^{2+}][CO_3^{2-}] = X^2$.

At 25°C, obtain $K_{sp} = 5 \times 10^{-9}$ for $CaCO_3$ from Table 9.12.

$$X = (K_{sp})^{1/2} = (5 \times 10^{-9})^{1/2} = 7.1 \times 10^{-5} \text{ mole/L}$$

Concentration of Ca^{2+}	$= 7.1 \times 10^{-5}$ mole/L × 40 g/mole × 1000 mg/g = 2.8 mg/L
Concentration of CO_3^{2-}	$= 7.1 \times 10^{-5}$ mole/L × 60 g/mole × 1000 mg/g = 4.3 mg/L
Concentration of $CaCO_3$	$= 7.1 \times 10^{-5}$ mole/L × 100 g/mole × 1000 mg/g = 7.1 mg/L

Therefore, the solubility of $CaCO_3$ at 25°C = 7.1 mg/L as $CaCO_3$.

2. Calculate the concentration of soluble $Mg(OH)_2$.

   The concentration of soluble $Mg(OH)_2$ is also calculated from the solubility product constant given in Table 9.12.

   The chemical equation for dissolution of $Mg(OH)_2$ is expressed below.

$$Mg(OH)_2 \leftrightarrow Mg^{2+} + 2\,OH^-$$

Assume X is the number of moles per liter (mole/L) of Mg^{2+} resulting from the dissociation of Mg $(OH)_2$, then $OH^- = 2X$ mole/L.

At the equilibrium, $K_{sp} = [Mg^{2+}][OH^-]^2 = X(2X)^2 = 4X^3$.

At 25°C, obtain $K_{sp} = 7 \times 10^{-12}$ for $Mg(OH)_2$ from Table 9.12.

$$X = \left(\frac{K_{sp}}{4}\right)^{1/3} = \left(\frac{7 \times 10^{-12}}{4}\right)^{1/3} = 1.2 \times 10^{-4} \text{ mole/L}$$

Concentration of Mg^{2+}	$= 1.2 \times 10^{-4}$ mole/L × 24 g/mole × 1000 mg/g = 2.9 mg/L
Concentration of OH^-	$= 2 \times (1.2 \times 10^{-4}$ mole/L) × 17 g/mole × 1000 mg/g = 4.1 mg/L
Concentration of $Mg(OH)_2$	$= 1.2 \times 10^{-4}$ mole/L × 58 g/mole × 1000 mg/g = 7.0 mg/L

Therefore, the solubility of $Mg(OH)_2$ at 25°C = 7.0 mg/L as $Mg(OH)_2$.

3. Calculate the total equivalent alkalinity caused by the soluble $CaCO_3$ and $Mg(OH)_2$.

   Equivalent calcium carbonate alkalinity (CaCA) = 7.1 mg/L as $CaCO_3$

   Equivalent magnesium hydroxide alkalinity (MgHA)

$$= 7.0 \text{ mg/L as Mg (OH)}_2 \times \frac{\text{Eq. wt. of CaCO}_3}{\text{Eq. wt. of Mg (OH)}_2}$$

$$= 7.0 \text{ mg/L as Mg (OH)}_2 \times \frac{50 \text{ g/eq. as CaCO}_3}{29 \text{ g/eq. as Mg (OH)}_2}$$

$$= 12.1 \text{ mg/L as CaCO3}$$

Total equivalent alkalinity $= (7.1 + 12.1)$ mg/L as $CaCO_3$

$= 19.2$ mg/L as $CaCO_3$

EXAMPLE 9.25: SOLUBILITY OF $Al(OH)_3$ AND $AlPO_4$

Wastewater is coagulated using alum to precipitate phosphorus. Determine the saturation concentrations of aluminum hydroxide and aluminum phosphate. Ignore the competition of Al between two precipitation reactions

Solution

1. Calculate the saturation concentration of $Al(OH)_3$.

   The chemical equation for dissolution of $Al(OH)_3$ is expressed below.

$$\begin{array}{ccccc} Al(OH)_3 & \leftrightarrow & Al^{3+} & + & 3\,OH^- \\ X & & X & & 3\,X \end{array}$$

   Assume the molar concentration of soluble $Al(OH)_3$ is X, mole/L. When equilibrium is reached, there should be X mole/L of Al^{3+} and 3 X mole/L of OH^-.

   The corresponding K_{sp} expression is $K_{sp} = [Al^{3+}][OH^-]^3 = X\,(3\,X)^3 = 27\,X^4$.

   At 25°C, obtain $K_{sp} = 1 \times 10^{-32}$ for $Al(OH)_3$ from Table 9.12.

$$X = \left(\frac{K_{sp}}{27}\right)^{1/4} = \left(\frac{10^{-32}}{27}\right)^{1/4} = 4.4 \times 10^{-9}\text{ mole/L}$$

$$X_{Al} = X = 4.4\ 10^{-9}\text{ mole/L}$$

$$X_{OH} = 3X = 3 \times (4.4\ 10^{-9})\text{ mole/L} = 1.3\ 10^{-8}\text{ mole/L}$$

Concentration of Al^{3+} $= 4.4 \times 10^{-9}$ mole/L × 27 g/mole × 1000 mg/g = 1.2×10^{-4} mg/L

Concentration of OH^- $= 1.3 \times 10^{-8}$ mole/L × 17 g/mole × 1000 mg/g = 2.2×10^{-4} mg/L

Concentration of $Al(OH)_3$ $= 4.4 \times 10^{-9}$ mole/L × 78 g/mole × 1000 mg/g = 3.4×10^{-4} mg/L

2. Calculate the saturation concentration of $AlPO_4$.

   The chemical equation for dissolution of $AlPO_4$ is expressed below.

$$\begin{array}{ccccc} AlPO_4 & \leftrightarrow & Al^{3+} & + & PO_4^{3-} \\ X & & X & & X \end{array}$$

   Assume that the molar concentration of soluble $AlPO_4$ is X, mole/L. When equilibrium is reached, there should be X mole/L of Al^{3+} and X mole/L of PO_4^{3-}.

   The corresponding K_{sp} expression is $K_{sp} = [Al^{3+}][\ PO_4^{3-}] = X^2$.

   At 25°C, obtain $K_{sp} = 1 \times 10^{-22}$ for $AlPO_4$ from Table 9.12.

$$X = (K_{sp})^{1/2} = (10^{-22})^{1/2} = 10^{-11}\text{ mole/L}$$

$$X_{Al} = X = 10^{-11}\text{ mole/L}$$

$$X_{PO4} = X = 10^{-11}\text{ mole/L}$$

Concentration of Al^{3+} $= 10^{-11}$ mole/L × 27 g/mole × 1000 mg/g = 2.7×10^{-7} mg/L

Concentration of PO_4^{3-} $= 10^{-11}$ mole/L × 95 g/mole × 1000 mg/g = 9.5×10^{-7} mg/L

Concentration of $AlPO_4$ $= 10^{-11}$ mole/L × 122 g/mole × 1000 mg/g = 1.2×10^{-6} mg/L

EXAMPLE 9.26: SOLUBILITY OF PRECIPITATE PRODUCED FROM CHEMICAL COAGULATION AND PRECIPITATION OF PHOSPHORUS

It is important to estimate the achievable concentrations of many precipitates in chemical processes for coagulation of wastewater, removal or hardness, and chemical removal of phosphorus. Tabulate the solubility values of common compounds of such reactions.

Solution

The solubility values of many common compounds are provided in Table 9.17.

TABLE 9.17 Solubility Values of Many Products of Coagulation and Precipitation Reactions (Example 9.26)

Precipitate	Chemical Formula	K_{sp} at 25°C	Mole Concentration, mole/L	Molecular Weight	Solubility, mg/L
Calcium carbonate	$CaCO_3$	5×10^{-9}	7.1×10^{-5}	100	7.1
Tricalcium phosphate	$Ca_3(PO_4)_2$	1×10^{-27}	1.6×10^{-6}	310	0.48
Dicalcium phosphate	$CaHPO_4$	3×10^{-7}	5.5×10^{-4}	136	75
Hydroxyapatite	$Ca_5OH(PO_4)_3$	1×10^{-56}	1.7×10^{-7}	502	8.5×10^{-2}
Aluminum hydroxide	$Al(OH)_3$	1×10^{-32}	4.4×10^{-9}	78	3.4×10^{-4}
Aluminum phosphate	$AlPO_4$	1×10^{-22}	1.0×10^{-11}	122	1.2×10^{-6}
Ferric hydroxide	$Fe(OH)_3$	6×10^{-38}	2.2×10^{-10}	107	2.3×10^{-5}
Ferric phosphate	$FePO_4$	1×10^{-22}	1.0×10^{-11}	151	1.5×10^{-6}
Ferrous hydroxide	$Fe(OH)_2$	5×10^{-15}	1.1×10^{-5}	90	0.97
Ferrous phosphate	$Fe_3(PO_4)_2$	1×10^{-36}	2.5×10^{-8}	357	8.8×10^{-3}
Magnesium hydroxide	$Mg(OH)_2$	7×10^{-12}	1.3×10^{-4}	58	7.0
Magnesium ammonium phosphate (MAP)	$MgNH_4PO_4$	3×10^{-13}	6.7×10^{-5}	137	9.2

Note: i = insoluble. Magnesium ammonium phosphate (MAP)
Source: Adapted in part from References 3, 4, 26, 27, 34, and 35.

EXAMPLE 9.27: LIME, ALUM, AND FERRIC CHLORIDE USAGE FOR PERCENT PHOSPHORUS REMOVAL

The average design flow of a municipal wastewater treatment plant is 3000 m^3/d. The average phosphorus concentration in the influent is 6 mg/L as P. Chemical removal of phosphorus is evaluated for the following chemicals: (a) lime (b) alum ($Al_2(SO_4)_3{\cdot}14\,H_2O$), and (c) ferric chloride ($FeCl_3$). Estimate the potential chemical usage (kg/d) for 95% phosphorus removal. Use the typical chemical dosages given in Table 9.13.

Solution

1. Estimate the lime usage.

   For 95% phosphorus removal, the suggested lime dose is 175 mg/L as $Ca(OH)_2$ in Table 9.13.

   Lime usage $= 175\ g/m^3 \times 3000\ m^3/d \times 10^{-3}\ kg/g = 525\ kg/d$ as $Ca(OH)_2$

2. Estimate the alum ($Al_2(SO_4)_3{\cdot}14\,H_2O$) usage.

   For 95% phosphorus removal, the desired dose ratio (Al^{3+}:P) is 2.0:1 by weight in Table 9.13.

$$\text{Al}^{3+}\text{ dose} = \frac{2.0\text{ mg as Al}^{3+}}{1\text{ mg as P}} \times 6\text{ mg/L as P} = 12\text{ mg/L as Al}^{3+}$$

$$\text{Alum dose} = 12\text{ mg/L as Al}^{3+} \times \frac{594\text{ g/mole as alum}}{2 \times 27\text{ g/mole as Al}^{3+}} = 132\text{ mg/L as alum}$$

$$\text{Alum usage} = 132\text{ g/m}^3\text{ as alum} \times 3000\text{ m}^3\text{/d} \times 10^{-3}\text{ kg/g} = 396\text{ kg/d as alum}$$

3. Estimate the ferric chloride ($FeCl_3$) usage.
   For 95% phosphorus removal, the expected dose ratio (Fe^{3+}:P) is 3.1:1 by weight in Table 9.13.

$$\text{Fe}^{3+}\text{ dose} = \frac{3.1\text{ mg as Fe}^{3+}}{1\text{ mg as P}} \times 6\text{ mg/L as P} = 18.6\text{ mg/L as Fe}^{3+}$$

$$\text{Ferric chloride dose} = 18.6\text{ mg/L as Fe}^{3+} \times \frac{162\text{ g/mole as FeCl}_3}{56\text{ g/mole as Fe}^{3+}} = 53.8\text{ mg/L as FeCl}_3$$

$$\text{Ferric chloride usage} = 53.8\text{ g/m}^3\text{ as FeCl}_3 \times 3000\text{ m}^3\text{/d} \times 10^{-3}\text{ kg/g} = 161\text{ kg/d as FeCl}_3$$

EXAMPLE 9.28: LIQUID FERRIC CHLORIDE USAGE FOR PHOSPHORUS REMOVAL AND STORAGE REQUIREMENT

A municipal wastewater treatment plant has an average design capacity of 10,000 m^3/d. The average phosphorus concentration in the influent is 6 mg/L as P in the influent. The standard jar test confirmed that 1.9 mole of Fe is needed for the precipitation of each mole of P. The liquid ferric chloride solution has chemical formula $FeCl_3 \cdot 13.1\ H_2O$. The $FeCl_3$ content is 41% by weight, and density is 1.3 kg/L. Calculate daily usage of liquid ferric chloride and the storage volume required for a 30-day supply.

Solution

1. Calculate the available Fe^{3+} per liter of liquid ferric chloride solution.

$$\text{Weight of FeCl}_3\text{ per liter of solution} = \frac{0.41\text{ kg FeCl}_3}{\text{kg solution}} \times \frac{1.3\text{ kg solution}}{\text{L}} = 0.53\text{ kg/L as FeCl}_3$$

$$\text{Molecular weight of FeCl}_3 = 162\text{ g/mole as FeCl}_3\text{ (Table 9.10)}$$

$$\text{Fe}^{3+}\text{ available per liter solution} = 0.53\text{ kg/L as FeCl}_3 \times \frac{56\text{ g/mole as Fe}^{3+}}{162\text{ g/mole as FeCl}_3} = 0.18\text{ kg/L as Fe}^{3+}$$

2. Calculate the amount of ferric chloride solution required per kg of P.
   Theoretical dose of Fe^{3+} = 1 mole Fe per mole of P (Table 9.13).
   Theoretical dose of Fe^{3+} required on weight basis from Equation 9.23a

$$= \frac{1\text{ mole Fe}^{3+}}{1\text{ mole P}} \times \frac{\text{MW of Fe}^{3+}}{\text{MW of P}}$$

$$= \frac{1\text{ mole Fe}^{3+}}{1\text{ mole P}} \times \frac{56\text{ g/mole as Fe}^{3+}}{31\text{ g/mole as P}}$$

$$= 1.8\text{ g Fe}^{3+}\text{/g P} \quad \text{or} \quad 1.8\text{ kg Fe}^{3+}\text{/kg P}$$

Actual dose of Fe^{3+} based on jar test results

$$= \frac{1.9\,\text{mole Fe}^{3+}}{1\,\text{mole P}} \times \frac{\text{MW of Fe}^{3+}}{\text{MW of P}}$$

$$= \frac{1.9\,\text{mole Fe}^{3+}}{1\,\text{mole P}} \times \frac{56\,\text{g/mole as Fe}^{3+}}{31\,\text{g/mole as P}}$$

$$= 3.4\,\text{g Fe}^{3+}/\text{g P} \quad \text{or} \quad 3.4\,\text{kg Fe}^{3+}/\text{kg P}$$

Liquid $FeCl_3$ dose

$$= \frac{3.4\,\text{kg Fe}^{3+}}{1\,\text{kg P}} \times \frac{\text{L liquid solution}}{0.18\,\text{kg Fe}^{3+}}$$

$$= 19\ \text{L liquid solution/kg P}$$

3. Calculate the quantity of liquid ferric chloride solution required per day.
   Mass of phosphorus targeted $= 6\ \text{g/m}^3 \times 10{,}000\ \text{m}^3\text{/d} \times 10^{-3}\,\text{kg/g} = 60\ \text{kg/d as P}$
   Liquid $FeCl_3$ utilized $= 60\ \text{kg P/d} \times 19\ \text{L/kg P} = 1140\ \text{L/d}$ as liquid solution
4. Determine 30-day storage requirement for liquid ferric chloride solution.
   Total volume required for 30-day supply $= 1140\ \text{L/d} \times 30\ \text{d} \times 10^{-3}\ \text{m}^3\text{/L} = 34\ \text{m}^3$

   The liquid ferric chloride is highly acidic solution with pH ranging from 0.1 to 1.5. The storage tanks shall be constructed of materials that are recommended for storage of corrosive solutions. The materials of choice for the storage tanks include fiberglass-reinforced polyester (FRP), or high-density cross-link polyethylene (HDXLPE), or polyvinyl chloride (PVC). The transfer of chemical from storage tank to the point of application shall utilize specialized chemical feed pumps. All piping, fittings, and any equipment that come in direct contact with liquid ferric chloride shall also be constructed of acid-resistant materials such as chlorinated polyvinyl chloride (CPVC), PVC, or Teflon. Common metal materials such as stainless steel (SS), carbon steel, or aluminum shall not be used for any parts that may come in direct contact with acid solution.
5. Determine dimensions of bulk storage tanks.
   Provide two storage cylindrical tanks.
   Volume required for each tank $= 34\ \text{m}^3/2\ \text{tanks} = 17\ \text{m}^3$ or 4500 gal per tank

   Provide flat bottom and domed top bulk tanks with a normal capacity of 4500 gal (17 m^3) per tank. Assume that the capacity of a chemical/acid tank trailer used by the local chemical supplier is 3000 gal (11 m^3). Therefore, the effective volume of each tank is sufficient for holding a regular truck load.

   The dimensions of commercially available tanks are provided by the tank manufacturers. In this example, the tank dimensions are 3.6 m (142 in) in diameter and ~1.7 m (66.5 in) effective depth. The overall height of the tank is 2.3 m (91 in) to include ~0.6 m (24 in) head space. Each tank shall also be provided with U-vent, overflow outlet, top-access manway, and tie-down/lifting lugs.
6. Prepare the general layout plan of the tank farm.

   Provide two tanks in a symmetrical grid inside a square-shape tank farm. Each tank is placed on a 4.5 m × 4.5 m square base. There is a minimum of 1.5-m (5-ft) clearance between the bases and from sidewalls of the tank farm.

   For safety precaution, provide a containment at the tank farm to contain the chemical from the tanks in case of accidental spill, leakage, or tank rupture. Consider a containment volume of at least 125% of the largest tank volume if the piping system is designed to prevent a combined release from the manifolded tanks. The containment area is 14 m long × 9 m wide and surrounded by 0.40-m-tall containment walls. This arrangement provides a 0.27-m effective depth and an additional 0.13-m freeboard. A chemical unloading station (2 m × 1.5 m) is also located inside the containment area.

Total effective containment area	$= (14\,\text{m} \times 9\,\text{m}) - 2 \times (4.5\,\text{m})^2 - 2\,\text{m} \times 1.5\,\text{m}$
	$= (126 - 40.5 - 3)\,\text{m}^2 = 82.5\,\text{m}^2$
Total containment volume	$= 82.5\,\text{m}^2 \times 0.27\,\text{m} = 22.3\,\text{m}^3$

The volume provided in the containment area is large enough to hold the normal capacity of one bulk tank.

$$\text{Safety factor} = \frac{22.3\,\text{m}^3}{17\,\text{m}^3} = 1.3\ (> 1.25)$$

A sump is also provided in the containment area for draining uncontaminated cleaning or stormwater by a manually operated submersible pump.

The layout plan is shown in Figure 9.26.

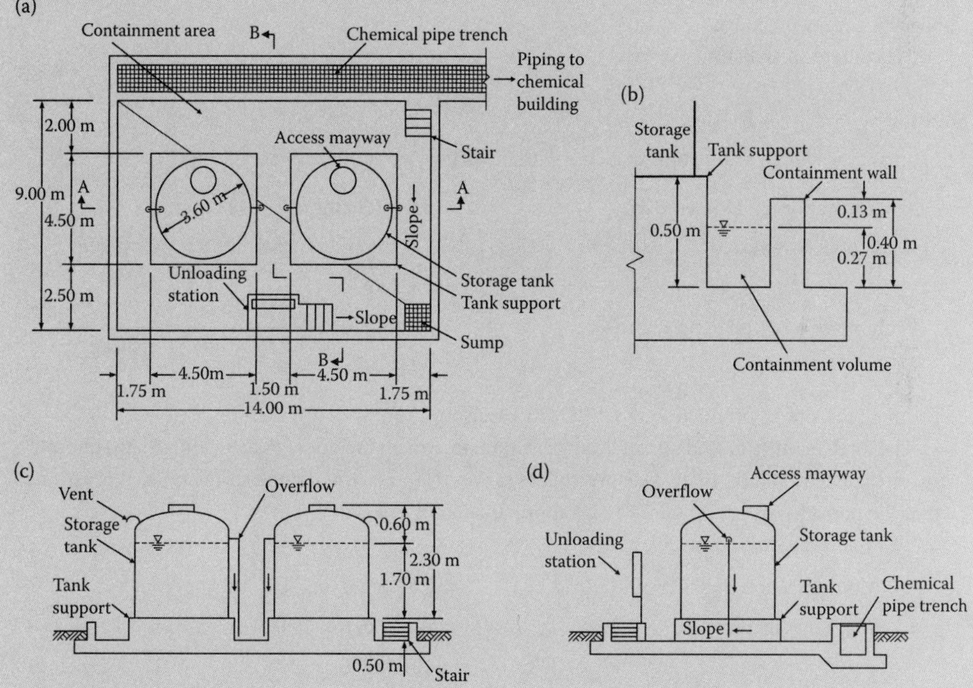

FIGURE 9.26 Ferric chloride storage facility: (a) plan, (b) details of containment sidewall, (c) Section AA, and (d) Section BB (Example 9.28).

EXAMPLE 9.29: SLUDGE PRODUCTION IN CONVENTIONAL PRIMARY TREATMENT, AND CHEMICALLY ENHANCED PRIMARY TREATMENTS (CEPTS) WITH ALUM AND FERRIC CHLORIDE

Prepare and compare the mass and volume of sludge produced by (1) a conventional primary treatment process and (2) chemically enhanced primary treatment (CEPT) processes using (a) alum ($Al_2(SO_4)_3{\cdot}14\,H_2O$) and (b) ferric chloride ($FeCl_3$). Use the typical chemical dosages given in Table 9.13. To be conservative, it is assumed that the phosphorous is removed completely by chemical precipitations.

The characteristics of raw wastewater and performances of conventional primary treatment and CEPT processes are given below:

Raw wastewater characteristics

Average design flow $= 10{,}000\ m^3/d$
TSS $= 220\ mg/L$
Total alkalinity $= 140\ mg/L$ as $CaCO_3$
Total phosphorus $= 6\ mg/L$ as P

Performance of conventional primary treatment process

TSS removed $= 60\%$
Specific gravity of sludge $= 1.02$
Solids content of sludge $= 4.5\%$

Performance of CEPT processes using alum and ferric chloride

TSS removed $= 85\%$
Desired P removed $= 95\%$
Specific gravity of sludge $= 1.03$
Solids content of sludge $= 6\%$

Solution

1. Estimate the sludge production in a conventional primary treatment process.

Concentration of TSS removed $= 0.6 \times 220\ mg/L = 132\ mg/L$

Mass of TSS removed $= 132\ mg/L \times 10{,}000\ m^3/d \times 10^{-3}\ kg/g$
$= 1320\ kg/d$

Volume of wet sludge produced $= 1320\ kg/d \times \dfrac{100\ kg\ \text{as wet}}{4.5\ kg\ \text{as dry}} \times \dfrac{1}{1020\ kg/m^3}$
$= 29\ m^3/d$

2. Estimate the sludge production in CEPT process using alum.

The total quantity of sludge produced from alum treatment includes (a) TSS removed, (b) phosphorus precipitates, and (c) hydroxide precipitate as $Al(OH)_3$. To simplify the calculations, it is assumed that the phosphorus precipitates can be approximated by $AlPO_4$.

a. Estimate TSS removed.

Concentration of TSS removed $= 0.85 \times 220\ mg/L = 187\ mg/L$

Mass of TSS removed $= 187\ mg/L \times 10{,}000\ m^3/d \times 10^{-3}\ kg/g$
$= 1870\ kg/d$

b. Estimate the quantity of $AlPO_4$ produced.

i. Concentrations of P removed in and remaining after the CEPT process.

Concentration of P removal required $= 0.95 \times 6\ mg/L = 5.7\ mg/L$ as P
Concentration of P remaining in the effluent $= (6 - 5.7)\ mg/L = 0.3\ mg/L$ as P

ii. Alum dose required to precipitate $AlPO_4$ (Equation 9.22a).

In accordance with Equation 9.22a, 1 mole of Al^{3+} stoichiometrically reacts with 1 mole of PO_4^{3-} (or P) to yield 1 mole of $AlPO_4$.

Al^{3+} dose required to precipitate $AlPO_4$ $= 5.7\ mg/L$ as P removed $\times \dfrac{\text{MW of } Al^{3+}}{\text{MW of P}}$
$= 5.7\ mg/L$ as P removed $\times \dfrac{27\ g/\text{mole as } Al^{3+}}{31\ g/\text{mole as P}}$
$= 5.0\ mg/L$ as Al^{3+}

iii. Quantity of $AlPO_4$ produced (Equation 9.22a).

Concentration of $AlPO_4$ in sludge $= 5.0\ \text{mg/L as}\ Al^{3+} \times \dfrac{122\ \text{g/mole as}\ AlPO_4}{27\ \text{g/mole as}\ Al^{3+}}$

$= 23\ \text{mg/L as}\ AlPO_4$

Mass of $AlPO_4$ in sludge $= 23\ \text{mg/L as}\ AlPO_4 \times 10{,}000\ \text{m}^3/\text{d} \times 10^{-3}\ \text{kg/g}$

$= 230\ \text{kg/d as}\ AlPO_4$

c. Estimate the quantity of $Al(OH)_3$ produced.

i. Overall alum dose required for chemical removal of P (Equations 9.22a and 9.22b).

For 95% phosphorus removal, the desired dose ratio (Al^{3+}:P) is 2.0:1 by weight in Table 9.13.

$$\text{Overall}\ Al^{3+}\text{dose for P removal} = \frac{2.0\ \text{mg/L as}\ Al^{3+}}{1\ \text{mg/L as P}} \times 6\ \text{mg/L P} = 12\ \text{mg/L as}\ Al^{3+}$$

ii. Alum dose available to precipitate $Al(OH)_3$.

Al^{3+} dose available to precipitate $Al(OH)_3 = (12 - 5.0)\ \text{mg/L} = 7.0\ \text{mg/L as}\ Al^{3+}$

iii. Quantity of $Al(OH)_3$ produced.

In accordance with Equation 9.22b, 1 mole of Al^{3+} stoichiometrically reacts with 3 moles of OH^- to yield 1 mole of $Al(OH)_3$.

Concentration of $Al(OH)_3$ in sludge $= 7.0\ \text{mg/L as}\ Al^{3+} \times \dfrac{78\ \text{g/mole as}\ Al(OH)_3}{27\ \text{g/mole as}\ Al^{3+}}$

$= 20\ \text{mg/L as}\ Al(OH)_3$

Mass of $Al(OH)_3$ in sludge $= 20\ \text{mg/L as}\ Al(OH)_3 \times 10{,}000\ \text{m}^3/\text{d} \times 10^{-3}\ \text{kg/g}$

$= 200\ \text{kg/d as}\ Al(OH)_3$

d. Estimate the total quantity of sludge produced in CEPT process using alum.

Mass of TSS in sludge $= 1870\ \text{kg/d}$

Mass of $AlPO_4$ in sludge $= 230\ \text{kg/d}$

Mass of $Al(OH)_3$ in sludge $= 200\ \text{kg/d}$

Total mass of alum-treated sludge $= 2300\ \text{kg/d}$

Volume of alum-treated sludge $= 2300\ \text{kg/d} \times \dfrac{100\ \text{kg as wet}}{6\ \text{kg as dry}} \times \dfrac{1}{1030\ \text{kg/m}^3}$

$= 37\ \text{m}^3/\text{d}$

e. Check the residual alkalinity.

Alkalinity is consumed during hydroxide precipitation (Equation 9.22b). Therefore, it is necessary to verify if residual alkalinity is sufficient to maintain the desired reaction pH.

Equivalent alum concentration that reacted to remove alkalinity

$= 7.0\ \text{mg/L as}\ Al^{3+} \times \dfrac{594\ \text{g/mole as alum}}{2 \times 27\ \text{g/mole as}\ Al^{3+}}$

$= 77\ \text{mg/L as alum}\ (Al_2(SO_4)_3{\cdot}14\ H_2O)$

Alkalinity of 0.51 mg as $CaCO_3$ is consumed per mg of alum for hydroxide precipitation (Table 9.11).

Alkalinity removed $= 0.51\ \text{mg CaCO}_3/\text{mg alum} \times 77\ \text{mg/L as alum}$

$= 39\ \text{mg/L as CaCO}_3$

Alkalinity remaining $= \text{Total alkalinity} - \text{Alkalinity removed}$

$= (140 - 39)\ \text{mg/L as CaCO}_3$

$= 101\ \text{mg/L as CaCO}_3$

The alkalinity remaining is sufficient to maintain the desired pH for the reactions during the CEPT process.

Note: The remaining alkalinity shall also be checked if the nitrification process is required to treat the effluent from the CEPT process (see more detailed information about alkalinity consumption during nitrification process in Chapter 10).

3. Estimate the sludge production in CEPT process using ferric chloride.

The total quantity of sludge produced from ferric chloride treatment includes (a) TSS removed, (b) phosphorus precipitates, and (c) hydroxide precipitate as $Fe(OH)_3$. To simplify the calculations, it is assumed that the phosphorus precipitates can be approximated by $FePO_4$.

The procedure to calculate the sludge and volume with ferric chloride treatment is similar to that with alum. Therefore, many steps are not repeated.

a. Estimate TSS removed.

The sludge mass produced from TSS is calculated in Step 2. The mass value is:

Mass of TSS removed $= 1870\ \text{kg/d}$

b. Estimate the quantity of $FePO_4$ produced.

i. Concentrations of P removed in and remaining after the CEPT process.

The concentrations of P removed and remaining in the effluent are calculated in Step 2. These values are:

Concentration of P removal required $= 5.7$ mg/L as P

Concentration of P remaining in the effluent $= 0.3$ mg/L as P

ii. Ferric chloride dose required to precipitate $FePO_4$ (Equation 9.23a).

In accordance with Equation 9.23a, 1 mole of Fe^{3+} stoichiometrically reacts with 1 mole of PO_4^{3-} (or P) to yield 1 mole of $FePO_4$.

Fe^{3+} dose required to precipitate $FePO_4$

$$= 5.7\ \text{mg/L as P removed} \times \frac{\text{MW of Fe}^{3+}}{\text{MW of P}}$$

$$= 5.7\ \text{mg/L as P removed} \times \frac{56\ \text{g/mole as Fe}^{3+}}{31\ \text{g/mole as P}}$$

$$= 10.3\ \text{mg/L as Fe}^{3+}$$

iii. Quantity of $FePO_4$ produced (Equation 9.23a).

Concentration of $FePO_4$ in sludge

$$= 10.3\ \text{mg/L as Fe}^{3+} \times \frac{151\ \text{g/mole as FePO}_4}{56\ \text{g/mole as Fe}^{3+}}$$

$$= 28\ \text{mg/L as FePO}_4$$

Mass of $FePO_4$ in sludge

$$= 28\ \text{mg/L as FePO}_4 \times 10{,}000\ \text{m}^3/\text{d} \times 10^{-3}\ \text{kg/g}$$

$$= 280\ \text{kg/d as FePO}_4$$

c. Estimate the quantity of $Fe(OH)_3$ produced.

i. Overall ferric chloride dose required for chemical removal of P (Equations 9.23a and 9.23b).

For 95% phosphorus removal, the desired dose ratio (Fe^{3+}:P) is 3.1:1 by weight in Table 9.13.

$$\text{Overall } Fe^{3+} \text{dose for P removal} = \frac{3.1\,\text{mg/L as } Fe^{3+}}{1\,\text{mg/L as P}} \times 6\,\text{mg/L P} = 18.6\,\text{mg/L as } Fe^{3+}$$

ii. Ferric chloride dose available to precipitate $Fe(OH)_3$.

Fe^{3+} dose available to precipitate $Fe(OH)_3 = (18.6 - 10.3)$ mg/L = 8.3 mg/L as Fe^{3+}

iii. Quantity of $Fe(OH)_3$ produced.

In accordance with Equation 9.23b, 1 mole of Fe^{3+} stoichiometrically reacts with 3 moles of OH^- to yield 1 mole of $Fe(OH)_3$.

Concentration of $Fe(OH)_3$ in sludge $= 8.3\,\text{mg/L as } Fe^{3+} \times \dfrac{107\,\text{g/mole as Fe (OH)}_3}{56\,\text{g/mole as } Fe^{3+}}$

$= 16\,\text{mg/L as } Fe(OH)_3$

Mass of $Fe(OH)_3$ in sludge $= 16\,\text{mg/L as } Fe(OH)_3 \times 10{,}000\,\text{m}^3/\text{d} \times 10^{-3}\,\text{kg/g}$

$= 160\,\text{kg/d as } Fe(OH)_3$

d. Estimate the total quantity of sludge produced in CEPT process using ferric chloride.

Mass of TSS in sludge = 1870 kg/d

Mass of $FePO_4$ in sludge = 280 kg/d

Mass of $Fe(OH)_3$ in sludge = 160 kg/d

Total mass of alum-treated sludge = 2310 kg/d

Volume of alum-treated sludge $= 2310\,\text{kg/d} \times \dfrac{100\,\text{kg as wet}}{6\,\text{kg as dry}} \times \dfrac{1}{1030\,\text{kg/m}^3}$

$= 37\,\text{m}^3/\text{d}$

e. Check the residual alkalinity.

Equivalent alum concentration that reacted to remove alkalinity

$= 8.3\,\text{mg/L as } Fe^{3+} \times \dfrac{162\,\text{g/mole as } FeCl_3}{56\,\text{g/mole as } Al^{3+}}$

$= 24\,\text{mg/L as } FeCl_3$

Alkalinity of 0.93 mg as $CaCO_3$ is consumed per mg of $FeCl_3$ for hydroxide precipitation (Table 9.11).

Alkalinity removed $= 0.93\,\text{mg } CaCO_3/\text{mg alum} \times 24\,\text{mg/L as } FeCl_3$

$= 22\,\text{mg/L as } CaCO_3$

Alkalinity remaining = Total alkalinity − Alkalinity removed

$= (140 - 22)\,\text{mg/L as } CaCO_3$

$= 118\,\text{mg/L as } CaCO_3$

The alkalinity remaining is sufficient to maintain the desired reaction pH during the CEPT process.

4. Compare the mass and volume of sludge with conventional primary treatment and CEPT processes. The comparative values are summarized below.

Treatment Process	Sludge Production				
	Mass, kg/d (% of total by weight)				Volume, m^3/d
	TSS Removal	Phosphorus Precipitation	Hydroxide Precipitation	Total	
Conventional primary treatment	1320 (100)	N/A	N/A	1320 (100)	29
CEPT with alum	1870 (81.3)	230 (10.0)	200 (8.7)	2300 (100)	37
CEPT with ferric chloride	1870 (81.0)	280 (12.1)	160 (6.9)	2310 (100)	37

Note: The sludge mass from the CEPT processes using alum or ferric chloride is about 1.7 times of that from the conventional primary treatment process with an approximately 30% increase in sludge volume.

EXAMPLE 9.30: SLUDGE PRODUCTION IN CONVENTIONAL PRIMARY TREATMENT AND CEPT WITH LIME

The chemical sludge produced by lime treatment is significantly higher than that produced with alum or ferric chloride treatment. Calculate the mass and volume of sludge produced by conventional primary treatment and by CEPT using lime. The raw wastewater data given in Example 9.29 also applies to this example. Also, compare the mass and volume of sludge produced with alum and ferric chloride calculated in Example 9.29. To simplify the calculations, it is assumed that the phosphorous is removed completely by the chemical precipitations. Additional data on raw influent and lime treatment performance are also given below:

Raw wastewater characteristics

Carbonate hardness (CH) = 140 mg/L as $CaCO_3$
Calcium = 42 mg/L as Ca^{2+}

Performance of CEPT process using lime

TSS removed = 85%
Desired P removed = 95%
Specific gravity of sludge = 1.025
Solids content of sludge = 5%

Solution

1. Estimate the sludge production in a conventional primary treatment process.
   The sludge mass and volume produced from TSS are calculated in Example 9.29. These values are:
   Mass of TSS in sludge = 1320 kg/d
   Volume of wet sludge produced = 29 m^3/d
2. Estimate the sludge production in CEPT process using lime.
   The total quantity of sludge produced from lime treatment includes (a) TSS removed, (b) phosphorus precipitates, and (c) carbonate hardness removed as $CaCO_3$ and $Mg(OH)_2$. To simplify the calculations, it is assumed that the phosphorus precipitates can be approximated by $Ca_5OH(PO_4)_3$.
   a. Estimate TSS removed.
      The sludge mass produced from TSS is calculated in Example 9.29. The mass value is:
      Mass of TSS removed = 1870 kg/d

b. Estimate the quantity of $Ca_5OH(PO_4)_3$ produced.

i. Concentrations of P removed and remaining after the CEPT process.

The concentrations of P removed and remaining in the effluent are calculated in Example 9.29. These values are:

Concentration of P removal required = 5.7 mg/L as P

Concentration of P remaining in the effluent = 0.3 mg/L as P

ii. Lime dose required to precipitate $Ca_5OH(PO_4)_3$ (Equation 9.21a).

In accordance with Equation 9.21a, 5 moles of Ca^{2+} stoichiometrically reacts with 1 mole of PO_4^{3-} (or P) to yield 1 mole of $Ca_5OH(PO_4)_3$.

Concentration of Ca^{2+} required to precipitate $Ca_5OH(PO_4)_3$

$$= 5.7\ \text{mg/L as P removed} \times \frac{5 \times \text{MW of } Ca^{2+}}{3 \times \text{MW of P}}$$

$$= 5.7\ \text{mg/L as P removed} \times \frac{5 \times 40\ \text{g/mole as } Ca^{2+}}{3 \times 31\ \text{g/mole as P}}$$

$$= 12.3\ \text{mg/L as } Ca^{2+}$$

Concentration of $Ca(OH)_2$ consumed

$$= 12.3\ \text{mg/L as } Ca^{2+} \times \frac{\text{MW of } Ca(OH)_2}{\text{MW of } Ca^{2+}}$$

$$= 12.3\ \text{mg/L as } Ca^{2+} \times \frac{74\ \text{g/mole as } Ca(OH)_2}{40\ \text{g/mole as } Ca^{2+}}$$

$$= 23\ \text{mg/L as } Ca(OH)_2$$

iii. Quantity of $Ca_5OH(PO_4)_3$ produced.

Concentration of $Ca(OH)_2$ consumed to precipitate $Ca_5OH(PO_4)_3$ for removal of P is calculated from Equation 9.21a.

Concentration of $Ca_5OH(PO_4)_3$ precipitate

$$= 12.3\ \text{mg/L as } Ca^{2+} \times \frac{\text{MW of } Ca_5OH(PO_4)_3}{5 \times \text{MW of } Ca^{2+}}$$

$$= 12.3\ \text{mg/L as } Ca^{2+} \times \frac{502\ \text{g/mole as } Ca_5OH(PO_4)_3}{5 \times 40\ \text{g/mole as } Ca^{2+}}$$

$$= 31\ \text{mg/L as } Ca_5OH(PO_4)_3$$

Mass of $Ca_5OH(PO_4)_3$ in sludge

$$= 31\ \text{mg/L as } Ca_5OH(PO_4)_3 \times 10{,}000\ \text{m}^3\text{/d} \times 10^{-3}\ \text{kg/g}$$

$$= 310\ \text{kg/d as } Ca_5OH(PO_4)_3$$

c. Estimate the quantity of $CaCO_3$ and $Mg(OH)_2$ produced.

i. Quantity of $CaCO_3$ produced due to removal of calcium carbonate hardness (CaCH).

Calcium carbonate alkalinity (CaCA) or calcium carbonate hardness (CaCH) is formed by combining calcium ion with alkalinity.

Equivalent CaCA or CaCH in the influent

$$= 42\ \text{mg/L as } Ca^{2+} \times \frac{\text{Eq. wt. of } CaCO_3}{\text{Eq. wt. of } Ca^{2+}}$$

$$= 42\ \text{mg/L as } Ca^{2+} \times \frac{50\ \text{g/eq. as } CaCO_3}{20\ \text{g/eq. as } Ca^{2+}}$$

$$= 105\ \text{mg/L as } CaCO_3$$

Concentration of $Ca(OH)_2$ consumed to precipitate $CaCO_3$ for complete removal of CaCH is calculated from Equation 9.16b.

Concentration of $Ca(OH)_2$ consumed

$$= 105\,\text{mg/L CaCA as } CaCO_3 \times \frac{\text{MW of } Ca(OH)_2}{\text{MW of CaCA}}$$

$$= 105\,\text{mg/L CaCA as } CaCO_3 \times \frac{74\,\text{g/mole as } Ca(OH)_2}{100\,\text{g/mole CaCA as } CaCO_3}$$

$$= 78\,\text{mg/L as } Ca(OH)_2$$

Concentration of $CaCO_3$ precipitate

$$= 105\,\text{mg/L CaCA as } CaCO_3 \times \frac{2 \times 100\,\text{g/mole as } Ca(OH)_2}{100\,\text{g/mole CaCA as } CaCO_3}$$

$$= 210\,\text{mg/L as } CaCO_3$$

Assume the solubility of $CaCO_3$ at 25°C = 7.1 mg/L as $CaCO_3$ is applicable (see calculations in Example 9.24).

Actual concentration of $CaCO_3$ in sludge

$$= (210 - 7.1)\,\text{mg/L as } CaCO_3$$

$$= 203\,\text{mg/L as } CaCO_3$$

Mass of $CaCO_3$ in sludge due to removal of CaCH

$$= 203\,\text{mg/L as } CaCO_3 \times 10{,}000\,\text{m}^3/\text{d} \times 10^{-3}\,\text{kg/g}$$

$$= 2030\,\text{kg/d as } CaCO_3$$

ii. Quantity of $Mg(OH)_2$ and $CaCO_3$ produced from removal of magnesium carbonate hardness (MgCH).

Magnesium carbonate alkalinity (MgCA) or magnesium carbonate hardness (MgCH) is formed by combining magnesium ion with alkalinity.

Concentration of	= Carbonate hardness (CH) − CaCH (or CaCA)
MgCA or MgCH in	= (140 − 105) mg/L as $CaCO_3$
the influent	= 35 mg/L as $CaCO_3$

Concentration of $Ca(OH)_2$ consumed to precipitate $Mg(OH)_2$ and $CaCO_3$ for complete removal of MgCH is calculated from Equation 9.16e.

Concentration of $Ca(OH)_2$ consumed

$$= 35\,\text{mg/L MgCA as } CaCO_3 \times \frac{2 \times \text{MW of } Ca(OH)_2}{\text{MW of MgCA}}$$

$$= 35\,\text{mg/L MgCA as } CaCO_3 \times \frac{2 \times 74\,\text{g/mole as } Ca(OH)_2}{100\,\text{g/mole MgCA as } CaCO_3}$$

$$= 52\,\text{mg/L as } Ca(OH)_2$$

Concentration of $Mg(OH)_2$ precipitate

$$= 35\,\text{mg/L MgCA as } CaCO_3 \times \frac{\text{MW of } Mg(OH)_2}{\text{MW of MgCA}}$$

$$= 35\,\text{mg/L MgCA as } CaCO_3 \times \frac{58\,\text{g/mole as } Mg(OH)_2}{100\,\text{g/mole MgCA as } CaCO_3}$$

$$= 20\,\text{mg/L as } Mg(OH)_2$$

Assume the solubility of $Mg(OH)_2$ at 25°C = 7 mg/L as $Mg(OH)_2$ is applicable (see calculations in Example 9.24).

Actual concentration of $Mg(OH)_2$ in sludge

$$= (20 - 7)\ \text{mg/L as CaCO}_3$$

$$= 13\ \text{mg/L as CaCO}_3$$

Mass of $Mg(OH)_2$ in sludge due to removal of MgCH

$$= 13\ \text{mg/L as Mg(OH)}_2 \times 10{,}000\ \text{m}^3\text{/d} \times 10^{-3}\ \text{kg/g}$$

$$= 130\ \text{kg/d as Mg(OH)}_2$$

Concentration of $CaCO_3$ precipitate

$$= 35\ \text{mg/L MgCA as CaCO}_3 \times \frac{2 \times \text{MW of CaCO}_3}{\text{MW of MgCA}}$$

$$= 35\ \text{mg/L MgCA as CaCO}_3 \times \frac{2 \times 100\ \text{mg/L as CaCO}_3}{100\ \text{mg/L MgCA as CaCO}_3}$$

$$= 70\ \text{mg/L as CaCO}_3$$

Since solubility of $CaCO_3$ is already considered for removal of CaCH, it is assumed that the $CaCO_3$ precipitate from removal of MgCH is insoluble.

Mass of $CaCO_3$ in sludge due to removal of MgCH

$$= 70\ \text{mg/L as CaCO}_3 \times 10{,}000\ \text{m}^3\text{/d} \times 10^{-3}\ \text{kg/g}$$

$$= 700\ \text{kg/d as CaCO}_3$$

d. Estimate the potentially maximum concentration of $Ca(OH)_2$ remaining as excess lime in the effluent.

i. Overall $Ca(OH)_2$ dose required for chemical removal of P (Equations 9.21a through 9.21c).

Overall lime dose for 95% P removal = 175 mg/L as $Ca(OH)_2$ (Table 9.13).

ii. Total $Ca(OH)_2$ consumption required for chemical removal of P (Equations 9.21a through 9.21c).

$Ca(OH)_2$ consumed for removal of P = 23 mg/L as $Ca(OH)_2$
$Ca(OH)_2$ consumed for removal of CaCH = 78 mg/L as $Ca(OH)_2$
$Ca(OH)_2$ consumed for removal of MgCH = 52 kg/d
Total $Ca(OH)_2$ consumed in reactions = 153 kg/d

iii. Maximum $Ca(OH)_2$ remaining as excess lime in the effluent.

Excess lime in effluent = Overall $Ca(OH)_2$ dose − Total $Ca(OH)_2$ consumed

$$= (175 - 153)\ \text{mg/L as Ca(OH)}_2$$

$$= 22\ \text{mg/L as Ca(OH)}_2$$

Note: The lime may also be consumed in the unknown reactions with many other constituents in the influent. Therefore, the actual concentration of $Ca(OH)_2$ remaining may be lower than the maximum value estimated above.

e. Estimate the total quantity of sludge produced in CEPT process using lime.

The concentration and mass of chemical precipitates, and lime consumed are summarized below.

Sludge Production	Concentration, mg/L	Mass, kg/d (% of Total by Weight)	Concentration of Lime Consumed, mg/L
TSS removed	187	1870	–
$Ca_5OH(PO_4)_3$	31	310	23
$CaCO_3$ precipitate from CaCH	203	2030	78
$Mg(OH)_2$ precipitate from MgCH	13	130	52
$CaCO_3$ precipitate from MgCH	70	700	
Total		5040	153

$$\text{Volume of lime-treated sludge} = 5040\,\text{kg/d} \times \frac{100\,\text{kg as wet}}{5\,\text{kg as dry}} \times \frac{1}{1025\,\text{kg/m}^3}$$
$$= 98\,\text{m}^3/\text{d}$$

f. Estimate the residual alkalinity in the effluent.

i. Alkalinity remaining from raw wastewater.

Since TA = CH = 140 as $CaCO_3$, all alkalinity in raw wastewater are consumed due to hardness removal.

ii. Equivalent alkalinity due to soluble $CaCO_3$ and $Mg(OH)_2$ in the effluent.

Equivalent alkalinity due to soluble $CaCO_3$ = 7.1 mg/L as $CaCO_3$

Equivalent alkalinity due to soluble $Mg(OH)_2$

$$= 7\,\text{mg/L as Mg(OH)}_2 \times \frac{\text{Eq. wt. of CaCO}_3}{\text{Eq. wt. of Mg(OH)}_2}$$
$$= 7\,\text{mg/L as Mg(OH)}_2 \times \frac{50\,\text{g/eq. as CaCO}_3}{29\,\text{g/eq. as Mg(OH)}_2}$$
$$= 12\,\text{mg/L as CaCO}_3$$

iii. Equivalent alkalinity due to excess $Ca(OH)_2$ in the effluent.

Equivalent alkalinity due to excess $Ca(OH)_2$

$$= 22\,\text{mg/L as Ca(OH)}_2 \times \frac{\text{Eq. wt. of CaCO}_3}{\text{Eq. wt. of Ca(OH)}_2}$$
$$= 22\,\text{mg/L as Ca(OH)}_2 \times \frac{50\,\text{g/eq. as CaCO}_3}{37\,\text{g/eq. as Ca(OH)}_2}$$
$$= 30\,\text{mg/L as CaCO}_3$$

iv. Total residual alkalinity in the effluent.

$$\text{Total alkalinity in the effluent} = (7.1 + 12 + 30)\,\text{mg/L as CaCO}_3$$
$$= 49\,\text{mg/L CaCO}_3$$

The alkalinity remaining is sufficient to maintain the desired reaction pH during the CEPT process.

Notes: The pH in the lime-treated primary effluent may not cause problems in an activated sludge process. Carbon dioxide generated in biological process may lower the pH back to a range near the neutral point. However, the alkalinity remaining in the effluent is not sufficient for activated sludge process with nitrification requirement. Therefore, alkalinity addition is required in the effluent from the CEPT process with lime (see more detailed information about alkalinity consumption during nitrification process in Chapter 10). This goal is typically achieved by adding sodium bicarbonate or sodium carbonate (soda ash) (see additional information of these chemicals Table 9.14). A combination of recarbonation process and additional lime may also be an effective approach for supplementing alkalinity.

3. Compare the mass and volume of sludge with conventional primary treatment and CEPT processes with alum, ferric chloride, and lime.

The comparative values are summarized below.

Treatment Process	Sludge Production				
	Mass, kg/d (% of total by weight)				Volume, m^3/d
	TSS Removal	Phosphorus Precipitation	Hydroxide Precipitation or Removal of Hardness	Total	
Conventional primary treatment (Example 9.29)	1320 (100)	N/A	N/A	1320 (100)	29
CEPT with alum (Example 9.29)	1870 (81.3)	230 (10.0)	200 (8.7)	2300 (100)	37
CEPT with ferric chloride (Example 9.29)	1870 (81.0)	280 (12.1)	160 (6.9)	2310 (100)	37
CEPT with lime (Example 9.30)	1870 (37.1)	310 (6.2)	2860 (56.7)	5040 (100)	98

Notes: The sludge mass from the CEPT process using lime is about 3.8 times than that from the conventional primary treatment process. In addition, the lime sludge has poor settling characteristics, causes scaling, and is difficult to handle. The sludge volume is also increased in more than 3 times in comparison with that from the conventional primary treatment process.

EXAMPLE 9.31: NEUTRALIZATION OF EFFLUENT AFTER LIME PRECIPITATION OF PHOSPHORUS

In a CEPT process, lime is used to precipitate phosphorus from wastewater. The primary effluent has excess lime and pH is 10.2. The total alkalinity of the effluent is 210 mg/L as $CaCO_3$. The average design flow is 2.5 MGD. Calculate the acid feed rate (lb/d) to maintain the pH of neutralized primary effluent around 7.5. The acid used for neutralization is 93% (by weight) industrial grade H_2SO_4. Also, describe the equipment to maintain and monitor the effluent pH.

Note: Recarbonation by diffusing CO_2 gas is also a common process for neutralization after a lime precipitation process.

Solution

1. Establish the pH and net alkalinity relationship.

   Alkalinity of a water sample is a measure of its acid-neutralizing capacity. It is measured by titrating the sample with a standard acid solution to an endpoint at pH 4.5. Likewise, acidity is a measure of base-neutralizing capacity of the effluent, and it is measured by titrating the sample with a base standard to pH 8.3. A sample can have both alkalinity and acidity. Net alkalinity of a sample = total alkalinity − total acidity (all expressed as mg/L $CaCO_3$).

   A relationship between pH and net alkalinity of wastewater effluent samples at different pH was developed in the laboratory. This relationship is shown in Figure 9.27. It has a wide application for neutralization of lime-treated effluent.

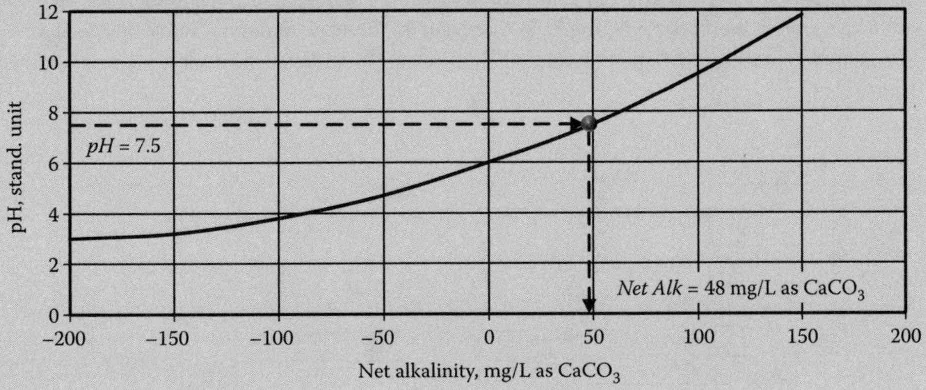

FIGURE 9.27 Titration result of pH and net alkalinity in mg/L $CaCO_3$ (Example 9.31).

2. Determine the net alkalinity at pH 7.5.

   The net alkalinity corresponding to pH 7.5 (Figure 9.27) = 48 mg/L as $CaCO_3$

3. Determine the sulfuric acid dose.

Net alkalinity $= \text{Total alkalinity} - \text{Acidity}$

Acidity $= \text{Total alkalinity} - \text{Net alkalinity}$

$= (210 - 48)\ \text{mg/L as CaCO}_3$

$= 162\ \text{mg/L as CaCO}_3$

Calculate the dose of H_2SO_4 that will produce equivalent acidity.

Dose of H_2SO_4 $= 162\ \text{mg/L as CaCO}_3 \times \dfrac{\text{Eq. wt. of H}_2\text{SO}_4}{\text{Eq. wt. of CaCO}_3}$

$= 162\ \text{mg/L as CaCO}_3 \times \dfrac{49\ \text{g/eq. as H}_2\text{SO}_4}{50\ \text{g/eq. as CaCO}_3}$

$= 159\ \text{mg/L as H}_2\text{SO}_4$

Dose of 93% H_2SO_4 solution $= 159\ \text{mg/L as H}_2\text{SO}_4 \times \dfrac{100\ \text{g as solution}}{93\ \text{g as H}_2\text{SO}_4}$

$= 171\ \text{mg/L as solution}$

Daily acid solution feed rate $= 171\ \text{mg/L as solution} \times \dfrac{8.34\ \text{lb/d}}{\text{Mgal}\cdot\text{mg/L}} \times 2.5\ \text{MGD}$

$= 3565\ \text{lb/d as solution}$

4. Describe the equipment for pH control.

   To maintain a desired effluent pH, an acid-feed system should be designed for variable flow condition. This means that the acid feed must be adjusted to maintain the effluent pH of nearly 7.5. At small installations, manual control is used. The operator measures the pH of neutralized effluent and then adjusts the feed rate of acid solution. Simple orifice-controlled constant-head arrangement or low-capacity proportioning pumps are used. Often, constant-speed feed pumps are programmed by time clock arrangement to start and stop the pump at desired intervals.

   At large installations, a complex automatic control loop with a feedback element is utilized. The control system consists of a (a) flow meter, (b) reaction vessel with mixer, (c) pH probe and transmitter, (d) controller, and (e) acid solution storage container. Signals from flow meter and pH meter are transmitted to a controller which senses these readings, integrates them, and then compares them with the preset pH value. It then signals the acid feeder to adjust the feed rate to maintain the pH near the preset value at all flow rates. A sketch of such system is shown in Figure 9.28.

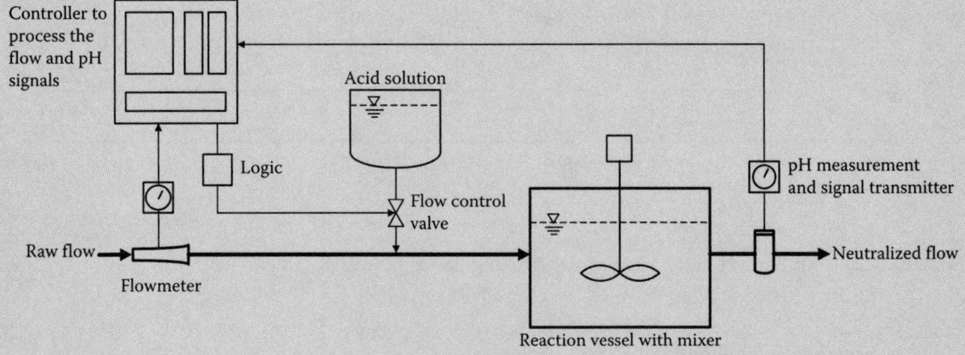

FIGURE 9.28 Process flow diagram of a typical automatic control loop with feedback element (Example 9.31).

9.5.5 Rapid Mixing

Mixing is a unit operation used to blend, disperse, or mingle chemicals to create a nearly homogeneous single-phase or multiple-phase system. Rapid dispersion of chemical in the stream is essential for coagulation or precipitation processes. Rapid mixing or flash mixing is accomplished by using (a) mechanical mixer or (b) static mixer. The mechanical mixers utilize a high-speed rotating impeller that causes agitation and mixing in a vessel. In the static mixer, the turbulence is created by vanes, orifice plates, direction change, obstruction, hydraulic jump, or by other means in the path of flowing liquid. Both types of devices are discussed below. The advantages and disadvantages of mechanical and static mixers are given in Table 9.18.[2,3]

Mechanical Rapid Mixer: The mechanical mixers are the most commonly used mixing devices for wastewater applications. They operate like a centrifugal pump without a casing. The impeller or propeller creates turbulence in the mixing chamber. The flow pattern or flow regime is one of three components: radial, axial, and tangential. The radial component is at right angle to the axis of rotation; the axial flow is parallel to the axis of rotation; and the tangential or rotational component act in a direction tangent to a circular path around the shaft. The rotating impeller mixers used for rapid mixing are divided into two groups: (a) turbine, and (b) propeller. The *turbine mixers* have impellers with straight, flat, pitched*, curved, vertical or vanned blades, and may be open, semiclosed, or shroud. The vertical impeller blades are more common. The principal currents are radial and tangential, or rotational. The *propeller mixers* are high-speed mixers. The impeller may be straight, curved, pitched, or vertical. These mixers generate currents that are primarily axial. Usually, the pitch is 1.0–2.0. In deep tanks, two or more impellers may be mounted in the same shaft, moving the liquid in the same direction.

Vortexing reduces mixing by reducing the velocity of impeller relative to the liquid. Baffles are used to reduce vortexing effectively without interfering with radial or tangential flow. Baffling is required in turbine and paddle mixers to minimize the vortexing and rotational flow except at very slow speeds. Baffles are also used with propeller impellers in large tanks. In small tanks, the propeller may be mounted off-center to avoid rotational flow. The turbine and propeller mixers are shown in Figure 9.29.[42]

Design Factors: The basic design factors of a rapid-mix basin are hydraulic retention time, basin geometry, type of impeller, dimensions and installation, velocity gradient G (describes the degree of mixing), power requirements, and laboratory scale-up. Basic design parameters and typical values are provided with additional comments in Table 9.19.

Power Requirements: The power imparted to the water by impeller is given by Equations 9.25a and 9.25b. Equations 9.25a is used for turbulent flow range (Reynolds number $> 10^4$). Mechanical mixers are designed to induce turbulent mixing in rapid-mix basin. Equations 9.25b is used for laminar flow range (Reynolds number < 10). Laminar mixing is encountered with some static mixers. The Reynolds number

TABLE 9.18 Comparison of Mechanical and Static Mixers

Type	Advantages	Disadvantage
Mechanical mixers	• Adjustable agitation • Not affected by flow rate • High flexibility in operation • Low head loss	• High energy cost • Requiring routine maintenance • High risk of mechanical failure • High O&M costs
Static mixer	• Low or no energy requirement • Little or no maintenance • Very reliable	• High dependence on flow rate • High head loss • Low flexibility in operation

* Pitch is defined as the distance the liquid moves axially during one revolution divided by the propeller diameter.

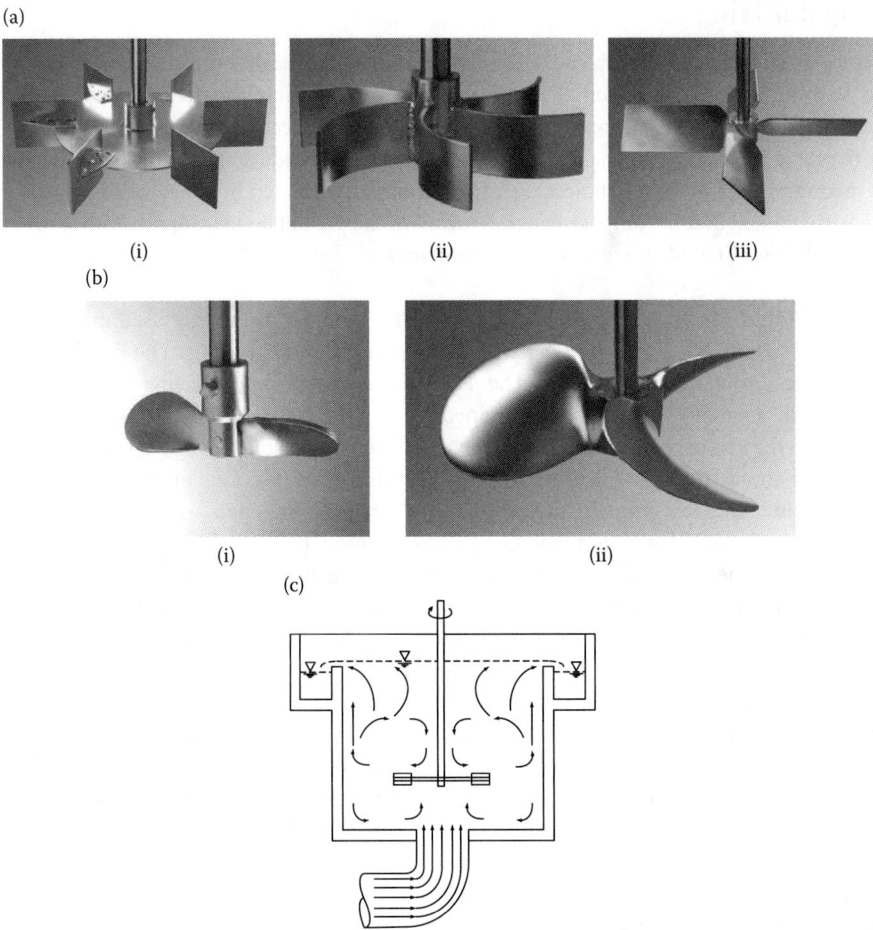

FIGURE 9.29 Impeller mixers and mixing arrangements: (a) turbine-type impeller (Courtesy SPX Flow, Inc.): (i) flat-blade radial-flow impeller, (ii) curved-blade radial-flow impeller, and (iii) pitched-blade axial-flow impeller; (b) propeller-type (Courtesy SPX Flow, Inc.): (i) fully open axial-flow impeller and (ii) marine-type axial-flow impeller; and (c) typical flow pattern in a mixing chamber using turbine radial-flow impeller.

is expressed by Equation 9.26.

$$P = N_T \rho n^3 d^5 \quad \text{or} \quad P = N_T n^3 d^5 \frac{\gamma}{g} \quad \text{(turbulent)} \tag{9.25a}$$

$$P = N_L \mu n^2 d^3 \quad \text{(laminar)} \tag{9.25b}$$

$$N_R = \frac{nd^2 \rho}{\mu} \quad \text{or} \quad N_R = \frac{nd^2 \gamma}{\mu g} \tag{9.26}$$

where

P = power imparted to the water, W or N·m/s (lb·ft/s)
N_T = power number or impeller constant for turbulent flow
N_L = power number or impeller constant for laminar flow
n = impeller speed, revolutions per second (rps)
d = impeller diameter, m (ft)

TABLE 9.19 Rapid-Mix Basin Basic Design Parameters and Typical Values

Design Parameter	Typical Value	Comment
Hydraulic retention time (HRT), min	0.5–2	HRT as small as 10 s and as long as 5 min have been used
Tank geometry	N/A	Square, rectangular, or circular mixing basins are used. Circular basins with baffles are more efficient than square or rectangular basins.
Mixing requirement		
Velocity gradient (G), s^{-1}	300–1500	As G increases, the degree of mixing increases. Rapid-mix basins for wastewater application have mixing in turbulent range ($N_R > 10^4$).
Power dissipation function (P/V), $N/m^2{\cdot}s$	120–3000	Power dissipation function can be used to estimate the approximate mixing effectiveness in a rapid-mixing unit
Mixing opportunity parameter (Gt), dimensionless	30,000–90,000	
Mixer speed, rpm	1750 (for small mixers) 400–800 (for large mixers)	Mixer speed depends upon power input, and propeller size and pitch. High-speed mixers are best suited for domestic wastewater. Low-speed mixers are used for viscous fluids.
Turbine mixer	56–125	These are relatively low-speed mixers and have high energy requirement. Flow is primarily radial and tangential.
Propeller mixer	400–1750	These are high-speed mixers and require low power. Flow is primarily axial. Best suited for thick solutions, and rotates at full motor speed.
Impeller diameter, % of tank diameter or width	25–40	The impeller diameter depends upon the tank size and mixer speed. High-speed mixers have smaller diameters.
Turbine mixer	30–50	
Propeller mixer	Rarely exceeds 460 mm (18 in)	
Baffles and projection, % of tank diameter or side width	< 10	Baffles reduce vortexing in turbine and paddle mixers, and in large tanks with propeller mixer. Circular basins without baffles have severe vortexing. Baffles are vertical strips installed perpendicular to the wall throughout the liquid depth. Four baffles are normally sufficient.
Turbine mixer	8–9	Projections are made to break the rotational movement
Propeller mixer	5–6	In small tanks, the propeller may be mounted off center to avoid vortexing
Mixer mounting	N/A	Mixer mounting is installation of mixer into the tank
Top mounting or top-entering mixer	N/A	Top mounting is used in small open tanks of <4000 L (1000 gal) in capacity.
Wall mounting or side-entering mixer	N/A	These mixers are mounted horizontally along the tank axis or off-center. Side-entry mixers are preferred for large basins to avoid extremely long shaft with top-entry mixers. The liquid flow is horizontal along the long axis of the tank and mixer shaft.
Impeller height above floor for vertical installation, % of mixer diameter		The impeller of top-mounting mixer is suspended above the floor and anchored in position. The shaft is held in vertical position.
Turbine mixer	50–100	Mixers are top mounted and always installed vertically in the center or off-center. Baffles are essential for top-mounted centrally located impeller. The impeller is mounted off-center in unbaffled tank.

(*Continued*)

TABLE 9.19 (*Continued*) Rapid-Mix Basin Basic Design Parameters and Typical Values

Design Parameter	Typical Value	Comment
Propeller mixer	50–100	Propeller mixers are mounted vertically at an angle, or horizontally along the center or off-center.
Multimixers or multipropeller mixer		Multipropeller mixers are considered where the liquid depth and tank width or diameter ratio is greater than unity. The impellers are installed in the shaft at several levels.

Source: Partly adapted from References 1 through 4, 26, 28 through 30, and 43.

μ = dynamic viscosity of the fluid, $N{\cdot}s/m^2$ ($lb{\cdot}s/ft^2$)
ρ = mass density of fluid ($\rho=\gamma/g$), kg/m^3 ($slug/ft^3$)
γ = specific weight of water, N/m^3 (lb/ft^3)
g = acceleration due to gravity, 9.81 m/s^2 (32.2 ft/s^2)
N_R = Reynolds number

Note: $\rho = \gamma/g = (N/m^3)/g = (kg{\cdot}m/s^2)/(m/s^2)\times (1/m^3)/(m/s^2) = kg/m^3$; N = mass × acceleration ($kg{\cdot}m/s^2$); Watt (W) = N·(m/s) = ($kg{\cdot}m/s^2$)× m/s = $kg{\cdot}m^2/s^2$; and $\mu = n{\cdot}s/m^2 = ((kg{\cdot}m/s)\ s)m^2 = kg/m{\cdot}s$.

The power number of an impeller depends upon the shape and size, the number of blades, number of baffles used to eliminate vortexing, and other variables not included in the power equation. Typical values of N_T are provided in Table 9.20. Typical values of N_L are in the range of 33–71. The manufacturers provide the power number of their equipment. For turbulent flow, the baffled square and circular tanks have same power requirements if tank width is equal to the diameter. The power imparted in an unbaffled tank is 75% of that of a baffled tank. Therefore, the value of N_T is reduced by this amount. The performance of a rapid-mix unit is described by several empirical parameters. These are (1) power dissipation function, (2) mixing opportunity parameter, and (3) mixing loading parameter.

Power Dissipation Function: Power dissipation function is an approximate measure of mixing effectiveness. It is power input per unit volume of the basin. Power input creates turbulence that leads to mixing. The degree of mixing is the rate of collision of particles, and is proportional to the velocity gradient G. Velocity gradient is also related to the shear forces. High-velocity gradient will break the floc in the

TABLE 9.20 Power Numbers of Various Rapid-Mix Impellers

Impeller Type	Power Number, N_T
Turbine-type	
Flat-blade	2.6
4 blade ($w/d = 0.15$)	3.3
4 blade ($w/d = 0.2$)	5.1
Disk	6.2
4 blade ($w/d = 0.25$)	
6 blade ($w/d = 0.25$)	
Propeller-type	0.3
1 : 1 pitch	0.7
1.5 : 1 pitch	
45° pitched blade	1.36
4 blade ($w/d = 0.15$)	1.94
4 blade ($w/d = 0.2$)	

Note: w/d = blade width-to-impeller diameter ratio.

Source: Adapted in part from References 2 through 4, and 26.

flocculation basin. In rapid-mix unit, the G value is typically in the range of 300–1500 s^{-1}, and power dissipation function is in the range of 120–3000 $N/m^2 \cdot s$. The power dissipation function with associated relationships is expressed by Equation 9.27.

$$\frac{P}{V} = \mu G^2 \quad or \quad G = \sqrt{\frac{P}{\mu V}} \quad or \quad P = \mu V G^2 \quad \text{or} \quad P = \left(\frac{P}{V}\right)V \tag{9.27}$$

where

P/V = power dissipation function, W/m^3 or $N/m^2 \cdot s$ ($lb/ft^2 \cdot s$)
G = velocity gradient, s^{-1} or mps/m (fps/ft)
V = volume of the basin, m^3 (ft^3)
P = power imparted to the water, N·m/s or W (lb·ft/s)

Mixing Opportunity Parameter: The total number of interparticles collisions is proportional to the product of velocity gradient G and detention time t. Thus, the value of Gt (mixing opportunity parameter) is important in design. The typical value of Gt for rapid mixing ranges from 30,000 to 90,000 (dimensionless). The mixing opportunity parameter Gt is also a ratio of power-induced rate of flow to hydraulic-induced rate of flow.[4] It is expressed by Equation 9.28.

$$Gt = \frac{1}{Q}\sqrt{\frac{PV}{\mu}} \tag{9.28}$$

where

Gt = mixing opportunity parameter, dimensionless
t = detention time of rapid-mixing basin, s
Q = flow, m^3/s (ft^3/s)

Mixing Loading Parameter: The hydraulic loading of a mixing unit is not only a function of hydraulic retention time but is also a function of power input and viscosity. The value of mixing loading parameter for a rapid-mix unit is expressed by Equation 9.29. For the municipal wastewater, the value of mixing loading parameter typically ranges from 0.0075 to 0.03 s^{-1}.

$$L_{\text{Mix}} = \frac{1}{t} = \frac{Q}{V} = QG\sqrt{\frac{\mu}{PV}} \tag{9.29}$$

where L_{Mix} = mixing loading parameter, s^{-1}

All other variables in the above equation have been defined earlier.

Static Mixer: The static (or hydraulic) mixer creates turbulence in flow stream by obstruction, contraction, or enlargement of flow area; sudden change in head loss and velocity patterns due to baffles; as well as momentum reversal. Common types of static mixers are static in-line mixer, baffled channel, pneumatic mixing by injection of compressed air, water jet, pump suction and discharge lines, hydraulic jump, Parshall flume, and weir.[1,3,30] Static mixers may be used for either rapid or slow mixing purposes. Several types of static mixers are shown in Figure 9.30.[44,45]

Head Loss or Pressure Drop: The degree of mixing in static mixer is related to head loss or pressure drop. The head loss and power dissipated by a static mixer are calculated from Equations 9.30a through 9.30c.[1,24,26,30]

$$P = \gamma Q h \quad \text{or} \quad P = \rho g Q h \tag{9.30a}$$

$$h = k\frac{v^2}{2g} = K_{\text{m}} v^2 \tag{9.30b}$$

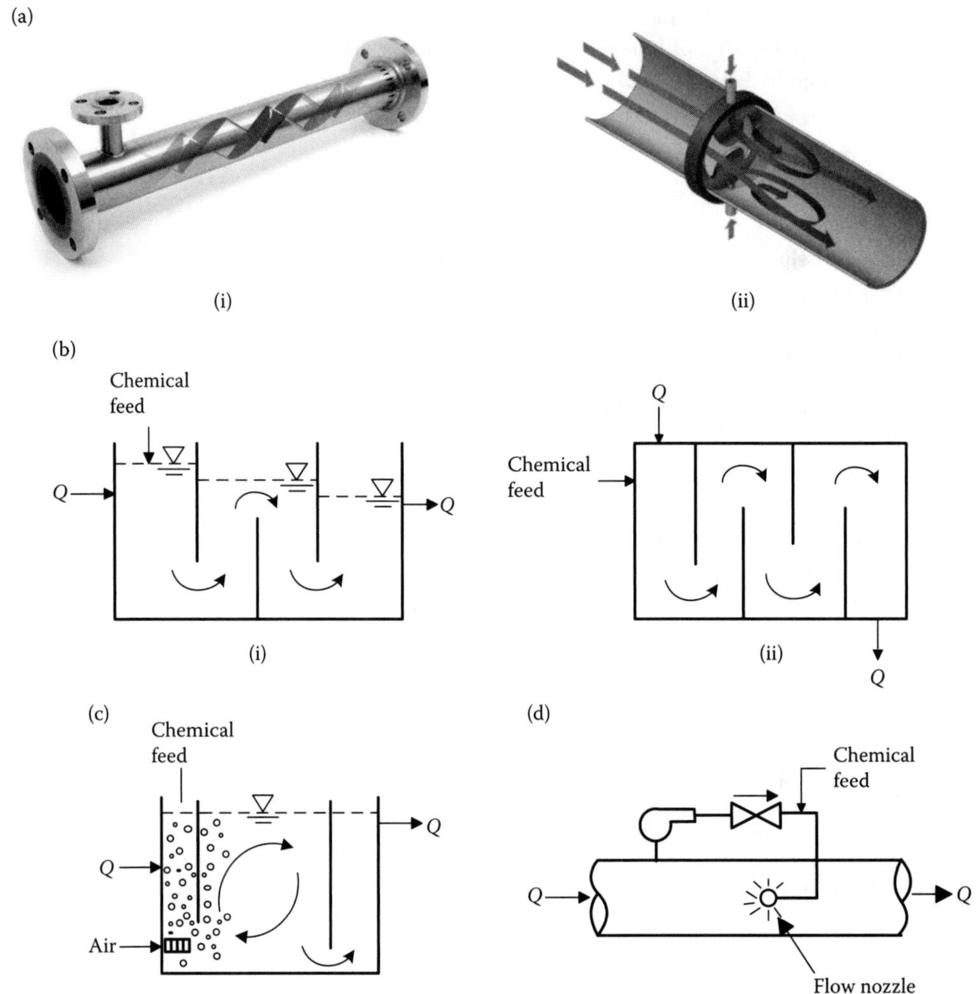

FIGURE 9.30 Examples of static mixers: (a) in-line static mixer: (i) internal vanes (Courtesy Koflo Corp.) and (ii) orifice plate (Courtesy Westfall Manufacturing Company); (b) baffled channel: (i) sectional view with over and under flow and (ii) plan view with around the end flow; (c) pneumatic mixer; and (d) pressurized water jet in pipe.

$$G = \sqrt{\frac{\gamma h}{\mu t}} \quad or \quad G = \sqrt{\frac{\rho g h}{\mu t}} \quad or \quad G = \sqrt{\frac{g h}{\nu t}} \tag{9.30c}$$

where

P = power dissipated in the device, W or N·m/s (lb·ft/s)
h = head loss encountered through the device, m (ft)
ν = kinematic viscosity, m^2/s (ft^2/s)
v = velocity, m/s (ft/s)
k = constant for mixing characteristic
K_m = overall mixing coefficient of the device, s^2/m (s^2/ft)

All other variables in the above equations have been defined earlier.

The typical value of K_m is in the range of 1–4 s^2/m. It varies with the type of the device. The manufacturers provide head loss or pressure drop curves for their equipment. At a hydraulic jump in a channel, t of 2 s is usually assumed if velocity in the downstream channel is 0.5 m/s or larger.[23,24]

Pneumatic mixer: In pneumatic mixer, the turbulence is caused by compressed air or oxygen gas induced near the bottom of the tank (Figure 9.30c). The velocity gradient may range from $G = 200$ to 820 s^{-1}. The power dissipated by a pneumatic mixer is expressed by Equation 9.31.[1,3,4,23]

$$P = p_a q_a \ln\left(\frac{p_d}{p_a}\right) \quad (9.31a)$$

$$P = K q_a \ln\left(\frac{h_d + 10.33}{10.33}\right) \quad \text{(SI unit)} \quad (9.31b)$$

$$P = K\, q_a \ln\left(\frac{h_d + 33.9}{33.9}\right) \quad \text{(U.S. customary unit)} \quad (9.31c)$$

where

p_a = atmospheric pressure N/m^2 (lb/ft^2)
p_d = air pressure at the discharge point of air, N/m^2 (lb/ft^2)
q_a = air flow at atmospheric pressure, m^3/s (ft^3/s)
h_d = submerged head at the discharge point of air, m (ft) of water
K = proportionality constant, N/m^2 (lb/ft^2). $K = 1689\ N/m^2$ in SI unit and $K = 35.28\ lb/ft^2$ in the U.S. customary units.

P has been defined earlier.

The velocity gradient G in pneumatic mixing is calculated by substituting P from Equation 9.31 in Equation 9.27.

9.5.6 Slow Mixing or Flocculation

The chemically destabilized colloidal particles after coagulation are gently stirred to promote the growth of floc so that they may readily settle in a clarifier. This process is known as a *flocculation* process. Flocculation is also used in the precipitation process to promote the growth of precipitates for settling. Generally, coagulation and precipitation are achieved in a rapid-mix basin, and gentle mixing is carried out in a flocculation basin for microfloc to aggregate into lager floc for settling in a clarifier.

Types of Flocculators: The principal types of mixers used for flocculation (also called flocculators) are (1) static and (2) paddle. The turbine- and propeller-type flocculators can also be used for flocculation. The common static flocculators are hydraulic types as shown in Figure 9.30. Several other types of mixers used for flocculation are shown in Figure 9.31. The *paddle mixers* are similar to turbine mixers but are not as efficient. The paddle-type flocculators consist of paddles or flats mounted over a horizontal or vertical shaft (Figures 9.31a and b). The mixing regime is principally radial and tangential, but they do not produce as much turbulence and shear forces as turbine or propeller. That is why paddle mixers are mostly used for flocculent mixing. These mixers have two or four blades. The tangential component promotes *rotational movement* or *vortexing around the impeller*. In addition to these paddle types, the walking beam flocculators may also be used (Figure 9.31c).[46]

Design Factors: The floc formation and characteristics of floc depend upon velocity gradient G, and the value of Gt of the flocculation process. If the G value is too high, the shear forces will break up the floc into

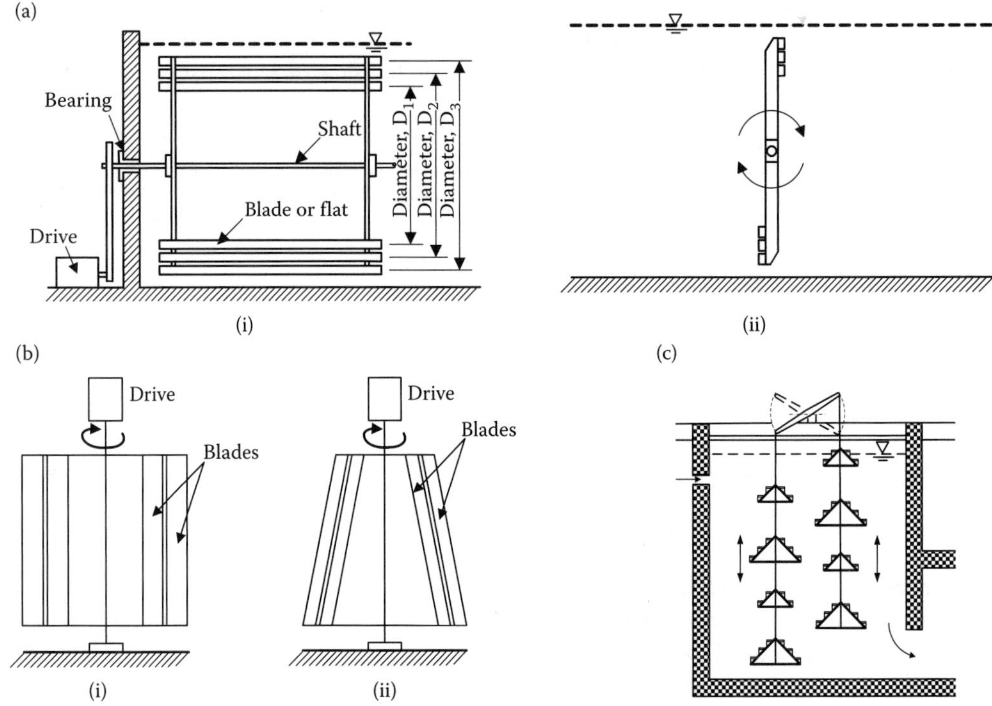

FIGURE 9.31 Flocculation equipments: (a) horizontal-shaft flocculation paddle wheel: (i) section and (ii) end view; (b) vertical-shaft paddle wheel: (i) vertical blade and (ii) sloping blades; and (c) walking beam flocculator.

smaller particles. If the G value is insufficient, adequate interparticulate collisions will not occur and proper floc will not be formed. Raw wastewater coagulates readily, and high-strength floc can normally be formed in a single-stage unit. Flocculators are frequently designed with tapered-stage flocculation in which the flow is subjected to decreasing value of G. A high value of Gt indicates a large number of interparticulate collisions during aggregation. Basic design parameters and typical values with additional comments are provided in Table 9.21.

Power Requirements: The power imparted to water by paddle wheels is calculated from the following equation (Equation 9.32):

$$P = \frac{C_D \gamma A v^3}{2g} \quad or \quad P = \frac{C_D \rho A v^3}{2} \tag{9.32}$$

where

C_D = coefficient of drag of flocculator paddles moving perpendicular to the fluid, dimensionless. For rectangular paddles, C_D varies with the length-to-width ratio of the paddle blade. The C_D values for different L/W ratios of paddle are summarized in Table 9.22.[1,26] A guidance curve is also provided in Figure 9.32 for a probable estimate of the C_D value for design of a paddle wheel flocculator when the L/W ratio is known.

A = area of the paddle(s), m^2 (ft^2)

v = velocity of the paddle relative to the rotating fluid, m/s (ft/s). Usually it is 0.6–0.75 times the paddle absolute velocity.

Variables of γ, ρ, and g in the above equations have been defined earlier.

TABLE 9.21 Flocculation Basin Basic Design Parameters and Typical Values

Design Parameter	Typical Value	Comment
Hydraulic retention time (HRT), min	5–30	HRT may vary in a range from 5 to 30 min
Tank geometry	N/A	Rectangular or square basins are usually used in single- or multistage arrangement
Mixing requirement		Slow mixing with relatively low energy intensity is required to promote interparticulate collisions for proper formation of floc
G value, s^{-1}	20–60	The typical value of G ranges from 20 to 60 s^{-1} in a flocculator
Single-stage	20–30	The typical value of G is narrowed between 20 and 30 s^{-1}.
Multistage	15–60	A typical series of G commonly used are 60, 30, and 15 s^{-1} in the first, second, and third stages, respectively.
Mixing opportunity parameter (Gt), dimensionless	10,000–150,000	Values of Gt are much higher than that for rapid mixing.
Basin		
Depth, m	3–5.5	Depth is typically 0.6–1 m greater than the wheel diameter.
Depth-to-width ratio	0.8:1–1.1:1	It is typically 0.8:1–1:1 for paddle wheel flocculators, and 0.9:1–1.1:1 for vertical turbine flocculators
Clearance at end or between stages, m	0.6–1	
Baffles, % of side width	~10	Baffles stop effectively rotational motion and induce turbulence
Paddle wheel flocculator		Paddle wheels are typically in wall-mounted mounting arrangement
Diameter, m	2.4–5	The diameter of the paddle wheel is typically 70–80% of basin width
Speed		
Shaft rotation, rpm	0.5–10	These are low-speed mixers with large area. Flow is primarily radial or tangential.
Absolute tip speed, m/s	<1	Low speed is suitable for promoting large floc formation. It is typical in front of clarifiers.
Paddle peripheral velocity relative to water, m/s	0.1–0.7	The peripheral velocity of paddle relative to the water is typically 60–75% of the absolute velocity of the paddle.
Paddle dimension		
Ratio of length to width (L/W)	10–25	
Length, m	2–3.5	
Width, m	0.1–0.2	Width of the paddle is 1/10–1/25 of the paddle diameter.
Vertical turbine flocculator		The flocculators are in top-mounted mounting arrangement
Diameter, % of the basin width	30–60	It is preferred 40–50% of the basin width
Rotation speed, rpm	10–30	They are typically faster than the paddle wheel flocculators. Flow is either radial or axial.
Absolute tip speed, m/s	<2	Relatively high speed is typically used for formation of pinpoint-sized floc. It is typical in front of filters.
Impeller mounting height above the floor, % of the water depth	30–50	It is typically at two-thirds of the depth

Source: Partly adapted from References 1 through 4, 26, 28 through 30, and 43.

TABLE 9.22 Recommended C_D Values for L/W Ratios of Paddle Blades

L/W Ratio	C_D
1	1.16
5	1.2
20	1.5
>>20	1.9

Source: Partly adapted from References 1 and 26.

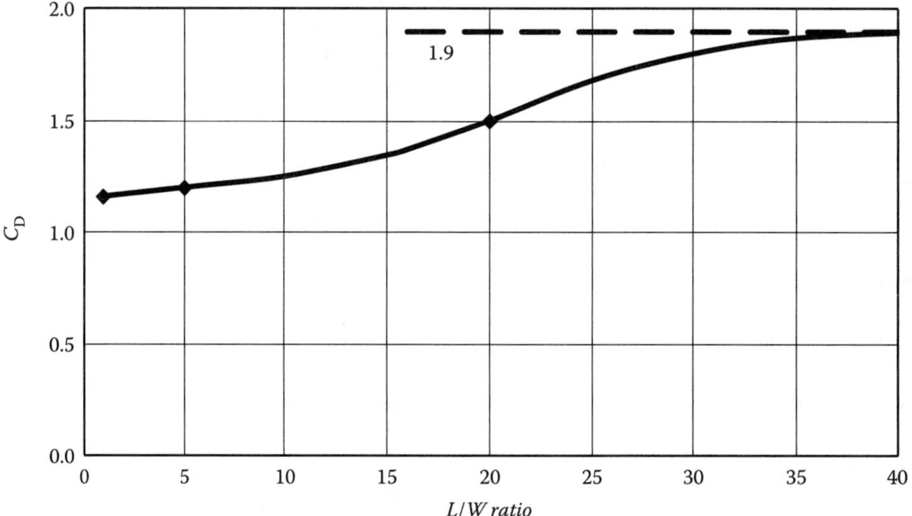

FIGURE 9.32 Estimated Drag Coefficient (C_D) from the ratio of length to width (L/W) for flat-paddle blade flocculator design.

EXAMPLE 9.32: VELOCITY GRADIENT

In a rapid-mix unit, two particles are 1.5 mm apart. Their velocity relative to each other is 1.2 m/s. Calculate the velocity gradient G.

Solution

1. Calculate the distance between the particles in m.

$$\text{Distance} = 1.5\,\text{mm} \times \frac{\text{m}}{10^3\,\text{m}} = 0.0015\,\text{m}$$

2. Calculate the velocity gradient G.

$$G = \frac{\text{Velocity}}{\text{Distance}} = \frac{1.2\,\text{m/s}}{0.0015\,\text{m}} = 800\,\text{s}^{-1}$$

EXAMPLE 9.33: RELATIONSHIP BETWEEN MIXING POWER DISSIPATION FUNCTION (*P*/*V*) AND VELOCITY GRADIENT (*G*)

Rapid mixers are used to disperse the coagulating chemicals into the flowing water. Develop a curve between the power requirement per unit volume (P/V, also called power dissipation function) and velocity gradient G within the typical range of G from 300 to 1500 s^{-1} at a wastewater temperature of 10°C.

Solution

1. Develop the relationship between P/V and G.

   The relationship is developed by taking a logarithm (base of 10) on each side of Equation 9.27.

$$\log\left(\frac{P}{V}\right) = \log(\mu G^2) = \log(\mu) + 2\log(G)$$

At 10°C, the dynamic viscosity $\mu = 1.307 \times 10^{-3}$ N·s/m^2 (Table B.2 in Appendix B).

$$\log\left(\frac{P}{V}\right) = \log(1.307 \times 10^{-3}) + 2\log(G) = -2.88 + 2\log(G)$$

2. Tabulate the values of $\log(G)$, $\log(P/V)$, and P/V at $G = 300\text{–}1500\ s^{-1}$.

G, s^{-1}	$\log(G)$	$\log(P/V)$	P/V, N/m^2·s
300	2.48	2.07	119
600	2.78	2.68	475
900	2.95	3.03	1068
1200	3.08	3.28	1898
1500	3.18	3.47	2966

3. Plot the P/V values with respect to G.

   The results of $\log(P/V)$ versus $\log(G)$, and P/V versus G on log-log paper are shown in Figures 9.33a and b, respectively.

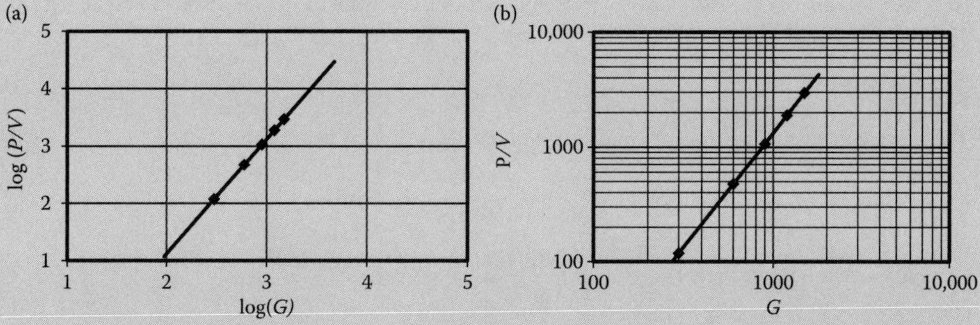

FIGURE 9.33 Relationship between power dissipation function (P/V) and velocity gradient (G): (a) log (P/V) versus log(G), and (b) P/V versus G on log-log paper (Example 9.33).

EXAMPLE 9.34: VELOCITY GRADIENT AND POWER IMPARTED BY A TURBINE IMPELLER

A turbine impeller has diameter of 2.0 m and speed of 30 rpm. The tank is square with four baffles. The power number N_T for the turbulent flow is 6.3. Calculate the power imparted by the impeller, and velocity gradient if the basin volume is 80 m^3. Assume $\rho = 995.7\ \text{kg/m}^3$ and $\mu = 0.798 \times 10^{-3}\ \text{N s/m}^2$ (kg/m·s).

Solution

1. Check for mixing regime.

$$n = \frac{30\ \text{rev/min}}{60\ \text{s/min}} = 0.5\ \text{rps}$$

Calculate N_R from Equation 9.26.

$$N_R = \frac{nd^2\rho}{\mu} = \frac{0.5\ \text{rps} \times (2.0\ \text{m})^2 \times 995.7\ \text{kg/m}^3}{0.789 \times 10^{-3}\ \text{kg/m·s}} = 2.5 \times 10^6$$

$N_R > 10^5$, therefore, turbulent mixing occurs.

2. Calculate the energy imparted by the mixer, P.

Calculate P from Equation 9.25a.

$$P = N_T\rho n^3 d^5 = 6.3 \times 995.7\ \text{kg/m}^3 \times (0.5\ \text{rps})^3 \times (2.0\ \text{m})^5 = 25{,}092\ \text{kg·m}^2/\text{s}^3$$
$$= 25{,}092\ \text{N·m/s} = 25{,}092\ \text{W} \quad \text{or} \quad 25.1\ \text{kW}$$

3. Calculate the velocity gradient, G.

Calculate G is calculated by Equation 9.27.

$$G = \sqrt{\frac{P}{\mu V}} = \left(\frac{25{,}092\ \text{N·m/s}}{0.798 \times 10^{-3}\ \text{N·s/m}^2 \times 80\ \text{m}^3}\right)^{1/2} = 627\ \text{s}^{-1}$$

EXAMPLE 9.35: MIXING OPPORTUNITY PARAMETER, VELOCITY GRADIENT, POWER DISSIPATION FUNCTION, POWER IMPARTED, AND MIXING LOADING PARAMETER OF A RAPID MIXER

A rapid-mix unit has a square basin of 45 m^3. It is designed to receive an average flow of 64,800 m^3/d. The Gt is 50,000 at an average wastewater temperature of 10°C. Calculate the power dissipation function (P/V), power imparted by the impeller (P), and mixing loading parameter (L_{Mix}).

Solution

1. Calculate the detention time of rapid-mix unit (t).

$$\text{Average flow, } Q = 64{,}800\ \text{m}^3/\text{d} \times \frac{d}{(60 \times 60 \times 24)\ \text{s}} = 0.75\ \text{m}^3/\text{s}$$

$$\text{Average detention time, } t = \frac{V}{Q} = \frac{45\ \text{m}^3}{0.75\ \text{m}^3/\text{s}} = 60\ \text{s}$$

2. Determine the mixing opportunity parameter (Gt).

Mixing opportunity parameter, Gt = 50,000 (Given)

3. Calculate the velocity gradient G.

   Calculate G from Gt.

$$\text{Velocity gradient, } G = \frac{Gt}{t} = \frac{50{,}000}{60\,\text{s}} = 833\,\text{s}^{-1}$$

4. Calculate the power dissipation function (P/V).

   At 10°C, the dynamic viscosity $\mu = 1.307 \times 10^{-3}\,\text{N}\cdot\text{s/m}^2$ (Table B.2 in Appendix B).

   Calculate P/V from Equation 9.27.

$$\text{Power dissipation function, } \frac{P}{V} = \mu G^2 = 1.307 \times 10^{-3}\,\text{N}\cdot\text{s/m}^2 \times (833\,\text{s}^{-1})^2 = 907\,\text{N/m}^2\cdot\text{s}$$

$$= 907(\text{N}\cdot\text{m/s})/\text{m}^3 \quad \text{or} \quad 907\,\text{W/m}^3$$

5. Calculate the power imparted by the impeller (P).

   Calculate P from Equation 9.27.

$$\text{Power imparted by the impeller, } P = \left(\frac{P}{V}\right) \times V = 907\,\text{W/m}^3 \times 45\,\text{m}^3 = 40815\,\text{W} = 40.8\,\text{kW}$$

6. Calculate the mixing loading parameter, L_{Mix}.

   Calculate L_{Mix} from Equation 9.29.

$$\text{Mixing loading parameter, } L_{\text{Mix}} = \frac{1}{t} = \frac{1}{60\,\text{s}} = 0.0167\,\text{s}^{-1}$$

EXAMPLE 9.36: DESIGN OF A RAPID-MIX BASIN COMPLEX

A rapid-mix basin complex with four process trains is used for CEPT process at a wastewater treatment plant. Each process train has equal flow distribution. Design one process train for a flow of 28,400 m^3/d (7.5 MGD). The mixing period is 30 s and velocity gradient is 950 s^{-1}. Select turbine mixer with four blades and $w/d = 0.16$. The wastewater temperature ranges from 5°C to 28°C. Develop the complete design of the rapid-mix unit and describe the design features and equipment selection.

Solution

1. Describe the rapid-mix basin arrangement and process trains.

   Chemical handling and delivery are among the major maintenance problems at a wastewater treatment plant. In this design example, the rapid-mix basin complex is located adjacent to the chemical building to minimize the potential problems by reducing the length of chemical delivery piping, and by using gravity feed where feasible. The raw wastewater enters a 4-m deep influent division chamber. Flow from the influent division chamber is divided equally into four exit pipes. Each pipe feeds an individual process train. In this example, each pipe carries a flow of 28,400 m^3/d to each process train headed by a rapid-mix unit. Coagulant is added and mixed in the rapid-mix chamber. The rapid mixer has turbine-type impeller with four blades that create turbulent mixing. Mixers may also be used in the influent division chamber in which lime or acid may be added for pH adjustment. The rapid-mix basin details are shown in Figure 9.34.

2. Describe the details of each rapid-mix unit.

   The major components of the rapid-mix unit are (a) reaction chamber, (b) rapid mixer, (c) influent structure, and (d) effluent structure. The reaction chamber is a square basin. The influent pipe is connected underneath the chamber to deliver the flow at the bottom directly below the turbine impeller.

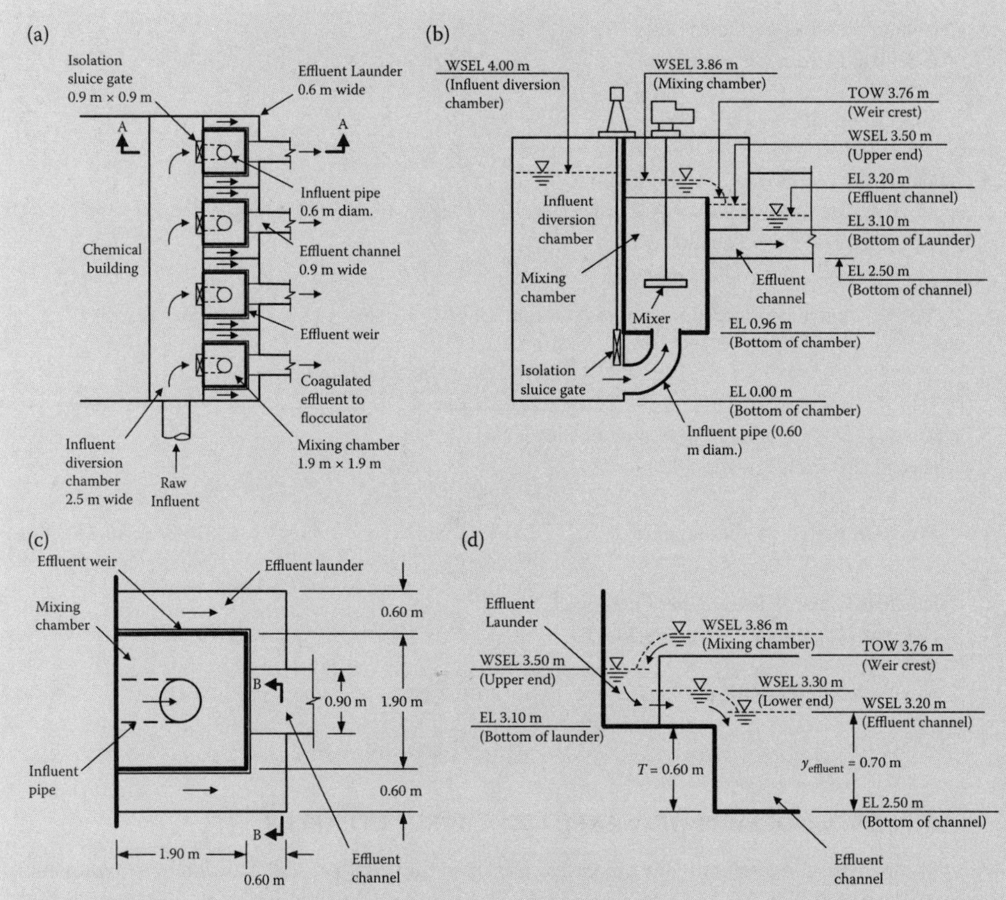

FIGURE 9.34 Design details of rapid-mix units: (a) sketch layout, (b) Section AA (hydraulic profile and major design elevations), (c) details of rapid-mixing chamber, and (d) Section BB showing water surface elevations in effluent launder and channel (Example 9.36).

An influent sluice gate at the entrance of the influent pipe in the division chamber is provided to isolate the rapid- mix unit from the other process trains if necessary. The effluent structure consists of a straight free-falling weir on three sides, effluent collection launder, and an exit channel. The exit channel carries the flow into a *Parshall flume* at the head of a flocculation basin.

The major instrumentation provided for control of coagulation process includes:

a. Flow measurement and recording at each Parshall flume
b. Temperature, pH, TSS (or turbidity), and TOC monitoring and recording in the influent diversion chamber
c. pH and oxidation–reduction potential (ORP) monitoring and recording in each rapid-mix unit
d. Ortho-P monitor and recording in the influent diversion chamber (optional if chemical precipitation of phosphorus is required)

3. Calculate the dimensions of the rapid-mix chamber.

$$\text{Design flow, } Q = 28{,}400\,\text{m}^3/d \times \frac{d}{(60 \times 60 \times 24)\text{s}} = 0.329\,\text{m}^3/\text{s}$$

Calculate the rapid-mix chamber volume required at the detention time of $t = 30$ s.

Volume, $V = Qt = 0.329\,\text{m}^3/\text{s} \times 30\,\text{s} = 9.86\,\text{m}^3$

Provide a depth-to-width ratio (H:W) of 1.5:1 in the square chamber.

$$V = W^2 \times H = W^2 \times (1.5\,W) = 1.5W^3 = 9.86\,\text{m}^3$$

Calculate the required width of the chamber.

$$W = \left(\frac{9.86\,\text{m}^3}{1.5}\right)^{1/3} = 1.87\,\text{m}$$

Proved width of the chamber, $W = 1.9$ m.

The required depth of the chamber, $H = 1.5\,W = 1.5 \times 1.9\,\text{m} = 2.85\,\text{m}$

Proved depth of the chamber, $H = 2.9$ m. The reaction chamber dimensions are 1.9 m × 1.9 m × 2.9 m (deep).

Actual volume of the chamber, $V = 1.9\,\text{m} \times 1.9\,\text{m} \times 2.9\,\text{m} = 10.5\,\text{m}^3$

Dimensions of the rapid-mixing chamber are illustrated in Figure 9.34c.

4. Design the rapid mixer.
   a. Determine the mixing requirements.

      Velocity gradient, $G = 950\,\text{s}^{-1}$ (given)

      Mixing opportunity parameter, $Gt = 950\,\text{s}^{-1} \times 30\,\text{s} = 28{,}500$
   b. Calculate the mixer water power.

      The minimum temperature of 5°C will present the critical condition for mixer design. At this temperature, the dynamic viscosity $\mu = 1.519 \times 10^{-3}\,\text{N·s/m}^2$ (Table B.2 in Appendix B).

      At $G = 950\,\text{s}^{-1}$ and $V = 10.5\,\text{m}^3$, calculate the required water power P_w from Equation 9.27.

      Water power, $P_w = \mu VG^2 = 1.519 \times 10^{-3}\,\text{N·s/m}^2 \times 10.5\,\text{m}^3 \times (950\,\text{s}^{-1})^2$

      $$= 1.44 \times 10^4\,\text{N·m/s} = 1.44 \times 10^4\,\text{W or } 14.4\,\text{kW}$$
   c. Calculate the motor output (or mixer input) power requirement.

      In general, input power requirement of a piece of equipment is calculated from the output requirement and the overall efficiency of the device. The relationship is expressed by Equation 9.33.

$$P_{input} = \frac{P_{output}}{E_{device}} \quad or \quad P_{output} = E_{device}P_{input} \tag{9.33}$$

where

P_{input} = input power requirement of a piece of equipment, N·m/s or W

P_{output} = output power requirement of a piece of equipment, N·m/s or W

E_{device} = overall efficiency of the equipment, percent

For the rapid mixer, the mixer output power is equal to the water power ($P_{output} = P_w$), and the overall efficiency is mainly determined by that of the gear box, which is typically around 90% ($E_{device} = E_{mixer} = 0.9$). Calculate the motor output (or mixer input) power requirement ($P_{input} = P_m$) from Equation 9.33.

$$\text{Motor output power, } P_m = \frac{P_w}{E_{mixer}} = \frac{14.4\,\text{kW}}{0.9} = 16.0\,\text{kW}(21.4\,\text{hp})$$

Note: 1 kW = 1.34 hp.

d. Calculate the motor wire (or input) power requirement.

Similar to the calculation for the rapid mixer, the motor wire (or input) power requirement ($P_{input} = P_{mw}$) is also calculated from Equation 9.33 at the motor output power ($P_{output} = P_m$). Assume a normal motor efficiency of 85% ($E_{device} = E_{motor} = 0.85$).

$$\text{Motor weir power, } P_{mw} = \frac{P_m}{E_{motor}} = \frac{16.0\,\text{kW}}{0.85} = 18.8\,\text{kW}$$

e. Determine the impeller size.

The rapid-mix chamber will have an *upflow* regime (Figure 9.29c). Experience has shown that the radial-flow mixers perform better than axial-flow mixer in a vertical-flow basin. The turbine mixer has four blades having $w/d = 0.16$. Provide mixer diameter ½ the width of the basin ($d =$ 0.95 m), and locate the mixer blade one diameter above the basin floor. The selected configuration will provide good mixing currents.

f. Calculate the impeller rotational speed.

Calculate the rotational speed by rearranging Equation 9.25a. The value of N_T for turbine mixer with four straight blades having $w/d = 0.16$ is interpolated from data provided in Table 9.20, $N_T =$ 2.74. At temperature of 5°C, the density of water $\rho = 999.9\,\text{g/cm}^3 \approx 1000\,\text{kg/m}^3$ (Table B.2 in Appendix B).

$$\text{Rotational speed, } n = \left(\frac{P}{\rho N_T d^5}\right)^{1/3} = \left(\frac{14.4\,\text{kN·m/s} \times 1000\,\text{N/kN} \times \text{kg·m/s}^2\text{·N}}{1000\,\text{kg/m}^2 \times 2.74 \times (0.95\,\text{m})^5}\right)^{1/3}$$

$$= 1.89\,\text{rps} = 113\,\text{rpm}$$

Check the Reynolds number for mixing regime (Equation 9.26).

$$N_R = \frac{nd^2\rho}{\mu} = \frac{1.89\,\text{s}^{-3} \times (0.95\,\text{m})^2 \times 1000\,\text{kg/m}^3}{1.518 \times 10^{-3}\,\text{N·s/m}^2 \times \text{kg.m/s}^2\text{·N}} = 1.1 \times 10^6 > 10^5$$

Since the Reynolds number is $>10^5$, turbulent mixing occurs and use of Equation 9.21a is valid.

g. Determine the gear reducer and motor size.

The mixer drive unit has a combination of speed and gear reducer with an alternating-current squirrel-cage motor. The gear reducer will regulate the mixer shaft speed to the required 113 rpm.

The output power of each motor shall be at least 26 hp (19.4 kW) to provide a safety factor of 1.2 (minimum). Each motor has contacts in the motor control center to monitor the operating status and malfunctions.

Note: The motor should also has a minimum service factor (SF) of 1.15. This means that the 26-hp motor could deliver 30-hp (22.4 kW) output for a short-term operation.

5. Design the influent and effluent structures.

a. Influent structure.

A design velocity $v = 1.2\,\text{m/s}$ is provided in the influent pipe leading to each rapid-mix unit.

$$\text{Diameter of the influent pipe, } D = \sqrt{\frac{4Q}{\pi v}} = \sqrt{\frac{4 \times 0.329\,\text{m}^3/\text{s}}{\pi \times 1.2\,\text{m/s}}} = 0.59\,\text{m} \approx 0.6\,\text{m}$$

The sluice gate has dimensions of 0.9 m × 0.9 m, and is operated from the top of the basin.

b. Effluent structure.

Total length of the effluent weir is three times the width of the chamber.

$$L_{weir} = 3W = 3 \times 1.9\,\text{m} = 5.7\,\text{m}$$

The width of the effluent launder $b = 0.6$ m
The width of exit channel $B = 0.9$ m
The length of effluent launder and weir on *each side* of the exit channel. (Figure 9.34c).

$$L_{\text{launder}} = W + 0.5(W - B) = 1.9\,\text{m} + 0.5 \times (1.9 - 0.9)\,\text{m} = 2.4\,\text{m}$$

The invert of the effluent launder is 0.6 m above the invert of the exit channel, $T = 0.6$ m.

6. Calculate the head losses through the rapid-mix unit.

The total head loss in the rapid-mix unit is the difference between water surface elevations in the influent division chamber and that in the effluent exit channel. Therefore, the hydraulic calculations are started from the effluent channel and ended at the influent division chamber. The head loss calculations are separately prepared for (a) exit point of the effluent collection launder, (b) effluent collection launder, (c) effluent rectangular weir, and (d) influent pipe.

a. Calculate the head loss at the exit point of the effluent collection launder.

Calculate the discharge per unit weir length from the rapid-mix chamber.

$$q_{\text{weir}} = \frac{Q}{L_{\text{weir}}} = \frac{0.329\,\text{m}^3/\text{s}}{5.7\,\text{m}} = 0.058\,\text{m}^3/\text{s·m}$$

Calculate the flow on each side of the launder.

$$Q_{\text{launder}} = L_{\text{launder}} q_{\text{weir}} = 2.4\,\text{m} \times 0.058\,\text{m}^3/\text{s·m} = 0.14\,\text{m}^3/\text{s·m}$$

The water depth in the exit channel leading to the flocculation basin should be regulated by the downstream Parshall flume. It is assumed that the water depth in the effluent exit channel $y_{\text{effluent}} = 0.7$ m.

Under a free-fall condition, the water depth at the exit point of the launder can normally be determined by the critical depth from Equation 7.10 (see Example 9.7). As shown in Figure 9.34d, the water depth in the effluent channel is higher than the floor of the effluent launder ($y_{\text{effluent}} > T$). The launder is submerged and free-fall condition does not exist at the exit point of the launder. The submergence will restrict the exit flow from the launder. The Villenmonte relationship has been developed to describe the impact of submerged head on discharge through various types of weirs (Figure 9.35), and is expressed by Equation 9.34.[47]

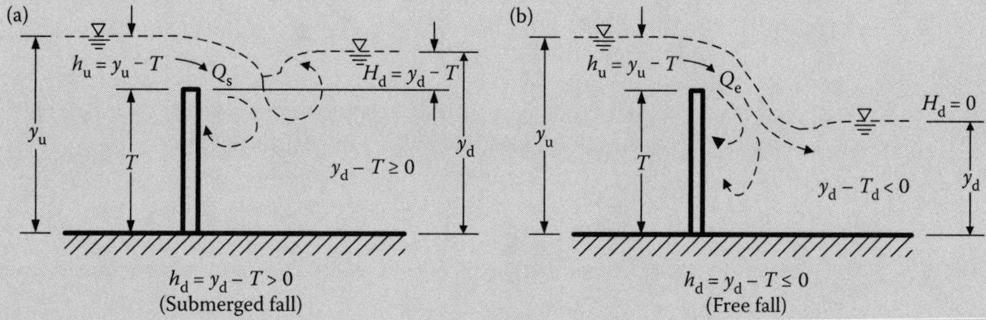

FIGURE 9.35 Discharge over submerged weir: (a) equivalent weir discharge and (b) submerged weir discharge (Example 9.36).

$$F_s = \frac{Q_s}{Q_e} \quad or \quad Q_s = F_s Q_e \quad or \quad Q_e = \frac{Q_s}{F_s} \tag{9.34a}$$

$$F_s = \left[1 - \left(\frac{h_d}{h_u}\right)^n\right]^{0.385} \tag{9.34b}$$

where

F_s = flow reduction factor, dimensionless

Q_s = submerged weir discharge at h_u under submerged fall ($h_d = y_d - T > 0$), m^3/s (ft^3/s)

Q_e = equivalent weir discharge at h_u under free (unsubmerged) fall ($h_d = y_d - T \leq 0$), m^3/s (ft^3/s). For rectangular and 90°V-notch weirs, Q_e are calculated from Equations 8.10 and 9.5b, respectively.

h_d = downstream (submerged) head over the weir, m (ft). $h_d = y_d - T$ when $y_d - T \geq 0$, and $h_d = 0$ when $y_d - T < 0$.

h_u = upstream head over the weir ($h_u = y_u - T$), m (ft)

n = constant, dimensionless

$n = 1.44$ for contracted rectangular weir

$n = 1.50$ for suppressed rectangular weir that stretches across the width of the channel

$n = 2.50$ for 90° V-notch weir

y_d = downstream water depth, m (ft)

y_u = upstream water depth, m (ft)

T = weir height above the channel bottom, m (ft)

At the exit point of the launder, critical depth equation under free-fall condition (Equation 7.10) is modified by using the Villenmonte relationship to estimate the actual outfall flow against the submerged head.

As shown in Figure 9.34d, the submerged head at the exit $h_d = y_{effluent} - T = (0.7 - 0.6)\text{ m} = 0.1\text{ m}$. The upstream head h_u is determined through trial-and-error procedure.

i. First iteration.

Calculate the critical depth under free-fall condition d_c from Equation 7.10 by assuming $C_w = 1.0$.

$$d_c = \left(\frac{Q_{launder}}{C_w b \sqrt{g}}\right)^{2/3} = \left(\frac{0.14\text{ m}^3/\text{s}}{1.0 \times 0.6\text{ m} \times \sqrt{9.81\text{ m/s}^2}}\right)^{2/3} = 0.177\text{ m}$$

Start the calculations by assuming $h_u = d_c = 0.177$ m.

Calculate the flow reduction factor from Equation 9.34b and using $n = 1.50$.

$$F_s = \left[1 - \left(\frac{h_d}{h_u}\right)^n\right]^{0.385} = \left[1 - \left(\frac{0.1\text{ m}}{0.177\text{ m}}\right)^{1.50}\right]^{0.385} = 0.808$$

Apply the flow reduction factor to calculate the equivalent free-fall flow Q_e to achieve $Q_s = Q_{launder} = 0.14\text{ m}^3/\text{s}$ against the submerged head h_d from Equation 9.34a.

$$Q_e = \frac{Q_s}{F_s} = \frac{0.14\text{ m}^3/\text{s}}{0.808} = 0.173\text{ m}^3/\text{s}$$

Calculate the critical depth at $Q_e = 0.173\text{ m}^3/\text{s}$ under free-fall condition from Equation. 7.10.

$$d_c = \left(\frac{0.173\text{ m}^3/\text{s}}{1.0 \times 0.6\text{ m} \times \sqrt{9.81\text{ m/s}^2}}\right)^{2/3} = 0.204\text{ m}$$

After the first iteration, an upstream head $h_u = d_c = 0.204$ m is required to achieve a submerged outfall flow $Q_s = Q_{launder} = 0.14\ m^3/s$.

ii. Second iteration.

Assume $h_u = 0.204$ m and repeat the same calculation steps in the first iteration.

$$F_s = \left[1 - \left(\frac{0.1\ m}{0.204\ m}\right)^{1.50}\right]^{0.385} = 0.851$$

$$Q_e = \frac{Q_s}{F_s} = \frac{0.14\ m^3/s}{0.851} = 0.165\ m^3/s$$

$$d_c = \left(\frac{0.165\ m^3/s}{1.0 \times 0.6\ m \times \sqrt{9.81\ m/s^2}}\right)^{2/3} = 0.198\ m$$

After the second iteration, $h_u = d_c = 0.198$ m.

iii. Third iteration.

Assume $h_u = 0.198$ m and repeat the same calculation steps in the first iteration.

$$F_s = \left[1 - \left(\frac{0.1\ m}{0.198\ m}\right)^{1.50}\right]^{0.385} = 0.843$$

$$Q_e = \frac{Q_s}{F_s} = \frac{0.14\ m^3/s}{0.843} = 0.166\ m^3/s$$

$$d_c = \left(\frac{0.166\ m^3/s}{1.0 \times 0.6\ m \times \sqrt{9.81\ m/s^2}}\right)^{2/3} = 0.198\ m$$

After the third iteration, $h_u = d_c = 0.198$ m. It is the same value obtained from the second iteration. Therefore, an approximate upstream head of 0.20 m is required to push a flow of $0.14\ m^3/s$ against a submerged head of 0.1 m at the exit point of the launder. The overall head loss at the exit point $\Delta h_{exit} = h_u - h_d = (0.20 - 0.1)\ m = 0.1$ m.

b. Calculate the head loss in the effluent collection launder.

Water surface profile in a launder receiving flow from a free-falling weir is estimated from Equation 8.13a. Assuming $y_2 = h_u = 0.20$ m, calculate the water depth y_1 at the upper end of the trough without channel friction loss.

$$y_1 = \sqrt{y_2^2 + \frac{2Q^2}{gb^2y_2}} = \sqrt{(0.20\ m)^2 + \frac{2 \times (0.14\ m^3/s)^2}{9.81\ m/s^2 \times (0.6\ m)^2 \times 0.20\ m}} = 0.31\ m$$

Add 30% for losses due to friction, turbulence, and 90° bend in the launder.

Total depth of water upstream of the launder $y_1' = 0.31\ m \times 1.3 = 0.40$ m

Add freeboard $FB_{weir} = 0.26$ m from the weir crest to the water surface upstream of the launder.

The total height of effluent launder $H_{launder} = y_1' + FB_{weir} = (0.40 + 0.26)\ m = 0.66$ m

The overall head loss in the launder $\Delta h_{launder} = y_1' - y_2 = (0.40 - 0.20)\ m = 0.20$ m.

c. Calculate the head loss at the effluent rectangular weir.

The head loss at the effluent weir consists of (a) head over the weir and (b) free fall after the weir into the effluent launder.

Calculate the head over the weir h_{weir} from Equation 8.10 at $Q = 0.329\ m^3/s$, $C_d = 0.6$, $L_{weir} = 5.7$ m, and $n = 0$.

$$L' = L_{weir} - 0.1\,nH = L_{weir} = 5.7\,\text{m}$$

$$h_{weir} = \left(\frac{3}{2} \times \frac{Q}{C_d L' \sqrt{2g}}\right)^{2/3} = \left(\frac{3}{2} \times \frac{0.329\,\text{m}^3/\text{s}}{0.6 \times 5.7\,\text{m} \times \sqrt{2 \times 9.81\,\text{m/s}^2}}\right)^{2/3} = 0.10\,\text{m}$$

Head over the effluent weir $h_{weir} = 0.10$ m. The weir crest elevation will be 0.10 m lower than the water surface in the rapid-mix chamber.

The overall head loss at the effluent weir $\Delta h_{weir} = h_{weir} + FB_{weir} = (0.10 + 0.26)\ \text{m} = 0.36$ m.

d. Calculate the head loss through the influent pipe.

$$\text{Actual velocity through the influent pipe, } v = \frac{Q}{\pi/4 \times D^2} = \frac{0.329\,\text{m}^3/\text{s}}{\pi/4 \times (0.60\,\text{m})^2} = 1.16\,\text{m/s}$$

Friction loss in the influent pipe is small because of small length and is ignored. Only minor losses are considered. The minor losses are calculated from Equation 6.15b. The minor losses are due to (a) entrance that is controlled by a sluice gate ($K = 0.7$), (b) one 90° elbow ($K = 0.4$), and (c) an exit loss ($K = 1.0$). The overall minor head loss coefficient $K = 0.7 + 0.4 + 1.0 = 2.1$.

$$h_m = K\frac{v^2}{2g} = 2.1 \times \frac{(1.16\,\text{m/s})^2}{2 \times 9.81\,\text{m/s}^2} = 0.14\,\text{m}$$

The head loss through the influent pipe $\Delta h_{influent} = h_m = 0.14$ m. The water surface elevation in the rapid-mix chamber will be 0.14 m lower than that in that in the influent division chamber.

7. Determine major elevations and prepare hydraulic profile through the rapid-mix unit.

Assume the reference elevation is the bottom of the influent division chamber, $EL_{influent} = 0.00$ m.

a. Prepare major water surface elevations (WSELs).

WSEL in the influent diversion chamber $y_{influent} = 4.00$ m

WSEL in the influent diversion chamber

$WSEL_{influent} = EL_{influent} + y_{influent} = (0.00 + 4.00)\ \text{m} = 4.00$ m

Head loss through the influent pipe $\Delta h_{influent} = 0.14$ m

WSEL in the rapid-mix chamber

$WSEL_{mix} = WSEL_{influent} - \Delta h_{influent} = (4.00 - 0.14)\ \text{m} = 3.86$ m

Overall head loss at the effluent weir $\Delta h_{weir} = 0.36$ m

WSEL at upper end of the effluent launder

$WSEL_{launder,upper} = WSEL_{mix} - \Delta h_{weir} = (3.86 - 0.36)\ \text{m} = 3.50$ m

Overall head loss in the launder $\Delta h_{launder} = 0.20$ m

WSEL at lower end of the effluent launder

$WSEL_{launder,lower} = WSEL_{launder,upper} - \Delta h_{launder} = (3.50 - 0.20)\ \text{m} = 3.30$ m

Overall head loss at the exit point of the effluent launder $\Delta h_{exit} = 0.10$ m

WSEL in the effluent channel

$WSEL_{effluent} = WSEL_{launder,lower} - \Delta h_{exit} = (3.30 - 0.10)\ \text{m} = 3.20$ m

b. Determine elevations (ELs) of the major components.

Water depth in the rapid-mix chamber $H = 2.90$ m.

EL of the rapid-mix chamber bottom

$EL_{mix} = WSEL_{mix} - H = (3.86 - 2.90)\ \text{m} = 0.96$ m

Head over the effluent weir $h_{weir} = 0.10$ m

Top of weir elevation TOW = $\text{WSEL}_{mix} - h_{weir} = (3.86 - 0.10)$ m = 3.76 m

Total height of effluent launder $H_{launder} = 0.66$ m

EL of the effluent launder invert

$\text{EL}_{launder} = \text{TOW} - H_{launder} = (3.76 - 0.66)$ m = 3.10 m

Water depth in the effluent exit channel $y_{effluent} = 0.70$ m

EL of the effluent channel invert

$\text{EL}_{effluent} = \text{WSEL}_{effluent} - y_{effluent} = (3.20 - 0.70)$ m = 2.50 m

c. Draw the hydraulic profile and show the important design elevations of the rapid-mix unit.

The hydraulic profile and important design elevations in rapid-mix basins are shown in Figures 9.34b and d.

EXAMPLE 9.37: POWER REQUIREMENT AND PADDLE DIMENSIONS OF A PADDLE FLOCCULATOR

A paddle flocculator has a total of four paddles and two paddles on each side of the shaft. The paddles rotate on a horizontal plane at a speed of 3.6 rpm. The mid-width distances of the paddles from the center of the shaft on both sides of the shaft are 2.5 m and 1.5 m, respectively. The flocculation basin volume is 150 m^3, and the velocity gradient G is 45 s^{-1}. The coefficient of drag C_D of paddles moving perpendicular to the fluid is 1.35. The experience has shown that the velocity of fluid is 0.75 times the paddle velocity. Calculate the power requirement, and paddle area and dimensions. Assume that the dynamic viscosity and bulk density of the fluid are 1.139×10^{-3} N·s/m^2 and 999.1 kg/m^3, respectively.

Solution

1. Draw the definition sketch.

   The definition sketch is shown in Figure 9.36.

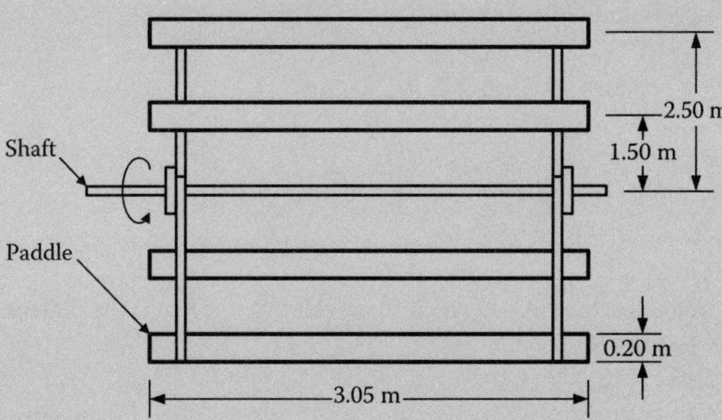

FIGURE 9.36 Definition sketch of flocculator (Example 9.37).

2. Calculate the water power requirement.

   Calculate P from Equation 9.27 at $\mu = 1.139 \times 10^{-3}$ N·s/m^2, $V = 150$ m^3, and $G = 45$ s^{-1}.

   Water power, $P = \mu V G^2 = 1.139 \times 10^{-3}\ \text{N·s/m}^2 \times 150\ \text{m}^3 \times (45\ \text{s}^{-1})^2 = 346\ \text{N·m/s}$ or $346\ \text{kg·m}^2/\text{s}^3$

3. Calculate the paddle area.

There are two paddles on each sides of the shaft at $r_1 = 1.50$ m and $r_2 = 2.50$ m. Assume the area of each paddle $= a$. The total paddle area $A = 2a$ at each radius. Calculate the paddle area from Equation 9.32.

$$P = \sum_{i=1}^{2} P_i = \sum_{i=1}^{2} \frac{C_D \rho A v_i^3}{2} = \frac{1}{2} C_D \rho (2a)\left(v_1^3 + v_2^3\right)$$

At radius of r, the absolute paddle velocity at the shaft rotational speed n, $v_{abs} = 2\pi r n$

The velocity of paddle relative to the water, $v = 0.75 v_{abs} = 0.75 \times (2\pi r n)$

$$P = \frac{1}{2} C_D \rho (2a)[(0.75(2\pi r_1)n)^3 + (0.75(2\pi r_2)n)^3]$$

$$= \frac{1}{2} C_D \rho (2a) \times (0.75(2\pi)n)^3 \times (r_1^3 + r_2^3)$$

Substitute $C_D = 1.35$, $\rho = 999.1$ kg/m^3, $n = 3.6$ rev/min/60 s/min $= 0.06$ rps , $r_1 = 2.50$ m, and $r_2 = 1.50$ m into the above equation and solve a at $P = 346$ kg·m^2/s^3.

$$P = \frac{1}{2} \times 1.35 \times 999.1\,\text{kg/m}^3 \times 2a \times (0.75 \times 2 \times \pi \times 0.06\,\text{s}^{-1})^3 \times ((2.50\,\text{m})^3 + (1.50\,\text{m})^3)$$

$$= 1349\,\text{kg/m}^3 \times a \times 0.0226\,\text{s}^{-3} \times 19.0\,\text{m}^3$$

$$= 579\,\text{kg/s}^3 \times a$$

$$346\,\text{kg·m}^2/\text{s}^3 = 579\,\text{kg/s}^3 \times a$$

$$a = 0.60\,\text{m}^2$$

4. Determine the paddle dimension.

At $C_D = 1.35$, a L/W ratio of 15 is estimated from Figure 9.32.

$$a = LW = (15W) \times W = 15W^2 = 0.60\,\text{m}^2$$

$$\text{Desired paddle width, } W = \sqrt{\frac{0.60\,\text{m}^2}{15}} = \sqrt{0.04\,\text{m}^2} = 0.2\,\text{m}$$

Desired paddle length, $L = 15\ W = 15 \times 0.2$ m $= 3$ m

Provide four paddles each 3.05 m long ×0.20 m wide (10 ft × 8 in). The dimensions of the paddle are also shown in Figure 9.36.

EXAMPLE 9.38: DESIGN OF A PADDLE FLOCCULATION BASIN

Design a flocculation basin that is an integral part of a primary clarifier. The separating partition is a 16.7-m-wide concrete diffusion wall. The flocculation basin has three stages. Each stage has 10-min detention time (total 30 min), and velocity gradient G in the first, second, and third stages are 60, 30, and 15 s^{-1}. The design flow to the basin is 28,400 m^3/d (7.5 MGD). A Parshall flume is located at the head of the influent distribution channel of the flocculation basin. The purposes of the Parshall flume are to measure the flow of coagulated water and mix a cationic polymer (coagulant aid) in the turbulence zone at the flume. Use μ

$= 1.519 \times 10^{-3}$ N·s/m^2 and $\rho \approx 1000$ g/cm^3 at the critical temperature of 5°C under the field condition (Table B.2 in Appendix B).

Solution

The design of a flocculation basin involves (a) determination of basin volume and dimensions, (b) selection of flocculators, (c) design of influent and effluent structures, (d) performing hydraulic calculations, and (e) preparing unit layout and hydraulic profile.

1. Determine the basin volume and dimensions.
   a. Calculate the volume of each stage of the flocculation basin.

$$\text{Total design flow to the basin, } Q = 28{,}400\,\text{m}^3/\text{d} \times \frac{\text{d}}{(60 \times 60 \times 24)\,\text{s}} = 0.329\,\text{m}^3/\text{s}$$

Calculate the flocculation basin volume required at the detention time of $t = 30$ min.

$$\text{Total volume of the basin, } V_{\text{basin}} = Qt = 0.329\,\text{m}^3/\text{s} \times (30\,\text{s} \times 60\,\text{s/min}) = 592\,\text{m}^3$$

Calculate the volume required in each stage.

$$\text{Volume of each stage, } V_{\text{stage}} = \frac{1}{3} \times V_{\text{basin}} = \frac{1}{3} \times 592\,\text{m}^3 = 197\,\text{m}^3$$

   b. Calculate the dimension of each stage.

The length of the flocculation basin is equal to the width of the diffusion wall with the clarifier.

Length of the flocculation basin (or stage), $L_{\text{basin}} = L_{\text{stage}} = 16.7$ m (perpendicular to the flow direction)

Assume an approximate depth-to-width ratio of 0.9:1 ($H_{\text{stage}} = 0.9\ W_{\text{stage}}$) in each stage of the flocculation basin.

$$\begin{aligned}\text{Volume of each stage, } V_{\text{stage}} &= L_{\text{stage}} \times W_{\text{stage}} \times H_{\text{stage}} = 16.7\,\text{m} \times W_{\text{stage}} \times (0.9W_{\text{stage}}) \\ &= 15.0\,\text{m} \times (W_{\text{stage}})^2 = 197\,\text{m}^3\end{aligned}$$

$$\text{Desired stage width, } W_{\text{stage}} = \sqrt{\frac{197\,\text{m}^3}{15.0\,\text{m}}} = 3.62\,\text{m}$$

$$\text{Desired water depth, } H_{\text{stage}} = 0.9 \times W_{\text{stage}} = 0.9 \times 3.62\,\text{m} = 3.26\,\text{m}$$

Provide each flocculator stage with the following dimensions: 16.7 m long × 3.65 m wide × 3.25 m deep (minimum).

Total volume of each stage,

$$V_{\text{stage}} = L_{\text{stage}} \times W_{\text{stage}} \times H_{\text{stage}} = 16.7\,\text{m} \times 3.65\,\text{m} \times 3.25\,\text{m} = 198\,\text{m}^3$$

The adjacent two flocculator stages are partitioned by a slotted stainless steel (SS) or fiberglass-reinforced plastic (FRP) baffle. Reserve a spacing of 15 cm (6 in) for installation of each baffle wall ($S_{\text{baffle}} = 0.15$ m).

Total width of the flocculation basin,

$$W_{\text{basin}} = 3 \times W_{\text{stage}} + 2 \times S_{\text{baffle}} = 3 \times 3.65\,\text{m} + 2 \times 0.15\,\text{m} = 11.25\,\text{m}$$

The bottom floor is slopped at ~1.5% from the first to third stage.,

Total floor elevation drop at the desired slope, $h_{slope} = 1.5\% \times W_{basin} = 0.015 \times 11.25\,m = 0.17\,m$

Provide a total floor elevation drop $\Delta h_{basin} = 0.20$ m, which gives drop of ~0.07 m per stage. The minimum water depth of 3.25 m is provided at the beginning of the first stage. The maximum water depth of 3.45 m is therefore provided at the end of the third stage. The volume increase due to the slope is small and ignored in process calculations. General layout of the flocculator is shown in Figure 9.37a.

2. Selection of the flocculators.

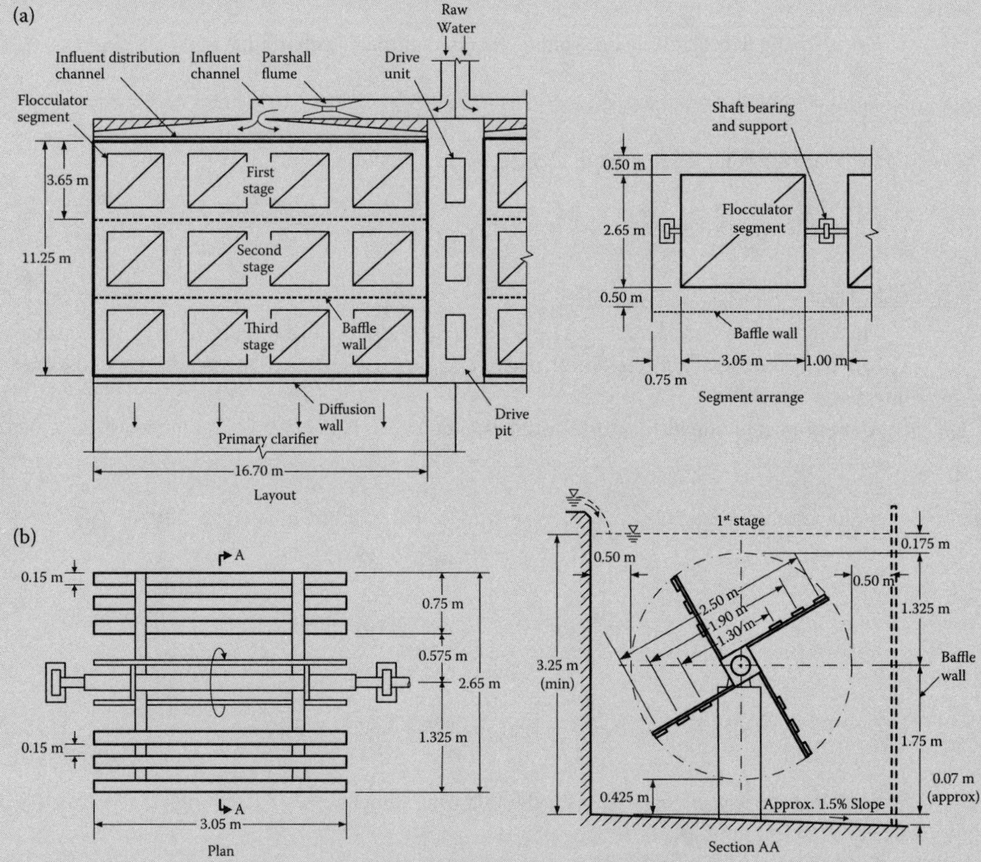

FIGURE 9.37 Details of flocculation basin arrangement and paddle wheel design: (a) flocculator layout and segment arrangement and (b) paddle wheel design details with dimensions (Example 9.38).

a. Determine the mixing requirements.
   i. Determine the power requirement in each stage

      The velocity gradients of $G_1 = 60\ s^{-1}$, $G_2 = 30\ s^{-1}$, and $G_3 = 15\ s^{-1}$ are given in the first, second, and third stages, respectively.
   ii. Calculate the mixing opportunity parameter in each stage.

Calculate the mixing opportunity parameter at the detention time of 10 min in each stage.

$$\text{Mixing opportunity parameter in the first stage, } Gt_1 = 60\,\text{s}^{-1} \times (10\,\text{min} \times 60\,\text{s/min}) = 36{,}000$$

$$\text{Mixing opportunity parameter in the second stage, } Gt_2 = 30\,\text{s}^{-1} \times (10\,\text{min} \times 60\,\text{s/min}) = 18{,}000$$

$$\text{Mixing opportunity parameter in the third stage, } Gt_3 = 15\,\text{s}^{-1} \times (10\,\text{min} \times 60\,\text{s/min}) = 9000$$

iii. Calculate the overall mixing requirements in the flocculation basin.

$$\text{Overall mixing opportunity parameter, } Gt = 36{,}000 + 18{,}000 + 9000 = 63{,}000$$

$$\text{Overall velocity gradients,} \quad G = \frac{63{,}000}{30\,\text{min} \times 60\,\text{s/min}} = 35\,\text{s}^{-1}$$

b. Calculate the flocculator power requirement.

The power requirement for each stage is calculated separately because the values of velocity gradient are gradually reduced from the first to third stage.

i. Calculate the power requirements for the first stage.

The power imparted to the water in the first stage P_w is calculated from Equation 9.27 at $\mu = 1.519 \times 10^{-3}\,\text{N·s/m}^2$, $V_{stage} = 198\,\text{m}^3$, and $G_1 = 60\,\text{s}^{-1}$.

$$\text{Water power, } P_w = \mu\, V_{stage}(G)^2 = 1.519 \times 10^{-3}\,\text{N·s/m}^2 \times 198\,\text{m}^3 \times (60\,\text{s}^{-1})^2$$
$$= 1083\,\text{N·m/s} = 1083\,\text{W or } 1.1\,\text{kW}$$

The flocculator output power is equal to the water power ($P_{output} = P_w$). For the flocculator, the overall efficiency can be estimated from the efficiencies of the gear box (E_{gear}) and the flocculator bearings ($E_{bearings}$).

$$\text{Overall efficiency of the flocculator, } E_{device} = E_{gear} \times E_{bearings}$$

Assume a gear box efficiency of 90% ($E_{gear} = 0.9$) and an efficiency of 70% for the bearings ($E_{bearings} = 0.7$). Calculate the motor output (or flocculator input) requirement ($P_{input} = P_m$) from Equation 9.33.

$$\text{Motor output power, } P_m = \frac{P_w}{E_{gear} \times E_{bearings}} = \frac{1.1\,\text{kW}}{0.9 \times 0.7} = 1.7\,\text{kW}(2.3\,\text{hp})$$

Assume a motor efficiency of 85% ($E_{device} = E_{motor} = 0.85$) and calculate the motor wire (or input) power requirement ($P_{input} = P_{mw}$) from Equation 9.33 at $P_{output} = P_{mw}$.

$$\text{Motor weir power, } P_{mw} = \frac{P_m}{E_{motor}} = \frac{1.7\,\text{kW}}{0.85} = 2.0\,\text{kW}$$

ii. Calculate the power requirements for the second and third stages.

The power requirements for the second and third stages are calculated in similar steps used for the first stage. The calculation results are summarized in Table 9.23.

TABLE 9.23 Mixing Requirements, Power Requirements, and Rotational Speed and Peripheral Velocity of the Paddle Wheels (Example 9.38)

Design Parameters	Design Value		
	Stage 1	Stage 2	Stage 3
Mixing requirements			
Velocity gradient G, s^{-1}	60	30	15
Mixing opportunity parameter Gt, dimensionless	36,000	18,000	9000
Power requirements			
Power imparted to water P_w, kN·m/s (or kW)	1.1	0.27	0.068
Motor output power requirement P_m, kN·m/s (or kW)	1.7	0.43	0.11
Motor wire power requirement P_{mw}, kN·m/s (or kW)	2.0	0.50	0.13
Flocculator rotational requirements			
Shaft rotational speed n, rpm	5.1	3.2	2.0
Tip speed v_{tip}, m/s	0.71	0.45	0.28
Range of peripheral velocity of the paddles v, m/s	0.26–0.50	0.16–0.31	0.10–0.20

c. Determine the number of segments and paddle wheel dimensions.

i. Determine the number of segments.

The paddle wheels are built in segments to facilitate construction, shipping, and installation. Select four segments (or wheels) to be accommodated in 16.7 m length of the flocculation basin. Provide a clearance of 0.75 m (30 in) between the end segment and the side wall of the basin, and a clearance of 1 m between two adjacent segments (Figure 9.37a). Calculate the length of the paddle blade.

$$L_{segment} = \frac{1}{4} \times (L_{stage} - 2 \times 0.75\,\text{m} - 3 \times 1\,\text{m}) = \frac{1}{4} \times (16.7\,\text{m} - 2 \times 0.75\,\text{m} - 3 \times 1\,\text{m})$$
$$= 3.05\,\text{m}$$

ii. Select the paddle blade dimensions.

Select paddle blade dimensions of 3.05 m (10 ft) long ($L = L_{segment} = 3.05$ m), and 15 cm (6 in) wide ($W = 0.15$ m) at a length-to-wide (L/W) ratio of approximately 20:1.

iii. Determine the paddle wheel dimensions.

Provide a clearance of 0.5 m (20 in) on both sides from either a sidewall or baffle wall and calculate the paddle wheel outer diameter at the tip (Figure 9.37a).

$$d_{outer} = W_{stage} - 2 \times 0.5\,\text{m} = (3.65 - 2 \times 0.5)\,\text{m} = 2.65\,\text{m}$$

Select 12 paddles per paddle wheel (6 on 2 perpendicular planes) and 3 paddles on each side of the shaft. Provide 15 cm (6 in) clearance between the paddles. Segment arrangement in the flocculator is also illustrated in Figure 9.37a. Calculate the dimension between the edge of the inner most paddle.

Dimension between the inner edges, $d_{inner} = d_{outer} - (6 \times 0.15\,\text{m} + 4 \times 0.15\,\text{m})$

$$= (2.65 - 1.5)\,\text{m} = 1.15\,\text{m}$$

Calculate the diameter (or center to center distance) between each pair of paddles:

Diameter of outer paddles, $d_1 = d_{\text{outer}} - 0.15\,\text{m} = (2.65 - 0.15)\,\text{m} = 2.50\,\text{m}$

Diameter of middle paddles, $d_2 = d_1 - 4 \times 0.15\,\text{m} = (2.50 - 4 \times 0.15)\,\text{m} = 1.90\,\text{m}$

Diameter of inner paddles, $d_3 = d_2 - 4 \times 0.15\,\text{m} = (1.90 - 4 \times 0.15)\,\text{m} = 1.30\,\text{m}$

Provide a minimum height of 1.75 m for the shaft above the floor of the flocculation basin. Calculate the minimum clearance below the bottom of the paddle wheel above the basin floor.

$$\text{Minimum clearance below the paddle wheel} = 1.75\,\text{m} - \frac{1}{2} \times d_{\text{tip}} = 1.75\,\text{m} - \frac{1}{2} \times 2.65\,\text{m}$$
$$= 0.425\,\text{m}\,(\text{about } 17\,\text{in})$$

Calculate the submergence at the top of the paddle wheel below the design water surface.

$$\begin{aligned}\text{Submergence of the paddle wheel,} &= H_{\text{stage}} - (d_{\text{outer}} + 0.425\,\text{m})\\ &= 3.25\,\text{m} - (2.65 + 0.425)\,\text{m}\\ &= 0.175\,\text{m or (about 7 in)}\end{aligned}$$

The paddle wheel layout and dimensions are shown in Figure 9.37b.

d. Calculate the rotational requirements of the paddle wheel.

In each stage, there are four paddle wheel segments and four paddle blades at each diameter per segment. Calculate the total area (A) at each diameter per stage.

$$\begin{aligned}A &= (4\,\text{segments/stage}) \times (4\,\text{blades/segment}) \times LW\\ &= 16\,\text{blades/stage} \times (3.05\,\text{m} \times 0.15\,\text{m})\,\text{per blade}\\ &= 7.32\,\text{m}^2\,\text{per stage}\end{aligned}$$

At the L/W ratio of 20:1, C_D of 1.5 is obtained from Table 9.22.

At each diameter, the absolute paddle velocity at the shaft rotational speed n, $v_{\text{abs}} = \pi dn$

Experience has shown that the velocity of paddle relative to the water is 75% of the absolute rotational velocity of the paddle, $v = 0.75\,v_{\text{abs}} = 0.75 \times (\pi dn) = 0.75 \times (\pi n)d$

The rotational speed in each stage is calculated from Equation 9.32 at $d_1 = 2.50$ m, $d_2 = 1.90$ m, and $d_3 = 1.30$ m.

$$\begin{aligned}P_w &= \sum_{i=1}^{3} P_{w,i} = \sum_{i=1}^{3} \frac{C_D \rho A v_i^3}{2} = \frac{1}{2} C_D \rho A \left(v_1^3 + v_2^3 + v_3^3\right)\\ &= \frac{1}{2} C_D \rho A (0.75 \times \pi n)^3 \times \left(d_1^3 + d_2^3 + d_3^3\right)\end{aligned}$$

Substitute $C_D = 1.5, \rho = 1000\,\text{kg/m}^3, A = 7.32\,\text{m}^2, d_1 = 2.50\,\text{m}, d_2 = 1.90\,\text{m}$, and $d_3 = 1.30$ m into the above equation.

$$\begin{aligned}P_w &= \frac{1}{2} \times 1.5 \times 1000\,\text{kg/m}^3 \times 7.32\,\text{m}^2 \times (0.75 \times \pi \times n)^3 \times ((2.50\,\text{m})^3 + (1.90\,\text{m})^3 + (1.30\,\text{m})^3)\\ &= 71{,}813\,\text{kg/m} \times n^3 \times 24.7\,\text{m}^3\\ &= 1{,}770{,}000\,\text{kg·m}^2 \times n^3\end{aligned}$$

Similar to the calculations of the power requirement, the rotation speed is separately calculated for each stage.

i. Calculate the rotational requirements in the first stage.

Substitute $P_w = 1083$ N·m/s $= 1083$ kg·m^2/s^3 into the above equation and solve for rotation speed n in the first stage.

$$1083\,\text{kg·m}^2/\text{s}^3 = 1{,}770{,}000\,\text{kg·m}^2 \times n^3$$

$$n = \left(\frac{1083\,\text{kg·m}^2/\text{s}^3}{1{,}770{,}000\,\text{kg·m}^2}\right)^{1/3} = 0.085\,\text{s}^{-1} \quad \text{or} \quad 0.085\,\text{rps}$$

$$n = 0.085\,\text{rps} \times 60\,\text{s/min} = 5.1\,\text{rpm}$$

Calculate the tip speed of the paddle wheel.

$$v_{\text{tip}} = \pi d_{\text{outer}}\, n = \pi \times 2.65\,\text{m} \times 0.085\,\text{rps} = 0.71\,\text{m/s}$$

Calculate the peripheral velocity of paddle relative to the water at each diameter.

$$v_1 = 0.75(\pi d_1)n = 0.75 \times \pi \times 2.50\,\text{m} \times 0.085\,\text{rps} = 0.50\,\text{m/s}$$

$$v_2 = 0.75(\pi d_2)n = 0.75 \times \pi \times 1.90\,\text{m} \times 0.085\,\text{rps} = 0.38\,\text{m/s}$$

$$v_3 = 0.75(\pi d_3)n = 0.75 \times \pi \times 1.30\,\text{m} \times 0.085\,\text{rps} = 0.26\,\text{m/s}$$

ii. Calculate the rotational requirements for the second and third stages.

The powers imparted to the water (P_w) are 271 and 68 N·m/s in the second and third stages. The rotational requirements for the second and third stages are calculated in similar steps used in the first stage. The calculation results are summarized in Table 9.23.

3. Design of influent and effluent structures.

a. Influent structure.

The influent structure of the flocculation basin consists of a Parshall flume, an influent channel, an influent distribution channel, and influent distribution side-weirs. The coagulated water is conveyed through a channel from the rapid-mix basin to the Parshall flume. The flow rate is metered through the flume and a free-flow condition is provided at the flume. The turbulence created by the hydraulic jump at the free fall of the flume mixes the cationic polymer (coagulant aid) with the coagulated water. The purpose of the polymer is to enlarge and toughen the pinhead floc in the coagulated water. The metered water flows into a short influent channel that is 0.85 m wide and ~0.9 m deep. The influent channel has one 90° turn and a flow split for additional polymer mixing prior to interring the influent distribution channel. The influent distribution channel is tapered with initial and final widths of 1 and 0.2 m, respectively. It provides a nearly constant velocity in the distribution channel. Four 3.8-m long straight side-weirs parallel to the channel distribute water evenly into the first stage of the paddle wheel segment. The flocculator layout and details are also shown in Figure 9.36a.

b. Effluent structure.

A simple concrete baffle wall with circular ports (called also *diffusion wall*) is provided at the end of the third stage of flocculation basin. It separates the flocculation basin from the clarifier. The purpose of the wall is to distribute the flocculated water evenly and to dissipate the kinetic energy. The port diameter is generally less than the thickness of the wall. To prevent the breakup of the floc, the optimum head loss through the ports is usually 2–3 mm. The allowable head loss through the orifice may be calculated from the kinetic energy exerted by the flocculator paddle in the third stage of the flocculator. A maximum velocity of 0.15 m/s through the ports may also be used to size the orifice.[14,18]

c. Diffusion wall.

The tip speed of the paddle wheel $v_{tip} = 0.28\,\text{m/s}$ in the third stage.

Calculate the kinetic energy h_k exerted by the flocculator paddle from Equation 6.15a.

$$h_k = \frac{(v_{tip})^2}{2g} = \frac{(0.28\,\text{m/s})^2}{2 \times 9.81\,\text{m/s}^2} = 0.004\,\text{m or } 4\,\text{mm}$$

Assume that the head loss through the orifice is equal to the kinetic energy $h_{port} = h_k = 0.004$ m. Calculate the velocity through the port (v_{port}) from orifice equation (Equation 7.4b) and assume $C_d = 0.6$.

$$v_{port} = C_d\sqrt{2gh_{port}} = 0.6 \times \sqrt{2 \times 9.81\,\text{m/s}^2 \times 0.004\,\text{m}} = 0.17\,\text{m/s}$$

To be conservative, use the maximum allowable velocity 0.15 m/s (<0.17 m/s) through the port for sizing the orifices on the diffusion wall, $v_{port} = 0.15$ m/s.

Calculate the total port area at the total design flow to the basin $Q = 0.329\ \text{m}^3/\text{s}$.

$$\text{Total port area, } A_{port} = \frac{Q}{v_{port}} = \frac{0.329\,\text{m}^3/\text{s}}{0.15\,\text{m/s}} = 2.19\,\text{m}^2$$

Provide 15 cm (6 in) diameter ports constructed by HDPE wall sleeves. Assume $d_{port} = 0.15$ m and calculate the area of each port.

$$\text{Area of each port, } a_{port} = \frac{\pi}{4}(d_{port})^2 = \frac{\pi}{4} \times (0.15\,\text{m})^2 = 0.0177\,\text{m}^2\text{per port}$$

$$\text{Total number of ports, } = N_{port} = \frac{A_{port}}{a_{port}} = \frac{2.19\,\text{m}^2}{0.0177\,\text{m}^2/\text{port}} = 124\ \text{ports}$$

Provide 125 ports in 5 rows and 25 ports on each row. The design details of the diffusion wall are provided in Figure 9.38.

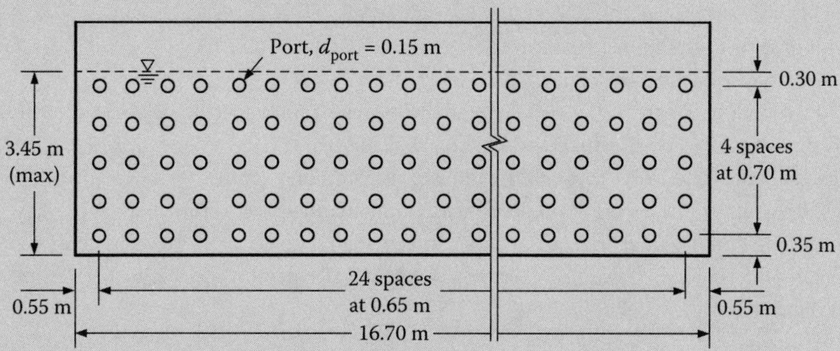

FIGURE 9.38 Design details of diffusion wall separating flocculation basin and clarifier (Example 9.38).

4. Perform hydraulic calculations of influent and effluent structures.

The head loss calculations are separately prepared for (a) effluent diffusion wall, (b) flocculation basin, (c) influent distribution weirs, (d) influent distribution channel, (e) influent channel, and (f) Parshall flume.

a. Calculate the head loss across the ports on the effluent diffusion wall.

The head loss is calculated from Equation 7.4b at $C_d = 0.6$ and $v_{port} = 0.15$ m/s.

$$\Delta h_{port} = \frac{1}{2g}\left(\frac{v_{port}}{C_d}\right)^2 = \frac{1}{2 \times 9.81\ \mathrm{m/s^2}} \times \left(\frac{0.15\ \mathrm{m/s}}{0.6}\right)^2 = 0.0032\ \mathrm{m\ or\ 3.2\ mm}$$

The head loss is small and ignored for hydraulic profile, $\Delta h_{port} = 0.00$ m.

b. Calculate the head losses through the flocculation basin.

In the flocculation basin, the head losses may be caused by the baffle walls separating the stages. They are normally small and ignorable, $\Delta h_{basin} = 0.00$ m.

c. Calculate the head loss at the influent distribution weirs.

The head loss at the influent distribution weirs consists of (a) head over the weir and (b) free fall after the weirs into the first stage of the flocculation basin.

Four straight weirs are provided on the sidewall of the influent distribution channel. The length of each weir plate $L_{weir} = 3.8$ m.

Calculate the discharge over each weir.

$$q_{weir} = \frac{Q}{N_{weir}} = \frac{0.329\ \mathrm{m^3/s}}{4\ \mathrm{weir\ plate}} = 0.0823\ \mathrm{m^3/s\ per\ weir\ plate}$$

Calculate the head over the weirs h_{weir} from Equation 8.10 at $C_d = 0.6$, and $n = 2$.

$$q_{weir} = \frac{2}{3} \times C_d L' \sqrt{2g(h_{weir})^3}$$

Assume $L' = 3.79$ m.

$$h_{weir} = \left(\frac{3}{2} \times \frac{q_{weir}}{C_d L' \sqrt{2g}}\right)^{2/3} = \left(\frac{3}{2} \times \frac{0.0823\ \mathrm{m^3/s}}{0.6 \times 3.79\ \mathrm{m} \times \sqrt{2 \times 9.81\ \mathrm{m/s^2}}}\right)^{2/3} = 0.05\ \mathrm{m}$$

Check: $L' = L_{weir} - 0.1\, n h_{weir} = 3.80\ \mathrm{m} - 0.1 \times 2 \times 0.05\ \mathrm{m} = 3.79\ \mathrm{m}$

It is same as the initial assumption.

Head over the effluent weir $h_{weir} = 0.05$ m.

Provide a freeboard $FB_{weir} = 0.20$ m after the weir. The overall head loss at the effluent weir $\Delta h_{weir} = h_{weir} + FB_{weir} = (0.05 + 0.20)\ \mathrm{m} = 0.25$ m.

d. Calculate the head losses through the influent distribution channel.

The width of the tapering influent distribution channel varies from 1 m ($w_{distyribution}$) at the beginning and 0.2 m at the end in order to provide relatively even-flow distribution along the length of the channel. Assume a relatively constant water depth $y_{distribution} = 0.9$ m is maintained in the channel.

$$\text{Initial velocity in influent distribution channel, } v_{distribution} = \frac{Q}{w_{distribution} \times y_{distribution}}$$

$$= \frac{0.329\ \mathrm{m^3/s}}{1\ \mathrm{m} \times 0.9\ \mathrm{m}} = 0.37\ \mathrm{m/s}$$

Both frictional and minor head losses are small and ignorable because of the low velocity in a channel without turns, $\Delta h_{distribution} = 0.00$ m.

e. Calculate the head losses through the influent channel.

The influent channel is 0.85-m wide ($w_{influent}$) and 0.9-m deep ($y_{influent,lower}$) at the lower end of the channel. The frictional head loss is small and ignorable in a short channel. The head losses may be caused by the turbulences at one 90° turn and the flow split at the entrance of the influent distribution channel (Figure 9.37a).

$$\text{Velocity in influent channel } v_{influent} = \frac{Q}{w_{influent} \times y_{influent,lower}} = \frac{0.329\,\text{m}^3/\text{s}}{0.85\,\text{m} \times 0.9\,\text{m}} = 0.43\,\text{m/s}$$

Calculate the minor head losses at the turns from Equation 6.15b using minor head loss coefficients: $K_m = 1.5$ at the 90° turn and $K_m = 2.0$ for the flow split.

$$h_m = K_m \frac{(v_{influent})^2}{2g} = (1.5 + 2.0) \times \frac{(0.43\,\text{m/s})^2}{2 \times 9.81\,\text{m/s}^2} = 0.03\,\text{m}$$

Therefore, the overall head loss in the influent channel $\Delta h_{influent} = 0.03$ m.

The water depth at the upper end of the channel is $y_{influent,upper} = y_{influent,lower} + \Delta h_{influent} = (0.90 + 0.03)\,\text{m} = 0.93$ m.

f. Calculate the head losses at the Parshall flume.

Flow through a Parshall flume is expressed in the U.S. customary unit by Equation 7.9.

$$Q = CH_a^n$$

Convert the design flow from m^3/s to ft^3/s.

$$Q = 0.329\,\text{m}^3/\text{s} \times \frac{\text{ft}^3/\text{s}}{0.0283\,\text{m}^3/\text{s}} = 11.6\,\text{ft}^3/\text{s}$$

Provide a standard Parshall flume with throat width $W = 1$ ft (~0.3 m) and obtain $C = 3.95$ and $n = 1.55$ are obtained from Table C.2 in Appendix C.

Substitute the variables in Equation 7.9 and calculate the measured head H_a at the throat.

$$11.6 = 3.95 \times H_a^{1.55}$$

$$H_a = \left(\frac{11.6}{3.95}\right)^{1/1.55} = 2.0\,\text{ft} \quad \text{or} \quad H_a = 2.0\,\text{ft} \times \frac{0.305\,\text{m}}{\text{ft}} = 0.61\,\text{m}$$

Under a free-flow condition at the Parshall flume, the maximum submergence head H_b allowed is 70% of the measured head H_a, that is, $H_b/H_a < 0.7$ (Table C.4). Also see Figure C.1 for the locations for H_a and H_b.

Calculate the maximum submergence head H_b.

$$H_b = 0.7 \times H_a = 0.7 \times 0.61\,\text{m} = 0.43\,\text{m}$$

Calculate the minimum head loss required under a free-flow condition.

$$\Delta h_{flume,min} = h_a - h_b = (0.61 - 0.43)\,\text{m} = 0.18\,\text{m}$$

Provide an overall head loss of 0.25 m at the Parshall flume $\Delta h_{flume} = 0.25\text{ m} > \Delta h_{flume,min}$.

Standard dimensions of a 1-ft Parshall flume are determined from Appendix C (Table C.3). The design details are provided in SI unit in Figure 9.39.

5. Determine major elevations and prepare hydraulic profile through the flocculation basin.

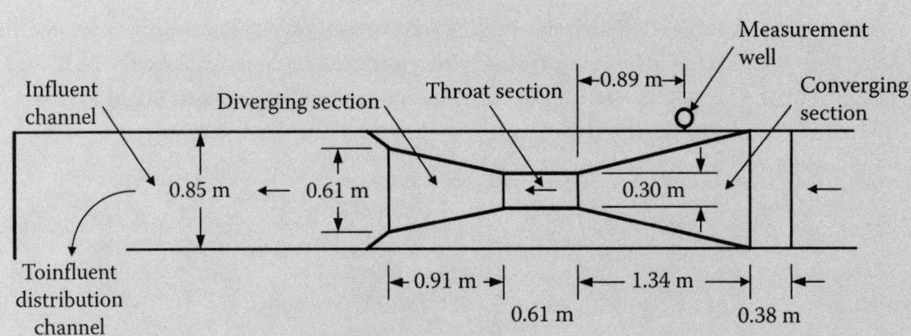

FIGURE 9.39 Design details of Parshall flume (Example 9.38).

Assume the reference elevation is the water surface elevation (WSEL) in the flocculation basin.

a. Prepare major WSELs.

WSEL in the flocculation basin $WSEL_{flocculation} = 100.00$ m (Reference)

Note: WSEL remains constant in the basin since the head losses through the slotted baffles are ignorable, $\Delta h_{basin} = 0.00$ m (Step 4b).

Head loss through the ports on diffusion wall $\Delta h_{port} = 0.00$ m (Step 4a)

WSEL in the influent primary clarifier

$WSEL_{clarifier} = WSEL_{flocculation} - \Delta h_{port} = (100.00 - 0.00)\text{ m} = 100.00\text{ m}$

Overall head loss at the influent distribution weirs $\Delta h_{weir} = 0.25$ m (Step 4c)

WSEL in the influent distribution channel

$WSEL_{distribution} = WSEL_{flocculation} + \Delta h_{weir} = (100.00 + 0.25)\text{ m} = 100.25\text{ m}$

Note: WSEL remains constant in the channel since the head losses through the channel are ignorable, $\Delta h_{distribution} = 0.00$ m (Step 4d).

WSEL at the lower end of the influent channel

$WSEL_{influent,lower} = WSEL_{distribution} = 100.25\text{ m}$ (Step 4e)

Overall head loss in the influent channel $\Delta h_{influent} = 0.03$ m (Step 4e)

WSEL at the upper end of the influent channel

$WSEL_{influent,upper} = WSEL_{influent,lower} + \Delta h_{influent} = (100.25 + 0.03)\text{ m} = 100.28\text{ m}$

Overall head loss at the Parshall flume $\Delta h_{flume} = 0.25$ m (Step 4f)

WSEL at the measurement well of the Parshall flume

$WSEL_{flume} = WSEL_{influent} + \Delta h_{flume} = (100.28 + 0.25)\text{ m} = 100.53\text{ m}$

b. Determine elevations (ELs) of the major components.

Provide a bottom slope from the first to third stage by a bottom elevation drop of 0.2 m.

Water depth in the first stage after the influent distribution weir, $H_{1st} = H_{stage} = 3.25\text{ m}$ (Step 1b)

Water depth in the third stage prior to the diffusion wall, $H_{3rd} = 3.45\text{ m}$ (Step 1b)

EL of the bottom in the first stage after the influent distribution weir

$EL_{1st} = WSEL_{flocculation} - H_{1st} = (100.00 - 3.25)\text{ m} = 96.75\text{ m}$

EL of the bottom in the third stage prior to the diffusion wall

$EL_{3rd} = WSEL_{flocculation} - H_{3rd} = (100.00 - 3.45)\text{ m} = 96.55\text{ m}$

Freeboard after the influent distribution weir $FB_{weir} = 0.20$ m (Step 4c)

Top of weir elevation $TOW = WSEL_{flocculation} + FB_{weir} = (100.00 + 0.20)\text{ m} = 100.20\text{ m}$

Water depth in the influent distribution channel $y_{distribution} = 0.90$ m and the heal loss through the channel $\Delta h_{distribution} = 0.00$ m (Step 4.d)

Water depth at the lower end of the influent channel $y_{influent,\ lower} = 0.90$ m (Step 4.d)

EL of the influent distribution channel

$EL_{distribution} = WSEL_{influent,lower} - y_{distribution} = (100.25 - 0.90)\ m = 99.35\ m$

EL at the lower end of the influent channel

$EL_{influent,lower} = WSEL_{influent,lower} - y_{influent,lower} = (100.25 - 0.90)\ m = 99.35\ m$

Water depth at the upper end of the influent channel $y_{influent,\ upper} = 0.93$ m (Step 4.e)

EL at the upper end of the influent channel

$EL_{influent,upper} = WSEL_{influent,upper} - y_{influent,upper} = (100.28 - 0.93)\ m = 99.35\ m$

Measured head at the throat of the Parshall flume $H_a = 0.61$ m (Step 4f)

EL of the flume invert at the throat

$EL_{flume} = WSEL_{flume} - H_a = (100.53 - 0.61)\ m = 99.92\ m$

Bottom elevation of the channel prior to the Parshall flume is 0.10 m lower than the throat invert. It is estimated from the dimensions in Table C.3 in Appendix C.

EL of the channel prior to the Parshall flume

$EL_{channel} = EL_{flume} - H_a = (99.92 - 0.10)\ m = 99.82\ m$

c. Draw the hydraulic profile and show the important design elevations of the flocculation basin.

The design details and hydraulic profile are shown for the flocculation basin in Figure 9.40a, and the Parshall flume in Figure 9.40b.

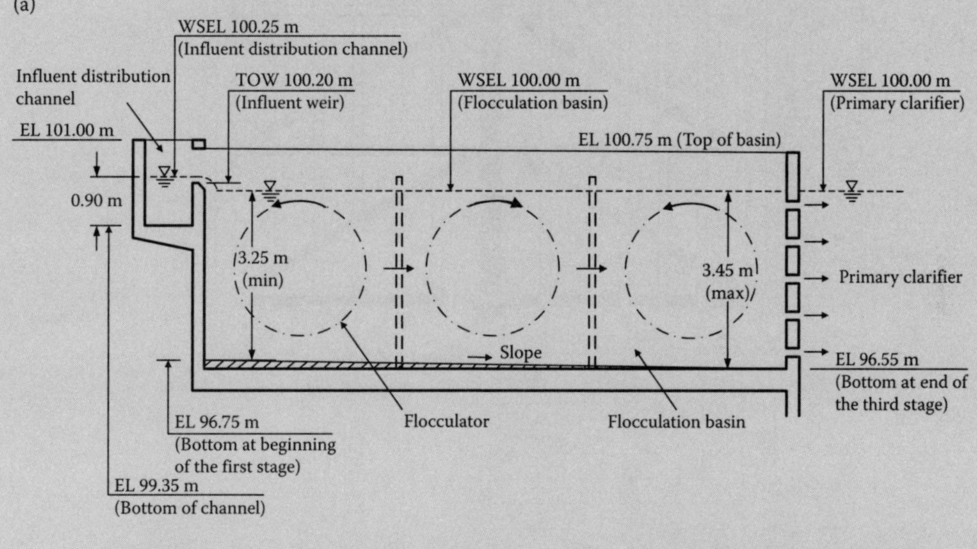

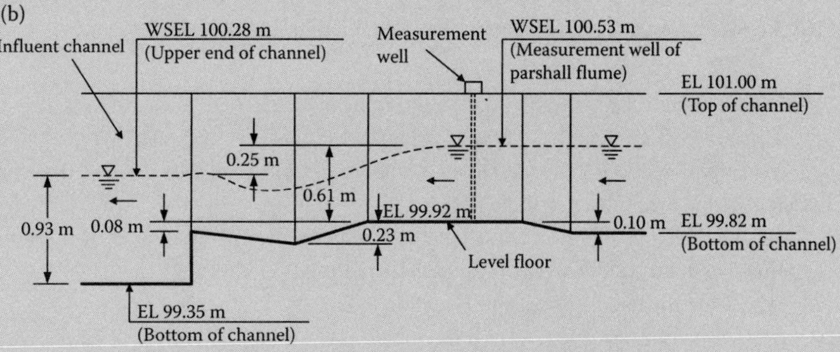

FIGURE 9.40 Hydraulic profiles: (a) through flocculation basin (Example 9.38), and (b) through Parshall flume (Example 9.38).

EXAMPLE 9.39: PADDLE AREA OF A VERTICAL FLOCCULATOR

Determine the area of a vertical paddle flocculator that has 2.5-m-long 12 redwood paddles (6 on each side of the vertical column). The distances to the middle of the paddles from the center of the column are 3.60, 3.00, 2.40, 1.80, 1.20, and 0.60 m. The velocity gradient G is 60 s^{-1} and the paddle rotational speed is 0.06 rps. The dynamic viscosity and mass density of coagulated wastewater are 1.002×10^{-3} N·s/m^2 and 998.2 kg/m^3 at 20°C, respectively. The volume V of the circular flocculation basin is 158 m^3. Assume that the relative velocity of the fluid is 75% of the paddle speed.

Solution

1. Draw the definition sketch.

   The design details of the vertical flocculator paddle are shown in Figure 9.41.

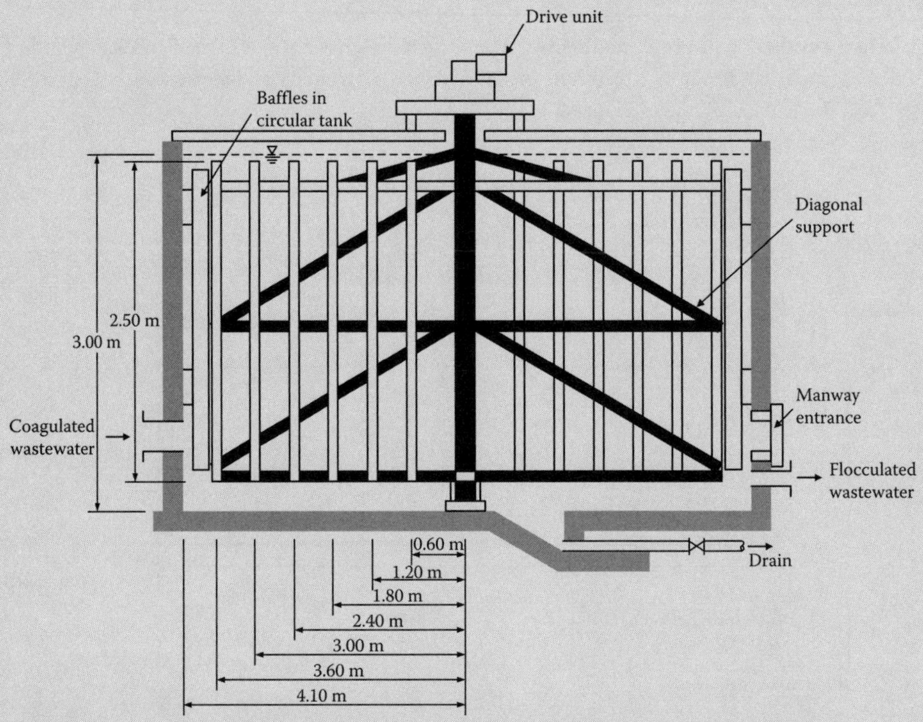

FIGURE 9.41 Design details of vertical flocculation basin (Example 9-39).

2. Calculate the flocculator water power requirement.

   Calculate P from Equation 9.27 at $\mu = 1.002\times10^{-3}$ N·s/m^2, $G = 60$ s^{-1}, and $V = 158$ m^3.

   Water power, $P = \mu VG^2 = 1.002 \times 10^{-3}\,\text{N·s/m}^2 \times 158\,\text{m}^3 \times (60\,\text{s}^{-1})^2 = 570\,\text{N·\ m/sor}570\text{kg·m}^2/\text{s}^3$

3. Calculate the area of each paddle flat.

   Assume the area of each paddle flat $= a$. There are two paddles on each sides of the shaft at each radius. The total paddle area $A = 2a$ at each radius.

   Calculate the paddle area from Equation 9.32.

$$P = \sum_{i=1}^{6} P_i = \sum_{i=1}^{6} \frac{C_D \rho A v_i^3}{2} = \frac{1}{2} C_D \rho(2a)(v_1^3 + v_2^3 + v_3^3 + v_4^3 + v_5^3 + v_6^3)$$

At radius of r, the absolute paddle velocity at the shaft rotational speed n, $v_{abs} = 2\pi rn$

Assume that the velocity of paddle relative to the water is 75% of the absolute paddle velocity and calculate the velocity of paddle relative to the water. $v = 0.75\, v_{abs} = 0.75 \times (2\pi\ r\ n)$

$$P = \frac{1}{2} C_D \rho(2a)\,[(0.75(2\pi r_1)n)^3 + (0.75(2\pi r_2)n)^3 + (0.75(2\pi r_3)n)^3 + (0.75(2\pi r_4)n)^3$$
$$+ (0.75(2\pi r_5)n)^3 + (0.75(2\pi r_6)n)^3]$$
$$= \frac{1}{2} C_D \rho(2a) \times (0.75(2\pi)n)^3 \times (r_1^3 + r_2^3 + r_3^3 + r_4^3 + r_5^3 + r_6^3)$$

Assume the coefficient of drag C_D of paddles is 1.9 and substitute $\rho = 998.2\ \text{kg/m}^3$, $r_1 = 3.60$ m, $r_2 = 3.00$ m, $r_3 = 2.40$ m, $r_4 = 1.80$ m, $r_5 = 1.20$ m, and $r_6 = 0.60$ m into the above equation and solve a at $P = 570\ \text{kg·m}^2/\text{s}^3$.

$$P = \frac{1}{2} \times 1.9 \times 998.2\,\text{kg/m}^3 \times 2a \times (0.75 \times 2 \times \pi \times 0.06\,\text{s}^{-1})^3 \times ((3.60\,\text{m})^3 + (3.00\,\text{m})^3$$
$$+ (2.40\,\text{m})^3 + (1.80\,\text{m})^3 + (1.20\,\text{m})^3 + (0.60\,\text{m})^3)$$
$$= 1897\,\text{kg/m}^3 \times a \times 0.0226\,\text{s}^{-3} \times 95.3\,\text{m}^3$$
$$= 4086\,\text{kg/s}^3 \times a$$

$$570\,\text{kg·m}^2/\text{s}^3 = 4086\,\text{kg/s}^3 \times a$$

$$a = 0.14\,\text{m}^2$$

4. Determine the paddle dimensions.
   Calculate the required paddle width at a paddle length $L = 2.5$ m.

$$\text{Desired paddle width, } W = \frac{a}{L} = \frac{0.14\,\text{m}^2}{2.5\,\text{m}} = 0.056\,\text{m} \quad \text{or} \quad 5.6\,\text{cm}$$

Provide paddles each 2.5 m long ×5.6 cm wide.

$$\text{Calculate the ratio } L/W = \frac{2.5\,\text{m}}{0.056\,\text{m}} = 45$$

At L/W ratio of 45 >> 20, $C_D = 1.9$ is estimated from Table 9.22 or Figure 9.32. Therefore, the assumed C_D value is suitable for use in determining the paddle dimensions.

EXAMPLE 9.40: VELOCITY GRADIENT CREATED BY HEAD LOSS AT A WEIR

A rectangular weir is constructed across a channel that is 1.2 m wide. The height of the weir is 1.5 m above the channel bottom. It carries a flow of 1.18 m^3/s. Calculate the velocity gradient G created by the turbulence at the weir at a water temperature of 10°C. Assume that the water depth in the channel downstream of the weir is 1.3 m, the weir discharge coefficient C_d is 0.62, and the mixing efficiency is 100% in the turbulence zone.

Solution

1. Draw the definition sketch (Figure 9.42).
2. Calculate the head over the weir.

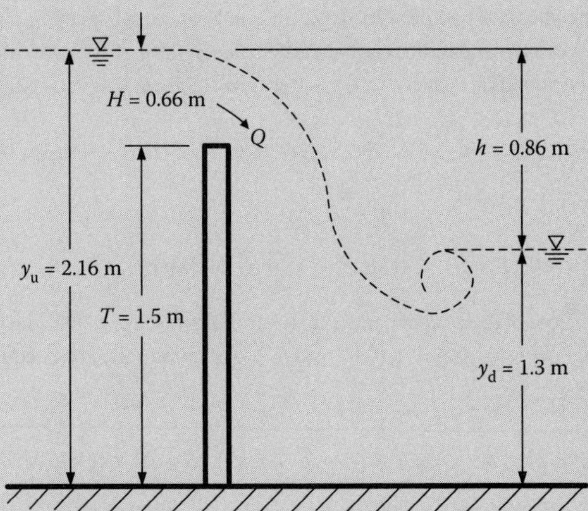

FIGURE 9.42 Definition sketch of hydraulic jump after a weir (Example 9.40).

The head over the weir (H) is calculated from Equation 8.10 at $Q = 1.18\ m^3/s$, $C_d = 0.62$, $L = w$ (channel width) = 1.2 m, and $n = 0$.

$$L' = L - 0.1\,nH = L = 1.2\,\text{m}$$

$$H = \left(\frac{3}{2} \times \frac{Q}{C_d L' \sqrt{2g}}\right)^{2/3} = \left(\frac{3}{2} \times \frac{1.18\,\text{m}^3/\text{s}}{0.62 \times 1.2\,\text{m} \times \sqrt{2 \times 9.81\,\text{m/s}^2}}\right)^{2/3} = 0.66\,\text{m}$$

3. Calculate the head loss at the weir.

   The head loss at the weir (h) is determined from the difference between the upstream and downstream water surface elevations.

   Weir height, $T = 1.5$ m (Given)

   Upstream water depth, $y_u = T + H = 1.5\ \text{m} + 0.66\ \text{m} = 2.16\ \text{m}$

   Downstream water depth, $y_d = 1.3$ m (Given)

   Head loss through the weir, $h = y_u - y_d = 2.16\ \text{m} - 1.3\ \text{m} = 0.86\ \text{m}$

4. Calculate the water power imparted by the turbulence at the weir.

   At 10°C, $\rho = 999.7\ \text{kg/m}^3 \approx 1000\ \text{kg/m}^3$ is obtained from Table B.2 in Appendix B.

   Calculate the power imparted into the water (P_w) by the turbulence from Equation 9.30a.

$$P_w = \rho g Q h = 1000\,\text{kg/m}^3 \times 9.81\,\text{m/s}^2 \times 1.18\,\text{m}^3/\text{s} \times 0.86\,\text{m} = 9955\,\text{kg·m}^2/\text{s}^3 \quad \text{or} \quad 9955\,\text{N·m/s}$$

5. Determine the mixing volume after the weir.

   Calculate the velocity in the downstream channel after the weir.

$$\text{Velocity in downstream channel, } v = \frac{Q}{w y_d} = \frac{1.18\,\text{m}^3/\text{s}}{1.2\,\text{m} \times 1.3\,\text{m}} = 0.76\,\text{m/s}$$

A mixing time $t = 2$ s is normally assumed when the velocity in the downstream channel is 0.5 m/s or larger.[14]

Calculate mixing volume at $Q = 1.18\ \text{m}^3/\text{s}$.

Mixing volume, $V = Qt = 1.18\ \text{m}^3/\text{s} \times 2\ \text{s} = 2.36\ \text{m}^3$

6. Calculate the velocity gradient G created by the turbulence at the weir.
   At 10°C, $\mu = 1.307 \times 10^{-3}\ \text{N·s/m}^2$ is obtained from Table B.2 in Appendix B.
   G is calculated from Equation 9.27.

$$G = \sqrt{\frac{P_w}{\mu V}} = \left(\frac{9955\ \text{N·m/s}}{1.307 \times 10^{-3}\ \text{N·s/m}^2 \times 2.36\ \text{m}^3}\right)^{1/2} = 1796\ \text{s}^{-1}$$

Note: Due to concerns of the effectiveness of hydraulic energy for rapid mixing, $G \approx 1500\ \text{s}^{-1}$ may be acceptable for design after applying a mixing effective factor of approx. 0.85 for G. A similar approach may also be applied to determine the power requirement (P) or velocity gradient (G) in Examples (9.41) to (9.44).

EXAMPLE 9.41: WEIR DESIGN TO CREATE A GIVEN VELOCITY GRADIENT

Design a weir to create a velocity gradient G of $1000\ \text{s}^{-1}$. The weir is used for rapid mixing of chemicals into a rectangular channel designed to carry wastewater flow. The weir is built across a rectangular channel. The velocity in the downstream channel must be >0.5 m/s at an average flow of 30,000 m^3/d. The average temperature of wastewater is expected to be around 20°C. Assume that the weir discharge coefficient C_d is 0.62 and the mixing efficiency is 100% in the turbulence.

Solution

1. Determine the optimum width and water depth in the channel.

   The optimum hydraulic section of a rectangular channel is achieved when the following optimum hydraulic conditions are met: (a) a width-to-depth ratio (w/d) of 2, and (b) a velocity of 0.5 m/s or greater.[12–14]

Design flow in the channel, $Q = 30{,}000\ \text{m}^3/d \times \dfrac{d}{(60 \times 60 \times 24)\ \text{s}} = 0.347\ \text{m}^3/\text{s}$

Determine the maximum velocity under the optimum hydraulic conditions, $w/d = 2$ and $v \geq 0.5$ m/s.

$$v = \frac{Q}{wd} = \frac{Q}{2d^2} = \frac{0.347\ \text{m}^3/\text{s}}{2d^2} \geq 0.5\ \text{m/s}$$

Solve v from the above relationship.

$$d \leq \sqrt{\frac{0.347\ \text{m}^3/\text{s}}{2 \times 0.5\ \text{m/s}}} = 0.59\ \text{m}$$

Provide d =0.50 m.

$w = 2d = 2 \times 0.50\ \text{m} = 1.00\ \text{m}$

2. Determine the mixing time in the channel downstream of the weir.

   The mixing time t in a channel downstream of a weir is around 2 s if the velocity in the channel is 0.5 m/s or larger.[14]

$$\text{Actual velocity, } v = \frac{0.347\,\text{m}^3/\text{s}}{0.50\,\text{m} \times 1.00\,\text{m}} = 0.69\,\text{m/s} > 0.5\,\text{m/s}$$

   Therefore, $t = 2$ s is acceptable.

3. Calculate the head loss required at the weir to meet the G value.

   At 20°C, $\mu = 1.002 \times 10^{-3}\,\text{N·s/m}^2$ (or kg/m·s) and $\rho = 998.2\,\text{kg/m}^3$ are obtain from Table B.2 in Appendix B.

   Rearrange Equation 9.30c and calculate head loss requirement (h) at $G = 1000\,\text{s}^{-1}$, and $t = 2$ s.

$$\text{Head loss, } h = \frac{\mu G^2 t}{\rho g} = \frac{1.002 \times 10^{-3}\,\text{kg/m·s} \times (1000\,\text{s}^{-1})^2 \times 2\,\text{s}}{998.2\,\text{kg/m}^3 \times 9.81\,\text{m/s}^2} = 0.20\,\text{m}$$

4. Calculate the head over the weir.

   Assume that free fall is achieved at the weir and calculate the head (H) over the rectangular weir across the full width of the channel from Equaiton 8.10 at $C_d = 0.62$, $Q = 0.347\,\text{m}^3/\text{s}$, $L = w$ (channel width) $= 1.00$ m, and $n = 0$.

$$L' = L - 0.1\,nH = L = 1.00\,\text{m}$$

$$H = \left(\frac{3}{2} \times \frac{Q}{C_d L' \sqrt{2g}}\right)^{2/3} = \left(\frac{3}{2} \times \frac{0.347\,\text{m}^3/\text{s}}{0.62 \times 1.00\,\text{m} \times \sqrt{2 \times 9.81\,\text{m/s}^2}}\right)^{2/3} = 0.33\,\text{m}$$

   Since $H > h$, a submerged weir condition is provided to achieve the desired $G = 1000\,\text{s}^{-1}$. The discharge over a submerged weir is expressed by Equaiton 9.34. Determine the upstream head (h_u) from Equaiton 9.34 to meet the head loss requirement $h = h_u - h_d = 0.20$ m through trial-and-error procedure.

   a. First iteration.

      Start the calculations by assuming upstream $h_u = H = 0.33$ m.

      Calculate the downstream head (submergence).

      $h_d = h_u - h = (0.33 - 0.20)\,\text{m} = 0.13\,\text{m}$

      Calculate the flow reduction factor from Equaiton 9.34b and at $n = 1.50$ for the rectangular weir across the width of the channel.

$$F_s = \left[1 - \left(\frac{h_d}{h_u}\right)^n\right]^{0.385} = \left[1 - \left(\frac{0.13\,\text{m}}{0.33\,\text{m}}\right)^{1.50}\right]^{0.385} = 0.896$$

      Apply the flow reduction factor to calculate the equivalent free-fall flow Q_e to achieve $Q_s = 0.347\,\text{m}^3/\text{s}$ from Equation 9.34a.

$$Q_e = \frac{Q_s}{F_s} = \frac{0.347\,\text{m}^3/\text{s}}{0.896} = 0.387\,\text{m}^3/\text{s}$$

      Calculate the upstream head h_u required at $Q_e = 0.387\,\text{m}^3/\text{s}$ under free-fall condition from Equation 8.10.

$$H = \left(\frac{3}{2} \times \frac{0.389\,\text{m}^3/\text{s}}{0.62 \times 1.00\,\text{m} \times \sqrt{2 \times 9.81\,\text{m/s}^2}}\right)^{2/3} = 0.355\,\text{m}$$

After the first iteration, an upstream head $h_u = H = 0.355$ m is required to achieve a submerged weir discharge flow $Q_s = 0.347\ m^3/s$.

b. Second through forth iterations.

Assume $h_u = 0.355$ m and repeat the same calculation steps in the first iteration. After the second iteration, $h_u = 0.360$ m. Repeat the same calculations in the third and fourth iterations. After the forth iteration, $h_u = 0.361$ m. It is the same value obtained from the third iteration. Therefore, an approximate upstream head of 0.36 m is required to push a flow of 0.347 m^3/s with an overall head loss of 0.2 m. The submerged head $h_d = (0.36 - 0.2)\ m = 0.16$ m.

5. Calculate the height of weir above the channel bottom.

The depth of water downstream of the weir, $y_d = d = 0.5$ m

The height of weir above the channel bottom, $T = y_d - h_d = (0.5 - 0.16)\ m = 0.34$ m

Therefore, the top of weir is 0.34 m above the downstream channel floor.

6. Draw the sketch.

The plan and longitudinal section of the channel are shown in Figure 9.43.

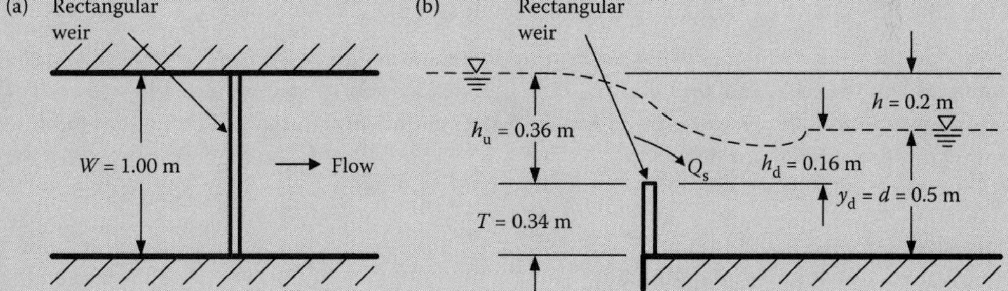

FIGURE 9.43 Rapid mixing by hydraulic jump after a weir: (a) plan and (b) hydraulic profile (Example 9.41).

EXAMPLE 9.42: VELOCITY GRADIENT CREATED BY FRICTION HEAD LOSS IN A PIPELINE

Coagulation chemicals are injected in a pipeline for mixing and to carry out coagulation. The pipeline is 0.75 m in diameter and 50 m long. The pipe discharge is 0.663 m^3/s. Calculate the velocity gradient G at a water temperature of 20°C. Assume that the mixing efficiency is 100% through the pipeline and the mixing caused by the diffusor at the injection point is ignored.

Solution

1. Calculate the head loss through the pipeline.

$$\text{Velocity through the pipeline, } v = \frac{Q}{A} = \frac{4Q}{\pi d^2} = \frac{4 \times 0.663\ m^3/s}{\pi \times (0.75\ m)^2} = 1.50\ m/s$$

The head loss through pipeline is calculated from Equation 6.13a (Darcy–Weisbach equation). The most commonly used value of f is 0.02–0.03 in water and wastewater applications. Assume f is 0.03 and calculate the friction head loss (h) at $D_h = d = 0.75$ m (for circular pipe flowing full), $L = 50$ m, and $v =$ 1.50 m/s

$$\text{Friction head loss, } h = \frac{fLv^2}{2gd} = \frac{0.03 \times 50\ m \times (1.5\ m/s)^2}{2 \times 9.81\ m/s^2 \times 0.75\ m} = 0.23\ m$$

2. Determine the mixing time in the pipeline.

$$\text{Mixing time, } t = \frac{L}{v} = \frac{50\,\text{m}}{1.5\,\text{m/s}} = 33.3\,\text{s}$$

3. Calculate the velocity gradient G created by the friction head loss through the pipeline.

At 20°C, $\mu = 1.002 \times 10^{-3}$ N·s/m^2 (or kg/m·s) and $\rho = 998.2$ kg/m^3 are obtain from Table B.2 in Appendix B.

Calculate velocity gradient G from Equation 9.27 at $h = 0.23$ m, and $t = 33.3$ s.

$$\text{Velocity gradient, } G = \sqrt{\frac{\rho g h}{\mu t}} = \sqrt{\frac{998.2\,\text{kg/m}^3 \times 9.81\,\text{m/s}^2 \times 0.23\,\text{m}}{1.002 \times 10^{-3}\,\text{kg/m·s} \times 33.3\,\text{s}}} = 260\,\text{s}^{-1}$$

EXAMPLE 9.43: VELOCITY GRADIENT CREATED BY AN ORIFICE DISCHARGE

Water is discharging from a head box through an orifice into a 2-m^3 basin. The driving head over the orifice is 2 m. The diameter of the orifice is 0.25 m. Calculate the velocity gradient G created by the orifice discharge at a water temperature of 20°C. Assume that the coefficient of discharge C_d is 0.65 and the mixing efficiency is 100% in the turbulence.

Solution

1. Draw the definition sketch (Figure 9.44).

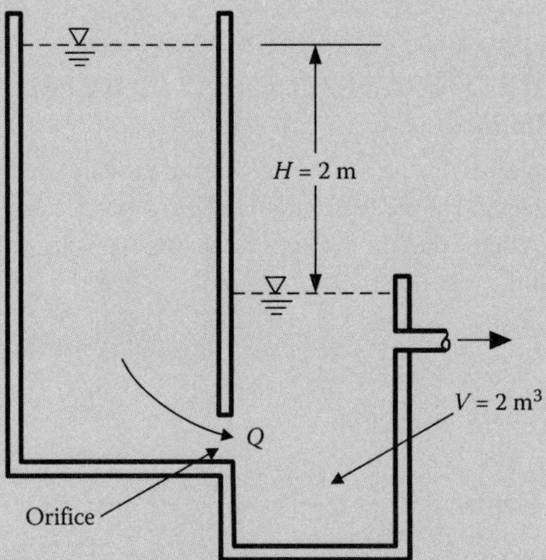

FIGURE 9.44 Definition sketch of hydraulic mixing by orifice discharge (Example 9.43).

2. Calculate the discharge through the orifice.

$$\text{Cross area of the orifice, } A = \frac{\pi}{4} d^2 = \frac{\pi}{4} \times (0.25\,\text{m})^2 = 0.049\,\text{m}^2$$

The discharge flow through an orifice is calculated from Equation 7.4a at $C_d = 0.65$ and $H = 2$ m.

$$Q = C_d A\sqrt{2gH} = 0.65 \times 0.049\,\text{m}^2 \times \sqrt{2 \times 9.81\,\text{m/s}^2 \times 2\,\text{m}} = 0.20\,\text{m}^3/\text{s}$$

3. Calculate the power imparted into the water (P_w) by the discharge through the orifice.

   At 20°C, $\rho = 998.2$ kg/m^3 is obtained from Table B.2 in Appendix B.

   The water power P_w imparted by the discharge at a head loss $H_L = 2$ m is calculated from Equation 9.30a.

$$P_w = \rho g Q H = 998.2\,\text{kg/m}^3 \times 9.81\,\text{m/s}^2 \times 0.20\,\text{m}^3/\text{s} \times 2\,\text{m} = 3917\,\text{kg·m}^2/\text{s}^3 \quad \text{or} \quad 3917\,\text{N·m/s}$$

4. Calculate the velocity gradient G created by the discharge through the orifice.

   At 20°C, $\mu = 1.002 \times 10^{-3}$ N·s/m^2 is obtained from Table B.2 in Appendix B.

   Calculate G from Equation 9.27 at $V = 2$ m^3.

$$G = \sqrt{\frac{P_w}{\mu V}} = \left(\frac{3917\,\text{N·m/s}}{1.002 \times 10^{-3}\,\text{N·s/m}^2 \times 2\,\text{m}^3}\right)^{1/2} = 1398\,\text{s}^{-1}$$

EXAMPLE 9.44: VELOCITY GRADIENT CREATED BY NOZZLE MIXING (OR EJECTOR)

Nozzle mixing is used for pH adjustment in a 8-ft-wide channel with a water depth of 8 ft. The design flow is 100 MGD in the channel. Water is pumped from the channel and discharged back into the channel through nozzle(s). Calculate the velocity gradient G created by a single nozzle at a water temperature of 60°F. Assume the mixing efficiency is 100% in the turbulence zone. Also evaluate the improved G values when two or four nozzles are used. Assume a discharge flow of 600 gpm at a discharge pressure of 10 psi remains the same at each nozzle in all scenarios.

Solution

1. Draw the definition sketch (Figure 9.45).

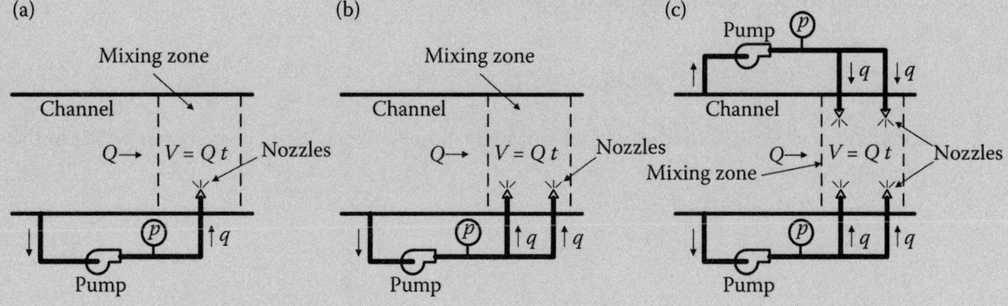

FIGURE 9.45 Definition sketches of hydraulic mixing by nozzle mixing: (a) single nozzle, (b) two nozzles, and (c) four nozzles (Example 9.44).

2. Determine the mixing volume in the channel.

$$\text{Design flow in the channel, } Q = 100\,\text{MGD} \times \frac{10^6\,\text{gal}}{1\,\text{Mgal}} \times \frac{1\,\text{ft}^3}{7.48\,\text{gal}} \times \frac{\text{d}}{(60 \times 60 \times 24)\,\text{s}} = 155\,\text{ft}^3/\text{s}$$

$$\text{Velocity in the channel, } v = \frac{155\,\text{ft}^3/\text{s}}{8\,\text{ft} \times 8\,\text{ft}} = 2.42\,\text{ft/s(orfps)}$$

Since $v = 2.42$ ft/s $= 0.74$ m/s > 0.5 m/s, assume $t = 2$ s and calculate the mixing volume at design flow $Q = 0.155\ \text{ft}^3/\text{s}$.

$$\text{Mixing volume, } V = Qt = 155\,\text{ft}^3/\text{s} \times 2\,\text{s} = 310\,\text{ft}^3$$

3. Calculate the water power imparted by a single nozzle.

$$\text{Nozzle discharge flow, } q = 600\,\text{gpm} \times \frac{1\,\text{ft}^3}{7.48\,\text{gal}} \times \frac{1\,\text{min}}{60\,\text{s}} = 1.34\,\text{ft}^3/\text{s(orcfs)}$$

At 60°F, the approximate water density $\gamma = 62.4\ \text{lb/ft}^3$ is obtained from Table B.3 in Appendix B. Calculate the head loss through the nozzle at the discharge pressure $p = 10$ psi.

$$\text{Head loss through the nozzle, } h = \frac{p}{\gamma} = 10\,\text{psi} \times \frac{144\,\text{in}^2}{1\,\text{ft}^2} \times \frac{1}{62.4\,\text{lb/ft}^3} = 10\,\text{psi} \times \frac{2.31\,\text{ft}}{1\,\text{psi}} = 23.1\,\text{ft}$$

Calculate the power imparted into the water (P_w) by the nozzle discharge from Equation 9.30a.

$$P_w = \gamma Q h = 62.4\,\text{lb/ft}^3 \times 1.34\,\text{ft}^3/\text{s} \times 23.1\,\text{ft} = 1932\,\text{lb·ft/s}$$

4. Calculate the velocity gradient G created by a single nozzle.
At 60°F, $\mu = 2.36 \times 10^{-5}\ \text{lb·s/ft}^2$ is obtained from Table B.3 in Appendix B.
Calculate G_1 with single nozzle from Equation 9.27.

$$G_1 = \sqrt{\frac{P_w}{\mu V}} = \left(\frac{1932\,\text{lb·ft/s}}{2.36 \times 10^{-5}\,\text{lb·s/ft}^2 \times 310\,\text{ft}^3}\right)^{1/2} = 514\,\text{s}^{-1}$$

5. Evaluate the improved G values using two or four nozzles.
Assume the mixing volume remains the same and the water power is doubled when two nozzles are used. Calculate G_2 with two nozzles from Equation 9.27.

$$G_2 = \sqrt{\frac{2P_w}{\mu V}} = \sqrt{2}G_1 = 1.41 \times 514\,\text{s}^{-1} = 725\,\text{s}^{-1}$$

Similarly, calculate G_4 with four nozzles from Equation 9.27 at four-time large water power in the same mixing volume.

$$G_4 = \sqrt{\frac{4P_w}{\mu V}} = 2G_1 = 2 \times 514\,\text{s}^{-1} = 1028\,\text{s}^{-1}$$

In comparison with the G value with single nozzle, it will be respectively increased by ~40% or 100% when two or four nozzles are installed and if the mixing volume remains the same. It can be further expected that the G value will be doubled when four nozzles are used.

EXAMPLE 9.45: BAFFLED FLOCCULATOR WITH AROUND-THE-END FLOW PATH

A mixing tank has around-the-end baffles to achieve flocculation of coagulated wastewater. Coagulating chemicals are added upstream of the influent weir at the entrance of the flocculator. The effluent end of the flocculator has a proportional weir to maintain a relatively constant velocity in the flocculator. The average flow through the flocculator is 3.0 MGD. The flocculator is divided into two similar sections in series. Each section consists of channels that are connected by around-the-end (180°) turns. The average wastewater temperature is 60°F. Calculate or prepare information for (a) velocity and detention time, (b) dimensions of the channel, (c) total number of around-the-end channels, (d) dimensions of the flocculator, and influent and effluent structures, (e) head losses through the channels, (f) the velocity gradient G, and (g) important design elevations and hydraulic profile through the flocculator.

Solution

1. Draw the definition sketch.

   The plan view of the around-the-end baffled flocculator is shown in Figure 9.45.

2. Determine the velocity and detention time in the flocculator.

   a. Determine the design velocity through the channel.

      The design velocity through the channel is typically in a range of 0.3–1 ft/s (0.1–0.3 m/s). A design velocity $v = 0.7$ ft/s is used.

      $$\text{Design flow in the channel, } Q = 3\,\text{MGD} \times \frac{10^6\,\text{gal}}{1\,\text{Mgal}} \times \frac{1\,\text{ft}^3}{7.48\,\text{gal}} \times \frac{d}{(60 \times 60 \times 24)\,\text{s}} = 4.64\,\text{ft}^3/\text{s}$$

      $$\text{Desired flow cross area, } A = \frac{Q}{v} = \frac{4.64\,\text{ft}^3/\text{s}}{0.7\,\text{ft/s}} = 6.63\,\text{ft}^2$$

   b. Determine the detention time.

      The hydraulic detention time is usually 15–20 min. A detention time $t = 20$ min is used.

3. Determine the dimensions of each channel.

   a. Determine the channel width and water depth.

      It is a practical design recommendation that the water depth ranges from one to three times the clear width of the channel between the baffle, that is, d/w ratio = 1–3.[48] Assume $d/w = 3$ and calculate the desired channel width (w) at the desired cross area $A = 6.63$ ft^2.

      $$\text{Desired flow cross area, } A = wd = w \times (3w) = 3w^2$$

      $$\text{Desired channel width, } w = \sqrt{\frac{A}{3}} = \sqrt{\frac{6.63\,\text{ft}^2}{3}} = 1.49\,\text{ft}$$

      Provide a design channel width $w = 1.50$ ft.

      $$\text{Design water depth, } d = 3w = 3 \times 1.50\,\text{ft} = 4.50\,\text{ft}$$

      Note: A minimum channel width of 0.45 m (1.5 ft) and minimum water depth of ~1 m (3 ft) are normally recommended.[49]

      $$\text{Actual average velocity in the channel, } v = \frac{4.64\,\text{ft}^3/\text{s}}{1.5\,\text{ft} \times 4.5\,\text{ft}} = 0.69\,\text{ft/s}$$

   b. Determine the width of clearance at the baffle end.

      It is a design practice to keep the around-the-end clearance (the clearance between the end of the baffle and the basin sidewall) to be 0.9–1.5 times the channel clear width; that is, w_{end}/w

ratio = 0.9–1.5.[48,49] Assume $w_{end}/w = 1.2$ and calculate the desired clearance width (w_{end}).

$$w_{end} = 1.2\,w = 1.2 \times 1.5\,\text{ft} = 1.80\,\text{ft}$$

c. Determine the overlap length of baffles.

The overlap length is also important in design of the around-the-end baffle flocculator. Practically, the overlap length is normally four to five times the channel clear width; that is, $l_{overlap}/w$ ratio = 4–5.[48] Assume the length of each channel $l = 10$ ft and calculate the overlap length ($l_{overlap}$).

$$l_{overlap} = l - 2w_{end} = 10\,\text{ft} - 2 \times 1.80\,\text{ft} = 6.40\,\text{ft}$$

$$\frac{l_{overlap}}{w} = \frac{6.40\,\text{ft}}{1.5\,\text{ft}} = 4.3\,\text{(within the practical range)}$$

d. Determine the baffle thickness.

Fiberglass baffles are used in the flocculator. Assume the thickness of the baffle wall $t_{baffle} = 0.25$ ft (3 in).

4. Calculate the number of around-the-end channels.

Calculate the total volume of the flocculator required at the hydraulic detention time $t = 20$ min and design capacity $Q = 4.64\ \text{ft}^3/\text{s}$.

$$\text{Total volume required, } V_{flocculator} = Qt = 4.64\,\text{ft}^3/\text{s} \times 20\,\text{min} \times 60\,\text{s/min} = 5570\,\text{ft}^3$$

Calculate the volume of each channel of the flocculator from $l = 10$ ft, $w = 1.5$ ft, and $d = 4.5$ ft.

$$\text{Volume of each channel, } V_{channel} = l \times w \times d = 10\,\text{ft} \times 1.5\,\text{ft} \times 4.5\,\text{ft} = 67.5\,\text{ft}^3\ \text{perchannel}$$

Calculate the total number of channels required in the flocculator.

$$\text{Total number of channels required, } N_{channel,total} = \frac{V_{flocculator}}{V_{channel}} = \frac{5570\,\text{ft}^3}{67.5\,\text{ft}^3/\text{channel}} = 83\ \text{channels}$$

Provide a total of 86 channels in the flocculator ($N_{channel,total} = 86$ channels).
There are two sections in the flocculator, $N_{section} = 2$ sections.

$$\text{Number of channels per section, } N_{channel,section} = \frac{N_{channel,total}}{N_{section}} = \frac{86\ \text{channels}}{2\ \text{sections}} = 43\ \text{channels per section}$$

$$\text{Number of baffles per section, } N_{baffle,section} = N_{channel,section} - 1 = 43 - 1 = 42\,\text{baffles per section}$$

There is a 180° turn at one end of each baffle.

$$\text{Number of } 180^\circ \text{ turns per section, } N_{turn,section} = N_{baffle,section} = 42\ \text{turns per section}$$

Total number of 180° turns in the flocculator,

$$N_{turn,total} = N_{section} \times N_{turn,section} = 2\,\text{sections} \times 42\,\text{turns/section} = 84\,\text{turns}$$

5. Design the basin, influent, and effluent structures.

a. Determine dimensions of the flocculator.

There are two sections in the flocculator. Assume the common wall between sections $t_{\text{wall}} = 1$ ft.

Overall width of the flocculator, $W = N_{\text{section}} \times l + t_{\text{wall}} = 2 \times 10\,\text{ft} + 1\,\text{ft} = 21\,\text{ft}$

$$\text{Overall length of the flocculator, } L = N_{\text{channel,section}} \times w + N_{\text{baffle,section}} \times t_{\text{baffle}}$$
$$= 43 \times 1.5\,\text{ft} + 42 \times 0.25\,\text{ft} = 75\,\text{ft}$$

Total volume provided, $V = N_{\text{channel,total}} V_{\text{channel}} = 86\,\text{channels} \times 67.5\,\text{ft}^3/\text{section} = 5800\,\text{ft}^3$

Note: The total extra volume at turns and between the sections is small and ignorable.

$$\text{Actual hydraulic retention time, } t = \frac{V}{Q} = \frac{5800\,\text{ft}^3}{4.64\,\text{ft}^3/\text{s} \times 60\,\text{s/min}} = 21\,\text{min}$$

b. Describe the basin.
The channel bottom is stepped downward to give allowance for head losses at the baffles. Ideally, the floor of each channel should be stepped down equal to the head loss encountered due to the flow turbulence at the end of the channel. In practice, the floor is dropped once and equally at every 4–5 channels.

c. Describe the influent structure.
The influent structure of the around-the-end baffled flocculator consists of a rectangular weir across the 1.5-ft-wide influent channel. The chemicals are fed into the upstream channel leading to the weir. The weir provides the needed turbulence at the free fall for dispersion of chemicals at the entry of the flocculator.

d. Describe the effluent structure.
The effluent structure consists of a 1.5-ft-wide channel and a proportional weir designed as an integral part of the primary clarifier. After the proportional weir, the coagulated water discharges into the primary clarifier. The design procedure of the proportional weir is similar to that presented in Examples 7.10 and 8.12.

6. Calculate the head losses through the flocculator.
Hydraulic calculations are performed for (a) the total minor head losses due to 180° turns at the around-the-end baffles, (b) the total friction loss through the total length of flow path in the flocculator, (c) the total head losses through the flocculator, and (d) the head over the influent rectangular weir. Calculations for the effluent proportional weir are not included in this example. See Examples 7.10 and 8.12 for detailed calculations of the proportional weir.

a. Calculate the minor head losses at the around-the-end baffles in the flocculator due to 180° turn.
The minor head loss due to a 180° turn at the end of each around-the-end baffle is calculated from Equation 6.15b using flow-through velocity $v = 0.69$ ft/s. Usually, the minor loss coefficient K_m for a 180° turn is 2–3.5. Assume $K_m = 3$ for each turn.

$$\text{Head loss at each 180° turn, } h_{\text{turn}} = K_m \frac{v^2}{2g} = 3 \times \frac{(0.69\,\text{ft/s})^2}{2 \times 32.2\,\text{ft/s}^2} = 0.022\,\text{ft per turn}$$

$$\text{Total minor head losses through the flocculator, } h_m = N_{\text{turn,total}} \times h_{\text{turn}}$$
$$= 84\,\text{turns} \times 0.022\,\text{ft/turn} = 1.85\,\text{ft}$$

b. Calculate the total friction loss through the total length of flow path in the flocculator.

The friction loss is calculated from the Manning's equation in the U.S. customary units (Equation 6.11c).

$$v = \frac{1.486}{n} R^{2/3} S^{1/2} = \frac{1.486}{n} R^{2/3} \left(\frac{h_f}{l_{\text{path,total}}} \right)^{1/2} \quad or \quad h_f = \left(\frac{nv}{1.486 R^{2/3}} \right)^2 l_{\text{path,total}}$$

Estimated length of flow path per channel, $l_{\text{path,channel}} = l - 2 \times \left(\frac{1}{2} \times w_{\text{end}} \right)$

$$= \left[10 - 2 \times \left(\frac{1}{2} \times 1.80 \right) \right] ft$$

$$= 8.20 \text{ ft per channel}$$

Total length of flow path in the flocculator,

$$l_{\text{path,total}} = N_{\text{channel,total}} \times l_{\text{path,channel}} = 86 \text{ channels} \times 8.20 \text{ ft/channel} = 705 \text{ ft}$$

Calculate the hydraulic mean radius from $d = 4.5$ ft and $w = 1.5$ ft.

$$R = \frac{dw}{2d + w} = \frac{4.5 \text{ ft} \times 1.5 \text{ ft}}{2 \times 4.5 \text{ ft} + 1.5 \text{ ft}} = 0.64 \text{ ft}$$

Assume $n = 0.013$ for the fiberglass baffles and calculated h_f at $v = 0.69$ ft/s from the above rearranged Equation 6.11c.

$$h_f = \left(\frac{0.013 \times 0.69 \text{ ft/s}}{1.486 \times (0.64 \text{ ft})^{2/3}} \right)^2 \times 705 \text{ ft} = 0.05 \text{ ft}$$

c. Calculate the total losses through the flocculator.

Provide extra head loss $h_{\text{extra}} = 0.18$ ft and calculate the total head losses through the flocculator.

$$h_{\text{flocculator}} = h_m + h_f + h_{\text{extra}} = 1.85 \text{ ft} + 0.05 \text{ ft} + 0.18 \text{ ft} = 2.08 \text{ ft} \quad \text{or} \quad 25 \text{ in}$$

Therefore, the total head loss through the flocculator $\Delta h_{\text{flocculator}} = 2.08$ ft and the head loss per section, $\Delta h_{\text{section}} = 1.04$ ft.

Provide a total 20 floor drops at 1¼ in per drop ($\Delta h_{\text{drop}} =$ 1¼ in or 0.104 ft) in the flocculator. The total drop in floor elevation is 25 in (2.08 ft) that matches total head losses through the flocculator.

d. Calculate the head over the influent rectangular weir.

Assume $C_d = 0.62$ and calculate the head (H) over the rectangular weir across the full width of the influent channel from Equation 8.10 at $Q = 4.64 \text{ ft}^3/\text{s}$, $L_{\text{weir}} = w$ (channel width) $= 1.5$ ft, and $n = 0$.

$$L' = L_{\text{weir}} - 0.1\, nH = L_{\text{weir}} = 1.5 \text{ ft}$$

$$H = \left(\frac{3}{2} \times \frac{Q}{C_d L' \sqrt{2g}} \right)^{2/3} = \left(\frac{3}{2} \times \frac{4.64 \text{ ft}^3/\text{s}}{0.62 \times 1.5 \text{ ft} \times \sqrt{2 \times 32.2 \text{ ft/s}^2}} \right)^{2/3} = 0.95 \text{ ft}$$

Provide a freeboard $FB_{\text{weir}} = 0.55$ ft after the weir. The overall head loss at the influent weir $\Delta h_{\text{weir}} = H + FB_{\text{weir}} = (0.95 + 0.55)$ ft $= 1.50$ ft. The free fall of 1.50 ft creates the turbulence for dispersion of chemicals at the entrance of the flocculator.

7. Calculate the velocity gradient G.

   Turbulence is provided at the influent weir for initial mixing of chemicals. Assume the mixing time is about 2-4 s after the weir, and a velocity gradient between 900–1000 s^{-1} is expected. It meets the typical requirement for rapid mixing of chemicals. See Examples 9.40 and 9.41 for the calculations of velocity gradient related to the head loss at a weir. The hydraulic gradient G for flocculation through the flocculator is calculated below.

   a. Calculate the water power imparted by the friction head loss through the flocculator.

      At 60°F, the approximate water density $\gamma = 62.4\,\text{lb/ft}^3$ is obtained from Table B.3 in Appendix B.

      The water power P_w imparted by the head losses in the flocculator is calculated from Equation 9.30a at $h_{\text{flocculator}} = 2.08$ ft.

      $$P_w = \gamma Q h_{\text{flocculator}} = 62.4\,\text{lb/ft}^3 \times 4.64\,\text{ft}^3\text{/s} \times 2.08\,\text{ft} = 602\,\text{lb·ft/s}$$

   b. Calculate the velocity gradient G created by the friction head loss through the flocculator.

      At 60°F, $\mu = 2.36 \times 10^{-5}$ lb·s/ft^2 is obtained from Table B.3 in Appendix B.

      Assume the mixing volume in the entire flocculator $V = 5800\,\text{ft}^3$. Calculate the G from Equation 9.27.

      $$G = \sqrt{\frac{P_w}{\mu V}} = \left(\frac{602\,\text{lb·ft/s}}{2.36 \times 10^{-5}\,\text{lb·s/ft}^2 \times 5800\,\text{ft}^3}\right)^{1/2} = 66\,\text{s}^{-1}$$

      The velocity gradient of 66 s^{-1} is within the typical range of 50–70 s^{-1} in an around-the-end baffled hydraulic flocculator.

8. Determine the major elevations and prepare hydraulic profile through the flocculator.

   Assume the reference elevation at the floor elevation in the first channel in the first section of the flocculator. $EL_{\text{1st,upper}} = 100.00$ ft (Reference).

   a. Prepare the major WSELs.

      WSEL in the first channel (upper end) of the first section of the flocculator

      $$WSEL_{\text{1st,upper}} = EL_{\text{1st,upper}} + d = (100.00 + 4.50)\,\text{ft} = 104.50\,\text{ft}$$

      Head loss through the influent weir $\Delta h_{\text{weir}} = 1.50$ ft

      WSEL in the influent channel prior to the weir

      $$WSEL_{\text{influent}} = EL_{\text{1st,upper}} + \Delta h_{\text{weir}} = (104.50 + 1.50)\,\text{ft} = 106.00\,\text{ft}$$

      Head loss through the first section of the flocculator $\Delta h_{\text{section}} = 1.04$ ft

      WSEL in the last channel (lower end) of the first section of the flocculator

      $$WSEL_{\text{1st,lower}} = WSEL_{\text{1st,upper}} - \Delta h_{\text{section}} = (104.50 - 1.04)\,\text{ft} = 103.46\,\text{ft}$$

      Assume there is no head loss between the sections.

      WSEL in the first channel (upper end) of the second section of the flocculator

      $$WSEL_{\text{2nd,upper}} = WSEL_{\text{1st,lower}} = 103.46\,\text{ft}$$

      Head loss through the second section of the flocculator $\Delta h_{\text{section}} = 1.04$ ft

      WSEL in the last channel (lower end) of the second section of the flocculator

      $$WSEL_{\text{2nd,lower}} = WSEL_{\text{2nd,upper}} - \Delta h_{\text{section}} = (103.46 - 1.04)\,\text{ft} = 102.42\,\text{ft}$$

Note: The maximum WSEL of 102.42 ft is used in design of the effluent proportional weir.

b. Determine the elevations (ELs) of the major components.

Freeboard after the influent weir $FB_{weir} = 1.05$ ft

Top of weir elevation $TOW_{influent} = EL_{1st,upper} + FB_{weir} = (104.50 + 0.55)\ ft = 105.05\ ft$

EL of the bottom in the last channel (lower end) of the first section of the flocculator

$$EL_{1st,lower} = WSEL_{1st,lower} - d = (103.46 - 4.50)\ ft = 98.96\ ft$$

Note: The floor elevation remains the same in every four consecutive channels and then drops $\Delta h_{drop} = 1\frac{1}{4}$ in (0.104 ft) at the fifth channel among the first 40 channels in the first section. There are a total of 10 drops for a total elevation drop $\Delta h_{section} = 1.04$ ft in the first section.

$$EL_{2nd,upper} = EL_{1st,lower} = 98.96\ ft$$

Note: The floor elevation for the first three channels of the second section remains the same as that for the last three channels of the first section.

EL of the bottom in the last channel (lower end) of the second section of the flocculator

$$EL_{2nd,lower} = WSEL_{2nd,lower} - d = (102.42 - 4.50)\ ft = 97.92\ ft$$

Note: The floor elevation remains the same in the first three channels and then drops $\Delta h_{drop} = 0.104$ ft at the forth channel in the second section. Among the last 40 channels, floor elevation remains the same in every four consecutive channels and then drops $\Delta h_{drop} = 0.104$ ft at the fifth channel. There are a total of 10 drops for a total elevation drop $\Delta h_{section} = 1.04$ ft in the second section.

c. Draw the design sketches showing the important design elevations and hydraulic profile of the flocculator.

The conceptual design sketch is shown in Figure 9.46. The design details are illustrated in Figure 9.47.

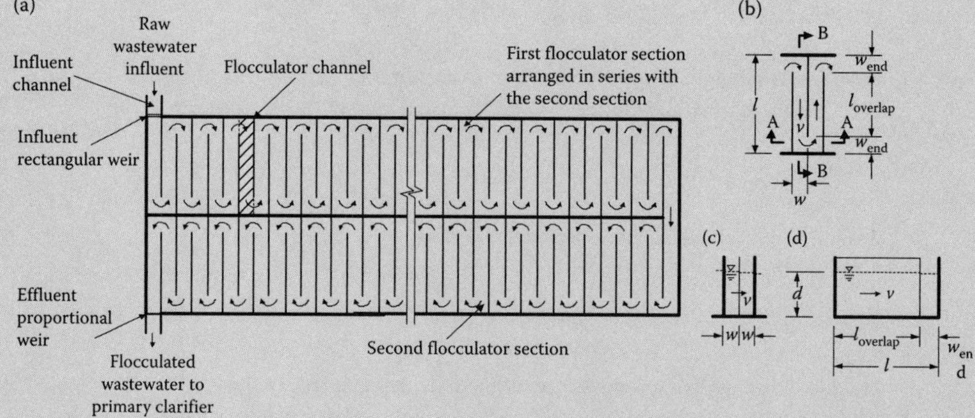

FIGURE 9.46 Definition sketch of the around-the-end baffled flocculator: (a) basin layout, (b) channel layout, (c) channel Section AA, and (d) channel Section BB (Example 9.45).

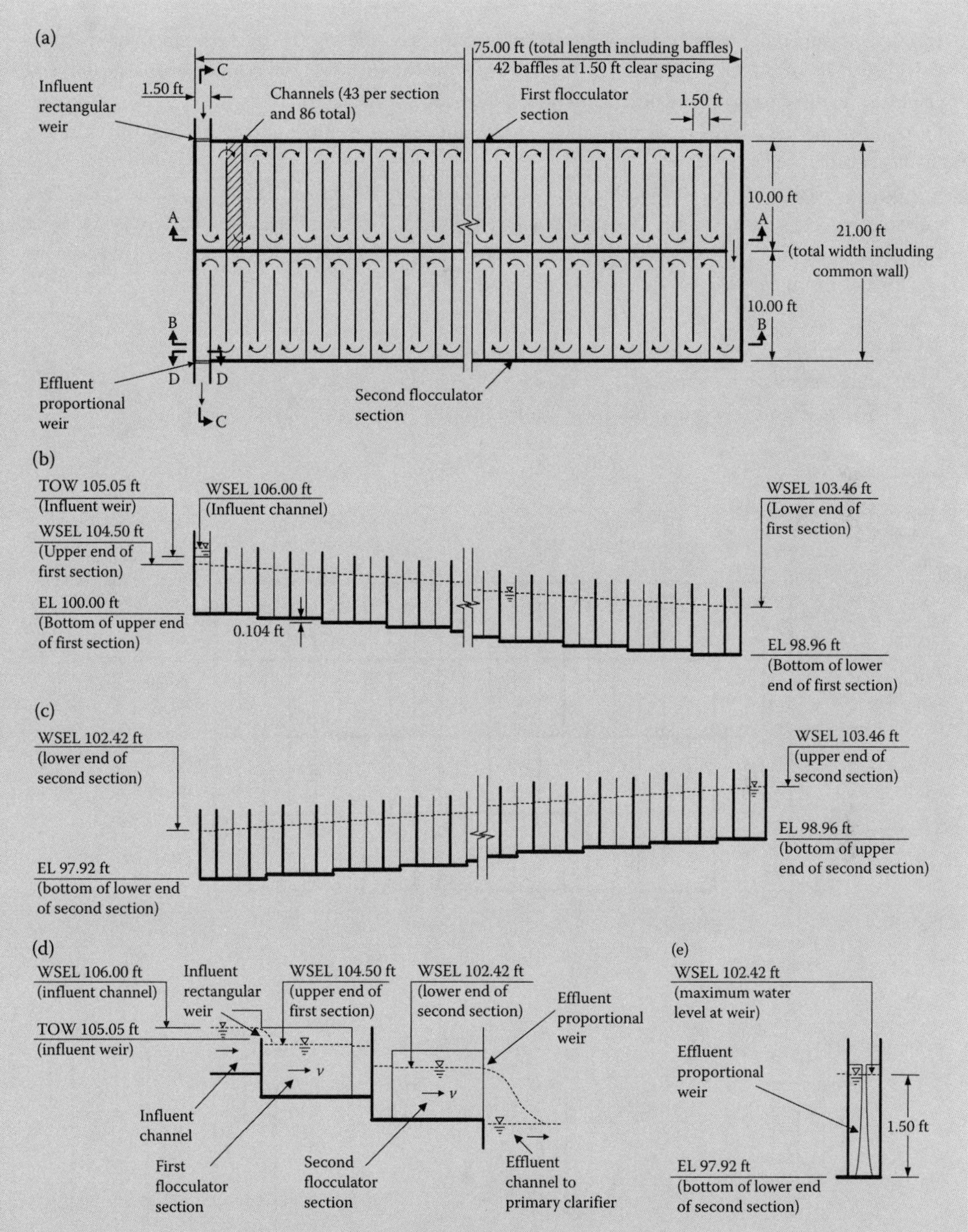

FIGURE 9.47 Design details of around-the-end flocculator: (a) plan view, (b) Section AA, (c) Section BB, (d) Section CC, and (e) Section DD (Example 9.45).

EXAMPLE 9.46: OVER-AND-UNDER BAFFLED CHANNEL FLOCCULATOR

Design an over-and-under baffled channel flocculator for a flow of 11,400 m^3/d (3 MGD). The design velocity in the flow-through basin is 0.25 m/s (0.8 fps). The detention time in the flocculator is 20 min. The flocculator is divided into four similar sections in series. Each section consists of channels that are connected by openings near the bottom or submerged weirs on the top. The water temperature is 20°C. Calculate or prepare information for (a) dimensions of the channel, (b) total number of around-the-end channels, (c) dimensions of the flocculator, and influent and effluent structures, (d) head losses through the channels, (e) the velocity gradient G, and (f) important design elevations and hydraulic profile through the flocculator.

Note: The basic design conditions for Examples 9.45 and 9.46 are similar except that the units for these examples are, respectively, in the U.S. customary and SI units. A similar design approach is used in both examples. Therefore, some design dimensions in this example are obtained from the calculation results obtained in Example 9.45.

Solution

1. Draw the definition sketch.

   The plan view of the over-and-under baffled flocculator is shown in Figure 9.48.

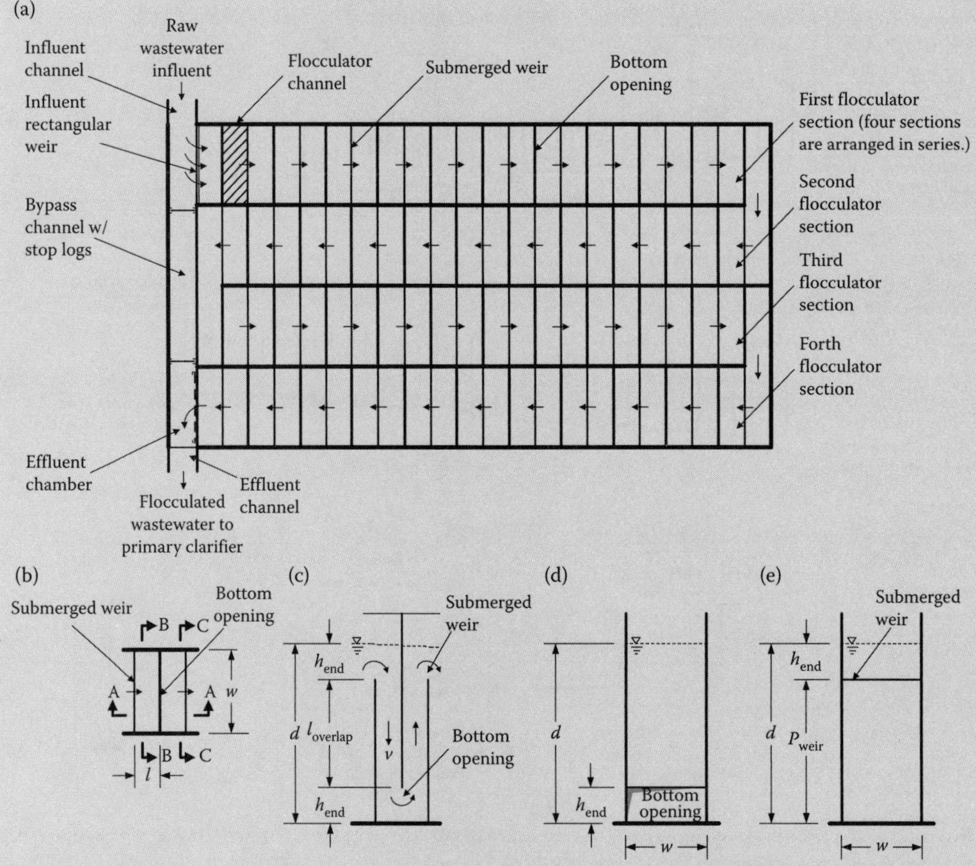

FIGURE 9.48 Definition sketch of the over-and-under baffled flocculator: (a) general layout, (b) channel layout, (c) channel Section AA, (d) channel Section BB, and (e) channel Section CC (Example 9.46).

2. Determine the dimensions of each channel.
   a. Determine the channel width and water depth.

      Provide a channel length $l = 0.45$ m.

Assume a w/l ratio of 3 and calculate the desired channel width (w).

Desired channel width, $w = 3l = 3 \times 0.45\,\text{m} = 1.35\,\text{m}$

Provide a channel width $w = 1.35$ m.

$$\text{Design flow in the channel, } Q = 11{,}400\,\text{m}^3/\text{d} \times \frac{\text{d}}{(60 \times 60 \times 24)\,\text{s}} = 0.132\,\text{m}^3/\text{s}$$

$$\text{Actual average velocity in the channel, } v = \frac{Q}{lw} = \frac{0.132\,\text{m}^3/\text{s}}{0.45\,\text{m} \times 1.35\,\text{m}} = 0.22\,\text{m/s}$$

The actual velocity is close to the design velocity of 0.25 m/s.

b. Determine the over-and-under clearance at the baffle end.

Assume the over-and-under clearance (the height of the opening near the bottom or the submergence over the weir on the top) is 1.3 times the channel length, $h_{\text{end}}/l = 1.3$.

Desired over-and-under clearance, $h_{\text{end}} = 0.60$ m.

c. Determine the overlap length of baffles.

Assume the ratio of baffle overlap length (l_{overlap}) to the channel length (l) is 4 and calculate the desired overlap length (l_{overlap}).

$$l_{\text{overlap}} = 4l = 4 \times 0.45\,\text{m} = 1.80\,\text{m}$$

d. Determine the water depth in the channel.

Water depth (d) is calculated by adding two over-and-under clearances (2 h_{end}) to the baffle overlap length (l_{overlap}).

$$d = l_{\text{overlap}} + 2h_{\text{end}} = 1.80\,\text{m} + 2 \times 0.60\,\text{m} = 3.00\,\text{m}$$

e. Determine the weir height in the channel.

Weir height (P_{weir}) is calculated by subtracting an over-and-under clearances (h_{end}) from the water depth (d).

$$P_{\text{weir}} = d - h_{\text{end}} = (3.00 - 0.60)\,\text{m} = 2.40\,\text{m}$$

f. Determine the baffle thickness.

Fiberglass baffles are used in the flocculator. Assume the thickness of the baffle wall, $t_{\text{baffle}} =$ 0.08 m or 8 cm (~3 in).

3. Calculate the number of around-the-end channels.

Calculate the total volume of the flocculator required at the hydraulic detention time $t = 20$ min and design capacity $Q = 0.132$ m^3/s.

$$\text{Total volume required, } V_{\text{flocculator}} = Qt = 0.132\,\text{m}^3/\text{s} \times 20\,\text{min} \times 60\,\text{s/min} = 158\,\text{m}^3$$

Calculate the volume of each channel of the flocculator from $l = 0.45$ m, $w = 1.35$ m, and $d =$ 3.00 m.

$$\text{Volume of each channel, } V_{\text{channel}} = l \times w \times d = 0.45\,\text{m} \times 1.35\,\text{m} \times 3.00\,\text{m}$$
$$= 1.82\,\text{m}^3\text{per channel}$$

Calculate the total number of channels required in the flocculator.

$$\text{Total number of channels required, } N_{\text{channel,total}} = \frac{V_{\text{flocculator}}}{V_{\text{channel}}} = \frac{158\,\text{m}^3}{1.82\,\text{m}^3/\text{channel}} = 87\text{ channels}$$

Provide a total of 88 channels in the flocculator ($N_{\text{channel,total}} = 88$ channels).

There are four sections in the flocculator, $N_{section} = 4$ sections.

$$\text{Number of channels per section, } N_{channel,section} = \frac{N_{channel,total}}{N_{section}} = \frac{88 \text{ channels}}{4 \text{ sections}} = 22 \text{ channels per section}$$

$$\text{Number of baffles per section, } N_{baffle,section} = N_{channel,section} - 1 = 22 - 1 = 21 \text{ baffles per section}$$

There is a 180° turn at one end of each baffle that is either an opening near the bottom or a submerged weir on the top.

$$\text{Number of } 180°\text{turns per section, } N_{turn,section} = N_{baffle,section} = 21 \text{ turns per section}$$

$$\text{Total number of } 180°\text{turns in the flocculator, } N_{turn,total} = N_{section} \times N_{turn,section} = 4 \text{ sections} \times 21 \text{ turns/section} = 84 \text{ turns}$$

There is an opening near the bottom at every two 180° turns.

$$\text{Total number of openings in the flocculator, } N_{opening,total} = \frac{1}{2} \times N_{turn,total} = \frac{1 \text{ opening}}{2 \text{turns}} \times 84 \text{turns} = 42 \text{ openings}$$

There is a submerged weir at every two 180° turns.

$$\text{Total number of weirs in the flocculator, } N_{weir,total} = \frac{1}{2} \times N_{turn,total} = \frac{1 \text{ weir}}{2 \text{ turns}} \times 84 \text{ turns} = 42 \text{ weirs}$$

4. Design the basin, influent, and effluent structures.
   a. Determine dimensions of the flocculator.

      There are two sections in the flocculator. Assume the common wall between sections $t_{wall} =$ 0.3 m.

$$\text{Overall width of the flocculator, } W = N_{section} \times w + (N_{section} - 1)t_{wall} = 4 \times 1.35 \text{ m} + (4 - 1) \times 0.3 \text{ m} = 6.30 \text{ m}$$

$$\text{Overall length of the flocculator, } L = N_{channel,section} \times l + N_{baffle,section} \times t_{baffle} = 22 \times 0.45 \text{ m} + 21 \times 0.08 \text{ m} = 11.58 \text{ m}$$

$$\text{Total volume provided, } V = N_{channel,total} V_{channel} = 88 \text{ channels} \times 1.82 \text{ m}^3/\text{section} = 160 \text{ m}^3$$

      Note: The total extra volume at turns and between the sections is small and ignorable.

$$\text{Actual hydraulic retention time, } t = \frac{V}{Q} = \frac{160 \text{ m}^3}{0.132 \text{ m}^3/\text{s} \times 60 \text{ s/min}} = 20 \text{ min}$$

   b. Describe the basin.

      The channel bottom is stepped downward to give allowance for head losses at the baffles. Ideally, the floor of each channel should be stepped down equal to the head loss encountered occurred due to the flow turbulence at one end of each baffle that is either an opening near the bottom or a submerged weir on the top. In practice, the floor is dropped equally at every four to five channels. To prevent accumulation of solids in the channels ahead of the baffle wells supported over the basin floor. Three small square sweep orifices (5 cm × 5 cm), are provided near the bottom of the baffle walls. The locations of these orifices are one in the center and two near the ends of the wall.

c. Describe the influent structure.

The influent structure consists of a rectangular weir constructed across the influent channel 0.50 m wide with a water depth of 1.20 m. A free fall of 0.45 m is required at the influent weir to provide the turbulence for rapid dispersion of chemicals at the entrance of the flocculator.

d. Describe the effluent structure.

The effluent structure consists of an effluent opening with stop gate an effluent chamber, and an effluent channel. The flocculated effluent flows from the last channel of the fourth section into the chamber through the 1.20 m × 1.20 m submerged opening. Stop log can be used to block the opening when bypass operation is desired. The effluent channel is 0.50 m wide with a water depth of 1.20 m. It leads to the influent distribution channel of the primary clarifier.

e. Describe the bypass channel.

A bypass channel is provided between the influent channel and the effluent chamber. The channel is normally closed by using isolation stop logs. The raw wastewater can flow into the primary clarifier directly when the channel is opened and the flocculator is taken out of service.

5. Calculate the head losses through the flocculator.

Hydraulic calculations are performed for (a) the total minor head losses at the openings due to 180° turns, (b) the total minor head losses at the submerged weirs, (c) the total friction loss through the total length of flow path in the flocculator, (d) the total head losses through the flocculator, and (e) the head over the influent rectangular weir. The head loss through the effluent opening is ignored.

a. Calculate the total minor head losses at the openings underneath the baffles in the flocculator.

The minor head loss at each opening due to 180° turn is calculated from Equation 6.15b using flow-through velocity $v = 0.22$ m/s and assuming $K_m = 3$ for each opening.

$$\text{Head loss at each opening, } \Delta h_{\text{opening}} = K_m \frac{v^2}{2g} = 3 \times \frac{(0.22\,\text{m/s})^2}{2 \times 9.81\,\text{m/s}^2} = 0.007\,\text{m per opening}$$

$$\begin{aligned}\text{Total minor head losses at the openings, } h_{\text{opening}} &= N_{\text{opening,total}} \times \Delta h_{\text{opening}} \\ &= 42\ \text{openings} \times 0.007\,\text{m/turn} = 0.29\,\text{m}\end{aligned}$$

b. Calculate the total minor head losses at the submerged weirs on top of the baffles in the flocculator.

A submerged weir condition is encountered when the coagulated water makes a 180° turn on top of a baffle. The impact of submergence on the discharge over the weir is evaluated to ensure the minor head loss calculation is used properly at the weir. The discharge over the weir under free-fall condition (Equation 8.10) is modified using the Villenmonte relationship (Equation 9.34) to estimate the submerged flow under the submerged weir head. The upward vertical velocity in the channel prior to the weir has been considered in the minor head loss calculation. It is therefore ignored in this evaluation.

Submerged head after the weir $h_d = h_{\text{end}} = 0.6$ m.

The upstream head h_u is evaluated through trial-and-error procedure. Start the first iteration by assuming $h_u = h_d = 0.6$ m and calculate the equivalent free-fall (unsubmerged) weir discharge Q_e over the rectangular weir across the channel from Equation 8.10 ($C_d = 0.62$, $L_{\text{weir}} = w$ (channel width) $= 1.35$ m, and $n = 0$).

$$L' = L_{\text{weir}} - 0.1\,nH, \text{ nd } L_{\text{weir}} = 1.35\,\text{ft}$$

$$Q_e = \frac{2}{3} \times C_d L' \sqrt{2gh_u^3} = \frac{2}{3} \times 0.62 \times (1.35\,\text{m}) \times \sqrt{2 \times 9.81\,\text{m/s}^2 \times (0.6\,\text{m})^3} = 1.15\,\text{m}^3/\text{s}$$

Calculate the flow reduction factor from Equation 9.34a for the submerged flow $Q_s = Q = 0.132\ \text{m}^3/\text{s}$ and $n = 1.50$.

$$F_s = \frac{Q_s}{Q_e} = \frac{0.132\ \text{m}^3/\text{s}}{1.15\ \text{m}^3/\text{s}} = 0.115$$

Rearrange Equation 9.34a and calculate h_u using $n = 1.50$.

$$h_u = \frac{h_d}{\left[1 - F_s^{1/0.385}\right]^{1/n}} = \frac{0.6\ \text{m}}{\left[1 - (0.115)^{1/0.385}\right]^{1/1.5}} = 0.601\ \text{m}$$

After the first iteration, an upstream head $h_u = 0.601$ m is required to achieve a submerged outfall flow $Q_s = Q = 0.132\ \text{m}^3/\text{s}$. Start the second iteration using $h_u = 0.601$ m and repeat the calculation steps in the first iteration. The same value of $h_u = 0.601$ m is obtained after the second iteration. Assume a factor of 2.0 to calculate the head loss due to turn prior to the weir and the minor head losses due to turbulences.

$\Delta h_{weir} = 2 \times (h_u - h_d) = 2 \times (0.601 - 0.6)\ \text{m} = 0.002$ m at each weir.

Total minor head losses at the submerged weirs,

$h_{weir} = N_{weir,total} \times \Delta h_{weir} = 42\ \text{weirs} \times 0.002\ \text{m/weir} = 0.08\ \text{m}$

c. Calculate the total friction loss through the total length of flow path in the flocculator.

The friction head loss through the channels is calculated from the Darcy–Weisbach equation (Equation 6.13a).

$$h_f = \frac{f l_{path,total} v^2}{2gD_h}$$

Estimated length of flow path per channel,

$$l_{path,channel} = d - 2 \times \left(\frac{1}{2} \times h_{end}\right)$$

$$= \left[3 - 2 \times \left(\frac{1}{2} \times 0.60\right)\right]\ \text{m} = 2.40\ \text{m per channel}$$

Total length of flow path in the flocculator, $l_{path,channel} = N_{channel,total} \times N_{channel,total}$

$= 88\ \text{channels} \times 2.40\ \text{m/channel} = 211\ \text{m}$

Calculate the hydraulic mean diameter of the rectangular channel from Equation 6.14e with $l = 0.45$ m and $w = 1.35$ m.

Hydraulic mean radius, $R = \frac{A}{P} = \frac{lw}{2(l + w)} = \frac{0.45\ \text{m} \times 1.35\ \text{m}}{2 \times (0.45\ \text{m} + 1.35\ \text{m})} = 0.17\ \text{m}$

Hydraulic mean diameter, $D_h = 4R = 4 \times 0.17\ \text{m} = 0.68\ \text{m}$

Assume $f = 0.03$ for the baffles. Calculate the friction head loss at $v = 0.22$ m/s.

$$h_f = \frac{0.03 \times 211\ \text{m} \times (0.22\ \text{m})^2}{2 \times 9.81\ \text{m/s}^2 \times 0.68\ \text{m}} = 0.02\ \text{m}$$

d. Calculate the total losses through the flocculator.

Provide extra head loss $h_{extra} = 0.11$ m and calculate the total head losses through the flocculator.

$$h_{flocculator} = h_{opening} + h_{weir} + h_f + h_{extra} = (0.29 + 0.08 + 0.02 + 0.11)\,\text{m} = 0.50\,\text{m}$$

Therefore, the total head loss through the flocculator $\Delta h_{section} = 0.50$ m, and the head loss per section $\Delta h_{section} = 0.5$ m/4 = 0.125 m.

Provide a total 20 floor drops at 2.5 cm per drop ($\Delta h_{drop} = 2.5$ cm or 0.025 m) in the flocculator. The total drop in floor elevation is 50 cm (0.50 m) that matches total head losses through the flocculator.

e. Calculate the head over the influent rectangular weir.

Assume $C_d = 0.62$ and calculate the head (H) over the rectangular weir across the full width of the influent channel from Equation 8.10 at $Q = 0.132$ m^3/s, $L_{weir} = w$ (channel width) = 1.35 m, and $n = 0$.

$$L' = L_{weir} - 0.1nH = L_{weir} = 1.35\,\text{m}$$

$$H = \left(\frac{3}{2} \times \frac{Q}{C_d L' \sqrt{2g}}\right)^{2/3} = \left(\frac{3}{2} \times \frac{0.132\,\text{m}^3/\text{s}}{0.62 \times 1.35\,\text{ft} \times \sqrt{2 \times 9.81\,\text{m/s}^2}}\right)^{2/3} = 0.14\,\text{m}$$

Calculate the freeboard (FB_{weir}) required at the weir to provide an assumed free fall (overall head loss) $\Delta h_{weir} = 0.45$ m.

$$FB_{weir} = \Delta h_{weir} - H = (0.45 - 0.14)\,\text{m} = 0.31\,\text{m}$$

6. Calculate the velocity gradient G.

A velocity gradient between 900–1000 s^{-1} is estimated for a weir under the similar flow condition (see Example 9.45). The hydraulic gradient G for flocculation through the flocculator is calculated below.

a. Calculate the water power imparted by the friction head loss through the flocculator.

At 20°C, $\rho = 998.2$ kg/m^3 is obtained from Table B.2 in Appendix B.

The water power P_w imparted by the head losses in the flocculator is calculated from Equation 9.30a at $h_{flocculator} = 0.50$ m.

$$P_w = \rho g Q h_{flocculator} = 998.2\,\text{kg/m}^3 \times 9.81\,\text{m/s}^2 \times 0.132\,\text{m}^3/\text{s} \times 0.50\,\text{m}$$
$$= 646\,\text{kg·m}^2/\text{s}^3 \quad \text{or} \quad 646\,\text{N·m/s}$$

b. Calculate the velocity gradient G created by the friction head loss through the flocculator.

At 20°C, $\mu = 1.002 \times 10^{-3}$ N·s/m^2 is obtained from Table B.2 in Appendix B.

Assume the mixing volume is achieved in the entire flocculator $V = 160$ m^3, and calculate the G from Equation 9.27.

$$G = \sqrt{\frac{P_w}{\mu V}} = \left(\frac{646\,\text{N·m/s}}{1.002 \times 10^{-3}\,\text{N·s/m}^2 \times 160\,\text{m}^3}\right)^{1/2} = 63\,\text{s}^{-1}$$

The velocity gradient of 63 s^{-1} is acceptable range for an over-and-under baffled hydraulic flocculator for mixing requirement while minimizing the settling of floc in the channel.

7. Determine major elevations and prepare hydraulic profile through the flocculator.

Assume the reference elevation is the floor elevation in the first channel in the first section of the flocculator. $EL_{1st,upper} = 100.00$ m (Reference).

a. Prepare major WSELs.

The procedure for determining WSELs through the flocculator is similar to that presented in Example 9.45.

WSEL in the first channel (upper end) of the first section of the flocculator

$WSEL_{1st,upper} = EL_{1st,upper} + d = (100.00 + 3.00)\ m = 103.00\ m$

Head loss through the influent weir $\Delta h_{weir} = 0.45$ m

WSEL in the influent channel prior to the weir

$WSEL_{influent} = EL_{1st,upper} + \Delta h_{weir} = (103.00 + 0.45)\ m = 103.45\ m$

Other WSELs at different locations through the flocculator are calculated and summarized below.

Location	Design WSELs, m		Head Loss, m
	Upper End	Lower End	
First section	103.00	102.88	$\Delta h_{section} = 0.125$ m
Second section	102.88	102.75	$\Delta h_{section} = 0.125$ m
Third section	102.75	102.63	$\Delta h_{section} = 0.125$ m
Fourth section	102.63	102.50[a]	$\Delta h_{section} = 0.125$ m

[a] The head loss through the bottom opening after the last channel of fourth section is small and ignored. Therefore, this is also the WSEL at the upper end of the effluent channel of the flocculator, $WSEL_{effluent} = 102.50$ m. The elevation is used in design of the influent distribution channel of the primary clarifier.

b. Determine elevations (ELs) of the major components.

The procedure for determining ELs of the major components of the flocculator is similar to that presented in Example 9.45.

Freeboard after the influent weir $FB_{weir} = 0.31$ m

Top of weir elevation $TOW_{influent} = EL_{1st,upper} + FB_{weir} = (103.00 + 0.31)\ m = 103.31\ m$

Water depth in the influent channel $d_{influent} = 1.20$ m

Bottom EL of the influent channel, $EL_{influent} = WSEL_{influent} - d_{influent} = (103.45 - 1.20)\ m = 102.25\ m$

Water depth in the effluent channel $d_{effluent} = 1.20$ m

Bottom EL of the effluent channel, $EL_{effluent} = WSEL_{effluent} - d_{effluent} = (102.50 - 1.20)\ m = 101.30\ m$

Other major design ELs at different locations through the flocculator are calculated and summarized below.

Notes: In each section, the floor elevation remains horizontal in the first three channels and then drops $\Delta h_{drop} = 0.025$ m at the fourth channel. The floor elevation remains the same in the last three channels. This sequence is continued in the floors of all four sections in the basin.

Location	Floor ELs, m		Top of Weir ELs, m	
	Upper End	Lower End	Upper End	Lower End
First section	100.00	99.88[a]	102.40	102.28[b]
Second section	99.88	99.75	102.28	102.15
Third section	99.75	99.63	102.15	102.03
Fourth section	99.63	99.50	102.03	101.90

[a] $EL_{First,lower} = WSEL_{First,lower} - d = (102.88 - 3.00)\ m = 99.88\ m$

[b] $TOW_{influent} = EL_{First,lower} + P_{weir} = (99.88 + 2.40)\ m = 102.28\ m$

c. Draw the design sketches showing the important design elevations and hydraulic profile of the flocculator.

The conceptual design sketches are shown in Figure 9.48. The design details are illustrated in Figure 9.49.

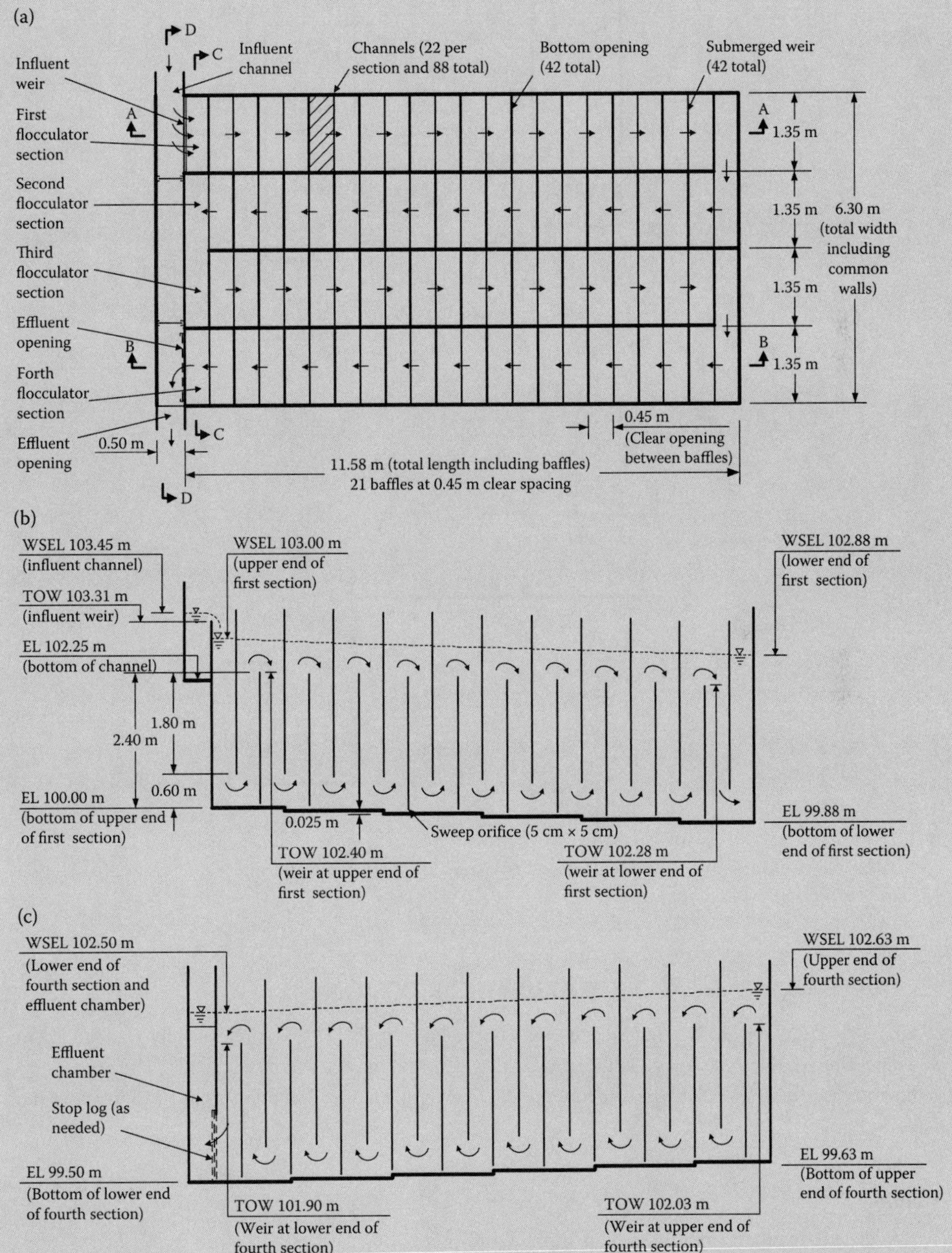

FIGURE 9.49 Design details and hydraulic profile of over-and-under flocculator: (a) plan, (b) Section AA, and (c) Section BB (Example 9.46). *(Continued)*

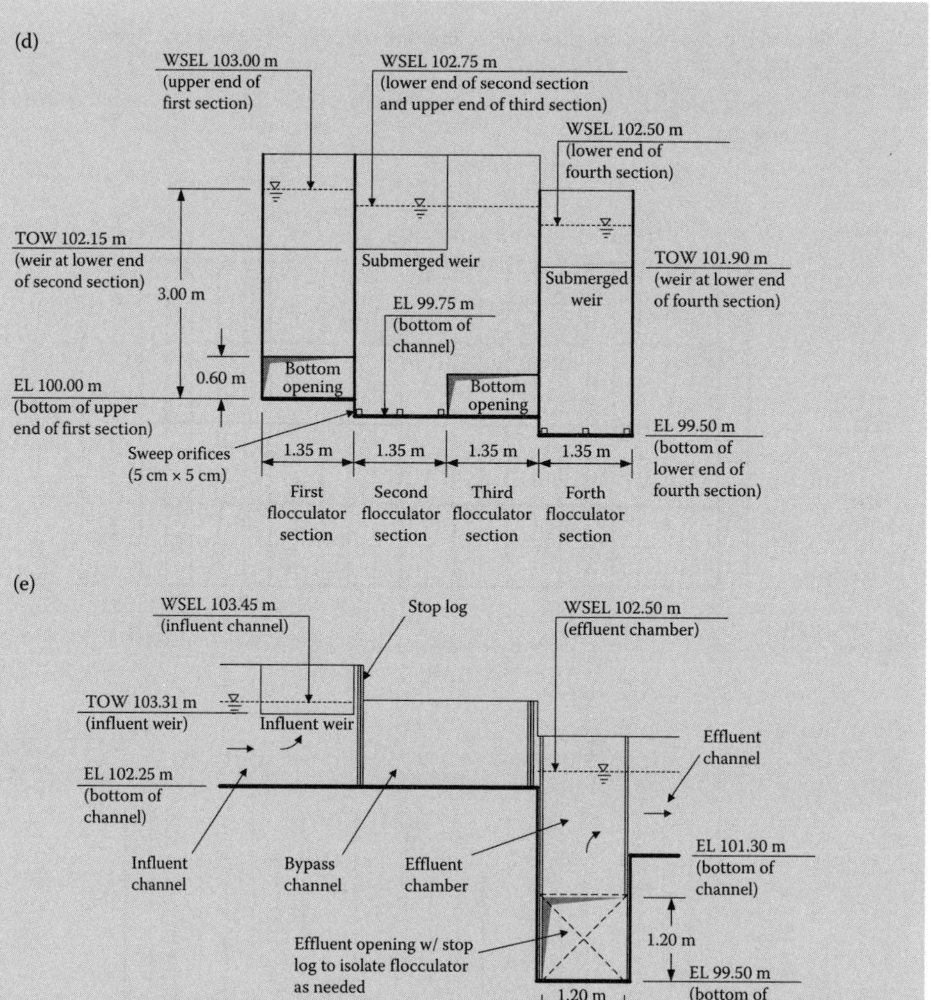

FIGURE 9.49 (Continued) Design details and hydraulic profile of over-and-under flocculator: (d) Section CC, and (e) Section DD (Example 9.46).

EXAMPLE 9.47: PNEUMATIC RAPID MIXER DESIGN

A square rapid-mixing basin is designed to treat a flow of 8000 m^3/d. The required velocity gradient is 700 s^{-1}. The depth of the square basin is 1.25 times the width, and the detention time is 45 s. The wastewater temperature is 10°C. Calculate the air required for pneumatic mixing. The diffusers are 0.15 m above the tank bottom.

Solution

1. Calculate the volume and dimensions of the rapid-mix basin.

$$Q = 8000\,\text{m}^3/\text{d} \times \frac{1}{(24 \times 60 \times 60)\,\text{s/d}} = 0.0926\,\text{m}^3/\text{d}$$

Calculate the basin volume at hydraulic retention time $t = 45$ s.

$$V = Qt = 0.0926\,\text{m}^3/\text{d} \times 45\,\text{s} = 4.17\,\text{m}^3$$

Calculate basin width required in the square basin ($l = w$) at the depth-to-width ratio of 1.25 ($d =$ 1.25 w).

$$V = dl\,w = 1.25\,w^3$$

$$w = \left(\frac{V}{1.25}\right)^{1/3} = \left(\frac{4.17\,\text{m}^3}{1.25}\right)^{1/3} = 1.49\,\text{m}$$

Provide $l = w = 1.5$ m and calculate the desired water depth d.

$$d = 1.25\,w = 1{,}25 \times 1.5\,\text{m} = 1.9\,\text{m}$$

Actual basin volume, $V = dlw = 1.90\,\text{m} \times (1.50\,\text{m})^2 = 4.28\,\text{m}^3$

2. Calculate the water power required for providing the desired G.
   At 10°C, $\mu = 1.307 \times 10^{-3}$ N·s/m^2 is obtained from Table B.2 in Appendix B.
   Calculate the water power requirement for providing $G = 700\,\text{s}^{-1}$ in the rapid-mixing basin $V =$ 4.28 m^3 from Equation 9.27.

$$\text{Water power, } P_w = \mu V G^2 = 1.307 \times 10^{-3}\,\text{N·s/m}^2 \times 4.28\,\text{m}^3 \times (700\,\text{s}^{-1})^2 = 2741\,\text{N·m/s} \quad \text{or} \quad 2741\,\text{W}$$

3. Calculate the air flow rate for pneumatic mixing.
   Calculate submerged head at the air discharge point that is 0.15 m above the tank bottom.

$$h_d = d - 0.15\,\text{m} = (1.90 - 0.15)\,\text{m} = 1.75\,\text{m}$$

Air supply rate is calculated from Equation 9.31b at $K = 1689$ N/m^2.

$$q_a = \frac{P_w}{K \ln(h_d + 10.33/10.33)} = \frac{2741\,\text{N·m/s}}{1689\,\text{N/m}^2 \ln(1.75\,\text{m} + 10.33\,\text{m}/10.33\,\text{m})} = 10.4\,\text{m}^3/\text{s}$$

9.6 High-Rate Clarification

The high-rate clarification (HRC) process involves rapid settling of settleable solids. To be efficient, it is typically applied to properly flocculated suspensions after chemical addition. However, it may be applied to nonflocculated particles at a reduced surface overflow rate. The HRC facility features a compact design with highly clarified effluent. It is also applied to retrofit existing overloaded clarifiers. HRC process can be achieved in a (1) solids contact clarifier, (2) inclined plate or tube settler, (3) ballasted flocculation and sedimentation, and (4) combined flocculator–clarifier facility. These systems are briefly discussed below.[3,5,30,43,50–55]

9.6.1 Solids Contact Clarifiers

Solids contact clarifiers, also called reactor clarifiers utilize the principle of solids contact. The influent is brought in contact with a sludge layer that is maintained near the tank bottom. The layer acts as a

TABLE 9.24 Process Design Parameters and Potential Removal Efficiencies of Using Solid Contact Clarifiers

Design Parameter	Value or Range
Design Parameters	
Detention time (h)	1–2
Surface overflow rate, $m^3/m^2 \cdot d$ (gpd/ft^2)	50–75 (1200–1850)
Weir loading rate, $m^3/m \cdot d$ (gpd/ft)	175–350 (14,000–28,000)
Solids recirculation rate, times of influent flow rate	3–5
Solids content of sludge, % dry solids[a]	1–3
Process Performance[b]	
TSS (%)	94
BOD or COD (%)	55
Ortho-P (%)	97

[a] Solids content shall be properly maintained. High organic contents in the sludge may create undesirable septic condition due to long sludge-holding time in the sludge blanket.

[b] The removal efficiencies are summarized for industrial wastewater treatment using solids contact clarifier after lime precipitation.[30]

Source: Summarized from information contained in References 2, 26, 28–30 and 55.

blanket, and the incoming solids agglomerate and remain enmeshed with the blanket. The liquid rises upward, while a distinct interface retains the solids below. Circular reactors are ideally suited to combine the influent central well with a flocculation compartment with paddles or mixer. The concentric outer compartment is the clarification zone. These clarifiers have better hydraulic performance, and have reduced detention time for equivalent solids removed in the horizontal-flow clarifiers. Solids contact clarifiers can be applied effectively in CEPT, tertiary treatment, and industrial wastewater treatment, especially when lime precipitation is desired. General process design parameters and potential process performance are summarized from publicly available data and presented in Table 9.24. The flow schematic and major components of example solids contact clarifiers are shown in Figure 9.50.[56,57] The manufacturer of the solid contact clarifier usually provide detailed basin design recommendations to accommodate their equipment.[56–58]

9.6.2 Inclined Surface Clarifiers

Process Description: The inclined surface clarifiers utilize inclined trays arranged inside a circular or rectangular tank (Figure 9.51a). Since the flow moves between the plates that are in an overlaying arrangement, the effective settling area is equal to the sum of each plate's area projected on the horizontal surface. It can be up to 10 times larger than the physical surface area of the basin. Therefore, the footprint of the clarifier, the falling depth of particles, and the settling time between the plates are all significantly reduced. There is also no wind effect, and the flow is laminar with less short-circuiting. The inclined surface concept is widely used to upgrade the existing overloaded primary and secondary clarifiers. It is possible to increase the treatment capacity up to 75%. The inclined surface clarifiers can also be used efficiently in wet weather flow management. The major drawbacks of inclined surface clarifiers include: (1) long period of sludge deposit on inner walls may cause septic conditions, (2) effluent quality may deteriorate when sludge deposits slide down, (3) there may be clogging of the inner tubes and channels, and (4) serious short-circuiting may occur when influent is warmer than the basin temperature.

Types and Configuration: There are two design variations to the inclined surface clarifiers: (a) tube settlers and (b) parallel plate or lamella separators. The tube settlers are constructed using thin-wall tubes

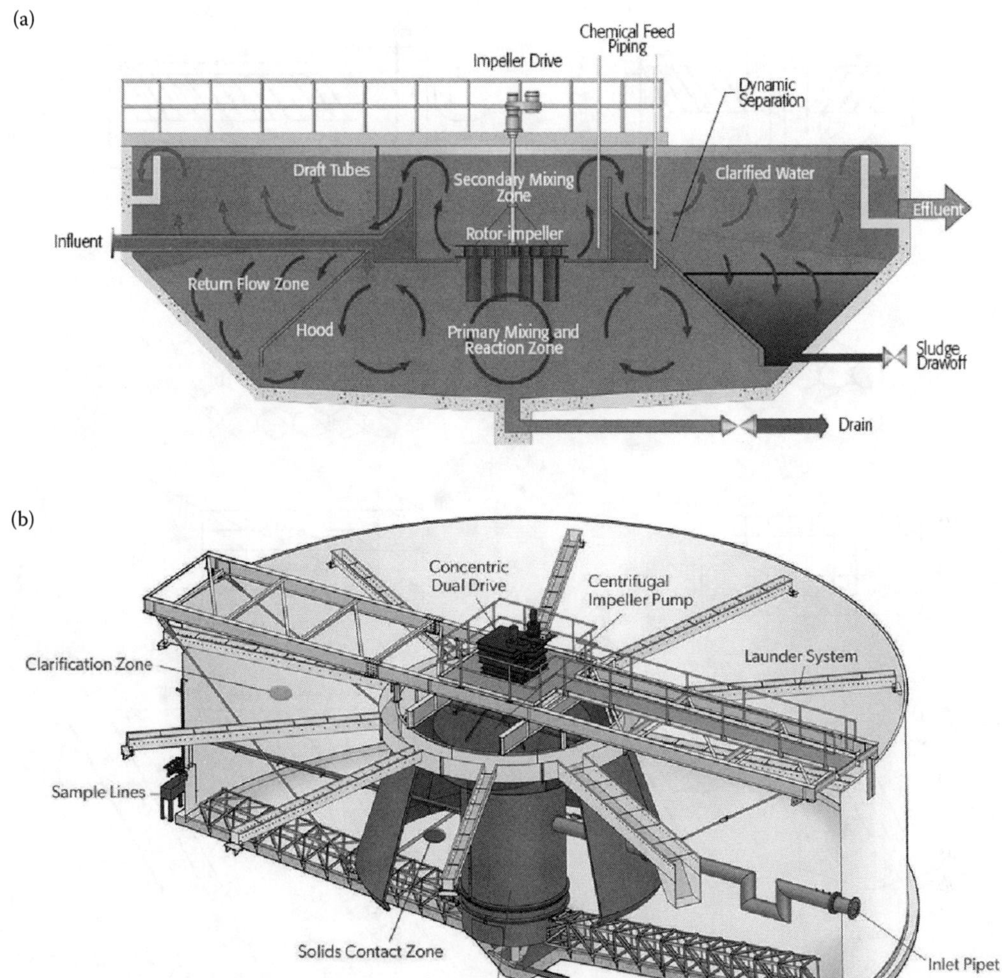

FIGURE 9.50 Solids contact clarifiers: (a) Accelator® Clarifier (Courtesy SUEZ), and (b) solids CONTACT CLARIFIER™ (Courtesy WesTech Engineering, Inc.).

in an inclined position within a basin. These tubes are circular, square, hexagonal, or any other geometric shape (Figure 9.51b). The incoming flow passes through these tubes. The solids settle on the inside of the tubes and slide down into a hopper. The parallel plate or lamella separators have parallel trays covering the entire width of the tank (Figure 9.51c).

Flow Patterns: The inclined plate or tube settler is usually designed to have one of the following three alternative flow patterns: (a) countercurrent, (b) cocurrent, and (c) cross-flow. These flow patterns are shown in Figure 9.51c. The flow pattern description, application, and design equations of inclined plate and tube settlers (Equations 9.34 through 9.37) are summarized in Table 9.25.

Design Recommendations and Applications: Both plate and tube settlers may have detention time <20 min, but they still have settling efficiency comparable to a rectangular basin of 2-h detention time. The operational principles of parallel plate separators are the same as those for the tube settlers. The self-cleaning plates or tubes are usually inclined between 55° and 60° to the horizontal. Steeper inclination reduces the efficiency, while lower inclination tends to accumulate solids within

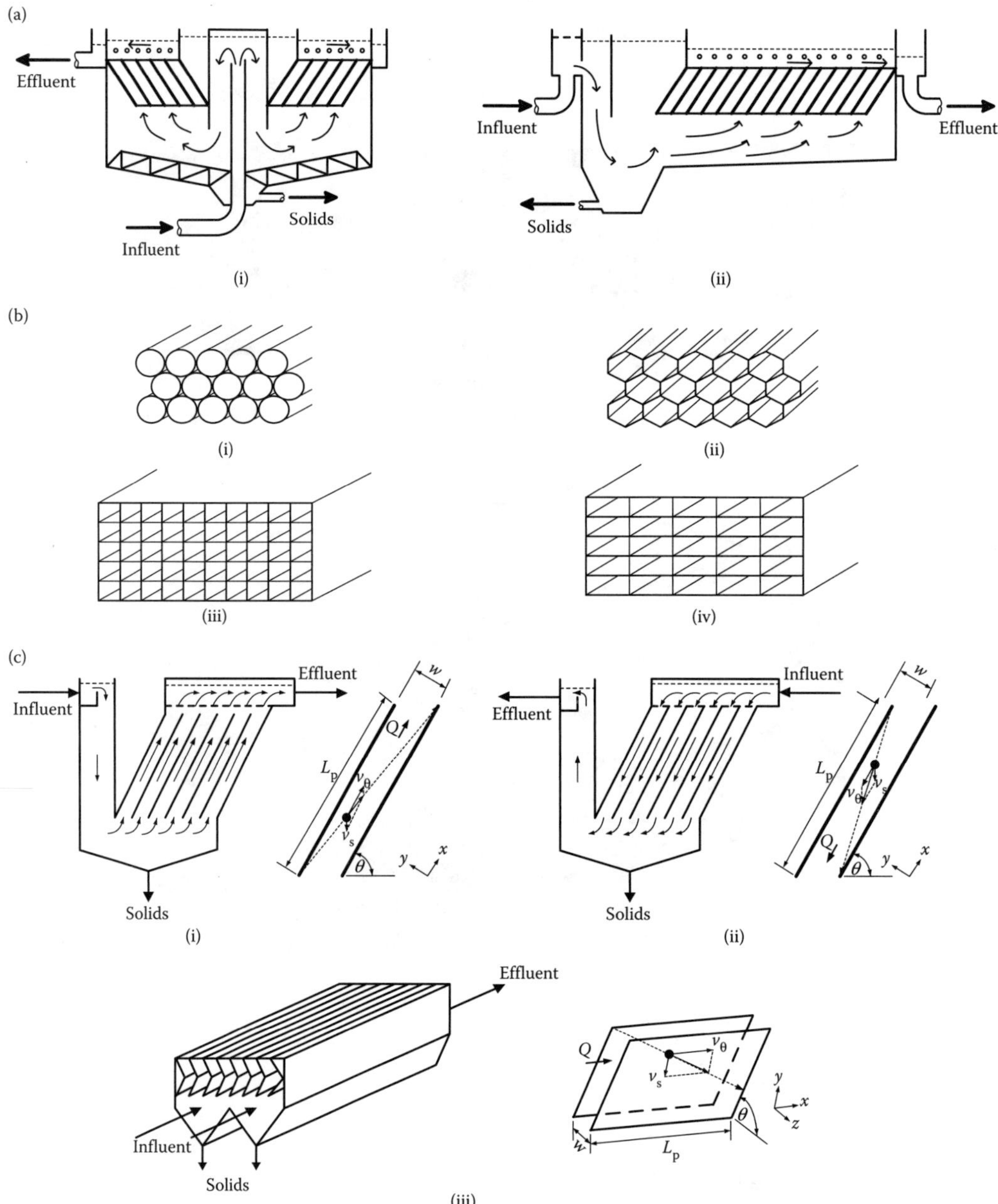

FIGURE 9.51 Inclined surface clarifier arrangements, configurations, and flow regimes: (a) general arrangements: (i) circular and (ii) rectangular; (b) typical configurations of tube settler: (i) circular, (ii) hexagonal, (iii) square, and (iv) rectangular; and (c) alternative flow patterns, (i) countercurrent, (ii) cocurrent, and (iii) cross-flow.

the plates or tubes. The normal spacing between the plates is 5–7.5 cm (2–3 in) at inclined length of 1–2 m (3–6 ft). The fraction of tank covered by plates or tubes is normally <75% in horizontal rectangular clarifiers. General design recommendations and potential process performance are summarized from publicly available data and presented in Table 9.26. In wastewater treatment practice, the inclined surface clarifiers is sized based on an overall surface overflow rate that is determined

TABLE 9.25 Flow Patterns in Inclined Plate and Tube Settlers

Flow Pattern	Description	Application	Design Equation	
Countercurrent	Flow enters from the bottom of tubes or plates, moves upward and exists from the top. Solids settle and slide downward (Figure 9.51c(i)).	Plates and tubes	$t = w/v_s \cos\theta$	(9.34a)
			$v_\theta = Q/Nwb$	(9.34b)
			$v_s \geq v_\theta w/L_p \cos\theta + w \sin\theta$	(9.35a)
			$L_p = w(v_\theta - v_s \sin\theta)/v_s \cos\theta$	(9.35b)
Cocurrent	Flow enters the tubes or plates from the top and moves downward. Solids settle and slide downward (Figure 9.51c(ii)).	Plates and tubes	$t = w/v_s \cos\theta$	
			$v_\theta = Q/Nwb$	
			$v_s \geq v_\theta w/L_p \cos\theta - w \sin\theta$	(9.36a)
			$L_p = w(v_\theta + v_s \sin\theta)/v_s \cos\theta$	(9.36b)
Cross-flow	The flow enters the space between the plates and moves horizontally along the longitudinal axis of the plate (Figure 9.51c(iii))	Plates only	$t = w/v_s \cos\theta$	
			$v_\theta = Q/Nwb$	
			$v_s \geq v_\theta w/L_p \cos\theta$	(9.37a)
			$L_p = v_\theta w/v_s \cos\theta$	(9.37b)

Note: t = settling time of particle to settle the vertical distance between two parallel inclined surfaces(s); w = perpendicular distance between parallel surfaces, m (ft); v_s = settling velocity of particles that are fully removed, m/s (ft/s); θ = angle of inclination to the horizontal, degrees; v_θ = liquid velocity in parallel to Q between the surfaces, m/s (ft/s); L_p = length of surface to travel in time t, m (ft); Q = flow rate, m^3/s (ft^3/s); N = number of channels made by (N+1) plates or tubes; and b = dimension of the surface at right angles to w and Q, m (ft).

Source: Partly adapted from References 3 through 5, 28 through 30, 43, and 55.

using the entire clarifier area. In water treatment, the surface overflow rate may be applicable to the portion of the clarifier area covered by the inclined surface settlers. Some settler manufacturers may also prefer using the specific surface overflow rates based on the total projected effective plate area within its settler cartridge. Other design considerations include: (1) equal-flow distribution through each plate or tube, (2) effective solid removal without resuspension, and (3) periodic flushing and cleaning of the narrow space between the plates and tubes.[3–5,28–30,43,55] Pilot test results with properly considered safety factors are widely used to establish the design criteria. HRC systems using inclined plate and tube clarifiers are commercially supplied as packaged systems.[59–61] An example of these systems is illustrated in Figure 9.52.[59]

9.6.3 Micro-Sand Ballasted Flocculation Process

Process Description: This is a physical–chemical treatment process that combines coagulation, flocculation, and sedimentation into a unique HRC system with the aid of polymer and micro-sand as a ballast material. The micro-sand acts as a seed and ballast for formation of high-density floc, while the polymer coats the sand particles and provides a medium to bind the floc onto the micro-sand. The process offers many advantages over the conventional coagulation/flocculation process.[5,50–54,62–65] Some of these are:

1. The density of floc is increased.
2. Coefficient of drag is decreased and Reynolds number is increased.
3. The shape factor is decreased as the sand and floc particles are more dense and spherical. They offer high surface-area-to-volume ratio to serve as effective seed.
4. The sand–floc particles exhibit discrete settling behavior.
5. Micro-sand does not react with the process chemistry, allowing it to be effectively separated from chemical sludge and reused.
6. The process has small footprint (10% space of conventional process), quick start-up, and high performance.

TABLE 9.26 General Design Parameters and Potential Removal Efficiencies of Inclined Surface Clarifiers

Design Parameters	Value or Range	
Detention time in clarifier (settling time in settler), min		
Primary treatment	10–30 (3–8)	
CEPT	5–8 (1–3)	
Wet weather flow treatment	2–6 (1–2.5)	
Surface overflow rate in the clarifier at peak flow, $m^3/m^2{\cdot}h$ (gpm/ft^2)		
Primary treatment	10–15 (4–6)	
CEPT	30–40 (12–16)	
Wet weather flow treatment	38–75 (15–30)	
Flow velocity between plates or through tubes, m/min (ft/min)		
Primary treatment	0.25–0.5 (0.8–1.6)	
CEPT	0.75–1.25 (2.5–4)	
Wet weather flow treatment	1–2.4 (3–8)	
Angle of inclination,°	55–60	
Normal spacing between the plates, cm (in)	5–10 (2–4)	
Length of the plates, m (in)	1–2 (3–6)	
Fraction of clarifier covered by the inclined surface settlers, %	75–100	
Side water depth, m (ft)	3–4 (10–13)	
Weir loading rate, $m^3/m{\cdot}d$ (gpd/ft)	100–360 (8,000–30,000)	
Chemical dosage in CEPT followed by inclined surface clarifier, mg/L		
Coagulant	60	
Polymer	1.5	
Solids content of sludge, % dry solids	~2.5–3	
Process Performance[a], Percent Removal	*Stormwater*	*Raw wastewater*
TSS	50–70	80
BOD	40–55	60
TKN	20–40	25
Phosphorus	70–75	80

[a] Summarized from the pilot study results of treating simulated stormwater by Parkson Lamella Clarifier in Reference 52.

Source: Summarized from information contained in References 3 through 5, 28 through 30, 43, 51, 52, and 55.

Major Components: The micro-sand ballasted flocculation process (MSBFP) system is typically arranged in four stages in series: (a) a coagulation chamber where a metal coagulant is added and flash-mixed with the influent that is normally screened by fine screen (3–6 mm) and degritted to remove large grits and particulates, (b) an injection chamber where micro-sand is added after injection of a polymer, (c) a maturation chamber where flocculation process is provided to enhance floc formation under gentle mixing, and (d) a clarifier where inclined plate or tube settlers are used for enhanced clarification. Settled solids are withdrawn from the clarifier. Micro-sand is separated from the residual solids using a hydrocyclone and reintroduced into the maturation chamber. MSBFP systems can be provided by equipment manufacturers. An example of MSBFP system called RapiSand™ process supplied by WesTech Engineering, Inc., is illustrated in Figure 9.53a.[62]

Design Recommendations: The MSBFP is supplied by the manufacturer provided as an integrated proprietary system. The system manufacturer guarantees the process performance, and is usually responsible for providing special design recommendations for each application on case-by-case basis. Pilot-scale study is frequently used to demonstrate the process performance and validate the process parameters. General

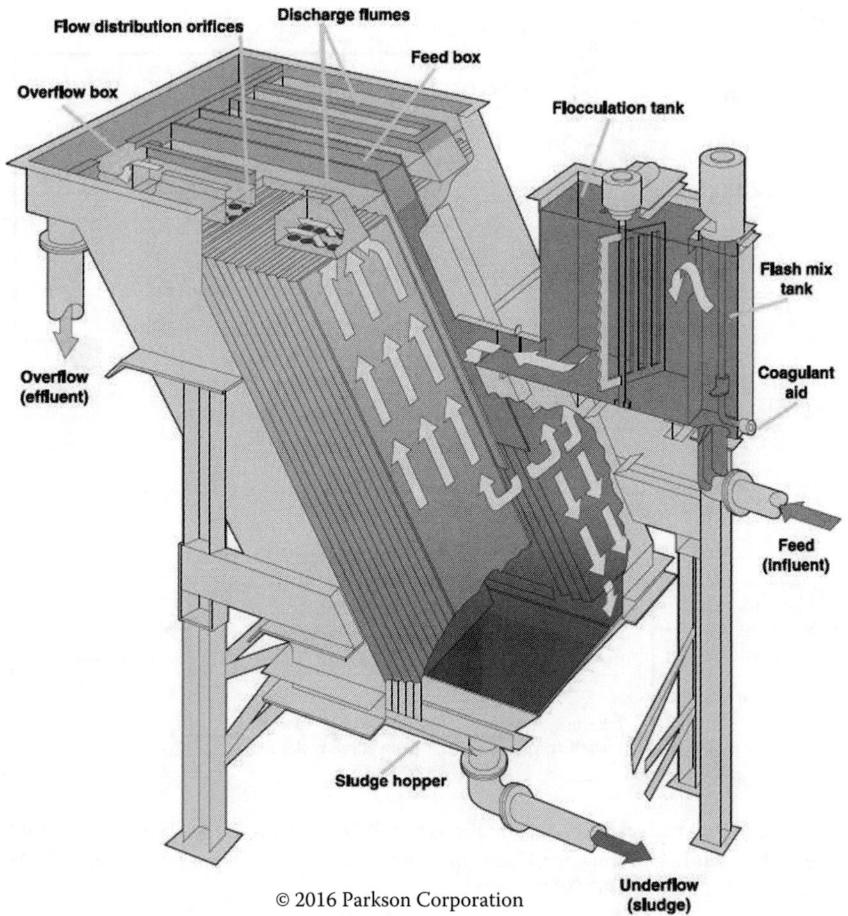

FIGURE 9.52 Lamella® Gravity Settler HRC System Using Inclined Surface Clarifier (Courtesy Parkson Corporation).

ranges of process design parameters of MSBFP systems are summarized from publicly available data and presented in Table 9.27.[5,50–54,62–65] The selected MSBFP system manufacturers should be required to provide the actual design recommendations of their systems.

Applications: The MSBFP has been used in Europe since 1991 in many drinking water and wastewater treatment applications. Other applications of MSBFP are treatment of (1) excess wet weather flow and combined sewer overflows (CSOs), (2) CEPT, (3) filter backwash water, and (4) side streams from biosolids processing areas. The potential process performance of MSBFP systems for different applications are summarized from publicly available data and presented in Table 9.28.[5,50–54,62–65] If acceptable, a parallel MSBFP unit can be used for treatment of excess wet weather flow at a conventional secondary wastewater treatment plant.[51,66,67] When the partial treatment by MSBFP alone is not enough to meet the requirement for secondary treatment, a biologically enhanced MSBFP (BE-MSBFP) system may be used to improve soluble BOD removal. In this system, activated sludge is brought from the secondary treatment into a biological solid contact chamber prior to the standard MSBFP unit to improve the TSS and BOD removal efficiencies up to 90% and 85%, respectively.[51] A conceptual diagram for using a parallel BE-MSBFP train to treat excess wet weather flow is shown in Figure 9.53b.

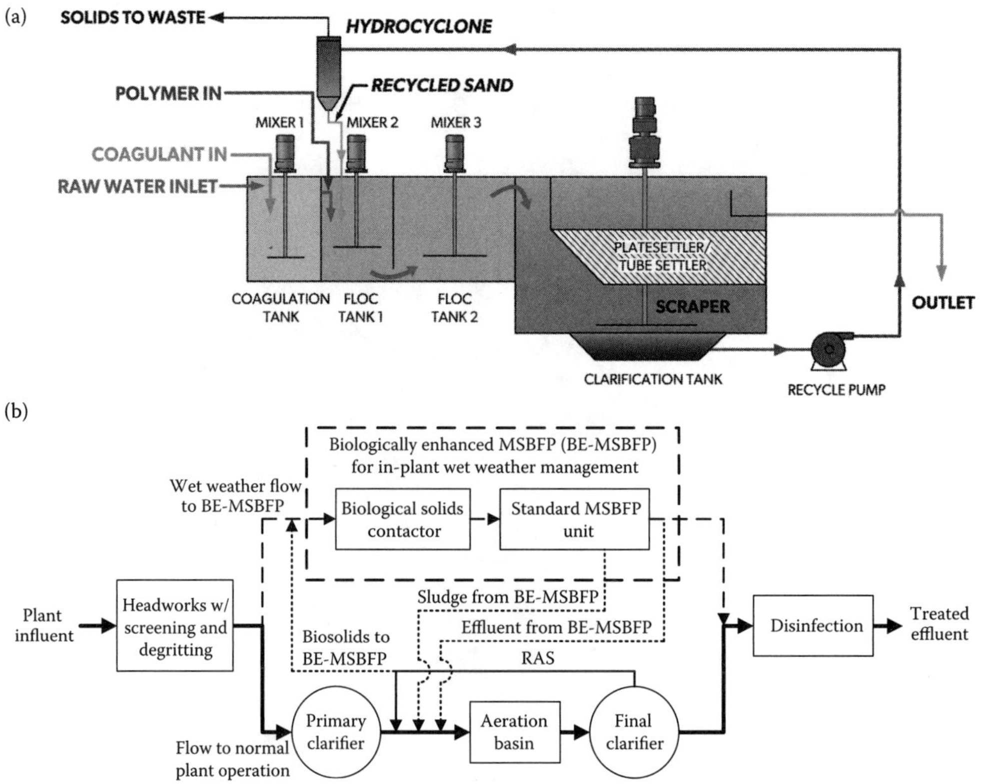

FIGURE 9.53 Micro-Sand Ballasted Flocculation Process (MSBFP) Systems: (a) RapiSand™ ballasted flocculation system (Courtesy WesTech Engineering, Inc.), and (b) conceptual process diagram for a parallel BE-MSBFP train to treat excess wet weather flow at a POTW.

9.6.4 Solids-Ballasted Flocculation Process

Process Description: Similar to the MSBFP, a ballasted material is also used in this HRC process. However, unlike the MSBFP using micro-sand, the solids-ballasted flocculation process (SBFP) recycles some chemically conditioned granules that act as ballasted material to improve the settling velocity of flocculated solids.[5,50–54,68]

Major Components: The SBFP is typically an integrated system with multiple functional zones. For example, DensaDeg® by SUEZ Treatment Solutions, Inc. has six different zones in series: (1) a rapid-mix zone where a metal coagulant is added and mixed with the screened and degritted influent for coagulation by a mechanical mixer, (2) a draft tube reaction zone, where a flocculant aid polymer is added with recycled solids under proper mixing by a low-share, high pumping axial-flow turbine to ensure dense and homogeneous solids formation, (3) a transition zone where flocculated wastewater flows through an upflow channel for effective separation of grease and scum, (4) a clarification stage where dense solids falls rapidly to the bottom, (5) a solids-thickening zone where the settled solids are stored and thickened, and (6) a polishing/collection zone where the clarified effluent is further polished in the inclined tube settler modules and collected by the collection launders from the surface. From the solids-thickening zone, part of the dense solids is recirculated back to the reaction zone and the residual solids are periodically discharged for disposal. Since micro-sand is not involved, use of a hydrocyclone is not required in the SBFP system. The schematic of DensaDeg is shown in Figure 9.54.[68]

TABLE 9.27 General Ranges of MSBFP System Design Parameters

Design Parameter	Value or Range
Single unit capacity, MGD	0.3–50
Detention time, min	
Coagulation	1–2
Injection	1–2
Maturation	3–5
Clarifier	4–7
Overall	~10–12
Velocity gradient (G), s^{-1}	
Coagulation	300
Injection	300
Maturation	200–250
Surface overflow rate in clarifier, $m^3/m^2 \cdot h$ (gpm/ft^2)	
CEPT/wet weather flow treatment/tertiary treatment	100–200 (40–80)
High strength industrial effluent/municipal wastewater	50–100 (20–40)
Start-up time, min	20–30
Micro-sand	
Effective size, μm	140–160
Specific gravity (water as 1)	$\geq$ 2.65
Dosage, g/L	1–12
Makeup, mg/L (lb/Mgal) treated	1–3 (8–25)
Chemical dosage, mg/L	
Coagulant	40–150
Polymer	0.5–1.25
Solids recirculation rate, times of influent flow rate	~6
Solids content of sludge, % dry solids	0.15–0.3
VSS/TSS ratio in sludge	0.4–0.6

Source: Summarized from information contained in References 5, 50 through 54, and 62 through 65.

Design Recommendations and Applications: General design recommendations and potential process performance of using DensaDeg process are summarized from publicly available data and presented in Table 9.29.[5,50–54] The general procedures for design and selection of the SBFP system are similar to those presented before for the MSBFP unit. The SBFP manufacturers shall be consulted to obtain the actual design recommendations of their systems.

TABLE 9.28 Process Performance of MSBFP Systems for Different Treatment Applications

Application	Removal Efficiency(%)							Log-Reduction of Fecal Coliforms
	TSS	COD	BOD	TKN	TP	Ortho-P	Metal	
Typical wastewater treatment	90–95	50–80	65–80	10–40	85–95	–	50–90	1.3
Primary treatment	30–90	55–80	–	–	50–95	50–98	–	1–1.5
Tertiary treatment	50–80	20–50	–	–	50–95	50–98	–	1–1.5
Stormwater	80–98	65–90	–	–	50–95	50–98	–	1–1.5
Biofilter backwash/biological sludge	75–99	55–80	–	–	50–95	50–98	–	1–1.5

Source: Summarized from information contained in References 5, 50 through 54, and 62 through 65.

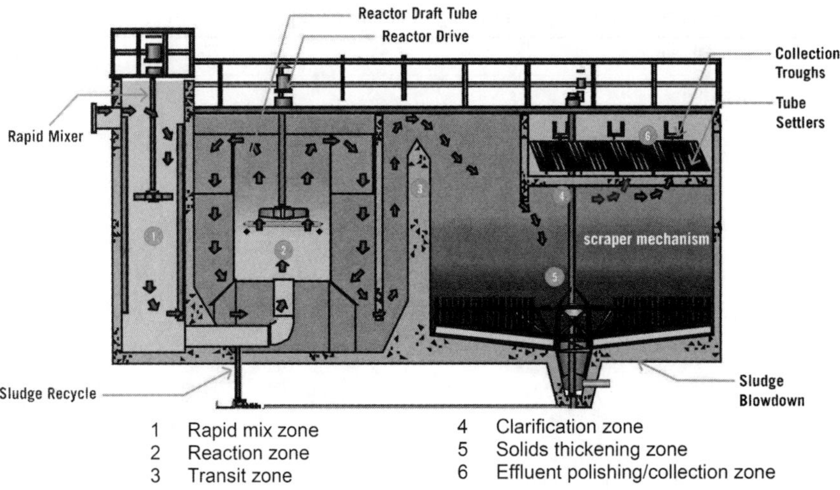

FIGURE 9.54 Solids-ballasted flocculation process (SBFP) system: DensaDeg® system (Courtesy SUEZ).

TABLE 9.29 General Ranges of DensaDeg® Process Design Parameters and Potential Removal Efficiencies

Design Parameters	Value or Range
Single unit capacity, MGD	8–100
Detention time, min	
Rapid mixing	1–2
Reaction/transit	4–6
Clarifier/thickener	7-10
Overall	13–18
Velocity gradient (G), s^{-1}	
Rapid mixing	100–250
Reaction	30–50
Surface overflow rate in clarifier, $m^3/m^2{\cdot}h$ (gpm/ft^2)	
Primary/CEPT	24–36 (10–15)
Wet weather flow treatment	60–150 (25–60)
Start-up time, min	15–30
Chemical dosage, mg/L	
Coagulant	30–150
Polymer	0.75–1.5
Solids recirculation rate, percent of influent flow rate	2–6
Solids content of sludge, % dry solids	4–6
VSS/TSS ratio in sludge	0.4–0.6
Process Performance[a], Percent Removal	
TSS	80–90
BOD	35–65
TKN	25–40
Phosphorus	85–95

[a] Summarized from the pilot study results using SUEZ DensaDeg® 4D in Reference 52.

Source: Summarized from information contained in References 5, and 50 through 54.

EXAMPLE 9.48: SOLIDS CONTACT CLARIFIER

A solids contact clarifier receives a flow of 14,000 m^3/d. The design surface overflow rate (SOR) is 75 $m^3/m^2{\cdot}d$. The side water depth (SWD) is 5.5 m. Calculate the hydraulic retention time (HRT).

Solution

1. Method 1: Determining HRT from side water depth.

   Calculate the HRT from the side water depth.

$$\text{Hydraulic retention time, } HRT = \frac{SWD}{SOR} = \frac{5.5\,\text{m}}{75\,\text{m}^3/\text{m}^2{\cdot}\text{d}} \times \frac{24\,\text{h}}{\text{d}} = 1.8\,\text{h}$$

2. Method 2: Determining HRT from clarifier volume.

   a. Calculate the surface area.

$$\text{Surface area, } A = \frac{Q}{SOR} = \frac{14{,}000\,\text{m}^3/\text{d}}{75\,\text{m}^3/\text{m}^2{\cdot}\text{d}} = 187\,\text{m}^2$$

   b. Calculate the clarifier volume.

$$\text{Clarifier volume, } V = A \times SWD = 187\,\text{m}^2 \times 5.5\,\text{m} = 1029\,\text{m}^3$$

   c. Calculate the HRT from the clarifier volume.

$$\text{Hydraulic retention time, } HRT = \frac{V}{Q} = \frac{1029\,\text{m}^3}{14{,}000\,\text{m}^3/\text{d}} \times \frac{24\,\text{h}}{d} = 1.8\,\text{h}$$

EXAMPLE 9.49: PERFORMANCE OF INCLINED SURFACE SETTLERS IN HRC

Inclined plate settlers are used in an HRC. The settler modules have a plate length of 2 m and clear space of 5 cm. The plates are inclined at 60° to the horizontal. The design surface overflow rate is 10 $m^3/m^2{\cdot}h$ (4 gpm/ft^2) in the HRC clarifier. For each of the three alternative flow patterns: countercurrent, cocurrent and cross-flow, (a) calculate terminal velocity v_s of the smallest particle that is removed, (b) calculate settling time t provided for particles to settle the vertical distance between the plates, and (c) verify the length of surface L_p traveled in time t. Compare the theoretical performance of the HRC in three different flow configurations, and the same clarifier without plate settlers at the same surface overflow rate. Assume that 75% of the basin area is covered by the plate settler modules with an efficiency factor of 0.8.

Solution

1. Determine the flow velocity between the plates.

   a. Determine the flow velocity in countercurrent and cocurrent flow patterns.

   The flow velocity between the plates is calculated from the design surface overflow rate in the clarifier. A definition sketch is shown in Figure 9.55.

   i. Calculate the surface overflow rate in the area covered by the inclined plate settlers.

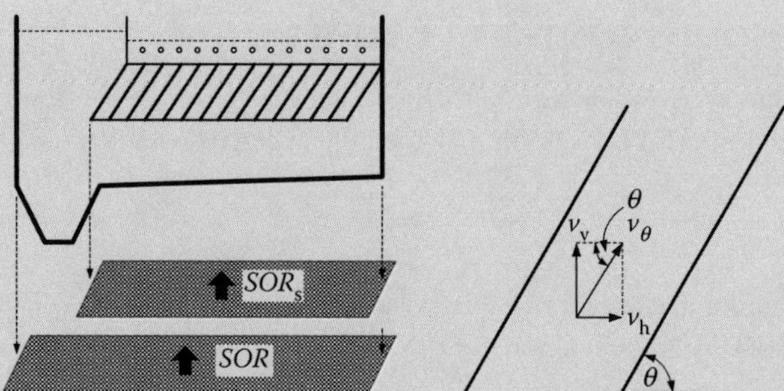

FIGURE 9.55 Definition sketch of overflow rates, vertical velocity, and flow velocity between plates (Example 9.49).

The surface overflow rate over the portion of the clarifier area covered by the inclined surface settlers is calculated from the following equation (Equation 9.38a):

$$SOR_s = \frac{SOR}{f_s} \tag{9.38a}$$

where

SOR = design overflow rate in the clarifier, m^3/s (ft^3/s)
SOR_s = overflow rate in the area covered by the inclined surface settlers, m^3/s (ft^3/s)
f_s = fraction of the basin area covered by the inclined surface settlers ($f_s \leq 0.75$), dimensionless

Calculate the surface overflow rate SOR_s from Equation 9.38a at $SOR = 10\ m^3/m^2{\cdot}h$ and $f_s = 0.75$.

$$SOR_s = \frac{SOR}{f_s} = \frac{10\ m^3/m^2{\cdot}h}{0.75} = 13.3\ m^3/m^2{\cdot}h$$

ii. Calculate the vertical vector of the flow velocity between the plates.

The vertical vector of the flow velocity is calculated from the following equation (Equation 9.38b):

$$v_v = \frac{SOR_s}{f_b} \tag{9.38b}$$

where

v_v = vertical vector of the flow velocity between the plates, m/s (ft/s)
f_b = area loss due to settler material thickness, settler support reinforce frames, angle of inclination, etc., as a fraction of the total area covered by the settlers ($f_b = 0.7$–1), dimensionless

Calculate the vertical vector of the flow velocity v_v from Equation 9.38b at $f_b = 0.8$.

$$v_v = \frac{SOR_s}{f_b} = \frac{13.3\ m^3/m^2{\cdot}h}{0.8} \times \frac{h}{3600\ s} = \frac{13.3\ m^3/m^2{\cdot}h}{0.8} \times \frac{h}{3600\ s} = 0.00462\ m/s$$

iii. Calculate the flow velocity in the direction parallel to the plates.

When the vertical vector is known, the flow velocity is calculated from the following equation (Equation 9.38c):

$$v_\theta = \frac{v_v}{\sin\theta} \tag{9.38c}$$

All variables have been defined before.

Calculate the flow velocity v_θ from Equation 9.38c at the angle $\theta = 60^\circ$.

$$v_\theta = \frac{v_v}{\sin\theta} = \frac{0.00462\,\text{m/s}}{\sin 60^\circ} = \frac{0.00462\,\text{m/s}}{0.866} = 0.00533\,\text{m/s} \quad \text{or} \quad 0.32\,\text{m/min}$$

This flow velocity will be used in evaluations of HRC in countercurrent and cocurrent flows.

b. Determine the flow velocity in cross-flow pattern.

In cross-flow pattern, the flow velocity v_θ is in horizontal direction. Theoretically, it is independent of the design overflow rate. It can be estimated from the width of the settler module. For evaluation purpose only in this example, the above flow velocity $v_\theta = 0.00533$ m/s is also used in evaluation of HRC in cross-flow pattern.

2. Evaluate the performance of HRC in countercurrent-flow pattern.

a. Calculate the terminal velocity.

Calculate the settling velocity v_s from Equation 9.35a at the flow velocity $v_\theta = 0.00533$ m/s, the length of surface $L_p = 2$ m, and clear space $w = 5$ cm $= 0.05$ m.

$$v_s = \frac{v_\theta w}{L_p \cos\theta + w \sin\theta} = \frac{0.00533\,\text{m/s} \times 0.05\,\text{m}}{2\,\text{m} \times \cos 60^\circ + 0.05\,\text{m} \times \sin 60^\circ} = \frac{0.00533\,\text{m/s} \times 0.05\,\text{m}}{2\,\text{m} \times 0.5 + 0.05\,\text{m} \times 0.866}$$

$$= 0.000255\,\text{m/s}$$

The theoretical terminal velocity is 0.000255 m/s, which is equivalent to a surface overflow rate of 22 $m^3/m^2{\cdot}d$ (0.38 gpm/ft^2).

b. Calculate the settling time in the settlers.

The settling time t is calculated from Equation 9.34a.

$$t = \frac{w}{v_s \cos\theta} = \frac{0.05\,\text{m}}{0.000255\,\text{m/s} \times \cos 60^\circ} = \frac{0.05\,\text{m}}{0.000255\,\text{m/s} \times 0.5} = 392\,\text{s} \quad \text{or} \quad 6.5\,\text{min}$$

c. Verify the length of surface in the settlers.

The length of surface L_p traveled in time t is calculated from Equation 9.35b.

$$L_p = \frac{w(v_\theta - v_s \sin\theta)}{v_s \cos\theta} = \frac{0.05\,\text{m} \times (0.00533 - 0.000255 \times \sin 60^\circ)\,\text{m/s}}{0.000255\,\text{m/s} \times \cos 60^\circ}$$

$$= \frac{0.05\,\text{m} \times (0.00533 - 0.000255 \times 0.866)\,\text{m/s}}{0.000255\,\text{m/s} \times 0.5} = 2.0\,\text{m}$$

The calculated length is exactly the same of the design length of 2 m.

3. Evaluate the performance of HRC in concurrent-flow pattern.

a. Calculate the terminal velocity.

Calculate the settling velocity v_s from Equation 9.36a.

$$v_s = \frac{v_\theta w}{L_p \cos\theta - w \sin\theta} = \frac{0.00533\,\text{m/s} \times 0.05\,\text{m}}{2\,\text{m} \times \cos 60^\circ - 0.05\,\text{m} \times \sin 60^\circ} = \frac{0.00533\,\text{m/s} \times 0.05\,\text{m}}{2\,\text{m} \times 0.5 - 0.05\,\text{m} \times 0.866}$$

$$= 0.000279\,\text{m/s}$$

The theoretical terminal velocity is 0.000279 m/s. It is equivalent to a surface overflow rate of 24 $m^3/m^2{\cdot}d$ (0.41 gpm/ft^2).

b. Calculate the settling time in the settlers.

The settling time t is calculated from Equation 9.34a.

$$t = \frac{w}{v_s \cos\theta} = \frac{0.05\,\text{m}}{0.000279\,\text{m/s} \times \cos 60^\circ} = \frac{0.05\,\text{m}}{0.000279\,\text{m/s} \times 0.5}$$
$$= 358\,\text{s} \quad \text{or} \quad 6.0\,\text{min}$$

c. Verify the length of surface in the settlers.

The length of surface L_p traveled in time t is calculated from Equation 9.36b.

$$L_p = \frac{w(v_\theta + v_s \sin\theta)}{v_s \cos\theta} = \frac{0.05\,\text{m} \times (0.00533 + 0.000279 \times \sin 60^\circ)\,\text{m/s}}{0.000279\,\text{m/s} \times \cos 60^\circ}$$
$$= \frac{0.05\,\text{m} \times (0.00533 + 0.000279 \times 0.866)\,\text{m/s}}{0.000279\,\text{m/s} \times 0.5} = 2.0\,\text{m}$$

The calculated length is exactly the same of the design length of 2 m.

4. Evaluate the performance of HRC in cross-flow pattern.

a. Calculate the terminal velocity.

Assuming that the length of surface remains at $L_p = 2$ m, calculate the settling velocity v_s from Equation 9.37a.

$$v_s = \frac{v_\theta w}{L_p \cos\theta} = \frac{0.00533\,\text{m/s} \times 0.05\,\text{m}}{2\,\text{m} \times \cos 60^\circ} = \frac{0.00533\,\text{m/s} \times 0.05\,\text{m}}{2\,\text{m} \times 0.5} = 0.000267\,\text{m/s}$$

The theoretical terminal velocity is 0.000267 m/s. It is equivalent to a surface overflow rate of 23 $\text{m}^3/\text{m}^2\cdot\text{d}$ (0.39 gpm/ft^2).

b. Calculate the settling time in the settlers.

The settling time t is calculated from Equation 9.34a.

$$t = \frac{w}{v_s \cos\theta} = \frac{0.05\,\text{m}}{0.000267\,\text{m/s} \times \cos 60^\circ} = \frac{0.05\,\text{m}}{0.000267\,\text{m/s} \times 0.5} = 375\,\text{s or } 6.3\,\text{min}$$

c. Verify the length of surface in the settlers.

The length of surface L_p traveled in time t is calculated from Equation 9.37b.

$$L_p = \frac{w v_\theta}{v_s \cos\theta} = \frac{0.05\,\text{m} \times 0.00533\,\text{m/s}}{0.000267\,\text{m/s} \times \cos 60^\circ} = \frac{0.05\,\text{m} \times 0.00377\,\text{m/s}}{0.000251\,\text{m/s} \times 0.5} = 2.0\,\text{m}$$

The calculated length is exactly the same as the design length of 2 m.

5. Determine the terminal velocity in the plain clarifier.

In the plain clarifier, the terminal velocity is theoretically equal to the design surface overflow rate. Calculate the theoretical terminal velocity at $SOR = 10\,\text{m}^3/\text{m}^2\cdot\text{h}$.

$$v_s = SOR = 10\,\text{m}^3/\text{m}^2\cdot\text{h} \times \frac{\text{h}}{3600\,\text{s}} = 0.00278\,\text{m/s}$$

This will be used in comparison with the terminal velocities determined for the different flow patterns of the HRC.

6. Compare the theoretical performance of the HRC in three flow configurations, and plain clarifier without plate settlers.

   The performance of a basin is compared based on its theoretical terminal velocity of the smallest particles that can be fully removed in the basin. The terminal velocities at the same design surface overflow rate of 10 $m^3/m^2{\cdot}h$ in the plain clarifier, and inclined surface clarifiers in countercurrent-, concurrent- and cross-flow configurations are summarized below.

Clarifier or Flow Configuration	Terminal Velocity v_s, m/s	Equivalent Surface Overflow Rate, $m^3/m^2{\cdot}d$
HRC with inclined plate settlers		
Countercurrent flow	0.000255	22
Cocurrent flow	0.000279	24
Cross-flow	0.000267	23
Plain clarifier without plate settlers	0.00278	240

   a. Comparison among three flow configurations.

      Based on the terminal velocity, the countercurrent flow configuration seems the most efficient because it provides the lowest terminal velocity of 0.000255 m/s for removing the smallest settling particles. The differences in performance of countercurrent, cocurrent, and cross-flow are so small that they may be virtually same. Practically, cocurrent- and cross-flow configurations are rarely used due to potentially serious operational difficulties.[49]

   b. Comparison between the HRC and plain clarifiers.

      Theoretically, a terminal velocity of 0.00278 m/s is achieved for removing the smallest particles in the plain clarifier. It is over 10 times greater than the terminal velocities in the inclined plate clarifiers at the same design overflow rate. Therefore, an inclined plate or tube settler is capable of enhancing the performance of a plain clarifier by nearly 10-fold. See Example 9.50 for estimating the range of capacity improvement after retrofitting an existing primary clarifier into an HRC.

EXAMPLE 9.50: CAPACITY IMPROVEMENT BY RETROFITTING A CONVENTIONAL PRIMARY CLARIFIER INTO HRC WITH INCLINED SURFACE SETTLERS

Use the information in Example 9.49 to estimate the clarifier capacity improvement after converting an original conventional clarifier into an HRC.

Solution

The conventional primary clarifiers usually have the design surface overflow rates within the range of 1–2 $m^3/m^2{\cdot}h$ (24–48 $m^3/m^2{\cdot}d$), while the design surface overflow rate of an HRC clarifier is 10 $m^3/m^2{\cdot}h$. The potential capacity improvement is evaluated based on the surface overflow rate ratios of HRC to conventional primary clarifier.

1. Calculate the overflow ratio at a surface overflow rate of 1 $m^3/m^2{\cdot}h$.

   At a surface overflow rate of 1 $m^3/m^2{\cdot}h$ in the conventional primary clarifier, the ratio is calculated below.

$$\frac{10\,m^3/m^2{\cdot}h}{1\,m^3/m^2{\cdot}h} = 10 \text{ (times)}$$

2. Calculate the ratio again at a surface overflow rate of 2 $m^3/m^2{\cdot}h$.

$$\frac{10\ m^3/m^2{\cdot}h}{2\ m^3/m^2{\cdot}h} = 5\ \text{(times)}$$

In practice, the capacity of a primary clarifier can be increased five to ten times by retrofitting it into an HRC with inclined surface settlers.

EXAMPLE 9.51: PROCESS ANALYSIS OF INCLINED SURFACE CLARIFIER

A clarifier has been retrofitted with 2-m (6.6-ft) square inclined plates spaced 50 mm (2 in) apart. Assume that the plates can be inclined at different angles between 10° and 80° and the clarifier can be operated for countercurrent-, concurrent-, or cross-flow configuration. At a given surface overflow rate (*SOR*), determine (a) theoretical performance for particle removal in three flow configurations and (b) theoretical effect of inclination angle of plates on the clarifier performance. Ignore any hydraulic problems that may arise as a result of flow distribution and sludge resuspension. To simplify the example, assume that the entire basin area is covered by the plate settler modules ($f_s = 1$) and the efficiency factor $f_b = 1$.

Solution

1. Develop the equations of the ratio of settling velocity to surface overflow rate.

   To evaluate the theoretical performance, the ratio of settling velocity to the surface overflow rate (v_s/SOR) is developed for each of the three flow configurations. Assuming laminar flow or constant velocity across the plate, the ratios are developed in two steps.

   a. Derive an equation to determine v_θ from *SOR*.

      The new equation (Equation 9.38d) is derived from Equations 9.38a, 9.38b, and 9.38c.

$$v_\theta = \frac{v_v}{\sin\theta} = \frac{1}{\sin\theta} \times \left(\frac{SOR_s}{f_b}\right) = \frac{1}{f_b \sin\theta} \times \left(\frac{SOR}{f_s}\right) = \frac{SOR}{f_s f_b \sin\theta} \tag{9.38d}$$

   b. Develop the ratio of settling velocity to surface overflow rate.

      Substitute Equation 9.38d for v_θ in Equations 9.35a, 9.36a, and 9.37a, and rearrange these equations. Using square plates, $L_p = 2$ m and $w = 0.05$ m in all three configurations. At $f_s = 1$ and $f_b = 1$, the ratios for three flow configurations are:

Countercurrent flow:
$$\frac{v_s}{SOR} = \frac{1}{f_s f_b \sin\theta} \times \frac{w}{L_p \cos\theta + w \sin\theta} = \frac{1}{\sin\theta} \times \frac{0.05\text{m}}{2\text{ m} \times \cos\theta + 0.05\text{ m} \times \sin\theta}$$

Cocurrent flow:
$$\frac{v_s}{SOR} = \frac{1}{f_s f_b \sin\theta} \times \frac{w}{L_p \cos\theta - w \sin\theta} = \frac{1}{\sin\theta} \times \frac{0.05\text{ m}}{2\text{ m} \times \cos\theta - 0.05\text{ m} \times \sin\theta}$$

Cross-flow:
$$\frac{v_s}{SOR_{\text{settler}}} = \frac{1}{f_e \sin\theta} \times \frac{w}{L_p \cos\theta} = \frac{1}{0.9 \times \sin\theta} \times \frac{0.05\text{ m}}{2\text{ m} \times \cos\theta}$$

2. Calculate the ratios of settling velocity to surface overflow rate at various plate angles.

Using the equations developed in Step 1, the ratio of v_s/SOR is calculated at $\theta = 10°, 20°, 30°, 40°, 45°, 50°, 60°, 70°$, and $80°$. The calculation results are summarized in the table below. These ratios are also shown in Figure 9.56.

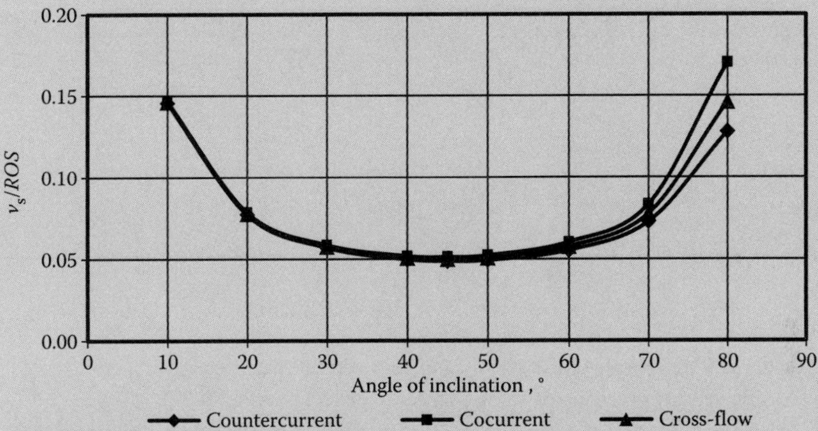

FIGURE 9.56 Ratio of settling velocity to surface overflow rate at various plate angles for three different flow patterns in an inclined surface clarifier (Example 9.51).

Flow Configuration	v_s/SOR at a Given Inclination Angle								
	10°	20°	30°	40°	45°	50°	60°	70°	80°
Countercurrent flow	0.15	0.077	0.057	0.050	0.049	0.049	0.055	0.073	0.13
Cocurrent flow	0.15	0.079	0.059	0.052	0.051	0.052	0.060	0.084	0.17
Cross-flow	0.15	0.078	0.058	0.051	0.050	0.051	0.058	0.078	0.15

3. Compare the ratios of settling velocity to surface overflow rate.

   The calculation results indicate the ratios are relatively low for all three flow configurations at the angles in the range between 30° and 60°. Therefore, the inclined surface clarifiers are most likely be efficient with very little difference among three flow configurations within this inclination angle range. In theory, the optimal ratios for particle settling are observed at an angle of 45°. In practice, the most efficient operations are achieved in the countercurrent-flow pattern at angles between 55° and 60°. These angles will not only allow efficient settlement of smaller particles but also provide effective removal of settled solids from the steeper slopes.

EXAMPLE 9.52: SETTLING VELOCITY OF BALLASTED PARTICLES IN MSBFP

Settling velocity of coagulated and flocculated particles is significantly smaller than that of ballasted particles. In an MSBFP unit, fine silica sand is used as a weighting agent. In the presence of a polymer, the coagulated particles stick to the surface of sand particle. As a result, the ballasted floc particles settle considerably faster than unballasted particles. Calculate the settling velocity of micro-sand ballasted floc particles. Also compare it with the settling velocities of (a) ideal micro-sand particles (spherical), (b) actual micro-sand particles (angular), and (c) particles naturally carried in raw wastewater. The parameters used in the evaluations of these particles are given below. Assume that the settling behavior of these particles is similar to that of discrete particles. Wastewater temperature = 10°C.

Parameters	Experimental Data			
	Ballasted Floc Particles in MSBFP	Ideal Micro-Sand (Spherical)	Actual Micro-Sand (Angular)	Particles in Raw Wastewater
Average diameter, μm	200	160	160	550
Specific gravity of particles (S_g), dimensionless	2.4	2.65	2.65	1.2
Shape factor (ϕ), dimensionless	2.5	1.0	1.9	2.5

Solution

1. Calculate the settling velocity of the micro-sand ballasted floc particles.
   The settling velocity is calculated using procedure given in Section 8-3-2.
   The kinematic viscosity ν of wastewater at 10°C is obtained from Table B.2 in Appendix B.

   $\nu = 1.307 \times 10^{-6}\ \text{m}^2/\text{s}$

   Assume that Stokes' equation (Equation 8.4) is applicable, and calculate the terminal settling velocity ($S_g = 2.4$ and $d = 200\ \mu\text{m} = 200 \times 10^{-6}$ m).

$$v_t = \frac{g(S_s - 1)d^2}{18\nu} = \frac{9.81\text{m/s}^2 \times (2.4 - 1) \times (200 \times 10^{-6}\ \text{m})^2}{18 \times (1.307 \times 10^{-6}\ \text{m}^2/\text{s})} = 0.0234\ \text{m/s}$$

   Calculated Reynolds number from Equation 8.3.

$$N_R = \frac{v_t d}{\nu} = \frac{0.0234\ \text{m/s} \times 200 \times 10^{-6}\ \text{m}}{1.307 \times 10^{-6}\ \text{m}^2/\text{s}} = 3.58$$

   N_R is in the transition range ($1.0 < N_R < 10^4$) and Newton's equation is applicable. Use trial-and-error procedure.
   Calculate C_D from Equation 8.2.

$$C_D = \frac{24}{N_R} + \frac{3}{\sqrt{N_R}} + 0.34 = \frac{24}{3.58} + \frac{3}{\sqrt{3.58}} + 0.34 = 8.63$$

   Calculate terminal settling velocity v_t from Equation 8.1b for nonspherical particles at $\phi = 2.5$.

$$v_t = \sqrt{\frac{4g(S_s - 1) \times d}{3C_D\phi}} = \sqrt{\frac{4 \times 9.81\ \text{m/s}^2 \times (2.4 - 1) \times (200 \times 10^{-6}\ \text{m})}{3 \times 8.63 \times 2.5}} = 0.0130\ \text{m/s}$$

   This is smaller than $v_t = 0.0234$ m/s calculated previously from Stokes' equation (Equation 8.4). Repeat the trial-and-error procedure until v_t does not change from that obtained in the previous trial. This objective is achieved after the fourth trial. The results of N_R, C_D, and v_t of the micro-sand ballasted floc particles at 10°C wastewater temperature are given below:

   $N_R = 1.37$, $C_D = 20.5$, and $v_t = 0.008$ m/s

2. Calculate the settling velocity of the other three particles.
   Using the same procedure, the settling velocities are also calculated for the other three particles. The calculation results are summarized below.

Parameters	Final Calculation Results		
	Ideal Micro-Sand (Spherical)	Actual Micro-Sand (Angular)	Particles in Raw Wastewater
Number of trial performed before reaching the final results	3	4	4
N_R, dimensionless	1.87	1.08	3.64
C_D, dimensionless	15.3	25.6	8.51
v_t, m/s	0.015	0.008	0.008

3. Compare the settling velocity of the micro-sand ballasted floc particles with the other particles.
   The settling velocities calculated for these particles are summarized below:

Micro-sand ballasted floc particles	$v_t = 0.008$ m/s
Ideal micro-sand (spherical)	$v_t = 0.015$ m/s
Actual micro-sand (angular)	$v_t = 0.009$ m/s
Particles in raw wastewater	$v_t = 0.003$ m/s

Observations:

a. At the same shape factor (ϕ) of 2.5, the micro-sand ballasted floc particles ($d = 200$ μm) may settle at a similar velocity to the much larger untreated particles ($d = 550$ μm) in the raw wastewater. The settling velocity is mainly enhanced due to an increase in specific gravity (S_g) from 1.2 to 2.4.

b. Even at a higher ϕ of 2.5 and lower S_g of 2.4, the micro-sand ballasted floc particles may settle at a similar velocity to the actual micro-sands used in the MSBFP. The settling velocity is mainly improved due to an increase in floc particle size (d) from 160 to 200 μm.

c. The smallest ideal micro-sands ($d = 160$ μm) will theoretically settle at the fastest rate than all other particles compared in this example. It is mainly because of the ideal spherical shape (ϕ of 1.0) and the highest S_g of 2.65.

9.7 Fine Mesh Screens for Primary Treatment

The conventional primary clarifiers occupy a large land area. They are also a major source of odors and other environmental problems at the plant site. A need to replace primary clarifier by a compact and enclosed process has been long realized. High-performance fine screen and microscreen technologies have been developed in recent years to replace primary clarifiers. They offer compact design, small footprint, can be installed inside a building, and provide suspend solids removal efficiencies equivalent or better than primary clarifiers. Fine screens have been presented and discussed in Section 7.2.3. The use of these fine screens has been extended to remove solid equivalent of primary treatment. They are preceded by coarse screens. They fall into two general categories: stationary and mobile screen types. General information such as type of screens, range of opening sizes, operational features, and application in wastewater treatment provided by equipment manufacturers are summarized in Table 7.6. Most of these screens are used for preliminary treatment but many provide TSS removal efficiency equivalent of primary treatment. A fine mesh screen-based filtration system can practically meet the BOD_5 requirements and TSS removal if a proper filter mat is formed and maintained on top of the screening material. Based on these process concepts, the innovative Salsnes Filter™ process, a rotating belt sieve (RBS) filtration system was developed with the capability to replace the primary clarifier.[69–72]

Process Descriptions: The Salsnes Filter process is illustrated in Figure 9.57.[72] The raw wastewater is fed into a feed chamber where it comes in contact with a belt-type endless filter cloth. As wastewater is filtered through the filter cloth, a filter mat of solids is formed on the cloth surface and the liquid level rises in the chamber due to an increase in head loss through the filter cloth. The liquid level in the feed chamber is measured by a pressure transmitter and used by the control system to adjust the moving speed of the belt, maintain an ideal filter mat thickness, and optimize the performance of the filtration process. Filtered wastewater is collected behind the filter cloth and the effluent is discharged through an outlet from the filter module. The solids accumulated on the cloth surface are conveyed upward by the moving belt through the sludge-thickening process. A compressed air-based, air-knife cleaning device is used to blow the solids off the filter surface into the sludge compartment once the belt rotates around a rotor located near the top of the filter module. As necessary, hot-water flush is also used to remove oil and grease (O&G) that accumulate on the filter cloth media. As an option, the collected sludge can also be pressed through an integrated dewatering screw unit or a standalone dewatering device to increase the dry solids content prior to further processing and final disposal.

Basic Features: Currently, the Salsnes Filter system is available for a mesh size from 0.1 to 1 mm (0.004 to 0.04 in), covering a wide range for fine screens and microscreens. However, the most commonly used mesh sizes for effective primary treatment are within a microscreening range between 0.15 and 0.5 mm

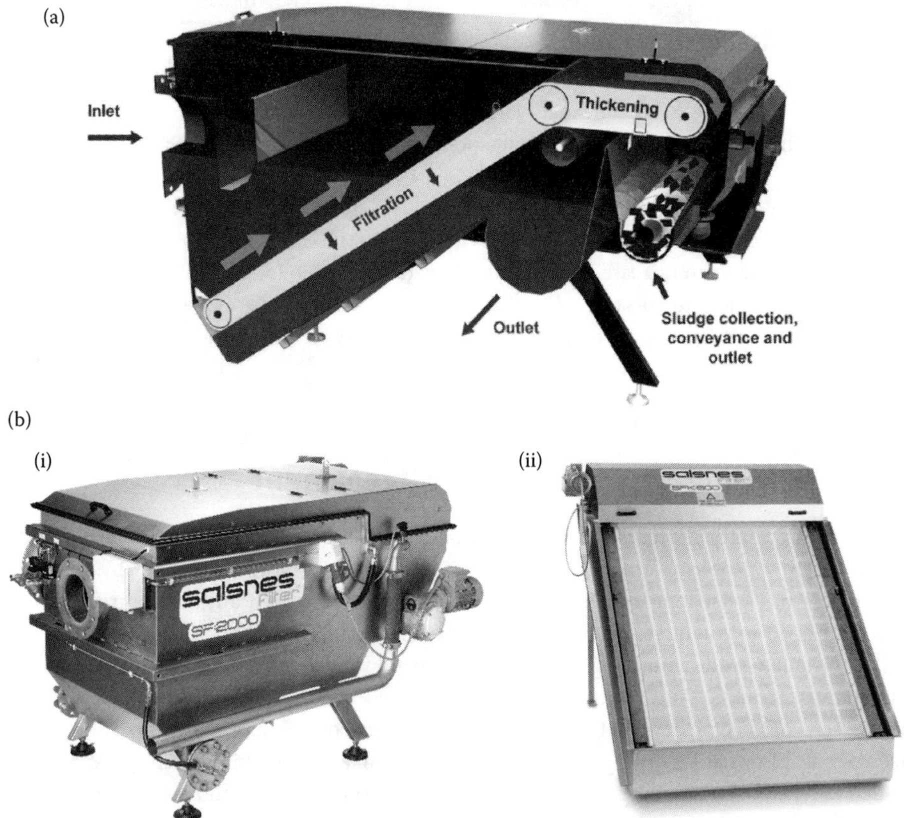

FIGURE 9.57 Salsnes Filter System (courtesy Salsnes Filter AS): (a) process illustration; and (b) installation options: (i) enclosed free-stand module and (ii) open filter for channel installation.

(0.006 and 0.02 in). When a filter mat with a thickness of >2 mm is properly developed on the belt with a typical mesh size of 0.35 mm, this process is capable of providing effective filtration of 0.02–0.1-mm particles.[51,72] The filter belt material is made of nylon with Kevlar support for a typical life time of 1–5 years. Other basic features of this fine mesh RBS filtration system are:

1. It is capable of providing pretreatment required by most secondary treatment processes, including MBR/MBBR.
2. It is capable of increasing secondary treatment capacity through enhanced primary treatment without adding chemicals.
3. Due to a small footprint, that is ~10% of the land required by a conventional primary clarifier, it is suitable for plant expansion where extra land may be unavailable.
4. It is suitable for retrofitting existing primary clarifiers with minimal structural modifications.
5. Enclosed design is effective for odor control.
6. The material of frame/enclosure construction is stainless steel.
7. Modular arrangement can be expanded easily in the future.
8. It is flexible for both free stand unit with or without manifolding or frame for channel-mounted installation [Figure 9.57b] in either indoor or outdoor applications.
9. Integrated thickening/dewatering option is available.
10. It is fully automated with low labor and moderate O&M requirements.
11. It is also suitable for CSO and stormwater treatment applications.

TABLE 9.30 General Ranges of Salsnes Filter System Design Parameters and Potential Removal Efficiencies

Design Parameters	Value or Range	
	Free-Stand Module	Channel Installation
Mesh size, mm (in)		
Available	0.1–1.0 (0.004–0.04)	0.1–1.0 (0.004–0.04)
Primary treatment[a]	0.15–0.5 (0.006–0.02)	0.15–0.5 (0.006–0.02)
Submerged sieve area, m^2 (ft^2)	0.25–2.2 (2.7–24)	0.25–2.2 (2.7–24)
Surface hydraulic loading rate over submerged sieve area[a], $m^3/m^2 \cdot h$ (gpm/ft^2)	20–260 (8–106)	20–260 (8–106)
Single unit capacity[a], m^3/h (MGD)	30–390 (0.2–2.5)	80–390 (0.5–2.5)
Overall dimensions, m (ft)		
Length	1.4–2.8 (5–9.1)	2–2.4 (6.6–8)
Width	1–1.8 (3.3–5.9)	1.3–2.5 (4.5–8.1)
Height	1.5–1.8 (5–5.9)	1.4–1.8 (4.7–6)
Single unit weight, kg (lb)	415–1120 (914–2469)	300–700 (661–1543)
Maximum head loss, m (in)	0.3–0.35 (12–14)	0.4 (16)
Power consumption per unit[a], kW (hp)	3.4–6.8 (4.6–9.1)	4.4–6.8 (5.9–9.1)
Solids content of sludge[a], % dry solids		
After thickening zone	3–8	3–8
After integrated dewatering device	20–30	20–30
After stand-alone dewatering device	20–40	20–40
Process Performance[a], Percent Removal		
TSS	40–80	40–80
BOD	20–35	20–35
TKN	10–15	10–15

[a] Raw wastewater characteristics and design preference dependent.

Source: Summarized from information contained in References 51, and 69 through 72.

Design Recommendations: General design recommendations and process performance of Salsnes Filter systems are summarized from publicly available data and presented in Table 9.30.[69–72] The most important design considerations are: (1) mesh size, (2) submerged sieve area, and (3) surface hydraulic loading rate over the submerged sieve area. These parameters are critical for proper development of the filter mat. In general, the alternative mesh sizes are first selected through bench-scale filter tests with or without the presence of a filter mat. Different testing procedures are developed by the manufacturer. A pilot study is usually recommended for finalizing the mesh size and determining other important design parameters that may be affected by the raw wastewater characteristics. The equipment manufacturer should be contacted to obtain detailed design recommendations.

EXAMPLE 9.53: SELECTING SIEVE MESH SIZE FOR SALSNES FILTER FROM PILOT TEST

Bench-scale filter test was performed on a raw wastewater sample with an initial TSS concentration of 276 mg/L. TSS concentration was measured in the filtrate from a filter column after passing a constant 1000-mL volume of raw sample prior to forming any filter mat. Five different sieve mesh sizes were used in the test. The test results are summarized below. Select the initial sieve mesh size to start the pilot test based on the filter test results. Assume that a filter mat can be properly formed if at least 25% particles in the raw wastewater can be captured by the filter cloth without the presence of a filter mat.

Sieve Mesh Size, mm	TSS Concentration in the Filtrate, mg/L
0.08	165
0.15	171
0.25	162
0.35	187
0.50	225

Solution

1. Calculate the TSS removal efficiency at each sieve mash size.

   From an initial TSS concentration of 276 mg/L, TSS removal efficiencies are calculated and tabulated below.

Sieve Mesh Size, mm	TSS Concentration in the Filtrate, mg/L	TSS Removed, mg/L	TSS Removal Efficiency, %
0.08	165	111	40
0.15	171	105	38
0.25	162	114	41
0.35	187	89	32
0.50	239	37	13

2. Plot the TSS removal efficiency curve.

   The TSS removal efficiency is plotted against the sieve mesh size in Figure 9.58.
3. Select the initial sieve mesh size for the pilot test.

   As shown in Figure 9.58, a mesh size of 0.41 mm is required to remove at least 25% of TSS (or catch at least 25% of particles) from the raw wastewater sample without the presence of a filter mat.

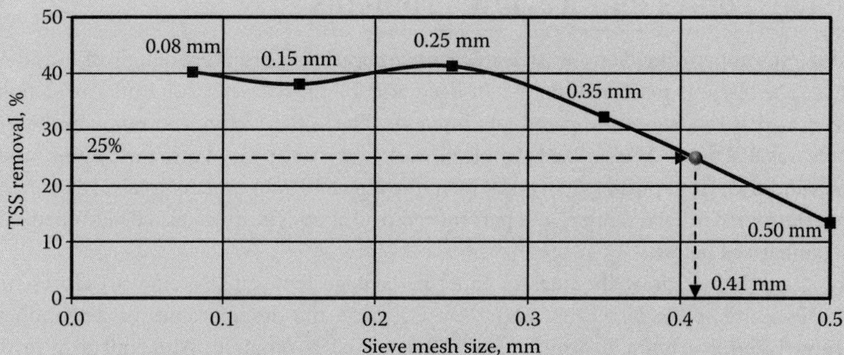

FIGURE 9.58 TSS removal efficiency versus sieve mesh size (Example 9.53).

Therefore, standard sieve mesh size of 0.35 mm is selected as initial size to start the pilot test. It is estimated that at least 40–50% removal of TSS can be expected when a filter mat is properly developed in the pilot test.

EXAMPLE 9.54: SURFACE HYDRAULIC LOADING RATE IN SALSNES FILTER UNIT

A pilot test was performed at a wastewater treatment plant to evaluate the potential application of Salsnes Filter systems for primary treatment. The pilot test results indicate the average flow allowed through the pilot unit is 450 gpm to achieve an average TSS removal of 60%. The TSS in raw wastewater is 220 mg/L. Estimate the average surface hydraulic loading rate during the pilot test. Also estimate the sludge quantity at a solid content of 5% and bulk specific gravity is 1.02. Assume the submerged sieve area is ~10.8 ft^2.

Solution

1. Calculate the surface hydraulic loading rate.

   At the average flow $Q = 450$ gpm over the submerge sieve area $A = 10.8/\ \text{ft}^2$.

   $$\text{Surface hydraulic loading rate, } L_r = \frac{Q}{A} = \frac{450\,\text{gpm}}{10.8\,\text{ft}^2} = 42\,\text{gpm/ft}^2 \quad \text{or} \quad 103\,\text{m}^3/\text{m}^2{\cdot}\text{h}$$

2. Calculate the quantity of sludge produced.

   $$\text{Solids removed in the sludge} = 0.6 \times 220\,\text{mg/L} = 132\,\text{mg/L}$$

   $$\text{Dry solids removed} = 132\,\text{mg/L} \times \frac{8.34\,\text{lb}}{\text{mg/L}{\cdot}\text{Mgal}} \times 450\,\text{gpm} \times 10^{-6}\,\text{Mgal/gal}$$
   $$= 0.5\,\text{lb/min dry solids}$$

   $$\text{Sludge volume removed} = 0.5\,\text{lb/ddrysolids} \times \frac{100\,\text{lb wet sludge}}{5\,\text{lb dry solids}}$$
   $$\times \frac{1}{(8.34 \times 1.02)\,\text{lb/gal as wet sludge}}$$
   $$= 1.2\,\text{gal/min of wet sludge}$$

Discussion Topics and Review Problems

9.1 A pilot test was performed on a wastewater sample containing a mixture of industrial and municipal wastes. The test was performed in a 5-m-deep and 18-cm-diameter column. The sampling ports were located at 1 m from the top and 1-m intervals. The initial TSS concentration in the well-mixed sample was 256 mg/L. Well-mixed sample of wastewater was poured into the column, and samples were withdrawn from ports 1, 2, 3, and 4 from the top at 10-min time intervals. TSS concentrations were determined in each sample, and percent removal at each port was also determined. The results are summarized below.

By interpolations draw the lines of equal percentage removal (isoremoval) representing 10%, 20%, 30%, 40%, 50%, 60%, 70% and 80%. Calculate the design values of detention time and overflow rate for a basin in which 85% TSS removed is required. Also, calculate (a) TSS concentration remaining in the sample withdrawn from 4.5-m-deep port and at a detention time of 60 min and (b) the water depth in the tank for 70% TSS removal at settling time of 80 min.

Pilot flocculant settling column test data on percent TSS removal in samples withdrawn from different ports.

Depth of Port from Top, m	Percent TSS Removal at Different Time (min), %							
	10	20	30	40	50	60	70	80
1.0	28	45	55	65	73	77	83	90
2.0	19	33	44	54	60	67	75	83
3.0	15	27	38	46	54	60	66	74
4.0	13	25	34	44	50	56	63	67

9.2 A 30-m-diameter sedimentation basin has a side water depth of 3.0 m. It is treating a wastewater flow of 0.3 m^3/s. Compute overflow rate and detention time.

9.3 A sedimentation basin has an overflow rate of 80 $m^3/m^2{\cdot}d$. What fraction of the particles that have a velocity of 0.02 m/min will be removed in this tank? Assume discrete settling.

9.4 Design a circular clarifier to treat coagulated wastewater at a flow rate of 0.22 m^3/s. The design overflow rate is 36 $m^3/m^2{\cdot}d$ and water depth is 3.1 m. The effluent launder is 0.5 m wide and is installed around the circumference of the basin, and 1 m away from the concrete wall. Eight -centimeter-deep 90°-V notches are provided on both sides of the launder. The effluent outlet box is 1 m × 1 m, and the depth of flow in the effluent box at peak design flow is 1 m. The invert of the effluent launder is 0.46 m above the invert of the effluent box. Also, calculate weir loading, total number of V-notches, and the depth of the effluent launder. Provide 15% loss for friction and turbulence, and 15 cm to allow for free fall. The peaking factor for design of effluent structure is 1.5.

9.5 A weir launder is 10 m long and 1 m wide. The weir crest is on one side of the trough and covers the entire length of the launder. Calculate the depth of effluent launder if discharge through the basin is 0.3 m^3/s. Depth of flow at the lower end of the launder is 0.9 m. Assume that the friction loss is 15% of the depth of water at the upper end, and free-fall allowance is 6 cm.

9.6 Determine the total head loss through a 4-m-wide sedimentation basin. Draw the hydraulic profile. The influent channel is 1.5 m wide and has one submerged orifice 1.5 m × 0.5 m. The details of the basin are as follows:

a. The invert of the influent channel is 0.5 m above the floor of the chamber. The water depth in the basin is 3.0 m.
b. The head loss in the basin is small and can be ignored.
c. The influent and effluent baffles occupy (respectively) 65% and 60% of the cross-sectional area of the basin, and head losses due to obstruction are 0.001 m at each baffle.
d. The flow through the basin is 1.6 m^3/s.

e. There is 0.4 m head loss into the effluent structure. This head loss is the difference between the water surface elevations in the basin at the effluent weir and in the outlet box of the basin.

9.7 A sedimentation facility was designed to treat an average flow of 0.6 m^3/s. The design overflow rate and the detention time are 45 $m^3/m^2 \cdot d$ and 2.5 h, respectively. The length-to-width ratio of the rectangular basin is 4.3:1. Calculate the dimensions of each basin if two basins in parallel are provided. Also compute the weir loading rate if the weir launder is constructed 1 m from the basin wall. The launder has weir plates on both sides. One outlet channel serves both basins and the width of the channel is 1 m. How the weir loading can be reduced to below 390 $m^3/m \cdot d$?

9.8 Conventional coagulation was carried out to polish the effluent from a POTW. The effluent contains 12 mg/L TSS, and total alkalinity (TA) and total hardness (TH) are, respectively, 150 and 130 mg/L as $CaCO_3$. The optimum alum dose ($Al_2(SO_4)_3 \cdot 14\ H_2O$) to coagulate, flocculate, and remove TSS is 75 mg/L. Calculate (a) TA remaining in the effluent, (b) carbonate hardness (CH) and noncarbonate hardness (NCH) in the coagulated effluent, and (c) solids concentration in the sludge if 95% TSS and 90% chemical precipitates are removed in the clarifier.

9.9 Phosphorus is precipitated in a primary sedimentation basin by the addition of ferric chloride. The total phosphorus and TSS concentrations in the raw municipal wastewater are 7.0 mg/L as P and 200 mg/L, respectively. The average design wastewater flow is 12,000 m^3/d. Ferric chloride is used to precipitate 95% P. TSS removal in the primary basin is 85%. Calculate (a) ferric chloride dose applied daily and (b) quantity of sludge produced daily.

9.10 Phosphorus is precipitated in a primary sedimentation basin by the addition of lime. The total phosphorus and TSS concentrations in raw wastewater, design flow, and total phosphorus and TSS removals are same as given in Example 9.9. Additional data for raw wastewater are: TA = 140 mg/L as $CaCO_3$, TH = 160 mg/L as $CaCO_3$, and Ca = 45 mg/L. Calculate (a) daily lime dose, (b) excess lime applied mg/L, and (c) quantity of sludge produced per day.

9.11 The effluent from a phosphorus precipitation facility has high pH and alkalinity due to excess lime. The design flow is 10,000 m^3/d. The alkalinity is 190 mg/L as $CaCO_3$. Industrial grade sulfuric acid of 90% purity is used to neutralize the effluent to pH 7.2. Calculate (a) the quantity of sulfuric acid required per day and (b) the average volume of 10% acid solution consumed per day to achieve the neutralization. The specific gravity of 10% sulfuric solution is 1.015.

9.12 Define velocity gradient *G*. In a rapid-mix basin, two pinhead floc particles are 0.8 mm apart, and they are moving at a relative velocity of 1.1 m/s. Calculate the velocity gradient *G*.

9.13 Calculate the dimension of a rapid-mix basin. The wastewater flow is 18,900 m^3/d (5 MGD) and is divided into four basins. The detention time is 20 s. All basins are square with a depth-to-width ratio of 1.6:1. Draw the unit configuration.

9.14 Calculate (a) the energy requirement, (b) rotation speed, and (c) Reynolds number of turbine mixer. The rapid-mix basin is 0.88 m × 0.88 m × 1.41 m deep. Provide a velocity gradient of 1000 s^{-1}. Use a four-blade turbine mixer with an impeller diameter of 0.7 m. The gear efficiency is 0.9 and the ratio of blade width to blade diameter is 0.2. Wastewater temperature is 5°C.

9.15 A flocculation basin is designed to treat a flow of 18,900 m^3/d (5 MGD). The flocculator has three stages. Each stage has 10-min detention time. The basin dimensions are 12 m perpendicular to flow direction, and three times the basin depth along the flow direction. Calculate (a) the flocculator dimensions and (b) volume of each stage.

9.16 The flocculator dimensions of a three-stage flocculation basin are calculated in Problem 9-15. The dimensions of each stage are 12 m × 3.31 m × 3.31 m deep. The flocculator is an integral part of a rectangular sedimentation basin that is 12 m wide. There are 10 segments and the length of paddle blades in each segment is 1 m. The absolute velocity of paddle exceeds the desired velocity relative to water by 33% (or water velocity relative to paddle velocity is 75%). The efficiencies of gear box and bearings are 90% and 70%, respectively. Basic requirements of paddle wheels are summarized below.

Stage	G, s^{-1}	Paddle Width, cm	Paddle Length, m	Number of Paddles on Each Segment	Number of Segments	Midpoint Paddle-Blade Radius from Shaft
1	45	15.2	1	8	10	4 at 1.25 m and 4 at 1.0 m
2	30	15.2	1	6	10	2 at 1.5 m, 2 at 1.0 m, and 2 at 0.75 m
3	15	15.2	1	4	10	2 at 1.25 m and 2 at 1.0 m

9.17 A flocculation basin is built as an integral part of a rectangular sedimentation basin. The flocculator is separated by a 12-m- long outlet baffle wall that has multiples ports each of 20 cm diameter. The dimensions of the third stage of flocculation basin are 12 m × 3.31 m × 3.31 m deep. The velocity gradient in the third stage is 30 s^{-1}. The design flow-through flocculator is 0.219 m^3/s. The absolute rotational speed of the outer paddle which is 1.25 m from the shaft is 2.5 rpm, and $C_D = 1.8$. The operating temperature is 20°C. Calculate (a) head loss through the port (b) number of ports, (c) third stage flocculator power requirement, and (d) third-stage total paddle area.

9.18 A rapid-mix and flocculation unit is designed to treat a design flow of 10,000 m^3/d. The rapid-mix unit is square with depth-to-width ratio of 1.5. The rapid mixer has four blades turbine type with width to diameter ratio of 0.18, and the mixer diameter is half the basin width. The flocculation basin is 9.0 m wide and has three stages. The flow length in each stage is equal to depth. Each stage has symmetrical flocculator paddles. There are eight segments of paddle wheels in each stage. Each segment contains paddle blades 0.8 m long (L) and 0.12 m wide (W). Two sets of blades are arranged at each diameters of 1.5 m, 1.0 m, and 0.75 m. The gear box and bearing efficiencies are 90% and 70%, respectively. The wastewater temperature is 5°C. The design information on rapid mix and flocculator are given below.

Rapid-mix detention time 30 s
Rapid-mix velocity gradient G 800 s^{-1}
Flocculation detention time, Three stages at 12 min each
Flocculation velocity gradient G
First stage 50 s^{-1}
Second stage 35 s^{-1}
Third stage 20 s^{-1}

Calculate (a) the dimensions of the rapid-mix basin, (b) motor power, and (c) rotational speed of the mixer. Also, calculate (d) the dimensions of the flocculation basin, (e) motor power requirement in each stage, and (f) rotational speed of the paddles in each stage.

9.19 A square rapid-mix basin has mixing opportunity parameter $G\,t = 30{,}000$. The power dissipation function $P/V = 1000\ (N/s{\cdot}m^2)/m^3$. Calculate (a) mixing loading parameter, (b) basin volume, (c) power imparted to water, and (d) mixing period t. The design flow to the basin is 9000 m^3/d, and wastewater temperature is 10°C.

9.20 Discharge over a rectangular weir across a 1-m-wide channel is 0.8 m^3/s. The weir height above the channel bottom is 1.05 m. The water depth in the rectangular channel downstream of the weir is 0.7 m. Calculate the velocity gradient G created at the weir due to free fall. It is assumed that the mixing time at the free fall is 2 s if the velocity in the downstream channel is >0.5 m/s. The coefficient of discharge at the weir is 0.62. Wastewater temperature is 10°C.

9.21 Wastewater is pumped into a rapid-mix basin that has a volume of 45 ft^3. The pump delivers 300 gpm flow through a 6-in discharge pipe. A coagulating chemical is added into the rapid-mix basin for reaction to occur. The wastewater temperature is 50°F and coefficient of discharge is 0.64. Calculate the velocity gradient G.

9.22 A wastewater treatment facility is using around-the-end baffled flocculator to achieve phosphate precipitation. The design flow is 5700 m^3/d (1.5 MGD). The flow velocity through the flocculator is 0.24 m/s and detention time is 15 min. The clear spacing of the baffle is 0.45 m and around-the-end clearance is 0.68 m. The average temperature of wastewater is 10° C. The basin width is

8 m. Calculate (a) total number of around-the-end channels, and number of baffles, (b) length of flocculator and total flow length, and (c) velocity gradient G. Also, draw the plan, longitudinal section, and hydraulic profile. Use the following information: width of baffles = 0.08 m, minor loss coefficient K_m around each baffle = 2.0, and Mannings $n = 0.013$.

9.23 A primary overload rectangular sedimentation basin is 10 m wide and 4.0 m long. The average side water depth is 4.5 m. It receives a flow of 48,000 m^3/d. This basin is modified by installing counter current parallel plates 5 cm clear. The length of plates are 2 m and are inclined as 60° to the horizontal. Determine the following (a) hydraulic loading and performance of existing basin, (b) settling time t and velocity v_s of particles to settle the vertical distance between the surfaces, (c) check the length of surface L_p traveled in time t, and (d) compare the overflow rate of the basin before and after the modification.

9.24 Micro-sand ballasted flocculation process is applied to treat the peak wet weather overflows at a POTW. The historical data show that the peak wet weather flow is 3.5 times the average design flow of 15000 m^3/d. Calculate (a) the dimensions of settling basin and (b) detention time. The length-to-width ratio is 4 and side water depth is 4.0 m. Use the following information. The average diameter, specific gravity, and shape factor ϕ of the coagulate and ballasted floc particles are 95 μm, 2.65, and 2.3, respectively. The critical wastewater temperature is 10°C.

References

1. Reynolds, T. D. and P. A. Richards, *Unit Operations and Processes in Environmental Engineering*, 2nd ed., PWS Publishing Co., Boston, MA, 1996.
2. Qasim, S. R., *Wastewater Treatment Plants: Planning, Design, and Operation*, 2nd ed., CRC Press, Boca Raton, FL, 1999.
3. Metcalf & Eddy AECOM, *Wastewater Engineering: Treatment and Resource Recovery*, 5th ed., McGraw-Hill, New York, NY, 2014.
4. Water Environment Federation, *Design of Municipal Wastewater Treatment Plants*, vol. 2, WEF Manual of Practice No. 8, ASCE Manual and Report on Engineering Practice No. 76, Water Environment Federation, Alexandria, VA, 1998.
5. Water Environment Federation, *Clarifier Design, WEF Manual of Practice No. FD-8*, 2nd ed., Water Environment Federation, Alexandria, VA, 2005.
6. Water Environment Federation, *Operation of Water Resource Recovery Facilities*, WEF Manual of Practice No. 11, 7th ed., McGraw Hill Education, New Nork, NY, 2016.
7. Randall, C. W., J. L. Barnard, and H. D. Stensel, *Design and Retrofit of Wastewater Treatment Plants for Biological Nutrient Removal, Water Quality Management Library Volume V*, 2nd ed., CRC Press, Boca Raton, FL, 1998.
8. Brentwood Industries, Inc., *Scum Removal System*, http://www.brentwoodindustries.com, November 2014.
9. Brentwood Industries, Inc., *Polychem®, Chain & Flight Scraper Systems*, http://www.brentwoodindustries.com, 2014.
10. Transdynamics Engineering Ltd., *Wastewater Treatment Process Equipment*, http://www.transdynamics.ca, November 2014.
11. Monroe Environmental, *Water & Wastewater Treatment*, http://www.mon-env.com, 2012.
12. WesTech Engineering, Inc., *Municipal Sedimentation Experts—COP™ Clarifier*, http://www.westech-inc.com, 2009.
13. Evoqua Water Technologies, *RIM-FLO® Secondary Clarifier with TOW-BRO Unitube Sludge Removal System*, http://www.evoqua.com, 2014.
14. Matsunaga, K., Design of multistory settling tanks in Osaka, *Journal Water Pollution Control Federation*, 52(5), 1980, 950–554.

15. Kelly, K., New clarifiers help save history, *Civil Engineering—American Society of Civil Engineering*, 58(10), 1988, 54–56.
16. Yuki, Y., Multi-story treatment in Osaka, Japan, *Water Environment and Technology*, 3(7), 1991, 72–76.
17. Nakai, K., Unique design features of Kobe's wastewater treatment plant, *Journal Water Pollution Control Federation*, 52(5), 1980, 955–960.
18. Kelly, K. F., D. P. O'Brien, W. McConnell, and J. Morris, Two tray clarifiers yield positive results, *Water Environment and Technology*, 7(12), 1995, 35–39.
19. Great Lakes—Upper Mississippi River Board of State and Provincial Public Health and Environmental Managers, *Recommended Standards for Wastewater Facilities*, Health Education Service, Albany, NY, 2004.
20. Camp, T. R., Sedimentation and design of settling tanks, *Transactions of the American Society of Civil Engineering*, 111(1), 1946, 895–958.
21. Daugherty, R. L. and J. B. Franzini, *Fluid Mechanics with Engineering Application*, 8th ed., McGraw-Hill Book Co., New York, NY, 1985.
22. Benefield, L. D., J. F. Judkins, and A. D. Parr, *Treatment Plant Hydraulics for Environmental Engineers*, Prentice-Hall, Inc., Englewood Cliffs, NJ, 1984.
23. Droste, R. L., *Theory and Practice of Water and Wastewater Treatment*, John Wiley & Sons, Inc., New York, NY, 1997.
24. Hudson, H. E., *Water Clarification Processes: Practical Design and Evaluation*, von Nostrand Reinehold Co., New York, NY, 1981.
25. Hudson, H. E., R. B. Uhler, and R. W. Barley, Dividing-flow manifolds with square-edged laterals, *Journal Environmental Engineering Division*, ASCE, 104(EE4), 1979, 254–258.
26. Qasim, S. R., E. M. Motely, and G. Zhu, *Water Works Engineering: Planning, Design, and Operation*, Prentice Hall PTR, Upper Saddle River, NJ, 2000.
27. Sawyer, C. N., P. L. McCarty, and G. F. Perkin, *Chemistry for Environmental Engineering*, McGraw-Hill, Inc., New York, NY, 1994.
28. American Water Works Association, *Water Quality and Treatment: A Hand Book of Community Water Supplies*, 5th ed., McGraw-Hill, Inc., New York, NY, 1999.
29. Montgomery Watson Harza, *Water Treatment: Principles and Design*, 2nd ed., John Wiley & Sons, Inc., Hoboken, NJ, 2005.
30. Hendricks, D., *Water Treatment Unit Processes: Physical and Chemical*, CRC Press, a Member of Taylor & Francis Group, Boca Raton, FL, 2006.
31. U.S. EPA, *Wastewater Technology Fact Sheet: Chemical Precipitation*, EPA 832-F-00-018, Office of Water, Washington, DC, September 2000.
32. U.S. EPA, *Nutrient Control Design Manual*, EPA/600/R-10/100, Office of Research and Development/National Risk Management Research Laboratory, Cincinnati, OH, August 2010.
33. U.S. EPA, *Design Manual: Phosphorus Removal*, EPA/625/1-87-001, Office of Research and Development/Center for Environmental Research Information/Water Engineering Research Laboratory, Cincinnati, OH, 1987.
34. U.S. EPA, *Phosphorus Precipitation with Ferrous Iron, Water Pollution Control Research Series*, 17010-EKI-09-71, Office of Research and Monitoring, Washington, DC, September 1971.
35. Speight, J., *Chemical Process and Design Handbook*. McGraw Hill Professional, New Nork, NY, 2002.
36. U.S. EPA, *Phosphorus Removal by Ferrous Iron and Lime, Water Pollution Control Research Series*, 110-10-EGO-01-71, Office of Research and Monitoring, Washington, DC, January 1971.
37. U.S. EPA, *Phosphorus Removal and Disposal from Municipal Wastewater, Water Pollution Control Research Series*, 17010-DYB-02/71, Office of Research and Monitoring, Washington, DC, February 1971.

38. U.S. EPA, *Calcium phosphorus precipitation in wastewater treatment, Environmental Protection Technology Series*, EPA-R2-72-064, Office of Research and Monitoring, Washington, DC, December 1972.
39. U.S. EPA, General information on phosphorus removal, *Presentation at Technology Transfer Design Seminar*, Chicago, IL, November 28–30, 1972.
40. U.S. EPA, *Municipal Nutrient Removal Technologies Reference Document, Vol. 1—Technical Report*, EPA 832-R-08-006, Office of Wastewater Management, Washington, DC, September 2008.
41. U.S. EPA, *Municipal Nutrient Removal Technologies Reference Document, Vol. 2—Appendices*, EPA 832-R-08-006, Office of Wastewater Management, Washington, DC, September 2008.
42. SPX Flow, Inc., *Water and Wastewater Treatment, Lighnin Mixers*, http://www.spxflow.com, 2015.
43. Howe, K. J., D. W. Hand, J. C. Crittenden, R. R. Rrussell, and G. Tchobanoglous, *Montgomery Watson Harza, Principles of Water Treatment*, John Wiley & Sons, Inc., Hoboken, NJ, 2012.
44. Koflo Corporation, *Static Mixers*, http://www.koflo.com, November 2014.
45. Westfall Manufacturing Company, *2800 Static Mixers*, http://westfallmfg.com, November 2014.
46. JDV Equipment Corporation, *JDV Walking Beam Flocculator*, http://www.jdvequipment.com, October 2009.
47. Gupta, R. S., *Hydrology and Hydraulic Systems*, 2nd. ed., Waveland Press, Inc., Prospect Heights, IL., 2001.
48. Haarhoff, J. and J. J. van der Walt, Towards optimal design parameters for around-the-end hydraulic flocculators, *Journal of Water Supply: Research and Technology—AQUA*, 50(3), 2001, 149–159.
49. Vigneswaran, S. and C. Visvanathan, *Water Treatment Processes: Simple Options*, CRC Press, Boca Raton, FL, 1995.
50. U.S. EPA, *Wastewater Technology Fact Sheet: Ballasted Flocculation*, EPA 832-F-03-010, Office of Water, Washington, DC, June 2003.
51. U.S. EPA, *Emerging Technologies for Wastewater Treatment and In-Plant Wet Weather Management*, EPA 832-R-12-011, Office of Wastewater Management, Washington, DC, March 2013.
52. Electric Power Research Institute (EPRI), Inc., *High-Rate Clarification for the Treatment of Wet Weather Flows*, EPRI, Inc., Palo Alto, CA, 1999.
53. Frank, D. A. and T. F., Smith III, Side By Side By Side, The evaluation of three high rate process technologies for wet weather treatment, *Proceedings of the 79th Annual WEFTEC®*, Dallas, TX, October 2006, pp. 6723–6747.
54. Jolis, D. and M. Ahmad, Evaluation of high-rate clarification for wet-weather-only treatment facilities, *Water Environmental Research*, 76(5), 2004, 474–489.
55. Kawamura, S. *Integrated Design and Operation of Water Treatment Facilities*, 2nd ed., Wiley, New York, NY, 2000.
56. SUEZ Treatment Solutions, Inc., *Suez ACCELATOR® Clarifier/Softener*, http://www.suez-na.com, 2013.
57. WesTech Engineering, Inc., *Contact Clarifier—A True Solids Contact Clarifier*, http://www.westech-inc.com, 2014.
58. Ovivo, *Reactor Clarifier™ Solids Contact Clarifiers—Solids Contact and Flocculating Clarifiers for Water & Wastewater Treatment*, http://www.ovivowater.com, 2010.
59. Parkson Corporation, *Lamella® Gravity Settler: Inclined Plate Settler*, http://www.parkson.com, 2009.
60. Parkson Corporation, *Lamella® EcoFlow™: Inclined Plate Settler*, http://www.parkson.com, 2010.
61. WesTech Engineering, Inc., *SuperSettler™ High Rate Sedimentation Technology*, http://www.westech-inc.com, 2010.
62. WesTech Engineering, Inc., *RapidSand™ Ballasted Flocculation System*, http://www.westech-inc.com, 2015.

63. Qasim, S. R. and C. Ramaswamy, *A Surrogate Test for Determination of BOD_5 in the Effluent from Microsand Ballasted Flocculation Facility, Final Report submitted to the City of Fort Worth, Texas, Department of Civil Engineering, The University of Texas at Arlington*, May 2003.
64. Kruger/Veolia Water, *ACTIFLO® The Ultimate Clarifier*, http://www.veoliawater.com
65. Veolia Water Technologies, *ACTIFLO® Microsand Ballasted Clarification Process*, http://www.krugerusa.com, 2007.
66. Crumb, S. F. and R. West, Blended flow process alleviates wet weather cost effectively for Fort Worth, Texas, *Journal of Water Environment and Technology*, 12(4), 2000, 40–45.
67. Reardon, R. D., Clarification concepts for treating peak wet weather wastewater flows, *Proceedings of the 78th Annual WEFTEC®*, Washington, D.C., October, 2005.
68. SUEZ Treatment Solutions, Inc., *DensaDeg® High-Rate Clarifier & Thickener for CSO/SSO*, http://www.suez-na.com, 2011.
69. Nussbaum, B. L., A. Soros, A. Mroz, and B. Rusten, Removal of particulate and organic matter from municipal and industrial wastewater Using Fine Mesh Rotating Belt Sieves, *Proceedings of the 79th Annual WEFTEC®*, Dallas, TX, October 2006, pp. 3052–3056.
70. Rusten, B. and A. Lundar, How a simple bench-scale test greatly improved the primary treatment performance of fine mesh sieves, *Proceedings of the 79th Annual WEFTEC®*, Dallas, TX, October 2006, pp. 1919–1935.
71. Sutton, P. M., B. Rusten, A. Ghanam, R. Dawson, and H. Kelly, Rotating belt screens: An attractive alternative for primary treatment of municipal wastewater, *Proceedings of the 81st WEFTEC®*, Chicago, IL, October 2008, pp. 1671–1687.
72. Salsnes Filter AS, *Eco-Efficient Solids Separation*, http://www.salsnes-filter.com, 2017.

10

Biological Waste Treatment

10.1 Chapter Objectives

The purpose of a secondary wastewater treatment is to remove soluble organics and suspended solids that escape the primary treatment. These removals are typically achieved by biological treatment processes. Although, conventional biological treatment may remove 85–90% BOD_5 and TSS, it does not achieve significant amounts of nitrogen and phosphorus removals. Recent advancements in the biological waste treatment technology are capable of providing enhanced organics and nutrients removals. The objectives of this chapter are to provide an overview of:

- Biological waste treatment
- Role of microorganisms in biological waste treatment
- Microbial growth kinetics and substrate removal
- Suspended, attached, and combined suspended and attached growth reactors
- Oxygen and other environmental requirements
- Aerobic, anoxic, and anaerobic reactors
- Biological nutrient removal
- Theory and design examples of biological waste treatment

10.2 Fundamentals of Biological Waste Treatment

10.2.1 Biological Growth and Substrate Utilization

Substrate is the term used to denote organic matter, nutrients, and other substances that are present in wastewater. Thus, the microorganisms are used to consume organics, nitrify ammonia, denitrify nitrate, and release and uptake phosphorus. Recent advancements in biological waste treatment technology provide economical method to achieve enhanced removal of organics and substantial amounts of nitrogen and phosphorus. In this section, the basic concepts of microbiology, growth phases, microbial metabolism and enzymes, and nutritional and environmental requirements are covered.

Microbiology: A mixed culture of microorganisms plays a vital role in biological waste treatment. In a mixed culture, the number of species and their population depends upon the characteristics of wastewater and the environmental conditions. A mixed culture of microorganisms includes bacteria (both single and multicellular), protozoa, fungi, rotifers, and sometimes nematodes. The principal organisms involved in the bio-oxidation of organics in wastewater are the single-celled bacteria. Although some individual microbial cells are present in suspended growth reactors, the majority are present as zooglea biomass or *floc*. The floc consists of mixed species of cells embedded in masses of polysaccharide gums released from the slime layers of living and lysed bacterial cells.[1] The protozoa, rotifers, and nematodes do not provide stabilization of organic matter in the wastewater. Instead, they feed on the bacterial population and other fragments of zooglea biomass. Their presence is simply an indication of healthy operation with appreciable aeration time. The classification of microorganisms and basic concepts of microbiology are covered in Chapter 5.

Basic Requirements: The basic requirements of aerobic biological waste treatment are: (1) presence of mixed population of active microorganisms, (2) good contact between microorganisms and substrates, (3) availability of oxygen, (4) availability of nutrients, and maintenance of other favorable environmental conditions such as temperature and pH, (5) sufficient contact time between food and microorganisms, and (6) effective separation of microorganisms from the treated effluent. These conditions are achieved in a wide variety of ways. Some of these are as follows:[2–4]

- Proper equipment and reactor design, and selection and assembly of hardware
- Effective ways to bring contact between the food and microorganisms
- Innovative ways to supply oxygen and maintain mixing
- Efficient ways to separate microorganisms and release the treated effluent

Details of these conditions are provided in later sections of this chapter.

Growth Phase: The bacteria reproduce by *binary fission* (division of original cell into two new cells). Time required for each fission is called *generation time*, which may vary from less than 20 minutes to several days. The growth of microorganisms in a batch reactor may be expressed in terms of time versus viable bacterial number or biomass. The viable bacterial *number curve* has six distinct phases.

As soon as food is brought in contact with the microorganisms, the growth starts. Initially, the growth is slow because the microorganisms become adjusted to the food and to the new environment. This is called the *lag growth phase*. After the initial adjustment phase, there is a rapid growth in number. This phase is referred to the *loggrowth phase* (or *geometric growth*) and is typical of microorganisms when there is an excess of food. In this phase, the rate of growth of cells and their subsequent division is limited by the ability of the microorganisms to process the substrate. During this phase, the microorganisms grow at their maximum rate, and consequently remove organic matter from the solution at the maximum rate.

As food supply is depleted, the rate of new cell production slows down and *declining growth phase* is reached. A further decrease in food concentration inhibits the microbial growth. This results in a *stationary phase*. At this phase, the population of the viable cells is balanced by the new and dead cells. As the biodegradable portion of the substrate is totally consumed, the death phase starts. In this phase, the bacterial death rate exceeds the production of new cells. Initially, the number declines slowly, then exponentially. These are called *lag death phase* and *increasing death phase*, respectively. The typical growth curves in terms of number of viable cells and the substrate utilization curves with respect to time are shown in Figure 10.1a.

The *biomass curve* is different from viable cell number curve because both live and dead cells are included in the cell mass. It has four distinct phases. The *lag-, log-, and declining-growth phases* are dependent upon the availability and utilization of substrate. At the end of declining growth phase, the cell mass reaches a maximum value. Beyond that time the *endogenous phase* predominates. During this phase, the microorganisms are forced to metabolize their own protoplasm, and decay rate predominates resulting in decrease in biological mass. The typical microbial cell mass and substrate utilization curves with respect to time are shown in Figure 10.1b.[1–3]

Microbial Metabolism: The microbial metabolism involves source of cell carbon, electron donor, electron acceptor, and end products. The carbon sources are organic compounds and CO_2. Different microorganisms can use a wide range of electron donors and acceptors. The electron donors are the reducing compounds, such as NH_3, NO_2^-, Fe^{2+}, H_2S, S, $S_2O_3^{2-}$, and others. The electron acceptors are oxidants, such as O_2, NO_2^-, NO_3^-, SO_4^{2-}, Fe^{3+}, organic compounds, and CO_2.

Microorganisms obtain energy for their growth from very complex and intricate biochemical reactions by substrate utilization. Special enzymes are involved in a series of reactions forming a sequence of enzyme–substrate complexes, which are then converted to the end products and the original enzymes. These enzymes are proteins that act as catalysts. They are specific to each substrate and have a high degree

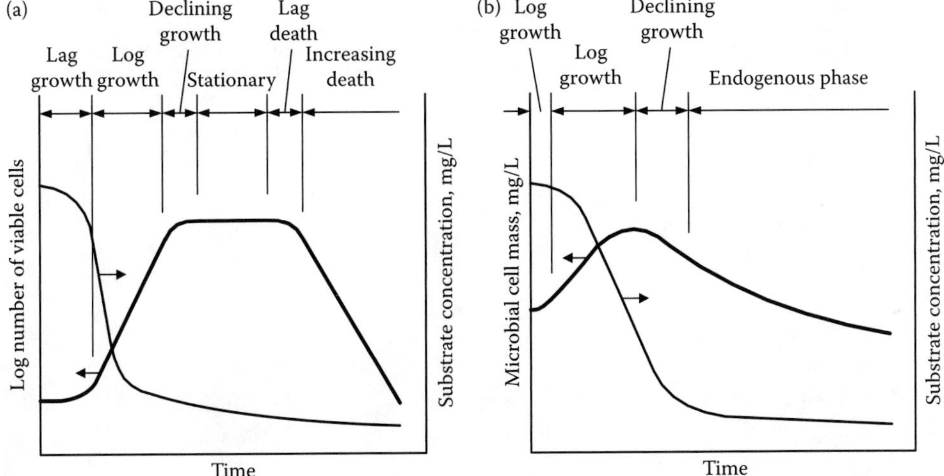

FIGURE 10.1 Typical microbial growth and substrate utilization curves: (a) viable number and (b) microbial cell mass.

of efficiency in converting the substrate to the end products. The enzymes can be *extracellular* or *intracellular*.[5–7]

Energy, Nutritional, and Growth Factors: For growth and survival of microorganisms, there must be (1) a source of energy, (2) sufficient carbon for synthesis of new cells, and (3) inorganic elements or nutrients. The energy needed for cell synthesis may be supplied by light or by chemical reactions. The energy-producing chemical reactions by *chemotrophs* are oxidation–reduction reactions: The electron donor is oxidized and the electron acceptor is reduced. If oxygen is the electron acceptor, the reaction is *aerobic.* Reactions involving any other electron acceptors are considered *anaerobic.* When nitrite and nitrate serve as designated electron acceptors, the reaction is distinguished as *anoxic.* Depending upon the electron acceptor requirements, the microorganisms are called *obligate aerobes* that utilize oxygen only and die in the absence of oxygen, *obligate anaerobes* that utilize SO_4^{2-}, Fe^{3+}, organic compounds, or CO_2 and die in the presence of oxygen, and *facultative anaerobes* that utilize oxygen if oxygen is present and switch to other electron acceptors in the absence of oxygen.

The principal inorganic nutrients needed by microorganisms are N, P, S, K, Na, Ca, Mg, Fe, and Cl. Minor nutrients of importance include trace elements such as Zn, Mn, Mo, Co, Cu, Ni, V, and W. In addition to inorganic nutrients, organic nutrients or growth factors are also needed by some microorganisms. The major growth factors are (1) amino acids, (2) purines and pyrimidines, and (3) vitamins.

Biomass Composition: The microbial cells consist of carbon, hydrogen, oxygen, nitrogen, and many minerals. The commonly used empirical formula to express the organic fraction of cells is $C_5H_7O_2N$. The empirical formula of cellular matter including phosphorus is $C_{60}H_{87}O_{23}N_{12}P$. In these formulations, the percent weight of carbon is 53% and 52%, respectively.[2,7] When biomass is generated using synthetic substrate, the organic fraction is ~95% of the cell mass generated. The typical compositions of bacterial cell are provided in Table 10.1.[2,8] These empirical formulations and values in Table 10.1 are considered typical as they may vary with different influent characteristics and process designs.[8]

Environmental Factors: The principal environmental factors affecting microbial activity are (1) temperature, (2) dissolved solids (salt content), (3) presence or absence of oxygen, and (4) mixing or contact between food and microorganisms. The temperature also affects the microbial activity. On an average, a temperature increase of 10°C approximately doubles the rate of the activity. The dissolved solids or salt

TABLE 10.1 Typical Composition of Bacterial Cell Mass

Constituent of Cell	Composition, % as Dry Weight
Carbon	50
Hydrogen	6
Oxygen	25
Nitrogen	12
Phosphorus	2
Total organic fraction	95%
Sulfur	1
Potassium	1
Sodium	1
Calcium	0.5
Magnesium	0.5
Chloride	0.5
Iron	0.2
Other trace elements	0.3
Total inorganic fraction	5%
Total cell mass	100%

Source: Adapted in part from References 2 and 8.

content affect the pH, buffering capacity, oxidation and reduction potential, and osmotic pressure of the aqueous solutions. Normally, microorganisms can only function within a narrow range of these factors. Heavy metal ions are toxic in relatively low concentrations. The toxicity in general increases with an increase in the atomic weight. The heavy metals that may be encountered in wastewaters are arsenic, barium, cadmium, chromium, copper, lead, mercury, nickel, silver, and zinc. Certain organic chemicals are also quite toxic to microorganisms.[1–3]

EXAMPLE 10.1: POPULATION OF MICROBIAL CELLS IN LOG-GROWTH PHASE

In the log-growth phase, the growth rate (dN/dt) of microorganisms at any time is proportional to the number present (N). Develop the log growth rate equation. Determine the number of cells at the end of 12 h of aeration period. Assume that the initial microbial number is 10^{18} and the generation time is 45 min.

Solution

1. Develop the log-growth rate equation.

   The log growth is expressed by Equation 10.1a.

$$\frac{dN}{dt} = kN \tag{10.1a}$$

where

N = number of cells present at time t, number of cells
t = reaction time, variable time unit
k = growth rate constant to the base e, t^{-1}

Rearrange and integrate the equation in the limits N_0 at $t = 0$, and N at time t to obtain the log-growth rate equation in Equation 10.1b.

$$\int_{N_0}^{N} \frac{dN}{N} = k \int_{0}^{t} dt$$

$$\ln\left(\frac{N}{N_0}\right) = k\,t \quad \text{or} \quad \frac{N}{N_0} = e^{kt} \cdots N = N_0 e^{kt} \tag{10.1b}$$

where N_0 = initial number at $t = 0$, number of cells

2. Develop the generation time equation (Equation 10.1c) from Equation 10.1b at the end of generation while the population is doubled or $N/N_0 = 2$.

$$\ln\left(\frac{N}{N_0}\right) = \ln(2) = k\,t_{\text{generation}}$$

$$t_{\text{generation}} = \frac{\ln(2)}{k} = \frac{0.693}{k} \tag{10.1c}$$

3. Calculate the growth rate constant (k) from Equation 10.1c at $t_{\text{generation}} = 45$ min or 0.75 ma h (given).

$$k = \frac{0.693}{t_{\text{generation}}} = \frac{0.693}{0.75\text{ h}} = 0.92\text{ h}^{-1}$$

4. Calculate the number of cells (N) from Equation 10.1b at the end of aeration period $t = 12$ h.

$$N = N_0 e^{kt} = 10^{18} \times e^{((0.92\text{h}^{-1}) \times 12\text{ h})} = 10^{18} \times 62{,}318 = 6.2 \times 10^{22}\text{cells}$$

EXAMPLE 10.2: MASS OF MICROORGANISMS IN A BIOLOGICAL REACTOR

Determine the cell biomass in a biological reactor. Total number of microorganisms is 8.8×10^{15}. Assume that the bacteria have spherical shape with the average diameter d of 1 μm and the specific gravity of 1.02.

Solution

1. Estimate the mass of each bacterium.
   a. Calculate the average volume of one spherical cell (V_{cell}) at $d = 1\ \mu\text{m} = 10^{-6}$ m.

   $$V_{\text{cell}} = \frac{\pi}{6} d^3 = \frac{\pi}{6} \times (10^{-6}\text{m})^3 = 5.24 \times 10^{-19}\text{m}^3/\text{cell}$$

   b. Calculate the average biomass of one cell (m_{cell}) using the specific gravity of 1.02 and a water density of 10^3 kg/m^3.

   $$\begin{aligned} m_{\text{cell}} &= 1.02 \times 10^3\text{ kg/m}^3 \times V_{\text{cell}} = 1.02 \times 10^6\text{ g/m}^3 \times 5.24 \times 10^{-19}\text{ m}^3/\text{cell} \\ &= 5.34 \times 10^{-13}\text{ g/cell} \end{aligned}$$

2. Determine the total biomass of cells in the reactor (m_{reactor}) at total cell number $N_{\text{reactor}} = 8.8 \times 10^{15}$ cells/reactor.

$$\begin{aligned} m_{\text{cell}} &= m_{\text{cell}} \times N_{\text{reactor}} = 5.34 \times 10^{-13}\text{ g/cell} \times 8.8 \times 10^{15}\text{ cells/reactor} \\ &= 4.7 \times 10^3\text{ g/reactor} = 4.7\text{ kg/reactor} \end{aligned}$$

EXAMPLE 10.3: CELL MASS GROWTH IN LOG GROWTH PHASE

In a log growth phase, the cell mass grows exponentially, and the cell growth is measured by the increase in the concentration of mixed liquor volatile suspended solids (MLVSS). The volume of a plug-flow aeration basin is 1 million gallon (MG), and the aeration period is 4 h. The initial concentration of MLVSS is 2000 mg/L and growth rate constant is 0.04 h^{-1} (base e). Calculate the mass of biosolids generated during the aeration period.

Solution

1. Calculate the MLVSS concentration (X) from Equation 10.1b at $X_0 = 2000$ mg/L, $k = 0.04\ h^{-1}$, and $t = 4$ h.

$$X = X_0 e^{kt} = 2000\,\text{mg/L} \times e^{((0.04\text{h}^{-1})\times 4\text{h})} = 2000\,\text{mg/L} \times 1.17 = 2340\,\text{mg/L}$$

2. Calculate the mass of biosolids generated during the aeration period.
   a. Calculate the increase in MLVSS concentration (ΔX) during the aeration period.

$$\Delta X = X - X_0 = (2340 - 2000)\,\text{mg/L} = 340\,\text{mg/L}$$

   b. Calculate the total biomass growth (Δm) during the aeration period in the aeration basin with a volume $V = 1$ MG.

$$\Delta m = 8.34\frac{\text{lb}}{\text{mg/L·MG}} \times \Delta X \times V = 8.34\frac{\text{lb}}{\text{mg/L·MG}} \times 340\,\text{mg/L} \times 1\,\text{MG} = 2800\,\text{lbs}$$

EXAMPLE 10.4: SUBSTRATE CONSUMPTION IN LOG GROWTH PHASE

The substrate is normally consumed exponentially in the log growth phase. Calculate the concentration of BOD_5 remaining after 10 h of reaction time. The BOD_5 consumption rate constant is 0.28 h^{-1} and initial BOD_5 concentration is 350 mg/L.

Solution

1. Set up the substrate consumption rate equation.

   The substrate consumption in the log growth phase is expressed by Equation 10.1d.

$$\frac{dS}{dt} = -kS \tag{10.1d}$$

where

S = substrate concentration at time t, mg/L

k = substrate consumption rate constant to the base e, t^{-1}

Note: The minus sign indicates depletion.

Rearrange and integrate the equation in the limits S_0 at $t = 0$, and S at time $t = t$ to obtain the substrate consumption rate equation in Equation 10.1e.

$$\ln\left(\frac{S}{S_0}\right) = -k\,t \quad \text{or} \quad \frac{S}{S_0} = e^{-kt} \quad \text{or} \quad S = S_0 e^{-kt} \tag{10.1e}$$

where S_0 = initial substrate concentration at $t = 0$, mg/L

2. Calculate the BOD_5 concentration (C) from Equation 10.1e $C_0 = 350$ mg/L, $k = 0.28\ h^{-1}$, and $t = 10$ h.

$$C = C_0 e^{-kt} = 350\ \text{mg/L} \times e^{-((0.28\ \text{h}^{-1})\times 10\ \text{h})} = 350\ \text{mg/L} \times 0.061 = 21\ \text{mg/L}$$

EXAMPLE 10.5: BIOMASS COMPOSITION WITHOUT PHOSPHORUS

Chemical analysis was performed on a biomass sample from an aerobic biological reactor. The following results of chemical analysis were obtained. Develop the chemical formula.

Element	Composition, % as Dry Weight
Carbon	42.5
Hydrogen	5.0
Oxygen	22.6
Nitrogen	9.9
Total volatile fraction	80.0
Total nonvolatile fraction	20.0

Solution

1. Describe the procedure for developing the chemical formula.

   A tabulation procedure is used to develop the chemical formula. A brief description is provided below for each step of the procedure as shown in Table 10.2.

   a. Obtain the percent of each element on the basis of total biomass in column (2).
   b. Calculate percent of each element on the basis of the volatile fraction only in column (3).
   c. Summarize the atomic weight of each element in column (4).
   d. Calculate the ratio of the percent volatile fraction to the atomic weight for each element in column (5).
   e. Identify the lowest ratio from the ratios for all elements in column (5).

TABLE 10.2 Calculation Results for Developing the Chemical Formula (Example 10.5)

Element	Percent in Total Biomass, %	Percent Volatile Fraction, %	Atomic Weight	Ratio of Percent Volatile Fraction to Atomic Weight	Relative Ratio or Number of Atoms
(1)	(2)	(3)	(4)	(5)	(6)
C	42.5[a]	53.1[b]	12[c]	4.43[d]	5[f]
H	5.0	6.2	1	6.20	7
O	22.6	28.3	16	1.77	2
N	9.9	12.4	14	0.89[e]	1
Total volatile fraction	80.0	100.0	N/A	N/A	N/A

[a] The percent of C in the total biomass is 42.5%.
[b] The percent of C is 42.5 ÷ 80 × 100% = 53.1% on the basis of the volatile fraction.
[c] The atomic weight of C is 12.
[d] The ratio of the percent volatile fraction to the atomic weight is 531 ÷ 12 = 4.43 for C.
[e] The lowest ratio of 0.89 for N is identified among the ratios in column (5).
[f] The relative ratio to the lowest ratio is 4.42 ÷ 0.89 ≈ 5 for C.

f. Calculate the relative ratio of each element to the lowest ratio in column (6). The relative ratio gives the number of atoms of each element in the chemical formula.

2. Determine the chemical formula of the biomass.

Calculation results are also tabulated in Table 10.2. The relative ratio in column (6) is the number of atoms of C, H, O, and N in the chemical formula. The chemical formula of the biomass is therefore $C_5H_7O_2N$.

EXAMPLE 10.6: BIOMASS COMPOSITION INCLUDING NITROGEN AND PHOSPHORUS

The chemical analysis of microbial cells from a biological reactor gave the following result. Develop the chemical formula.

Element	Composition, % as Dry Weight
Carbon	45.0
Hydrogen	5.5
Oxygen	27.0
Nitrogen	10.5
Phosphorus	2.0
Total volatile fraction	90.0
Total nonvolatile fraction	10.0

Solution

A tabulation procedure is described step by step for developing the chemical formula in Example 10.5. The same procedure is also used for this example. Calculation results are summarized in Table 10.3. Based on the relative ratios in column (6), the chemical formula of the biological cells is determined as $C_{60}H_{87}O_{27}N_{12}P$.

TABLE 10.3 Calculation Results for Developing the Chemical Formula (Example 10.6)

Element	Percent in Total Biomass, %	Percent Volatile Fraction, %	Atomic Weight	Ratio of Percent Volatile Fraction to Atomic Weight	Relative Ratio or Number of Atoms
(1)	(2)	(3)	(4)	(5)	(6)
C	45.0	50.0	12	4.17	60
H	5.5	6.1	1	6.10	87
O	27.0	30.0	16	1.88	27
N	10.5	11.7	14	0.84	12
P	2.0	2.2	31	0.07	1
Total volatile fraction	90.0	100.0	N/A	N/A	N/A

10.2.2 Types of Biological Treatment Processes

The common biological waste treatment processes fall into three major groups. These are aerobic, anaerobic, and biological nutrient removal processes. These processes are briefly described below:

Major Groups and Subgroups: The major groups and subgroups of biological waste treatment processes and their applications are presented in Table 10.4. The discussion on these processes is divided into the following categories:

a. Aerobic suspended growth
b. Aerobic attached growth
c. Combined suspended and attached growth
d. Anaerobic suspended growth
e. Anaerobic attached growth and combinations of these
f. Biological nutrient removal systems that include combination of anaerobic, anoxic, and aerobic processes

Common Terms and Definitions: Biological waste treatment involves removal of many constituents in the waste stream using live microbial culture. Although many useful terms and definitions have been presented and discussed in the earlier chapters, it is appropriate to review many of them again. This will provide a new perspective for the materials covered in this chapter. These common terms and definitions are provided in Table 10.5. The basic reactor types, definition terms, and flow schematic are shown in Figure 10.2.

TABLE 10.4 Major Groups and Subgroups of Biological Waste Treatment Processes and Their Applications

Major Group	Process Description	Major Subgroup	Applications
Aerobic process	It occurs in the presence of oxygen.	• Suspended growth • Attached growth • Combined suspended and attached growth	• Removal of carbonaceous BOD_5 • Nitrification
Anaerobic process	It occurs in the absence of oxygen, nitrate, and nitrite.	• Suspended growth • Attached growth	• Removal of carbonaceous BOD_5 • Solids stabilization
Biological nutrient removal	It occurs in a single or multiple reactors where anaerobic, anoxic, and/or aerobic conditions are created in a special sequence. Denitrification, and phosphorus release and uptake occur in the presence of suitable carbon source.	• Suspended growth • Combined suspended and attached growth	• Removal of carbonaceous BOD_5 • Nitrification • Denitrification • Phosphorus removal

TABLE 10.5 Common Terms and Definitions in Biological Waste Treatment Processes

Common Terms	Definitions
Acclimated culture or acclimated activated sludge	A microbial culture that has been cultivated to metabolize a particular substrate or wastewater. Mixed culture is cultivated by gradually increasing the desired dose of substrate until microbes metabolize in higher concentrations.
Activate sludge	The active biomass that is maintained in an aeration basin.
Activated sludge process	It consists of an aeration basin, a secondary (or final) clarifier, and a pumping system for returning and wasting activated sludge. The mixed liquor is aerated in the aeration basin for bio-oxidation of organic matter. It is then settled in the clarifier. The supernatant is discharged and the settled sludge is either returned to the aeration basin or wasted for further disposal.

(*Continued*)

TABLE 10.5 (*Continued*) Common Terms and Definitions in Biological Waste Treatment Processes

Common Terms	Definitions
Advanced wastewater treatment	Physical, chemical, and biological treatment processes used to upgrade secondary treatment, and to remove nutrients.
Aeration process	A physical process that dissolves oxygen into wastewater in order to create a desired condition for aerobic processes. It may be achieved by using air diffusers or mechanical surface (or brush) aerators.
Aerobic or oxic process	Treatment occurs in the presence of free molecular oxygen. Oxygen serves as an electron acceptor. This process is mainly used for removal of organic matter or nitrification.
Anaerobic process	Organic waste is stabilized in the absence of oxygen. The electron acceptors are PO_4^{3-}, NO_2^-, NO_3^-, SO_4^{2-}, Fe^{3+}, organic compounds, and CO_2. The decomposition products are odorous compounds. This process is mainly used for removal of organic matter or solids stabilization. It may also be an integral part of biological nutrient removal process.
Anoxic process	Treatment occurs in the absence of oxygen but in the presence of NO_3^-, NO_2^-, and a carbon source. NO_3^- and NO_2^- serve as electron acceptors. Nitrate nitrogen is biologically converted to nitrogen gas (N_2). This process is also called denitrification. It is an integral part of biological nutrient removal process.
Attached growth reactor	In an attached growth or fixed-film reactor, the biomass remains attached to some inert or fixed media. It can be used in either aerobic or anaerobic process. Example is a trickling filter.
Autotrophic bacteria	Bacteria that use inorganic materials for both energy and growth.
Basal metabolism	Basal metabolism occurs when microorganisms consume very small amount of substrate for survival without any appreciable growth in cell mass.
Batch reactor	A reactor that does not have continuous flow entering or leaving. The reactants are added, the reaction occurs, then the products are discharged with or without settling.
Bench-scale reactor	A laboratory-scale operation, process, or combination thereof.
Binary fission	Division of original cell into two cells.
Biological nutrient removal	A biological waste treatment process in which nitrogen and phosphorous are removed. The process utilizes anaerobic, anoxic, and aerobic sequences in different combinations to achieve phosphorous release and uptake, nitrification, and denitrification.
Biomass	Biomass represents the living microbiological culture responsible for biological waste treatment. It is measured and expressed by total suspended solids (TSS), volatile suspended solids (VSS), mixed liquor suspended solids (MLSS), and mixed liquor volatile suspended solids (MLVSS). Both TVSS and MLVSS are better representatives of active biomass than TSS and MLSS. Normally, the ratio of MLVSS to MLSS is 0.7–0.85 in biomass.
Biomass yield[a]	It is the ratio of the amount of biomass produced to the amount of substrate consumed or utilized. Most common expression of biomass yield is: $Y =$ g VSS-produced/g BOD_5 or COD (substrate)-consumed The actual observed biomass yield (Y_{obs}) is normally less than the above yield because a portion of microorganisms is reduced due to endogenous decay.
Bulking sludge (sludge bulking)	An activated sludge that settles poorly because of low bulk density. The main cause of sludge bulking is the extensive growth of filamentous bacteria.
Combined attached and suspended growth	A fixed-film reactor is used in conjunction with an activated sludge process in a dual system.
Completely mixed reactor	The fluid elements upon entry into the reactor are dispersed almost immediately throughout the entire reactor content.

(*Continued*)

TABLE 10.5 (*Continued*) Common Terms and Definitions in Biological Waste Treatment Processes

Common Terms	Definitions
Continuous flow reactor with or without sludge recycle	A reactor that has continuous flow entering and leaving the reactor. In reactors with sludge recycle, the biomass is settled in a separate clarifier and returned to the reactor. The example is an activated sludge plant. In reactors with no recycle, the biomass is not settled. Example is an aerated lagoon.
Dispersed plug-flow reactor	A reactor that has significant longitudinal or axial mixing of fluid elements throughout its length.
Endogenous decay	The biological oxidation occurs with limited food supply. The microbial cell tissue is consumed.
Enzymes	Organic catalysts that are proteins and are produced by living cells.
Excess activated sludge (waste activated sludge)	The microbial growth that is in excess of that needed for plant operation. The excess growth must be wasted for balanced operation.
Facultative process	The biological decomposition occurs in the presence or absence of molecular oxygen.
Food-to-microorganism (F/M) ratio[a]	The ratio of substrate (BOD_5, COD, or TOC) mass entering the reactor to the total mass of biological solids (MLVSS) maintained in the reactor: kg BOD_5/kg VSS·d or kg COD/kg VSS·d.
Gasification	A conversion of soluble organic matter into gases during anaerobic bio-oxidation.
Heterotrophic bacteria	Bacteria that feed only on organic matter.
Hydraulic retention time (HRT or θ)[a]	HRT or detention time is the average time liquid stays in a reactor. It is equal to reactor volume (V) divided by the flow (Q): $\theta = V/Q$.
Integrated fixed-film media in aeration basin	The biomass remains in suspension and also attaches to fixed or floating media. The examples of these hybrid systems include integrated fixed-film activated sludge (IFAS) and moving bed biofilm reactor (MBBR).
Mass organic loading (MOL)	The mass loading rate of organics (BOD_5, COD, or TOC) that is utilized by the microorganisms in the aeration basin: kg organics/kg biomass per day.
Mass transfer	The transfer of a substance from one phase to the other. For example, transfer of oxygen from gaseous phase into soluble phase in a liquid.
Mean cell residence time (MCRT or θ_c)[a]	MCRT is the average time a microbial cell remains in a reactor. It is also called solids retention time (SRT) or sludge age. The MCRT is equal to the mass of biological solids (XV) in the reactor divided by the rate of biomass loss from the system (ΔX): $\theta_c = XV/\Delta X$.
Metabolic functions	These functions are necessary for the maintenance of living organisms. Based on metabolic functions, the biological processes are classified as aerobic, anaerobic, anoxic, facultative and combined processes.
Mixed liquor	It is a mixture of raw or settled wastewater and activated sludge contained in an aeration basin.
Nitrobacter	A genus of bacteria that oxidizes nitrite (NO_2^-) to nitrate (NO_3^-) in the presence of oxygen.
Nitrosomonas	A genus of bacteria that oxidizes ammonia (NH_3) to nitrite (NO_2^-) in the presence of oxygen.
Photosynthesis	The synthesis of complex organic compounds from carbon dioxide, water and inorganic salts. The sunlight is used as the energy source in the presence of catalyst such as chlorophyll.
Plug-flow reactor	A reactor in which all fluid elements that enter and flow through the reactor exit in the same sequence as they enter. The travel time of the fluid elements is equal to the theoretical detention time, and there is no longitudinal mixing. Long narrow tank and pipe flow are examples of such reactor.
Precursor	A substance from which another substance is later formed.

(*Continued*)

TABLE 10.5 (*Continued*) Common Terms and Definitions in Biological Waste Treatment Processes

Common Terms	Definitions
Return sludge or recycle ratio[a]	The ratio of returned activated sludge (RAS) flow (Q_r) to the influent flow to the aeration basin (Q): $R_{ras} = Q_r/Q$.
Shock (or slug) loading	A sudden change in either organic or hydraulic load to a waste treatment plant.
Sludge volume index (SVI)[a]	It indicates the settling behavior of mixed liquor in an aeration basin. It is the volume (mL) occupied by 1 g of settled sludge. It is also defined as the ratio of percent volume occupied by settled sludge in a 1-L graduated cylinder after 30 min settling to the MLSS concentration in percent.
Substrate	Substrate (S) is organic matter, nutrients, or other substances that are present in wastewater and are consumed by microorganisms for growth and survival. BOD_5, COD, and TOC are the commonly used nonspecific measures of organic matter.
Suspended growth reactor	The biomass in the reactor remains in suspension within the liquid. It can be used in either aerobic or anaerobic process. The example is an activated sludge process.
Volumetric organic loading (VOL)	The mass loading rate of organics (BOD_5, COD, or TOC) that is fed into unit volume of the aeration basin: kg organics/m^3 of aeration basin volume per day.
Wastewater reuse (or reclamation) process	The wastewater treatment process for intended reuse of the effluent.
Zooglea	The gelatinous material resulting from attrition of bacterial slime layer. It is an important constituent of activated sludge floc and biological growth in trickling filter.

[a] These are important parameters used in design and operation of biological waste treatment process.
Source: Adapted in part from References 2, 7, and 8.

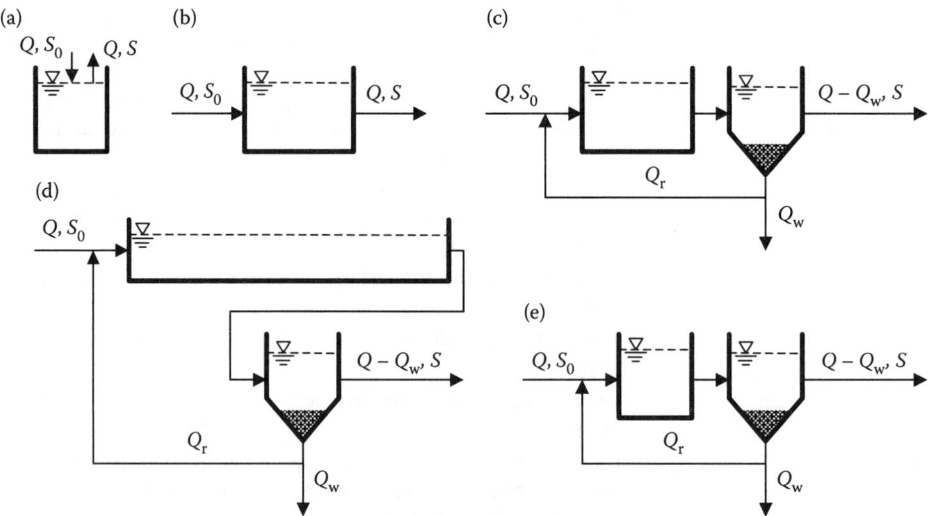

FIGURE 10.2 Basic reactor types and definitions of terms: (a) a batch reactor, (b) a continuous flow reactor with no sludge recycle, (c) a continuous flow reactor with sludge recycle, (d) a plug-flow reactor, and (e) a completely mixed reactor.

EXAMPLE 10.7: ESTIMATION OF OBSERVED BIOMASS YIELD Y_{obs} FROM BATCH EXPERIMENTAL DATA

A batch reactor was operated. The initial COD and VSS concentrations were 350 and 560 mg/L, respectively. The COD and VSS concentrations after 4-h operation were 110 and 660 mg/L. Estimate the observed biomass yield.

Solution

1. Calculate the increase in VSS (ΔX) over 4-h period from $X_0 = 560$ mg/L at $t = 0$, and $X_e = 660$ mg/L at $t = 4$ h.

$$\Delta X = X_e - X_0 = (660 - 560)\,\text{mg} - \text{VSS/L} = 100\,\text{mg} - \text{VSS/L}$$

2. Calculate the COD consumption (ΔS) in 4-h period from $S_0 = 350$ mg/L at $t = 0$, and $S = 110$ mg/L at $t = 4$ h.

$$\Delta S = S_0 - S = (350 - 110)\,\text{mg COD/L} = 240\,\text{mg COD/L}$$

3. Estimate the observed biomass yield (Y_{obs}) from $\Delta X = 100$ mg VSS/L and $\Delta S = 240$ mg COD/L.

$$Y_{obs} = \frac{\Delta X}{\Delta S} = \frac{100\,\text{mg VSS/L}}{240\,\text{mg COD/L}} = 0.42\,\text{mg VSS/mg COD} \quad \text{or} \quad 0.42\ \text{g VSS/g COD}$$

Therefore, the observed biomass yield is about 0.42. It indicates that 0.42 g of biomass as VSS is produced per gram of substrate as COD is consumed.

EXAMPLE 10.8: ESTIMATION OF OBSERVED BIOMASS YIELD Y_{obs} FROM A CONTINUOUS FLOW REACTOR WITH NO RECYCLE

A continuous flow biological reactor is operated without return sludge. The flow to the reactor is 1580 m^3/d, and influent and effluent soluble COD are 450 and 15 mg/L, respectively. The effluent VSS concentration is 180 mg/L. Calculate the observed biomass yield Y_{obs}.

Solution

1. Calculate the amount of soluble COD consumed from $S_0 = 450$ mg/L (or g/m^3) and $S = 15$ mg/L (or g/m^3) at $Q = 1580\ m^3$/d.

$$\text{COD consumed} = \Delta S Q = (S_0 - S) \times Q = (450 - 15)\,\text{g/m}^3 \times 1580\,\text{m}^3\text{/d} \times 10^{-3}\,\text{kg/g} = 687\,\text{kg COD/d}$$

2. Calculate the amount of VSS produced from $X_0 \approx 0$ g/m^3 (due to no recycle) and $X_e = 180$ mg/L (or g/m^3) at $Q = 1580\ m^3$/d.

$$\text{VSS produced} = \Delta X\, Q = (X_e - X_0) \times Q = (180 - 0)\,\text{g/m}^3 \times 1580\,\text{m}^3\text{/d} \times 10^{-3}\,\text{kg/g} = 284\,\text{kg VSS/d}$$

3. Calculate the observed biomass yield Y_{obs} from VSS produced = 284 mg VSS/d, and COD consumed = 687 mg COD/d.

$$Y_{obs} = \frac{\text{VSS produced}}{\text{COD consumed}} = \frac{284\,\text{mg VSS/d}}{687\,\text{mg COD/d}} = 0.41\,\text{mg VSS/mg COD} \quad \text{or} \quad 0.41\ \text{g VSS/g COD}$$

Therefore, the observed biomass yield is about 0.41.

EXAMPLE 10.9: ESTIMATION OF BIOMASS YIELD Y FROM STOICHIOMETRY

A biological reactor is operated using glucose ($C_6H_{12}O_6$) as a substrate. The biomass is represented as $C_5H_7O_2N$. Estimate the biomass yield from the stoichiometric relationship between: (a) biomass and the substrate and (b) biomass and COD.

Solution

1. Write the stoichiometric relationship between the substrate utilized and biomass yield.

   By neglecting nutrients other than nitrogen, the stoichiometric relationship is expressed by Equation 10.2.

$$\underset{3\times180}{3\,C_6H_{12}O_6} + \underset{8\times32}{8\,O_2} + 2\,NH_3 \rightarrow \underset{2\times113}{2\,C_5H_7O_2N} + 8\,CO_2 + 14\,H_2O \tag{10.2}$$

2. Calculate the biomass yield (Y) from Equation 10.2 due to the consumption of substrate ($C_6H_{12}O_6$).

$$Y = \frac{\text{Cell mass yield}}{\text{Glucose consumed}} = \frac{2 \times 113\,\text{g/mole as cell mass}}{3 \times 180\,\text{g/mole as glucose}} = 0.42\,\text{g cell mass/g glucose consumed}$$

   Therefore, 0.42 g cell mass is produced per gram of glucose consumed for cell synthesis.

3. Write the balanced stoichiometric reaction for the oxidation of glucose.

   The stoichiometric relationship is expressed in Equation 10.3.

$$\underset{180}{C_6H_{12}O_6} + \underset{6\times32}{6\,O_2} \rightarrow 6\,CO_2 + 6\,H_2O \tag{10.3}$$

4. Calculate the COD equivalent of glucose from Equation 10.3.

$$\text{COD equivalent of glucose} = \frac{O_2\ \text{required}}{\text{Glucose consumed}} = \frac{6 \times 32\,\text{g/mole as}\ O_2}{180\,\text{g/mole as glucose}}$$
$$= 1.07\,\text{g}\ O_2\text{/g glucose consumed or}\ 1.07\,\text{g COD/g glucose consumed}$$

5. Calculate the biomass yield (Y) from Equation 10.2 and the COD equivalent of glucose.

$$Y = \frac{\text{Cell mass yield}}{(\text{Glucose consumption}) \times (\text{COD equivalent of glucose})}$$
$$= \frac{2 \times 113\,\text{g/mole as cell mass}}{(3 \times 180\,\text{g/mole as glucose}) \times (1.07\,\text{g COD/g glucose consumed})}$$
$$= 0.39\,\text{g cell mass/g COD consumed}$$

   Therefore, 0.39 g cell mass is produced per gram of COD consumed for cell synthesis.

EXAMPLE 10.10: COD EXERTED OR OXYGEN CONSUMED FOR OXIDATION OF BIOMASS

Oxygen is consumed when biomass is oxidized. Calculate the amount of oxygen equivalent or COD exerted by 15 mg/L biomass (VSS) in the effluent from a wastewater treatment plant. The biomass is expressed by the chemical formula $C_5H_7O_2N$.

Solution

1. Write the stoichiometric equation of biomass oxidation.

The stoichiometric relationship is expressed in Equation 10.4.

$$\underset{113}{C_5H_7O_2N} + \underset{5\times 32}{5\ O_2} \rightarrow 5\ CO_2 + NH_3 + 2H_2O \qquad (10.4)$$

2. Calculate the COD equivalent of biomass oxidized from Equation 10.4.

$$\text{COD equivalent of biomass} = \frac{O_2 \text{ required}}{\text{Biomass oxidized}} = \frac{5 \times 32\ \text{g/mole as } O_2}{113\ \text{g/mole as biomass}}$$

$$= 1.42\ \text{g } O_2/\text{g biomass oxidized} \quad \text{or} \quad 1.42\ \text{g COD/g VSS oxidized}$$

The oxygen required or COD, or ultimate BOD exerted for oxidation of microbial cell (VSS) is 1.42 g COD/g VSS oxidized.

3. Calculate the amount of oxygen required for oxidation of 15 mg/L as VSS.

$$\text{COD exerted} = (15\ \text{mg/L as VSS}) \times (1.42\ \text{g COD/g VSS removed}) = 21.3\ \text{mg/L as COD}$$

EXAMPLE 10.11: OXYGEN REQUIRED FOR OXIDATION OF SUBSTRATE WITH CELL SYNTHESIS

Calculate the oxygen required for oxidation of substrate with cell synthesis expressed by Equation 10.2.

Solution

Direct method:

In this method, the oxygen requirement is directly calculated from Equation 10.2.

$$O_2 \text{ required} = \frac{O_2 \text{ required}}{\text{Glucose consumption}} = \frac{8 \times 32\ \text{g/mole as } O_2}{3 \times 180\ \text{g/mole as glucose}}$$

$$= 0.47\ \text{g } O_2/\text{g glucose consumed}$$

The oxygen required for oxidation of substrate with cell synthesis is 0.47 g O_2/g glucose consumed.

Indirect method:

The net oxygen required can also be calculated using the relationships expressed in Equations 10.2 through 10.4. The calculation steps are given below.

a. Calculate the cell mass yield from consumption of the substrate (Equation 10.2).

$$Y = 0.42\ \text{g cell mass/g glucose consumed}$$

b. Calculate the total oxygen consumed for complete oxidation of the substrate (Equation 10.3):

$$\text{COD equivalent of glucose} = 1.07\ \text{g } O_2/\text{g glucose consumed}$$

c. Calculate the equivalent O_2 required to oxidize cellular mass (Equation 10.4):

$$\text{COD equivalent of biomass} = 1.42\ \text{g } O_2/\text{g cell mass oxidized}$$

d. Calculate the net oxygen required for oxidation of the substrate with cell synthesis:

$$\begin{aligned}\text{Net oxygen required} &= \text{COD equivalent of glucose} - Y \times (\text{COD equivalent of biomass})\\ &= 1.07\ \text{g } O_2/\text{g glucose consumed} - (0.42\ \text{ng cell mass/g glucose consumed})\\ &\quad \times (1.42\ \text{g } O_2/\text{g cell mass oxidized})\\ &= 0.47\ \text{g } O_2/\text{g glucose consumed}\end{aligned}$$

EXAMPLE 10.12: DETERMINATION OF RETURN ACTIVATED SLUDGE FLOW Q_r

An activated sludge plant receives a flow of 28,000 m^3/d. The MLVSS concentration in the aeration basin is 2100 mg/L and the volatile suspended solids concentration in the return sludge is 7000 mg/L. Calculate the return flow and return flow ratio.

Solution

1. Draw the definition sketch.

   The definition sketch of an activated sludge process is shown in Figure 10.3.

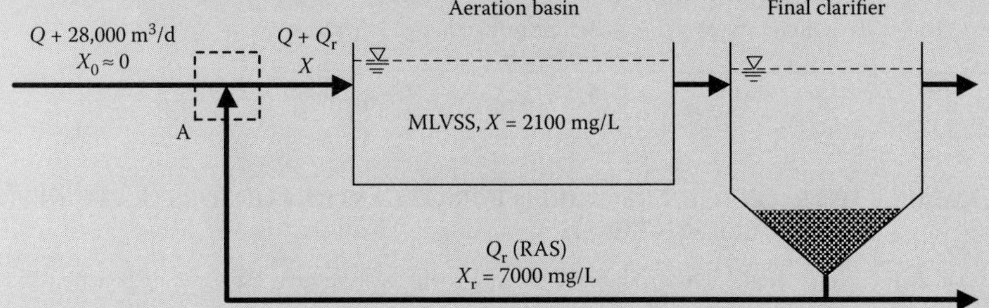

FIGURE 10.3 Definition sketch of an activated sludge process (Example 10.12).

2. Calculate the required RAS flow, Q_r.

   Conduct mass balance at point A.

$$X_0 \times Q + Q_r \times X_r = (Q + Q_r) \times X$$

Concentration of biomass X_0 in the influent is assumed to be zero to obtain Equation 10.5a.

$$Q_r \times X_r = Q \times X + Q_r \times X$$

$$Q_r \times (X_r - X) = Q \times X$$

$$Q_r = \frac{X}{X_r - X} Q \tag{10.5a}$$

where

Q_r = return activated sludge (RAS) flow, m^3/d (MGD)
Q = aeration basin influent flow, m^3/d (MGD)
X = MLVSS (or biomass) concentration in the aeration basin, mg/L
X_r = RAS volatile suspended solids (VSS_{ras}) concentration, mg/L

Calculate the return flow Q_r from Equation 10.5a at $X = 2100$ mg/L, $X_r = 7000$ mg/L, and $Q =$ 28,000 m^3/d.

$$Q_r = \frac{2100\ \text{mg/L}}{7000\ \text{mg/L} - 2100\ \text{mg/L}} \times 28{,}000\ \text{m}^3/\text{d} = 12{,}000\ \text{m}^3/\text{d}$$

3. Calculate the required RAS flow ratio, R_{ras} from Equation 10.5b.

$$R_{ras} = \frac{Q_r}{Q} = \frac{X}{X_r - X} \tag{10.5b}$$

where R_{ras} = RAS or recycle ratio, dimensionless

Calculate the RAS flow ratio R_{ras} from Equation 10.5b.

$$R_{ras} = \frac{2100\ \text{mg/L}}{7000\ \text{mg/L} - 2100\ \text{mg/L}} = 0.43;$$

$$\text{Also, } R_{ras} = \frac{12{,}000\ \text{m}^3/\text{d}}{28{,}000\ \text{m}^3/\text{d}} = 0.43$$

The required RAS flow is 43% of the flow received by the aeration basin.

EXAMPLE 10.13: DETERMINATION OF AERATION PERIOD, Q_r/Q, F/M RATIOS, REACTOR LOADING, AND MEAN CELL RESIDENCE TIME

An aeration basin receives 25,000 m^3/d flow, and influent BOD_5 concentration is 150 mg/L. The volume of aeration basin is 6250 m^3 and MLSS concentration is 2800 mg/L. The effluent soluble BOD_5 and TSS concentrations are 4 and 10 mg/L, respectively. The suspended solids concentration in the underflow (TSS_{uf}) is 10,000 mg/L, and volume of waste activated sludge (WAS) $Q_{wr} = 150\ \text{m}^3/\text{d}$. The VSS portion is 78% of TSS in the biosolids. Calculate the following: (a) aeration period, (b) return flow and ratio, (c) mass organic loading, (d) volumetric organic loading, (e) mean cell residence time, and (f) observed biomass yield.

Solution

1. Draw the definition sketch.

   The definition sketch of an activated sludge process is shown in Figure 10.4.

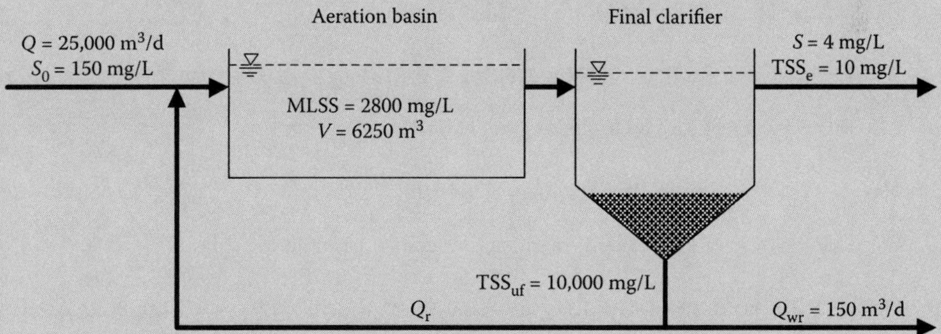

FIGURE 10.4 Definition sketch of an activated sludge process (Example 10.13).

2. Calculate the aeration period.

   The aeration period is expressed in Equation 10.5c.

$$\theta = \frac{V}{Q} \times \frac{24\,\text{h}}{\text{d}} \tag{10.5c}$$

where

θ = aeration period or hydraulic retention time (HRT or θ), h

V = aeration basin volume, m^3 (MG)

Calculate the aeration period from Equation 10.5c at $V = 6250\ \text{m}^3$ and $Q = 25{,}000\ \text{m}^3/\text{d}$.

$$\theta = \frac{6250\ \text{m}^3}{25{,}000\ \text{m}^3/\text{d}} \times \frac{24\ \text{h}}{\text{d}} = 6\ \text{h}$$

3. Calculate the return flow and ratio.

   Assume the VSS/TSS ratio of 0.78 is relatively constant in the mixed liquor in the aeration basin, and effluent and underflow from the final clarifier. Therefore, X and X_r can be substituted by MLSS and TSS_{ras} (return activated sludge suspended solids), respectively, in both Equation 10.5a and 10.5b to obtain Equation 10.5d and 10.5e.

$$Q_r = \frac{\text{MLSS}}{\text{TSS}_{\text{ras}} - \text{MLSS}} Q \qquad (10.5\text{d})$$

$$R_{\text{ras}} = \frac{Q_r}{Q} = \frac{\text{MLSS}}{\text{TSS}_{\text{ras}} - \text{MLSS}} \qquad (10.5\text{e})$$

   where TSS_{ras} = TSS concentration in return sludge, mg/L

   Calculate the return flow Q_r from Equation 10.5d at MLSS = 2800 mg/L, $\text{TSS}_{\text{ras}} = \text{TSS}_{\text{uf}} = 10{,}000$ mg/L, and $Q = 25{,}000\ \text{m}^3/\text{d}$.

$$Q_r = \frac{\text{MLSS}}{\text{TSS}_{\text{ras}} - \text{MLSS}} \times Q = \frac{2800\ \text{mg/L}}{10{,}000\ \text{mg/L} - 2800\ \text{mg/L}} \times 25{,}000\ \text{m}^3/\text{d} = 9722\ \text{m}^3/\text{d}$$

   Calculate the return flow ratio R_{ras} from Equation 10.5e.

$$R_{\text{ras}} = \frac{\text{MLSS}}{\text{TSS}_{\text{ras}} - \text{MLSS}} = \frac{2800\ \text{mg/L}}{10{,}000\ \text{mg/L} - 2800\ \text{mg/L}} = 0.39$$

4. Calculate the mass organic loading.

   Calculate the concentration of MLVSS in the aeration basin.

$$X = \text{VSS/TSS ratio} \times \text{MLSS} = 0.78 \times 2800\ \text{mg/L} = 2184\ \text{mg/L or } 2184\ \text{g/m}^3$$

   Calculate the mass of MLVSS in the aeration basin.

$$\text{Mass of MLVSS in aeration basin} = XV = 2184\ \text{g/m}^3 \times 6250\ \text{m}^3 \times 10^{-3}\ \text{kg/g} = 13{,}650\ \text{kg VSS}$$

   Calculate the mass of BOD_5 (food) reaching the aeration basin during a day at $S_0 = 150\ \text{g/m}^3$.

$$\text{Mass of BOD}_5 \text{ per day} = S_0 Q = 150\ \text{g/m}^3 \times 25{,}000\ \text{m}^3/\text{d} \times 10^{-3}\ \text{kg/g} = 3750\ \text{kg BOD}_5/\text{d}$$

   Calculate the mass organic loading to the biomass or food-to-microorganism (F/M) ratio.

$$\text{F/M ratio} = \frac{\text{Mass of BOD}_5 \text{ per day}}{\text{Mass of MLVSS in aeration basin}} = \frac{3750\ \text{kg BOD}_5/\text{d}}{13{,}650\ \text{kg VSS}}$$
$$= 0.27\ \text{kg BOD}_5/\text{kg VSS·d}$$

5. Calculate the volumetric organic (BOD_5) loading (*VOL*) in the aeration basin.

$$VOL = \frac{\text{Mass of BOD}_5 \text{ per day}}{V} = \frac{3750\ \text{kg BOD}_5/\text{d}}{6250\ \text{m}^3}$$
$$= 0.60\ \text{kg BOD}_5/\text{m}^3\text{·d}$$

6. Calculate the mean cell residence time.
   Calculate the VSS concentration in the effluent from the clarifier.

$$X_e = \text{VSS/TSS} \times \text{TSS}_e = 0.78 \times 10\,\text{mg/L} = 7.8\,\text{mg/L or } 7.8\,\text{g/m}^3$$

Calculate the mass of VSS lost in the effluent.

$$\text{Mass of VSS in effluent} = X_e(Q - Q_{wr}) = 7.8\,\text{g/m}^3 \times (25{,}000 - 150)\,\text{m}^3/\text{d} \times 10^{-3}\,\text{kg/g} = 194\,\text{kg VSS/d}$$

Calculate the VSS concentration in the WAS (X_{wr}) from the clarifier.

$$X_{wr} = \text{VSS/TSS ratio} \times \text{TSS}_{uf} = 0.78 \times 10{,}000\,\text{mg/L} = 7800\,\text{mg/L or } 7800\,\text{g/m}^3$$

Calculate the mass of VSS wasted in WAS.

$$\text{Mass of VSS in WAS} = X_{wr}Q_{wr} = 7800\,\text{g/m}^3 \times 150\,\text{m}^3/\text{d} \times 10^{-3}\text{kg/g} = 1170\,\text{kg VSS/d}$$

Calculate the total biomass lost from the system.

$$\text{Total mass of VSS (biomass) lost} = \text{Mass of VSS in WAS} + \text{Mass of VSS in effluent} = (1170 + 194)\,\text{kg VSS/d} = 1364\,\text{kg VSS/d}$$

Calculate the mean cell residence time (MCRT or θ_c) maintained in the activated sludge process.

$$\text{F/M ratio} = \frac{\text{Mass of MLVSS in aeration basin}}{\text{Total mass of VSS (biomass) lost}} = \frac{13{,}650\,\text{kg VSS}}{1364\,\text{kg VSS/d}} = 10\text{ days}$$

7. Calculate the observed biomass yield.
   Calculate BOD_5 consumed from $S_0 = 150\,\text{g/m}^3$ and $S = 4\,\text{g/m}^3$ at $Q = 25{,}000\,\text{m}^3/\text{d}$.

$$\text{BOD}_5\text{ consumed} = \Delta S\,Q = (S_0 - S) \times Q = (150 - 4)\,\text{g/m}^3 \times 25{,}000\,\text{m}^3/\text{d} \times 10^{-3}\text{kg/g} = 3650\,\text{kg BOD}_5/\text{d}$$

In a balanced biological system, the biomass produced is equal the total biomass of VSS lost from the system. Calculate Y_{obs} from VSS produced = total biomass lost = 1364 kg VSS/d.

$$Y_{obs} = \frac{\text{VSS produced}}{\text{BOD}_5\text{ consumed}} = \frac{1364\,\text{kg VSS/d}}{3650\,\text{kg BOD}_5/\text{d}} = 0.37\,\text{kg VSS/kg BOD}_5$$

10.3 Suspended Growth Aerobic Treatment Processes

Biological waste treatment involves dynamics of substrate utilization and microbial growth. In the past, the design of biological systems was based on empirical parameters developed by experience. Examples of these parameters are organic loading, hydraulic loading (HL), aeration period, and others. Today, however, the design utilizes empirical, as well as rational parameters based on biological kinetics equations.

These equations express the biomass growth and substrate utilization rates in terms of biological kinetic coefficients, food-to-microorganism ratio, the mean cell residence time (MCRT), and more. Using these equations, the design parameters such as reactor volume, substrate utilization, biomass growth, and effluent quality can be calculated.

In a suspended growth biological process, the biomass is maintained in suspension by appropriate mixing mechanism. The substrate and microorganisms remain in close contact in the presence of oxygen to carry out the substrate utilization. The biological kinetic equations for batch and continuous flow reactors are given below.

10.3.1 Microbial Growth Kinetics in Batch Reactor

In a batch reactor the contents are added, reaction occurs, and then the products are discharged. The design details and assembly of a single and multiple batch reactors are shown in Figure 10.5.[9] To develop the kinetic coefficients, the batch reactor is inoculated by well-acclimated microbial cells, and biomass growth and substrate utilization are monitored with respect to time. There are two extreme conditions: (1) the substrate concentration is relatively high and biological kinetics follows a pseudo zero-order reaction, and (2) substrate concentration is limiting and the kinetics follows a pseudo first-order reaction. The reaction rates and order of reactions, including irreversible or reversible, zero- or first-order, saturation-type or enzymatic, and consecutive reactions are discussed in Chapters 2 and 3. Readers may refer to Sections 2.3 and 2.5, and Examples 2.4 through 2.35, and Section 3.4, and Examples 3.14 through 3.23 on these topics. Many of these reactions are repeated here in the context of biological waste treatment.

Pseudo Zero-Order Substrate Utilization: The substrate concentration is relatively high, and the specific rate of substrate utilization is expressed by Equation 10.6.[10–12]

$$-\frac{1}{X}\frac{dS}{dt} = k \tag{10.6}$$

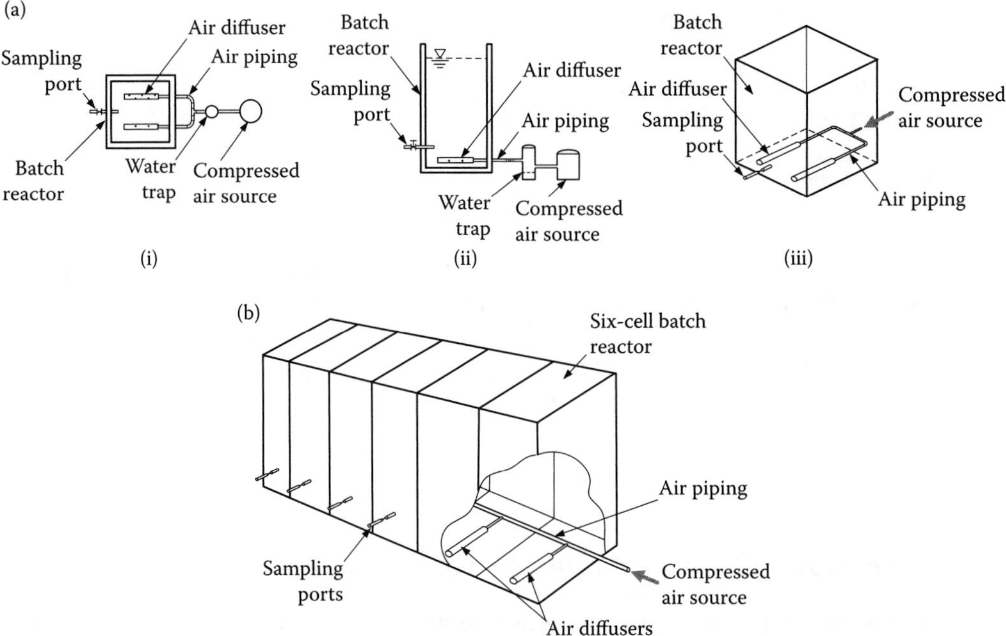

FIGURE 10.5 Design details and assembly of a batch reactor: (a) details of a single batch reactor, (i) plan, (ii) side view, and (iii) isometric view; and (b) assembly of a six-cell batch reactor unit.

The negative sign indicates that the substrate is decreasing with respect to time. Theoretically, the cell mass X is not a constant. It increases with time. To simplify the integration procedure, an average concentration of cell mass ($\bar{X}$) during the biological reaction period from $t = 0$ to $t = \mathrm{t}$ is used to replace X in Equation 10.6 that is then rearranged and integrated to yield Equation 10.7.

$$dS = -k\bar{X}dt$$

$$\int_{S_0}^{S} dS = -k\bar{X}\int_{0}^{t} dt$$

$$S = S_0 - k\bar{X}\,t \tag{10.7}$$

where

(dS/dt)	= substrate utilization rate per unit of reactor volume, mg substrate/L·h
$(1/\bar{X})(dS/dt)$	= specific substrate utilization rate per unit mass of microorganisms, mg substrate/mg VSS·h (mg substrate/mg VSS·d)
$\bar{X}$	= average concentration of cell mass during the biological reaction period from $t = 0$ to $t = t$, that is $\bar{X} = (X_0 + X_t)/2$, mg-VSS/L
k	= pseudo zero-order reaction rate constant, h^{-1} (d^{-1})
S	= substrate concentration remaining at any time t, mg substrate/L
S_0	= initial substrate concentration at $t = 0$, mg substrate/L

Note: The substrate is the organics that can be consumed as food by the microorganisms. It could be BOD_5 as well as biodegradable portion of COD or TOC, or any other organic measurements.

Equation 10.7 represents a linear relationship between S and $\bar{X}t$ with a slope of $-k$, and an intercept S_0.

Pseudo First-Order Substrate Utilization: When the substrate concentration is limiting, the specific rate of substrate utilization is expressed by Equation 10.8.

$$-\frac{1}{X}\frac{dS}{dt} = kS \tag{10.8}$$

After substituting X with $\bar{X}$, rearranging and integrating Equation 10.8 yields Equation 10.9.

$$\frac{1}{S}dS = -k\bar{X}dt$$

$$\int_{S_0}^{S} \frac{1}{S}dS = -k\bar{X}\int_{0}^{t} dt$$

$$\ln\left(\frac{S}{S_0}\right) = -k\bar{X}t$$

$$\ln(S) = \ln(S_0) - k\bar{X}t \quad \text{or} \quad S = S_0 e^{-k\bar{X}t} \tag{10.9}$$

where k = pseudo first-order reaction rate constant to the base e, L/mg VSS·h (L/mg VSS·d)

Equation 10.9 expresses a linear relationship between $\ln(S)$ and $\bar{X}t$ on semilog graph. The intercept is ln S_0, and the slope is rate constant $-k$. The value of k varies for different wastewaters and should be determined experimentally. Some of these values based on TOC data for different wastewaters are summarized in Table 10.6.

TABLE 10.6 First-Order Reaction Rate Constant k Based on TOC Data for Selected Wastewaters

Wastewater	k at 25°C, L/g-VSS·h
Pulp and paper	0.38–0.53
Chemical manufacturing	0.48–0.60
Oil refinery	0.50–0.66
Petrochemical manufacturing	0.59–1.33
Municipal (domestic)	1.15–1.95

Source: Adapted in part from Reference 1.

The first-order reaction rate constant for municipal wastewater based on BOD_5 may range from 0.1 to 1.25 L/g VSS·h. In absence of experimental data, the k value should be carefully selected for design purposes. A reasonable value may range from 0.20 to 0.40 L/g VSS·h.[1]

Biological Growth Kinetics: The rate of biomass production and substrate utilization based on kinetic coefficients is expressed by Equation 10.10a.

$$\frac{1}{\bar{X}}\frac{\Delta X}{\Delta t} = Y\left(-\frac{1}{\bar{X}}\frac{\Delta S}{\Delta t}\right) - k_d \qquad (10.10a)$$

Note: See Example 10.31 for derivation.

where

$\frac{1}{\bar{X}}\frac{\Delta X}{\Delta t}$ = specific rate of biomass production per unit mass of microorganisms during the time period (Δt), mg VSS/mg VSS·h (mg VSS/mg VSS·d)

$-\frac{1}{\bar{X}}\frac{\Delta S}{\Delta t}$ = specific rate of substrate utilization per unit mass of microorganisms during the time period (Δt), mg substrate/mg VSS·h (mg substrate/mg VSS·d)

Y = biomass yield coefficient, mg VSS/mg substrate

k_d = specific endogenous decay coefficient, h^{-1} (d^{-1})

$\bar{X}$ = average concentration of cell mass during the time period (Δt) between two consecutive time intervals of t_{t-1} and t_t, that is, $\bar{X}_t = (X_{t-1} + X_t)/2$, mg VSS/L

Equation 10.10a represents a linear relationship of the specific rate of cell production with respect to the specific rate of substrate utilization. The slope of the line is the biomass yield coefficient Y, and the intercept is the negative value of endogenous decay coefficient $-k_d$.

Relationship between COD and BOD_5: The ultimate biochemical oxygen demand (BOD_L) is a measure of total biodegradable portion of the organic matter in the wastewater. Chemical oxygen demand (COD), on the other hand, is a measure of total organic content in the wastewater including BOD_L. It is a common practice to conduct COD analysis for routine operation of a wastewater treatment plant as the results are obtained in 2 hours or less (some COD measurement methods can give reliable results in 15 minutes).[13] To correlate the COD to BOD_5 or COD to BOD_L data, normally a series of COD and time-dependent BOD tests are run on the same sample to establish COD/BOD_5 and COD/BOD_L relationships. Some of this information has been presented in Sections 5.4.1 and 5.4.3. Determination of biodegradable portion of COD is given in Example 10.14.

Relationship between VSS and Active Biomass in Biological Growth Kinetics: The traditional unit to express the active biomass X in a biological reactor is VSS. In reality, the active biomass is only a portion of the VSS. The other components of VSS in a bioreactor include nonbiodegradable portion that comes from (a) cell wall of microorganisms that have undergone lysing, and (b) inert organic fraction contained in the municipal wastewater.

EXAMPLE 10.14: DETERMINATION OF BIODEGRADABLE PORTION OF COD

An industrial wastewater was treated in a batch reactor using the acclimated seed. The time series BOD and COD tests were conducted on raw and treated effluent samples. The soluble BOD_5 and COD results and BOD_L values obtained from time series BOD data are given below. Calculate the nonbiodegradable component of COD in the influent and effluent samples.

Parameter	Influent Sample	Effluent Sample
Soluble COD, mg/L	455	35.0
Soluble BOD_5, mg/L	250	4.4
BOD_5/BOD_L	0.63	0.34

Solution

1. Calculate the nonbiodegradable portion of influent COD.

$$\text{Ultimate BOD of influent} = \frac{\text{Soluble BOD}_5 \text{ in influent}}{\text{BOD}_5/\text{BOD}_\text{L} \text{ in influent}} = \frac{250\,\text{mg BOD}_5/\text{L}}{0.63\ \text{BOD}_5/\text{BOD}_\text{L}} = 397\,\text{mg/L}$$

$$\text{Biodegradable COD of influent} = \text{Ultimate BOD in influent} = 397\,\text{mg/L}$$

$$\begin{aligned}\text{Nonbiodegradable COD of influent} &= \text{Soluble COD in influent} \\ &\quad - \text{Biodegradable COD of influent} \\ &= (455 - 397)\,\text{mg/L} = 58\,\text{mg/L}\end{aligned}$$

2. Calculate the nonbiodegradable portion of effluent COD.

$$\text{Ultimate BOD of effluent} = \frac{\text{Soluble BOD}_5 \text{in effluent}}{\text{BOD}_5/\text{BOD}_\text{L} \text{in effluent}} = \frac{4.4\,\text{mg BOD}_5/\text{L}}{0.34\ \text{BOD}_5/\text{BOD}_\text{L}} = 12.9\,\text{mg/L}$$

$$\begin{aligned}\text{Nonbiodegradable COD of effluent} &= \text{Soluble COD in effluent} - \text{Ultimate BOD in effluent} \\ &= (35.0 - 12.9)\,\text{mg/L} = 22.1\,\text{mg/L}\end{aligned}$$

EXAMPLE 10.15: DETERMINATION OF REACTION RATE CONSTANT OF A PSEUDO ZERO-ORDER REACTION

A bench-scale batch reactor study was conducted on an industrial wastewater using well-acclimated biological seed. The initial soluble COD of the wastewater and VSS in the reactor were 500 and 100 mg/L, respectively. The effluent samples were withdrawn hourly, and the soluble COD tests were conducted. The results are summarized below. At the end of 5 h, the VSS concentration was 166 mg/L. Determine the reaction rate constant k.

Time, h	0	1	2	3	4	5
Soluble COD (S), mg/L	500	455	425	383	330	300

Solution

1. Calculate the average biomass growth ($\bar{X}$) during 5-h reaction period at $X_0 = 100$ mg/L and $X_5 = 166$ mg/L.

$$\bar{X} = \frac{X_0 + X_5}{2} = \frac{(100 + 166)\ \text{mg/L}}{2} = 133\ \text{mg/L}$$

2. Calculate the $\bar{X}t$ value at each time step.

Time, h	0	1	2	3	4	5
$\bar{X}t$, mg·h/L	0	133	266	399	532	665

3. Plot the soluble COD data with respect to $\bar{X}t$.

   The arithmetic plot of soluble COD (S) data with respect to $\bar{X}t$ is shown in Figure 10.6.

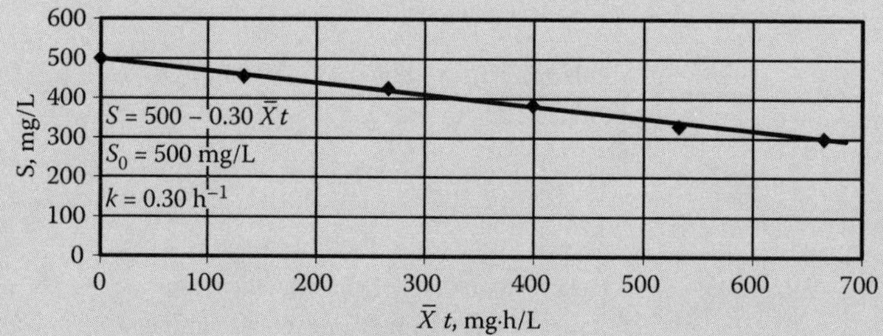

FIGURE 10.6 Arithmetic plot of soluble COD data with respect to reaction time (Example 10.15).

   The arithmetic plot shows a linear relationship. Therefore, the pseudo zero-order reaction rate applies.

4. Calculate the pseudo zero-order reaction rate constant.

   The reaction rate constant k is calculated from Equation 10.7 at $S_0 = 500$ mg/L at $t = 0$, and $S = 300$ mg/L at $t = 5$ h, and $\bar{X} = 133$ mg/L.

$$S = S_0 - k\bar{X}\,t$$

$$300\ \text{mg/L} = 500\ \text{mg/L} - k \times 133\ \text{mg/L} \times 5\ \text{h}$$

$$k = \frac{(500 - 300)\ \text{mg/L}}{133\ \text{mg/L} \times 5\text{h}} = 0.30\ \text{h}^{-1}$$

EXAMPLE 10.16: DETERMINATION OF REACTION RATE CONSTANT OF A PSEUDO FIRST-ORDER REACTION

A batch reactor was operated on an industrial wastewater to determine the reaction rate constant. The biomass used in this study was well-acclimated seed developed on the industrial wastewater. Hourly samples were withdrawn from the batch reactor. The biodegradable soluble COD and MLVSS concentrations were measured on each sample. Determine (1) whether the reaction rate follows a pseudo first-order reaction, and (2) the reaction rate constant k. The results are given below.

Time, h	0	1	2	3	4	5
Biodegradable soluble COD (S), mg/L	900	470	220	102	55	22
MLVSS (X), mg/L	2000	2000	2080	2120	2140	2160

Solution

1. Tabulate the experimental data.

   The values of biodegradable soluble COD (S), MLVSS (X), $\bar{X}$, and $\bar{X}t$ at each sampling time (t) are summarized in Table 10.7.

TABLE 10.7 Calculation Steps for Pseudo First-Order Reaction Rate (Example 10.16)

t, h	S, mg/L	X, mg/L	$\bar{X}$, mg/L	$\bar{X}t$, mg/L·h
0	900	2000	–	0
1	470	2000	2000	2000
2	220	2080	2040	4080
3	102	2120	2060	6180
4	55	2140	2070	8280
5	22	2160	2080	10,400

2. Plot the S versus $\bar{X}t$ values on a semilog paper.

   The plot is shown in Figure 10.7.

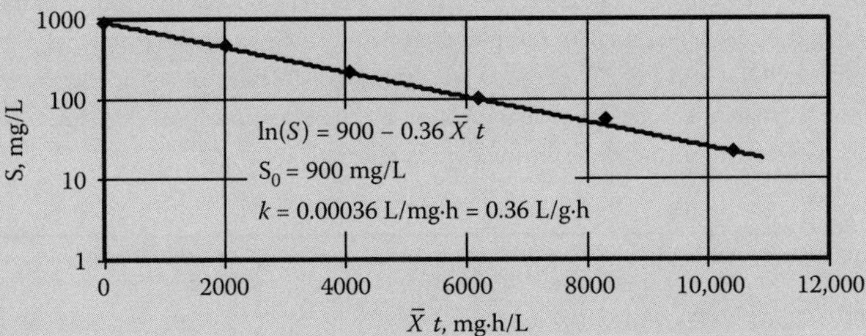

FIGURE 10.7 Semilog plot of batch reactor data (Example 10.16).

3. Check the validity of pseudo first-order reaction.

   The semilog plot of the experimental data (S versus $\bar{X}t$) has a linear relationship. Therefore, the pseudo first-order reaction rate applies.

4. Determine the reaction rate constant.

   The reaction rate constant k is calculated from Equation 10.9 at $S_0 = 900$ mg/L at $t = 0$, $S = 22$ mg/L at $t = 5$ h, and $\bar{X} = 133\,\text{mg/L}$.

$$\ln(S) = \ln(S_0) - k\bar{X}t$$

$$\ln(22\,\text{mg/L}) = \ln(900\,\text{mg/L}) - k \times 10{,}400\,\text{mg·h/L}$$

$$k = \frac{\ln[(900\,\text{mg/L})/(22\,\text{mg/L})]}{10{,}400\,\text{mg·h/L}} = \frac{\ln(40.9)}{10{,}400\,\text{mg·h/L}} = \frac{3.71}{10{,}400\,\text{mg·h/L}}$$

$$= 0.00036\,\text{L/mg·h}$$

The slope of the line is $k =$ 0.0003 L/mg·h or 0.36 L/g·h.

EXAMPLE 10.17: ESTIMATE OF EFFLUENT QUALITY FROM PSEUDO FIRST-ORDER REACTION RATE CONSTANT k

Estimate the effluent COD from an activated sludge plant to be designed to treat an industrial waste stream. The influent COD is 450 mg/L. The average MLVSS concentration in the aeration basin and the aeration period are 1800 mg/L and 4 h, respectively. The pseudo first-order reaction coefficient is 0.4 L/g·h.

Solution

Calculate the effluent COD (S) from Equation 10.9 at $S_0 = 450$ mg/L, $\bar{X} = 1800$ mg/L, $k = 0.4$ L/g·h $=$ 0.0004 L/mg·h, and $t = 4$ h.

$$S = S_0 e^{-k\bar{X}t}$$

$$S = 450\,\text{mg/L} \times e^{-0.00040\,\text{L/mg·h} \times 1800\,\text{mg/L} \times 4\,\text{h}} = 450\,\text{mg/L} \times e^{-2.88} = 450\,\text{mg/L} \times 0.0561$$

$$= 25\,\text{mg/L}$$

EXAMPLE 10.18: DETERMINATION OF BIOLOGICAL KINETIC COEFFICIENTS OF Y AND k_d FROM BATCH REACTOR EXPERIMENTAL DATA

A batch biological reactor unit was operated in the laboratory. The reactor contained an industrial waste and well-acclimated biological seed. The samples were withdrawn from the reactor at 1-h interval and analyzed for soluble BOD_5 and VSS. The results are summarized below. Describe the experimental steps and calculate the biological kinetic coefficients Y and k_d. See also Example 10.30 for determination of kinetic coefficients k and K_s and μ_{max} from the data give in this example.

Aeration Period (t), h	0	1	2	3	4
Soluble BOD_5 (S), mg/L	310	218	130	54	9.5
VSS X, mg/L	1000	1034	1069	1095	1111

Solution

1. Describe batch reactor assembly and experimental procedure.
   The design details and arrangement of a single-cell batch reactor are shown in Figure 10.5a. The assembly consists of a reaction chamber with a sampling port and air diffusers, and an air supply source.[9]
2. Establish the concentrations of VSS and substrate for the batch reactor operation.
   The concentrations of soluble BOD_5 and VSS in the batch reactor should be nearly the same as those expected in the full-scale treatment plant. The procedure to prepare the substrate and VSS concentrations in the batch reactor is (a) develop a concentrated seed sample for reactor operation, (b) conduct

mass balance calculations based on the volumes and concentrations of soluble BOD_5 in the industrial waste and those in the developed seed.*

3. Describe the experimental procedure.
   a. Place a known volume of concentrated and well-acclimated seed in the reactor and start aeration.
   b. Pour a known volume of wastewater sample into the reactor. The total volume of seed and sample should reach a predetermined height in the reactor.
   c. Purge the sample collection line then collect a sample immediately ($t = 0$) and other samples at predetermined reaction times.
   d. Conduct MLSS and MLVSS tests on well-mixed samples, and BOD_5 on centrifuged supernatant samples. The measured concentration of soluble BOD_5 in the mixture at time zero may be slightly lower than that calculated because the biological seed may uptake a portion of soluble BOD_5 upon initial contact with the substrate causing an instantaneous drop in the concentration.
   e. The aeration may cause foaming in the reactor, and a layer of solid may build up on the reactor wall near the liquid surface. Scrape the surface with a mixer blade and push the solids into the reactor periodically.
   f. Tabulate the experimental data (reaction time, MLVSS, and soluble BOD_5).
4. Calculate the specific substrate utilization and biomass growth rates.
   The relationship between $(-\Delta S/\Delta t)/\bar{X}$ and $(\Delta x/\Delta t)/\bar{X}$ is provided in Equation 10.10a.

$$\frac{1}{\bar{X}}\frac{\Delta X}{\Delta t} = Y\left(-\frac{1}{\bar{X}}\frac{\Delta S}{\Delta t}\right) - k_d$$

   Prepare $(-\Delta S/\Delta t)/\bar{X}$ and $(\Delta X/\Delta t)/\bar{X}$ from the experimental data so the linear relationship can be developed. The experimental data of soluble BOD_5 (S) and VSS (X), and the calculated values of $-\Delta S/\Delta t$, $\Delta X/\Delta t$, $\bar{X}$, $(-\Delta S/\Delta t)/\bar{X}$, and $(\Delta X/\Delta t)/\bar{X}$ are summarized at each reaction time (t) in Table 10.8.

TABLE 10.8 Calculated Values from the Experimental Data (Example 10.18)

t, h	S, mg/L	X, mg/L	$-\Delta S/\Delta t$, mg/L·h	$\Delta X/\Delta t$, mg/L·h	$\bar{X}$, mg/L	$(-\Delta S/\Delta t)/\bar{X}$, h^{-1}	$(\Delta X/\Delta t)/\bar{X}$, h^{-1}
0	310	1000	–	–	–	–	–
1	218	1034	92	34	1017	0.0905	0.0334
2	130	1069	88	35	1052	0.0837	0.0333
3	54	1095	76	26	1082	0.0702	0.0240
4	9.5	1111	44.5	16	1103	0.0403	0.0145

5. Plot the graphical relationship.
   The plot of $(\Delta X/\Delta t)/\bar{X}$ versus $(-\Delta S/\Delta t)/\bar{X}$ values is shown in Figure 10.8.
6. Determine the biomass growth yield coefficient and endogenous decay coefficient.

---

* $C = \frac{C_1V_1 + C_2V_2}{V_1 + V_2}$, and $M = \frac{M_1V_1 + M_2V_2}{V_1 + V_2}$, where C and M are soluble BOD^5 and VSS concentrations in the mixture; C_1, C_2 and M_1, M_2 are respectively soluble BOD_5 and VSS concentrations in the industrial waste and prepared seed; and V_1 and V_2 are volumes of industrial waste and seed in the mixture. In most cases C_2 and M_1 are negligible and considered zero.

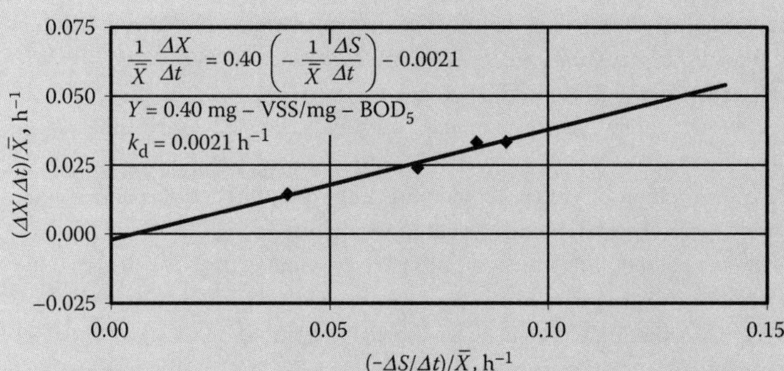

FIGURE 10.8 Plot of specific biological growth versus substrate utilization rate from the batch reactor data (Example 10.18).

Using MS Excel, a linear trend line is obtained below:

$$\frac{1}{\bar{X}}\frac{\Delta X}{\Delta t} = 0.40\left(-\frac{1}{\bar{X}}\frac{\Delta S}{\Delta t}\right) - 0.0021$$

In accordance with Equation 10.10a, the slope of this linear relationship is the biomass growth yield coefficient $Y = 0.40$ mg VSS/mg BOD_5.

At $(-\Delta S/\Delta t)/\bar{X} = 0$, the intercept is the negative value of endogenous decay coefficient $-k_d = -0.0021\ h^{-1}$. It gives $k_d = 0.0021\ h^{-1} = 0.0021\ h^{-1} \times 24$ h/d $= 0.050\ d^{-1}$.

Note: See Example 10.30 for a procedure to determine other kinetic coefficients k, K_s, and μ_{max}.

7. Describe the important features of the plot in Figure 10.8.
   a. At $(\Delta X/\Delta t)/\bar{X} = 0$, the $(-\Delta S/\Delta t)/\bar{X}$ intercept is ~0.005 mg BOD_5/mg VSS·h. At this critical point, the microorganisms live at a substrate utilization rate for survival without any appreciable growth. It is called basal metabolism. When the $(-\Delta S/\Delta t)/\bar{X}$ is below this point, the cell mass growth reaches the endogenous phase and a reduction in cell mass occurs. At $(-\Delta S/\Delta t)/\bar{X} = 0$, the endogenous activity reaches the maximum decay rate, that is expressed by the negative value of endogenous decay coefficient $-k_d$. The negative sign indicates that the biomass is decreasing with respect to time.
   b. In accordance with the experimental data, the actual consumption of soluble BOD_5, $\Delta S_{max} = (310 - 9.5)$ mg BOD_5/L $= 300$ mg BOD_5/L during 4-h aeration period. At $Y = 0.40$ mg VSS/mg BOD_5, the maximum cell mass yield $\Delta X'_{max} = 0.40$ mg VSS/mg $BOD_5 \times 300$ mg BOD_5/L $=$ 120 mg VSS/L. The potentially maximum cell mass concentration $X'_{max} = (1000 + 120)$ mg VSS/L $= 1120$ mg VSS/L in the batch reactor.
   c. The experimental data also show that the maximum cell mass concentration $X = 1111$ mg VSS/L at the end of 4-h aeration period. The actual cell mass yield $\Delta X_{max} = (1111 - 1000)$ mg VSS/L $= 111$ mg VSS/L. The observed biomass yield $Y_{obs} = 111$ mg VSS/L–300 mg BOD_5/L $= 0.37$ mg VSS/mg BOD_5. Y_{obs} is only 92.5% of Y due to basal metabolism.
   d. At $X'_{max} = 1120$ mg VSS/L and $k_d = 0.0021\ h^{-1}$, the maximum endogenous decay rate $\Delta X_d/\Delta t = 0.0021\ h^{-1} \times 1120$ mg VSS/L $= 2.4$ mg VSS/L·h due to basal metabolism.
   e. At $X'_{max} = 1120$ mg VSS/L and $(-\Delta S/\Delta t)/\bar{X} = 0.005$ mg BOD_5/mg VSS·h, the potentially minimum substrate utilization rate by the microorganisms for survival $(-\Delta S/\Delta t)_{min} = 0.005$ mg BOD_5/mg VSS·h $\times$ 1120 mg VSS/L $= 5.6$ mg BOD_5/L·h.

EXAMPLE 10.19: APPLICATION OF BATCH REACTOR DATA FOR ESTIMATION OF SLUDGE GROWTH RATE IN A CONTINUOUS FLOW REACTOR

A continuous flow biological reactor is receiving settled municipal wastewater from a primary clarifier. The flow (Q) and influent BOD_5 (S_0) are 18,150 m^3/d and 150 mg/L, respectively. The aeration period (θ) is 5 h. The biomass yield coefficient (Y), endogenous decay coefficient (k_d), effluent soluble BOD_5 (S), and average MLVSS concentration ($\bar{X}$) are 0.5 g VSS/g BOD_5, 0.04 d^{-1}, 6 mg/L, and 1800 mg/L, respectively. Calculate the biomass growth in the reactor and mean cell residence time (θ_c).

Solution

1. Develop the relationship between the average biomass (VSS) growth rate and average BOD_5 consumption rate.

   Rearrange Equation 10.10a to obtain Equation 10.10b that describes the relationship between the average biomass (VSS) growth rate ($\Delta X/\Delta t$) and average BOD_5 consumption rate per unit of reactor volume.

$$\frac{\Delta X}{\Delta t} = Y\frac{\Delta S}{\Delta t} - k_d\bar{X} \tag{10.10b}$$

   where

   $\Delta X/\Delta t$ = average biomass (VSS) growth rate per unit of reactor volume, mg VSS/L·d (g VSS/m^3·d)

   $\Delta S/\Delta t$ = average BOD_5 consumption rate per unit of reactor volume, mg BOD_5/L·d (g BOD_5/m^3·d)

2. Calculate the BOD_5 consumption rate ($\Delta S/\Delta t$) in the aeration basin from $S_0 = 150$ mg/L, $S = 6$ mg/L, and $\Delta t = \text{HRT} = 5$ h.

$$\frac{\Delta S}{\Delta t} = \frac{S_0 - S}{\Delta t} = \frac{(150-6)\text{ mg BOD}_5\text{/L}}{5\text{ h}} \times \frac{24\text{ h}}{\text{d}} = 691\text{ mg BOD}_5\text{/L·d} \quad \text{or} \quad 691\text{ g BOD}_5\text{/m}^3\text{·d}$$

3. Calculate the biomass (VSS) growth rate ($\Delta X/\Delta t$) from Equation 10.10b at $\bar{X} = 1800$ mg/L (or g/m^3), $Y = 0.5$ g VSS/g BOD_5, $k_d = 0.04\ d^{-1}$.

$$\frac{\Delta X}{\Delta t} = 0.5\text{ g VSS/g BOD}_5 \times 691\text{ g BOD}_5\text{/m}^3\text{·d} - 0.04\text{ d}^{-1} \times 1800\text{ g VSS/m}^3$$

$$= (346 - 72)\text{ g VSS/m}^3\text{·d} = 274\text{ g VSS/m}^3\text{·d}$$

4. Calculate the aeration basin volume.

   Rearrange 10.5c and calculate aeration volume (V) at $Q = 18{,}150\ m^3/d$ at $\theta = 5$ h.

$$V = \theta\, Q \times \frac{\text{d}}{24\text{ h}} = 5\text{ h} \times 18{,}150\text{ m}^3\text{/d} \times \frac{\text{d}}{24\text{ h}} = 3781\text{ m}^3$$

5. Calculate the bio mass growth per day (Δm).

$$\Delta m = V\frac{\Delta X}{\Delta t} = 3781\text{ m}^3 \times 274\text{ g VSS/m}^3\text{·d} \times 10^{-3}\text{ kg/g} = 1036\text{ kg VSS/d}$$

6. Calculate the mean cell residence time (θ_c).

Calculate the biomass (VSS) in aeration basin, $V\bar{X} = 3781\ \text{m}^3 \times 1800\ \text{g/m}^3 \times 10^{-3}\ \text{kg/g} = 6806\ \text{kg VSS}$

$$\theta_c = \frac{V\bar{X}}{\Delta m} = \frac{6806\ \text{kg}}{1036\ \text{VSS/d}} = 6.6 \text{ or } 7\ \text{d}$$

10.3.2 Microbial Growth Kinetics in Continuous Flow Reactor

In a continuous flow suspended growth reactor, the flow enters and leaves the reactor on a continuous basis. The biomass grows while the substrate is consumed.

Cell Growth Rate: In general, the growth rate of microbiological cells can be expressed by Equation 10.11a.

$$r_g = \frac{dX}{dt} = \mu X \quad \text{or} \quad \mu = \frac{1}{X}\frac{dX}{dt} = \frac{r_g}{X} \tag{10.11a}$$

where

r_g or dX/dt = rate of microbial growth per unit of reactor volume, mg VSS/L·d (g VSS/m^3·d)
μ = specific growth rate per unit mass of microorganisms, mg VSS/mg VSS·d (d^{-1})
X = concentration of biomass in the reactor, normally expressed as mg VSS/L (g VSS/m^3)

In a biological reactor, the growth of microorganisms can be described by enzymatic-based saturation-type reaction. Such biomass growth under substrate limitation is defined by an expression proposed by Monod (Equation 10.11b).[14,15] The inverse of this equation gives a linear relationship (Equation 10.11c).

$$\mu = \frac{\mu_{max} S}{K_s + S} \tag{10.11b}$$

$$\frac{1}{\mu} = \frac{K_s}{\mu_{max} S} + \frac{1}{\mu_{max}} \tag{10.11c}$$

where

μ_{max} = maximum specific growth rate per unit mass of microorganisms, mg VSS/mg VSS·d (d^{-1})
K_s = half-velocity constant or the substrate concentration at one-half the maximum specific growth rate, mg substrate/L (g substrate/m^3)
S = concentration of growth-limiting substrate in solution, mg substrate/L (g substrate/m^3)

The relationship between specific growth rate and substrate concentration in the reactor (Equation 10.11b) is shown in Figure 10.9a. As discussed in Section 2.3.2, the saturation-type reactions can be linearized by inverting Equation 10.11b. The linear relationship expressed by Equation 10.11c is shown in Figure 10.9b.

The growth of microorganisms under a limited-substrate environment is obtained by substituting the value of μ from Equation 10.11a into Equation 10.11b. This relationship is expressed by Equation 10.11d.

$$r_g = \frac{\mu_{max} X S}{K_s + S} \tag{10.11d}$$

The growth of microorganisms in a biological reactor may occur in two extreme conditions, limited or excess substrate as described below.

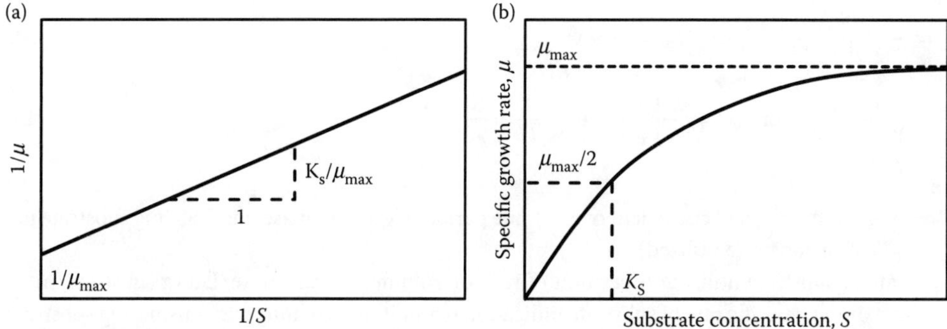

FIGURE 10.9 Relationship between substrate concentration remaining and specific growth rate: (a) effect of limiting nutrient on specific growth rate, and (b) $1/\mu$ versus $1/S$ plot to develop a linear relationship.

Cell Growth Rate with Excess Substrate ($S \gg K_s$): In excess substrate situation, the substrate concentration (S) in the bioreactor is significantly larger than value of K_s and $K_s + S \approx S$. Therefore, Equation 10.11b can be simplified to Equation 10.11e.

$$\mu = \frac{\mu_{max} S}{K_s + S} \approx \mu_{max} \frac{S}{S} = \mu_{max} \quad \text{or} \quad \mu \approx \mu_{max} \tag{10.11e}$$

When the substrate is unlimited, the growth follows the zero-order reaction, in which the specific growth rate (μ) reaches the maximum value of μ_{max} and independent of the substrate concentration (S) in the reactor.

Cell Growth Rate with Low Substrate ($S \ll K_s$): When the substrate concentration (S) in the bioreactor is much lower than the value of K_s and $K_s + S \approx K_s$, and Equation 10.11b can be modified to Equation 10.11f.

$$\mu = \frac{\mu_{max} S}{K_s + S} \approx \frac{\mu_{max} S}{K_s} = \frac{\mu_{max}}{K_s} S \quad \text{or} \quad \mu \approx k' S \tag{10.11f}$$

where k' = specific growth rate constant simplified when the substrate is limited $\left(k' = \frac{\mu_{max}}{K_s}\right)$,

$$\frac{\text{mg VSS/mg VSS·d}}{\text{mg substrate/L}}$$

When the substrate is limited, the growth follows the first-order reaction, in which the specific growth rate (μ) is proportional to the substrate concentration (S) in the reactor.

Substrate Utilization Rate: In a biological reactor, many simultaneous and interdependent reactions occur. These are as follows: (1) microorganisms grow, (2) substrate is consumed, (3) a portion of substrate is oxidized for energy, and (4) a portion of substrate is converted into new cells. As a result, the biomass concentration increases and the substrate concentration decreases. The relationships between the substrate utilization and growth of microorganisms are expressed by Equations 10.12a through 10.12e.

$$Y = -\frac{r_g}{r_{su}} \quad \text{or} \quad r_{su} = -\frac{r_g}{Y} \quad \text{or} \quad r_g = -Y r_{su} \tag{10.12a}$$

$$k = \frac{\mu_{max}}{Y} \quad \text{or} \quad \mu_{max} = Yk \tag{10.12b}$$

$$r_{su} = \frac{dS}{dt} = -\frac{\mu_{max} XS}{Y(K_s + S)} \quad \text{or} \quad r_{su} = -\frac{kXS}{K_s + S} \tag{10.12c}$$

$$U = -\frac{1}{X}\frac{dS}{dt} = -\frac{r_{su}}{X} \quad \text{or} \quad r_{su} = -UX \tag{10.12d}$$

$$U = \frac{\mu_{max} S}{Y(K_s + S)} \quad \text{or} \quad U = \frac{kS}{K_s + S} \quad \text{or} \quad \frac{1}{U} = \frac{K_s}{kS} + \frac{1}{k} \tag{10.12e}$$

where

Y = maximum cell yield coefficient over a finite period of growth phase, mg VSS/mg substrate utilized (g VSS/g substrate utilized)

r_{su} = rate of substrate utilization per unit of reactor volume, mg substrate/L·d (g substrate/m^3·d)

k = maximum specific rate of substrate utilization per unit mass of microorganisms, mg substrate/mg VSS·d (d^{-1})

U = specific substrate utilization rate per unit mass of microorganisms, which may also be called mass organic loading (MOL), mg substrate/mg VSS·d (d^{-1})

Endogenous Metabolism: The growth of mixed culture biomass is normally accompanied by death, predation, cell lysing, and decay. The natural decrease in cell mass is called endogenous decay, and is normally assumed proportional to the concentration of biomass present. The net growth and net specific growth of biomass are expressed by Equations 10.13a and 10.13b.

$$r_g' = r_g - k_d X = -Y r_{su} - k_d X \quad \text{or} \quad r_g' = YUX - k_d X \quad \text{or} \quad r_g' = \frac{\mu_{max} XS}{K_s + S} - k_s X \tag{10.13a}$$

$$\mu' = \frac{r'_g}{X} \quad \text{or} \quad \mu' = \mu - k_d \quad \text{or} \quad \mu' = YU - k_d \quad \text{or} \quad \mu' = \frac{\mu_{max} S}{K_s + S} - k_d \tag{10.13b}$$

where

r_g' = net growth rate of biomass per unit of reactor volume, mg VSS/L·d (g VSS/m^3·d)

μ' = specific net biomass growth per unit mass of microorganisms, mg VSS/mg VSS·d or d^{-1}

The endogenous respiration reduces the biomass and has an effect on net cell yield. Therefore, the observed biomass yield Y_{obs} is expressed by Equation 10.13c.

$$Y_{obs} = \frac{r_g'}{-r_{su}} \quad \text{or} \quad Y_{obs} = \frac{YU - k_d}{U} \tag{10.13c}$$

where Y_{obs} = observed biomass yield to consider the impact of endogenous metabolism on net biomass growth in a biological reactor, mg VSS/mg substrate consumed (g VSS/g substrate consumed)

10.3.3 Continuous Flow Completely Mixed Reactor

There are two limiting cases of design and operation of a continuous flow suspended growth reactor: (1) without recycle and (2) with recycle. In the first case, the high concentration of biomass leaves the reactor without settling in a clarifier. In the second case, the biomass is settled in a clarifier and the solids are returned to maintain high concentration of biomass in the reactor, and the effluent leaving the clarifier has low suspended solids. Different types of continuous flow completely mixed reactors, with and without recycle, are illustrated in Figure 10.10.

Many useful process design relationships can be developed from the preceding discussion. Also, mass balance approach similar to that presented in Chapter 3 can be used to develop many of these relationships. Important process design relationships are presented below.

General Process Design Equations: The important relationships for organic loading, removal efficiency, substrate utilization, and oxygen utilization in a continuous flow completely mixed reactor

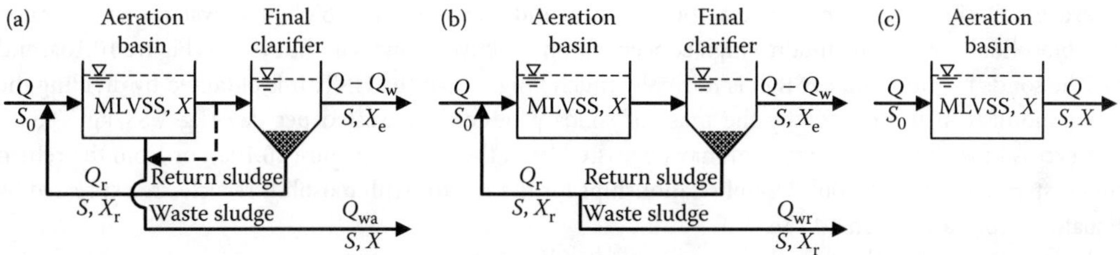

FIGURE 10.10 Continuous flow complete-mixed reactors: (a) returning solids from final clarifier and wasting sludge directly from the aeration basin, (b) returning solids and wasting sludge from the return sludge line, and (c) no solids return.

are expressed by Equations 10.14a through 10.14f.

$$E = \frac{S_0 - S}{S_0} \times 100\% \tag{10.14a}$$

$$\text{F/M} = \frac{QS_0}{VX} = \frac{S_0}{\theta X} \quad \text{or} \quad \text{F/M} = \frac{S_0}{S_0 - S} \times U = \frac{100\%}{E} \times U \tag{10.14b}$$

$$r_{su} = -\frac{Q(S_0 - S)}{V} = -\frac{S_0 - S}{\theta} \quad \text{or} \quad VOL = -r_{su} = \frac{Q(S_0 - S)}{V} = \frac{S_0 - S}{\theta} \tag{10.14c}$$

$$U = \frac{Q(S_0 - S)}{VX} = \frac{S_0 - S}{\theta X} \quad \text{or} \quad MOL = U = \frac{S_0 - S}{S_0} \times \text{F/M} = \frac{E}{100\%} \times \text{F/M} \tag{10.14d}$$

$$\text{MLSS} = \frac{X}{\text{VSS/TSS ratio}} \quad \text{or} \quad X = \text{VSS/TSS ratio} \times \text{MLSS} \tag{10.14e}$$

$$O_2 = \frac{Q(S_0 - S)}{\text{BOD}_5/\text{BOD}_\text{L}} - 1.42\ P_\text{x} \tag{10.14f}$$

Note: See Example 10.34 for derivation. It is a simplified oxygen utilization rate for carbonaceous oxygen demand from assumption that the biodegradable solids ≈ biomass (VSS). For more precise calculation of carbonaceous and nitrogenous oxygen demands, see Sec 10.3.8 (Example 10.70).

where

E	= substrate removal efficiency through the biological process, percent
F/M	= food to microorganisms ratio, g substrate/g VSS·d or d^{-1}
Q	= influent flow to the reactor, m^3/d
V	= volume of the reactor, m^3
S_0	= substrate concentration in the influent (soluble BOD_5 or COD), mg/L (g/m^3)
S	= substrate concentration in the effluent (soluble BOD_5 or COD), mg/L (g/m^3)
θ	= hydraulic retention time (HRT), d
VOL	= volumetric organic loading in the reactor, mg substrate/L·d (g substrate/m^3·d)
MOL	= mass organic loading to the microorganisms, mg substrate/mg VSS·d
MLSS	= mixed liquor suspended solids concentration in the reactor, mg TSS/L
VSS/TSS ratio	= ratio of volatile to total suspended solids, mg VSS/mg TSS
O_2	= oxygen utilization rate for carbonaceous oxygen demand, g O_2/d
BOD_5/BOD_L	= ratio of 5-d BOD to ultimate carbonaceous BOD (see Chapter 5), dimensionless
P_x	= biosolids (VSS) growth rate in the reactor, g VSS/d

The biomass growth in reactors with and without recycle is separately described below.

Process Design Equations for Reactor with Biosolids Recycle: Most biological systems are operated with biosolids recycle to maintain a high concentration of active biomass in the system (Figure 10.10a, and b). The solids retention time (SRT) is normally much longer than the HRT. It is obtained by dividing the active biomass in the reactor by the mass of solids produced or wasted per day (kg VSS/kg VSS·d). The excess biosolids from the system may be wasted from the reactor (Figure 10.10a), or from the return sludge stream (Figure 10.10b). Useful relationships for the reactor with biosolids recycle are expressed by Equations 10.15a through 10.15o.

Mean cell residence time or solids retention time (SRT or θ_c)

$$\theta_c = \frac{VX}{Q_{wa}X + (Q - Q_{wa})X_e} = \frac{VX}{Q_{wr}X_r + (Q - Q_{wr})X_e} \tag{10.15a}$$

$$\theta_c = \frac{VX}{YQ(S_0 - S) - k_d VX} \quad \text{or} \quad \theta_c = \frac{VX}{P_x} = \frac{\theta X}{p_x} = \frac{VX}{Y_{obs}Q(S_0 - S)} \tag{10.15b}$$

$$\frac{1}{\theta_c} = \frac{YQ(S_0 - S)}{VX} - k_d \quad \text{or} \quad \frac{1}{\theta_c} = \frac{P_x}{VX} = \frac{p_x}{\theta X} = \frac{Y_{obs}Q(S_0 - S)}{VX} \tag{10.15c}$$

$$\frac{1}{\theta_c} = YU - k_d = \frac{YkS}{K_s + S} - k_d = \mu - k_d = \mu' \tag{10.15d}$$

Note: See Example 10.20 for derivation.

Minimum mean cell residence time (θ_c^{min})

$$\frac{1}{\theta_c^{min}} = \frac{YkS}{K_s + S} - k_d \tag{10.15e}$$

Since $S_0 \gg K_S$, Equation 10.15f is obtained.

$$\frac{1}{\theta_c^{min}} \approx Yk - k_d = \mu_{max} - k_d \quad \text{or} \quad \theta_c^{min} \approx \frac{1}{Yk - k_d} = \frac{1}{\mu_{max} - k_d} \tag{10.15f}$$

Hydraulic retention time (HRT or θ)

$$\theta = \frac{Y}{1 + k_d\theta_c}\frac{(S_0 - S)}{X}\theta_c \quad \text{or} \quad \theta = \frac{V}{Q} = \frac{P_x}{QX}\theta_c = \frac{p_x}{X}\theta_c = \frac{Y_{obs}(S_0 - S)}{X}\theta_c \tag{10.15g}$$

Note: See Example 10.21 for derivation.

$$\theta = \frac{Y}{1 + k_d\theta_c}\frac{(S_0 - S)}{X}\theta_c$$

Effluent substrate concentration (S)

$$S = \frac{K_s(1 + k_d\theta_c)}{\theta_c(Yk - k_d) - 1} = \frac{K_s(1 + k_d\theta_c)}{\theta_c(\mu_{max} - k_d) - 1} \approx \frac{\theta_c^{min}K_s(1 + k_d\theta_c)}{\theta_c - \theta_c^{min}} \tag{10.15h}$$

Note: See Example 10.22 for derivation.

$$S = S_0 - \frac{\theta X}{\theta_c Y_{obs}} = S_0 - \frac{P_x}{Y_{obs} Q} = S_0 - \frac{p_x}{Y_{obs}} \tag{10.15i}$$

$$S \approx \frac{U}{k''} = \frac{Q(S_0 - S)}{k'' VX} = \frac{S_0 - S}{k'' \theta X} \quad \text{or} \quad S \approx \frac{S_0}{1 + k'' \theta X} \tag{10.15j}$$

Note: This equation is simplified from Equation 10.12e when $S \ll K_s$. See Example 10.23 for derivation.
Mixed liquor solids concentration (X)

$$X = \frac{Y}{1 + k_d \theta_c} \frac{\theta_c}{\theta} (S_0 - S) \quad \text{or} \quad X = \frac{\theta_c P_x}{V} = \frac{\theta_c}{\theta} p_x = Y_{obs} \frac{\theta_c}{\theta} (S_0 - S) \tag{10.15k}$$

Note: See Example 10.24 for derivation.
Solids yield and production

$$Y_{obs} = Y - k_d \frac{VX}{Q(S_0 - S)} = \frac{Y}{1 + k_d \theta_c} \tag{10.15l}$$

Note: See Example 10.25 for derivation.

$$P_x = YQ(S_0 - S) - k_d VX \quad \text{or} \quad P_x = Y_{obs} Q(S_0 - S) \tag{10.15m}$$

$$p_x = Y(S_0 - S) - k_d \theta X \quad \text{or} \quad p_x = \frac{P_x}{Q} = Y_{obs}(S_0 - S) \tag{10.15n}$$

Reactor volume (V)

$$V = \frac{Y}{1 + k_d \theta_c} \frac{\theta_c Q(S_0 - S)}{X} \quad \text{or} \quad V = Q\theta = \frac{\theta_c P_x}{X} = Y_{obs} \frac{\theta_c Q(S_0 - S)}{X} \tag{10.15o}$$

Note: See Example 10.26 for derivation.

where
- θ_c = mean cell residence time or solids retention time (SRT), d
- θ_c^{min} = minimum mean cell residence time below which substrate utilization and cell growth does not occur, d
- Q_{wa} = sludge wasted from the reactor, m^3/d
- Q_{wr} = sludge wasted from the return sludge line, m^3/d
- X_r = concentration of VSS in the return sludge, mg VSS/L (g VSS/m^3)
- X_e = concentration of VSS in the effluent, mg VSS/L (g VSS/m^3)
- k'' = specific growth rate constant simplified when the substrate is limited ($k'' = k/K_s$), L/mg VSS·d
- p_x = biosolids (VSS) growth rate per unit flow through the reactor, mg VSS/L (g VSS/m^3)

Process Design Equations for Reactor without Biosolids Recycle: A steady-state condition is reached when the biomass concentration in the reactor reaches a constant value. Under this condition, the loss of biosolids from the system is equal to the net growth, and the solids retention time θ_c is equal to HRT θ. Such condition normally exists in aerated lagoons. The flow diagram and variables in these relationships are shown in Figure 10.10c. The important relationships for the reactor without biosolids recycle are expressed by Equations 10.16a through 10.16e.

Mean cell residence time or solids retention time (SRT or θ_c) and hydraulic retention time (HRT or θ)

$$\theta_c = \theta \quad \text{and} \quad \frac{1}{\theta} = YU - k_d = \frac{YkS}{K_s + S} - k_d \tag{10.16a}$$

Minimum hydraulic retention time (θ^{min})

$$\frac{1}{\theta^{min}} \approx Yk - k_d = \mu_{max} - k_d \quad \text{or} \quad \theta^{min} \approx \frac{1}{Yk - k_d} = \frac{1}{\mu_{max} - k_d} \tag{10.16b}$$

Effluent substrate concentration (S)

$$S = \frac{K_s(1 + k_d\theta)}{\theta(Yk - k_d) - 1} \approx \frac{\theta^{min} K_s(1 + k_d\theta)}{\theta - \theta^{min}} \quad \text{or} \quad S = S_0 - \frac{X}{Y_{obs}} \tag{10.16c}$$

Mixed liquor solids concentration (X), and solids yield production

$$Y_{obs} = \frac{Y}{1 + k_d\theta} \quad \text{or} \quad Y_{obs} = \frac{X}{S_0 - S} \tag{10.16d}$$

$$X = \frac{Y(S_0 - S)}{(1 + k_d\theta)} \quad \text{or} \quad X = Y_{obs}(S_0 - S) \quad \text{or} \quad px = X \quad \text{or} \quad P_x = QX \tag{10.16e}$$

Temperature Effects on Reaction Rates: The microbial metabolism and growth involve enzymatic reactions. These enzyme-catalyzed reactions are temperature dependent. Normally the reaction rates or reaction velocity doubles for each 10°C rise in temperature. Additionally, the generation time (cell division or fission time) decreases as temperature increases.

The reaction rate constant as a function of temperature is given by Equation 2.8 in Section 2.4. The general form of this equation for maximum specific substrate utilization rate constant in the activated sludge process is given by Equation 10.17a.

$$k_2 = k_1 \theta_T^{(T_2 - T_1)} \tag{10.17a}$$

where

k_2 = maximum specific substrate utilization rate constant at mixed liquor temperature T_2 (°C), mg substrate/mg VSS·d or d^{-1}

k_1 = maximum specific substrate utilization rate constant at mixed liquor temperature T_1 (°C), mg substrate/mg VSS·d or d^{-1}

θ_T = temperature correction coefficient for maximum specific substrate utilization rate constant, dimensionless. The value of θ_T varies from 1.03 to 1.09 depending upon the type of substrate and activated sludge system. θ_T increases with increase in F/M ratio.[1,11] The typical value of θ_T is 1.047.

The temperature also has effect upon the endogenous decay coefficient, and is expressed by Equation 10.17b.[1]

$$k_{d2} = k_{d1} \theta_T^{(T_2 - T_1)} \tag{10.17b}$$

where

k_{d2} = endogenous decay coefficient at mixed liquor temperature T_2 (°C), d^{-1}

k_{d1} = endogenous decay coefficient at mixed liquor temperature T_1 (°C), d^{-1}

θ_T = temperature correction coefficient for endogenous decay coefficient, dimensionless. The value of θ_T may range from 1.065 to 1.085 depending upon the type of activated sludge system.

EXAMPLE 10.20: DERIVATION OF EQUATION EXPRESSING RELATIONSHIP BETWEEN θ_c AND U

Equation 10.15d expresses important relationship between the mean cell residence time (θ_c) and specific substrate utilization rate (U). Give step by step derivation of this equation.

Solution

1. Write the general word statement of mass balance relationship of microorganisms in the reactor (Equation 10.18a).

$$\begin{bmatrix}\text{Rate of accumulation}\\ \text{of microorganisms}\end{bmatrix} = \begin{bmatrix}\text{Rate of inflow of}\\ \text{microorganisms}\end{bmatrix} - \begin{bmatrix}\text{Rate of outflow of}\\ \text{microorganisms}\end{bmatrix} + \begin{bmatrix}\text{Net growth of}\\ \text{microorganisms}\end{bmatrix} \tag{10.18a}$$

2. Develop the mass balance equation.
   Assume the sludge is wasted from the return sludge line and give the symbolic representation (Equation 10.18b) based on the statement in Equation 10.18a.

$$V\frac{dX}{dt} = QX_0 - [Q_{wr}X_r + (Q - Q_{wr})X_e] + r'_g V \tag{10.18b}$$

   where
   dX/dt = rate of change of biomass concentration in the reactor, g VSS/m^3
   X_0 = biomass concentration in the influent, g VSS/m^3
   Note: Other variables have been defined previously in Equations 10.11 through 10.15.

3. Simplify the equation.
   Under steady-state condition, $dX/dt = 0$ and the biomass concentration in the influent is negligible, $X_0 = 0$.

$$0 = -[Q_{wr}X_r + (Q - Q_{wr})X_e] + r'_g V$$

   Rearrange the equation to obtain Equation 10.18c.

$$r'_g = \frac{Q_{wr}X_r + (Q - Q_{wr})X_e}{V} \tag{10.18c}$$

4. Develop the relationship between θ_c and U.
   Equate Equations 10.18c and 10.13a to obtain Equation 10.18d.

$$\frac{Q_{wr}X_r + (Q - Q_{wr})X_e}{V} = YUX - k_d X \tag{10.18d}$$

   Inverse of Equation 10.18d yields Equation 10.18e.

$$\frac{VX}{Q_{wr}X_r + (Q - Q_{wr})X_e} = \frac{1}{YU - k_d} \tag{10.18e}$$

   In accordance with Equation 10.15a, the left side of Equation 10.18e equals to θ_c. Therefore, θ_c is expressed by Equation 10.18f.

$$\theta_c = \frac{1}{YU - k_d} \tag{10.18f}$$

Inverse of Equation 10.18f to obtain Equation 10.15d.

$$\frac{1}{\theta_c} = YU - k_d$$

EXAMPLE 10.21: DERIVATION OF EXPRESSION FOR θ IN THE REACTOR

Equation 10.15g expresses relationship between the hydraulic retention time (θ) in the reactor in terms of kinetic coefficients of Y, k_d, and Y_{obs}. Give step-by-step derivation of this equation.

Solution

1. Write the basic relationship between θ_c and U.
   Start derivation from Equation 10.19a that is copied from the expression in Equation 10.15d.

$$\frac{1}{\theta_c} = YU - k_d \tag{10.19a}$$

2. Develop the relationship between θ_c and θ.
   Substitute the expression of U given by Equation 10.14d into Equation 10.19a to yield Equation 10.19b.

$$\frac{1}{\theta_c} = \frac{YQ(S_0 - S)}{VX} - k_d = Y\left(\frac{Q}{V}\right)\frac{(S_0 - S)}{X} - k_d \tag{10.19b}$$

Obtain Equation 10.19c by substituting $\frac{Q}{V} = \frac{1}{\theta}$ in Equation 10.19b.

$$\frac{1}{\theta_c} = \frac{Y}{\theta}\frac{(S_0 - S)}{X} - k_d \tag{10.19c}$$

3. Develop the expression of θ by Y_{obs} at a given θ_c.
   Rearranging Equation 10.19c yields Equation 10.19d.

$$\theta = \frac{Y}{1 + k_d\theta_c}\frac{(S_0 - S)}{X}\theta_c \tag{10.19d}$$

Substitute $Y_{obs} = \frac{Y}{1 + k_d\theta_c}$ from Equation 10.15k into Equation 10.19d to yield Equation 10.19e.

$$\theta = \frac{Y_{obs}(S_0 - S)}{X}\theta_c \tag{10.19e}$$

Both Equations 10.19d and 10.19e are part of Equation 10.15g.

EXAMPLE 10.22: DERIVATION OF EXPRESSION FOR S IN THE REACTOR

Equation 10.15h expresses the substrate concentration remaining (S) in the reactor in terms of kinetic coefficients K_s, k, Y, k_d, and Y_{obs}. Give step by step derivation of this equation.

Solution

1. Write basic relationship between θ_c and U.

Start derivation from Equation 10.20a that is copied from the expression in Equation 10.15d.

$$\frac{1}{\theta_c} = YU - k_d \tag{10.20a}$$

2. Develop the expression of S in terms of kinetic constants.

Substitute the expression of U given by Equation 10.12e into Equation 10.20a to yield Equation 10.20b.

$$\frac{1}{\theta_c} = Y\frac{kS}{K_s + S} - k_d \tag{10.20b}$$

Rearrange Equation 10.20b to obtain Equation 10.20c.

$$\frac{1}{\theta_c} = \frac{YkS - k_d(K_s + S)}{K_s + S} = \frac{YkS - K_s k_d - k_d S}{K_s + S}$$

$$K_s + S = \theta_c YkS - K_s k_d \theta_c - k_d \theta_c S$$

$$K_s + K_s k_d \theta_c = \theta_c YkS - k_d \theta_c S - S = S[\theta_c(Yk - k_d) - 1]$$

$$S = \frac{K_s(1 + k_d\theta_c)}{\theta_c(Yk - k_d) - 1} \tag{10.20c}$$

Substitute the expression of $(1/\theta_c^{\min}) \approx Yk - k_d$ from Equation 10.15f into Equation 10.20c to yield Equation 10.20d.

$$S \approx \frac{K_s(1 + k_d\theta_c)}{(\theta_c/\theta_c^{\min}) - 1} = \frac{\theta_c^{\min}}{\theta_c - \theta_c^{\min}} K_s(1 + k_d\theta_c) \tag{10.20d}$$

Both Equations 10.20c and 10.20d are part of Equation 10.15h.

3. Discussion of the potential impact of θ_c^{design} on S.

In Equation 10.20d, the expression of S indicates clearly that S would increase quickly when θ_c is too close to $\theta_c^{\min}$. Therefore, it is a common design practice of using a safety factor $f_{\theta_c}^{\text{design}}$ as expressed by Equation 10.20e.

$$\theta_c^{\text{design}} = f_{\theta_c}^{\text{design}} \theta_c^{\min} \tag{10.20e}$$

where

θ_c^{design} = design θ_c, d

$f_{\theta_c}^{\text{design}}$ = design safety factor ($f_{\theta_c}^{\text{design}} \geq 2$ normally), dimensionless

Use Equation 10.20e to further simplify Equation 10.20d and obtain Equation 10.20f.

$$S \approx \frac{\theta_c^{\min} K_s(1 + \theta_c^{\text{design}} k_d)}{\theta_c^{\text{design}} - \theta_c^{\min}} = \frac{\theta_c^{\min} K_s(1 + \theta_c^{\text{design}} k_d)}{f_{\theta_c}^{\text{design}} \theta_c^{\min} - \theta_c^{\min}} = \frac{K_s(1 + \theta_c^{\text{design}} k_d)}{f_{\theta_c}^{\text{design}} - 1} \tag{10.20f}$$

EXAMPLE 10.23: DERIVATION OF A SIMPLIFIED EXPRESSION FOR LOW *S* IN THE REACTOR

Equation 10.15j provides a simplified expression of the substrate concentration remaining (S) in the reactor when S is low. Give step-by-step derivation of this equation.

Solution

1. Identify and write the specific substrate utilization rate equation.
   Start derivation from Equation 10.21a that is copied from Equation 10.12e.

$$U = \frac{kS}{K_s + S} \tag{10.21a}$$

2. Develop the expression of low S in the effluent.
   In limited- substrate situation, the $S \ll K_s$ in the bioreactor and $K_s + S \approx K_s$, and Equation 10.21a can be modified to Equation 10.21b, which is a first-order reaction.

$$U = \frac{kS}{K_s + S} \approx \frac{kS}{K_s} = \frac{k}{K_s}S \quad \text{or} \quad U = k''S \tag{10.21b}$$

   Substitute the expression of $U = \dfrac{S_0 - S}{\theta X}$ from Equation 10.14d into Equation 10.21b to obtain Equation 10.21c.

$$\frac{S_0 - S}{\theta X} = k''S \tag{10.21c}$$

   Rearranging Equation 10.21c yields Equation 10.15j.

$$S = \frac{S_0 - S}{k''\theta X} \quad \text{or} \quad S = \frac{S_0}{1 + k''\theta X}$$

EXAMPLE 10.24: DERIVATION OF EXPRESSION FOR X (BIOMASS OR MLVSS CONCENTRATION) IN THE REACTOR

The biomass is constantly generated in a bioreactor. It is lost in the effluent, and is also wasted to maintain a desired MLVSS concentration (X) in the reactor. Equation 10.15k expresses X in terms of kinetic coefficients of Y, k_d, and Y_{obs}. Provide step-by-step derivation of this equation.

Solution

1. Develop the relationship between θ_c and X.
   Follow the first two steps in Example 10.21 to obtain Equation 10.22a.

$$\frac{1}{\theta_c} = \frac{Y}{\theta}\frac{(S_0 - S)}{X} - k_d \tag{10.22a}$$

2. Develop the expression of X in terms of kinetic constants.
   Rearrange Equation 10.22a to obtain Equation 10.22b.

$$X = \frac{Y}{1 + k_d\theta_c}\frac{\theta_c}{\theta}(S_0 - S) \tag{10.22b}$$

   Substitute $Y_{obs} = \dfrac{Y}{1 + k_d\theta_c}$ from Equation 10.15l into Equation 10.22b to yield Equation 10.22c.

$$X = Y_{obs}\frac{\theta_c}{\theta}(S_0 - S) \tag{10.22c}$$

   Both Equations 10.22b and 10.22c are part of Equation 10.15k.

EXAMPLE 10.25: DERIVATION OF EXPRESSION FOR Y_{obs}

The endogenous respiration reduces the biomass, and has an effect on the net cell yield. Equation 10.15l expresses the observed biomass yield (Y_{obs}) in terms of kinetic coefficients of Y, and k_d. Give step-by-step derivation of this equation.

Solution

1. Write the basic relationship between θ_c and U.

   Start derivation from Equation 10.23a that is copied from the expression in Equation 10.15d.

$$\frac{1}{\theta_c} = YU - k_d \tag{10.23a}$$

   Rearrange Equation 10.23a to obtain Equation 10.23b.

$$U = \frac{1 + k_d\theta_c}{Y\theta_c} \tag{10.23b}$$

2. Develop the expression of Y_{obs} in terms of kinetic constants.

   Rearranging Equation 10.13c yields Equation 10.23c.

$$Y_{obs}U = YU - k_d \tag{10.23c}$$

   Equation 10.23d is obtained by comparing the right sides in both Equations 10.23a and 10.23c.

$$\frac{1}{\theta_c} = Y_{obs}U \tag{10.23d}$$

   Substitute the expression of U given by Equation 10.23b into Equation 10.23d to yield Equation 10.23e.

$$\frac{1}{\theta_c} = Y_{obs}\frac{1 + k_d\theta_c}{Y\theta_c} \tag{10.23e}$$

   Rearrange Equation 10.23e to obtain Equation 10.15l.

$$Y_{obs} = \frac{Y}{1 + k_d\theta_c}$$

EXAMPLE 10.26: DERIVATION OF EXPRESSION FOR V OF THE REACTOR

Proper volume (V) in the reactor is also important to meet the design θ_c at an operational MLVSS concentration. Equation 10.15o expresses V in terms of kinetic coefficients of Y, k_d, and Y_{obs}. Provide step-by-step derivation of this equation.

Solution

1. Develop the relationship between θ_c and V.

   Follow the first two steps in Example 10.21 to obtain Equation 10.24a.

$$\frac{1}{\theta_c} = \frac{YQ(S_0 - S)}{VX} - k_d \tag{10.24a}$$

2. Develop the expression of V in terms of kinetic constants.
   Rearrange Equation 10.24a to obtain Equation 10.24b.

$$V = \frac{Y}{1 + k_d\theta_c}\frac{\theta_c Q(S_0 - S)}{V} \tag{10.24b}$$

   Substitute $Y_{obs} = \dfrac{Y}{1 + k_d\theta_c}$ from Equation 10.15l into Equation 10.24b to yield Equation 10.24c.

$$X = Y_{obs}\frac{\theta_c Q(S_0 - S)}{V} \tag{10.24c}$$

   Both Equations 10.24b and 10.24c are part of Equation 10.15o.

EXAMPLE 10.27: DEPENDENT AND INDEPENDENT VARIABLES IN BIOLOGICAL TREATMENT PROCESS

Following is a list of parameters that are used for biological process design. Identify the primary (independent) and dependent (designers' choice) variables.

S_0, S, Y, K_s, k_d, k, θ, θ_c, V, X, X_r, X_{wa}, X_{wr}

Solution

1. List of primary variables.

   S_0, Y, K_s, k_d, and k

2. List of dependent variables.

   S, θ_c, θ, V, X, X_r, X_{wa}, and X_{wr}

EXAMPLE 10.28: SATURATION OR ENZYMATIC REACTION

In a biological treatment facility, it was found that 1 g of microorganisms would decompose glucose at a maximum rate of 1.5 mg per day when the concentration of glucose was high. It was also found that in the reactor at glucose concentration of 15 mg/L, the reaction rate is half of maximum rate. Calculate the rate of glucose decomposition (dS/dt) when glucose concentration remaining is 5 mg/L and microorganism concentration is 2 g/L.

Solution

1. Identify and write the specific substrate utilization rate equation.
   The specific substrate utilization rate equations are given in Equations 10.12d and 10.12e. Combine these two equations to obtain Equation 10.25.

$$U = -\frac{r_{su}}{X} = -\frac{1}{X}\frac{dS}{dt} = \frac{kS}{K_s + S} \quad \text{or} \quad -\frac{dS}{dt} = \frac{kSX}{K_s + S} \tag{10.25}$$

2. Calculate the rate of glucose decomposition.
   Substitute the given data in Equation 10.25 and determine dS/dt.
   $k = 1.5$ mg glucose/mg VSS·d
   $K_s = 15$ mg glucose/L

$S = 5$ mg glucose/L
$X = 2$ g/L $= 2000$ mg VSS/L

$$-\frac{dS}{dt} = \frac{1.5 \text{ mg glucose/mg VSS·d} \times 5 \text{ mg glucose/L} \times 2000 \text{ mg VSS/L}}{(15 + 5) \text{ mg glucose/L}}$$

$$= 750 \text{ mg glucose/L·d}$$

or

$$\frac{dS}{dt} = -0.75 \text{ g glucose/L·d} \quad \text{or} \quad -0.75 \text{ kg glucose/m}^3\text{·d}$$

The negative sign indicates that the glucose concentration is decreasing with respect to time.

EXAMPLE 10.29: DETERMINATION OF K AND K_S FROM EXPERIMENTAL DATA

A batch biological reactor was operated with mixed microorganisms. The specific waste utilization rate and substrate concentration remaining at various time intervals are given below. Calculate k and K_s.

$U\left(\text{or } -\frac{1}{X}\frac{dS}{dt}\right)$, d^{-1}	1.9	1.8	1.7	1.4	1.1	0.9	0.6	0.35	0.15
S, mg/L	243	150	97	44	26	16	9	5	2

Solution

The values of k and K_s can be determined by using either of two methods: (a) saturation plot, or (b) inverse linear plot. Both methods are presented below.

Saturation Plot Method

1. Write the relationship between U and S.
   The relationship is expressed by Equation 10.12e.

$$U = \frac{kS}{K_s + S}$$

2. Plot the values of U versus S.
   Typically, K_s has a relatively small value in comparison with S_0. It will be easy to read if the plot is made on a semilog paper as shown in Figure 10.11a.
3. Determine k and K_s.
   A potentially maximum value of 2.2 d^{-1} is estimated from the forward extension of the curve. This value gives the maximum specific rate of substrate utilization, $k = 2.2\ d^{-1}$.
   $K_s = 30$ mg/L is then estimated from the curve at $U = k/2 = 1.1\ d^{-1}$.

Inverse Linear Plot Method

1. Develop a linear form of the equation.
   Equation 10.12e is inversed to reduce it to a linear form.

$$\frac{1}{U} = \frac{K_s}{k}\frac{1}{S} + \frac{1}{k}$$

The linear relationship is obtained by plotting $1/U$ versus $1/S$. The slope of the line is K_s/k and intercept is $1/k$. Tabulate the data for $1/U$ and $1/S$.

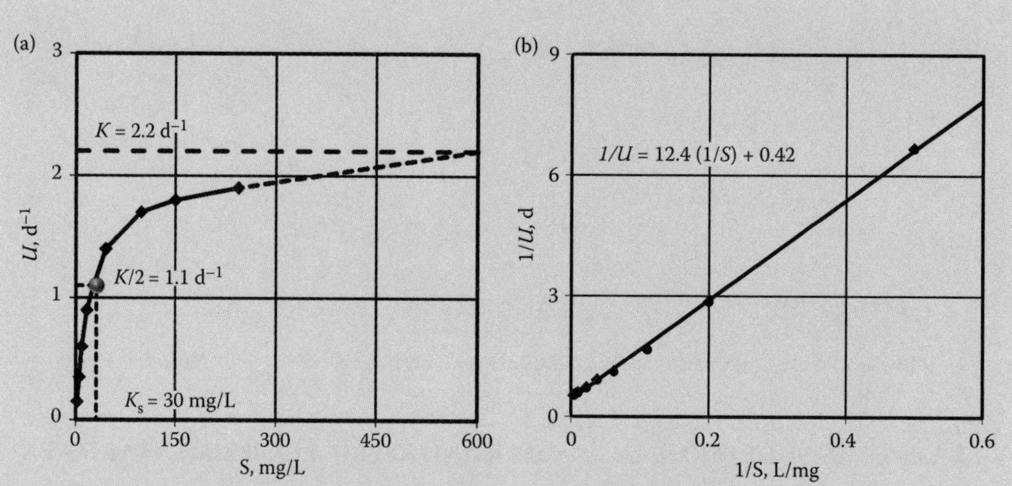

FIGURE 10.11 Determination of kinetic coefficients k and K_s: (a) saturation plot, and (b) inverse linear plot (Example 10.29).

$\frac{1}{U}\left(\text{or} - X\frac{dt}{dS}\right)$, d	0.526	0.556	0.588	0.714	0.909	1.11	1.67	2.86	6.67
$1/S$, L/mg	0.00412	0.00667	0.0103	0.0227	0.0385	0.0625	0.111	0.200	0.500

2. Prepare the linear plot.
   The inverse linear plot is shown in Figure 10.11b.
3. Determine k and K_s.
   k and K_s are obtained from this plot.
   The intercept of the line is 0.42 d.

$$k = \frac{1}{0.42 \text{ d}} = 2.4 \text{ d}^{-1}$$

The slope of the line is 12.4 mg·d/L.

$$K_s = 12.4 \text{ mg·d/L} \times 2.4 \text{ d}^{-1} = 30 \text{ mg/L}$$

Therefore, the results from both methods are very close. The results from *saturation plot method* are not precise. They are completely estimated by visual judgment.

EXAMPLE 10.30: DETERMINATION OF BIOLOGICAL KINETIC COEFFICIENTS OF k AND K_S FROM BATCH REACTOR EXPERIMENTAL DATA

The kinetic coefficients Y and k_d were developed from the batch reactor study presented in Example 10.18. Determine the kinetic coefficients k and K_s from the relationship developed between the specific substrate utilization rate and substrate concentrations remaining. Also, determine the kinetic coefficient μ_{max}. Utilize the reactor data and results of Example 10.18. Summarize all kinetic coefficients developed in Examples 10.18 and 10.30.

Solution

1. Apply the batch reactor data given in Example 10.18 to determine the kinetic coefficients.

   There are two methods that can be used to estimate the biological coefficients as presented in Example 10.29. In this example, the inverse linear plot method is used.

2. Calculate the inverted specific substrate utilization and biomass growth rates.

   The relationship between $\bar{X}/(-\Delta S/\Delta t)$ and $1/\bar{S}$ (Equation 10.26 is developed from Equation 10.12e).

$$\bar{X}\left(-\frac{\Delta t}{\Delta S}\right) = \frac{K_s}{k}\frac{1}{\Delta S} + \frac{1}{k} \tag{10.26}$$

   Prepare the relationship $\bar{X}/(-\Delta S/\Delta t)$ and $1/\bar{S}$ from the experimental data given in Example 10.18 so that a linear relationship can be developed. The experimental data of soluble BOD_5 (S) and VSS (X), and the calculated values of $-\Delta S/\Delta t$, $\bar{X}$, $\bar{S}$, $(-\Delta S/\Delta t)/\bar{X}$, $\bar{X}/(-\Delta S/\Delta t)$, and $1/\bar{S}$ are summarized at different reaction times (t), in Table 10.9.

TABLE 10.9 Calculated Values from the Experimental Data (Example 10.30)

t, h	S, mg/L	X, mg/L	$-\Delta S/\Delta t$, mg/L·h	$\bar{X}$, mg/L	$\bar{S}$, mg/L	$(-\Delta S/\Delta t)/\bar{X}$, h^{-1}	$1/\bar{S}$, L/mg	$\bar{X}/(-\Delta S/\Delta t)$, h
0	310	1000	–	–	–	–	–	–
1	218	1034	92	1017	264	0.0905	0.00379	11.1
2	130	1069	88	1052	174	0.0837	0.00575	11.9
3	54	1095	76	1082	92	0.0702	0.0109	14.2
4	9.5	1111	44.5	1103	32	0.0403	0.0315	24.8

3. Plot the graphical relationship.

   The plot of $(\Delta X/\Delta t)/\bar{X}$ versus $(-\Delta S/\Delta t)/\bar{X}$ values is shown in Figure 10.12.

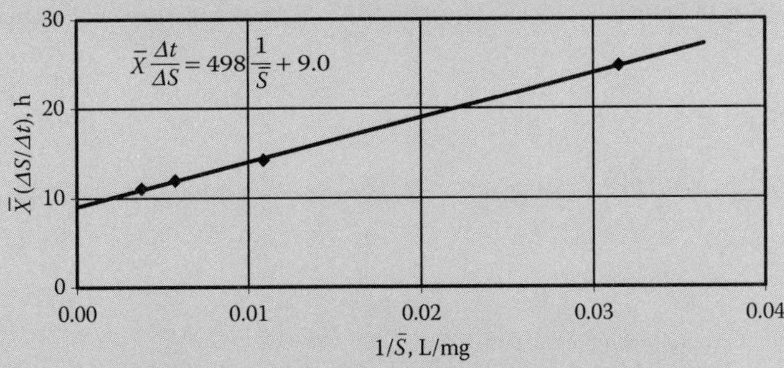

FIGURE 10.12 Determination of kinetic coefficients k and K_s (Example 10.30).

4. Determine k and K_s.

   k and K_s are obtained from this plot.

   The intercept of the line on the Y axis is 9.0 h.

$$k = \frac{1}{9\,\text{h}} = 0.11\ \text{h}^{-1} = 0.11\ \text{h}^{-1} \times 24\ \text{h/d} = 2.6\ \text{d}^{-1} \quad \text{or} \quad 2.6\ \text{mg BOD}_5/\text{mg VSS·d}$$

The slope of the line is 298 mg·h/L.

$K_s = 498 \text{ mg·h/L} \times 0.11 \text{ h}^{-1} = 55 \text{ mg/L}$

5. Calculate μ_{max}.

The maximum specific growth rate μ_{max} is calculated from Equation 10.12b at $k = 2.6$ mg BOD_5/mg VSS·d and $Y = 0.40$ mg VSS/mg BOD_5. The kinetic coefficients k_d and Y are obtained in Example 10.18.

$$\mu_{max} = Yk = 0.40 \text{ mg VSS/mg BOD}_5 \times 2.6 \text{ mg BOD}_5\text{/mg VSS·d}^{-1}$$
$$= 1.0 \text{ mg VSS/mg VSS·d}$$

6. Summarize all kinetic coefficients obtained from the batch reactor study in Examples 10.18 and 10.30.

$Y = 0.40$ mg VSS/mg BOD_5, $k_d = 0.050$ mg VSS/mg VSSd^{-1}, $k = 2.6$ mg BOD_5/mg VSS·d, $K_s = 55$ mg BOD_5/L, and $\mu_{max} = 1.0$ mg VSS/mg VSS·d.

EXAMPLE 10.31: RATE OF CHANGE IN SUBSTRATE AND CELL CONCENTRATIONS WITH TIME IN A BATCH REACTOR

Develop an expression that can be used to determine the relationship between the rate of change of substrate and cell concentrations for a batch reactor.

Solution

1. Write the general word statement of mass balance relationship of microorganisms in the batch reactor (Equation 10.27a).

$$\begin{bmatrix} \text{Rate of accumulation of} \\ \text{microorganisms or substrate} \end{bmatrix} = \begin{bmatrix} \text{Rate of generation of microorganisms} \\ \text{or utilization of substrate} \end{bmatrix} \quad (10.27a)$$

2. Develop the substrate mass balance.

A substrate mass balance equation in Equation 10.27b is developed based on the statement in Equation 10.27a.

$$V\frac{dS}{dt} = V\, r_{su} \quad (10.27b)$$

Substitute $r_{su} = -\dfrac{kXS}{K_s + S}$ from Equation 10.12c into Equation 10.27b to yield Equation 10.27c.

$$\frac{dS}{dt} = -\frac{kXS}{K_s + S} \quad (10.27c)$$

3. Develop the microorganism mass balance.

The microorganism mass balance equation is developed based on the statement in Equation 10.27a and expressed by Equation 10.27d.

$$V\frac{dX}{dt} = Vg' \quad (10.27d)$$

Substitute $r'_g = \dfrac{\mu_{max}XS}{K_s + S} - k_dX$ from Equation 10.13a into Equation 10.27d to obtain Equation 10.27e.

$$\frac{dX}{dt} = \frac{\mu_{max}XS}{K_s + S} - k_dX \quad (10.27e)$$

Equation 10.27f is obtained from Equation 10.27e since $\mu_{max} = Yk$ (Equation 10.12b).

$$\frac{dX}{dt} = \frac{YkXS}{K_s + S} - k_d X = Y\frac{kXS}{K_s + S} - k_d X \tag{10.27f}$$

4. Develop the relationship between the rate of change of substrate and cell concentrations.
   Substitute Equation 10.27c in Equation 10.27f to obtain Equation 10.27g.

$$\frac{dX}{dt} = Y\left(-\frac{dS}{dt}\right) - k_d X \quad \text{or} \quad \frac{1}{X}\frac{dX}{dt} = Y\left(-\frac{1}{X}\frac{dS}{dt}\right) - k_d \tag{10.27g}$$

5. Compare the relationships in Equation 10.27g and Equation 10.10a.
   Both relationships are identical. Procedure for determination of Y and k_d from a batch reactor is given in Example 10.18.

EXAMPLE 10.32: DETERMINATION OF DESIGN PARAMETERS FROM THE KINETIC COEFFICIENTS

An activated sludge plant is treating municipal wastewater. The kinetic constants Y, k_d, k, and K_s are 0.50, 0.05 d^{-1}, 5.0 d^{-1}, and 50 mg BOD$_5$/L, respectively. The influent BOD$_5$ (S_0) is 200 mg/L, aeration period (θ) is 6 h, and solids retention time (θ_c) is 5 days. Calculate the following: (1) effluent soluble BOD$_5$ (S), (2) removal efficiency (E) of BOD$_5$, (3) concentration of microorganism (X) in the aeration basin, (4) MLSS concentration in the aeration basin, (5) the specific BOD$_5$ utilization rate (U), and (6) food to microorganism (F/M) ratio. Assume that VSS/TSS ratio in the mixed liquor is 0.78.

Solution

1. Calculate the soluble BOD$_5$ (S) in the effluent from Equation 10.15h.

$$S = \frac{K_s(1 + k_d\theta_c)}{\theta_c(Yk - k_d) - 1} = \frac{50\,\text{mg BOD}_5/\text{L} \times (1 + 0.05\,\text{d}^{-1} \times 5\,\text{d})}{5\,\text{d} \times (0.50 \times 5.0\,\text{d}^{-1} - 0.05\,\text{d}^{-1}) - 1} = 5.6\,\text{mg BOD}_5/\text{L}$$

2. Calculate the removal efficiency (E) of BOD$_5$ in the effluent from Equation 10.14a.

$$E = \frac{S_0 - S}{S_0} \times 100\% = \frac{(200 - 5.6)\,\text{mg BOD}_5/\text{L}}{200\,\text{mg BOD}_5/\text{L}} \times 100\% = 97.2\%$$

3. Calculate the biomass concentration (X) in the reactor from Equation 10.15k.

$$X = \frac{Y}{1 + k_d\theta_c}\frac{\theta_c}{\theta}(S_0 - S) = \frac{0.50\,\text{mg VSS/mg BOD}_5}{1 + 0.05\,\text{d}^{-1} \times 5\,\text{d}} \times \frac{5\,\text{d} \times 24\,\text{h/d}}{6\,\text{h}} \times (200 - 5.6)\,\text{mg BOD}_5/\text{L}$$

$$= 1555\,\text{mg VSS/L}$$

4. Calculate the MLSS concentration in the reactor from Equation 10.14e.

$$\text{MLSS} = \frac{X}{\text{VSS/TSS}} = \frac{1555\,\text{mg VSS/L}}{0.78\,\text{mg VSS/mg TSS}} = 1994\,\text{mg TSS/L} \approx 2000\,\text{mg TSS/L}$$

5. Calculate the specific BOD_5 utilization rate (U) from Equation 10.14d.

$$U = \frac{S_0 - S}{\theta X} = \frac{(200 - 5.6)\text{ mg BOD}_5/\text{L} \times 24\text{ h/d}}{6\text{ h} \times 1555\text{ mg VSS/L}} = 0.50\text{ mg BOD}_5/\text{mg VSS·d}$$

6. Calculate the F/M ratio from Equation 10.14b.

$$\text{F/M} = \frac{S_0}{\theta X} = \frac{200\text{ mg BOD}_5/\text{L} \times 24\text{ h/d}}{6\text{ h} \times 1555\text{ mg VSS/L}} = 0.51\text{ mg BOD}_5/\text{mg VSS·d}$$

EXAMPLE 10.33: BIOMASS PRODUCTION AND OXYGEN UTILIZATION

An aeration basin receives 3800 m^3/d primary settled wastewater. The influent and effluent soluble BOD_5 concentrations are 150 and 6 mg/L, respectively. The process-related data include $Y = 0.50$, $k_d = 0.06$ d^{-1}, $\theta_c = 10$ d, and $BOD_5/BOD_L = 0.68$. Calculate (1) biomass growth and (2) oxygen utilization.

Solution

1. Calculate the observed yield coefficient (Y_{obs}) from Equation 10.15l.

$$Y_{obs} = \frac{Y}{1 + k_d\theta_c} = \frac{0.50\text{ mg VSS/mg BOD}_5}{1 + 0.06\text{ d}^{-1} \times 10\text{ d}} = 0.31\text{ mg VSS/mg BOD}_5 = 0.31\text{ g VSS/g BOD}_5$$

2. Calculate the biomass growth or VSS generation rate (P_x) in the aeration basin from Equation 10.15m.

$$P_x = Y_{obs}Q(S_0 - S) = 0.31\text{ g VSS/g BOD}_5 \times 3800\text{ m}^3/\text{d} \times (150 - 6)\text{ g BOD}_5/\text{m}^3 \times 10^{-3}\text{ kg/g}$$
$$= 170\text{ kg VSS/d}$$

3. Calculate the oxygen utilization rate (O_2) from Equation 10.14f.

$$O_2 = \frac{Q(S_0 - S)}{BOD_5/BOD_L} - 1.42P_x$$
$$= \frac{3800\text{ m}^3/\text{d} \times (150 - 6)\text{ g BOD}_5/\text{m}^3 \times 10^{-3}\text{ kg/g}}{0.68\text{ g BOD}_5/\text{g BOD}_L} - 1.42\text{ g BOD}_L/\text{g VSS} \times 170\text{ kg VSS/d}$$
$$= (805 - 241)\text{ kg BOD}_L/\text{d} = 564\text{ kg BOD}_L/\text{d} \quad \text{or} \quad 564\text{ kg O}_2/\text{d}$$

EXAMPLE 10.34: OXYGEN UTILIZATION EQUATION

There are two parts in Equation 10.14f. Explain each part.

Solution

1. Write Equation 10.14f.

$$O_2 = \frac{Q(S_0 - S)}{BOD_5/BOD_L} - 1.42P_x$$

2. Explain the first part $\frac{Q(S_0 - S)}{BOD_5/BOD_L}$ of this equation,

   The term $(S_0 - S)$ expresses the 5-day carbonaceous BOD concentration removed in the reactor and is equal to the difference between the influent and effluent BOD_5 concentrations. Multiplying it by the flow to the reactor gives the BOD_5 removal rate in the reactor. Dividing the product by the ratio of BOD_5 to BOD_L yields the BOD_L removed rate in the reactor. Therefore, an equivalent oxygen supply rate must be provided in order to oxidize the ultimate carbonaceous BOD_L into CO_2 and water.

$$BOD_L + O_2 \rightarrow CO_2 + H_2O$$

3. Explain the second part $-1.42P_x$ of this equation.

   Since a portion of BOD_L is converted into biomass, this amount of BOD_L is not oxidized and oxygen is not needed. Therefore, the amount of oxygen required to oxidize the biomass produced is subtracted from the total oxygen required to oxidize the incoming BOD_L. In Example 10.10, a factor of 1.42 g O_2/g VSS has been developed based on the molecular ratio of the oxygen utilized to cellular mass produced from the stoichiometric equation for carbonaceous oxidation of cellular biomass, expressed by $C_5H_7O_2N$. This amount is subtracted from the total oxygen required for oxidation of total carbonaceous BOD_L expressed by the first term in the Equation 10.14f.

EXAMPLE 10.35: PROFILES OF SUBSTRATE CONCENTRATION IN THE EFFLUENT AND PROCESS REMOVAL EFFICIENCY

The biological kinetic coefficients are used to calculate the effluent concentration of substrate remaining (S), and the efficiency of the treatment process (E). Develop the profiles of effluent concentration and process efficiency with respect to mean cell residence time (θ_c). The process data are given below: $S_0 = 200$ mg BOD_5/L, $Y = 0.5$, $k_d = 0.05\ d^{-1}$, $k = 5\ d^{-1}$, and $K_s = 60$ mg BOD_5/L.

Solution

1. Calculate the minimum mean cell residence time (θ_c^{min}).

   In a biological rector, there is a minimum mean cell residence time (θ_c^{min}) at or below which the cell mass growth rate is lower than the washout rate of biomass in the effluent. For this reason, a stable MLVSS concentration cannot be maintained in the reactor when $\theta_c \leq \theta_c^{min}$. Therefore, there will be eventually $S = S_0$ if $\theta_c \leq \theta_c^{min}$. Calculate θ_c^{min} from Equation 10.15e.

$$\frac{1}{\theta_c^{min}} = \frac{YkS_0}{K_s + S_0} - k_d = \frac{0.50 \times 5.0\ d^{-1} \times 200\ mg\ BOD_5/L}{(60 + 200)\ mg\ BOD_5/L} - 0.05\ d^{-1} = 1.87\ d^{-1}$$

$$\theta_c^{min} = \frac{1}{1.87\ d^{-1}} = 0.53\ d$$

2. Identify the equations used to calculate the effluent concentration (S) and the process removal efficiency (E) from Equations 10.15h and 10.14a, respectively

$$S = \frac{K_s(1 + k_d\theta_c)}{\theta_c(Yk - k_d) - 1} \quad \text{and} \quad E = \frac{S_0 - S}{S_0} \times 100\%$$

3. Calculate S and E.

Assume different values of $\theta_c > 0.53$ d and calculate S and E. These values are calculated below.

θ_c, d	0.00	0.53	0.55	0.65	0.75	1	2	4	6	8	10
S, mg BOD$_5$/L	500	500	177[a]	105	74	43	17	8.2	5.7	4.5	3.8
E, %	0	0	12[b]	48	63	79	92	96	97	98	98

[a] $S = \dfrac{60 \text{ mg BOD}_5/\text{L} \times (1 + 0.05 \text{ d}^{-1} \times 0.55 \text{ d})}{0.55 \text{ d} \times (0.5 \times 5 \text{ d}^{-1} - 0.05 \text{ d}^{-1}) - 1} = 177 \text{ mg BOD}_5/\text{L}$

[b] $E = \dfrac{(200 - 177) \text{ mg BOD}_5/\text{L}}{200 \text{ mg BOD}_5/\text{L}} \times 100\% = 12\%$

4. Plot the results.

The profiles of S and E with respect to θ_c are shown in Figure 10.13.

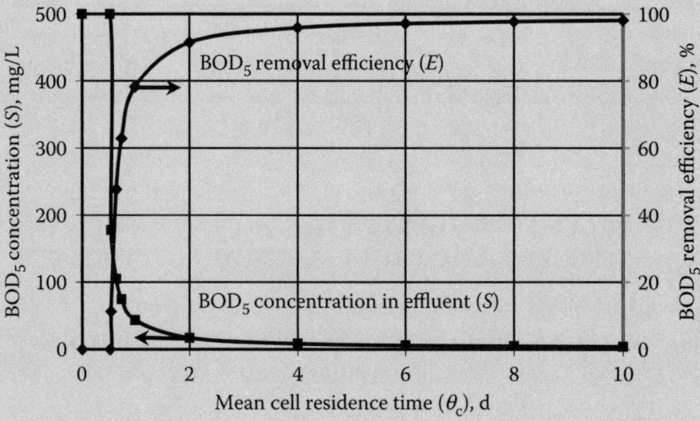

FIGURE 10.13 Performance efficiency of a continuous flow complete-mixed biological reactor (Example 10.35).

EXAMPLE 10.36: DESIGN OF BIOLOGICAL PROCESS

A suspended growth biological reactor receives primary settled wastewater at an average flow rate (Q) of 3800 m^3/d. Calculate the following: (a) effluent soluble BOD$_5$ (S), (b) design solids retention time (θ_c), (c) biosolids produced (P_x), (d) aeration period (θ), (e) volume of aeration basin (V), (f) daily volume of waste activated sludge (WAS) (Q_{wr}), (g) flow rate of return activated sludge (RAS) (Q_r), (h) food to microorganism (F/M) ratio, (i) mass organic loading (MOL), (j) volumetric organic loading (VOL), and (k) theoretical oxygen required (O_2).

The following data may be used: influent BOD$_5$ concentration (S_0) = 150 mg/L, effluent total BOD$_5$ concentration = 10 /L, effluent TSS concentration (TSS$_e$) = 10 mg/L, MLVSS = 2000 mg/L, $Y = 0.50$, $k_d = 0.06$ d^{-1}, $k = 5.0$ d^{-1}, and $K_s = 50$ mg BOD$_5$/L, VSS/TSS ratio in the mixed liquor = 0.8, BOD$_5$/BOD$_L$ = 0.68, and TSS concentration in RAS (TSS$_{ras}$) = 10,000 mg/L. Assume 0.65 g biodegradable solids in 1 g TSS, and 1.42 g BOD$_L$ in 1 g biodegradable solids. The WAS is withdrawn from the return sludge line. To simplify the calculations, the impact of influent TSS on solids production is ignored in this example.

Solution

1. Calculate soluble BOD_5 (S) in the effluent from the aeration basin.

$$\text{Effluent BOD}_5 \text{ exerted by TSS} = \text{TSS}_e \times \frac{0.65\text{ g biodegradable solids}}{\text{g TSS}}$$

$$\times \frac{1.42\text{ g BOD}_L}{\text{g biodegradable solids}} \times \frac{0.68\text{ g BOD}_5}{\text{g BOD}_L}$$

$$= 10\text{ mg TSS/L} \times \frac{0.63\text{ g BOD}_5}{\text{g TSS}} = 6.3\text{ mg BOD}_5\text{/L}$$

$$\text{Effluent soluble BOD}_5\ (S) = \text{Effluent total BOD}_5 - \text{Effluent BOD}_5 \text{ exerted by TSS}$$

$$= (10 - 6.3)\text{ mg BOD}_5\text{/L} = 3.7\text{ mg BOD}_5\text{/L}$$

Calculate soluble BOD_5 removal efficiency (e) from Equation 10.14a.

$$E = \frac{(150 - 3.7)\text{ mg BOD}_5\text{/L}}{150\text{ mg BOD}_5\text{/L}} \times 100\% = 97.5\%$$

2. Calculate the solids retention time (θ_c) required in the aeration basin.
The minimum θ_c for effective cell reproduction (θ_c^{min}) is calculated from Equation 10.15e.

$$\frac{1}{\theta_c^{min}} = \frac{YkS_0}{K_s + S_0} - k_d = \frac{0.50 \times 5.0\text{ d}^{-1} \times 150\text{ mg BOD}_5\text{/L}}{(50 + 150)\text{ mg BOD}_5\text{/L}} - 0.06\text{ d}^{-1} = 1.82\text{ d}^{-1}$$

$$\theta_c^{min} = \frac{1}{1.82\text{ d}^{-1}} = 0.55\text{ d}$$

The required θ_c to achieve S in the effluent ($\theta_c^{required}$) is calculated from Equation 10.15d.

$$\frac{1}{\theta_c^{required}} = \frac{YkS}{K_s + S} - k_d = \frac{0.50 \times 5.0\text{ d}^{-1} \times 3.7\text{ mg BOD}_5\text{/L}}{(50 + 3.7)\text{ mg BOD}_5\text{/L}} - 0.06\text{ d}^{-1} = 0.11\text{ d}^{-1}$$

$$\theta_c^{required} = \frac{1}{0.11\text{ d}^{-1}} = 9.1\text{ d}$$

Select a design $\theta_c = 10$ d, which is slightly greater than $\theta_c^{required}$ and much larger than θ_c^{min}.

3. Calculate the biomass growth rate (P_x) in the aeration basin.
The observed yield coefficient (Y_{obs}) is calculated from Equation 10.15l.

$$Y_{obs} = \frac{Y}{1 + k_d\theta_c} = \frac{0.50\text{ mg VSS/mg BOD}_5}{1 + 0.06\text{ d}^{-1} \times 10\text{ d}} = 0.31\text{ mg VSS/mg BOD}_5 = 0.31\text{ g VSS/g BOD}_5$$

Calculate the P_x from Equation 10.15m.

$$P_x = Y_{obs}Q(S_0 - S) = 0.31\text{ g VSS/g BOD}_5 \times 3800\text{ m}^3\text{/d} \times (150 - 3.7)\text{g BOD}_5\text{/m}^3 \times 10^{-3}\text{ kg/g}$$
$$= 172\text{ kg VSS/d}$$

$$\text{Total TSS produced} = \frac{P_x}{\text{VSS/TSS ratio}} = \frac{172\text{ kg VSS/d}}{0.8\text{ kg VSS/kg TSS}} = 215\text{ kg TSS/d}$$

4. Calculate the aeration period (θ).

The required θ in the aeration basin ($\theta_c^{\text{required}}$) is calculated from Equation 10.15g at MLVSS = 2000 mg VSS/L = 2000 g VSS/m^3.

$$\theta^{\text{required}} = \frac{P_x}{QX}\theta_c = \frac{172\,\text{kg VSS/d}}{3800\,\text{m}^3/\text{d} \times 2000\,\text{g VSS/m}^3 \times 10^{-3}\,\text{kg/g}} \times 10\,\text{d} \times 24\,\text{h/d} = 5.4\,\text{h}$$

Provide a design hydraulic retention time $\theta = 6$ h in the aeration basin.

5. Calculate the aeration basin volume (V) from Equation 10.15o.

$$V = Q\theta = 3800\,\text{m}^3/\text{d} \times 6\,\text{h} \times \frac{\text{d}}{24\,\text{h}} = 950\,\text{m}^3$$

6. Calculate the daily volume of WAS (Q_{wr}).
Since the WAS flow Q_{wr} flow is much smaller than the influent flow Q, the effluent flow $Q_e = Q - Q_{wr} \approx Q$. Calculate effluent VSS (X_e) from the clarifier.

$$X_e = \text{VSS/TSS ratio} \times \text{TSS}_e = 0.8\,\text{mg VSS/mg TSS} \times 10\,\text{mg TSS/L} = 8\,\text{mg VSS/L}$$

Estimate weights of TSS and VSS lost in the effluent.

$$\text{TSS lost in the effluent} = Q \times \text{TSS}_e = 3800\ \text{m}^3/\text{d} \times 10\ \text{g TSS/m}^3 \times 10^{-3}\,\text{kg/g} = 38\ \text{kg TSS/d}$$

$$\text{VSS lost in the effluent} = Q \times X_e = 3800\ \text{m}^3/\text{d} \times 8\ \text{g VSS/m}^3 \times 10^{-3}\,\text{kg/g} = 30\ \text{kg TSS/d}$$

7. Calculate the dry weights of TSS and VSS wasted with WAS.

$$\begin{aligned}\text{Quantity of total dry solids (TSS) in WAS} &= \text{Total TSS produced} - \text{TSS lost in the effluent}\\ &= (215 - 38)\ \text{kg TSS/d} = 177\,\text{kg TSS/d}\end{aligned}$$

$$\begin{aligned}\text{Quantity of VSS in WAS} &= P_x - \text{VSS lost in the effluent}\\ &= (172 - 30)\,\text{kg VSS/d} = 142\,\text{kg VSS/d}\end{aligned}$$

Calculate the WAS flow (Q_{wr}) at $\text{TSS}_{wr} = \text{TSS}_{ras} = 10{,}000\ \text{g/m}^3$.

$$Q_{wr} = \frac{\text{Total TSS produced}}{\text{TSS}_{wr}} = \frac{177\,\text{kg VSS/d}}{10{,}000\,\text{g/m}^3 \times 10^{-3}\ \text{kg/g}} = 18\,\text{m}^3/\text{d}$$

The effluent flow rate from the final clarifier, $Q_e = Q - Q_{wr} = (3800 - 18)\,\text{m}^3/\text{d} = 3782\,\text{m}^3/\text{d}$

Note: Q_e is ~99.5% of Q.

8. Calculate the flow rate of return activated sludge (RAS) (Q_r).
Calculate VSS concentration (X_r) in the RAS.

$$\begin{aligned}X_r &= \text{VSS/TSS ratio} \times \text{TSS}_{ras} = 0.8\,\text{mg VSS/mg TSS} \times 10{,}000\,\text{mg TSS/L}\\ &= 8000\,\text{mg VSS/L or 8000 g VSS/m}^3\end{aligned}$$

Calculate the return flow Q_r from Equation 10.5a.

$$Q_r = \frac{X}{X_r - X} \times Q = \frac{2000\,\text{mg VSS/L}}{8000\,\text{mg VSS/L} - 2000\,\text{mg VSS/L}} \times 3800\,\text{m}^3/\text{d} = 1267\,\text{m}^3/\text{d}$$

Calculate MLSS concentration in the aeration basin from Equation 10.14e.

$$\text{MLSS} = \frac{X}{\text{VSS/TSS ratio}} = \frac{2000\,\text{mg TSS/L}}{0.8\,\text{mg VSS/mg TSS}} = 2500\,\text{mg TSS/L or } 2500\,\text{g TSS/m}^3$$

Calculate the return flow ratio R_{ras} from Equation 10.5e.

$$R_{ras} = \frac{\text{MLSS}}{\text{TSS}_{ras} - \text{MLSS}} = \frac{2500\,\text{mg TSS/L}}{10{,}000\,\text{mg TSS/L} - 2500\,\text{mg TSS/L}} = 0.33$$

A mass balance around the biological system is illustrated in Figure 10.14.

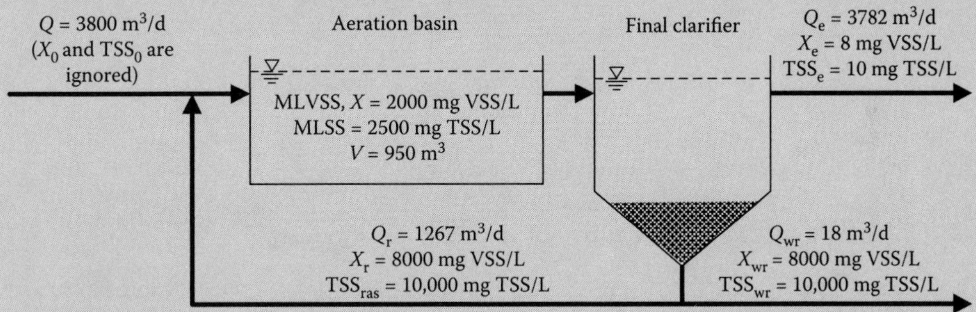

FIGURE 10.14 Mass balance sketch (Example 10.36).

9. Calculate the food to microorganism (F/M) ratio from Equation 10.14b.

$$\text{F/M} = \frac{S_0}{\theta X} = \frac{150\,\text{mg BOD}_5\text{/L} \times 24\,\text{h/d}}{6\,\text{h} \times 2000\,\text{mg VSS/L}} = 0.30\,\text{mg BOD}_5\text{/mg VSS·d}$$

10. Calculate the mass organic loading (MOL) to the aeration basin from Equation 10.14d.

$$MOL = \frac{E}{100\%} \times \text{F/M} = \frac{97.5\%}{100\%} \times 0.30\text{ mg BOD}_5\text{/mg VSS·d} = 0.29\text{ mg BOD}_5\text{/mg VSS·d}$$

11. Calculate the volumetric organic loading (VOL) in the aeration basin.

Since the soluble BOD reaching the aeration basin from Q_r is very small, it is usually ignored when the VOL is calculated from Equation 10.14c.

$$VOL = \frac{S_0 - S}{\theta} = \frac{(150 - 3.7)\,\text{g BOD}_5\text{/m}^3 \times 10^{-3}\,\text{kg/g} \times 24\,\text{h/d}}{6\,\text{h}} = 0.59\,\text{kg BOD}_5\text{/m}^3\text{·d}$$

12. Calculate the theoretical oxygen required (O_2) from Equation 10.14f.

$$O_2 = \frac{Q(S_0 - S)}{\text{BOD}_5/\text{BOD}_L} - 1.42P_x$$

$$= \frac{3800\,\text{m}^3\text{/d} \times (150 - 3.7)\,\text{g BOD}_5\text{/m}^3 \times 10^{-3}\,\text{kg/g}}{0.68\,\text{g BOD}_5\text{/g BOD}_L} - 1.42\,\text{g BOD}_L\text{/g VSS} \times 172\,\text{kg VSS/d}$$

$$= (818 - 244)\,\text{kg BOD}_L\text{/d} = 574\,\text{kg BOD}_L\text{/d} \quad \text{or} \quad 574\,\text{kg O}_2\text{/d}$$

EXAMPLE 10.37: ESTIMATE THE EFFECT OF TEMPERATURE ON *S*

The temperature effect on the performance of biological treatment process is expressed by Equation 10.17a. Calculate the effluent COD concentration S at mixed liquor temperatures of 20°C and 25°C. The aeration period is 10 h, influent COD is 480 mg/L and MLVSS concentration is 2000 mg VSS/L. Equation 10.15j can be used to estimate S from a completely mixed aeration basin since S is relatively small in comparison with K_s. Assume the temperature correction coefficient $\theta_T = 1.047$ and the simplified specific growth rate constant $k''_{20} = 0.48$ L/mg VSS·h.

Solution

1. Write the expression for S with respect to S_0 when S is low.

   The effluent substrate concentration (S) is expressed by Equation 10.15j when S is relatively low.

$$S \approx \frac{S_0}{1 + k''\theta X}$$

2. Calculate the effluent COD concentration at 20°C (S_{20}) from Equation 10.15j and $k''_{20} = 0.48$ L/mg VSS·h.

$$S_{20} = \frac{480\text{ mg COD/L}}{1 + 0.48\text{ L/g VSS/L·h} \times 10\text{ h} \times 2500\text{ mg VSS/L} \times 10^{-3}\text{ g/mg}} = 37\text{ mg COD/L}$$

3. Calculate the simplified specific growth rate constant at 25°C (k''_{25}) from Equation 10.17a and $\theta_T = 1.047$.

$$k''_{25} = k''_{20}\theta_T^{(25-20)} = 0.48\text{ L/g VSS/L·h} \times 1.047^{(25-20)} = 0.48\text{ L/gVSS/L·h} \times 1.26$$
$$= 0.60\text{ L/g VSS/L·h}$$

4. Calculate the effluent COD concentration at 25°C (S_{25}) from Equation 10.15j and $k''_{25} = 0.60$ L/mg VSS·h.

$$S_{25} = \frac{480\text{ mg COD/L}}{1 + 0.60\text{ L/g VSS/L·h} \times 10\text{ h} \times 2500\text{ mg VSS/L} \times 10^{-3}\text{ g/mg}} = 30\text{ mg COD/L}$$

Note: The effluent quality improves at higher temperature.

10.3.4 Determination of Biological Kinetic Coefficients in a Continuous Flow Completely Mixed Reactor

In recent years, the design of biological wastewater treatment plants utilizes rational equations based on biological kinetic coefficients Y, k_d, k, and K_s. These coefficients greatly influence the biological treatment process. The numerical values of these coefficients depend upon the characteristics of wastewater, and environmental conditions such as temperature, nutrients, mixing intensity, and air supply. Therefore, these coefficients must be developed from bench or pilot-plant studies for most wastewaters especially if they contain notable industrial wastes. The procedure to conduct bench-scale study may be found in many references.[1–3,9,16,17] Brief discussion on bench-scale study to determine biological kinetic coefficients is given below.

Reactor Design Details: There are many reactor designs that are used for continuous flow bench-scale studies. One such design is described in this section. The reactor assembly has aeration and settling zones separated by an adjustable sliding baffle, constant-speed influent feed pump, effluent piping, and influent and effluent collection containers. The volume of aeration zone in which active biomass carries out substrate stabilization ranges from 60% to 80% of the total reactor assembly (settling plus aeration zones). The design details of the biological reactor are given in Figure 10.15. The adjustable sliding baffle is used to adjust a desired opening at the base so that the settled solids may move from the settling zone back to the aeration zone.[1,2,9,16] The diffuser location is such that the settled solids are circulated into the aeration zone.

Ideally, four (preferably five) identical reactor assembly consisting of reactors, diffused aerators and pumps should be used. The feed tubes should draw the influent sample from a common feed reservoir. Each reactor assembly should be operated at different target MLSS concentrations. This will give variable SRT (θ_c), F/M ratio, and specific substrate utilization rate (U). Authors' experience is that with such an assembly, the experimental program can be completed in about 2–3 weeks, although intense daily work load will be required to operate the reactors and conduct the analytical work. Also, large volume of feed will be needed.

Alternative to a multireactor assembly is a one-reactor assembly. However, the experimental duration will be quite long. The data will be collected at different target MLSS concentrations, and may take up to 2 weeks at each setting. Authors suggest changing the target MLSS concentration from higher to lower

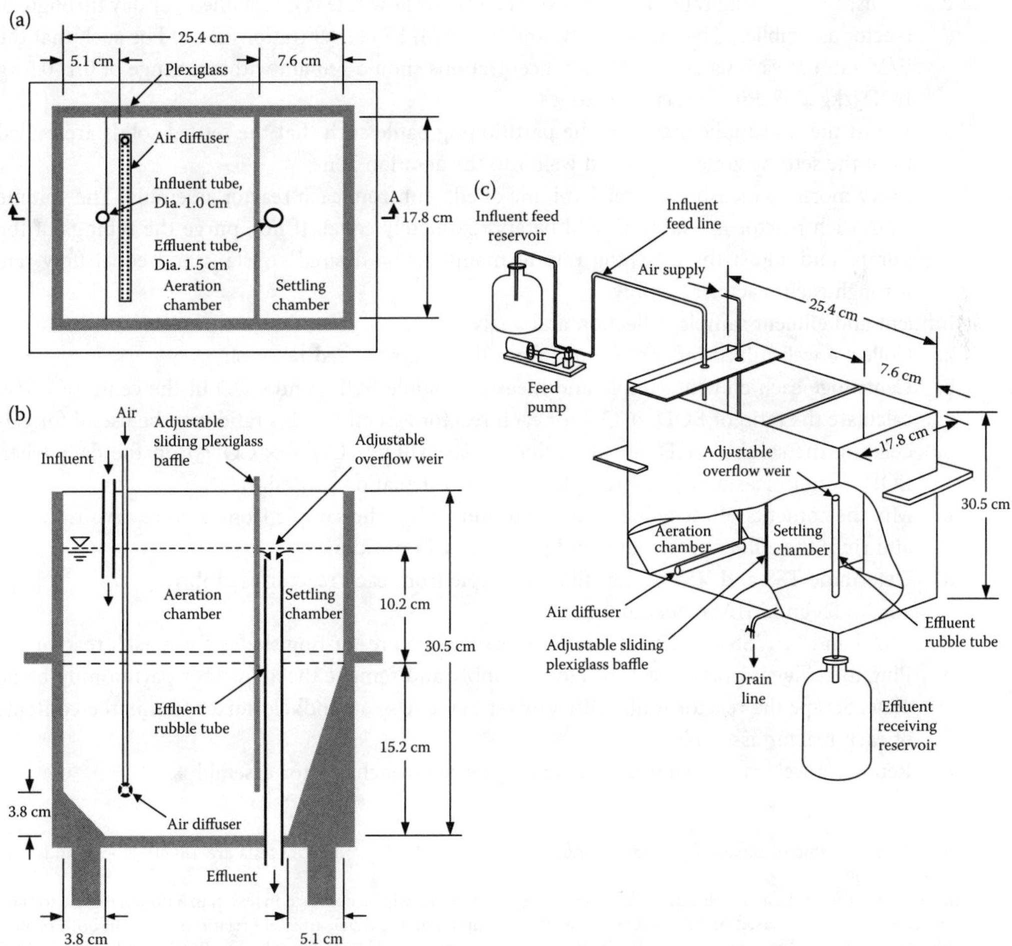

FIGURE 10.15 Design details of a one-reactor continuous flow complete-mixed biological reactor assembly: (a) plan, (b) Section AA, and (c) isometric view.

values. This gives a smoother transition from higher to lower target MLSS concentration and provides steady-state operation in a relatively shorter time. The data collection period will also be reduced at each target MLSS setting.

Experimental Procedure: Following is a detailed procedure for developing kinetic coefficients from a four-reactor assembly.

1. Installation of reactor assembly.

   Install four identical reactor assemblies with separate feed pumps and effluent reservoirs, and a common influent reservoir. Ideally a variable speed pump with four identical rotor heads arranged in parallel over a common shaft with identical tubings provides a constant and equal inflow into each reactor assembly. This arrangement gives ideal parallel operation.
2. Reactor operation.
   a. Estimate the volumetric fraction factor ($f_{aeration}$) of the aeration zone with respect to the entire volume of the reactor assembly. This factor is usually in the range of 0.6–0.8.
   b. Fill the reactor assemblies with the mixed liquor obtained from a local activated sludge plant to achieve the target MLSS concentration in the reactors. The MLSS concentration in each reactor assembly is calculated from the target MLSS concentration in the aeration zone and the volumetric fraction factor ($MLSS_{reactor} = f_{aeration} \times MLSS_{target}$).
   c. Adjust the pumping rate so that a desired constant flow rate is maintained per day through all reactor assemblies. The influent flow and target MLSS concentrations should be such that the F/M ratio at various target MLSS concentrations should remain within a range of 0.1–0.5 kg BOD_5/kg VSS·d in the aeration zones.
   d. Adjust the air supply and open the partitioning baffle such that the settled solids are pulled from the settling zone and mixed well into the aeration zone.
   e. Every morning measure the total volume of effluent from each reactor assembly. The volume from each reactor assembly should be approximately equal. If not, purge the tubings of the pumps and adjust the pumping rate to maintain the desired constant and equal flow rate through each reactor assembly.
3. Influent and effluent sample collection and analyses.*
   a. Collect a well-mixed influent sample from the common feed reservoir.
   b. Centrifuge each effluent sample and measure soluble BOD_5 and COD in the centrate.† Also calculate the ratio of BOD_5/COD for each reactor assembly. This ratio may be useful for calculating the soluble BOD_5 in the influent ($S_0 = (BOD_5/COD) \times COD_0$) on the days when BOD_5 is not measured (see Example 10.39 for calculation procedures).
   c. Mix the contents of the effluent reservoir and collect the sample from each reservoir. Empty and rinse the effluent reservoirs and place them in position.
   d. Determine TSS and VSS in the effluent sample from each reactor assembly.
4. Reactor MLSS and MLVSS measurement.
   a. MLSS and MLVSS concentrations are measured before wasting sludge from each reactor.
   b. Plug the effluent port of each reactor assembly and remove the adjustable partitioning baffle plate. Scrape the reactor walls with a mixer blade. Use a handheld mixer to mix the contents of each reactor assembly.‡
   c. Remove a well-mixed sample of mixed liquor from each reactor assembly.

---

* Normally the volume of mixed liquor sample needed for solids, BOD_5, and COD tests are 15–40, 20–50, and 5–20 mL, respectively.

† The BOD_5 results will be available after 5 days, while the COD results will be obtained in less than 3 hours. It is customary to operate the reactors based on the COD data, while develop a reliable BOD_5 to COD relationship from both results. Afterwards, only COD tests are run daily while BOD_5 tests are conducted less frequently. The BOD_5 results are obtained from the relationship thus developed. This procedure is presented in Examples 5.31 and 5.32.

‡ Additionally, Steps 4(b) and (d) should be performed several times a day to assure that the settled solids are circulated and MLSS in the reactor assembly is well mixed.

d. Place the partitioning baffle back. Wait for 15 min prior to removing the plug from the effluent port to resume normal operation of the reactor (see footnote).
e. Mix each mixed liquor sample and determine the MLSS and MLVSS concentrations in each reactor assembly. Convert the MLVSS measured in each reactor assembly to equivalent MLVSS concentration expected in the aeration zone (X_{aeration}). Also calculate the ratio of MLVSS/MLSS for each reactor assembly. This ratio may be used to calculate (i) the target MLVSS concentrations in the aeration zones (X_{target}), and (ii) the MLVSS concentrations in the reactor assemblies (X_{measured}) on the days when MLVSS is not measured (see Example 10.38).
f. Return any leftover of mixed liquor back into the respective reactor assembly after the lab tests of the sample are completed and data are analyzed.

5. Daily net increase in MLVSS concentration.

   The daily net increase in MLVSS concentration (ΔX_{net}) includes two parts: (a) the increase in MLVSS concentration measured directly in the aeration zone ($\Delta X_{\text{increase}}$), and (b) the equivalent MLVSS concentration (ΔX_{lost}) due to VSS lost in the effluent (see Example 10.38).
6. Waste excess sludge at the end of each day.
   a. Based on the net increase in MLVSS concentration from Step 5, calculate the volume of excess mixed liquor to be wasted from each reactor assembly to maintain the target MLSS concentration in the respective aeration zone (see Example 10.38).
   b. Plug the effluent port of each reactor assembly and remove the adjustable partitioning baffle plate. Scrape the reactor walls with a mixer blade. Use a handheld mixer to mix the contents of the reactor assembly.
   c. Remove the calculated volume of excess mixed liquor from each reactor assembly.
   d. Place the partitioning baffle back. Wait for 15 min prior to removing the plug from the effluent port to resume normal operation of the reactor assembly.
7. Reach the steady-state condition.

   Continue the reactor operation until the steady-state condition is reached. The steady-state condition is reached when daily operation data show: (a) a constant daily percent substrate removal, and (b) a constant daily growth of biomass (sludge). Normally it takes 3–7 days to reach the steady-state operation.
8. Data collection.

   After reaching the steady-state conditions in all reactor assemblies, the data collection is started. All reactors are then operated for 4–7 days to obtain sufficient data points. Average the values recorded during the steady-state operation in each reactor assembly. Use the average value of the data points obtained from each reactor assembly for determination of the kinetic coefficients.

Experimental Data Summary: Develop a summary table to include the experimental data collected under the steady-state operation from each reactor assembly in a respective row. The table should contain the following columns: reactor number; target MLSS concentration maintained in the aeration zone; measured MLVSS concentration in the reactor assembly; MLVSS/MLSS ratio; influent soluble COD and BOD_5, and BOD_5/COD ratio; effluent VSS, soluble COD and BOD_5, and BOD_5/COD ratio; and total daily effluent volume and measured MLVSS concentration prior to wasting sludge at the end of each day (see Example 10.39 and Table 10.11). Also, organize the influent COD, BOD_5, and BOD_5/COD ratio (see Example 10.39).

Procedure to Determine Kinetic Coefficients from Experimental Data: The experimental data collected under the steady-state condition from all reactors is used to develop the kinetic coefficients (see Example 10.39, Tables 10.11 and 10.12, and Figures 10.16 and 10.17 for detailed procedures). The general procedure is given below:

1. Calculate solids retention time (θ_c) in each reactor.

TABLE 10.10 Typical Values of Kinetic Coefficients for Activated Sludge Process

Coefficient	Unit	Values	
		Range	Typical
Y	mg VSS/mg BOD_5	0.3–0.7	0.5
	mg VSS/mg COD	0.2–0.5	0.45
k_d	d^{-1}	0.03–0.07	0.05
μ_{max}	mg VSS/mg VSS	–	2.5
k	mg BOD_5/mg VSS·d	2–8	5
	mg COD/mg VSS·d	5–20	12
K_s	mg BOD_5/L	40–120	60
	mg COD /L	20–80	40

θ_c is obtained by dividing the active biomass in the aeration zone by net mass of solids lost in the effluent and wasted as excess sludge per day. It is calculated from Equation 10.15a.

2. Calculate specific substrate utilization rate (U) in each reactor.

   The specific substrate utilization rate U is the substrate consumption rate per unit mass of the biomass. It is calculated from Equation 10.14d.

3. Plot $1/\theta_c$ versus U using the data from all four reactors.

   From Equation 10.15d, the plot of $1/\theta_c$ versus U gives a straight line (Figures 10.16a and 10.17a). The slope of the straight line is the value of Y and the intercept is k_d.

4. Plot $1/U$ versus $1/S$ using the data from all four reactors.

   From Equation 10.12e, the plot of $1/U$ versus $1/S$ also gives a linear relationship (Figures 10.16b and 10.17b). The slope of the straight line is K_s/k, and the intercept is $1/k$.

5. The range and typical values of the kinetic coefficients.

   The range and typical values of Y, k_d, k, and K_s for activated sludge process are summarized on Table 10.10. It may be noted that the values of kinetic coefficients are different for BOD_5 and COD.

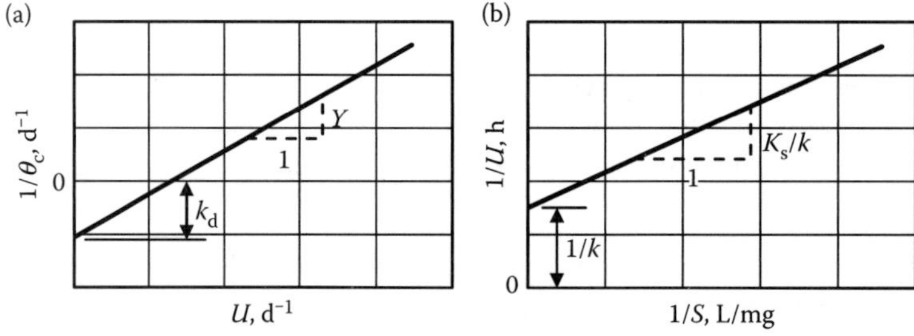

FIGURE 10.16 Procedure to determine the biological kinetic coefficients: (a) $1/\theta_c$ versus U to give Y and k_d, and (b) $1/U$ versus $1/S$ to give k and K_s.

EXAMPLE 10.38: NET BIOMASS GROWTH AND SLUDGE WASTING

The total volume of a reactor assembly is 15.2 L, and the average effluent volume collected from the reactor is 38 L/ day. The average concentration of VSS in the effluent is 6 mg/L, and the measured MLVSS concentration in the reactor assembly prior to wasting sludge is 1345 mg/L. Calculate: (1) volume of MLSS wasted from the reactor assembly each day to maintain the average operating MLSS concentration of 2250 mg/L in the aeration zone, and (2) net increase in MLVSS concentration in the aeration zone per day. Assume the average MLVSS/MLSS ratio is 0.77 and the volume of aeration zone is 75% of the reactor assembly.

Solution

1. Calculate the target MLVSS concentration in the aeration zone.

$$X_{target} = \text{MLVSS/MLSS ratio} \times \text{MLSS} = 0.77\,\text{mg VSS/mg TSS} \times 2250\,\text{mg TSS/L} = 1733\,\text{mg VSS/L}$$

2. Calculate the MLVSS concentration increase ($\Delta X_{increase}$) in the aeration zone after 1 day.

   Convert the MLVSS concentration measured in the entire volume of the reactor assembly into the MLVSS concentration expected in the aeration zone using the aeration chamber fraction factor $f_{aeration} = 0.75$.

$$X_{aeration} = \frac{X_{measured}}{f_{aeration}} = \frac{1345\,\text{mg VSS/L}}{0.75} = 1793\,\text{mg VSS/L}$$

$$\Delta X_{increase} = \frac{X_{aeration} - X_{target}}{t} = \frac{(1793 - 1733)\,\text{mg VSS/L}}{1\text{d}} = 60\,\text{mg VSS/L·d}$$

3. Calculate the mixed liquor to be wasted (V_{was}) to maintain the operating MLSS in the aeration zone.

   The mass of VSS wasted from the reactor assembly.

$$M_{was} = \Delta X_{increase} \times (f_{aeration} \times V_{reactor}) = 60\,\text{mg VSS/L·d} \times (0.75 \times 15.2\text{L}) = 684\,\text{mg VSS/d}$$

   The volume of sludge wasted from the reactor at the end of the day.

$$V_{was} = \frac{M_{was}}{X_{measured}} = \frac{684\,\text{mg VSS/d}}{1345\,\text{mg VSS/L}} = 0.51\,\text{L/d} \quad \text{or} \quad 510\,\text{mL/d}$$

4. Calculate the equivalent MLVSS concentration (ΔX_{lost}) in the aeration zone due to VSS lost in the effluent during 1 day of operation.

   Calculate the mass of VSS lost in the effluent per day.

$$M_{lost} = X_e \times V_{effluent} = 6\,\text{mg VSS/L} \times 38\,\text{L/d} = 228\,\text{mg VSS/d}$$

   Calculate the equivalent X_{lost} in the reactor.

$$\Delta X_{lost} = \frac{M_{lost}}{f_{aeration} \times V_{reactor}} = \frac{228\,\text{mg VSS/d}}{0.75 \times 15.2\text{L}} = 20\,\text{mg VSS/L·d}$$

5. Calculate the net increase in MLVSS concentration (ΔX_{net}) in the aeration zone after one day of operation.

$$\Delta X_{net} = \Delta X_{increase} + \Delta X_{lost} = (60 + 20)\,\text{mg VSS/L·d} = 80\,\text{mg VSS/L·d}$$

   Note: If solids lost in the effluent and sludge wasted from the reactor were both contained in the aeration zone, the MLVSS concentration expected in the aeration zone after one day of operation can be calculated from the following two approaches:

$$X_{total} = X_{target} + \Delta X_{net} \times t = 1733\,\text{mg/L} + 80\,\text{mg/L·d} \times 1\,\text{d} = 1813\,\text{mg VSS/L}, \quad \text{or}$$

$$X_{total} = X_{aeration} + \Delta X_{lost} \times t = 1793\,\text{mg/L} + 20\,\text{mg/L·d} \times 1\,\text{d} = 1813\,\text{mg VSS/L}.$$

EXAMPLE 10.39: DETERMINATION OF KINETIC COEFFICIENTS FROM CONTINUOUS FLOW COMPLETELY MIXED BENCH-SCALE REACTOR STUDY

Four completely mixed bench-scale reactor assemblies were operated. The reactor assemblies were identical in design and operation. The only difference was in the concentration of mixed liquor suspended solids (MLSS) that was maintained in each reactor. When the reactors reached the steady-state condition, the data collection began. Each day the following measurements were made:

a. Soluble COD and BOD_5 concentrations in the common feed reservoir,
b. MLVSS and MLSS concentrations in each reactor assembly after the day of operation,
c. VSS, and soluble COD and BOD_5 concentrations in the effluent from each reactor, and
d. Total volume of effluent from each reactor during the day.

The 7-day average experimental results at four target MLSS concentrations are summarized in Table 10.11. The average soluble COD concentration in the common feed reservoir (COD_0) is 151 mg/L with an average soluble BOD_5/COD ratio of 0.71. The total volume of each reactor assembly ($V_{reactor}$) including the section that is partitioned for settling is 13.6 L with a volumetric fraction factor ($f_{aeration}$) of 0.8 for the aeration zone. Determine the biological kinetic coefficients Y, k_d, k, and K_s based on experimental results.

TABLE 10.11 Summary of Experimental Data of Example 10.39

Reactor No.	Mixed Liquor			Effluent			
	MLSS[a], mg TSS/L	$X_{measured}$[b], mg VSS/L	MLVSS/MLSS ratio[c]	Xe, mg/L	CODe, mg/L	BOD_5/COD ratio	$V_{effluent}$[d], L
1	2450	1458	0.73	4.2	20.2	0.20	37.0
2	1700	1034	0.73	5.0	22.5	0.27	37.0
3	1250	780	0.72	6.2	29.0	0.28	36.4
4	1000	646	0.74	7.4	30.6	0.30	35.6

[a] Target MLSS concentration in each aeration zone during 7-day operation under steady-state condition.
[b] Average MLVSS concentration measured in each reactor assembly (aeration plus settling zones) at the end of each day.
[c] The MLVSS/MLSS ratio is calculated from the MLVSS and MLSS concentrations measured in each reactor assembly.
[d] Average effluent volume collected at the end of each day from each reactor.

Solution

1. Calculate θ_c and $1/\theta_c$ from the experimental data.

   Calculate the average BOD_5 concentrations in the common feed reservoir.

$$S_0 = (BOD_5/COD\ ratio) \times COD_0 = (0.71\ mg\ BOD_5/mg\ COD) \times 151\ mg\ COD/L = 107\ mg\ BOD_5/L$$

   The calculated values of the targeted MLVSS concentration in the aeration zone (X_{target}), expected MLVSS concentration in the aeration zone ($X_{aeration}$), MLVSS concentration increase in the aeration zone ($\Delta X_{increase}$), the equivalent MLVSS concentration in the aeration zone due to VSS lost in the effluent (ΔX_{lost}), the net MLVSS concentration increase in the aeration zone (ΔX_{net}), solids retention time (θ_c), and $1/\theta_c$ are summarized for each reactor in Table 10.12. The hydraulic retention time (θ) for all reactors are also calculated and included in Table 10.12.
2. Calculate U and 1/U from the experimental data.

TABLE 10.12 Calculation Results for Determination of θ_c and $1/\theta_c$ (Example 10.39)

Reactor No.	X_{target}, mg VSS/L	$X_{aeration}$, mg VSS/L	$\Delta X_{increase}$, mg VSS/L	ΔX_{lost}, mg VSS/L	ΔX_{net}, mg VSS/L	θ_c, d	$1/\theta_c$, d^{-1}	θ, d
1	1789[a]	1823[b]	34[c]	14.3[d]	48.3[e]	37[f]	0.027[g]	7.1[h]
2	1241	1293	52	17.0	69.0	18	0.056	7.1
3	900	975	75	20.7	95.7	9.4	0.106	7.2
4	740	808	68	24.2	92.2	8.0	0.125	7.3

[a] $X_{desired} = (\text{MLVSS/MLSS}) \times \text{MLSS} = 0.73\,\text{mg VSS/mg TSS} \times 2450\,\text{mg TSS/L} = 1789\,\text{mg VSS/L}$

[b] $$X_{aeration} = \frac{X_{measured}}{f_{aeration}} = \frac{1458\,\text{mg VSS/L}}{0.8} = 1823\,\text{mg VSS/L}$$

[c] $$\Delta X_{increase} = \frac{X_{aeration} - X_{target}}{t} = \frac{(1823 - 1789)\,\text{mg VSS/L}}{1\text{d}} = 34\ \text{mg VSS/L·d}$$

[d] $$\Delta X_{lost} = \frac{X_e \times V_{effluent}}{f_{aeration} \times V_{reactorassembly}} = \frac{4.2\,\text{mg VSS/L} \times 37\text{L/d}}{0.8 \times 13.6\text{L}} = 14.3\,\text{mg VSS/L·d}$$

[e] $\Delta X_{net} = \Delta X_{increase} + \Delta X_{lost} = (34 + 14.3)\,\text{mg VSS/L·d} = 48.3\ \text{mg VSS/L·d}$

[f] $$\theta_c = \frac{VX_{target}}{VX_{net}} = \frac{X_{target}}{\Delta X_{net}} = \frac{1789\,\text{mg VSS/L}}{48.3\,\text{mg VSS/L·d}} = 37\,\text{d}$$

[g] $$\frac{1}{\theta_c} = \frac{1}{37.0\text{d}} = 0.027\text{d}^{-1}$$

[h] $$\theta = \frac{f_{aeration} \times V_{reactor}}{V_{effluent}} = \frac{0.8 \times 13.6\,\text{L}}{37\,\text{L/d}} \times \frac{24\,\text{h}}{\text{d}} = 7.1\,\text{h}$$

The calculated values of the BOD_5 concentration in the effluent (S), reduction of BOD_5 concentration in the reactor (ΔS), specific substrate utilization rate (U), and $1/U$ are summarized for each reactor in Table 10.13. The F/M ratios for all reactors are also calculated and included in Table 10.13.

TABLE 10.13 Calculation Results for Determination of U, $1/U$, and $1/S$ (Example 10.39)

Reactor No.	S, mg BOD_5/L	$\Delta S = S_0 - S$, mg BOD_5/L	U, mg BOD_5/mg VSS·d	$1/U$, d^{-1}	$1/S$, L/mg BOD_5	F/M Ratio, kg BOD_5/kg VSS·d
1	4.04[a]	103[b]	0.195[c]	5.1[d]	0.25[e]	0.20[f]
2	6.08	101	0.275	3.6	0.16	0.29
3	8.12	99	0.367	2.7	0.12	0.40
4	9.18	98	0.436	2.3	0.11	0.47

[a] $S = BOD_5/COD \times COD_e = 0.20\,\text{mg}\ BOD_5/\text{mg COD} \times 20.2\,\text{mg COD/L} = 4.04\,\text{mg}\ BOD_5/\text{L}$

[b] $\Delta S = S_0 - S = (107 - 4.04)\,\text{mg}\ BOD_5/\text{L} = 103\,\text{mg}\ BOD_5/\text{L}$

[c] $$U = \frac{S_0 - S}{\theta X_{desired}} = \frac{\Delta S}{\theta X_{desired}} = \frac{103\,\text{mg}\,BOD_5/\text{L}}{7.1\,\text{h} \times 1789\,\text{mg VSS/L}} \times \frac{24\text{h}}{\text{d}} = 0.195\,\text{d}^{-1}$$

[d] $$1/U = \frac{1}{0.195\ \text{d}^{-1}} = 5.1\text{d}$$

[e] $$1/S = \frac{1}{4.04\,\text{mg}\ BOD_5/\text{L}} = 0.25\text{L/mg}\ BOD_5$$

[f] $$\text{F/M ratio} = \frac{S_0}{\theta X_{desired}} = \frac{107\,\text{mg}\ BOD_5/\text{L}}{7.1\,\text{h} \times 1789\,\text{mg VSS/L}} \times \frac{24\,\text{h}}{\text{d}} = 0.20\,\text{kg}\ BOD_5/\text{kg VSS·d}$$

Note: Due to the changes in MLVSS concentration, different θ_c values as well as F/M ratios are achieved in four aeration zones at a constant influent BOD_5 concentration and near-constant θ. The experimental F/M ratios cover the typical range of 0.1–0.5 kg BOD_5/kg VSS·d.

3. Plot the graphical relationships.

   The plot of $1/\theta_c$ versus U, and $1/U$ versus $1/S$ values are shown in Figures 10.17a and b, respectively.

4. Determine Y and k_d.

   The relationship between $1/\theta_c$ and U (Figure 10.17a) is expressed by Equation 10.15d. The slope of this linear relationship is $Y = 0.42$ mg VSS/mg BOD_5. At $U = 0$, the intercept is the negative value of endogenous decay coefficient $-k_d = -0.057\ d^{-1}$. It gives $k_d = 0.057\ d^{-1}$.

5. Determine k and K_s.

   The relationship between $1/U$ and $1/S$ (Figure 10.17b) is expressed by Equation 10.12e. The intercept of the line is 0.33 d that is $1/k$.

$$k = \frac{1}{0.33\ \text{d}} = 3.0\ \text{d}^{-1} \quad \text{or} \quad 3.0\ \text{mg BOD}_5/\text{mg VSS}\cdot\text{d}$$

   The slope of this linear relationship is 19.3 mg·d/L that is K_s/k.

$$K_s = 19.3\ \text{mg}\cdot\text{d/L} \times 3.0\ \text{d}^{-1} = 58\ \text{mg/L} \quad \text{or} \quad 58\ \text{mg BOD}_5/\text{L}$$

6. Calculate μ_{max}.

   The maximum specific growth rate μ_{max} is calculated from Equation 10.12b at $Y = 0.42$ mg VSS/mg BOD_5 and $k = 3.0$ mg BOD_5/mg VSS·d.

$$\mu_{max} = Yk = 0.42\ \text{mg VSS/mg BOD}_5 \times 3.0\ \text{mg BOD}_5/\text{mg VSS}\cdot\text{d}^{-1} = 1.3\ \text{mg VSS/mg VSS}\cdot\text{d}$$

7. Summarize the biological kinetic coefficients from the completely mixed bench-scale reactor study.

   The biological kinetic coefficients that are developed from the experimental data are summarized below:

   $Y = 0.42$ mg VSS/mg BOD_5, $k_d = 0.057\ d^{-1}$, $k = 3.0$ mg BOD_5/mg VSS·d, $K_s = 58$ mg BOD_5/L, and $\mu_{max} = 1.3$ mg VSS/mg VSS·d.

   Note: See Examples (10.18) and (10.30) also for procedures to determine the kinetic coefficients of Y, k_d, k, K_s, and μ_{max} from the batch reactor experimental data.

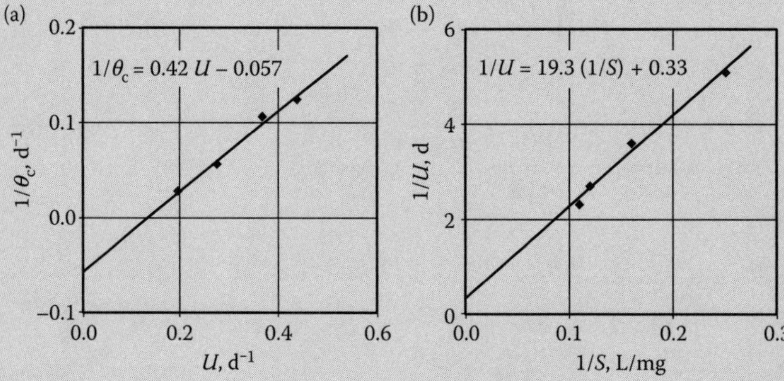

FIGURE 10.17 Determination of kinetic coefficients from experimental data: (a) $1/\theta_c$ versus U to determine Y and k_d, and (b) $1/U$ versus $1/S$ to determine K_s and k (Example 10.39).

10.3.5 Continuous Plug-Flow Reactor

True PFR: In a true plug-flow reactor (PFR), all fluid elements exit in the same sequence as they enter. The travel time is equal to the theoretical hydraulic detention time, and there is no longitudinal mixing. In-depth discussions on plug-flow reactor regime in comparison with continuous flow stirred-tank reactor (CFSTR), and many solved examples have been presented in Chapter 3. Readers may review Sections 3.4 and 3.5, and Examples 3.14 through 3.51 for background information about different types of reactors.

True plug-flow reactors are difficult to obtain because of the presence of some longitudinal dispersion. In a biological reactor because of sludge recycle some particles may make many passes through the reactor, making the mathematical modeling for the plug-flow reactor difficult. The most accepted and useful kinetic model of the plug-flow reactor for the activated sludge process was proposed by Lawrence and McCarty.[18,19] This model is based on two simplifying assumptions:

1. The concentration of microorganisms X in the influent and effluent of the reactor are approximately the same. This assumption is valid only if $\theta_c/\theta > 5$, and
2. Equation 10.12c applies to substrate utilization rate r_{su} as the wastewater flows through the aeration basin. Integration of Equation 10.12c over the aeration period and substituting Equation 10.15k for X and simplification give Equation 10.28a. This equation gives solids residence time in the plug-flow activated sludge process. The influent substrate concentration to the reactor after mixing with the return sludge is given by Equation 10.28b. If R_{ras} approaches 0 (or $R_{ras} \ll 1$ and $S_i \approx S_0$), Equation 10.28a reduces to Equation 10.28c.

$$\frac{1}{\theta_c} = \frac{Yk(S_0 - S)}{(S_0 - S) + (1 + R_{ras})K_s \ln(S_i/S)} \tag{10.28a}$$

$$S_i = \frac{S_0 + R_{ras}S}{1 + R_{ras}} \tag{10.28b}$$

$$\frac{1}{\theta_c} = \frac{Yk(S_0 - S)}{(S_0 - S) + K_s \ln(S_0/S)} \tag{10.28c}$$

where

S_0 = influent substrate concentration, mg/L
S = effluent substrate concentration, mg/L
S_i = influent substrate concentration to the reactor after mixing with the return sludge, mg/L
R_{ras} = RAS or recycle flow ratio (Equation 10.5b or 10.5e), dimensionless
All other terms are defined earlier.

Equation 10.28c is applicable to a PFR, similar to Equation 10.15d or 10.16a developed for a completely mixed (or CFSTR) biological reactor with or without recycle.

Dispersed PFR: Most biological reactors exhibit some degree of partial mixing between plug-flow and completely mixed conditions because of the presence of longitudinal or axial mixing. The longitudinal mixing is characterized by the dispersion number d (Equation 3.41). The ranges of dispersion number applying to the types of biological reactors are summarized in Table 10.14.

TABLE 10.14 Ranges of Dispersion Number for Different Types of Biological Reactors

Dispersion Number (*d*)	Type of Reactor
0–0.2	Plug-flow reactor (PFR) such as a long narrow tank or flow in a pipe (see Section 3.4.3, and Examples 3.24 through 3.27)
0.2–4	Dispersed plug-flow reactor such as a conventional aeration basin
4–∞	Completely mixed reactor such as a CFSTR (see Section 3.4.2, and Examples 3.16 through 3.23)

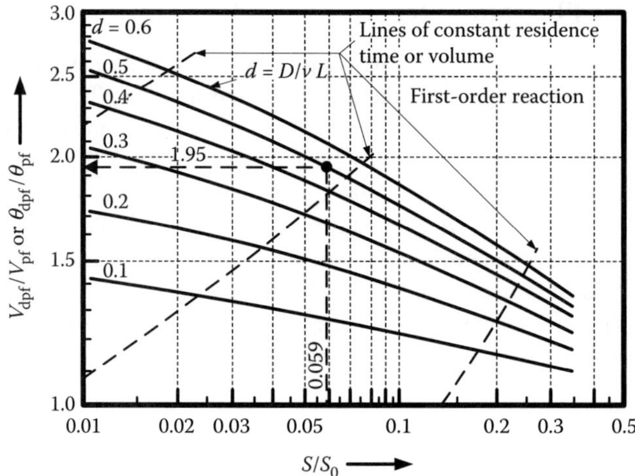

FIGURE 10.18 Performance of ideal plug-flow and dispersed plug-flow reactors achieving first-order reaction. *Note:* 1. Subscripts pf and dpf are for PFR and dispersed PFR, respectively. 2. S_0 and S are for influent and effluent concentrations, respectively.

As expressed in Equation 3.41, the dispersion number can be determined from the longitudinal axial dispersion coefficient. In a biological reactor, the mixing is created by diffused aeration or mechanical mixer. For a spiral flow rectangular basin, Murphy and Boyko proposed Equation 10.29 for dispersion coefficient as a function of basin width and air supply.[20]

$$D = CW^2(Q_a)^{0.346} \tag{10.29}$$

where

D = dispersion coefficient, m^2/h (ft^2/h)
C = constant, h^{-1}. $C = 3.118\ h^{-1}$ for both SI and U.S. customary units
W = width of the basin, m (ft)
Q_a = air flow under standard conditions (at 20°C and 1 atm.), $sm^3/min \cdot 1000\ m^3$ ($scfm/1000\ ft^3$). In a biological reactor, Q_a is normally 20–30 $sm^3/min \cdot 1000\ m^3$ (20–30 $scfm/1000\ ft^3$).

In a truly plug-flow biological reactor exhibiting a pseudo first-order reaction, the substrate removal is identical to a batch reactor. This relationship is given by Equation 10.9. However, the reaction time θ required in a dispersed PFR is increased to achieve the same level of treatment as in a true PFR. Levenspiel and Dischoff compared the performance of ideal and dispersed plug-flow reactors achieving first-order reaction. This relationship is given in Figure 10.18.[1,21,22]

EXAMPLE 10.40: PERFORMANCE COMPARISON OF BIOLOGICAL CFSTR AND PFR

Equations 10.15d and 10.28c express the kinetic performance in a biological CFSTR and PFR, respectively. Calculate the effluent substrate concentration S from a CFSTR and PFR, and compare the results. Use the following data: $Y = 0.6$, $k_d = 0.05\ d^{-1}$, $k = 4\ d^{-1}$, $K_s = 80$ mg/L, $\theta_c = 0.75$ d, and $S_0 = 250$ mg/L.

Solution

1. Determine S and the percent removal in a CFSTR.

a. Calculate S from Equation 10.15h.

$$S = \frac{K_s(1 + k_d\theta_c)}{\theta_c(Yk - k_d) - 1} = \frac{80\,\text{mg/L} \times (1 + 0.05\,\text{d}^{-1} \times 0.75\,\text{d})}{0.75\,\text{d} \times (0.6 \times 4\,\text{d}^{-1} - 0.05\,\text{d}^{-1}) - 1} = 109\,\text{mg/L}$$

b. Calculate the removal efficiency from Equation 10.14a.

$$E = \frac{S_0 - S}{S_0} \times 100\% = \frac{(250 - 109)\,\text{mg/L}}{250\,\text{mg/L}} \times 100\% = 56\%$$

2. Determine S and the percent removal in a PFR.
   a. Calculate S from Equation 10.28c.

$$\frac{1}{\theta_c} = \frac{Yk(S_0 - S)}{(S_0 - S) + K_s \ln(S_0/S)}$$

$$\frac{1}{0.75\,\text{d}} = \frac{0.6 \times 4\,\text{d}^{-1} \times (250 - S)\,\text{mg/L}}{(250\,\text{mg/L} - S) + 80\,\text{mg/L} \times \ln[(250\,\text{mg/L})/S]}$$

$$(250\,\text{mg/L} - S) + 80\,\text{mg/L} \times \ln[(250\,\text{mg/L})/S] = 1.8 \times (250 - S)\,\text{mg/L}$$

Solve for S by trial and error.

$$S = 26\,\text{mg/L}$$

   b. Calculate the removal efficiency from Equation 10.14a.

$$E = \frac{S_0 - S}{S_0} \times 100\% = \frac{(250 - 26)\,\text{mg/L}}{250\,\text{mg/L}} \times 100\% = 90\%$$

3. Compare the results.
   The results are summarized below. The PFR is significantly more efficient than the CFSTR.

Reactor Type	Effluent Concentration (S), mg/L	Percent Removal (E), %
CSTR	109	56
PFR	26	90

EXAMPLE 10.41: COMPARATIVE PERFORMANCE CURVES OF A BIOLOGICAL CFSTR AND PFR

Prepare the substrate concentration remaining in the effluent (S) and percent substrate removal (E) curves with respect to θ_c for a CFSTR and PFR. Compare the curves. Both reactors are treating the same wastewater. Use the following data: $Y = 0.6$, $k_d = 0.05\,\text{d}^{-1}$, $k = 4\,\text{d}^{-1}$, $K_s = 80\,\text{mg/L}$, and $S_0 = 250\,\text{mg/L}$. Assume $R_{ras} < 1$.

Solution

1. Develop the generalized equation for CFSTR and PFR, respectively, to express S with respect to θ_c.

Substitute the given data in Equations 10.15h and 10.28c to develop the equations between S and θ_c.

$$S = \frac{80\,\text{mg/L} \times (1 + 0.05\,\text{d}^{-1} \times \theta_c)}{\theta_c \times (0.6 \times 4\,\text{d}^{-1} - 0.05\,\text{d}^{-1}) - 1} = \frac{80\,\text{mg/L} + 4\,\text{mg/L·d} \times \theta_c}{\theta_c \times (2.35\,\text{d}^{-1}) - 1} \quad \text{for CFSTR}$$

$$\frac{1}{\theta_c} = \frac{0.6 \times 4\,\text{d}^{-1} \times (250 - S)\,\text{mg/L}}{(250\,\text{mg/L} - S) + 80\,\text{mg/L} \times \ln[(250\,\text{mg/L})/S]} \quad \text{for PFR } (R_{ras} < 1)$$

2. Develop data table to prepare comparative performance curves for CFSTR and PFR.

   Assume different values of θ_c and calculate S using the generalized equations obtained in Step 1 for CFSTR and PFR. Using Equation 10.14a, the removal efficiency (E) is then calculated for CFSTR and PFR, respectively. The calculation results are summarized below.

θ_c, d	CFSTR		PFR	
	S, mg/L	E, %	S, mg/L	E, %
0.00	250	0	250	0
0.55	250	0	250	0
0.60	201	20	127	49
0.65	157	37	71	72
0.70	128	49	43	83
0.75	109	56	26	90
0.85	84	67	11	96
1.0	62	75	3.3	99
1.2	47	81	0.71	100
1.5	34	86	0.07	100
2.0	24	90	0.00	100
3.0	15	94	0.00	100
6.0	8	97	0.00	100
10	5	98	0.00	100

3. Draw S and E versus θ_c curves for CFSTR and PFR.

   The substrate concentration remaining in the effluent and removal efficiency of CFSTR and PFR are shown in Figure 10.19.

4. Compare the curves for CFSTR and PFR.

   The plug-flow biological reactor is significantly more efficient than a complete-mix biological reactor. At θ_c of 1 d, the substrate removal efficiency of PFR is 99% while that of CFSTR is 75%.

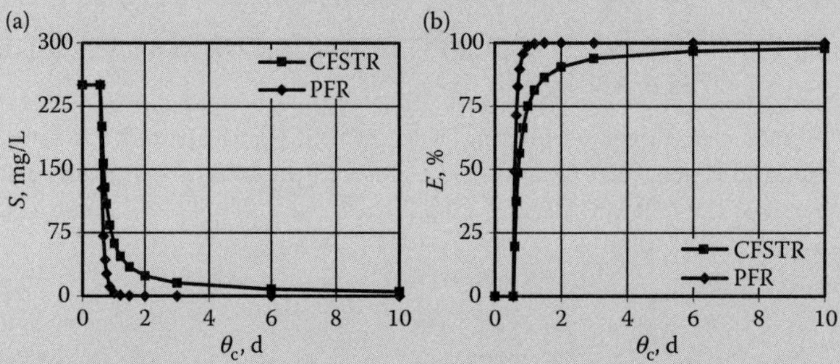

FIGURE 10.19 Comparative performance curves with respect to solids retention time (θ_c) for biological CFSTR and PFR: (a) effluent substrate concentration remaining (S), and (b) removal efficiency (Example 10.41).

EXAMPLE 10.42: COMPARISON BETWEEN TRUE AND DISPERSED PFRs

Compare the hydraulic retention time (θ) and reactor volume (V) of a true PFR and dispersed PFR. The design flow to the plant (Q) is 13,300 m^3/d (3.5 MGD) and the design MLVSS concentration (X) is 2000 mg/L in the reactors. The influent and effluent BOD_5 concentrations (S_0 and S) are 135 and 8 mg/L, respectively. Assume the reaction is pseudo first-order with the reaction rate constant k of 0.3 L/g VSS·h. The dispersion number d of dispersed plug-flow reactor is 0.5.

Solution

1. Calculate θ and V of a true PFR.

   Equation 10.9 applies to a batch reactor undergoing pseudo first-order reaction. The same equation applies to a true PFR.

$$S = S_0 e^{-kX\theta}$$

   Rearrange Equation 10.9 to obtain Equation 10.30.

$$\theta = \frac{1}{kX}\ln\left(\frac{S_0}{S}\right) \tag{10.30}$$

   Calculate θ_{pf} from Equation 10.30 at $k = 0.3$ L/g VSS·h and $X = 2000$ mg VSS/L $= 2$ g VSS/L.

$$\theta_{pf} = \frac{1}{0.3\,\text{L/g VSS·h} \times 2\,\text{g VSS/L}}\ln\left(\frac{8\,\text{mg BOD}_5\text{/L}}{135\,\text{mg BOD}_5\text{/L}}\right) = \frac{2.83}{0.6\,\text{h}^{-1}} = 4.7\,\text{h}$$

The volume of a true PFR at $Q = 13{,}300\,\text{m}^3/\text{d}$, $V_{pf} = \theta_{pf}Q = 4.7\,\text{h} \times 13{,}300\,\text{m}^3/\text{d} \times \frac{\text{d}}{24\,\text{h}} = 2605\,\text{m}^3$

2. Calculate θ and V of a dispersed PFR (or DPFR).

   The ratio of S to S_0, $\frac{S}{S_0} = \frac{8\,\text{mg BOD}_5\text{/L}}{135\,\text{mg BOD}_5\text{/L}} = 0.059$

   Obtain approximately θ_{dpf}/θ_{pf} or V_{dpf}/V_{pf} from the curve corresponding to $d = 0.5$ at $S/S_0 = 0.059$ on Figure 10.18.

$$\frac{\theta_{dpf}}{\theta_{pf}} = \frac{V_{dpf}}{V_{pf}} = 1.95$$

   Calculate θ_{dpf} and V_{dpf} using the above ratio.

$$\theta_{dpf} = \frac{\theta_{dpf}}{\theta_{pf}} \times \theta_{pf} = 1.95 \times 4.7\,\text{h} = 9.2\,\text{h}$$

$$V_{dpf} = \frac{V_{dpf}}{V_{pf}} \times \theta_{pf}Q = 1.95 \times 2605\,\text{m}^3 = 5080\,\text{m}^3$$

EXAMPLE 10.43: DISPERSION NUMBER OF A DISPERSED PLUG-FLOW AERATION BASIN

A long narrow path is created by baffles in aeration basin. The dimensions of the path are: total length = 75 m, width = 8 m, and depth = 5.5 m. The design flow is 10,000 m^3/d and air supply is 80 m^3/min. Determine the dispersion number and verify if it is valid for a dispersed PFR.

Solution

1. Calculate the aeration basin volume (V) and unit air supply to the aeration basin (Q_a).

$$V = LWH = 75\,\text{m} \times 8\,\text{m} \times 5.5\,\text{m} = 3300\,\text{m}^3$$

$$Q_a = \frac{q_a}{V} = \frac{80\,\text{m}^3/\text{min}}{3300\,\text{m}^3} \times \frac{1000\,\text{m}^3}{1000\,\text{m}^3} = 24.2\,\text{m}^3/\text{min}\cdot 1000\,\text{m}^3$$

2. Calculate the dispersion coefficient (D) from Equation 10.29.

$$D = CW^2(Q_a)^{0.346} = 3.118\,\text{h}^{-1} \times (8\,\text{m})^2 \times (24.2)^{0.346} = 601\,\text{m}^2/\text{h}$$

3. Calculate the dispersion number d.

$$\text{Fluid axial velocity } v,\ v = \frac{Q}{WH} = \frac{10{,}000\,\text{m}^3/\text{d}}{8\,\text{m} \times 5.5\,\text{m}} \times \frac{\text{d}}{24\,\text{h}} = 9.5\ \text{m/h}$$

$$\text{Value of } d \text{ from Equation 3.41, } d = \frac{D}{vL} = \frac{601\,\text{m}^2/\text{h}}{9.5\,\text{m/h} \times 75\,\text{m}} = 0.84$$

Note: The dispersion number d of 0.84 is within the typical range of 0.2–4 for a dispersed PFR.

10.3.6 Operational Parameters of Activated Sludge Process

The most important parameters of an activated sludge plant are oxygen or air supply, mixing, food to microorganism ratio (or organic loading), and return activated sludge flow. Additionally, some nontheoretical indexes such as sludge blanket, volume of settled sludge, sludge volume index (SVI), and sludge density index (SDI) are also valuable tools for daily operation and control of an activated sludge process.

Sludge Blanket: The depth of sludge blanket in the final clarifier indicates the depth of thickening zone, and the approximate mass of solids retained in the clarifier. Plant operator measures the depth of sludge blanket by a sampling device called *sludge judge*. It is designed to measure accurate depth of settled solids up to 5% concentration.

Volume of Settled Sludge: A quick method to determine the return sludge ratio (Q_r/Q) is to settle mixed liquor for 30 min in a 1-L graduated cylinder. The return sludge ratio R_{ras} is roughly determined from Equation 10.31.

$$R_{ras} = \frac{Q_r}{Q} = \frac{\text{SV}}{1000\,\text{mL/L} - \text{SV}} \tag{10.31}$$

where

SV = settled sludge volume, mL/L

R_{ra} = RAS or recycle flow ratio (Equation 10.5b or 10.5e), dimensionless

Sludge Volume Index: A more accurate method of determining RAS flow is based on SVI. The SVI is defined as the volume (mL) occupied by 1 g of settled sludge (Equation 10.32a). It is also defined as the ratio of percent volume occupied by settled sludge in a 1-L graduated cylinder in 30 min (SV_p) to the MLSS percent concentration ($MLSS_p$) (Equation 10.32b). SV_p and MLSSp are calculated by Equations 10.32c and 10.32d, respectively. The RAS ratio (R_{ras}) is related to SVI, MLSS percent concentration, and SV_p (Equation 10.32e). The maximum TSS concentration in RAS (TSS_{ras}^{max}) depends upon SVI, and occurs when the volume of settled mixed liquor in 1-L graduated cylinder in 30 min remains 1000 mL.

This relationship is expressed by Equation 10.32f.

$$SVI = \frac{\text{Volume occupied in mL}}{\text{One gram of settled sludge}} = \frac{SV \times 1000\,\text{mg/g}}{MLSS} \quad (10.32a)$$

$$SVI = \frac{SV_p}{MLSS_p \times 1\,\text{g/mL}} \quad (10.32b)$$

$$SV_p = \frac{SV}{1000\,\text{mL/L}} \times 100\% \quad (10.32c)$$

$$MLSS_p = \frac{MLSS}{1000\,\text{g/L} \times 1000\,\text{mg/g}} \times 100\% \quad \text{or} \quad MLSS_p = \frac{MLSS}{10^6\,\text{mg/L}} \times 100\% \quad (10.32d)$$

$$R_{ras} = \frac{1}{(100\%/(SVI \times MLSS_p \times 1\,\text{g/mL}) - 1} \quad \text{or} \quad R_{ras} = \frac{SV_p}{100\% - SV_p} \quad (10.32e)$$

$$TSS_{ras}^{max} = \frac{1000\,\text{mL/L} \times 1000\,\text{mg/g}}{SVI} = \frac{10^6(\text{mL/g})\cdot(\text{mg/L})}{SVI} \quad (10.32f)$$

Note: See Example 10.46 for derivation.

where

SVI = sludge volume index, mL/g
MLSS = mass concentration of MLSS, mg/L
SV_p = percent of settled sludge volume, percent by volume
$MLSS_p$ = percent concentration of MLSS, percent by weight with respect to water density (10^6 mg/L)
TSS_{ras}^{max} = maximum TSS concentration in RAS, mg/L
Note: 1000 g/L or 1 g/mL is the approximate density of water (sp.gr. = 1.0).

Other alternative methods for SVI tests are diluted SVI (DSVI) and stirred SVI test.[23,24]

SVI test has been traditionally used to determine the settling behavior of MLSS. It is an important factor in process control. It limits the MLSS concentration in the aeration basin and determines the return sludge flow rate. Typically, the value of SVI is 80–150 for the municipal wastewater treatment plants operating at MLSS concentrations of 2000–3500 mg/L. An SVI below 100 is an indication of excellent settling sludge. Poor settling is characterized by high SVI. At SVI values above 200, the mixed liquor does not settle well, fills up the clarifier with high risk of being washed out with the effluent.

Sludge Density Index: Similar to SVI, SDI is an operational parameter. It is the weight in gram of 1 mL of sludge after settling for 30 min. It is calculated from Equation 10.33.

$$SDI = \frac{1}{SVI \times 1\,\text{g/mL}} \times 100\% \quad \text{or} \quad SDI = \frac{100}{SVI \times 1\,\text{g/mL}} \quad (10.33)$$

where SDI = sludge density index, percent of sludge density with respect to water density (1 g/mL)

Note: 1 g/mL is the density of water (sp. gr. = 1.0).

Typically, an SDI between 1 and 2 is an indicator of good settling sludge. SDI of 0.5 indicates a poor settling sludge.

EXAMPLE 10.44: DETERMINATION OF SVI

An aeration basin is operating at an MLSS concentration of 2500 mg/L. To determine SVI, the mixed liquor is settled in a 1-L graduated cylinder for 30 min. The volume of settled sludge (SV) is 200 mL/L. Calculate SVI from Equations 10.32a and 10.32b. Also calculate sludge density index (SDI).

Solution

1. Calculate SVI from Equation 10.32a at SV = 200 mL/L and MLSS = 2500 mg/L.

$$SVI = \frac{SV \times 1000\,mg/g}{MLSS} = \frac{200\,mL/L \times 1000\,mg/g}{2500\,mg/L} = 80\,mL/g$$

2. Calculate SVI from Equation 10.32b.

$$SVp \text{ from Equation 10.32c, } SV_p = \frac{SV}{1000\,mL/L} \times 100\% = \frac{200\,mL/L}{1000\,mL/L} \times 100\% = 20\%$$

$$MLSSp \text{ from Equation 10.32d, } MLSS_p = \frac{MLSS}{10^6\,mg/L} \times 100\% = \frac{2500\,mg/L}{10^6\,mg/L} \times 100\% = 0.25\%$$

$$SVI \text{ from Equation 10.32b, } SVI = \frac{SV_p}{MLSS_p \times 1\,g/mL} = \frac{20\%}{0.25\% \times 1\,g/mL} = 80\,mL/g$$

3. Calculate SDI from Equation 10.33.

$$SDI = \frac{1}{SVI \times 1\,g/mL} \times 100\% = \frac{1}{80\,mL/g \times 1\,g/mL} \times 100\% = 1.25\% \text{ or } 1.25$$

EXAMPLE 10.45: SVI, SETTLED VOLUME, AND RETURN SLUDGE RELATIONSHIPS

An activated sludge plant is operating at a MLSS concentration of 3000 mg/L. The settled volume of sludge (SV) is 300 mL after the mixed liquor was settled in a 1-L graduated cylinder for 30 min. The TSS concentration in the return sludge is 10,000 mg/L. Calculate SVI from Equations 10.32a and 10.32b. Also calculate the percent return sludge (R_{ras}) from Equations 10.31, 10.32e, and 10.5e. Validate if TSS_{ras} is achievable in the RAS from Equation 10.5e.

Solution

1. Calculate SVI from Equation 10.32a at SV = 300 mL/L and MLSS = 3000 mg/L.

$$SVI = \frac{SV \times 1000\,mg/g}{MLSS} = \frac{300\,mL/L \times 1000\,mg/g}{3000\,mg/L} = 100\,mL/g$$

2. Calculate SVI.

$$SV_p \text{ Equation 10.32c, } SV_p = \frac{SV}{1000\,mL/L} \times 100\% = \frac{300\,mL/L}{1000\,mL/L} \times 100\% = 30\%$$

$$MLSS_p \text{ from Equation 10.32d, } MLSS_p = \frac{MLSS}{10^6\,mg/L} \times 100\% = \frac{3000\,mg/L}{10^6\,mg/L} \times 100\% = 0.3\%$$

$$\text{SVI from Equation 10.32b, SVI} = \frac{\text{SV}_\text{p}}{\text{MLSS}_\text{p} \times 1\,\text{g/mL}} = \frac{30\%}{0.3\% \times 1\,\text{g/mL}} = 100\,\text{mL/g}$$

3. Calculate R_{ras} from Equation 10.31.

$$R_{ras} = \frac{\text{SV}}{1000\ \text{mL/L} - \text{SV}} = \frac{300\,\text{mL/L}}{1000\,\text{mL/L} - 300\,\text{mL/L}} = 0.43 \text{ or } 43\%$$

4. Calculate R_{ras} from Equation 10.32e.
   R_{ras} can be calculated from SVI and MLSS_p.

$$R_{ras} = \frac{1}{(100\%/(\text{SVI} \times \text{MLSS}_\text{p} \times 1\ \text{g/mL}) - 1} = \frac{1}{(100\%/(100\ \text{mL/g} \times 0.3\% \times 1\ \text{g/mL}) - 1}$$

$$= 0.43 \text{ or } 43\%$$

   R_{ras} can also be calculated directly from SV_p.

$$R_{ras} = \frac{\text{SV}_\text{p}}{100\% - \text{SV}_\text{p}} = \frac{30\%}{100\% - 30\%} = 0.43 \text{ or } 43\%$$

5. Calculate R_{ras} from Equation 10.5e using MLSS = 3000 mg/L and TSS_{ras} = 10,000 mg/L.

$$R_{ras} = \frac{\text{MLSS}}{\text{TSS}_{ras} - \text{MLSS}} = \frac{3000\,\text{mg/L}}{10{,}000\,\text{mg/L} - 3000\,\text{mg/L}} = 0.43 \text{ or } 43\%$$

6. Validate the maximum TSS concentration in the RAS allowed by the SVI.
   The maximum achievable TSS concentration in RAS (TSS_{ras}^{max}) can be calculated from Equation 10.32f.

$$\text{TSS}_{ras}^{max} = \frac{1000\,\text{mL/L} \times 1000\,\text{mg/g}}{\text{SVI}} = \frac{1000\,\text{mL/L} \times 1000\,\text{mg/g}}{100\,\text{mL/g}} = 10{,}000\,\text{mg/L}$$

   The calculated TSS_{ras}^{max} indicates the TSS concentration of 10,000 mg/L is achievable in the RAS based on the sludge SVI.

EXAMPLE 10.46: MAXIMUM SOLIDS CONCENTRATION IN RAS BASED ON SVI

The maximum TSS concentration in RAS (TSS_{ras}^{max}) depends upon the SVI (Equation 10.32f). Develop this relationship. In the typical SVI range from 80 to 150 mL/g for activated sludge plants, estimate (a) the potential range of TSS_{ras} in the RAS, (b) the potential range of R_{ras} of RAS at an MLSS concentration of 2000 mg/L, and (c) the potential range of MLSS concentration in the aeration basin at an RAS ratio $R_{ras} = 1/3$.

Solution

1. Develop the relationship between TSS_{ras}^{max} and SVI.

   When the sludge sample is settled in a 1-L graduated cylinder, the maximum possible SV is 1000 mL. Assume MLSS $= TSS_{ras}^{max}$ is achieved at SV = 1000 mL/L, the following expression is obtained from Equation 10.32a.

$$SVI = \frac{SV \times 1000\,\text{mg/g}}{TSS_{ras}^{max}} = \frac{1000\,\text{mL/L} \times 1000\,\text{mg/g}}{TSS_{ras}^{max}}$$

   Rearranging the above expression yields Equation 10.32f.

$$TSS_{ras}^{max} = \frac{1000\,\text{mL/L} \times 1000\,\text{mg/g}}{SVI} \quad \text{or} \quad TSS_{ras}^{max} = \frac{10^6(\text{mL/g})\cdot(\text{mg/L})}{SVI}$$

2. Estimate the potential range of TSS_{ras}^{max} in RAS.

   The lower and upper end of TSS_{ras}^{max} range is estimated from Equation 10.32f at SVI = 150 and 80 mL/g, respectively.

$$TSS_{ras}^{lower} = \frac{10^6\,(\text{mL/g})\cdot(\text{mg/L})}{SVI} = \frac{10^6\,(\text{mL/g})\cdot(\text{mg/L})}{150\,\text{mL/g}} = 6700\,\text{mg/L}$$

$$TSS_{ras}^{upper} = \frac{10^6\,(\text{mL/g})\cdot(\text{mg/L})}{80\,\text{mL/g}} = 12\,500\,\text{mg/L}$$

   Therefore, a range from ~6000 to 12,000 mg/L is estimated for the TSS concentration in the RAS.

3. Estimate the potential range of R_{ras} required at a given MLSS concentration.

   Use Equation 10.5e to estimate the lower and upper end of R_{ras} range at MLSS = 2000 mg/L when TSS_{ras} concentration is 12,000 and 6000 mg/L in the RAS, respectively.

$$R_{ras}^{lower} = \frac{MLSS}{TSS_{ras} - MLSS} = \frac{2000\,\text{mg/L}}{12{,}000\,\text{mg/L} - 2000\,\text{mg/L}} = 0.2 \text{ or } 20\%$$

$$R_{ras}^{upper} = \frac{2000\,\text{mg/L}}{6000\,\text{mg/L} - 2000\,\text{mg/L}} = 0.5 \text{ or } 50\%$$

   Therefore, the estimated R_{ras} range is ~20–50% for maintaining an MLSS of 2000 mg/L.

4. Estimate the potential range of MLSS concentration at a given R_{ras}.

   Rearrange Equation 10.5e and substitute $R_{ras} = 1/3$ to obtain the following expression.

$$MLSS = \frac{R_{ras}}{1 + R_{ras}} \times TSS_{ras} = \frac{1/3}{1 + 1/3} \times TSS_{ras} = \frac{1}{4} \times TSS_{ras}$$

   Estimate the lower and upper end of MLSS range at $R_{ras} = 1/3$ when TSS_{ras} concentration is 6000 and 12,000 mg/L in the RAS, respectively.

$$MLSS^{lower} = \frac{1}{4} \times TSS_{ras} = \frac{1}{4} \times 6000\,\text{mg/L} = 1500\,\text{mg/L}$$

$$MLSS^{upper} = \frac{1}{4} \times 12{,}000\,\text{mg/L} = 3000\,\text{mg/L}$$

Therefore, the estimated MLSS concentration is within a range between 1500 and 3000 mg/L at $R_{ras} = 1/3$.

EXAMPLE 10.47: RAS FLOW ADJUSTMENT

The morning shift operator conducts several routine tests for plant operation and process control. These tests are sludge blanket in the clarifier, settling of MLSS in a 1-L graduated cylinder, MLSS concentration, and TSS concentration in RAS (TSS_{ras}). The results of these tests are given below. Calculate SVI, and RAS flow (Q_r). The daily average plant flow is 8960 m^3/d. Settled volume in 1-L graduated cylinder, SV = 220 mL/L, MLSS = 2500 mg/L, and $TSS_{ras} = 11{,}370$ mg/L. The evening shift operator finds $TSS_{ras} = 11{,}900$ mg/L. What should be the adjusted RAS flow to maintain MLSS = 2500 mg/L? Estimate SV and SVI in the settling test. Compare the results of both shifts. Assume that there is no change in plant flow during both shifts.

Solution

1. Determine the SVI in the morning shift from Equation 10.32a.

$$\text{SVI} = \frac{\text{SV} \times 1000\,\text{mg/g}}{\text{MLSS}} = \frac{220\,\text{mL/L} \times 1000\,\text{mg/g}}{2500\,\text{mg/L}} = 88\,\text{mL/g}$$

2. Determine the RAS flow (Q_r) in the morning shift from Equation 10.31.

$$R_{ras} = \frac{\text{SV}}{1000\,\text{mL/L} - \text{SV}} = \frac{220\,\text{mL/L}}{1000\,\text{mL/L} - 220\,\text{mL/L}} = 0.282 \text{ or } 28.2\%$$

Calculate Q_r from $Q = 8960\,\text{m}^3/\text{d}$, $Q_r = R_{ras} \times Q = 0.282 \times 8960\,\text{m}^3/\text{d} = 2527\,\text{m}^3/\text{d}$
Verify R_{ras} from Equation 10.5e,

$$R_{ras} = \frac{\text{MLSS}}{\text{TSS}_{ras} - \text{MLSS}} = \frac{2500\,\text{mg/L}}{11{,}370\,\text{mg/L} - 2500\,\text{mg/L}} = 0.282 \text{ or } 28.2\%$$

Therefore, the R_{ras} ratios calculated from two different methods are consistent.

3. Determine the RAS flow to be adjusted by the evening shift operator.
Calculate the required R_{ras} from 10.5e to maintain MLSS = 2500 mg/L.

$$R_{ras} = \frac{\text{MLSS}}{\text{TSS}_{ras} - \text{MLSS}} = \frac{2500\,\text{mg/L}}{11{,}900\,\text{mg/L} - 2500\,\text{mg/L}} = 0.266 \text{ or } 26.6\%$$

Calculate Q_r from $Q = 8960\,\text{m}^3/\text{d}$, $Q_r = R_{ras} \times Q = 0.266 \times 8960\,\text{m}^3/\text{d} = 2383\,\text{m}^3/\text{d}$

4. Estimate the operating parameters values of evening shift.
Rearrange Equation 10.31 and calculate SV from $R_{ras} = 0.266$.

$$\text{SV} = \frac{R_{ras}}{1 + R_{ras}} \times 1000\,\text{mL/L} = \frac{0.266}{1 + 0.266} \times 1000\,\text{mL/L} = 210\,\text{mL/L}$$

Calculate SVI from Equation 10.32a, $SVI = \frac{SV \times 1000\,mg/g}{MLSS} = \frac{210\,mL/L \times 1000\,mg/g}{2500\,mg/L} = 84\,mL/g$

Calculate MLSSp from Equation 10.32d, $MLSS_p = \frac{MLSS}{10^6\,mg/L} \times 100\% = \frac{2500\,mg/L}{10^6\,mg/L} \times 100\% = 0.25\%$

Verify Q_r from Equation 10.32e.

$$R_{ras} = \frac{1}{(100\%/(SVI \times MLSS_p \times 1\,g/mL) - 1} = \frac{1}{(100\%/(84\,mL/g \times 0.25\% \times 1\,g/mL) - 1}$$

$$= 0.266 \text{ or } 26.6\%$$

Therefore, the R_{ras} ratios calculated from two different methods are consistent.

5. Compare the results of both shifts.

The results of lab tests and estimated values are summarized below.

Parameter	Morning Shift	Evening Shift	Expectation in the Evening Shift
MLSS, mg/L	2500	2500	Remain the same as an operating goal
TSS_{ras}, mg/L	11,370	11,900	
SV, mL/L	220	210	Sludge settling property is improved
SVI (mL/g)	88	84	
R_{ras}	0.282	0.266	RAS flow is lowered
Q_r, m^3/d	2527	2383	

10.3.7 Activated Sludge Process Modifications

The suspended growth biological treatment processes have many modifications. These modifications differ from each other in the manner in which the influent is applied, microorganisms are utilized, and hardware is assembled. The major modifications of suspended growth biological treatment process include (1) conventional activated sludge; (2) completely mixed aeration; (3) tapered aeration; (4) step-feed aeration; (5) high-rate aeration; (6) extended aeration (oxidation ditch); (7) single-stage nitrification; (8) separate-stage nitrification; (9) deep-shaft technology; (10) contact stabilization process; (11) Kraus process; (12) pure oxygen activated sludge process; (13) combined anaerobic, anoxic, and aerobic process; (14) sequencing batch reactor (SBR); (15) continuous flow sequencing batch reactor (CF-SBR); (16) counter current aeration system; (17) Biolac™ process; (18) fixed-film media in suspended growth reactor; and (19) combined activated sludge and powdered activated carbon (AS-PAC) system; and (20) membrane bioreactor (MBR) process. Many of these alternatives have been widely applied in biological nutrient removal (BNR) processes. For background information, review the following sections and related examples: Section 10.6 for biological nitrogen removal, Section 10.7 for biological phosphorus removal, and Section 9.5 for phosphorus removal by chemical precipitation. Application of membrane technology in wastewater treatment is presented in Section 15.4.10. These process modifications with brief descriptions, important design parameters, and simple definition sketches are summarized in Table 10.15.

TABLE 10.15 Description and Design Parameters of Various Activated Sludge Process Modifications

Process Modification Alternative with Brief Description	Reactor Type[a]	SRT (θ_c), d	F/M Ratio[b], d^{-1}	VOL[c], kg $BOD_5/m^3 \cdot d$	MLSS[d], mg TSS/L	HRT (θ), h	RAS Ratio ($R_{ras} = Q_r/Q$)
1. Conventional activated sludge (CAS) The influent and returned activated sludge enter the tank at the head end of the basin and are mixed by the aeration system (Figure 10.20).	PF	5–15	0.2–0.4	0.3–0.7	1000–3000	4–8	0.25–0.5
2. Completely mixed aeration The influent and the returned sludge are mixed and applied at several points along the length and width of the basin. The contents are mixed, and the mixed liquor flows across the tank to the effluent channel. The oxygen demand and organic loading are uniform along the entire length of the basin Figure 10.21, and Example 10.49.	CM	3–15	0.2–0.6	0.3–1.6	1500–4000	3–6	0.25–1

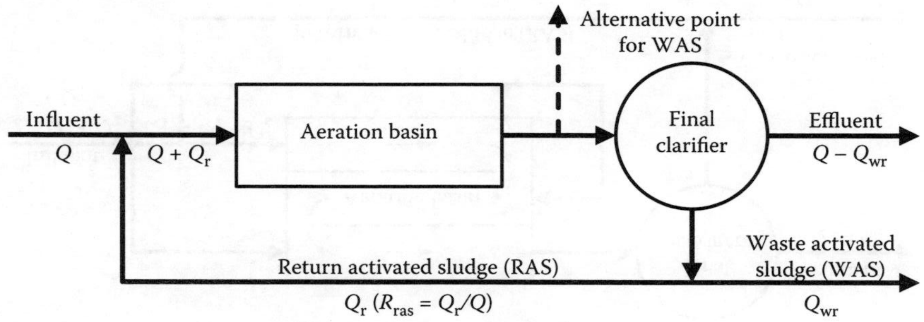

FIGURE 10.20 Definition sketch of conventional activated sludge.

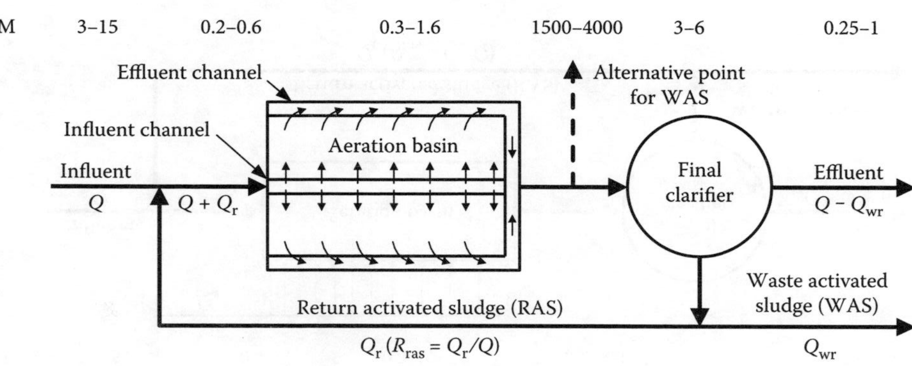

FIGURE 10.21 Definition sketch of complete-mixed aeration.

(*Continued*)

TABLE 10.15 (*Continued*) Description and Design Parameters of Various Activated Sludge Process Modifications

Process Modification Alternative with Brief Description	Reactor Type[a]	SRT (θ_c), d	F/M Ratio[b], d^{-1}	VOL[c], kg $BOD_5/m^3 \cdot d$	MLSS[d], mg TSS/L	HRT (θ), h	RAS Ratio ($R_{ras} = Q_r/Q$)
3. Tapered aeration The tapered aeration system is similar to conventional activated sludge process. The major difference is in the arrangement of air diffusers. The diffusers are close together at the influent end where more oxygen is needed. The spacing of diffusers is increased toward the effluent end of the aeration basin (Figure 10.22).	PF	3–15	0.2–0.4	0.3–0.6	1500–3000	4–8	0.25–0.5
FIGURE 10.22 Definition sketch of tapered aeration.							
4. Step-feed aeration The influent is applied at several points in the aeration basin. Generally, the tank is subdivided into three or more parallel channels with around the end baffles, and the influent is applied at separate channels or steps. The oxygen demand is uniformly distributed. The MLSS concentration is increased in the first chamber to reduce the initial F/M ratio effectively (Figure 10.23).	PF	3–15	0.2–0.4	0.7–1	1500–4000	3–5	0.25–0.75

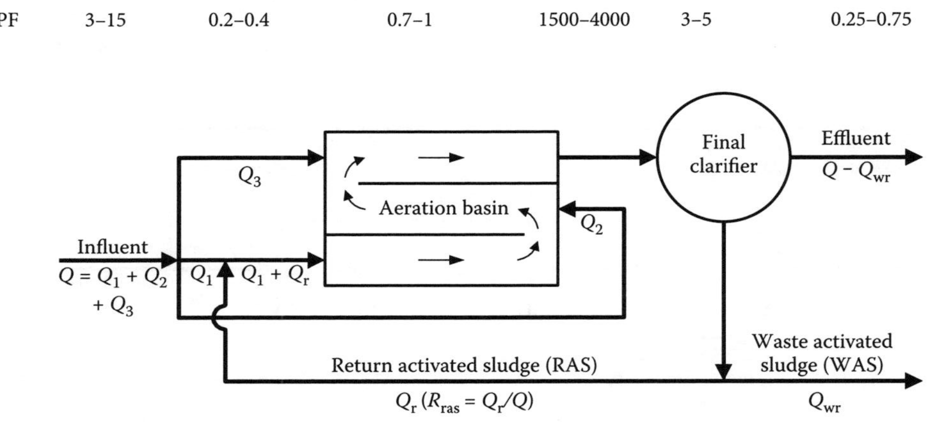

FIGURE 10.23 Definition sketch of step-feed aeration.

(*Continued*)

TABLE 10.15 (*Continued*) Description and Design Parameters of Various Activated Sludge Process Modifications

Process Modification Alternative with Brief Description	Reactor Type[a]	SRT (θ_c), d	F/M Ratio[b], d^{-1}	VOL[c], kg $BOD_5/m^3 \cdot d$	MLSS[d], mg TSS/L	HRT (θ), h	RAS Ratio ($R_{ras} = Q_r/Q$)
5. High-rate aeration This process is similar to the conventional treatment process. Low MLSS concentration and high VOL are applied. Low SRT and high F/M ratio are maintained. Aeration period is relatively short (Figure 10.24).	PF	0.5–2	1.5–2	1–2.5	500–1500	1–2	1–2
6. Extended aeration (oxidation ditch) The extended aeration process utilizes a large aeration basin in which high MLSS concentration is maintained. It is normally used for small applications. Prefabricated package plants utilize this process extensively. Oxidation ditch is a variation of extended aeration. It has a channel in the shape of a race track. Mechanical surface aerators are used to supply oxygen and maintain circulation. The Orbal™ or Carrousel® process is a variation of oxidation ditch [Figure 10.25, and Examples (10.50) and (10.51)].[1,2,17,25,26] See additional information in Tables 10.55 and 10.57.	PF	15–40	0.04–0.1	0.1–0.3	2000–5000	15–30	0.5–2

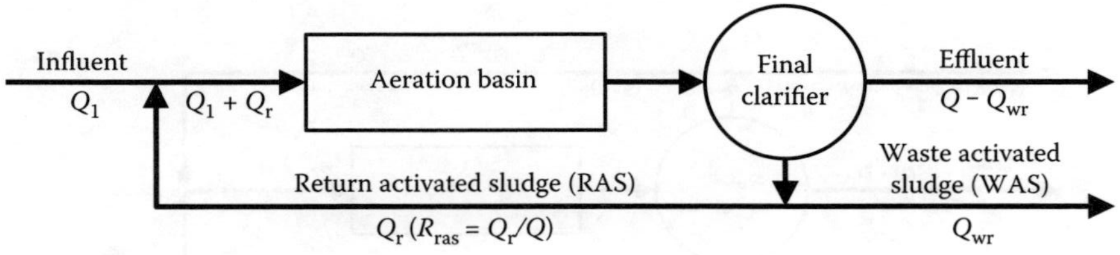

FIGURE 10.24 Definition sketch of high-rate aeration.

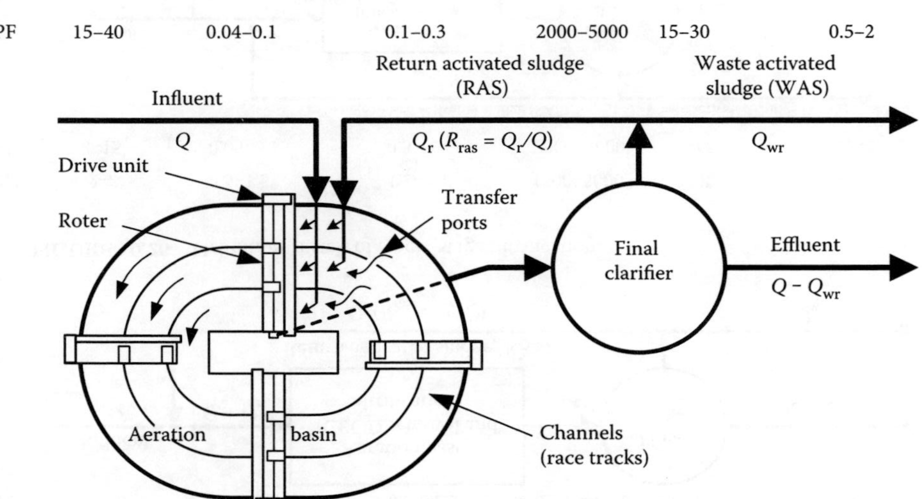

FIGURE 10.25 Definition sketch of extended aeration (Orbal™ Process).

(*Continued*)

TABLE 10.15 (*Continued*) Description and Design Parameters of Various Activated Sludge Process Modifications

Process Modification Alternative with Brief Description	Reactor Type[a]	SRT (θ_c), d	F/M Ratio[b], d^{-1}	VOL[c], kg $BOD_5/m^3 \cdot d$	MLSS[d], mg TSS/L	HRT (θ), h	RAS Ratio ($R_{ras} = Q_r/Q$)
7. Single-stage nitrification Organic carbon oxidation and nitrification are carried out in one aeration basin. High SRT and low F/M ratio are utilized. HRT is relatively long (Figure 10.26).	PF	10–20	0.05–0.2	0.08–0.3	2500–4500	6–15	0.5–1.5
8. Separate-stage nitrification Two-stage aeration is applied. In the first stage CBOD is reduced. The reactor is similar to high-rate aeration. In the second reactor, high SRT is maintained and effective nitrification is achieved. The process may commonly be an add-on to an existing activated sludge plant to enhance nitrification. A portion of influent wastewater may be bypassed around the first stage to provide carbon source which may promote flocculation and solids capture in the nitrification tank (Figure 10.27).	PF[e] PF[f]	2–8[e] 8–15[f]	0.5–1.5[e] 0.05–0.15[f]	1.2–2.4[e] 0.3–0.8[f]	1500–3000[e] 2000–4000[f]	2–4[e] 3–6[f]	0.2–0.5[e] 0.2–0.5[f]

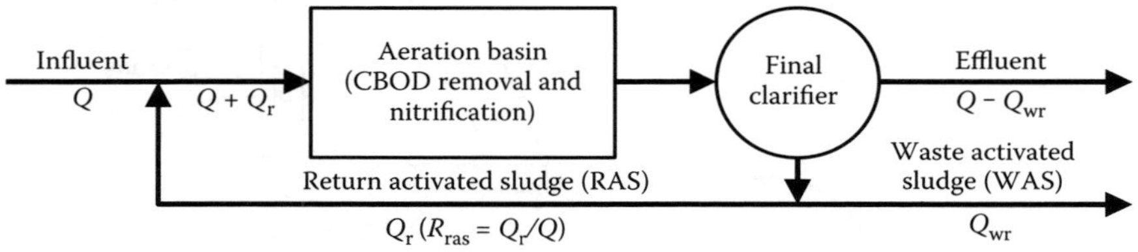

FIGURE 10.26 Definition sketch of single-stage nitrification.

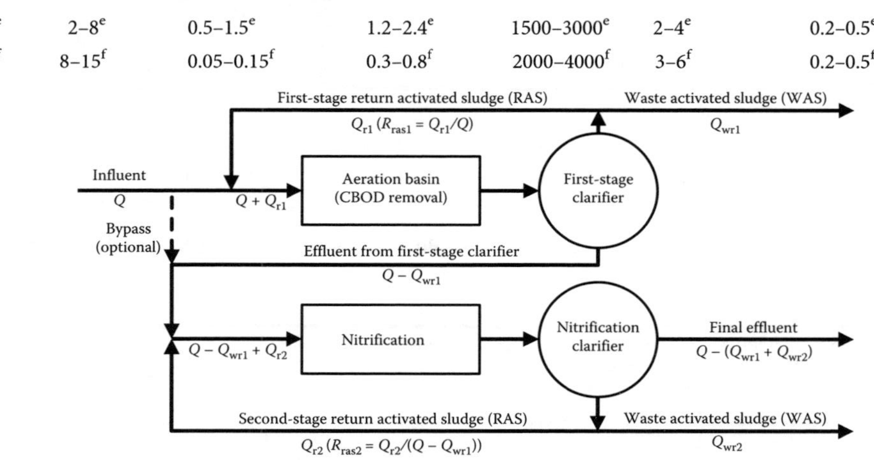

FIGURE 10.27 Definition sketch of separate-stage nitrification.

(*Continued*)

TABLE 10.15 (*Continued*) Description and Design Parameters of Various Activated Sludge Process Modifications

Process Modification Alternative with Brief Description	Reactor Type[a]	SRT (θ_c), d	F/M Ratio[b], d^{-1}	VOL[c], kg $BOD_5/m^3 \cdot d$	MLSS[d], mg TSS/L	HRT (θ), h	RAS Ratio ($R_{ras} = Q_r/Q$)
9. Deep-shaft technology It is an aerobic biological subsurface wastewater treatment process. A vertical shaft about 30–150 m (100–500 ft) deep is drilled and lined with steel shell. A concentric pipe forms an annular reactor. MLSS and air are forced down the center of the shaft. The contents move upward into the riser section. The deep-shaft reactor replaces the primary clarifier and aeration basin. The solids/liquid separation is achieved in an air flotation tank (Figure 10.28, and Example 10.52).[27,28]	PF	2–4	0.75–1.25	5.6–8.0	7000–12,000	0.5–0.75	0.2–0.5
10. Contact stabilization process Contact stabilization process uses two separate tanks. The activated sludge is mixed with influent in the contact tank in which the organics are adsorbed by microorganisms. The mixed liquor is settled in the clarifier, and effluent is released. The returned sludge is aerated in the reaeration basin to stabilize the organics. The process requires 30–50% less aeration volume in comparison to conventional aeration (Figure 10.29 and Example 10.53).	PF[g] PF[h]	5–10	0.2–0.6	1–1.3	1000–3000[g] 5000–10,000[h]	0.5–1[g] 3–4[h]	0.5–1

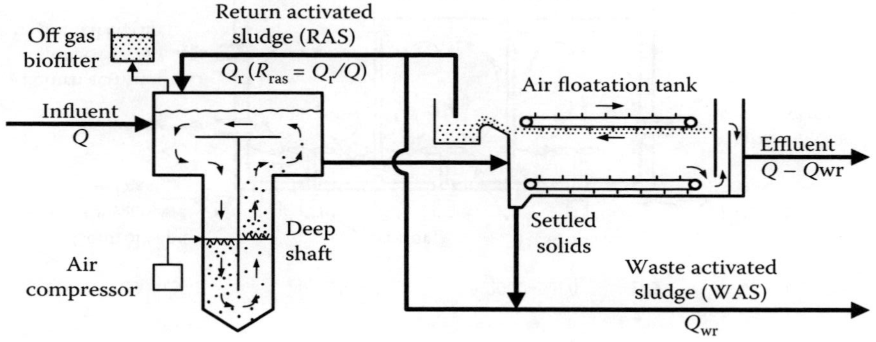

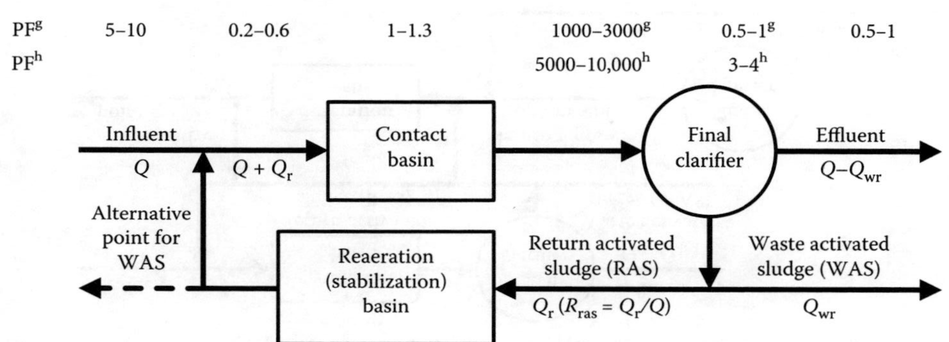

FIGURE 10.28 Definition sketch of deep-shaft technology.

FIGURE 10.29 Definition sketch of contact stabilization.

(*Continued*)

TABLE 10.15 (*Continued*) Description and Design Parameters of Various Activated Sludge Process Modifications

Process Modification Alternative with Brief Description	Reactor Type[a]	SRT (θ_c), d	F/M Ratio[b], d^{-1}	VOL[c], kg $BOD_5/m^3 \cdot d$	MLSS[d], mg TSS/L	HRT (θ), h	RAS Ratio ($R_{ras} = Q_r/Q$)
11. Kraus process This is a variation of step aeration process. Used commonly for industrial wastes low in nitrogen. In a separate aeration basin, the digester supernatant is mixed with a portion of waste activated sludge, and is aerated to provide needed nutrients. A portion of mixed liquor is combined with returned sludge and is added into the aeration basin at desired locations. The remaining mixed liquor is returned to influent line (Figure 10.30 and Example 10.53).	PF	5–15	0.3–1	0.5–1.5	2000–3000	4–8	0.5–1
12. High-purity oxygen (HPO) activated sludge process Oxygen is diffused into a covered aeration tank that has three or four stages. A portion of gas is wasted from the tank to reduce the concentration of CO_2. The process is suitable for high-strength industrial wastes where space may be limited and/or oxygen may be available as a by-product (Figure 10.31 and Example 10.54).[29]	CFST	1–4	0.5–1	1.5–3	2000–5000	1–3	0.25–0.5

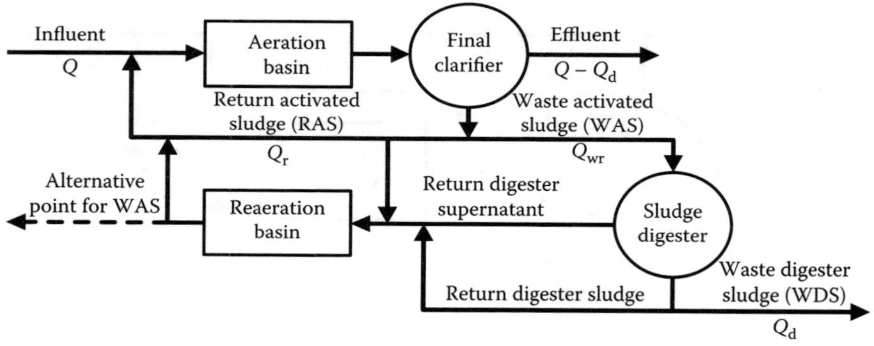

FIGURE 10.30 Definition sketch of Kraus process.

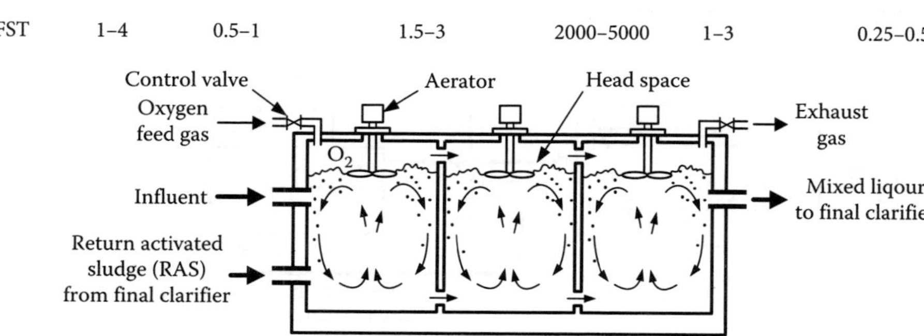

FIGURE 10.31 Definition sketch of high-purity oxygen (HPO) activated sludge process.

(*Continued*)

TABLE 10.15 (*Continued*) Description and Design Parameters of Various Activated Sludge Process Modifications

Process Modification Alternative with Brief Description	Reactor Type[a]	SRT (θ_c), d	F/M Ratio[b], d^{-1}	VOL[c], kg $BOD_5/m^3 \cdot d$	MLSS[d], mg TSS/L	HRT (θ), h	RAS Ratio ($R_{ras} = Q_r/Q$)
13. Combined anaerobic, anoxic, and aerobic process The process uses multiple zones and alternatives for recirculation. There are anaerobic, anoxic, and aerobic zones that are arranged in many configurations. The process has capabilities to enhance biological phosphorus and nitrogen removal while keeping effective CBOD removal and nitrification. There are various process modifications using three, four, or five zones and several recirculation patterns. One such process is Virginia Initiative Plant (VIP) (Figure 10.32).[30]	CM[i] CM[j] PF[k]	5–10	0.1–0.2	0.4–1	2000–4000	1–2[i] 1–2[j] 4–6[k]	0.8–1 1–3[l] 1–2[m]
14. Conventional sequencing batch reactor (SBR) This is a batch type reactor that acts as aeration basin and a final clarifier in a single basin. It operates in a five-stage cycle. The mixed liquor remains in the reactor all the time. Anaerobic, anoxic, and aerobic periods are created during the reaction stage with mixing and/or aeration. The mixed liquor is settled and effluent is decanted. SBRs are extensively used for the treatment of small and medium flows (Figure 10.33, and Examples 10.55 and 10.56).[31,32]	BOCM	15–30	0.04–0.1	0.1–0.3	2000–5000	15–40 4–9[n]	NA

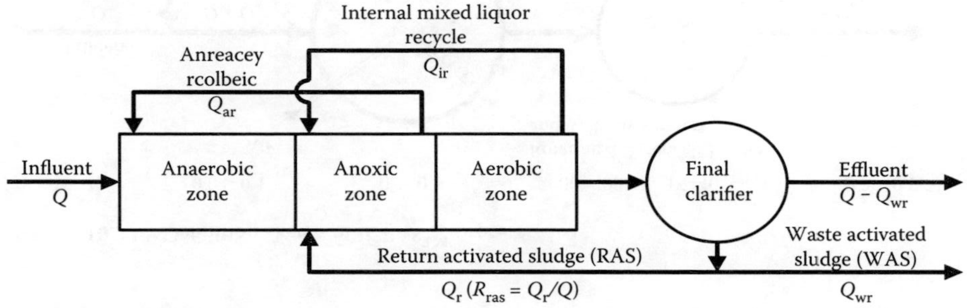

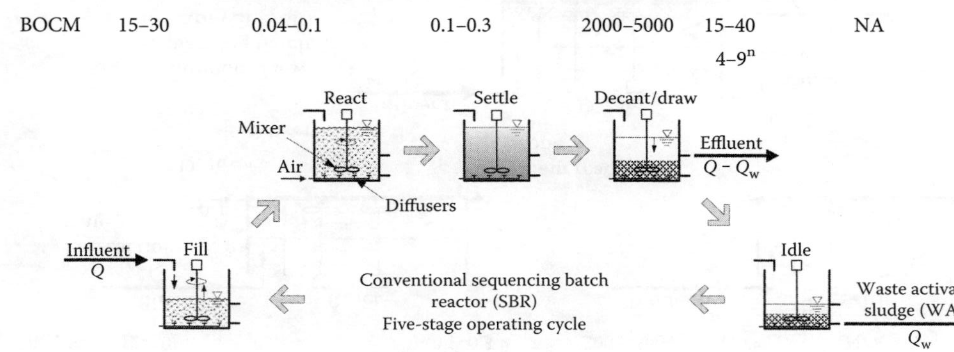

FIGURE 10.32 Definition sketch of Virginia Initiative Plant (VIP) process.

FIGURE 10.33 Definition sketch of conventional sequencing batch reactor (SBR).

(*Continued*)

TABLE 10.15 (*Continued*) Description and Design Parameters of Various Activated Sludge Process Modifications

Process Modification Alternative with Brief Description	Reactor Type[a]	SRT (θ_c), d	F/M Ratio[b], d^{-1}	VOL[c], kg $BOD_5/m^3 \cdot d$	MLSS[d], mg TSS/L	HRT (θ), h	RAS Ratio ($R_{ras} = Q_r/Q$)
15. Continuous flow sequencing batch reactor (CF-SBR) The process is also called intermittent cycle extended aeration system (ICEAS™). It consists of prereaction and main reaction zones. The screened and degritted influent is fed continuously into the prereaction zone and directed through the submerged ports into the main reaction zone.[33,34] Anaerobic, anoxic, and aerobic sequences can be arranged for use in biological nutrient removal. The mixed liquor is settled and effluent is decanted (Figure 10.34).	BOPF	12–30	0.04 –0.1	0.1–0.3	2000–5000	4–9[n]	N/A
16. Counter current aeration system The system utilizes a circular aeration basin and a final clarifier. The diffusers are mounted at the bottom of a revolving bridge to maintain low level of DO in the basin. When the air is turned off, the circular motion created by the rotating bridge keeps the suspended solids in suspension. Alternation between aerobic and anoxic conditions encourages nitrification and denitrification (Figure 10.35).	IOCM	15–30	0.04–0.1	0.1–0.3	2000–4000	15–40	0.25–0.75

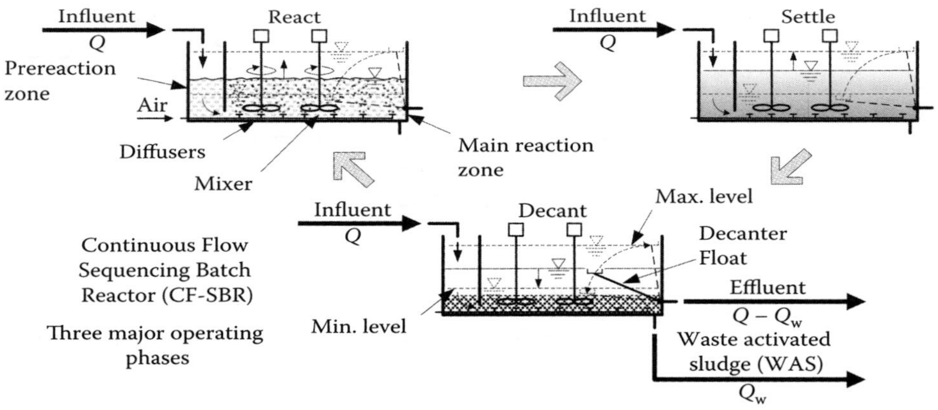

FIGURE 10.34 Definition sketch of ICEAS™ process.

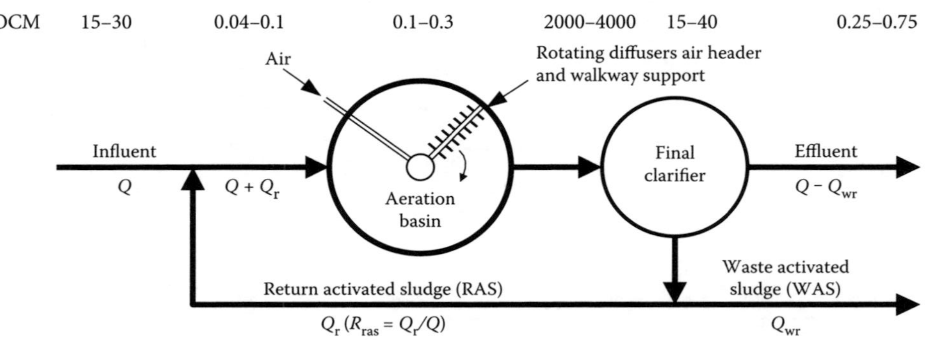

FIGURE 10.35 Definition sketch of counter-current aeration system.

(*Continued*)

TABLE 10.15 (*Continued*) Description and Design Parameters of Various Activated Sludge Process Modifications

Process Modification Alternative with Brief Description	Reactor Type[a]	SRT (θ_c), d	F/M Ratio[b], d^{-1}	VOL[c], kg $BOD_5/m^3 \cdot d$	MLSS[d], mg TSS/L	HRT (θ), h	RAS Ratio ($R_{ras} = Q_r/Q$)
17. Biolac™ process It is a proprietary process. Earthen or lined aeration basins use moving aeration chains. The process operates at high SRT and low F/M ratio. Cyclic aeration is used for simultaneous nitrification and denitrification. Anaerobic zone may also be created for enhanced phosphorus removal. Internal or external clarifier may be used (Figure 10.36).	PF	30–70	0.04–0.1	0.1–0.3	1500–4000	15–40	0.5–1

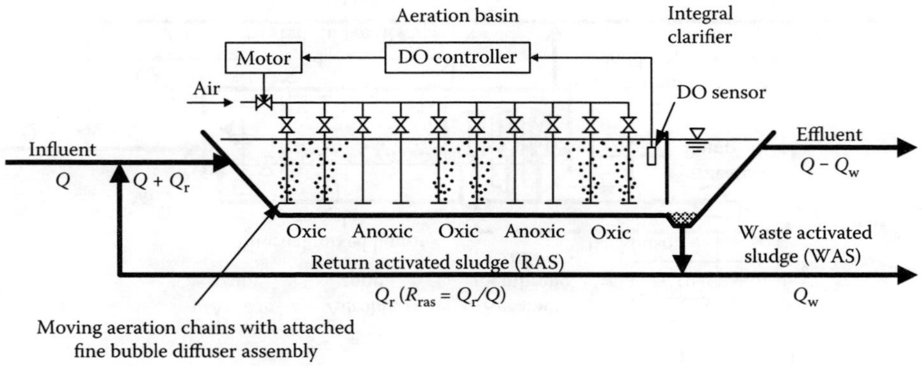

FIGURE 10.36 Definition sketch of Biolac® process.

Process Modification Alternative with Brief Description	Reactor Type[a]	SRT (θ_c), d	F/M Ratio[b], d^{-1}	VOL[c], kg $BOD_5/m^3 \cdot d$	MLSS[d], mg TSS/L	HRT (θ), h	RAS Ratio ($R_{ras} = Q_r/Q$)
18. Fixed-film media in suspended growth reactor It utilizes fixed-film media carriers submerged or dispersed in a suspended growth reactor. There are two major types: (1) integrated fixed -film activated sludge (IFAS), and (2) moving bed biofilm reactor (MBBR). It facilitates high MLSS, low F/M ratio, and high SRT due to large biofilm developed on the media. Hybrid systems are also available for biological nutrient removal [Figure 10.37, and Examples 10.109 and 10.110].[35–39]	CM[j]	10–20	0.1–0.3	1–6	1500–3000	1–2[j]	0–0.5
	PF[k]				2500–6000[o]	2–4[k]	1–4[l]

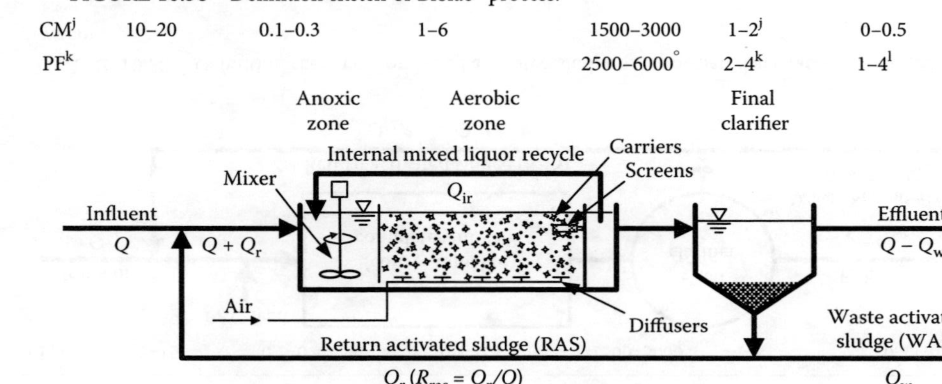

FIGURE 10.37 Definition sketch of IFAS process with nitrification.

(*Continued*)

TABLE 10.15 (*Continued*) Description and Design Parameters of Various Activated Sludge Process Modifications

Process Modification Alternative with Brief Description	Reactor Type[a]	SRT (θ_c), d	F/M Ratio[b], d^{-1}	VOL[c], kg $BOD_5/m^3 \cdot d$	MLSS[d], mg TSS/L	HRT (θ), h	RAS Ratio ($R_{ras} = Q_r/Q$)
19. Combined activated sludge and powdered activated carbon (AS-PAC) system Addition of PAC in an activated sludge process enhances the removal of organics (BOD and COD), and priority pollutants.[40] Both the commercially available PAC and the waste biological sludge converted to an adsorbent by chemical activation produce similar results in AS-PAC system (Figure 10.38).[41]	PF/CM	5–15	0.2–0.4	0.3–0.6	1500–3000	4–8	0.25–0.5
20. Membrane bioreactor (MBR) process MBR process is an innovative biological treatment system where final clarifiers are replaced by membrane modules for effective separation of the effluent from the biosolids. High SRT and lower F/M ratio are achieved at high MLSS. Superior effluent quality with low TSS and turbidity is suitable for reuse. MBR is often an integral component of many biological nutrient removal processes (Figure 10.39).[42–47]	CM[j]	10–25	0.1–0.3	0.5–4	5000–15,000	1–2[j]	0.25–0.5
	PF[k]					2–4[k]	1–4[l]

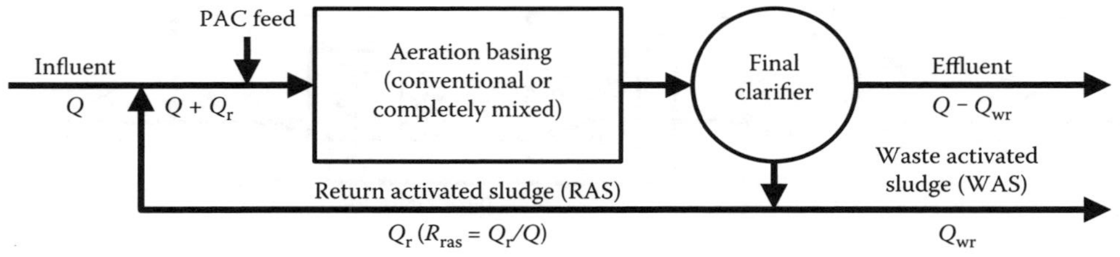

FIGURE 10.38 Definition sketch of combined activated sludge and powdered activated carbon (AS-PAC) system.

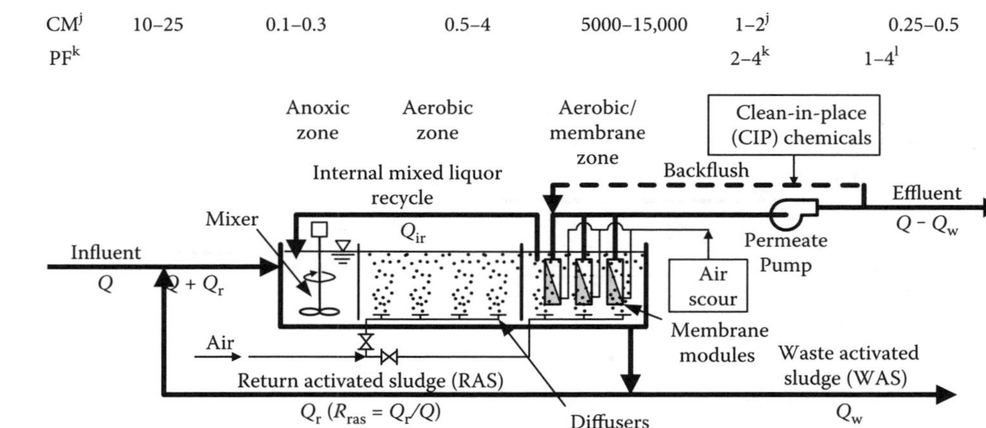

FIGURE 10.39 Definition sketch of membrane bioreactor (MBR) activated sludge process with nitrification.

(Continued)

TABLE 10.15 (*Continued*) Description and Design Parameters of Various Activated Sludge Process Modifications

[a] Reactor type: PF = plug flow, CM = completely mixed, CFST = continuous-flow stirred-tank, BOCM = batch operated completely mixed, BOPF = batch operated plug flow, IOCM = intermittently operated completely mixed.

[b] Food-to-microorganisms ratio: kg BOD_5 applied per day per kg of MLVSS in the reactor.

[c] Volumetric organic loading: kg of BOD_5 applied per day per cubic meter of the reactor.

[d] The ratio of MLVSS to MLSS is normally 0.70–0.85.

[e] Applicable to CBOD removal.

[f] Applicable to nitrification basin based on influent $BOD_5 = 50$ mg/L.

[g] Applicable to contact basin.

[h] Applicable to reaeration (stabilization) basin.

[i] Applicable to anaerobic zone.

[j] Applicable to anoxic zone.

[k] Applicable to aerobic zone.

[l] Internal recycle (Q_{ir}/Q) for denitrification.

[m] Anaerobic recycle (Q_{ar}/Q) for phosphorus release.

[n] Total cycle time.

[o] Total biomass concentration, including suspended and attached growths in the reactor.

Source: Adapted in part from References 1, 2, 7, 17, and 25 through 47.

EXAMPLE 10.48: OXYGEN DEMAND AND SUPPLY IN ACTIVATED SLUDGE PROCESS MODIFICATIONS

The oxygen demand and supply vary greatly in conventional, completely mixed, tapered aeration, and step-feed aeration activated sludge processes. Draw the typical oxygen demand and supply curves in these processes with respect to tank length.

Solution

The oxygen demand and supply curves of conventional, completely mixed, tapered aeration, and plug-flow activated sludge processes with respect to tank length are shown with process diagrams in Figure 10.40.

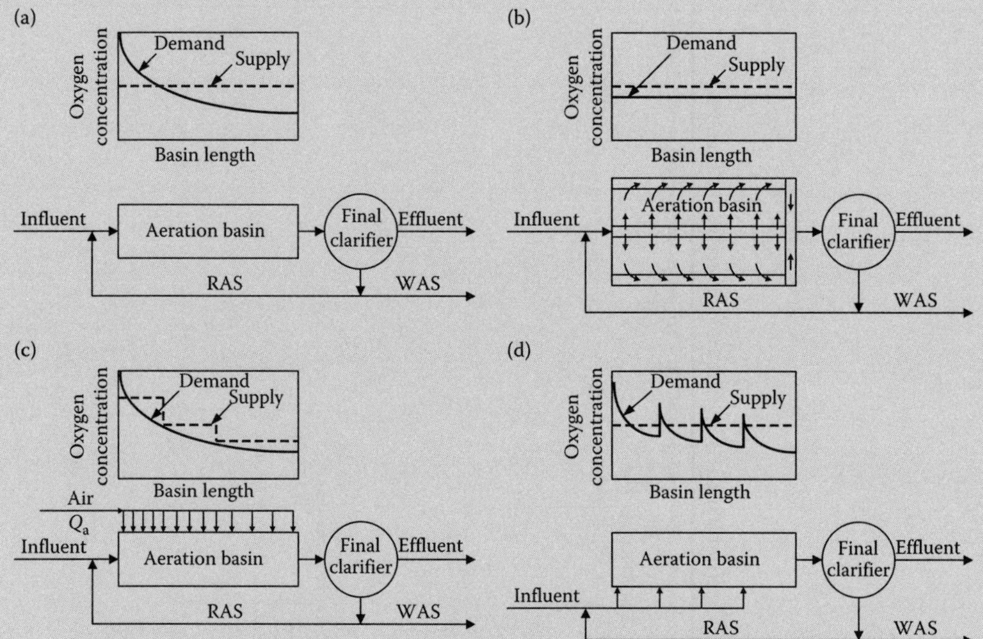

FIGURE 10.40 Oxygen demand and supply with respect to tank length: (a) conventional, (b) completely mixed, (c) tapered aeration, and (d) step-feed aeration (Example 10.48).

EXAMPLE 10.49: ADVANTAGES OF A COMPLETELY MIXED ACTIVATED SLUDGE PROCESS

The performance of completely mixed activated sludge process is less efficient than plug-flow or dispersed plug-flow systems. From operational point of view, however, it has many advantages. List these operational advantages.

Solution

The advantages of a completely mixed activated sludge process are:

1. A slug of organic load (BOD, COD or other substrates) are quickly dispersed throughout the entire aeration, thus the effect is dampened.

2. A slug of toxic substance is quickly distributed in the entire reactor volume, thus the concentration of toxic substance is reduced and the potential impact on biological treatment is lowered.
3. The oxygen uptake by the microbial activities is uniform throughout the basin. The aeration system can be operated efficiently.
4. Distribution and neutralization of CO_2 produced from bio-oxidation of carbonaceous BOD is maximized.
5. The environmental conditions such as nutrients, mixing, oxygen level, pH, and temperature are uniform for growth of active biomass.
6. Operational flexibility is much higher than that of using other process modifications.

EXAMPLE 10.50: MLVSS CONCENTRATION IN AN OXIDATION DITCH (EXTENDED AERATION SYSTEM) WITHOUT SLUDGE WASTING

Most oxidation diches (extended aeration systems) are designed at very low F/M ratios. The biomass undergoes endogenous decay and sludge production is reduced significantly. Develop a relationship to express MLVSS concentration maintained in an oxidation ditch so that the net sludge production reaches zero.

Solution

1. Identify the general expression for sludge production in the oxidation ditch without sludge wasting.

In Example 10.20, the following expression (Equation 10.18e) is obtained in Step 4 from mass balance analysis of a biological treatment system.

$$\frac{VX}{Q_{wr}X_r + (Q - Q_{wr})X_e} = \frac{1}{YU - k_d}$$

When no sludge is wasted from the oxidation ditch, $Q_{wr} = 0$. Simplifying Equation 10.18e yields Equation 10.34a.

$$\frac{VX}{QX_e} = \frac{1}{YU - k_d} \quad \text{or} \quad \frac{\theta X}{X_e} = \frac{1}{YU - k_d} \quad \text{or} \quad YU\theta X - k_d\theta X = X_e \tag{10.34a}$$

2. Develop the expression of MLVSS concentration (X) with respect to TSS in the effluent (TSS_e).

Substitute the expression $U = (S_0 - S)/\theta X$ from Equation 10.14d into Equation 10.34a to obtain Equation 10.34b.

$$Y\frac{S_0 - S}{\theta X}\theta X - k_d\theta X = X_e \quad \text{or} \quad Y(S_0 - S) - k_d\theta X = X_e \tag{10.34b}$$

Rearrange Equation 10.34b to obtain Equation 10.34c.

$$X = \frac{Y(S_0 - S) - X_e}{k_d\theta} \tag{10.34c}$$

Assume that $f_{VSS/TSS}$ is the fraction of volatile solids in suspended solids in the effluent from the oxidation ditch, $X_e = f_{VSS/TSS} \times TSS_e$. Therefore, the MLVSS concentration (X) in the oxidation ditch can be expressed by Equation 10.34d with respect to the TSS concentration (TSS_e) in the effluent.

$$X = \frac{Y(S_0 - S) - f_{VSS/TSS} \times TSS_e}{k_d \theta} \tag{10.34d}$$

where $f_{VSS/TSS}$ = fraction of volatile solids in suspended solids in the effluent ($f_{VSS/TSS} = 0.5–0.7$), dimensionless or mg VSS/mg TSS

Note: Because of long aeration period (θ) and SRT (θ_c), and high TSS in the effluent from an oxidation ditch process, the value of $f_{VSS/TSS}$ is typically lower than those in other modifications of activated sludge process.

Other terms in Equation 10.34d have been defined earlier.

EXAMPLE 10.51: DESIGN OF OXIDATION DITCH WITH NO SLUDGE WASTING

An oxidation ditch unit is designed to treat municipal wastewater from a small community. The following data apply: average design flow (Q) = 4200 m^3/d, influent BOD_5 concentrtion (S_0) is 180 mg/L, $Y = 0.65$ mg VSS/mg BOD_5, $k_d = 0.05$ d^{-1}, $\theta = 18$ h, and $f_{VSS/TSS} = 0.55$ mg VSS/mg TSS. Effluent soluble BOD_5 concentration (S) and TSS_e are not to exceed 4 and 10 mg/L, respectively. The ditch has a rectangular cross section. The channel width (W), water depth (D), and freeboard (H) are 8, 2, and 0.75 m, respectively. Assume that the thickness (d) is 0.3 m for all internal partition and baffle walls within the basin. A horizontal velocity in the range of 0.25–0.4 m/s should be provided to maintain the solids in suspension. Determine the following:

a. Major dimensions of the basin,
b. Operating MLSS concentration without sludge wasting,
c. Operating solids retention time (θ_c),
d. Average flow ($Q_{channel}$) required to keep the solids in suspension in the channel.

Solution

1. Draw a definition sketch.

Definition sketch of oxidation ditch process is shown in Figure 10.41.

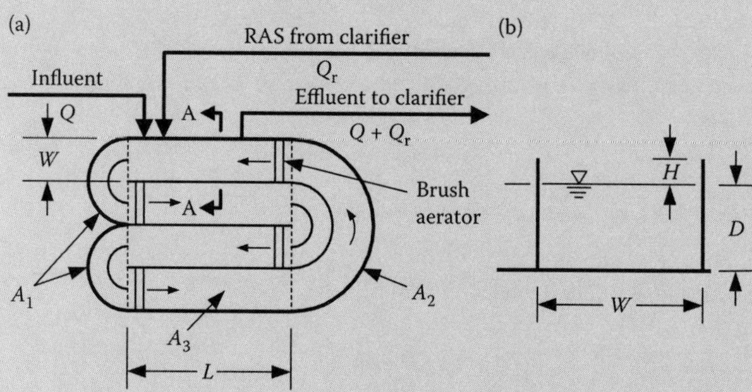

FIGURE 10.41 Definition sketch of oxidation ditch: (a) plan and (b) Section AA (Example 10.51).

2. Determine the major dimensions of the oxidation ditch.

Total volume required in the oxidation ditch, $V = \theta Q = 18\,\text{h} \times 4200\,\text{m}^3/\text{d} \times \frac{\text{d}}{24\,\text{h}} = 3150\,\text{m}^3$

Total surface area required at a water depth $D = 2$ m, $A = \frac{V}{D} = \frac{3150\,\text{m}^3}{2\,\text{m}} = 1575\,\text{m}^2$

As shown in Figure 10.41, the total surface area can be broken into three basic areas that are divided by the dashed lines: two half-circular areas ($2A_1$), one half-circular area (A_2), and four rectangular areas (A_3). The relationship among these subareas is: $A = 2A_1 + A_2 + 4A_3$.

Calculate the area $2A_1$.

$$\begin{aligned}2A_1 &= 2 \times (\text{half circular area} - \text{area occupied by the curved baffle wall})\\ &= 2 \times \left(\frac{1}{2} \times \frac{\pi}{4} \times (2W + d)^2 - \frac{1}{2} \times \pi \times (W + d) \times d\right)\\ &= 2 \times \left(\frac{1}{2} \times \frac{\pi}{4} \times (2 \times 8\,\text{m} + 0.3\,\text{m})^2 - \frac{1}{2} \times \pi \times (8\,\text{m} + 0.3\,\text{m}) \times 0.3\,\text{m}\right)\\ &= 2 \times (104\,\text{m}^2 - 4\text{m}^2) = 200\,\text{m}^2\end{aligned}$$

Calculate the area A_2.

$$\begin{aligned}A_2 &= 2 \times (\text{half circular area} - \text{area occupied by the partition wall}\\ &\qquad - \text{area occupied by the curved baffle walls})\\ &= \frac{1}{2} \times \frac{\pi}{4} \times (4W + 3d)^2 - \frac{1}{2} \times \pi \times (2W + d) \times d) - \frac{1}{2} \times \pi \times (W + d) \times d)\\ &= \frac{1}{2} \times \frac{\pi}{4} \times (4 \times 8\,\text{m} + 3 \times 0.3\,\text{m})^2 - \frac{1}{2} \times \pi \times (2 \times 8\,\text{m} + 0.3\,\text{m}) \times 0.3\,\text{m})\\ &\quad - \frac{1}{2} \times \pi \times (8\,\text{m} + 0.3\,\text{m}) \times 0.3\,\text{m}\\ &= 425\,\text{m}^2 - 8\,\text{m}^2 - 4\text{m}^2 = 413\,\text{m}^2\end{aligned}$$

Calculate the area $4A_3$, $4A_3 = 4 \times W \times L = 4 \times 8\,\text{m} \times L = 32\,\text{m} \times L$

Total surface area, $A = 2A_1 + A_2 + 4A_3 = 200\,\text{m}^2 + 413\,\text{m}^2 + 32\,\text{m} \times L = 1575\,\text{m}^2$

Solve the expression for L.

$$L = \frac{(1575 - 200 - 413)\,\text{m}^2}{32\,\text{m}} = 30\,\text{m}$$

Major dimensions of the basin are shown in Figure 10.42.

The provided oxidation ditch has overall dimensions of 54.3 m (length inside walls) by 32.9 m (width inside walls) by 2.75 m (height, including a freeboard of 0.75 m). The actual volume, $V \approx 3150\,\text{m}^3$.

Note: In accrual design, multiple oxidation ditches may be considered as an alternative for operational flexibility.

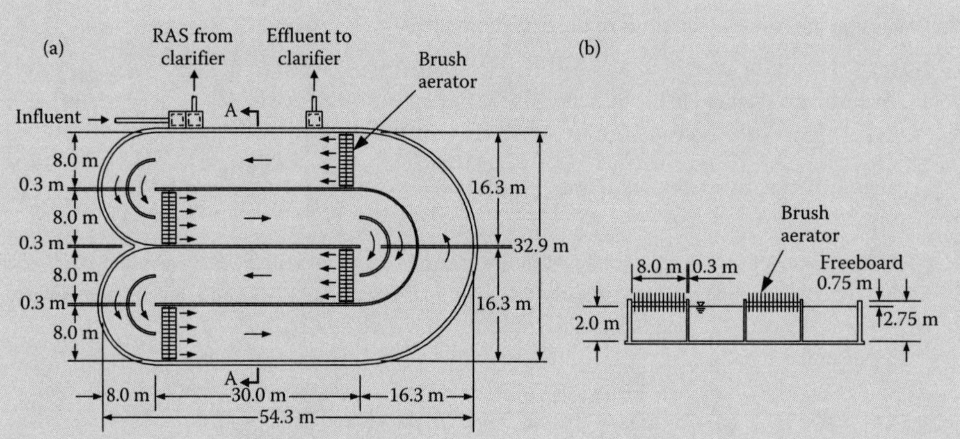

FIGURE 10.42 Design details of oxidation ditch: (a) plan and (b) Section AA (Example 10.51).

3. Calculate the operating MLVSS concentration (X) from Equation 10.34d.

$$X = \frac{Y(S_0 - S) - f_{\mathrm{VSS/TSS}} \times \mathrm{TSS_e}}{k_d \theta}$$

$$= \frac{0.65\ \mathrm{mg\ VSS/mg\ BOD_5}(180 - 4)\ \mathrm{mg\ BOD_5/L} - 0.55\ \mathrm{mg\ VSS/mg\ TSS} \times 10\ \mathrm{mg\ TSS/L}}{0.05\ \mathrm{d^{-1}} \times \mathrm{d/24\,h} \times 18\,\mathrm{h}}$$

$$= 2904\ \mathrm{mg\ VSS/L}$$

4. Calculate the operating MLSS concentration from Equation 10.14e.

$$\mathrm{MLSS} = \frac{X}{f_{\mathrm{VSS/TSS}}} = \frac{2904\ \mathrm{mg/VSS/L}}{0.55\ \mathrm{mg\ VSS/mg\ TSS}} = 5280\ \mathrm{mg/L}$$

5. Calculate the operating solids retention time (θ_c).
   Calculate total biomass maintained in the oxidation ditch at MLSS = 5280 mg/L = 5280 g/m^3.

$$W_{\mathrm{MLSS}} = V \times \mathrm{MLSS} = 3150\ \mathrm{m^3} \times 5280\ \mathrm{g\ TSS/m^3} \times 10^{-3}\ \mathrm{kg/g} = 16{,}630\ \mathrm{kg\ TSS}$$

   Calculate the TSS lost in the effluent at $\mathrm{TSS_e}$ = 10 mg/L = 10 g/m^3.

$$\Delta\mathrm{TSS_e} = Q \times \mathrm{TSS_e} = 4200\ \mathrm{m^3/d} \times 10\ \mathrm{g\ TSS/m^3} \times 10^{-3}\ \mathrm{kg/g} = 42\ \mathrm{kg\ TSS/d}$$

$$\theta_c = \frac{W_{\mathrm{MLSS}}}{\Delta TSS_e} = \frac{16{,}630\ \mathrm{kg\ TSS}}{42\ \mathrm{kg\ TSS/d}} = 396\ \mathrm{d} \approx 400\ \mathrm{d}$$

6. Calculate the required average channel flow (Q_{channel}).
   Assume a design horizontal velocity, $v_h = 0.35$ m/s and calculate the required Q_{channel}.

$$Q_{\mathrm{channel}} = v_h WD = 0.35\ \mathrm{m/s} \times 8\ \mathrm{m} \times 2\ \mathrm{m} \times 86{,}400\ \mathrm{s/d} = 484{,}000\ \mathrm{m^3/d}$$

   The required channel flow is 115 times of the influent flow. The brush aerators will be used to provide oxygen for microbial activities and maintain the required channel flow.

EXAMPLE 10.52: DESIGN OF DEEP-SHAFT BIOLOGICAL REACTOR

Develop the preliminary design of a deep-shaft biological reactor. The design flow is 13,600 m^3/d, and influent BOD_5 concentration is 200 mg/L. The reactor receives raw influent without primary treatment. Apply the following design and operating criteria that are summarized partially in Table 10.15: hydraulic retention time = 0.5–0.75 h; MLSS = 7000–12,000 mg/L; VSS/TSS ratio = 0.6–0.7; F/M ratio = 0.75–1.25 kg BOD_5/kg VSS·d; volumetric organic loading (*VOL*) = 5.6–8 kg BOD_5/m^3·d; sludge retention time = 2–4 d; sludge production = 0.4–0.5 kg VSS/kg BOD_5 removed; sludge recycle ratio = 0.2–0.5; oxygen consumption ratio for BOD_5 removal = 2–2.4 kg O_2/kg BOD_5 removed; oxygen transfer efficiency (*OTE*) = 40–90%; and surface overflow rate (*SOR*) and mass loading rate in flotation units for solids separation are 20–29 m^3/m^2·d and 293–439 kg TSS/m^2·d, respectively. Assume BOD_5 removal efficiency = 90% in the reactor.

Solution

1. Determine the volume of the deep-shaft reactor.

   Calculate the amount of BOD_5 in the influent from $S_0 = 200$ mg BOD_5/L = 200 g BOD_5/m^3.

$$\Delta S_0 = Q \times S_0 = 13{,}600\,\text{m}^3/\text{d} \times 200\,\text{g BOD}_5/\text{m}^3 \times 10^{-3}\,\text{kg/g} = 2720\,\text{kg BOD}_5/\text{d}$$

   Assume the volumetric organic loading (*VOL*) = 6.8 kg $BOD_5/\ m^3$·d.

$$\text{Total reactor volume } (V) \text{ required, } V = \frac{\Delta S_0}{VOL} = \frac{2720\,\text{kg BOD}_5/\text{d}}{6.8\,\text{kg BOD}_5/\text{m}^3\cdot\text{d}} = 400\,\text{m}^3$$

2. Check the hydraulic retention time (θ).

$$\theta = \frac{V}{Q} = \frac{400\,\text{m}^3/\text{d}}{13{,}600\,\text{m}^3/\text{d}} \times \frac{24\,\text{h}}{\text{d}} = 0.7\,\text{h}$$

   The retention time is within the design criteria (0.5–0.75 h) given in the problem.

3. Determine the dimensions of the reactor.

   Assume the shaft diameter, $D = 2.5$ m.

$$\text{Required depth } (L) \text{ of the reactor, } L = \frac{4V}{\pi d^2} = \frac{4 \times 400\,\text{m}^3}{\pi \times (2.5\,\text{m})^2} = 81\,\text{m}$$

   Note: It is within the typical range of 30–150 m.

4. Determine the air supply rate.

   Assume a BOD_5 removal efficiency $E = 90\%$ is achieved and calculate the mass of BOD_5 removed in the reactor.

$$\Delta S_{\text{reactor}} = E \times \Delta S_0 = 0.9 \times 2720\,\text{kg BOD}_5/\text{d} = 2448\,\text{kg BOD}_5/\text{d}$$

   Assume the consumption ratio, $O_{2,\text{consumed}} = 2.2$ kg O_2 consumed/kg BOD_5 removed, and calculate the amount of oxygen consumption for BOD_5 removal.

$$\Delta O_{2,\text{consumed}} = O_{2,\text{consumed}} \times \Delta S_0 = 2.2\,\text{kg O}_2\text{ consumed/kg BOD}_5\text{ removed} \times 2448\,\text{kgBOD}_5/\text{d}$$
$$= 5386\,\text{kg O}_2\text{ consumed/d}$$

Assume oxygen transfer efficiency (*OTE*) of 60% (kg oxygen consumed per kg oxygen supplied) and calculate the amount of oxygen that needs to be delivered by the air supply system.

$$\Delta O_{2,\text{supplied}} = \frac{O_{2,\text{consumed}}}{OTE} = \frac{5386\,\text{kg O}_2\ \text{consumed/d}}{0.6\,\text{kg O}_2\ \text{consumed/kg O}_2\ \text{supplied}} = 8977\,\text{kg O}_2\ \text{supplied/d}$$

Calculate the air supply rate q_{air} at an oxygen content of 0.232 kg O_2/kg air, and a specific weight (or density) of air of 1.23 kg air/sm^3 (standard cubic meter).

$$q_{air} = \frac{O_{2,\text{supplied}}}{0.232\,\text{kg O}_2/\text{kg air} \times 1.23\,\text{kg air/sm}^3} = \frac{8977\,\text{kg O}_2\ \text{supplied/d}}{0.232\,\text{kg O}_2/\text{kg air} \times 1.23\,\text{kg air/sm}^3}$$

$$= 31{,}500\,\text{sm}^3/\text{d}$$

$$\text{or} \quad q_{air} = 31{,}500\,\text{sm}^3/\text{d} \times \frac{\text{d}}{1440\,\text{min}} = 22\,\text{sm}^3/\text{min}$$

5. Determine the dimensions of the flotation units for solids separation.
   Assume surface over flow rate (*SOR*) = 25 $m^3/m^2{\cdot}d$ and calculate the total area required in the flotation units.

$$A_{\text{flotation}} = \frac{Q}{SOR} = \frac{13{,}600\,\text{m}^3/\text{d}}{25\,\text{m}^3/\text{m}^2{\cdot}\text{d}} = 544\,\text{m}^2$$

Provide two flotation units with the following dimensions for each unit:
Length, $l = 30$ m, width $w = 10$ m, and depth $d = 6$ m. Calculate the total surface area of the flotation units.

$$A_{\text{flotation}} = 2lw = 2 \times 30\,\text{m} \times 10\,\text{m} = 600\,\text{m}^2$$

6. Check the solids loading rate in the flotation units.
   Assume the return flow ratio $R_{ras} = 0.3$ and calculate the total flow reaching the solids separation facility.

$$Q_{\text{flotation}} = (1 + R_{ras}) \times Q = (1 + 0.3) \times 13{,}600\,\text{m}^3/\text{d} = 17{,}680\,\text{m}^3/\text{d}$$

Assume the MLSS concentration in the reactor = 10,000 mg/L = 10,000 g/m^3 and calculate the total amount of TSS reaching the flotation units.

$$\Delta TSS_{\text{flotation}} = Q \times \text{MLSS} = 17{,}680\,\text{m}^3/\text{d} \times 10{,}000\,\text{g TSS/m}^3 \times 10^{-3}\,\text{kg/g} = 176{,}800\,\text{kg TSS/d}$$

Calculate the solids loading rate (*SLR*) in the flotation units.

$$SLR = \frac{\Delta\text{TSS}_{\text{reactor}}}{A_{\text{flotation}}} = \frac{176{,}800\,\text{kg TSS/d}}{600\,\text{m}^2} = 295\,\text{kg TSS/m}^2{\cdot}\text{d}$$

The solid loading rate is at the lower end of the criteria (293 kg TSS/m^2·d) given in the problem.

7. Check the F/M ratio.
   Calculate the total mass of TSS maintained in the deep-shaft reactor.

$$W_{\text{MLSS}} = V \times \text{MLSS} = 400\,\text{m}^3 \times 10{,}000\,\text{g TSS/m}^3 \times 10^{-3}\,\text{kg/g} = 4000\,\text{kg TSS}$$

Assume VSS/TSS ratio = 0.65 and calculate the amount of MLVSS in the reactor.

$$W_X = \text{VSS/TSS ratio} \times W_{MLSS} = 0.65\,\text{kg VSS/kg TSS} \times 4000\,\text{kg TSS} = 2600\,\text{kg VSS}$$

$$\text{F/M ratio in the reactor, F/M ratio} = \frac{\Delta S_0}{W_X} = \frac{2720\,\text{kg BOD}_5/\text{d}}{2600\,\text{kg VSS}} = 1.0\,\text{kg BOD}_5/\text{kg VSS·d}$$

The F/M ratio is within the design criteria (0.75–1.25 kg BOD_5/kg VSS·d) given in the problem.

8. Check the sludge retention time (θ_c).

   Assume an observed sludge production coefficient $Y_{obs} = 0.45$ kg VSS/kg BOD_5 removed and calculate the total mass of VSS produced from the reactor.

$$P_x = Y_{obs}\Delta S_{reactor} = 0.45\,\text{kg VSS/kg BOD}_5 \times 2448\,\text{kg BOD}_5/\text{d} = 1102\,\text{kg VSS/d}$$

$$\theta_c = \frac{W_X}{P_x} = \frac{2600\,\text{kg VSS}}{1102\,\text{kg VSS/d}} = 2.4\,\text{d}$$

The sludge retention time $\theta_c = 2.4$ d is within the design criteria (2–4 d) given in the problem.

EXAMPLE 10.53: COMPARISON OF BASIN CAPACITY AND F/M RATIO OF CONTACT STABILIZATION AND CONVENTIONAL ACTIVATED SLUDGE PROCESSES

A contact stabilization activated sludge plant is designed for a flow of 6800 m^3/d. The aeration period of contact and reaeration basin are 45 min and 5 h, respectively. The MLVSS concentration in the contact basin is 2500 mg/L, and return flow is 60% of influent flow. The influent BOD_5 concentration is 150 mg/L. Compare the combined volume of contact and reaeration basins with that of conventional activated sludge plant that has an aeration period of 6 h. Also compare the F/M ratio of both processes. The MLVSS concentration, and influent flow and BOD_5 concentration for conventional plant are same as those for the contact basin.

Solution

1. Calculations for the contact stabilization (CS) process plant.
   a. Calculate the total volume required.

$$\text{Volume of contact basin, } V_{contact} = Q \times \theta_{contact} = 6800\,\text{m}^3/\text{d} \times 45\,\text{min} \times \frac{\text{d}}{1440\,\text{min}} = 213\,\text{m}^3$$

Calculate the return sludge flow to the reaeration basin at $R_{ras} = 0.6$.

$$Q_r = R_{ras} \times Q = 0.6 \times 6800\,\text{m}^3/\text{d} = 4080\,\text{m}^3/\text{d}$$

$$\text{Volume of reaeration basin, } V_{reaeration} = Q_r \times \theta_{reaeration} = 4080\,\text{m}^3/\text{d} \times 5\,\text{h} \times \frac{\text{d}}{24\,\text{h}} = 850\,\text{m}^3$$

$$\text{Total volume required using CS process, } V_{CS} = V_{contact} + V_{reaeration} = (213 + 850)\,\text{m}^3 = 1063\,\text{m}^3$$

   b. Calculate the amounts of VSS.

   Calculate the mass of VSS in the contact basin at $X = 2500$ mg VSS/L = 2500 g VSS/m^3.

$$W_{X,contact} = V_{contact} \times X = 213\,\text{m}^3 \times 2500\,\text{g VSS/m}^3 \times 10^{-3}\,\text{kg/g} = 533\,\text{kg VSS}$$

Rearrange Equation 10.5b and calculate the VSS concentration (X_r) in the return flow.

$$X_r = \frac{1 + R_{ras}}{R_{ras}} X = \frac{1 + 0.6}{0.6} \times 2500 \text{ mg VSS/L} = 6667 \text{ mg VSS/L} = 6667 \text{ g VSS/m}^3$$

Calculate the mass of VSS in the reaeration basin.

$$W_{X,\text{reaeration}} = V_{\text{reaeration}} \times X_r = 850 \text{ m}^3 \times 6667 \text{ g VSS/m}^3 \times 10^{-3} \text{ kg/g} = 5667 \text{ kg VSS}$$

Calculate the total mass of VSS in the CS process.

$$W_{X,CS} = W_{X \text{ contact}} + W_{X,\text{reaeration}} = (533 + 5667) \text{ kg VSS} = 6200 \text{ kg VSS}$$

c. Calculate the F/M ratios.

Calculate the amount of BOD_5 in the influent from $S_0 = 150$ mg BOD_5/L $= 150$ g BOD_5/m^3.

$$\Delta S_0 = Q \times S_0 = 6800 \text{ m}^3/\text{d} \times 150 \text{ g BOD}_5/\text{m}^3 \times 10^{-3} \text{ kg/g} = 1020 \text{ kg BOD}_5/\text{d}$$

Calculate the F/M ratio in the contact basin.

$$\text{F/M ratio}_{\text{contact}} = \frac{\Delta S_0}{W_{X,\text{contact}}} = \frac{1020 \text{ kg BOD}_5/\text{d}}{533 \text{ kg VSS}} = 1.9 \text{ kg BOD}_5/\text{kg VSS·d}$$

Calculate the F/M ratio in the CS process.

$$\text{F/M ratio}_{CS} = \frac{\Delta S_0}{W_{XCS}} = \frac{1020 \text{ kg BOD}_5/\text{d}}{6200 \text{ kg VSS}} = 0.16 \text{ kg BOD}_5/\text{kg VSS·d}$$

2. Calculations for the conventional activated sludge (CAS) process plant.
   a. Calculate the volume of aeration basin.

$$V_{CAS} = Q \times \theta_{CAS} = 6800 \text{ m}^3/\text{d} \times 6 \text{ h} \times \frac{\text{d}}{24 \text{ h}} = 1700 \text{ m}^3$$

   b. Calculate the mass of VSS in the aeration basin.

$$W_{X,CAS} = V_{CAS} \times X = 1700 \text{ m}^3 \times 2500 \text{ g VSS/m}^3 \times 10^{-3} \text{ kg/g} = 4250 \text{ kg VSS}$$

   c. Calculate the F/M ratio in the aeration basin.

$$\text{F/M ratio}_{CAS} = \frac{\Delta S_0}{W_{X\,CAS}} = \frac{1020 \text{ kg BOD}_5/\text{d}}{4250 \text{ kg VSS}} = 0.24 \text{ kg BOD}_5/\text{kg VSS·d}$$

3. Comparison of aeration volume and F/M ratio.
   a. Summarize the results.

Treatment Process	Aeration Basin Volume, m^3	F/M Ratio, kg BOD_5/kg VSS·d
Contact stabilization (CS) process		
Contact basin	213	1.9
Stabilization basin	850	N/A
Overall	1063	0.16
Conventional activated sludge (CAS) process	1700	0.24

b. Comment on the results.

The combined volume of contact and stabilization basins of the CS plant is nearly 40% less than that of the CAS plant. Also, the overall F/M ratio in the contact basin is much higher than that of the CAS plant. Since the organics in the contact basin are mainly kept by the microorganisms as food storage at high F/M ratio, the overall removal efficiency is usually low in a contact stabilization process. Also, the organic and toxic slugs are poorly equalized in the contact basin. Therefore, it is only suitable for the application where a high-quality effluent is not required.

EXAMPLE 10.54: DESIGN OF HIGH-PURITY OXYGEN (HPO) ACTIVATED SLUDGE PROCESS REACTOR

A high-purity oxygen reactor is treating 8000 m^3/d primary settled municipal wastewater. The influent BOD_5 concentration is 140 mg/L. The reactor volume is 500 m^3. The MLSS concentration in the reactor is 4000 mg/L and VSS/TSS ratio is 0.75. Oxygen consumption ratio for BOD_5 removal is 0.82 kg O_2 consumed/kg of BOD_5 removed. BOD_5 removal is 90%. The oxygen transfer efficiency (*OTE*) is 92%. Assume the density and purity of oxygen gas is ~1.4 kg/sm^3 and 100%, respectively. Calculate the following: (a) hydraulic retention time (θ); (b) volumetric organic loading (*VOL*); (c) F/M ratio; (d) oxygen supply rate; and (e) mean cell residence time (θ_c) based on sludge production of 0.6 kg TSS/kg BOD_5 removed. Compare the results with those given in Table 10.15.

Solution

1. Calculate the hydraulic retention time.

   Calculate the retention time (θ) from the reactor volume $V = 500\ m^3$ at the influent flow $Q = 8000\ m^3/d$.

$$\theta = \frac{V}{Q} = \frac{500\,m^3/d}{8000\,m^3/d} \times \frac{24\,h}{d} = 1.5\,h$$

2. Calculate the volumetric organic loading (*VOL*) received by the reactor.

   Calculate the mass of BOD_5 in the influent.

$$\Delta S_0 = Q \times S_0 = 8000\,m^3/d \times 140\,g\ BOD_5/m^3 \times 10^{-3}\,kg/g = 1120\,kg\ BOD_5/d$$

$$VOL = \frac{\Delta S_0}{V} = \frac{1120\,kg\ BOD_5/d}{500\,m^3} = 2.2\,kg\ BOD_5/m^3{\cdot}d$$

3. Calculate the F/M ratio in the reactor.

   Calculate MLVSS concentration (X) in the reactor from MLSS = 4000 mg/L = 4000 g/m^3.

$$X = VSS/TSS\ ratio \times MLSS = 0.75\,g\ VSS/g\ TSS \times 4000\,g\ TSS/m^3 = 3000\,g\ VSS/m^3$$

   Calculate the total mass of TSS maintained in the reactor.

$$W_X = V \times X = 500\,m^3 \times 3000\,g\ VSS/m^3 \times 10^{-3}\,kg/g = 1500\,kg\ VSS$$

$$F/M\ ratio = \frac{\Delta S_0}{W_X} = \frac{1120\,kg\ BOD_5/d}{1500\,kg\ VSS} = 0.75\,kg\ BOD_5/kg\ VSS{\cdot}d$$

4. Calculate the oxygen supply rate.

Calculate the mass of BOD_5 removed in the reactor at the BOD_5 removal efficiency $E = 90\%$.

$$\Delta S_{removal} = E \times \Delta S_0 = 0.9 \times 1120\,\text{kg BOD}_5/\text{d} = 1008\,\text{kg BOD}_5/\text{d}$$

Calculate the mass of oxygen consumed for BOD_5 removal.

$$\begin{aligned}\Delta O_{2,consumed} &= O_{2,consumed} \times \Delta S_{removal} = 0.82\,\text{kg O}_2\text{ consumed/kg BOD}_5\text{ removed} \times 1008\,\text{kg BOD}_5/\text{d}\\ &= 827\,\text{kg O}_2\text{ consumed/d}\end{aligned}$$

Calculate the mass of oxygen that needs to be delivered.

$$\Delta O_{2,supplied} = \frac{O_{2,consumed}}{OTE} = \frac{827\,\text{kg O}_2\text{ consumed/d}}{0.92\,\text{kg O}_2\text{ consumed/kg O}_2\text{ supplied}} = 899\,\text{kg O}_2\text{ supplied/d}$$

Calculate the oxygen supply rate q_{oxygen} at the oxygen gas density of 1.4 kg air/ sm^3 with a purity of ~100%.

$$q_{oxygen} = \frac{O_{2,supplied}}{1.00 \times 1.4\,\text{kg O}_2/\text{sm}^3} = \frac{899\,\text{kg O}_2\text{ supplied/d}}{1.00 \times 1.4\,\text{kg O}_2/\text{sm}^3} = 642\,\text{sm}^3/\text{d}$$

$$\text{or}\quad q_{oxygen} = 642\,\text{sm}^3/\text{d} \times \frac{\text{d}}{24\,\text{h}} = 27\,\text{sm}^3/\text{h}$$

5. Calculate the mean cell residence time (θ_c).

   Calculate the total mass of TSS in the waste sludge and effluent from the reactor at the sludge production yield of 0.6 kg TSS/kg BOD_5 removed.

$$\begin{aligned}P_{TSS} &= 0.6\,\text{kg TSS/kg BOD}_5 \times \Delta S_{removal} = 0.6\,\text{kg TSS/kg BOD}_5 \times 1008\,\text{kg BOD}_5/\text{d}\\ &= 605\,\text{kg TSS/d}\end{aligned}$$

   Calculate the total mass of VSS produced in the reactor at VSS/TSS ratio = 0.75.

$$P_x = \text{VSS/TSS ratio} \times P_{TSS} = 0.75\,\text{kg VSS/kg TSS} \times 605\,\text{kg TSS/d} = 454\,\text{kg VSS/d}$$

$$\theta_c = \frac{W_X}{P_X} = \frac{1500\,\text{kg VSS}}{454\,\text{kg VSS/d}} = 3.3\,\text{d}$$

6. Compare the results with those in Table 10.15.

   The calculated values of θ, *VOL*, F/M ratio, and θ_c fall within the range given in Table 10.15.

EXAMPLE 10.55: DESCRIBE OPERATIONAL DETAILS OF THE SEQUENCING BATCH REACTOR (SBR) PROCESS

The use of SBR units for biological waste treatment is increasing rapidly. The reported benefits are improved process stability in removal of traditional pollutants (BOD_5 or COD), and enhanced removal of nutrients (nitrogen and phosphorus) (see Figures 10.33, 10.94, and 10.114, and Sections 10.6, 10.7, and 10.8). Describe in detail the operations of conventional SBR units for removal of traditional contaminants with nitrification, and in different modes for biological nutrient removal processes.

Solution

1. Describe the operating stages of conventional SBR units for removal of traditional contaminants with nitrification.

   Refer to Figure 10.33. The operation of a conventional SBR unit is normally divided into a *fill* cycle followed by four operating stages or cycles: *react*, *settle*, *decant/draw*, and *idle*. The total cycle time includes fill and operational cycle times. For the purpose of removing traditional contaminants with nitrification, aerobic condition is provided during fill and react cycles. The settle, decant, and idle cycles are mainly operated as physical process for liquid–solids separation. Aerobic condition is desired even though anoxic condition may occur during decant and/or idle cycles. Anaerobic condition shall be avoided in any cycles of a traditional SBR process. Reaeration may be required if the effluent DO concentration is too low to meet the DO requirement in the discharge permit.

   a. Fill cycle.

      At the beginning of the fill cycle, about 25% of the reactor volume is normally occupied by the settled biosolids. The influent is fed to provide the organic substrate into the basin while aeration and/or mixing are provided continuously until the basin is 100% full. With two SBR units in service, the influent flows into one tank while the other tank completes its operational cycles. The total fill time is about 50% of the total cycle time. With more units in service, the fill time may be reduced to 20–50% of the total cycle time. A fast increase in biomass may be seen due to rapid consumption of readily biodegradable organic substrates under aerobic condition during the fill cycle.

   b. Operational cycles.

      i. React cycle: It starts when the tank is 100% full. Aeration is continuously provided for further stabilization of organic substrates (BOD_5 or COD) and nitrification of ammonia nitrogen (see Section 10.6.1) during the entire react cycle that is about 15–25% of the total cycle time.

      ii. Settle cycle: It provides calm condition for biosolids settling while aeration and mixing is completely stopped. The tank remains 100% full during the cycle when the biosolids settle to about one-third tank depth leaving a clarified supernatant layer above the solid–liquid interface. The settle time is about 20–30% of the total cycle time.

      iii. Decant/draw cycle: The supernatant is removed by a decanter during the decant cycle. As the decant arm lowered slowly, the supernatant flows over the effluent weir and into the exit pipe for discharge. The liquid level drops from 100% to about 30% full. The total decant period is about 10–20% of the total cycle time.

      iv. Idle cycle: It provides flexibility in process operation. The idle time varies from 0% to 5% of the total cycle time. It may include the time required while waiting for completion of fill cycle in another unit. Sludge wasting may also be performed during settle or idle stage. The liquid level is normally lowered to about 25% of the basin depth at the end of the idle time. The tank is then ready for being switched to fill cycle again.

2. Describe the operation of using conventional SBR units in biological nutrient removal (BNR) process.

   Many operational procedures have been developed for using SBR units in BNR processes. Two of the mostly used procedures are presented below.

   a. Procedure I: This procedure is normally used when moderate removal of phosphorus is required.

      Refer to Figure 10.33. In this procedure, the operation of the SBR unit is similar to that described above in Step 1. Major difference is to provide anoxic and anaerobic phases in sequence during the fill cycle to accommodate both denitrification and release of phosphorus.

      i. Fill cycle: In the early stage of fill cycle, gentle stirring by submerged mixer is provided without aeration and an anoxic phase occurs due to the presence of nitrate produced during the last operational cycles. The nitrate is reduced to nitrogen gas along with consumption of the readily biodegradable organics such as short-chain volatile fatty acids (SCVFAs) under anoxic condition (Section 10.6.2). After the nitrate in the settled sludge is completely denitrified, anaerobic phase starts and allows the phosphorus-accumulating organisms (PAOs) to release phosphorus and uptake the remaining SCVFAs (Section 10.7.1). The air supply is typically turned on when the tank reaches about 75% full and an aerobic condition is maintained for the rest of the fill cycle.

ii. Operational cycles: The aerobic microbial activity occurs while the tank is still filling and continues in the react cycle after tank is full. During this time period, the organic substrate is oxidized rapidly, PAOs uptake orthophosphate (Ortho-P) (Section 10.7.2), and ammonia nitrogen is nitrified (see Section 10.6.1). The reactor will be maintained under aerobic condition during the entire react and most settle cycles. Partial denitrification may occur in decant and idle cycles. Reaeration is normally required to meet the effluent DO requirement in the discharge permit.

Note: In this procedure, chemical precipitation is required if high removal of phosphorus is desired. Chemical can be fed during react cycle so that the phosphorus that precipitates settles with the biosolids during the settle cycle. Post chemical phosphorus removal process can also be used if filtration is provided after the SBR process. The chemical usages and sludge production for precipitation of phosphorus from different chemicals are covered in Examples 9.27 to 9.30.

b. Procedure II: This procedure is primarily developed for enhanced biological phosphorus removal (EBPR) process.

Refer to Figures 10.94 and 10.114, and Tables 10.55 and 10.57, and Sections 10.6 and 10.8. In this procedure, the basic fill and operational cycles are still the same as those described in Step 1. Major difference is to create a predominately anaerobic phase to favor the release of phosphorus than denitrification during the fill cycle.

i. Fill cycle: During the entire fill cycle, air supply is turned off and the basin contents are stirred only by the submerged mixer. Since the denitrification is nearly complete around the end of the react cycle, there is insignificant nitrate remaining in the settled sludge to compete for the SCVFAs. Therefore, an anaerobic phase can be maintained in nearly the entire fill phase and initial react cycle until the aeration is turned on. This will maximize the release of phosphorus and uptake of SCVFAs by the PAOs to achieve EBPR.

ii. Operational cycles: By effective control of the aeration system, the following operating phases are created during the react cycle:

A. Anaerobic phase: If necessary, this phase is provided in the initial react cycle so the PAOs continue to release phosphorus and temporarily store SCVFAs.

B. Aerobic phase: The aerobic phase is started by turning on the air supply. The organic substrate is oxidized, PAOs uptake and store soluble Ortho-P, and ammonia nitrogen is nitrified into nitrate.

C. Cyclic aerobic/anoxic phase: From some point during the react cycle, operation may be switched between aerobic and anoxic conditions back and forth to maximize nitrification/denitrification or stimulate simultaneous nitrification–denitrification. Additional carbon source such as methanol may be required for synthesis of new cells during the denitrification cyclic phase. At the beginning of the settle phase, minimum nitrate concentration should be reached. Anoxic condition is usually maintained to enhance denitrification in the remaining three cycles, i.e., settle, decant, and idle cycles. Reaeration is normally required to meet the effluent DO requirement in the discharge permit.

Note: The concentration profiles of BOD_5, Ortho-P, and total nitrogen in a BNR facility are developed in Example 10.140. The cycle times and many operating parameters of an SBR unit used in BNR process are summarized in Table 10.57.

3. Summary of the SBR operations.

   The conceptual operation of conventional SBR units in three different SBR treatment modes is illustrated in Figure 10.43.

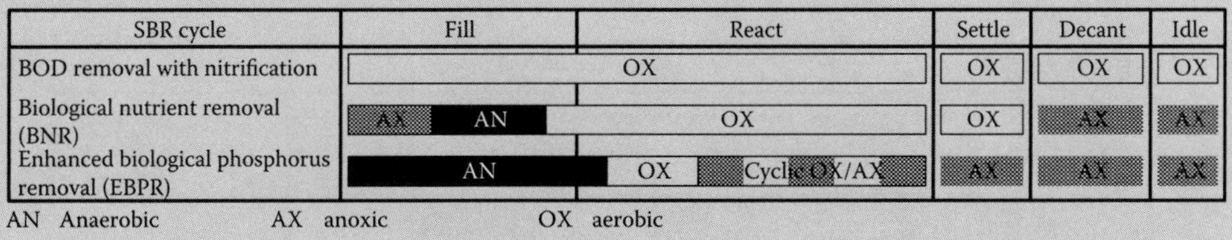

SBR cycle	Fill	React	Settle	Decant	Idle
BOD removal with nitrification	OX	OX	OX	OX	OX
Biological nutrient removal (BNR)	AX, AN	OX	OX	AX	AX
Enhanced biological phosphorus removal (EBPR)	AN	AN, OX, Cyclic OX/AX	AX	AX	AX

AN Anaerobic AX anoxic OX aerobic

FIGURE 10.43 Conceptual operation of conventional SBR unit in three treatment modes (Example 10.55).

EXAMPLE 10.56: CYCLE TIME AND VOLUME OF A SEQUENCING BATCH REACTOR (SBR)

An SBR facility is designed to treat a municipal wastewater flow of 3785 m^3/d (1 MGD). The SBR influent and effluent BOD_5 concentrations are 180 and 5 mg/L, respectively. The SVI is 120 mL/g. There are two tanks in parallel operation. The operation cycle times after filling are as follows: react or aeration time (t_R) = 2 h; settle time (t_S) = 0.5 h; decant time (t_D) = 0.4 h; and idle time (t_I) = 0.1 h. Assume the sludge yield is 0.68 kg TSS/kg BOD_5 removed and the VSS/TSS ratio is 0.75. Use a design F/M ratio of 0.07 kg BOD_5/kg VSS·d for the SBR units and a design weir loading rate $WLR_{decanter} = 100\ m^3/m{\cdot}h$ for the effluent decanters. The MLVSS concentration at full volume is 2500 mg/L and maximum water depth is 6 m. Calculate (a) volume requirement, (b) volumetric organic loading, equivalent hydraulic retention time, and solids retention time, (c) major dimensions, (d) operating cycles, (e) operating levels, (f) requirements of the effluent decanters, and (g) requirement of waste activated sludge pumps.

Solution

1. Draw the definition sketch.

   Definition sketch of a conventional SBR process with design operating levels are shown in Figure 10.44a and b.

2. Determine the total volume required in the SBR units.

   Calculate the amount of BOD_5 in the influent, $S_0 = 180\ mg\ BOD_5/L = 180\ g\ BOD_5/m^3$.

$$\Delta S_0 = Q \times S_0 = 3785\ m^3/d \times 180\ g\ BOD_5/m^3 \times 10^{-3}\ kg/g = 681\ kg\ BOD_5/d$$

   Calculate the total mass of VSS required at the design F/M ratio of 0.07 kg BOD5/kg VSS·d.

$$W_X = \frac{\Delta S_0}{F/M\ ratio} = \frac{681\ kg\ BOD_5/d}{0.07\ kg\ BOD_5/kg\ VSS{\cdot}d} = 9729\ kg\ VSS$$

   Calculate the total volume required in the SBR units at $X = 2500$ mg VSS/L or 2500 g VSS/m^3.

$$V = \frac{W_X}{X} = \frac{9729\ kg\ VSS}{2500\ g\ VSS/m^3 \times 10^{-3}\ kg/g} = 3892\ m^3$$

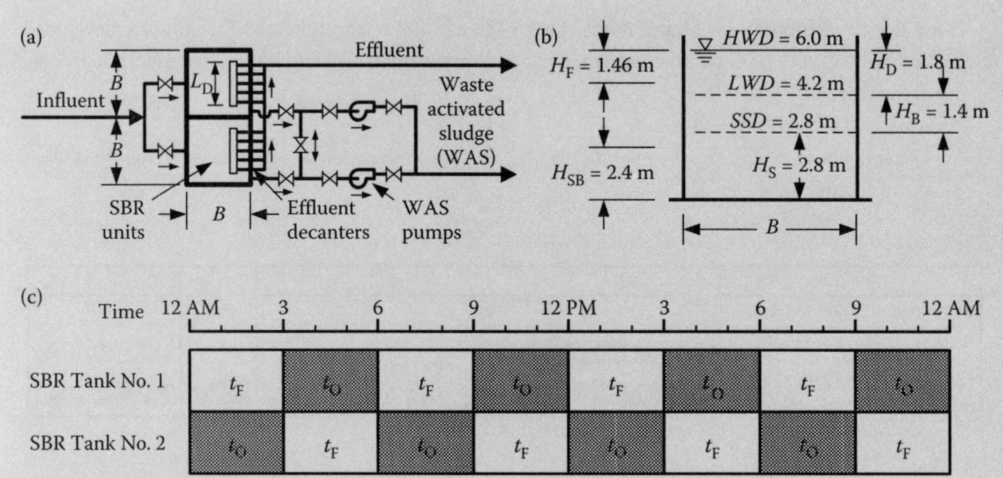

FIGURE 10.44 Definition sketch of conventional SBR units: (a) conceptual layout, (b) operating levels, and (c) operational cycles (Example 10.56).

3. Verify the volumetric organic loading (*VOL*) and equivalent hydraulic retention time (θ) in the SBR units.

$$VOL \text{ in the SBR units, } VOL = \frac{\Delta S_0}{V} = \frac{681\,\text{kg BOD}_5/\text{d}}{3892\,\text{m}^3} = 0.17\,\text{kg BOD}_5/\text{m}^3{\cdot}\text{d}$$

$$\text{Equivalent } \theta \text{ in the SBR units, } \theta = \frac{V}{Q} = \frac{3892\,\text{m}^3}{3872\,\text{m}^3/\text{d}} \times \frac{24\,\text{h}}{\text{d}} = 25\,\text{h}$$

The calculated values of *VOL* and θ are both within the range given in Table 10.15.

4. Verify the solids retention time in the SBR units.

Calculate the total mass of BOD_5 removed in the SBR units at the effluent BOD_5 concentration $S = 5$ mg BOD_5/L $= 5$ g BOD_5/m^3.

$$\Delta S_{SBR} = Q \times (S_0 - S) = 3785\,\text{m}^3/\text{d} \times (180 - 5)\,\text{g BOD}_5/\text{m}^3 \times 10^{-3}\,\text{kg/g} = 662\,\text{kg BOD}_5/\text{d}$$

Calculate the total mass of TSS produced in the SBR units at the sludge production yield of 0.68 kg TSS/kg BOD_5 removed.

$$P_{TSS} = 0.68\,\text{kg TSS/kg BOD}_5 \times \Delta S_{SBR} = 0.68\,\text{kg TSS/kg BOD}_5 \times 662\,\text{kg BOD}_5/\text{d}$$
$$= 450\,\text{kg TSS/d}$$

Calculate the total mass of VSS produced in the SBR units at VSS/TSS ratio $= 0.75$.

$$P_x = \text{VSS/TSS ratio} \times P_{TSS} = 0.75\,\text{kg VSS/kg TSS} \times 450\,\text{kg TSS/d} = 338\,\text{kg VSS/d}$$

$$\text{Sludge retention time } (\theta_c) \text{ in the SBR units, } \theta_c = \frac{W_X}{P_X} = \frac{9792\,\text{kg VSS}}{338\,\text{kg VSS/d}} = 29\,\text{d}$$

The calculated θ_c is slightly below the upper end of the range given in Table 10.15.

5. Determine major dimensions of the SBR units.

$$\text{Volume required in each SBR unit, } V_{unit} = \frac{1}{N} \times V = \frac{1}{2} \times 3892\,\text{m}^3 = 1946\,\text{m}^3$$

Calculate the surface area required in each SBR unit at the high water depth $HWD = 6$ m.

$$A = \frac{V_{\text{unit}}}{HWD} = \frac{1946\,\text{m}^3}{6\,\text{m}} = 324\,\text{m}^2$$

Length required for each side of a square basin, $B = \sqrt{A} = \sqrt{324\,\text{m}^2} = 18\,\text{m}$

Provide two 18 m × 18 m square SBR basins.

6. Determine the operating cycles of the SBR process.

In an SBR system with two tanks, the balance operation is reached when one tank is in the fill cycle while the other tank is undergoing operating cycle. The fill time required can be calculated from Equation 10.35.

$$t_{\text{F}} = \frac{1}{N_{\text{unit}} - 1} t_{\text{O}} \tag{10.35}$$

where

t_{F} = fill cycle time, h
t_{O} = total operating cycle time, h
N_{unit} = number of SBR units, tank

Calculate the required t_{O} from the separate operating cycle times.

$$t_{\text{O}} = t_{\text{R}} + t_{\text{S}} + t_{\text{D}} + t_{\text{I}} = (2 + 0.5 + 0.4 + 0.1)\,\text{h} = 3\,\text{h}$$

Required t_{F} from Equation 10.35, $t_{\text{F}} = \dfrac{1}{N_{\text{unit}} - 1} t_{\text{O}} = \dfrac{1}{2-1} \times 3\,\text{h} = 3\,\text{h}$

Total cycle time (t_{T}) is the sum of t_{F} and t_{O}, $t_{\text{T}} = t_{\text{F}} + t_{\text{O}} = (3+3)\,\text{h} = 6\,\text{h}$ or 6 h/cycle

Number of cycles per SBR unit per day (or 24 h), $N_{\text{cycle,unit}} = \dfrac{24\,\text{h/d}}{t_{\text{cycle}}} = \dfrac{24\,\text{h/d}}{6\,\text{h/cycle}} = 4\,\text{cycles/d}$

Total number of cycles in two SBR units, $N_{\text{cycle}} = 2\,\text{units} \times 4\,\text{cycles/unit} = 8\,\text{cycles/d}$

The SBR normal operation during a period of 24 h is illustrated conceptually in Figure 10.44c.

7. Determine the operating levels in the SBR units.

Fill volume reached in each SBR unit during the fill cycle, $V_{\text{F}} = Qt_{\text{F}} = 3785\,\text{m}^3/\text{d} \times \dfrac{\text{d}}{24\,\text{h}} \times 3\,\text{h}$ $= 473\,\text{m}^3$

Calculate the minimum operating depth required in the basin to contain the fill volume.

$$H_{\text{F}} = \frac{V_{\text{F}}}{A} = \frac{473\,\text{m}^3}{324\,\text{m}^2} = 1.46\,\text{m}$$

Provide a maximum decant depth $H_{\text{D}} = 1.8\,\text{m} > H_{\text{F}}$. Calculate the low water depth (LWD) at the end of decanting operation in each cycle.

$$LWD = HWD - H_{\text{D}} = (6 - 1.8)\text{m} = 4.2\,\text{m}$$

Calculate the maximum achievable TSS concentration after settling from Equation 10.32f.

$$TSS_{settle} = \frac{1000\,\text{mL/L} \times 1000\,\text{mg/g}}{\text{SVI}} = \frac{1000\,\text{mL/L} \times 1000\,\text{mg/g}}{120\,\text{mL/g}}$$

$$= 8333\,\text{mg TSS/L or } 8333\,\text{g TSS/m}^3$$

Calculate the total mass of MLSS maintained in the SBR units.

$$W_{MLSS} = \frac{W_X}{\text{VSS/TSS ratio}} = \frac{9729\,\text{kg VSS}}{0.75\,\text{kg VSS/kg TSS}} = 12{,}972\,\text{kg TSS}$$

The mass of TSS in each SBR unit, $W_{MLSS,unit} = \frac{1}{2} \times W_{MLSS} = \frac{1}{2} \times 12{,}972$ kg TSS $= 6486$ kg TSS

Calculate the volume of the settled solids at $TSS_{settle} = 8333\,\text{g/m}^3$ in each SBR unit.

$$V_{SB} = \frac{W_{MLSS,\,unit}}{TSS_{settle}} = \frac{6486\,\text{kg TSS} \times 10^3\,\text{g/kg}}{8333\,\text{g TSS/m}^3} = 778\,\text{m}^3$$

Settled solids blanket depth, $H_{SB} = \frac{V_{SB}}{A} = \frac{778\,\text{m}^3}{324\,\text{m}^2} = 2.4\,\text{m}$

Provide a design settled sludge depth $SSD = H_S = 2.8\,\text{m} > H_{SB}$ and calculate the buffer zone depth H_B that provides a distance to avoid drawing solids from the sludge blanket.

$$H_B = LWD - SSD = (4.2 - 2.8)\,\text{m} = 1.4\,\text{m}$$

The operating levels in SBR reactor are illustrated in Figure 10.44b.

8. Design of the effluent decanters.
   Calculate the potential volume decanted during the decant cycle.

$$V_D = A\,H_D = 324\,\text{m}^2 \times 1.8\,\text{m} = 583\,\text{m}^3$$

Calculate the maximum decant flow that needs to be collected by the decanter during $t_D = 0.4$ h.

$$Q_D = \frac{V_D}{t_D} = \frac{583\,\text{m}^3}{0.4\,\text{h}} = 1458\,\text{m}^3/\text{h}$$

Calculate the maximum surface overflow rate due to withdrawal of decant flow during the decant cycle.

$$SOR_D = \frac{Q_D}{A} = \frac{1458\,\text{m}^3/\text{h}}{324\,\text{m}^2} \times \frac{24\,\text{h}}{\text{d}} = 108\,\text{m}^3/\text{m}^2{\cdot}\text{d}$$

Calculate the required effluent weir length at the design weir loading rate $WLR_{decanter} = 100\,\text{m}^3/\text{m}{\cdot}\text{h}$.

$$L_D = \frac{Q_{decanter}}{WLR_{decanter}} = \frac{1458\,\text{m}^3/\text{h}}{100\,\text{m}^3/\text{m}{\cdot}\text{h}} = 14.6\,\text{m}$$

Provide one decanter with a rectangular weir length of 15 m.

9. Design of the waste activated sludge pumps.

Calculate the total volume of wasted activated sludge (WAS) from the SBR process at $TSS_{settle} = 8333\ g/m^3$.

$$V_w = \frac{p_{TSS}}{TSS_{settled}} = \frac{450\ kg\ TSS \times 10^3\ g/kg}{8333\ g\ TSS/m^3} = 54\ m^3/d$$

The volume of sludge wasted during each cycle, $V_{w,cycle} = \frac{V_w}{N_{cycle}} = \frac{54\ m^3/d}{8\ cycles/d} = 6.75\ m^3/cycle$

Calculate the required pump flow to waste the sludge in $t_{was} = 5$ min during each idle cycle ($t_I = 0.1\ h = 6$ min).

$$Q_w = \frac{V_{w,cycle}}{t_{was}} = \frac{6.75\ m^3/cycle}{5\ min/cycle} = 1.35\ m^3/min \text{ or } 81\ m^3/h$$

Provide two pumps, including one standby unit. Each has a capacity of 90 m^3/h (400 gpm).

10.3.8 Oxygen Transfer

In a suspended growth aerobic reactor, sufficient oxygen supply and mixing must be provided to meet the respiration and mixing requirements of biomass. These requirements are provided by submerged diffusion system of compressed air, by mechanical surface aeration, or by pure oxygen with mechanical agitation or mixing.[48] In this section, theory of oxygen transfer and different types of aeration systems are presented.

Saturation Concentration of Oxygen in Water: The solubility of gases in natural waters depends upon four factors: (1) temperature, (2) pressure or elevation, (3) gas fraction in overlying atmosphere, and (4) salt concentration or TDS. Solubility increases with increase in pressure and gas fraction; and decreases with increase in temperature and salinity.[49] The saturation concentrations of dissolved oxygen (DO) in water at different temperatures, elevations, and salinity are given in Table 10.16.[50,51] Many equations based on Henry's law constants and partial pressure of gases, and empirical relationships are also used to calculate the solubilities of gases at different temperatures, pressures, and salinities. These equations and useful tables along with many solved examples on this subject are provided in Sections 10.5.3, 11.6.2, 15.4.5, and Appendix B.

Two-Film Theory and Oxygen Transfer Rate: The transfer of a solute gas from a gas mixture into a contact liquid is described by two-film theory. The two-film theory involves two phases in contact with each other. Figure 10.45 shows the schematics.

General discussion about the two-film theory is provided below.

1. The partial pressures of solute gas in the bulk gas, and at the gas–liquid interface are P_b and P_i, respectively.
2. The concentrations of the solute gas at the gas–liquid interface and in the bulk liquid are C_i and C_b, respectively.
3. The solute gas must diffuse through the gas film (laminar layer), pass through the interface, and then diffuse through the liquid film (laminar layer). The interface layer offers no resistance to the solute gas transfer.
4. The rate-limiting step for very soluble gases is the diffusion of the solute gas through the gas film. For gases that are slightly soluble in liquid (such as oxygen in water), the rate-limiting step is the diffusion of the solute gas through the liquid film.
5. The diffusion transfer coefficient K_L for oxygen in water is given by $K_L = D_L/y_L$.

TABLE 10.16 Saturation Concentration of Oxygen at Various Temperatures, Elevations (Pressures), and Total Dissolved Solids (TDS) Concentrations

Temperature, °C (°F)	DO Saturation at Sea Level, mg/L				DO Saturation at TDS Concentration = 0 mg/L, mg/L		
	TDS Concentration, mg/L				Elevation above Sea Level, m (ft)		
	0	800	1500	2500	610 (2000)	1219 (4000)	1829 (6000)
0 (32)	14.62	–	–	–	13.6	12.6	11.7
2 (35.6)	13.83	13.68	13.58	13.42	12.8	11.9	11.1
4 (39.2)	13.11	12.98	12.89	12.75	12.2	11.4	10.5
6 (42.8)	12.45	12.38	12.29	12.15	11.6	10.8	10.0
8 (46.4)	11.84	11.80	11.70	11.58	11.0	10.2	9.5
10 (50)	11.29	11.20	11.12	11.00	10.5	9.8	9.1
12 (53.6)	10.78	10.71	10.64	10.52	10.1	9.4	8.6
14 (57.2)	10.31	10.32	10.25	10.15	9.6	8.9	8.3
16 (60.8)	9.87	9.92	9.85	9.75	9.2	8.6	8.0
18 (64.4)	9.47	9.43	9.36	9.27	8.9	8.2	7.6
20 (68)	9.09	9.13	9.06	8.97	8.5	7.9	7.3
22 (71.6)	8.74	8.73	8.68	8.60	8.2	7.6	7.1
24 (75.2)	8.42	8.43	8.38	8.30	7.9	7.3	6.8
26 (78.8)	8.11	8.13	8.08	8.00	7.6	7.1	6.6
28 (82.4)	7.83	7.83	7.78	7.70	7.4	6.8	6.3
30 (86)	7.56	7.53	7.48	7.40	7.1	6.6	6.1

Note: Equations, solved examples, and tabulated data on partial pressure of gases, Henry's law, and solubility of gases may be found in Appendix B.

Source: Adapted in part from References 1, 2, 50 and 51.

6. The overall mass transfer coefficient K_La is the product of K_L and a, the interfacial bubble area per unit volume of water ($a = A/V$).
7. Since the liquid resistance is controlling, $P_b = P_i$ and $C_i = C_e$, where C_e is the ultimate concentration of oxygen in equilibrium with the partial pressure P_{Gb} of oxygen in air bubbles. This is one form of the Henry's law, $P_{Gb} = (\text{constant}) \times C_s$.
8. The driving force for mass transfer is $(C_{Li} - C_L)$.

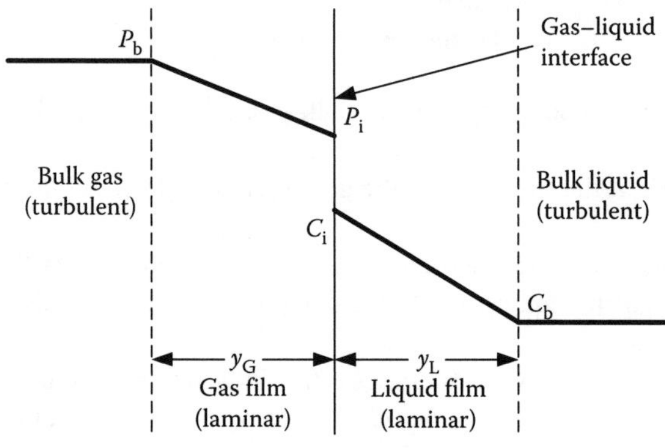

FIGURE 10.45 Schematic of two-film theory.

9. Since the rate of oxygen mass transfer is equal to the mass transfer coefficient times the driving force, the mass transfer is expressed by Equation 10.36a or 10.36b.

$$\frac{dM}{dt} = K_{L}aV(C_{e} - C) \quad \text{or} \quad \frac{1}{V}\frac{dM}{dt} = K_{L}a(C_{e} - C) \tag{10.36a}$$

$$\frac{dC}{dt} = K_{L}a(C_{e} - C) \quad \text{or} \quad OTR = K_{L}a(C_{e} - C) \tag{10.36b}$$

where

$\frac{dM}{dt}$ = mass oxygen transfer rate (OTR) in water, mass/time, for example, kg O_2/h (lb O_2/h)
$\frac{dC}{dt}$ = rate of mass oxygen transferred into unit volume of water, mg O_2/L·h (kg O_2/m^3·h)
$K_{L}a$ = overall oxygen mass transfer coefficient for water, h^{-1}
C_e = equilibrium concentration of dissolved oxygen (DO) in water, mg/L
C = actual DO concentration in water, mg/L
OTR = oxygen transfer rate in water, kg O_2/h (lb O_2/h)
a = interfacial bubble area per unit volume of water ($a = A/V$), m^{-1}(ft^{-1})
A = total bubble surface area, m^2 (ft^2)
V = volume of water, m^3 (ft^3)
K_L = diffusion rate coefficient for oxygen in water ($K_L = D_L/y_L$), m/h (ft/h)
D_L = diffusivity coefficient for oxygen in water, m^2/h (ft^2/h)
y_L = thickness of liquid film, m (ft)

Rearrange Equation 10.36b and integrate within the proper limits to obtain Equation 10.37.

$$\int_{C_0}^{C} \frac{-dC}{C_e - C} = -K_{L}a \int_{0}^{t} dt \tag{10.37a}$$

$$\ln(C_e - C) = \ln(C_e - C_0) - K_{L}a \cdot t \quad \text{or} \quad \ln\left(\frac{C_e - C}{C_e - C_0}\right) = -K_{L}a \cdot t \tag{10.37b}$$

$$C = C_e - (C_e - C_0)e^{-K_{L}a \cdot t} \tag{10.37c}$$

where

C_0 = initial concentration of dissolved oxygen in water at $t = 0$, mg/L
C = concentration of dissolved oxygen in water at t, mg/L

Overall Oxygen Mass Transfer Coefficient in Clean Water: Manufacturers commonly report the data on their aeration devices in clean water at 20°C and at 1 atm. The test is conducted on deoxygenated water (DO ≈ 0). The nonsteady-state aeration procedure has been adopted as a standard.[1,48,52–54] Deoxygenation is achieved by addition of *sodium sulfite* (Na_2SO_3) at 1.25–1.75 times of the stoichiometric requirement. Cobalt in the form of *cobalt chloride* ($CoCl_2$) is added to catalyze the deoxygenation reaction. Only 0.05–0.5 mg/L cobalt (as $CoCl_2$) is needed. At higher concentrations such as 2 mg/L or higher, the cobalt interferences with Winkler test for DO measurement has been reported.

The general procedure for determination of $K_{L}a$ using nonsteady-state aeration system is suggested below.

1. Fill the aeration test vessel with tap water. Measure the temperature and DO using a DO probe in small tanks. In large tanks, multiple probes are used to compensate for concentration gradient. The DO probes shall be fast-responding type and precalibrated.

2. Remove the DO by adding sodium sulfite and cobalt chloride solutions (see Example 10.57 for calculations of the required concentrations). Mix thoroughly the tank content by a low-speed mechanical mixer or diffused nitrogen or air. The deoxygenation reaction is fast. Usually 1–2 min will be sufficient to reach near-zero DO.
3. The aeration device is normally operated during the deoxygenation and reaeration periods. It may take several minutes for DO being established after all sodium sulfite is exhausted. Check and maintain the aeration at the desired air flow rate. Measure the DO at appropriate time intervals until 98% of the expected DO saturation concentration is reached. Measure the temperature at the end of the test.
4. Establish the value of K_La from the experimental data for each DO probe using an appropriate method, and correct the value for standard temperature of 20°C (see Example 10.58). When multiple DO probes are used, the system K_La is the average value calculated from the results estimated for all probes.

Based on experimental data obtained from the nonsteady-state clean water test, the K_La may be developed using either *linear* or *nonlinear* regression method. The linear regression method is the traditional approach to establish the K_La using Equation 10.37b. It is simple and easy for use but may cause considerable bias due to potential error transformation when the log deficit is used. In the nonlinear regression method, the values of K_La, C_0, and C_e are solved simultaneously from Equation 10.37c. This method produces more accurate data since the log deficit is not involved at all. It is a more complicated procedure that requires using a computer or programmable calculator. The latter method is used in the latest ASCE standard for measurement of oxygen transfer in clean water.[52,53]

In the linear regression approach, the value of C_e is estimated first so Equation 10.37b will give a linear relationship of the form $y = mx + c$. A plot of $\ln(C_e - C)$ versus time t gives a straight line. The slope of the line is $-K_La$ for the aeration device tested in the liquid, and the intercept on the Y axis is $\ln(C_e - C)$. Also, an arithmetic plot of $\ln[(C_e - C)/(C_e - C_0)]$ versus t gives a linear relationship. The line passes through the origin and has a slope of $-K_La$. The value of C_e may be estimated experimentally from a separate aeration test in a small column. As alternatives, multiple methods including *surface saturation*, *mid-depth saturation*, *bottom saturation*, and *mid-depth corrected saturation* have also been developed for estimating the value of C_e from the saturation concentration of oxygen under standard conditions.[52] For aeration design, the value of C_e is usually corrected to account for the changes in solubility of oxygen, hydrostatic pressure, and the oxygen mole fraction in the air bubbles. See further discussions later and Examples 10.58 through 10.65, 10.68, and 10.70 for calculation procedure.

Overall Oxygen Mass Transfer Coefficient and Oxygen Uptake Rate in Wastewater: In actual applications, the liquid could be tap water, raw wastewater, treated wastewater, or the mixed liquor in aeration basin. The oxygen transfer in wastewater containing mixed liquor is affected due to oxygen uptake by the microorganisms. The procedure for determination of K_La is modified to include oxygen uptake (or utilization) rate (γ_m) of the active biomass. The relationship is given by Equation 10.38a.[2]

$$\frac{dC}{dt} = K_La_{ww}(C_{e,ww} - C_c) - \gamma_m \tag{10.38a}$$

where

K_La_{ww} = overall oxygen mass transfer coefficient for wastewater at operating condition, h^{-1}

$C_{e,ww}$ = equilibrium concentration of DO for wastewater in the aeration basin, mg/L

C_c = concentration of oxygen maintained in the reactor, mg/L. Typically, C_c is the operating DO concentration that is normally maintained constant in a range of 1−3 mg/L.

γ_m = oxygen uptake rate (OUR), that is, rate of DO consumed by the microorganisms, mg O_2/L·h. The OUR is usually determined by respirometry or reactor study.

Under a steady-state condition, the DO concentration (C_c) is relatively a constant with $dC/dt \approx 0$ in the aeration basin. Equation 10.38a is therefore reduced to Equation 10.38b.

$$K_L a_{ww}(C_{e,ww} - C_c) - \gamma_m = 0 \tag{10.38b}$$

The oxygen mass transfer coefficient of an aerator in the aeration basin is determined from Equation 10.38c.

$$K_L a_{ww} = \frac{\gamma_m}{C_{e,ww} - C_c} \tag{10.38c}$$

The $K_L a_{ww}$ may be determined by two procedures: (a) *steady-state*, or (b) *nonsteady-state*.

Steady-State Procedure: In this procedure, the test facility is operated at desired MLVSS concentration. The operating DO concentration C_c is maintained at a relatively constant value in the basin. Equation 10.38c is used to calculate $K_L a_{ww}$.

Nonsteady-State Procedure: The nonsteady-state procedure is similar to that utilized for clean water. No chemicals are added. Instead, the DO in the test basin is allowed to approach zero by microbial respiration. Aeration device is then turned on, and DO concentration (C) is recorded with respect to time. Equation 10.38a is modified and rearranged to give a linear relationship expressed by Equation 10.38d.

$$\frac{dC}{dt} = K_L a_{ww}(C_{e,ww} - C) - \gamma_m \quad \text{or} \quad \frac{dC}{dt} = (K_L a_{ww},\, C_{e,ww} - \gamma_m) - K_L a_{ww}\, C \tag{10.38d}$$

In this equation, the factor ($K_L a_{ww}\, C_{e,ww} - \gamma_m$) is a constant under a specific operating condition. A plot of (dC/dt) versus C gives a linear relationship. The slope of the line is $-K_L a_{ww}$, and the intercept on the Y axis is ($K_L a_{ww}\, C_{e,ww} - \gamma_m$).

Once the aeration device in the nonsteady-state procedure is turned off, the DO level in the basin will drop gradually due to microbial respiration. A plot of DO concentration versus elapsed time gives a linear relationship. The slope of the line is $-\gamma_m$ (negative value).

In recent years, an alternative off-gas test procedure has also been developed for direct measurement of the system oxygen transfer rate in the wastewater under field operating conditions.[55]

Factors Influencing Oxygen Mass Transfer Coefficient: The oxygen transfer coefficient $K_L a$ depends upon many physical and chemical factors. These are (1) aeration device, (2) temperature, (3) mixing and basin geometry, (4) liquid depth, and (5) nature of wastewater (dissolved organics and minerals).[52,54,56] The factors influencing $K_L a$ are presented below.

Aeration Device: The aeration device brings air and liquid in close contact. The bubble size, intimate contact of gas–liquid film, and thickness of the film influence the value of $K_L a$. Manufacturers conduct test on their devices and specify $K_L a$ under standard operating conditions.

Temperature: The change in temperature affects the size of bubble and the liquid film coefficient. The change in $K_L a$ due to temperature effect is expressed by Equation 10.39a.

$$K_L a_T = K_L a_{20} \theta_T^{(T-20)} \tag{10.39a}$$

where

$K_L a_T$ = oxygen transfer coefficient in clean water at operating temperature T, h^{-1}

$K_L a_{20}$ = standard oxygen transfer coefficient in clean water at 20°C, h^{-1}

T = operating temperature, °C

θ_T = temperature correction coefficient for $K_L a$, dimensionless. The value of θ_T depends upon the test condition and is in the range of 1.015–1.040. The typical value for both diffused and mechanical aeration devices is 1.024.

Mixing and Basin Geometry: The intensity of mixing and basin geometry affect the contact time and surface area of gas–liquid film. In general, K_La increases with turbulent mixing and tank depth.

Liquid Depth: The liquid depth in the basin increases the contact time of oxygen in the gas–liquid film. For diffused aeration, the relationship of K_La with depth is given by Equation 10.39b.[50]

$$\frac{K_La_{H_1}}{K_La_{H_2}} = \left(\frac{H_1}{H_2}\right)^n \tag{10.39b}$$

where

$K_La_{H_1}$ and $K_La_{H_2}$ = overall oxygen mass transfer coefficients at liquid depths of H_1 and H_2, h^{-1}
H_1 and H_2 = liquid depths where the air is released, respectively, m (ft)
n = exponent, dimensionless. The typical value of n is 0.7 for most systems.

Wastewater Characteristics: The wastewater characteristics affect the K_La. The presence of surface active agents, organics, and inorganics has profound effect on K_L and a (or A/V). Two correction factors α and β are used to make proper adjustments for wastewater characteristics at a given temperature. These correction factors are expressed by Equations 10.39c and 10.39d.[1,2,12,17,48]

$$\alpha = \frac{K_La_{ww}\,(\text{wastewater})}{K_La\,(\text{clean water})} \tag{10.39c}$$

$$\beta = \frac{C_{e,ww}\,(\text{wastewater})}{C_e\,(\text{clean water})} \tag{10.39d}$$

The value of α varies with the wastewater characteristics as well as the type of aeration device, tank geometry, and degree of mixing. The value of α for diffused and mechanical aeration system is in the range of 0.4–0.8 and 0.6–1.12, respectively. The correction factor β is used to adjust the solubility of oxygen in the wastewater due to the presence of significant dissolved and suspended solids, and surface active substances. It is more important for diffused aeration where the β value is in the range of 0.8 to 1 with a typical value of 0.95. For mechanical aerators, $\beta = 1$ is usually assumed. Manufacturers provide the standard value of K_La_{20} for their aeration devices in clean water at 20°C. They may also suggest the value of α and β for different wastewaters. It is desirable to determine K_La, K_La_{20}, α, β, and γ_m from bench-, pilot-, or full-scale testing. Procedures to determine these coefficients are given in Example 10.63. Scale-up models should be considered to project the data for field conditions.

The K_La_{ww} for wastewater under field conditions are generally expressed by Equation 10.39e.

$$K_La_{ww} = K_La_{20}(1.024)^{(T-20)}\alpha \tag{10.39e}$$

Equilibrium Concentation of DO in Wastewater ($C_{e,ww}$): Similar to the general oxygen transfer rate (OTR) in Equation 10.36b, the OTR for wastewater under field condition (OTR_f) can be expression by Equation 10.39f.

$$OTR_f = K_La_{ww}(C_{e,ww} - C_c)V \quad \text{or} \quad OTR_f = K_La_{20}(C_{e,ww} - C_c)V(1.024)^{(T-20)}\alpha \tag{10.39f}$$

where ORT_f = field oxygen transfer rate for wastewater in the aeration basin under field operating conditions, kg O_2/h (lb O_2/h)

The factor ($C_{e,ww} - C_c$) represents the driving force, and the term ($1.024^{(T-20)}\,\alpha$) is the conversion factor from clean water to wastewater under the field operating condition.

In an aeration basin, $C_{e,ww}$ is the average concentration of DO for wastewater at equilibrium. For design purpose, it can be estimated from the surface saturation concentration of DO in clean water

($C_{s,20}$) at 20°C and 1 atm.* Multiple correction factors are used to obtain the relationship between $C_{e,ww}$ and $C_{s,20}$ in Equation 10.39g. These factors are expressed by Equations 10.39d, and 10.39h through 10.39j. The procedure to obtain these terms is covered in many solved examples later.

$$C_{e,ww} = \left(\frac{C_{e,ww}}{C_e}\right) \times \left(\frac{C_e}{C_{s,f}}\right) \times \left(\frac{C_{s,f}}{C_{s,t}}\right) \times \left(\frac{C_{s,t}}{C_{s,20}}\right) \times C_{s,20} \quad \text{or} \quad C_{e,ww} = \beta f_a f_p f_t C_{s,20} \tag{10.39g}$$

$$f_a = \frac{C_e}{C_{s,f}} \tag{10.39h}$$

$$f_p = \frac{C_{s,f}}{C_{s,t}} \tag{10.39i}$$

$$f_t = \frac{C_{s,t}}{C_{s,20}} \tag{10.39j}$$

where

$C_{e,ww}$ = average concentration of DO for wastewater at equilibrium in the aeration basin at the water temperature, atmospheric pressure, and altitude in the field, mg/L

C_e = average concentration of DO for clean water at equilibrium in the aeration basin at the water temperature, atmospheric pressure, and altitude in the field, mg/L

$C_{s,f}$ = surface saturation concentration of oxygen in clean water at the water temperature, atmospheric pressure, and altitude in the field, mg/L

$C_{s,t}$ = surface saturation concentration of oxygen in clean water at the field water temperature and 1 atm, mg/L. The value of $C_{s,t}$ can be found in Table 10.16. For TDS concentration ≤ 800 mg/L, the impact of TDS on solubility of oxygen is ignorable at sea level.

$C_{s,20}$ = surface saturation concentration of oxygen in clean water under the standard conditions (20°C and 1 atm), 9.09 mg/L

f_a = aeration process correction factor for depth and/or oxygen content, dimensionless. This correction factor is primarily used for the diffused aeration systems. The calculation procedures used to determine the value of f_a are presented later.

f_p = pressure correction factor for field altitude, dimensionless. By rearranging Equation B.2 in Appendix B, the value of f_b is calculated from Equation 10.39k.

$$f_p = \frac{P_f - P_{v,t}}{P_s - P_{v,t}} \tag{10.39k}$$

where

P_f = absolute atmospheric pressure at field altitude, kPa (psia)

P_s = absolute atmospheric pressure at sea level (1 atm), 101.325 kPa (14.7 psia)

$P_{v,t}$ = saturated vapor pressure of water at the field water temperature, kPa (psia)

For field altitudes <600 m (2000 ft) and field temperature <35°C (95°F), $P_{v,t} < 0.06\,P_s$. Therefore, Equation 10.39m can be simplified to Equation 10.39l.

$$f_p = \frac{P_f}{P_s} \quad \text{or} \quad f_p = \frac{H_f}{H_s} \tag{10.39l}$$

where

f_t = temperature correction factor for oxygen solubility, dimensionless

H_f = absolute atmospheric pressure at the field altitude above sea level measured in head of water, m (ft)

H_s = absolute atmospheric pressure at sea level (1 atm) measured in head of water, 10.33 m (33.9 ft)

* 1 atm = 760 mm Hg = 10.33 m (33.9 ft) water = 101.325 kPa (kN/m^2) = 14.7 psia.

Equation 10.39m is obtained by substituting Equation 10.39g into Equation 10.39f.

$$\begin{aligned} OTR_f &= K_La_{ww}(\beta f_a f_p f_t C_{s,20} - C_c)V \\ &= K_La_{20}(\beta f_a f_p f_t C_{s,20} - C_c)V(1.024)^{(T-20)}\alpha \end{aligned} \tag{10.39m}$$

Standard and Field Oxygen Transfer Rates: By rearranging Equation 10.39m and applying a fouling factor (F), an important relationship between the OTR_f and standard oxygen transfer rate ($SOTR$) is obtained and expressed by Equations 10.40a through 10.40c.

$$OTR_f = SOTR\left(\frac{\beta f_a f_p f_t C_{s,20} - C_c}{C_{e,20}}\right)(1.024)^{(T-20)}\alpha F \tag{10.40a}$$

$$SOTR = K_La_{20}C_{e,20}V \tag{10.40b}$$

$$F = \frac{K_La_{ww}\text{ (used device)}}{K_La_{ww}\text{ (new device)}} \tag{10.40c}$$

where

$SOTR$ = standard oxygen transfer rate of a new aeration device in clean water at 20°C and zero DO, kg O_2/h (lb O_2/h)

$C_{e,20}$ = equilibrium concentration of oxygen in clean water under standard conditions (20°C and 1 atm) during the nonsteady state clean water test, mg/L. The value of $C_{e,20}$ for the aeration device shall be obtained from non-linear regression method and provided by the manufacturer. For design purpose, it may be estimated from Equation 10.40d.[2,57]

$$C_{e,20} = C_{s,20}\left(1 + d_e\frac{H_b}{H_s}\right) \tag{10.40d}$$

where

H_b = depth of aeration basin, m (ft)

d_e = effective depth correction factor for the aeration device, dimensionless. For the diffused aeration systems, the value of d_e may be in a range of 0.25–0.45 when the aeration device is submerged to more than 90% of the basin depth. A typical d_e of 0.4 may be used.

The value of $C_{e,20}$ is typically higher than $C_{s,20}$ of 9.09 mg/L for diffused aerators. It may be considered that $C_{e,20} = C_{s,20}$ for surface aerators if $d_e \approx 0$ is assumed.

V = aeration basin volume, m^3 (ft^3)

F = fouling factor due to growth of slime and diffuser clogging from impurities in air, dimensionless. Typical value of F is 0.65–0.9 for diffused aerators, and 1 for surface aerators.

All other terms have been defined previously.

K_La_{20}, $SOTR$, and $C_{e,20}$ are all device-specific parameters and shall be provided as the clean water test results by the aeration device manufacturers.[52,57]

Standard and Field Oxygen Transfer Efficiencies: The standard oxygen transfer efficiency ($SOTE$) of a new aeration device is the ratio of $SOTR$ and the standard oxygen supply rate ($SOSR$). It is expressed by

Equations 10.41a and 10.41b.[57]

$$SOTE = \frac{\text{weight of oxygen transferred per unit time}}{\text{weight of oxygen supplied per unit time}} \times 100\% \quad (10.41a)$$

$$SOTE = \frac{SOTR}{SOSR} \times 100\% \quad \text{or} \quad SOTE = \frac{SOTR}{\rho_a w_{O2} Q_a} \times 100\% \quad (10.41b)$$

where

$SOTE$ = standard oxygen transfer efficiency of a new aeration device in clean water at 20°C and zero DO, %

$SOSR$ = standard (or specified) oxygen supply rate, kg O_2/h (lb O_2/h)

ρ_a = density or specific weight of air under standard conditions (20°C and 1 atm), 1.20 kg/sm^3 for SI units, and 0.075 lb/scf for U.S. customary units

w_{O2} = weight fraction of oxygen in air, 0.232 kg O_2/kg air (0.232 lb O_2/lb air)

Q_a = specified air flow under standard conditions, sm^3/h (scfm × 60 min/h)

The values of *SOTE* are calculated from the experimental data obtained at the air flow rate specified for the clean water test and should be provided by the manufacturers of aeration device.[52,57] See Tables 10.18 and 10.19 for the *SOTE* values for different aeration devices.

The correction from *SOTR* to OTR_f in Equation 10.40a is also applicable for converting *SOTE* to the field oxygen transfer efficiency (OTE_f) under process operating conditions. It is expressed by Equation 10.41c.[52,57]

$$OTE_f = SOTE\left(\frac{\beta f_a f_p f_t C_{s,20} - C_c}{C_{e,20}}\right)(1.024)^{(T-20)}\alpha F \quad (10.41c)$$

where OTE_f = field oxygen transfer efficiency for wastewater in the aeration basin under process operating conditions, %

The value of OTE_f may also be estimated from Equation 10.41d when the field operating information is available while the field oxygen transfer rate (OSR_f) for wastewater is expressed by Equation 10.41e.

$$OTE_f = \frac{OTR_f}{OSR_f} \times 100\% \text{ or } OTE_f = \frac{OTR_f}{\rho_{a,f} w_{O2} Q_{a,f}} \times 100\% \quad (10.41d)$$

$$OSR_f = \rho_{a,f} w_{O2} Q_{a,f} \quad (10.41e)$$

where

OSR_f = field oxygen supply rate for wastewater, %

$\rho_{a,f}$ = density or specific weight of air under field conditions, kg/sm^3 (lb/scf)

$Q_{a,f}$ = air flow under field conditions, m^3/h (cfm × 60 min/h)

Standard and Field Aeration Efficiencies: The standard aeration efficiency (*SAE*) is also one of the most important operating parameters for the aeration devices. It is the standard mass oxygen transfer rate achieved per unit of power input as expressed by Equation 10.42a.

$$SAE = \frac{SOTR}{P_a} \quad (10.42a)$$

where

SAE = standard aeration efficiency of a new aeration device in clean water at 20°C and zero DO, kg O_2/kW·h (lb O_2/hp·h)

P_a = standard (or specified) power input, kW (hp). P_a may be specified as shaft (or delivered), brake, wire, or total wire horsepower.

The values of SAE are also determined from the clean water test results and shall be provided by the manufacturers.[52,57] See Tables 10.18 and 10.19 for the SAE values for different aeration devices.

Similar to oxygen transfer efficiencies, the relationship between AE_f and SAE can also be expressed by Equation 10.42b.[52,57]

$$AE_f = SAE\left(\frac{\beta f_a f_p f_t C_{s,20} - C_c}{C_{e,20}}\right)(1.024)^{(T-20)}\alpha F \tag{10.42b}$$

where AE_f = field aeration efficiency for wastewater in the aeration basin under process operating conditions, kg O_2/kW·h (lb O_2/hp·h)

Oxygen Transfer by Diffused Aeration: Diffused aerators utilize air bubbles released near the bottom of the tank. These bubbles are formed at the orifice from which they break and rise through the liquid column, and burst at the liquid surface.

Aeration Process Correction Factor (f_a): For diffused aeration devices, one of the most challenging steps in oxygen transfer calculations is to estimate the value of f_a, a comprehensive oxygen solubility correction factor for aeration process. It is the ratio of C_e to $C_{s,f}$. It involves corrections for the operating depth of aeration device and/or oxygen content in the air bubbles. The most commonly used formulas for this purpose are summarized in Equations 10.43a through 10.43f.[2,7,50,53,57]

a. *Mid-depth model*

$$f_a = 1 + \gamma_w \frac{H_r}{2P_f} \quad \text{or} \quad f_s = 1 + \frac{H_r}{2H_f} \tag{10.43a}$$

b. *Mid-depth model with oxygen content correction*

$$f_a = \frac{1}{2}\left(1 + \gamma_w \frac{H_r}{P_f}\right) + \frac{O_{ex}}{2O_{in}} \quad \text{or} \quad f_a = \frac{1}{2}\left(1 + \frac{H_r}{H_f}\right) + \frac{O_{ex}}{2O_{in}} \tag{10.43b}$$

$$O_{ex} = \frac{0.21 \times (1 - (OTE_f/100\%))}{0.79 + 0.21 \times \left(1 - (OTE_f/100\%)\right)} \times 100\% \tag{10.43c}$$

$$O_{ex} \approx \frac{0.21 \times (1 - (SOTE/100\%))}{0.79 + 0.21 \times (1 - (SOTE/100\%))} \times 100\% \tag{10.43d}$$

c. *Mid-depth model based on oxygen transfer efficiency*

$$f_a = \frac{1}{2}\left(1 + \gamma_w \frac{H_r}{P_f}\right) + \frac{1}{2}\left(1 - \frac{OTE_f}{100\%}\right) \quad \text{or} \quad f_a = \frac{1}{2}\left(1 + \frac{H_r}{H_f}\right) + \frac{1}{2}\left(1 - \frac{OTE_f}{100\%}\right) \tag{10.43e}$$

$$f_a \approx \frac{1}{2}\left(1 + \gamma_w \frac{H_r}{P_f}\right) + \frac{1}{2}\left(1 - \frac{SOTE}{100\%}\right) \quad \text{or} \quad f_a \approx \frac{1}{2}\left(1 + \frac{H_r}{H_f}\right) + \frac{1}{2}\left(1 - \frac{SOTE}{100\%}\right) \tag{10.43f}$$

where

γ_w = specific weight of water at the field water temperature, kN/m^3 (lb/in^3)

H_r = hydrostatic pressure caused by the submerged depth of aeration device in the aeration basin measured in head of water, m (in). It may be substituted by the aeration basin depth (H_b) when the aeration device is submerged to more than 90% of the basin depth ($H_r \approx H_b$).

O_{ex} = oxygen content in the air bubbles at the exhaust, percent by volume

O_{in} = oxygen content in the air supply flow, percent by volume. The oxygen content of atmospheric air is 20.95% or ~21% by volume.

All other terms have been defined previously.

In Equation 10.43a, the change in oxygen content in the air bubbles is ignored while Equation 10.43b is used when the oxygen content correction is considered. Equations 10.43e and 10.43f are simplified expressions using directly the oxygen transfer efficiencies. It is a conservative practice of using *SOTE* (Equation 10.43f) instead of OTE_f (Equation 10.43e).

Empirical Equations for Oxygen Transfer: Empirical equations have also been developed to estimate the oxygen transfer from the diffused aeration devices. The interfacial bubble area per unit volume of water is a or A/V. It is expressed by Equation 10.44a. The standard K_La_{20} for a diffused aeration device is expressed by Equation 10.44b, and the K_La_{ww} for wastewater at operating conditions is expressed by Equation 10.44c.[50]

$$a = \frac{A}{V} = \frac{6Q_aH}{d_Bv_BV} \tag{10.44a}$$

$$K_La_{20} = \frac{C_1H^{2/3}Q_a^{1-n}}{V} \quad \text{or} \quad K_La_{20}V = C_1H^{2/3}Q_a^{1-n} \tag{10.44b}$$

$$K_La_{ww} = \frac{C_1H^{2/3}Q_a^{1-n}}{V}(1.024)^{(T-20)}\alpha$$

$$\text{or} \quad K_La_{ww}V = C_1H^{2/3}Q_a^{1-n}(1.024)^{(T-20)}\alpha \tag{10.44c}$$

where

Q_a = air flow per diffuser under standard condition (at 20°C and 1 atm), sm³/s (scfm)
d_B = bubble diameter, m (ft)
v_B = bubble velocity, m/s (ft/s)
H = liquid depths where the air is released, m (ft)
C_1 and n = constants provided by the manufacturer for the diffused aeration device. The values of C_1 are 0.04233 and 0.0017 in the SI and U.S. customary units, respectively; and n is 0.1 in both units.[1]

All other terms have been defined previously.

Therefore, the OTR_f per diffuser for wastewater in an aeration basin is expressed by Equation 10.44d.[50]

$$OTR_f = C_1H^{2/3}Q_a^{1-n}(C_{e,ww} - C_c)(1.024)^{(T-20)}\alpha \tag{10.44d}$$

Equation 10.44d can also be modified to include the effect of basin width. This relationship is expressed by Equation 10.44e.[11]

$$OTR_f = C_2\frac{H^m}{W^p}Q_a^n(C_{e,ww} - C_c)(1.024)^{(T-20)}\alpha \tag{10.44e}$$

where

OTR_f = oxygen transfer rate under field conditions, lb O_2/h
W = width of the basin, ft
C_2 = constant that may vary upon W
m, n, and p = exponents. The values depend upon the characteristics of diffused aeration device.

All other terms have been defined previously.

Power Requirements: One of the equations that can be used to calculate the shaft horsepower (SHP) requirement by a blower is expressed by Equation 10.45a.[2,58]

$$P_{\text{shp,blower}} = \frac{(Q_a/60)\rho_a R T_i}{C(k-1/k)(\eta_b/100\%)}\left[\left(\frac{P_d}{P_i}\right)^{\frac{k-1}{k}} - 1\right] \tag{10.45a}$$

where

$P_{\text{shp,blower}}$ = blower SHP requirement, kW (hp)
Q_a = air flow under standard conditions, sm^3/min (scfm)
R = universal gas constant for air, 8.314 J/mole·°K in SI units, and 53.3 ft-lb/lb·°R in U.S. customary units
T_i = blower inlet air temperature, °K (°R). °K = °C + 273, and °R = °F + 460
k = ratio of specific heats of air, C_p/C_v, dimensionless
k = $C_p/C_v = 1.395$, and $(k-1)/k = 0.283$
η_b = blower efficiency, %. The blower efficiency is usually 70–90%.
P_d = absolute blower discharge pressure, atm (psia)
psia = psig + barometric pressure (psia)
P_i = absolute blower inlet pressure, atm (psia)
C = constant, 28.97 kg/k mole in SI units, and 550 ft-lb/s·hp in U.S. customary units
60 = conversion factor, 60 s/min

The brake horsepower (BHP) requirement for the drive unit is expressed by Equation 10.45b.

$$P_{\text{bhp}} = \frac{P_{\text{shp}}}{E_T} \times 100\% \tag{10.45b}$$

where

P_{bhp} = BHP requirement, hp
P_{shp} = SHP requirement, hp
E_T = efficiency of gear box or drive belt, %

Oxygen Transfer by Mechanical Aeration: Mechanical aerators can be either surface or submerged devices. The surface mechanical aerators create turbulence near the surface. The *turbine aerators* are submerged mechanical aerators that shear coarse bubbles from a diffuser into fine bubbles and disperse the bubbles by the pumping action of the rotating impeller.

Aeration Process Correction Factor (f_a): For surface mechanical aerators, the value of $f_a = 1$ is typically assumed. It is a conservative consideration since the OTR may be enhanced when a portion of air bubbles created by the turbulence near the surface is actually carried down into the basin by the rolling action. The calculations summarized in Equation 10.43 for diffused aerators may also be used for submerged mechanical aerators with engineering judgment.

Empirical Equations for Oxygen Transfer: For a turbine aerator, the standard $K_L a_{20}$ is expressed by Equation 10.46a, while the $K_L a_{ww}$ for wastewater at operating conditions is expressed by Equation 10.46b.[11]

$$K_L a_{20} = \frac{C_3 R_i^x Q_a^n d_i^m}{V} \quad \text{or} \quad K_L a_{20} V = C_3 R_i^x Q_a^n d_i^m \tag{10.46a}$$

$$K_L a_{ww} = \frac{C_3 R_i^x Q_a^n d_i^m}{V}(1.024)^{(T-20)}\alpha$$

$$\text{or} \quad K_L a_{ww} V = C_3 R_i^x Q_a^n d_i^m (1.024)^{(T-20)}\alpha \tag{10.46b}$$

where

R_i = impeller peripheral speed, m/s (fps). Typical value of R_i is 3–5.5 m/s (10–18 fps).
d_i = impeller diameter, ft (m). Typical ratio of d_i to equivalent tank diameter is 0.1–0.2.
C_3 = constant. Typical value of C_3 is 20.
x, n, m = exponents. The typical values are: $x = 1.2–2.4$, $n = 0.4–0.9$, and $m = 0.6–1.8$.
All other terms have been defined previously.

The OTR_f of a turbine aerator is expressed by Equation 10.46c.

$$OTR_f = C_3 R_i^x G_s^n d_i^m (C_{e,ww} - C_c)(1.024)^{(T-20)}\alpha \quad (10.46c)$$

Power Requirements: The power requirement delivered to a turbine aerator is expressed by Equation 10.47.[11]

$$P_{shp,turbine} = C_4 d_i^n r_i^m \quad (10.47)$$

where

$P_{shp,turbine}$ = turbine shaft power, hp
d_i = impeller diameter, ft
r_i = impeller rotation speed, rps
n, m = exponents. The typical values are: $n = 4.8–5.3$, and $m = 2–2.5$.
C_4 = constant. The typical value of C_4 is 0.02.

The BHP requirement for the drive unit can be calculated from Equation 10.45b.

Theoretical and Standard Oxygen Requirements: The theoretical oxygen requirement (*ThOR*) in an aerobic suspended growth reactor is determined by the microbial activities involved in the wastewater treatment processes. It is calculated from the removal of ultimate oxygen demand (UOD) minus CBOD and ammonia nitrogen fixed in the cell mass. It is calculated from Equation 10.48a or 10.48b. The derivation of these equations may be found in Example 10.70.

$$UOR = UOD_0 - UOD_e - \left(\frac{0.92}{\text{VSS/TSS ratio}} + 0.56\right)p_x \quad (10.48a)$$

$$ThOR = Q(UOD_0 - UOD_e) - \left(\frac{0.92}{\text{VSS/TSS ratio}} + 0.56\right)P_x$$

$$\text{or} \quad ThOR = Q \times UOR \quad (10.48b)$$

where

UOR = ultimate oxygen concentration requirement in the aeration basin under process operating conditions, mg O_2/L
UOD_0 = ultimate oxygen demand concentration in the influent to the aeration basin, mg/L (g/m^3)
UOD_e = ultimate oxygen demand concentration in the effluent from the aeration basin, mg/L (g/m^3)
p_x = biosolids (VSS) concentration increase during the biological treatment process, mg VSS/L
$ThOR$ = theoretical oxygen requirement in the aeration basin under process operating conditions, kg O_2/d
Q = flow rate of influent to the aeration basin, m^3/d
P_x = growth rate of biosolids (VSS) in the aeration basin, kg VSS/d

The value of *UOD* is expressed by Equation 5.18a and P_x is calculated from Equation 10.15m.

In process design, the OTR by the aeration device under field condition (OTR_f) must be equal to *ThOR*. The relationship between *SOTR* and OTR_f in Equation 10.40a is also used to calculate the standard oxygen requirement (*SOR*) from *ThOR*. It is expressed by Equation 10.48c.[7] The *SOR* provides a basis for design of air supply system.

$$SOR = \frac{ThOR}{((\beta f_a f_p f_t C_{s,20} - C_c)/C_{e,20})(1.024)^{(T-20)}\alpha F}$$

or

$$SOR = \frac{ThOR}{((C_{e,ww} - C_c)/C_{e,20})(1.024)^{(T-20)}\alpha F} \quad (10.48c)$$

where SOR = standard oxygen requirement, kg O_2/d

All other terms have been defined previously.

EXAMPLE 10.57: SODIUM SULFITE DOSE FOR DEOXYGENATION

Deoxygenation dose of sodium sulfite is slightly higher than the stoichiometric dose of sodium sulfite that is required to deoxygenate the water for determination of K_La in clean water. Calculate the stoichiometric dose, and quantity of sodium sulfite added in a test tank. The volume of the tank is 30 m^3, and the initial DO concentration in the water is 8.3 mg/L.

Solution

1. Calculate the stoichiometric dose of sodium sulfite required for deoxygenation.

   The deoxygenation reaction with sodium sulfite is carried out in the presence of 0.5 mg/L cobalt as a catalyst. The reaction with sodium sulfite is given below.

$$\underset{2\times 126}{2\,Na_2SO_3} + \underset{32}{O_2} \xrightarrow{\text{Co (Catalyst)}} 2\,Na_2SO_4$$

   Calculate the stoichiometric dose of Na_2SO_3 required (D_{stoi}) for deoxygenation per mg of DO.

$$D_{stoi} = \frac{2 \times 126\ \text{mg/L as}\ Na_2SO_3}{32\ \text{mg/L as}\ O_2} = 7.9\ \text{mg}\ Na_2SO_3/\text{mg}\ O_2$$

2. Calculate the quantity of Na_2SO_3 required for deoxygenation in the tank.

   Slightly higher than the stoichiometric dose of Na_2SO_3 is added in the tank. The most practically used dose (D_{deox}) for the clean water test is 8 mg Na_2SO_3/mg O_2. The actual dosage of Na_2SO_3 (C_{deox}) required for deoxygenation of the initial DO concentration DO = 8.3 mg/L in the tank is calculated below.

$$C_{deox} = D_{deox}DO = 8\ \text{mg}\ Na_2SO_3/\text{mg}O_2 \times 8.3\ \text{mg/L as}\ O_2 = 66.4\ \text{mg/L or}\ 66.4\ \text{g/m}^3\ \text{as}\ Na_2SO_3$$

   Calculate the quantity of Na_2SO_3 (W_{deox}) required for deoxygenation in the tank with volume V = 30 m^3.

$$W_{deox} = C_{deox}V = 66.4\ \text{g}\ Na_2SO_3/\text{m}^3 \times 30\ \text{m}^3 \times 10^{-3}\ \text{kg/g} = 1.99\ \text{kg} \approx 2\ \text{kg as}\ Na_2SO_3$$

EXAMPLE 10.58: DETERMINATION OF STANDARD K_La OF A SURFACE AERATOR IN CLEAN WATER

A surface aerator was tested in clean water in a tank of 250 m^3 capacity. The water temperature was 7.5°C, and was deoxygenated with sodium sulfite and cobalt chloride as a catalyst. The dissolved oxygen data with respect to elapsed time is given below. Determine the standard K_La_{20} using linear regression method and calculate standard oxygen transfer rate (*SOTR*) in clean water under standard conditions (20°C and 1 atm). The average power input was 2.9 kW during the 60-min testing period. Also calculate the standard aeration efficiency (*SAE*). Assume the field altitude is ignorable.

Elapsed Time (t), min	0	6	12	18	24	30	36	42	48	54	60
DO concentration (C), mg/L	2.3	3.7	4.9	5.9	6.8	7.5	8.1	8.7	9.2	9.6	9.9

Solution

1. Determine the linear relationship used for determination of K_La.

   Both relationships in Equation 10.37b can be used in the linear regression method. These expressions are summarized below.

   $$\ln(C_e - C) = \ln(C_e - C_0) - K_La{\cdot}t$$

   The linear relationship is obtained by plotting $\ln(C_e - C)$ versus t.

   $$\ln\left(\frac{C_e - C}{C_e - C_0}\right) = -K_La{\cdot}t$$

   The linear relationship is obtained by plotting $\ln[(C_e - C)/(C_e - C_0)]$ versus t.
2. Determine the equilibrium concentration of DO in clean water at 7.5°C.

   Since the field altitude is ignorable, C_e can be estimated from the surface saturation concentration of DO (C_s) in clean water, and $C_e = C_s$ for a surface aerator.

   At 1 atm (sea level), $C_s = 12.45$ and 11.84 mg/L at 6°C and 8°C, respectively, as obtained from Table 10.16. Interpolate C_s at 7.5°C.

   $$C_s = 12.45\,\text{mg/L} + \frac{(7.5-6)^\circ\text{C}}{(8-6)^\circ\text{C}} \times (11.84 - 12.45)\text{mg/L} = 12.0\,\text{mg/L}$$

   $$C_e = C_s = 12.0\,\text{mg/L}$$

3. Tabulate the test data to prepare linear plots.

   Calculate the difference of C from C_e and tabulate the terms expressed for different linear relationships.
4. Plot linear relationships for determination of K_La at 7.5°C.

   Two linear relationships identified in Step 1 are plotted in Figure 10.46. The plots of $\ln(C_e - C)$ and $\ln[(C_e - C)/(C_e - C_0)]$ with respect to t are shown in Figure 10.46a and b, respectively.
5. Determine the value of K_La at 7.5°C from both plots.
   a. From the linear relationship between $\ln(C_e - C)$ and t.

      Slope of the linear relationship, $-K_La = -0.026\,\text{min}^{-1}$.

      It gives $K_La = 0.026\,\text{min}^{-1}$ or $1.6\,\text{h}^{-1}$.

Intercept of the line, $\ln(C_e - C_0) = 2.27$.

$$(C_e - C_0) = e^{2.27} = 9.68 \text{ mg/L} \quad \text{or} \quad C_0 = (C_e - 9.68) \text{ mg/L} = (12.0 - 9.68) \text{ mg/L} = 2.3 \text{ mg/L}$$

t, min	C, mg/L	$C_e - C$, mg/L	$\ln(C_e - C)$	$(C_e - C)/(C_e - C_0)$	$\ln[(C_e - C)/(C_e - C_0)]$
0	2.3	9.7[a]	2.27	1.000	0.00
6	3.7	8.3[b]	2.12[c]	0.856[d]	−0.156[e]
12	4.9	7.1	1.96	0.732	−0.312
18	5.9	6.1	1.81	0.629	−0.464
24	6.8	5.2	1.65	0.536	−0.623
30	7.5	4.5	1.50	0.464	−0.768
36	8.1	3.9	1.36	0.402	−0.911
42	8.7	3.3	1.19	0.340	−1.08
48	9.2	2.8	1.03	0.289	−1.24
54	9.6	2.4	0.875	0.247	−1.40
60	9.9	2.1	0.742	0.216	−1.53

[a] $C_e - C_0 = (12.0 - 2.3)\text{mg/L} = 9.7 \text{ mg/L}$

[b] $C_e - C = (12.0 - 3.7)\text{mg/L} = 8.3 \text{ mg/L}$

[c] $\ln(C_e - C) = \ln(8.3) = 2.12$

[d] $\dfrac{C_e - C}{C_e - C_0} = \dfrac{8.3 \text{ mg/L}}{9.7 \text{ mg/L}} = 0.856$

[e] $\ln\left(\dfrac{C_e - C}{C_e - C_0}\right) = \ln(0.856) = -0.156$

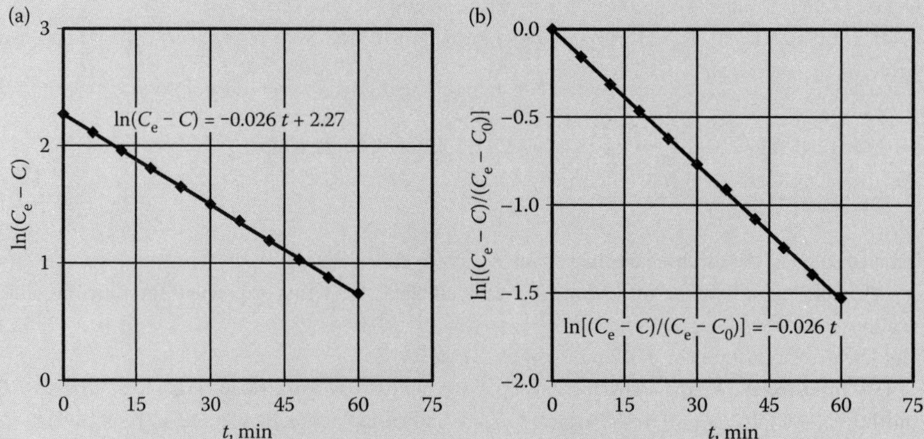

FIGURE 10.46 Plots for determination of K_La: (a) $\ln(C_e - C)$ versus t, and (b) $\ln[(C_e - C)/(C_e - C_0)]$ versus t (Example 10.58).

The calculated C_0 is verified with the measured DO concentration at $t = 0$. The calculated and measured values are identical.

b. From the linear relationship between $\ln[(C_e - C)/(C_e - C_0)]$ versus t.
Slope of the linear relationship, $-K_La = -0.026\ \text{min}^{-1}$. It also gives $K_La = 1.6\ \text{h}^{-1}$.
Therefore, the results obtained from both expressions are the same at 7.5°C: $K_La_{7.5} = 1.6\ \text{h}^{-1}$.

6. Determine the standard K_La_{20} at 20°C.
Rearrange Equation 10.39a and calculate the standard K_La_{20} assuming $\theta_T = 1.024$.

$$K_La_{20} = K_La_T\theta_T^{(20-T)} = K_La_{7.5}\theta_T^{(20-7.5)} = 1.6\ \text{h}^{-1} \times 1.024^{(20-7.5)} = 1.6\ \text{h}^{-1} \times 1.35 = 2.2\ \text{h}^{-1}$$

7. Calculate the standard oxygen transfer rate (*SOTR*).
For a surface aerator, assume $C_{e,20} = C_{s,20}$.
$C_{e,20} = 9.09\ \text{mg}\ O_2/\text{L} = 9.09\ \text{g}\ O_2/\text{m}^3$ is obtained from Table 10.16 at 20°C and 1 atm.
Calculate *SORT* from Equation 10.40b at $V = 250\ \text{m}^3$.

$$SOTR = K_La_{20}C_{e,20}\ V = 2.2\ \text{h}^{-1} \times 9.09\ \text{g}\ O_2/\text{m}^3 \times 250\ \text{m}^3 \times 10^{-3}\ \text{kg/g} = 5.0\ \text{kg}\ O_2/\text{h}$$

8. Calculate the standard aeration efficiency (*SAE*).
SAE is calculated from Equation 10.42a at $P_a = 2.9\ \text{kW}$.

$$SAE = \frac{SOTR}{P_a} = \frac{5.0\ \text{kg}\ O_2/\text{h}}{2.9\ \text{kW}} = 1.7\ \text{kg}\ O_2/\text{kW·h}$$

EXAMPLE 10.59: EQUILIBRIUM CONCENTRATION OF OXYGEN FOR WASTEWATER IN AERATION BASIN

The equilibrium concentration of oxygen in aeration basin is needed to calculate the oxygen transfer rate under field conditions. An aeration basin is designed to operate at an elevation of 610 m (2000 ft) above sea level. Calculate the equilibrium concentration of oxygen for wastewater in the basin from Equation 10.39g. Use all three models expressed in Equations 10.43a, 10.43b, and 10.43e to calculate the pressure correction factor (f_a) and compare the results. Use the following data: $\beta = 0.9$, operating temperature $T = 26°\text{C}$, depth of wastewater above the diffusers = 5 m, diffuser $OTE_f = 10\%$.

Solution

1. Write the equation to estimate the equilibrium DO concentration for wastewater in the aeration basin ($C_{e,ww}$).
Equation 10.39g is used to calculate $C_{e,ww}$.

$$C_{e,ww} = \beta f_a f_p f_t C_{s,20}$$

$C_{e,20} = 9.09\ \text{mg}\ O_2/\text{L} = 9.09\ \text{g}\ O_2/\text{m}^3$ is obtained from Table 10.16 at 20°C and 1 atm.
$\beta = 0.9$ is given in the statement. The other factors, f_t, f_p, and f_a are calculated below.

2. Calculate the temperature correction factor (f_t).
The factor f_t is calculated from Equation 10.39j. The surface saturation concentration of oxygen $C_{s,t} = 8.11\ \text{mg/L}$ is obtained from Table 10.16 at 26°C and sea level.

$$f_t = \frac{C_{s,t}}{C_{s,20}} = \frac{8.11\ \text{mg/L}}{9.09\ \text{mg/L}} = 0.89$$

3. Calculate the pressure correction factor (f_p).

The barometric pressures $P_s = 101.325$ kPa or 760 mm Hg at sea level. $P_f = 706$ mm Hg at an altitude of 610 m is obtained from Table B.7 in Appendix B. Calculate P_f in kPa.

$$P_f = \frac{706 \text{ mm Hg}}{760 \text{ mm Hg}} \times 101.325 \text{ kPa} = 94.13 \text{ kPa or } 94.13 \text{ kN/m}^2$$

Water vapor pressure $P_{v,t} = 25.21$ mm Hg is obtained at 26°C from Table B.5 in Appendix B. Calculate $P_{v,t}$ in kPa.

$$P_{v,t} = \frac{25.21 \text{ mm Hg}}{760 \text{ mm Hg}} \times 101.325 \text{ kPa} = 3.36 \text{ kPa}$$

Calculate factor f_p from Equation 10.39k.

$$f_p = \frac{P_f - P_{v,t}}{P_s - P_{v,t}} = \frac{(94.13 - 3.36) \text{ kPa}}{(101.325 - 3.36) \text{ kPa}} = 0.93$$

4. Calculate the aeration process correction factor (f_a).
   The factor f_a is calculated using three methods as presented below.
   a. Mid-depth model in Equation 10.43a.
      The specific weight (γ_w) of 9.777 and 9.764 kN/m^3 is obtained for water at temperature 25°C and 30°C, respectively, from Table B.2 in Appendix B. The γ_w of water at 26°C is interpolated from these values to ~9.77 kN/m^3.
      Calculate f_a from Equation 10.43a at $H_r = 5$ m

$$f_a = 1 + \gamma_w \frac{H_r}{2P_f} = 1 + 9.77 \text{ kN/m}^3 \frac{5 \text{ m}}{2 \times 94.13 \text{ kN/m}^2} = 1.26$$

   b. Mid-depth model with oxygen content correction in Equation 10.43b.
      Assume $O_{in} = 21\%$. Calculate the oxygen content in the air bubbles at the exhaust O_{ex} from Equation 10.43c at $OTE_f = 10\%$.

$$O_{ex} = \frac{0.21 \times (1 - (OTE_f/100\%))}{0.79 + 0.21 \times (1 - (OTE_f/100\%))} \times 100\% = \frac{0.21 \times (1 - (10\%/100\%))}{0.79 + 0.21 \times (1 - (10\%/100\%))} \times 100\% = 19.3\%$$

      Calculate f_a from Equation 10.43b at $H_r = 5$ m.

$$f_a = \frac{1}{2}\left(1 + \gamma_w \frac{H_r}{P_f}\right) + \frac{O_{ex}}{2O_{in}} = \frac{1}{2}\left(1 + 9.77 \text{ kN/m}^3 \frac{5 \text{ m}}{94.13 \text{ kN/m}^2}\right) + \frac{19.3\%}{2 \times 21\%} = 0.76 + 0.46 = 1.22$$

   c. Mid-depth model based on oxygen transfer efficiency in Equation 10.43e.

$$f_a = \frac{1}{2}\left(1 + \gamma_w \frac{H_r}{P_f}\right) + \frac{1}{2}\left(1 - \frac{OTE_f}{100\%}\right) = \frac{1}{2}\left(1 + 9.77 \text{ kN/m}^3 \frac{5 \text{ m}}{94.13 \text{ kN/m}^2}\right) + \frac{1}{2}\left(1 - \frac{10\%}{100\%}\right)$$

$$= 0.76 + 0.45 = 1.21$$

   d. Compare the results obtained from Equations 10.43a, 10.43b, and 10.43e.

The factor f_a calculated using Equation 10.43a is higher than those calculated using other two models (Equations 10.43b and 10.43e). The procedure for calculation of f_a from Equation 10.43e is much shorter and simpler than that from Equation 10.43b. Equation 10.43a is, however, the simplest to use among three models.

5. Calculate the equilibrium DO concentration ($C_{e,ww}$).

   The value of $C_{e,ww}$ is calculated from Equation 10.39g using all the correction factors obtained above. The calculation results are summarized below in accordance with the models used for calculating f_a factor.

   a. Mid-depth model (Equation 10.43a).

   $$C_{e,ww} = \beta f_a f_p f_t C_{s,20} = 0.90 \times 1.26 \times 0.93 \times 0.89 \times 9.09\,\text{mg/L} = 0.94 \times 9.09\,\text{mg/L} = 8.5\,\text{mg/L}$$

   b. Mid-depth model with oxygen content correction (Equation 10.43b).

   $$C_{e,ww} = \beta f_a f_p f_t C_{s,20} = 0.90 \times 1.22 \times 0.93 \times 0.89 \times 9.09\,\text{mg/L} = 0.91 \times 9.09\,\text{mg/L} = 8.3\,\text{mg/L}$$

   c. Mid-depth model based on oxygen transfer efficiency (Equation 10.43e).

   $$C_{e,ww} = \beta f_a f_p f_t C_{s,20} = 0.90 \times 1.21 \times 0.93 \times 0.89 \times 9.09\,\text{mg/L} = 0.90 \times 9.09\,\text{mg/L} = 8.2\,\text{mg/L}$$

EXAMPLE 10.60: DETERMINATION OF K_La_{ww} OF A DIFFUSED AERATOR IN PRIMARY SETTLED WASTEWATER

A diffused aerator was tested in primary settled wastewater. The temperature of wastewater and the depth of diffusers below the water surface were 26°C and 4.5 m. The test result indicates that the surface saturation concentration of DO in the wastewater was 7.30 mg/L at temperature of 26°C and operating pressure of 1 atm. The average volume of the aeration basin during testing was 72 m^3. The wastewater was deoxygenated to $C_0 = 0$. The aeration test data are tabulated below. Calculate β for the wastewater tested and $K_La_{ww,20}$ of the diffused aerator in the wastewater at 20°C.

Elapsed Time (t), min	0	5	10	20	30	40	50	60
DO concentration in the wastewater (C), mg/L	0	0.8	1.4	2.8	3.8	4.7	5.3	6.0

Solution

1. Calculate the β coefficient for the wastewater.

   The surface saturation concentration of oxygen in wastewater at 26°C and 1 atm, $C_{s,t,ww} = 7.30$ mg/L.

   The surface saturation concentration of oxygen in clean water at 26°C and 1 atm, $C_{s,t} = 8.11$ mg/L (Table 10.16). Assume $C_{e,ww} = C_{s,t,ww} = 7.30$ mg/L. Calculate from Equation 10.39d.

   $$\beta = \frac{C_{e,ww}}{C_e} = \frac{C_{s,t,ww}}{C_{s,t}} = \frac{7.30\,\text{mg/L}}{8.11\,\text{mg/L}} = 0.90$$

2. Calculate the equilibrium DO concentration for wastewater in the aeration basin ($C_{e,ww}$).

   $C_{e,ww}$ is calculated from Equation 10.39g. Since the operating pressure is 1 atm, the pressure correction factor $f_p = 1$ and $C_{e,ww} = \beta f_a f_t C_{s,20}$. Substitute Equation 10.39j for f_t to obtain the following expression.

$$C_{e,ww} = \beta f_a \left(\frac{C_{s,t}}{C_{s,20}}\right) C_{s,20} = \beta f_a C_{s,t}$$

Use the mid-depth model (Equation 10.43a) to calculate the aeration process correction factor f_a. At 26°C, γ_w of water is ~ 9.77 kN/m³ (see Example 10.59). Calculate f_a from Equation 10.43a at $H_r = 4.5$ m and $P_f = 101.325$ kPa or 101.325 kN/m² at 1 atm.

$$f_a = 1 + \gamma_w \frac{H_r}{2P_f} = 1 + 9.77\,\text{kN/m}^3 \frac{4.5\,\text{m}}{2 \times 101.325\,\text{kN/m}^2} = 1.22$$

$$C_{e,ww} = 0.90 \times 1.22 \times 8.11\,\text{mg/L} = 8.9\,\text{mg/L}$$

3. Tabulate the test data to prepare linear plot.
   The linear relationship in Equation 10.37b is used for determination of $K_L a_{ww}$. Calculate and tabulate the values of $(C_{e,ww} - C)$, and $\ln(C_{e,ww} - C)$ at each elapsed time t.

t, min	C_{ww}, mg/L	$C_{e,ww} - C$, mg/L	$\ln(C_{e,ww} - C)$
0	0	8.9	2.19
5	0.8	8.1	2.09
10	1.4	7.5	2.01
20	2.8	6.1	1.81
30	3.8	5.1	1.63
40	4.7	4.2	1.44
50	5.3	3.6	1.28
60	6.0	2.9	1.06

4. Plot the data.
   The values of $\ln(C_{e,ww} - C)$ versus t are plotted in Figure 10.47.

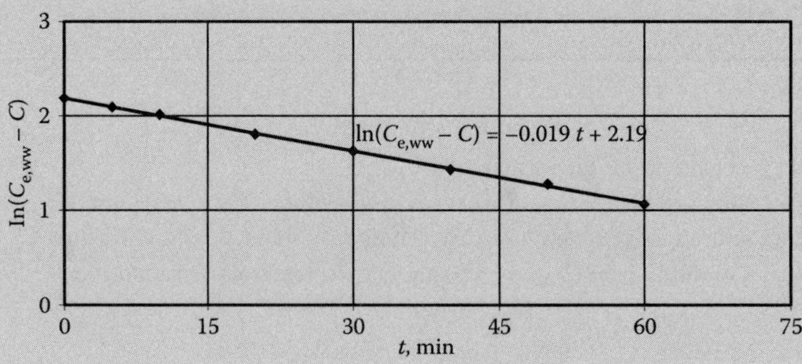

FIGURE 10.47 Plot of $\ln(C_{e,ww} - C)$ versus t for determination of $K_L a_{ww}$ of diffused aerator in primary settled wastewater (Example 10.60).

5. Determine $K_L a_{ww}$ in wastewater at 26°C.
   For the linear relationship between $\ln(C_{e,ww} - C)$ and t, the slope of $-0.019\,\text{min}^{-1}$ is $-K_L a_{ww}$.

$$K_L a_{ww} = 0.019\,\text{min}^{-1} \times 60\,\text{min/h} = 1.1\,\text{h}^{-1}$$

The intercept of 2.19 is $\ln(C_{e,ww} - C_0)$.

$$(C_{e,ww} - C_0) = e^{2.19} = 8.9 \text{ mg/L or } C_0 = C_{e,ww} - 8.9 \text{ mg/L} = (8.9 - 8.9) \text{ mg/L} = 0 \text{ mg/L}$$

This means that the DO concentration is zero at $t = 0$. Therefore, $K_La_{ww} = 1.1\ \text{h}^{-1}$.

6. Determine $K_La_{ww,20}$ in wastewater at 20°C.
   Rearrange Equation 10.39a and calculate the standard $K_La_{ww,20}$ assuming $\theta_T = 1.024$.

$$K_La_{ww,20} = K_La_{ww,T}\theta_T^{(20-T)} = K_La_{ww,26}\theta_T^{(20-26)} = 1.1\,\text{h}^{-1} \times 1.024^{(20-26)} = 1.1\,\text{h}^{-1} \times 0.87 = 1.0\,\text{h}^{-1}$$

EXAMPLE 10.61: ESTIMATION OF K_La AND OXYGEN TRANSFER EFFICIENCY OF A DIFFUSER SYSTEM

A completely mixed aeration basin uses fine-bubble aeration device at 1 atm. The test data on diffuser device in clean water at 20°C is given below. Air flow $Q_a = 1.05\ \text{sm}^3/\text{min}$ at 20°C and 1 atm, volume of the test basin $V = 30\ \text{m}^3$, basin depth $H_b = 4.5$ m, depth of water above the air release $H = 4.2$ m, bubble diameter $d^B = 0.8$ mm, average bubble rise velocity $v_B = 15$ cm/s, and oxygen transfer coefficient $K_L = 0.061$ m/h at 20°C. Estimate the following: (a) standard K_La_{20} using the empirical equation given in Equation 10.44a, (b) standard oxygen transfer rate (*SOTR*) and standard oxygen transfer rate (*SOTE*) in tap water, (c) OTE_f and OTR_f in the wastewater under the following operating condition: an aeration basin with same dimensions and volume at sea level, temperature of mixed liquor $T = 30$°C, $\alpha = 0.85$, $\beta = 0.90$, DO maintained in the aeration basin $C_c = 1.5$ mg/L, and fouling factor $F = 0.9$.

Solution

1. Calculate the ratio of interfacial area to volume (A/V).
   The ratio is calculated from Equation 10.44a at $Q_a = 1.05\ \text{sm}^3/\text{min}$ at 20°C and 1 atm, $H = 4.2$ m, $d_B = 0.8\ \text{mm} \times 10^{-3}\ \text{m/mm} = 0.8 \times 10^{-3}$ m, $v_B = 15\ \text{cm/s} \times 10^{-2}\ \text{m/cm} \times 60\ \text{s/min} = 9$ m/min, and $V = 30\ \text{m}^3$.

$$a = \frac{A}{V} = \frac{6Q_aH}{d_Bv_BV} = \frac{6 \times 1.05\,\text{m}^3/\text{min} \times 4.2\,\text{m}}{0.8 \times 10^{-3}\,\text{m} \times 9\,\text{m/min} \times 30\,\text{m}^3} = 123\,\text{m}^2/\text{m}^3$$

2. Calculate the standard K_La_{20} at $K_{L,20} = 0.061$ m/h at 20°C.

$$K_La_{20} = K_{L20} \times a = 0.061\,\text{m/h} \times 123\,\text{m}^2/\text{m}^3 = 7.50\,\text{h}^{-1}$$

3. Calculate the equilibrium concentration of oxygen in clean water ($C_{e,20}$) under standard conditions.
   $C_{s,20} = 9.09\ \text{mg O}_2/\text{L} = 9.09\ \text{g O}_2/\text{m}^3$ is obtained from Table 10.16 at 20°C and 1 atm.
   Calculate the ratio of H to H_b.

$$\frac{H}{H_b} = \frac{4.2\,\text{m}}{4.5\,\text{m}} = 0.93 \text{ or } 93\% > 90\%$$

Since $H/H_b > 90\%$, use the typical value of $d_e = 0.4$ and calculate $C_{e,20}$ from Equation 10.40d at $H_s = 10.33$ m at 20°C and 1 atm.

$$C_{e,20} = C_{s,20}\left(1 + d_e\frac{H_b}{H_s}\right) = 9.09\,\text{mg O}_2/\text{L} \times \left(1 + 0.4 \times \frac{4.5\,\text{m}}{10.33\,\text{m}}\right) = 10.7\,\text{mg/L or } 10.7\,\text{g O}_2/\text{m}^3$$

4. Calculate *SOTR* from Equation 10.40b.

$$SOTR = K_La_{20}C_{e,20}\,V = 7.50\,\text{h}^{-1} \times 10.7\,\text{g O}_2/\text{m}^3 \times 30\,\text{m}^3 \times 10^{-3}\,\text{kg/g} = 2.4\,\text{kg O}_2/\text{h}$$

5. Calculate *SOTE* from Equation 10.41b.

$$SOTE = \frac{SOTR}{\rho_a w_{O2} Q_a} \times 100\%$$
$$= \frac{2.4\,\text{kg O}_2/\text{h}}{1.05\,\text{sm}^3/\text{min} \times 60\,\text{min/h} \times 0.232\,\text{kg O}_2/\text{kg air} \times 1.23\,\text{kg air/sm}^3} \times 100\% \quad = 13.3\%$$

6. Calculate the equilibrium DO concentration in the wastewater under operating conditions ($C_{e,ww}$).

$C_{e,ww}$ is calculated from Equation 10.39g. $\beta = 0.9$ is given in the statement and $f_p = 1$ at sea level. The other correction factors, f_t and f_a are calculated below.

a. Calculate the temperature correction factor.

The factor f_t is calculated from Equation 10.39j. The surface saturation concentration of oxygen $C_{s,t} = 7.54$ mg/L is obtained from Table 10.16 at 30°C and sea level.

$$f_t = \frac{C_{s,t}}{C_{s,20}} = \frac{7.54\,\text{mg/L}}{9.09\,\text{mg/L}} = 0.83$$

b. Calculate the aeration process correction factor.

A trial and error procedure is required to obtain f_a since OTE_f is unknown. Assuming $OTE_f =$ 7.7%, calculate f_a from Equation 10.43e at sea level.

$$f_a = \frac{1}{2}\left(1 + \frac{H_r}{H_s}\right) + \frac{1}{2}\left(1 - \frac{OTE_f}{100\%}\right) = \frac{1}{2}\left(1 + \frac{4.2\,\text{m}}{10.33\,\text{m}}\right) + \frac{1}{2}\left(1 - \frac{7.7\%}{100\%}\right) = 0.70 + 0.46 = 1.16$$

c. Calculate $C_{e,ww}$.

$$C_{e,ww} = \beta f_a f_p f_t C_{s,20} = 0.90 \times 1.16 \times 1 \times 0.83 \times 9.09\,\text{mg/L} = 0.87 \times 9.09\,\text{mg/L} = 7.91\,\text{mg/L}$$

7. Calculate OTE_f under field conditions.

Calculate OTE_f from Equation 10.41c.

$$OTE_f = SOTE\left(\frac{\beta f_a f_p f_t C_{s,20} - C_c}{C_{e,20}}\right)(1.024)^{(T-20)}\alpha F = SOTE\left(\frac{C_{e,ww} - C_c}{C_{e,20}}\right)(1.024)^{(T-20)}\alpha F$$
$$= 13.3\% \times \left(\frac{(7.91 - 1.5)\,\text{mg/L}}{10.7\,\text{mg/L}}\right)(1.024)^{(30-20)} \times 0.85 \times 0.9$$
$$= 7.7\%$$

The calculated OTE_f is same as the assumed value in Step 6.b. If it is not close to the assumed value, carry out another iteration using Steps 6.b through 7.

8. Calculate OTR_f under field conditions.

Calculate OTR_f from Equation 10.40a at $T = 30°C$, $C_c = 1.5$ mg/L, $\alpha = 0.85$, and $F = 0.9$.

$$OTR_f = SOTR\left(\frac{\beta f_a f_p f_t C_{s,20} - C_c}{C_{e,20}}\right)(1.024)^{(T-20)}\alpha F = SOTR\left(\frac{C_{e,ww} - C_c}{C_{e,20}}\right)(1.024)^{(T-20)}\alpha F$$

$$= 2.4\,\text{kgO}_2/\text{h} \times \left(\frac{(7.91 - 1.5)\ \text{mg/L}}{10.7\ \text{mg/L}}\right)(1.024)^{(30-20)} \times 0.85 \times 0.9$$

$$= 1.4\,\text{kgO}_2/\text{h}$$

EXAMPLE 10.62: DIFFUSED OXYGEN TRANSFER RATE AT HIGHER ELEVATION

A completely mixed activated sludge plant has diffused aerators. The air flow per diffuser is 0.45 sm^3/min at 20°C and 1 atm. The depth of aeration tank $H_b = 5.0$ m, and diffuser is located 0.5 m above the floor. Estimate the oxygen transfer rate using the empirical equation given in Equation 10.44d under the following operating condition: $\alpha = 0.75$, $\beta = 0.92$, $C_1 = 0.04233$, $n = 0.1$, operating temperature $T = 25°C$, operating DO concentration in the basin $C_c = 1.5$ mg/L, and diffuser $OTE_f =$ 6%. The elevation of the plant is 610 m (2000 ft). Diffuser $OTE_f = 6\%$.

Solution

1. Calculate the equilibrium DO concentration in the wastewater under operating conditions ($C_{e,ww}$).

   Calculate $C_{e,ww}$ from Equation 10.39g. $\beta = 0.92$ is given in the statement. The other correction factors, f_t, f_p, and f_a are calculated below.

   a. Calculate the temperature correction factor (f_t).

      $C_{s,t} = 8.42$ and 8.11 mg/L at 24°C and 26°C are obtained at 1 atm (at sea level) from Table 10.16. Interpolate C_s at 25°C and 1 atm, $C_{s,25} = 8.27$ mg/L.

      $C_{e,20} = 9.09$ mg O_2/L = 9.09 g O_2/m^3 is obtained at 20°C and 1 atm from Table 10.16. The factor f_t is calculated from Equation 10.39j.

$$f_t = \frac{C_{s,t}}{C_{s,20}} = \frac{C_{s,25}}{C_{s,20}} = \frac{8.27\ \text{mg/L}}{9.09\ \text{mg/L}} = 0.91$$

   b. Calculate the pressure correction factor (f_p).

      The factor at an elevation of 610 m has been calculated from Equation 10.39k in Step 3 of Example 10.59. The calculated valu e of $f_p = 0.93$.

   c. Calculate the aeration process correction factor (f_a).

      Assume $O_{in} = 21\%$. Calculate the oxygen content in the air bubbles at the exhaust O_{ex} from Equation 10.43c at $OTE_f = 6\%$.

$$O_{ex} = \frac{0.21 \times (1-(OTE_f/100\%))}{0.79+0.21 \times (1-(OTE_f/100\%))} \times 100\% = \frac{0.21 \times (1-(6\%/100\%))}{0.79+0.21 \times (1-(6\%/100\%))} \times 100\% = 20.0\%$$

      $H_r = H_b - 0.5\ \text{m} = (5 - 0.5)\ \text{m} = 4.5$ m and $H_f = 9.6$ m at the field altitude of 610 m from Table B.7 of Appendix B. Calculate f_a from Equation 10.43b.

$$f_a = \frac{1}{2}\left(1 + \frac{H_r}{H_f}\right) + \frac{O_{ex}}{2O_{in}} = \frac{1}{2}\left(1 + \frac{4.5\ \text{m}}{9.6\ \text{m}}\right) + \frac{20.0\%}{2 \times 21\%} = 0.73 + 0.48 = 1.21$$

d. Calculate $C_{e,ww}$.

$$C_{e,ww} = \beta f_a f_p f_t C_{s,20} = 0.92 \times 1.21 \times 0.93 \times 0.91 \times 9.09\,\text{mg/L} = 0.94 \times 9.09\,\text{mg/L}$$
$$= 8.5\,\text{mg/L or } 8.5\,\text{g } O_2/\text{m}^3$$

2. Determine the oxygen transfer rate under operating conditions.

Calculate OTR_f from Equation 10.44d at $C_1 = 0.04233$, $n = 0.1$, $Q_a = 0.45\,\text{sm}^3/\text{min}$, $H = H_r = 4.5\,\text{m}$, $C_c = 1.5\,\text{mg/L}$ or $1.5\,\text{g}/\text{m}^3$, $T = 25°\text{C}$, and $\alpha = 0.75$.

$$OTR_f = C_1 H^{2/3} Q_a^{1-n}(C_{e,ww} - C_c)(1.024)^{(T-20)}\alpha$$
$$= 0.04233 \times (4.5\,\text{m})^{2/3} \times (0.45\,\text{m}^3/\text{min})^{(1-0.1)} \times (8.5 - 1.5)\,\text{g } O_2/\text{m}^3 \times (1.024)^{(25-20)} \times 0.75$$
$$= 0.33\,\text{kg } O_2/\text{h per diffuser}$$

EXAMPLE 10.63: DETERMINATION OF DO SATURATION COEFFICIENT β, OXYGEN UTILIZATION RATE γ_m, AND $K_L a_{WW}$

A sample of mixed liquor was aerated in a biological reactor at 20°C. The DO profile was recorded by a DO meter with respect to the elapsed time. After reaching nearly 90% of equilibrium concentration, the aeration device was turned off. The DO measurement was continued as the DO dropped due to microbial respiration. The results of this study are summarized below. Calculate (a) DO equilibrium concentration of the mixed liquor ($C_{e,ww}$) and β coefficient, (b) oxygen utilization rate (γ_m) by the microorganisms, and (c) $K_L a_{ww}$ of diffuser device using nonsteady-state procedure. Assume the test was performed at sea level and the effect of water depth on the equilibrium concentration in the clean water is ignored.

Aeration data

Elapsed Time (t), min	0	2	5	10	15	20	30	40	50	60	70
Measured DO (C), mg/L	0	1.5	3.0	4.6	5.6	6.2	6.8	7.1	7.4	7.6	7.7

Microbial respiration data

Elapsed Time (t), min	0	2	5	10	15	20	25	30	40
Measured DO (C), mg/L	7.4	7.1	6.9	6.0	5.5	4.8	4.3	3.7	2.4

Solution

1. Plot the DO profiles for aeration and microbial respiration.

The DO concentration (C) profiles for aeration and microbial respiration with respect to elapsed time (t) are shown in Figure 10.48a and b.

2. Determine the DO equilibrium concentration of the mixed liquor.

The DO concentration in the mixed liquor follows first-order reaction during aeration. The DO saturation equilibrium concentration $C_{e,ww}$ is obtained from Fujimoto method (refer to Section 5.4.1, Examples (5.22) and (5.28)). Plot a line between C_{t+1} versus C_t. Also, draw a line with slope = 1 from the origin. These plots are shown in Figure 10.49. The DO equilibrium concentration $C_{e,ww}$ of 8.6 mg/L is obtained from the intersection of the two lines.

3. Calculate the β coefficient.

Calculate $C_{e,ww}$ from Equation 10.39g. Since the test was performed at sea level and the effect of water depth on the equilibrium concentration in the clean water is ignored, $f_p = 1$ and $f_a = 1$. At $T = 20°\text{C}$, $C_{s,t} = C_{s,20}$. From Equation 10.39j, there is also $f_t = C_{s,t}/C_{s,20} = 1$. Therefore, $C_{e,ww}$ is

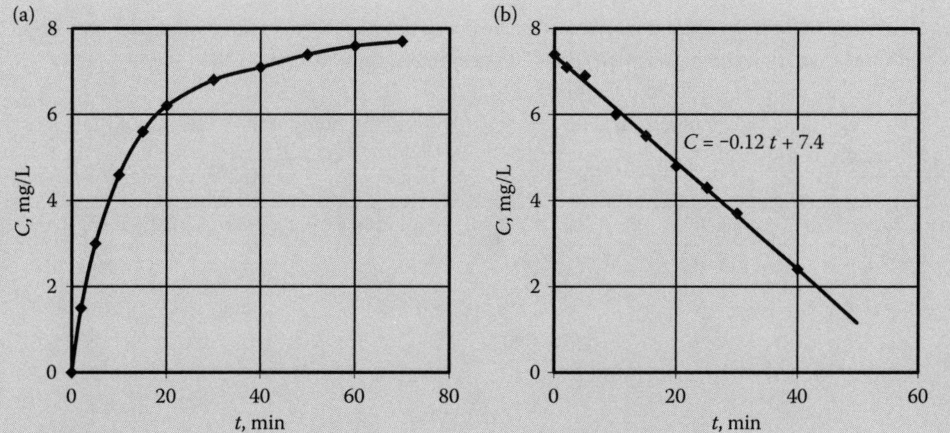

FIGURE 10.48 DO concentration profiles with respect to elapsed time: (a) aeration data and (b) microbial respiration data (Example 10.63).

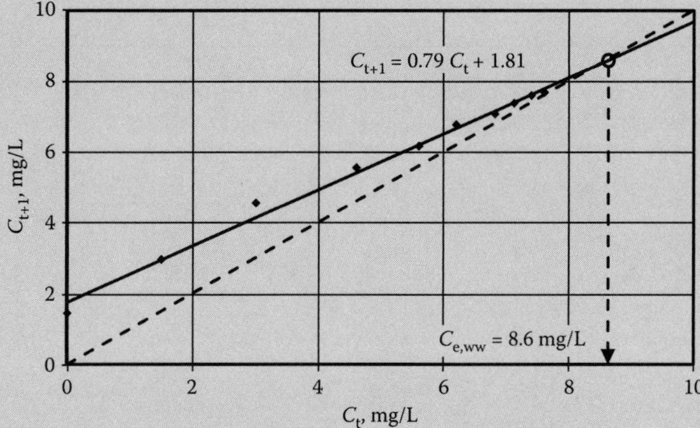

FIGURE 10.49 Determination of DO equilibrium concentration of wastewater from Fujimoto method (Example 10.63).

described by the expression below.

$$C_{e,ww} = \beta f_a f_p f_t C_{s,20} = \beta C_{s,20} \text{ or } \beta = \frac{C_{e,ww}}{C_{s,20}}$$

Obtain $C_{e,20} = 9.09$ mg/L at 20°C and 1 atm from Table 10.16 and calculate β coefficient.

$$\beta = \frac{C_{e,ww}}{C_{s,20}} = \frac{8.6\,\text{mg/L}C_{e,ww}}{9.09\,\text{mg/L}} = 0.95$$

4. Determine the oxygen uptake rate γ_m of microbial respiration.

   The respiration curve is prepared by plotting C with respect to t (Figure 10.48b). The negative slope of the linear relationship is the value of $-\gamma_m$.

$$-\gamma_m = -0.12\,\text{mg/L·min or } \gamma_m = 0.12\,\text{mg/L·min} \times 60\,\text{min/h} = 7.2\,\text{mg/L·h}$$

The negative slope indicates that DO is consumed due to respiration.

5. Determine the oxygen mass transfer coefficient for wastewater K_La_{ww} using nonsteady-state data. A linear relationship between dC/dt and t is expressed by Equation 10.38d.
   a. Prepare data for the linear relationship.
      Calculate values of $\Delta C/\Delta t$ with respect to t from the nonsteady-state data and tabulate the results below.

t, min	C_t, mg/L	$\Delta C = (C_t - C_{t-1})$, mg/L	$\Delta t = (t_t - t_{t-1})$, min	$\Delta C/\Delta t$, mg/L·min
0	0.0[a]	N/A	N/A	N/A
2	1.5[b]	1.5[c]	2[d]	0.75[e]
5	3.0	1.5	3	0.50
10	4.6	1.6	5	0.32
15	5.6	1.0	5	0.20
20	6.2	0.6	5	0.12
30	6.8	0.6	10	0.06
40	7.1	0.3	10	0.03
50	7.4	0.3	10	0.03
60	7.6	0.2	10	0.02

[a] $C_1 = 0$ mg/L
[b] $C_2 = 1.5$ mg/L
[c] $\Delta C = (C_2 - C_1) = (1.5 - 0.0)$ mg/L $= 1.5$ mg/L
[d] $\Delta t = t_2 - t_1 = (2 - 0)$min $= 2$ min
[e] $\frac{\Delta C}{\Delta t} = \frac{1.5\text{ mg/L}}{2\text{ min}} = 0.75$ mg/L·min

   b. Plot the linear relationship.
      Tabulated values of $\Delta C/\Delta t$ are plotted with respect to C_t in Figure 10.50.
   c. Determine the K_La_{ww} from the slope of the line.
      From Equation 10.38d, the slope of the line is $-K_La_{ww}$. It is -0.12 min^{-1} from the plot.

$$-K_La_{ww} = -0.12\,\text{min}^{-1} \text{ or } K_La_{ww} = 0.12\,\text{min}^{-1} \times 60\,\text{min/h} = 7.2\,\text{h}^{-1}$$

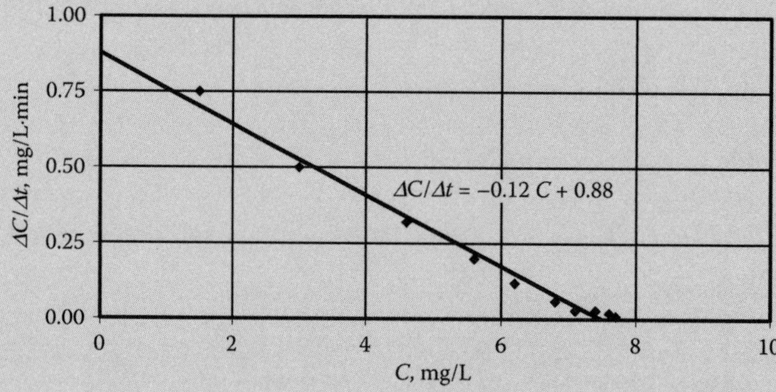

FIGURE 10.50 Determination of K_La_{ww} from nonsteady-state procedure (Example 10.63).

   d. Check the γ_m from the intercept of the line on the Y axis.
      From Equation 10.38d, the intercept of the line is $(K_La_s \times C_{sww} - \gamma_m)$.

From the plot, $(K_La_{ww} \times C_{e,ww} - \gamma_m) = 0.88$ mg/L·min.

Solve the value of γ_m from the expression.

$$\gamma_m = K_La_{ww} \times C_{e,ww} - 0.88\,\text{mg/L·min} = 0.12\,\text{min}^{-1} \times 8.6\,\text{mg/L} - 0.88\,\text{mg/L·min}$$
$$= (1.0 - 0.88)\,\text{mg/L·min} = 0.12\,\text{mg/L·min}$$

The calculated γ_m is same as obtained in Step 4.

EXAMPLE 10.64: OXYGEN UTILIZATION RATE OF ACTIVATED SLUDGE

Determine the rate of oxygen utilization by the active microorganisms in an aeration basin holding 85,000 gallons mixed liquor. A mechanical aerator is provided to meet the oxygen demand. The operating DO concentration is 1.5 mg/L. Use the following data: $\alpha = 0.87$, $K_La_{20} = 14\,\text{h}^{-1}$, operating temperature = 26°C, TDS concentration of the mixed liquor = 800 mg/L, $F = 1.0$ for the mechanical aerator. Assume the basin is located at sea level.

Solution

1. Determine the DO equilibrium concentration of the mixed liquor.

   $C_{e,ww}$ is calculated from Equation 10.39g. Since the basin is at sea level and a mechanical aerator is used for aeration, no correction is required for either pressure or process ($f_p = 1$ and $f_a = 1$).

   Assume that the solubility of oxygen in wastewater is mainly affected by the TDS concentration. From Table 10.16, $C_{s,26,TDS} = 8.13$ mg/L at 26°C and TDS of 800 mg/L and $C_{s,20} = 9.09$ mg/L at 20°C and TDS = 0 mg/L.

   Calculate f_t from Equation 10.39j.

$$f_t = \frac{C_{s,t}}{C_{s,20}} = \frac{C_{s,26,TDS}}{C_{s,20}} = \frac{8.13\,\text{mg/L}}{9.09\,\text{mg/L}} = 0.89$$

$$C_{e,ww} = \beta f_a f_p f_t C_{s,20} = \beta f_t C_{s,20} = 0.89 \times 9.09\,\text{mg/L} = 8.1\,\text{mg/L}$$

2. Determine the oxygen transfer coefficient K_La_{ww}.

   K_La_{ww} is calculated from Equations 10.39a and 10.39c at $T = 26$°C, $K_La_{20} = 14\,\text{h}^{-1}$and $\alpha = 0.87$.

$$K_La_{ww} = K_La_{20}\theta_T^{(T-20)}\alpha = 14\,\text{h}^{-1} \times 1.024^{(26-20)} \times 0.87 = 14\,\text{h}^{-1} \times 1.15 \times 0.87 = 14\,\text{h}^{-1}$$

3. Determine the oxygen utilization rate γ_m.

   Rearrange Equation 10.38b and calculate γ_m at a constant DO concentration $C_c = 1.5$ mg/L .

$$\gamma_m = K_La_{ww}(C_{e,ww} - C_c) = 14\,\text{h}^{-1} \times (8.1 - 1.5)\,\text{mg O}_2\text{/L} = 92\,\text{mg/L·h}$$

4. Determine the mass oxygen utilization rate (OUR).

   Calculate mass oxygen utilization rate in the basin volume $V = 85{,}000$ gal = 0.085 Mgal.

$$OUR = \gamma_m V = 92\,\text{mg O}_2\text{/L·h} \times 0.085\,\text{Mgal} \times 8.34\,\text{lb/Mgal·(mg/L)} = 65\,\text{lb O}_2\text{/h}$$

EXAMPLE 10.65: IMPACT OF WATER DEPTH ON K_La

A diffused aeration device is used in an aeration basin that has a water depth of 5.5 m. The same device was tested in a pilot plant facility. The water depth was 4.5 m, and the K_La_{20} was 14.2 h^{-1}. Estimate the K_La_{20} in the shallower basin using Equation 10.39b and describe impact of the water depth on the oxygen transfer coefficient. Assume that the device is mounted 0.3 m above the basin floor in both cases. Use the typical value of $n = 0.7$.

Solution

1. Estimate K_La_{20} in the aeration basin.

   Rearrange Equation 10.39b and calculate $K_La_{H_2}$ from $H_1 = (4.5 - 0.3)$ m = 4.2 m, $H_2 = (5.5 - 0.3)$ m = 5.2 m, $K_La_{H_1} = 14.2\ h^{-1}$, and $n = 0.7$.

$$K_La_{H_2} = K_La_{H_1}\left(\frac{H_2}{H_1}\right)^n = 14.2\,h^{-1} \times \left(\frac{5.2\,m}{4.2\,m}\right)^{0.7} = 14.2\,h^{-1} \times 1.16 = 16.5\,h^{-1}$$

2. Describe the impact of the water depth on K_La.

   The value of K_La increases about 15% per meter increase in water depth in the aeration basin. Typically, both higher oxygen transfer rate and efficiency are also expected for the same diffused aeration device in deeper water depth.

EXAMPLE 10.66: FIELD OXYGEN TRANSFER EFFICIENCY OF DIFFUSED AERATOR

Compute the field oxygen transfer rate (OTR_f) and oxygen transfer efficiency (OTE_f) of an air diffusion system operating in an aeration basin. The operating DO and volume of the basin are 1 mg/L and 1.5 million gallons (MG). $K_La_{ww,20} = 5\ h^{-1}$. The total air supply rate is 3500 cfm; the operating water temperature in the basin is 79°F; and the ambient air temperature is 73°F. The absolute atmospheric pressure at field altitude is 13.7 psia. The air temperature rise is 77°F after compression to reach an air supply pressure of 8.0 psig. The average concentration of DO for wastewater at equilibrium $C_{e,ww}$ is 8.95 mg/L at mid-depth of the aeration basin. Assume fouling factor $F = 0.9$.

Solution

1. Develop a simplified expression for calculating the field oxygen transfer rate (OTR_f).

   Equation 10.49a is developed by combining Equations 10.39c, 10.39g, 10.40a, and 10.40b.

$$OTR_f = K_La_{ww,20}V(C_{e,ww} - C_c)(1.024)^{(T-20)}F \quad (10.49a)$$

   This is a simplified equation for direct calculation of OTR_f from $K_La_{ww,20}$ and $C_{e,ww}$ when they are both known.

2. Calculate the oxygen transfer rate at operating conditions.

   Equation 10.49b is used to calculate the temperature in Celsius (°C) from Fahrenheit (°F).

$$T_C = \frac{5}{9} \times (T_F - 32) \quad (10.49b)$$

   where

   T_C = temperature in Celsius, °C

   T_C = temperature in Fahrenheit, °F

Convert the operating water temperature $T_F = 79°F$ to °C from Equation 10.49b.

$$T = T_c = \frac{5}{9} \times (79°F - 32) = 26.1°C$$

Calculate the OTR_f from Equation 10.49a at $K_La_{ww,20} = 5\ h^{-1}$, $V = 1.5$ MG, $C_{e,ww} = 8.95$ mg/L, $C_c = 1$ mg/L and $F = 0.9$.

$$OTR_f = 5\,h^{-1} \times 1.5\,MG \times (8.95 - 1)mg\ O_2/L \times (1.024)^{(26.1-20)} \times 0.9 \times 8.34\,lb/(mg/L{\cdot}MG)$$

$$= 5\,h^{-1} \times 1.5\,MG \times 7.95\ mg\ O_2/L \times 1.16 \times 0.9 \times 8.34\,lb/(mg/L{\cdot}MG)$$

$$= 519\,lb\ O_2/h$$

3. Calculate the density of air at operating conditions.
   The air density at operating conditions is expressed by Equation 10.49c.

$$\rho_{a,b} = (0.0808\ lb\ air/ft^3) \times \left(\frac{P_d}{P_s} \times \frac{273}{T_d}\right) \qquad (10.49c)$$

where
$\rho_{a,b}$ = density of the air discharged by the blower, lb/ft^3
P_d = absolute pressure of the air discharged by the blower, psia
P_s = absolute atmospheric pressure at sea level (1 atm), 14.7 psia
T_d = temperature of the air discharged by the blower, K

The absolute pressure of the air at supply pressure can be calculated from Equation 10.49d.

$$P_d = P_g + P_i \qquad (10.49d)$$

where
P_i = absolute atmospheric pressure at the inlet of the blower, psia
P_g = gauge pressure of the air discharged by the blower, psig

The temperature in Kelvin (°K) can be calculated from Fahrenheit (°F) using Equation 10.49e.

$$T_K = \frac{5}{9} \times (T_F + 460) \qquad (10.49e)$$

where T_K = temperature in Kelvin, °K
Calculate the air temperature after compression from the ambient air temperature of 73°F and the air temperature rise of 77°F.

$$T_d = (73 + 77)°F = 150°F$$

Convert the air temperature after compression $T_d = 150°F$ to °K from Equation 10.49e.

$$T_d = T_K = \frac{5}{9} \times (150°F + 460) = 339°K$$

The absolute pressure of the air at supply pressure is calculated from Equation 10.49d at $P_g = 8$ psig and $P_i = 13.7$ psia.

$$P_d = 8\ psig + 13.7\ psia = 21.7\ psia$$

Calculate the air density $\rho_{a,b}$ at operating conditions from Equation 10.49c.

$$\rho_{a,b} = (0.0808 \text{ lb air/ft}^3) \times \left(\frac{21.7 \text{ psia}}{14.7 \text{ psia}} \times \frac{273°\text{K}}{339°\text{K}}\right) = 0.096 \text{ lb air/ft}^3$$

4. Calculate the oxygen supply rate under the field operating conditions.
   Calculate the field oxygen supply rate (OSR_f) from Equation 10.41e at the field air density of $\rho_{a,f} = \rho_{a,b} = 0.096$ lb air/ft^3, oxygen content of $w_{O,2} = 0.232$ lb O_2/lb air, and air flow $Q_{a,f} = 3500$ cfm.

$$OSR_f = \rho_{a,f} \times w_{O2} \times Q_{a,f} = 0.096 \text{ lb air/ft}^3 \times 0.232 \text{ lb } O_2\text{/lb air} \times 3500 \text{ ft}^3\text{/min} \times 60 \text{ min/h} = 4677 \text{ lb } O_2\text{/h}$$

5. Calculate the oxygen transfer efficiency under the field operating conditions.
   The field oxygen transfer efficiency (OTE_f) is calculated from Equation 10.41d.

$$OTE_f = \frac{OTR_f}{OSR_f} \times 100\% = \frac{519 \text{ lb } O_2\text{/h}}{4677 \text{ lb } O_2\text{/h}} \times 100\% = 11.1\%$$

EXAMPLE 10.67: AERATION EFFICIENCY OF DIFFUSED AERATOR

Calculate (a) standard oxygen transfer rate ($SOTR$), (b) standard oxygen transfer efficiency ($SOTE$), (c) the total horsepower requirement, and (d) standard and field aeration efficiencies (SAE and AE_f) of the air diffusion system given in Example 10.66. Use the raw data and calculation results from Example 10.66. Additional data are given below: $K_La_{20} = 5.9 \text{ h}^{-1}$, $C_{e,20} = 10.6$ mg/L, and blower efficiency $\eta_b = 70\%$.

Solution

1. Calculate the standard oxygen transfer rate ($SOTR$).
   Calculate from Equation 10.40b at $K_La_{20} = 5.9 \text{ h}^{-1}$, $C_{e,20} = 10.6$ mg/L, and $V = 1.5$ MG (Example 10.66).

$$SOTR = K_La_{20}C_{e,20}V = 5.9 \text{h}^{-1} \times 10.6 \text{mg } O_2\text{/L} \times 1.5\text{MG} \times 8.34 \text{lb/(mg/L·MG)} = 782 \text{lb } O_2\text{/h}$$

2. Calculate the ratio of OTR_f to $SOTR$ at $OTR_f = 519$ lb O_2/h.

$$\frac{OTR_f}{SOTR} = \frac{519 \text{lb } O_2\text{/h}}{782 \text{lb } O_2\text{/h}} = 0.66$$

3. Calculate the standard oxygen transfer efficiency ($SOTE$).
   Combine Equations 10.40a, 10.41c, and 10.42b to obtain Equation 10.49f.

$$\frac{OTR_f}{SOTR} = \frac{OTE_f}{SOTE} = \frac{AE_f}{SAE} \tag{10.49f}$$

Rearrange Equation 10.49f and calculate $SOTE$ from $OTE_f = 11\%$ (from Example 10.66).

$$SOTE = \frac{OTE_f}{\left(\frac{OTR_f}{SOTR}\right)} = \frac{11\%}{0.66} = 17\%$$

4. Calculate the total horsepower requirement for the air blowers.

Convert the air flow $Q_{a,b} = 3500$ cfm under the field operating conditions to the standard air flow Q_a at 1 atm and 0°C.

$$Q_a = Q_{a,b} \times \left(\frac{P_d}{P_s} \times \frac{273}{T_d}\right) = 3500 \text{ cfm} \times \left(\frac{21.7 \text{ psia}}{14.7 \text{ psia}} \times \frac{273°\text{K}}{339°\text{K}}\right) = 3500 \text{ cfm} \times 1.19 = 4165 \text{ scfm}$$

The temperature in Rankine (°R) is converted from Fahrenheit (°F) using Equation 10.49g.

$$T_R = T_F + 460 \tag{10.49g}$$

where T_R = temperature in Rankine, °R

The blower discharge air temperature $T_d = 150$°F, and assume that the inlet air temperature is equal to the ambient air temperature $T_i = 73$°F.

Convert T_i and T_d to °R from Equation 10.49g.

$$T_i = 73°\text{F} + 460 = 533°\text{R}$$

$$T_d = 150°\text{F} + 460 = 610°\text{R}$$

Calculate the total shaft power requirement $P_{blowers}$ from Equation 10.45a at $\eta_b = 70\%$ and $\gamma_a = 0.075$ lb/scf, $R = 53.3$ ft-lb/lb·°R, $(k-1)/k = 0.283$, and $C = 550$ ft-lb/s·hp.

$$P_{blowers} = \frac{(Q_a/60)\rho_a R T_i}{C((k-1)/k)(\eta_b/100\%)}\left[\left(\frac{P_d}{P_i}\right)^{(k-1)/k} - 1\right]$$

$$= \frac{(4165 \text{ scfm}/(60 \text{ s/min})) \times 0.075 \text{ lb/scf} \times 53.3 \text{ ft-lb/lb·°R} \times 533°\text{R}}{550 \times 0.283 \times (70\%/100\%)}$$

$$\left[\left(\frac{21.7 \text{ psia}}{13.7 \text{ psia}}\right)^{0.283} - 1\right] = 1357 \text{ hp} \times 0.139 = 189 \text{ hp}$$

5. Calculate the standard and field aeration efficiencies (SAE and AE_f).

The SAE specified as shaft horsepower is calculated from Equation 10.42a.

$$SAE = \frac{SOTR}{P_{blowers}} = \frac{782 \text{ lb } O_2/\text{h}}{189 \text{ hp}} = 4.1 \text{ lb } O_2/\text{hp·h as shaft horsepower}$$

Rearrange Equation 10.49f and calculate AE_f.

$$AE_f = \left(\frac{OTR_f}{SOTR}\right) \times SAE = 0.66 \times 4.1 \text{ lb } O_2/\text{hp·h} = 2.7 \text{ lb } O_2/\text{hp·h as shaft horsepower}$$

EXAMPLE 10.68: TOTAL HORSEPOWER AND NUMBER OF TURBINE AERATORS

An activated sludge system is treating wastewater from a community of 3000 population. The aeration basin receives primary settled wastewater at an average rate of 100 gpcd with influent BOD_5 concentration of 130 mg/L. Assume that the basin is located at sea level. Determine the number of turbine aerators and total horsepower required in the aeration basin. The field condition and pilot plant data on turbine aerators are given below.

Field conditions

Aeration basin depth	= 15 ft
Operating water temperature	= 15°C
BOD_L removed	= 90%
α	= 0.9
β	= 0.8

DO maintained in the basin $C_c = 1$ mg/L

Pilot plant data on turbine aerator

Square test tank dimensions	= 26 ft × 26 ft × 10 ft (deep)
Testing water temperature	= 20°C
Aerator diameter d_i	= 30 in
Impeller peripheral speed R_i	= 12 fps
Air flow Q_a	= 11 scfm

Equations developed

DO transfer in clean water (Equation 10.46a)	$K_La_{20} = \dfrac{20R_i^{1.5}Q_a^{0.45}d_i^{1.8}}{V}$
Horsepower per turbine (Equation 10.47)	$P_{turbine} = 0.02d_i^{5.25}r_i^{2.75}$
BOD_5 concentration (Equation 5.7)	$BOD_5 = BOD_L(1 - e^{-0.23t})$

Solution

1. Calculate the K_La_{20} for 10-ft deep test tank.

Calculate the volume of the test tank, $V = 26\text{ ft} \times 26\text{ ft} \times 10\text{ ft} = 6760\text{ ft}^3$

Apply pilot plant data to calculate K_La_{20,H_t} in the test tank.

$$K_La_{20,H_t} = \frac{20 \times (12\text{ fps})^{1.5} \times (11\text{ scfm})^{0.45} \times (2.5\text{ ft})^{1.8}}{6760\text{ ft}^3} = 1.88\text{ h}^{-1}$$

2. Calculate the K_La_{20} for 15 ft deep aeration basin.

Assume that the air is released near the bottom of both aeration basin and the test tank. Rearrange Equation 10.39b and calculate K_La_{20,H_a} using $H_a = 15$ ft, $H_t = 10$ ft, $K_La_{H_t} = 1.88\text{ h}^{-1}$, and $n = 0.7$.

$$K_La_{20,H_a} = K_La_{20,H_t}\left(\frac{H_a}{H_t}\right)^n = 1.88\text{ h}^{-1} \times \left(\frac{15\text{ ft}}{10\text{ ft}}\right)^{0.7} = 1.88\text{ h}^{-1} \times 1.33 = 2.50\text{ h}^{-1}$$

3. Calculate the horsepower per aerator.

The impeller rotation speed, $r_i = \dfrac{R_i}{\pi d_i} = \dfrac{12\text{ fps}}{\pi \times 2.5\text{ ft}} = 1.53\text{ rps}$

Apply pilot plant data to calculate $P_{turbine}$.

$$P_{turbine} = 0.02 \times (2.5\text{ ft})^{5.25} \times (1.53\text{ rps})^{2.75} = 7.91\text{ hp per turbine aerator}$$

4. Calculate the standard oxygen transfer rate (*SOTR*) and standard aeration efficiency (*SAE*).

For a turbine aerator, assume $C_{e,20} \approx C_{s,20} = 9.09\text{ mg }O_2/L$ (see Table 10.16 at 20°C and 1 atm). Convert the test tank volume from ft^3 to Mgal.

$$V = 6760\text{ ft}^3 \times 7.48\text{ gal/ft}^3 \times \text{Mgal}/10^6\text{ gal} = 0.051\text{ Mgal}$$

Calculate *SOTR* from Equation 10.40b.

$$SOTR = K_L a_{20,H_a} C_{e,20} V = 2.50\ \text{h}^{-1} \times 9.09\ \text{mg O}_2/\text{L} \times 0.051\ \text{Mgal} \times 8.34\ \text{lb/(mg/L·Mgal)}$$
$$= 9.67\ \text{lbO}_2/\text{h per turbine aerator}$$

Calculate *SAE* from Equation 10.42a, $SAE = \dfrac{SOTR}{P_{\text{turbine}}} = \dfrac{9.67\ \text{lb O}_2/\text{h}}{7.91\ \text{hp}} = 1.22\ \text{lb O}_2/\text{hp·h}$

5. Calculate the oxygen transfer rate (OTR_f) and standard aeration efficiency (AE_f) under field conditions.

   The equilibrium DO concentration of wastewater in the aeration basin ($C_{e,ww}$) is calculated from Equation 10.39g. Since the basin is at sea level, no correction is required for the altitude pressure ($f_p = 1$). For a turbine aerator, assume the correction factor for process $f_a \approx 1$. Substitute Equation 10.39j for f_t to obtain the following expression.

$$C_{e,ww} = \beta f_a \left(\frac{C_{s,t}}{C_{s,20}}\right) C_{s,20} = \beta C_{s,t}$$

   At 1 atm (sea level), $C_s = 9.87$ and 10.31 mg/L are obtained at 16°C and 14°C, respectively, from Table 10.16. Use interpolation method to obtain $C_s = 10.1$ mg/L at 15°C.

   $C_{e,ww} = \beta C_{s,t} = 0.80 \times 10.1$ mg/L $= 8.08$ mg/L. Assume $F = 1.0$ for the turbine aerators, calculate OTR_f from Equation 10.40a at $T = 15°\text{C}$, $C_c = 1$ mg/L, and $\alpha = 0.9$.

$$OTR_f = SOTR\left(\frac{\beta f_a f_p f_t C_{s,20} - C_c}{C_{e,20}}\right)(1.024)^{(T-20)}\alpha F = SOTR\left(\frac{C_{e,ww} - C_c}{C_{e,20}}\right)(1.024)^{(T-20)}\alpha F$$
$$= 9.67\ \text{lb O}_2/\text{h} \times \left(\frac{(8.08-1)\ \text{mg/L}}{9.09\ \text{mg/L}}\right)(1.024)^{(15-20)} \times 0.9 \times 1.0$$
$$= 6.02\ \text{lb O}_2/\text{h per turbine aerator}$$

   Rearrange Equation 10.49f and calculate AE_f.

$$AE_f = \frac{OTR_f}{SOTR} \times SAE = \frac{6.02\ \text{lb O}_2/\text{h}}{9.67\ \text{lb O}_2/\text{h}} \times 1.22\ \text{lb O}_2/\text{hp·h} = 0.76\ \text{lb O}_2/\text{hp·h}$$

6. Calculate the amount of BOD_L removed.

Influent flow, $Q = 3000\ \text{persons} \times \dfrac{100\ \text{gal}}{\text{person·d}} \times \dfrac{10^{-6}\text{Mgal}}{\text{gal}} = 0.3\ \text{MGD}$

BOD_5 loading, $\Delta S_{0,\text{BOD5}} = Q \times S_0 = 0.3\ \text{MGD} \times 130\ \text{mg BOD}_5/\text{L} \times 8.34\ \text{lb/(mg/L·Mgal)} = 325\ \text{lb BOD}_5/\text{d}$

$$\text{BOD}_L\text{loading}, \Delta S_{0,\text{BODL}} = \frac{\text{BOD}_L}{\text{BOD}_5} \times \Delta S_{0,\text{BOD5}} = \frac{\Delta S_{0,\text{BOD5}}}{1 - e^{-0.23t}} \times \frac{\text{lb BOD}_L}{\text{lb BOD}_5} = \frac{325\ \text{lb BOD}_5/\text{d}}{1 - e^{-0.23\ \text{d}^{-1} \times 5\ \text{d}}} \times \frac{\text{lb BOD}_L}{\text{lb BOD}_5}$$
$$= \frac{325\ \text{lb BOD}_L/\text{d}}{0.68} = 478\ \text{lb BOD}_L/\text{d}$$

Calculated the mass of BOD_L removed in the aeration basin at the BOD_L removal efficiency E = 90%.

$$\Delta S_{\text{removal}} = E \times \Delta S_{0,\text{BODL}} = 0.9 \times 478\ \text{lb BOD}_L/\text{d} = 430\ \text{lb BOD}_L/\text{d}$$
$$\text{or}\quad \Delta S_{\text{reactor}} = 430\ \text{lb BOD}_L/\text{d} \times (\text{d}/24\ \text{h}) = 18\ \text{lb BOD}_L/\text{h or } 18\ \text{lb O}_2/\text{h}$$

7. Calculate the number of aerators and total horsepower.

   Calculate the number of aerators required.

$$N_{\text{turbine}} = \frac{\Delta S_{\text{removal}}}{OTR_{\text{f}}} = \frac{18\,\text{lb O}_2/\text{h}}{6.02\,\text{lb O}_2/\text{h}\cdot\text{turbine aerator}} = 3 \text{ turbine aerators}$$

   Provide 5 aerators to meet the peak oxygen demand and provide redundancy under average flow conditions.

   Calculate the total horsepower requirement under average flow conditions.

$$P_{\text{total}} = \frac{\Delta S_{\text{removal}}}{AE_{\text{f}}} = \frac{18\text{ lb O}_2/\text{h}}{0.76\text{ lb O}_2/\text{hp}\cdot\text{h}} = 24\,\text{hp}$$

EXAMPLE 10.69: DIFFUSERS IN A BASIN OF KNOWN WIDTH

An aeration basin has width and depth of 30 and 15 ft. Calculate the amount of oxygen transferred per diffuser and the total number of diffusers required at a total oxygen requirement of 120 lb O_2/h under field conditions. Use the following data: operating temperature = 30°C, $C_{e,ww} = 7.94$ mg/L, $\alpha = 0.85$, $C_c = 1.5$ mg/L, $Q_a = 7.5$ scfm per diffuser tube, and diffuser tube height above the basin floor $d_d = 1.5$ ft. Use Equation 10.44e with the exponents *m*, *n*, *p*, and constant C_2 of 0.72, 1.02, 0.35, and 0.0081, respectively.

Solution

1. Calculate the oxygen transfer rate (OTR_f) per diffuser tube under field conditions.

   Calculate the water depth above the diffuser from the basin depth $H_b = 15$ ft and the diffuser height above the floor $d_d = 1.5$ ft.

$$H = H_b - d_d = (15 - 1.5)\text{ ft} = 13.5\,\text{ft}$$

   Calculate the OTR_f per diffuser from Equation 10.44e.

$$\begin{aligned} OTR_f &= C_2 \frac{H^m}{W^p} Q_a^n (C_{e,ww} - C_c)(1.024)^{(T-20)}\alpha \\ &= 0.0081 \times \frac{(13.5\,\text{ft})^{0.72}}{(30\,\text{ft})^{0.35}} \times (7.5\,\text{scfm})^{1.02} \times (7.94 - 1.5) \times (1.024)^{(30-20)} \times 0.85 \\ &= 0.0081 \times \frac{6.51}{3.29} \times 7.81 \times 6.44 \times 1.27 \times 0.85 \\ &= 0.87\,\text{lb O}_2/\text{h per tube diffuser} \end{aligned}$$

2. Calculate total number of diffuser tubes required to meet the total oxygen requirement, $\Delta S_{\text{basin}} = 120$ lb O_2/h.

$$N_{\text{tube}} = \frac{\Delta S_{\text{basin}}}{OTR_f} = \frac{120\,\text{lb O}_2/\text{h}}{0.87\,\text{lb O}_2/\text{h}\cdot\text{tube}} = 138 \text{ tube diffusers}$$

   Provide 150 tube diffusers in the aeration basin.

EXAMPLE 10.70: THEORETICAL AND STANDARD OXYGEN REQUIREMENTS IN A SINGLE-STAGE NITRIFICATION REACTOR

A single-stage suspended growth biological reactor is designed to achieve carbonaceous and nitrogenous BOD removal. The characteristics of influent to the aeration basin are: design average flow = 42,000 m^3/d (0.486 m^3/s), BOD_5 = 200 mg/L, ON (organic nitrogen) = 15 mg/L, and AN (ammonia nitrogen) = 20 mg/L. The maximum effluent limits are: soluble BOD_5 = 10 mg/L, TSS = 10 mg/L, and AN = 1.0 mg/L as N. Other design parameters are: $\theta_c = 12$ d, biomass yield coefficients $Y_{BOD5} = 0.6$ g VSS/g BOD_5 for removal of BOD_5, and $Y_N = 0.2$ g VSS/g AN as N for nitrification, specific endogenous decay coefficients $k_{d,BOD5} = 0.06\ d^{-1}$ for removal of BOD_5 and $k_{d,N} = 0.05\ d^{-1}$ for nitrification, MLVSS/MLSS ratio = 0.8, $\alpha = 0.75$, $\beta = 0.9$, $F = 0.95$, $OTE_f = 8\%$, $C_c = 2$ mg/L, basin depth $H_b = 5$ m, depth of water above the air release $H = 4.6$ m, the elevation of treatment plant is 450 m above sea level, and the operating temperature is 24°C. Assume $BOD_5/BOD_L = 0.68$ (see Example 5.18), 65% biomass is biodegradable, ultimate BOD_L exerted by biomass is 1.42 g BOD_L/g biodegradable biomass (see Example 10.10), and ON exerted by biomass is 12.2% of VSS. Calculate the theoretical oxygen requirement (*ThOR*) and standard oxygen requirement (*SOR*).

Solution

1. Develop the equations for calculating the theoretical oxygen requirement (*ThOR*).

   In a single-stage suspended growth biological reactor, the theoretical oxygen requirement (*ThOR*) consists of (a) oxygen required for removal of ultimate carbonaceous BOD ($CBOD_L$), and (b) oxygen required for removal of ultimate nitrogenous BOD ($NBOD_L$).

   a. Oxygen required for the removal of $CBOD_L$.

   Write a general mass balance statement for the oxygen required for removal of $CBOD_L$.

$$\begin{bmatrix}\text{Oxygen required}\\ \text{for CBOD}_L\\ \text{removal, mg O}_2/\text{L}\end{bmatrix} = \begin{bmatrix}\text{CBOD}_L\\ \text{in influent,}\\ \text{mg BOD}_L/\text{L}\end{bmatrix} - \begin{bmatrix}\text{Soluble CBOD}_L\\ \text{in effluent,}\\ \text{mg BOD}_L/\text{L}\end{bmatrix} - \begin{bmatrix}\text{CBOD}_L\text{fixed}\\ \text{in biomas,}\\ \text{mg BOD}_L/\text{L}\end{bmatrix}$$

   A symbolic representation of the above statement is expressed by Equation 10.50a.

$$r_{CBODL} = S_{0,CBODL} - S_{CBODL} - p_{x,CBODL} \tag{10.50a}$$

where

r_{CBODL} = oxygen concentration required for removal of $CBOD_L$, mg O_2/L
$S_{0,CBODL}$ = $CBOD_L$ concentration in the influent, mg O_2/L
S_{CBODL} = soluble $CBOD_L$ concentration in the effluent, mg O_2/L
$p_{x,CBODL}$ = $CBOD_L$ concentration exerted in the biomass, mg O_2/L

Using the fraction factors of 1.42 g BOD_L/g biodegradable biomass, and 0.65 g biodegradable biomass/g VSS, the ultimate BOD_L exerted in the biomass is calculated below.

$$p_{x,CBODL} = \left[\left(\frac{1.42\text{ g BOD}_L}{\text{g biodegradable solids}}\right) \times \left(\frac{0.65\text{ g biodegradable solids}}{\text{g TSS}}\right) \times \left(\frac{1}{\text{VSS/TSS ratio}}\right)\right] p_x$$

$$= \left(\frac{0.92\text{ g BOD}_L/\text{g TSS}}{\text{VSS/TSS ratio}}\right) \times p_x \quad \text{or} \quad \left(\frac{0.92\text{ g O}_2/\text{g TSS}}{\text{VSS/TSS ratio}}\right) \times p_x$$

The VSS concentration increase due to biomass growth (p_x) is calculated from Equation 10.15n. Substitute the above expression into Equation 10.50a to obtain Equation 10.50b.

$$r_{CBODL} = S_{0,CBODL} - S_{CBODL} - \frac{0.92\ p_x}{VSS/TSS\ ratio} \qquad (10.50b)$$

b. Oxygen required for the removal of $NBOD_L$.

Write a similar general mass balance statement for the oxygen required for removal of $NBOD_L$ (Equation 10.50c).

$$\begin{bmatrix} \text{Oxygen required} \\ \text{for } NBOD_L \\ \text{removal, mg } O_2/L \end{bmatrix} = \begin{bmatrix} NBOD_L \\ \text{in influent,} \\ \text{mg } BOD_L/L \end{bmatrix} - \begin{bmatrix} \text{Soluble } NBOD_L \\ \text{in effluent,} \\ \text{mg } BOD_L/L \end{bmatrix} - \begin{bmatrix} NBOD_L \text{ fixed} \\ \text{in biomas,} \\ \text{mg } BOD_L/L \end{bmatrix}$$

Above statement can also be expressed by Equation 10.50c.

$$r_{NBODL} = S_{0,NBODL} - S_{NBODL} - p_{x,NBODL} \qquad (10.50c)$$

where

r_{NBODL} = oxygen concentration required for removal of $NBOD_L$, mg O_2/L
$S_{0,CBODL}$ = $NBOD_L$ concentration in the influent, mg O_2/L
S_{CBODL} = soluble $NBOD_L$ concentration in the effluent, mg O_2/L

Using the fraction factors of 4.57 g BOD_L/g N (see Examples 5.26 and 5.27) and 0.122 g N/g VSS the ultimate BOD_L exerted in the biomass is obtained below, and r_{NBODL} is expressed by Equation 10.50d.

$$p_{x,NBODL} = \left[\left(\frac{4.57\,\text{g } BOD_L}{\text{g N}}\right) \times \left(\frac{0.122\,\text{g N}}{\text{g VSS}}\right)\right] p_x$$

$$= (0.56\,\text{g } BOD_L/\text{g VSS}) \times p_x \quad \text{or} \quad (0.56\,\text{g } O_2/\text{g VSS}) \times p_x$$

$$r_{NBODL} = S_{0,NBODL} - S_{CNODL} - 0.56 p_x \qquad (10.50d)$$

c. Theoretical oxygen requirement (*ThOR*) for the removal of $CBOD_L$ and $NBOD_L$.

The *ThOR* for the overall removal of $CBOD_L$ and $NBOD_L$ is derived by combining Equations 10.50b and 10.50d.

$$r_{BODL} = r_{CBODL} + r_{NBODL}$$

$$= \left(S_{0,CBODL} - S_{CBODL} - \frac{0.92 p_x}{VSS/TSS\ ratio}\right) + (S_{0,NBODL} - S_{NBODL} - 0.56 p_x)$$

$$= (S_{0,CBODL} + S_{0,NBODL}) - (S_{CBODL} + S_{NBODL}) - \left(\frac{0.92}{VSS/TSSratio} + 0.56\right) p_x$$

where r_{BODL} = total oxygen concentration required for removal of $CBOD_L$ and $NBOD_L$, mg O_2/L

Assume $UOD = S_{CBODL} + S_{NBODL}$ (from Equation 5.17), and $UOR = r_{BODL}$ to obtain Equation 10.48a.

$$UOR = UOD_0 - UOD_e - \left(\frac{0.92}{VSS/TSS\ ratio} + 0.56\right) p_x$$

Multiply Q on both sides of the above expression (Equation 10.48a) to obtain Equation 10.48b.

$$ThOR = Q(UOD_0 - UOD_e) - \left(\frac{0.92}{\text{VSS/TSS ratio}} + 0.56\right)P_x$$

2. Calculate the UOD concentration (UOD_0) in the influent.
   a. Calculate the $CBOD_L$ concentration ($S_{0,CBODL}$) in the influent.

$$S_{0,CBODL} = \frac{1}{0.68 \text{ g BOD}_5/\text{g BOD}_L} \times S_0$$
$$= \frac{1}{0.68 \text{ g BOD}_5/\text{g BOD}_L} \times 200 \text{ mg BOD}_5/\text{L} = 294 \text{ mg BOD}_L/\text{L}$$

   b. Calculate the $NBOD_L$ concentration ($S_{0,NBODL}$) in the influent.

$$TKN_0 = ON_0 + AN_0 = (15 + 20) \text{ mg/L as N} = 35 \text{ mg/L as N}$$

$$S_{0,NBODL} = 4.57 \text{ g BOD}_L/\text{g N} \times TKN_0 = 4.57 \text{ g BOD}_L/\text{g N} \times 35 \text{ mg/L as N} = 160 \text{ mg BOD}_L/\text{L}$$

   c. Calculate the UOD concentration (UOD_0) in the influent.
   The UOD_0 in the influent is calculated from Equation 5.17.

$$UOD_0 = S_{0,CBODL} + S_{0,NBODL} = (294 + 160) \text{ mg BOD}_L/\text{L} = 454 \text{ mg BOD}_L/\text{L or } 454 \text{ mg O}_2/\text{L}$$

3. Calculate the UOD concentration (UOD_e) in the effluent.
   a. Calculate the soluble $CBOD_L$ concentration (S_{CBODL}) in the effluent.

$$S_{BOD5inTSS} = \left[\left(\frac{0.68 \text{ g BOD}_5}{\text{g BOD}_L}\right) \times \left(\frac{1.42 \text{ g BOD}_L}{\text{g biodegradable solids}}\right) \times \left(\frac{0.65 \text{ g biodegradable solids}}{\text{g TSS}}\right)\right] TSS_e$$
$$= 0.63 \text{ g BOD}_5/\text{g TSS} \times 10 \text{ mg TSS/L} = 6.3 \text{ mg BOD}_5/\text{L}$$

Calculate the soluble BOD_5 concentration in the effluent.

$$S = (10 - 6.3) \text{ mg BOD}_5/\text{L} = 3.7 \text{ mg BOD}_5/\text{L}$$

$$S_{CBODL} = \frac{1}{0.68 \text{ g BOD}_5/\text{g BOD}_L} \times S = \frac{1}{0.68 \text{ g BOD}_5/\text{g BOD}_L} \times 3.7 \text{ mg BOD}_5/\text{L} = 5.4 \text{ mg BOD}_L/\text{L}$$

   b. Calculate the soluble $NBOD_L$ concentration (S_{NBODL}) in the effluent.

$$S_{NBODL} = 4.57 \text{ g BOD}_L/\text{g N} \times AN_e = 4.57 \text{ g BOD}_L/\text{gN} \times 1.0 \text{ mg/L as N}$$
$$= 4.6 \text{ mg BOD}_L/\text{L}$$

   c. Calculate the soluble UOD concentration (UOD_e) in the effluent.

$$UOD_e \text{ from Equation 5.17} = S_{CBODL} + S_{NBODL} = (5.4 + 4.6) \text{ mg BOD}_L/\text{L}$$
$$= 10 \text{ mg BOD}_L/\text{L or } 10 \text{ mg O}_2/\text{L}$$

4. Calculate the biosolids (VSS) concentration increase (p_x) in the reactor.
   a. Calculate the increase in the concentration of biomass ($p_{x,BOD5}$) due to removal of BOD_5.

The observed yield coefficient ($Y_{obs,BOD5}$) is calculated from Equation 10.15l.

$$Y_{obs,BOD5} = \frac{Y_{BOD5}}{1 + k_{d,BOD5}\theta_c} = \frac{0.6 \text{ mg VSS/mg BOD}_5}{1 + 0.06 \text{ d}^{-1} \times 12 \text{ d}} = 0.35 \text{ mg VSS/mg BOD}_5$$

The $p_{x,BOD5}$ is calculated from Equation 10.15n.

$$p_{x,BOD5} = Y_{obs,BOD5}(S_0 - S) = 0.35 \text{ g VSS/g BOD}_5 \times (200 - 3.7) \text{ mg BOD}_5\text{/L} = 69 \text{ mg VSS/L}$$

b. Calculate the increase in the concentration of biomass ($p_{x,N}$) due to nitrification.
The observed yield coefficient ($Y_{obs,N}$) is calculated from Equation 10.15l.

$$Y_{obs,N} = \frac{Y_N}{1 + k_{d,N}\theta_c} = \frac{0.2 \text{ mg VSS/mg AN as N}}{1 + 0.05 \text{ d}^{-1} \times 12 \text{ d}} = 0.13 \text{ mg VSS/mg AN as N}$$

The $p_{x,N}$ is calculated from Equation 10.15n while $p_{x,BOD5}$ is used to estimate the ON exerted by biomass.

$$p_{x,N} = Y_{obs,N}\left(TKN_0 - \left(\frac{0.122 \text{ g N}}{\text{g VSS}}\right) \times p_{x,BOD5} - AN_e\right)$$

$$= 0.13 \text{ g VSS/g AN as N} \times \left(35 \text{ mg/L as N} - \left(\frac{0.122 \text{ mg N}}{\text{mg VSS}}\right) \times 69 \text{ mg VSS/L} - 1.0 \text{ mg/L as N}\right)$$

$$= 3.3 \text{ mg VSS/L}$$

c. Calculate the total biomass concentration increase (p_x) due to removal of $CBOD_L$ and $NBOD_L$.

$$p_x = p_{x,BOD5} + p_{x,N} = (69 + 3.3) \text{ mg VSS/L} = 72 \text{ mg VSS/L}$$

Note: Since $p_{x,BOD5}$ (69 mg/L) is more than 95% of p_x (72 mg/L), it is valid to use $p_{x,BOD5}$ for estimating the ON exerted in biomass.

5. Calculate the theoretical oxygen requirement (*ThOR*) for the removal of $CBOD_L$ and $NBOD_L$.
The *UOR* is calculated from Equation 10.48a.

$$UOR = 454 \text{ mg O}_2\text{/L} - 10 \text{ mg O}_2\text{/L} - \left(\frac{0.92 \text{ mg O}_2\text{/mg TSS}}{0.8 \text{ mg VSS/mg TSS}} + 0.56 \text{ mg O}_2\text{/mg VSS}\right) \times 72 \text{ mg VSS/L}$$

$$= 321 \text{ mg O}_2\text{/L or } 321 \text{ g O}_2\text{/m}^3$$

The amount of *ThOR* is calculated from Equation 10.48b.

$$ThOR = Q \times UOR = 42{,}000 \text{ m}^3\text{/d} \times 321 \text{ g O}_2\text{/m}^3 \times 10^{-3} \text{ kg/g} = 13{,}480 \text{ kg O}_2\text{/d}$$

6. Calculate equilibrium concentration of oxygen in clean water ($C_{e,20}$) under standard conditions.
$C_{s,20} = 9.09 \text{ mg O}_2\text{/L} = 9.09 \text{ g O}_2\text{/m}^3$ is obtained from Table 10.16 at 20°C and 1 atm.
Calculate the ratio of H to H_b.

$$\frac{H}{H_b} = \frac{4.6 \text{ m}}{5 \text{ m}} = 0.92 \text{ or } 92\% > 90\%$$

Since $H/H_b > 90\%$, use the typical value of $d_e = 0.4$ to calculate $C_{e,20}$ from Equation 10.40d at $H_s =$ 10.33 m at 20°C and 1 atm.

$$C_{e,20} = C_{s,20}\left(1 + d_e\frac{H_b}{H_s}\right) = 9.09\,\text{mg}\ O_2/\text{L} \times \left(1 + 0.4 \times \frac{5\,\text{m}}{10.33\,\text{m}}\right) = 10.8\,\text{mg/L or } 10.8\,\text{g}O_2/\text{m}^3$$

7. Calculate equilibrium DO concentration in the wastewater ($C_{e,ww}$) under operating conditions.

   $C_{e,ww}$ is calculated from Equation 10.39g. $\beta = 0.9$ is given in the statement. The other correction factors, f_t, f_p, and f_a are calculated below.

   a. Calculate temperature correction factor.

      The factor f_t is calculated from Equation 10.39j. The surface saturation concentration of oxygen $C_{s,t} = 8.42$ mg/L is obtained from Table 10.16 at 24°C and sea level.

$$f_t = \frac{C_{s,t}}{C_{s,20}} = \frac{8.42\,\text{mg/L}}{9.09\,\text{mg/L}} = 0.93$$

   b. Calculate pressure correction factor.

      The barometric pressures $P_s = 101.325$ kPa or 760 mm Hg at sea level. $P_f \approx 721$ mm Hg at the field altitude of 450 m is obtained from Table B.7 in Appendix B.

$$P_f = \frac{721\,\text{mm Hg}}{760\,\text{mm Hg}} \times 101.325\,\text{kPa} = 96.13\,\text{kPa}$$

      Water vapor pressure $P_{v,t} = 22.38$ mm Hg is obtained at 24°C from Table B.5 in Appendix B.

$$P_{v,t} = \frac{22.38\,\text{mm Hg}}{760\,\text{mm Hg}} \times 101.325\,\text{kPa} = 2.98\,\text{kPa}$$

      The factor f_p is calculated from Equation 10.39k,

$$f_p = \frac{P_f - P_{v,t}}{P_s - P_{v,t}} = \frac{(96.13 - 2.98)\,\text{kPa}}{(101.325 - 2.98)\,\text{kPa}} = 0.95$$

   c. Calculate aeration process correction factor.

      The absolute atmospheric pressure at the field altitude $H_f \approx 9.8$ m at the field altitude of 450 m from Table B.7 of Appendix B.

      The factor f_a is calculated from Equation 10.43e at $OTE_f = 8\%$ and $H_r = H = 4.6$ m.

$$f_a = \frac{1}{2}\left(1 + \frac{H_r}{H_f}\right) + \frac{1}{2}\left(1 - \frac{OTE_f}{100\%}\right) = \frac{1}{2}\left(1 + \frac{4.6\,\text{m}}{9.8\,\text{m}}\right) + \frac{1}{2}\left(1 - \frac{8\%}{100\%}\right) = 0.73 + 0.46 = 1.19$$

   d. Calculate $C_{e,ww}$.

$$C_{e,ww} = \beta f_a f_p f_t C_{s,20} = 0.90 \times 1.19 \times 0.95 \times 0.93 \times 9.09\,\text{mg/L} = 0.95 \times 9.09\,\text{mg/L} = 8.64\,\text{mg/L}$$

8. Calculate the standard oxygen requirement (SOR) for the removal of $CBOD_L$ and $NBOD_L$.

   The amount of SOR is calculated from Equation 10.48c.

$$SOR = \frac{ThOR}{((C_{e,ww} - C_c)/C_{e,20})(1.024)^{(T-20)}\alpha F} = \frac{13{,}480\,\text{kg}O_2/\text{d}}{(((8.64-2)\,\text{mg/L})/10.8\,\text{mg/L})(1.024)^{(24-20)} \times 0.75 \times 0.95}$$
$$= 28{,}000\,\text{kg}O_2/\text{d}$$

Note: The SOR of 28,000 kg O_2/d at 20°C and 1 atm (standard operating conditions) will satisfy both carbonaceous and nitrogenous oxygen demands.

10.3.9 Aeration Device, Equipment, and Hardware Assembly

Aeration system is one of the most important components in a wastewater treatment plant. It is used for transferring oxygen from air into the wastewater to meet the oxygen demand of biomass and provide mixing in the aeration basin. The aeration systems are primarily categorized into either *diffused-air* or *mechanical* aeration system. Commonly used aeration devices under these categories are illustrated in Figure 10.51.[2,7,57,59]

Diffused-Air Aeration: In a typical submerged diffused-air aeration system, oxygen transfer and mixing occurs as air bubbles rise to the surface from the porous or nonporous diffusers that are placed near the bottom of the tank. Many other types of diffused-air aeration devices are also commercially available. In these devices, compressed air or air–water mixture is fed by air jet or porous diffusers. Also, single vertical tube, V-tube, and tubular devices produce airlift pumping action. Aspirator draws air through hollow tube and propeller action near the tip creates turbulence and fine bubbles. In a diffused-air surface aeration system, oxygen transfer occurs due to turbulence when water in thin sheets comes in contact with atmospheric air. Good example is a cascade aeration device.[59,60] Many types of diffused-air aeration devices, their advantage and disadvantages, and applications are provided in Table 10.17.[2,7,57,60,61]

The factors affecting the oxygen transfer are bubble sizes and air supply rate; diffuser arrangement and submergence depth; and turbulence and surface tension of the surrounding liquid. The SOTE and SAE values of fine-bubble diffusers are usually in the range of 25–35%, and 2.5–10 kg O_2/kw·h (4–16 lb O_2/hp·h), respectively. These two values for coarse bubble diffusers and other diffused-air aeration devices are in the range of 7–15% and 1.2–2.5 kg O_2/kw·h (2–4 lb O_2/hp·h), respectively.[61–64] Brief description, technical and performance data, diffuser layout, airflow rate, SOTE, and SAE of many types of air diffusion devices are summarized in Table 10.18.[1,2,7,27,57,59–76] The manufacturers provide the technical and performance data in their devices. Many of these aeration devices are shown in Figure 10.52.[57,62,64,66,69]

Mechanical Aeration: The mechanical aerators fall into two major groups: (1) aerators with horizontal axis, and (2) aerators with vertical axis (Figure 10.51). Both groups are further divided into surface and submerged aerators. The performance of a mechanical aerator is usually characterized by the SAE value.

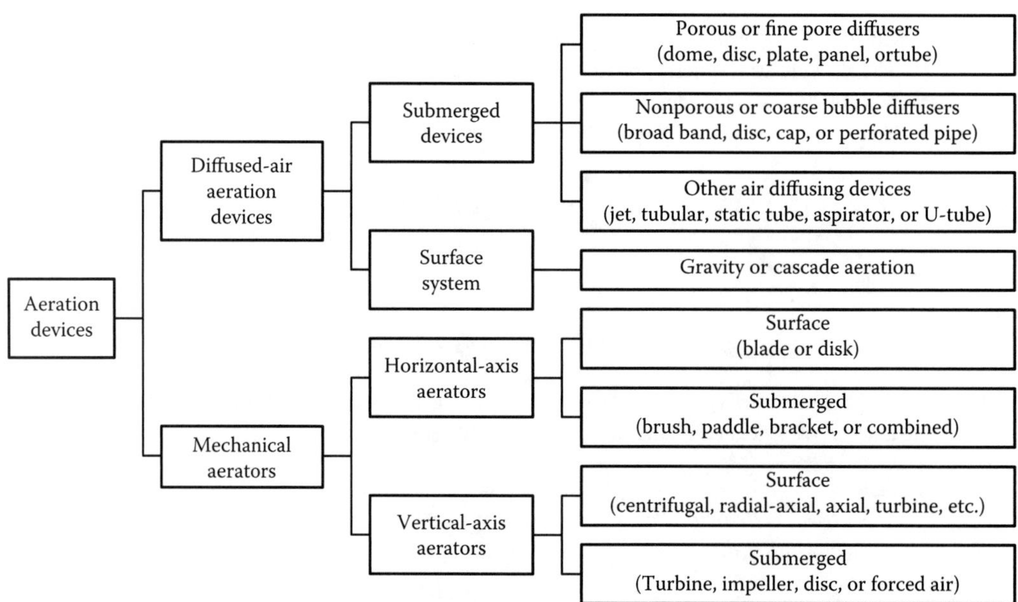

FIGURE 10.51 Commonly used aeration devices.

TABLE 10.17 General Information about Commonly Used Diffused-Air Aeration Devices

Aeration Category	Description	Advantage	Disadvantage	Application
Submerged devices				
Porous or fine-pore diffusers	The diffusers are provided in domes, disks, panels, plates, or tubes made of porous materials such as ceramic, plastic, or membrane. Typical bubble sizes are: micro bubbles ≤0.5 (≤0.02 in), ultrafine bubbles of 0.5–1 mm (0.02–0.04 in); and fine bubbles of 1–2 mm (0.04–0.08 in), (Figure 10.52a)	High OTR and SAE suitable for uniform oxygen supply throughout the basin in a grid configuration, and provide operational flexibility in air flow rate	High initial costs, need for air filter, high head loss, high maintenance, and potential clogging issues	Most activated sludge process modifications
Nonporous or coarse bubble diffusers	The diffusers may be supplied as (a) broad or narrow bands, disks, or caps constructed of ABS, PVC, CPVC, (b) disks covered with perforated membrane diaphragm, or (c) perforated stainless steel or plastic pipes. Typical bubble sizes are medium bubbles of 2–10 mm (0.08–0.4 in), and coarse bubbles of 10 to > 50 mm (0.4 to > 2 in), (Figure 10.52b)	Low initial cost, high operational reliability, good for mixing and scouring applications, low maintenance, nonclogging, no need for air filter, diaphragm with check-valve feature, and may be suitable for top or bottom mounting	Low OTR, OTE, and SAE	Most activated sludge process modifications especially in RBC, IFAS, MBBR, and MBR, aerated grit chamber, flow equalization (EQ) basin, aerobic digester, sludge holding tank, and channel mixing
Other air diffusing devices				
Jet aerators	Compressed air injected into liquid are mixed and discharged horizontally. The rising plume of fine bubbles and water mixture produces oxygen transfer and mixing (Figure 10.52c)	High OTR, moderate to high OTE, moderate initial cost, and suited for use in deep tanks	Requires blower and pumping equipment, and nozzle clogging	Most activated sludge process modifications, flow equalization (EQ) basin, sludge-holding tank, and deep tank
Tubular aerators	Air flows upward through a tortuous pathway within a tube that may be of different height. Alternately placed deflection plates increase air–wastewater contact. Efficient mixing and oxygen transfer is accomplished because the tube aerator acts as an airlift pump (Figure 10.52d)	Low initial cost, moderate to high OTE, and low maintenance	Low mixing	Most activated sludge process modifications, and channel mixing
Static tubes	Static tubes are stationary vertical tubes mounted on basin floor. A series of diffuser membranes or internal baffles may be arranged. Static tubes also function like an airlift pump (Figure 10.52e)	Low initial cost, medium OTE, and low maintenance	Low mixing	Most activated sludge process modifications, and aerated lagoons
Aspirators	The high-speed inclined propeller draws air through hollow tubes and injects it underwater. Air velocity and propeller action creates turbulence and fine	High OTE, and moderate to high mixing	Requires propeller, needs mounting on fixed structure or on pontoons,	Aerated lagoons, flow equalization (EQ) basin, and sludge-holding tank

(*Continued*)

TABLE 10.17 (*Continued*) General Information about Commonly Used Diffused-Air Aeration Devices

Aeration Category	Description	Advantage	Disadvantage	Application
	bubbles (~2 mm) near the tip (Figure 10.52f)		and high maintenance	
U-tubes	It consists of downdraft and updraft tubes. The mixed liquor flows from the downcomer to the bottom and returns to the surface through the riser. Compressed air is mixed in the downcomer and bubbles rise into the updraft tube. The system acts as an airlift pump (see Figure 10.15)	High OTE due to high solubility of oxygen under high static pressure	Requires high-pressure blower	Used specifically in a deep-shaft process
Surface system				
Cascade aerators	Oxygen transfer occurs due to turbulences in sheet flow when water flows over a series of steps by gravity in the open channel-type aerator	Effective oxygen transfer	High head loss, cost-effective only if hydraulic conditions allow gravity flow	Post aeration

Note: OTE = oxygen transfer efficiency, OTR = oxygen transfer rate, SAE = standard aeration efficiency.

It is usually in the range of 1.2–2.4 kg O_2/kW·h (2–4 lb O_2/hp·h). Brief description and typical performance data of commonly used mechanical aerators are provided in Table 10.19.[2,4,7,11,12,59,77–83] Equipment details and installation features are shown in Figure 10.53.[79–83]

10.3.10 Aeration System Design

Aeration system is one of the most important unit processes at a wastewater treatment plant. Its operation usually accounts up to 60–70% of the total energy consumed at the plant. The design of a diffused-air aeration system involves sizing of aeration basin, selection of diffuser layout, sizing air piping, and selection of air blowers or compressors. For a mechanical aeration system the design may involve an aerator platform. When the aeration system is also used for mixing purpose, the mixing intensity may need to be verified. Normally, spray nozzles are also involved for foam control.

Aeration Basin: The aeration basin is usually a reinforced concrete structure. A prefabricated steel tank may also be used for a small wastewater treatment package system. The influent and effluent sections are the integral part of the basin. The aeration device is installed in the basin with special consideration of operation and maintenance requirements without draining the basin.

Diffuser Layouts: The typical layouts used for arranging the air diffusers in an aeration basin are shown in Figure 10.54. In general, the micro- and fine-bubble diffusers are suitable for use in the grid layout. Medium to coarse bubble aeration devices are most likely to be arranged in cross or dual spiral roll layouts. Mid-width or single-roll layout is normally used for coarse bubble diffusion devices.[2,57]

Normally, air diffusion devices are installed on fixed grids or drop pipes that are anchored to the floor or to the walls. Shutdown and dewatering of aeration basin are required for maintenance or service. When access of aeration devices without draining the basin is required, use of pullable, retrievable, or liftable design must be considered. For a small plant, a flexible rubber hose that is connected to an individual diffuser may be used. However, special retrievable aeration assemblies must be used in large aeration basins. An example of retrievable aeration system is illustrated in Figure 10.55.[84,85]

Air piping: The air piping consists of large and small diameter pipes, valves, flow meters, elbows and fittings, manifolds, and air diffusers. Typically, for good distribution of air in the basin, the head loss in the air piping between the last flow split and farthest diffuser should be less than the head loss across the

TABLE 10.18 Brief Descriptions with Typical Technical and Performance Data of Selected Air Diffusion Devices

Type of Device	General Description	Technical Data	Diffuser Layout	Airflow Rate/Diffuser[a], Nm^3/h (scfm)	SOTE[b], %	SAE[c], kg O_2/kW·h (lb O_2/hp·h)
Ceramic diffusers	Mostly made of ceramic porous media, bonded grains of fused crystalline aluminum oxide, vitreous-silicate-bonded grains or resin-bonded grains of pure silica with typical pore sizes of 0.1–0.4 mm					
Dome	Mounted through a center bolt or a lock ring on a baseplate with rubber gasket to seal (Figure 10.52a(i))	Diameter ~180 mm (7 in), and height ~40 mm (1.5 in)	Grid	0.8–4 (0.5–2.5)	27–37	–
Disk	Mounted on a saddle-type baseplate that is either solvent-cemented or mechanically attached to the header	Diameter 180–230 mm (7–9 in)	Grid	0.5–12 (0.3–7.5)	25–35	2.5–6 (4–10)
Plate	Ceramic media mounted on a plenum (Figure 10.52a(ii))	Width ~0.3 m (1 ft), and length of 0.3–1.2 m (1–4 ft)	Grid	35–90 (2–5)[d]	26–33	–
Tube	Self-supported ceramic media mounted on one or both sides of lateral header	Diameter 65–100 mm (2.5–4 in), and length 500–600 mm (20–24 in)	Grid	0.55–5.5 (0.35–3.5)	25–35	–
Rigid plastic diffusers	Porous media of high-density polyethylene (HDPE) or styrene-acrylonitrile (SAN) with typical pore sizes of 0.02–0.12 mm					
Disk	Mounted on a saddle-type baseplate that is either solvent-cemented or mechanically attached to the header	Diameter 180–230 mm (7–9 in)	Grid	1.6–7.1 (1–4.5)	25–35	–
Tube	Self-supported ceramic media connected on one or both sides of lateral header	Diameter 12.5–100 mm (0.5–4 in), and length 500–1200 mm (20–48 in)	Grid, and single or dual spiral roll	4–6.3 (2.5–4)	28–32	–
Nonrigid plastic diffusers	Porous media of rubber-HDPE with typical pore sizes of 0.02–0.04 mm					

(Continued)

TABLE 10.18 (*Continued*) Brief Descriptions with Typical Technical and Performance Data of Selected Air Diffusion Devices

Type of Device	General Description	Technical Data	Diffuser Layout	Airflow Rate/Diffuser[a], Nm^3/h (scfm)	SOTE[b], %	SAE[c], kg O_2/kW·h (lb O_2/hp·h)
Tube	Flexible media connected to lateral header on one end with a rubber check-valve on the opposite end	Diameter 25–50 mm (1–2 in), and length 600–900 mm (24–36 in)	Grid, and single spiral roll	1.6–11 (1–7)	26–36	–
Membrane diffusers	Mostly made of perforated polyurethane or ethylene-propylene dimers (EPDMs) membranes with typical pore sizes of 0.25–5 mm					
Disk	Membrane material placed over a support plate and secured to the plate around the periphery by different types of rings (Figure 10.52a(iii))	Diameter 180–300 mm (7–12 in), and upward ~6–64 mm (0.24–2.6 in) when air supply is turned on	Grid	0.8–21 (0.5–13)	27–35	2.5–6 (4–10)
Tube	Membrane material covering an internal support tube with slots or openings (Figure 10.52a(iv))	Diameter 65–90 mm (2.5–3.5 in), and length 500–1000 mm (20–40 in)	Grid, single spiral roll, or quarter points	1.6–21 (1–13)	20–35	–
Panel	Membrane material stretched over a baseplate anchored on the basin floor (Figure 10.52a(v))	Width ~1.2 m (4 ft), and length 1.8–3.6 m (6–12 ft)	Grid	9–90 (0.5–5)[d]	≈ 40	Up to 8 (Up to 13)
Strip	Membrane material stretched over a long opening on top of a holding pipe anchored on the basin floor (Figure 10.52a(vi))	Width 100–150 mm (4–6 in), and length 0.5–4 m (1.6–13 ft)	Grid	3.5–35 (0.2–2)[d]	30–45	5–8 (8–13)
Coarse bubble broad or narrow band diffusers	Saddle-mounted stainless steel or plastic broad band devices containing perforations and slots (Figure 10.52b(i))	Length 300–600 mm (12–24 in)	Single or dual spiral roll, or mid width	8–63 (5–40)	7–17	0.7–2 (1.1–3.5)
Coarse bubble disk, cap, or plate diffusers	Disks or caps with fixed orifices made of molded plastic with NPT threads for connection on lateral pipes or spring-loaded metal plates(Figures 10.52b(ii–iv))	Diameter 75–150 mm (3–6 in)	Side header, or single spiral roll	4.8–9.5 (3–6)	7–14	–
Perforated pipes	Stainless steel or plastic pipes with perforations	Length 300–600 mm (12–24 in)	Side header	8–80 (5–50)	5–15	–
Jet aerators	Dual nozzle jet device fabricated from fiberglass reinforced plastic (Figure 10.52c)	–	Side header	85–475 (54–300)	10–25	1.2–2.5 (2–4)
Tubular aerators	A vertical tube with a tortuous pathway with alternately placed deflection plates (Figure 10.52d)	–	–	–	7–10	1.2–1.5 (2–2.5)

(*Continued*)

TABLE 10.18 (*Continued*) Brief Descriptions with Typical Technical and Performance Data of Selected Air Diffusion Devices

Type of Device	General Description	Technical Data	Diffuser Layout	Airflow Rate/Diffuser[a], Nm^3/h (scfm)	SOTE[b], %	SAE[c], kg O_2/kW·h (lb O_2/hp·h)
Static tubes	Stationary vertical tubes with a series of diffuser membranes or internal baffles (Figure 10.52e)	–	–	–	10–12	1.4–1.7 (2.3–2.8)
Aspirators	Stainless steel jet device assembly including a high-speed propeller mounted on the lower end of an inclined hollow shaft and an air intake at the top end of the shaft (Figure 10.52f)	–	N/A	N/A	N/A	1.5-2.5 (2.5–4)
U-tubes	Concrete or steel downdraft and updraft tubes with diffusers and compressor	Diameters 1–6 m (3–20 ft), and length 30–150 m (100–500 ft)	N/A	N/A	40–90[e]	1.8–2.7 (3–4.5)

[a] For aeration devices, the typical air flow rate units are Nm^3/h at 0°C and 1 atm or sm^3/h at 20°C and 1 atm in SI units, and scfm at 20°C and 14.7 psi in U.S. customary units.
[b] SOTE: Standard oxygen transfer efficiency (see definition and description in Section 10.3.8 and Equation 10.41). The approximate values are mostly 4.5 m (12) submergence.
[c] SAE: Standard aeration efficiency (see definition and description in Section 10.3.8 and Equation 10.42). The values will vary with change in blower efficiency.
[d] The air flow rate unit for this device is $sm^3/min \cdot m^2$ ($scfm/ft^2$).
[e] The values depend upon the shaft diameter and depth.

Note: 1 Nm^3/h = 1.07 sm^3/h = 0.631 scfm, 1 $sm^3/h \cdot m^2$ = 0.0547 $scfm/ft^2$, and 1 kg O_2/kW·h = 1.64 lb O_2/hp·h.

Source: Adapted in part from References 1, 2, 7, 27, 57, and 59 through 76.

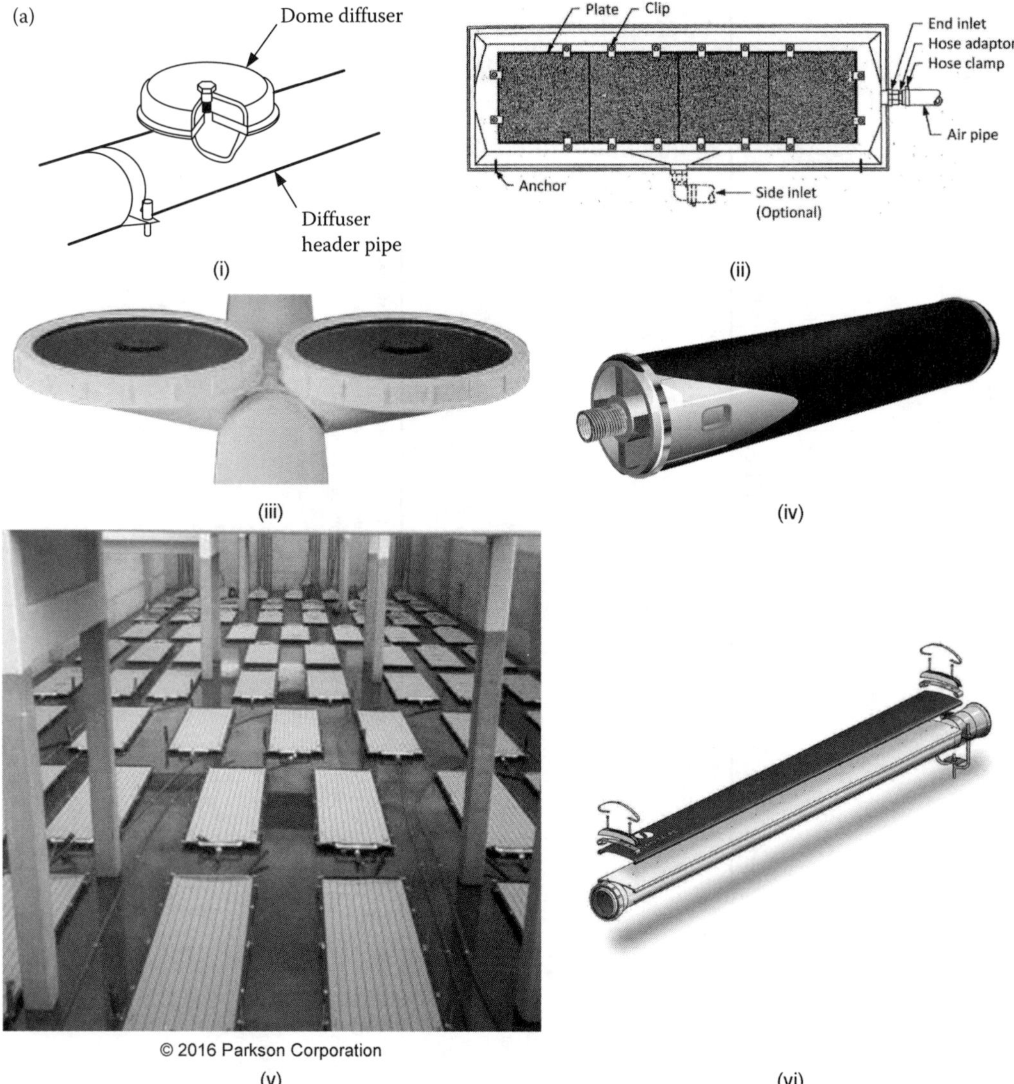

FIGURE 10.52 Various types of diffused-air aeration devices: (a) porous diffusers: (i) ceramic dome, (ii) ceramic plate (Adapted in part from Reference 57), (iii) fine bubble dual-disk system (Courtesy Evoqua Water Technologies), (iv) membrane tubular (Courtesy EDI), (v) membrane panel (Courtesy Parkson Corporation), and (vi) membrane strip (Courtesy Xylem). *(Continued)*

diffuser. The total head loss in pipe headers is generally 5–20 cm (2–8 in) of water. The head loss across fine-bubble diffusers generally range 25–50 cm (10–20 in) of water. Manufacturers normally provide the rating curves for the diffusers and allowance for clogging. The range of air velocities in pipe headers are provided in Table 10.20.[2,7,17] The friction factor and head loss for a straight steel pipe can be estimated from Equations 10.51a through 10.51e.

$$h_L = f\frac{L}{D}h_v \tag{10.51a}$$

$$f = \frac{0.029D^{0.027}}{Q_d^{0.148}} \tag{10.51b}$$

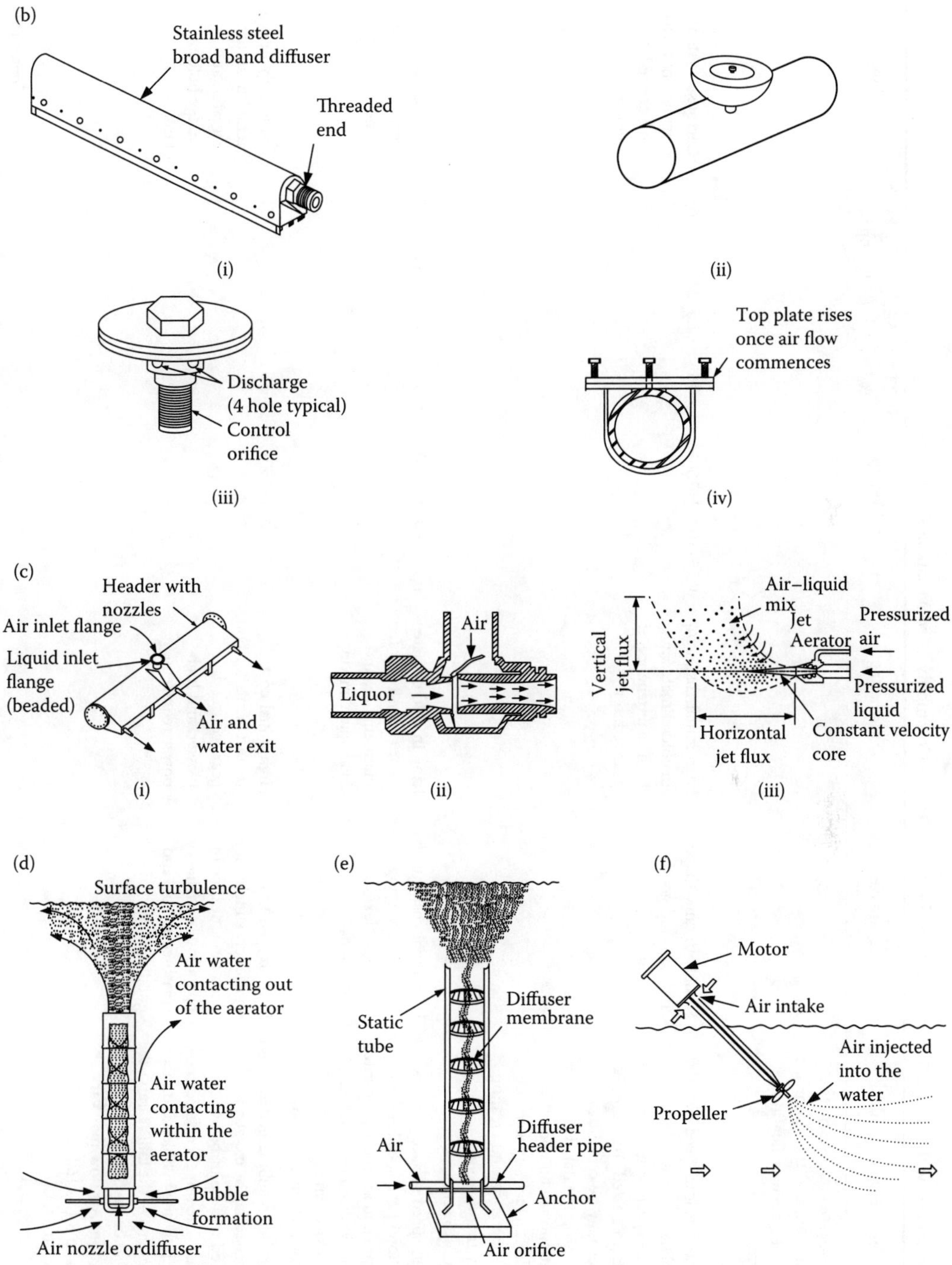

FIGURE 10.52 (Continued) Various types of diffused-air aeration devices: (b) nonporous diffusers: (i) stainless steel broadband, (ii) orifice, (iii) cap, and (iv) spring-loaded plate; (c) jet aerators: (i) manifold arrangement, (ii) sectional view, and (iii) air–water plume; (d) tubular aerators; (e) static tube; and (f) aspirator.

TABLE 10.19 Brief Descriptions with Typical Technical and Performance Data of Commonly Used Mechanical Aerators

Type of Aerator	General Description	Advantage	Disadvantage	Power, kW, hp	SOTE[a], %	SAE[b], kg O_2/kW·h (lb O_2/hp·h)	Application
Vertical axis	The blades are attached on the vertical shaft, and the motor sits on top of a fixed or floating platform.						
Surface aerator							
Low speed	The aerator is low speed (20–100 rpm). The motor with a gear box to reduce speed is mounted on the float or on a fixed structure. The impeller may be centrifugal, radial-axial, or axial. The circulation may be updraft or downdraft. Dual impellers are used for aeration and creating velocity in a carrousel basin (Figure 10.53a).	Flexible in tank shape and size, and good mixing	High initial cost, icing in cold climate, and high maintenance of gear reducer	0.75–100 (1–150)	10–15	1.5–2.8 (2.5–4.6)	Carrousel system (oxidation ditch), and aerated lagoon
High speed	The high-speed surface aerators (axial flow) have speeds of 300–1200 rpm and are mostly mounted on floats with mooring cable. The impeller is submerged or partly submerged (Figure 10.53b).	Low initial cost, suitable for use on varying water level, and flexible operations	Icing in cold climate, poor accessibility for maintenance, and inadequate mixing	0.75–100 (1–150)	10–15	1.1–1.4 (1.8–2.3)	Aerated lagoon, and stabilization pond
Submerged or turbine aerator	The impeller is submerged and compressed air or oxygen is delivered to a point below the impeller. The impeller disperses the air into fine bubbles and mixes the contents of the tank. Draft tube may also be used to increase circulation. Air flow may vary from 4 to 8 L/s (60–120 gpm) (Figure 10.53c).	High oxygen transfer efficiency, suitable for deep tank, flexible in operation, no icing or splash concerns, and good mixing	High initial cost, requirements for both gear reducer and blower, and high power requirement	0.75–100 (1–150)	15–35	1.1–2.1 (1.8–3.5)	Completely mixed aeration, aerobic digester, and sludge holding tank

(Continued)

TABLE 10.19 (*Continued*) Brief Descriptions with Typical Technical and Performance Data of Commonly Used Mechanical Aerators

Type of Aerator	General Description	Advantage	Disadvantage	Power, kW, hp	SOTE[a], %	SAE[b], kg O_2/kW·h (lb O_2/hp·h)	Application
Horizontal axis	The aerator has a horizontal axis. A cylinder or drum either exposed or submerged provides aeration and forward movement of the liquid.						
Surface or brush aerator	Consists of cylinder or drum with bristles of steel protruding from the perimeter into wastewater, provides aeration and moves the liquid forward (Figure 10.53d)	Moderate initial cost, provides aeration and circulation, and good maintenance accessibility	Limitation in tank geometry need gear reducer, and have low efficiency	0.1–0.75 (0.15–1)	–	1–2 (1.6–3.3)	Oxidation ditch
Submerged disk aerator	Consists of disks that are submerged in the liquid approximately one-eighth to three-eighth of the diameter. The recess in the disks introduces entrapped air into the submerged section. The disk spacing and submergence can vary depending upon the oxygen requirement (Figure 10.53e)	Same as surface aerator	Same as surface aerator	0.1–0.75 (0.15–1)	–	1.2–2.4 (2–4)	Oxidation ditch

[a] SOTE: Standard oxygen transfer efficiency (see definition and description in Section 10.3.8 and Equation 10.41). The approximate values are mostly 4.5 m (12 ft) submergence.

[b] SAE: Standard aeration efficiency (see definition and description in Section 10.3.8 and Equation 10.42). The values will vary with change in blower efficiency.

Note: 1 kW = 1.34 hp, and kg O_2/kW·h × 1.64 = lb O_2/hp·h.

Source: Adapted in part from References 2, 4, 7, 11, 12, 59, and 77 through 83.

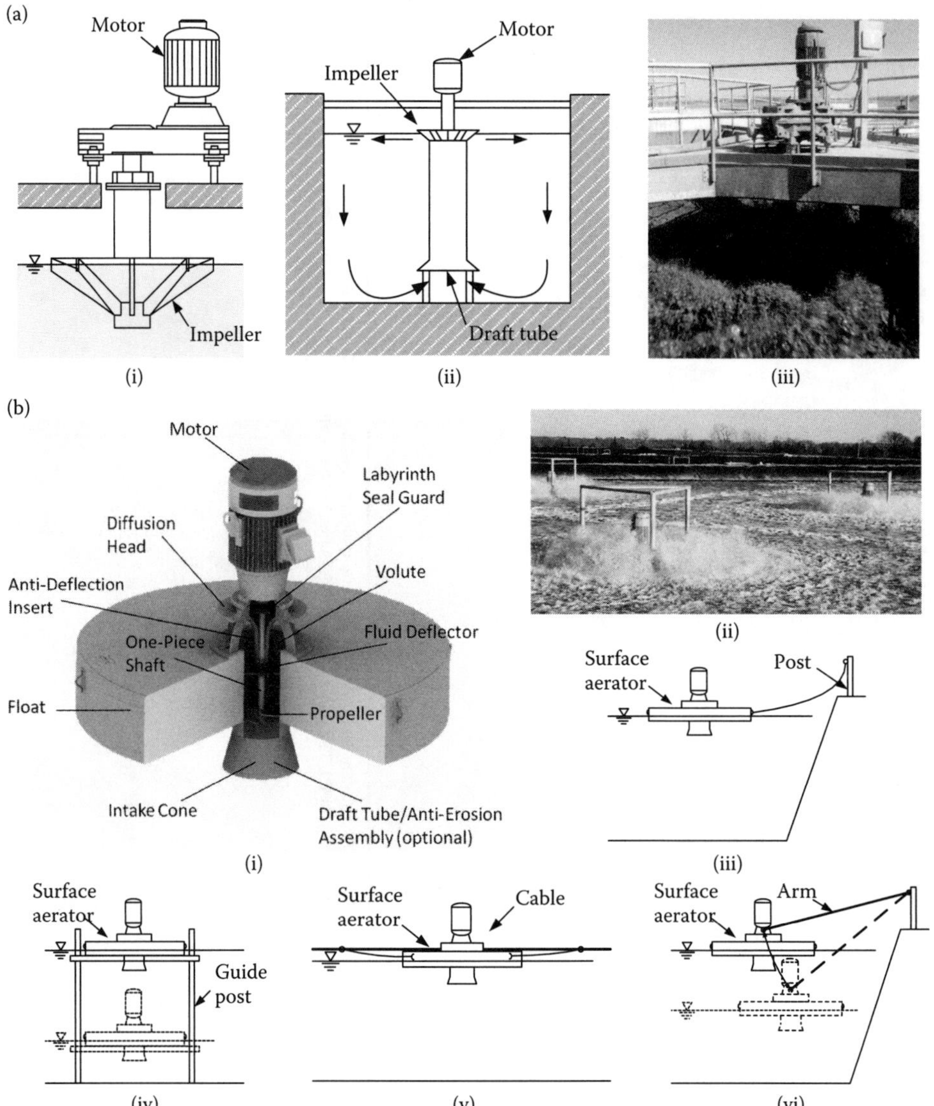

FIGURE 10.53 Various types of mechanical aerators: (a) vertical-axis low-speed surface aerator: (i) impeller and motor on fixed bases, (ii) surface aerator with draft tube, and (iii) impeller in operation; (b) floating surface aerator components and typical installations: (i) major components (Courtesy Aqua-Aerobic System, Inc.), (ii) aerator in operation (Courtesy Evoqua Water Technologies), (iii) shore post mooring, (iv) restrained mooring, (v) cable anchor mooring, and (vi) pivoting arm mooring. *(Continued)*

$$h_L = 9.82 \times 10^{-8} \times \left(\frac{fT_d Q_d^2}{P_d D^5}\right) L \tag{10.51c}$$

$$Q_d = \left(\frac{1\ \text{atm}}{P_d} \times \frac{T_d}{293°\text{K}}\right) Q_a \tag{10.51d}$$

$$T_d = T_0 \left(\frac{P_d}{P_0}\right)^{0.283} \tag{10.51e}$$

FIGURE 10.53 (Continued) Various types of mechanical aerators: (c) vertical-axis submerged aerator (Courtesy Invent): (i) hyperboloid-shape mixer/aerator and (ii) aerator in operation; (d) horizontal-axis brush-type surface aerator (Courtesy Envirodyne Systems Inc.): (i) installation in ditch and (ii) aerator in operation; and (e) horizontal axis disk type submerged aerator (Courtesy Evoqua Water Technologies): (i) close view of disk and (ii) aerator in operation.

where

h_L = friction head loss in the straight pipe, mm (in) of water

f = friction factor, dimensionless. The friction factor is obtained from Moody diagram based on relative roughness. It is recommended that f be increased by at least 10% to allow for an increase in friction factor as the pipe ages

L = equivalent length of the straight pipe, m (ft)

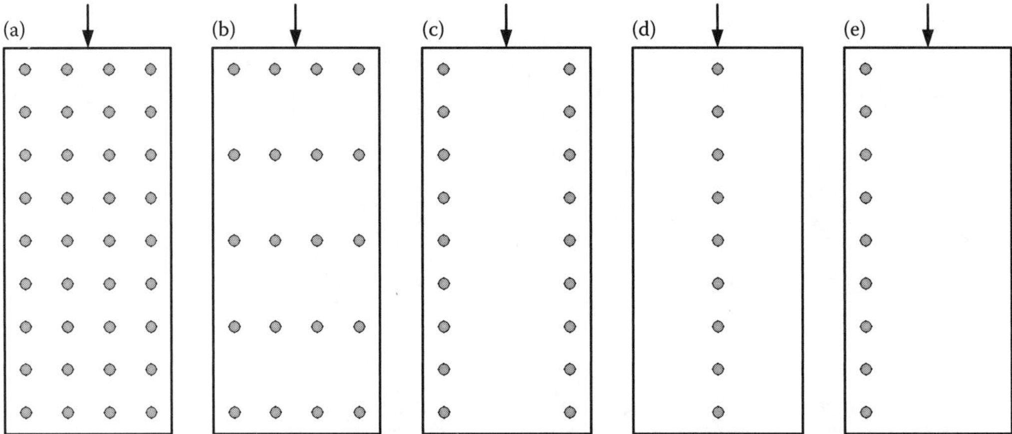

FIGURE 10.54 Typical diffuser layouts: (a) grid, (b) cross roll, (c) dual spiral roll, (d) mid-width roll, and (e) single spiral roll.

D = pipe diameter, m (ft)
h_v = velocity head of air, mm (in) of water
Q_d = air flow under blower discharge pressure and temperature, sm^3/min (scfm)
P_d = blower discharge pressure, atm
P_0 = ambient barometric pressure, atm
T_d = blower discharge temperature in air piping, °K
T_0 = ambient air temperature, °K. Usually, the maximum sustained air temperature in the summer is used

The equivalent pipe length for pipe fittings (elbows, tees, valves, meters, etc.) is calculated from Equation 10.51f.[2,7,17]

$$L = 55.4\ CD^{1.2} \tag{10.51f}$$

FIGURE 10.55 Retrievable aeration system (Courtesy Xylem): (a) retrievable section and (b) installation in basin.

TABLE 10.20 Typical Velocities in Aeration Header Pipes

Pipe Diameter, cm (in)	Velocity at Standard Conditions, m/min (ft/min)
2.5–7.5 (1–3)	350–550 (1200–1800)
10–25 (4–10)	500–900 (1600–3000)
30–60 (12–24)	800–1200 (2700–4000)
75–150 (30–60)	1050–2000 (3400–6500)

Source: Adapted in part from References 2, 7, and 17.

where C = coefficient for converting air pipe fittings to the equivalent pipe length, dimensionless. It is determined from Table 10.21.

The submergence over the diffusers is normally the largest pressure to overcome. All air pipes used for fine bubble diffusers are of noncorroding and nonscaling material such as stainless steel, galvanized metal, or plastic pipes. The head loss in piping, diffusers, and other accessories is usually 20–40% of the diffusers submergence.

Blowers: Blowers move air or gases under pressure. They develop a pressure differential between the inlet and discharge point. Blowers are usually divided into two groups: (1) positive displacement (PD), and (2) centrifugal. The centrifugal blowers are further categorized as multistage (MSC), single-stage (SSC), or high-speed turbo (HST). Brief descriptions with typical performance data of commonly used mechanical aerators are provided in Table 10.22. Examples of different blowers are shown in Figure 10.56.[86,87]

The PD or HST blowers are more suitable for use at small plants, while SSC blowers are mostly used for large applications. MSC blowers have been used for wide range of flows. Centrifugal blowers are most commonly used for installations in larger than 425 m^3/min (15,000 scfm) in a pressure range of 50–70 kPa (7–10 psi). These blowers have head-capacity (H-Q) curve similar to low specific-speed centrifugal pumps. In air supply system design, the operating point is determined from the intersection of H-Q curve and the piping system curve. The air flow may be adjusted by throttling the inlet or discharge. Throttling of outlet for low volume may cause surging of the machine.* HST blowers are the latest innovative development in blower design that emerged into the North American market around 2007 and have been installed at many plants during recent years. Using aerodynamically designed gearless impellers with low-friction bearings, they are typically 10–20% more energy efficient than other types of centrifugal or PD blowers. Only a slight decrease in efficiency may be seen at a turndown flow ratio less than 50%.[86–92]

TABLE 10.21 Coefficient for Converting Typical Fittings to Equivalent Pipe Length

Fitting	*C* Coefficient
Long-radius elbow or run of standard tee	0.33
Medium-radius elbow or run of tee reduced by 25% in size	0.42
Standard elbow or run of tee reduced by 50% in size	0.67
Tee through side outlet	1.33
Angle valve	0.90
Gate valve	0.25
Globe valve	2.00

Source: Adapted in part from References 2, 7, and 17.

* Surging is a phenomenon in which a blower runs alternately at zero and full capacity. This causes vibration and overheating.

TABLE 10.22 Brief Descriptions with Typical Technical and Performance Data of Commonly Used Air Blowers

Type of Blower	General Description	Advantage	Disadvantage	Capacity, sm^3/min (scfm)	Pressure, kN/m^2 (psi)	Nominal Efficiency, %	Nominal Turndown, % of Rated Flow
Positive displacement (PD)	Delivers a fixed volume of air over a broad range of discharge pressure for each shaft revolution. It uses two- or three-lobed impellers on each shaft at constant or variable frequency drive (VFD).	Low capital costs, available for high discharge pressure, good turndown ratio, and simple control scheme	Least energy efficient, high power consumption, requires silencers for noise control, requires VFD for variable flow, limited to small plants, and maintenance intensive	<425 (15,000)	>55 (8)	45–65	50
Multistage centrifugal (MSC)	Delivers a wide range of flows at high discharge pressure by using a series of enclosed impellers operating at low speed (3600–4400 rpm); uses antifriction roller-type bearings; uses inlet or discharge throttling for control.	More efficient than PD blower, lower capital costs than SSC blower, good to fair efficiency, lower noise than SSC blowers, and simple control scheme	Less energy efficient than SSC blower, not efficient at low turndown flows, need for monitoring bearing temperature and vibration, and need for grease or oil lubrication	>35–700 (1250–25,000)	7–175 (1–25)	50–70	50–60
Single-stage centrifugal (SSC)	Delivers a wide range of flows over a broad range of discharge pressure by using a single open face-type impeller operating at high speed (typically 10,000–14,000 rpm). It also uses gearing and journal type bearings, inlet guide vanes, and discharge variable diffuser vanes for control.	More efficient than MSC blower, efficient at low turndown flows, easy for installation, and good protection of power system from surge	Higher capital costs than PD and SSC blowers, high maintenance, complex control scheme, requires enclosure for noise control, needs monitoring of power consumption, and grease or oil lubrication	>425 (15,000)	50–70 (7–10)	70–80	50–65
High-speed turbo (HST)	Delivers moderate range of flows at a moderate discharge pressure. It uses a single or twin impellers operating at super high speed (typically 40,000–75,000 rpm); uses air-foil or magnetic bearings, and single- or dual-point control.	High efficiency at low flows, efficient at low turndown flows, small footprint, easy to install, low maintenance, low noise, low vibration, integrated-VFD, no requirement for lubrication and air cooling	Potentially highest capital costs, requirement of filters for preventing power surge, and lack of long-term operating experience	<567 (20,000)	100–140 (15–20)	70–80	50

Source: Adapted in part from References 2, 7, 58, and 86.

FIGURE 10.56 Examples of air blowers used for diffused-air aeration systems: (a) positive displacement (Courtesy Aerzen USA Corporation), (b) multistage centrifugal (Courtesy Gardner Denver Nash, LLC), and (c) high-speed turbo (Courtesy Sulzer).

In a complete air supply system, the blower requires other accessories such as air filter, silencer, check valve, and other fittings. Head losses are also associated with these accessories. The values of head losses in Table 10.23 may be used as a design guide. The designer shall consult with the equipment manufacturers to obtain the proper design values.

The blower power requirements are estimated from the air flow, discharge and inlet pressures, and air temperature. Equation 10.45a is used to calculate the blower shaft horsepower based on the assumption of adiabatic conditions.[1,2,7,58] The normal procedure involves calculation of the maximum head loss that may potentially occur in the air piping system. The sum of all head losses in piping, diffuser, blower accessories, and submergence gives the discharge pressure desired at the blower. Since the head losses in piping and diffusers depend on the supply pressure and temperature of the air, an iterative procedure is normally required in design.

Mixing: Mixing requirement in an aeration basin is normally expressed by airflow for a given aerator volume or power intensity. The requirements for good mixing are given in Table 10.24, and may be used as a design guide.

TABLE 10.23 Typical Head Losses for Devices Associated with Air Blowers

Device	Head Loss, mm (in)
Air filter	13–75 (0.5–3)
Silencer	
Centrifugal	13–38 (0.5–1.5)
Positive displacement	150–200 (6–8)
Check valve	20–200 (0.8–8)

Source: Adapted in part from Reference 7.

TABLE 10.24 Mixing Requirements in Aeration Basins

Tank Type	Requirement for Good Mixing
Diffused-air aeration	
Spiral roll patterns (Figure 10.54c–e)	20–30 m^3/min per 1000 m^3 (20–30 cfm/1000 ft^3) tank volume
Grid pattern (Figure 10.54a)	10–15 m^3/min per 1000 m^3 (10–15 cfm/1000ft^3) tank volume
Fine buddle diffusers in grid pattern (Figure 10.54a)	2.2 m^3/h per m^2 (0.12 cfm/ft^2) tank surface area
Mechanical aeration	
Deep tank	30–40 kW/1000 m^3 (1.1–1.5 hp/1000 ft^3) tank volume
Shallow tank	20–30 kW/1000 m^3 (0.75–1.1 hp/1000 ft^3) tank volume

Source: Adapted in part from References 1, 2, 57, and 59.

Spray Nozzles: The effluent spray is used to break the foam on the surface of aeration basin. Nozzles should produce a hard, flat spray flow of nearly 10 L/min (2.5 gpm) at a nozzle pressure of 103 kN/m^2 (15 psi). The pump and piping should be capable of pumping total flow of all nozzles at the required nozzle pressure.

EXAMPLE 10.71: DIFFUSER CONFIGURATION AND AERATION BASIN DESIGN

The primary settled wastewater enters four aeration basins. The average and peak design flows are 42,000 and 126,000 m^3/d. The total standard oxygen requirement (*SOR*) under the field conditions to carry out BOD_5 stabilization and nitrification in a single-stage aerobic reactor is 28,000 kgO_2/d (see Example 10.70 for calculation procedure). The aeration period (or HRT) is 6 h. The diffusers are standard tube 76 cm long × 6.6 cm diameter and air rating is 0.15 Nm^3 per minute per tube (5.7 scfm). The standard oxygen transfer efficiency (*SOTE*) of diffuser tubes is 20%. Air weighs 1.20 kg/m^3 and contains 23.2% oxygen by weight. The diffuser and piping shall be capable of delivering 150% of theoretical average air requirement under field conditions. Calculate the number of diffusers. Design the influent and effluent structures, diffuser assembly, and piping arrangement.

Solution

1. Determine the dimensions of the aeration basin.

$$\text{Average design flow, } Q_{\text{total}} = 42{,}000\,\text{m}^3/\text{d} \times \frac{\text{d}}{24\,\text{h}} = 1750\,\text{m}^3/\text{h}$$

$$\text{Total volume of the basins, } V_{\text{total}} = \theta \times Q_{\text{total}} = 6\,\text{h} \times 1750\,\text{m}^3/\text{h} = 10{,}500\,\text{m}^3$$

Provide four aeration basins in process trains 1 through 4.

$$\text{Volume of each basin, } V = \frac{1}{4\text{ basins}} \times V_{\text{Total}} = \frac{1}{4\text{ basins}} \times 10{,}500\,\text{m}^3 = 2625\,\text{m}^3 \text{ per basin}$$

Provide an effluent weir crest of 5 m above the floor and assume the liquid depth $H_b \approx 5$ m.

$$\text{Surface area of each basin, } A = \frac{V}{H_b} = \frac{2625\,\text{m}^3}{5\,\text{m}} = 525\,\text{m}^2$$

Provide basin width $W_b = 12$ m, and length $L_b = 45$ m.

$$\text{Actual volume per basin, } V = L_b \times W_b \times H_b = 45\,\text{m} \times 12\,\text{m} \times 5\,\text{m} = 2700\,\text{m}^3$$

A general layout of aeration basins with influent and effluent structures is shown in Figure 10.57.

2. Design the influent structure at the peak design flow.
The influent structure consists of a 0.75-m wide channel along the width of the basin and an influent junction box (1.5-m long × 1-m wide) at the middle of the channel. The primary settled effluent enters through a 0.75-m diameter influent pipe into the junction box. Also, the returned activated sludge (RAS) is brought by a 0.5-m diameter pipe into the junction box. The primary effluent and RAS flows are combined and then split into the influent channels on both sides of the basin. At peak design flow, the water depth is approximately 3.5 m in the junction box and a water depth of about 1 m is maintained in the channel. The influent channel has four 0.25 m × 0.25 m submerged ports on the basin wall on each side of the influent channel. The design details are shown in Figure 10.58a.

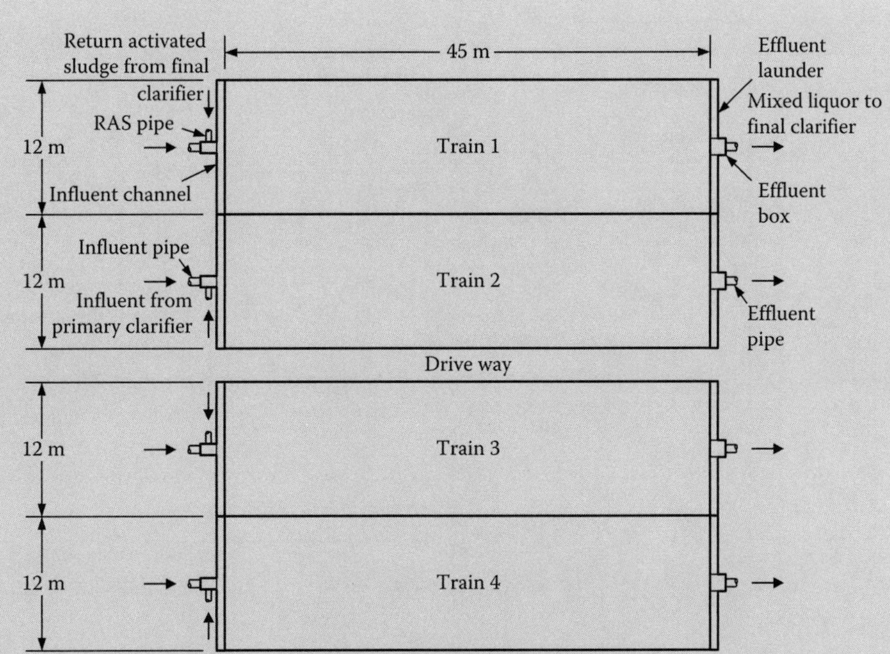

FIGURE 10.57 General layout of aeration basins with major components (Example 10.71).

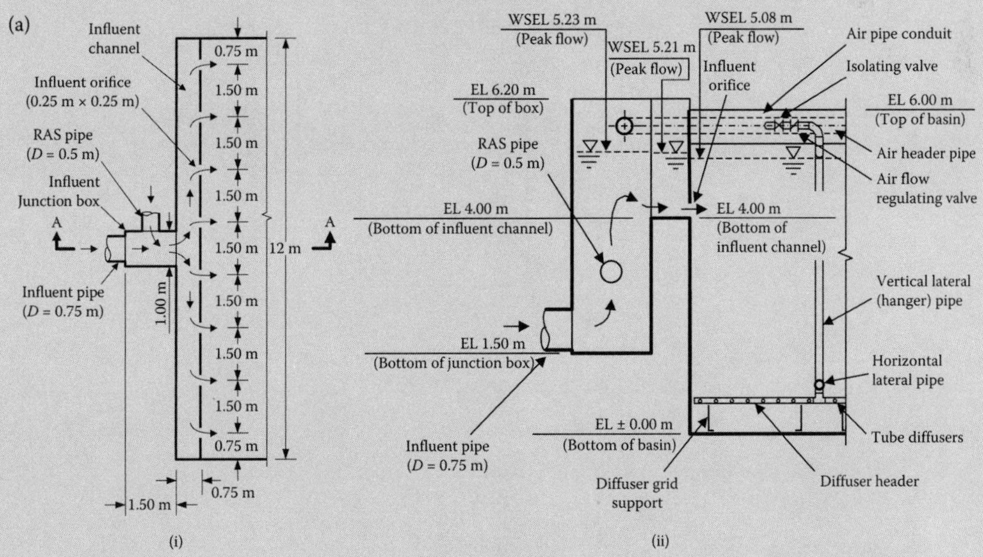

FIGURE 10.58 Aeration basin design details: (a) influent structure: (i) plan and (ii) Section AA (Example 10.71). (*Continued*)

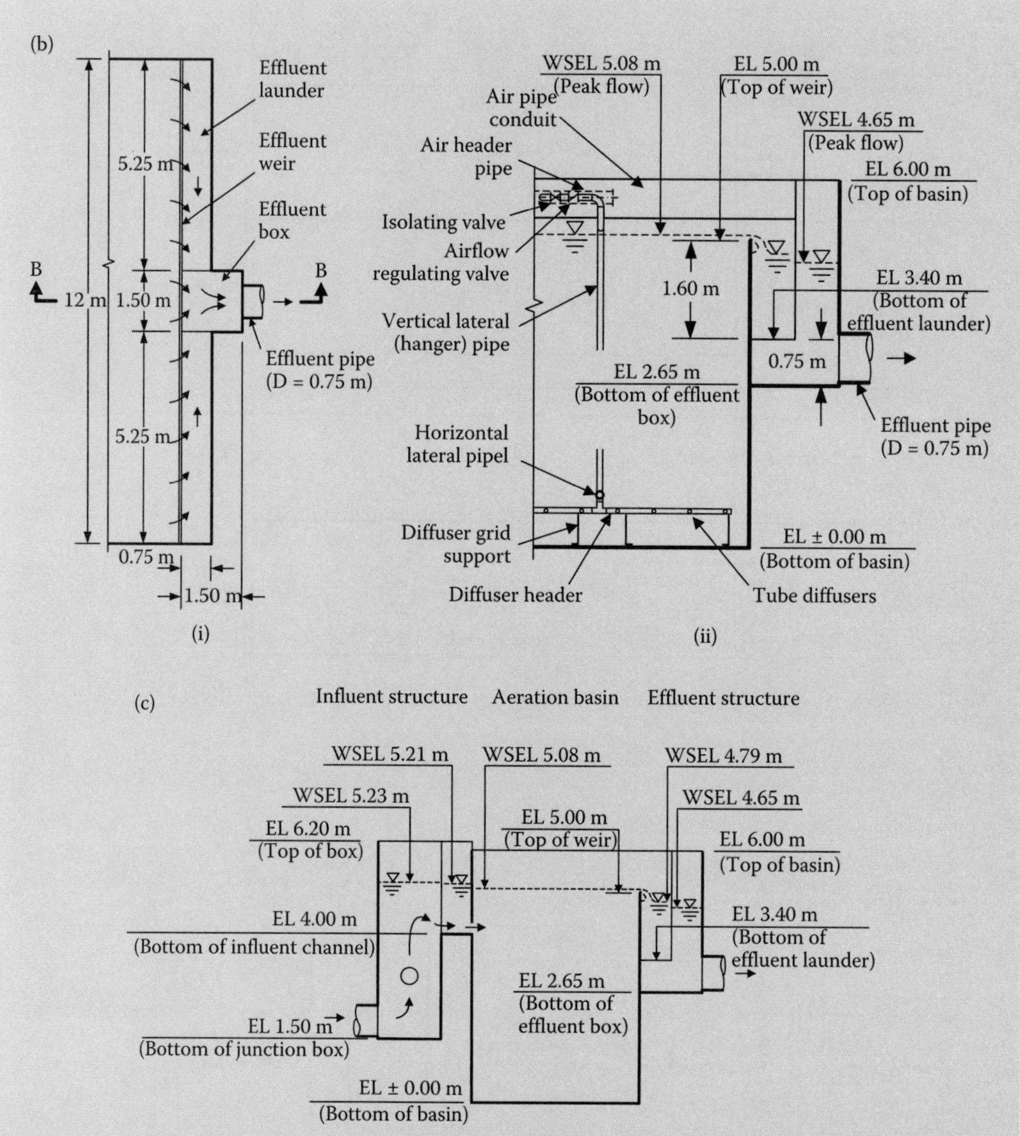

FIGURE 10.58 (Continued) Aeration basin design details: (b) effluent structure: (i) plan and (ii) Section BB; and (c) hydraulic profile at peak design flow (Example 10.71).

a. Calculate the total influent flow to each basin.

$$\text{Peak design flow, } Q_{\text{Total,peak}} = 126{,}000\ \text{m}^3/\text{d} \times \frac{\text{d}}{24\ \text{h}} \times \frac{\text{h}}{60\ \text{min}} \times \frac{\text{min}}{60\text{s}} = 1.46\ \text{m}^3/\text{s}$$

$$\text{Peak design flow to each basin, } Q_{\text{peak}} = \frac{1}{4\ \text{basins}} \times Q_{\text{Total,peak}}$$

$$= \frac{1}{4\ \text{basins}} \times 1.46\ \text{m}^3/\text{s} = 0.365\ \text{m}^3/\text{s per basin}$$

Assume that the design RAS flow ratio $R_{ras,peak}$ is one-third of the peak design flow.

$$\text{RAS flow to each basin, } Q_{r,peak} = R_{ras}Q_{peak} = \frac{1}{3} \times 0.365\,\text{m}^3/\text{s} = 0.122\,\text{m}^3/\text{s per basin}$$

$$\text{Total peak influent flow to each basin, } Q_{inf,peak} = Q_{peak} + Q_{ras,\,peak} = (0.365 + 0.122)\,\text{m}^3/\text{s}$$
$$= 0.487\,\text{m}^3/\text{s per basin}$$

Assume the R_{ras} is 0.5 under the average design flow condition.

$$\text{Total design influent flow to each basin, } Q_{inf} = \frac{(1+R_{ras})}{4\,\text{basins}} Q_{total}$$
$$= \frac{(1+0.5)}{4\,\text{basins}} \times 42{,}000\,\text{m}^3/\text{d} \times \frac{\text{d}}{24\,\text{h}} \times \frac{\text{h}}{60\,\text{min}} \times \frac{\text{min}}{60\,\text{s}}$$
$$= 1.5 \times 0.122\,\text{m}^3/\text{s} = 0.183\,\text{m}^3/\text{s per basin}$$

b. Calculate the head loss at the influent junction box at peak design flow.
Provide a 1-m wide exit opening to connect the junction box to the influent channels. Assume that the water depth at the exit opening in the influent channel is 1 m.

$$\text{Exit velocity at peak design flow, } v_{exit,peak} = \frac{Q_{inf,peak}}{w_{exit} \times d_{exit}} = \frac{0.487\,\text{m}^3/\text{s}}{1\,\text{m} \times 1\,\text{m}} = 0.487\,\text{m/s}$$

Minor head loss is expected in the junction box due to the turbulences caused by (a) changing flow direction, (b) combining primary effluent and RAS flows, and (c) splitting flow between channels. Calculate the head loss from Equation 6.15b using a minor head loss coefficient $K = 2.0$ at the exit velocity.

$$h_{m,peak} = K\frac{(v_{exit,peak})^2}{2g} = 2.0 \times \frac{(0.487\text{ m})^2}{2 \times 9.81\text{ m/s}^2} = 0.02\text{ m}$$

The head loss in the influent junction box, $\Delta h_{junction,peak} = \text{h}_{m,peak} = 0.02$ m at peak flow.
Similarly calculate the head loss at the average design flow of 0.183 m^3/s, $\Delta h_{junction,avg} \approx$ 0.00 m.

c. Calculate the head loss through the influent channel.
Because of short length, friction loss in the influent channel is small and ignored under both peak and average design flow conditions.

d. Calculate the head loss across the submerged ports at peak design flow.

$$\text{Peak design flow across each port, } q_{port,\,peak} = \frac{Q_{inf,peak}}{2\text{ channels} \times 4\text{ ports per channel}}$$
$$= \frac{0.487\text{ m}^3/\text{s}}{8\text{ ports}} = 0.0609\text{ m}^3/\text{s per port}$$

$$\text{Peak velocity across the ports, } v_{port,peak} = \frac{q_{port,peak}}{w_{port} \times h_{port}} = \frac{0.0609\text{ m}^3/\text{s}}{0.25\text{ m} \times 0.25\text{ m}} = 0.974\text{ m/s}$$

Rearrange Equation 7.4b and calculate head loss through the influent orifices using $C_d = 0.61$.

$$h_{port,peak} = \frac{1}{2g}\left(\frac{v_{port,peak}}{C_d}\right)^2 = \frac{1}{2 \times 9.81\text{ m}^2/\text{s}} \times \left(\frac{0.974\text{ m/s}}{0.61}\right)^2 = 0.13\text{ m}$$

The head loss across the influent ports $\Delta h_{port,peak} = h_{port,peak} = 0.13$ m at the peak design flow.

Similarly calculate the head loss under the average design flow condition, $\Delta h_{port,avg} \approx 0.02$ m.

3. Design the effluent structure at the peak design flow.

The effluent structure consists of a rectangular weir, a horizontal effluent launder, and an effluent box in the middle of the effluent launder. Provide a straight weir across the basin so the length of the weir is equal to the basin width $L = W_b = 12$ m. The mixed liquor flows over the weir and drops into the 0.75-m wide launder. The effluent box is 1.5 m × 1.5 m. The length of the effluent launder on both sides of the outlet box is 5.25 m. An effluent pipe of 0.75-m diameter carries the mixed liquor from the effluent box to a junction box for the final clarifiers. A liquid depth of 2 m is maintained in the effluent box at the peak design flow. The differential elevation between inverts of the launder and effluent box is 0.75 m. The design details are shown in Figure 10.58b.

a. Calculate the head over the rectangular effluent weir at the peak design flow.

The head over the weir is calculated from Equation 8.10 at $C_d = 0.6$, and $n = 0$.

$$L' = L - 0.1nH = L = 12 \text{ m}$$

Calculate the head over the effluent weir under the peak design flow condition.

$$h_{weir,peak} = \left(\frac{3}{2} \times \frac{Q_{inf,peak}}{C_d L' \sqrt{2g}}\right)^{2/3} = \left(\frac{3}{2} \times \frac{0.487 \text{ m}^3/\text{s}}{0.6 \times 12 \text{ m} \times \sqrt{2 \times 9.81 \text{ m/s}^2}}\right)^{2/3} = 0.08 \text{ m}$$

The head loss over the effluent weir $\Delta h_{weir,peak} = h_{weir,peak} = 0.08$ m at the peak design flow.

Similarly, the head over the weir is also calculated at the average design flow, $\Delta h_{weir,avg} \approx 0.04$ m.

b. Calculate the depth of effluent launder at peak design flow.

The water surface profile in the effluent launder receiving flow from a free-falling weir is estimated from Equation 8.13a.

The water depth at the exit point ($y_{2,peak}$) at the peak design flow.

$$y_{2,peak} = (2 - 0.75) \text{ m} = 1.25 \text{ m}$$

Calculate the peak design flow collected in the effluent launder on each side of the effluent box (launder width $b = 0.75$ m).

$$Q_{launder,peak} = \frac{Q_{inf,peak}}{2 \text{ sides}} = \frac{0.487 \text{ m}^3/\text{s}}{2 \text{ sides}} = 0.244 \text{ m}^3/\text{s per side of launder}$$

Calculate the water depth $y_{1,peak}$ at the upper end of the launder without channel friction loss from Equation 8.13a.

$$y_{1,peak} = \sqrt{(y_{2,peak})^2 + \frac{2(Q_{launder,peak})^2}{gb^2 y_{2,peak}}} = \sqrt{(1.25 \text{ m})^2 + \frac{2 \times (0.244 \text{ m}^3/\text{s})^2}{9.81 \text{ m/s}^2 \times (0.75 \text{ m})^2 \times 1.25 \text{ m}}}$$
$$= 1.26 \text{ m}$$

Add 10% for losses due to friction, turbulence, and 90° bend in the launder.

Total depth of water upstream of the launder, $y_{1,peak} = 1.26 \text{ m} \times 1.1 = 1.39$ m

Add a weir drop (freeboard) at peak design flow $FB_{weir,peak} = 0.21$ m from the weir crest to the water surface upstream of the launder.

The total height of effluent launder $H_{launder} = y_{1,peak} + FB_{weir,peak} = (1.39 + 0.21)\ m = 1.60\ m$.

The overall head loss in the launder $\Delta h_{launder,peak} = y_{1,peak} - y_{2,peak} = (1.39 - 1.25)\ m = 0.14\ m$ at the peak design flow.

Similarly, the maximum water depth and overall head loss in the launder are also calculated for the average design flow. Assume that the liquid depth is 1.8 m in the effluent box and obtain the following results at the average design flow: $y_{2,avg} = 1.05\ m$, $y_{1,avg} = 1.16\ m$, $FB_{weir,avg} = 0.44\ m$, and $\Delta h_{launder,avg} = 0.11\ m$.

4. Prepare the hydraulic profile at the peak design flow.

   A hydraulic profile through the aeration basin, including influent and effluent structures is prepared for the peak design flow in Figure 10.58c.

5. Calculate the number of diffuser tubes.

   The oxygen requirement under standard conditions $SOR = 28{,}000\ kg\ O_2/d$ (see Step 7 of Example 10.70).

   Rearrange Equation 10.41b and calculate the air requirement under standard conditions while $SOTR = SOR$.

$$Q_{air,total} = \frac{SOTR}{\rho_a w_{O2} SOTE} \times 100\% = \frac{SOR}{\rho_a w_{O2} SOTE} \times 100\%$$

$$= \frac{28{,}000\ kg\ O_2/d}{1.20\ kg\ air/m^3 \times 0.232\ g\ O_2/g\ air \times 20\%} \times 100\% = 502{,}900\ sm^3 air/d$$

$$\text{or } Q_{air,total} = 502{,}900\ sm^3\ air/d \times \frac{d}{1440\ min} = 349\ sm^3/min \approx 350\ sm^3/min$$

Convert the air rating from Nm^3 (0°C and 1 atm) to sm^3 (20°C and 1 atm) per minute per tube.

$$Q_a = \frac{(273+20)^\circ K}{273^\circ K} \times 0.15\ Nm^3/min{\cdot}tube = 1.07 \times 0.15\ Nm^3/min{\cdot}tube = 0.16\ sm^3/min{\cdot}tube$$

The diffuser system shall deliver 150% air required under standard condition.

$$\text{Total number of diffuser tubes required, } N_{tube,total} = \frac{Q_{air,total}}{Q_a} = \frac{1.5 \times 350\ sm^3/min}{0.16\ sm^3/min{\cdot}tube} = 3280\ \text{tubes}$$

$$\text{Number of diffuser tubes required per basin } N = \frac{1}{4\ \text{basins}} \times N_{tube,total} = \frac{1}{4\ \text{basins}} \times 3280\ \text{tubes}$$

$$= 820\ \text{tubes per basin}$$

6. Select the diffuser tube arrangement.

   Open grid type arrangement is used in each aeration zone. The overall arrangement of the aeration system in Train 1 aeration basin is shown in Figure 10.59a. The general design considerations are summarized below:

   a. Provide three equally divided aeration zones (Zones A, B and C) along the length of each aeration basin. Each zone is 15 m long (Figure 10.59a).
   b. Provide 12 identical aeration grids in each aeration zone. There are 32, 24, and 16 tube diffusers mounted on the diffuser header of each grid in Zones A, B, and C, respectively. A total of 864 diffusers are provided in each aeration basin. The tapered number of air diffusers are provided to match approximately the oxygen demand profile through the aeration basin (Figure 10.59b).

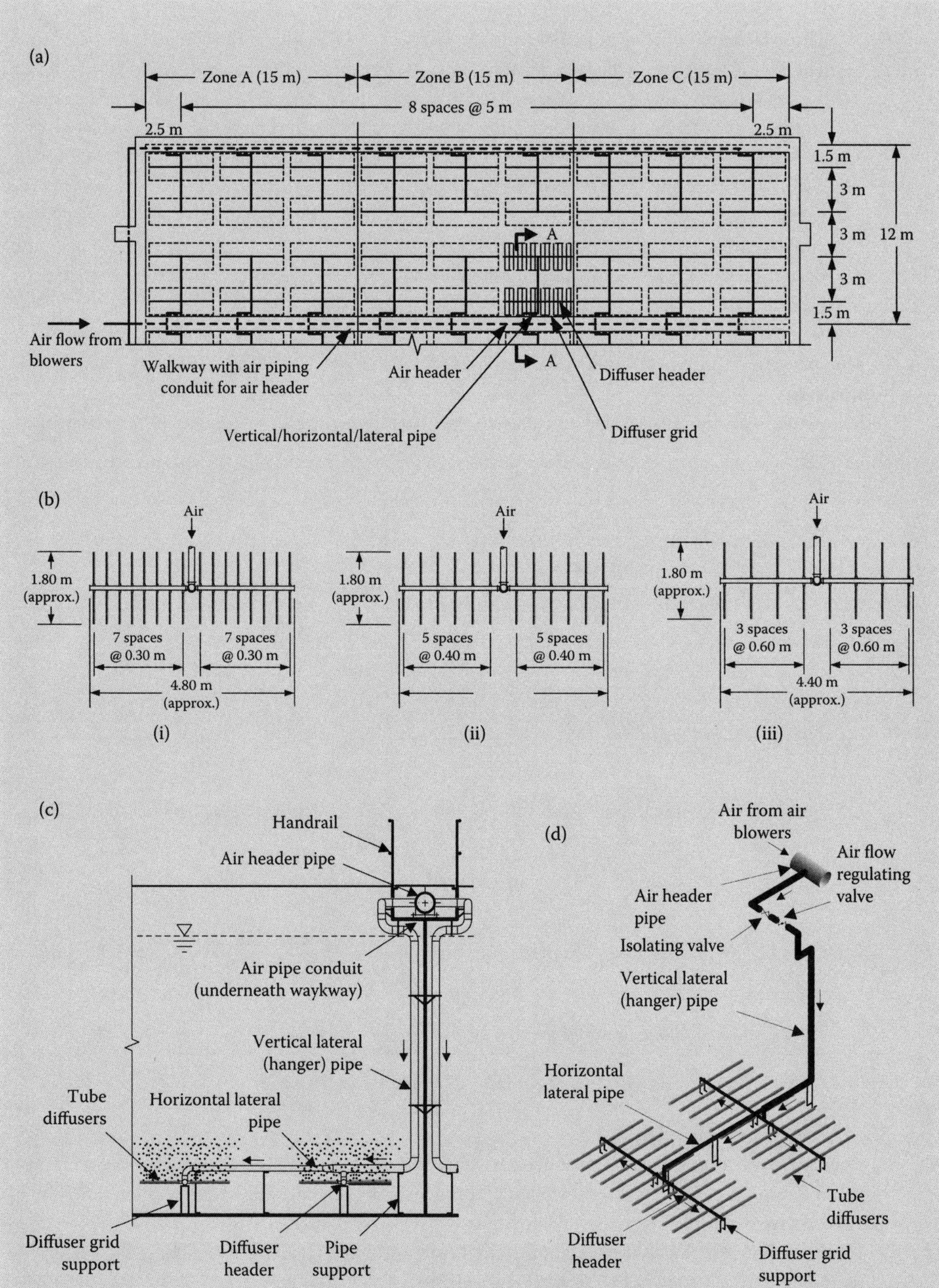

FIGURE 10.59 Aeration system design details: (a) general layout of aeration system; (b) diffuser grid design: (i) 32 diffusers per grid for Zone A, (ii) 24 diffusers per grid for Zone B, and (iii) 16 diffusers per grid for Zone C; (c) Section AA to illustrate lateral air piping; and (d) isometric view of lateral piping with diffuser grids (Example 10.71).

c. Arrange the air grids in four rows. Air headers are provided along both sides of the basin. A lateral pipe is used to deliver air to two diffuser grids from the header. Therefore, there are six lateral air pipes in each zone. Provide an isolation valve and a butterfly regulating valve on each lateral pipe (Figure 10.59c and d).

The detailed design information about the diffusers, air grids, and aeration zones is also summarized in Table 10.25. In accordance with the design guidance summarized in Table 10.24, a minimum air flow of 10–15 m^3/min per 1000 m^3 tank volume is recommended for mixing in grid pattern. In this example, the average air flow provided in the aeration basin is over 50 sm^3/min per 1000 m^3 tank volume (see Table 10.25). It is 3-5 times of the minimum air flow requirement. Therefore, the air flow is considered sufficient for good mixing under design conditions.

TABLE 10.25 Summary of Diffuser Arrangement with Airflow (Example 10.71)

Aeration Zone	Number of Grids in Each Zone	Number of Diffusers on Each Grid	Number of Diffusers in Each Zone	Airflow in Each Zone		Air Intensity for Mixing, m^3/min per 1000 m^3 Tank Volume
				sm^3/min	% of Total	
A	12	32	384	61.4	45	68
B	12	24	288[a]	46.1[b]	33[c]	51[d]
C	12	16	192	30.7	22	34
Average	12	24	288	46.1	33	51
Total	–	–	864[e]	138	100	–

[a] 12 grids/zone × 24 diffusers/grid = 288 diffusers/zone.
[b] 288 diffusers/zone × 0.16 sm^3/min per diffuser = 46.1 sm^3/min per zone.
[c] (46.1 sm^3/min per zone) ÷ (138 sm^3/min per zone) × 100% = 33%.
[d] $\frac{46.1\ \text{sm}^3/\text{min} \times 1000\ \text{m}^3}{(15\ \text{m} \times 12\ \text{m} \times 5\ \text{m}) \times 1000\ \text{m}^3} = 51\ \text{sm}^3/\text{min per } 1000\ \text{m}^3$
[e] Total number of diffusers can be determined in three different approaches: (a) (384 + 288 + 192) diffusers = 864 diffusers. (b) 3 zones × 12 grids/zone × 24 diffusers/grid (average) = 864 diffusers, or (c) 3 zones × 288 diffusers/zone (average) = 864 diffusers.

Note: The piping arrangement used in this example is the open (or open end) grid type. An alternate arrangement would be to provide loop around the basins to connect lateral pipes on both ends. Such an arrangement would give improved air flow, but the network analysis is complex. Readers are referred to References 93 through 95 for design information.

EXAMPLE 10.72: SIZING OF MECHANICAL AERATOR

A completely mixed activated sludge plant is serving a city of 10,000 population. The average wastewater flow is 120 gpcd and the influent BOD_5 after primary treatment is 150 mg/L. The aeration period HRT = 4.5 h and there are four aerators in the basin. Calculate the nameplate horsepower of the mechanical aerators and determine the aeration basin dimensions. The manufacturer's certified standard aeration efficiency (*SAE*) is 2.3 lb/hp·h. The general operating parameters are provided below: $\alpha = 0.85$, $\beta = 0.9$, temperature $T = 25°C$, the operating DO concentration $C_c = 2$ mg/L, and $BOD_5/BOD_L = 0.68$. Assume that the aeration basin is located at sea level ($P = 1$ atm) and nitrification is not achieved at the plant.

Solution

1. Calculate the theoretical oxygen requirement (*ThOR*) in the wastewater.

$$\text{Total wastewater flow, } Q = 10{,}000 \text{ persons} \times \frac{120 \text{ gal}}{\text{person·d}} \times \frac{10^{-6}\text{Mgal}}{\text{gal}} = 1.2 \text{ MGD}$$

$$\text{BOD5 loading } \Delta S_{0,\text{BOD5}} = Q \times S_0 = 1.2 \text{ MGD} \times 150 \text{ mg BOD}_5/\text{L} \times 8.34 \text{ lb/(mg/L·Mgal)} = 1501 \text{ lb BOD}_5/\text{d}$$

$$\text{CBODL loading } \Delta S_{0,\text{CBODL}} = \frac{\text{BOD}_\text{L}}{\text{BOD}_5} \times \Delta S_{0,\text{BOD5}} = \frac{1 \text{ lb BOD}_\text{L}}{0.68 \text{ lb BOD}_5} \times 1501 \text{ lb BOD}_5/\text{d} = 2207 \text{ lb BOD}_\text{L}/\text{d}$$

To be conservative, the CBOD_L lost in the effluent and fixed into the biomass are ignored. The oxygen requirement is estimated from the influent CBOD_L loading.

$$ThOR \approx \Delta S_{0,\text{CBODL}} = 2207 \text{ lb BOD}_\text{L}/\text{d} \quad \text{or} \quad 2207 \text{ lb O}_2/\text{d} \times \frac{\text{d}}{24 \text{ h}} = 92.0 \text{ lb O}_2/\text{h}$$

2. Calculate the equilibrium DO concentration in the wastewater under operating conditions ($C_{e,ww}$).

   $C_{e,ww}$ is calculated from Equation 10.39g. $\beta = 0.9$ is given in the statement, $f_p = 1$ at sea level, and $f_a = 1$ for the mechanical surface aerator. The factor f_t is calculated from Equation 10.39j.

   At 1 atm (sea level), $C_{s,t} = 8.42$ and 8.11 mg/L at 24°C and 26°C, respectively, as obtained from Table 10.16. Interpolate C_s at 25°C,$C_{s,25} = 8.27$ mg/L.

   $C_{s,20} = 9.09$ mg/L is also obtained from Table 10.16 at 20°C and 1 atm.

$$f_t = \frac{C_{s,t}}{C_{s,20}} = \frac{C_{s,25}}{C_{s,20}} = \frac{8.27 \text{ mg/L}}{9.09 \text{ mg/L}} = 0.91$$

$$C_{e,ww} = \beta f_a f_p f_t C_{s,20} = 0.90 \times 1 \times 1 \times 0.91 \times 9.09 \text{ mg/L} = 0.82 \times 9.09 \text{ mg/L} = 7.45 \text{ mg/L}$$

3. Calculate the standard oxygen requirement (*SOR*) for the removal of CBOD_L loading.

   Calculate the total amount of *SOR* from Equation 10.48c at $\alpha = 0.85$ and $C_c = 2$ mg/L. For a surface aerator, $F = 1$ and assuming $C_{e,20} \approx C_{s,20} = 9.09$ mg/L.

$$SOR = \frac{ThOR}{((C_{e,ww} - C_c)/C_{e,20})(1.024)^{(T-20)}\alpha F} = \frac{92.0 \text{lbO}_2/\text{h}}{(((7.45-2)\text{mg/L})/9.09\text{mg/L})(1.024)^{(25-20)} \times 0.85 \times 1} = 160 \text{lbO}_2/\text{h}$$

$$SOR_\text{aerator} = \frac{160 \text{ lb O}_2/\text{h}}{4 \text{ aerators}} = 40 \text{ lb O}_2/\text{h per aerator}$$

4. Determine the nameplate horsepower of each aerator.

   Rearrange Equation 10.42a and calculate the power requirement (P_a) at $SOTR = SOR_\text{aerator}$.

$$P_a = \frac{SOR_\text{aerator}}{SAE} = \frac{40 \text{ lb O}_2/\text{h}}{2.3 \text{ lb O}_2/\text{hp·h}} = 17.4 \text{ hp per aerator}$$

   Provide nameplate horsepower of the aerator = 18 hp.

   Assume a motor efficiency of 91%.

   The estimated horsepower of the motor = (18 hp/0.91) ≈ 20 hp per motor

5. Determine the dimensions of the aeration basin.

Volume of the basin, $V = \theta \times Q = 4.5\ \text{h} \times 1.2\ \text{MGD} \times \frac{\text{d}}{24\ \text{h}} = 0.225\ \text{Mgal}$

$$V = 0.225 \times 10^6 \text{gallons} \times \frac{\text{ft}^3}{7.48\ \text{gallon}} = 30{,}100\ \text{ft}^3$$

Provide a square basin with a liquid depth $H_b = 16$ ft.

$$L = W = \sqrt{\frac{30{,}100\ \text{ft}^3}{16\ \text{ft}}} = 43\ \text{ft}$$

Provide the following dimensions of the basin are $L = 44$ ft, $W = 44$ ft, and $H_b = 16$ ft.

6. Arrange the aerators and show the installation details.

The aerators arrangement and installation details are shown in Figure 10.60. The design procedures for influent and effluent structures are similar to those given in Example 10.71.

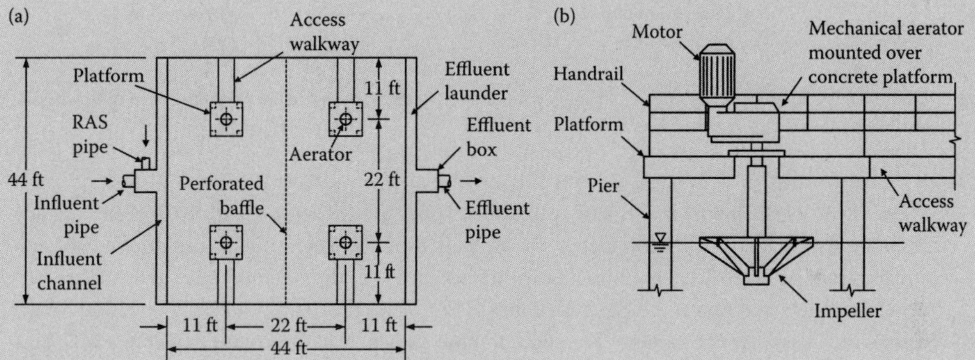

FIGURE 10.60 General layout of aeration basins with mechanical aerators: (a) plan and (b) platform and installation of the aerator (Example 10.72).

EXAMPLE 10.73: BLOWER AND AIR PIPING SIZING

Blowers supply the required amount of air against the operating water head. Determine the power requirement and number of blowers to supply the air to a given open grid type diffuser arrangement in an aeration basin. The blowers shall be capable of delivering the maximum design air requirements considering the largest single unit out of service. The piping layout and open grid type diffuser arrangement are shown in Figure 10.61. The diffusers and air piping are designed for providing 150% of the theoretical air requirement. Use the following data: design diffuser air flow = 0.25 sm^3/min delivery; and diffuser submergence = 4.5 m. The head losses through air filter, silencer, compressor, piping, fittings, valves and specials, and diffusers are supplied by the manufacturers. The operating temperature in summer is 28°C at an altitude of 457 m (1500 ft) above sea level ($P = 1$ atm).

Solution

1. Describe the open grid type diffuser arrangement, piping, layout, and aeration equipment.

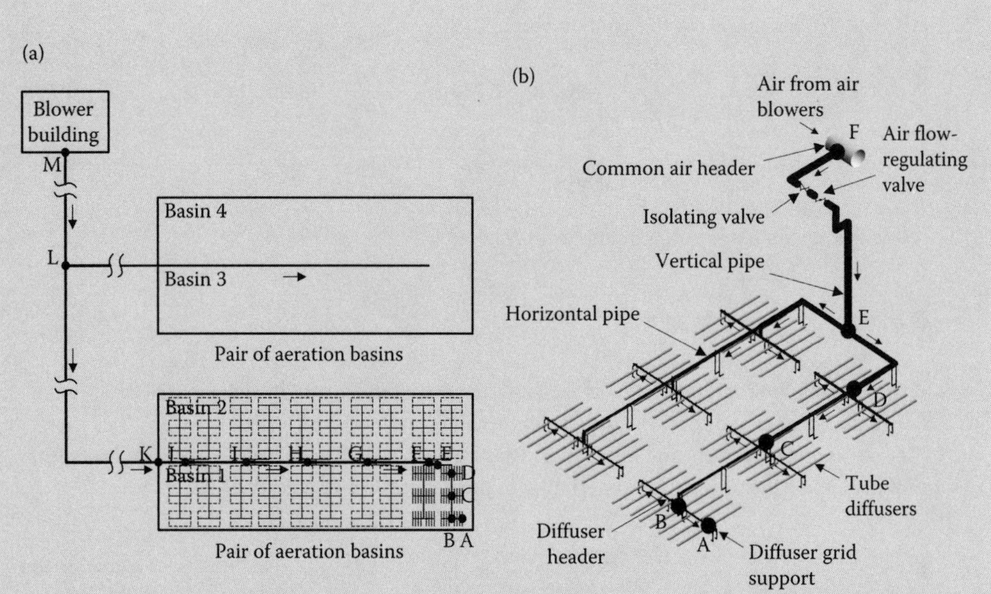

FIGURE 10.61 Air supply system: (a) piping layout and (b) isometric view of lateral piping with diffuser grids (Example 10.73).

The details of the piping layout and open diffuser grids are shown in Figure 10.61. Four aeration basins are arranged in pairs of two basins. The air main from the blower building splits into two common air headers. These headers and walkways are installed over the common walls of the paired aeration basins. The common air header from Points K to E serves the Basins 1 and 2. Five vertical hanger pipes in each basin receive air flows from the common air header at five connecting Points F, G, H, I, and J and distribute the air into the diffuser grids in each basin. An isolating valve and a flow-regulating valve are provided on each vertical hanger pipe. Two horizontal lateral pipes are connected to each vertical hanger pipe near the bottom of the tank. One horizontal lateral pipe in Basin 1 is from Points E to B. Three diffuser grids are connected to each horizontal lateral pipe. Each diffuser grid consists of a diffuser header and 16 tube diffusers, eight tubes on each side of the header. There are 8 diffuser tubes on the header from Points A to B, four on each side of pipe. There are a total of 30 diffuser grids in each aeration basin that contains a total of 480 diffuser tubes [(5 vertical hangers pipes/basin) × (2 horizontal lateral pipes/vertical hanger pipe) × (3 diffuser header pipes/horizontal lateral pipe) × (16 diffuser tubes/diffuser header pipe)].

2. Calculate the air flows in air piping system.

   Air flow per diffuser tube, $Q_a = 0.25\ \text{sm}^3/\text{min}$ per tube

   Total air flow in each diffuser header (Line AB), $Q_{air,AB} = 0.25\ \text{sm}^3/\text{min per tube} \times 8\ \text{tubes} = 2\ \text{sm}^3/\text{min}$

   Total air flow for one diffuser grid (Line BC), $Q_{air,BC} = 2 \times 2\ \text{sm}^3/\text{min} = 4\ \text{sm}^3/\text{min}$

   Total air flow in one horizontal lateral pipe (Line DE), $Q_{air,DE} = 3 \times 4\ \text{sm}^3/\text{min} = 12\ \text{sm}^3/\text{min}$

   Total air flow in each vertical lateral pipe (Line EF), $Q_{air,EF} = 2 \times 12\ \text{sm}^3/\text{min} = 24\ \text{sm}^3/\text{min}$

   Total air flow for Basins 1 and 2 (Line JK), $Q_{air,JK} = 10 \times 24\ \text{sm}^3/\text{min} = 240\ \text{sm}^3/\text{min}$

   Total air flow for Basins 1 through 4 (Line LM), $Q_{air,LM} = 2 \times 240\ \text{sm}^3/\text{min} = 480\ \text{sm}^3/\text{min}$

   Detailed flow rates in the air piping system are summarized in Table 10.26.

3. Determine the diameters of the pipes.

TABLE 10.26 Air Pipe Sizing, Airflows, and Head Loss Calculations for Blower Design (Example 10.73)

Line[a]	Description	D^b, cm (in)	$Q_a{}^c$, sm^3/min	V^d, m/min	$L_{pipe}{}^e$, m	Overall C Coefficient[f]	$L_{fittings}{}^g$, m	L^h, m	$Q_d{}^i$, m^3/min	f^j	$h_L{}^k$, mm
(a)	(b)	(c)	(d)	(e)	(f)	(g)	(h)	(i)	(j)	(k)	(l)
AB	Diffuser header pipe containing 8 diffusers	7.5 (3)	2	453	2.2	2.32	5.74	7.94	1.56	0.0253	4.6
BC	Horizontal lateral pipe supplying air to one diffuser grid with 16 diffusers	10 (4)	4	509	3.2	1.09	3.81	7.01	3.12	0.0222	3.4
CD	Horizontal lateral pipe supplying air to two diffuser grids with 32 diffusers	15 (6)	8	453	3.0	0.67	3.81	6.81	6.23	0.0203	1.6
DE	Horizontal lateral pipe supplying air to three diffuser grids with 48 diffusers	15 (6)	12	679	3.7	0.42	2.39	6.09	9.35	0.0191	3.0
EF	Vertical hanger pipe supplying air to two horizontal lateral pipes with 96 diffusers	20 (8)	24	764	7.5	7.01	56.3	63.8	18.7	0.0173	27
FG	Horizontal common main air header pipe supplying air to two vertical hanger pipes with 192 diffusers	30 (12)	48	679	10.0	1.33	17.4	27.4	37.4	0.0158	5.6
GH	Horizontal common main air header pipe supplying air to four vertical hanger pipes with 384 diffusers	40 (16)	96	764	10.0	0.42	7.75	17.7	74.8	0.0144	3.1
HI	Horizontal common main air header supplying to six vertical header hanger pipes with 576 diffusers	50 (20)	144	733	10.0	0.42	10.1	20.1	112	0.0136	2.5
IJ	Horizontal common main air header supplying air to eight vertical hanger pipes with 768 diffusers	50 (20)	192	978	10.0	0.42	10.1	20.1	150	0.0131	4.2
JK	Horizontal common main air header supplying air to ten vertical hanger pipes with 960 diffusers	60 (24)	240	849	3.5	2.00	60.0	63.5	187	0.0127	8.1
KL	Yard air main pipe from the first pair of aeration basins to Junction L, supplying air in Basins 1 and 2 with 960 diffusers	60 (24)	240	849	40.0	0.33	9.90	49.9	187	0.0127	6.4
LM	Yard air main header from Junction L to blower, supplying air to 1920 diffusers in Basins 1 through 4	60 (24)	480	1698	30.0	0.66	19.8	49.8	374	0.0115	23
										Total	93

[a] Junction points are marked in Figure 10.61.
[b] Pipe diameter is selected such that the velocity in Column (e) is within the range given in Table 10.20.
[c] Air flow is based on 150% of the theoretical air requirement under standard condition.
[d] Velocity = air flow/area of pipe.
[e] Linear length of the physical pipe segment.
[f] Overall C coefficient is estimated from Table 10.21.
[g] Equivalent length of all fittings estimated from Equation 10.51f.
[h] Total equivalent length of the pipe (L) is the sum of values in columns (f) and (h).
[i] Air flow (Q_d) under discharge pressure and temperature is calculated from Equation 10.51d.
[j] Value of f is calculated from Equation 10.51b.
[k] Head loss (h_L) is calculated from Equation 10.51c.

The diameter of a pipe is selected such that the velocity in the pipe is within the range given in Table 10.20.

a. Sample calculations for sizing Line AB (Figure 10.61).

At the standard conditions, the air flow (Q_a) in the diffuser header varies from 0.5 (for 2 outer tubes) to 2 sm^3/s (for 8 tubes in the diffuser header). The maximum flow of 2 sm^3/min is used in this calculation for a conservative design.

Select a diameter $D = 7.5$ cm $= 0.075$ m and verify the velocity at standard conditions.

$$V = \frac{Q_a}{((\pi/4)D^2)} = \frac{2\ \text{sm}^3/\text{min}}{((\pi/4)(0.075\ \text{m})^2)} = 453\ \text{m/min}$$

The velocity is within the range of 350–550 m/min (see Table 10.20).

b. Calculations for sizing lines in entire air piping system.

Follow the above procedure, to determine the pipe diameters in the entire air piping system. The selections of pipe diameter and velocity verification results are summarized in Table 10.26.

4. Calculate the head loss in the air piping system.

The head loss (h_L) in the air piping system is calculated from Equation 10.51c. The steps involve an iterative procedure. The air supply pressure P at the blower is first assumed and h_L is then calculated for each pipe segment. After completing calculations for all pipes, the final value of P is obtained. It must match the assumed value. If not, the procedure is repeated until the desired results are obtained. The calculation steps for the final iteration are given below.

a. Sample calculations for h_L in Line AB.

The equivalent length is determined by adding the equivalent lengths of fittings ($L_{fittings}$) obtained from Equation 10.51f and Table 10.21 to the actual pipe length (L_{pipe}).

The actual length of diffuser header $L_{pipe} = 2.2$ m. On this segment of pipe, there are three runs of standard tee ($C = 0.33$ for each tee from Table 10.21) and one tee through side outlet ($C = 1.33$ from Table 10.21). Therefore, the overall $C = 3 \times 0.33 + 1.33 = 2.32$. Calculate equivalent length of these fittings from Equation 10.51f.

$$L_{fittings} = 55.4\ CD^{1.2} = 55.4 \times 2.32 \times (0.075\ \text{m})^{1.2} = 5.74\ \text{m}$$

Total equivalent length, $L = L_{pipe} + L_{fittings} = (2.2 + 5.74)\ \text{m} = 7.94\ \text{m}$

The barometric pressure of 721 mm Hg at an elevation of 457 m above sea level is obtained from Table B.7 of Appendix B.

$$\text{Ambient barometric pressure, } P_0 = 721\ \text{mm Hg} \times \frac{1\ \text{atm}}{760\ \text{mm Hg}} = 0.95\ \text{atm}$$

Ambient air temperature, $T_0 = 273 + 28°\text{C} = 301°\text{K}$

Assume the absolute air pressure at the blower discharge and in the air piping system, $P_d \approx$ 1.5 atm.

Calculate the air temperature in air piping from Equation 10.51e.

$$T = T_0\left(\frac{P_d}{P_0}\right)^{0.283} = 301\ \text{K}\left(\frac{1.5\ \text{atm}}{0.95\ \text{atm}}\right)^{0.283} = 343°\text{K}$$

Calculate the air flow Q_d in Line AB at $P_d = 1.5$ atm and $T_d = 343°$K from Equation 10.51d.

$$Q_d = \left(\frac{1\ \text{atm}}{P_d} \times \frac{T_d}{293°\text{K}}\right)Q_a = \left(\frac{1\ \text{atm}}{1.5\ \text{atm}} \times \frac{343°\text{K}}{293°\text{K}}\right) \times 2\ \text{sm}^3/\text{min} = 1.56\ \text{m}^3/\text{min}$$

Calculate f from Equation 10.51b.

$$f = \frac{0.029\ D^{0.027}}{Q_d^{0.148}} = \frac{0.029 \times (0.075\ \text{m})^{0.027}}{(1.56\ \text{m}^3/\text{min})^{0.148}} = 0.0253$$

Calculate h_L from Equation 10.51c.

$$h_L = 9.82 \times 10^{-8} \times \left(\frac{fT_dQ_d^2}{P_dD^5}\right)L = 9.82 \times 10^{-8} \times \left(\frac{0.0253 \times 343°\text{K} \times (1.56\ \text{m}^2/\text{min})^2}{1.5\ \text{atm} \times (0.075\ \text{m})^5}\right)$$

$$\times\ 7.94\ \text{m} = 4.6\ \text{mm H}_2\text{O}$$

b. Calculate the head losses in the entire air piping system.

Follow the above procedure to calculate the values of Q_d, f, and h_L for each pipe segment in the entire air piping system. The calculation results are summarized in Table 10.26.

5. Determine the discharge pressure P_d at the blower.

The discharge pressure at the blower is the sum of pipe losses, air filter, silencer, wiring around the blower for parallel combinations, valves and fittings, diffuser losses, and submergence. Many of these losses are provided by the manufacturers. The discharge pressure P_d is calculated below.

Total head losses between Points A and M in the air piping system (Table 10.26)	= 93 mm H_2O
Unaccounted for head losses in the air piping system that is calculated from 10% of the total head losses in Table 10.26	= 9 mm
Air filter head loss that is estimated from the data in Table 10.23	= 44 mm
Head losses in the silencers of centrifugal blowers that is estimated from the data in Table 10.23	= 26 mm
Estimated additional head losses in the blower building that include piping, valves, and connections	= 250 mm
Diffuser head losses	= 300 mm
Extra head losses due to diffuser clogging, and miscellaneous	= 450 mm
Diffuser submergence depth	= 4500 mm
Total head losses	= 5672 mm or 5.67 m H_2O

The absolute discharge pressure, $P_d = 0.95\ \text{atm} + 5.67\ \text{m H}_2\text{O} \times \dfrac{1\ \text{atm}}{10.33\ \text{m H}_2\text{O}} \approx 1.5\ \text{atm}$

The estimated pressure is same as the assumed value.

Note: The calculated head losses in the air piping system are less than 2% of the total head losses estimated for the blowers. Also, ~80% of total discharge pressure at the blower is caused by the submergence of the diffusers in this example.

6. Compute the design air flow and number of blowers.

Under the standard conditions, the total air flow supplied by the blowers is 480 sm^3/min in Line LM (see Table 10.26).

Provide five blowers in parallel, each rated at a capacity of 120 sm^3/min.

Four blowers will provide the maximum air flow required with one unit out of service. It is expected that three blowers will meet the average air requirement.

7. Calculate the power requirement.

Blower power is calculated from Equation 10.45a. Assume a machine efficiency $\eta_b = 75\%$.

$\rho_a = 1.20\ \text{kg/sm}^3$, $R = 8.314\ \text{kJ/k mole}\cdot{}^\circ\text{K}$, $(k-1)/k = 0.283$, $C = 28.97\ \text{kg/k mole}$, $Q_a = 120\ \text{sm}^3/\text{min}$, $T_i = T_0 = 301^\circ\text{K}$, and $P_i = P_0 = 0.95$ atm.

$$P_{\text{blowers}} = \frac{(Q_a/60)\rho_a RT_i}{C(k-1k)(\eta_b/100\%)}\left[(P_d/P_i)^{(k-1)/k} - 1\right]$$

$$= \frac{((120\ \text{sm}^3/\text{min})/(60\ \text{s/min})) \times 1.20\ \text{kg/m}^3 \times 8.314\ \text{kJ/k mole}\cdot{}^\circ\text{K} \times 301^\circ\text{K}}{28.97\ \text{kg/k mole} \times 0.283 \times (75\%/100\%)}$$

$$\times \left[\left(\frac{1.5\ \text{atm}}{0.95\ \text{atm}}\right)^{0.283} - 1\right] = 135\ \text{kW}\ (181\ \text{hp})$$

The power requirement of each blower is 135 kW (181 hp).
Total power requirement of five blowers = 5 × 135 kW = 675 kW (905 hp)

10.3.11 Aerated Lagoon

Aerated lagoons are large suspended growth flow-through biological reactors built in earthen- or synthetic membrane-lined earthen basins. They are relatively inexpensive to build and simple to operate. They are mostly used by small and/or rural communities where minimum secondary treatment requirements may be acceptable and land is available at low cost. The basic design and operational features are listed below while design procedure is covered in several solved examples later.[96–98]

- The aerated lagoons can be either partially or completely mixed systems.
- Liquid depth is 2–6 m (6.5–20 ft).
- The aeration period (or HRT) is normally 3–10 days. In a completely mixed lagoon, the sludge residence time (SRT) is theoretically equal to the HRT. However, the SRT is normally longer than the HRT due to incomplete mixing conditions, especially in the partially mixed system.
- The design is similar to that of an activated sludge process with no sludge recycle. Aerated lagoons with sludge recycle are less common.
- As a general rule, the oxygen requirement is around 0.7–1.6 kg O_2 per kg BOD_5 removed with the largest single aeration unit in the lagoon system out of service. The energy requirement for keeping the solids in suspension is significantly larger than that for oxygen supply.
- The threshold energy required for keeping the solids in suspension is 1.6–2 kW/1000 m^3 (8–10 hp/Mgal).* When the mixing energy is below 0.8–1.6 kW/1000 m^3 (4–8 hp/Mgal), accumulation of settleable solids begins on the lagoon bottom where anaerobic decompositions may occur. High mixing energy of 5–8 kW/1000 m^3 (25–40 hp/Mgal) and 16–20 kW/1000 m^3 (80–100 hp/Mgal) are typically required for partially and completely mixed lagoons, respectively.
- Mechanical aerators (floating or fixed base type) are commonly used to provide mixing and oxygen demand. In few cases, diffused aeration has also been used. In spite of mixing, a certain degree of settling does occur in different parts of the basin.
- Aerated lagoons produce effluent with high suspended solids in a range 80–250 mg/L. The suspended solids contain biological solids, small amounts of algae, and inorganic matter. The total BOD_5 concentration may be high even though the soluble BOD_5 may be low in the effluent. The BOD_5 removal efficiencies are normally within the range of 80–95%. Therefore, the aerated lagoons may only be considered as either the primary treatment or to meet the minimum secondary treatment requirement by many regulatory agencies. Disinfection of effluent may also be required. To meet the secondary effluent standards, clarifiers both externally and

* 1 kW/1000 m^3 = 5.05 hp/Mgal = 0.038 hp/1000 ft^3.

internally have been added with success. Return sludge is also included to increase active biomass in the basin.

- Special arrangement in piping and valving are normally required so that the lagoon system can be operated in either parallel or series to meet the potentially seasonal requirements.
- Long aeration period under low organic loading at high temperature may encourage nitrification.

Aerated lagoons may be classified as aerobic and facultative. In aerobic lagoons, aeration and mixing are sufficient to maintain solids in suspension and DO throughout the basin depth. Clarification of solids in a settling basin is needed. The facultative lagoons maintain DO in the upper depths while oxygen is absent at lower depths. Settled solids undergo anaerobic decomposition.

Temperature: Aerated lagoons occupy large surface area. The operating temperature, therefore, depends upon air and influent temperatures. Equation 10.52 is used to calculate the lagoon temperature.[99]

$$T_w = \frac{AfT_a + QT_i}{Af + Q} \tag{10.52}$$

where

T_w = water temperature in the lagoon, °C
T_a = ambient air temperature, °C
T_i = influent wastewater temperature, °C
A = surface area of lagoon, m^2
Q = m^3/d
f = proportionality factor, m/d. The value of f for the central and eastern part of the United States is 0.5 m/d, and is 0.8 m/d for the Gulf states

BOD Removal: The aerated lagoon is essentially an extended aeration system with no sludge recycle. Therefore, the biochemical kinetic equations presented earlier can be used for process design. If pseudo first-order reaction rate is assumed for substrate removal, the concentration equation under steady-state condition is expressed by Equation 10.53. This topic has been presented in Chapter 3 (see Section 3.4.2). Equation 3.11 for steady-state operation of a nonconservative substance with first-order decay was developed and application is shown in Example 3.18.

$$\frac{S}{S_0} = \frac{1}{1 + k(V/Q)} \quad \text{or} \quad \frac{S}{S_0} = \frac{1}{1 + k\theta} \quad \text{or} \quad \theta = \frac{1}{k}\left(\frac{S_0}{S} - 1\right) \tag{10.53}$$

where

S = effluent soluble BOD_5 concentration, mg/L
S_0 = influent soluble BOD_5 concentration, mg/L
k = overall first-order reaction rate constant for BOD_5 removal, d^{-1}
θ = hydraulic retention time ($\theta = (V/Q)$), d

Kinetic model based on complete-mix hydraulics and first-order reaction rate has been proposed by Marias and Shaw for a series of aerated lagoons and stabilization ponds. This topic has been presented in Chapter 3 (see Section 3.4.5). Equations 3.29 and 3.30 have been developed for nonconservative substance with first-order decay. The application is shown in Example 3.40. These relationships are expressed by Equations 10.54a and 10.54b.[100–106]

$$\frac{S_n}{S_0} = \frac{1}{(1 + k_c\theta)^n} \quad \text{or} \quad \theta = \frac{1}{k_c}\left[\left(\frac{S_0}{S_n}\right)^{1/n} - 1\right] \tag{10.54a}$$

$$\frac{S_n}{S_0} = \left(\frac{1}{1 + k_{c1}\theta_1}\right)\left(\frac{1}{1 + k_{c2}\theta_2}\right)\cdots\left(\frac{1}{1 + k_{ci}\theta_i}\right)\cdots\left(\frac{1}{1 + k_{cn}\theta_n}\right) \tag{10.54b}$$

where

S_n = effluent BOD_5 concentration from the nth lagoon in series, mg/L

S_0 = influent BOD_5 concentration in the first lagoon, mg/L

k_c = uniform first-order reaction rate constant for BOD_5 removal in each equal lagoon in series, d^{-1}. The typical value of $k_{c20} = 0.3\ d^{-1}$ for partially mixed lagoons, and 0.5–2.5 d^{-1} for completely mixed lagoons. For soluble BOD_5, the values of k_{c20} may be considerably higher than the above values in some applications. The value of k_c for operating temperature T is obtained from Equation 10.17a and $\theta_T = 1.036–1.06$.

θ = hydraulic retention time in each equal-volume lagoon, d

n = number of lagoons in series

$k_{c1}, k_{c2}, \ldots, k_{ci}, \ldots, k_{cn}$ = the first-order reaction rate constants for uniform complete-mix 1st, 2nd, …, ith, …, nth lagoons, d^{-1}

$\theta_1, \theta_2, \ldots, \theta_i, \ldots, \theta_n$ = unequal hydraulic retention times in 1st, 2nd, …, ith, …, nth lagoons, d

Thirumurthi applied the complete-mix chemical reactor equations developed by Wehner and Wilhelm for facultative pond design.[107–109] This topic has been presented in detail in Chapter 3 (Section 3.5.3). Equation 3.40 and Figure 3.34 express the Wehner-Wilhelm equation, and simplified graphical solution suggested by Thirumurthi. Examples 3.45 through 3.51, and 10.79 are developed from Wehner-Wilhelm equation and Thirumurthi's graphical solution for design of reactors, aerated lagoons, and stabilization ponds.

Construction Considerations: The aerated lagoon is normally a rectangular facility unless available land area does not allow such geometry. Basic construction considerations are listed below:

- Rectangular lagoons usually have length to width ratio of 2–3:1.
- At least two lagoons should be provided for O&M flexibility. The influent and effluent piping should be arranged such that parallel or series operation can be achieved.
- The influent structure should distribute the flow outward into the lagoon. The effluent structure should have baffles to prevent the floating matter from going over the weir.
- The side slope of the lagoon and dike should be 1V:3H or less. The slope should be protected by riprap or other means from wave erosion.
- The outer slope of dike should be sodded to protect from erosion from surface runoff.
- The freeboard should be at least 0.9 m (3 ft).
- The bottom and side slopes should be lined with clay liner or flexible membrane liner if soil is pervious.
- To prevent erosion, gabions or a concrete scour pad should be provided in the areas where the velocity is equal to or greater than 1.0 ft/s. These may include the areas around the influent pipe, or underneath the surface aerators.
- The aerated lagoon should have provision for installation of mooring cables for floating aerators, and fixed platform for turbine aerators.
- To enhance the effluent quality, provision should be made for internal or external clarifier, stabilization pond, or intermittent sand filter.

Solids Separation Facility: The suspended solids concentration in the effluent is high. Therefore, solids separation is essential prior to effluent discharge. Earthen basins of one to two days detention time have been suggested.[98,102,103] Additionally, separate facultative lagoons are provided to bioflocculate, separate, stabilize, and store the settled suspended solids for dredging and disposal. Such arrangements may cause serious environmental problems, and such an option has been rejected for long-term use. In the authors' judgement, the desirable option is to provide a conventional sedimentation facility with associated sludge dewatering and biosolids reuse. To reduce the cost of construction, lined earthen basins with deep hopper for pumping out the sludge have been used with success for small communities.

A design example of a large earthen sedimentation basin (lagoon) for a suspended growth flow-through aeration basin may be found in Reference 98. Design of a sedimentation basin in conjunction with a flow-through aerated lagoon is given in Example 10.81.

EXAMPLE 10.74: LIQUID TEMPERATURE IN AN AERATED LAGOON

An aerated lagoon receives municipal wastewater at an average flow of 4000 m^3/d. The average wastewater temperature during the winter is 16°C and air temperature is 4°C. The surface area of the lagoon is 0.5 ha. Calculate the average temperature of the lagoon content. Use the proportionality factor of 0.5 for the eastern part of the United States.

Solution

1. Determine the surface area of the lagoon.
   Surface area, $A = 0.5 \text{ ha} \times 10{,}000 \text{ m}^2/\text{ha} = 5000 \text{ m}^2$
2. Calculate the average temperature of the lagoon content.
   The average water temperature (T_w) in the lagoon during winter months is calculated from Equation 10.52 using $f = 0.5$ m/d for the eastern part of U.S.

$$T_w = \frac{AfT_a + QT_i}{Af + Q} = \frac{5000 \text{ m}^2 \times 0.5 \text{ m/d} \times 4°\text{C} + 4000 \text{ m}^3/\text{d} \times 16°\text{C}}{5000 \text{ m}^2 \times 0.5 \text{ m/d} + 4000 \text{ m}^3/\text{d}} = \frac{74{,}000°\text{C}}{6500} = 11.4°\text{C}$$

EXAMPLE 10.75: COMPARISON OF A SINGLE AERATED LAGOON WITH THREE LAGOONS IN SERIES

Three aerated lagoons of equal detention time are arranged in series. The critical operating temperature of lagoons is 7°C. The first-order reaction rate constant is 2.5 d^{-1} at 20°C. The desired soluble effluent BOD_5 concentration = 20 mg/L. The influent BOD_5 concentration is 200 mg/L and the temperature coefficient $\theta_T = 1.06$. Compare the total detention time of the three-lagoon system with that of a single lagoon. Assume that the reaction rate constant remains the same in all applications.

Solution

1. Calculate the first-order reaction rate constant at 7°C (k_7) from Equation 10.17a using $\theta_T = 1.06$.

$$k_7 = k_{20}\theta_T^{(T-20)} = 2.5 \text{ d}^{-1} \times 1.06^{(7-20)} = 1.2 \text{ d}^{-1}$$

2. Calculate the detention time (θ_{single}) of the one-lagoon system from Equation 10.53.

$$\theta_{single} = \frac{1}{k_7}\left(\frac{S_0}{S} - 1\right) = \frac{1}{1.2 \text{ d}^{-1}}\left(\frac{200 \text{ mg/L}}{20 \text{ mg/L}} - 1\right) = 7.5 \text{ d}$$

3. Calculate the total detention time (θ_{total}) of the three-lagoon system.
   The detention time θ_{cell} of each lagoon is calculated from Equation 10.54a at $k_{c7} = k_7 = 1.2 \text{ d}^{-1}$ and $n = 3$.

$$\theta_{cell} = \frac{1}{k_{c7}}\left[\left(\frac{S_0}{S_n}\right)^{1/n} - 1\right] = \frac{1}{1.2 \text{ d}^{-1}}\left[\left(\frac{200 \text{ mg/L}}{20 \text{ mg/L}}\right)^{\frac{1}{3}} - 1\right] = 0.96 \text{ d}$$

   Total detention time of three lagoons, $\theta_{total} = n\theta_{cell} = 3 \times 0.96 \text{ d} = 2.9 \text{ d}$

4. Compare the results.

   If the effluent quality is comparable from both arrangements, the detention time of a single aerated lagoon is about 2.6 times the combined detention times of three equal lagoons arranged in series.

EXAMPLE 10.76: COMPARISON OF A SINGLE AERATED LAGOON WITH THREE LAGOONS WITH UNEQUAL DETENTION TIMES IN SERIES

Three aerated lagoons are operating in series. The detention time of the first cell is two times that of the second and third cells. The influent and effluent BOD_5 concentrations are 200 and 20 mg/L, respectively. The reaction rate constant at the critical operating temperature is 2.5 d^{-1}. Compare the total detention time of three-lagoon system with the detention time of a single lagoon.

Solution

1. Determine the detention time (θ_{single}) of the single lagoon.

   The detention time θ_{single} is calculated from Equation 10.53 at $k = 2.5\ d^{-1}$.

$$\theta_{single} = \frac{1}{k}\left(\frac{S_0}{S} - 1\right) = \frac{1}{2.5\ d^{-1}}\left(\frac{200\ mg/L}{20\ mg/L} - 1\right) = 3.6\ d$$

2. Calculate the total detention time (θ_{total}) of the three lagoons in series.

   Assume the detention times are θ_1, θ_2, and θ_3 of lagoons 1, 2, and 3, respectively. The relationships of θ_1, θ_2, and θ_3 are $\theta_1 = 2\theta_2$ and $\theta_3 = \theta_2$. Apply Equation 10.54b with $k_{c1} = k_{c2} = k_{c3} = k = 2.5\ d^{-1}$ and $n = 3$.

$$\frac{S_3}{S_0} = \left(\frac{1}{1 + k_{c1}\theta_1}\right)\left(\frac{1}{1 + k_{c2}\theta_2}\right)\left(\frac{1}{1 + k_{c3}\theta_3}\right)$$

$$\frac{20\ mg/L}{200\ mg/L} = \left(\frac{1}{1 + (2.5\ d^{-1}) \times 2\theta_2}\right)\left(\frac{1}{1 + (2.5\ d^{-1}) \times \theta_2}\right)\left(\frac{1}{1 + (2.5\ d^{-1}) \times \theta_2}\right)$$

$$(1 + 5\,\theta_2)(1 + 2.5\,\theta_2)^2 = 10$$

$$31.25(\theta_2)^3 + 31.25(\theta_2)^2 + 10(\theta_2) - 9 = 0$$

   Solving by trial and error, the detention time of lagoon 2, $\theta_2 = 0.36$ d.

   The detention time of lagoon 1, $\theta_1 = 2 \times \theta_2 = 2 \times 0.36\ d = 0.72$ d.

   The detention time of lagoon 3, $\theta_1 = \theta_2 = 0.36$ d.

   The total detention time of three lagoons, $\theta_{total} = \theta_1 + \theta_2 + \theta_3 = (0.72 + 0.36 + 0.36)\ d = 1.44$ d.

3. Compare the results.

   The detention time of a single lagoon is 2.5 times the combined detention time of three lagoons in series. The three-lagoon system is more efficient.

EXAMPLE 10.77: AERATED LAGOONS WITH UNEQUAL REACTION RATE CONSTANTS IN SERIES

Three aerated lagoons of equal detention time are arranged in series. The first-order reaction rate constant of the first lagoon is 1.25 d^{-1} at 20°C, which is twice that of the second and third cell. Calculate the detention time of each cell. The operating temperature is 7°C, and the influent and effluent BOD_5 concentrations are 200 mg/L and 20 mg/L, respectively. The temperature coefficient $\theta_T = 1.06$.

Solution

1. Determine the first-order reaction rate constant (k_{c20}) of three lagoons.
   The first-order reaction rate constant at 20°C for the first lagoon is $k_{c20,1} = 1.25\ \text{d}^{-1}$, and for the second and third lagoons is $k_{c20,2} = k_{c20,3} = 0.625\ \text{d}^{-1}$.
2. Calculate the first-order reaction rate constant (k_{c7}) at critical operating temperature of 7°C.
   Calculate k_{c7} from Equation 10.17a at $\theta_T = 1.06$.

   In the first lagoon, $k_{c7,1} = k_{c20,1}\theta_T^{(T-20)} = 1.25\ \text{d}^{-1} \times 1.06^{(7-20)} = 0.59\ \text{d}^{-1}$

   In the second and third lagoons, $k_{c7,2} = k_{c7,3} = 0.625\ \text{d}^{-1} \times 1.06^{(7-20)} = 0.29\ \text{d}^{-1}$

3. Calculate the total detention time (θ_{total}) of the aerated lagoons in series.
   Assume the equal retention time θ and apply Equation 10.54b at $n = 3$.

$$\frac{S_3}{S_0} = \left(\frac{1}{1 + k_{c7,1}\theta}\right)\left(\frac{1}{1 + k_{c7,2}\theta}\right)\left(\frac{1}{1 + k_{c7,3}\theta}\right)$$

$$\frac{20\ \text{mg/L}}{200\ \text{mg/L}} = \left(\frac{1}{1 + (0.59\ \text{d}^{-1}) \times \theta}\right)\left(\frac{1}{1 + (0.29\ \text{d}^{-1}) \times \theta}\right)\left(\frac{1}{1 + (0.29\ \text{d}^{-1}) \times \theta}\right)$$

$$(1 + 0.59\,\theta)(1 + 0.29\,\theta)^2 = 10$$

$$0.05\,\theta^3 + 0.43\,\theta^2 + 1.17\,\theta - 9 = 0$$

Solving by trial and error, the equal detention time of each aerated lagoon, $\theta = 3.1$ d.
Total detention time of three lagoons, $\theta_{total} = 3\theta = 3 \times 3.1\ \text{d} = 9.3$ d.

EXAMPLE 10.78: AERATED LAGOON AND TEMPERATURE RELATIONSHIP

Equations 10.52 and 10.53 indicate that the surface area of an aerated lagoon, detention time, and first-order reaction rate constant have dependence upon mean air and influent temperatures. An aerated lagoon is designed in a region where the lowest mean air temperature is 5°C and wastewater temperature is 16°C. The first-order reaction rate constant k_{20} is $2.8\ \text{d}^{-1}$ at 20°C. The wastewater flow is 4000 m^3/d and influent BOD_5 concentration is 220 mg/L. The effluent BOD_5 concentration is 20 mg/L. Calculate the detention time and surface area of the lagoon. Use a proportionality factor $f = 0.5$ m/d and a temperature coefficient $\theta_T = 1.06$.

Solution

The surface area (A_s), detention time (θ), first-order reaction rate constant (k_T), and temperature of lagoon content (T_w) are dependent variables. For this reason, an iterative procedure is required to solve this problem.

1. First Iteration
   a. Calculate the critical temperature of lagoon content (T_w).

Assume that the water surface area of the lagoon $A_s = 0.8$ ha $= 8000$ m^2.
Apply Equation 10.52 to calculate the water temperature T_w in the lagoon.

$$T_w = \frac{AfT_a + QT_i}{Af + Q} = \frac{8000\ \text{m}^2 \times 0.5\ \text{m/d} \times 5°\text{C} + 4000\ \text{m}^3/\text{d} \times 16°\text{C}}{8000\ \text{m}^2 \times 0.5\ \text{m/d} + 4000\ \text{m}^3/\text{d}} = \frac{84{,}000°\text{C}}{8000} = 10.5°\text{C}$$

b. Calculate the first order reaction rate constant at 10.5°C ($k_{10.5}$) from Equation 10.17a using $\theta_T = 1.06$.

$$k_7 = k_{20}\theta_T^{(T-20)} = 2.8\ \text{d}^{-1} \times 1.06^{(10.5-20)} = 1.6\ \text{d}^{-1}$$

c. Calculate the detention time (θ) from Equation 10.53.

$$\theta = \frac{1}{k_{10.5}}\left(\frac{S_0}{S} - 1\right) = \frac{1}{1.6\ \text{d}^{-1}}\left(\frac{220\ \text{mg/L}}{20\ \text{mg/L}} - 1\right) = 6.3\ \text{d}$$

d. Determine the water surface and base area dimensions of the lagoon.
Provide inner side slope of the lagoon at 1V to 3H, and ratio of $L_s/W_s = 2$ at the water surface.

$$A_s = L_s W_s = 2W_s \times W_s = 2W_s^2 = 8000\ \text{m}^2$$

$$W_s = \sqrt{\frac{8000\ \text{m}^2}{2}} = 63.2\ \text{m}$$

Provide $W_s = 63$ m and $L_s = 126$ m.
Actual water surface area, $A_s = 63\ \text{m} \times 126\ \text{m} = 7938\ \text{m}^2$
Assume a water depth $D = 4.5$ m and calculate the dimensions of the base area at a slope $S = 1/3$ (1V to 3H).

$$W_b = W_s - 2\,D \times \frac{1}{S} = 63\ \text{m} - 2 \times 4.5\ \text{m} \times 3 = 36\ \text{m}$$

$$L_b = L_s - 2\,D \times \frac{1}{S} = 126\ \text{m} - 2 \times 4.5\ \text{m} \times 3 = 99\ \text{m}$$

Base area, $A_b = 36\ \text{m} \times 99\ \text{m} = 3564\ \text{m}^2$

e. Calculate the volume of the lagoon.
The volume of the lagoon V is calculated from Equation 10.55.

$$V = \frac{D}{3}[A_b + A_s + \sqrt{A_b \times A_s}] \tag{10.55}$$

where
V = volume of the lagoon, m^3 (ft^3)
D = water depth, m (ft)
A_b and A_s = areas of the base and water surface, respectively, m^2 (ft^2)

$$V = \frac{4.5\ \text{m}}{3} \times [3564\ \text{m}^2 + 7938\ \text{m}^2 + \sqrt{3564\ \text{m}^2 \times 7938\ \text{m}^2}] = 25{,}231\ \text{m}^3$$

2. Second Iteration

Verify the hydraulic detention time of the lagoon.

$$\text{Hydraulic retention time, } \theta = \frac{25{,}231 \text{ m}^3}{4000 \text{ m}^3/\text{d}} = 6.3 \text{ d}$$

The calculated detention time of $\theta = 6.3$ d is same as that in Step 1c. Therefore, further iteration is not needed. The calculated dimensions of aerated lagoon in Steps 1d and 1e are final.

Note: If, however, the calculated θ in Step 2 was significantly different from that in Step 1c, more iterations are carried out by adjusting the water depth (D) in Step 1d and repeating Steps 1d through 2 until close agreement is reached. If the adjusted water depth is out of the desired range of 2–5 m, further adjustment of the water surface area (A_s) in Step 1a is required. Steps 1 and 2 are repeated until the difference between the calculated and assumed θ is within an acceptable range.

EXAMPLE 10.79: WEHNER-WILHELM EQUATION, AND THIRUMURTHI GRAPHICAL PROCEDURE FOR AERATED LAGOONS

An aerated lagoon is provided for treatment of municipal wastewater from a community of 4000 residents. The average wastewater generation rate is 380 Lpcd and BOD_5 concentration is 220 mg/L. The lagoon is square in shape and water depth is 4 m. Mechanical aerators provide aeration and mixing. The dispersion number d approaches 2.0. The first-order reaction rate constant at the critical temperature are 1.25 d^{-1} and the desired detention time is 4.7 d. Calculate the BOD_5 concentration in the effluent and the dimensions of lagoon. The surface side slope is 1V to 3H. Assume baffles are provided in the lagoon to prevent short circuiting.

Solution

1. Calculate the BOD_5 concentration of the effluent from Wehner-Wilhelm equation.

   Wehner-Wilhelm equation is Equation 3.40. Calculate coefficient a.

$$a = \sqrt{1 + 4\,k\theta d} = \sqrt{1 + 4 \times 1.25 \text{ d}^{-1} \times 4.7 \text{ d} \times 2.0} = 6.93$$

   Calculate C/C_o from Equation 3.40.

$$\frac{C}{C_0} = \frac{4ae^{1/2d}}{(1+a)^2 e^{a/2d} - (1-a)^2 e^{-a/2d}} = \frac{4 \times 6.93 \times e^{\frac{1}{2\times 2.0}}}{(1+6.93)^2 e^{6.93/2\times 2.0} - (1-6.93)^2 e^{-6.93/2\times 2.0}}$$

$$= \frac{27.7 \times e^{0.25}}{62.9 \times e^{1.73} - 35.2 \times e^{-1.73}} = \frac{35.6}{355 - 6.24} = 0.10$$

   BOD_5 concentration in effluent, $C = 0.10 \times C_0 = 0.10 \times 220 \text{ mg/L} = 22 \text{ mg/L}$

2. Calculate and check the effluent BOD_5 concentration from Thirumurthi's graphical solution.

   The Thirumurthi's graphical solution is given in Figure 3.34.

$$k\theta = 1.25 \text{ d}^{-1} \times 4.7 \text{ d} = 5.9$$

$C/C_0 = 0.105$ is obtained from Figure 3.34 at $d = 2.0$.

$$C = 0.105 \times 220 \text{ mg/L} = 23 \text{ mg/L}$$

This is very close to that obtained in Step 1.

3. Determine the dimensions of the lagoon.

Wastewater flow, $Q = 380 \text{ L/person·d} \times 4000 \text{ people} \times 10^{-3}\text{m}^3/\text{L} = 1520 \text{ m}^3/\text{d}$

Volume of the lagoon, $V = Q\theta = 1520 \text{ m}^3/\text{d} \times 4.7 \text{ d} = 7144 \text{ m}^3$

Assume $L_b = W_b = b$ of the bottom dimensions of a square basin.

Calculate the water surface dimensions at a water depth $D = 4$ m and a horizontal to vertical side slope ratio of 1V to 3H.

$$L_s = W_s = b + 2 \times 3 \times D = b + 2 \times 3 \times 4 \text{ m} = b + 24 \text{ m}$$

Calculate the volume of the lagoon from Equation 10.55.

$$V = \frac{D}{3}\left[A_b + A_s + \sqrt{A_b \times A_s}\right] = \frac{4 \text{ m}}{3} \times \left[b^2 + (b + 24 \text{ m})^2 + \sqrt{b^2 \times (b + 24 \text{ m})^2}\right]$$

$$= \frac{4}{3} \times (b^2 + b^2 + 48b + 576 + b^2 + 24b) = \frac{4}{3} \times (3b^2 + 72b + 576) = 4b^2 + 96b + 768$$

$$4b^2 + 96b - 6376 = 7144 \quad \text{or} \quad 4b^2 + 96b - 6376 = 0$$

Solving by trial and error, $b = 29.7$ m.

Provide $b = 30$ m.

$$L_b = W_b = b = 30 \text{ m} \quad \text{and} \quad L_s = W_s = b + 24 \text{ m} = (30 + 24) \text{ m} = 54 \text{ m}$$

These dimensions are shown in Figure 10.62. The baffles are not shown in this figure.

The actual volume of the lagoon, $V = \frac{4 \text{ m}}{3} \times \left[(30 \text{ m})^2 + (54 \text{ m})^2 + \sqrt{(30 \text{ m})^2(54 \text{ m})^2}\right] = 7248 \text{ m}^3$

Calculate the actual detention time, $\theta = \frac{7248 \text{ m}^3}{1520 \text{ m}^3/\text{d}} = 4.8 \text{ d}$

The actual detention time of $\theta = 4.8$ d is very close to the desired value of 4.7 d.

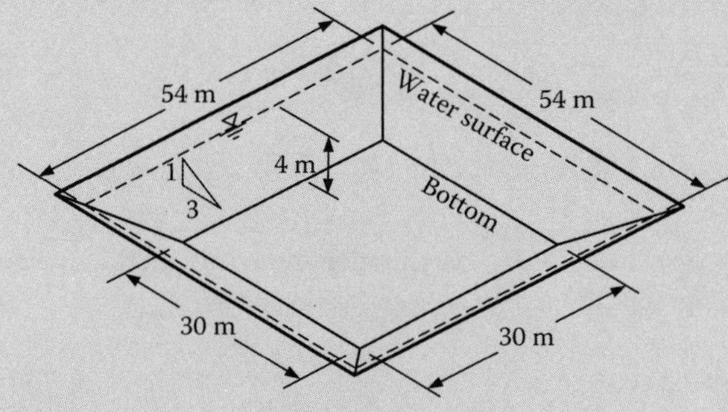

FIGURE 10.62 Dimensions of the square aerated lagoon (Example 10.79).

EXAMPLE 10.80: DESIGN OF AN AERATED LAGOON SYSTEM USING A KINETIC-BASED METHOD

Design a flow-through aerated lagoon system to treat wastewater from a community of 17,000 residents. The average wastewater generation rate is 380 L/capita·d. The following data apply:

Wastewater: The influent BOD_5 and TSS concentrations are 160 and 200 mg/L, respectively. The influent temperatures are 18°C and 23°C in the winter and summer, respectively. The biodegradable TSS in the influent is 45% of the influent TSS, and BOD_5/BOD_L is 0.68.

Aerated lagoons: Two aerated lagoons that can be operated in either parallel or series. The aeration period of each lagoon is 4 d. The average ambient air temperature in the winter and summer months are 8°C and 30°C, respectively. A DO level of 2.0 mg/L is maintained in the lagoons that are 4-m deep with a freeboard of 1.5 m. Use a proportionality factor $f = 0.5$ m/d.

Biological kinetic coefficients: Yield coefficient $Y = 0.62$ mg VSS/mg BOD_5; $k_d = 0.05$ mg VSS/mg VSS·d; maximum specific rate of substrate utilization $k = 2.5$ mg BOD_5/mg VSS·d; $K_s = 60$ mg BOD_5/L; first-order soluble BOD_5 removal rate constant $k_{20} = 2.5$ mg BOD_5/mg VSS·d at 20°C; VSS/TSS ratio = 0.75; and temperature coefficient $\theta_T = 1.04$.

Aerators: Surface aerators are used. The standard aeration efficiency (*SAE*) is 1.2 kg O_2/kW·h (2 lb O_2/hp·h), $\alpha = 0.8$, and $\beta = 0.9$. The elevation is 610 m above sea level.

Solution

1. Calculate the average wastewater flow.

   Calculate the total average wastewater influent flow.

$$Q_{total} = 380 \text{ L/person·d} \times 17{,}000 \text{ people} \times 10^{-3} \text{ m}^3/\text{L} = 6460 \text{ m}^3/\text{d}$$

$$\text{Average flow per aerated lagoon, } Q = \frac{Q_{total}}{2} = \frac{1}{2} \times 6460 \text{ m}^3/\text{d} = 3230 \text{ m}^3/\text{d}$$

2. Calculate the volume, dimensions, and surface area of the aerated lagoon.

   Volume of the lagoon $V = Q\theta = 3230 \text{ m}^3/\text{d} \times 4 \text{ d} = 12{,}920 \text{ m}^3$

   Provide an L_b/W_b ratio = 3, side slope is 1 vertical to 3 horizontal. If bottom width $W_b = a$, then the other bottom and water surface dimensions are: $L_b = 3a$, $W_s = W_b + 2 \times 3 \times D = a + 2 \times 3 \times 4 \text{ m} = a + 24$ m, and $L_s = L_b + 24 \text{ m} = 3a + 24$ m.

   Area of the base, $A_b = a \times 3a = 3a^2$, and area of water surface,

$$A_s = (a + 24) \times (3a + 24) = 3a^2 + 96a + 576.$$

The volume of the lagoon is equated to the volume in Equation 10.55.

$$V = \frac{D}{3}[A_b + A_s + \sqrt{A_b \times A_s}] = \frac{4}{3} \times [3a^2 + 3a^2 + 96a + 576 + \sqrt{3a^2 \times (3a^2 + 96a + 576)}]$$

$$= 12{,}920 \text{ m}^3$$

Solving by trial and error, $a = 24.9$ m. Provide $a = 25$ m.

Dimensions of the bottom: $W_b = a = 25$ m, and $L_b = 3W_b = 2 \times 25 \text{ m} = 75$ m.

Dimensions of the water surface: $W_s = (25 + 24)\text{m} = 49$ m, and $L_s = (75 + 24)\text{m} = 99$ m

Lagoon bottom area, $A_b = L_b \times W_b = 75\ m \times 25\ m = 1875\ m^2$

Water surface area, $A_s = L_s \times W_s = 99\ m \times 49\ m = 4851\ m^2$ (0.49 ha)

The major dimensions of one aerated lagoon are shown in Figure 10.63.

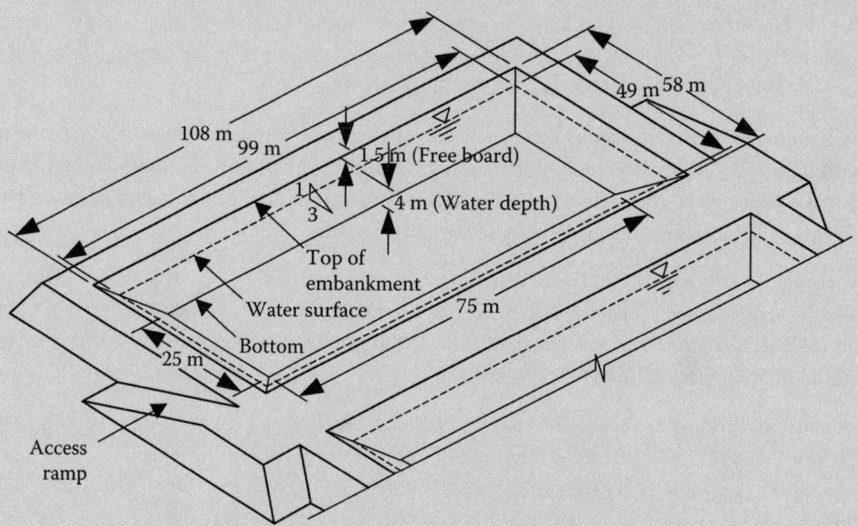

FIGURE 10.63 Major dimensions of the aerated lagoon (Example 10.80).

The actual volume of the lagoon,

$$V = \frac{4\ m}{3} \times [1875\ m^2 + 4851\ m^2 + \sqrt{1875\ m^2 \times 4851\ m^2}] = 13{,}000\ m^3$$

$$\text{Actual detention time, } \theta = \frac{13{,}000\ m^3}{3230\ m^3/d} = 4\ d$$

3. Describe the aerated lagoon details.

The design concepts are shown in Figure 10.64. The general layout of the lagoon system with special piping arrangement is illustrated in Figure 10.64a. There are two aerated lagoons: Lagoons I and II. The bar-screened influent is measured by a Parshall flume prior to a flow splitting box in which a weir gate (Weir Gate A or B) is provided for flow control to the designated lagoon. There are two influent pipes: Pipes A and B. Each pipe has an influent box at the exit point in a lagoon (Figure 10.64b). An additional Pipe C with a sluice gate (Sluice Gate C) is also provided between the effluent box in Lagoon I and the influent box of Lagoon II. There are also two effluent pipes: Pipes D and E. Each pipe collects effluent from the effluent box in aerated lagoon. At each effluent box, an adjustable weir gate (Weir Gate D or E) is also provided and used for operating the lagoon system in either parallel or series modes (Figure 10.64c). The conceptual arrangement of the weir and sluice gates is illustrated in Figure 10.64d). By opening, closing, or adjusting these gates, the lagoon system is operated in either parallel or series mode, or to isolate one lagoon for

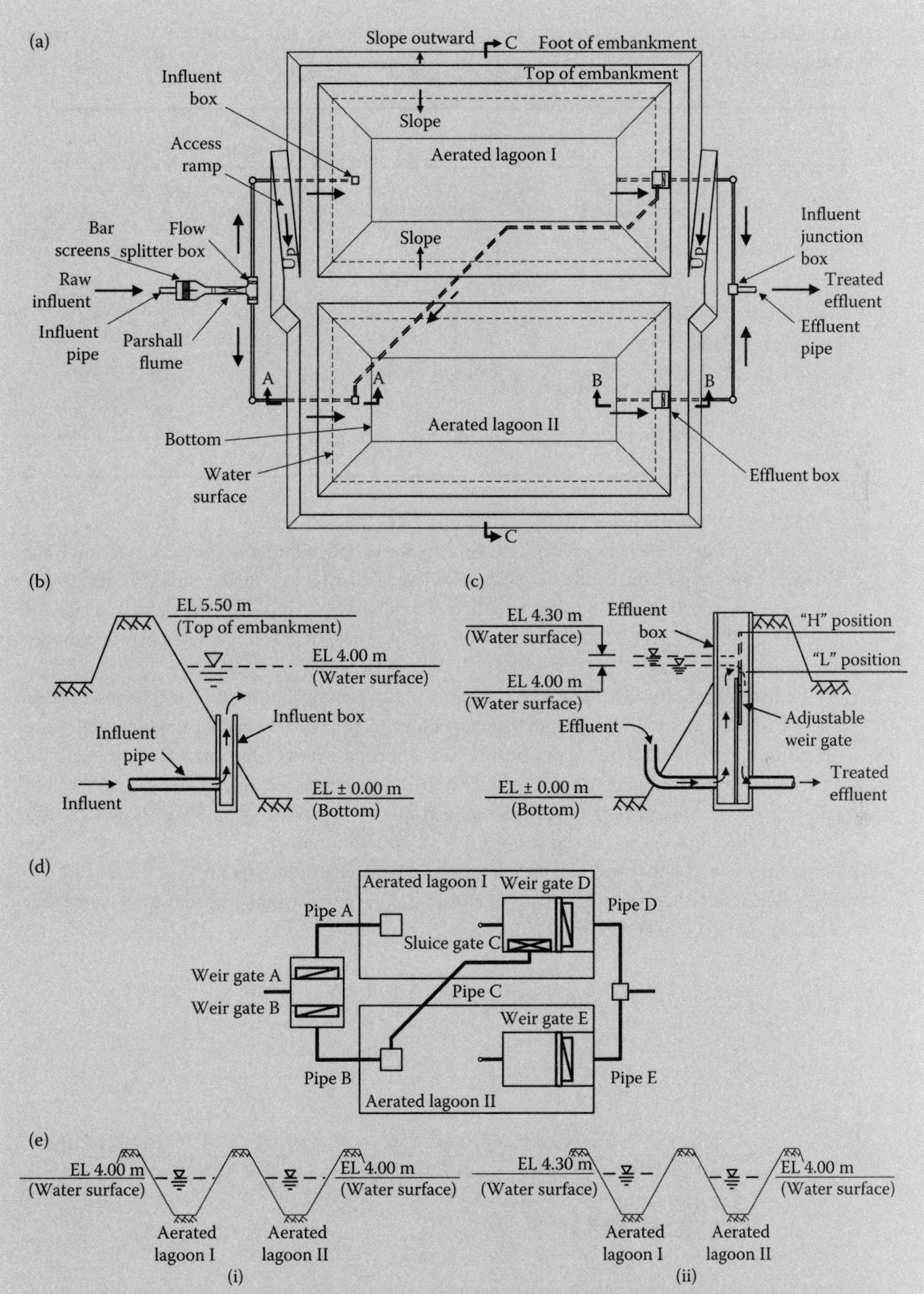

FIGURE 10.64 Design details of the aerated lagoons: (a) plan of the lagoon system; (b) Section AA showing influent box; (c) Section BB showing effluent box; (d) conceptual arrangement of pipes and gates used for lagoon operation; and (e) Section CC showing water surface levels in different operation modes: (i) operation in parallel and (ii) operation in series (Example 10.80).

maintenance. The gates operations and weir position settings are summarized below for different operating modes:

Operating Mode	Gate Operation					Effluent Weir Gate Position		Water Depth, m	
	Weir Gate A	Weir Gate B	Sluice Gate C	Weir Gate D	Weir Gate E	Weir Gate D	Weir Gate E	Lagoon I	Lagoon II
Both lagoons in service									
Parallel operation	Open	Open	Close	Open	Open	L	L	4.0	4.0
Series operation	Open	Close	Open	Close	Open	H	L	4.3	4.0
Isolation of one lagoon									
Lagoon I out of service	Close	Open	Close	Close	Open	H	L	Out	4.0
Lagoon II out of service	Open	Close	Close	Open	Close	L	H	4.0	Out

a. Parallel operation mode.

   In this mode, both Weir Gates A and B are open and both adjustable Weir Gates D and E are placed at the "L" position. Sluice Gate C is also closed. In either lagoon, the effluent is discharged over the discharge weir and the water depth is maintained at 4.0 m since both weirs are placed at the "L" position (Figure 10.64e(i)). Therefore, the lagoon system is in parallel operation mode.

b. Series operation mode.

   In this mode, both Weir Gate A and Sluice Gate C are open while Weir Gate B is closed. Weir Gate E remains at the "L" position to maintain a water depth of 4.0 m in the Lagoon II. Weir Gate D is adjusted to the "H" position for closure. The effluent discharge from Lagoon I is stopped and the water depth is raised to 4.3 m (Figure 10.64e(ii)). The higher depth gives the needed head to push the flow through Pipe C into the influent box of Lagoon II when Sluice Gate C is also opened. Therefore, the lagoon system is in series operation mode.

4. Calculate the average winter and summer temperatures of water in the lagoon.

   Apply Equation 10.52 to calculate $T_{\text{w,winter}}$ and $T_{\text{w,summer}}$ in the lagoon, respectively. Assume that the lagoon system is operated in parallel mode.

$$T_{\text{w,winter}} = \frac{A_s f T_{\text{a,winter}} + Q T_{\text{i,winter}}}{Af + Q} = \frac{4851\ \text{m}^2 \times 0.5\ \text{m/d} \times 8^\circ\text{C} + 3230\ \text{m}^3/\text{d} \times 18^\circ\text{C}}{4851\ \text{m}^2 \times 0.5\ \text{m/d} + 3230\ \text{m}^3/\text{d}}$$

$$= \frac{77{,}544^\circ\text{C}}{5656} = 13.7^\circ\text{C} \quad \text{or} \quad 14^\circ\text{C}$$

$$T_{\text{w,summer}} = \frac{A_s f T_{\text{a,summer}} + Q T_{\text{i,summer}}}{Af + Q} = \frac{4851\ \text{m}^2 \times 0.5\ \text{m/d} \times 30^\circ\text{C} + 3230\ \text{m}^3/\text{d} \times 23^\circ\text{C}}{4851\ \text{m}^2 \times 0.5\ \text{m/d} + 3230\ \text{m}^3/\text{d}}$$

$$= \frac{147{,}055^\circ\text{C}}{5656} = 26.0^\circ\text{C} \quad \text{or} \quad 26^\circ\text{C}$$

   Note: In winter or summer, the water temperature in the lagoon system remain the same no matter which operating mode is selected.

5. Estimate the soluble BOD_5 concentration in the effluent from the aerated lagoon system.

   The soluble BOD_5 concentration in the effluent is evaluated based on the first-order BOD_5 removal rate constant $k_{20} = 2.5$ mg BOD_5/mg VSS·d at 20°C.

At the average winter and summer water temperatures in the lagoon, calculate k_{14} and k_{26} from Equation 10.17a at $\theta_T = 1.04$.

$$k_{14} = k_{20}\theta_T^{(T_{w,winter}-20)} = 2.5\ d^{-1} \times 1.04^{(14-20)} = 2.0\ d^{-1}$$

$$k_{26} = k_{20}\theta_T^{(T_{w,summer}-20)} = 2.5\ d^{-1} \times 1.04^{(26-20)} = 3.2\ d^{-1}$$

Calculate the soluble BOD_5 in the effluent from Equation 10.53 when the lagoon system is operated in parallel ($\theta_{parallel} = \theta = 4$ d) during winter months.

$$S_{winter,parallel} = \frac{S_0}{1 + k_{14}\theta_{parallel}} = \frac{160\ mg/L}{1 + 2.0\ d^{-1} \times 4\ d} = 18\ mg/L$$

Calculate the soluble BOD_5 in the effluent from Equation 10.54a when the lagoon system is operated in series ($\theta_{in\ series} = \theta/2 = 2$ d, $k_{c14} = k_{14} = 2.0\ d^{-1}$, and $n = 2$) during winter time.

$$S_{winter,\ in\ series} = \frac{S_0}{(1 + k_{c14}\theta_{in\ series})^n} = \frac{160\ mg/L}{(1 + 2.0\ d^{-1} \times 2\ d)^2} = 6.4\ mg/L$$

The soluble BOD_5 concentration in the effluent is also calculated under the summer temperature in both operating modes. All calculation results are summarized below.

Operating Mode	Soluble BOD_5 Concentration in the Effluent, mg/L	
	Winter	Summer
Parallel operation	18	12
Series operation	6.4	2.9

The soluble BOD_5 concentration in the effluent during the winter is ~50–100% higher than that in the summer time. However, the soluble BOD_5 concentration in the effluent is reduced by ~30–45% in both seasons when the lagoon operation is switched from the parallel to series mode. Therefore, the plant operator may select different operation mode based on the seasonal effluent BOD_5 limitations in the discharge permit.

6. Estimate the TSS concentration in the effluent from the aerated lagoon system.
   a. Calculate the soluble BOD_5 concentration (S) in the effluent.

      Use kinetic equation Equation 10.15h or 10.16c to calculate the soluble BOD_5 concentration (S) in the effluent assuming $\theta_c = \theta = 4$ d when the lagoon system is operated in parallel.

$$S = \frac{K_s(1 + k_d\theta_c)}{\theta_c(Yk - k_d) - 1} = \frac{60\ mg\ BOD_5/L \times (1 + 0.05\ d^{-1} \times 4\ d)}{4\ d \times (0.62 \times 2.5\ d^{-1} - 0.05\ d^{-1}) - 1} = 14\ mg\ BOD_5/L$$

      Note: The soluble BOD_5 concentration is between the values estimated from the first-order BOD_5 removal rate constant at winter and summer water temperatures when the lagoons are operated in parallel.

      Check the lagoon volume from Equation 10.15o.

$$V = \frac{Y}{1 + k_d\theta_c}\frac{\theta_c Q(S_0 - S)}{X} = \frac{0.62}{(1 + 0.05\ d^{-1} \times 4\ d)}\frac{4\ d \times 3230\ m^3/d \times (160 - 14)\ mg/L}{75\ mg/L} = 13{,}000\ m^3$$

      This is exactly the same volume calculated and provided in Step 2.

b. Calculate the TSS concentration in the effluent due to biomass generation.
Calculate the biomass yield (X) from Equation 10.15k or 10.16e.

$$X = \frac{Y}{1 + k_d\theta_c}\frac{\theta_c}{\theta}(S_0 - S) = \frac{0.62 \text{ mg VSS/mg BOD}_5}{1 + 0.05 \text{ d}^{-1} \times 4 \text{ d}} \times \frac{4 \text{ d}}{4 \text{ d}} \times (160 - 14) \text{ mg BOD}_5\text{/L}$$

$$= 75 \text{ mg VSS/L} \quad \text{or} \quad 75 \text{ g VSS/m}^3$$

The TSS concentration due to biomass generation is calculated from the VSS/TSS ratio of 0.75.

$$TSS_{e,biomass} = \frac{X}{\text{TSS/VSS ratio}} = \frac{75 \text{ mg VSS/L}}{0.75 \text{ mg VSS/mg TSS}} = 100 \text{ mg TSS/L or } 100 \text{ g TSS/m}^3$$

c. Calculate the TSS concentration in the effluent due to nonbiodegradable TSS in the influent.
The biodegradable to total TSS ratio, $TSS_{i,bd}/TSS_i$ ratio = 45% of the influent TSS. Calculate the nonbiodegradable TSS remaining in the influent ($TSS_{e,nbd}$).

$$TSS_{e,nbd} = TSS_i \times (1 - TSS_{i,biodeg}/TSS_i \text{ ratio})$$
$$= 200 \text{ mg TSS/L} \times (1 - 0.45) = 110 \text{ mg TSS/L}$$

d. Calculate the TSS concentration in the effluent ($TSS_{e,lagoon}$).
Assume that the remaining of biodegradable TSS ($TSS_{i,bd}$) is ignorable. If no settling occurs in the lagoons, the TSS concentration in the effluent is calculated below.

$$TSS_{e,lagoon} = TSS_{e,biomass} + TSS_{e,nbd} = (100 + 110) \text{ mg TSS/L} = 210 \text{ mg TSS/L}$$

Note: The TSS concentration in the aerated lagoon effluent is too high and clarification will be required (see Example 10.81).

7. Determine the theoretical oxygen requirement (*ThOR*).
   a. Calculate the biomass growth or VSS generation rate (P_x) in the aerated lagoon.
   Apply Equation 10.15m,

$$P_x = XQ = 75 \text{ g TSS/m}^3 \times 3230 \text{ m}^3\text{/d} \times 10^{-3} \text{ kg/g} = 242 \text{ kg VSS/d}$$

   b. Calculate the $CBOD_L$ loading.
   The oxygen requirement (O_2) is calculated from Equation 10.14f.

$$O_2 = \frac{Q(S_0 - S)}{BOD_5/BOD_L} - 1.42P_x$$

$$= \frac{3230 \text{ m}^3\text{/d} \times (160 - 14) \text{ g BOD}_5\text{/m}^3 \times 10^{-3} \text{ kg/g}}{0.68 \text{ g BOD}_5\text{/g BOD}_L} - 1.42 \text{ g BOD}_L\text{/g VSS}$$

$$\times 242 \text{ kg VSS/d} = (694 - 344) \times \text{kg BOD}_L\text{/d} = 350 \text{ kg BOD}_L\text{/d}$$

   c. Determine the *ThOR*.

$$ThOR = O_2 = 350 \text{ kg BOD}_L\text{/d} \quad \text{or} \quad 350 \text{ kg O}_2\text{/d} \times \frac{\text{d}}{24 \text{ h}} = 15 \text{ kg O}_2\text{/h}$$

d. Determine the ratio of oxygen required to BOD_5 removed.

$$\frac{O_2 \text{ required}}{BOD_5 \text{ removed}} = \frac{350 \text{ kg } O_2/\text{d}}{3230 \text{ m}^3/\text{d} \times (160 - 14) \text{ g } BOD_5/\text{m}^3 \times 10^{-3} \text{ kg/g}} = \frac{350 \text{ kg } O_2/\text{d}}{472 \text{ kg } BOD_5/\text{d}}$$

$$= 0.74 \text{ kg } O_2/\text{kg } BOD_5$$

8. Calculate equilibrium DO concentration in the wastewater under operating conditions ($C_{e,ww}$).

$C_{e,ww}$ is calculated from Equation 10.39g. $\beta = 0.9$ is given in the statement, and $f_a = 1$ is typically for the mechanical surface aerator. The factors f_t and f_p are determined below.

a. Calculate the temperature correction factor.

The factor f_t is calculated from Equation 10.39j. The critical temperature for aeration occurs during summer time.

The surface saturation concentrations of oxygen are obtained from Table 10.16.

$C_{s,t} = 8.11$ mg/L is obtained at 26°C and sea level and $C_{e,20} = 9.09$ mg/L is also obtained at 20°C and 1 atm.

$$f_t = \frac{C_{s,t}}{C_{s,20}} = \frac{8.11 \text{ mg/L}}{9.09 \text{ mg/L}} = 0.89$$

b. Calculate the pressure correction factor.

The barometric pressures $P_s = 101.325$ kPa or 760 mm Hg at sea level. $P_f = 706$ mm Hg at an altitude of 610 m is obtained from Table B.7 in Appendix B. Calculate P_f in kPa.

$$P_f = \frac{706 \text{ mm Hg}}{760 \text{ mm Hg}} \times 101.325 \text{ kPa} = 94.13 \text{ kPa}$$

Water vapor pressure $P_v = 25.21$ mm Hg is obtained at 26°C from Table B.5 in Appendix B. Calculate P_v in kPa.

$$P_v = \frac{25.21 \text{ mm Hg}}{760 \text{ mm Hg}} \times 101.325 \text{ kPa} = 3.36 \text{ kPa}$$

Calculate factor f_p from Equation 10.39k.

$$f_p = \frac{P_f - P_{v,t}}{P_s - P_{v,t}} = \frac{(94.13 - 3.36) \text{ kPa}}{(101.325 - 3.36) \text{ kPa}} = 0.93$$

c. Calculate the equilibrium DO concentration ($C_{e,ww}$).

The value of $C_{e,ww}$ is calculated from Equation 10.39g using all the correction factors obtained above.

$$C_{e,ww} = \beta f_a f_p f_t C_{s,20} = 0.90 \times 1 \times 0.93 \times 0.89 \times 9.09 \text{ mg/L} = 0.74 \times 9.09 \text{ mg/L}$$
$$= 6.73 \text{ mg/L}$$

9. Calculate the standard oxygen requirement (*SOR*).

The total amount of *SOR* is calculated from Equation 10.48c at $\alpha = 0.8$ and $C_c = 2$ mg/L. For a surface aerator, $F = 1$ and assuming $C_{e,20} \approx C_{s,20} = 9.09$ mg/L.

$$SOR = \frac{ThOR}{((C_{e,ww} - C_c)/C_{e,20})(1.024)^{(T-20)} \alpha F}$$

$$= \frac{15 \text{ kg } O_2/\text{h}}{(((6.73 - 2) \text{ mg/L})/9.09 \text{ mg/L})(1.024)^{(26-20)} \times 0.8 \times 1} = 31 \text{ kg } O_2/\text{h}$$

10. Calculate the energy required in the aerated lagoon system.
   a. Calculate the energy requirement for oxygen transfer.
      Rearrange Equation 10.42a and calculate the power requirement ($P_{aeration}$) at $SOTR = SOR$.

$$P_{aeration} = \frac{SOR}{SAE} = \frac{31\ \text{kgO}_2/\text{h}}{1.2\ \text{kg O}_2/\text{kW·h}} = 26\ \text{kW}(34\,\text{hp})\ \text{for oxygen transfer}$$

   b. Calculate the energy requirement for mixing.
      The energy required for partially mixed regime is 5–8 kW/1000 m^3. Assume $p_{mixing} = 6.5$ kW/1000 m^3 and calculate the energy requirement for mixing (P_{mixing}).

$$P_{mixing} = V \times p_{mixing} = \frac{13{,}000\ \text{m}^3}{1000\ \text{m}^3} \times 6.5\ \text{kW}/1000\ \text{m}^3 = 85\ \text{kW}\ (113\text{hp})\ \text{for mixing}$$

      Note: the energy required for mixing is over three times the energy required for oxygen transfer.
   c. Determine the energy requirement for sizing the aerators.
      Provide total energy $P_{design} = 85$ kW to meet the energy requirement for mixing in each aerated lagoon.
      Provide eight floating surface aerators in each lagoon and calculate the power requirement for each aerator (P_a).

$$P_a = \frac{P_{design}}{8\ \text{aerators}} = \frac{85\ \text{kW}}{8\ \text{aerators}} = 11\ \text{kW}\ (15\ \text{hp})\ \text{per aerator}$$

      Provide nameplate horsepower of the aerator = 11 kW.
      Assume a motor efficiency of 92%.
      The estimated horsepower of the motor = 11 kW/0.92 = 12 kW (16 hp) per motor
      Provide 15 kW (20 hp) motor on each surface aerator. The total power requirement is 120 kW (160 hp) per lagoon and 240 kW (320 hp) in both lagoons.
11. Select the surface aerators and their arrangement in the lagoons.
    The floating surface aerators are provided in two rows and kept in place by cable anchor mooring method (see Figure 10.53b(v)). Details about the arrangement of aerators in the lagoon are shown in Figures 10.65a and b.
12. Verify the mixing condition provided in the lagoon.
    Example performance data are given below for the selected surface aerator:
    a. Impingement mixing (white water) diameter (IMD), $D_{im} = 8$ m (~25 ft)
    b. Diameter of complete mix (DCM), $D_{cm} = 20$ m (~65 ft)
    c. Diameter of complete oxygen dispersion (DCO), $D_{O2} = 60$ m (~200 ft)
    The performance ranges are illustrated by the IMD, DCM, and DCO circles in Figure 10.65c. The partial mixing condition is therefore confirmed.
13. Describe the construction requirements.
    The aerated lagoons should be lined with flexible membrane liners to control percolation. Others but less desirable methods are pond sealers such as clay, cement, and chemical sealants. The influent pipe should be discharged in a concrete box to prevent erosion. The slope facing the wave action should have breakwaters to dissipate wave energy. Rock revetment may be considered for slope protection. Different types of impermeable liners are presented in Example 10.91.

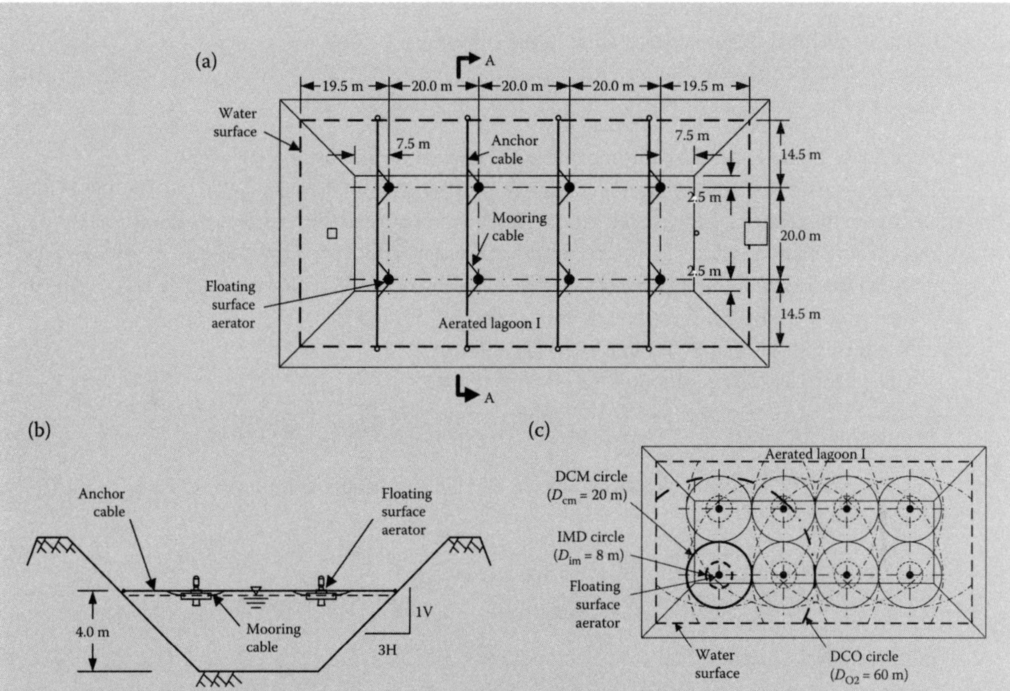

FIGURE 10.65 Arrangement and performance ranges of the floating surface aerators: (a) plan, (b) conceptual Section AA, and (c) aerator performance ranges (Example 10.80).

EXAMPLE 10.81: SOLIDS SEPARATION FACILITY FOR AN AERATED LAGOON

The TSS concentration in the effluent from a flow-through aerated lagoon is high and an external solids separation facility is required to improve the effluent quality. In Example 10.80, the effluent from the aerated lagoon has $TSS_{e,lagoon} = 210$ mg/L and soluble $BOD_5 = 12$ mg/L when the lagoons are operated in parallel in summer. Select and design a solids separation facility to lower the TSS concentration in the effluent. Calculate (a) total BOD_5 concentration in the effluent from the lagoons, (b) TSS and total BOD_5 in the effluent after the improvements, and (c) total quantity of dry solids removed from the solids separation facility at the total design average flow $Q_{total} = 6460\ m^3/d$. Assume that the TSS in the lagoon effluent is 30% biodegradable, the ratio of ultimate BOD_L exerted in the biomass to the biodegradable solids = 1.42, and $BOD_5/BOD_L = 0.68$.

Solution

1. Calculate the total BOD_5 concentration in the effluent.

   The total BOD_5 in the effluent includes soluble BOD_5 and that exerted by TSS. Calculate the BOD_5 exerted by TSS in the lagoon effluent.

$$S_{TSS,lagoon} = \left[\left(\frac{0.68\text{ g BOD}_5}{\text{g BOD}_L}\right) \times \left(\frac{1.42\text{ g BOD}_L}{\text{g biodegradable solids}}\right) \times \left(\frac{0.30\text{ g biodegradable solids}}{\text{g TSS}}\right)\right] TSS_{e,lagoon}$$

$$= 0.29\text{ g BOD}_5/\text{g TSS} \times 210\text{ mg TSS/L} = 61\text{ mg BOD}_5/\text{L}$$

Calculate the total BOD_5 concentration in the effluent.

$$S_{e,lagoon} = S + S_{TSS,laggon} = (12 + 61)\ \text{mg BOD}_5/\text{L} = 73\ \text{mg BOD}_5/\text{L}$$

2. Select a solids separation facility for improving the effluent quality from the lagoons.
   Provide an external circular clarifier for removing the solids from the lagoon effluent. The TSS in the effluent from the aerated lagoons is 210 mg/L. This is comparable to the raw municipal wastewater to a primary sedimentation facility. Therefore, the typical design surface overflow rate of 40 $m^3/m^2{\cdot}d$ is recommended for design of the external clarifier (see Table 9.6). A TSS removal efficiency $E_{TSS} = 60\%$ is estimated from Figure 9.12. Provide a side water depth $SWD = 5$ m.
3. Calculate the TSS and total BOD_5 in the clarifier effluent.
   Calculate the TSS concentration in the clarifier effluent.

$$\text{TSS}_{e,clarifier} = (1 - 0.6) \times \text{TSS}_{e,lagoon} = (1 - 0.6) \times 210\ \text{mg TSS/L} = 84\ \text{mg TSS/L}$$

Calculate total BOD_5 concentration in the clarifier effluent using the ration of 0.29 g BOD_5/g TSS obtained from Step 1.

$$\begin{aligned} S_{e,clarifier} &= S + 0.29\ \text{g BOD}_5/\text{g TSS} \times \text{TSS}_{e,clarifier} \\ &= 12\ \text{mg BOD}_5/\text{L} + 0.29\ \text{g BOD}_5/\text{g TSS} \times 84\ \text{mg TSS/L} = 36\ \text{mg BOD}_5/\text{L} \end{aligned}$$

BOD_5 removal efficiency,

$$E_{BOD5} = \left(1 - \frac{S_{e,clarifier}}{S_{e,lagoon}}\right) \times 100\% = \left(1 - \frac{36\ \text{mg BOD}_5/\text{L}}{73\ \text{mg BOD}_5/\text{L}}\right) \times 100\% = 51\%$$

4. Calculate the total quantity of dry solids removed from the clarifier.
   Dry solids removed,

$$W_{TSS} = 0.60 \times \text{TSS}_{e,lagoon} \times Q_{total} = 0.60 \times 210\ \text{g/m}^3 \times 6460\ \text{m}^3/\text{d} \times \frac{\text{kg}}{10^3\ \text{g}} = 814\ \text{kg/d}$$

5. Calculate the dimensions of the clarifier.
   Calculate the required surface area at the total design average flow of 6460 m^3/d, and design surface overflow rate of 40 $m^3/m^2{\cdot}d$.

$$A' = \frac{Q_{total}}{SOR_{design}} = \frac{Q_{total}}{40\ \text{m}^3/\text{m}^2{\cdot}\text{d}} = \frac{6460\ \text{m}^3/\text{d}}{40\ \text{m}^3/\text{m}^2{\cdot}\text{d}} = 162\ \text{m}^2$$

The required diameter of the clarifier, $D' = \sqrt{\frac{4}{\pi}A'} = \sqrt{\frac{4}{\pi} \times 162\ \text{m}^2} = 14.4\ \text{m}$

Provide a standard diameter of basin $D = 14.5$ m.

Actual surface area of the clarifier, $A = \frac{\pi}{4}D^2 = \frac{\pi}{4} \times (14.5\ \text{m})^2 = 165\ \text{m}^2$

Actual surface overflow rate, $SOR = \frac{Q_{total}}{A} = \frac{6460\ \text{m}^3/\text{d}}{165\ \text{m}^2} = 39\ \text{m}^3/\text{m}^2{\cdot}\text{d}$

Volume of the clarifier, $V = A \times SWD = 165\ \text{m}^2 \times 5\ \text{m} = 825\ \text{m}^3$

Hydraulic retention time, $HRT = \frac{V}{Q_{total}} = \frac{825\ \text{m}^3}{6460\ \text{m}^3/\text{d}} \times \frac{24\ \text{h}}{\text{d}} \approx 3\ \text{h}$

6. Develop the design of the external clarifier.

   The design steps are similar to those given in Example 9.8. More detailed design approach is also presented in Example 9.21 for a rectangular primary sedimentation basin, and in Example 10.166 for circular final clarifier.

10.3.12 Stabilization Ponds

Stabilization ponds* are earthen basins with relatively shallow body of water. They are of different shapes, and are designed to treat wastewater from small communities and industries. They have the advantage of low construction and maintenance costs. The major disadvantages are large land area, odors and insects, potential ground water contamination, and poor effluent quality.[1,7,98]

Process Description: A wide variety of microscopic plants and animals finds the environment a suitable habitat. Bacteria and protozoa are the primary feeders that metabolize the organic matters. The secondary feeders are protozoa and higher animals, including rotifers, crustaceans and nemotoids. The nutrients released from the sediments are utilized by algae and other aquatic plants. The accumulated solids layer at the bottom is decomposed by anaerobic bacteria. The major sources of oxygen supply are natural reaeration and oxygen produced from photosynthesis. The schematic representation of a facultative waste stabilization pond is shown in Figure 10.66.[12,97,110,111]

Types of Stabilization Ponds: The stabilization ponds are classified as (a) *aerobic*, (b) *facultative*, and (c) *anaerobic*. This classification is based on oxygen profile, biological activity, organic loading, and physical factors such as depth, detention time, and temperature. Effluent quality also varies greatly among three types of ponds. A brief description of each type of pond is given below.

Aerobic Pond: Aerobic condition prevails in the entire depth. There is a diurnal variation in DO concentration. During the day DO is high due to photosynthesis, and the maximum concentration reaches around the midday. In the evening DO concentration decreases as photosynthesis ceases, and there is oxygen demand by microbial and algae population. The DO concentration may vary from about 4–6 mg/L during the day to around 2–3 mg/L during the night.

Facultative (Aerobic–Anaerobic) Pond: The upper zone of the pond is aerobic while the lower zones are anaerobic. On a sunny day the entire depth may be aerobic while during overcast days anaerobic condition may occur throughout the pond depth. The bottom layer undergoes anaerobic decomposition, and the sludge accumulates and decomposes slowly at the bottom.

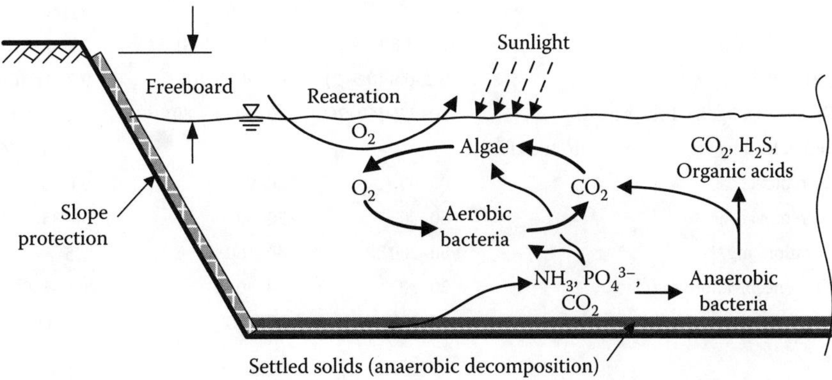

FIGURE 10.66 Schematic of a facultative stabilization pond showing the biochemical reactions of bacteria and algae.

* The terms stabilization pond, *oxidation pond*, and *lagoon* are interchangeably used in the literature.

Anaerobic Pond: Anaerobic condition prevails throughout the pond. They are deep, and are designed as an anaerobic digester without mixing. Normally these ponds are used for treatment of concentrated industrial wastes from dairies, slaughterhouses, and meat processing, and canning industries.

Effluent Quality: The effluent from stabilization ponds is poor and does not meet the EPA secondary discharge permit. The total suspended solids concentration is high because of algae and other microorganisms. The soluble BOD_5 and nutrients are low in the effluent.

If some type of solids removal system is applied for effluent polishing, a high-quality effluent is obtained. Common effluent polishing systems are (1) coagulation-flocculation and clarification, (2) dissolved air flotation, (3) microscreening, (4) sand and rock filtration, and (5) specialized algae harvesting devices. Detailed discussion on effluent polishing from stabilization ponds may be found in References 1, 97, 102, 106, and 112 through 114.

Design Considerations: The design of stabilization pond is based on organic loading, detention time, first-order reaction kinetics, and dispersion factor (varying from ideal plug-flow to complete-mix reactor), or by equating the oxygen resources of the pond to the applied organic loading. The basic design parameters of three types of stabilization ponds are summarized in Table 10.27.

Process design is important but careful considerations must also be given to the physical factors and facility layout to ensure that the optimum performance of the stabilization ponds is achieved. Some of the basic factors are listed below.[97,102,106,112–114]

- Influent and effluent piping should be located to minimize short-circuiting and maximize use of the entire pond area.
- Multiple inlet arrangement should be considered for better hydraulic distribution in large ponds. All gravity influent pipes should discharge horizontally onto a discharge aprons for erosion control while force main influent pipes should discharge vertically up at a minimum outlet submergence of at least 0.6 m (2 ft).

TABLE 10.27 Design Parameters for Aerobic, Facultative, and Anaerobic Stabilization Ponds

Parameter	Value		
	Aerobic (High Rate)	Facultative (Aerobic–Anaerobic)	Anaerobic
Mixing	Intermittent	Mixed surface layer	Stagnant
Operation	Series or parallel	Series or parallel	Series
Detention time, d	2–6	20–180	5–180
Water depth, m (ft)	0.6–1.8 (2–6)	1.2–2.4 (4–8)	2.4–6 (8–20)
Area, ha (acre)	0.2–0.8 (0.5–2)	0.8–4 (2–10)	0.2–0.8 (0.5–2)
Surface BOD_5 loading, kg/ha·d (lb/acre·d)	70–225 (60–200)	20–90 (15–80)	225–1100 (200–1000)
Volumetric BOD_5 loading, kg/1000 m^3·d (lb/1000 ft^3·d)	–	–	100–400 (6–25)
Soluble BOD_5 removal, %	80–95	80–95	80–95
Overall BOD_5 removal, %	60–80	70–90	50–85
Algae concentration, mg/L	80–200	40–160	<5
Effluent BOD_5 concentration, mg/L	20–30	20–50	100–300[a]
Effluent TSS concentration, mg/L	150–300	30–400	80–160
Temperature, °C			
Range	5–30	0–50	15–50
Optimum	20	20	30

[a] Normally used for the treatment of high strength of industrial wastewater.

Note: 1m = 3.281 ft; 1 ha = 2.471 acre; 1 kg/ha·d = 0.8922 lb/acre·d; and 1 kg/1000 m^3·d = 0.06243 lb/1000 ft^3·d.

Source: Adapted in part from References 1, 7, 97, 102, 106, 112, and 115 through 119.

- The effluent structure should be located as far as possible from the influent release. The effluent structure should have valved draw-off lines arranged at multiple levels or use adjustable draw-off weir. The effluent weir should be protected from floating debris. The effluent structure should be accessible from the shore. Multiple effluent structures are desirable to protect from flooding.
- The pond depth is an important design parameter. A minimum depth of 3 ft (1.0 m) is required to control the potential growth of emergent vegetation. Deeper ponds have inadequate surface area to support photosynthetic action, mixing, and natural aeration.
- Multiple ponds should have transfer structure and interconnection piping to divert flow from one pond to the other. These structures should be designed to minimize the change in flow pattern. The head loss should not exceed 7.5–10 cm (3–4 in).
- All discharge pipes should have valves or adjustable overflow devices to regulate and divide the flow.
- The freeboard should be at least 0.9 m (3 ft) above maximum water surface in the pond.
- The dike should have a top width of 3 m (10 ft). The inner slope should be 1 vertical to 3–6 horizontal. Outer slope should be 1 vertical to 3 horizontal. Vegetation cover should be provided to minimize erosion.
- The slope facing the wave action should have breakwaters to dissipate wave energy. They should be fixed or floating type. Rock revetment should be considered for slope protection.
- The pond bottom should be well compacted, smooth, and level. Pond sealer or liner may be provided to reduce percolation. The sealer should be earthen or cement liners, chemical sealant, or flexible membrane liners. Readers should refer to References 106, and 112 through 114 for pond sealers and liners, and slope protection. Impermeable liners for stabilization ponds are presented in Example 10.91.

Design Equations: Many equations for process design of stabilization ponds have been proposed over the past 50 years but no equation can be universally applied. Many factors such as wind mixing, solar energy, temperature, and turbidity influence the results. These relationships include the ideal equations of plug-flow and complete-mix models, as well as empirical equations. Some of these equations have been proposed by Gloyna, Marais, Oswald, and Thirumurthi. Many of these equations are presented below.

Gloyna Equation: Gloyna developed equations for determining volume and area of a facultative stabilization pond treating municipal wastewater. The effective depth of the pond is 3–5 ft (1–1.6 m). The hydraulic residence time of the facultative pond is calculated from Equation 10.56a. The volume and area of the pond may be calculated from Equations 10.56b and 10.56c, respectively.[110,111,114,120]

$$\theta = 0.035\ \mathrm{BOD_L}\theta_\mathrm{T}^{(35-T)}ff' \tag{10.56a}$$

$$V = CQ\ \mathrm{BOD_L}\theta_\mathrm{T}^{(35-T)}ff' \tag{10.56b}$$

$$A = C'Q\ \mathrm{BOD_L}\theta_\mathrm{T}^{(35-T)}ff' \tag{10.56c}$$

where

θ = hydraulic residence time at operating temperature, d
V = volume of pond, m^3 (ft^3)
Q = influent flow rate, m^3/d (MGD)
C = constant, $C = 0.035$ for SI units, and $C = 4680$ for U.S. customary units
$\mathrm{BOD_L}$ = ultimate BOD concentration in the influent, mg/L
θ_T = temperature correction coefficient, $\theta_\mathrm{T} = 1.085$ for an effective depth $D = 1.52$ m (5 ft)
T = water temperature in the pond, °C
f = algal toxicity factor, $f = 1$ for municipal wastewater
f' = sulfide correction factor, $f' = 1$ for SO_4^{2-} equivalent concentration less than 500 mg/L
A = area of the pond acre, ha (acre)
C' = constant, $C' = 2.30 \times 10^{-6}$ for SI units, and $C' = 0.0215$ for U.S. customary units

Note: It is applicable to a pond effective depth $D = 1.52$ m (5 ft) plus 0.30 m (1 ft) allowance for sludge storage. This is a typical depth for the locations with significant seasonal variation in temperature.

Oswald Equation: Oswald equation (Equation 10.57a) is used to design aerobic ponds. It is based on oxygen resources of the pond equal to the applied organic load. The principal source of oxygen in aerobic pond is photosynthesis which is governed by solar radiation.[121,122]

$$Y_{O2} = 0.25\,F\,SR \tag{10.57a}$$

where

Y_{O2} = oxygen yield by photosynthesis, lb O_2/acre·d
F = oxygenation factor, g O_2 produced/g algal cell synthesized
SR = solar radiation, calories (cal)/cm^2·d

The term oxygenation factor represents the ratio of the net weight of oxygen produced daily (g/d) to the net weight of algal cell synthesized daily (g/d). The oxygenation factor is calculated from the stoichiometric relationship, or solar energy conversation efficiency. The stoichiometric relationship of photosynthesis is expressed by Equation 10.57b and F can be estimated by Equation 10.57c.[113]

$$a\,CO_2 + (0.5b - 1.5d)\,H_2O + d\,NH_3 \rightarrow C_aH_bO_cN_d + (a + 0.25b - 0.5c - 0.75d)\,O_2 \tag{10.57b}$$

$$F = \frac{\text{Oxygen produced by photosynthesis}}{\text{Algal cell yield by photosynthesis}}$$

$$= \frac{(2a + 0.5b - c - 1.5d)AW_O}{aAW_C + bAW_H + cAW_O + dAW_N} \tag{10.57c}$$

where

a, b, c, and d = the atoms of carbon, hydrogen, oxygen and nitrogen in algal cellular mass ($C_aH_bO_cN_d$) produced
AW_C, AW_H, AW_O, and AW_N = the atomic weights of C, H, O and N

The solar radiation (SR) varies with time of the year and latitude. The probable solar radiation as a function of months and latitude is provided in Table 10.28.[123] Hourly solar radiation data are also available on 10 km grids for all states in the United States excluding Alaska in the National Solar Radiation Database (NSRDB) that is updated periodically by a team led by the National Renewable Energy Laboratory (NRWL).[124]

The actual solar radiation value is obtained from Equation 10.57d after a correction for cloudiness.

$$SR_c = SR_{min} + p(SR_{max} - SR_{min}) \tag{10.57d}$$

where

SR_c, SR_{max}, and SR_{min} = actual, maximum, and minimum solar radiations in a day, cal/cm^2·d
p = total actual hours of sunlight divided by total possible hours of sunlight, hr per hr or dimensionless

Elevation correction may also be considered if the ponds are located at a significantly high elevation. The correction of solar radiation for an elevation up to 3000 m (10,000 ft) above sea level is expressed by Equation 10.57e.[113,117]

$$SR_e = SR_c(1 + C''E) \tag{10.57e}$$

where

SR_e = solar radiation corrected for site elevation, cal/cm^2·d
C'' = constant, $C'' = 0.0033$ m^{-1} for SI units, and $C'' = 0.001$ ft^{-1} for U.S. customary units
E = site elevation, m (ft)

TABLE 10.28 Probable Values of Visible Solar Energy as a Function of Latitude and Month

Latitude, °		Solar Radiation (*SR*) by Month, cal/cm^2·d											
		Jan	Feb	Mar	Apr	May	Jun	Jul	Aug	Sep	Oct	Nov	Dec
0	Max	255	266	271	266	249	236	238	252	269	265	256	253
	Min	210	219	206	188	182	103	137	167	207	203	202	195
10	Max	223	244	264	271	270	262	265	266	266	248	228	225
	Min	179	184	193	183	192	129	158	176	196	181	176	162
20	Max	183	213	246	271	284	284	282	272	252	224	190	182
	Min	134	140	168	170	194	148	172	177	176	150	138	120
30	Max	136	176	218	261	290	296	289	271	231	192	148	126
	Min	76	96	134	151	184	163	178	166	147	113	90	70
40	Max	80	130	181	181	286	298	288	258	203	152	95	66
	Min	30	53	95	125	162	173	172	147	112	72	42	24
50	Max	28	70	141	210	271	297	280	236	166	100	40	26
	Min	10	19	58	97	144	176	155	125	73	40	15	7
60	Max	7	32	107	176	249	294	268	205	126	43	10	5
	Min	2	4	33	79	132	174	144	100	38	26	3	1

Source: Adapted in part from Reference 123.

The algal cell yield is related to light conversion efficiency of solar radiation, and is expressed by Equation 10.57f. The organic surface loading in ponds is calculated from Equation 10.57g.

$$Y_c = 0.25\ SR \tag{10.57f}$$

$$L_0 = 2.71\left(\frac{D}{\theta}\right)\ \text{BOD}_\text{L} \tag{10.57g}$$

where
Y_c = algal cell yield, lb algae/acre·d
L_0 = organic loading, lb BOD_L/acre·d
D = depth of the pond, ft
θ = hydraulic residence time at operating temperature, d
BOD_L = ultimate BOD concentration in the influent, mg/L

If the oxygen yield (Y_{O2}) in Equation 10.57a is equated to the organic surface loading (L_0) in Equation 10.57g, the design equation (Equation 10.57h) is obtained.

$$0.25\ F\ SR = 2.71\left(\frac{D}{\theta}\right)\text{BOD}_\text{L} \quad \text{or} \quad \frac{D}{\theta} = \frac{0.0922\ F\ SR}{\text{BOD}_\text{L}} \tag{10.57h}$$

In Equation 10.57h, use $SR = SR_e$ after corrections per Equations 10.57d and 10.57e, if applicable.

Marais and Shaw Equation: In addition to aerated lagoons, the kinetic model based on complete-mix hydraulics and first-order reaction rate has also been proposed for pond design.[100,101] The pond performance is also expressed by Equations 10.53 through 10.54b. The temperature correction coefficient θ_T and first-order reaction rate constant for facultative ponds are 1.085 and 1.2 d^{-1} at 35°C, respectively.

Based on long-term behavior data on pond performance, an equation was also developed by empirical relationships to express the effluent BOD_5. This relationship is expressed by Equation 10.58.

$$S_{e,max} = \frac{600}{0.6D + 8} \quad (10.58)$$

where

$S_{e,max}$ = maximum BOD_5 concentration in the pond effluent, mg/L

D = depth of the pond, ft

Equation 10.58 has several limitations. These are: (1) applicable to U.S. customary units, (2) applicable to aerobic and complete-mix ponds, (3) reaction rate constant $k = 0.17\ d^{-1}$, and (4) lower limit of (A/D) ratio of 1000 ft (~300 m).

Plug-Flow Equation: Kinetic model based on plug-flow hydraulics is also used in pond design. In depth discussion on plug-flow reactors is provided in Section 3.4.3. Equations 3.16b through 3.16d and Examples 3.24 through 3.27 cover the performance of plug-flow reactors. The first-order reaction rates for stabilization ponds are modified to reflect the BOD_5 loading rate on pond surface and corrected to the operating temperature using a temperature coefficient $\theta_T = 1.09$. The standard values of first-order reaction rate constant k_p at 20°C for plug-flow at different BOD_5 loading rates are summarized in Table 10.29.[102]

Wehner-Wilhelm Equation and Thirumurthi Application: Thirumurthi pointed out that neither the complete-mix nor the plug-flow kinetic model fully applies to the rational design of stabilization ponds. Rather, the facultative ponds exhibit a nonideal flow pattern that is between the complete-mix and plug-flow regimes.[107,108] On that basis, he suggested the use of chemical reactor equation developed by Wehner and Wilhelm for pond design (Equation 3.40).[109] Wehner-Wilhelm equations have been discussed in detail for nonideal flow reactors in Chapter 3 (Section 3.5.3), and for aerated lagoons in Section 10.3.11, and Examples 3.45 through 3.51, and 10.79. It is further discussed below in the context of stabilization ponds. To facilitate the use of Equation 3.40, Thirumurthi developed Figure 3.34. In this equation, the term $k\theta$ is plotted against BOD_5 concentration remaining in the effluent (S) for different dispersion numbers d.[107–109] The dispersion number d for stabilization ponds seldom exceeds 1.0. Equation 3.40 has two constants: the reaction rate k and dispersion number d. For a desired S, an iterative solution is required to obtain the hydraulic residence time (θ). Also, the selected values of d and k greatly influence the value of θ required to produce a given quality effluent S.[102] Thirumurthi further suggested a simplified equation for stabilization pond design (Equation 10.59).[107,108] The temperature correction coefficient θ_T is 1.09 and the first-order reaction rate constant for facultative ponds is 0.15 d^{-1} at 20°C.

$$\frac{S}{S_0} = \frac{4ae^{(1-a)/2d}}{(1+a)^2} \quad (10.59)$$

where

S_0, S = influent and effluent BOD_5 concentrations, mg/L

a = coefficient ($a = \sqrt{1 + 4k\theta d}$, see Equation 3.40), dimensionless

d = dispersion number, dimensionless

k = first-order reaction rate constant, d^{-1}

θ = detention time, d

TABLE 10.29 Reaction Rate Constant with Respect to Organic Loading Rate in a Plug-Flow Stabilization Pond

Parameter	Value				
BOD_5 loading rate, kg/ha·d (lb/acre·d)	22 (20)	45 (40)	67 (60)	90 (80)	112 (100)
First-order reaction rate constant k_p, d^{-1} at 20°C	0.045	0.071	0.083	0.096	0.129

Source: Adapted in part from Reference 102.

TABLE 10.30 Effluent Characteristics of Various Types of Stabilization Ponds

Type of Pond	Concentration Ratio				
	Suspended Solids			BOD_5	
	Algae $\frac{}{S_0}$	Microorganisms $\frac{}{S_0}$	Other $\frac{}{TSS_i}$	$\frac{S}{S_0}$	$\frac{S_P}{TSS_0}$
Aerobic	0.5–1.2	0.2–0.5	Low (≈ 0)	0.02–0.1	0.3–1.5
Aerobic-anaerobic	0.02–0.1	0.2–0.5	0.1–0.4	≤0.1	0.3–1.0
Anaerobic	–	0.1–0.3	0.3–0.5	0.04–0.2	0.3–0.8

Note: S_0 = influent BOD_5 concentration, mg/L; TSS_i = influent TSS concentration, mg/L; S = effluent soluble BOD_5 concentration, mg/L; S_p = effluent particulate BOD_5 concentration, mg/L; and TSS_e = effluent TSS concentration, mg/L.

Source: Adapted in part from Reference 111.

Stabilization Pond Efficiency: Many factors affect the performance of a stabilization pond. These are solar radiation, temperature, organic load, toxicity, depth, mixing, and pond shape. The estimated effluent characteristics of aerobic, aerobic–anaerobic, and anaerobic ponds in terms of influent BOD_5 and effluent suspended solids are given in Table 10.30.[111]

EXAMPLE 10.82: FACULTATIVE POND DESIGN USING GLOYNA EQUATION

A facultative stabilization pond is designed for a population of 10,000 residents in the south central region of the United States. The average wastewater flow is 100 gpcd (380 Lpcd). The influent BOD_5 concentration is 200 mg/L, and $BOD_5/BOD_L = 0.68$. The mean temperature of pond content during summer and winter are approximately 297°C and 7°C, respectively. The data for Gloyna equation are $f = 1$, $f' = 1$, water depth $D = 5$ ft (1.52 m). Provide two ponds in parallel operation. Calculate (a) hydraulic residence time, (b) total surface area, and volume of the pond, and (c) applied surface BOD_5 loading. The effluent soluble BOD_5 concentration (S) under critical condition may range from 20 to 55 mg/L.

Solution

1. Calculate the average flow per pond (Q).

$$\text{Average flow from the community, } Q_{total} = 10{,}000 \text{ persons} \times \frac{100 \text{ gal}}{\text{person·d}} \times \frac{10^{-6}\text{Mgal}}{\text{gal}}$$

$$= 1.0 \text{ MGD (or } 3800 \text{ m}^3/\text{d)}$$

$$\text{Design flow per pond, } Q = \frac{Q_{total}}{2 \text{ ponds}} = \frac{1 \text{ MGD}}{2 \text{ ponds}} = 0.5 \text{ MGD per pond (or } 1900 \text{ m}^3/\text{d)}$$

2. Calculate the ultimate BOD concentration (BOD_L).

$$BOD_L = \frac{BOD_5}{BOD_5/BOD_L} = \frac{200 \text{ mg/L}}{0.68} = 294 \text{ mg/L}$$

3. Calculate the hydraulic residence time (θ).
   Winter condition is critical for the pond design. Calculate θ from Equation 10.56a at $T = 7°C$ and $f = 1$ and $f' = 1$.

$$\theta = 0.035\, BOD_L \theta_T^{(35-T)} ff' = 0.035 \times 294 \text{ mg/L} \times 1.085^{(35-7)} \times 1 \times 1 = 101 \text{ d}$$

4. Calculate the volume of each pond (V).

   Calculate V from Equation 10.56b in U.S. customary units.

$$V = CQ\text{BOD}_\text{L}\theta_\text{T}^{(35-T)}ff' = 4680 \times 0.5 \text{ MGD} \times 294 \text{ mg/L} \times 1.085^{(35-7)} \times 1 \times 1 = 6{,}800{,}000 \text{ ft}^3$$

   $V = 190{,}000 \text{ m}^3$ is also obtained from Equation 10.56b in SI units.

   Note: Same value of V can also be calculated from $V = \theta\, Q$.

5. Calculate the area of each pond (A).

   At $D = 5$ ft, calculate A from Equation 10.56c in U.S. customary units.*

$$A = C'Q\text{BOD}_\text{L}\theta_\text{T}^{(35-T)}ff' = 0.0215 \times 0.5 \text{MGD} \times 294 \text{mg/L} \times 1.085^{(35-7)} \times 1 \times 1 = 31 \text{ acres or } 1{,}400{,}000 \text{ ft}^2$$

   $A = 13$ ha or $130{,}000 \text{ m}^2$ is also obtained from Equation 10.56c in SI units.

   Note: Same value of A can also be calculated from $A = V/D$.

6. Calculate the applied mass BOD_5 loading ($\Delta S_{0,\text{BOD5}}$).

$$\text{Mass BOD}_5 \text{ loading, } \Delta S_{0,\text{BOD5}} = Q \times S_0 = 0.5 \text{ MGD} \times 200 \text{ mg BOD}_5/\text{L} \times 8.34 \text{ lb/(mg/L·Mgal)}$$
$$= 835 \text{ lb BOD}_5/\text{d}$$

   $\Delta S_{0,\text{BOD5}} = 380 \text{ kg BOD}_5/\text{d}$ is obtained in SI units.

7. Calculate the applied surface BOD_5 loading.

$$\text{Surface BOD}_5 \text{ loading} = \frac{\Delta S_{0,\text{BOD5}}}{A} = \frac{835 \text{ lb BOD}_5/\text{d}}{31 \text{ acres}} = 27 \text{ lb BOD}_5/\text{acre·d}$$

   Surface BOD_5 loading $= 30 \text{ kg BOD}_5/\text{ha·d}$ is also obtained in SI units.

* 1 acre = 43,560 ft^2 = 0.4047 ha; 1 ha = 10,000 m^2 = 2.471 acres; and 1 ft^2 = 0.0929 m^2.

EXAMPLE 10.83: AEROBIC POND DESIGN USING OSWALD EQUATION

Oswald equation is used to design an aerobic pond. Design two stabilization ponds in parallel operation for the data given in Example 10.82. Additional data for Oswald equation are: $D = 3$ ft (0.91 m), the site latitude = 30°, and the site elevation $E = 655$ ft (~200 m). Assume that the actual hours of sunlight is 8 hours over a possible 10 hours ($p = 8\text{ h}/10\text{ h} = 0.8$) in December and algal cell composition equation is $C_{6.14}H_{10.3}\,O_{2.24}N$.

Solution

1. Determine the average solar radiation *(SR)*.

   The average solar radiation SR_c in critical month is calculated from Equation 10.57d using the probable values of visible solar radiation ($SR_{min} = 70 \text{ cal/cm}^2\text{·d}$ and $SR_{max} = 126 \text{ cal/cm}^2\text{·d}$) obtained from Table 10.28 at latitude 30° for the month of December.

$$SR_c = SR_{min} + p(SR_{max} - SR_{min}) = 70 \text{ cal/cm}^2\text{·d} + 0.8 \text{ hr/hr} \times (126 - 70) \text{ cal/cm}^2\text{·d}$$
$$= 115 \text{ cal/cm}^2\text{·d}$$

Altitude correction is obtained from Equation 10.57e.

$$SR_e = SR_c(1 + C''E) = 115\ \text{cal/cm}^2\cdot\text{d} \times (1 + 0.001\ \text{ft}^{-1} \times 655\ \text{ft}) = 190\ \text{cal/cm}^2\cdot\text{d}$$

2. Determine the oxygenation factor (F).

The oxygenation factor F is determined from the stoichiometric relationship or conversion efficiency.

The stoichiometric relationship applied to determine F is expressed by Equation 10.57b. Determine F from Equation 10.57c at the algal cell composition of $C_{6.14}H_{10.3}O_{2.24}N$.

$a = 6.14$, $b = 10.3$, $c = 2.24$, $d = 1$, $AW_C = 12$, $AW_H = 1$, $AW_O = 16$, and $AW_N = 14$.

$$F = \frac{(2a + 0.5b - c - 1.5d)AW_O}{aAW_C + bAW_H + cAW_O + bAW_N} = \frac{(2 \times 6.14 + 0.5 \times 10.3 - 2.24 - 1.5 \times 1) \times 16}{6.14 \times 12 + 10.3 \times 1 + 2.24 \times 16 + 1 \times 14} = \frac{219}{134} = 1.63$$

3. Calculate the hydraulic residence time (θ).

Rearrange Equation 10.57h and calculate θ at $D = 3$ ft, $BOD_L = 294$ mg/L (from Step 2 of Example 10.82), and $SR = SR_e = 190\ \text{cal/cm}^2\cdot\text{d}$.

$$\theta = \frac{D\ BOD_L}{0.0922\ F\ SR} = \frac{3\ \text{ft} \times 294\ \text{mg/L}}{0.0922 \times 1.63 \times 190\ \text{cal/cm}^2\cdot\text{d}} = 31\ \text{d}$$

4. Calculate the volume (V) and surface area (A) of each pond.

Design flow per pond, $Q = 0.5$ MGD (from Step 1 of Example 10.82).

$$V = \theta\, Q = 31\ \text{d} \times 0.5\ \text{MGD} \times \frac{10^6\ \text{gal}}{\text{Mgal}} \times \frac{1\ \text{ft}^3}{7.48\ \text{gal}} = 2{,}100{,}000\ \text{ft}^3\ (59{,}000\ \text{m}^3)$$

$$A = \frac{V}{D} = \frac{2{,}100{,}000\ \text{ft}^3}{3\ \text{ft}} = 700{,}000\ \text{ft}^2\ (65{,}000\ \text{m}^2) \quad \text{or} \quad A = \frac{700{,}000\ \text{ft}^2}{43{,}560\ \text{ft}^2/\text{acre}} = 16\ \text{acres}\ (6.5\ \text{ha})$$

5. Calculate the applied surface BOD_5 loading.

Mass BOD_5 loading in each pond = 835 lb BOD_5/d (from Step 6 of Example 10.82).

$$\text{Surface BOD}_5\ \text{loading} = \frac{\Delta S_{0,BOD5}}{A} = \frac{835\ \text{lbBOD}_5/\text{d}}{16\ \text{acres}} = 52\ \text{lb BOD}_5/\text{acre}\cdot\text{d}\ (59\ \text{kg BOD}_5/\text{ha}\cdot\text{d})$$

EXAMPLE 10.84: FACULTATIVE POND DESIGN USING MARAIS AND SHAW EQUATION

Marias and Shaw equation was originally developed for complete-mix ponds in series arrangement. Design two stabilization ponds in parallel operation. Use the data given in Example 10.82. Additional data for Marias and Shaw equation are: $S_{e,max} = 55$ mg/L and $k_{c35} = 1.2\ \text{d}^{-1}$.

Solution

1. Calculate the water depth (D) in the pond.

Rearrange Equation 10.58 and calculate D.

$$D = \frac{1}{0.6}\left(\frac{600}{S_{e,max}} - 8\right) = \frac{1}{0.6}\left(\frac{600}{55\ \text{mg/L}} - 8\right) = 4.8\ \text{ft}$$

Use a water depth $D = 5$ ft (1.52 m).

2. Calculate the first-order reaction rate constant at 7°C (k_{c7}) during winter from Equation 10.17a at $k_{c35} = 1.2\ \text{d}^{-1}$ and $\theta_T = 1.085$.

$$k_{c7} = k_{c35}\theta_T^{(T-35)} = 1.2\ \text{d}^{-1} \times 1.085^{(7-35)} = 0.12\ \text{d}^{-1}$$

3. Calculate the hydraulic residence time (θ) from Equation 10.53.

$$\theta = \frac{1}{k_{c7}}\left(\frac{S_0}{S_{e,\,max}} - 1\right) = \frac{1}{0.12\ \text{d}^{-1}}\left(\frac{200\ \text{mg/L}}{55\ \text{mg/L}} - 1\right) = 22\ \text{d}$$

4. Calculate the volume (V) and surface area (A) of each pond.

Design flow per pond, $Q = 0.5$ MGD (from Step 1 of Example 10.82).

$$V = \theta\ Q = 22\ \text{d} \times 0.5\ \text{MGD} \times \frac{10^6\ \text{gal}}{\text{Mgal}} \times \frac{1\ \text{ft}^3}{7.48\ \text{gal}} = 1{,}500{,}000\ \text{ft}^3\ (42{,}000\ \text{m}^3)$$

$$A = \frac{V}{D} = \frac{1{,}500{,}000\ \text{ft}^3}{5\ \text{ft}} = 300{,}000\ \text{ft}^2\ (28{,}000\ \text{m}^2) \quad \text{or} \quad A = \frac{300{,}000\ \text{ft}^2}{43{,}560\ \text{ft}^2/\text{acre}} = 6.9\ \text{acres}\ (2.8\ \text{ha})$$

5. Calculate the applied surface BOD_5 loading.

Mass BOD_5 loading in each pond = 835 lb BOD_5/d (from Step 6 of Example 10.82).

$$\text{Surface BOD}_5\ \text{loading} = \frac{\Delta S_{0,BOD5}}{A} = \frac{835\ \text{lb BOD}_5/\text{d}}{6.9\ \text{acres}} = 121\ \text{lb BOD}_5/\text{acre·d}\ (136\ \text{kg BOD}_5/\text{ha·d})$$

EXAMPLE 10.85: FACULTATIVE POND DESIGN USING PLUG-FLOW EQUATION

Plug-flow equation is used when mixing is low due to limited wave action. Design two stabilization ponds in parallel operation for the data given in Example 10.82. The process parameters for plug-flow equation are: $D = 6$ ft (1.83 m), soluble BOD_5 in the effluent is 40 mg/L and k_p at 20°C varies with BOD_5 loading rate as given in Table 10.29.

Solution

1. Describe the design procedure.

   Since the reaction rate k_p at 20°C varies with the BOD_5 loading rate, an iterative solution is required. Assume a reaction rate constant or BOD_5 loading rate for first iteration. Calculate the corresponding k_p or BOD_5 loading rate from Table 10.29. Repeat the step until the values converge.

2. First iteration.

   a. Determine the reaction rate constant (k_{p20}).

Assume $k_{p20} = 0.08\ d^{-1}$ to start the iteration procedure. Calculate the reaction rate constant at the critical water temperature of 7°C (k_{p7}) during winter from Equation 10.17a at $\theta_T = 1.09$.

$$k_{p7} = k_{p20}\theta_T^{(T-20)} = 0.08\ d^{-1} \times 1.09^{(7-20)} = 0.026\ d^{-1}$$

b. Calculate the detention time (θ), volume (V), surface area (A), and surface BOD_5 loading for each pond.

Design flow per pond, $Q = 0.5$ MGD (from Step 1 of Example 10.82).

Calculate θ from Equation 10.1e.

$$\theta = -\frac{1}{k_{p7}}\ln\left(\frac{S}{S_0}\right) = -\frac{1}{0.026\ d^{-1}} \times \ln\left(\frac{40\ mg/L}{200\ mg/L}\right) = 61.9\ d$$

$$V = \theta\, Q = 61.9\ d \times 0.5\ MGD \times \frac{10^6\ gal}{Mgal} \times \frac{1\ ft^3}{7.48\ gal} = 4{,}140{,}000\ ft^3\ (117{,}000\ m^3)$$

$$A = \frac{V}{D} = \frac{4{,}140{,}000\ ft^3}{6\ ft} = 690{,}000\ ft^2 (64{,}000\ m^2) \quad \text{or} \quad A = \frac{690{,}000\ ft^2}{43{,}560\ ft^2/acre}$$

$$= 15.8\ \text{acres}\ (6.4\ \text{ha})$$

Mass BOD_5 loading in each pond = 835 lb BOD_5/d (from Step 6 of Example 10.82).

$$\text{Surface } BOD_5 \text{ loading} = \frac{\Delta S_{0,BOD5}}{A} = \frac{835\ lb\ BOD_5/d}{15.8\ cres} = 52.8\ lb\ BOD_5/acre{\cdot}d\ (59.3\ kg\ BOD_5/ha{\cdot}d)$$

3. Second iteration.

a. Determine k_{p20}.

Reaction rate constants k_{p20} of 0.071 and 0.083 d^{-1} are obtained, respectively, at surface BOD_5 loadings of 40 and 60 lb/acre·d from Table 10.29. Calculate k_{p20} at the surface BOD_5 loading of 52.8 lb/acre·d.

$$k_{p20} = 0.071\ d^{-1} + \frac{(52.8 - 40)\ lb\ BOD_5/acre{\cdot}d}{(60 - 40)\ lb\ BOD_5/acre{\cdot}d} \times (0.083 - 0.071)\ d^{-1} = 0.079\ d^{-1}$$

$$k_{p7} = 0.079\ d^{-1} \times 1.09^{(7-20)} = 0.026\ d^{-1}$$

b. Calculate θ, V, A, and surface BOD_5 loading for each pond.

$$\theta = -\frac{1}{0.026\ d^{-1}} \times \ln\left(\frac{40\ mg/L}{200\ mg/L}\right) = 61.9\ d$$

$$V = 61.9\ d \times 0.5\ MGD \times \frac{10^6\ gal}{Mgal} \times \frac{1\ ft^3}{7.48\ gal} = 4{,}140{,}000\ ft^3\ (117{,}000\ m^3)$$

$$A = \frac{4{,}140{,}000\ ft^3}{6\ ft} = 690{,}000\ ft^2\ (64{,}000\ m^2) \quad \text{or} \quad A = \frac{690{,}000\ ft^2}{43{,}560\ ft^2/acre}$$

$$= 15.8\ \text{acres}\ (6.4\ \text{ha})$$

$$\text{Surface } BOD_5 \text{ loading} = \frac{835\ lbBOD_5/d}{15.8\ acres} = 52.8\ lb\ BOD_5/acre{\cdot}d\ (59.3\ kg\ BOD_5/ha{\cdot}d)$$

4. Third iteration.

The following results are obtained after the third iteration:

$k_{p20} = 0.079\ d^{-1}$

$k_{p7} = 0.026\ d^{-1}$

$\theta = 61.9\ d$

$V = 4{,}140{,}000\ ft^3\ (117{,}000\ m^3)$

$A = 690{,}000\ ft^2\ (64{,}000\ m^2)$ or 15.8 acres (6.4 ha)

Surface BOD_5 loading = 52.8 lb BOD_5/acre·d (59.3 kg BOD_5/ha·d)

These iterative results are same as that in Step 2. Therefore further iteration is not needed.

Note: If the calculated values in Step 3 were significantly different from that in Step 2, more iterations are carried out by adjusting the reaction rate constants (k_{p20}) and repeating Steps 2 until close agreement is reached.

EXAMPLE 10.86: THIRUMURTHI APPLICATION OF WEHNER-WILHELM EQUATION FOR FACULTATIVE POND DESIGN

Thirumurthi's procedure for stabilization pond design has been well accepted. Design two stabilization ponds in parallel operation. The basic information is given in Example 10.82. The data for Wehner-Wilhelm equation are: $k_{20} = 0.15\ d^{-1}$ at 20°C, $\theta_T = 1.09$, dispersion number $d = 0.5$, water depth $D =$ 6 ft (1.83 m), and soluble BOD_5 in the effluent is 40 mg/L.

Solution

1. Calculate the reaction rate constant (k_7) at the critical temperature of 7°C from Equation 10.17a at $\theta_T = 1.09$.

$$k_7 = k_{20}\theta_T^{(T-20)} = 0.15\ d^{-1} \times 1.09^{(7-20)} = 0.049\ d^{-1}$$

2. Determine $k\theta$.

$$\frac{S}{S_0} = \frac{40\ mg/L}{200\ mg/L} = 0.2$$

$k\theta = 2.3$ is obtained at $S/S_0 = 0.2$ from the curve plotted for $d = 0.5$ in Figure 3.34.

3. Determine the hydraulic residence time (θ).

$$k_7\theta = 0.049\ d^{-1} \times \theta = 2.3 \quad \text{or} \quad \theta = \frac{2.3}{0.049\ d^{-1}} = 47\ d$$

4. Calculate the volume (V), surface area (A), and surface BOD_5 loading for each pond.

$$V = \theta\ Q = 47\ d \times 0.5\ MGD \times \frac{10^6\ gal}{Mgal} \times \frac{1\ ft^3}{7.48\ gal} = 3{,}100{,}000\ ft^3\ (88{,}000\ m^3)$$

$$A = \frac{V}{D} = \frac{3{,}100{,}000\ ft^3}{6\ ft} = 520{,}000\ ft^2\ (48{,}000\ m^2) \quad \text{or} \quad A = \frac{520{,}000\ ft^2}{43{,}560\ ft^2/acre} = 12\ acres\ (4.8\ ha)$$

Mass BOD_5 loading in each pond = 835 lb BOD_5/d (from Step 6 of Example 10.82).

$$\text{Surface } BOD_5 \text{ loading} = \frac{\Delta S_{0,BOD5}}{A} = \frac{835 \text{ lb } BOD_5/\text{d}}{12 \text{ acres}} = 70 \text{ lb } BOD_5/\text{acre·d} \, (78 \text{ kg } BOD_5/\text{ha·d})$$

EXAMPLE 10.87: THIRUMURTHI'S SIMPLIFIED EQUATION FOR FACULTATIVE POND DESIGN

Equation 10.59 is a simplified form of Wehner-Wilhelm equation for pond design. Design two stabilization ponds in parallel operation. The basic information is given in Example 10.82. The data for Thirumurthi's simplified equation are: $\theta_T = 1.09$, $k_{20} = 0.15 \text{ d}^{-1}$ at 20°C, dispersion number $d =$ 0.5, and water depth $D = 6$ ft (1.83 m). Assume $S/S_0 = 0.2$ is required.

Solution

1. Calculate the reaction rate constant (k_7) at the critical temperature of 7°C from Equation 10.17a at $\theta_T = 1.09$.

$$k_7 = k_{20}\theta_T^{(T-20)} = 0.15 \text{ d}^{-1} \times 1.09^{(7-20)} = 0.049 \text{ d}^{-1}$$

2. Calculate the hydraulic residence time (θ).

   Since detention time θ is not known, a trial and error solution is required.

   a. First iteration.

   Assume $\theta = 45$ d and calculate a from Equation 3.40.

$$a = \sqrt{1 + 4k\theta d} = \sqrt{1 + 4 \times 0.049 \text{ d}^{-1} \times 45 \text{ d} \times 0.5} = 2.33$$

   Calculate S/S_0 from Equation 10.59.

$$\frac{S}{S_0} = \frac{4ae^{(1-a)/2d}}{(1+a)^2} = \frac{4 \times 2.33 \times e^{(1-2.33)/(2\times 0.5)}}{(1+2.33)^2} = \frac{9.32 \times 0.264}{11.1} = 0.22$$

   The S/S_0 of 0.22 is slightly higher than the required value of 0.2.

   b. Second iteration.

   Increase θ to 50 d.

$$a = \sqrt{1 + 4 \times 0.049 \text{ d}^{-1} \times 50 \text{ d} \times 0.5} = 2.43$$

$$\frac{S}{S_0} = \frac{4 \times 2.43 \times e^{(1-2.43)/(2\times 0.5)}}{(1+2.43)^2} = \frac{9.72 \times 0.239}{11.8} = 0.20$$

   S/S_0 of 0.2 reaches the required value of 0.2.

3. Calculate the volume (V), surface area (A), and surface BOD_5 loading for each pond.

$$V = \theta\, Q = 50 \text{ d} \times 0.5 \text{ MGD} \times \frac{10^6 \text{ gal}}{\text{Mgal}} \times \frac{1 \text{ ft}^3}{7.48 \text{ gal}} = 3{,}300{,}000 \text{ ft}^3 \, (93{,}000 \text{ m}^3)$$

$$A = \frac{V}{D} = \frac{3{,}300{,}000 \text{ ft}^3}{6 \text{ ft}} = 550{,}000 \text{ ft}^2 \, (51{,}000 \text{ m}^2) \text{ or } A = \frac{550{,}000 \text{ ft}^2}{43{,}560 \text{ ft}^2/\text{acre}} = 13 \text{ acres} \, (5.1 \text{ ha})$$

Mass BOD_5 loading in each pond = 835 lb BOD_5/d (from Step 6 of Example 10.82).

$$\text{Surface BOD}_5\text{ loading} = \frac{\Delta S_{0,BOD5}}{A} = \frac{835\text{ lb BOD}_5/\text{d}}{13\text{ acres}} = 64\text{ lb BOD}_5/\text{acre·d (72 kg BOD}_5/\text{ha·d)}$$

EXAMPLE 10.88: COMPARISON OF DESIGN PARAMETERS OF STABILIZATION POND CALCULATIONS FROM DIFFERENT EQUATIONS

Two stabilization ponds in parallel operation were designed using six different methods in Examples 10.82 through 10.87. These methods included: (1) Gloyna equation, (2) Oswald equation, (3) Marais and Shaw equation, (4) Plug-flow reactor equation, (5) Wehner-Wilhelm equation with Thirumurthi application, and (6) Thirumurthi's simplified equation. Compare and comment on the results obtained from these methods.

Solution

1. Summarize the results.

   The design parameters and information developed of the same stabilization pond are compared in Table 10.31.

TABLE 10.31 Summary of Design Parameters Developed by Using Six Different Equations for Stabilization Pond Design (Example 10.88)

Parameter	Gloyna Equation	Oswald Equation	Marais and Shaw Equation	Plug-Flow Equation	Wehner-Wilhelm Equation with Thirumurthi Application	Thirumurthi Simplified Equation
	(Example 10.82)	(Example 10.83)	(Example 10.84)	(Example 10.85)	(Example 10.86)	(Example 10.87)
Pond type	Facultative	Aerobic	Facultative	Facultative	Facultative	Facultative
No. of ponds in parallel operation	2	2	2	2	2	2
Total flow (Q_{total}), MGD (m^3/d)	1 (3800)	1 (3800)	1 (3800)	1 (3800)	1 (3800)	1 (3800)
Flow per pond (Q), MGD (m^3/d)	0.5 (1900)	0.5 (1900)	0.5 (1900)	0.5 (1900)	0.5 (1900)	0.5 (1900)
Influent BOD_5 (S_0), mg/L	200	200	200	200	200	200
Effluent soluble BOD_5 (S), mg/L	20–55	–	55	40	40	40
S/S_0	0.1–0.3	–	0.28	0.20	0.20	0.20
Pond depth (D), ft (m)	5 (1.52)	3 (0.91)	5 (1.52)	6 (1.83)	6 (1.83)	6 (1.83)
Detention time (θ), d	101	31	22	62	47	50
Volume of each pond (V), 10^6 ft^3 (10^3 m^3)	6.8 (190)	2.1 (59)	1.5 (42)	4.1 (117)	3.1 (88)	3.3 (93)
Surface area of each pond (A), acres (ha)	31 (13)	16 (6.5)	6.9 (2. 8)	16 (6.4)	12 (4.8)	13 (5.1)
Surface BOD_5 loading, lb/acre·d (kg/ha·d)	27 (30)	52 (59)	121 (136)	53 (59)	70 (78)	64 (72)

2. Comments on the results.

The results from various design methods vary greatly. The reason is the varying conditions, assumptions, and limitations under which each method is developed. Therefore, direct comparison is difficult to make. Some general comments are provided below:

a. Gloyna method is applicable for facultative ponds, and the BOD_5 removal efficiency ranges 70–90%. Depth of the pond varies with climate from 1 m for tropical to 2.5 m for severe climate.
b. Oswald method relies upon photosynthetic oxygenation and is only applicable to aerobic ponds. Liquid depth is limited to 3 ft (~1 m).
c. The Marias and Shaw method is based on complete-mix hydraulics and first-order reaction kinetics. Complete-mix hydraulics is not approached in facultative ponds.
d. Plug-flow hydraulics and first-order kinetic equations best describe the performance of facultative ponds in series. In a single pond, the plug-flow hydraulics is not reachable due to short circuiting.
e. Wehner-Wilhelm equation requires knowledge of both the reaction rate and dispersion factor. This method will yield satisfactory results if hydraulic characteristics and reaction rate are estimated properly.
f. In summary, all methods will provide a valid design if proper design parameters are selected under the field conditions.

EXAMPLE 10.89: DESIGN AND LAYOUT OF TWO AEROBIC STABILIZATION PONDS

An aerobic stabilization pond system is designed in south central region of the country. The population served is 2000 residents. The average wastewater flow is 400 Lpcd, and average soluble BOD_5 concentration (S_0) is 180 mg/L. The critical average temperature of the pond is 12°C. Develop the dimensions and configuration of two pond system. Arrange the piping so that parallel or series operation is achieved. Also, one lagoon may be taken out of service for maintenance. The surface BOD_5 loading must not exceed 50 kg/ha·d (45 lb/acre·d). Soluble BOD_5 removal must be 90% or greater. Justify all assumptions.

Solution

1. Make the following assumptions.

a. For aerobic pond system, the first-order soluble BOD_5 removal rate constant $k_{20} = 0.25\ d^{-1}$ at 20°C. This is consistent with the residential wastewater flow.
b. The influent BOD_5 concentration of 180 mg/L is mainly soluble. Assume that the total BOD_5 contribution by residents is 95 g BOD_5/capita·d (see Table 5.15). At 400 Lpcd wastewater flow rate, the average influent total BOD_5 concentration should be ~240 mg/L.* The difference is suspended BOD_5 concentration (60 mg/L) which is not included in the given data.
c. The pond depth $D = 1$ m to maintain aerobic condition throughout the entire depth of the pond.
d. The dispersion number $d = 1.0$ because there is sufficient wave action to warrant mixing and dispersion of pond contents in shallow ponds.
e. The inner and outer slopes are 1V to 3H. The top width of the embankment is 3 m. These dimensions are such that mowing and maintenance equipment can be operated.
f. The length to width ratio is 2. This ratio will provide good flow distribution and mixing.
g. The freeboard $H_{fb} = 1$ m to give sufficient allowance for wave action.

* BOD_5 concentration $= \dfrac{95\ \text{g}}{\text{capita·d}} \times \dfrac{1000\ \text{mg}}{\text{g}} \times \dfrac{\text{capita·d}}{400\ \text{L}} \approx 240$ mg/L

2. Calculate the reaction rate constant (k_{12}) at the critical pond temperature of 12°C from Equation 10.17a at $\theta_T = 1.09$.

$$k_{12} = k_{20}\theta_T^{(T-20)} = 0.25 \text{ d}^{-1} \times 1.09^{(12-20)} = 0.125 \text{ d}^{-1}$$

3. Determine the hydraulic residence time (θ) in the pond.
   Thirumurthi's application of Wehner-Wilhelm equation is used to determine $k\theta$.
   $k\theta = 5$ is obtained at $S/S_0 = 0.1$ (at 90% removal) and $d = 1.0$ in Figure 3.34.
   Determine the θ.

$$k_{12}\theta = 0.125 \text{ d}^{-1} \times \theta = 5 \quad \text{or} \quad \theta = \frac{5}{0.125 \text{ d}^{-1}} = 40 \text{ d}$$

4. Calculate the volume (V) and surface area (A) of the pond.

$$\text{Total average wastewater influent flow, } Q_{\text{total}} = 400 \text{ L/person·d} \times 2000 \text{ people} \times 10^{-3} \text{ m}^3/\text{L} = 800 \text{ m}^3/\text{d}$$

   Since there are two ponds in parallel operation, calculate the average flow per pond (Q).

$$Q = \frac{Q_{\text{total}}}{2} = \frac{1}{2} \times 800 \text{ m}^3/\text{d} = 400 \text{ m}^3/\text{d}$$

$$V = \theta Q = 40 \text{ d} \times 400 \text{ m}^3/\text{d} = 16{,}000 \text{ m}^3 \text{ per pond}$$

$$A = \frac{V}{D} = \frac{16{,}000 \text{ m}^3}{1 \text{ m}} = 16{,}000 \text{ m}^2 \quad \text{or} \quad A = \frac{16{,}000 \text{ m}^2}{10{,}000 \text{ m}^2/\text{ha}} = 1.6 \text{ ha per pond}$$

5. Verify the surface BOD_5 loading.

$$\text{Mass BOD5 loading, } \Delta S_{0,\text{BOD5}} = Q \times S_0 = 400 \text{ m}^3/\text{d} \times 180 \text{ g BOD}_5/\text{m}^3 \times 10^{-3} \text{ kg/g} = 72 \text{ kg BOD}_5/\text{d}$$

$$\text{Surface BOD}_5 \text{ loading} = \frac{\Delta S_{0,\text{BOD5}}}{A} = \frac{72 \text{ kg BOD}_5/\text{d}}{1.6 \text{ ha}} = 45 \text{ kg BOD}_5/\text{ha·d } (40 \text{ lb BOD}_5/\text{acre·d})$$

   The design surface BOD_5 loading rate does exceed the criterion of the 50 kg/ha·d (45 lb/acre·d).

6. Determine the water surface dimensions at the length to width ratio (L_s/W_s) of 2.
   $L_s = 2W_s$ and $A = L_sW_s = 2W_s^2 = 16{,}000 \text{ m}^2$.

$$W_s = \sqrt{\frac{16{,}000 \text{ m}^2}{2}} = 89 \text{ m, provide } W_s = 92 \text{ m and then } L_s = 2W_s = 2 \times 92 \text{ m} = 184 \text{ m.}$$

7. Determine the bottom dimensions at the water depth $D = 1$ m.

$$W_b = W_s - 2 \times 3 \times D = 92 \text{ m} - 2 \times 3 \times 1 \text{ m} = 86 \text{ m}$$

$$L_b = L_s - 2 \times 3 \times D = 184 \text{ m} - 2 \times 3 \times 1 \text{ m} = 178 \text{ m}$$

8. Determine the top dimensions of embankment at freeboard $H_f = 1$ m.

$$W_{top} = W_s + 2 \times 3 \times H_f = 92 \text{ m} + 2 \times 3 \times 1 \text{ m} = 98 \text{ m}$$

$$L_{top} = L_s + 2 \times 3 \times H_f = 184 \text{ m} - 2 \times 3 \times 1 \text{ m} = 190 \text{ m}$$

9. Calculate the actual volume and detention time provided.

   Area of base, $A_b = L_b \times W_b = 176\ \text{m} \times 86\ \text{m} = 15{,}308\ \text{m}^2$

   Area of water surface, $A_s = L_s \times W_s = 184\ \text{m} \times 92\ \text{m} = 16{,}928\ \text{m}^2$

   Calculate the actual pond volume from Equation 10.55.

$$V=\frac{D}{3}[A_b+A_s+\sqrt{A_b\times A_s}]=\frac{1\,\text{m}}{3}\times[15{,}308\,\text{m}^2+16{,}928\,\text{m}^2+\sqrt{15{,}308\,\text{m}^2\times 16{,}928\,\text{m}^2}]=16{,}111\,\text{m}^3$$

$$\text{Detention time, } \theta = \frac{V}{Q} = \frac{16{,}111\ \text{m}^3}{400\ \text{m}^3/\text{d}} = 40\ \text{d}$$

   It is same as the desired residence time obtained in Step 3.
10. Develop the ponds layout, piping arrangement, and influent and effluent structures.
    The ponds are arranged longitudinally with a common dike. The layout, piping arrangement, and influent and effluent structures are very similar to the aerated lagoon system that has been presented in Example 10.80 and Figures 10.63 and 10.64.
11. Parallel and series operation.
    In parallel operation, the influent is split equally over two weir gates into two basins. The water depth in both basins is maintained at the design water depth of 1 m. In series operation, the entire flow enters the Lagoon I. The effluent structure of this lagoon is designed and operated such that the water depth in the lagoon is raised to 1.2 m by closing the adjustable weir. By opening an isolation sluice gate the effluent from Lagoon I is diverted by the extra head to Lagoon II which has a design water depth of 1 m. The pipings, adjustable weir gates, and sluice gate arrangements are similar to those of aerated lagoons designed in Example 10.80. Design details are shown in Figure 10.64. Readers should consult Example 10.80 for specific information.

EXAMPLE 10.90: POLISHING OF EFFLUENT FROM A STABILIZATION POND

Effluent from stabilization ponds contains high concentration of suspended solids. List different methods used for polishing the effluent from a stabilization pond.

Solution

The effluent polishing systems are listed below:

1. Sedimentation lagoon (maturation pond) or sedimentation pond
2. Coagulation, flocculation, and sedimentation
3. Dissolved air flotation
4. Microscreens
5. Intermittent sand filters
6. Rock filters
7. Rapid sand filters
8. Natural systems to include aquaculture (water hyacinth and constructed wetlands), and land treatment such as slow irrigation, overland flow, and rapid infiltration

EXAMPLE 10.91: IMPERMEABLE LINERS IN STABILIZATION PONDS

Impermeable liners are installed along the bottom and on the sides of the stabilization basins to reduce the migration of water or contaminants to groundwater. List various methods that are used to impede the flow of water and pollutants from stabilization ponds.

Solution

The impervious liners are broadly divided into two types: (1) liners that are constructed on site, and (2) liners that are prefabricated of different synthetic membranes, and are installed on site. Both types of liners are briefly presented below.[7,125]

Liners constructed on site

1. Native clays or engineered soils formed on-site.
2. Admixed liners are formed on site from hot mixture of asphalt cement and mineral aggregate, or soil and liquid asphalt. They fall into three categories: (a) asphalt concrete, (b) soil cement, and (c) soil asphalt.
3. Sprayed on liners formed in the field by spraying chemical solutions on prepared surface. They are of four types: (a) air-blow asphalt, (b) emulsified asphalt, (c) urethane modified asphalt, and (d) rubber or plastic latex.
4. Soil sealants are also used to reduce the permeability of soil by application of sealant by spraying or pressure injection.

Prefabricated liners installed on site

The liners are of two types: (a) flexible membrane liners (FMLs) or geomembranes, and (b) geosynthetic clay liners. Both types are presented below:

1. FMLs also called polymeric membranes or geomembranes are manufactured from different polymers and are available in different thicknesses. They come in rolls of different length and width, and are installed on prepared base. A list of commercial FMLs are listed below:
   a. Butyl rubber,
   b. Chlorinated polyethylene (CPE),
   c. Chlorosulfonated polyethylene (CSPE),
   d. Elasticized polyolefin (ELPO),
   e. Elasticized polyvinyl chloride (PVC-E),
   f. Epichlorohydrin rubber (ECH),
   g. Ethylene propylene rubber (EPDM),
   h. High-density polyethylene (HDPE),
   i. Neoprene (chloroprene rubber, CR),
   j. Polyethylene (PE),
   k. Polyvinyl chloride (PVC),
   l. Very low density polyethylene (VLDPE),
   m. Thermoplastic elastomer (TPE).
2. Geosynthetic clay liners are fabricated from sodium bentonite layer that is sandwiched between a woven and a nonwoven geotextile. Under moist conditions, the bentonite forms an impermeable barrier.

10.4 Fixed-Film or Attached Growth Aerobic Treatment Processes

Fixed-film or attached growth biological reactors utilize inert packing material on which microorganisms grow and form a *biofilm*. The organic material and nutrients are removed from the wastewater as they pass over the attached biofilm. The support material includes rocks, redwood, or a wide variety of synthetic media.

In the attached growth processes, the substrate is consumed within the biofilm. The thickness of the biofilm may range from 0.1 to 10 mm. A stagnant liquid layer (diffusion layer) separates the biofilm from the bulk liquid that passes over it. Oxygen and substrate diffuse through the stagnant liquid and biofilm, while the products of metabolic degradation diffuse in the opposite direction. In fixed-film reactors, the biomass remains attached to the media. As a result, high population of microorganisms and thus high sludge age can be maintained in these reactors. The attached growth processes are divided into three major groups: (1) nonsubmerged attached growth, (2) combined attached and suspended growth, (3) integrated attached and suspended growth, and (4) submerged aerobic attached growth. These processes are briefly described in Table 10.32. An in-depth discussion is provided below.

10.4.1 Nonsubmerged Attached Growth Processes

In nonsubmerged attached growth processes, air freely flows through the air space in the media. The two most common fixed-film aerobic processes are the *trickling filter* and *rotating biological contactor (RBC)*. In a trickling filter the packaging material remains stationary, whereas in an RBC unit the packaging material rotates in the tank. Both systems are discussed below.

Trickling Filter: A trickling filter consists of a shallow bed filled with natural or synthetic media. Wastewater is applied on the surface by means of a self-propelled or mechanical rotary distributor arms. The organics are removed by the biofilm that develops over the media. The underdrain system collects the trickled liquid that also contains the biological solids that are detached (or sloughed off) from the media. The air circulates through the voids in the media due to natural draft developed by thermal gradient. The trickled liquid is settled in a clarifier. A portion of flow is recycled. Recirculation helps to prevent ponding in the filter, maintains uniform HL, dilutes the influent, and reduces the nuisance from odors and flies. The typical process flow diagrams with recirculation patterns are shown in Figure 10.67.[2,7]

TABLE 10.32 Attached Growth Aerobic Treatment Processes

Process	Description	Reference
Nonsubmerged attached growth	There is a free flow of liquid and gasses through the large air space in the packing media. Common types are (1) trickling filters and (2) rotating biological contractor (RBC).	Section 10.4.1
Combined attached and suspended growth	Combined attached and suspended growth (dual) systems have been developed to use a fixed-film reactor (usually a trickling filter) in conjunction with an activated sludge process to reduce the overall organic loading on the suspended growth.	Section 10.4.2
Integrated fixed-film media in aeration basin	Several types of fixed-film media are used in aeration basins. The purpose is to provide greater biomass that may enhance the performance of the activated sludge process, and reduce the size of the aeration basin. These hybrid systems include integrated fixed-film activated sludge (IFAS) and moving bed biofilm reactor (MBBR).	Section 10.4.3
Submerged attached growth	The process uses submerged packing material. The oxygen is supplied by diffused aeration. Common types are (1) downflow packed-bed reactor, (2) upflow packed-bed reactor, and (3) upflow fluidized-bed reactor. Submerged RBC with diffused aeration is also included. The major advantage is a relatively small footprint and the ability to treat diluted wastewater.	Section 10.4.4

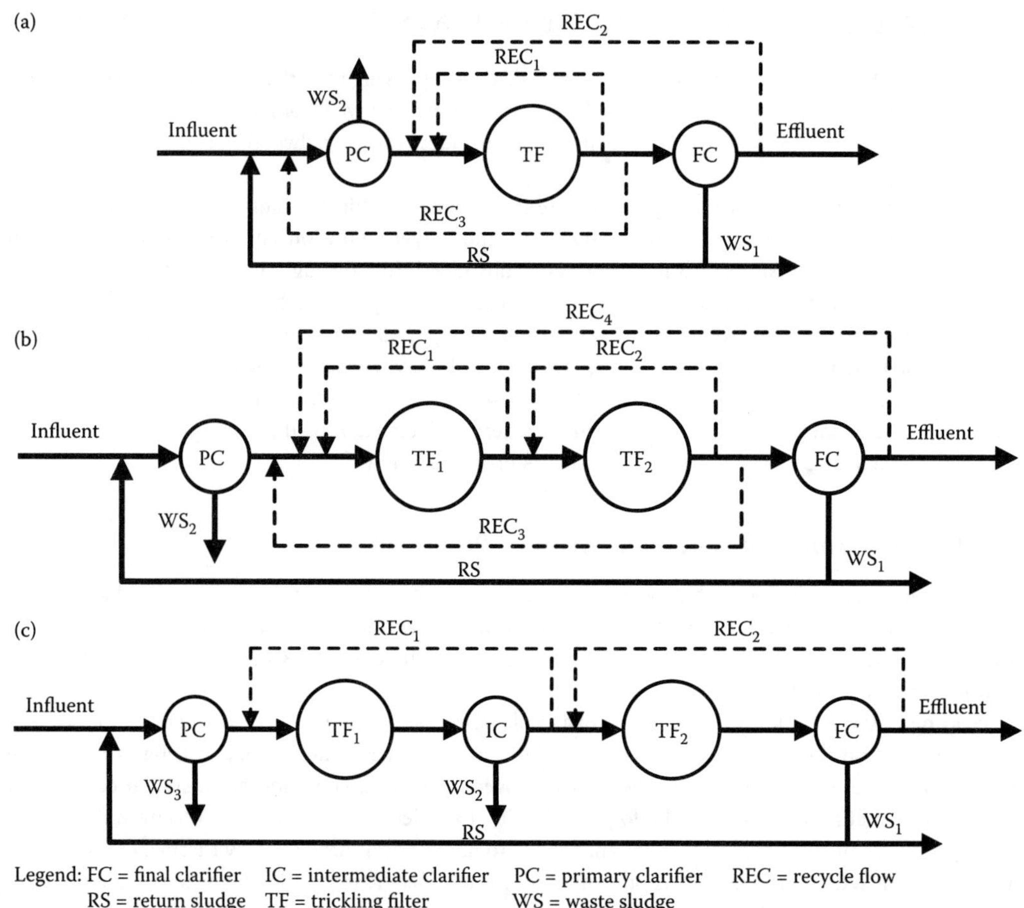

FIGURE 10.67 Trickling filter process flow diagrams: (a) intermediate and high-rate single-stage filter, (b) two-stage filter with one clarifier, and (c) two-stage trickling filter with intermediate and final clarifiers.
Note: Various recirculation and sludge wasting options are numbered. They can be used in different schematics of trickling filter processes.

The principal advantages of a trickling filter over the activated sludge process are (1) low energy requirement, (2) simple and reliable operation, no return or waste sludge, no food to microorganism (F/M) ratio to maintain, or sludge bulking to deal with, (3) low equipment maintenance, (4) better sludge thickening, (5) requiring moderate level of skill to operate the system, and (6) better process recovery from toxic shock. The considerable disadvantages of a trickling filter over activated sludge process are (1) poorer effluent quality, (2) larger land area requirement, (3) odors, fly, and vector problems, (4) sensitivity to low temperatures, (5) requiring regular operator attention, (6), clogging and snail problems, and (7) inability to provide biological nitrogen and phosphorus removal.[2,7,126]

Types of Trickling Filters: Based on organic and HLs, the filters are classified into (1) low-rate, (2) intermediate-rate, (3) high-rate, and (4) super-rate (or roughing) filters. Two-stage trickling filters are often used for treating high-strength industrial wastes or achieving nitrification. Tall towers packed with synthetic media and forced air ventilation reduce the land requirement. They are called activated biofilters. The process description, basic design parameters as well as major operation features of different types of trickling filters are summarized in Table 10.33.[2,7,127]

TABLE 10.33 Process Description, Basic Design Parameters, and Major Operation Features of Trickling Filters

Item	Low Rate with Nitrification	Standard Low Rate	Intermediate to High Rate	Super Rate	Roughing
Process description	• Two filters arranged in series • With or without intermediate clarifier • Recirculation at low to medium flow • Use of many recirculation patterns • Full nitrification	• Shallow depth in single-stage only • Large area • Without or with recirculation at low flow	• Shallow depth • Single- or two-stage operation • Recirculation at low to medium flow • Use of many recirculation patterns • Dependable operation	• Shallow depth but high hydraulic loading • Single- or two-stage operation • Recirculation at low to medium rate • Use of many recirculation patterns • Consistent effluent quality	• Deep bed • Similar to a packed bed reactor • High organic and hydraulic loadings • Without or with recirculation at low to medium flow • Synthetic packing is random or in modules • Mainly used ahead of other trickling filters or aeration basins to treat strong industrial waste
BOD_5 loading, kg/m^3·d (lb/10^3 ft^3·d)	0.08–0.48 (5–30)	0.08–0.32 (5–20)	0.24–2.4 (15–150)	0.8–4.8 (50–300)	1.6–6 (100–375)
Hydraulic loading, m^3/m^2·d (gpd/ft^2)	5–16 (125–400)	1–4 (25–100)	4–40 (100–1000)	15–80 (350–2000)	40–100 (1000–2500)
Recirculation ratio	1–2	0–1	0.1–2	1–12	0–2
Depth, m (ft)	2.5–12 (8–40) for plastic, 1–2.5 (3–8) for rock	1–2.5 (3–8)	2–2.5 (6–8)	2.5–12 (8–40)	1–10 (3–30)
BOD_5 removal, %	85–95	80–90	80–90 single-stage 90–95 two-stage	70–80 single-stage 80–90 two-stage	40–70
Effluent quality					
BOD_5, mg/L	<20	<25	<30	<30	<40
NH_4 (mg/L as N)	<3	<5	<5	<5	No nitrification
Filter media	Plastic or rock	Rock or slag	Rock or slag	Plastic	Plastic
Ventilation	Forced air	Natural	Forced air	Forced air	Forced air
Sloughing	Intermittent	Intermittent	Intermittent to continuous	Intermittent	Continuous

(Continued)

TABLE 10.33 (*Continued*) Process Description, Basic Design Parameters, and Major Operation Features of Trickling Filters

Item	Low Rate with Nitrification	Standard Low Rate	Intermediate to High Rate	Super Rate	Roughing
Dosing rate[a], mm/pass (in/pass)					
Normal	15–100 (0.6–4)	15–60 (0.6–2.5)	45–450 (1.8–18)	50–900 (2–36)	100–1000 (4–40)
Flushing	100–150 (4–6)	100–125 (4–5)	300–450 (12–18)	150–900 (6–36)	300–1000 (12–40)
Power, $kW/10^3\ m^3$ ($hp/10^3\ ft^3$)	—	2–4	2–8	6–10	10–20
Filter flies (*Psychoda*)	Few to none	Many	Varies to few	Few to none	Few to none

[a] It is the amount of liquid applied for each pass of each distributor arm.

Note: $1\ kg/m^3{\cdot}d = 62.4\ lb/10^3\ ft^3{\cdot}d$; and $1\ m^3/m^2{\cdot}d = 24.5\ gpd/ft^2$.

Source: Adapted in part from References 2, 7, and 127.

Trickling Filter Performance Equations: A number of trickling filter performance equations have been proposed over the past 60 years. Most of these equations were proposed by Velz, Eckenfelder, and the National Research Council (NRC). These classical equations are discussed below.

1. *Velz Equation:* One of the earliest equations was proposed by Velz in 1948.[128] This equation, as expressed by Equation 10.60, is a modification of the first-order reaction with respect to filter depth.

$$-\frac{dL}{dD} = KL \quad \text{or} \quad \frac{L_D}{L_i} = e^{-kD} \quad \text{or} \quad \frac{L_D}{L_i} = 10^{-KD} \tag{10.60}$$

where

L_D = ultimate carbonaceous BOD concentration at the depth D, mg/L
L_i = ultimate carbonaceous BOD concentration applied over the filter surface, mg/L
k = the first-order reaction rate constant (base e) based on ultimate BOD, m^{-1} (ft^{-1})
K = the first-order reaction rate constant (base 10, $K = (k/2.3)$) based on ultimate BOD, m^{-1} (ft^{-1})
D = total filter depth, m (ft)

The reaction rate constant depends upon HL. The reaction rate constants provided by Velz are summarized in Table 10.34. The temperature correction coefficient θ_T for reaction rate constant in Equation 10.17a is 1.047.

2. *Eckenfelder Equations:* Eckenfelder proposed a filter performance equation based on specific rate of substrate removal for pseudofirst-order reaction (Equation 10.61a).[129] Equation 10.61b is further developed by integration of Equation 10.61a.

$$-\frac{1}{\overline{X}}\left(\frac{dS}{dt}\right) = kS \tag{10.61a}$$

$$\frac{S}{S_i} = e^{-k\overline{X}t} \tag{10.61b}$$

where

$-\frac{1}{\overline{X}}\left(\frac{dS}{dt}\right)$ = specific rate of substrate utilization, g substrate/g biomass·d
$\overline{X}$ = average cell mass concentration, mg/L
k = first-order reaction rate constant, mg/L·d
t = mean contact time, d
S = substrate concentration after contact time t, mg/L
S_i = substrate concentration applied to the filter surface, mg/L

It has been reported that average cell mass concentration $\overline{X}$ is proportional to the specific surface area of the packing media (or surface area per unit volume) (m^2/m^3 (ft^2/ft^3)).[129] Also, the flow through a filter is tortuous. For this reason, the mean contact time t in a filter is related to (1) filter depth, (2) media geometry, specific surface area and characteristics, and (3) surface HL. All these

TABLE 10.34 Reaction Rate Constants Used in Design of Trickling Filters

Filter Type	Hydraulic Loading, m^3/m^2·d (MG/ac·d)	Reaction Rate Constant		Removal, %
		k, m^{-1} (ft^{-1})	K, m^{-1} (ft^{-1})	
Low-rate	1.8–5.4 (45–135)	1.35 (0.41)	0.59 (0.18)	90
High-rate	18 (450)	1.15 (0.35)	0.49 (0.15)	79

Note: 1 m^3/m^2·d = 24.54 gpd/ft^2.
Source: Adapted in part from References 2 and 7.

relationships are combined in Equations 10.61c and 10.61d. Substitution of these equations in Equation 10.61b gives the trickling filter performance equation (Equation 10.61e).[130]

$$\overline{X} = C_1 A_s \tag{10.61c}$$

$$t = \frac{C_2 D}{Q_L^n} \tag{10.61d}$$

$$\frac{S}{S_i} = e^{-kC_1 A_s^m C_2 D/Q_L^n} \quad \text{or} \quad \frac{S}{S_i} = e^{-k_1 A_s^m D/Q_L^n} \tag{10.61e}$$

where

A_s = specific surface area of the packing media (area/volume), m^2/m^3 (ft^2/ft^3) The nominal size and specific surface area of different trickling filter media are given in Table 10.35.[1,2,54]

k_1 = first-order reaction rate constant ($k_1 = k \times C_1 \times C_2$), mg/L·d

C_1 and C_2 = constants

Q_L = total hydraulic loading rate (or wetting rate), including recirculation flow, $m^3/m^2{\cdot}d$ (gpm/ft^2)

m and n = exponential constants

Other terms have been previously defined.

The factors k_1 and A_s in Equation 10.61e are dependent upon wastewater characteristics and filter media. For a given wastewater and packing material, these factors may be combined and Equation 10.61e is simplified by Equation 10.61f.

$$\frac{S}{S_i} = e^{-k_r D/Q_L^n} \tag{10.61f}$$

where k_r = experimentally determined rate constant ($k_r = k_1 \times A_s^m$), $(m^3/d)^n/m^2$, $(gpm)^n/ft^2$. The temperature correction coefficient θ_T in Equation 10.17a is 1.035 for k_r in Equation 10.61f. The value of n depends upon the flow characteristics and the packing media. The most accepted value of $n = 0.5$. Therefore, k_r is commonly expressed in the unit of $(m^3/d)^{0.5}/m^2$, $(gpm)^{0.5}/ft^2$.

German reported the values of k_r and n from pilot testing of municipal wastewater in filters packed with synthetic media.[131] The specific surface area of the media (A_s) was

TABLE 10.35 Physical Characteristics of Fixed-Film Media Used in Trickling Filter

Medium	Nominal Size, cm (in)	Bulk Unit Weight, kg/m^3 (lb/ft^3)	Specific Surface Area, m^2/m^3 (ft^2/ft^3)	Void Volume, %
River rock	7.5–10 (3–4)	800–1440 (50–90)	40–70 (12–21)	45–55
Slag	5–9 (2–3.5)	800–1200 (50–75)	45–70 (14–21)	45–55
Redwood	120 × 120 × 50 (48 × 48 × 20)	32–96 (2–6)	40–50 (12–15)	70–80
Plastic, random packing				
Cylindrical	7.5–18.5 dia. × 5 (3–7.3 dia. × 2)	27–40 (1.7–2.5)	100–115 (30–35)	95–96
Spherical	9 dia. (3.5 dia.)	53 (3)	125 (38)	95
Plastic, modular packing media				
Cross-flow	61 × 61 × 122 (24 × 24 × 48)	25–45 (1.6–2.8)	100–225 (30–69)	>95
Vertical flow	61 × 61 × 122 (24 × 24 × 48)	25–45 (1.6–2.8)	100–130 (30–40)	>94

Note: 1 kg/m^3 = 0.0624 lb/ft^3; and 1 m^2/m^3 = 0.305 ft^2/ft^3.
Source: Adapted in part from References 1, 2, and 54.

89 m^2/m^3 (27 ft^2/ft^3). The filter depth (D) and total hydraulic loading (Q_L) were 1.83 m (6 ft) and 8 $m^3/m^2{\cdot}d$ (0.136 gpm/ft^2), respectively. The experimental values of n and k_r were 0.5 and 2.0 $(m^3/d)^{0.5}/m^2$ or 0.22 $(L/s)^{0.5}/m^2$ (0.08 $(gpm)^{0.5}/ft^2$), respectively. The conversion steps between the U.S. customary and SI units may be found in Example 10.99.

The most commonly used kinetic equation for the trickling filter performance is also proposed by Eckenfelder. It is given by Equation 10.62a.[132]

$$\frac{S}{S_i} = \frac{1}{1 + C(D^{0.67}/Q_L^{0.5})} \tag{10.62a}$$

where

S = BOD_5 concentration in the effluent from the filter, mg/L
S_i = BOD_5 concentration in the filter influent, including the recycle flow, mg/L
C = constant, 5.4 for SI units, and 2.5 for U.S. customary units
Q_L = total hydraulic loading, $m^3/m^2{\cdot}d$ (MG/ac·d)
Other terms have been previously defined.

The BOD_5 concentration applied to the filter surface (S_i) is obtained from the mass–balance relationship, and is given by Equation 10.62b.

$$S_i = \frac{QS_0 + RQS}{Q + RQ} = \frac{S_0 + RS}{(1 + R)} \tag{10.62b}$$

where

Q = influent flow to the trickling filter, m^3/d (MGD)
S_0 = BOD_5 concentration in the influent to the filter, not including the recycle flow, mg/L
S = BOD_5 concentration in the effluent from the filter, mg/L
R = recirculation ratio of recycle flow rate to influent flow rate, dimensionless
Other terms have been previously defined

Normally, an experimental program is conducted to determine k_r and n in Equation 10.61f. A pilot filter is operated under normalized depth conditions. If the depth of the pilot filter is different from full-scale design, Equation 10.63 is applied to make the depth correction.[2] This correction is applicable to the k_r in Equation 10.61f.

$$k_{r,s} = k_{r,p}\left(\frac{D_p}{D}\right)^{0.5}\left(\frac{S_p}{S_i}\right)^{0.5} \tag{10.63}$$

where

$k_{r,s}$ = rate constant corresponding to site-specific filter depth D and influent BOD_5 concentration S_i, $(m^3/d)^{0.5}/m^2$, $(gpm)^{0.5}/ft^2$
$k_{r,p}$ = rate constant corresponding to the normalized pilot test filter depth D_n and influent BOD_5 concentration S_n, $(m^3/d)^{0.5}/m^2$, $(gpm)^{0.5}/ft^2$
D = site-specific filter depth, m (ft)
D_p = pilot test filter depth, m (ft). A normalized filter depth of 6.1 m (20 ft) is typically used in the pilot tests.
S_i = site-specific filter influent BOD_5 concentration, mg/L
S_p = pilot test filter influent BOD_5 concentration, mg/L. A normalized filter influent BOD_5 concentration of 150 mg/L is typically used in the pilot tests

3. *NRC's Equations:* The NRC developed in 1946 many empirical performance equations for trickling filters using stone media (Equations 10.64a through 10.64d).[133] These equations utilized data

collected from numerous trickling filters treating domestic wastewater at military bases. These equations apply for (1) one trickling filter and its final clarifier, and (2) two trickling filters in series with an intermediate clarifier between the filters. These equations are given below (Equation 10.64).

$$E_1 = \frac{100\%}{1 + C\sqrt{\dfrac{W_1}{V_1 F_1}}} \tag{10.64a}$$

$$E_2 = \frac{100\%}{1 + \dfrac{C}{1 - E_1'}\sqrt{\dfrac{W_2}{V_2 F_2}}} \tag{10.64b}$$

$$F_1 = \frac{1 + R_1}{(1 + 0.1R_1)^2} \quad \text{or} \quad F_2 = \frac{1 + R_2}{(1 + 0.1R_2)^2} \tag{10.64c}$$

$$E_1 = \frac{S_0 - S_1}{S_0} \times 100\% \quad \text{or} \quad E_2 = \frac{S_1 - S}{S_1} \times 100\% \tag{10.64d}$$

where

E_1 = BOD removal efficiency through single- or the first-stage filter and clarifier, percent
E_2 = BOD removal efficiency through the second-stage filter and clarifier, percent
E_1' = BOD removal efficiency through the first-stage filter and clarifier, fraction. It is only used for two trickling filters in series.
C = constant, 0.443 for SI units, and 0.0085 for U.S. customary units
W_1 = BOD_5 applied to single- or the first-stage filter, kg/d (lb/d)
W_2 = BOD_5 applied to the second-stage filter, if applicable, kg/d (lb/d)

$$W_2 = W_1(1 - E_1')$$

V_1 = volume of stone media in single- or the first-stage filter, m^3 (acre·ft)
V_2 = volume of stone media in the second-stage filter, m^3 (acre·ft)
F_1 = recycle factor applied to single- or the first-stage filter, dimensionless
F_2 = recycle factor applied to the second-stage filter, dimensionless
R_1 and R_2 = recycle ratio (recycle flow to influent flow), dimensionless
S_1 = BOD_5 concentration in the effluent from the first-stage filter, mg/L

Other terms have been previously defined

The amounts of BOD_5 carried in the recirculation flows are not included in the calculations of E_1, E_2, E'_1, W1, and W_2 in Equations 10.64a through 10.64d.

Components of Trickling Filter: Several components of trickling filters must be considered carefully in the design of filters. These are (1) filter media, (2) dosing control, (3) distribution system, (4) underdrains, and (5) air circulation (or draft). These components are illustrated in Figure 10.68a. Each of these components is briefly described below.

1. *Filter Media:* The common media used are gravel, crushed stone, and slag in a nominal size between 5 and 10 cm (2 and 4 in). Soft stone such as limestone should not be used. Other media include redwood and plastic media made of polyvinylchloride (PVC), high-density polyethylene (HDPE), and others. The plastic modular packing media have a much larger specific surface area than stone (Table 10.35). Bio-Pac SF#30, Bio-Rings®, FLEXIRING®, and AccuPac® are examples of commercially available plastic media of different types.[134–139] Grating panels are typically provided on top of the plastic modular packing media to (1) provide an interlocked and nonskid surface for safe operator access, (2) minimize the hydraulic impact on the media, and (3) protect the plastic media from damages by foot traffics or falling objects.[140]

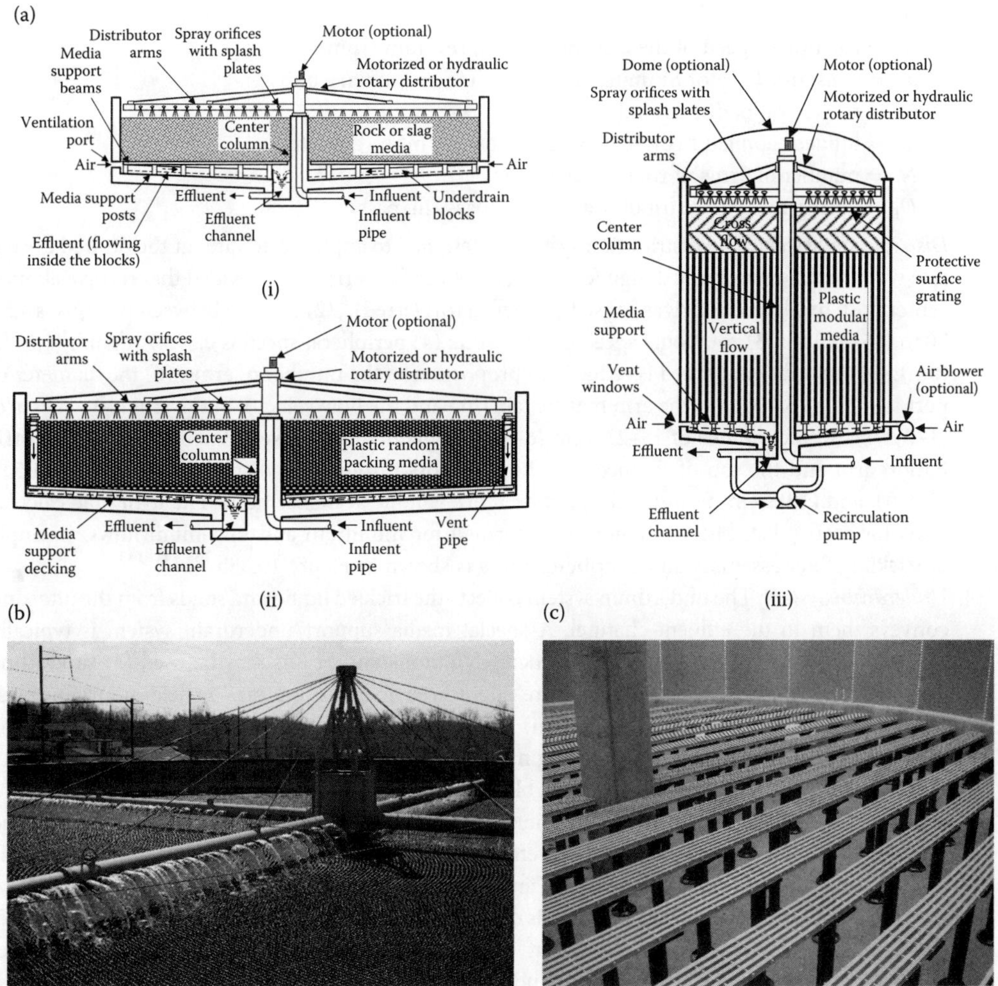

FIGURE 10.68 Trickling filter details: (a) major components: (i) rock media, (ii) plastic random packing media, and (iii) plastic modular packing media; (b) trickling filter assembly and distributor arms (Courtesy Brentwood Industries, Inc.); and (c) plastic modular packing media support/underdrain system (Courtesy Brentwood Industries, Inc).

2. *Dosing Control:* The dosing rate is the application of wastewater in inches (mm) per pass. The required dosing rate depends upon the organic loading used in the trickling filter process (Table 10.35). The normal dosing rate is a sustained low dose used on daily basis to maintain the desired organic loading rate while the flushing dosing rate is an intermittent high dose applied once a day for control of biofilm thickness and solids inventory. Typically, a normal dosing rate of 190 mm (7.5 in) per pass of each distributor arm is required for each kg $BOD_5/m^3{\cdot}d$ (62.4 lb $BOD_5/10^3\ ft^3{\cdot}d$). For a standard low-rate trickling filter, the organic loading is below 0.32 $kg/m^3{\cdot}d$ (20 $lb/10^3\ ft^3{\cdot}d$). The normal dosing rate is typically less than 60 mm (2.5 in). Uniform dosing rate is also desired for uniform biofilm growth and sloughing. The rotational speed of the distributor arm depends upon the dosing rate and the number of arms. Equation 10.65 is used to calculate the rotational speed of the distributor arm.

$$n = \frac{a(1+R)Q}{ND_R} \tag{10.65}$$

where

n = rotational speed of the distributor arm, rev/min (rpm)
a = constant, 16.7 for SI units, and 1.6 for U.S. customary units
R = recycle ratio, dimensionless
Q = influent applied hydraulic loading rate, $m^3/m^2 \cdot h$ (gpm/ft^2)
N = number of arms of rotary distributor
D_R = dosing rate per distributor arm, mm/pass (in/pass)

3. *Distribution System:* The distribution system is designed to apply wastewater at the desired dosing rate with the following basic design features: (1) two or four arms are provided that revolve about a center post; fixed, motor-driven, or self-propelled (preferred);[5] (2) the arm between two tips is 4.5–70 m (15–230 ft); (3) rotational speed is 0.5–2 rpm; (4) peripheral speed is 0.5–3.7 m/min (1.5–12 fpm); (5) the rotational speed is reduced in proportion to the number of arms; (6) the diameter of port is 95 mm (3.7 in); (7) the arm may be tapering and the flow velocity in the arm is 0.3–0.6 m/s (1–2 ft/s); (8) a clearance of 15–22.5 cm (6–9 in) is maintained between the bottom of the distribution arm and the top of the bed; (9) the head loss through the distributor arm is 0.6–1.5 m (2–5 ft); and (10) twin dosing tanks provide more volume at higher head. The hydraulic calculations for rotary distributors should be performed for minimum and maximum flows. Example of trickling filter assembly and distributor arms is shown in Figure 10.68b.[139,141–143]
4. *Underdrain System:* The underdrain system collects the trickled liquid and solids from the filter and conveys them to the effluent channel. A special media support/underdrain system is typically required underneath the filter media. Concrete columns and beams may be used for supporting heavy media such as rocks or slags while pre-engineered fiberglass gratings on field-adjustable plastic supporting piers (25–100 cm (9–42 in) tall) are typically used for plastic modular packing media.[144] For the plastic random packing media, HDPE decking modules with 15-cm (6-in) legs may be used for the underdrain.[134] Precast blocks of vitrified clay may also be laid on a sloping reinforced concrete floor to form an underdrain system. The channels of underdrain blocks are aligned to collect the flow and convey it to the effluent channel. Example of the trickling filter media support/underdrain system is shown in Figure 10.68c.[144]
5. *Air Circulation:* Airflow in trickling filters is due to natural draft induced by temperature differences between the ambient air and wastewater. Warmer wastewater than the ambient air decreases air density causing air to flow upward. Under the reverse condition the air flows downward. Air enters and exits through the underdrains and collecting channels. These passageways should be designed to flow no more than half full. Ventilation ports around the periphery of the filter base and the ventilating manholes and vent stacks with open gratings should be provided at ends of the main collection channel. Total ventilating area should be at least 0.4% of the filter surface area. The top opening of the underdrain block should be at least 15% of the filter surface area. The pressure head draft from temperature difference is given by Equation 10.66. In heavily loaded filters, if forced ventilation is provided, the minimum air flow should be 0.3 $m^3/m^2 \cdot min$ (1 $ft^3/ft^2 \cdot min$). In extremely cold weather, it is desirable to have low air flow to keep the filter from freezing.

$$P_{air} = 353\left(\frac{1}{T_m} - \frac{1}{T_h}\right)Z \quad \text{(SI units)} \tag{10.66a}$$

$$P_{air} = 7.64\left(\frac{1}{T_m} - \frac{1}{T_h}\right)Z \quad \text{(U.S. customary units)} \tag{10.66b}$$

where

P_{air} = natural air draft or pressure head in column of water, mm (in)
T_m = log-mean pore air temperature, °K (°R)
T_h = wastewater temperature which is warmer, °K (°R)
Z = filter depth, m (ft)

The value of K can be calculated from °C using Equation 10.49e while the value of °R can be calculated from °F using Equation 10.49g. T_m is further calculated from Equation 10.66c.

$$T_m = \frac{(T_c - T_h)}{\ln(T_c/T_h)} \tag{10.66c}$$

where T_c = air temperature which is colder, °K (°R)

The vertical velocity of natural draft is approximately given by Equation 10.66d.[145]

$$V_a = 0.135\,\Delta T - 0.46 \tag{10.66d}$$

where

V_a = velocity of air through the filter media by the natural draft, ft/min
The negative and positive signs of V_a indicate upward and downward air flow, respectively.
ΔT = temperature difference between the atmosphere and wastewater, °F

In tall filters, the natural draft cannot be relied upon for oxygen needs. Forced-air system is used to provide reliable oxygen supply. The required oxygen supplies for BOD removal, and BOD removal plus nitrification is given by Equations 10.67a and 10.67b, respectively.[2]

$$O_2 = (20 \text{ kg } O_2/\text{kg BOD}_5)(0.80\, e^{-9\,L_b} + 1.2\, e^{-0.17\,L_b})\, PF \tag{10.67a}$$

$$O_2 = (40 \text{ kg } O_2/\text{kg BOD}_5)(0.80 e^{-9\,L_b} + 1.2\, e^{-0.17\,L_b} + 4.6(\,N_x/\text{BOD}_5))\, PF \tag{10.67b}$$

where

O_2 = oxygen supply, kg O_2/kg BOD_5 applied
L_b = BOD_5 loading to the filter, kg $BOD_5/m^3{\cdot}d$
N_x/BOD = ratio of influent nitrogen oxidized to influent BOD_5, mg N/mg BOD_5
PF = peaking factor of maximum to average load for the loading rate, dimensionless

The pressure drop through packing material depends upon the air velocity and tower resistance. The air velocity and tower resistance are given by Equations 10.68a and 10.68b.[2,17]

$$\Delta P = N_p \frac{v_s^2}{2g} \tag{10.68a}$$

$$N_P = 10.33 \quad k_m\, k_i\, De^{(bL_w/A)} \tag{10.68b}$$

where

ΔP = total head loss, m of air
N_p = tower resistance as number of velocity head, dimensionless
v_s = superficial velocity, m/s
D = packing depth, m
L_w = liquid mass loading rate, kg/h
A = filter surface area, m^2
b = constant, $b = 1.36 \times 10^{-5}\ m^2{\cdot}h/kg$
k_m = correction factor for packing material, dimensionless. Typical values of k_m are: 2 for rock media, 1.6 for plastic random packing media, and 1.3–1.6 for plastic cross-flow media.
k_i = correction factor for inlet condition, underdrain system, and other minor losses, dimensionless. Typical value of k_i is 1.5.

Operational Problems: Trickling filters have fewer operational problems than the activated sludge plants. Common problems include (1) odors in warm weather, particularly when wastewater is septic and organic loading is high. Also, if the temperature difference between the ambient air and wastewater is small, the air circulation due to natural draft does not occur and septic condition may develop, (2) the filter fly (Psychoda fly) may be a nuisance in the vicinity of the filter. These flies breed in the filter beds. The most effective control measure is to flood the filter to kill the fly larvae, (3) the detached biomass settle rapidly in the clarifier. If these solids are not removed quickly by the scraper mechanism, they may turn anaerobic causing floating sludge, (4) trickling filters are associated with poorer effluent quality, and (5) in cold weather, freezing may be a problem.[1,2,146]

Clarifier Design: The solids in trickling filter settle rapidly. The following considerations should be given for the intermittent or final clarifier design: (1) influent to the clarifier is the plant flow plus recirculation, (2) return sludge for biomass is not needed, and (3) solids withdrawn from the clarifier are entirely waste sludge. The design criteria of the final clarifier are given in Section 10.9.

EXAMPLE 10.92: MICROBIOLOGICAL GROWTH AND MOVEMENT OF SUBSTRATE AND END PRODUCTS

Draw the cross section of an attached growth biofilm. Show the movement of sorbed organic materials, oxygen, and end products. Also, draw the concentration profile of substrate remaining in the effluent with respect to filter depth.

Solution

The organics and oxygen are sorbed through the fixed water layer and the microbial film. The inorganic end products of decomposition are released. The biofilm nearest to the surface of the media does not get enough oxygen and may turn anaerobic. The end products of anaerobic decomposition are organic compounds. The biofilm grows until it sloughs off.

The concentration of organics in the percolating water is largest on the surface and lowest near the bottom of the filter. The cross section of a microbial film and the concentration profile of the organics in the filter bed are shown in Figure 10.69.

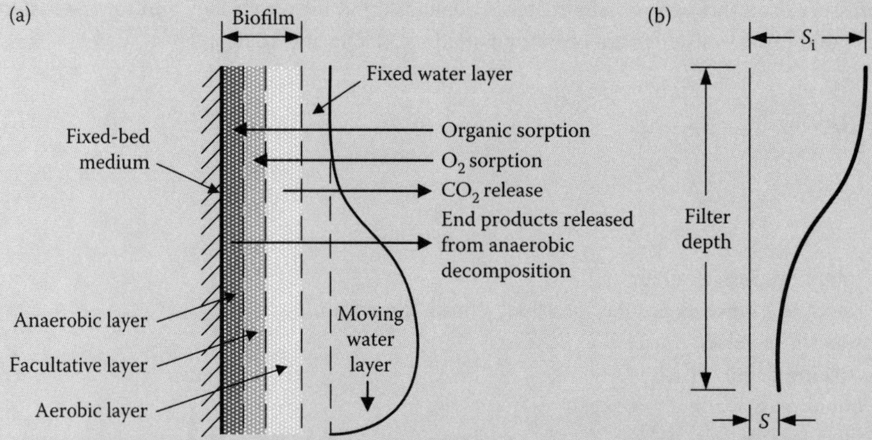

FIGURE 10.69 Fixed-film microbiological growth and substrate concentration profile: (a) cross section of fixed-film microbial growth, sorbed and released products, and (b) profile of organic concentration remaining in the effluent with respect to filter depth (Example 10.92).

EXAMPLE 10.93: BOD_5 REMOVAL EFFICIENCY OF A SINGLE-STAGE TRICKLING FILTER

A high-rate trickling filter receives primary settled wastewater that has a BOD_5 concentration (S_0) of 135 mg/L. The effluent BOD_5 concentration (S) is 20 mg/L and recirculation ratio (R) is 1.5. Calculate the influent BOD_5 concentration applied to the filter surface (S_i) and determine the overall BOD_5 removal efficiency (E) of the filter.

Solution

1. Draw the definition sketch.
   The definition sketch is drawn in Figure 10.70.

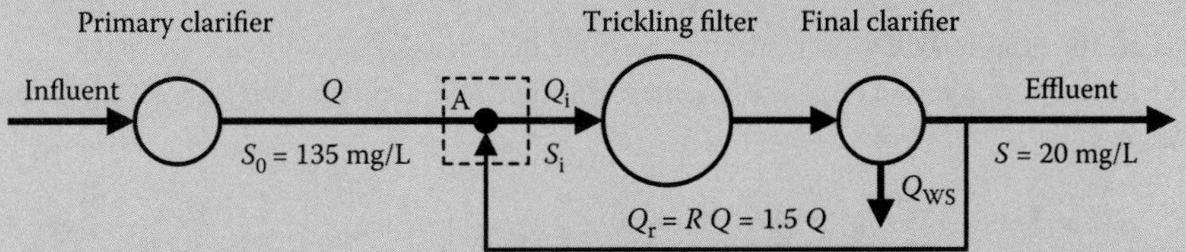

FIGURE 10.70 Definition sketch of a tricking filter process (Example 10.93).

2. Conduct a mass balance analysis at point A and calculate the BOD_5 concentration applied to the filter surface (S_i).

$$S_0 \times Q + S \times (RQ) = S_i \times (Q + RQ)$$

$$S_i = \frac{S_0 + RS}{1 + R}$$

This is the same relationship as expressed by Equation 10.62b. Substitute the data to obtain S_i.

$$S_i = \frac{135 \text{ mg/L} + 1.5 \times 20 \text{ mg/L}}{1 + 1.5} = \frac{165 \text{ mg/L}}{2.5} = 66 \text{ mg/L}$$

3. Calculate the overall BOD_5 removal efficiency of the filter (E).
   The efficiency E is calculated based on the BOD_5 concentration in the primary settled wastewater (S_0).

$$\text{BOD}_5\text{removalEfficiency, E} = \frac{S_0 - S}{S_0} \times 100\% = \frac{(135 - 20) \text{ mg/L}}{135 \text{ mg/L}} = 85\%$$

EXAMPLE 10.94: APPLICATION OF VELZ EQUATION FOR FILTER PERFORMANCE

A single-stage intermediate-rate trickling filter is designed. The depth of the filter is 2 m. The flow and BOD_5 concentration in the primary settled wastewater are 4000 m^3/d and 120 mg/L, respectively. The recirculation ratio is 1. The ratio of BOD_5 to ultimate BOD (BOD_L) is 0.68. The reaction rate constant K (base 10) is 0.23 m^{-1} at 20°C. The design hydraulic loading rate (HLR) is 6 $m^3/m^2{\cdot}d$ and operating temperature is 25°C. Calculate (a) filter effluent BOD_5 concentration, (b) the filter area, and (c) organic loading to the filter.

Solution

1. Calculate the BOD_L concentration (L_0) of primary settled wastewater.

$$L_0 = \frac{S_0}{BOD_5/BOD_L} = \frac{120 \text{ mg/L}}{0.68} = 176 \text{ mg/L}$$

2. Calculate the reaction rate constant at 25°C (K_{25}) from Equation 10.17a at $\theta_T = 1.047$.

$$K_{25} = K_{20}\theta_T^{T-20} = (0.23 \text{ m}^{-1}) \times 1.047^{(25-20)} = 0.23 \text{ m}^{-1} \times 1.26 = 0.290 \text{ m}^{-1}$$

3. Calculate the influent BOD_L concentration reaching the filler (L_i).
   A trial and error solution is required. Assume effluent BOD_L concentration $L_D = 30$ mg/L and apply Equation 10.62b.

$$L_i = \frac{L_0 + RL_D}{(1+R)} = \frac{176 \text{ mg/L} + 1 \times 30 \text{ mg/L}}{1+1} = 103 \text{ mg/L}$$

4. Calculate L_D from the Velz equation (Equation 10.60).

$$\frac{L_D}{L_i} = 10^{-K_{25}D} = 10^{-(0.290 \text{ m}^{-1})\times 2 \text{ m}} = 10^{-0.58} = 0.263$$

$$L_D = 0.263 \times L_i = 0.263 \times 103 \text{ mg/L} = 27 \text{ mg/L}$$

   The assumption for $L_D = 30$ mg/L is invalid. Repeat Steps 2 to 4, and $L_D = 27$ mg/L is finally obtained.
   Calculate the BOD_5 concentration in the effluent.

$$S = (BOD_5/BOD_L)L_D = 0.68 \times 27 \text{ mg/L} = 18 \text{ mg/L}$$

5. Determine the diameter of the filter (d).

   Total flow to the filters, $Q_i = (1+R)Q = (1+1) \times 4000 \text{ m}^3/\text{d} = 8000 \text{ m}^3/\text{d}$

   Total hydraulic loading rate from the HLR $= 6 \text{ m}^3/\text{m}^2\cdot\text{d}$,

$$Q_L = (1+R)HLR = (1+1) \times 6 \text{ m}^3/\text{m}^2\cdot\text{d} = 12 \text{ m}^3/\text{m}^2\cdot\text{d}$$

   Required filter area, $A = \frac{Q_i}{Q_L} = \frac{8000 \text{ m}^3/\text{d}}{12 \text{ m}^3/\text{m}^2\cdot\text{d}} = 667 \text{ m}^2$

   Two identical filters are to be provided. Therefore, the required filter area for each filter is 334 m^2.

   Required diameter of each filter, $d' = \sqrt{\frac{4 \times 334 \text{ m}^2}{\pi}} = 20.6$ m. Provide a diameter $d = 21$ m.

   Volume of each filter, $V = \frac{\pi}{4}d^2D = \frac{\pi}{4} \times (21 \text{ m})^2 \times 2 \text{ m} = 693 \text{ m}^3$

6. Calculate the volumetric organic loading (VOL) applied to the filter.

Calculate mass BOD_5 loading applied to each filter (ΔS_i).

$$\Delta S_i = \frac{1}{2} \times (S_0 + RS)Q = \frac{1}{2} \times (120 + 1 \times 18)\ \text{g/m}^3 \times 4000\ \text{m}^3/\text{d} \times 10^{-3}\text{kg/g} = 276\ \text{kg/d}$$

$$VOL = \frac{\Delta S_i}{V} = \frac{276\ \text{kg/d}}{693\ \text{m}^3} = 0.40\ \text{kg/m}^3{\cdot}\text{d}$$

EXAMPLE 10.95: SIZING A SUPER-RATE TRICKLING FILTER USING THE ECKENFELDER EQUATION

Two super-rate trickling filters are designed to operate in parallel to treat municipal wastewater. The influent flow rate and BOD_5 concentration to the plant are 4000 m^3/d and 210 mg/L. The BOD_5 removal efficiency of primary sedimentation basin is 35%. The critical operating temperature in the winter is 16°C. The filter depth is 6.1 m and recycle ratio is 2. The pilot plant data with a plastic media at a similar filter depth gave the BOD_5 removal rate constant $k_{r,20} = 1.3\ (m^3/d)^{0.5}/m^2$ at 20°C, exponent $n = 0.5$ and effluent BOD_5 concentration = 30 mg/L. Calculate (a) filter hydraulic loading, (b) filter organic loading, and (c) filter diameter. Use the simplified filter equation proposed by Eckenfelder (Equation 16.61f).

Solution

1. Determine the BOD_5 concentration applied to the filter (S_i).

   Calculate the BOD_5 concentration in the primary settled wastewater.

   $$S_0 = (1 - 0.35) \times 210\ \text{mg/L} = 137\ \text{mg/L}$$

   $$BOD_5 \text{ removal efficiency in the filter, } E = \frac{S_0 - S}{S_0} \times 100\% = \frac{(137\text{–}30)\ \text{mg/L}}{137\ \text{mg/L}} = 78\%$$

   Note: It is within the typical range of 70–80% (see Table 10.33).

   Calculate S_i from Equation 10.62b.

   $$S_i = \frac{S_0 + RS}{(1 + R)} = \frac{137\ \text{mg/L} + 2 \times 30\ \text{mg/L}}{1 + 2} = 66\ \text{mg/L}$$

2. Calculate the reaction rate constant at 16°C ($k_{r,16}$) from Equation 10.17a at $\theta_T = 1.035$.

   $$k_{r,16} = k_{r,20}\theta_T^{T-20} = 1.3\ (\text{m}^3/\text{d})^{0.5}/\text{m}^2 \times 1.035^{(16-20)} = 1.3\ (\text{m}^3/\text{d})^{0.5}/\text{m}^2 \times 0.87 = 1.13\ (\text{m}^3/\text{d})^{0.5}/\text{m}^2$$

3. Calculate the total hydraulic loading rate (Q_L) from Equation 10.61f.

   $$\frac{S}{S_i} = e^{-k_{r,16}D/Q_L^n}$$

   $$\frac{30\ \text{mg/L}}{66\ \text{mg/L}} = e^{-1.13\ (\text{m}^3/\text{d})^{0.5}/\text{m}^2 \times 6.1\ \text{m}/Q_L^{0.5}}$$

   $$0.455 = e^{-6.89\ (\text{m}^3/\text{m}^2{\cdot}\text{d})^{0.5}/Q_L^{0.5}}$$

   $$\ln(0.455) = \frac{-6.89\ (\text{m}^3/\text{m}^2{\cdot}\text{d})^{0.5}}{Q_L^{0.5}}$$

   $$Q_L = \left(\frac{-6.89\ (\text{m}^3/\text{m}^2{\cdot}\text{d})^{0.5}}{\ln(0.455)}\right)^2 = \left(\frac{-6.89}{-0.787}\right)^2 \text{m}^3/\text{m}^2{\cdot}\text{d} = 76.6\ \text{m}^3/\text{m}^2{\cdot}\text{d}$$

   Note: It is within the typical hydraulic loading range of 15–80 $m^3/m^2{\cdot}d$ (see Table 10.33).

4. Calculate the required filter area.

   Total flow to the filters, $Q_i = (1 + R)Q = (1 + 2) \times 4000\ m^3/d = 12{,}000\ m^3/d$

   Required filter area, $A = \dfrac{Q_i}{Q_L} = \dfrac{12{,}000\ m^3/d}{76.6\ m^3/m^2{\cdot}d} = 157\ m^2$

5. Determine the diameter of the filter (d).
   There are two trickling filters. The required filter area for each filter is 79 m^2.

   Required diameter of each filter, $d' = \sqrt{\dfrac{4 \times 79\ m^2}{\pi}} = 10$ m. Provide a diameter $d = 10$ m.

6. Calculate the volumetric organic loading (VOL) on the filter.
   Mass BOD_5 loading applied to both filter (ΔS_i).

   $\Delta S_i = (S_0 + RS)Q = (137 + 2 \times 30)\ g/m^3 \times 4000\ m^3/d \times 10^{-3}\ kg/g = 788\ kg/d$

   Total volume of both filters, $V = 2 \times \dfrac{\pi}{4} d^2 D = 2 \times \dfrac{\pi}{4} \times (10\ m)^2 \times 6.1\ m = 958\ m^3$

   $VOL = \dfrac{\Delta S_i}{V} = \dfrac{788\ kg/d}{958\ m^3} = 0.82\ kg/m^3{\cdot}d$

   Note: It is within the typical VOL loading range of 0.8–4.8 $kg/m^3{\cdot}d$ (see Table 10.33).

EXAMPLE 10.96: DERIVATION OF ECKENFELDER EQUATION

Derive the Eckenfelder equation (Equation 10.62a) for a trickling filter performance.

Solution

Apply the second-order kinetic equation to obtain Equation 10.69a.

$$\frac{1}{\overline{X}} = \frac{dS}{dt} = kS^2 \qquad (10.69a)$$

Integrate this equation to obtain Equation 10.69b.

$$\int_{S_i}^{S} \frac{1}{S^2}\, dS = \int_{0}^{t} \overline{X}\, k\, dt$$

$$\frac{S}{S_i} = \frac{1}{1 + S_i k \overline{X} t} \qquad (10.69b)$$

Assume that the mean contact time t is expressed by Equation 10.69c and c is a constant.

$$t = c\left(\frac{D^{0.67}}{Q_L^{0.5}}\right) \qquad (10.69c)$$

Substitute t in Equation 10.69c to yield Equation 10.69d.

$$\frac{S}{S_i} = \frac{1}{1 + S_i k \overline{X} c (D^{0.67}/Q_L^{0.5})} \tag{10.69d}$$

Assume that both S_i and $\overline{X}$ are maintained at constant values, combine constants S_i, k, $\overline{X}$, and c into one overall constant C to obtain Equation 10.69e. The resulting expression is the Eckenfelder equation (Equation 10.62a).

$$\frac{S}{S_i} = \frac{1}{1 + C(D^{0.67}/Q_L^{0.5})} \tag{10.69e}$$

EXAMPLE 10.97: APPLICATION OF THE ECKENFELDER EQUATION IN FILTER DESIGN

Two low-rate trickling filters in parallel are designed for a city. The average wastewater flow rate and average BOD_5 concentration of primary settled wastewater are 1.0 MGD (3780 m^3/d) and 130 mg/L, respectively. The average effluent BOD_5 concentration is 20 mg/L and filter depth is 5 ft (1.52 m). Since the recycle is intermittent for a typical low-rate trickling filter process, it is safe to assume that $S_i \approx S_0$ and $Q_i \approx Q$ at $R \approx 0$. Calculate the hydraulic loading rate (HLR), diameter of the filters and volumetric organic loading (VOL) in U.S. customary units and SI units.

Solution

1. Calculate the hydraulic loading rate (*HLR*) in U.S. customary and SI units.

   Rearrange Equation 10.62a to yield the following expression, $Q_L = \left(\frac{CD^{0.67}S}{S_0 - S}\right)^2$

   Calculate Q_L using $C = 2.5$ in U.S. customary units.

$$Q_L = \left(\frac{2.5 \times (5\text{ ft})^{0.67} \times 20\text{ mg/L}}{(130 - 20)\text{ mg/L}}\right)^2 = 1.8\text{ Mgal/ac·d} \quad \text{or} \quad Q_L = 41\text{ gpd/ft}^2$$

$$HLR = Q_L = 41\text{ gpd/ft}^2 \text{ (since } R \approx 0)$$

   Q_L using $C = 5.4$ in SI units, $Q_L = \left(\frac{5.4 \times (1.52\text{ m})^{0.67} \times 20\text{ mg/L}}{(130 - 20)\text{ mg/L}}\right)^2 = 1.7\text{ m}^3/\text{m}^2\cdot\text{d}$

$$HLR = Q_L = 1.7\text{ m}^3/\text{m}^2\cdot\text{d}$$

2. Calculate the required area of the two filters (A).

   Calculate the area of both filters (A) in U.S. customary units.

$$A = \frac{Q}{Q_L} = \frac{1\text{ MGD}}{1.8\text{ Mgal/ac·d}} = 0.56\text{ acre} \quad \text{or} \quad A = 0.56\text{ acre} \times 43{,}560\text{ ft}^2/\text{acre} = 24{,}400\text{ ft}^2$$

   A in SI units, $A = 24{,}400\text{ ft}^2 \times 0.0929\text{ m}^2/\text{ft}^2 = 2270\text{ m}^2$

3. Calculate the diameter of each filter (d).

$$\text{Area of each filter, } A' = \frac{A}{2} = \frac{24{,}400\ \text{ft}^2}{2} = 12{,}200\ \text{ft}^2$$

$$\text{Required filter diameter, } d' = \sqrt{\frac{4 \times 12{,}200\ \text{ft}^2}{\pi}} = 125\ \text{ft}$$

Provide a diameter $d = 125$ ft (38.1 m).

4. Calculate the volumetric organic loading (VOL).

Calculate the mass BOD_5 loading reaching the filters (ΔS_0) in U.S. customary units.

$$\Delta S_0 = Q \times S_0 = 1\ \text{MG} \times 130\ \text{mg/L} \times 8.34\ \text{lb/(mg/L·MG)} = 1084\ \text{lb/d}$$

$$\text{Total volume of both filters, } V = 2 \times \frac{\pi}{4} d^2 D = 2 \times \frac{\pi}{4} \times (125\ \text{ft})^2 \times 5\ \text{ft} = 123{,}000\ \text{ft}^3$$

$$VOL\ \text{in U.S. customary units, } VOL = \frac{\Delta S_0}{V} = \frac{1084\ \text{lb/d}}{123{,}000\ \text{ft}^3} = 8.8\ \text{lb}/10^3\ \text{ft}^3\text{·d or } VOL = 385\ \text{lb/acre-ft·d}$$

$$VOL\ \text{in SI units, } VOL = 8.8\ \text{lb}/10^3\ \text{ft}^3\text{·d} \times \frac{0.016\ \text{kg/m}^3\text{·d}}{\text{lb}/10^3\ \text{ft}^3\text{·d}} = 0.14\ \text{kg/m}^3\text{·d}$$

EXAMPLE 10.98: DETERMINATION OF KINETIC COEFFICIENTS

A pilot plant study was conducted using a high-rate trickling filter. Four filter depths (D), and four total hydraulic loading rates (Q_L) at each depth were investigated. The results such as filter effluent to influent ratio (S/S_i) are summarized below. Determine the rate constant k_r and exponent n in Equation 10.61f.

Filter Depth (D), m	BOD_5 Remaining Ratio (S/S_i) at Different Total Hydraulic Loading (Q_L), dimensionless			
	$Q_L = 2\ \text{m}^3/\text{m}^2\text{·d}$	$Q_L = 4\ \text{m}^3/\text{m}^2\text{·d}$	$Q_L = 6\ \text{m}^3/\text{m}^2\text{·d}$	$Q_L = 10\ \text{m}^3/\text{m}^2\text{·d}$
0.5	0.41	0.54	0.60	0.67
1.0	0.17	0.29	0.36	0.45
1.5	0.07	0.15	0.22	0.31
2.0	0.03	0.08	0.13	0.21

Solution

1. Develop a linear relationship between BOD_5 remaining ratio (S/S_i) and filter depth (D).

A nonlinear expression between S/S_i and D is expressed by Equation 10.61f.

$$\frac{S}{S_i} = e^{-k_r D/Q_L^n}$$

Take the natural log on both sides to yield Equation 10.70a.

$$\ln\left(\frac{S}{S_i}\right) = -k_r D/Q_L^n \qquad (10.70a)$$

A linear relationship between $\ln(S/S_i)$ and D is therefore obtained and expressed by Equations 10.70b and 10.70c.

$$\ln\left(\frac{S}{S_i}\right) = -mD \quad (10.70b)$$

$$m = k_r/Q_L^n \quad (10.70c)$$

A plot of $\ln(S/S_i)$ versus D on arithmetic graph (or S/S_i versus D on semilog paper) gives a linear relationship that passes through the origin and has a slope of $-m$. The variable m is in a unit of m^{-1}.

2. Prepare data and determine the slope ($-m$) of each linear relationship from the plots.
The values of $\ln(S/S_i)$ and D are prepared at each Q_L. The data are summarized below and are plotted in Figure 10.71. Using the MS Excel program, a best-fit trend line is developed for each data set. The value of m is obtained for each linear relationship as shown in the figure.

Filter Depth (D), m	$Q_L = 2\ m^3/m^2{\cdot}d$		$Q_L = 4\ m^3/m^2{\cdot}d$		$Q_L = 6\ m^3/m^2{\cdot}d$		$Q_L = 10\ m^3/m^2{\cdot}d$	
	S/S_i	$\ln(S/S_i)$	S/S_i	$\ln(S/S_i)$	S/S_i	$\ln(S/S_i)$	S/S_i	$\ln(S/S_i)$
0.5	0.41	−0.89	0.54	−0.62	0.60	−0.51	0.67	−0.40
1.0	0.17	−1.77	0.29	−1.24	0.36	−1.02	0.45	−0.80
1.5	0.07	−2.66	0.15	−1.90	0.22	−1.51	0.31	−1.17
2.0	0.03	−3.51	0.08	−2.53	0.13	−2.04	0.21	−1.56

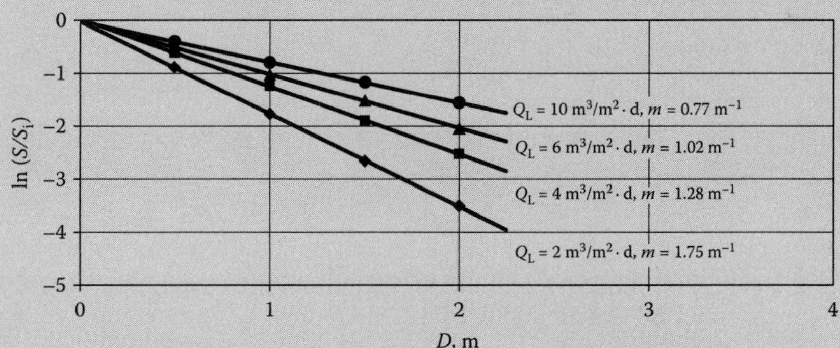

FIGURE 10.71 Linear plots of $\ln(S/S_i)$ versus D at different Q_L values (Example 10.98).

3. Develop a linear relationship to determine the constant k_r and exponent n.
In Equation 10.70b, the slope of the linear relationship is $-m$, and $m = k_r/Q_L^n$.
Take the natural log on both sides of Equation 10.70c to yield Equation 10.70d.

$$\ln(m) = \ln(k_r) - n\ln(Q_L) \quad (10.70d)$$

The slope of the line is $-n$ and the intercept on the Y axis is $\ln(k_r)$. Therefore, n and k_r can be determined from the linear relationship between $\ln(m)$ and $\ln(Q_L)$.

4. Prepare data and determine the exponent n and constant k_r.

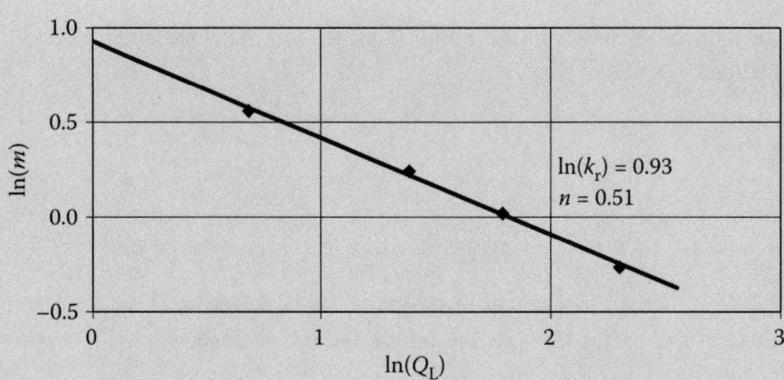

FIGURE 10.72 Determination of k_r and n from linear plot (Example 10.98).

The values of $\ln(Q_L)$ and $\ln(m)$ are prepared and summarized below. The data are then plotted in Figure 10.72.

Q_L, m^3/m^2d	$Ln(Q_L)$	m, m^{-1}	$\ln(m)$
2	0.69	1.75	0.56
4	1.39	1.28	0.25
6	1.79	1.02	0.02
10	2.30	0.77	−0.26

From the plot, the intercept $\ln(k_r) = 0.93$ and slope $-n = -0.51$ are obtained, respectively.

$$k_r = e^{0.93} = 2.5\ (m^3/m^2{\cdot}d)^{0.5}/m \quad \text{or} \quad 2.5\,(m^3/d)^{0.5}/m^2$$

$$n = 0.51 \approx 0.5.$$

Write Equation 10.61f with the experimental values of k_r and n.

$$\frac{S}{S_i} = e^{-2.5\,D/Q_L^{0.5}}$$

EXAMPLE 10.99: CONVERSION OF REACTION RATE CONSTANT FROM U.S. CUSTOMARY UNITS TO SI UNITS

The reaction rate constant k_r in U.S. customary units is 0.08 $(gpm)^{0.5}/ft^2$. Convert values of k_r in SI units (a) $(m^3/d)^{0.5}/m^2$ and (b) $(L/s)^{0.5}/m^2$.

Solution

1. Develop the conversion factor from $(gpm)^{0.5}/ft^2$ to $(m^3/d)^{0.5}/m^2$.

$$1\ (gal/min)^{0.5}/ft^2 = 1 \times \left(\frac{gal \times (3.78\,L/gal) \times (10^{-3}\ m^3)/L}{min \times (d/(60 \times 24\ min))}\right)^{0.5} \bigg/ \left(ft^2 \times \frac{0.3048^2\ m^2}{ft^2}\right)$$

$$= \frac{1 \times (3.78 \times 10^{-3} \times 60 \times 24)^{0.5}}{0.3048^2}\ (m^3/d)^{0.5}/m^2$$

$$= \frac{1 \times 2.33}{0.0929}\ (m^3/d)^{0.5}/m^2 = 25\ (m^3/d)^{0.5}/m^2$$

The conversion factors are:

$$(\text{gal/min})^{0.5}/\text{ft}^2 \times 25 = (\text{m}^3/\text{d})^{0.5}/\text{m}^2 \quad \text{or} \quad (\text{m}^3/\text{d})^{0.5}/\text{m}^2 \times 0.04 = (\text{gal/min})^{0.5}/\text{ft}^2$$

2. Convert 0.08 $(\text{gpm})^{0.5}/\text{ft}^2$ in $(\text{m}^3/\text{d})^{0.5}/\text{m}^2$.

$$0.08\ (\text{gal/min})^{0.5}/\text{ft}^2 = 0.08 \times 25\ (\text{m}^3/\text{d})^{0.5}/\text{m}^2 = 2.0\ (\text{m}^3/\text{d})^{0.5}/\text{m}^2$$

3. Develop the conversion factor from $(\text{gpm})^{0.5}/\text{ft}^2$ to $(\text{L/s})^{0.5}/\text{m}^2$.

$$\begin{aligned} 1\ (\text{gal/min})^{0.5}/\text{ft}^2 &= 1 \times \left(\frac{\text{gal} \times (3.78\ \text{L/gal})}{\text{min} \times (60\ \text{s/min})}\right)^{0.5} \Big/ \left(\text{ft}^2 \times \frac{0.3048^2\ \text{m}^2}{\text{ft}^2}\right) \\ &= \frac{1 \times (3.78/60)^{0.5}}{0.3048^2}\ (\text{L/s})^{0.5}/\text{m}^2 \\ &= \frac{1 \times 0.251}{0.0929}\ (\text{L/s})^{0.5}/\text{m}^2 \\ &= 2.7\ (\text{L/s})^{0.5}/\text{m}^2 \end{aligned}$$

The conversion factors are:

$$(\text{gal/min})^{0.5}/\text{ft}^2 \times 2.7 = (\text{L/s})^{0.5}/\text{m}^2 \quad \text{or} \quad (\text{L/s})^{0.5}/\text{m}^2 \times 0.37 = (\text{gal/min})^{0.5}/\text{ft}^2$$

4. Convert 0.08 $(\text{gpm})^{0.5}/\text{ft}^2$ in $(\text{L/s})^{0.5}/\text{m}^2$.

$$0.08\ (\text{gal/min})^{0.5}/\text{ft}^2 = 0.08 \times 2.7\ (\text{L/s})^{0.5}/\text{m}^2 = 0.22\ (\text{L/s})^{0.5}/\text{m}^2$$

EXAMPLE 10.100: FILTER PERFORMANCE EQUATION WITH CORRECTIONS FOR TEMPERATURE, FILTER DEPTH, AND INFLUENT BOD$_5$ CONCENTRATION

A pilot plant study was conducted with cross-flow plastic modular packing media to determine the kinetic coefficients k_r and n. The pilot plant data are: $k_{r,20,p} = 1.9\ (\text{m}^3/\text{d})^{0.5}/\text{m}^2$, $n = 0.5$, $\theta_T = 1.035$, $D_p = 6.1$ m, $S_p = 150$ mg/L, $R = 0.5$, and $Q_L = 40\ \text{m}^3/\text{m}^2\text{·d}$. Two trickling filters are designed that has the same packing material, recirculation ratio R, and total hydraulic loading Q_L as those of the pilot plant. The depth of the designed filter is 6.5 m and the critical operating temperature is 14°C. Calculate the effluent BOD$_5$ concentration S, and the filter area. The BOD$_5$ concentration of primarily settled wastewater = 135 mg/L, and flow $Q = 3780\ \text{m}^3/\text{d}$. Apply Equation 10.61f.

Solution

1. Correct the rate constant $k_{r,20,p}$ for temperature, filter depth, and influent BOD$_5$ concentration. Apply Equation 10.63 to correct $k_{r,20}$ for $D = 6.5$ m and $S_0 = 135$ mg/L.

$$\begin{aligned} k_{r,20,s} &= k_{r,20,p}\left(\frac{D_p}{D}\right)^{0.5}\left(\frac{S_p}{S_i}\right)^{0.5} = 1.9\ (\text{m}^3/\text{d})^{0.5}/\text{m}^2\left(\frac{6.1\ \text{m}}{6.5\ \text{m}}\right)^{0.5}\left(\frac{150\ \text{mg/L}}{135\ \text{mg/L}}\right)^{0.5} \\ &= 1.9\ (\text{m}^3/\text{d})^{0.5}/\text{m}^2 \times 0.969 \times 1.05 = 1.93\ (\text{m}^3/\text{d})^{0.5}/\text{m}^2 \end{aligned}$$

Calculate $k_{r,14,s}$ at the temperature of 14°C from Equation 5.8 at $\theta_T = 1.035$.

$$k_{r,14,s} = k_{r,20,s}\theta_T^{T-20} = 1.93\ (\text{m}^3/\text{d})^{0.5}/\text{m}^2 \times 1.035^{(14-20)} = 1.93\ (\text{m}^3/\text{d})^{0.5}/\text{m}^2 \times 0.814$$
$$= 1.57\ (\text{m}^3/\text{d})^{0.5}/\text{m}^2$$

2. Calculate S/S_i ratio for the designed filters from Equation 10.61f.

$$\frac{S}{S_i} = e^{-k_{r,16,s}D/Q_L^n} = e^{-1.57(\text{m}^3/\text{d})^{0.5}/\text{m}^2 \times 6.5\,\text{m}/(40\,\text{m}^3/\text{m}^2\cdot\text{d})^{0.5}} = e^{-1.61} = 0.20 \quad \text{or} \quad S = 0.20\ S_i$$

3. Calculate the effluent BOD_5 concentration (S).
   Apply $S = 0.20\ S_i$ to Equation 10.62b.

$$S_i = \frac{S_0 + RS}{(1+R)} = \frac{S_0 + R \times 0.20\ S_i}{(1+R)}$$

$$S_i = \frac{S_0}{(1 + R - 0.20R)} = \frac{135\ \text{mg/L}}{(1 + 0.5 - 0.20 \times 0.5)} = 96\ \text{mg/L}$$

$$S = 0.20\ S_i = 0.20 \times 96\,\text{mg/L} = 19\ \text{mg/L}$$

4. Calculate the required total filter area.

Total flow to the filters, $Q_i = (1+R)Q = (1+0.5) \times 3780\ \text{m}^3/\text{d} = 5670\ \text{m}^3/\text{d}$

Required filter area at $Q_L = 40\,\text{m}^3/\text{m}^2\cdot\text{d}$, $A = \dfrac{Q_i}{Q_L} = \dfrac{5670\,\text{m}^3/\text{d}}{40\ \text{m}^3/\text{m}^2\cdot\text{d}} = 142\ \text{m}^2$

For two filters, the required filter area for each filter is 71 m^2.

Required diameter of each filter, $d' = \sqrt{\dfrac{4 \times 71\ \text{m}^2}{\pi}} = 9.5\ \text{m}$.

Provide two filters with a diameter of 10 m for each.

EXAMPLE 10.101: APPLICATION OF NRC EQUATION

A city of 15,000 residents is considering a two-stage high-rate trickling filter system. The BOD_5 concentration of primarily settled wastewater is 135 mg/L and per capita flow is 100 gpcd. The filter diameter and depth are 50 and 5.5 ft, respectively. The recycle ratio = 1. Both filters are equal in diameter and depth, and have equal recycle ratios. Calculate the effluent BOD_5 concentration. Use the National Research Council (NCR) equation (Equation 10.64).

Solution

1. Calculate the daily average flow rate (Q).

$$Q = 100\ \text{gpcd} \times 15{,}000\ \text{people} \times 10^{-6}\,\text{Mgal/gal} = 1.5\,\text{MGD}$$

2. Calculate the BOD_5 applied to the first-stage filter (W_1).

$$W_1 = Q \times S_0 = 1.5\,\text{MGD} \times 135\,\text{mg/L} \times 8.34\,\text{lb/(mg/L·Mgal)} = 1689\,\text{lb/d}$$

3. Calculate the filter volume of the first-stage filter (V_1).

$$\text{Filter area, } A_1 = \frac{\pi}{4} \times d_1^2 = \frac{\pi}{4} \times (50\text{ ft})^2 \times \frac{\text{acre}}{43{,}560\text{ ft}^2} = 0.045\text{ acre}$$

$$\text{Filter volume, } V_1 = A_1 \times D_1 = 0.045\,\text{acre} \times 5.5\text{ ft} = 0.25\,\text{acre-ft}$$

4. Calculate the recycle factor of the first-stage filter (F_1).

F_1 is calculated from Equation 10.64c, $F_1 = \dfrac{1+R_1}{(1+0.1R_1)^2} = \dfrac{1+1}{(1+0.1\times 1)^2} = 1.65$

5. Calculate the efficiency of the first-stage filter (E_1).
E_1 is calculated from Equation 10.64a,

$$E_1 = \frac{100\%}{1 + C\sqrt{W_1/V_1F_1}} = \frac{100\%}{1 + 0.0085\sqrt{1689/(0.25 \times 1.65)}} = 65\% \quad \text{or} \quad E_1' = 0.65$$

6. Calculate the efficiency of the second-stage filter (E_2).
The E_2 is calculated by the steps similar to that used for calculating the E_1 in Steps 2 through 5. The BOD_5 applied to the second-stage filter,

$W_2 = W_1(1 - E_1') = 1689\text{ lb/d} \times (1 - 0.65) = 591\,\text{lb/d}$

$V_2 = V_1 = 0.25$ acre-ft and $F_2 = F_1 = 1.65$ since $d_2 = d_1 = 50$ ft, $D_2 = D_1 = 5.5$ ft, and $R_2 = R_1 = 1$.

E_2 is calculated from Equation 10.64b,

$$E_2 = \frac{100\%}{1 + (C/(1 - E'_1))\sqrt{W_2/(V_2F_2)}} = \frac{100\%}{1 + (0.0085/(1 - 0.65))\sqrt{591/(0.25 \times 1.65)}} = 52\%$$

7. Calculate the effluent BOD_5 concentration after the second-stage filter (S).
Rearrange Equation 10.64d to calculate S_1 and S.

$$S_1 = S_0\left(1 - \frac{\text{E}_1}{100\%}\right) = 135\,\text{mg/L} \times \left(1 - \frac{65\%}{100\%}\right) = 47\,\text{mg/L}$$

$$S = S_1\left(1 - \frac{E_2}{100\%}\right) = 47\,\text{mg/L} \times \left(1 - \frac{52\%}{100\%}\right) = 23\,\text{mg/L}$$

EXAMPLE 10.102: ROTATIONAL SPEED OF ROTARY DISTRIBUTOR AND NATURAL DRAFT

A trickling filter is 6-ft deep and is designed for an organic loading of 35 lb $BOD_5/1000\text{ ft}^3$·d. The total hydraulic loading rate (including recycle), $Q_L = 0.45\text{ gpm/ft}^2$. There are two distributor arms. The wastewater and ambient air temperatures during the critical summer period are 82°F and 64°F, respectively. Calculate (a) the rotational speed of the distributor arm, (b) the natural air draft or pressure head available for air draft, and (c) velocity of the air. Also, calculate the volume of air draft if the filter diameter $d = 50$ ft. The void volume λ_{void} of packing media is 45%, and 30% of void space (λ_{bio}) is filled with biomass.

Solution

1. Calculate the dosing rate.

   Use the typical unit dosing rate $D'_R = 190$ mm/pass per kg $BOD_5/m^3{\cdot}d$ (7.5 in/pass per 62.4 $lb/10^3\ ft^3{\cdot}d$) per distribution arm and convert it in the U.S. customary units.

$$\text{Unit dosing rate, } D'_R = \frac{7.5\ \text{in/pass}{\cdot}\text{arm}}{62.4\ \text{lb}/10^3\ \text{ft}^3{\cdot}\text{d}} = 0.12\ \text{in/pass}{\cdot}\text{arm per lb}/10^3\ \text{ft}^3{\cdot}\text{d}$$

   Calculate the design dosing rate at the organic loading 35 lb $BOD_5/10^3\ ft^3{\cdot}d$.

$$D_R = 35\ \text{lb}/10^3\ \text{ft}^3{\cdot}\text{d} \times \frac{0.12\ \text{in/pass}{\cdot}\text{arm}}{\text{lb}/10^3\ \text{ft}^3{\cdot}\text{d}} = 4.2\ \text{in/pass}{\cdot}\text{arm}$$

2. Calculate the rotational speed of the distributor arms (n).

   Apply Equation 10.65 at $(1+R)\ Q = Q_L = 0.45$ gpm/ft^2, $N = 2$ arm, $D_R = 4.2$ in/pass·arm, and $a = 1.6$.

$$n = \frac{a(1+R)Q}{ND_R} = \frac{1.6 \times 0.45}{2 \times 4.2} = 0.09\ \text{rev/min(rpm)} \quad \text{or} \quad 11.6\ \text{min per revolution}$$

3. Calculate the pressure head available for natural air draft (P_{air}).

   Convert the wastewater temperature $T_h = 82°F$ and air temperature $T_c = 64°F$ in °R from Equation 10.49g.

$$T_h = (82 + 460)°\text{R} = 542°\text{R} \quad \text{and} \quad T_c = (64 + 460)°\text{R} = 524°\text{R}$$

   Calculate the log-mean pore air temperature T_m from Equation 10.66c.

$$T_m = \frac{(T_c - T_h)}{\ln(T_c/T_h)} = \frac{(524 - 542)°\text{R}}{\ln(524°\text{R}/542°\text{R})} = \frac{-18°\text{R}}{-0.0338} = 533°\text{R}$$

   P_{air} is then calculated from Equation 10.66b at $Z = 6$ ft.

$$P_{air} = 7.64\left(\frac{1}{T_m} - \frac{1}{T_h}\right)Z = 7.64 \times \left(\frac{1}{533°\text{R}} - \frac{1}{542°\text{R}}\right) \times 6\ \text{ft} = 7.64 \times (0.00188 - 0.00185) \times 6\ \text{ft}$$

$$= 0.0014\ \text{in (0.036 mm) of water}$$

   The air draft is sufficient for the natural air flow through the pores, if all the underdrain channels are partially full and sufficient air ports are provided along the periphery of the filter base.

4. Calculate the air flow velocity due to natural draft (V_a) from Equation. 10.66d.

$$V_a = 0.135\ \Delta T - 0.46 = 0.135 \times (64 - 82)°\text{F} - 0.46 = -2.9\ \text{ft/min}$$

   Note: The negative value indicates a velocity in upward direction.

5. Calculate the volume of the air draft (Q_a).

   Calculate the area of void space available for air flow (A_a).

$$A_a = \frac{\pi}{4}d^2 \times \lambda_{void} \times (1 - \lambda_{bio}) = \frac{\pi}{4}(50\ \text{ft})^2 \times 0.45 \times (1 - 0.3) = 620\ \text{ft}^2$$

$$\text{Volume of air draft, } Q_a = V_aA_a = 2.9\ \text{ft/min} \times 620\ \text{ft}^2 = 1800\ \text{ft}^3/\text{min (cfm)}$$

EXAMPLE 10.103: FORCED VENTILATION REQUIREMENT IN A FILTER TOWER

A tower filter with cross-flow plastic modular packing media is designed to stabilize carbonaceous BOD_5 from primary settled wastewater. The influent BOD_5 concentration and flow rate are 135 mg/L and 6400 m^3/d, respectively. The critical wastewater and air temperatures are 25°C and 15°C, respectively. The BOD_5 loading (L_b) and organic loading peaking factor (PF) are 0.57 kg BOD_5/m^3·d and 1.6, respectively. The head loss correction factors for the packing material, and the inlet and underdrain system are $k_m = 1.3$ and $k_i = 1.5$, respectively. The tower diameter (d) is 18 m and depth (D or Z) is 6 m. Assume the packing media void volume $\lambda_{void} = 95\%$, and biofilm void volume (λ_{bio}) = 20%. Calculate (a) forced air-ventilation requirement and (b) pressure drop through the plastic packing media.

Solution

1. Calculate the oxygen supply requirement for BOD_5 stabilization (OSR).

   Apply Equation 10.67a to calculate the oxygen supply required for carbonaceous BOD removal only.

$$\begin{aligned} O_2 &= 20 \text{ kg } O_2/\text{kg BOD}_5 \times (0.80\, e^{-9L_b} + 1.2e^{-0.17L_b})PF \\ &= 20 \text{ kg } O_2/\text{kg BOD}_5 \times (0.80\, e^{-9\times0.57} + 1.2\, e^{-0.17\times0.57}) \times 1.6 \\ &= 20 \text{ kg } O_2/\text{kg BOD}_5 \times (0.0047 + 1.0892) \times 1.6 \\ &= 35.0 \text{ kg } O_2/\text{kg BOD}_5 \text{ applied} \end{aligned}$$

   Calculate the BOD_5 reaching the filter (ΔS_0).

$$\Delta S_0 = S_0 Q = 135 \text{ g BOD}_5/\text{m}^3 \times 6400 \text{ m}^3/\text{d} \times 10^{-3} \text{ kg/g} = 864 \text{ kg BOD}_5/\text{d}$$

$$OSR = O_2 \Delta S_0 = 35.0 \text{ kg } O_2/\text{kg BOD}_5 \text{ applied} \times 864 \text{ kg BOD}_5/\text{d} = 30{,}240 \text{ kg } O_2/\text{d}$$

2. Calculate the standard draft-air requirement ($SDAR$).

   At 20°C and 1 atm, the air contains 23.2% O_2 by weight, and density of air $\rho_{a20} = 1.204 \text{ kg/m}^3$.

$$SDAR = \frac{30{,}240 \text{ kg} O_2/\text{d}}{1.204 \text{ kg air/m}^3 \text{ air} \times 0.232 \text{ kg} O_2/\text{kg air} \times 1440 \text{ min/d}} = 75 \text{ m}^3 \text{ air/min at } 20^\circ\text{C and 1 atm}$$

3. Calculate the draft-air requirement under field conditions.

   Apply the ideal gas law to correct for temperature and pressure. Assume the air pressure is 1 atm and calculate the draft-air requirement (DAR) at 15°C.

$$DAR = \frac{(273 + 15)\text{K}}{(273 + 20)\text{K}} \times 75 \text{ sm}^3 \text{ air/min} = 0.98 \times 75 \text{ sm}^3 \text{ air/min} = 74 \text{ m}^3 \text{ air/min}$$

   Apply correction for low solubility of oxygen at higher temperature. The air supply requirement (ASR) under field conditions is increased by 1% for each degree of temperature increase in wastewater over 20°C. Calculate the ASR at the wastewater temperature of 25°C.

$$\begin{aligned} ASR &= (1 + 0.01 \times \Delta T)\, DAR = (1 + 0.01 \times (25 - 20)^\circ\text{C}) \times 74 \text{ m}^3 \text{ air/min} \\ &= 1.05 \times 74 \text{ m}^3 \text{ air/min} = 78 \text{ m}^3 \text{ air/min} \end{aligned}$$

4. Calculate the superficial velocity through tower packing media (v_s).

   The overall surface area of the filter, $A = (\pi/4)d^2 = (\pi/4) \times (18 \text{ m})^2 = 254 \text{ m}^2$

The area of void space available for air flow, $A_a = A \times \lambda_{void} \times (1-\lambda_{biod}) = 254\ m^2 \times 0.95 \times (1-0.2) = 193\ m^2$

Calculate the superficial velocity of air through the filter media.

$$v_s = \frac{ASR}{A_a} = \frac{78\ m^3\ air/min}{193\ m^2} = 0.40\ m/min\ or\ 0.0067\ m/s$$

5. Calculate the tower resistance (N_p).

   Assume the density of wastewater $\rho_w = 10^3\ kg/m^3$ and calculate the weight of liquid loading rate L_w.

$$L_w = Q\rho_w = 6400\ m^3/d \times \frac{d}{24\ h} \times \frac{10^3\ kg}{m^3} = 2.67 \times 10^5\ kg/h$$

   N_p is calculated from Equation 10.68b at $D = 6$ m, $k_m = 1.3$ and $k_i = 1.5$.

$$N_P = 10.33\ k_m k_i D e^{(bL_W/A)} = 10.33 \times 1.3 \times 1.5 \times 6 \times e^{(1.36\times10^{-5}\times2.67\times10^5/254)}$$
$$= 10.33 \times 1.3 \times 1.5 \times 6 \times e^{0.0143} = 121 \times e^{0.0143} = 121 \times 1.01 = 122$$

   Therefore, $N_p = 122$ number of air velocity head.

6. Calculate the pressure drop through packing material due to air flow (ΔP).

   Calculate ΔP from Equation 10.68a at $v_s = 0.0067$ m/s.

$$\Delta P = N_P \frac{v_s^2}{2g} = 122 \times \frac{(0.0067\ m/s)v_s^2}{2 \times 9.81\ m/s^2} = 0.00028\ m\ of\ air$$

   Convert the pressure drop to kPa using the density of air $\rho_{a15} = 1.225\ kg/m^3$.

$$\Delta P = 0.00028\ m \times 1.225\ kg/m^3\ air \times 9.81\ m/s^2 = 0.0034\ N/m^2 = 0.0034\ Pa = 3.4 \times 10^{-6}\ kPa$$

   Convert the pressure drop to m of water column using the density of water $\rho_{w15} = 999.1\ kg/m^3$.

$$\Delta P = 0.00028\ m \times \frac{1.225\ kg/m^3}{999.1\ kg/m^3} = 0.00034\ mm\ of\ water$$

7. Calculate the pressure head available for the natural air (P_{air}).

   Convert the wastewater temperature $T_h = 25°C$ and air temperature $T_c = 15°C$ in °K from Equation 10.49e.

$$T_h = (25 + 273)°K = 298°K\ and\ T_c = (15 + 273)°K = 288°K$$

   Calculate the log-mean pore air temperature T_m from Equation 10.66c.

$$T_m = \frac{(T_c - T_h)}{\ln(T_c/T_h)} = \frac{(288 - 298)°K}{\ln(288°K/298°K)} = \frac{-10°K}{-0.0341} = 293°K$$

   P_{air} is then calculated from Equation 10.66b at $Z = D = 6$ m.

$$P_{air} = 353\left(\frac{1}{T_m} - \frac{1}{T_h}\right)Z = 353 \times \left(\frac{1}{293°K} - \frac{1}{298°K}\right) \times 6\ m = 353 \times (0.00341 - 0.00336) \times 6\ m$$
$$= 0.12\ mm\ (0.005\ in)\ of\ water$$

8. Compare the air pressure drop due to forced ventilation and natural air draft available.

   The natural air draft available (0.12 mm H_2O) is significantly higher than the pressure drop due to forced ventilation (0.00034 mm H_2O). There is sufficient air draft available for the supplied air. The required draft air of 78 m^3/min must be divided and supplied at several points at the base of the tower. This is necessary for uniform distribution of air through the filter media.

Rotating Biological Contactor (RBC): Rotating biological contractors (also called bio-disks) consists of a series of circular plastic plates or disks mounted over a shaft that slowly rotates. The disks remain ~ 30–50% submerged in a tank that has contoured bottom. The disks are *spaced* to allow the biological growth to develop over the surfaces on both side of the disk. As the disks rotate, the biofilm is exposed alternately to wastewater and air. In wastewater, the biofilm sorbs organics and grow in addition to losing excess growth. In air, the oxygen is absorbed to keep the biofilm aerobic. The active biofilm thickness is typically 1.3–5 mm (0.05–0.2 in) with a solids content of about 5–10%. The detached biomass is removed in a clarifier and recycling of biomass is not employed.

Properly designed and operated RBC provides superior performance compared to other fixed-film systems. They are simple to operate and have relatively low energy requirements.

Components of RBC: An RBC assembly consists of (a) disks, (b) shaft, (c) drive unit, (d) housing, and (e) settling basin. The *disks* are circular plates made of high-density polyethylene, or PVC. The disks are mounted over a shaft passing through the center of the disks. The cross section of the shaft may be round, square, or octagonal. The RBC process flow diagram and typical components of a RBC assembly are shown in Figure 10.73a and b, respectively.

The drive units are mechanical drive or compressed air-drive. The mechanical drive unit has an electrical motor attached indirectly to the shaft through a gear box. The compressed air-driven units have a series of cups fixed along the periphery of the disks. Air cups capture the air jet released at the bottom of the basin causing rotation.

The disk housing consists of a tank and enclosure. The tank is contoured and provides a detention time of 1–1.5 h. Fiberglass reinforced plastic (FRP) covers are normally provided over each segment. The purpose of the enclosure is to (1) protect the assembly against cold and damaging weather, (2) protect against UV light, (3) control algae growth, (4) control odors, and (5) improve aesthetics. Clarifier is provided to remove solids. The solids are pumped to the sludge handling facility or to the primary clarifier. The solids generation rate is 0.6–1.1 kg TSS/kg BOD_5 removed (0.6–1.1 lb TSS/lb BOD_5 removed).[147]

In an RBC system, the oxygen is transferred in two mechanisms: (a) directly from the atmosphere to the nonsubmerged biofilm and (b) from the DO in the bulk liquid of wastewater to the submerged biofilm. The oxygen is transferred into the bulk liquid by (a) the turbulence generated by the rotation of the disks, (b) the wastewater flowing back by gravity across the media surface, and/or (c) optional diffused-air supply. Due to the potential limitation by the oxygen availability, a maximum organic loading rate is normally applied in the first stage of a RBC system.[148,149]

Multistage RBC: Multistage RBC consists of several units or stages connected in series. The stages may be created by baffling a single tank or using separate tanks. The number of stages depends on the organic loading rate required by the treatment goals. These goals may be pretreatment of industrial wastewater, BOD removal, nitrification/denitrification, and removal of refractory organics. As wastewater flows through different stages, each stage receives influent of reduced organics and substrate of different nature. Varieties of microorganisms selectively grow in subsequent stages and consume different contaminants. Step feed may be applied to different stages or tapered stages may be utilized to reduce cost. It is customary to provide a minimum of two parallel trains. This is necessary to isolate one train for routine maintenance. The typical flow patterns, staging, baffling, step feeding, and tapered arrangements are shown in Figure 10.73c. Illustrations of RBC equipment details and installation are shown in Figure 10.73d.[98,150,151]

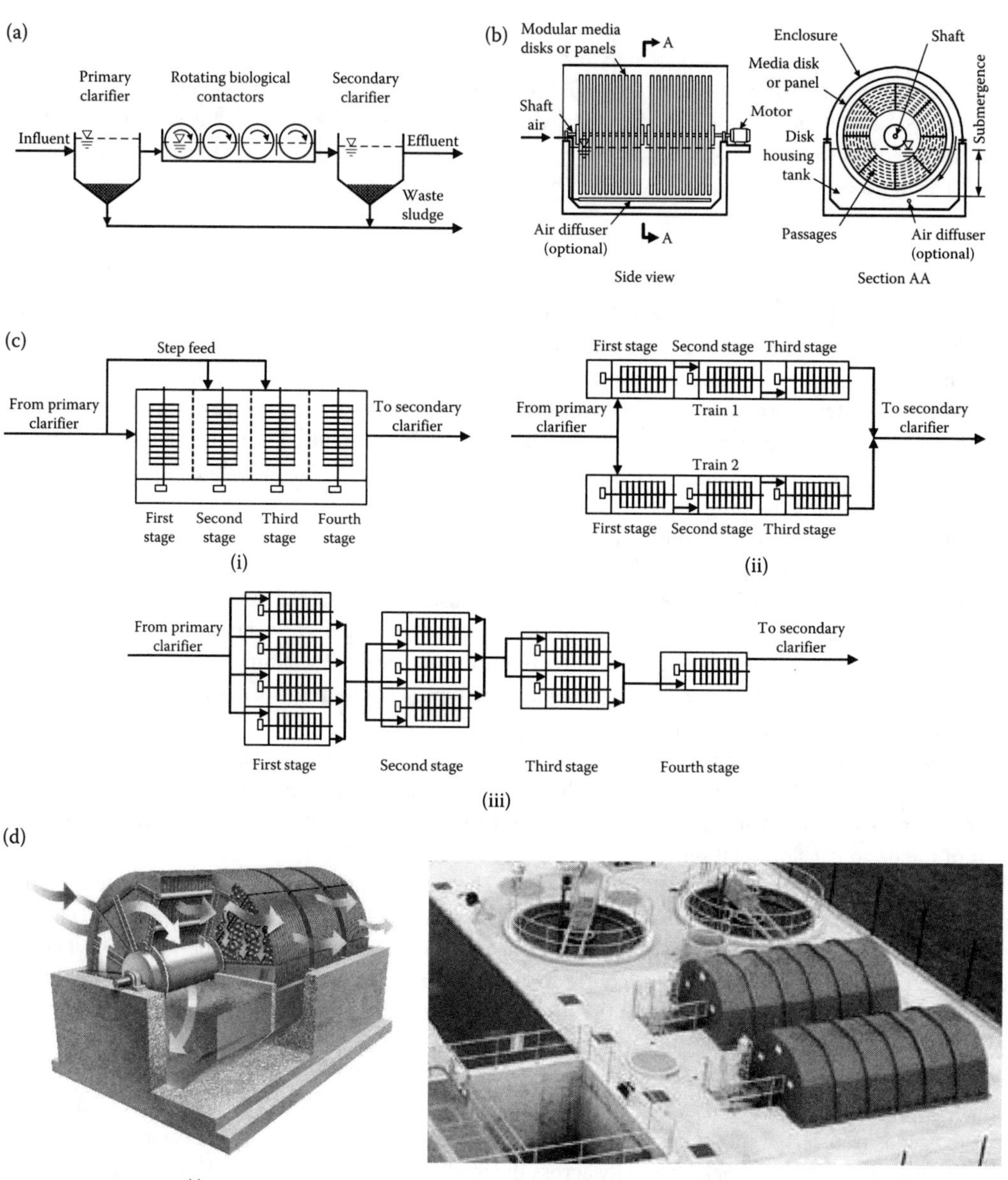

FIGURE 10.73 Details of RBC systems: (a) process flow diagram; (b) typical components; (c) arrangement alternatives: (i) flow perpendicular to shaft in four-stage RBC in a baffled single tank with step feed, (ii) flow parallel to shaft in two parallel trains with three-stage RBC in separate tanks in each train, and (iii) tapered four-stage RBC in separate tanks in series; (d) RBC systems: (i) process operation (Courtesy Walker Process Equipment), and (ii) installation with enclosure (Courtesy Ovivo).

Basic Design Considerations: The principal process design parameters of RBC are (a) hydraulic loading (HL) based on the total disk surface area, (b) organic and/or ammonia nitrogen loading on the total disk surface area, (c) HRT in the disk housing tank, and (e) effluent qualities. The typical process design parameters are summarized in Table 10.36. Other equipment parameters are major disk, shaft and tank dimensions, rotational speed, submergence, and energy consumption. The typical values of these parameters are summarized in Table 10.37.[1,17,54,98,147–149]

TABLE 10.36 Typical Process Design Parameters for RBC Systems[a]

Parameter	BOD Removal	BOD Removal and Nitrification	Separate Nitrification
Hydraulic loading (HL), $m^3/m^2{\cdot}d$ (gpd/ft^2)	0.08–0.16 (2–4)	0.03–0.08 (0.75–2)	0.04–0.10 (1–2.5)
Surface organic loading (SOL)[b], g $BOD_5/m^2{\cdot}d$ (lb $BOD_5/10^3$ ft$^2{\cdot}$d)			
Overall	8–20 (1.6–4)	5–16 (1–3.3)	1–2 (0.2–0.4)
Maximum in the first stage	24–30 (5–6)	24–30 (5–6)	–
Surface ammonia nitrogen loading (SANL), g NH_3-N/$m^2{\cdot}d$ (lb NH_3-N/10^3 ft$^2{\cdot}$d)	–	0.75–1.5 (0.15–0.3)	–
Hydraulic detention time (θ), h	0.75–1.5	1.5–4	1.2–3
No. of stages, stage	1–2	4	2–4
Effluent quality			
BOD_5 concentration, mg/L	15–30	7–15	7–15
NH_3-N concentration, mg/L	–	<2	1–2

[a] These process design parameters are applicable for a wastewater temperature above 13°C (55°F).
[b] The SOL value should be reduced to half if a soluble BOD_5 concentration is used.
Note: 1 $m^3/m^2{\cdot}d$ = 24.5 gpd/ft^2; and 1 g/$m^2{\cdot}d$ = 0.205 lb/10^3 ft$^2{\cdot}$d.
Source: Adapted in part from References 1, 17, 54, 98, 148, and 149.

TABLE 10.37 Typical Equipment Parameters of RBCs

Parameter	Typical Range
Disk dimensions	
Diameter, m (ft)	2–3.65 (6.5–12)
Thickness, cm (in)	1.3 (0.5)
Spacing center to center, cm (in)	3.4 (1.3)
Total disk area on standard shaft length, m^2 (ft^2)	9300–16,700 (100,000–180,000)
Shaft length, m (ft)	
Standard	8.2 (27)
Short	1.5–7.6 (5–25)
Disk housing tank requirement	
Volume for each stage, m^3/m^2 (gal/ft^2)	0.005 (0.12)
Speed requirement	
Rotational speed, rev/min	1–2
Peripheral speed, m/min (ft/min)	6–18 (20–60)
Submergence requirement, %	
Mechanically driven unit	35–40
Air-driven unit	70–90
Rotational energy or airflow requirement	
Power required by mechanically driven unit, kW (hp)	3.7–5.6 (5–7.5)
Airflow required for air-driven standard-density shaft, sm^3/min (scfm)	5.4 (190)

Note: 1 m^3/m^2 = 24.5 gpd/ft^2; 1 kW = 1.34 hp; and 1 sm^3/min = 35.3 scfm.
Source: Adapted in part from References 1, 17, 54, 98, 148, and 149.

Design Equations: Many empirical and kinetic equations have been developed for the design of RBC. The process variables are total disk surface area, and hydraulic and organic loadings.[147–149,152] A kinetic equation has been proposed by Eckenfelder.[50] This equation uses specific rate of substrate removal in a pseudo first-order reaction. This relationship is expressed by Equation 10.71a and is called the Eckenfelder equation. The derivation of this equation is covered in Example 10.105.

$$\frac{Q}{A}(S_0 - S) = kS \tag{10.71a}$$

where
Q = influent flow rate, m^3/d or L/d (gpd)
A = total disk surface area, m^2 (ft^2)
S_0 = BOD_5 reaching the RBC, mg/L (g/m^3)
S = effluent BOD_5, mg/L (g/m^3)
k = reaction rate constant, $m^3/m^2{\cdot}d$ or $L/m^2{\cdot}d$ (gpd/ft^2)

The term $(Q/A)(S_0 - S)$ is the rate of reaction (r) that is expressed by Equation 10.71b. The term Q/A is the *HL* that is expressed by Equation 10.71c. The term $(Q/A)S_0$ is the surface organic loading (*SOL*) that is expressed by Equation 10.71d.

$$r = \frac{Q}{A}(S_0 - S) \quad \text{or} \quad r = kS \tag{10.71b}$$

$$HL = \frac{Q}{A} \quad \text{or} \quad HL = \frac{r}{S_0 - S} \quad \text{or} \quad HL = \frac{kS}{S_0 - S} \tag{10.71c}$$

$$SOL = \frac{Q}{A}S_0 \quad \text{or} \quad SOL = \frac{\Delta S_0}{A} \quad \text{or} \quad SOL = HL\,S_0 \tag{10.71d}$$

where
r = rate of reaction based on BOD_5 removed, g $BOD_5/m^2{\cdot}d$ (lb $BOD_5/ft^2{\cdot}d$)
HL = hydraulic loading, $m^3/m^2{\cdot}d$ (gpd/ft^2)
SOL = surface organic loading (*SOL*), g $BOD_5/m^2{\cdot}d$ (lb $BOD_5/ft^2{\cdot}d$)
ΔS_0 = BOD_5 reaching the RBC system, g BOD_5/d (lb BOD_5/d)

Equation 10.72b gives a linear relationship. A plot of r versus S gives a straight line passing through the origin and the slope is k. The reaction rate constant k reaches a maximum value of r_{max} when r is high and oxygen becomes a limiting factor for biofilm growth. This relationship is shown in Figure 10.74.

Equation 10.71a is rearranged for application with multistage RBC system. This relationship is given by Equation 10.71e.

$$S = \left(\frac{1}{1 + k/(Q/A)}\right)S_0 \tag{10.71e}$$

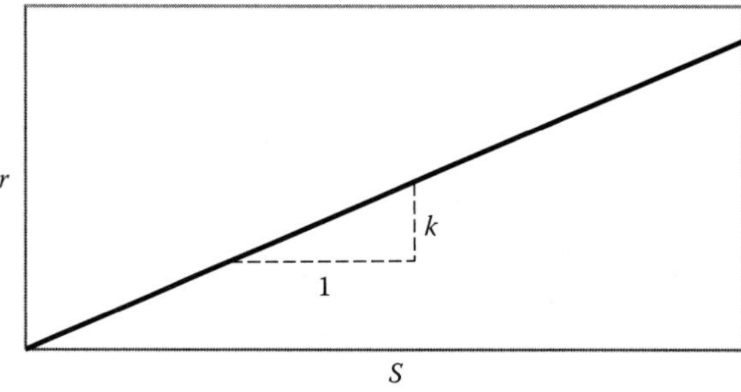

FIGURE 10.74 Linear plot of r (reaction rate) versus S (effluent BOD_5 concentration).

Since there is no recycle in an RBC system, the BOD_5 concentration in the effluent from the first, second, … , and nth stages may be expressed by $S_1, S_2, \ldots$, and S_n. The BOD_5 concentrations from different stages are given by Equations 10.71f through 10.71h. These equations are applicable when the total disk area remains the same in all stages.

$$S_1 = \left(\frac{1}{1 + k/(Q/A)}\right)S_0 \quad \text{(Effluent BOD}_5\text{ from the first stage)} \tag{10.71f}$$

$$S_2 = \left(\frac{1}{1 + k/(Q/A)}\right)S_1 \quad \text{or} \quad S_2 = \left(\frac{1}{1 + k/(Q/A)}\right)^2 S_0 \tag{10.71g}$$

(Effluent BOD_5 from the second stage)

$$S_n = \left(\frac{1}{1 + k/(Q/A)}\right)^n S_0 \quad \text{(Effluent BOD}_5\text{ from the } n\text{th stage)} \tag{10.71h}$$

A graphical solution of Equation 15.71h is also obtained. The procedure is similar to that of a series of CFSTRs given in Section 3.4.6 and Examples 3.41 and 3.42. The graphical procedure for a series of RBCs is provided in Example 10.107.

Another model based on second-order reaction is given in References 2, 148, and 149. It can be used to estimate the soluble BOD_5 concentration in the nth stage of RBC assembly, and is expressed by Equation 10.71i.

$$S_n = \frac{-1 + \sqrt{1 + (4)(0.00974)(A_s/Q)S_{n-1}}}{(2)(0.00974)(A_s/Q)} \tag{10.71i}$$

where

S_n = soluble BOD_5 concentration in stage n, mg/L. The soluble BOD_5 concentration is normally 50–75% of the total BOD_5 concentration.
A_s = disk surface area per stage, m^2
Q = influent flow rate, m^3/d

EXAMPLE 10.104: DISK LOADINGS AND DIMENSIONS OF DISK HOUSING

A three-stage RBC is designed to treat primary settled wastewater. The disk area per 7.5 m shaft length is 9000 m^2 in each stage. The diameter of disks and submergence are 3.5 m and 40%. The influent flow and BOD_5 concentration are 2000 m^3/d and 130 mg/L, respectively. The flow is perpendicular to the shaft. The clearances between the disks in adjacent stages and at the ends from the walls are 0.75 m. The bottom clearance between the disks and the floor is 0.3 m. Assume the required volume of disk housing is at least 0.005 m^3/m^2. Calculate (a) hydraulic loading, (b) surface organic loading, (c) dimensions, volume, and retention time of the disk housing.

Solution

1. Calculate the hydraulic loading (HL) from Equation 10.71c.

$$HL = \frac{Q}{A} = \frac{Q}{N_{\text{stage}}A_{\text{stage}}} = \frac{2000\ \text{m}^3/\text{d}}{3\ \text{stages} \times 9000\ \text{m}^2/\text{stage}} = 0.074\ \text{m}^3/\text{m}^2{\cdot}\text{d}$$

2. Calculate the surface organic loading (SOL).

Calculate the total BOD_5 reaching the RBC (ΔS_0).

$$\Delta S_0 = S_0 Q = 130\ \text{g}\,BOD_5/\text{m}^3 \times 2000\ \text{m}^3/\text{d} = 260{,}000\ \text{g}\,BOD_5/\text{d}$$

Calculate *SOL* over the entire three stages of RBC from Equation 10.71d using ΔS_0.

$$SOL = \frac{\Delta S_0}{A} = \frac{\Delta S_0}{N_{stage}\,A_{stage}} = \frac{260{,}000\ \text{g}\,BOD_5/\text{d}}{3\ \text{stages} \times 9000\ \text{m}^2/\text{stage}} = 9.6\ \text{g}\,BOD_5/\text{m}^2{\cdot}\text{d}$$

Alternatively, *SOL* can also be calculated from Equation 10.71d using *HL*.

$$SOL = HL\,S_0 = 0.074\ \text{m}^3/\text{m}^2{\cdot}\text{d} \times 130\ \text{g}\ BOD_5/\text{m}^3 = 9.6\ \text{g}\,BOD_5/\text{m}^2{\cdot}\text{d}$$

SOL over the first stage, $SOL_1 = \dfrac{\Delta S_0}{A_{stage}} = \dfrac{260{,}000\ \text{g}\ BOD_5/\text{d}}{9000\ \text{m}^2/\text{stage}} = 29\ \text{g}\ BOD_5/\text{m}^2{\cdot}\text{d}$

3. Calculate the dimensions of the RBC housing.

The layout of the RBC is similar to that in Figure 10.73a.

The total length (L) of the tank is equal to three disk diameters and total four clearances between the disks and at the ends.

$$L = 3\ \text{stages} \times 3.5\ \text{m (disk diameter)} + 4\ \text{clearances} \times 0.75\ \text{m/space} = 13.5\ \text{m}$$

The total width (W) of the tank is equal to the shaft length plus two clearances at the ends.

$$W = 7.5\ \text{m (shaft length)} + 2\ \text{clearances} \times 0.75\ \text{m/space} = 9\ \text{m}$$

The water depth (h) is equal to the 40% submergence of the disk plus the bottom clearance.

$$h = 0.4 \times 3.5\ \text{m (disk diameter)} + 0.3\ \text{m (bottom clearance)} = 1.7\ \text{m}$$

The dimensions of the RBC housing are: 13.5 m (L) × 9 m (W) × 1.7 m (h)

4. Calculate the total volume (V) and hydraulic detention time (HRT) of RBC housing.

Two baffle walls are provided between the RBC disk assembles. The loss of volume due to the baffles is small and ignored in volume calculation.

$$V = L \times W \times h = 13.5\ \text{m} \times 9\ \text{m} \times 1.7\ \text{m} = 207\ \text{m}^3$$

Calculate the V/A ratio and verify if the minimum volume requirement is met.

$$\frac{V}{A} = \frac{V}{N_{stage}\,A_{stage}} = \frac{207\ \text{m}^3}{3\ \text{stages} \times 9000\ \text{m}^2/\text{stage}}$$
$$= 0.008\ \text{m}^3/\text{m}^2 > 0.005\ \text{m}^3/\text{m}^2 \quad \text{(minimum requirement)}$$

$$HRT = \frac{V}{Q} = \frac{207\ \text{m}^3}{2000\ \text{m}^3/\text{d}} \times \frac{24\ \text{h}}{\text{d}} = 2.5\ \text{h}$$

EXAMPLE 10.105: DERIVATION OF ECKENFELDER EQUATION

The Eckenfelder equation (Equation 10.71a) is developed based on the specific rate of substrate removal in a pseudo first-order reaction. Develop this equation.

Solution

The specific rate of substrate removal for a pseudo first-order reaction is given by Equation 10.72a.

$$\frac{1}{X}\left(\frac{dS}{dt}\right) = k'S \tag{10.72a}$$

where

dS/dt = rate of substrate utilized in unit volume of RBC disk housing tank, mg/L·d or g/m^3·d
X = equivalent biomass concentration in the RBC disk housing tank, mg/L or g/m^3
k' = rate constant, L/mg·d or m^3/g·d
S = substrate concentration after time t, mg/L or g/m^3·d

The substrate utilization rate is given by Equation 10.72b.

$$\frac{dS}{dt} = \frac{Q}{V}(S_0 - S) \tag{10.72b}$$

where

S_0 = influent substrate concentration, mg/L or g/m^3
Q = influent flow rate, m^3/d
V = volume of RBC disk housing tank, m^3

Substitute dS/dt from Equation 10.72b in Equation 10.72a to obtain Equation 10.72c.

$$\frac{1}{X}\frac{Q}{V}(S_0 - S) = -k'S \quad \text{or} \quad \frac{1}{XV}Q(S_0 - S) = -k'S \tag{10.72c}$$

The term $X\ V$ is the cell mass weight that is proportional to the disk area A and is expressed by Equation 10.72d.

$$XV = k''A \tag{10.72d}$$

where

k'' = unit area biomass coefficient, g/m^2
A = total surface area of disks, m^2

Substitute XV from Equation 10.72d in Equation 10.72c to obtain Equation 10.72e.

$$\frac{1}{k''A}Q(S_0 - S) = k'S \quad \text{or} \quad \frac{Q}{A}(S_0 - S) = k'k''S \quad \text{or} \quad \frac{Q}{A}(S_0 - S) = kS \tag{10.72e}$$

where k = reaction rate constant ($k = k'\ k''$), m^3/m^2·d or L/m^2·d

Note: 1 m^3/m^2·d = 10^3 L/m^2·d.

EXAMPLE 10.106: EFFLUENT QUALITY, DISK AREA, AND NUMBER OF SHAFTS IN A MULTIPLE-STAGE RBC SYSTEM

A four-stage RBC system is used to treat municipal wastewater. The influent BOD_5 concentration and flow rate are 135 mg/L and 3000 m^3/d, respectively. The effluent BOD_5 concentration from the fourth stage is 20 mg/L and $k = 55$ L/m^2·d. Total disk area is 9200 m^2 per standard shaft. Calculate the following: (a) Q/A ratio on each stage, (b) effluent BOD_5 concentration from each stage, (c) number of disks per stage, (d) number of shafts per stage, and (e) draw the layout plan.

Solution

1. Develop the generalized performance equation.

   Assume that the total disk area per stage (A) is same for all stages in this design. A generalized performance equation is obtained from Equation 10.71h.

$$S_n = \left(\frac{1}{1 + k/(Q/A)}\right)^n S_0$$

2. Calculate the hydraulic loading (Q/A) on each stage.

   Substitute $S_4 = 20$ mg/L, $S_0 = 135$ mg/L, $n = 4$, and $k = 55$ L/m^2·d into the expression above and solve for Q/A.

$$20\text{ mg/L} = \left(\frac{1}{1 + (55\text{ L/m}^2\cdot\text{d})/(Q/A)}\right)^4 \times 135\text{ mg/L}$$

$$Q/A = \frac{55\text{ L/m}^2\cdot\text{d}}{(135\text{ mg/L}/20\text{ mg/L})^{1/4} - 1} = 90\text{ L/m}^2\cdot\text{d}$$

   Q/A ratio per stage = 90 L/m^2·d or 0.09 m^3/m^2·d.

3. Determine the effluent BOD_5 concentration from each stage.

   The effluent BOD_5 concentration from the first stage (S_1) is calculated from the generalized performance equation obtained in Step 1.

$$S_1 = \left(\frac{1}{1 + (55\text{ L/m}^2\cdot\text{d})/(90\text{ L/m}^2\cdot\text{d})}\right) \times 135\text{ mg/L} = 0.62 \times 135\text{ mg/L} = 84\text{ mg/L}$$

   The same equation is used to determine the effluent BOD_5 concentrations from the other three stages: $S_2 = 52$ mg/L, $S_3 = 32$ mg/L, and $S_4 = 20$ mg/L (same as given).

4. Calculate the total disk area per stage (A).

$$A = \frac{Q}{Q/A\text{ ratio}} = \frac{3000\text{ m}^3/\text{d}}{0.09\text{ m}^3/\text{m}^2\cdot\text{d}} = 33{,}300\text{ m}^2\text{ per stage}$$

5. Determine the number of shafts per stage (N_{shaft}).

$$N_{shaft} = \frac{A}{A_{shaft}} = \frac{33{,}000\text{ m}^2/\text{stage}}{9200\text{ m}^2/\text{shaft}} = 3.6\text{ shafts per stage}$$

   Provide 4 standard shafts in each stage.

6. Draw the layout plan.

   The layout plan is shown in Figure 10.75.

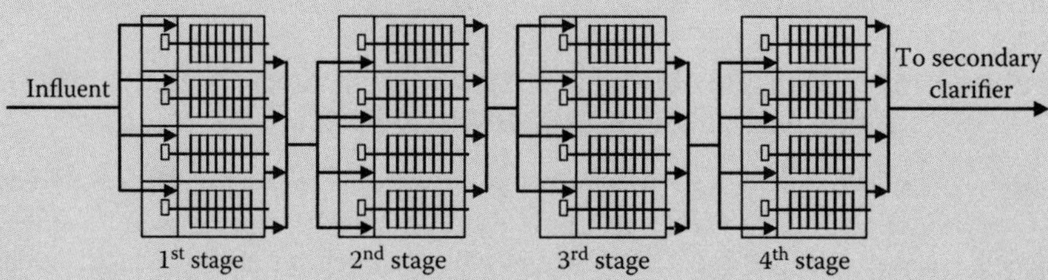

FIGURE 10.75 Conceptual layout of a four-stage RBC system (Example 10.106).

EXAMPLE 10.107: GRAPHICAL SOLUTION OF ECKENFELDER EQUATION

Graphical solution is often used for multi-stage RBC system. Explain the graphical procedure and solve Example 10.106. Compare the results.

Solution

1. State the Eckenfelder equation.

   Eckenfelder equation (Equation 10.71b) gives a linear relationship $r = k\,S$. The plot of r versus S is therefore a straight line passing through the origin.

2. Description of general procedure.

   a. Develop Line r from rate equation.

$$k = 55\,\text{L/m}^2\text{·d} = 0.055\,\text{m}^3/\text{m}^2\text{·d} \ \text{ is given in Example 10.106.}$$

   Assume $S = 100\,\text{mg/L} = 100\,\text{g/m}^3$, calculate

$$r = kS = 0.055\,\text{m}^3/\text{m}^2\text{·d} \times 100\,\text{g/m}^3 = 5.5\,\text{g/m}^2\text{·d}$$

   Plot a Point P of coordinates (100, 5.5). Draw a straight line from the origin (0, 0) and passing through Point P. This is the reaction rate line r developed from the rate equation and is shown in Figure 10.76a.

   b. Develop Line L_n from Q/A ratio.

   To determine the effluent BOD_5 concentration from the nth stage, a straight line L_n needs to be developed first. The intercept of L_n on the X axis is S_{n-1}. The intercept of L_n on the Y axis (r'_n) is calculated from Equation 10.73 for an assumed Q/A ratio.

$$r'_n = (Q/A)S_{n-1} \qquad (10.73)$$

   Graphically, the intercepting point of lines r and L_n gives S_n and r_n for the assumed Q/A ratio.

3. Determine the effluent BOD_5 concentration from each stage.

   A trial and error procedure is needed to obtain a solution. The procedure is presented below.

   a. Trial 1.

   Assume $Q/A = 0.1\,\text{m}^3/\text{m}^2\text{·d}$ to initiate the procedure. Calculate r'_1 from Equation 10.73.

$$r'_1 = (Q/A)S_0 = 0.1\,\text{m}^3/\text{m}^2\text{·d} \times 135\,\text{g/m}^3 = 13.5\,\text{g/m}^2\text{·d}$$

   Draw a straight line L_1 from (135 mg/L, 0) through (0, 13.5 g/m^2·d) in Figure 10.76a. Determine the reaction rate r_1 at the point where line L_1 intercepts the rate line r. Draw a perpendicular from r_1 to intercept X axis at $S_1 = 87$ mg/L.

   Calculate the reaction rate r_1 from Equation 10.71b.

$$r_1 = kS_1 = 0.055\,\text{m}^3/\text{m}^2\text{·d} \times 87\,\text{g/m}^3 = 4.8\,\text{g/m}^2\text{·d}$$

   The above steps used to determine S_1 are shown by the dotted lines in Figure 10.76a. The same steps are repeated to determine the effluent BOD_5 concentration from the other stages: $S_2 = 56$ mg/L, $S_3 = 36$ mg/L, and $S_4 = 23$ mg/L. These steps are also shown by the dotted lines in Figure 10.76a.

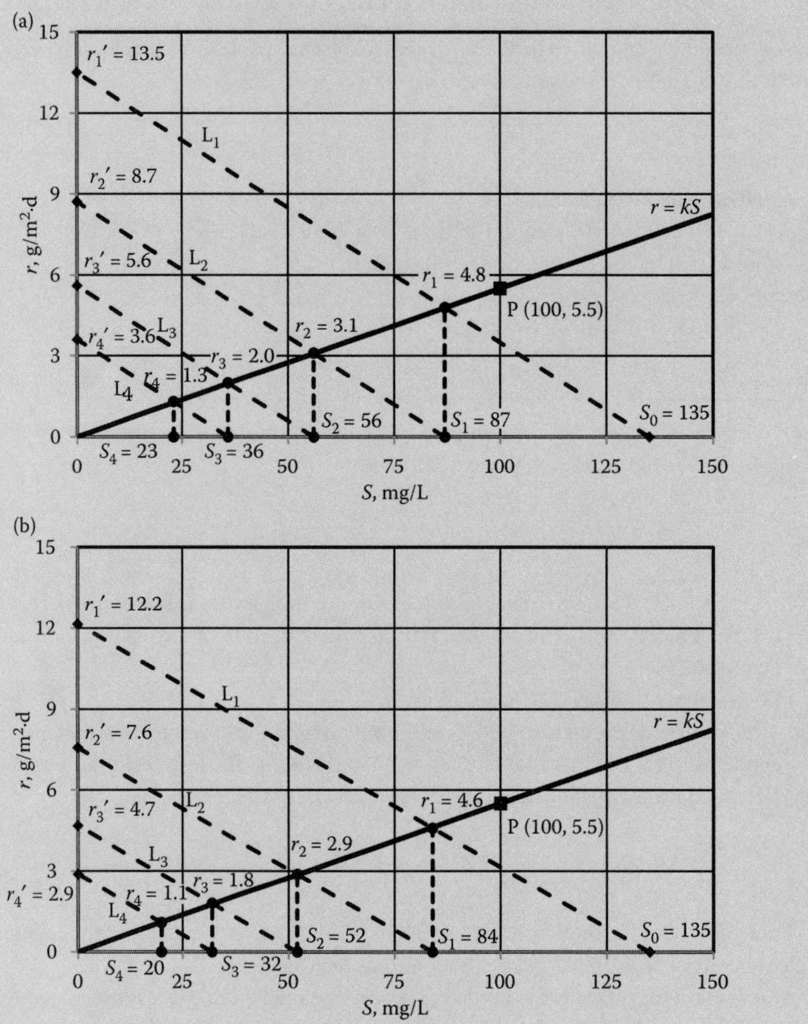

FIGURE 10.76 Graphical solutions of effluent BOD_5 concentrations for each stage in a four-stage RBC system: (a) results from Trial 1 and (b) results from Trial 2 (Example 10.107).

The final effluent BOD_5 concentration $S_4 = 23$ mg/L is higher than the desired value of 20 mg/L. It indicates the assumed Q/A ratio of 0.1 $m^3/m^2{\cdot}d$ seems high. Repeat the procedure at a lowered Q/A ratio.

b. Trial 2.

Assume $Q/A = 0.09\ m^3/m^2{\cdot}d$ to repeat the procedure. The results from Trial 2 are summarized below:

$S_1 = 84$ mg/L $r_1 = 4.6\ g/m^2{\cdot}d$

$S_2 = 52$ mg/L $r_2 = 2.9\ g/m^2{\cdot}d$

$S_3 = 32$ mg/L $r_3 = 1.8\ g/m^2{\cdot}d$

$S_4 = 20$ mg/L $r_4 = 1.1\ g/m^2{\cdot}d$

The results from Trial 2 are illustrated in Figure 10.76b.

The final effluent BOD_5 concentration $S_4 = 20$ mg/L is exactly the desired value. Therefore, and the assumed Q/A ratio of 0.09 $m^3/m^2{\cdot}d$ and the effluent BOD_5 concentrations obtained from Trial 2 are the final resolutions.

4. Compare the results.

   The values of S_1, S_2, S_3, S_4, and Q/A ratio are the same as those obtained in Example 10.106. Therefore, the graphic procedure is valid to develop solutions for a series RBC using Eckenfelder equation.

Notes: It may be noted that the data $S_4 = 20$ mg/L given in Example 10.106 is not used. If this information is used in this example then the trial and error solution will not be needed. Perform Step 1 to obtain line r from the rate equation ($r = k\,S$). Also, calculate $Q/A = 0.09\ m^3/m^2{\cdot}d$ (step 2 in Example 10.106) and then $r_1' = (Q/A)\ S_0 = 0.09\ m^3/m^2{\cdot}d \times 135\ g/m^3 = 12.2\ g/m^2{\cdot}d$. Draw line L_1, and then lines L_2, L_3, and L_4 parallel to line L_1. The values of S_1, S_2, and S_3 are obtained from the intercepts of these lines with line r (Figure 10.76b).

EXAMPLE 10.108: APPLICATION OF SECOND-ORDER MODEL IN RBC DESIGN

An RBC plant is designed to treat municipal wastewater. The influent flow rate (Q_{total}) and $sBOD_5$ concentration (S_0) of primary settled wastewater are 3800 m^3/d and 100 mg/L, respectively. The plant has standard shaft disk area (A_{shaft}) of 9290 m^2. There are three parallel trains and three stages in each train. Each stage contains a single standard shaft. Determine the effluent $sBOD_5$ concentration and organic loadings on each stage. Also, calculate overall hydraulic and organic loading. Use the second-order model (Equation 10.71i).

Solution

1. Draw the conceptual sketch.

   The conceptual sketch is given in Figure 10.77.

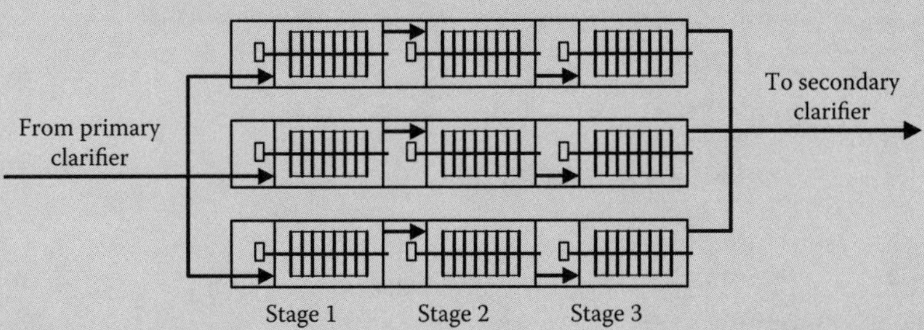

FIGURE 10.77 Conceptual sketch of a three-stage and three-shaft RBC system (Example 10.108).

2. Determine the flow rate to each process train (Q).

$$Q = \frac{Q_{total}}{N} = \frac{3800\ m^3/d}{3\ trains} = 1267\ m^3/d$$

3. Determine the effluent $sBOD_5$ concentration from each stage.

   Since there is only one shaft per stage, the disk surface area per stage is equal to the disk area per standard shaft ($A_s = A_{shaft} = 9290\ m^2$). Calculated the A_s/Q ratio.

$$\frac{A_s}{Q} = \frac{9290\ m^2}{1267\ m^3/d} = 7.33\ (m^3/m^2{\cdot}d)^{-1}$$

   Apply Equation 10.71i to determine the effluent $sBOD_5$ concentration from each stage (S_1, S_2, and S_3).

$$S_1 = \frac{-1+\sqrt{1+(4)(0.00974)(A_s/Q)S_0}}{(2)(0.00974)(A_s/Q)} = \frac{-1+\sqrt{1+4\times 0.00974\times 7.33\times 100}}{2\times 0.00974\times 7.33} = 31\ mg/L$$

$$S_2 = \frac{-1+\sqrt{1+(4)(0.00974)(A_s/Q)S_1}}{(2)(0.00974)(A_s/Q)} = \frac{-1+\sqrt{1+4\times 0.00974\times 7.33\times 31}}{2\times 0.00974\times 7.33} = 15\ mg/L$$

$$S_3 = \frac{-1+\sqrt{1+(4)(0.00974)(A_s/Q)S_2}}{(2)(0.00974)(A_s/Q)} = \frac{-1+\sqrt{1+4\times 0.00974\times 7.33\times 15}}{2\times 0.00974\times 7.33} = 9\ mg/L$$

4. Determine the surface organic loading (SOL) on each stage.

   Apply Equation 10.71d to determine SOL_1, SOL_2, and SOL_3.

$$SOL_1 = \frac{Q}{A_s}S_0 = \frac{S_0}{(A_s/Q)} = \frac{100\ g/m^3}{7.33\ (m^3/m^2{\cdot}d)^{-1}} = 14\ g/m^2{\cdot}d$$

$$SOL_2 = \frac{S_1}{(A_s/Q)} = \frac{31\ g/m^3}{7.33\ (m^3/m^2{\cdot}d)^{-1}} = 4.2\ g/m^2{\cdot}d$$

$$SOL_3 = \frac{S_2}{(A_s/Q)} = \frac{15\ g/m^3}{7.33\ (m^3/m^2{\cdot}d)^{-1}} = 2.0\ g/m^2{\cdot}d$$

5. Determine the $sBOD_5$ loading over the entire RBC system (SOL_{total}).

   The SOL_{total} is also calculated from Equation 10.71d.

$$SOL_{total} = \frac{Q_{total}}{A_{total}}S_0 = \frac{3800\ m^3/d}{3\ \text{trains}\times 3\ \text{stages/train}\times 1\ \text{shaft/stage}\times 9290\ m^2/\text{shaft}}\times 100\ g/m^3$$
$$= 4.5\ g/m^2{\cdot}d$$

6. Determine the hydraulic loading (HL_{total}) over the entire RBC system.

   The HL_{total} is calculated from Equation 10.71c.

$$HL_{total} = \frac{Q_{total}}{A_{total}} = \frac{3800\ m^3/d}{3\ \text{trains}\times 3\ \text{stages/train}\times 1\ \text{shaft/stage}\times 9290\ m^2/\text{shaft}} = 0.045\ m^3/m^2{\cdot}d$$

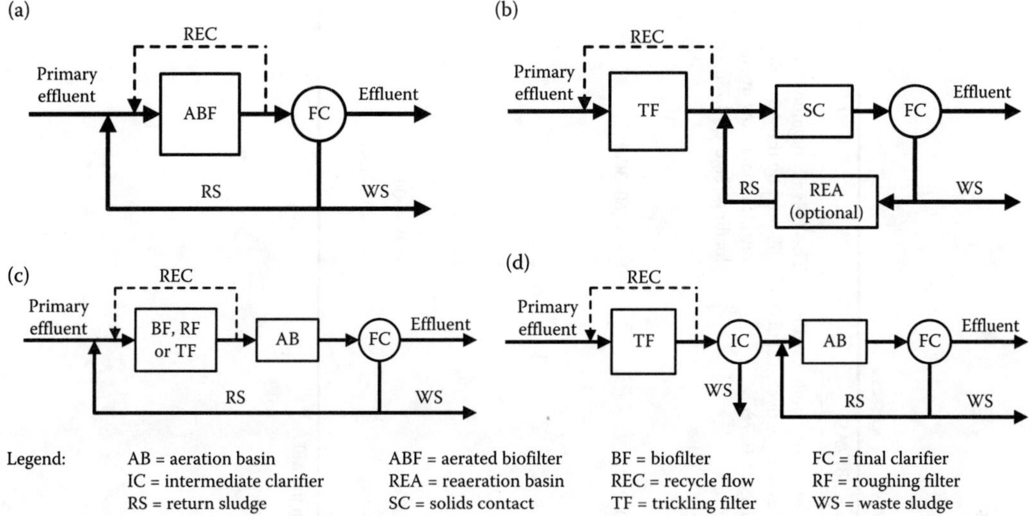

FIGURE 10.78 Combined attached and suspended growth systems: (a) activated biofilter (ABF), (b) trickling filter/solids contact (TF/SC), (c) roughing, biofilter, or trickling filter/activated sludge (BF/AS, RF/AS, TF/AS), and (d) trickling filter with intermediate clarifier/activated sludge (TF-IC/AS).

Note: (1) The ABF does not have an aeration basin but has recycle and return sludge to the filter; (2) The TF/SC may have sludge reaeration, final clarifier may need flocculating center feed well, and has recycle to the filter and return sludge to the AB, respectively; (3) The BF/AS, RF/AS, or TF/AS has recycle to the biotower and return sludge to the AB; and (4) The TF-IC/AS has recycle to filter, no return sludge from the intermediate clarifier, and return sludge from the final clarifier to the AB.

10.4.2 Combined Attached and Suspended Growth Processes

Process Description: The process uses a combination of fixed-film (trickling filter) and suspended growth (activated sludge with or without modification) reactors as dual aerobic treatment systems. The advantages of this process over suspended growth are (1) improved process stability, (2) reduced shock loading effects, (3) better sludge settling and effluent quality, and (4) lower energy requirement. Also, fixed-film reactor ahead of aeration basin helps to upgrade an overloaded aeration basin. The dual processes have many names based on the modifications. These are (a) ABF, (b) trickling filter/solid contact (TF/SC), or (c) roughing filter, biofilter, or trickling filter/activated sludge (RF/AS, BF/AS, or TF/AS). The major differences or process modifications are in return sludge either to the fixed-film reactor or aeration basin, and with or without intermediate clarifier. The schematic process diagrams are shown in Figure 10.78.[2,17,98] The process description and basic design criteria are summarized in Table 10.38.

Process Design: The design of dual system utilizes design equations of trickling filter and aeration basin. The organic loading and biodegradable and nonbiodegradable solids reaching the aeration basin is dependent upon trickling filter performance. The biomass produced and oxygen consumed in fixed-film reactor and aeration basin depend upon BOD load and θ_c (MCRT). The estimate of θ_c and particulate BOD removal rate as functions of BOD loading are provided in Table 10.39.[2] The theory and design procedure of combined fixed-film reactor and activated sludge system are provided in Example 10.109.

TABLE 10.38 Description and Typical Process Design Parameters for Dual Fixed-Film Reactor and Suspended Growth Processes

Parameter	ABF[a]	TF/SC[b]	BF/AS[c]	TF-IC/AS
Process description	High-rate fixed-film reactor with plastic or redwood media is used for treating medium- to high-strength wastes (Figure 10.78a).	Fixed-film reactor is followed by a solids contact channel (Figure 10.78b).	The processes are mainly used to upgrade existing activated sludge plants (Figure 10.78c).	The TF with organic loading is followed by the AS with nitrification at low organic loading (Figure 10.78d).
Design parameters				
Fixed-film reactor process				
Organic loading, kg $BOD_5/m^3{\cdot}d$ (lb $BOD_5/10^3\ ft^3{\cdot}d$)	0.24–1.2 (15–75)	0.4–1.6 (25–100)	1.4–4 (90–250)	1–4.8 (60–300)
Suspended growth process				
HRT (θ), h	–	0.75–2	2–8	2–8
SRT (θ_c) (d)		1–2	2–10[d]	2–10[d]
F/M ratio, kg BOD_5/kg VSS·d (lb BOD_5/lb VSS·d)	–	–	0.5–1.2	0.2–0.5
MLSS, mg/L	–	1500–3000	2500–5000	1500–4000
Final clarifier surface overflow rate (SOR), $m^3/m^2{\cdot}h$ (gpm/ft^2)	2–3 (0.8–1.2)	1–2 (0.4–0.8)	1–2 (0.4–0.8)	2–3.5 (0.8–1.4)[e]

[a] The ABF does not have an aeration basin. It has recycle and sludge return to ABF. SRT (θ_c) = 0.5–2 d, and MLSS = 1500–4000 mg/L in the ABF.

[b] The solids contact chamber may require 10–15% of the volume of an aeration basin.

[c] It may also be applicable to other combinations, including RF/AS and TF/AS. The RF may require 10–20% of the volume of high-rate TF.

[d] An SRT below 6 d may be used unless nitrification is desired.

[e] It is applicable to both the intermediate and final clarifiers in the TF-IC/AS process.

Note: 1 $kg/m^3{\cdot}d$ = 62.4 $lb/10^3\ ft^3{\cdot}d$; and 1 $m^3/m^2{\cdot}h$ = 0.409 gpm/ft^2.

Source: Adapted in part from References 2, 7, 98, and 153 through 157.

TABLE 10.39 Equivalent SRT (θ_c) and Particulate BOD_5 Removed in Fixed-Film Reactor

Organic Loading, kg $BOD_5/m^3 \cdot d$	Equivalent SRT (θ_c), d	Particulate BOD_5 Removed, %
0.25	6.8	67
0.5	3.9	63
1	2.3	55
1.5	1.6	47
2	1.3	39
3	0.9	22

Source: Adapted in part from References 2, 155, and 157.

EXAMPLE 10.109: DESIGN OF A COMBINED TF/AS SYSTEM

A dual TF/AS system is designed to treat primary-settled combined municipal and industrial wastewater. The wastewater, trickling filter, and activated sludge process data are given below:

Primary Settled Wastewater Data	Trickling Filter Data	Activated Sludge Data	Kinetic Coefficients (Applicable for Entire TF = AS Process)
Flow = 3000 m^3/d	Media = plastic packing media	SRT, $\theta_{c,AB} = 4$ d	Yield coefficient $Y = 0.5$ g VSS/g BOD_5
$TBOD_5 = 140$ mg/L	No. of filter = 1	MLVSS = 2000 mg/L	Specific endogenous decay coefficient $k_d = 0.06\ d^{-1}$
$sBOD_5/TBOD_5 = 0.7$	Depth = 8 m	BOD_5 removed = 90%	
Critical temperature = 15°C	Recirculation ratio = 1		
$BOD_5/BOD_L = 0.68$	Volumetric organic loading (VOL) = 1 kg $BOD_5/m^3 \cdot d$		
	Rate constant $k_{20} = 1.8\ (m^3/d)^{0.5}/m^2$		

Determine the following: (1) the soluble and particulate BOD_5 concentrations in the primary effluent, (2) the soluble and particulate BOD_5 concentrations in the TF effluent, (3) BOD_5 removal efficiencies of the TF, (4) diameter and hydraulic loading of the filter, (5) biomass produced and oxygen utilized in the TF, (6) biomass produced and oxygen required in the aeration basin (AB), (7) total biomass produced and oxygen requirements in the TF/AS process, (8) the AB volume and hydraulic retention time. The process flow schematic is shown in Figure 10.78c.

Solution

1. Calculate the total flow to the TF (Q_i).

$$Q_i = (1 + R)Q = (1 + 1) \times 3000\ m^3/d = 6000\ m^3/d$$

2. Calculate the soluble BOD_5 ($S_{s,0}$) and particulate BOD_5 ($S_{p,0}$) in the primary effluent.

$$S_{s,0} = (sBOD_5/TBOD_5) \times S_{t,0} = 0.7 \times 140\,mg/L = 98\,mg/L$$

$$S_{p,0} = S_{t,0} - S_{s,0} = (140 - 98)\,mg/L = 42\,mg/L$$

3. Calculate the reaction rate constant at the critical temperature 15°C ($k_{r,15}$) from Equation 10.17a at $\theta_T = 1.035$.

$$k_{r,15} = k_{r,20}\theta_T^{T-20} = 1.8\,(m^3/d)^{0.5}/m^2 \times 1.035^{(15-20)} = 1.8\,(m^3/d)^{0.5}/m^2 \times 0.842$$
$$= 1.52\,(m^3/d)^{0.5}/m^2$$

4. Calculate the total BOD_5 (S_t) in the TF effluent.
   A trial and error procedure is needed to obtain S_t and presented below.
   Assume $S_t = 20$ mg/L to initiate the trial and error procedure. Calculate S_i from Equation 10.62b.

$$S_{t,i} = \frac{S_{t,0} + RS_t}{(1+R)} = \frac{140\,mg/L + 1 \times 20\,mg/L}{1+1} = 80\,mg/L \text{ or } 80\,g/m^3$$

   Calculate the mass of total BOD_5 applied to the TF ($\Delta S_{t,i}$).

$$\Delta S_{t,i} = S_{t,i} \times Q_i = 80\,g/m^3 \times 6000\,m^3/d \times 10^{-3}\,kg/g = 480\,kg/d$$

   The TF volume required at the design $VOL = 1$ kg $BOD_5/m^3{\cdot}d$,

$$V_{TF} = \frac{\Delta S_i}{VOL} = \frac{480\,kg/d}{1\,kg/m^3{\cdot}d} = 480\,m^3$$

   The TF area required at the design filter depth $D = 8$ m, $A = \dfrac{V_{TF}}{D} = \dfrac{480\,m^3}{8\,m} = 60\,m^2$

   The hydraulic loading applied to the TF area,

$$Q_L = \frac{Q_i}{A_{TF}} = \frac{6000\,m^3/d}{60\,m^2} = 100\,m^3/m^2{\cdot}d$$

   Calculate the $S_t/S_{t,i}$ ratio from Equation 10.61f.

$$\frac{S_t}{S_{t,i}} = e^{-k_{r,16,s}D/Q_L^n} = e^{-1.52\,(m^3/d)^{0.5}/m^2 \times 8m/(100m^3/m^2{\cdot}d)^{0.5}} = 0.296 \text{ or } S_t = 0.296S_{t,i}$$

$$S_t = 0.296S_{t,i} = 0.296 \times 80\,mg/L = 23.7\,mg/L$$

   The assumption for $S_t = 20$ mg/L is invalid. Repeat the trial procedure assuming $S_t = 23.7$ mg/L. The final value $S_t = 23.9$ mg/L $\approx$ 24 mg/L is obtained in the second trial.

5. Calculate the particulate and soluble BOD_5 concentrations in the TF effluent.
   Particulate BOD_5 removal efficiency of 55% is obtained at the VOL of 1 kg $BOD_5/m^3{\cdot}d$ from Table 10.39. Calculate particulate BOD_5 concentration in the TF effluent (S_p).

$$S_p = (1 - \text{particulate } BOD_5 \text{ removal efficiency})S_{p,0} = (1 - 0.55) \times 42\,mg/L = 19\,mg/L$$

   Soluble BOD_5 concentration in the TF effluent, $S = S_t - S_p = (24 - 19)$ mg/L $= 5$ mg/L

6. Calculate the BOD_5 removal efficiencies of the TF.

   The BOD_5 removal efficiencies are calculated from Equation 10.14a.

   Total BOD_5 removal efficiency,

$$E_{\text{total}} = \frac{S_{t,0} - S_t}{S_{t,0}} \times 100\% = E = \frac{(140 - 24)\,\text{mg/L}}{140\,\text{mg/L}} \times 100\% = 83\%$$

   Soluble BOD_5 removal efficiency,

$$E_{\text{soluble}} = \frac{S_{s,0} - S}{S_{s,0}} \times 100\% = E = \frac{(98 - 5)\,\text{mg/L}}{98\,\text{mg/L}} \times 100\% = 95\%$$

   Particulate BOD_5 removal efficiency, $E_{\text{particulate}} = 55\%$ (from Table 10.39)

7. Determine the hydraulic loading (Q_L) and TF diameter (d).

   $S_t = 24$ mg/L is obtained from Step 4. Calculate $S_{t,i}$ from Equation 10.62b.

$$S_{t,i} = \frac{S_{t,0} + RS_t}{(1 + R)} = \frac{140\,\text{mg/L} + 1 \times 24\,\text{mg/L}}{1 + 1} = 82\,\text{mg/L or } 82\,\text{g/m}^3$$

$$\Delta S_{t,i} = S_{t,i} \times Q_i = 82\,\text{g/m}^3 \times 6000\,\text{m}^3/\text{d} \times 10^{-3}\,\text{kg/g} = 492\,\text{kg/d}$$

$$V_{TF} = \frac{\Delta S_i}{VOL} = \frac{492\,\text{kg/d}}{1\,\text{kg/m}^3\cdot\text{d}} = 492\,\text{m}^3$$

$$A = \frac{V_{TF}}{D} = \frac{492\,\text{m}^3}{8\,\text{m}} = 61.5\,\text{m}^2$$

$$Q_L = \frac{Q_i}{A_{TF}} = \frac{6000\,\text{m}^3/\text{d}}{61.5\,\text{m}^2} = 98\,\text{m}^3/\text{m}^2\cdot\text{d}$$

   Calculate the required filter diameter, $d' = \sqrt{\dfrac{4 \times 61.5\,\text{m}^2}{\pi}} = 8.8\,\text{m}$

   Provide the trickling filter diameter $d = 9$ m.

8. Determine the biomass produced in the TF.

   Equivalent $\theta_{c,TF}$ in the TF depends upon the VOL applied to the filter. At VOL of 1 kg $BOD_5/\text{m}^3\cdot\text{d}$, $\theta_{c,TF} = 2.3$ d is obtained from Table 10.39. Calculated the observed biomass yield ($Y_{obs,TF}$) for the TF from Equation 10.15l.

$$Y_{obs,TF} = \frac{Y}{1 + k_d\theta_{c,TF}} = \frac{0.50\,\text{mg VSS/mg BOD}_5}{1 + 0.06\,\text{d}^{-1} \times 2.3\,\text{d}} = 0.44\,\text{mg VSS/mg BOD}_5 = 0.44\,\text{g VSS/g BOD}_5$$

   Apply Equation 10.15n to calculate the biomass produced in the TF ($p_{x,TF}$).

$$\begin{aligned} p_{x,TF} &= Y_{obs,TF}(S_{t,0} - S) = 0.44\,\text{mg VSS/mg BOD}_5 \times (140 - 5)\,\text{mg BOD}_5/\text{L} \\ &= 59\,\text{mg VSS/L or } 59\,\text{g VSS/m}^3 \end{aligned}$$

$$P_{x,TF} = Qp_{x,TF} = 3000\,\text{m}^3/\text{d} \times 59\,\text{g VSS/m}^3 \times 10^{-3}\,\text{kg/g} = 177\,\text{kg VSS/d}$$

9. Determine the amount of oxygen utilized in the TF ($O_{2,TF}$) from Equation 10.14f.

$$O_{2,TF} = \frac{Q(S_{t,0} - S)}{BOD_5/BOD_L} - 1.42P_{x,TF}$$

$$= \frac{3000\,m^3/d \times (140-5)\,g\,BOD_5/m^3 \times 10^{-3}\,kg/g}{0.68\,g\,BOD_5/g\,BOD_L} - 1.42\,g\,BOD_L/g\,VSS \times 177\,kg\,VSS/d$$

$$= (596 - 251) \times kg\,BOD_L/d = 345\,kg\,BOD_L/d \text{ or } 345\,kg\,O_2/d$$

$$o_{2,TF} = \frac{P_{x,TF}}{Q} = \frac{345\,kg\,O_2/d \times 10^3\,g/kg}{3000\,m^3/d} = 115\,g\,O_2/m^3 \text{ or } 115\,mg\,O_2/L$$

10. Determine the biomass produced in the aeration basin (AB).
Calculate the BOD_5 concentration in the effluent from the AB.

$$S = (1 - BOD_5 \text{ removal efficiency})\,S_t = (1-0.9) \times 24\,mg/L = 2.4\,mg/L$$

Calculated the $Y_{obs,AS}$ due to removal of BOD_5 in the activated sludge process (AS) from Equation 10.15l.

$$Y_{obs,AS} = \frac{Y}{1 + k_d\theta_{c,AB}} = \frac{0.50\,mg\,VSS/mg\,BOD_5}{1 + 0.06\,d^{-1} \times 4\,d} = 0.40\,mg\,VSS/mg\,BOD_5 = 0.40\,g\,VSS/g\,BOD_5$$

The biomass produced in the AS ($p_{x,AS}$) is calculated from Equation 10.15n.

$$p_{x,AS} = Y_{obs,AS}(S_t - S) = 0.40\,mg\,VSS/mg\,BOD_5 \times (24 - 2.4)\,mg\,BOD_5/L$$
$$= 8.6\,mg\,VSS/L \text{ or } 8.6\,g\,VSS/m^3$$

$$P_{x,AS} = Qp_{x,AS} = 3000\,m^3/d \times 8.6\,g\,VSS/m^3 \times 10^{-3}\,kg/g = 25.8\,kg\,VSS/d$$

The biomass produced in the TF will undergo endogenous decay under aerobic condition in the AB. The biomass remaining after the endogenous decay is expressed by Equation 10.74.

$$R_{x,TF/AB} = \frac{1}{1 + k_d\theta_{c,AB}} \tag{10.74}$$

where $R_{x,TF/AB}$ = TF biomass (VSS) fraction remaining fraction after endogenous decay in the AB, g VSS remaining/g VSS produced in TF

Calculate the TF biomass (VSS) remaining fraction from Equation 10.73.

$$R_{x,TF/AB} = \frac{1}{1 + k_d\theta_{c,AB}} = \frac{1}{1 + 0.06\,d^{-1} \times 4\,d} = 0.81$$

$$p_{x,TF/AB} = R_{x,TF/AB}p_{x,TF} = 0.81 \times 59\,mg\,VSS/L = 48\,mg\,VSS/L \text{ or } 48\,g\,VSS/m^3$$

$$P_{x,TF/AB} = Qp_{x,TF/AB} = 3000\,m^3/d \times 48\,g\,VSS/m^3 \times 10^{-3}\,kg/g = 144\,kg\,VSS/d$$

Calculate the total biomass produced in the AB.

$$p_{x,AB} = p_{x,TF/AB} + p_{x,AS} = (48 + 8.6)\,mg\,VSS/L = 56.6\,mg\,VSS/L \text{ or } 56.6\,g\,VSS/m^3$$

$$P_{x,AB} = Qp_{x,AB} = 3000\,m^3/d \times 56.6\,g\,VSS/m^3 \times 10^{-3}\,kg/g = 170\,kg\,VSS/d$$

11. Determine the amount of oxygen required in the AS ($O_{2,AS}$) from Equation 10.14f.

$$O_{2,AS} = \frac{Q(S_t - S)}{BOD_5/BOD_L} - 1.42P_{x,AS}$$

$$= \frac{3000\,m^3/d \times (24 - 2.4)g\,BOD_5/m^3 \times 10^{-3}\,kg/g}{0.68\,g\,BOD_5/g\,BOD_L} - 1.42\,g\,BOD_L/g\,VSS \times 25.8\,kg\,VSS/d$$

$$= (95.3 - 36.6) \times kg\,BOD_L/d = 58.7\,kg\,BOD_L/d \text{ or } 58.7\,kg\,O_2/d$$

$$o_{2,AS} = \frac{P_{x,TF}}{Q} = \frac{58.7\,kg\,O_2/d \times 10^3\,g/kg}{3000\,m^3/d} = 19.6\,g\,O_2/m^3 \text{ or } 19.6\,mg\,O_2/L$$

12. Determine the total oxygen requirement in the TF/AS ($O_{2,TF/AS}$).

$$o_{2,TF/AS} = o_{2,TF} + o_{2,AS} = (115 + 19.6)\,mg\,O_2/L = 135\,mg\,O_2/L \text{ or } 135\,g\,O_2/m^3$$

$$O_{2,TF/AS} = Qo_{2,F/AB} = 3000\,m^3/d \times 135\,g\,O_2/m^3 \times 10^{-3}\,kg/g = 405\,kg\,O_2/d$$

13. Determine the volume (V_{AB}) and hydraulic retention time (θ) of the aeration basin.
The required V_{AB} is calculated from Equation 10.15o.

$$V_{AB} = \frac{\theta_{c,AB}P_{x,AB}}{X} = \frac{4\,d \times 150\,kg\,VSS/d \times 10^3\,g/kg}{2000\,g\,VSS/m^3} = 300\,m^3$$

$$\theta_{AB} = \frac{V_{AB}}{Q} = \frac{300\,m^3}{3000\,m^3/d} \times \frac{24\,h}{d} = 2.4\,h$$

14. Summarize the results.
   a. BOD_5 concentrations and removal efficiencies of the TF.
   The primary effluent BOD_5 concentrations, TF effluent BOD_5 concentrations and removal efficiencies are summarized in the table below:

Parameter	BOD_5 Concentration, mg/L		TF BOD_5 Removal Efficiency, %
	Primary Effluent	TF Effluent	
Soluble	42	5	95
Particulate	98	19	55
Total	140	24	83

   b. Diameter of the filter, $d = 9$ m and hydraulic loading, $Q_L = 98\,m^3/m^2{\cdot}d$.
   c. Biomass productions and oxygen requirements in the TF, AB, and TF/AS process are summarized below:

Parameter	Value		
	TF	AS	TF/AS
Biomass production			
p_x, mg VSS/L	59	8.6	56.6[a]
P_x, kg VSS/d	177	25.8	170[a]
Oxygen requirement			
o_2, mg O_2/L	115	19.6	135
O_2, kg O_2/d	345	58.7	405

[a] The net biomass production after endogenous decay of the TF biomass in the aeration basin (see the last two lines of Step 10).

   d. Aeration basin volume, $V_{AB} = 300\,m^3$ and hydraulic retention time, $\theta_{AB} = 2.4$ h.

10.4.3 Integrated Fixed-Film Media in Aeration Basin

Proprietary processes have been developed that utilize fixed-film packing media in aeration basin. These packing materials hold attached growth within a suspended growth reactor.

A brief description of the process is given in Table 10.15. This process offers the following advantages over conventional activated sludge: (1) increasing biomass in aeration basin, (2) reducing volume requirement and improving treatment capacity, (3) increasing SRT for same aeration volume, (4) eliminating concerns of bulking, (5) enhancing nitrification and denitrification in one reactor, and (6) reducing solid loading to the final clarifier since the attached growth stays in the aeration basin. The major disadvantages include: (1) increasing energy demand at elevated operating DO concentration of 4–6 mg/L, (2) use of proprietary media, (3) requiring pretreatment processes such as fine screening, degritting, and primary sedimentation, and (4) additional head loss up to 0.15 m (6 in) through the media screens at the effluent end of the aeration basin.[2] Different types of fixed-film media have been used to support the attached growth. The important physical parameters of these media are summarized in Table 10.40. The suspended media include sponges and random plastic biofilm carriers (Figure 10.79a).[38] The effluent end of the aeration basin has screens to retain the suspended media. The commercial screens are available in either horizontal or vertical installation (Figure 10.79b). The screens are continuously cleaned and media is recirculated to the influent end. The fixed media modules may utilize fabric rope- or web-type media attached to rigid frames (Figure 10.79c).[158] In early development, submerged rotating biological contactors (SRBCs) were also used to provide fixed media at a submergence of ~85% (Figure 10.79d). There are two variations of this process: (1) *integrated fixed-film activated sludge (IFAS) process* and (2) *moving bed biofilm reactor (MBBR) process.* Brief descriptions of both variations are presented below.

IFAS Process: The IFAS process is commercially available as many proprietary systems. These systems use either suspended media or fixed packing modules to support attached growth in the aeration basin. The suspended growth biomass in the IFAS process is typically returned to maintain the desired biomass within the biological reactor. Coarse- to medium-bubble aerators may be used to achieve proper mixing and maintain the desire DO concentration in the aeration basin. The general design and operating parameters are summarized in Table 10.41. The process schematic for an IFAS process with nitrification is shown in Figure 10.37.

MBBR Process: The MBBR process is similar to the IFAS system. They are also available as proprietary systems in the market. The MBBR process is different from the IFAS process because of the following major features: (1) majority substrate removal by attached growth, (2) no RAS is needed due to a low MLSS concentration (<250 mg/L), (3) use only cylindrical-shaped polyethylene perforated plates, and (4) high fill volume of 60–70% for the plastic biofilm carriers. The general design and operating parameters for achieving different process goals are summarized in Table 10.42. An example MBBR reactor process schematic is shown in Figure 10.80.

TABLE 10.40 Important Physical Parameters of Media for Attached Growth

Parameter	Value for Media Type			
	Sponge	Plastic Wheels	Plastic Chips	Fabric
Dimension, mm (in)	15 (0.6) × 15 (0.6) × 12 (0.5) depth	4–9 (0.16–0.35) × 10–25 (0.4–1) dia.	2–3 (0.08–0.12) × 45–48 (1.8–1.9) dia.	45 (1.8) dia.
Specific gravity of media, g/cm^3	0.95	0.96–0.98	0.96–1.02	–
Surface area, m^2/m^3 of media volume	850	500–800	900–1200	2.85 m^2/m

Source: Adapted in part from Reference 2.

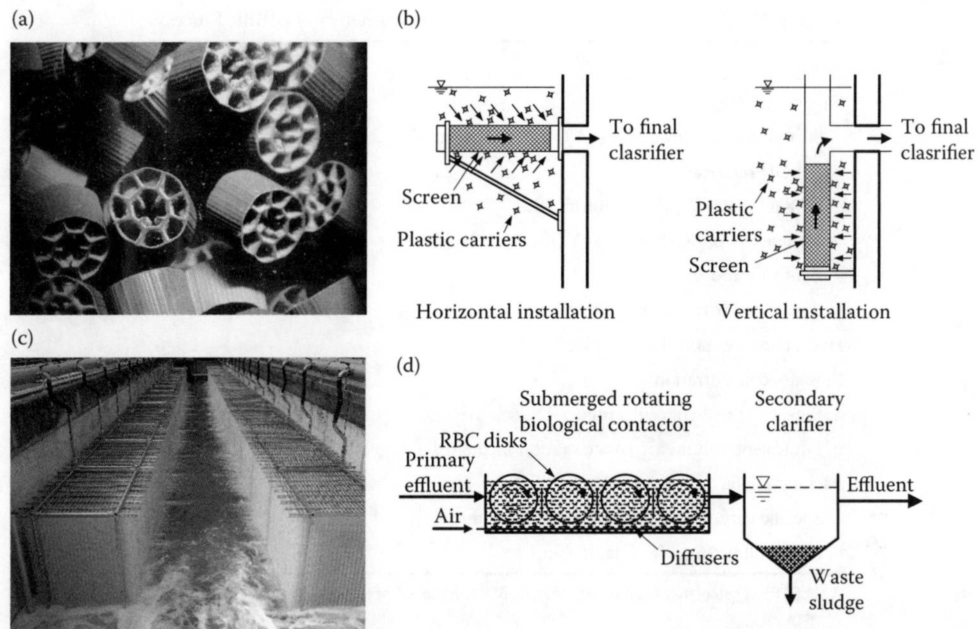

FIGURE 10.79 Fixed-film media commonly used in aeration basin: (a) suspended plastic biofilm carriers (Courtesy SUEZ), (b) effluent media screens, (c) fixed fabric media (Courtesy Ovivo), and (d) submerged RBC with diffused aeration.

Process Design Approaches: Three basic approaches are utilized for the design of IFAS and MBBR processes: (a) calculations based on the process performance parameters, (b) semiempirical equations, and (c) advanced process simulation software. Two of the most commonly used process performance parameters are (a) equivalent MLSS concentration and (b) surface area substrate removal flux (g substrate/m^2·d). Both parameters are usually approximated based on experimental data obtained from bench-, pilot-, or full-

TABLE 10.41 General Design and Operating Parameters of IFAS Process

Parameter	Value for Media Type			
	Sponge	Plastic Wheels	Plastic Chips	Fabric
Organic loading				
Surface area loading rate (SALR)[a], g BOD_5/m^2·d	10	10	10	10
Volumetric organic loading (VOL), kg BOD_5/m^3·d	1.5–3	1.5–3	1.5–3	1.5–3
Hydraulic retention time (HRT), h	3.5–5	3.5–5	3.5–5	4–6
Biomass concentration (g TSS/L)				
Activated sludge MLSS	2.5	2.5	2.5	3
Biosolids in the media volume	15–20	15–20	15–20	9–11
Equivalent MLSS concentration in the reactor	6–8	8–10	8–10	10–12
Packing media				
Specific surface area, m^2/m^3 of total tank volume	150–250	150–400	300–600	50–100
Media fill volume, % of total tank volume	15–20	20–40	20–35	45–55

[a] SALR is calculated from the amount of BOD_5 applied per unit surface area of the media per day.

Note: 1 g/m^2·d = 0.205 lb/10^3 ft^2·d; 1 kg/m^3·d = 62.4 lb/10^3 ft^3·d; and 1 m^2/m^3 = 0.305 ft^2/ft^3.

Source: Adapted in part from Reference 2.

TABLE 10.42 General Design and Operating Parameters of MBBR Process

Parameter	Value
Organic loading	
Surface area removal flux (SARF)[a], g BOD_5/m^2·d	
BOD removal	5–20
BOD removal prior to nitrification	2–4
Volumetric organic loading (VOL), kg BOD_5/m^3·d	
BOD removal	1.5–6
BOD removal prior to nitrification	1–1.2
Hydraulic retention time (HRT), h	1.5–6
Biomass concentration	
Biosolids areal concentration, g TSS/m^2	15–25
Equivalent volumetric concentration in the reactor, g TSS/m^3	4–8
Packing media	
Specific surface area, m^2/m^3 of tank volume	250–500
Media fill volume, % of tank volume	50–60

[a] SARF is calculated from the amount of BOD_5 removed per unit surface area of the media per day.

Source: Adapted in part from Reference 2.

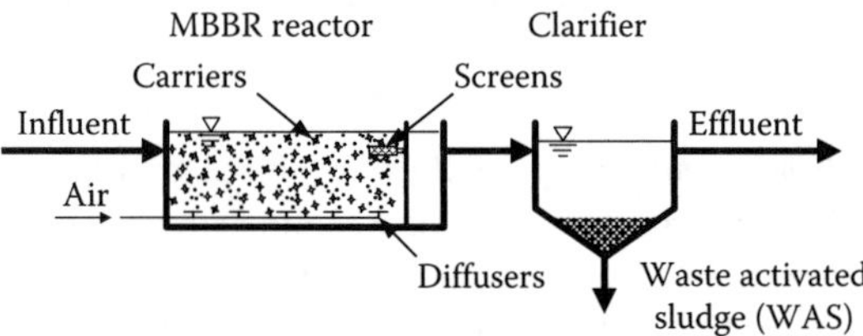

FIGURE 10.80 Schematic of an MBBR process.

scale tests. Process calculations for IFAS and MBBR processes based on the process performance parameters are provided in Examples 10.110 and 10.111. Other kinetic-based semiempirical design approaches have been developed in recent years for the fixed-film media in aeration basin with nitrification. Use of mass-balance-based process modeling software is also a valid approach for the design of both processes.[2]

EXAMPLE 10.110: DESIGN OF INTEGRATED FIXED-FILM ACTIVATED SLUDGE (IFAS) PLANT FOR BOD REMOVAL

A conventional activated sludge plant is modified into an IFAS process by adding plastic biofilm carriers into the existing aeration basins. Total volume of the basins is 2500 m^3. The current treatment capacity is 12,000 m^3/d at the design F/M ratio of 0.35 kg BOD_5/kg VSS·d and volumetric organic loading (VOL) of 0.7 kg BOD_5/m^3·d. The influent BOD_5 concentration is 140 mg/L. The media fill volume is 40% of the total basin volume. Assume that the MLSS concentration in the aerated sludge volume and solids concentration in the media fill volume are 2500 and 17,000 mg/L, respectively. The design F/M ratio is 0.25 kg BOD_5/kg VSS·d, and the overall VSS/TSS ratio is ~ 0.8 with respect to total biomass that includes both attached and suspended growths. Calculate (a) equivalent biomass, (b) treatment capacity improvement after the modification, (c) aeration period, and (d) *VOL* at improved treatment capacity after the modification.

Solution

1. Calculate the equivalent biomass after the modification.

   The equivalent MLSS concentration in the IFAS reactor ($MLSS_{eq}$) can be calculated from Equation 10.75a.

$$MLSS_{eq} = \frac{1}{V}(V_{as}\ MLSS + V_m\ MS) \text{ or } MLSS_{eq} = (1 - f_m)\ MLSS + f_m\ MS \qquad (10.75a)$$

where

$M_{MLSS,\ eq}$ = equivalent MLSS concentration in the IFAS reactors, mg TSS/L (g TSS/m^3)
$MLSS$ = MLSS concentration in the activated sludge volume, mg TSS/L (g TSS/m^3)
MS = average solids concentration due to attached growth within the media fill volume, mg TSS/L (g TSS/m^3)
V_{as} = activated sludge volume, m^3
V_m = media fill volume, m^3
V = total IFAS reactor volume ($V = V_{as} + V_m$), m^3
f_m = fraction of the media fill volume in the total IFAS reactor volume ($f_m = V_m/V$), dimensionless

Apply Equation 10.75a to calculate the equivalent MLSS concentration ($MLSS_{eq}$) in the IFAS reactors.

$$MLSS_{eq} = (1 - f_m)MLSS + f_m\ MS = (1 - 0.4) \times 2500\ \text{g TSS/m}^3 + 0.4 \times 17{,}000\ \text{g TSS/m}^3$$
$$= 8300\ \text{g TSS/m}^3$$

Calculate the total mass of equivalent MLSS ($M_{MLSS,\ eq}$) in the reactors.

$$M_{MLSS,eq} = V\ MLSS_{eq} = 2500\ \text{m}^3 \times 8300\ \text{g TSS/m}^3 \times 10^{-3}\ \text{kg/g} = 20{,}750\ \text{kg TSS}$$

Calculated the total equivalent biomass ($M_{X,eq}$) in the reactors.

$$M_{X,eq} = \text{VSS/TSS ratio} \times MLSS_{eq} = 0.8\ \text{kg VSS/kg TSS} \times 20{,}750\ \text{kg TSS} = 16{,}600\ \text{kg VSS}$$

2. Calculate the treatment capacity improvement after the modification.

   Calculate the acceptable mass BOD_5 loading from the total equivalent biomass at the design F/M ratio of 0.25 kg BOD_5/kg VSS·d.

$$\Delta S_0 = M_{X,total}\text{F/M ratio} = 16{,}600\ \text{kg VSS} \times 0.25\ \text{kg BOD}_5\text{/kg VSS·d} = 4150\ \text{kg BOD}_5\text{/d}$$

Calculate the influent flow to deliver the given mass BOD_5 loading at the influent BOD_5 concentration of 140 mg/L.

$$Q = \frac{\Delta S_0}{S_0} = \frac{4150\ \text{kg BOD}_5\text{/d}}{140\ \text{g BOD}_5/\text{m}^3 \times 10^{-3}\ \text{kg/g}} = 29{,}600\ \text{m}^3\text{/d}$$

The average treatment capacity will be 29,600 m^3/d after the modification. This is nearly 2.5 times of the current average treatment capacity of 12,000 m^3/d. Therefore, significant improvement can be achieved after the modification of the existing conventional activated sludge process to the IFAS process.

Note: Retrofitting existing air supply system may be needed to meet the elevated air supply requirement at increased BOD_5 loading to the biological reactors. Existing aeration devices may also be evaluated for proper mixing of the content after the modification.

3. Calculate the aeration period after the modification.

$$\theta = \frac{V}{Q} = \frac{2500\,\text{m}^3}{29{,}600\,\text{m}^3/\text{d}} \times \frac{24\,\text{h}}{\text{d}} = 2.0\,\text{h}$$

Note: The current aeration period is 5 h before the improvement.

4. Calculate the volumetric organic loading at improved treatment capacity after the modification.

$$VOL = \frac{\Delta S_0}{V} = \frac{4150\,\text{kg BOD}_5/\text{d}}{2500\,\text{m}^3} = 1.7\,\text{kg BOD}_5/\text{m}^3\cdot\text{d}$$

Note: A much higher VOL is achieved in the IFAS process even though the F/M ratio is lowered after the improvement.

EXAMPLE 10.111: SIZING OF MBBR REACTOR FOR BOD REMOVAL

The MBBR process is used for secondary treatment. The primary settled wastewater flow and BOD_5 concentration are 4000 m^3/d and 150 mg/L, respectively. The MBBR reactor has two compartments with same volume in series. Use the plastic biofilm carriers with a media specific surface area of 500 m^2/m^3. Assume that the media fill volume is 50% in each compartment; a 75% BOD_5 removal at a removal flux of 15 g $BOD_5/m^2\cdot d$ in the first compartment; and a 90% BOD_5 removal at a removal flux of 5 g $BOD_5/m^2\cdot d$ in the second compartment. Determine (a) volume of each compartment, (b) overall volume and hydraulic retention time, and (c) total volume and surface area of the media in the MBBR reactor. Also calculate the volumetric organic loading (VOL) in the first compartment and the entire MBBR rector.

Note: See Example 10.136 for use of MBBR reactor for tertiary nitrification.

Solution

1. Determine the volume required in the first compartment of the MBBR reactor.
   a. Determine the media area requirement.

   The media area required in a compartment (A_m) can be calculated from Equation 10.75b.

$$A_m = \frac{Q(S_0 - S)}{SARF} \text{ or } A_m = \frac{QS_0}{SARF}\frac{E}{100\%} \text{ or } A_m = \frac{\Delta S_0}{SARF}\frac{E}{100\%} \tag{10.75b}$$

where

A_m = total media area required in the compartment, m^2

Q = influent flow applied to the compartment, m^3/d

S_0 = BOD_5 concentration in the influent to the compartment, mg BOD_5/L or g BOD_5/m^3

S = BOD_5 concentration in the effluent from the compartment, mg BOD_5/L or g BOD_5/m^3

$SARF$ = surface area removal flux, g $BOD_5/m^2\cdot d$

E = BOD_5 removal efficiency achieved in the compartment ($E = ((S_0 - S)/S_0)$ 100%), percent

Apply Equation 10.75b to calculate the media area required in the first compartment.

$$A_{m,1} = \frac{QS_{0,1}}{SARF_1} \times \frac{E_{BOD,1}}{100\%} = \frac{4000\,\text{m}^3/\text{d} \times 150\,\text{g BOD}_5/\text{m}^3}{15\,\text{g BOD}_5/\text{m}^2\cdot\text{d}} \times \frac{75\%}{100\%} = 30{,}000\,\text{m}^2$$

b. Determine the media volume requirement.

The media volume required in a compartment (V_m) can be calculated from Equation 10.75c.

$$V_m = \frac{A_m}{SA_m} \tag{10.75c}$$

where

V_m = total media volume required in the compartment, m^3

SA_m = specific surface area of the plastic biofilm carrier media, m^2/m^3

Apply Equation 10.75c to calculate the media volume required in the first compartment.

$$V_{m,1} = \frac{A_{m,1}}{SA_m} = \frac{30{,}000\ m^2}{500\ m^2/m^3} = 60\ m^3$$

c. Determine the required compartment volume.

The required compartment volume (V_c) can be calculated from Equation 10.75d.

$$V_c = \frac{V_m}{f_m} \tag{10.75d}$$

where V_c = total volume required in the compartment, m^3

Apply Equation 10.75d to calculate the total volume required in the first compartment.

$$V_{c,1} = \frac{V_{m,1}}{f_{m,1}} = \frac{60\ m^3}{0.5} = 120\ m^3$$

Provide a volume $V_{c,1} = 125\ m^3$ in the first compartment.

2. Determine the volume required in the second compartment of the MBBR reactor.

Calculate the BOD_5 concentration in the influent applied to the second compartment.

$$S_{0,2} = S_{0,1}\left(\frac{1 - E_{BOD,1}}{100\%}\right) = 150\ g\ BOD_5\ m^3 \times \left(1 - \frac{75\%}{100\%}\right) = 37.5\ g\ BOD_5 m^3$$

Follow the same procedure presented in Step 1 to calculate the total volume required in the second compartment.

$$A_{m,2} = \frac{QS_{0,2}}{SARF_2}\frac{E_{BOD,2}}{100\%} = \frac{4000\ m^3/d \times 37.5\ g\ BOD_5/m^3}{5\ g\ BOD_5/m^2{\cdot}d}\frac{90\%}{100\%} = 27{,}000\ m^2$$

$$V_{m,2} = \frac{A_{m,2}}{SA_m} = \frac{27{,}000\ m^2}{500\ m^2/m^3} = 54\ m^3$$

$$V_{c,2} = \frac{V_{m,2}}{f_{m,2}} = \frac{54\ m^3}{0.5} = 108\ m^3$$

Use the same volume $V_{c,2} = 125\ m^3$ in the second compartment.

3. Calculate the total volume of the MBBR reactor.

Calculate the total volume of two 125-m^3 compartments in series.

$$V = V_{c,1} + V_{c,2} = (125 + 125)\ m^3 = 250\ m^3$$

4. Calculate the overall hydraulic retention time (θ) in the MBBR reactor.

$$\theta = \frac{V}{Q} = \frac{250\,\text{m}^3}{4000\,\text{m}^3/\text{d}} \times \frac{24\,\text{h}}{\text{d}} = 1.5\,\text{h}$$

   The HRT is 0.75 h in each compartment.

5. Calculate the total media volume and surface area in the MBBR reactor.

   Calculate the total media volume.

$$V_{\text{m}} = f_{\text{m}} V = 0.5 \times 250\,\text{m}^3 = 125\,\text{m}^3$$

   Therefore, the media volume is 62.5 m^3 in each compartment.

   Calculate the total media volume in each compartment.

$$A_{\text{m}} = SA_{\text{m}} V_{\text{m}} = 500\,\text{m}^3/\text{m}^2 \times 125\,\text{m}^3 = 62{,}500\,\text{m}^2$$

   The media surface area is 31,250 m^2 in each compartment.

6. Calculate the overall volumetric organic loading (VOL).

   Calculate the mass BOD_5 loading applied to the MBBR reactor.

$$\Delta S_0 = QS_0 = 4000\,\text{m}^3/\text{d} \times 150\,\text{g BOD}_5/\text{m}^3 \times 10^{-3}\,\text{kg/g} = 600\,\text{kg BOD}_5/\text{d}$$

$$VOL \text{ in the first compartment, } VOL_{\text{c},1} = \frac{\Delta S_0}{V_{\text{c},1}} = \frac{600\,\text{kg BOD}_5/\text{d}}{125\,\text{m}^3} = 4.8\,\text{kg BOD}_5/\text{m}^3\cdot\text{d}$$

$$VOL \text{ in the MBBR reactor, } VOL_{\text{MBBR}} = \frac{\Delta S_0}{V} = \frac{600\,\text{kg BOD}_5/\text{d}}{250\,\text{m}^3} = 2.4\,\text{kg BOD}_5/\text{m}^3\cdot\text{d}$$

Note: The VOL in the first compartment is twice of the overall VOL in the MBBR reactor.

10.4.4 Submerged Attached Growth Systems

The submerged attached growth processes utilize a housing packed with granular media where the solids remain trapped. For this reason, clarifier may not be needed. However, periodic backwashing may be necessary to flush out these trapped solids. The traditional submerged biofilm reactor may be (1) upflow, (2) downflow, or (3) fluidized-bed reactor. During recent years, emerging biofilm reactors have also been under development. These systems may include membrane biofilm reactors (MBfRs), biofilm airlift suspension reactors (BASRs), aerobic granular sludge membrane bioreactors (AGMBRs), and granular sludge sequencing batch reactors (GSBRs). Brief descriptions and additional information about these processes can be found in References 2, 159 through 165.

Brief descriptions about the traditional biofilm processes are presented below. Most of these processes are proprietary systems. Equipment manufacturers shall be consulted for system specific design information. The general design and operating parameters are summarized in Table 10.43.

Submerged Upflow Reactor: The major process components are similar to those of downflow reactor, except that the flow is upward. Either *sunken* or *floating* media can be used in the upflow reactors. The sunken media are the same as those used by the downflow reactors. The floating media are usually polystyrene or polyethylene beads that are lighter than water. When the floating media are used a perforated plate is placed on top of the bed for collecting filtrate while restraining the media. The backwashing is normally done daily. The process is used for BOD removal, or combined BOD removal, nitrification, and denitrification. The process schematics are shown in Figure 10.81a.

Submerged Downflow Reactor: The process is also called *biological aerated filter (BAF)*. The reactor is similar to a typical gravity filter, except that air supply is provided at about 30 cm above the underdrain.

TABLE 10.43 General Design and Operating Parameters of Submerged Attached Growth Systems

Parameter	Value for Reactor Type			
	Submerged Upflow Reactor		Submerged Downflow Reactor (BAF)	Upflow Fluidized-Bed Bioreactor (FBBR)
	Sunken Media	Floating Media		
Volumetric organic loading (VOL), kg $BOD_5/m^3{\cdot}d$				
BOD removal	5–6	4–5	3.5–5	5–10
BOD removal with nitrification	2.5–3	2–2.5	1.8–2.5	2.5–5
Hydraulic loading, $m^3/m^2{\cdot}h$ ($ft^3/ft^2{\cdot}h$)	4–6 (13–20)	4–6 (13–20)	2.4–4.8 (8–16)	10–40 (33–130)
Empty bed contact time (EBCT), min	20–60	30–60	20–60	5–20
Typical media				
Material	Expanded clay or anthracite	Plastic beads	Expanded clay or anthracite	Silica sand or activated carbon
Diameter, mm	3.5–4.5	2.3–6	2–5	0.3–0.7 (silica sand) 0.6–1.4 (activated carbon)
Specific gravity (water = 1)	1.6	0.5–0.95	1.6	2.4–2.6 (silica sand) 1.4–1.5 (activated carbon)
Depth, m	2–4	3–4	1.6–2	3–4
Backwash				
Air scour, $m^3/m^2{\cdot}h$ (gpm/ft^2)	100 (41)	10 (4)	90 (37)	N/A
Water, $m^3/m^2{\cdot}h$ (gpm/ft^2)	20 (8)	50 (20)	15 (6)	N/A

Note: 1 $kg/m^3{\cdot}d = 62.4\ lb/10^3\ ft^3{\cdot}d$; 1 $m^3/m^2{\cdot}h = 3.28\ ft^3/ft^2{\cdot}h$; and 1 $m^3/m^2{\cdot}h = 0.409\ gpm/ft^2$.
Source: Adapted in part from Reference 2.

The major components are (1) underdrain system, (2) air header nozzles for diffused aeration, (3) backwash piping, (4) air scour system, and (5) housing. The typical packing media are expanded clay or anthracite. The head loss through the media is monitored and backwashing is normally performed on daily basis or when the head loss reaches about 1.8 m. The effluent BOD_5 and TSS concentrations are normally <10 mg/L. Partial or complete nitrification can be achieved by lowering the organic loading. The schematic process diagram is shown in Figure 10.81b.

Upflow Fluidized-Bed Bioreactor: In a fluidized-bed bioreactor (FBBR) or expanded bed reactor, the bacteria are grown on granular media. A recirculation flow at 200–500% of the influent flow is typically required and pumped upward causing suspension of the media through the entire reactor. A high rate of mass transfer is obtained due to high liquid velocity. The fluid shears the bacterial film to small thickness, thus the diffusional resistance to mass transfer of substrate, nutrients and oxygen is less than those in thicker biofilm. Also, the bacterial end products are removed rapidly. The typical fluidized bed media are silica sand or activated carbon. Substantial loss of media in the effluent may occur due to aeration. Therefore, a media screen may be required on the top to prevent the media loss. Also, an oxygenation tank may be considered to predissolve oxygen in the recycled effluent to lower the aeration requirement.

In a fluidized-bed reactor, large masses of dense solids accumulate on the upper layer. This increases SRT substantially. Eventually, the excess solids are washed out due to media agitation. The advantages of FBBR over other processes are (1) there is a high SRT that help to disperse and degrade shock loadings of toxic compounds, (2) the media with smaller diameter provides more surface area for bacteria to grow

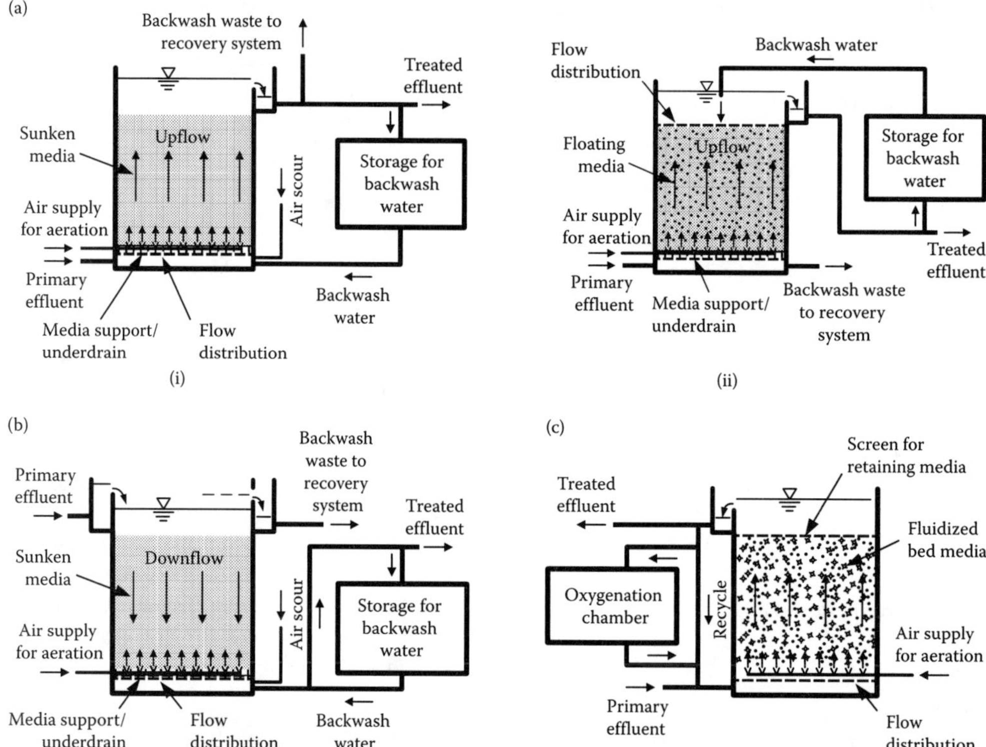

FIGURE 10.81 Submerged attached growth processes: (a) upflow reactor: (i) sunken media, and (ii) floating media; (b) downflow reactor or biological aerated filter (BAF); and (c) fluidized-bed bioreactor (FBBR).

thus more treatment capability per unit volume, (3) improved effluent quality due to more uniform flow distribution through the reactor, and (4) the operation is simple and reliable. The major drawback of using FBBR process includes high energy consumption, high oxygen supply requirement, and operational issues with biomass control.[2] The process schematic is shown in Figure 10.81c.

EXAMPLE 10.112: DESIGN OF SUBMERGED DOWNFLOW REACTOR

A submerged downflow reactor receives 1500 m^3/d primary settled wastewater. The influent total BOD_5 concentration is 150 mg/L with a fraction of 0.7 for soluble BOD_5. The influent TSS concentration is 80 mg/L with a VSS/TSS ratio of 0.8. Assume that 60% of VSS in the influent is degraded in the filter and TSS concentration from the reactor is 5 mg/L. The design media depth and the volumetric organic loading are 2 m and 4.5 kg $BOD_5/m^3{\cdot}d$. The granular media is expanded clay with a diameter of 4 mm. The filter is backwashed once per day at an initial air scour flowrate of 90 $m^3/m^2{\cdot}h$ followed by a water backwash of 18 $m^3/m^2{\cdot}h$ for 16 min. Determine (a) dimensions of reactor, (b) hydraulic loading, (c) media surface area loading rate (SALR), (d) solids generation, (e) air and water requirements during backwash, and (f) oxygen requirement.

Note: Theory and design of gravity filters are covered in detail in Chapter 15. Reader may review Examples 15.22, and 15.26 through 15.31 for filter unit sizing and calculating backwash velocity required to achieve a desired expansion of a given media layer.

Solution

1. Determine the dimensions of the reactor.

Mass BOD_5 loading in influent,

$$\Delta S_0 = QS_0 = 1500\,\text{m}^3/\text{d} \times 150\,\text{g BOD}_5/\text{m}^3 \times 10^{-3}\,\text{kg/g} = 225\,\text{kg BOD}_5/\text{d}$$

Volume of the reactor, $V = \frac{\Delta S_0}{VOL} = \frac{225\,\text{kg BOD}_5/\text{d}}{4.5\,\text{kg BOD}_5/\text{m}^3{\cdot}\text{d}} = 50\,\text{m}^3$

Area of the reactor, $A = \frac{V}{H} = \frac{50\,\text{m}^3}{2\,\text{m}} = 25\,\text{m}^2$

Provide a square filter with the length of the side, $a = \sqrt{A} = \sqrt{25\,\text{m}^2} = 5\,\text{m}$

2. Determine the hydraulic loading.

$$HL = \frac{Q}{A} = \frac{1500\,\text{m}^3/\text{d}}{25\,\text{m}^2} = 60\,\text{m}^3/\text{m}^2{\cdot}\text{d} \text{ or } HL = 60\,\text{m}^3/\text{m}^2{\cdot}\text{d} \times \frac{\text{d}}{24\,\text{h}} = 2.5\,\text{m}^3/\text{m}^2{\cdot}\text{h}$$

3. Estimate the media surface area loading rate (*SALR*).

The specific surface area of bulk spherical media can be estimated from Equation 10.76a.[2]

$$SA_\text{m} = \frac{6000}{d_\text{m}} \quad (10.76\text{a})$$

where

SA_m = specific surface area in bulk volume of spherical media, m^2/m^3

d_m = diameter of the media, mm

Apply Equation 10.76a to calculate the specific surface area of the media.

$$SA_\text{m} = \frac{6000}{d_\text{m}} = \frac{6000}{4\,\text{mm}} = 1500\,\text{m}^2/\text{m}^3$$

Calculate the total surface area of the biofilm attached to the media surface.

$$A_\text{m} = SA_\text{m}V = 1500\,\text{m}^2/\text{m}^3 \times 50\,\text{m}^3 = 75{,}000\,\text{m}^2$$

$$SALR = \frac{\Delta S_0}{A_\text{m}} = \frac{225\,\text{kg BOD}_5/\text{d}}{75{,}000\,\text{m}^2} = 0.003\,\text{kg BOD}_5/\text{m}^2{\cdot}\text{d} \text{ or } 3\,\text{g BOD}_5/\text{m}^2{\cdot}\text{d}$$

4. Estimate the solids generation.

The solids generation through the submerged attached growth reactor can be estimated from Equations 10.76b through 10.76d.[2]

$$p_\text{VSS} = 0.60\,(\text{SF}_{\text{BOD5}})S_0 + (1 - \text{BF}_\text{VSS})(\text{VSS/TSS ratio})TSS_0 \quad (10.76\text{b})$$

$$p_\text{NVSS} = (1 - \text{VSS/TSS ratio})TSS_0 \quad (10.76\text{c})$$

$$p_\text{TSS} = p_\text{VSS} + p_\text{NVSS}$$
$$\text{or } p_\text{TSS} = 0.60\,(\text{SF}_{\text{BOD5}})S_0 + [1 - (\text{BF}_\text{VSS})(\text{VSS/TSS ratio})]TSS_0 \quad (10.76\text{d})$$

where

p_VSS = volatile solids generated in the reactor, mg VSS/L (g VSS/m^3)

p_NVSS = nonvolatile solids generated in the reactor, mg NVSS/L (g NVSS/m^3)

p_TSS = total solids generated in the reactor, mg TSS/L (g TSS/m^3)

S_0 = total BOD_5 concentration in the influent, mg BOD_5/L (g BOD_5/m^3)

TSS_0 = TSS concentration in the influent, mg BOD_5/L (g BOD_5/m^3)
SF_{BOD5} = fraction of soluble BOD_5 in total BOD_5 in the influent, g $sBOD_5$/g $TBOD_5$
BF_{VSS} = fraction of VSS degraded in VSS in the influent, g VSS-degraded/g VSS
VSS/TSS ratio = ratio of VSS to TSS in the influent, g VSS/g TSS

Apply Equation 10.76d to calculate the TSS generated in the reactor.

$$\begin{aligned} p_{TSS} &= 0.60\,(SF_{BOD5})S_0 + [1 - (BF_{VSS})(VSS/TSS\ ratio)]TSS_0 \\ &= 0.60 \times 0.70 \times 150\,mg/L + (1 - 0.6 \times 0.8) \times 80\,mg/L \\ &= 105\,mg\ TSS/L \end{aligned}$$

Calculated the total solids captured in the reactor.

$$TSS_{accumulated} = p_{TSS} - TSS_e = (105 - 5)\ mg\ TSS/L = 100\,mg\ TSS/L \text{ or } 100\,g\ TSS/m^3$$

Calculated the amount of total solids accumulated in the reactor per day.

$$\Delta TSS = Q\ TSS_{accumulated} = 1500\,m^3/d \times 100\,g\ TSS/m^3 \times 10^{-3}\,kg/g = 150\,kg\ TSS/d$$

Calculated the concentration of total solids accumulated in the reactor if the backwash is performed once per day.

$$TSS_{reactor} = \frac{\Delta TSS \times 1\,d}{V} = \frac{150\,kg\ TSS/d \times 1\,d}{50\,m^3} = 3\,kg\ TSS/m^3$$

Note: A maximum solids storage capacity of 4 kg TSS/m^3 filter volume is typically used in filter design.[2] Therefore, the backwash on daily basis is acceptable.

5. Estimate the alir and water requirements during backwash.

Calculate the backwash air flow.

$$q_a = q'_a A = 90\,m^3/m^2{\cdot}h \times 25\,m^2 = 2250\,m^3/h \text{ or } 37.5\,m^3/min$$

Calculate the backwash water flow.

$$q_w = q'_w A = 18\,m^3/m^2{\cdot}h \times 25\,m^2 = 450\,m^3/h \text{ or } 7.5\,m^3/min$$

Calculate the amount of backwash water used during 16 min backwash operation per day.

$$\Delta q_w = q_w t_{bw} = 7.5\,m^3/min \times 16\,min/d = 120\,m^3/d$$

Calculate the percent of treated flow used for backwash.

$$P_{bw} = \frac{\Delta q_w}{Q} \times 100\% = \frac{120\,m^3/d}{1500\,m^3/d} \times 100\% = 8\%$$

Note: The backwash water is normally 5–15% of the influent applied to the filter.[2]

Calculate the TSS concentration in the backwash waste.

$$TSS_{bw} = \frac{\Delta TSS}{\Delta q_w} = \frac{150\,kg\ TSS/d}{120\,m^3/d} = 1.25\,kg\ TSS/m^3 = 1250\,g\ TSS/m^3 \text{ or } 1250\,mg\ TSS/L$$

Note: The TSS concentration in the backwash waste is normally 500–1500 mg/L.[2]

6. Estimate the oxygen requirements.

The oxygen demand can be estimated from Equation 10.76e.[2]

$$q_{O2} = [0.82\,(SF_{BOD5}) + 1.6\,BF_{VSS}(VSS/TSS\ ratio)]S_0 \qquad (10.76e)$$

where q_{O2} = oxygen demand in the influent applied to the reactor, mg O_2/L (g O_2/m^3)

Other parameters have been defined previously.

Apply Equation 10.76e to calculate the oxygen demand in the influent.

$$q_{O2} = [0.82\,(SF_{BOD5}) + 1.6\,BF_{VSS}(\text{VSS/TSS ratio})]S_0 = (0.82 \times 0.70 + 1.6 \times 0.6 \times 0.8) \times 150\,\text{mg/L}$$
$$= 1.34 \times 150\,\text{mg/L} = 201\,\text{mg}\,O_2/\text{L} \approx 200\,\text{mg}\,O_2/\text{L or } 200\,\text{g}\,O_2/\text{m}^3$$

Calculated the amount of oxygen required by the reactor per day.

$$O_2 = Q q_{O2} = 1500\,\text{m}^3/\text{d} \times 200\,\text{g}\,O_2/\text{m}^3 \times 10^{-3}\,\text{kg/g} = 300\,\text{kg}\,O_2/\text{d}$$

10.5 Anaerobic Treatment Processes

Anaerobic process is one of the oldest methods of municipal wastewater treatment. Anaerobic lagoon, cesspool, septic tank, Imhoff tank, and sludge digestion are the early examples of anaerobic treatment processes. In recent decades, the advancements in anaerobic treatment technologies have greatly increased the attractiveness and enhanced the advantages of anaerobic processes. It is well recognized that the waste is treated more quickly and reliably in the newly developed anaerobic reactors in which the net production of methane is also significantly increased.

10.5.1 Capabilities of Anaerobic Treatment Processes

These days, the anaerobic process is looked upon as a viable alternative to aerobic system for treatment of medium to high strength wastes. Although the growth rate of anaerobes is relatively slow, it does not mean that their rate of processing substrate is also slow. A comparison of both processes indicates the following advantages of anaerobic process over aerobic process: (1) biomass growth is 5–20 times lesser, therefore the sludge production is reduced significantly; (2) organic loading rate is 5–10 times higher, thus reactor size is reduced significantly; (3) nutrient requirements are 5–20 times lesser; (4) no energy requirement for aeration; (5) methane (CH_4) is produced as a source for energy recovery; (6) no release of volatile organic compounds (VOCs) and odorous compounds since the facility has airtight cover; (7) refractory organics is 10 times less; and (8) anaerobic biomass can stay dormant for a long time without feed, thus it is very attractive for seasoned industries.[2,7,54,166,167]

10.5.2 Fundamentals of Anaerobic Process

In anaerobic process, oxygen is not present or consumed. The energy-rich organic compounds are oxidized, and many other organic compounds and carbon dioxide (CO_2) are used as electron (hydrogen) acceptors. Transfer of electrons releases small amounts of chemically bound energy (reduction in free energy). This energy is used by anaerobes. The reduction in COD occurs by conversion of most organics into gases that contains mainly CH_4 and CO_2 with very small amounts of hydrogen (H_2) and other gasses. Anaerobic decomposition involves a consortium of microorganisms that execute a complex process in a series of interdependent steps. In general, the anaerobic biochemical reactions involve in three basic steps: (1) *hydrolysis*, (2) *acidogenesis (fermentation)/acetogenesis*, and (3) *methanogenesis*. A simplified schematic of overall anaerobic decomposition process is shown in Figure 10.82.[2,3,7,54,98] These steps are briefly discussed below.

Hydrolysis: Hydrolysis is a solubilization process of complex soluble and insoluble organic compounds (carbohydrates, proteins, and lipids) into simpler and smaller molecules (monosaccharides, amino acids,

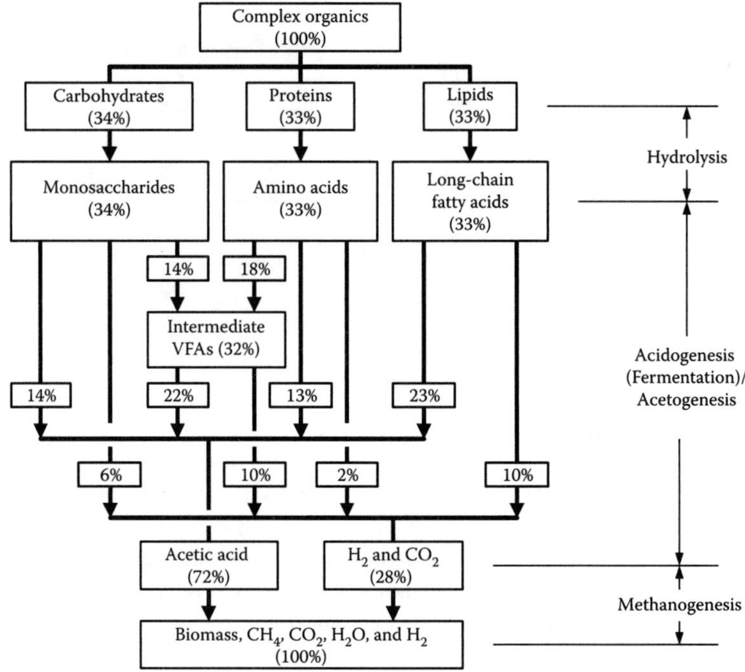

FIGURE 10.82 Simplified schematic of the anaerobic decomposition process.

and long-chain fatty acids (LCFAs)). These compounds are later transported into the cells and are metabolized.

Acidogenesis (Fermentation)/Acetogenesis: The microorganisms that cause hydrolysis also carry out fermentation. The major products of fermentation include organic volatile fatty acids (VFAs), other lower molecular weight compounds, CO_2, and H_2. The primary intermediate product is acetic acid (CH_3COOH) at the end of the step. The group of microorganisms responsible for the fermentation is facultative and obligate anaerobes; collectively called *acetogens* or *acid formers*. There is only a little change in organic content, although some lowering of pH may occur.

Methanogenesis: The final step involves gasification or conversion of metabolic intermediates, that is, acetic acid, hydrogen, and carbon dioxide into methane. The group of responsible microorganisms is strict anaerobes and is called *methanogens*. The conversion process is complex and a chain of many rate-limiting reactions occur. Several overall reactions for conversion of CO_2 and H_2 to CH_4 are given by Equation 10.77.

$$CH_3COOH \rightarrow CH_4 + CO_2 \tag{10.77a}$$

$$CH_3COOH + 4\ H_2 \rightarrow 2\ CH_4 + 2\ H_2O \tag{10.77b}$$

$$4\ H_2 + CO_2 \rightarrow CH_4 + 2\ H_2O \tag{10.77c}$$

10.5.3 Environmental Factors

Many environmental conditions are essential for anaerobic decomposition as the methane production by methanogens is normally the rate-limiting step. Some of these factors are listed below.[54]

Oxygen: The DO must be absent. The oxidation–reduction potential (ORP) is usually below −350 mV for maintaining the anaerobic condition.

pH and Alkalinity: The optimum pH for acid formers is 5–6 while that for methane formers is around 7 (near neutral condition). The acid formers tend to lower the pH while methane formers are sensitive to low pH. If the pH drops below 6.2, the methane formers are adversely affected while acids continue to accumulate. This brings the digestion process to a standstill. For proper operation, the alkalinity should be 1000–5000 mg/L as $CaCO_3$, and volatile acid should remain below 250 mg/L in the reactor.

Temperature: The anaerobic process has two optimum temperature ranges: (1) mesophilic range at 30–38°C (85–100°F), and (2) thermophilic range at 50–57°C (122–135°F). Most anaerobic processes are operated in the mesophilic range. Operation in thermophilic range is normally not practical because of high heating energy requirements unless an economical alternative heat source is available.

Mixing: Mixing is essential to maintain uniform environmental condition. Mixing distributes buffering agents, nutrients, intermediate metabolic products, toxic, compounds, and slug loads.

Nutrients: Proper ratio of nitrogen, phosphorus and sulfur should be present. Since the anaerobic sludge production is nearly one-fifth of that from an aerobic system, the nutrient requirements are only 5–20% of that required in the aerobic system. Typical *N*, *P*, and *S* requirements are typically 11.5, 2.3, and 1.5 mg per 100 mg of biomass depending upon the type of waste and organic loadings. Micronutrients (or trace metals) such as cobalt, iron, nickel, and zinc promote methane production.

Ammonia and Sulfide Control: Both free ammonia and sulfides at high concentration are inhibitory to methanogens. Sulfate under anaerobic condition is reduced to sulfides. The toxicity threshold concentrations of NH_3-N and H_2S may reach 250 mg/L.

10.5.4 Process Analysis

Anaerobic processes have high efficiency for conversion of COD into methane with a minimal biomass production. Also, high organic loading results in a smaller reactor volume when compared with aerobic processes. A discussion on process details is provided below.

Solids Yield and SRT: The biomass yield in anaerobic treatment is very small. In low- to medium-strength wastewater, the biomass growth is often less than the amount of solids lost in the effluent. For this reason, it may be difficult to maintain a proper SRT. Good solids capture and return is an essential part of anaerobic treatment. Following methods may be used to reduce the loss of biosolid in the effluent: (1) add a clarifier to separate solids and return them to the reactor, (2) use fixed-film to retain the biomass in the reactor; (3) develop dense sludge blanket to trap biomass in the reactor; (4) provide filtration or membrane separation of solids; and (5) provide a long HRT to settle solids in the reactor.[2,54,166]

Methane Production Rate: The methane production potential is the maximum quantity of methane that can be produced from a wastewater sample. The standard test is called *biochemical methane potential (BMP)*.[54,166,167] The methane production rate is calculated for Equation 10.78a.

$$Q_M = Q(S_0 - S)M \text{ or } Q_M = QES_0M \qquad (10.78a)$$

where

Q_M = volumetric methane production rate, m^3/d
Q = influent wastewater flow, m^3/d
S_0 = total COD (including soluble and particulate) in the influent, mg/L
S = soluble COD in the effluent, mg/L
M = volume of methane produced per unit of COD removed, m^3 CH_4/kg COD (ft^3 CH_4/lb COD). The value M ranges from 0.10–0.35 m^3/kg COD (1.6–5.6 ft^3/lb COD).
E = COD removal efficiency through the anaerobic process, dimensionless. The value of E may range from 0.6 to 0.9.

The volumetric methane generation rate for anaerobic wastewater treatment processes can be estimated from the kinetic equations. These equations are given in Equations 10.78b through 10.78e.[7,98]

$$\Delta S_M = QES_0 - 1.42P_x \text{ or } \Delta S_M = E\Delta S_0 - 1.42P_x \quad (10.78b)$$

$$Q_M = f_v[QES_0 - 1.42P_x] \text{ or } Q_M = f_v[E\Delta S_0 - 1.42P_x] \text{ or } Q_M = f_v\Delta S_M \quad (10.78c)$$

$$P_x = Y_{obs}QES_0 \text{ or } P_x = \frac{Y}{1 + k_d\theta_c}QES_0 \text{ or } P_x = \frac{Y}{1 + k_d\theta_c}E\Delta S_0 \quad (10.78d)$$

$$E_M = \frac{\Delta S_M}{\Delta S_0} \text{ or } \Delta S_M = E_M\Delta S_0 \quad (10.78e)$$

where

ΔS_M = COD consumption or stabilization rate, kg COD/d
P_x = net mass of biosolids (VSS) produced, kg VSS/d
Y_{obs} = observed biomass yield, g VSS/g COD utilized
ΔS_0 = mass COD loading in the influent, kg COD/d
Y = biomass yield coefficient, g VSS/g COD utilized
k_d = specific endogenous decay coefficient, d^{-1}
θ_c = mean cell residence time, d. It is normally equal to the reaction period or hydraulic retention time (HRT) since a reactor without recycle is usually utilized in anaerobic wastewater treatment processes.
f_v = volumetric conversion factor for the amount of methane produced from the COD consumption, m^3 CH_4/kg COD. The theoretical value of f_v at the standard temperature and pressure (STP) conditions (0°C and 1 atm) is 0.35 m^3 CH_4/kg COD (Example 10.113).
1.42 = conversion factor for the amount of COD inserted in the biomass (VSS), g COD/g VSS
E_M = observed COD removal (or stabilization) efficiency through the anaerobic process, g COD/g VSS or dimensionless

All other terms have been defined previously.

Note: This is a simplified approach to estimate the methane production rate by assuming that the biodegradable solids ≈ biomass (VSS).

The mass methane generation rate (Equation 10.79) has also been developed from the substrate utilization rate (Equation 10.12e).[7,54]

$$W_M = f_m\,UXV \text{ or } W_M = f_m\frac{kS}{K_s + S}XV \text{ or } W_M = f_m\frac{kS}{K_s + S}X\theta Q \quad (10.79)$$

where

W_M = mass methane generation rate, g CH_4/d
U = specific substrate utilization rate per unit mass of microorganisms, g COD/g VSS·d or d^{-1}
f_m = theoretical mass conversion factor of COD to methane, g CH_4/g COD. The theoretical value of f_m is 0.25 g CH_4/g COD (Example 10.113).
X = biomass concentration in the reactor, mg VSS/L (g VSS/m^3)
V = volume of reactor, m^3
k = maximum specific rate of substrate-utilization per unit mass of microorganisms, g COD/g VSS·d or d^{-1}
K_s = half-velocity constant or the substrate concentration at one-half the maximum specific growth rate, mg COD/L (g COD/m^3)
S = COD concentration in the effluent, mg COD/L (g COD/m^3)

The amounts of CH_4, CO_2, NH_3, and H_2S are also estimated if the composition of waste is known. The generalized composition with and without sulfur are $C_aH_bO_cN_dS_e$ and $C_aH_bO_cN_d$. The reaction of a compound without sulfur, and products produced are presented in Example 10.116.

Kinetic Coefficients: The biological kinetic coefficients for the anaerobic process are summarized in Table 10.44.

Properties of Methane: Methane is a colorless and odorless gas. It is combustible and a 5–15% mixture with air is explosive. Methane is very soluble in water. It is not toxic when inhaled, but it can produce suffocation by reducing the concentration of oxygen inhaled. It is an important greenhouse gas. The properties of methane are summarized in Table 10.45.[2,98,166,167]

Simplified Model for an Anaerobic Process: In a completely mixed suspend growth reactor, without recycle the substrate and biomass balance is expressed by Equations 10.80a and 10.80b, respectively.

$$\text{Substrate balance:} \quad V\frac{dS}{dt} = QS_0 - QS - \frac{kS}{K_s + S}XV \tag{10.80a}$$

$$\text{Biomass balance:} \quad V\frac{dX}{dt} = -QX + \frac{YkS}{K_s + S}XV - k_dXV \tag{10.80b}$$

Note: Both equations are developed from the mass-balance relationship given by Equation 3.1. The derivation steps of Equations 10.80a and 10.80b are given in Example 10.118.

Advanced Models for Anaerobic Process: Anaerobic process involves hydrolysis, acid formation, and methane generation. In a simplified approach, the biomass production and substrate utilization in three stages are usually modeled by the first-order or Monod-type reaction. However, many interactive equations are involved in the advanced modeling programs. These equations are complex and must be solved simultaneously using standard numerical procedures. More information on these models may be found in References 54 and 168 through 170.

TABLE 10.44 Biological Kinetic Coefficients for Anaerobic Treatment Processes

Kinetic Coefficient	Value	
	Range	Typical
Acidogenesis phase		
Y, g VSS/g COD	0.06–0.12	0.1
k_d, d^{-1}	0.02–0.06	0.04
k, g COD/g VSS·d	2.3–3.7	2.5
K_s, mg COD/L	–	200
Methanogenesis phase		
Y, g VSS/g COD	0.02–0.06	0.04
k_d, d^{-1}	0.01–0.04	0.02
k, g COD/g VSS·d	4.7–11	6.3
K_s, mg COD/L	–	50
Overall process		
Y, g VSS/g COD	0.05–0.10	0.08
k_d, d^{-1}	0.02–0.04	0.03
k, g COD/g VSS·d	2–7	4.4
K_s, mg COD/L	60–1000	120

Source: Adapted in part from References 2, 54, 98, 166, and 167.

TABLE 10.45 Properties of Methane

Parameter	Value
Density, kg/m^3 ($lb/10^3\ ft^3$)	
At 0°C and 1 atm	0.71 (44)
At 35°C and 1 atm	0.63 (39)
Specific energy, kJ/kg (Btu/lb)	55,600 (23,900)
Energy content, kJ/m^3 (Btu/ft^3)	
At 0°C and 1 atm	39,700 (1060)
At 35°C and 1 atm	35,200 (940)
Production, m^3/kg COD (ft^3/lb)	
At 0°C and 1 atm	0.35 (5.6)
At 35°C and 1 atm	0.39 (6.3)
Mass content in biogas, % w/w	
Range	35–45
Average	40
Volumetric content in biogas, % v/v	
Range	60–70
Average	65
ΔH combustion, kJ/mole	−891
Greenhouse effect (times heat trapping compared to CO_2)	20
Solubility in water, g/kg	
At 0°C	40
At 20°C	23

Note: 1 kg/m^3 = 62.4 $lb/10^3\ ft^3$; 1 kJ/kg = 0.430 Btu/lb; 1 kJ/m^3 = 0.0268 Btu/ft^3, and 1 m^3/kg = 16.02 ft^3/lb.

Source: Adapted in part from References 2, 98, 166, and 167.

EXAMPLE 10.113: THEORETICAL CONVERSION FACTORS OF COD TO METHANE

Develop the theoretical mass and volumetric conversion factors of COD to methane at the standard temperature and pressure (STP) conditions (0°C and 1 atm) using the ideal gas law.

Solution

1. Write the balanced reaction for COD as O_2 from methane.

   The reaction between methane (CH_4) and COD (as O_2) is given by Equation 10.81.

$$\underset{16}{CH_4} + \underset{2\times32}{2\,O_2} \rightarrow CO_2 + 2\,H_2O \tag{10.81}$$

2. Develop the theoretical mass conversion factor (f_m).

   On stoichiometric basis calculate the value of f_m from Equation 10.81.

$$f_m = \frac{16\,\text{g/mole as CH}_4}{2\times 32\,\text{g/mole COD as O}_2} = 0.25\,\text{g CH}_4/\text{g COD or } 0.25\,\text{kg CH}_4/\text{kg COD}$$

   Therefore, one kg COD is equivalent to 0.25 kg CH_4 or 250 g CH_4.

3. Calculate the number of moles of methane equivalent to one kg COD.

$$n_{CH4} = \frac{0.25\,\text{g } CH_4/\text{g COD}}{16\,\text{g/mole as } CH_4} \times 10^3\,\text{g/kg} = 15.6\text{ mole } CH_4/\text{kg COD}$$

4. Write the expression of the ideal gas law.

The ideal gas law is expressed by Equation 10.82.

$$V = nV_m \text{ or } V = n\frac{RT}{P} \text{ or } nR = \frac{PV}{T} \tag{10.82}$$

where

V = volume occupied by the gas at temperature T and pressure P, L
n = the amount of the gas, mole
V_m = molar volume of the gas at STP, 22.4 L/mole
R = universal gas constant, 0.082056 at L·atm/mole·°K
T = average temperature of the gas, °K
P = absolute pressure of the gas, atm

5. Calculate the volume of methane equivalent to one kg COD (V_{CH4}).

a. Calculate V_{CH4} from the molar volume (V_m).

Use Equation 10.82 to calculate the volume of methane equivalent to one kg COD from $V_m =$ 22.4 L/mole of an ideal gas at STP.

$$V_{CH4} = n_{CH4}V_m = 15.6\,\text{g mole } CH_4/\text{kg COD} \times 22.4\,\text{L/g mole} = 350\,\text{L } CH_4/\text{kg COD}$$

b. Calculate V_{CH4} from the universal gas constant (R).

Use Equation 10.82 to calculate f_v from $R = 0.082057$ L·atm/mole·°K, $T = 237$°K, and $P =$ 1 atm.

$$V_{CH4} = n_{CH4}\frac{RT}{P} = 15.6\text{ mole } CH_4/\text{kg COD} \times \frac{0.082056\,\text{L atm/mole·°K} \times 273\text{°K}}{1\text{ atm.}}$$

$$= 350\,\text{L } CH_4/\text{kg COD}$$

The value of $V_{CH4} = 350$ L CH_4/kg COD is obtained from both methods.

6. Develop the theoretical volumetric conversion factor (f_v).

$$f_v = V_{CH4} = 350\,\text{L } CH_4/\text{kg COD or } 0.35\,\text{m}^3\ CH_4/\text{kg COD}$$

Therefore, one kg COD produces 0.35 m^3 CH_4 theoretically at STP.

Note: The f_v value of 0.35 m^3 CH_4 /kg COD is equal to the upper value of the typical range for M used in Equation 10.78a.

EXAMPLE 10.114: BIOGAS GENERATION FROM INDUSTRIAL WASTE

An anaerobic treatment facility receives industrial waste. The wastewater flow is 800 m^3/d and the total COD (soluble and particulate) is 2500 mg/L. Calculate (a) the volume of methane produced and (b) the total volume of biogas at STP. Assume 80% of the waste is degraded and the methane content in the biogas is 64%. Use $M = 0.35$ m^3 CH_4 /kg COD.

Solution

1. Calculate the volume of methane produced from the industrial waste at STP from Equation 10.78a.

$$Q_M = QES_0M = 800\,m^3/d \times 0.8 \times 2500\,g\,COD/m^3 \times 0.35\,m^3\,CH_4/kg\,COD \times 10^{-3}\,kg/g$$
$$= 560\,m^3/d$$

2. Calculate the total biogas volume at STP.

$$Q_{biogas} = \frac{Q_M}{64\%} \times 100\% = \frac{560\,m^3/d}{64\%} \times 100\% = 875\,m^3/d$$

EXAMPLE 10.115: METHANE GENERATION FROM MASS COD LOADING

An anaerobic reactor receives an average mass COD loading of 1500 kg COD/d from a local food processing plant. Calculate the volume of methane generated daily if efficiency of waste utilization is 80% in the reactor. Also estimate the amount of energy contained in the methane. Assume the biomass yield coefficient $Y = 0.08$ g VSS/g COD, the specific endogenous decay coefficient $k_d = 0.03\,d^{-1}$, and the average reaction period $\theta = 10$ d. Use $f_v = 0.35\,m^3$ CH_4/kg COD at STP.

Solution

1. Calculate the net biomass produced in the reactor (P_x).

   Since solids recirculation is not used, $\theta_c = \theta = 10$ d. Calculate P_x from Equation 10.78d.

$$P_x = \frac{Y}{1 + k_d\theta_c} E\Delta S_0 = \frac{l0.08\,kg\,VSS/kg\,COD}{1 + 0.03\,d^{-1} \times 10\,d} \times 0.8 \times 1500\,kg\,COD/d = 74\,kg\,VSS/d$$

2. Calculate the daily volume of methane produced (Q_M) from Equation 10.78c.

$$Q_M = f_v[E\Delta S_0 - 1.42P_x]$$
$$= 0.35\,m^3\,CH_4/kg\,COD \times (0.8 \times 1500\,kg\,COD/d - 1.42\,kg\,COD/kg\,VSS \times 74\,kg\,VSS/d)$$
$$= 383\,m^3\,CH_4/d$$

3. Estimate the potential energy contained in the methane (E_M).

   The energy content of methane is 39,700 kJ/m^3 CH_4 in Table 10.45. Calculate the potential daily energy recovery.

$$E_M = 383\,m^3\,CH_4/d \times 39{,}700\,kJ/m^3\,CH_4 = 15{,}200{,}000\,kJ/d = 15.2 \times 10^6\,kJ/d$$

EXAMPLE 10.116: BIOGAS GENERATION FROM A WASTE COMPOSITION

An organic waste is undergoing anaerobic decomposition. The chemical formula of the waste on dry weight basis is $C_{50}H_{100}O_{40}N$. The dry weight of organic waste decomposed in the reactor is 500 kg/d. Assume that the weight of biomass produced is ignorable. Calculate the total mass and volume of biogas produced and the mass and volumetric ratios of CH_4 to CO_2.

Solution

1. Write the generalized reaction equation.

The generalized reaction equation is given by Equation 10.83.

$$C_aH_bO_cN_d + \frac{1}{4}(4a - b - 2c + 3d)\ H_2O \rightarrow \frac{1}{8}(4a + b - 2c - 3d)\ CH_4 + \frac{1}{8}(4a - b + 2c + 3d)\ C_2O + d\ NH_3 \tag{10.83}$$

2. Write the stoichiometric reaction equation.

 From Equation 10.83 the stoichiometric reaction equation is developed for the organic waste $C_{50}H_{100}O_{40}N$ where a = 50, b = 100, c = 40, and d = 1.

$$C_{50}H_{100}O_{40}N + \frac{1}{4}(4 \times 50 - 100 - 2 \times 40 + 3 \times 1)\ H_2O \rightarrow \frac{1}{8}(4 \times 50 + 100 - 2 \times 40 - 3 \times 1)\ CH_4 + \frac{1}{8}(4 \times 50 - 100 + 2 \times 40 + 3 \times 1)\ CO_2 + NH_3$$

 Simplify the reaction equation.

$$C_{50}H_{100}O_{40}N + 5.75\ H_2O \rightarrow 27.125\ CH_4 + 22.875\ CO_2 + NH_3$$

3. Develop the balanced reaction equation.

 Calculate the weight of each component in a balanced reaction equation.

$$\begin{aligned}\text{Weigh of 1 mole of the waste } (C_{50}H_{100}O_{40}N) &= 1\,\text{g mole} \times (50 \times 12 + 100 \times 1 + 40 \times 16 + 1 \times 14)\ \text{g/g mole as waste}\\ &= 1354\,\text{g as waste}\end{aligned}$$

$$\text{Weigh of 7.57 moles of } H_2O = 5.75\ \text{mole} \times (2 \times 1 + 16)\ \text{g/mole} = 103.5\,\text{g}$$

$$\text{Weigh of 27.125 moles of } CH_4 = 27.125\ \text{mole} \times (1 \times 12 + 4 \times 1)\ \text{g/mole} = 434\,\text{g}$$

$$\text{Weigh of 22.875 moles of } CO_2 = 22.875\ \text{mole} \times (1 \times 12 + 2 \times 16)\ \text{g/mole} = 1006.5\,\text{g}$$

$$\text{Weigh of 1 mole of } NH_3 = 1\ \text{mole} \times (1 \times 14 + 3 \times 1)\ \text{g/mole} = 17\,\text{g}$$

 The balanced reaction equation is summarized below.

$$\underset{1354}{C_{50}H_{100}O_{40}N} + \underset{103.5}{5.75\ H_2O} \rightarrow \underset{434}{27.125\ CH_4} + \underset{1006.5}{22.875\ C_2O} + \underset{17}{NH_3}$$

4. Calculate the weights of CH_4, CO_2, and NH_3 produced from the waste.

 Calculate the mole of each product from the balanced reaction equation for 500 kg of waste decomposed per day.

$$\text{Weight of } CH_4,\ w_{CH4} = \frac{434\,\text{g}}{1354\,\text{g}} \times 500\,\text{kg/d} = 160\,\text{kg } CH_4/\text{d}$$

$$\text{Weight of } CO_2,\ w_{CO2} = \frac{1006.5\,\text{g}}{1354\,\text{g}} \times 500\,\text{kg/d} = 372\,\text{kg } CO_2/\text{d}$$

$$\text{Weight of } NH_3,\ w_{NH3} = \frac{17\,\text{g}}{1354\,\text{g}} \times 500\,\text{kg/d} = 6.3\,\text{kg } NH_3/\text{d}$$

5. Calculate the moles of CH_4, CO_2, and NH_3 produced from the waste.

$$\text{Moles of } CH_4,\ n_{CH4} = \frac{160\,\text{kg}\,CH_4/\text{d}}{16\,\text{g/mole as}\,CH_4} \times 10^3\,\text{g/kg} = 10{,}000\,\text{mole}\,CH_4/\text{d}$$

$$\text{Moles of } CO_2,\ n_{CO2} = \frac{372\,\text{kg}\,CO_2/\text{d}}{44\,\text{g/mole as}\,CO_2} \times 10^3\,\text{g/kg} = 8450\,\text{mole}\,CO_2/\text{d}$$

$$\text{Moles of } NH_3,\ n_{NH3} = \frac{6.3\,\text{kg}\,NH_3/\text{d}}{17\,\text{g/mole as}\,NH_3} \times 10^3\,\text{g/kg} = 371\,\text{mole}\,NH_3/\text{d}$$

6. Calculate the volumes of CH_4, CO_2, and NH_3 produced from the waste.
   The volume of an ideal gas at STP, $V_m = 22.4$ L/mole.

$$V_{CH4} = n_{CH4} V_m = 10{,}000\,\text{mole}\,CH_4/\text{d} \times 22.4\,\text{L/mole} \times 10^{-3}\,\text{m}^3/\text{L} = 224\,\text{m}^3\,CH_4/\text{d}$$

$$V_{CO2} = n_{CO2} V_m = 8450\,\text{mole}\,CO_2/\text{d} \times 22.4\,\text{L/mole} \times 10^{-3}\,\text{m}^3/\text{L} = 189\,\text{m}^3\,CO_2/\text{d}$$

$$V_{NH3} = n_{NH3} V_m = 371\,\text{mole}\,CO_2/\text{d} \times 22.4\,\text{L/mole} \times 10^{-3}\,\text{m}^3/\text{L} = 8.3\,\text{m}^3\,NH_3/\text{d}$$

7. Calculate the total mass and volume of the biogas produced from the waste.
   The ammonia produced from the waste is not considered in the biogas since its weight and volume are relatively small and most of it remains in solution as NH_4^+ at the typical pH in the anaerobic reactor.
   Calculate the total weight of biogas from the weights of CH_4 and CO_2 produced from the waste.

$$w_{biogas} = w_{CH4} + w_{CO2} = (160 + 372)\,\text{kg/d} = 532\,\text{kg biogas/d}$$

   Calculate the total volume of biogas from the volumes of CH_4 and CO_2 produced from the waste.

$$V_{biogas} = V_{CH4} + V_{CO2} = (224 + 189)\,\text{m}^3/\text{d} = 413\,\text{m}^3\,\text{biogas/d}$$

8. Calculate the mass and volumetric ratios of CH_4 to CO_2.
   Calculate the mass ratio of CH_4 to CO_2.

$$\text{Proportion of } CH_4 = \frac{160\,\text{kg}\,CH_4/\text{d}}{532\,\text{kg biogas/d}} \times 100\% = 30\%\ \text{w/w}$$

$$\text{Proportion of } CO_2 = \frac{372\,\text{kg}\,CO_2/\text{d}}{532\,\text{kg biogas/d}} \times 100\% = 70\%\ \text{w/w}$$

$$\text{Mass ratio of } CH_4 \text{ to } CO_2 = \frac{160\,\text{kg}\,CH_4/\text{d}}{372\,\text{kg}\,CO_2/\text{d}} = 0.43$$

   Calculate the volumetric ratio of CH_4 to CO_2.

$$\text{Volume fraction of } CH_4 = \frac{224\,\text{m}^3\,CH_4/\text{d}}{413\,\text{m}^3\,\text{biogas/d}} \times 100\% = 54\%\ \text{w/w}$$

$$\text{Volume fraction of } CO_2 = \frac{189\,\text{m}^3\,CO_2/\text{d}}{413\,\text{m}^3\,\text{biogas/d}} \times 100\% = 46\%\ \text{w/w}$$

$$\text{Volumetric ratio of } CH_4 \text{ to } CO_2 = \frac{224\,\text{m}^3\,CH_4/\text{d}}{189\,\text{m}^3\,CO_2/\text{d}} = 1.2$$

   Note: In a mixture of ideal gases, the mole fraction for a gas is equal to the volumetric fraction.

EXAMPLE 10.117: METHANE GENERATION FROM SUBSTRATE UTILIZATION

Methane is generated from the consumption of COD. An 800 m^3 anaerobic reactor is operated at a biomass concentration of 1500 g/m^3. The effluent COD concentration is 30 mg/L. The kinetic coefficient k and K_S are 4 g COD/g VSS·d and 120 mg COD/L, respectively. Calculate the volumetric methane generation rate (m^3/d) at the field conditions of 35°C and 1.2 atm. Use $f_m = 0.25$ kg CH_4/kg COD.

Solution

1. Calculate the mass generation rate of CH_4 (W_M) from Equation 10.79.

$$W_M = f_m \frac{kS}{K_s + S} XV$$

$$= 0.25\,\text{g CH}_4/\text{g COD} \times \frac{4\,\text{g COD/g VSS·d} \times 30\,\text{mg COD/L}}{(120+30)\,\text{mg COD/L}} \times 1500\,\text{g VSS/m}^3 \times 800\,\text{m}^3$$

$$= 240{,}000\,\text{g CH}_4/\text{d or } 240\,\text{kg CH}_4/\text{d}$$

2. Calculate the moles of CH_4 produced from the consumption of COD.
   Calculate the moles from the weight of CH_4 produced.

$$\text{Moles of CH}_4,\ n_{CH4} = \frac{240{,}000\,\text{g CH}_4/\text{d}}{16\,\text{g/mole as CH}_4} = 15{,}000\,\text{mole CH}_4/\text{d}$$

   Calculate the volume of methane generated at STP.
   Calculate the volume of methane from $V_m = 22.4$ L/mole of an ideal gas at STP.

$$V_{CH4,STP} = n_{CH4} V_m = 15{,}000\,\text{mole CH}_4/\text{d} \times 22.4\,\text{L/mole} \times 10^{-3}\,\text{m}^3/\text{L} = 336\,\text{m}^3\,\text{CH}_4/\text{d}$$

3. Estimate the volume of methane at the field conditions.
   The following expression is obtained from Equation 10.82 since nR remains as a constant under both STP and field conditions.

$$nR = \frac{P_1 V_1}{T_1} = \frac{P_2 V_2}{T_2} \text{ or } V_2 = \frac{T_2}{T_1}\frac{P_1}{P_2} V_1$$

   Calculate the volume of methane at the field conditions of 35°C and 1.2 atm.

$$V_{CH4} = \frac{T_2}{T_{STP}}\frac{P_{STP}}{P_2} V_{CH4,STP} = \frac{(273+35)^\circ\text{K}}{273^\circ\text{K}} \times \frac{1\,\text{atm.}}{1.2\,\text{atm.}} \times 336\,\text{m}^3/\text{d} = 316\,\text{m}^3/\text{d}$$

EXAMPLE 10.118: DERIVATION OF SUBSTRATE UTILIZATION AND BIOMASS GROWTH RATE EQUATIONS

Equations 10.80a and 10.80b express the substrate utilization and biomass growth rates. These equations are developed from mass balance relationships. Provide step-by-step derivations of these equations.

Solution

1. Write the mass balance relationship.

The generalized mass balance relationship of a bioreactor is prepared below from Equation 3.1.

$$\begin{bmatrix}\text{Accumulation}\\ \text{rate}\end{bmatrix} = \begin{bmatrix}\text{Input}\\ \text{rate}\end{bmatrix} - \begin{bmatrix}\text{Output}\\ \text{rate}\end{bmatrix} + \begin{bmatrix}\text{Utilization or}\\ \text{net growth}\end{bmatrix}$$

2. Provide step-by-step derivation of the expression for the substrate utilization.

Apply the generalized relationship to obtain the expression for the substrate utilization in mathematical terms.

$$V\frac{dS}{dt} = QS_0 - QS + r_{su}V$$

The substrate utilization rate (r_{su}) is given by Equation 10.12c.

$$r_{su} = -\frac{kXS}{K_s + S} \text{ or } r_{su} = -\frac{kS}{K_s + S}X$$

Substitute the r_{su} value in the above substrate utilization expression to obtain Equation 10.80a.

$$V\frac{dS}{dt} = QS_0 - QS - \frac{kXS}{K_s + S}XV$$

3. Provide step-by-step derivation of the expression for the biomass growth.

Apply the generalized relationship to obtain the expression for the biomass growth in mathematical terms.

$$V\frac{dX}{dt} = QX_0 - QX + r_g'V$$

The net biomass growth rate (r_g') is given by Equation 10.13a.

$$r_g' = \frac{\mu_{max}XS}{K_s + S} - k_dX$$

Apply the expression of $\mu_{max} = Yk$ in Equation 10.12b to obtain the following expression.

$$r_g' = \frac{Yk\,XS}{K_s + S} - k_dX \text{ or } r_g' = \frac{YkS}{K_s + S}X - k_dX$$

Since $X_0 << X$, it is assumed that $QX_0 - QX \approx -QX$.
Substitute the above expressions to obtain Equation 10.80b.

$$V\frac{dX}{dt} = -QX + \frac{YkS}{K_s + S}XV - k_dXV$$

EXAMPLE 10.119: ESTIMATE ALKALINITY REQUIREMENT IN ANAEROBIC PROCESS

An anaerobic reactor is treating an industrial wastewater flow of 1000 m^3/d with an alkalinity concentration of 450 mg/L as $CaCO_3$. An average pH of 7.2 is desired in the reactor at the operating temperature of 35°C. The average CO_2 content is 40% by volume in the biogas that is held at a pressure of 1.1 atm in the gas phase. Determine (1) the HCO_3^- concentration required in the reactor, (2) the alkalinity concentration required in the reactor, (3) additional alkalinity concentration required, and (4) the dosage and daily amount of sodium bicarbonate required to maintain the desired operating pH.

Solution

1. Calculate the saturation concentration of carbonic acid (H_2CO_3) in the liquid.

Alkalinity is required to neutralize the CO_2 or H_2CO_3 produced and maintain the desired pH in the reactor. In accordance with Henry's law, the mole fraction of a gas in the liquid at a given temperature is a function of the gas pressure and mole fraction of the gas in the gas phase. At the equilibrium this relationship is given in Equation 10.84a.[2]

$$x_{g,l} = \frac{P_b}{H} x_{g,b} \tag{10.84a}$$

where

$x_{g,l}$ = mole fraction of the gas in the liquid, mole gas/mole liquid
$X_{g,b}$ = mole fraction of the gas in the biogas, mole gas/mole biogas
P_b = total biogas pressure above the liquid surface, atm
H = mole fraction-based Henry's law constant, atm·(mole gas/mole air)/(mole gas/mole H_2O), atm·mole H_2O/mole air, or atm. The value of H can be estimated by Equation 10.84b.

$$H = 10^{(-(A/T)+B)} \tag{10.84b}$$

where

T = average temperature of the gas, °K
A and B = empirical constants. For CO_2, $A = 1012.40$ and $B = 6.606$. The constants are available for other dissolved gases in Reference 2.

Note: The mole fraction of a gas is proportional to its volume fraction in the gas phase. Also see Examples 11.18 through 11.20 for applications of Henry's law in dissolution of gaseous chlorine and Examples 15.10 through 15.14 for that during air stripping.

Estimate the Henry's law constant H for CO_2 from Equation 10.84b at $T = (273.15 + 35)°K = 308.15°K$.

$$H = 10^{(-(A/T)+B)} = 10^{(-(1012.40/308.15°K)+6.606)} = 2092 \text{ atm}$$

Calculate the saturation mole concentration of H_2CO_3 at the equilibrium with CO_2 at a mole (or volumetric) fraction of 40% in the biogas.

$$x_{H2CO3,l} = \frac{P_b}{H} x_{CO2,b} = \frac{1.1 \text{ atm}}{2092 \text{ atm}} \times 0.4 = 2.10 \times 10^{-4} \text{ mole } H_2CO_3/\text{mole liquid}$$

The mole fraction of a gas in the liquid can also be expressed by Equation 10.84c.[2]

$$x_{g,l} = \frac{n_g}{\sum n_g + n_{H2O}} \quad \text{or} \quad n_g = x_{g,l}\left(\sum n_g + n_{H2O}\right) \tag{10.84c}$$

where

n_g = mole concentration of the gas in the liquid, mole gas/L
$\sum n_g$ = mole concentration of dissolved gases in the liquid, mole gases/L
n_{H2O} = mole concentration of water in the liquid, mole H_2O/L

Since $\sum n_g << n_{H2O}$, it is reasonable to assume $\sum n_g + n_{H2O} \approx n_{H2O}$, and therefore, $n_g \approx x_{g,l} n_{H2O}$.

$$n_{H2O} \approx \frac{1000\,\text{g/L}}{18\,\text{g/mole as H}_2\text{O}} = 55.6 \text{ mole H}_2\text{O (or liquid)/L}$$

$$\begin{aligned} n_{H2CO3,l} &= x_{H2CO3,l} n_{H2O} = (2.10 \times 10^{-4} \text{ mole H}_2\text{CO}_3\text{/mole liquid}) \times 55.6 \text{ mole liquid/L} \\ &= 1.17 \times 10^{-2} \text{ mole H}_2\text{CO}_3\text{/L} \end{aligned}$$

2. Calculate the concentration of bicarbonate (HCO_3^-) dissociated at the desired operating pH value.
   The relationship between the pH and alkalinity can be established using the first step dissociation reaction of H_2CO_3. The dissociation reaction of H_2CO_3 is expressed by Equation 10.84d.

$$H_2CO_3 \leftrightarrow H^+ + HCO_3^- \tag{10.84d}$$

The corresponding relationship at the equilibrium is expressed by Equation 10.84e.

$$K_{a1} = \frac{[HCO_3^-][H^+]}{[H_2CO_3]} \text{ or } [HCO_3^-] = \frac{[H_2CO_3]}{[H^+]} K_{a1} \tag{10.84e}$$

where K_{a1} = first step dissociation constant, mole/L. The value of K_{a1} is generally a function of temperature and ionic strength. At 35°C, $K_{a1} = 4.85 \times 10^{-7}$ for H_2CO_3. The K_{a1} values at other temperatures are available in Reference 2.

At pH = 7, $\text{pH} = -\log[H^+] = 7$ or $[H^+] = 10^{-7}$ mole H^+/L.

Calculate the mole concentration of HCO_3^- from Equation 10.84e.

$$n_{HCO3,l} = \frac{[H_2CO_3]}{[H^+]} K_{a1} = \frac{1.17 \times 10^{-2} \text{ mole H}_2\text{CO}_3\text{/L}}{10^{-7} \text{ mole H}^+\text{/L}} \times 4.85 \times 10^{-7} = 0.0567 \text{ mole HCO}_3^-\text{/L}$$

Calculate the mass concentration of HCO_3^- required to maintain pH of 7 in the reactor.

$$C_{HCO3} = 0.0567 \text{ mole HCO}_3^-\text{/L} \times 61\,\text{g/mole as HCO}_3^- = 3.46\,\text{g HCO}_3^-\text{/L or } 3460\,\text{mg HCO}_3^-\text{/L}$$

3. Calculate the alkalinity concentration required in the reactor.
   Calculate the equivalents of HCO_3^-.

$$eq_{HCO3} = \frac{3.46\,\text{mg HCO}_3^-\text{/L}}{61\,\text{g/eq. as HCO}_3^-} = 0.0567\,\text{eq./L}$$

Calculate the alkalinity concentration required in the reactor.

$$\begin{aligned} Alk_{reactor} &= eq_{HCO3} \times 50\,\text{g/eq. as CaCO}_3 = 0.0567\,\text{eq./L} \times 50\,\text{g/eq. as CaCO}_3 \\ &= 2.84\,\text{g/L as CaCO}_3 \text{ or } 2840\,\text{mg/L as CaCO}_3 \end{aligned}$$

4. Calculate the additional alkalinity concentration required from the supplemental chemical.

The additional alkalinity concentration is determined from the difference between the alkalinity concentration required in the reactor and the alkalinity concentration contained in the influent.

$$Alk_{additional} = Alk_{ireactor} - Alk_{inf} = (2840 - 450)\text{mg/L as CaCO}_3$$
$$= 2390\text{ mg/L as CaCO}_3 \text{ or } 2390\text{ g/m}^3 \text{ as CaCO}_3$$

5. Determine the dosage and daily amount of sodium bicarbonate ($NaHCO_3$) required.
   Calculate the $NaHCO_3$ dosage required to provide additional alkalinity.

$$C_{NaHCO3} = Alk_{additional} \times \frac{84\text{ g/eq. as NaHCO}_3}{50\text{ g/eq. as CaCO}_3} = 2390\text{ mg/L as CaCO}_3 \times \frac{84\text{ g/eq. as NaHCO}_3}{50\text{ g/eq. as CaCO}_3}$$
$$= 4020\text{ mg/L as NaHCO}_3 \text{ or } 4020\text{ g/m}^3 \text{ as CaCO}_3$$

Calculate the daily amount of $NaHCO_3$ required at the wastewater influent flow of 1000 m^3/d.

$$\Delta m_{NaHCO3} = C_{NaHCO3}Q = 4020\text{ g/m}^3 \text{ as CaCO}_3 \times 1000\text{ m}^3/\text{d} \times 10^{-3}\text{ kg/g} = 4020\text{ kg NaHCO}_3/\text{d}$$

Therefore, a significant amount of sodium bicarbonate is theoretically required for the anaerobic treatment process.

10.5.5 Anaerobic Suspended Growth Processes

The major types of suspended growth anaerobic treatment processes are (1) anaerobic digestion, (2) anaerobic contact process (AnCP), (3) upflow anaerobic sludge blanket (UASB) process, and (4) anaerobic sequencing batch reactor (AnSBR). The process diagrams of these treatment processes are illustrated in Figure 10.83. The important design and operational parameters of these processes are summarized in Table 10.46.[2,7,54,98,171–175] Each process is briefly described below.

Anaerobic Digestion: Anaerobic digestion is one of the oldest processes used for treatment of sludge and wastewater with high volatile solids. The thickened sludge is heated and mixed in an air tight digestion tank. Anaerobic digesters are divided into different types based on hydraulic and organic loadings. These types are standard rate, high rate, two-stage, and two-phase digesters.[2,7] Residuals management systems including anaerobic sludge digestion are presented in Chapter 13.

Anaerobic Contact Process (AnCP): The AnCP consists of an airtight, mixed and heated reactor followed by a solids separation device. The solids are returned to the reactor to maintain high MLVSS. The solid separation device may be a sedimentation basin, lamella, or tube settler, filtration, membrane, or gas flotation unit. Gravity settling may have poor performance because of gas formation and uplift of biological floc. Gas stripping and vacuum degasification prior to settling is helpful. Flotation is achieved by dissolving the process gases under pressure into digestor overflow and releasing the content in a flotation tank at atmospheric pressure. The AnCP with membrane separation, also called anaerobic membrane bioreactor (AnMBR) features a high VOL of 5–15 kg COD/m^3·d (310–930 lb COD/10^3 ft^3), a long SRT, and a low HRT. The process diagrams of AnCP are shown in Figure 10.83a.

Upflow Anaerobic Sludge Blanket (UASB) Process: The UASB process was initially developed in the Netherlands to treat medium- to high-strength industrial wastes.[173] The flow moves upward through a sludge blanket composed of biologically formed granules. The granules are the most significant aspect of the reactor. They are 1–3 mm in size and may be dark brown, smooth, granular or flocculent. As the influent comes in contact with the dense sludge blanket of granules, the organics are degraded to CH_4 and CO_2. The gases produced cause internal recirculation and upward velocity which keep the granules in suspension. Some granules attached with gases rise to the top. They strike the degassing baffle and

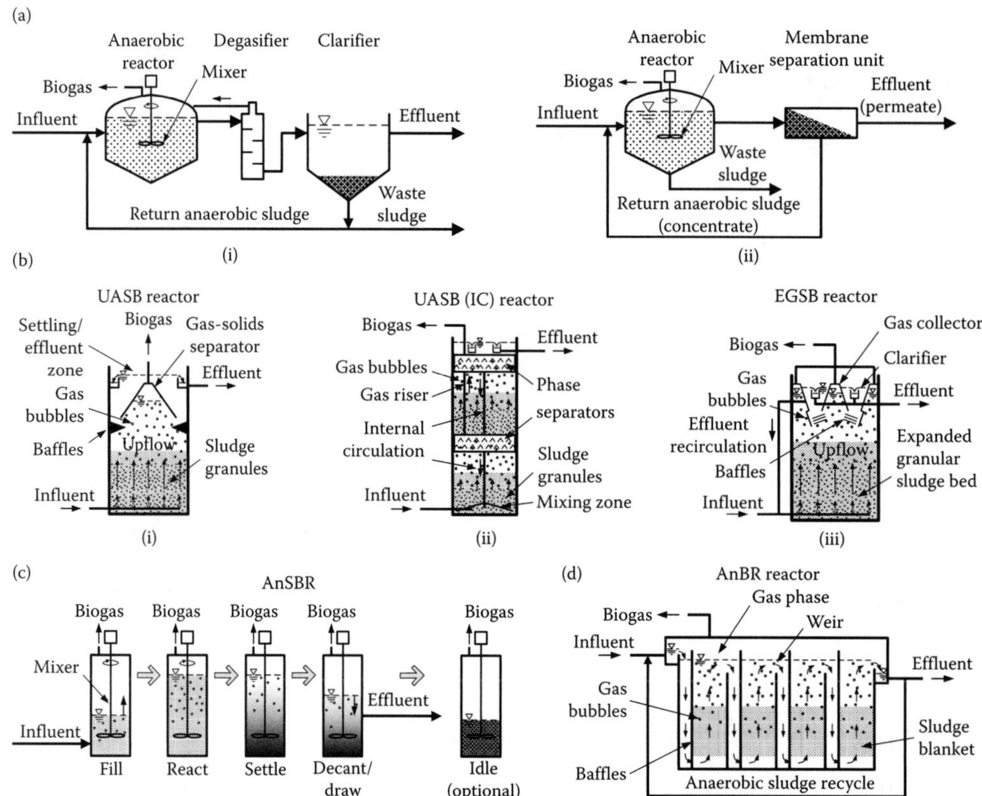

FIGURE 10.83 Anaerobic suspended growth processes: (a) anaerobic contact process (AnCP): (i) clarifier after degasifier, and (ii) membrane separation of solids; (b) upflow anaerobic sludge blanket (UASB) reactors: (i) conventional UASB, (ii) stacked UASB with internal circulation (UASB (CI)), and (iii) expanded granular sludge blanket (EGSB); (c) anaerobic sequencing batch reactor (AnSBR); and (d) anaerobic baffled reactor(AnBR).

return to the blanket. A settling zone is provided in the upper portion of the reactor. Solids separation also occurs in this zone and the granules move down to the blanket. The gas is collected from the top. The process diagrams of UASB and modofocations are shown in Figure 10.83b.

The biggest challenge in anaerobic treatment of weak- to medium-strength wastewater is in maintaining the desired concentration of biomass in the UASB reactor. Several improvements have been suggested to better capture and recycle the biomass. Some of the improvements are: (1) internal lamella clarifiers, (2) external clarification, (3) stacked UASB reactors with internal circulation (UASB (IC)), (4) expanded granular sludge blanket (EGSB) process, (5) anaerobic hybrid (AnHYB) process by adding fix-film media, and (6) membrane separation with return of concentrate.[2,174,175] For treatment of low-strength wastewater, the application of EGSB process provides higher upflow velocities (4–10 m/h) that expands the granules blanket to have a high VOL up to 35 kg COD/m^3·d (2200 lb COD/10^3 ft^3). This gives some benefits of fluidized-bed reactor such as better mixing and enhanced contact of wastewater with the surface area of granules. Some of these improvements are also shown in Figure 10.83b.

Anaerobic Sequencing Batch Reactor (AnSBR): Sequencing batch reactor is also used in anaerobic treatment. The process is much like an aerobic SBR. The main differences are a cover over the tank and no air supply. In a single covered tank, the process involves five sequential cycles are carried out very similar to those in an aerobic SBR (Table 10.15 and Examples 10.55). The feed and react cycles utilize mixing. Granulated sludge normally develops after sufficient operating time. Good settling property of granules allows a settling cycle as short as 30 min. A schematic of AnSBR process is shown in Figure 10.83c.

TABLE 10.46 Design and Operational Parameters for Anaerobic Suspended Growth Processes

Parameter	Typical Range for Process Type		
	AnCP	UASB	AnSBR
Volumetric organic loading (VOL), kg COD /m^3·d (lb COD/10^3 ft^3·d)	2–5 (125–310)	5–20 (310–1250)	1–2.4 (60–150)
COD removal efficiency, %	75–90	80–95	75–98
Solids retention time (SRT), d	15–30	>30	50–200
Hydraulic retention time (HRT), h	12–120	4–16	6–24
MLVSS or sludge content in blanket, g VSS/L (kg VSS/m^3)	4–8	35–40	2–4
Upflow hydraulic loading, m^3/m^2·h (ft^3/ft^2·h)			
Average	–	0.5–1.5 (1.5–5)	–
Peak	–	2–6 (6.6–20)	–
Surface overflow rate in clarifier, m^3/m^2·h (ft^3/ft^2·h)	0.5–1 (1.6–3.3)	–	–
Major dimensions and configurations			
Reactor process zone height, m (ft)	–	3–7 (10–23)	–
Gas-solids separator height, m (ft)	–	2–3 (6.6–10)	–
Gas-solids separator area, % of reactor area	–	80–85	–
Slope of gas collector plate, degree from horizontal	–	45–60	–

Abbreviation: AnCP = anaerobic contact process, UASB = upflow anaerobic sludge blanket process, and AnSBR = anaerobic sequencing batch reactor process.

Note: 1 kg/m^3·d = 62.4 lb/10^3 ft^3·d; and 1 m^3/m^2·h = 3.28 ft^3/ft^2·h.

Source: Adapted in part from References 2, 7, 54, 98, and 171 through 175.

Improvements in Anaerobic Sludge Blanket Processes: Several anaerobic sludge blanket processes are under development stages. One of them is the anaerobic baffled reactor (AnBR) process (Figure 10.83d). This process utilizes a series of baffled chambers in which the wastewater flows, rises, and falls while moving slowly through baffled chambers. The contact of wastewater with the sludge blanket occurs mainly in the upward paths. Gasses are collected from the top. Low-speed mixer may be used. The directions of influent and effluent may be reversed periodically to maintain uniform concentration of sludge blanket in each chamber. References 2 and 176 provide more details on this process.

EXAMPLE 10.120: SIZING OF AN ANAEROBIC CONTACT PROCESS REACTOR

An AnCP reactor is designed to treat industrial wastewater. The characteristics of influent are given below.

Average daily flow	= 800 m^3/d	Biodegradable VSS	= 90% of VSS
Total COD concentration	= 5000 mg/L	Nitrogen	= 10 mg/L as N
TSS concentration	= 400 mg TSS/L	Phosphorus	= 6 mg/L as P
VSS/TSS ratio	= 0.80		

The reactor is designed to remove 85% COD. Calculate (1) SRT, (2) sludge production, (3) volume and dimensions, (4) hydraulic retention time and volumetric organic loading, (5) gas production, (6) energy recovered, and (7) nutrient requirements.

Apply also the following design parameters:

Kinetic coefficients: $Y = 0.08$ g VSS/g COD, $k_d = 0.03\ d^{-1}$, $k = 2.5\ d^{-1}$, and $K_s = 800$ mg/L.
Process parameters: SRT safety factor $SF = 1.5$, MLVSS concentration = 6000 mg/L in the reactor, field conditions of 25°C and 1.2 atm, and CH_4 content by volume = 65% of total biogas.
Effluent quality: TSS concentration = 150 mg/L, and VSS/TSS ratio = 0.85. Assume 0.65 g biodegradable VSS/g VSS and 1.42 g COD/g biodegradable VSS.

Solution

1. Determine the amount of influent COD reaching the anaerobic reactor.

$$\Delta S_0 = Q \times S_0 = 800\ \text{m}^3/\text{d} \times 5000\ \text{g COD/m}^3 \times 10^{-3}\ \text{kg/g} = 4000\ \text{kg COD/d}$$

2. Determine the effluent total and soluble COD concentrations.
Effluent total COD concentration,

$$S_{t,e} = (1 - E_{COD}) \times S_0 = (1 - 0.85) \times 5000\ \text{mg COD/L} = 750\ \text{mg COD/L}$$

Effluent VSS concentration, $VSS_e = TSS_e \times$ VSS/TSS ratio $= 150$ mg TSS/L $\times$ 0.85 mg VSS/mg TSS $= 128$ mg VSS/L

Effluent particulate COD concentration in VSS,

$$\begin{aligned} S_{p,e} &= \text{VSS} \times \text{biodegradable VSS/VSS ratio} \times \text{COD/biodegradable VSS ratio} \\ &= 128\ \text{mg VSS/L} \times 0.65\ \text{mg biodegradable VSS/mg VSS} \times 1.42\ \text{mg COD/mg biodegradable VSS} \\ &= 118\ \text{mg COD/L} \approx 100\ \text{mg COD/L} \end{aligned}$$

Effluent soluble COD concentration, $S = S_{t,e} - S_{p,e} = (750 - 118)$ mg COD/L $= 632$ mg COD/L

3. Determine the design SRT.
Apply Equation 10.15d or 10.16a to calculate the SRT (θ_c)) required for process operation.

$$\frac{1}{\theta_c} = \frac{YkS}{K_s + S} - k_d = \frac{0.08 \times 2.5\ \text{d}^{-1} \times 632\ \text{mg COD/L}}{800\ \text{mg COD/L} + 632\ \text{mg COD/L}} - 0.03\ \text{d}^{-1} = (0.088 - 0.03)\ \text{d}^{-1} = 0.058\ \text{d}^{-1}$$

$$\theta_c = \frac{1}{0.058\ \text{d}^{-1}} = 17\ \text{d}$$

Design SRT, $\theta_c^{design} = SF\theta_c = 1.5 \times 17\ \text{d} = 26\ \text{d}$

Provide $\theta_c^{design} = 30$ d. The design SRT is within the acceptable range (Table 10.46).

4. Determine the sludge production.
The total volatile solids produced by an AnCP can be estimated from Equation 10.85a.

$$P_{x,t} = P_{x,bm} + P_{x,cd} + P_{x,nb} + P_{x,bd} \text{ or } p_{x,t} = p_{x,bm} + p_{x,cd} + p_{x,nb} + p_{x,bd} \qquad (10.85a)$$

where
$P_{x,t}$ or $p_{x,t}$ = total volatile solids produced, kg VSS/d (g/m^3)

$P_{x,bm}$ or $p_{x,bm}$ = biomass generated from COD removal (Equation 10.15m or 10.16e), kg VSS/d (g/m^3)

$P_{x,cd}$ or $p_{x,cd}$ = cell debris remained after endogenous decay in the reactor, mg VSS/L (g/m^3)

A relatively long SRT is typically required in an anaerobic treatment process. For this reason, the cell debris remained after endogenous decay may be considered in the calculations of sludge production. The cell debris can be estimated from Equation 10.85b.

$$P_{x,cd} = f_{cd}\, k_d\, \theta_c\, P_{x,bs} \quad \text{or} \quad p_{x,cd} = f_{cd}\, k_d\, \theta_c\, p_{x,bs} \tag{10.85b}$$

where

f_{cd} = cell debris factor, g VSS/g VSS. The typical value of f_{cd} is 0.15 g VSS/g VSS. This means 0.15 gram of cell debris will be left per gram of VSS destroyed by endogenous decay.

$P_{x,nb}$ or $p_{x,nb}$ = nonbiodegradable volatile solids carried in the influent, kg VSS/d (g/m^3)

$P_{x,bd}$ or $p_{x,bd}$ = remaining biodegradable volatile solids carried in the influent, kg VSS/d (g/m^3). In this example, the remaining biodegradable VSS is ignorable for treating high soluble organic industrial wastewater. However, this component should be analyzed when the biodegradable VSS in the wastewater influent is extremely high or the anaerobic digestion process is used for sludge stabilization (Example 13.29).

a. Calculate the amount of biomass generated from COD removal ($P_{x,bm}$) from Equation 10.15l or 10.16d.

$$Y_{obs} = \frac{Y}{1 + k_d\theta_c^{design}} = \frac{0.08\text{ mg VSS/mg COD}}{1 + 0.03\text{ d}^{-1} \times 30\text{ d}} = 0.042\text{ mg VSS/mg COD}$$
$$= 0.042\text{ g VSS/g COD}$$

Calculate $P_{x,bm}$ from Equation 10.15m.

$$P_{x,bm} = Y_{obs}Q(S_0 - S) = 0.042\text{ g VSS/g COD} \times 800\text{ m}^3\text{/d} \times (5000 - 632)\text{ g COD/m}^3 \times 10^{-3}\text{ kg/g}$$
$$= 147\text{ kg VSS/d}$$

b. Calculate the amount of cell debris produced ($P_{x,cd}$).
Use $f_{cd} = 0.15$ and apply Equation 10.5b to calculate $P_{x,cd}$.

$$P_{x,cd} = f_d\, k_d\, \theta_c^{design}\, P_{x,bm} = 0.15 \times 0.03\text{ d}^{-1} \times 30\text{ d} \times 147\text{ kg VSS/d} = 0.135 \times 147\text{ kg VSS/d}$$
$$= 20\text{ kg VSS/d}$$

c. Calculate the amount of nonbiodegradable VSS ($P_{x,nb}$).
Nonbiodegradable VSS concentration in the influent,

$$\text{VSS}_{nb,0} = \text{TSS}_0 \times \text{VSS/TSS ratio} \times (1 - \text{biodegradable fraction of VSS})$$
$$= 400\text{ mg TSS/L} \times 0.8\text{ mg VSS/mg TSS} \times (1 - 0.9)$$
$$= 32\text{ mg VSS/L or } 32\text{ g VSS/m}^3$$

The amount of $P_{x,nb}$ in the influent,

$$P_{x,nb} = Q\,\text{VSS}_{nb,0} = 800\text{ m}^3\text{/d} \times 32\text{ g VSS/m}^3 \times 10^{-3}\text{ kg/g} = 26\text{ kg VSS/d}$$

d. Calculate the amount of total volatile solids generated from the process ($P_{x,t}$).

Calculate the $P_{x,t}$ from Equation 10.85a.

$$P_{x,t} = P_{x,bm} + P_{x,cd} + P_{x,nb} = (147 + 20 + 26)\ \text{kg VSS/d} = 193\ \text{kg VSS/d}$$

e. Calculate the sludge production from the process.
   Total amount of TSS produced from the process,

$$\Delta TSS = \frac{P_{x,t}}{\text{VSS/TSS ratio}} = \frac{193\ \text{kg VSS/d}}{0.85\ \text{kg VSS/kg TSS}} = 227\ \text{kg TSS/d}$$

   Amount of TSS in the effluent,

$$\Delta TSS_e = Q\ \text{TSS}_e = 800\ \text{m}^3\text{/d} \times 150\ \text{g TSS/m}^3 \times 10^{-3}\ \text{kg/g} = 120\ \text{kg TSS/d}$$

   Calculate the amount of sludge wasted from the reactor,

$$\Delta WS = \Delta TSS - \Delta TSS_e = (227 - 120)\ \text{kg TSS/d} = 107\ \text{kg TSS/d}$$

5. Calculate the required reactor volume and dimensions.
   Calculate the volume from Equation 10.15o at $X = 6000$ mg VSS/L $= 6$ kg VSS/m^3.

$$V = \frac{\theta_c^{design} P_{x,t}}{X} = \frac{30\ \text{d} \times 193\ \text{kg VSS/d}}{6\ \text{kg VSS/m}^3} = 965\ \text{m}^3$$

   Provide two reactors with a total volume of 970 m^3.

$$\text{Volume of each reactor, } V_{reactor} = \frac{1}{2}V = \frac{1}{2} \times 970\ \text{m}^3 = 485\ \text{m}^3$$

   Each reactor is a square basin with a pyramidal hopper. Assume a side water depth $SWD = 5$ m and a hopper depth $D = 1.5$ m. Write the reactor volume as a function of the basin width (B).

$$V_{reactor} = SWD \times B^2 + \frac{2}{3} \times D \times B^2 = 5\ \text{m} \times B^2 + \frac{2}{3} \times 1.5\ \text{m} \times B^2 = 6\ \text{m} \times B^2$$

   Solve 6 m $\times B^2 = 485$ m^3 and $B = 9$ m.
   Provide two square reactors; each has 9 m × 9 m cross section, 5-SWD, and 1.5-m deep pyramidal hopper.

6. Calculate the hydraulic detention time and the volumetric organic loading.

$$\text{Hydraulic retention time, } \theta = \frac{V}{Q} = \frac{970\ \text{m}^3 \times 24\ \text{h/d}}{800\ \text{m}^3\text{/d}} = 29\ \text{h}$$

$$\text{Volumetric organic loading, } V = \frac{\Delta S_0}{VOL} = \frac{4000\ \text{kg COD/d}}{970\ \text{m}^3} = 4.1\ \text{kg COD/m}^3\cdot\text{d}$$

   Both design HRT and VOL are within the acceptable ranges (Table 10.46).

7. Determine the volume of CH_4 and total biogas generated.
   The volume of CH_4 generated is calculated from Equation 10.78a. Use $M = f_v = 0.35$ m^3 CH_4/kg COD at STP (Example 10.113).

$$Q_{CH4,STP} = Q(S_0 - S)M = 800\,\text{m}^3\text{/d} \times (5000 - 632)\,\text{g COD/m}^3 \times 0.35\,\text{m}^3\ \text{CH}_4\text{/kg COD} \times 10^{-3}\,\text{kg/g}$$
$$= 1223\,\text{m}^3\ \text{CH}_4\text{/d}$$

Calculate the volume of methane at the field conditions of 25°C and 1.2 atm from Equation 10.82.

$$Q_{CH4,field} = \frac{T_{field}}{T_{STP}}\frac{P_{STP}}{P_{field}} V_{CH4,STP} = \frac{(273+25)^\circ K}{273^\circ K} \times \frac{1\ atm.}{1.2\ atm.} \times 1223\ m^3\ CH_4/d = 1112\ m^3\ CH_4/d$$

Calculate the total biogas volume at the field conditions.

$$Q_{biogas,field} = \frac{Q_{CH4,field}}{65\%} \times 100\% = \frac{1112\ m^3/d}{65\%} \times 100\% = 1711\ m^3/d$$

8. Determine the energy contained in CH_4.
   Use $f_m = 0.25$ kg CH_4/kg COD (from Example 10.113) to calculate the mass CH_4 generated (W_M) from Equations 10.14d and 10.79.

$$\begin{aligned} W_{CH4} &= f_m\ UXV = f_m\ Q(S_0 - S) \\ &= 0.25\ kg\ CH_4/kg\ COD \times 800\ m^3/d \times (5000 - 632)\ g\ COD/m^3 \times 10^{-3}\ kg/g \\ &= 874\ kg\ CH_4/d \end{aligned}$$

   Calculate the energy contained in CH_4 using the specific energy $SE_{CH4} = 55{,}600$ kJ/kg obtained from Table 10.45.

$$E_{CH4} = W_{CH4}\ SE_{CH4} = 874\ kg\ CH_4/d \times 55{,}600\ kJ/kg\ CH_4 = 48{,}600{,}000\ kJ/d = 48.6 \times 10^6\ kJ/d$$

9. Determine the nutrients requirements.
   a. Calculate the nitrogen and phosphorus contents of biomass.
      Mixed biological growth is normally expressed by formula $C_{60}H_{87}O_{23}N_{12}P$. The nitrogen and phosphorus contents of biomass are estimated from this formula, respectively.

$$N_{bvss} = \frac{12 \times \text{molar mass of N}}{\text{molar mass of } C_{60}H_{87}O_{23}N_{12}P} = \frac{12 \times 14\ g/mole\ as\ N}{1374\ g\ VSS/mole\ as\ C_{60}H_{87}O_{23}N_{12}P} = 0.12\ g\ N/g\ VSS$$

$$P_{bvss} = \frac{1 \times \text{molar mass of P}}{\text{molar mass of } C_{60}H_{87}O_{23}N_{12}P} = \frac{1 \times 31\ g/mole\ as\ P}{1374\ g\ VSS/mole\ as\ C_{60}H_{87}O_{23}N_{12}P} = 0.023\ g\ P/g\ VSS$$

   b. Calculate the amount of VSS yield ($P_{x,bvss}$) due to biological growth.
      The amount of VSS yield due to biological growth include the biomass ($P_{x,bm}$) generated from COD removed and the cell debris ($P_{x,cd}$) remained in the reactor.

$$P_{x,bvss} = P_{x,bm} + P_{x,cd} = (147 + 20)\ kg\ VSS/d = 167\ kg\ VSS/d$$

   c. Calculate the nitrogen and phosphorus requirements by the biological growth.
      Nitrogen requirement,

$$w_{N,bvss} = N_{bvss}\ P_{x,bvss} = 0.12\ kg\ N/kg\ VSS \times 167\ kg\ VSS/d = 20\ kg/d\ as\ N$$

      Phosphorus requirement,

$$w_{P,bvss} = P_{bvss}\ P_{x,bvss} = 0.023\ kg\ P/kg\ VSS \times 167\ kg\ VSS/d = 3.8\ kg/d\ as\ P$$

   d. Calculate the amount of nitrogen and phosphorus available in the influent.

Nitrogen available,

$$w_{N,0} = N_0\,Q = 10\,g/m^3 \text{ as N} \times 800\,m^3/d \times 10^{-3}\,kg/g = 8.0\,kg/d \text{ as N}$$

Phosphorus available,

$$w_{P,0} = P_0\,Q = 6\,g/m^3 \text{ as P} \times 800\,m^3/d \times 10^{-3}\,kg/g = 4.8\,kg/d \text{ as P}$$

e. Evaluate the nutrient requirements.

Since $w_{N,0} < w_{N,bvss}$, nitrogen is not sufficient and must be supplied.

Nitrogen supply needed, $w_{N,s} = w_{N,bvss} - w_{N,0} = (20 - 8)\text{ kg/d as N} = 12\,kg/d \text{ as N}$

Since $w_{P,0} > w_{P,bvss}$, there is sufficient phosphorus available in the influent. No additional phosphorus is required.

Note: The alkalinity requirements may be calculated from the procedure given in Example 10.119.

EXAMPLE 10.121: CHARACTERISTICS OF GRANULES

The granular-sludge blanket in a UASB reactor is most significant aspect of the reactor. List some of the important characteristics of dense sludge blanket that is produced in a USAB.

Solution

The UASB is a high rate anaerobic treatment process. Contact with dense granular sludge allows rapid breakdown of organic matters. The characteristics of granular sludge are listed below.[2,54,98,173–175]

1. The formation of granules depends upon the characteristics of wastewater, upflow velocity, and nutrients. High carbohydrates, pH around 7, sufficient NH_4-N, and high hydrogen concentration encourage formation of granules.
2. High protein waste encourages formation of fluffy floc or flocculant sludge.
3. Granule formation is slow and may take several months.
4. The COD:N:P ratio initially should be 300:5:1 and 600:5:1 under steady-state operation.
5. Formation of granular sludge is accelerated and enhanced when the required trace elements are available.
6. The granules may reach pea size but normally range 1–3 mm with densities in the range of 1.0–1.05 g/L.
7. The granular sludge blanket is dense. The solids concentration may range from 50 to100 g/L at the bottom of the blanket and 10 to 30 g/L in the sludge layer above the blanket.
8. The granules have high sludge-thickening property; and SVI may be <20 mL/g.
9. The operation and performance of UASB is not affected whether the sludge blanket is granular or flocculant.
10. The granules can stay dormant for a long time without substrate.

EXAMPLE 10.122: DESIGN OF A UASB REACTOR

Two UASB reactors are designed to treat the same industrial wastewater that is given in Example 10.120. Assume the wastewater contains mainly the carbohydrate waste. Use the wastewater characteristic and kinetic coefficients given in Example 10.120. Additional process design parameters are given below:

Process parameters: design volumetric organic loading = 10 kg COD/m^3·d, height of the process volume (h_{pv}) = 8 m, and design average solids content in the sludge blanket = 40 kg VSS/m^3. Provide a SRT safety factor SF = 1.5. A COD removal efficiency of 90% is desired. The UASB reactor is operated at 25°C.

Effluent qualities: TSS concentration = 240 mg/L and VSS/TSS ratio = 0.85. Assume 0.65 g biodegradable VSS/g VSS and 1.42 g COD/g biodegradable VSS.

Determine (1) the required process volume, (2) effluent COD concentrations, (3) SRT, (4) solids generation, (5) average solids concentration in the sludge blanket, (6) HRT in the process volume, (7) dimensions of the reactor, and (8) upflow velocity in the reactor.

Solution

1. Determine the total process volume requirement.

   Mass COD loading in the influent,

$$\Delta S_0 = QS_0 = 800\,\text{m}^3/\text{d} \times 5000\,\text{g COD/m}^3 \times 10^{-3}\,\text{kg/g} = 4000\,\text{kg COD/d}$$

   Required total reactor process volume, $V_t = \dfrac{\Delta S_0}{VOL} = \dfrac{4000\,\text{kg COD/d}}{10\,\text{kg COD/m}^3\cdot\text{d}} = 400\,\text{m}^3$

   Provide two UASB reactors, each has a process volume of 200 m^3.

2. Determine the effluent total and soluble COD concentrations.

   Effluent total COD concentration,

$$S_{t,e} = (1 - E_{COD}) \times S_0 = (1 - 0.9) \times 5000\,\text{mg COD/L} = 500\,\text{mg COD/L}$$

$$\text{Effluent VSS concentration, } X_e = \text{TSS}_e \times \text{VSS/TSS ratio} = 240\,\text{mg TSS/L} \times 0.85\,\text{mg VSS/mg TSS} = 204\,\text{mg VSS/L}$$

   Effluent particulate COD concentration in VSS,

$$\begin{aligned} S_{p,e} &= X_e \times \text{biodegradable VSS/VSS ratio} \times \text{COD/biodegradable VSS ratio} \\ &= 204\,\text{mg VSS/L} \times 0.65\,\text{mg biodegradable VSS/mg VSS} \times 1.42\,\text{mg COD/mg biodegradable VSS} \\ &= 188\,\text{mg COD/L} \end{aligned}$$

   Effluent soluble COD concentration, $S = S_{t,e} - S_{p,e} = (500 - 188)\,\text{mg COD/L} = 312\,\text{mg COD/L}$

3. Determine the design SRT.

   Calculate the SRT (θ_c)) required by the process from Equation 10.15d.

$$\frac{1}{\theta_c} = \frac{YkS}{K_s + S} - k_d = \frac{0.08\,\text{d}^{-1} \times 2.5\,\text{d} \times 312\,\text{mg COD/L}}{800\,\text{mg COD/L} + 312\,\text{COD/L}} - 0.03\,\text{d}^{-1} = (0.056 - 0.03)\,\text{d}^{-1} = 0.026\,\text{d}^{-1}$$

$$\theta_c = \frac{1}{0.026\,\text{d}^{-1}} = 38\,\text{d}$$

   Design SRT, $\theta_c^{\text{design}} = SF\,\theta_c = 1.5 \times 38\,\text{d} = 57\,\text{d}$

   Provide $\theta_c^{\text{design}} = 60\,\text{d}$.

4. Determine the solids generated.

   a. Calculate the amount of biomass ($P_{x,bm}$) generated from COD removed.

Calculate Y_{obs} from Equation 10.15l.

$$Y_{obs} = \frac{Y}{1 + k_d\theta_c^{design}} = \frac{0.08\,\text{mg VSS/mg COD}}{1 + 0.03\,\text{d}^{-1} \times 60\,\text{d}} = 0.029\,\text{mg VSS/mg COD} = 0.029\,\text{g VSS/g COD}$$

Calculate $P_{x,bm}$ from Equation 10.15m.

$$P_{x,bm} = Y_{obs}Q(S_0 - S) = 0.029\,\text{g VSS/g COD} \times 800\,\text{m}^3/\text{d} \times (5000 - 312)\,\text{g COD/m}^3 \times 10^{-3}\,\text{kg/g}$$
$$= 109\,\text{kg VSS/d}$$

b. Calculate the amount of debris ($P_{x,cd}$) remained in the reactor.
   Use $f_{cd} = 0.15$ and apply Equation 10.15m to calculate $P_{x,cd}$.

$$P_{x,cd} = f_d k_d \theta_c^{design} P_{x,bs} = 0.15 \times 0.03\,\text{d}^{-1} \times 60\,\text{d} \times 109\,\text{kg VSS/d} = 0.27 \times 109\,\text{kg VSS/d}$$
$$= 29\,\text{kg VSS/d}$$

c. Calculate the amount of nonbiodegradable VSS ($P_{x,nb}$).
   The value of $P_{x,nb} = 26$ kg VSS/d is obtained from Step 3.c in Example 10.120.
d. Calculate the amount of total volatile solids ($P_{x,t}$) generated from the process.
   The $P_{x,t}$ is calculated from Equation 10.85a.

$$P_{x,t} = P_{x,bm} + P_{x,cd} + P_{x,nb} = (109 + 29 + 26)\,\text{kg VSS/d} = 164\,\text{kg VSS/d}$$

e. Calculate the amount of TSS produced from the process (ΔTSS).

$$\Delta TSS = \frac{P_{x,t}}{\text{VSS/TSS ratio}} = \frac{164\,\text{kg VSS/d}}{0.85\,\text{kg VSS/kg TSS}} = 193\,\text{kg TSS/d}$$

Without wasting sludge the total amount of TSS generated from the process is discharged in the effluent.

Calculate the expected TSS concentration in the effluent.

$$TSS_e = \frac{\Delta TSS}{Q} = \frac{193\,\text{kg VSS/d}}{800\,\text{m}^3/\text{d}} \times 10^3\,\text{g/kg} = 241\,\text{g TSS/m}^3 \text{ or } 241\,\text{mg TSS/L}$$

This is essentially the same TSS concentration given in the problem statement. Therefore, additional waste of sludge is not required.

5. Estimate the solids concentration in the reactor.

Rearrange Equation 10.15a, and calculate the average solids concentration in the reactor process volume without wasting sludge ($Q_{wa} = 0$).

$$X_{pv} = \frac{\theta_c^{design}}{V_t}(Q_{wa}X + (Q - Q_{wa})X_e) = \frac{\theta_c^{design}QX_e}{V_t} = \frac{60\,\text{d} \times 800\,\text{m}^3/\text{d} \times 204\,\text{mg VSS/L}}{400\,\text{m}^3}$$
$$= 24{,}500\,\text{mg VSS/L}$$

Assume that the majority of the solids are maintained as sludge granules in the sludge blanket that occupies approximately two-thirds of the total reactor process volume. Estimate the solids concentration in the sludge blanket.

Height of the sludge blanket,

$$h_{sb} = \text{sludge blanket/reactor process volume ratio} \times h_{pv} = \frac{2}{3} \times 8\,\text{m} = 5.3\,\text{m}$$

$$X_{sb} = \frac{X_{pv}}{\text{sludge blanket/reactor process volume ratio}} = \frac{3}{2} \times 24{,}500\,\text{mg VSS/L} = 36{,}750\,\text{mg VSS/L}$$

The solids concentration is ~ 37 kg/m^3 in the sludge blanket. It is within the acceptable range for UASB reactor process (Table 10.46).

6. Determine the HRT in the process volume (θ).

$$\theta = \frac{V_t}{Q} = \frac{400\,\text{m}^3 \times 24\,\text{h/d}}{800\,\text{m}^3/\text{d}} = 12\,\text{h}$$

It is within the acceptable range in Table 10.46

7. Determine the dimensions of the UASB reactors.

Provide a height of clearance zone $h_{cz} = 0.5$ m between the reactor process volume and the gas-solids separator, a height of the gas-solids separator $h_{gs} = 2.5$ m, and a height of settling/effluent collection zone $h_{ez} = 0.9$ m.

The total water depth in the reactor,

$$H_w = h_{pv} + h_{cz} + h_{gs} + h_{ez} = (8 + 0.5 + 2.5 + 0.9)\ \text{m} = 11.9\,\text{m}$$

Adding a freeboard $FB = 0.6$ m, the total height of the reactor,

$$H = H_w + FB = (11.9 + 0.6)\ \text{m} = 12.5\,\text{m}$$

Calculate the total surface area required by the UASB process.

$$\text{Required total surface area, } A_t = \frac{V_t}{h_{pz}} = \frac{400\,\text{m}^3}{8\,\text{m}} = 50\,\text{m}^2$$

$$\text{Provide two circular reactors and the surface area of each reactor, } A = \frac{1}{2}A_t = \frac{1}{2} \times 50\,\text{m}^2 = 25\,\text{m}^2$$

$$\text{Diameter of each reactor, } D = \sqrt{\frac{4A}{\pi}} = \sqrt{\frac{4 \times 25\,\text{m}^2}{\pi}} = 5.6\,\text{m}$$

Assume the surface area of the gas-solids separator is 80% of reactor area.

Diameter of the gas-solids separator,

$$d_{gs} = \sqrt{\frac{4}{\pi} \times 0.8 \times \frac{\pi}{4}(5.6\,\text{m})^2} = \sqrt{0.8} \times 5.6\,\text{m} = 0.89 \times 5.6\,\text{m} = 5.0\,\text{m}$$

The slope of gas collector plate is 45°.

Assume that the baffle-ring is extended 0.2 m inside the gas-solids separator surface area.

Diameter of the opening area at the baffle,

$$d_b = d_{gs} - 2 \times 0.2\,\text{m} = (5 - 2 \times 0.2)\ \text{m} = 4.6\,\text{m}$$

The conceptual sectional view of the UASB reactor is given in Figure 10.84.

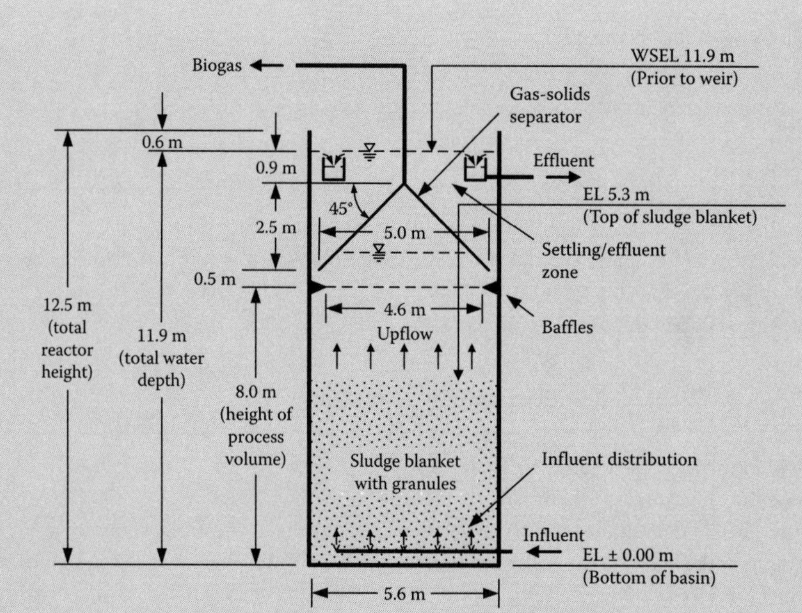

FIGURE 10.84 Conceptual sectional view of the UASB reactor (Example 10.122).

8. Verify the upflow velocity.

Calculate the upflow velocity, $V_h = \dfrac{Q}{A_t} = \dfrac{800\,\text{m}^3/\text{d}}{50\,\text{m}^2 \times 24\,\text{h/d}} = 0.7\,\text{m/h}$

At design flow, the upflow velocity is within the typical range (Table 10.46). Proper upflow velocity is one of the critical factors for formation of sludge granules. Therefore, additional external recirculation may be required to achieve the desired upflow velocity during start-up period and/or under low flow conditions.

Note: The gas production, energy contained in CH_4, and nutrient and alkalinity requirements may be calculated following the procedures given in Examples 10.119 and 10.120.

10.5.6 Anaerobic Attached Growth Processes

The anaerobic attached growth process utilizes biological reactors with packed media to support biological growth. These anaerobic reactors are: (1) packed-bed reactor (AnPBR), (2) expanded-bed reactor (AnEBR), (3) fluidized-bed reactor (AnFBR), and (4) rotating biological contactor (AnRBC). High microbiological population develops over large surface area of the media. Low-strength wastewaters have been effectively treated in relatively short HRT with COD removal efficiencies in a typical range of 50–90%. These processes are shown in Figure 10.85. The design and operational parameters for these anaerobic attached growth processes are summarized in Table 10.47. A brief description of these processes is given below.

Anaerobic Packed-Bed Reactor (AnPBR): The AnPBR utilizes circular or rectangular housing. The packing materials are similar to those used in the aerobic submerged attached growth processes (Section 10.4.4). The flow may be in either upflow or downflow direction. In an upflow anaerobic packed-bed reactor (AnUPBR), a sedimentation basin may be integrated on the top that is similar to the general arrangement in a UASB reactor. Recent developments provide 30–50% empty space at the bottom. Suspended

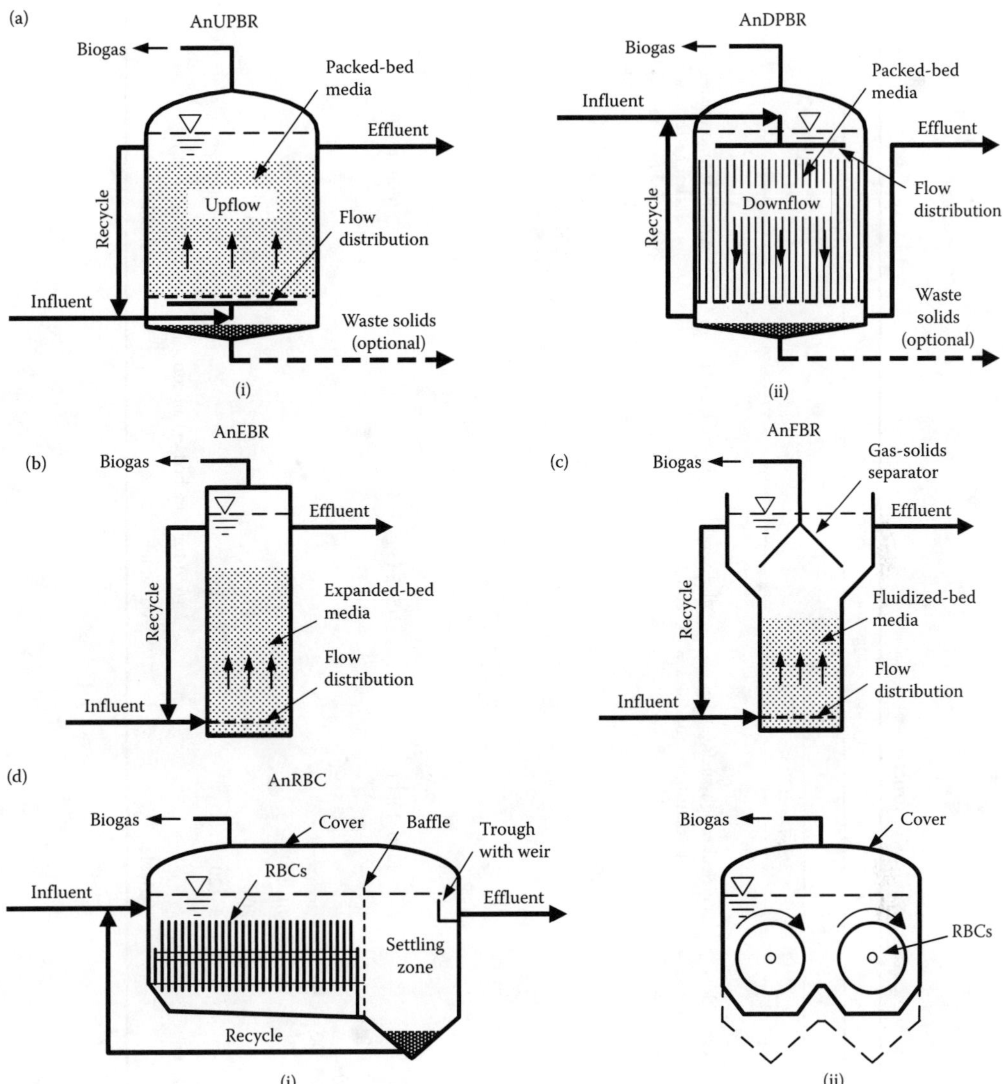

FIGURE 10.85 Anaerobic attached growth reactors: (a) packed-bed reactors (AnPBRs): (i) upflow (AnUPBR), and (ii) downflow (AnDPBR); (b) expanded-bed reactor (AnEBR); (c) fluidized-bed reactor (AnFBR); and (d) anaerobic rotating biological contactor (AnRBR): (i) longitudinal section view, and (ii) cross-section view.

growth accumulates in this zone to give a hybrid design.[2,98,166,167] A downflow packed-bed reactor (AnDPBR) uses vertically orientated modular surface media.[54,98,177] These reactors are shown in Figure 10.85a.

Anaerobic Expanded-Bed Reactor (AnEBR): The AnEBR is an upflow reactor in which the media is partially expanded. The media is usually the fine sands. High population of microorganisms is supported on large surface area over the media. The process is applied for treatment of municipal wastewater. The bed expansion reduces bed plugging.[98,166,171] The process is shown in Figure 10.85b.

Anaerobic Fluidized-Bed Reactor (AnFBR): In an AnFBR, the media is fully expanded due to high upflow velocity. Effluent recycle is necessary for desired upflow velocity. To prevent media wash out particularly with low-density media, various types of devices have been applied (tube settlers, screens,

TABLE 10.47 Design and Operational Parameters for Anaerobic Attached Growth Processes

Parameter	Typical Range for Process Type			
	AnPBR		AnEBR	AnFBR
	Upflow	Downflow		
Media				
Type and size	Corrugated plastic cross-flow or tubular modules; raschig, flexi- or pall rings; and rock or plastic balls	Vertical rough-surface clay and plastic tubular modules	Silica sand with dia. of 0.2–0.5 mm and sp. gr. of 2.65	Sand, diatomaceous earth, or ion-exchange resin with dia. of 0.1–0.7 mm; and GAC with dia. of 0.6–0.8 mm
Void ratio or pore volume, %	40–95	60–90	40–50	40–50
Specific surface area, m^2/m^3	85–200	70–100	10,000	5000–10,000
Expansion ratio, L_{fb}/L	–	–	1–1.3	1.5–2
Reactor dimensions				
Height, m	3–13	2–4	–	4–6
Diameter or width, m	2–8	–	–	–
Volumetric organic loading, kg $COD/m^3 \cdot d$ (lb $COD/10^3\ ft^3$)	0.1–1.2 (6–75) at COD 200–300 mg/L 1.5–15 (95–950) at COD 2500–25,000 mg/L	5–20 (300–1250)	0.5–5 (30–300)	10–40 (600–1500)
SRT, d	15–40	10–40	10–35	10–15
Solids concentration, g VSS/L	5–15	20	40–60	15–20
HRT, h	20–96	8–120	–	9–24
Upflow velocity, $m^3/m^2 \cdot h$ ($ft^3/ft^2 \cdot h$)	0.5–0.8 (1.5–2.5)	–	2 (6.5) for 20% bed expansion	10–20 (30–60)
Recycle ratio (Q_r/Q)	0–10	Variable	Variable	Variable

Source: Adapted in part from References 2, 7, 54, 98, 166, 167, and 171.

membranes).[54,98,166] The advantages of FBR are (1) high biomass, (2) high organic loading, (3) high mass transfer, (4) reduced shock loading, and (5), reduced media plugging. The biomass concentration in reactor may reach 15–20 g/L. The FBRs are best suited for low-strength wastes.[166,171,178] The process schematic is shown in Figure 10.85c.

Anaerobic Rotating Biological Contactor (AnRBC): The AnRBC reactors have submerged disks, use fixed covers, and may be under slight pressure.[179] The gases produced are collected. The effluent is settled and solids are returned. The AnRBC process is illustrated in Figure 10.85d. More design parameters and design details of RBC systems are covered in Sections 10.4.3 and 10.4.4.

EXAMPLE 10.123: DESIGN OF AN ANAEROBIC FLUIDIZED-BED REACTOR (AnFBR)

An AnFBR is designed to treat industrial wastewater. The media is uniform-sized GAC with diameter = 0.6 mm, sp. gr. = 1.4, and porosity ratio = 0.48. The fluidized bed ratio (L_{fb}/L) = 1.6. The COD concentration and flow rate of the influent wastewater are 5000 mg/L and 180 m^3/d, respectively. A design volumetric organic loading of 12 kg COD/m^3·d is desired at the operating temperature of 25°C. Estimate (1) the required fluidized-bed volume, (2) empty bed HRT in the fluidized-bed volume, (3) the dimensions of the reactor, (4) settling velocity of the clean media, (5) porosity of the fluidized bed, (6) upflow velocity, and (7) required recycle flow.

Solution

1. Determine the volume of the fluidized-bed media.

   Mass COD loading in influent, $\Delta S_0 = QS_0 = 180\,m^3/d \times 5000\,g\,COD/m^3 \times 10^{-3}\,kg/g = 900\,kg\,COD/d$

   Volume of the fluidized-bed media, $V_{fb} = \dfrac{\Delta S_0}{VOL} = \dfrac{900\,kg\,COD/d}{12\,kg\,COD/m^3{\cdot}d} = 75\,m^3$

2. Determine the empty bed HRT in the fluidized-bed volume (θ).

$$\theta = \frac{V_{fb}}{Q} = \frac{24\,m^3 \times 75\,h/d}{180\,m^3/d} = 10\,h$$

   It is within the typical range (Table 10.47).

3. Determine the dimensions of the fluidized-bed reactor.

   Provide a circular reactor with a design fluidized-bed depth (h_{fb}) to reactor diameter (D) ratio of 1.5 (h_{fb}:D = 1.5 or h_{fb} = 1.5 D. Calculate the diameter of the reactor.

$$V_{fb} = \frac{\pi}{4}D^2h_{fb} = \frac{\pi}{4}D^2 \times 1.5D = 0.375\pi D^3 \text{ or } D = \left(\frac{V_{fb}}{0.375\pi}\right)^{1/3} = \left(\frac{75\,m^3}{0.375\pi}\right)^{1/3} = 4.0\,m$$

   Depth of fluidized bed, $h_{fb} = 1.5D = 1.5 \times 4.0\,m = 6.0\,m$

   Note: It is within the acceptable range (Table 10.47).

   Height of unfluidized bed, $h_{ufb} = \dfrac{h_{fb}}{L_{fb}/L \text{ ratio}} = \dfrac{6\,m}{1.6} = 3.75\,m$

   Surface area of the reactor, $A_{fb} = \dfrac{\pi}{4}D^2 = \dfrac{\pi}{4}(4\,m)^2 = 12.6\,m^2$

Assume that the diameter of gas-solids separator is 0.5 m larger than the diameter of the reactor.

Diameter of the gas-solids separator, $d_{gs} = D + 0.5\,\text{m} = (4.0 + 0.5)\,\text{m} = 4.5\,\text{m}$

Surface area of the gas-solids separator, $A_{gs} = \dfrac{\pi}{4} D_{gs}^2 = \dfrac{\pi}{4} \times (4.5\,\text{m})^2 = 15.9\,\text{m}^2$

Provide a clearance $h_{cz} = 0.6\,\text{m}$ between the bottom of the settling zone and the top of the fluidized bed.

Provide a settling area that is equal to the surface area of the gas-solids separator, plus the surface area of the reactor.

Total surface area of the settling zone, $A_{sz} = A_{gs} + A_{fb} = (15.9 + 12.6)\,\text{m}^2 = 28.5\,\text{m}^2$

Diameter of the settling zone, $d_{sz} = \sqrt{\dfrac{4}{\pi} \times A_{sz}} = \sqrt{\dfrac{4}{\pi} \times 28.5\,\text{m}^2} = 6.0\,\text{m}$

Height of the hopper frustum, $h_{hf} = \dfrac{1}{2}(d_{sz} - d) = \dfrac{1}{2} \times (6 - 4)\,\text{m} = 1\,\text{m}$

Provide a gas-solids separator height, $h_{gs} = 2.25$ m at a slope of gas collector plate of 45°.
Provide an additional settling/effluent collection zone $h_{ez} = 1.05$ m.
Calculate the total water depth in the AnFBR.

$$H_w = h_{fb} + h_{cz} + h_{hf} + h_{gs} + h_{ez} = (6 + 0.6 + 1 + 2.25 + 1.05)\,\text{m} = 10.9\,\text{m}$$

Adding a freeboard $FB = 0.6$ m and calculate the total height of the reactor.

$$H = H_w + FB = (10.9 + 0.6)\,\text{m} = 11.5\,\text{m}$$

The conceptual sectional view of the AnFBR unit is given in Figure 10.86.

4. Determine the settling velocity of the clean media.

The kinematic viscosity $\nu = 0.893 \times 10^{-6}\,\text{m}^2/\text{s}$ for the wastewater temperature of 25°C is obtained from Table B.2 in Appendix B.

Assume that Stokes' equation (Equation 8.4) is applicable and calculate the terminal settling velocity of particle diameter $d = 0.6\,\text{mm} = 6 \times 10^{-4}\,\text{m}$, and $S_s = 1.4$.

$$v_t = \frac{g(S_s - 1)d^2}{18\nu} = \frac{9.81\,\text{m/s}^2 \times (1.4 - 1) \times (6 \times 10^{-4}\,\text{m})^2}{18 \times (0.893 \times 10^{-6}\,\text{m}^2/\text{s})} = 0.088\,\text{m/s}$$

Check N_R from Equation 8.3,

$$N_R = \frac{v_t d}{\nu} = \frac{0.088\,\text{m/s} \times (6 \times 10^{-4}\,\text{m})}{0.893 \times 10^{-6}\,\text{m}^2/\text{s}} = 59.1$$

Since $1 < N_R < 10^4$, the settling velocity is in the transition zone, and trial and error procedure will apply.

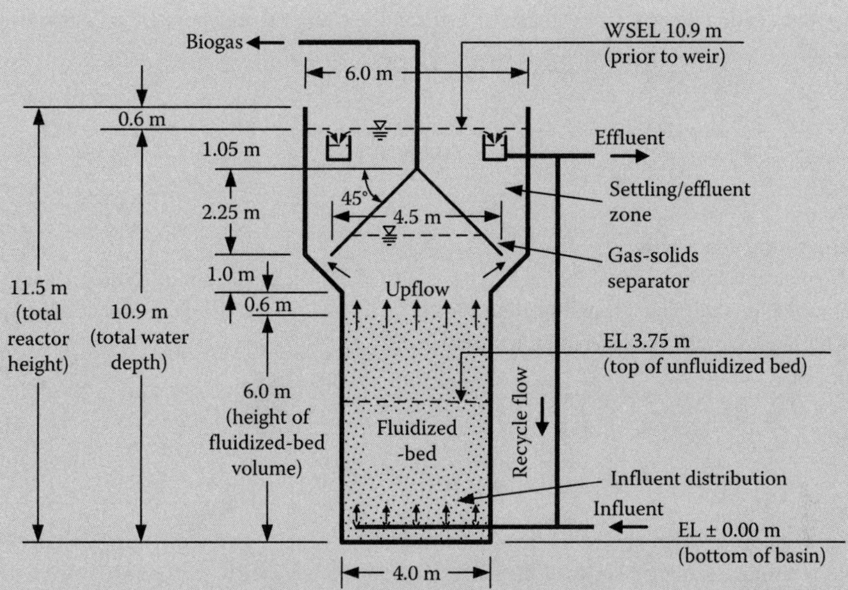

FIGURE 10.86 Conceptual sectional view of the AnFER (Example 10.123).

Calculate C_D from Equation 8.2.

$$C_D = \frac{24}{N_R} + \frac{3}{\sqrt{N_R}} + 0.34 = \frac{24}{59.1} + \frac{3}{\sqrt{59.1}} + 0.34 = 1.14$$

Calculate terminal settling velocity v_t from Equation 8.1a.

$$v_t = \sqrt{\frac{4g(S_s - 1)d}{3C_D}} = \sqrt{\frac{4 \times 9.81\,\text{m/s}^2 \times (1.4 - 1) \times (6 \times 10^{-4}\,\text{m})}{3 \times 1.14}} = 0.052\,\text{m/s}$$

Continue iterative solution until a stable value is reached. After fifth iterations, final value of $v_t =$ 0.043 m/s is reached.

5. Determine the porosity of the fluidized bed.

Since the volume of media before and after fluidization remain the same, the porosity of fluidized bed can be estimated from the porosity of unfluidized bed at the design fluidized-bed ratio.[180] The relationship is expressed by Equation 10.86a. Additional information on this topic may be found in Section 15.4.6.

$$(1 - e)L = (1 - e_{fb})L_{fb} \text{ or } \frac{L_{fb}}{L} = \frac{1 - e}{1 - e_{fb}} \quad (10.86a)$$

where

L and L_{fb} = depth of unfluidized and fluidized beds, m (ft)

e and e_{fb} = porosity of unfluidized and fluidized beds, dimensionless

Apply Equation 10.86a at the fluidized bed ratio $L_{fb}/L = 1.6$ and porosity of unfluidized bed $e = 0.48$.

$$1.6 = \frac{1-0.48}{1-e_{fb}} \text{ or } (1-e_{fb}) = \frac{1-0.48}{1.6} = 0.33$$

Solve for e_{fb}. $= 0.67$.

6. Determine the upflow velocity.

   The upflow velocity is calculated from the porosity of fluidized bed and the settling velocity of the clean media. It is calculated from Equation 10.86b.[1,98,180] Theory and design of expanded bed gravity filter during backwash are covered in detail in Chapter 15.

$$e_{fb} = \left(\frac{v_{uf}}{v_t}\right)^{0.22} \text{ or } v_{uf} = v_t(e_{fb})^{4.5} \tag{10.86b}$$

   where

   v_{uf} = upflow velocity, m/s or m^3/m^2·s (ft/s or ft^3/ft^2·s)

   v_t = terminal settling velocity of the media particle after backwash, m/s (ft/s)

   Apply Equation 10.86b to calculate v_{uf}.

$$v_{uf} = 0.043\,\text{m/s} \times (0.67)^{4.5} = 0.0071\,\text{m/s or } 25.6\,\text{m/h}$$

   It is slightly higher than the upper end of the typical range (Table 10.47).

   It should be noted that the calculated v_{uf} of 25.6 m/h is for clean media. The attached biological growth on the surface of the GAC media may reduce the upflow velocity required to achieve the desired fluidized ratio from that calculated for the clean media.

7. Determine the recycle flow.

   The upflow velocity calculated for the clean media is only used as a general guide for estimating the recycle flow rate.

   Upflow rate,

$$Q_{uf} = A_{fb}v_{uf} = 12.6\,\text{m}^2 \times 25.6\,\text{m/h} = 323\,\text{m}^3/\text{h or } 7752\,\text{m}^3/\text{d}$$

   Recycle flow,

$$Q_r = Q_{uf} - Q = (7752 - 180)\,\text{m}^3/\text{d} = 7572\,\text{m}^3/\text{d}$$

   The recycle flow ratio,

$$\frac{Q_r}{Q} = \frac{7572\,\text{m}^3/\text{d}}{180\,\text{m}^3/\text{d}} \approx 42$$

   The recycle ratio of 42 is calculated from the general procedure for gravity filter. It is recommended that the actual upflow velocity and recycle ratio should be validated from a pilot study.

   Note: The SRT, sludge production, solids concentration, gas production, energy contained in CH_4, and nutrient and alkalinity requirements may be calculated following the procedures given in Examples 10.119, 10.120, and 10.122.

10.6 Biological Nitrogen Removal

Nitrogen in aquatic environment may exist in many forms. These forms are *organic* nitrogen (bound in proteins), dissolved *ammonia* gas (NH_3) or *ammonium* ion (NH_4^+), *nitrite* NO_2^-, *nitrate* NO_3^-, and *nitrogen* gas (N_2). Atmospheric nitrogen (N_2) may undergo fixation by microorganisms into proteins and denitrification may cause volatilization of N_2 into atmosphere. Significance of different forms of nitrogen in aquatic environment is summarized in Table 10.48.[2,7]

Nitrogen removal from wastewater is essential to protect the quality of receiving waters and for beneficial reuse of effluent. Biological nitrogen removal processes have received much interest. The benefits of biological process over physical and chemical methods are (1) less sludge quantity, (2) enhanced BOD_5 and TSS removal, and added process stability and reliability, and (3) savings of chemicals and associated costs. The theoretical and practical aspects of biological nitrogen removal are well understood. It is a two-step process: nitrification and denitrification. Both steps are discussed below.

10.6.1 Nitrification

Nitrification is the biological oxidation of ammonia (or ammonium) to nitrate. Prior to the nitrification, a heterotrophic hydrolysis is required for microbial decomposition, converting organic nitrogen contained in the TKN into ammonia (or ammonium). The two-step oxidation process then involves: (1) oxidation of ammonia to nitrite and (2) oxidation of nitrite to nitrate. A simplified schematic for a complete nitrification is shown in Figure 10.87.

Microbiology: The biological nitrification is carried out by special aerobic chemoautotrophic bacteria that are called *nitrifiers*. They are normally categorized in two distinct groups of bacteria: (1) the ammonia-oxidizing bacteria (AOB) that are Nitroso-organisms, and they oxidize ammonia to nitrite, and (2) the nitrite-oxidizing bacteria (NOB) that are Nitro-organisms, and they oxidize nitrite to nitrate. The AOB population is dominated by the genera *Nitrosomonas*, *Nitrosospira*, and *Nitrosococcus* while the NOB population includes the genera *Nitrobactor*, *Nitrospira*, *Nitrococcus*, and *Nitrospina*.[2,7,181] The important information about the substrate utilization, biomass growth and environmental factors for the nitrification are summarized below.

TABLE 10.48 Significance of Different Forms of Nitrogen in Aquatic Environment

Form of Nitrogen	Significance
Organic nitrogen (ON)	ON is bound in complex form in proteins, amino acids, and amino sugars. It may be soluble or particulate. Microorganisms convert ON into ammonia. Typical concentration of ON in raw municipal wastewater is 15 mg/L as N. Presence of ON in natural waters indicates recent pollution.
Ammonia (ammonium) nitrogen (NH_3-N (NH_4-N))	Decomposition of ON by microorganisms produces ammonia. Depending upon the pH, ammonia may exist as gas (NH_3) or ammonium ion (NH_4^+). Ammonia is toxic to aquatic life and exerts nitrogenous oxygen demand (NOD). High levels of NH_3 in receiving waters indicate active decomposition. Typical concentration of ammonia in municipal wastewater is 30 mg/L, and it is totally soluble. Total Kjeldahl nitrogen (TKN) is the sum of organic and ammonia nitrogen.
Nitrite and nitrate nitrogen (NO_2-N and NO_3-N)	Nitrite nitrogen is an intermediate state of ammonia oxidation. Its presence in natural waters is an indication of ongoing recovery from sewage pollution. It is highly toxic to fish and increases chlorine demand for effluent disinfection. Nitrate is the final stage of ammonia oxidation. Its presence may indicate full recovery from pollution. Well-nitrified effluent may contain nitrate nitrogen in the range 15–20 mg/L. High concentrations of nitrate nitrogen is associated with eutrophication of receiving waters. Both nitrite and nitrate in drinking water are associated with severe health effects to infants. Their MCLs in drinking water are 1 and 10 mg/L, respectively.

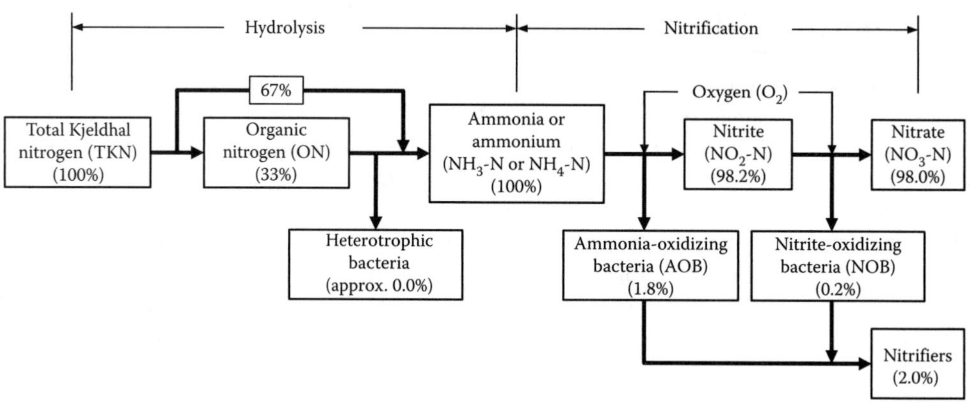

FIGURE 10.87 Simplified schematic of the biological nitrification process.

1. Nitrification can be carried out in conjunction with carbonaceous BOD removal or separately. The organisms responsible for BOD removal are heterotrophic bacteria while nitrifiers are autotrophic organisms and grow much slower than the heterotrophs. In a combined system, nitrification normally starts after the heterotrophs have consumed most of the carbonaceous BOD.
2. Like carbonaceous BOD removal, nitrification is carried out in suspended, attached, or hybrid systems.
3. The concentration of total Kjeldahl nitrogen (TKN) in wastewater is significantly lower than the organic matter (TKN $= 40$ mg/L and $BOD_5 = 180$ mg/L). Also, cell mass produced from nitrification is much less than that from heterotrophs. For these reasons, the fraction of nitrifiers in the MLVSS is typically 2–5%.
4. The growth of nitrifiers in a single-stage reactor depends upon BOD_5/TKN ratio. At a ratio of 5 and 0.5 the fractions of nitrifiers in MLVSS are 0.054 and 0.35, respectively.[182] As the ratio decreases, the nitrification rate increases.
5. The growth rate of nitrifier at DO level below 0.2 mg/L is insignificant. In general, NOB is more sensitive than AOB to low DO condition. For rapid and complete nitrification the DO level should be above 2 mg/L.
6. Nitrifiers are sensitive to lower pH. The optimum pH is in the range of 7.2–8.6. Nitrification practically stops below pH 6.3. Since alkalinity is consumed during nitrification and pH is lowered, an alkalinity concentration over 200 mg/L as $CaCO_3$ in the influent may be required for stable and complete nitrification if there is insignificant denitrification.
7. The activity of nitrifiers is adversely affected by lower temperatures. Temperature correction for nitrification rate constant must be applied in designing the reactor.

Deammonification process can also be performed by autotrophic AOB and the *anammox* bacteria through anaerobic ammonium oxidation. More detailed information about this process is covered in Section 13.10.4.

Stoichiometric Reactions: The biological oxidation reactions of ammonia to nitrite and then to nitrate are given by Equation 10.87. The biomass synthesis from nitrification are expressed by Equation 10.88.[2,7,181]

Oxidation of ammonia to nitrite by AOB:

$$NH_4^+ + 1.5\,O_2 + 2\,HCO_3^- \xrightarrow{\text{AOB}} NO_2^- + 2\,CO_2 + 3\,H_2O \tag{10.87a}$$

Oxidation of nitrite to nitrate by NOB:

$$NO_2^- + 0.5\,O_2 \xrightarrow{\text{NOB}} NO_3^- \tag{10.87b}$$

Overall ammonia oxidation reaction:

$$NH_4^+ + 2\,O_2 + 2\,HCO_3^- \xrightarrow{\text{AOB\&NOB}} NO_3^- + 2\,CO_2 + 3\,H_2O \tag{10.87c}$$

Biomass synthesis of AOB:

$$NH_4^+ + 1.38\,O_2 + 1.98\,HCO_3^- \rightarrow 0.018\,C_5H_7NO_2 + 0.98\,NO_2^- + 1.89\,CO_2 + 2.93\,H_2O \tag{10.88a}$$

Biomass synthesis of NOB:

$$NH_4^+ + 400\,NO_2^- + 195\,O_2 + HCO_3^- + 4\,CO_2 + H_2O \rightarrow C_5H_7NO_2 + 400\,NO_3^- \tag{10.88b}$$

Overall biomass synthesis of AOB and NOB from nitrification:

$$NH_4^+ + 1.86\,O_2 + 1.98\,HCO_3^- \rightarrow 0.021\,C_5H_7NO_2 + 0.98\,NO_3^- + 1.88\,CO_2 + 2.92\,H_2O \tag{10.88c}$$

The stoichiometric reactions represented above provide the following useful relationships:

1. The stoichiometric oxygen requirement for oxidation of ammonia to nitrate is 4.57 g O_2/g NH_4-N oxidized (Equation 10.87c). For cell synthesis, the consumption of oxygen is 4.25 g O_2/g NH_4-N utilized (Equation 10.88c). To be conservative, it is a common practice to ignore the biomass synthesis and use only the oxidation value of 4.57 g O_2/g NH_4-N consumed in the calculations of oxygen requirement. The calculation procedures for these ratios are shown in Example 10.124.
2. An alkalinity consumption of 7.14 g as $CaCO_3$ is required per gram of NH_4-N oxidized (Equation 10.87c). It is slightly higher than the alkalinity consumption of 7.07 g alkalinity as $CaCO_3$/g NH_4-N utilized by biomass synthesis (Equation 10.88c). Similar to the approach for dealing with the oxygen requirement, the higher value of 7.14 g alkalinity as $CaCO_3$/g NH_4-N consumed is typically used to estimate the alkalinity requirement by the nitrification (see Example 10.124).
3. The biomass synthesis are 0.15 g VSS/g NH_4-N utilized and 0.02 g VSS/g NO_2-N utilized for AOB and NOB, respectively. The overall biomass yield is ~ 0.17 g VSS/g NH_4-N utilized by nitrification. For a complete conversion of ammonia to nitrate, AOB is nearly 90% of total population for the nitrifiers. A simplified nitrogen mass balance through biological nitrification process is also shown in Figure 10.87. The calculation procedures for these constants are shown in Example 10.125.

Suspended Growth Kinetics: In a suspended growth aerobic reactor, the biomass is aerated and kept in suspension. Liquid and solids separation is achieved in a clarifier. These reactors are discussed in Section 10.3.2. Most biological kinetic equations developed in Sections 10.3.2 and 10.3.3 for aerobic suspended growth reactors also apply to AOB and NOB for nitrification (Equations 10.11 through 10.15). Since nitrification is sensitive to many environmental constraints, several safety or correction factors are applied. Also, the kinetic coefficients for nitrification are significantly different from those for organic substrate (BOD or COD) removal.

Specific Growth and Substrate Utilization Rates: The specific growth rate of biomass, specific substrate utilization rate, and specific net growth rate of biomass in an aerobic suspended growth reactor are expressed by Equations 10.11b, 10.12e, and 10.13b. These important equations in the context of nitrification kinetics are expressed by Equation 10.89.[2,3,7,181–185]

$$\mu_N = \frac{\mu_{max,N} N}{K_N + N} \tag{10.89a}$$

$$\mu'_N = \mu_N - k_{d,N} \text{ or } \mu'_N = \frac{\mu_{max,N} N}{K_N + N} - k_{d,N} \tag{10.89b}$$

$$k_N = \frac{\mu_{max,N}}{Y_N} \text{ or } \mu_{max,N} = Y_N k_N \tag{10.89c}$$

$$U_N = \frac{\mu_{max,N} N}{Y_N(K_N + N)} \text{ or } U_N = \frac{k_N N}{K_N + N} \tag{10.89d}$$

where

- μ_N = specific growth rate per unit mass of either AOB or NOB, mg VSS/mg VSS·d or d^{-1}
- $\mu_{max,N}$ = maximum specific growth rate per unit mass of either AOB or NOB, mg VSS/mg VSS·d or d^{-1}
- N = substrate concentration (NH_4-N or NO_2-N) in the effluent, mg/L (g/m^3) as N
- K_N = half-velocity constant for substrate (NH_4-N or NO_2-N), mg/L (g/m^3) as N
- μ'_N = specific net growth rate per unit mass of either AOB or NOB, mg VSS/mg VSS·d or d^{-1}
- $k_{d,N}$ = specific endogenous decay coefficient for either AOB or NOB, mg VSS/mg VSS·d or d^{-1}
- U_N = specific substrate utilization rate of NH_4-N or NO_2-N per unit mass of either AOB or NOB, mg N-utilized/mg VSS·d or d^{-1}
- Y_N = maximum cell yield coefficient over a finite period of growth phase, mg VSS/mg N-utilized as NH_4-N or NO_2-N
- k_N = maximum specific rate of substrate-utilization for NH_4-N or NO_2-N per unit mass of either AOB or NOB, mg N-utilized/mg VSS·d or d^{-1}

Kinetic Coefficients with Correction Factors: The nitrification kinetic coefficients commonly used for reactor design are summarized in Table 10.49.[2,7,181–185]

Correction Factors for Inhibitory Conditions: The nitrification process is sensitive to temperature, DO concentration, pH, and many other inhibitory conditions. The practically used corrections of nitrification

TABLE 10.49 Common Kinetic Coefficients for Nitrification

Parameter	Value[a]			
	AOB[b]		NOB[c]	
	Range	Typical	Range	Typical
Y_N, mg VSS/mg N-utilized	0.08–0.33	0.15	0.02–0.1	0.05
$k_{d,N}$, d^{-1}	0.15–0.2	0.17	0.15–0.2	0.17
$\mu_{max,N}$, mg VSS/mg VSS·d	0.32–4.55	0.9	0.65–1.8	1
k_N[d], mg N-utilized/mg VSS·d	–	6	–	20
K_N, mg/L as N	0.14–5	0.5	0.05–0.3	0.2

[a] These are the kinetic coefficients estimated at the temperature of 20°C and pH of 7.2. Use the correction factors presented below to obtain the design values under the field conditions.
[b] The substrate (N) is NH_4-N for AOB.
[c] The substrate (N) is NO_2-N for NOB.
[d] The value of k_N is calculated from Equation 10.89c.
Source: Adapted in part from References 2, 7, and 181 through 185.

kinetic coefficients for some of these factors are given in Equation 10.90.[2,7,181–186]

$$\mu_{max,N,F} = F_T F_{DO} F_{pH} \mu_{max,N} \text{ or } k_{N,F} = F_T F_{DO} F_{pH} k_N \quad (10.90a)$$

$$k_{d,N,F} = F_T k_{d,N} \quad (10.90b)$$

$$F_T = \theta_T^{T-20} \quad (10.90c)$$

$$F_{DO} = \frac{DO}{K_{DO} + DO} \quad (10.90d)$$

$$F_{pH} = 0.0004017\, e^{1.0946pH} \quad (pH < 7.0) \quad (10.90e)$$

$$F_{pH} = 2.35^{pH-7.2} \quad (pH \leq 7.2) \quad (10.90f)$$

$$F_{pH} = \frac{1.13(9.5 - pH)}{9.8 - pH} \quad (7.2 < pH \leq 9.5) \quad (10.90g)$$

where

$\mu_{max,N,F}$ = maximum specific growth rate per unit mass of either AOB or NOB under the field operating conditions, mg VSS/mg VSS·d or d^{-1}

$k_{N,F}$ = maximum specific rate of substrate-utilization per unit mass of either AOB or NOB under the field operating conditions, mg N-utilized/mg VSS·d or d^{-1}

F_T = operating temperature correction factor, dimensionless. The factor of F_T is applicable to either $\mu_{max,N}$ or $k_{d,N}$.

F_{DO} = operating DO concentration correction factor, dimensionless

F_{pH} = operating pH correction factor, dimensionless

$k_{d,N,F}$ = specific endogenous decay coefficient for either AOB or NOB under the field operating conditions, mg VSS/mg VSS·d or d^{-1}

θ_T = temperature correction coefficient, dimensionless. The typical values of θ_T for both AOB and NOB are summarized in Table 10.50. At a given temperature, same value of F_T is applicable for either $\mu_{max,N}$ or $k_{d,N}$ for NOB. However, the values of F_T may be different for AOB since the value of θ_T is different for $\mu_{max,N}$ and $k_{d,N}$.

T = critical temperature, °C

K_{DO} = DO half-saturation concentration for either AOB or NOB, mg/L or g/m^3. The typical values of K_{DO} for both AOB and NOB are also summarized in Table 10.50.

DO = operating dissolved oxygen (DO) concentration, mg/L or g/m^3

pH = operating pH value, standard pH unit

Note: The $\mu_{max,N}$ and $k_{d,N}$ in Equation 10.90 represent those at temperature of 20°C and pH of 7.2. Their ranges and typical values are summarized in Table 10.49. The procedure for correction of $\mu_{max,N}$ and $k_{d,N}$ from Equation 10.90 for the field conditions is given in Example 10.126.

High concentrations of free ammonia (NH_3-N), nitrous acid (HNO_2), and salinity (or TDS) may have potentially inhibitory effects on nitrification. Certain organic compounds and heavy metals

TABLE 10.50 Typical Kinetic Correction Coefficients for Nitrification

Parameter	Kinetic Coefficient to Be Corrected	Typical Value AOB	Typical Value NOB
θ_T, dimensionless	$\mu_{max,N}$	1.072	1.063
	$k_{d,N}$	1.029	1.063
K_{DO}, mg DO/L	$\mu_{max,N}$	0.50	0.90

Source: Adapted in part from Reference 181.

were also reported to be toxic to nitrifiers. Additional information on the topic may be found in References 2, 98, and 181.

Mean Cell Residence Time with Design Safety Factor: The general expression and minimum requirement of the MCRT (or SRT) in an aerobic suspended growth reactor are given by Equations 10.15d and 10.15f. Similar equations (Equations 10.91a and 10.91b) can also be used for nitrification.[2,7,181]

$$\frac{1}{\theta_{c,N}} = Y_N U_N - k_{d,N} \text{ or } \theta_{c,N} = \frac{1}{Y_N U_N - k_{d,N}} \text{ or } \theta_{c,N} = \frac{1}{\mu'_N} \tag{10.91a}$$

$$\theta_{c,N}^{min} = \frac{1}{\mu_{max,N} - k_{d,N}} \tag{10.91b}$$

where

$\theta_{c,N}$ = theoretical SRT determined from nitrification kinetics, d
$\theta_{c,N}^{min}$ = minimum SRT below which nitrification does not occur, d

There are many uncertainties associated with plant operation. These are flow and concentration fluctuations, slug and sustained loadings, and inhibitory effects of many organic and inorganic compounds under different environmental conditions. Therefore, safety factors are normally applied to determine the design SRT ($\theta_{c,N}^{design}$) from either of the two methods given by Equations 10.91c and 10.91d (Example 10.127 for calculation the design SRT from these two methods).

$$\theta_{c,N}^{design} = SF_P \theta_{c,N} \quad \text{(Theoretical SRT method)} \tag{10.91c}$$

$$\theta_{c,N}^{design} = SF_P SF_K \theta_{c,N}^{min} \quad \text{(Minimum SRT method)} \tag{10.91d}$$

where

$\theta_{c,N}^{design}$ = design SRT with safety considerations, d
SF_P = safety factor to account for uncertainty against process design and operation, dimensionless. The suggested values for SF_P are in the range of 1.5 to 2.5.
SF_K = safety factor to compensate for the sensitivity of nitrification kinetics under field conditions, dimensionless. The suggested values for SF_K are in the range of 2 to 5.

The suggested values for SF_K are in the range of 2–5.

During the normal operation of a properly designed nitrification process, the nitrifier population is normally dominated by AOB. Therefore, the design practice is to use the kinetic coefficients of AOB for estimating the overall substrate utilization and growth rates of the nitrification process. When the $\theta_{c,N}^{design}$ is determined, the specific substrate utilization rate (U_N), HRT (θ), reactor volume (V), biomass concentration in the reactor (X_N), and effluent ammonia concentration (N) can be estimated from applicable Equations 10.12 through 10.15 (see the calculations of using these equations in Example 10.128). The biomass growth rate and oxygen requirement for the nitrification are briefly discussed below.

Biomass Growth Rate. The observed biomass yield coefficient and biomass growth rate in an aerobic suspended growth reactor are expressed by Equations 10.15l, 10.15m, and 10.15n. They are modified to obtain the general equations (Equations 10.92a and 10.92b) for estimating the autotrophic biomass growth in a separate second-stage nitrification reactor. On account of the dominance of AOB for a complete oxidation of NH_4-N to NO_3-N, the overall biomass growth rate from nitrification is typically estimated from Equation 10.92c or 10.92d by substituting the kinetic coefficients for AOB with $N_0 = N_{1,NH4\text{-}N}$ and $N = N_{NH4\text{-}N}$.

$$Y_{obs,N} = \frac{Y_N}{1 + k_{d,N}\theta_{c,N}^{design}} \tag{10.92a}$$

$$P_{x,N} = Y_{obs,N}Q(N_0 - N)Q \text{ or } P_{x,N} = \frac{Y_N Q(N_0 - N)}{1 + k_{d,N}\theta_{c,N}^{design}} \quad (10.92b)$$

$$P_{x,N} = Y_{obs,AOB}Q(N_{1,NH4\text{-}N} - N_{NH4\text{-}N}) \text{ or } P_{x,N} = \frac{Y_{AOB}Q(N_{1,NH4\text{-}N} - N_{NH4\text{-}N})}{1 + k_{d,AOB}\theta_{c,N}^{design}} \quad (10.92c)$$

$$p_{x,N} = Y_{obs,AOB}(N_{1,NH4\text{-}N} - N_{NH4\text{-}N}) \text{ or } p_{x,N} = \frac{Y_{AOB}(N_{1,NH4\text{-}N} - N_{NH4\text{-}N})}{1 + k_{d,AOB}\theta_{c,N}^{design}} \quad (10.92d)$$

where

$Y_{obs,N}$ = observed biomass yield for either AOB or NOB due to consumption of substrate (NH_4-N or NO_2-N), g VSS/g N-consumed

$P_{x,N}$ = biosolids (VSS) growth rate of either AOB or NOB in the reactor, g VSS/d

Q = influent flow rate to the reactor, m^3/d

N_0 = initial substrate concentration (NH_4-N or NO_2-N), mg/L (g/m^3)

N = effluent substrate concentration (NH_4-N or NO_2-N), mg/L (g/m^3)

$N_{1,NH4\text{-}N}$ = NH_4-N concentration in the effluent from the first-stage, mg/L or g/m^3. The value of $N_{1,NH4\text{-}N}$ is calculated from Equation 10.92e.

$$N_{1,NH4\text{-}N} = N_{0,TKN} - \Delta N_{1,NH4\text{-}N} \quad (10.92e)$$

where

$N_{0,TKN}$ = influent TKN concentration to the first stage, mg/L (g/m^3)

$\Delta N_{1,NH4\text{-}N}$ = ammonia nitrogen utilized by the heterotrophs for cell synthesis in the first stage, mg/L (g/m^3)

$N_{NH4\text{-}N}$ = NH_4-N concentration in the effluent from the second-stage, mg/L (g/m^3)

$p_{x,N}$ = biosolids (VSS) growth rate per unit flow through the reactor, mg VSS/L (g VSS/m^3)

In a single aerobic reactor, the suspended growths of heterotrophs and autotrophs are added to give the combined biomass growth. Therefore, Equation 10.15m is combined with either Equation 10.92c or 10.92d after two modifications: (1) the design solid retention time (θ_c^{design}) is the larger value between $\theta_{c,BOD}^{design}$ for CBOD removal and $\theta_{c,N}^{design}$ for nitrification and (2) the amount of ammonia nitrogen utilized by the heterotrophs for cell synthesis is not available to the consumption by the nitrifiers and therefore it is subtracted. The relationship is expressed by Equation 10.92f.

$$P_{x,N} = \frac{Y_{AOB}Q}{1 + k_{d,AOB}\theta_c^{design}}(N_{0,TKN} - N_{NH4\text{-}N} - f_N P_{x,BOD}) \quad (10.92f)$$

The combined biomass growth rate from both CBOD removal and nitrification in a single reactor is expressed by Equation 10.92g or 10.92h.

$$P_{x,BOD+N} = P_{x,BOD} + P_{x,N} = P_{x,BOD} + \frac{Y_{AOB}Q}{1 + k_{d,AOB}\theta_c^{design}}(N_{0,TKN} - N_{NH4\text{-}N} - f_N P_{x,BOD}) \text{ or}$$

$$P_{x,BOD+N} = \frac{Y_{BOD}Q(S_0 - S)}{1 + k_{d,BOD}\theta_c^{design}} + \frac{Y_{AOB}Q}{1 + k_{d,AOB}\theta_c^{design}}\left(N_{0,TKN} - N_{NH4\text{-}N} - f_N\frac{Y_{BOD}(S_0 - S)}{1 + k_{d,BOD}\theta_c^{design}}\right) \quad (10.92g)$$

$$p_{x,BOD+N} = p_{x,BOD} + p_{x,N} = p_{x,BOD} + \frac{Y_{AOB}}{1 + k_{d,AOB}\theta_c^{design}}(N_{0,TKN} - N_{NH4\text{-}N} - f_N p_{x,BOD}) \text{ or}$$

$$p_{x,BOD+N} = \frac{Y_{BOD}(S_0 - S)}{1 + k_{d,BOD}\theta_c^{design}} + \frac{Y_{AOB}}{1 + k_{d,AOB}\theta_c^{design}}\left(N_{0,TKN} - N_{NH4\text{-}N} - f_N \frac{Y_{BOD}(S_0 - S)}{1 + k_{d,BOD}\theta_c^{design}}\right) \quad (10.92h)$$

where

$P_{x,BOD+N}$ = combined biosolids (VSS) growth rate for both CBOD removal and nitrification in the single reactor, g VSS/d

S_0 = BOD_5 concentration in the influent, mg BOD_5/L (g BOD_5/m^3)

S = soluble BOD_5 concentration in the effluent, mg BOD_5/L (g BOD_5/m^3)

$Y_{obs,BOD}$ = observed biomass yield for heterotrophs due to CBOD removal, g VSS/g BOD_5

$k_{d,BOD}$ = specific endogenous decay coefficient for heterotrophs, mg VSS/mg VSS·d or d^{-1}

f_N = fraction of nitrogen synthesized in the biomass (VSS), mg N/mg VSS. The typical value of f_N of 0.12 g N/g VSS is calculated based on the chemical formula of cellular mass $C_{60}H_{87}O_{27}N_{12}P$ (Table 10.3 in Example 10.6).

$p_{x,BOD+N}$ = combined biosolids (VSS) growth rate per unit flow through the single reactor, mg VSS/L or g VSS/m^3

Oxygen Requirement: The ultimate oxygen demand (*UOR*) for nitrification only is calculated from Equation 10.93a. The combined *UOR* for CBOD removal and nitrification in a single reactor is estimated from Equation 10.93b. From the *UOR*, the theoretical oxygen requirement (*ThOR*) is further calculated from Equation 10.48b.

$$UOR_N = 4.57(N_{0,TKN} - N_{NH4\text{-}N}) - 1.42 p_{x,N} \quad (10.93a)$$

$$UOR_{BOD+N} = \frac{(S_0 - S)}{BOD_5/BOD_L} + 4.57(N_{0,TKN} - N_{NH4\text{-}N} - f_N p_{x,BOD+N}) - 1.42 p_{x,BOD+N} \quad (10.93b)$$

where

UOR_N = ultimate oxygen concentration requirement for nitrification only, mg O_2/L (g O_2/m^3)

UOR_{BOD+N} = total ultimate oxygen concentration requirement for CBOD removal and nitrification in the single reactor, mg O_2/L (g O_2/m^3)

Note: The constant of 4.57 g O_2/g NH_4-N is the oxygen required for oxidation of ammonia to nitrate (Examples 5.26 and 10.124) for derivation of this factor). The constant of 1.42 g O_2/g VSS is the oxygen consumed for oxidation of biomass (VSS) (Example 10.10 for derivation of the constant). In a different arrangement, the general equation for *UOR* is also expressed by Equation 10.48a and derived in Example 10.70.

Attached Growth Kinetics: The types of fixed-film or attached growth reactors that may be used for nitrification include trickling filter (Section 10.4.1), ABF (Section 10.4.2), and RBC (Section 10.4.1). As an integrated fixed-film media and suspended growth process (Section 10.4.3), the IFAS or MBBR are more efficient than suspended growth systems for nitrification, especially under cold weather conditions.[2,98,165,181] Design information of these reactors for nitrification is given below.

TF and ABF. Nitrification in TF and ABF reactors is difficult to predict because of many uncertainties associated with the process. Some of these are: (1) biofilm thickness and density, (2) biofilm area covering the media surface, (3) wetted area of the biofilm, and (4) effectiveness of CBOD removal over nitrification. Most design equations therefore are based on semi empirical relationships. Some rules of thumb are (a) nitrification may not occur if the soluble CBOD loading is above 9 g/m^2·d (1.8 lb/1000 ft^2·d) or the soluble CBOD concentration is above 30 mg/L in the influent, (b) nearly complete nitrification typically occurs if soluble CBOD loading rate is below 2 g/m^2·d (0.4 lb/1000 ft^2·d) or the soluble CBOD concentration is below 5 mg/L in the influent, and (c) the nitrification is inhibited proportionately as soluble CBOD concentration increases from 5 to 30 mg/L or soluble CBOD loading rate increases from 2 to 9 g/m^2·d.[2,184]

For combined CBOD removal and nitrification, both organic and nitrogen loadings may need to be considered in the process design. Several design approaches have been developed by using the following design parameters: (1) BOD-based volumetric organic loading (VOL), (2) oxygen requirement based on volumetric oxidation rate (VOR), or (3) nitrogen-based specific nitrogen oxidation rate (R_N). A typical value of 0.2 kg BOD_L/m^3·d (12.5 lb BOD_L/10^3 ft^{3}d) is suggested for the cross-flow plastic modular packing media. A range of 0.08–0.5 kg BOD_5/m^3·d (5–30 lb BOD_5/10^3 ft^{3}d) is applied over trickling filters for low-rate CBOD removal with nitrification. This is given in Table 10.33. The VOR is expressed by Equation 10.94a and its value may be in a range of 0.4–1.3 kg O_2/m^3·d. The R_N is calculated from Equation 10.94b. A relationship between the R_N and ratio of BOD_5 to TKN in the influent is given by Equation 10.94c. When the applied organic, nitrogen, and HLs are known, the NH_4-N concentration in the effluent may also be estimated for a given effluent temperature from the semi empirical expression given by Equation 10.94d.[2,187,188]

$$VOR = \frac{(S_0 + 4.57N_{0,\text{TKN}})Q}{V_\text{m}} \tag{10.94a}$$

$$R_\text{N} = \frac{(N_{0,\text{TKN}} - N_{\text{NH4-N}})Q}{A_\text{m}} \text{ or } R_\text{N} = \frac{(N_{0,\text{TKN}} - N_{\text{NH4-N}})Q}{SA_\text{m} V_\text{m}} \tag{10.94b}$$

$$R_\text{N} = 0.82\left(\frac{S_0}{N_{0,\text{TKN}}}\right)^{-0.44} \tag{10.94c}$$

$$N_{\text{NH4-N}} = 20.81(SLR_\text{BOD})^{1.03}(SLR_\text{N})^{1.52}(SLR_\text{Q})^{-0.36}T^{-0.12} \tag{10.94d}$$

where

VOR = volumetric oxidation rate, kg O_2/m^3·d
R_N = specific nitrogen oxidation rate, g N-oxidied/m^2·d
V_m = volume of packing material, m^3
A_m = total surface area of packing material, m^2
SA_m = specific surface area of packing material, m^2/m^3
SLR_{BOD} = specific BOD_5 surface loading rate on the packing media, g BOD_5-applied/m^2·d
SLR_N = specific nitrogen surface loading rate on the packing media, g N-applied/m^2·d
SLR_Q = specific hydraulic surface loading rate on the packing media, L/m^2·d
T = effluent temperature, °C

Other terms have been previously defined.

The general expression for substrate removal flux by attached growth under substrate-limited conditions is provided in Equation 10.94e. This relationship may be used to determine the BOD_5 removal by heterotrophic growth in the biofilm under the low BOD_5 concentration in the media fill volume.

$$J_\text{BOD} = J_\text{BOD,max}\left(\frac{S}{K_\text{BOD,m} + S}\right) \tag{10.94e}$$

where

J_{BOD} = BOD_5 removal flux, g BOD_5 consumed/m^2·d
$J_{BOD,max}$ = maximum BOD_5 removal flux, g BOD_5 consumed/m^2·d. The value of $J_{BOD,max}$ may be affected by the applied substrate loading flux. Proper adjustment and/or correction may be required based on the pilot- and full-scale testing data.
$K_{BOD,m}$ = half-velocity constant for the BOD_5 utilization in the biofilm, mg BOD_5/L (g BOD_5/m^3).
S = BOD_5 concentration in the effluent from the media fill volume, mg BOD_5/L (g BOD_5/m^3)

Note: The values of $J_{BOD,max} = 5$ g $BOD_5/m^2{\cdot}d$ and $K_{BOD,m} = 60$ mg BOD_5/L are normally used for municipal wastewater.

For tertiary nitrification, a separate nitrification reactor is utilized after CBOD removal. Since low BOD loading applied to the filter is expected, the nitrogen removal flux (J_N) is the key parameter in the process design. It is described by Equation 10.94f.[2]

$$J_{N,T} = J_{N,max,T}\left(\frac{N_{NH4\text{-}N}}{K_{NH4\text{-}N} + N_{NH4\text{-}N}}\right) \tag{10.94f}$$

where

$J_{N,T}$ = nitrogen removal flux and bulk liquid temperature T, g N oxidized/$m^2{\cdot}d$

$J_{N,max,T}$ = maximum nitrogen removal flux at temperature T, g N oxidized/$m^2{\cdot}d$. The value of $J_{N,max,T}$ may vary broadly in a range from 1.2 to 2.9 g N/$m^2{\cdot}d$. When the bulk liquid temperature is in the range of 10–25°C, $J_{N,max,T} \approx J_{N,max,10}$ and remains almost unchanged. When the temperature is below 10°C, temperature correction is performed using Equation 10.94g.

$$J_{N,max,T} = J_{N,max,10}(1.045)^{T-10} \tag{10.94g}$$

where $J_{N,max,10}$ = maximum nitrogen removal flux at 10°C, g N-removed/$m^2{\cdot}d$

$K_{NH4\text{-}N}$ = half-velocity constant for NH_4-N, mg/L or g/m^3. The value of $K_{NH4\text{-}N}$ is typically in the range of 0.5–2 mg NH_4-N/L for design of trickling filter and ABFs for tertiary nitrification.

Other terms have been previously defined.

Using Equation 10.94f a mass balance is performed across an incremental depth of the packing media at a given filter depth. By integrating the expression within the boundary conditions, a general expression is obtained and expressed by Equation 10.94h. A trial and error approach is required to solve this equation.[2]

$$(N_i - N) + K_{NH4\text{-}N}\ln\left(\frac{N_i}{N}\right) = J_{N,max,T}\frac{Z\,SA_m}{Q_L} \tag{10.94h}$$

where N_i = NH_4-N concentration in the total influent flow applied to the filter, mg NH_4-N/L or g NH_4-N/m^3. When recirculation is used, N_i is calculated from Equation 10.94i.

$$N_i = \frac{N_0 + RN}{1 + R} \tag{10.94i}$$

where

R = recirculation ratio of recycle flow rate to influent flow rate, dimensionless

N_0 = NH_4-N concentration in the flow received by the filter from the secondary treatment process, mg/L

Z = plastic packing media depth, m

Q_L = total hydraulic surface loading rate applied to the filter, $m^3/m^2{\cdot}d$. When recirculation is used, Q_L is calculated from Equation 10.94j.

$$Q_L = (1 + R)q \tag{10.94j}$$

where q = surface hydraulic loading rate (HLR) on the filter (not including the recirculation), $m^3/m^2{\cdot}d$

Other terms have been previously defined.

Rotating Biological Contactor (RBC): Design details of multistage RBC for CBOD removal are covered in Section 10.4.3. In a combined carbon and nitrogen oxidation system, both organic and nitrogen loadings must be considered and the growth of nitrifiers is inhibited until the soluble BOD (sBOD) concentration drops below 10 mg/L. The common application of RBC for nitrification is to provide additional stages. Typical design parameters for combined CBOD removal with nitrification and separate nitrification are provided in Table 10.36. The effect of sBOD on nitrification can be estimated using Equation 10.95a.[98,184]

$$R_N = F_N R_{N,max} \tag{10.95a}$$

where

R_N = specific nitrogen oxidation rate, g NH_4-N/m^2·d

$R_{N,max}$ = maximum specific nitrogen oxidation rate, g NH_4-N/m^2·d. The typical value of $R_{N,max}$ is 1.5 g NH_4-N/m^2·d

F_N = reduction in nitrogen oxidation rate as a fraction of the maximum nitrification rate ($R_{N,max}$) without the influence of mass organic loading, dimensionless.
The F_N is calculated from Equation 10.95b.

$$F_N = 1.0 - 0.1\ SOL \tag{10.95b}$$

where SOL = soluble organic surface loading rate based on soluble BOD_5, g $sBOD_5/m^2$·d. At $SOL \geq 10$ g sBOD/m^2·d, the nitrification does not occur ($F_N = 0$ or $R_N = 0$).

At low organic loading the nitrification rate in a RBC may follow a first-order relationship with respect to the effluent NH_4-N concentration. The first-order plot of nitrogen oxidation rate (g NH_4-N/m^2·d) versus effluent nitrogen concentration (mg NH_4-N/L) has been reported.[184] Equation 10.95c is developed for the first-order nitrification rate in a RBC.

$$R_N = R_{N,max}\left(1 - e^{-k'_N N_{NH4\text{-}N}}\right) \tag{10.95c}$$

where k'_N = NH_4-N utilization rate constant, L/mg NH_4-N. The average NH_4-N utilization reaction-rate constant (k'_N) is normally 0.44 L/mg NH_4-N.

At a $N_{0,TKN} \geq 5$ mg/L, the reaction rate approaches zero order and the nitrogen oxidation rate (R_N) reaches $R_{N,max} = 1.5$ g NH_4-N/m^2·d. The application of Equation 10.95 is shown in Example 10.128.

IFAS and MBBR. Design details of IFAS and MBBR processes for CBOD removal are covered in Section 10.4.3. Similar to design of trickling filter and ABF for tertiary nitrification, the nitrogen removal flux (J_N) is also a key design parameter to determine the media surface area required by the attached growth nitrification in IFAS or MBBR systems. The value of J_N is determined from the DO and NH_4-N concentrations in the bulk liquid at the critical operating temperature. At a given design operating DO concentration, there is a critical NH_4-N concentration (N_c) for the zero-order nitrification flux. By comparing the NH_4-N concentration in the reactor ($N_{NH4\text{-}N}$) with the value of N_c, the nitrification rate can be identified as either DO or NH_4-N limited. The values of N_c in the DO concentration range of 2–6 mg/L are summarized in Table 10.51. More detailed information about the NH_4-N and DO limited conditions may be found in Reference 2.

TABLE 10.51 Criteria for NH_4-N Concentration for Zero-Order Nitrogen Removal Flux

Parameter	Value				
Operating DO concentration, mg/L	2	3	4	5	6
Critical NH_4-N concentration for DO-limited conditions (N_c), mg/L	0.5	0.8	1	1.3	1.65

Source: Developed from information provided in Reference 2.

Depending on the limiting conditions, the value of J_N can be expressed by either Equation 10.96a or 10.96b.[2]

$$J_{N,T} = J_{N,max,15}\left(\frac{N_c}{K_{NH4\text{-}N} + N_c}\right)(1.058)^{T-15} \quad \text{(DO limited conditions)} \qquad (10.96a)$$

$$J_{N,T} = J_{N,max,15}\left(\frac{N_{NH4\text{-}N}}{K_{NH4\text{-}N} + N_{NH4\text{-}N}}\right)(1.098)^{T-15} \quad (NH_4\text{-N limited conditions}) \qquad (10.96b)$$

where

$J_{N,T}$ = nitrogen removal flux at critical operating temperature T, g NH_4-N/m^2·d

$J_{N,max,15}$ = maximum nitrogen removal flux at bulk liquid temperature of 15°C, g NH_4-N/m^2·d. The typical value of $J_{N,max,15}$ is 3.3 g NH_4-N/m^2·d for attached growth nitrification in IFAS and MBBR systems.

N_c = critical NH_4-N concentration at a given operating DO concentration (Table 10.51), mg/L or g/m^3

$N_{NH4\text{-}N}$ = bulk liquid NH_4-N concentration in the reactor, mg/L or g/m^3

$K_{NH4\text{-}N}$ = half-velocity constant for NH_4-N, mg/L or g/m^3. The typical value of $K_{NH4\text{-}N}$ is 2.2 mg NH_4-N/L for design of attached growth nitrification in IFAS and MBBR system.

The application of Equation 10.96 to determine the design nitrogen removal flux in IFAS and MBBR is presented in Examples 10.135 and 10.136.

EXAMPLE 10.124: OXYGEN AND ALKALINITY REQUIREMENTS IN NITRIFICATION

Determine the oxygen and alkalinity requirements for ammonia oxidation and biomass synthesis in nitrification process.

Solution

1. Determine the oxygen requirements in the nitrification process.
   a. Determine the oxygen requirement due to ammonia oxidation.
      The ratio of oxygen consumed to nitrogen oxidized (g O_2/g NH_4-N) due to converting ammonia to nitrate by AOB and NOB is determined from Equation 10.87c.

$$NH_4^+ + 2\,O_2 + 2\,HCO_3^- \xrightarrow{\text{AOB \& NOB}} NO_3^- + 2\,CO_2 + 3\,H_2O$$

      Two moles of O_2 are theoretically needed per mole of NH_4-N oxidized.

$$O_{2,oxidation} = \frac{2 \text{ moles } O_2 \times 32\,g\,O_2/\text{mole } O_2}{1 \text{ mole } NH_4^+\text{-N} \times 14\,g\,NH_4^+\text{-N/mole } NH_4^+\text{-N}} = 4.57\,g\,O_2/g\,NH_4^+\text{-N oxidized}$$

   b. Determine the oxygen requirement due to biomass synthesis.
      The ratio of oxygen consumed to nitrogen utilized (g O_2/g NH_4-N) due to biomass synthesis of AOB and NOB is determined from Equation 10.88c.

$$NH_4^+ + 1.86\,O_2 + 1.98\,HCO_3^- \rightarrow 0.021\,C_5H_7NO_2 + 0.98\,NO_3^- + 1.88\,CO_2 + 2.92\,H_2O$$

Theoretically, 1.86 moles of O_2 are needed per mole of NH_4-N utilized.

$$O_{2,\text{synthesis}} = \frac{1.86 \text{ moles } O_2 \times 32 \text{ g } O_2/\text{mole } O_2}{1 \text{ mole } NH_4^+\text{-N} \times 14 \text{ g } NH_4^+\text{-N/mole } NH_4^+\text{-N}} = 4.25 \text{ g } O_2/\text{g } NH_4^+\text{-N utilized}$$

2. Determine the alkalinity requirements in the nitrification process.
   a. Determine the alkalinity requirement due to ammonia oxidation.
      The ratio of alkalinity consumed to nitrogen oxidized (g $CaCO_3$/g NH_4-N) from oxidation of ammonia to nitrate by AOB and NOB is determined from Equation 10.87c. Two moles of HCO_3^- are theoretically needed per mole of NH_4-N oxidized. Calculate the equivalents of two moles of HCO_3^-.

$$eq_{HCO3,\text{oxidation}} = \frac{2 \text{ moles } HCO_3^- \times 61 \text{ g } HCO_3^-/\text{mole } HCO_3^-}{61 \text{ g } HCO_3^-/\text{eq. as } HCO_3^-} = 2 \text{ eq.}$$

Calculate the alkalinity consumed.

$$w_{CaCO3,\text{oxidation}} = eq_{HCO3,\text{oxidation}} \times 50 \text{ g/eq. as } CaCO_3 = 2 \text{ eq.} \times 50 \text{ g/eq. as } CaCO_3 = 100 \text{ g as } CaCO_3$$

Therefore, 100 g of alkalinity as $CaCO_3$ are needed per mole of NH_4-N oxidized.

$$\text{Alk}_{\text{oxidation}} = \frac{100 \text{ g as } CaCO_3}{1 \text{ mole } NH_4^+\text{-N} \times 14 \text{ g } NH_4^+\text{-N/mole } NH_4^+\text{-N}}$$
$$= 7.14 \text{ g Alk as } CaCO_3/\text{g } NH_4^+\text{-N oxidized}$$

   b. Determine the alkalinity requirement due to biomass synthesis.
      The ratio of alkalinity consumed to nitrogen utilized (g $CaCO_3$/g NH_4-N) due to biomass synthesis of AOB and NOB is determined from Equation 10.88c. Theoretically, 1.98 moles of HCO_3^- are needed per mole of NH_4-N utilized. Calculate the equivalents of 1.98 moles of HCO_3^-.

$$eq_{HCO3,\text{synthesis}} = \frac{1.98 \text{ moles } HCO_3^- \times 61 \text{ g } HCO_3^-/\text{mole } HCO_3^-}{61 \text{ g } HCO_3^-/\text{eq. as } HCO_3^-} = 1.98 \text{ eq.}$$

Calculate the alkalinity consumed.

$$w_{CaCO3,\text{synthesis}} = eq_{HCO3,\text{synthesis}} \times 50 \text{ g/eq. as } CaCO_3 = 1.98 \text{ eq.} \times 50 \text{ g/eq. as } CaCO_3$$
$$= 99 \text{ g as } CaCO_3$$

Therefore, 99 g of alkalinity as $CaCO_3$ are needed per mole of NH_4-N utilized.

$$\text{Alk}_{\text{synthesis}} = \frac{99 \text{ g as } CaCO_3}{1 \text{ mole } NH_4^+\text{-N} \times 14 \text{ g } NH_4^+\text{-N/mole } NH_4^+\text{-N}}$$
$$= 7.07 \text{ g Alk as } CaCO_3/\text{g } NH_4^+\text{-N utilized}$$

3. Summarize the calculation results.
   The calculation results of the oxygen and alkalinity requirements in the nitrification process are summarized below.

Parameter	Nitrification Reaction	
	Oxidation of Ammonia	Biomass Synthesis
Oxygen requirement, g O_2/g NH_4-N	4.57	4.25
Alkalinity requirement, g Alk as $CaCO_3$/g NH_4-N	7.14	7.07

EXAMPLE 10.125: CELL SYNTHESIS YIELDS IN NITRIFICATION

Determine the cell synthesis yields of AOB, NOB, and nitrifiers as a whole in nitrification process.

Solution

1. Determine the cell yield of AOB.

   The cell yield of AOB (g VSS/g NH_4-N) from biomass synthesis during oxidation of ammonia to nitrite is determined from Equation 10.88a.

$$NH_4^+ + 1.38\,O_2 + 1.98\,HCO_3^- \rightarrow 0.018\,C_5H_7NO_2 + 0.98\,NO_2^- + 1.89\,CO_2 + 2.93\,H_2O$$

   From the stoichiometric relationship, 0.018 mole of biomass VSS as $C_5H_7NO_2$ is produced per mole of NH_4-N utilized. The molecular weight of biomass, $w_{bVSS} = 113$ g VSS/mole $C_5H_7NO_2$.

$$Y_{AOB} = \frac{0.018 \text{ moles } C_5H_7NO_2 \times 113\,\text{g VSS/mole } C_5H_7NO_2}{1 \text{ mole } NH_4^+\text{-N} \times 14\,\text{g } NH_4^+\text{-N/mole } NH_4^+\text{-N}} = 0.15\,\text{g VSS/g } NH_4^+\text{-N utilized}$$

2. Determine the cell yield of NOB.

   The cell yield of NOB (g VSS/g NO_2-N) from biomass synthesis during oxidation of nitrite to nitrate is determined from Equation 10.88b.

$$NH_4^+ + 400\,NO_2^- + 195\,O_2 + HCO_3^- + 4\,CO_2 + H_2O \rightarrow C_5H_7NO_2 + 400\,NO_3^-$$

   Theoretically, one mole of biomass VSS as $C_5H_7NO_2$ is produced per 400 moles of NO_2-N utilized.

$$Y_{NOB} = \frac{1 \text{ mole } C_5H_7NO_2 \times 113\,\text{g VSS/mole } C_5H_7NO_2}{400 \text{ mole } NO_2^-\text{-N} \times 14\,\text{g } NO_2^-\text{-N/mole } NO_2^-\text{-N}} = 0.02\,\text{g VSS/g } NO_2^-\text{-N utilized}$$

3. Determine the overall cell yield of nitrifiers.

   The overall cell yield of nitrifier (g VSS/g NH_4-N) from biomass synthesis of AOB and NOB in a complete oxidation of ammonia to nitrate is determined from Equation 10.88c.

$$NH_4^+ + 1.86\,O_2 + 1.98\,HCO_3^- \rightarrow 0.021\,C_5H_7NO_2 + 0.98\,NO_3^- + 1.88\,CO_2 + 2.92\,H_2O$$

   The stoichiometric relationship indicates that 0.021 mole of biomass VSS as $C_5H_7NO_2$ is generated per mole of NH_4-N utilized.

$$Y_{nitrifier} = \frac{0.021 \text{ moles } C_5H_7NO_2 \times 113\,\text{g VSS/mole } C_5H_7NO_2}{1 \text{ mole } NH_4^+\text{-N} \times 14\,\text{g } NH_4^+\text{-N/mole } NH_4^+\text{-N}} = 0.17\,\text{g VSS/g } NH_4^+\text{-N utilized}$$

4. Summarize the calculation results.
   The calculation results of the cell synthesis yields in the nitrification process are summarized below.

Parameter	Value
Y_{AOB}, g VSS/g NH_4-N	0.15
Y_{NOB}, g VSS/g NO_2-N	0.02
$Y_{nitrifier}$, g VSS/g NH_4-N	0.17

EXAMPLE 10.126: DETERMINATION OF NITRIFICATION KINETIC COEFFICIENTS UNDER FIELD OPERATING CONDITIONS

Determine the following nitrification kinetic coefficients under the field operating conditions: (a) the maximum specific growth rate and (b) specific endogenous decay coefficient. The DO concentration in the reactor and critical operating temperature are 2.0 mg/L and 12°C, respectively. An operating pH of 7.2 is expected based on the buffering capacity of wastewater. Use typical kinetic coefficients for AOB given in Tables 10.49 and 10.50.

Solution

1. Determine the typical nitrification kinetic coefficients.
   The typical nitrification kinetic coefficients for AOB at 20°C and pH of 7.2 are obtained from Table 10.49.

$$\mu_{max,AOB} = 0.9\,\text{mg VSS/mg VSS·d}$$

$$k_{d,AOB} = 0.17\,\text{d}^{-1}$$

2. Determine the maximum specific growth rate under the field operating conditions.
   Equation 10.90a is used for correction of $\mu_{max,AOB}$ for the field operating conditions.

$$\mu_{max,AOB,F} = F_T F_{DO} F_{pH} \mu_{max,AOB}$$

   a. Determine the operating temperature correction factor (F_T).
      Calculate F_T from Equation 10.90c using $\theta_T = 1.072$ for AOB from Table 10.50.

$$F_T = \theta_T^{T-20} = 1.072^{12-20} = 0.573$$

   b. Determine the operating DO concentration correction factor (F_{DO}).
      Calculate F_{DO} from Equation 10.90d using $K_{DO,AOB} = 0.5$ mg/L for AOB from Table 10.50.

$$F_{DO} = \frac{DO}{K_{DO,AOB} + DO} = \frac{2.0\,\text{mg/L}}{(0.50 + 2.0)\,\text{mg/L}} = 0.80$$

   c. Determine the operating pH correction factor (F_{pH}).
      Calculate F_{pH} from Equation 10.90f for pH ≤ 7.2.

$$F_{pH} = 2.35^{pH-7.2} = 2.35^{7.2-7.2} = 1.00$$

   d. Calculate the maximum specific growth rate ($\mu_{max,AOB.F}$).

Calculate $\mu_{max,AOB.F}$ from Equation 10.90a.

$$\mu_{max,AOB,F} = F_T F_{DO} F_{pH} \mu_{max,AOB} = 0.573 \times 0.8 \times 1 \times 0.90 \text{ mg VSS/mg VSS·d}$$
$$= 0.458 \times 0.90 \text{ mg VSS/mg VSS·d} = 0.412 \text{ mg VSS/mg VSS·d}$$

3. Determine the specific endogenous decay coefficient under the field operating conditions.
   Calculate F_T from Equation 10.90c using $\theta_T = 1.029$ for AOB from Table 10.50.

$$F_T = \theta_T^{T-20} = 1.029^{12-20} = 0.796$$

Equation 10.90b is used for correction of $k_{d,AOB,F}$ for the field operating conditions.

$$k_{d,AOB,F} = F_T k_{d,AOB} = 0.796 \times 0.17 \text{ d}^{-1} = 0.135 \text{ d}^{-1}$$

EXAMPLE 10.127: DESIGN SRT FOR NITRIFICATION

The design SRT for nitrification is determined by applying safety factors. Two methods are used for determination of the design SRT for nitrification: (1) *Theoretical SRT* and (2) *Minimum SRT*. Determine the design SRT for nitrification by both methods. An effluent NH_4-N concentration of 0.5 mg/L is desired. Use the kinetic coefficients given for AOB in Table 10.49, and the maximum specific growth rate that is corrected for the field operating conditions in Example 10.126.

Solution

1. Determine the specific growth and substrate utilization rates for AOB under field conditions.
   The specific growth rate for AOB under field conditions ($\mu_{AOB.F}$) is calculated from Equation 10.89a using $N = 0.5$ mg NH_4-N/L, $\mu_{max,AOB.F} = 0.412$ mg VSS/mg VSS·d that is calculated in Step 2, d of Example 10.126 and $K_{NH4\text{-}N} = 0.50$ mg NH_4-N/L for AOB given in Table 10.49.

$$\mu_{AOB,F} = \frac{\mu_{max,AOB,F} N}{K_{NH4\text{-}N} + N} = \frac{0.412 \text{ mg VSS/mg VSS·d} \times 0.5 \text{ mg NH}_4\text{-N/L}}{(0.5 + 0.5) \text{ mg NH}_4\text{-N/L}}$$
$$= 0.206 \text{ mg VSS/mg VSS·d}$$

Calculate the specific net growth rate for AOB under field conditions ($\mu'_{AOB,F}$) from Equation 10.89b using $k_{d,AOB.F} = 0.135 \text{ d}^{-1}$ from Example 10.126.

$$\mu'_{AOB,F} = \mu_{AOB,F} - k_{d,AOB,F} = (0.206 - 0.135) \text{ d}^{-1} = 0.071 \text{ d}^{-1}$$

2. Determine the design SRT from the theoretical SRT.
   Calculate the theoretical SRT ($\theta_{c,N}$) from Equation 10.91a.

$$\theta_{c,N} = \frac{1}{\mu'_{AOB,F}} = \frac{1}{0.071 \text{ d}^{-1}} = 14 \text{ d}$$

Provide $SF_P = 1.5$ for the uncertainty in process design and operation and calculate the design SRT $\left(\theta_{c,N}^{design}\right)$ from Equation 10.91c.

$$\theta_{c,N}^{design} = SF_P \theta_{c,N} = 1.5 \times 14 \text{ d} = 21 \text{ d}$$

3. Determine the design SRT from the minimum SRT.

Calculate the minimum SRT ($\theta_{c,N}^{min}$) from Equation 10.91b.

$$\theta_{c,N}^{min} = \frac{1}{\mu_{max,AOB,F} - k_{d,AOB,F}} = \frac{1}{(0.412 - 0.135)\,d^{-1}} = 3.6\,d$$

Use $SF_P = 1.5$, provide $SF_K = 3.5$ for the sensitivity of nitrification kinetics, and calculate the $\theta_{c,N}^{design}$ from Equation 10.91d.

$$\theta_{c,N}^{design} = SF_P SF_K \theta_{c,N}^{min} = 1.5 \times 3.5 \times 3.6\,d = 5.25 \times 3.6\,d = 19\,d$$

4. Comment on the calculated results from two methods.

The design SRTs determined from two methods are 21 and 19 d. A design SRT $\theta_c^{design} = 21$ d is recommended for nitrification achieving a desired effluent NH4-N concentration of 0.5 mg/L.

Notes: The *Theoretical SRT* method provides the designer a more reliable design SRT due to less uncertainties. The *Minimum SRT* method has some uncertainties. Therefore, more experience-based judgment from the designer may be needed.

EXAMPLE 10.128: DESIGN OF A SINGLE-STAGE SUSPENDED GROWTH REACTOR FOR COMBINED CBOD REMOVAL AND NITRIFICATION

A suspended growth biological reactor is designed to achieve BOD_5 removal and nitrification. The characteristics of primary settled influent to the reactor are given below: flow = 4000 m^3/d, BOD_5 concentration = 150 mg/L, and TKN concentration = 35 mg/L. The design MLSS concentration is 2500 mg/L and VSS/TSS ratio is 0.8 in the mixed liquor. An effluent soluble BOD_5 concentration <3 mg/L is desired. Apply the data and results obtained for the nitrification process in Examples 10.126 and 10.127. For BOD_5 removal, use the typical kinetic coefficients given in Table 10.10. Calculate the following: (a) concentrations of soluble BOD_5 and NH_4-N in the effluent, (b) specific substrate utilization rates for BOD_5 removal and nitrification, (c) approximate fraction of nitrifiers in the mixed liquor, (d) hydraulic retention time, and (e) reactor volume. Ignore the amount of ammonia nitrogen utilized by the heterotrophs for cell synthesis.

Solution

1. List the kinetic parameters used in the calculations.
   a. Apply the calculated results of nitrification obtained in Examples 10.126 and 10.127:

$$\mu_{max,AOB.F} = 0.412\,mg\,VSS/mg\,VSS{\cdot}d,\ k_{d,AOB.F} = 0.135\,d^{-1},\ \mu'_{AOB,F} = 0.071\,d^{-1},\ and\ \theta_c^{design} = 21\,d$$

   b. Typical kinetic coefficients for AOB obtained from Table 10.49:

$$Y_{AOB} = 0.15\,mg\,VSS/mg\,NH_4\text{-}N,\ \ and\ \ K_{NH4\text{-}N} = 0.50\,mg\,NH_4\text{-}N/L$$

   c. Typical kinetic coefficients for BOD_5 removal obtained from Table 10.10:

$$Y_{BOD} = 0.5\,mg\,VSS/mg\,BOD_5,\ \ k_{d,BOD} = 0.05\,d^{-1},$$
$$k_{BOD} = 5\,mg\,BOD_5/mg\,VSS\,d,\ \ and\ \ K_{S,BOD5} = 60\,mg\,BOD_5/L$$

2. Calculate the concentrations of BOD_5 and NH_4-N in the effluent.

   Apply Equation 10.15h to calculate the BOD_5 concentration (S) and NH_4-N concentration (N) in the effluent.

$$S = \frac{K_{s,BOD}(1 + k_{d,BOD}\theta_c^{design})}{\theta_c^{design}(Y_{BOD}k_{BOD} - k_{d,BOD}) - 1} = \frac{60\text{ mg BOD}_5/\text{L} \times (1 + 0.05\text{ d}^{-1} \times 21\text{ d})}{21\text{ d} \times (0.5 \times 5\text{ d}^{-1} - 0.05\text{ d}^{-1}) - 1} = 2.44\text{ mg BOD}_5/\text{L}$$

   Note: The effluent soluble BOD_5 concentration of 2.44 mg/L is below the desired value of 3 mg/L. Therefore, the θ_c^{design} of 21 d is proper for the CBOD removal. If $S > 3$ mg/L, it indicates the design SRT required by CBOD removal is longer than that by the nitrification. Additional calculations are required to determine a design SRT that will meet the CBOD removal requirement (see the calculation procedure presented in Example 10.36).

$$N = \frac{K_{NH4\text{-}N}(1 + k_{d,AOB,F}\theta_c^{design})}{\theta_c^{design}(\mu_{max,AOB,F} - k_{d,AOB,F}) - 1} = \frac{0.5\text{mg NH}_4\text{-N/L} \times (1 + 0.135\text{d}^{-1} \times 21\text{d})}{21\text{d} \times (0.412\text{d}^{-1} - 0.135\text{d}^{-1}) - 1} = 0.40\text{mg NH}_4\text{-N/L}$$

   Note: The effluent NH_4-N concentration of 0.4 mg/L is below the desired value of 0.5 mg/L (from Example 10.127).

3. Calculate the specific substrate utilization rates for BOD_5 removal and nitrification.

   The specific substrate utilization rates for BOD_5 removal (U_{BOD}) and nitrification (U_{AOB}) are calculated from Equation 10.12e.

$$U_{BOD} = \frac{k_{BOD}S}{K_{s,BOD} + S} = \frac{5\text{ mg BOD}_5/\text{mg VSS·d} \times 2.44\text{ mg BOD}_5/\text{L}}{(60 + 2.44)\text{ mg BOD}_5/\text{L}} = 0.195\text{ mg BOD}_5/\text{mg VSS·d}$$

$$U_{AOB} = \frac{\mu_{max,AOB,F}N}{Y_{AOB}(K_{NH4\text{-}N} + N)} = \frac{0.412\text{ mg NH}_4\text{-N/mg VSS·d} \times 0.4\text{ mg NH}_4\text{-N/L}}{0.15\text{ d}^{-1} \times (0.5 + 0.4)\text{ mg NH}_4\text{-N/L}}$$
$$= 1.22\text{ mg NH}_4\text{-N/mg VSS·d}$$

4. Estimate the approximate fraction of nitrifiers in the mixed liquor.

   The biomass growth from BOD_5 removal and nitrification can be estimated from Equation 10.15n.

$$p_{BOD} = Y_{obs,BOD}(S_0 - S)$$

$$p_{AOB} = Y_{obs,AOB}(N_{0,TKN} - N)$$

   The approximate fraction of nitrifiers in the mixed liquor (f_{AOB}) is expressed by Equation 10.97a.

$$f_{AOB} = \frac{p_{AOB}}{p_{BOD} + p_{AOB}} = \frac{Y_{obs,AOB}(N_{0,TKN} - N)}{Y_{obs,BOD}(S_0 - S) + Y_{obs,AOB}(N_{0,TKN} - N)} \quad (10.97a)$$

   Calculate the observed biomass yields for heterotrophs due to BOD_5 removal ($Y_{obs,BOD}$) from Equation 10.15l.

$$Y_{obs,BOD} = \frac{Y_{BOD}}{1 + k_{d,BOD}\theta_c^{design}} = \frac{0.5\text{ mg VSS/mg BOD}_5}{1 + 0.05\text{ d}^{-1} \times 21\text{ d}} = 0.244\text{ mg VSS/mg BOD}_5$$

   Calculate the observed biomass yields for AOB ($Y_{obs,AOB}$) from Equation 10.92a.

$$Y_{obs,AOB,F} = \frac{Y_{AOB}}{1 + k_{d,AOB,F}\theta_c^{design}} = \frac{0.15\text{ mg VSS/mgNH}_4\text{-N}}{1 + 0.135\text{ d}^{-1} \times 21\text{ d}} = 0.039\text{ mg VSS/mg NH}_4\text{-N}$$

Estimate the fraction factor (f_{AOB}) from Equation 10.97a.

$$f_{AOB} = \frac{0.039 \times (35 - 0.4)\ \text{mg/L}}{0.244 \times (150 - 2.44)\ \text{mg/L} + 0.039 \times (35 - 0.4)\ \text{mg/L}} = 0.036$$

Note: The nitrifiers count for about 3.6% of the total biomass in the mixed liquor. It is within in the typical range of 2–5%.

Calculate the concentration of biomass as MLVSS (X) in the mixed liquor.

$$X = \text{VSS/TSS ratio} \times \text{MLSS} = 0.8\ \text{mg VSS/mg TSS} \times 2500\ \text{mg TSS/L} = 2000\ \text{mg VSS/L}$$

Calculate the concentration of AOB (X_{AOB}) in the mixed liquor.

$$X_{AOB} = f_{AOB}X = 0.036 \times 2000\ \text{mg VSS/L} = 72\ \text{mg VSS/L}$$

Calculate the concentration of heterotrophs (X_{BOD}) in the mixed liquor.

$$X_{BOD} = X - X_{AOB} = (2000 - 72)\ \text{mg VSS/L} = 1928\ \text{mg VSS/L}$$

5. Determine the design hydraulic retention time in the reactor.

The required hydraulic retention time can be calculated from either BOD_5 removal (θ_{BOD}) or nitrification (θ_{AOB}) using rearranged Equation 10.14d. Both calculations are shown below.

$$\theta_{BOD} = \frac{S_0 - S}{U_{BOD}X_{BOD}} = \frac{(150 - 2.44)\ \text{mg BOD}_5\text{/L}}{0.195\ \text{mg BOD}_5\text{/mg VSS·d} \times 1928\ \text{mg VSS/L}} \times \frac{24\ \text{h}}{\text{d}} = 9.4\ \text{h}$$

$$\theta_{AOB} = \frac{N_{0,TKN} - N}{U_{AOB}X_{AOB}} = \frac{(35 - 0.4)\ \text{mg NH}_4\text{-N/L}}{1.22\ \text{mg NH}_4\text{-N/mg VSS·d} \times 72\ \text{mg VSS/L}} \times \frac{24\ \text{h}}{\text{d}} = 9.4\ \text{h}$$

Provide a design $\theta^{design} = 10\ \text{h}$.

6. Determine the volume of the reactor.

$$V = Q\theta^{design} = 4000\ \text{m}^3\text{/d} \times 10\ \text{h} \times \frac{\text{d}}{24\ \text{h}} = 1670\ \text{m}^3$$

EXAMPLE 10.129: HETEROTROPHIC AND AUTOTROPHIC BIOMASS GROWTH

A biological reactor is designed to carry out CBOD removal and nitrification. Determine the growth of heterotrophic and autotrophic organisms. Also check the fraction of autotrophs in the combined biomass. Use $f_N = 0.12$ g N/g VSS. The characteristics of primary settled influent and secondary effluent, and kinetic coefficients are given below.

Primary settled influent: $Q = 3000\ \text{m}^3\text{/d}$, $BOD_5 = 140$ mg/L, and TKN = 35 mg/L

Secondary effluent: $BOD_5 = 10$ mg/L, TSS = 10 mg/L, NH_4-N = 1 mg/L, VSS/TSS = 0.8, biodegradable solids/TSS = 0.65, BOD_L/biodegradable solids = 1.42, and $BOD_5/BOD_L = 0.68$.

Kinetic coefficients: $Y_{BOD} = 0.6$ mg VSS/mg BOD_5, $k_{d,BOD} = 0.06\ \text{d}^{-1}$, $Y_{AOB} = 0.15$ mg VSS/mg NH_4-N, $k_{d,AOB,F} = 0.14\ \text{d}^{-1}$, and $\theta_c = 12$ d.

Solution

1. Determine the soluble BOD_5 (S) in the effluent.

Calculate the concentration of particulate BOD_5 exerted by TSS (S_p) in the effluent.

$$S_p = \frac{\text{biodegradable solids}}{\text{TSS}} \times \frac{BOD_L}{\text{biodegradable solids}} \times \frac{BOD_5}{BOD_L} \times TSS_e$$

$$= (0.65 \times 1.42 \times 0.68 \times 10) \text{ mg } BOD_5/L = 6.3 \text{ mg } BOD_5/L$$

Calculate the soluble BOD_5 concentration (S_p) from the total BOD_5 concentration $S_t = 10$ mg BOD_5/L.

$$S = S_t - S_p = (10 - 6.3) \text{ mg } BOD_5/L = 3.7 \text{ mg } BOD_5/L$$

2. Determine the combined biomass growth.
   Calculate the growth of heterotrophic, autotrophic, and combined organisms separately from Equations 10.15n, 10.92f and 10.92g.
   a. Calculate the growth of heterotrophic organisms ($p_{x,BOD}$) from Equation 10.15n.

$$p_{x,BOD} = \frac{Y_{BOD}(S_0 - S)}{1 + k_{d,BOD}\theta_c} = \frac{0.6 \text{ mg VSS/mg } BOD_5 \times (140 - 3.7) \text{ mg } BOD_5/L}{1 + 0.06 \text{ d}^{-1} \times 12 \text{ d}} = 47.5 \text{ mg VSS/L}$$

   b. Calculate the growth of autotrophic organisms ($p_{x,N}$) from Equation 10.92f.

$$p_{x,N} = \frac{Y_{AOB}}{1 + k_{d,AOB,F}\theta_c^{design}}(N_{0,TKN} - N_{NH4\text{-}N} - f_N p_{x,BOD})$$

$$= \frac{0.15 \text{ mg VSS/mg } NH_4\text{-N}}{1 + 0.14 \text{ d}^{-1} \times 12 \text{ d}} \times ((35 - 1) \text{ mg } NH_4\text{-N/L} - 0.12 \text{ mg N/mg VSS}$$

$$\times 47.5 \text{ mg VSS/L}) = 1.6 \text{ mg VSS/L}$$

   c. Calculate the combined growth of heterotrophic and autotrophic organisms ($p_{x,BOD+N}$).

$$p_{x,BOD+N} = p_{x,BOD} + p_{x,N} = (47.5 + 1.6) \text{ mg VSS/L} = 49.1 \text{ mg VSS/L or } 49.1 \text{ g VSS/m}^3$$

3. Check the fraction of autotrophs in the combined biomass.

$$f_{autotrophs} = \frac{p_{x,N}}{p_{x,BOD+N}} = \frac{1.6 \text{ mg/L}}{49.1 \text{ mg/L}} = 0.033$$

   Note: The amount of autotrophic organisms is about 3.3% of the combined biomass. It is within in the typical range of 2–5%.
4. Determine the total mass of the sludge produced per day.
   Calculate the total biosolids ($P_{x,BOD+N}$) produced per day.

$$P_{x,BOD+N} = Qp_{x,BOD+N} = 3000 \text{ m}^3/\text{d} \times 49.1 \text{ g VSS/m}^3 \times 10^{-3} \text{ kg/g} = 147 \text{ kg VSS/d}$$

   Note: $P_{x,BOD+N}$ can also be calculated directly from Equation 10.92f.
   Calculate the total mass of sludge (W_{BOD+N}) produced per day.

$$W_{BOD+N} = \frac{P_{x,BOD+N}}{\text{VSS/TSS ratio}} = \frac{147 \text{ kg VSS/d}}{0.8 \text{ kg VSS/kg TSS}} = 184 \text{ kg TSS/d}$$

EXAMPLE 10.130: THEORETICAL OXYGEN REQUIREMENT (ThOR) IN A SINGLE REACTOR WITH CBOD REMOVAL AND NITRIFICATION

Determine the total theoretical oxygen requirement (*ThOR*) for combined CBOD removal and nitrification in a reactor. Apply the data and results obtained in Example 10.129.

Solution

1. List the data developed in Example 10.129 and used in the calculations.

$$Q = 3000\,\text{m}^3/\text{d},\ S_0 = 140\,\text{mg/L},\ S = 3.7\,\text{mg/L},\ N_{0,\text{TKN}} = 35\,\text{mg/L},\ N_{\text{NH4-N}} = 1\,\text{mg/L},$$
$$p_{\text{x,BOD+N}} = 49.1\,\text{g VSS/m}^3,\ f_\text{N} = 0.12\,\text{g N/g VSS, and BOD}_5/\text{BOD}_\text{L} = 0.68$$

2. Determine the ultimate oxygen concentration demand (*UOR*) for CBOD removal and nitrification. Substitute the data identified in Step 1 in Equation 10.93b.

$$UOR_{\text{BOD+N}} = \frac{(S_0 - S)}{\text{BOD}_5/\text{BOD}_\text{L}} + 4.57(N_{0,\text{TKN}} - N_{\text{NH4-N}} - f_\text{N} p_{\text{x,BOD+N}}) - 1.42 p_{\text{x,BOD+N}}$$
$$= \frac{(140 - 3.7)\,\text{mg BOD}_5/\text{L}}{0.68\,\text{mg BOD}_5/\text{mg BOD}_\text{L}}$$
$$+ 4.57\,\text{mg O}_2/\text{mg NH}_4\text{-N} \times ((35-1)\,\text{mg NH}_4\text{-N/L} - 0.12\,\text{mg N/mg VSS} \times 49.1\,\text{mg VSS/L})$$
$$- 1.42\,\text{mg O}_2/\text{mg VSS} \times 49.1\,\text{mg VSS/L}$$
$$= (200 + 128 - 70)\,\text{mg O}_2/\text{L} = 258\,\text{mg O}_2/\text{L or } 258\,\text{g O}_2/\text{m}^3$$

3. Determine the theoretical oxygen requirement (*ThOR*) for CBOD removal and nitrification. Calculate the *ThOR* from Equation 10.48b.

$$ThOR_{\text{BOD+N}} = Q \times UOR_{\text{BOD+N}} = 3000\,\text{m}^3/\text{d} \times 258\,\text{g O}_2/\text{m}^3 \times 10^{-3}\,\text{kg/g} = 774\,\text{kg O}_2/\text{d}$$

Note: See also Example 10.70 for derivation and calculation of *UOR*, *ThOR* and standard oxygen requirement (*SOR*) from Equation 10.48.

EXAMPLE 10.131: BIOMASS GROWTH AND SUPPLEMENTAL ALKALINITY FOR NITRIFICATION

A second-stage reactor is designed to carry out nitrification. The design influent characteristics are: flow = 8500 m^3/d, TKN = 35 mg/L and alkalinity = 140 mg/L as $CaCO_3$. An amount of ammonia nitrogen of 6 mg/L is utilized by the heterotrophs for cell synthesis in the first stage. The final effluent ammonia = 1 mg NH_4-N/L. The reactor operation conditions are: critical temperature = 10°C, DO = 2.5 mg/L and pH is in the range of 7.2–7.6. The typical kinetic coefficients are given for AOB in Tables 10.49 and 10.50 and applied to this problem. Use a safety factor $SF_\text{P} = 2$ for design SRT. Determine (a) design SRT, (b) biomass growth, (c) alkalinity consumed by the nitrification, (d) supplemental alkalinity, and (e) dosage and quantity of sodium bicarbonate ($NaHCO_3$) required. It is necessary to maintain a minimum residual alkalinity of 50 mg/L as $CaCO_3$ for pH control. Nitrogen fixed in the biomass is 12% ($f_\text{N} = 0.12$ g N/g VSS). Assume that the change in alkalinity concentration is minimal in the first stage of the treatment process.

Solution

1. Determine the typical nitrification kinetic coefficients used in the calculations.
   The nitrification kinetic coefficients for AOB at 20°C and pH of 7.2 are obtained from Table 10.49.

$$\mu_{\text{max,AOB}} = 0.9\,\text{mg VSS/mg VSS·d},\ k_{\text{d,AOB}} = 0.17\,\text{d}^{-1},$$
$$Y_{\text{AOB}} = 0.15\,\text{mg VSS/mg NH}_4\text{-N, and } K_{\text{NH4-N}} = 0.5\,\text{mg NH}_4\text{-N/L.}$$

2. Determine the maximum specific growth rate under the field operating conditions.
   a. Determine the operating temperature correction factor (F_T).
      Calculate F_T from Equation 10.90c using $\theta_T = 1.072$ for AOB from Table 10.50.

$$F_T = \theta_T^{T-20} = 1.072^{10-20} = 0.499$$

   b. Determine the operating DO concentration correction factor (F_{DO}).
      Calculate F_{DO} from Equation 10.90d using $K_{DO,AOB} = 0.5$ mg/L for AOB from Table 10.50.

$$F_{\text{DO}} = \frac{\text{DO}}{K_{\text{DO,AOB}} + \text{DO}} = \frac{2.5\,\text{mg/L}}{(0.50 + 2.5)\,\text{mg/L}} = 0.833$$

   c. Determine the operating pH correction factor (F_{pH}).
      Calculate F_{pH} from Equation 10.90g for $7.2 < \text{pH} \le 9.5$.

$$F_{\text{pH}} = \frac{1.13(9.5 - \text{pH})}{9.8 - \text{pH}} = \frac{1.13 \times (9.5 - 7.6)}{9.8 - 7.6} = 0.976$$

   d. Calculate the maximum specific growth rate ($\mu_{\text{max,AOB.F}}$).
      Calculate $\mu_{\text{max,AOB.F}}$ from Equation 10.90a.

$$\mu_{\text{max,AOB,F}} = F_T F_{\text{DO}} F_{\text{pH}} \mu_{\text{max,AOB}} = 0.499 \times 0.833 \times 0.976 \times 0.90\,\text{mg VSS/mg VSS·d}$$
$$= 0.406 \times 0.90\,\text{mg VSS/mg VSS·d} = 0.365\,\text{mg VSS/mg VSS·d}$$

3. Determine the specific endogenous decay coefficient under the field operating conditions.
   Calculate F_T from Equation 10.90c using $\theta_T = 1.029$ for AOB from Table 10.50.

$$F_T = \theta_T^{T-20} = 1.029^{10-20} = 0.751$$

   Calculate $k_{\text{d,AOB,F}}$ from Equation 10.90b.
   Equation 10.90b is used for correction of $k_{\text{d,AOB,F}}$ for the field operating conditions.

$$k_{\text{d,AOB,F}} = F_T k_{\text{d,AOB}} = 0.751 \times 0.17\,\text{d}^{-1} = 0.128\,\text{d}^{-1}$$

4. Determine the specific growth and substrate utilization rates for AOB under field conditions.
   The specific growth rate for AOB under field conditions ($\mu_{\text{AOB.F}}$) is calculated from Equation 10.89a.

$$\mu_{\text{AOB,F}} = \frac{\mu_{\text{max,AOB,F}} N}{K_{\text{NH4-N}} + N} = \frac{0.365\,\text{mg VSS/mg VSS·d} \times 1\,\text{mg NH}_4\text{-N/L}}{(0.5 + 1)\,\text{mg NH}_4\text{-N/L}}$$
$$= 0.243\,\text{mg VSS/mg VSS·d}$$

   Calculate the specific net growth rate for AOB under field conditions ($\mu'_{\text{AOB,F}}$) from Equation 10.89b.

$$\mu'_{\text{AOB,F}} = \mu_{\text{AOB,F}} - k_{\text{d,AOB,F}} = (0.243 - 0.128)\,\text{d}^{-1} = 0.115\,\text{d}^{-1}$$

5. Determine the design SRT from the theoretical SRT.
   Calculate the theoretical SRT ($\theta_{c,N}$) from Equation 10.91a.

$$\theta_{c,N} = \frac{1}{\mu'_{AOB,F}} = \frac{1}{0.115\,d^{-1}} = 8.7\,d$$

   Apply $SF_P = 2$ and calculate the design SRT ($\theta_{c,N}^{design}$) from Equation 10.91c.

$$\theta_{c,N}^{design} = SF_P\theta_{c,N} = 2 \times 8.7\,d = 17.4\,d$$

   Provide $\theta_{c,N}^{design} = 18\,d$.
6. Verify the design SRT based on the minimum SRT.
   Calculate the minimum SRT ($\theta_{c,N}^{min}$) from Equation 10.91b.

$$\theta_{c,N}^{min} = \frac{1}{\mu_{max,AOB,F} - k_{d,AOB,F}} = \frac{1}{(0.365 - 0.128)\,d^{-1}} = 4.2\,d$$

   Rearrange Equation 10.91d and calculate the SF_K from $\theta_{c,N}^{design} = 18\,d$.

$$SF_K = \frac{\theta_{c,N}^{design}}{SF_P\theta_{c,N}^{min}} = \frac{18\,d}{2 \times 4.2\,d} = 2.1$$

   A SF_K of 2.1 is provided. It is within the suggested range of 2–5. Therefore, $\theta_{c,N}^{design} = 18\,d$ is acceptable.
7. Determine the biomass growth.
   Apply Equation 10.92e to calculate the effluent TKN concentration from the first-stage treatment.

$$N_{1,NH4\text{-}N} = N_{0,TKN} - \Delta N_{1,NH4\text{-}N} = (35 - 6)\text{ mg NH}_4\text{-N/L} = 29\text{ mg NH}_4\text{-N/L}$$

   Calculate the growth of autotrophic organisms ($p_{x,N}$) from Equation 10.92d.

$$p_{x,N} = \frac{Y_{AOB}(N_{1,NH4\text{-}N} - N_{NH4\text{-}N})}{1 + k_{d,AOB}\theta_{c,N}^{design}} = \frac{0.15\text{ mg VSS/mg NH}_4\text{-N} \times (29 - 1)\text{ mg NH}_4\text{-N/L}}{1 + 0.128\,d^{-1} \times 18\,d} = 1.27\text{ mg VSS/L}$$

8. Determine the total nitrogen nitrified ($N_{nitrified}$).

$$N_{nitrified} = N_{1,NH4\text{-}N} - N_{NH4\text{-}N} - f_N p_{x,N} = (29 - 1)\text{ mg NH}_4\text{-N/L} - 0.12\text{ mg N/mg VSS} \times 1.27\text{ mg VSS/L} = 27.8\text{ mg NH}_4\text{-N/L}$$

9. Determine the total alkalinity consumed during nitrification ($Alk_{consumed}$).

$$Alk_{consumed} = Alk_{oxidation}N_{nitrified} = 7.14\text{ mg Alk as CaCO}_3\text{/mg NH}_4\text{-N} \times 27.8\text{ mg NH}_4\text{-N/L} = 198\text{ mg/L as CaCO}_3$$

   Note: See Example 10.124 for calculation of the ratio of 7.14 mg Alk as $CaCO_3$/mg NH_4-N.
10. Determine the supplemental alkalinity ($Alk_{supplemental}$) required.

A remaining alkalinity concentration $Alk_{residual} = 50$ mg/L as $CaCO_3$ is required to maintain a pH in the desired range of 7.2–7.6.

$$Alk_{supplemental} = Alk_{consumed} + Alk_{residual} - Alk_{inf} = (198 + 50 - 140)\ \text{mg/L as } CaCO_3$$
$$= 108\ \text{mg/L as } CaCO_3$$

11. Determine the dosage and quantity of sodium bicarbonate ($NaHCO_3$) required.
    Calculate the $NaHCO_3$ dosage required to provide additional alkalinity.

$$C_{NaHCO3} = Alk_{supplemental} \times \frac{84\,\text{g/eq. as } NaHCO_3}{50\,\text{g/eq. as } CaCO_3} = 108\,\text{mg/L as } CaCO_3 \times \frac{84\,\text{g/eq. as } NaHCO_3}{50\,\text{g/eq. as } CaCO_3}$$
$$= 181\,\text{mg/L as } NaHCO_3 \text{ or } 181\,\text{g/m}^3 \text{ as } CaCO_3$$

Calculate the daily amount of $NaHCO_3$ required under the design flow.

$$\Delta m_{NaHCO3} = C_{NaHCO3}Q = 181\,\text{g/m}^3 \text{ as } CaCO_3 \times 8500\,\text{m}^3/\text{d} \times 10^{-3}\,\text{kg/g} = 1540\,\text{kg } NaHCO_3/\text{d}$$

Note: The amount of sodium bicarbonate can be reduced considerably if denitrification process is implemented. See Example 10.138 for calculation of alkalinity recovery in denitrification.

EXAMPLE 10.132: COMBINED CBOD REMOVAL AND NITRIFICATION IN AN ATTACHED GROWTH REACTOR

A circular trickling filter with plastic packing media is designed to provide combined CBOD removal and nitrification of primary settled wastewater. The influent characteristics and design parameters are provided below.

Influent: $Q = 4000\ \text{m}^3/\text{d}$, $BOD_5 = 150$ mg/L, and TKN = 30 mg/L.

Design parameters: volumetric organic loading $VOL = 0.2\ \text{kg/m}^3{\cdot}\text{d}$, volumetric oxidation rate $VOR = 0.4\ \text{kg/m}^3{\cdot}\text{d}$, specific surface area of the packing media $SA_m = 100\ \text{m}^2/\text{m}^3$, depth of the packing media $D = 6.1$ m, and the desired effluent NH_4-N < 3 mg/L at an operating temperature 18°C.

Determine (a) the required volume of packing media based on the following three parameters: (i) *VOL*, (ii) *VOR*, and (iii) specific nitrogen oxidation rate (R_N) and (b) surface area and diameter of the filter. Also verify the hydraulic loading rate and estimate the expected effluent NH_4-N concentration.

Solution

1. Determine the required volume of packing media (V_m) using three design parameters.
   a. Volume required from the design *VOL*.
      Calculate the mass BOD_5 loading applied to each filter without recirculation (ΔS).

$$\Delta S = S_0Q = 150\,\text{g/m}^3 \times 4000\,\text{m}^3/\text{d} \times 10^{-3}\,\text{kg/g} = 600\,\text{kg/d}$$

$$V_m = \frac{\Delta S}{VOL} = \frac{600\,\text{kg/d}}{0.2\,\text{kg/m}^3{\cdot}\text{d}} = 3000\,\text{m}^3$$

   b. Volume required from the design *VOR*.

Rearrange Equation 10.94a and calculate the volume required at the design *VOR*.

$$V_m = \frac{(S_0 + 4.57N_{0,TKN})Q}{VOR} = \frac{(150\,g/m^3 + 4.57 \times 30\,g/m^3) \times 4000\,m^3/d \times 10^{-3}\,kg/g}{0.4\,kg/m^3{\cdot}d}$$
$$= 2870\,m^3$$

c. Volume required from the design R_N.

Apply Equation 10.94c to determine specific nitrogen oxidation rate (R_N) at $S_0 = 150$ mg/L and $N_{0,TKN} = 30$ mg/L.

$$R_N = 0.82\left(\frac{S_0}{N_{0,TKN}}\right)^{-0.44} = 0.82 \times \left(\frac{150\,mg/L}{30\,mh/L}\right)^{-0.44} = 0.40\,g\,N/m^2{\cdot}d$$

Rearrange Equation 10.94b and calculate the required total surface area of packing media A_m.

$$A_m = \frac{(N_{0,TKN} - N_{NH4\text{-}N}) \times Q}{R_N} = \frac{(30 - 3)\,g\,N/m^3 \times 4000\,m^3/d}{0.40\,g\,N/m^2{\cdot}d} = 270{,}000\,m^2$$

Calculate the required volume using $SA_m = 100\,m^2/m^3$.

$$V_m = \frac{A_m}{SA_m} = \frac{270{,}000\,m^2}{100\,m^2/m^3} = 2700\,m^3$$

d. Select the design volume of the packing media.

Select the design volume required from the design VOL, $V_{m,design} = 3000\,m^3$.

2. Determine the surface area and diameter of the filter.

$$\text{Filter area, } A = \frac{V_{m,design}}{D} = \frac{3000\,m^3}{6.1\,m} = 492\,m^2$$

Calculate the required diameter of each filter.

$$d = \sqrt{\frac{4 \times 492\,m^2}{\pi}} = 25\,m. \quad \text{Provide a diameter } d = 25\,m.$$

3. Verify the hydraulic loading rate (*HLR*).

$$\text{The } HLR \text{ without recirculation, } HLR = \frac{Q}{A} = \frac{4000\,m^3/d}{492\,m^2} = 8.1\,m^3/m^2{\cdot}d$$

Note: The *HLR* or *q* is within the typical range of 5 –16 $m^3/m^2{\cdot}d$ with recirculation (Q_L) of a low-rate trickling filter with nitrification (Table 10.33). Therefore, a recirculation rate up to 100% ($R = 1$) may be required in this case.

4. Estimate the potential effluent NH_4-N concentration ($N_{NH4\text{-}N,potential}$).

Calculate the design surface area of the packing media.

$$A_{m,design} = SA_m V_{m,design} = 100\,m^2/m^3 \times 3000\,m^3 = 300{,}000\,m^2$$

Calculate the specific BOD_5 surface loading rate (SLR_{BOD}).

$$SLR_{BOD} = \frac{S_0Q}{A_{m,design}} = \frac{150\,g\,BOD_5/m^3 \times 4000\,m^3/d}{300{,}000\,m^2} = 2\,g\,BOD_5/m^2{\cdot}d$$

Calculate the specific nitrogen surface loading rate (SLR_N).

$$SLR_N = \frac{N_{0,TKN}Q}{A_{m,design}} = \frac{30\,g/m^3 \times 4000\,m^3/d}{300{,}000\,m^2} = 0.4\,g\,N/m^2{\cdot}d$$

Calculate the specific hydraulic surface loading rate (SLR_Q).

$$SLR_Q = \frac{Q}{A_{m,design}} = \frac{4000\,m^3/d \times 10^3\,L/m^3}{300{,}000\,m^2} = 13.3\,L/m^2{\cdot}d$$

Apply Equation 10.94d to estimate the expected effluent NH_4-N concentration.

$$\begin{aligned} N_{NH4\text{-}N,expected} &= 20.81(SLR_{BOD})^{1.03}(SLR_N)^{1.52}(SLR_Q)^{-0.36}T^{-0.12} \\ &= 20.81 \times (2)^{1.03} \times (0.4)^{1.52} \times (13.3)^{-0.36} \times (18)^{-0.12} \\ &= 20.81 \times 2.04 \times 0.248 \times 0.394 \times 0.707 \\ &= 2.94\,mg\,NH_4\text{-}N/L \end{aligned}$$

Therefore, the expected effluent NH_4-N concentration of 2.94 mg/L meets the desired effluent NH_4-N concentration < 3 mg/L at an operating temperature of 18°C.

EXAMPLE 10.133: TERTIARY NITRIFICATION IN A FIXED-FILM REACTOR

A fixed-film reactor is designed to provide nitrification after CBOD removal. The design influent flow and ammonia nitrogen concentration are 2500 m^3/d and 25 mg/L, respectively. The design parameters are: critical operating temperature = 15°C, the NH_4-N concentration in the effluent < 1.0 mg/L, $J_{N,max} = 2\,g/m^2{\cdot}d$ at 15°C, $K_{NH4\text{-}N} = 1.5$ mg/L, specific surface area of the cross-flow plastic modular packing media $SA_m = 120\,m^2/m^3$, the height of the modular packing media $h_m = 1.2$ m (48 inches) and total hydraulic loading rate $Q_L = 72\,m^3/m^2{\cdot}d$ at a recirculation ratio $R = 1$. Determine (a) required layers of the packing media, (b) surface area and volume of the packing media, and (c) dimensions of the filter.

Note: Initially the nitrification rate will be zero order due to high NH_4-N concentrations at the top part of the packing media. The NH_4-N concentration drops gradually through the media under a constant nitrification rate. As the NH_4-N concentration reaches a certain lower level at a critical depth (Z_c) the nitrification rate becomes first order. Assume that the limiting NH_4-N concentration for zero-order reaction (N_c) is ~ 6 mg/L and the value of $J_{N,max}$ is decreased by 0.1 $g/m^2{\cdot}d$ per meter incremental media depth when the NH_4-N concentration is below 6 mg/L.

Solution

1. Determine the NH_4-N concentration in the flow with recirculation fed at the top of the filter (N_i).

   Calculated q from Equation 10.94i at the desired effluent NH_4-N concentration $N = 1.0$ mg/L and $R = 1$.

$$N_i = \frac{N_0 + RN}{1+R} = \frac{(25 + 1 \times 1)\,mg\,NH_4\text{-}N/L}{1+1} = 13\,mg\,NH_4\text{-}N/L$$

2. Determine the media depth (Z_c) required for the zero-order reaction.

   Rearrange Equation 10.94h and calculate the Z_c for reaching $N_c = 6$ mg/L.

$$Z_c = \frac{Q_L}{J_{N,max} S A_m}\left((N_i - N_c) + K_{NH4\text{-}N} \ln\left(\frac{N_i}{N_c}\right)\right) = \frac{72\ m^3/m^2{\cdot}d}{2\ g\ N/m^2{\cdot}d \times 120\ m^2/m^3}$$

$$\times \left((13 - 6)\ mg\ NH_4\text{-}N/L + 1.5\ mg/L \times \ln\left(\frac{13\ mg\ NH_4\text{-}N/L}{6\ mg\ NH_4\text{-}N/L}\right)\right) = 2.45\ m$$

   Therefore, the nitrification remains zero-order reaction within the 2.45 m depth of the filter media from the top. This depth is approximately the total height of two packing media.

3. Determine the NH_4-N concentration at the bottom of the second layer (N_2).

   Apply Equation 10.94h to determine N_2 using $J_{N,max,1-2} = J_{N,max} = 2\ g/m^2{\cdot}d$ at the media depth $Z_{1-2} = 2h_m = 2 \times 1.2\ m = 2.4\ m$.

$$(N_i - N_2) + K_{NH4\text{-}N} \ln\left(\frac{N_i}{N_2}\right) = \frac{J_{N,max,1-2} S A_m Z_{1-2}}{Q_L}$$

$$(13\ mg\ NH_4\text{-}N/L - N_2) + 1.5\ mg/L \times \ln\left(\frac{13\ mg\ NH_4\text{-}N/L}{N_2}\right) = \frac{2\ g\ N/m^2{\cdot}d \times 120\ m^2/m^3 \times 2.4\ m}{72\ m^3/m^2{\cdot}d}$$

   A trial and error procedure is used to obtain $N_2 = 6.13$ mg NH_4-N/L.

4. Determine the additional layers of packing media required to meet the desired effluent NH_4-N concentration.

   Since N_2 is close to and slightly higher than N_c, it is assumed that the first-order reaction occurs in the packing media in the third layer and below. The reaction rate drops continuously as the filter depth is increased. Therefore, the evaluation is performed at a depth increment that is equal to the height of the media ($h_m = 1.2$ m). To be conservative, the value of $J_{N,max}$ at the bottom of each layer is used for that increment.

   a. Determine the NH_4-N concentration at the bottom of the third layer (N_3).

      Determined the value of $J_{N,max,3}$ for the third layer.

$$J_{N,max,3} = J_{N,max,1-2} - 0.1h_m = 2\ g\ N/m^2{\cdot}d - (0.1\ g\ N/m^2{\cdot}d{\cdot}m) \times 1.2\ m = 1.88\ g\ N/m^2{\cdot}d$$

      Use the similar approach to solve N_3.

$$(N_2 - N_3) + K_{NH4\text{-}N} \ln\left(\frac{N_2}{N_3}\right) = \frac{J_{N,max,3} S A_m Z_3}{Q_L}$$

$$(6.13\ mg\ NH_4\text{-}N/L - N_3) + 1.5\ mg/L \times \ln\left(\frac{6.13\ mg\ NH_4\text{-}N/L}{N_3}\right)$$

$$= \frac{1.88\ g\ N/m^2{\cdot}d \times 120\ m^2/m^3 \times 1.2\ m}{72\ m^3/m^2{\cdot}d}$$

      $N_3 = 3.30$ mg NH_4-N/L.

   b. Determine the NH_4-N concentration at the bottom of each additional layer.

      Repeat the procedure in Step 4.a for two more layers. The calculation results are: $N_4 = 1.24$ mg NH_4-N/L and $N_5 = 0.27$ mg NH_4-N/L.

   c. Packing media layer design.

      A total of five (5) layers of packing media are provided to produce an effluent NH_4-N concentration below the desired value of 1 mg NH_4-N/L. The estimated nitrification rates and NH_4-N concentrations through the media layers are summarized below.

Layer	Nitrogen removal flux ($J_{N,max}$), g N/m^2·d	Ammonia concentration, mg NH_4-N/L
Applied to the filter	2	13
After Layer 1	2	9.48
After Layer 2	2	6.13
After Layer 3	1.88	3.30
After Layer 4	1.76	1.24
After Layer 5	1.64	0.27

5. Determine the dimensions of the filter.
   Total height of the plastic packing media, $H_m = 5h_m = 5 \times 1.2\ \text{m} = 6\ \text{m}$.
   Provide a design media height $H_{design} = 6\ \text{m}$.
   Calculate the hydraulic loading rate (q) from Equation 10.94j.

$$q = \frac{Q_L}{1+R} = \frac{72\ \text{m}^3/\text{m}^2\cdot\text{d}}{1+1} = 36\ \text{m}^3/\text{m}^2\cdot\text{d}$$

Surface area of the filter, $A = \dfrac{Q}{q} = \dfrac{2500\ \text{m}^3/\text{d}}{36\ \text{m}^3/\text{m}^2\cdot\text{d}} = 69.4\ \text{m}^2$

Diameter of the filter, $d = \sqrt{4A/\pi} = \sqrt{4 \times 69.4\ \text{m}^2/\pi} = 9.4\ \text{m}$. Provide a diameter $d_{design} = 9.5\ \text{m}$.

6. Determine the volume and surface area of the packing media.

$$\text{Total volume of the packing media, } V_{design} = \frac{\pi}{4}(d_{design})^2 H_{design} = \frac{\pi}{4} \times (9.5\ \text{m})^2 \times 6\ \text{m} = 425\ \text{m}^3$$

$$\text{Total surface area of the packing media, } A_{design} = SA_m V_{design} = 120\ \text{m}^2/\text{m}^3 \times 425\ \text{m}^3 = 51{,}000\ \text{m}^2$$

EXAMPLE 10.134: TERTIARY NITRIFICATION IN A RBC SYSTEM

Nitrification in a RBC system occurs at a reduced organic loading. In Example 10.108, three parallel trains with three stages were designed. For complete nitrification additional stages are considered. The influent NH_4-N concentration is 25 mg/L and the required effluent NH_4-N concentration is 1.0 mg/L or less. Use the problem data and results of Example 10.108. It is assumed that at NH_4-N concentration of 5 mg/L or higher the reaction rate approaches zero-order and $R_{N,max} = 1.5$ g NH_4-N/m^2·d.

Solution

1. List the selected data and results of Example 10.108.
   Influent flow = 3800 m^3/d, number of stages = 3, number of parallel shafts = 3, standard shaft disk area (A_{shaft}) of 9290 m^2, and soluble BOD_5 loadings ($SOLs$) on three stages: $SOL_1 = 14$ g sBOD$_5$/m^2·d, $SOL_2 = 4.2$ g sBOD$_5$/m^2·d, and $SOL_3 = 2.0$ g sBOD$_5$/m^2·d (Example 10.108 Step 4).
2. Determine the reduction of nitrogen oxidation rate (F_N).
   Calculate the F_N from Equation 10.95b for each of the three stages.

$$F_{N,1} = 1.0 - 0.1\ SOL_1 = 1.0 - 0.1 \times 14 = -0.4 \text{ or } 0$$

No nitrification occurs in the first stage since $SOL_1 \geq 10$ g sBOD$_5$/m^2·d.

$$F_{N,2} = 1.0 - 0.1SOL_2 = 1.0 - 0.1 \times 4.2 = 0.58 \text{ or } 58\% \text{ nitrogen oxidation rate } (R_N)$$

$$F_{N,3} = 1.0 - 0.1SOL_3 = 1.0 - 0.1 \times 2.0 = 0.8 \text{ or } 80\% \ \ R_N$$

3. Determine the nitrogen oxidation rate (R_N).
   Apply Equation 10.95a to calculate the R_N in each of the three stages.

$$R_{N,1} = F_{N,1}R_{N,\max} = 0 \times 1.5\,\text{g NH}_4\text{-N/m}^2\cdot\text{d} = 0 \quad \text{(No nitrification occurs.)}$$

$$R_{N,2} = F_{N,2}R_{N,\max} = 0.58 \times 1.5\,\text{g NH}_4\text{-N/m}^2\cdot\text{d} = 0.87\,\text{g NH}_4\text{-N/m}^2\cdot\text{d}$$

$$R_{N,3} = F_{N,3}R_{N,\max} = 0.8 \times 1.5\,\text{g NH}_4\text{-N/m}^2\cdot\text{d} = 1.2\,\text{g NH}_4\text{-N/m}^2\cdot\text{d}$$

4. Determine the concentration of NH$_4$-N removed (ΔN).
   Calculate the total disk surface area in each stage (A_{stage}).

$$A_{stage} = N_{shaft}A_{shaft} = 3\text{ shaft/stage} \times 9290\,\text{m}^2\text{/shaft} = 27{,}870\,\text{m}^2$$

Calculate the ΔN in each of the three stages.

$$\Delta N_1 = \frac{R_{N,1}A_{stage}}{Q} = 0 \quad \text{(No NH}_4\text{-N removal occurs.)}$$

$$\Delta N_2 = \frac{R_{N,2}A_{stage}}{Q} = \frac{0.87\,\text{g NH}_4\text{-N/m}^2\cdot\text{d} \times 27{,}870\,\text{m}^2}{3800\,\text{m}^3\text{/d}} = 6.4\,\text{g NH}_4\text{-N/m}^3 \text{ or } 6.4\,\text{mg NH}_4\text{-N/L}$$

$$\Delta N_3 = \frac{R_{N,3}A_{stage}}{Q} = \frac{1.2\,\text{g NH}_4\text{-N/m}^2\cdot\text{d} \times 27{,}870\,\text{m}^2}{3800\,\text{m}^3\text{/d}} = 8.8\,\text{g NH}_4\text{-N/m}^3 \text{ or } 8.8\,\text{mg NH}_4\text{-N/L}$$

Calculate the overall removal ΔN in three stages.

$$\Delta N = \Delta N_1 + \Delta N_2 + \Delta N_3 = (0 + 6.4 + 8.8)\,\text{mg NH}_4\text{-N/L} = 15.2\,\text{mg NH}_4\text{-N/L}$$

5. Determine the NH$_4$-N concentration (N).
   Calculate N from the third stage.

$$N_3 = N_{0,\text{NH4-N}} - \Delta N = (25 - 15.2)\,\text{mg NH}_4\text{-N/L} = 9.8\text{ mg NH}_4\text{-N/L}$$

Since N_3 is much higher than the required NH$_4$-N concentration in the effluent, a fourth stage is therefore needed for providing additional nitrification.

6. Determine the number of shafts ($N_{shaft,4}$) required in the fourth stage.
   Calculate the minimum removal of NH$_4$-N concentration required in the fourth stage to meet the required NH$_4$-N concentration in the effluent.

$$\Delta N_4 = N_3 - N = (9.8 - 1.0)\,\text{mg NH}_4\text{-N/L} = 8.8\,\text{mg NH}_4\text{-N/L or } 8.8\,\text{g NH}_4\text{-N/m}^3$$

Since BOD$_5$ remaining is small it is fair to assume that the nitrogen oxidation rate reaches the maximum value in the fourth stage, $R_{N,4} = R_{N,\max} = 1.5$ g NH$_4$-N/m^2·d.

Calculate the total disk surface area required in the fourth stage.

$$A_{stage,4} = \frac{\Delta N_4 Q}{R_{N,4}} = \frac{8.8\,\text{g NH}_4\text{-N/m}^3 \times 3800\,\text{m}^3\text{/d}}{1.5\,\text{g NH}_4\text{-N/m}^2\cdot\text{d}} = 22{,}300\,\text{m}^2$$

Calculate the $N_{shaft,4}$ in the fourth stage.

$$N_{shaft,4} = \frac{A_{stage,4}}{A_{shaft}} = \frac{22{,}300\,\text{m}^2}{9290\,\text{m}^2\text{/shaft}} = 2.4\text{ shafts}$$

Assume that there are still three trains in the fourth stage and calculate the number of shafts required in each train.

$$N_{shaft/train,4} = \frac{N_{shast,4}}{N_{train}} = \frac{2.4\text{ shafts}}{3\text{ trains}} = 0.8\text{ shaft/train}$$

Provide one shaft in each train of the fourth stage.

7. Comment on the results.

The first stage provides mainly CBOD removal. The second and third stages provide both CBOD removal and nitrification. The fourth stage is added exclusively for nitrification. Since the provided total disk surface area is more than the required area, the effluent NH_4-N concentration from the fourth stage will meet the required NH_4-N concentration of 1.0 mg NH_4-N/L or below.

EXAMPLE 10.135: COMBINED CBOD REMOVAL AND NITRIFICATION IN AN IFAS SYSTEM

An IFAS system is selected to enhance nitrification by providing attached growth biofilm in the downstream section of the aeration basin. Apply the data and results obtained for the nitrification process in Examples 10.126 and 10.127. For BOD_5 removal, use the typical kinetic coefficients given in Table 10.10. Use $f_N = 0.12$ g N/g VSS. Additional influent and effluent characteristics and design parameters are provided below.

Influent: $Q = 4000\,\text{m}^3\text{/d}$, $BOD_5 = 150$ mg/L, and TKN = 35 mg/L.
Effluent: desired effluent quality of $sBOD_5 \leq 2.5$ mg/L and $NH_4\text{-N} \leq 0.5$ mg/L.
General process operating parameters: DO concentration = 4.0 mg/L and critical operating temperature = 12°C.
Design parameters of the activated sludge volume: MLSS = 2500 mg/L and mixed liquor VSS/TSS ratio = 0.8.
Design parameters of the media fill volume: specific surface area of the plastic film carriers $SA_m = 500\,\text{m}^2/\text{m}^3$, the media fill volume fraction $f_m \leq 0.4$, and VSS/TSS ratio = 0.7 in the biofilm.

Determine (a) design SRT in the activated sludge volume, (b) suspended growth and fraction of heterotrophic biomass in the mixed liquor, (c) hydraulic retention times (HRTs) required in the activated sludge volume, media fill volume, and the IFAS system, (d) total volume, fraction of the media fill volume, and total surface area required in the IFAS system, and (e) biomass surface concentration and fraction of autotropic biomass in the biofilm.

Solution

1. List the kinetic parameters used in the calculations.
   a. The nitrification results from Examples 10.126 and 10.127.
      $k_{d,AOB.F} = 0.135\,\text{d}^{-1}$, and $\theta_c^{design} = 21$ d for nitrification in the medial fill volume achieving an effluent NH_4-N concentration ≤ 0.5 mg/L.

b. Typical kinetic coefficients for BOD_5 removal from Table 10.10.

$$\mu_{max,BOD} = 2.5\,\text{mg VSS/mg VSS·d},\ \ Y_{BOD} = 0.5\,\text{mg VSS/mg BOD}_5,\ \ k_{d,BOD} = 0.05\,\text{d}^{-1}$$
$$k_{BOD} = 5\,\text{mg BOD}_5/\text{mg VSS d},\ \ \text{and}\ K_{S,BOD5} = 60\,\text{mg BOD}_5/\text{L}.$$

c. Typical kinetic coefficients for AOB from Table 10.49.

$$Y_{AOB} = 0.15\,\text{mg VSS/mg NH}_4\text{-N and } K_{NH4\text{-}N} = 0.50\,\text{mg NH}_4\text{-N/L}$$

2. List the general assumptions to simplify the complex kinetic calculations.
Complicated modeling approach is required to solve the complex kinetic equations simultaneously for IFAS or other similar process with integrated fixed-film media in aeration basin. The following mass-balance-based ammunitions are necessary to simplify the calculations.
   a. The IFAS system consists of two functional sections arranged in series. The upstream section is an *activated sludge volume* (V_{as}) with suspended growth only. The downstream section is a *media fill volume* (V_m) with suspended growth predominantly on plastic biofilm carriers.
   b. The overall effluent quality requirements of BOD_5 and NH_4-N are achieved by biological growth in both sections of the IFAS system.
   c. The suspended growth section primarily removes CBOD in excess of 95% of the total removal of BOD_5 from the IFAS system. Small nitrification is also achieved in this section.
   d. The attached growth section is responsible for nitrification of NH_4-N in excess of 85%. The attached growth also removes the residual soluble BOD_5.
3. Determine the design SRT in the activated sludge volume ($\theta_{c,as}^{design}$).
   a. Determine the specific and net growth rates for heterotrophs in the activated sludge volume.
Calculate the concentration of total BOD_5 removed in the IFAS system at the desired effluent BOD_5 concentration of 2.5 mg BOD_5/L.

$$\Delta S = S_0 - S = (150 - 2.5)\,\text{mg BOD}_5/\text{L} = 147.5\,\text{mg BOD}_5/\text{L}$$

Calculate the concentration of BOD_5 removed at the targeted removal efficiency of 95% in the activated sludge volume.

$$\Delta S_{as} = E_{as}\Delta S = 0.95 \times 147.5\,\text{mg BOD}_5/\text{L} = 140\,\text{mg BOD}_5/\text{L}$$

The concentration of $sBOD_5$ remaining after the activated sludge volume,

$$S_{as} = S_0 - \Delta S_{as} = (150 - 140)\,\text{mg BOD}_5/\text{L} = 10\,\text{mg BOD}_5/\text{L}$$

Calculate the specific growth rate (μ_{BOD}) from Equation 10.11b.

$$\mu_{BOD} = \frac{\mu_{max,BOD} S_{as}}{K_{S,BOD} + S_{as}} = \frac{2.5\,\text{mg VSS/mg VSS·d} \times 10\,\text{mg BOD}_5/\text{L}}{(60 + 10)\,\text{mg BOD}_5/\text{L}}$$
$$= 0.357\,\text{mg VSS/mg VSS·d}$$

Calculate the specific net growth rate (μ'_{BOD}) from Equation 10.13b.

$$\mu'_{BOD} = \mu_{BOD} - k_{d,BOD} = (0.357 - 0.05)\,\text{d}^{-1} = 0.307\,\text{d}^{-1}$$

   b. Determine the design SRT in the activated sludge volume ($\theta_{c,as}^{design}$).

Rearrange Equation 10.91a and calculate the theoretical SRT ($\theta_{c,BOD}$).

$$\theta_{c,BOD} = \frac{1}{\mu'_{BOD}} = \frac{1}{0.307\,d^{-1}} = 3.3\,d$$

Provide a $SF_P = 1.5$ for the uncertainty in the process design and operation to calculate the SRT required ($\theta_{c,as}$).

$$\theta_{c,as} = SF_P\theta_{c,BOD} = 1.5 \times 3.3\,d = 5\,d$$

Provide a design SRT, $\theta_{c,as}^{design} = 5\,d$ in the activated sludge volume.

4. Determine the heterotrophic and autotrophic biomass growth in the activated sludge volume.
   a. Estimate the heterotrophic growth in the activated sludge volume ($p_{x,BOD,as}$).

   Apply Equation 10.15h to calculate the $sBOD_5$ concentration remaining (S_{as}) after BOD_5 removal by the heterotrophs at $\theta_{c,as}^{design}$.

$$S_{as} = \frac{K_{s,BOD}(1 + k_{d,BOD}\theta_{c,as}^{design})}{\theta_{c,as}^{design}(Y_{BOD}k_{BOD} - k_{d,BOD}) - 1} = \frac{60\,mg\,BOD_5/L \times (1 + 0.05\,d^{-1} \times 5\,d)}{5\,d \times (0.5 \times 5\,d^{-1} - 0.05\,d^{-1}) - 1} = 6.7\,mg\,BOD_5/L$$

Calculate the $p_{x,BOD,as}$ from Equation 10.92g.

$$p_{x,BOD,as} = \frac{Y_{BOD}(S_0 - S_{as})}{1 + k_{d,BOD}\theta_{c,as}^{design}} = \frac{0.5\,mg\,VSS/mg\,BOD_5 \times (150 - 6.7)\,mg\,BOD_5/L}{1 + 0.05\,d^{-1} \times 5\,d} = 57.3\,mg\,VSS/L$$

   b. Estimate the autotrophic growth in the activated sludge volume ($p_{x,AOB,as}$).

   The total nitrogen concentration available for nitrification ($N_{0,NH4\text{-}N}$),

$$N_{0,NH4\text{-}N} = N_{0,TKN} - f_N p_{x,BOD,as} = 35\,mg\,TKN/L - 0.12\,mg\,N/mg\,VSS \times 57.3\,mg\,VSS/L = 28.1\,mg\,NH_4\text{-}N/L$$

Note: The amounts of nitrogen utilized for cell synthesis from the autotrophic growth in the activated sludge volume and the heterotrophic and autotrophic growths in the biofilm in the media fill volume are not included in the above calculation.

Calculate the concentration of NH_4-N nitrified (ΔN).

$$\Delta N = N_{0,NH4\text{-}N} - N = (28.1 - 0.5)\,mg\,NH_4\text{-}N/L = 27.6\,mg\,NH_4\text{-}N/L$$

Calculate the concentration of NH_4-N removed in the media fill volume at the targeted removal efficiency of 85%.

$$\Delta N_m = E_m \Delta N = 0.85 \times 27.6\,mg\,NH_4\text{-}N/L = 23.5\,mg\,NH_4\text{-}N/L$$

Calculate the concentration of NH_4-N removed in the activated sludge volume.

$$\Delta N_{as} = \Delta N - \Delta N_m = (27.6 - 23.5)\,mg\,NH_4\text{-}N/L = 4.1\,mg\,NH_4\text{-}N/L$$

Nitrification (or NH_4-N removal) efficiency in the activated sludge volume,

$$E_{as} = \frac{\Delta N_m}{\Delta N} \times 100\% = \frac{4.1\,mg\,NH_4\text{-}N/L}{27.6\,mg\,NH_4\text{-}N/L} \times 100\% = 15\%$$

The concentration of NH_4-N remaining after the nitrification by the suspended autotrophs in the activated sludge volume,

$$N_{NH4\text{-}N,as} = N_{0,NH4\text{-}N} - \Delta N_{as} = (28.1 - 4.1)\,\text{mg NH}_4\text{-N/L} = 24\,\text{mg NH}_4\text{-N/L}$$

Calculate the observed biomass yields for AOB ($Y_{obs,AOB}$) from Equation 10.92a.

$$Y_{obs,AOB,as} = \frac{Y_{AOB}}{1 + k_{d,AOB,F}\theta_{c,as}^{design}} = \frac{0.15\,\text{mg VSS/mg NH}_4\text{-N}}{1 + 0.135\,\text{d}^{-1} \times 5\,\text{d}} = 0.090\,\text{mg VSS/mg NH}_4\text{-N}$$

The autotrophic growth $p_{x,AOB,as}$ is calculated from Equation 10.92g.

$$p_{x,AOB,as} = \frac{Y_{AOB}}{1 + k_{d,AOB,F}\theta_{c,as}^{design}}(N_{0,TKN} - N_{NH4\text{-}N,as} - f_N p_{x,BOD,as})$$

$$= \frac{0.15\,\text{mg VSS/mg NH}_4\text{-N}}{1 + 0.135\,\text{d}^{-1} \times 5\,\text{d}} \times ((35 - 24)\,\text{mg NH}_4\text{-N/L}$$

$$-0.12\,\text{mg N/mg VSS} \times 57.3\,\text{mg VSS/L})\,\text{mg VSS/L}$$

$$= 0.37\,\text{mg VSS/L}$$

The combined growth of heterotrophic and autotrophic organisms ($p_{x,as}$) in the activated sludge volume,

$$p_{x,as} = p_{x,BOD,as} + p_{x,AOB,as} = (57.3 + 0.37)\,\text{mg VSS/L} = 57.67\,\text{mg VSS/L or } 57.67\,\text{g VSS/m}^3$$

Estimate the fraction of heterotrophs in the combined biomass in the mixed liquor ($f_{BOD,as}$).

$$f_{BOD,as} = \frac{p_{x,BOD,as}}{p_{x,as}} = \frac{57.3\,\text{mg/L}}{57.67\,\text{mg/L}} = 0.9936 \text{ or } 99.36\%$$

Note: The total biomass growth is dominated by the heterotrophic organisms in the activated sludge volume. The nitrifiers count is $<0.5\%$ of the total biomass in the mixed liquor. This is much lower than the typical range of 2–5% for nitrifiers in the activated sludge process for CBOD removal with nitrification.

5. Determine the design HRT (θ_{as}^{design}) in the activated sludge volume.

The concentration of biomass as MLVSS (X_{as}) in the mixed liquor,

$$X_{as} = \text{VSS/TSS ratio} \times \text{MLSS} = 0.8\,\text{mg VSS/mg TSS} \times 2500\,\text{mg TSS/L} = 2000\,\text{mg VSS/L}$$

The concentration of heterotrophs ($X_{BOD,as}$) in the mixed liquor,

$$X_{BOD,as} = f_{BOD,as}X = 0.9936 \times 2000\,\text{mg VSS/L} = 1987.2\,\text{mg VSS/L}$$

The concentration of AOB (X_{AOB}) in the mixed liquor,

$$X_{AOB,as} = X_{as} - X_{BOD,as} = (2000 - 1987.2)\,\text{mg VSS/L} = 12.8\,\text{mg VSS/L}$$

The required HRT can be calculated from either BOD_5 removal ($\theta_{BOD,as}$) or nitrification ($\theta_{AOB,as}$) from Equation 10.15g. Both calculations are shown below.

$$\theta_{BOD,as} = \frac{p_x}{X_{BOD,as}}\theta_{c,as}^{design} = \frac{57.3\ \text{mg VSS/L}}{1987.2\ \text{mg VSS/L}} \times 5\ \text{d} \times \frac{24\ \text{h}}{\text{d}} = 3.5\ \text{h}$$

$$\theta_{AOB,as} = \frac{p_{x,AOB,as}}{X_{BOD,as}}\theta_{c,as}^{design} = \frac{0.37\ \text{mg VSS/L}}{12.8\ \text{mg VSS/L}} \times 5\ \text{d} \times \frac{24\ \text{h}}{\text{d}} = 3.5\ \text{h}$$

Same HRT is obtained from both calculations. Provide a design $\theta_{as}^{design} = 3.6$ h, giving a small margin for safety.

6. Determine the design HRT (θ_m^{design}) in the media fill volume.
   a. Determine the HRT required for BOD_5 removal ($\theta_{BOD,m}$) in the media fill volume.

      The BOD_5 removal flux (J_{BOD}) is calculated from Equation 10.94e using $J_{BOD,max} = 5$ g $BOD_5/m^2{\cdot}d$, and $K_{BOD,m} = 60$ mg BOD_5/L.

$$J_{BOD} = J_{BOD,max}\left(\frac{S}{K_{BOD,m} + S}\right) = 5\ \text{g BOD}_5/\text{m}^2{\cdot}\text{d} \times \left(\frac{2.5\ \text{mg BOD}_5/\text{L}}{(60 + 2.5)\ \text{mg BOD}_5/\text{L}}\right)$$
$$= 0.2\ \text{g BOD}_5/\text{m}^2{\cdot}\text{d}$$

The hydraulic retention time in the media fill volume can be calculated from Equation 10.97b.

$$\theta_m = \frac{S_0 - S}{SA_m J_s} \qquad (10.97b)$$

where
θ_m = hydraulic retention time in the media fill volume, d
S_0 = substrate concentration in the influent to the media fill volume, mg/L (g/m^3)
SA_m = specific surface area of the plastic biofilm carrier media, m^2/m^3. The typical ranges of SA_m for different media materials are given in Table 10.40.

Calculate the $\theta_{BOD,m}$ is calculated from Equation 10.97b.

$$\theta_{BOD,m} = \frac{S_{as} - S}{SA_m J_{BOD}} = \frac{(6.7 - 2.5)\ \text{g BOD}_5/\text{m}^3}{500\ \text{m}^2/\text{m}^3 \times 0.2\ \text{g BOD}_5/\text{m}^2{\cdot}\text{d}} \times \frac{24\ \text{h}}{\text{d}} = 1.0\ \text{h}$$

   b. Determine the HRT required for nitrification ($\theta_{AOB,m}$) in the media fill volume.

      At the design operating DO concentration of 4 mg/L, the critical NH_4-N concentration N_c = 1 mg/L is obtained from Table 10.51. Since the $N = 0.5\ \text{mg/L} < N_c$, the nitrogen removal flux is NH_4-N limited. Calculate the $J_{N,12}$ from Equation 10.96b using $J_{N,max,15} = 3.3$ g NH_4-N/$m^2{\cdot}d$ and $K_{NH4\text{-}N} = 2.2$ mg NH_4-N/L.

$$J_{N,12} = J_{N,max,15}\left(\frac{N}{K_{NH4\text{-}N} + N}\right)(1.098)^{T-15}$$
$$= 3.3\ \text{g NH}_4\text{-N/m}^2{\cdot}\text{d} \times \left(\frac{0.5\ \text{mg NH}_4\text{-N/L}}{(2.2 + 0.5)\ \text{mg NH}_4\text{-N/L}}\right) \times (1.098)^{12-15}$$
$$= 0.462\ \text{g NH}_4\text{-N/m}^2{\cdot}\text{d}$$

Calculate the $\theta_{AOB,m}$ from Equation 10.97a.

$$\theta_{AOB,m} = \frac{N_{NH4\text{-}N,as} - N}{SA_m J_N} = \frac{(24 - 0.5)\,g\,NH_4\text{-}N/m^3}{500\,m^2/m^3 \times 0.462\,g\,NH_4\text{-}N/m^2\cdot d} \times \frac{24\,h}{d} = 2.4\,h$$

c. Determine the design HRT (θ_m^{design}) in the media fill volume.

The HRT required by nitrification is longer that that by BOD_5 removal. Provide a design $\theta_m^{design} = 2.4\,h$.

7. Determine the design HRT (θ^{design}) in the IFAS system.

Total HRT in the IFAS basin, $\theta_{design} = \theta_{as}^{design} + \theta_m^{design} = (3.6 + 2.4)\,h = 6\,h$

8. Determine the volumes in the IFAS system.

Activated sludge volume, $V_{as} = Q\theta_{as}^{design} = 4000\,m^3/d \times 3.6\,h \times \frac{d}{24\,h} = 600\,m^3$

Media fill volume, $V_m = Q\theta_m^{design} = 4000\,m^3/d \times 2.4\,h \times \frac{d}{24\,h} = 400\,m^3$

Total IFAS basin volume, $V = V_{as} + V_m = (600 + 400)\,m^3 = 1000\,m^3$

Fraction of the media fill volume in the total IFAS basin volume,

$$f_m = \frac{V_m}{V} = \frac{400\,m^3}{1000\,m^3} = 0.4 \text{ or } 40\%$$

Note: The design f_m of 0.4 is the maximum fill fraction typically allowed for plastic film carriers in an IFAS system.

9. Determine the total surface area required in the IFAS system.

Total media surface area, $A_m = SA_m V_m = 500\,m^2/m^3 \times 400\,m^3 = 200{,}000\,m^2$

10. Estimate the biomass surface concentration in the biofilm (X_m).

The biomass concentration in the biofilm on the media surface can be estimated from either Equation 10.97c or Equation 10.97d at the design substrate removal flux, SRT, and microbial kinetic coefficients.

$$X_m = \frac{Y(S_0 - S)Q}{A_m(1 + k_d\theta_{c,m})}\theta_{c,m} \text{ or } X_m = \frac{P_x}{A_m}\theta_{c,m} \text{ or } X_m = \frac{p_x}{SA_m\theta_m}\theta_{c,m} \quad (10.97c)$$

$$X_m = \frac{Y}{1 + k_d\theta_{c,m}}J_s\theta_{c,m} \text{ or } X_m = Y_{obs}J_s\theta_{c,m} \quad (10.97d)$$

where

X_m = biomass surface concentration in the biofilm, g VSS/m^2

$\theta_{c,m}$ = design SRT in the biofilm, d

Other terms have been previously defined.

At $\theta_{c,m}^{design} = 21$ d (from Example 10.127), estimate the biomass surface concentrations for heterotrophic, autotrophic, and combined organisms from Equation 10.97c, separately.

$$X_{BOD,m} = \frac{Y_{BOD}}{1 + k_{d,BOD}\theta_{c,m}^{design}} J_{BOD}\theta_{c,m}^{design} = \frac{0.5 \text{ mg VSS/mgBOD}_5}{1 + 0.05 \text{ d}^{-1} \times 21 \text{ d}} \times 0.2 \text{ g BOD}_5/\text{m}^2\text{·d} \times 21 \text{ d}$$

$$= 1.0 \text{ g VSS/m}^2$$

$$X_{AOB,m} = \frac{Y_{AOB}}{1 + k_{d,AOB,F}\theta_{c,m}^{design}} J_N\theta_{c,m}^{design} = \frac{0.15 \text{ mg VSS/mg NH}_4\text{-N}}{1 + 0.135 \text{ d}^{-1} \times 21 \text{ d}}$$

$$\times 0.462 \text{ g NH}_4\text{-N/m}^2\text{·d} \times 21 \text{ d} = 0.38 \text{ g VSS/m}^2$$

$$X_m = X_{BOD,m} + X_{AOB,m} = (1.0 + 0.38) \text{ g VSS/m}^2 = 1.38 \text{ g VSS/m}^2$$

Calculate the surface concentration of total solids in the biofilm (TSS_m).

$$TSS_m = \frac{X_m}{\text{VSS/TSS ratio}} = \frac{1.38 \text{ g VSS/m}^2}{0.7 \text{ g VSS/g TSS}} = 2.0 \text{ g VSS/m}^2$$

11. Estimate the fraction of autotropic biomass in the biofilm ($f_{AOB,m}$).

Fraction of autotropic biomass in the biofilm, $f_{AOB,m} = \frac{X_{AOB,m}}{X_m} = \frac{0.38 \text{ g VSS/m}^2}{1.38 \text{ g VSS/m}^2} = 0.28$ or 28%

Note:

a. The reported biomass surface concentration is in the range of 5–50 g TSS/m^2 over the attached growth on a plastic media.[2] The biomass concentration over the surface of the media calculated in this example is 2 g TSS/m^2.
b. The reported COD removal flux is in the range of 0.5–5 g COD/m^2·d over an attached growth media.[2] In this example, $J_{BOD} = 0.2$ g BOD_5/m^2·d.
c. The reported nitrogen removal flux is in the range of 0.05–0.5 g NH_4-N/m^2·d at temperature of 15°C.[2] In this example, $J_N = 0.46$ g NH_4-N/m^2·d.
d. Nearly complete nitrification is expected if the soluble CBOD loading rate is below 2 g BOD_5/m^2·d, or the concentration of soluble CBOD in the influent to the attached media is below 5 mg/L.
e. The design value of SRT $\theta_{c,as}^{design} = 5$ d in the activated sludge volume is determined from heterotrophic growth to achieve the targeted removal efficiency of 95% with a safety factor of 1.5 for the uncertainty in the process design and operation. A nitrification efficiency of 15% in the activated sludge volume is then estimated at $\theta_{c,as}^{design}$. This is within the normal range of 10–20% in the activated sludge volume.[2]
f. The design substrate removal fluxes used in the example may be conservative for both BOD removal and nitrification. Actual concentrations of $sBOD_5$ and NH_4-N in the effluent from the IFAS system may be lower than those calculated in this example.

Based on the information developed in this example, it is concluded that the attached autotrophic growth in an IFAS system has significantly improved nitrification, and enhanced performance and reliability over the suspended growth reactor with nitrification.

EXAMPLE 10.136: TERTIARY NITRIFICATION IN A MBBR REACTOR

The CBOD removal in a MBBR reactor has been presented in Example 10.111. Design a new second MBBR reactor for tertiary nitrification after the CBOD removal in the first reactor. The new reactor consists of two compartments of equal volume arranged in series. Use the data and results from Examples 10.111 and 10.126. For BOD_5 removal, use the typical kinetic coefficients given in Table 10.10. Use $f_N =$ 0.12 g N/g VSS. Assume the SRT $\theta_c = 12$ d for the biofilm in the first MBBR reactor. Ignore the potential nitrification in the first MBBR reactor and BOD_5 removal in the second reactor. Additional influent and effluent characteristics, and design parameters are provided below.

Influent: TKN = 35 mg/L.
Effluent: NH_4-N $\leq$ 0.5 mg/L.
Design parameters: DO concentration = 4.0 mg/L, critical operating temperature = 12°C, specific surface area of the plastic film carriers $SA_m = 500\ m^2/m^3$, the media fill volume fraction $f_m = 0.6$ in both compartments, and VSS/TSS ratio = 0.7 in the biofilm.

Determine (a) total nitrogen nitrified in the second MBBR reactor, (b) nitrogen removal flux in each compartment of the second MBBR reactor, (c) NH_4-N concentration in the effluent from the first compartment of the second MBBR reactor, (d) nitrogen removal efficiencies in the MBBR system, and (e) HRTs, reactor volumes, media volumes, and media surface areas in the second MBBR reactor. Also, summarize the calculation results from Examples 10.111 and 10.136 for CBOD removal and nitrification in the complete MBBR system.

Solution

1. List the kinetic parameters used in the calculations.
   a. Influent BOD_5 concentration and removal efficiencies from Example 10.111.
      Influent BOD_5 concentration $S_0 = 150$ mg/L, and BOD_5 removal efficiencies of $E_{BOD,1} = 75$ and $E_{BOD,2} = 90\%$ in the first and second compartments of the first MBBR reactor, respectively.
   b. The kinetic coefficients for nitrification from Example 10.126.

$$\mu_{max,AOB.F} = 0.412\ \text{mg VSS/mg VSS·d,and}\ k_{d,AOB.F} = 0.135\ \text{d}^{-1}.$$

   c. Typical kinetic coefficients for BOD_5 removal obtained from Table 10.10:

$$Y_{BOD} = 0.5\ \text{mg VSS/mg BOD}_5\text{, and}\ k_{d,BOD} = 0.05\ \text{d}^{-1}.$$

2. Determine the total nitrogen removal by nitrification in the second MBBR reactor (ΔN).
   Calculate BOD_5 concentration in the effluent from the second MBBR reactor.

$$S = S_0\left(1-\frac{E_{BOD,1}}{100\%}\right)\left(1-\frac{E_{BOD,2}}{100\%}\right) = 150\,\text{g BOD}_5/\text{m}^3 \times \left(1-\frac{75\%}{100\%}\right)\left(1-\frac{90\%}{100\%}\right) = 3.75\,\text{g BOD}_5/\text{m}^3$$

   Calculate the overall BOD_5 removal efficiency in the MBBR system.

$$E_{BOD,MBBR} = \left(1-\frac{S}{S_0}\right) \times 100\% = \left(1-\frac{3.75\ \text{mg BOD}_5/\text{L}}{150\ \text{mg BOD}_5/\text{L}}\right) \times 100\% = 97.5\%$$

   Combine Equations 10.15l and 10.15m to calculate the heterotrophic growth ($p_{x,BOD}$) in the first MBBR reactor.

$$p_{x,BOD} = \frac{Y_{BOD}(S_0 - S)}{1 + k_{d,BOD}\theta_c} = \frac{0.5\ \text{mg VSS/mg BOD}_5 \times (150 - 3.75)\ \text{mg BOD}_5/\text{L}}{1 + 0.05\ \text{d}^{-1} \times 12\ \text{d}} = 45.7\ \text{mg VSS/L}$$

Estimate the total nitrogen concentration available for nitrification ($N_{0,\text{NH4-N}}$).

$$N_{0,\text{NH4-N}} = N_{0,\text{TKN}} - f_N p_{x,\text{BOD,as}} = 35\text{ mg TKN/L} - 0.12\text{ mg N/mg VSS} \times 45.7\text{ mg VSS/L}$$
$$= 29.5\text{ mg NH}_4\text{-N/L}$$

Calculate the total NH_4-N concentration nitrified in the second MBBR reactor (ΔN).

$$\Delta N = N_{0,\text{NH4-N}} - N = (29.5 - 0.5)\text{ mg NH}_4\text{-N/L} = 29\text{ mg NH}_4\text{-N/L}$$

3. Determine the nitrogen removal flux (J_N) in each compartment.
   The design J_N is determined from Equation 10.97a for each compartment.
   a. Determine the design $J_{N,1}$ in the first compartment.
      At the design operating DO concentration of 4 mg/L, the critical NH_4-N concentration $N_c =$ 1 mg/L for zero-order nitrification flux (Table 10.51). Assume the NH_4-N concentration in the first compartment is higher than 1 mg/L. Therefore, the nitrification removal flux is under DO limited conditions. Calculate $J_{N,1}$ from Equation 10.96a using $J_{N,\text{max},15} = 3.3\text{ g NH}_4\text{-N/m}^2\cdot\text{d}$ and $K_{\text{NH4-N}} = 2.2\text{ mg NH}_4\text{-N/L}$.

$$J_{N,1} = J_{N,\text{max},15}\left(\frac{N_c}{K_{\text{NH4-N}} + N_c}\right)(1.058)^{T-15}$$
$$= 3.3\text{ g NH}_4\text{-N/m}^2\cdot\text{d} \times \left(\frac{1\text{ mg NH}_4\text{-N/L}}{(2.2+1)\text{ mg NH}_4\text{-N/L}}\right) \times (1.058)^{12-15}$$
$$= 0.871\text{ g NH}_4\text{-N/m}^2\cdot\text{d}$$

   b. Determine the design $J_{N,2}$ in the second compartment.
      Assume the NH_4-N concentration in the second compartment is equal to the desired effluent NH_4-N concentration $N = 0.5$ mg/L. Since the $N < N_c$, the nitrogen removal flux is NH_4-N limited. Calculate the $J_{N,2}$ from Equation 10.96b using $J_{N,\text{max},15} = 3.3\text{ g NH}_4\text{-N/m}^2\cdot\text{d}$ and $K_{\text{NH4-N}} =$ 2.2 mg NH_4-N/L.

$$J_{N,2} = J_{N,\text{max},15}\left(\frac{N}{K_{\text{NH4-N}} + N}\right)(1.098)^{T-15}$$
$$= 3.3\text{ g NH}_4\text{-N/m}^2\cdot\text{d} \times \left(\frac{0.5\text{ mg NH}_4\text{-N/L}}{(2.2+0.5)\text{ mg NH}_4\text{-N/L}}\right) \times (1.098)^{12-15}$$
$$= 0.462\text{ g NH}_4\text{-N/m}^2\cdot\text{d}$$

4. Determine the NH_4-N concentration in the effluent from the first compartment.
   The hydraulic retention time required by the attached growth nitrification in a MBBR reactor (θ_N) is expressed by Equation 10.97e.

$$\theta_N = \frac{1}{f_m}\theta_m \text{ or } \theta_N = \frac{S_0 - S}{f_m S A_m J_s} \quad (10.97\text{e})$$

   where θ_N = hydraulic retention time in the MBBR reactor, d

   Other parameters have been defined previously.
   Assume that the NH_4-N concentration removed in the first and second compartments are ΔN_1 and ΔN_2, respectively. The following expressions in Equation 10.97f can be obtained

using Equation 10.97e.

$$\theta_{N,1} = \frac{\Delta N_1}{f_m SA_m J_{N,1}} \text{ and } \theta_{N,2} = \frac{\Delta N_2}{f_m SA_m J_{N,2}} \tag{10.97f}$$

Since the two compartments are of equal volume, $\theta_{N,1} = \theta_{N,2}$. Equate the above two expressions in Equation 10.97f and obtain the following relationship after proper rearrangement.

$$\frac{\Delta N_1}{J_{N,1}} = \frac{\Delta N_2}{J_{N,2}} \text{ or } \Delta N_2 = \frac{J_{N,2}}{J_{N,1}}\Delta N_1 = \frac{0.462 \text{ g NH}_4\text{-N/m}^2\text{·d}}{0.871 \text{ g NH}_4\text{-N/m}^2\text{·d}}\Delta N_1 = 0.53\Delta N_1$$

From the mass balance of nitrogen concentration in the second MBBR reactor, the following relationship is further obtained.

$$\Delta N = \Delta N_1 + \Delta N_2 = \Delta N_1 + 0.53\Delta N_1 = 1.53\Delta N_1 = 29 \text{ mg NH}_4\text{-N/L}$$

Solve the relationship to obtain $\Delta N_1 = \frac{29 \text{ mg NH}_4\text{-N/L}}{1.53} = 18.95 \text{ mg NH}_4\text{-N/L}$.

$$\Delta N_2 = \Delta N - \Delta N_1 = (29 - 18.95) \text{ mg NH}_4\text{-N/L} = 10.05 \text{ mg NH}_4\text{-N/L}$$

Assume the NH_4-N concentration in the effluent from the first compartment is N_1.

$$N_1 = N_{0,\text{NH4-N}} - \Delta N_1 = (29.5 - 18.95) \text{ mg NH}_4\text{-N/L} = 10.55 \text{ mg NH}_4\text{-N/L}$$

Note: $N_1 > N_c$. The previous assumption of DO limited conditions in the first compartment is acceptable.

5. Determine the nitrogen removal efficiencies in the MBBR system.

Calculate the overall nitrogen removal efficiency due to biomass synthesis in the first MBBR reactor.

$$E_{\text{N,biomass synthesis}} = \left(1 - \frac{N_{1,\text{NH4-N}}}{N_{0,\text{TKN}}}\right) \times 100\% = \left(1 - \frac{29.5 \text{ mg NH}_4\text{-N/L}}{35 \text{ mg TKN/L}}\right) \times 100\% = 15.7\%$$

Calculate the NH_4-N removal efficiency in the first compartment.

$$E_{N,1} = \left(1 - \frac{N_1}{N_{0,\text{NH4-N}}}\right) \times 100\% = \left(1 - \frac{10.55 \text{ mg NH}_4\text{-N/L}}{29.5 \text{ mg NH}_4\text{-N/L}}\right) \times 100\% = 64.2\%$$

Calculate the NH_4-N removal efficiency in the second compartment.

$$E_{N,2} = \left(1 - \frac{N}{N_1}\right) \times 100\% = \left(1 - \frac{0.5 \text{ mg NH}_4\text{-N/L}}{10.55 \text{ mg NH}_4\text{-N/L}}\right) \times 100\% = 95.3\%$$

Calculate the overall NH_4-N removal efficiency by the attached growth nitrification in the second MBBR reactor.

$$E_{\text{N,nitrification}} = \left(1 - \frac{N}{N_{0,\text{NH4-N}}}\right) \times 100\% = \left(1 - \frac{0.5 \text{ mg NH}_4\text{-N/L}}{29.5 \text{ mg NH}_4\text{-N/L}}\right) \times 100\% = 98.3\%$$

Calculate the overall NH_4-N removal efficiency in the MBBR system.

$$E_{N,MBBR} = \left(1 - \frac{N}{N_{0,TKN}}\right) \times 100\% = \left(1 - \frac{0.5\,\text{mg NH}_4\text{-N/L}}{35\,\text{mg NH}_4\text{-N/L}}\right) \times 100\% = 98.6\%$$

6. Determine the overall HRT in the second MBBR reactor.

   Calculate the HRTs required by nitrification in the first and second compartments ($\theta_{r,1}$ and $\theta_{r,2}$) from Equation 10.97d, respectively.

$$\theta_{N,1} = \frac{N_{0,NH4\text{-}N} - N_1}{f_m\,SA_m\,J_{N,1}} = \frac{(29.5 - 10.55)\,\text{g NH}_4\text{-N/m}^3}{0.6 \times 500\,\text{m}^2/\text{m}^3 \times 0.871\,\text{g NH}_4\text{-N/m}^2\cdot\text{d}} \times \frac{24\,\text{h}}{\text{d}} = 1.74\,\text{h}$$

$$\theta_{N,2} = \frac{N_1 - N}{f_m\,SA_m\,J_{N,2}} = \frac{(10.55 - 0.5)\,\text{g NH}_4\text{-N/m}^3}{0.6 \times 500\,\text{m}^2/\text{m}^3 \times 0.462\,\text{g NH}_4\text{-N/m}^2\cdot\text{d}} \times \frac{24\,\text{h}}{\text{d}} = 1.74\,\text{h}$$

   Provide a HRT $\theta_N = 1.75$ h in each compartment of the second MBBR reactor. The overall HRT in the second MBBR reactor is $\theta = 3.5$ h.

7. Determine the volume of the second MBBR reactor.

   The volume of each compartment, $V_c = \theta_N Q = 1.75\,\text{h} \times 4000\,\text{m}^3/\text{d} \times \frac{\text{d}}{24\,\text{h}} = 292\,\text{m}^3$

   Provide a volume $V_c = 300\,\text{m}^3$ in each compartment of the second MBBR reactor.

   The total volume of the second MBBR reactor is 600 m^3.

8. Determine the media volume in the second MBBR reactor.

   The media volume in each compartment, $V_m = f_m V_r = 0.6 \times 300\,\text{m}^3 = 180\,\text{m}^3$

   Therefore, the total media volume in the second MBBR reactor is 360 m^3.

9. Determine the surface area required in the second MBBR reactor.

   The media volume in each compartment, $A_m = SA_m V_{\_m} = 500\,\text{m}^2/\text{m}^3 \times 180\,\text{m}^3 = 90{,}000\,\text{m}^2$.

   The media surface area in the second MBBR reactor is 180,000 m^2.

10. Summarize the design information in the complete MBBR system.

    The calculation results for CBOD removal and nitrification in the entire MBBR system obtained from Examples 10.111 and 10.136 are summarized below.

Parameter	First Reactor			Second Reactor			MBBR System
	Compartment		Subtotal	Compartment		Subtotal	
	First	Second		First	Second		
Process performance							
BOD_5 removal, %	75	90	97.5	–	–	–	97.5
NH_4-N removal, %	–	–	15.7	64.2	95.3	98.3	98.6
HRT, h	0.75	0.75	1.5	1.75	1.75	3.5	5
Volume, m^3	125	125	250	300	300	600	850
Media volume, m^3	62.5	62.5	125	180	180	360	485
Media surface area, m^2	31,250	31,250	62,500	90,000	90,000	180,000	242,500

10.6.2 Denitrification

Denitrification is a biological conversion of nitrate or nitrite to nitrogen. This conversion takes place through a series of reduction reactions under anaerobic or anoxic condition in presence of a suitable carbon source. The alternative sources of carbon are methanol, ethanol, acetate, raw or primary settled wastewater, molasses and sugars, and others. The nitrate reduction reaction involves several steps. These steps are conceptually illustrated in Figure 10.88. Biological nitrogen removal may also be achieved by converting ammonia to nitrogen gas directly through deammonification (an anaerobic ammonium oxidation also called partial nitritation/anammox (PN/A)) process. Information about the PN/A process is presented in Section 13.10.4.

Microbiology: The denitrification process is primarily carried out by heterotrophic bacteria called *denitrifiers*. Most of them are facultative aerobic microorganisms that can utilize either oxygen, nitrate or nitrite as their electron acceptors for oxidizing organic materials.[2,7,181] The basic features of denitrification process with affecting factors are listed below.

1. The basic requirements for denitrification by *denitrifiers* are (a) anoxic condition that requires absence of oxygen and presence of nitrate and/or nitrite and (b) presence of a suitable carbon source.
2. Based on organic carbon source denitrification may be classified as (a) preanoxic system that utilizes the incoming raw wastewater as the sole organic carbon source (Figure 10.89a) and (b) postanoxic system that may use (i) exogenous organic carbon source such as *methanol* in substrate driven mode or (ii) organic substrate provided by the biomass endogenous decay in endogenous driven mode (Figure 10.89b).
3. In a preanoxic system with internal recycle, facultative aerobic heterotrophic population is alternated between anoxic and aerobic conditions. For this reason, a relatively lower fraction of total heterotrophs than that in a postanoxic system would be involved in the biological denitrification. The lower denitrification rate in a preanoxic system is adjusted by applying a fraction of total population of heterotrophs that actually carries out denitrification (Example 10.144).
4. The denitrification reaction can be carried out in a suspended or attached growth reactor.
5. Denitrification can also be achieved in a single biological reactor without distinct zones or designated sequences. These processes include (a) alternative nitrification/denitrification (AltNdN) processes, and (b) simultaneous nitrification/denitrification (SNdN).
6. The desirable range of pH for denitrification is 6.5–8.

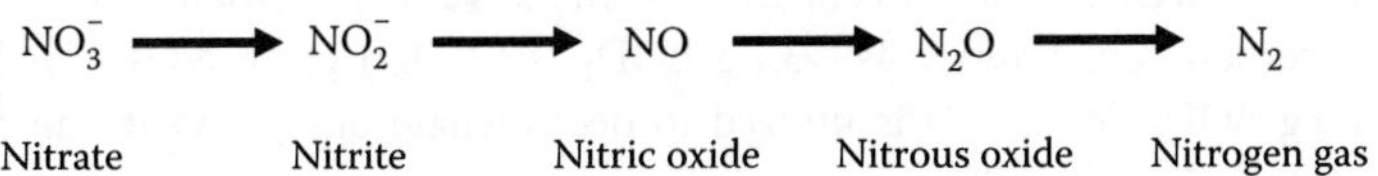

FIGURE 10.88 Reduction reaction steps of the denitrification process.

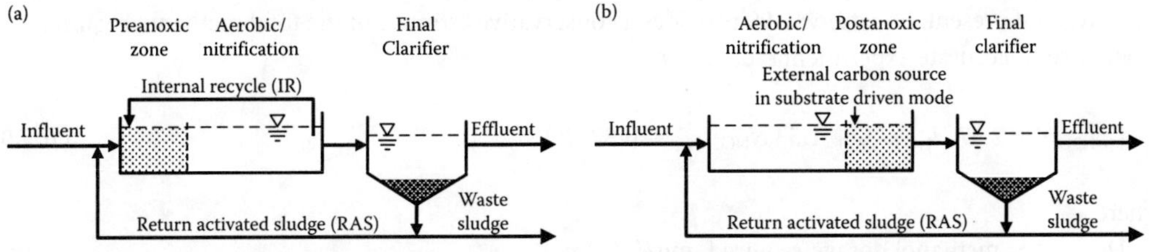

FIGURE 10.89 Schematics of denitrification processes: (a) preanoxic system, and (b) postanoxic system.

7. The denitrification rate is sensitive to DO level in the anoxic zone. Denitrification is inhibited at a DO concentration above 0.1 mg/L. A maximum DO level of 0.1 mg/L in the anoxic zone is normally assumed for design of denitrification process.
8. Denitrifiers are also sensitive to low temperatures. Temperature correction of the denitrification rate is needed in the anoxic zone.

Stoichiometric Reactions: Simplified conceptual denitrification reactions with methanol (CH_3OH) as an organic carbon source are expressed by Equation 10.98. Similar reactions with other carbon sources are presented in Example 10.137.

Energy reaction for reduction of NO_3^- to NO_2^-:

$$3\,NO_3^- + CH_3OH \rightarrow 3\,NO_2^- + CO_2 + 2\,H_2O \tag{10.98a}$$

Energy reaction for reduction of NO_2^- to N_2 (nitrogen gas):

$$2\,NO_2^- + CH_3OH + CO_2 \rightarrow N_2 + H_2O + 2\,HCO_3^- \tag{10.98b}$$

Overall energy reaction for reduction of NO_3^- to N_2:

$$6\,NO_3^- + 5\,CH_3OH + CO_2 \rightarrow 3\,N_2 + 7\,H_2O + 6\,HCO_3^- \tag{10.98c}$$

Biomass synthesis from denitrification:

$$3\,NO_3^- + 14\,CH_3OH + 4\,CO_2 \rightarrow 3\,C_5H_7NO_2 + 16\,H_2O + 3\,HCO_3^- \tag{10.98d}$$

Overall energy and synthesis reaction for denitrification:

$$NO_3^- + 1.08\,CH_3OH + 0.24\,CO_2 \rightarrow 0.47\,N_2 + 0.065\,C_5H_7NO_2 + 1.43\,H_2O + HCO_3^- \tag{10.98e}$$

Following relationships are obtained from Equation 10.98e. The calculations for these ratios are provided in Example 10.138.

1. The cell yield from cell synthesis reaction is 0.52 g VSS per g NO_3-N reduced.
2. The methanol requirement for conversion to biomass is 2.47 g methanol per g NO_3-N reduced.
3. The organic carbon consumption is ~3.7 g BOD_L (or COD) per g NO_3-N reduced to nitrogen gas. Also, 1.3 g BOD_L (or COD) is utilized to deoxygenate one g DO in the nitrate containing influent or recycle flow.[97,183,184]
4. Approximately 3.57 g of alkalinity as $CaCO_3$ is recovered per g NO_3-N reduced.

Methanol Requirement: In Equation 10.98e, methanol is demonstrated as an exogenous carbon source for reducing nitrate to nitrogen gas during denitrification. It may also be required for reduction of nitrite and oxygen if present. Equation 10.99 provides a conservative estimate of the total methanol requirement in absence of accurate experimental data.

$$D_{CH3OH} = 2.47\,N_{NO3\text{-}N,r} + 1.53\,N_{NO2\text{-}N,r} + 0.87\,DO_r \tag{10.99}$$

where

D_{CH3OH} = methanol dosage required, mg/L (g/m^3)
$N_{NO3\text{-}N,r}$ = nitrate nitrogen removed, mg/L (g/m^3)

$N_{NO2\text{-}N,r}$ = nitrite nitrogen removed, mg/L (g/m^3)
DO_r = DO removed, mg/L (g/m^3)

Suspended Growth Biological Kinetics: The general kinetic equations for the aerobic suspended growth biological reactors also apply for biological denitrification. These equations are provided in Sections 10.3.2 and 10.3.3. The substrate utilization rate during denitrification is still controlled by the soluble organic substrate concentration (as electron donor) as described by the Monod kinetics. Under anoxic conditions, nitrate serves as an electron acceptor in the same role as oxygen plays under aerobic conditions. The nitrate nitrogen utilization rate is proportional to the substrate utilization rate. Therefore, substrate utilization rate for soluble BOD_5 or biodegradable soluble COD ($sBOD_5$ or bsCOD) is typically used as the basis for developing denitrification kinetics.

Specific Growth and Substrate Utilization Rates. The most important equations for denitrification kinetics are expressed by Equations 10.100a through 10.100c. The kinetic coefficients for biological denitrification are given in Table 10.52.[2,98,181–184]

$$\mu_H = \frac{\mu_{max,H} S}{K_{s,H} + S} \tag{10.100a}$$

$$r_{su,H} = -\frac{\mu_H}{Y_H} f_{AN} X_H \quad \text{or} \quad r_{su,H} = -\frac{\mu_{max,H} S}{K_{s,H}+S} f_{AN} X_H \tag{10.100b}$$

$$r_{NO} = -\frac{r_{su,H}}{CR_{NO}} f_{AN} X_H \quad \text{or} \quad r_{NO} = \frac{\mu_H}{CR_{NO} Y_H} f_{AN} X_H \quad \text{or}$$

$$r_{NO} = \frac{\mu_{max,H} S}{CR_{NO} Y_H (K_{s,H} + S)} f_{AN} X_H \tag{10.100c}$$

TABLE 10.52 Common Kinetic Coefficients for Denitrification

Parameter	Value for Process Type[a]			
	Preanoxic Zone		Postanoxic Zone[b]	
	Range	Typical	Range	Typical
Y_H,				
BOD_5-based, mg VSS/mg BOD_5-utilized	–	0.67	0.25–0.5	0.48
bCOD-based, mg VSS/mg bCOD-utilized	–	0.42	0.16–0.33	0.3
$k_{d,H}$, d^{-1}	0.09–0.17	0.12	0.04–0.25	0.05
$\mu_{max,H}$, mg VSS/mg VSS·d	1.1–4.5	3.2	0.94–1.86	1.2
k_H[c], mg bCOD-utilized/mg VSS·d	–	6.8	6.7–10.3	4
$K_{s,H}$, mg bCOD/L	–	5	1.5–32	1.5
θ_T, dimensionless				
Temperature correction of $\mu_{max,H}$	–	1.063	1.09–1.12	1.1
Temperature correction of $k_{d,H}$	–	1.029	–	1.04
$K_{NO,H}$, mg NO_3-N/L	–	0.1	–	0.1
$K_{DO,H}$, mg DO/L	0.1–0.2	0.2	0.02–0.2	0.2

[a] These are the kinetic coefficients estimated for heterotrophic growth at 20°C. Use the correction factors presented in this table to obtain the design values under the field conditions.
[b] These are the kinetic coefficients estimated for heterotrophic growth with methanol.
[c] The value of k_H is calculated from Equation 10.89c.
Source: Adapted in part from References 2, 98, and 181 through 184.

where

μ_H = specific growth rate per unit mass of heterotrophs under anoxic conditions, mg VSS/mg VSS·d or d^{-1}

$\mu_{max,H}$ = maximum specific growth rate per unit mass of heterotrophs under anoxic conditions, mg VSS/mg VSS·d or d^{-1}

S = substrate concentration ($sBOD_5$ or bsCOD) in the effluent, mg/L (g/m^3)

$K_{s,H}$ = half-velocity constant for substrate $sBOD_5$ or bsCOD), mg/L (g/m^3)

$r_{su,H}$ = rate of substrate ($sBOD_5$ or bsCOD) utilization per unit of reactor by heterotrophs under anoxic conditions, mg substrate-utilized/L·d (g substrate-utilized/m^3·d)

Y_H = maximum cell yield coefficient over a finite period of growth phase, mg VSS/mg substrate ($sBOD_5$ or bsCOD) utilized

f_{AX} = fraction of heterotrophic bacteria that can utilize NO_3-N in lieu of O_2 under anoxic condition, mg VSS/mg VSS. The value of f_{AX} varies from 0.2 to 1 with a typical value of 0.8 for preanoxic system. In a postanoxic system, $f_{AX} \approx 1$ may be assumed.

X_H = heterotrophic bacteria concentration in the reactor, mg VSS/L (g VSS/m^3)

r_{NO} = NO_3-N reduction rate by heterotrophs, mg N-utilized/L·d (g N-utilized/m^3·d)

CR_{NO} = net oxygen consumption ratio, mg substrate ($sBOD_5$ or bsCOD)/mg electron acceptor (NO_3^-, NO_2^- or O_2) removed. The value of CR_{NO} is calculated from either Equation 10.100d or 10.100e.

$$CR_{NO} = \frac{OE_{NO}}{1.6 - 1.42Y_H} \quad \text{(for BOD}_5\text{-based } Y_H) \tag{10.100d}$$

$$CR_{NO} = \frac{OE_{NO}}{1 - 1.42Y_H} \quad \text{(for bCOD-based } Y_H) \tag{10.100e}$$

where

OE_{NO} = theoretical oxygen consumption ratio, mg O_2/mg electron acceptor (NO_3^-, NO_2^- or O_2) completely reduced

$OE_{NO3\text{-}N}$ = 2.86 mg O_2/mg NO_3-N completely reduced

$OE_{NO2\text{-}N}$ = 1.71 mg O_2/mg NO_2-N completely reduced

OE_{O2} = 1.0 mg O_2/mg O_2 completely reduced

The calculation procedures of these ratios for NO_3^- and NO_2^- are shown in Example 10.140.

Correction Factors for Inhibitory Conditions: The denitrification process is sensitive to many critical operating conditions such as temperature, DO, substrate (carbon source) concentration, of NO_3^- to NO_2^- concentration, and presence of other inhibitors. Many factors are used for corrections of denitrification kinetic coefficients. Some of these factors are given in Equation 10.101.[2,98,181]

$$\mu_{max,H,F} = F_{T,H}F_{NO,H}F_{DO,H}\mu_{max,H} \quad \text{or} \quad k_{H,F} = F_{T,H}F_{NO,H}F_{DO,H}k_H \tag{10.101a}$$

$$k_{d,H,F} = F_{T,H}k_{d,H} \tag{10.101b}$$

$$F_{T,H} = \theta_T^{T-20} \tag{10.101c}$$

$$F_{NO,H} = \frac{N_{NO3\text{-}N}}{K_{NO,H} + N_{NO3\text{-}N}} \tag{10.101d}$$

$$F_{DO,H} = \frac{K_{DO,H}}{K_{DO,H} + DO} \tag{10.101e}$$

where

$\mu_{max,H,F}$ = maximum specific growth rate per unit mass of heterotrophs in anoxic zone under the field operating conditions, mg VSS/mg VSS·d or d^{-1}

$k_{H,F}$ = maximum specific rate of substrate (sBOD or bsCOD) utilization per unit mass of heterotrophs in anoxic zone under the field operating conditions, mg substrate-utilized/mg VSS·d or d^{-1}

$F_{T,H}$ = operating temperature correction factor, dimensionless
The factor of F_T is applicable to either $\mu_{max,N}$ or $k_{d,N}$.

$F_{NO,H}$ = operating NO_3-N concentration correction factor, dimensionless

$F_{DO,H}$ = operating DO concentration correction factor, dimensionless

$k_{d,N,F}$ = specific endogenous decay coefficient for heterotrophs in anoxic zone under the field operating conditions, mg VSS/mg VSS·d or d^{-1}

θ_T = temperature correction coefficient, dimensionless

T = critical temperature, °C

$K_{NO,H}$ = NO_3-N half-saturation concentration for heterotrophs in anoxic zone, mg/L (g/m^3)

$N_{NO3\text{-}N}$ = operating NO_3-N concentration, mg/L (g/m^3)

$K_{DO,H}$ = DO half-saturation concentration for heterotrophs in anoxic zone, mg/L (g/m^3)

DO = operating dissolved oxygen (DO) concentration, mg/L (g/m^3)

Note: The typical values of θ_T, $K_{NO,H}$, and $K_{DO,H}$ are also summarized in Table 10.52.

Specific Denitrification Rate: SDNR is the most commonly used design parameter to size the anoxic zone for denitrification. It is expressed by Equation 10.102.[2,98,181]

$$SDNR = \frac{R_{NO3\text{-}N}}{VX} \quad \text{or} \quad V = \frac{R_{NO3\text{-}N}}{SDNR\,X} \tag{10.102}$$

where

SDNR = specific denitrification rate per unit mass of heterotrophs in anoxic zone, mg NO_3-N/mg VSS·d. The typical ranges of *SDNR* for different denitrification systems are summarized in Table 10.53.

$R_{NO3\text{-}N}$ = nitrate nitrogen removal rate in anoxic zone, kg NO_3-N/d

V = volume of anoxic zone, m^3

X = MLVSS concentration in anoxic zone, mg VSS/L (g VSS/m^3)

SDNR for Preanoxic Zone: For sizing the preanoxic zone, several empirical expressions have been established between the SDNR and the F/M ratio (F/M_b) based on the active heterotrophic biomass in the anoxic zone.[2,98,181] Two useful expressions are given in Equations 10.103a and 10.103b. They can be used to estimate the SDNR at temperature of 20°C and an internal recycle ratio $IR = 1$ ($SDNR_{b,20}$) while the value of F/M_b is determined.[2]

$$SDNR_{b,20} = a + b\ln(F/M_b) \quad (\text{for } F/M_b > 0.5) \tag{10.103a}$$

$$SDNR_{b,20} = 0.24\,F/M_b \qquad (\text{for } F/M_b \le 0.5) \tag{10.103b}$$

TABLE 10.53 Typical Ranges of Specific Denitrification Rate (SDNR)

Type of Denitrification System	Type of Substrate	SDNR at 20°C, mg NO_3-N/mg VSS·d
Preanoxic zone	Wastewater influent	0.05–0.1
Postanoxic zone	Methanol addition	0.1–0.25
	Endogenous decay	0.01–0.04

Source: Adapted in part from References 2 and 181.

where

$SDNR_{b,20}$ = specific denitrification rate per unit mass of active heterotrophic biomass that utilize NO_3-N in anoxic zone at 20°C and $IR = 1$, mg NO_3-N/mg VSS·d

F/M_b = food to microorganisms ratio based on active heterotrophic biomass that utilize NO_3-N in anoxic zone, mg BOD_5/mg VSS·d

a *and* b = empirical constants, mg NO_3-N/mg VSS·d. The values of a and b are dependent of the ratio of readily biodegradable COD (rbCOD) to biodegradable COD (bCOD) in the preanoxic influent.[2] Using an approximate ratio of bCOD/BOD_5 = 1.6, these constants can be estimated from Equations 10.103c and 10.103d within an applicable range of rbCOD/BOD_5 = 0.2–0.6:

$$a = 0.12(\text{rb COD/BOD}_5) = 0.2 - 0.6 \quad (10.103c)$$

$$b = 0.053\ \ln(\text{rb COD/BOD}_5) + 0.18 \quad (10.103d)$$

where rbCOD/BOD_5 = ratio of rbCOD to BOD_5 in the influent to preanoxic zone, mg COD/mg BOD_5

The value of F/M_b may be estimated from Equation 10.103e.

$$F/M_b = \frac{1}{f_{AX} f_b}(F/M) \quad (10.103e)$$

where

F/M = food to microorganisms ratio based on MLVSS in anoxic zone, mg BOD_5/mg VSS·d. F/M is calculated from Equation 10.14b using the volume of anoxic zone.

f_b = active heterotrophic biomass fraction in MLVSS (X_H/X), mg VSS/mg VSS. The value of f_b can be estimated from Equation 10.103f.

$$f_b = \frac{Y_{obs,H,F}}{Y_{obs,H,F} + Y_{ivss}} \quad (10.103f)$$

where

$Y_{obs,H,F}$ = observed biomass yield due to heterotrophic growth in anoxic zone, mg VSS/mg BOD_5. $Y_{obs,H,F}$ is calculated from Equation 10.15l using Y_H (BOD_5-based) and $k_{d,H,F}$ (after temperature correction from Equation 10.101b) at design SRT (θ_c).

Y_{ivss} = fraction of inert (nonbiomass) VSS in the influent to the biological reactor, mg VSS/mg VSS. The values of Y_{ivss} range from 0.1 to 0.3 in primary effluent and 0.3 to 0.5 without primary treatment.

Normally, temperature correction of $SDNR_{b,20}$ is also required to obtain the SDNR at the critical operating temperature ($SDNR_{b,T}$) using Equation 10.103g.[2,181]

$$SDNR_{b,T} = SDNR_{b,20}\theta_T^{T-20} \quad (10.103g)$$

where

$SDNR_{b,T}$ = specific denitrification rate per unit mass of active heterotrophic biomass that utilize NO_3-N in anoxic zone at design operating temperature T and $IR = 1$, mg NO_3-N/mg VSS·d

θ_T = temperature correction coefficient, dimensionless. The typical value of θ_T is 1.026 for temperature correction of $SDNR_{b,20}$.

T = critical temperature, °C

When the values of IR and F/M_b are both greater than 1, an adjustment of the SDNR may be needed. The adjustment can be performed using either Equation 10.103h or 10.103i depending on the value of IR.

No adjustment is necessary when $F/M_b \leq 1$.[2]

$$SDNR_{b,T,IR} = SDNR_{b,T} - 0.0166 \ln(F/M_b) - 0.078 \quad (\text{for } IR = 2) \tag{10.103h}$$

$$SDNR_{b,T,IR} = SDNR_{b,T} - 0.029 \ln(F/M_b) - 0.12 \quad (\text{for } IR = 3\text{–}4) \tag{10.103i}$$

where

$SDNR_{b,T,IR}$ = specific denitrification rate per unit mass of active heterotrophic biomass that utilize NO_3-N in anoxic zone at design operating temperature T with adjustment, mg NO_3-N/mg VSS·d

IR = internal recycle ratio that is the ratio of recycle flow to influent flow, dimensionless

The value of *IR* can be estimated from either Equation 10.103j or 10.103k depending upon the concentration of NO_3-N in the effluent from the preanoxic zone. It is typically in a range of 1–4 to meet the desired NO_3-N concentration in the effluent.

$$IR = \frac{NO_x}{N_{NO3\text{-}N} - N_{NO3\text{-}N,AX}} - (1 + R_{RS}) \tag{10.103j}$$

$$IR \approx \frac{NO_x}{N_{NO3\text{-}N}} - (1 + R_{RS}) \quad (N_{NO3\text{-}N} >> N_{NO3\text{-}N,AX} \quad \text{or} \quad N_{NO3\text{-}N,AX} \approx 0) \tag{10.103k}$$

where

NO_x = NO_3-N concentration produced from nitrification in the aeration basin, mg/L (g/m^3). The value of NO_x should be approximately equal to the total NH_4-N concentration completely oxidized by nitrifiers during nitrification process (Section 10.6.1). Therefore, it is normally estimated from Equation 10.103l.

$$NO_X = N_{0,TKN} - N_{NH4\text{-}N} - f_N p_x \tag{10.103l}$$

$N_{NO3\text{-}N}$ = NO_3-N concentration in the effluent or internal recycle from aeration basin, mg/L or g/m^3

$N_{NO3\text{-}N,AX}$ = NO_3-N concentration in the effluent from preanoxic zone, mg/L or g/m^3

Note: Theoretically, the concentration of NO_3-N in the preanoxic zone must be present to create and maintain a true anoxic condition. However, it may be ignored if the NO_3-N concentration in the aeration basin is significantly higher than that in the preanoxic zone ($N_{NO3\text{-}N} \gg N_{NO3\text{-}N,AX}$ and $N_{NO3\text{-}N,AX} \approx 0$). This assumption simplifies the process calculations (Equation 10.103k).

R_{RS} = ratio of return sludge flow to influent flow, dimensionless

All other parameters have been defined previously.

The overall *SDNR* based on the MLVSS concentration in the anoxic zone (*SDNR*) is eventually converted from $SDNR_{b,T,IR}$ by using Equation 10.103m.

$$SDNR = f_{AX} f_b \, SDNR_{b,T,IR} \tag{10.103m}$$

where $SDNR$ = specific denitrification rate per unit mass of MLVSS in anoxic zone at design operating temperature T after IR adjustment, mg NO_3-N/mg VSS·d

The extra NO_3-N consumption due to endogenous decay of heterotrophic biomass in the preanoxic zone is small and is normally ignored for conservative consideration; and the nitrate nitrogen removal rate ($R_{NO3\text{-}N}$) can be estimated from Equation 10.103n or 10.103o.

$$R_{NO3\text{-}N} = Q[(IR + R_{RS})(N_{NO3\text{-}N} - N_{NO3\text{-}N,AX}) - N_{NO3\text{-}N,AX}] \quad (10.103n)$$

$$R_{NO3\text{-}N} = Q(IR + R_{RS})N_{NO3\text{-}N} \ (N_{NO3\text{-}N} >> N_{NO3\text{-}N,AX} \quad \text{or} \quad N_{NO3\text{-}N,AX} \approx 0) \quad (10.103o)$$

All parameters in Equation 10.103n or 10.103o have been defined previously.

The consumption of organics by heterotrophic denitrification in the preanoxic zone can be estimated from Equation 10.103p.

$$S_{r,AX} = CR_{NO}SDNR\,X\,V \quad \text{or} \quad S_{r,AX} = CR_{NO}R_{NO3\text{-}N} \quad (10.103p)$$

where $S_{r,AX}$ = organic ($sBOD_5$ or bsCOD) consumption rate by denitrification in preanoxic zone, kg BOD_5/d or kg COD/d

All other parameters in Equation 10.103p have been defined previously.

SDNR in Postanoxic Zone with Substrate Driven Mode: The denitrification kinetics are normally applied in the postanoxic zone with substrate driven mode. The *SDNR* under field operating conditions in Equation 10.104a is obtained by dividing the NO_3-N reduction rate in Equation 10.100c by the MLVSS concentration (X).[2]

$$SDNR = \frac{r_{NO,F}}{X} \quad \text{or} \quad SDNR = \frac{\mu_{max,H,F}\,S}{CR_{NO}\,Y_H(K_{s,H} + S)} f_{AX}\left(\frac{X_H}{X}\right) \quad \text{or}$$

$$SDNR = \frac{\mu_{max,H,F}\,S}{CR_{NO}\,Y_H(K_{s,H} + S)} f_{AX} f_b \quad (10.104a)$$

All parameters in Equation 10.104a have been defined previously.

To include the NO_3-N consumption by endogenous decay of heterotrophic biomass, the nitrate nitrogen removal rate ($R_{NO3\text{-}N}$) in the postanoxic zone can be estimated from Equation 10.104b. When the endogenous decay is excluded, a simplified but conservative expression is obtained from Equation 10.104c. All parameters in Equations 10.104b and 10.104c have been defined previously.

$$R_{NO3\text{-}N} = Q(NO_x - N_{NO3\text{-}N}) - \frac{1.42}{OE_{NO3\text{-}N}} k_{d,H,F} f_b X\,V \quad (10.104b)$$

$$R_{NO3\text{-}N} = Q(NO_x - N_{NO3\text{-}N}) \quad (10.104c)$$

When the concentration remaining of carbon source in the effluent after nitrification is not sufficient, the external carbon source (ECS) required for denitrification at a given *SDNR* in the postanoxic zone is also calculated from Equation 10.103p.

SDNR in Postanoxic Zone with Endogenous Driven Mode: The SDNR in the endogenous driven mode may be estimated from Equation 10.105a at operating temperature.[181]

$$SDNR = \frac{1.42}{OE_{NO3\text{-}N}} k_{d,H,F} f_b \quad (10.105a)$$

All parameters in Equation 10.105a have been defined previously. CR_{NO} is calculated from Equation 10.100d or Equation 10.100e using Y_H while $Y_{obs,H,F}$ is calculated from Equation 10.15l using Y_H and $k_{d,H,F}$ after temperature correction from Equation 10.101b at the design SRT (θ_c)

In the endogenous driven mode, the nitrate nitrogen removal rate ($R_{NO3\text{-}N}$) is completely controlled by the endogenous decay of heterotrophic biomass in the postanoxic zone. It is generally expressed by Equation 10.105b.[2]

$$R_{NO3\text{-}N} = \frac{1.42}{OE_{NO3\text{-}N}} k_{d,H,F} X_H V \quad \text{or} \quad R_{NO3\text{-}N} = \frac{1.42}{OE_{NO3\text{-}N}} k_{d,H,F} f_b X V \quad (10.105b)$$

All parameters in Equation 10.105b have been defined previously.

Release of ammonia due to the endogenous decay may need to be checked in the postanoxic zone with the endogenous driven mode. The potential NH_4-N release rate and concentration are calculated from Equations 10.105c and 10.105d, respectively.

$$R_{r,NH4\text{-}N} = f_{NH4\text{-}N} f_N k_{d,H,F} X_H V \quad \text{or} \quad R_{r,NH4\text{-}N} = f_{NH4\text{-}N} f_N k_{d,H,F} f_b X V \tag{10.105c}$$

$$N_{r,NH4\text{-}N} = \frac{R_{r,NH4\text{-}N}}{(1 + R_{RS})Q} \tag{10.105d}$$

where

$R_{r,NH4\text{-}N}$ = rate of NH_4-N release due to endogenous decay, kg NH_4-N/d

$f_{NH4\text{-}N}$ = fraction of NH_4-N released from total synthesized organic nitrogen destroyed due to endogenous decay, mg MH_4-N/mg N decayed

$N_{r,NH4\text{-}N}$ = concentration of NH_4-N release due endogenous decay, mg NH_4-N/L (mg NH_4-N/m^3)

All other parameters have been defined previously.

Attached Growth for Postanoxic Denitrification: Denitrification of nitrate containing effluent is also achieved in submerged attached growth processes. These processes include (1) denitrification filters (DNFs), (2) anoxic fluidized-bed bioreactors (AnoxFBBRs), and (3) suspended media denitrifications. The flow schematics of these processes are very similar to those presented in Sections 10.4.3 and 10.4.4 except for an additional arrangement for feeding an ECS such as methanol in the well nitrified influent is needed for denitrification in a reactor without aeration. The typical ratio of methanol to NO_3-N is 3–3.5 kg CH_3OH/kg NO_3-N that includes the requirement for reducing the DO concentration remaining in the nitrified influent.[2,98,183,184]

The DNFs may be either an upflow or downflow filter using floating or sunken media as shown in Figuares 10.81a and 10.81b. Continuous backwash denitrification filter (CBDNF) has also been developed for anoxic denitrification (Figure 10.90a).[98,181,189] The AnoxFBBR can also be used for postanoxic denitrification. The flow schematic is similar to that shown in Figure 10.81c. The anoxic suspended media reactors for tertiary denitrification may include the anoxic moving biofilm reactor (AnoxMBBR, Figure 10.90b) and the submerged denitrification RBC (Figure 10.79d). A solid separation device may be required for removal of TSS after these suspended media denitrification reactors. The important design and operational information of these reactors for denitrification is summarized in Table 10.54.[2,98,183,184] Brief description in the context of denitrification for these reactors is given below.

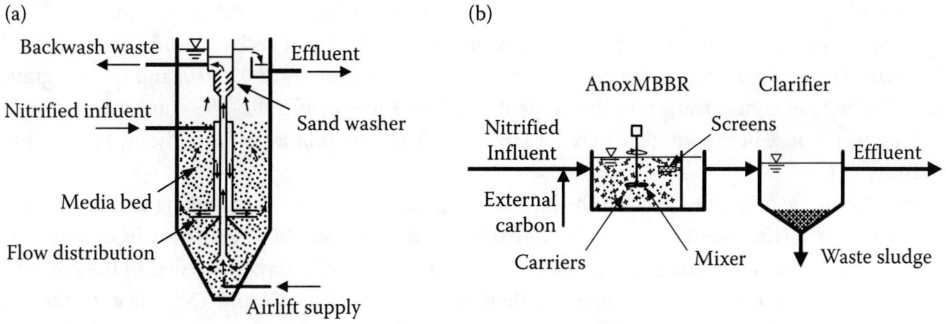

FIGURE 10.90 Schematics of selected attached growth postanoxic denitrification processes: (a) continuous backwash denitrification filter (CBDNF), and (b) anoxic MBBR (AnoxMBBR).

TABLE 10.54 Important Design and Operational Parameters of Submerged Attached Growth Systems for Denitrification

Parameter	Value for Reactor Type					
	DFDNF	UFDNF		CBDNF	AnoxFBBR	AnoxMBBR
		Sunken Media	Floating Media			
Volumetric nitrate loading, kg NO_3-N/m^3·d (lb/10^3 ft^3·d)	0.3–3 (20–200)	0.8–5 (50–300)	1.5–2 (95–125)	0.3–2 (19–125)	3–5 (200–300)	0.25–0.5 (15–30)
Hydraulic loading, m^3/m^2·h (ft^3/ft^2·h)	2.4–4.8 (8–16)	4–6 (13–20)	4–6 (13–20)	2.4–8 (8–26)	15–24 (50–80)[a]	N/A
HRT or empty bed contact time (EBCT), min	25	20	20	15	5–10	30[b]
Depth, m	1.2–1.8	2	2	3–4	1.5–2.5[c]	4–6
Backwash						
Air scour, m^3/m^2·h (gpm/ft^2)	90 (37)	100 (41)	12 (5)	Airlift	N/A	N/A
Water, m^3/m^2·h (gpm/ft^2)	18 (7.5)	20 (8)	55 (23)	0.4 (0.16)	N/A	N/A
Water wash duration, min	15	10	12	Continuous	N/A	N/A
Wash water volume, percent of feed flow (approximate)	4.5–5	3.5	10–15	10	N/A	N/A

[a] The recirculation ratio is 2–5.
[b] Use of a specific surface area SA_m = 500 m^2/m^3 and a fill volume f_m = 0.5.
[c] The bed expansion is 75–150% of reactor depth.
Note: 1 kg/m^3·d = 62.4 lb/10^3 ft^3·d; 1 m^3/m^2·h = 3.28 ft^3/ft^2·h; and 1 m^3/m^2·h = 0.409 gpm/ft^2.
Source: Adapted in part from Reference 2.

Upflow Denitrification Filter: The upflow denitrification filter (UFDNF) utilizes 2–6 mm natural or synthetic granular media. No *bumping* is needed for nitrogen release. Effluent concentration of NO_3-N below 3 mg/L is achieved.

Downflow Denitrification Filter: The downflow denitrification filter (DFDNF) utilizes expanded clay or anthracite media with effective size of 1.8–6 mm, uniformity coefficient of 1.3–1.7, sphericity of 0.7–0.9, and specific gravity of ~ 1.6. The reactor is similar to a gravity filter and provides denitrification and suspended solids removal. Filter head gradually increases and air scour followed by air-water backwash is needed to remove solid and excess biomass. Frequently the filter media is flushed or purged in an upward direction to remove accumulated nitrogen. This is called filter *bump*. The achievable TSS and NO_3-N concentrations in the effluent are >5 and 3 mg/L, respectively.

Anoxic Fluidized-Bed Bioreactors: The AnoxFBBR utilizes the silica sand that is expanded and fluidized under the upflow velocity created by the combined influent and effluent recirculation flow. The sand has effective size of 0.3–0.5 mm, uniformity coefficient = 1.25–1.5, sphericity of 0.8–0.9, and specific gravity of 2.4–2.6. The NO_3-N concentration in the effluent is below 3 mg/L, and the TSS concentration is 15–20 mg/L. The media lost in the effluent may be separated from the biomass, cleaned and returned to the AnoxFBBR.

Anoxic Moving Bed Biofilm Reactor: The basic configuration of the anoxic moving bed biofilm reactor (AnoxMBBR) for tertiary denitrification is similar to the aerobic MBBR systems for BOD removal and nitrification (Sections 10.4.3 and 10.6.1). Use of smaller plastic media may be needed in the AnoxMBBR to provide larger surface area for more effective denitrification. The NO_3-N and TSS concentrations below 3 and 10 mg/L are achievable in the effluent from the clarifier.

Submerged Denitrification RBC: The process provides good denitrification. The basic design features of RBC system are covered in Sections 10.4.1 and 10.4.3. The denitrification depends upon the surface NO_3-

N loading on disk area. The achievable effluent NO_3-N concentrations are 1 and 6 mg/L at a surface NO_3-N loading rate of 0.4 and 3.3 kg NO_3-N/10^3 m^2 d, respectively.

Attached Growth for Preanoxic Denitrification: In a preanoxic system, the BOD removal, nitrification, and denitrification are carried out with one clarifier and one sludge line. Soluble BOD_5 in the primary effluent provides organic carbon source for the denitrification. A packed-bed filter may be used for denitrification followed by nitrification sequence in a suspended or attached growth reactor. The well nitrified effluent is recycled through the attached growth DNFs. Example process diagrams of two attached growth preanoxic systems are shown in Figure 10.91.

Common Processes for Biological Nitrogen Removal: In a preanoxic system combined nitrification-denitrification can be achieved in a single or a series of reactors that create aerobic and anoxic conditions in different zones. Using proper configurations with recycle flows, raw wastewater can serve as the only organic carbon source. In the postanoxic systems, the SDNR is significantly higher (75–125 g NO_3-N/kg VSS·d) in the substrate driven mode with excess organic carbon source than that in the endogenous driven mode with limited carbon source (15–50 g NO_3-N/kg VSS·d). Many preanoxic and postanoxic process configurations have been integrated in the MBR systems. Examples of commonly used suspended growth processes for biological nitrogen removal are presented in Table 10.55. Biological nitrogen removal can also be achieved in aerobic and anoxic attached growth reactors. These reactors may include (a) nitrification filters and RBCs and (b) DNFs, fluidized-bed reactors and submerged RBCs.[190–192]

Alternated Nitrification/Denitrification (AltNdN): In these cyclic or phased processes, the operating conditions are changed between anoxic and aerobic conditions sequentially or spatially to achieve nitrification and denitrification. In general, the nitrification occurs when or where the aeration is provided while the denitrification happens when or where mixing is provided without aeration. The submersible mixers are typically used to provide the desired mixing energy under the anoxic conditions. Automatic process control with DO/ORP may be required to alternate operating conditions efficiently and precisely. Examples of these cyclic or phased processes are illustrated in Table 10.55.

Simultaneous Nitrification/Denitrification (SNdN): Under DO below 0.5 mg/L both aerobic and anoxic conditions may exist inside activated sludge floc. Due to a limitation on DO penetration the aerobic zone can only be maintained at the exterior of the floc while an anoxic zone may occur in the interior of the floc. As a result, nitrification and denitrification may take place simultaneously in the same tank. Under favorable conditions, total nitrogen removal may reach over 90%. Processes utilizing simultaneous nitrification-denitrification are presented in Table 10.55.

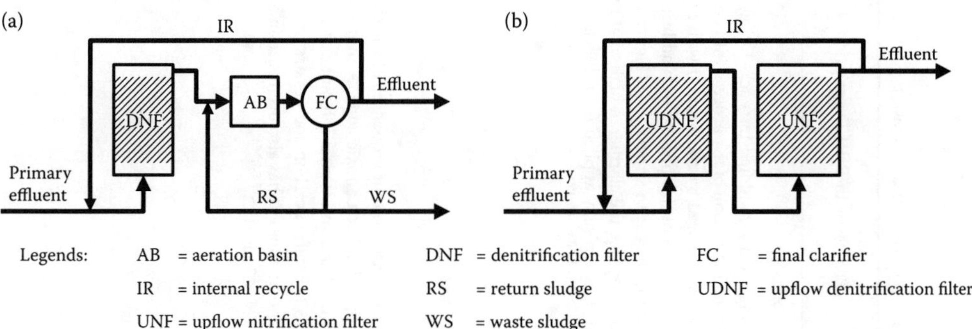

FIGURE 10.91 Schematics of selected attached growth preanoxic denitrification processes: (a) denitrification filter (DNF) prior to aeration basin for BOD removal and nitrification, and (b) sequential upflow denitrification-nitrification filters for combined BOD removal, nitrification and denitrification.

TABLE 10.55 Description and Design Parameters of Common Biological Nitrogen Removal Processes

Process with Brief Description	SRT (θ_c), d	HRT (θ), h			MLSS[a], mg TSS/L	Return/Recycle Ratio[b]		Effluent TN, mg/L as N
		Anoxic	Aerobic	Total		R_{ir} or R_{mlr}	R_{rs}	
1. Preanoxic systems								
a. Ludzack–Ettinger (LE)/modified Ludzack–Ettinger (MLE) LE is the basic configuration of the process with an anoxic–aerobic process sequence followed by clarification with return sludge. MLE is a modification of this process with addition of an internal recycle (IR) from aerobic to anoxic zone (Figure 10.92). When an MBR is used to replace the clarifier, a high total recycle ratio of 6 for RAS and IR is typically applied for achieving an effluent TN <6 mg/L.[2,7,17,98,181,183–185,193,194]	7–20	1–4	4–12	6–16	3000–4000[c]	1–2	0.5–1	<10
b. Step-feed The process uses three to four sequences of anoxic–aerobic zones. Influent is applied at each anoxic zone. A four-sequence system with RAS to the head and typical influent distributions is shown in Figure 10.93. External carbon sources are optional in the 3rd and/or 4th anoxic zones. It has also been used in MBR process for an effluent TN <3 mg/L.[2,17,98,181]	7–20	2–3[d]	6–10[e]	8–12	2000–6000[c]	N/A	0.25–0.75	<5

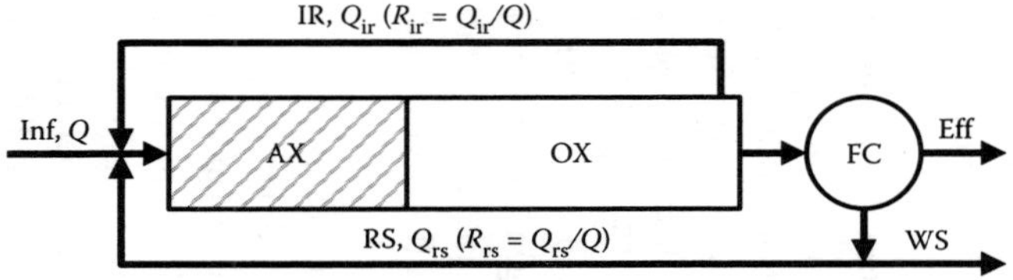

FIGURE 10.92 Definition sketch of MLE process.

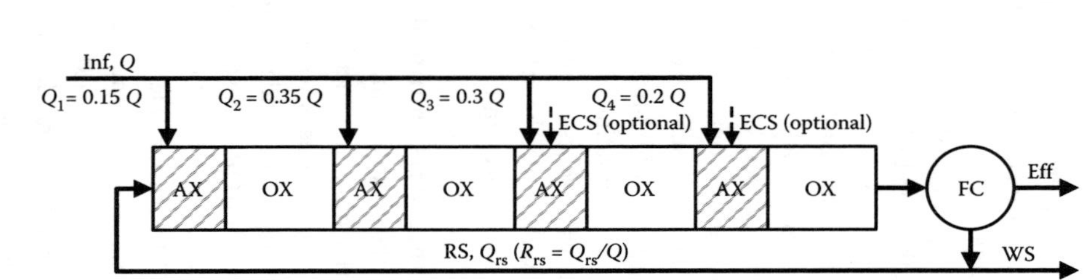

FIGURE 10.93 Definition sketch of four-stage step-feed process.

(*Continued*)

TABLE 10.55 (*Continued*) Description and Design Parameters of Common Biological Nitrogen Removal Processes

Process with Brief Description	SRT (θ_c), d	HRT (θ), h			MLSS[a], mg TSS/L	Return/Recycle Ratio[b]		Effluent TN, mg/L as N
		Anoxic	Aerobic	Total		R_{ir} or R_{mlr}	R_{rs}	
c. Sequencing batch reactor (SBR) The SBR process provides biological BOD and nitrogen removals in a single reactor (see Table 10.15). NO_3-N is mostly removed in the late period of the fill cycle due to anoxic/anaerobic environment. Nitrification occurs in the react cycle followed by partial denitrification in nonaerated settle, decant, and idle cycles (Figure 10.94).[2,7,17,29,98,181,195,196]	10–30	Variable	Variable	8–10[f]	3000–5000	N/A	N/A	<8
2. Postanoxic systems								
a. Wuhrmann process (aerobic–anoxic) The process uses aerobic–anoxic sequence followed by a clarifier. Returned sludge and influent are added at the head of the basin. External carbon source may be needed to improve denitrification rate and efficiency (Figure 10.95). An elevated effluent NH_4-N may be seen in the postanoxic zone due to release of NH_4-N during endogenous decay.[2,7,17,98,181,197]	7–20	1–4[g]	4–12	6–16	3000–4000	N/A	0.5–1	<3[h]

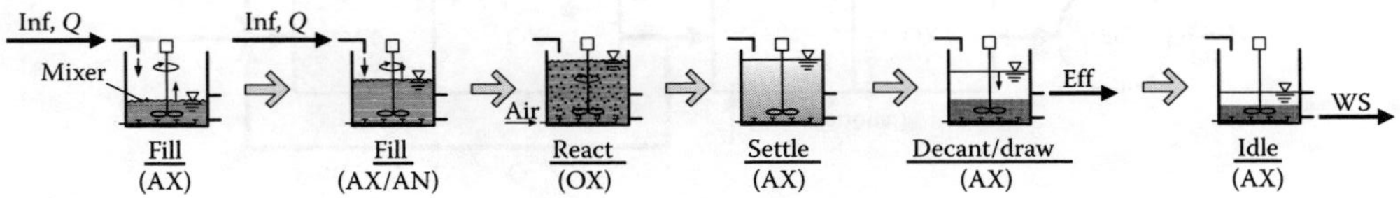

FIGURE 10.94 Definition sketch of SBR process.

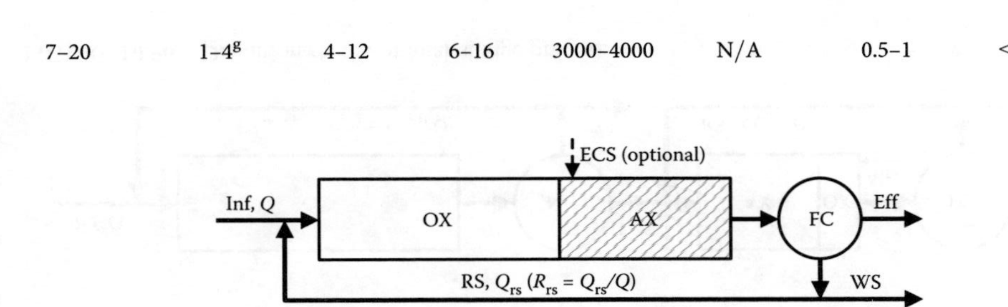

FIGURE 10.95 Definition sketch of Wuhrmann process.

(*Continued*)

TABLE 10.55 (*Continued*) Description and Design Parameters of Common Biological Nitrogen Removal Processes

Process with Brief Description	SRT (θ_c), d	HRT (θ), h			MLSS[a], mg TSS/L	Return/Recycle Ratio[b]		Effluent TN, mg/L as N
		Anoxic	Aerobic	Total		R_{ir} or R_{mlr}	R_{rs}	
b. Dual- or triple-sludge systems Dual- or triple-sludge system is also utilized to achieve nitrification–denitrification. In a dual-sludge system, an activated sludge process with nitrification is used in the 1st system. An anoxic zone is used for denitrification in front of the 2nd system and followed by a short aerobic zone for effluent polishing and degassing. Methanol is typically needed as external carbon source (Figure 10.96).[2,7,98,181,198,199]	10–20	1–4[g]	6–8	8–12	3000–4000	N/A	0.5–1	<3
c. Four-stage Bardenpho This is a proprietary process that uses anoxic, aerobic, anoxic, and aerobic zones. Sludge is returned from clarifier to the first anoxic zone. Mixed liquor return (MLR) from the first aerobic to the first anoxic zone is usually used to reduce effluent NO_3-N concentration. External carbon may be added into the second anoxic zone (Figure 10.97). A single MBR reactor can be used to replace both the last aerobic zone and the clarifier.[2,7,17,98,181,193,200]	10–20	2–6[i]	6–12[i]	8–18	3000–4000[c]	2–4	0.5–1	<3

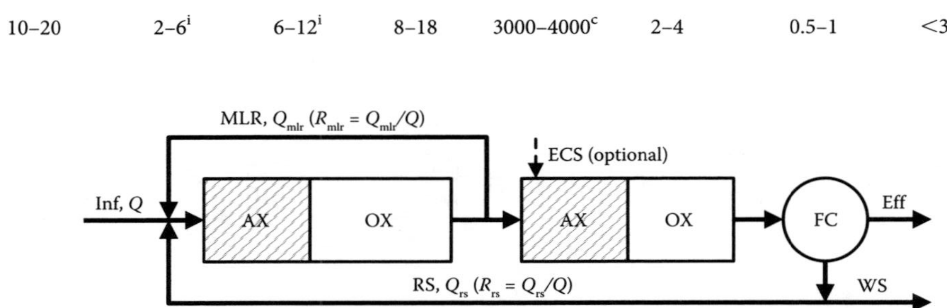

FIGURE 10.96 Definition sketch of dual-sludge process.

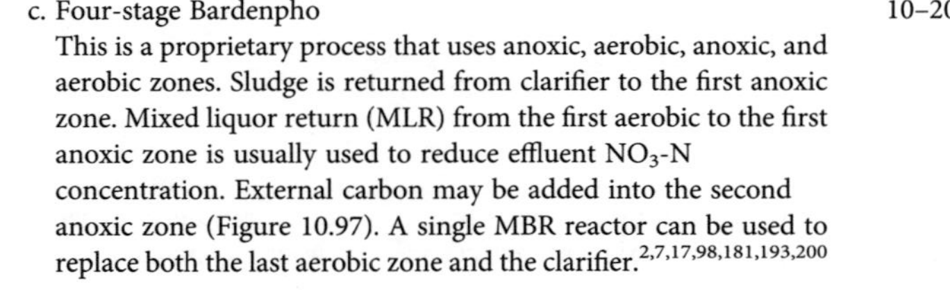

FIGURE 10.97 Definition sketch of four-stage Bardenpho process.

(*Continued*)

TABLE 10.55 (*Continued*) Description and Design Parameters of Common Biological Nitrogen Removal Processes

Process with Brief Description	SRT (θ_c), d	HRT (θ), h			MLSS[a], mg TSS/L	Return/Recycle Ratio[b]		Effluent TN, mg/L as N
		Anoxic	Aerobic	Total		R_{ir} or R_{mlr}	R_{rs}	
3. Alternated nitrification/denitrification (AltNdN) processes								
a. Oxidation ditch for nitrogen removal Oxidation ditch has a channel in the shape of a racetrack. Aerobic zone exists after the aerator and anoxic zone is created away from the aerator. The oxidation ditch has large tank volume and long SRT to accommodate several nitrification and denitrification zones, if needed (Figure 10.98).[2,7,17,98,181,201] See additional information in Tables 10.15 and 10.57.	20–30	4–6	12–24	18–30	2000–4000	N/A	0.5–1	<5
b. Nitrox™ or dNOx™ process These are oxidation ditch systems where operation is alternated between aerobic and anoxic conditions by turning on and off the aerators. The aerators are turned off at least once per day for 3–5 h in each anoxic period. Submerged mixers are utilized to maintain the desired velocity when the aerators are turned off. Oxidation–reduction potential (ORP) control is monitored to restart the aerator when the denitrification is completed (Figure 10.99).[2,98]	10–20	3–8	12–24	18–30	2000–3000	N/A	0.5–1	<10[j]

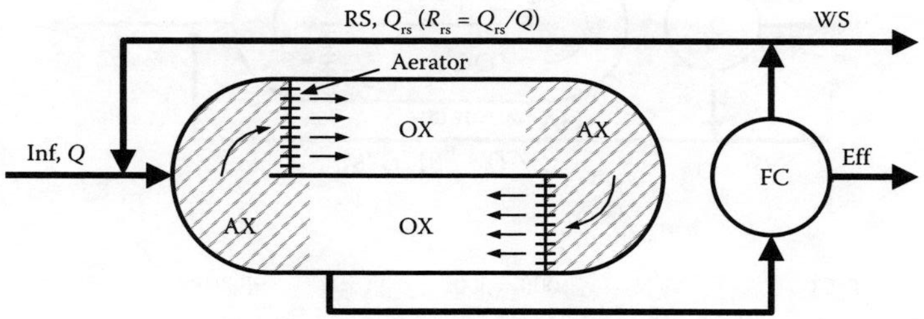

FIGURE 10.98 Definition sketch of oxidation ditch process.

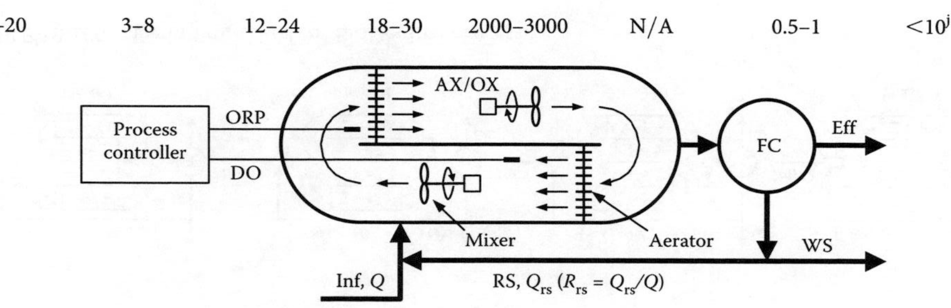

FIGURE 10.99 Definition sketch of dNOx process.

(*Continued*)

TABLE 10.55 (*Continued*) Description and Design Parameters of Common Biological Nitrogen Removal Processes

Process with Brief Description	SRT (θ_c), d	HRT (θ), h			MLSS[a], mg TSS/L	Return/Recycle Ratio[b]		Effluent TN, mg/L as N
		Anoxic	Aerobic	Total		R_{ir} or R_{mlr}	R_{rs}	
c. BioDenitro™ process for nitrogen removal This is a typical phased oxidation ditch technology. It uses at least two oxidation ditches that are operated in anoxic–aerobic, aerobic–aerobic, aerobic–anoxic, and aerobic–aerobic stages. The durations of these stages are 1.5, 0.5, 1.5, and 0.5 h, respectively. The RAS is returned to the head of the plant (Figure 10.100).[2,7,17,98,181,202]	20–40	9[k]	15[j]	20–30	3000–4000	N/A	0.5–1	<8
4. Simultaneous nitrification/denitrification (SNdN) processes								
a. Low DO process In this process, a DO of 0–0.3 mg/L is maintained to accommodate both nitrification and denitrification in the same basin. The desired operating condition can be controlled precisely using aerators with variable frequency drives (VFDs) under automatic DO control (Figure 10.101).[2,98] The ADH/NAD ratio[l] may also be monitored for process control in low DO oxidation ditch system.[98,203]	10–20	Variable	Variable	18–36	3000–4000	N/A	0.5–1	<8

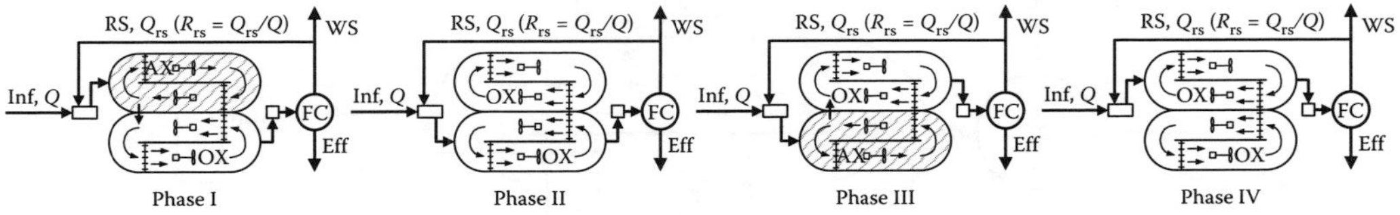

FIGURE 10.100 Definition sketch of BioDenitro process.

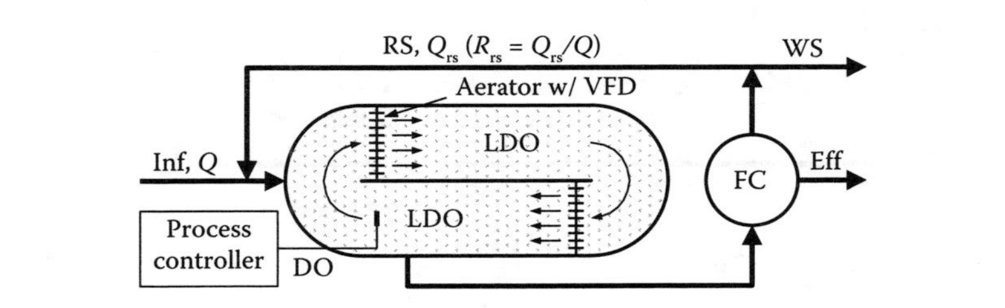

FIGURE 10.101 Definition sketch of low DO oxidation ditch process.

(*Continued*)

TABLE 10.55 (*Continued*) Description and Design Parameters of Common Biological Nitrogen Removal Processes

Process with Brief Description	SRT (θ_c), d	HRT (θ), h			MLSS[a], mg TSS/L	Return/Recycle Ratio[b]		Effluent TN, mg/L as N
		Anoxic	Aerobic	Total		R_{ir} or R_{mlr}	R_{rs}	
b. Orbal® process for nitrogen removal The Orbal® process has three concentric channels. DO below 0.3 mg/L in the 1st channel (which is the outer channel that receives influent and RAS), 0.5–1.5 mg/L in the 2nd channel, and 2–3 mg/L in the 3rd channel are maintained, respectively. The volumes of first, second, and third channels are 1/2, 1/3, and 1/6 of total volume (Figure 10.102).[2,7,98,181] See additional information in Tables 10.15 and 10.57.	10–30	6–10	8–12	10–20	2000–4000	1–2	0.5–1	<5
c. Low DO MBR process The system configuration is similar to an MLE process with MBR. However, a slightly higher DO of 0.3–0.7 mg/L is required in the aerobic zone to achieve the SNdN due to high volumetric oxygen uptake rate at higher MLVSS concentration (Figure 10.103).[2,7,98]	20–30	Variable	Variable	12–18	8000–12,000	N/A	0.5–1	<8

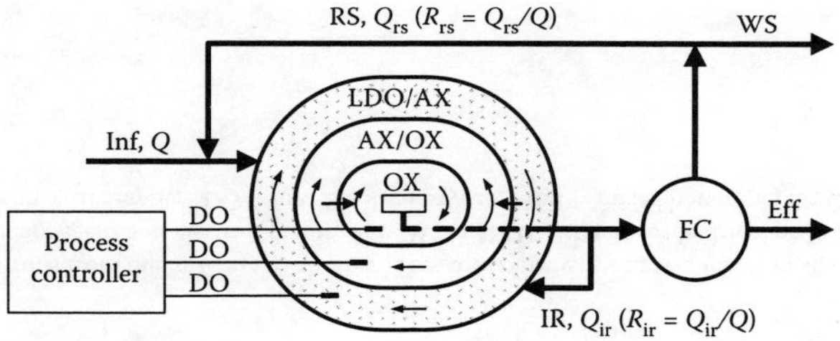

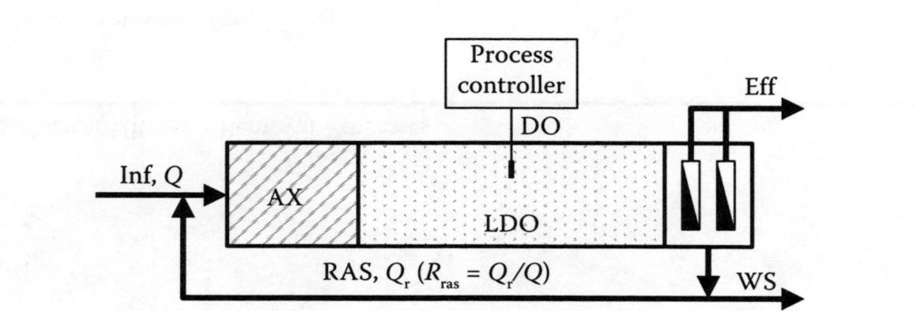

FIGURE 10.102 Definition sketch of Orbal® process.

FIGURE 10.103 Definition sketch of low DO MBR process.

(*Continued*)

TABLE 10.55 (*Continued*) Description and Design Parameters of Common Biological Nitrogen Removal Processes

[a] The ratio of MLVSS to MLSS is normally 0.70–0.85.

[b] It is the ratio to the influent flow *Q*.

[c] The MLSS may be in the range of 7500–12,000 mg/L when MBR is used to replace the clarifier for solids separation.

[d] The HRT ranges from 0.5 to 0.57 h in the anoxic zone in each pass.

[e] The HRT ranges from 1.5 to 2.5 h in the aerobic zone in each pass.

[f] A short aerobic zone with HRT <0.5 may be needed after the postanaerobic zone to strip the nitrogen gas and improve the mixed liquor settling in the final clarifier.

[g] The cycle times may be 3–4, 3–4, 0.5–1, 0.5, 0–0.5 h for fill, react, settle, decant and idle, respectively.

[h] External carbon source is typically required to achieve a low effluent TN concentration.

[i] The HRT range is for each anoxic or aerobic zone. The HRT values may be: 1–3 h in the 1st anoxic zone, 4–12 h in the 2nd aerobic zone, 2–4 h in the 3rd anoxic zone, and 0.5–1 h in the 4th aerobic zone.

[j] The process performance may be effective at low to medium influent TKN concentration.

[k] This is the overall HRT in each basin during a 24-h period.

[l] In metabolism, nicotinamide adenine dinucleotide (NAD) is involved in redox reactions, carrying electrons from one reaction to another. This coenzyme exists in two forms in all cells: NAD^+ is an oxidizing agent and it accepts electrons from other molecules and becomes reduced to form NADH, which is a reducing agent and electron donor. The "H" in NADH stands for hydride, or "high-energy hydrogen."

Legends: AX = anoxic; DO = dissolved oxygen; ECS = external carbon source; FC = final clarifier; IR = internal recycle; LDO = low dissolved oxygen; MLR = mixed liquor recycle; ORP = oxidation–reduction potential; OX = aerobic; RS = return sludge; WS = waste sludge.

Source: Adapted in part from References 2, 7, 17, 98, 181, 183 through 185, 189, and 193 through 203.

EXAMPLE 10.137: DENITRIFICATION REACTIONS WITH CARBON SOURCE OTHER THAN METHANOL

Nitrate reduction reaction (energy reaction) with methanol is given by Equation 10.98c. Write similar reactions with acetic acid, ethanol, sugar, and wastewater as a carbon source. The chemical formula for biodegradable portion of wastewater is $C_{10}H_{19}O_3N$.

Solution

1. Write the reduction reaction of NO_3^- to N_2 with acetic acid (CH_3COOH).

$$8\,NO_3^- + 5\,CH_3COOH \rightarrow 4\,N_2 + 10\,CO_2 + 6\,H_2O + 8\,OH^-$$

2. Write the reduction reaction of NO_3^- to N_2 with ethanol (C_2H_5OH).

$$2\,NO_3^- + C_2H_5OH \rightarrow N_2 + 2\,CO_2 + 3\,H_2O$$

3. Write the reduction reaction of NO_3^- to N_2 with sugar ($C_6H_{12}O_6$).

$$4\,NO_3^- + C_6H_{12}O_6 \rightarrow 2\,N_2 + 6\,CO_2 + 6\,H_2O$$

4. Write the reduction reaction of NO_3^- to N_2 with wastewater ($C_{10}H_{19}O_3N$).

$$10\,NO_3^- + C_{10}H_{19}O_3N \rightarrow 5\,N_2 + 10\,CO_2 + 3\,H_2O + NH_3 + 10\,OH^-$$

EXAMPLE 10.138: CELL SYNTHESIS, METHANOL AND EQUIVALENT BOD$_L$ CONSUMPTIONS, AND ALKALINITY RECOVERY FACTORS FOR DENITRIFICATION

The factors for cell synthesis, methanol, and equivalent BOD$_L$ consumptions as carbon source and alkalinity recovery in biological denitrification are, respectively, 0.52 g VSS, 2.47 g CH_3OH (methanol), 3.7 g BOD$_L$, and 3.57 g alkalinity as $CaCO_3$ per gram of NO_3-N reduced. Develop these factors. Also, determine methanolconsumption from deoxygenation.

Solution

1. Identify the reaction for denitrification.
   The overall energy and synthesis reaction for denitrification is given by Equation 10.98e.

$$NO_3^- + 1.08\,CH_3OH + 0.24\,CO_2 \rightarrow 0.47\,N_2 + 0.065\,C_5H_7NO_2 + 1.43\,H_2O + HCO_3^-$$

2. Determine the cell synthesis factor for reduction of NO_3-N to N_2.
   In accordance with the stoichiometric relationship, 0.065 moles of biomass VSS as $C_5H_7NO_2$ are produced per mole of NO_3-N reduced. The molecular weight of biomass $w_{bVSS} = 113$ g VSS/mole $C_5H_7NO_2$. Calculated the cell synthesis factor ($bVSS_{NO3\text{-}N\ reduced}$) for denitrification (g VSS/g NO_3-N) from Equation 10.98e.

$$bVSS_{NO3\text{-}N\ reduced} = \frac{0.065\text{ moles } C_5H_7NO_2 \times 113\text{ g VSS/mole } C_5H_7NO_2}{1\text{ mole } NO_3\text{-N} \times 14\text{ g } NO_3\text{-N/mole } NO_3\text{-N}}$$

$$= 0.52\text{ gVSS/g } NO_3\text{-N reduced}$$

3. Determine the methanol (CH_3OH) consumption from denitrification.
   From Equation 10.98e, 1.08 moles of CH_3OH are theoretically required per mole of NO_3-N reduced. The molecular weight of CH_3OH $w_{CH3OH} = 32$ g CH_3OH/mole CH_3OH. Calculate the CH_3OH consumption ($CH_3OH_{NO3\text{-}N\ reduced}$) during denitrification (g CH_3OH/g NO_3-N).

$$CH_3OH_{NO3\text{-}N\,reduced} = \frac{1.08\,\text{moles}\,CH_3OH \times 32\,\text{g}\,CH_3OH/\text{mole}\,CH_3OH}{1\,\text{mole}\,NO_3\text{-N} \times 14\,\text{g}\,NO_3\text{-N}/\text{mole}\,NO_3\text{-N}}$$

$$= 2.47\,\text{g}\,CH_3OH/\text{g}\,NO_3\text{-N reduced}$$

4. Determine the equivalent BOD_L consumption from denitrification.
   Assume methanol is completely biodegradable. The BOD_L equivalent of methanol is given by Equation 10.106.

$$\underset{32}{CH_3OH} + \underset{1.5\times 32}{1.5\,O_2} \rightarrow CO_2 + 2\,H_2O \tag{10.106}$$

   Determine the equivalent BOD_L consumption ($BOD_{L,CH3OH}$) from methanol for degradation (g BOD_L/g CH_3OH) from Equation 10.106.

$$BOD_{L,CH3OH} = \frac{1.5\,\text{moles}\,O_2 \times 32\,\text{g}\,O_2/\text{mole}\,O_2}{1\,\text{mole}\,CH_3OH \times 32\,\text{g}\,CH_3OH/\text{mole}\,CH_3OH} = 1.5\,\text{g}\,BOD_L/\text{g}CH_3OH\ \text{degraded}$$

   Calculated the organic carbon consumption as BOD_L from denitrification.

$$BOD_{L,NO3\text{-}N\,reduced} = BOD_{L,CH3OH} \times CH_3OH_{NO3\text{-}N\,reduced}$$

$$= 1.5\,\text{g}\,BOD_L/\text{g}\,CH_3OH\text{degraded} \times 2.47\,\text{g}\,CH_3OH/\text{g}\,NO_3\text{-N reduced}$$

$$= 3.7\,\text{g}\,BOD_L/\text{g}\,NO_3\text{-N reduced}$$

5. Determine the alkalinity recovery from denitrification.
   Calculate bicarbonate generation as HCO_3^- from Equation 10.98e that indicates 1 mole of HCO_3^- is theoretically recovered per mole of NO_3-N reduced.

$$\text{Bicarbonate}_{NO3\text{-}N\,reduced} = \frac{1\,\text{mole}\,HCO_3^- \times 61\,\text{g}\,HCO_3^-/\text{mole}\,HCO_3^-}{1\,\text{mole}\,NO_3\text{-N} \times 14\,\text{g}\,NO_3\text{-N}/\text{mole}\,NO_3\text{-N}}$$

$$= 4.36\,\text{g}\,HCO_3^-/\text{g}\,NO_3\text{-N reduced}$$

   Alkalinity recovery as $CaCO_3$ is calculated from the bicarbonate generation.

$$Alk_{NO3\text{-}N,recovered} = 4.36\,\text{g}\,HCO_3^-/\text{g}\,NO_3\text{-N reduced} \times \frac{50\,\text{g/eq. as}\,CaCO_3}{61\,\text{g/eq. as}\,HCO_3^-}$$

$$= 3.57\,\text{g}\,CaCO_3/\text{g}\,NO_3\text{-N reduced}$$

6. Determine the theoretical methanol consumption from deoxygenation.
   The theoretical methanol consumption from deoxygenation can be determined by inverting the value of $BOD_{L,CH3OH}$ obtained in Step 4.

$$CH_3OH_{deoxygenation} = \frac{1}{BOD_{L,CH3OH}} = \frac{1}{1.5\,\text{g}\,BOD_L/\text{g}\,CH_3OH} = 0.67\,\text{g}\,CH_3OH/\text{g}\,O_2\ \text{removed}$$

EXAMPLE 10.139: DETERMINE THEORETICAL OXYGEN CONSUMPTION RATIOS

Determine the theoretical oxygen consumption ratio for complete reduction of NO_3-N or NO_2-N.

Solution

1. Write the oxidation-reduction half reactions.

   The oxidation-reduction half reactions with one mole of electron being accepted by oxygen, nitrate, and nitrite are expressed by Equation 10.107.

$$\frac{1}{4}O_2 + H^+ + e^- \rightarrow \frac{1}{2}H_2O \qquad (10.107a)$$

$$\frac{1}{5}NO_3^- + \frac{6}{5}H^+ + e^- \rightarrow \frac{1}{10}N_2 + \frac{3}{5}H_2O \qquad (10.107b)$$

$$\frac{1}{3}NO_2^- + \frac{4}{3}H^+ + e^- \rightarrow \frac{1}{6}N_2 + \frac{2}{3}H_2O \qquad (10.107c)$$

   From Equation 10.107, the theoretical oxygen consumption ratio can be determined by comparing the equivalent capability of replacing oxygen by NO_3^- or NO_2^- for accepting electron.

2. Determine the theoretical oxygen consumption ratio for NO_3^-.

   Comparing Equations 10.107a and 10.107b, one-fifth mole of nitrate is equivalent to one-fourth of oxygen for accepting one mole of electron. From this relationship, determine the theoretical oxygen consumption ratio for NO_3^- ($OE_{NO3\text{-}N}$).

$$OE_{NO3\text{-}N} = \frac{\frac{1}{4}\text{ mole } O_2 \times 32\text{ g } O_2/\text{mole } O_2}{\frac{1}{5}\text{ mole } NO_3\text{-N} \times 14\text{ g } NO_3\text{-N}/\text{mole } NO_3\text{-N}} = \frac{5 \times 32\text{ g } O_2}{4 \times 14\text{ g } NO_3\text{-N}}$$

$$= 2.86\text{ g } O_2/\text{g } NO_3\text{-N completely reduced or } 2.86\text{ mg } O_2/\text{mg } NO_3\text{-N completely reduced}$$

3. Determine the theoretical oxygen consumption ratio for NO_2^-.

   Comparing Equations 10.107a and 10.107c, one-third mole of nitrite is equivalent to one-fourth of oxygen for accepting one mole of electron. From this relationship, determine the theoretical oxygen consumption ratio for NO_2^- ($OE_{NO2\text{-}N}$).

$$OE_{NO2\text{-}N} = \frac{\frac{1}{4}\text{ mole } O_2 \times 32\text{ g } O_2/\text{mole } O_2}{\frac{1}{3}\text{ mole } NO_2\text{-N} \times 14\text{ g } NO_2\text{-N}/\text{mole } NO_2\text{-N}} = \frac{3 \times 32\text{ g } O_2}{4 \times 14\text{ g } NO_2\text{-N}}$$

$$= 1.71\text{ g } O_2/\text{g } NO_2\text{-N completely reduced or } 1.71\text{ mg } O_2/\text{mg } NO_2\text{-N completely reduced}$$

EXAMPLE 10.140: ESTIMATE METHANOL REQUIREMENT FOR POSTANOXIC DENITRIFICATION

A complete-mixed biological plant is designed to achieve nitrogen removal. The separate-stage denitrification system uses methanol as the external carbon source. The design data for denitrification facility are as follows: flow = 4800 m^3/d, NO_3-N = 30 mg/L, NO_2-N = 4 mg/L, and DO = 3 mg/L. Estimate the methanol feed as mg/L and kg/d.

Solution

The methanol dose (mg/L) required for denitrification is estimated from Equation 10.99.

$$D_{CH3OH} = 2.47 N_{NO3\text{-}N,r} + 1.53 N_{NO2\text{-}N,r} + 0.87 DO_r$$
$$= 2.47 \times 30\,\text{mg}\,NO_3\text{-N/L} + 1.53 \times 4\,\text{mg}\,NO_2\text{-N/L} + 0.87 \times 3\,\text{mg}\,O_2\text{/L}$$
$$= 83\,\text{mg}\,CH_3OH\text{/L or } 83\,\text{g}\,CH_3OH/m^3$$

Calculate the daily amount of methanol requirement (kg/d) for denitrification.

$$W_{CH3OH} = 83\,\text{g}\,CH_3OH/m^3 \times 4800\,m^3/d \times 10^{-3}\,\text{kg/g} = 398\,\text{kg}\,CH_3OH/d$$

Approximately 400 kg methanol is required as the external carbon source for denitrification in the separate-stage denitrification system.

EXAMPLE 10.141: DENITRIFICATION KINETIC COEFFICIENTS UNDER FIELD OPERATING CONDITIONS

Determine the following denitrification kinetic coefficients under the field operating conditions: (a) the maximum specific growth rate and (b) specific endogenous decay coefficient. The NO_3-N concentration in the anoxic zone is 2 mg/L. The average DO concentration is 0.05 mg/L in the anoxic zone and the critical operating temperature are 12°C. Use typical kinetic coefficients for postanoxic denitrification in Table 10.52.

Solution

1. Determine the typical denitrification kinetic coefficients.
   The denitrification kinetic coefficients for denitrification at 20°C are obtained from Table 10.52. $\mu_{max,H} = 1.2$ mg VSS/mg VSS·d, $k_{d,H} = 0.05\,d^{-1}$, $K_{NO,H} = 0.1$ mg NO_3-N/L, $K_{DO,H} = 0.2$ mg DO/L. The values of θ_T are 1.1 and 1.04 for $\mu_{max,H}$ and $k_{d,H}$, respectively.
2. Determine the maximum specific growth rate ($\mu_{max,H,F}$) under the field operating conditions.
   Equation 10.101a is used for correction of $\mu_{max,H,F}$ for the field operating conditions.

$$\mu_{max,H,F} = F_{T,H} F_{NO,H} F_{DO,H} \mu_{max,H}$$

   a. Determine the operating temperature correction factor ($F_{T,H}$) from Equation 10.101c using $\theta_T = 1.1$ at $T = 12$°C.

$$F_{T,H} = \theta_T^{T-20} = 1.1^{12-20} = 0.467$$

   b. Determine the operating NO_3-N concentration correction factor ($F_{NO,H}$) from Equation 10.101d using $K_{NO,H} = 0.1$ mg/L at $N_{NO3\text{-}N} = 1$ mg/L.

$$F_{NO,H} = \frac{N_{NO3\text{-}N}}{K_{NO,H} + N_{NO3\text{-}N}} = \frac{2\,\text{mg/L}}{(0.1 + 2)\,\text{mg/L}} = 0.952$$

c. Determine the operating DO correction factor ($F_{DO,H}$) from Equation 10.101e using $K_{DO,H} =$ 0.2 mg/L at $DO = 0.05$ mg/L.

$$F_{DO,H} = \frac{K_{DO,H}}{K_{DO,H} + DO} = \frac{0.2\text{ mg/L}}{(0.2 + 0.05)\text{ mg/L}} = 0.8$$

d. Calculate the maximum specific growth rate ($\mu_{max,H.F}$) from Equation 10.101a.

$$\mu_{max,H,F} = F_{T,F}F_{NO,F}F_{DO,H}\mu_{max,AOB} = 0.467 \times 0.952 \times 0.8 \times 1.2\text{ mg VSS/mg VSS·d}$$
$$= 0.356 \times 1.2\text{ mg VSS/mg VSS·d} = 0.427\text{ mg VSS/mgVSS·d}$$

3. Determine the specific endogenous decay coefficient ($k_{d,H,F}$) under the field operating conditions.
   Calculate $F_{T,H}$ from Equation 10.101c using $\theta_T = 1.04$ at $T = 12°C$.

$$F_{T,H} = \theta_T^{T-20} = 1.04^{12-20} = 0.731$$

Calculate $k_{d,H,F}$ from Equation 10.101b.

$$k_{d,H,F} = F_{T,H}k_{d,H} = 0.731 \times 0.05\text{ d}^{-1} = 0.037\text{ d}^{-1}$$

EXAMPLE 10.142: KINETICS-BASED NITRATE REDUCTION AND SPECIFIC DENITRIFICATION RATES

Based on the kinetic coefficients determined in Example 10.141, calculate (a) the NO_3-N reduction rate by heterotrophs and (b) the specific denitrification rate under the field operating conditions. Assume the design SRT $\theta_c^{design} = 21$ d, $X = 2500$ mg VSS/L, bsCOD concentration = 9 mg/L, fraction of inert (nonbiomass) VSS in the influent $Y_{ivss} = 0.1$, and fraction of heterotrophic bacteria active for denitrification $f_{AX} = 0.9$.

Solution

1. Determine the net oxygen consumption ratio for NO_3-N ($CR_{NO3\text{-}N}$).
   Calculate $CR_{NO3\text{-}N}$ from Equation 10.100e using $OE_{NO3\text{-}N} = 2.86$ mg O_2/mg NO_3-N (Example 10.139) and $Y_H = 0.3$ mg VSS/mg bCOD (from Table 10.52).

$$CR_{NO3\text{-}N} = \frac{OE_{NO3\text{-}N}}{1 - 1.42Y_H} = \frac{2.86\text{ mg } O_2/\text{mg } NO_3\text{-N}}{1 - 1.42 \times 0.3\text{ mg VSS/mg bCOD}} = 4.98\text{ mg } O_2/\text{mg } NO_3\text{-N}$$

or $CR_{NO3\text{-}N} \approx 4.98$ mg bCOD/mg NO_3-N

2. Determine the observed biomass yields for heterotrophs.
   Calculate the observed biomass yields for heterotrophs ($Y_{obs,H,F}$) from Equation 10.151 using $Y_H =$ 0.3 mg VSS/mg bCOD (from Table 10.52) and $k_{d,H,F} = 0.037\text{ d}^{-1}$ (Example 10.141).

$$Y_{obs,H,F} = \frac{Y_H}{1 + k_{d,H.F}\theta_c^{design}} = \frac{0.3\text{ mg VSS/mg bCOD}}{1 + 0.037\text{ d}^{-1} \times 21\text{ d}} = 0.169\text{ mg VSS/mg bCOD}$$

3. Determine the active heterotrophic biomass concentration (X_H).
   Calculate the active heterotrophic biomass fraction (f_b) from Equation 10.103f using $Y_{ivss} = 0.1$.

$$f_b = \frac{Y_{obs,H,F}}{Y_{obs,H,F} + Y_{iVSS}} = \frac{0.169}{0.169 + 0.1} = 0.628$$

Calculate the active heterotrophic biomass concentration.

$$X_H = f_b X = 0.628 \times 2500 \text{ mg VSS/L} = 1570 \text{ mg VSS/L}$$

4. Determine the NO_3-N reduction rate by heterotrophs ($r_{NO3\text{-}N,F}$) under the field operating conditions.

   Calculate $r_{NO3\text{-}N}$ from Equation 10.104a using $\mu_{max,H,F} = 0.427$ mg VSS/mg VSS·d (Example 10.141), $K_{s,H} = 1.5$ mg bCOD/L (from Table 10.52), $S = 9$ mg bsCOD/L, $f_{AX} = 0.9$, and $X_H = 1570$ mg VSS/L.

$$r_{NO3\text{-}N,F} = \frac{\mu_{max,H,F} S}{CR_{NO} Y_H (K_{s,H} + S)} f_{AX} X_H$$

$$= \frac{0.427 \text{ mg VSS/mg VSS·d} \times 9 \text{ mg bsCOD/L}}{4.98 \text{ mg bCOD/mg } NO_3\text{-N} \times 0.3 \text{ mg VSS/mg bCOD} \times (1.5 + 9) \text{ mg bCOD/L}}$$

$$\times 0.9 \times 1570 \text{ mg/VSS/L}$$

$$= 346 \text{ mg } NO_3\text{-N/L·d or } 0.35 \text{ kg } NO_3\text{-N/m}^3\text{·d}$$

5. Determine the specific denitrification rate (*SDNR*) under the field operating conditions.

   Calculate *SDNR* from Equation 10.104a using $r_{NO3\text{-}N,F} = 346$ mg NO_3-N/L·d and $X = 2500$ mg VSS/L.

$$SDNR = \frac{r_{NO3\text{-}N,F}}{X} = \frac{346 \text{ mg } NO_3\text{-N/L·d}}{2500 \text{ mg VSS/L}} = 0.14 \text{ mg } NO_3\text{-N/mgVSS·d}$$

EXAMPLE 10.143: DENITRIFICATION IN PREANOXIC ZONE

A preanoxic zone is utilized as part of a MLE process for combined removal of BOD and nitrogen (see process diagram and additional information in Figure 10.92 and Table 10.55). The influent characteristics and design parameters are provided below.

Influent: Flow $Q = 4000$ m^3/d, $BOD_5 = 150$ mg/L, TKN = 35 mg/L, and rbCOD/$BOD_5 = 0.6$.
Effluent: Desired effluent NO_3-N = 6 mg/L and NH_4-N = 0.5 mg/L.
Design parameters: Operating temperature 12°C, design preanoxic zone $F/M = 0.45$ kg BOD_5/kg VSS·d, NO_3-N concentrations in the preanoxic zone = 1 mg NO_3-N/L, MLVSS concentration $X = 2500$ mg VSS/L, return sludge ratio $R_{RS} = 0.6$, approximate biomass growth $p_x = 57$ mg VSS/L, fraction of heterotrophic bacteria active for denitrification $f_{AX} = 0.8$, active heterotrophic biomass fraction $f_b = 0.55$, and $f_N = 0.12$ g N/g VSS. Use typical kinetic coefficients for preanoxic denitrification in Table 10.52.

Determine (a) foodtomicroorganisms ratio based on active heterotrophic biomass, (b) internal recycle rate, (c) specific denitrification rate, (d) nitrate nitrogen removal rate, (e) volume and HRT of the preanoxic zone, (f) consumption of BOD_5 in the preanoxic zone, and (g) alkalinity recovery from denitrification.

Solution

1. Determine the foodtomicroorganisms ratio in the preanoxic zone (F/M_b).

   Calculate F/M_b, the foodtomicroorganisms ratio based on active heterotrophic biomass that utilizes NO_3-N in the preanoxic zone from Equation 10.103e.

$$F/M_b = \frac{1}{f_{AX} f_b}(F/M) = \frac{1}{0.8 \times 0.55} \times 0.45 \text{ kg } BOD_5\text{/kg VSS·d} = 1.02 \text{ kg } BOD_5\text{/kg VSS·d}$$

2. Determine the internal recycle rate required from the aeration basin to the preanoxic zone (*IR*).

   Calculate the NO_3-N concentration produced from nitrification in the aeration basin (NO_x) from Equation 10.103l.

$$NO_X = N_{0,TKN} - N_{NH4\text{-}N} - f_N p_x = (35 - 0.5)\,\text{mg N/L} - 0.12\,\text{mg N/mg VSS} \times 57\,\text{mg VSS/L}$$
$$= 27.7\,\text{mg NO}_3\text{-N/L}$$

   IR is estimated from Equation 10.103j at $N_{NO3\text{-}N,AX} = 1$ mg NO_3-N/L.

$$IR = \frac{NO_x}{N_{NO3\text{-}N} - N_{NO3\text{-}N,AX}} - (1 + R_{RS}) = \frac{27.7\,\text{mg N/L}}{(6-1)\,\text{mg NO}_3\text{-N/L}} - (1 + 0.6) = 3.94 \approx 4$$

3. Determine the specific denitrification rate (*SDNR*).

   a. Determine the specific denitrification rate based on the active heterotrophic biomass that utilize NO_3-N in anoxic zone at 20°C ($SDNR_{b,20}$).

      Since $F/M_b > 0.5$, the $SDNR_{b,20}$ is calculated from Equation 10.103a after the empirical constants a and b are first estimated from Equations 10.103c and 10.103d at rbCOD/$BOD_5 = 0.6$, respectively.

$$a = 0.12\ (\text{rbCOD/BOD}_5) + 0.17 = 0.12 \times 0.6 + 0.17 = 0.24$$

$$b = 0.053\ \ln(\text{rbCOD/B OD}_5) + 0.18 = 0.053 \ln(0.6) + 0.18 = 0.15$$

$$SDNR_{b,20} = a + b \ln(F/M_b) = 0.24 + 0.15 \times \ln(1.02) = 0.243\,\text{mg NO}_3\text{-N/mg VSS·d}$$

   b. Determine the specific denitrification rate based on the active heterotrophic biomass that utilize NO_3-N in anoxic zone at 12°C ($SDNR_{b,12}$).

      Calculate $SDNR_{b,12}$ from Equation 10.103g using $\theta_T = 1.026$ at 12°C.

$$SDNR_{b,12} = SDNR_{b,20}\theta_T^{T-20} = 0.243\,\text{mg NO}_3\text{-N/mgVSS·d} \times 1.026^{12-20}$$
$$= 0.198\ \text{mg NO}_3\text{-N/mgVSS·d}$$

   c. Determine the specific denitrification rate based on the active heterotrophic biomass that utilize NO_3-N in anoxic zone at 12°C with adjustment for IR ($SDNR_{b,12,IR}$).

      At $IR = 4$, $SDNR_{b\text{-}12,IR}$ is adjusted using Equation 10.103i.

$$SDNR_{b,12,IR} = SDNR_{b,12} - 0.029 \ln(F/M_b) - 0.012 = 0.198 - 0.029 \ln(1.02) - 0.012$$
$$= 0.185\ \text{mg NO}_3\text{-N/mgVSS·d}$$

   d. Determine the specific denitrification rate based on MLVSS in anoxic zone (*SDNR*).

      Calculate *SDNR* from Equation 10.103m.

$$SDNR = f_{AX} f_b SDNR_{b,12,IR} = 0.8 \times 0.55 \times 0.185\,\text{mg NO}_3\text{-N/mg VSS·d}$$
$$= 0.081\,\text{mg NO}_3\text{-N/mg VSS·d}$$

4. Determine the nitrate nitrogen removal rate ($R_{NO3\text{-}N}$) from Equation 10.103n.

$$R_{NO3\text{-}N} = Q[(IR + R_{RS})(N_{NO3\text{-}N} - N_{NO3\text{-}N,AX}) - N_{NO3\text{-}N,AX}]$$
$$= 4000\,\text{m}^3/\text{d} \times [(4 + 0.6) \times (6 - 1)\ \text{g NO}_3\text{-N/m}^3 - 1\,\text{g NO}_3\text{-N/m}^3] \times 10^{-3} \times \text{kg/g}$$
$$= 88.0\,\text{kg NO}_3\text{-N/d}$$

5. Determine the volume and HRT of the preanoxic zone.

Calculate the volume of preanoxic zone (V) required for the desired denitrification from Equation 10.102.

$$V = \frac{R_{\text{NO3-N}}}{SDNRX} = \frac{88.0\,\text{kg}\,\text{NO}_3\text{-N/d} \times 10^3\,\text{g/kg}}{0.081\,\text{g}\,\text{NO}_3\text{-N/gVSS}\cdot\text{d} \times 2500\,\text{g VSS/m}^3} = 435\,\text{m}^3$$

Provide a design volume $V = 450\,\text{m}^3$ and calculate HRT of the preanoxic zone.

$$\theta = \frac{V}{Q} = \frac{450\,\text{m}^3}{4000\,\text{m}^3/\text{d}} \times \frac{24\,\text{h}}{\text{d}} = 2.7\,\text{h}$$

Note: Typically, a slightly larger volume of preanoxic zone is provided to achieve complete deoxygenation of IR to maximize denitrification.

6. Determine the removal of BOD_5 in the preanoxic zone.

Calculate $CR_{\text{NO3-N}}$ from Equation 10.100d using $OE_{\text{NO3-N}} = 2.86\,\text{mg O}_2/\text{mg NO}_3\text{-N}$ (Example 10.139) and $Y_H = 0.67\,\text{mg VSS/mg BOD}_5$ (from Table 10.52).

$$CR_{\text{NO3-N}} = \frac{OE_{\text{NO3-N}}}{1.6 - 1.42Y_H} = \frac{2.86\,\text{mg}\,\text{O}_2/\text{mg}\,\text{NO}_3\text{-N}}{1.6 - 1.42 \times 0.67\,\text{mg}\,\text{VSS/mg}\,\text{BOD}_5} = 4.41\,\text{mg}\,\text{O}_2/\text{mg}\,\text{NO}_3\text{-N}$$

or $CR_{\text{NO3-N}} \approx 4.41\,\text{mg}\,\text{BOD}_5/\text{mg}\,\text{NO}_3\text{-N}$

The amount of organic consumption due to denitrification in the preanoxic zone ($\Delta S_{\text{r,AX}}$) is calculated from Equation 10.103p.

$$\Delta S_{\text{r,AX}} = CR_{\text{NO}}R_{\text{NO3-N}} = 4.41\,\text{kg}\,\text{BOD}_5/\text{kg NO}_3\text{-N} \times 88.0\,\text{kg}\,\text{NO}_3\text{-N/d} = 388\,\text{kg}\,\text{BOD}_5/\text{d}$$

Calculate the BOD_5 concentration removal normalized on basis of influent $Q = 4000\,\text{m}^3/\text{d}$.

$$S_{\text{r,AX}} = \frac{\Delta S_{\text{r,AX}}}{Q} = \frac{388\,\text{kg}\,\text{BOD}_5/\text{d} \times 10^3\,\text{g/kg}}{4000\,\text{m}^3/\text{d}} = 97\,\text{g}\,\text{BOD}_5/\text{m}^3 \text{ or } 97\,\text{mg BOD}_5/\text{L}$$

Calculate the BOD_5 concentration remaining after preanoxic zone normalized on the basis of influent $Q = 4000\,\text{m}^3/\text{d}$.

$$S_{\text{AX}} = S_0 - S_{\text{r,AX}} = (150 - 97)\,\text{mg}\,\text{BOD}_5/\text{L} = 53\,\text{mg}\,\text{BOD}_5/\text{L}$$

$$\text{BOD}_5 \text{ removal in anoxic zone, } E_{\text{AX}} = \frac{S_{\text{r,AX}}}{S_0} \times 100\% = \frac{97\,\text{mg/L}}{150\,\text{mg/L}} \times 100\% = 65\%$$

Note: Significant amount of soluble organics is consumed by denitrification process. In this example, the removal is 65%. This indicates that sufficient amount of soluble organics is required in the influent to the preanoxic zone for successful denitrification. This results in a significant reduction of oxygen demand and air supply requirement for removal of remaining BOD_5 in the aerobic zone.

7. Determine the alkalinity recovery from denitrification.

Use the alkalinity recovery ratio of $3.57\,\text{g CaCO}_3/\text{g NO}_3\text{-N}$ derived in Example 10.138 to calculate the amount of alkalinity recovery from denitrification.

$$\begin{aligned}\Delta Alk_{\text{recovered}} &= Alk_{\text{NO3-N,recovered}}R_{\text{NO3-N}} = 3.57\,\text{kg}\,\text{CaCO}_3/\text{kg}\,\text{NO}_3\text{-N} \times 88.0\,\text{kg}\,\text{NO}_3\text{-N/d} \\ &= 314\,\text{kg}\,\text{CaCO}_3/\text{d}\end{aligned}$$

Calculate the alkalinity concentration normalized on the basis of influent $Q = 4000\ m^3/d$.

$$Alk_{recovered} = \frac{\Delta Alk_{recovered}}{Q} = \frac{314\ kg\ CaCO_3/d \times 10^3\ g/kg}{4000\ m^3/d}$$
$$= 78.5\ g\ CaCO_3/m^3 \text{ or } 78.5\ mg\ CaCO_3/L$$

Note: Significant amount of alkalinity is recovered from denitrification process in the preanoxic zone. It helps to compensate for the alkalinity consumption later during nitrification in aerobic zone.

EXAMPLE 10.144: DENITRIFICATION IN POSTANOXIC ZONE WITH SUBSTRATE DRIVEN MODE

In Wuhrmann process, a postanoxic zone is used for denitrification (see process diagram and additional information in Figure 10.95 and Table 10.55). Assume methanol is added as external carbon source. Determine (a) volume and HRT of the postanoxic zone and (b) requirement of methanol. Use the kinetic coefficients and calculation results obtained in Examples 10.141 and 10.142. Additional influent characteristics and design parameters are provided below.

Influent: Flow $Q = 4000\ m^3/d$ and TKN = 35 mg/L.
Effluent: Desired effluent NO_3-N = 3 mg/L and NH_4-N = 0.5 mg/L.
Design parameters: MLVSS concentration $X = 2500$ mg VSS/L, return sludge ratio $R_{RS} = 0.6$, DO concentration in the mixed liquor from the aerobic zone = 4 mg/L, approximate biomass growth $p_x = 57$ mg VSS/L, and $f_N = 0.12$ g N/g VSS.

Solution

1. Determine the nitrate nitrogen removal rate ($R_{NO3\text{-}N}$).

Calculate the NO_3-N concentration produced from nitrification in the aerobic zone (NO_x) from Equation 10.103l.

$$NO_X = N_{0,TKN} - N_{NH4\text{-}N} - f_N\, p_X = (35 - 0.5)\ mg\,N/L - 0.12\ mg\,N/mg\,VSS \times 57\ mg\,VSS/L$$
$$= 27.7\ mg\,N/L \quad \text{or} \quad 27.7\ g\,N/m^3$$

For conservative consideration, exclude the endogenous decay and calculate $R_{NO3\text{-}N}$ from Equation 10.104c at $Q = 4000\ m^3/d$.

$$R_{NO3\text{-}N} = Q(NO_X - N_{NO3\text{-}N}) = 4000\ m^3/d \times (27.7 - 3)\ g\,N/m^3 \times 10^{-3}\ kg/g = 98.8\ kg\,NO_3\text{-}N/d$$

2. Determine the volume and HRT of the postanoxic zone.

Calculate the volume of postanoxic zone (V) required for the desired denitrification from Equation 10.102 using $SDNR = 0.14$ mg NO_3-N/mg VSS·d (Example 10.142, Step 5).

$$V = \frac{R_{NO3\text{-}N}}{SDNR\,X} = \frac{98.8\ kg\ NO_3\text{-}N/d \times 10^3 g/kg}{0.14\ g\ NO_3\text{-}N/g\,VSS\cdot d \times 2500\ g\,VSS/m^3} = 282\ m^3$$

Consider potential deoxygenation requirement and provide a design volume $V = 300\ m^3$ and calculate HRT of the postanoxic zone.

HRT of the postanoxic zone,

$$\theta = \frac{V}{Q} = \frac{300\ m^3}{4000\ m^3/d} \times \frac{24\ h}{d} = 1.8\ h$$

Note: Typically, a slightly larger volume is provided for complete deoxygenation in the postanoxic zone.

3. Determine the external organic carbon required by denitrification in the postanoxic zone.

The amount of organic consumption due to denitrification in the postanoxic zone ($\Delta S_{r,NO3\text{-}N}$) is calculated from Equation 10.103p using $CR_{NO3\text{-}N} = 4.98$ mg O_2/mg NO_3-N (Example 10.142, Step 1).

$$\Delta S_{r,NO3\text{-}N} = CR_{NO}\, R_{NO3\text{-}N} = 4.98 \text{ kg bCOD/kg } NO_3\text{-N} \times 98.8 \text{ kg } NO_3\text{-N/d} = 492 \text{ kg bCOD/d}$$

Calculate the required external organic carbon concentration normalized over the influent $Q =$ 4000 m^3/d.

$$S_{r,NO3\text{-}N} = \frac{\Delta S_{r,NO3\text{-}N}}{Q} = \frac{492 \text{ kg bCOD/d} \times 10^3 \text{ g/kg}}{4000 \text{ m}^3\text{/d}} = 123 \text{ g bCOD/m}^3 \quad \text{or} \quad 123 \text{ mg } O_2\text{/L}$$

4. Determine the external organic carbon required by deoxygenation in the postanoxic zone.

The amount of organic consumption due to deoxygenation in the postanoxic zone ($\Delta S_{r,DO}$) is depending upon on the DO concentration in the mixed liquor from the aerobic zone. At $DO = 4$ mg/L, calculate the required external organic carbon concentration normalized on basis of influent $Q = 4000$ m^3/d.

$$S_{r,DO} = \frac{(1 + R_{RS})Q\,DO}{Q} = (1 + R_{RS})\,DO = (1 + 0.6) \times 4 \text{ mg } O_2\text{/L} = 6.4 \text{ mg } O_2\text{/L} \quad \text{or} \quad 6.4 \text{ g } O_2\text{/m}^3$$

5. Determine the requirement of methanol as external carbon source for denitrification in the postanoxic zone.

Calculate the total amount of organic consumption due to denitrification and deoxygenation in the postanoxic zone ($\Delta S_{r,AX}$).

$$S_{r,AX} = S_{r,NO3\text{-}N} + S_{r,DO} = (123 + 6.4) \text{ mg } O_2\text{/L} = 129.4 \text{ mg } O_2\text{/L} \approx 130 \text{ mg } O_2\text{/L} \quad \text{or} \quad 130 \text{ g } O_2\text{/m}^3$$

Calculate the methanol concentration required by the denitrification using the equivalent oxygen for methanol $O_{2,CH3OH} = BOD_{L,CH3OH} = 1.5$ mg O_2/mg CH_3OH (Example 10.138, Step 4).

$$D_{CH3OH} = \frac{S_{r.AX}}{O_{2,CH3OH}} = \frac{130 \text{ mg } O_2\text{/L}}{1.5 \text{ mg } O_2\text{/mg } CH_3OH} = 87 \text{ mg } CH_3OH\text{/L} \quad \text{or} \quad 87 \text{ g } CH_3OH\text{/m}^3$$

Provide a methanol dosage $D_{CH3OH} = 90$ mg/L and calculate the amount of methanol required at influent $Q = 4000$ m^3/d.

$$w_{CH3OH} = D_{CH3OH}\, Q = 90 \text{ g } CH_3OH\text{/m}^3 \times 4000 \text{ m}^3\text{/d} \times 10^{-3} \text{ kg/g} = 360 \text{ kg } CH_3OH\text{/d}$$

Note: Significant amount of methanol is required by postanoxic denitrification process. In actual design, additional safety factors may be also considered for sizing methanol feed and storage systems.

EXAMPLE 10.145: DENITRIFICATION IN POSTANOXIC ZONE WITH ENDOGENOUS DRIVEN MODE

A postanoxic zone is added after the aerobic zone of the MLE process presented in Example 10.143. The goal of adding the postanoxic zone is to reduce the effluent NO_3-N concentration from 6 to 3 mg/L without adding external carbon source. Determine (a) volume and HRT of the postanoxic zone, and (b) potential NH_4-N release due to endogenous decay in the postanoxic zone. Use the applicable kinetic coefficients and calculation results obtained in Example 10.143. Use $k_{d,H,F} = 0.037\ d^{-1}$ (Example 10.141). Assume the 50% of organic nitrogen released from endogenous decay is in the form of NH_4-N ($f_{NH4\text{-}N} = 0.5$ mg NH_4-N/mg N decayed).

Solution

1. Draw the definition sketch.
   The definition sketch of the process postanoxic zone is shown in Figure 10.104.

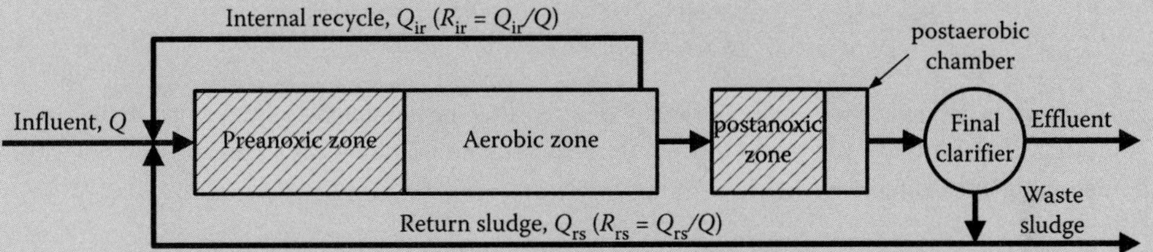

FIGURE 10.104 Definition sketch of MLE process followed by a postanoxic zone (Example 10.145).

2. Determine the nitrate nitrogen removal rate ($R_{NO3\text{-}N}$) in the postanoxic zone.
   Calculate $R_{NO3\text{-}N}$ from mass balance at $Q = 4000\ m^3/d$, $R_{RS} = 0.6$, $N_{NO3\text{-}N,OX} = 6$ mg/L, and $N_{NO3\text{-}N} = 3$ mg/L.

$$R_{NO3\text{-}N} = Q(1 + R_{RS})(N_{NO3\text{-}N,OX} - N_{NO3\text{-}N}) = 4000\ m^3/d \times (1 + 0.6) \times (6 - 3)\ g\ N/m^3 \times 10^{-3}\ kg/g = 19.2\ kg\ NO_3\text{-}N/d$$

3. Determine the specific denitrification rate (*SDNR*).
   Calculate *SDNR* from Equation 10.105a using $OE_{NO3\text{-}N} = 2.86$ mg O_2/mg NO_3-N (from Example 10.139, Step 2), $f_b = 0.55$ (Example 10.143), and $k_{d,H,F} = 0.037\ d^{-1}$.

$$SDNR = \frac{1.42}{OE_{NO3\text{-}N}} k_{d,H,F} f_b = \frac{1.42}{2.86\ mgO_2/mg\ NO_3\text{-}N} \times 0.037\ d^{-1} \times 0.55$$
$$= 0.0101\ kg\ NO_3\text{-}N/kg\ VSS{\cdot}d$$

4. Determine the volume and HRT of the postanoxic zone.
   Rearrange Equation 10.102 and calculate the required postanoxic zone volume (V) at $X = 2500$ mg VSS/L (Example 10.143).

$$V = \frac{R_{NO3\text{-}N}}{SDNR\,X} = \frac{19.2\ kg\ NO_3\text{-}N/d \times 10^3\ g/kg}{0.0101\ g\ NO_3\text{-}N/g\ VSS{\cdot}d \times 2500\ g\ VSS/m^3} = 760\ m^3$$

   Since deoxygenation is essential prior to reaching anoxic condition, provide a larger design volume $V = 800\ m^3$, and calculate HRT of the postanoxic zone.

$$\text{HRT of the postanoxic zone, } \theta = \frac{V}{Q} = \frac{800\ m^3}{4000\ m^3/d} \times \frac{24\ h}{d} = 4.8\ h$$

   Note: Significant reactor volume is required when slow endogenous decay is the only carbon source for postanoxic denitrification.

5. Determine the potential NH_4-N release rate ($R_{NH4\text{-}N}$).

   Calculate $R_{NH4\text{-}N}$ from Equation 10.105c at $Q = 4000\ m^3/d$, $f_{NH4\text{-}N} = 0.5$ mg NH_4-N/mg N decayed, $f_b = 0.55$, and $f_N = 0.12$ g N/g VSS (Example 10.143).

$$R_{r,NH4\text{-}N} = f_{NH4\text{-}N} f_N\, k_{d,H,F} f_b XV$$

$$= 0.5\,g\ NH_4\text{-}N/gN \times 0.12\,g\ N/g\ VSS \times \times 0.037\,d^{-1} \times 0.55 \times 2500\,g\ VSS/m^3 \times 800\,m^3 \times 10^{-3}\,kg/g = 2.44\,kg\ NH_4\text{-}N/d$$

6. Determine the NH_4-N concentration elevated after the postanoxic zone ($N_{r,NH4\text{-}N}$).

$$N_{r,NH4\text{-}N} = \frac{R_{r,NH4\text{-}N}}{(1 + R_{RS})Q} = \frac{2.44\,kg\ NH_4\text{-}N/d \times \times 10^3 g/kg}{(1 + 0.6) \times 4000\,m^3/d} = 0.38\,g/m^3 \quad \text{or} \quad 0.38\,mg/L$$

   Note: The NH_4-N concentration is slightly elevated after the postanoxic zone. A small aerobic nitrification chamber is typically added to lower the NH_4-N concentration while increasing DO concentration in the final effluent.

EXAMPLE 10.146: ATTACHED GROWTH DOWNFLOW DENITRIFICATION FILTER

The attachedgrowth downflow denitrification filter (DFDNF) process is utilized for denitrification of well nitrified wastewater. Determine (a) filter dimensions, (b) water flow rates required for filter backwash and nitrogen release bump, (c) filtered water storage requirement, (d) water volume requirements for filter backwash and filter bump, (e) methanol feed rate, and (f) sludge production. The DFDNF influent and effluent characteristics, and design data are provided below.

Influent: Flow $Q = 6000\ m^3/d$, NO_3-N = 27 mg/L, and TSS = 20 mg/L.

Effluent: The effluent TSS and NO_3-N concentrations are <5 and 2 mg/L, respectively.

Design parameters: Filter depth = 1.6 m; number of fillers = 4 (including one standby); hydraulic loading $HL = 80\ m^3/m^2{\cdot}d$ (or $3.33\ m^3/m^2{\cdot}h$); volumetric NO_3-N loading $VNL = 1.3$ kg NO_3-N/m^3·d; cell yield with methanol $Y_H = 0.18$ g VSS/g COD removed; and methanol requirement 3.2 kg CH_3OH/kg NO_3-N removed. Allowable solids storage in the bed between backwash = 4 kg TSS/m^3. Filter is backwashed once a day at backwash flow rate of 18 m^3/m^2·h for a duration of 15 min. Filter bump is activated every 2 h at a flow rate of 12 m^3/m^2·h for a duration of 3 min.

Notes: Design of submerged aerobic downflow reactor for BOD removal is presented in Example 10.112. The procedures for (a) head loss calculations, (b) bed expansion, (e) air flow rates required during backwash operation, (f) air volume requirements, and (g) component design of the granular media filters may be found in Section 15.4.6 and Examples 15.22, 15.26 through 15.31.

Solution

1. Determine the minimum total filter area.

   Calculate the total filter area ($A_{Total,HL}$) based on the hydraulic loading $HL = 80\ m^3/m^2{\cdot}d$.

$$A_{Total,HL} = \frac{Q}{HL} = \frac{6000\,m^3/d}{80\,m^3/m^2{\cdot}d} = 75\,m^2$$

   Calculate the nitrate nitrogen removal rate ($R_{NO3\text{-}N}$) at $Q = 6000\ m^3/d$, $N_{NO3\text{-}N,0} = 27$ mg/L, and $N_{NO3\text{-}N} = 2$ mg/L.

$$R_{NO3\text{-}N} = Q(N_{NO3\text{-}N,0} - N_{NO3\text{-}N}) = 6000\,m^3/d \times (27 - 2)\,g\ N/m^3 \times 10^{-3}\,kg/g = 150\,kg\ NO_3\text{-}N/d$$

Calculate the total filter volume ($V_{\text{Total,VNL}}$) based on the volumetric NO_3-N loading $VNL = 1.3$ kg NO_3-N/m^3·d.

$$V_{\text{Total,VNL}} = \frac{R_{\text{NO3-N}}}{VNL} = \frac{150\,\text{kg NO}_3\text{-N/d}}{1.3\,\text{kg NO}_3\text{-N/m}^3\cdot\text{d}} = 115\,\text{m}^3$$

Calculate the total filter area ($A_{\text{Total,VNL}}$) based on the VNL at the filter depth $D = 1.6$ m.

$$A_{\text{Total,HL}} = \frac{V_{\text{Total,VNL}}}{D} = \frac{115\,\text{m}^3}{1.6\,\text{m}} = 72\,\text{m}^2$$

A slightly larger filter area is required based on the hydraulic loading. Therefore, select $A_{\text{Total}} = 75\,\text{m}^2$.

2. Determine the dimensions of each filter.

Calculate the required area of each filter (A) with three filters in service at the design flow of 6000 m^3/d.

$$A = \frac{A_{\text{Total}}}{N_{\text{filter}} - 1} = \frac{75\,\text{m}^2}{(4-1)\,\text{filter}} = 25\,\text{m}^2 \text{ per filter}$$

Provide four square filters and filter dimensions = 5 m × 5 m × 1.6 m. The empty bed volume of each filter $V = 40\,\text{m}^3$.

3. Determine the filter backwash flow rate and volume.

The required filter backwash water flow per filter (q_{bw}) at the $q'_{\text{bw}} = 18\,\text{m}^3/\text{m}^2\cdot\text{h}$.

$$q_{\text{bw}} = q'_{\text{bw}} A = 18\,\text{m}^3/\text{m}^2\cdot\text{h} \times 25\,\text{m}^2 = 450\,\text{m}^3/\text{h} \quad \text{or} \quad 7.5\,\text{m}^3/\text{min}$$

The amount of backwash water required during each backwash operation (ΔV_{bw}) for $t_{\text{bw}} = 15$ min.

$$\Delta V_{\text{bw}} = q_{\text{bw}}\, t_{\text{bw}} = 7.5\,\text{m}^3/\text{min} \times 15\ \text{min/backwash} = 112.5\,\text{m}^3/\text{backeash}$$

4. Determine the filter bumping water flow rate and volume.

Calculate the required filter bump water flow per filter (q_{fb}) at the $q'_{\text{fb}} = 12\,\text{m}^3/\text{m}^2\cdot\text{h}$.

$$q_{\text{fb}} = q'_{\text{fb}} A = 12\,\text{m}^3/\text{m}^2\cdot\text{h} \times 25\,\text{m}^2 = 300\,\text{m}^3/\text{h} \quad \text{or} \quad 5\,\text{m}^3/\text{min}$$

Calculate the amount of filter bump water required during each bumping operation (ΔV_{fb}) for $t_{\text{fb}} = 3$ min.

$$\Delta V_{\text{fb}} = q_{\text{fb}}\, t_{\text{bw}} = 5\,\text{m}^3/\text{min} \times 3\ \text{min/bump} = 15\,\text{m}^3/\text{bump}$$

5. Determine the required filtered water storage volume.

The filtered water is used for backwashing and filter bumping. Since either the required backwash or filter bump flow rate is higher than the total filter design flow ($Q = 6000\,\text{m}^3/\text{d} = 4.2\,\text{m}^3/\text{min}$), a filtered water storage tank is required in order to meet the flow required for filter backwashing and bumping.

Assume the filtered water storage tank is large enough to hold the amount of water potentially required for (a) two filter backwashes, and (b) twelve filter bumping operations when three filters are in the service during a 6 h plant operation period. Calculate the required volume of the filtered water storage tank ($V_{storage}$).

$$V_{storage} = 2\Delta V_{bw} + 12\Delta V_{fb} = 2\,\text{backwashes} \times 112.5\,\text{m}^3/\text{backeash} + 12\,\text{bumps} \times 15\,\text{m}^3/\text{bump} = 405\,\text{m}^3$$

Provide a storage volume of 420 m^3 that is 7% of the total filter design flow.

6. Determine the daily total water volume required for filter backwash and bumping.

   Calculate the daily total amount of backwash water required for one backwash per filter per day when three filters are in service under design operating conditions.

$$\Delta q_{bw} = (N_{filter} - 1)\,N_{bw}\Delta V_{bw} = (4-1)\,\text{filters} \times \frac{1\,\text{backwash}}{\text{filter·d}} \times 112.5\,\text{m}^3/\text{backeash} = 338\,\text{m}^3/\text{d}$$

   Calculate the daily total amount of filtered water required for one bump every 2 h per filter when three filters are in service under design operating conditions.

$$\Delta q_{fb} = (N_{filter} - 1)\,N_{fb}\Delta V_{fb} = (4-1)\,\text{filters} \times \frac{1\,\text{bump}}{\text{filter·2 h}} \times \frac{24\,\text{h}}{\text{d}} \times 15\,\text{m}^3/\text{bump} = 540\,\text{m}^3/\text{d}$$

   Calculate the daily total amount of filtered water required for both filter backwash and bumping operations.

$$\Delta q_w = \Delta q_{bw} + \Delta q_{fb} = (338 + 540)\,\text{m}^3/\text{d} = 878\,\text{m}^3/\text{d}$$

   Calculate the percent of filtered flow used for filter backwash and bumping.

$$P_w = \frac{\Delta q_{fw}}{Q} \times 100\% = \frac{878\,\text{m}^3/\text{d}}{6000\,\text{m}^3/\text{d}} \times 100\% = 14.6\%$$

   Note: It is within the normal range of 5–15% of the influent applied to the filter.[2]

7. Determine the methanol feed rate.

   Calculate the daily methanol feed rate at the methanol requirement of $R_{CH3OH} = 3.2$ kg CH_3OH/kg NO_3-N removed.

$$w_{CH3OH} = R_{CH3OH}\,R_{NO3\text{-}N} = 3.2\,\text{kg CH}_3\text{OH/kg NO}_3\text{-N} \times 150\,\text{kg NO}_3\text{-N/d} = 480\,\text{kg CH}_3\text{OH/d}$$

8. Determine the sludge production from the filters.

   Biomass generation from the denitrification filters is estimated using the equivalent COD for methanol $COD_{CH3OH} = BOD_{L,CH3OH} = 1.5$ mg COD/mg CH_3OH (from Example 10.138) at the cell yield coefficient $Y_H = 0.18$ g VSS/g COD.

$$P_X = Y_H\,COD_{CH3OH}\,w_{CH3OH} = 0.18\,\text{kg VSS/kg COD} \times 1.5\,\text{kg COD/kg CH}_3\text{OH} \times 480\,\text{kg CH}_3\text{OH/d} = 130\,\text{kg VSS/d}$$

   Estimate the biomass generation rate based on nitrate nitrogen removal.

$$Y_{VSS/NO3\text{-}N} = \frac{P_X}{R_{NO3\text{-}N}} = \frac{130\,\text{kg VSS/d}}{150\,\text{kg NO}_3\text{-N/d}} = 0.87\,\text{kg VSS/kg NO}_3\text{-N}$$

   Note: The typical biomass production rate is 0.4–0.8 kg VSS/kg NO_3-N. The estimated biomass generation rate is slightly high and on the conservative side.

Assume the biodegradation of the influent VSS in the filters is ignorable and estimate the TSS removed by the filters from the influent and effluent TSS concentrations.

$$P_{\text{TSS}} = Q(TSS_{\text{inf}} - TSS_{\text{eff}}) = 6000\,\text{m}^3/\text{d} \times (20 - 5)\,\text{g TSS/m}^3 \times 10^{-3}\,\text{kg/g} = 90\,\text{kg TSS/d}$$

Calculate the daily total sludge produced from the denitrification filters.

$$P_{\text{sludge}} = P_X + P_{\text{TSS}} = (130 + 90)\,\text{kg/d} = 220\,\text{kg TSS/d}$$

Calculated the amount of total solids accumulated in each filter for one backwash per filter per day when three filters are in service under design operating conditions.

$$\Delta TSS = \frac{P_{\text{sludge}}}{(N_{\text{filter}} - 1)\,N_{\text{bw}}} = \frac{220\,\text{kg TSS/d}}{(4-1)\,\text{filters} \times \dfrac{1\,\text{backwash}}{\text{filter}\cdot\text{d}}} = 73\,\text{kg TSS/backwash}$$

Calculated the concentration of total solids accumulated in the filter if the backwash is performed once per day.

$$TSS_{\text{filter}} = \frac{\Delta TSS}{V} = \frac{73\,\text{kg TSS/d} \times 1\,\text{d}}{40\,\text{m}^3} = 1.8\,\text{kg TSS/m}^3$$

Note: The amount of solids accumulated in the filter beds is below the allowed maximum solids storage capacity of 4 kg TSS/m^3 filter volume. Therefore, the filters can be backwashed every other day if head loss is not excessive.

Calculate the TSS concentration in the combined backwash and bump waste flow.

$$TSS_{\text{bw}} = \frac{P_{\text{sludge}}}{\Delta q_{\text{w}}} = \frac{220\,\text{kg TSS/d}}{878\,\text{m}^3/\text{d}} = 0.25\,\text{kg TSS/m}^3 = 250\,\text{g TSS/m}^3 \quad \text{or} \quad 250\,\text{mg TSS/L}$$

Note: The TSS concentration in the backwash waste is normally 500–1500 mg/L. The relatively low TSS concentration in the combined waste is potentially because a large amount of filtered water for filter bump is included in the calculation.[2]

EXAMPLE 10.147: DENITRIFICATION IN AN ANOXIC FLUIDIZED-BED BIOREACTOR

Anoxic fluidized-bed bioreactors (AnoxFBBRs) are used for denitrification of well nitrified waste stream given in Example 10.146. The effluent characteristics and design parameters for the AnoxFBBR are given below.

Effluent: The effluent NO_3-N concentrations = 2 mg/L.

Design parameters: Reactor fluidized bed depth = 2.5 m, volumetric NO_3-N loading = 5 kg NO_3-N/m^3·d, recycle ratio = 4, and number of reactors = 4 (including one standby).

Determine (a) reactor volume and dimensions, (b) hydraulic loading (or flux rate) and empty bed contact time (EBCT) (or HRT), and (c) power requirements for fluidization.

Notes: The design procedure is provided for an anaerobic fluidized-bed reactor (AnFBR) in Example 10.123. The methanol feed rate and sludge production in the AnoxFBBRs are similar to that estimated in Example 10.146

Solution

1. Determine the reactor volume and dimensions.

   Calculate the nitrate nitrogen removal rate ($R_{NO3\text{-}N}$) at $Q = 6000\ m^3/d$, $N_{NO3\text{-}N,0} = 27\ mg/L$, and $N_{NO3\text{-}N} = 2\ mg/L$.

$$R_{NO3\text{-}N} = Q(N_{NO3\text{-}N,0} - N_{NO3\text{-}N}) = 6000\ m^3/d \times (27 - 2)\ g\ N/m^3 \times 10^{-3}\ kg/g = 150\ kg\ NO_3\text{-}N/d$$

   Calculate the total fluidized media volume (V_{Total}) based on the volumetric NO_3-N loading $VNL = 5$ kg NO_3-N/m^3·d.

$$V_{Total,VNL} = \frac{R_{NO3\text{-}N}}{VNL} = \frac{150\ kg\ NO_3\text{-}N/d}{5\ kg\ NO_3\text{-}N/m^3 \cdot d} = 30\ m^3$$

   Calculate the total reactor area (A_{Total}) at the fluidized bed depth $D = 2.5$ m.

$$A_{Total} = \frac{V_{Total}}{D} = \frac{30\ m^3}{2.5\ m} = 12\ m^2$$

   Calculate the required area of each reactor (A) with three reactors in service at the design flow of 6000 m^3/d.

$$A = \frac{A_{Total}}{N_{reactor} - 1} = \frac{12\ m^2}{(4 - 1)\ reactor} = 4\ m^2\ \text{per reactor}$$

   Calculate the diameter of the reactor (d).

$$d = \sqrt{\frac{4A}{\pi}} = \sqrt{\frac{4 \times 4\ m^2}{\pi}} = 2.3\ m$$

   Provide four circular reactors with dimensions = 2.3 m (diameter) × 2.5 m (fluidized bed depth). For each reactor, the actual surface area $A = 4.15\ m^2$ and empty bed volume $V = 10.4\ m^3$. The total surface area and volume are $A_{Total} = 12.45\ m^2$ and $V_{Total} = 31.2\ m^3$ when three reactors are in service.

2. Determine the hydraulic loading (HL) and EBCT.

$$HL = \frac{Q}{A_{Total}} = \frac{6000\ m^3/d}{12.45\ m^2} = 482\ m^3/m^2 \cdot d \quad \text{or} \quad 20\ m^3/m^2 \cdot h$$

   Note: The design HL is within the typical range of 15–24 m^3/m^2·d (Table 10.54).

$$EBCT = \frac{V_{Total}}{Q} = \frac{31.2\ m^3}{6000\ m^3/d} \times \frac{1440\ min}{d} = 7.5\ min$$

   Note: The design EBCT is within the typical range of 5–10 min (Table 10.54).

3. Determine the power requirement for fluidization.

   Provide five circulation pumps, one for each of four reactor and one as standby. Each pump is sized to pump the recycle flow (Q_r) that creates the desired upflow velocity for bed fluidization. Calculate the pump design flow (Q_p) at the recycle flow ratio of $R_r = 4$ when three reactors are in service.

$$Q_p = Q_r = \frac{R_r Q}{(N_{reactor} - 1) N_{pump}} = \frac{4 \times 6000\,m^3/d}{(4-1)\,reactor \times 1\,pump/reactor} \times \frac{d}{86400\,s} = 0.093\,m^3/s \text{ per pump}$$

Assume the total dynamic head (TDH) = 12.2 m at the design flow, specific weight of the fluid pumped$\gamma \approx 9.81$ kN/m^3, pump efficiency $E_p = 75\%$, and motor efficiency $E_m = 90\%$. Calculate the power output of the pump (P_w) from Equation 6.17a with $K' = 1$ kW/(kN·m/s).

$$P_w = K' Q_p (TDH)\gamma = 1\,kw/(kN{\cdot}m/s) \times 0.093\,m^3/s \times 12.2\,m \times 9.81\,kN/m^3 = 11.1\,kW \quad \text{or} \quad 15\,hp$$

The power input to the motor (P_m) is then calculated by rearranging (Equations 6.17b and 6.17c).

$$P_m = \frac{P_p}{E_m} = \frac{P_w}{E_p E_m} = \frac{11.1\,kW}{0.75 \times 0.9} = 16.4\,kW \quad \text{or} \quad 22\,hp$$

Provide a 20-kW (27 hp) motor for each of five pumps.

10.7 Enhanced Biological Phosphorus Removal

Phosphorus is essential for growth of microorganisms. It is a limiting nutrient in most natural waters. Excess of phosphorus is associated with eutrophication. Phosphorus concentration in raw municipal wastewater is 4–10 mg/L. It exists in inorganic and organic forms. Ortho- and polyphosphates constitute the major inorganic component, and account for 70% of total phosphorus in municipal wastewater. Inorganic phosphorus is used up by microorganisms for cell synthesis and energy transport. For biosynthesis, phosphorus is consumed based on the stoichiometric requirement. The typical formula for cell biomass is $C_{60}H_{87}O_{23}N_{12}P$ and the phosphorus content is 1.5–2.5% of VSS on a dry weight basis. Typical phosphorus removal in a conventional secondary wastewater treatment plant is 10–20% of influent phosphorus.[2,7]

10.7.1 Process Fundamentals

In the enhanced biological phosphorus removal (EBPR) processes, successive change between anaerobic and aerobic conditions encourages growth of special heterotrophic microorganisms that are capable of uptaking phosphorus beyond the stoichiometric requirement for growth and energy transfer. As a result, the average organic phosphorus (Org-P) levels of 20–30% of microbiological solids (VSS) may be achievable due to the heterotrophic biomass yield, and phosphorus removal up to 95% of influent concentration may take place. A simplified illustration of the EBPR process is shown in Figure 10.105a. The primary organisms associated with enhanced phosphorus removal belong to the genius *Acinetobacter*, and are known as phosphorus-accumulating organisms (PAOs). Under anaerobic stress the PAOs assimilate easily biodegradable organic matters such as SCVFAs within the cell biomass. SCVFAs are the intermediate products known as poly-β-hydroxyalkanoates (PHAs), of which the most common compound is poly-β-hydroxybutyrate (PHB) with concomitant release of orthophosphate (Ortho-P) from stored polyphosphates (or Poly-P). Under the aerobic environment the SCVFAs stored as PHA is oxidized to provide energy for cell growth, "luxury" uptake of Ortho-P, and storage of energy in Poly-P. A typical process configuration for achieving the EBPR goal is illustrated in Figure 10.105b.

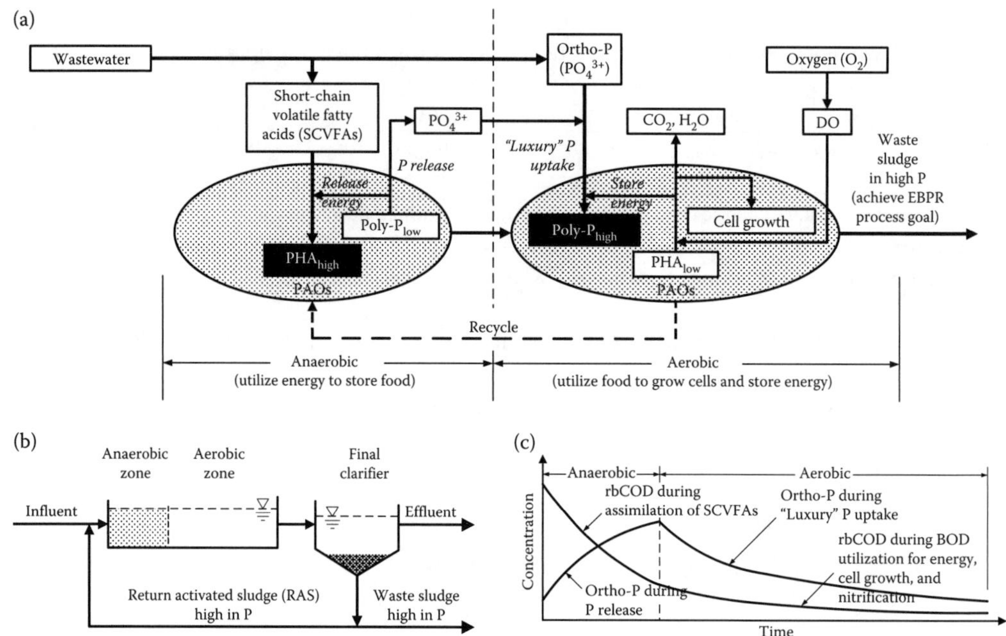

FIGURE 10.105 Enhanced Biological Phosphorus Removal (EBPR) process: (a) simplified mechanism, (b) example configuration (A/O or Pho-redox process), and (c) rbCOD and Ortho-P concentration profiles.

10.7.2 Biochemical Reactions in Anaerobic Zone

A series of complex biochemical reactions occur in the anaerobic zone. Basic information about some of these reactions with important relationships are provided below.

Reactions: The simplified biochemical reactions in anaerobic zones are expressed by Equation 10.108.

Energy storage reaction

$$\text{Poly-P}_n + \text{ADP} \xrightarrow{\text{PAOs}} \text{Poly-P}_{n-1} + \text{ATP} \tag{10.108a}$$

P-release reaction

$$\text{SCVFAs} + \text{ATP} \xrightarrow{\text{PAOs}} \text{PHA} + \text{ADP} + \text{PO}_4^{3-} + \text{Cations} \tag{10.108b}$$

ATP (adenosine triphosphate) is a coenzyme that is used as an intercellular energy carrier for metabolism while ADP (adenosine diphosphate) is the product of dephosphorization of ATP after releasing energy and Ortho-P.

SCVFA Uptake with Ortho-P Release: Using the energy stored in the cellular polyphosphates, PAOs assimilate the soluble readily biodegradable COD (rbCOD) that are mostly in the form of SCVFAs such as acetic and propionic acids and store them as PHB. The uptake of SCVFAs is a temporary storage that is finally stabilized in the aerobic zone. Concurrently with SCVFA uptake, Ortho-P is released when the polyphosphate bonds are broken to produce the needed energy. Cations that are bonded with Ortho-P in the Poly-P are also released. The SCVFA uptake and Ortho-P release curves under the anaerobic condition are conceptually illustrated in Figure 10.105c. The soluble Ortho-P concentration after release in anaerobic zone may reach 20–40 mg/L.[204–206]

TABLE 10.56 Recommended Minimum Organic Substrate to TP Ratios for EBPR

Ratio of Organic Substrate to TP	Value or Range[a]
COD	40–60
BOD_5	20–30
rbCOD	10–18[b]
VFA	4–16

[a] These are the minimum ratios recommended for achieving an effluent TP concentration less than 1 mg/L.
[b] This may be the most accurate ratio range.
Note: rbCOD = readily biodegradable chemical oxygen demand, VFA = volatile fatty acid.
Source: Adapted in part from References 2 and 181.

Glycogen may also be partially involved in energy production for both anaerobic SCVFA uptake and aerobic PHA oxidation by glycogen-accumulating organisms (GOAs). Since there is neither Ortho-P release nor subsequently uptake, GOAs do not contribute to EBPR. Lowering the growth of GOAs should improve the EBPR.[2,181]

rbCOD to TP Ratios: Many studies have shown that EBPR is clearly influenced by the availability of rbCOD (or SCVFAs). High enough rbCOD concentration in the influent to the anaerobic zone is essential to carry out both the phosphorus release in the anaerobic zones and uptake in aerobic zones. Many rules of thumb have been proposed to access the favorable conditions and actions needed to maximize the EBPR performance. Some of these relationships are given below.[204–206]

1. The reported minimum ratios of different organic substrates to TP to obtain an effluent P concentration <1 mg/L are summarized in Table 10.56.[181]
2. A relationship between effluent total phosphorus and the COD/TP ratio in the influent is given in Figure 10.106.[200] To achieve a TP concentration of one or less mg/L in the effluent, the COD/TP ratio in the influent should be 33 or higher.
3. More than 90% TP removal and effective nitrification/denitrification may be achieved if soluble BOD_5 concentration in the influent is above 100 mg/L.
4. Fermentation and hydrolysis of biodegradable organics in the influent and/or recycled biosolids are the typical sources for soluble rbCOD. Approximately 1.0 g acetate COD is produced per gram of bCOD fermented.
5. The COD/TP ratio greater than 40 or COD/BOD_5 ratio below 2 in the influent is an indication that sufficiently fermented SCVFAs are available. The major function of anaerobic zone is to allow PAOs to store SCVFAs as PHB or similar polymers. This step is rapid and is completed within

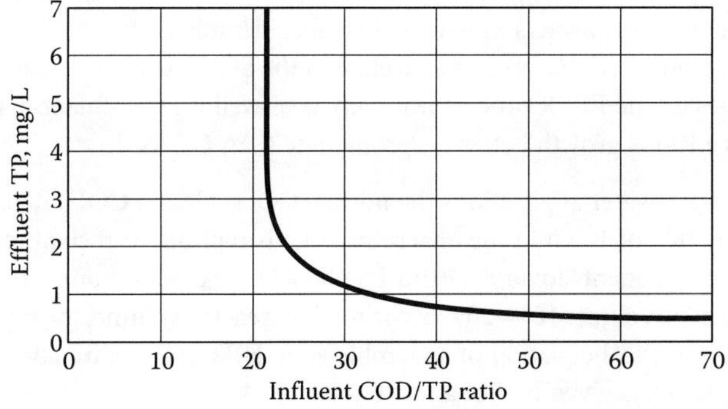

FIGURE 10.106 Relationship between effluent total phosphorus and COD/TP ratio in the influent (Adapted in part from Reference 200).

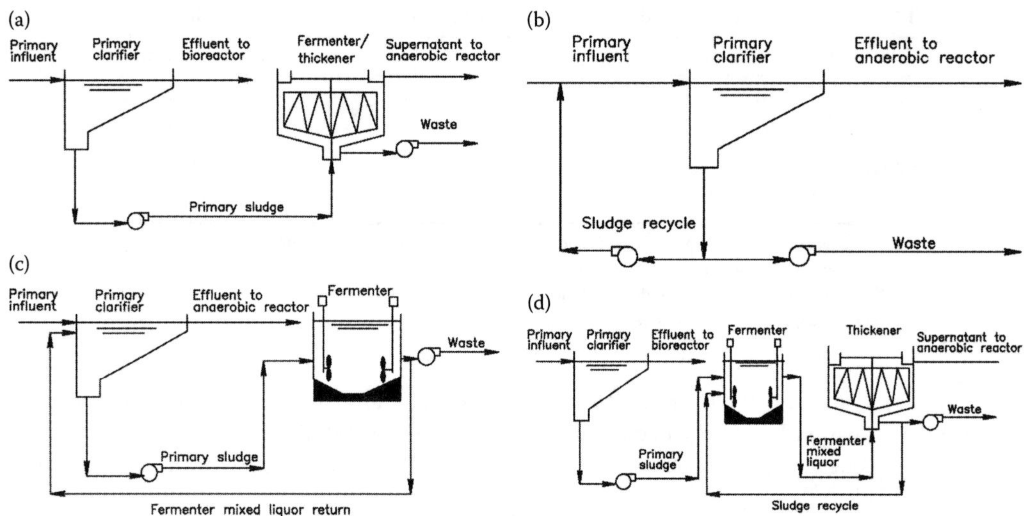

FIGURE 10.107 Primary and thickener fermenters: (a) static primary thickener fermenter to hydrolyze primary sludge, (b) activated primary fermenter to recirculate sludge for hydrolysis, (c) complete-mix primary with an external fermenter, and (d) complete-mix fermenter-thickener combination to hydrolyze primary sludge (Adapted from Reference 210, used with permission of the Author).

30–60 min. At a COD/TP ratio below 20 and a COD/BOD_5 ratio higher than 2, the influent may have insignificant fermentation products and P release may not occur. For this reason, an increase in the anaerobic zone volume, an additional fermentation facility, or an external source of VFAs may be needed.

6. On-site fermentation is achieved in an activated primary or thickener fermenter. Schematic flow diagrams of several fermentation facilities are shown in Figure 10.107.[207–210] The typical SRT range is 2–6 days in the fermentators. The HRTs are normally 6–12 h in the primary fermenters and 2–4 days in thickener fermenters. A low recycle ratio for elutriation is also required in a range of 0.05–0.1 for primary clarifiers and 0.1–0.4 for thickeners.[2]
7. When the supply of rbCOD is limited in the raw influent or after significant consumption of rbCOD by denitrification, phosphorus release may still be associated with insufficient PHA storage. Subsequently phosphorus uptake under aerobic condition will not occur since sufficient PHA is not available for oxidation and energy release. Under these conditions, the P release is called *secondary release*. The overall phosphorus removal will be considerably lower.
8. The ratio of propionate ($C_2H_5COO^-$) to acetate in the SCVFAs after fermentation also affects the EBPR performance. The EBPR process normally goes well using influent that has high percent of propionate since PAOs growth better on propionate than GOAs do.[2]

HRT: In municipal wastewater applications the normal COD/TP and COD/BOD_5 ratios are sufficient to carry out EBPR. An HRT of 1–2 h in the anaerobic zone is typically sufficient for fermentation and P release followed by the subsequent "luxury" Ortho-P uptake in the aerobic zone. This is usually equivalent to having an anaerobic zone volume 10%–20% of the total bioreactor volume, including anoxic, anaerobic, and aerobic zones. An empirical equation of anaerobic zone HRT (θ) as a function of the concentration removal of total phosphorus is given by Equation 10.109.[2,7,181]

$$\theta = 0.81(TP_0 - SP_e) - 2.11 \tag{10.109}$$

where

θ = hydraulic retention time of anaerobic zone, h

TP_0 = total phosphorus in the influent, mg/L (g/m^3)

SP_e = soluble phosphorus in the effluent, mg/L (g/m^3)

Mixing Energy: Sufficient mixing is needed in the anaerobic or anoxic zones to keep the contents in suspension while minimizing the surface turbulence. The minimum power requirement of the mixer is given by Equation 10.110.[7]

$$P/V = 0.0075\,\mu^{0.3}\,\text{MLSS}^{0.298} \tag{10.110}$$

where

P/V = power requirement for mixing, kW/m^3

μ = dynamic viscosity, N·s/m^2

MLSS = concentration of mixed liquor suspended solid, mg/L

DO and Nitrate: Presence of DO and NO_3-N in anaerobic zone will consume SCVFAs before PAOs are able to store it as PHA while releasing Ortho-P. Therefore, DO and NO_3-N in the return or recycle flows must be destroyed so a true anaerobic environment can be established for PAOs. Approximately 0.8–1.5 and 2.5–4 mg/L of BOD_5 concentration is consumed to remove, respectively, each mg/L of DO or NO_3-N (see Section 10.6.2 for the organics required for denitrification and deoxygenation).[7]

10.7.3 Biochemical Reactions in Aerobic Zone

The generalized biochemical reactions that occurred in the aerobic zone and some basic information are presented below.

Reactions: The simplified biochemical reactions in aerobic zones are expressed by Equation 10.111.

Energy release reaction

$$\text{Poly-P}_{n-1} + \text{ATP} \xrightarrow{\text{PAOs}} \text{Poly-P}_n + \text{ADP} \tag{10.111a}$$

P-uptake reaction

$$\text{PHA} + O_2 + \text{ADP} + PO_4^{3-} + \text{Cations} \xrightarrow{\text{PAOs}} \text{Biomass with high P} + \text{ATP} + CO_2 + H_2O \tag{10.111b}$$

Under the aerobic condition, the stored PHA is oxidized to provide carbon source for cell synthesis and energy to rebuild the polyphosphate bonds as energy storage in the cell. Through the “luxury” uptake process, soluble Ortho-P is removed from the liquid and high amounts of Poly-P are stored within the new and old cells. A notable amount of cations is also brought into the cell with Ortho-P uptake. The EBPR goal is eventually reached when the excess biomass with high P content is wasted from the biological treatment system.

DO Requirement: Sufficient DO supply in the aerobic reactor is essential for PAOs to completely metabolize PHA. In high rate EBPR systems, a DO concentration of 2 mg/L or higher in the effluent may be required, although satisfactory EBPR is achieved at a DO concentration below 0.5 mg/L in long SRT systems. Also, sufficient DO in the sludge zone is needed to prevent premature P release in the clarifier.

10.7.4 Overall Process Considerations

The stoichiometric information, operational considerations, and factors affecting environmental conditions for EBPR process are summarized below. Detailed information may be found in References 2, 7, and 181.

COD Consumptions: A minimum of 10 mg/L of soluble rbCOD is theoretically required to remove each mg/L of P by the biological storage mechanism. Practically, the rbCOD requirements range from 10 to 20 mg rbCOD/mg P due to competition between PAOs and GAOs. The overall rbCOD consumption from municipal wastewater is normally 20–50 mg/L for each mg/L P removed by the EBPR. Additional removal of P is also achieved by normal cell synthesis during BOD removal in the activated sludge systems.

Cations: The minimum requirements for Mg^{2+}, K^{+}, and Ca^{2+} are, respectively, 0.28–0.33, 0.26–0.33, and 0.09 mole per mole of P removed. In municipal wastewater, the concentrations of these cations are significantly higher than the minimum requirements.

Nitrate or Nitrite: Ortho-P uptake can also occur when oxygen is substituted by nitrate or nitrite as the final electron acceptor. In some BNR process configurations, the denitrifying PAOs carry out Ortho-P uptake in the anoxic zone between the anaerobic and aerobic zone.

SRT: An increase in SRT normally has an adverse effect on the EBPR process. An EBPR may only require a short SRT of 3–5 days (based on total reactor volume) for effective Ortho-P uptake under aerobic condition. Long SRT has been reported more favorable for growth of undesired GAO populations. Secondary release may also occur due to endogenous decay at long SRT.

Temperature: The performance of EBPR process improves at temperatures below 15°C because the growth of PAOs dominates over GAOs. The favorable temperature range for fast growth of GOAs is between 15°C and 30°C. At lower temperatures the fermentation is reduced significantly. For this reason the overall EBPR performance also drops significantly when the temperatures are below 5°C.

pH: The optimum pH range for an EBPR process is 7.0–7.4. The performance steadily declines below pH of 6.9 and practically stops below 5.5. Low pH may be encountered where high concentration of ammonia is nitrified without benefit of denitrification. Lower pH is also more favorable for growth of GAOs over PAOs. At pH above 7.25 the growth of GAOs is nearly prohibited.

Selector: The PAOs loaded with polyphosphates produce dense floc that settles rapidly. For this reason, activated sludge plants have been provided with an anaerobic zone ahead of the aeration basin to encourage growth of PAOs. This is called *selector technology*. Added benefits of selector technology are foaming and sludge bulking control.

10.7.5 Biosolids from EBPR Process

The biosolids from EBPR process have many important characteristics. Some of these are briefly summarized below. Additional information may be found in References 2, 7, and 181.

Solids Production: The biomass yield is normally 0.4–0.5 g VSS per g rbCOD (or 0.6–0.7 g VSS/g $sBOD_5$) consumed by PAOs. This is the typical heterotrophic cell yield rate of heterotrophs that carry out the conventional biological waste treatment. The typical range of PAOs in the total heterotrophs is 20–40%.

Org-P Content in VSS: The phosphorus content in the cell mass of PAOs may range from 0.2 to 0.3 g P/g VSS. The Org-P content in the other heterotrophic cell growth due to BOD removal is typically 0.02 g P/g VSS. Therefore, the P content in the PAOs is 10–15 times of the biomass produced from the activated sludge systems without phosphorus removal. The typical Org-P content in the combined biomass from the EBPR systems may range from 0.05 to 0.15 g P/g VSS.

VSS/TSS Ratio: The PAOs uptake cations which causes an increase in inorganics content of the cell mass. The inorganic content in biomass of PAOs is about 0.63 g SS/g P removed. As a result, the VSS/TSS ratio may be lowered to 0.60–0.65 g VSS/g TSS in PAOs in comparison with the typical VSS/TSS ratio of 0.7–0.85 in the WAS from the conventional activated sludge systems.

Solids Handling: The phosphorus rich WAS from an EBPR system releases phosphorus under anaerobic condition. For this reason, the sidestreams from gravity thickener, anaerobic digester, and dewatering facility will contain high concentration of soluble Ortho-P. Since the sidestreams are recycled through the plant, a buildup of phosphorus in the system will occur. The sCOD/TP ratio will decrease and effluent phosphorus requirement may not be met. For this reason, a phosphorus stripping process should be provided to chemically immobilize the phosphorus before the sidestreams are returned to the liquid treatment process (Section 10.8 and Example 10.160). Phosphorus recovery from the phosphorus rich sidestreams is also effective for preventing phosphorus release from the sludge processing and handling areas (Section 13.10.1 and Example 13.46).

EXAMPLE 10.148: EFFLUENT TP AND HRT OF ANAEROBIC ZONE

The anaerobic basin of an EBPR facility receives influent that contains COD = 260 mg/L, BOD_5 = 125 mg/L, ready biodegradable COD (rbCOD) = 80 mg/L, and TP = 6 mg/L. Estimate (a) the fermentation condition, (b) availability of SCVFAs, (c) concentration of effluent TP, and (d) HRT of anaerobic zone. Apply the rule of thumb ratios given in Section 10.7.2. Assume that $TSS_e = 10$ mg/L and VSS/TSS ratio = 0.75 in the final clarifier effluent. The overall P content ($f_{P,EBPR}$) = 0.07 g P/g VSS in the biomass.

Solution

1. Evaluate the fermentation condition in the influent.
   Determine the COD/BOD_5 ratio in the influent.

$$COD/BOD_5 = \frac{260 \text{ mg/L}}{125 \text{ mg/L}} = 2.1$$

   This ratio is slightly higher than 2. It indicates that the fermentation may be close to completion, but a slight larger anaerobic zone may be needed. Assume an efficiency $f_{fm} = 0.8$ for fermentation of rbCOD. Estimate the SCVFA concentration as rbCOD in the influent.

$$SCVFA = f_{fm} \times rbCOD = 0.8 \times 80 \text{ mg rbCOD/L} = 64 \text{ mg rbCOD/L}$$

2. Evaluate the sufficiency of SCVFAs for achieving efficient EBPR.
   Evaluate the sufficiency of SCVFAs based on the recommendations summarized in Table 10.56.
   a. Determine the COD/TP ratio in the influent.

$$COD/TP = \frac{260 \text{ mg/L}}{6 \text{ mg/L}} = 43 \quad (40 < COD/TP < 60)$$

   b. Determine the BOD_5/TP ratio in the influent.

$$BOD_5/TP = \frac{125 \text{mg/L}}{6 \text{ mg/L}} = 21 \quad (20 < BOD_5/TP < 30)$$

   c. Determine the rbCOD/TP ratio in the influent.

$$rbCOD/TP = \frac{80 \text{ mg/L}}{6 \text{ mg/L}} = 13 \quad (10 < rbCOD/TP < 18)$$

   d. Determine the SCVFA/TP ratio in the influent.

$$SCVFA/TP = \frac{64 \text{ mg/L}}{6 \text{ mg/L}} = 11 \quad (4 < VFA/TP < 16)$$

   e. Evaluate the availability of SCVFAs.

All calculated ratios are within or above the minimum ratios required for achieving TP < 1 mg/L in the effluent (see the recommendations in Table 10.56). Therefore, the fermentation products in the influent should be sufficient and efficient P removal is expected.

3. Determine the phosphorus concentrations in the effluent.

An effluent total phosphorus concentration (TP_e) of 0.8 mg/L is estimated at the COD/TP ratio of 43 from Figure 10.106. It is consistent with TP < 1 mg/L expected in Step 2. The total phosphorous concentration in the effluent is expressed by Equation 10.112a while the particulate phosphorus can be estimated from Equation 10.112b.

$$TP_e = SP_e + PP_e \quad \text{or} \quad SP_e = TP_e - PP_e \qquad (10.112a)$$

$$PP_e = f_{P,EBPR} \times \text{VSS/TSS ratio} \times TSS_e \qquad (10.112b)$$

where

TP_e = total phosphorus in the effluent, mg/L (g/m^3)

PP_e = particulate phosphorus in the effluent, mg/L (g/m^3)

$f_{P,EBPR}$ = overall P content in the biomass, including PAOs and other heterotrophs, g P/g VSS

Note: SP_e has been defined in Equation 10.109.

All other parameters in Equation 10.112 have been defined previously.

Calculate the particulate phosphorus concentration PP_e in the effluent from Equation 10.112b.

$$PP_e = 0.07\,\text{mg P/mg VSS} \times 0.75\,\text{mg VSS/mg TSS} \times 10\,\text{mg TSS/L} = 0.53\,\text{mg P/L}$$

Determine the SP_e from Equation 10.112a.

$$SP_e = TP_e - PP_e = (0.8 - 0.53)\,\text{mg P/L} = 0.27\,\text{mg P/L}$$

Note: To meet an effluent TP < 0.5 mg/L is possible when TSS concentration < 4 mg/L is achieved after a proper tertiary treatment such as a filtration process (see more information about the tertiary and advanced treatment processes presented in Chapter 15).

4. Determine the HRT of anaerobic zone.

The HRT is estimated from Equation 10.109.

$$\theta = 0.81(TP_0 - SP_e) - 2.11 = 0.81\,\text{h·L/mg} \times (6 - 0.27)\,\text{mg/L} - 2.11\,\text{h} = 2.5\,\text{h}$$

5. Select the HRT for anaerobic zone.

The organic substrate to TP ratios obtained in Step 2 indicate that a HRT of 2 h or less in anaerobic zone may be considered sufficient. Since the COD/BOD_5 ratio is exceeding 2, a HRT of 2.5 h is considered more desirable for improving fermentation of rbCOD, storage of SCVFAs as PHAs and release of Ortho-P.

EXAMPLE 10.149: PHOSPHORUS RELEASE AND UPTAKE, AND COD CONSUMPTION IN ANAEROBIC ZONE

An EBPR plant is providing excellent phosphorus removal. The biological system consists of anaerobic and aerobic zones. Develop the approximate Ortho-P release and uptake, and rbCOD consumption curves. The influent and effluent qualities and process operating parameters are given below.

Influent: COD = 280 mg/L, biodegradable COD (bCOD) = 220 mg/L, rbCOD = 85 mg/L, TP = 6 mg/L, and Ortho-P = 4.2 mg/L.

Effluent: bsCOD = 15 mg/L, and $SP_e = 0.3$ mg/L.

Operating parameters: The MLVSS = 1400 mg/L and return sludge ratio $R_{ras} = 0.25$. The HRTs are 2 and 10 h in the anaerobic zone and the entire EBPR system, respectively. Assume that an uptake of ~12 mg/L of SCVFAs as rbCOD for each mg/L of P removed by "luxury" P uptake; overall P content in the biomass from EBPR system is 0.08 g P/g VSS; Org-P content due to cell synthesis in biomass is 0.02 g P/g VSS; 40% of the excess P is released in the anaerobic zone; the observed biomass yield for heterotrophic growth is 0.3 g VSS/g bCOD under field conditions; and the fermentation is nearly completed with a fermentation efficiency $f_{fm} = 0.8$ g SCVFA as rbCOD/g rbCOD. Ignore the consumption of rbCOD by denitrification/deoxygenation.

Solution

1. Determine the COD concentrations in the flow applied to the anaerobic zone.

   The concentration of a soluble substrate or nutrient applied to the anoxic zone after mixing of return sludge with influent is calculated by Equation 10.112c.

$$C_{AN,0} = \frac{C_0 + R_{ras}C}{1 + R_{ras}} \qquad (10.112c)$$

where

$C_{AN,0}$ = concentration of substrate or nutrient applied to the anoxic zone, mg/L (g/m^3)
C_0 = concentration of substrate or nutrient in the influent, mg/L (g/m^3)
C = concentration of substrate or nutrient in the effluent from the aeration zone or return sludge, mg/L (g/m^3)
R_{ras} = ratio of return sludge to influent flow, dimensionless

The following concentrations are calculated from Equation 10.112c.

a. Determine the bCOD concentration ($S_{AN,0}$).

$$S_{AN,0} = \frac{S_0 + R_{ras}S}{1 + R_{ras}} = \frac{220\text{ mg rbCOD/L} + 0.25 \times 15\text{ mg bsCOD/L}}{1 + 0.25} = 179\text{ mg bCOD/L}$$

b. Determine the rbCOD concentration ($rbCOD_{AN,0}$).

$$rbCOD_{AN,0} = \frac{rbCOD_0 + R_{ras}S}{1 + R_{ras}} = \frac{85\text{ mg rbCOD/L} + 0.25 \times 15\text{ mg bsCOD/L}}{1 + 0.25} = 71\text{ mg rbCOD/L}$$

c. Estimate the SCVFA concentration ($SCVFA_{AN,0}$).

   Calculate the SCVFA concentration as rbCOD ($SCVFA_0$) in the influent.

$$SCVFA_0 = f_{fm} \times rbCOD = 0.8 \times 85\text{ mg rbCOD/L} = 68\text{ mg rbCOD/L}$$

   Assume $SCVFA_e \approx 0$ in the effluent and calculate the $SCVFA_{AN,0}$.

$$SCVFA_{AN,0} = \frac{SCVFA_0 + R_{ras}SCVFA_e}{1 + R_{ras}} = \frac{68\text{ mg rbCOD/L} + 0.25 \times 0}{1 + 0.25} = 54\text{ mg rbCOD/L}$$

2. Determine the phosphorus concentrations in the flow applied to the anaerobic zone.

   Calculate the phosphorus concentrations from Equation 10.112c.

a. Determine the TP concentration ($TP_{AN,0}$), excluding the particulate P content in the biomass.

$$TP_{AN,0} = \frac{TP_0 + R_{ras}SP_e}{1 + R_{ras}} = \frac{6\,\text{mg P/L} + 0.25 \times 0.3\,\text{mg P/L}}{1 + 0.25} = 4.86\,\text{mg P/L}$$

b. Determine the Ortho-P concentration ($Ortho\text{-}P_{AN,0}$).

$$Ortho\text{-}P_{AN,0} = \frac{Ortho\text{-}P_0 + R_{ras}SP_e}{1 + R_{ras}} = \frac{4.2\,\text{mg P/L} + 0.25 \times 0.3\,\text{mg P/L}}{1 + 0.25} = 3.42\,\text{mg P/L}$$

3. Determine the Ortho-P concentration profile through the EBPR system.
   a. Determine the Ortho-P release concentration in the anaerobic zone.
   The total P concentration the MLVSS,

$$P_{EBPR} = f_{P,EBPR} \times MLVSS = 0.08\,\text{mg P/mg VSS} \times 1400\,\text{mg VSS/L} = 112\,\text{mg P/L}$$

The Org-P concentration from cell synthesis in the MLVSS,

$$P_{synthesis} = f_{p,synthesis} \times MLVSS = 0.02\,\text{mg P/mg VSS} \times 1400\,\text{mg VSS/L} = 28\,\text{mg P/L}$$

The excess P concentration available for release in the anaerobic zone,

$$P_{excess} = P_{EBPR} - P_{synthesis} = (112 - 28)\,\text{mg P/L} = 84\,\text{mg P/L}$$

The Ortho-P release concentration at a release rate of 40% ($f_{P,release} = 0.4$),

$$P_{release} = f_{P,release} \times P_{excess} = 0.4 \times 84\,\text{mg P/L} = 33.6\,\text{mg P/L}$$

   b. Determine the Ortho-P concentration in the effluent from the anaerobic zone after the P release.

$$Ortho\text{-}P_{AN} = Ortho\text{-}P_{AN,0} + P_{release} = (3.42 + 33.6)\,\text{mg P/L} = 37.0\,\text{mg P/L}$$

Note: The estimated soluble Ortho-P concentration after release in the anaerobic zone is within the typical range of 20–40 mg P/L.

4. Determine the rbCOD concentration profile through the EBPR system.
   a. Determine the "luxury" P uptake by PAOs.
   The TP concentration removed in the EBPR system based on the combined influent to the anaerobic zone, $P_{r,EBPR} = TP_{AN,0} - SP_e = (4.86 - 0.3)\,\text{mg P/L} = 4.56\,\text{mg P/L}$
   The overall biomass production concentration ($p_{x,EBPR}$) due to bCOD removal in the EBPR is calculated from Equation 10.15n.

$$p_{X,EBPR} = Y_{obs,H,F}(S_{AN,0} - S) = 0.3\,\text{mg VSS/mg bCOD} \times (179 - 15)\,\text{mg bCOD/L} = 49.2\,\text{mg VSS/L}$$

The P concentration removed due to cell synthesis in the biomass,

$$P_{r,synthesis} = f_{P,synthesis} \times p_{X,EBPR} = 0.02\,\text{mg P/mg VSS} \times 49.2\,\text{mgVSS/L} = 0.98\,\text{mg P/L}$$

The P concentration removed due to "luxury" P uptake by PAOs,

$$P_{r,uptake} = P_{r,EBPR} - P_{r,synthesis} = (4.56 - 0.98)\,\text{mg P/L} = 3.58\,\text{mg P/L}$$

b. Determine the rbCOD concentration uptake in the anaerobic zone.
The rbCOD concentration uptaken by PAOs,

$$S_{uptake} = f_{rbCOD/P} \times P_{r,uptake} = 12\,\text{mg rbCOD/mg P} \times 3.58\,\text{mg P/L} = 43.0\,\text{mg rbCOD/L}$$

Note: The estimated SCVFA concentration uptake of 43 mg rbCOD/L is less than the SCVFA concentration of 54 mg rbCOD/L available in the anaerobic zone. There is an additional SCVFA concentration of 11 mg/L available in the anaerobic zone. It is needed to carry out denitrification/deoxygenation in RAS flow.

c. Determine the rbCOD concentration in the effluent from the anaerobic zone after the SCVFA uptake.

$$rbCOD_{AN} = rbCOD_{AN,0} - S_{uptake} = (71 - 43)\,\text{mg P/L} = 28\,\text{mg rbCOD/L}$$

Note: The actual rbCOID concentration may be higher than 28 mg/L because the additional conversion of bCOD to rbCOD through hydrolysis and fermentation processes in the anaerobic zone.

5. Draw the Ortho-P and rbCOD concentration profiles.

The conceptual Ortho-P and rbCOD concentration curves in the anaerobic and aerobic zones are shown in Figure 10.108.

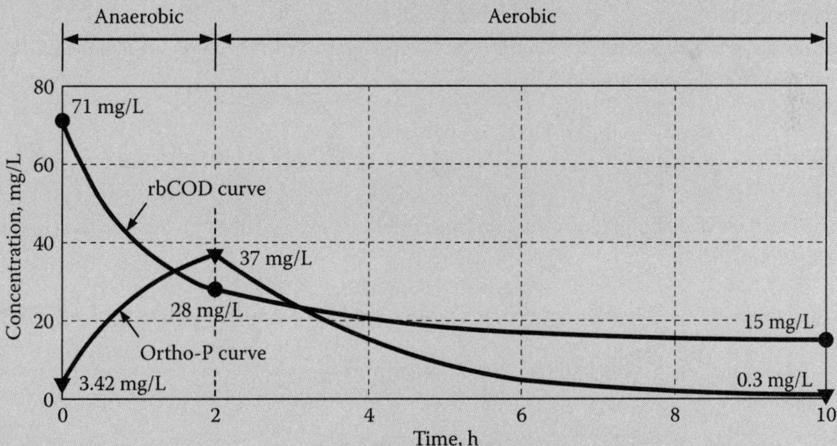

FIGURE 10.108 Approximate ortho-P and rbCOD concentration profiles through anaerobic and aerobic zones (Example 10.149).

EXAMPLE 10.150: PHOSPHORUS REMOVAL IN AN EBPR PROCESS

Excellent phosphorus removal is achieved by an EBPR process where sufficient SCVFAs are available in the influent. Determine (a) the overall P concentration removed, (b) P concentration removed due to cell synthesis, (c) P concentration removal by the "luxury" P uptake, (d) fraction of PAOs in the total biomass, (e) P content in PAOs, and (f) fermentation requirement. The influent, effluent, and process data are given below.

Influent: $BOD_5 = 140$ mg/L, soluble BOD_5 ($sBOD_5$) = 100 mg/L, and TP = 6 mg/L.
Effluent: Effluent $sBOD_5 = 3$ mg/L, TSS = 5 mg/L, TP = 0.5 mg/L, and VSS/TSS ratio = 0.7.
Process design data: $\theta_c = 10$ d, $Y_H = 0.65$ mg VSS/mg BOD_5, $k_{d,H,F} = 0.04\ d^{-1}$, 10 mg $sBOD_5$/L uptake from SCVFAs for each mg/L of P removed by "luxury" P uptake, and Org-P content due to cell synthesis in biomass = 0.02 g P/g VSS.

Solution

1. Determine the overall P concentration removed in the EBPR system.
   Assume that the overall P content in the biomass from the EBPR system $f_{P,EBPR} = 0.09$ g P/g VSS. The particular phosphorus concentration PP_e is calculated from Equation 10.112b.

$$PP_e = f_{P,EBPR} \times \text{VSS/TSS ratio} \times TSS_e = 0.09\,\text{mg P/mg VSS} \times 0.7\,\text{mg VSS/mg TSS} \times 5\text{mg TSS/L} = 0.32\,\text{mg P/L}$$

   The soluble phosphorus concentration SP_e is calculated from Equation 10.112a.

$$SP_e = TP_e - PP_e = (0.5 - 0.32)\,\text{mg P/L} = 0.18\,\text{mg P/L}$$

   The overall P concentration removed in the EBPR system based on the influent,

$$P_{r,EBPR} = TP_{AN,0} - SP_e = (6 - 0.18)\,\text{mg P/L} = 5.82\,\text{mg P/L}$$

2. Determine the P concentration removed due to cell synthesis.
   The observed biomass yields for heterotrophic cell growth due to BOD_5 removal in the EBPR system is calculated from Equation 10.15l.

$$Y_{obs,H,F} = \frac{Y_H}{1 + k_{d,H,F}\theta_c} = \frac{0.65\,\text{mg VSS/mg BOD}_5}{1 + 0.04\,\text{d}^{-1} \times 10\,\text{d}} = 0.46\,\text{mg VSS/mg BOD}_5$$

   Calculate the overall biomass production concentration ($p_{x,EBPR}$) due to BOD_5 removal in the EBPR from Equation 10.15n.

$$p_{x,EBPR} = Y_{obs}(S_0 - S) = 0.46\,\text{mg VSS/mg BOD}_5 \times (140 - 3)\,\text{mg BOD}_5\text{/L} = 63.0\,\text{mg VSS/L}$$

   The overall P content in the biomass from the EBPR system,

$$f_{P,EBPR} = \frac{P_{r,EBPR}}{p_{x,EBPR}} = \frac{5.82\,\text{mg P/L}}{63.0\,\text{mg VSS/L}} = 0.09\,\text{mg P/mg VSS}$$

   The calculated overall P content is same as the assumed value in Step 1.

3. Determine the P concentration removed by the "luxury" P uptake.
   The P concentration removed due to cell synthesis in the biomass,

$$P_{r,synthesis} = f_{P,synthesis} \times p_{x,EBPR} = 0.02\,\text{mg P/mg VSS} \times 63.0\,\text{mg VSS/L} = 1.26\,\text{mg P/L}$$

   The P concentration removed due to "luxury" P uptake by PAOs,

$$P_{r,uptake} = P_{r,EBPR} \times p_{r,synthesis} = (5.82 - 1.26)\,\text{mg P/L} = 4.56\,\text{mg P/L}$$

The fraction of P removed due to "luxury" P uptake of PAOs on the basis of total P removal in the EBPR system, $f_{\text{P,uptake}} = \dfrac{P_{\text{r,uptake}}}{P_{\text{r,EBPR}}} = \dfrac{4.56\,\text{mg P/L}}{5.82\,\text{mg P/L}} = 0.78$ or 78%

Note: Approximately 75–80% of P removal is achieved by the "luxury" P uptake of PAOs in the EBPR system.

4. Determine the fraction of PAOs in the total biomass.

The uptake of sBOD_5 concentration by PAOs,

$$S_{\text{uptake}} = f_{\text{sBOD5/P}} \times P_{\text{r,uptake}} = 10\,\text{mg sBOD}_5/\text{mg P} \times 4.56\,\text{mg P/L} = 45.6\,\text{mg sBOD}_5/\text{L}$$

The biomass production concentration due to growth of PAOs is calculated from Equation 10.15n.

$$p_{\text{x,PAO}} = Y_{\text{obs}} S_{\text{uptake}} = 0.46\,\text{mg VSS/mg BOD}_5 \times 45.6\,\text{mg sBOD}_5/\text{L} = 21.0\,\text{mg VSS/L}$$

The fraction of PAOs in the total biomass, $f_{\text{PAO}} = \dfrac{p_{\text{x,PAO}}}{p_{\text{x,EBRP}}} = \dfrac{21.0\,\text{mg VSS/L}}{63.0\,\text{mg VSS/L}} = 0.33$ or 33%

Note: PAOs count is approximately one-third of the total biomass in the EBPR system.

5. Determine the overall P removed by PAOs.

The P concentration removed due to cell synthesis in PAOs,

$$P_{\text{r,synthesis, PAO}} = f_{\text{P,synthesis}} \times p_{\text{x,PAO}} = 0.02\,\text{mg P/mg VSS} \times 21.0\,\text{mg VSS/L} = 0.42\,\text{mg P/L}$$

The total P concentration removed by PAOs ($P_{\text{r,PAO}}$) through cell synthesis and "luxury" P uptake,

$$P_{\text{r,PAO}} = P_{\text{r,synthesis, PAO}} + P_{\text{r,uptake}} = (0.42 + 4.56)\,\text{mg P/L} = 4.98\,\text{mg P/L}$$

The fraction of P removed by PAOs ($f_{\text{r,PAO}}$) on the basis of total P removed in the EBPR system,

$$f_{\text{r,PAO}} = \frac{P_{\text{r,PAO}}}{P_{\text{r,EBRP}}} = \frac{4.98\,\text{mg P/L}}{5.82\,\text{mg P/L}} = 0.86 \text{ or } 86\%$$

Note: Over 85% of P removal is achieved by PAOs in the EBPR system.

6. Determine the P content in PAOs.

$$f_{\text{P,PAO}} = \frac{P_{\text{r,PAO}}}{p_{\text{x,PAO}}} = \frac{4.98\,\text{mg P/L}}{21.0\,\text{mg VSS/L}} = 0.24\,\text{mg P/mg VSS}$$

Note: The calculated P content of 0.24 g P/g VSS is within the typically range of 0.2–0.3 g P/g VSS for PAOs.

7. Determine the fermentation requirement in the influent for effective EBPR performance.

$$f_{\text{fm,required}} = \frac{S_{\text{uptake}}}{S_{\text{0,sBOD5}}} = \frac{45.6\,\text{mg sBOD}_5/\text{L}}{100\,\text{mg sBOD}_5/\text{L}} = 0.46\,\text{mg SCVFA as sBOD}_5/\text{sBOD}_5 \text{ or } 46\% \text{ of sBOD}_5$$

Note: Normally, the expected fermentation efficiency in the influent $f_{\text{fm}} = 60\%–70\%$. Therefore, the influent will have sufficient SCVFA concentration for efficient P removal by the EBPR system.

EXAMPLE 10.151: MIXER SIZING FOR ANAEROBIC OR ANOXIC BASINS

A mixer is typically provided for anaerobic or anoxic zone to keep the mixed liquor in suspension while minimizing the surface turbulence. An anaerobic zone has a volume of 250 m^3 and the MLSS concentration is 3500 mg/L. Determine the mixer and motor power requirements for the mixer. The efficiencies of mixer and motor are 80% and 92%, respectively. The design operating temperature is 10°C.

Solution

1. Determine the water power requirement.
   At 10°C, the dynamic viscosity $\mu = 1.307 \times 10^{-3}\,\text{N·s/m}^2$ (Table B.2 in Appendix B).
   Calculate P/V from Equation 10.110.

$$P/V = 0.0075\,\mu^{0.3}\,\text{MLSS}^{0.298} = 0.0075 \times (1.307 \times 10^{-3}\,\text{N·s/m}^2)^{0.3} \times (3500\,\text{mg/L})^{0.298}$$
$$= 0.0116\,\text{kW/m}^3$$

   Calculate water power (P_w) or output power from the mixer.

$$P_W = (P/V)V = 0.0116\,\text{kW/m}^3 \times 250\,\text{m}^3 = 2.9\,\text{kW (3.9 hp)}$$

2. Determine the mixer power requirement.
   Calculate the mixer input (or motor output) power requirement (P_m) from Equation 9.33 at $E_{mixer} = 0.80$.

$$P_m = \frac{P_W}{E_{mixer}} = \frac{2.9\,\text{kW}}{0.8} = 3.6\,\text{kW (4.8 hp)}$$

3. Determine the motor power requirement.
   Calculate the motor input (or wire) power requirement (P_{mw}) from Equation 9.33 at $E_{motor} = 0.92$.

$$P_{mw} = \frac{P_m}{E_{motor}} = \frac{3.6\,\text{kW}}{0.92} = 3.9\,\text{kW (5.2 hp)}$$

   Provide the mixer a 4-kW motor with variable speed drive (VFD).

10.8 Biological Nutrient Removal

Excess discharge of nutrients (primarily, nitrogen and phosphorus) in waterways contributes to significant water quality problems. These problems include harmful algae blooms and aquatic plants growth, low DO, and decline in wildlife and their habitats. The EPA has laid out a roadmap for states to adopt numeric nutrient water quality standards. The nutrients abatement programs involve enhanced nutrient reduction at the conventional wastewater treatment plants, and watershed-based strategies.[211]

The BNR processes provide integrated design for enhanced removal of nitrogen and phosphorus in conjunction with BOD and TSS removal. The benefits of BNR systems include (1) relatively low cost of nitrogen and phosphorus removal, (2) no chemical cost, (3) reduced sludge quantity, (4) reduced aeration, (5) enhanced BOD_5 and TSS removal, (6) increased process stability and reliability, and (7) possibly for phosphorus recovery. In depth theory and design of biological nitrogen removal processes (nitrification and denitrification) and EBPR have been covered in Sections 10.6 and 10.7, respectively. In the following section, the theory and design of BNR processes are covered.

10.8.1 General Description of BNR Processes

Several proprietary BNR systems are currently available. The most common processes for combined removal of nitrogen, phosphorus, BOD_5, and TSS include (1) A^2/O, (2) modified Bardenpho, (3) standard and modified UCT, (4) VIP, (5) Johannesburg (JHB), (6) sequencing batch reactor (SBR) with EBPR, (7) Orbal® with EBPR, (8) PhoStrip, and (9) Anoxic/anaerobic/aerobic. Most of these processes are MBR compatible. Brief descriptions, important design parameters, and simple sketches are presented for these processes in Table 10.57.[2,7,17,30,98,181,205,212–224]

10.8.2 Performance of BNR Processes

The BNR processes are capable of removing significant amounts of TP, TN, BOD_5, and TSS. The degree of removal depends upon the characteristics of wastewater, operating temperature, and many other facility design and operational features. The achievable limit of technology (LOT) for the BNR processes at large facilities are 3 and 0.1 mg/L for TN and TP, respectively. The typical LOTs for different forms of nutrients are summarized in Table 10.58.[2,7,181,225,226] Under the optimum conditions the BNR system in assistance by tertiary treatment normally produces an effluent quality of total BOD_5, TSS, TP, and NH_4-N below 5, 5, 0.5, and 0.2 mg/L, respectively. Typical soluble Org-N and NO_3-N concentrations are in the range of 1–1.5 and 3–6 mg/L. A more economically achievable TN concentration is 6 mg/L. Important information on the technologies, case studies, and capital and O&M costs for the full-scale BNR applications at municipal wastewater treatment facilities in the United States may be found in References 225 and 226.

10.8.3 General Design Considerations for BNR Facilities

The design and operation of a BNR facility requires thorough understanding of influent characteristics, environmental factors, and design parameters such as kinetic coefficients, SRT, number and configuration of reactors, and recirculation patterns. The following basic design steps are useful for understanding the design procedure:

1. Develop the influent waste characteristics reaching the BNR facility. A material-mass balance analysis may be required to establish flow and concentrations of many constituents (BOD_5, COD, TKN, and TP).
2. Evaluate the fermentation conditions with expected SCVFAs and the ratios of organics (COD, BOD_5, rbCOD, or VFA) to TP. If sufficient SCVFAs are not present, consider including a fermentation facility or adding ECS.
3. Establish the kinetic coefficients for carbonaceous BOD removal, nitrification, and denitrification. These coefficients are well developed for typical municipal wastewater. If the influent is not of a typical municipal quality the kinetic coefficients must be developed based on experience from bench scale or pilot testing.
4. Determine the process configuration, and number and size of different functional zones: anaerobic, anoxic, and aerobic (or oxic). Using kinetic equations and critical winter conditions, determine HRT (θ) and SRT (θ_c) for these zones. The volume of anoxic zone will depend upon the recycle ratio and extent of denitrification desired. The volume of anaerobic zone will depend upon extent of hydrolysis and availability of SCVFAs. The volume of aerobic zone should be large enough to consume the leftover carbonaceous BOD and provide nitrification.
5. Determine the recycle ratios for denitrification in anoxic zones using total nitrogen balance through the system. Total NO_3-N destroyed by denitrification is equal to TKN in the influent minus the particulate Org-N fixed in the biomass, and soluble NH_4-N, Org-N, and NO_3-N lost in the effluent.

TABLE 10.57 Description and Design Parameters of Common BNR Processes

Process with Brief Description	SRT (θ_c), d	F/M Ratio[a], d^{-1}	HRT (θ), h	MLSS, mg TSS/L	Return/Recycle Ratio[b]	Effluent[c], mg/L
1. A^2/O process A^2/O process (called also three-stage Pho-redox) is a modification of A/O and provides both N and P removal.[213] The process diagram utilizes anaerobic, anoxic, and aerobic sequence with RAS to the anaerobic zone and IR to the anoxic zone (Figure 10.109). It has stability to produce high effluent quality.	5–25	0.1–0.25	AN: 0.5–1.5 AX: 1–3 OX: 4–8 Total: 6–12	3000–4000[d]	R_{ir}: 1–4 R_{rs}: 0.25–1	TN < 6 TP < 2
2. Modified Bardenpho process The proprietary Bardenpho process for N removal was modified to provide combined N and P removal in five stages (Figure 10.110).[212] P release occurs in the 1st anaerobic zone and most denitrification occurs in the 1st anoxic zone. Nitrification and P uptake are carried out in the 1st oxic zone. The 2nd anoxic zone adds additional denitrification. The final aerobic zone strips N_2 and adds DO to minimize P release in the final clarifier.	10–20	0.1–0.2	AN: 0.5–1.5 AX_1: 1–3 OX_1: 4–12 AX_2: 2–4 OX_2: 0.5–1 Total: 8–20	3000–4000[d]	R_{ir}: 2–4 R_{rs}: 0.5–1	TN < 6 TP < 2

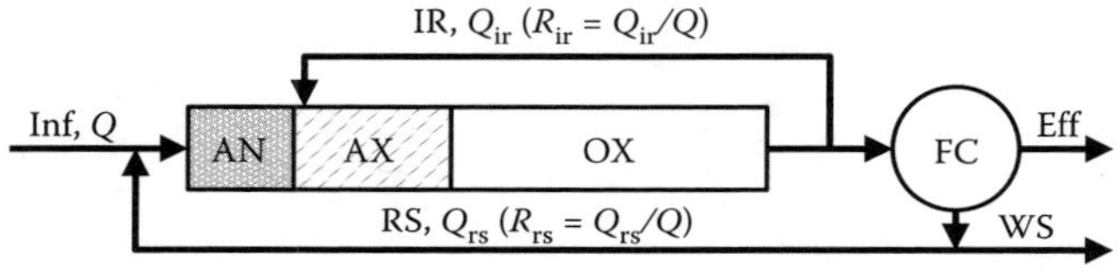

FIGURE 10.109 Definition Sketch of A^2/O Process.

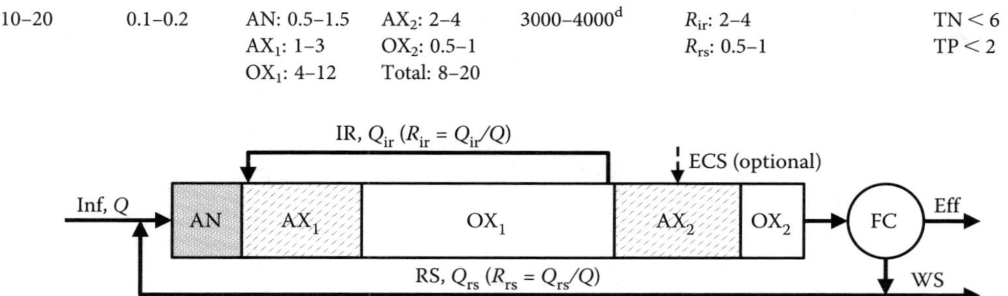

FIGURE 10.110 Definition Sketch of Modified 5-Stage Bardenpho Process.

(*Continued*)

TABLE 10.57 (*Continued*) Description and Design Parameters of Common BNR Processes

Process with Brief Description	SRT (θ_c), d	F/M Ratio[a], d^{-1}	HRT (θ), h		MLSS, mg TSS/L	Return/Recycle Ratio[b]		Effluent[c], mg/L
3. Standard and modified UCT processes The University of Cape Town (UCT) process utilizes anaerobic, anoxic, and aerobic sequence for both N and P removal.[214,215] Both the RAS and IR are returned from the end of aeration basin into the anoxic zone. The contents of the anoxic zone are then recycled into the anaerobic zone. The modified UCT process has anaerobic, anoxic, anoxic, and aerobic sequence (Figure 10.111). The 2nd anoxic tank receives IR for enhanced denitrification.	10–25	0.1–0.2	AN: 1–2 AX_1: 1–2 AX_2: 2–3	OX: 4–12 Total: 8–18	3000–4000[d]	R_{mlr}: 2–4 R_{ir}: 1–3 R_{rs}: 0.8–1		TN < 10 TP < 2
4. VIP process The zone sequence of VIP (named for Virginia Initiative Plant in Norfolk, VA) is similar to the A^2/O and UCT processes except for the method of sludge recycle (Figure 10.112).[30] At least two stages are created using baffles in the anoxic and aerobic zones to ensure that the desired operating conditions are firmly met.	5–10	0.1–0.2	AN: 1–2 AX: 1–3	OX: 4–6 Total: 6–10	2000–4000[d]	R_{mlr}: 1–2 R_{ir}: 1–3	R_{rs}: 0.8–1	TN < 6 TP < 2

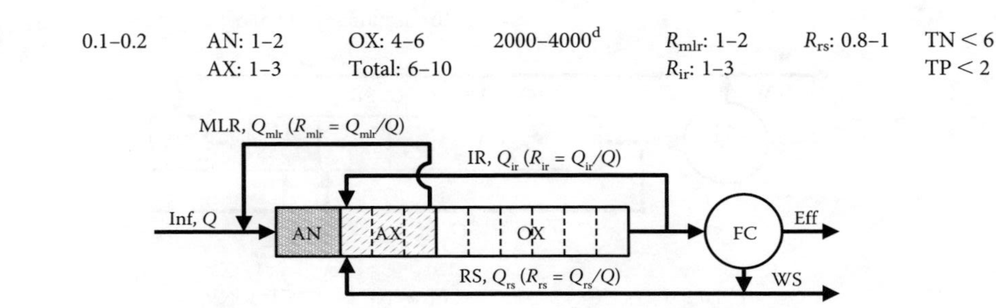

FIGURE 10.111 Definition Sketch of Modified UCT Process.

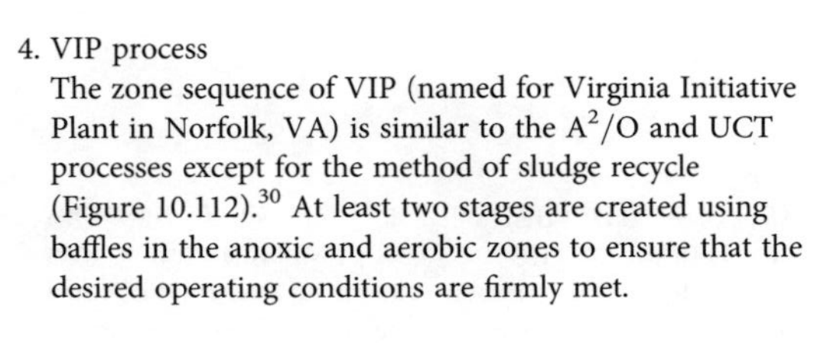

FIGURE 10.112 Definition Sketch of VIP Process.

(*Continued*)

TABLE 10.57 (*Continued*) Description and Design Parameters of Common BNR Processes

Process with Brief Description	SRT (θ_c), d	F/M Ratio[a], d^{-1}	HRT (θ), h		MLSS, mg TSS/L	Return/Recycle Ratio[b]	Effluent[c], mg/L
5. Johannesburg (JHB) process This process originated in Johannesburg, South Africa, and is an alternative to UCT or modified UCT processes.[2] An anoxic chamber is used to reduce nitrate in the RAS prior to returning it into the anaerobic zone. This modification is intended to improve P removal for weak wastewater influent that is low in SCVFAs (Figure 10.113).	5–10	0.1–0.2	AX_1: 0.5–1 AN: 1–1.5 AX_2: 1–2	OX: 4–6 Total: 8–10	3000–4000[d]	R_{ir}: 3–4 R_{rs}: 0.5–1	TN < 6 TP < 2
6. SBR with EBPR process Enhanced N and P removal can be achieved in an SBR. P release and SCVFA uptake occur during the anaerobic react (stir) operation after fill. P uptake, BOD reduction, and nitrification occur during the aerobic cycle. Denitrification is achieved during the anoxic stir and settling/decant cycles (Figure 10.114).[2] Additional information about SBR systems is provided in Tables 10.15 and 10.55, and Examples 10.55 and 10.56.	20–40	0.1–0.2	AX_1: 1–2 AN: 1.5–3 OX: 2–4	AX_2: 0.5–2 AX_3: 1.5–2 Total: 8–12	3000–4000	N/A	TN < 6 TP < 2

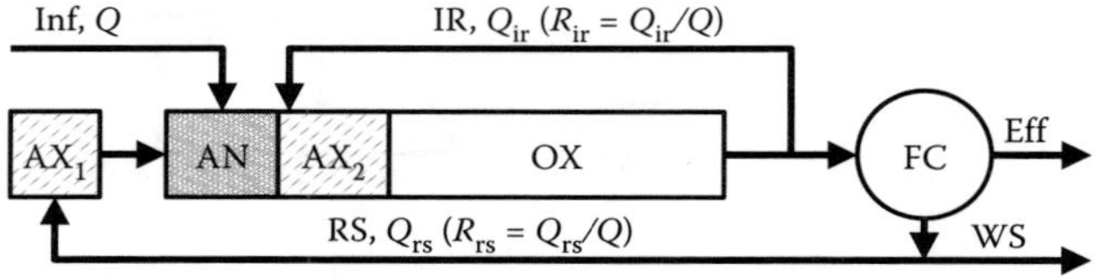

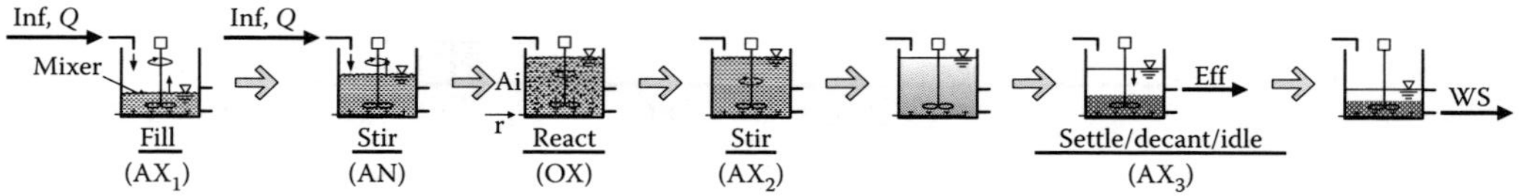

FIGURE 10.113 Definition Sketch of Johannesburg Process.

FIGURE 10.114 Definition Sketch of SBR with EBPR Process

(*Continued*)

TABLE 10.57 (*Continued*) Description and Design Parameters of Common BNR Processes

Process with Brief Description	SRT (θ_c), d	F/M Ratio[a], d^{-1}	HRT (θ), h	MLSS, mg TSS/L	Return/Recycle Ratio[b]	Effluent[c], mg/L
7. Orbal® with EBPR process The Orbal® process typically consists of three concentric channels operating in series. The outer channel receives influent and RAS and maintains an oxygen-deficient environment by intermittent aeration for simultaneous nitrification–denitrification. The 2nd and 3rd channels maintain higher levels of DO.[216–219] These systems are designed for 80% TN removal. Above 90% removal is possible by recycling MLSS from the 3rd to the 1st channel. For EBPR, a strong oxygen deficit is maintained in the 1st channel, a minor deficit in the 2nd, and a DO of 2–3 mg/L in the 3rd channel (Figure 10.115). See additional information in Tables 10.15 and 10.55.	10–30	0.1–0.2	AN: 3–6 OX: 8–12 AX: 4–8 Total: 16–24	3000–6000	R_{ir}: 1–4 R_{rs}: 0.75–1	TN < 5 TP < 3
8. PhoStrip process This proprietary process has essentially an anaerobic/aerobic sequence. It has a stripper tank in which a portion of the return sludge is diverted. The sludge in the stripper tank may be elutriated with primary effluent to provide the necessary carbon source for P release.[220,221] The P-stripped biological solids are separated and returned to the aeration basin. The P-rich supernatant is coagulated to precipitate phosphorus (Figure 10.116). Lime dose is a function of alkalinity and not the amount of P present. Alum and ferric salts may also be used but the dose is related to the amount of P released.	5–20	0.1–0.5	AN: 8–12	1000–3000	R_{ir}: 0.1–0.2 R_{rs}: 0.5–1	TN < 8 TP < 1

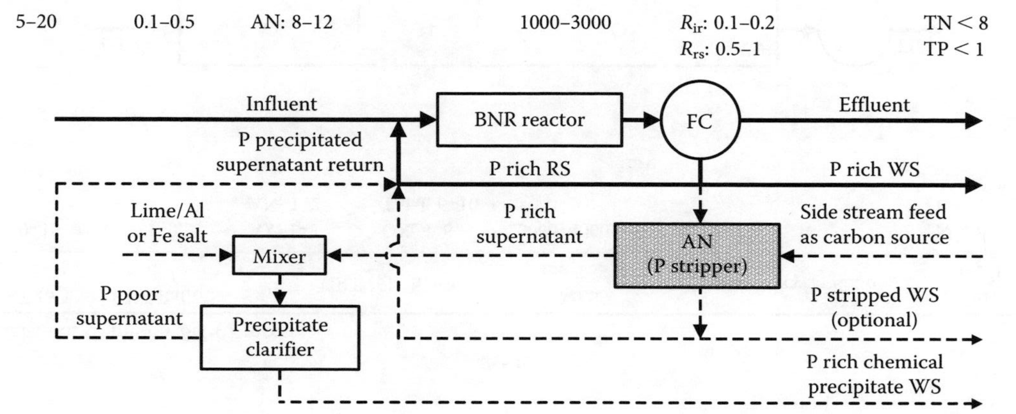

FIGURE 10.115 Definition Sketch of Oral® Process with EBPR Process.

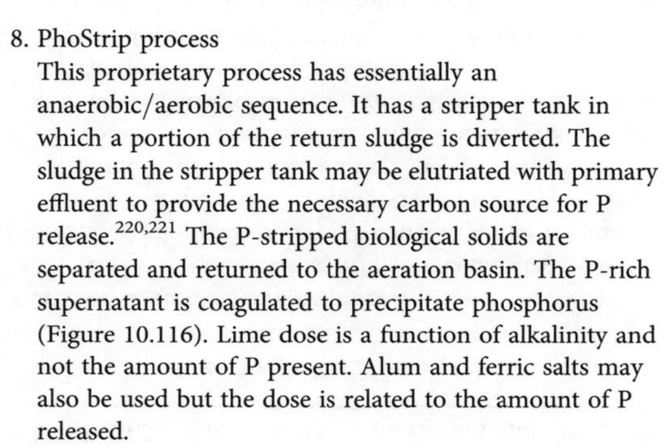

FIGURE 10.116 Definition Sketch of PhoStrip Process.

(*Continued*)

TABLE 10.57 (*Continued*) Description and Design Parameters of Common BNR Processes

Process with Brief Description	SRT (θ_c), d	F/M Ratio[a], d^{-1}	HRT (θ), h	MLSS, mg TSS/L	Return/Recycle Ratio[b]	Effluent[c], mg/L
9. Anoxic/anaerobic/aerobic process The University of Texas at Arlington conducted bench-scale and pilot plant studies to enhance nutrient removal from an existing activated sludge plant that would require a minimum unit process addition and use a single recycle line common to most activated sludge designs. A reactor with anoxic, anaerobic, and aerobic sequence with sludge recycle from the clarifier to anoxic zone was used (Figure 10.117). Significant TN and TP removals are achieved. High return sludge ratio would increase nitrate recycle and enhance N removal. However, it would also (a) increase the amount of DO returned, (b) increase the consumption of rbCOD for deoxygenation and denitrification, (c) reduce the availability of rbCOD in the effluent from the anoxic zone, and (d) affect eventually the P release in the anaerobic zone, and subsequent P uptake in the aerobic zone. This process is more effective for wastewater high in rbCOD and without primary treatment.[205,222–224]	10–15	0.1–0.2	AX: 1–3 AN: 1–2 OX: 4–6 Total: 6–10	3000–4000[d]	R_{rs}: 1–3	TN < 8 TP < 2

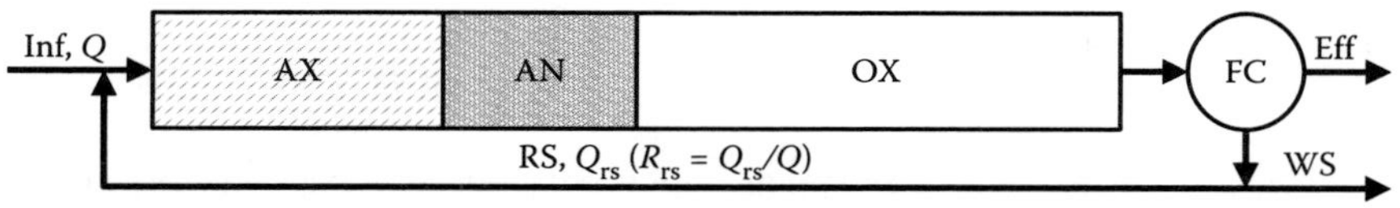

FIGURE 10.117 Definition Sketch of Anoxic/Anaerobic/Aerobic Process.

[a] The unit of F/M ratio is g BOD_5/g VSS·d.

[b] It is the ratio to the influent flow Q.

[c] All processes are capable of achieving BOD_5 and TSS concentrations below 5 mg/L.

[d] It indicates that the process is MBR compatible. The MLSS may be in the range of 7500–12,000 mg/L when MBR is used to replace the final clarifier.

Legends: AN = anaerobic; AX = anoxic; DO = dissolved oxygen; ECS = external carbon source; FC = final clarifier; R_{ir} = internal recycle ratio; LDO = low dissolved oxygen; R_{mlr} = mixed liquor recycle ratio; ORP = oxidation–reduction potential; OX = aerobic or oxic; R_{rs} = return sludge ratio; WS = waste sludge.

Source: Adapted in part from References 2, 7, 17, 30, 98, 181, 205, and 212 through 224.

TABLE 10.58 Limits of Technology (LOT) for Nutrients in BNR Processes

Parameter	Removal Process	Limit of Technology, mg/L	Reference
Nitrogen			
NH_4-N	Nitrification or BNR	<0.5	Sections 10.6.1 and 10.8
NO_3-N	Denitrification or BNR	1–2	Sections 10.6.2 and 10.8
Particulate Org-N	Solids separation	<1	Chapters 9 and 15, and Section 10.9
Soluble Org-N	None[a]	0.5–1.5	N/A
TN	Nitrification/denitrification or BNR	3	Sections 10.6.1, 10.6.2, and 10.8
Phosphorus			
Ortho-P	Chemical precipitation, EBPR or BNR	0.1	Chapter 9, and Sections 10.7 and 10.8
Particulate Org-P	Solids separation	<0.05	Chapters 9 and 15, and Section 10.9
TP	Chemical precipitation, EBPR or BNR	0.1	Chapter 9, and Sections 10.7 and 10.8

[a] In general, biological treatment processes are not effective for removal of soluble Org-N. Effective removal of soluble Org-N may require advanced treatment processes such as carbon absorption, ion exchange, membrane process, or advanced oxidation processes. See Chapter 15 for detailed discussion on these processes.

Source: Adapted in part from References 2, 7, 181, 225, and 226.

6. Determine the organics required in the post anoxic zones, and provide sufficient ECS for effective denitrification.
7. Check the alkalinity of the system by conducting alkalinity balance based on the alkalinity concentrations in influent, destroyed by nitrification, recovered by denitrification, and lost in the effluent. An alkalinity buffer of 50–60 mg/L as $CaCO_3$ in the effluent must be maintained to keep favorable pH (above 6.5).
8. A minimum DO of 2.0 mg/L must be maintained in the aerobic zone. DO should be absent in the anoxic and anaerobic zones. Deoxygenation by rbCOD may be used to destroy the DO in the return or recycle flows if desired.

EXAMPLE 10.152: CONCENTRATION PROFILES OF sBOD$_5$, ORTHO-P, AND TN IN A BNR FACILITY

A BNR facility uses A^2/O process as described in Table 10.57 and shown in Figure 10.109. Draw the conceptual concentration profiles of biodegradable organic matter (sBOD$_5$ or rbCOD), Ortho-P, and TN through the entire process. Explain the components of each curve in anaerobic, anoxic, and aerobic zones.

Solution

1. Draw the conceptual concentration profile curves of sBOD$_5$, Ortho-P, and TN.

   The conceptual concentration profile curves of sBOD$_5$, Ortho-P, and TN with respect to nominal HRT based on influent flow are shown in Figure 10.118.
2. Describe the sBOD$_5$, Ortho-P, and TN curves.

   The general comments on these curves are provided in Table 10.59.

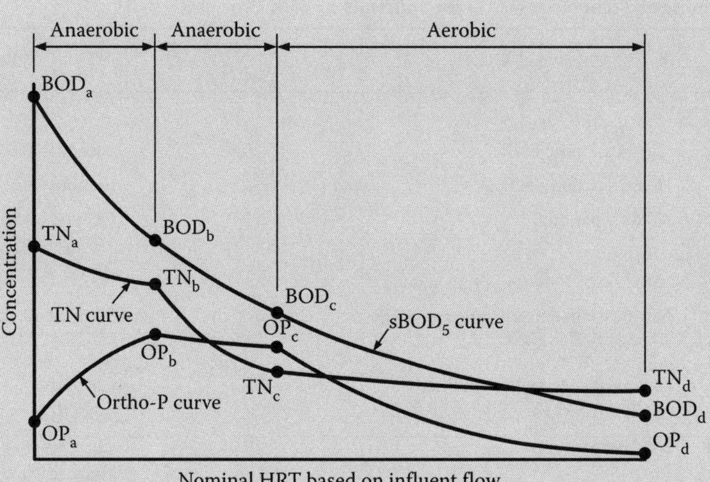

FIGURE 10.118 Typical concentration profiles of BOD_5, Ortho-P, and TN in a BNR facility (Example 10.152).

TABLE 10.59 Conceptual $sBOD_5$, Ortho-P, and TN Concentration Profiles through A^2/O Process (Example 10.152)

Curve	Anaerobic Zone	Anoxic Zone	Aerobic Zone
$sBOD_5$ curve	*Between BOD_a and BOD_b:* $sBOD_5$ is utilized for denitrification of NO_3-N in RAS. PAOs temporarily store ready biodegradable organics (SCVFAs) as PHA within the cells. Small amount of organics is also consumed for deoxygenation of DO.	*Between BOD_b and BOD_c:* $sBOD_5$ is utilized for denitrification of NO_3-N in the internal recycle (IR) flow. Organics are also consumed for energy and cell synthesis. Small amount of organics is consumed for deoxygenation of DO. Small amount of stored PHA may also be consumed by denitrification.	*Between BOD_c and BOD_d:* The remaining BOD_5 and stored PHA are utilized for energy, cell growth, "luxury" P uptake, and storage of Poly-P as energy. The BNR effluent BOD_5 is typically less than 5–10 mg/L and is lost in the discharge from final clarifier.
Ortho-P curve	*Between OP_a and OP_b:* Ortho-P is released from the stored Poly-P while the uptake of SCVFAs occurs. The Ortho-P concentration may reach 20–40 mg/L in the anaerobic zone. Small amounts of Ortho-P are fixed in the biomass.	*Between OP_b and OP_c:* Relatively high concentration of Ortho-P is maintained in the anoxic zone. Small amount of Ortho-P is fixed in the biomass.	*Between OP_c and OP_d:* Extra Ortho-P is consumed by "luxury" P uptake and storage as Poly-P while organics are oxidized for energy. The TP concentration in the BNR effluent is low, normally less than 0.5–1 mg/L with Ortho-P less than 0.5 mg/L.
TN curve	*Between TN_a and TN_b:* NO_3-N carried in RAS is denitrified prior to true anaerobic condition is established and small amount of alkalinity is recovered. Nitrogen is removed due to loss of N_2 into atmosphere. Some NH_4-N is fixed in the biomass.	*Between TN_b and TN_c:* Large amount of NO_3-N in the recycle flow is denitrified and N_2 is lost into the atmosphere. Large amount of alkalinity is recovered in this process. Some NH_4-N is fixed in the biomass.	*Between TN_c and TN_d:* The organic nitrogen and NH_4-N are oxidized to NO_3-N in nitrification, and a large amount of alkalinity is destroyed. Some NH_4-N is also fixed in the biomass. Typically, the BNR effluent contains TN below 6–8 with 3–6 mg/L NO_3-N, <0.5–1 mg/L NH_4-N, and 1 mg/L of soluble and 0.5–1.5 mg/L of particulate Org.-N.

EXAMPLE 10.153: PHOSPHORUS CONTENT OF BIOMASS AND TP IN THE EFFLUENT

A BNR facility uses A^2/O process. The influent and effluent characteristics and process details are given below. Determine (a) the P content of biomass and (b) TP concentration in the effluent.

Influent: $BOD_5 = 180$ mg/L, TKN = 35 mg/L, and TP = 6 mg/L.
Effluent: $sBOD_5 = 3$ mg/L, TSS = 4 mg/L, NH_4-N = 0.5 mg/L, Ortho-P = 0.3 mg/L, and VSS/TSS ratio = 0.7.
Kinetic coefficients: $Y_H = 0.65$ mg VSS/mg BOD_5, $k_{d,H,F} = 0.04\ d^{-1}$, $Y_N = 0.17$ mg VSS/mg NH_4-N, $k_{d,N,F} = 0.12\ d^{-1}$, and $\theta_c = 10$ d. Use $f_N = 0.12$ g N/g VSS and ignore inert (nonbiomass) VSS in the influent. Org-P content due to cell synthesis in biomass $f_{P,synthesis} = 0.02$ g P/g VSS.

Note: The subscripts H, N, and F express, respectively, heterotrophs, nitrifiers, and field condition.

Solution

1. Determine the combined biomass growth.
   Calculate the growth of heterotrophic, autotrophic, and combined organisms separately from Equation 10.92g.
   a. Calculate the growth of heterotrophic organisms ($p_{x,H}$) due to consumption of BOD_5.

$$p_{x,H} = \frac{Y_H(S_0 - S)}{1 + k_{d,H,F}\theta_c} = \frac{0.65\ \text{mg VSS/mg BOD}_5 \times (180 - 3)\ \text{mg BOD}_5\text{/L}}{1 + 0.04\ \text{d}^{-1} \times 10\ \text{d}} = 82.2\ \text{mg VSS/I}$$

   Note: The growth of heterotrophic organisms ($p_{x,H}$) can occur due to (a) consumption of BOD_5 by active denitrifying heterotrophs in the anoxic zone and (b) consumption of PHAs stored by phosphorus-accumulating organisms (PAOs), and (c) removal of stored and remaining BOD_5 by all heterotrophs in the aerobic zone.
   b. Calculate the growth of autotrophic organisms ($p_{x,N}$) due to nitrification in the aerobic zone.

$$p_{x,N} = \frac{Y_N}{1 + k_{d,N,F}\,\theta_c^{design}}(N_{0,TKN} - N_{NH4\text{-}N} - f_N\,p_{x,BOD})$$

$$= \frac{0.17\ \text{mg VSS/mg NH}_4\text{-N}}{1 + 0.12\ \text{d}^{-1} \times 10\ \text{d}} \times ((35 - 0.5)\ \text{mg NH}_4\text{-N/L} - 0.12\ \text{mg N/mg VSS}$$

$$\times\ 82.2\ \text{mg VSS/L}) = 1.90\ \text{mg VSS/L}$$

   c. Calculate the combined growth of heterotrophic and autotrophic organisms ($p_{x,H+N}$) in the BNR system.

$$P_{x,BNR} = p_{x,H} + p_{x,N} = (82.2 + 1.90) = 84.1\ \text{mg VSS/L}$$

2. Determine the P removal in the BNR system.
   a. Calculate the overall P concentration removed by both "luxury" P uptake and cell synthesis in the BNR system ($P_{r,BNR}$).

$$P_{r,BNR} = TP_0 - SP_e = (6 - 0.3)\ \text{mg P/L} = 5.7\ \text{mg P/L}$$

   b. Calculate the P concentration removed by autotrophic organisms in the BNR system ($P_{r,N}$).

$$P_{r,N} = f_{P,synthesis} \times p_{x,N} = 0.02\ \text{mg P/mgVSS} \times 1.90\ \text{mg VSS/L} = 0.038\ \text{mg P/L}$$

c. Calculate the P concentration removed by heterotrophic organisms in the BNR system ($P_{r,H}$).

$$P_{r,H} = P_{r,BNR} - P_{r,N} = (5.7 - 0.038)\,\text{mg P/L} = 5.66\,\text{mg P/L}$$

3. Determine the overall P content in the biomass from the BNR system.
   a. Calculate the overall P content in the biomass from the BNR system ($f_{P,BNR}$).

$$f_{P,BNR} = \frac{P_{r,BNR}}{p_{x,BNR}} = \frac{5.7\,\text{mg P/L}}{84.1\,\text{mg VSS/L}} = 0.068\,\text{mg P/mg VSS}$$

Note: The overall P content is ~ 7% of the biomass (VSS) from the BNR system. It is within the typical range 0.05–0.15 g P/g VSS.

   b. Calculate the P content in the total heterotrophs from the BNR system ($f_{P,H}$).

$$f_{P,H} = \frac{P_{r,H}}{p_{x,H}} = \frac{5.66\,\text{mg P/L}}{82.2\,\text{mg VSS/L}} = 0.069\,\text{mg P/mg VSS}$$

4. Determine the TP concentration in the effluent.
   Calculated the effluent particular phosphorus concentration PP_e from Equation 10.112b.

$$PP_e = f_{P,BNR} \times \text{VSS/TSS ratio} \times TSS_e = 0.068\,\text{mg P/mg VSS} \times 0.7\,\text{mg VSS/mg TSS} \times 4\,\text{mg TSS/L} = 0.19\,\text{mg P/L}$$

Determine the TP concentration TP_e from Equation 10.112a.

$$TP_e = SP_e + PP_e = (0.3 + 0.19)\,\text{mg P/L} = 0.49\,\text{mg P/L}$$

The TP concentration is ~0.5 mg/L in the effluent from the BNR system.

EXAMPLE 10.154: HETEROTROPHIC AND AUTOTROPHIC GROWTHS IN BNR SYSTEM

Apply the influent, effluent, and process design data given and the results developed in Example 10.153 to estimate the following different functional biomass growths in the BNR system: (a) nitrifying autotrophic (ammonia-oxidizing bacteria, AOB) growth, (b) phosphorus-accumulating organisms (PAOs) heterotrophic growth, (c) denitrifying heterotrophic growth, and (d) non-PAO aerobic heterotrophic growth in the BNR system. Also determine the fermentation requirement in the influent for effective BNR performance. Additional process design data includes: influent $sBOD_5 = 150$ mg/L, effluent nitrate concentration $N_{NO3\text{-}N} = 4$ mg/L, and SCVFA uptake $f_{sBOD5/P} = 10$ g $sBOD_5$/g P uptake.

Solution

1. Summarize calculation results for the combined biomass growth.
   The biomass growths are calculated in Example 10.153 and the results are summarized below.
   $p_{x,H} = 82.2$ mg VSS/L for the heterotrophic organisms, $p_{x,N} = 1.90$ mg VSS/L for the autotrophic ($p_{x,N}$), and $p_{x,BNR} = 84.1$ mg VSS/L for the combined biomass growth in the BNR system.
2. Estimate the fractions of autotrophs ($f_{x,N}$) and heterotrophs ($f_{x,H}$) in the combined biomass growth.

$$\text{Fraction of autotrophs, } f_{x,N} = \frac{p_{x,N}}{p_{x,BNR}} = \frac{1.90\,\text{mg/L}}{84.1\,\text{mg/L}} = 0.023 \text{ or } 2.3\%;$$

Fraction of heterotrophs, $f_{x,H} = 1 - 0.023 = 0.977$ or 97.7%

Note: The AOB count is about 2% of the total biomass in the mixed liquor. It is at the lower end of the typical range of 2–5%.

3. Estimate the fraction of PAO heterotrophic biomass growth.
   a. Calculate the P concentration removed ($P_{r,\text{synthesis}}$) due to cell synthesis of the biomass.

$$P_{r,\text{synthesis}} = f_{P,\text{synthesis}} \times p_{x,\text{BNR}} = 0.02\,\text{mg P/mgVSS} \times 84.1\,\text{mg VSS/L} = 1.68\,\text{mg P/L}$$

   b. Calculate the P concentration removed due to "luxury" P uptake ($P_{r,\text{uptake}}$) by PAOs.
   Total P concentration removed in the BNR system $P_{r,\text{BNR}} = 5.7$ mg P/L (Example 10.153)

$$P_{r,\text{uptake}} = f_{r,\text{BNR}} - P_{r,\text{synthesis}} = (5.7 - 1.68)\,\text{mg P/L} = 4.02\,\text{mg P/L}$$

   c. Calculate the SCVFA uptake ($S_{r,\text{AN}}$) by PAOs.

$$S_{r,\text{AN}} = f_{\text{sBOD5/P}} \times P_{r,\text{uptake}} = 10\,\text{mg sBOD}_5/\text{mg P} \times 4.02\,\text{mg P/L} = 40.2\,\text{mg sBOD}_5/\text{L}$$

   d. Calculate the heterotrophic biomass production concentration ($p_{x,\text{H,PAO}}$) due to growth of PAOs from Equation 10.15n.

$$p_{x,\text{H,PAO}} = \frac{Y_H S_{r,\text{AN}}}{1 + k_{d,\text{H,F}}\theta_c} = \frac{0.65\,\text{mg VSS/mg BOD}_5 \times 40.2\,\text{mg BOD}_5/\text{L}}{1 + 0.04\,\text{d}^{-1} \times 10\,\text{d}} = 18.7\,\text{mg VSS/L}$$

   Note: The actual growth of PAOs occurs in aerobic zone. It is part of the overall heterotrophic aerobic growth calculated later in Step 5.

   e. Estimate the fraction of PAOs ($f_{x,\text{H,PAO}}$) in the heterotrophic growth.

$$f_{x,\text{H,PAO}} = \frac{p_{x,\text{H,PAO}}}{p_{x,\text{H}}} = \frac{18.7\,\text{mg/L}}{82.2\,\text{mg/L}} = 0.23 \text{ or } 23\%;$$

   Note: The PAOs are about one-fourth of the total heterotrophic biomass growth. It is within the typical range of 20–30%.

4. Estimate the P content in PAOs.
   a. Calculate the P concentration removed ($P_{r,\text{synthesis}}$) due to cell synthesis of PAOs.

$$f_{r,\text{synthesis,PAO}} = P_{P,\text{synthesis}} \times P_{x,\text{H,PAO}} = 0.02\,\text{mg P/mg VSS} \times 18.7\,\text{mg VSS/L} = 0.37\,\text{mg P/L}$$

   b. Calculate the total P concentration removed ($P_{r,\text{PAO}}$) by PAOs through cell synthesis and "luxury" P uptake.

$$f_{r,\text{PAO}} = P_{r,\text{synthesis,PAO}} + P_{r,\text{uptake}} = (0.37 + 4.02)\,\text{mg P/L} = 4.39\,\text{mg P/L}$$

   c. Calculate the fraction of P removed by PAOs on the basis of total P removed ($f_{r,\text{PAO}}$) in the BNR system.

$$f_{r,\text{PAO}} = \frac{P_{r,\text{PAO}}}{p_{r,\text{BNR}}} = \frac{4.39\,\text{mg P/L}}{5.7\,\text{mg P/L}} = 0.77 \text{ or } 77\%;$$

   Note: Over three quarters of P removal is achieved by PAOs in the BNR system.

   d. Calculate the P content in PAOs ($f_{P,\text{PAO}}$).

$$f_{P,\text{PAO}} = \frac{P_{r,\text{PAO}}}{p_{x,\text{PAO}}} = \frac{4.39\,\text{mg P/L}}{18.7\,\text{mg VSS/L}} = 0.23\,\text{mg P/mg VSS}$$

Note: The calculated P content of 0.23 g P/g VSS is within the typically range of 0.2–0.3 g P/g VSS for PAOs.

5. Estimate the fraction of denitrifying heterotrophic biomass growth.
   a. Calculate from Equation 10.103l the NO_3-N concentration produced from nitrification (NO_x) in the aerobic zone. Use $N_{0,TKN} = 35$ mg/L, $N_{NH4\text{-}N} = 0.5$ mg/L, and $f_N = 0.12$ g N /g VSS (Example 10.153).

$$NO_x = N_{0,TKN} - N_{NH4\text{-}N} - f_N p_{x,BNR} = (35 - 0.5)\,\text{mg NH}_4\text{-N/L} - 0.12\,\text{mg N/mg VSS} \times 84.1\,\text{mg VSS/L} = 24.4\,\text{mg NH}_4\text{-N/L or } 24.4\text{ mg NO}_3\text{-N/L}$$

   b. Calculate from Equation 10.103l the NO_3-N concentration removed by denitrification ($R'_{NO3\text{-}N}$) in the anoxic zone.

$$R'_{NO3\text{-}N} = NO_x - N_{NO3\text{-}N} = (24.4 - 4)\,\text{mg NO}_3\text{-N/L} = 20.4\,\text{mg NO}_3\text{-N/L}$$

   c. Calculate from Equation 10.100d the net oxygen consumption ratio for NO_3-N ($CR_{NO3\text{-}N}$). Use the theoretical oxygen consumption ratio for NO_3^-, $OE_{NO3\text{-}N} = 2.86$ mg O_2/mg NO_3-N (Example 10.139), and $Y_H = 0.65$ mg VSS/mg BOD_5 (Example 10.153).

$$CR_{NO3\text{-}N} = \frac{OE_{NO3\text{-}N}}{1.6 - 1.42Y_H} = \frac{2.86\,\text{mg O}_2\text{/mg NO}_3\text{-N}}{1.6 - 1.42 \times 0.65\,\text{mg VSS/mg BOD}_5}$$
$$= 4.22\,\text{mg O}_2\text{/mg NO}_3\text{-N or } CR_{NO3\text{-}N} \approx 4.22\,\text{mg sBOD}_5\text{/mg NO}_3\text{-N}$$

   The concentration of organic consumption due to denitrification in the preanoxic zone ($S_{r,AX}$) is calculated from Equation 10.103p.

$$S_{r,AX} = CR_{NO} R'_{NO3\text{-}N} = 4.22\,\text{mg sBOD}_5\text{/mg NO}_3\text{-N} \times 20.4\,\text{mg NO}_3\text{-N/L} = 86.1\,\text{mg sBOD}_5\text{/L}$$

   d. Calculate the growth of heterotrophic organisms ($p_{x,H}$) due to consumption of BOD_5 during denitrification.

$$p_{x,H,AX} = \frac{Y_H S_{r,AX}}{1 + k_{d,H,F}\theta_c} = \frac{0.65\,\text{mg VSS/mg BOD}_5 \times 86.1\,\text{mg BOD}_5\text{/L}}{1 + 0.04\,\text{d}^{-1} \times 10\,\text{d}} = 40.0\,\text{mg VSS/L}$$

   e. Estimate the fraction of ng ($f_{x,H,AX}$) in the heterotrophic growth.

$$f_{x,H,AX} = \frac{p_{x,H,AX}}{p_{x,H}} = \frac{40.0\,\text{mg/L}}{82.2\,\text{mg/L}} = 0.49 \text{ or } 49\%;$$

   Note: Nearly half of the overall heterotrophic growth occurs during denitrification in the anoxic zone.

6. Determine the fermentation requirement in the influent for effective BNR performance.
   a. Calculate the overall consumption of sBOD5 ($S_{r,AN/AX}$) in both anaerobic and anoxic zones.

$$S_{r,AN/AX} = S_{r,AN} + S_{r,AX} = (40.2 + 86.1)\,\text{mg sBOD}_5\text{/L} = 126\,\text{mg sBOD}_5\text{/L}$$

   b. Calculate the fermentation efficiency ($f_{fm,required}$) required to produce the desired SCVFA concentration.

$$f_{fm,required} = \frac{S_{uptake}}{S_{0,sBOD5}} = \frac{126\,\text{mg sBOD}_5\text{/L}}{150\,\text{mg sBOD}_5\text{/L}} = 0.84 \text{ or } 84\%;$$

Note: Fermentation efficiency over 80% in the influent is required to produce the desired SCVFA concentration as $sBOD_5$ for organics uptake in the anaerobic zone and denitrification in the anoxic zone.

7. Estimate the fraction of non-PAO aerobic heterotrophic biomass growth.
   a. Calculate the overall BOD_5 removal ($S_{r,OX}$), including the SCVFAs stored as PHA in the cell mass of PAOs in the aerobic zone.

$$S_{r,OX} = S_0 - S_{r,AX} - S = (180 - 86.1 - 3)\,\text{mg}\,BOD_5/\text{L} = 90.9\,\text{mg}\,BOD_5/\text{L}$$

   b. Calculate the overall aerobic heterotrophic growth ($p_{x,H,OX}$) in the aerobic zone.

$$p_{x,H,OX} = \frac{Y_H S_{r,OX}}{1 + k_{d,H,F}\theta_c} = \frac{0.65\,\text{mg VSS/mg}\,BOD_5 \times 90.9\,\text{mg}\,BOD_5/\text{L}}{1 + 0.04\,\text{d}^{-1} \times 10\,\text{d}} = 42.2\,\text{mg VSS/L}$$

   c. Calculate the non-PAO aerobic heterotrophic growth ($p_{x,H,OX}$) in the aerobic zone.

$$p_{x,H,\text{non-PAO}} = p_{x,H,OX} - p_{x,H,\text{non-PAO}} = (42.2 - 18.7)\,\text{mg VSS/L} = 23.5\,\text{mg VSS/L}$$

   d. Estimate the fraction of non-PAO heterotrophs ($f_{x,H,\text{non-PAO}}$) in the heterotrophic growth.

$$f_{x,H,\text{non-PAO}} = \frac{p_{x,H,\text{non-PAO}}}{p_{x,H}} = \frac{23.5\,\text{mg/L}}{82.2\,\text{mg/L}} = 0.29 \text{ or } 29\%;$$

   e. Calculate the P concentration removed ($P_{r,\text{synthesis,non-PAO}}$) due to cell synthesis of non-PAOs.

$$P_{r,\text{synthesis,non-PAO}} = P_{P,\text{synthesis}} \times p_{X,H,\text{non-PAO}} = 0.02\,\text{mg P/mg VSS} \times 23.5\,\text{mg VSS/L} = 0.47\,\text{mg P/L}$$

8. Summarize the calculation results.

   The calculation results for heterotrophic and autotrophic growths with P contents in the BNR system are summarized in Table 10.60.

TABLE 10.60 Heterotrophic and Autotrophic Growths in BNR System (Example 10.154)

Parameter	Heterotrophic Growth				Autotrophic (AOB) Growth	Overall
	PAO	Non-PAO	Denitrifier	Total		
Biomass growth						
mg VSS/L	18.7[a]	23.5[b]	40.0[c]	82.2[d]	1.90[e]	84.1[f]
% of heterotrophic growth	22.7	28.6	48.7	100	–	–
% of overall growth	22.2	27.9	47.6	97.7	2.3	100
P content in biomass, g P/g VSS	0.23[g]	0.02[h]	0.02[h]	0.069[i]	0.02[h]	0.068[j]

[a] See Step 3.d.
[b] See Step 7.c.
[c] See Step 5.d.
[d] See Example 10.153, Step 1.a.
[e] See Example 10.153, Step 1.b.
[f] See Example 10.153, Step 1.c.
[g] See Step 4.d.
[h] It is the Org-P content in the biomass due to cell synthesis that is given in the problem statement of Example 10.153.
[i] See Example 10.153, Step 3.b.
[j] See Example 10.153, Step 3.a.

EXAMPLE 10.155: SUBSTRATE REMOVALS UNDER ANAEROBIC, ANOXIC, AND AEROBIC CONDITIONS IN BNR SYSTEM

Develop a simplified breakdown of substrate removals under anaerobic, anoxic, and aerobic conditions in a BNR facility. These substrates are BOD5, nitrogen (NH_4-N, NO_2-N, and NO_3-N), and phosphorus (Ortho-P and Org-P). Ignore the soluble organic nitrogen in the effluent. Apply the influent, effluent, and process design data given in Examples 10.153 and 10.154. Also, develop the additional information on the breakdown of substrate removals that is not covered in these examples.

Solution

1. Prepare a table to summarize the breakdown of substrate removal data.

   Substrates BOD_5, nitrogen (N), and phosphorus (P) removal data under anaerobic, anoxic, and aerobic conditions are summarized in in Table 10.61. Brief discussion and procedure used to develop these numbers are presented below.

TABLE 10.61 Substrate Removals under Anaerobic, Anoxic, and Aerobic Conditions in BNR System (Example 10.155)

Parameter[a]	Influent	Substrate Removal in BNR System				Effluent[b]
		Anaerobic	Anoxic	Aerobic	Overall	
BOD_5						
mg BOD_5/L	180	40.2[c]	86.1[d]	50.7[e]	177	3
% of influent concentration	100	22.3	47.8	28.2	98.3	1.7
N						
mg N/L	35	0[f]	25.2[g]	5.3[h]	30.5	4.5
% of influent concentration	100	0	72.0	15.1	87.1	12.9
P						
mg P/L	6	0[i]	0.8[j]	4.9[k]	5.7	0.3
% of influent concentration	100	0	13.3	81.7	95.0	5.0

[a] The substrate removal is expressed on the basis of influent flow reaching the BNR system.

[b] Only soluble concentration of different parameters is included in the table. Particulate substrates such as VSS, Org-N, and Org-P are discharged as part of TSS in the effluent. The loss of these substrates in the effluent are considered substrate removals by the BNR system.

[c] See Example 10.154, Step 3.c.

[d] See Example 10.154, Step 5.c.

[e] See Step 4.a.

[f] See Step 2.b.

[g] See Step 3.b.

[h] See Step 4.b.

[i] See Step 2.b.

[j] See Step 3.c.

[k] See Step 4.c.

2. Summary of the substrate removals under anaerobic conditions.
   a. BOD_5 removal.

      The $sBOD_5$ consumption is associated with the SCVFA uptake and simultaneously release of Ortho-P by PAOs under the anaerobic condition. The stored organics are utilized later for cell synthesis and P uptake by APOs in the aerobic condition. The amount of $sBOD_5$ consumed, $S_{r,AN} =$ 40.2 mg $sBOD_5$/L.
   b. N and P removals.

Both N and P in the influent are not removed under the anaerobic condition. Actually, soluble P concentration is increased due to Ortho-P release by APOs. NH_4-N consumed by APOs is included in the overall nitrogen requirement for heterotrophic growth under aerobic condition.

3. Summary of the substrate removals under anoxic conditions.
   a. BOD_5 removal:

   The $sBOD_5$ is consumed by the denitrifying heterotrophic bacteria (or denitrifiers) in the denitrification process. The amount of $sBOD_5$ removed, $S_{r,AX} = 86.1$ mg $sBOD_5$/L.

   b. N removal:

   The NO_3-N is reduced to N_2 as an electron acceptor by the denitrifiers for oxidizing organics. A removal of 20.4 mg NO_3-N/L is estimated in Example 10.154. NH_4-N is also used for new cell synthesis of the denitrifiers. The NH_4-N consumption for cell synthesis ($N_{r,synthesis,AX}$) is calculated from $p_{x,H,AX} = 40.0$ mg VSS/L and $f_N = 0.12$ g N/g VSS.

$$N_{r,synthesis,\,AX} = f_N p_{X,H,AX} = 0.12\,\text{mg N/mg VSS} \times 40.0\,\text{mg VSS/L} = 4.80\,\text{mg N/L}$$

   The overall nitrogen removal ($N_{r,AX}$) under the anoxic condition is given below.

$$N_{r,AX} = R'_{NO3\text{-}N} + N_{r,synthesis,AX} = (20.4 + 4.8)\,\text{mg N/L} = 25.2\,\text{mg N/L}$$

   c. P removal:

   Assume no P uptake occurs under the anoxic condition, P is only removed by cell synthesis ($P_{r,synthesis,AX}$), and it is calculated from $p_{x,H,AX} = 40.0$ mg VSS/L and $f_{P,synthesis} = 0.02$ g P/g VSS.

$$P_{r,synthesis,AX} = P_{P,synthesis} \times p_{X,H,AX} = 0.02\,\text{mg P/mg VSS} \times 40.0\,\text{mg VSS/L} = 0.80\,\text{mg P/L}$$

4. Summary of the substrate removals under aerobic conditions.
   a. BOD_5 removal:

   The consumption of $sBOD_5$ due to SCVFA uptake by PAOs has been considered under anaerobic condition. Therefore, the removal of BOD_5 ($S_{r,OX}$) is determined only from the utilization of remaining BOD_5 by the non-PAOs heterotrophic aerobic growth.

$$S_{r,OX} = S_0 - S_{r,AN} - S_{r,AX} - S = (180 - 40.2 - 86.1 - 3)\,\text{mg BOD}_5\text{/L} = 50.7\text{mg BOD}_5\text{/L}$$

   b. N removal:

   NH_4-N is oxidized to NO_3-N by AOB in nitrification process. No direct removal of nitrogen occurs since both NH_4-N and NO_3-N are in soluble forms. Therefore, N removal is only caused by cell synthesis of both heterotrophic and autotrophic organisms under aerobic conditions. The overall nitrogen removal ($N_{r,OX}$) is calculated from $p_{x,H,OX} = 42.2$ mg VSS/L, $p_{x,N} = 1.90$ mg VSS/L and $f_N = 0.12$ g N/g VSS.

$$N_{r,OX} = f_N(p_{x,H,OX} + p_{x,N}) = 0.12\,\text{mg N/mg VSS} \times (42.2 + 1.90)\,\text{mg VSS/L} = 5.3\,\text{mg N/L}$$

   c. P removal:

   P removal is achieved by "luxury" P uptake by PAOs, and for cell synthesis by both heterotrophic and autotrophic growths. The overall nitrogen removal ($P_{r,OX}$) is calculated from $P_{r,PAO} = 4.39$ mg P/L, $p_{x,H,non\text{-}PAO} = 23.5$ mg VSS/L, $p_{x,N} = 1.90$ mg VSS/L and $f_{P,synthesis} = 0.02$ g P/g VSS.

$$\begin{aligned} P_{r,OX} &= P_{r,PAO} + f_{P,synthesis}(p_{x,H,non-PAO} + p_{x,N}) \\ &= 4.39\,\text{mg P/L} + 0.02\,\text{mg P/mg VSS} \times (23.5 + 1.90)\,\text{mg VSS/L} = 4.90\,\text{mg VSS/L} \end{aligned}$$

5. General comments on the substrate removals in the BNR system.
   a. BOD_5 removal:
   Removal of organics (BOD_5) is achieved under all three operating conditions. An uptake of about 20–25% of total BOD_5 in the influent occurs when the PAOs store the $sBOD_5$ as food in the form of PHA in the cell mass under the anaerobic condition. Almost 40–50% of the total BOD_5 removal is achieved under the anoxic condition. Over 50% of total BOD_5 is utilized under the aerobic condition, including utilizations of stored organics by PAOs, and remaining BOD_5 by non-PAO heterotrophic growth (nearly 30%). A high BOD_5 removal efficiency above 95–98% may be expected in the BNR system.
   b. N removal:
   A nitrogen removal over 70–75% is achieved by denitrifying heterotrophs due to denitrification of NO_3-N and consumption of NH_4-N for cell synthesis under the anoxic condition. High recycle flow is required for return of NO_3-N from the aerobic to anoxic zone after effective oxidation of NH_4-N. Additionally, about 15–20% of total nitrogen in the form of NH_4-N is fixed by heterotrophic and autotrophic growths under the aerobic condition. An overall N removal of 85–90% may be achieved in the BNR system.
   c. P removal:
   Over 80% of P removal is achieved by "luxury" P uptake by PAOs and cell synthesis by all heterotrophic and autotrophic growths under the aerobic condition. Another 10–15% of P is fixed by denitrifying heterotrophic growth under the anoxic condition. The P release by PAOs under the anaerobic condition is necessary for a successful EBPR process. The overall P removal may be 90–95% in a BNR system.

EXAMPLE 10.156: DESIGN SRT OF A BNR FACILITY

A BNR facility uses A^2/O process for removals of phosphorus, nitrogen, BOD, and TSS. Calculate SRT (θ_c) for the BNR facility based on the autotrophic growth required for nitrification. Also, verify if the requirement for soluble BOD_5 and NH_4-N concentrations are met in the effluent. The biological kinetic coefficients under the field operating conditions and desired effluent quality data are given below.

Kinetic coefficients for nitrifiers: $\mu_{max,N.F} = 0.55$ mg VSS/mg VSS·d, and $k_{d,N.F} = 0.12\ d^{-1}$, $Y_N = 0.17$ mg VSS/mg NH_4-N, and $K_{NH4\text{-}N} = 0.50$ mg NH_4-N/L.
Kinetic coefficients for aerobic heterotrophs: $Y_{BOD} = 0.65$ mg VSS/mg BOD_5, $k_{d,BOD} = 0.04\ d^{-1}$, $k_{BOD} =$ 5 mg BOD_5/mg VSS·d, and $K_{S,BOD5} = 60$ mg BOD_5/L.
Desired effluent quality: $S = 3$ mg $sBOD_5$/L and $N = 0.5$ mg NH_4-N/L.

Note: Procedures for obtaining nitrification kinetic coefficients under the field operating conditions are presented in Example 10.126.

Solution

1. Determine the theoretical SRT ($\theta_{c,N}$).
   Calculate the specific growth rate for nitrifiers ($\mu_{N.F}$) under the field operating condition from Equation 10.89a. Use $\mu_{max,N.F} = 0.55$ mg VSS/mg VSS·d, $K_{NH4\text{-}N} = 0.50$ mg NH_4-N/L (from Table 10.49), and the desired effluent NH_4-N concentration $N = 0.5$ mg NH_4-N/L.

$$\mu_{N,F} = \frac{\mu_{max,N,F} N}{K_{NH4\text{-}N} + N} = \frac{0.55\ \text{mg VSS/mg VSS·d} \times 0.5\ \text{mg NH}_4\text{-N/L}}{(0.5 + 0.5)\ \text{mg NH}_4\text{-N/L}} = 0.275\ \text{mg VSS/mg VSS·d}$$

Calculate the specific net growth rate for nitrifiers ($\mu'_{AOB,\,F}$) from Equation 10.89b. Use $k_{d,N,F} = 0.12\ d^{-1}$.

$$\mu'_{N,F} = \mu_{N,F} - k_{d,N,F} = (0.275 - 0.12)\ d^{-1} = 0.155\ d^{-1}$$

2. Determine the design SRT from the theoretical SRT.

Calculate the theoretical SRT ($\theta_{c,N}$) from Equation 10.91a.

$$\theta_{c,N} = \frac{1}{\mu'_{N,F}} = \frac{1}{0.155\ d^{-1}} = 6.5\ d$$

Provide $SF_P = 1.5$ for the uncertainty in process design and operation. Calculate the design SRT ($\theta_{c,N}^{design}$) from Equation 10.91c.

$$\theta_{c,N}^{design} = SF_P\,\theta_{C,N} = 1.5 \times 6.5\ d = 9.75\ d$$

Provide a design SRT $\theta_c^{design} = 10\ d$

3. Calculate the concentrations of soluble BOD_5 and NH_4-N expected in the effluent.

Apply Equation 10.15h to calculate the soluble BOD_5 concentration (S) and NH_4-N concentration (N) in the effluent.

$$S = \frac{K_{s,BOD}(1 + k_{d,BOD}\theta_c^{design}}{\theta_c^{design}(Y_{BOD}k_{BOD} - k_{d,BOD}) - 1} = \frac{60\ \text{mg BOD}_5/\text{L} \times (1 + 0.04\ d^{-1} \times 10\ d)}{10\ d \times (0.65 \times 5\ d^{-1} - 0.04\ d^{-1}) - 1} = 2.7\ \text{mg BOD}_5/\text{L}$$

$$N = \frac{K_{NH4\text{-}N}(1 + k_{d,N,F}\theta_c^{design})}{\theta_c^{design}(\mu_{max,N,F} - k_{d,N,F}) - 1} = \frac{0.5\text{mg NH}_4\text{-N/L} \times (1 + 0.12\ d^{-1} \times 10\ d)}{10\ d \times (0.55\ d^{-1} - 0.12\ d^{-1}) - 1}$$
$$= 0.33\ \text{mg NH}_4\text{-N/L}$$

Both the effluent soluble BOD_5 concentration of 2.7 mg/L and NH_4-N concentration of 0.33 mg/L are below the required values.

Note: See Example 10.127 for information about different methods that may be used to determine the design SRT.

EXAMPLE 10.157: HRT AND VOLUME OF ANAEROBIC, ANOXIC, AND AEROBIC ZONES

A BNR facility uses A^2/O process. Calculate the HRTs, volumes, and percent volumes of anaerobic, anoxic, and aerobic zones. Apply the influent, effluent, and process design data given and developed in Examples 10.153 through 10.156. Additional data applicable to this example are given below.

Influent: Average design flow = 4000 m^3/d, COD = 320 mg/L.

Process design data: MLSS = 4000 mg/L, return sludge ratio $R_{RS} = 0.6$, and specific denitrification rate $SDNR = 0.09$ mg NO_3-N/mg VSS·d.

Note: Procedures for determining *SDNR* are presented in Example 10.143.

Solution

1. Determine the HRT and volume of the anaerobic zone.
   a. Evaluate the fermentation condition in the influent.

$$\text{The COD/BOD}_5\text{ ratio in the influent,COD/BOD}_5 = \frac{320\ \text{mg/L}}{180\ \text{mg/L}} = 1.8$$

A ratio <2 indicates that the influent is well fermented. Assume that the fermentation efficiency $f_{fm} = 0.9$ for $sBOD_5 = 150$ mg/L (Example 10.154). Estimate the SCVFA concentration from $sBOD_5$ in the influent.

$$SCVFA = f_{fm} \times sBOD_5 = 0.9 \times 150\,\text{mg sBOD}_5/\text{L} = 135\,\text{mg sBOD}_5/\text{L}$$

Note: The estimated $SCVFA = 135$ mg $sBOD_5$/L is higher than the desired SCVFA concentration ($S_{r,AN/AX}$) of 126 mg $sBOD_5$/L for organics uptake in the anaerobic zone and denitrification in the anoxic zone (Example 10.154, Step 6).

b. Estimate the TP concentrations in the effluent.

The COD/TP ratio in the influent,

$$\text{COD/TP} = \frac{320\,\text{mg/L}}{6\,\text{mg/L}} = 53 \quad (40 < \text{COD/TP} < 60)$$

An effluent TP ≈ 0.5 mg/L is estimated from Figure 10.106 at the COD/TP ratio of 53. It is the same as that estimated in Example 10.153, Step 4.

c. Estimate the HRT required in the anaerobic zone (θ_{AN}).

The HRT is estimated from Equation 10.109 at $TP_0 = 6$ mg/L and $SP_e = 0.3$ mg/L (Example 10.153).

$$\theta_{AN} = 0.81\,(TP_0 - SP_e) - 2.11 = 0.81\,\text{h} \cdot \text{L/mg} \times (6 - 0.3)\,\text{mg/L} - 2.11\,\text{h} = 2.5\,\text{h}$$

The calculated HRT of 2.5 h from Equation 10.109 seems over conservative. In Authors' judgement, a $\theta_{AN} = 1.5$ h is sufficient for anaerobic zone under the fermentation condition estimated for the influent. It equals the upper end of the typical range of 0.5–1.5 h for A^2O process (Table 10.57).

d. Determine the volume of anaerobic zone (V_{AN}).

$$V_{AN} = \theta_{AN} Q = 1.5\,\text{h} \times 4000\,\text{m}^3/\text{d} \times \frac{1\,\text{d}}{24\,\text{h}} = 250\,\text{m}^3$$

2. Determine the HRT and volume of the anoxic zone.

a. Determine the nitrate nitrogen removal rate ($R_{NO3\text{-}N}$).

The NO_3-N concentration removed by denitrification in the anoxic zone ($R'_{NO3\text{-}N} = 20.4$ mg NO_3-N/L) is obtained from Example 10.154, Step 5.b. Determine $R_{NO3\text{-}N}$ at influent flow $Q = 4000\,\text{m}^3/\text{d}$.

$$R_{NO3\text{-}N} = QR'_{NO3\text{-}N} = 4000\,\text{m}^3/\text{d} \times 20.4\,\text{g NO}_3\text{-N/m}^3 \times 10^{-3} \times \text{kg/g} = 81.6\,\text{kg NO}_3\text{-N/d}$$

b. Determine the concentration of biomass as MLVSS (X) in the mixed liquor.

X is calculated from VSS/TSS = 0.7 mg VSS/mg TSS and MLSS = 4000 mg/L.

$$X = \text{VSS/TSS ratio} \times \text{MLSS} = 0.7\,\text{mg VSS/mg TSS} \times 4000\,\text{mg TSS/L} = 2800\,\text{mg VSS/L}$$

c. Determine the volume of anoxic zone (V_{AX}).

V_{AX} required for the desired denitrification is calculated from Equation 10.102. Use $SDNR = 0.09$ mg NO_3-N/mg VSS·d.

$$V_{AX} = \frac{R_{NO3\text{-}N}}{SDNR\,X} = \frac{81.6\,\text{kg NO}_3\text{-N/d} \times 10^3\,\text{g/kg}}{0.09\,\text{g NO}_3\text{-N/g VSS·d} \times 2800\,\text{g VSS/m}^3} = 324\,\text{m}^3$$

Provide a design volume $V_{AX} = 333\,\text{m}^3$.

d. Determine the HRT in anoxic zone (θ_{AX}).

Calculate HRT of the anoxic zone.

$$Q_{AX} = \frac{V_{AX}}{Q} = \frac{333\ \text{m}^3}{4000\ \text{m}^3} \times \frac{24\text{h}}{\text{d}} = 2\ \text{h}$$

Note: The HRT of 2 h is within the typical range of 1–3 h for the anoxic zone in A^2O process (Table 10.57).

e. Estimate the internal recycle rate required from the aeration basin to the anoxic zone (*IR*).

Assume $N_{NO3\text{-}N} >> N_{NO3\text{-}N,AX}$ or $N_{NO3\text{-}N,AX} \approx 0$ and determine *IR* from Equation 10.103k using the NO_3-N concentration produced from nitrification in the aerobic zone, $NO_x = 24.4$ mg NO_3-N/L (Example 10.154, Step 5.a).

$$IR \approx \frac{NO_X}{N_{NO3\text{-}N}} - (1 + R_{RS}) = \frac{24.4\ \text{mg N}/L}{4\ \text{mg NO}_3\text{-N}/L} - (1 + 0.6) = 4.5$$

Provide a design $IR = 5$. It is slightly higher than the typical range of 1–4 for the anoxic zone in A^2O process (Table 10.57).

3. Determine the HRT of the aerobic zone.

The aerobic zone should be large enough to provide (a) desired nitrification and (b) maximum BOD_5 removal. The longer HRT required by either of two processes is used for design.

a. Determine the concentrations of nitrifiers (X_N) and heterotrophs (X_H) in the mixed liquor.

X_N in the mixed liquor is calculated using $f_{x,N} = 2.3\%$ (Example 10.154, Step 2).

$$X_N = f_N X = 0.023 \times 2800\ \text{mg VSS/L} = 64.4\ \text{mg VSS/L}$$

X_H in the mixed liquor, $X_H = X - X_N = (2800 - 64.4)$ mg VSS/L $= 2736$ mg VSS/L

b. Determine the HRT required for the desired nitrification (θ_N) in the aerobic zone.

The required θ_N is calculated from Equation 10.15g using $p_{x,N} = 1.90$ mg VSS/L (Example 10.153, Step 1) and the design SRT $\theta_c^{\text{design}} = 10$ d (Example 10.156, Step 2).

$$\theta_N = \frac{p_{x,N}}{X_N}\theta_c^{\text{design}} = \frac{1.90\ \text{mg VSS/L}}{64.4\ \text{mgVSS/L}} \times 10\ \text{d} \times \frac{24\ \text{h}}{1\ \text{d}} = 7.1\ \text{h}$$

c. Determine the HRT required for removal of BOD_5 (θ_{BOD}) in the aerobic zone.

The required θ_{BOD} is calculated from Equation 10.15g using $p_{x,H,OX} = 42.2$ mg VSS/L (Example 10.154, Step 7.b) and $\theta_c^{\text{design}} = 10$ d.

$$\theta_{BOD} = \frac{p_{x,H,OH}}{X_H}\theta_c^{\text{design}} = \frac{42.2\ \text{mg VSS/L}}{2736\ \text{mg VSS/L}} \times 10\ \text{d} \times \frac{24\ \text{h}}{1\ \text{d}} = 3.7\ \text{h}$$

d. Select the design HRT (θ_{OX}) for the aerobic zone.

The HRT of aerobic zone should be longer than 7.1 h. Provide a design $\theta_{OX} = 7.5$ h.

Note: See Example 10.128 for information about a different approach that may also be used to determine the design HRT.

4. Determine the volume of the aerobic zone.

Determine the volume of aerobic zone (V_{OX}).

$$V_{OX} = \theta_{OX}\, Q = 7.5\ \text{h} \times 4000\ \text{m}^3/\text{d} \times \frac{1\ \text{d}}{24\ \text{h}} = 1250\ \text{m}^3$$

5. Determine the total volume and overall HRT of the BNR facility.

   The total volume (V_{BNR}) of BNR facility, $V_{BNR} = V_{AN} + V_{AX} + V_{OX} = (250 + 333 + 1250)\ m^3 = 1833\ m^3$

   The overall HRT (θ_{BNR}) of BNR facility, $\theta_{BNR} = \theta_{AN} + \theta_{AX} + \theta_{OX} = (1.5 + 2 + 7.5)\ h = 11\ h$

6. Determine the percent volume of anaerobic, anoxic, and aerobic zones in the BNR facility.

   Calculate the percent of each zone in the BNR facility.

$$\text{Percent of anaerobic zone,} \quad f_{V,AN} = \frac{250\ m^3}{1833\ m^3} \times 100\% = 14\%$$

$$\text{Percent of anoxic zone,} \quad f_{V,AX} = \frac{333\ m^3}{1833\ m^3} \times 100\% = 18\%$$

$$\text{Percent of aerobic zone,} \quad f_{V,OX} = \frac{1250\ m^3}{1833\ m^3} \times 100\% = 68\%$$

EXAMPLE 10.158: THEORETICAL OXYGEN REQUIREMENT (ThOR) IN A BNR FACILITY

Determine the ThOR in a BNR facility using A^2/O process. The influent and effluent, and process design data are as presented or developed in Examples 10.153, 10.154, and 10.157. Use this information to develop ThOR. Assume $BOD_5/BOD_L = 0.68$.

Solution

1. Determine the ultimate oxygen concentration demand (*UOR*) in the aerobic zone.

   The UOR in the aerobic zone is due to (a) BOD_5 removal by the aerobic heterotrophic growth, (b) nitrification by autotrophic growth, and (c) oxygen consumption by endogenous respiration of biomass produced. Apply Equation 10.93b to calculate UOR. Use (a) the overall BOD_5 removal in the aerobic zone, $S_{r,OX} = 90.9$ mg BOD_5/L (Example 10.154, Step 7.a), (b) the NO_3-N concentration produced from nitrification in the aerobic zone, $NOx = 24.4$ mg NH_4-N/L (Example 10.154, Step 5.a), and (c) the combined growth of heterotrophic and autotrophic organisms in the BNR facility, $p_{x,BNR} = 84.1$ mg VSS/L (Example 10.153, Step 1.c).

$$UOR_{OX} = \frac{S_{r,OX}}{BOD_5/BOD_L} + 4.57\ NO_X - 1.42 p_{X,BNR}$$

$$= \frac{90.9\ \text{mg}\ BOD_5/L}{0.68\ \text{mg}\ BOD_5/\text{mg}\ BOD_L} + 4.57\ \text{mg}\ O_2/\text{mg}\ NH_4\text{-N} \times 24.4\ \text{mg}\ NH_4\text{-N/L}$$

$$- 1.42\ \text{mg}\ O_2/\text{mg VSS} \times 84.1\ \text{mg VSS/L}$$

$$= (134 + 112 - 119)\ \text{mg}\ O_2/L = 127\ \text{mg}\ O_2/L \text{ or } 127\ g\ O_2/m^3$$

2. Determine the theoretical oxygen requirement (*ThOR*) in the aerobic zone.

   Calculate the *ThOR* from Equation 10.48b.

$$ThOR_{OX} = Q \times UOR_{OX} = 4000\ m^3/d \times 127\ g\ O_2/m^3 \times 10^{-3} kg/g = 508\ kg\ O_2/d$$

Note: See also Examples 10.70 and 10.130 for calculations of *UOR*, *ThOR* and standard oxygen requirement (*SOR*).

EXAMPLE 10.159: ALKALINITY BALANCE IN A BNR FACILITY

The alkalinity in a BNR facility is destroyed by nitrification but partly recovered by denitrification. Use the data developed in Example 10.154, and determine the net alkalinity concentration remaining in the effluent. Also, comment on pH of the effluent. The initial alkalinity concentration of the influent is 200 mg/L as $CaCO_3$.

Solution

1. Determine the alkalinity destroyed by nitrification in the aerobic zone.

   The total alkalinity consumed by nitrification ($Alk_{consumed}$) is calculated from (a) the NO_3-N concentration produced from nitrification in the aerobic zone, $NOx = 24.4$ mg NH_4-N/L (Example 10.154, Step 5.a) and (b) the stoichiometric alkalinity destruction ratio of 7.14 g alkalinity as $CaCO_3$ per g NH_4-N oxidized (Example 10.124).

$$\begin{aligned}Alk_{consumed} &= Alk_{oxidation} NO_X = 7.14\,\text{mg Alk as CaCO}_3/\text{mg NH}_4\text{-N} \times 24.4\,\text{mg NH}_4\text{-N/L}\\ &= 174\,\text{mg/L asCaCO}_3\end{aligned}$$

   Note: See the calculations of the ratio of 7.14 mg Alk as $CaCO_3$/mg NH_4-N in Example 10.124 and supplemental alkalinity for nitrification in Example 10.131.

2. Determine the amount of alkalinity recovered by denitrification.

   The stoichiometric alkalinity recovery ratio is 3.57 g alkalinity as $CaCO_3$/g NO_3-N (Example 10.138). The NO_3-N concentration removed by denitrification in the anoxic zone, $R'_{NO3\text{-}N} = 20.4$ mg NO_3-N/L (Example 10.154, Step 5.b.). The alkalinity recovery from denitrification is calculated is calculated below.

$$\begin{aligned}Alk_{recovered} &= Alk_{NO3\text{-}N,recovered} R'_{NO3\text{-}N} = 3.57\,\text{mg CaCO}_3/\text{mg NO}_3\text{-N} \times 20.4\,\text{mg NO}_3\text{-N/L}\\ &= 73\,\text{mg CaCO}_3/\text{L}\end{aligned}$$

   Note: See the calculations of the ratio of 3.57 mg Alk as $CaCO_3$/mg NO_3-N in Example 10.138 and alkalinity recovery from denitrification in Example 10.143.

3. Determine the alkalinity remaining in the effluent.

   The alkalinity concentration remaining in the effluent ($Alk_{remaining}$) from the BNR facility is calculated from the alkalinity balance.

$$Alk_{remaining} = Alk_{inf} + Alk_{recovered} - Alk_{consumed} = (200 + 73 - 174)\,\text{mg CaCO}_3/\text{L} = 99\,\text{mg CaCO}_3/\text{L}$$

4. Comment on the pH of effluent.

   The remaining alkalinity concentration of 99 mg/L as $CaCO_3$ is sufficient to maintain the desired bicarbonate buffer. The pH will remain in the desirable range of around 7.2.

EXAMPLE 10.160: PRECIPITATION OF PHOSPHORUS FROM WAS

The phosphorus rich WAS from a BNR facility will release phosphorous in the gravity thickening, anaerobic digestion, and other solids handling facilities. This will cause a buildup of high Ortho-P level in the BNR facility influent due to recycle of sidestreams. For this reason, the released Ortho-P must be stripped from the WAS before sending the biosolids for processing in the solids handing area. A phosphorus stripping system similar to that shown in Figure 10.116 is designed to release and precipitation of Ortho-P by alum coagulation. The stripper tank receives P-rich WAS from a BNR facility and supernatant from a gravity thickener to provide the needed SCVFAs. The WAS contains 1200 kg TSS/d and the VSS/TSS ratio is 0.7. The stripped P is captured in the stripper supernatant and then treated with liquid alum to precipitate P that is separated in a clarifier. The commercial alum solution contains 50% $Al_2(SO_4)_3{\cdot}14\,H_2O$ by weight and the bulk density (specific gravity) is 1320 kg/m^3. The alum dose is determined at an Al^{3+}/P molar ratio of 2.5 to precipitate 98% of Ortho-P released from the WAS. The average P contents of P-rich WAS and P-stripped solids are 12% and 3% of VSS by weight, respectively. Ignore the solids carried in the stripper supernatant. To simplify the calculations, it is also assumed that 90% of the Ortho-P released is captured in the supernatant from the stripper, and 95% of the precipitated solids are removed from the clarifier. Determine (a) the Ortho-P released from the stripper, (b) amount of Ortho-P in the supernatant stripper, (c) the amount of commercial alum solution required to precipitate the Ortho-P released, (d) capacity of alum feed system, and (e) total precipitated solids removed in the clarifier.

Solution

1. Determine the amount of Ortho-P released in the anaerobic stripper.

Amount of Ortho-P released = (0.12 – 0.03) kg P/Kg VSS × 0.7 kg VSS/kg TSS × 1200 kg TSS/d
= 75.6 kg P/d

Amount of Ortho-P in the supernatant = 0.9 × 75.6 kg P/d = 68.0 kg P/d

Amount of P-stripped biosolids = (1200−75.6) kg TSS/d = 1124 kg TSS/d

2. Determine the daily requirement of commercial alum solution at the desired Al^{3+}/P molar ratio of 2.5. The phosphorus precipitation reaction by alum is given by Equation 9.22a.

$$Al^{3+}PO_4^{3-} \leftrightarrow AlPO_4(s)$$

Amount of P precipitated as $AlPO_4$ = 0.98 × amount of Ortho-P in supernatant

= 0.98 × 68.0 kg P/d = 66.6 kg P/d

$$\text{Desired } Al^{3+}/P \text{ mass ratio} = Al^{3+}/P \text{ molar ratio} \times \frac{\text{MW of } Al^{3+}}{\text{MW of P}}$$

$$= \frac{2.5 \text{ moles } Al^{3+}}{1 \text{ mole P}} \times \frac{27\text{ g } Al^{3+}/\text{mole}}{31\text{ g P/mole}} = 2.2\text{ g}Al^{3+}/\text{g P}$$

Amount of Al^{3+} desired = Al^{3+}/P mass ratio × amount of P precipitated as $AlPO_4$

$$= 2.2\text{ kg } Al^{3+}/\text{kg P} \times 66.6\text{ kg P/d} = 147\text{ kg } Al^{3+}/\text{d}$$

$$\text{Amount of alum } (Al_2(SO_4)_3\,14H_2O) \text{ required} = \text{amount of } Al^{3+} \text{ required} \times \frac{\text{MW of } Al_2(SO_4)_3{\cdot}14H_2O}{2 \times \text{MW of } Al^{3+}}$$

$$= 147\text{ kg } Al^{3+}/\text{d} \times \frac{594\text{ g alum/mole}}{2 \times 27\text{ g } Al^{3+}/\text{mole}} = 1617\text{ kg alum/d}$$

$$\text{Amount of the commercial solution required} = \frac{\text{amount of alum}}{0.5\,\text{kg alum/kg alum solution}}$$

$$= \frac{1617\,\text{kg alum/d}}{0.5\,\text{kg alum/kg alum solution}} = 3234\,\text{kg alum solution/d}$$

$$\text{Daily volume of commercial solution required} = \frac{\text{amount of solution required}}{1320\,\text{kg alum soluction/m}^3}$$

$$= \frac{3234\,\text{kg alum/d}}{1320\,\text{kg alum solution/m}^3}$$

$$= 2.5\,\text{m}^3\,\text{alum solution/d}$$

3. Determine the capacity of alum feed system.

   Provide the design capacity of alum feed system at 150% excess.

   $$\text{Design capacity of alum feed system} = \text{safety factor} \times \text{daily volume of solution required}$$

   $$= 1.5 \times 2.5\,\text{m}^3/\text{d} = 3.75\,\text{m}^3/\text{d or } 2.6\,\text{L/min}$$

4. Determine the total solids generated by the precipitation processes.

   The total solids are produced due to (a) precipitations of $AlPO_4$ (Equation 9.22a and 9.22b precipitation of $Al(OH)_3$ (Equation 9.22b).

   a. Estimate the amount of $AlPO_4$ produced.

      One mole of Al^{3+} stoichiometrically reacts with one mole of PO_4^{3-}(or P) to yield one mole of $AlPO_4$ (Equation 9.22a).

$$\text{Amount of Al}^{3+}\text{used for precipitation of AlPO}_4 = \text{amount of P precipitated as AlPO}_4 \times \frac{\text{MW of Al}^{3+}}{\text{MW of P}}$$

$$= 66.6\,\text{kg P/d} \times \frac{27\,\text{g Al}^{3+}/\text{mole}}{31\,\text{g P/mole}} = 58\,\text{kg Al}^{3+}/\text{d}$$

$$\text{Amount of AlPO}_4\text{ precipitated} = \text{amount of Al}^{3+}\text{ required} \times \frac{\text{MW of AlPO}_4}{\text{MW of Al}^{3+}}$$

$$= 58\,\text{kg Al}^{3+}/\text{d} \times \frac{122\,\text{g AlPO}_4/\text{mole}}{27\,\text{g Al}^{3+}/\text{mole}}$$

$$= 262\,\text{kg AlPO}_4/\text{d or } 262\,\text{kg TSS/d}$$

   b. Estimate the amount of $Al(OH)_3$ produced.

      In accordance with Equation 9.22b, one mole of Al^{3+} stoichiometrically reacts with three mole of OH^- to yield one mole of $Al(OH)_3$.

$$Al^{3+} + 3\,OH^- \leftrightarrow Al(OH_3)(s)$$

$$\text{Amount of Al}^{3+}\text{ for precipitation of Al(OH)}_3 = \text{amount of Al}^{3+}\text{ desired amount of Al}^{3+}\text{precipitated as AlPO}_4$$

$$= (147 - 58)\,\text{kg Al}^{3+}/\text{d} = 89\,\text{kg Al}^{3+}/\text{d}$$

$$\text{Amount of Al(OH)}_3\text{ precipitated} = \text{amount of Al}^{3+}\text{ required} \times \frac{\text{MW of Al(OH)}_3}{\text{MW of Al}^{3+}}$$

$$= 89\,\text{kg Al}^{3+}/\text{d} \times \frac{78\,\text{g Al(OH)}_3/\text{mole}}{27\,\text{g Al}^{3+}/\text{mole}}$$

$$= 257\,\text{kg Al(OH)}_3/\text{d or } 257\,\text{kg TSS/d}$$

c. Estimate the total amount of precipitates produced.

$$\text{Total amount of TSS produced} = \text{amount of } AlPO_4 + \text{amount of } Al(OH)_3$$

$$= (262 + 257)\,\text{kg TSS/d} = 519\,\text{kg TSS/d}$$

$$\text{Amount of TSS removed by clarifier} = 0.95 \times \text{total amount of TSS produced}$$

$$= 0.95 \times 519\,\text{kg TSS/d} = 493\,\text{kg TSS/d}$$

Notes:

a. See also Section 9.6.2, and Examples 9.27 through 9.31 for detailed calculations of sludge generation, phosphorus removal, and alkalinity consumption using different chemicals in the chemically enhanced primary treatment (CEPT).

b. Phosphorus recovery is achieved by precipitation of struvite from the phosphorus stripped supernatant. Detailed discussion on this subject may be found in Section 13.10.3 and Examples 13.57, 13.58, and 13.61.

10.8.4 Computer Application for BNR Facility Design

The biological waste treatment facilities are designed to remove BOD, TSS, and phosphorus, and achieve nitrification and denitrification. Microorganisms such as PAOs, heterotrophs, and autotrophs (AOB and NOB) carry out these functions. Heterotrophs compete for carbonaceous organic substrates that may exist as soluble, insoluble, biodegradable, and nonbiodegradable. For these reasons, complex equations are needed to describe the microbial kinetics and various reactions. Computer modeling is needed to utilize large number of variables and equations to express reaction rates for substrate utilization, oxygen consumption, and biomass production.

ASM Models: International Water Association (IWA) formerly International Association on Water Quality (IAWQ) or International Association on Water Pollution Research and Control (IAWPRC) has proposed a series of Activated Sludge Models (ASMs) for suspended growth biological reactors with and without simulations for BNR. These models include ASM1, ASM2, ASM2d, and ASM3.[227–230] Some of the basic features of these models are (1) specific growth rate kinetic equations are used instead of substrate utilization rate, (2) COD is used as a common measure of oxygen consumption, substrate utilization, and biomass growth, (3) endogenous respiration is expressed by lysis-regrowth model instead of endogenous coefficient, (4) the wastewater characteristics are represented by 13 components such as inert and biodegradable soluble and particulate organic matter; slowly and readily biodegradable organics; debris from biomass death and lysis; soluble, particulate, and biodegradable organic nitrogen, ammonia, and nitrate nitrogen; and many others, and (5) matrix format is used to express model expression instead of large number of mathematical expressions involved. Model details, wastewater components, relationships, default values, and application may be found in References 2, 98, 182, 229 through 235.

Commercial Models: Several computer programs are commercially available that have the capability to design and analyze the system components of wastewater treatment facilities. Two commercially available computer programs are briefly discussed in this section. The most commonly used programs are Biowin and GPS-X™.[236,237] These programs include not only the basic and modified ASMs but also their own special models to provide comprehensive analysis platforms for design of wastewater treatment facilities. Generalized process configurations of many unit operation and processes are provided in a package that can be used to build specific and complex treatment plant configurations. These computer programs have capability to assist designers with a wide range of activities such as (a) troubleshooting and "what if" analyses, (b) plant design and retrofitting, (c) plant capacity analysis and re-rating, (d) control system design and tuning, (e) data and sensor validation, and (f) plant auditing and scheduling.

The modeling and simulation expertise available in these programs can be used to investigate many diverse treatment systems and components such as (a) BNR system design and operation, (b) CFSTR and PF reactors, and step-feed, (c) aeration system setup and control, (d) sludge recycle and wasting control, (e) internal recycle, material mass balance analyses and sidestreams, (f) settleability analysis, and (g) plant energy management. These programs have the capability to interface with Microsoft® Office applications and allow to export data, tables, and charts in word or spreadsheet files to create attractive professional reports.

Model Calibration: Model calibration is an important step before its application. Model calibration is achieved by comparing the model predicted values of many process parameters with the actual operating or desired design values. The input data include the wastewater characteristics, kinetic coefficients, and rates for different reactions; and stoichiometric ratios for different substrates. The comparisons may be made with (1) effluent quality parameters, (2) intermediate soluble substrate concentration data in anaerobic, anoxic, and aerobic zones, and (3) oxygen uptake rates at different stages.

Model Applications: Computer models provide a valuable tool to obtain useful information regarding activated-sludge processes. Some benefits of modeling are to (a) incorporate large number of biological kinetic coefficients, reactions, and wastewater characteristics that are impossible by manual calculations, (b) evaluate process performance in both steady-state and dynamic simulations, (c) design a wastewater treatment plant with a single-stage reactor as well as multiple-staged reactors with complete-mix to plug-flow regimes, (d) evaluate the treatment capacity of an existing facility, and (e) apply as a research tool to better understand the significance of process parameters such as oxygen consumption, substrate utilization, sludge production, and biodegradable, inert, soluble, suspended solids, and biomass debris upon the plant performance.

EXAMPLE 10.161: WASTEWATER CHARACTERIZATION FOR COMPUTER MODEL INPUT

Computer simulation models require large number of wastewater characterization data, stoichiometric parameters, and kinetic coefficients to predict accurately the performance of a suspended growth biological reactor. The influent constituents based on analytical results are given in Table 10.62. $BOD_5/BOD_L =$ 0.68. Calculate and list other important constituents that are used as input in the model simulation. The terminology and calculation procedures are adapted from References 2 and 98.

Solution

1. Determine the BOD parameters.

   Determine the BOD_L from Equation 10.113.

$$BOD_L = \frac{BOD_5}{0.68} = \frac{190\ \text{mg/L}}{0.68} = 279\ \text{mg/L} \qquad (10.113)$$

2. Determine the COD parameters.

   a. Determine the bCOD, nbCOD, and SbCOD from Equations 10.114a through 10.114c.

$$\text{bCOD} = 1.6\,BOD_5 = 1.6 \times 190\ \text{mg/L} = 304\ \text{mg/L} \qquad (10.114a)$$

$$\text{nbCOD} = \text{COD} - \text{bCOD} = (460 - 304)\ \text{mg/L} = 156\ \text{mg/L} \qquad (10.114b)$$

$$\text{SbCOD} = \text{bCOD} - \text{rbCOD} = (304 - 80)\ \text{mg/L} = 224\ \text{mg/L} \qquad (10.114c)$$

TABLE 10.62 Input of Influent Measured and Calculated Constituents for Computer Model Simulation (Example 10.161)

Constituent	Definition	Input Value, mg/L		Reference
		Measured	Calculated	
BOD parameters				
BOD_5	Total 5-day biochemical oxygen demand	190		
$sBOD_5$	Soluble 5-day biochemical oxygen demand	92		
BOD_L	Ultimate biochemical oxygen demand		279	Equation 10.113
COD parameters				
COD	Total chemical oxygen demand	460		
sCOD	Soluble chemical oxygen demand	168		
rbCOD	Readily biodegradable chemical oxygen demand	80		
bCOD	Biodegradable chemical oxygen demand		304	Equation 10.114a
nbCOD	Nonbiodegradable chemical oxygen demand		156	Equation 10.114b
SbCOD	Slowly biodegradable chemical oxygen demand		224	Equation 10.114c
rbsCOD	Readily biodegradable soluble chemical oxygen demand		29	Equation 10.114d
rbpCOD	Readily biodegradable particulate chemical oxygen demand		51	Equation 10.114e
bsCOD	Biodegradable soluble COD		111	Equation 10.114f
nbsCOD	Nonbiodegradable soluble chemical oxygen demand		57	Equation 10.114g
nbpCOD	Nonbiodegradable particulate chemical oxygen demand		99	Equation 10.114h
pCOD	Particulate chemical oxygen demand		292	Equation 10.114i
bpCOD	Biodegradable particulate chemical oxygen demand		193	Equation 10.114j
Suspended solids parameters				
TSS	Total suspended solids	230		
VSS	Volatile suspended solids	200		
nbVSS	Nonbiodegradable volatile solids		68	Equation 10.115a
iTSS	Inert total suspended solids		30	Equation 10.115b
Nitrogen parameters				
TKN	Total Kjeldahl nitrogen	45		
NH_4-N	Ammonia nitrogen (soluble)	30		
ON	Organic nitrogen	15		
sON	Soluble (filtered) organic nitrogen	1.2		
pON	Particulate organic nitrogen		13.8	Equation 10.116a
nbpON	Nonbiodegradable particulate organic nitrogen		4.7	Equation 10.116b
nbsON	Nonbiodegradable soluble organic nitrogen		0.4	Equation 10.116c
nbON	Nonbiodegradable organic nitrogen		5.1	Equation 10.116d
bON	Biodegradable organic nitrogen		9.9	Equation 10.116e
bTKN	Biodegradable total Kjeldahl nitrogen		39.9	Equation 10.116f
sTKN	Soluble (filtered) total Kjeldahl nitrogen		31.2	Equation 10.116g
Phosphorus parameters				
TP	Total phosphorus	6		
iP	Inorganic phosphorus (ortho and polyphosphates)	5		
OP	Organic phosphorus (bound phosphorus in cell mass)		1	Equation 10.117

b. Determine the rbsCOD and rbpCOD from Equations 10.114d and 10.114e.

$$\text{rbsCOD} = \text{rbCOD} \times \frac{\text{sCOD}}{\text{COD}} = 80\,\text{mg/L} \times \frac{168\,\text{mg/L}}{460\,\text{mg/L}} = 29\,\text{mg/L} \tag{10.114d}$$

$$\text{rbpCOD} = \text{rbCOD} - \text{rbsCOD} = (80 - 29)\,\text{mg/L} = 51\,\text{mg/L} \tag{10.114e}$$

c. Determine the bsCOD and nbsCOD from Equations 10.114f through 10.114h.

$$\text{bsCOD} = \text{sCOD} \times \frac{\text{bCOD}}{\text{COD}} = 168\,\text{mg/L} \times \frac{304\,\text{mg/L}}{460\,\text{mg/L}} = 111\,\text{mg/L} \tag{10.114f}$$

$$\text{nbsCOD} = \text{sCOD} - \text{bsCOD} = (168 - 111)\,\text{mg/L} = 57\,\text{mg/L} \tag{10.114g}$$

$$\text{nbpCOD} = \text{nbCOD} - \text{nbsCOD} = (156 - 57)\,\text{mg/L} = 99\,\text{mg/L} \tag{10.114h}$$

d. Determine the pCOD and bpCOD from Equations 10.114i and 10.114j.

$$\text{pCOD} = \text{COD} - \text{sCOD} = (460 - 168)\,\text{mg/L} = 292\,\text{mg/L} \tag{10.114i}$$

$$\text{bpCOD} = \text{pCOD} - \text{nbpOD} = (292 - 99)\,\text{mg/L} = 193\,\text{mg/L} \tag{10.114j}$$

3. Determine the suspended solids parameters.

a. Determine the nbVSS from Equation 15.115a.

$$\text{nbVS} = \frac{\text{nbpCOD}}{\text{pCOD}} \times \text{VSS} = \frac{99\,\text{mg/L}}{292\,\text{mg/L}} \times 200\,\text{mg/L} = 68\,\text{mg/L} \tag{10.115a}$$

b. Determine the iTSS from Equation 15.115b.

$$\text{iTSS} = (\text{TSS} - \text{VSS}) = (230 - 200)\,\text{mg/L} = 30\,\text{mg/L} \tag{10.115b}$$

4. Determine the nitrogen parameters.

a. Determined the pON, nbpON, nbsON, nbON, and bON from Equations 10.116a through 10.116e.

$$\text{pON} = \text{ON} - \text{sON} = (15 - 12)\,\text{mg/L} = 13.8\,\text{mg/L} \tag{10.116a}$$

$$\text{nbpON} = \text{pON}\frac{\text{nbVSS}}{\text{VSS}} = 13.8\,\text{mg/L} \times \frac{68\,\text{mg/L}}{200\,\text{mg/L}} = 4.7\,\text{mg/L} \tag{10.116b}$$

$$\text{nbsON} = \text{sON}\frac{\text{nbVSS}}{\text{VSS}} = 1.2\,\text{mg/L} \times \frac{68\,\text{mg/L}}{200\,\text{mg/L}} = 0.4\,\text{mg/L} \tag{10.116c}$$

$$\text{nbON} = \text{nbsON} + \text{nbsON} = (0.4 + 4.7)\,\text{mg/L} = 5.1\,\text{mg/L} \tag{10.116d}$$

$$\text{bON} = \text{ON} + \text{nbON} = (15 - 5.1)\,\text{mg/L} = 9.9\,\text{mg/L} \tag{10.116e}$$

b. Determine the bTKN and sTKN from Equations 10.116f and 10.116g.

$$\text{bTKN} = \text{TKN} - \text{nbON} = (45 - 5.1)\,\text{mg/L} = 39.9\,\text{mg/L} \quad (10.116f)$$

$$\text{sTKN} = \text{NH}_4\text{-N} + \text{sON} = (30 + 1.2)\,\text{mg/L} = 31.2\,\text{mg/L} \quad (10.116g)$$

5. Determine the phosphorus parameters.
Determine the OP from Equation 10.117.

$$\text{OP} = \text{TP} - \text{iP} = (6 - 5)\,\text{mg/L} = 1\,\text{mg/L} \quad (10.117)$$

10.9 Secondary Clarification

The MLSS from a biological reactor is settled to produce well-clarified effluent. The secondary clarifier is designed to perform two major functions: (1) provide gravity settling of solids and clarification of effluent and (2) produce thickened underflow. To achieve both these functions, sufficient depths of clarified zone and sludge zone are needed. In deeper clarified zone, the settled solids remain considerably below the weir and are not lost in the effluent. The deeper sludge zone provides storage for settled solids for thickening, and maintaining adequate sludge blanket. If sufficient sludge blanket is not maintained, unthickened sludge will be returned to the aeration basin. The basic design considerations for primary sedimentation basins have been presented in Chapter 9. These considerations include (1) surface overflow rate or surface settling rate, (2) detention period, (3) weir loading rate (WLR) and weir length, (4) tank shape and dimensions, (5) solid loading rate, (6) influent structure, (7) effluent structure, and (8) sludge collection and removal.

10.9.1 Design Considerations for Secondary Clarifier

The secondary clarifiers are designed for solids separation. Most common approach is to use surface overflow and solids loading rates. Allowance should be made for fluctuations in influent and return sludge flows and MLSS concentration. Typical values of these parameters for the secondary clarifiers used for different biological treatment processes are given in Table 10.63.[2,7,98,238] The surface overflow rate is based on influent flow because return sludge is withdrawn from the bottom and does not contribute to upward velocity.

10.9.2 Zone or Hindered Settling (Type III)*

At high solids concentrations, forces between the particles become significant. The settling is hindered due to additional resistance to movement caused by surrounding particles. The suspension tends to settle as a *blanket*. An interface develops that leaves a clarified zone in the upper portion of the clarifier. Below the interface the suspension moves downward concentrating toward a bottom layer where sludge slowly compacts.

Batch Single Settling Column Test: The zone settling data are usually gathered from a batch column settling test. A one- or two-liter graduated cylinder may be used, although the latter is preferred to minimize the wall effects and to prevent bridging. Gentle stirring at a speed around 1 rpm is recommended.

* Discussion on discrete settling (Type I), and flocculent settling (Type II) and mathematical equations may be found in Sections 8.3 and 9.2, respectively.

TABLE 10.63 Typical Design Surface Overflow and Solids Loading Rates for Secondary Clarifiers

Treatment Process	Surface Overflow Rate (SOR), $m^3/m^2·d$ (gpd/ft^2)	Solid Loading Rate (SLR), $kg/m^2·h$ (lb/$ft^2·h$)
	Suspended Growth[a]	
Activated sludge[b]		
Average flow	16–24 (400–600)	4–6 (0.8–1.2)
Peak flow	40–48 (1000–1200)	8–10 (1.6–2)
Extended aeration		
Average flow	8–16 (200–400)	1–5 (0.2–1)
Peak flow	24–32 (600–800)	7–8 (1.4–1.6)
High-purity oxygen (HPO)		
Average flow	16–28 (400–700)	5–7 (1–1.4)
Peak flow	40–64 (1000–1600)	9 (1.8)
Enhanced biological phosphorus removal (EBPR) with chemical addition		
TP = 2 mg/L at average flow	24–32 (600–800)	–
TP = 1 mg/L at average flow	16–24 (400–600)	–
TP = 0.2–0.5 mg/L at average flow	12–20 (300–500)	–
Biological nutrient removal (BNR) or selector technologies		
Average flow	24–32 (600–800)	5–8 (1–1.6)
Peak flow	40–64 (1000–1600)	9–10 (1.8–2)
	Attached Growth	
Trickling filter[c]		
Average flow	32 (800)	–
Peak flow	64 (1600)	–
Activated biofilter (ABF)		
Peak flow	40–72 (1000–1800)	–
	Fixed-film Reactor and Suspended Growth	
Biofilter/activated sludge (BF/AS)	48–84 (1200–2100)	–

[a] The typical side water depth (SWD) is in the range of 4–5.5 m (13–18).
[b] Conventional activated sludge processes and modifications except extended aeration.
[c] At an SWD of 5 m (16 ft).
Note: 1 $m^3/m^2·d$ = 24.5 gpd/ft^2; and 1 $kg/m^2·h$ = 0.205 lb/$ft^2·h$.
Source: Adapted in part from References 2, 7, and 98.

The initial concentration of the suspension is measured and subsidence of the top interface is monitored with time. Several important observations can be made and are summarized below.

1. A batch column test is started with uniform initial concentration of C_0 at time t_0. Two interfaces develop as the suspension settles from the initial height of the interface at Point A. A relatively clear water zone exists above the top interface, and a concentrated sludge zone develops below the bottom interface.
2. As time progresses, two interfaces propagate downward and upward, respectively. At time t_c, the two interfaces meet to form a single interface. After this time, compaction or compression of sludge starts and continues at a relatively slow pace until reaching ultimate compression or compaction. The progression of zone settling in a batch column is shown in Figure 10.119a.
3. The graphical procedure for obtaining clarifier area is shown in Figure 10.119b, and calculations are provided in Example 10.162.

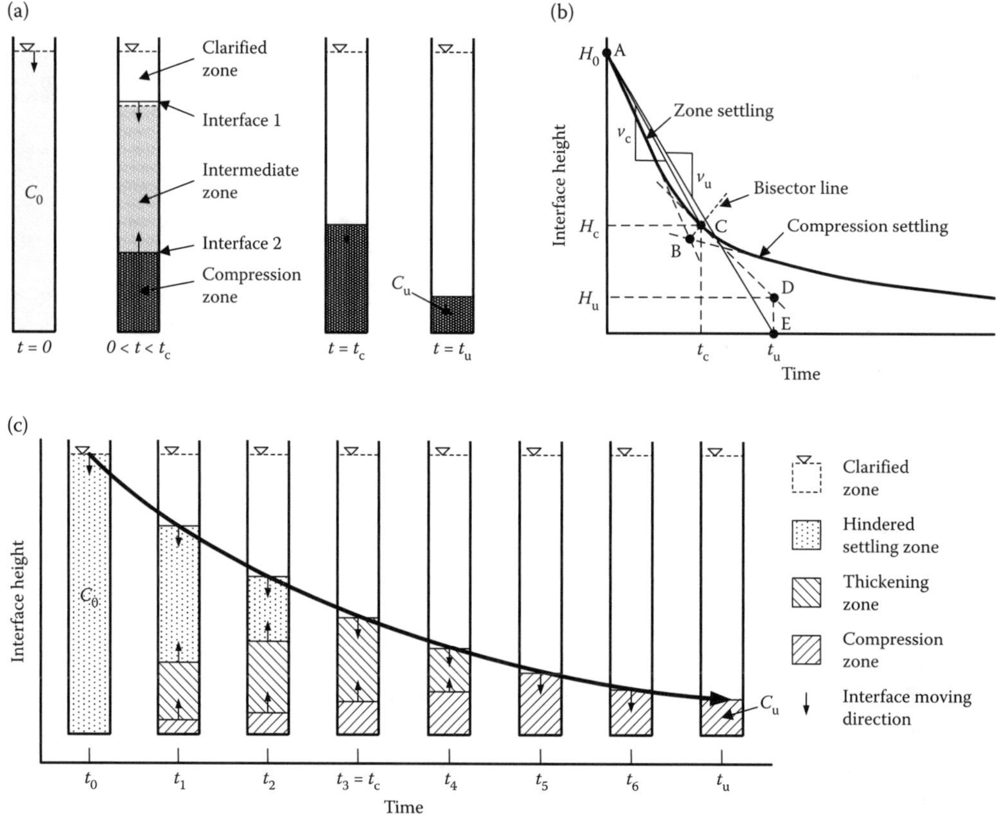

FIGURE 10.119 Zone or hindered settling in single column test: (a) progression of zone settling column test, (b) graphical procedure for obtaining clarifier area, and (c) subsidence of sludge blanket with respect to time of settling.

4. The settling rate of the upper interface is a function of solids concentration and their characteristics. The settled depth of upper interface with time in a column test is shown in Figure 10.119c. The settling rate decreases with time, and eventually it stops subsiding when the maximum compaction of settled sludge is reached at time t_u.

The settling behavior of solids in a continuous flow clarifier is different from that in a settling column. The following behavior of solids may be noted in a continuous flow clarifier: (a) a dynamic move of solids occurs due to settling and underflow through the sludge blanket into the concentrated sludge zone; (b) in the upper sludge blanket (or the hindered settling) zone there is an upflow velocity caused by fluid movement toward the effluent weir; (c) in the concentrated sludge (or the thickening) zone the net movement of fluid is downward; (d) the settling velocity of the suspension in the upper zone governs the design of clarifier; (e) the thickening of sludge in the concentrated sludge zone is not influenced by the draw off of clarified effluent at the top; and (f) the detention time of sludge in the sludge zones is independent of the detention time in the upper zone. Three different approaches are used to develop the design data from the column settling tests: (i) batch single column settling test, (ii) solid flux approach, and (iii) state point analysis. These approaches are discussed in the following sections.

Design Equations Based on Batch Single Settling Column Test Data: The hindered settling column test data are used to obtain the clarifier area. The clarifier area is calculated to achieve (a) effluent clarification (A_c) and (b) sludge compression (A_u). The clarifier area is selected from the larger of the two areas. The surface overflow and solids loading rates are then calculated from the selected design clarifier area.

The zone settling is expressed by Equation 10.118. The graphical procedure for determination of design variable is described briefly below.

$$A_c = \frac{Q}{v_c} \tag{10.118a}$$

$$v_c = \frac{H_0 - H_2}{t_2} \tag{10.118b}$$

$$A_u = \frac{Q_0}{v_u} \tag{10.118c}$$

$$H_u = \frac{C_0 H_0}{C_u} \tag{10.118d}$$

$$v_u = \frac{H_0}{t_u} \tag{10.118e}$$

$$Q_0 = \left(\frac{H_0}{H_0 - H_u}\right) \quad \text{or} \quad Q_0 = Q\left(\frac{C_u}{C_u - C_0}\right) \tag{10.118f}$$

where

A_c = area required for effluent clarification, m^2 (ft^2)
A_u = area required for sludge compression, m^2 (ft^2)
Q = clarified effluent flow, m^3/s (ft^3/s)
Q_0 = total influent flow, including recycled underflow, m^3/s (ft^3/s)
v_c = zone settling velocity of the interface, m/s (ft/s)
v_u = average settling velocity of the compressed sludge, m/s (ft/s)
C_0 = concentration of MLSS, g/L (kg/m^3)
C_u = the desired underflow concentration, g/L (kg/m^3)
H_0 = initial height of the interface, m
H_c = virtual height of the single interface at settling time t_c, m
H_u = virtual depth at which the underflow concentration C_u is reached, m (ft)
t_c = time to reach the height of the virtually single interface, s
t_u = time to reach the desired underflow concentration, s

The values of v_c and v_u are obtained by a graphical method using Figure 10.119b. Two tangents are drawn on the zone settling and compression regions of the settling curve. These tangents meet at Point B. A bisector of the angel between these tangents is drawn that cuts the curve at Point C to yield H_c and t_c. The zone settling velocity of the interface (v_c) is calculated from Equation 10.118b by dividing the vertical distance between H_0 and H_c by the time t_c. Another tangent is then drawn on the curve at Point C. This point is considered an imaginary dividing point between the zone and compression settling regions. This point is important because depth and solids concentration corresponding to this point govern the design of sludge storage and removal systems. The time corresponding to the intersection point (Point D) of this tangent, and the horizontal line through H_u is t_u. The virtual settling velocity of the compressed sludge (v_u) is eventually estimated from H_0 and t_u at Point E. This procedure may be found in References 1, 2, 7, 15, 17, 97, and 98, and explained in Example 10.162.

Design Based on Solids Flux Analysis: The solids flux analysis is used more widely to design a clarifier. Solids flux is defined as the mass of solids per unit volume passing through a unit area perpendicular to the direction of the flow. In a secondary clarifier, it is the product of the solids concentration (g/L or kg/m^3) and the velocity (m/h). Thus, the preferred unit for solids flux is kg/m^2·h. The downward movement of solids in a secondary clarifier is brought about by (1) the hindered gravity settling of the solids relative to

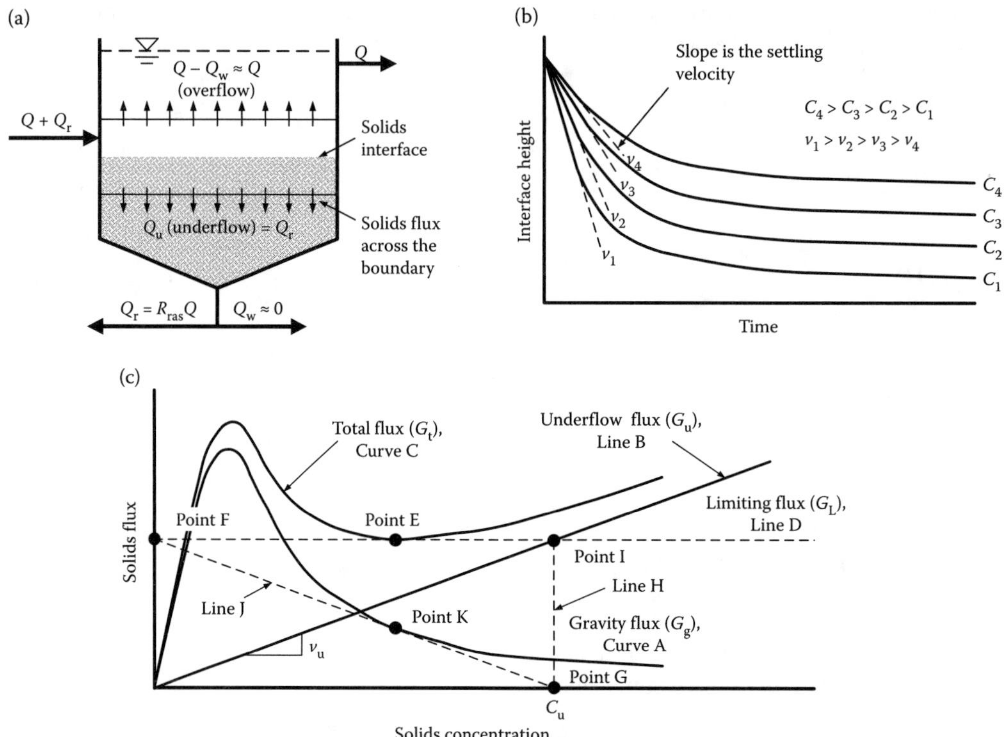

FIGURE 10.120 Solids flux analysis to develop design data for clarifier design: (a) definition sketch for a clarifier operating at steady state, (b) zone settling column test results at different solids concentrations, and (c) definition sketch for the solids flux analysis method.

the water and (2) the transport velocity caused by the underflow (withdrawal of returned sludge). The definition sketch for a settling basin operating at a steady state is given by Figure 10.120a.

The combined solids flux caused by gravity and underflow are expressed by Equation 10.119.

$$G_g = v_g C_i \tag{10.119a}$$

$$G_u = v_u C_i \quad \text{or} \quad G_u = C_i \frac{Q_u}{A} \tag{10.119b}$$

$$G_t = G_g + G_u \tag{10.119c}$$

where

G_g = solids flux caused by gravity, kg/m^2·h (lb/ft^2·h)
G_u = solids flux caused by underflow, kg/m^2·h (lb/ft^2·h)
G_t = combined flux caused by gravity and underflow, kg/m^2·h (lb/ft^2·h)
C_i = concentration of solids, g/L (kg/m^3)
Q_u = underflow flow rate, m^3/h (ft^3/h)
v_g = hindered settling velocity, m/h (ft/h)
v_u = downward velocity due to underflow, m/h (ft/h)

The hindered settling velocity at different solids concentration is determined by settling column tests. Several column tests at different solids concentrations are conducted and interface settling depth with respect to time for each column are recorded. The results of hindered settling column tests are shown in Figure 10.120b. The slope of the initial portion of the curve is the hindered settling velocity. The velocity caused by underflow is the downward movement resulting from withdrawal of the returned sludge (Q_r) and area of the clarifier (A).

The procedure for determination of the limiting solids flux for clarifier design is given below and shown in Figure 10.120c.

1. The gravity solids flux G_g (Curve A) caused by gravity is prepared from the hindered settling tests. This test is conducted at different solids concentrations as shown in Figure 10.119c.
2. The underflow solids flux G_u (Line B) is prepared from the downward velocity resulting from the return sludge flow and the clarifier area. It is a linear function of the concentration with a slope of v_u.
3. The total solids flux G_t (Curve C) is drawn by adding the G_g and G_u values.
4. A tangent (Line D) is drawn horizontally from the lowest point (Point E) after the hump on the total solids flux curve that give the value G_L of limiting solids flux or maximum allowable solids loading to the clarifier (Point F). If the quantity of solids discharged to the clarifier is greater than the limiting solids flux G_L, the solids will build up in the clarifier and ultimately washed out with the clarifier effluent from the top.
5. The maximum achievable underflow solids concentration C_u, corresponding to the limiting solids flux G_L, is obtained at the intersection on X axis (Point G) that is obtained by drawing a vertical line (Line H) from the intersection of Lines B and D at Point I. The clarifier must be designed for a desired underflow solids concentration ($C_c \leq C_u$). The G_L can also be found at the intersection on Y axis (Point F) by drawing a tangent (Line J) on the Curve A at Point K from Point F.

Using the limiting solid flux G_L value, the required area derived from the material mass balance is given by Equation 10.120.

$$A = \frac{(Q + Q_u)C_i}{G_L} \tag{10.120a}$$

$$A = \frac{(1 + \alpha)QC_i}{G_L} \tag{10.120b}$$

where

A = area of the clarifier in plan, m^2 (ft^2)

$Q + Q_u$ = total flow rate to the clarifier (overflow + underflow), m^3/d (ft^3/d)

α = Q_u/Q ratio, dimensionless

The procedure for solids flux analyses may be found in References 2, 7, 17, and 98. Design of final clarifiers using limiting solids flux is presented in Examples 10.163, 10.164, and 10.166.

Design Based on State Point Analysis: The state point analysis provides a convenient method to evaluate the performance of a clarifier under different operating condition relative to the limiting solid flux. The clarifier operating parameters such as MLSS concentrations and overflow rate are used to determine the optimal thickening of underflow and return sludge recycle ratio for a given influent flow condition and within the solids-flux limitation.[239,240]

The coordinates of the state point as shown in Figure 10.121 are the aeration basin MLSS concentration and clarifier overflow solids flux rate. The clarifier overflow solids flux rate is calculated from Equation 10.121.

$$G_0 = \frac{QC_0}{A} \quad \text{or} \quad G_0 = SOR\, C_0 \tag{10.121}$$

where

G_0 = clarifier overflow solids flux rate, $kg/m^2{\cdot}d$ ($lb/ft^2{\cdot}h$)

C_0 = MLSS concentration in the aeration basin, g/L (kg/m^3)

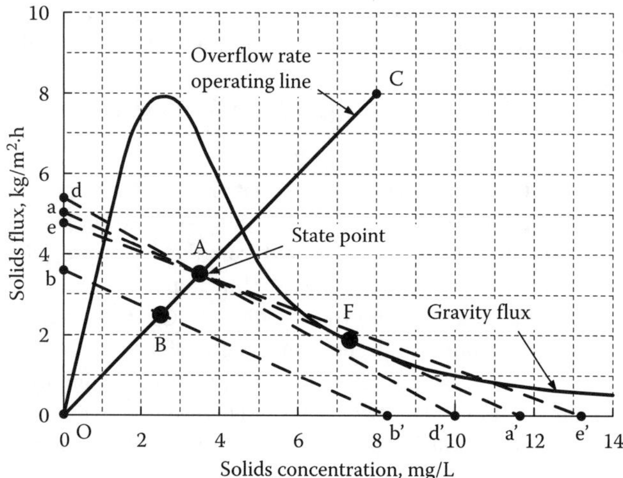

FIGURE 10.121 State point analysis to determine clarifier operating conditions for underloaded, critically loaded, and overloaded solids.

A = clarifier surface area, m^2 (ft^2)
SOR = clarifier surface overflow rate, $m^3/m^2{\cdot}d$ (gpd/ft^2)

The line joining the origin and state point is the overflow rate operating line or overflow solids flux line. The line joining the state point and the underflow solids concentration represents the underflow rate operating line. The basic features of state point analysis with reference to Figure 10.121 are summarized below.

1. Line OBAC is overflow rate operating line. The aeration basin MLSS concentration at any point along the overflow rate operating line is found by constructing a vertical line to the X axis and the Y intercept of the horizontal line drawn from the point of intersection of the vertical line and operating line. The Y intercept is the total solids flux (G_x) to the clarifier.
2. Point A is the state point corresponding to a MLSS concentration $C_0 = 3.5\ kg/m^3$ and clarifier overflow solids flux rate of $G_0 = 3.5\ kg/m^2{\cdot}h$ on Line OBAC.
3. Point B is the state point corresponding to MLSS concentration = $2.5\ kg/m^3$ and clarifier overflow solids flux rate $G_0 = 2.5\ kg/m^2{\cdot}h$ on Line OBAC also.
4. Lines aAa′, bBb′, dAd′, and eAe′ are the underflow operating lines. They have negative slopes and represent the clarifier underflow velocity.
5. The state point and underflow solids flux lines can be compared to the gravity flux curve to determine the operational condition of the clarifier.
   a. The underflow solids flux line aAa′ is tangent to gravity flux curve and represents the limiting solids flux condition since the line is a tangent at Point F on the gravity flux curve. The clarifier is critically loaded for this underflow solids flux and MLSS concentration.
   b. Lines bBb′ and dAd′ represent underloaded condition as they are below the gravity flux curve. The underflow solids flux is high and solids concentration in the clarifier is below the critical MLSS concentration.
   c. Line eAe′ represents higher underflow solids concentration. The underflow line crosses the lower limb of the gravity flux curve. The limiting solids flux will be exceeded and clarifier blanket will rise to the effluent weir.

The clarifier operating conditions are explained based on the state point analysis in Example 10.165.

EXAMPLE 10.162: CLARIFIER DESIGN BASED ON BATCH SINGLE SETTLING COLUMN TEST DATA

A settling column 0.75 m high was used to determine the zone settling of mixed liquor from an extended aeration basin that receives an average influent of 4000 m^3/d (~1 MGD). The MLSS concentration is 4000 mg/L. Determine the final clarifier area required to produce an underflow sludge concentration C_u of 12,000 mg/L, and verify the surface overflow and solids loading rates developed from the settling column test data. The subsidence rate of the interface with time is given in Figure 10.122.

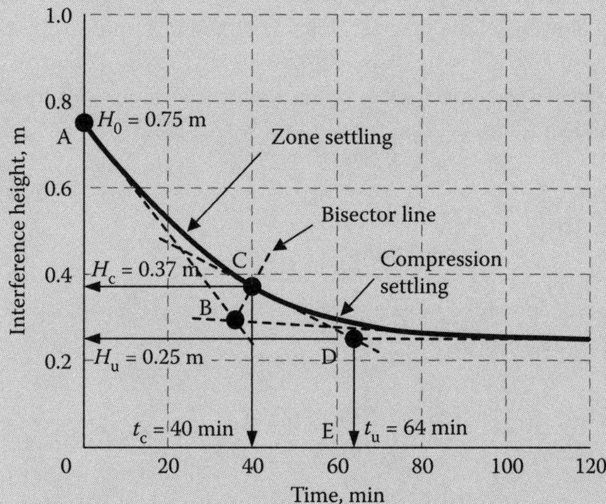

FIGURE 10.122 Subsidence of interface with time in a settling column test (Example 10.162).

Solution

1. Determine the area required for effluent clarification.
   a. Determine the zone settling velocity (v_c).

      Draw a tangent at the initial interface height H_0 to the zone settling curve. Draw another tangent to the compression settling region of the lower settling curve. Draw a bisector line of the angel between two tangents from Point B and find the interception (Point C) on the settling curve. Determine $H_c = 0.37$ m and $t_c = 40$ min at Point C. Calculate the velocity v_c from Equation 10.181b.

$$v_c = \frac{H_0 - H_c}{t_c} = \frac{(0.75 - 0.37)\ \text{m}}{40\ \text{min}} = 0.0095\ \text{m/min}$$

   b. Determine the area required for effluent clarification (A_c).

      The surface area requirement shall be determined from the effluent flow through the clarified zone. Exclude the waste sludge volume and the effluent flow (or upflow) shall be equal to the flow received by the extended aeration basin, $Q = 4000\ \text{m}^3/\text{d}$. calculate the Ac from Equation 10.118a.

$$A_c = \frac{Q}{v_c} = \frac{4000\ \text{m}^3/\text{d}}{0.0095\ \text{m/min} \times 1440\ \text{min/d}} = 292\ \text{m}^2$$

2. Determine the area required for sludge compression.
   a. Determine the depth for desired underflow concentration (H_u) from Equation 10.118d at $C_0 = 4000$ mg/L and $C_u = 12{,}000$ mg/L.

$$H_u = \frac{C_0 H_0}{C_u} = \frac{4000\,\text{mg/L} \times 0.75\,\text{m}}{12{,}000\,\text{mg/L}} = 0.25\,\text{m}$$

   b. Determine the time to reach the desired underflow concentration (t_u).
      Draw another tangent on the settling curve at Point C. Determine the time $t_u = 64$ min at $H_u = 0.25$ m (Point D) on the tangent. Point E is then determined from the vertical line from Point D.
   c. Determine the average settling velocity of the compressed sludge (v_u) from Equation 10.118e using height and time values relative to Points A and E.

$$v_u = \frac{H_0}{t_u} = \frac{0.75\,\text{m}}{64\,\text{min}} = 0.012\,\text{m/ min}$$

   d. Determine the total influent flow received by the clarifier (Q_0), including effluent and recycled flows from Equation 10.181f.

$$Q_0 = Q\left(\frac{H_0}{H_0 - H_u}\right) = 4000\,\text{m}^3/\text{d} \times \frac{0.75\,\text{m}}{(0.75 - 0.25)\text{m}} = 4000\,\text{m}^3/\text{d} \times 1.5 = 6000\,\text{m}^3/\text{d}$$

$$\text{Return sludge ratio,}\quad R_{ras} = \frac{Q_0 - Q}{Q} = \frac{(6000 - 4000)\,\text{m}^3/\text{d}}{4000\,\text{m}^3/\text{d}} = 0.5 \quad \text{or} \quad Q_{ras} = 2000\,\text{m}^3/\text{d}$$

   Note: The calculation results indicate that a return sludge ratio of 0.5 is desired. This is same as the R_{ras} calculated from Equation 10.5e.

   e. Determine the area required for sludge compression (A_u) from Equation 10.118c.

$$A_u = \frac{Q_0}{v_u} = \frac{6000\,\text{m}^3/\text{d}}{0.012\,\text{m/min} \times 1440\,\text{min/d}} = 347\,\text{m}^2$$

3. Determine the design area for the final clarifier (A).
   The clarifier area is governed by the compression settling based on the settling column test data. Provided a total final clarifier area $A = 350\,\text{m}^2$.
4. Verify the design surface overflow and solids loading rates.

$$\text{The surface overflow rate, } SOR = \frac{Q}{A} = \frac{4000\,\text{m}^3/\text{d}}{350\,\text{m}^2} = 11\,\text{m}^3/\text{m}^2{\cdot}\text{d}\,(270\,\text{gpd/ft}^2)$$

$$\text{The solids loading rate, } SLR = \frac{C_0 Q_0}{A} = \frac{4000\,\text{g/m}^3 \times 6000\,\text{m}^3/\text{d}}{350\,\text{m}^2 \times 10^3\,\text{g/kg} \times 24\,\text{h/d}} = 2.9\,\text{kg/m}^2{\cdot}\text{h}\,(0.59\,\text{lb/ft}^2{\cdot}\text{h})$$

Both design values are within the typical ranges of design SOR (8–16 $\text{m}^3/\text{m}^2{\cdot}\text{d}$) and SLR (1–5 $\text{kg/m}^2{\cdot}\text{h}$) for the extended aeration process (Table 10.63).

EXAMPLE 10.163: FINAL CLARIFIER AREA AT DESIGN RETURN SLUDGE RATIO BASED ON SOLIDS FLUX DATA

A conventional activated sludge process facility is designed to treat mixed municipal and industrial wastewater. The influent flow is 8000 m^3/d. Final clarifiers are operating in parallel to remove biological solids, and produce effluent of high quality. See Figure 10.120a for a definition sketch for the clarifier operating at steady state. A series of hindered settling column tests were conducted at different solids concentrations to establish the settling behavior and initial settling velocity of the biomass. The initial settling velocity of different solids concentrations are provided below. The return sludge is 50% of the influent flow. Determine (a) design solid loading rate to the clarifier, (b) the solid concentration in the returned sludge, (c) MLSS concentration maintained in the aeration basin, and (d) clarifier surface area. The waste sludge flow is small and is ignored ($Q_w \approx 0$).

Solids concentration (C_i), kg/m^3	1	1.5	2	2.5	3	3.5	4	4.5	5	6	7	8	10	12	15
Initial settling velocity (v_g), m/h	5	4.7	4	3.2	2.3	1.6	1.2	0.9	0.7	0.45	0.3	0.2	0.1	0.07	0.05

Note: 1 kg/m^3 = 1000 g/m^3 = 1000 mg/L.

Solution

1. Determine the underflow velocity.

   A typical surface overflow rate (*SOR*) in the range of 16–24 $m^3/m^2{\cdot}d$ is recommended for operation of a conventional activated sludge facility under the average flow (Table 10.63). Select a design $SOR = 20\ m^3/m^2{\cdot}d$ and calculate the underflow velocity from the design return sludge ratio $R_{ras} = 0.5$.

$$v_u = R_{ras}SOR = 0.5 \times 20\,m^3/m^2{\cdot}d \times \frac{1\,d}{24\,h} = 0.42\,m/h$$

2. Develop the computational table.

   The solid flux data are provided in Table 10.64. The calculations of the solid fluxes (G_g, G_u, and G_t) at different solids concentrations are summarized in the table.
3. Plot the solids flux data.

   The gravity, underflow, and total solids flux data are plotted with respect to solids concentration in Figure 10.123.
4. Determine the design solid loading rate.

   After the hump, the lowest point on the G_t curve is at Point A. Draw a horizontal tangent at Point A to find the limiting solids flux $G_L = 5.0\ kg/m^2{\cdot}h$ at Point B. This is the maximum allowable solids flux. Select the design solids loading rate, $SLR = G_L = 5.0\ kg/m^2{\cdot}h$. It is within the typical range of 4–6 $kg/m^2{\cdot}h$ for the conventional activated sludge process (Table 10.63).
5. Determine the design return sludge solid concentration.

   Extend the horizontal tangent line developed in Step 4 to intercept the G_u line at Point C. Draw a vertical line from Point C and find the Point D on X axis. This is the maximum allowable underflow soluble concentration $C_u = 11.8\ kg/m^3$. Select the return sludge suspended solids concentration, $TSS_{ras} = C_u = 11.8\ kg/m^3$ or 11,800 g/m^3 or 11,800 mg/L.

   Note: The value of C_u can also be determined by a line from Point B and tangent to the G_g curve at Point E. The tangent intercepts the X axis at Point D that gives $C_u = 11.8\ kg/m^3$.

TABLE 10.64 Computational Table for Determination of Solids Fluxes (Example 10.163)

MLSS Concentration (C_i), mg/L	Initial Settling Velocity (v_g), m/h	Solids Flux, kg/m²·h		
		Gravity (G_g)	Underflow (G_u)	Total (G_t)
(1)	(2)	(3)	(4)	(5)
1000[a]	5.0[a]	5.0[b]	0.42[c]	5.42[d]
1500	4.7	7.05	0.63	7.68
2000	4.0	8.00	0.84	8.84
2500	3.2	8.00	1.05	9.05
3000	2.3	6.90	1.26	8.16
3500	1.6	5.60	1.47	7.07
4000	1.2	4.80	1.68	6.48
4500	0.9	4.05	1.89	5.94
5000	0.7	3.50	2.10	5.60
6000	0.45	2.70	2.52	5.22
7000	0.3	2.10	2.94	5.04
8000	0.2	1.60	3.36	4.96
10,000	0.1	1.00	4.20	5.20
12,000	0.07	0.84	5.04	5.88
15,000	0.05	0.75	6.30	7.05

[a] The initial hindered settling velocities are obtained at different solids concentrations from the settling column tests (see Figure 10.120b).

[b] $G_g = v_g C_i = 5.0\ \text{m/h} \times 1000\ \text{g/m}^3 \times 10^{-3}\ \text{kg/g} = 5.0\ \text{kg/m}^2\cdot\text{h}$ (from Equation 10.119a).

[c] $G_u = v_u C_i = 0.42\ \text{m/h} \times 1000\ \text{g/m}^3 \times 10^{-3}\ \text{kg/g} = 0.42\ \text{kg/m}^2\cdot\text{h}$ (from Equation 10.119b).

[d] $G_t = G_g + G_u = (5.0 + 0.42)\ \text{m/h} = 5.42\ \text{kg/m}^2\cdot\text{h}$ (from Equation 10.119c).

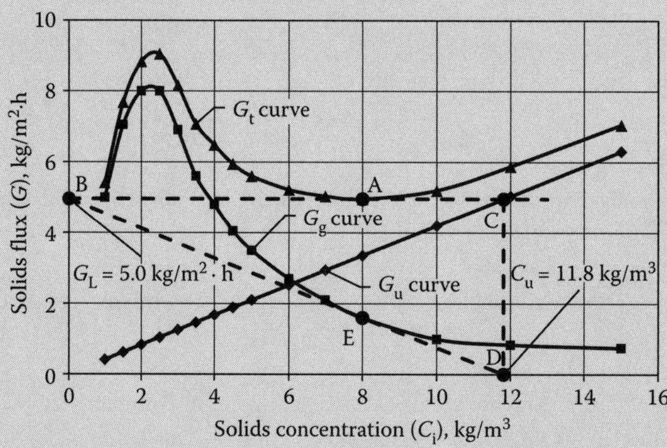

FIGURE 10.123 Plot of solids flux data for determination of final clarifier area at design return sludge ratio (Example 10.163).

6. Determine the MLSS concentration from the rearranged Equation 10.5e.

$$\text{MLSS} = \frac{R_{ras}}{1 + R_{ras}} \text{TSS}_{ras} = \frac{0.5}{1 + 0.5} \times 11{,}800\ \text{mg/L} = 3930\ \text{mg/L}$$

7. Determine the total clarifier surface area.

a. Determine the clarifier surface area (A_{SLR}) required from the design SLR.

The total solids reaching the clarifier, $\Delta TSS = (1 + R_{ras})\,Q\,\text{MLSS}$

$$= (1 + 0.5) \times 8000\,\text{m}^3/\text{d} \times 3930\,\text{g/m}^3 \times 10^{-3}\,\text{kg/g} = 47{,}200\,\text{kg/d}$$

The clarifier surface area required at the design SLR, $A_{SLR} = \dfrac{\Delta TSS}{SLR} = \dfrac{47{,}200\,\text{kg/d}}{5.0\,\text{kg/m}^2\cdot\text{h} \times 24\,\text{h/d}} = 393\,\text{m}^2$

b. Determine the clarifier surface area (A_{SOR}) required from the design SOR.

$$A_{SOR} = \frac{Q}{SOR} = \frac{8000\,\text{m}^3/\text{d}}{20\,\text{m}^3/\text{m}^2\cdot\text{d}} = 400\,\text{m}^2$$

c. Determine the total design clarifier surface area.

The clarifier surface areas required by SLR and SOR are virtually the same. Provide a design clarifier surface area $A = 400\,\text{m}^2$.

EXAMPLE 10.164: FINAL CLARIFIER AREA AT A DESIGN RETURN SLUDGE CONCENTRATION BASED ON SOLIDS FLUX DATA

A series of column settling tests were conducted at different MLSS concentrations. The testing results are summarized below. Determine the final clarifier surface area that will produce underflow concentration of 10,000 mg/L. The MLSS concentration and influent flow are 3000 mg/L and 8000 m^3/d, respectively.

MLSS concentration (C_i), mg/L	1400	2200	3000	3700	4500	5200	6500	8200
Initial settling velocity (v_g), m/h	3.00	1.85	1.21	0.76	0.45	0.30	0.16	0.089

Solution

1. Tabulate the solids flux calculation results for different solids concentrations.

Gravity solids flux (G_g) is calculated from Equation 10.119a. The solids flux at different solids concentrations are summarized below.

Solids concentration (C_i), mg/L	1400	2200	3000	3700	4500	5200	6500	8200
Initial settling velocity (v_g), m/h	3.00	1.85	1.21	0.76	0.45	0.30	0.16	0.089
Gravity solids flux (G_g), kg/m²·h	4.20[a]	4.07	3.63	2.81	2.03	1.56	1.04	0.73

[a] $G_g = v_g C_i = 3.00\,\text{m/h} \times 1400\,\text{g/m}^3 \times 10^{-3}\,\text{kg/g} = 4.2\,\text{kg/m}^2\cdot\text{h}$ (from Equation 10.119a).

2. Plot the gravity solids flux (G_g) curve.

The gravity solids flux data are plotted with respect to solids concentration in Figure 10.124.

3. Determine the limiting solids flux (G_L).

Draw a line from the underflow solids concentration $C_u = 10\,\text{kg/m}^3$ (Point A) that is tangent to the G_g curve at Point B. The limiting solids flux $G_L = 2.9\,\text{kg/m}^2\cdot\text{h}$ is obtained at Point C.

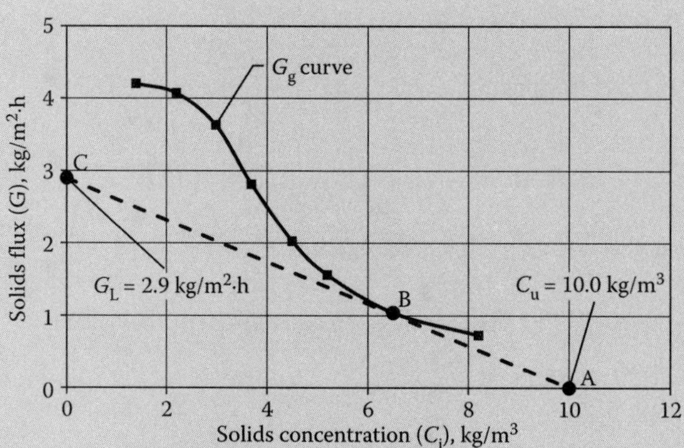

FIGURE 10.124 Plot of solids flux data for determination of final clarifier area at design return sludge concentration (Example 10.164).

4. Determine the return activated sludge ratio (R_{ras}) from Equation 10.5e.

$$R_{ras} = \frac{\text{MLSS}}{\text{TSS}_{ras} - \text{MLSS}} = \frac{3000\ \text{mg/L}}{10{,}000\ \text{mg/L} - 3000\ \text{mg/L}} = 0.43$$

5. Determine the clarifier area (A_{SLR}) based on the solids loading rate (SLR).

The total solids reaching the clarifier, $\Delta TSS = (1 + R_{ras})\, Q\, \text{MLSS}$

$$= (1 + 0.43) \times 8000\ \text{m}^3/\text{d} \times 3000\ \text{g/m}^3 \times 10^{-3}\ \text{kg/g} = 34{,}320\ \text{kg/d}$$

Select a design solids flux, $SLR = G_L = 2.9\ \text{kg/m}^2{\cdot}\text{h}$. It is below the typical range of 4–6 kg/m²·h for a conventional activated sludge process (Table 10.63).

The clarifier surface area, $$A_{SLR} = \frac{\Delta TSS}{SLR} = \frac{34{,}320\ \text{kg/d}}{2.9\ \text{kg/m}^2{\cdot}\text{h} \times 24\ \text{h/d}} = 493\ \text{m}^2$$

Provide a design total clarifier surface area $A = 500\ \text{m}^2$ and check the surface overflow rate (SOR).

$$SOR = \frac{Q}{A} = \frac{8000\ \text{m}^3/\text{d}}{500\ \text{m}^2} = 16\ \text{m}^3/\text{m}^2{\cdot}\text{d}$$

The design SOR is at the lower end of the typical range of 16–28 m³/m²·d for a conventional activated sludge process (Table 10.63).

EXAMPLE 10.165: EVALUATE CLARIFIER OPERATING CONDITIONS FROM STATE POINT ANALYSIS

A BNR facility is operating at a MLSS concentration of 3500 mg/L. There are multiple clarifiers operating in parallel under the same surface overflow rate of 24 m³/m²·d. Determine (a) the solids flux loadings, (b) underflow velocities, (c) return sludge ratios, and (d) operating conditions if the underflow solids concentrations are 10, 12, and 14 kg/m^3, respectively. Also, comment on the operational conditions of the clarifiers at three underflow solids concentrations and limiting solids flux. The solids settling and gravity flux data are given below.

Solids concentration (C_i), kg/m^3	0.5	1	1.5	2	2.5	3.5	5	7	9	12	15
Initial settling velocity (v_g), m/h	12	7	4.9	3.6	2.6	1.4	0.68	0.30	0.17	0.10	0.07
Gravity solids flux (G_g), kg/m^2·h	6.0	7.0	7.4	7.2	6.5	4.9	3.4	2.1	1.5	1.2	1.1

Solution

1. Plot the gravity solids flux curve.

   The gravity solids flux (G_g) curve is shown in Figure 10.125.

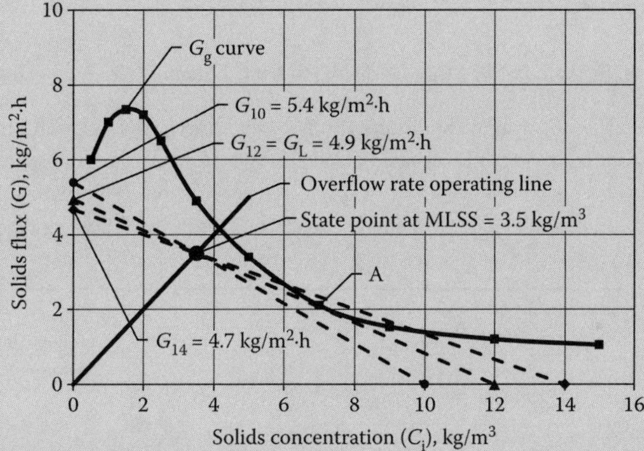

FIGURE 10.125 State point analysis for evaluation of clarifier operating conditions (Example 10.165).

2. Determine the coordinates of the state point.

   $SOR = 24$ m^3/m^2·d $= 1$ m/h is maintained constantly at all three underflow solids concentrations. Calculate the overflow solids flux at the state point (G_0) from Equation 10.121 at $C_0 = 3500$ mg/L $=$ 3.5 kg/m^3.

$$G_0 = SORC_0 = 1\,\text{m/h} \times 3.5\,\text{kg/m}^3 = 3.5\,\text{kg/m}^2\cdot\text{h}$$

   Plot the state point at the coordinates of X = 3.5 kg/m^3 and Y = 3.5 kg/m^2·h in Figure 10.125.

3. Determine the clarifier operating conditions at the underflow solids concentration of 10 kg/m^3.
   a. Determine the solids flux loading (G_{10}).

      Draw a line starting from the $C_{u,10} = 10$ kg/m^3 on X axis, passing the state point, and meeting the point $G_{10} = 5.4$ kg/m^2·h on Y axis (Figure 10.125).
   b. Determine the underflow velocity ($v_{u,10}$) from the negative slope of the line.

$$v_{u,10} = -\frac{0 - G_{10}}{C_{u,10} - 0} = \frac{G_{10}}{C_{u,10}} = \frac{5.4\,\text{kg/m}^2\cdot\text{h}}{10\,\text{kg/m}^3} = 0.54\,\text{m/h}$$

c. Determine the return sludge flow ratio ($R_{rs,10}$).

Assume the waste sludge flow is small and ignorable ($Q_w \approx 0$), the return sludge flow should be equal to the underflow rate, $Q_r = Q_u$. A relationship is derived below for calculating the R_{ras} (Equation 10.122).

$$R_{ras} = \frac{Q_r}{Q} = \frac{Q_u}{Q} = \frac{Q_u/A}{Q/A} = \frac{v_u}{SOR} \tag{10.122}$$

Calculate the ratio from Equation 10.122 at $SOR = 1$ m/h and $v_{u,10} = 0.54$ m/h.

$$R_{ras,10} = \frac{v_{u,10}}{SOR} = \frac{0.54\,\text{m/h}}{1\,\text{m/h}} = 0.54$$

Validate the R_{ras} from Equation 10.5e.

$$R_{ras,10} = \frac{C_0}{C_{u,10} - C_0} = \frac{3500\,\text{mg/L}}{10{,}000\,\text{mg/L} - 3500\,\text{mg/L}} = 0.54$$

Therefore, the value of R_{ras} that is calculated from Equation 10.122 is valid.

4. Determine the clarifier operating conditions at other two underflow solids concentrations.

The procedure used in Step 3 is repeated for evaluation of the clarifier operating conditions at the underflow solids concentrations of 12 and 14 kg/m^3. The evaluation results with additional comments are summarized for all three underflow solids concentrations in the following table.

	Clarifier Operating Condition			
Underflow Solids Concentration (C_u), kg/m^3	Solids Flux Loading (G), kg/m^2·h	Underflow Velocity (v_u), m/h	Return Sludge Ratio (R_{ras})	Comment on Clarifier Operating Condition
10	5.4	0.54	0.53	The underflow line is below the G_g curve. The solids flux loading of 5.4 kg/m^2·h is higher than the value of G_L. The clarifier operation is at underloaded condition.
12	4.9	0.41	0.41	The underflow line is tangent to the G_g curve at Point A. This gives the limiting solids flux $G_L = 4.9$ kg/m^2·h. Therefore, the clarifier operation is at critically loaded condition at the limiting solids flux.
14	4.7	0.33	0.33	The underflow line crosses the lower limb of the G_g curve. The solids flux loading of 4.7 kg/m^2·h is lower than the value of G_L. The clarifier operation is at overloaded condition. The clarifier blanket will rise to the effluent weir.

10.9.3 Design of Secondary Clarifiers

The purpose of a secondary clarifier is to remove biological solids and produces well-clarified effluent that is low in suspended solids and BOD. The design of plain and chemically enhanced primary sedimentation basins has been presented in Chapter 9. Much of this information is also applicable to the secondary clarifiers. Therefore, only selected information applicable to zone or hindered settling solids with high solids concentrations are presented below. Readers may review Chapter 9 for background information about the clarification process and clarifier design.

Types: The secondary clarifiers are normally circular or rectangular in shape (Figures 9.6 and 9.9). Square tanks perform poorly due to accumulation of solids in the corners and disturbance of settled solids by sludge collectors. The length of rectangular tank should normally not exceed 10 times the side water depth (SWD). Circular tanks are 3–60 m (10–200 ft) in diameter (more frequently 10–35 m (30–120 ft)). The tank radius preferably should not exceed five times the SWD.

SORs and SLRs: The recommended SORs and SLRs at average and design peak flows for different biological treatment processes are summarized in Table 10.63.

Side Water Depth: The SWD of a clarifier is in the range of 4–6 m (13 –20 ft). The depth should be sufficient to allow clear water and settling zones, and to maintain sufficient sludge blanket at the bottom. Deeper tanks provide greater flexibility of operation and a larger margin of safety for low-density activated sludge.

Inlet Structure: The inlet structure should provide even distribution of flow, dissipate influent energy, mitigate density currents, minimize disturbance of sludge blanket, and promote flocculation. Circular basins may utilize center or peripheral feed. The diameter of the center feed well should be 25–35% of the basin diameter. The design of influent channel for rectangular basins and details of center and peripheral feed clarifiers are covered in Section 9.4.2.

Effluent Structure: The effluent structure should evenly collect effluent flow, minimize the upflow velocity and minimize the loss of suspended solids in the effluent. The upflow velocity in the vicinity of effluent weir should be 3.5–7 m/h. The WLR in large basins should be below 250 m^3/m·d (20,000 gpd/ft) at the peak flow. In small basins, the WLR should not exceed 125 m^3/m·d and 250 m^3/m·d (10,000 and 20,000 gpd/ft) at average and peak flows, respectively. The effluent weir configuration and flow regime in the basin are covered in Section 9.4.5.

Scum Control: Scum generated from the secondary treatment contain significant amount of filamentous and/or foam generating microorganisms. Effective removal of scum is typically required to avoid recirculating these unhealthy microorganisms back to the biological system. Different scum collection and removal systems have been described in Section 9.4.

Sludge Collection: The sludge collection system for rectangular clarifiers are traveling flights, and traveling bridge similar to those for primary sedimentation tank. In circular clarifiers, revolving scraper or plows move the sludge into a central hopper for pumping. Suction type sludge collector and removal systems utilize suction orifices at reduced static head on suction pipes or on manifolds. Details may be found in Section 9.4.

Design and Performance Optimization: Utilization of advanced computational fluid dynamics (CFD) models has been popular in recent years. Powerful two- or three-dimensional (2D or 3D) CFD software packages are now commercially available for assisting engineers in clarifier design.[241] The potential applications of using CFD models have been reported in (a) validation of one-dimensional design criteria (SOR and SLR), (b) evaluation of inlet/feed well arrangement alternatives for efficient energy dissipation, uniform flow distribution and ideal flocculation, (c) assistance in developing proper baffle arrangement for reducing short circuiting due to density currents, and (d) study of balancing between sludge wasting and solids inventory.[242–246] Many equipment manufacturers may also offer proprietary package clarifier systems for optimum clarification. As an example, Clarifier Optimization Package (COP™) is delivered with key design parameters established by the manufacturer based on the specific site conditions.[247]

Pilot test is typically required to validate the claims and effectiveness of these systems. See Section 11.10 for additional information about applications of CFD modeling in disinfection reactor design.

EXAMPLE 10.166: DESIGN OF A SECONDARY CLARIFIER

A series of process descriptions and calculations on a BNR facility using A^2/O process have been provided in Examples 10.153 through 10.159. Design final clarifiers for this BNR facility to produce the desired effluent quality. Apply the influent, effluent, and process design data given and developed in these examples. Additional clarifier design criteria are given below.

a. Provide two circular clarifiers with each unit serving one of the two parallel BNR process trains.
b. The 2-h peaking factor is 2.8 based on the design average flow.
c. Design surface overflow rate (SOR) shall not exceed 24 $m^3/m^2{\cdot}d$ at the design average flow. The SOR at the design peak flow shall be within the typical range given for BNR process in Table 10.63.
d. The design of the clarifier shall be based on the gravity solids flux (G_g) curve shown in Figure 10.126. (*Note*: Detailed procedure for developing G_g cure from the laboratory settling tests data has been provided in Examples 10.163 through 10.165). The design solids loading rates shall not exceed 8 $kg/m^2{\cdot}h$ at the design average flow (Table 10.63).

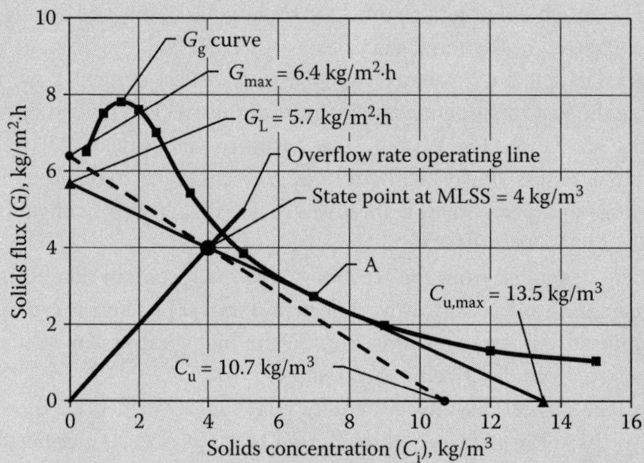

FIGURE 10.126 Determination of design solids flux from gravity solids flux curve (Example 10.166).

e. The clarifier shall meet the clarification as well as the thickening requirements for the effluent and underflow. The underflow solids concentration shall not exceed 12,000 mg/L. The selected depth shall provide adequate clear water zone, thickening zone, and sludge storage zone.
f. The effluent weir loading rates (WLRs) shall not exceed 125 and 250 $m^3/m{\cdot}d$ (10,000 and 20,000 gpd/ft) at average and design peak flows, respectively.

Solution

1. Determine the flow conditions for each clarifier.

The design average total flow to the BNR facility is 4000 m^3/d (Example 10.157). The design average flow for each clarifier is $Q = 2000\ m^3/d$ or $0.023\ m^3/s$ when two clarifier are operating in parallel. Apply the 2-h peaking factor of 2.8 to calculate the peak design flow Q_{peak}.

$$Q_{peak} = PF_{2h}Q = 2.8 \times 2000\ m^3/d = 5600\ m^3/d \text{ or } Q_{peak} = 5600\ m^3/d \times \frac{d}{86{,}400\ s} = 0.065\ m^3/s$$

2. Determine the design solids loading rate (*SLR*).

Apply the state point analysis method to determine the flux to determine the value of *SLR*. Detailed procedure is presented in Example 10.165.

a. Determine the limiting solids flux (G_L).

The design MLSS concentration $C_0 = 4000\ mg/L = 4\ kg/m^3$ (Example 10.157). Apply the state point analysis the flux $G_L = 5.7\ kg/m^2{\cdot}h$ and $C_{u,max} = 13.5\ kg/m^3$ (Figure 10.126 at the design $SOR = 24\ m^3/m^2{\cdot}d = 1\ m/h$).

b. Determine the maximum solids flux (G_{max}).

The return sludge ratio $R_{RS} = 0.6$ is given in Example 10.157. Rearrange Equation 10.5e and calculate the desired solids concentration in the underflow (or return sludge).

$$C_u = \frac{1 + R_{RS}}{R_{RS}} C_0 = \frac{1 + 0.6}{0.6} \times 4\ kg/m^3 = 10.7\ kg/m^3$$

Note: C_u is lower than $C_{u,max}$. Therefore, clarifier operation at $C_u = 10.7\ kg/m^3$ is achievable. It is also $<12\ kg/m^3$ (required in problem statement).

Corresponding to a design $C_u = 10.7\ kg/m^3$, the flux $G_{max} = 6.4\ kg/m^2{\cdot}h$ (Figure 10.126). The clarifier operation is clearly underloaded since G_{max} is higher than the limiting solids flux $G_L = 5.7\ kg/m^2{\cdot}h$ or C_u is lower than the maximum allowable underflow soluble concentration $C_{u,max} = 13.5\ kg/m^3$.

c. Determine the value of *SLR*.

Use the design $SLR = G_L = 5.7\ kg/m^2{\cdot}h$ for the design of the final clarifier. This will give a conservative design since G_L is less than G_{max}.

3. Determine the surface area and diameter required per clarifier.

a. Determine the required surface area based on the *SOR*, $A_{SOR} = \dfrac{Q}{SOR} = \dfrac{2000\ m^3/d}{24\ m^3/m^2{\cdot}d} = 83.3\ m^2$

b. Determine the required surface area based on the *SLR* (A_{SLR}).

The solids reaching the clarifier, $\Delta TSS = (1 + R_{RS})\,Q\,C_0$

$$= (1 + 0.6) \times 2000\ m^3/d \times 4\ kg/m^3 = 12{,}800\ kg/d$$

The clarifier surface area required at the design *SLR*,

$$A_{SLR} = \frac{\Delta TSS}{SLR} = \frac{12{,}800\ kg/d}{5.7\ kg/m^2{\cdot}h \times 24\ h/d} = 93.6\ m^2$$

c. Determine the surface area and diameter of the clarifier.

Select the surface area of the final clarifier $A = 93.6\ m^2$ based on SLR requirement. The corresponding diameter of each clarifier is 11 m (36 ft). The clarifier plan and section views with design details of influent and effluent structures are shown in Figure 10.127.

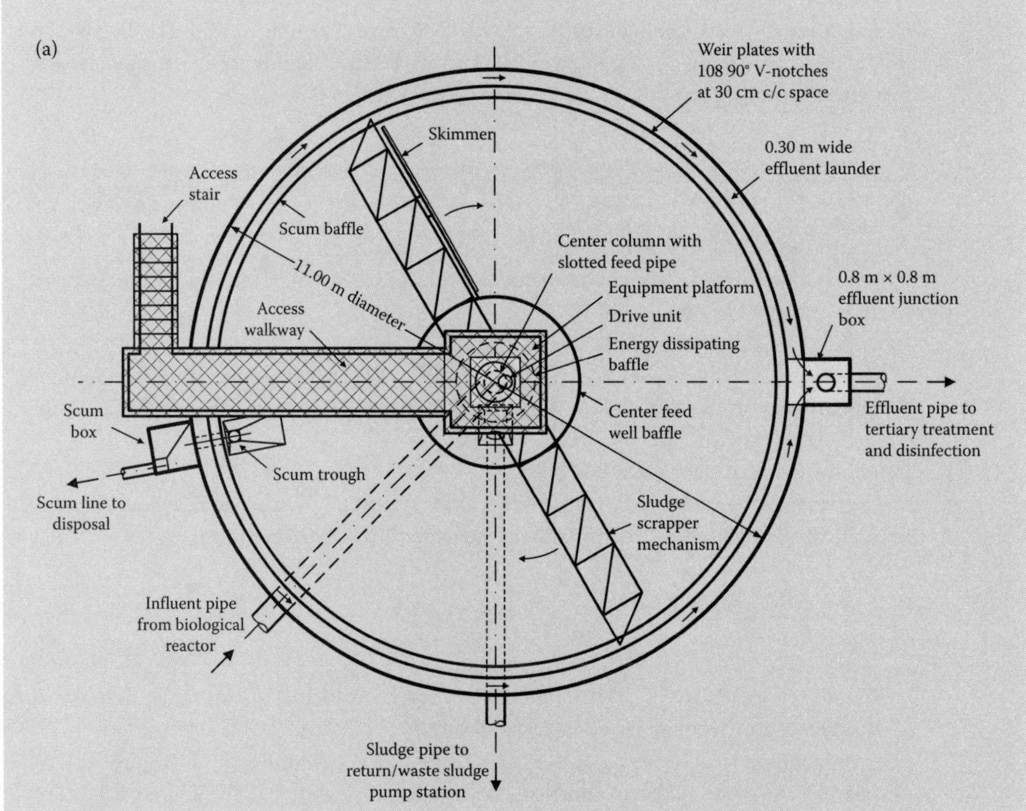

FIGURE 10.127 Design details of final clarifier: (a) plan view (Example 10.166) (*Continued*)

4. Verify the design SOR and SLR.

The actual surface area of each clarifier, $A = \frac{\pi}{4}D^2 = \frac{\pi}{4} \times (11\text{ m})^2 = 95.0\text{ m}^2$

$$SOR = \frac{Q}{A} = \frac{2000\text{ m}^3/\text{d}}{95\text{ m}^2} = 21\text{ m}^3/\text{m}^2{\cdot}\text{d} < 24\text{ m}^3/\text{m}^2{\cdot}\text{d} \quad \text{(See the problem statement)}$$

$$SOR_{\text{peak}} = \frac{Q_{\text{peak}}}{A} = \frac{5600\text{ m}^3/\text{d}}{95\text{ m}^2} = 59\text{ m}^3/\text{m}^2{\cdot}\text{d}$$

(It is within the typical range of 40–64 $\text{m}^3/\text{m}^2{\cdot}\text{d}$ for a BNR process in Table 10.63.)

$$SLR = \frac{\Delta TSS}{A} = \frac{12{,}800\text{ kg/d}}{95\text{ m}^2 \times 24\text{ h/d}} = 5.6\text{ kg/m}^2{\cdot}\text{h} < 8\text{ kg/m}^2{\cdot}\text{h} \quad \text{(See the problem statement)}$$

5. Determine the depth of clarifier.
   a. Select the depth for clear water and settling zones.

      The clear water zone is normally 1.0–1.5 m. The settling zone is normally 1.5–2 m. Provide a total depth $H_{\text{csz}} = 3.5$ m for the clear water and settling zones.
   b. Determine the depth of thickening zone.

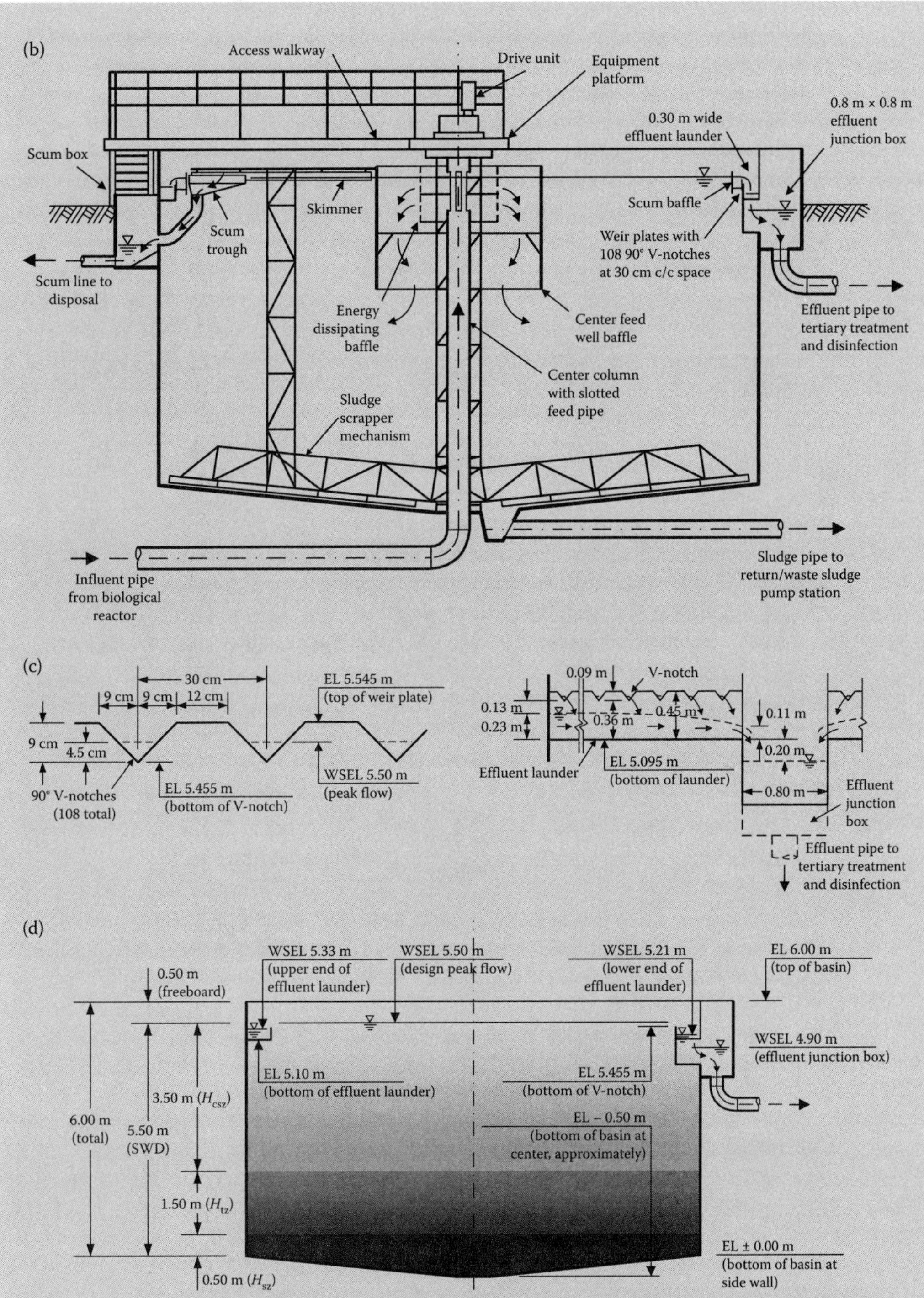

FIGURE 10.127 **(Continued)** Design details of final clarifier: (b) conceptual section view, (c) effluent structure design details with water surface elevations at the design peak flow condition, and (d) hydraulic profile, depths of functional zones, and other key elevations through final clarifier (Example 10.166).

The depth of thickening sludge zone is calculated using the procedure given in References 2, 7, 17, 97, and 98.

The reactor volume for each BNR process train is half of the total volume $V_{BNR} = 1833\ m^3$ determined in Example 10.157. The reactor volume served by each clarifier $V = 917\ m^3$.

The biomass in each biological reactor, $\Delta m_{BNR} = C_0 \times V = 4\ kg/m^3 \times 917\ m^3 = 3668\ kg\ TSS$

Assume that 30% of total biomass in the suspended growth biological reactor is retained in the thickening zone.

The biomass in thickening zone of each clarifier, $\Delta m_{tz} = 0.3 \times \Delta m_{BNR}$

$$= 0.3 \times 3668\ kg\ TSS = 1100\ kg\ TSS$$

Assume that the average solids concentration of the sludge blanket, $C_{tz} = 7.5\ kg/m^3$ (It is approximately the average value of C_0 and C_u).

The required depth of the thickening zone, $H_{tz} = \dfrac{\Delta m_{tz}}{C_{tz}A} = \dfrac{1100\ kg}{7.5\ kg/m^3 \times 95\ m^2} = 1.5\ m$

c. Determine the depth of sludge storage zone.

The biomass produced in a BNR facility is due to BOD_5 consumption, nitrification, and denitrification. The combined heterotrophic and autotrophic biomass growth in the BNR system, $p_{x,BNR} = 84.1$ mg VSS/L (Example 10.153).

The total amount of biomass yield in each BNR process train at the design average flow,

$$\Delta m_{x,BNR} = p_{x,BNR} \times Q = 84.1\ g/m^3 \times 2000\ m^3/d \times 10^{-3}\ kg/g = 168\ kg\ VSS/d$$

Calculate the total solids generated ($\Delta m_{TSS,BNR}$) using a VSS/TSS ratio of 0.7 (Example 10.153).

$$\Delta m_{TSS,BNR} = \frac{\Delta m_{x,BNR}}{\text{VSS/TSS ratio}} = \frac{168\ kg\ VSS/d}{0.7\ kg\ VSS/kg\ TSS} = 240\ kg\ TSS/d$$

Long sludge storage in the clarifier may create anaerobic condition and phosphorus release may occur. In a well-designed and operated system a 1-day storage capacity may be sufficient. A safety factor $SF_{BOD} = 2$ is normally applied for the sustained high BOD_5 loading.

The maximum amount of solids in the storage zone, $\Delta m_{sz} = SF_{BOD} t_{sz} \Delta m_{TSS,BNR}$

$$= 2 \times 1\ d \times 240\ kg\ TSS/d = 480\ kg\ TSS$$

Assume the underflow solids concentration $C_u = 10.7\ kg/m^3$ is reached in the storage zone and calculate the required depth of the storage zone (H_{sz}).

$$H_{sz} = \frac{\Delta m_{sz}}{C_u A} = \frac{480\ kg}{10.7\ kg/m^3 \times 95\ m^2} = 0.5\ m$$

d. Determine the total side water depth (SWD) and total clarifier depth.

$$SWD = H_{csz} + H_{tz} + H_{sz} = (3.5 + 1.5 + 0.5)\ m = 5.5\ m$$

Provide a freeboard of 0.5 m. The total side wall height of the clarifier is 6 m.

6. Determine the design hydraulic retention times (HRTs) under design average and peak flows.

   The effective volume of each clarifier, $V = SWD \times A = 5.5\,\text{m} \times 95.0\,\text{m}^2 = 523\,\text{m}^3$

   At the design average flow, $HRT = \dfrac{V}{Q} = \dfrac{523\,\text{m}^3 \times 24\,\text{h/d}}{2000\,\text{m}^3/\text{d}} = 6.3\,\text{h}$

   At the design peak flow, $HRT_{\text{peak}} = \dfrac{V}{Q_{\text{peak}}} = \dfrac{523\,\text{m}^3 \times 24\,\text{h/d}}{5600\,\text{m}^3/\text{d}} = 2.2\,\text{h}$

   Note: The HRT for final clarifier under design peak flow condition is normally in the range of 1.8–2.2 h.
7. Select the influent structure.

   The influent structure consists of a center feed well in which properly designed ports with baffles are provided for dissipating energy and distributing the flow evenly throughout the tank. The center feed well is normally designed, fabricated, and supplied as a package with center draft pipe and/or support column by the clarifier manufacturer. A conceptual sectional view of the final clarifier with influent structure and equipment details is shown in Figure 10.127b. Basic design considerations of influent structures for sedimentation basin are provided in Section 9.4.5. Different types of influent structurers for rectangular and circular clarifiers are shown in Figures 9.13 and 9.14. Review Example 9.21 for detailed design calculations of influent structure of a rectangular primary sedimentation basin.
8. Develop the design of effluent structure.

   The effluent structure consists of an effluent baffle around the periphery, V-notches, effluent launder, effluent box, and an exit pipe. Basic design considerations of effluent structures for sedimentation basin are provided in Section 9.4.5. Detailed design calculations of effluent structure are presented for a rectangular primary sedimentation basin in Example 9.21. Similar procedure is used to calculate the effluent weir with 90° V-notches, and effluent launder with free fall. Important calculation results are briefly summarized below.

   a. Calculate the total number of 90° V-notches per clarifier.

      Provide a 0.3-m wide effluent launder with weir plates installed on the inner side of the launder wall. The total length of the weir plates per clarifier, $L_{\text{weir}} = \pi(D - 2b) = \pi \times (11\text{ m} - 2 \times 0.3\text{ m}) = 32.7\text{ m}$

      The total number of 90° V-notches per clarifier at a center-to-center space of 30 cm,

      $$N_{\text{notch}} = \frac{L_{\text{weir}}}{L_{\text{space}}} - 1 = \frac{32.7\,\text{m}}{0.3\,\text{m/notch}} - 1 = 108\text{ notches}$$

   b. Calculate the height of the weir plate.

      The height of weir plate is determined under the design peak flow condition.

      The flow per 90° V-notch at design peak flow,

      $$q_{\text{peak}} = \frac{Q_{\text{peak}}}{N_{\text{notch}}} = \frac{0.065\,\text{m}^3/\text{s}}{108\text{ notches}} = 0.00060\,\text{m}^3/\text{s per notch}$$

      The head over 90° V-notch is calculated from Equation 9.5b.

      $$H_{\text{peak}} = \left(\frac{15}{8} \times \frac{0.0006\,\text{m}^3/\text{s}}{0.6 \times \sqrt{2 \times 9.81\,\text{m/s}^2}}\right)^{2/5} = 0.045\,\text{m or } 4.5\,\text{cm}$$

Similar procedure is used to obtain 3 cm head over the weir under the design average flow condition.

Provide 4.5 cm freeboard at the design peak flow. Total depth of V-notch weir plate $D_{plate} =$ 4.5 cm + 4.5 cm (free board) = 9 cm. The hydraulic profile through the effluent weir is shown in Figure 10.127c.

c. Check the weir loading rates (WLRs) under both design average and peak flow conditions.

$$\text{At the design average flow, } WLR = \frac{Q}{L_{weir}} = \frac{2000\,\text{m}^3/\text{d}}{32.7\,\text{m}} = 61\,\text{m}^3/\text{m·d} < 125\,\text{m}^3/\text{m·d}$$

$$\text{At the design peak flow, } WLR = \frac{Q_{peak}}{L_{weir}} = \frac{5600\,\text{m}^3/\text{d}}{32.7\,\text{m}} = 171\,\text{m}^3/\text{m·d} < 250\,\text{m}^3/\text{m·d}$$

Note: The design WLRs under both design average and peak flow conditions are within the acceptable range (see problem statement).

d. Calculate the depth of the effluent launder.

Provide a common junction box 0.8 m × 0.8 m. The box receives effluent from the effluent launders on both sides. The flow in each launder is one half of the total peak flow from the clarifier.

$$Q_{peak,\,launder} = \frac{1}{2} Q_{peak} = \frac{1}{2} \times 0.065\,\text{m}^3/\text{s} = 0.033\,\text{m}^3/\text{s}$$

The combined effluent exits through a 0.3-m diameter outlet pipe. The water depth in the junction box at design peak flow is maintained 0.2 m below the bottom of the effluent launder. Therefore, a free fall exists at the end of the effluent launder. The critical depth at design peak flow is calculated from rearranged Equation 7.10 with $C_w = 1.0$.

$$y_c = \left(\frac{Q_{peak,\,launder}}{C_w b \sqrt{g}}\right)^{2/3} = \left(\frac{0.033\,\text{m}^3/\text{s}}{1.0 \times 0.3\,\text{m} \times \sqrt{9.81\,\text{m/s}^2}}\right)^{2/3} = 0.11\,\text{m}$$

The water depth y_1 at the upper end of the launder is calculated from Equation 8.13a using $y_2 = y_c = 0.09$ m at the lower end of the launder.

$$y_1 = \sqrt{y_2^2 + \frac{2(Q_{peak,\,launder})}{gb^2 y_2}} = \sqrt{(0.11\,\text{m})^2 + \frac{2 \times (0.033\,\text{m}^3/\text{s})^2}{9.81\,\text{m/s}^2 \times (0.3\text{m})^2 \times 0.11\,\text{m}}} = 0.19\,\text{m}$$

Add 20% additional depth for friction losses and turbulence. Add a clearance of 0.13 m for free fall after the V-notches. Total depth of effluent launder = 1.2 × 0.19 m + 0.13 m + 0.09 m (height of V-notch) = 0.45 m. The water depths through the effluent launder is also shown in Figure 10.127c.

At the design average flow, the critical depth is 0.055 and the water depth at the upper end of the launder is 0.11 m.

9. Develop the hydraulic profile.

An overall hydraulic profile showing major design water surface elevations, depths of functional zones, and other key elevations through the final clarifier is shown in Figure 10.127d.

10. Describe the sludge collection system and skimmer.

The sludge collection system consists of a rotating rake with scraper blades. The settled sludge will be collected into a hopper near the center of the clarifier. The RAS is recycled back to the BNR biological reactor and waste activated sludge (WAS) is sent to solids handling area. The fixed access bridge with a walkway will house the drive unit, and will be supported by a central column. The skimmer will move the scum into a scum trough for removal.

11. Describe the RAS/WAS pumps.

 Provide three nonclog, recessed impeller sludge pumps with variable speed drive (VFD), one pump for each clarifier and one identical standby pump. Each pump shall be capable of pumping 150% of average return flow (1800 m^3/d). A magnetic flow meter shall be provided in each force main to monitor and control the RAS flow from each clarifier. A sonic sludge blanket meter shall be provided in the sludge wet well to control the pumping operation. A WAS line with magnetic flow meter shall be provided from each RAS force main to remove and monitor the WAS flow. Basic information about pumping equipment and flow measuring devices are briefly presented in Chapter 6. The pump design is similar to that of a dry well horizontal centrifugal pump with flooded suction pipe. The readers may review References 2, 7, 17, 97, and 98 for details of pump design.

Discussion Topics and Review Problems

10.1 In the log-growth phase, the biomass grows and substrate is consumed. Both biomass growth and substrate reduction follow exponential relationship with respect to aeration period. A plug-flow reactor has aeration period of 5 h. The reaction rates of biomass growth and substrate utilization are 0.05 and 0.61 h^{-1}, respectively. The concentrations of influent soluble BOD_5 and MLVSS are 980 and 2000 mg/L. Determine the following: (a) VSS growth and concentration at the end of aeration period, (b) effluent BOD_5 concentration, (c) yield coefficient based on VSS and BOD_5, and (d) theoretical aeration period when biomass concentration is doubled and substrate concentration is reduced to half.

10.2 The chemical formula of volatile solids of a mixed-culture biomass is given by $C_{60}H_{90}O_{30}N_{10}P$. Determine the percent composition of each element.

10.3 The pseudo first-order reaction rate constant was determined from a batch reactor study. An industrial wastewater sample was aerated using the acclimated biomass. Samples of MLSS were withdrawn on the hour. The soluble COD and VSS concentrations were measured in each hourly sample. The results are summarized below. Determine the reaction rate constant k.

Time, h	0	1	2	3	4	5
Biodegradable COD (S_{bd}), mg/L	700	400	230	130	80	42
MLVSS, mg/L	1980	2020	2040	2060	2080	2090

10.4 A batch reactor was operated to determine the biological kinetic coefficients of an industrial wastewater. A mixture of wastewater and acclimated biological seed was aerated. Hourly samples were withdrawn and analyzed for soluble COD and VSS. The results are summarized below. Determine the biological kinetic coefficients Y and k_d and substrate utilization rate for basal metabolism. Also, determine from the same data the maximum specific growth rate μ_{max} and kinetic coefficients K_s and k.

Aeration period, h	0	1	2	3	4
Soluble BOD_5, mg/L	372	204	120	66	36
VSS, mg/L	960	1010	1035	1047	1056

10.5 A bench-scale batch reactor study was conducted on an industrial wastewater. The operational data used were biodegradable soluble COD and VSS. The kinetic coefficients Y and k_d were 0.4 and 0.05 d^{-1}, respectively. Determine the volume of aeration basin to treat a flow of 3800 m^3/d having influent biodegradable soluble COD of 300 mg/L. The MLVSS concentration in the reactor was 2000 mg/L. The biomass growth in the reactor was 20 mg VSS/L·h.

10.6 A flow-through aerated lagoon is designed to pretreat an industrial wastewater. The aeration period is 3 d and influent soluble $BOD_5 = 400$ mg/L. The biodegradable kinetic coefficients are: $K_s = 140$ mg/L, $k = 5\ d^{-1}$, $Y = 0.5$, and $k_d = 0.05\ d^{-1}$. The TSS is 65% biodegradable, and each gram of biodegradable solids = 1.42 g BOD_L. Determine the effluent TSS, soluble BOD_5, and total BOD_5. Also, determine the theoretical oxygen requirements. Assume $BOD_5/BOD_L = 0.68$ and VSS/TSS = 0.8.

10.7 A flow-through aerated lagoon is designed to treat wastewater that has soluble BOD_5 of 400 mg/L. The aeration period is 3 days. Determine the pseudo first-order reaction rate constant *K*. The biological kinetic coefficients are: $K_s = 112$ mg/L, $k = 5\ d^{-1}$, $k_d = 0.05\ d^{-1}$, and $Y = 0.5$.

10.8 Equation 10.53 is normally used to estimate the effluent BOD from an aerated lagoon at the operating temperature. The equation is based on first-order reaction kinetics under steady-state condition. Develop this equation.

10.9 A bench-scale study using a complete-mix reactor without clarification was conducted to determine the reaction rate constant *k*. The results are summarized below. Write the procedure and determine reaction rate constant *k*. The reactor volume is 15.5 L, and influent $BOD_5 = 200$ mg/L.

Flow, L/h	3.1	3.9	5.2	7.8	15.5
Effluent soluble BOD_5, mg/L	72.7	83.3	97.6	117.6	148.2

10.10 A bench-scale continuous flow completely mixed biological reactor study was conducted to determine the biological kinetic coefficients. Four identical biological reactors were installed and operated over a 4-week period. The total volume of reactor assembly was 10.9 L. The clarifier section was separated from the reactor assembly by a removable partition wall. The volume of clarifier section was 18% of total volume of the reactor assembly. The MLSS concentration maintained in each reactor was different. After the steady-state condition was reached, the daily data collection began. The average data over 1-week period are summarized in the table below. Determine the kinetic coefficients K_s, *k*, *Y*, and k_d.

Reactor NO.	Target MLSS Concentration Maintained in Aeration Zone, mg/L	MLVSS/MLSS Ratio	MLVSS Maintained in Aeration Zone (X_m), mg/L	MLVSS Measured in the Total Reactor Volume after 1 Day, mg/L	TSS in Effluent, mg/L	Influent BOD_5 (S_0), mg/L	Effluent BOD_5 (S_e), mg/L	Effluent Volume (Q), mg/L
1	2050	0.73	1500	1249	4	148	5	30.0
2	1800	0.72	1300	1100	4	156	6	31.0
3	1500	0.73	1100	953	6	180	9	31.5
4	1250	0.72	900	797	7	187	11	32.5

10.11 A small dairy industry is treating its wastewater in an activated sludge plant. The flow and soluble BOD_5 to the aeration basin are 2000 m^3/d and 300 mg/ L, respectively. The kinetic coefficients and design parameters are $\theta_c = 10$ d, $Y = 0.55$, $k = 4.0\ d^{-1}$, $K_s = 95$ mg/L, $k_d = 0.04\ d^{-1}$, $BOD_5 = 0.7$ BOD_L, effluent soluble $BOD_5 = 30$ mg/L, TSS in the effluent = 30 mg/L, biological solids/TSS = 0.72, BOD_L/biodegradable solids = 1.42, MLVSS maintained = 2800 mg/L, and TSS in return sludge = 12,000 mg/L. Determine the following (a) volume of aeration basin, (b) aeration period, (c) return sludge flow and Q_r/Q ratio, (d) food to MO ratio, (e) biomass produced, (f) oxygen requirement, and (g) volume of waste activated sludge at sp. gr. of 1.025.

10.12 The MLSS concentration in an aeration basin is 3000 mg/L. The MLSS was settled in a 1-L graduated cylinder for 30 min. The settled volume is 250 mL. Calculate SVI from Equations 10.30d and 10.32a. Also, calculate the sludge density index (SDI).

10.13 Compare the performance of a CFSTR and PFR treating municipal wastewater that has influent soluble BOD_5 of 150 mg/L. Use the following design data: $Y = 0.6$, $k = 6\ d^{-1}$, $K_s = 95$ mg/L, $k_d = 0.05\ d^{-1}$, and $\theta_c = 1.0$ d. Assume that the return sludge ratio $R_r = 0.25$, and waste activated sludge volume Q_{was} is small and is ignored.

10.14 A biological reactor is designed for an average flow of 10,000 m^3/d. Compare detention times and reactor volumes for three reactor configurations: CFSTR, PFR, and dispersed plug-flow reactor (DPFR). The dispersion number $d = 0.3$, influent and effluent substrate concentrations are 100 and 5 mg/L, respectively, and the first-order reaction kinetic coefficient is 0.40 h^{-1}.

10.15 A package-extended aeration plant is designed to treat domestic wastewater from a small community. The plant receives a flow of 0.5 MGD. The influent BOD_5 and TSS are 210 and 250 mg/L. The aeration period, Y, and k_d are 24 h, 0.65, and 0.05 d^{-1}, respectively. The fraction f of volatile solids in aeration basin is 0.5. The aeration and mixing is provided by turbine aerator. The soluble BOD_5 and TSS in the effluent are 4 and 10 mg/L, respectively. There is no sludge wasting. Calculate (a) volume of aeration basin, (b) MLSS concentration in aeration basin, (c) solids retention time (*SRT*), (d) F/M ratio, and (e) theoretical biomass concentration in the aeration basin.

10.16 A contact stabilization plant is designed for an average flow of 8000 m^3/d. The contact time is 50 min, and the volume of reaeration basin is twice that of contact basin. The reaeration period is 6 h and MLSS concentration in the contact basin is 2400 mg/L. Determine the following: (a) volumes of contact and reaeration basin, (b) the return flow Q_r, percent ratio of Q_r/Q, TSS concentration in the return activated sludge (RAS), and (c) the design capacity of an equivalent conventional plant that has aeration basin volume equal to the combined volumes of contact basin and reaeration basin. The aeration period is 8 h and MLSS concentration is same as in the contact basin.

10.17 A surface aerator was tested in tap water and in primary settled wastewater. The test temperatures in both cases were 24°C, and testing was done at 1 atm. The average volume of the aeration basin during testing was 75 m^3. The DO of the deoxygenated tap water and primary settled wastewater was zero ($C_0 = 0$). The DO saturation of wastewater at 20°C was 7.9 mg/L. The measured DO concentrations in aeration basins at different time intervals for both cases are summarized below. Calculate the K_La of aerator in tap water and settled wastewater at 20°C. The temperature coefficient θ for K_La is 1.024. Also, calculate the correction factor α and β.

Elapsed time, min	0	5	10	20	40	50	60
DO concentration in tap water (C_w), mg/L	0	2.88	4.64	6.74	8.00	8.25	8.34
DO concentration in wastewater (C_{ww}), mg/L	0	2.45	3.95	5.85	7.35	7.58	7.78

10.18 A diffused aerator was tested in tap water. The depth of aerator below the surface and operating temperatures and pressure were 4.5 m, 24°C, and 1 atm, respectively. The tap water was deoxygenated to $C_0 = 0$ and $\theta = 1.024$. The aeration test data are tabulated below. Determine the standard $K_La_{,20}$ of the aerator in tap water.

Elapsed time (t), min	0	5	10	20	40	50	60
DO concentration (C), mg/L	0	0.85	1.65	3.35	5.76	6.57	7.21

10.19 A turbine aerator is used in an aeration basin for oxygen supply and to provide mixing. The operating DO and temperature under field conditions are 2.0 mg/L and 24°C. A small bench-scale laboratory column was used to determine the DO saturation and oxygen uptake rate by the microorganisms. Diffused air was supplied in the column containing the MLSS at 24°C. The DO concentration increased to a maximum level of 8.5 mg/L. At that time the air supply was stopped and the DO concentration was measured in the column at different time intervals. The results are summarized below. Determine (a) the oxygen uptake γ_m by the microorganisms at 24°C, and (b) standard oxygen transfer coefficient $K_La_{,w}$ at 20°C in tap water. The correction factor α and θ given by the manufacturer are 0.85 and 1.024.

Time elapsed, min	0	5	10	15	20	25	30	35	40	45
DO concentration, mg/L	8.5	8.1	7.7	7.2	6.9	6.5	6.0	5.7	5.2	4.8

10.20 A fine-bubble diffuser device is tested in clean water at 20°C and at 1 atm. The aeration basin is 5 m deep, and air flow rate, bubble diameter and rise velocity are 1.0 sm^3/min, 0.75 mm, and 13 cm/s, respectively. The tank volume is 35 m^3 and depth of water above diffuser is 4.5 m. The oxygen transfer coefficient K_L at 20°C = 0.07 m/h. Determine the following (a) standard $K_La_{,w}$ at 20°C, (b) standard oxygen transfer rate in tap water at 20°C, (c) diffuse efficiency, and (d) oxygen transfer rate in the wastewater under field conditions. The initial DO in deoxygenated tap water = 0, $\alpha = 0.85$, $\beta =$ 0.90, temperature of MLSS under field conditions = 25°C, DO maintained in aeration basin = 1.0 mg/L, and fouling factor $F = 0.90$.

10.21 A surface aerator is used in a square completely mixed aeration basin that has a volume of 250 m^3. The aeration period is 5 h. The standard K_La_{20} given by the manufacturer is 2.1 d^{-1}. The coefficients α, β, θ, and F are 0.9, 0.85, 1.024, and 1.0, respectively. The power consumption after 24 -h operation is 71.5 kWh. Determine (a) standard oxygen transfer rate (*SOTR*), (b) standard power and oxygen transfer rating of the aerator, (c) oxygen transfer rate (*OTR*) under field conditions, and (d) aerator efficiency. The operating temperature and DO in the aeration basin are 24°C and 1.5 mg/L.

10.22 An aeration basin receives 0.265 MGD (1000 m^3/d) primary treated effluent that has a $BOD_5 = 120$ mg/L. The volume of the aeration basin is 6321 ft^3 (179 m^3). The critical operating temperature in the field is 15°C. The aeration and mixing are achieved by a turbine aerator. The diameter, peripheral speed, and air flow are 2.5 ft (0.762 m), 15 ft/s (4.57 m/s), and 16 scfm (0.45 sm^3/min). The other operating factors are: $\alpha = 0.9$, $\beta = 0.8$, and $F = 1$. BOD_5 in the removed aeration basin = 90% and DO maintained in the aeration basin = 1.0 mg/L. Determine (a) standard oxygen transfer coefficient $K_La_{w,20}$, (b) aerator horse power, (c) oxygen transfer rate in the field (OTR_f), and (d) number of aerators required.

10.23 A completely mixed aeration basin is designed to remove carbonaceous and nitrogenous BOD in a single-stage nitrification process. The characteristics of influent and effluent, and many designed parameters are summarized below:

Influent: flow = 25,000 m^3/d, $BOD_5 = 180$ mg/L, ammonia and organic nitrogen = 25 and 20 mg/L as N.

Effluent: effluent standards are $BOD_5 = 10$ mg/L, TSS = 10 mg/L, and ammonia nitrogen = 1.0 mg/L as N.

Other design parameters: $\theta = 15$ d, $Y_{BOD5} = 0.5$ g VSS/g BOD_5, Y_N for nitrification = 0.2 g VSS/g AN as N, $k_{d,BOD5} = 0.05\ d^{-1}$, $k_{d,n}$ for nitrification = 0.06 d^{-1}, MLVSS/MLSS = 0.75, BOD_5/BOD_L = 0.68, biodegradability of biomass = 0.68 %, BOD_L exerted by each g of biodegradable solids = 14.2 g, organic N in biomass = 12.2%, $\alpha = 0.80$, $\beta = 0.85$, $F = 0.9$, DO maintained in the aeration basin $C_e = 2.0$ mg/L, operating temperature $T = 25$°C, plant is located at 610 m above sea level, liquid depth in the aeration basin = 5 m, and depth of water above diffusers = 4.5 m, and oxygen transfer efficiency of diffuser (OTE_f) = 8%. Calculate (a) theoretical oxygen requirement (*ThOR*), (b) standard oxygen requirement (*SOR*), (c) air supplied to the diffusers, and (d) number of diffuser tubes if diffusers are rated at 0.2155 sm^3/min per tube.

10.24 A blower is designed to supply air in a diffused air system of a BNR facility located 450 m above sea level. An air supply header delivers air from the compressor to the swing-type diffuser pipings and medium fine diffuser tubes. The equivalent length and diameter of the pipe header are 130 and 45 cm. The inlet summer temperature and design air flow are 30°C and 250 m^3/min, respectively.

The air supply pressure at the blower includes total head loss in (a) pipe header, (b) other pipings, valves, elbows, tees, meters, and specials at 30% of head loss in pipe header, (c) accessory losses in head of water for air filter = 50 mm, silencers = 30 mm, compressor pipings, valves and connections = 210 mm, allowance for clogging of diffuser and miscellaneous head losses under emergency condition = 550 mm. Calculate total head loss in pipe header, and the absolute supply pressure at the blower.

10.25 An aeration basin has diffused aeration system. The oxygen requirement is 1200 kg O_2/d. The oxygen transfer efficiency of diffuser is 8% and water depth over diffusers is 4.5 m. The compressor supplies 150% of required air, and has an efficiency of 75% at an inlet temperature and pressure of 30°C and 1 atm. The head loss in piping, air filter, silence, diffuser losses, and allowance for diffuser clogging add up to 1.7 m of water. Determine the power requirement of the compressor.

10.26 A completely mixed aerated lagoon is designed to treat wastewater from a small community. The influent soluble BOD_5 and aeration period are 150 mg/L and 4 d, respectively. Estimate the concentration of soluble and total BOD_5 and TSS in the effluent. Use the following data: $K_s = 70$ mg/L, $k = 5\ d^{-1}$, $Y = 0.45$, $k_d = 0.05\ d^{-1}$, $BOD_5/BOD_L = 0.68$, VSS/TSS = 0.8, biomass is 65% biodegradable, and each gram of biodegradable solids exerts 1.42 g BOD_L. Also, calculate the first-order reaction rate constant for BOD_5 removal in a completely mixed aerated lagoon.

10.27 An aerated lagoon is designed to treat 3000 m^3/d wastewater that has a soluble BOD_5 of 150 mg/L. The operating field temperature is 25°C and pressure of 1 atm. The soluble effluent BOD_5 is not to exceed 15 mg/L. The lagoon length to width ratio = 2:1, water depth = 4.5 m, and side slope = 1:1. The kinetic coefficients are: $K_s = 100$ mg/L, $Y = 0.6$, $k_d = 0.05\ d^{-1}$, and $k = 5.0\ d^{-1}$. The available standard oxygen transfer rate of aeration in water at 20°C and at zero DO given by the manufacturer is 1.2 kg O_2/kwh. The available aerator sizes are 6, 12, and 18 kw. The other information for aerator sizing are: the operating oxygen maintained in the reactor = 2.5 mg/L, $\alpha = 0.9$, and $\beta = 0.9$. Determine the following: (a) dimensions of the aerator lagoon, (b) power requirement of the aerators under field condition, and (c) number of aerators. Show the aerators arrangement.

10.28 Two equal aerobic stabilization ponds are designed to treat wastewater from a residential subdivision. The average total flow is 1000 m^3/d and influent BOD_5 = 180 mg/L. The first-order soluble BOD_5 removal rate constant at 20°C = 0.34 d^{-1} and $\theta = 1.06$. The pond depth is 1.0 m and critical lagoon temperature is 12°C. The side slope is 1V to 3H. The length to width ratio is 2:1. The Thirumurthi's application of Wehner–Wilhelm equation (Equation 3.35) is used to determine the detention time. The dispersion number $d = 1.0$. The soluble BOD_5 removal is 90%. The freeboard = 1.0 m. Determine for each lagoon (a) detention time and volume, (b) the water surface and base dimension, (c) the organic loading, and (d) effluent total BOD_5. The expected maximum algae concentration in the effluent is 120 mg/L. The algae is 68% biodegradable, and each gram of biodegradable matter exerts 1.42 g BOD_L, and $BOD_5/BOD_L = 0.68$. Draw the plan and sectional view showing the influent and effluent structures. The lagoons are designed to operate in parallel and series. Describe the operational features.

10.29 A high-rate trickling filter is designed to treat primary settled municipal wastewater. The flow and BOD_5 concentration are 4000 m^3/d and 120 mg/L. The filter depth is 2.5 m, recycle ratio $R = 2$, and the required effluent BOD_5 is 20 mg/L. The critical operating temperature is 15°C. The BOD_5 removal rate constant in Eckenfelder equation $k_{r,20} = 2.5\ (m^3/d)^{0.5}/m^2$ at 20°C, $n = 0.5$ and $\theta = 1.035$. Calculate (a) filter hydraulic loading, (b) filter area, and (c) volumetric organic loading. Check the required effluent concentration from Velz equation. $BOD_5/BOD_L = 0.68$, the reaction rate constant K (base 10) at 20°C = 0.97 m^{-1}, and $\theta = 1.047$.

10.30 An RBC system is designed to treat primary settled wastewater from a residential community. The average wastewater flow and BOD_5 are 2800 m^3/d and 125 mg/L. The required effluent BOD_5 is 20 mg/L, and Q/A ratio on each stage is 100 L/m^2·d. The reaction rate constant k and total disk area per standard shaft are 60 L/m^2·d. Determine (a) concentration from each stage, (b) number of shafts per stage, (c) hydraulic loading (*HL*), and (d) surface organic loading (*SOL*) over the entire RBC, and SOL over each stage.

10.31 A conventional activated sludge was modified into an integrated fixed-film activated sludge (IFAS) plant to increase the plant capacity. Plastic biofilm carriers were placed in the existing aeration basin to support the attached growth. The volume of aeration basin = 2000 m^3, average plant capacity = 10,000 m^3/d, the BOD_5 of primary settled effluent = 150 mg/L. The media fill volume = 35% of total basin volume. The expected MLSS concentration in aerated basin volume and media fill volume are 3000 and 18,000 mg/L, respectively. The overall VSS/TSS ratio is 0.75. The design F/M ratio is 0.25 kg BOD_5/kg VSS·d. Calculate (a) equivalent biomass, (b) treatment capacity improvement after modification, (c) aeration period before and after modification, and (d) volumetric organic loading (*VOL*) at improved capacity after modification.

10.32 A high-rate completely mixed activated sludge plant has hydraulic retention time of 2.6 h and MLSS concentration of 1500 mg/L. It receives 3000 m^3/d primary settled wastewater having BOD_5 = 160 mg/L. The *SRT* (θ_c) is 2.5 d. The biological kinetic coefficients Y, k_d, and VSS/TSS ratio are 0.5 mg VSS per mg BOD_5 consumed, 0.05 d^{-1}, and 0.73, respectively. The aeration basin was modified into a moving bed biofilm reactor (MBBR) by adding plastic biofilm carriers. The media film reactor volume is 40% of aeration basin volume. The specific surface area of the media is 500 m^2/m^3, and the expected BOD_5 removal flux is 6.8 g BOD_5/m^2·d.

10.33 A submerged downflow reactor is designed to treat primary settled wastewater that has a flow, BOD_5 and TSS of 2000 m^3/d, 140 mg/L, and 100 mg/L, respectively. The soluble fraction of BOD_5 in the influent is 70%, VSS/TSS ratio is 0.75, and 60% VSS is biodegraded in the filter. The TSS concentration in the filter effluent is 5 mg/L. The designed depth of the downflow reactor is 2.5 m, and volumetric organic loading is 4.0 kg BOD_5/m^3·d. The granular filter media is of expanded clay with an average diameter of 3.7 mm. The filter is designed for daily backwash. Determine (a) filter diameter, (b) hydraulic loading, (c) media surface area loading rate, (d) solids generated and accumulated between backwash, and (e) oxygen requirement.

10.34 An anaerobic reactor is designed to treat industrial wastewater that contains 1000 kg COD per day. The volume of the reactor and biomass concentration maintained in the reactor are 590 m^3 and 1500 g/m^3. The biological kinetic coefficients are Y, k_d, k, and K_s are 0.07 g VSS/g COD, 0.03 d^{-1}, 5 g COD/g VSSd, and 125 mg/L, respectively. The waste utilization efficiency E of the reactor is 80%. The stabilization period in the reactor is 10 days, and effluent COD concentration is 25 mg/L. Estimate the volume of methane generated daily from kinetic coefficients Y and k_d, and from kinetic coefficients k and K_s. Compare the results.

10.35 Anaerobic contact process is designed to treat industrial wastewater. The influent characteristics are: average flow = 1000 m^3/d, COD = 4000 mg/L, TSS = 400 mg/L, VSS/TSS = 0.75, biodegradable VSS = 90%. The reactor is designed to remove 85% COD. The expected effluent characteristics are: TSS concentration = 140 mg/L, VSS/TSS ratio = 0.8, biodegradable solids/TSS = 0.65 g biodegradable VSS/g TSS, and COD/biodegradable solids = 1.42 g COD/g biodegradable VSS. The kinetic coefficients and process design parameters are: Y = 0.07 g VSS/g COD, k_d = 0.03 d^{-1}, k = 3.0 d^{-1}, K_s = 700 mg COD/L, SRT safety factor *SF* = 1.5, MLVSS concentration in the reactor (X) = 5000 mg/L, operating temperature and pressure are 35°C and 1.3 atm. The methane content in biogas is 65%. Calculate (a) *SRT* (θ_c), (b) sludge production and waste sludge, (c) reactor volume, and (d) gas production under field condition.

10.36 An UASB reactor is designed to treat industrial wastewater containing high carbohydrates. The influent flow = 900 m^3/d, COD concentration = 4500 mg/L, TSS = 300 mg/L, VSS/TSS = 0.8, biodegradable VSS = 90%. The effluent characteristics required are: TSS concentration = 210 mg/L, VSS/TSS ratio = 0.85, biodegradable solids/TSS = 0.65 g biodegradable VSS/g TSS, and COD/biodegradable solids = 1.42 g COD/g biodegradable VSS. The process design parameters are: design volumetric organic loading (*VOL*) = 10 kg COD/m^3·d, the desired COD removal efficiency of the reactor = 90%, the SRT safety factor *SF* = 1.5, the operating temperature = 30°C.

The biological kinetic coefficients are: $Y = 0.08$ g VSS/g COD, $k_d = 0.03\ d^{-1}$, $k = 2.4\ d^{-1}$, $K_s =$ 800 mg COD/L. Calculate the following: (a) reactor volume, (b) total and soluble COD concentration in the effluent, (c) *SRT* (θ_c), (d) total solids generated and effluent TSS, (e) volatile solids concentration in the reactor and sludge blanket, and (f) HRT.

10.37 An anaerobic fluidized-bed reactor is designed to treat industrial wastewater that has a COD of 3500 mg/L and flow of 200 m^3/d. The fluidized-bed reactor has uniform-sized GAC media of 0.58 mm, sp. gr. of 1.4, and porosity ratio of 0.45. The fluidized-bed ratio (L_{fb}/L) = 1.6. The design volumetric organic loading is 12 kg COD/m^3·d at operating temperature of 25°C. The depth of the reactor is 1.5 times the diameter. Estimate the following: (a) volume of fluidized-bed and empty-bed contact time, (b) settling velocity of clean media, (c) porosity ratio of fluidized bed, (d) upflow velocity, and (e) recycle flow.

10.38 A single-stage suspended growth biological treatment plant is designed to carry out BOD_5 removal and nitrification. The influent flow, BOD_5, and TKN are respectively 2000 m^3/d, 150 mg/L, and 35 mg/L. The cell yield coefficients for BOD_5 removal and nitrification are $Y_{BOD} = 0.5$ mg VSS/mg BOD_5 and $Y_N = 0.15$ mg VSS/mg NH_4-N, and endogenous decay coefficients are $k_{d,BOD} = 0.05$ d^{-1} and $k_{d,N} = 0.15\ d^{-1}$. The effluent BOD_5, TSS, and NH_4-N are 5, 5, and 0.5 mg/L. The ratios of BOD_L/biodegradable solids = 1.42 g BOD_L/g biodegradable VSS, biodegradable solids/TSS = 0.65 g biodegradable VSS/g TSS, $BOD_5/BOD_L = 0.68$, and VSS/TSS = 0.7. Determine the growth of heterotrophic and autotrophic organisms, their ratios, and total solids produced per day. The fraction of nitrogen synthesized in the biomass, $f_N = 0.12$ mg N/mg VSS. The *SRT* (θ_c) = 10 d.

10.39 A suspended growth nitrification facility is designed to achieve less than 0.5 mg/L NH_4-N concentration in the effluent. The operating temperature and DO in the reactor are 10°C and 2.5 mg/L. An operating pH of 7.4 is expected based on buffering capacity of the wastewater. At 20°C the maximum specific growth rate $\mu_{max.AOB} = 0.95$ mg VSS/mg VSS·d, the decay coefficient $k_{d,AOB} = 0.20$ d^{-1}, and the half-saturation concentrations for DO ($K_{DO,AOB}$), and half-velocity constant for NH_4-N (K_N) are 0.55 mgO_2/L and 0.65 mg NH_4-N/L, respectively. The temperature correction coefficients θ_T for $\mu_{max.N}$ and $k_{d,N}$ are 1.072 and 1.029, respectively. Determine the design *SRT* for nitrification of NH_4-N ($\theta_{c,N}^{design}$) from two methods: (a) theoretical *SRT*, and (b) minimum *SRT*. Two safety factors to account for uncertainty against process design and operation SF_p, and safety factor to compensate for nitrification kinetic under field conditions SF_K are 1.5 and 4.55, respectively.

10.40 A single-stage suspended growth biological treatment plant is designed for BOD_5 removal and nitrification. The influent BOD_5 and TKN are 160 mg/L and 36 mg/L, respectively. The maximum specific growth rate for ammonia-oxidizing bacteria under field conditions $\mu_{max.AOB.F} = 0.412$ mg VSS/mg VSS·d are: $Y_{AOB} = 0.13$ mg VSS/mgBOD_5, $k_{d,BOD} = 0.05\ d^{-1}$, $k_{BOD} = 5$ mg BOD/mg VSS·d, $K_{s,BOD5} = 70$ mg BOD/L. Calculate (a) effluent concentration of soluble BOD_5 and NH_4-N, (b) autotrophic and heterotrophic biomass in the aeration basin, and (c) ratio of autotrophs and heterotrophs with respect to total biomass. The MLSS concentration in the aeration basin = 3000 mg TSS/L, design *SRT*, $\theta_c^{design} = 15$ d, and VSS/TSS = 0.75.

10.41 A single-stage suspended growth biological facility is designed to remove CBOD and achieve nitrification. The design flow is 2500 m^3/d. The influent BOD_5 (S_0) and TKN ($N_{0,TKN}$) are 150 and 38 mg/L, respectively. The effluent soluble BOD_5 (S) and NH_4-N concentrations ($N_{NH4\text{-}N}$) are 3.5 and 0.5 mg/L, respectively. The total biomass growth $p_{x,BOD+N} = 51$ mgVSS/L, nitrogen fixed in the biomass is 12% ($f_N = 0.12$ g N/g VSS), and $BOD_5/BOD_L = 0.68$. The operating temperature and DO in the aeration basin are 24°C and 2 mg/L. The DO saturation concentration $C_{e,ww}$ in the wastewater at operating temperature and pressure is 8.64 mg/L. The liquid depth in the basin and diffusers are 5 and 4.6 m. The correction factors α and F for the diffused aeration system are 0.85 and 0.9, respectively. The oxygen transfer efficiency of the diffusers is 15%. Calculate the total theoretical oxygen requirement (*ThOR*) for combined CBOD removal and nitrification. Also, calculate the standard air requirement.

10.42 A two-stage suspended growth biological facility is designed for CBOD removal and nitrification. The influent flow = 3500 m^3/d, TKN = 38 mg/L, and alkalinity = 160 mg/L as $CaCO_3$. In the first stage, CBOD is removed, and 5 mg/L NH_4-N is fixed in the heterotrophic biomass. In the second stage, the remaining NH_4-N is nitrified. In the final effluent, the ammonia concentration is 0.5 mg NH_4-N/L. The autotrophic biomass growth $p_{x,N} = 1.5$ mg VSS/L and nitrogen fixed in the biomass is 12% ($f_N = 0.12$ g N/g VSS). The alkalinity consumption in nitrification process is 7.14 mg Alk as $CaCO_3$ per mg NH_4-N nitrified. Determine (a) total alkalinity consumed in the second stage, (b) supplemental alkalinity need, and (c) dosage and quantity of sodium bicarbonate ($NaHCO_3$) required to maintain a residual alkalinity of 50 mg/L as $CaCO_3$ for pH control.

10.43 A circular trickling filter is designed to provide combined CBOD removal and nitrification. The trickling filter has a diameter of 20 m and filled with plastic media to an effective depth of 7 m. The specific surface area (SA_m) of the media is 100 m^2/m^3. The influent flow = 3000 m^3/d, BOD_5 = 140 mg/L, TKN = 30 mg/L, and recirculation ratio $R = 1.0$. Determine the specific BOD_5 surface loading rate (SLR_{BOD}), specific nitrogen surface loading rate (SLR_N), specific hydraulic surface loading rate (HLR_Q) and the NH_4-N concentration in the effluent at critical operating temperature of 12°C. Also, verify the hydraulic loading rate (*HLR*) and expected recirculation ratio *R*.

10.44 A fixed-film nitrification reactor has cross-flow plastic media modules. The specific surface area SA_m is 110 m^2/m^3. The height of the packing module is 1.4 m. The influent flow and NH_4-N concentration are 2000 m^3/d and 20 mg/L, respectively. The NH_4-N concentration in the effluent = 1 mg/L or less. The hydraulic loading rate $Q_L = 75$ m^3/m^2·d, recirculation ratio $R = 1$, and $J_{N,\,max}$ at critical temperature of 12°C = 2.4 g/m^2·d. Initially the nitrification is zero order to a limiting NH_4-N concentration of 6 mg/L. After that, the nitrification rate becomes first order and the average $J_{N,\,max}$ drops to 1.5 g/m^2·d. Calculate the number of packing modules needed to achieve the required NH_4-N concentration in the effluent $K_{NH4\text{-}N} = 1.5$ mg/L.

10.45 An RBC assembly is designed to carry out CBOD removal and nitrification. The RBC assembly has three parallel trains and three stages. Each parallel shaft in three stages has standard disk area (A_{shaft}) of 9500 m^2. The soluble BOD_5 loading rates (*SOLs*) on three stages are: $SOL_1 = 12$ g sBOD_5/m^2·d, $SOL_2 = 4.0$ g sBOD_5/m^2·d, and $SOL_3 = 1.5$ g sBOD_5/m^2·d. The influent flow is 3000 m^3/d and influent NH_4-N concentration is 22 mg/L. If NH_4-N concentration is 5 mg/L or higher, the reaction rate approaches zero order, $R_{N,max} = 1.5$ g NH_4-N/m^2d. Determine the NH_4-N concentration in the effluent from the third stage.

10.46 A suspended growth biological reactor is converted to an integrated fixed-film activated sludge (IFAS) process to achieve BOD removal and nitrification. The influent flow, BOD_5, and TKN are, respectively, 3000 m^3/d, 150 mg/L, and 38 mg/L. The MLSS concentration in the aeration basin = 2800 mg/L, aeration period = 6 h. The fraction of nitrogen f_N in the biomass = 0.12 g N/g VSS, and VSS/TSS = 0.75. The design parameters of IFAS plant are: the media fill volume fraction $f_m = 0.4$, the specific surface area of the plastic-film carriers $SA_m = 500$ m^2/m^3, and VSS/TSS in the biofilm = 0.7. The desired effluent quality from the IFAS plant sBOD_5 is <2.5 mg/L, and NH_4-N ≤ 0.5 mg/L. The DO concentration = 4.0 mg/L and critical temperature is 12°C. The suspended growth volume removes CBOD in excess of 95% of the influent BOD_5. Nitrification in this section is in the range of 10%–20%. The media fill volume provides nitrification of NH_4-N in excess of 85%. The biological kinetic coefficients for BOD_5 removal are: $Y_{BOD} = 0.6$ mg VSS/mg BOD_5, $k_{d,\,BOD} = 0.05$ d^{-1}, $k_{BOD} = 5$ mg BOD_5/mg VSS·d, $K_{BOD5} = 65$ mg BOD/L, and $\mu_{max,BOD} = 2.5$ mg VSS/mg VSS·d. The biological kinetic coefficients for nitrification (AOB) are: $Y_{AOB} = 0.15$ mg VSS/ mg NH_4-N, $K_{NH4\text{-}N} = 0.05$ mg NH_4-N/L, $k_{d,AOB,F} = 0.135$ d^{-1}, and θ_c^{design} for biofilm = 21 d. The maximum BOD removal flux $J_{BOD,max}$ and maximum nitrogen removal flux $J_{N,max,12}$ in the media fill volume are 6 g BOD_5/m^2·d and 3.5 g NH_4-N/m^2·d. The half-velocity constant for BOD_5 ($K_{m,\,BOD}$) and nitrogen ($K_{m,NH4\text{-}N}$) in the media fillvolume are 65 mg BOD_5/L and 2.5 mg NH_4-N/L, respectively. Determine (a) the design *SRT* in the activated sludge volume $\theta_{c,as}^{design}$, and (b) the ratios of heterotrophic and autotrophic biomass growth in activated sludge and media fill volumes.

10.47 A suspended growth denitrification facility is designed for a flow of 5000 m^3/d. The influent NO_3-N and DO concentrations are 35 and 4 mg/L. The NO_2-N concentration is small and is ignored. Estimate the methanol feed rate as kg/d. Also, calculate the alkalinity recovery as kg $CaCO_3$/d.

10.48 A suspended growth denitrification facility is designed to provide NO_3-N reduction by heterotrophs. The *SRT* $\theta_c^{design} = 25\ d^{-1}$, and MLVSS concentration = 3000 mg/L. The fraction of heterotrophic bacteria active for denitrification $f_{AX} = 0.85$. The bsCOD concentration = 8 mg/L, and fraction of inert (nonbiomass) VSS in the influent $Y_{iVSS} = 0.12$. The theoretical bound oxygen consumption ratio $OE_{NO3\text{-}N} = 2.86$ mg O_2/mg NO_3-N reduced, maximum cell yield coefficient $Y_H =$ 0.3, specific endogenous decay coefficient for heterotrophs under field conditions $k_{d,H,F} = 0.04$ d^{-1}, the maximum specific growth rate per unit mass of heterotrophs under field conditions, $\mu_{max,H,F} = 0.45$ mg VSS/mg VSS·d, and half-velocity constant for sbCOD $K_{S,H} = 0.18$ mg/L. Calculate (a) the NO_3-N reduction rate by heterotrophs ($r_{NO3\text{-}N,F}$), and (b) specific denitrification rate under field operating conditions (*SDNR*).

10.49 An attached growth downflow denitrification filter (DFDNF) is designed for post denitrification of treated effluent. The influent flow, NO_3-N, and TSS concentrations are 3000 m^3/d, 32 mg/L, and 15 mg/L, respectively. The effluent NO_3-N and TSS concentrations in the effluent are 2 and 5 mg/L. Two identical filters are provided. The depth of filters, hydraulic loading (HL), and volumetric NO_3-N loadings (VNL) are 2 m, 75 m^3/m^2·d, and 1.2 kg NO_3-N/m^3·d. Cell yield from methanol $Y_H =$ 0.18 g VSS/g COD. The equivalent COD for methanol utilization is 1.5 mg COD/mg CH_3OH. Filter is backwashed once a day, and backwash flow and duration are 18 m^3/m^2·h and 15 min. Filter bump frequency, flow rate, and duration are 2 h, 12 m^3/m^2·h, and 3 min. Determine (a) filter diameter, (b) methanol feed rate, (c) sludge production, (d) filter backwash water flow rate, and bumping water.

10.50 An enhanced biological phosphorus removal (EBPR) facility is designed. The influent COD = 350 mg/L, BOD_5 = 150 mg/L, rbCOD = 140 mg/L, and total P = 6 mg/L. The expected TSS concentration in the effluent (TSS_e) is less than 8 mg/L. The fermentation efficiency factor $f_{fm} = 0.82$, the overall P content $f_{P,EBPR}$ and VSS/TSS ratio in the effluent TSS are, respectively, 0.03 mg P/mg VSS and 0.79. Determine (a) fermentation condition and availability of SCVFAs, (b) concentration of effluent TP, and (c) HRT of the anoxic zone.

10.51 An EBPR facility has anaerobic, anoxic, and aerobic zones. The influent COD = 280 mg/L, biodegradable soluble COD (bsCOD) = 220 mg/L, rbCOD = 85 mg/L, TP = 6 mg/L, and Ortho-P = 4.5 mg/L. The effluent bsCOD = 20 mg/L and soluble P, $SP_e = 0.3$ mg/L. The MLVSS concentrations = 1500 mg/L, and return sludge ratio $R_{ras} = 0.3$. The fermentation efficiency $f_{fm} = 0.8$. The uptake of SCVFAs as rbCOD is ~12 mg/L for each mg/L of P removed by "luxury" P uptake. The P content in the EBPR biomass is 0.08 g P/g VSS. The Org-P content of biomass from cell synthesis = 0.02 g P/g VSS, and 45% of the excess P is released. The observed biomass yield for heterotrophs is 0.3 g VSS/g bCOD under field conditions. Determine the maximum Ortho-P concentration in the anaerobic zone after P release.

10.52 A BNR facility uses A^2/O process. The influent BOD_5 and TKN are 160 and 38, respectively. The effluent $sBOD_5$ = 35 mg/L and NH_4-N = 0.5 mg/L. The cell yield coefficient and the field endogenous decay coefficients for heterotrophs and nitrifiers are: $Y_H = 0.6$ mg VSS/mg BOD_5, $k_{d,H,F} =$ 0.05 d^{-1}, $Y_N = 0.15$ mg VSS/mg NH_4-N, and $k_{d,N,F} = 0.10\ d^{-1}$. The SRT $\theta_c = 10$ d, and fraction of nitrogen fixed in biomass $f_N = 0.12$ g N/g VSS. Estimate the biomass growth of heterotrophs and autotrophs, and fractional growth of each with respect to total biomass growth.

10.53 A BNR facility has anaerobic, anoxic, and oxic zones similar to A^2/O process train. The heterotrophic $p_{x,H}$ and autotrophic $p_{x,N}$ biomass growth are 62.8 and 2.2 mg VSS/L, respectively. Determine (a) the P content of biomass and (b) TP concentration in the effluent. Additional data are: influent P concentration = 6 mg P/L, and effluent Ortho-P concentration = 0.3 mg P/L. The Org-P content from cell synthesis in the biomass $f_{P,synthesis} = 0.02$ g P/g VSS. TSS concentration in the effluent = 5 mg TSS/L and VSS/TSS = 0.75.

10.54 A BNR facility is designed to carry out BOD_5, nitrogen, and phosphorus removal. The influent BOD_5 = 160 mg/L, effluent nitrate concentration NO_3-N = 5 mg/L and SCVFA uptake $f_{sBOD5/P} = 10$ g $sBOD_5$/g P uptake. The Org-P content of biomass from cell synthesis $f_{P,synthesis} = 0.02$ g P/g VSS. The heterotrophic and autotrophic biomass growths $p_{x,H}$ and $p_{x,N}$ are 62.8 and 2.2 mg VSS/L, respectively. The influent P and effluent Ortho-P concentrations are 6 and 0.3 mg P/L. The cell yield coefficient of heterotrophs $Y_H = 0.6$ mg VSS/mg BOD_5, The field endogenous decay coefficient for heterotrophs $k_{d,H,F} = 0.05\ d^{-1}$ and the design *SRT* $\theta_c = 10$ d.

10.55 A BNR facility is designed to remove BOD_5, TSS, nitrogen, and phosphorus. The P concentration in the influent is 6 mg/L and Ortho-P concentration in effluent is 0.3 mg/L. The total concentrations of autotrophic and heterotrophic biomass $p_{x,H}$ and $p_{x,BNR}$ are 2.2 and 87.2 mg VSS/L. The SCVFA uptake ratio by PAO_5 ($f_{sBOD5/P}$) = 10 mg $sBOD_5$/mg P, and P content of biomass synthesized ($f_{P,synthesis}$) is 2%. The cell yield coefficient Y_H for heterotrophs under field conditions $k_{d,H,F} = 0.05\ d^{-1}$ and $\theta_c = 10$ d. Estimate the heterotrophic biomass growth due to PAO_5 and the fraction of PAO_5 in heterotrophic growth.

10.56 A BNR facility is achieving 0.3 mg Ortho P/L in the effluent. The influent concentration is 6 mg P/L. The heterotrophic biomass production $p_{x,H,PAO}$ with PAOs = 19.2 mg VSS/L. The fraction of P synthesized in the biomass, $f_{P,synthesis} = 0.02$ mg P/mg VSS. The heterotrophic biomass growth in the BNR system $p_{x,BNR} = 85$ mg VSS/L. Determine the fraction of P removed by PAOs, and fraction of P content in PAOs.

10.57 A BNR facility is removing nitrate by nitrification and denitrification. The concentration of TKN in the influent is 38 mg N/L. The NH_4-N and NO_3-N concentrations in the effluent are 0.5 and 4 mg/L, respectively. The fraction of nitrogen fixed in biomass $f_N = 0.12$ mg N/mg VSS. The heterotrophic ($p_{x,N}$) and total biomass growth ($p_{x,BNR}$) in the BNR system are 82.4 and 85 mg VSS/L, respectively. The theoretical oxygen consumption for NO_3-N ($OE_{NO3\text{-}N}$) = 2.86 mg O_3/mg NO_3-N. The cell yield coefficient for heterotrophs $Y_N = 0.6$ mg VSS/mg BOD_5. The endogenous decay coefficient for heterotrophs under field conditions $k_{d,H,F} = 0.05\ d^{-1}$ and design *SRT* $\theta_c = 10$ d.

10.58 In a biological nitrification–denitrification facility, 21 mg/L NH_4-N is nitrified and then 50% nitrate is denitrified. Calculate the alkalinity after nitrification and after denitrification. Assume that the alkalinity of raw wastewater is 200 mg/L as $CaCO_3$.

10.59 A graduated cylinder 0.50 m high is used for zone settling test. A mixed sample of MLSS obtained from a BNR facility was placed in the cylinder. The initial solids concentration C_0 was 4000 mg/L. The height of the interface from the base was measured with respect to time. The data are summarized below. Determine the area of clarifier required to yield a thickened sludge concentration of 20,000 mg/L. The inflow to the basin is 5000 m^3/d. Also, determine the returned sludge ratio and verify the surface overflow and solids loading rates.

Settling time, min	0	2	4	6	10	14	20	30	40	50	60
Height of interface from base, m	0.50	0.43	0.36	0.29	0.18	0.13	0.09	0.06	0.05	0.05	0.05

10.60 Five calibrated and identical columns were used for hindered settling test. Five MLSS samples were prepared from a conventional activated sludge plant. The initial concentrations of MLSS samples were 1000, 2000, 3000, 5000, 10,000, and 15,000 mg/L. Well-mixed samples were placed in five columns. The settled depths of the interface in each column were measured with respect to time. The settling data are summarized in the table provided below. Determine (a) the percent recycle rate, (b) clarifier surface area that will produce an underflow concentration of 12,500 mg/L, and (c) the surface overflow rate (*SOR*). The MLSS concentration in the clarifier is 3000 mg/L, and influent flow to the plant is 8000 m^3/d.

Settling Time, min	Settled Depth of Interface of Different MLSS Concentrations from the Top of Column, m					
	1000	2000	3000	5000	10,000	15,000
10	1.25	0.85	0.42	0.14	0.05	0.03
20	1.85	1.65	0.82	0.27	0.11	0.06
30	1.89	1.80	1.25	0.40	0.16	0.09
40	1.90	1.81	1.42	0.52	0.25	0.11
50	1.90	1.82	1.51	0.64	0.30	0.14
60	1.91	1.82	1.60	0.78	0.34	0.17
80	1.92	1.84	1.66	1.03	0.39	0.21
100	1.92	1.86	1.70	1.17	0.50	0.27
120	1.93	1.88	1.70	1.18	0.53	0.31

A circular clarifier is designed to settle MLSS from a BNR facility. The design solids flux of 4.0 $kg/m^2 \cdot h$ is determined from extensive hindered settling column tests. The average influent flow to the plant is 2500 m^3/d and average MLSS concentration in the BNR facility and return sludge are 4000 and 10,000 mg/L. The hydraulic loading rates at average and 2-h peak design flow should not exceed 24 and 65 $m^3/m^2 \cdot d$. The 2-h peaking factors are 2.5 and 2.4 for flow and solids, respectively. The average solids concentration in the return sludge = 12,000 mg/L. Determine (a) the surface area of the clarifier, (b) check the solids and hydraulic loading rates at average and peak flow, and (c) clear water zone, thickening zone, sludge storage zone, and total side water depth of the clarifier. The HRT in the BNR reactor = 12 h, the average heterotrophic and autotrophic biomass growth is 85 mg VSS/L, and VSS/TSS = 0.7.

References

1. Reynolas, T. D. and P. A. Richards, *Unit Operations and Processes in Environmental Engineering*, 2nd ed, PWS Publishing Co., Boston, MA, 1996.
2. Metcalf & Eddy AECOM, *Wastewater Engineering: Treatment and Resource Recovery*, 5th ed., McGraw-Hill, New York, NY, 2014.
3. Grady, C. P. L., Jr., G. T. Daigger, N. G. Love, and C. D. M. Filipe, *Biological Wastewater Treatment*, 3rd ed., CRC Press, Boca Raton, FL, 2011.
4. Lawson, A. E., *Biology: An Inquiry Approach*, 3rd ed., Kendall/Hunt, Dubuque, Iowa, 2011.
5. Bitton, G. (Editor), *Encyclopedia of Environmental Microbiology*, Wiley, New York, NY, 2002.
6. Mathews, C. K., K. E. van Holde, D. R. Appling, and S. J. Anthony-Cahill, *Biochemistry*, 4th ed., Prentice Hall, Upper Saddle River, NJ, 2013.
7. Qasim, S. R., *Wastewater Treatment Plants: Planning, Design, and Operation*, 2nd ed., CRC Press, Boca Raton, FL, 1999.
8. Madigan, M. T., J. M. Martinko, K. S. Bender, D. H. Buckley, D. A. Stahl, and T. Brock, *Brock Biology of Microorganisms*, 14th ed., Prentice-Hall, Upper Saddle River, NJ, 2014.
9. Eckenfelder, W. W., Jr. and D. L. Ford, *Water Pollution Control: Experimental Procedures for Process Design*, Pemberton Press, Jenkins Publishing Co., Austin and New York, 1970.
10. Benefield, L. D. and C. W Randall, *Biological Process Design for Wastewater Treatment*, Prentice Hall, Englewood Cliffs, NJ, 1980.
11. Eckenfelder, W. W., Jr., *Industrial Water Pollution Control*, 3rd ed., McGraw-Hill, New York, NY, 1999.

12. Tchobanoglous, G. and E. D. Schroeder, *Water Quality: Characteristics Modeling and Modification*, Addison-Welsey, Publishing Co., Reading, MA, 1985.
13. Ramaswamy, C., Qasim, S. R., and J., Zabolio, III, A surrogate test for BOD_5 in the effluent from microsand ballasted flocculation facility, *Proceedings of the Texas Water 2004, Joint Conference of Water Environment Association of Texas and Texas Section of American Water Works Association*, April 5–8, 2004, Arlington, Texas.
14. Monod, J., The growth of bacterial cultures, *Annual Review of Microbiology*, III, 1949, 371–394.
15. Sunstorm, D. W. and H. E. Klei, *Wastewater Treatment*, Prentice-Hall, Englewood Cliffs, NJ, 1979.
16. Qasim, S. R. and M. L. Stinehelfer, Effect of a bacterial culture product on biological kinetics, *Journal Water Pollution Control Federation*, 54(3), March 1982, 255–260.
17. Task Force of the Water Environment Federation and the American Society of Civil Engineers/ Environmental and Water Resources Institute, *Design of Municipal Wastewater Treatment Plant*, WEF Manual of Practice No. 8 and ASCE Manuals and Reports on Engineering Practice No. 76, 5th ed., McGraw-Hill, New York, NY, 2009.
18. Lawrence, A. W. and P. L. McCarty, Kinetics of methane fermentation in anaerobic treatment, *Journal Water Pollution Control Federation*, 41(2), 1969, R1–R17.
19. Lawrence, A. W. and P. L. McCarty, Unified basis for biological treatment design and operation, *Journal of the Sanitary Engineering Division, ASCE*, 96(SA3), 1970, 757–778.
20. Murphy K. L. and B. I. Boyko, Longitudinal mixing in spiral flow aeration tanks, *Journal of the Sanitary Engineering Division, ASCE*, 96(2), 1971, 211–221.
21. Levenspiel O. and K. B. Bischoff, Backmixing in the design of chemical reactors, *Industrial & Engineering Chemistry*, 51(12), 1959, 1431–1434.
22. Szekely, J., O. Levenspiel, and K. B. Bischoff, Correspondence. Reaction rate constant may modify the effects of backmixing, *Industrial & Engineering Chemistry*, 53(4), 1961, 313–314.
23. Jenkins, D., M. G. Richards, and G. T. Daigger, *Manual on the Causes and Control of Activated Sludge Bulking, Foaming, and Other Solids Separation Problems*, 3rd ed., Lewis Publishers, Ann Arbor, MI, 2003.
24. Palm, J. C., D. Jenkins, and D. S. Parker, Relationship between organic loading, dissolved oxygen concentration and sludge settleability in the completely-mixed activated sludge process, *Journal Water Pollution Control Federation*, 52(10), 1980, 2484–2506.
25. Insel, G., D. Russell, B. Beck, and P. A. Vanrolleghem, Evaluation of nutrient removal performance for an Orbal plant using ASM2d Model, *Proceedings of the 76th Annual WEFTEC®, Water Environment Federation*, Los Angeles, CA, October 2003.
26. Evoqua Water Technologies, *Orbal® Multichannel Oxidation Ditch for Wastewater Treatment*, http://www.evoqua.com, 2014.
27. U.S. Environmental Protection Agency, *Technology Assessment of the Deep Shaft Biological Reactor*, EPA 600/2-82-002, Office of Research and Development, Cincinnati, OH, 1982.
28. Forster, C., *Wastewater Treatment and Technology*, Thomas Telford, London, 2003.
29. McWhirter, J. R., *The Use of High-purity Oxygen in the Activated Sludge Process*, Vol. 2, CRC Press, West Palm Beach, FL, 1978.
30. Daigger, G. T., G. D. Waltrip, E. D. Romm, and L. M. Morales, Enhanced secondary treatment incorporating biological nutrient removal, *Journal Water Pollution Control Federation*, 60(10), 1988, 1833–1842.
31. Irvine, R. L., L. H. Ketchum, R. Breyfogle, and E. F. Barth, Municipal application of sequencing batch treatment, *Journal Water Pollution Control Federation*, 55(5), 1983, 484–488.
32. U.S. Environmental Protection Agency, *Wastewater Technology Fact Sheet: Sequencing Batch Reactors*, EPA 832-F-99-073, Office of Water, Washington, DC, September 1999.
33. Mahvi, A. H., A. R. Mesdaghinia, and F. Karakani, Feasibility of continuous flow sequencing batch reactor in domestic wastewater treatment, *American Journal of Applied Sciences*, 1(4), 2004, 348–353.

34. Xylem, Inc., *Sanitaire ICEAS^{TM} Advanced SBR*, http://www.xylemwatersolutions.com, 2012.
35. Sriwiriyarat, T. and C. W. Randall, Evaluation of integrated fixed film activated sludge wastewater treatment processes at high mean cell residence time and low temperatures, *Journal of Environmental Engineering, ASCE*, 131(11), 2005, 1550–1556.
36. Randall, C. W. and D. Sen, Full-scale evaluation of an integrated fixed-film activated sludge (IFAS) process for enhanced nitrogen removal, *Water Science and Technology*, 33(12), 1996, 155–162.
37. HeadworksBIO, Inc., *ACTIVECELL® MBBR/IFAS Wastewater Treatment*, http://www.headworksbio.com, 2012.
38. SUEZ Treatment Solutions, Inc., *Meteor® IFAS/MBBR Process*, http://www.suez-na.com, 2011.
39. Kruger/Veolia Water Technologies, *AnoxKaldnes^{TM} Moving Bed Biofilm Reactor (MBBR) and Integrated Fixed-Film Activated Sludge (IFAS)*, http://www.krugerusa.com, 2013.
40. Hegg, B. A., L. DeMers, and J. Barber, *Retrofitting Publicly-Owned Treatment Works for Compliance*, Noyes Data Corporation, Park Ridge, NJ, 1990.
41. Martin M. J., A. Artola, M. D. Balaguer, and M. Rigola, Enhancement of activated sludge process by activated carbon produced from surplus biological sludge, *Biotechnology Letters*, 24(3), 2002, 163–168.
42. U.S. Environmental Protection Agency, *Wastewater Management Fact Sheet: Membrane Bioreactors*, September 2007.
43. Aqua-Aerobic Systems, Inc., *Aqua-Aerobic MBR® Membrane Bioreactor System*, http://www.aqua-aerobic.com, 2013.
44. General Electric Company, *LEAPmbr, taking ZeeWeed* MBR technology to the next level*, http://www.ge-energy.com, 2011.
45. Koch Membrane Systems, Inc., *PURON® Solutions, Low Energy, Cost-Effective Wastewater Treatment*, http://www.kochmembrane.com, 2014.
46. Kruger/Veolia Water Technologies, *NEOSEP^{TM} MBR*, http://www.krugerusa.com, 2014.
47. WesTech Engineering, Inc., *ClearLogic® Membrane Bioreactor*, http://www.westech-inc.com, 2010.
48. Kalinske, A. A., Economics of aeration in waste treatment, *Proceedings of the 23rd Purdue Industrial Waste Conference*, 1968, 338–397.
49. Elmore, HY. L., and T. W. Hayes, Solubility of atmospheric oxygen in water, 29th Progress Report of the Committee on Sanitary Engineering Research, *Journal of the Sanitary Engineering Division, ASCE*, 86(AS4), 1960, 41–53.
50. Eckenfelder, W. W., D. L. Ford, and A. J. Englande Jr., *Industrial Water Quality*, 4th ed., McGraw-Hill, New York, NY, 2009.
51. APHA, AWWA, and WEF, *Standard Methods for Examination of Water and Wastewater*, 22nd ed., American Public Health Association, Washington, DC, 2012.
52. ASCE, *Measurement of Oxygen Transfer in Clean Water*, ASCE/EWRI Standard 2-06, Reston, VA, 2007.
53. Stenstrom, M., S. Leu, and P. Jiang, Theory to practice: oxygen transfer and the new ASCE Standard, *Proceedings of the 79th Annual WEFTEC®*, Water Environment Federation, Dallas, TX, October 2006.
54. Droste, R. L., *Theory and Practice of Water and Wastewater Treatment*, John Wiley & Sons, Inc., New York, NY, 1997.
55. ASCE, Standard guidelines for in-process oxygen transfer testing, ASCE Standard 18-96, Reston, VA, 1997.
56. Graves, K., G. T. Daigger, T. J. Simpkin, D. T. Redman, and L. Ewing, Evaluation of oxygen transfer efficiency and alpha factor on variety of diffused aeration systems, *Water Environment Research*, 64(5), 1992, 691–698.

57. U.S. Environmental Protection Agency, *Fine Pore Aeration Systems*, EPA/625/1-89/023, Center for Environmental Research Information/Risk Reduction Engineering Laboratory, Cincinnati, OH, September 1989.
58. Jenkins, T. E., *Aeration Control System Design: A Practical Guide to Energy and Process Optimization*, John Wiley & Sons, Inc., Hoboken, NJ, 2014.
59. ASCE and WPCF, *Aeration*, ASCE Manuals and Reports on Engineering Practice, No. 68 and WPCF Manual of Practice, No. FD-13, American Society of Civil Engineering and Water Pollution Control Federation, 1988.
60. U.S. Environmental Protection Agency, *Wastewater Technology Fact Sheet: Fine Bubble Aeration*, September 1999.
61. Mueller, J. A., W. C. Boyle, and H. J. Popel, *Aeration: Principles and Practice*, CRC Press LLC, Boca Raton, FL, 2002.
62. Xylem, *Aeration Products for Energy-Efficient Biological Treatment*, http://www.xyleminc.com, 2011.
63. Evoqua Water Technologies, *Flexdome™ Fine Bubble Membrane Diffuser*, http://www.evoqua.com, 2014.
64. Evoqua Water Technologies, *Dualair™ Fine Bubble Ceramic or Membrane Diffuser System*, http://www.evoqua.com, 2014.
65. Bilfinger Water Technologies, *Passavant® Fine Bubble Aeration Roeflex®/Bioflex®*, http://www.water.bilfinger.com, 2014.
66. Environmental Dynamics Inc. (EDI), *EDI FlexAir® T-Series Tube Diffuser, Fine Bubble Flexible Membrane Technology*, http://www.wastewater.com, 2006.
67. Aqua-Aerobic Systems, Inc., *Aqua Diffused Aeration Products*, http://www.aqua-aerobic.com, 2001.
68. Ovivo, *Aerostrip® System*, http://www.ovivowater.com.
69. Parkson Corporation, *HiOx® Aeration System*, http://www.parkson.com, 2012.
70. Evoqua Water Technologies, *Wideband™ Stainless Steel Coarse Bubble Diffuser*, http://www.evoqua.com, 2014.
71. Stamford Scientific International Inc., *SSI™ Coarse Bubble Diffusers*, http://www.StamfordScientific.com, 2014.
72. Evoqua Water Technologies, *Snap-Cap™ Diffuser*, http://www.evoqua.com, 2014.
73. Ashbrook Simon-Hartley Corporations LP, *Dura-Disc/CB™ Coarse Bubble Diffuser*, http://www.as-h.com, 2010.
74. Evoqua Water Technologies, *Vari-Cant® Jet Aeration Systems*, http://www.evoqua.com, 2014.
75. RWL Water, *Tornado® Surface Aspirating Aerator*, http://www.rwlwater.com, 2014.
76. Aeration Industries® International, *Aire-O_2® Surface Aspirating Aerator*, http://www.aireo2.com, 2012.
77. Ovivo, *Carrousel® Systems*, http://www.ovivowater.com.
78. WesTech Engineering, Inc., *OxyStream™ Oxidation Ditch Technology*, http://www.westech-inc.com, 2011.
79. Aqua-Aerobic Systems, Inc., *Aqua-Jet® Surface Mechanical Aerator*, http://www.aqua-aerobic.com, 2013.
80. Evoqua Water Technologies, *Aqua-Lator® Surface Aerator*, http://www.evoqua.com, 2014.
81. Invent, *HyperClassic® Mixing and Aeration System*, http://www.invent-et.com, 2014.
82. Envirodyne Systems Inc., *Brush Rotor Oxidation Ditches*, http://www.envirodynesystems.com.
83. Evoqua Water Technologies, *Torque Tube Disc Aerator—An Addition to the Disc Aerator Family of Product Offerings*, http://www.evoqua.com, 2014.
84. Xylem, *Eco-Lift™ Retrievable Aeration Grid for Sanitaire Systems*, http://www.xyleminc.com, 2014.
85. Environmental Dynamics Inc. *(EDI), EDI ModuleAir™ System*, http://www.wastewater.com.

86. Gardner Denver Nash, LLC, *Multistage Centrifugal Blowers/Exhausters*, http://www.HoffmanandLamson.com, 2017.
87. Sulzer, *Meeting Today's Challenges within Wastewater Treatment, Products and Solutions that Save Energy and Increase Reliability*, http://www.sulzer.com, 2013.
88. GE, *Roots Blowers, Compressors and Controls*, http://www.ge.com, 2015.
89. HSi, Inc./Siemens, *High Speed Turbo Blowers*, http://www.hsiblowers.com, 2011.
90. APG-Neuros/APGN, *High Efficiency Air Bearing Turbo Blower*, http://www.apg-neuros.com.
91. ABS Group, *ABS High Speed Turbocompressor*, http://www.absgroup.com, 2011.
92. Spellman, F. R.: *Handbook of Water and Wastewater Treatment Plant Operations*, 3rd ed., CRC Press, Boca Raton, 2014.
93. Jeppson, R.W., *Analysis of Flow in Pipe Networks*, Ann Arbor Science, Ann Arbor, MI, 1976.
94. Walski, T. M., *Analysis of Water Distribution Systems*, Van Nostrand Reinhold, New York, NY, 1984.
95. Mays, L. W., *Water Distribution Systems Handbook*, McGraw-Hill, New York, NY, 2000.
96. U.S. Environmental Protection Agency, *Wastewater Technology Fact Sheet: Aerated, Partial Mix Lagoons*, September 2002.
97. Metcalf & Eddy, Inc., *Wastewater Engineering: Treatment, Disposal and Reuse*, 3rd ed., McGraw-Hill, Inc., New York, NY, 1991.
98. Metcalf & Eddy, Inc., *Wastewater Engineering, Treatment and Reuse*, 4th ed., McGraw-Hill Companies, Inc., New York, NY, 2003.
99. Mancini, J. L. and E. L. Barnhart, Industrial waste treatment in Aerated Lagoon, In *Advances in Water Quality Improvement*, E. F. Gloyna and W.W., Eckenfelder, Jr. (eds), University of Texas Press, Austin, TX, 1971.
100. Marais, G. V. R. and V. A. Shaw, A rational theory for the design of sewage stabilization ponds in central and south Africa, *Transactions of the South Africa Institute of Civil Engineers*, Johannesburg, South Africa, 3, 1961, 205–227.
101. Marais, G. V. R., Dynamic behaviour of oxidation ponds, *Proceedings of the Second International Symposium for Waste Treatment Lagoons*, University of Kansas, Kansas City, MO, June 23–25, 1970, pp. 15–46.
102. U.S. Environmental Protection Agency, *Design Manual, Municipal Wastewater Stabilization Ponds*, EPA-625/1-83-015, Office of Water Program Operations, Washington, DC, October 1983.
103. Adams, C. E. and W. W., Eckenfelder, Jr. (eds), *Process Design Techniques for Industrial Waste Treatment*, Enviro Press, Nashville, TN, 1974.
104. Rich, L. G., Design approach to dual-power aerated lagoons, *Journal of the Environmental Engineering Division, ASCE*, 108(3), 1982, 532–548.
105. Rich, L. G., Modification of design approach to aerated lagoons, *Journal of the Environmental Engineering Division, ASCE*, 122(2), 1996, 149–153.
106. U.S. Environmental Protection Agency, *Principles of Design and Operations of Wastewater Treatment Pond Systems for Plant Operators, Engineers, and Managers*, EPA-600/R-11/088, Office of Research and Development/National Risk Management Research Laboratory, Cincinnati, OH, August 2011.
107. Thirumurthi, D., Design principles of waste stabilization ponds, *Journal of the Sanitary Engineering Division, ASCE*, 95(2), 1969, 311–332.
108. Thirumurthi, D., Design criteria for waste stabilization ponds, *Journal Water Pollution Control Federation*, 46(9), 1974, 2094–2106.
109. Wehner, J. F. and R. H. Wilhelm, Boundary conditions of flow reactor, *Chemical Engineering Science*, 6(2), 1956, 89–93.
110. Chiang, W. W. and E. F. Gloyna, *Biodegradation in Waste Stabilization Ponds*, Technical Report EHE-70-24 CRWR-74, The University of Texas at Austin, Austin, TX, December 1970.

111. Aquirre, J. and E. F. Gloyna, Waste Stabilization Pond Performance, *Design Guides for Biological Wastewater Treatment Processes*, Technical Report EHE-71-3 and CRWR-77, The University of Texas at Austin, Austin, TX, May 1970.
112. Middlebrooks, E. J., C. H. Middlebrooks, J. H. Reynolds, G. Z. Walters, C. S. Reed, and D. B. George, *Wastewater Stabilization Lagoon Design, Performance, and Upgrading*, MacMillan Publishing Co., Inc., New York, NY, 1982.
113. U.S. Environmental Protection Agency, EPA design information report: Protection of wastewater lagoon interior slopes, *Journal Water Pollution Control Federation*, 58(10), 1986, 1010–1014.
114. Gloyna, E. F., Facultative waste stabilization pond design, In *Ponds as a Wastewater Treatment Alternative*, E. F. Gloyna, J. F., Malina, Jr., and E. M. Davis (eds), University of Texas Press, Austin, TX, 1976.
115. U.S. Environmental Protection Agency, *Wastewater Technology Fact Sheet: Facultative Lagoons*, September 2002.
116. U.S. Environmental Protection Agency, *Wastewater Technology Fact Sheet: Anaerobic Lagoons*, September 2002.
117. Wang, L. K., N. C. Pereira, and Y. Hung, Biological treatment processes, In *Handbook of Environmental Engineering*^{TM}, Vol. 8, N. K. Shammas (eds), Humana Press, New York, NY, 2009.
118. Celenza, G. J., Biological processes, In *Industrial Waste Treatment Process Engineering*, Vol. II, Technomic Publishing Co., Inc., Lancaster, PA, 2000.
119. Salvato, J. A., N. L. Nemerow, and F. J. Agardy, *Environmental Engineering*, 5th ed., John Wiley & Sons, Inc., New York, NY, 2003.
120. Crites, R. W., E. J. Middlebrooks, R. K. Bastian, and S. C. Reed, *Natural Wastewater Treatment Systems*, CRC Press, Boca Raton, FL, 2014.
121. Oswald, W. J., Fundamental factors in stabilization pond design, In *Advances in Waste Treatment*, Pergamon, New York, NY, 1963.
122. Oswald, W. J. and H. B. Gotaas, Photosynthesis in sewage treatment, *Transactions ASCE*, 122(1), 1957, 73–79.
123. Metcalf & Eddy, Inc., *Wastewater Engineering: Collection, Treatment, and Disposal*, McGraw-Hill Book Co., New York, NY, 1972.
124. Wilcox, S., *National Solar Radiation Database 1991–2010 Update: User's Manual*, Technical Report NREL/TP-5500-54824, National Renewable Energy Laboratory (NRWL), Golden, CO, August 2012.
125. Qasim, S. R. and W. W. Chiang, *Sanitary Landfill Leachate: Generation, Control and Treatment*, CRC Press, Boca Raton, FL, 1994.
126. U.S. Environmental Protection Agency, *Wastewater Technology Fact Sheet, Trickling Filters*, EPA 832-F-00-014, Office of Water, Washington, DC, September 2000.
127. Brentwood Industries, *Design & Application, Trickling Filters*, http://www.brentwoodindustries.com, 2009.
128. Velz, C. J., A basic law for performance of biological beds, *Sewage Works Journal*, 20(4), 1948, 607–617.
129. Eckenfelder, W. W., Jr., *Water Quality Engineering for Practicing Engineers*, Barnes and Noble, Inc., New York, NY, 1970.
130. Howland, W. E., Flow over porous media as in a trickling filter, *Proceedings of the 12th Industrial Waste Conference*, Purdue University, 1957, pp. 435–465.
131. Germain, J. E., Economical treatment of domestic waste by plastic-medium trickling filters, *Journal Water Pollution Control Federation*, 38(2), 1966, 192–203.
132. Eckenfelder, W. W., Jr., Trickling filter design and performance, *Journal of the Sanitary Engineering Division, ASCE*, 87(SA 4, Part 1), 1961, 33–46.
133. National Research Council (NRC), Sewage treatment at military installations (subcommittee report), *Sewage Works Journal*, 18(5), 1946, 787–1028.

134. Jaeger Environmental, *Bio-Pac SF#30 Media for Trickling Filters, Submerged Beds and Anaerobic Reactors*, http://www.jaeger.com
135. Jaeger Environmental, *Plastic Jaeger Bio-Rings®*, Product Bulletin 850, http://www.jaeger.com
136. Koch-Glitsch, *Plastic Random Packing*, http://www.koch-glitsch.com, 2010.
137. Brentwood Industries, *Accu-Pac®, Cross Flow Media*, http://www.brentwoodindustries.com.
138. Brentwood Industries, Inc., *Accu-Pac®, Vertical Flow Media*, http://www.brentwoodindustries.com.
139. Brentwood Industries, Inc., *Biological Treatment, Attached Growth Processes*, http://www.brentwoodindustries.com, 2014.
140. Brentwood Industries, Inc., *AccuGrid® Bio-Grating*, http://www.brentwoodindustries.com, 2014.
141. WesTech Engineering, Inc., *HydroDoc™ Rotary Distributors*, http://www.westech-inc.com, 2012.
142. Walker Process Equipment, *Rotoseal Rotary Distributors*, http://www.walker-process.com
143. Kusters-Water/Kusters Zima, *Trickling Filter Rotary Distributors*, http://www.kusterswater.com
144. Brentwood Industries, Inc., *AccuPier® Support System*, http://www.brentwoodindustries.com, 2014.
145. Fair, G. M., J. C. Geyer, and D. A. Okun, *Water and Wastewater Engineering, Volume 2: Water Purification and Wastewater Treatment and Disposal*, John Wiley & Sons Inc., New York, NY, 1968.
146. Culp, G. L. and N. F. Heim, *Field Manual for Performance Evaluation and Troubleshooting at Municipal Wastewater Treatment Facilities*, EPA-430/9-78-001, U.S. Environmental Protection Agency, Washington, DC, January 1978.
147. U.S. Environmental Protection Agency, *Onsite Wastewater Treatment Systems Manual*, EPA/625/R-00/008, Office of Water/Office of Research and Development, Washington, DC, February 2002.
148. U.S. Environmental Protection Agency, *Summary of Design Information on Rotating Biological Contactors*, EPA-430/9-84-008, Office of Water Program Operations (WH-546), Washington, DC, September 1984.
149. U.S. Environmental Protection Agency, *Design Information on Rotating Biological Contactors*, EPA-600/2-84-106, Municipal Environmental Research Laboratory, Cincinnati, OH, June 1984.
150. Walker Process Equipment, *EnviroDiscTM Rotating Biological Contactor*, http://www.walker-process.com
151. Ovivo, *Ovivo® RBC*, http://www.ovivowater.com, 2014.
152. U.S. Environmental Protection Agency, *Project Summary: Review of Current RBC Performance and Design Procedures*, EPA-600/2-85-033, Water Engineering Research Laboratory, Cincinnati, OH, June 1985.
153. Water Pollution Control Federation, *O&M of Trickling Filters, RBCs, and Related Processes*, Manual of Practice OM-10, Water Pollution Control Federation, Alexandria, VA, 1988.
154. Arora, M. L. and M. B. Umphres, Evaluation of activated biofiltration and Active biofiltration/activated sludge technologies, *Journal Water Pollution Control Federal*, 59(4), 1987, 183–190.
155. Harrison, J. R. and P. L. Timpany, Design considerations with the trickling filter solids contact process, *Proceedings of the Joint Canadian Society of Civil Engineers and American Society of Civil Engineers National Conference*, Environmental Engineering, Vancouver, B. C., Canada, 1988, pp. 753–762.
156. Matasci, R. N., D. L. Clark, J. A. Heidman, D. S. Parker, B. Petrik, and D. Richards, Trickling filter/solids contact performance with rock filters at high organic loadings, *Journal Water Pollution Control Federation*, 60(1), 1988, 68–76.
157. Harrison, J. R., G. T. Daigger, and J. W. Filbert, A survey of combined trickling filter and activated sludge processes, *Journal Water Pollution Control Federation*, 56(10), 1984, 1073–1079.
158. Ovivo, *Cleartec® Systems*, http://www.ovivowater.com
159. Stensel, H. D., R. C. Brenner, K. M. Lee, H. Melcer, and K. Rackness, Biological aerated filter evaluation, *Journal of the Environmental Engineering Division, ASCE*, 114(3), 1988, 655–671.

160. Martin, K. J. and R. Nerenberg, The membrane biofilm reactor (MBfR) for water and wastewater treatment: Principles, applications, and recent developments, *Bioresource Technology*, 122(October), 2012, 83–94.
161. Lin, H., S. L. Ong, and W. J. Ng, Performance of a biofilm airlift suspension reactor for synthetic wastewater treatment, *Journal of the Environmental Engineering Division, ASCE*, 130(1), 2004, 26–36.
162. Tay, J. H., P. Yang, W. Q. Zhang, S. T. L. Tay, and Z. H. Pan, Reactor performance and membrane filtration in aerobic granular sludge membrane bioreactor, *Journal of Membrane Science*, 304 (1–2), 2007, 24–32.
163. Adav, S. S., D. J., Lee, K. Y. Show, and J. H. Tay, Aerobic granular sludge: Recent advances, *Biotechnology Advances*, 26(5), 2008, 411–423.
164. de Bruin, L. M. M., M. K. de Kreuk, H. F. R. van der Roest, C. Uijterlinde, and M. C. M. van Loosdrecht, Aerobic granular sludge technology: An alternative to activated sludge?, *Water Science and Technology*, 49(11–12), 2004, 1–7.
165. U.S. Environmental Protection Agency, *Emerging Technologies for Wastewater Treatment and In-Plant Wet Weather Management*, EPA 832-R-06-006, Office of Wastewater Management, Washington, DC, February 2008.
166. Spreece, R. E., *Anaerobic Biotechnology and Odor/Corrosion Control for Municipalities and Industries*, Archae Press, Nashville, TN, 2008.
167. Spreece R. E., *Anaerobic Biotechnology for Industrial Wastewaters*, Archae Press, Nashville, TN, 1996.
168. Shin, H. S. and Y. C. Song, A model for evaluation of anaerobic degradation characteristics of organic waste: Focusing on kinetics, rate-limiting step, *Environmental Technology*, 16(8), 1995, 775–784.
169. Jeyaseelam, S., A simplified mathematical model for anaerobic digestion process, *Water Science and Technology*, 35(8), 1997, 185–191.
170. Husain A., Mathematical models of the kinetics of anaerobic digestion—A selected review, *Biomass and Bioenergy*, 14(5–6), 1998, 561–571.
171. Malina, J. F. and F. G. Pohland, *Design of Anaerobic Processes for Treatment of Industrial and Municipal Wastewaters*, Water Quality Management Library, vol. 7, CRC Press, Boca Raton, FL, 1992.
172. Water Pollution Control Federation, *Anaerobic Sludge Digestion, MOP 16*, 2nd ed., Water Pollution Control Federation, Alexandria, VA, 1987.
173. Lettinga, G., and L. H. Pol, Advanced reactor design, operation and economy, *Water Science and Technology*, 18(12), 1986, 99–108.
174. Lettinga, G. and L. W. Hulshoft Pol, UASB-process design for various types of wastewaters, *Water Science and Technology*, 24(8), 1991, 87–107.
175. Lettinga, G. and J. N. Vinken, Feasibility of the upflow anaerobic sludge blanket (UASB) process for the treatment of low-strength wastes, *Proceedings of the 35th Industrial Waste Conference*, Purdue University, Lafayette, IN, May 13–15 1980, pp. 625–634.
176. Barber, W. P. and D. C. Stuckey, The use of the anaerobic baffled reactor (ABR) for wastewater treatment: A review, *Water Research*, 33(7), 1999, 1559–1578.
177. Kennedy, K. J. and R. L. Dorste, Anaerobic wastewater treatment in downflow stationary fixed film reactors, *Water Science and Technology*, 24(8), 1991, 157–177.
178. Iza, J., Fluidized bed reactors for anaerobic wastewater treatment, *Water Science and Technology*, 24(8), 1991, 109–132.
179. Laquidara, M. J., F. C. Blanc, and J. C. O'Shaughnessy, Development of biofilm, operating characteristics, and operational control in the anaerobic rotating biological contactor process, *Journal Water Pollution Control Federation*, 58(2), 1986, 107–114.
180. Qasim, S. R., E. M. Motely, and G. Zhu, *Water Works Engineering: Planning, Design, and Operation*, Prentice Hall PTR, Upper Saddle River, NJ, 2000.

181. U.S. Environmental Protection Agency, *Nutrient Control Design Manual*, EPA/600/R-10/100, Office of Research and Development/National Risk Management Research Laboratory, Cincinnati, OH, August 2010.
182. Grady, C. P. L., G. T. Daigger, N. G. Love, and C. D. M. Filipe, *Biological Wastewater Treatment*, 3rd ed., CRC Press, Boca Raton, FL, 2011.
183. U.S. Environmental Protection Agency, *Process Design Manual for Nitrogen Control*, EPA/625/1-77/007, Office of Technology Transfer, Washington, DC, October 1975.
184. U.S. Environmental Protection Agency, *Manual Nitrogen Control*, EPA/625/R-93/010, Office of Research and Development, Washington, DC, September 1993.
185. U.S. Environmental Protection Agency, *Nutrient Control Design Manual: State of Technology Review Report*, EPA/600/R-09/012, Office of Research and Development/National Risk Management Research Laboratory, Cincinnati, OH, January 2009.
186. Hartley, K., *Tuning Biological Nutrient Removal Plants*, IWA Publishing, London, UK, 2013.
187. Daigger, G. T., T. A. Heinemann, G. Land, and R. S. Watson, Practical experience with combined carbon oxidation and nitrification in plastic media trickling filters, *Water Science and Technology*, 29(10–11), 1994, 189–196.
188. Pearce, P. and W. Edward, A design model for nitrification unstructured cross-flow plastic packing media trickling filters, *Water Environment Journal*, 25(2), 2011, 257–265.
189. Parkson Corporation, *DynaSand® Continuous, Upflow, Granular Media Filter*, http://www.parkson.com, 2010.
190. Iida, Y. and A. Teranishi, Nitrogen removal from municipal wastewater by a single submerged filter, *Journal Water Pollution Control Federation*, 56(3 Part I), 1984, 251–258.
191. Shieh, W. K., P. M. Sutton, and P. Kos, Predicting reactor biomass concentration in a fluidized-bed system, *Journal Water Pollution Control Federation*, 53(11), 1981, 1574–1584.
192. Freeney, P. K. and C. E. Hucks, Denitrification using submerged RBCs: A case history, In *Biological Nitrogen and Phosphorous Removal: The Florida Experience*, TREED Center, University of Florida, Gainesville, FL, 1987.
193. Barnard, J. L., Biological denitrification, *Journal International Water Pollution Control Federation*, 72(6), 1973, 705–717.
194. Ludzack, F. J. and M. B. Ettinger, Controlling operation to minimize activated sludge effluent nitrogen, *Journal Water Pollution Control Federation*, 34(9), 1962, 920–931.
195. Abufayed, A. A. and E. D. Schroeder, Kinetics and stoichiometry of SBR/denitrification with primary sludge carbon source, *Journal Water Pollution Control Federation*, 58(5), 1986, 398–405.
196. Arora, M. L., E. F. Barth, and M. B. Umphres, Technology evaluation of sequencing batch reactors, *Journal Water Pollution Control Federation*, 57(8), 1985, 867–875.
197. Wuhrmann, K., High-rate activated sludge treatment and its relation to stream sanitation: II. Biological river tests of plant effluents, *Sewage and Industrial Wastes*, 26(2), 1954, 212–220.
198. Barth, E. F. and H. D. Stensel, International nutrient control technology for municipal effluents, *Journal Water Pollution Control Federation*, 53(12), 1981, 1691–1701.
199. Mulbarger, M. C., The three sludge system for nitrogen and phosphorus removal, *Proceedings of the 44th Annual Conference of the Water Pollution Control Federation*, San Francisco, CA, 1970.
200. Randall, C. W., J. L. Barnard, and H. D. Stensel, *Design and Retrofit of Wastewater Treatment Plants for Biological Nutrient Removal*, vol V, 2nd ed., CRC Press, Boca Raton, FL, 1998.
201. Barnes, D., C. F. Forster, and D. W. M. Johnstone, *Oxidation Ditches in Wastewater Treatment*, Pitman Publishing Inc., Marshfield, MA, 1983.
202. Stensel, H. D. and T. E. Coleman, *Technology Assessments: Nitrogen Removal Using Oxidation Ditches*, Water Environment Research Foundation, Water Environment Federation, Alexandria, VA, 2000.
203. Trivedi, H. and N. Heinen, Simultaneous nitrification/denitrification by monitoring NADH fluorescence in activated sludge, *Proceedings of the 73th Annual WEFTEC®*, Water Environment Federation, Anaheim, CA, 2000.

204. Wentzel, M. C., R. E. Loewenthal, G. A. Ekama, and G. v. R. Marais, Enhanced polyphosphate organism cultures in activated sludge systems—Part I: Enhanced culture development, *Water SA*, 14(2), 1988, 81–92.
205. Qasim, S. R., W. Chiang, G. Zhu, and R. Miller, Effect of biodegradable carbon on biological phosphorus removal, *Journal of the Environmental Engineering Division, ASCE*, 122(9), 1996, 875–878.
206. Ekama, G. A. and G. V. R. Marais, The Influence of Wastewater Characteristics on Process Design, Chapter 3, In *Theory, Design and Operation of Nutrient Removal Activated Sludge Process*, Water Research Commission, Pretoria, South Africa, 1984.
207. Dawson, R. N., S. S. Jeyanayagam, K. Abraham, and C. L. Wallis-Lage, The importance of primary sludge fermentation in the BNR process, *Proceedings of the Water Environment Federation 67th Annual Conference and Exposition*, vol. 1, October 1994, pp. 607–612.
208. Barnard, J. L., Activated primary tanks for phosphorus removal, *Water SA*, 10(3), 1984, 121–126.
209. Oldham, W. K., Full-scale optimization of biological phosphorus removal at Kelowna, Canada, *Water Science and Technology*, 17, 1985, 243–257.
210. Box, T. F., Truly Soluble COD: *A Surrogate for Volatile Fatty Acids in an Activated Primary Clarifier*, MS thesis, The University of Texas at Arlington, December 1996.
211. U.S. Environmental Protection Agency, *State Adoption of Numeric Nutrient Standards (1998--2008)*, EPA-821-F-08-007, Office of Water, Washington, DC, December 2008.
212. Barnard, J. L., Background to biological phosphorus removal, *Water Science and Technology*, 15 (3–4), 1983, 1–13.
213. Deakyne, C. W., M. A. Patel, and D. J. Krichten, Pilot plant demonstration of biological phosphorus removal, *Journal Water Pollution Control Federation*, 56(7), 1984, 867–873.
214. Ekama, G. A., P. Siebritz, and G. V. R. Marais, Considerations in the process design of nutrient removal activated sludge processes, *Water Science and Technology*, 15(3–4), 1983, 283–318.
215. Wentzel, M. C., P. L. Dold, G. A. Ekama, G. V. R. Marais, Kinetics of biological phosphorus release, *Water Science and Technology*, 17(11–12), 1985, 57–71.
216. Bundgaard, E., G. H. Kristensen, and E. Arvin, Full scale experience with phosphorus removal in an alternating system, *Water Science and Technology*, 15(3–4), 1983, 197–217.
217. U.S. Environmental Protection Agency, *Evaluation of Oxidation Ditches for Nutrient Removal*, EPA-832-R-92-003, Office of Wastewater Enforcement and Compliance, Washington, DC, September 1992.
218. Applegate, C. S., B. Wilder, and J. R. DeShaw, Total nitrogen removal in a multi-channel orbal plant, Huntsville, Texas, *Technical paper presented at the 51st Annual Conference of Water Pollution Control Federation*, Anaheim, CA, October 1978, pp. 1–6.
219. Evoqua Water Technologies, *The Orbal® System for TrueSND™ Biological Treatment, Simultaneous Nitrification-Denitrification from Envirex*, http://www.evoqua.com, 2015.
220. Matsch, L. C. and R. F. Drnevich, PhoStrip: A biological system for removing phosphorus, In *Advances in Water and Wastewater Treatment: Biological Nutrient Removal*, M. P. Wanielista and W. W. Eckenfelder (eds), Ann Arbor Science Publishers, Ann Arbor, MI, 1978.
221. Upton, J., J. Churchley, and A. Fergusson, Pilot scale comparison of mainstream BNR and sidestream phostrip systems for biological removal nutrients, *Proceedings of the 66th Annual Conference*, Water Environment Federation, Anaheim, CA, October 1993, pp. 261–272.
222. Qasim, S. R., S. Y. Chang, and C. E. Parker, Biological nutrient removal in three-stage suspended growth reactor system, *International Journal of Environmental Studies*, 46(1), 1994, 21–30.
223. Qasim, S. R., C. E. Parker, and R. T. McMillon, Biological nutrient removal: A pilot plant study, *Proceedings of the 45th Purdue Industrial Waste Conference*, West Lafayette, IN, May 8–10, 1990, pp. 303–310.
224. Qasim, S. R. and K. Udomsinrot, Biological nutrient removal in anoxic-anaerobic-aerobic treatment process, *International Journal of Environmental Studies*, 30(4), 1986, 257–270.

225. U.S. Environmental Protection Agency, *Municipal Nutrient Removal Technologies Reference Document*, EPA-832-R-08-006, Office of Wastewater Management, Municipal Support Division, Washington, DC, September 2008.
226. U.S. Environmental Protection Agency, *Biological Nutrient Removal Processes and Costs*, EPA-832-R-07-002, Office of Water, Washington, DC, June 2007.
227. Henze, M., C. P. L., Grady, Jr., W. Gujer, G. V. R. Marais, and T. Matsuo, *Activated Sludge Model No. 1 (ASMI)*, Scientific and Technical Report, No. 1, International Association on Water Pollution Research and Control (IAWPRC), London, 1987.
228. Henze, M., W. Gujer, T. Mino, T. Matsuo, M. C. Wentzel, and G. V. R. Marais, *Activated Sludge Model No. 2 (ASM2)*, Scientific and Technical Report No. 3, International Association on Water Quality (IAWQ), London, 1995.
229. Henze, M., W. Gujer, T. Mino, T. Matsuo, M. C. Wentzel, G. V. R. Marais, and M. v. Loosdrecht, Activated sludge model No. 2D, ASM2D, *Water Science Technology*, 39(1), 1999, 165–182.
230. Henze, M., W. Gujer, T. Mino, and M. V. Loosdrecht, *Activated Sludge Models ASM1, ASM2, ASM2d and ASM3*, Scientific and Technical Report No. 9, IWA (International Water Association), London, 2000.
231. Henze, M., C. P. L., Grady, Jr., W. Gujer, G. V. R. Marais, and T. Matsuo, A general model for single-sludge wastewater treatment systems, *Water Research*, 21(5), 1987, 505–515.
232. Baker, P. S. and P. L. Dold, General model for biological nutrient removal activated-sludge systems: Model presentation, *Water Environment Research*, 69(5), 1997, 969–984.
233. Baker, P. S. and P. L. Dold, General model for biological nutrient removal activated-sludge systems: Model application, *Water Environment Research*, 69(5), 1997, 985–991.
234. Grady, C. P. L., , Jr., W. Gujer, G. V. R. Marais, and T. Matsuo, A model for single-sludge wastewater treatment system, *Water Science Technology*, 18(6), 1986, 47–61.
235. Matinia, J., *Mathematical Modelling and Computer Simulation of Activated Sludge Systems*, IWA Publishing, London, 2010.
236. EnviroSim Association, Ltd., *BioWin*, http://envirosim.com
237. Hydromantis Environmental Software Solutions, Inc., *GPS-X*^{TM}, http://www.hydromantis.com
238. Water Environment Federation, *Clarifier Design*, WEF Manual of Practice No. FD-8, 2nd ed., Water Environment Federation, Alexandria, VA, 2005.
239. Keinath, T. M., D. A. Hofer, C. H. Dana, and M. D. Ryckman, Activated sludge-unified system design and operation, *Journal Environmental Engineering Division, ASCE*, 103(5), 1977, 829–849.
240. Keinath, T. M., Operational dynamics and control of secondary clarifiers, *Journal Water Pollution Control Federation*, 57(7), 1985, 770–776.
241. Ansys, Inc., *Fluid Dynamics*, http://www.ansys.com, 2011.
242. Ekama, G. A. and P. Marais, Assessing the applicability of the 1D flux theory to full-scale secondary settling tank design with a 2D hydrodynamic model, *Water Research*, 38(3), 2004, 495–506.
243. Griborio, A., P. Pitt, and J. A. McCorquodale, Next-generation modeling tool helps you get the most from your clarifier, *Water Environment and Technology*, 20(10), 2008, 1986, 52–58.
244. Wicklein, E. A. and R. W. Samstag, Comparing commercial and transport CFD models for secondary sedimentation, *Proceedings of the 82th Annual WEFTEC®*, Water Environment Federation, Orlando, FL, 2009.
245. Samstag, R. W., S. Zhou, R. L. Chan, C. Royer, and K. Brown, Comprehensive evaluation of secondary sedimentation performance, *Proceedings of the 83th Annual WEFTEC®*, Water Environment Federation, New Orleans, LA, 2010.
246. Xanthos, S., K. Ramalingam, S. Lipke, B. McKenna, and J. Fillos, Implementation of CFD modeling in the performance assessment and optimization of secondary clarifiers: The PVSC case study, *Water Science and Technology*, 68(9), 2013, 1901–1913.
247. WesTech Engineering, Inc., *Clarifier Optimization Package™, Advanced Clarifier Technology*, http://www.westech-inc.com, 2015.

Appendix A: Abbreviations and Symbols, Basic Information about Elements, Useful Constants, Common Chemicals Used in Water and Wastewater Treatment, and the U.S. Standard Sieves and Size of Openings

In this appendix, many abbreviations and symbols, atomic numbers and weights, and valances of selected elements are listed in Tables A.1 and A.2. Many useful constants for unit conversions in different fields of engineering are summarized in Table A.3. Additionally, the alphabetical listing of commonly used chemicals in water and wastewater treatment and their properties are summarized in Table A.4. Also, the opening sizes of the U.S. Standard Sieves are summarized in Table A.5.[1–7]

TABLE A.1 Abbreviations and Symbols

Acceleration due to gravity (LT^{-2})	g	Gallon per minute (L^3T^{-1})	gal/min or gpm
Area (L^2)	A	Gallon per second (L^3T^{-1})	gal/s or gps
Atmosphere	atm	Horsepower	hp
British thermal unit	Btu	Horsepower-hour	hp·h
Calorie	cal	Hour (T)	h
Centimeter (L)	cm	Inch	in
Centipoise	cp	Joule	J
Cubic centimeter (L^3)	cm^3	Kilocalorie	kcal
Cubic feet per minute (L^3T^{-1})	cfm or ft^3/min	Kilogram (M)	kg
Cubic feet per second (L^3T^{-1})	cfs or ft^3/s	Kilonewton per cubic meter	kN/m^3
Cubic meter (L^3)	m^3	Kilowatt-hour	kWh
Cubic meter per day (L^3T^{-1})	m^3/d	Liter (L^3)	L
Cubic meter per second (L^3T^{-1})	m^3/s	Milligram (M)	mg
Degree Celsius	°C	Million gallons per day (L^3T^{-1})	MGD
Degree Fahrenheit	°F	Pound force	lb_f
Degree Kelvin	°K	Pound mass	lb_m
Degree Rankine	°R	Pounds per square inch	lb_f/in^2 or psi
Density (ML^{-3})	ρ	Power	P
Discharge or flow (L^3T^{-1})	Q	Second (T)	s
Energy	E	Square meter (L^2)	m^2
Feet per minute (LT^{-1})	fpm or ft/min	Standard temperature and pressure	STP (20°C and 760 mm Hg)
Feet per second (LT^{-1})	fps or ft/s	Volume (L^3)	V
Gallon (L^3)	gal	Watt	W

Note: M = mass, L = length, T = time.

Source: Adapted in part from References 1 through 5.

TABLE A.2 Basic Information of Selected Elements[a]

Element	Symbol	Atomic Number	Atomic Weight	Valence
Aluminum	Al	13	26.982	3
Antimony	Sb	51	121.760	3, 5
Argon	Ar	18	39.948	–
Arsenic	As	33	74.922	3, 5
Barium	Ba	56	137.327	2
Beryllium	Be	4	9.012	–
Bismuth	Bi	83	208.980	3, 5
Boron	B	5	10.811	3
Bromine	Br	35	79.904	1, 3, 5, 7
Cadmium	Cd	48	112.411	2
Calcium	Ca	20	40.078	2
Carbon	C	6	12.011	4

(*Continued*)

TABLE A.2 (*Continued*) Basic Information of Selected Elements[a]

Element	Symbol	Atomic Number	Atomic Weight	Valence
Chlorine	Cl	17	35.453	1, 3, 5, 7
Chromium	Cr	24	51.996	2, 3, 6
Cobalt	Co	27	58.933	2, 3
Copper	Cu	29	63.546	1, 2
Fluorine	F	9	18.998	1
Gold	Au	79	196.966	1, 3
Helium	He	2	4.003	–
Hydrogen	H	1	1.008	1
Iodine	I	53	126.904	1, 3, 5, 7
Iron	Fe	26	55.845	2, 3
Lead	Pb	82	207.190	2, 4
Lithium	Li	3	6.941	1
Magnesium	Mg	12	24.305	2
Manganese	Mn	25	54.938	2, 3, 4, 6, 7
Mercury	Hg	80	200.592	1, 2
Molybdenum	Mo	42	94.940	3, 4, 6
Neon	Ne	10	20.180	–
Nickel	Ni	28	58.693	2, 3
Nitrogen	N	7	14.007	3, 5
Oxygen	O	8	15.999	2
Phosphorus	P	15	30.974	3, 5
Platinum	Pt	78	195.084	2, 4
Potassium	K	19	39.098	1
Radium	Ra	88	226.05	–
Silicon	Si	14	28.086	4
Silver	Ag	47	107.868	1
Sodium	Na	11	22.990	1
Strontium	Sr	38	87.62	2
Sulfur	S	16	32.064	2, 4, 6
Tin	Sn	50	118.710	2, 4
Titanium	Ti	22	47.867	–
Tungsten	W	74	183.84	3, 4
Uranium	U	92	238.029	4, 6
Zinc	Zn	30	65.38	2

[a] For a complete list, consult a handbook of chemistry and physics.

TABLE A.3 Useful Constants

Acceleration due to gravity, $g = 9.807\ m/s^2$ ($32.174\ ft/s^2$)

pi, $\pi = 3.14159265$

$e = 2.71828$

U.S. standard atmosphere = 101.325 kN/m^2 (14.696 lb_f/in^2 or psi) = 101.325 kPa (1.013 bars) = 10.333 m (33.899 ft) of water = 760 mm Hg = 29.9213 in of Hg

Molecular weight of air (0.21 molar fraction of oxygen and 0.79 molar fraction of nitrogen) = 29.1 g/mole

Density of air (15°C and 1 atm) = 1.23 kg/m^3 (0.0765 lb_m/ft^3)

Dynamic viscosity of air, μ (15°C and 1 atm) = 17.8×10^{-6} $N{\cdot}s/m^2$ (0.372×10^{-6} $lb_f{\cdot}s/ft^2$)

Kinematic viscosity of air, ν (15°C and 1 atm) = 14.4×10^{-6} m/s^2 (155×10^{-6} ft^2/s)

1 centipoise = 0.01 g/cm·s

1 stoke = 1 cm^2/s

Specific heat of air = 1005 J/kg·K (0.24 Btu/lb_m·°R)

Universal gas constant, R = 1.987 cal/mole·°K = 0.08206 L·atm/mole·°K = 8.314 J/mole·°K (R = 1.986 Btu/lb mole·°R = 0.7302 atm·ft3/lb mole·°R)

1 N = 1 kg·m/s^2 = 0.2248 lb_f

1 pascal (Pa) = 1 N/m^2

1 J = 1 N·m

1 J/s = 1 N·m/s = 1 w

1 kJ = 0.9478 Btu

1 amp = 1 coulomb/s

1 cal (20°C) = 0.003966 Btu

1 cal = 4.184 J

1°C = 1.8°F

$°C = \frac{5}{9}(°F - 32)$ or $°F = \frac{9}{5}°F + 32$

°K = °C + 273.15 or °R = °F + 446

1 Faraday = 96,485 coulombs/eq = 23,061 cal/volt eq

1 fathom = 1.829 m

1 grain = 0.06480 g

1 ha = 10,000 m^2 = 2.471 acres

1 knot = 1.852 km/h = 1.151 mph

1 hp = 745.7 w = 745.7 J/s = 550 ft·lb_f/s

1 lumen (at 5550 A) = 0.001471 W

1 lumen/cm^2 = 1 Lambert

1 lumen/m^2 = 0.0929 foot-candle

1 ton (U.S.) = 2000 lb

1 tonne (metric) = 1000 kg = 2204.6 lb

1 ton (U.K.) = 2240 lb

1 U.S. gallon = 3.78541 L = 8.34 lb of water

1 Imperial gallon (U.K.) = 4.546 L = 10 lb of water

1 ft^3 = 7.481 U.S. gallon = 62.43 lb_m of water

Source: Adapted in part from References 1 through 5.

TABLE A.4 Common Chemicals Used in Water and Wastewater Treatment

Chemical Name and Formula	Common or Trade Name	Shipping Containers, and Suitable Storage Material[a]	Available Forms, Description, and Density	Solubility, lb/gal	Commercial Strength, %	Additional Characteristics and Properties
Activated carbon	Powdered carbon, PAC, Aqua Nuchor, Hydrodarco, Herite	Bags or bulk; dry: iron, steel; wet: rubber and silicon linings, or type 316 stainless steel	Black granules, powder, 5–30 lb/ft^3	Insoluble (suspension used)		1 lb/gal suspension used for storage and handling
Aluminum oxide, Al_2O_3	Activated alumina	Bags or drums; iron or steel	Powder granules (up to 1 in. in diameter); 95 lb/ft^3	Insoluble	100	
Aluminum sulfate, $Al_2(SO_4)_3{\cdot}14H_2O$ (dry)	Alum, filter alum, sulfate of alumina	100–200-lb bags, 300–400-lb barrels, bulk (carloads), or tank cars and trucks; dry: iron, or steel; wet: stainless steel, or rubber plastic	Ivory colored; powder (38–45 lb/ft^3), granule (60–63 lb/ft^3), or lump (62–70 lb/ft^3)	6.2 (60°F)	17 (Al_2O_3), or ~9 (Al)	pH of 1% solution: 3.5
Aluminum sulfate, $Al_2(SO_4)_3{\cdot}49\ H_2O$ (45-55% solution)	Alum solution	Tank cars and trucks; FRP, PE, type 316 stainless steel, rubber linings	Liquid; 10.8-11.2 lb/gal	–	8.5 (Al_2O_3), or 4.1-5 (Al)	pH: 1.9-2.3; freezing point: 4°F
Ammonium aluminum sulfate, $(NH_4)Al(SO_4)_2{\cdot}12H_2O$	Ammonium alum, crystal alum	100-lb bags, barrels, or bulk; FRP, PE, type 316 stainless steel, or rubber linings	Colorless crystals or white powder; 65-75 lb/ft^3	0.3 (32°F) 1.3 (68°F)	99	pH of 1% solution: 3.5
Ammonium hydroxide, NH_4OH	Ammonia water, ammonium hydrate, aqua ammonia	Carboys, 750-lb drums, or bulk; glass lining, steel, iron, FRP, or PE	Colorless liquid; 7.48 lb/gal	Complete	29.4 (NH_3) max 26° (Baumé)	pH 14; freezing point: −107°F
Ammonium silicofluoride, $(NH_4)_2SiF_6$	Ammonium fluorsilicate	100- and 400-lb drums; steel, iron, FRP, or PE	White crystals; 65-70 lb/ft^3	1.7 (63°F)	100	White, free flowing solid
Ammonium sulfate, $(NH_4)_2SO_4$	Sulfate of ammonia	50- and 100-lb bags, or 725-lb drums; FRP, PE, ceramic and rubber linings; or iron (dry)	White or brown crystal; 70 lb/ft^3	6.3 (68°F)	>99	Cakes in dry feed; add $CaSO_4$ for free flow
Anhydrous ammonia, NH_3	Ammonia	50-, 100-, and 150-lb cylinders, tank cars, trucks, or bulk; shipping containers	Colorless gas; 5.1 lb/gal (liquid at 68°F), or 0.048 lb/ft^3 (gas at 32°F and 1 atm)	3.9 (32°F) or 3.1 (60°F)	>99.9 (NH_3)	

(Continued)

TABLE A.4 (*Continued*) Common Chemicals Used in Water and Wastewater Treatment

Chemical Name and Formula	Common or Trade Name	Shipping Containers, and Suitable Storage Material[a]	Available Forms, Description, and Density	Solubility, lb/gal	Commercial Strength, %	Additional Characteristics and Properties
Bentonite	Colloidal clay, volclay, wilkinite	100-lb bags or bulk; iron, steel, FRP, or PE	Powder, pellet, or mixed sizes; 40-60 lb/ft^3	Insoluble (colloidal solution used)		Free flowing; nonabrasive
Calcium fluoride, CaF_2	Fluorspar	Bags, drums, barrels, hopper cars, or trucks; steel, iron, FRP, or PE	White crystalline solid or powder; 200 lb/ft^3 (crystal)	Very slight	85 (CaF_2) or < 5 (SiO_2)	
Calcium hydroxide, $Ca(OH)_2$	Hydrated lime, slaked lime	50-lb bags or bulk; FRP, PE iron, steel, or rubber lining	White powder, light, or dense; 28-36 lb/ft^3	0.14 (68°F), or 0.12 (90°F)	85-99 ($Ca(OH)_2$), or 63-73 (CaO)	Hopper agitation required for dry feed of light form
Calcium hypochlorite, $Ca(OCl)_2$	HTH, perchloron, pittchlor, bleaching power	5-lb cans; or 100-, 300-, and 800-lb drums; glass, plastic, and rubber linings, FRP, or PE	White granule, powder, or tablet; 50 lb/ft^3 (powder)	1.5 (25°F)	65 (available Cl_2)	1–3 (available Cl_2 solution used)
Calcium oxide, CaO	Burnt lime, chemical lime, quicklime, unslaked lime	80- and 100-lb bags, or bulk; FRP, PE, iron, steel, or rubber	Lump, pebble, or granule; 55-60 lb/ft^3	Slaked to form hydrated lime	75-95 (CaO)	pH of saturated solution depending on detention time and temperature; amount of water critical for efficient slaking
Carbon dioxide, liquid CO_2	Carbonic anhydride	Bulk; carbon steel (dry); or type 316 stainless steel (solution)	Liquid	0.012 (25°C)	99.5	Solution is acid
Chlorinated lime, CaO, $2CaOCl_2{\cdot}3H_2O$	Bleaching powder, chloride of lime	100-, 300-, and 800-lb drums; glass and rubber lining, FRP, PE	White powder; 48 lb/ft^3		25-37 (available Cl_2)	Deteriorates
Chlorine, Cl_2	Chlorine gas, liquid chlorine	100-lb and 150-lb cylinders, ton containers, or 16-, 30-, and 55-ton tank cars; shipping containers	Greenish-yellow, liquefied gas under pressure; 11.8 lb/gal (liquid at 68°F), or 0.201 lb/ft^3 (gas at 32°F and 1 atm)	0.07 (60°F), or 0.04 (100°F)	99.8 (Cl_2)	Forms HCl and HOCl when mixed with water

(*Continued*)

TABLE A.4 (*Continued*) Common Chemicals Used in Water and Wastewater Treatment

Chemical Name and Formula	Common or Trade Name	Shipping Containers, and Suitable Storage Material[a]	Available Forms, Description, and Density	Solubility, lb/gal	Commercial Strength, %	Additional Characteristics and Properties
Chlorine dioxide, ClO_2	Chlorine dioxide	Generated at site of application; glass, PVC, and rubber linings; FRP, or PE	Greening-yellow gas; 0.150 lb/ft^3 (gas at 32°F and 1 atm)	0.02 (30 mu)	263 (available Cl_2)	Explosive under certain conditions
Copper sulfate, $CuSO_4{\cdot}5H_2O$	Blue vitriol, blue stone	100-lb bags, 450-lb barrels, or drums; FRP, PE, silicon lining, iron, or stainless steel	Crystal, lump, or powder; 60-90 lb/ft^3	1.6 (32°F), 2.2 (68°F), or 2.6 (86°F)	99 ($CuSO_4$)	pH of 25% solution: ~3
Disodium phosphate, anhydrous, $Na_2HPO_4{\cdot}12H_2O$	Basic sodium phosphate, DSP, secondary sodium phosphate	100- and 300-lb drums, or 50- and 100-lb bags; cast iron, steel, FRP, or PE	White crystal, granular, or powder; 60-64 lb/ft^3	0.4 (32°F), 0.65 (86°F), or 1 (77°F)	64.3 (PO_4), or 48 (P_2O_5)	Precipitates with Ca and Mg; pH of 1% solution: 9.1
Ferric chloride, $FeCl_3$	Anhydrous ferric chloride	500-lb caska; 100-, 300-, 400-lb kegs; or 65-, 135-, and 250-lb drums; keep in original containers	Greenish-black powder, or crystals; 45-60 lb/ft^3		98 ($FeCl_3$), or 31-34 (Fe)	
Ferric chloride, $FeCl_3$ (solution)	Ferrichlor, iron chloride	55-gal drums, or bulk; Glass, PVC and rubber linings; FRP, or PE	Dark brown syrupy liquid; 10.5-12.4 lb/gal (25-47%)	Complete	37-45 ($FeCl_3$), or 8.6-16.3 (Fe)	
Ferric chloride, $FeCl_3{\cdot}6H_2O$	Crystal ferric chloride	300-lb barrels; keep in original containers	Yellow-brown lump; 60-65 lb/ft^3		59-61 ($FeCl_3$), or 20-21 (Fe)	Hygroscopic (store lumps and powder in tight container), no dry feed; optimum pH: 4-11
Ferric sulfate, $Fe_2(SO_4)_3{\cdot}9H_2O$	Ferrifloc, ferrisulfate	100–175-lb bags, or 400–425-lb drums; glass, plastic and rubber linings, FRP, PE, or type 316 stainless steel	Red-brown powder, or granule; 60–70 lb/ft^3	Soluble in 2–4 parts cold water	90-94 (Fe $(SO_4)_3$), or 25-26 (Fe)	Mildly hygroscopic coagulant at pH 3.5-11
Ferrous chloride, $FeCl_2{\cdot}4H_2O$	Iron(II) chloride tetrahydrate, ferrous dichloride tetrahydrate, ferrous chloride tetrahydrate	Bags, barrels, or bulk; glass, plastic, rubber linings, FRP, PE, or type 316 stainless steel	Light green crystal or lump; 56 lb/ft^3	~13 (50°F)	27-28 (Fe)	pH of 1% solution: <7

(*Continued*)

TABLE A.4 (*Continued*) Common Chemicals Used in Water and Wastewater Treatment

Chemical Name and Formula	Common or Trade Name	Shipping Containers, and Suitable Storage Material[a]	Available Forms, Description, and Density	Solubility, lb/gal	Commercial Strength, %	Additional Characteristics and Properties
Ferrous sulfate, $FeSO_4 \cdot 7H_2O$	Copperos, green vitriol	Bags, barrels, or bulk; Glass, plastic, rubber linings, FRP, PE, or type 316 stainless steel	Green crystal, granule, or lump; 45–75 lb/ft^3		55 ($FeSO_4$), or 20 (Fe)	Hygroscopic; cakes in storage; optimum pH: 8.5–11
Fluorosilicic acid, H_2SiF_6	Hexafluoro-silicic acid	Rubber-lined drums, trucks, or railroad tank cars; rubber-lined steel, or PE	Liquid; 11.5 lb/gal	~1.2 (68°F)	35 (approx.)	Corrosive, etches glass
Hydrogen fluoride, HF	Hydrofluoric acid	Steel drums, or tank cars; steel, FRP, PE	Liquid; 8.3 lb/gal		70 (HF)	Below 60% steel cannot be used
Oxygen, liquid	LOX	Dewars, cylinders, or truck and rail tankers; steel	Pale blue cryogenic liquid; 0.089 lb/ft^3 (gas at 68°F and 1 atm)	3.16% by volume (77°F)	>99.5	Prevent LOX from contacting grease, oil, asphalt, or other combustibles
Ozone, O_3	Ozone	Generated at site of application	Colorless gas; 0.125 lb/ft^3 (gas at 68°F and 1 atm)	12 mg/L (68°C)	1–20	Feed gas for ozone generation: air, oxygen-generated on-site, or liquid oxygen (LOX)
Phosphoric acid, H_3PO_4	Orthophosphoric acid, phosphoric(V) acid	Polyethylene drums or bulk; FRP, epoxy, rubber linings, polypropylene, or type 316 stainless steel	Odorless watery white liquid; 13–14 lb/gal		75, 80, or 85	Freezing point: 0.5°F at 75%, 40.2°F at 80%, and 70.0°F at 85%
Polyaluminum chloride, $Al_n(OH)_m Cl_{(3n-m)}$	PACl, SternPac	55-gal drums and bulk; FRP, PE, type 316 stainless steel, or rubber linings	Pale amber liquid; 10–11 lb/gal (bulk)	Soluble	8.5–18 (Al_2O_3)	Freezing point: −12°C; pH of 5% solution: 3–5
Potassium aluminum sulfate, $KAl(SO_4)_2 \cdot 12H_2O$	Potash alum, potassium alum	Bags, or lead-lined bulk (carloads); FRP, PE, or ceramic and rubber linings	Lump, granule, or powder; 60–67 lb/ft^3	0.5 (32°F), 1 (68°F), or 1.4 (86°F)	10–11 (Al_2O_3)	Low or even solubility; pH of 1% solution: 3.5
Potassium permanganate, $KMnO_4$	Purple salt	Bulk, barrels, or drums; iron, steel, FRP, or PE	Purple crystals; 170 lb/ft^3	0.54 (68°F)	100	Danger of explosion in contact organic matters

(*Continued*)

TABLE A.4 (*Continued*) Common Chemicals Used in Water and Wastewater Treatment

Chemical Name and Formula	Common or Trade Name	Shipping Containers, and Suitable Storage Material[a]	Available Forms, Description, and Density	Solubility, lb/gal	Commercial Strength, %	Additional Characteristics and Properties
Sodium aluminate, $Na_2OAl_2O_3$	Sodium aluminum oxide, Sodium metaaluminate, soda alum	100- and 150-lb bags, 250- and 440-lb drums, or 55-gal solution drums; iron, FRP, PE, rubber, or steel	Brown powder, or liquid; 50-65 lb/ft^3 (powder), or 12.7 lb/gal (30-45% solution)	3 (68°F), or 3.3 (86°F)	30-45 ($Na_2OAl_2O_3$), or 27° (Baumé)	Hopper agitation required for dry feed; very hygroscopic; and pH of 1% solution: 11.5-14
Sodium carbonate, Na_2CO_3	Soda ash, soda crystals, washing soda	Bags, barrels, bulk (carloads), or trucks; iron, rubber linings, steel, FRP, or PE	White powder; 31-56 lb/ft^3 (extra light or light), or 56-69 lb/ft^3 (dense)	1.5 (68°F) or 2.3 (86°F)	99.4 (Na_2CO_3) or 57.9 (Na_2O)	Hopper agitation required for dry feed of light and extra-light forms; pH of 1% solution: 11.3
Sodium chloride, NaCl	Common salt, salt	Bags, barrels, or bulk (carloads); bronze, FRP, PE, or rubber	Rock (50-60 lb/ft^3), or fine (58-78 lb/ft^3)	2.9 (32°F) or 3 (68°F)	98 (NaCl)	Absorbs moisture
Sodium chlorite, $NaClO_2$	ADOX dry	100-lb drums; metals (avoid cellulose materials)	Light orange powder, flake, or crystals; 53-56 lb/ft^3	3.5 (68°F)	80 ($NaClO_2$), or 30 (available Cl_2)	Generates ClO_2 at pH 3; explosive
Sodium fluoride, NaF	Fluoride	Bags, barrels, fiber drums, or kegs; iron, steel, FRP, or PE	Nile blue or white powder, light, or dense; 50-75 lb/ft^3	0.3-0.4 (32-68°F)	90-100 (NaF)	pH of 4% solution: 6.6
Sodium fluorosilicate, Na_2SiF_6	Sodium silicofluoride, sodium hexafluoro-silicate	Bags, barrels, fiber drums; cast iron, rubber lining, steel, FRP, PE	Nile blue or yellowish white powder; 72 lb/ft^3	0.03 (32°F), 0.06 (68°F), or 0.12 (140°F)	100	pH of 1% solution: 5.3
Sodium hexametaphosphate, $(NaPO_3)_6$	Calgon, glassy phosphate, metaphosphoric acid, sodium polyphosphate, SHMP	100-lb bags; rubber linings, plastics, or type 316 stainless steel	Crystal, flake, or powder; 47 lb/ft^3	1–4.2	90-100	pH of 1% solution: 7 (neutral)
Sodium hydroxide, NaOH	Caustic soda, soda lye, lye solution	100–700-lb drums, or bulk; carbon, steel, polypropylene, FRP, or rubber lining	Flake, lump, or liquid; 95.5 lb/ft^3 (solids), 9.6 lb/gal (25% solution), or 12.8 lb/gal (50% solution)	2.4 (32°F), 4.4 (68°F), or 4.8 (104°F)	25-100 (NaOH) or 74-76 (NaO_2)	Solid, hygroscopic; pH of 1% solution: 12.9; and freezing point of 50% solution: 54°F

(*Continued*)

TABLE A.4 (*Continued*) Common Chemicals Used in Water and Wastewater Treatment

Chemical Name and Formula	Common or Trade Name	Shipping Containers, and Suitable Storage Material[a]	Available Forms, Description, and Density	Solubility, lb/gal	Commercial Strength, %	Additional Characteristics and Properties
Sodium hypochlorite, NaOCl (6-12 trade %)	Sodium hypochlorite solution, bleach	5-, 13-, and 50-gal carboys; or 1300–2000-gal tank trucks; ceramic, glass, plastic, and rubber linings; FRP, or PE	Light yellow liquid; 8.8-10.1 lb/gal		3.8-13.2 (available Cl_2)	Unstable; pH of 1% solution: 7 (neutral)
Sodium pyrosulfite, $Na_2S_2O_5$	Sodium metabisulfite	Bags, drums, or barrels; iron, steel, FRP, or PE	White to yellow crystalline powder;	5.5 (68°F)	67 (SO_2 dry), or 33.3 (SO_2 solution)	Sulfurous odor
Sodium silicate, $Na_2O(SiO_2)_n$	Liquid glass, water glass	Drums or bulk (tank trucks and cars); cast iron, rubber lining, steel, FRP, or PE	Opaque, viscous liquid, 9.6 lb/gal (40% solution)	1.9 (77°F)	40, or 40° (Baumé)	Variable ratio of Na_2O to SiO_2; pH of 1% solution: 12.3
Sodium sulfite, Na_2SO_3	Sulfite	Bags, drums, or barrels; cast iron, rubber lining, steel, FRP, or PE	White crystalline powder, 80–90 lb/ft^3	2.3 (68°F)	23 (SO_2)	Sulfurous taste and odor
Sulfur dioxide, SO_2	Sulfurous acid anhydride	100–150-lb steel cylinders, ton-containers, or tank cars and trucks; shipping container	Colorless gas; 11.5 lb/gal (liquid at 68°F), or 0.201 lb/ft^3(gas at 32°F and 1 atm)	1.7 (68°F)	99 (SO_2)	Irritating gas
Sulfuric acid, H_2SO_4	Hydrogen sulfate, oil of vitriol, green vitriol	Bottles, carboys, drums, trucks, or tank cars; FRP, PE, porcelain, glass, or rubber linings	Solution; 15.4 lb/gal	Complete	95-98	pH of 0.5% solution: ~1.2
Tetrasodium pyrophosphate, $Na_4P_2O_7{\cdot}10H_2O$	Sodium pyrophosphate, tetrasodium phosphate, TSPP	125-lb kegs, 200-lb bags, or 300-lb barrels; cast iron, steel, or plastics	White powder, 68 lb/ft^3	0.6 (77°F), or 3.5 (212°F)	53 (P_2O_5)	pH of 1% solution: 10.8
Tricalcium phosphate, $Ca_{10}(OH)_2(PO_4)_6$	Tribasic calcium phosphate (TCP), bone phosphate of lime (BPL),	Bags, drums, bulk, or barrels; cast iron, steel, or plastics	Granular; ~200 lb/ft^3	Insoluble	100	Also available as white powder
Trisodium phosphate, $Na_3PO_4{\cdot}12H_2O$	Normal sodium phosphate, tertiary sodium phosphate, trisodium orthophosphate, TSP	125-lb kegs, 200-lb bags, or 325-lb barrels; cast iron, steel, or plastics	Crystal-course, medium; 56-61 lb/ft^3	1 (68°F) or 1.9 (104°F)	19 (P_2O_5)	pH of 1% solution: 11.9

[a] Always contact chemical suppliers to select best materials for handling.

Note: 1 lb = 0.4536 kg, 1 lb/ft^3 = 16.02 kg/m^3, 1 lb/gal = 119.8 kg/m^3, FRP = fiberglass-reinforced plastic, PE = polyethylene.

Source: Adapted in part from References 1 through 7.

TABLE A.5 U.S. Standard Sieves and Size of Openings

Sieve Size or Number	Size of Opening	
	in	mm
3/8	0.375	9.53
1/4	0.250	6.35
4	0.187	4.76
6	0.132	3.35
8	0.0937	2.36
10	0.0787	2.00
12	0.0661	1.68
14	0.0555	1.41
16	0.0469	1.19
18	0.0394	1.00
20	0.0331	0.841
25	0.0278	0.710
30	0.0232	0.595
35	0.0197	0.500
40	0.0165	0.420
45	0.0136	0.345
50	0.0117	0.300
60	0.0098	0.250
70	0.0083	0.210
80	0.0070	0.177
100	0.0059	0.149
120	0.0049	0.125
140	0.0041	0.105
170	0.0035	0.088
200 (75 μ)	0.0029	0.075
230 (63 μ)	0.0024	0.063
270 (53 μ)	0.0021	0.053
325 (45 μ)	0.0017	0.045
400 (38 μ)	0.0015	0.038
450 (32 μ)	0.0012	0.032
500 (25 μ)	0.0010	0.025
635 (20 μ)	0.0008	0.020
850 (10 μ)	0.0004	0.010

Source: Adapted in part from Reference 7.

References

1. Qasim, S. R., *Wastewater Treatment Plants: Planning, Design, and Operation*, 2nd ed., CRC Press, Boca Raton, FL, 1999.
2. Qasim, S. R., E. M. Motley, and G. Zhu, *Water Works Engineering: Planning, Design, and Operation*, Prentice Hall PTR, Upper Saddle River, NJ, 2000.

3. Droste, R. L., *Theory and Practice of Water and Wastewater Treatment*, John Wiley of Sons, Inc., New York, NY, 1997.
4. Metcalf & Eddy | AECOM, *Wastewater Engineering: Treatment and Resource Recovery*, 5th ed., McGraw-Hill, New York, NY, 2014.
5. Metcalf & Eddy, Inc., *Wastewater Engineering: Treatment and Reuse*, 4th ed., McGraw-Hill Companies, Inc., New York, NY, 2003.
6. AWWA and ASCE, *Water Treatment Plant Design*, 5th ed., McGraw-Hill Education, New York, NY, 2012.
7. Chen, W. F. and J. Y. Richard Liew (editors), *The Civil Engineering Handbook*, 2nd ed., CRC Press, Boca Raton, FL, 2002.

Appendix B: Physical Constants and Properties of Water, Solubility of Dissolved Gases in Water, and Important Constants for Stability and Sodicity of Water

The standard values of many useful physical constants and important properties of water, including specific weight, density, dynamic and kinetic viscosities, surface tension, and vapor pressure are provided in Section B.1. The definitions of these constants and properties, as well as the solubility and Henry's law constants of selected dissolved gases are provided in Section B.2. The relationships of these constants are also illustrated in many solved examples in Section B.2. Langelier saturation index (LSI) and Ryznar Stability Index (RSI) are used to determine the stability of feed water in design of membrane processes. The $(pK_2 - pK_s)$ values that are used to calculate these values are summarized in Section B.3. Also, the sodium adsorption ratio (SAR) is used to evaluate the sodicity of reclaimed effluent for irrigation. The adjusted concentrations of calcium ion $[Ca_X^{2+}]$ that are used to determine the SAR are listed in Section B.4.

B.1 Physical Constants

The useful physical constants of water are summarized in Tables B.1.[1–5] The principal physical properties of water are summarized in SI and the U.S. customary units in Tables B.2 and B.3, respectively.[3,4,6–8]

B.2 Definitions and Physical Relationships

The definitions and relationships of many physical constants and properties most commonly used in wastewater treatment process calculations are provided below.[3,4,6–8] The solubility of oxygen and other gases depends upon temperature, pressure, and TDS concentration. Many solved examples are included to illustrate these relationships and solubility calculations from Henry's law constants.

TABLE B.1 Useful Physical Constants of Water

Constant	Value
Molecular formula	$= H_2O$
Molecular weight	= 18 g/mole
Ionization constant (K_w) at 25°C	$= 10^{-14}$
Specific weight, γ at 4°C	= 9.81 kN/m^3 (62.4 lb$_f$/ft^3)
Weigh of 1 L of water (average)	= 1 kg of water (2.205 lb of water)
Weigh of 1 U.S. gal of water (average)	= 8.345 lb of water
1 U.S. gal	= 3.785 L = 0.003785 m^3 (0.1337 ft^3)
1 Imperial U.K. gal	= 4.546 L = 4.546 kg (10.02 lb) of water
1 ft^3	= 28.32 L (7.481 U.S. gal)
1 mg/L	= 1 g/m^3 $\approx$ 1 ppm
1 μg/L	= 1 mg/m^3 $\approx$ 1 ppb
1 g/U.S. gal	= 264.2 mg/L
1 lb/d	= 8.34 × 1 mg/L × 1 MGD
Density at 4°C, ρ_w	= 1 g/cm^3 = 1000 kg/m^3 (1.94 slug/ft^3)
Dynamic viscosity at 20°C (68°F), μ	= 1.002 × 10^{-3} N·s/m^2 (2.089 × 10^{-5} lb$_f$·s/ft^2)
Kinematic viscosity at 20°C (68°F), ν	= 1.003 × 10^{-6} m^2/s (1.078 × 10^{-5} ft^2/s)
1 atm	= 10.33 m (33.9 ft) of water = 760 mm Hg (14.7 psi)
1 psi	= 2.31 ft (0.703 m) of water = 0.0680 atm (51.71 mm Hg)
1 m head of water at 20°C	= 9.790 kN/m^2 (1.420 lb$_f$/in^2)
Boiling point at 1 atm	= 100°C (212°F)
Melting (or freezing) point at 1 atm	= 0°C (32°F)
Specific heat	= 1 cal/g·°C = 4200 J/kg·°C (1 Btu/lb·°F)
Heat of fusion at 0°C	= 80 cal/g = 335 kJ/kg (144 Btu/lb)
Heat of vaporization at 100°C	= 540 cal/g = 2260 kJ/kg (973 Btu/lb)
Modulus of elasticity at 15°C	= 2.15 kN/m^2 (312 × 10^3 lb$_f$/in^2)

Source: Adapted in part from References 1 through 5.

B.2.1 Density

The density (ρ) of a fluid is mass per unit volume. In SI units, density is expressed in g/cm^3 or kg/m^3. At 4°C, the density of water (ρ_w) is 1000 kg/m^3. In the U.S. customary units, the density is expressed in slug/ft^3. At 4°C, the density of water is 1.94 slug/ft^3.

B.2.2 Dynamic Viscosity

The dynamic viscosity (μ) is a measure of the fluid resistance to tangential or shear stress. Dynamic viscosity in SI units is expressed in Newton-second per square meter (N·s/m^2). In the U.S. customary units, it is expressed as lb$_f$·s/ft^2. The typical values of μ in both units at 10°C is 1.307 × 10^{-3} N·s/m^2 and 2.735 × 10^{-5} lb$_f$·s/ft^2.

B.2.3 Kinematic Viscosity

Kinematic viscosity (ν) is the ratio of dynamic viscosity and the density ($\nu = \mu/\rho$). The units for kinematic viscosity are m^2/s in SI units and ft^2/s in U.S. customary units. The typical values of kinematic viscosity at 10°C in SI and the U.S. customary units are 1.307 × 10^{-6} m^2/s and 1.410 × 10^{-5} ft^2/s.

TABLE B.2 Physical Properties of Water in SI Units

Temperature (T)		Specific Weight (γ),	Density (ρ_w),	Dynamic Viscosity (μ),	Kinematic Viscosity (ν),	Surface Tension against Air (σ),	Vapor Pressure (P_v)	
°C	°F	kN/m^3	kg/m^3	10^{-3} N·s/m^2	10^{-6} m^2/s	10^{-3} N/m	mm Hg	kN/m^2
0	32	9.805	999.8	1.787	1.786	75.6	4.579	0.61
3.98	39.2	9.806	1000	1.568	1.568	75.0	6.092	0.82
5	41	9.807	999.9	1.519	1.519	74.9	6.543	0.87
10	50	9.804	999.7	1.307	1.307	74.2	9.209	1.23
15	59	9.798	999.1	1.139	1.140	73.5	12.79	1.70
20	68	9.789	998.2	1.002	1.003	72.8	17.54	2.34
25	77	9.777	997.0	0.890	0.893	72.0	23.76	3.17
30	86	9.764	995.7	0.798	0.801	71.2	31.82	4.24
40	104	9.730	992.2	0.653	0.658	69.6	55.32	7.38
50	122	9.689	988.1	0.547	0.553	97.9	92.51	12.3
60	140	9.642	983.2	0.466	0.474	66.2	149.6	19.9
70	158	9.589	977.8	0.404	0.413	64.4	233.9	31.2
80	176	9.530	971.8	0.354	0.364	62.6	355.4	47.3
90	194	9.466	965.3	0.315	0.326	60.8	526.3	70.1
100	212	9.399	958.4	0.282	0.294	58.9	760.0	101

Source: Adapted in part from References 1 through 5.

TABLE B.3 Physical Properties of Water in U.S. Customary Units

Temperature (T)		Specific Weight (γ),	Density (ρ_w),	Dynamic Viscosity (μ),	Kinematic Viscosity (ν),	Surface Tension against Air (σ),	Vapor Pressure (P_v)	
°F	°C	lb$_f$/ft^3	slug/ft^3	10^{-5} lb$_f$·s/ft^2	10^{-5} ft^2/s	lb$_f$/ft	in Hg	lb$_f$/in^2
32	0	62.42	1.940	3.746	1.931	0.00518	0.18	0.09
40	4	62.43	1.940	3.229	1.664	0.00614	0.24	0.12
50	10	62.41	1.940	2.735	1.410	0.00509	0.37	0.18
60	16	62.37	1.938	2.359	1.217	0.00504	0.53	0.26
70	21	62.30	1.936	2.050	1.059	0.00498	0.73	0.36
80	27	62.22	1.934	1.799	0.930	0.00492	1.04	0.51
90	32	62.11	1.931	1.595	0.826	0.00486	1.43	0.70
100	38	62.00	1.927	1.424	0.739	0.00480	1.93	0.95
110	43	61.86	1.923	1.284	0.667	0.00473	2.59	1.27
120	44	61.71	1.918	1.168	0.609	0.00467	3.44	1.69
130	54	61.55	1.913	1.069	0.558	0.00460	4.51	2.22
140	60	61.38	1.908	0.981	0.514	0.00454	5.88	2.89
150	66	61.20	1.902	0.905	0.476	0.00447	7.57	3.72
160	71	61.00	1.896	0.838	0.442	0.00441	9.65	4.74
170	77	60.80	1.890	0.780	0.413	0.00434	12.2	5.99
180	82	60.58	1.883	0.726	0.385	0.00427	15.3	7.51
190	88	60.36	1.876	0.678	0.362	0.00420	19.0	9.34
200	93	60.12	1.868	0.637	0.341	0.00413	23.5	11.5
212	100	59.83	1.860	0.593	0.319	0.00404	29.9	14.7

Source: Adapted in part from References 2 through 4.

B.2.4 Specific Gravity

The specific gravity (sp. gr.) is the relative density of a substance with respect to water at a given temperature. The density of water and other liquids change with temperature. The density of 1 g/cm^3 (or 1 g/mL, 1000 kg/m^3) is commonly used as the basis of specific gravity (water = 1).

B.2.5 Specific Weight

The specific weight (γ) of a fluid is its weight per unit volume. In SI units, it is expressed in kilonewton per cubic meter (kN/m^3). The commonly used relationship between specific weight (γ), density (ρ), and acceleration due to gravity (g) is $\gamma = \rho \cdot g$. The commonly used value of γ in SI units is 9.81 $kN/m^3(kg/m^2{\cdot}s^2)$. In the U.S. customary, the value of γ is 62.4 lb_f/ft^3.

B.2.6 Surface Tension

Surface tension is a physical property that is created through the attraction of the molecules to each other in a liquid. Due to surface tension, a glass filled with water has surface slightly above the brim before it spills; a metal needle floats in the surface; or a drop of water is held in suspension at the tip of a pipette. The units of expression of surface tension are N/m (lb/ft). The surface tension of water at 10°C is 74.2 N/m (0.00509 lb_f/ft). The surface tension decreases as the temperature increases.

B.2.7 Vapor Pressure

Liquid molecules constantly change phase from liquid to gas due to sufficient kinetic energy they possess. The vapor pressure (p_v) is the pressure exerted by the vapors at the free water surface. The vapor pressure is normally expressed in mm mercury (Hg) or in Hg. In SI and the U.S. customary units, the vapor pressure is expressed in kilonewtons per square meter (kN/m^2) and lb_f/in^2. The vapor pressure of water at 10°C is 9.209 mm Hg, 1.23 kN/m^2, 0.37 in Hg, and 0.18 lb_f/in^2.

B.2.8 Henry's Law Constants

The saturation concentration of a dissolved gas in a liquid depends upon the partial pressure of the gas exerted over the liquid surface. Henry's law expresses the relationship between the partial pressure of the gas in the gas phase above the liquid and the concentration of gas in the liquid phase. The Henry's law constant may be expressed in a number of different ways and units. Some commonly used relationships are given by Equations 10.84a, 10.84b, 11.10, and 15.6, and explained in Example 10.119, and Sections 11.6.2 and 15.4.5. The values of Henry's law constants H (mole fraction based) and H_s (volumetric fraction based) for many gases are given in Tables B.4 and B.5.[2–4,6–8]

B.2.9 Solubility of Gases

The solubility of gases in natural waters depends upon four factors: (1) *temperature*, (2) *pressure*, (3) *gas content* in overlying atmosphere, and (4) salt concentration.[7,8]

Temperature: The solubility of any gas in water decreases with rising temperature. The solubility of gases including oxygen can be determined by Henry's law constant given in Table B.6. The saturation

TABLE B.4 Mole Fraction-Based Henry's Law Constant (H) for Common Gases

Temperature (T), °C	H (10^4 atm)							
	Air	O_2	CO_2	CO	H_2	H_2S	CH_4	N_2
0	4.32	2.55	0.0728	3.52	5.79	0.0268	2.24	5.29
10	5.49	3.27	0.104	4.42	6.36	0.0367	2.97	6.68
20	6.64	4.01	0.142	5.36	6.83	0.0483	3.76	8.04
30	7.71	4.75	0.186	6.20	7.29	0.0609	4.49	9.24
40	8.70	5.35	0.233	6.96	7.51	0.0745	5.20	10.4
50	9.46	5.88	0.283	7.61	7.65	0.0884	5.77	11.3
60	10.1	6.29	0.341	8.21	7.65	0.1030	6.26	12.0

Source: Adapted in part from Reference 6.

TABLE B.5 Volumetric Concentration-Based Henry's Law Constant (H_v) for Common Gases

Temperature (T)		Water Vapor Pressure (P_v), mm Hg	H_v at 0°C and 1 atm[a], mL/L·atm			
°C	°F		Air	O_2	CO_2	N_2
0	32	4.58	29.18	48.89	1713	23.54
2	36	5.29	27.69	46.33	1584	22.41
4	39	6.10	26.32	43.97	1473	21.35
6	43	7.01	25.06	41.80	1377	20.37
8	46	8.05	23.90	39.83	1282	19.45
10	52	9.21	22.84	38.02	1194	18.61
12	54	15.52	21.87	36.37	1117	17.86
14	57	11.99	20.97	34.86	1050	17.17
16	61	13.63	20.14	33.48	985	16.54
18	64	15.48	19.38	32.20	928	15.97
20	68	17.54	18.68	31.02	878	15.45
22	72	19.83	18.01	29.88	829	14.98
24	75	22.38	17.38	28.81	781	14.54
26	79	25.21	16.79	27.83	738	14.13
28	82	28.35	16.21	26.91	699	13.76
30	86	31.82	15.64	26.08	665	13.42
35	95	42.18	–	24.40	592	12.56
40	104	55.32	–	23.06	530	11.84
45	113	71.88	–	21.87	479	11.30
50	122	92.51	–	20.90	436	10.88
60	140	156.4	–	19.46	359	10.23
70	158	233.7	–	18.33	–	9.77
80	176	355.1	–	17.61	–	9.58
90	194	525.8	–	17.20	–	9.50
100	212	760.0	–	17.00	–	9.50

[a] It is also called the absorption *coefficient of gas* at 0°C and 1 atm.
Source: Adapted in part from References 6 through 8.

TABLE B.6 Solubility of Common Gases in Water

Temperature (T)		Solubility of Gases (C_s), mg/L			
°C	°F	Air	O_2	CO_2	N_2
0	32	37.50	14.62	1.00	22.81
5	41	32.94	12.80	0.83	20.16
10	50	29.27	11.33	0.70	17.93
15	59	26.25	10.15	0.59	16.15
20	68	23.74	9.17	0.51	14.72
25	77	21.58	8.38	0.43	13.55
30	86	19.60	7.63	0.38	12.53

Source: Adapted in part from References 6 through 8.

concentration of oxygen in water at 1 atm can also be calculated from several empirical equations (Equation B.1).[7,8]

$$C_s = 14.652 - 0.41022T + 0.00799T^2 - 0.0000777T^3 \quad \text{(B.1a)}$$

$$C_s = \left(0.68 - 6 \times 10^{-4}T\right)\left(\frac{760 - P_v}{T + 35}\right) \quad \text{(B.1b)}$$

$$C_s = 14.652 - 10.53(1 - e^{-0.03896T}) \quad \text{(B.1c)}$$

where

C_s = saturation concentration of oxygen in freshwater at 1 atm, mg/L
T = water temperature, °C
P_v = vapor pressure of water, mm Hg

The saturation concentrations of oxygen, nitrogen, carbon dioxide, and air in freshwater with respect to temperature and at 1 atm are summarized in Table B.6.

Pressure: The solubility of any gas increases at higher pressures and vice versa. The solubility of oxygen at any given pressure is calculated from Equation B.2.[9–11]

$$C_{s,P} = \frac{P - P_v}{760 - P_v} \times C_s \quad \text{(B.2)}$$

where

$C_{s,P}$ = solubility of oxygen at any given pressure (P), mg/L
C_s = solubility of oxygen at 1 atm, mg/L
P = prevailing barometric pressure, mm Hg
p_v = vapor pressure of water at 1 atm, mm Hg

At higher altitudes, the barometric pressure is lower, as a result the solubility of oxygen is also lower. The barometric pressure with altitude is given in Table B.7.

Gas Content in Air: The solubility of a gas in water increases if its content in the atmosphere increases because of its partial pressure above water–air interface. The oxygen concentration in dry air is 21% by volume, and 23.2% by weight.[9–11]

Salt Concentration: The solubility of a gas in water is inversely proportional to the salt concentration. The degree of salinity is normally expressed in concentration (mg/L) of total dissolved solid (TDS), chloride ions (Cl^-), or common salt (NaCl). The North Atlantic Ocean has chloride content of nearby 18,000 mg/L, and its DO saturation is about 82% of that in fresh water. The DO saturation of domestic wastewater is about 95% of that in fresh water. The TDS concentrations in strong, medium, and weak

TABLE B.7 Barometric Pressure with Altitude

Elevation above Sea Level		Absolute Pressure in Head of Water (H_{abs})		Barometric Pressure, mm Hg
m	ft	m	ft	
0	0	10.3	33.9	760
305	1000	10.0	32.8	736
457	1500	9.8	32.1	721
610	2000	9.6	31.5	706
1219	4000	8.9	29.2	655
1829	6000	8.3	27.2	611
2438	8000	7.7	25.2	566
3048	10,000	7.1	23.4	522
4572	15,000	5.9	19.2	434

Source: Adapted in part from References 1 and 12.

domestic wastewaters are 850, 500, and 250 mg/L, respectively. The saturation concentration of oxygen in saline water of different chloride concentrations and other gases in wastewater with respect to temperature are summarized in Table B.8.

The saturation concentrations of oxygen in saline water at different temperatures can be calculated from several methods. Three such methods are: (a) Equation B.3a based on the correction factor given in Table B.8, (b) the empirical relationship given by Equation B.3b based on chloride concentration, and (c) the empirical equation (Equation B.3c) based on chloride content, barometric pressure, and temperature.[12,13]

$$C_s'' = C_s - \frac{f_{Cl}}{100} S_{Cl} \tag{B.3a}$$

$$C_s'' = C_s(1 - 9 \times 10^{-6} \times S_{Cl}) \qquad (T = 0\text{–}35°C) \tag{B.3b}$$

$$C_s'' = \left(\frac{475 - 2.65 \times 10^{-3} \times S_{Cl}}{33.5 + T}\right)\left(\frac{P}{760}\right) \tag{B.3c}$$

TABLE B.8 Saturation Concentration of Oxygen at Different Chloride Concentrations and Temperatures

Temperature (T)		Dissolved Oxygen Saturation (C_s), mg/L — Chloride Concentration, mg/L					Decrease in Oxygen Concentration, mg O_2 per 100 mg Chloride
°C	°F	0	5000	10,000	15,000	20,000	
0	32	14.62	13.79	12.97	12.14	11.32	0.017
5	41	12.80	12.09	11.39	10.70	10.01	0.014
10	50	11.33	10.73	10.13	9.55	8.98	0.012
15	59	10.15	9.65	9.14	8.63	8.14	0.010
20	68	9.17	8.73	8.30	7.86	7.42	0.009
25	77	8.38	7.96	7.56	7.15	6.74	0.008
30	86	7.63	7.25	6.86	6.49	6.13	0.008

Source: Adapted in part from References 1, 3, 4, 8, 10 and 11.

where

C_s'' = saturation concentration of oxygen in saline water, mg/L
C_s = saturation concentration of oxygen in freshwater, mg/L
f_{Cl} = correction factor, mg/100 mg chloride. This factor is given in the last column of Table B.8.
S_{Cl} = salinity of chloride (Cl^-), mg/L
P = prevailing barometric pressure, mm Hg
T = temperature of saline water, °C

The solubility of oxygen at various temperatures, TDS concentrations and elevations are summarized in Table 10.16. The values in this table are extensively used in oxygen transfer relationships and solved examples in Section 10.3.8.

EXAMPLE B.1: SPECIFIC GRAVITY MEASUREMENT

An empty 100-mL graduated cylinder weighs 110.675 g. The cylinder filled with water at 20°C weighs 210.495 g. The same graduated cylinder filled with oil at 20°C weighs 209.475 g. Calculate the density of the oil and its specific gravity.

Solution

1. Calculate the density of water at 20°C.

$$\text{Weight of water} = (210.495 - 110.675)\text{ g} = 99.820\text{ g}$$

$$\text{Density of water at }20°\text{C} = \frac{\text{Mass}}{\text{Volume}} = \frac{99.820\text{ g}}{100\text{ mL}} = 0.9982\text{ g/mL or }0.9982\text{ g/cm}^3.$$

2. Calculate the density of oil at 20°C.

$$\text{Weight of oil} = (209.475 - 110.675)\text{ g} = 98.800\text{ g}$$

$$\text{Density of oil at }20°\text{C} = \frac{98.800\text{ g}}{100\text{ mL}} = 0.9880\text{ g/mL or }0.9880\text{ g/cm}^3.$$

3. Calculate the specific gravity of oil.

$$\text{The specific gravity of oil at }20°\text{C} = \frac{\text{Density of oil}}{\text{Density of water}} = \frac{0.9880\text{ g/cm}^3}{0.9982\text{ g/cm}^3} = 0.990.$$

Note: The specific gravity of oil can also be calculated directly from the weights.

$$\text{Specific gravity of oil} = \frac{\text{wt. of 100 mL of oil}}{\text{wt. of 100 mL of water}} = \frac{98.800\text{ g}}{99.820\text{ g}} = 0.990.$$

EXAMPLE B.2: WEIGHT OF A GIVEN VOLUME OF SOLUTION

The specific gravity of a salt solution is 1.15. How many lbs are in 55 gallons of the solution?

Solution

Calculate weight of 50 gal of the salt solution by assuming density of water is 1 g/cm^3.
Weight of 1 gallon of water = 8.345 lb/gal (Table B.1)
Density of salt solution = 1.15 × 8.345 lb/gal = 9.60 lb/gal of solution
Weight of 50 gal of salt solution = 55 gal × 9.60 lb/gal = 528 lbs

EXAMPLE B.3: RELATIVE WEIGHT OF A SUBSTANCE

The specific gravity of seawater is 1.25. How many percent it is heavier than that of water?

Solution

Calculate the relative weight of seawater heavier than the distilled water of a specific gravity of 1.

$$\begin{aligned}\text{Relative weight of seawater heavier than distilled water} &= \frac{\text{sp. gr. of seawater} - \text{sp. gr. of distilled water}}{\text{sp. gr. of distilled water}} \times 100\% \\ &= \frac{1.25 - 1.00}{1.00} \times 100\% = 25\%\end{aligned}$$

EXAMPLE B.4: EXPRESSION OF γ, μ, AND P_V IN g, s, AND m UNITS

The common unit for specific weight (γ) is N/m^3 (or kN/m^3); dynamic viscosity (μ) is N·s/m^2; and vapor pressure (P_v) is N/m^2 (or kN/m^2). Express these units in g, s, and m units.

Solution

1. Convert the unit of specific weight (γ).

   Unit for $\gamma = \text{N/m}^3$

   N is a force and is equal to mass × acceleration, $\text{N} = \text{g} \times \dfrac{\text{m}}{\text{s}^2} = \dfrac{\text{g}\cdot\text{m}}{\text{s}^2}$

$$\gamma = \frac{\text{N}}{\text{m}^3} = \frac{\text{g}\cdot\text{m}}{\text{s}^2} \times \frac{1}{\text{m}^3} = \text{g/m}^2\cdot\text{s}^2$$

2. Convert the unit of dynamic viscosity (μ)

$$\mu = \frac{\text{N}\cdot\text{s}}{\text{m}^2} = \frac{\text{g}\cdot\text{m}}{\text{s}^2} \times \frac{\text{s}}{\text{m}^2} = \text{g/m}\cdot\text{s}$$

3. Convert the units of vapor pressure (P_v)

$$P_v = \frac{\text{N}}{\text{m}^2} = \frac{\text{g}\cdot\text{m}}{\text{s}^2} \times \frac{1}{\text{m}^2} = \text{g/m}\cdot\text{s}^2$$

EXAMPLE B.5: SOLUBILITY OF OXYGEN FROM MOLE-FRACTION-BASED HENRY'S LAW CONSTANT (H)

Determine the DO saturation concentration C_s in water at 10°C and 1 atm using the mole-fraction-based Henry's law constant, H (atm/mole fraction).

Solution

1. Select the applicable equation.

   Apply Equation 15.6e, $H = \dfrac{P_g}{x_g}$.

   The Henry's law constant, H at 10°C for oxygen = 3.27×10^4 atm/mole fraction (Table B.4).
2. Determine the partial pressure of oxygen (P_g) at 10°C and 1 atm.

   Vapor pressure of water on the surface of water at 10°C and 1 atm, $P_v = 9.21$ mm Hg (Table B.4).

Partial pressure of a gas can be calculated from Equation B.4, while the content of the gas in the air is known.

$$P_g = \frac{760 - P_v}{760} \times \frac{V_g}{100\%} \tag{B.4}$$

where

P_g = partial pressure of the gas in air, mm Hg
P_v = vapor pressure of water, mm Hg
V_g = content of gas in the air, % by volume

$$\text{Partial pressure of oxygen at 21\% (from Equation B.4)} = \frac{(760 - 9.21)\,\text{mm Hg}}{760\,\text{mm Hg/atm}} \times \frac{21\%}{100\%} = 0.207\,\text{atm}$$

3. Determine the mole fraction of oxygen in water (x_g) from rearranging Equation 15.6e.

$$x_g = \frac{P_g}{H} = \frac{0.207\,\text{atm}}{3.27 \times 10^4\,\text{atm/mole fraction}} = 6.33 \times 10^{-6}\,\text{mole fraction.}$$

4. Determine the mole concentration of O_2 in water (n_a).
   Since $n_a \ll n_{H_2O}$ calculate n_a from Equation 15.6l using mole concentration of water $n_{H_2O} = 55.6$ mole H_2O/L.

$$n_a \approx x_g n_{H_2O} = 6.33 \times 10^{-6}\,\text{mole fraction} \times 55.6\,\text{mole}\,H_2O/L = 3.52 \times 10^{-4}\,\text{mole}\,O_2/L$$

5. Determine the saturation concentration of oxygen (C_s) in mg/L.

$$C_s = 3.52 \times 10^{-4}\,\text{mole}\,O_2/L \times 32\,\text{g/mole} \times 103\,\text{mg/g} = 11.3\,\text{mg/L}$$

EXAMPLE B.6: SOLUBILITY OF OXYGEN FROM VOLUMETRIC SATURATION-CONCENTRATION-BASED HENRY'S LAW CONSTANT (H_V)

Determine the DO saturation C_s in water at 10°C and 1 atm using the volumetric saturation-concentration-based Henry's law constant at 0°C and 1 atm, H_v (mL/L·atm). Compare the result obtained in Example B.5.

Solution

1. Select the applicable equation.
   Apply Equation 15.6r, $C'_s = H_v P_g$
2. Determine the partial pressure of oxygen (P_g) at 10°C and 1 atm.
   Partial pressure of oxygen $P_g = 0.207$ atm (Example B.5, Step 2.)
3. Determine the absorption coefficient H_v at 10°C.
   From Table B.5, $H_v = 38.02$ mL/L·atm is obtained at 10°C after reducing to the standard condition, 0°C and 1 atm.
4. Determine the solubility of oxygen in water (volumetric saturation concentration, C'_s) at 10°C and 1 atm from Equation 15.6r at $P_g = 0.207$ atm.

$$C'_s = H_v P_g = 38.02\,\text{mL/L·atm} \times 0.207\,\text{atm} = 7.87\,\text{mL/L}$$

5. Convert the solubility of oxygen from volumetric saturation concentration (C'_s) in mL/L to mass concentration C_s in mg/L.

   By substituting Equations 15.6p and 15.6r into Equation 15.6s the relationship between these two concentrations is obtained in Equation B.5.

$$C'_s = \frac{22.4}{mw_g} C_s \quad \text{or} \quad C_s = \frac{mw_g}{22.4} C'_s \tag{B.5}$$

   where

   C'_s = volumetric saturation concentration of gas in liquid phase, mL/L
   C_s = solubility or saturation concentration of gas in liquid phase, mg/L
   mw_g = molecular weight of gas, g/mole
   22.4 = mL·g/mole·mg

   Mass concentration from Equation B.5, $C_s = \frac{32\,\text{g/mole}}{22.4\,\text{mL}\cdot\text{g/mole}\cdot\text{mg}} \times 7.87\,\text{mL/L} = 11.2\,\text{mg/L}$

6. Compare the result with that obtained in Example B.5.

   The C_s values obtained in Examples B.5 and B.6 are 11.3 and 11.2 mg/L, respectively.

EXAMPLE B.7: SATURATION CONCENTRATION OF OXYGEN FROM EMPIRICAL EQUATIONS

Determine the solubility of oxygen in distilled water at 10°C and 1 atm from Equations B.1a, B.1b, and B.1c, respectively.

Solution

1. Calculate the saturation concentration of oxygen (C_s) at 10°C and 1 atm from Equation B.1a.

$$\begin{aligned} C_s &= 14.652 - 0.41022T + 0.00799T^2 - 0.0000777T^3 \\ &= 14.652 - 0.41022 \times 10 + 0.00799 \times 10^2 - 0.0000777 \times 10^3 \\ &= 14.652 - 4.102 + 0.799 - 0.078 = 11.3\,\text{mg/L} \end{aligned}$$

2. Calculate the saturation concentration of oxygen (Cs) at 10°C and 1 atm from Equation B.1b.

   Vapor pressure at 10°C, $P_v = 9.21$ mm Hg (Table B.4).

$$\begin{aligned} C_s &= (0.68 - 6 \times 10^{-4}T)\left(\frac{760 - P_v}{T + 35}\right) = (0.68 - 6 \times 10^{-4} \times 10) \times \left(\frac{760 - 9.21}{10 + 35}\right) \\ &= 0.674 \times \frac{751}{45} = 11.2\,\text{mg/L}. \end{aligned}$$

3. Calculate the saturation concentration of oxygen (C_s) at 10°C and 1 atm from Equation B.1c.

$$\begin{aligned} C_s &= 14.652 - 10.53(1 - e^{-0.03896T}) = 14.652 - 10.53(1 - e^{-0.03896 \times 10} - 1) \\ &= 14.652 - 10.53 \times (1 - 0.677) = 11.3\,\text{mg/L}. \end{aligned}$$

EXAMPLE B.8: PREVAILING PRESSURE CORRECTION

Determine the saturation concentration of oxygen in surface water at an altitude of 6000 ft above sea level. The temperature of water is 15°C.

Solution

At an altitude of 6000 ft above sea level, the barometric pressure, $P = 611$ mm Hg (Table B.7).
At 15°C, the vapor pressure at 1 atm, $P_v = 12.788$ mm Hg (Table B.2).
The saturation concentration of oxygen at 15°C and 1 atm, $C_s = 10.15$ mg/L (Table B.6).
Calculate the saturation concentration of oxygen ($C_{s,P}$) at 15°C and 6000 ft above sea level from Equation B.2.

$$C_{s,P} = \frac{P - P_v}{760 - P_v} \times C_s = \frac{(611 - 12.788)\,\text{mm Hg}}{(760 - 12.788)\,\text{mm Hg}} \times 10.15\,\text{mg/L} = \frac{598\,\text{mm Hg}}{747\,\text{mm Hg}} \times 10.15\,\text{mg/L} = 8.13\,\text{mg/L}.$$

EXAMPLE B.9: SATURATION CONCENTRATION OF OXYGEN IN SALINE WATER

The saturation concentration of oxygen decreases at higher salt concentration. Calculate the DO saturation in saline water having chloride concentration of 5000 mg/L. The temperature and barometric pressure are 15°C and 1 atm. Use Equations B.3a, B.3b, and B.3c, respectively. Compare the results with the value given in Table B.8.

Solution

1. Calculate the DO saturation concentration using Equation B.3a.
   The saturation concentration of oxygen at 15°C and 1 atm, $C_s = 10.15$ mg/L (Table B.6).
   Correction factor, $f_{Cl} = 0.010$ mg O_2/100 mg chloride (Table B.8).
   DO saturation from Equation B.3a,

$$C_s'' = C_s - f_{Cl}\frac{S_{Cl}}{100} = 10.15\,\text{mg/L} - 0.010\,\text{mg/100 mg chloride} \times \frac{5000\,\text{mg chloride/L}}{100\,\text{mg chlorine}} = 9.65\,\text{mg/L}$$

2. Calculate the DO saturation concentration using Equation B.3b.

$$\begin{aligned} C_s'' &= C_s(1 - 9 \times 10^{-6} \times S_{Cl}) = 10.15\,\text{mg/L} \times (1 - 9 \times 10^{-6} \times 5000) \\ &= 10.15\,\text{mg/L} \times (1 - 0.045) = 9.69\,\text{mg/L}. \end{aligned}$$

3. Calculate the DO saturation concentration using Equation B.3c.

$$C_s'' = \left(\frac{475 - 2.65 \times 10^{-3} \times S_{Cl}}{33.5 + T}\right)\left(\frac{P}{760}\right) = \left(\frac{475 - 2.65 \times 10^{-3} \times 5000\,\text{mg/L}}{33.5 + 15}\right) \times \left(\frac{760}{760}\right) = 9.52\,\text{mg/L}$$

4. Compare the results.
   The saturation concentration of oxygen in saline water containing 5000 mg/L chloride at 15°C and 1 atm is 9.65 mg/L (Table B.8). The values calculated from Equations B.3a and B.3b are close to the value given in the Table B.8 and acceptable. The value obtained from Equation B.3c is about 13% lower that obtained from Table B.8. Therefore, Equation B.3c should be used with caution.

B.3 Stability of Feed Water for Membrane Systems

The scaling or corrosive nature of feed water for treatment by membrane processes may be evaluated from LSI or RSI. These indexes are calculated by using ($pK_2 - pK_s$) values. These values with respect to temperature and TDS are summarized in Table B.9.[14] The application procedure is shown in Example 15.58.

TABLE B.9 Values of $pK_2 - pK_s$ with Respect to Temperature and Total Dissolved Solid (TDS)

TDS, mg/L	$pK_2 - pK_s$						
	0°C	10°C	20°C	30°C	40°C	50°C	80°C
0	2.45	2.23	2.02	1.86	1.68	1.52	1.08
40	2.58	2.36	2.15	1.99	1.81	1.65	1.21
80	2.62	2.40	2.19	2.03	1.85	1.69	1.25
120	2.66	2.44	2.23	2.07	1.89	1.73	1.29
160	2.68	2.46	2.25	2.09	1.91	1.75	1.31
200	2.71	2.49	2.28	2.12	1.94	1.78	1.34
240	2.74	2.52	2.31	2.15	1.97	1.81	1.37
280	2.76	2.54	2.33	2.17	1.99	1.83	1.39
320	2.78	2.56	2.35	2.19	2.01	1.85	1.41
360	2.79	2.57	2.36	2.20	2.02	1.86	1.42
400	2.82	2.59	2.38	2.22	2.04	1.88	1.44
440	2.83	2.61	2.40	2.24	2.06	1.90	1.46
480	2.84	2.62	2.41	2.25	2.07	1.91	1.47
520	2.86	2.64	2.43	2.27	2.09	1.93	1.49
560	2.87	2.65	2.44	2.28	2.10	1.94	1.50
600	2.88	2.66	2.45	2.29	2.11	1.95	1.51
640	2.90	2.68	2.47	2.31	2.13	1.97	1.53
680	2.91	2.69	2.48	2.32	2.14	1.98	1.54
720	2.92	2.70	2.49	2.33	2.15	1.99	1.55
760	2.92	2.70	2.49	2.33	2.15	1.99	1.55
800	2.93	2.71	2.50	2.34	2.16	2.00	1.56

Source: Adapted in part from Reference 14.

B.4 Salinity of Reclaimed Water for Irrigation

Sodicity is a condition when Na^+ ions build up in soil matrix from use of reclaimed water for irrigation. This condition reduces movement of water and air causing water logging.[15] The adjusted calcium concentration $[Ca_X^{2+}]$ value is needed to evaluate this situation. The $[Ca_X^{2+}]$ values based on different HCO_3^-/Ca^{2+} ratios and salinities are given in Table B.10. The procedure for determining this condition from $[Ca_X^{2+}]$ data is explained in Example 12.25.

TABLE B.10 Values of $[Ca_X^{2+}]$ as a Function of the HCO_3^--to-Ca^{2+} Ratio and Salinity

Ratio of HCO_3^-/Ca^{2+}, meq/L	Value of $[Ca_X^{2+}]$, meq/L											
	Salinity of Applied Water (EC_w), dS/m or mmhos/cm											
	0.1	0.2	0.3	0.5	0.7	1	1.5	2	3	4	6	8
0.05	13.2	13.6	13.9	14.4	14.8	15.3	15.9	16.4	17.3	18.0	19.1	19.9
0.1	8.31	8.57	8.77	9.07	9.31	9.62	10.0	10.4	10.9	11.3	12.0	12.6
0.15	6.34	6.54	6.69	6.92	7.11	7.34	7.65	7.90	8.31	8.64	9.17	9.58
0.2	5.24	5.40	5.52	5.71	5.87	6.06	6.31	6.52	6.86	7.13	7.57	7.91
0.25	4.51	4.65	4.76	4.92	5.06	5.22	5.44	5.62	5.91	6.15	6.52	6.82
0.3	4.00	4.12	4.21	4.36	4.48	4.62	4.82	4.98	5.24	5.44	5.77	6.04

Continued

TABLE B.10 (*Continued*) Values of $[Ca_X^{2+}]$ as a Function of the HCO_3^--to-Ca^{2+} Ratio and Salinity

Ratio of HCO_3^-/Ca^{2+}, meq/L	Value of $[Ca_X^{2+}]$, meq/L											
	Salinity of Applied Water (EC_w), dS/m or mmhos/cm											
	0.1	0.2	0.3	0.5	0.7	1	1.5	2	3	4	6	8
0.35	3.61	3.72	3.80	3.94	4.04	4.17	4.35	4.49	4.72	4.91	5.21	5.45
0.4	3.30	3.40	3.48	3.60	3.70	3.82	3.98	4.11	4.32	4.49	4.77	4.98
0.45	3.05	3.14	3.22	3.33	3.42	3.53	3.68	3.80	4.00	4.15	4.41	4.61
0.5	2.84	2.93	3.00	3.10	3.19	3.29	3.43	3.54	3.72	3.87	4.11	4.30
0.75	2.17	2.24	2.29	2.37	2.43	2.51	2.62	3.70	2.84	2.95	3.14	3.28
1	1.79	1.85	1.89	1.96	2.01	2.09	2.16	2.23	2.35	2.44	2.59	2.71
1.25	1.54	1.50	1.63	1.68	1.73	1.78	1.86	1.92	2.02	2.10	2.23	2.33
1.5	1.37	1.41	1.44	1.49	1.53	1.58	1.65	1.70	1.79	1.86	1.97	2.07
1.75	1.23	1.27	1.30	1.35	1.38	1.43	1.49	1.54	1.62	1.68	1.78	1.86
2	1.13	1.16	1.19	1.23	1.26	1.31	1.36	1.40	1.48	1.54	1.63	1.70
2.25	1.04	1.08	1.10	1.14	1.17	1.21	1.26	1.30	1.37	1.42	1.51	1.58
2.5	0.97	1.00	1.02	1.06	1.09	1.12	1.17	1.21	1.27	1.32	1.40	1.47
3	0.85	0.89	0.91	0.94	0.96	1.00	1.04	1.07	1.13	1.17	1.24	1.30
3.5	0.78	0.80	0.82	0.85	0.87	0.90	0.94	0.97	1.02	1.06	1.12	1.17
4	0.71	0.73	0.75	0.78	0.80	0.82	0.86	0.88	0.93	0.97	1.03	1.07
4.5	0.66	0.68	0.69	0.72	0.74	0.76	0.79	0.82	0.86	0.90	0.95	0.99
5	0.61	0.63	0.65	0.67	0.69	0.71	0.74	0.76	0.80	0.83	0.88	0.93
7	0.49	0.50	0.52	0.53	0.55	0.57	0.59	0.61	0.64	0.67	0.71	0.74
10	0.39	0.40	0.41	0.42	0.43	0.45	0.47	0.48	0.51	0.53	0.56	0.58
20	0.24	0.25	0.26	0.26	0.27	0.28	0.29	0.30	0.32	0.33	0.35	0.37

Source: Adapted in part from Reference 15.

References

1. Qasim, S. R., *Wastewater Treatment Plants: Planning, Design, and Operation*, 2nd ed., CRC Press, Boca Raton, FL, 1999.
2. Droste, R. L., *Theory and Practice of Water and Wastewater Treatment*, John Wiley of Sons, Inc., New York, NY, 1997.
3. Metcalf & Eddy | AECOM, *Wastewater Engineering: Treatment and Resource Recovery*, 5th ed., McGraw-Hill, New York, NY, 2014.
4. Metcalf & Eddy, Inc., *Wastewater Engineering: Treatment and Reuse*, 4th ed., McGraw-Hill Companies, Inc., New York, NY, 2003.
5. Qasim, S. R., E. M. Motley, and G. Zhu, *Water Works Engineering: Planning, Design, and Operation*, Prentice Hall PTR, Upper Saddle River, NJ, 2000.
6. Tchobanoglous, G., H. Theisen, and S. Vigil, *Integrated Solid Waste Management: Engineering Principles and Management Issues*, McGraw-Hill, Inc., New York, NY, 1993.
7. Lange, N. A. and J. A. Dean, *Lange's Handbook of Chemistry*, 13th ed., McGraw-Hill, Inc., New York, NY, 1985.
8. Tchobanoglous, G. and E. D. Schroeder, *Water Quality: Characteristics Modeling and Modification*, Addison-Welsey, Publishing Co., Reading, MA, 1985.
9. Elmore, H. L. and T. W. Hayes, Solubility of atmospheric oxygen in water, No. 29 Report of the Committee on Sanitary Engineering Research, *Journal of the Sanitary Engineering Division*, ASCE, 86(SA 4), 1960, 41–53.

10. Gameson, A. H. and K. G. Robertson, The solubility of oxygen in pure water and sea water, *Journal Chemical Technology and Biotechnology*, 5(9), 1955, 502.
11. APHA, AWWA, and WEF, *Standard Methods for Examination of Water and Wastewater*, 22nd ed., American Public Health Association, Washington, DC, 2012.
12. Hunter, J. S. and J. C. Ward, The effects of water temperature and elevation upon aeration, *Proceedings of International Symposium on Wastewater Treatment in Cold Climates*, Institute of Northern Studies, University of Saskatchewan, Canada, 1974.
13. Fair, G. M., J. C. Geyer, and D. A. Okum, *Water and Wastewater Engineering*, Vol. 2, John Wiley and Sons Inc., New York, NY, 1968.
14. Viessman, W., Jr., M. J. Hammer, E. M. Perez, and P. A. Chadik, *Water Supply and Pollution Control*, 8th ed., Prentice Hall PTR, Upper Saddle River, NJ, 2008.
15. Pettygrove, G. S. and T. Asano (editors), *Irrigation with Reclaimed Municipal Wastewater—A Guide Manual*, Lewis Publishers, Inc., Chelsea, MI, 1985.

Appendix C: Minor Head Loss Coefficients for Pressure Conduits and Open Channels, Normal Commercial Pipe Sizes, and Design Information of Parshall Flume

In this appendix, the constants commonly used to calculate the minor head losses in pressure conduits and open channels are provided in Sections C.1 and C.2. Standard commercial pipe sizes are listed in Table C.1 of Section C.3. Additionally, the basic information needed to design standard Parshall flumes are provided in Section C.4. The information provided in this appendix may be found in many hydraulics texts and handbooks. Many of these references are cited at the end of this appendix.

C.1 Minor Head Loss Coefficients (*K*) for Pressure Conduits

In a pressure conduit, the minor head losses are created at fittings, valves, bends, entrance, exit, etc. Each of these losses is calculated from Equation 6.15b by applying a proper minor head loss coefficient (K) to the velocity head at the application point. The typical values of K used in plant hydraulic calculations are presented below.[1–7]

C.1.1 Gate Valve

Full Open	One-Fourth Closed	One-Half Closed	Three-Fourths Closed	Typical Value	Equation 6.15b
0.19	1.15	5.6	24	1.0	$h_m = K\frac{V^2}{2g}$

Note: V = velocity in the pipe upstream or downstream of the valve.

C.1.2 Butterfly Valve

Open Full	Angle Closed			Typical Value	Equation 6.15b
	0°	40°	60°		
0.3	1.4	10	94	1.2	$h_m = K\frac{V^2}{2g}$

Note: V = velocity in the pipe upstream or downstream of the valve.

C.1.3 Other Valves

Check Valve	Plug Globe Valve (Full Open)	Diaphragm Valve (Full Open)	Equation 6.15b
1.5–2.5	1.0–4.0	2.3	$h_m = K\frac{V^2}{2g}$

Note: V = velocity in the pipe upstream or downstream of the valve.

C.1.4 Entrance and Exit

Entrance				Exit from Conduit to Still Water	Equation 6.15b
Pipe Projecting into Tank	End of Pipe Flushed with Tank	Slightly Rounded	Bell-Mouthed		
0.83	0.50	0.23	0.04	1.0	$h_m = K\frac{V^2}{2g}$

Note: V = velocity in the pipe after the entrance or before the exit.

C.1.5 Elbow (45–61 cm diameter)

22.5°	45°	90° Regular	90° Long	Equation 6.15b
0.1–0.2	0.15–0.3	0.20–0.40	0.14–0.23	$h_m = K\frac{V^2}{2g}$

Note: V = velocity in the pipe upstream or downstream of the elbow.

C.1.6 Tee

Run-to-Run	Branch-to-Run	Run-to-Branch	Equation 6.15b
0.25–0.6	0.6–1.8	0.6–1.8	$h_m = K\frac{V^2}{2g}$

Note: V = velocity in the pipe upstream or downstream of the elbow.

C.1.7 Increaser and Reducer

Increaser[a]			Reducer[b]			Equation 6.15b
$d/D = 0.25$	$d/D = 0.5$	$d/D = 0.75$	$d/D = 0.25$	$d/D = 0.5$	$d/D = 0.75$	
0.42	0.33	0.19	0.92	0.56	0.19	$h_m = K\frac{V^2}{2g}$

[a] d/D = ratio of pipe diameter upstream and downstream of increaser and V = velocity in the pipe upstream of the increaser.
[b] d/D = ratio of pipe diameters downstream and upstream of the reducer and V = velocity in the pipe downstream of the reducer.

C.2 Minor Head Loss Constants (K) for Open Channels

In an open channel, the minor head losses occur due to turbulences caused by changing the flow direction and cross sectional area. The minor head loss equation for the pressure conduit (Equation 6.15b) may also be applied to appurtenances such as manholes, junction and diversion boxes, syphons, and sluice gates (submerged or nonsubmerged). This equation is also modified to Equation 7.3b for enlargement (or outlet) losses, and Equation 7.3c for contraction (or inlet) losses, respectively. The typical minor head loss coefficients for these expressions are presented below.[1–7]

C.2.1 Manholes, and Junction and Diversion Boxes

Head losses in manholes, and junction and diversion boxes depend on the size of the appurtenances, change in direction, and the contour of the bottom.

1. Manhole, junction box, or diversion box with no change in channel or pipe size or direction.

Flow-Through Box	Terminal Box	Equation 6.15b
0.05	1.0	$h_m = K\frac{v^2}{2g}$

Note: v = velocity in the upstream or downstream channel or pipe of the manhole or box.

2. Manhole, junction box, or diversion box with no change in channel or pipe size, but a change in channel or pipe direction.

45° turn		90° turn		Equation 6.15b
Without Shaping	With Shaping	Without Shaping	With Shaping	
0.3–0.4	0.2–0.3	1.2–1.4	1.0–1.2	$h_m = K\frac{v^2}{2g}$

Note: v = velocity in the upstream or downstream channel or pipe of the manhole or box.

3. Large junction or diversion box in which the velocity is small.

Entrance	Exit	Equation 6.15b
0.5	1.0	$h_m = K\frac{v^2}{2g}$

Note: v = velocity in the channel or pipe after the entrance or before the exit.

C.2.2 Syphon

Syphon	Equation 6.15b
2.78	$h_m = K\frac{v^2}{2g}$

Note: v = velocity through the syphon.

C.2.3 Sluice Gate

Sluice Gate	Equation 6.15b
0.2–0.8	$h_m = K\frac{v^2}{2g}$

Note: v = velocity through the full gate opening (submerged or nonsubmerged).

C.2.4 Sudden Enlargement or Outlet Losses

Sharp-Cornered	Bell-Mouthed	Equation 7.3b
0.2–1.0	0.1	$h_L = \frac{K_e}{2g}(v_1^2 - v_2^2)$

Note: v_1 and v_2 = velocities upstream and downstream of the enlargement or outlet.

C.2.5 Sudden Contraction or Inlet Losses

Sharp-Cornered	Round-Cornered	Bell-Mouthed	Equation 7.3c
0.5	0.25	0.05	$h_L = \frac{K_c}{2g}(v_2^2 - v_1^2)$

Note: v_1 and v_2 = the velocities upstream and downstream of the contraction or inlet.

C.3 Standard Commercial Pipe Sizes

The normal diameters of pipes that may be commercially available in the U.S. market, and equivalent normal pipe sizes in SI units are summarized in Table C.1. Certain normal size may not be applicable for some pipe materials. The actual inside diameter of a pipe will also vary with the *pipe material*, and the *wall thickness* determined by the *pipe schedule* or *pipe class*. Readers should consult the pipe manufacturers to obtain the detailed technical data of the pipes used in the design.

C.4 Standard Design of Parshall Flume

A Parshall flume is a fixed hydraulic structure used for the flow measurement of water and wastewater. It has a converging section, throat, and the diverging section.[8–10] Under the free-flow condition, the depth of water at a specified location upstream of the throat is converted to flow rate by Equation 7.9. The constant C is a free-flow coefficient, and the exponent n varies with flume size. These constants for different throat width (W) are summarized in Table C.2.[9] The design components of a standard Parshall flume are shown in Figure C.1.[8] The standard dimensions are given in Table C.3.[9]

The free-flow condition is applied until the submergence ratio (submergence head to measuring head) is exceeded by a certain criteria (Table C.4).[9] Design procedures of Parshall flumes are presented in Examples 7.8, 8.13, and 9.38. The hydraulic profiles through these flumes are also illustrated in

Figures 7.14, 9.39, and 9.40. When a Parshall flume is submerged, the head loss through the flume depends upon the submergence.[8–10] The head losses through standard Parshall flumes for different percent submergences may be read from the graphics contained in Reference 9.

TABLE C.1 Commonly Available Commercial Pipe Sizes (4–240 in Diameter)

Normal Pipe Size Available in the U.S. Market		Equivalent Normal Pipe Size in SI Units, mm
in	mm	
4	101.6	100
5	127.0	–
6	152.4	150
8	203.2	200
10	254.0	250
12	304.8	300
14	355.6	350
15	381.0	–
16	406.4	400
18	457.2	450
20	508.0	500
21	533.4	–
24	609.6	600
27	685.8	700
30	762.0	750
36	914.4	900
42	1067	1050
48	1219	1200
54	1372	1400
60	1524	1500
66	1676	–
72	1829	1800
78	1981	2000
84	2134	2100
90	2286	2250
96	2438	2400
102	2591	–
108	2743	2700
114	2896	–
120	3048	3000
–	–	3300
144	3658	3600
180	4572	–
204	5182	–
240	6096	–

TABLE C.2 Coefficients (C) and Exponents (n) Used for Design of Parshall Flumes

Throat Width (W)	Coefficient (C)	Exponent (n)
1 in	0.338	1.55
2 in	0.676	1.55
3 in	0.992	1.55
6 in	2.06	1.58
9 in	3.07	1.53
1 ft	3.95	1.55
2 ft	8.00	1.55
3 ft	12.00	1.57
4 ft	16.00	1.58
5 ft	20.00	1.59
6 ft	24.00	1.59
7 ft	28.00	1.60
8 ft	32.00	1.61
10 ft	39.38	1.60
12 ft	46.75	1.60
15 ft	57.81	1.60
20 ft	76.25	1.60
25 ft	94.69	1.60
30 ft	113.13	1.60
40 ft	150.00	1.60
50 ft	186.88	1.60

Source: Adapted in part from Reference 9.

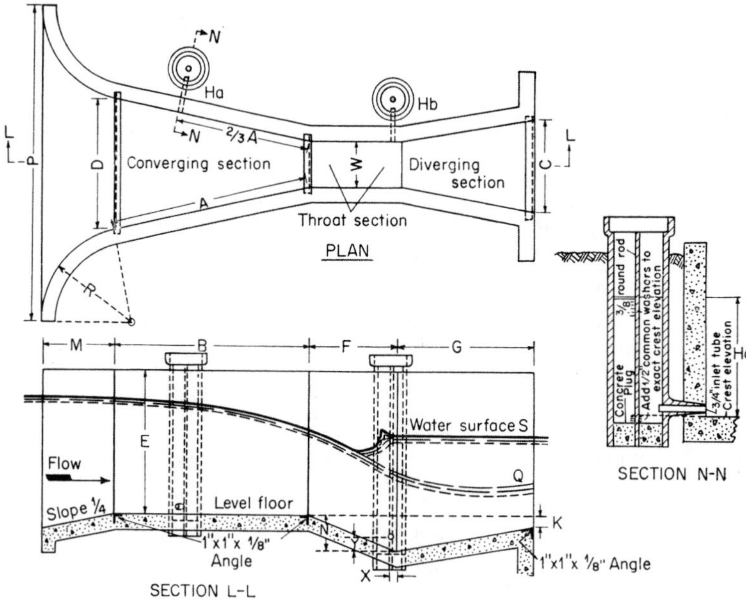

FIGURE C.1 Standard design details of Parshall flume. (Adapted in part from Reference 8. (Courtesy U.S. Department of the Interior, Bureau of Reclamation.))

TABLE C.3 Standard Dimensions of Selected Parshall Flumes ($W = 0.5$–6 ft)

Dimension and Capacity Parameter	Throat Size (W) in ft, Dimension in ft-in, and Capacity Parameter in ft^3/s									
	$W = 0.5$		$W = 1$		$W = 2$		$W = 4$		$W = 6$	
	ft	in	ft	in	ft	in	ft	in	ft	in
A	2	7/16	4	6	5	0	6	0	7	0
2/3 A	1	4 5/16	3	0	3	4	4	0	4	8
B	2	0	4	4 7/8	4	10 7/8	5	10 5/8	6	10 3/8
C	1	3½	2	0	3	0	5	0	7	0
D	1	3 5/8	2	9¼	3	11½	6	4¼	8	9
E	2	0	3	0	3	0	3	0	3	0
F	1	0	2	0	2	0	2	0	2	0
G	2	0	3	0	3	0	3	0	3	0
K	0	3	0	3	0	3	0	3	0	3
M	0	0	1	3	1	3	1	6	1	6
N	0	4¼	0	9	0	9	0	9	0	9
P	2	11½	4	10¾	6	1	8	10¾	11	3½
R	1	4	1	8	1	8	2	0	2	0
X	0	2	0	2	0	2	0	2	0	2
Y	0	3	0	3	0	3	0	3	0	3
Minimum capacity	0.05		0.11		0.42		1.3		2.6	
Maximum capacity	3.9		16.1		33.1		67.9		103.5	

Source: Adapted in part from Reference 9.

TABLE C.4 Determination of Free-Flow Condition for Parshall Flumes

Throat Width (W)	Submergence Ratio (H_b/H_a) Allowed under Free-Flow Condition[a], %
1–3 in	≤50
6–9 in	≤60
1–8 ft	≤70
10–50 ft	≤80

[a] H_a = water depth at the throat and H_b = submergence head downstream of the throat (Figure C.1).

Source: Adapted in part from Reference 9.

References

1. Qasim, S. R., E. M. Motley, and G. Zhu, *Water Works Engineering: Planning, Design, and Operation*, Prentice Hall PTR, Upper Saddle River, NJ, 2000.
2. Qasim, S. R., *Wastewater Treatment Plants: Planning, Design, and Operation*, 2nd ed., CRC Press, Boca Raton, FL, 1999.
3. FMC Corporation, *Hydraulics and Useful Information*, Chicago Pumps, Chicago, IL, 1963.
4. Benefield, L. D., J. F. Judkins, and A. D. Parr, *Treatment Plant Hydraulics for Environmental Engineers*, Prentice-Hall Inc., Englewood Cliffs, NJ, 1984.
5. Brater, E. F., H. W. King, J. E. Lindell, and C. Y. Wei, *Handbook of Hydraulics*, 7th ed., MaGraw-Hill, New York, NY, 1996.

6. Heald, C. C. (editor), *Cameron Hydraulic Data*, 18th ed., Ingersoll-Dresser Pumps, Liberty Corner, NJ, 1995.
7. Zipparro, V. J. (editor), *Davis's Handbook of Applied Hydraulics in Engineering*, 4th ed., McGraw-Hill Book Co., New York, 1993.
8. U.S. Department of the Interior, *Water Measurement Manual*, 2nd ed., Bureau of Reclamation, Denver, CO, 1967.
9. U.S. Department of the Interior, Bureau of Reclamation, *Water Measurement Manual, A Water Resources Technical Publication*, 3rd ed., 1997 (Revised Reprint, 2001), Bureau of Reclamation, http://www.usbr.gov, 2001.
10. Parshall, R. L., *Measuring Water in Irrigation Channels with Parshall Flumes and Small Weirs*, U.S. Soil Conservation Service, Circular 843, U.S. Department of Agriculture, Washington, DC, 1950.

Appendix D: Unit Conversions

In this appendix, the conversion factors between SI units and the U.S. customary units are provided in Table D.1. The conversion factors for mixed units are summarized in Table D.2.[1-3]

TABLE D.1 Conversion Factors between SI Units and the U.S. Customary Units

SI Units	Factor *f* Multiply by Factor *f* to Obtain → ← Divide by Factor *f* to Obtain	U.S. Customary Units
Length		
cm	0.3937	in
m	3.281	ft (1 yd = 3 ft)
km (1000 m)	0.6214	mile, mi (1760 yd or 5280 ft)
Area		
cm^2	0.1550	in^2
m^2	10.76	ft^2
m^2	1.196	yd^2 (9 ft^2)
Hectare, ha (10,000 m^2)	2.471	acre (4840 yd^2 or 43,560 ft^2)
km^2	0.3861	mi^2
km^2	247.1	acre
Volume		
cm^3	6.102×10^{-2}	in^3
L	0.2642	gal
m^3 (1000 L)	35.31	ft^3 (7.481 gal)
m^3	1.308	yd^3 (27 ft^2)
m^3	264.2	gal
m^3	8.13×10^{-4}	acre-ft (43,560 ft^3 or 325,900 gal)
Mass		
g	3.527×10^{-2}	oz
g	2.205×10^{-3}	lb (1 lb = 16 oz)
g	15.43	grain (1 lb = 7000 grain)
kg	2.205	lb
Tonne (1000 kg)	2205	lb
Force		
N	0.2248	lb_f

(*Continued*)

TABLE D.1 (*Continued*) Conversion Factors between SI Units and the U.S. Customary Units

SI Units	Factor f Multiply by Factor f to Obtain → ← Divide by Factor f to Obtain	U.S. Customary Units
Velocity		
m/s (1 m/s = 86.4 km/d)	3.281	ft/s (16.36 mi/d)
m/s	2.237	mi/h
km/h	0.6214	mi/h
km/h	2.237	mi/h
Flow rate		
m^3/s	35.31	ft^3/s (7.481 gal/s or 448.9 gal/min)
m^3/d	264.2	gal/d
m^3/d	2.642×10^{-4}	Mgal/d
m^3/s	22.82	Mgal/d
L/s	22,820	gal/d
Acceleration		
m/s^2 ($g = 9.81\ m/s^2$)	3.281	ft/s^2 ($g = 32.2\ ft/s^2$)
Temperature		
°C	1.8 × (°C) + 32 → ← 0.0555 × (°F) − 32	°F
°k	1.8 × (°k) − 459.7 → ← 0.0555 × (°F) + 459.7	°F
Power		
kW	0.9478	Btu/s
kW	1.341	hp (550 ft·lb_f/s)
W	0.7376	ft·lb/s
Pressure		
kPa (Pascal)	0.1450	lb_f/in^2
Pa (N/m^2)	1.450×10^{-4}	lb_f/in^2
Pa (N/m^2)	2.089×10^{-2}	lb_f/in^2
Pa (N/m^2)	2.961×10^{-4}	in Hg
Pa (N/m^2)	4.019×10^{-3}	in of water
Pa (N/m^2)	9.869×10^{-3}	atm (14.7 lb/in^2 or 1 psi)
Dynamic viscosity (μ)		
N·s/m^2 (k/m·s)	0.021	lb_f·s/ft^2 (32.15 lb_m/ft·s)
N·s/m^2 (k/m·s)	0.671	lb_m/ft·s
Kinematic viscosity (ν)		
Centistoke	1.076×10^{-5}	ft^2/s
m^2/s	10.76	ft^2/s
Energy		
kW·h	1.341	hp·h
kW·h	3412	Btu

Source: Adapted in part from References 1 through 3.

TABLE D.2 Conversion Factors, Mixed Units

Length (L)

Mile (mi)	yd	ft	in	m	cm
1	1760	5280	6.336×10^4	1.609×10^3	1.609×10^5
5.68×10^{-4}	**1**	3	36	0.9144	91.44
1.894×10^{-4}	0.333	**1**	12	0.3048	30.48
1.578×10^{-5}	0.028	0.083	**1**	0.0254	2.54
6.214×10^{-4}	1.094	3.281	39.37	**1**	100
6.214×10^{-6}	0.01094	0.03281	0.3937	0.01	**1**

Area (A)

mi^2	acre	yd^2	ft^2	in^2	m^2
1	640[a]	3.098×10^6	2.788×10^7	4.014×10^9	2.59×10^6
1.563×10^{-3}	**1**	4840	43,560	6.27×10^6	4047
3.228×10^{-7}	2.066×10^{-4}	**1**	9	1296	0.836
3.587×10^{-8}	2.3×10^{-5}	0.111	**1**	144	0.093
2.491×10^{-10}	1.59×10^{-7}	7.716×10^{-4}	6.944×10^{-3}	**1**	6.452×10^{-4}
3.861×10^{-7}	2.5×10^{-4}	1.196	10.764	1550	**1**

Volume (V)

acre·ft	U.S. gal	ft^3	in^3	L	m^3	cm^3
1	325,851	43,560	75.3×10^6	1.23×10^6	1230	1.23×10^9
3.07×10^{-6}	**1**	0.134	231.552	3.785	3.875×10^{-3}	3875.412
2.3×10^{-5}	7.481	**1**	1728	28.317	0.028	28,316.846
1.33×10^{-8}	4.329×10^{-3}	5.787×10^{-4}	**1**	0.016	1.639×10^{-5}	16.387
8.1×10^{-7}	0.264	0.035	61.024	**1**	1×10^{-3}	1000
8.13×10^{-4}	264.2	35.31	6.10×10^4	1000	**1**	10^6
8.13×10^{-10}	2.642×10^{-4}	3.531×10^{-5}	6.10×10^{-2}	10^{-3}	10^{-6}	**1**

Time (T)

Year	Months	Days	h	min	s
1	12	365	8760	525,600	3.154×10^7

Velocity (L/T)

ft/s	ft/min	m/s	m/min	cm/s
1	60	0.3048	18.29	30.48
0.017	**1**	5.08×10^{-3}	0.3048	0.5080

(*Continued*)

TABLE D.2 (*Continued*) Conversion Factors, Mixed Units

Velocity (L/T)				
ft/s	ft/min	m/s	m/min	cm/s
3.281	196.8	**1**	60	100
0.055	3.28	0.017	**1**	1.70
0.032	1.969	0.01	0.588	**1**

Discharge (L^3/T)					
mgd	gpm	ft^3/s	ft^3/min	L/s	m^3/d
1	694.4	1.547	92.82	43.75	3.78×10^3
1.44×10^{-3}	**1**	2.228×10^{-3}	0.134	0.063	5.45
0.646	448.9	**1**	60	28.32	2447
0.011	7.481	0.017	**1**	0.472	40.78
0.023	15.850	0.035	2.119	**1**	86.41
2.64×10^{-4}	0.183	4.09×10^{-4}	0.025	0.012	**1**

Mass (M)					
Ton	lb_m	Grain	Ounce (oz)	kg	g
1	2000	1.4×10^7	32,000	907.2	907,200
0.0005	**1**	7000	16	0.454	454
7.14×10^{-8}	1.429×10^{-4}	**1**	2.29×10^{-3}	6.48×10^{-5}	0.065
3.125×10^{-5}	0.0625	437.6	**1**	0.028	28.35
1.10×10^{-3}	2.205	1.54×10^4	35.27	**1**	1000
1.10×10^{-6}	2.20×10^{-3}	15.43	0.035	10^{-3}	**1**

Temperature (T)			
°F	°C	°K	°R
°F	$\frac{5}{9} \times (°F - 32)$	$\frac{5}{9} \times (°F - 32) + 273.2$	°F + 459.7
1.8 × (°C + 32)	**°C**	°C + 273.2	1.8 × (°C + 273.2)
1.8 × °K − 459.7	°K − 273.2	**°K**	1.8 × °K
°R − 459.7	$\frac{5}{9} \times (°R - 491.7)$	$\frac{5}{9} \times °R$	**°R**

Density (M/L^3)			
lb/ft^3	lb/gal (U.S.)	kg/m^3 (g/L)	kg/L (g/cm^3)
1	0.1337	16.02	0.01602
7.48	**1**	119.8	0.1198

(*Continued*)

TABLE D.2 (*Continued*) Conversion Factors, Mixed Units

Density (M/L^3)			
lb/ft^3	lb/gal (U.S.)	kg/m^3 (g/L)	kg/L (g/cm^3)
0.06243	8.345×10^{-3}	**1**	0.001
62.43	8.345	1000	**1**

Pressure (F/L^2)						
lb/in^2	ft water	in Hg	atm	mm Hg	kg/cm^2	N/m^2
1	2.307	2.036	0.068	51.71	0.0703	6895
0.4335	**1**	0.8825	0.0295	22.41	0.0305	2989
0.4912	1.133	**1**	0.033	25.40	0.035	3386.44
14.70	33.93	29.92	**1**	760	1.033	1.013×10^5
0.019	0.045	0.039	1.30×10^{-3}	**1**	1.36×10^{-3}	133.3
14.23	32.78	28.96	0.968	744.7	**1**	98,070
1.45×10^{-4}	3.35×10^{-4}	2.96×10^{-4}	9.87×10^{-6}	7.50×10^{-3}	1.02×10^{-5}	**1**

Force (F)		
lb$_f$	N	dyne
1	4.448	4.448×10^5
0.225	**1**	10^5
2.25×10^{-6}	10^{-5}	**1**

Energy (E)					
kW·h	hp·h	Btu	J	kJ	Calorie (cal)
1	1.341	3412	3.6×10^6	3600	8.6×10^5
0.7457	**1**	2545	2.684×10^6	2684	6.4×10^5
2.930×10^{-4}	3.929×10^{-4}	**1**	1054.8	1.055	252
2.778×10^{-7}	3.72×10^{-7}	9.48×10^{-4}	**1**	0.001	0.239
2.778×10^{-4}	3.72×10^{-4}	0.948	1000	**1**	239
1.16×10^{-6}	1.56×10^{-6}	3.97×10^{-3}	4.186	4.18×10^{-3}	**1**

Power (P)					
kW	Btu/min	hp	ft·lb/s	kg·m/s	cal/min
1	56.89	1.341	737.6	102	14,330
0.018	**1**	0.024	12.97	1.793	252
0.746	42.44	**1**	550	76.09	10,690

(*Continued*)

TABLE D.2 (*Continued*) Conversion Factors, Mixed Units

Power (P)					
kW	Btu/min	hp	ft·lb/s	kg·m/s	cal/min
1.35×10^{-3}	0.077	1.82×10^{-3}	**1**	0.138	19.43
9.76×10^{-3}	0.558	0.013	7.233	**1**	137.6
6.977×10^{-5}	3.97×10^{-3}	9.355×10^{-5}	0.0514	7.12×10^{-3}	**1**

Dynamic (or Absolute) Viscosity (μ)				
cp	$lb_f \cdot s/ft^2$	$lb_m/ft \cdot s$	g/cm·s	$N \cdot s/m^2$ (kg/m·s or dp)
1	2.09×10^{-5}	6.72×10^{-4}	0.01	1×10^{-3}
4.78×10^4	**1**	32.15	478.5	47.85
1488	0.031	**1**	14.88	1.488
100	2.09×10^{-3}	0.0672	**1**	0.10
1000	0.021	0.672	10	**1**

Kinematic Viscosity (ν)			
Centistoke	ft^2/s	cm^2/s	m^2/s (Myriastoke)
1	1.076×10^{-5}	0.01	10^{-6}
9.29×10^4	**1**	929.4	0.093
100	1.076×10^{-3}	**1**	10^{-4}
10^6	10.76	10^4	**1**

[a] 1 acre = 0.4047 hectare (ha), and 1 ha = 10,000 m^2.
Source: Adapted in part from References 1 through 3.

References

1. Qasim, S. R., *Wastewater Treatment Plants: Planning, Design, and Operation*, 2nd ed., CRC Press, Boca Raton, FL, 1999.
2. Qasim, S. R., E. M. Motley, and G. Zhu, *Water Works Engineering: Planning, Design, and Operation*, Prentice Hall PTR, Upper Saddle River, NJ, 2000.
3. Mechtly, E. A., *The International System of Units, Physical Constants and Conversion Factors*, 2nd rev., Scientific and Technical Information Office, National Aeronautics and Space Administration (NASA), Washington, D.C., 1973.

Appendix E: Summary of Design Parameters for Wastewater Treatment Processes

In this appendix, the practical ranges of selected major design parameters for wastewater treatment processes expressed in both SI and U.S. customary units, and the conversion factors between two units, are summarized in Table E.1.[1–3]

TABLE E.1 Design Parameter and Units of Expressions for Wastewater Treatment Processes

Design Parameter	Typical Range	Conversion Factor
	Coarse Screen, Pretreatment	
Opening size, mm (in)		
Manually cleaned	25–75 (1–3)	1 mm = 0.03937 in
Mechanically cleaned	6–50 (0.25–2)	1 mm = 0.03937 in
Velocity through clean screen, m/s (ft/s)		
Manually cleaned	0.3–0.6 (1–2)	1 m/s = 3.28 ft/s
Mechanically cleaned	0.6–1 (2–3.3)	1 m/s = 3.28 ft/s
Volume of screenings, $m^3/10^6\ m^3$ (ft^3/Mgal)	20–75 (2.5–10)	$1\ m^3/10^6\ m^3 = 0.1337\ ft^3$/Mgal
	Fine Screen, Pretreatment	
Opening size, mm (in)	1–6 (0.04–0.25)	1 mm = 0.03937 in
	Grit Removal	
Velocity in horizontal flow channel, m/s (ft/s)		1 m/s = 3.28 ft/s
Velocity controlled channel	0.25–0.4 (0.8–1.3)	
Aerated grit chamber	0.6–0.8 (2–2.5)	
Surface overflow rate (SOR) in horizontal flow chamber, m^3 flow/m^2 surface area·d (gpd/ft^2)	500–1000 (12,500–25,000)	$1\ m^3/m^2{\cdot}d = 24.54$ gpd/ft^2
Detention time, min		
Velocity controlled channel	0.75–1.5	
Aerated grit chamber	2–5	
Horizontal flow chamber	4–6	
Volume of grits, m^3 grit/$10^6\ m^3$ raw influent (ft^3/Mgal)	5–50 (0.7–7)	$1\ m^3/10^6\ m^3 = 0.1337\ ft^3$/Mgal

(Continued)

TABLE E.1 (*Continued*) Design Parameter and Units of Expressions for Wastewater Treatment Processes

Design Parameter	Typical Range	Conversion Factor
	Flow Equalization	
Mixing power requirement, kW/10^3 m^3 volume (hp/Mgal)	4–8 (20–40)	1 kW/10^3 m^3 = 5.2 hp/Mgal
Air supply, m^3 air/10^3 m^3 tank volume·min (ft^3/Mgal·min)	10–20 (1350–2700)	1 m^3/10^3 m^3·min = 133.7 ft^3/Mgal·min
	Primary Clarifier	
Surface overflow rate (SOR), m^3 flow/m^2 surface area·d (gpd/ft^2)		1 m^3/m^2·d = 24.54 gpd/ft^2
Average flow	24–48 (600–1200)	
Peak 2-h flow	40–100 (1000–2500)	
Detention time, h	1.5–4	
Weir loading rate, m^3 flow/m weir length·d (gpd/ft)	125–375 (10,000–30,000)	1 m^3/m·d = 80.52 gpd/ft
Volume of sludge, m^3 sludge/10^3 m^3 (ft^3/Mgal)	2.2–10 (300–1300)	1 m^3/10^3 m^3 = 133.7 ft^3/Mgal
Mass of sludge, kg dry solids/10^3 m^3 (lb/Mgal)	100–170 (850–1400)	1 kg/10^3 m^3 = 8.345 lb/Mgal
Solids content of sludge, % total solids	1–6	
	Chemically Enhanced Primary Treatment (CEPT) for Phosphorus Removal	
Chemical dosage, practical molar ratio (mole/mole)		
Lime (Ca:P)	1.3:1–2:1	
Aluminum (Al:P)	1:1–2.5:1	
Iron, ferric (Fe:P)	1.5:1–3.5:1	
Iron, ferrous (Fe:P)	1.5:1–1.7:1	
	High Rate Clarification for Primary Treatment	
Surface overflow rate (SOR), m^3 flow/m^2 surface area·h (gpm/ft^2)		1 m^3/m^2·h = 0.409 gpm/ft^2
Solids contact	2–3 (0.8–1.2)	
Inclined surface clarification	10–15 (4–6)	
Micro-sand ballasted flocculation	100–200 (40–80)	
Solids ballasted flocculation	24–36 (10–15)	
Overall detention time, h		
Solids contact	1–2	
Inclined surface clarification	0.15–0.5	
Micro-sand ballasted flocculation	0.15–0.2	
Solids ballasted flocculation	0.2–0.3	
	Fine Mesh Screen for Primary Treatment	
Mesh size, mm (in)	0.15–0.5 (0.006–0.02)	1 mm = 0.03937 in
Surface overflow rate, m^3 flow/m^2 surface area·h (gpm/ft^2)	20–260 (8–106)	1 m^3/m^2·h = 0.409 gpm/ft^2
	Rapid Mixing for Coagulation	
Detention time (t), min	0.5–2	
Hydraulic gradient (G), s^{-1}	300–1500	
Mixing opportunity parameter (Gt), dimensionless	30,00–90,000	

(*Continued*)

TABLE E.1 (*Continued*) Design Parameter and Units of Expressions for Wastewater Treatment Processes

Design Parameter	Typical Range	Conversion Factor
Flocculation after Coagulation		
Detention time (t), min	5–30	
Hydraulic gradient (G), s^{-1}	15–60	
Mixing opportunity parameter (Gt), dimensionless	10,000–150,000	
Suspended Growth Biological Treatment, Activated Sludge and Modifications		
Volumetric organic loading rate, kg BOD_5/m^3 aeration basin volume·d (lb/10^3 ft^3·d)		1 kg/m^3·d = 62.43 lb/10^3 ft^3·d
Conventional with operational modification	0.3–1 (20–60)	
High-rate aeration/high pure oxygen (HPO)	1–3 (60–185)	
Extended aeration/oxidation ditch/single nitrification/sequencing bioreactor (SBR)	0.08–0.3 (5–20)	
Fixed-film media	0.5–6 (30–375)	
Membrane bioreactor (MBR)	0.8–1.5 (50–95)	
F/M ratio, kg BOD_5/kg VSS·d (lb/lb·d)		1 kg/kg·d = 1 lb/lb·d
Conventional with operational modification	0.2–0.6 (0.2–0.6)	
High-rate aeration/HPO	0.5–2 (0.5–2)	
Extended aeration/oxidation ditch/single nitrification/SBR	0.04–0.2 (0.04–0.2)	
Fixed-film media	0.1–0.3 (0.1–0.3)	
MBR	<0.1 (0.1)	
Solids retention time (SRT), d		
Conventional with operational modification	3–15	
High-rate aeration/HPO	0.5–4	
Extended aeration/oxidation ditch/single nitrification/SBR	15–40	
Fixed-film media	10–25	
MBR	10–30	
Return activated sludge ratio (return flow/influent flow), dimensionless	0.2–1	
Air requirement by BOD_5 removal, m^3 air/kg BOD_5 removed (ft^3/lb)	20–100 (300–1500)	1 m^3/kg = 16.02 ft^3/lb
Air requirement by flow treated, m^3 air/m^3 wastewater treated (ft^3/gal)	3.75–15 (0.5–2)	1 m^3/m^3 = 0.1337 ft^3/gal
Oxygen requirement		
Organic, kg O_2/kg BOD_5 applied (lb/lb)	0.6–2.4 (0.6–2.4)	1 kg/kg = 1 lb/lb
Ammonia, kg O_2/kg NH_3-N applied·d (lb/lb·d)	4.57 (4.57)	
Aeration System		
Standard oxygen transfer efficiency (SOTE), %		
Fine bubble diffuser	25–35	
Coarse bubble diffuser/mechanical aerator	7–15	
Standard aeration efficiency (SAE), kg O_2 transferred/kW·h (lb/hp·h)		1 kg/kW·h = 1.644 lb/hp·h
Fine bubble diffuser	2.5–10 (4–16)	
Coarse bubble diffuser/mechanical aerator	1.2–2.5 (2–4)	

(*Continued*)

TABLE E.1 (*Continued*) Design Parameter and Units of Expressions for Wastewater Treatment Processes

Design Parameter	Typical Range	Conversion Factor
	Final Clarifier	
Surface overflow rate (SOR) at average flow, m^3 flow/m^2 surface area·d (gpd/ft^2)		1 $m^3/m^2·d$ = 24.54 gpd/ft^2
Activated sludge	16–24 (400–600)	
Extended aeration/oxidation ditch	8–16 (200–400)	
High pure oxygen (HPO)	16–28 (400–700)	
Enhanced biological phosphorus removal (EBPR) with chemical addition/biological nutrient removal (BNR) or selector technologies	12–32 (300–800)	
Surface overflow rate (SOR) at peak 2-h flow, m^3 flow/m^2 surface area·d (gpd/ft^2)	40–64 (1000–1600)	1 $m^3/m^2·d$ = 24.54 gpd/ft^2
Solids loading rate, kg solids/m^2 surface area·h (lb/ft^2·h)		1 kg/m^2·h = 0.2048 lb/ft^2·h
Average flow	1–8 (0.2–1.6)	
Peak 2-h flow	7–10 (1.4–2)	
	Stabilization Pond System	
Surface organic loading rate, kg BOD_5/ha pond surface area·d (lb/ac·d)		1 kg/ha·d = 0.8922 lb/ac·d
Aerobic (high rate)	70–225 (60–200)	
Facultative (aerobic–anaerobic)	20-90 (15–80)	
Anaerobic	225–1120 (200–1000)	
Volumetric organic loading rate for anaerobic, kg BOD_5/m^3 pond volume·d (lb/10^3 ft^3·d)	0.1–0.4 (6–25)	1 kg/m^3·d = 62.43 lb/10^3 ft^3·d)
Detention time, d		
Aerobic (high rate)	2–6	
Facultative (aerobic–anaerobic)	20–180	
Anaerobic	5–180	
	Attached Growth Biological Treatment, Trickling Filter (TF) and Rotating Biological Contactor (RBC)	
Volumetric organic loading rate for TF, kg BOD_5/m^3 reactor volume·d (lb/10^3 ft^3·d)		1 kg/m^3·d = 62.43 lb/10^3 ft^3·d
Low rate/standard rate with nitrification TF	0.08–0.5 (5–30)	
Intermediate-rate/high-rate TF	0.24–2.4 (15–150)	
Super-rate/roughing TF	0.8–6 (50–375)	
Surface organic loading rate for RBC, g BOD_5/m^2·d (lb BOD_5/10^3 ft^2·d)		1 g/m^2·d = 0.2048 lb/10^3 ft^2·d
For BOD removal	8–20 (1.6–4)	
For BOD removal and nitrification	5–16 (1–3.3)	
For separate nitrification	1–2 (0.2–0.4)	
Hydraulic loading rate, m^3 flow/m^2 surface area·d (gpd/ft^2)		1 $m^3/m^2·d$ = 24.54 gpd/ft^2
Low rate/standard rate with nitrification TF	1–16 (25–400)	
Intermediate-rate/high-rate TF	4–40 (100–1000)	
Super-rate/roughing TF	15–100 (350–2500)	
RBC for BOD removal	0.08–0.16 (2–4)	

(*Continued*)

TABLE E.1 (*Continued*) Design Parameter and Units of Expressions for Wastewater Treatment Processes

Design Parameter	Typical Range	Conversion Factor
RBC for BOD removal and nitrification	0.03–0.08 (0.75–2)	
RBC for separate nitrification	0.04–0.10 (1–2.5)	
Hydraulic retention time (HRT) in RBC, h		
BOD removal	0.75–1.5	
BOD removal and nitrification	1.5–4	
Separate nitrification	1.2–3	
Recirculation rate for TF, % of influent flow		
Low-rate/standard-rate with nitrification	0–2	
Intermediate rate/high rate	0.1–2	
Super rate/roughing	0–12	
Combined Attached and Suspended Growth Biological Treatment		
Organic loading rate in fixed film reactor, kg $BOD_5/m^3{\cdot}d$ ($lb/10^3\ ft^3{\cdot}d$)		$1\ kg/m^3{\cdot}d = 62.43\ lb/10^3\ ft^3{\cdot}d$
Activated biofilter (ABF)	0.24–1.2 (15–75)	
Trickling filter/solids contact (TF/SC)	0.4–1.6 (25–100)	
Biofilter/activated sludge (BF/AS)	1.4–4 (90–250)	
Trickling filter with intermediate clarifier/activated sludge (TF-IC/AS)	1–4.8 (60–300)	
F/M ratio for suspended growth, kg BOD_5/kg VSS·d (lb/lb·d)		1 kg/kg = 1 lb/lb
BF/AS	0.5–1.2 (0.5–1.2)	
TF-IC/AS	0.2–0.5 (0.2–0.5)	
Solids retention time (SRT), d	15–200	
Hydraulic retention time (HRT) for suspended growth, h		
TF/SC	0.75–2	
BF/AS, or TF-IC/AS	2–8	
Final clarifier surface overflow rate (SOR), $m^3/m^2{\cdot}h$ (gpm/ft^2)		$1\ m^3/m^2{\cdot}h = 0.409\ gpm/ft^2$
ABF	2–3 (0.8–1.2)	
TF/SC or BF/AS	1–2 (0.4–0.8)	
TF-IC/AS	2–3.5 (0.8–1.4)	
Integrated Fixed-Film Media Biological Treatment		
Volumetric organic loading rate, kg $BOD_5/m^3{\cdot}d$ (lb $BOD_5/10^3\ ft^3{\cdot}d$)		$1\ kg/m^3{\cdot}d = 62.43\ lb/10^3\ ft^3{\cdot}d$
Integrated fixed-film activated sludge (IFAS)	1.5–3 (95–190)	
Moving bed biofilm reactor (MBBR)	1–6 (60–380)	
Surface organic loading rate, g BOD_5/m^2 surface area·d ($lb/10^3\ ft^2{\cdot}d$)		$1\ g/m^2{\cdot}d = 0.2048\ lb/10^3\ ft^2{\cdot}d$
Surface area loading rate (SALR) for IFAS	10 (2)	
Surface area removal flux (SARF) for MBBR	5–20 (1–4)	
Hydraulic retention time (HRT), h	1.5–6	

(*Continued*)

TABLE E.1 (*Continued*) Design Parameter and Units of Expressions for Wastewater Treatment Processes

Design Parameter	Typical Range	Conversion Factor
Submerged Attached Growth Biological Treatment		
Volumetric organic loading rate, kg BOD_5/m^3·d (lb/10^3 ft^3·d)	2–10 (120–600)	1 kg/m^3·d = 62.43 lb/10^3 ft^3·d
Hydraulic loading rate, m^3 flow/m^2 reactor surface area·h (ft^3/ft^2·h)	2.4–40 (8–130)	1 m^3/m^2·h = 3.281 ft^3/ft^2·h
Empty bed contact time (EBCT), min	5–60	
Anaerobic Suspended Growth Biological Treatment		
Volumetric organic loading rate, kg COD/m^3·d (lb/10^3 ft^3·d)	1–20 (60–1200)	1 kg/m^3·d = 62.43 lb/10^3 ft^3·d
Solids retention time (SRT), d	15–200	
Hydraulic loading rate, m^3/m^2·h (ft^3/ft^2·h)		
Upflow velocity in upflow anaerobic sludge blanket process (UASB)	0.5–1.5 (1.5–5)	1 m^3/m^2·h = 3.281 ft^3/ft^2·h
Surface overflow rate (SOR) in clarifier	0.5–1 (1.6–3.3)	
Hydraulic retention time (HRT), h	4–120	
Anaerobic Attached Growth Biological Treatment		
Volumetric organic loading rate, kg COD/m^3·d (lb/10^3 ft^3·d)	0.1–40 (6–1500)	1 kg/m^3·d = 62.43 lb/10^3 ft^3·d
Solids retention time (SRT), d	10–40	
Hydraulic loading, m^3/m^2 reactor surface area·h (ft^3/ft^2·h)	0.5–20 (1.5–60)	1 m^3/m^2·h = 3.281 ft^3/ft^2·h
Hydraulic retention time (HRT), h	8–120	
Suspended Growth for Biological Nitrogen Removal		
Solids retention time (SRT), d	7–40	
Hydraulic retention time (HRT), h		
Anoxic	1–8	
Aerobic	4–24	
Overall	6–36	
Mixed liquor internal recycle ratio (recycle flow/influent flow), dimensionless	1–4	
Return activated sludge ratio (return flow/influent flow), dimensionless	0.5–1	
Suspended Growth for Biological Nutrient Removal (BNR)		
Solids retention time (SRT), d	5–40	
F/M ratio, kg BOD_5/kg VSS·d (lb/lb·d)	0.1–0.25 (0.1–0.25)	1 kg/kg·d = 1 lb/lb·d
Hydraulic retention time (HRT), h		
Anaerobic (AN)	0.5–6	
Anoxic (AX)	1–8	
Oxic (OX)	4–12	
Overall	6–24	
Mixed liquor internal recycle ratio (recycle flow/influent flow), dimensionless	1–4	
Return activated sludge ratio (return flow/influent flow), dimensionless	0.25–1	

(*Continued*)

TABLE E.1 (*Continued*) Design Parameter and Units of Expressions for Wastewater Treatment Processes

Design Parameter	Typical Range	Conversion Factor
Submerged Attached Growth for Biological Denitrification		
Volumetric organic loading rate, kg NO_5-N/m^3·d (lb/10^3 ft^3·d)	0.3–5 (20–300)	1 kg/m^3·d = 62.43 lb/10^3 ft^3·d
Hydraulic loading rate, m^3 flow/m^2 reactor surface area·h (ft^3/ft^2·h)	2.4–24 (1–10)	1 m^3/m^2·h = 3.281 ft^3/ft^2·h
Hydraulic retention time (HRT) or empty bed contact time (EBCT), min	5–30	
Filtration with Granular and Surface Filters		
Filtration rate, m^3 flow/m^2 surface area·h (gpm/ft^2)		1 m^3/m^2·h = 0.409 gpm/ft^2
Average flow	5–20 (2–8)	
Peak flow	14–40 (6–16)	
Land Treatment System		
Annual volumetric loading rate, m^3/m^2·yr (gal/ft^2·yr)		1 m^3/m^2·yr = 24.54 gal/ft^2·yr
Slow rate (SR)	0.5-6 (12–150)	
Rapid infiltration (RI)	6-125 (150-3000)	
Overland flow (OF)	3-20 (75–500)	
Organic loading rate[d], kg BOD_5/ha·d (lb/ac·d)		1 kg/ha·d = 0.8922 lb/ac·d
Slow rate (SR)	50-500 (45–450)	
Rapid infiltration (RI)	145–1000 (130–900)	
Overland flow (OF)	40–110 (35–100)	
Constructed Wetland Systems		
Hydraulic loading rate, m^3/ha·d (gpm/ac)		1 m^3/ha·d = 0.07424 gpm/ac
Free water surface (FWS)	135–900 (10–70)	
Subsurface flow (SF)	200–1800 (15–140)	
Adsorption by Granular Activated Carbon		
Operating velocity, L/s·m^2 (gpm/ft^2)	1.4-4.2 (2–6)	1 L/s·m^2 = 1.472 gpm/ft^2
Empty bed contact time (EBCT), min	5–30	
Ion Exchange Column		
Operating velocity, L/s·m^2 (gpm/ft^2)	0.7-5.5 (1-8)	1 L/s·m^2 = 1.472 gpm/ft^2
Empty bed contact time (EBCT), min	10–30	
Membrane System		
Flux rate, L/m^2·d (gpd/ft^2)		1 L/m^2·d = 0.0245 gpd/ft^2
Microfiltration (MF)	600–1400 (15–35)	
Ultrafiltration (UF)	960–1400 (25–35)	
Nanofiltration (NF) or reverse osmosis (RO)	340–480 (8–12)	
Recovery factor, %		
MF/UF	85–95	
NF/RO	80–90	
Disinfection		
Chlorine dosage, mg Cl_2 applied/L (lb/Mgal)	4–40 (33–330)	1 mg/L = 8.345 lb/Mgal

(*Continued*)

TABLE E.1 (*Continued*) Design Parameter and Units of Expressions for Wastewater Treatment Processes

Design Parameter	Typical Range	Conversion Factor
UV dose for inactivation of bacteria and virus in filtered secondary effluent, mJ/cm^2		
1- to 2-log	10–70	
3- to 4-log	20–190	
Gravity Sludge Thickening		
Solids loading rate, kg dry solids/m^2 surface area·d ($lb/ft^2·d$)	20–150 (4–30)	1 $kg/m^2·d$ = 0.2048 $lb/ft^2·d$
Hydraulic loading rate, m^3 sludge/m^2 surface area·d (gpd/ft^2)	4–30 (100–750)	1 $m^3/m^2·min$ = 24.54 gpm/ft^2
Chemical dosage		
Lime, mg/L as CaO (lb/Mgal)	6–9 (50–75)	1 mg/L = 8.345 lb/Mgal
Ferric chloride, mg/L as $FeCl_3$ (lb/Mgal)	1–3 (8–25)	
Potassium permanganate, mg/L $KMnO_4$ (lb/Mgal)	10–40 (80–330)	
Polymer, g polymer/kg dry solids (lb/ton)	2.5–6 (5–12)	1 g/kg = 2 lb/ton
Dissolved Air Flotation (DAF) for Sludge Thickening		
Solids loading rate with polymer addition, kg dry solids/m^2 surface area·d ($lb/ft^2·d$)	220–300 (45–60)	1 $kg/m^2·d$ = 0.2048 $lb/ft^2·d$
Polymer dosage, g polymer/kg dry solids (lb/ton)	1–5 (2–10)	1 g/kg = 2 lb/ton
Gravity Belt Thickener (GBT)		
Solids loading rate, kg dry solids/h (lb/h) per m belt width	300–1350 (650–3000)	1 kg/h = 2.205 lb/h
Hydraulic loading rate, L wet sludge/min (gpm) per m belt width	400–1000 (100–250)	1 L/min = 0.2642 gpm
Polymer dosage, g polymer/kg dry solids (lb/ton)	1–7 (2–14)	1 g/kg = 2 lb/ton
Wash water requirement, L/min (gpm) per m belt width	60–80 (15–20)	1 L/min = 0.2642 gpm
Anaerobic Sludge Digestion		
Per capita capacity requirement (PCCR), m^3 (ft^3) per capita	0.03–0.18 (1–6)	1 m^3 = 35.31 ft^3
Volumetric solids loading rate, kg VSS/m^3 digester volume·d ($lb/10^3\ ft^3·d$)		1 $kg/m^3·d$ = 62.43 $lb/10^3\ ft^3·d$
Standard rate or staged mesophilic	0.5–1.6 (30–100)	
High rate	1.6–4.8 (100–300)	
Staged thermophilic, acid/gas phased (A/GAnD), or temperature-phased (TPAnD)	4.8–6.4 (300–400)	
Solids retention time (SRT), d		
Standard rate mesophilic	30–60	
High rate mesophilic	15–20	
Staged mesophilic, 1st reactor (2nd reactor)	7–10 (variable)	
Staged thermophilic, 1st reactor (2nd or 3rd reactor)	17–22 (1.5–2)	
A/GAnD, 1st reactor (2nd reactor)	1–3 (>10)	
TPAnD, 1st reactor (2nd reactor)	3–10 (5–15)	

(*Continued*)

TABLE E.1 (*Continued*) Design Parameter and Units of Expressions for Wastewater Treatment Processes

Design Parameter	Typical Range	Conversion Factor
Digester heating loss, J/s·m^2 surface area·°C (Btu/h·ft^2·°F)	0.7–2.8 (0.12–0.5)	1 J/s·m^2·°C = 0.1763 Btu/ft^2·h·°F
Digester mixing requirement		
Gas injection, m^3 gas/10^3 m^3 volume·min (ft^3/10^3 ft^3·min)	4.5–7 (4.5–7)	1 m^3/10^3 m^3·min = 1 ft^3/10^3 ft^3·min
Mechanical mixing, kW/10^3 m^3 volume (hp/10^3 ft^3)	5–8 (0.2–0.3)	1 KW/10^3 m^3 = 0.03795 hp/10^3 ft^3
Hydraulic mixing (G and T), s^{-1} and min	50–80 and 20–30	
Aerobic Sludge Digestion		
Per capita capacity requirement (PCCR), m^3 (ft^3) per capita	0.06–0.2 (2–7)	1 m^3 = 35.31 ft^3
Volumetric solids loading rate, kg VSS/m^3 digester volume·d (lb/10^3 ft^3·d)		1 kg/m^3·d = 62.43 lb/10^3 ft^3·d
Conventional aerobic digestion	1.6–4.8 (100–300)	
Autothermal thermophilic aerobic digestion (ATAD)	3–4 (200–250)	
Solids retention time (SRT), d		
Conventional	10–60	
ATAD	4–15	
Oxygen requirement, kg O2/kg VSS reduced (lb/lb)	1.6–2.3 (1.6–2.3)	1 kg/kg = 1 lb/lb
Sludge Drying Beds		
Area requirement per capita for conventional sludge drying bed, m^2 bed area/person (ft^2/person)	0.09–0.23 (1–2.5)	1 m^2/person = 10.76 ft^2/person
Mass sludge applying rate, kg dry solids/m^2 bed area·yr (lb/ft^2·yr)		1 kg/m^2·yr = 0.2048 lb/ft^2·yr
Conventional sand drying bed	50–200 (10–40)	
Artificial-media drying bed	750–1800 (150–360)	
Solar drying bed	50–900 (10–180)	
Reed drying bed	30–100 (6–20)	
Mass sludge applying rate for vacuum assisted drying bed, kg dry solids/m^2 bed area·h (lb/ft^2·h)	6–60 (1.2–12)	1 kg/m^2·yr = 0.2048 lb/ft^2·yr
Belt-Filter Press (BFP) Dewatering		
Solids loading rate, kg dry solids/h (lb/h) per m belt width	200–1500 (450–3300)	1 kg/h = 2.205 lb/h
Hydraulic loading rate, L wet sludge/min (gpm) per m belt width	100–800 (25–200)	1 L/min = 0.2642 gpm
Polymer dosage, g polymer/kg dry solids (lb/ton)	1.5–12.5 (3–25)	1 g/kg = 2 lb/ton
Wash water requirement, L/min (gpm) per m belt width	150–300 (40–80)	1 L/min = 0.2642 gpm
Plate and Frame Filter Press Dewatering		
Solids loading rate, kg dry solids/m^2 press area·h (lb/ft^2·h)	~5 (1)	1 kg/m^2·h = 0.2048 lb/ft^2·h
Chemical dosage, g chemical/kg dry solids (lb/ton)		1 g/kg = 2 lb/ton
Ferric chloride	50 (100)	
CaO	100 (200)	

(*Continued*)

TABLE E.1 (*Continued*) Design Parameter and Units of Expressions for Wastewater Treatment Processes

Design Parameter	Typical Range	Conversion Factor
Vacuum Filter Dewatering		
Solids loading rate kg dry solids/m^2 filter area·h (lb/ft^2·h)	10–25 (2–5)	1 kg/m^2·h = 0.2048 lb/ft^2·h
Chemical dosage, g chemical/kg dry solids (lb/ton)		1 g/kg = 2 lb/ton
Ferric chloride	20–40 (40–80)	
CaO	100–150 (200–300)	
Rotary Press for Dewatering		
Solids loading rate, kg dry solids/m^2 filter area·h (lb/ft^2·h)	~250 (50)	1 kg/m^2·h = 0.2048 lb/ft^2·h
Volumetric sludge feed rate, m^3 feed wet sludge/m^2 filter area·h (ft^3/ft^2·h)	2.5–3.5 (1–1.5)	1 m^3/m^2·h = 0.409 gpm/ft^2
Polymer dosage, g polymer/kg dry solids (lb/ton)	5–15 (10–30)	1 g/kg = 2 lb/ton
Land Application of Biosolids		
Mass application rate, metric ton (mt) biosolids/ha field area·yr (ton/ac·yr)		1 mt/ha·yr = 0.4461 ton/ac·yr
Agricultural land utilization	2–70 (1–30)	
Forest land utilization	10–225 (4–100)	
Land reclamation utilization	7–450 (3–200)	
Dedicated land disposal	12–2250 (5–1000)	

Source: Adapted in part from References 1 through 3.

References

1. Qasim, S. R., *Wastewater Treatment Plants: Planning, Design, and Operation*, 2nd ed., CRC Press, Boca Raton, FL, 1999.
2. Metcalf & Eddy | AECOM, *Wastewater Engineering: Treatment and Resource Recovery*, 5th ed., McGraw-Hill, New York, NY, 2014.
3. Metcalf & Eddy, Inc., *Wastewater Engineering: Treatment and Reuse*, 4th ed., McGraw-Hill Companies, Inc., New York, NY, 2003.

Appendix F: List of Examples and Solutions

In this book, over 700 design examples with in-depth solutions are presented to cover the complete spectrum of wastewater treatment and reuse. A listing of these illustrative examples is provided in Table F.1.

Note: In this table, the examples of Chapters 1–10 are presented in Volume 1: Principles and Basic Treatment while the examples of Chapters 11–15 are presented in Volume 2: Post-Treatment, Reuse, and Disposal.

TABLE F.1 Summary of Examples

(Continued)

TABLE F.1 (*Continued*) Summary of Examples

(*Continued*)

TABLE F.1 (*Continued*) Summary of Examples

No	Title	Page
Example 3.26	Conversion Equation of Reactive Substance in a PFR	3-37
Example 3.27	Residence Time in a PFR for a Given Conversion	3-38
Example 3.28	Residence Time Equations for CFSTR and PFR	3-39
Example 3.29	Performance Evaluation of First-Order Reaction in CFSTR and PFR for a given Residence Time	3-40
Example 3.30	Second-Order Residence Time Equations of CFSTR and PFR	3-41
Example 3.31	Performance Evaluation of Second-Order Reaction in a CFSTR and PFR for a given Residence Time	3-42
Example 3.32	Comparison of Residence-Times of CFSTR and PFR for a given Removal by Saturation-Type Reaction	3-43
Example 3.33	Performance Evaluation of Saturation-Type Reaction in a CFSTR and PFR for a given Hydraulic Residence Time	3-44
Example 3.34	Slug Conservative Dye Tracer in Series CFSTRs	3-44
Example 3.35	Effluent Concentration of a Conservative Tracer through Three CFSTRs in Series	3-46
Example 3.36	Conservative Tracer Concentration Profile from Three CFSTRs in Series	3-47
Example 3.37	Effluent Quality from CFSTRs in Series with First-Order Reaction and under Steady State	3-49
Example 3.38	Performance of Five CFSTRs in Series Treating a Nonconservative Substance	3-50
Example 3.39	Total Volume of CFSTRs in Series to Achieve a given Removal Ratio of a Nonconservative Substance	3-50
Example 3.40	Comparison of HRTs in a Single and Three Equal CFSTRs with that of a PFR	3-51
Example 3.41	Graphical Solution Procedure of CFSTRs in Series	3-52
Example 3.42	Graphical Solution of Five CFSTRs in Series	3-54
Example 3.43	Performance of a Sedimentation Basin with Conservative Dye Tracer	3-58
Example 3.44	Dead Volume in a CFSTR	3-60
Example 3.45	Effluent Quality from a Reactor with Dispersion	3-63
Example 3.46	Dispersion Coefficient	3-64
Example 3.47	Observed *E.coli* Die-off in a Series of Stabilization Ponds	3-65
Example 3.48	Dispersion Factor and Performance of a Chlorine Contact Basin	3-66
Example 3.49	Comparison of Residence Times for Ideal and Dispersed PFRs Treating a Nonconservative Substance	3-67
Example 3.50	Comparison of Fraction Remaining of a Nonconservative Substance in Ideal and Dispersed PFRs	3-68
Example 3.51	Reactors in Series to Achieve a given Removal of a Nonconservative Substance	3-68
Example 3.52	Process Train with Equalization Basins	3-70
Example 3.53	Storage Volume Based on Flow Diagram and Change in Volume	3-72
Example 3.54	Volume of an In-line Equalization Basin	3-74
Example 3.55	Equalization of Mass Loading	3-76
Example 3.56	Equalization of Mass Loading over 24-h Cycle	3-78
Chapter 4	**Sources and Flow Rates of Municipal Wastewater**	**4-1**
Example 4.1	Relationship between Diurnal Water Demand and Waste Water Flow	4-2
Example 4.2	Determination of Average Flow from Diurnal Flow Data	4-3
Example 4.3	Municipal Water Demand and Components	4-7
Example 4.4	Water Demand and Wastewater Flow	4-7
Example 4.5	Water Saving from Supply Pressure Reduction	4-10

(*Continued*)

TABLE F.1 (*Continued*) Summary of Examples

(*Continued*)

TABLE F.1 (*Continued*) Summary of Examples

(*Continued*)

TABLE F.1 (*Continued*) Summary of Examples

(*Continued*)

TABLE F.1 (*Continued*) Summary of Examples

(*Continued*)

TABLE F.1 (*Continued*) Summary of Examples

(Continued)

TABLE F.1 (*Continued*) Summary of Examples

(*Continued*)

TABLE F.1 (*Continued*) Summary of Examples

(*Continued*)

TABLE F.1 (*Continued*) Summary of Examples

(*Continued*)

TABLE F.1 (*Continued*) Summary of Examples

(*Continued*)

TABLE F.1 (*Continued*) Summary of Examples

(*Continued*)

TABLE F.1 (*Continued*) Summary of Examples

(*Continued*)

TABLE F.1 (*Continued*) Summary of Examples

(*Continued*)

TABLE F.1 (*Continued*) Summary of Examples

(*Continued*)

TABLE F.1 (*Continued*) Summary of Examples

(*Continued*)

TABLE F.1 (*Continued*) Summary of Examples

(*Continued*)

TABLE F.1 (*Continued*) Summary of Examples

Index

Note: Volume 1—Chapters 1–10
Volume 2—Chapters 11–15

A

B

C

F

H

N

O

P

S

T

U

V

W

Y

Z